GEOGRAPHY
AND
ANTHROPOLOGY

WIDENER LIBRARY SHELFLIST

Volumes in Print

5–6.	LATIN AMERICA. 1966. 2 VOLS. ISBN 0–674–51247–2
9–13.	AMERICAN HISTORY. 1967. 5 VOLS. ISBN 0–674–02400–1
15.	PERIODICAL CLASSES. 1968. 758 PP. ISBN. 0–674–66300–4
16–17.	EDUCATION. 1968. 2 VOLS. ISBN 0–674–23800–1
18.	LITERATURE: GENERAL AND COMPARATIVE. 1968. 189 PP. ISBN 0–674–53650–9
19.	SOUTHERN ASIA. 1968. 543 PP. ISBN 0–674–82500–4
20.	CANADIAN HISTORY AND LITERATURE. 1968. 411 PP. ISBN 0–674–09351–8
21.	LATIN AMERICAN LITERATURE. 1969. 498 PP. ISBN 0–674–51251–0
22.	GOVERNMENT. 1969. 263 PP. ISBN 0–674–35786–8
23–24.	ECONOMICS. 1970. 2 VOLS. ISBN 0–674–23125–2
25.	CELTIC LITERATURES. 1970. 192 PP. ISBN 0–674–10480–3
26–27.	AMERICAN LITERATURE. 1970. 2 VOLS. ISBN 0–674–02535–0
28–31.	SLAVIC HISTORY AND LITERATURES. 1971. 4 VOLS. ISBN 0–674–81090–2
32.	GENERAL EUROPEAN AND WORLD HISTORY. 1970. 959 PP. ISBN 0–674–34420–0
33.	REFERENCE COLLECTIONS. 1971. 130 PP. ISBN 0–674–75201–5
34.	AFRICAN HISTORY AND LITERATURES. 1971. 600 PP. ISBN 0–674–00780–8
35–38.	ENGLISH LITERATURE. 1971. 4 VOLS. ISBN 0–674–25663–8
40.	FINNISH AND BALTIC HISTORY AND LITERATURES. 1972. 250 PP. ISBN 0–674–30205–2
41.	SPANISH HISTORY AND LITERATURE. 1972. 770 PP. ISBN 0–674–83095–4
42–43.	PHILOSOPHY AND PSYCHOLOGY. 1973. 2 VOLS. ISBN 0–674–66486–8
44.	HUNGARIAN HISTORY AND LITERATURE. 1974. 186 PP. ISBN 0–674–42700–9
45–46.	SOCIOLOGY. 1973. 2 VOLS. ISBN 0–674–81625–0
47–48.	FRENCH LITERATURE. 1973. 2 VOLS. ISBN 0–674–32215–0
49–50.	GERMAN LITERATURE. 1974. 2 VOLS. ISBN 0–674–35070–7
51–52.	ITALIAN HISTORY AND LITERATURE. 1974. 2 VOLS. ISBN 0–674–46955–0
53–54.	BRITISH HISTORY. 1975. 2 VOLS. ISBN 0–674–08240–0
55.	ANCIENT HISTORY. 1975. 363 PP. ISBN 0–674–03312–4
56.	ARCHAEOLOGY. 1979. 442 PP. ISBN 0–674–04318–9
57.	CLASSICAL STUDIES. 1979. 215 PP. ISBN 0–674–13461–3
58.	ANCIENT GREEK LITERATURE. 1979. 638 PP. ISBN 0–674–03310–8
59.	LATIN LITERATURE. 1979. 610 PP. ISBN 0–674–51295–2
60.	GEOGRAPHY AND ANTHROPOLOGY. 1979. 270 PP. ISBN 0–674–34855–9

HARVARD UNIVERSITY LIBRARY

WIDENER LIBRARY SHELFLIST, 60

GEOGRAPHY AND ANTHROPOLOGY

CLASSIFICATION SCHEDULES
CLASSIFIED LISTING BY CALL NUMBER
CHRONOLOGICAL LISTING
AUTHOR AND TITLE LISTING

Published by the Harvard University Library
Cambridge, Massachusetts
Distributed by the Harvard University Press
Cambridge, Massachusetts and London, England
1979

Library of Congress Cataloging in Publication Data

Harvard University. Library.
  Geography and anthropology : classification schedules, classified listing by call number, chronological listing, author and title listing.

  (Widener Library shelflist ; 60)
  1. Geography—Bibliography—Catalogs. 2. Anthropology—Bibliography—Catalogs. 3. Classification—Books—Geography. 4. Classification—Books—Anthropology. 5. Harvard University. Library—Catalogs. I. Title. II. Series: Harvard University Library. Widener Library shelflist ; 60.
  Z6009.H37 1979 [G116] 016.91   79-617
  ISBN 0-674-34855-9

Copyright © 1979
By the President and Fellows of Harvard College
international standard book number: 0-674-34855-9
printed in u.s.a.

Foreword

The shelflists of libraries not having classified catalogs have long been used by librarians and knowing readers as implements for systematically surveying holdings in a particular subject. Examining a shelflist, one can see all the titles that have been classified in a given area, and not merely those with legible spine lettering which happen to be on the shelves. However, the potential bibliographical usefulness of the shelflist has been difficult to exploit because it usually exists in only one copy kept in a relatively inaccessible location. Computer technology has made it possible to enlarge the concept of the shelflist and to expand its usefulness and accessibility while improving the techniques of maintaining it.

The Harvard University Library is converting to machine-readable form the shelflist and classification schedules of the Harry Elkins Widener Memorial Library, which houses Harvard's central research collections for the humanities and social sciences.* After each class or group of related classes is converted, it is published in the *Widener Library Shelflist* series.

SCOPE OF THIS VOLUME

Each new volume in the series is comprehensive for materials housed in the Widener Library book stacks and cataloged before July 1976, when the Library's unique system of classifying materials was discontinued and the Library of Congress classification was adopted for new acquisitions. Works cataloged since that date are not included. Some books which were formerly shelved in Widener but subsequently moved to other locations are listed here, but complete coverage of this type of material has not been attempted. In particular, many rare or early books housed in the Houghton Library do not appear in this shelflist.

This volume, *Geography and Anthropology*, the sixtieth in the published series, includes 9000 shelflist entries comprising the Widener *Geog* and *An* classes. Detailed information on the scope and arrangement of this material may be found in the Classification Schedule section of this book, beginning on page 3. The *An* class, containing only 1600 entries, has been included in this volume with *Geog* because of the close association of these two subject areas both in the Library of Congress classification and in their physical situation in the stacks. Harvard's major research collection in anthropology is in the Tozzer Library, whose holdings are represented by the *Catalogue of the Library of the Peabody Museum of Archaeology and Ethnology, Harvard University.* (Boston: Hall, 1963) in two parts, *Authors* (26 v.) and *Subjects* (27 v.), each of which has three supplements.

ORGANIZATION OF THIS VOLUME

This catalog is arranged in four parts. The classification schedule is the first of these. It serves as an outline of the second part, which presents the entries in shelflist order; that is, in order by call number, as the books are arranged on the shelves. Together these two parts form a classified catalogue and browsing guide.

The third section lists the same items (excluding periodicals and other serials) in chronological order by date of publication. In addition to its obvious reference use, this list yields information on the quantity and rate of publication in the field. It can be helpful in determining patterns of collection development and in identifying existing strengths and weaknesses.

Access to the collection by author and by title is provided by the alphabetical list which constitutes the fourth part of the catalog. Computer-generated entries are included for titles of works listed elsewhere by author. This section equips the reader with a subject-oriented subset of the card catalog — a finding list which offers substantial advantages of conciseness and portability over the card catalog as a whole.

BIBLIOGRAPHIC LIMITATIONS OF THE SHELFLIST

A shelflist has traditionally served as an inventory record of the books in the library and as an indispensable tool for assigning call numbers to books as they are added to the collection. Since the Widener Shelflist was designed and maintained to fulfill these two intramural functions, its bibliographic and syndetic standards are not equal to those that prevail in the public card catalogs. Shelflists entries are less complete than the public catalog entries and may contain errors and inconsistencies which have not been eliminated during the conversion process. Cross references and name added entries are not provided. Entries for serials rarely reflect changes in title, and serial holding statements are terse.

The reader must also bear in mind that the design of any classification schedule, or variations in its interpretation or application, may result in a book's not being classified in the first place one would expect to find it.

INTERPRETING THE ENTRIES

As a general rule, entries in the this catalog include call number, author, title, place and date of publication, and, if more than one, the number of physical volumes held by the Library. Title added entries generated for the Author and Title Listing have the author in parentheses following the title. For works of classical authors and certain other works, the "author" portion of the entry may include information about the edition; to wit, language, date, and editor.

For serials, a summary holdings statement giving the

---

* For information on the relationship of the Widener Library to other units of the Harvard University Library, consult the Library's occasional publication, *The Research Services of the Harvard College Library*, or Edwin E. Williams, "Harvard University Library," *The Encyclopedia of Library and Information Science*, X (New York: Dekker, 1973), 317–373. A more extended description of the Widener Library Shelflist conversion project is provided by Richard De Gennaro, "Harvard University's Widener Library Shelflist Conversion and Publication Program," *College and Research Libraries*, XXXI (September 1970), 318–331.

year or volume number of the first and last volumes in the Library replaces the publication date. Gaps are not specified. A plus sign following the beginning date or volume number indicates that the title is currently being received. The conventional serial records in the Widener Library should be consulted for detailed holdings information.

The letter *A* following a call number in the chronological or alphabetical lists indicates that the Library holds more than one copy of the book on this number. The Classified Listing shows all copies.

A book in the shelflist but not in the Widener stacks is identified either by a *V* preceding the call number or by one of the marginal notations listed below. In the latter case, the call number given in this volume may be obsolete. The current call number should be obtained from the card catalogs before requesting the book from the Widener Circulation Division.

Symbol	Location
Htn	Houghton Library (Rare books and special collections)
NEDL	New England Deposit Library
X Cg	Protective storage
RRC	Russian Research Collection (apply to Coolidge Hall Library for use)

Contents

Foreword v

Statistical Summaries ix

Classification Schedules 3

Classified Listing

    Geog 15

    An 70

Chronological Listing 85

Author and Title Listing 149

Statistical Summaries of Classes in this Volume
August 1976

Analysis of Shelflist Entries by Language

Entries in the *Geog* Class   7,364

English	4,344	Russian	307
German	793	Ukrainian	6
Dutch	120	Polish	56
Swedish	64	Yugoslav languages	29
Danish	138	Czech & Slovak	17
Norwegian	40	Bulgarian	12
Icelandic	23	Greek	3
Latin	81	Hungarian	7
French	745	Turkish	5
Italian	223	Finnish	8
Spanish	197	Estonian	3
Portuguese	84	Lithuanian	1
Catalan	1	Uncoded	57

Entries in the *An* Class   1,614

English	1,011	Portuguese	21
German	219	Russian	62
Dutch	14	Ukrainian	3
Swedish	4	Polish	9
Danish	5	Yugoslav languages	4
Norwegian	3	Czech & Slovak	1
Icelandic	1	Bulgarian	1
Latin	11	Hungarian	2
French	164	Finnish	4
Italian	31	Lithuanian	1
Spanish	36	Uncoded	7

Count of Titles

		Widener	Elsewhere	Total
Geog	Monographs	6,447	327	6,774
	Serials	463	46	509
	Pamphlets in Tract Volumes	220	12	232
	Pamphlet Boxes	44	2	46
	Total	7,174	387	7,561
An	Monographs	1,574	20	1,594
	Serials	4	0	4
	Pamphlets in Tract Volumes	70	2	72
	Pamphlet Boxes	5	0	5
	Total	1,653	22	**1,675**
Total Titles		8,827	409	9,236

Count of Volumes

		Widener	Elsewhere	Total
Geog	Monograph	8,039	573	8,612
	Serial	5,338	962	6,300
	Tract	30	5	35
	Total	13,407	1,540	14,947
An	Monograph	1,784	19	1,803
	Serial	22	0	22
	Tract	10	1	11
	Total	1,816	20	1,836
Total Volumes		15,223	1,560	16,783

WIDENER LIBRARY SHELFLIST, 60

GEOGRAPHY AND ANTHROPOLOGY

CLASSIFICATION SCHEDULES

Geography

NOTE ON THE CLASSIFICATION

The Geog class provides for general works relating to geography. It contains works on the theory and methods of geography as a science, general history of the growth of geographical knowledge, descriptive geographies of the whole world, and general voyages and travels covering the whole world. It also includes descriptive geographies and voyages and travels covering Europe in general, the Arctic and Antarctic regions, and regions (e.g. the tropics in general) not included in one of the history classes. Descriptive geographies and voyages and travels covering other areas are provided for in the various history classes.

Special types of geography are contained in other classes. Physical geography is in the S class, and Mathematical geography goes with Geodesy in the Eng class. General anthropogeography is in the An class, general historical and political geography is in the H class, and general commercial and economic geography goes in the Econ class. All world maps and atlases are in the Map Room.

The accompanying Outline shows the general arrangement of the class. The position of the section for the History of geography seems awkward, but otherwise the order is fairly straightforward.

In the sections for Voyages and travels and Arctic and Antarctic, it should be noted that individual journeys and exploring expeditions are arranged by the date of the beginning of the voyage rather than by the date of publication.

During the present revision of the scheme, no major changes have been made. But many notes and references have been added to clarify the situation for the classifiers. It has not been possible as yet to reclassify all the books which had been wrongly classified.

Bartol Brinkler
Classification Specialist
April 1975

OUTLINE

1-585	Periodicals; Bibliographies; Dictionaries; etc.
600-626	Biographies of geographers
650-665	Theory, methods, etc.
750-960	General geographies
	Guidebook collections
1500-1660	Baedeker
1700-1860	Murray
	Others. See various history classes
3000-3699	History of geography
4000-4029	Adventure and adventurers in general
4100-4480	Voyages and travels in general
4500-4521	Mountaineering in general
4550-4710	Ocean life in general
4750-4760	Tropics in general
	Arctic and Antarctic
5050-5130	General
5150-5560	Arctic regions
5600-5898	Greenland
5900-6100	Antarctic regions

Geog   Geography   Geog

	Geography in general
1-296	Periodicals and Societies (Special Table)
	[Include here general geographical periodicals and the serial publications of geographical societies. For a society, include its general serials and proceedings, annual reports, membership lists, and histories of the society. Include also the proceedings of annual or continuing congresses. See also Geog 5057, 5153, 5606, and 5907; also Geog 500.
	The works are arranged in approximate alphabetical order - by a distinctive title, by the name of a society (sometimes by its original name), or by the place where a society is located.]
500	Bibliographies
	[Include bibliographical periodicals. See also Geog 4005, 4500, 4750, 5050, 5150, 5600, and 5900.]
510	General pamphlet volumes
[520]	Collections [Discontinued. See Geog 665, 4154-4182, etc.]
535	Directories
540	Museums
557-560	Exhibitions (By date)
574-580	Dictionaries and Gazetteers (By date)
585	Tables
	Biographies of geographers
600	Collected
601-626	Individual (A-Z by person)
	[Include a person's autobiography, journals, letters, etc.; but his collected writings go in Geog 665-820, etc.]
	History of geography. See Geog 3000-3699
656-660	Theory, aims, methods of geography (By date)
	[Include works on the principles of geography and on the techniques of geography. For works on map-making and map-making apparatus, see Eng 525-535.]
	Study and teaching of geography in elementary and secondary schools. See Educ 2230
665	General miscellany about geography
	[Include miscellaneous essays, congresses, Festschriften, etc.; also works on the natural "wonders of the world".]
	Geographical terms. See Geog 574-580
	Geographical names. See 1252 (Linguistics)
	Physical, Mathematical, and Descriptive geographies
	Periodicals. See Geog 1-296
	Bibliographies. See Geog 500
	Dictionaries. See Geog 574-580
755-760	General works (By date)
	[N.B. - Include here only works which contain all types of geography together.]
	Physical geography. See S 6000-6499
	Mathematical geography. See Eng 496-500 and Astr 3306-3310
	Descriptive geographies
	[N.B. - See also Geog 4264-4270 for Europe]
	Periodicals. See Geog 1-296
	Bibliographies. See Geog 500
	Dictionaries. See Geog 574-580
	General works
805-810	Multi-volume sets (By date)
815-820	Others (By date)
	Views, pictorial works
953	Pamphlet volumes; Bibliographies
956-960	General works (By date)

	Descriptive geographies (cont.)
	Maps and atlases. Send to Map Room
	Gazetteers. See Geog 574-580
	Voyages and travels. See Geog 4100-4480
	Guidebook collections
	Baedeker
1500	Worldwide
	Europe
1505	General
1506	Northern Europe in general
1507	Central Europe in general
	Great Britain
1508	General and England
1509.1-.39	Counties of England (Special Table)
1510	London
1511	Other cities of England
1512	Scotland
1513	Ireland
1514	Wales
	France
1515	General
1516	Northern France in general
1517	Southern France in general
1518	Other regions
1519	Paris
1520	Other cities
	Corsica. See Geog 1544
	Germany
1522	General
1523	Northern Germany in general
1524	Southern Germany in general
1525	Other regions
1526	Berlin
1527	Other cities
	Austria-Hungary
1530	General
1531	Austria
1532	Hungary
1533	Eastern Alps
	[See also Geog 1536]
	Switzerland
1535	General
1536	Alps
	Italy
1538	General
1539	Northern Italy in general
1540	Central Italy in general
1541	Rome
1542	Southern Italy in general
1543	Sicily
1544	Sardinia and Corsica
	Spain and Portugal
1545	General
1548	Spain
1550	Portugal
	Low Countries
1555	General
1558	Belgium and Luxemburg
1560	Holland, Netherlands
	Scandinavia
1563	General
1565	Denmark
1568	Norway and Sweden
1570	Norway
1573	Sweden
1575	Russia
1580	Other Slavic countries
1582	Ionian Islands

Geog   Geography   Geog

	Guidebook collections (cont.)			**Guidebook collections (cont.)**
	Baedeker (cont.)			Murray (cont.)
1585	Greece			Switzerland
1590	Other Europe		1735	General
	[See also Geog 1580 and 1603-1604.]		1736	Alps
	Asia			Italy
1600	General		1738	General
1601	Asia Minor		1739	Northern Italy in general
	Turkey		1740	Central Italy in general
1603	General		1741	Rome
1604	Constantinople, Istanbul		1742	Southern Italy in general
1605	Palestine and Syria		1743	Sicily
	India		1744	Sardinia and Corsica
1607	General			
1608	Large regions			Spain and Portugal
1609	Provinces, States, Cities		1745	General
1610	China		1748	Spain
1611	Japan		1750	Portugal
1613	Other Asia			
				Low Countries
	Africa		1755	General
1615	General		1758	Belgium and Luxemburg
	Egypt		1760	Holland, Netherlands
1616	General			
1617	Upper Egypt			Scandinavia
1618	Lower Egypt		1763	General
1619	Cities		1765	Denmark
1620	Algeria and Tunisia		1768	Norway and Sweden
1622	Other Africa		1770	Norway
1625	Australia		1773	Sweden
1628	New Zealand		1775	Russia
1630	Pacific Islands		1780	Other Slavic countries
1640	Canada		1782	Ionian Islands
1645	United States		1785	Greece
1650	Mexico		1790	Other Europe
1655	Central America			[See also Geog 1780 and 1803-1804.]
1660	South America			
				Asia
	Murray		1800	General
1700	Worldwide		1801	Asia Minor
				Turkey
	Europe		1803	General
1705	General		1804	Constantinople, Istanbul
1706	Northern Europe in general		1805	Palestine and Syria
1707	Central Europe in general			India
			1807	General
	Great Britain		1808	Large regions
1708	General and England		1809	Provinces, States, Cities
1709.1-.39	Counties of England (Special Table)		1810	China
1710	London		1811	Japan
1711	Other cities of England		1813	Other Asia
1712	Scotland			
1713	Ireland			Africa
1714	Wales		1815	General
				Egypt
	France		1816	General
1715	General		1817	Upper Egypt
1716	Northern France in general		1818	Lower Egypt
1717	Southern France in general		1819	Cities
1718	Other regions		1820	Algeria and Tunisia
1719	Paris		1822	Other Africa
1720	Other cities		1825	Australia
	Corsica. See Geog 1744		1828	New Zealand
			1830	Pacific Islands
	Germany		1840	Canada
1722	General		1845	United States
1723	Northern Germany in general		1850	Mexico
1724	Southern Germany in general		1855	Central America
1725	Other regions		1860	South America
1726	Berlin			
1727	Other cities			Other guidebooks. See with the appropriate history classes.
	Austria-Hungary			
1730	General			
1731	Austria			
1732	Hungary			
1733	Eastern Alps			
	[See also Geog 1736]			

Geography

	Historical and political geography. See H 1930-1969
	Anthropogeography. See An 375-380
	Commercial and economic geography. See Econ 6917-6920
	History of geography
	[Include here works on the history and growth of geographical knowledge, the history of geographical discovery and exploration, etc.]
3000	Pamphlet volumes
	Periodicals and Societies. See Geog 1-296
	Bibliographies. See Geog 500
	General history
	[Include here works covering all periods and works on the history of discoveries since 1600.]
3015-3020	General works (By date)
3025	Commentary on maps
3030	Special topics
	Local history. See Geog 3400-3699
	Mythical geography, Geographical myths
3045-3050	General works (By date)
3055	Atlantis
3057	Ophir
3060	Other special
	[Include Isle of Pines, Mu, Lemuria, etc.]
	History of cartography
3070	General works
	Local history. See Geog 3400-3699
	Commentaries on specific maps. See Geog 3025, 3135, 3235, and 3251
	Treatises on mapmaking. See Eng 525-535
	Ancient geography, to ca.300 A.D.
3105-3110	Description of the ancient world (By date)
	[N.B. - See also AH, etc.]
	History of ancient geography
3125-3130	General works (By date)
	[N.B. - See also AH]
3135	Commentary on maps
3140	Special topics
	Medieval and Renaissance geography, 300-1600
	Description of the medieval world
3199	Before 1399
3204-3210	Later works (By date)
	History of medieval geography
3225-3230	General works (By date)
3235	Commentary on maps
3240	Special topics
	Local history. See Geog 3400-3699
	History of renaissance geography
3245-3250	General works (By date)
3251	Commentary on maps
3255	Special topics
	Local history. See Geog 3400-3699
	Modern geography, since 1600
	Description of the modern world. See Geog 755-760 and 805-960
	History of modern geography
	General works. See Geog 3015-3030
	Local history. See Geog 3400-3699
	Local history
	[N.B. - For each area or country, include histories of its geographical research and activities; also works on the geographical explorations and discoveries made by the people of that area or country. For biographies of geographers, see Geog 600-626.]

	History of geography (cont.)
	Local history (cont.)
	North America
3400	General works
3405	Canada
3410	United States
	West Indies
3415	General works
3416	Cuba
3417	Puerto Rico
3418	Jamaica
3421-3446	Other islands (A-Z by island)
3448	Mexico
	Central America
3450	General works
3451	British Honduras
3452	Costa Rica
3453	Guatemala
3454	Honduras
3455	Nicaragua
	Panama. See Geog 3470
3457	Salvador
	South America
3460	General works
3461	Argentina
3462	Bolivia
3463	Brazil
3464	Chile
3465	Colombia
3466	Ecuador
3467	British Guiana, Guyana
3468	Dutch Guiana, Surinam
3469	French Guiana
3470	Panama
3471	Paraguay
3472	Peru
3473	Uruguay
3474	Venezuela
	Oceania
3480	General works
3481	Australia
3482	New Zealand
3483	Indonesia
3485	Philippines
3489	Other islands
	Europe
	General works. See Geog 3015-3030, 3070, 3125-3140, 3225-3255
	Ancient Greece and Rome. See Geog 3125-3140 and AH
	Low countries
3505	General works
3506	Belgium
3508	Luxemburg
3510	Holland
	Great Britain
3520	General works, England
3522	Wales
3524	Scotland
3526	Ireland
3530	France
3535	Switzerland
3540	Germany
3542	Austria
3544	Hungary
3546	Rumania
3550	Italy
	Spain and Portugal
3560	General works
3570	Portugal
3580	Spain
3590	Russia, Soviet Union
	[See also Geog 3610-3616]
3594	Poland

Geography

History of geography (cont.)
    Local history (cont.)
        Europe (cont.)

3596	Czechoslovakia
	Scandinavia
3600	General works
3602	Norway
3604	Sweden
3606	Denmark
3608	Iceland
	Baltic States
3610	General works
3612	Lithuania
3614	Latvia
3616	Estonia
3618	Finland
3620	Yugoslavia
3622	Bulgaria
3624	Albania
3626	Greece
	Turkey. See Geog 3632
	Asia
3630	General works
	[Include also Near East in general. See also Geog 3240 for the history of medieval Muslim geography in general.]
3632	Turkey
3633	Syria
3634	Lebanon
3635	Palestine, Israel
3636	Jordan
3637	Arabia, Saudi Arabia
3638	Persian Gulf States
3640	Iraq
3642	Iran
	Russian Asia. See Geog 3590
3648	Afghanistan
3650	India
3652	Pakistan
3654	Ceylon
3655	Burma
3656	Siam, Thailand
3657	Malaya
	Indonesia. See Geog 3483
3660	Indochina, Vietnam
3662	Laos
3663	Cambodia
3665	China
3668	Korea
3670	Japan
3672	Formosa, Taiwan
	Africa
3680	General works
	[Include North Africa in general]
3681	Morocco
3682	Algeria
3683	Tunisia
3684	Libya
3685	Egypt
3687	Ethiopia
3690	South Africa (Union and Republic)
3699	Other Africa

Adventure and adventurers in general
    [N.B. - Include here only very general works. See also Voyages and travels, Mountaineering, and Ocean life in Geog below and in the various history classes.]

	Periodicals. See Geog 1-296
4005	Bibliographies
4015-4020	General works (By date)
	[Include general history of adventure]
	Biographies of adventurers
	[Include only such works as cannot be classed elsewhere. See also Geog 4135-4145, 4520-4521, etc.]
4028	Collected
4029	Individual
	[Include also a person's memoirs]

Voyages and travels in general
    [N.B. - Include here only voyages and travels which cannot be placed in one of the country history classes.]

	Periodicals. See Geog 1-296; also Geog 4155-4182
	Bibliographies. See Geog 500
4100	General pamphlet volumes
	Art and history of travel
4105	Travelling instructions, handbooks
4110	General guidebooks
4125-4130	General works (By date)
4131	Tourism, tourist trade (By date, e.g. .65 for 1965)
4132	Other special topics
	Biographies of travellers
4135-4140	Collected (By date)
4145	Individual (99 scheme, A-Z by person)
	Collections of travels
4155-4160	Multivolume works (By date)
4165-4170	Smaller works (By date)
	Hakluyt Society
4179	Miscellany by and about the Society
4180	Publications, Series I
4181	Publications, Series II
4182	Linschoten Vereeniging Werken
4184	Hakluyt Society Publications, Extra Series
	Individual travels
	[N.B. - When such works are in a date arrangement, the date of the beginning of the voyage or travel is to be used. Editions are to be kept together.]
4205-4210	Miscellaneous travels (By date of travel)
	[Include here travels which take in various parts of the world.]
4215-4220	Circumnavigations (By date of travel)
	[Include voyages around the world by ship or on land. For Magellan's voyage around the world, see US 2412. Airplane flights around the world go in Eng 5508-5510, or in Geog 4205-4210 if the travel aspect is emphasized.]
	Americas in general. See US and SA
4225-4230	America and other regions (By date of travel)
	[Travel books which touch on America but do not devote much space to it should go in Geog 4205-4210.]
4235-4240	Atlantic Ocean (By date of voyage)
	[Include also works on the history of the Atlantic Ocean in general. Works on individual islands or groups of islands go in the appropriate history class.]
4245-4250	Pacific Ocean (By date of voyage)
	[Include also works on the history of exploration of the Pacific Ocean as a whole. See also Oc 1-23 for general histories of the Pacific Ocean.]
	Europe in general
	Bibliographies. See Geog 500
4260	Pamphlet volumes
4264-4270	Descriptive geographies (By date)
	[N.B. - See also Geog 4301-4310 for travel books]
4275-4280	History of geography and travel in Europe (By date)
4300	Guidebooks and handbooks
4301-4310	Travels (By date of travel)

Geog — Geography

Voyages and travels in general (cont.)
  Individual travels (cont.)
    Europe in general (cont.)
      Special regions
        [N.B. - Prefer the appropriate history class whenever possible]

4311	Danube River
4312	North Sea
4313	Baltic Sea
4314	Black Sea
4321-4330	Mediterranean Sea (By date of travel)

          [Include also works dealing with all the islands of the Mediterranean and with the lands surrounding the Mediterranean. Works on the Levant in its broadest sense, i.e. the Mediterranean lands east of Italy, go here. See also Ott for works on the Levant in its more restricted sense.]

4332	Adriatic Sea
	Aegean Sea. See MG

    Europe and other regions

4341-4350	Europe and Asia (By date of travel)
	[See also Asia and Ott]
4354-4360	Europe and Africa (By date of travel)
4364-4370	Europe and Oceania (By date of travel)
	Europe and America. See Geog 4225-4230

  Africa in general. See Afr

  Africa and other regions
    Africa and America. See Geog 4225-4230
    Africa and Europe. See Geog 4354-4360
    Africa and Asia. See Geog 4411-4420

  Asia in general. See Asia

  Asia and other regions

4411-4420	Asia and Africa, Asia and Oceania (By date of travel)
	Asia and Europe. See Geog 4341-4350
	Asia and America. See Geog 4225-4230
4424-4430	Indian Ocean (By date of travel)
	Red Sea. See Geog 4411-4420
	Pacific Ocean. See Geog 4245-4250
4450	Voyages to unidentified places
4474-4480	Imaginary voyages (By date)

    [N.B. - For geographical myths, see Geog 3045-3060.]

Mountaineering in general
  Periodicals. See Geog 1-296

4500	Bibliographies
4502	Gesellschaft Alpiner Bücherfreunde publications
	[N.B. - This is a special series.]
4506-4510	General works (By date)

    [Include general works on methods of mountain climbing and general histories of mountaineering.]

  Literary works

4512.1-.90	Anthologies, collections
4512.100-.899	Individual works (800 scheme, A-Z)
4515	Special topics

    [Include use of ropes, skiing in the mountains, etc.]

  Biographies of mountaineers

4520	Collected
4521	Individual

  Local mountaineering. See the appropriate history class, e.g. SA, Swi, US

Ocean life in general
  Periodicals. See Geog 1-296

4550	Bibliographies
4556-4560	General works (By date)

    [Include general works about and histories of seafaring life and ocean travel; also accounts of voyages not limited to a particular ocean or sea and general reminiscences of life at sea. Prefer to class works about whaling in F.]

  Biographies of seamen

4570	Collected
4571	Individual

    [N.B. - Autobiographies and diaries have gone in Geog 4556-4560. See also the appropriate history classes.]

  Local voyages. See Geog 4235-4250, 4311-4332, 4424-4430, etc.

  Shipwrecks and other marine disasters
    Bibliographies. See Geog 4550

4652	Dictionaries, etc.
4656-4660	General history (By date)
	Local history in general. See Geog 4656-4660
4672	Pamphlet volumes; Collected narratives
4675-4680	Individual wrecks, etc. (By date of event)

    [N.B. - Shipwrecks followed by accounts of adventures on shore are classed locally, e.g. Afr, SA, etc.]

  Piracy and treasure trove in general
    Bibliographies. See Geog 4550

4706-4710	General works (By date)

    Local history. See the appropriate history classes, e.g. Afr, SA, etc.

Tropics in general
  Periodicals. See Geog 1-296

4750	Bibliographies
4756-4760	General works (By date)

  Special regions. See Afr, SA, etc.

Arctic and Antarctic Regions in general
  [N.B. - Class here only works dealing with both the Arctic and Antarctic regions together. Include historical and geographical works and works on exploration in general.]

5050	Bibliographies
5055	Pamphlet volumes
5057	Periodicals and Societies
5060	History of polar exploration
	Biographies of explorers. See Geog 5310-5336 and 5980-5985
5070	General treatises

    [Include geographical works, general works on methods of polar exploration, and collected voyages.]

  Special topics

5080	Sovereignty
5100	Essays, addresses
5110	Miscellaneous speculations
5120	Instructions for explorers

    [Include works on equipment, hygiene, etc.]

5130	Aerial exploration

Arctic Regions
  [Include the North Polar area]

5150	Bibliographies
5153	Periodicals and Societies
5155	Pamphlet volumes
5160	General works, geography and description

  Special topics

5170	Theory of North Polar exploration
5180	Miscellaneous essays, etc.

	Arctic Regions (cont.)
	Special topics (cont.)
	Eskimos
5182	General works
	Local. See US, Can, and Geog 5855
	History of exploration and discoveries
5208-5210	General works
	Special periods
5220	Before 1800
5225	19th century
5235	20th century
5300	By special countries
	Biographies of explorers
	[See also Geog 5980 and 5985]
5310	Collected
5311-5336	Individual (A-Z by person)
	[N.B. - For Sir John Franklin, see Geog 5530. Narratives of their expeditions go with the expeditions in Geog 5375-5400 and 5515-5560.]
5340	Collected Arctic voyages
5350	Northeast and Northwest Passages together
	Eastern hemisphere
	Bibliographies. See Geog 5150
	Northeast Passage
5365-5370	General works (By date)
5375-5380	Special expeditions (By date of voyage)
5395-5400	Polar expeditions (By date of expedition)
	[Include here arctic and polar explorations made in the eastern hemisphere, i.e. above Europe and Asia. See also Geog 5555-5560.]
	Local description, etc.
	Franz Joseph Land. See Slav 3280
	Novaya Zemlya. See Slav 3280
	Spitzbergen, Svalbard. See Scan 2180-2189
	Other local. See Slav 3280 and 3650
	Western hemisphere
	Bibliographies. See Geog 5150
	Northwest passage
5505-5510	General works (By date)
5515-5520	Special expeditions (By date of voyage)
	Sir John Franklin's voyage of 1845-1846
5530	General works; Biography of Franklin
	Search for Franklin
5535	History
5538	Special voyages (By date of voyage)
5555-5560	Polar expeditions (By date of expedition)
	[Include here arctic and polar explorations made in the western hemisphere, i.e. above Alaska and Canada. See also Geog 5395-5400.]
	Local description, etc. See US and Can
	Greenland
5600	Bibliographies
5603	Pamphlet volumes
5606	Periodicals and Societies
5610	Collected source materials
5615	History of archives
5617	Historiography
	Government and administration
5620	General works
5623	Special topics
	History
5636-5640	General works (By date)
5645	General special
	Special periods
5650	Before 1600
5660	1600-1900
5670	1900-

	Arctic Regions (cont.)
	Greenland (cont.)
5700	Religion, Missions
5835-5840	Geography, description and travel (By date)
5845	General economic conditions
	[See also Econ for special topics]
5850	Civilization, Social life and conditions
5855	Races
5861-5886	Local history, etc. (A-Z by place)
5892	Genealogies
	Biographies
5896	Collected
5898	Individual (299 scheme, A-Z by person)
	Antarctic Regions
	[Include the South Polar area]
5900	Bibliographies
5905	Pamphlet volumes
5907	Periodicals and Societies
5918-5920	General works, geography and description (By date)
5925	Special topics
	[Include question of sovereignty, theory and methods of antarctic exploration, etc.]
	History of exploration and discoveries
5938-5940	General works (By date)
	Special periods. See Geog 5938-5940
5970	By special countries
	Biographies of explorers
5980	Collected
5985	Individual (363 scheme, A-Z by person)
	[N.B. - Narratives of their expeditions go with the expeditions in Geog 6008-6010.]
6005	Collected expeditions
6008-6010	Special expeditions (By date of voyage)
6100	Local history, description, etc. (99 scheme, A-Z by place)
	[Include areas on the antarctic continent and isolated islands or island groups in the Antarctic Ocean, such as Crozet, Heard, and Kerguelen. See also Oc and SA.]

An Anthropology An

NOTE ON THE CLASSIFICATION

The An class provides for books of a general nature relating to Anthropology and Ethnology. It should be noted at once that the Tozzer Library of the Peabody Museum is Harvard's major research library for anthropology. Periodicals and the serial publications of anthropological societies are in the Sci class. Works on special races and local ethnology are in the H class and the various special history classes, such as Afr, Br, Ger, Slav, US.

The accompanying Outline shows the three major divisions of the An class. The second section, Prehistoric and primitive man, is primarily for general works on cultural and social anthropology, but with the emphasis on the primitive aspects. The Arc class contains books on the more developed prehistoric civilizations and especially on local prehistoric civilizations, and also works on the excavations of prehistoric remains. The section on somatology again is largely made up of general material.

Most topics in this class have a numbering which allows for a date arrangement of the books. In contrast to the more usual practice of keeping editions together by the date of the first edition, the An class follows the practice in the other science classes (Astr, Chem, Math, Phys) of arranging the books strictly by the date of issue, on the theory that later editions would have more up-to-date contents. Thus, editions of a book are not kept together here, except in the case of photomechanical reproductions.

During the present revision of the scheme, no major changes have been made. But many notes have been added to clarify the situation for the classifiers. It has not been possible as yet to reclassify all the books which had been wrongly classified.

Bartol Brinkler
Classification Specialist
March 1975

OUTLINE

	Anthropology and ethnology in general
5-40	Bibliographies; Dictionaries
96-150	History; Biographies
170-360	General works
375-380	Anthropogeography
606-630	Museums; Research methods
	Prehistoric and primitive man
2052	Bibliographies
2056-2060	Antiquity and origin of man
2106-2410	Primitive customs and institutions
2600	Wild men
	Somatology, Physical anthropology
3202	Bibliographies
3206-3210	General works
3305-3410	Anthropometry; Fingerprints
3506-3560	Osteology; Craniology

Anthropology and ethnology in general
    Periodicals and Societies. See Sci 3050-3125
5   Bibliographies
        [Include bibliographical periodicals. See also
        An 125-150 for bibliographies of a person.]
36-40   Dictionaries (By date of issue)
96-100   History (By date of issue)
        [Include general histories, histories of special
        periods, and histories of anthropological theory
        and activities in a particular country.]

    Biographies of anthropologists
124   Collected
125-150   Individual (A-Z by person)
        [Include a person's autobiography, memoirs,
        journals, bibliographies, etc. But a person's
        collected writings go in An 175-200.]

    Collected essays, etc.
170   Several authors (299 scheme, A-Z)
        [Include congresses, miscellaneous readings,
        etc.; also Festschriften, alphabetized by
        person honored.]
175-200   Individual authors (A-Z)
        [Include a person's collected and
        selected works, volumes of his essays, etc.
        For biographies, see An 125-150. For
        Festschriften, see An 170.]
205   General pamphlet volumes

    General treatises
[346-350]   Folios [Discontinued]
355-360   Monographs (By date of issue)
        [Include here general works on race
        and racial characteristics, general
        ethnology and ethnography. See also
        H 1970-1989 for purely historical works on
        ethnography. See also the various history
        classes for works on the races of a country.
        See also Phil 5893 for race psychology; also
        Soc 636 for race problems in general.]

    Anthropogeography
        [N.B. - Include here general works only.
        See the various history classes for works on
        the races of a country.]
375   Pamphlet volumes
376-380   General works (By date of issue)
        [Include human geography, social geography in
        general.]

    Local. See Br, Ger, Slav, US, etc.
606-610   Anthropological museums (By date of issue)
        [Include general works about museums; also
        catalogues and histories of museums.]
626-630   Research methods and apparatus (By date of issue)
    Local ethnology, Special races. See H and
        the appropriate history classes

Prehistoric and primitive man
    Periodicals. See Sci 3050-3125
2052   Bibliographies
2056-2060   Antiquity and origin of man (By date of issue)
        [Include works on theories of the origins of
        man, works on early hominids, etc. For general
        works on prehistoric man, prefer An 2106-2110
        and Arc.]

    Primitive customs and institutions
        [Include here primarily works on the primitive
        aspects of prehistoric culture and on the
        culture of modern primitive societies. For
        works on the more developed prehistoric
        civilizations, prefer Arc. All local material
        goes in Arc or the appropriate history class.]
2106-2110   General works (By date of issue)
        [Include social and cultural anthropology in
        general; also works on primitive economic,
        political, and social customs and institutions.]
2156-2160   Science and technology (By date of issue)

Prehistoric and primitive man (cont.)
 Primitive customs and institutions (cont.)
2206-2210	Medicine (By date of issue)
	Religion. See R
2306-2310	Burial (By date of issue)
2336-2340	Arms and armor, tools (By date of issue)
2356-2360	Cannibalism (By date of issue)
2406-2410	Law (By date of issue)
[2493]	Early man [Discontinued. See An 2056-2060]
2600	Wild men, Wolf children

Somatology, Physical anthropology
	Periodicals. See Sci 3050-3125
3202	Bibliographies
3206-3210	General works (By date of issue)
3305-3310	Anthropometry (By date of issue)
	[Include also works relating to a particular country.]
3406-3410	Fingerprints (By date of issue)
	[Include also palm and sole prints]
3506-3510	Osteology, Skeleton (By date of issue)
	Craniometry, Skull
[3538-3539]	Folios [Discontinued]
3556-3560	Monographs (By date of issue)

 Local, Special races. See H and the appropriate history classes.

WIDENER LIBRARY SHELFLIST, 60

GEOGRAPHY
AND
ANTHROPOLOGY

CLASSIFIED LISTING BY CALL NUMBER

Classified Listing

Geog 1 - 296 Geography in general - Periodicals and Societies (Special Table)

	Geog 3.1	Acta arctica. København. 1-10,1943-1958 3v.
	Geog 3.25	Acta geographica. Kaapstad. 1,1967+
	Geog 3.30	Acta geographica Lovaniensia. Louvain. 1,1961+ 9v.
	Geog 3.35	Acta geographica. Paris. 1970+ 2v.
	Geog 3.35.2	Acta geographica. Tables générales, anciennes séries, 1947-1969. Paris, n.d.
	Geog 10.2	Akademiia Nauk SSSR. Institut Nauchnoi Informatsii. Gidrologiia rushi. Moskva. 1963+
	Geog 10.4	Akademiia Nauk SSSR. Institut Nauchnoi Informatsii. Itogi nauki: geografiia SSSR. Moskva. 1,1965+ 4v.
	Geog 10.5	Itogi nauki: geomorfologiia. Moskva. 2,1971+ 3v.
	Geog 10.6	Akademiia Nauk SSSR. Institut Geografii. Geograficheskie soobshcheniia. Moskva. 2,1961+
	Geog 10.7	Itogi nauki i tekhniki. Geografiia zarubezhnykh stran. Moskva. 1,1972+
NEDL	Geog 12.1	Allgemeine geographische Ephemeriden. Weimar. 1-51 51v.
	Geog 12.2	Alpine journal. London. 1,1863+ 46v.
	Geog 12.2.3	Alpine journal. Index. v.39-58, 1927-52. London, 1954.
	Geog 12.3	Alpine Club of Canada. The gazette. Banff. 1,1921
	Geog 12.4	Die Alpen. Monatsschrift des Schweitzer Alpenclub. Bern. 1,1915+ 50v.
	Geog 12.4.5	Die Alpen; Zeitschrift des SAC. Bern. 33,1957+ 14v.
X Cg	Geog 12.5	Almanach géographique, ou Petit atlas élémentaire. Paris. 1770
	Geog 12.6	Almanacco del turista. Roma. 1951-1960 4v.
	Geog 13.1	American Geographical Society of New York. Bulletin. 1-2,1852-1856
	Geog 13.2	Pamphlet vol. American Geographical Society of New York. Charter and by-laws. 15 pam.
	Geog 13.2.3	American Geographical Society of New York. Charter and by-laws. 1857
	Geog 13.2.5	American Geographical Society of New York. Charter, by-laws and list of members. N.Y. 1870
	Geog 13.2.10	American Geographical Society of New York. The role of geography in the modern world. N.Y., 195-.
	Geog 13.3	American Geographical Society of New York. Proceedings. 1-2
	Geog 13.4	American Geographical Society of New York. Journal. 1-47,1895-1915 45v.
	Geog 13.4.4	American Geographical Society of New York. Index to the Bulletin of the American Geographical Society. 1852-1910. N.Y., n.d.
	Geog 13.4.20	Wright, J.K. Geography in the making; the American Geographical Society, 1851-1951. N.Y., 1952.
	Geog 13.9	Association of American Geographers. Annals. Albany, N.Y. 1+ 47v.
	Geog 13.9.5	Association of American Geographers. Annals. Index for volumes 1-25. Lancaster, Pa.? n.d.
	Geog 13.10	Association of American Geographers. Handbook-directory. Washington. 1956
	Geog 13.12	American Bureau of Geography. Bulletin. Winona. 1900-1901
	Geog 13.200	Nederlandsch Aardrijkskundig Genootschap, Amsterdam. Tijdschrift. Amsterdam. 1,1900+ 72v.
NEDL	Geog 13.200	Nederlandsch Aardrijkskundig Genootschap, Amsterdam. Tijdschrift. Amsterdam. 1-17,1876-1899
	Geog 13.200.5	Nederlandsch Aardrijkskundig Genootschap, Amsterdam. Systematisch register. Leiden. 1876-1960 3v.
	Geog 13.201	Nederlandsch Aardrijkskundig Genootschap, Amsterdam. Bijbladen. Amsterdam. 1879-1883 3v.
	Geog 14.10	Annali di richerche e studi di geografia. Genova. 5,1949+ 11v.
	Geog 14.200	Annales de géographie. Paris. 1,1892+ 76v.
	Geog 14.200.15	Annales de géographie. Table décennale, 1891-1901. Paris, 1902- 3v.
NEDL	Geog 14.201	Annales maritimes. 99v.
NEDL	Geog 14.204	Nouvelles annales de la marine. 1849-1864 15v.
NEDL	Geog 14.205	Revue maritime. 1861-1899
	Geog 14.205	Revue maritime. 1900-1971//
NEDL	Geog 14.206	Revue maritime. Table. 1861-1888 2v.
NEDL	Geog 14.207	Annales des voyages. Paris. 1808-1814 25v.
NEDL	Geog 14.208	Annales des voyages. Table. Paris. 1813
NEDL	Geog 14.209	Nouvelles annales des voyages, géographiques, historiques, et archéologiques. Paris. 1819-1854 124v.
NEDL	Geog 14.210	Nouvelles annales des voyages, géographiques, historiques, et archéologiques. Paris. 1855-1870 42v.
	Geog 14.500	L'année cartographique. 1-23
	Geog 14.501	L'année geographique. 1862-1878 16v.
	Geog 14.550	Annuaire des voyages et de la géographie. Paris. 1844-1845 2v.
	Geog 14.600	Société Royale de Géographie d'Anvers. Bulletin. Anvers. 1-75,1877-1964 30v.
NEDL	Geog 14.600	Société Royale de Géographie d'Anvers. Bulletin. Anvers. 1-23,1877-1899 23v.
	Geog 15.2	Antarctic; a news bulletin published quarterly by the New Zealand Antarctic Society. Wellington. 1,1956+ 5v.
	Geog 15.5	Antarctica, the last frontier; the annual report of the officer in charge, United States Antarctic Programs. Washington.
	Geog 16.1	Appalachia. Boston. 1,1879+ 39v.
	Geog 16.1.2	Appalachia. Index, v.1-10. Boston, 1906.
	Geog 16.1.5	Appalachian Mountain Club. Register. 1888-1955 13v.
	Geog 16.1.7	Appalachian Mountain Club. Bulletin. 1-27,1907-1934 27v.
	Geog 16.1.8	Appalachian Mountain Club. Bulletin. 1935+ 26v.
	Geog 16.1.9	Pamphlet vol. Appalachian Mountain Club.
	Geog 16.1.9.5F	Pamphlet vol. Appalachian Mountain Club.
	Geog 16.1.10	Appalachian Mountain Club. Annual report.
	Geog 16.2	Appalachian Trail Conference. Publication. Washington. 4,1942 3v.
	Geog 16.2.5	Appalachian trailway news. Washington. 25,1964+ 4v.
	Geog 16.2.9	Pamphlet vol. Appalachian Trail Conference. Washington. 2 pam.
NEDL	Geog 18.1	Archiv für Geographie, Historie. Wien. 1-19,1810-1828 19v.
NEDL	Geog 18.2	Neues Archiv für Geschichte. Wien. 1-2,1829-1830 2v.
	Geog 18.3	Österreichisches Archiv für Geschichte. Wien. 1-3,1831-1833 3v.
	Geog 18.4	Österreichisches Zeitschrift für Geschichte. Wien. 1-3,1835-1837 3v.
	Geog 18.5	Arktis; Vierteljahrsschrift der Internationalen Studiengesellschaft zur Erforschung der Arktie. Gotha. 1-4,1928-1931 2v.

Geog 1 - 296 Geography in general - Periodicals and Societies (Special Table) - cont.

	Geog 18.10	Academia...Buenos Aires. Academia Argentina de Geografia, Buenos Aires. Anales. 1-6,1957-1962// 3v.
	Geog 18.150	Arctic Institute of North America. Bulletin. Montreal.
	Geog 18.155	Arctic; journal of the Arctic Institute of North America. Montreal. 1,1948+ 21v.
	Geog 18.160	Arctic Institute of North America. Special publication. Washington. 1-4,1952-1962//? 3v.
	Geog 19.2	Associated Mountaineering Clubs of North America. Bulletin. N.Y. 1918-1921
	Geog 19.3	Associação dos Geografos Brasileiros. Anais. 1,1945+ 8v.
	Geog 19.5	Association of Indian Geographers. Bulletin. New Delhi. 1-2,1956-1957
	Geog 19.5.10	Association of Indian Geographers. The Indian geographer. New Delhi. 1957
	Geog 21.1	Royal Geographical Society of Australasia. Proceedings and transactions. Queensland. 1-62,1885-1964 9v.
	Geog 21.1.40	Royal Geographical Society of Australasia. Victorian Branch. Transactions. Melbourne. 15-19,1898-1901
	Geog 21.15	The Australian geographer. Sydney. 8,1960+
	Geog 21.25	The Australian geographical record. Armidale. 1-7
	Geog 21.26	Australian geographical studies. Melbourne. 1,1963+ 5v.
	Geog 28.1	Berne. Geographische Gesellschaft. Jahresbericht. Bern. 1,1879+ 14v.
NEDL	Geog 28.2	Berlin. Gesellschaft für Erdkunde. Verhandlungen. Berlin. 1-27,1873-1896 25v.
NEDL	Geog 28.2.5	Berlin. Gesellschaft für Erdkunde. Übersicht der Aufsatze...in den Monatsberichten über die Verhandlungen. Berlin. 1863
NEDL	Geog 28.3	Berlin. Gesellschaft für Erdkunde. Übersicht der Aufsatze. Zeitschrift. Berlin. 1884-1921 15v.
	Geog 28.3.40	Berlin. Freie Universität. Geographisches Institut. Abhandlungen. 1-16 8v.
	Geog 28.3.50	Berlin. Universität. Verein der Studierenden der Geographie. Mitteilungen. Berlin. 1-2,1915-1918
	Geog 28.3.60	Berliner geographische Arbeiter. Berlin.
	Geog 28.4	Jonard, E.F. Notice sur l'etablissement géographique de Bruxelles. v.1-4. Paris, n.d.
	Geog 28.4.3	Lettre sur l'etablissement géographique de Bruxelles. Bruxelles, 1836.
	Geog 28.4.5	Brussels. Université Nouvelle. Institut Géographique. Publication. Bruxelles.
	Geog 28.6	Berg und Buch. München. 1,1928
	Geog 28.7	Bersteiger Almanach. München.
	Geog 28.8	Frankische Geographische Gesellschaft. Mitteilungen. Erlangen. 1,1954+ 12v.
	Geog 30.2	Paschin, O. Bibliotheca geographica. Berlin. 1-19,1891-1912 19v.
	Geog 30.5	Boletin de estudios geograficos. Mendoza. 2,1950+ 6v.
	Geog 30.7	Boletin paulista de geografia. Sao Paulo, Brazil. 1,1949+ 8v.
	Geog 32.5	Bombay Geographical Society. Transactions. Bombay. 1-19,1836-1873 19v.
	Geog 32.5.5	Bombay Geographical Society. Transactions. Index, v.1-17. Edinburgh, 1866.
	Geog 32.10	Bombay geographical magazine. Bombay.
	Geog 32.50	Brazil. Diretoria do Serviço Geográfico do Exército. Anuário. 1962+
	Geog 32.60	British Columbia geographical series. Vancouver. 3,1965+ 2v.
	Geog 32.90	Budapest. Tudomány-Egyetem. Annales. Sectio geographica. Budapest. 1,1965+ 2v.
	Geog 33.1	Bulletin des sciences géographiques. Paris. 1-28 28v.
	Geog 33.10	Bulgarska Akademiia na Naukite, Sofia. Geografska Institut. Izvestiia. 1,1951+ 8v.
	Geog 34.5	Bulgarsko Geografsko Druzhestvo, Sofia. Izvestiia. 1,1933+ 11v.
	Geog 34.15	Bruenn. Universita. Přérodovědecká Fakulta. Geographia. Praha. 5,1971+
	Geog 35.2	Pamphlet vol. Société Khédivienne de Géographie.
	Geog 35.3	California. Geographical Society. Bulletin. 1-2
	Geog 35.4	California. University. Publications in geography. Berkeley, Calif. 1,1913+ 19v.
	Geog 35.6	Les cahiers d'Outre-mer. Bordeaux. 1,1948+ 21v.
	Geog 35.8	Cahiers de géographie; publication de l'institut d'histoire et de géographie, Université Laval. Québec.
	Geog 35.9	Cahiers de géographie de Québec. Québec. 1,1955+ 13v.
	Geog 36.2	The Canadian Alpine journal. Banff. 1,1907+ 23v.
	Geog 36.3	Canadian geographical journal. Ottawa. 1,1930+ 46v.
	Geog 36.3.5	Canadian geographical journal. Regional index of articles, 1930-59. Ottawa, 1960.
	Geog 36.15	Pamphlet box. Canadian Geographical Society.
	Geog 36.20	Revue canadienne de géographie, organe de la Société de geographie de Montréal et de l'Institut de géographie de l'université de Montréal. Montréal. 1,1947+ 13v.
	Geog 36.25	Canadian geographer. Ottawa? 1,1951+ 13v.
	Geog 36.25.2	Canadian Association of Geographers. Canadien geographer. Author and subject index. Montreal. 1951+
	Geog 36.26	Canadian Association of Geographers. Occasional papers in geography. Vancouver. 1-7,1960-1965//
NEDL	Geog 39.1	Club Alpin Française. Annuaire. Paris. 1-30,1874-1903 30v.
NEDL	Geog 39.1.3	Club Alpin Française. Annuaire. Table générale. Paris. 1874-1888
	Geog 39.1.5	Pamphlet vol. Club Alpin Française. 2 pam.
	Geog 39.1.10	La montagne. Paris. 1904+ 38v.
NEDL	Geog 39.2	Club Alpino Italiano. Bollettino. Torino. 1-41 13v.
NEDL	Geog 39.2.5	Club Alpino Italiano. Indice generale dei cinquanta primi numeri (dal 1865 al 1884). Torino, 1885.
	Geog 39.2.10	Club Alpino Italiano. Revista mensile publicata per cura de consiglio direttivo della sede centrale. Torino. 1-60,1882-1941 37v.
NEDL	Geog 39.2.10	Club Alpino Italiano. Revista mensile publicata per cura de consiglio direttivo della sede centrale. Torino. 1-18,1882-1899 18v.
	Geog 39.2.11	Club Alpino Italiano. I cento anni del club alpino italiano. 2. ed. Milano, 1964.
	Geog 39.2.16	Club Alpino Italiano. Sezione di Roma. Annuario. Roma. 1-3,1886-1889
	Geog 39.2.18	Centro Alpinistico Italiano. Sezione di Biella. Annuario. Biella.
	Geog 39.2.20	Club Alpino Italiano. Sezione Antonio Locatelli, Bergamo. Annuario. 1951+ 4v.
	Geog 40.2	Congresso Geografico Italiano. Atti. Genova. 1892-1907 39v.

Classified Listing

Geog 1 - 296 Geography in general - Periodicals and Societies (Special Table) - cont.

Geog 40.9 Congresso Brasiliero de Geografia. Miscellaneous publications.

Geog 40.10 Congresso Brasileiro de Geografia. 4 annaes. Pernambuco. 1-2 3v.

Geog 40.15 Colton's journal of geography and collateral sciences. N.Y.

Geog 40.17F Cook's excursionist. English edition. London.

Geog 40.17.5F Cook's excursionist. American edition. N.Y.

Geog 42.1 Die Erde; Zeitschrift der Gesellschaft für Erdkunde zu Berlin. Berlin. 1,1949+ 20v.

Geog 45.1 Deutsche Geographentage. Verhandlungen. Berlin. 1-39 25v.

Geog 45.2 Deutsche geographische Blätter. Bremen. 13

Geog 45.5 Deutscher und Österreichischer Alpenvereins. Mittheilungen. Salzburg. 1875-1929 24v.

NEDL Geog 45.5 Deutscher und Österreichischer Alpenvereins. Mittheilungen. Salzburg. 1-25,1875-1899 25v.

Geog 45.6 Deutscher und Österreichischer Alpenvereins. Zeitschrift. Salzburg. 1-74,1869-1949 40v.

NEDL Geog 45.6 Deutscher und Österreichischer Alpenvereins. Zeitschrift. Salzburg. 1-30,1869-1899 30v.

Geog 45.6.7 Deutscher und Österreichischer Alpenvereins. Zeitschrift. Beilagen. 1869-1910

Geog 45.9 Pamphlet box. Deutscher und Oesterreichischer Alpenvereins.

Geog 45.10 Deutscher und Österreichischer Alpenvereins. Register zu den Vereinsschriften. Innsbruck. 1906

Geog 45.12 Deutscher Geographentag, 38th, Erlangen and Nuremberg, 1971. Tagungsbericht und wissenschaftliche Abhandlungen. Wiesbaden, 1972.

Geog 48.1 Dresden. Vereins für Erdkunde. Jahresbericht. Dresden. 1-27 12v.

Geog 48.1.3 Dresden. Vereins für Erdkunde. Mitteilungen. Dresden. 1905-1935 5v.

Geog 48.1.7 Dresden. Vereins für Erdkunde. Mitgleider-Verzeichnis. Dresden.

Geog 48.1.9 Dresden. Vereins für Erdkunde. Festschrift. Dresden. 1888

Geog 51.5 Durham, England. University. Durham Colleges. Department of Geography. Occasional papers series.

Geog 51.5.10 Durham, England. University. Department of Geography. Occasional publications. Durham. 1,1973+

Geog 55.1 Edinburgh. University. Department of Geography. Papers. 1960-1962

Geog 60.3 Erdkunde; Archiv für wissenschaftliche Geographie. Bonn. 5,1951+ 17v.

Geog 60.3.5 Erdkunde; Archiv für wissenschaftliche Geographie. Gesamtregister, v.1-17, 1947-63. Bonn, 1965.

Geog 62.2 L'esploratore. Milano. 1-10,1876-1886 10v.

Geog 62.3 Esplorazione commerciale. Milano. 2-32,1887-1917 31v.

Geog 63.5 Estudios geograficos. Madrid. 1,1940+ 30v.

Geog 63.5.2 Estudios geograficos. Madrid. 22-77,1950-1961 2v.

Geog 64.2 Europa. Kjøbenhavn. 1-3,1895-1902 3v.

Geog 64.3 L'exploration. Paris. 1-18,1876-1884 16v.

Geog 64.4F L'explorateur. Paris. 1-4,1875-1876 4v.

Geog 64.5 The explorers' journal. N.Y. 36,1958+ 9v.

Geog 68.2F Fels und Firn. München. 1925

Geog 69.5 Finisterra. Lisboa. 1,1966+ 5v.

Geog 70.2 Florida. University. Publication. Geography series. Gainesville.

Geog 72.5 Forschungen zur theoretischen Kartographie. Wien. 1,1971+ 2v.

Geog 75.1 Frankfurter Vereins. Jahresbericht - Geographie und Statistik. Frankfurt. 1905-1919

Geog 75.1.3 Frankfurter geographische Hefte. Frankfurt. 1,1927+ 14v.

Geog 78.1 Fennia. 1+ 52v.

Geog 78.1F Fennia. 16-66 3v.

Geog 78.1PF Fennia. 48,1929

Geog 78.2 Vetenskapliga. 1-11,1892-1920 6v.

Geog 79.2 Folia geographica. Series geographica-oeconomica. Kraków. 1,1968+ 2v.

Geog 81.1 France. Ministère de l'Instruction Publique. Bulletin de géographie historique et descriptive. Paris. 1886+ 58v.

Geog 81.1.3 France. Comité des Travaux Historiques et Scientifiques. Section de Géographie. Actes du congrès national des sociétés savantes. Paris. 84,1959+ 11v.

Geog 81.5 Union Geographique du France. Congrès national compte-rendu. 1879-1904 22v.

Geog 83.5 Focus, by the American Geographical Society. N.Y. 1,1950+ 3v.

Geog 84.5 Genoa. Università. Istituto di Scienze Geografiche. Pubblicazione. Genova. 7,1968+

Geog 85.1 Geneva. Société de Géographie. Mémoires. Paris. 1-104 43v.

Geog 85.1.2 Le globe. Table des matières. Genève. 71-90,1932-1951

Geog 85.2 Geografia. Karachi. 3,1964+

Geog 85.78 Geographica; collana di sussidi didattici e bibliografici. Roma.

Geog 85.80 Geographica. Uppsala. 1-38,1936-1968// 15v.

Geog 85.90 Geographica helvetica. Bern. 1,1946+ 17v.

Geog 85.95 Geographica slovaca. Bratislava. 1,1949

Geog 85.100 Geografisk tidsskrift. Kjøbenhavn. 29+ 20v.

Geog 85.100F Geografisk tidskrift. Kjøbenhavn. 1-28 28v.

Geog 85.100.5F Geografisk tidskrift. Index, 1-20. Kjøbenhavn, 1910.

Geog 85.105 Geografiska annaler. Stockholm. 1,1919+ 19v.

Geog 85.135 Geografiska Sällskspet i Finland, Helsingfors. Acta geographica. Helsinki. 1-27,1927-1972// 11v.

Geog 85.135.5 Terra; geografiska sällskapets i Finland tidskrift. Helsingfors. 1,1888+ 32v.

Geog 85.180.15 Pamphlet vol. Geographical Association, Great Britain.

Geog 85.195 The geographical magazine. London. 1,1935+ 53v.

Geog 85.195.2F The geographical magazine. Index, 1950-1972. London, 1972.

Geog 85.200 Geographical magazine. London. 1-5,1874-1878 5v.

Geog 85.201 Geographical review published by the American Geographical Society. N.Y. 1,1916+ 55v.

Geog 85.201.10 Geographical review. Index, 1-45, 1926-1957. N.Y., n.d. 4v.

Geog 85.202 Geographical Teacher. Geography; the magazine of the Geographical Association. London. 2,1902+ 41v.

Geog 85.203 Geographical teacher. Supplement. London. 1,1925

Geog 85.205 Geographical Society of the Pacific. Transcriptions and proceedings. San Francisco. 1902

Geog 85.206 Geographical Society of the Pacific. Bulletin. San Francisco. 1905

Geog 1 - 296 Geography in general - Periodicals and Societies (Special Table) - cont.

Geog 85.225 Geographische Gesellschaft in Hamburg. Mittheilungen. Hamburg. 1-61 47v.

Geog 85.225.2 Geographische Gesellschaft in Hamburg. Mittheilungen. Register. 1-60, 1873-1972. Hamburg, 1973.

Geog 85.228 Geographisches Jahrbuch. Gotha. 1-3,1850-1851

Geog 85.300 Geographisches Jahrbuch. Gotha. 1-61 60v.

Geog 85.301 Geographische Abhandlungen. Wien. 1886-1936 15v.

Geog 85.301PF Geographische Abhandlungen. Wien. 6 2v.

Geog 85.302 Geographische Zeitschrift. Leipzig. 1,1895+ 55v.

Geog 85.302.3 Erdkundliches Wissen. Wiesbaden. 1-19 9v.

Geog 85.303 Geographische Zeitschrift. Register. Leipzig. 1-20,1895-1914 2v.

Geog 85.303.2 Geographische Zeitschrift. Register. v.1-50, 1895-1944. Wiesbaden, 1970. 2v.

Geog 85.309F Genoa. Civico Istituto Colombiano. Bollettino. 1-4,1953-1956

Geog 85.310 Geographisch-Ethnographische Gesellschaft in Basel. Mitteilungen der geographisch-Ethnographischen Gesellschaft in Basel. Basel. 1-8 4v.

Geog 85.315 Geographical Society of Ireland. Bulletin. Dublin. 1,1944+ 5v.

Geog 85.350 Istituto Geografico de Agostini. Communicazione dell...la geografia. Novara. 1-18,1912-1930 9v.

Geog 85.365 Geographical studies. London. 1-5,1954-1958 3v.

Geog 85.370 Geografski glasnik. Zagreb. 8-28 7v.

Geog 85.372 Zagreb. Univerzitet. Geografski Institut. Radovi. Travaux. 7,1968+

Geog 85.373 Geographische Berichte; Mitteilungen der geographischen Gesellschaft in der Deutschen Demokratischen Republik. Berlin. 1,1956+ 9v.

Geog 85.375 The geographical digest. London. 1963+ 3v.

Geog 85.377 Belgrade. Geografski Institut. Zbornik radova. 18+ 5v.

Geog 85.379 Geographia Polonica. Warszawa. 1,1964+ 13v.

Geog 85.379.2 Geographia Polonica. Index, 1-32,1964-1975. Warszawa, n.d.

Geog 85.380 Geografski pregled. Sarajevo. 2,1958+ 3v.

Geog 85.382 Geografski zbornik. Ljubljana. 10,1967+ 6v.

Geog 85.383 Geografski razgledi. Skopje. 4,1966+ 5v.

Geog 85.384 Belgrade. Univerzitet. Prirodno-Matematičhi Fakultet. Geografski Zavod. Zbornik radova. Beograd. 13,1966+ 2v.

Geog 85.385 Geografsko Društvo, Belgrade. Podnežnica, Kragujevac. Geografski godišnjak. Kragujevac. 4,1968+

Geog 85.385.5 Spomenica o pedecetogodišnjici Srpskog geógrafskog društva, 1910-1960. Beograd, 1961.

Geog 85.386 Geograficheskii sbornik. Kazan. 2,1967+

Geog 85.387 Geograficheskii sbornik. Moskva. 1,1963+ 4v.

Geog 85.388 Geograficheskii sbornik. L'vov. 7,1963+

Geog 85.389 Zagreb. Univerzitet. Geografski Institut. Geographical papers. Zagreb. 1,1970+

Geog 85.390 Geographical report. Umeå. 2,1971+

Geog 88.25 Giornale popolare di viaggi. Weekly. Milano. 1-8,1871-1874 8v.

Geog 89.1F Globus; illustrirte Zeitschrift für Länder- und Völkerkunde. Braunschweig. 1-98,1862-1910 76v.

Geog 89.4 Der Globusfreund. Wien. 1-17 4v.

Geog 90.10 Goldthwaite's geographical magazine. N.Y. 2v.

Geog 90.15 Göttinger geographische Abhandlungen. Göttingen. 1,1948+ 19v.

Geog 91.1 Greifswald. Geographische Gesellschaft. Jahresbericht. Greifswald. 1-60,1882-1942 16v.

Geog 91.1.10 Braun, Gustav. Geographische Gesellschaft zu Greifswald, 1882-1927. Greifswald, 1932.

Geog 91.4 Grønlandske Selskab. Aarskrift. Kjøbenhavn. 1910-1952 10v.

Geog 91.5 The long trail news. Brandon, Vt. 2v.

Geog 95 Geographische Gesellschaft zu Hannover. Jahrbuch. Hannover. 1926+ 8v.

Geog 95.5 Geographische Gesellschaft zu Hannover. Jahrbuch. Sonderheft. Hannover. 2,1968+ 4v.

Geog 97.2 Harvard mountaineering. Cambridge, Mass. 10-14,1951-1959

Geog 97.10 Harvard Travellers' Club. Yearbook.

Geog 97.12 Pamphlet vol. Harvard Travellers' Club.

Geog 105.15 The geographer. Aligarh, India.

Geog 109.5 Institute of British Geographers. Publications. London. 1933-1975// 20v.

Geog 109.6 Institute of British Geographers. Transactions. New series. London. 1,1976+

Geog 110.01.8 Pamphlet vol. International Geographical Congress, 8th.

Geog 110.1 International Geographical Congress. v.1-8. Anvers, 1872. 23v.

Geog 110.1.13 International Geographical Congress, 13th, 1931. Livret-guide du congressiste. Paris, 1931.

Geog 110.1.13.5 International Geographical Congress. pt. A-B. Paris, 1931. 2v.

Geog 110.1.14 International Geographical Congress, 14th, Warsaw, 1934. Comptes rendus. v.1-7. Varsovie, 1934.

Geog 110.1.15 International Geographical Congress, 15th, Amsterdam, 1938. Comptes rendus du Congrès international de géographie, Amsterdam, 1938. v.1-5,21. Leiden, 1938. 6v.

Geog 110.1.16 International Geographical Congress. Comptes rendus du Congrès international de géographie. Lisbonne, 1950. 2v.

Geog 110.1.18 International Geographical Congress, 18th, Rio de Janeiro. Abstracts of papers. Rio de Janeiro, 1956.

Geog 110.1.18.5 International Geographical Congress, 18th, Rio de Janeiro, 1956. Comptes rendus du XVIIIe congrès international de géographie. Rio de Janeiro, 1959. 4v.

Geog 110.1.20 International Geographical Congress, 20th, London, 1964. Abstracts of papers. London, 1964.

Geog 110.1.20.2 International Geographical Congress, 20th, London, 1964. Abstracts of papers. Supplement. London, 1964.

Geog 110.1.20.5 International Geographical Congress, 20th, London, 1964. Congress programme. London, 1964.

Geog 110.1.20.10 International Geographical Congress, 20th, London, 1964. Congress proceedings. London, 1967.

Geog 110.1.20.15 Natsional'nyi Komitet Sovetskikh Geografov. XX mezhdunarodnyi geograficheskii kongress, London, iiul 1964 gg. Moskva, 1966.

Geog 110.1.21 International Geographical Congress, 21st, Delhi, 1968. Abstracts of papers. Supplement. Calcutta, 1968.

Geog 110.1.22 International Geographical Congress, 22nd, Montreal, 1972. Doklady k XXII mezhdunarodnomu geograficheskomu kongress (Kanada, avgust, 1972). Leningrad, 1972.

Geog 110.1.25 International Geographical Congress, 20th, London, 1964. Sovremennye problemy geografii. London, 1964.

Classified Listing

17

Geog 1 - 296 Geography in general - Periodicals and Societies (Special Table) - cont.

	Geog 110.1.500	Pamphlet vol. International Geographical Congress.
	Geog 110.2.3	United States. War Department. Corps of Engineers. Report of 3rd International Geographical Congress and Exhibition at Venice, Italy, 1881. Washington, 1885.
	Geog 110.2.17	International Geographical Congress, 7th, Berlin. Miscellaneous papers. v.1-6. Berlin, 1899.
	Geog 110.2.18	Pamphlet box. International Geographical Congress, 8th, Berlin.
	Geog 110.3	Congrès des Géographes et Ethnographes Slaves, 3rd, 1930. Zbornik radova III Kongresa slovenskih geografa i ethnografa u Kraljevini Jugoslaviji, 1930. Beograd, 1933.
	Geog 110.3.10	Congrès des Géographes et Ethnographes Slaves, 2nd, Krakow, 1927. Pamiętnik II zjardu słowiańskich geografów i etnografów odbytego w Polsce w roku, 1927. Kraków, 1929-30. 2v.
	Geog 110.3.25	Congrès des Géographes et Ethnographes Slaves, 4th, Sofia, 1936. Sbornik na IV Kongress na slavianskite geografi i etnografi. Sofiia, 1938.
	Geog 112.5	Istanbul. Üniversite. Cografya Ensitüsü. Coğrafi araştirmalar. n.p., n.d.
	Geog 113.5	International Geographical Union. The IGU newsletter. N.Y. 1-13,1950-1962
	Geog 114.5	Istituto Cartografico Italiano, Rome. Annuario dell'Istituto. Roma. 1-4,1884-1889
	Geog 114.7.85	Coën, A. Venticinque anni di Lavoro dell'Istituto Geographico Militare. Firenze, 1898.
	Geog 115.5	Istoriia geograficheskikh znanii i istoricheskaia geografiia. Etnografiia. Moskva. 2,1967+
	Geog 120.5	Jerusalem studies in geography. Jerusalem. 1,1970+
	Geog 121.1	The journal of school geography. Lancaster, 1897. 5v.
	Geog 121.1.2	Journal of geography. Lancaster. 1,1902+ 63v.
	Geog 121.1.6	The journal of geography. Index. 1897-1956. N.Y., 1922-58. 2v.
	Geog 121.2F	Journal des voyages et des aventures de terre et de mer. Paris. 1877-1909 30v.
	Geog 126.10	Kiel. Universität. Geographisches Institut. Schriften. Kiel. 1,1932+ 36v.
	Geog 128.2	Konigsberg. Universität. Geographisches Institut. Veröffentlichungen. Hamburg. 10v.
	Geog 128.2.5	Konigsberg. Universität. Geographisches Institut. Veröffentlichungen. N.F. Reihe Geographie. Konigsberg. 1-10,1931-1937 10v.
	Geog 128.2.8	Konigsberg. Universität. Geographisches Institut. Veröffentlichungen. N.F. Reihe Ethnographie. Neudamm. 1-2,1931-1932 2v.
	Geog 129.5	Krymskii Garnyi Klub, Odessa. Zapiski. Odessa. 1895-1912 2v.
	Geog 132.1	Kulturgeografi; tidsskrift. Kjøbenhavn. 5,1953+ 9v.
	Geog 132.4	Kiev. Universitet. Visnyk. Ser. heohrafiï. Kiev. 9,1967+
	Geog 133.1	Krakow. Wyższa Szkoła Pedagogiczna. Prace geograficzne. Kraków. 3,1964+ 2v.
	Geog 135.1	Gesellschaft für Erdkunde zu Leipzig. Verein für Erdkunde. Jahresbericht. 1884-1941 24v.
	Geog 135.2	Gesellschaft für Erdkunde zu Leipzig. Beiträge zur Geographie des festen wassers. v.1. Leipzig, 1891.
	Geog 135.2.2	Ratzel, Friedrich. Anthropogeographische Beiträge. v.2. Leipzig, 1895.
	Geog 135.2.3	Baumann, Oskar. Der Sansibar-Archipel. v.3. Leipzig, 1896.
	Geog 135.2.4	Ratzel, Friedrich. Beiträge zur Geographie des mittleren Deutschland. v.4. Leipzig, 1899.
	Geog 135.2.5	Ule, Willi. Der Würmsee (Starnbergersee) in Oberbayern. v.5. Leipzig, 1901.
	Geog 135.2.6	Beiträge zur Biogeographie und Morphologie der Alpen. v.6. Leipzig, 1904.
	Geog 135.2.7	Hauthal, Rudolf. Reisen in Bolivien und Peru. v.7. Leipzig, 1911.
	Geog 137.2	Lisbon. Sociedade de Geographia. Boletin. 1900+ 48v.
NEDL	Geog 137.2	Lisbon. Sociedade de Geographia. Boletin. 1-17,1877-1899
	Geog 137.2.5	Geographica. Lisboa. 1,1965+ 8v.
	Geog 137.2.12	Lisbon. Sociedade de Geographia. Catalogo de Vendas. 1900-1916
	Geog 137.5	Linschoten. Vereinigung. Jaarveslag. 1-31,1908-1938 2v.
	Geog 137.10	Pamphlet vol. Linschoten. Vereinigung. Minor publications.
	Geog 139.1	London. Royal Geographical Society. Journal. London. 1831-1880 50v.
	Geog 139.2	London. Royal Geographical Society. Journal. Index. v.1-20,21-40,41-50. London, n.d. 3v.
	Geog 139.2.9	London. Royal Geographical Society. Proceedings. London. 2-22 16v.
	Geog 139.3	London. Royal Geographical Society. Proceedings. London. 1-14,1879-1894 14v.
	Geog 139.3.5	London. Royal Geographical Society. Proceedings. Index. 1879-1892. London, 1896.
	Geog 139.4	London. Royal Geographical Society. Geographical journal. London. 1,1893+ 128v.
	Geog 139.4.5	London. Royal Geographical Society. Geographical journal. Index. 1893-1902. London, 1906. 3v.
	Geog 139.4.12	London. Royal Geographical Society. Recent geographical literature, maps. London. 19-41,1926-1932 11v.
	Geog 139.4.13	London. Royal Geographical Society. Recent geographical literature, maps. Index. 1-4, 1918-1932. London, 1936.
	Geog 139.4.15F	London. Royal Geographical Society. Library. Library series.
	Geog 139.5	London. Royal Geographical Society. Supplementary papers. v.1-3. Photoreproduction. 1951-1955 6v.
	Geog 139.5.5	Playfair, R.L. Supplement to the bibliography of Algeria, 1895. London, 1898.
	Geog 139.6	London. Royal Geographical Society. Yearbook record. London. 1898-1905 2v.
	Geog 139.7	Smyth, W.H. Address at anniversary meeting of the Royal Geographical Society. London. 1851-1868
	Geog 139.8	Pamphlet vol. Royal Geographical Society.
	Geog 139.9	London. Royal Geographical Society. Technical series. London. 1-5,1920-1929
	Geog 139.10	London. Royal Geographical Society. Its foundation and history. London, 1930.
	Geog 139.25	Mill, Hugh R. The record of the Royal Geographical Society, 1830-1930. London, 1930.
	Geog 139.30	London. University. Queen Mary College. Department of Geography. Occasional papers. London. 1,1974+
	Geog 141.5	Lund. Universitet Geografiska Institution. Meddelanden. Avhandlingar. Lund. 1-60 32v.
	Geog 141.5F	Lund. Universitet Geografiska Institution. Meddelanden. Avhandlingar. Lund.

Geog 1 - 296 Geography in general - Periodicals and Societies (Special Table) - cont.

	Geog 141.10	Lund studies in geography. Series A: Physical geography. Lund. 1-37 4v.
	Geog 141.10.2	Lund studies in geography. Series B: Human geography. Lund. 1-31 6v.
	Geog 141.10.4	Lund studies in geography. Series C: General and mathematical geography. Lund. 1-4
	Geog 142.1	Madrid. Instituto de Geográfico y Estadistico. Memorias. Madrid. 1925-1927 4v.
NEDL	Geog 142.1	Madrid. Instituto de Geográfico y Estadistico. Memorias. Madrid. 1-11,1875-1899 11v.
	Geog 142.2	Sociedad Geografico de Madrid. Boletin. Madrid. 1876-1901 32v.
NEDL	Geog 142.2	Sociedad Geografico de Madrid. Boletin. Madrid. 1-44,1876-1900
	Geog 142.2.2	Sociedad Geografico de Madrid. Repertorio, 1901-1910. Madrid, 1911.
	Geog 142.2.3	Sociedad Geografico de Madrid. Boletin. Revista geografica colonial y mercantil. Actas. Madrid. 1-21,1899-1924 19v.
	Geog 142.2.5	Pamphlet vol. Sociedad Geografico de Madrid.
	Geog 142.2.10F	Madrid. Museo Naval. Publicaciones. Madrid. 1,1932
	Geog 142.3	Malayan journal of tropical geography. Singapore. 1,1953+ 10v.
	Geog 142.3.2	The journal of tropical geography. Index. 1-39,1953-1974. Singapore, n.d. 2v.
	Geog 142.4	Marburger geographischer Schrifter. Marburg. 1+ 43v.
	Geog 142.5	Marseilles. Société de géographie. Marseilles. 1-65,1877-1954 18v.
NEDL	Geog 142.5	Marseilles. Société de géographie. Marseilles. 1-23,1877-1899 23v.
	Geog 142.6	Marseilles. Exposition Coloniale Nationale, 1922. Semaine internationale des géographes, des explorateur et des ethnologues, 22-28 Sept. 1922. Marseilles, 1923.
	Geog 142.7F	Marco Polo; turismo scolastico del touring club italiano. Milano. 5-10,1954-1959 7v.
	Geog 142.12	Manchester Geographical Society. Journal. Manchester. 1-7 6v.
	Geog 142.12.5	Brown, Theodore Nigel Leslie. The history of the Manchester Geographical Society, 1884-1950. Manchester, 1971.
	Geog 143.5	México (City). Universidad Nacional. Facultad de Filosofiá y Letras. Anuario de geografía. 1,1961 4v.
	Geog 143.10	México (City). Universidad Nacional. Instituto de Geografía. Publicaciones. México. 1,1965+
	Geog 143.10.5	México (City). Universidad Nacional. Instututo de Geografía. Boletín. México. 1,1969+ 2v.
	Geog 143.70	Milan. Università Cattolica del Sacro Cuore. Pubblicazione. Serie 10. Scienze geografico. Milano.
NEDL	Geog 144.1	Mittheilungen. Berlin. 1-36,1888-1929 36v.
NEDL	Geog 144.1.2F	Mittheilungen. Berlin, 1908. 6v.
NEDL	Geog 144.1.3	Mittheilungen. v.2. Berlin, 1915.
	Geog 144.9	The mountain world. London. 1,1953+ 9v.
	Geog 144.10	Montagnes du monde. Genève. 1,1953+
	Geog 145.1	The mountaineer. Seattle. 1-9,1907-1916 2v.
	Geog 145.1.10	The mountaineer.
	Geog 145.2F	Le mouvement geographique. Bruxelles. 1884-1905 23v.
	Geog 145.4	Monographs in geography. Enfield, Eng. 1,1973//
	Geog 146.1	Munich. Geographische Gesellschaft Jahresbericht. 1869-1902 7v.
	Geog 146.1.5	Munich. Mittheilungen der geographische Gesellschaft in München. München. 1,1906+ 33v.
	Geog 146.3	Meunchener geographische Hefte. Kallmunz. 1,1953+ 17v.
	Geog 146.5	Murray Park College of Advanced Education. Department of Geography. Occasional papers in geography series. Magill. 1,1974+
	Geog 147.5	National Council for Geographic Education. Yearbook. Palo Alto, Calif. 1,1970+ 2v.
	Geog 148.1	National geographic magazine. 1915+ 170v.
Htn	Geog 148.1*	National geographic magazine. 1-6 6v.
	Geog 148.1.6	National geographic magazine. Cumulative index, 1899-1936. Washington, 1937.
	Geog 148.1.8	Buxbaum, Edwin Clarence. Collecting national geographic magazines. Milwaukee, 1935.
	Geog 148.1.9	Buxbaum, Edwin Clarence. Collectors guide to the national geographic magazine. Wilmington, 1962.
	Geog 148.1.15	National Geographic Society. Contributed technical papers. Katmai series. Washington. 1,1923
	Geog 148.1.18	National Geographic Society. Contributed technical papers. Stratosphere series. Washington.
	Geog 148.1.30	Pamphlet vol. National Geographic Society.
	Geog 148.5	National Geographical Society of India. Bulletin. Benares.
	Geog 148.5.5	National geographical journal of India. Benares. 1,1955+ 7v
Htn	Geog 149.1*	Pallas, Peter. Neue nordische Beyträge. St. Petersburg, 1781. 4v.
	Geog 150.1	Neue allgemeine geographische Ephemeriden. Weimar.
	Geog 150.5	New Zealand geographer. 3,1947+ 12v.
	Geog 150.5.2	New Zealand geographer. Cumulative index, v.1-25, 1945-1969. Wellington, 1971.
	Geog 150.10	New Zealand Geographical Society. Record of proceedings of the society and its branches. Christchurch. 1,1946+ 3v.
	Geog 151.1	Norske geografisk selskabs årbog. 1-32 20v.
	Geog 151.1.2	Norsk geografisk tidsskrift. Oslo. 18,1961+ 18v.
	Geog 151.1.3	Norsk geografisk tidsskrift. Register. Oslo.
	Geog 151.2	Den Norske turistforenings årbog. Kristiania. 1868-1913 23v.
	Geog 151.4F	North German Lloyd bulletin. N.Y.
	Geog 151.5	North Western University studies in geography. Evanston. 1+ 11v.
	Geog 151.6	Norois; revue géographique de l'Ouest et des pays de l'Atlantique nord. Poitiers. 1,1954+ 20v.
	Geog 151.6.5	Norois. Table décennale, 1954-73. Poitiers, n.d. 2v.
	Geog 152.1	Société neuchateloise de géographie. Bulletin. Neuchatel. 1,1885+ 16v.
	Geog 152.5	Nottingham, Eng. University. Geographical Society. Magazine. 1-5,1963-1967//?
	Geog 152.10	The northern universities' geographical journal. Leicester, Eng. 1-9,1960-1968//?
	Geog 156.1	Ocean highways. 1-12
	Geog 157.50	Österreichischer Alpen-Verein, Vienna. Mittheilungen. Wien. 1-2,1863-1864 2v.
NEDL	Geog 158.1	Österreichische Touristen-Zeitung. Wien. 1-56,1881-1936 28v.
	Geog 158.1.25	Österreichische Touristen-Zeitung. Sondernummer anläslich des 60 jähren Bestande. Innsbruck, 1929.

Classified Listing

Geog 1 - 296 Geography in general - Periodicals and Societies (Special Table) - cont.

Geog 160.5	Ocherki po fizicheskoi geografii. Riga. 6,1966+
Geog 161.1	Ohio State University. Contributions in geographical exploration. Columbus. 1,1920
VGeog 162.5	Okhrana prirody i vosproizvodstvo prirodnykh resursov. Moskva. 1,1968+
Geog 170.3	Ottawa, Ont. University. Department of Geography and Regional Planning. Travaux du Département de géographie et d'aménagement régional, Université d'Ottawa. Ottawa. 1,1971+
Geog 174.5F	Oxford. University. School of Geography. Research papers. Oxford. 1,1972+
Geog 175.5	Pan American Institute of Geography and History. Commission on Cartography. Progress report on the cartographic activities of the United States. St. Louis. 1946-1952 3v.
Geog 178.1	Pamphlet vol. Pacific Geographic Society.
Geog 178.5	Pacific geographic magazine. Beverly Hills, Calif.
Geog 178.7	Pamphlet vol. Pacific Crest Trail Conference, Pasadena, California.
Geog 178.7.5	Pamphlet vol. Pacific Crest Trail Conference, Pasadena, California.
Geog 179.2	France. Centre de Documentation Cartographique et Géographique. Mémoires et documents. 1-10 22v.
Geog 179.2.5	France. Centre de Documentation Cartographique et Géographique. Centre de documentation cartographique et géographique. Paris, 1956?
Geog 180.1	Geographische Mittheilungen. 1+ 113v.
Geog 180.2	Geographische Mittheilungen. Ergänzungshefte. 2+ 73v.
Geog 180.3	Geographische Mittheilungen. Inhaltsverz. 1855-1934 6v.
Geog 180.4	Geographischen Anzeiger. Gotha. 1-41 40v.
Geog 180.10	Perspectives in geography. Dekalb. 1,1971+
Geog 182.1	Philadelphia Geographical Club. Bulletin. Philadelphia. 1-36 16v.
Geog 182.2	Philadelphia Geographical Society. Charter. Philadelphia.
Geog 183.2	Philippine geographical journal. Manila. 1,1953+ 4v.
Geog 183.5	Places. Indiana, Pa. 1,1974+
Geog 184.1	The polar record. Cambridge. 3,1941+ 14v.
Geog 185.1	Portugal. Ministerio dos Negocios da Marinha e Ultramar. Annaes da commissão central permanente de geographia. Lisboa. 1,1876
Geog 185.5	Polska Akademia Umiejętności. Komisja Geograficzna. Prace. Kraków. 1-2 2v.
Geog 185.5PF	Polska Akademia Umiejętności. Komisja Geograficzna. Prace. Kraków. 2
Geog 185.6	Polska Akademia Nauk. Instytut Geografii. Prace geograficzne. Warszawa. 1,1954+ 40v.
Geog 185.7	Polski przegląd kartograficzny. Lwów.
Geog 185.8	Polskie Towarzystwo Geograficzne. Polskie towarzystwo geograficzne w 50 rocznicę dzialalności. Warszawa, 1968.
Geog 185.10	Portugal. Agrupamento de Estudos de Cartografia Antiga. Secção de Lisboa. Publicacion. Ser. separatas. Lisboa. 1-21 5v.
Geog 185.12	Portugal. Agrupamento de Estudos de Cartografia Antiga. Secção de Coimbra. Publicacion. Ser. monografias. Lisboa. 1,1963+ 12v.
Geog 185.15	Polsko-Czeskie Seminarium Geograficzne, 3rd, Warsaw, 1967. Polsko-czeskie seminarium geograficzne. Wyd. 1. Warszawa, 1968.
Geog 185.15.10	Polsko-Czeskie Seminarium Geograficzne, 5th, Warsaw, 1972. V. Czesko-Polskie seminarium geograficzne. Wyd. 1. Warszawa, 1975.
Geog 185.16	Polska Akademia Nauk. Institut Geografii. Streszczenia prac habilitacyjnych i dobtorskich. Warszawa. 1967+
Geog 186.2	The professional geographer. Washington. 1,1949+ 14v.
Geog 186.3	Prague. Universita Karlova. Acta Universitis Carolinae. Geographica. Praha. 1,1966+ 3v.
Geog 186.5	Presidente Prudente, Brazil. Faculdade de Filosofia, Ciências e Letras. Departamento de Geografia. Boletim do Departamento de Geografia. Presidente Prudente. 4,1972+
Geog 186.10	Progress in geography; international reviews of current research. London. 1,1969+ 8v.
Geog 186.15	Problemi na paleogeomorfolozhko to razvitie na Bulgariia. Sofiia. 1,1971+
Geog 186.20	Problemi na geografiiata. Sofiia. 1,1975+
Geog 187.1	Quaderni geografici. Roma. 1-11
Geog 187.5	Quaderns de geografia. Barcelona. 1,1949 2v.
Geog 187.10	Quaestiones geographicae. Poznań. 1,1974+
Geog 189.2	Quebec Geographical Society. Transactions. Quebec. 1880-1879 3v.
Geog 189.2.3	Quebec Geographical Society. Bulletin. Quebec. 3-23,1908-1929 10v.
Geog 189.2.4	Quebec Geographical Society. Tables des matieres contenus dans le Bulletin. Quebec, 1913.
Geog 189.2.5	Société de Géographie de Québec. Bulletin. 2v.
Geog 189.2.6	Société de Géographie de Québec. Bulletin. Index. 1880-1934. Québec, 1969.
Geog 189.2.8	Morissonneau, Christian. La société de géographie de Québec, 1877-1970. Québec, 1971.
Geog 190.5	Revista geográfica americana. Buenos Aires. 1-43,1933-1959 41v.
Geog 190.5.5	Ricossa, J.A. Pueblos. Buenos Aires, 195-? 6v.
Geog 190.10	Pam American Institute of Geography and History. Revista geográfica del instituto panamericano de geográfica e historia. México. 9,1949+ 11v.
Geog 190.15	Revista uruguaya de geografia. Montevideo. 1-6,1950-1952
Geog 190.20	Regio Basiliensis. Hefte für jurassische und oberrheinische Landeskunde. Basel. 1,1959+ 8v.
Geog 190.20.2	Regio Basiliensis. Hefte für jurassische und oberrheinische Landeskunde. Register. 1-10,1959-1969. Basel, n.d.
Geog 192.1	Revue de geographie. Annuelle. Paris. 1-12,1906-1924 65v.
Geog 192.2	Revue geographique internationale. 89-112,1883-1885
Geog 192.3	Revue française de l'étranger et des colonies et exploration. Paris. 1-39,1886-1914 39v.
Geog 192.4	Grenoble. Université. Institut de Géographie Alpine. Revue de géographie alpine. 1,1913+ 61v.
Geog 192.4.5	Grenoble. Université. Institut de Géographie Alpine. Table décennale. v.2 (1923-32), v.5 (1953-62). Grenoble, 1932-63. 3v.
Geog 192.5	Revue géographique des pyrénées et du sud-ouest. Toulouse. 1,1930+ 33v.

Geog 1 - 296 Geography in general - Periodicals and Societies (Special Table) - cont.

Geog 192.5.5	Revue géographique des pyrénées et du sud-ouest. Table décennale, 1950-1959. Toulouse, n.d.
Geog 193.1	Rennes. Université. Laboratoire de Géographie. Travaux. Rennes. 1-6,1903-1909 2v.
Geog 193.10	Reims. Université. Institut de Géographie. Travaux. Reims. 3,1970+ 3v.
Geog 194.1	Rivista geografica italiana. Roma. 1,1894+ 37v.
Geog 194.1.5	Memorie geografiche. Firenze. 1-13,1907-1919 11v.
Geog 194.1.6	Rivista geografia italiana. Firenze. Reprint ed. Amsterdam, 1970. 8v.
Geog 194.2	Rivista di geografia didattica. Firenze. 1-13,1917-1933 3v.
Geog 194.5	Russkoe Obshchestvo Parokhodstva i Torgovli. Putevoditel' po Krymu, Kavkazu, i Blizhnemu Vostoku. Odessa. 1913
Geog 195.2	Memoire geografiche. Roma. 1-9 7v.
Geog 195.3	Rome, Italy. Università. Istituto di Geografia. Pubblicazioni. Serie B (Geostorica). Roma. 1,1969+
Geog 195.4	Rome, Italy. Università. Istituto di Geografia. Pubblicazioni. Serie C (Miscellanea). Roma. 2,1969+
Geog 197.5	Sao Paulo, Brazil. Universidade. Faculdade de Filosofia, Ciências e Letras. Geografia. 5-7
Geog 197.10	Saarbruecken. Universität des Saarlandes. Geographisches Institut. Arbeiten. 1,1956+ 4v.
NEDL Geog 199.1	Schweizer Alpenclub. Jahrbuch. Bern. 1-35 23v.
Geog 199.1	Schweizer Alpenclub. Jahrbuch. Bern. 36-58,1900-1923 23v.
Geog 199.1.3	Schweizer Alpenclub. Beilagen zum Jahrbuch. Bern. 1-46 46v.
Geog 199.1.4	Pamphlet vol. Schweizer Alpenclub. Clubhütten. Ergänzungsblätter.
Geog 199.1.5	Schweizer Alpenclub. Repertorium und Ortsregister. Bern. 1-44,1886-1910 2v.
Geog 199.1.18	Schweizer Alpenclub. Die ersten 25 Jahre des Schweizer Alpenclub. Glarus, 1889.
Geog 199.1.20	Dübi, Heinrich. Die ersten fünfzig Jahre des Schweizer Alpenclub. Bern, 1913.
Geog 199.2	Schweizer Alpen-Zeitung. Zürich. 1-11 3v.
Geog 199.3	Alpina; Mitteilungen des Schweizer Alpen-Club. Zürich. 1-32 18v.
NEDL Geog 199.10	Schweizer Alpenclub. Section Genevoise. Bulletin. Genève.
Geog 199.20	Der Schweizer Geograph; Zeitschrift des Vereins schweizerischer Geographialehrer sowie der geographischen Besellschaften von Basel, Bern, St. Gallen und Zürich. Bern. 1-22,1923-1945 4v.
Geog 200.1	Scottish geographical magazine. Edinburgh. 1+ 63v.
Geog 200.1.3	Scottish geographical magazine. Index. v.1-50, 1885-1934. Edinburgh, n.d.
Geog 205.1	Don; the journal of Sheffield University Geographical Society. Sheffield, Eng. 1,1957+
Geog 207.1	Sierra club bulletin. San Francisco. 2,1897+ 30v.
Geog 207.5	Sierra club circular.
Geog 209.1	Slutskaia, Raisa D. Geograficheskoe obshchestvo globus. Moskva, 1972.
Geog 212.1	Sociedad Geográfica de la Paz. Boletin. La Paz. 27-66,1909-1943 3v.
Geog 212.9	Sociedad Geográfica de la Paz. Estatutos de la "Sociedad geográfica de la Paz." 2. ed. La Paz, 1907.
Geog 212.15	Sociedad Geográfica, Madrid. Publicaciones. Serie B. Madrid. 455,1966+
Geog 212.20	Sociedad Geográfica de Lima. La reorganización de la Sociedad Geográfica de Lima. Lima, 1944.
Geog 212.30	Urteaga, H.H. Memoria del presidente de la Sociedad Geográfica de Lima. Lima, 1942.
Geog 212.40	Sociedad Geográfica de Colombia. Boletin. Bogotá. 14,1956+ 5v.
Geog 212.50	Sociedad Mexicana de Geográfica y Estadística. Informe sobre los trabajos cartográficos. México. 5-7,1938-1947 4v.
NEDL Geog 212.100	Societa Geografica Italiana. Bollettino. Firenze. 1868-1899 12v.
Geog 212.100	Societa Geografica Italiana. Bollettino. Firenze. 1900+ 82v.
NEDL Geog 212.100.2	Societa Geografica Italiana. Bollettino. Indice generale della serie II-III. Roma, n.d.
Geog 212.105	Societa Geografica Italiana. Memorie. Roma. 1-30 27v.
Geog 212.109	Societa Geografica Italiana. Catalogo della biblioteca sociale. Roma, 1903.
Geog 212.110	Vedova, G. La Societa Geografica Italiana e l'opera sua nel secolo XIX. Roma. 5-6,1940-1941 3v.
Geog 212.110.10	Brunialti, A. Relazioni sulla fondazione e sull'ordinamento della sezione di geografia commerciale della Societa italiana. Roma, 1879.
Geog 212.112	La terra e la vita. Roma. 2v.
NEDL Geog 212.200	Société Royale Belge de Géographie. Bulletin. Bruxelles. 1-23,1877-1899 23v.
Geog 212.200	Société Royale Belge de Géographie. Bulletin. Bruxelles. 24,1900+ 46v.
NEDL Geog 212.200.5	Société Royale Belge de Géographie. Compte rendu. Tables des matières des v.1-25, 1876-1901. Bruxelles, n.d.
Geog 212.200.20	Société Belge d'Études Coloniales. Bulletin. Bruxelles. 1894-1925 22v.
NEDL Geog 212.201	Société Normande de Géographie. Bulletin. Rouen. 1-21,1879-1899 21v.
Geog 212.201	Société Normande de Géographie. Bulletin. Rouen. 22-43,1900-1928 14v.
NEDL Geog 212.202	Société de Géographie de Lyon. Bulletin. Lyon. 1-16,1875-1900 11v.
Geog 212.202	Société de Géographie de Lyon. Bulletin. Lyon. 1901-1929 6v.
Geog 212.202.5	Société de Géographie de Lyon. Bulletin du cinquantenaire, 1922-23. Lyon, 1923.
NEDL Geog 212.203	Société de Géographie. Bulletin. Paris. 1-20 73v.
Geog 212.203.5	Société de Géographie. Bulletin. Table, series V-VII. Paris, n.d.
NEDL Geog 212.204	Société de Géographie. Comptes rendus. Paris. 1882-1899 18v.
Geog 212.205	Société de Géographie. La geographie; bulletin de la société de géographie. Paris. 1-72,1925-1939 60v.
Geog 212.206	Société de Géographie. Liste des membres. Paris. 1868-1897 11v.
NEDL Geog 212.215	Société de Géographie de l'Est. Bulletin. Nancy. 1879-1912 34v.

Classified Listing

Geog 1 - 296 Geography in general - Periodicals and Societies (Special Table) - cont.

Geog 212.216F Société Royale de Géographie d'Egypte. Mémoires. Le Caire. 1,1919+ 17v.
Geog 212.216.10 Société Royale de Géographie d'Egypte. Bulletin. Le Caire. 2+ 27v.
Geog 212.216.11 Société Royale de Géographie d'Egypte. Bulletin. Tables. 16-30,1928-1957
Geog 212.216.95 Foucart, G. La société Sultanich de géographie du Caire. Le Caire, 1921.
Geog 212.218 Société de Géographie du Maroc. Bulletin. Casablanca. 1916-1933 6v.
NEDL Geog 212.225 Société de Géographie de Toulouse. Bulletin. Toulouse. 1-18,1882-1899 18v.
Geog 212.225 Société de Géographie de Toulouse. Bulletin. Toulouse. 19-50 18v.
NEDL Geog 212.235 Société de Géographie de Lille. Bulletin. Lille. 1-32,1882-1899 17v.
Geog 212.235 Société de Géographie de Lille. Bulletin. Lille. 1900-1962 20v.
Geog 212.235.15 Société de Géographie de Lille. Publications. 1943-1953
Geog 212.240 Société Geographique de Liège. Bulletin année. Liège. 1,1965+ 2v.
NEDL Geog 212.245 Société Languedocieme de Géographie, Montpellier. Bulletin. Montpellier. 1-34,1878-1911 33v.
Geog 213.5 The South Hampshire geographer. Portsmouth, Eng. 3,1970+
Geog 214.5 Soviet geography: review and translation. N.Y. 1,1960+ 9v.
Geog 214.10 Studia geograficzno-fizyczne z obszasu opolszocyzny. Opole. 1,1968+
Geog 214.20 Studien zur Kartographie. Berlin. 1-2
Geog 214.25 Studia geographica. Brno. 1,1969+ 4v.
Geog 215.1 Gesellschaft für Erdkunde und Kolonialwesen zu Strassburg. Mitteilungen. Strassburg.
Geog 218.2 Svensk geografisk årsbok. Lund. 39,1963+ 23v.
Geog 218.2.2 Svensk geografisk årsbok. Register. Lund.
Geog 220.5 Terrae incognitae. Amsterdam. 1,1969+
Geog 220.10 Itogi nauki: Teoreticheskie i obshchie voprosy geografii. Moskva. 1,1974+
Geog 224.3F Tijdschrift voor economische en sociale geografie. Rotterdam. 40,1935+ 21v.
Geog 224.10 Tydskrif vir aardrykskunde; tydskrif van die vereniging vir aardrykskunde-onderwys. Stellenbosch. 2,1962+ 2v.
Geog 225.5 Tokyo Metropolitian University. Department of Geography. Geographical reports. Tokyo. 1,1966+ 2v.
Geog 226.1 Toscanelli. 1,1893
NEDL Geog 226.2F Tour du monde. Paris. 1861-1899
Geog 226.2F Tour du monde. Paris. 1900-1914 15v.
Geog 226.2.5 Tour du monde. Table alphabetique, 1860-1910. Paris, n.d.
Geog 226.3 Touring Club de France. Revue mensuelle. Paris. 36-47,1926-1937 3v.
Geog 227.2 Trident; magazine of the sea. London. 17-18,1955-1956 2v.
Geog 228.1 Travel and exploration; monthly illustrated magazine. London. 1-4,1909-1910 4v.
Geog 228.5F Travel and camera. N.Y.
Geog 228.25 Trieste. Università. Istituto di Geografia. Notiziario.
Geog 240.5 Ukrains'ke Heohrafichne Tovarystvo. Heohrafichnyi zbirnyk. Kyïv. 1-5 5v.
Geog 243.2 Union Geographique du Nord de la France. Bulletin. Lille. 1-34,1880-1913 17v.
Geog 243.5 United States. Department of State. Office of the Geographer. Geographic bulletin.
Geog 243.15 L'universo. Firenze. 1,1920+ 62v.
Geog 243.15.5 L'universo. Index, 1920-40. Firenze, n.d.
VGeog 249.5 Venture. Des Moines. 3,1967+ 10v.
Geog 250.5 Vereniging vir Aardrykskunde-Onderwys. Special publication. Stellenbosch.
Geog 252.2 Viaggi e scoperte di navigatori ed esploratori italiani. Milano. 1-18,1929-1932 18v.
Geog 252.3 Viaggi esplorazioni e scoperte. Milano. 1-11 11v.
Geog 258.1 Vienna. K.K. Geographische Gesellschaft. Mittheilungen. Wien. 1,1857+ 81v.
Geog 258.1.5 Österreichische Geographische Gesellschaft. Registerband, 1908-59. Wien, 1960.
Geog 258.2 Vienna. K.K. Geographische Gesellschaft. Abhandlungen. Wien. 1-18,1899-1959 11v.
Geog 260.5 Voprosy geografii. Moskva. 1-98,1946-1975 68v.
Geog 265.1F World traveler. N.Y.
Geog 265.2 Warsaw. Universytet. Instytut Geograficzny. Katedra Klimatologii. Prace i studia. 1,1964+
Geog 265.3 Warsaw. Universytet. Instytut Geograficzny. Katedra Geografii Fizycznej. Prace i studia. 1,1967+
Geog 266.2 Westfälische geographische Studien. Münster. 1,1949+ 8v.
Geog 275.1 Ymer; tidskrift utgifven af svenska sällskapet för antropologi och geografi. Stockholm. 1+ 61v.
Geog 275.2 Ymer; tidskrift utgifven af svenska sällskapet för antropologi och geografi. Person-och ämnesregister; 1-70, 1881-1950. n.p., n.d. 2v.
Geog 287.2 Zeitschrift für wissenschaftliche Geographie. Lahr. 1-3,1880-1882 3v.
Geog 287.4 Zeitschrift für Erdkunde. Frankfurt am Main. 4
Geog 290.2 Zemlia i liudi. Moskva. 9,1966+ 9v.
Geog 290.4 Danzig. Wyższa Szkoła Pedagogiczna. Wydział Geograficzny. Zeszyty geograficzne. 1-11,1959-1969// 3v.
Geog 290.5 Danzig. Universytet. Wydział Biologii i Nauk o Ziemi. Zeszyty naukowe. Geografia. Gdańsk. 1,1970+

Geog 500 Geography in general - Bibliographies

Geog 500.1 Pamphlet box. Geography. Bibliography.
Geog 500.5 Pamphlet vol. Geography. Bibliography. 15 pam.
Geog 500.7 Boucher, G. Bibliothèque universelle des voyages. Paris, 1808. 6v.
Geog 500.9 Rome, Italy. Bibliotheca Collegio Romano. Catalogo ragionato. Roma, 1876.
Geog 500.11 Beckmann, J. Literatur der...Reisebeschreibungen. Göttingen, 1808-09. 2v.
Geog 500.13 Hagers, J.G. Geographischer Buchersaal. Chemnitz, 1766-78. 3v.
Geog 500.15 Jomard, E.F. De la collection geographique. Paris, 1848.
Geog 500.17 Ternaux-Compaus, H. Bibliothèque asiatique et africaine. Paris, 1841.
Geog 500.17.2 Ternaux-Compaus, H. Bibliothèque asiatique et africaine. Notes. Paris, 1841.
Geog 500.19 Engelmann, W. Bibliotheca geographica. Leipzig, 1858.
Geog 500.21 London. Royal Geographical Society. Catalogue of the library. London, 1852.

Geog 500 Geography in general - Bibliographies - cont.

Geog 500.21.3 London. Royal Geographical Society. Catalogue of the library. London, 1865.
Geog 500.21.5 London. Royal Geographical Society. Catalogue of the library. London, 1895.
Geog 500.22 Amat di San Filippo, Pietro. Bibliografia di viaggiatori italiani ordinata cronologicamenti. Roma, 1874.
Geog 500.23 Cardon, F. Publicazioni geografiche. Roma, 1892.
Geog 500.25 Georg, Carl. Die Reiseliteratur Deutschlands. Leipzig, 1877.
Geog 500.27A Jackson, J. Liste provisoire de bibliographies geographiques. Paris, 1881.
Geog 500.27B Jackson, J. Liste provisoire de bibliographies geographiques. Paris, 1881.
Geog 500.27.2A Jackson, J. Liste provisoire de bibliographies geographiques. Paris, 1881.
Geog 500.27.2B Jackson, J. Liste provisoire de bibliographies geographiques. Paris, 1881.
Geog 500.29 Stuck, G.H. Verzeichnis von...Land und Reisebeschreibungen. Halle, 1784.
Geog 500.31 Ersch, J.S. Repertorium über...allgemeinern...Journale. Lemgo, 1790. 3v.
Geog 500.32 Muller, F. Topographie ancienne - catalogue a prix marquès de cartes anciennes. Amsterdam, 1896.
Geog 500.33 Asher, George Michael. Viro venerabili Friderico Laurentio Hoffmann. Berolinenses, 1860.
Geog 500.35 Petherick, E.A. Catalogue of the York Gate Library. London, 1886.
Geog 500.40 Gesellschaft für Erdkunde zu Berlin. Bibliothek. Katalog der Bibliothek der Gesellschaft für Erdkunde zu Berlin. Berlin, 1903.
Geog 500.55 Newark, N.J. Free Public Library. Foreign countries. Washington, 1918.
Geog 500.56 Amsterdam. Universiteit. Bibliotheek. Catalogus geographie en reizen. Amsterdam, 1923.
Geog 500.57 Almagia, Roberto. La geografia. Roma, 1919.
Geog 500.58 Milan. Bibliotheca. Catalogo ragionato della geografica. Milano, 1927.
Htn Geog 500.59F* Atkinson, G. La litterature géographique française de la renaissance. Paris, 1927.
Htn Geog 500.59.2F* Atkinson, G. La litterature géographique française de la renaissance. Supplement. Paris, 1936.
Geog 500.60 Maggs Bros., London. Bibliotheca asiatica et africana. pt.4-5. London, 1929.
Geog 500.61.5 Société de Géographie de Lille. Bibliothèque. Supplément au catalogue...paru en décembre, 1887. 1,1889
Geog 500.62 Stevenson, Edward L. Publications of Edward L. Stevenson. n.p., 191-.
Geog 500.63 Pan American Union. Columbus Memorial Library. Books and magazine articles on geography in the Columbus Memorial Library. Washington, 1935.
Geog 500.64 Current geographical publications. N.Y. 1,1938+ 35v.
Geog 500.65 Verein für Erdkunde, Dresden. Bibliothek. Bücherei-Verzeichnis des Vereins für Erdkunde zu Dresden. Dresden, 1905.
Geog 500.66 Bolles, E.C. The literature of sea travel since the introduction of steam, 1830-1930. Philadelphia, 1943.
Geog 500.67 Curtiss, Frederic H. A little book on travel books. Boston, 1936.
Geog 500.68 International Committee of Historical Sciences. Committee on the History of Great Voyages and Great Discoveries. Travaux de la Commission pour l'histoire des grands voyages et des grandes decouvertes. Paris, 1936.
Geog 500.70 Anderson, Ernst. Bok-katalog omfattande geografi och resor. v.1-2. Stockholm, 1929-31.
Geog 500.71 Perthes, Justus, publishers. Wandkarten, Cottanten, Bücher, Zeitschriften für den geographischen Unterricht. Gotha, 1930.
Geog 500.72 American School of Classic Studies at Athens. Voyages and travels in the Near East made during the nineteenth century. Princeton, N.J., 1952.
Geog 500.72.5 American School of Classical Studies at Athens. Voyages and travels in Greece. Princeton, N.J., 1953.
Geog 500.73 Ossa Varela, P. Catálogo alfabético de algunos géografos y exploradores. Bogotá, 1951.
Geog 500.75A Baranskii, N.N. Istoricheskii obzor uchebniko geografii, 1876-1934. Moskva, 1954.
Geog 500.75B Baranskii, N.N. Istoricheskii obzor uchebniko geografii, 1876-1934. Moskva, 1954.
Geog 500.78 Hanover, S. Geographie. Hannover, 1955.
Geog 500.80 Migliorini, Elio. Giuda bibliografica allo studio della geografia. Napoli, 1945.
Geog 500.83 L'Information Geographique. La géographie française au milleu du XX. siècle. Paris, 1957.
Geog 500.85 Moscow. Gosudarstvennyi Biblioteka SSSR imeni V.I. Lenin. Glazami sovetskikh liudei. Moskva, 1956.
Geog 500.88 Fossati Bellani, Luigi Vittorio. I libri di viaggio e le guide della raccolta Luigi Vittorio. Roma, 1957. 3v.
Geog 500.90 Harris, Chauncy Donnison. A union list of geographical serials. 2. ed. Chicago, 1950.
Geog 500.91 Harris, Chauncy Donnison. International list of geographical serials. Chicago, 1960.
Geog 500.95 Bibliografia geografii polskiej. Warszawa. 1936+ 8v.
Geog 500.97 Coronelli, Marco V. Il catalogo degli autori; una biobibliografia geografica del '600. Firenze, 1957.
Geog 500.98F Kosack, Hans P. Die Kartographie, 1943-1954. Lahr, 1955.
Geog 500.100 Istituto Veneto di Scienza, Lettre ed Arte. Carte geografiche cinquecentesche a stampa della Biblioteca Marciana e della Biblioteca del museo correr di Venezia. Venezia, 1954.
Geog 500.105 Barras de Aragón, F. Los ultimos escritores de Indias. Madrid, 1949.
Geog 500.110 Powell, Lawrence. Around the world in sixty books. Los Angeles, 1960.
Geog 500.115F American Geographical Society of New York. Research catalogue of the American Geographical Society. Boston, 1962. 15v.
Geog 500.115.1F American Geographical Society of New York. Research catalogue. First supplement. Boston, 1974. 2v.
Geog 500.115.2F American Geographical Society of New York. Research catalogue. Map supplement. Boston, 1962.
Geog 500.116 Harris, Chauncy Donnison. Geographic bibliography. Chicago, 1961.
Geog 500.116.5 Harris, Chauncy Donnison. Annotated world list of selected current geographical serials in English. 2. ed. Chicago, 1964.
Geog 500.117 Kaufman, Isaak M. Geograficheskie slovari; bibliografiia. Moskva, 1964.

Classified Listing

Geog 500   Geography in general - Bibliographies - cont.
- Geog 500.118   Maggs Bros., London. Voyages and travels. v.1,4-5. London, 1962- 3v.
- Geog 500.119   Church, Martha. A basic geographical library. Washington, 1966.
- Geog 500.120   Polska Akademia Nauk. Institut Geografii. Katalog rękopisów geograficznych w zbiorach polskich. Warszawa, 1965- 2v.
- Geog 500.121   Josuweit, Werner. Studienbibliographie Geographie. Wiesbaden, 1973.
- Geog 500.125   Geographical abstracts. A: Geomorphology. London. 1966+ 9v.
- Geog 500.126   Geographical abstracts. B: Biogeography, climatology and cartography. London. 1966-1974 10v.
- Geog 500.127   Geographical abstracts. C: Economic geography. London. 1966+ 9v.
- Geog 500.128   Geographical abstracts. D: Social geography. London. 1966+ 7v.
- Geog 500.129   Geographical abstracts. Index: section A-D. London. 1966-1971// 6v.
- Geog 500.130   Geographical abstracts. F: Regional and community planning. Norwich, Eng. 1972+ 3v.
- Geog 500.130.3   Geographical abstracts. G: Remote sensing and cartography. Norwich. 1974+
- Geog 500.131   Geographical abstracts. Annual index. Norwich, Eng. 1972+ 4v.
- Geog 500.135   Schmidt, Rolf Dietrich. Verzeichnis der geographischen Zeitschriften, periodischen Veröffentlichungen und Schriftreihen Deutschlands. Bad Godesberg, 1964.
- Geog 500.140   Wystawa pt. Rozwój Historyczny Geografii Polskiej i Pismicnnictwo Polskie o Zakresu Historii Geografii, Warsaw, 1965. Catalogue of literature on the history of geography at the exposition. Warsaw, 1965.
- Geog 500.145   Documentatio geographica; geographische Zeitschriften- und Serien-Literatur. Jahresband. Berlin. 1966+ 9v.
- Geog 500.150   Browning, Clyde Eugene. A bibliography of dissertations in geography, 1901 to 1969; American and Canadian universities. Chapel Hill, 1970.
- Geog 500.155   Smith, Harold F. American travellers abroad. Carbondale, 1969.
- Geog 500.160   Arnim, Hlmuth. Bibliographie der geographischen Literatur in deutscher Sprache. 1. Aufl. Baden Baden, 1970.
- Geog 500.165   Trecento tesi di laurea in geografia. Padova, 1969.
- Geog 500.170   Schwickerath, Hildegard. Inhaltsverzeichnis der Festschriften zur Ehrung und Würdigung deutscher. Bad Godesberg, 1969.
- Geog 500.171   Tolchinskaia, L.I. Geograficheskaia literatura. Moskva, 1971.
- Geog 500.175   Wright, John K. Aids to geographical research. N.Y., 1923.
- Geog 500.180   International Committee of Historical Sciences. Commission Internationale d'Histoire Maritime. Compte rendu des travaux de la commission internationale d'histoire maritime, 1965. Paris? 1965?
- Geog 500.182   Vinge, Clarence L. United States government publications for research and teaching in geography. Totowa, N.J., 1967.

Geog 510   Geography in general - General pamphlet volumes
- Geog 510.1   Pamphlet vol. Geography. 10 pam.
- Geog 510.3F   Pamphlet vol. Geography.
- Geog 510.5   Pamphlet vol. Geography. 3 pam.
- Geog 510.6   Pamphlet vol. Geography. 4 pam.
- Geog 510.7   Pamphlet vol. Geography. 7 pam.
- Geog 510.8   Pamphlet vol. Geography and maps. 5 pam.
- Geog 510.9   Pamphlet vol. Geography. Islands. 2 pam.
- Geog 510.10   Pamphlet vol. Geography. Islands.
- Geog 510.11   Pamphlet vol. Geography. Islands. 24 pam.
- Geog 510.12   Pamphlet vol. Geography. Islands. 15 pam.
- Geog 510.13   Pamphlet vol. Geographical papers. 20 pam.
- Geog 510.20   Reden over geografie. v.1-2. Groningen, 1958.
- Geog 510.25   Pamphlet vol. Geografie. 3 pam.
- Htn   Geog 510.50PF*   Pamphlet box. Geography. Broadsides on geographical subjects.

Geog 520   Geography in general - Collections [Discontinued]
- Geog 520.1   Harrisse, H. Jean et Sébastien Cabot. Paris, 1882.
- Geog 520.2   Le voyage de la Sainte Cyté de Hierusalem. Paris, 1882.
- Geog 520.3   Harrisse, H. Les Corte-Real et leurs voyages. Paris, 1883.
- Geog 520.3.5   Harrisse, H. Gaspar Corte-Real. Paris, 1883.
- Geog 520.4   Parmentier, J. Le discours de la navigation. Paris, 1883.
- Geog 520.5   Thenaud, J. Le voyage d'Outremer. Paris, 1884.
- Geog 520.6   Harrisse, H. Christophe Colomb. Paris, 1884. 2v.
- Geog 520.8   Chesneau, J. Le voyage de M. d'Aramon. Paris, 1887.
- Geog 520.9   Varthema, L. di. Les voyages de L. di Varthema. Paris, 1888.
- Geog 520.10   Odoric de Pardenone. Les voyages en Asie. Paris, 1891.
- Geog 520.11   Possot, D. Le voyage de la Terre Sainte. Paris, 1890.
- Geog 520.12   La Broquière, Bertrandon de. Le voyage d'Outremer. Photoreproduction. Paris, 1892.
- Geog 520.13   Leone, G. Description de l'Afrique. v.2-3. Paris, 1896-98. 2v.
- Geog 520.16   Du Fresne-Canayl, P. Le voyage du Levant. Paris, 1897.
- Geog 520.17   Maurand, J. Itineraire de Jerome Maurand. Paris, 1901.
- Geog 520.18   Vignaud, Henry. La lettre et la carte. Paris, 1901.
- Geog 520.19   Codex Ramirez. Histoire de l'origine des Indiens. Paris, 1903.
- Geog 520.20   Fonteneau, J. La cosmographie. Paris, 1904.
- Geog 520.21   Anghiera, P.M. De orbe novo. Paris, 1907.
- Geog 520.22   Le Bouvier, Gilles. Le livre de la description des pays. Paris, 1908.
- Geog 520.23   Vignaud, Henry. Americe Vespuce, 1451-1512. Paris, 1917.
- Geog 520.24   Denucé, J. Pigafetta; relation du premier voyage...par Magellan. Paris, 1923.
- NEDL   Geog 520.25   Conder, Josiah. The modern traveller. London, 1830. 30v.
- Geog 520.30   Turistresor och forskningsfärder. v.3-12. Helsingfors, 1918. 5v.

Geog 535   Geography in general - Directories
- Geog 535.5   Geographen Kalendar. Gotha. 1903-1914 12v.
- Htn   Geog 535.7*   Étrennes intéressants des quatre parties du monde. Paris. 1788
- Geog 535.10   Orbis geographicus; world directory of geography. Wiesbaden. 1952 3v.
- Geog 535.15   Denis, Jacques. Guide de la recherche géographique en Belgique. Namur, 1970.

Geog 540   Geography in general - Museums
- Geog 540.15   Frabetti, Pietro. La collezione della antiche carte geografiche, a cura di Pietro Frabetti, Il museo delle navi, a cura di Amedio Rizzi. Bologna, 1959.

Geog 557 - 560   Geography in general - Exhibitions (By date)
- Geog 557.62   Succinta descrizione. Venezia, 1763.
- Geog 559.52   United States. National Archives. Geographical exploration and topographic mapping by the United States government. Washington, 1952.

Geog 574 - 580   Geography in general - Dictionaries and Gazetteers (By date)
- Geog 576.67   Poyares, P. de. Diccionario lusitanico-latino de nomes proprios de regioens, reinos, provincias, cidades. Lisboa, 1667.
- Geog 576.70F   Ferrari, Filippo. Lexicon geographicum. Parisiis, 1670.
- Geog 576.80   Fondeur, F. Urbium insularum regionum. Londuni, 1680.
- Geog 576.82F   Baudrand, M.A. Parisini geographia. Parisiis, 1681-82. 2v.
- Htn   Geog 576.93F*   Bohun, E. Geographical dictionary. London, 1693.
- Geog 576.97F   Ferrari, P. Novum lexicon geographicum. Patavii, 1697.
- Geog 577.01   Baudrand, M.A. Dictionaire geographique universal. Amsterdam, 1701.
- Htn   Geog 577.04*   The gazetteer's or newsman's interpreter. pt.2. London, 1707.
- Htn   Geog 577.09*   Eachard, Laurence. Gazetteer's geographical index of Europe. London, 1709.
- Geog 577.09.15   Eachard, Laurence. The gazetteer's or news-man's interpreter. London, 1751.
- Geog 577.09.18   Ladvocat, J.B. Dictionnaire géographique portatif. Paris, 1770.
- Geog 577.13F   Savonarola, R. Universus terrarum orbis. Patavii, 1713. 2v.
- Geog 577.26F   Bruzen de la Martiniere, A.A. Le grand dictionnaire géographique. Haye, 1726-39. 10v.
- Geog 577.31   Hederich, M.B. Reales Schul-Lexicon. Leipzig, 1731.
- Geog 577.58   Salmon, T. Modern gazeteer. London, 1758.
- Geog 577.58.2   Salmon, T. Modern gazeteer. London, 1759.
- Geog 577.58.4   Salmon, T. Modern gazeteer. London, 176-.
- Geog 577.58.6   Salmon, T. Modern gazeteer. London, 1766.
- Geog 577.59F   New geographical dictionary of the known world. London, 1759. 2v.
- Geog 577.60F   Brice, A. Grand gazetteer, or topographical dictionary. n.p., n.d.
- Geog 577.60.3   The universal gazetteer. 2. ed. London, 1760.
- Geog 577.62   Brookes, R. The general gazetteer. London, 1762.
- Geog 577.68F   Bruzen de la Martiniere, A.A. Le grand dictionnaire géographique. Paris, 1768. 6v.
- Htn   Geog 577.73*   Macbean, A. Dictionary of ancient geography. London, 1773.
- Geog 577.76   Johnson, R. New gazeteer. London, 1776.
- Geog 577.91   Brookes, R. General gazetteer. London, 1791.
- Geog 577.91.3   Brookes, R. General gazetteer. London, 1796.
- Geog 577.91.4   Brookes, R. General gazetteer. London, 1823.
- Geog 577.91.5   Brookes, R. General gazetteer. Boston, 1816.
- Geog 577.91.6   Brookes, R. A new universal gazetteer containing a description of the principal nations. N.Y., 1832.
- Geog 577.91.6.5   Brookes, R. A new universal gazetteer. Philadelphia, 1844.
- Geog 577.91.7   Brookes, R. A new universal gazetteer. Philadelphia, 1847.
- Geog 577.91.10   Brookes, R. New universal gazetteer. Boston, 1850.
- Htn   Geog 577.97*   Malham, J. Naval gazetteer. Boston, 1797. 2v.
- Geog 577.98   Cruttwell, C. New universal gazetteer. London, 1798. 3v.
- Htn   Geog 578.02*   Morse, Judith. A new gazeteer of the Eastern continent. Charlestown, 1802.
- Geog 578.10.2   Ladvocat, J.B. Dictionnaire géographique. 2. ed. Paris, 1812.
- Geog 578.17   Worcester, J.E. Geographical dictionary or universal gazetteer. Andover, 1817. 2v.
- Geog 578.17.2   Worcester, J.E. Geographical dictionary or universal gazetteer. 2. ed. Boston, 1823. 2v.
- Geog 578.21   Morse, Jedidiah. New universal gazeteer of the known world. 3. ed. New Haven, 1821.
- Geog 578.22   Edinburgh gazetteer. v.2-6. Edinburgh, 1822. 4v.
- Geog 578.23.5   Dictionaire géographique. Paris, 1823. 10v.
- Geog 578.27   Darby, William. Darby's universal gazetteer. 2. ed. Philadelphia, 1827.
- Geog 578.27.3   Hawkes, P. The American companion. Philadelphia, 1827.
- Geog 578.29   Bischoff, F.H.T. Vergleichendes Wörterbuch. Gotha, 1829.
- Geog 578.31   Müller, Johann Wilhelm. Lexicon manuale. Lipsiae, 1831.
- Geog 578.34   Wright, G.N. New and comprehensive gazeteer. London, 1834-37. 4v.
- Geog 578.40   Landmann, G. Universal gazetteer. London, 1840.
- Geog 578.41   McCulloch, J.R. Dictionary geographical, statistical. London, 1841-42. 2v.
- Geog 578.41.2   McCulloch, J.R. Dictionary geographical, statistical. N.Y., 1843-44. 2v.
- Geog 578.41.3   McCulloch, J.R. Dictionary geographical, statistical. N.Y., 1844-45. 2v.
- Geog 578.41.4   McCulloch, J.R. Dictionary geographical, statistical. N.Y., 1844-47. 2v.
- Geog 578.41.5   McCulloch, J.R. Dictionary geographical, statistical. N.Y., 1849. 2v.
- Geog 578.41.6   McCulloch, J.R. Dictionary geographical, statistical. London, 1851. 2v.
- Geog 578.41.7   McCulloch, J.R. Dictionary geographical, statistical. N.Y., 1852. 2v.
- Geog 578.45   Baldwin, Thomas. Universal pronouncing gazetteer. Philadelphia, 1845.
- Geog 578.47   Ritter, Karl. Geographisches-statistisches Lexikon. 3. Aufl. Leipzig, 1847.
- Geog 578.47.9   Ritter, Karl. Geographisches-statistisches Lexikon. Leipzig, 1874. 2v.
- Geog 578.49.9   Baldwin, Thomas. A pronouncing gazetteer. 9. ed. Philadelphia, 1851.
- Geog 578.51   Johnstone, A.K. Dictionary of geography. London, 1851.
- Geog 578.53   Gazetteer of the world. London, 1853. 14v.
- Geog 578.54A   Knight, Charles. The English cyclopaedia. London, 1854-55. 4v.
- Geog 578.54B   Knight, Charles. The English cyclopaedia. v.3. London, 1854-55.
- Geog 578.55   Lippincott, J.B. and Co. Complete pronouncing gazetteer. Philadelphia, 1855.
- Geog 578.55.3   Lippincott, J.B. and Co. Lippincott's pronouncing gazetteer. Philadelphia, 1857.

Classified Listing

Geog 574 - 580 Geography in general - Dictionaries and Gazetteers (By date) - cont.

	Geog 578.55.7	Lippincott, J.B. and Co. Lippincott's pronouncing gazetteer. Philadelphia, 1873.
	Geog 578.55.9	Lippincott, J.B. and Co. Lippincott's gazetteer of the world. Philadelphia, 1880.
	Geog 578.55.11	Lippincott, J.B. and Co. Supplementary tables of population. Philadelphia, 1883.
	Geog 578.55.13A	Lippincott, J.B. and Co. Gazetteer of the world. Philadelphia, 1893.
	Geog 578.55.15	Smith, John Calvin. Harper's statistical gazetteer of the world. N.Y., 1855.
	Geog 578.61	Graesse, J.G.T. Orbis latinus. Dresden, 1861.
Htn	Geog 578.61.5*	Graesse, J.G.T. Orbis latinus. Dresden, 1861.
	Geog 578.68	Beeton, S.O. Dictionary of geography. London, 1868.
	Geog 578.70	Deschamps, P. Dictionnaire de géographie, ancienne et moderne. Paris, 1870.
Htn	Geog 578.71*A	Rosser, William H. The Bijou gazetteer of the world. London, 1871.
Htn	Geog 578.71*B	Rosser, William H. The Bijou gazetteer of the world. London, 1871.
	Geog 578.75	Post-Lexicon; ein Verzeichniss der wichtigeren Verkehrs Orte. Berlin, 1875.
	Geog 578.76	Blackie, W.G. The imperial gazetteer. London, 1876. 2v.
	Geog 578.77	Johnston, A.K. A general dictionary of geography. London, 1877.
	Geog 578.79F	Vivien de St. Martin, L. Nouveau dictionnaire de géographie universelle contenant...la géographie physique. Paris, 1879-95. 7v.
	Geog 578.79.3F	Vivien de St. Martin, L. Nouveau dictionnaire de géographie universelle contenant...la géographie physique. Paris, 1879-95. 7v.
	Geog 578.79.4F	Vivien de St. Martin, L. Nouveau dictionnaire de géographie universelle contenant...la géographie physique. Supplement. Paris, 1897-1900. 2v.
	Geog 578.94	Chamber's consise gazetteer of the world, topographical, statistical, historical. London, 1894.
	Geog 578.98	Garollo, G. Dizionario geografico universale. 4. ed. Milano, 1898.
	Geog 579.07	Demangeon, A. Dictionnaire, manuel, illustré. Paris, 1907.
	Geog 579.26	Vergara y Martin, Gabriel Maria. Diccionario de voces y términos geográficos. Madrid, 1926.
	Geog 579.36A	Schmidt, A.J. Kleines deutsch-portugiesisches...Verzeichnis geographischer Eigennamen. Rio de Janeiro, 1938.
	Geog 579.36B	Schmidt, A.J. Kleines deutsch-portugiesisches...Verzeichnis geographischer Eigennamen. Rio de Janeiro, 1938.
	Geog 579.41	Bonacker, Wilhelm. Karten-Wörterbuch eine Verdeutschung fremdsprachiger Kartensignatur-Bezeichnungen. Berlin, 1941.
	Geog 579.43	Kosack, H.P. Wörterverzeichnis für russische Karten. Berlin, 1943.
	Geog 579.44	Bargilliot, A. Vocabulaire pratique anglais-français. Paris, 1944.
	Geog 579.49	Webster's geographical dictionary. Springfield, Mass., 1960.
	Geog 579.49.5	Webster's geographical dictionary. Springfield, Mass., 1962.
	Geog 579.50F	Lagoa, J.A.M.J. Glossário taponimico da antiga historiografia. Lisboa, 1950- 4v.
	Geog 579.52	Zavatti, Silvio. Dizionario geografico. Catania, 1952.
	Geog 579.52.5F	The Columbia Lippincott gazetteer of the world. N.Y., 1952.
	Geog 579.53	Villalba y Rubio, F. Diccionario geografico universal. Madrid, 1953.
	Geog 579.54A	Bodnarskii, M.S. Slovar' geograficheskikh nazvanii. Moskva, 1954.
	Geog 579.54B	Bodnarskii, M.S. Slovar' geograficheskikh nazvanii. Moskva, 1954.
	Geog 579.54.5	Bodnarskii, M.S. Slovar' geograficheskikh nazvanii. Moskva, 1958.
	Geog 579.55A	Volostnova, M.B. Slovar' russkoi...geograficheskikh nazvanii. Moskva, 1955. 2v.
	Geog 579.55B	Volostnova, M.B. Slovar' russkoi...geograficheskikh nazvanii. v.1. Moskva, 1955.
	Geog 579.55.5	Webster's geographical dictionary. Springfield, 1955.
	Geog 579.56	Grigson, G. Places. N.Y., 1956?
	Geog 579.56.5	Swayne, James Colin. A concise glossary of geographical terms. London, 1956.
	Geog 579.56.7	Swayne, James Colin. A concise glossary of geographical terms. 2. ed. London, 1962.
	Geog 579.59	Öngör, Sami. Coğrafya sözlüğü. Fasc.1-5. İstanbul, 1959.
	Geog 579.59.5	Öngör, Sami. Coğrafya terimleri sözlüğü. İstanbul, 1975.
	Geog 579.60	Kratkaia geograficheskaia entsiklopecha. Moskva, 1960. 5v.
	Geog 579.60.5	The worldmark encyclopedia of the nations. N.Y., 1960.
	Geog 579.61	British Association for the Advancement of Science. A glossary of geographical terms. London, 1961.
	Geog 579.63	Hustich, I. Tämän päivän maailmaa. 4. ed. Helsinki, 1963.
	Geog 579.65.5	Stolitsy stran mira. Moskva, 1965.
	Geog 579.66	Longman's dictionary of geography. London, 1966.
	Geog 579.66.5	Soto Mora, Consuelo. Glosario de términos geográficos. 1. ed. México, 1966.
	Geog 579.68	Kalesnik, Stanislaw W. Entsiklopedicheskii slovar' geograficheskikh terminov. Moskva, 1968.
	Geog 579.68.5	Westermann Lexikon der Geographie. Braunschweig, 1968-72. 5v.
	Geog 579.70	Dictionnaire de la géographie. Paris, 1970.
	Geog 579.71	Fochler-Hauze, Gustav. Allgemeine Geographie. Frankfurt am Main, 1971.
	Geog 579.73	Entsiklopedicheskii slovar' geograficheskikh nazvanii. Moskva, 1973.
	Geog 579.73.5	Pietkiewicz, Stanisław. Słownik pojęć geograficznych. Wyd. 1. Warszawa, 1973.

Geog 585 Geography in general - Tables

Geog 585.5	A geographical instructor. London, 1825.
Geog 585.7	Coulier, P.J. Tables des principales positions geonomiques. Paris, 1828.
Geog 585.9	Levasseur, E. Extrait de l'Annuaire in Bureau de Longitudes. n.p., 1905.
Geog 585.11	Crump, William H. The world in a pocket-book. Philadelphia, 1841.
Geog 585.11.2	Crump, William H. The world in a pocket-book. Philadelphia, 1842.

Geog 585 Geography in general - Tables - cont.

Geog 585.11.5A	Crump, William H. The world in a pocket-book. Philadelphia, 1845.
Geog 585.11.5B	Crump, William H. The world in a pocket-book. Philadelphia, 1845.
Geog 585.11.10	Crump, William H. The world in a pocket-book, or Universal popular statistics. 12th ed. Phialdelphia, 1860.
Geog 585.12F	Borbstaedt, A. Allgemeine geographische und statistische Verhältnisse. Berlin, 1846.
Geog 585.13	Tavole sinottiche di geografia. Livorno, 1834.
Geog 585.15.3	Albrecht, Theodor. Formeln und Hülfstafeln für geographische Ortsbestemmungen. 3. Aufl. Leipzig, 1894.

Geog 600 Geography in general - Biographies of geographers - Collected

Geog 600.5	Abulhasan Mansur. Arabische Schriftsteller über die Geographie Indiens. Inaug. Diss. Berlin, 1919.
Geog 600.6	Literary record of Cleveland Abbe. Ithaca, 1931.
Geog 600.7	Ciampi, I. Oltre l'Alpe e il mare. Roma, 1865.
Geog 600.10	Baranskii, N.N. Otechestvennye ekonomiko-geografy XVIII-XX vv. Moskva, 1957.
Geog 600.11	Baranskii, N.N. Otechestvennye fiziko-geografy i puteshestvenniki. Moskva, 1959.
Geog 600.12	Ekonomicheskaia geografiia v SSSR, istoriia i sovremennoe razvitie. Moskva, 1965.
Geog 600.15	Carcie, Giuseppe. Tre Fiorentini del Rinascimento. v.7. Roma, 1959?
Geog 600.20	Harms, Hans. Künstler des Kartenbildes. Oldenburg, 1962.
Geog 600.21	Olszewicz, Bolesław. Dziewięć wieków geografii polskiej. Warszawa, 1967.
Geog 600.22	Matveeva, T.P. Russkie geografy i puteshestvenniki. Leningrad, 1971.
Geog 600.24	Gilbert, Edmund William. British pioneers in geography. Newton Abbot, 1972.

Geog 601 - 626 Geography in general - Biographies of geographers - Individual (A-Z by person)

Geog 601.5	Pertish, E.N. K.I. Arsen'ev i ego raboty po Raiomirovamiiu Rossii. Moskva, 1960.
Geog 602.1	Baudet, P.J.H. Leven en werken van Williem Janszoon Blaeu. Utrecht, 1871.
Geog 602.1.5	Stevenson, E.L. Willem Janszoon Blaeu. N.Y., 1914.
Geog 602.1.10	Amsterdam. Nederlansh Historisch Schlepvaart Museum. De Blaeu's beschrijvers van land. Amsterdam, 1952.
Geog 602.3	Blanchard, Raoul. Ma jeunese sous l'aile de Péguy. Paris, 1961.
Geog 602.3.5	Blanchard, Raoul. Je découvre l'université. Paris, 1963.
Geog 602.3.10	Association des Amis de l'Université de Grenoble. In memoriam Raoul Blanchard, 1877-1965. Grenoble, 1966.
Geog 603.1	Graubner, Paul. Fr. Cannabich (1777-1859) sein Leben und sein Werke. Koniejeberg, 1913.
Geog 603.2	Partsch, J. Philipp Clüver der Begründer der historischen Landerkunde. Wien, 1891.
Geog 603.4	Briceno, Alfonso. Augustin Codazzi. Tesis. Caracas, 1928.
Geog 603.5	Jean-Baptiste Charcot, 1867-1936. Paris, 1936.
Geog 603.5.5	Oulié, M. Jean Charcot. 14. éd. Paris, 1937.
Geog 603.5.8	Emmanuel, Marthe. J.B. Charcot. Paris, 1945.
Geog 603.5.8.5	Emmanuel, Marthe. Tel fut Charcot, 1867-1936. Paris, 1967.
Geog 603.5.10	Oulié, M. Charcot of the Antartic. London, 1938.
Geog 603.6	Armao, Ermanno. Vincenzo Coronelli. Firenze, 1944.
Geog 603.6.5	Catalogo dei globi antici conservati in Italia. pt.1-2. Firenze, 1957.
Geog 603.7	Constantini, Otto. Leben und Wirken eines österreichischen Geographieprofessors und Erwachsenenbildners. Linz, 1969.
Geog 603.8	Kleopov, Igor' L'. Aleksandr Lavrent'evich Chekanovskii, 1833-1876. Leningrad, 1972.
Geog 603.9	Brown, Lloyd A. Jean Domenique Cassini and his world map of 1696. Ann Arbor, 1941.
Geog 604.1	Anthiaume. L'abbé Guillaume Denys de Dieppe, 1624-1689. Paris, 1927.
Geog 604.2	Anthiaume. Pierre Desceliers. Rouen, 1926.
Geog 604.3	Couto, Gustavo. O cosmografo Fernam Vaz Dourado, fronteiro da India e a sua obra. Lisboa, 1928.
Geog 604.4	Wagner, H.R. George Davidson, geographer of the northwest coast of America. n.p., 1932.
Geog 604.4.5	Lewis, Oscar. George Davidson. California, 1954.
Geog 604.5	Hoff, Bert van. Jacob van Deventer. 's-Gravenhage, 1953.
Geog 606.1	Gallois, L. De Crontio finaeo gallico geographo. Paris, 1890.
Geog 606.2	Coma Soley, V. Jaime Ferrer...y el descubimiento de America. 1. ed. Barcelona, 1952.
Geog 606.3	Uzbekistan. Tsentral'nyi Gosudarstvennyi Arkhiv. Otdel Dorevoliutsionnykh Fondov. A.P. Fedchenko. Tashkent, 1956.
Geog 606.3.5	Leonov, Nikolai I. Aleksei Pavlovich Fedchenko, 1844-1873. Moskva, 1972.
Geog 607.1	North, S.N.D. Henry Gannett, president of the National Geographic Society, 1910-1914. n.p., 1915.
Geog 607.5	Fraerman, R.I. Zhizn' i prikl. K.L. Golovnina. Moskva, 1946.
Geog 607.5.5	Fraerman, R.I. Zhizn' i neobyknovennye prikl. K.L. Golovnina. Moskva, 1957.
Geog 607.5.10	Ivashchenko, M.M. Admiral Golovnin. Moskva, 1946.
Geog 607.5.15	Davydov, Iurii Vl. Golóvnin. Moskva, 1968.
Geog 607.10	Galorm'a, R.M. Sochineniia. Moskva, 1949.
Geog 607.11	Vlora, Gribaudi. L'uomo e lo studioso. Bari, 1971.
Geog 607.15	Bonasera, F. Un Gobo Terrestre di Matteo Greuter conservato nella Biblioteca. Camarino? 1959.
Geog 607.16	Gunchev, Guncho S. Guncho Gunchev. Sofiia, 1941.
Geog 607.20	Schweizinscher Zofinguverein. Souvenir de l'inauguration du monument élevé à Arnold Guyot par la Société de Zofingue à l'Académie de Neuchâtel le 6 mai 1892. Neuchâtel, 1892.
Geog 608.2	Hubbard, Gardiner Greene. Contents of box in corner stone of Hubbard Memorial. n.p., n.d.
Geog 608.5	Almagià, Roberto. L'opera geografica di Luca Holstenio. Citta del Vaticano, 1942.
Geog 608.10	Martin, Geoffrey J. Ellsworth Huntington; his life and thought. Hamden, Conn., 1973.
Geog 608.11	Drake, Fred W. China charts the world. Hsu, Chi-Yü and his geography of 1848. Cambridge, 1975.
Geog 608.15	Alfred Hettner. 6.8.1859 Gedenkschrift zum 100. Geburtstag. Heidelberg, 1960.
Geog 610.1	Cortambert, R. Notice sur la vie et les oeuvres de M. Jomard. Paris, 1863.
Geog 610.5	Martin, Geoffrey J. Mark Jefferson, geographer. Ypsilanti, Mich., 1968.
Geog 611.1	Dupouy, A. Le Briton Yves de Kerguelen. Paris, 1928.

Classified Listing

Geog 601 - 626 Geography in general - Biographies of geographers - Individual (A-Z by person) - cont.

Geog 611.1.5	Brossard, Maurice Raymond de. Kerguelen, le découvreur et ses îles. Paris, 1970-1971. 2v.
Geog 611.2	Heinrich Kiepert bei seiner Rückkehr November 1886 von Freunden gewidmet. Berlin, 1886.
Geog 611.3	Meier-Lemgo, K. Engelbert Kampfer, der erste deutsche Forschungsreisende, 1651-1716. Stuttgart, 1937.
Geog 611.5	Novlianskaia, Mariia G. I.K. Kirilov i ego. Atlas Vserossisskoi imperii. Moskva, 1958.
Geog 611.5.5	Novlianskaia, Mariia G. Ivan Kirilovich Kirilov, geograf XVIII veka. London, 1964.
Geog 612.1	Materialien zu einer Biographie des Fr. J.M. Liechtenstern. Schneeberg, 1823.
Geog 612.2	Mazuel, J. L'oeuvre géographique de Linant de Bellefonds. Thèse. Le Caire, 1937.
Geog 612.3	Allen, Edward W. Jean François Galaup de Lapérouse. San Francisco, 1941.
Geog 612.3.2	Allen, Edward W. The vanishing Frenchman. Rutland, 1959.
Geog 612.3.5	Maine, René. Lapérouse. Paris, 1946.
Geog 612.3.8	Bellessort, André. La Pérouse. Paris, 1926.
Geog 612.3.10	Scott, Ernest. Laperouse. Sydney, 1912.
Geog 612.3.15	Dondo, M.M. La Perouse in Maui. Wailuku, 1959.
Geog 612.4.1	Fradkin, Naum G. Akademik I.I. Lepekhin i ego puteshestviia po Rossii v 1768-1773 gg. Moskva, 1950.
Geog 612.4.2	Fradkin, Naum G. Akademik I.I. Lepekhin i ego puteshestviia po Rossii v 1768-1773 gg. 2. izd. Moskva, 1953.
Geog 612.4.4	Fradkin, Naum G. Puteshestviia I.I. Lepekhina, N. Ia. Ozeretskovskogo, V.F. Zueva. Moskva, 1948.
Geog 612.4.5	Lukina, Tat'iana A. Ivan Ivanovich Lepekhin. Leningrad, 1965.
Geog 612.5	Kondracki, Jerzy. Stanisław Lencewicz. Warszawa, 1966.
Geog 612.6	Alekseev, Aleksandr G. R. Fedor Petrovich Litke. Moskva, 1970.
Geog 613.1	Pamphlet vol. Gerardus Mercator.
Geog 613.1.3	Raemdonck, J. van. Gérard Mercator. St. Nicolas, 1869.
Geog 613.1.5	Raemdonck, J. van. Gérard de Cremer ou Mercator. St. Nicolas, 1870.
Geog 613.1.7	Cologne, Germany. Stadtbibliothek. Katalog einer Mercator-Ausstellung. Koeln, 1894.
Geog 613.1.11	Breusing, A. Gerhard Kremer...Mercator. Duisburg, 1869.
Geog 613.1.13	Raemdonck, J. van. Relations commerciales entre G. Mercator et C. Plantin. Anvers, 1880.
Geog 613.1.15	Dinse, Paul. Zum Gedächtnis Gerhard Mercator's. Berlin, 1894.
Geog 613.1.18	Mercator, Gerardus. Correspondance mercatorienne. Anvers, 1959.
Geog 613.2.5	Frenzel, R. Malthe Conrad Bruun. Crimmitschau, 1908.
Geog 613.3.5	Murray, John. John Murray III, 1808-1892; memoir. London, 1919.
Geog 613.4.5	Markham, A.H. The life of Sir Clements R. Markham. London, 1917.
Geog 613.4.15	Bernstein, H. Sir Clements R. Markham as a translator. n.p., 1937?
Geog 613.4.20	Olivas, A. Contribución a la bibliotheca de C.R. Markham. Lima, 1924.
Geog 613.5	Borodajkez, Taras. Konrad Millers Lebenswerk. Salzburg, 19- .
Geog 613.5.5	Festschrift zum 70...Konrad Miller. Bremen, 1949.
Geog 613.6	Espinosa Cordero, N. Pedro Vicente Maldonado, y la Misión geodésica del siglo XVIII. Cuenca, 1936.
Geog 613.7	Mill, H.R. An autobiography. London, 1951.
Geog 613.8	Gilbert, Edmund W. Sir Halford MacKinder, 1861-1947, an appreciation of his life and work. London, 1961.
Geog 613.9	Novlianskaia, Mariia G. Daniil Gotlib Messershmidt i ego raboty po issledovaniiu Sibiri. Leningrad, 1970.
Geog 613.10	Markov, Konstantin K. Vospominaniia i razmyshleniia geografa. Moskva, 1973.
Geog 614.5F	Marinelli, G. Cristoforo Negri. Torino, 1897.
Geog 614.6	Kupferschmidt, F. Karl Neumann. Inaug. Diss. Leipzig, 1935.
Geog 614.7	Nałkowska, Z. Moj ojciec. Warszawa, 1953.
Geog 614.8	Nałkowska, Z. Moja ojciec. Wyd. 2. Warszawa, 1955.
Geog 614.9	Olszewicz, B. Wacław Nałkowski. Warszawa, 1962.
Geog 614.10	Dontsova, Zoia N. Sergei Semenovich Neustruev, 1874-1928. Moskva, 1967.
Geog 614.15	Barcinski, Florian. Stanisław Nowakowski. Warszawa, 1965.
Geog 616.1.3	Gravier, G. Notice sur Jean Parmentier. Rouen, 1902.
Geog 616.2	Spano, Benito. Gli atlanti corografici del cavaliere C.G. Pocelli. Bari, 1958.
Geog 616.3.5	Peattie, R. The incurable romantic. N.Y., 1941.
Geog 616.5	Eavenson, H.N. Map maker and Indian traders...John Patten. Pittsburgh, 1949.
Geog 616.25	Weller, E. August Petermann. Leipzig, 1911.
Geog 616.30	Wyder, Samuel. Die Schaffhauser Karten von Hauptmann Heinrich Peyer (1621-1690). Zürich, 1951.
Geog 616.31	Olszewicz, Bolesław. Stanisław Pawłowski. Warszawa, 1968.
Geog 616.32	Talyzin, Fedor P. Puteshestviia za nevidimym vragom. Moskva, 1974.
Geog 618.1.5	Gage, W.L. Life of Carl Ritter. N.Y., 1867.
Geog 618.1.7	Guyot, A. Carl Ritter. Princeton, 1860.
Geog 618.1.9	Richter, O. Der teleologische Zug im Denken C. Ritters. Borna, 1905.
Geog 618.2	Markham, C.R. Major Jame Rennell. London, 1895.
Geog 618.2.2	Markham, C.R. Major James Rennell. N.Y., 1895.
Geog 618.2.5	Frenzel, C.A. Major James Rennell. Leipzig, 1904.
Geog 618.3	Reclus, Élisée. Correspondance. v.1-2,3. Paris, 1911-25. 2v.
Geog 618.3.3	Ishill, Joseph. Élisée and Elie Reclus, in memoriam. Berkeley Heights, N.J., 1927.
Geog 618.3.5	Greef, Guillaume J. de. Discours prononcé par Monsieur le recteur Guillaume de Greef. Gand, 1906.
Geog 618.3.7	Nettlau, Max. Elisée Reclus. Berlin, 1928.
Geog 618.3.8	Nettlau, Max. Eliseo Reclus, la vida de un sabio justo. Barcelona, 1929. 2v.
Geog 618.4	Romer, E. Wybór prac. Warszawa, 1960. 3v.
Geog 618.4.5	Mazurkiewicz-Herzowa, Kucja. Eugeniusz Romer. Warszawa, 1966.
Geog 618.5	Wanklyn, H.G. Friedrich Ratzel. Cambridge, Eng., 1961.
Geog 618.5.5F	Babicz, J. Nauka o ludakh Fryderyka Ratzla. Wrocław, 1962.
Geog 618.5.10	Ratzel, Fridrich. Jugenderinnerungen. München, 1966.
Geog 618.6	Gol'denberg, Leonid A. Semen Ul'ianovich Remezov. Moskva, 1965.
Geog 618.7	Mil'kov, Fedor N. P.I. Rychkov. Moskva, 1953.
Geog 619.1	Gedenkboek ter herinnering aan den 70sten verjaardag van R. Schuiling, 27 mei, 1924. Groningen, 1924.

Geog 601 - 626 Geography in general - Biographies of geographers - Individual (A-Z by person) - cont.

Geog 619.5	Shokal'skaia, Z. Iu. Zhiznennyi put' Iu. M. Shokal'skogo. Moskva, 1960.
Geog 619.6	Rowley, V.M. J. Russell Smith, geographer, educator and conservationist. Philadelphia, 1964.
Geog 619.7	Schultén, N.G. Levnadsteckning. Helsingfors, 1964.
Geog 619.8	Novlianskaia, Mariia G. Filipp Iogann Stralenberg. Leningrad, 1966.
Geog 619.9	Gol'denberg, Leonid A. Fedor Ivanovich Soimonov, 1692-1780. Moskva, 1966.
Geog 619.10	Dobrowolska, Maria. Ludomir Sawicki. Warszawa, 1968.
Geog 619.11	Schmieder, Oskar. Lebenserinnerungen und Tagebuchblätter eines Geographen. Kiel, 1972.
Geog 619.12	Termer, Franz. Karl Theodor Sapper, 1866-1945. Leipzig, 1966.
Geog 622.1	Ciampi, Ignatius. Della vita e...opere di Pietro della Valle il Pellegrino. Roma, 1880.
Geog 622.1.5	Blunt, Wilfred. Pietro's pilgrimage. London, 1953.
Geog 622.2	Shostiu, N.A. M.P. Vronchenko. Moskva, 1956.
Geog 622.5	Bonapace, Umberto. Luigi Visintin. Novara, 1958.
Geog 623.1	Gebhard, J.F. Het leven van Mr. Nicolaas C. Witsen. Utrecht, 1881. 3v.
Geog 623.2	Wright, John K. Human nature in geography. Cambridge, 1966.
Geog 623.3	Babicz, Józef. Teoria Moritza Wagnera o powstawaniu gatunków. Wrocław, 1966.
Geog 624.1	Konnyü, Leslie. John Xantus. Köln, 1965.

Geog 656 - 660 Geography in general - Theory, aims, methods of geography (By date)

Geog 658.81	Cora, G. Cenni intorno all'attuale. Torino, 1881.
Geog 659.19	Schrader, Franz. The foundations of geography in the 20th century. Oxford, 1919.
Geog 659.28.1	Spethman, Hans. Dynamische Länderkunde. Kiel, 1972.
Geog 659.39.5A	Hartshorne, Richard. Perspective on the nature of geography. Chicago, 1959.
Geog 659.39.5B	Hartshorne, Richard. Perspective on the nature of geography. Chicago, 1959.
Geog 659.42	Cholley, André. Guide de l'étudiant en géographie. 1. éd. Paris, 1942.
Geog 659.42.5	Cholley, André. La géographie. 2. éd. Paris, 1951.
Geog 659.53	Sestini, Aldo. Avviamento allo studio della geografia. 1. ed. Firenze, 1953.
Geog 659.53.5	Spain. Consejo Superior de Investigaciones Cientificas. Iniciación a la geografia local. Zaragoza, 1953.
Geog 659.55	Jong, Guben. Het karakter van de geografische totaliteit. Groningen, 1955.
Geog 659.57	Filchner, W. Route-mapping and position locating in unexplored regions. Basel, 1957.
Geog 659.58	Ackerman, Ed. Geography as a fundamental research discipline. Chicago, 1958.
Geog 659.59	Scotti, Pietro. Elementi di geografia. Genova, 1959.
Geog 659.59.5	Birot, Pierre. Précis de géographie physique générale. Paris, 1959.
Geog 659.60	Phlipponneau, Michel. Geographie et action. Paris, 1960.
Geog 659.62	Jong, Guben. Chonological differentiation as the fundamental principle of geography. Groningen, 1962.
Geog 659.63.2	Gregory, Stanley. Statistical methods and the geographer. 2nd ed. London, 1971.
Geog 659.65	Overbeck, Hermann. Kulturlandschaftsforschung und Landeskunde. Heidelberg, 1965.
Geog 659.65.7	Chorley, Richard J. Frontiers in geographical teaching. 2nd ed. London, 1970.
Geog 659.67	Storkebaum, Werner. Zum Gegenstand und zer Methode der Geographie. Darmstadt, 1967.
Geog 659.69	King, Leslie J. Statistical analysis in geography. Englewood Cliffs, 1969.
Geog 659.69.5	Conference on Quantitative Methods in Geography, New York, 1969. Quantitative methods in geography; a symposium. N.Y., 1969.
Geog 659.70	Harvey, David. Explanation in geography. N.Y., 1970.
Geog 659.71	Hampl, Martin. Teorie komplexity a diferenciace světa se zolástním zřetelem na diferenciaci geograficksu. 1. vyd. Praha, 1971.
Geog 659.71.5	Chorley, Richard J. Models in geography. London, 1971.
Geog 659.71.10	Santos, Milton. Le métier de géographe en pays sous-développé. Paris, 1971.
Geog 659.71.15	Beaujeu-Garnier, Jacqueline. La géographie: méthods et perspectives. Paris, 1971.
Geog 659.73	Racine, Jean Bernard. L'analyse quantitative en géographie. 1. éd. Paris, 1973.
Geog 659.73.5	Simpozium po teoreticheskim problemam geografii, Riga, 1973. Teoreticheskaia geografiia. Riga, 1973.
Geog 659.73.10	Entwicklungstendenzen der Geographie. Berlin, 1973.
Geog 659.73.15	Hard, Gerhard. Die Geographie. Berlin, 1973.
Geog 659.74	Hammond, Robert. Quantitative techniques in geography: an introduction. Oxford, 1974.
Geog 659.74.5	Mukitanov, Naurzbai K. Problema tselostnosti v fizicheskoi geografii. Alma-Ata, 1974.
Geog 659.74.10	Baker, Laurie. A selection of geographical computer programs. London, 1974.
Geog 659.75	Amedeo, Douglas. An introduction to scientific reasoning in geography. N.Y., 1975.

Geog 665 Geography in general - General miscellany about geography

Geog 665.4	Umlauft, F. Die Pflege der Erdkunde in Oesterreich...Festschrift...Franz Josef I. Wien, 1898.
Geog 665.5	Ratzel, F. Zu Friedrich Ratzels Gedächtnis. Leipzig, 1904.
Geog 665.7	Strachey, R. Lectures on geography. London, 1888.
Geog 665.9A	Ritter, Karl. Geographical studies. Boston, 1863.
Geog 665.9B	Ritter, Karl. Geographical studies. Boston, 1863.
Geog 665.9.3	Ritter, Karl. Geographical studies. Cincinnati, 1861.
Geog 665.12	Tagsberichte über die Forstschritte. Weimar, 1852.
Geog 665.15	Geographical, commercial...essays. London, 1812.
Geog 665.17	Günther, S. Geographische Studien. Stuttgart, 1907.
Geog 665.18	Peschel, O.F. Abhandlungen zur Erd- und Völkerkunde. v.1,2-3. Leipzig, 1877. 2v.
Geog 665.20	Historisch-statistisch-geographische Belustigungen. Leipzig, 1782.
Geog 665.21	Scenes in foreign lands. N.Y., 1853.
Geog 665.22	Phillips, R. The hundred wonders of the world. 1st American ed. New Haven, 1821.
Geog 665.23	Sherwood, M.E. Here and there and everywhere. Chicago, 1898.
Geog 665.24	Ingwood of Westchester. Transatlantic souvenirs. N.Y., 1868.

Classified Listing

Geog 665 Geography in general - General miscellany about geography - cont.

	Geog 665.26	Famous islands. Boston, n.d.
	Geog 665.27	Aufsätze Prof. Dr. Eugen Oberhummer gewidmet. Brünn, 1919.
Htn	Geog 665.28*	Pamphlet box. Greeley, A.W. Essays on geographical subjects.
	Geog 665.29F	Santarem, M.F. de B. Inéditos (miscellanea). Lisboa, 1914.
	Geog 665.30F	Santarem, M.F. de B. Opusculos e esparsos. Lisboa, 1910. 2v.
	Geog 665.31	Miller, Émile. Pour qu'on aime la géographie. Montréal, 1921.
	Geog 665.32	Vallaux, Camille. Les sciences géographiques. Paris, 1925.
	Geog 665.34	Hoel. Geographische Charakter-Bilder für Schule und Haus. pt.1-10. Supplement. Wien, 1886.
	Geog 665.35	Mélanges de géographie offerts par ses collègues et amis de l'étranger à M. Václav Svambera. Praha, 1936.
	Geog 665.36	Freeman, Henry. Wonders of the world. Boston, 1873.
	Geog 665.37	Singleton, E. Wonders of nature. N.Y., 1900.
	Geog 665.37.5	Singleton, E. Wonders of nature. N.Y., 1911.
	Geog 665.37.6	Singleton, E. The wonders of nature as seen and described by Alexandre Dumas. Washington, 1962.
	Geog 665.37.10	Singleton, E. Greatest wonders of the world. N.Y., 1906.
	Geog 665.38	A geographical present. N.Y., 1831.
	Geog 665.39	Ricchieri, G. Dopo il viaggio d'istruzione. Firenze, 1914.
	Geog 665.40F	Hachette, firm, publishers, France. Les merveilles du monde. Paris, 192-?
	Geog 665.40.5F	Hachette, firm, publishers, France. Les merveilles du monde. Paris, 1957.
	Geog 665.42	Ward, F.K. Modern exploration. London, 1945.
	Geog 665.43	Errera, Carlo. Scritti geografici scelti e ordinati a cura del Comitato nazionale. Bologna, 1937.
	Geog 665.45	Kühn, Arthur. Die Neugestaltung der deutschen Geographie im 18. Jahrhundert. Leipzig, 1939.
	Geog 665.46	Arden-Close, Charles. Geographical by-ways and some other geographical essays. London, 1947.
	Geog 665.47	Scritti di geografia e di storia della geografia consernenti l'Italia pubblicati in onore di Giuseppe della Vedova. Firenze, 1908.
	Geog 665.49	Mélanges géographiques offerts par ses élèves à Raoul Blanchard. Grenoble, 1932.
	Geog 665.50	Van Loon, H.W. Van Loon's geography, the story of the world we live in. N.Y., 1932.
	Geog 665.50.5	Van Loon, H.W. Van Loon's geography. Garden City, N.Y., 1940.
	Geog 665.52	Mogey, J.M. The study of geography. London, 1950.
	Geog 665.53	Chicago. University. Norman Wait Harris Foundation. Reports of round tables, 1937. Geographic aspects of international relations. n.p., 1937.
	Geog 665.54	Livre jubilaire offert à Maurice Zimmermann. Lyon, 1949.
	Geog 665.56	Stamp, L. London essays in geography. Cambridge, 1951.
	Geog 665.58A	Taylor, Griffith. Geography in the twentieth century; a story of growth, fields, techniques, aims and trends. N.Y., 1951.
	Geog 665.58B	Taylor, Griffith. Geography in the twentieth century; a story of growth, fields, techniques, aims and trends. N.Y., 1951.
	Geog 665.58.2	Taylor, Griffith. Geography in the twentieth century. 2nd ed. N.Y., 1953.
	Geog 665.58.3	Taylor, Griffith. Geography in the twentieth century. 3rd ed. N.Y., 1957.
	Geog 665.60	Mélanges de géographie et d'orientalisme offerts à E.-F. Gautier. Tours, 1937.
	Geog 665.62	Karper, Kurt. Landschaft und Lund. Remagen, 1951.
	Geog 665.64	Dardel, E. L'homme et la terre. Paris, 1952.
	Geog 665.68	Geographische Studien. Wien, 1951.
	Geog 665.69	Geographische Gesellschaft in Wien. Festschrift zur Hundertjahrfeier der geographischen Gesellschaft in Wien, 1856-1956. Wien, 1957.
	Geog 665.70	Bologna. Università. Istituto di Geografia. Studi geografici in onore di Antonio Renato Toniolo. Milano, 1952.
	Geog 665.72	Mélanges géographiques offerts au Doyen ernest Benevent. Gap, 1954.
	Geog 665.73	Geograficheskoe Obshchestvo SSSR. Essais de géegraphie. Moscou, 1956.
	Geog 665.80	Koegel, Ludwig. Geographische Plaudereien. Bonn, 1957.
	Geog 665.81	International Geographical Union. Regional Conference in Japan, Tokyo and Nara, 1957. Proceedings of the IGU Regional Conference in Japan, August 28-September 3, 1957. Tokyo, 1959.
	Geog 665.82	Sorre, Maximilien. Rencontres de la géographie et de la sociologie. Paris, 1957.
	Geog 665.84	Bulgarska Akademiia na Naukite, Sofia. Geografski Institut. Sbornik v chest na akademik Anastas Stoianov Beshkov. Sofiia, 1959.
	Geog 665.86	Miller, Ronald. Geographical essays in memory of Alan G. Ogilvie. London, 1959.
	Geog 665.88	Demangeon, Albert. Problèmes de géographie humaine. 4. éd. Paris, 1952.
	Geog 665.92	Plattie, Roderick. College geography. Boston, 1926.
	Geog 665.93	France. Centre National de la Recherche Scientifique. Colloque national de géographie appliquée, Strasbourg, 20-22 avril 1961. Paris, 1962.
	Geog 665.94	McCashill, Murray. Land and livelihood. Christchurch, 1962.
	Geog 665.95	Mori, Assunto. Scritti geografici, scelti e ordinati. Pisa, 1960.
	Geog 665.96	Hafemann, D. Mainzer geographische Studien. Braunschweig, 1961.
	Geog 665.97	Leidlmar, A. Herman von Wissmann-Festschrift. Tübingen, 1962.
	Geog 665.98	Bulgarska Akademiia na Naukite, Sofia. Geografski Institut. Sbornik v chestna chlen-korespondent Iordan Zakhariev. Sofiia, 1964.
	Geog 665.99	Akademiia Nauk SSSR. Institut Geografii. Razvitie i preobrazovanie geografii predy. Moskva, 1964.
	Geog 665.100	National Research Council. Ad Hoc Committee on Geography. The science of geography. Washington, 1965.
	Geog 665.101	Whittow, John Byron. Essays in geography for Austin Miller. Reading, Eng., 1965.
	Geog 665.102	Schickel, Joachim. Terra incognita. Bergisch Gladbach, 1965.
	Geog 665.103	House, John. The frontiers of geography. New Castle upon Tyne, 1965.
	Geog 665.104	Neue Fragen der allgemeinen Geographie. Wurzburg, 1964.

Geog 665 Geography in general - General miscellany about geography - cont.

Geog 665.105	Weigt, Ernst. Angewandte Geographie; Festschrift für Professor Dr. Erwin Scheu. Nürnberg, 1966.
Geog 665.107	Andrews, John. Frontiers and men; a volume in memory of Griffith Taylor (1880-1963). Melbourne, 1966.
Geog 665.110	Chile. Universidad, Santiago. Facultad de Filosofia y Educacion. Estudios geográficos. Homenaje de la Facultad de Filosofia y Educacion a Don Huberto Fuenzalida Villegas. Santiago, 1966.
Geog 665.112	Colloque International de Géographie Appliquée, Liège, 1967. Comptes rendus. Liège, 1968.
Geog 665.115	Sauer, Carl Ortwin. Land and life. Berkeley, 1965.
Geog 665.116	Lulovac, Milisav. Cvijićev zbornik u spomen 100. godišnjice njegovog rodjeuja. Beograd, 1968.
Geog 665.116.5	Cvijić, Jovan. Opšta geografija; antropogeografija. Beograd, 1969.
Geog 665.118	Johnston, William. Dynamic relationships in physical geography. Christchurch, 1967.
Geog 665.120	Bowen, Emzys George. Geography at Aberystwyth: essays written on the occasion of the departmental jubilee 1917-1918-1967-1968. Cardiff, 1968.
Geog 665.122	Festschrift für Hans Kinzl zum siebzigsten Geburtstag. Innsbruck, 1968.
Geog 665.124	Symposium über Fragen der Naturräumlichen Gliederung. Probleme der landschaftsökologischen Erkundung und naturräumlichen Gliederung. Leipzig, 1967.
Geog 665.126	Leipziger geographische Beiträge. Text and Atlas. Leipzig, 1965. 2v.
Geog 665.130	Steering Committee for Celebration of the Sixtieth Year of Prof. S.P. Chatterjee. Essays in geography. Calcutta, 1965.
Geog 665.132	Settlement and encounter; geographical studies presented to Sir Grenfell Price. Melbourne, 1968.
Geog 665.134	Geographers in government: a series of papers given at meetings of the Geography Section of the AAAS in New York. N.Y., 1968.
Geog 665.136	Geographical essays in honour of K.C. Edwards. Nottingham, 1970.
Geog 665.138	Festkolloquim: 100 Jahre Geographie in Giessen. Giessen, 1965.
Geog 665.142	Mass, Walther Gerhard Eduard. Menschen und Landschaften. Hildesheim, 1967.
Geog 665.144	Trends in geography. 1st ed. Oxford, 1969.
Geog 665.146	Gottmann, Jean. The renewal of the geographic environment: an inaugural lecture delivered before the University of Oxford on 11 February 1969. Oxford, 1969.
Geog 665.148	Beiträge zur Geographie der Tropen und Subtropen; Festschrift zum 60. Geburtstag von Herbert Wilhelmy. Tübingen, 1970.
Geog 665.150	Argumenta geographica; Festschrift Carl Troll zum 70. Geburtstag. Bonn, 1970.
Geog 665.152	Mélanges de géographie physique. Gembloux, 1967. 2v.
Geog 665.154	Eesti Geograafia Selts. Orazvitii geografii v Estonskoi SSR 1960-1968. Tallin, 1970.
Geog 665.156	Davis, William Morris. Geographical essays. N.Y., 1954.
Geog 665.158	Riva, Ambrogio. Una piccola biblioteca. Milano, 1970.
Geog 665.162	Raeumliche und zeitliche und Bewegungen. Würzburg, 1972.
Geog 665.164	James, Preston Everett. On geography: Selected writings of Preston E. James. 1st ed. Syracuse, 1971.
Geog 665.165	Forschungen zur allgemeinen und regionalen Geographie; Festschrift für Kurt Kayser. Wiesbaden, 1971.
Geog 665.166	La pensée géographique française contemporaine; Mélanges offerts à André Meynier. Saint-Brieuc, 1972.
Geog 665.167	Prague. Universita Karlova. Sborník prací Geografických kateder UK k 75. narozeninám prof. dr. Jaromíra Korčáka, DrSc. Praha, 1970.
Geog 665.168	Gould, Peter R. Mental maps. Harmondsworth, 1974.
Geog 665.169	Hans Graul-Festschrift. Heidelberg, 1974.

Geog 755 - 760 Physical, Mathematical, and Descriptive geographies - General works (By date)

Htn	Geog 755.19*	Introductio in Ptolomei. n.p., n.d.
Htn	Geog 755.27*	Glareanus, H. De geographia liber unus. Basileae, 1527.
Htn	Geog 755.27.2*	Glareanus, H. De geographia liber unus. Friburgum, 1530.
	Geog 755.27.10	Fritzsche, O.F. Glarean sein Leben und seine Schriften. Frauenfeld, 1890.
Htn	Geog 755.63.2*	Postel, Guillaume. De universitate liber. 2a ed. pt.1-2. Parisiis, 1563.
Htn	Geog 755.71*	Nores, Jason de. Breve trattato del mondo e delle sue parti. Venetia, 1571.
Htn	Geog 756.28*	Velazquez Minaya, F. Esfera forma del mundo. Madrid, 1628.
Htn	Geog 756.36*	Postelli, G. Cosmographica disciplina. Lugduni, 1636.
	Geog 756.44	Herigone, P. Cursus mathematicus. v.4. n.p., 1644.
Htn	Geog 756.91*	Abraham ben Mordecai Farissol. Itinera mundi. v.1-2. Oxonii, 1691.
Htn	Geog 757.42*	Maupertius, Pierre L. Elements de geographie. Paris, 1742.
	Geog 757.55	Vaissete, Joseph. Geographie historique, ecclesiastique et civile. Paris, 1755. 4v.
	Geog 757.64F	Fenning, Daniel. A new system of geography: or A general description of the world. n.p., 17- . 2v.
	Geog 757.70	Buy de Mornas, C. Cosmographie methodique. Paris, 1770.
	Geog 757.93	Brookes, R. The general gazetteer. London, 1795.
	Geog 758.07	Dicuil. Liber de mensura orbis terrae. Paris, 1807.
	Geog 758.07.2	Dicuil. Liber de mensura orbis terrae. Paris, 1807.
	Geog 758.07.5	Dicuil. Liber de mensura orbis terrae. Dublin, 1967.
	Geog 758.45.25	Daniel, H.A. Lehrbuch der Geographie für höhere Unterrichtsanstalten. Halle, 1891
	Geog 758.48	Raumer, K.G. Lehrbuch der allgemeinen Geographie. Leipzig, 1848.
	Geog 758.50	Milner, Thomas. A universal geography in four parts. London, 1850.
	Geog 758.70	Dicuil. Liber de mensura orbis terra a Gustavo Parthey recognitus. Bercolini, 1870.
	Geog 758.79	Bevan, W.L. Students' manual of modern geography mathematical, physical and descriptive. London, 1879.
	Geog 758.90	Günther, S. Handbuch der Mathematischen Geographie. Stuttgart, 1890.
	Geog 759.11	Giamnitrapani, D. Geografia mathematica, geografia generale. Firenze, 1911.
	Geog 759.19	Beltran, Juan G. Lo inerte y lo vital. Buenos Aires, 1919.
	Geog 759.21.5	Herbertson, Andrew J. The senior geography. 5th ed. Oxford, 1921.
	Geog 759.23	Gribandi, P. Il mondo e l'Italia. 3. éd. v.1-2. Torino, 1923.
	Geog 759.32.5	Case, E.C. College geography. N.Y., 1932.

Classified Listing

Geog 755 - 760 Physical, Mathematical, and Descriptive geographies - General works (By date) - cont.

Geog 759.33	Allgemeine Geographie. Potsdam, 1933. 2v.
Geog 759.35	Mitteleuropa. Potsdam, 1935.
Geog 759.36.4	Finch, V.C. Elements of geography. 4th ed. N.Y., 1957.
Geog 759.38	West- und Nordeuropa in Natur. Potsdam, 1938.
Geog 759.40	Costa Pereira, J.V. da. Geographia humana. Rio de Janeiro, 194-.
Geog 759.42F	Lawrence, C.H. New world horizons; geography for the air age. N.Y., 1942.
Geog 759.43	Howe, E.L. Air world. Denver, 1943.
Geog 759.45	Almagia, R. Fondamenti di geografia generale. v.1-2. Roma, 1946-48.
Geog 759.46	Ward, Francis K. About this earth. London, 1946.
Geog 759.49	James, P.E. A geography of man. Boston, 1949.
Geog 759.49.2A	James, P.E. A geography of man. 2nd ed. Boston, 1959.
Geog 759.49.2B	James, P.E. A geography of man. 2nd ed. Boston, 1959.
Geog 759.55	Schmieder, Oscar. Geografía del viejo mundo. México, 1955.
Geog 759.56	Lobeck, Armin Kohl. Things maps don't tell us; an adventure into map interpretation. N.Y., 1968.
Geog 759.57	Fiziko-geograficheskie raionirovanie Kitaia; sbornik statei. Moskva, 1957.
Geog 759.58	Rebagliato, F. Geografia universal. Barcelona, 1958.
Geog 759.59	Doerr, Arthur H. Principles of geography. N.Y., 1968.
Geog 759.59.7	Obst, Erich. Lehrbuch der allegemeinen Geographie. 3. Aufl. v.1,6,7. Berlin, 1966- 3v.
Geog 759.65	Grigor'ev, Andrei A. Razvitie teoretichekikh problem sovetskoi fizicheskoi geografii, 1917-1934 gg. Moskva, 1965.
Geog 759.66.5	Journaux, André. Géographie générale. Paris, 1966.
Geog 759.68	Cole, John Peter. Quantitative geography. London, 1968.
Geog 759.70	Birot, Pierre. Les régions naturelles du globe. Paris, 1970.
Geog 759.70.5	Voprovy geografii. Kaliningrad, 1970.
VGeog 759.71	Thuchkevich, Vadim A. Geografiia v tsifrakh i sravneniiakh. Minsk, 1971.
Geog 759.71.5	Prirodnye resursy i kulturnye landshafty materikov. Moskva, 1971.
Geog 759.72	Haggeh, Peter. Geography: a modern synthesis. N.Y., 1972.
Geog 759.72.5	Predsrazhenskii, Vladimir S. Besedy v sovremennoi fisicheskoi geografii. Moskva, 1972.
Geog 759.73	Eramov, R.A. Fizicheskaia geografiia zarubezhnoi Evropy. Moskva, 1973.
Geog 759.74	Juillard, Étienne. La région, contributions à une géographie générale des espaces régionaux. Paris, 1974.
Geog 759.75	Géographie régionale. Paris, 1975-
Geog 759.75.5	Hart, John Fraser. The look of the land. Englewood Cliffs, 1975.

Geog 805 - 810 Descriptive geographies - General works - Multi-volume sets (By date)

Htn	Geog 805.67F*	Franck, S. Erst Theil dieses Weltbuchs von neuen Erfundnen Landtschafften Warhafftige Beschreibunge aller Theil der Welt. Franckfurt, 1567.
Htn	Geog 805.75F*	Thevet, A. La cosmographie universelle. Paris, 1575. 2v.
	Geog 806.87	Happelius, E.G. Mundus mirabilis tripartitus. Ulm, 1687. 3v.
	Geog 807.00F	Middleton, C.T. System of geography. v.1-2. London, 17- 4v.
	Geog 807.03	Scherer, R.P.H. Atlas novus exhibens orbem. pt.1-7. Augustae, 1710. 4v.
	Geog 807.16F	Notoras, C. Introductio ad geographia et spheram. Paris, 1716.
	Geog 807.29	Berckenmeier, P. Le curieux antiquaire. Leide, 1729. 3v.
	Geog 807.39.5F	Salmon, Thomas. Modern history. 3rd ed. London, 1744-1746. 3v.
NEDL	Geog 807.40	Salmon, Thomas. Lo stato presente di tutti i paesi e popoli del mondo naturale, politico e morale. 2a ed. v.1-26. Venezia, 1738-66. 27v.
	Geog 807.42	Lenglet, N. Methode pour etudier la geographie. Paris, 1742. 7v.
Htn	Geog 807.51F*	Sanson, N. La France,...les Isles Britanniques. Paris, 1651.
	Geog 807.62	Büsching, Anton Friedrich. New system of geography. London, 1762. 6v.
	Geog 807.68	The wonders of nature and art. 2nd ed. London, 1768. 6v.
	Geog 807.85	Büsching, Anton Friedrich. Auszug aus seiner Erdbeschreibung. 6. Aufl. Hamburg, 1785.
	Geog 807.85.5	Handbuch der alten Erdbeschreibung zum Gebrauch der Eilf. v.1-2, pt.1-2. Nürnberg, 1785-1793. 3v.
	Geog 807.87	Büsching, Anton Friedrich. Erdbeschreibung. Hamburg, 1787-1792. 10v.
	Geog 807.87.2	Büsching, Anton Friedrich. Grosse Erdbeschreibung. Troppau, 1784-1787. 24v.
	Geog 807.87.3	Bohn, C.E. Ankündigung einer neuen Ausgabe von Büschings Erdbeschreibung welche auf Pränumeration gedruckt wird. Hamburg, 1802.
	Geog 807.87.5	Bankes, T. A new, royal, authentic and complete system of universal geography. London, 1787-1810? 2v.
	Geog 808.04.3	Blomfield, E. A general view of the world. Bungay, 1807. 2v.
	Geog 808.07	System of geography. Glasgow, 1807. 4v.
	Geog 808.08	Playfair, J. System of geography. Edinburgh, 1808. 6v.
	Geog 808.11	Bigland, J. Geographical...view of the world. Boston, 1811. 5v.
	Geog 808.12.5	Malte-Brun, Conrad. Précis de la geographie universelle. Paris, 1812-1829. 8v.
	Geog 808.12.5F	Malte-Brun, Conrad. Précis de la geographie universelle. Atlas. Paris, 1812.
	Geog 808.12.7	Malte-Brun, Conrad. Précis de la geographie universelle. v.5. Paris, 1817.
	Geog 808.12.9	Malte-Brun, Conrad. Universal geography. Boston, 1824. 8v.
	Geog 808.12.10	Malte-Brun, Conrad. Universal geography. v.1-14. Boston, 1824-1829. 7v.
	Geog 808.12.11	Malte-Brun, Conrad. Universal geography. Philadelphia, 1827-1832. 6v.
	Geog 808.12.12	Malte-Brun, Conrad. A system of universal geography. Boston, 1834. 3v.
	Geog 808.12.13	Malte-Brun, Conrad. System of universal geography. v.1-2. Boston, 1844.
	Geog 808.12.15F	Malte-Brun, Conrad. Geographie universelle. v.1-6, Atlas. Paris, 1841-1847. 7v.

Geog 805 - 810 Descriptive geographies - General works - Multi-volume sets (By date) - cont.

Geog 808.12.16F	Malte-Brun, Conrad. A description of all parts of the world. Boston, 1859. 3v.
Geog 808.12.17	Malte-Brun, Conrad. Traité elementaire de géographie. Bruxelles, 1832.
Geog 808.17	Ritter, Carl. Die Erkunde. Berlin, 1817-1818. 2v.
Geog 808.17.5	Ritter, Carl. Die Erkunde. v.1-19. Berlin, 1822-1859. 21v.
Geog 808.17.6	Ritter, Carl. Namen- und Sach-Verzeichniss. Berlin, 1841-1849. 2v.
Geog 808.25	Casado Giraldes, J.P.C. Tratado completo de cosmographia. Paris, 1825-1828. 4v.
Geog 808.33	Montenegro Colon, Feliciano. Geografia general para el uso de la juventud de Venezuela. Caracás, 1833-1837. 4v.
Geog 808.33.2	Blanc, L.G. Handbuch...der Natur und Geschichte der Erde. Halle, 1833-1834. 3v.
Geog 808.34.10	Murray, Hugh. The encyclopaedia of geography. Philadelphia, 1845. 3v.
Geog 808.36	Bell, J. System of geography. Glasgow, 1836. 6v.
Geog 808.36.5	Bell, J. System of geography. London, 1850. 6v.
Geog 808.37	Murray, H. Encyclopedia of geography. Philadelphia, 1837.
Geog 808.50.9	Grube, A.W. Geographische Charakterbilder in algerundeten Gemälden aus der Länder- und Völkerkunde. Leipzig, 1868. 3v.
Geog 808.59	Klöden, G.A. Handbuch der Erdkunde. Berlin, 1859-1862. 3v.
Geog 808.59.3	Klöden, G.A. Handbuch der Erdkunde. Berlin, 1873. 5v.
Geog 808.66	Daniel, H.A. Handbuch der Geographie. Leipzig, 1866-1868. 4v.
Geog 808.66.5	Daniel, H.A. Handbuch der Geographie. Leipzig, 1874. 4v.
Geog 808.66.7	Daniel, H.A. Handbuch der Geographie. Leipzig, 1895. 4v.
Geog 808.75F	Colange, Leo de. The picturesque world, or Scenes in many lands. pt.1,3-6,8-13,15-16,18,20,26. Boston, 1875.
NEDL Geog 808.76	Reclus, J.J.E. Nouvelle geographie universelle. Paris, 1876-1894. 19v.
Geog 808.76.5	Reclus, J.J.E. Earth and its inhabitants - Europe. N.Y., 1885. 5v.
Geog 808.76.6	Reclus, J.J.E. Earth and its inhabitants - Asia. N.Y., 1891. 4v.
Geog 808.76.7	Reclus, J.J.E. Earth and its inhabitants - Africa. N.Y., 1886-1890. 4v.
Geog 808.76.8	Reclus, J.J.E. Earth and its inhabitants - Oceanica. N.Y., 1890.
Geog 808.76.9	Reclus, J.J.E. Earth and its inhabitants - North America. N.Y., 1890-1893. 3v.
Geog 808.76.10	Reclus, J.J.E. Earth and its inhabitants - South America. N.Y., 1894-1895. 2v.
Geog 808.77	Brown, R. The countries of the world. London, 1877-1880? 4v.
Geog 808.83	Marinelli, G. La terra. Milano, 1883-1885. 7v.
Geog 808.87	Kirchoff, A. Landerkunde des Erdteils Europa. v.1-3. Wien, 1887-1907. 5v.
Geog 808.90F	DePuy, W.H. The universal guide and gazetteer. N.Y., 1890.
Geog 808.96	Kerp, H. Methodisches Lehrbuch...Erdkund. Bonn, 1896-1904. 3v.
Geog 809.07	Hettner, A. Grundzüge der Landerkunde. Leipzig, 1907-1925. 2v.
Geog 809.10	Land og folk. Geografi i skildringer og livsbilleder. Kjøbenhavn, 1910. 2v.
Geog 809.19	Camena d'Almeida, P. La tierra; geografia general. 2a ed. Barcelona, 1919.
Geog 809.19.2	Camena d'Almeida, P. Europa. Barcelona, 1914.
Geog 809.19.3	Blásquez y Delgado Aguilera, Antonio. Peninsula Ibérica. 2a ed. Barcelona, 1921.
Geog 809.19.4	Camena d'Almeida, P. Asia, India insular, Africa. Barcelona, 1914.
Geog 809.19.5	Camena d'Almeida, P. América Septentrional, América Central, Las Antillas, Alaska, Canada, Estados Unidos. Barcelona, 1916.
Geog 809.19.6	Blásquez y Delgado Aguilera, Antonio. América Meridional, Oceania. Barcelona, 1916.
Geog 809.21	Bader, G. Erläuterungen zu 938 ausgewählten Lichtbilder zur Länderkunde. Stuttgart, 1921. 3v.
Geog 809.22F	Grauper, Ernest. Nouvelle géographie universelle. pt.1-10. Paris, 1922. 2v.
Geog 809.22.15	Vahl, Martin. Jorden og menneskelivet. København, 1922-1927. 4v.
Geog 809.27A	Geographie universelle. v.1-15. Paris, 1927-1946. 22v.
Geog 809.27B	Geographie universelle. v.11, pt.1-2. Paris, 1927-1946. 2v.
Geog 809.27.2	Vidal de La Blache, P. Géographie universelle. 2. éd. v.6, pt.1. Paris, 1947.
Geog 809.28	Allgemeine Länderkunde der Erdteile. v.3-4,6. Hannover, 1928-1935.
Geog 809.36	Ozonf, R. (Mme.). Lectures géographiques. v.1-2. Paris, 1936-1938. 4v.
Geog 809.51	Gutersloh, H. Die Erde. v.1-13. Bern, 1951. 2v.
Geog 809.52	Teran, M. de. Imago mundi. Madrid, 1952. 2v.
Geog 809.59	Obst, Erich. Lehrbuch der allgemeinen Geographie. v.1-4,6-8,10-11. Berlin, 1959. 9v.
Geog 809.59.5	Vooys, Adriaan. Panorama der Wireld. Roermond, 1959. 3v.
Geog 809.59.10F	Istituto Geografico de Agostini. Il milione; enciclopedia di geografia. Novara, 1959-1965. 15v.

Geog 815 - 820 Descriptive geographies - General works - Others (By date)

Htn	Geog 815.19*	Denciso, M.F. Suma de geografia. Seville, 1519.
Htn	Geog 815.34*	Watt, J. von. Epitome trium terrae. Tiguri, 1534.
	Geog 815.35.4	Servetus, Michael. Descripciones geograficas del estado moderno de las regiones. Madrid, 1932.
	Geog 815.35.5	Servetus, Michael. Descripciones geograficas del estado moderno de las regiones. Madrid, 1932.
	Geog 815.35.10	Servetus, Michael. Michael Servetus, a translation of his geographical, medical, and astrological writings. Philadelphia, 1953.
Htn	Geog 815.38*	Rithaymer, Georg. Georgii Rithaymeri De orbis terrarvm sitr compendium. Norimbergae, 1538.
Htn	Geog 815.50*	Hondius, Jodocus. Thresor de chartes contenant les tableaux de tous les pays du monde. Franckfort? 1602.
Htn	Geog 815.59.3*	Botero, G. Le relationi universali...divise in quattro parti. Venetia, 1597.

Classified Listing

Geog 815 - 820 Descriptive geographies - General works - Others (By date) - cont.

Htn	Geog 815.59.5*	Botero, G. The travellers breviat, or An historical description of the most famous kingdoms in the world. London, 1601.
Htn	Geog 815.59.7*	Botero, G. Le relationi universali...divise in quattro parti. Venetia, 1622.
	Geog 815.59.10	Botero, G. Descripcion de todas las provincias. Gerona, 1748.
Htn	Geog 816.00*	Abbott, George. A briefe description of the whole world. London, 1600.
Htn	Geog 816.00.3*	Abbott, George. A briefe description of the whole world. 3d ed. London, 1608. 2 pam.
Htn	Geog 816.00.5*	Abbott, George. A briefe description of the whole world. 5th ed. London, 1620.
Htn	Geog 816.21.7*	Heylyn, Peter. Mikrokosmos: a little description of the great world. 7th ed. Oxford, 1636.
Htn	Geog 816.33.5*	Heylyn, Peter. Mikrokosmos; little description. Oxford, 1633.
Htn	Geog 816.33.7*	Heylyn, Peter. Mikrokosmos; little description. Oxford, 1639.
Htn	Geog 816.33.8*	Heylyn, Peter. Mikrokosmos; a little description of the great world. 8th ed. Oxford, 1639.
Htn	Geog 816.35*	Carpenter, N. Geographie delineated forth in two books. Oxford, 1635.
	Geog 816.36	Carvalho Da Costa, Antonio. Compendio geographico. Lisboa, 1636.
Htn	Geog 816.40*	Bertius, Petrus. Livre premier [-septiesme] des tables geographiques auquel est traité du monde en general. n.p., n.d.
	Geog 816.46.13	Labbé, Philippe. La geographie royalle. 2e éd. Paris, 1653.
Htn	Geog 816.52*	Francois, Jean. La science de la geographie divisée en trois parties. Rennes, 1652.
Htn	Geog 816.55*	Linda, Lucas de. Descriptio orbis et omniumejus rerumpublicarum. Lugdunum Batavorum, 1655.
Htn	Geog 816.57.5*	Clarke, Samuel. A geographical description of all the countries in the known world. London, 1671.
	Geog 816.61	Cluverius, P. Introductionis in...geographiam. Amsterdam, 1661.
NEDL	Geog 816.61.3	Cluverius, P. Introduction into geography. Oxford, 1657.
Htn	Geog 816.61.4*	Cluverius, P. Introductionis in universam geographiam. Amsterdam, 1659.
	Geog 816.61.5	Cluverius, P. Introductionis in...geographiam. Brunsvigae, 1672.
	Geog 816.61.6	Cluverius, P. Introductionis in...geographiam. Amsterdam, 1686.
	Geog 816.61.7	Cluverius, P. Introductionis in universam geographiam. Amsterdam, 1697.
Htn	Geog 816.61.9*	Cluverius, P. Introductionis in...geographiam. Londini, 1711.
Htn	Geog 816.61.13*	Cluverius, P. Introductionis in universam geographiam. Amsterdam, 1729.
Htn	Geog 816.61.16*	Cluverius, P. Introductionis in universam geographiam. Lugdunum Batavorum, 1627.
	Geog 816.61.19	Cluverius, P. Introduction à la geographie universelle. Paris, 1631.
	Geog 816.65	Linda, Lucas de. Descriptio orbis. Amsterdam, 1665.
	Geog 816.70	Joosten, J. De kleyne wonderlijcke werelt. Amsterdam, 1670.
Htn	Geog 816.80.4*	Morden, Robert. Geography rectified. 4th ed. London, 1700.
	Geog 816.82	Duval, Pierre. La geographie du temps. pt.1-2. Paris, 1682. 2v.
Htn	Geog 816.85*	Suval, P. Geographia universalis. London, 1685.
	Geog 816.93	Sanson, N. Introduction à la geographie. Paris, 1693.
Htn	Geog 816.94*	Seller, John. A new system of geography. n.p., 1694?
	Geog 816.96	El atlas abreviado ô compendiosa geografia. Amberes, 1696.
Htn	Geog 817.00*	Sanson, N. Description de tout universe. Amsterdam, 1700.
Htn	Geog 817.00.5*	Bion. L'usages des globes celestes et terrestes. Adam, 1700.
Htn	Geog 817.01F*	Moll, H. System of geography. London, 1701.
	Geog 817.01.5F	Moll, H. The compleat geographer. London, 1723.
	Geog 817.04	Gordon, P. Geography anatomized. 4th ed. London, 1704.
Htn	Geog 817.04.6*	Gordon, P. Geography anatomized. 6th ed. London, 1711.
	Geog 817.04.8	Gordon, P. Geography anatomized. 8th ed. London, 1719.
	Geog 817.04.12	Gordon, P. Geography anatomized. 12th ed. London, 1730.
	Geog 817.04.19	Gordon, P. Geography anatomized. 19th ed. London, 1749.
	Geog 817.04.20	Gordon, P. Geography anatomized. 20th ed. London, 1754.
	Geog 817.26	Kolb, P.G. Compendium totius orbis. Rottwike, 1726.
	Geog 817.51.2	Salmon, T. A new geographical...grammar. London, 1749.
	Geog 817.51.3	Salmon, T. New geographical...grammar. London, 1751.
	Geog 817.51.5	Salmon, T. New geographical...grammar. London, 1764.
	Geog 817.51.7	Salmon, T. New geographical...grammar. London, 1766.
	Geog 817.51.8	Salmon, T. New geographical...grammar. London, 1769.
	Geog 817.51.9	Salmon, T. Geographical and astronomical grammar. London, 1785.
	Geog 817.73	Jones, E. The young geographer and astronomer's best companion. London, 1773.
Htn	Geog 817.74*	André, Noël. Description et usages de la Mappemonde. Paris, 1774.
	Geog 817.79F	Carver, J. New universal traveller. London, 1779.
	Geog 817.82	Guthrie, William. A new geographical, historical, and commercial grammar and present state of several kingdoms of the world. 7th ed. London, 1732.
	Geog 817.83	Guthrie, William. A new geographical, historical, and commercial grammar and present state of the several kingdoms of the world. 8th ed. London, 1783.
	Geog 817.88	Guthrie, William. A new geographical, historical and commercial grammar. 11th ed. London, 1788.
	Geog 817.89	Gordon, William. New geographical grammar. Edinburgh, 1789.
	Geog 817.92	Guthrie, William. A new geographical, historical and commercial grammar. 13th ed. London, 1792.
	Geog 817.94	Guthrie, William. New system of modern geography. Philadelphia, 1794-1795. 2v.
	Geog 817.94.5F	Guthrie, William. New system of modern geography. London, 1795.
	Geog 817.94.10	Guthrie, William. New geographical, historical...grammar. London, 1796.
	Geog 817.94.13	Guthrie, William. New geographical, historical...grammar. Montreal, 1810.
	Geog 817.96	Walker, J. Elements of geography. London, n.d.
	Geog 818.00.5	LaCroix, L.A.N. Geographie moderne et universelle. Paris, 1800. 2v.
	Geog 818.02	Adam, A. A summary of geography and history. 3rd ed. London, 1802.
	Geog 818.02.90	Pinkerton, J. Modern geography. London, 1802. 2v.

Geog 815 - 820 Descriptive geographies - General works - Others (By date) - cont.

	Geog 818.04	Pinkerton, J. Modern geography. Philadelphia, 1804. 2v.
	Geog 818.04.5	Pinkerton, J. Modern geography. London, 1811. 2v.
Htn	Geog 818.05*	Davies, Benjamin. A new system of modern geography. Philadelphia, 1805.
	Geog 818.07	Aikin, J. Geographical delineations. Philadelphia, 1807.
	Geog 818.09	Guthrie, William. A new geographical, historical and commercial grammar. 1st American ed. Philadelphia, 1809.
	Geog 818.10	Phillips, R. General view of manners...of nations. Philadelphia, 1810. 2v.
	Geog 818.10.3	Phillips, R. Geographical view of the world. N.Y., 1826.
	Geog 818.10.5	Evans, J. New system of geography. London, 1810. 2v.
	Geog 818.13	Dickinson, R. Elements of geography. Boston, 1813.
NEDL	Geog 818.16	Adam, A. Summary of geography and history. London, 1816.
NEDL	Geog 818.16.20	Cannabich, T.G.F. Lehrbuch der Geographie. 17e Aufl. Weimar, 1855.
	Geog 818.19	Riise, J. Haanbog i geographien. Kjobenhavn, 1819-1820. 2v.
	Geog 818.20	The traveller; or, An entertaining journey round the habitable globe. 3rd ed. London, 182-.
	Geog 818.21	Oddsson, G. Almenn landaskipunarfraede. Kaupmannahøfn, 1821-1827. 2v.
	Geog 818.21.3	Oddsson, G. Almenn jardarfrädi og landskipun edur geographia. Kaupmannahøfn, 1822. 4v.
	Geog 818.22	Morse, S.E. New system of modern geography. Boston, 1822.
	Geog 818.23	Worcester, J.E. Sketches of the earth. Boston, 1823. 2v.
Htn	Geog 818.24*	Engelmann, G. Porte-feuille géographique et ethonographique. 2e éd. v.1-2, Atlas. Mulhouse, 1824. 2v.
	Geog 818.26	Blake, J.L. Geographical, chronological...atlas. N.Y., 1826.
	Geog 818.28	Venning, I.A. (Mrs.). A geographical present. 1st American ed. N.Y., 1829.
	Geog 818.30	Hale, N. Epitome of universal geography. Boston, 1830.
	Geog 818.32	Goodrich, S.G. A system of universal geography. Boston, 1832.
	Geog 818.33	Depping, G.B. Evening entertainments...manners and customs of nations. Philadelphia, 1833.
	Geog 818.33.5	Goodrich, S.G. A system of universal geography. 2d ed. Boston, 1833.
	Geog 818.34	Murray, H. An encyclopaedia of geography. London, 1834.
	Geog 818.36	Cannabich, J.G.F. Lehrbuch der Geographie. 14e Aufl. Weimar, 1836.
	Geog 818.36.10	Perkins, Samuel. The world as it is. 5th ed. n.p., 1839.
	Geog 818.36.11	Perkins, Samuel. The world as it is. 5th ed. n.p., 1840.
	Geog 818.36.14	Perkins, Samuel. The world as it is. 6th ed. n.p., 1842.
	Geog 818.39	Mitchell, S.A. Accompaniment to map of world. Philadelphia, 1839.
	Geog 818.40	Murray, H. Encyclopedia of geography. London, 1840.
	Geog 818.40.4	Goodrich, S.G. A pictorial geography of the world. Boston, 1840.
	Geog 818.40.5	Goodrich, S.G. Pictorial geography of the world. 2nd ed. Boston, 1840. 2v.
	Geog 818.40.6	Goodrich, S.G. Pictorial geography of the world. 3rd ed. Boston, 1840.
	Geog 818.40.7	Goodrich, S.G. Pictorial geography of the world. Boston, 1856. 2v.
	Geog 818.42	Laurie, J. System of universal geography. Edinburgh, 1842.
	Geog 818.42.5	Mitchell, S.A. An accompaniment to Mitchell's map. Philadelphia, 1842.
	Geog 818.43	St. Platou, L. Stutt landaskipunarfraedi. n.p., 1843.
	Geog 818.43.5	Murray, Hugh. The encyclopaedia of geography. Philadelphia, 1843. 3v.
	Geog 818.44	Mitchell, S.A. An accompaniment to Mitchell's map. Philadelphia, 1844.
	Geog 818.45	Mitchell, S.A. An accompaniment to Mitchell's map. Philadelphia, 1845.
	Geog 818.46	Mitchell, S.A. An accompaniment to Mitchell's map. Philadelphia, 1846.
	Geog 818.51	Geelmuyden, J. Loerebogi geografien. Christiania, 1851.
	Geog 818.51.14	Arendts, Carl. Leitfaden für den ersten Wissenschaftlichen Unterricht in der Geographie. 14. Aufl. Regensburg, 1874.
	Geog 818.52.5	Nicolay, Charles G. A manuel of geographical science. v.11. London, 1859.
	Geog 818.53	Savage, C.C. The world; geographical, historical, statistical. N.Y., 1853.
	Geog 818.54	Ingerslev, C.F. Stutt kennslubok i landafroedinni. Reykjavik, 1854.
	Geog 818.54.10	Condaminas, S. Nociones generales de geografía astronómica. Matanzas, 1854.
	Geog 818.59	Pütz, Wilhelm. Charakteristiken zur Erd und Völkerkunde. Köln, 1859-1860. 2v.
	Geog 818.59.6	Pütz, Wilhelm. Grundriss der Geographie und Geschichte der...Zeit. v.1-3. 10. Aufl, 12. Aufl. Koblenz, 1865-1867.
	Geog 818.59.7	Pütz, Wilhelm. Grundriss der Geographie und Geschichte. Koblenz, 1866.
	Geog 818.59.9	Young, Francis. Elementary geography. London, 1859.
	Geog 818.61	Mackay, A. Manual of modern geography. Edinburgh, 1861.
	Geog 818.61.10	Bohn, Henry G. A pictorial hand-book of modern geography. 1st ed. London, 1865.
	Geog 818.62	Harris, A. Geographical hand-book. Lancaster, Pa., 1862.
	Geog 818.62.5	Cortambert, P.F.E. Cours de geographie. Paris, 1862.
	Geog 818.64	Milner, Thomas. The gallery of geography, a pictorial and descriptive tour of the world. London, 1864. 2v.
	Geog 818.68	Lavallée, T.S. Physical, historical and military geography. London, 1868.
	Geog 818.73.3	Reclus, O. Geographie. La terre a vol d'oiseau. 3e ed. Paris, 1877. 2v.
	Geog 818.78	Erslev, Ed. Agrip af landafraedi. Reykjavik, 1878.
	Geog 818.80	Cañas Pinochet, A. El estudio de la jeografía por el dibujo de las cartas jeográficas. Santiago de Chile, 1880.
	Geog 818.80.5	Johnston, Alexander K. A physical, historical, political, and descriptive geography. London, 1880.
	Geog 818.80.6	Johnston, Alexander K. A physical, historical, political, and descriptive geography. 2nd ed. London, 1881.
	Geog 818.82.5	Guthe, H. Lehrbuch der Geographie. Hannover, 1882-1883. 2v.
	Geog 818.82.7	Wagner, H. Lehrbuch der Geographie. Hannover, 1903. 2v.
	Geog 818.86	Pequeña geografia. Asunción, 1886.
	Geog 818.92	Reclus, O. A bird's-eye view of the world. Boston, 1892.
	Geog 818.93	Gilbert, Frank. The world, historical and actual. Chicago, 1893.
	Geog 818.97.5	Hellwald, F. von. Die Erde und ihre Völker. Stuttgart, 1897.

Classified Listing

Geog 815 - 820		Descriptive geographies - General works - Others (By date) - cont.
	Geog 819.00	Beltrán y Rózpide, R. La geografía en 1898. Madrid, 1899.
	Geog 819.01	Mill, H.R. The international geography. N.Y., 1901.
	Geog 819.01.6	Mill, H.R. The international geography. N.Y., 1920.
	Geog 819.01.7	Mill, H.R. The international geography. N.Y., 1907.
	Geog 819.01.10	Mill, H.R. International geography. 2nd ed. N.Y., 1900.
	Geog 819.02	Seydlitz, Ernst von. Grosses Lehrbuch der Geographie. Breslau, 1902.
	Geog 819.03	Seydlitz, Ernst von. Kleines Lehrbuch der Geographie. Breslau, 1903.
	Geog 819.09	Scobel, A. Geographisches Handbuch. Bielefeld, 1909-1910. 2v.
	Geog 819.10F	Rand, McNally and Company. The world and its peoples photographed and described. Chicago, 1910.
	Geog 819.11	Newbigin, Marion I. Modern geography. London, 1911.
	Geog 819.11.1	Newbigin, Marion I. Modern geography. N.Y., 1911.
	Geog 819.11.3	Busson, Henri. Les principales puissances du monde. Paris, 1911.
	Geog 819.11.5	Busson, Henri. Les principales puissances d'aujourd'hui. 5. éd. Paris, 1924.
	Geog 819.13	Andrews, A.W. A text-book of geography. London, 1913
	Geog 819.13.5	Salisbury, Rollin D. Modern geography for high schools. N.Y., 1913.
	Geog 819.14	Seydlitz, Ernst von. Handbuch der Geographie. Breslau, 1914.
	Geog 819.16	Andrews, A.W. A text book of geography. London, 1916.
	Geog 819.18	Santa Cruz, Alonso de. Islario general de todas las isles del mundo. Atlas. Madrid, 1918. 2v.
	Geog 819.19	Banse, E. Illustrierte Länderkunde. Berlin, 1919.
	Geog 819.21	Bowman, I. The new world. Yonkers-on-Hudson, 1921.
	Geog 819.21.3	Wilmore, Albert. The groundwork of modern geography. London, 1921.
	Geog 819.22	Bowman, I. The new world; problems in political geography. Yonkers-on-Hudson, 1922.
	Geog 819.22.2	Bowman, I. Supplement to The new world. Yonkers-on-Hudson, 1923-1924. 2v.
	Geog 819.27.5	Hettner, A. Die Geographie; ihre Geschichte. Breslau, 1927.
	Geog 819.27.10	Paquet, A. Städte, Landschaften und ewige Bewegung. Hamburg, 1927.
	Geog 819.28	Mitchell, J. Leslie. Hauno, or Future of exploration. London, 1928.
	Geog 819.28.5A	Bowman, I. The new world. 4th ed. Yonkers-on-Hudson, 1928.
	Geog 819.28.5B	Bowman, I. The new world. 4th ed. Yonkers-on-Hudson, 1928.
	Geog 819.30	Newbigin, Marion I. A new regional geography of the world. N.Y., 193-?
	Geog 819.34	Kalmár, Gusztáv. Négy Világrésg Földje és Népei. Budapest, 1934-1935.
	Geog 819.35	James, Preston E. An outline of geography. Boston, 1935.
	Geog 819.38F	Globus geograficheskii eksegodnik dlia detei. Moskva, 1938.
	Geog 819.42.2	Davis, D.H. The earth and man. N.Y., 1942.
	Geog 819.43	Engelhardt, N.L. Toward new frontiers of our global world. N.Y., 1943.
	Geog 819.44	Renner, G.T. Global geography. N.Y., 1944.
	Geog 819.44.5F	Hankins, G.C. Our global world. N.Y., 1944.
	Geog 819.45	Pickles, Thomas. The work of men. 1st ed. London, 1945.
	Geog 819.45.10	George, Pierre. Géographie sociale du monde. 6. éd. Paris, 1964.
	Geog 819.47	Davis, D.H. The earth and man. N.Y., 1947.
	Geog 819.47.5	Kinkead, Eugene. Our own Baedeker, from the New Yorker. N.Y., 1947.
	Geog 819.50	Jones, S.B. Geography and world affairs. Chicago, 1953.
	Geog 819.51	Krebs, N. Vergleichende Landerkunde. Stuttgart, 1951.
	Geog 819.52F	Visintin, L. Continenti e poesi. Novara, 1952.
	Geog 819.57	Geografía y Atlas Universal. Geografía y atlas universal. Barcelona, 1957.
	Geog 819.58F	Unsere Erde. Heidelberg, 1958.
	Geog 819.58.5	Larousse, firm, publishers. Geographie universelle Larousse. Paris, 1958. 3v.
	Geog 819.58.10	Larousse encyclopedia of geography. N.Y., 1961.
	Geog 819.59.4	Länder der Erde. 4. Aufl. Berlin, 1967.
	Geog 819.61	Manley, Gordon. Geography; our planet, its peoples and resources. London, 1961.
	Geog 819.73	McCormick, Donald. How to buy an island. N.Y., 1973.
Geog 953		Descriptive geographies - Views, pictorial works - Pamphlet volumes; Bibliographies
	Geog 953.1	Pamphlet box. Descriptive geographies. Views. Bibliographies.
	Geog 953.5FA	Bachmann, Friedrich. Die alten Städtebilder. Leipzig, 1939.
	Geog 953.5FB	Bachmann, Friedrich. Die alten Städtebilder. Leipzig, 1939.
	Geog 953.5.2F	Bachmann, Friedrich. Die alten Städtebilder. 2. Aufl. Stuttgart, 1965.
	Geog 953.10	Stockholm. Biblioteket. Magnus Gabriel de la Gordie's samling af öldre stadsvger. Stockholm, 1915.
	Geog 953.12	Ellis, Jessie (Croft). Travel through pictures; references to pictures, in books and periodicals. Boston, 1935.
	Geog 953.14	Kinauer, Rudolf. Lexikon geographischer Bilbände. Wien, 1966.
Geog 956 - 960		Descriptive geographies - Views, pictorial works - General works (By date)
Htn	Geog 956.78*	Meissner, D. Sciagraphia cosmica. v.1-8. Nürnberg, 1678. 2v.
Htn	Geog 958.10*	Künstliche Erdkugel zur Uerbreitung gemeinnütziger Kentnisse über die Eintheilung. n.p., 18- ?
	Geog 958.33F	Meyer, J. Meyer's Universum. v.1-6,9. Hildburghausen, 1833- 3v.
	Geog 958.33.3	Meyer, J. Meyer's Universum. v.1,4. N.Y., 1850. 2v.
	Geog 958.33.5	Meyer, J. Meyer's Universum, or Views of...all contries. N.Y., 1852.
	Geog 958.40	Our Globe. Philadelphia, 184-.
	Geog 958.90F	La panorama, merveilles de France, Belgique. Paris, 189-?
	Geog 958.91	Shepp, J.W. Shepp's photographs of the world. Philadelphia, 1891.
	Geog 958.92F	Stoddard, John L. Glimpses of the world. Chicago, 1892.
	Geog 958.92.5F	Stoddard, John L. Portfolio of photographs of famous scenes, cities and paintings. Chicago, 189-?
	Geog 958.94	Photographic views of the world. Boston, 1894.
	Geog 958.96	Dubois, M. Album géographique. Paris, 1896-1906. 5v.
	Geog 958.98F	Hirt, F. Geographische Bildertafeln. v.1-3. Breslau, 1886-98. 5v.

Geog 956 - 960		Descriptive geographies - Views, pictorial works - General works (By date) - cont.
	Geog 959.00	Photograph album containing views of Alaska about 1900, and mountaineering in the Swiss Alps. n.p., n.d
	Geog 959.07	Grosvenor, G.H. Scenes from every land. Washington, D.C., 1907.
	Geog 959.07.5	Grosvenor, G.H. Scenes from every land. 2. ser. Washington, D.C., 1909.
	Geog 959.07.15	Grosvenor, G.H. Scenes from every land. Washington, D.C., 1918.
	Geog 959.12F	Raymond, E.L. Marvelous scenes of the world. Chicago, 1902.
Geog 1508		Guidebook collections - Baedeker - Great Britain - General and England
	Geog 1508.5	Baedeker, publishers. Great Britain. Leipsic, 1887.
	Geog 1508.10	Baedeker, publishers. Great Britain. 2. ed. Leipsic, 1890.
	Geog 1508.11	Baedeker, publishers. Great Britain. 2. ed. Leipsic, 1890.
NEDL	Geog 1508.12	Baedeker, publishers. Great Britain. 3. ed. Leipsic, 1894.
	Geog 1508.13	Baedeker, publishers. Great Britain. 3. ed. Leipsic, 1894.
	Geog 1508.15	Baedeker, publishers. Great Britain. 4. ed. Leipsic, 1897.
	Geog 1508.16	Baedeker, publishers. Great Britain. 4. ed. Leipsic, 1897.
	Geog 1508.20	Baedeker, publishers. Great Britain. 5. ed. Leipsic, 1901.
	Geog 1508.25	Baedeker, publishers. Great Britain. 6. ed. Leipsic, 1906.
	Geog 1508.30	Baedeker, publishers. Great Britain. 7. ed. Leipzig, 1910.
	Geog 1508.32	Baedeker, publishers. Great Britain. 8. ed. Leipzig, 1927.
	Geog 1508.34	Baedeker, publishers. Great Britain. N.Y., 1937.
Geog 1510		Guidebook collections - Baedeker - Great Britain - London
	Geog 1510.2	Baedeker, publishers. London und Umgebung. 15. Aufl. Leipzig, 1905.
	Geog 1510.3	Baedeker, publishers. Londres, ses environs: le Sud de l'Angleterre. 3. éd. Leipzig, 1875.
	Geog 1510.5	Baedeker, publishers. London and its environs. Leipsic, 1878. 2v.
	Geog 1510.8	Baedeker, publishers. London and its environs. 2. ed. Leipsic, 1879.
	Geog 1510.10	Baedeker, publishers. London and its environs. 3. ed. Leipsic, 1881.
	Geog 1510.15	Baedeker, publishers. London and its invirons. 4. ed. Leipsic, 1883.
	Geog 1510.20	Baedeker, publishers. London and its invirons. 5. ed. Leipsic, 1885.
	Geog 1510.25	Baedeker, publishers. London and its environs. 6. ed. Leipsic, 1887.
	Geog 1510.30	Baedeker, publishers. London and its environs. 7. ed. Leipsic, 1889.
	Geog 1510.35	Baedeker, publishers. London and its environs. 8. ed. Leipsic, 1892.
	Geog 1510.36	Baedeker, publishers. London and its environs. 8. ed. Leipsic, 1892.
	Geog 1510.40	Baedeker, publishers. London and its environs. 9. ed. Leipsic, 1894.
	Geog 1510.41	Baedeker, publishers. London and its environs. 9. ed. Leipsic, 1894.
	Geog 1510.42	Baedeker, publishers. London und Umgebungen. 11. Aufl. Leipzig, 1894.
	Geog 1510.45	Baedeker, publishers. London and its environs. 10. ed. Leipsic, 1894.
	Geog 1510.46	Baedeker, publishers. London and its environs. 10. ed. Leipsic, 1896.
	Geog 1510.47	Baedeker, publishers. London and its environs. 11. ed. Leipsic, 1898.
	Geog 1510.47.12	Baedeker, publishers. London and its environs. 12. ed. Leipsic, 1900.
	Geog 1510.48	Baedeker, publishers. London and its environs. 13. ed. Leipsic, 1902. 2v.
	Geog 1510.49	Baedeker, publishers. London and its environs. 13. ed. Leipsic, 1902.
	Geog 1510.50	Baedeker, publishers. London and its environs. 13. ed. Leipsic, 1902.
	Geog 1510.55	Baedeker, publishers. London and its environs. 14. ed. Leipsic, 1905.
	Geog 1510.56	Baedeker, publishers. London and its environs. 15. ed. Leipsic, 1908.
	Geog 1510.57	Baedeker, publishers. London and its environs. 15. ed. Leipsic, 1908.
	Geog 1510.59A	Baedeker, publishers. London and its invirons. 16. ed. Leipsic, 1911.
	Geog 1510.59B	Baedeker, publishers. London and its invirons. 16. ed. Leipsic, 1911.
	Geog 1510.61	Baedeker, publishers. London and its environs. 18. ed. Leipzig, 1923.
	Geog 1510.62	Baedeker, publishers. London and its environs. 18. ed. Leipzig, 1923.
	Geog 1510.62.2	Baedeker, publishers. London and its environs. 20. ed. Hamburg, 1951.
Geog 1516		Guidebook collections - Baedeker - France - Northern France in general
	Geog 1516.3	Baedeker, publishers. Le Nord de la France. Leipzig, 1884.
	Geog 1516.4	Baedeker, publishers. Le Nord de la France. 2. éd. Leipzig, 1887.
	Geog 1516.5	Baedeker, publishers. Northern France. Leipsic, 1889.
	Geog 1516.10	Baedeker, publishers. Northern France. 2. ed. Leipsic, 1894.
	Geog 1516.10.2	Baedeker, publishers. Northern France. 2. ed. Leipsic, 1894.
	Geog 1516.15	Baedeker, publishers. Northern France. 3. ed. Leipsic, 1899.
	Geog 1516.15.3	Baedeker, publishers. Northern France. 3. ed. Leipsic, 1899.
	Geog 1516.20.4	Baedeker, publishers. Northern France. 4. ed. Leipzig, 1905.
	Geog 1516.25	Baedeker, publishers. Northern France. 5. ed. Leipzig, 1909.

Classified Listing

Geog 1517	Guidebook collections - Baedeker - France - Southern France in general	
Geog 1517.5	Baedeker, publishers. Southern France...including Corsica. Leipzig, 1897.	
Geog 1517.17	Baedeker, publishers. Southern France. 4. ed. Leipsic, 1902.	
Geog 1517.19	Baedeker, publishers. Southern France. 5. ed. Leipzig, 1907.	
Geog 1517.19.5	Baedeker, publishers. Southern France. 5. ed. Leipzig, 1907.	
Geog 1517.21A	Baedeker, publishers. Southern France. 6. ed. Leipzig, 1914.	
Geog 1517.21B	Baedeker, publishers. Southern France. 6. ed. Leipzig, 1914.	

Geog 1518 Guidebook collections - Baedeker - France - Other regions
- Geog 1518.3 Baedeker, publishers. South eastern France. 3. ed. Leipsic, 1898.
- Geog 1518.26 Baedeker, publishers. South-western France from the Loire and the Rhone to the Spanish frontier. 2. ed. Leipsic, 1895.
- Geog 1518.55 Baedeker, publishers. Le Nord-est de la France. 5. éd. Leipzig, 1895.
- Geog 1518.80 Baedeker, publishers. Le Nord-ouest de la France. 5. éd. Leipzig, 1895.
- Geog 1518.89 Baedeker, publishers. Le Nord-ouest de la France. 9. éd. Leipzig, 1913.

Geog 1519 Guidebook collections - Baedeker - France - Paris
- Geog 1519.10 Baedeker, publishers. Paris, Rouen, Havre, Dieppe. 5. Aufl. Coblenz, 1864.
- Geog 1519.13 Baedeker, publishers. Paris, Rouen, Havre, Dieppe. Coblenz, 1865.
- Geog 1519.15 Baedeker, publishers. Paris and northern France. 2. ed. Coblenz, 1867.
- Geog 1519.19 Baedeker, publishers. Paris, nebst einigen Routen. 16. Aufl. Leipzig, 1905.
- Geog 1519.20 Baedeker, publishers. Paris and northern France. 3. ed. Coblenz, 1872.
- Geog 1519.21 Baedeker, publishers. Paris and northern France. 3. ed. Coblenz, 1872.
- Geog 1519.21.3 Baedeker, publishers. Paris and northern France. 3. ed. Leipsic, 1872.
- Geog 1519.23 Baedeker, publishers. Paris and its environs. 4. ed. Leipsic, 1874.
- Geog 1519.24 Baedeker, publishers. Paris, ses environs et les principaux itinéraires. 4. éd. Leipzig, 1876.
- Geog 1519.25 Baedeker, publishers. Paris and its environs. 5. ed. Leipsic, 1876.
- Geog 1519.26 Baedeker, publishers. Paris and its environs. 5. ed. Leipsic, 1876.
- Geog 1519.27 Baedeker, publishers. Paris et ses environs. 5. éd. Leipzig, 1878.
- Geog 1519.30 Baedeker, publishers. Paris and its environs. 6. ed. Leipsic, 1878.
- Geog 1519.33 Baedeker, publishers. Paris et ses environs. 7. éd. Leipzig, 1881.
- Geog 1519.33.2 Baedeker, publishers. Paris and its invirons. 7. ed. Leipzig, 1881.
- Geog 1519.36 Baedeker, publishers. Paris and its environs. 8. ed. Leipsic, 1884.
- Geog 1519.36.2 Baedeker, publishers. Paris et ses environs. 7. éd. Leipzig, 1884.
- Geog 1519.37 Baedeker, publishers. Paris et ses environs. 8. éd. Leipzig, 1887.
- Geog 1519.39 Baedeker, publishers. Paris and its environs. 9. ed. Leipsic, 1888.
- Geog 1519.42 Baedeker, publishers. Paris et ses environs. 9. éd. Leipzig, 1889.
- Geog 1519.45 Baedeker, publishers. Paris et ses environs. 10. éd. Leipzig, 1891.
- Geog 1519.46 Baedeker, publishers. Paris and its environs. 10. ed. Leipsic, 1891.
- Geog 1519.46.2 Baedeker, publishers. Paris and its environs. 10. ed. Leipsic, 1891.
- Geog 1519.48 Baedeker, publishers. Paris and its environs. 11. ed. Leipsic, 1894.
- Geog 1519.49 Baedeker, publishers. Paris et ses environs. 11. éd. Leipzig, 1894.
- Geog 1519.50 Baedeker, publishers. Paris and its environs. 12. ed. Leipsic, 1896.
- Geog 1519.50.2 Baedeker, publishers. Paris and its environs. 12. ed. Leipsic, 1896.
- Geog 1519.51 Baedeker, publishers. Paris et ses environs. 12. éd. Leipzig, 1896.
- Geog 1519.54 Baedeker, publishers. Paris and its environs. 14. ed. Leipsic, 1900.
- Geog 1519.55 Baedeker, publishers. Paris and its environs. 14. ed. Leipsic, 1900.
- Geog 1519.58 Baedeker, publishers. Paris and its environs. 15. ed. Leipzig, 1904.
- Geog 1519.60 Baedeker, publishers. Paris and its environs. 16. ed. Leipzig, 1907.
- Geog 1519.63 Baedeker, publishers. Paris and its environs. 17. ed. Leipzig, 1910.
- Geog 1519.68 Baedeker, publishers. Paris and its environs. 18. ed. Leipzig, 1913.
- Geog 1519.70 Baedeker, publishers. Paris and its environs. 19. ed. Leipzig, 1924.
- Geog 1519.75 Baedeker, publishers. Paris und Umgebung und Supplement. 20. Aufl. Leipzig, 1931-37. 2v.

Geog 1522 Guidebook collections - Baedeker - Germany - General
- Geog 1522.15 Baedeker, publishers. Handbuch für Reisende in Deutschland. 2. Aufl. Coblenz, 1846.
- Geog 1522.22 Baedeker, publishers. Handbuch für Reisende in Deutschland. 5. Aufl. Coblenz, 1854.
- Geog 1522.23 Baedeker, publishers. Handbuch für Reisende in Deutschland. pt.1-2. 6. Aufl. Coblenz, 1855.
- Geog 1522.25 Baedeker, publishers. Deutschland. 7. Aufl. Coblenz, 1857.
- Geog 1522.30 Baedeker, publishers. Deutschland. 8. Aufl. Coblenz, 1858.
- Geog 1522.33 Baedeker, publishers. Deutschland. v.2. 10. Aufl. Coblenz, 1862.
- Geog 1522.35 Baedeker, publishers. Deutschland. 12. Aufl. Coblenz, 1865.
- Geog 1522.36 Baedeker, publishers. Deutschland und Österreich. 14. Aufl. Coblenz, 1869.

Geog 1522 Guidebook collections - Baedeker - Germany - General - cont.
- Geog 1522.38 Baedeker, publishers. Deutschland in einem Bande. 4. Aufl. Leipzig, 1925.
- Geog 1522.39.5 Baedeker, publishers. Autoführer, Deutsches Reich (Grossdeutschland). 2. Aufl. Leipzig, 1939.
- Geog 1522.42 Baedeker, publishers. Das Deutsche Reich. 6. Aufl. Leipzig, 1936.

Geog 1523 Guidebook collections - Baedeker - Germany - Northern Germany in general
- Geog 1523.15 Baedeker, publishers. Rhine and northern Germany. 3. ed. Coblenz, 1868.
- Geog 1523.17 Baedeker, publishers. Rhine and northern Germany. 4. ed. Coblenz, 1970.
- Geog 1523.20 Baedeker, publishers. Northern Germany. 5. ed. Coblenz, 1873.
- Geog 1523.25 Baedeker, publishers. Northern Germany. 6. ed. Leipsic, 1877.
- Geog 1523.26 Baedeker, publishers. Northern Germany. 6. ed. Leipsic, 1877.
- Geog 1523.30 Baedeker, publishers. Northern Germany. 7. ed. Leipsic, 1881.
- Geog 1523.35 Baedeker, publishers. Northern Germany. 8. ed. Leipsic, 1884.
- Geog 1523.36 Baedeker, publishers. Northern Germany. 8. ed. Leipsic, 1884.
- Geog 1523.40 Baedeker, publishers. Northern Germany. 9. ed. Leipsic, 1886.
- Geog 1523.41 Baedeker, publishers. Northern Germany. 9. ed. Leipsic, 1886.
- Geog 1523.45 Baedeker, publishers. Northern Germany. 10. ed. Leipsic, 1890.
- Geog 1523.46 Baedeker, publishers. Northern Germany. 10. ed. Leipsic, 1890.
- Geog 1523.50 Baedeker, publishers. Northern Germany. 11. ed. Leipsic, 1893.
- Geog 1523.55 Baedeker, publishers. Northern Germany. 12. ed. Leipsic, 1897.
- Geog 1523.60 Baedeker, publishers. Northern Germany. 13. ed. Leipsic, 1900.
- Geog 1523.65 Baedeker, publishers. Northern Germany. 14. ed. Leipzig, 1904.
- Geog 1523.66 Baedeker, publishers. Northern Germany. 14. ed. Leipzig, 1904.
- Geog 1523.67A Baedeker, publishers. Northern Germany. 15. ed. Leipzig, 1910.
- Geog 1523.67B Baedeker, publishers. Northern Germany. 15. ed. Leipzig, 1910.
- Geog 1523.70 Baedeker, publishers. Northern Germany. 16. ed. Leipzig, 1913.
- Geog 1523.73 Baedeker, publishers. Northern Germany. 17. ed. Leipzig, 1925.
- Geog 1523.75 Baedeker, publishers. Nordost-Deutschland. 25. Aufl. Leipzig, 1896.
- Geog 1523.80 Baedeker, publishers. Nordwest-Deutschland. 26. Aufl. Leipzig, 1899.
- Geog 1523.81 Baedeker, publishers. Nordwestdeutschland. 27. Aufl. Leipzig, 1902.

Geog 1524 Guidebook collections - Baedeker - Germany - Southern Germany in general
- Geog 1524.5A Baedeker, publishers. Southern Germany and Austria. 2. ed. Coblenz, 1871.
- Geog 1524.5B Baedeker, publishers. Southern Germany and Austria. 2. ed. Coblenz, 1871.
- Geog 1524.10 Baedeker, publishers. Southern Germany and Austria. 3. ed. Coblenz, 1873.
- Geog 1524.14 Baedeker, publishers. L'Allemagne, l'Autriche. 6. éd. Leipzig, 1878.
- Geog 1524.15 Baedeker, publishers. Süd-Deutschland und Oesterreich. 8. Aufl. Leipzig, 1879.
- Geog 1524.19 Baedeker, publishers. Süd-Deutschland. 22. Aufl. Leipzig, 1888.
- Geog 1524.19.2 Baedeker, publishers. Süd-Deutschland. 23. Aufl. Leipzig, 1890.
- Geog 1524.30A Baedeker, publishers. Southern Germany and Austria. 4. ed. Leipsic, 1880.
- Geog 1524.30B Baedeker, publishers. Southern Germany and Austria. 4. ed. Leipsic, 1880.
- Geog 1524.32 Baedeker, publishers. Süd-Deutschland und Oesterreich. 19. Aufl. Leipzig, 1882.
- Geog 1524.33 Baedeker, publishers. Southern Germany and Austria. 5. ed. Leipzig, 1883.
- Geog 1524.35 Baedeker, publishers. Southern Germany and Austria. 6. ed. Leipzig, 1887.
- Geog 1524.36 Baedeker, publishers. Southern Germany and Austria. 6. ed. Leipzig, 1887.
- Geog 1524.38 Baedeker, publishers. Allemagne du sud et Autriche. 9. éd. Leipzig, 1888.
- Geog 1524.40 Baedeker, publishers. Southern Germany and Austria. 7. ed. Leipsic, 1891.
- Geog 1524.41 Baedeker, publishers. Southern Germany and Austria. 7. ed. Leipsic, 1891.
- Geog 1524.45 Baedeker, publishers. Southern Germany. 8. ed. Leipsic, 1895.
- Geog 1524.46 Baedeker, publishers. Southern Germany. 8. ed. Leipsic, 1895.
- Geog 1524.51 Baedeker, publishers. Southern Germany. 9. ed. Leipsic, 1902.
- Geog 1524.55 Baedeker, publishers. Southern Germany. 10. ed. Leipzig, 1907.
- Geog 1524.56 Baedeker, publishers. Southern Germany. 11. ed. Leipzig, 1910.
- Geog 1524.69A Baedeker, publishers. Southern Germany. 13. ed. Leipzig, 1929.
- Geog 1524.69B Baedeker, publishers. Southern Germany. 13. ed. Leipzig, 1929.
- Geog 1524.72 Baedeker, publishers. Northern Bavaria. Hamburg, 1951.
- Geog 1524.73 Baedeker, publishers. Südbayern: Alpenvorland, Alpen, österreichische Gunzgebiete. 41. Aufl. Hamburg, 1953.
- Geog 1524.75 Baedeker, publishers. Südbaiern, Tirol und Salzburg. 18. Aufl. Leipzig, 1878.
- Geog 1524.77 Baedeker, publishers. Südbaiern, Tirol und Salzburg. 22. Aufl. Leipzig, 1886.

Classified Listing

Geog 1525   Guidebook collections - Baedeker - Germany - Other regions
- Geog 1525.15   Baedeker, publishers. Die Rheinlande. 9. Aufl. Coblenz, 1856.
- Geog 1525.20   Baedeker, publishers. A handbook for travellers on the Rhine. 2. ed. Coblenz, 1864.
- Geog 1525.22   Baedeker, publishers. Die Rheinlande von der Schweize zu holländisch Grenze. 14. Aufl. Coblenz, 1866.
- Geog 1525.25   Baedeker, publishers. The Rhine from Rotterdam to Constance. 5. ed. Leipsic, 1873.
- Geog 1525.25.3   Baedeker, publishers. The Rhine from Rotterdam to Constance. 5. ed. Leipsic, 1873.
- Geog 1525.27   Baedeker, publishers. Les bords du Rhin de la frontière suisse à la frontière de Hollande. 9. éd. Leipzig, 1875.
- Geog 1525.29   Baedeker, publishers. The Rhine from Rotterdam to Constance. 6. ed. Leipsic, 1878.
- Geog 1525.30   Baedeker, publishers. The Rhine from Rotterdam to Constance. 7. ed. Leipsic, 1880.
- Geog 1525.35   Baedeker, publishers. The Rhine from Rotterdam to Constance. 8. ed. Leipsic, 1882.
- Geog 1525.35.3   Baedeker, publishers. The Rhine from Rotterdam to Constance. 8. ed. Leipsic, 1882.
- Geog 1525.37   Baedeker, publishers. Les bords du Rhin. 12. éd. Leipzig, 1882.
- Geog 1525.40A   Baedeker, publishers. The Rhine from Rotterdam to Constance. 10. ed. Leipsic, 1886.
- Geog 1525.40B   Baedeker, publishers. The Rhine from Rotterdam to Constance. 10. ed. Leipsic, 1886.
- Geog 1525.41   Baedeker, publishers. Die Rheinlande von der Schweize zu holländisch Grenze. 28. Aufl. Leipzig, 1889.
- Geog 1525.41.3   Baedeker, publishers. The Rhine from Rotterdam to Constance. 11. ed. Leipsic, 1889.
- Geog 1525.45   Baedeker, publishers. The Rhine from Rotterdam to Constance. 12. ed. Leipsic, 1892.
- Geog 1525.45.2   Baedeker, publishers. The Rhine from Rotterdam to Constance. 12. ed. Leipsic, 1892.
- Geog 1525.45.5   Baedeker, publishers. Die Rheinlande von der Schweize. 26. Aufl. Leipzig, 1892.
- Geog 1525.47   Baedeker, publishers. The Rhine from Rotterdam to Constance. 13. ed. Leipsic, 1896.
- Geog 1525.50   Baedeker, publishers. The Rhine from Rotterdam to Constance. 14. ed. Leizsic, 1900.
- Geog 1525.50.2   Baedeker, publishers. The Rhine from Rotterdam to Constance. 14. ed. Leizsic, 1900.
- Geog 1525.52.5   Baedeker, publishers. The Rhine from Rotterdam to Constance. 15. ed. Leipsic, 1903.
- Geog 1525.53   Baedeker, publishers. The Rhine from Rotterdam to Constance. 16. ed. Leipsic, 1906.
- Geog 1525.55A   Baedeker, publishers. The Rhine including the Black Forest and Vosges. 17. ed. Leipzig, 1911.
- Geog 1525.55B   Baedeker, publishers. The Rhine including the Black Forest and Vosges. 17. ed. Leipzig, 1911.
- Geog 1525.55.15   Baedeker, publishers. Schwarzwald. 3. Aufl. Leipzig, 1936.
- Geog 1525.55.16   Baedeker, publishers. Schwarzwald. 4. Aufl. Malente, 1956.
- Geog 1525.56   Baedeker, publishers. Die Rheinland, Schwarzwald, Vogesen. 32. Aufl. Leipzig, 1912.
- Geog 1525.66   Baedeker, publishers. The Rhine from the Dutch to the Alsatian frontier. 18. ed. Leipzig, 1926.
- Geog 1525.70   Baedeker, publishers. Schleswig-Holstein und Hamburg. Hamburg, 1949.
- Geog 1525.72   Baedeker, publishers. Köln und der Rheinland zwischen Köln und Mainz. Hamburg, 1953.
- Geog 1525.75   Baedeker, publishers. Wiesbaden, Mainz, Rheingau, Rheinhessen. Malente, 1956.

Geog 1526   Guidebook collections - Baedeker - Germany - Berlin
- Geog 1526.10   Baedeker, publishers. Berlin und Umgebungen. 7. Aufl. Leipzig, 1891.
- Geog 1526.12   Baedeker, publishers. Berlin und Umgebungen. 8. Aufl. Leipzig, 1894.
- Geog 1526.17   Baedeker, publishers. Berlin und Umgebungen. 11. Aufl. Leipzig, 1900.
- Geog 1526.20   Baedeker, publishers. Berlin and its environs. Leipsic, 1903.
- Geog 1526.25   Baedeker, publishers. Berlin and its environs. 2. ed. Leipsic, 1905.
- Geog 1526.30   Baedeker, publishers. Berlin and its environs. 3. ed. Leipsic, 1908.
- Geog 1526.35   Baedeker, publishers. Berlin und its environs. 5. ed. Leipzig, 1912.
- Geog 1526.36   Baedeker, publishers. Berlin und its environs. 6. ed. Leipzig, 1923.
- Geog 1526.40   Baedeker, publishers. Berlin. 23. Aufl. Freiburg, 1964.

Geog 1527   Guidebook collections - Baedeker - Germany - Other cities
- Geog 1527.2   Baedeker, publishers. München und Südbayern. 39. Aufl. Leipzig, 1935.
- Geog 1527.5   Baedeker, publishers. München und Umgebung. Augsburg. Leipzig, 1935.
- Geog 1527.10   Baedeker, publishers. Munich and its environs. Hamburg, 1950.
- Geog 1527.12   Baedeker, publishers. Munich and its environs. 2. ed. Hamburg, 1956.
- Geog 1527.15   Baedeker, publishers. München. Leipzig, 1921.
- Geog 1527.18   Baedeker, publishers. München und Umgebung. 4. Aufl. Freiburg, 1960.
- Geog 1527.50   Baedeker, publishers. Leipzig. Leipzig, 1948.
- Geog 1527.60   Baedeker, publishers. Köln und Umgebung. 2. Aufl. Freiburg, 1960.
- Geog 1527.70   Baedeker, publishers. Hamburg und die Niederelbe. Hamburg, 1953.

Geog 1530   Guidebook collections - Baedeker - Austria-Hungary - General
- Geog 1530.8   Baedeker, publishers. Austria. 8. ed. Leipsic, 1896.
- Geog 1530.9   Baedeker, publishers. Austria. 9. ed. Leipsic, 1900.
- Geog 1530.9.5   Baedeker, publishers. Austria. 9. ed. Leipsic, 1900.
- Geog 1530.10A   Baedeker, publishers. Austria-Hungary. 10. ed. Leipzig, 1905.
- Geog 1530.10B   Baedeker, publishers. Austria. 10. ed. Leipzig, 1905.
- Geog 1530.13   Baedeker, publishers. Austria. 11. ed. Leipzig, 1911.
- Geog 1530.15   Baedeker, publishers. Österreich-Ungarn. 29. Aufl. Leipzig, 1913.
- Geog 1530.18   Baedeker, publishers. Oesterreich-Ungarn. 20. Aufl. Leipzig, 1884.

Geog 1530   Guidebook collections - Baedeker - Austria-Hungary - General - cont.
- Geog 1530.69   Baedeker, publishers. Austria, together with Budapest. 12. ed. Leipzig, 1929.

Geog 1531   Guidebook collections - Baedeker - Austria-Hungary - Austria
- Geog 1531.3   Baedeker, publishers. Tirol: Vorarlberg und Teile von Salzburg und Kärnten. 37. Aufl. Leipzig, 1923.
- Geog 1531.4   Baedeker, publishers. Tirol, Vorarlberg, Etschland. 39. Aufl. Leipzig, 1929.
- Geog 1531.5   Baedeker, publishers. Tirol, Vorarlberg, westliche Salzburg, Hochkärnten. 40. Aufl. Leipzig, 1938.
- Geog 1531.6   Baedeker, publishers. Tirol, Vorarlberg, westliche Salzburg, Hochkärnten. 41. Aufl. Leipzig, 1943.
- Geog 1531.10   Baedeker, publishers. Tyrol and Salzburg. 14. ed. Freiburg, 1961.
- Geog 1531.15   Baedeker, publishers. Wien und Niederdonau. Leipzig, 1943.

Geog 1533   Guidebook collections - Baedeker - Eastern Alps
- Geog 1533.4   Baedeker, publishers. Eastern Alps. 4. ed. Leipsic, 1879.
- Geog 1533.6   Baedeker, publishers. Eastern Alps. 6. ed. Leipsic, 1888.
- Geog 1533.7   Baedeker, publishers. Eastern Alps. 7. ed. Leipsic, 1891.
- Geog 1533.8   Baedeker, publishers. Eastern Alps. 8. ed. Leipsic, 1895.
- Geog 1533.10   Baedeker, publishers. Eastern Alps. 10. ed. Leipsic, 1903.
- Geog 1533.12   Baedeker, publishers. Eastern Alps. 11. ed. Leipzig, 1907.
- Geog 1533.13   Baedeker, publishers. Eastern Alps. 12. ed. Leipzig, 1911.
- Geog 1533.14A   Baedeker, publishers. Tyrol and the Dolomites including the Bavarian Alps. 13. ed. Leipzig, 1927.
- Geog 1533.14B   Baedeker, publishers. Tyrol and the Dolomites including the Bavarian Alps. 13. ed. Leipzig, 1927.

Geog 1535   Guidebook collections - Baedeker - Switzerland - General
- Geog 1535.10   Baedeker, publishers. Die Schweiz. 5. Aufl. Coblenz, 1853.
- Geog 1535.13   Baedeker, publishers. Die Schweiz. 5. Aufl. Coblenz, 1854.
- Geog 1535.16   Baedeker, publishers. Die Schweiz, die italienischen Seen. 6. Aufl. Coblenz, 1856.
- Geog 1535.20   Baedeker, publishers. La Suisse. 7. éd. Coblenz, 1867.
- Geog 1535.22   Baedeker, publishers. Die Schweiz. 12. Aufl. Coblenz, 1869.
- Geog 1535.23   Baedeker, publishers. La Suisse. 8. éd. Coblenz, 1869.
- Geog 1535.23.9   Baedeker, publishers. Die Schweiz. 24. Aufl. Leipzig, 1891.
- Geog 1535.24   Baedeker, publishers. Switzerland and the adjacent portions of Italy. 2. ed. Coblenz, 1864.
- Geog 1535.24.25   Baedeker, publishers. Switzerland and the adjacent portions of Italy. 3. ed. Coblenz, 1867.
- Geog 1535.25   Baedeker, publishers. Switzerland. 3. ed. Coblenz, 1869.
- Geog 1535.26   Baedeker, publishers. Switzerland. 4. ed. Coblenz, 1869.
- Geog 1535.30   Baedeker, publishers. Switzerland...Italy, Savoy...Tyrol. 5. ed. Coblenz, 1872.
- Geog 1535.31   Baedeker, publishers. Switzerland...Italy, Savoy...Tyrol. 6. ed. Coblenz, 1873.
- Geog 1535.33   Baedeker, publishers. Switzerland...Italy, Savoy...Tyrol. 7. ed. Leipsic, 1877.
- Geog 1535.35   Baedeker, publishers. Switzerland...Italy, Savoy...Tyrol. 8. ed. Leipsic, 1879.
- Geog 1535.40   Baedeker, publishers. Switzerland...Italy, Savoy...Tyrol. 9. ed. Leipsic, 1881.
- Geog 1535.42   Baedeker, publishers. Switzerland...Italy, Savoy...Tyrol. 10. ed. Leipsic, 1883.
- Geog 1535.43   Baedeker, publishers. Switzerland...Italy, Savoy...Tyrol. 11. ed. Leipsic, 1885.
- Geog 1535.44   Baedeker, publishers. La Suisse. 14. éd. Leipzig, 1885.
- Geog 1535.45   Baedeker, publishers. Switzerland...Italy, Savoy...Tyrol. 12. ed. Leipsic, 1887.
- Geog 1535.46   Baedeker, publishers. Switzerland...Italy, Savoy...Tyrol. 13. ed. Leipsic, 1889.
- Geog 1535.47   Baedeker, publishers. Switzerland...Italy, Savoy...Tyrol. 13. ed. Leipsic, 1889.
- Geog 1535.48   Baedeker, publishers. Switzerland...Italy, Savoy...Tyrol. 14. ed. Leipsic, 1891.
- Geog 1535.50   Baedeker, publishers. Switzerland...Italy, Savoy...Tryol. 15. ed. Leipsic, 1893.
- Geog 1535.55   Baedeker, publishers. Switzerland...Italy, Savoy...Tyrol. 16. ed. Leipsic, 1895.
- Geog 1535.57   Baedeker, publishers. Switzerland...Italy, Savoy...Tyrol. 17. ed. Leipsic, 1897.
- Geog 1535.58   Baedeker, publishers. Switzerland and the adjacent portions of Italy, Savoy, and Tyrol. 18. ed. Leipsic, 1899.
- Geog 1535.59   Baedeker, publishers. Switzerland and the adjacent portions of Italy, Savoy, and Tyrol. 19. ed. Leipsic, 1901.
- Geog 1535.59.50   Baedeker, publishers. Switzerland and the adjacent portions of Italy, Savoy, and Tyrol. 20. ed. Leipsic, 1903.
- Geog 1535.60   Baedeker, publishers. Switzerland and the adjacent portions of Italy, Savoy, and Tyrol. 21. ed. Leipsic, 1905.
- Geog 1535.61   Baedeker, publishers. Switzerland and the adjacent portions of Italy, Savoy, and Tyrol. 21. ed. Leipsic, 1905.
- Geog 1535.65   Baedeker, publishers. Switzerland and the adjacent portions of Italy, Savoy and Tyrol. 22. ed. Leipzig, 1907.
- Geog 1535.68   Baedeker, publishers. Switzerland and the adjacent portions of Italy, Savoy, and Tyrol. 23. ed. Leipsic, 1909.
- Geog 1535.70   Baedeker, publishers. Switzerland and the adjacent portions of Italy, Savoy, and Tyrol. 24. ed. Leipzig, 1911.
- Geog 1535.72   Baedeker, publishers. Switzerland and the adjacent portions of Italy, Savoy, and Tyrol. 25. ed. Leipsic, 1913.
- Geog 1535.74   Baedeker, publishers. Switzerland together with Chamonix and the Italian lakes. 26. ed. Leipzig, 1922.
- Geog 1535.76   Baedeker, publishers. Switzerland together with Chamonix. 27. ed. Leipzig, 1928.

Classified Listing

Geog 1538		Guidebook collections - Baedeker - Italy - General
	Geog 1538.20	Baedeker, publishers. Italien. Coblenz, 1866-68. 2v.
	Geog 1538.30	Baedeker, publishers. Italien. 9. Aufl. Leipzig, 1879-89. 2v.
	Geog 1538.35	Baedeker, publishers. Italien. 14. Aufl. Leipzig, 1894.
	Geog 1538.37	Baedeker, publishers. Italien. v.3. 11. Aufl. Leipzig, 1895.
	Geog 1538.45	Baedeker, publishers. Italien von den Alpen bis Neapel. 3. Aufl. Leipzig, 1895.
	Geog 1538.50	Baedeker, publishers. Italy from the Alps to Naples. Leipzig, 1904.
	Geog 1538.56A	Baedeker, publishers. Italy from the Alps to Naples. 2. ed. Leipzig, 1909.
	Geog 1538.56B	Baedeker, publishers. Italy from the Alps to Naples. 2. ed. Leipzig, 1909.
	Geog 1538.57A	Baedeker, publishers. Italy from the Alps to Naples. 3. ed. Leipzig, 1928.
	Geog 1538.57B	Baedeker, publishers. Italy from the Alps to Naples. 3. ed. Leipzig, 1928.
	Geog 1538.61	Baedeker, publishers. Italy, including Sicily and Sardinia. Freiburg, 1961.
Geog 1539		Guidebook collections - Baedeker - Italy - Northern Italy in general
	Geog 1539.5	Baedeker, publishers. L'Italie...septentrionale. 3. éd. Coblenz, 1865.
	Geog 1539.8	Baedeker, publishers. Northern Italy, as far as Leghorn. Coblentz, 1868.
	Geog 1539.10	Baedeker, publishers. L'Italie...septentrionale. 5. éd. Coblenz, 1870.
	Geog 1539.11	Baedeker, publishers. L'Italie...septentrionale. 12. éd. Leipzig, 1889.
	Geog 1539.12	Baedeker, publishers. Italy...northern Italy. 2. ed. Coblenz, 1870.
	Geog 1539.17	Baedeker, publishers. Italy...northern Italy. 3. ed. Leipsic, 1874.
	Geog 1539.17.5	Baedeker, publishers. Italy...northern Italy. 3. ed. Leipsic, 1874.
	Geog 1539.18	Baedeker, publishers. L'Italie...septentrionale. 7. éd. Leipzig, 1876.
	Geog 1539.19	Baedeker, publishers. L'Italie...septentrionale. 8. éd. Leipzig, 1878.
	Geog 1539.21	Baedeker, publishers. Italy...northern Italy. 5. ed. Leipzig, 1879.
	Geog 1539.22	Baedeker, publishers. L'Italie...septentrionale. 9. éd. Leipzig, 1880.
	Geog 1539.23	Baedeker, publishers. Italy...northern Italy. 6. ed. Leipsic, 1882.
	Geog 1539.25	Baedeker, publishers. Italy...northern Italy. 7. ed. Leipsic, 1886.
	Geog 1539.30	Baedeker, publishers. Italy...northern Italy. 8. ed. Leipsic, 1889.
	Geog 1539.35	Baedeker, publishers. Italy...northern Italy. 9. ed. Leipsic, 1892.
	Geog 1539.36	Baedeker, publishers. Italy...northern Italy. 9. ed. Leipsic, 1892.
	Geog 1539.38	Baedeker, publishers. Italy...northern Italy. 10. ed. Leipsic, 1895.
	Geog 1539.40	Baedeker, publishers. Italy...northern Italy. 11. ed. Leipsic, 1899.
	Geog 1539.45	Baedeker, publishers. Italy...northern Italy. 12. ed. Leipsic, 1903.
	Geog 1539.46	Baedeker, publishers. Italy...northern Italy. 12. ed. Leipsic, 1903.
	Geog 1539.50	Baedeker, publishers. Italy...northern Italy. 13. ed. Leipsic, 1906-
	Geog 1539.51	Baedeker, publishers. Italie septentrionale. 17. éd. Leipzig, 1908.
	Geog 1539.55	Baedeker, publishers. Northern Italy. 14. ed. Leipzig, 1913.
	Geog 1539.60	Baedeker, publishers. Northern Italy. 15. ed. Leipzig, 1930.
Geog 1540		Guidebook collections - Baedeker - Italy - Central Italy in general
	Geog 1540.5	Baedeker, publishers. Italy. Pt. 2: Central Italy and Rome. Coblenz, 1867.
	Geog 1540.10	Baedeker, publishers. Italy. Pt. 2: Central Italy and Rome. 2. ed. Coblenz, 1869.
	Geog 1540.12	Baedeker, publishers. Italy. Pt. 2: Central Italy and Rome. 3. ed. Coblenz, 1872.
	Geog 1540.15	Baedeker, publishers. Italy. Pt. 2: Central Italy and Rome. 4. ed. Leipsic, 1875.
	Geog 1540.17	Baedeker, publishers. Italie. Pt. 2: Italie centrale et Rome. 4. éd. Leipzig, 1875.
	Geog 1540.17.25	Baedeker, publishers. Italie. Pt. 2: Italie centrale et Rome. 5. éd. Leipzig, 1877.
	Geog 1540.18	Baedeker, publishers. Italy. Pt. 2: Central Italy and Rome. 6. ed. Leipsic, 1879.
	Geog 1540.18.25	Baedeker, publishers. Italy. Pt. 2: Central Italy and Rome. 7. ed. Leipsic, 1881.
	Geog 1540.19	Baedeker, publishers. Italy. Pt. 2: Central Italy and Rome. 8. ed. Leipsic, 1883.
	Geog 1540.19.5	Baedeker, publishers. Italy. Pt. 2: Central Italy and Rome. 8. ed. Leipsic, 1883.
	Geog 1540.20	Baedeker, publishers. Italy. Pt. 2: Central Italy and Rome. 9. ed. Leipsic, 1886.
	Geog 1540.21	Baedeker, publishers. Italy. Pt. 2: Central Italy and Rome. 9. ed. Leipsic, 1886.
	Geog 1540.25	Baedeker, publishers. Italy. Pt. 2: Central Italy and Rome. 10. ed. Leipsic, 1890.
	Geog 1540.26	Baedeker, publishers. Italy. Pt. 2: Central Italy and Rome. 10. ed. Leipsic, 1890.
	Geog 1540.30	Baedeker, publishers. Italy. Pt. 2: Central Italy and Rome. 11. ed. Leipzig, 1893.
	Geog 1540.31A	Baedeker, publishers. Italy. Pt. 2: Central Italy and Rome. 11. ed. Leipzig, 1893.
	Geog 1540.31B	Baedeker, publishers. Italy. Pt. 2: Central Italy and Rome. 11. ed. Leipzig, 1893.
	Geog 1540.37	Baedeker, publishers. Italy. Pt. 2: Central Italy and Rome. 12. ed. Leipsic, 1897.
	Geog 1540.40	Baedeker, publishers. Italy. Pt. 2: Central Italy and Rome. 13. ed. Leipsic, 1900.
	Geog 1540.45	Baedeker, publishers. Italy. Pt. 2: Central Italy and Rome. 14. ed. Leipzig, 1904.
	Geog 1540.47	Baedeker, publishers. Central Italy and Rome. 15. ed. Leipzig, 1909.
	Geog 1540.48	Baedeker, publishers. Central Italy and Rome. 15. ed. Leipzig, 1909.
Geog 1540		Guidebook collections - Baedeker - Italy - Central Italy in general - cont.
	Geog 1540.50	Baedeker, publishers. Rome and central Italy. 16. ed. N.Y., 1930.
Geog 1542		Guidebook collections - Baedeker - Italy - Southern Italy in general
	Geog 1542.5	Baedeker, publishers. Italy. Pt. 3: Southern Italy, Sicily. Coblenz, 1867.
	Geog 1542.9	Baedeker, publishers. Italy. Pt. 3: Southern Italy, Sicily. 2. ed. Coblenz, 1869.
	Geog 1542.10	Baedeker, publishers. Italy. Pt. 3: Southern Italy, Sicily. 2. ed. Coblenz, 1869.
	Geog 1542.13	Baedeker, publishers. Italy. Pt. 3: Southern Italy, Sicily. 4. ed. Leipsic, 1873.
	Geog 1542.14	Baedeker, publishers. Italie. Pt. 3: Italie du sud et la Sicile. 4. éd. Leipzig, 1875.
	Geog 1542.15	Baedeker, publishers. Italy. Pt. 3: Southern Italy, Sicily. 6. ed. Leipsic, 1876.
	Geog 1542.16	Baedeker, publishers. Italie. Pt. 3: Italie méridionale et la Sicile. 5. éd. Leipzig, 1877.
	Geog 1542.20	Baedeker, publishers. Italy. Pt. 3: Southern Italy, Sicily. 8. ed. Leipsic, 1883.
	Geog 1542.21	Baedeker, publishers. Italy. Pt. 3: Southern Italy, Sicily. 8. ed. Leipsic, 1883.
	Geog 1542.25	Baedeker, publishers. Italy. Pt. 3: Southern Italy, Sicily. 9. ed. Leipsic, 1887.
	Geog 1542.27	Baedeker, publishers. Italy. Pt. 3: Southern Italy, Sicily. 10. ed. Leipsic, 1890.
	Geog 1542.30	Baedeker, publishers. Italy. Pt. 3: Southern Italy, Sicily. 11. ed. Leipsic, 1893.
	Geog 1542.33	Baedeker, publishers. Italy. Pt. 3: Southern Italy, Sicily. 12. ed. Leipsic, 1896.
	Geog 1542.35	Baedeker, publishers. Italy. Pt. 3: Southern Italy, Sicily. 13. ed. Leipsic, 1900.
	Geog 1542.37	Baedeker, publishers. Southern Italy and Sicily. 15. ed. Leipzig, 1908.
	Geog 1542.39	Baedeker, publishers. Southern Italy and Sicily. 17. ed. Leipzig, 1930.
Geog 1545		Guidebook collections - Baedeker - Spain and Portugal - General
	Geog 1545.5	Baedeker, publishers. Spain and Portugal. Leipsic, 1898.
	Geog 1545.10	Baedeker, publishers. Spain and Portugal. 2. ed. Leipsic, 1901.
	Geog 1545.11	Baedeker, publishers. Spain and Protugal. 2. ed. Leipsic, 1901.
	Geog 1545.15	Baedeker, publishers. Spain and Portugal. 3. ed. Leipzig, 1908.
	Geog 1545.18A	Baedeker, publishers. Spain and Portugal. 4. ed. Leipzig, 1913.
	Geog 1545.18B	Baedeker, publishers. Spain and Portugal. 4. ed. Leipzig, 1913.
	Geog 1545.25	Baedeker, publishers. Espagne et Portugal. 3. éd. Leipzig, 1920.
Geog 1555		Guidebook collections - Baedeker - Low Countries - General
	Geog 1555.10	Baedeker, publishers. Belgique et Hollande. 4. éd. Coblenz, 1866.
	Geog 1555.13	Baedeker, publishers. Belgium and Holland. Coblenz, 1869.
	Geog 1555.15	Baedeker, publishers. Belgique et Hollande. 6. éd. Coblenz, 1871.
	Geog 1555.16	Baedeker, publishers. Belgium and Holland. 2. ed. Coblenz, 1871.
	Geog 1555.17	Baedeker, publishers. Belgium and Holland. 3. ed. Leipsic, 1874.
	Geog 1555.18	Baedeker, publishers. Belgique et Hollande. 8. éd. Leipzig, 1875.
	Geog 1555.20	Baedeker, publishers. Belgium and Holland. 5. ed. Leipsic, 1878.
	Geog 1555.21	Baedeker, publishers. Belgium and Holland. 5. ed. Leipsic, 1878.
	Geog 1555.27	Baedeker, publishers. Belgium and Holland. 5. ed. Leipsic, 1878.
	Geog 1555.29	Baedeker, publishers. Belgien und Holland. 23. Aufl. Leipzig, 1904.
	Geog 1555.30	Baedeker, publishers. Belgium and Holland. 6. ed. Leipsic, 1881.
	Geog 1555.35	Baedeker, publishers. Belgium and Holland. 7. ed. Leipsic, 1884.
	Geog 1555.36	Baedeker, publishers. Belgium and Holland. 7. ed. Leipsic, 1884.
	Geog 1555.38	Baedeker, pbulishers. Belgium and Holland. 8. ed. Leipsic, 1885.
	Geog 1555.40	Baedeker, publishers. Belgium and Holland. 9. ed. Leipzig, 1888.
	Geog 1555.43	Baedeker, publishers. Belgien und Holland. 18. Aufl. Leipzig, 1888.
	Geog 1555.45	Baedeker, publishers. Belgium and Holland. 10. ed. Leipsic, 1891.
	Geog 1555.50	Baedeker, publishers. Belgium and Holland. 11. ed. Leipsic, 1894.
	Geog 1555.51	Baedeker, publishers. Belgium and Holland. 11. ed. Leipsic, 1894.
	Geog 1555.55	Baedeker, publishers. Belgium and Holland. 12. ed. Leipsic, 1897.
	Geog 1555.60	Baedeker, publishers. Belgium and Holland. 13. ed. Leipsic, 1901.
	Geog 1555.65	Baedeker, publishers. Belgium and Holland. 14. ed. Leipsic, 1905.
	Geog 1555.70	Baedeker, publishers. Belgium and Holland. 15. ed. Leipsic, 1910.
Geog 1558		Guidebook collections - Baedeker - Low Countries - Belgium and Luxemburg
	Geog 1558.10	Baedeker, publishers. Belgien. 5. Aufl. Coblenz, 1855.
	Geog 1558.20	Baedeker, publishers. Belgique et Luxembourg. 20. éd. Leipzig, 1928.
	Geog 1558.30	Baedeker, publishers. Belgium and Luxembourg. 16. ed. Leipzig, 1931.
Geog 1560		Guidebook collections - Baedeker - Low Countries - Holland, Netherlands
	Geog 1560.67	Baedeker, publishers. Holland; Hanbuch für Reisende. 26. Aufl. Leipzig, 1927.
Geog 1568		Guidebook collections - Baedeker - Scandinavia - Norway and Sweden
	Geog 1568.5	Baedeker, publishers. Norway and Sweden. Leipzig, 1879.
	Geog 1568.15	Baedeker, publishers. Norway and Sweden. 3. ed. Leipsic, 1885.
	Geog 1568.16	Baedekr, publishers. Norway and Sweden. 3. ed. Leipsic, 1885.

Classified Listing

Geog 1568 Guidebook collections - Baedeker - Scandinavia - Norway and Sweden - cont.
- Geog 1568.18A Baedeker, publishers. Schweden und Norwegen. 3. Aufl. Leipzig, 1885.
- Geog 1568.18B Baedeker, publishers. Schweden und Norwegen. 3. Aufl. Leipzig, 1885.
- Geog 1568.24   Baedeker, publishers. Norway, Sweden and Denmark. 5. ed. Leipsic, 1892.
- Geog 1568.25A Baedeker, publishers. Norway, Sweden and Denmark. 6. ed. Leipsic, 1895.
- Geog 1568.25B Baedeker, publishers. Norway, Sweden and Denmark. 6. ed. Leipsic, 1895.
- Geog 1568.30   Baedeker, publishers. Norway, Sweden and Denmark. 7. ed. Leipzig, 1899.
- Geog 1568.35   Baedeker, publishers. Norway, Sweden and Denmark. 8. ed. Leipsic, 1903.
- Geog 1568.40   Baedeker, publishers. Norway, Sweden and Denmark. 9. ed. Leipzig, 1909.
- Geog 1568.45A Baedeker, publishers. Norway, Sweden and Denmark. 10. ed. Leipzig, 1912.
- Geog 1568.45B Baedeker, publishers. Norway, Sweden and Denmark. 10. ed. Leipzig, 1912.

Geog 1575 Guidebook collections - Baedeker - Russia
- Geog 1575.5   Baedeker, publishers. La Russie. 2. éd. Leipzig, 1897.
- Geog 1575.10   Baedeker, publishers. Russland. Handbuch für Reisende. 5. Aufl. Leipzig, 1901.
- Geog 1575.15A Baedeker, publishers. Russia. N.Y., 1914.
- Geog 1575.15B Baedeker, publishers. Russia. N.Y., 1914.
- Geog 1575.16   Baedeker, publishers. Handbook for travellers: Russia. N.Y., 1970.

Geog 1580 Guidebook collections - Baedeker - Other Slavic countries
- Geog 1580.5   Baedeker, publishers. Das General gouvernement. Leipzig, 1943.

Geog 1585 Guidebook collections - Baedeker - Greece
- X Cg Geog 1585.5   Baedeker, publishers. Griechenland. Leipzig, 1883.
- Geog 1585.10   Baedeker, publishers. Greece. Leipsic, 1889.
- Geog 1585.15   Baedeker, publishers. Greece. 2. ed. Leipsic, 1894.
- Geog 1585.25   Baedeker, publishers. Greece. 4. ed. Leipzig, 1909.
- Geog 1585.26   Baedeker, publishers. Greece. 4. ed. Leipzig, 1909.

Geog 1590 Guidebook collections - Baedeker - Other Europe
- Geog 1590.25   Orlowicz, M. Illustrierter Führer durch Galizien. Wien, 1914.
- Geog 1595.5A   Baedeker, publishers. Mediterranean...Madeira, Canary Islands. Leipzig, 1911.
- Geog 1595.5B   Baedeker, publishers. Mediterranean...Madeira, Canary Islands. Leipzig, 1911.
- Geog 1595.7   Baedeker, publishers. Mittelmeer. 2. Aufl. Leipzig, 1934.

Geog 1604 Guidebook collections - Baedeker - Asia - Turkey - Constantinople, Istanbul
- Geog 1604.5   Baedeker, publishers. Konstaninopel und das westliche Kleinasien. Leipzig, 1905.
- Geog 1604.15   Baedeker, publishers. Konstantinopel und Kleinasien, Archipel, Cypern. 2. Aufl. Leipzig, 1914.

Geog 1605 Guidebook collections - Baedeker - Asia - Palestine and Syria
- Geog 1605.5   Baedeker, publishers. Palestine and Syria. Leipzig, 1876.
- Geog 1605.10   Baedeker, publishers. Palestine and Syria. 2. ed. Leipsic, 1894.
- Geog 1605.18   Baedeker, publishers. Palästina und Syrien. 6. Aufl. Leipzig, 1904.
- Geog 1605.25A Baedeker, publishers. Palestine and Syria. 5. ed. Leipzig, 1912.
- Geog 1605.25B Baedeker, publishers. Palestine and Syria. 5. ed. Leipzig, 1912.

Geog 1616 Guidebook collections - Baedeker - Africa - Egypt - General
- Geog 1616.15   Baedeker, publishers. Egypt. 4. ed. Leipsic, 1898.
- Geog 1616.18   Baedeker, publishers. Egypt. 4. ed. Leipsic, 1898.
- Geog 1616.20   Baedeker, publishers. Egypt. 5. ed. Leipsic, 1902.
- Geog 1616.20.2 Egyptian Museum, Cairo. Gratis supplement to the fifth edition of Baedeker's Egypt. Leipzig, 1904.
- Geog 1616.21   Baedeker, publishers. Egypt. 6. ed. Leipzig, 1908.
- Geog 1616.22   Baedeker, publishers. Egypt and the Sudan. 7. ed. Leipzig, 1914.
- Geog 1616.24A Baedeker, publishers. Egypt and the Sudan. 8. ed. Leipzig, 1929.
- Geog 1616.24B Baedeker, publishers. Egypt and the Sudan. 8. ed. Leipzig, 1929.

Geog 1617 Guidebook collections - Baedeker - Africa - Egypt - Upper Egypt
- Geog 1617.5   Baedeker, publishers. Egypt. Leipsic, 1892.

Geog 1618 Guidebook collections - Baedeker - Africa - Egypt - Lower Egypt
- Geog 1618.5   Baedeker, publishers. Egypt...lower Egypt. Leipsic, 1878.
- Geog 1618.10A Baedeker, publishers. Egypt...lower Egypt. 2. ed. Leipsic, 1885.
- Geog 1618.10B Baedeker, publishers. Egypt...lower Egypt. 2. ed. Leipsic, 1885.
- Geog 1618.15   Baedeker, publishers. Egypt. Leipsic, 1892-95. 2v.

Geog 1640 Guidebook collections - Baedeker - Canada
- Geog 1640.5   Baedeker, publishers. Dominion of Canada. Leipzig, 1894.
- Geog 1640.11   Baedeker, publishers. Dominion of Canada. 2. ed. Leipzig, 1900.
- Geog 1640.15A Baedeker, publishers. Dominion of Canada with Newfoundland and an excursion to Alaska. 3. ed. Leipzig, 1907.
- Geog 1640.15B Baedeker, publishers. Dominion of Canada with Newfoundland and an excursion to Alaska. 3. ed. Leipzig, 1907.
- Geog 1640.15.2 Baedeker, publishers. The dominion of Canada. 4. ed. Leipzig, 1922.

Geog 1645 Guidebook collections - Baedeker - United States
- Htn Geog 1645.5*   Baedeker, publishers. United States...Mexico. Leipsic, 1893.
- Geog 1645.5.5 Baedeker, publishers. United States...Mexico. Leipsic, 1893.
- Geog 1645.7   Baedeker, publishers. United States...Mexico. 2. ed. Leipsic, 1899.
- Geog 1645.7.5 Baedeker, publishers. United States...Mexico. 2. ed. Leipsic, 1899.
- Geog 1645.10A Baedeker, publishers. United States...Mexico. 3. ed. Leipzig, 1904.
- Geog 1645.10B Baedeker, publishers. United States...Mexico. 3. ed. Leipzig, 1904.

Geog 1645 Guidebook collections - Baedeker - United States - cont.
- Geog 1645.10C Baedeker, publishers. United States...Mexico. 3. ed. Leipzig, 1904.
- Geog 1645.11   Baedeker, publishers. Nordamerika. Die Vereinigten Staaten...Mexiko. 2. Aufl. Leipzig, 1904.
- Geog 1645.15A Baedeker, publishers. United States...Mexico. 4. ed. Leipzig, 1909.
- Geog 1645.15B Baedeker, publishers. United States...Mexico. 4. ed. Leipzig, 1909.

Geog 1706 Guidebook collections - Murray - Europe - Northern Europe in general
- Geog 1706.5   Murray, John, publisher, London. Handbook for northern Europe. London, 1849. 2v.

Geog 1707 Guidebook collections - Murray - Europe - Central Europe in general
- Geog 1707.3   Murray, John, publisher, London. Handbook for travellers on the continent. 2. ed. London, 1838.
- Geog 1707.5   Murray, John, publisher, London. Handbook the travellers on the continent. London, 1840.
- Geog 1707.6   Murray, John, publisher, London. Handbook for travellers on the continent. 5. ed. London, 1845.
- Geog 1707.7   Murray, John, publisher, London. Handbook for travellers on the continent. 7. ed. London, 1850.
- Geog 1707.8   Murray, John, publisher, London. Handbook for travellers on the continent. 8. ed. London, 1951.
- Geog 1707.9   Murray, John, publisher, London. Handbook for travellers on the continent. 9. ed. London, 1853.
- Geog 1707.10   Murray, John, publisher, London. Handbook for travellers on the continent. 10. ed. London, 1854.
- Geog 1707.11   Murray, John, publisher, London. Handbook for travellers on the continent. 11. ed. London, 1856.
- Geog 1707.12   Murray, John, publisher, London. Handbook for travellers on the continent. 12. ed. London, 1858.
- Geog 1707.13   Murray, John, publisher, London. Handbook for travellers on the continent. 13. ed. London, 1860.
- Geog 1707.15   Murray, John, publisher, London. Handbook for travellers on the continent. 15. ed. London, 1865.
- Geog 1707.16   Murray, John, publisher, London. Handbook for travellers on the continent. 16. ed. London, 1868.
- Geog 1707.17   Murray, John, publisher, London. Handbook for travellers on the continent. 17. ed. London, 1870.
- Geog 1707.17.5 Murray, John, publisher, London. Handbook for travellers on the continent. 17. ed. London, 1871.
- Geog 1707.18   Murray, John, publisher, London. Handbook for travellers on the continent. 18. ed. London, 1873-74. 2v.
- Geog 1707.19   Murray, John, publisher, London. Handbook for travellers on the continent. 19. ed. London, 1875. 2v.

Geog 1708 Guidebook collections - Murray - Great Britain - General and England
- Geog 1708.1   Murray, John, publisher, London. Handbook for England and Wales. London, 1878.
- Geog 1708.15   Murray, John, publisher, London. Handbook for England and Wales. 2. ed. London, 1890.

Geog 1709.1 - .39 Guidebook collections - Murray - Great Britain - Counties of England (Special Table)
- Geog 1709.1   Murray, John, publisher, London. Handbook for...Berks, Bucks and Oxfordshire. London, 1860.
- Geog 1709.1.3 Murray, John, publisher, London. Handbook for...Berks, Bucks and Oxfordshire. 3. ed. London, 1882.
- Geog 1709.1.10 Murray, John, publisher, London. Handbook for Berkshire. London, 1902.
- Geog 1709.2   Murray, John, publisher, London. Handbook for Buckinghamshire. London, 1903.
- Geog 1709.5.10 Murray, John, publisher, London. Handbook for...Cornwall. 10. ed. London, 1882.
- Geog 1709.6.25 Murray, John, publisher, London. Handbook to the English lakes...Cumberland, Westmorland, and Lancashire. London, 1889.
- Geog 1709.7.2 Murray, John, publisher, London. Handbook for...Derbyshire, Nottinghamshire. 2. ed. London, 1874.
- Geog 1709.7.3 Murray, John, publisher, London. Handbook for Derbyshire, Nottinghamshire. 3. ed. London, 1904.
- Geog 1709.8.4 Murray, John, publisher, London. Handbook for Devon and Cornwall. Newton Abbot, Eng., 1971.
- Geog 1709.8.5 Murray, John, publisher, London. Handbook for Devon and Cornwall. 5. ed. London, 1863.
- Geog 1709.8.8 Murray, John, publisher, London. Handbook for Devon and Cornwall. 8. ed. London, 1872.
- Geog 1709.8.10A Murray, John, publisher, London. Handbook for...Devonshire. 10. ed. London, 1887.
- Geog 1709.8.10B Murray, John, publisher, London. Handbook for...Devonshire. 10. ed. London, 1887.
- Geog 1709.10.5 Murray, John, publisher, London. Handbook for...Durham and Northumberland. London, 1873.
- Geog 1709.11.2 Murray, John, publisher, London. Handbook for Essex, Suffolk and Norfolk. 2. ed. London, 1875.
- Geog 1709.12   Murray, John, publisher, London. Handbook for...Gloucestershire, Worcestershire. London, 1867.
- Geog 1709.12.3 Murray, John, publisher, London. Handbook for...Gloucestershire, Worcestershire. 3. ed. London, 1884.
- Geog 1709.12.4 Murray, John, publisher, London. Handbook for...Gloucestershire. 4. ed. London, 1895.
- Geog 1709.13.5 Murray, John, publisher, London. Handbook for...Hampshire. 5. ed. London, 1898.
- Geog 1709.15   Murray, John, publisher, London. Handbook for Hertfordshire, Bedfordshire. London, 1895.
- Geog 1709.17   Murray, John, publisher, London. Handbook for...Kent and Sussex. London, 1858.
- NEDL Geog 1709.17.3 Murray, John, publisher, London. Handbook for...Kent and Sussex. 3. ed. London, 1868.
- Geog 1709.17.4 Murray, John, publisher, London. Handbook for travellers in Kent. 4. ed. London, 1877.
- Geog 1709.18.3 Murray, John, publisher, London. Handbook for Lancashire. London, 1880.
- Geog 1709.20   Murray, John, publisher, London. Handbook for Lincolnshire. London, 1890.
- Geog 1709.20.2 Murray, John, publisher, London. Handbook for Lincolnshire. 2. ed. London, 1903.
- Geog 1709.24   Murray, John, publisher, London. Handbook for...Northamptonshire. London, 1878.
- Geog 1709.24.2 Murray, John, publisher, London. Handbook for...Northamptonshire. 2. ed. London, 1901.
- Geog 1709.27   Murray, John, publisher, London. Handbook for...Oxfordshire. London, 1894.
- Geog 1709.29   Murray, John, publisher, London. Handbook for Shropshire. London, 1870.

Classified Listing

Geog 1709.1 - .39 Guidebook collections - Murray - Great Britain - Counties of England (Special Table) - cont.

Geog 1709.29.3   Murray, John, publisher, London. Handbook for Shropshire and Cheshire. London, 1879.
Geog 1709.29.5   Murray, John, publisher, London. Handbook for Shropshire and Cheshire. 3. ed. London, 1897.
Geog 1709.30.5   Murray, John, publisher, London. Handbook for Somerset. 5. ed. London, 1899.
Geog 1709.33   Murray, John, publisher, London. Handbook for...Surrey, Hampshire. London, 1858.
Geog 1709.33.4   Murray, John, publisher, London. Handbook for...Surrey, Hampshire. 4. ed. London, 1888.
Geog 1709.33.5   Murray, John, publisher, London. Handbook for travellers in Surrey. 5. ed. London, 1898.
Geog 1709.34.4   Murray, John, publisher, London. Handbook for Sussex. 4. ed. London, 1877.
Geog 1709.34.5   Murray, John, publisher, London. Handbook for Sussex. 5. ed. London, 1905.
Geog 1709.35   Murray, John, publisher, London. Handbook of Warwickshire. London, 1899.
Geog 1709.36.2   Murray, John, publisher, London. Handbook for Westmorland...and the Lakes. 2. ed. London, 1869.
Geog 1709.37   Murray, John, publisher, London. Handbook for...Wiltshire, Dorsetshire and Somersetshire. London, 1856.
Geog 1709.37.2   Murray, John, publisher, London. Handbook for...Wiltshire, Dorsetshire and Somersetshire. London, 1859.
Geog 1709.37.10 Murray, John, publisher, London. Handbook for...Wiltshire, Dorsetshire and Somersetshire. 4. ed. London, 1882.
Geog 1709.37.12 Murray, John, publisher, London. Handbook for...Wilts and Dorset. 5. ed. London, 1899.
Geog 1709.38.4   Murray, John, publisher, London. Handbook for...Worcestershire. 4. ed. London, 1894.
Geog 1709.39   Murray, John, publisher, London. Handbook for Yorkshire. London, 1867.
Geog 1709.39.3   Murray, John, publisher, London. Handbook for...Yorkshire. 3. ed. London, 1882.
Geog 1709.39.4   Murray, John, publisher, London. Handbook for...Yorkshire. 4. ed. London, 1904.

Geog 1710 Guidebook collections - Murray - Great Britain - London

Geog 1710.5   Cunningham, Peter. Handbook for London, past and present. London, 1849. 2v.
Geog 1710.7   Cunningham, Peter. Handbook for London, past and present. London, 1850.
Geog 1710.9A   Cunningham, Peter. Handbook for modern London. London, 1851.
Geog 1710.9B   Cunningham, Peter. Handbook for modern London. London, 1851.
Geog 1710.15   Cunningham, Peter. London. London, 1856.
Geog 1710.20   Murray, John, publisher, London. Handbook to London. London, 1869?
Geog 1710.22   Murray, John, publisher, London. Handbook to London. London, 1870.
Geog 1710.24   Cunningham, Peter. London, as it is. London, 185-.
Geog 1710.26   Murray, John, publisher, London. Handbook to London. London, 1871.
Geog 1710.28   Murray, John, publisher, London. Handbook to London. London, 1873.
Geog 1710.29   Murray, John, publisher, London. Handbook to London. London, 1874.
Geog 1710.30   Murray, John, publisher, London. Handbook to London. London, 1876.
Geog 1710.35   Murray, John, publishers, London. Handbook to London. London, 1879.

Geog 1711 Guidebook collections - Murray - Great Britain - Other cities of England

Geog 1711.1   Thorne, James. Handbook to environs of London. London, 1876. 2v.

Geog 1712 Guidebook collections - Murray - Great Britain - Scotland

Geog 1712.1   Murray, John, publisher, London. Handbook for...Scotland. London, 1867.
Geog 1712.2   Murray, John, publisher, London. Handbook for...Scotland. 2. ed. London, 1868.
Geog 1712.4   Murray, John, publisher, London. Handbook for...Scotland. 4. ed. London, 1875.
Geog 1712.7.25 Murray, John, publisher, London. Handbook for...Scotland. 8. ed. London, 1903.
Geog 1712.8   Murray, John, publisher, London. Handbook for...Scotland. 8. ed. London, 1907.
Geog 1712.9A   Murray, John, publisher, London. Handbook for...Scotland. 9. ed. London, 1913.
Geog 1712.9B   Murray, John, publisher, London. Handbook for...Scotland. 9. ed. London, 1913.

Geog 1713 Guidebook collections - Murray - Great Britain - Ireland

Geog 1713.2.5   Murray, John, publisher, London. Handbook for travellers in Ireland. 2. ed. London, 1866.
Geog 1713.4   Murray, John, publisher, London. Handbook for travellers in Ireland. 4. ed. London, 1878.
Geog 1713.5   Murray, John, publisher, London. Handbook for travellers in Ireland. 5. ed. London, 1896.
Geog 1713.6   Murray, John, publisher, London. Handbook for travellers in Ireland. 6. ed. London, 1902.
Geog 1713.6.75 Murray, John, publisher, London. Handbook for travellers in Ireland. 7. ed. London, 1906.
Geog 1713.7   Murray, John, publisher, London. Handbook for travellers in Ireland. 7. ed. London, 1906.
Geog 1713.8   Murray, John, publisher, London. Handbook for travellers in Ireland. 7. ed. London, 1906.

Geog 1714 Guidebook collections - Murray - Great Britain - Wales

Geog 1714.1.2   Murray, John, publisher, London. Handbook for...North Wales. 2. ed. London, 1864.
Geog 1714.1.5   Murray, John, publisher, London. Handbook for...North Wales. 5. ed. London, 1885.
Geog 1714.2.4   Murray, John, publisher, London. Handbook for...South Wales. 4. ed. London, 1890.

Geog 1715 Guidebook collections - Murray - France - General

Geog 1715.2   Murray, John, publisher, London. Handbook for travellers in France. London, 1843.
Geog 1715.2.5 Murray, John, publisher, London. Handbook for travellers in France. London, 1844.
Geog 1715.3   Murray, John, publisher, London. Handbook for travellers in France. 3. ed. London, 1848.
Geog 1715.4   Murray, John, publisher, London. Handbook for travellers in France. 4. ed. London, 1853.

Geog 1715 Guidebook collections - Murray - France - General - cont.

Geog 1715.7   Murray, John, publisher, London. Handbook for travellers in France. 7. ed. London, 1859.
Geog 1715.9   Murray, John, publisher, London. Handbook for travellers in France. 9. ed. London, 1864.
Geog 1715.10   Murray, John, publisher, London. Handbook for travellers in France. 10. ed. London, 1867.
Geog 1715.11   Murray, John, publisher, London. Handbook for travellers in France. 11. ed. London, 1869.
Geog 1715.12   Murray, John, publisher, London. Handbook for travellers in France. 12. ed. London, 1873. 2v.
Geog 1715.13   Murray, John, publisher, London. Handbook for travellers in France. 13. ed. London, 1875.
Geog 1715.16   Murray, John, publisher, London. Handbook for travellers in France. 16. ed. London, 1882-84. 2v.
Geog 1715.18   Murray, John, publisher, London. Handbook for travellers in France. 18. ed. London, 1892. 2v.

Geog 1719 Guidebook collections - Murray - France - Paris

Geog 1719.1   Murray, John, publisher, London. Handbook for visitors to Paris. London, 1864.
Geog 1719.3   Murray, John, publisher, London. Handbook for visitors to Paris. 3. ed. London, 1867.
Geog 1719.5   Murray, John, publisher, London. Handbook for visitors to Paris. 5. ed. London, 1872.
Geog 1719.8   Murray, John, publisher, London. Handbook for visitors to Paris. 8. ed. London, 1876.
Geog 1719.10   Murray, John, publisher, London. Handbook for visitors to Paris. 10. ed. London, 1879.

Geog 1723 Guidebook collections - Murray - Germany - Northern Germany in general

Geog 1723.19   Murray, John, publisher, London. Handbook for north Germany. 19. ed. London, 1877.

Geog 1724 Guidebook collections - Murray - Germany - Southern Germany in general

Geog 1724.2   Murray, John, publisher, London. Handbook for...southern Germany. London, 1837.
Htn Geog 1724.5*   Murray, John, publisher, London. Handbook for...southern Germany. 5. ed. London, 1850.
Geog 1724.5.5   Murray, John, publisher, London. Handbook for...southern Germany. 5. ed. London, 1851.
Geog 1724.6   Murray, John, publisher, London. Handbook for...southern Germany. 6. ed. London, 1853.
Geog 1724.7   Murray, John, publisher, London. Handbook for...southern Germany. 7. ed. London, 1855.
Geog 1724.7.2   Murray, John, publisher, London. Handbook for...southern Germany. 7. ed. London, 1855.
Geog 1724.8   Murray, John, publisher, London. Handbook for...southern Germany. 8. ed. London, 1858.
Geog 1724.9   Murray, John, publisher, London. Handbook for...southern Germany. 9. ed. London, 1863.
Geog 1724.10   Murray, John, publisher, London. Handbook for...southern Germany. 10. ed. London, 1867.
Geog 1724.11   Murray, John, publisher, London. Handbook for...southern Germany. 11. ed. London, 1871.
Geog 1724.12   Murray, John, publisher, London. Handbook for...southern Germany. 12. ed. London, 1873.
Geog 1724.14   Murray, John, publisher, London. Handbook for...south Germany and Austria. 14. ed. London, 1881.

Geog 1733 Guidebook collections - Murray - Eastern Alps

Geog 1733.5   Murray, John, publisher, London. The knapsack guide...Tyrol and the eastern Alps. London, 1867.

Geog 1735 Guidebook collections - Murray - Switzerland - General

Geog 1735.1   Murray, John, publisher, London. Handbook for...Switzerland. London, 1838.
Geog 1735.3   Murray, John, publisher, London. Handbook for...Switzerland. London, 1839.
Geog 1735.7   Murray, John, publisher, London. Handbook for...Switzerland. 3. ed. London, 1846.
Geog 1735.9   Murray, John, publisher, London. Handbook for...Switzerland. 4. ed. London, 1851.
Geog 1735.11   Murray, John, publisher, London. Handbook for...Switzerland. 5. ed. London, 1852.
Geog 1735.13   Murray, John, publisher, London. Handbook for...Switzerland. 6. ed. London, 1854.
Geog 1735.15   Murray, John, publisher, London. Handbook for...Switzerland. 7. ed. London, 1856.
Geog 1735.17   Murray, John, publisher, London. Handbook for...Switzerland. 8. ed. London, 1858.
Geog 1735.22   Murray, John, publisher, London. Knapsack guide for...Switzerland. London, 1864.
Geog 1735.23A   Murray, John, publisher, London. Handbook for...Switzerland. 11. ed. London, 1865.
Geog 1735.23B   Murray, John, publisher, London. Handbook for...Switzerland. 11. ed. London, 1865.
Geog 1735.25   Murray, John, publisher, London. Handbook for...Switzerland. 12. ed. London, 1867.
Geog 1735.26   Murray, John, publisher, London. Knapsack guide for...Switzerland. London, 1867.
Geog 1735.30   Murray, John, publisher, London. Handbook for...Switzerland. 15. ed. London, 1874.
Geog 1735.35   Murray, John, publisher, London. Handbook for...Switzerland. 17. ed. London, 1886.

Geog 1739 Guidebook collections - Murray - Italy - Northern Italy in general

Geog 1739.4   Murray, John, publisher, London. Handbook for...northern Italy. 4. ed. London, 1852.
Geog 1739.6   Murray, John, publisher, London. Handbook for...northern Italy. 6. ed. London, 1856. 2v.
Geog 1739.8   Murray, John, publisher, London. Handbook for...northern Italy. 8. ed. London, 1860.
Geog 1739.9   Murray, John, publisher, London. Handbook for...northern Italy. 9. ed. London, 1863.
Geog 1739.10   Murray, John, publisher, London. Handbook for...northern Italy. 10. ed. London, 1866.
Geog 1739.11   Murray, John, publisher, London. Handbook for...northern Italy. 11. ed. London, 1867.
Geog 1739.14   Murray, John, publisher, London. Handbook for...northern Italy. 14. ed. London, 1877.

Geog 1740 Guidebook collections - Murray - Italy - Central Italy in general

Geog 1740.1   Murray, John, publisher, London. Handbook for...central Italy. London, 1843.
Geog 1740.3   Murray, John, publisher, London. Handbook for...central Italy. 3. ed. London, 1853. 2v.

Classified Listing

Geog 1740 Guidebook collections - Murray - Italy - Central Italy in general - cont.
- Geog 1740.4 — Murray, John, publisher, London. Handbook for...central Italy. 4. ed. London, 1857.
- Geog 1740.5 — Murray, John, publisher, London. Handbook for...central Italy. 5. ed. London, 1861.
- Geog 1740.6 — Murray, John, publisher, London. Handbook for...central Italy. 6. ed. London, 1864.
- Geog 1740.7.5 — Murray, John, publisher, London. Handbook for...central Italy. 7. ed. London, 1867.
- Geog 1740.9 — Murray, John, publisher, London. Handbook for...central Italy. 9. ed. London, 1875.
- Geog 1740.11 — Murray, John, publisher, London. Handbook for...central Italy. 11. ed. pt.2. London, 1889.

Geog 1741 Guidebook collections - Murray - Italy - Rome
- Geog 1741.5 — Murray, John, publisher, London. Handbook of Rome and its environs. 5. ed. London, 1858.
- Geog 1741.7 — Murray, John, publisher, London. Handbook of Rome and its environs. 7. ed. London, 1864.
- Geog 1741.8 — Murray, John, publisher, London. Handbook of Rome and its environs. 8. ed. London, 1867.
- Geog 1741.9 — Murray, John, publisher, London. Handbook of Rome and its environs. 9. ed. London, 1869.
- Geog 1741.10 — Murray, John, publisher, London. Handbook of Rome and its environs. 10. ed. London, 1871.
- Geog 1741.12 — Murray, John, publisher, London. Handbook of Rome and its environs. 12. ed. London, 1875.

Geog 1742 Guidebook collections - Murray - Italy - Southern Italy in general
- Geog 1742.1 — Murray, John, publisher, London. Handbook for...southern Italy. London, 1853.
- Geog 1742.2 — Murray, John, publisher, London. Handbook for...southern Italy. 2. ed. London, 1855.
- Geog 1742.3 — Murray, John, publisher, London. Handbook for...southern Italy. 3. ed. London, 1858.
- Geog 1742.4 — Murray, John, publisher, London. Handbook for...southern Italy. 4. ed. London, 1862.
- Geog 1742.6 — Murray, John, publisher, London. Handbook for...southern Italy. 6. ed. London, 1868.
- Geog 1742.8 — Murray, John, publisher, London. Handbook for...southern Italy. 8. ed. London, 1878.
- Geog 1742.9 — Murray, John, publisher, London. Handbook for...southern Italy and Sicily. 9. ed. London, 1892.

Geog 1743 Guidebook collections - Murray - Italy - Sicily
- Geog 1743.1 — Murray, John, publisher, London. Handbook for...Sicily. London, 1864.

Geog 1744 Guidebook collections - Murray - Italy - Sardinia and Corsica
- Geog 1744.1 — Murray, John, publisher, London. Handbook for...Corsica and Sardinia. London, 1860.

Geog 1748 Guidebook collections - Murray - Spain and Portugal - Spain
- Geog 1748.1 — Murray, John, publisher, London. A handbook for travellers in Spain. London, 1845. 2v.
- Geog 1748.2 — Murray, John, publisher, London. Handbook for travellers in Spain. 2. ed. London, 1847.
- Geog 1748.3A — Murray, John, publisher, London. Handbook for...Spain. 3. ed. London, 1855. 2v.
- Geog 1748.3B — Murray, John, publisher, London. Handbook for...Spain. 3. ed. London, 1855.
- Geog 1748.5 — Murray, John, publisher, London. Handbook for...Spain. 5. ed. London, 1878.
- Geog 1748.6 — Murray, John, publisher, London. Handbook for...Spain. 6. ed. London, 1882. 2v.
- Geog 1748.7 — Murray, John, publisher, London. Handbook for...Spain. 7. ed. London, 1890. 2v.

Geog 1750 Guidebook collections - Murray - Spain and Portugal - Portugal
- Geog 1750.1 — Murray, John, publisher, London. Handbook for...Portugal. London, 1855.
- Geog 1750.2 — Murray, John, publisher, London. Handbook for...Portugal. 2. ed. London, 1856.
- Geog 1750.3 — Murray, John, publisher, London. Handbook for...Portugal. 3. ed. London, 1875.

Geog 1763 Guidebook collections - Murray - Scandinavia - General
- Geog 1763.3 — Murray, John, publisher, London. Handbook for...Denmark, Norway, Sweden. 3. ed. London, 1858.
- Geog 1763.6 — Murray, John, publisher, London. Handbook for...Denmark, Norway and Sweden. 3. ed. London, 1871.

Geog 1765 Guidebook collections - Murray - Scandinavia - Denmark
- Geog 1765.4 — Murray, John, publisher, London. Handbook for...Denmark. 4. ed. London, 1875.

Geog 1770 Guidebook collections - Murray - Scandinavia - Norway
- Geog 1770.5 — Murray, John, publisher, London. Handbook for Norway. 5. ed. London, 1874.
- Geog 1770.7 — Murray, John, publisher, London. Handbook for Norway. 7. ed. London, 1880.
- Geog 1770.8 — Murray, John, publisher, London. Handbook for...Norway. 8. ed. London, 1892.

Geog 1773 Guidebook collections - Murray - Scandinavia - Sweden
- Geog 1773.4 — Murray, John, publisher, London. Handbook for...Sweden. 4. ed. London, 1875.
- Geog 1773.6 — Murray, John, publisher, London. Handbook for...Sweden. 6. ed. London, 1883.

Geog 1775 Guidebook collections - Murray - Russia
- Geog 1775.2 — Murray, John, publisher, London. Handbook for...Russia, Poland and Finland. 2. ed. London, 1868.
- Geog 1775.3 — Murray, John, publisher, London. Handbook for...Russia, Poland and Finland. 3. ed. London, 1875.
- Geog 1775.5 — Murray, John, publisher, London. Handbook for...Russia, Poland and Finland. 5. ed. London, 1893.
- Geog 1775.10 — Murray, John, publisher, London. Hand-book for northern Europe: Finland and Russia. N.Y., 1970.

Geog 1782 Guidebook collections - Murray - Ionian Islands
- Geog 1782.1 — Murray, John, publisher, London. Handbook for...Ionian islands. London, 1840.
- Geog 1782.2 — Murray, John, publisher, London. Handbook for...Ionian islands. London, 1845.

Geog 1785 Guidebook collections - Murray - Greece
- Geog 1785.2 — Murray, John, publisher, London. Handbook for...Greece. London, 1854.
- Geog 1785.4 — Murray, John, publisher, London. Handbook for...Greece. 4. ed. Lonon, 1872.
- Geog 1785.5 — Murray, John, publisher, London. Handbook for...Greece. 5. ed. London, 1884. 2v.
- Geog 1785.6 — Murray, John, publisher, London. Handbook for...Greece. 6. ed. London, 1896. 2v.
- Geog 1785.7 — Murray, John, publisher, London. Handbook for...Greece. 7. ed. London, 1900.

Geog 1790 Guidebook collections - Murray - Other Europe
- Geog 1795.3 — Murray, John, publisher, London. Handbook to the Mediterranean. 3. ed. London, 1892. 2v.

Geog 1801 Guidebook collections - Murray - Asia - Asia Minor
- Geog 1801.1 — Murray, John, publisher, London. Handbook for Asia Minor. London, 1895.

Geog 1803 Guidebook collections - Murray - Asia - Turkey - General
- NEDL Geog 1803.3 — Murray, John, publisher, London. Handbook for...Turkey. London, 1854.
- Geog 1803.4 — Murray, John, publisher, London. Handbook for Turkey. 4. ed. London, 1878.

Geog 1804 Guidebook collections - Murray - Asia - Turkey - Constantinople, Istanbul
- Geog 1804.1A — Murray, John, publisher, London. Handbook for...Constantinople. London, 1900.
- Geog 1804.1B — Murray, John, publisher, London. Handbook for...Constantinople. London, 1900.

Geog 1805 Guidebook collections - Murray - Asia - Palestine and Syria
- Geog 1805.1 — Murray, John, publisher, London. Handbook for...Syria and Palestine. London, 1858. 2v.
- Geog 1805.1.10 — Murray, John, publisher, London. Handbook for...Syria and Palestine. London, 1875.
- Geog 1805.2 — Murray, John, publisher, London. Handbook for...Syria and Palestine. London, 1892.

Geog 1807 Guidebook collections - Murray - Asia - India - General
- Geog 1807.5 — Murray, John, publisher, London. Handbook for...India. London, 1859. 2v.
- Geog 1807.10 — Murray, John, publisher, London. Handbook for...India and Ceylon. London, 1892.
- Geog 1807.13 — Murray, John, publisher, London. Handbook for...India, Burma and Ceylon. 3. ed. London, 1898.
- Geog 1807.17 — Murray, John, publisher, London. Handbook for...India, Burma and Ceylon. 5. ed. London, 1905.
- Geog 1807.17.5 — Murray, John, publisher, London. Handbook for...India, Burma and Ceylon. 5. ed. London, 1906.
- Geog 1807.18 — Murray, John, publisher, London. Handbook for...India, Burma and Ceylon. 6. ed. London, 1907.
- Geog 1807.19 — Murray, John, publisher, London. Handbook for...India, Burma and Ceylon. 8. ed. London, 1911.
- Geog 1807.21 — Murray, John, publisher, London. Handbook for...India, Burma and Ceylon. 10. ed. London, 1920.
- Geog 1807.25 — Murray, John, publisher, London. Handbook for...India, Burma and Ceylon. 13. ed. London, 1929.
- Geog 1807.25.2 — Murray, John, publisher, London. Handbook for...India, Burma and Ceylon. 13. ed. N.Y., 1929.
- Geog 1807.27 — Murray, John, publisher, London. Handbook for...India, Burma and Ceylon. 14. ed. London, 1933.
- Geog 1807.29 — Murray, John, publisher, London. Handbook for...India, Pakistan, Burma and Ceylon. London, 1949.
- Geog 1807.30 — Murray, John, publisher, London. The Imperial guide to India. London, 1904.

Geog 1809 Guidebook collections - Murray - Asia - India - Provinces, States, Cities
- Geog 1809.5 — Murray, John, publisher, London. Handbook of the Bengal presidency. London, 1882.
- Geog 1809.30 — Murray, John, publisher, London. Handbook of the Bombay presidency. London, 1881.
- Geog 1809.55 — Murray, John, publisher, London. Handbook of the Madras presidency. London, 1879.
- Geog 1809.80 — Murray, John, publisher, London. Handbook of the Punjab...Kashmir. London, 1883.

Geog 1811 Guidebook collections - Murray - Asia - Japan
- Geog 1811.2 — Chamberlain, B.H. Handbook for...Japan. 3. ed. London, 1891.
- Htn Geog 1811.3* — Chamberlain, B.H. Handbook for...Japan. N.Y., 1893.
- Geog 1811.4 — Murray, John, publisher, London. Handbook for...Japan. 4. ed. London, 1896.
- Geog 1811.5 — Murray, John, publisher, London. Handbook for Japan. 5. ed. London, 1899.
- Geog 1811.6 — Murray, John, publisher, London. Handbook for...Japan. 6. ed. London, 1901.
- Geog 1811.8A — Murray, John, publisher, London. Handbook for...Japan. 8. ed. London, 1907.
- Geog 1811.8B — Murray, John, publisher, London. Handbook for...Japan. 8. ed. London, 1907.
- Geog 1811.8C — Murray, John, publisher, London. Handbook for...Japan. 8. ed. London, 1907.
- Geog 1811.9A — Murray, John, publisher, London. Handbook for...Japan. 9. ed. London, 1913.
- Geog 1811.9B — Murray, John, publisher, London. Handbook for...Japan. 9. ed. London, 1913.
- Geog 1811.9C — Murray, John, publisher, London. Handbook for...Japan. 9. ed. London, 1913.

Geog 1816 Guidebook collections - Murray - Africa - Egypt - General
- Geog 1816.5 — Murray, John, publisher, London. Handbook for...Egypt. London, 1847.
- Geog 1816.10 — Murray, John, publisher, London. Handbook for...Egypt. London, 1858.
- Geog 1816.15 — Murray, John, publisher, London. Handbook for...Egypt. 4. ed. London, 1873.
- Geog 1816.17 — Murray, John, publisher, London. Handbook for...Egypt. 5. ed. London, 1875.
- Geog 1816.19 — Murray, John, publisher, London. Handbook for...Egypt. 7. ed. London, 1888.
- Geog 1816.20A — Murray, John, publisher, London. Handbook for...Egypt. 8. ed. London, 1891.
- Geog 1816.20B — Murray, John, publisher, London. Handbook for...Egypt. 8. ed. London, 1891.

Classified Listing

Geog 1816 Guidebook collections - Murray - Africa - Egypt - General - cont.

Geog 1816.25 Murray, John, publisher, London. Handbook for...Egypt and the Sudan. 11. ed. London, 1907.

Geog 1820 Guidebook collections - Murray - Africa - Algeria and Tunisia

Geog 1820.2 Murray, John, publisher, London. Handbook for...Algeria and Tunis. 2. ed. London, 1878.
Geog 1820.4 Murray, John, publisher, London. Handbook for...Algeria and Tunis. 4. ed. London, 1891.
Geog 1820.5 Murray, John, publisher, London. Handbook for...Algeria and Tunis. 5. ed. London, 1895.

Geog 1828 Guidebook collections - Murray - New Zealand

Geog 1828.1 Murray, John, publisher, London. Handbook for...New Zealand. London, 1893.

Geog 3000 History of geography - Pamphlet volumes

Geog 3000.1 Pamphlet box. History of geography.
Geog 3000.1.2 Pamphlet box. History of geography.
Geog 3000.3 Pamphlet vol. Geography. Almagia. 10 pam.

Geog 3015 - 3020 History of geography - General history - General works (By date)

Geog 3017.55 Robert de Vaugondy, D. Essai sur l'histoire de géographie. Paris, 1755.
Htn Geog 3017.62* Bollan, William. Colonae anglicanae illustratae. Londini, 1762.
Geog 3017.85 Pluche, N.A. Concorde de la géographie des différents ages. Paris, 1785.
Geog 3018.03 Clarke, J.S. Progress of maritime discovery. London, 1803.
Geog 3018.28 Larenaudière. Histoire abrégée de l'origine...de géographie. Paris, 1828.
Geog 3018.30 Cooley, W.D. The history of maritime and inland discovery. London, 1830-31. 3v.
Geog 3018.30.5 Cooley, W.D. The history of maritime and inland discovery. London, 1833. 2v.
Geog 3018.36.2 Bajot, L.M. Abrégé historique...des...voyages. Paris, 1836.
Geog 3018.37 Historical account of circumnavigation. N.Y., 1837.
Geog 3018.37.2 Historical account of circumnavigation. N.Y., 1837.
Geog 3018.56 Cortambert, E. Coup d'oeil...sur...les progrès de géographie. Langny, n.d.
Geog 3018.56.5 Taylor, B. Cyclopedia of modern travel. Cincinnati, 1856.
Geog 3018.56.10 Taylor, B. Cyclopaedia of modern travel. N.Y., 1860. 2v.
Geog 3018.58 Goodrich, F.B. Man upon the sea. Philadelphia, 1858.
Geog 3018.61 Ritter, C. Geschichte der Erdkunde. Berlin, 1861.
Geog 3018.74 Tiele, Pieter Anton. De ontdekkingsreizen sedert de vijftiende eeuw. Leiden, 1874.
Geog 3018.76 Bruniatti, A. I progressi della generale e della geografia esploratrice in Europa. Vicenza, 1877.
Geog 3018.77.2 Peschel, O. Geschichte der Erdkunde. München, 1877.
Geog 3018.79 Verne, Jules. Exploration of the world. N.Y., 1879-81. 3v.
Geog 3018.79.3 Verne, Jules. Jardens op da gelseshistorie. Kristiania, 1879-83.
Geog 3018.79.5 Jurien de la Gravière, J.B.E. Les marins du XVe et du XVIe siècle. Paris, 1879. 2v.
Geog 3018.81 Low, C.R. Maritime discovery. London, 1881. 2v.
Geog 3018.82.2 Kingsley, H. Tales of old travel. London, 1882.
Geog 3018.82.5 Dussieux, L.E. Les grands faits de l'histoire de géographie. Paris, 1882-83. 5v.
Geog 3018.83 Rubiner, W. Die Entdeckungsreisen. Glogau, 1883.
Geog 3018.83.10 Crotambert, R. Nouvelle histoire des voyages. Paris, 1883-84.
Geog 3018.88 Ruge, S. Abhandlungen...zur Geschichte der Erdkunde. Dresden, 1888.
Geog 3018.93 Rainaud, A. Lę continent austral. Paris, 1893.
Geog 3018.96 Hamy, J.T.E. Études historiques et géographiques. Paris, 1896.
Geog 3018.97 Wisotzki, E. Zeitströmungen in der Geographie. Leipzig, 1897.
Geog 3018.99 Jacobs, J. Story of geographical discovery. London, 1899.
Geog 3018.99.5 Nystrom, J.F. Geografiens och de...historia. Stockholm, 1899.
Geog 3018.99.10 Partsch, J. Die geographische Arbeit des 19. Jahrhunderts. Breslau, 1899.
Geog 3018.99.15 Keane, John. Evolution of geography. London, 1899.
Geog 3019.00 Johnson, W.H. The world's discoveries. Boston, 1900.
Geog 3019.02 Günther, S. Entdeckungsgeschichte. Berlin, 1902.
Geog 3019.04 Günther, S. Geschichte der Erdkunde. Leipzig, 1904.
Geog 3019.04.3 Böhme, Max. Die grossen Reisesammlungen des 16. Jahrhunderts und ihre Bedeutung. Strassburg, 1904.
Geog 3019.05 Athlenius, K. Landkonturer och hafsvidder. Stockholm, 1905.
Geog 3019.06 Roberts, C.G.D. Discoveries and explorations in the century. Toronto, 1906.
Geog 3019.12.2 Heawood, Edward A. History of geographical discovery in the seventeenth and eighteenth centuries. N.Y., 1965.
Geog 3019.13 Keltie, J.S. History of geography. N.Y., 1913.
Geog 3019.17 Teleki, Pal. A foldrajzi gondolat torténete. Budapest, 1917.
Geog 3019.20 Synge, M.B. A book of discovery. N.Y., 1920.
Geog 3019.22 Ispizúa, Segundo de. Historia de la geografía y de la cosmografía. Madrid, 1922-26. 2v.
Geog 3019.24 Capasso, C. Le scoperte geografiche e i viaggi d'esplorazione. Messina, 1924.
Geog 3019.26 Parlss, George B. The forerunners of Hakluyt. n.p., 1926?
Geog 3019.27.2 Iorga, N. Les voyageurs français dans l'Orient européen. Paris, 1928.
Geog 3019.27.5 Iorga, N. Les voyageurs orientaux en France. Paris, 1927.
Geog 3019.27.10 Iorga, N. Une vingtaine de voyageurs dans l'Orient européen. Paris, 1928.
Geog 3019.29.5 Olsen, Ørjan. La conquête de la terre. Paris, 1933-36. 5v.
Geog 3019.31 Baker, John Norman Leonard. A history of geographical discovery and exploration. London, 1931.
Geog 3019.31.2 Baker, John Norman Leonard. A history of geographical discovery and exploration. N.Y., 1967.
Geog 3019.32F Morrison, E.R. Explorographs. Cleveland Heights, 1932.
Geog 3019.33 Dickinson, R.E. The making of geography. Oxford, 1933.
Geog 3019.33.5 Plischke, Hans. Entdeckungsgeschichte von Altertum bis zur Neuzeit. Leipzig, 1933.
Geog 3019.33.10 Banse, Ewald. Grosse Forschungsreisende. München, 1933.
Geog 3019.34 Sykes, P.M. A history of exploration from the earliest times to the present day. N.Y., 1934.

Geog 3015 - 3020 History of geography - General history - General works (By date) - cont.

Geog 3019.34.2 Sykes, P.M. A history of exploration. Westport, Conn., 1975.
Geog 3019.34.10 Mitchell, J.L. Earth conquerors. N.Y., 1934.
Geog 3019.35 Outhwaite, L. Unrolling the map. N.Y., 1935.
Geog 3019.35.5 Spilhaus, M.N. (Mrs.). The background of geography. Philadelphia, 1935.
Geog 3019.35.10 Rosh, J.H. Man and the sea. Cambridge, Eng., 1935.
Geog 3019.35.15 Kábmár, G. Régi népak, ujvilagok. Budapest, 1936.
Geog 3019.35.20 Ellsworth, L. Exploring today. N.Y., 1935.
Geog 3019.35.25 Sanchez, Pedro C. Evolución de la geografía. México, 1935.
Geog 3019.37 Olschki, L. Storia letteraria delle scoperte geografiche. Firenze, 1937.
Geog 3019.38 Key, Charles E. The story of twentieth-century exploration. N.Y., 1938.
Geog 3019.38.3 St. Croix de la Roncière, G. À la conquête des mers. Paris, 1938.
Geog 3019.38.5F La Roncière, Charles. Histoire de la découverte de la terre. Paris, 1938.
Geog 3019.38.10 Toschi, Umberto. Schemi e notizie di storia delle esplorazione geografiche. 5. ed. Firenze, 1953?
Geog 3019.40 Balen, Willem Julius van. Pioniers (De ontdekking van de wereld). Amsterdam, 1940. 2v.
Geog 3019.40.5 Dainville, François de. La géographie des humanistes. Paris, 1940.
Geog 3019.42.4 Clozier, René. Histoire de la géographie. 4. éd. Paris, 1967.
Geog 3019.43 Mexico (City). Biblioteca Benjamin Franklin. La era de las exploraciones. México, 1943.
Geog 3019.46 Darby, Henry C. The theory and practice of geography. Liverpool, 1946.
Geog 3019.47 Beckman, Leif. Vår väg genom världen. Stockholm, 1947-51. 3v.
Geog 3019.48 Schwarz, G. Die Entwicklung der geographischen Wissenschaft seit dem 18. Jahrhunderts. Berlin, 1948.
Geog 3019.49A Hanson, Earl P. New worlds emerging. 1. ed. N.Y., 1949.
Geog 3019.49B Hanson, Earl P. New worlds emerging. 1. ed. N.Y., 1949.
Geog 3019.50 Dainelli, G. La conquista della terra. Torino, 1950.
Geog 3019.51 Wooldridge, S. William. The spirit and purpose of geography. London, 1951.
Geog 3019.52 Herrmann, Paul. Sieben vorbei und acht verweht. 2. Aufl. Hamburg, 1952.
Geog 3019.52.2 Herrmann, Paul. Conquest by man. N.Y., 1954.
Geog 3019.53 Bonse, Ewald. Entwicklung und Aufgabe der Geographie. Stuttgart, 1953.
Geog 3019.53.5 Leithaeuser, J.G. Worlds beyond the horizon. 1. American ed. N.Y., 1955.
Geog 3019.53.10 Fochler-Hanke, Gustav. Introducción a la historia de la geografía. Tucumán, 1953.
Geog 3019.54.5 Le Gentil, Georges. Découverte du monde. Paris, 1954.
Geog 3019.55 Parias, L.H. Histoire universelle des explorations. Paris, 1955- 4v.
Geog 3019.55.5 Samhaber, Ernst. Knaurs Geschichte der Entdeckungsreisen. München, 1955.
Geog 3019.56 Herrmann, Paul. Zeigt mir Adams Testament. Hamburg, 1956.
Geog 3019.56.2 Herrmann, Paul. The great age of discovery. N.Y., 1958.
VGeog 3019.56.4 Herrmann, Paul. Historia de los descubrimientos geográficos. 2. ed. Barcelona, 1967.
Geog 3019.56.5 Colamonico, Carmelo. Compendio di storia della geografia e delle esplorazioni geografiche. Napoli, 1956.
Geog 3019.57.5 Magidovich, I.P. Ocherki po istorii geograficheskii otkrytii. Moskva, 1957.
Geog 3019.58 Albertini, Renzo. Storia delle esplorazioni geografiche. Venezia, 1958.
Geog 3019.59 Fradkin, N.G. Roshidenie karty. Moskva, 1959.
Geog 3019.59.5 Herrmann, Paul. Traumen, Wagen und Vollbringen. Hamburg, 1959.
Geog 3019.59.10 Codazzi, Angelo. Storia della geografia. Milano, 1959.
NEDL Geog 3019.59.15 Cabal, Juan. Grandes exploradores, en la mar. Barcelona, 1959.
Geog 3019.60.5F Bettex, A.W. Welten der Entdecker. München, 1960.
Geog 3019.61.5 Almagia, Roberto. Scritti geografici. Roma, 1961.
Geog 3019.62 Grenville, J.A.S. The coming of the Europeans. London, 1962.
Geog 3019.62.5 Freeman, Thomas. A hundred years of geograpy. Chicago, 1962.
Geog 3019.63 Parry, John Horace. The age of reconnaissance. London, 1963.
Geog 3019.63.1 Parry, John Horace. The age of reconnaissance. 1. ed. Cleveland, 1963.
Geog 3019.63.5 Baker, J.N.L. The history of geography. Oxford, 1963.
Geog 3019.63.10 Sharaf, A. Torayah. A short history of geographical discovery. Alexandria, 1963.
Geog 3019.64 Landström, B. The quest for India. London, 1964.
Geog 3019.65 Belov, Mikhail I. Puteshestviia i geograficheskia otkrytiia v XV-XIX vv. Leningrad, 1965.
Geog 3019.67 Baker, John Norman Leonard. A history of geographical discovery and exploration. N.Y., 1967.
Geog 3019.69 Chaunu, Pierre. L'expansion européene du XIIIe au XVe siècle. 1. éd. Paris, 1969.
Geog 3019.69.5 Fuson, Robert Henderson. A geography of geography; origins and development of the discipline. Dubuque, Iowa, 1969.
Geog 3019.70 Schmithuesen, Josef. Geschichte der geographischen Wissenschaft von den ersten Anfägen bis zur Ende des 18. Jahrhunderts. Mannheim, 1970.
Geog 3019.71 Isachenko, Anatolii G. Razvitie geograficheskikh idei. Moskva, 1971.
Geog 3019.72 James, Preston Everett. All possible worlds; a history of geographical ideas. Indianapolis, 1972.
Geog 3019.72.5 Fradkin, Naum G. Geografickeskie otkrytiia i nauchnoe poznanie Zemli. Moskva, 1972.
Geog 3019.72.15 Anuchin, V.A. Teoreticheskie osnovy geografii. Moskva, 1972.
Geog 3019.72.20 Langley, Michael. When the pole star shone. London, 1972.
Geog 3019.72.25 Aleksandrovskaia, Ol'ga A. Frantsvzskaia geograficheskaia shkola kontsa deviatnadtsatogo nachala dvadtsatogo veka. Moskva, 1972.
Geog 3019.72.30 Deschamps, Hubert J. Les Européens hors d'Europe de 1434 à 1815. 1. éd. Paris, 1972.
Geog 3019.72.35 Claral, Paul. La pensée géographique. Paris, 1972.
Geog 3019.73 Studia z dziejów geografii i kartografii. Wrocław, 1973.
Geog 3019.73.5 Voprosy istoricheskoi geografii i istorii geografii. Moskva, 1973.
Geog 3019.73.10 Büttner, Manfred. Die Geographie generalis vor Varenius. Wiesbaden, 1973.

Classified Listing

Geog 3015 - 3020 History of geography - General history - General works (By date) - cont.

Geog 3019.73.15 Beck, Hanno. Geographie; europäische Entwicklung in Texten und Erläuterungen. Freiburg, 1973.
Geog 3019.74 Fradkin, Naum G. Obraz Zemli. Moskva, 1974.
Geog 3019.75F Newby, Eric. The Mitchell Beazley world atlas of exploration. London, 1975.

Geog 3025 History of geography - General history - Commentary on maps

Geog 3025.5 Barbier, J.V. Rapport sur les travaux cartographiques. Nancy, 1884.
Geog 3025.7 International Geographical Congress. Über...Herstellung...Erdkarte im Mafestabe. Wien, n.d.
Geog 3025.8 Portolan charts of XVth, XVIth, XVIIth centuries collected by Dr. Theodore J.E. Hamy. N.Y., 1912.
Geog 3025.8.2F Portolan charts of XVth, XVIth, and XVIIth centuries. N.Y., 1912.

Geog 3030 History of geography - General history - Special topics

Geog 3030.3F Guénin, E. La route de l'Inde. Paris, 1903.
Geog 3030.4 Honigmann, Ernst. Die sieben Klimata. Heidelberg, 1929.
Geog 3030.7 Magidovich, I.P. Istoriia otkrytiia i issledovaniia Evropy. Moskva, 1970.
Geog 3030.7.1 Magidovich, I.P. Historiia poznania Europy. Wyd 1. Warszawa, 1974.
Geog 3030.8 Baker, Alan R.H. Progress in historical geography. N.Y., 1972.

Geog 3045 - 3050 History of geography - Mythical geography, Geographical myths - General works (By date)

Geog 3048.73 Moreau de Jonnes, A.C. L'ocean des anciens. Paris, 1873.
Geog 3048.83.1 Donnelly, Ignatius. Ragnarok: the age of fire and gravel. N.Y., 1970.
Geog 3049.52 DeCamp, L.S. Lands beyond. N.Y., 1952.
Geog 3049.59 Correa-Calderon, E. Floria de la Atlantida y otras historias fabulosas. Madrid, 1959.
Geog 3049.62 Cozzi, Piero. Los paises legendarios de la mitologia. Buenos Aires, 1962.
Geog 3049.62.5 Hutin, Serge. Les civilisations inconnues; mythes ou réalités. Paris, 1962.
Geog 3049.68 Kolosimo, Peter. Timeless earth. Secaucus, N.J., 1974.
Geog 3049.70 Kohlenberg, Karl Friedrich. Enträtselte Vorzeit. München, 1970.

Geog 3055 History of geography - Mythical geography, Geographical myths - Atlantis

Geog 3055.9 Block, R. de. Quelques mots sur l'Atlantide. n.p., n.d.
Geog 3055.11 Hoernes, M. Atlantis. Wien, 1884.
Geog 3055.13.2 Baer, F.C. Essai sur l'Atlantique des anciens. Avignon, 1835.
Geog 3055.15 Knötel, A.F.R. Atlantis und das Volk der Atlanten. Leipzig, 1893.
Geog 3055.19 Unger, F.X. Die versunkene Insel Atlantis. Wien, 1860.
Geog 3055.21 Unger, F.X. The sunken island of Atlantis. n.p., n.d.
Geog 3055.23 Norof, A.S. Die Atlantis. St. Petersburg, 1854.
Geog 3055.25 Jolibois. Dissertation sur l'Atlantide. Lyon, 1846.
Geog 3055.27 Baour-Lormian, Pierre Marie F.L. L'Atlantide. Paris, n.d.
Geog 3055.29 Nicaise, A. Les terres disparues. Chalons sur Marne, 1885.
Geog 3055.31 Roisel, G. de. Les Atlantes. Paris, 1874.
Htn Geog 3055.33* Bailly, J.S. Lettres sur l'Atlantide de Platon. Paris, 1779.
Geog 3055.33.5 Bailly, J.S. Lettres sur l'Atlantide de Platon. Paris, 1804.
Geog 3055.33.10 L'antiquité dévoilée par les principles de la magie naturelle. Paris? 1799-1800.
Htn Geog 3055.37* Tomasi, T. La spinalba antica historia. Venetia, 1647.
Geog 3055.39.1 Donnelly, Ignatius. Atlantis: the antediluvian world. Blauvelt, N.Y., 1971.
Geog 3055.39.10 Donnelly, Ignatius. Atlantis, the antediluvian world. 7th ed. N.Y., 1882.
Geog 3055.39.20 Donnelly, Ignatius. Atlantis, the antediluvian world. 1st ed. N.Y., 1949.
Geog 3055.39.25 Donnelly, Ignatius. Atlantis, the antediluvian world. London, 1970.
Geog 3055.40 Buelua, E. La Atlantida y la ultima tule. Mexico, 1895.
Geog 3055.41 Manzi, M. Le livre de l'Atlantide. Paris, 1923.
Geog 3055.42 Spence, Lewis. Atlantis in America. London, 1925.
Geog 3055.42.5 Spence, Lewis. The history of Atlantis. London, 1926.
Geog 3055.42.7 Spence, Lewis. The history of Atlantis. 4th ed. London, 1927.
Geog 3055.43 Le Cour, Paul. A la recherche d'un monde perdu. Paris, 1926.
Geog 3055.44 Bjorkman, E. The search for Atlantis. N.Y., 1927.
Geog 3055.45 Dévigné, R. Un continent disparu l'Atlantide. Paris, 1923.
Geog 3055.45.5 Dévigné, R. Un continent disparu l'Atlantide. Paris, 1924.
Geog 3055.46 Moreux, T. L'Atlantide a-t-elle existé? Paris, 1924.
Geog 3055.47 Gattefossé, R.M. La vérité sur l'Atlantide. Lyon, 1923.
Geog 3055.48 Gattefossé, Jean. Bibliographie de l'Atlantide et des questions connexes. Lyon, 1926.
Geog 3055.49 Rosmy, Léon de. L'Atlantide historique. Paris, 1902.
Geog 3055.50 Whishaw, M. Atlantis in Andalucia; a study of folk memory. London, 1929.
Geog 3055.51 Barroso, Gustavo. Aquem da Atlantida. São Paulo, 1931.
Geog 3055.52 Bessmertny, A. Das Atlantisrätsel. Leipzig, 1932.
Geog 3055.53 Karst, Josef. Atlantis und der Liby-athiopische Kulturkreis. Heidelberg, 1931.
Geog 3055.54A Bramwell, James. Lost Atlantis. N.Y., 1938.
Geog 3055.54B Bramwell, James. Lost Atlantis. N.Y., 1938.
Geog 3055.56 Bragbine, A. The shadow of Atlantis. N.Y., 1940.
Geog 3055.57 Requena, Rafael. Vestigios de la Atlántida. Caracas, 1932.
Geog 3055.58 Rodriguez Prampolini, Ida. La Atlántida de Platón en los cronistas del siglo XVI. Mexico, 1947.
Geog 3055.60 Spanuth, Jürgen. Das enträtselte Atlantis. Stuttgart, 1953.
Geog 3055.60.5 Spanuth, Jürgen. Und doch. Stuttgart, 1959.
Geog 3055.60.10 Spanuth, Jürgen. Atlantis; Heimat. Tübingen, 1965.
Geog 3055.62 Weyl, R. Atlantis enträtselt? Kiel, 1953.
Geog 3055.64 Spence, Lewis. The problem of Atlantis. 2d ed. N.Y., 1925?
Geog 3055.65 Saint-Michel, Léonard. Aux sources de l'Atlantide. Bourges, 1953.
Geog 3055.66 Saurat, Denis. L'Atlantide et la règne des géants. Paris, 1954.
Geog 3055.67 Muck, O.H. Atlantis-gefunden. Stuttgart, 1954.

Geog 3055 History of geography - Mythical geography, Geographical myths - Atlantis - cont.

Geog 3055.68 Gadow, Gerhard. Der Atlantis-Streit. Frankfurt am Main, 1973.
Geog 3055.70 Rousseau-Liessens, A. Les colonnes d'Hercule et l'Atlantide. Bruxelles, 1955.
Geog 3055.72 DeCamp, L.S. Lost continents; the Atlantis theme in history, science and literature. 1st ed. N.Y., 1954.
Geog 3055.75 Cordeau, Catherine L. Poséidones. Paris, 1957.
Geog 3055.76 Lehmann, Einar. Atlantis. København, 1934.
Geog 3055.78 Cayce, Edgar Evans. Edgar Cayce in Atlantis. N.Y., 1968.
Geog 3055.80 Galanopoulos, Angelos Georgiou. Atlantis, the truth behind the legend. Indianapolis, 1969.
Geog 3055.82 Berlitz, Charles Frambach. The mystery of Atlantis. N.Y., 1969.
Geog 3055.84 Bergquist, Nils Olof. Ymdogat-Atlantis. Solna, 1971.
Geog 3055.85 Falk, Bertil. Atlantis och svenskarna. Stockholm, 1974.
Geog 3055.86 Le Cour, Paul. L'Atlantide atlantique. Bordeaux, 1971.
Geog 3055.87 Gleich, Sigismund von. Der Mensch der Eiszeit und Atlantis. Stuttgart, 1969.

Geog 3057 History of geography - Mythical geography, Geographical myths - Ophir

Htn Geog 3057.3* Lipenius, M. Navigatio Salomonis Ophirifica. n.p., 1660.
Geog 3057.4 Quatremère. Memoire sur le pays d'Ophir. Paris, 1845.
Geog 3057.5 Peters, Karl. Ophir nach dem neuesten Forschungen. Berlin, 1908.
Geog 3057.5.5 Peters, Karl. King Solomon's golden Ophir. N.Y., 1969.
Geog 3057.6 Keane, Augustus Henry. The gold of Ophir. London, 1901.
Geog 3057.7 Der wohleingerichtete Staat des Bishero von vielen gesuchten aber nicht gefundenen Königreichs Ophir. Photoreproduction. Leipzig, 1699. 2v.

Geog 3060 History of geography - Mythical geography, Geographical myths - Other special

Geog 3060.5 Gaffarel, P. Les isles fantastiques de l'Atlantique. n.p., 1883.
Geog 3060.6A Babcock, William H. Legendary islands of the Atlantic. N.Y., 1922.
Geog 3060.6B Babcock, William H. Legendary islands of the Atlantic. N.Y., 1922.
Geog 3060.6C Babcock, William H. Legendary islands of the Atlantic. N.Y., 1922.
Geog 3060.7 Hofmann, C. Setzungsberichte. n.p., 1865.
Geog 3060.8 Ford, Worthington Chauncey. The isle of pines, 1668. Boston, 1920.
Htn Geog 3060.8* Ford, Worthington Chauncey. The isle of pines, 1668. Boston, 1920.
Htn Geog 3060.9* Das verdächtiger Pineser-Eyland. Hamburg, 1668.
Geog 3060.10 Firestone, C.B. The coasts of illusion. N.Y., 1924.
Geog 3060.11 Nunn, G.E. Origin of Strait of Asian concept. Philadelphia, 1929.
Geog 3060.12.25 Churchward, J. The lost continent of Mu. N.Y., 1934.
Geog 3060.12.45A Churchward, J. The sacred symbols of Mu. N.Y., 1934.
Geog 3060.12.45B Churchward, J. The sacred symbols of Mu. N.Y., 1934.
Geog 3060.12.50 Churchward, J. The children of Mu. N.Y., 1937.
Geog 3060.12.55 Churchward, J. The sacred symbols of Mu. London, 1960.
Geog 3060.12.70 Vincent, Louis Claude. Le paradis perdu de Mu. Paris? 1969. 2v.
Geog 3060.13.5 Cervé, W.S. Lemuria. 6th ed. San Jose, Calif., 1954.
Geog 3060.13.10 Spence, Lewis. The problem of Lemuria. Philadelphia, 1933.
Geog 3060.14 Carta em resposta a hum amigo. Lisboa, 1815.
Geog 3060.15 Relação que trata de como em cincoenta e oito gráos do sul fay descuberta huma ilha. pt.2. Lisboa, 17- ?
Htn Geog 3060.16* Notica certa do descobrimento de huma nova terra. Lisboa, 1757.
Geog 3060.18 Barreto, Costa. A lenda das Sete Cidades. Porto, 1949.
Geog 3060.19 Adams, Percy G. Travellers and travel liars, 1660-1800. Berkeley, 1962.

Geog 3070 History of geography - History of cartography - General works

Geog 3070.01 Pamphlet vol. Cartography.
Geog 3070.02 Pamphlet vol. Erwin Raisz.
Geog 3070.2 Desimoni, C. Atlante idrografico. Genova, 1867.
Geog 3070.4 Breusing, A. Leitfaden durch das Wiegenalter. Frankfurt, 1883.
Geog 3070.6 Fischer, T. Sammlung mittelalterlicher Welt. Venedig, 1886.
Geog 3070.8 D'Avezac, M.A.P. Coup d'oeil historique sur projection. Paris, 1863.
Geog 3070.12 Wauwermans, H. Histoire de l'ecole cartographique. Bruxelles, 1895. 2v.
Geog 3070.14 Mayer, E. Die Entwicklung der Seekarten. Wien, 1877.
Geog 3070.16 Dröber, W. Kartographie bei den Naturvölkern. Erlangen, 1903.
Geog 3070.22 Matkovic, P. Alte handschriftliche Schiffer-Karten. Wien, 1863.
Geog 3070.24 Desimoni, C. Elenco di carte ed Atlanti nautici. Genova, n.d.
Geog 3070.25 Stevenson, E.L. Maps reproduced as glass transparencies. N.Y., 1913.
Geog 3070.26 Thomas, G.M. Der Periplus des Pontus Euxinus nach Münchener Handschriften. München, 1864.
Geog 3070.27 Denucé, J. Oud-Nederlandsches kaartmakers in betrekking met Plantijn. Antwerpe, 1912. 2v.
Geog 3070.29 Mager, Henri. De la lecture des cartes étrangères. Paris, 1883.
Geog 3070.30 Anthiaume, A. Cartes marines, constructions navales, 1500-1650. Paris, 1916. 2v.
Geog 3070.31 Fordham, Herbert G. Maps. Cambridge, Eng., 1921.
Geog 3070.33.2A Holman, Louis A. Old maps and their makers. 2d ed. Boston, 1926.
Geog 3070.33.2B Holman, Louis A. Old maps and their makers. 2d ed. Boston, 1926.
Geog 3070.34 Baulig, Henri. Exercices cartographiques. Paris, 1927.
Geog 3070.35 Hayes, Gerald R. The production of an admiralty chart. London, 1929.
Geog 3070.36 Barker, W.H. The history of cartography. Manchester, 1926.
Geog 3070.39 Jervis, W.W. The world in maps. N.Y., 1937.
Geog 3070.39.5 Jervis, W.W. The world in maps. 2d ed. N.Y., 1938.
Geog 3070.40 Bagrow, Leo. A. Ortelii Catalogus cartographorum bearbeitet. v.1-2. Gotha, 1928-30.
Geog 3070.40.5 Bagrow, Leo. Die Geschichte der Kartographie. Berlin, 1951.
Geog 3070.42 Wroth, Lawrence C. The early cartography of the Pacific. N.Y., 1944.
Geog 3070.45 Salishahev, K.A. Osnovy kartovesheniia. Izd. 2. Moskva, 1943.

Classified Listing

Geog 3070   History of geography - History of cartography - General works - cont.

Geog 3070.47.5	Durand, D.B. The Vienna-Klosterneuberg map corpus of the 15th century. Leiden, 1952.
Geog 3070.48	Brown, Lloyd A. The story of maps. 1st ed. Boston, 1949.
Geog 3070.48.5	Brown, Lloyd A. The story of maps. Boston, 1950.
Geog 3070.50A	Lynam, E. The mapmaker's art. London, 1953.
Geog 3070.50B	Lynam, E. The mapmaker's art. London, 1953.
Geog 3070.56	Codazzi, Angela. Storia delle carte geografiche. Milano, 1952-
Geog 3070.56.5	Codazzi, Angela. Storia delle carte geografiche. Milano, 1958.
Geog 3070.80	Conseil Scientifique pour l'Afrique au Sud du Sahara. Mapping and surveying of Africa south of the Sahara. London, 1954.
Geog 3070.85	Raisz, E.J. Mapping the world. N.Y., 1956.
Geog 3070.90	Lauf, G.B. The origin and development of cartography. Johannesburg, 1955.
Geog 3070.95	Guarnieri, Gino. Geografia e cartografia nautica nella loro evoluzione storica e scientifica. Genova, 1956.
Geog 3070.105	Leithäuser, J.G. Mappae Mundi. Berlin, 1958.
Geog 3070.115	Internationaler Kurz für Kartendruck. Fortschrittsberichte auf dem Gebiet des Kartendrucks. Hamburg, 1958.
Geog 3070.135	Brown, Lloyd. Map making; the art that became a science. 1st ed. Boston, 1960.
Geog 3070.155	Salinari Emiliani, M. Nozioni di cartografia. Roma, 1959.
Geog 3070.157	Bagrow, Leo. History of cartography. Cambridge, 1964.
Geog 3070.165	Kremling, Helmut. Die Beziehungsgrundlage in thematischen Karten in ihrem Verhältnis zum Kartengegenstand. München, 1970.
Geog 3070.166	Ditmar, Andrei B. Rubezh oikumeny. Moskva, 1973.

Geog 3105 - 3110   History of geography - Ancient geography, to ca.300 A.D. - Description of the ancient world (By date)

Htn	Geog 3106.48*	Brietio, O. Parallela geographiae. Paris, 1648-49. 3v.
	Geog 3107.26	Wells, Edward. Treatise of antient and present geography. London, 1726.
	Geog 3107.31	Cellarius, C. Notitia orbis antiqui. Lipsiae, 1731-32. 2v.
	Geog 3107.94	Adam, A. Summary of geography and history. Edinburgh, 1794.
	Geog 3107.94.5	Adam, A. Geographical index. Edinburgh, 1795.
	Geog 3108.25	Butler, S. Sketch of modern and ancient geography. London, 1825.
	Geog 3108.37.2	Schirlitz, S.C. Handbuch der alten Geographie. Halle, 1837.
	Geog 3108.39	Arrowsmith, A. Grammar of ancient geography. London, 1839.
	Geog 3108.39.3	Arrowsmith, A. Compendium of ancient and modern geography. London, 1839.
	Geog 3108.42	Forbiger, A. Handbuch der alten Geographie. Leipzig, 1842-48. 3v.
	Geog 3108.42.5	Forbiger, A. Handbuch der alten Geographie. Hamburg, 1877. 3v.
	Geog 3108.50	Anthon, C. System of ancient and mediaeval geography. N.Y., 1850.
	Geog 3108.71	Anthon, C. System of ancient and mediaeval geography. N.Y., 1871.
	Geog 3108.74	Moreau de Jonnés, A. Estudios prehistóricos. Sevilla, 1874.
	Geog 3108.78	Kiepert, J.S.H. Lehrbuch der alten Geographie. Berlin, 1878.
	Geog 3108.82	Hahn, H. Leitfaden der alten Geographie. Leipzig, 1882.

Geog 3125 - 3130   History of geography - Ancient geography, to ca.300 A.D. - History of ancient geography - General works (By date)

Geog 3127.98	Gosselin, P.F.J. Recherches sur la géographie...des anciens. Paris, 1798. 2v.
Geog 3128.07	Dureau de la Malle, Adolphe. Géographie physique de la Mer Noire, de l'intérieur de l'Afrique et de la Méditerranée. Paris, 1807.
Geog 3129.32A	Burton, H.E. The discovery of the ancient world. Cambridge, 1932.
Geog 3129.32B	Burton, H.E. The discovery of the ancient world. Cambridge, 1932.
Geog 3129.34	Kahlo, G. Die Keuntnis der Erde im Altertum. München, 1934.
Geog 3129.36	Saa, Mario. Evudania. Lisboa, 1936.
Geog 3129.48A	Thomson, J.O. History of ancient geography. Cambridge, 1948.
Geog 3129.48B	Thomson, J.O. History of ancient geography. Cambridge, 1948.
Geog 3129.49	Almagià, Roberto. Storia dell'esplorazione e della scienza geografica: l'eta greca. Roma, 1949.
Geog 3129.55	Codazzi, Angela. La geografia dei Greci e dei Romani. 2. ed. Milano, 1955.
Geog 3129.57	Paassen, Christian van. The classical tradition of geography. Groningen, 1957.
Geog 3129.58	Paassen, Christian van. The classical tradition of geography. Groningen, 1957.
Geog 3129.68	Guarnieri, Giuseppe Gino. Le correnti del pensiero geografico nell'antichità classica e il lorocontributo alla cartografia nautica medioevale. Pisa, 1968-69. 2v.

Geog 3135   History of geography - Ancient geography, to ca.300 A.D. - History of ancient geography - Commentary on maps

Geog 3135.5A	Heidel, William A. The frame of the ancient Greek maps. N.Y., 1937.
Geog 3135.5B	Heidel, William A. The frame of the ancient Greek maps. N.Y., 1937.

Geog 3140   History of geography - Ancient geography, to ca.300 A.D. - History of ancient geography - Special topics

Geog 3140.5	Ruge, S. Der Chaldäer Seleukos. Dresden, 1865.
Geog 3140.7	Warren, W.F. True key to ancient cosmology. Boston, 1882.
Geog 3140.9	Blau, J. Memoirs sur deux monuments geographiques. Nancy, 1836.
Geog 3140.11F	Rylands, T.G. Geography of Ptolemy elucidated. Dublin, 1893.
Geog 3140.13	Mer, A. Memoire sur le Periple d'Hannon. Paris, 1885.
Geog 3140.15	Gosselin, P.Z.J. De l'évaluation et de l'emploi des mesures itinéraires grecques et romaines. Paris, 1813.
Geog 3140.16	Antichan, P.N. Grands voyages de découvertes des anciens. Paris, 1888.
Geog 3140.17	Ninck, M. Die Entdeckung von Europa durch die Griechen. Basel, 1945.

Geog 3199   History of geography - Medieval and Renaissance geography, 300-1600 - Description of the medieval world - Before 1399

Geog 3199.5	Manitius, M. Anonymi de situ orbis. Stuttgardiae, 1884.
Geog 3199.10	al-Idrisi, Muhammad ibn Muhammad. Deutschland und seine Nachbarländer nach der grossen Geographie des Idrisi. Stuttgart, 1937.

Geog 3204 - 3210   History of geography - Medieval and Renaissance geography, 300-1600 - Description of the medieval world - Later works (By date)

Htn	Geog 3205.34F*	Franck, S. Weltbuch: Spiegel...in Asiam, Aphrica, Europam und America. n.p., 1534.
Htn	Geog 3205.40.5*	Boehme, J. Omnium gentium, mores, leges. Friburgi, 1540.
Htn	Geog 3205.40.7*	Boehme, J. Omnium gentium, mores, leges. Antverpiae, 1542.
Htn	Geog 3205.40.11*	Boehme, J. Mores, leges et ritus omnium gentium. Lugduni, 1561.
Htn	Geog 3205.40.15*	Boehme, J. Mores, leges et ritus omnium gentium. Lugduni, 1576.
Htn	Geog 3205.40.17*	Boehme, J. Mores, leges et ritus omnium gentium. Lugduni, 1582.
Htn	Geog 3205.40.21*	Boehme, J. Gli costumi, le leggi et lusanze. Venetia, 1560.
Htn	Geog 3205.40.25*	Boehme, J. Fardle of facious...ancient manners. London, 1555. 3v.
	Geog 3205.50F	Pacheco Pereira. Esmeraldo de situ orbis. Lisboa, 1892.
	Geog 3205.50.5	Pacheco Pereira. Esmeraldo de situ orbis. Lisboa, 1905.
Htn	Geog 3205.95*	Romanus, A. Parvum theatrum urbium. Frankoforti, 1595.
	Geog 3209.13	En Islandsk Vejviser for pilgrimme fra 12 årh. København, 1913.

Geog 3225 - 3230   History of geography - Medieval and Renaissance geography, 300-1600 - History of medieval geography - General works (By date)

Geog 3228.76	Beltrán y Rózpide, Ricardo. Viajes y descubrimientos, efectuados en la edad media. Madrid, 1876.
Geog 3228.80	Gravier, S. La cosmographie avant la decouverte de l'Amerique. Paris, 1880.
Geog 3228.80.5	Bullo, C. La vera patria di Nicolò de Conti e di Giovanni Caboto. Chioggia, 1880.
Geog 3228.82	Marinelli, G. La geografia. Roma, 1882.
Geog 3228.82.3	Marinelli, G. Die Erdkunde bei den Kirchenvätern. Leipzig, 1884.
Geog 3228.97A	Beazley, C.R. Dawn of modern geography. London, 1897. 3v.
Geog 3228.97B	Beazley, C.R. Dawn of modern geography. London, 1897. 2v.
Geog 3228.97.2	Beazley, C.R. Dawn of modern geography. London, 1897-1906. 3v.
Geog 3229.06	Sensburg, W. Poggio Bracciolini und Nicolò de Conti. Wien, 1906.
Geog 3229.10	Errera, C. L'epoca delle grandi scoperte geografiche. Milano, 1910.
Geog 3229.20	Dark, Richard. The quest of the Indies. N.Y., 1920.
Geog 3229.25A	Wright, John K. The geographical lore of the time of the Crusades. N.Y., 1925.
Geog 3229.25B	Wright, John K. The geographical lore of the time of the Crusades. N.Y., 1925.
Geog 3229.29	Mžik, Hans von. Beiträge zur historischen Geographie, Kulturgeographie, Ethnographie und Kartographie. Leipzig, 1929.
Geog 3229.38	Kimble, George H.T. Geography in the Middle Ages. London, 1938.
Geog 3229.39	Rohr, Heinz. Die Entwicklung des Kartenbildes. Inaug. Diss. Borna, 1939.
Geog 3229.61	Roux, Jean Paul. Les explorateurs au Moyen Âge. Paris, 1961.
Geog 3229.66	Alavi, S.M. Ziauddin. Geography in the Middle Ages. 1st ed. Delhi, 1966.

Geog 3235   History of geography - Medieval and Renaissance geography, 300-1600 - History of medieval geography - Commentary on maps

Geog 3235.5	Bevan, W.L. Mediaeval geography. London, 1873.
Geog 3235.7F	De Luca, G. Carte nautiche del medio evo. Napoli, 1866.
Geog 3235.9	Wuttke, J.K.H. Über Erdkunde und Karten. Leipzig, 1853.
Geog 3235.11	Cortambert, P.F.E. Tros...monuments geographiques. Paris, 1877.
Geog 3235.13	Longhena, M. Atlanti e carte nautiche. Parma, 1907.
Geog 3235.17	Pezzana, A. De l'anciennete de la mappemonde. Genes, 1808.
Geog 3235.19	Miller, K. Die Ebstorfkarte. Stuttgart, 1900.
Geog 3235.21	Crivellari, G. Alcuni cimeli della cartografia. Firenze, 1903.
Geog 3235.27F	Berchet, G. Il planisfero di Giovanni Leardo. Venezia, 1880.
Geog 3235.29	Buchon, J.A.C. Notice sur un atlas en langue catalane. Paris, 1838.
Geog 3235.29.3	Buchon, J.A.C. Notices et extraits des manuscrits. Paris, n.d.
Geog 3235.31	Amat, P. Del planisferio. Roma, 1878.
Geog 3235.33	Hany, E.T. La mappemonde. Paris, 1887.
Geog 3235.35	Matkovic, P.P. Alte handschriftliche Schifferkarten. Agram, 1860.
Geog 3235.37	Pezzana, A. Estratto di una nota pesta a.f. 365-66...la carta nautica. Berlin, 1881.
Geog 3235.39	Berchet, G. Portolani, esistenti nelle...bibliotheque. Venezia, 1866.
Geog 3235.41	Odorici, F. Carte geografiche. Milano, 1877.
Geog 3235.43	Schweder, E. Uber die Weltkarte des Kosmographen. Kiel, 1886.
Geog 3235.45	Avezac, M.A.P. Un digression géographique...la mappemonde. Paris, 1870.
Geog 3235.47	Jomard, E.F. Introduction à l'atlas des monuments de la géographie. Paris, 1879.
Geog 3235.47.9	Santarem. Examen des assertions contenues...des monuments de la géographie. n.p., n.d.
Geog 3235.49	Avezac, M.A.P. Note sur un atlas hydrographique. v.1-2. Paris, 1850.
Geog 3235.50	Kretschmer, K. Die italienischen Portolane des Mittelalters. Berlin, 1909.
Geog 3235.50.9	Errera, Carlo. I portolani italiani del medioevo secondo l'opera di K. Kretschmer. Firenze, 1911.
Geog 3235.51	Avezac, M.A.P. Note sur la mappemonde historiée de la cathedral de Hereford. Paris, 1862.
Geog 3235.52	Avezac, M.A.P. Deux notes sur d'anciennes cartes historiées. Paris, 1844.
Geog 3235.53F	Gaffarel, Paul. Etude sur un portolan inédit. Dijon, 1876.
Geog 3235.55	Desimoni, C. Nuovi studi sull'atlante luxoro. Geneva, 1869.

35

36

Classified Listing

Geog 3235 History of geography - Medieval and Renaissance geography, 300-1600 - History of medieval geography - Commentary on maps - cont.

Geog 3235.56A Stevenson, E.L. Portolan charts. N.Y., 1911.
Geog 3235.56B Stevenson, E.L. Portolan charts. N.Y., 1911.
Htn Geog 3235.57* Marcel, G. Note sur une carte catalane de Dulceri. Paris, 1887.
Geog 3235.58 Beans, G.H. A collection of maps. Jenkintown, Pa., 1943.
Geog 3235.61 Koeman, Cornelis. The history of Lucas Janszoon Wazehnaer and his Spieghel der Zeevaerdt. Lausanne, 1964.
Geog 3235.65 Gross, Hans. Zur Entstehungs-Geschichte der Tabula Purtingeriana. Diss. Bonn, 1913.
Geog 3235.66 Vinland Map Conference, Smithsonian Institution. Proceedings. Chicago, 1971.

Geog 3240 History of geography - Medieval and Renaissance geography, 300-1600 - History of medieval geography - Special topics

Geog 3240.5 Bernard, A. De Adamo Bremensi geographo. Parisiis, 1895.
Geog 3240.6 Lönborg, S. Adam af Bremen...skildring af Nordeuropas länder. Uppsala, 1897.
Geog 3240.7 Thomassy, M.J.R. Les papes géographes. Paris, 1852.
Geog 3240.9 Boek, C.P. Lettres...intitulé: liber guldonis. n.p., n.d.
Geog 3240.17 Devic, L. Marcel. Coup d'oeil sur la litterature géographique arabe au moyen âge. Paris, 1882.
Geog 3240.18 Ortroy, F.G. van. L'oeuvre cartographique de Gérard et de Corneille de Jode. Gand, 1914.
Htn Geog 3240.18.5* Orion, booksellers, ltd., London. Description of a rare and precious atlas. London, 1935?
Geog 3240.19A Malone, Kemp. King Alfred's North. Cambridge, 1930.
Geog 3240.19B Malone, Kemp. King Alfred's North. Cambridge, 1930.
Geog 3240.20 Andriani, G. Giacomo Bracelli nella storia della geografia. Pontremoli, n.d.
Geog 3240.22 Ahmad, Nafia. Muslim contribution to geography. Lahore, 1947.
Geog 3240.25 Garcia Franco. Le legua nautica en la edad media. Madrid, 1957.
Geog 3240.26 Jacob, Georg. Studien in arabischen Geographen. Berlin, 1891-92.
Geog 3240.26.5 Miquel, André. La géographie humaine du monde musalman jusqu'au milieu du 11e siècle. Paris, 1967.
Geog 3240.28 Strzelczyk, Jerzy. Gerwazy z Tilbury. Wrocław, 1970.

Geog 3245 - 3250 History of geography - Medieval and Renaissance geography, 300-1600 - History of renaissance geography - General works (By date)

Geog 3248.58 Peschel, O.F. Geschichte des Zeitalters. Stuttgart, 1858.
Geog 3248.58.5 Peschel, O.F. Geschichte des Zeitalters. Stuttgart, 1877.
Geog 3248.81 Ruge, Sophus. Geschichte des Zeitalters der Entdeckungen. Berlin, 1881.
Geog 3248.82 Cat, E. Les grands decouvertes maritimes. Paris, 1882.
Geog 3249.03 Hugues, Luigi. Cronologia delle scoperte e delle esplorazioni geografiche dall'anno 1492 a tutto il secolo XIX. Milano, 1903.
Geog 3249.25 Pereyra, Carlos. La conquête des routes océaniques d'Henri le navigateur à Magellan. Paris, 1925.
Geog 3249.28.3 Bullon y Fernández, Eloy. Miguel Servet y la geographia del renacimiento. 3. ed. Madrid, 1945.
Geog 3249.30 Jacome Correa, Ayres. Discussão historica das medidas geographicas no seculo XVI. Lisboa, 1930.
Geog 3249.52A Penrose, B. Travel and discovery in the Renaissance. Cambridge, 1952.
Geog 3249.52B Penrose, B. Travel and discovery in the Renaissance. Cambridge, 1952.
Geog 3249.54 Nowell, C.E. The great discoveries and the first colonial empires. Ithaca, 1954.
Geog 3249.64 Goldstein, Thomas. Fifteenth century geography against the background of medieval science. Salem, Mass., 1964.
Geog 3249.68 Hale, John Higby. Renaissance exploration. London, 1968.
Geog 3249.70 Wright, Louis Booker. Gold, glory, and the gospel. 1. ed. N.Y., 1970.
Geog 3249.72 Krämer, Walter. Neue Horizonte. 1. Aufl. Leipzig, 1972.

Geog 3251 History of geography - Medieval and Renaissance geography, 300-1600 - History of renaissance geography - Commentary on maps

Geog 3251.7 Nordenskiold, A.E. Om en märklig globkarta. Stockholm, 1884.
Geog 3251.9 Brenner, O.K. Die ächte Karte des Olaus Magnus vom Jahre 1539. Christiania, 1886.
Geog 3251.13 Ceradini, G. A proposito dei due globi mercatoriani. Milano, 1894.
Geog 3251.19 Wieser, F. Der Portulan des Infanten. n.p., n.d.
Geog 3251.21 Behrmann, W. Über die niederdeutschen Seebücher. Hamburg, 1906.
Geog 3251.23 Stevenson, E.L. Map of the world. N.Y., 1907.
Geog 3251.23.3 Stevenson, E.L. Genovese world map, 1457. Facsimile. N.Y., 1912.
Geog 3251.23.5 Stevenson, E.L. Marine world chart, 1502. N.Y., 1908.
Geog 3251.27F Nunn, George E. World map of Francesco Roselli. Philadelphia, 1928.
Geog 3251.31 Magnagni, A. D'Anamia e Botero. Ciriè, 1914.
Geog 3251.32 Lehmann, Edgar. Alte deutsche Landkarten. Leipzig, 1935.

Geog 3400 History of geography - Local history - North America - General works

Geog 3400.5A James, Preston E. American geography; inventory and prospect. Syracuse, 1954.
Geog 3400.5B James, Preston E. American geography; inventory and prospect. Syracuse, 1954.

Geog 3410 History of geography - Local history - North America - United States

Geog 3410.5 Pfeifer, Gottfried. Regional geography in the United States since the war. N.Y., 1938.
Geog 3410.7 Problems and trends in American geography. N.Y., 1967.

Geog 3506 History of geography - Local history - Europe - Low countries - Belgium

Geog 3506.5 Saint-Genois, Jules de. Les voyageurs belges. v.1-2. Bruxelles, 1846-47?
Geog 3506.7 Hennequin, Emile. Etude historique sur l'exécution de la carte de Ferraris et l'évolution de la cartographie topographique en Belgique. Bruxelles, 1891.
Geog 3506.8 Smet, Antoine de. Album Antoine de Smet. Bruxelles, 1974.

Geog 3510 History of geography - Local history - Europe - Low countries - Holland

Geog 3510.7 Van Loon, H.W. The golden book of the Dutch navigators. N.Y., 1916.
Geog 3510.11 Belen, Willem Julius. Nederlands voorhoede. 2. druk. Amsterdam, 1946.
Geog 3510.13 Smet, Antoine de. La cartographie hollandaise. Bruxelles, 1971.

Geog 3520 History of geography - Local history - Europe - Great Britain - General works, England

Geog 3520.5 Jurien de la Gravière, J.B.E. Les Anglais et les Hollandais. Paris, 1890. 2v.
Geog 3520.7F Aa, P. van der. De wijd beroemde voyagien...der Engelsen. Leyden, 17- . 2v.
Geog 3520.8 Warmer, Oliver. English maritime writing. London, 1958.
Htn Geog 3520.10F* Hakluyt, Richard. Principall navigations, voiages. London, 1589.
Htn Geog 3520.10.5F* Hakluyt, Richard. Principal navigations, voyages. v.1-3. London, 1599. 2v.
Geog 3520.10.10F Hakluyt, Richard. Collection of early voyages, travels. London, 1809. 5v.
Geog 3520.10.15 Hakluyt, Richard. A selection of the principal voyages. N.Y., 1926.
Geog 3520.10.20 Hakluyt, Richard. Principal navigations, voyages. v.1-12,14-16. Edinburgh, 1885-90. 15v.
NEDL Geog 3520.10.20 Hakluyt, Richard. Principal navigations, voyages. v.13. Edinburgh, 1885-90.
Geog 3520.10.25A Hakluyt, Richard. Voyages of Drake and Gilbert. Oxford, 1909.
Geog 3520.10.25B Hakluyt, Richard. Voyages of Drake and Gilbert. Oxford, 1909.
Geog 3520.10.35 Hakluyt, Richard. Principal navigations, voyages. London, 1907-13. 8v.
Geog 3520.10.45 Hakluyt, Richard. Principal navigations, voyages...of the English nation. London, 1927. 8v.
Geog 3520.10.45.5 Hakluyt, Richard. The principal navigations. Cambridge, 1965. 2v.
Geog 3520.10.50.2 Hakluyt, Richard. Voyages and documents. London, 1963.
Geog 3520.10.55 Hakluyt, Richard. They told Mr. Hakluyt. London, 1964.
Geog 3520.10.75A Hakluyt, Richard. Fighting merchant men. Boston, 1927.
Geog 3520.10.75B Hakluyt, Richard. Fighting merchant men. Boston, 1927.
Geog 3520.11A Parks, George Bruner. Richard Hakluyt and the English voyages. N.Y., 1928.
Geog 3520.11B Parks, George Bruner. Richard Hakluyt and the English voyages. N.Y., 1928.
Geog 3520.11C Parks, George Bruner. Richard Hakluyt and the English voyages. N.Y., 1928.
Geog 3520.11.2 Parks, George Bruner. Richard Hakluyt and the English voyages. 2. ed. N.Y., 1961.
Geog 3520.11.10 Quinn, David Beers. A study of the facsimile edition of Richard Hakluyt's Divers voyages. Amsterdam, 1968. 2v.
Geog 3520.12 Penrose, Boies. Tudor and early Stuart voyaging. Washington, 1962.
Geog 3520.13 Moorehead, Alan. The fatal impact. London, 1966.
Geog 3520.14 Pennington, Loren E. Hakluytus posthumus; Samuel Purches and the promotion of English overseas expansion. Emporia, 1966.
Geog 3520.15 Lynam, Edward. British maps and map-makers. London, 1944.
Geog 3520.16 Robinson, Adrian. Marine cartography in Britain. Leicester, 1962.
Geog 3520.17 Taylor, Eva G. Tudor geography, 1485-1583. London, 1930.
Geog 3520.17.6 Taylor, Eva G. Late Tudor and early Stuart geography, 1583-1650. N.Y., 1968.
Geog 3520.18 Haender, Wilhelmina. De Engelsche geographie in de 20ste eeuw. Utrecht, 1934.

Geog 3530 History of geography - Local history - Europe - France

Geog 3530.5 Dahlgren, E.W. De franska sjöfärderna. Stockholm, 1900.
Geog 3530.7 Margery, P. Les navigations françaises. Paris, 1867.
Geog 3530.7.2 Margery, P. Les navigations françaises. Paris, 1867.
Geog 3530.9 Barré, H. Voyageurs et explorateurs provençaux. Marseille, 1905.
Geog 3530.11 Estancelin, L. Recherches sur les voyages et découvertes des...Normands. Paris, 1832.
Geog 3530.13 Jolly, Raoul. Les missions françaises; causeries géographiques. Paris, 1894-96. 2v.
Geog 3530.14 Deschamps, Léon. De Rasilliis Gabriel, Isaac et Claudio proenominatis Richelii adjutoribus. Paris, 1898.
Geog 3530.15 Dupic, Jeanne. La Normandie exploratrice et colonisatrice du XVe au XVIIIe siècle. Rouen, 1932.
Geog 3530.16 Martonne, E. de. La science géographique. Paris, 1915.
Geog 3530.17 Julien, C.A. Les voyages de découverte et les premiers établissements. 1. éd. Paris, 1948.
Geog 3530.18 Lauga, Henri. De la banquise à la jungle. Paris, 1952.
Geog 3530.20 Meynier, André. Histoire de la pensée géographique en France, 1872-1969. Paris, 1969.

Geog 3540 History of geography - Local history - Europe - Germany

Geog 3540.7 Gallois, Lucien. Les géographes allemands de la renaissance. Paris, 1890.
Geog 3540.10 Pannwitz, Max. Deutsche Pfadfinder des 16. Jahrhunderts in Afrika, Asien und Südamerika. Stuttgart, 1911-12.
Geog 3540.11 Beck, Carl. Deutsches Reisen im Wandel. Berlin, 1936.
Geog 3540.12 Wotte, Herbert. In blaver Ferne lag Amerika. 3. Aufl. Leipzig, 1974.
Geog 3540.13 Timpte, Helmut. Typologische Studien zur historischen Kartographie in Westfalen. Düsseldorf, 1961.
Geog 3540.15 Schulte-Althoff, Franz Josef. Studien zur politischen Wissenschaftsgeschichte der deutschen Geographie im Zeitalter des Imperialismus. Paderborn, 1971.

Geog 3542 History of geography - Local history - Europe - Austria

Geog 3542.5 Hassinger, H. Österreichs Anteil an der Erforschung der Erde. Wien, 1949?

Geog 3550 History of geography - Local history - Europe - Italy

Geog 3550.2 Venice. Biblioteca Nazionale Marciana. Mostra dei navigatori veneti del quattrocento e del cinquecento. Venezia, 1957.
Geog 3550.3 Studi bibliografici e biografici sulla storia della geografia in Italia. Roma, 1875.
Geog 3550.4 Gubernatis, A. de. Memoria ai viaggiatori italiani. Firenze, 1867.
Geog 3550.5 Gubernatis, A. de. Storia dei viaggiatori italiani. Livorno, 1875.
Geog 3550.6 Fischer, Theobald. Über italienischen Seekarten und Kartographen des Mittelalters. Berlin, 1882.
Geog 3550.7 Branca, G. Storia dei viaggiatori italiani. Roma, 1873.
Geog 3550.9 Canale, M.G. Indicazioni di opere...sopra i viaggi, le navigazioni, le scoperte...degl'Italiani nel medio evo. Lucca, 1861.
Htn Geog 3550.10F* Morelli, Iacopo. Dissertazione intorno ad alcuni viaggiatori eruditi veneziani. Venezia, 1803.
Geog 3550.11 Bertacchi, C. Geografi ed esploratori italiani contemporanei. Milano, 1929.
Htn Geog 3550.12* Franciulli, G. I grandi navigatori italiani. Roma, 1930.

Classified Listing

Geog 3550 History of geography - Local history - Europe - Italy - cont.

Geog 3550.13 Ricchieri, G. Il contributo degli Italiani alla conoscenza della terra ed agli studi cinquatennio. Roma, 1912.
Geog 3550.14 Revelli, P. La casa di Savoia e gli studii geografici. Milano, 1903.
Geog 3550.16 Scarin, Maria Luisa. Viaggi ed esplorazioni di capitani marittimi della Riviera di Levante nella prima metà del secolo XIV. Genova, 1968.
Geog 3550.17 Miscellanea di storia delle esplorazioni. Genova, 1975.

Geog 3560 History of geography - Local history - Europe - Spain and Portugal - General works

Geog 3560.5 Great Britain. India Office. Map of the world, commonly known as the second Borgian map. n.p., 1889.
Geog 3560.10 Pereyra, C. La conquista de las rutas oceánicas. Madrid, 1940.
Geog 3560.15 Farinelli, Arturo. Viajes por España y Portugal. Roma, 1942-44. 3v.
Geog 3560.20 Perez Emlied, F. Los descubrimientos en el Atlántico y la rivalidad castellano-portuguesa hosta el tratado de Tordesillas. 1. ed. Sevilla, 1948.
Geog 3560.25 Dória, A.A. Los descubrimientos en el Atlántico y la rivalidad castellano-portuguesa. Braga, 1951.
Geog 3560.26 Sanz, Carlos. La huella de España en el mundo. Madrid, 1971-73. 3v.

Geog 3570 History of geography - Local history - Europe - Spain and Portugal - Portugal

Geog 3570.2 Consiglieri-Pedroso, L. Catalogo bibliographico das publicacões. Lisboa, 1912.
Htn Geog 3570.5* Sonsa Viterbo, F.M. Trabalhos nauticos dos Portuguezes nos seculos XVI e XVII. Lisboa, 1890.
Geog 3570.5.3F Sonsa Viterbo, F.M. Trabalhos nauticos dos Portuguezes. Lisboa, 1884-1900. 2v.
Geog 3570.7 Oliveira Marlins, J.P. de. Les explorations des Portugais. Paris, 1893.
Geog 3570.8 Marcondes de Souza, T.O. Novas achegos à historia dos descobrimentos maritimos. São Paulo, 1963.
Geog 3570.9 Bersaude, J. Les légendes allemandes. Genève, 1917-20. 2v.
Geog 3570.9.5 Bersaude, J. The attacks against Portuguese history. Lisbon, 1950.
Geog 3570.10 Maix de Sori, A.F. Descobrimentos dos Portuguezes nos seculos XV e XVI. Lisboa, 1867.
Htn Geog 3570.11* Barrós, J. de. De alder erste scheepo-tog ten der Portugesen. Leyden, 1700?
Geog 3570.15 Colleccão de opusculos...relativos a historia das navegacões. pt.1-4. Lisboa, 1875.
Geog 3570.16 Frazão de Vasconcellos, J. Os pilotos dos seculos XV e XVI e a nobreza do reino. Lisboa, 1932.
Geog 3570.16.5 Frazão de Vasconcellos, J. Pilobas das navegacões portuguesas das seculos XVI e XVII. Lisboa, 1942.
Geog 3570.17 Prestage, Edgar. The Portuguese pioneers. London, 1933.
Geog 3570.18 Prestage, Edgar. Descobridores portugueses. Porto, 1934.
Geog 3570.18.5 Prestage, Edgar. Descobridores portugueses. 2. ed. Lisboa, 1943.
Geog 3570.20 Peres, Damião. Historia dos descobrimentos portugueses. pt.1-15. Lisboa, 1943-45.
Geog 3570.20.2 Peres, Damião. Historia dos descobrimentos portugueses. Lisboa, 1959.
Geog 3570.20.4 Peres, Damião. A history of the Portuguese. Lisbon, 1960.
Geog 3570.21 Amzalah, Moses. La Méditerranée et les découvertes maritimes des Portugais. Lisbonne, 1951.
Geog 3570.22 Goncalões Niana, M. As viagens terrestres dos Portuguezes. Porto, 1945.
Geog 3570.23F Cortesão, Jaime. Os descrobrimentos portugueses. Lisboa, 195-? 2v.
Geog 3570.24 Costa Brochado. Historia de uma polemica. Lisboa, 1944.
Geog 3570.25 Castro Soromenho. A maravilhosa viagem dos exploradores portugueses. v.1-12. Lisboa, 1946.
Geog 3570.26 Bandeira Ferreira, F. As viagens de descobrimento de iniciativa particular no tempo. Lisboa, 1946.
Geog 3570.27 Pina Manique, Luiz da. Subsidios para a história de cartografia portuguesa. Lisboa, 1943.
Geog 3570.28 Lima, Manuel C. Deux voyages portuguèses de découverte dans l'Atlantique occidental. Lisbonne, 1946.
Geog 3570.30A Hart, Henry H. Sea road to the Indies. N.Y., 1950.
Geog 3570.30B Hart, Henry H. Sea road to the Indies. N.Y., 1950.
Geog 3570.31 Fontoura da Costa, Abel. Roteiros portugueses ineditos. Lisboa, 1940.
Geog 3570.31.5 Fontoura da Costa, Abel. Descobrimentos maritimos africanos dos Portugueses com D. Henrique. Lisboa, 1938.
Geog 3570.33 Sa, Ayres de. Frei Goncalo Velho. Lisboa, 1899-1900. 2v.
Geog 3570.34 Rogers, F.M. Valentim Fernandes. Lisboa, 1957?
Geog 3570.34.5 Congresso Internacional de Historia dos Descobrimentos. Actas. Lisboa, 1960- 6v.
Geog 3570.36 Renault-Roulier, Gilbert. The caravels of Christ. N.Y., 1959.
Geog 3570.45 Tracey, H. Antonio Fernandes. Lourenco Marques, 1940.
Geog 3570.50 Albuquerque, Louis. Introducão a historia dos descobrimentos. Coimbra, 1962.
Geog 3570.51 O seculo dos descobrimentos. São Paulo, 1961.
Geog 3570.52 Cortesão, Jaime. Os descobrimentos pre-colombinos dos Portugueses. Lisboa, 1966.
Geog 3570.53 Crone, Gerald. The discovery of the East. London, 1972.
Geog 3570.55 Campos, Viriato. Viagens de Diogo Cão e de Bartolomeu Dias. Lisboa, 1966.
Geog 3570.58 Heleno, Manuel Domingues. Colaboração portuguesa nos descobrimentos nauticos das outras nações. Lisboa, 1932.
Geog 3570.60 Marjay, Frederico Pedro. Navegadores portugueses, herois do mar. Lisboa, 1970.
Geog 3570.61 Ferro, Gaetano. Le conoscenze geografiche del Medioevo. Genova, 1972.

Geog 3580 History of geography - Local history - Europe - Spain and Portugal - Spain

Geog 3580.5 Fernandes de Navaretti, Martin. Coleccion de los viajes y descubrimientos. 2. ed. v.1-3. Madrid, 1958. 5v.
Geog 3580.5.5 Fernandes de Navaretti, Martin. Coleccion de los viajes y descubrimientos. Buenos Aires, 1945-46. 5v.
Geog 3580.10 Anghiera, P.M. d'. Lettres relatives aux découvertes maritimes des Espagnols. Paris, 1885.
Geog 3580.15 Pamphlet box. Spain. Consejo Superior de Investigaciones Cientificas. Instituto Historico de Marina.
Geog 3580.20 Ballesteros Gaibrois, Manuel. España en los mares. 2. ed. Madrid, 1943.
Geog 3580.25A Vicens Vives, J. Rumbos oceanicos. Barcelona, 1946.

Geog 3580 History of geography - Local history - Europe - Spain and Portugal - Spain - cont.

Geog 3580.25B Vicens Vives, J. Rumbos oceanicos. Barcelona, 1946.
Geog 3580.30 Garcia de Herreros, E. Quatre voyageurs espagnols à Alexandria d'Egypt. Alexandria, 1923.
Geog 3580.35 Gavira, José. La ciencia geografica española del siglo XVI. Madrid, 1931.
Geog 3580.40 Colbrecht, Jozsf. De vleminfen en de Spansche. Antwerp, 1927.
Geog 3580.45 Diaz-Trechuelo Spinola, Maria Lourdes. Navegantes y conquistadores vascos. Madrid, 1965.
Geog 3580.46 Prieto, Carlos. El Oceano pacifico. Navegantes española del siglo XVI. Madrid, 1972.

Geog 3590 History of geography - Local history - Europe - Russia, Soviet Union

Geog 3590.5 Nozikov, N. Russian voyages round the world. London, 1947.
Geog 3590.5.5 Nozikov, N. Russkie krugosvetiye moreplavateli. Izd. 2. Moskva, 1947.
Geog 3590.10 Lialina, M.A. Russkie moreplavateli, arkticheckie krulosvetnye. Sankt Peterburg, 1892.
Geog 3590.15 Berg, L.S. Ocherki po istorii russk. geogr. otkrytii. Moskva, 1946.
Geog 3590.15.2 Berg, L.S. Ocherki po istorii russk. geogr. otkrytii. 2. izd. Moskva, 1949. 2v.
Geog 3590.15.5 Doklady na ezhegodnykh chteniiakh pamiati L.S. Berga. Moskva. 1-7
Geog 3590.15.10 Berg, L.S. Belikie russkie iute estvenniki. Moskva, 1950.
Geog 3590.15.15 Berg, L.S. Geschichte der russischer geographischer Entdeckungen. Leipzig, 1954.
Geog 3590.15.20 Berg, L.S. Izbrannye trudy. Moskva, 1956. 5v.
Geog 3590.15.40 Berg, R.L. Po ozeram Sibiri i Srednei Azii. Moskva, 1955.
Geog 3590.17 Doklady na ezhegodnykh chteniiakh pamiati V.A. Obrucheva. Moskva. 1-5
Geog 3590.20 Pallas, P.S. Bering's successors, 1745-1780. Seattle, 1948.
Geog 3590.25 Bodnarskii, M.S. Ocherkii po istorii russk. zemleved. Moskva, 1947.
Geog 3590.30 Lebedev, Dimitrii M. Geogr. v Rossii XVII v. ocherkii. Moskva, 1949.
Geog 3590.30.5 Lebedev, Dimitrii M. Geog. v Rossii petrovsk. vremeni. Moskva, 1950.
Geog 3590.30.10 Lebedev, Dimitrii M. Plavanie A.I. Chirukova. Moskva, 1951.
Geog 3590.30.15 Divin, Vasilii A. Belikii russkii moreplavateli A.I. Chirukova. Moskva, 1953.
Geog 3590.35 Efimov, Aleksei V. Iz istorii russkii geogr. otkrytii v sever ledov i tiloke okeanov. Moskva, 1950.
Geog 3590.35.2 Efimov, Aleksei V. Iz istorii velikikh russkikh geograf. otkrytii. Moskva, 1971.
Geog 3590.35.3 Efimov, Aleksei V. Iz istorii velikikh russkikh geograf. otkrytii. Moskva, 1949.
Geog 3590.35.5 Efimov, Aleksei V. Iz istorii russkikh aksped. na Tikhom okeane. Moskva, 1948.
Geog 3590.35.10 Efimov, Aleksei V. Otkrytiia russk. zemleprov. na ser. Vost. Asii. Moskva, 1951.
Geog 3590.40 Adamov, Arkadii. Pervye russkie issledovateli Aliaski. Moskva, 1950.
Geog 3590.42 Zogosken, L.A. Puteshestviia i issledovaniu...v Russkoi Amerike v 1802-04 gg. Moskva, 1956.
Geog 3590.45 Lupach, V.S. Russkie moraplavateli. Moskva, 1953.
Geog 3590.47 Zubov, N.N. Otecheatvennye moraplavateli-issledovanii morel i okeanov. Moskva, 1954.
Geog 3590.49 Zabrodskaia, M.R. Russkie puteshestvenniki po Afrike. Moskva, 1955.
Geog 3590.51 Moscow. Gosudarstvennyi Biblioteka SSSR Imeni V.I. Lenin. Russkie geografi i puteshestvenniki. Moskva, 1955.
Geog 3590.55 Lebedev, Dmitri M. Ocherki po istorii geografii v Rossii XV i XVI veko. Moskva, 1956.
Geog 3590.55.5 Lebedev, Dmitri M. Ocherki po istorii geografii v Rossii XVIII veka. Moskva, 1957.
Geog 3590.60 Murator, M.V. Navatrechu apasnostiam. Moskva, 1956.
Geog 3590.65 Gvozdetskii, Nikolai Andreevich. Sorok let issledovanii i otkrytii. Moskva, 1957.
Geog 3590.67 Gesellschaft für Deutsch-Sowjetische Freundschaft. Beiträge aus der sowjetische Kartographie. Berlin, 1953.
Geog 3590.70 Sovetskie ekspeditisii god. 1959. Moskva, 1962.
Geog 3590.71.3 Akademiia Nauk SSSR. Soviet geography, accomplishments and tasks. N.Y., 1962.
Geog 3590.72 Kaulbars, N.A. Aperçu des travaux géographiques en Russie. St. Pétersbourg, 1889.
Geog 3590.73 Esakov, V.A. Russkie geograficheskie issledovaniia Evropeiskoi Rossii i Vrala v XIX-nachale XX v. Moskva, 1964.
Geog 3590.74 Alekseev, Mikhail P. Kolumby rosskie. Magadan, 1966.
Geog 3590.75 Gvozdetskii, Nikolai Andreevich. Sovetskie geograficheskie issledovaniia i otkrytiia. Moskva, 1967.
Geog 3590.75.1 Gvozdetskii, Nikolai Andreevich. Soviet geographical explorations and discoveries. Moscow, 1974.
Geog 3590.76 Wotte, Herbert. Kurs auf Unerforscht. Leipzig, 1967.
Geog 3590.77 Geograficheskoe Obshchestvo SSSR. Geograficheskoe obshchestvo za 125 let. Leningrad, 1970.
Geog 3590.78 Alekseev, Aleksandr Ivanovich. Syny otvazhnye Rossii. Magadan, 1970.
Geog 3590.79 Divin, Vasilii A. Russkie moreplavaniia na Tikhom okeane r XVIII neke. Moskva, 1971.
Geog 3590.80 Lebedev, Dmitrii M. Russkie geograficheskie otkrytiia i issledovaniia s drevnikh vremen do 1917 goda. Moskva, 1971.
Geog 3590.81 Akademiia Nauk SSSR. Sovetskaia geografiia. Moskva, 1960.
Geog 3590.82 Moscow. Universitet. Sovetskaia geografiia v period stroitel'stva kommunizma. Moskva, 1963.
Geog 3590.83 Novokshanova, Zinaida K. Kartograficheskie i geodezicheskie raboty v Rossii v XIX-nachale XX v. Moskva, 1967.
Geog 3590.84 Krempol'skii, Viktor F. Istoriia razvitiia kartoizdaniia v Rossii i v SSSR. Moskva, 1959.
Geog 3590.85 Tikhomirov, Georgii S. Bibliograficheskii ocherk istorii geografii v Rossii XVIII veka. Moskva, 1968.
Geog 3590.86 Tel', Sergei E. Kartografiia Rossii XVIII veka. Moskva, 1960.
Geog 3590.87 Problemy heohrafichnoï nauky v Ukraïns'kii RSR. Kyïv. 1,1972+ 2v.
Geog 3590.88 Krupenikov, Igor' A. Istoriia geograficheskoi mysli v Moldavii. Kishinev, 1974.

Classified Listing

Geog 3594 History of geography - Local history - Europe - Poland
Geog 3594.5 Pertek, Jerzy. Polacy ner szlakach morskich świata. Gdańsk, 1957.
Geog 3594.10 Kuźmiński, Bolesław. Polskie nazwy na mapie śceiata. Wyd. 1. Warszawa, 1967.
Geog 3594.15 Turley, Tomasz J. Polacy badacze Ameryki. Chicago, 1968.

Geog 3596 History of geography - Local history - Europe - Czechoslovakia
Geog 3596.2 Vitásek, František. Výroj moravské geografie. 1. vyd. Praha, 1973.

Geog 3602 History of geography - Local history - Europe - Scandinavia - Norway
Geog 3602.5 Richter, Sørea. Great Norwegian expeditions. Oslo, 1955.

Geog 3604 History of geography - Local history - Europe - Scandinavia - Sweden
Geog 3604.5 Selander, Sten. Linnélärjungar i främmande länder. Stockholm, 1960.

Geog 3618 History of geography - Local history - Europe - Finland
Geog 3618.3 Saellskapit för Finlands Geografi, Helsingfors. Exposé des travaux géographiques executés en Finlande jusqu'en 1895. Helsingfors, 1895.
Geog 3618.10 Mead, William R. The geographical tradition in Finland. London, 1963.

Geog 3620 History of geography - Local history - Europe - Yugoslavia
Geog 3620.2 Radosčić, Nikola. Geografsko znanje o Srbiji početkom 19 veka. Beograd, 1927.

Geog 3650 History of geography - Local history - Asia - India
Geog 3650.5 Dube, Beehan. Geographical concepts in ancient India. Varanasi, 1967.

Geog 4015 - 4020 Adventure and adventurers in general - General works (By date)
Geog 4019.52 Lamb, Geoffrey F. Modern action and adventure. London, 1952.
Geog 4019.57 Wie sie entkamen; mit einer Einleitung von Kasimir Edschmid. Düsseldorf, 1957.
Geog 4019.60 Mier, Waldo de. Escogieron la inquietud. Madrid, 1960.
Geog 4019.63 National Geographic Society, Washington. Great adventures with NationalGeographic. Washington, 1963.
Geog 4019.68 Petro, W. Triple commission. London, 1968.
Geog 4019.74 Zweig, Paul. The adventure. London, 1974.

Geog 4028 Adventure and adventurers in general - Biographies of adventurers - Collected
Geog 4028.2 Green, Timothy. The adventures; four profiles of contemporary travellers. London, 1970.
Geog 4028.3 Cartier, Jean Pierre. Explorateurs et explorations. Paris, 1974.
Geog 4028.4.2 Bridges, Thomas C. Heroes of modern adventure. Boston, 1928.

Geog 4029 Adventure and adventurers in general - Biographies of adventurers - Individual
Geog 4029.2 Mämpel, J.C. The young rifleman's comrade. London, 1826.
Geog 4029.4 Franco, Ramon. Aguilas y garras. Madrid, 1929.
Geog 4029.6 Borden, Norman E. Dear Sarah. Freeport, Me., 1966.
Geog 4029.8 Dean, Harry. The Pedro Gorino. Boston, 1929.
Geog 4029.9 Aresty, Miguel de. Los papeles de Juan de Aresty. Barcelona, 1972.
Geog 4029.10 Anderton, Russ. Tic-polonga. 1. ed. Garden City, 1953.
Geog 4029.12 Loeffler, Johann Friedrich. Abenteur in drei Erdteilen. Heidenheim, 1971.
Geog 4029.14 Ridgway, John M. Journey to Ardmore. London, 1971.
Geog 4029.15 Monfreid, Henri de. Le feu de Saint-Elme. Paris, 1974.
Geog 4029.16 Root, Jonathan. Halliburton, the magnificient myth; a biography. N.Y., 1965.
Geog 4029.17 Seering, Ruth. Mein tödliches Risiko. Bergisch Gladbach, 1973.

Geog 4100 Voyages and travels in general - General pamphlet volumes
Geog 4100.01F Pamphlet vol. Travel.
Geog 4100.1 Pamphlet box. Travel.
Geog 4100.2 Pamphlet vol. Travel. 3 pam.
Geog 4100.3 Pamphlet vol. Travel.
Geog 4100.4 Pamphlet vol. Travel. 9 pam.
Geog 4100.5 Pamphlet vol. Travel.

Geog 4105 Voyages and travels in general - Art and history of travel - Travelling instructions, handbooks
Geog 4105.2 Jackson, J.R. What to observe. London, 1841.
Geog 4105.4 Noyes, E.H. Steamship notes...a handbook. N.Y., 1874.
Geog 4105.6 Verax, V. (pseud.). Cautions for the first tour. London, 1863.
Geog 4105.10 Buffum, E.G. Pocket guide for Americans going to Europe. N.Y., 1859.
Htn Geog 4105.12* Gratarolo, G. De regimine iter agentium. Basileae, 1561.
Geog 4105.13A Harvard Travel Club. Handbook of travel. Cambridge, 1917.
Geog 4105.13B Harvard Travel Club. Handbook of travel. Cambridge, 1917.
Geog 4105.13.3 Harvard Travel Club. Handbook of travel. Cambridge, 1935.
Geog 4105.14 Kitchiner, William. The travellers' oracle. 2. ed. London, 1827. 2v.
Geog 4105.15 Darde, J.B. The travellers' handbook. London, 1842.
Geog 4105.16 Brouwer, H.A. Practical hints to scientific travellers. Leyden, 1922-29. 6v.
Geog 4105.17.5 Tatchell, Frank. The happy traveller. 5. ed. London, 1927.
Geog 4105.18.4 Luce, Robert. Going abroad? 4. ed. Boston, 1906.
Htn Geog 4105.19* Palmer, Thomas. An essay of the meanes how to make...travailes. London, 1606.

Geog 4110 Voyages and travels in general - Art and history of travel - General guidebooks
Geog 4110.5 King, M. Where to stop. Boston, 1893.
Geog 4110.5.2 King, M. Where to stop. Boston, 1894.
Geog 4110.7 Thorpe, D. Universal guide of standard routes. Boston, 1907.
Geog 4110.9 Rand, McNally and Co. Pocket atlas of the world. N.Y., 1887.
Geog 4110.9.2 Rand, McNally and Co. Pocket atlas of the world. Chicago, n.d.
Geog 4110.11F Cuward Steamship co. Official guide and album. London, 1877.
Geog 4110.13 Loftie, W.J. Orient line guide. London, 1890.
Geog 4110.16 Dempsey, J.M. Our ocean highways. London, 1871.
Geog 4110.17 Burns, Philp and Co. Picturesque travel. Sydney, 1914.

Geog 4110 Voyages and travels in general - Art and history of travel - General guidebooks - cont.
Geog 4110.18 Phillips, Morris. Abroad and at home. N.Y., 1892.
Geog 4110.19 Hall, E.H. The picturesque tourist. N.Y., 1877.
Geog 4110.20 Peninsular and Oriental Steam Navigation Company. Pocket book, 1890. London, 1890.
Geog 4110.20.3 Peninsular and Oriental Steam Navigation Company. The P. and O. pocket book. London, 1908.
Geog 4110.21 Goodrich, Charles A. The universal traveller. 2. ed. Hartford, 1836.
Geog 4110.22 Pacific Mail Steamship Co. A sketch of the route to California, China. San Francisco, 1867.
Geog 4110.23 Sherriff's illustrated route charts and travellers' hand book. v.1-4. London, 1887.
Geog 4110.25 Coon, Horace. 100 vacations costing from $50.00 to $500.00. N.Y., 1939.
Geog 4110.27 Meyer, H.J. Weltreise. Leipzig, 1907.
Geog 4110.30 Ford, Norman D. Bargain paradises of the world. 4. ed. Greenlawn, N.Y., 1957.
Geog 4110.30.5 Ford, Norman D. How to travel without being rich. Greenlawn, N.Y., 1957.
Geog 4110.35 Jewish travel guide. London, 1961+ 3v.
Geog 4110.36 Sheraton, Mimi. City portraits; a guide to 60 of the world's great cities. 1. ed. N.Y., 1964.
Geog 4110.37 Croft-Cooke, Rupert. Cities. London, 1951.
Geog 4110.38 Gunther, John. Twelve cities. 1. ed. N.Y., 1969.
Geog 4110.40 Travel routes around the world. N.Y. 23,1957
Geog 4110.42 Let's go II. Cambridge, Mass. 1968

Geog 4125 - 4130 Voyages and travels in general - Art and history of travel - General works (By date)
Geog 4125.75.1 Turler, H. The traveller, 1575. Gainesville, Fla., 1951.
Htn Geog 4125.91* De arte peregrinandi. Libri II. Noribergae, 1591.
Htn Geog 4126.14* Lithgow, W. Peregrination from Scotland. Lond, 1614.
Htn Geog 4126.33* Essex, Robert. Profitable instrucions. London, 1633.
Htn Geog 4126.43* Neale, Thomas. A treatise of direction, how to travell safely, and profitably into forraigne countries. London, 1643.
Geog 4127.05 Schroeter, J.C. Diatriba...peregrinationum eruditarum. Jenae, 1705.
Geog 4127.77.2 Schloezer, August Ludwig von. Vorlesungen über Land- und Seereisen. Göttingen, 1962.
Geog 4128.56 Galton, Francis. The art of travel. London, 1856.
Geog 4128.80.3 Knox, Thomas W. How to travel. N.Y., 1881.
Geog 4128.80.5 Knox, Thomas W. How to travel. N.Y., 1888.
Geog 4128.91 Hunt, R. Steamship lines of the world. N.Y., 1891.
Geog 4128.97 Ludwig, Friedrich. Untersuchungen...Reise...Itineraire der deutsech Königer. Inaug. Diss. Berlin, 1897.
Geog 4128.97.1 Ludwig, Friedrich. Untersuchungen über die Reise und Marschegesehwindigkeit in XII. und XIII. Jahrhundert. Berlin, 1897.
Geog 4129.03 Shand, A.I. Old-time travel. London, 1903.
Geog 4129.06 Raleigh, Walter. The English voyages of the sixteenth century. Glasgow, 1906.
Geog 4129.10 Hopkins, A.A. Scientific American handbook of travel. N.Y., 1910.
Geog 4129.11 Hedin, Sven. Van pool tot pool. Amsterdam, 194-?
Geog 4129.11.5 Hedin, Sven. Von Pol zu Pol. 81. Aufl. Leipzig, 1942.
Geog 4129.14 Howard, Clare. English travellers of the Renaissance. London, 1914.
Geog 4129.14.5 Mead, William E. The grand tour in the 18th century. Boston, 1914.
Geog 4129.21 Roget, S.R. Travel in the two last centuries of three generations. N.Y., 1921.
Geog 4129.23 Hungerford, E. Planning a trip abroad. N.Y., 1923.
Geog 4129.23.5 Hungerford, E. Planning a trip abroad. N.Y., 1923.
Geog 4129.25 Gorce, Denys. Les voyages l'hospitalité...dans le monde chrétien des IVe et Ve siècles. Wépion-sur-Meuse, 1925.
Geog 4129.25.2 Gorce, Denys. Les voyages l'hospitalité...dans le monde chrétien des IVe et Ve siècles. Thèse. Wépion-sur-Meuse, 1925.
Geog 4129.31 Titayna (pseud.). Mademoiselle against the world. N.Y., 1931.
Geog 4129.37 Barraud, G. Touristes de jadis. Paris, 1937.
Geog 4129.47 Carrington, D. The traveller's eye. N.Y., 1947.
Geog 4129.47.5 Cook, H.K. Over the hills and far away. London, 1947.
Geog 4129.50 Michael, M. Traveller's quest. London, 1950.
Geog 4129.57 Hughes, Spike. The art of coarse travel. London, 1957.
Geog 4129.58 Greenen, E. Reisen seit Anno dazumal. Hamburg, 1958.
Geog 4129.58.5 Randall, C.B. International travel. Washington, 1958.
Geog 4129.59 Schadendorf, Wulf. Zu Pferde, im Wagen, zu Fuss. München, 1959.
Geog 4129.70 Anderson, John Richard Lane. The Ulysses factor: the exploring instinct in man. London, 1970.
Geog 4129.71 Bauer, Hans. Wenn einer eine Reise tat. Leipzig, 1971.
Geog 4129.74 Parry, John Horace. The discovery of the sea. N.Y., 1974.

Geog 4131 Voyages and travels in general - Art and history of travel - Tourism, tourist trade (By date, e.g. .65 for 1965)
Geog 4131.2 Organization for European Economic Cooperation. Trends in economic sectors: tourism in Europe. Paris. 1953-1961
Geog 4131.4 Tourism in O.E.C.D. member countries. Paris. 1961+ 6v.
Geog 4131.5 Rae, W.F. The business of travel. London, 1891.
Geog 4131.10 Ogilvie, F.W. The tourist movement. London, 1933.
Geog 4131.34 Grünthal, Adolf. Probleme der Fremdenverkehrsgeographie. Diss. Berlin, 1934.
Geog 4131.38 Trimbach, André. Le tourisme international. Thèse. Paris, 1938.
Geog 4131.38.5 Leveillé-Nizerolle, Claude. Le tourisme dans l'economie contemporaine. Thèse. Paris, 1938.
Geog 4131.48 Winble, Ernest W. European recovery, 1948-1951, and the tourist industry. London, 1948.
Geog 4131.51 International Touring Association. Scientific Commission. Publication de la Commission scientifique de l'alliance internationale de tourisme. Berne. 1-6,1951-1956 3v.
Geog 4131.53 Pudney, J. The Thomas Cook story. London, 1953.
Geog 4131.55 Ignacio de Arrillaga, José. Sistema de política turistíca. Madrid, 1955.
Geog 4131.58 Tusci, Leonida. Elementi di tecnica professionale turistica. Roma, 1958.
Geog 4131.59 Schweizerischer Fremdenverkehrsverband. Festschrift für Walter Hienzeker zum 60. Geburtstag. Bern, 1959.
Geog 4131.60 Knebel, Hans J. Soziologische Strukturwandlungen im Mardernen. Stuttgart, 1960.
Geog 4131.60.5 Fuss, K. Geschichte der Reisebüros. Darmstadt, 1960.
Geog 4131.61 Cheechi and Company, Washington, D.C. The future of tourism in the Pacific and Far East. Washington, 1961.

Classified Listing

Geog 4131 Voyages and travels in general - Art and history of travel - Tourism, tourist trade (By date, e.g. .65 for 1965) - cont.

	Geog 4131.62.5	Poeschl, A.E. Fremdenverkehr und Fremdenverkehrspolitik. Berlin, 1962.
	Geog 4131.66	Ritter, Wigand. Fremdenverkehr in Europa. Leiden, 1966.
	Geog 4131.67	Spain. Comisaría del Plan de Desarrollo Economico y Social. Comisión de Turismo. Turismo. Madrid, 1967?
	VGeog 4131.67.5	Seminàr o Ekonomickej Efektívnosti Iuvesticii Cestovného Ruchu, Piešťany, 1966. Efektivnost' investicí cestovného ruchu. Bratislava, 1967.
	VGeog 4131.68	Anan'ev, Mikhail A. Mezhdunarodruji turizm. Moskva, 1968.
	Geog 4131.68.5	Meinke, Hans. Tourismus und wirtschaftliche Entwicklung. Göttingen, 1968.
	Geog 4131.68.10	Aeschlimann, Jean Louis. Structure et tâches d'un organisme national de tourisme. Inaug. Diss. Bern, 1968.
	Geog 4131.68.15	Internationale Informationstagung zur Geographie des Fremdenverkehrs. Dresden, 1965. Probleme der Geographie des Fremdenverkehrs der Deutschen Demokratischen Republik und anderer Staaten. Leipzig, 1968.
	Geog 4131.68.20	Corna Pellegrini, Giacomo. Studi e ricerche sulla regione turistica. Milano, 1968.
	Geog 4131.69	Frentrup, Klaus. Die ökonomische Bedeutung des internationalen Tourismus fur die Ehtwicklungsländer. Hamburg, 1969.
	Geog 4131.70	Ruppert, Karl. Zur Geographie des Freizeitverhaltens. Kallmünz, 1970.
	Geog 4131.71	Patev, Iliia. Statistika na turizma. Varne, 1971.
	Geog 4131.73	Keller, Peter. Soziologische Probleme in modernen Tourismus. Bern, 1973.
	Geog 4131.74	International Geographical Union. Working Group, Geography of Tourism and Recreation. Studies in the geography of tourism. Frankfurt, 1974.
	Geog 4131.74.5	Haulot, Arthur. Tourisme et environnement. Verviers, 1974.

Geog 4132 Voyages and travels in general - Art and history of travel - Other special topics

	Geog 4132.5	Silverberg, Robert. The longest voyage; circumnavigators in the age of discovery. Indianapolis, 1972.
	Geog 4132.6	Trantina, V. Staří Čechové na cestách. Praha, 1941.

Geog 4135 - 4140 Voyages and travels in general - Biographies of travellers - Collected (By date)

Htn	Geog 4136.76*	Bosch, L. Leeven en daden...zee-helden. Amsterdam, 1676.
	Geog 4136.76.5	Bosch, L. Leben...der...See-Helden. Nürnberg, 1681.
	Geog 4138.30	Brendon, J.A. Great navigators and discoverers. N.Y., 1930.
	Geog 4138.31	Lives and voyages of Drake. Edinburgh, 1831.
	Geog 4138.36	Lives and voyages of Drake. N.Y., 1836.
	Geog 4138.37	St. John, J.A. Lives of celebrated travellers. N.Y., 1837. 3v.
	Geog 4138.37.5	Johnstone, C. (Mrs.). Lives and voyages of Drake. Edinburgh, 1837.
	Geog 4138.54	Charton, E.T. Voyageurs anciens et modernes. Paris, 1854-57. 4v.
	Geog 4138.82.5	Werner, R. Berühmte Seeleute. Berlin, 1882.
	Geog 4138.82.7	Embacher, F. Lexikon der Reisen und Entdeckungen. Leipzig, 1882.
	Geog 4138.93	Greely, A.W. Explorers and travellers. N.Y., 1893.
	Geog 4139.10	Johnstone, C. (Mrs.). Buccaneers of America. Akron, 1910.
	Geog 4139.25A	Beston, Henry B. The book of gallant vagabonds. N.Y., 1925.
	Geog 4139.25B	Beston, Henry B. The book of gallant vagabonds. N.Y., 1925.
	Geog 4139.42	Penrose, Boies. Urbane travelers, 1591-1635. Philadelphia, 1942.
	Geog 4139.46	Majó Framis, R. Vida de los navegantes y conquistadores españoles del siglo XVI. 1. ed. Madrid, 1946.
	Geog 4139.61	Kunský, J. Čeští cestovatelé. Praha, 1961. 2v.
	Geog 4139.64	Mirsky, J. The great Chinese travelers, an anthology. N.Y., 1964.
	Geog 4139.68	Wertheim, Willem Frederik. Ketters en kwezels, regenten en rebellen. Drachten, 1968.
	Geog 4139.69	Histoire générale des grands aventuriers de la mer. v.1,4-15,17-18. Paris, 1969. 16v.
	Geog 4139.73	Sławezyński, Wacław. Polscy podróznicy i odkrywey. 1. wyd. Warszawa, 1973.
	Geog 4139.73.5	Kuźmiński, Bolesław. Przygody polskich obieżyświatów na morzach i lądach. 1. wyd. Gdeńsk, 1973.

Geog 4145 Voyages and travels in general - Biographies of travellers - Individual (99 scheme, A-Z by person)

	Geog 4145.3	Allen, Katharine M. Foreign service diary. Washington, 1967.
	Geog 4145.4	Lewis, Warren. Levantine adventurer...Chevalier d'Arvieux. London, 1962.
	Geog 4145.7	Bertrand, Alfred. Alfred Bertrand. London, 1926.
	Geog 4145.9	Thiery, Maurice. Bougainville. London, 1932.
	Geog 4145.9.5	Bodrick, Alan Houghton. Casual change. London, 1961.
	Geog 4145.10	Burkov, Boris S. Ustrechi na piati kontinentakh. Moskva, 1973.
	Geog 4145.11	Peres, Damião. Diogo Cão. Lisboa, 1957.
	Geog 4145.11.10	Fernandez de Navarrete, Eustaquio. Historia de Juan Sebastian del Cano. Vitoria, 1872.
	Geog 4145.14	Lopes, Francisco Fernandes. The brothers Corte Real. Lisboa, 1957.
	Geog 4145.14.5	Fernandes, José dos Santos. Miguel Corte Real. Coimbra, 1960.
	Geog 4145.14.10	Brazão, Eduardo. Os Corte Reais e o novo mundo. Lisboa, 1965.
	Geog 4145.14.10.5	Brazão, Eduardo. Les Corte-Real et le nouveau monde. Lisbonne, 1967.
	Geog 4145.14.100	Swinglehurst, Edmund. The romantic journey. London, 1974.
	Geog 4145.15	Chapman, F.S. Living dangerously. N.Y., 1953.
	Geog 4145.15.2	Chapman, F.S. Living dangerously. London, 1953.
	Geog 4145.17	Connolly, J.B. Master mariner...Amasa Delano. Garden City, 1943.
	Geog 4145.20	Day, George. Dumont d'Urville. Paris, 1947.
	Geog 4145.23	Ely, Edward. The wanderings of Edward Ely. N.Y., 1954.
	Geog 4145.24	Eschels, Jeus Jacob. Das obenteuerliche Leben des Jeus Jacob Eschels. Hamburg, 1966.
	Geog 4145.25.5	Wilson, P.W. An explorer of changing horizons. N.Y., 1927.
	Geog 4145.27	Fink, O. Auf dem Kirs der Raben. Hamburg, 1963.
	Geog 4145.28	Filchner, Wilhelm. Ein Forscherleben. 3. Aufl. Wiesbaden, 1953.

Geog 4145 Voyages and travels in general - Biographies of travellers - Individual (99 scheme, A-Z by person) - cont.

	Geog 4145.28.5	Filchner, Wilhelm. In China, auf Asiens Hochsteppen. Freiburg, 1930.
	Geog 4145.29	Fournier-Aubry, Fernand. Mon metier l'aventure. Paris, 1953.
	Geog 4145.29.5	Fournier-Aubry, Fernand. Don Fernando. Paris, 1972.
	Geog 4145.29.6	Fournier-Aubry, Fernand. Don Fernando. 1st American ed. N.Y., 1974.
	Geog 4145.34.2	Grum-Grzhimailo, A.G. Dela i dni G.E. Grum-Grzhimailo. Moskva, 1947.
	Geog 4145.34.25	Guzanov, Vitalii G. Odissei s Beloi Rusi. Minsk, 1969.
	Geog 4145.37	Jacoby, Arnold. Señor Kon-Tiki; the biography of Thor Heyerdahl. Chicago, 1967.
	Geog 4145.39	Baum, Jiří. Holub a Mašukulumbové. Praha, 1955.
	Geog 4145.39.5	Hoyt, Jo Wasson. For the love of Mike. N.Y., 1966.
	Geog 4145.46	Pasetskii, Vasilii M. Ivan Fedorovich Kruzenshtern. Moskva, 1974.
	Geog 4145.48	Lantzsch, W. Die Welt in allen Zonen. München, 1961.
	Geog 4145.50	Parr, Charles. Jan van Linschaten. N.Y., 1964.
	Geog 4145.52	Dainelli, Giotto. Il duca degli Abruzzi. Torino, 1967.
	Geog 4145.53	Willers, U. Xavier Marmier och Sverige. Stockholm, 1949.
	Geog 4145.53.5	Davydov, I.V. V moriakh i stranstviakh. Moskva, 1956.
	Geog 4145.56	Cordier, Stéphane. Balthazar de Monconys. Bruxelles, 1967.
	Geog 4145.61	Noailles, Loise de. Souvenirs de quatre horizons. Fribourg, 1958.
	Geog 4145.68	Mueller, Martin. Julius von Payer. Stuttgart, 1958.
	Geog 4145.68.5	Gnevusheva, E.I. Zaky tyi pute Shestvennik. Moskva, 1958.
	Geog 4145.70	Collis, Maurice. The grand peregrination; being the life of...Fernão Mendes Pinto. London, 1949.
	Geog 4145.71	Obruchev, V.A. Grigorii Nikolaenin Potanin; zhizn' i deiatel'nost'. Moskva, 1947.
	Geog 4145.71.5	Obruchev, V.A. Puteshchestviia Potanina. Moskva, 1953.
	Geog 4145.71.500	Zariu, V.M. Puteshchestviia A.V. Potaninoi. Moskva, 1950.
	Geog 4145.72	Price, W. I cannot rest from travel. N.Y., 1951.
	Geog 4145.78	Touring Club de France. De l'Himalaya aux Pyrénées. Paris? 1959.
	Geog 4145.79	Sack, John. Report from practically nowhere. N.Y., 1959.
	Geog 4145.80	Schindler, Fritz. Meine schönste Autoreise. Wien, 1958.
	Geog 4145.81	Slocum, Victor. Captain Joshua Slocum. N.Y., 1950.
	Geog 4145.81.5	Teller, Walter Magnes. The search for Captain Slocum. N.Y., 1956.
	Geog 4145.81.6	Teller, Walter Magnes. Joshua Slocum. New Brunswick, 1971.
	Geog 4145.81.10	Slocum, Joshua. The voyages of Joshua Slocum. New Brunswick, 1958.
	Geog 4145.81.50	Siegfried, André. Geographie poétique des cinq continents. Paris, 1952.
	Geog 4145.83	Stark, Freya. Perseus in the wind. London, 1948.
	Geog 4145.83.5	Szumańska-Grossowa, Hanna. Podróze Stefana Srolca Rogozinskiego. Warszawa, 1967.
	Geog 4145.84	Stark, Freya. Traveller's prelude. London, 1950.
	Geog 4145.84.5	Stark, Freya. Beyond Euphrates. 1. ed. London, 1951.
	Geog 4145.84.10	Stark, Freya. The coast of incense. 1. ed. London, 1953.
	Geog 4145.84.15	Stark, Freya. Dust in the lion's paw. London, 1961.
	Geog 4145.84.20	Stark, Freya. Letters. Compton Chamber, 1974-
	Geog 4145.87	Tschiffely, A. Bohemia junction. London, 1950.
	Geog 4145.87.5	Kharitanovskii, Aleksandr A. Chelovek s zheleznym olenem. Moskva, 1965.
	Geog 4145.89	Arteche, J. de. Urdaneta. Madrid, 1945.
	Geog 4145.89.5	Mitchell, Mairin. Friar Andrés de Urdaneta, O.S.A. London, 1964.
	Geog 4145.91	Parr, Charles. The voyages of David de Vries. N.Y., 1969.
	Geog 4145.94	Willis, Wiiliam. The hundred lives of an ancient mariner. London, 1967.
	Geog 4145.95	Wollschläger, Alfred. Menschen an meinen Wegen. München, 1973.
	Geog 4145.98	Seaver, G. Francis Younghusband. London, 1952.

Geog 4155 - 4160 Voyages and travels in general - Collections of travels - Multivolume works (By date)

	Geog 4157.04F	Churchill, J. Collection of voyages and travels. London, 1704-07. 8v.
	Geog 4157.04.3F	Churchill, J. Collection of voyages and travels. 3. ed. London, 1744. 6v.
Htn	Geog 4157.05.2F*	Harris, J. Navigantium...compleat collection. London, 1705.
	Geog 4157.05.10F	Harris, J. Navigantium...compleat collection. London, 1744. 2v.
	Geog 4157.05.15F	Harris, J. Navigantium...voyages and travels. London, 1764. 2v.
NEDL	Geog 4157.07	Aa, P.V. Naaukeurige...der zee en land-reysen. v.1-28. Leyden, 1707. 29v.
	Geog 4157.29	Dampier, W. Collection of voyages. London, 1729. 4v.
	Geog 4157.31	Bernard, J.F. Recueil de voyages au nord. Amsterdam, 1731-37. 9v.
	Geog 4157.45	Green, J. New collection of voyages and travels. London, 1745-47. 4v.
	Geog 4157.46	Prevost, A.T. Histoire generale des voyages. Paris, 1746-89. 20v.
	Geog 4157.47	Allgemeine Historie der Reisen. v.1-19,21. Leipzig, 1747-74. 20v.
	Geog 4157.59	World displayed. London, 1759-61. 20v.
	Geog 4157.59.3	World displayed. v.3-4,5-6,8,15-16,17-18. 3.-4. ed. London, 1760-88. 5v.
	Geog 4157.66	Barrow, J. Abrégé...histoire des decouvertes. Paris, 1766. 12v.
	Geog 4157.66.5	Callander, J. Terra Australis cognita, or Voyages. Edinburgh, 1766. 3v.
	Geog 4157.67	Knox, J. New collection of voyages. London, 1767. 7v.
	Geog 4157.74	Henry, David. Historical account of voyages. London, 1773-74. 4v.
	Geog 4157.80	LaHarpe, J.F. Abrégé de l'histoire generale des voyages. v.1-32, atlas. Paris, 1780-1801. 33v.
	Geog 4157.88	Berenger, J.P. Collection de tous les voyages. Lausanne, 1788-91. 9v.
	Geog 4157.96	Mavor, William. Historical account of...voyages. v.1-2,4-22. London, 1796. 19v.
	Geog 4157.96.3	Mavor, William. Historical account of...voyages...Columbus to present. Philadelphia, 1802.
	Geog 4157.96.5	Mavor, William. Historical account of most celebrated voyages, travels, and discoveries from time of Columbus. New Haven, 1802-03. 24v.
	Geog 4158.03F	Burney, J. Chronological history...discoveries..South Sea. London, 1803-1817. 5v.
	Geog 4158.05	Phillips, R. Collection of...voyages and travels. London, 1805-08. 7v.

Classified Listing

Geog 4155 - 4160   Voyages and travels in general - Collections of travels - Multivolume works (By date) - cont.

	Geog 4158.05.2	Phillips, R. Collection of...voyages and travels. London, 1810. 3v.
	Geog 4158.05.5	Pouqueville, F.C.H. Voyage en Morée. v.1,2-3. Paris, 1805. 2v.
	Geog 4158.08	Pinkerton, John. General collection of voyages. London, 1808-14. 17v.
NEDL	Geog 4158.10F	Pinkerton, John. General collection of voyages. Philadelphia, 1811-14. 17v.
	Geog 4158.11	Kerr, R. General history...of voyages. Edinburgh, 1811-24. 18v.
	Geog 4158.19	Phillips, R. New voyages and travels. London, 1819-20. 9v.
	Geog 4158.22	Eyries, J.B.B. Abrégé des voyages modernes. Paris, 1822-24. 14v.
	Geog 4158.24	Paris. Société de Geographie. Recueil de voyages et de memoires. Paris, 1824-1864. 7v.
NEDL	Geog 4158.33	Montemont, A. Histoire...des voyages. Paris, n.d. 46v.
	Geog 4158.42	Smith, William. Nouvelle bibliotheque des voyages. Paris, 1842. 12v.
	Geog 4158.60	Galton, F. Vacation tourists...in 1860, 1861, 1862, 1863. Cambridge, 1861-64. 3v.
	Geog 4159.10.10	Northcliffe, A.C.W.H. The world's greatest books. v.19. n.p., 1910.
	Geog 4159.21	Viajes clasicos. Madrid. 1-34 23v.
	Geog 4159.58	Les marins à la découverte. Paris, 1958. 12v.
	Geog 4159.62	Fruehe Reisen und Seefahrten in Originalberichten. Graz. 1,1962+ 7v.

Geog 4165 - 4170   Voyages and travels in general - Collections of travels - Smaller works (By date)

	Geog 4166.64F	Thevenot, M. de. Relations de divers voyages. Paris, 1664-66. 2v.
	Geog 4166.64.5F	Thevenot, M. de. Relation de divers voyages. Paris, 1696. 2v.
	Geog 4166.78.5	Dyck, J. Seer gedenckwaerdige voyagien. Amsterdam, 1678.
	Geog 4167.07.2	Bellegarde, J.B.M. General history of all voyages...old and new world. London, 1708.
Htn	Geog 4167.11*	Stevens, J. New collection of voyages. London, 1711. 2v.
	Geog 4167.51	Reisen nach Peru, Acadien und Egypten. Göttingen, 1751.
	Geog 4167.65	Barrow, J. A collection of authentic...voyages. v.2. London, 1765.
	Geog 4167.75	Dalrymple, A. A collection of voyages, chiefly in the S. Atlantic. London, 1775.
	Geog 4167.92	Adams, J. Flowers of modern travels. London, 1792. 2v.
Htn	Geog 4167.92.3*	Adams, J. Flowers of modern travels. Boston, 1797. 2v.
	Geog 4167.92.10	Adams, J. Flowers of celebrated travellers. Baltimore, 1834.
	Geog 4167.97	Heron, Robert. A collection of late voyages and travels. Edinburgh, 1797.
	Geog 4168.07	Campe, Joachim Heinrich. Voyages anecdotiques. Paris, 1807.
	Geog 4168.23	Collection de relations de voyages. Paris, 1823.
	Geog 4168.29	Bennet, Roelof Gabriel. Nederlandsch zeeseizen in het laatst der z estiende. v.1-4. Wijk, 1828-29. 3v.
	Geog 4168.31	Adams, W. The modern voyager and traveller. London, 1831. 4v.
	Geog 4168.34	Dumont d'Urville, Jules. Voyage pittoresque autour du monde. Paris, 1834-35. 2v.
	Geog 4168.36	Atlas zur Kunde fremder Weltheile. Leipzig. 1-2 2v.
	Geog 4168.40	Ternaux-Compans, H. Archives des voyages. pt.1-4. Paris, 1840. 4v.
	Geog 4168.43.4	Voyages round the world from the death of Captain Cook to the present time. 4. ed. London, 1850.
	Geog 4168.50	Hombron, Bernard. Aventures...des voyageurs. Paris, n.d.
	Geog 4168.54	Charton, Edouard T. Voyageurs anciens et modernes. v.1-4. Paris, 1854-57. 4v.
	Geog 4168.75	Adventures of famous travellers in many lands. N.Y., 18- .
	Geog 4168.84.5	Buel, James W. The worlds wonders. St. Louis, 1884.
	Geog 4168.86	Documentos para la historia de la nautica en Chile. Santiago de Chile, 1886.
	Geog 4168.87F	Colange, Leo de. Voyages and travels. Boston, 1887. 2v.
	Geog 4168.87.2F	Colange, Leo de. Voyages and travels. Boston, 1887.
	Geog 4168.89	Travel, adventure and sport. v.2-4. N.Y., 1889.
	Geog 4168.90	Griswold, W.M. Travel, series of narratives of...visits. v.1-106. Cambridge, 1890. 2v.
	Geog 4168.95	Bonnaffé, E. Voyages et voyageurs de la renaissance. Paris, 1895.
	Geog 4169.01	Singleton, Lesther. Romantic castles and palaces, as seen and described by famous writers. N.Y., 1901.
	Geog 4169.02A	Voyages and travels, 16th and 17th centuries. N.Y., 1902. 2v.
	Geog 4169.02B	Voyages and travels, 16th and 17th centuries. N.Y., 1902. 2v.
	Geog 4169.02C	Voyages and travels, 16th and 17th centuries. N.Y., 1902. 2v.
	Geog 4169.04	Autour du monde. Paris, 1904.
	Geog 4169.12	Boer, M.G. de. Van oude voyagien. Amsterdam, 1912-13. 3v.
	Geog 4169.12.5	Boer, M.G. de. Van oude voyagien. 3. druk. Amsterdam, 1939.
	Geog 4169.19A	Newbolt, Henry. The book of the long trail. N.Y., 1919.
	Geog 4169.19B	Newbolt, Henry. The book of the long trail. N.Y., 1919.
	Geog 4169.28	Chatterton, E.K. Ventures and voyages. London, 1928.
	Geog 4169.30	Adler, Elkan N. Jewish travellers. London, 1930.
	Geog 4169.31	Dulles, F.R. Eastward ho! London, 1931.
	Geog 4169.31.2	Dulles, F.R. Eastward ho! The 1st English adventures to the Orient. Boston, 1931.
	Geog 4169.31.15	Blossom, F.A. Told at the Explorers Club. N.Y., 1931.
	Geog 4169.33	Ward, Edward. Five travel scripts commonly attributed to Edward Ward. N.Y., 1933.
	Geog 4169.36	Hennig, R. Terrae incognitae. Leiden, 1936-39. 4v.
	Geog 4169.36.3	Hennig, R. Terrae incognitae. Leiden, 1944- 4v.
	Geog 4169.36.5A	Ingram, B.S. Three sea journals of Stuart times. London, 1936.
	Geog 4169.36.5B	Ingram, B.S. Three sea journals of Stuart times. London, 1936.
	Geog 4169.38	Compton, Ray. The open road. N.Y., 1938.
	Geog 4169.41	Explorers Club, N.Y. Thru hell and high water. N.Y., 1941.
	Geog 4169.45	Bailey, Leslie. Travellers' tales, a series of BBC programmes broadcast throughout the world. London, 1945.

Geog 4165 - 4170   Voyages and travels in general - Collections of travels - Smaller works (By date) - cont.

Geog 4169.45.5	Stood, Frederick T. Modern travel. London, 1946.
Geog 4169.47	Stefansson, V. Great adventures and explorations. N.Y., 1947.
Geog 4169.47.5	Ley, Charles D. Portuguese voyages, 1498-1663. London, 1947.
Geog 4169.51	Van Thal, H. Victoria's subjects travelled. London, 1951.
Geog 4169.55	Cooper, G. Forbidden lands. N.Y., 1955.
Geog 4169.59	Dillon, Richard. Embarcadero. N.Y., 1959.
Geog 4169.59.5	Spagnol, Mario. Avventure e viaggi di mare. Milano, 1959.
Geog 4169.62	Vincenti, Leonello. Viaggiatori del Settecento. Torino, 1962.
Geog 4169.64	Jahn, Janheinz. Wir nannten sie Wilde. München, 1964.

Geog 4179   Voyages and travels in general - Collections of travels - Hakluyt Society - Miscellany by and about the Society

Geog 4179.5	Markham, C.R. Richard Hakluyt: his life and work. London, 1896.
Geog 4179.5.5	Markham, C.R. Address...on fiftieth anniversary. London, 1911.
Geog 4179.7	Hakluyt Society, London. Prospectus and list of members. London. 1850-1918 2v.
Geog 4179.10	Hakluyt Society, London. A list of the publications of the Hakluyt Society. London, 1946.

Geog 4180   Voyages and travels in general - Collections of travels - Hakluyt Society - Publications, Series I

	Geog 4180.1	Hawkins, R. Observations...in voyage...South Sea. London, 1847.
	Geog 4180.2	Columbus, C. Letters...four voyages to new world. London, 1847.
	Geog 4180.3	Raleigh, W. Discovery of...empire of Guiana. London, 1848.
	Geog 4180.4	Maynarde, T. Sir Francis Drake, his voyage 1595. London, 1849.
	Geog 4180.5	Rundall, T. Narratives of voyages...north-west. London, 1849.
	Geog 4180.6	Strachey, W. Historie of travaile. London, 1849.
	Geog 4180.7A	Hackluyt, R. Divers voyages...discovery of America. London, 1850.
	Geog 4180.7B	Hackluyt, R. Divers voyages...discovery of America. London, 1850.
	Geog 4180.8	Rundall, T. Memorials of the empire of Japan. N.Y., 1964?
	Geog 4180.9	Relacam Verdadeira dos Trabalhos. Discovery...of Terra Florida by De Soto. London, 1851.
	Geog 4180.10A	Herberstein, Sigmund. Notes upon Russia. N.Y., 1963? 2v.
	Geog 4180.10B	Herberstein, Sigmund. Notes upon Russia. N.Y., 1963? 2v.
X Cg	Geog 4180.10	Herberstein, Sigmund. Notes upon Russia. London, 1851-52. 2v.
	Geog 4180.11	Coats, W. Geography of Hudson's Bay. London, 1852.
	Geog 4180.13	Veer, G. de. True description...voyages by the North-East. London, 1852-53.
	Geog 4180.14	Gonzalez de Mendoza, J. History of...kingdom of China. London, 1853-54. 2v.
	Geog 4180.16	Feltcher, F. The world encompassed by Francis Drake. N.Y., 1964?
	Geog 4180.17	Orleans, P.J. d'. History of...tartar conquerors of China. N.Y., 1964?
	Geog 4180.18	Martens, F. Collection of documents on Spitzbergen and Greenland. London, 1855.
	Geog 4180.19	Middleton, H. Voyage...to Bantum and Maluco Islands. London, 1855.
	Geog 4180.20	Fletcher, G. Russia at close of sixteenth century. N.Y., 1964?
	Geog 4180.21	Benzoni, G. History of the new world. N.Y., 1964?
	Geog 4180.22	Major, R.H. India in the fifteenth century. London, 1857.
	Geog 4180.23	Champlain, S. Narrative of voyages to West Indies and Mexico. London, 1859.
	Geog 4180.24	Markham, C.R. Expeditions into valley of the Amazons. N.Y., 1964.
	Geog 4180.25	Major, R.H. Early voyages to Terra Australis. London, 1859.
	Geog 4180.26	Clavijo, R.G. de. Narrative of embassy...to court of Timar. London, 1859.
	Geog 4180.27A	Asher, G.M. Henry Hudson the navigator. London, 1860.
	Geog 4180.27B	Asher, G.M. Henry Hudson the navigator. London, 1860.
	Geog 4180.28	Simon, P. Expedition of...Ursua and...Aguirre. N.Y., 1964?
	Geog 4180.29	Guzman, A.E. de. Life and acts of...Guzman. N.Y., 1970.
	Geog 4180.30A	Galvano, A. Discoveries of the world. London, 1862.
	Geog 4180.30B	Galvano, A. Discoveries of the world. London, 1862.
	Geog 4180.31	Jordanus de Saxonia. Mirabilia descripta...wonders of the East. N.Y., 1964?
	Geog 4180.32	Varthema, S. di. Travels of Varthema. N.Y., 1964?
	Geog 4180.33	Cieza de Leon, P. de. Travels of Piedro Cieza de Leon. London, 1864.
	Geog 4180.34	Andagoya, P. de. Narrative of proceedings of Pedrarias Davila. London, 1865.
	Geog 4180.35	Barbosa, D. Description of...East Africa and Malabar. London, 1866.
	Geog 4180.36	Yule, H. Cathay and the way thether. London, 1866.
	Geog 4180.38	Best, G. Three voyages of Martin Frobisher. London, 1867.
	Geog 4180.39	Morga, A. de. Phillipine Islands, Moluccas. London, 1868.
	Geog 4180.40	Cortes, H. Fifth letter of...Cortes to Emperor Charles V. London, 1868.
	Geog 4180.41	Vega, G. de la. First part of royal commentaries of the Yncas. N.Y., 1964. 2v.
	Geog 4180.42	Correa, G. Three voyages of Vasco da Gama. Photoreproduction. London, 1869.
	Geog 4180.43	Columbus, C. Letters...four voyages to the new world. London, 1870.
	Geog 4180.44	Salîl-Ibn-Razîk. History of the Inâms...of Omân. N.Y., 1964?
	Geog 4180.46	Bontier, P. The Canarien...conquest and conversion. N.Y., 1964?
	Geog 4180.47	Markham, C.R. Reports on the discovery of Peru. N.Y., 1964.
	Geog 4180.48	Markham, C.R. Narrative of the rites and laws of the Yncas. N.Y., 1964?
	Geog 4180.49	Barbaro, J. Travels to Tana and Persia. N.Y., 1964?
	Geog 4180.50	Zeno, N. Voyages of...Nicolo and Antonio Zeno. London, 1873.
	Geog 4180.51	Stade, H. Captivity of Hans Stade of Hesse. London, 1874.

Classified Listing

Geog 4180 Voyages and travels in general - Collections of travels - Hakluyt Society - Publications, Series I - cont.

	Geog 4180.52A	Stanley, E.J. First voyage round the world by Magellan. London, 1874.
	Geog 4180.52B	Stanley, E.J. First voyage round the world by Magellan. London, 1874.
	Geog 4180.53	Alboquerque, A. d'. Commentaries of Dalboquerque. N.Y., 1964?
	Geog 4180.54	Veer, G. de. Three voyages of...William Barents. London, 1876.
	Geog 4180.56	Markham, C.R. Voyages of Sir James Lancaster. N.Y., 1964?
	Geog 4180.57	Markham, C.R. The Hawkins voyages. London, 1878.
	Geog 4180.58A	Schiltberger, J. Bondage and travels of J. Schiltberger. London, 1879.
	Geog 4180.58B	Schiltberger, J. Bondage and travels of J. Schiltberger. London, 1879.
	Geog 4180.59	Davis, J. Voyages and works of J. Davis. London, 1880.
	Geog 4180.59.3	Maps (1600). Map of the world A.D. 1600. London, 1880.
	Geog 4180.60	Acosta, J. de. National and moral history of the Indies. v.2. London, 1880.
	Geog 4180.60.2	Acosta, J. de. National and moral history of the Indies. N.Y., 1964? 2v.
	Geog 4180.63	Baffin, W. Voyages of William Baffin 1612-22. N.Y., 1964?
	Geog 4180.64	Alvares, F. Narrative of Portuguese Embassy. London, 1881.
	Geog 4180.65	Butler, N. Historye of the Bermudaes. London, 1882.
	Geog 4180.66	Cocks, R. Diary of Richard Cocks...1615-22. N.Y., 1964. 2v.
	Geog 4180.68	Cieza de Leon, P. de. Second part of chronicle of Peru. N.Y., 1964?
	Geog 4180.70	Linschoten, J.H. Voyage of Linschoten to East Indies. N.Y., 1964? 2v.
X Cg	Geog 4180.70	Linschoten, J.H. Voyage of Linschoten to East Indies. London, 1885. 2v.
	Geog 4180.72	Jenkinson, A. Early voyages...to Russia and Persia. N.Y., 1964? 2v.
	Geog 4180.74	Hedges, W. Diary of William Hedges...during agency in Bengal. London, 1887-89. 3v.
	Geog 4180.76	Pyrard, F. Voyage of F. Pyrard...to East Indies. v.1-2. N.Y., 1964? 3v.
NEDL	Geog 4180.76	Pyrard, F. Voyage of F. Pyrard...to East Indies. v.1-2. London, 1887. 3v.
	Geog 4180.79	Hues, R. Tractatus de globis et eorum usu. London, 1889.
	Geog 4180.81	Dominguez, L.L. Conquest of the river plate. London, 1891.
	Geog 4180.82	Leguat, F. Voyage of F. Leguat...to Rodriguez. London, 1891. 2v.
	Geog 4180.84A	Valle, P. della. Travels of...Valle in India. London, 1892. 2v.
	Geog 4180.84B	Valle, P. della. Travels of...Valle in India. London, 1892. 2v.
	Geog 4180.86A	Markham, C.R. Journal of Christopher Columbus. London, 1893.
	Geog 4180.86B	Markham, C.R. Journal of Christopher Columbus. London, 1893.
	Geog 4180.87A	Bent, J.T. Early voyages and travels in the Levant. London, 1893.
	Geog 4180.87B	Bent, J.T. Early voyages and travels in the Levant. London, 1893.
	Geog 4180.88	Christy, M. Voyages of Captain Luke Foxe and Captain T. James. London, 1894. 2v.
	Geog 4180.90	Markham, C.R. Letters of Amerigo Vespucci. London, 1894.
	Geog 4180.91	Sarmiento de Gamboa, Pedro. Narrative of voyages of...Sarmiento. London, 1895.
	Geog 4180.92	Leo Africanus, J. History and description of Africa. London, 1896. 3v.
	Geog 4180.92.2	Leo Africanus, J. History and description of Africa. N.Y., 1964? 3v.
	Geog 4180.95	Azurara, G.E. de. Chronicle of discovery...of Guinea. London, 1896-99. 2v.
	Geog 4180.96A	Gosch, C.C.A. Danish Arctic expeditions, 1605-20. London, 1897.
	Geog 4180.96B	Gosch, C.C.A. Danish Arctic expeditions, 1605-20. London, 1897.
	Geog 4180.98A	Cosmas. Christian topography of Cosmas. London, 1897.
	Geog 4180.98B	Cosmas. Christian topography of Cosmas. London, 1897.
	Geog 4180.99A	Gama, V. da. Journal of first voyage of V. da Gama. London, 1898.
	Geog 4180.99B	Gama, V. da. Journal of first voyage of V. da Gama. London, 1898.

Geog 4181 Voyages and travels in general - Collections of travels - Hakluyt Society - Publications, Series II

Geog 4181.1	Roe, T. Embassy of Sir T. Roe to court of great mogul. London, 1899. 2v.
Geog 4181.3	Dudley, R. Voyage of R. Dudley...to West Indies. London, 1899.
Geog 4181.4	Rubruquis, G. de. Journey of William of Rubruck...and John of Pian de Carpine. London, 1900.
Geog 4181.5	Saris, J. Voyage of Captain John Saris to Japan, 1613. London, 1900.
Geog 4181.6	Battell, A. Strange adventures of A. Battell. London, 1901.
Geog 4181.7	Amherst of Hackney. Discovery of Solomon Islands by...Mendaña. London, 1901. 2v.
Geog 4181.9	Teixeira, P. Travels of Teixeira with his "kings of Harmuz". London, 1902.
Geog 4181.10	Castanhoso. Portugese expedition to Abyssinia. London, 1902.
Geog 4181.11	Conway, W.M. Early Dutch and English voyages. London, 1904.
Geog 4181.12	Bowrey, T. Geographical account...countries...Bengal. Cambridge, 1905.
Geog 4181.13	Corney, B.G. Voyage of Captain Don Felippe Gonzalez. Cambridge, 1908.
Geog 4181.14	Belmonte Bermudez, L. de. Voyages of Pedro Fernandez de Quiros. London, 1904. 2v.
Geog 4181.16	Jourdain, J. Journal of John Jourdain, 1608-17. Cambridge, 1905.
Geog 4181.17	Mundy, P. Travels of Peter Mundy in Europe and Asia, 1608-67. v.1-5. Cambridge, 1907-36. 6v.
Geog 4181.18	Spillbergen, J. van. East and West Indian mirror. London, 1906.
Geog 4181.19	Fryer, J. New account of East India and Persia. London, 1909.
Geog 4181.21	Espinosa, A. de. The guanches of Tenerife. London, 1907.
Geog 4181.22A	Sarmiento de Gamboa, Pedro. History of the Incas and execution of the Inca Tupac Amaru. Cambridge, 1907.

Geog 4181 Voyages and travels in general - Collections of travels - Hakluyt Society - Publications, Series II - cont.

Geog 4181.22B	Sarmiento de Gamboa, Pedro. History of the Incas and execution of the Inca Tupac Amaru. Cambridge, 1907.
Geog 4181.23A	Diaz del Castillo, B. True history of conquest of New Spain. London, 1908-10. 5v.
Geog 4181.23B	Diaz del Castillo, B. True history of conquest of New Spain. London, 1908-10. 4v.
Geog 4181.23C	Diaz del Castillo, B. True history of conquest of New Spain. London, 1908.
Geog 4181.26	Storm van 's Gravesande, L. Rise of British Guiana. London, 1911. 2v.
Geog 4181.28	Markham, C.R. Early Spanish voyages to the Strait of Magellan. v.28. London, 1911.
Geog 4181.29	Markham, C.R. Book of knowledge of all kingdoms, lands. London, 1912.
Geog 4181.31	Cieza del Leon, P. de. The war of Quito. London, 1913.
Geog 4181.32	Corney, B.G. The quest and occupation of Tahiti by emissaries of Spain 1772-76. London, 1913-19. 3v.
Geog 4181.33	Yule, Henry. Cathay and the way thither. London, 1913-14. 4v.
Geog 4181.33.1	Yule, Henry. Cathay and the way thither. v.1-4. Taipei, 1966. 2v.
Geog 4181.34	Mittall, Zelia. New light on Drake. London, 1914.
Geog 4181.35	Cieza de Leon, P. de. The war of Chupas. London, 1918.
Geog 4181.36	Barbosa, Duarte. The book. London, 1918-21. 2v.
Geog 4181.48	Montesinos, Fernando. Memorias antiguas historiales del Peru. London, 1920.
Geog 4181.51	Edmundson, George. Journal of...Father Samuel Fritz. London, 1922.
Geog 4181.52	Lockerby, William. The journal of William Lockerby. London, 1925.
Geog 4181.53	Life of the Icelander Jón Ólafsson. London, 1923-31. 2v.
Geog 4181.54	Cieza de Leon, P. de. The war of Las Salinas. London, 1823.
Geog 4181.56	Harlow, V.T. Colonizing expeditions to the West Indies and Guiana, 1623-67. London, 1925.
Geog 4181.57	Mortoft, F. Francis Mortoft; his book...1658-59. London, 1925.
Geog 4181.58	Bowry, T. The papers. London, 1927.
Geog 4181.59A	Luard, C.E. Travels of Fray Sebastian Manrique. London, 1927. 2v.
Geog 4181.59B	Luard, C.E. Travels of Fray Sebastien Manrique. v.2. London, 1927.
Geog 4181.60	Farcourt, R. Relation of a voyage to Guiana. London, 1928.
Geog 4181.62	Wright, L.A. Spanish documents concerning English voyages to the Caribbean. n.p., 1929.
Geog 4181.63	Carruthers, D. The desert route to India. London, 1929.
Geog 4181.64	Stevens, H.N. New light on the discovery of Australia. London, 1930.
Geog 4181.65	Jane, Cecil. Select documents illustrating the 4 voyages of Columbus. London, 1930-33. 2v.
Geog 4181.66	Moreland, W.H. Relations of Golconda in the early 17th century. London, 1931.
Geog 4181.67	Foster, William. Travels of John Sanderson in the Levant. London, 1931.
Geog 4181.69	Barlow, Roger. A brief summe of geographie. London, 1932.
Geog 4181.71	Wright, I.A. Documents concerning English voyages to the Spanish Main, 1569-1580. London, 1933.
Geog 4181.72	Burnell, John. Bombay in the days of Queen Anne. London, 1933.
Geog 4181.73	Joyce, L.E. Elliott. A new voyage and description of the Isthmus of America. Oxford, 1934.
Geog 4181.74	Moreland, W.H. Peter Floris. London, 1934.
Geog 4181.75	Foster, William. The voyage of Thomas Best to the East Indies. London, 1934.
Geog 4181.76	Taylor, E.G.R. The original writings and correspondance of the two Richard Hakluyts. London, 1935. 2v.
Geog 4181.79	Pacheco Pereira, Duarte. Esmeraldo de situ orbis. London, 1937.
Geog 4181.80	Crone, G.R. The voyages of Cadamosto. London, 1937.
Geog 4181.81	Greenlee, William B. The voyage of Pedro Alvares Cabral to Brazil and India. London, 1938.
Geog 4181.82	Foster, William. The voyage of Nicholas Downton to the East Indies, 1614-15. London, 1939.
Geog 4181.83	Quinn, David B. The voyages...of Sir Humphrey Gilbert. London, 1940. 2v.
Geog 4181.85	Foster, William. The voyages of Sir James Lancaster to Brazil and the East Indies, 1591-1603. London, 1940.
Geog 4181.86A	Blake, John W. Europeans in West Africa, 1450-1560. London, 1942. 2v.
Geog 4181.86B	Blake, John W. Europeans in West Africa, 1450-1560. London, 1942. 2v.
Geog 4181.88	Foster, William. The voyage of Sir Henry Middleton to the Moluccas. London, 1943.
Geog 4181.89	Cortesão, Armando. Suma Oriental of Tomé Pires...book of Francisco Rodrigues. London, 1944. 2v.
Geog 4181.91	Debenham, Frank. Voyage of Captain Bellingshausen to Antarctic seas, 1819-21. London, n.d. 2v.
Geog 4181.93	Lynam, E. Richard Hakluyt and his successors. London, 1946.
Geog 4181.94	Harff, Arnold. The pilgrimage of Arnold von Harff. London, 1946.
Geog 4181.95	Carre. The travels of the Abbé Carre in India and the Near East. London, 1947-48. 3v.
Geog 4181.98A	Robertson, George. The discovery of Tahiti. London, 1948.
Geog 4181.98B	Robertson, George. The discovery of Tahiti. London, 1948.
Geog 4181.99	Spain. Archivo General de Indias, Seville. Further English voyages to Spanish America. London, 1951.
Geog 4181.100	Foster, William. The Red Sea and adjacent countries of the close of the 17th century as described by Joseph Pitts. London, 1949.
Geog 4181.101	Mandeville, J. Mandeville's travels. London, 1953. 2v.
Geog 4181.103	Strachey, William. The historie of travel into Virginia Britania. London, 1953.
Geog 4181.104	Quinn, D.B. The Roanoke voyages, 1584-1590. London, 1955. 2v.
Geog 4181.106	Boxer, C.R. South China in the sixteenth century. London, 1953.
Geog 4181.107	Beckingham, C.F. Some records of Ethiopia. London, 1954.
Geog 4181.108	Letts, M.H.I. The travels of Leo of Rozmital through Germany. Cambridge, 1957.
Geog 4181.109	Crawford, O.G.S. Ethiopian itineraries circa 1400-1524. Cambridge, Eng., 1958.

Classified Listing

Geog 4181 Voyages and travels in general - Collections of travels - Hakluyt Society - Publications, Series II - cont.

Geog 4181.110A Ibn Batuta. The travels of Ibn Battuta. Cambridge, 1958. 2v.
Geog 4181.110B Ibn Batuta. The travels of Ibn Battuta. Cambridge, 1958. 2v.
Geog 4181.111 Andrews, Kenneth. English privateering voyages to the West Indies. Cambridge, 1959.
Geog 4181.112A Gomes de Brito, Bernardo. The tragic history of the sea. Cambridge, Eng., 1959.
Geog 4181.112B Gomes de Brito, Bernardo. The tragic history of the sea. Cambridge, Eng., 1959.
Geog 4181.113 Taylor, Eva. The troublesome voyage of Captain Edward Fenton. Cambridge, Eng., 1959.
Geog 4181.114 Alvares, Francisco. The Prester John of the Indies. Cambridge, 1961. 2v.
Geog 4181.116 Davies, J. The history of the Tahitian Mission, 1799-1830. Cambridge, Eng., 1961.
Geog 4181.118 Navarrete, Domingo Fernández de. The travels and controversies of Friar Domingo Navarrete. Cambridge, 1962. 2v.
Geog 4181.120 Williamson, J.A. The Cabot voyages and Bristol discovery under Henry VII. Cambridge, 1962.
Geog 4181.121 Bourne, William. A regiment for the sea. Cambridge, Eng., 1963.
Geog 4181.122 Byron, John. Byron's journal of his circum navigation, 1764-1766. Cambridge, 1964.
Geog 4181.123 Bovill, E.W. Missions to the Niger. Cambridge, Eng., 1964. 4v.
Geog 4181.124 Carteret, Philip. Carteret's voyage round the world, 1766-69. Cambridge, 1965. 2v.
Geog 4181.126 Kelly, Celsus. La Austrialia del Espiritu Santo. Cambridge, 1966. 2v.
Geog 4181.131 Bishop, Charles. The journal and letters of Captain Charles Bishop on the north-west of America. Cambridge, 1967.
Geog 4181.132 Gomes de Brito, Bernardo. Further selections from the tragic history of the sea, 1559-1565. Cambridge, 1968.
Geog 4181.133 Leichhardt, Ludwig. The letters of F.W. Ludwig Leichhardt. Cambridge, Eng., 1968. 3v.
Geog 4181.136 Barbour, Philip L. The Jamestown voyages under the first charter, 1606-1609. Cambridge, Eng., 1969. 2v.
Geog 4181.138 Allen, William Edward David. Russian embassies to the Georgian kings, 1589-1605. Cambridge, 1970. 2v.
Geog 4181.140 Morga, Antonio de. Sucesos de las Islas Filipinas. Cambridge, Eng., 1971.
Geog 4181.142A Andrews, Kenneth Raymond. The last voyage of Drake and Hawkins. Cambridge, 1972.
Geog 4181.142B Andrews, Kenneth Raymond. The last voyage of Drake and Hawkins. Cambridge, 1972.
Geog 4181.143 Peard, George. To the Pacific and Arctic with Beechey. Cambridge, 1973.

Geog 4182 Voyages and travels in general - Collections of travels - Linschoten Vereeniging Werken

Geog 4182.05 Linschoten. Vereeniging. Werken. Register. v.1-25, 26-50. 's-Gravenhage, 1939-57. 2v.
Geog 4182.1 Muller, S. De reis van Jan C. May. 's-Gravenhage, 1909.
Geog 4182.2 Linschoten, Jan H. van. Itinerario voyage. 's-Gravenhage, 1910. 5v.
Geog 4182.3 Colenbrander, H.T. Korte historiael...voyagiens...David P. de Vries. 's-Gravenhage, 1911.
Geog 4182.4 Mulert, F.E. De reis van Mr. Jacob Roggeveen. 's-Gravenhage, 1911.
Geog 4182.5 Marees, P. de. Beschrijving...van het Gout Koninckrijek van Gunea. 's-Gravenhage, 1912.
Geog 4182.6 Naber, S.P. Toortse der zee-vaert...D. Ruiters 1623 en Samuel Brun's schiffarten (1624). 's-Gravenhage, 1913.
Geog 4182.7 Rouffaer, G.P. De eerste schipvaart der nederlanders naar Ost-Indie. 's-Gravenhage, 1915-29. 3v.
Geog 4182.8 Linschoten, Jan H. van. Reizen van...naar het noorden 1594-95. 's-Gravenhage, 1914.
Geog 4182.9 Ijzermann, J.W. Dirck Gerritsz Pomp. 's-Gravenhage, 1915.
Geog 4182.10 Rogerius, Abraham. De open-deure tot het verborgen heydendom. 's-Gravenhagen, 1915.
Geog 4182.11 Godeé Molsbergen, E.C. Reizen in Zuid-Afrika. 's-Gravenhage, 1916-32. 4v.
Geog 4182.12 Ottsen, Hendrick. Journael van de reis naar Zuid-Amerika. 's-Gravenhage, 1918.
Geog 4182.14 Hamel, Hendrik. Verhaal van het vergaan van het jacht De Sperwer. 's-Gravenhage, 1920.
Geog 4182.15 Ledyard, Gari. The Dutch come to Korea. Seoul, 1971.
Geog 4182.19 Juet, Robert. Henry Hudson's reize onder Nederlander vlag. 's-Gravenhage, 1921.
Geog 4182.21 Wieder, F.C. De reis van Mahu en de Cordes. 's-Gravenhage, 1923-25. 3v.
Geog 4182.23 Naber, S.P. Hessel Gerritsz, samoyeden land ten Spitsberghe. 's-Gravenhage, 1924.
Geog 4182.26 Wieder, F.C. Die stichting van New York in Juli 1625. 's-Gravenhage, 1925.
Geog 4182.27 Ijzermann, J.W. De reis om de wereld door Olivier van Noort. 's-Gravehage, 1926. 2v.
Geog 4182.29 Van Nouhuys, J.W. De eerste nederlandsche. 's-Gravenhage, 1927-51. 2v.
Geog 4182.30 Woard, C. de. Zeeuwsche expedite...Cornelis Evertsen. 's-Gravenhage, 1928.
Geog 4182.31 Caland, W. Die remonstrantie von W. Geleynssen de Jongh. 's-Gravenhage, 1929.
Geog 4182.33 Warnsinck, J.E.M. Reisen van Nicolaus de Graaff. 's-Gravenhage, 1930.
Geog 4182.34 Naber, S.P. Johannes de Laet. Iaerlyck Verhael van der verichtinghen der...Compagnie. 's-Gravenhage, 1931-37. 4v.
Geog 4182.38 De reis van Voris van Spilbergen naar Ceylon, Atjeh en Bantam, 1601-1604. 's-Gravenhage, 1933.
Geog 4182.41 Vogel, J.P. Journal van J.J. Ketelaar's hofreis...1711-1713. 's-Gravenhage, 1937.
Geog 4182.42 Keuning, J. De tweede schipvaart der Nederlanders. v.1-5. 's-Gravenhage, 1938-51. 8v.
Geog 4182.45 Adrichem, D. van. Journaal van Dircq Van Adrichem hofreis. 's-Gravenhage, 1941.
Geog 4182.47 Spilbergen, J. van. De reis om de wereld. 's-Gravenhage, 1943. 2v.
Geog 4182.49 LeMaire, Jacob. De ontdekkingsreis van Jacob le Maire. 's-Gravenhage, 1945. 2v.
Geog 4182.51 Unger, W. De oudste reizen van de Zieuwen naar Oost-Indie. 's-Gravenhage, 1948.

Geog 4182 Voyages and travels in general - Collections of travels - Linschoten Vereeniging Werken - cont.

Geog 4182.52 Broecke, Pieter van den. Reizen naar West-Afrika van Pieter van den Broecke. 's-Gravenhage, 1950.
Geog 4182.54 Bontekae, W.Y. Journalen van de gedenckenaerdige reijsen. 's-Gravenhage, 1952.
Geog 4182.55 Ratelband, K. Vijf dagregisters van het Kasteel São Jorge da Maria. 's-Gravenhage, 1953.
Geog 4182.56 Quast, Mathijs H. De reis van Mathijs Hendriksz. 's-Gravenhage, 1954.
Geog 4182.57 Linschoten, Jan H. van. Itinerario. 's-Gravenhage, 1955-57. 3v.
Geog 4182.59 Goens, Rijklof van. De vijf gezantschapsreizen van Rijklof van Goens naar Hethof van Mataram. 's-Gravenhage, 1956.
Geog 4182.61 Hein, Pieter P. De Westafrikaanse reis van Piet Heyn. 's-Gravenhage, 1959.
Geog 4182.62 Ruyter, M.A. De reis van Michiel Adriaanszoom de Ruyter in 1664-1665. 's-Gravenhage, 1961.
Geog 4182.63 Broecke, Pieter van den. Pieter van den Broecke in Azie. 's-Gravenhage, 1962. 2v.
Geog 4182.65 De reis om de wereld van de nassausche vloot, 1623-1926. 's-Gravenhage, 1964.
Geog 4182.66 Witsen, Nicolaas. Moscovische reyse 1664-1665. 's-Gravenhage, 1966-67. 3v.
Geog 4182.69 Schagen, Adriaen. Reijse gadaen bij Adriaen Schagen. 's-Gravenhage, 1968.
Geog 4182.70 Booy, A. de. De derde reis van de V.O.C. naar Oost-Indië onder het beleid van Admiraal van Caerden. 's-Gravenhage, 1968.
Geog 4182.72 Briel, Johan Jurgen. De expeditie van Anthonio Hurdt. 's-Gravenhage, 1971.
Geog 4182.75 Pijnacker, Cornelis. Historysch verhael van den steden Thunes. 's-Gravenhage, 1975.
Geog 4182.76 Kreekel, Willem. De reis van Z.M. De Vlieg. 's-Gravenhage, 1975.

Geog 4184 Voyages and travels in general - Collections of travels - Hakluyt Society Publications, Extra Series

Geog 4184.1A Hakluyt, R. Principal navigations...voyages traffiques. Glasgow, 1903-05. 12v.
Geog 4184.1B Hakluyt, R. Principal navigations...voyages traffiques. v.8. Glasgow, 1903-05.
Geog 4184.13 Hakluyt, R. Texts and versions of...Carpini and...Rubruquis. London, 1903.
Geog 4184.14A Purchas, Samuel. Hakluytus posthumus, or Purchas, his pilgrimes. Glasgow, 1905-07. 20v.
Geog 4184.14B Purchas, Samuel. Hakluytus posthumus, or Purchas, his pilgrimes. Glasgow, 1905-07. 20v.
Geog 4184.14C Purchas, Samuel. Hakluytus posthumus, or Purchas, his pilgrimes. v.16,18. Glasgow, 1905-07. 2v.

Geog 4205 - 4210 Voyages and travels in general - Individual travels - Miscellaneous travels (By date of travel)

Geog 4205.03 Avezac, A. de. Campagne du navire l'Espoir de Honfleur, 1503-1505. Paris, 1869.
Geog 4205.96.1 Ultzheimer, Andreas J. Warhaffte Beschreibung ettlicher Reisen in Europa, Africa, Asien und America 1596-1610. Tübingen, 1971.
Geog 4205.96.9F Linschoten, Jan H. van. Itinerarium oste schip-vaert naer. Amsterdam, 1644.
Htn Geog 4205.98F* Linschoten, Jan H. van. Discours of voyages...East and West Indies. London, 1598.
Htn Geog 4205.98.2F* Linschoten, Jan H. van. Discours of voyages...East and West Indies. London, 1598.
Geog 4205.98.10 Nijhoff, Wouter. Bibliographie...die voyagie om den geheelen werelt...door Olivier van Noort. 's-Gravenhage, 1926.
Htn Geog 4205.99F* Linschoten, Jan H. van. Navigatio ac itinerarium. Hagae-Comitis, 1599.
Htn Geog 4206.10.5* Linschoten, Jan H. van. Histoire de la navigation. 2e éd. v.1-3. Amsterdam, 1619.
Htn Geog 4206.10.6F* Linschoten, Jan H. van. Histoire de la navigation. Amsterdam, 1638.
Htn Geog 4206.13F* Purchas, Samuel. Purchas, his pilgrimage. London, 1613.
Htn Geog 4206.13.5F* Purchas, Samuel. His pilgrimage, or Relations of the world. 3rd ed. London, 1617.
Htn Geog 4206.13.9F* Purchas, Samuel. His pilgrimes. London, 1625- 4v.
Geog 4206.13.12 Purchas, Samuel. Samuel Purchas pelgrimagir. Amsterdam, 1655.
Geog 4206.13.25 Amman, H.J. Reiss ins gelobte Land...Servian...Aegypten. Zürich, 1630.
Geog 4206.13.27 Amman, H.J. Reiss in das gelobte Land. Berlegung, 1678.
Geog 4206.13.30F Amman, H.J. Hans Jakob Ammann genannt der Thalwyler Schärer und seineReise ins gelobte Land. Zürich, 1919.
Geog 4206.13.35 Purchas, Samuel. Narratives from Purchas, his pilgrimes. Cambridge, 1931.
Htn Geog 4206.43.5* La Boullaye, C. Goux. Les voyages et observations. Paris, 1653.
Htn Geog 4206.45* Mocquet, J. Voyages en Afrique, Asie. Rouen, 1645.
Geog 4206.45.5 Mocquet, J. Travels and voyages into Africa. London, 1696.
Geog 4206.54 Bäckhoff. Anhang zwer Reisen. Berlin, n.d.
Geog 4206.58 Welsch, Hier. Warhafftige Reiss-Beschreibung. v.1-2. Stuttgart, 1664.
Htn Geog 4206.59* Account of several late voyages and discoveries to south and north. London, 1694.
Geog 4206.59.5 Account of several late voyages and discoveries to north and south. London, 1711.
Htn Geog 4206.60* LeBlanc, V. The world surveyed. London, 1660.
Geog 4206.62 Valle, Pietro. Les fameux voyages de Pietro della Valle. Paris, 1663-70. 4v.
Htn Geog 4206.65* Desboys du Chastelet, R. L'odyssee ou diversite d'avantures...en Europe, Asie et Afrique. Fleche, 1665.
Geog 4206.65.5 Piolin, Paul. René Desboys du Chastelet. n.p., 1885.
Geog 4206.65.9 Journal des voyages de Monsieur de Monconys. Lyon, n.d.
Geog 4206.65.11 Les voyages de Balthasar de Monconys. Paris, 1887.
Geog 4206.81 Melton, E. (pseud.). Eduward Meltons, engelsch edelmans, zeldzaame en gedenkwaardige zee- en land reizen. Amsterdam, 1681.
Htn Geog 4206.82* Glanius. A new voyage to the East Indies. 2d ed. London, 1682.
Htn Geog 4206.99* Hacke, William. A collection of original voyages. London, 1699.
Geog 4207.41 Kuhns, J.M. Lebens und Reise Beschreibung. Gotha, 1741.
Geog 4207.53.3 Hanway, Jonas. Reize van London, door Rusland, nae en in Persie. Amsterdam, 1758. 2v.
Geog 4207.53.5F Hanway, Jonas. Historical account of British trade. 3. ed. London, 1762. 2v.

Classified Listing

Geog 4205 - 4210 Voyages and travels in general - Individual travels - Miscellaneous travels (By date of travel) - cont.

	Geog 4207.55	Morris, Drake. Travels of...voyage at sea. London, 1775.
	Geog 4207.55.5	Brelin, Johan. En äfventyrlig resa til och ifrån Ost-Indien, Södra America och en del af Europa åren 1755, 56 och 57. Stockholm, 1973.
	Geog 4207.68	Viaggi da Parma in varie partie del mondo. Parma, 1768.
	Geog 4207.69.5	Wallenberg, Jacob. Min son på galejan. Stockholm, 1835.
	Geog 4207.69.9	Wallenberg, Jacob. Min son på galejan. 2. uppl. Stockholm, 1913.
	Geog 4207.76A	Sparks, Jared. The life of John Ledyard. Cambridge, 1828.
	Geog 4207.76B	Sparks, Jared. The life of John Ledyard. Cambridge, 1828.
	Geog 4207.76.5	Sparks, Jared. The life of John Ledyard. 2. ed. Cambridge, 1829.
	Geog 4207.76.9	Sparks, Jared. Life of John Ledyard...American traveller. Boston, 1847.
	Geog 4207.76.10	Sparks, Jared. Life of John Ledyard, the American traveller. Boston, 1864.
	Geog 4207.76.12	Sparks, Jared. Leben des...Americanischen Reisenden J. Ledyard. Leipzig, 1829.
	Geog 4207.76.15	Munford, K. John Ledyard: an American Marco Polo. Portland, 1939.
	Geog 4207.76.20	Augur, Helen. Passage to glory; John Ledyard's America. 1. ed. Garden City, 1946.
	Geog 4207.79	Evers. Journal kept on a journey from Bassora to Bagdad...1779. Horsham, 1784.
	Geog 4207.80.2	Baranshchikov, V. Neshchastiia prikliucheniia...Amerike, Azii i Evrope s" 1780 ro 1787 god". 2. izd. Sankt Peterburg, 1787.
	Geog 4207.99	Nevens, William. Forty years at sea. Portland, 1846.
	Geog 4208.06	General history of voyages and travels. Glasgow, 1806.
	Geog 4208.10	Swaving, J.G. Swavings reizen en logevallen, doorhem. v.1-2. Dordrecht, 1827.
	Geog 4208.12	Bacardí Moreau, E. Hacia tierras viejas. Valencia, 1913.
	Geog 4208.13	Dunham, Jacob. Journal of voyages. N.Y., 1850.
Htn	Geog 4208.14*	Bunnell, D.C. Travels and adventures. Palmyra, N.Y., 1831.
	Geog 4208.17F	Fitzclarence. Journal of route across India...to England. London, 1819.
	Geog 4208.19	Lumsden, T. A journey from Merut in India to London. London, 1822.
	Geog 4208.21.5	Ames, Nathaniel. Nautical reminiscences. Providence, 1832.
	Geog 4208.22	Rezo, José Luiz do. Viagens à China. Porto, 1822.
	Geog 4208.22.6	Nicol, John. The life and adventures of John Nicol, mariner. London, 1937.
	Geog 4208.25	Howison, J. Foreign scenes and travel recreations. 2. ed. Edinburgh, 1825. 2v.
	Geog 4208.27	Haus, Rosalie. Voyage of the Caroline, 1827-28. London, 1927.
	Geog 4208.28	Butterworth, William. Three years adventures of a minor in England Africa, South Carolina and Georgia. Leeds, 1831.
	Geog 4208.29.3	Prinsep, A. (Mrs.) The journal of a voyage from Calcutta. London, 1833.
	Geog 4208.29.5	Roberts, Jane. Two years at sea. London, 1834.
	Geog 4208.29.7	Roberts, Jane. Two years at sea. 2. ed. London, 1837.
	Geog 4208.30	Holbrook, S.P. Sketches by a traveller. Boston, 1830.
	Geog 4208.30.5	Ames, N. A mariner's sketches. Providence, 1830.
	Geog 4208.30.7	Lefavor, William. Captain Lefavor's forty years' travels. Philadelphia, 1877.
	Geog 4208.31	Hall, B. Fragments of voyages and travels. Philadelphia, 1831. 2v.
	Geog 4208.31.2	Hall, B. Fragments of voyages and travels. v.1-2. 2d ed. Edinburgh, 1832.
	Geog 4208.31.5	Morrell, B. Narrative of four voyages. N.Y., 1832.
	Geog 4208.31.7	Morrell, Abby J. Narrative of voyage to Ethiopia. N.Y., 1833.
	Geog 4208.34	Rapelje, George. A narrative of excursions, voyages and travels. N.Y., 1834.
	Geog 4208.37	Lofé, G. Cartas á mis hijos. N.Y., 1839.
	Geog 4208.38	Jenkins, J.S. United States exploring expeditions...voyage...1838-1842. New Orleans, 1854.
	Geog 4208.38.1	Jenkins, J.S. United States exploring expeditions. N.Y., 1855.
	Geog 4208.38.3	Rose, W.G. Three months' leave, or Military reminiscences. London, 1838.
	Geog 4208.39	Were, J.B. Voyage from Plymonth to Melbourne in 1839. Melbourne, 1964.
	Geog 4208.40	Cleveland, R.J. Voyages and commercial enterprises. N.Y., n.d.
Htn	Geog 4208.40.3*	Cleveland, R.J. Narrative of voyages and commercial enterprises. Cambridge, 1842. 2v.
	Geog 4208.40.5	Cleveland, R.J. Narrative of voyages and commercial enterprises. Boston, 1850.
	Geog 4208.40.10	Cleveland, R.J. Voyages of a merchant navigator. N.Y., 1886.
	Geog 4208.41	Oliphant, L. Episodes in a life of adventure. N.Y., 1887.
	Geog 4208.41.2	Oliphant, L. Episodes in a life of adventure. Edinburgh, 1887.
	Geog 4208.41.4	Oliphant, L. Episodes in a life of adventure. 4. ed. Edinburgh, 1887.
	Geog 4208.43	Yvan, M. Voyages and recits. v.1-2. Bruxelles, 1853.
	Geog 4208.44	Capobianco, R. Breve racconto delle cose. Napoli, 1846.
	Geog 4208.44.3	Smith, T.W. Narrative of life...of T.W. Smith. Boston, 1844.
	Geog 4208.44.5	Santos, Eusebio. Diario del viaje desde Madrid a Manila. Madrid, 1851.
	Geog 4208.45	Bustamante, J. Viaje. Lima, 1845.
	Geog 4208.45.2	Bustamante, J. Viaje al antiguo mundo. 2. ed. Lima, 1959.
	Geog 4208.47	Clark, J.G. Lights and shadows of sailor life. Boston, 1848.
	Geog 4208.47.5	Clark, J.G. Lights and shadows of sailor life. Boston, 1848.
	Geog 4208.48.2	Cöllen, Franz A. Reise Album vom 15ten bis zum 22ten Lebensjahre. Hamburg, 1852.
	Geog 4208.48.3	Cöllen, Franz A. Reisen und Dichtungen. Berlin, 1864.
	Geog 4208.48.5	Barbeida, Claudio Lagrange. Huma viagem de duas mil legoas. Nova-Goa, 1848.
	Geog 4208.49	Donald, C.S. Adventures of a Greek lady. London, 1849. 2v.
	Geog 4208.49.5	Barra, E.I. A tale of two oceans. San Francisco, 1893.
	Geog 4208.50.3	Prince, Nancy. Narrative of life and travels. 2. ed. Boston, 1853.
	Geog 4208.50.5	Prince, Nancy. Narrative of life and travels. 3. ed. Boston, 1856.
	Geog 4208.51	Coggeshall, G. Voyages to various parts of the world...1799-1844. N.Y., 1851.

Geog 4205 - 4210 Voyages and travels in general - Individual travels - Miscellaneous travels (By date of travel) - cont.

	Geog 4208.51.5	Blaney, Henry. Journal of voyages to China and return. Boston, 1913.
	Geog 4208.52	The world here and there. N.Y., 1852.
	Geog 4208.52.5	Colvocoresses, G.M. Four years in a government exploring expedition. N.Y., 1852.
	Geog 4208.52.6	Colvocoresses, G.M. Four years in the government exploring expedition. 5. ed. N.Y., 1855.
	Geog 4208.52.7	Lamson, Joseph. Round Cape Horn...in 1852. Bangor, 1878.
	Geog 4208.53.2	Coggeshall, G. Voyages to various parts...1800-1831. N.Y., 1853.
	Geog 4208.53.3	Ungewitter, D.H. Portfolio für Iländer und Völkerkunde. Pest, 1853.
	Geog 4208.53.5	Mather, Frank J. A clipper ship and her commander. n.p., 1904.
	Geog 4208.54A	Pumpelly, R. My reminiscences. N.Y., 1918. 2v.
	Geog 4208.54B	Pumpelly, R. My reminiscences. N.Y., 1918. 2v.
	Geog 4208.54.5	Warren, E. A doctor's experiences in three continents. Baltimore, 1885.
	Geog 4208.54.10	Coffin, Robert. The last of the Logan. Ithaca, 1941.
	Geog 4208.54.15	Romance of travel...by an old traveler. Philadelphia, 1854.
	Geog 4208.55	King, Paul. Voyaging to China in 1855 and 1904. London, 1936.
	Geog 4208.57	Nordhoff, Charles. Nine years a sailor. Cincinnati, 1857.
	Geog 4208.58	Davenport, H.E. Rovings on land and sea. Boston, 1858.
	Geog 4208.58.5	Coggeshall, G. Thirty-six voyages to various parts of the world, 1799-1841. 3. ed. N.Y., 1858.
	Geog 4208.58.9	Petano y Mazariegos, Gorgonio. Viajes por Europa y América de Don Gorgonio Petano y Mazariegos. Paris, 1858.
	Geog 4208.59	Johnes, M. Boys' book of modern travel. N.Y., 1859.
	Geog 4208.60.3	Colquohoun, A.R. Dan to Beersheba. London, 1908.
	Geog 4208.60.5	Coggeshall, G. An historical sketch of commerce and navigation. N.Y., 1860.
	Geog 4208.60.7	Bray, M.M. A sea trip in clipper ship days. Boston, 1920.
	Geog 4208.63	Massett, S.C. Drifting about. N.Y., 1863.
	Geog 4208.63.7F	Tillotson, John. The overland route to India. London, 1863?
	Geog 4208.64	Mompou, José. De la Habana a Madrid. Habana, 1865.
	Geog 4208.64.5	Root, Sidney. Exotic leaves, gathered by a wanderer. London, 1865.
	Geog 4208.66.3	Cruise of "The Wave". N.Y., 1866.
	Geog 4208.67	Watt, Robert. Kjøbenhavn. Kjøbenhavn, 1867.
	Geog 4208.67.5	Pacific Mail Steamship Co. A sketch of the new route to China and Japan. San Francisco, 1867.
	Geog 4208.68	Vogel, H.W. Vom indischen Ocean bis zum Goldlande. Berlin, 1877.
X Cg	Geog 4208.70	Willis, G.R. The cruise of the Colorado. N.Y., 1873.
	Geog 4208.70.2	Willis, G.R. The cruise of the Colorado. N.Y., 1873.
	Geog 4208.70.5	Around the world. n.p., n.d.
	Geog 4208.72	Revere, J.W. Keel and saddle, a retrospect. Boston, 1872.
	Geog 4208.72.2	Revere, J.W. Keel and saddle, a retrospect. Boston, 1873.
	Geog 4208.72.7	Barker, Mary Ann B. Travelling about over new and old ground. London, 1872.
	Geog 4208.72.9	Smith, G.A. Correspondence of Palestine tourists...while traveling in Europe, Asia and Africa. Salt Lake City, 1875.
	Geog 4208.73	Wood, C.F. Yachting cruise in the South Seas. London, 1875.
	Geog 4208.73.5	Chapman, Charles. The ocean waves. London, 1875.
	Geog 4208.78	Forbes, R.L. Personal reminiscences. Boston, 1878.
X Cg	Geog 4208.78.2	Forbes, R.L. Personal reminiscences. 2. ed. Boston, 1882.
Htn	Geog 4208.78.3*	Forbes, R.B. Personal reminiscences. 2. ed. Boston, 1882.
	Geog 4208.78.4A	Connolly, J.B. Canton captain. Garden City, N.Y., 1942.
	Geog 4208.78.4B	Connolly, J.B. Canton captain. Garden City, N.Y., 1942.
	Geog 4208.78.5	Holden, J.W. A wizard's wanderings from China to Peru. London, 1886.
	Geog 4208.78.7	Wakeman, Edgar. The log of an ancient marine. San Francisco, 1878.
	Geog 4208.78.9	Browne, Ernest A. Round the north hemisphere. London, 1880.
	Geog 4208.80	Lopez, Lucio V. Recuerdos de viaje. Buenos Aires, 1915.
	Geog 4208.80.3	Briggs, L.V. Around Cape Horn to Honolulu on the bark "Amy Turner" 1880. Boston, 1926.
	Geog 4208.80.5	Boyle, Frederick. Chronioles of no-man's land. London, 1880.
	Geog 4208.81	Cané, Miguel. En viaje, 1881-82. Buenos Aires, 1917.
	Geog 4208.81.5	Cané, Miguel. En viaje. Buenos Aires, 1949.
	Geog 4208.81.10	Haeckel, Ernst. Tropenfahrten Reiseschilderungen aus Ceylon, Java und den Mittelmeergebieten. Leipzig, 1969.
	Geog 4208.82	Helms, L.V. Pioneering in the Far East. London, 1882.
Htn	Geog 4208.82.3*	Meyer, Hans. Blätter aus mienem Reisetagbuch 1881-1883. Leipzig, 1882.
	Geog 4208.82.6	Coote, W. Wanderings, south and east. London, 1882.
	Geog 4208.83	Lucy, H.W. East by West. London, 1885. 2v.
	Geog 4208.83.6	Tangye, Richard. Reminiscences of travel in Australia, America and Egypt. 2. ed. London, 1884.
	Geog 4208.83.10	Wilkinson, H. Sunny lands and seas. London, 1883.
	Geog 4208.84.5	McCarthy, W.J. Cruise of the U.S. flagship Lancaster. n.p., 1887.
	Geog 4208.85	Frachebourd, J.J. Monséjour en France...mes pelerinages a Rome. Fribourg, 1885.
	Geog 4208.85.3	Forbes, Archibald. Souvenirs of some continents. London, 1885.
	Geog 4208.85.5	Burton, R.G. Tropics and snows. London, 1898.
	Geog 4208.86	Böckmann, W. Reise nach Japan aus Briefen und Tagebüchern zusammengestellt. Berlin, 1886.
	Geog 4208.87A	Kipling, Rudyard. From sea to sea. N.Y., 1899. 2v.
	Geog 4208.87B	Kipling, Rudyard. From sea to sea. N.Y., 1899. 2v.
	Geog 4208.87.3	Kipling, Rudyard. From sea to sea. N.Y., 1909.
	Geog 4208.87.10	Palgrove, William G. Ulysses; or Scenes and studies in many lands. London, 1887.
	Geog 4208.87.15	Chapin, J.H. From Japan to Granada. N.Y., 1889.
	Geog 4208.88	Howells, W.D. Library of universal adventure by sea and land. N.Y., 1888.
	Geog 4208.88.5	Keane, J.F. Three years of a wanderer's life. London, 1888.
	Geog 4208.90	Platen, Carl von. Impressions of travels. Stockholm, 1890.
	Geog 4208.91	Cimon, Henri. Impressions de voyage. Québec, 1895-1902. 2v.
	Geog 4208.91.5	Carrasco, Gabriel. Del Atlántico al Pacífico y un Argentino en Europa. 2a ed. Buenos Aires, 1891.
	Geog 4208.92	Great streets of the world. London, 1892.
	Geog 4208.94	Arnold, E. Wandering words. London, 1894.
	Geog 4208.95	Ballou, M.M. Foot-prints of travel. Boston, 1895.
	Geog 4208.95.5	Nordhoff, C. The merchant vessel. N.Y., 1895.

Classified Listing

Geog 4205 - 4210 Voyages and travels in general - Individual travels - Miscellaneous travels (By date of travel) - cont.

Geog 4208.95.5.5 Nordhoff, C. The merchant vessel. N.Y., 1895.

Geog 4208.96 Arnold, E. East and West. London, 1896.

Geog 4208.96.10 Bond, C. Goldfields and chrysantemums. London, 1898.

Geog 4208.97 Davis, R.H. A year from a reporter's note-book. N.Y., 1897.

Geog 4208.97.3 Stoddard, J.L. Lectures. Boston, 1899-1900. 10v.

Geog 4208.97.4 Stoddard, J.L. Lectures. Supplement. n.p., 1901-05. 4v.

Geog 4208.97.5 Stoddard, J.L. Lectures. v.8. Boston, 1905.

Geog 4208.97.7 Schanz, M. Ein Zug nach Osten. Hamburg, 1897. 2v.

Geog 4208.98 Voss, J.C. Venturesome voyages of Captain Voss. Tokyo (Kanda), Japan. Yokohama, 1913.

Geog 4208.98.2 Voss, J.C. The venturesome voyages of Captain Voss. 2. ed. London, 1949.

Geog 4208.98.5 Schweitzer, Georg. Eine Reise um die Welt. 2e Aufl. Berlin, 1899.

Geog 4208.98.11 Temple, E.L. Old world memories. Boston, 1899. 2v.

Geog 4208.98.18 Mackin, S.M.A. (Mrs.). A society woman on two continents. N.Y., 1898.

Geog 4208.99.2 Antin, M. From Plotzk to Boston. Boston, 1899.

Geog 4209.00 Boyer, Bessie B. Chief Officer Brown. Chicago? 1900.

Geog 4209.01 Creelman, J. On the great highway. Boston, 1901.

Geog 4209.01.5A Burton, Richard F. Wanderings in three continents. N.Y., 1901.

Geog 4209.01.5B Burton, Richard F. Wanderings in three continents. N.Y., 1901.

Geog 4209.02 Prado, Eduardo. Viagens. v.1, 2a ed; v.2, 1a ed. São Paulo, 1902. 2v.

Geog 4209.02.3 Colquhoun, E.C. (Mrs.). Two on their travels. London, 1902.

Geog 4209.02.5 Colquhoun, E.C. (Mrs.). Two on their travels. N.Y., 1902.

Geog 4209.02.10 Hellen, Gustav. Aus meiner Wanderzeit, 1902-1906 und 1908-1909. Hamburg, 1966.

Geog 4209.03 Simpson, William. Autobiography. London, 1903.

Geog 4209.04 Arthur, Richard. Ten thousand miles in a yacht. N.Y., 1906.

Geog 4209.05 Herboso, F.J. Reminiscencias de viajes. Caracas, 1905- 3v.

Geog 4209.05.5 Williams, Archibald. The romance of modern exploration. Philadelphia, 1905.

Geog 4209.06 Gambier, J.W. Links on my life on land and sea. N.Y., 1906.

Geog 4209.07 Mallik, M.C. Impressions of a wanderer. London, 1907.

Geog 4209.07.5 Russell, J.W. The romance of an old time shipmaster. N.Y., 1907.

Geog 4209.07.10 Nicholson, T.R. Adventurer's road. N.Y., 1958.

Geog 4209.08 Podmore, P.S.M. Rambles and adventures in Australasia. London, 1909.

Geog 4209.09 Burge, C.O. Adventures of a civil engineer. London, 1909.

Geog 4209.09.6 Ewers, Hanns H. "Mit meinen Augen"; Fahrten durch die lateinische Welt. München, 1914.

Geog 4209.09.10 Gordon, F. Listy z podrozy. Chicago, 1910.

Geog 4209.10 Beltramelli, Antonio. Il diario di un viandante, dal deserto al mar glaciale. Milano, 1911.

Geog 4209.10.5 Fraser, Mary Crawford. A diplomatist's wife in many lands. N.Y., 1911. 2v.

Geog 4209.10.6 Fraser, Mary Crawford. A diplomatist's wife in many lands. London, 1911. 2v.

Geog 4209.10.9 Fraser, Mary Crawford. Reminiscences of a diplomat's wife. N.Y., 1913.

Geog 4209.11.2 Fraser, Mary Crawford. Further reminiscences of a diplomat's wife. London, 1912.

Geog 4209.11.5 Lowther, H.C. From pillar to post. London, 1911.

Geog 4209.11.6 Lowther, H.C. From pillar to post. London, 1912.

Geog 4209.11.7 Jacobs, A.H. Reisbrieven uit Afrika en Azie. Almelo, 1913. 2v.

Geog 4209.13 Casey, R.H. Notes made during a cruise around the world in 1913. N.Y., 1914.

Geog 4209.13.7 Kawabata, A. A hermit turned loose. 2. ed. Tokio, 1915.

Geog 4209.14 Aldao, Carlos A. A través del mundo. 5. ed. Buenos Aires, 1914.

Geog 4209.15 Rohrbach, P. Weltpolitisches Wanderbuch 1897-1915. Königstein, 1916.

Geog 4209.15.5 Rohrbach, P. Weltpolitisches Wanderbuch 1897-1915. Königstein, 1916.

Geog 4209.15.10 Lubbock, A.B. Round the Horn before the most. London, 1915.

Geog 4209.17 Anderson, I. (Mrs.). Odd corners. N.Y., 1917.

Geog 4209.19 Mackay, John. The ten islands and Ireland. Dublin, 1919.

Htn Geog 4209.19.5* Romero de Terreros, Juan. Apuntaciones de viaje en 1849. México, 1919.

Geog 4209.20 Curle, Richard. Wanderings. n.p., n.d.

Geog 4209.21 Lucas, E.V. Roving east and roving west. London, 1921.

Geog 4209.21.2 Lucas, E.V. Roving east and roving west. N.Y., 1921.

Geog 4209.21.3 Hamilton, Frederick. Here, there, and everywhere. London, 1921?

Geog 4209.21.4 Hamilton, Frederick. Here, there, and everywhere. N.Y., 1921.

Geog 4209.22 Glass, James. Chats over a pipe; a tale of two brothers. London, 1922.

Geog 4209.23 Buchan, John. The last secrets. London, 1923.

Geog 4209.23.10 Wehde, Albert. Seit ich die heimat Verliess. Berlin, 1924.

Geog 4209.23.15 Balmont, K.D. Visions solaires Mexique, Égypte, Inde, Japon, Océanie. Paris, 1923.

Geog 4209.23.20 Bryce, J.B. Memories of travel. N.Y., 1923.

Geog 4209.23.25 Phelan, James D. Travel and comment. San Francisco, 1923.

Geog 4209.24A Landor, A.H.S. Everywhere; the memoirs of an explorer. N.Y., 1924. 2v.

Geog 4209.24B Landor, A.H.S. Everywhere; the memoirs of an explorer. N.Y., 1924. 2v.

Geog 4209.24.5A Burton, Richard F. Selected papers on anthropology, travel and exploration. London, 1924.

Geog 4209.24.5B Burton, Richard F. Selected papers on anthropology, travel and exploration. London, 1924.

Geog 4209.25A Coolidge, J.G. Random letters from many countries. Boston, 1924.

Geog 4209.25B Coolidge, J.G. Random letters from many countries. Boston, 1924.

Geog 4209.25.5 Bridgman, H.B. Conquering the world. N.Y., 1925.

Geog 4209.25.10 Harper, H.H. Highlights of foreign travel. N.Y., 1925.

Geog 4209.25.15 Jacques, N. Im Kaleidoskop der Weltteile. Berlin, 1925.

Geog 4209.26 Hall, James Norman. On the stream of travel. Boston, 1926.

Geog 4209.27 Hoare, Samuel. India by air. London, 1927.

Geog 4209.28 Forbes, Rosita. Adventure. Boston, 1928.

Geog 4205 - 4210 Voyages and travels in general - Individual travels - Miscellaneous travels (By date of travel) - cont.

Geog 4209.28.5 Cust, Nina. Wanderers. London, 1928.

Geog 4209.28.10 Benson, Stella. Worlds within worlds. London, 1928.

Geog 4209.28.20A Cameron, J. John Cameron's odyssey. N.Y., 1928.

Geog 4209.28.20B Cameron, J. John Cameron's odyssey. N.Y., 1928.

Geog 4209.28.25 Bloem, Walter. Weldgesicht; ein Buch von heutiger und kommender Menscheit. Leipzig, 1928.

Geog 4209.29 Goldring, Douglas. People and places. Boston, 1929.

Geog 4209.29.5 Buxton, Noel. Travels and reflections. London, 1929.

Geog 4209.29.10 Bruce, Michael. Peaks of hazard. Indianapolis, 1929.

Geog 4209.30.2 Newton, A.E. A tourist in spite of himself. Boston, 1930.

Geog 4209.30.3 Newton, A.E. A tourist in spite of himself. Boston, 1933.

Geog 4209.30.5 Fairchild, David G. Exploring for plants. N.Y., 1930.

Geog 4209.30.10 Bremer, M. Memoirs of a Ceylon planter's travels, 1851 to 1921. London, 1930.

Geog 4209.30.15 Mather, Norman C. Travel notes, 1929-30. Chicago, 1930.

Geog 4209.30.20 Chable, Jacques Edouard. Jazz, boomerang et kimonos. Paris, 1930.

Geog 4209.31 Hyne, C.J.C. People and places. London, 1931.

Geog 4209.31.5 Desmond, Alice C. Far horizons. N.Y., 1931.

Geog 4209.31.10 Leslie, L.A.D. Wilderness trails in three continents. London, 1931.

Geog 4209.31.15 Rastron, A.H. Home from the sea. N.Y., 1931.

Geog 4209.31.20 Cornish, Naughan. The poetic impression of natural scenery. London, 1931.

Geog 4209.31.25 Keilpflug, Erich R. An den Rändern dreier Erdteile. Berlin, 1931.

Geog 4209.32 Murchie, Guy. Men on the horizon. Boston, 1932.

Geog 4209.32.5 Hering, O.C. Down the world. N.Y., 1932.

Geog 4209.32.10 Powell, Edward A. Yonder lies adventure! N.Y., 1932.

Geog 4209.32.15 Mordaunt, E. (pseud.). Rich tapestry. N.Y., 1932.

Geog 4209.32.20 Hauser, H. Fair winds and foul. N.Y., 1932.

Geog 4209.32.25 Johnson, Irving. Round the Horn in a square rigger. Springfield, 1932.

Geog 4209.33 Howard, S. Thames to Tahiti. London, 1951.

Geog 4209.34 Coyne, William M. A sailor's log of facts: not fables. Boston, 1934.

Geog 4209.34.7 Fuchs, Hans. Heimkehr ins Dritte Reich. Dresden, 1934.

Geog 4209.34.10 Gutierrez Zamora, I. Contando un viaje. n.p., 1934.

Geog 4209.35.5 Foley, Arthur. Breezy adventure. Boston, 1935.

Geog 4209.35.15 Matisse, Maarten. Wanderer from sea to sea. N.Y., 1936.

Geog 4209.35.20A Halliburton, Richard. Seven league boots. Indianapolis, 1935.

Geog 4209.35.20B Halliburton, Richard. Seven league boots. Indianapolis, 1935.

Geog 4209.35.23 Halliburton, Richard. Seven league boots. 3. ed. London, 1941.

Geog 4209.35.25 Harris, P.P. Peregrinations. v.2-3. n.p., 1935-37. 2v.

Geog 4209.35.33 Aubert de la Rue, Edgar. L'homme et les iles. 2. ed. Paris, 1935.

Geog 4209.36 Macdonald, J.R. At home and abroad; essays. London, 1936.

Geog 4209.36.5 Davis, R.H. People, people, everywhere. N.Y., 1936.

Geog 4209.36.10 Howland, Charles. Travel essays. n.p., 1936?

Geog 4209.36.15 Wøller, Johan. Zest for life; recallections of a philosophic traveller. London, 1936.

Geog 4209.36.30 Durtain, L. Le globe sous le bras. Paris, 1936.

Geog 4209.36.35 Lissner, Ivar. Völker und Kontinente. Hamburg, 1936.

Htn Geog 4209.37F* LaCombe, Jean de. A compendium of the East; being an account of voyages to the Grand Indies. London, 1937.

Geog 4209.37.5 Vrázová, Vlasta. Život a cesty E. St. Vráze. Praha, 1937.

Geog 4209.37.7 Prague. Městská Lidová Knihovna. Život a délo E. St. Vráze. Praha, 1960.

Geog 4209.37.10 Wechsberg, J. Die grosse Mauer; das Buch einer Weltreise. Leipzig, 1937.

Geog 4209.38 Hart, M.R. Who called that lady a skipper? N.Y., 1938.

Geog 4209.38.1 Long, Dwight. Seven seas on a shoestring. N.Y., 1939.

Geog 4209.38.5A Hallet, R.M. The rolling world. Boston, 1938.

Geog 4209.38.5B Hallet, R.M. The rolling world. Boston, 1938.

Geog 4209.38.10 Dighy, G. Goose feathers. N.Y., 1938.

Geog 4209.38.15A Dos Passos, John. Journeys between wars. N.Y., 1938.

Geog 4209.38.15B Dos Passos, John. Journeys between wars. N.Y., 1938.

Geog 4209.38.15C Dos Passos, John. Journeys between wars. N.Y., 1938.

Geog 4209.38.16 Dos Passos, John. Journeys between wars. London, 1938.

Geog 4209.38.21 Beddington, C. We sailed from Brixham. London, 1938.

Geog 4209.38.30 Craig, John D. Danger is my business. N.Y., 1938.

Geog 4209.38.35 Bonnard, A. Le bouquet du monde. Paris, 1938.

Geog 4209.38.40 Sigvaldson, J. Ferdasaga Fritz Liebig. Reykjavik, 1938.

Geog 4209.38.45 Mielche, H. Let's see if the world is round. London, 1950.

Geog 4209.39 Forbes, Rosita T. These are real people. N.Y., 1939.

Geog 4209.39.5 O'Malley, C.J. It was news to me. Boston, 1939.

Geog 4209.39.10 Maury, Richard. The saga of "Cimba". N.Y., 1939.

Geog 4209.39.15 Reinius, I. Journal hållen på resan till Canton i China. Helsingfors, 1939.

Geog 4209.39.20A Harris, N.D. Moving on; the romance of travel. Chicago, 1939.

Geog 4209.39.20B Harris, N.D. Moving on; the romance of travel. Chicago, 1939.

Geog 4209.39.25 Larigaudie, Guy de. La route aux aventures. Paris, 1948.

Geog 4209.40 Woodward, C. Lanterns alight; journeys to far places. Chicago, 1940.

Geog 4209.40.3 Forsyth, W.J. Journal. Boston, 1940.

Geog 4209.40.10 Davis, R.H. Let's go with Bob Davis to India, Ceylon, Venezuela. N.Y., 1940.

Geog 4209.40.15 Baarslag, K. Islands of adventure. N.Y., 1940.

Geog 4209.40.20 Halliburton, Richard. Richard Halliburton, his story of his life's adventures as told in letters to his mother and father. Indianapolis, 1940.

Geog 4209.40.25 Dennis, W.H. Glittering horizons. N.Y., 1940.

Geog 4209.40.30 Hurst, Ida. Dare to live. London, 1940.

Geog 4209.40.35 Weidman, J. Letter of credit. N.Y., 1940.

Geog 4209.40.40 Hogben, L. Author in transit. N.Y., 1940.

Geog 4209.40.45 Haven, V.S. Many ports of call. N.Y., 1940.

Geog 4209.40.50 Simon, C.R. Gran'ma goes by freight. Placerville, Calif., 1940.

Geog 4209.40.55 Tangye, Derek. The time was mine. London, 1940.

Geog 4209.40.60 Martyr, W. The wandering years. N.Y., 1940.

Geog 4209.41 Lynch, Kathleen M. Travellers must be content. N.Y., 1941.

Geog 4209.41.3 Buck, F. All in a lifetime. N.Y., 1941.

Geog 4209.41.5 Baerlein, Henry. Travels without a passport. London, 1941.

Geog 4209.41.6 Baerlein, Henry. Travels without a passport. London, 1943.

Geog 4209.41.10 DeValda, F.W. Adventure is my business. London, 1941.

Geog 4209.41.15 Low, G.W. Gold rush by sea. Philadelphia, 1941.

Classified Listing

Geog 4205 - 4210 Voyages and travels in general - Individual travels - Miscellaneous travels (By date of travel) - cont.

Geog 4209.41.20 Maury, A.F. Intimate Virginiana. Richmond, 1941.
Geog 4209.42 Domville-Fife, C.W. I tell of the seven seas. London, 1942.
Geog 4209.42.5 Considine, J.J. Across a world. Toronto, 1942.
Geog 4209.42.10 Barrett, C.L. On the Wallaby. Melbourne, 1942.
Geog 4209.42.15 Birnbaum, M. Vanishing Eden. N.Y., 1942.
Geog 4209.42.20 Maillart, E.K. Gypsy afloat. London, 1942.
Geog 4209.42.25 Vejarano, J.R. Rutas del mundo. Bogota, 1942.
Geog 4209.43 Thatcher, T.C. Travel letters. Yarmouth Port, 1943.
Geog 4209.43.5 Collins, Grace. Chin takes and talks, an absurd travel diary. Chicago, 1943.
Geog 4209.43.10 Medern, W.E.A.A. Blick in die weite Welt. Berlin, 1943.
Geog 4209.44 Allan, D. Lightning strikes once. N.Y., 1944.
Geog 4209.44.5 Armstrong, W. Saltwater tramp. London, 1944.
Geog 4209.44.10 Baerlein, Henry. The caravan rolls on. London, 1944.
Geog 4209.44.15 Gagnon, H.J. Blanc et noir. Montréal, 1944.
Geog 4209.45 Caldwell, E.N. Satin skirts of commerce. N.Y., 1945.
Geog 4209.45.5 Orcutt, R. Merchant of alphabets. Garden City, N.Y., 1945.
Geog 4209.45.10 Allan, Doug. Gamblers with fate. 1. ed. N.Y., 1945.
Geog 4209.45.15 Galimir, Mosco. Half a century of world travel; impressions and reflections. N.Y., 1945.
Geog 4209.45.20 Retuerto, Marcial. Sur les traces de Magellan. Paris, 1945.
Geog 4209.46 Baerlein, Henry. Leaves in the wind. London, 1946.
Geog 4209.46.2 Baerlein, Henry. So many roads. London, 1946?
Geog 4209.46.5 Shaw, Frank H. White sails and spendthrift. London, 1946.
Geog 4209.46.10 George, Albert J. The cap'n's wife. Syracuse, 1946.
Geog 4209.46.15 Etherton, P.T. All over the world. London, 1946.
Geog 4209.46.20 Wever, Jan. Journal der merkwaardige reizen van Jan Wever. 's-Gravenhage, 1946.
Geog 4209.46.28 Waugh, Evelyn. When the going was good. London, 1946.
Geog 4209.47 Colam, Lance. Death over my shoulder. London, 194-.
Geog 4209.47.5 Wethered, Herbert Newton. The four paths of pilgrimage. London, 1947.
Geog 4209.48 Mackenzie, C. All over the place. London, 1948.
Geog 4209.50 Riddell, J. Flight of fancy. London, 1950.
Geog 4209.51 Harris, F.R. Itchin' feet. Greenfield, Ohio, 1951.
Geog 4209.51.5 Rosenberg, Halger. Jorden rundt med Halger Rosenberg. København, 1951.
Geog 4209.51.10 Barreto, Paulo. A alma encantadora das ruas. Rio de Janeiro, 1951.
Geog 4209.52 Warden, W.R. Vale enchanting. London, 1952.
Geog 4209.52.5 Beonio-Brocchieri, V. Il Marcopolo. Milano, 1952.
Geog 4209.52.10 Szos Kies, Henryk J. Your world and mine. N.Y., 1952.
Geog 4209.53 Meiss-Teuffen, Hans. Wanderlust. N.Y., 1953.
Geog 4209.53.5 Davenport, P. The voyage of Waltzing Matilda. London, 1953.
Geog 4209.54 Osborn, C. He lived for adventure. Washington, 1954.
Geog 4209.54.5 Kivikoski, Olavi. Yksin yli Atlantin. Helsinki, 1954.
Geog 4209.55 Woodcock, A.W.W. To Hong Kong and return. Boston, 1955.
Geog 4209.55.5 Starobin, J.R. Paris to Peking. 1. ed. N.Y., 1955.
Geog 4209.55.10 Praz, Mario. Viaggi in Occidente. Firenze, 1955.
Geog 4209.55.15 Pineda Yañez, R. La isla y Colón. Buenos Aires, 1955.
Geog 4209.56A Wilson, Edmund. Red, black, blond, and olive. N.Y., 1956.
Geog 4209.56B Wilson, Edmund. Red, black, blond, and olive. N.Y., 1956.
Geog 4209.56.5 Elmberg, J.E.O. Islands of to-morrow. London, 1956.
Geog 4209.56.10 Gouy, Robert. Terres de l'amitié. Neuchâtel, 1956.
Geog 4209.56.15 Svenson, Sven. Varlden ar så stor, så stor. Stockholm, 1956.
Geog 4209.56.20 Slessor, Tim. First overland. London, 1957.
Geog 4209.56.25 Uberti, Roberto degli. Soli attraverso gli oceani. 2. ed. Brescia, 1956.
Geog 4209.56.30 Gaudio, Attilio. A la recherche des îles ignorees. Paris, 1956.
Geog 4209.57 Gosset, R.P. Retour du bout du monde. Paris, 1957.
Geog 4209.57.5 Raitchewitch, M. Biographien und Autogramme beruhmter Staatsmanner und anderer Personlichkeiten des 20. Jahrhundert. Bielefeld, 1957.
Geog 4209.57.10 Jara Peralta, José. La ciudad de Toledo. Madrid, 1957.
Geog 4209.58 Hadley, Leila. Give me the world. N.Y., 1958.
Geog 4209.58.5 Korshunov, V. Khodili my pokhodami. Moskva, 1958.
Geog 4209.58.10 Toynbee, A.J. East to west. N.Y., 1958.
Geog 4209.58.15 Olszewicz, B. Ze wspomnień podróżników. Warszawa, 1958.
Geog 4209.58.20 La Croix, Robert de. Mysteres des îles. Paris, 1958.
Geog 4209.58.25 Knies, Donald. Walk the wide world. N.Y., 1958.
Geog 4209.59 Pizzinelli, Corrado. Viaggio nel mondo. Bologna, 1959.
Geog 4209.59.5 Trolliet, Héli. Sur les routes du monde. Lausanne, 1959.
Geog 4209.59.10 Tey, José M. Hong Kong-Barcelona en el junco Rubia. Barcelona, 1959.
Geog 4209.60 Hammond-Innes, Ralph. Harvest of journeys. N.Y., 1960.
Geog 4209.60.5 Nicholson, Timothy. Five roads to danger. London, 1960.
Geog 4209.60.10 Nikhailov, Nikolai I. From pole to pole. Moscow, 1960.
Geog 4209.60.15 Baisen-Moeller, Axel. Globetrotter og hovedjaeger. Vordingborg, 1960.
Geog 4209.60.20 Burden, W.D. Look to the wilderness. Boston, 1960.
Geog 4209.60.25 Moncharmont, Simone. La croisière du Ly-Kau. Paris, 1960. 2v.
Geog 4209.60.30 Roy, Claude. Le journal des voyages. Paris, 1960.
Geog 4209.60.35 Portes Gil, Emilio. El mundo a través de sus grandes estadistas. México, 1960.
Geog 4209.60.40 Jelling Kristoffersen, Frode. Jorden rundt på tommelfinger. Sirius, 1960.
Geog 4209.60.45 Firsé, Adolf. Reise-Journal. Gütersloh, 1967.
Geog 4209.60.50 Basso, Hamilton. A quota of seaweed. Garden City, 1960.
Geog 4209.61 Kahn, E.J. A reporter here and there. N.Y., 1961.
Geog 4209.61.5 Kazakova, M.S. Goroda i vstrechi. Riga, 1961.
Geog 4209.61.10 Drachev, B.S. K beregam Vostoka. Moskva, 1961.
Geog 4209.61.15 Hentzel, R. Aventyr jorden runt. Stockholm, 1961.
Geog 4209.62 Shakirianov, N.A. Atlanticheskii dnevnik. Riga, 1962.
Geog 4209.62.5 Koroteev, V.I. Ia eto videl; ocherki raznykh let. Moskva, 1962.
Geog 4209.62.10 Lattmann, D. Mit einem deutschem Pass. München, 1964.
Geog 4209.62.15 Thorlacius, Sigriðm. Feroabók. Akureyri, 1962.
Geog 4209.63 Nasedkin, V.N. Piatnadtsat' let skitanii po zemnomu sharu. Moskva, 1963.
Geog 4209.63.5 Zhukov, Iu.A. Eti semnadtsat' let. Moskva, 1963.
Geog 4209.63.10 Morris, J. Cities. London, 1963.
Geog 4209.63.15 Ajala, O. An African abroad. London, 1963.
Geog 4209.63.20 Diez del Corval, L. Del nuevo al viejo mundo. Madrid, 1963.
Geog 4209.63.25 Turri, Eugenio. Viaggio a Samacanda. Novara, 1963.
Geog 4209.63.30 Bouvier, Nicolas. L'usage du monde. Genève, 1963.
Geog 4209.63.35 Green, Lawrence George. A decent fellow doesn't work. Cape Town, 1963.
Geog 4209.64 Knowles, J. Double vision. N.Y., 1964.

Geog 4205 - 4210 Voyages and travels in general - Individual travels - Miscellaneous travels (By date of travel) - cont.

Geog 4209.64.5 Pamphlet vol. Voyages and travel. 2 pam.
Geog 4209.64.10 Benchley, P. Time and a ticket. Boston, 1964.
Geog 4209.64.15 Brooks, P. Roadles area. 1. ed. N.Y., 1964.
Geog 4209.64.20 Muhlethader, J. Toutes voiles dehors. Genève, 1964.
Geog 4209.64.25 Sakhnin, Arkadii Ia. Vot liudi. Moskva, 1964.
Geog 4209.64.30 Kozlovs'kyi, Mykola F. Cherez 15 morei i 2 okeana; puteshestviia. Kiev, 1964.
Geog 4209.64.35 Eckert, Gerhard. Richtig Reisen. Bergisch Gladbach, 1964.
Geog 4209.64.40A Fleming, Ian. Thrilling cities. N.Y., 1964.
Geog 4209.64.40B Fleming, Ian. Thrilling cities. N.Y., 1964.
Geog 4209.65 Rand, Christopher. Mountains and water. N.Y., 1965.
Geog 4209.65.5 Shelekhov, Grigorii I. Puteshestvie s 1783 po 1790 god iz okhotska po vostoch. okeanu. Ann Arbor, 1965. 2v.
Geog 4209.65.10 Efimov, Gerontii V. Strany i liudi. Leningrad, 1965.
Geog 4209.65.15 Pleshakov, Leonid P. Vokrug sveta s Zarei. Moskva, 1965.
Geog 4209.65.20 Horvat, Joža. Besa-brodski dnevnik. Zagreb, 1973.
Geog 4209.65.25 Vinogradov, Aleksandr A. Gde shumiat chuzhie goroda. Moskva, 1974.
Geog 4209.66 Mikhailov, Nikolai N. S planetoi vmestve. Moskva, 1966.
Geog 4209.66.5 Kovanov, Vladimir V. Meridiany, sobytiia, vstrechi. Moskva, 1966.
Geog 4209.66.10 Pogačnik, Bogdan. Povsod so ljudje. Maribor, 1966.
Geog 4209.66.15 Monteil, Vincent. Soldat de fortune. Paris, 1966.
Geog 4209.66.20 Quilici, Folco. Giramare. Roma, 1966.
Geog 4209.66.25 Scott, Jack Denton. Passport to adventure. N.Y., 1966.
Geog 4209.67 Pochivalov, Leonid V. My za granitsei. Moskva, 1967.
Geog 4209.67.5 Mikosha, Vladislav V. Gody i strany. Moskva, 1967.
Geog 4209.67.10 Díaz Plaja, Guillermo. Con variado rumbo. 1. ed. Barcelona, 1967.
Geog 4209.67.15 Fiore, Ilario. L'italiano di Ponte Cayumba. Firenze, 1967.
Geog 4209.68 Maevskii, Viktor V. Polmilliona kilometrov pozadi. Moskva, 1968.
Geog 4209.68.5 Cole, Jean. Trimaran against the trades. Wellington, 1968.
Geog 4209.68.10 Gritskevich, Valentin P. Puteshestviia nashikh zemliakov. Minsk, 1968.
Geog 4209.68.15 Ogni dalekikh gorodov. Moskva, 1968.
Geog 4209.68.20 Expédition France-Inde. Expédition France-Inde; mission culturelle 1968-1969 sous le patronage de la Fédération régionale speleologique, Nancy. Saint Max, 1970.
Geog 4209.68.25 Rodin, Leonid Efimovich. Po iuzhnym stranam. Moskva, 1968.
VGeog 4209.69 Ogrin, Miran. Širine sveta. Ljubljana, 1969.
VGeog 4209.69.5 Karlin, Alma M. Samotno potovanje. Ljubljana, 1969.
Geog 4209.69.10 Sytin, Viktor A. Puteshestviia. Moskva, 1969.
Geog 4209.69.15 Paleckis, Justas. Kelioniu, Knyga. Vilnius, 1969.
Geog 4209.69.20 Boldyrev, Ivan I. Na raznykh shirotakh. Riga, 1969.
Geog 4209.69.25 Stoianovich, Ivan. Bregovete na khorata. Sofiia, 1969.
Geog 4209.70 Wollschaeger, Alfred. Weltreise auf den Spuren der Unruhe. Gütersloh, 1970.
Geog 4209.70.5 Kiknadze, Aleksandr Vasil'evich. Ot Madrida do Tokio. Baku, 1970.
Geog 4209.70.10 Kozima, Marjan. Veseli potopisi. Maribor, 1970.
Geog 4209.70.15 Rubissow, Helen. Ozni na dorogakh chetyrekh chastei sveta. Parizh, 1970.
Geog 4209.71 Šulentić, Zeatko. Ljudi krajevi Beskraj. Karlovac, 1971.
Geog 4209.71.5 Górnicki, Wiesław. Opowieści zdyszane. 1. wyd. Warszawa, 1971.
Geog 4209.71.10 Šteinbergs, Valentins. Filosofiia i dzhentel'meny. Riga, 1971.
Geog 4209.71.15 Pamphlet vol. "Voyages and travels". 6 pam.
Geog 4209.71.20 Shelekhov, Grigorii I. Rossiiskogo kuptsa Grigoriia Shelekhova stranstvoraniia iz Okholska po Vostochnomu okeanu k amerikanskiam beregam. Khabarovsk, 1971.
Geog 4209.71.25 Kuleshov, Aleksandr P. Shest' gorodov piati kontinentov. Moskva, 1971.
Geog 4209.71.30 Nikolaev, Vladimir D. Maiak Zemli. Moskva, 1971.
Geog 4209.71.35 Kearns, Des. World wanderer; 100,000 miles under sail. Sydney, 1971.
Geog 4209.71.40 Knudsen, Kim. Grenseløs. Oslo, 1971.
Geog 4209.71.45 Swale, Rosie. Children of Cape Horn. London, 1974.
Geog 4209.72 Hammond Innes, Dorothy. Occasions. London, 1972.
Geog 4209.72.5 Mon'ko, Aleksei M. Potoratiny. Volgograd, 1972.
Geog 4209.72.10 Morris, James. Places. London, 1972.
Geog 4209.72.15 Bamans, Godfried. Op reis rond de wereld en op Rottumerplaat. Amsterdam, 1972.
Geog 4209.73 Matveev, Vikentii A. Sud'ia-vremia. Moskva, 1973.

Geog 4215 - 4220 Voyages and travels in general - Individual travels - Circumnavigations (By date of travel)

Geog 4215.19 Alexander, Philip. Earliest voyages round the world, 1519-1617. Cambridge, 1916.
Geog 4215.19.5 Kölliker, O. Die erste Umseglung der Erde...1519-1522. München, 1908.
Geog 4215.19.10 Mitchell, Mairin. Elcano. London, 1958.
Htn Geog 4215.86* Neaudro, M. Orbis terrae partium. Lipsiae, 1586.
Geog 4216.19 Schouten, W.C. Relación diaria del viaje de J. Le Maire y G.C. Shouten. Santiago de Chile, 1897.
Geog 4216.21 Fernberger von Egenberg, C.M. Unfreiwillige Reise um die Welt, 1621-1628. Leipzig, 1928.
Geog 4216.99 Careri, G. Giro del mondo. Napoli, 1699-1700. 6v.
Geog 4216.99.10 Nunnari, F.A. Un viaggiatore calabrese. Mesina, 1901.
Geog 4217.03 Funnell, William. Voyage round the world. London, 1707.
Geog 4217.08.10 MacLiesh, Archibald F. The privateers. N.Y., 1962.
Geog 4217.21 Bree, Levinus W. de. Jacob Roggeveen en zijn reis naar het zuidland, 1721-1722. Amsterdam, 1942.
Geog 4217.28 La Barbinais le Gentil. Nouveau voyage autour du monde. Amsterdam, 1728. 2v.
Geog 4217.37 Behrens, E.F. Sud-Lander und um die Balt. Leipzig, 1737.
Geog 4217.40.5 Heaps, Leo. Log of the centurion. London, 1973.
Geog 4217.64 Byron, John V. Viaje al rededor del mundo hecho en 1764, 65 y 66. Madrid, 1833.
Geog 4217.64.7 Ortega, D.C. de. Viage del Comandante Byron al rededor del mundo. Madrid, 1769.
Geog 4217.64.10 Byron, John V. Viaggio intorno al mondo. Firenze, 1768.
Geog 4217.66 Bougainville, L. de. Voyage autour du monde. Paris, 1771.
Geog 4217.66.3A Bougainville, L. de. Voyage round the world...1766. London, 1772.
Geog 4217.66.3B Bougainville, L. de. Voyage round the world...1766. London, 1772.
Geog 4217.66.5 Bougainville, L. de. Voyage autour du monde. Paris, 1772. 3v.
Geog 4217.66.11 Bougainville, L. de. Reis rondom de weereldt. Dordrecht, 1772.

Classified Listing

Geog 4215 - 4220 Voyages and travels in general - Individual travels - Circumnavigations (By date of travel) - cont.

	Geog 4217.67	Pagès, P.M.F. de. Voyage autour du monde. Paris, 1782. 2v.
	Geog 4217.67.5	Pagès, P.M.F. de. Travels round the world. London, 1791. 3v.
	Geog 4217.72	Sparrman, Anders. Voyage au Cap de Bonne-Esperance. Paris, 1787.
	Geog 4217.72.3	Sparrman, Anders. Reise nach dem Vorgebirge der guten Hoffnung. Berlin, 1784.
	Geog 4217.72.5	Sparrman, Anders. Voyage to Cape of Good Hope. Perth, 1789.
	Geog 4217.72.10F	Sparrman, Anders. A voyage round the world with Captain James Cook. London, 1944.
Htn	Geog 4217.85.3F*	Portlock, N. Voyage round the world. London, 1789.
	Geog 4217.85.5	Burney, J. Memoir on the voyage...in search of La Pérouse. London, 1820.
	Geog 4217.85.13	Lapérouse, Jean François de Galaup. Voyage de Lapérouse autour du monde. Paris, 1965.
	Geog 4217.85.15	Gordon, Mona. The mystery of La Pérouse. Christchurch, N.Z., 1961.
	Geog 4217.88	Pagès, P.M.F. de. Nouveau voyage...en 1788. Paris, 1797. 3v.
	Geog 4217.89	Viana, F.J. de. Diario del viaje explorador...en 1789-94. Cerrito de la Victoria, 1849.
Htn	Geog 4217.90*	Vancouver, G. Voyage of discovery...in 1790-1792. London, 1798. 3v.
	Geog 4217.90.3	Vancouver, G. A voyage of discovery of the North Pacific Ocean. London, 1801. 6v.
	Geog 4217.90.7	Fleurieu, Charles P. Claret de. Voyage autour du monde...1790-1792. Paris, 1798-1800. 4v.
	Geog 4217.90.15	Fleurieu, Charles P. Claret de. A voyage round the world. London, 1801. 3v.
	Geog 4217.90.18	Fleurieu, Charles P. Claret de. Die neuste Reise um die Welt in den Jahren 1790, 1791 und 1792 von Etienne Marchand. Leipzig, 1801? 2v.
	Geog 4217.91.2	Marshall, James S. Vancouver's voyage. 2. ed. Vancouver, 1967.
Htn	Geog 4217.97*	Milet-Mureau, M.L.A. Voyage de La Pérouse. Paris, 1797. 4v.
Htn	Geog 4217.97PF*	Milet-Mureau, M.L.A. Voyage de La Pérouse. Atlas. Paris, 1797.
Htn	Geog 4218.03*	Langsdorff, G.H. von. Voyages and travels...in 1803-1805. London, 1813. 2v.
Htn	Geog 4218.03.3*	Krusenstern, A.J. von. Reise um die Welt in den Jahren 1803-1806 auf Befehl seiner Kaisere. v.1-2. Berlin, 1811-12. 3v.
Htn	Geog 4218.03.5*	Krusenstern, A.J. Voyage around the world...1803-1806. London, 1813. 2v.
Htn	Geog 4218.03.7*	Krusenstern, A.J. Voyage autour du monde...1803-06. Paris, 1821. 2v.
Htn	Geog 4218.03.8F*	Krusenstern, A.J. Voyage autour du monde. Atlas. Paris, 1821.
Htn	Geog 4218.03.9*	Lisianskii, I. Voyage round the world...1803-06. London, 1814.
	Geog 4218.03.10	Lisianskii, I. Putesh. vokrug sveta na kor. ""Neva", 1803-1806. Moskva, 1947.
	Geog 4218.03.11	Camus, A.G. Rapport à l'Institut national. Paris, 1803.
	Geog 4218.03.12	Nevskii, Vladimir V. Perv. putesh. rossii. vokrug sveta, 1803-1806. Moskva, 1951.
	Geog 4218.03.13	Lupach, V.S. I.F. Kruzenshtern. Moskva, 1953.
	Geog 4218.03.14	Nevskii, Vladimir V. Vokrug sveta pod russkim flagom. Moskva, 1953.
Htn	Geog 4218.06*	Campbell, A. Voyage round the world from 1806-12. Edinburgh, 1816.
Htn	Geog 4218.12*	Langsdorff, G.H. von. Bemerkungen auf einer Reise um die Welt in den Jahren 1803-1807. Frankfurt, 1812. 3v.
	Geog 4218.16	Roquefeuil, C. de. Voyage autour du monde. Paris, 1823. 2v.
	Geog 4218.17	Arago, J. Narrative of a voyage round the world. London, 1823.
	Geog 4218.17.3	Arago, J. Souvenirs...voyage...du monde. Paris, 1839.
	Geog 4218.17.8	Delano, Amasa. Narrative of voyages and travels. N.Y., 1970.
	Geog 4218.17.10	Bassett, M.M. Realms and islands. London, 1962.
	Geog 4218.24	Memoires du Capitaine Péron. Paris, 1824. 2v.
	Geog 4218.24.5PF	La Touanne, E. de. Album pittoresque de la frégate la Thétis et de la corvette l'Esperance. Paris, 1828.
Htn	Geog 4218.26*	Duhaut-Cilly, A. Voyage autour du monde. Paris, 1834.
	Geog 4218.26.5	Duhaut-Cilly, A. Viaggio intorno al globo. Napoli, 1842.
	Geog 4218.26.10	Le Netrel, Edmond. Voyage of the Lerós. Los Angeles, 1951.
	Geog 4218.26.17	Luetke, Fedor P. Puteshestvne vokrug sveda na volunom shliupe "Seniavin" 1826-1829. 2. izd. Moskva, 1948.
	Geog 4218.27	Allyn, Gurdon L. The old sailor's story. Norwich, Conn., 1879.
	Geog 4218.30	Holman, James. A voyage round the world. v.4. London, 1835.
	Geog 4218.30.5	Meyen, F.J.F. Reise um die Erde. Berlin, 1834-35. 2v.
	Geog 4218.31	Reynolds, J.N. Voyage of the United States frigate Potomac. N.Y., 1835.
	Geog 4218.31.2	Reynolds, J.N. Voyage of the United States frigate Potomac. N.Y., 1835.
	Geog 4218.31.3	Warriner, F. Cruise of the United States frigate Potomac round the world. N.Y., 1835.
	Geog 4218.33.10A	Fanning, E. Voyages and discoveres in the South Seas, 1792-1832. Salem, 1924.
	Geog 4218.33.10B	Fanning, E. Voyages and discoveries in the South Seas, 1792-1832. Salem, 1924.
	Geog 4218.33.25	Laplace, Cyril P.T. Voyage autour du monde. Paris, 1833-39. 4v.
	Geog 4218.33.27PF	Laplace, Cyril P.T. Voyage autour du monde. Atlas. Paris, 1835.
	Geog 4218.35	Belcher, Edward. Narrative of a voyage round the world...1836-42. London, 1843. 2v.
	Geog 4218.35.3	Ruschenberger, W.S.W. A voyage round the world. Philadelphia, 1838.
	Geog 4218.37.5	Dumont d'Urville, J. Malerische Reise um die Welt. Leipzig, 1837.
	Geog 4218.38	Erskine, C. Twenty years before the mast. Boston, 1890.
	Geog 4218.38.2	Taylor, F.W. A voyage around the world. v.1-2. 9th ed. New Haven, 1850.
	Geog 4218.38.3	Taylor, F.W. A voyage around the world. v.1-2. 9th ed. New Haven, 1847.
	Geog 4218.38.4	Murrell, William M. Cruise frigate Columbia around the world...1838-1840. Boston, 1840.
	Geog 4218.40	Lafond, G. Quinze ans de voyages. Paris, 1840. 2v.

	Geog 4218.40.5F	Lafond, G. Fragments de voyages autour du monde. Paris, 1861.
	Geog 4218.40.10	Blok, G.K. Dva goda iz zhizni russkago moriaka. Sanktpeterburg, 1854. 2v.
Htn	Geog 4218.41*	Simpson, George. Overland journey round the world. Philadelphia, 1847.
	Geog 4218.41.2	Simpson, George. Overland journey round the world. Philadelphia, 1847.
	Geog 4218.41.3	Simpson, George. Narrative of a journey round the world. London, 1847. 2v.
	Geog 4218.45	Bille, C.S.A. Beretning om corvetten Galathea's reise omkring jorden i 1845. 2. udg. Kjøbenhavn, 1853.
	Geog 4218.45.5	Seemann, Berthold. Narrative of the voyage of H.M.S. Herald during the years 1845-51...being a circumnavigation of the globe. London, 1853.
NEDL	Geog 4218.46.3	Pfeiffer, I. A woman's journey round the world. London, 1846.
	Geog 4218.46.5	Pfeiffer, I. A lady's voyage round the world. N.Y., 1852.
	Geog 4218.46.7	Pfeiffer, I. A lady's second journey...world. N.Y., 1856.
	Geog 4218.48	Beck, Christian. Reise um die Welt. 11e Aufl. Dresden, 1907.
	Geog 4218.49	Coffin, George. Pioneer voyage to California and round the world, 1849-1852. Chicago? 1908.
	Geog 4218.49.5	Davis, R.C. Reminiscences of a voyage around the world. Ann Arbor, 1869.
	Geog 4218.50	Gerstaecker, F. Narrative of a journey round the world. N.Y., 1853.
	Geog 4218.50.3	Marmier, Xavier. Voyage d'une femme autour du monde. v.1-2. Bruxelles, 1853.
	Geog 4218.51	Stogman, C.D. Fregatten Eugenies resa omkring jorden, aren 1851-1853. v.1-2. Stockholm, 1855?
	VGeog 4218.51.5	Svenska Vetenskaps-Akademien, Stockholm. Voyage autour du monde sur la frégate suédoise l'Eugénie executé pendant des années 1851-1853. Stockholm, 1858-74.
	Geog 4218.52	Spalding, J.W. The Japan expedition. N.Y., 1855.
	Geog 4218.55.5	Nordhoff, Charles. Man-of-war life. Cincinnati, 1856.
	Geog 4218.55.9	Nordhoff, Charles. Man-of-war life. N.Y., 1883.
	Geog 4218.57	Scherzer, Karl. Narrative of circumnavigation of globe...1857-58. London, 1861-63. 3v.
	Geog 4218.57.3	Miller, Thomas. Over five seas and oceans. N.Y., 1894.
	Geog 4218.57.10	Wallisch, Friedrich. Sein Schiff hiess Novara. Wien, 1966.
	Geog 4218.58	Dorr, D.F. A colored man round the world. Cleveland? 1858.
	Geog 4218.59	Hoffman, William. The Monitor; or, jottings of a N.Y. merchant during a trip around the globe. N.Y., 1863.
Htn	Geog 4218.61*	D'Wolf, J. Voyage to the North Pacific. Cambridge, 1861.
	Geog 4218.61.3	Ainsworth, W.F. All round the world. London, 1861. 2v.
	Geog 4218.66	Weppner, M. The North Star and the Southern Cross. 3. American ed. Albany, 1880.
	Geog 4218.66.5	Weppner, M. The North Star and the Southern Cross. Albany, 1876. 2v.
	Geog 4218.68	Smiles, S. Round the world. N.Y., 1871.
	Geog 4218.68.5	Bell, William M. Other countries. London, 1872. 2v.
	Geog 4218.68.10	Jentsch, August. Ein Mann und ein Traktor auf Weltreise. v.2. Wien, 1968-69.
	Geog 4218.69	Coffin, C.C. Our new way round the world. Boston, 1869.
	Geog 4218.70	Longworth, M.T. Teresina peregrina. London, 1874. 2v.
	Geog 4218.70.3	Peebles, James. Around the world. 2. ed. Boston, 1876.
	Geog 4218.70.4	Seward, William H. William H. Seward's travels around the world. N.Y., 1873.
	Geog 4218.70.5	Seward, William H. William H. Seward's travels around the world. N.Y., 1873.
	Geog 4218.70.6	Seward, William H. William H. Seward's travels around the world. N.Y, 1874.
	Geog 4218.70.8	Beaumont, Thomas E. Pencilling by the way: a constitutional voyage round the world: 1870 and 1871. London, 1971.
	Geog 4218.71	Hübner, J.A. Ramble round the world. N.Y., 1874.
	Geog 4218.71.2	Hübner, J.A. Promenade autour du monde, 1871. 2e éd. Paris, 1873. 2v.
	Geog 4218.71.3	Hübner, J.A. Ramble round the world. London, 1874. 2v.
	Geog 4218.71.7	Hübner, J.A. Promenade autour du monde, 1871. Paris, 1877. 2v.
	Geog 4218.71.9F	Hübner, J.A. Passeggiata intorno al mondo. Milano, 1879.
	Geog 4218.71.15	Adams, N. A voyage around the world. Boston, 1871.
	Geog 4218.72	Simpson, William. Meeting the sun; journey all round the world. London, 1874.
	Geog 4218.72.2	Simpson, William. Meeting the sun; journey all round the world. London, 1877.
	Geog 4218.72.5	Hildebrandt, E.W. Reise um die Erde. 3. Aufl. Boston, 1872.
	Geog 4218.72.15	Brooks, James. A seven month's run up and down and around the world. N.Y., 1872.
	Geog 4218.73.5	Walworth, E.H. An old world...travels around the world. N.Y., 1877.
	Geog 4218.73.9	Guillemard, A.C. Over land and sea. London, 1875.
	Geog 4218.73.10	Dupuy de Lôme, E. De Madrid a Madrid. Madrid, 1877.
	Geog 4218.73.15	Cook, Thomas. Letters from the sea and from foreign lands. London, 1873.
	Geog 4218.73.20	Bastian, Adolf. Geographische und ethnologische Bilder. Jena, 1873.
	Geog 4218.74	Prime, E.D.C. Around the world, sketches of travel. N.Y., 1874.
	Geog 4218.75	Wieting, M.E. Prominent incidents in life of John M. Wieting...around the world. N.Y., 1889.
	Geog 4218.76	Curtis, B.R. Dottings round the circle. Boston, 1876.
	Geog 4218.76.2	Curtis, B.R. Dottings round the circle. 3. ed. Boston, 1877.
	Geog 4218.76.3	Hendrix, E.R. Around the world. St. Louis, 1881.
	Geog 4218.76.5	Campbell, J.F. My circular notes. v.1-2. London, 1876.
	Geog 4218.76.9	Bedinello, Ugo. Diario del viaggio intorno al globo. Trieste, 1876.
	Geog 4218.76.15	Fenzi, S. Gita intorno alla terra dal gennaio al settembre dell'anno 1876. Firenze, 1877.
	Geog 4218.78	Brassey, A. (Mrs.). Around the world in the "Sunbeam". N.Y., 1878.
	Geog 4218.78.3	Brassey, A. (Mrs.). A voyage in the "Sunbeam". London, 1879.
	Geog 4218.78.5	Brassey, A. (Mrs.). Around the world in the "Sunbeam". N.Y., 1879.
	Geog 4218.78.7A	Brassey, Thomas. The "Sunbeam". London, 1918.
	Geog 4218.78.7B	Brassey, Thomas. The "Sunbeam". London, 1918.

Classified Listing

Geog 4215 - 4220 Voyages and travels in general - Individual travels - Circumnavigations (By date of travel) - cont.

	Geog 4218.78.8	Brassey, A. (Mrs.). A voyage in the "Sunbeam". Chicago, 1881.
	Geog 4218.78.9	Parry, S.H. Jones. My journey round the world. London, 1881. 2v.
	Geog 4218.78.11	Carnegie, Andrew. Notes of a trip round the world. N.Y., 1879.
	Geog 4218.78.15	Wernich, Agathon. Geographisch-medicinische Studien. Berlin, 1878.
	Geog 4218.78.20	Nichols, E.P. The Ocean Chronicle, 1878-1891. N.Y., 1941.
NEDL	Geog 4218.79	Wylie, A.H. Chatty letters from the East and West. London, 1879.
	Geog 4218.79.10A	Carnegie, Andrew. Round the world. Garden City, 1933.
	Geog 4218.79.10B	Carnegie, Andrew. Round the world. Garden City, 1933.
	Geog 4218.80	Coop, Timothy. A trip around the world. Cincinnati, 1882.
	Geog 4218.80.5	Vetromile, E. A tour in both hemispheres. N.Y., 1880.
	Geog 4218.80.7	Packard, J.F. Grant's tour around the world. Cincinnati, 1880.
	Geog 4218.81	Ludwig, S. Um die Welt ohne zu wollen. Prag, 1881.
	Geog 4218.81.5	Glass, C. The world, round it and over it. Toronto, 1881.
	Geog 4218.81.9	Bennett, D.M. A truth seeker around the world. N.Y., 1882. 4v.
	Geog 4218.82.3	Pitt, George. The collected remarkable travels of G. Pitt. 3. ed. Glasgow, 1887.
	Geog 4218.83.2	Pidgeon, D. An engineer's holiday...trip from 0 degrees to 0 degrees. 2. ed. London, 1883.
	Geog 4218.83.5	Bridges, F.D. Journal of a lady's travels. London, 1883.
	Geog 4218.83.10	Lambert, C. Voyage of the Wanderer. London, 1883.
	Geog 4218.84	Ballou, M.M. Due west, or Round the world. Boston, 1884.
	Geog 4218.84.2	Ballou, M.M. Due west, or Round the world in ten months. 2. ed. Boston, 1884.
	Geog 4218.84.5	Ballou, M.M. Due west, or Round the world. 3. ed. Boston, 1885.
	Geog 4218.84.9	Ballou, M.M. Due west, or Round the world. 9. ed. Boston, 1891.
	Geog 4218.84.13	Ballou, M.M. Foot-prints of travel. Boston, 1889.
	Geog 4218.84.14	Ballou, M.M. Foot-prints of travel. Boston, 1889.
	Geog 4218.84.15	Ballou, M.M. Foot-prints of travel. Boston, 1892.
	Geog 4218.85	Richardson, D.N. A girdle round the earth. Chicago, 1888.
	Geog 4218.85.5	Mitford, R.C.W.M. Orient and Occident. London, 1888.
	Geog 4218.85.10	Arme-Webb, A. (Mrs.). A glimpse of two hemispheres. Hertford, 1885.
	Geog 4218.86	Raum, G.E. A tour around the world. N.Y., 1886.
	Geog 4218.87	Stevens, Thomas. Around the world on a bicycle. N.Y., 1887-88. 2v.
	Geog 4218.87.7	M'Collester, S.H. Round the globe. 3. ed. Boston, 1890.
	Geog 4218.87.12	Cecil, Evelyn. Notes of my journey round the world. London, 1889.
	Geog 4218.87.13	Caine, William S. A trip round the world in 1887-88. London, 1891.
	Geog 4218.87.14	Caine, W.S. A trip round the world in 1887-88. London, 1892.
	Geog 4218.87.15	Paterson, Florence E. Reminiscences of a skipper's wife. London, 19- .
	Geog 4218.87.20	Floyd Jones. Letters from the Far East. N.Y., 1887.
	Geog 4218.88	Lewis, I.N. Pleasant hours in sunny lands. Boston, 1888.
	Geog 4218.89.5	Gillis, C.J. Around the world in seven months. N.Y., 1891.
	Geog 4218.89.7	Wetmore, E.B. A flying trip around the world. N.Y, 1891.
	Geog 4218.90	Leland, Lillian. Travelling alone, a woman's journey. N.Y., 1890.
	Geog 4218.90.5	Cotes, Sara J.D. A social departure. N.Y., 1890.
	Geog 4218.90.6	Cotes, Sara J.D. A social departure. London, 1890.
	Geog 4218.91	Cotes, Sara J.D. A social departure. How Orthodocia and I went round the world by ourselves. N.Y., 1891.
	Geog 4218.91.4	Vassar, J.G. Twenty years around the world. N.Y., 1891.
	Geog 4218.92	Esmonde, T.H.G. Round the world with the Irish delegates. Dublin, 1892.
	Geog 4218.92.3	Peters, G.H. Impressions of a journey round the world. London, 1897.
	Geog 4218.92.5F	Radde, G.F.R. 23.000 mil'na iakhte "Tamara". Sanktpeterburg, 1892-93. 2v.
	Geog 4218.92.7	Scott, Clement. Pictures of the world. London, 1894.
	Geog 4218.92.15	Aubertin, J.J. Wanderings and wonderings. London, 1892.
	Geog 4218.92.20	MacGregor, John. Toil and travel. London, 1892.
	Geog 4218.93	Thompson, F.D. In the track of the sun. N.Y., 1893.
	Geog 4218.93.5	Stephens, A.A. A Queenslander's travel notes. Sydney, 1894.
	Geog 4218.93.10	Egerton, Alice A. Glimpses of four continents. London, 1894.
	Geog 4218.94	Dunn, S.H. The world's highway. London, 1894.
	Geog 4218.95	Raum, G.E. A tour around the world. N.Y., 1895.
	Geog 4218.95.5	Brewster, F. Carroll. From Independence Hall around the world. Philadelphia, 1895.
	Geog 4218.96	Fraser, J.F. Round the world on a wheel. N.Y., 1899.
	Geog 4218.97	Barrows, John D. A world-pilgrimage. Chicago, 1897.
	Geog 4218.98	Albrey, J. d'. Du Tonkin au Havre. Paris, 1898.
	Geog 4218.98.5	McMahon, William. Journey with the sun around the world. Cleveland, 1900.
	Geog 4218.98.10	Mikhailovskii, N.G. Iz dnevnikov krugosvetnogo puteshestviia. Moskva, 1952.
	Geog 4219.00	Slocum, Joshua. Sailing alone around the world. N.Y., 1900.
	Geog 4219.00.5	Slocum, Joshua. Sailing alone around the world. N.Y., 1900.
	Geog 4219.00.10	Slocum, Joshua. Sailing alone around the world. London, 1948.
	Geog 4219.01	Loewenbach, Lothaire. Promenade autour du globe. Paris, 1903.
	Geog 4219.04	Pearson, D.G. Trip to the Phillipines...in 1904. n.p., n.d.
	Geog 4219.05.3	Treves, F. The other side of the lantern. London, 1905.
	Geog 4219.05.4	Treves, F. The other side of the lantern. London, 1905.
	Geog 4219.05.10	Treves, F. The other side of the lantern. London, 1928.
	Geog 4219.06	Huber, Max. Tagebuchblätter aus Siberien, Japan. Zürich, 1906.
	Geog 4219.06.7	Sebok, I. Öt világrészun heresztül. Budapest, 1934.
	Geog 4219.07	Aldao, C.A. A través del mundo. Buenos Aires, 1907.
	Geog 4219.07.5	Stephens, Edwin William. Around the world; a narrative in letter form of a trip around the world, from October 1907 to July 1908. Columbia, 1909.
	Geog 4219.08.2	Treves, F. Other side of the lantern. London, 1908.
	Geog 4219.08.5	Carter, James. In the wake of the setting sun. London, 1908.
	Geog 4219.08.10	Bertrand, A. Quelques notes sur les conférences de Tokyo et de Shanghai. Genève, 19- .

Geog 4215 - 4220 Voyages and travels in general - Individual travels - Circumnavigations (By date of travel) - cont.

	Geog 4219.09	Bidwell, Daniel D. As far as the East is from the West. Hartford, 1910.
	Geog 4219.09.5	Pageot, G. A travers les pays jaunes. Paris, 1909.
	Geog 4219.10	Franck, H.A. Vagabond journey around the world. N.Y., 1910.
	Geog 4219.10.2	Franck, H.A. A vagabond journey around the world. N.Y., 1911.
	Geog 4219.11	Pearse, Albert W. Recent travel. Sydney, 1914.
	Geog 4219.11.5	Bertram, C. A magician in many lands. N.Y., 1911.
	Geog 4219.11.7	Dittmar, J. Eine Fahrt um die Welt. Berlin, 1911.
	Geog 4219.12	Zobeltitz, Fedor von. Ein Bummel um die Welt. Berlin, 1912.
	Geog 4219.20F	Kincaid, Earle H. History and cruises of the United States ship Whipple. Constantinople, n.d.
	Geog 4219.20.5	Chevalier, Haakon Maurice. The last voyage of the schooner Rosamond. London, 1970.
	Geog 4219.21	Büchler, E. Rund um die Erde. Bern, 1921.
	Geog 4219.23	Northcliffe, A.C.W.H. My journey round the world. Philadelphia, 1923.
	Geog 4219.23.5	O'Brien, Conor. Across three oceans. London, 1949.
	Geog 4219.24	Clements, Rex. A gipsy of the Horn. London, 1924.
	Geog 4219.24.2	Clements, Rex. A gipsy of the Horn. 2. ed. Boston, 1925.
	Geog 4219.24.5	Blasco Ibáñez, Vicente. La vuelta al mundo de un novelista. Valencia, 1924-25. 3v.
	Geog 4219.25	Dennison, C.S. Around the world with Texaco. Houston, 1925.
	Geog 4219.25.3	Knight, L.L. Trucking the sunset. Atlanta, 1925.
	Geog 4219.25.5	Reinhardt, Walther. Querweltein. Berlin, 1925.
	Geog 4219.25.10A	Halliburton, Richard. The royal road to romance. Indianapolis, 1925.
	Geog 4219.25.10B	Halliburton, Richard. The royal road to romance. Indianapolis, 1925.
	Geog 4219.25.15	Halliburton, Richard. The royal road to romance. Garden City, 1925.
	Geog 4219.26	Wells, Linton. Around the world in twenty-eight days. Boston, 1926.
	Geog 4219.27	Lunn, Henry. Round the world with a dictaphone. London, 1927.
	Geog 4219.27.2	Lunn, Henry. Round the world with a dictaphone. London, 1928.
	Geog 4219.27.5	Martin, Lillien J. Round the world with a psychologist. San Francisco, 1927.
	Geog 4219.28	Cook, Thomas, firm, publishers, London. The supreme travel-adventure: around the world in the "Franconia" 1929. N.Y., 1928.
	Geog 4219.28.5	Kircheiss, C. Meine Weltumsegelung mit dem Fischkutter Hamburg. 2. Aufl. Berlin, 1928.
	Geog 4219.28.10	Costes Dieudonné. La grande croisière de Costes et Le Brie. Paris, 1928.
	Geog 4219.29.2	Mann, Erik. Rundherm. 3.-4. Aufl. Berlin, 1929.
	Geog 4219.29.5	Baus, Thomas J. 30,000 miles around the world. Philadelphia, 1929.
Htn	Geog 4219.29.10F*	Vanderbilt, William K. Taking one's own ship around the world. N.Y., 1929.
	Geog 4219.29.15	Lewis, Aswold. Because I've not been there before. London, 1929.
	Geog 4219.30	Nelson, Robert. World beaters. Boston, 1930.
	Geog 4219.32	Fletcher, A.C.B. Keep moving. Chiago, 1932.
	Geog 4219.32.5	Gilmore, A.F. Yes, 'tis round. Boston, 1932.
	Geog 4219.32.10A	Halliburton, Richard. The flying carpet. Indianapolis, 1932.
	Geog 4219.32.10B	Halliburton, Richard. The flying carpet. Indianapolis, 1932.
	Geog 4219.32.11	Halliburton, Richard. The flying carpet. London, 1933.
	Geog 4219.32.15	Dinglreiter, Senta. Deutsches Mädel auf Fahrt um die Welt. Leipzig, 1932.
	Geog 4219.32.20	Katz, Richard. Ernte. Zürich, 1935.
	Geog 4219.32.25	Katz, Richard. Ein Bummel um die Welt. Zürich, 1936.
	Geog 4219.32.35.5	Pidgeon, H. Around the world single-handed. N.Y., 1932.
Htn	Geog 4219.33F*	Vanderbilt, William K. West made East with the loss of a day. N.Y., 1933.
	Geog 4219.33.15	Gezork, H. So sah ich die Welt. 12e Aufl. Kassel, 1938?
	Geog 4219.34.5	Atkinson, B. The Cingalese prince. Garden City, 1934.
	Geog 4219.34.10	Unwin, S. Two young men see the world. London, 1934.
	Geog 4219.35	Benn, Wedgwood. Beckoning horizon. London, 1935.
	Geog 4219.36	Merrick, Lesley. A good time. N.Y., 1936.
	Geog 4219.36.5	Johnson, Irving. Westward bound in the schooner Yankee. N.Y., 1936.
	Geog 4219.36.10	Hurja, E.E. Westward ho - fare paid. Juneau, Alaska, 1936.
	Geog 4219.36.15	Bernicot, Louis. The voyage of Anahita; single-handed round the world. London, 1953.
	Geog 4219.37	Belfrage, C. Away from it all. N.Y., 1937.
	Geog 4219.37.10	Villiers, A.J. Cruise of the Conrad...1934-1936. London, 1937.
	Geog 4219.37.12	Marine Historical Association, Mystic, Conn. The Joseph Conrad. Mystic, 195-.
	Geog 4219.38	Hypes, J.L. Knights of the road. Washington, 1938.
	Geog 4219.38.6	Bradshaw, M.J. Third class world. Alliance, 1939.
	Geog 4219.39	Alford, B. Around the world on $100.00 a month. Emmaus, 1939.
	Geog 4219.39.5	Seligman, A. The voyage of the Cap Pilar. London, 1939.
	Geog 4219.40.10	Michael, Rudolf. Roman einer Weltreise. Hamburg, 1940.
	Geog 4219.41	Parsons, C. Vagabondage. London, 1941.
	Geog 4219.41.5	Wiener, P. Last man around the world. N.Y., 1941.
	Geog 4219.41.10	Faber, Kurt. Tausend und ein Abenteuer. Berlin, 1941.
	Geog 4219.43	Metzelaar, A.C. Europa ahoy! De geschiedenis van een zeilschip. Amsterdam, 1943.
	Geog 4219.47	Petterson, H. Westward ho with the Albatross. 1. ed. N.Y., 1953.
	Geog 4219.51	Skolov, A.V. Tri prugosvetiykh plavaniia M.P. Lazareva. Moskva, 1951.
	Geog 4219.51.5	Tregaskis, R.W. Seven leagues to paradise. 1. ed. Garden City, 1951.
	Geog 4219.51.10	Mulhauser, G.H.P. The cruise of the Amaryllis. London, 1951.
	Geog 4219.53	Le Toumelin, J.I. Kurun around the world. N.Y., 1955.
	Geog 4219.55	Vocino, Michele. Marinai italiani e iberici sulle vie delle Indie. Roma, 1955.
	Geog 4219.55.5	Crove, Bill. Heaven, hell and salt water. London, 1957.
	Geog 4219.56	Wollschläger, A. Grosse Weltreise mit A.E Johann (pseud.). 1. Aufl. Gutersloh, 1956.
	Geog 4219.58.5	Fenyvessy, J. Mein Traum wurde Wirklichkeit. 10. Aufl. Köln, 1961.
	Geog 4219.58.10	Bedford, Jimmy. Around the world on a nickel. New Delhi, 1967.

48

Classified Listing

Geog 4215 - 4220 Voyages and travels in general - Individual travels - Circumnavigations (By date of travel) - cont.

Geog 4219.59 Hiscock, Eric C. Beyond the west horizon. London, 1963.
Geog 4219.60 Baty, Eben Neal. Citizen abroad. N.Y., 1960.
Geog 4219.60.5 Uittenbogaard, Leo. De wereld. Den Haag, 1958-60. 3v.
Geog 4219.62 Gandon, Yves. A la recherche de l'Eden. Paris, 1962.
Geog 4219.64 Menon, Kumara P.S. Journey round the world. Bombay, 1966.
Geog 4219.64.5 Lewis, David Henry. Children of three oceans. London, 1969.
Geog 4219.65 Graham, Robin Lee. Dove. 1. ed. N.Y., 1972.
Geog 4219.66 Marti-Ibáñez, Félix. Journey around myself. N.Y., 1966.
Geog 4219.66.5 Simpson, Colin. Sir Francis Chichester: voyage of the century. London, 1967.
Geog 4219.66.10 Rowland, John. Lone adventurer: the story of Sir Francis Chichester. London, 1968.
Geog 4219.67 Chichester, Francis Charles. Gipsy Moth circles the world. N.Y., 1967.
Geog 4219.69 Price, Willard. Odd way around the world. N.Y., 1969.
Geog 4219.69.5 Knox-Johnston, Robin. A world of my own: the single-handed, non-stop circumnavigation of the world in Suahili. London, 1969.
VGeog 4219.69.10 Ogrin, Miran. Na juga sveta. Ljubljana, 1969.
Geog 4219.70 Blyth, Chay. The impossible voyage. London, 1971.
Geog 4219.72.5 Cluzel, Magdeleine. Au fil de l'eau, autour de la terre. Paris, 1973.

Geog 4225 - 4230 Voyages and travels in general - Individual travels - America and other regions (By date of travel)

Geog 4227.35.1 Atkins, John. A voyage to Guinea, Brazil, and the West Indies. London, 1970.
Geog 4227.46 Goelet, Francis. The voyages and travels of Francis Goelet, 1746-1758. N.Y., 1970.
Geog 4227.85F Strange, James. Journal and narrative of the...expedition from Bombay to the N.W. coast of America. Madras, 1928.
Geog 4227.89 Langstedt, F.L. Reisen nach Sudamerika, Asien. Hildesheim, 1789.
Geog 4227.90 Ingraham, Joseph. Journal of the brigantine Hope on the voyage to the northwest coast of North America, 1790-92. Barre, 1971.
Geog 4228.10 Reynolds, Stephen. The voyage of the New Hazard...1810-1813. Salem, 1938.
Geog 4228.16 Montulé, E. de. Voyage en Amerique. Paris, 1821. 2v.
Geog 4228.17 Montulé, E. de. A voyage to North America and West Indies in 1817. London, 1821.
Geog 4228.17.5 Montulé, E. de. Travels in America. 1st ed. Bloomington, 1951.
Geog 4228.17.10 Golovnin, Vasilii M. Puteshestvie vokrug sveta. Moskva, 1965.
Geog 4228.23 Landolphe, Jean François. Mémoires du capitaine Landolphe. Paris, 1823. 2v.
Geog 4228.26 Boelen, Jacobus. Reize naar de oost- en westkust van Zuid-Amerika. Amsterdam, 1835-36. 3v.
Geog 4228.28 Combier, C. Voyage au Golfe de Californie. Paris, 1864.
Geog 4228.28.5 Davis, S.H. Journal of Captain S.H. Davis, a Gloucester sea-captain, 1828-1846. Norwood, 1922.
Geog 4228.31 Compendious view of...travels. London, 1831.
Geog 4228.35 Ferrer, Antonio C. Pases por Europa y América, en 1835 y 1836. Madrid, 1838.
Geog 4228.40 Fitch, F.Y. Life, travels...of an American wanderer...Alonzo P. De Milt. N.Y., 1883.
Geog 4228.45A Van Denburgh, E.D. My voyage in the U.S. frigate "Congress". N.Y., 1913.
Geog 4228.45B Van Denburgh, E.D. My voyage in the U.S. frigate "Congress". N.Y., 1913.
Geog 4228.49 Lewis, Oscar. Sea routes to the gold fields. 1st ed. N.Y., 1949.
Htn Geog 4228.50* Bryant, W.C. Letters of a traveller. N.Y., 1850.
Geog 4228.50.2 Bryant, W.C. The picturesque souvenir. N.Y., 1851.
Geog 4228.50.5 Bryant, W.C. Letters of a traveller. 4. ed. N.Y., 1855.
Geog 4228.51 Bryant, W.C. Letters of a traveller. 3. ed. N.Y., 1851.
Geog 4228.56A Ossoli, M.F. At home and abroad. 3. ed. Boston, 1856.
Geog 4228.56B Ossoli, M.F. At home and abroad. 3. ed. Boston, 1856.
Geog 4228.56.5 Ossoli, M.F. At home and abroad. Boston, 1860.
Geog 4228.57 Ennemoser, F.J. Reise vom Mittelrhein...nach dem Nordamerikanischen Freistaaten. Kaiserlautern, 1864.
Geog 4228.59 Warren, T.R. Dust and foam, or Three oceans and two continents. N.Y., 1859.
Geog 4228.61 Bradford, Ruth. Maskee! The journal and letters of Ruth Bradford, 1861-1872. Hartford, 1938.
Geog 4228.70 Pumpelly, Raphael. Across America and Asia. N.Y., 1870.
Geog 4228.71.5 Pumpelly, Raphael. Across America and Asia. N.Y., 1871.
Geog 4228.73 Barra de Cobo, Maipina de la. Mis impresiones y mis vicisitudes en mi viaje a Europa. Buenos Aires, 1878.
Geog 4228.79 MacMichael, M. A landlubber's log of a voyage round the "Horn". Philadelphia, 188-?
Geog 4228.85 Beehler, W.H. The cruise of the Brooklyn. Philadelphia, 1885.
Geog 4228.86 Molinari, M.G. Au Canada et aux montagnes. Paris, 1886.
Geog 4228.86.10 Jackson, Helen F.H. Glimpses of three coasts. Boston, 1891.
Geog 4228.87 Francis, Harriet E. Across the meridians. N.Y., 1887.
Geog 4228.89 Barneby, W.H. New Far West and old Far East. London, 1889.
Geog 4228.89.5 Faucher de Sant-Maurice, N.H.E. Loin du Pays. Souvenirs d'Europe, d'Afrique, et d'Amérique. Quebec, 1889. 2v.
Geog 4228.97 Riesenberg, Felix. Under sail. N.Y., 1918.
Geog 4228.98 Trevelyan, Charles Philips. Letters from North America and the Pacific, 1898. London, 1969.
Geog 4228.99 Riseis, G. de. Dagli stati uniti alle Indie. Roma, 1899.
Geog 4229.00.5 Winter, James M. New York to Alaska...May to July, 1900. Middletown, N.Y, 1943.
Geog 4229.04 Armstrong, W.N. Around the world with a king. London, 1904.
Geog 4229.05 Macdonald, A. In search of El Dorado. London, 1905.
Geog 4229.11 Wilmore, C. Ray. Square rigger round the Horn. Camden, Me., 1972.
Geog 4229.25.5 Kleinschmitt, Edmund. Durch Werkstätten und Gassen dreier Erdteile. 2. Aufl. Hamburg, 1928.
Geog 4229.26 Huxley, Aldous. Jesting pilate; an intellectual holiday. N.Y., 1926.
Geog 4229.28 Log of the ketch Seven Bells. n.p., 1928?
Geog 4229.30 Toller, Ernest. Querdurch; Reisebilder und Reden. Berlin, 1930.
Geog 4229.35 Carré, J.M. Promenades dans trois continents. Paris, 1935.

Geog 4225 - 4230 Voyages and travels in general - Individual travels - America and other regions (By date of travel) - cont.

Geog 4229.42 Faber, Kurt. Der göttliche Vagabund. Stuttgart, 1942.
Geog 4229.47 Vargas Ugarte, Ruben. Relaciones de viajes. Lima, 1947.
Geog 4229.71 Shur, Leonid A. K beregam Novogo Sveta. Moskva, 1971.
Geog 4229.72 Struve, Wolfgang. Unglaubliche Wirklichkeit. Salzburg, 1972.
Geog 4229.73 Cousteau, Jacques Yves. Trois aventures de la Calypso: Galapago, Titicaca, Trous Bleus. Paris, 1973.

Geog 4235 - 4240 Voyages and travels in general - Individual travels - Atlantic Ocean (By date of voyage)

Geog 4235.70.2 Sousa, F. de. Tratado das ilhas novas. Ponta Delgada, 1884.
Geog 4237.00 Stearns, R.P. The course of Captain Edmond Halley in the year 1700. n.p., 1936.
Htn Geog 4237.52.5* Pierce, Nathaniel. Account of great dangers...and remarkable deliverance of N. Pierce. n.p., n.d.
Geog 4237.56 Pierce, Nathaniel. Account of the great dangers...of Captain N. Pierce. Boston, 1756.
Geog 4237.61 An account of the voyages and cruizes of Captain Walker. Boston, 1761.
Geog 4237.78 Alexandre e Silva, Elias. Relação, ou Noticia particular da infeliz viajem da nas de Nassa Senhora da Ajuda. Lisboa, 1778.
Geog 4237.83 Burall, Paul. Cornwall to America in 1783 from the journal of Paul Burall (1755-1826). n.p., 19- .
Htn Geog 4238.03* Sutherland, D. A diary kept by Reverend David Sutherland on a voyage from Greenock Scotland to New York. Woodsville, 1910.
Geog 4238.32.10 Roebling, John A. Diary of my journey from Muehlhausen in Thurangia via Bremen to the United States...1831. Trenton, 1931.
Geog 4238.45 D'Avezac. Notice des découvertes...dans l'Ocean. Paris, 1845.
Geog 4238.47 Rosenbaum, S.E. A voyage to America 90 years ago...Hamburg to New York in 1847. N.Y., 1939.
Geog 4238.48 Whaley, Thomas. Consignments to El Dorado. 1st ed. N.Y., 1972.
Geog 4238.49F Jenkins, F.H. Journal of a voyage to San Francisco, 1849. Northridge? 1975.
Geog 4238.53 Hauch, J.C. von. En Rejse til Amerika. Kjøbenhavn, 1853.
Geog 4238.72 Wells, Theodore. Narrative of life and adventures of Captain Wells...voyages. Biddeford, 1874.
Geog 4238.73 Thomson, C.W. The Atlantic. N.Y., 1878. 2v.
Geog 4238.77 Thomson, C.W. Voyage of the Challenger. The Atlantic. v.1-2, atlas. London, 1877. 3v.
Geog 4238.78 Benjamin, S.G.W. Atlantic Islands. N.Y., 1878.
Geog 4238.79 Moulton, Francis D. At sea in the Celtic from January 23rd to February 11th, 1879. n.p., 1879.
Geog 4238.80 Andrews, William Albert. Dangerous voyages of Captain William Andrews. N.Y., 1966.
Geog 4238.83.5 Brassey, Thomas. 11,506 knots in the "Sunbeam". London, 1884.
NEDL Geog 4238.83.7 Brassey, Anna. 14,000 miles in the Sunbeam in 1883. London, 1885.
Geog 4238.83.8F Brassey, Anna. In the trades, the tropics, and the 'roaring forties. London, 1886.
Geog 4238.92 Ward, Artemus. Columbus outdone, an exact narrative of the voyage of the Yankee skipper Captain N.A. Andrews. N.Y., 1893.
Geog 4239.01 Lindsey, N. Allen. Cruising in the Madiana. Boston, 1901.
Geog 4239.21 Craig, Gavin. Boy aloft. Lymington, 1971.
Geog 4239.23 Gerbault, Alain. Seul a travers l'Atlantique. Paris, 1924.
Geog 4239.24 Wells, F. de W. The last cruise of the Shanghai. N.Y., 1925.
Geog 4239.27 Ihering, H. von. Die Geschichte des Atlantischen Ozeans. Jena, 1927.
Geog 4239.29 Franceschi, F. Odisea del yate "Mary" de Puerto Rico a España. Madrid, 1930.
Geog 4239.29.5 England, George Allan. Isles of romance. N.Y., 1929.
Geog 4239.29.13 Andersen, Lis. Lis sails the Atlantic. London, 1953.
Geog 4239.38 Moyne, W.E.G. Atlantic circle. London, 1938.
Geog 4239.47 Durand-Couppel de St. Trent, M.N.P. Wind aloft, wind alone. N.Y., 1947.
Geog 4239.47.5 Turen, T. The Tuntsa. Chicago, 1961.
Geog 4239.49 Wightman, Frank Armstrong. The wind is free. N.Y., 1949.
Geog 4239.50 Barton, H.D.E. Westward crossing. 1st American ed. N.Y., 1951.
Geog 4239.51 Uriburu, E.C. Seagoing Gaucho. N.Y., 1951.
Geog 4239.51.5 Buhler, Jean. Sur les routes de l'Atlantique. Lausanne, 1951.
Geog 4239.52 Veedam, V. Sailing to freedom. N.Y., 1952.
Geog 4239.54 Mitchell, Carleton. Passage east. London, 1954.
Geog 4239.55 Barton, H.D.E. Atlantic adventurers. N.Y., 1955?
Geog 4239.55.5 Mastrangelo, F. Atlantica dall'intuizione alla scoperta del nuovo mondo. Napoli, 1955.
Geog 4239.57 Villiers, Alan John. Wild ocean. N.Y., 1957.
Geog 4239.57.5 Outhwaite, L. The Atlantic. N.Y., 1957.
Geog 4239.57.10 Villiers, Alan John. The new Mayflower. Leicester, 1959.
Geog 4239.58 Costa Brochado, José de. Descobrimento do Atlântico. Lisboa, 1958.
Geog 4239.58.5 Costa Brochado, José de. The discovery of the Atlantic. Lisboa, 1960.
Geog 4239.62 Nikiforovskii, V.A. Ekspeditsiia na "Sedove" v Atlanticheskii okean. Moskva, 1962.
Geog 4239.63 Marx, R.F. The voyage of the Niña II. 1st ed. Cleveland, 1963.
Geog 4239.65 Manry, Robert. Tinkerbelle. N.Y., 1966.
Geog 4239.65.5 Hiscock, Eric C. Atlantic cruise in Wanderer III. London, 1968.
Geog 4239.66 Snaith, William. Across the western ocean. 1st ed. N.Y., 1966.
Geog 4239.66.5 Ridgway, John M. A fighting chance. London, 1967.
Geog 4239.67 Volovich, Vitalii G. Tridtsatyi meridian. Moskva, 1967.
Geog 4239.68 Cassidy, Vincent H. The sea around them; the Atlantic Ocean, A.D. 1250. Baton Rouge, 1968.
Geog 4239.68.5 Sauer, Carl Ortwin. Northern mists. Berkeley, 1968.
Geog 4239.68.10 Naydler, Merton. The penance way; the mystery of Puffin's Atlantic voyage. London, 1968.
Geog 4239.68.15 Haward, Peter. High latitude crossing. London, 1968.
Geog 4239.68.20 Tomalin, Nichols. The strange voyage of Donald Crowhurst. London, 1970.
Geog 4239.68.25 Svirin, Vladimir P. Vzemliakh blizkikh i dal'nikh. Stavropol', 1970.
Geog 4239.69 Wharran, James. Two girls, two catamarans. London, 1969.

Classified Listing

Geog 4235 - 4240 Voyages and travels in general - Individual travels - Atlantic Ocean (By date of voyage) - cont.

Geog 4239.69.5 Castelbajac, Bertrand de. De Plymouth à Newport par le sud de Nantucket. Paris, 1969.
Geog 4239.69.10 McClean, Tom. I had to dare; rouring the Atlantic in seventy days. London, 1971.
Geog 4239.69.15 Woolass, Peter. Stelda, George and I. London, 1971.
Geog 4239.70 Hills, Lawrence Donegan. Lands of the morning. London, 1970.
Geog 4239.70.6 Heyerdahl, Thor. The Ra expeditions. London, 1971.
Geog 4239.70.10 Chichester, Francis Charles. The romantic challenge. London, 1971.
Geog 4239.70.15 Genovés Tarazaga, Santiago. Ra, una balsa de papyrus a través del Atlantico. 1. ed. Mexico, 1972.
Geog 4239.71 Vihlen, Hugo. April Fool or, How I sailed from Casablanca to Florida in a six-foot boat. Chicago, 1971.
Geog 4239.71.5 Brinnin, John M. The sway of the grand saloon: a social history of the North Atlantic. N.Y., 1971.
Geog 4239.73 Senkevich, Iurii A. Na "RA" cherez Atlantiku. Leningrad, 1973.

Geog 4245 - 4250 Voyages and travels in general - Individual travels - Pacific Ocean (By date of voyage)

Geog 4247.69 Dunmore, John. The fateful voyage of the St. Jean Baptiste. Christchurch, 1969.
Geog 4247.89 Colnett, J. The journal of Captain James Colnett aboard the Argonaut from Apr. 26, 1789 to Nov. 3, 1791. Toronto, 1940.
Geog 4247.96 Smith, William. Journal of a voyage in the Duff to Pacific. N.Y., 1813.
Htn Geog 4248.08* Patterson, S. Narrative of adventures...of S. Patterson, experienced in Pacific Ocean. Palmer, 1817.
Geog 4248.12 Porter, David. Voyage in the South Seas in...1812-1814. London, 1823.
Geog 4248.13 Corney, Peter. Voyages in the northern Pacific, 1813-1818. Honolulu, 1896.
Geog 4248.37 Rovings in the Pacific from 1837 to 1849. London, 1851.
Geog 4248.39 Mercier, H.J. Life in a man-of-war...during cruise in Pacific. Philadelphia, 1841.
Geog 4248.53 Jenkins, J.S. Recent exploring expeditions to the Pacific. London, 1853.
Geog 4248.54 Niboyet, P. Les mondes nouveaux. Paris, 1854.
Geog 4248.61 Jones. Life and adventures in the south Pacific. N.Y., 1861.
Geog 4248.73 Eardley-Wilmot, S. Our journal in the Pacific. London, 1873.
Geog 4248.82 Gilbay, Bernard. A voyage of pleasure. Cambridge, 1956.
Geog 4248.86 Guillemard, F.H.H. Cruise of the Marchesa. London, 1886. 2v.
Geog 4248.86.5 Guillemard, F.H.H. Cruise of the Marchesa to Kamschatka and New Guinea. 2nd ed. London, 1889.
Geog 4249.01 Silva, Abeillard. A través da Malasia. Coimbra, 1901.
Geog 4249.26 Akademiia Nauk, Petrograd. The Pacific Russian scientific investigation. Leningrad, 1926.
Geog 4249.36 Allen, Edward Weber. North Pacific: Japan, Siberia, Alaska, Canada. N.Y., 1936.
Geog 4249.37 Lissner, Ivar. Menschen und Mächte am Pazifik. Hamburg, 1937.
Geog 4249.38 Fahnestvak, B. Stars to windward. N.Y., 1938.
Geog 4249.44A Medrs, Eliot G. Pacific Ocean handbook. Stanford, Calif., 1944.
Geog 4249.44B Medrs, Eliot G. Pacific Ocean handbook. Stanford, Calif., 1944.
Geog 4249.46 Robson, R.W. Where the trade-winds blow. Sydney, 1946.
Geog 4249.50 Heyerdahl, T. Kon-Tiki. Chicago, 1950.
Geog 4249.50.2 Heyerdahl, T. Kon-Tiki; across the Pacific by raft. Chicago, 1964.
Geog 4249.50.5 Barrett, C.L. The Pacific. Melbourne, 1950?
Geog 4249.50.10A Hesselberg, E. Kon-Tiki and I. N.Y., 1950.
Geog 4249.50.10B Hesselberg, E. Kon-Tiki and I. N.Y., 1950.
Geog 4249.51 Clune, Frank. Hands across the Pacific. Sydney, 1951.
Geog 4249.54 Willis, William. The gods were kind. 1st ed. N.Y., 1955.
Geog 4249.56 Raitt, Helen. Exploring the deep Pacific. 1st ed. N.Y., 1956.
Geog 4249.58 Baker, DeVere. The raft Lehi IV. Long Beach, 1959.
Geog 4249.58.5 Bisschop, Eric de. Tahiti-Nui. London, 1959.
Geog 4249.59 Van Sinderen, Adrian. The other half of the earth. N.Y., 1959. 2v.
Geog 4249.60 Sharp, C.A. The discovery of the Pacific Islands. Oxford, 1960.
Geog 4249.61 Hoever, Otto. Alt-Asiaten unter Segel in Indischen und Pazifischen. Braunschweig, 1961.
Geog 4249.61.5 United States. National Archives. United States scientific geographical exploration of the Pacific Basin, 1783-1899. Washington, 1961.
Geog 4249.65 Dainelli, Giotto. L'esplorazione del Grande Oceano. Torino, 1965.
Geog 4249.67 Friis, Herman Ralph. The Pacific Basin; a history of its geographical exploration. N.Y., 1967.
Geog 4249.67.5 Downs, Hugh. A shoal of stars. Garden City, 1967.
Geog 4249.67.10 Sandvad, Jørgen. Rejsernes rytme. Kronikker. København, 1967.

Geog 4260 Voyages and travels in general - Individual travels - Europe in general - Pamphlet volumes

Geog 4260.3 Pamphlet vol. Europe. 11 pam.
Htn Geog 4260.5F* Zeiller, M. Topographia Italiae. Franckfurt, 1688.
Geog 4260.7 Pamphlet vol. Panoramas. 7 pam.
Geog 4260.8 Pamphlet vol. European geography.

Geog 4264 - 4270 Voyages and travels in general - Individual travels - Europe in general - Descriptive geographies (By date)

Htn Geog 4265.61* Barreiros, G. Chorographia. Coimbra, 1561.
Htn Geog 4265.76* Verstegen, R. The post of the world. London, 1576.
Geog 4265.95.1 Saur, Abraham. Theatrum urbium. Unterschneidheim, 1971.
Geog 4266.09 Ens, Gaspar. Deliciarum germaniae...viatorius...itinera ad omnes ciirtates. Coloniae, 1608.
Htn Geog 4266.31* Gölnitz, A. Ulysses belgico-gallicus...per Belgium hispan. Lugdunum Batavorum, 1631.
Geog 4266.36 Mervla, P. Cosmographiae generalis libri tres. Amsterdami, 1636.
Htn Geog 4266.41.7F* Zeiller, Martin. Sämtliche...Topographias. v.1-31. Frankfurt, 1677-1736. 10v.
Htn Geog 4266.41.8F* Zeiller, Martin. Sämtliche...Topographias. Haupt-Register. Frankfurt, 1726.
Geog 4266.41.10F Zeiller, Martin. Topographia Sueviae. Kassel, 1960.
Geog 4266.41.80 Schuchhard, C. Die Zeiller-Merianschen Topographien bibliographisch Beschrieben. Hamburg, 1960.

Geog 4264 - 4270 Voyages and travels in general - Individual travels - Europe in general - Descriptive geographies (By date) - cont.

Htn Geog 4266.65* Gerbier, B. Subsidium peregrinatibus. Oxford, 1665.
Geog 4267.18 Pockh, J.J. Der politische catholische Passagier. Augsburg, 1718.
Geog 4267.26 Breval, John D. Remarks on several parts of Europe. v.1-2. London, 1726.
Geog 4267.38 Breval, John D. Remarks on several parts of Europe. v.1-2. London, 1738.
Geog 4267.63 Expilly, J.J. d'. Le geographe manuel. 6e éd. Paris, 1763.
Geog 4267.71.3 Gutierrez de la Hacera, P.R. Descripción general de la Europa. Madrid, 1782. 2v.
Geog 4267.79 Raff, G.C. Geographie für Kinder. Tübingen, 1779.
Geog 4267.92 Meusel, J.G. Lehrbuch der Statistik. Leipzig, 1792.
Geog 4267.92.5 Meusel, J.G. Lehrbuch der Statistik. Leipzig, 1804.
Geog 4268.00F Boetticher, J.G. Geographical, historical...description. London, 1800.
Geog 4268.30 Bertaut, Jules. Villégiatures romantiques. Paris, 1927.
Geog 4268.31.7 Hofland, B. (Mrs.). Panorama of Europe. 7th ed. London, n.d.
Geog 4268.48 Maps. Europe. Description...of countries of Europe. Hartford, 1848.
Geog 4268.48.5 Description of Bayne's gigantic panorama of a voyage to Europe. Boston, 1848.
Geog 4268.50 Lavallée, T. Military topography of...Europe. London, 1850.
Geog 4268.52 Hughes, W. Manual of geography. pt.1. London, 1852.
Geog 4268.63 Ritter, K. Europa. Berlin, 1863.
Geog 4268.73.2 Levasseur, E. L'Europe...geographie. Paris, 1873.
Geog 4268.73.4 Levasseur, E. Precis de la geographie...de l'Europe. v.1, atlas. Paris, 189-. 2v.
Geog 4268.75F Picturesque Europe. N.Y., 1875-1879. 3v.
Geog 4268.77F Sherer, J. Europe...picturesque scenes. London, 1877. 2v.
Geog 4268.77.5 Faucher, J. Vergleichende Kultur Gilder. Hannover, 1877.
Geog 4268.83.6 Dutens, Louis. Itinéraire des routes les plus fréquentées. 6. éd. Paris, 1788.
Geog 4268.85 Rudler, F.W. Europe. London, 1885.
Geog 4268.85.5A Chisholm, G.G. Europe. London, 1899-1902. 2v.
Geog 4268.85.5B Chisholm, G.G. Europe. London, 1899-1902. 2v.
Geog 4268.86.4 Lanier, L. L'Europe...choix de lectures de geographie. 4. ed. Paris, 1888.
Geog 4268.89.5 Vidal de la Blache, Paul. Etats et nations de l'Europe, autour de la France. 4. éd. Paris, 189-?
Geog 4268.94 Philippson, A. Europa...eine...Landeskunde. Leipzig, 1894.
Geog 4268.94.5 Philippson, A. Europa. Leipzig, 1906.
Geog 4269.02 Carpenter, F.G. Europe. N.Y., 1902.
Geog 4269.03 Partsch, J. Central Europe. London, 1903.
Geog 4269.03.5 Symons, A. Cities. London, 1903.
Geog 4269.03.10 Herbertson, (Mrs.). Descriptive geography of Europe from original sources. London, 1919.
X Cg Geog 4269.07.7PF Hunnewell, H.H. Nineteen thousand miles in 1907. Boston, 1908. 8v.
Geog 4269.10 Sothriados, G. Chōrai kai ladi tēs Evrōpēs. Athēnai, 1910.
Geog 4269.10.5 Bartholomew, J.G. Literary and historical atlas of Europe. London, 1910.
Geog 4269.13 Lyde, L.W. The continent of Europe. London, 1913.
Geog 4269.13.2 Lyde, L.W. The continent of Europe. London, 1920.
Geog 4269.13.4 Lyde, L.W. The continent of Europe. 2nd ed. London, 1924.
Geog 4269.13.10 Fitch, G.H. Critic in the occident. San Francisco, 1913.
Geog 4269.14 Couperus, Louis. Van en over alles en iedereen. pt.1-5. Amsterdam, 1915. 5v.
Geog 4269.16 Sievers, Wilhelm. Die geographischen Grenzen Mitteleuropas. Giessen, 1916.
Geog 4269.18A National Research Council. Division of Geology and Geography. The geography of Europe. New Haven, 1918.
Geog 4269.18B National Research Council. Division of Geology and Geography. The geography of Europe. New Haven, 1918.
Geog 4269.19 Beltrán y Rozpide, Ricardo. Nuevas nacionalidades en Europa. 2a ed. Madrid, 1919.
Geog 4269.23 Bartholomew, J.G. A literary and historical atlas of Europe. London, 1923.
Geog 4269.31 Lyde, L.W. Peninsular Europe. London, 1931.
Geog 4269.31.10 Suedost- und Südeuropa in Natur. v.1-7,8-14,18. Potsdam, 1931-1936. 3v.
Geog 4269.32 Hausenstein, W. Europäische Hauptstädte. Erlenbach, 1932.
Geog 4269.34 Bogardus, J.F. Europe; a geographical survey. N.Y., 1934.
Geog 4269.34.7 Shackleton, Margaret Reid. Europe: a regional geography. 7th ed. N.Y., 1965.
Geog 4269.35 Blanchard, Raoul. A geography of Europe. N.Y., 1935.
Geog 4269.35.4 Blanchard, Raoul. Géographie de l'Europe. Paris, 1936.
Geog 4269.35.10A Van Valkenburg, S. Europe. N.Y., 1935.
Geog 4269.35.10B Van Valkenburg, S. Europe. N.Y., 1935.
Geog 4269.36 Price, Lucien. We Northmen. Boston, 1936.
Geog 4269.37 Hubbard, George D. The geography of Europe. N.Y., 1937.
Geog 4269.39 Gandy, Roxana Smith. Are people more a like than different. Cape May Court House, 1939.
Geog 4269.46 Antonovych, Roman. Burlats'kym shliakom. Vynnypeg, 1946.
Geog 4269.47A Rahv, Philip. Discovery of Europe. Boston, 1947.
Geog 4269.47B Rahv, Philip. Discovery of Europe. Boston, 1947.
Geog 4269.50 Gottmann, J. A geography of Europe. N.Y., 1950.
Geog 4269.50.5 Gottmann, J. A geography of Europe. N.Y., 1961.
Geog 4269.52F Hürlimann, Martin. Europe in photographs. London, 1952.
Geog 4269.52.5F Hürlimann, Martin. Europe in photographs. London, 1951.
Geog 4269.54.5 George, Pierre. L'Europe centrale. 1. éd. v.1-2. Paris, 1954.
Geog 4269.54.10 Fremantle, Anne (Jackson). Europe: a journey with pictures. N.Y., 1954.
Geog 4269.55 Lehmann, H. Europa. 16. Aufl. Frankfurt, 1955.
Geog 4269.57 Ogilvie, A.G. Europe and its borderlands. Edinburgh, 1957.
Geog 4269.58 Chabot, Georges. L'Europe du Nord et du Nord-Ouest. Paris, 1958. 3v.
Geog 4269.58.5 Koeppen, Wolfgang. Nach Russland und anderswohin. Stuttgart, 1958.
Geog 4269.58.15 Egli, Emil. Flugbild Europas. Zürich, 1958.
Geog 4269.59.5 Egli, Emil. Europe from the air. London, 1959.
Geog 4269.59.10 Monkhouse, Francis. A regional geography of western Europe. London, 1959.
Geog 4269.60 Kuleshov, A.P. 500,000 [i.e. Piat'sot tysiach] kolometrov v puti. Moskva, 1960.
Geog 4269.60.5 Nejedlý, O. Malířovy toulky po Evropě, Eejlonu a Indii. Praha, 1960.
Geog 4269.61.5 Mutton, Alice. Central Europe; a regional and human geography. London, 1961.

Geog 4264 - 4270 Voyages and travels in general - Individual travels - Europe in general - Descriptive geographies (By date) - cont.

Geog 4269.61.10 Hoffman, George W. A geography of Europe including Asiatic USSR. N.Y., 1961.
Geog 4269.62 Kukanov, A.A. Doroga k serdtsu; reportazh. Tallin, 1962.
Geog 4269.62.5 Salvadori, M. Western ports in Europe. N.Y., 1962.
Geog 4269.62.10 Koen, Albert. Putishta i spirki. Sofiia, 1962.
Geog 4269.63 Glushchenk, I. Strany, strechi, uchenye. Moskva, 1963.
Geog 4269.64.10 George, Pierre. Géographie de l'Europe centrale slave et danubienne. Paris, 1964.
Geog 4269.64.15 Bonasera, Francesco. Gli stati d'Europa. Palermo, 1964.
Geog 4269.64.20 Phloros, Paulos. Staurodromia tēs Eurōpēs. Athēnai, 1964.
Geog 4269.65 Filipenko, Aloiz A. Na chuzhikh ulitsakh. Moskva, 1965.
Geog 4269.65.12 Kares, Olavi. Kallaveden rannalta, päiväkirja valpurista 1964 valpuriin 1965. 3. painos. Porvoo, 1966.
Geog 4269.66.5 Markovic, Slobodan. Putovanja grešnog Elefterija. Beograd, 1971.
Geog 4269.67 Maevskii, Viktor. V. Evropa bez dzhentl'menov. Moskva, 1967.
Geog 4269.67.5 Le Lannou, Maurice. Le déménagement du territoire. Paris, 1967.
Geog 4269.67.10 Houter, F. den. Vluchtige verlenningen. Laven, 1967.
VGeog 4269.68 Lah, Avguštin. Naše sosedne države. Ljubljana, 1968.
Geog 4269.68.5 Phloros, Paulos. Eurōpaikē suphmōnia taxidia. Athēnai, 1968.
Geog 4269.70 Coghill, Ian G. Western Europe: geographical studies. London, 1970.
Geog 4269.72 Lisakovskii, Igor' N. Gamlet ne prishel na svidanie. Odessa, 1972.
Geog 4269.73 Budrewicz, Olgierd. Ścieżkami Starego Świàta. Wyd. 1. Warszawa, 1973.
Geog 4269.74 Agapov, Boris N. Shest' zagranits. Moskva, 1974.

Geog 4275 - 4280 Voyages and travels in general - Individual travels - Europe in general - History of geography and travel in Europe (By date)

Geog 4278.93 Fernandez Duro, Cesario. Viajes regios por mar en el trancurso de quimentos años. Madrid, 1893.
Geog 4279.11A Bates, Ernest S. Touring in 1600. Boston, 1911.
Geog 4279.11B Bates, Ernest S. Touring in 1600. Boston, 1911.
Geog 4279.52 Stoye, J.W. English travellers abroad. London, 1952.
Geog 4279.64 Dulles, F.R. Americans abroad. Ann Arbor, 1964.
Geog 4279.67 Trease, Geoffrey. The grand tour. London, 1967.
Geog 4279.69 Anderson, Patrick. Over the Alps. London, 1969.
Geog 4279.69.5 Hibbert, Christopher. The grand tour. London, 1969.
Geog 4279.71 Bocca, Geoffrey. The great resorts. N.Y., 1971.

Geog 4300 Voyages and travels in general - Individual travels - Europe in general - Guidebooks and handbooks

Geog 4300.4 Fodor's Europe. N.Y. 1967+ 3v.
Geog 4300.5 Fetridge, W.P. Harper's handbook for...Europe. N.Y. 1-17,1862-1878 7v.
Geog 4300.6 Satchel guide for the vacation tourist in Europe. N.Y. 1872-1929 27v.
Geog 4300.8 Osgood, J.R. Osgood's pocket guide to Europe. Boston, 1882.
Geog 4300.8.5A Osgood, J.R. Osgood's pocket guide to Europe. Boston, 1883.
Geog 4300.8.5B Osgood, J.R. Osgood's pocket guide to Europe. Boston, 1883.
Geog 4300.10 Frager, M.D. Practical European guide. Boston, 1907.
Geog 4300.12 Sargent, H.W. Skeleton tours. N.Y., 1870.
Geog 4300.12.2 Sargent, H.W. Skeleton tours. N.Y., 1870.
Geog 4300.12.3 Sargent, H.W. Skeleton tours. N.Y., 1871.
VGeog 4300.16 Guerber, H.A. How to prepare for Europe. N.Y., 1906.
Geog 4300.17 Practical general continental guide. London, 1868.
Geog 4300.18.5 Starke, M. Information and directions for travellers. Paris, 1826.
Geog 4300.20.8 Reichard, H.A.O. Reichard's Passagier. Berlin, 1834.
Geog 4300.22 Audin, J.M.V. Guide classique du voyageur. Paris, 1828-1829. 2v.
Geog 4300.23 Illustrated Europe. Zürich.
Geog 4300.24 Cyclist's touring club handbook and guide. London, 1900.
Geog 4300.25.3 Engelmann, J.B. Manuel pour voyageurs en Allemagne et dans les Payslimitrophes. 3.ed. Francfort, 1827.
Geog 4300.26 Union Steamship Company, Ltd. Homeward bound. 3rd ed. London, 1892.
Geog 4300.27 Morford, Henry. Morford's short-trip guide to Europe. N.Y., 1872.
Geog 4300.28.7 Presbrey, F. Presbrey's information guide to transatlantic travelers. 7th ed. N.Y., 1911.
Geog 4300.29 Schoonmaker, F. Through Europe on two dollars a day. N.Y., 1927.
Geog 4300.30.4 Hungerford, E. Planning a trip abroad. N.Y., 1927.
Geog 4300.30.8 Hungerford, E. Planning a trip abroad. N.Y., 1931.
Geog 4300.31.8 Touring Club Suisse. Europa touring; guide automobile d'Europe. Berne, n.d.
Geog 4300.32 Cook, T., firm. Tourist handbook for Holland, Belgium, the Rhine and Black Forest. London, 1895.
Geog 4300.33.3 Palmer, J.E. Palmer's European pocket guide. N.Y., 188-.
Geog 4300.34 New York Herald Tribune. European Edition. Guide to Europe. Paris?
Geog 4300.35 Continental touring, 1947-1948 for the British motorist. Leeds, 1947.
Geog 4300.36 Nederlandsch-Amerikaansche Stoomvaart Maatschappij, N.V. Guide for Europe. The Hague, 1925.
Geog 4300.37 Pastene, J. Auto guide to Europe. N.Y., 1954.
Geog 4300.40 Joseph, Richard. Guide to Europe and the Mediterranean. 1st ed. Garden City, N.Y., 1956.
Geog 4300.42 Nagel, publishers. Europe. Geneva, 1956.
Geog 4300.44 Ross, Janet. Budgetouring Europe. N.Y., 1933.
Geog 4300.46 Olson, H.S. Aboard and abroad. 3d ed. Philadelphia, 1956.
Geog 4300.47 Vanderbilt, C. European travel directory. N.Y., 1954.
Geog 4300.48 Vidari, Giovanni Maria. Il viaggio in pratica. Venezia, 1718.
Geog 4300.50 Bertelsmann, C. Ferien einmal anders. Gütersloh, 1965.
Geog 4300.51 Owen, Charles. Britons abroad; a report on the package tour. London, 1968.

Geog 4301 - 4310 Voyages and travels in general - Individual travels - Europe in general - Travels (By date of travel)

Geog 4303.08 Anonymi descriptio Europae orientalis. Cracoviae, 1916.
Geog 4303.99 Lannoy, Guillebert de. Voyages et ambassades...1399-1450. Mons, 1840.
Geog 4303.99.2 Lannoy, Guillebert de. Oeuvres. Louvain, 1878.
Geog 4303.99.3 Sachet, E. Examen critique des voyages et ambassades de G. de Lannoy, 1399-1450. Bruxelles, 1843.
Geog 4303.99.5 Lelewel, J. Guillebert de Lannoy et ses voyages en 1413, 1414 et 1421. Bruxelles, 1844.

Geog 4301 - 4310 Voyages and travels in general - Individual travels - Europe in general - Travels (By date of travel) - cont.

Geog 4303.99.20A Klimas, Petras. Ghillebert de Lannoy in medieval Lithuania. N.Y., 1945.
Geog 4303.99.20B Klimas, Petras. Ghillebert de Lannoy in medieval Lithuania. N.Y., 1945.
Geog 4304.65 Schaschek. Des böhmischen Herrn Leo's von Rozmital...Reise...Abendlande 1465-1467. Stuttgart, 1844.
Geog 4304.65.5 Cust, Nina W.G. Gentlemen errant; being the journeys and adventures of four noblemen in Europe during the fifteenth and sixteenth centuries. London, 1909.
Geog 4304.65.10 Slavík, František A. Cesta pana Lva z Rožmitála po západní Europě roku 1465-1467. Telči, 1890.
Geog 4304.90 Martyr. Relation d'un voyage fait en Europe. Paris, 1827.
Geog 4304.95 Muenzer, Hieronymus. Voyage aux Pays-Bas, 1495. Bruxelles, 1942.
Geog 4305.17 Antonio de Beatis, Don. Voyage du Cardinal d'Aragon en Allemagne. Paris, 1913.
Geog 4305.91A Moryson, Fynes. An itinerary containing his 10 yeeres travell. Glasgow, 1907. 4v.
Geog 4305.91B Moryson, Fynes. An itinerary containing his 10 yeeres travell. Glasgow, 1907. 4v.
Geog 4305.93.5 Esprinchard, Jacques. Vie. Paris, 1957.
Htn Geog 4305.96* Hentznero, P. Itinerarium Germaniae, Galliae. Norinbergae, 1612.
VGeog 4306.00 Rohan de. Voyage...en l'an 1600. Amsterdam, 1646.
Htn Geog 4306.17F* Moryson, Fynes. An itinerary. London, 1617.
Geog 4306.17.1F Moryson, Fynes. An itinerary. Amsterdam, 1971.
Geog 4306.17.3 Moryson, Fynes. Shakespeare's Europe. London, 1903.
Geog 4306.17.4 Moryson, Fynes. Shakespeare's Europe: a survey of the condition of Europe at the end of the 16th century. 2nd ed. N.Y., 1967.
Geog 4306.17.7 Küsel Saxon, S. Itinerarium Germaniae, Siciliae. Erphordiae, 1617.
Htn Geog 4306.56* Ogier, Charles. Ephemerides sive iter Danicum, Suecicum, Polonicum. Lutetiae Parisiorum, 1656.
Geog 4306.56.5 Hiltebrandt, C.J. Dreifache schwedische Gesandtschaftsreise nach Siebenbürgen. Leiden, 1937.
Geog 4306.56.10 Ogier, Charles. Oziennik podrózy do Polski, 1635-1636. Gdańsk, 1950-1953. 2v.
Htn Geog 4306.60.2* Loménie, Louis Henri. Itinerarium. Paris, 1662.
Geog 4306.63.3 Payen. Les voyages de Monsieur Payen. 2e éd. Paris, 1667.
Htn Geog 4306.68* Birken, S. von. Hoch fürstlicher Brandenburgischer Ulysses. Bayreuth, 1668.
Geog 4306.71.5 Patin, Charles. Relations historiques...de voyages en Allemagne. 2e éd. Lyon, 1676.
Htn Geog 4306.73* Brown, E. Brief account of some travels. London, 1673.
Htn Geog 4306.73.5F* Brown, E. Brief account of some travels. London, 1685.
Geog 4306.81.3 Regnard, J.F. Voyage de...en Flandre, Hollanke, Danemark. Paris, 1874.
Geog 4306.83.2 Melle, Jakob von. Beschreibung einer Reise durch das Nordwestliche Deutschland nach den Niederländen. Lübeck, 1891.
Geog 4306.85 Pacichelli, G.B. Memorie de viaggi per l'Europa. Napoli, 1685. 3v.
Htn Geog 4306.86*A Burnet, Gilbert. Some letters...Switzerland, Italy. Rotterdam, 1686.
Htn Geog 4306.86*B Burnet, Gilbert. Some letters...Switzerland, Italy. Rotterdam, 1686.
Htn Geog 4306.86.3* Burnet, Gilbert. Dr. Burnet's travels or letters. Amsterdam, 1687.
Geog 4306.86.5 Burnet, Gilbert. Some letters...Switzerland, Italy. n.p., 1708.
Geog 4306.86.6 Burnet, Gilbert. Some letters containing an account of what seemed most remarkable in Switzerland. Menston, Eng., 1972.
Geog 4306.86.10 Burnet, Gilbert. Travels through France, Italy. London, 1750.
Geog 4306.87 Regnard, J.F. Voyages. n.p., n.d.
Geog 4306.91.1 Retcher, Wilhelm. Der sächsische Robinson. Leipzig, 1722.
Htn Geog 4306.92* Voyages historiques de l'Europe. v.1-8. Paris, 1692. 2v.
Geog 4306.92.6 Bromley, William. Remarks in the grand tour of France and Italy. 2d ed. London, 1705.
Geog 4306.95 Patin, Charles. Relations historiques...de voyages. Amsterdam, 1695.
Geog 4306.97F Sheremt'ev, B.P. Zapiska puteshestviia. Moskva, 1773.
Geog 4307.03 Broekhuizen, G. Die nieuwe Béreisde wereld. Amsterdam, n.d. 2 pam.
Geog 4307.22 Careri, G. Viaggi per Europa. Napoli, 1722. 2v.
Geog 4307.28 Remarques d'un voyageur. Haye, 1728.
Geog 4307.35.3 Pöllnitz, K.L. von. Memoires de Charles-Louis, Baron de Pöllnitz. Amsterdam, 1735. 2v.
Geog 4307.35.5 Pöllnitz, K.L. von. Nouveaux memoires du Baron de Pöllnitz. Amsterdam, 1737. 2v.
Geog 4307.37 Pöllnitz, K.L. von. Memoirs of...travels. London, 1737. 4v.
Geog 4307.37.2 Pöllnitz, K.L. von. Memoirs of Charles-Lewis, Baron de Pöllnitz. 2nd ed. London, 1739. 4v.
Geog 4307.37.9 Pöllnitz, K.L.. von. A vagabond courtier. London, 1913. 2v.
VGeog 4307.39F Sagramoso, M.E. Lettera...al E. Ignazio Zanardi di Mantova. Verona, 1877.
Geog 4307.43 Blainville, J. de. Travels through Holland, Germany, Switzerland, and other parts of Europe. London, 1743-1745. 3v.
Geog 4307.50 Clancy, Michael. The memoirs. v.1-2. Dublin, 1750.
Htn Geog 4307.53* Uffenbach, Z.C. Merkwürdige Reisen. Ulm, 1753-1754. 3v.
Geog 4307.56 Nugent, T. Grand tour...through Netherlands. London, 1756. 4v.
Geog 4307.65 Defeller. Itinéraire ou voyages. Liege, 1820. 2v.
Geog 4307.65.10 Bousquet, (Mrs.). Mrs. Bousquet's diary, 1765. Norwich, 1927.
Geog 4307.65.15 Pennant, T. Tour on the continent. London, 1948.
Geog 4307.70.10A Burney, Charles. Continental travels. London, 1927.
Geog 4307.70.10B Burney, Charles. Continental travels. London, 1927.
Geog 4307.72 Marshall, J. Travels through Holland. London, 1772. 3v.
Geog 4307.76 Sander, Heinrich. Beschreibung seiner Reisen durch Frankreich. Pt.1. Leipzig, 1783.
Geog 4307.77 Staszic, S. Dziennik podrózy, 1777-1791. v.1-2. Warszawa, 1903.
Geog 4307.78 Staszic, S. Dziennik podrózy, 1789-1805. Krakow, 1931.
Geog 4307.80 Björnstähls, J.J. Briefe auf seinem Auslandischen Reisen. Leipzig, 1780-1783. 6v.
Geog 4307.80.3 Beckford, William. Italy, with sketches of Spain and Portugal. Philadelphia, 1834.

Classified Listing

Geog 4301 - 4310 Voyages and travels in general - Individual travels - Europe in general - Travels (By date of travel) - cont.

	Geog 4307.80.5	Beckford, William. Italy, Spain and Portugal. London, 1840.
	Geog 4307.80.7	Beckford, William. Italy, Spain and Portugal. Pt.1-2. N.Y., 1845.
	Geog 4307.80.15	Beckford, William. The travel-diares of William Beckford of Fonthill. Cambridge, 1928. 2v.
	Geog 4307.82	Bruce, Peter Henry. Memoirs of Peter Henry Bruce, Esq. London, 1782.
	Geog 4307.82.3	Bruce, Peter Henry. Memoirs of Peter Henry Bruce, Esq. Reprint. London, 1970.
	Geog 4307.83	The American wanderer...Europe. London, 1783.
	Geog 4307.83.2	The American wanderer...Europe. Dublin, 1783.
	Geog 4307.84	Piozzi, Hester Lynch Thrale. Observations and reflections made in the course of a journey through France, Italy and Germany. Ann Arbor, 1967.
	Geog 4307.85.5	Matthisson, F. Letters...from various parts of the continent. London, 1799.
	Geog 4307.85.7	Ponz, A. Viage fuera de Espana. Madrid, 1785. 2v.
	Geog 4307.86	Smith, J.E. Sketch of a tour on the continent. London, 1793. 3v.
	Geog 4307.86.2	Smith, J.E. Sketch of a tour on the continent. 2nd ed. London, 1807. 3v.
	Geog 4307.86.6	Moore, J. View of society and manners. Boston, 1792.
	Geog 4307.86.7	Moore, J. View of society and manners. 5th ed. Dublin, 1793. 2v.
	Geog 4307.86.8	Moore, J. View of society and manners. 6th ed. London, 1786. 2v.
	Geog 4307.86.10	Moore, J. View of society and manners. Paris, 1803. 2v.
Htn	Geog 4307.86.12*	Ansbach, Elizabeth Craven. Journey thro' Crimea to Constantinople. London, 1789. 2 pam.
	Geog 4307.86.13	Ansbach, Elizabeth Craven. Journey thro' Crimea to Constantinople. London, 1789.
	Geog 4307.87.2	Watkins, T. Travels through Switzerland, Italy. London, 1794. 2v.
	Geog 4307.87.5	Walker, Adam. Bemerkungen auf einer Reise durch Flandern. Berlin, 1791.
	Geog 4307.91	Stolberg, F.L.G. Reise in Deutschland, der Schweiz. Königsberg, 1794. 4v.
	Geog 4307.91.2	Stolberg, F.L.G. Travels through Germany. London, 1796. 2v.
	Geog 4307.91.4	Stolberg, F.L.G. Travels through Germany, Switzerland, Italy. 2nd ed. London, 1797. 4v.
	Geog 4307.91.10	Gaultier, Pierre R.A. Séjour de mon grand-oncle, P. Gaultier, en Espagne...1791-1802. v.1-2. Angers, 1912.
	Geog 4307.92	Sketches and observations...tour. London, 1797.
Htn	Geog 4307.94*	Radcliffe, A. (Mrs.). A journey made in summer of 1794. London, 1795.
	Geog 4307.94.2	Radcliffe, A. (Mrs.). Journey...through Holland. 2. ed. London, 1795. 2v.
	Geog 4307.97.2	Pratt, S.J. Gleanings through Wales. Dublin, 1797. 2v.
	Geog 4307.99	A concise narrative of a tour through some parts of England, France, Holland, Switzerland, and Italy in the years 1799, 1800, 1801, and 1802. Philadelphia, 1821.
	Geog 4308.01	Wilmot, Catherine. An Irish peer on the continent. London, 1920.
	Geog 4308.01.5	Wakefield, Priscilla (Bell). The juvenile travellers. London, 1801.
	Geog 4308.04	Kotzebue, A. Travels from Berlin thro' Switzerland. London, 1804. 3v.
	Geog 4308.05.2	Silliman, B. Journal of travels. Boston, 1812. 2v.
	Geog 4308.05.4	Silliman, B. Journal of travels. 3rd ed. New Haven, 1820. 3v.
	Geog 4308.06	Cruz y Bahamonde, N. Viage de Espana, Francia. v.1-10,12,14. Madrid, 1806-1813. 8v.
	Geog 4308.06.5	Wakefield, Priscilla (Bell). Juvenile travellers. London, 1806.
	Geog 4308.07	Irving, Peter. Peter Irving's journals. N.Y., 1943.
	Geog 4308.08	Salvo, C. Travels in the year 1806, from Italy to England. Troy, N.Y., 1808.
Htn	Geog 4308.08.2*	Salvo, C. Travels in...1806 from Italy to England. Troy, N.Y., 1808.
	Geog 4308.16	Raumer, F.L.G. Die Herbstreise nach Benedig. Berlin, 1816. 2v.
	Geog 4308.17	Raffles, T. Letters during tour thro'...France. Liverpool, 1818.
	Geog 4308.17.3	Matthews, H. The diary of an invalid. 2nd ed. London, 1820.
	Geog 4308.17.3.5	Matthews, H. The diary of an invalid. 3rd ed. London, 1822. 2v.
	Geog 4308.17.4	Matthews, H. The diary of an invalid. Paris, 1825.
	Geog 4308.17.5	Matthews, H. Diary of an invalid...journey...1817. Paris, 1836.
	Geog 4308.17.6	Charleville, H.C.B. A journey to Florence in 1817. London, 1951.
	Geog 4308.18	Neale, A. Travels thro'...Germany, Poland. London, 1818.
	Geog 4308.18.3	Griscom, J. A year in Europe. N.Y., 1823.
	Geog 4308.18.4	Griscom, J. A year in Europe. 2nd ed. N.Y., 1824. 2v.
	Geog 4308.18.5	Baillie, M. First impressions on tour upon continent. London, 1819.
	Geog 4308.18.6	Russell, Jonathan. Journal, 1818-1819. Boston, 1918.
	Geog 4308.19	Beskow, B. Wandrings minnen. Stockholm, 1833-1834. 2v.
	Geog 4308.19.5	Plessis, Joseph O. Journal d'un voyage en Europe. Québec, 1903.
	Geog 4308.20	Niemeyer, A.H. Beobachtungen auf Reisen in und ausser Deutschland. v.2-4. Halle, 1822-1826. 4v.
	Geog 4308.21	Tennant, C. Tour thro'...Netherlands, Holland. London, 1824. 2v.
	Geog 4308.21.5	Scott, J. Sketches of manners...in French provinces. London, 1821.
	Geog 4308.22.5	Duppa, Richard. Miscellaneous observations and opinions on the continent. London, 1825.
	Geog 4308.23.2	Wilson, D. Letters from an absent brother. London, 1824.
	Geog 4308.23.5	Glagolev, A.G. Zapiski russkago puteshestve norika [s 1823 po 1826 g.]. Sankt Peterburg, 1845. 4v.
	Geog 4308.24	Moore, John. A journey from London to Odessa. Paris, 1833.
	Geog 4308.25	Carter, N.H. Letters from Europe. v.2. N.Y., 1827.
	Geog 4308.25.2	Carter, N.H. Letters from Europe. N.Y., 1829. 2v.
NEDL	Geog 4308.25.10	Pennington, Thomas. A journey into various parts of Europe. London, 1825. 2v.
	Geog 4308.26	Carus, Karl Gustav. Reisen und Briefe. Leipzig, 1926. 2v.
	Geog 4308.27	Delestre, P. De Paris a Varsovie. Paris, 1827.
	Geog 4308.28A	Topliff, S. Topliff's travels, letters from abroad. Boston, 1906.
	Geog 4308.28B	Topliff, S. Topliff's travels, letters from abroad. Boston, 1906.
	Geog 4308.28.5	Walter, Weever. Letters from the continent. Edinburgh, 1828.
	Geog 4308.30	Inglis, H.D. Switzerland...France and Pyrenees. Edinburgh, 1831. 2v.
	Geog 4308.30.5	Candolle, Augustin Pyramus de. L'Europe de 1830 vue à travers la correspondence de Augustin Pyremus de Candolle et Madame de Cércourt. Genève, 1966.
	Geog 4308.31.2A	Johnson, J. Change of air, or The diary. London, 1831.
	Geog 4308.31.2B	Johnson, J. Change of air, or The diary. London, 1831.
	Geog 4308.31.5	McLellan, H.B. Journal of residence in Scotland and tour. Boston, 1834.
	Geog 4308.32	Cooper, J.F. Residence in France with excursion up the Rhine to Switzerland. Paris, 1836.
	Geog 4308.32.2	Willis, N.P. Pencillings by the way. London, 1842.
Htn	Geog 4308.32.3*	Willis, N.P. Pencillings by the way. 1st ed. N.Y., 1844.
Htn	Geog 4308.32.5*	Briggs, Charles F. Working a passage: or Life in a liner. N.Y., 1844.
	Geog 4308.32.10	Ritchie, Leitch. Travelling sketches in the north of Italy, the Tyrol, and on the Rhine. London, 1832.
	Geog 4308.33	Ritchie, Leitch. Travelling sketches. London, 1833.
	Geog 4308.33.2	Dewey, O. Old world and the new. N.Y., 1836. 2v.
	Geog 4308.33.5	Marmier, X. Souvenirs de voyages. Paris, 1841.
	Geog 4308.33.10	Strombeck, Friedrich Karl von. Darstellungen aus meinem Leben und aus meiner Zeit. Braunschweig, 1833-1840. 4v.
	Geog 4308.34	Michelet, J. Sur les chemins de l'Europe. Paris, 1893.
	Geog 4308.34.5	Boddington, Mary. Slight reminiscences of the Rhine. London, 1834. 2v.
	Geog 4308.35	Roby, John. Seven wekks in Belgium, Switzerland, Lombardy. London, 1838. 2v.
	Geog 4308.35.5	Fisk, Wilbur. Travels in Europe. 4th ed. N.Y., 1838.
	Geog 4308.35.10	Pückler-Muskau, H. Vorletzter Weltgang von Semilasso. Stuttgart, 1835. 3v.
	Geog 4308.35.15	Hoppus, John. The continent in 1835. N.Y., 1837.
	Geog 4308.36	The tourist in Europe. N.Y.. 1838.
	Geog 4308.36.5	Hall, Fanny W. Rambles in Europe. N.Y., 1939. 2v.
	Geog 4308.36.10	Gross-Hoffinger, A.J. Empfindsame Reise eines expatriirten Schwärmers. Leipzig, 1836.
	Geog 4308.37	Mundt, T. Spaziergänge und Weltfarhrten. Altona, 1838-1839. 3v.
	Geog 4308.37.3	Cavalier auf Reisen im...1837. Leipzig, 1838.
	Geog 4308.37.5	Jewett, I.A. Paggages in foreign travel. Boston, 1838. 2v.
	Geog 4308.38	Wright, H.H. Desultory reminiscences of a tour. Boston, 1838.
	Geog 4308.38.5	Holmes, D. A ride to Florence thro' France. London, 1841. 2v.
Htn	Geog 4308.38.8*A	Stephens, J.L. Incidents of travel in Greece, Turkey and Poland.7th ed. N.Y., 1838. 2v.
Htn	Geog 4308.38.8*B	Stephens, J.L. Incidents of travel in Greece, Turkey and Poland.7th ed. N.Y., 1838. 2v.
	Geog 4308.38.14	Stephens, J.L. Incidents of travel in Greece, Turkey and Poland. 17th ed. N.Y., 1845. 2v.
	Geog 4308.39	Gibson, William. Rambles in Europe in 1839. Philadelphia, 1841.
	Geog 4308.39.3	Locmaria. Souvenirs des voyages de...duc de Bordeaux. Paris, 1846. 2v.
Htn	Geog 4308.39.5*	Sedgwick, C.M. Letters from abroad. N.Y., 1841. 2v.
	Geog 4308.39.7	Tuttolasso. Wanderungen durch Deutschland, Polen. Stuttgart, 1839.
	Geog 4308.39.9	Sreznevskii, I.I. Putevyia pis'ma...iz slavianskikh zemel, 1839-1842. Sankt Peterburg, 1895.
	Geog 4308.40.3	Hall, Basil. Patchwork. London, 1841. 3v.
	Geog 4308.40.5	Hall, Basil. Patchwork. 2nd ed. London, 1841. 3v.
	Geog 4308.41	Lindeberg, A. Betraktelser under en resa. Stockholm, 1841.
	Geog 4308.42	Faber, F.W. Sights and thoughts...foreign. Photoreproduction. London, 1842.
	Geog 4308.42.5	Baruffi, G.F. Pellegrinazioni autunnali ed opuscoli. Torino, 1842.
	Geog 4308.42.10	Rives, J.P.W. Tales and souvenirs of a residence in Europe. Philadelphia, 1842.
	Geog 4308.43	Dana, William C. A transatlantic tour. Philadelphia, 1845.
	Geog 4308.44	Durbin, J.P. Observations in Europe. N.Y., 1844. 2v.
	Geog 4308.44.5	Volkov, M. Otryuki iz zagranichnykh pisem. Sankt Peterburg, 1857.
	Geog 4308.45	Pedestrian and other reminiscences. London, 1846.
	Geog 4308.45.9	Nicholson, Asenath. Loose papers...residence in Ireland. N.Y., 1853.
	Geog 4308.45.15	Talfourd, T.N. Vacation rambles and thoughts. London, 1845. 2v.
	Geog 4308.45.19	Talfourd, T.N. Vacation rambles...1841-1843. 3rd ed. London, 1851.
	Geog 4308.45.25	Talfourd, T.N. Supplement to "Vacation rambles". London, 1854.
	Geog 4308.45.30	White, T.H. A pilgrim's reliquary. London, 1845.
	Geog 4308.46.2	Laing, S. Notes of a traveller. Philadelphia, 1846.
	Geog 4308.46.3	Laing, S. Observations on social...state of European people. London, 1850.
	Geog 4308.46.5	Smith, J.J. Summer's jaunt across the water. v.1-2. Philadelphia, 1846.
	Geog 4308.46.35	Taylor, Bayard. Views a-foot. 17th ed. N.Y., 1854.
	Geog 4308.47	Herzen, A.I. Tsis'ma iz Frantsii i Italii. London, 1858.
	Geog 4308.47.3	Herzen, A.I. Lettres de France et d'Italie (1847-1852). Genève, 1871.
	Geog 4308.47.5	Eames, J.A. A budget of letters. Boston, 1847.
	Geog 4308.47.7	Eames, J.A. The budget closed. Boston, 1860.
	Geog 4308.47.9	Dodge, Robert. Diary, sketches and reviews. N.Y., 1850.
	Geog 4308.48	Buckingham, J.S. Belgium, the Rhine, Switzerland. London, 1848. 2v.
	Geog 4308.48.2	Description of Bayne's gigantic panorama of a voyage to Europe. Philadelphia, 1849.
	Geog 4308.48.5	Heinzelmann, F. Reisen durch Belgien, Holland und Grossbritannien. Leipzig, 1848.
	Geog 4308.48.11	Ow, Max von. Mit dem jüngsten Sohn König Ludwigs I. als Reisebegleiter nach Spanien 1848-1849. München, 1967.
	Geog 4308.49	Colman, H. European life and manners. Boston, 1849. 2v.
	Geog 4308.49.2	Colman, H. European life and manners. Boston, 1850. 2v.

52

Classified Listing

Geog 4301 - 4310 Voyages and travels in general - Individual travels - Europe in general - Travels (By date of travel) - cont.

	Geog 4308.49.5	Kirkland, C.M. Holidays abroad. N.Y., 1849. 2v.
	Geog 4308.49.10	Baxter, William E. Impressions of central and southern Europe. London, 1850.
	Geog 4308.50	Lundeberg, A. La perle trouvée. Lausanne, 1850.
X Cg	Geog 4308.50.5	McFarland, A. Five months abroad...1850. Concord, 1851.
	Geog 4308.50.7	Bullard, A.T.J. (Mrs.). Sights and scenes in Europe. St. Louis, 1852.
	Geog 4308.50.9	Conway, George. Running sketches of men and places. N.Y., 1851.
	Geog 4308.50.12	George, W.C. A year abroad. Boston, 1852.
	Geog 4308.50.15	Taylor, J.B. Views a-foot or Europe. N.Y., 1850.
	Geog 4308.50.17	Dickinson, A. My first visit to Europe. 5th ed. N.Y., 1856.
	Geog 4308.50.25	Crockett, Henry C. The American in Europe. London, 184-.
	Geog 4308.51	Silliman, B. Visit to Europe in 1851. N.Y., 1853. 2v.
	Geog 4308.51.2	Silliman, B. Visit to Europe in 1851. N.Y., 1854. 2v.
	Geog 4308.51.4	Silliman, B. Visit to Europe in 1851. N.Y., 1854. 2v.
	Geog 4308.51.7	Cox, Samuel S. A Buckeye abroad. N.Y., 1852.
	Geog 4308.51.10	Huidskoper, A. Glimpses of Europe in 1851 and 1867-1868. Meadville, Pa., 1882.
	Geog 4308.51.15	McFarland, A. The escape or loiterings amid the scenes. Boston, 1851.
	Geog 4308.51.20	Greely, H. Glances at Europe. N.Y., 1851.
	Geog 4308.51.25	A summer's tour in Europe in 1851. Charleston, 1852.
	Geog 4308.52	Sewell, E.M. Diary...summer tour. N.Y., 1852.
	Geog 4308.52.3	Baxter, W.E. The Tagus and the Tiber. London, 1852. 2v.
	Geog 4308.52.5	Tappan, H.P. A step from the new world to the old. N.Y., 1852. 2v.
Htn	Geog 4308.52.7*	Calvert, G.H. Scenes and thoughts in Europe. 2d series. N.Y., 1852.
	Geog 4308.52.8	Calvert, G.H. Scenes and thoughts in Europe. 1st-2d series. Boston, 1863. 2v.
	Geog 4308.52.9	Trafton, M. Rambles in Europe. Boston, 1852.
	Geog 4308.52.10	Clarke, J.F. Eleven weeks in Europe. Boston, 1852.
	Geog 4308.52.11	Horwitz, O. Brushwood, picked up on the continent. Philadelphia, 1855.
	Geog 4308.52.20	Eddy, Daniel C. Europa, or Scenes and society in England, France. 50th ed. Boston, 1860.
	Geog 4308.53	Murray, N. Men and things...in Europe. N.Y., 1853.
	Geog 4308.53.3	Tripp, A. Crests from the ocean-world. Boston, 1853.
	Geog 4308.53.4	Tripp, A. Crests from the ocean-world. Boston, 1853.
	Geog 4308.53.6	Tripp, A. Crests from the ocean-world. Boston, 1861.
	Geog 4308.53.7	Tripp, A. Crests from the ocean-world. Boston, 1862.
	Geog 4308.53.8	Wild oats sown abroad. Philadelphia, 1853.
	Geog 4308.53.9	Young, Samuel. A Wall-street bear in Europe. N.Y., 1855.
	Geog 4308.53.12	Prime, Samuel I. Travels in Europe and the East. N.Y., 1855. 2v.
	Geog 4308.53.13	Prime, Samuel I. Travels in Europe and the East. N.Y., 1856. 2v.
	Geog 4308.53.14	Boucher de Crèrecoeur de Perthes. Voyage à Constantinople par l'Italie. v.2. Paris, 1855.
	Geog 4308.54	Lippincott, S.J. Haps and mishaps...in Europe. Boston, 1854.
	Geog 4308.54.2	Lippincott, S.J. Haps and mishaps...in Europe. Boston, 1854.
	Geog 4308.54.3	Lippincott, S.J. Haps and mishaps...in Europe. Boston, 1854.
	Geog 4308.54.4	Lippincott, S.J. Haps and mishaps...in Europe. Boston, 1854.
Htn	Geog 4308.54.5*	Stowe, H.B. Sunny memories of foreign lands. Boston, 1854. 2v.
	Geog 4308.54.10	Murray, E.C.G. The roving Englishman. London, 1854.
	Geog 4308.54.15	Maney, H. Memories over the water. Nashville, 1854.
	Geog 4308.54.20	Choules, J.O. Cruise of the steam yacht, North Star. Boston, 1854.
	Geog 4308.54.25	Choules, J.O. Cruise of the steam yacht, North Star. Boston, 1854.
	Geog 4308.55	Ochoa, Eugenio de. Paris, Londres y Madrid. Paris, 1861.
	Geog 4308.55.3A	Wallace, H.B. Art, scenery and philosophy. Philadelphia, 1855.
Htn	Geog 4308.55.3*B	Wallace, H.B. Art, scenery and philosophy. Philadelphia, 1855.
	Geog 4308.55.5	Bartol, C.A. Pictures of Europe...ideas. Boston, 1855.
	Geog 4308.55.7	Bartol, C.A. Pictures of Europe...ideas. Boston, 1856.
	Geog 4308.56	Channing, W. A physician's vacation. Boston, 1856.
	Geog 4308.56.3	Cuvillier-Heury, A.A. Voyages et voyageurs, 1837-1854. Paris, 1856.
	Geog 4308.56.7	Vicuña Mackenna, Benjamin. Pajinas de mi diario. Santiago, 1856.
	Geog 4308.56.7.10	Vicuña Mackenna, Benjamin. Páginas de mi diario durante tres años de viaje. Santiago de Chile, 1936. 2v.
	Geog 4308.56.9	Eddy, D.C. Europa, or Scenes and society. Boston, 1856.
	Geog 4308.56.13	Sigourney, L.H. Pleasant memories. Boston, 1856.
	Geog 4308.56.15	Haskins, G.F. Travels in England, France. Boston, 1856.
	Geog 4308.56.17	Edwards, J.E. Random sketches...travel in 1856. N.Y., 1857.
	Geog 4308.57	Doré. A stroller in Europe. N.Y., 1857.
	Geog 4308.57.3	LeVert, O.W. Souvenirs of travel. Mobile, 1857. 2v.
	Geog 4308.57.7	Fisk, Samuel. Mr. Dunn Browne's experiences in foreign parts. Boston, 1857.
	Geog 4308.57.8	Fisk, Samuel. Mr. Dunn Browne's experiences in foreign parts. Boston, 1857.
	Geog 4308.58	Letters from Europe...summer of 1858. Buffalo, 1859.
	Geog 4308.58.3	Glimpses of Europe. Cincinnati, 1859.
	Geog 4308.58.5	Samper, J.M. Viajes de un Colombiano en Europa. Paris, 1862. 2v.
	Geog 4308.59	Sweat, M.J.M. Highways of travel. Boston, 1859.
	Geog 4308.59.3	Field, H.M. Summer pictures. N.Y., 1859.
	Geog 4308.59.5	Tait, J.R. European life, legend. Philadelphia, 1859.
	Geog 4308.59.11	Pas Soldan y Unánue, Pedro. Memories de un viajero peruano; apuntes y recuerdos de Europa y Oriente (1859-1863). Lima, 1971.
Htn	Geog 4308.60*	Benedict, E.C. A run through Europe. N.Y., 1860.
	Geog 4308.61	Bremer, F. Two years in Switzerland and Italy. London, 1861. 2v.
	Geog 4308.61.5	Hale, E.E. Ninety days worth of Europe. Boston, 1861.
	Geog 4308.62	Bucher, L. Unterwegs. v.1-2. Berlin, 1862.
	Geog 4308.63	Travelling notes in France, Italy...of an invalid in search of health. Glasgow, 1863.
	Geog 4308.65	Felton,C.C. Familiar letters from Europe. Boston, 1865.
	Geog 4308.65.3	Felton, C.C. Familiar letters from Europe. Boston, 1866.
	Geog 4308.65.5	Haven, Gilbert. The pilgrim's wallet. N.Y., 1866.

Geog 4301 - 4310 Voyages and travels in general - Individual travels - Europe in general - Travels (By date of travel) - cont.

	Geog 4308.65.9	Morford, Henry. Over-sea. N.Y., 1867.
	VGeog 4308.65.15	Smith, Amy G. Letters from Europe, 1865-1866. Washington, 1948.
	Geog 4308.66	Calvert, G.H. First years in Europe. Boston, 1866.
	Geog 4308.66.5	Macgregor, J. Thousand miles in the Rob Roy canoe. London, 1866.
	Geog 4308.66.7	Macgregor, J. Voyage alone in the yawl "Rob Roy". London, 1868.
	Geog 4308.66.10	Macgregor, J. Voyage alone in the yawl "Rob Roy". London, 1880.
	Geog 4308.66.12	Watt, Robert. Tgjennen Europa. København, 1866.
	Geog 4308.67	Forney, J.W. Letters from Europe. Philadelphia, 1867.
	Geog 4308.67.3	Meissner, A. Unterwegs. Leipzig, 1867.
Htn	Geog 4308.67.5*	Sala, George A. From Waterloo to the Peninsula. London, 1867. 2v.
	Geog 4308.67.7	Haeseler, Charles H. Across the Atlantic. Philadelphia, 1868.
Htn	Geog 4308.67.9*	Stone, H.S. From Cleveland to Russia. n.p., 1867.
	Geog 4308.67.11F	Richardson, W.L. Europe, 1867-1869. n.p., n.d.
	Geog 4308.68	Peabody, A.P. Reminiscences of European travel. N.Y., 1868.
Htn	Geog 4308.68.3*	Howe, Julia W. From the oak to the olive. Boston, 1868.
	Geog 4308.68.7	Bellows, H.W. The old world in its new face. N.Y., 1868. 2v.
Htn	Geog 4308.68.15*	Darley, F.O.C. Sketches abroad with pen and pencil. N.Y., 1868.
	Geog 4308.68.20	Coghill, James Henry. Abroad. N.Y., 1868.
	Geog 4308.69	Taylor, B. By-ways of Europe. N.Y., 1869.
Htn	Geog 4308.69.2*	Taylor, B. Byeways of Europe. London, 1869. 2v.
	Geog 4308.69.3	Darley, F.O.C. Sketches abroad with pen and pencil. N.Y., 1869.
	Geog 4308.69.4	Darley, F.O.C. Sketches abroad with pen and pencil. Boston, 1878.
	Geog 4308.69.5	Buffum, E.G. Sights and sensations in France. N.Y., 1869.
	Geog 4308.69.7	Tousey, S. Papers from over the water. N.Y., 1869.
	Geog 4308.69.8	Macmillan, Hugh. Holidays on high lands. London, 1869.
	Geog 4308.69.9	Montgomery, J.E. Our admiral's flag abroad. N.Y., 1869.
	Geog 4308.69.15	Byers, S.H.M. Twenty years in Europe. Chicago, 1900.
	Geog 4308.70	Harding, W.M. Trans-Atlantic sketches. N.Y., 1870.
	Geog 4308.70.5	Spender, E. Fjord, Isle and Tor. London, 1870.
	Geog 4308.70.10	Beste, John R.D. Nowadays. London, 1870. 2v.
	Geog 4308.71A	Guild, C. Over the ocean. Boston, 1871.
	Geog 4308.71B	Guild, C. Over the ocean. Boston, 1871.
	Geog 4308.71.2	Guild, C. Over the ocean, or Sights and scenes in foreign lands. Boston, 1875.
	Geog 4308.71.3	Hoyt, James M. Glances on the wing at foreign lands. Cleveland, 1872.
	Geog 4308.71.5	Haskins, George F. Six weeks abroad in Ireland, England and Belgium. Boston, 1872.
	Geog 4308.71.9	McCray, R.H. Narrative of travels, scenes...in the Old World. 3rd ed. Philadelphia, 1871.
	Geog 4308.71.12	Morris, Caspar. Letters of travel. Philadelphia, 1896. 2v.
	Geog 4308.72	Jackson, H.H. Bits of travel. Boston, 1872.
	Geog 4308.72.2	Jackson, H.H. Bits of travel. Boston, 1873.
	Geog 4308.72.3	Trafton, A. An American girl abroad. Boston, 1872.
	Geog 4308.72.5	Warner, Charles Dudley. Saunterings. Boston, 1872.
	Geog 4308.72.5.2	Warner, Charles Dudley. Saunterings. Boston, 1900.
	Geog 4308.72.6	Warner, Charles Dudley. Saunterings. Boston, 1872.
	Geog 4308.72.7	Kennedy, David. Colonial travel. London, 1876.
	Geog 4308.72.9F	Pressed flowers and leaves; souvenirs of travel. n.p., n.d.
	Geog 4308.73.5	Prime, Samuel I. The Alhambra and the Kremlin. N.Y., 1873.
	Geog 4308.74	Forney, John W. A centennial commissioner in Europe 1874-1876. Philadelphia, 1876.
	Geog 4308.74.5	Hutton, William. Twelve thousand miles over land and sea. Philadelphia, 1878.
Htn	Geog 4308.75*	James, H. Transatlantic sketches. Boston, 1875.
	Geog 4308.75.6	Waring, G.E. A farmer's vacation. Boston, 1876.
	Geog 4308.76	Moulton, Louise. Random rambles. Boston, 1881.
	Geog 4308.77A	Guild, C. Abroad again. Boston, 1877.
	Geog 4308.77B	Guild, C. Abroad again. Boston, 1877.
	Geog 4308.77.5	Halsey, F.W. Two months abroad. Binghampton, 1878.
	Geog 4308.77.7	Field, Henry M. From Egypt to Japan. 13th ed. N.Y., 1886.
	Geog 4308.77.8	Field, Henry M. From Egypt to Japan. 13th ed. N.Y., 1877.
	Geog 4308.78	King, H. Sketches of travel...Europe. Washington, 1878.
Htn	Geog 4308.78.5*	King, Edward. A collection of newspaper articles, describing his experiences in Europe, 1878-1880. n.p., n.d.
	Geog 4308.78.10	Prentis, N.L. A Kansan abroad. Topeka, Kansas, 1878.
	Geog 4308.78.15	Stokes, F.A. College tramps. N.Y., 1880.
	Geog 4308.80	Bush, R. Personal impressions...of European travel. Photoreproduction. N.Y., 1880?
	Geog 4308.81	Colton, G. A Maryland editor abroad. Annapolis, 1881.
	Geog 4308.81.5	Butterworth, Hezekiah. Zigzag journeys in classic lands. Boston, 1881.
	Geog 4308.81.6	Butterworth, Hezekiah. Zigzag journeys in classic lands. Boston, 1882.
	Geog 4308.82	Pitman, M.J. European breezes. Boston, 1882.
	Geog 4308.82.5	Hale, E.E. A family flight thro' France, Germany, Norway and Switzerland. Boston, 1882.
	Geog 4308.83	Thacher, S.O. What I saw in Europe. Topeka, Kansas, 1883.
	Geog 4308.83.3	Aldrich, T.B. From Ponkapog to Pesth. Boston, 1883.
	Geog 4308.83.10	Stokes, F.A. A jolly summer. N.Y., 1883.
	Geog 4308.84	James, H. Portraits of places. Boston, 1884.
Htn	Geog 4308.84*	James, H. Portraits of places. Boston, 1884.
	Geog 4308.84.5	James, H. Portraits of places. 3rd ed. Boston, 1893.
	Geog 4308.84.7	Warner, C.D. A roundabout journey. Boston, 1884.
	Geog 4308.85	Johnson, E.C. On the track of the crescent. London, 1885.
	Geog 4308.85.3	Tyler, I. Waymarks, or Sola in Europe. Chicago, 1885.
	Geog 4308.85.5	Ninde, Mary L. We two-alone in Europe. Chicago, 1889.
Htn	Geog 4308.85.7*	Sala, George A. A journey due south, travels. London, 1885.
	Geog 4308.85.9	Gjellerup, Karl. Vandreaaret. Kjøbenhavn, 1885.
	Geog 4308.85.15	Johnston, Richard M. Two gray tourists. Baltimore, 1885.
	Geog 4308.85.20	Evans, M.L. Glimpses by sea and land. Philadelphia, 1885.
	Geog 4308.86.5	Preston, M.J. (Mrs.). A handfull of monographs, continental and English. N.Y., 1886.
	Geog 4308.86.10	Buckley, James M. The midnight sun; the tsar and the nihilist. Boston, 1886.
	Geog 4308.86.17	Smith, F.H. Well-worn roads of Spain, Holland and Italy. Boston, 1892.
	Geog 4308.86.20	Bartlett, David L. Letters from Europe. Baltimore, 1886.
	Geog 4308.87	Meriwether, Lee. A tramp trip. N.Y., 1887.
	Geog 4308.87.1	Meriwether, Lee. A tramp trip. 5th ed. N.Y., 1887.

Classified Listing

Geog 4301 - 4310 Voyages and travels in general - Individual travels - Europe in general - Travels (By date of travel) - cont.

Geog 4308.87.2 Morrison, Leonard A. Rambles in Europe. Boston, 1887.
Geog 4308.87.3 Morrison, Leonard A. Rambles in Europe. Boston, 1887.
Geog 4308.87.5 Morrison, Leonard A. Among the Scotch-Irish. Boston, 1891.
Geog 4308.87.7 Cust, R.H.H. How I spent my summer holidays. Eton, 1887.
Geog 4308.88 Kingston, W.B. A wanderer's notes. London, 1888. 2v.
Geog 4308.89 Child, T. Summer holidays...Europe. N.Y., 1889.
Geog 4308.89.3 Walker, B. Aboard and abroad. Lowell, 1889.
Geog 4308.89.5 Un paseo por Europa. Habana, 1891.
Htn Geog 4308.89.10* Stockton, F.R. Personally conducted. N.Y., 1889.
Geog 4308.90 Robinson, L.B. A bundle of letters from over the sea. Boston, 1890.
Geog 4308.90.5 Potter, V.M. To Europe on a stretcher. N.Y., 1890.
Geog 4308.91 Bates, J.H. Notes of foreign travel. N.Y., 1891.
Geog 4308.91.3F Richardson, W.L. Cornwall-Devon, 1891; [Collection of photographic views...of Germany, Holland, Switzerland, Paris and England]. n.p., n.d.
Geog 4308.91.14 Elson, L.C. European reminiscences. Philadelphia, 1914.
Geog 4308.92 Hayward, H.C. From Finland to Greece, or Three seasons in Eastern Europe. N.Y., 1892.
Geog 4308.93 Pennell, J. Our sentimental journey thro' France. London, 1893.
Geog 4308.93.3 Bishop, W.H. A house-hunter in Europe. N.Y., 1893.
Geog 4308.93.5 D'Apery, Tello J. Europe seen through a boy's eyes. N.Y., 1893.
Geog 4308.93.10 Papa, Dario. Viaggi. v.2. Milano, 1893.
Geog 4308.95 Magness, E. Tramp tales of Europe. Buffalo, 1895.
Geog 4308.96 Dudley, L.B. Letters to Ruth. N.Y., 1896.
Geog 4308.96.3 Zorrilla de San Martin, J. Resonancias del camino. Paris, 1896.
Geog 4308.96.5 Zorrilla de San Martin, J. Resonancias del camino. Barcelona, 190-?
Geog 4308.97 Widmann, J.V. Sommerwanderungen und Winterfahrten. Frauenfeld, 1897.
Geog 4308.97.3 Cassell and Company, Inc. Complete pocket guide to Europe. London, 1897.
Geog 4308.98 Ferro, Adersón. Minhas viagens. Ceara, 1898.
Geog 4308.99 Letchworth, J. Home letters from the continent. N.Y., 1902.
Geog 4308.99.10 Allen, Grant. The European tour. N.Y., 1909.
Geog 4309.00 Steevens, G.W. Glimpses of three nations. N.Y., 1900.
Geog 4309.00.10 Schück, Henrik. Ur en resandes anteckningar. v.1-3. Stockholm, 1900-1909.
Geog 4309.00.15 Taylor, Charles M. Odd bits of travel with brush and camera. Philadelphia, 1900.
Geog 4309.01 Vivian, O.W.H. (Mrs.). The romance of religion. N.Y., 1901.
Geog 4309.01.5 Dudley, L.B. (Mrs.). A royal journey. N.Y., 1901.
Geog 4309.01.9 Proctor, Rachel. To the land of the midnight sun. Utica, N.Y., 1901.
Geog 4309.02 Woodward, M. His last log. Chicago, 1903.
Geog 4309.02.3A Belloc, H. The path to Rome. 4th ed. London, 1916.
Geog 4309.02.3B Belloc, H. The path to Rome. 4th ed. London, 1916.
Geog 4309.02.4 Belloc, H. The path to Rome. N.Y., 1902.
Geog 4309.02.5 Some account of...travels of myself. N.Y., 1903.
Geog 4309.02.7 Bierbaum, Otto Julius. Mit der Kraft; Automobilia. Berlin, 1906.
Geog 4309.02.13 Bierbaum, Otto Julius. Die Yankeedoodle-Fahrt und andere Reisegeschichten. München, 1920.
Geog 4309.02.20 Bell, L.L. Abroad with the Jimmies. Boston, 1902.
Geog 4309.03 Bolton, C.E. Travels in Europe. N.Y., 1903.
Geog 4309.03.5 Higinbotham, John V. Three weeks in Europe. Chicago, 1905.
VGeog 4309.03.15 Deák, Imre. Zarandoklat Rómába. Fajsz? 1903.
Geog 4309.05 Cabrera, Raimundo. Cartas a Estevez. Habana, 1906.
Geog 4309.06 Cestero, T.M. Hombres y piedras. Madrid, 1915.
Geog 4309.07 Seitz, Don C. Discoveries in every-day Europe. N.Y., 1907.
Geog 4309.07.5 Bull, Edvard. Over lave og høie fjelde; breve fra en aeldre berre. Kristiania, 1907.
Geog 4309.07.10 Society recollections in Paris and Vienna, 1789-1904, by an English officer. London, 1907.
Geog 4309.07.13 More society recollections, by an English officer. London, 1908.
Geog 4309.08 Osborne, T.M. Adventures of a green dragon. Auburn, 1908.
Geog 4309.08.5 Darling, Kenneth G. My early life in California and my first American and European tours (1890-1908). Cambridge, Mass.? 1942?
Geog 4309.09 O'Laughlin, John. From the jungle through Europe with Roosevelt. Boston, 1910.
Geog 4309.10A Hale, Louise (Closser). Motor journeys. Chicago, 1912.
Geog 4309.10B Hale, Louise (Closser). Motor journeys. Chicago, 1912.
Geog 4309.10.5 Smith, Bertha W. Traveller's tales told in letters. N.Y., 1912.
Geog 4309.10.10 Matto de Turner, C. Viaje de recrea. Valencia, 1910?
Geog 4309.11 Meriwether, Lee. Seeing Europe by automobile. N.Y., 1911.
Geog 4309.12 Maraini, Y. A year of strangers. N.Y., 1912.
Geog 4309.12.5 Peçanha, Nilo. Impressões da Europa. Nice, 1912.
Geog 4309.12.10 Howell, Charles F. Around the clock in Europe. Boston, 1912.
Geog 4309.12.15 Van Allen, William H. Travel-pictures. 2nd series. Milwaukee, 1912.
Geog 4309.13 Vasconcelos, C. de. Notas da Europa. Rio de Janeiro, 1913.
Geog 4309.13.5 Dreiser, Theodore. A traveler at forty. N.Y., 1913.
Geog 4309.13.6 Dreiser, Theodore. A traveler at forty. N.Y., 1930.
Geog 4309.14 Vrooman, Carl S. The lure and the lore of travel. Boston, 1914.
Geog 4309.14.11A Cobb, I.S. Europe revised. N.Y., 1914.
Geog 4309.14.11B Cobb, I.S. Europe revised. N.Y., 1914.
Geog 4309.14.15 Mencken, Henry L. Europe after 8:15. N.Y., 1914.
Geog 4309.15 Gardiner, Sarah D. Pages in azure and gold. New Haven, 1915.
Geog 4309.16 Villaverde, J.R. Desde lajos. Habana, 1916.
Geog 4309.21 Unstead, J.V. Europe of to-day. London, 1921.
Geog 4309.21.10A Futera, Count Y. The crown prince's European tour (March 3-September 3, 1921). Osaka, 1926.
Geog 4309.21.10B Futera, Count Y. The crown prince's European tour (March 3-September 3, 1921). Osaka, 1926.
Geog 4309.21.15A Graham, S. Europe - whither bound? N.Y., 1922.
Geog 4309.21.15B Graham, S. Europe - whither bound? N.Y., 1922.
Geog 4309.22 Morse, William I. Seeing Europe backwards. Boston, 1922.
Geog 4309.23 Traz, Robert de. Dépaysements. Paris, 1923.
Geog 4309.23.5 Osborne, A.B. Finding the worth while in Europe. N.Y., 1923.

Geog 4301 - 4310 Voyages and travels in general - Individual travels - Europe in general - Travels (By date of travel) - cont.

Geog 4309.23.10 Osborne, A.B. Finding the worth while in Europe. N.Y., 1925.
Geog 4309.24 Morse, William I. Twisting trails in the Auvergnes, Cevennes, Alps of Provence. Boston, 1924.
Geog 4309.25.10 Hamilton, C.M. Wanderings. Garden City, 1925.
Geog 4309.25.11 MacDonald, J.R. Wanderings and excursions. Indianapolis, 1928.
Geog 4309.25.12 MacDonald, J.R. Wanderings and excursions. London, 1929.
Geog 4309.25.20 Shotwell, J.T. A Balkan mission. N.Y., 1949.
Geog 4309.26A Morse, William I. The diary of a musketeer. Boston, 1926.
Geog 4309.26B Morse, William I. The diary of a musketeer. Boston, 1926.
Geog 4309.26.10 Blanco-Fombona, R. Por los caminos del mundo. Madrid, 1926.
Geog 4309.26.20 Farson, N. Sailing across Europe. London, 1928.
Geog 4309.26.25 Reynolds, Bruce. A cocktail continentale. N.Y., 1926.
Geog 4309.27 Thomas, Lowell. European skyways. Boston, 1927.
Geog 4309.27.10 Petre, E.R. When you go to Europe. N.Y., 1933.
Geog 4309.27.15 Edschmid, Kasimir. Das grosse Reisebuch. Berlin, 1927.
Geog 4309.28 Marín Vicuna, Santiago. Viajando. Santiago, 1928.
Geog 4309.28.5 Brandt, R. Das Gesicht Europas. Hamburg, 1928.
Geog 4309.28.10 Zozulia, Efim D. Iz Moskvy na Korsiku i Obratno. Leningrad, 1928.
Geog 4309.29 Muller, Julius W. "Ever thine". N.Y., 1928.
Geog 4309.29.10 MacOrlan, Pierre. Villes: Rouen-Montmartre-Brest-Londres-Villes rhénanes-Rome. 4. éd. Paris, 1929.
Geog 4309.29.11 MacOrlan, Pierre. Villes, mémoires: Montmartre, Rouen, souvenirs de Picardie et d'Artois, Brest, Londres, Villes rhénanes. Paris, 1966.
Geog 4309.29.12 MacOrlan, Pierre. Villes, mémoires: Montmartre, Rouen, souvenirs de Picardie et d'Artois. 1. éd. Evreux, 1969.
Geog 4309.29.20 Davis, Robert H. With Bob Davis hither and yon. N.Y., 1931.
Geog 4309.30 Ferreira de Mira, M. Em viagem (na velha Europa). Lisboa, 1930.
Htn Geog 4309.30.5* Morse, W.I. Nordic trails. Boston, 1930.
Geog 4309.30.6 Morse, W.I. Nordic trails. Boston, 1930.
Geog 4309.30.15 Benedict, C.W. (Mrs.). "The Benedicts abroad". London, 1930.
Geog 4309.30.20 Terán, Juan B. Lo gótico, signo de Europa. Buenos Aires, 1930?
Geog 4309.30.25 Tully, J. Beggaro abroad. Garden City, 1930.
Geog 4309.30.30 Kushner, Boris A. 103 dnia na Zapade, 1924-1926 gg. 2. izd. Moskva, 1930.
Geog 4309.31 McGeehan, William O. Trouble in the Balkans. N.Y., 1931.
Geog 4309.31.5 Ponten, J. Zwischen Rhone und Wolga. Leipzig, 1931.
Geog 4309.31.10 Duhamel, Georges. Mon Europe. Paris, 1931.
Geog 4309.31.15 Eddelbüttel, H.F. Schön ist die Welt. Berlin, 1931.
Geog 4309.31.20 Osborne, A.B. Finding the worth while in Europe. N.Y., 1931.
Geog 4309.31.25 Iraizoz y de Villar, Antonio. Apuntes de un turista tropical. Habana, 1931.
Geog 4309.31.30 Powell, Gertrude C. The quiet side of Europe. Los Angeles, 1959.
Geog 4309.32 Landau, M.A. Eine unsentimentale Reise. München, 1932.
Geog 4309.32.5 Grisar, Erich. Mit Kamera und Schreibonaschine durch Europa. Berlin, 1932.
Geog 4309.33 American Automobile Association. Foreign. Motering abroad. N.Y., 1933.
Geog 4309.33.3 Halévy, Daniel. Courrier d'Europe. Paris, 1933.
Geog 4309.34 Patterson, Sara K. Out of the fog. Caldwell, 1934.
Geog 4309.34.5 Stackpole, Edward J. Land of the midnight sun, Scandinavian region, Russia, and Germany. Harrisburg, 1934.
Geog 4309.34.10 Kisch, Egon E. Eintritt verboten. Paris, 1934.
Geog 4309.35 Hull, M.L. The gangway to Europe. Boston, 1935.
Geog 4309.35.10 Paquet, Alfons. Fluggast über Europa. München, 1935.
Geog 4309.35.15 Ofaire, Cilette. The San Luca. N.Y., 1935.
Geog 4309.36 Harris, J.P. Traveling light (Europe between boats). Hutchinson, Kansas, 1936.
Geog 4309.36.5 Abella Caprile, Margarita. Geografias (notas de viaje). Buenos Aires, 1936.
Geog 4309.37A Ichikawa, H. Japanese lady in Europe. N.Y., 1937.
Geog 4309.37B Ichikawa, H. Japanese lady in Europe. N.Y., 1937.
Geog 4309.37.5A Dorgelès, R. Vive la liberté. Paris, 1937.
Geog 4309.37.5B Dorgelès, R. Vive la liberté. Paris, 1937.
Geog 4309.38 Hamburg-American Line. Your trip to Europe. N.Y., 1938.
Geog 4309.38.5 Fellman, A. Voyage en Orient du roi Erik Ejegod. Helsinki, 1938.
Geog 4309.38.10 Pinochet, T. Viaje, plebeys por Europa. 2. ed. Santiago, 1938.
Geog 4309.38.15 Fish, Helen D. Invitation to travel. N.Y., 1938.
Geog 4309.40 Abbe, P. No place like home. N.Y., 1940.
Geog 4309.41 Moody, M.N. Trails of two travelers. N.Y., 1941.
Geog 4309.41.5 McGee, A.E. Black America abraod. Boston, 1941.
Geog 4309.41.10 Belloc, Hilaire. Places. N.Y., 1941.
Geog 4309.41.15 Blanco-Fombona, R. Dos años y medio de Inquietad. Caracas, 1942.
Geog 4309.44 Cornette, Arthur H. Van Toledo tot Budapest. Antwerpen, 1944.
Geog 4309.44.5 Baretti, Giuseppe. Oga Magoga. Milano, 1944.
Geog 4309.44.10 Hiett, Helen. No matter where. N.Y., 1944.
Geog 4309.46 Jaunes, Elly. Mánnishor därute. Stockholm, 1946.
Geog 4309.47A Wilson, Edmund. Europe without Baedeker. 1st ed. Garden City, 1947.
Geog 4309.47B Wilson, Edmund. Europe without Baedeker. 1st ed. Garden City, 1947.
Geog 4309.47C Wilson, Edmund. Europe without Baedeker. 1st ed. Garden City, 1947.
Geog 4309.47.2 Wilson, Edmund. Europe without Baedeker. 2nd ed. N.Y., 1966.
Geog 4309.47.5 Thompson, R.W. Devil at my heels. London, 1947.
Geog 4309.47.12 Noguera Mora, Neftali. Alegría y llanto de Europe; verano de 1946. 2. ed. Mérida, 1965.
Geog 4309.48.5F Bemelmans, Ludwig. The best of times, an account of Europe revisited. N.Y., 1948.
Geog 4309.51.5 Hausenstein, Wilhelm. Abendländische Wanderungen. München, 1951.
Geog 4309.52 Lins do Rêgo, José. Bota de sete léguas. Rio de Janeiro, 1952.
Geog 4309.53 Lockley, R.M. Travels with a tent in western Europe. London, 1953.
Geog 4309.53.5 Middleton, E.E. The cruise of the Kate. London, 1953.
Geog 4309.53.11 Edschmid, Kasimir. Europäisches Reisebuch. Hamburg, 1954.
Geog 4309.55 Gibbings, R. Trumpets from Montparnasse. London, 1955.
Geog 4309.55.5 Cox, Alfred B. The other half. Adelaide, 1955?

54

Classified Listing

Geog 4301 - 4310 Voyages and travels in general - Individual travels - Europe in general - Travels (By date of travel) - cont.

Geog 4309.56 Haya de la Torre. Mensaje de la Europa nordica. Buenos Aires, 1956.
Geog 4309.57 Nothomb, Pierre. Pélerinages européens. Paris, 1957.
Geog 4309.58 Bell, Robert. By road to Turkey. London, 1958.
Geog 4309.59 Fortin Magaña, Romeo. Ajo otros cielos. San Salvador, 1959.
Geog 4309.60 Anjaneyulu, D. Window to the West. 1st ed. Madras, 1967.
Geog 4309.61.5 Malinowksi, P. Der synthetische Bazar. Wien, 1961.
Geog 4309.61.10 Giloteaux, Paulin. Voyage à Moscou et au-dela. Paris, 1961.
Geog 4309.61.15 Beadle, Muriel. These ruins are inhabited. N.Y., 1961.
Geog 4309.62 Johnson, Irving. Yankee sails across Europe. N.Y., 1962.
Geog 4309.62.5 Alimzhanov, An. Piat'desiat tysiach nil' po vode i sushe. Alma-Ata, 1962.
Geog 4309.63 Cluzel, M.E. Des brumes de la Mer du Nord. Paris, 1963.
Geog 4309.63.5 Levi, P. La tregua. Torino, 1963.
Geog 4309.63.10 Wićaz, Jurij. Z Kamjenskim nosom; što dóžiwi serbski nowinar we swěće. 2. wyd. Budyšin, 1963.
Geog 4309.64 Pritchett, V.S. The offensive traveller. 1. ed. N.Y., 1964.
Geog 4309.64.5 Cluzel, M.E. Impressions sur l'Europe. Paris, 1964.
Geog 4309.64.10 Pritchett, V.S. Foreign faces. London, 1964.
Geog 4309.64.15 Selucký, Radoslav. Západ je Západ. Praha, 1964.
Geog 4309.64.16 Selucký, Radoslav. Západ je Západ. 2. vyd. Praha, 1965.
Geog 4309.64.20 Streeter, Edward. Along the ridge; from northwestern Spain to southern Yugoslavia. N.Y., 1964.
Geog 4309.64.25 Linhares, Temístocles. Jornal da Europe. Brasil, 1964.
Geog 4309.64.31 Sansom, William. Away to it all. N.Y., 1966.
Geog 4309.65 MacShane, Frank. The American in Europe. 1st ed. N.Y., 1965.
Geog 4309.65.5 Cione, Edmondo. Questa Europa. Milano, 1965.
Geog 4309.66 Basnett, Fred. Travels of a capitalist lackey. South Brunswick, 1966.
Geog 4309.66.5 Andersch, Alfred. Aus einem römischen Winter. Olten, 1966.
Geog 4309.66.10 Utchenko, Sergei L. Glazami istorika. Moskva, 1966.
Geog 4309.67 Lorck, Carl von. Europa Privat. Frankfurt, 1967.
Geog 4309.68 Lisakovskii, Igor' N. Dalnie sosedi. Odessa, 1968.
Geog 4309.68.5 Garrido, Felipe. Viejo continente. 1. ed. México, 1973.
Geog 4309.69 Duffus, Robert Luther. The polar route to time gone by. 1st ed. N.Y., 1969.
Geog 4309.70 Lottman, Herbert R. Detours from the grand tour. Englewood Cliffs, N.J., 1970.
Geog 4309.71 Bąbiński, Gan. Noc w Soho. Wyd. 1. Łódź, 1971.
Geog 4309.71.5 Derruau, Max. L'Europe. Paris, 1971.
Geog 4309.71.10 Krueger, Horst. Fremde Vaterländer. München, 1971.
Geog 4309.71.15 Holmqvist, Lasse. Vitt och brett. Stockholm, 1971.
Geog 4309.72 Hillaby, John D. Journey through Europe. London, 1972.
Geog 4309.73 Piontek, Heinz. Helle Tage anderswo. München, 1973.
Geog 4309.74 Kujundžić, Miodrag. Tudje avlije. Novi Sad, 1974.
Geog 4309.74.5F Europas Hovedstaeder. Billeder og stemninger oplevet af Eva Bendix. København, 1974.

Geog 4311 Voyages and travels in general - Individual travels - Europe in general - Special regions - Danube River

Geog 4311.5 Götz, W. Das Donaugebeit mit Rücksicht. Stuttgart, 1882.
Geog 4311.7 Senkel, K.F. De istri ostiis dissertatio historico-geographica. Wratislaviae, 1820.
Geog 4311.9 Bigelow, P. Paddles and politics down the Danube. London, 1892.
Geog 4311.11 Schweiger-Lerchenfeld, Amand von. Die Donau als Völkeweg. Wien, 1896.
Geog 4311.12 Jerrold, Walter. The Danube. N.Y., 1911.
Htn Geog 4311.15F* Meyer's Donau-Ansichten. pt.1-2,5,7-11. Hildburghausen, 1838.
Geog 4311.17 Beattie, William. The Danube, its history, scenery. London, 1844?
Geog 4311.18 Feuerstein, Ernst. Deudonex. Dichtung. n.p., 1966.
Geog 4311.19 Jonák, Ján. 2200 kilo-metrov po Dunaiu. Moskva, 1960.
Geog 4311.20 Blairy, Jean. Le beau Danube gris. Paris, 1948.
Geog 4311.21 Buettner, R. Burgen und Schlösser and der Donau. Wein, 1964.
Geog 4311.22 Sergeeva, Nelli A. Dunai-reka druzhby. Odessa, 1966.
Geog 4311.23 Kazaka, Tenin. Plavashchi domovc. Plovdiv, 1965.

Geog 4312 Voyages and travels in general - Individual travels - Europe in general - Special regions - North Sea

Geog 4312.5 Raynes, Rozelle. North in a nutshell. Lymington, 1968.
Geog 4312.10 Pilkington, Roger. Small boat to Elsinore. N.Y., 1969.

Geog 4313 Voyages and travels in general - Individual travels - Europe in general - Special regions - Baltic Sea

Geog 4313.5 Catteau-Calleville, J.P. Tableau de la Mer Baltique. Paris, 1812. 2v.
Geog 4313.6 Schulz, Bruno. Die deutsche Ostsee. Bielefeld, 1931.
Geog 4313.7 Schlump, E. Die politisch-geographische Bedeutung der Ostsee. Inaug.-Diss. Königsberg, 1934.
Geog 4313.8 Egor'eva, A.V. Baltiiskoe more. Moskva, 1961.
Geog 4313.9 Kiecksee, Heinz. Die Ostsee-Sturmflut 1872. Heide, 1972.

Geog 4314 Voyages and travels in general - Individual travels - Europe in general - Special regions - Black Sea

Geog 4314.5 Das Schwarze Meer. Leipzig, 1854.
Geog 4314.6 Symonds, W. Extract from journal in the Black Sea. London, 184-?
Geog 4314.7 Zenkovich, V.P. Berega Chennogo i Azovskego morei. Moskva, 1958.
Geog 4314.8 Schweiger-Lerchenfeld, Amand von. Zwischen Donau und Kaukasus. Wien, 1887.

Geog 4321 - 4330 Voyages and travels in general - Individual travels - Europe in general - Special regions - Mediterranean Sea (By date of travel)

Geog 4321.83 Ibn Gubáyer. Viaggio in Espana, Sicilia, Siria e Palestina. Roma, 1906.
Geog 4323.35 Jacopo da Verona. Liber peregrinationis. Roma, 1950.
Htn Geog 4324.59* Ehingen, Georg von. Itinerarium. Augsburg, 1600.
Geog 4324.59.5 Pfeiffer, F. Georg von Ehingen Reisen. Stuttgart, 1842.
Geog 4324.59.10 Ehingen, Georg von. Diary. London, 1929.
Geog 4324.80 Brasca, Santo. Viaggio in Terrasanta di Santo Brasca, 1480. Milano, 1966.
Geog 4324.84 Neeffs, E. Un voyage au XVe siècle. Louvain, 1873.
Geog 4324.89 Illustrazioni in un anonimo Viaggiatore (1489). Livorno, 1785.
Geog 4325.73 Lubenau, R. Beschreiben der Reisen des Reinhold Lubenau. Königsberg, 1912-1920. 3v.

Geog 4321 - 4330 Voyages and travels in general - Individual travels - Europe in general - Special regions - Mediterranean Sea (By date of travel) - cont.

Htn Geog 4325.90.5* Webbe, Edward. Edward Webbe, his travailes. Edinburgh, 1885.
Geog 4325.90.6 Webbe, Edward. Edward Webbe, chief master gunner, his travailes. Edinburgh, 1885.
Geog 4325.90.10 Webbe, Edward. Edward Webbe, his travailes, 1590. Birmingham, 1868.
Geog 4325.90.12 Webbe, Edward. Edward Webbe, chief master gunner, his travailes, 1590. London, 1868.
Geog 4326.10 Ens, K. Deliciae transmari. Coloniae Agrippinae, 1610.
Htn Geog 4326.32* Lithgow, William. The totall discourse of...adventures. Lyon, 1632.
Htn Geog 4326.32.3* Lithgow, William. The totall discourse of...adventures. London, 1640.
Geog 4326.32.5 Lithgow, William. Travels and voyages thro' Europe, Asia and Africa. 12th ed. Leith, 1814.
Geog 4326.32.10A Lithgow, William. Travels and voyages thro' Eruope, Asia and Africa. Glasgow, 1906.
Geog 4326.32.10B Lithgow, William. Travels and voyages thro' Eruope, Asia and Africa. Glasgow, 1906.
Geog 4326.32.20 Lithgow, William. Rare adventures and painefull peregrinations. London, 1928.
Geog 4326.32.30 Lithgow, William. Discourse of a peregrination in Europe, Asia and Affricke. Facsimile. N.Y., 1971.
Geog 4326.74 Smith, T. Epistulae...moribus...turcarum. Oxonii, 1674.
Geog 4326.75 Teonge, Henry. The diary of H. Teonge...1675-1679. London, 1825.
Geog 4326.75.5A Teonge, Henry. The diary of H. Teonge, chaplain on board H.M.'s ships Assistance, Bristol and Royal Oak, 1675-1679. London, 1927.
Geog 4326.75.5B Teonge, Henry. The diary of H. Teonge, chaplain on board H.M.'s ships Assistance, Bristol and Royal Oak, 1675-1679. London, 1927.
Geog 4326.86 Neitzschitz, G.C. Sieben-Jährige...Welt-Beschauung. Nurnberg, 1686.
Geog 4327.22 Shaw, Thomas. Voyages...dans plusieurs provinces de la Barbarie et du Levant. La Haye, 1743. 2v.
Geog 4327.38 Montague, J. Voyage...round the Mediterranean. London, 1799.
Htn Geog 4327.39* Campbell, J. Travels and adventures of Eduard Brown. London, 1739.
Geog 4327.44 Thompson, C. Travels. Reading, 1744. 2v.
Geog 4327.44.3F Drummond, A. Travels thro' different cities of Germany, Italy, Greece and parts of Asia. London, 1754.
Geog 4327.88 Jardine, A. Letters from Barbary, France. London, 1788. 2v.
Geog 4327.88.10 Bisani, Alessandro. Lettres sur divers endroits de l'Europe. Londres, 1791.
Geog 4327.93 Bisani, Alessandro. Picturesque tour thro'...Europe, Asia. London, 1793.
Geog 4327.94 Morritt, John B.S. The letters of...journeys in Europe and Asia Minor. London, 1914.
Geog 4327.98F Willyams, Cooper. A voyage up the Mediterranean. London, 1802.
Geog 4328.01 Collins, F. Voyages...from 1796-1801. London, 1819.
Geog 4328.05 Griffiths, J. Travels in Europe, Asia Minor, and Arabia. London, 1805.
Geog 4328.10 Cockburn, G. A voyage to Cadiz and Gibraltar, up the Mediterranean, to Sicily and Malta, in 1810 and 1811. London, 1815. 2v.
Geog 4328.13.2 Bramsen, John. Travels in Egypt, Syria, Cyprus. 2nd ed. London, 1820. 2v.
Geog 4328.19 Noah, M.M. Travels in England, France and Spain. N.Y., 1819.
Geog 4328.19.10 James, John Thomas. Journal of a tour in Germany, Russia, Poland in 1813-1814. 3rd ed. London, 1819.
Geog 4328.20 Monson, W.J.M. Extracts from a journal. London, 1820.
Geog 4328.29 Jones, George. Sketches of naval life...on Mediterranean. New Haven, 1829. 2v.
Htn Geog 4328.32* Wines, E.C. Two years and a half...in Mediterranean and Levant. Philadelphia, 1832.
Geog 4328.34 Mott, V. Travels in Europe and the East. N.Y., 1842.
Geog 4328.35 Ship and shore...cruise to the Levant. N.Y., 1835.
Geog 4328.36.2 Cumming, W.F. Notes of a wanderer. London, 1840. 2v.
Geog 4328.36.5 Maritime scraps, or Scenes in the frigate United States during a cruise in the Mediterranean. Boston, 1838.
Geog 4328.37.2 Wilde, William R. Narrative voyage to Madeira...Cyprus and Greece. Dublin, 1852.
Geog 4328.38 Records of travel. Boston, 1838.
Geog 4328.38.5 Happoldt, Christopher. The Christopher Happoldt journal. Charleston, 1960.
Geog 4328.39 Maximilian, H. Wanderung nach dem Orient. München, 1839.
Htn Geog 4328.39.3* Torrey, F.P. Journal of a cruise of the U.S.S. Ohio in the Mediterranean. Boston, 1841.
Geog 4328.40.3 Wilde, William R. Narrative of a voyage to Madeira, Teneriffe. 2nd ed. Dublin, 1844.
Geog 4328.40.10 Wright, G.N. The shores and islands of the Mediterranean. London, 1840.
Geog 4328.42 Rockwell, C. Sketches of foreign travel. Boston, 1842. 2v.
Geog 4328.42.5 Damer, G.L.D. Diary of a tour in Greece, Turkey. London, 1842.
Geog 4328.43 Chenavard, Antoine N. Voyage en Grèce et dans le Levant fait en 1843-1844. Lyon, 1849.
Geog 4328.45 Borrer, D. Journey from Naples to Jerusalem. London, 1845.
Geog 4328.46 Schroeder, F. Shores of the Mediterranean. N.Y., 1846. 2v.
Geog 4328.47 Margoliouth, Moses. A pilgrimage to the land of my fathers. London, 1850. 2v.
Geog 4328.51 Colton, W. Ship and shore in Madeira. N.Y., 1851.
Geog 4328.51.5 Noel-Fearn, H. The shores and islands of the Mediterranean. London, 1851. 3v.
Geog 4328.52.2 Aiton, John. The lands of the Messiah, Mahomet, and the pope. 2nd ed. London, 1852.
Geog 4328.53 Willis, N.P. Summer cruise in the Mediterranean. Auburn, 1853.
Geog 4328.53.5 Paris. Services Maritimes. Les paquebots du Levant. Paris, 1853.
Geog 4328.54 Smyth, W.H. The Mediterranean. London, 1854.
Geog 4328.55 Taylor, B. Lands of the Saracew. N.Y., 1855.
Geog 4328.57 Wise, H.A. Scampavias from Gibel Tarek to Stamboul. N.Y., 1857.
Geog 4328.59 Taylor, Bayard. Travels in Greece and Russia. N.Y., 1859.
Geog 4328.59.5 Böttger, C. Das Mittelmeer. Leipzig, 1859.

Classified Listing

Geog 4321 - 4330 Voyages and travels in general - Individual travels - Europe in general - Special regions - Mediterranean Sea (By date of travel) - cont.

Geog 4328.61 Bennet, J.H. Winter and spring on the shores of the Mediterranean. 4th ed. London, 1870.
Geog 4328.62 Phelps, S.D. Holy Land with glimpses of Europe and Egypt. N.Y., 1864.
Geog 4328.66 Lycklama, A. Nijeholt. Voyage en Russie, au Caucase. Paris, 1872-1875. 4v.
Geog 4328.66.5 Sleeper, M.G.Z. (Mrs.). The Mediterranean Islands. Boston, 1866.
Geog 4328.69 Cox, S.S. Search for winter sunbeams. London, 1869.
Geog 4328.69.5 Cox, S.S. Search for winter sunbeams. N.Y., 1874.
Geog 4328.70 Fouché, William Grey (Mrs.). Dagbqk...resatill Egypten, Konstantinopel. Stockholm, 1870.
Geog 4328.70.5 Woerl, Leo. Erzherzog Ludwig Salvator aus dem Osterreichischen Kaiserhause als Forscher des Mittelmeeres. Leipzig, 1899.
Geog 4328.73 Nāsir, al-Dīn Shāh. Ein Harem in Bismarcks Reich. Tübingen, 1969.
Geog 4328.74.5 Aguilar, Federico C. Recuerdos de un viaje a Oriente en el año de 1874. Bogota, 1875.
Geog 4328.79 Jeréz Perchét, Augusto. El mediterráneo. 3a ed. Madrid, 1879.
Geog 4328.83 Lyautey, L.H.G. Lettres de jeunesse. Paris, 1931.
Geog 4328.88 Buckley, J.M. Travels in three continents. N.Y., 1895.
Geog 4328.91 Groome, P.L. Rambles...in three continents. Greensboro, 1891.
Geog 4328.92 Meriwether, Lee. Afloat and ashore on the Mediterranean. N.Y., 1892.
Geog 4328.93 Barber, J.L. Mediterranean mosaics. N.Y., 1895.
Geog 4328.94A Davis, R.H. Rulers of the Mediterranean. N.Y., 1894.
Geog 4328.94B Davis, R.H. Rulers of the Mediterranean. N.Y., 1894.
Geog 4328.94.5 Hoyt, W.M. A cruise on the Mediterranean. Chicago, 1894.
Geog 4328.94.7PF Rothschild, N. Skizzer aus dem Süden. Wien, 1894.
Geog 4328.95 Brooks, N. The Mediterranean trip. N.Y., 1895.
Geog 4328.96 Woolson, C.F. Mentone, Cairo and Corfu. N.Y., 1896.
Geog 4328.98 Domsthorpe, Wordsworth. Down the stream of civilization. London, 1898.
Geog 4328.99 Honeyman, Abraham Van Doren. From America to the Orient. Plainfield, 1899.
Geog 4329.00 Acker, F. Pen sketches. Philadelphia, 1900.
Geog 4329.02 McCready, R.H. Cruise of the Celtic...Mediterranean. N.Y., 1902.
Geog 4329.04 Loring, J.M. The old world thro' new world eyes. St. Louis, 1904.
Geog 4329.05.20 Lorenz, I.E. The new Mediterranean traveller. 13th ed. N.Y., 1927.
Geog 4329.06 Fischer, T. Mittelmeerbilder. Leipzig, 1906.
Geog 4329.06.2 Fischer, T. Mittelmeerbilder. Leipzig, 1908.
Geog 4329.07 Letters from a Tarheel traveler. Oxford, 1907.
Geog 4329.09 Bayne, S.G. A fantasy of Mediterranean travel. N.Y., 1909.
Geog 4329.09.5 Roberts, William C. The boys' account of it...foreign travel. N.Y., 1909.
Geog 4329.09.10 Naumann, Friedrich. Sonnenfahrten. Berlin, 1909.
Geog 4329.14F Januszewski, W. Dokoła morza śródziemnego. Wilnie, 1914.
Geog 4329.17.5 Cimon, Henri H.M. Aux vieux pays. 3. ed. Montréal, 1917.
Geog 4329.20 Bunge de Galvez, D. Tierras del mar Azul. Buenos Aires, 192-.
Geog 4329.21 Prioleau, John. The adventures of Imshi; a two-seater in search of the sun. Boston, 1923.
Geog 4329.22 Maurras, Charles. Anthénea. Paris, 1922.
Geog 4329.23 Hildebrand, A.S. Blue water. N.Y., 1923.
Geog 4329.25 McAllister, J.G. Borderlands of the Mediterranean. Richmond, 1926.
Geog 4329.26 Wilstach, Paul. Islands of the Mediterranean. London, 1926.
Geog 4329.27 Haliburton, R. The glorious adventure. Indianapolis, 1927.
Geog 4329.27.5 Oppenheim, E.P. The quest for winter sunshine. Boston, 1927.
Htn Geog 4329.27.10* Morse, William I. Sicilian days and other journeys round the Mediterranean and Adriatic. Boston, 1927.
Geog 4329.27.15 Ludwig, Emil. Am Mittelmeer. Berlin, 1927.
Geog 4329.27.20 Meier-Graefe, J. Pyramide und Tempel. Berlin, 1927.
Geog 4329.28 Rydh, Hanna. Kring medelhavets stränder. Stockholm, 1928.
Geog 4329.29 Spaini, Alberto. Viaggi di Bertoldo. Aquila, 1929.
Geog 4329.29.5 Chalmers, T.M. Under the olive trees. N.Y., 1931.
Geog 4329.29.10 Ludwig, Emil. On Mediterranean shores. Boston, 1929.
Geog 4329.30.3 Waugh, E. Labels; a Mediterranean journal. London, 1930.
Geog 4329.30.7 Anderson, I.P. (Mrs.). A yacht in Mediterranean seas. Boston, 1930.
Geog 4329.31 Wilstach, Paul. Islands of the Mediterranean. N.Y., 1931.
Geog 4329.32 Edschmid, K. Zauber und Grösse des Mittelmeers. Frankfurt, 1932.
Geog 4329.33 Rèpaci, L. Con la ciurma dell'Alessandro. Milano, 1933.
Geog 4329.35.3 Audisio, G. Jeunesse de la Méditerranée. Paris, 1935-1936.
Geog 4329.35.10 Bloomfield, Paul. The Mediterranean; an anthology. London, 1935.
Geog 4329.35.15 Laughlin, C.E. So you're going to the Mediterranean. Boston, 1935.
Geog 4329.36.2 Ervine, St. J.G. A journey to Jerusalem. London, 1936.
Geog 4329.36.6 Morton, H.V. In the steps of St. Paul. London, 1936.
Geog 4329.36.7 Morton, H.V. In the steps of St. Paul. N.Y., 1936.
Geog 4329.36.10 Parain, Charles. La Méditerranée; les hommes et leurs travaux. 4e éd. Paris, 1936.
Geog 4329.37F Hürlimann, M. Das Mittelmeer. Zürich, 1937.
Geog 4329.38 Miller, M.S. (Mrs.). Cruising the Mediterranean. N.Y., 1938.
Geog 4329.38.5 Duran, Faik Sabri. Akdenirde leir yor gerintisi. Istanbul, 1938.
Geog 4329.40 Pfuhl, E. Ostgriechische Reisen; Kleinasien, Kypros und Syrien. Basel, 1940.
Geog 4329.44 Fateili, A. Dea senza volto. Milano, 1944.
Geog 4329.47 Newman, B. Middle eastern journey. London, 1947.
Geog 4329.49 Newman, B. Mediterranean background. London, 1949.
Geog 4329.51 Thompson, D. The phoenix in the desert. London, 1951.
Geog 4329.53 Landery, C.F. Whistling for a wind. 1st American ed. N.Y., 1953.
Geog 4329.53.5F Dunand, M. De l'Amanus au Sinai. Beyrouth, 1953.
Geog 4329.53.10 Birot, Pierre. La Méditerranée et le Moyen-Orient. Paris, 1953. 2v.
Geog 4329.53.12 Birot, Pierre. La Méditerranée et le Moyen-Orient. 2. éd. Paris, 1964.
Geog 4329.53.15 Cordan, W. Das Mittelmeer. Düsseldorf, 1953.
Geog 4329.53.20 Méditerranée orientale. Paris, 1953.

Geog 4321 - 4330 Voyages and travels in general - Individual travels - Europe in general - Special regions - Mediterranean Sea (By date of travel) - cont.

Geog 4329.54 Méditerranée occidentale. Paris, 1954.
Geog 4329.55 Golding, Louis. Good-bye to Ithaca. London, 1955.
Geog 4329.57 Todd, Roberto. Viajando por Europa. Madrid, 1957.
Geog 4329.58F Beny, Roloff. The thrones of earth and heaven. London, 1958.
Geog 4329.58.5 Van Sinderen, Adrian. A voyage through the azure seas. N.Y., 1958.
Geog 4329.58.10F Beny, Roloff. Merveilles de la Méditerranée. Paris, 1958.
Geog 4329.63 Morpurgo, J.E. The road to Athens. London, 1963.
Geog 4329.64 Le Lannou, Maurice. Problèmes géographiques de la Méditerranée européenne. Paris, 1964.
Geog 4329.67 Houston, James M. The western Mediterranean world. N.Y., 1967.
Geog 4329.69 Lewis, Cecil. Turn right for Corfu. London, 1972.
Geog 4329.71 Chernishev, Slaveho. Po more. Sofiia, 1971.
Geog 4329.73 Isnard, Hildebert. Pays et paysages mediterranéens. 1. éd. Paris, 1973.
Geog 4329.74 Fries, Carl. Europas morgen. Stockholm, 1974.

Geog 4332 Voyages and travels in general - Individual travels - Europe in general - Special regions - Adriatic Sea

Geog 4332.5 Schweiger-Lerchenfeld, A.F. Die Adria. Wien, 1883.
Geog 4332.7F Yriarte, C. Les bords de l'Adriatique. Paris, 1878.
Geog 4332.8 Paton, A.A. Highlands and islands of the Adriatic. London, 1849. 2v.
Geog 4332.8.5 Paton, A.A. Researches on the Danube and the Adriatic. Leipzig, 1861. 2v.
Geog 4332.9 L'Adriatico. Milano, 1914.
Geog 4332.10 L'Adriatico; studio geografico. Milano, 1915.
Geog 4332.12 Cassi, Gellio. Il Mare Adriatico. Milano, 1915.
Geog 4332.14 Roncagli, Giovanni. Il problema militaire dell'Adriatico spiegatto a tutti. Roma, 1918.
Geog 4332.17 Goracuchi, J.A. Die Adria und ihre Küsten. Triest, 1863.
Geog 4332.19 Hodgkinson, H. The Adriatic Sea. N.Y., 1956.
Geog 4332.20 Novak, Grga. Naše more. 2. izd. Zagreb, 1932.
Geog 4332.21 Soljan, Antun. The thousand islands of the Adriatic. Beograd, 1965.
Geog 4332.22 Denham, Henry Mangles. The Adriatic; a sea guide to its coasts and islands. London, 1967.
Geog 4332.24 Košutić, Ivo. Gusari Jadranskoga mora. Zagreb, 1971.
Geog 4332.26 Kolar, Vladimir. Jadran - more mira. Ljubjana, 1970.

Geog 4341 - 4350 Voyages and travels in general - Individual travels - Europe and other regions - Europe and Asia (By date of travel)

Htn Geog 4341.73* Montano, B.A. Itinerarium beniamini tudelemsis. Antwerpiae, 1575.
Htn Geog 4341.73.3* Benjamin ben Jonah, of Tudela. Itinerarium Benjaminis. Lugdunum Batavorum, 1633.
Geog 4341.73.5 Benjamin ben Jonah, of Tudela. The itinerary of Rabbi Benjamin of Tudela. pt.1-2. London, 1840.
Geog 4341.73.6 Benjamin ben Jonah, of Tudela. Die Reisebeschreibungen. Jerusalem, 1903.
Geog 4341.73.7 Carmoly, E. Notice historique sur Benjamin de Tudèle. Bruxelles, 1852.
Geog 4341.73.10A Benjamin ben Jonah, of Tudela. The itinerary of Benjamin of Tudela. London, 1907.
Geog 4341.73.10B Benjamin ben Jonah, of Tudela. The itinerary of Benjamin of Tudela. London, 1907.
Geog 4341.73.15 Benjamin ben Jonah, of Tudela. Viajes de Benjamin de Tudela, 1160-1173. Madrid, 1918.
Geog 4341.73.16 Benjamin ben Jonah, of Tudela. Beniamini Tudelensis Itinerarium; ex versione Benedicti Ariae Montani. Bononiae, 1967.
Geog 4344.35.10 Tafur, Pero. Travels and adventures, 1435-1439. N.Y., 1926.
Geog 4344.35.12 Tafur, Pero. Travels and adventures, 1435-1439. London, 1926.
Geog 4345.73 Krafft, Hans U. Ein deutscher Kaufmann. Göttingen, 1862.
Geog 4345.99 Sherley, A. The three brothers or travels. London, 1825.
Geog 4345.99.2 Nixon, Anthony. The three English brothers. London, 1607.
Geog 4345.99.4 Penrose, B. The Sherleian odyssey, being a record of the travels of three famous brothers. Taunton, 1938.
Geog 4345.99.10 Davies, David William. Elizabethans errant; the strange fortunes of Sir Thomas Sherley and his three sons. Ithaca, N.Y., 1967.
Geog 4346.73.2 Ray, John. Travels thro' Low Countries, Germany. 2nd ed. London, 1738.
Htn Geog 4346.76.5* Struys, Jan J. Les voyages de Jean Struys. Amsterdam, 1681.
Geog 4346.76.25 Struys, Jan J. Tri puteshestviia. Moskva, 1935.
Geog 4346.83 Kaempfer, Engelbert. Die Reisetagebücher. Wiesbaden, 1968.
Geog 4346.85 Avril, P. Voyage en divers etats d'Europe et d'Asie. Paris, 1693.
Htn Geog 4346.93* Ray, John. Collection of curious travels and voyages...eastern countries. London, 1693.
Htn Geog 4347.01.5F* Bruyn, C. de. Reizen over Muskovie door Persie. Amsterdam, 1711.
Htn Geog 4347.01.6F* Bruyn, C. de. Voyage au Levant. Paris, 1725. 5v.
Geog 4347.01.7F Bruyn, C. de. Voyages...au Levant. Haye, 1732. 5v.
Htn Geog 4347.01.8F* Bruyn, C. de. Voyages...en Perse et aux Indes Orient. Amsterdam, 1718. 2v.
Geog 4347.01.10FA Bruyn, C. de. Travels into Muscovy, Persia, E. Indies. London, 1737. 2v.
Geog 4347.01.10FB Bruyn, C. de. Travels into Muscovy, Persia, E. Indies. London, 1737. 2v.
Htn Geog 4347.18* Struys, Jan J. Les voyages...en Moscovie. v.1-3. Amsterdam, 1718.
Geog 4347.27 La Montraye, A. de. Voyage du...en Europe, Asie et Afrique. Haye, 1727. 2v.
Geog 4347.27.3F La Montraye, A. de. Voyage en Anglois et en François. Haye, 1732.
Htn Geog 4347.27.5* La Montraye, A. de. Voyages and travels, Prussia, Russia and Poland. London, 1732. 3v.
Htn Geog 4347.85F* Witsen, N. Noord en Oost Tartaryen. Amsterdam, 1785. 2v.
Geog 4347.91.1 Steuke, Johann Kaspar. Von Amsterdam nach Temiswar. Berlin, 1969.
Geog 4347.98 Forster, G. Journey from Bengal to England. London, 1798.
Geog 4347.99 Abu Taleb Khan. Travels...in Asia,...Europe. London, 1810. 2v.
Geog 4347.99.5 Jackson, J. Journey from India. London, 1799.
Geog 4347.99.7 Taylor, J. Travels from England to India. London, 1799. 2v.

56

Classified Listing

Geog 4341 - 4350 Voyages and travels in general - Individual travels - Europe and other regions - Europe and Asia (By date of travel) - cont.

	Geog 4348.07	Campbell, D. Journey over land to India. Philadelphia, 1807.
	Geog 4348.08	Darwin, F.S. Travels in Spain and the East, 1808-1810. Cambridge, 1927.
	Geog 4348.10.5	Clarke, E.D. Travels in various countries of Europe, Asia and Africa. Edinburgh, 1811. 2v.
Htn	Geog 4348.10.10*	Clarke, E.D. Travels in various countries of Europe. 1st American ed. v.3-4. N.Y., 1815. 2v.
	Geog 4348.13.5	Clarke, E.D. Travels in various countries of Europe. N.Y., 1813. 4v.
	Geog 4348.13.10	Clarke, E.D. Travels in...Europe, Asia and Africa. London, 1816-1824. 11v.
	Geog 4348.13.20	Clarke, E.D. Travels in Russia, Tartary and Turkey. Edinburgh, 1839.
	Geog 4348.14	Almerté, T. Voyages de sa majesté la reine d'Angleterre. Paris, 1821.
	Geog 4348.17	Johnson, John. A journey from India to England thro' Persia, Georgia, Russia, Poland and Prussia, in 1817. London, 1818.
	Geog 4348.18	Journal of Jonathan Russell, 1818-1819. n.p., n.d.
	Geog 4348.32	Stocqueler, J.H. Fifteen months pilgrimage. London, 1832. 2v.
	Geog 4348.34	Marmont, Auguste Frédéric L. Voyage du...en Hongrie, en Transylvanie. Paris, 1837. 5v.
	Geog 4348.45	Barber, J. Overland guide-book. London, 1845.
	Geog 4348.47	Marmier, X. Du Rhin au Nil. Paris, 1847. 2v.
	Geog 4348.48	Terry, C. Scenes and thoughts in foreign lands. London, 1848.
	Geog 4348.49.2	Richardson, D.L. Anglo-Indian passage. London, 1849.
	Geog 4348.54	Hoppin, J.M. Notes of a theological student. N.Y., 1854.
	Geog 4348.60.10	Beluze, X. Les pérégrinations en Orient et en Occident. 21e éd. Citeaux, 1883. 2v.
	Geog 4348.65	Ussher, John. Journey from London to Persepolis. London, 1865.
	Geog 4348.67	Creagh, J. A scamper to Sebastopol and Jerusalem 1867. London, 1873.
	Geog 4348.81	Cuyler, T.L. From the Nile to Norway and homeward. N.Y., 1881.
	Geog 4348.87	McKenzie, A. Some things abroad. Boston, 1887.
	Geog 4348.93	Scollard, Clinton. On sunny shores. N.Y., 1893.
	Geog 4349.07	Barzini, Luigi. Da Pechino a Parigi in sessanta giorni. Milano, 1970.
	Geog 4349.10	The all-rail route between the Far East and Europe. London, 1910.
	Geog 4349.14	Young, Ernest. From Russia to Siam, with a voyage down the Danube. London, 1914.
	Geog 4349.23	Ross, Halford. By devious ways. London, 1927.
	Geog 4349.28	Brennan, John W. The student abroad. Boston, 1928.
	Geog 4349.29	Ellis, M.H. Express to Hindustan. London, 1929.
	Geog 4349.30	St.-Félix, Max de. A travers l'Orient (1930). Paris, 1931.
	Geog 4349.31	Toynbee, Arnold J. A journey to China; or, Things which are seen. London, 1931.
	Geog 4349.36	Nichols, Beverley. No place like home. London, 1936.
	Geog 4349.36.2	Nichols, Beverley. No place like home. 1st ed. Garden City, 1936.
	Geog 4349.36.3	Morand, Paul. La route des Indes. Paris, 1936.
	VGeog 4349.68	Bregman, Aleksander. Rubiezè woluorci. Londyn, 1968.
	Geog 4349.72	Ovcharenko, Aleksandr I. V etom bushuiushchem mire. Moskva, 1972.

Geog 4354 - 4360 Voyages and travels in general - Individual travels - Europe and other regions - Europe and Africa (By date of travel)

	Geog 4357.81	Rooke, Henry. Travels to the coast of Arabia Felix, and from thence by the Red Sea and Egypt to Europe. London, 1783.
	Geog 4358.90	Maupassant, G. de. La vie errante. 13. éd. Paris, 1890.
	Geog 4358.92	Blackburn, Henry. Artistic travel in Normandy, Brittany, the Pyrenees, Spain and Algeria. London, 1892.
	Geog 4359.25	Flambeau, V. Red letter days in Europe. With a glimpse of Northern Africa. N.Y., 1925.
	Geog 4359.27	Belloc, Hilaire. Towns of destiny. N.Y., 1931.
	Geog 4359.28	Belloc, Hilaire. Many cities. London, 1928.
	Geog 4359.40	Newman, Bernard. Savoy! Corsica! Tunis! Mussolini's dream lands. London, 1940.
	Geog 4359.70	Brandi, Cesare. A passo d'uomo. Milano, 1970.

Geog 4411 - 4420 Voyages and travels in general - Individual travels - Asia and other regions - Asia and Africa, Asia and Oceania (By date of travel)

	Geog 4412.91	Pertz, G.H. Der älteste Versuch zur Entdeckung des Seeweges nach Ostindien im Jahre 1291. Berlin, 1859.
	Geog 4412.91.5	Avezac-Macaya. Expédition génoise des frères Vivaldi...route maritime des Indes orientales XIII. siècle. Paris, 1859.
	Geog 4413.25.5A	Ibn Batoutah. Travels of Ibn Batuta. London, 1829.
	Geog 4413.25.5B	Ibn Batoutah. Travels of Ibn Batuta. London, 1829.
	Geog 4413.25.9	Ibn Batoutah. Voyages. Paris, 1874. 5v.
	Geog 4415.02	Varthema, L. de. Itinerário. Lisboa, 1949.
	Geog 4415.02.5	Bracciolini, Poggio. A traveler in disguise. Cambridge, 1963.
	Geog 4415.05	Den Öchten Weg Auss zu Faren von Lissbona gen Kallakuth. From Lisbon to Calicut. Minneapolis, 1956.
Htn	Geog 4415.11F*	Varthema, L. de. Itinerarium aethio piae. Milan, 1511.
	Geog 4416.21	Lobo, Jeronymo. Itinerário e outros escritos inéditos. Porto, 1971.
Htn	Geog 4416.26*	Herbert, Thomas. Some years travels into...Asia and Afrique. London, 1638.
	Geog 4416.26.5F	Herbert, Thomas. Some years travels into...Africa and Asia. London, 1677.
	Geog 4416.54	LeBlanc, V. De vermaarde reizen. Amsterdam, 1654.
	Geog 4416.54.5	LeBlanc, V. Les voyages fameux. Paris, 1658.
	Geog 4417.38F	Shaw, Thomas. Travels or observations. Oxford, 1738.
	Geog 4417.38.3	Shaw, Thomas. Travels or observations...Barbary...Levant. Edinburgh, 1808. 2v.
	Geog 4417.38.5	Shaw, Thomas. Reisen der Barbery und der Levante. Leipzig, 1765.
	Geog 4417.44	Lade, R. Voyages du capitaine...Lade. Paris, 1744. 2v.
	Geog 4417.44.5	Lade, R. Voyages du capitaine...Lade. Paris, 1810.
	Geog 4417.45	Poivre, Pierre. Un manuscrit inédit de Pierre Poivre: Les mémoires d'un voyageur. Paris, 1868.
	Geog 4417.64	Orleans, G. de. Travels of Father William Orleans a Jesuit. London, n.d.
	Geog 4417.68	Poivre, Pierre. Voyages d'un philosophe. Yverdon, 1768.

Geog 4411 - 4420 Voyages and travels in general - Individual travels - Asia and other regions - Asia and Africa, Asia and Oceania (By date of travel) - cont.

	Geog 4417.70	Thunberg, Karl P. Travels in Europe, Africa and Asia. London, 1793? 4v.
	Geog 4417.70.2	Thunberg, Karl P. Voyage en Afrique et en Asie. 1770-1779. Paris, 1794.
	Geog 4417.80	Irwin, E. Series of adventures...Red Sea, Arabia, Egypt. Dublin, 1780.
	Geog 4417.80.3	Irwin, E. Series of adventures...Red Sea, Arabia, Egypt. 3. ed. London, 1787. 2v.
	Geog 4417.80.5	Parsons, Abraham. Travels in Asia and Africa. London, 1808.
	Geog 4417.98F	Pennant, T. The view of Hindoostan. London, 1798-1800. 4v.
	Geog 4417.99	Abu Taleb Khan. Reise...durch Asien, Africa. Wien, 1813.
	Geog 4418.02F	Annesley, G. Voyages and travels to India. London, 1809. 3v.
	Geog 4418.03	Badia y Leblich. Voyages d'Ali Bey el Abbassi en Afrique et en Asie...1803...1807. v.1-3, atlas. Paris, 1814. 4v.
	Geog 4418.03.5	Badia y Leblich. Viajes de Ali Bay el Abbassi por Africa y Asia. Valencia, 1836. 3v.
	Geog 4418.03.9	Badia y Leblich. Viatjes de Ali Bey el Abbassi per Africa y Asia. v.1-3. Barcelona, 1888.
	Geog 4418.03.10	Badia y Leblich. Atlas dels viatjes d'Ali Bey el Abbassi. v.1-3. Barcelona, 1892.
	Geog 4418.03.15	Gonzalez y Rodriguez de la Peña, Hipolito. Viajes Ali Bey el Abbassi. Madrid, 1951.
	Geog 4418.22	Owen, William Fitz William. Narrative of voyages to explore the shores of Africa, Arabia, and Madagascar. Farnborough, 1968. 2v.
NEDL	Geog 4418.22	Owen, William Fitz William. Narrative of voyages to explore the shores of Africa, Arabia, and Madagascar. London, 1833. 2v.
	Geog 4418.29	Lushington, C. (Mrs.). Narrative of journey from Calcutta. London, 1829.
	Geog 4418.32	Tyerman, David. Journal of voyages and travels. Boston, 1832. 3v.
	Geog 4418.38	Sherer, M. The imagery of foreign travel. London, 1838.
	Geog 4418.39.4	Olin, S. Travels in Egypt, Arabia, Petraea and Holy Land. 4th ed. N.Y., 1844. 2v.
	Geog 4418.44	Davidson, G.F. Trade and travel in the Far East. London, 1846.
	Geog 4418.45	Frege, C.G. Erinnerungen aus dem Osten. Leipzig, 1845.
	Geog 4418.54	Osgood, J.B.F. Notes of travel or recollections. Salem, 1854.
	Geog 4418.54.3	Mortimer, F.L.B. (Mrs.). Far off, or Asia and Australia described with anecdotes and illustrations. N.Y., 1854.
	Geog 4418.55	Eames, J.A. Another budget...in the East. Boston, 1855.
	Geog 4418.57	Train, G.F. An American merchant. N.Y., 1857.
	Geog 4418.57.5	Train, G.F. Young American abroad. London, 1857.
	Geog 4418.61	Tilley, H.A. Japan, the Amoor and the Pacific. London, 1861.
	Geog 4418.61.3	Rand, E.A. All aboard for sunrise lands. 2. ed. Boston, 1861.
	Geog 4418.62	Brumund, J.F.G. Schetsen eener mail-reize van Batavia naar Maastricht op reis en thuis. Amsterdam, 1862.
	Geog 4418.64	Russell-Killough, H. Seize mille lieues a travers l'Asie et l'Oceanie. Paris, 1864.
	Geog 4418.70	Beauvior, L. Voyage round the world. London, 1870. 2v.
	Geog 4418.72	Fogg, W.P. "Round the world", letters from Japan, China, India, and Egypt. Cleveland, 1872.
	Geog 4418.73	Blyden, E.W. From West Africa to Palestine. Freetown, Sierra Leone, 1873.
	Geog 4418.73.5	Obligade, P.S. Viaje a Oriente. Paris, 1873.
	Geog 4418.75	Fogg, W.P. Arabistan, or Land of "Arabian Nights." London, 1875.
	Geog 4418.79	Farnham, J.M.W. Homeward or travels. Schenectady, 1879.
	Geog 4418.79.5	Ludwig, Salvator. Die Karawanen-Strasse von Aegypten nach Syrien. Prag, 1879.
	Geog 4418.89A	Brassey, A. The last voyage. London, 1889.
	Geog 4418.89B	Brassey, A. The last voyage. London, 1889.
	Geog 4418.90	Tchihatchef, P. de. Etudes de géographie et d'histoire naturelle. Florence, 1890.
	Geog 4418.95	Tiffany, F. This goodly frame, the earth. Boston, 1895.
	Geog 4418.95.2	Tiffany, F. This goodly frame, the earth. Boston, 1896.
	Geog 4418.95.3	Tiffany, F. This goodly frame, the earth. Boston, 1896.
	Geog 4419.02	Krausz, S. Zu Land und See im Orient. Chicago, 1902.
	Geog 4419.07	Hochberg, Fritz. An eastern voyage; travels thro' the British Empire...in East and Japan. London, 1910. 2v.
	Geog 4419.10	Pick, Emil G. Reisebriefe eines österreichischen Industriellen aus Abessinien, Indien und Ostasien. Prag, 1910?
	Geog 4419.12A	Gardner, G.P. Chiefly the Orient. Boston, 1912.
	Geog 4419.12B	Gardner, G.P. Chiefly the Orient. Boston, 1912.
	Geog 4419.25	Huntington, E. West of the Pacific. N.Y., 1925.
	Geog 4419.28	Bosch, Carl. Karawanen-Reisen...Ägypten, Mesopotamien, Persien und Abessinien. Berlin, 1928.
	Geog 4419.30	Monfreid, Henri de. Pearls, arms and hashish; pages from the life of a Red Sea navigator. N.Y., 1930.
	Geog 4419.33	Makin, William J. Red Sea nights. N.Y., 1933.
	Geog 4419.34	Monfreid, Henri de. Secrets of the Red Sea. London, 1934.
	Geog 4419.34.3	Monfreid, Henri de. Les secrets de la Mer Rouge. Paris, 1935.
	Geog 4419.34.4	Monfreid, Henri de. La poursuite du "Kaipan". Paris, 1934.
	Geog 4419.34.5	Rogers, C.K. (Mrs.). Journal letters from the Orient. Boston, 1934.
	Geog 4419.38	Morton, H.V. Through the lands of the Bible. London, 1938.
	Geog 4419.38.15	Monfreid, Henri de. Le trésor du pélerin; roman. 12. ed. Paris, 1938.
	Geog 4419.40	Cutting, S. The fire ox and other years. N.Y., 1940.
	Geog 4419.45	Ponder, S.E.G. A wanderer in khaki. London, 1945.
	Geog 4419.47	Goolid-Adams, R. Middle East journey. 1. ed. London, 1947.
	Geog 4419.49	Foster, William. The Red Sea and adjacent countries...seventeenth century. London, 1949.
	Geog 4419.51	Calder, Ritchie. Med against the desert. London, 1951.
	Geog 4419.59	Serstevens, Albert. Les précurseurs de Marco Polo. Paris, 1959.
	Geog 4419.68	Gotev, Goran. Probuzhdane to na Shekherazada. Sofiia, 1968.
	Geog 4419.69	Nolan, Cynthia. A sight of China. London, 1969.
	Geog 4419.70	Zemlia dobrykh liudei. Tashkent, 1970.

Classified Listing

Geog 4424 - 4430 Voyages and travels in general - Individual travels - Indian Ocean (By date of travel)

Geog 4426.08 Castro Daire, A. de Ataide. Viagens de Reino para a India e da India para o Reino. Lisboa, 1957-58. 3v.
Geog 4427.35 Schwartz, G.L. Utdrag af en Ost-Indisk rese-bescrifning. Westerås, 1784.
Geog 4427.80 Le Gentil de la Galaisière, G.J.H.J.B. Voyage dans les mers de l'Inde. v.1-2. Suisse, 1780.
Geog 4427.80.5 Le Gentil de la Galaisière, G.J.H.J.B. A voyage to the Indian Seas. Manila, 1964.
Geog 4427.82 Kerguélen. Relation de deux voyages. Paris, 1782.
Geog 4427.99 Henry, Pierre. Route de l'Inde. Paris, 1798-99.
Geog 4428.01 Avine, Grégoire. Les voyages du chirurgien Avine à l'Ile de France et dans la mer de Indes au début du XIXe siècle. Paris, 1961.
Geog 4428.68 Codine, J. Mémoire géographique sur mer des Indes. Paris, 1868.
Geog 4429.40 Villiers, Alan J. Sons of Sinbad. N.Y., 1940.
Geog 4429.52A Villiers, Alan J. Monsoon seas. N.Y., 1952.
Geog 4429.52B Villiers, Alan J. Monsoon seas. N.Y., 1952.
Geog 4429.52.5 Ommanney, F.D. The shoals of Capricorn. London, 1952.
Geog 4429.53A Doyle, A.C. Heaven has claws. N.Y., 1953.
Geog 4429.53B Doyle, A.C. Heaven has claws. N.Y., 1953.
Geog 4429.55 Auber, Jacques. Histoire de l'Ocean Indien. Tananarive, 1955.
Geog 4429.61 Toussaint, Auguste. History of the Indian Ocean. London, 1966.
Geog 4429.61.5 Toussaint, Auguste. History of the Indian Ocean. Chicago, 1966.
Geog 4429.72 Tendiuk, Leonid M. Al'batros-clukach moriv. Kyïv, 1972.

Geog 4474 - 4480 Voyages and travels in general - Individual travels - Imaginary voyages (By date)

Geog 4476.38.1 Godwin, Francis. The man in the moon, 1638. London, 1971.
Htn Geog 4476.71* Schooten, Henry (pseud.). The hairy-giants. London, 1671.
Htn Geog 4476.75* Hamilton, William (pseud.). O-Brazile. London, 1675.
Htn Geog 4476.92* Les avantures de Jaques Sadeur. Paris, 1692.
Htn Geog 4476.92.2* Foigny, Gabriel de. A new discovery of terra Incognita Australis. London, 1693.
Htn Geog 4477.08* Posos, Juan de (pseud.). Beschryoinge van het magtig. Amsterdam, 1708.
Geog 4477.09 Relation...du voyage du Prince de Montberaud. Merinde, 1709.
Htn Geog 4477.10* Voyages et avantures de Jaques Masse. Bourdeaux, 1710.
Geog 4477.21 Relation d'un voyage du Pole Arctique au Pole Antarctique. Amsterdam, 1721.
Geog 4477.21.3 Relation un voyage du Pole Arctique au Pole Antarctique. 2. ed. La Haye, 1734.
Htn Geog 4477.26* Boyle, R. (pseud.). Voyages and adventures of Captain Boyle. London, 1726.
Geog 4477.26.3 Boyle, R. (pseud.). The voyages and adventures of Captain Boyle. 1st American ed. Cooperstown, N.Y., 1796.
Geog 4477.26.5 Boyle, R. (pseud.). The voyages and adventures of Captain Boyle. Dublin, 1741.
Geog 4477.29 Les voyages de Glantzby. Paris, 1739.
Htn Geog 4477.35.3* Godwin, Francis. The strange voyage and adventures of Domingo Gonsales. 2. ed. London, 1768.
Geog 4477.51 Geschichte des Gaudenti di Lucca. Frankfurt, 1751.
Geog 4477.51.7 The adventures of Signore Gaudentio di Lucca. London, 1774.
Geog 4477.51.15 Berington, S. The adventures of Gaudentio di Lucca. London, 1763.
Geog 4477.51.25 Berington, S. The adventures of Gaudentio di Lucca. London, 1850.
Geog 4477.57 Histoire d'un peuple nouveau. Londres, 1757.
Geog 4477.59 Brachfelds, J.M. Begebenheisen...Sudlandern. Eisenach, 1759.
Geog 4477.61 Listonai. Le voyageur philosophe. Amsterdam, 1761. 2v.
Geog 4477.64 White, D.M. Zaccaria Seriman, 1709-1784, and the Viaggi di Enrico Wanton. Manchester, 1961.
Geog 4477.64.5 Seriman, Zaccaria. Viages de Enrique Wanton. Madrid, 1781-85. 4v.
Geog 4477.87 Wollap, G. Mémoires. v.1-4. Londres, 1787. 2v.
Geog 4477.87.7 Voyages imaginaires, songes. v.1-31,36. Amsterdam, 1787-89. 32v.
Geog 4478.78 Voyage to the North Pole. Cambridge, 1878.
Geog 4479.69 Darvère, Pierre. Le Baitoto. Paris, 1969.

Geog 4500 Mountaineering in general - Bibliographies

Geog 4500.5 Jeffers, LeRoy. Selected list of books on mountaineering. N.Y., 1916.
Geog 4500.7 Dreyer, Aloys. Bücherverzeichnis der Alpenvereinsbucherei, mit Verfasser- und Bergnamen-Verzeichnis. München, 1927.
Geog 4500.10 Istituto di Bibliografia Alpina, Biella. Opere antiche e rare sulla montagna. Biella, 1951.
Geog 4500.15 Smith, J.A. Mountaineering. Cambridge, 1955.
Geog 4500.16 Harvard University. Library. Mountaineering literature in the Harvard College Library. n.p., 1931.
Geog 4500.17 Scotland. National Library, Edinburgh. Shelf-catalogue of the Lloyd collection of Alpine books. Boston, 1964.

Geog 4502 Mountaineering in general - Gesellschaft Alpiner Bücherfreunde publications

Geog 4502.5 Saussure, H.B. de. Relation abrégés d'un voyage a la cime du Mont-Blanc en aout 1787. Genève, 1928? 2 pam.
Geog 4502.5.2 Rohrer, M. Berglieder der Völker. München, 1928.
Geog 4502.5.3 Steinberger, S. Leben und Schriften. München, 1929.
Geog 4502.5.4 Rombert, E. Das Murmeltier mit dem Halsband. München, 1929.
Geog 4502.5.5 Rickmers, W.R. Querschnitt durch mich. München, 1930.
Geog 4502.5.6 Schmidkunz, W. Der Berg des Herzens. München, 1930.
Geog 4502.5.7 Simmler, J. De alpibus Commentarius; die Alpen. München, 1931.
Geog 4502.5.8 Hoek, H. Der denkende Wanderer. München, 1932.
Geog 4502.5.9 Knorr, O. Der Grossvenediger in der Alpinismus. München, 1932.
Geog 4502.5.10 Schuster, K. Weise Berge - schwarze Zelte. München, 1932.
Geog 4502.5.11 Bühler, H. Alpine Bibliographie. München. 1932
Geog 4502.5.11.5 Bühler, H. Alpine Bibliographie. Gesamtregister, 1931-1938. München, 1949.
Geog 4502.5.12 Tscharner, J.B. von. Die Bernina. 1786. München, 1933.
Geog 4502.5.13 Lebwald, A. Damographia oder Gemsen-Beschreibung in zewy Theil Abgetheilet. Saltzburg, 1933.
Geog 4502.5.14 Gallhuber, J. Das Gesäuse und seine Berge. 2. Aufl. München, 1933.
X Cg Geog 4502.5.15 Schmidkunz, W. Auf der Alm. Erfurt, 1934.
Geog 4502.5.16 Hager, F. Das Chiemgaubuch. München, 19- .
Geog 4502.5.17 Maduschka, L. Junger Mensch im Gebirg. München, 1936.

Geog 4502 Mountaineering in general - Gesellschaft Alpiner Bücherfreunde publications - cont.

Geog 4502.5.18 Rudatis, D. Das Letzte im Fels. München, 1936.
Geog 4502.5.19 Kobell, F. von. Wildonger. München, 1936.
Geog 4502.5.20 Stephen, L. Der Tummelplatz Europas. München, 19- .
Geog 4502.5.21 Montis, R. Kampf um den Berg. München, 19- .
Geog 4502.5.22 Rohrer, M. Der Feuerberg. München, 19- .
Geog 4502.5.23 Schmitt, F. Mensch, Berg und Tod. München, 19- .
Geog 4502.5.25 Dreyer, A. Geschichte der Alpinen Literatur. München, 1938.
Geog 4502.5.26 Javelle, E. Erinnerungen eines Bergsteigers. München, 1938.
Geog 4502.5.27 Graber, A. Der Weg zum Berg. München, 1939?
Geog 4502.5.28 Stöger-Ostin, Georg. Georg Jennerwein, der Wildschütz. München, 1939?
Geog 4502.5.29 Flaig, Walther. Das Silvretta-Buch. München, 1940.
Geog 4502.5.30 Mason, Alfred E.W. Das Gesetz der Berge. München, 1940?
Geog 4502.5.31 Geissler, Paul. Um den Montblanc. München, 1940?
Geog 4502.10.2 Petrarca, F. Des F. Petrarca Sendschreiben die Besteigung des Mont Ventoux betreffend. München, 1936.
Geog 4502.10.3 Christ, Fritz. Die erste Ersteigung des Totenkirches durch den Christ Fick-Kamin. München, 1937.
Geog 4502.10.4 Vierthaler, Franz Michael. Die Reise auf den Grossglockner, 1800. München, 1938.
Geog 4502.10.15 Berge der Welt. Zürich. 1946-1957 10v.
Geog 4502.12 The American alpine journal. N.Y. 7,1948+ 13v.

Geog 4506 - 4510 Mountaineering in general - General works (By date)

Geog 4508.81 Reclus, Elisee. The history of a mountain. N.Y., 1881.
Geog 4508.89 Fiorio, C. I pericoli dell'Alpinismo e noime per vitarli. Torino, 1889.
Geog 4508.93 Balch, E.S. Mountain exploration. Philadelphia, 1893.
Geog 4508.93.5 Wilson, Claude. Mountaineering. London, 1893.
Geog 4508.93.10 Meurer, Julius. Der Bergsteiger in Hochgebirge. Wien, 1893.
Geog 4508.97 Wilson, E.L. Mountain climbing. N.Y., 1897.
Geog 4508.99A Gribble, F. The early mountaineers. London, 1899.
Geog 4508.99B Gribble, F. The early mountaineers. London, 1899.
Geog 4508.99.5 Zurbriggen, M. From the Alps to the Andes. London, 1899.
Geog 4509.07.1 Abraham, G.D. The complete mountaineer. N.Y., 1908.
Geog 4509.07.2 Abraham, G.D. The complete mountaineer. 2. ed. London, 1908.
Geog 4509.07.3 Abraham, G.D. The complete mountaineer. 3. ed. London, 1923.
Geog 4509.10 Abraham, G.D. Mountain adventures at home and abroad. London, 1910.
Geog 4509.11 Turner, S.F.R. My climbing adventures in 4 continents. London, 1913.
Geog 4509.13 Casella, G. L'alpinisme. Paris, 1913.
Geog 4509.14 Rey, G. Alpinismo acrobatico. Torino, 1914.
Geog 4509.16 Taüber, C. Auf fremden Bergfpfaden. Zürich, 1916.
Geog 4509.18 Schweizer Alpen Club. Le conseiller de l'ascensionniste. v.2. Genève, 1918.
Geog 4509.20 Young, G.W. Mountain craft. London, 1920.
Geog 4509.20.5 Raeburn, H. Mountaineering art. N.Y., 1920.
Geog 4509.20.10 Conway, W.M.C. Mountain memories. N.Y., 1920.
Geog 4509.22 Hamilton, H. Mountain madness. London, 1922.
Geog 4509.23 Dupin. Zum Wortschatze des Bergsteigers. Wien, 1923-
Geog 4509.23.10 Freshfield, D.M. Below the snow line. London, 1923.
Geog 4509.23.15 Collins, F.A. Mountain climbing. London, 1924.
Geog 4509.24 Flaig, W. Felsklettern in Bildern und Merkworten. Stuttgart, 1924.
Geog 4509.24.6 Finch, G.L. The making of a mountaineer. London, 1924.
Geog 4509.24.10 Schwartz, M. Vers l'idéal par la montagne. Paris, 1924.
Geog 4509.25 Lunn, A.H.M. The mountains of youth. London, 1925.
Geog 4509.25.6 Flaig, W. Eistechnik des Bergsteigers in Bildern und Merkworten. pt.1-4. 2. Aufl. Stuttgart, 1925.
Geog 4509.31 Robbins, L.H. Mountains and men. N.Y., 1931.
Geog 4509.31.5 Schwartz, M. Et la montagne conquit l'homme. Paris, 1931.
Geog 4509.31.10 Giussani, C. Chiacchiere di un alpinista. Milano, 1931.
Geog 4509.31.15 Smythe, F.S. Climbs and ski runs. Edinburgh, 1931.
Geog 4509.33 Abraham, G.D. Modern mountaineering. London, 1931.
Geog 4509.33.5A Pickman, D.L. Some mountain views. Boston, 1933.
Geog 4509.33.5B Pickman, D.L. Some mountain views. Boston, 1933.
Geog 4509.34 Club Alpin Francais. Manuel d'alpinisme. Chambéry, 1934. 2v.
Geog 4509.34.5 Spencer, S. Mountaineering. London, 19- .
Geog 4509.35.5 Richards, D.P. (Mrs.). Climbing days. London, 1935.
Geog 4509.40 Jahn, A. Exrusionismo y alpinismo; historia de su evolución. Caracas, 1940.
Geog 4509.40.5 Irving, R.L.G. Ten great mountains. N.Y., 1940.
Geog 4509.41 Ullman, J.R. High conquest, the story of mountaineering. Philadelphia, 1941.
Geog 4509.41.5 Peacocke, T.A.H. Mountaineering. London, 1941.
Geog 4509.41.7 Peacocke, T.A.H. Mountaineering. 3. ed. London, 1953.
Geog 4509.41.10 Rondet, C. Montagne: des terrasses aux arêtes. Paris, 1941.
Geog 4509.41.17 Smythe, F.S. The mountain vision. London, 1942.
Geog 4509.42 Henderson, K.A. The American Alpine Club's handbook of American mountaineering. Boston, 1942.
Geog 4509.43.5 Shipton, Eric E. Upon that mountain. London, 1943.
Geog 4509.46 Tilman, H.W. When man and mountains meet. Cambridge, Eng., 1946.
Geog 4509.47 Smythe, F.S. The mountain top. London, 1947.
Geog 4509.48.5 Kurz, Marcel. Fremde Berge-Ferne Ziele. Bern, 1948.
Geog 4509.48.13 Pause, Walter. Mit gleichlichen Augen. 3. Aufl. München, 1951.
Geog 4509.49 Allain, Pierre. Alpinisme et compétition. Grenoble, 1949.
Geog 4509.52 Negri, Carlo. Alpinismo. 3. ed. Milano, 1952.
Geog 4509.53 Petzoldt, P. On top of the world. N.Y., 1953.
Geog 4509.53.5 Chatellus, A. de. Alpiniste. Grenoble, 1953.
Geog 4509.53.10 Michel, Aime. Montagnes héroïques. Tours, 1953.
Geog 4509.53.15 Club Alpino Italiano. Alpinismo italiano nel mondo. Milano, 1953.
Geog 4509.53.20 Jerschak, Federico. L'alpinismo a Cortina dai suoi primordi ai giorni nostri. Roma, 1953.
Geog 4509.54 Jolis Felisart, A. La conquista de la montaña. Barcelona, 1954.
Geog 4509.54.5 Ullman, J.R. The age of mountaineering. Philadelphia, 1954.
Geog 4509.54.10 Buhl, H. Achttausend, drüber und drunter. München, 1954.
Geog 4509.56 Garobbio, A. Uomini del sesto grado. Milano, 1956.
Geog 4509.56.10 Herzog, Maurice. La montagne. Paris, 1956.
Geog 4509.56.15 Evans, Charles. On climbing. London, 1956.
Geog 4509.56.20 Rébuffat, Gaston. Mont Blanc to Everest. London, 1956.
Geog 4509.56.25 Underhill, Miriam. Give me the hills. London, 1956.
Geog 4509.57.5 Young, G.W. The influence of mountains upon the development of human intelligence. Glasgow, 1957.

58

Classified Listing

Geog 4506 - 4510 Mountaineering in general - General works (By date) - cont.
Geog 4509.58 Moravec, Fritz. Weisse Berge. Wien, 1958.
Geog 4509.63 Gudhjohnsen, P. Endurminningar fjallgöngumanns. Reykjavik, 1963.
Geog 4509.65 Boutron, Michel. La montagne et ses hommes. Paris, 1965.
Geog 4509.68 Potočnik, Miha. Srečanja z gorami. Ljubljana, 1968.
VGeog 4509.69 Planinsko berilo. Ljubljana, 1969.

Geog 4512.1 - .90 Mountaineering in general - Literary works - Anthologies, collections
Geog 4512.1 Irving, R.L.G. The mountain way. London, 1938.

Geog 4512.100 - .899 Mountaineering in general - Literary works - Individual works (800 scheme, A-Z)
Geog 4512.535 Lunn, A. Mountain jubilee. London, 1943.
Geog 4512.537 Luther, C.J. Wil der Schneelauf nach Deutschland kam. München, 1925.
Geog 4512.537.5 Luther, C.J. Skunterhaltungen. München, 1925.
Geog 4512.561 Mayer, T.H. Der Führer; Novelle. München, 1924.
Geog 4512.767 Salacrow, Armand. A pied, au-dessus des nuages. Paris, 1956.

Geog 4515 Mountaineering in general - Special topics
Geog 4515.14 Colò, Carlo. Attrezzature per soccorso alpino. Trento, 1958.
Geog 4515.25 Underhill, R.L.K. On the use and management of the rope in rock work. San Francisco, 1931.
Geog 4515.80 Serebrianye lyzhi. Uzhgorod, 1973.
Geog 4515.101 Niederl, Franz. Das gehen auf Eis und Schnee. 2. Aufl. München, 1927.
Geog 4515.102 Hoferer, Erwin. Winterliches Bergsteigen alpine Schilaustechnik. München, 1925.
Geog 4515.103.2 Lunn, A.H.M. Alpine ski-ing at all heights and seasons. 2d ed. London, 1926.
Geog 4515.104.5 Bilgeri, George. Colonel Bilgeri's handbook on mountain ski-ing. London, 1929.
Geog 4515.105.2 Faes, Henry. Le manuel du skieur. 2e éd. Lausanne, 1925.
Geog 4515.106.2 Kurz, Marcel. Alpinisme hivernal. Paris, 1928.
Geog 4515.107 Couttet, Alfred. L'enchantement du ski. Paris, 1930.
Geog 4515.108 Patani, Osvaldo. Soli con le montagne. Milano, 1955.
Geog 4515.201 Prusik, Karl. Gymnastik für Bergsteiger. München, 192-.
Geog 4515.301 Schmidkunz, W. Kochbuch für Bergsteiger. München, 1925.
Geog 4515.401 Saysse-Tobiczyk, K. W skałach i lodach swiata. Warszawa, 1959- 2v.

Geog 4520 Mountaineering in general - Biographies of mountaineers - Collected
Geog 4520.5 Ségogne, Henry de. Les alpinistes célèbres. Paris, 1956.

Geog 4521 Mountaineering in general - Biographies of mountaineers - Individual
Geog 4521.2 Banks, Mike. Commando climber. London, 1955.
Geog 4521.3 Clark, Ronald. An eccentric in the Alps. London, 1959.
Geog 4521.8 Hackel, Heinrich. Meine Berge, nuen Leben. Salzburg, 1960.
Geog 4521.10 Jackson, John. More than mountains. London, 1955.
Geog 4521.11 Kain, Conrad. Where the clouds can go. N.Y., 1935.
Geog 4521.23 Whymper, Edward. Edward Whymper, alpinist of the heroic age. Nashville, Tenn., 1914.
Geog 4521.23.5 Whymper, Edward. A letter addressed to the...Alpine Club. London, 1900.
Geog 4521.25 Lowe, George. Because it is there. London, 1959.
Geog 4521.26 Chapman, F.S. Memories of a mountaineer. London, 1951.
Geog 4521.27 Ullman, James Ramsey. Straight up; the life and death of John Harlin. Garden City, N.Y., 1968.

Geog 4556 - 4560 Ocean life in general - General works (By date)
Geog 4556.59 Barlow, E. Barlow's journal of his life at sea in king's ships, East and West Indiamen. London, 1934. 2v.
Geog 4556.94 Coxere, Edward. Adventures by sea of Edward Coxere. Oxford, 1945.
Geog 4557.85 Pontoppidam, C. Hval og robbesangsten. Risbenhavn, 1785.
Geog 4557.95 Kelly, Samuel. Samuel Kelly, an 18th century seaman. N.Y., 1925.
Htn Geog 4557.98* Colnett, J. Voyages to the South Atlantic. London, 1798.
Htn Geog 4558.05* Burton, C. Journal of voyage from London. London, 1805.
Geog 4558.11 Forbes, R.B. Voyage of the Niedas, 1811. Boston, 1885.
Geog 4558.20.3 Durand, J.R. The life and adventures. Rochester, 1820.
Htn Geog 4558.26F* Hunnewell, J. Journal of voyage of "Missionary Packet". Charlestown, 1880.
Geog 4558.31 Davis, George. Recollections of a sea-wanderer's life. N.Y., 1887.
Htn Geog 4558.33* Mariner's library. Boston, 1833.
Htn Geog 4558.40* Dana, Richard Henry. Two years before the mast. N.Y., 1840.
Geog 4558.40.5 Dana, Richard Henry. Two years before the mast. N.Y., 1840.
Htn Geog 4558.40.6* Dana, Richard Henry. Two years before the mast. Glasgow, 1842.
Htn Geog 4558.40.7* Dana, Richard Henry. Two years before the mast. London, 1841.
Geog 4558.40.8 Dana, Richard Henry. Two years before the mast. N.Y., 1854.
Geog 4558.40.9 Dana, Richard Henry. Two years before the mast. N.Y., 1847.
Htn Geog 4558.40.10* Dana, Richard Henry. Two years before the mast. Boston, 1869.
Geog 4558.40.11 Dana, Richard Henry. Two years before the mast. Boston, 1869.
Geog 4558.40.11.5 Dana, Richard Henry. Two years before the mast. Boston, 1869.
Geog 4558.40.12 Dana, Richard Henry. Two years before the mast. Boston, 1883.
Htn Geog 4558.40.13* Dana, Richard Henry. Two years before the mast. Boston, 1871.
Htn Geog 4558.40.14* Dana, Richard Henry. Two years before the mast. N.Y., 1892.
Geog 4558.40.15 Dana, Richard Henry. Two years before the mast. Cambridge, 1895.
Geog 4558.40.20A Dana, Richard Henry. Two years before the mast. N.Y., 1907.
Geog 4558.40.20B Dana, Richard Henry. Two years before the mast. N.Y., 1907.
Geog 4558.40.22 Dana, Richard Henry. Two years before the mast. N.Y., 1911.
Geog 4558.40.24A Dana, Richard Henry. Two years before the mast. Boston, 1911.
Geog 4558.40.24B Dana, Richard Henry. Two years before the mast. Boston, 1911.

Geog 4556 - 4560 Ocean life in general - General works (By date) - cont.
Geog 4558.40.25 Dana, Richard Henry. Two years before the mast. Boston, 1911.
Geog 4558.40.26 Dana, Richard Henry. Two years before the mast. N.Y., 1922.
Geog 4558.40.27 Dana, Richard Henry. Two years before the mast. Boston, 1911.
Geog 4558.40.28 Dana, Richard Henry. Two years before the mast. Boston, 1911.
Geog 4558.40.30 Dana, Richard Henry. Two years before the mast. N.Y., 1915.
Geog 4558.40.35A Dana, Richard Henry. Two years before the mast. N.Y., 1921.
Geog 4558.40.35B Dana, Richard Henry. Two years before the mast. N.Y., 1921.
Geog 4558.40.35C Dana, Richard Henry. Two years before the mast. N.Y., 1921.
Geog 4558.40.40 Dana, Richard Henry. Two years before the mast. N.Y., 1928.
Htn Geog 4558.40.50* Dana, Richard Henry. Two years before the mast. London, 1946?
Htn Geog 4558.40.52* Dana, Richard Henry. Two years before the mast. Sydney, 1946.
Geog 4558.40.60 Dana, Richard Henry. Two years before the mast. Los Angeles, 1964. 2v.
Geog 4558.44.2 Little, G. Life on the ocean. Boston, 1844.
Geog 4558.44.3 Little, G. Life on the ocean, or Twenty years at sea. Aberdeen, 1847.
Geog 4558.44.4 Little, G. Life on the ocean, or Twenty years at sea. 14. ed. N.Y., 1852.
Geog 4558.48 Hart, J.C. Romance of yachting. N.Y., 1848.
Geog 4558.50 Browne, J.R. Etchings of a whaling cruise. N.Y., 1850.
Geog 4558.50.5 Cheever, H.T. The whale and its captors. N.Y., 1850.
Geog 4558.51 Congar, Obadiah. The autobiography and memorials of Captain Obadiah Congar. N.Y., 1851.
Geog 4558.53 Holcomb, E. A wonderful providence in many incidents at sea. 8. ed. Boston, 1853.
Geog 4558.55 Ocean scenes. N.Y., 1855.
Geog 4558.56 Nordhoff, C. Whaling and fishing. Cincinnati, 1856.
Geog 4558.56.5 Abbey, Charles A. Before the mast in clippers...1856 to 1860. N.Y., 1937.
Geog 4558.61.5 Stevens, Charles. A sailor boy's experience. Napanee, Ont., 1892.
Geog 4558.65 Waites, Alfred. My diary from England to India around Cape of Good Hope. Calcutta, 1865.
Geog 4558.66 Smith, C.E. From the deep of the sea. London, 1922.
Geog 4558.73 Wogan, E. de. Du Far-West à Bornéo. Paris, 1873.
Geog 4558.74 Nordhoff, C. Life on the ocean. Cincinnati, 1874.
Geog 4558.74.3 Jewell, J.G. Among our sailors. N.Y., 1874.
Geog 4558.76 Lindsay, W.S. Recollections of a sailor. London, 1876.
Geog 4558.76.3 The tourist, a magazine of information for ocean travellers, 1876. Boston, 1876.
Geog 4558.77 Nelson, A.W. Yankee Swanson. N.Y., 1913.
Geog 4558.79 Adams, Robert C. On board the "Rocket". Boston, 1879.
Geog 4558.81 Wyman, Walter. A cruise on the United States practice ship, S.P. Chase. N.Y., 1910.
Geog 4558.84 Rideing, William H. Boys coastwise, or All along the shore. N.Y., 1884.
Geog 4558.87A Samuels, S. From the forecastle to the cabin. N.Y., 1887.
Geog 4558.87B Samuels, S. From the forecastle to the cabin. N.Y., 1887.
Geog 4558.88 Spear, P.S. The old sailor's story of his life. Portland, 1888.
Geog 4558.89 Aldrich, H.L. Arctic Alaska and Siberia. Chicago, 1889.
Geog 4558.90.5 Amicis, E. de. Sull'oceano. Milano, 1890.
Geog 4558.90.6 Amicis, E. de. Sull'oceano. Milano, 1913.
Geog 4558.90.7 Amicis, E. de. On blue water. N.Y., 1897.
Geog 4558.94 Stewart, J.A.E. En mer. n.p., 1895?
Geog 4558.98 Stevenson, P.E. A deep-water voyage. Philadelphia, 1898.
Geog 4558.98.5 Stevenson, P.E. By way of Cape Horn. Philadelphia, 1899.
Geog 4558.98.10 Bullen, F.T. Cruise of the Cachalot. London, 1898.
Geog 4558.98.11 Bullen, F.T. Cruise of the Cachalot. N.Y., 1899.
Geog 4558.98.12 Bullen, F.T. Cruise of the Cachalot. N.Y., 1899.
Geog 4558.98.13 Bullen, F.T. Cruise of the Cachalot. N.Y., 1899.
Geog 4558.98.14 Bullen, F.T. Cruise of the Cachalot. N.Y., 1899.
Geog 4558.98.15 Bullen, F.T. Cruise of the Cachalot. N.Y., 1920.
Geog 4558.99A Bullen, F.T. The log of a sea-waif. N.Y., 1899.
Geog 4558.99B Bullen, F.T. The log of a sea-waif. N.Y., 1899.
Geog 4559.02 Lubbock, A.B. Round the Horn before the mast. N.Y., 1902.
Geog 4559.10 Shaw, Frank H. The sea and its story from viking ship to submarine. London, 1910.
Geog 4559.12 Crutchley, William C. My life at sea. London, 1912.
Geog 4559.13 Burns, Walter. A year with a whaler. N.Y., 1913.
Geog 4559.16 Verrill, A.H. The real story of the whaler. N.Y., 1916.
Geog 4559.16.3 The thrilling adventures of the whaler Alcyone. Peabody, 1916.
Geog 4559.16.5 Andrews, R.C. Whale hunting with gun and camera. N.Y., 1916.
Geog 4559.17 Gelett, Charles N. A life on the ocean; autobiography. Honolulu, 1917.
Geog 4559.23 Angel, W.H. The clipper ship "Sheila". Boston, 1923.
Geog 4559.24 Chatterton, E.K. Seamen all. Boston, 1924.
Geog 4559.25 Hayes, Bertram. Hull down. N.Y., 1925.
Geog 4559.25.3 Perry, F. Fair winds and foul. Boston, 1925.
Geog 4559.25.10F Lubbock, A.B. Adventures by sea from art of old time. London, 1925.
Geog 4559.25.15 Brown, Charles W. My ditty bag. Boston, 1925.
Geog 4559.26 Hemy, Thomas M. Deep sea days. London, 1926.
Geog 4559.26.3 Farmer, H.F. The log of a shellback. N.Y., 1926.
Geog 4559.26.10 Crane, Mannin. Yarns from a windjammer. Boston, 1926.
Geog 4559.26.15 Boughton, George P. Seafaring. London, 1926.
Geog 4559.26.20 Beaumont, J.C.H. Ships and people. N.Y., 1926.
Geog 4559.27 Smith, Cicely Fox. A sea chest; an anthology of ships and sailormen. London, 1927.
Geog 4559.27.5A Senior, William. Naval history in the law courts. London, 1927.
Geog 4559.27.5B Senior, William. Naval history in the law courts. London, 1927.
Geog 4559.27.10 Cooper, F.G. Yarns of the seven seas. London, 1927.
Geog 4559.27.15 Chatterton, E.K. The brotherhood of the sea. London, 1927.
Geog 4559.28 Sorrell, George. The man before the mast. London, 1928.
Geog 4559.28.5 Harlow, Frederick Pease. The making of a sailor. Salem, Mass., 1928.
Htn Geog 4559.28.6* Harlow, Frederick Pease. The making of a sailor. Salem, Mass., 1928.
Geog 4559.28.10 Smith, Cicely Fox. Ancient mariners. London, 1928.
Geog 4559.28.15 Amicis, E. de. Sull'oceano. Milano, 1928.
Geog 4559.29 Munro, D.J. The roaring forties and after. London, 1929.

Classified Listing

Geog 4556 - 4560 Ocean life in general - General works (By date) - cont.

Geog 4559.29.3 Stanton, William H. The journey of...pilot, of Deal. London, 1929.
Geog 4559.29.15 Chatterton, E.K. On the high seas. Philadelphia, 1929.
Geog 4559.29.22 Rogers, S.R.H. Sea lore. London, 1934.
Geog 4559.29.30 Villiers, A.J. Falmouth for orders...Cape Horn. Garden City, N.Y., 1929.
Geog 4559.30.3A Villiers, A.J. By way of Cape Horn. N.Y., 1930.
Geog 4559.30.3B Villiers, A.J. By way of Cape Horn. N.Y., 1930.
Geog 4559.30.5 Barnes, William M. When ships were ships and not tin pots. N.Y., 1930.
Geog 4559.30.10 Randell, Jack. I'm alone, by Captain Jack Randell, as told to Meigs O. Frost. 1. ed. Indianapolis, 1930.
Geog 4559.30.15 Valéry, Paul. Mer, marines, marins. Paris, 1930.
Geog 4559.30.20 Perry, Joseph Malcolm. Cruising in many seas. Springfield, 1930.
Geog 4559.31 Chatterton, E.K. Sailing the seas. London, 1931.
Geog 4559.32 Villiers, A.J. Grain race. N.Y., 1933.
Geog 4559.33 McCulloch, J.H. A million miles in sail. N.Y., 1933.
Geog 4559.34 Grant, George H. Consigned to Davy Jones. Boston, 1934.
Geog 4559.35 Van Loon, H.W. Ships and how they sailed the seven seas. (5000 B.C.-A.D. 1935). N.Y., 1935.
Geog 4559.35.5 Stevers, M.D. Sea lanes; man's conquest of the ocean. N.Y., 1935.
Geog 4559.35.7 Stevers, M.D. Sea lanes; man's conquest of the ocean. Garden City, N.Y., 1938.
Geog 4559.35.10 Brown, Charles W. Journal and letters, 1876-84. n.p., 1935.
Geog 4559.35.15 Sheridan, Richard B. Heavenly hell. London, 1935.
Geog 4559.35.20 Devine, Eric. Midget Magellans; great cruises in small ships. N.Y., 1935.
Htn Geog 4559.37* Dana, Richard Henry. Cruelty to seamen; being the case of Nichols and Couch. Berkeley, 1937.
Geog 4559.37.3 Wead, Frank W. Gales, ice and men. N.Y., 1937.
Geog 4559.37.5 Making, V.L. In sail and steam...merchant service, 1902-1927. London, 1937.
Geog 4559.38 Tompkins, W.M. Fifty south to fifty south...around Cape Horn...Wander Bird. N.Y., 1938.
Geog 4559.38.5 Wead, F.W. Gales, ice and men; a biography of the Steam Barkentine Bear. London, 1938.
Geog 4559.38.12 Nixon, Laurence A. Vagabond voyaging; the story of freighter travel. Boston, 1939.
Geog 4559.39.5 Tompkins, W.M. Two sailors and their voyage around Cape Horn. N.Y., 1939.
Geog 4559.40 Nordhoff, C. In Yankee windjammers. N.Y., 1940.
Geog 4559.40.3 Gainard, J.A. Yankee skipper; the life story of Joseph A. Gainard. N.Y., 1940.
Geog 4559.40.5 Smale, R. There go the ships. Caldwell, 1940.
Geog 4559.40.7 Blain, W. Home is the sailor. N.Y., 1940.
Geog 4559.40.9 Bisschop, E. de. The voyage of the Kaimiloa. London, 1940.
Geog 4559.40.10 Pierrefeu, François de. Les confessions de Tatibouet. Paris, 1939.
Geog 4559.40.11 Wossidlo, R. Reise, Quartier in Gottesnaam. Seestadt Rostock, 1940.
Geog 4559.40.12 Jónsson, Gisli. Frekjan; aefintýralegt...frá Danmoëku. Reykjavik, 1940.
Geog 4559.44 Snow, A.R. Log of a sea captain's daughter. Boston, 1944.
Geog 4559.44.5 Green, Alfred. Jottings from a cruise. Seattle, 1944.
Geog 4559.45 Klitgaard, Kaj. Oil and deep water. Chapel Hill, 1945.
Geog 4559.45.5 Carter, G.G. Looming lights. London, 1945.
Geog 4559.45.10 Anson, P.F. Harbour Head. London, 1944.
Geog 4559.46 Chaveriat, R. Ceux de la voile. Paris, 1946.
Geog 4559.48 Öngör, Sami. Deniz cağrafyasį. Ankara, 1948.
Geog 4559.49 Villiers, A.J. The set of the sails. N.Y., 1949.
Geog 4559.52 Rasmussen, A.H. Sea fever. London, 1952.
Geog 4559.53 Mitchell, C. Beyond horizons. 1. ed. London, 1953.
Geog 4559.53.5 Hayet, Armand. Us et coutumes à bord des long-courriers. Paris, 1953.
Geog 4559.53.10 Freuchen, Peter. Vagrant viking. N.Y., 1953.
Geog 4559.53.16 Bombard, Alain. The voyage of the Hérétique. Ann Arbor, Mich., 1971.
Geog 4559.54 Neider, C. Man against nature. 1. ed. N.Y., 1954.
Geog 4559.54.5 Berge, Victor. Danger is my life. London, 1954.
Geog 4559.54.10 Edwards, H.W. Under four flags. London, 1954.
Geog 4559.54.15 Learmont, J.S. Master in sail. 2d ed. London, 1954.
Geog 4559.54.20 Eaddy, P.A. Sails beneath the Southern Cross. Wellington, 1954.
Geog 4559.55 Thomas, L.J. Great true adventures. N.Y., 1955.
Geog 4559.55.5 Cross, Beverley. Mars in Capricorn. 1. ed. Boston, 1955.
Geog 4559.55.10 Stenhouse, J.R. Cracker hash. London, 1955.
Geog 4559.56 Newby, Eric. The last grain race. Boston, 1956.
Geog 4559.56.5 Rutzebeck, Hjalmar. Mad sea. N.Y., 1956.
Geog 4559.56.10 Jones, William Herbert Sidney. The Cape Horn breed. N.Y., 1956.
Geog 4559.56.15 Noble, Arthur H. Three ships came sailing. London, 1956.
Geog 4559.56.20 Tunstall Behrens, Hilary. Pamir. London, 1956.
Geog 4559.57 Freuchen, Peter. Book of the seven seas. N.Y., 1957.
Geog 4559.57.5 Houter, F. den. Neptuania, maritieme curiosa. Amsterdam, 1957.
Geog 4559.58 Rehm, Arnold. Das fröhliche Logbuch. Bremerhaven, 1958.
Geog 4559.58.5 Noyce, Wilfrid. The springs of adventure. Cleveland, 1958.
Geog 4559.59 Carfi, Francesco. Geografia economica e sociale del mare. Livorno, 1959.
Geog 4559.59.5 Bradford, G. Yonder is the sea. Barre, Mass., 1959.
Geog 4559.60 Whipple, Addison. Tall ships and great captains. N.Y., 1960.
Geog 4559.60.5 O'Daniel, John. The nation that refused to starve. N.Y., 1960.
Geog 4559.62 Bradford, G. In with the sea wind. Barre, 1962.
Geog 4559.62.5 Villiers, A.J. Men, ships, and the sea. Washington, 1962.
Geog 4559.62.10 Dane, Peter. The seamen are down below. Ilfracombe, 1962.
Geog 4559.63 Asgeirsson, Rikki. Alltaf má fá annad skip. Reykjavik, 1963.
Geog 4559.64.3 Hanke, Helmut. Männer, Planken, Ozeane. 3. Aufl. Leipzig, 1966.
Geog 4559.65 West, Ellsworth Luce. Captain's papers. Barre, Mass., 1965.
Geog 4559.66 Karlsson, Elis. Pully-Haul: the story of a voyage. London, 1966.
Geog 4559.66.5 Randier, Jean. Hommes et navires au Cap Horn, 1616-1939. Paris, 1966.
VGeog 4559.67 Proface, Bruno. Nebo, more i ponekad kopno. Zagreb, 1967.
Geog 4559.67.5 Jupp, Ursula. Home port Victoria. Victoria, 1967.
Geog 4559.67.10 Heikkinen, Helge. Runt Kap Horn med Herzogin Cecilie. Ekenäs, 1967.
Geog 4559.67.15A Hugill, Stan. Sailortown. London, 1967.
Geog 4559.67.15B Hugill, Stan. Sailortown. London, 1967.
Geog 4559.68 Phillips-Burt, Douglas H.C. Reflections in the sea. Lymington, 1968.
Geog 4559.68.5 Armstrong, Richard. A history of seafaring. N.Y., 1968-69. 3v.
Geog 4559.69 Bradley, Wendell P. They live by the wind. 1. ed. N.Y., 1969.
Geog 4559.69.5 Fréminville, René Marie de La Poix de. La vie quotidienne des marins au Moyen Age. Paris, 1969.
Geog 4559.69.10 Spiers, George. The Wavertree; being an account of an ocean wanderer. N.Y., 1969.
Geog 4559.70 Bertino, Serge. Guide de la mer mystérieuse. Paris, 1970.
Geog 4559.71.5 Villiers, A.J. The war with Cape Horn. N.Y., 1971.
Geog 4559.71.10 Robinson, Cyril. Men against the sea. Windsor, N.S., 1971.
Geog 4559.72.1 Hassloef, Olof. Ships and shipyards, sailors and fishermen. Copenhagen, 1972.
Geog 4559.72.5 Throner, William R. Life at sea in the age of sail. London, 1972.
Geog 4559.72.10 Chapman, Gifford Desmond. Kangaroo Island shipwrecks. Canberra, 1972.

Geog 4571 Ocean life in general - Biographies of seamen - Individual

Geog 4571.2 Garland, Joseph E. Lone voyager. 1. ed. Boston, 1963.

Geog 4652 Ocean life in general - Shipwrecks and other marine disasters - Dictionaries, etc.

Geog 4652.5 Heden, Karl E. Direction of shipwrecks of the Great Lakes. Boston, 1966.
Geog 4652.7 Hocking, Charles. Dictionary of disasters at sea during the age of steam, 1824-1962. London, 1969. 2v.
Geog 4652.8 Boswell, David B. Loss list of Grimsby vessels, 1800-1960. Grimsby, 1969.
Geog 4652.9 Marx, Robert F. Shipwrecks of the Western Hemisphere, 1492-1825. N.Y., 1975.

Geog 4656 - 4660 Ocean life in general - Shipwrecks and other marine disasters - General history (By date)

Geog 4658.33 Redding, Cyrus. A history of shipwrecks, and disasters at sea. London, 1833. 2v.
Geog 4658.50 Rastoul de Mongeot, Alphonse. Histoire des naufrages depuis les temps ancien jusqu'au 1850. v.1-2. Bruxelles, 1850.
Geog 4658.90 Vidal Gormaz, Francisco. Algunos naufragios occuridos en las costas chilenes. Valparaiso, 1890.
Geog 4658.90.2 Vidal Gormaz, Francisco. Algunas naufragios occuridos en las costas chilenes. Santiago de Chile, 1901.
Geog 4658.93.2 Lussich, Antonio D. Naufragios célebres en el cabo Polonio, el banco Inglés, y el océano Atlántico. 2. ed. Montevideo, 1893.
Geog 4658.93.4 Lussich, Antonio D. Naufragios célebres en el cabo Polonio, banco Inglés, y océano Atlántico. Montevideo, 1938.
Geog 4659.36 Ingram, Charles W.N. Shipwrecks; New Zealand disasters, 1795-1936. Dunedin, 1936.
Geog 4659.36.3 Ingram, Charles W.N. New Zealand shipwrecks, 1795-1960. 3. ed. Wellington, 1961.
Geog 4659.36.10 Masters, David. S.O.S., a book of sea adventure. N.Y., 1936.
Geog 4659.52 Snow, Edgar R. Great gales and dire disasters. N.Y., 1952.
Geog 4659.54 Mackenzie, Margaret E. Shipwrecks; being the historical account of shipwrecks along the Victorian Coast, 1813-1914. Melbourne, 1954.
Geog 4659.55A Duffy, James E. Shipwrecks and empire; being an account of Portuguese maritime disasters in a century of decline. Cambridge, Mass., 1955.
Geog 4659.55B Duffy, James E. Shipwrecks and empire; being an account of Portuguese maritime disasters in a century of decline. Cambridge, Mass., 1955.
Geog 4659.55.5 O'May, Harry. Wrecks in Tasmanian waters, 1797-1950. Tasmania, 1955?
Geog 4659.57 Gibbs, James A. Shipwrecks of the Pacific Coast. Portland, Ore., 1957.
Geog 4659.60 Ratigan, William. Great Lakes shipwrecks and survivals. Grand Rapids, 1960.
Geog 4659.60.2 Ratigan, William. Great Lakes shipwrecks and survivals. 2. ed. Grand Rapids, 1969.
Geog 4659.63.4 Williams, Peter J. Shipwrecks at Port Phillips Heads since 1840. Melbourne, 1967.
Geog 4659.64 Noall, Cyril. Wreck and rescue round the Cornish coast. Truro, 1964-65. 3v.
Geog 4659.66 Farr, Grahame E. Wreck and rescue in the Bristol Channel. Truro, 1966-67. 2v.
Geog 4659.67 Burman, Jose. Great shipwrecks off the coast of southern Africa. Cape Town, 1967.
Geog 4659.68 Malster, Robert. Wreck and rescue on the Essex coast. Truro, 1968.
Geog 4659.68.5 Farr, Grahame E. Wreck and rescue on the coast of Devon. Truro, 1968.
Geog 4659.68.10 Loney, Jack Kenneth. Wrecks on the Gippsland coast. Geelong, 1968.
Geog 4659.69 Parry, Henry. Wreck and rescue on the coast of Wales. Truro, 1969-73. 2v.
Geog 4659.69.5 Larn, Richard. Cornish shipwrecks. Newton Abbot, 1969- 2v.
Geog 4659.74 Berlitz, Charles F. The Bermuda triangle. 1st ed. Garden City, 1974.
Geog 4659.75 Kusche, Lawrence D. The Bermuda triangle mystery - solved. N.Y., 1975.

Geog 4672 Ocean life in general - Shipwrecks and other marine disasters - Pamphlet volumes; Collected narratives

Geog 4672.1 Pamphlet vol. Geography. Ocean life. Shipwrecks.
Geog 4672.3 Pamphlet vol. Geography. Ocean life. Shipwrecks. 6 pam.
Geog 4672.4 Remarkable shipwrecks. Hartford, 1813.
Geog 4672.5 Ellms, Charles. The tragedy of the seas. Philadelphia, 1841.
Geog 4672.5.5 Ellms, Charles. Shipwrecks and disasters at sea. N.Y., 1844.
Geog 4672.6 Pamphlet vol. Geography. Ocean life. Shipwrecks.
Geog 4672.7 Book of shipwrecks and adventures on the ocean. N.Y., n.d.
Geog 4672.10 Paine, Ralph D. Lost ships and lonely seas. N.Y., 1922.
Geog 4672.12 Kitchin, Frederick H. Dead men's tales. Edinburgh, 1926.
Geog 4672.12.6 Kitchen, Frederick H. Tales of S.O.S. and T.T.T. Edinburgh, 1927.
Geog 4672.14 O'Donnell, Elliott. Strange sea mysteries. London, 1926.

60

Classified Listing

Geog 4672 Ocean life in general - Shipwrecks and other marine disasters - Pamphlet volumes; Collected narratives - cont.

Geog 4672.16 Lockhart, John Gilbert. Strange adventures of the sea. N.Y., 1931.
Geog 4672.16.1 Lockhart, John Gilbert. Strange adventures of the sea. N.Y., 1926.
Geog 4672.16.5 Lockhart, John Gilbert. Strange tales of the seven seas. London, 1929.
Geog 4672.16.10 Lockhart, John Gilbert. Mysteries of the sea. London, 1924.
Geog 4672.16.15 Lockhart, John Gilbert. The "Mary Celeste" and other strange tales of the sea. London, 1952.
Geog 4672.18 Neider, Charles. Great shipwrecks and castaways; authentic accounts of adventures at sea. N.Y., 1952.
Geog 4672.20.4 Gomes de Brito, Bernardo. História trágico-marítima. v.1-5. Pôrto, 1936-37.
Geog 4672.20.10 Gomes de Brito, Bernardo. Quadros da História trágico-marítima. Lisboa, 1944.
Geog 4672.21.2 Deperthes, Jean L.H.S. Histoire des naufrages, ou Recueil des relations les plus intéressantes des naufrages. Paris, 1790. 3v.
Geog 4672.21.4 Deperthes, Jean L.H.S. Histoire des naufrages, ou Recueil des relations les plus intéressantes des naufrages. Paris, 1828. 3v.
Geog 4672.22 Shipwrecks and disasters at sea, or Historical narratives of the most noted calamities. Edinburgh, 1812. 3v.
Geog 4672.23 Huntress, Keith G. Narrative of shipwrecks and disasters, 1586-1860. Ames, 1974.
Geog 4672.24 Baldwin, Hanson W. Sea fights and shipwrecks. Garden City, 1955.
Geog 4672.25 Dyall, Valentine. Famous sea tragedies. London, 1955.
Geog 4672.26 Snow, Edward Rowe. The vengeful sea. N.Y., 1956.

Geog 4675 - 4680 Ocean life in general - Shipwrecks and other marine disasters - Individual wrecks, etc. (By date of event)

Htn Geog 4676.02* Godinho Cardozo, M. Relaçam do naufragio da nao Santiago e itinerario da gente que delle se salvou. Lisboa, 1602. 3 pam.
Geog 4676.61 Kerckhoven, J. van. Wijtloopig...beschrijvinge van d. onzel voyage van't schip Arnhem, 1661. Amsterdam, 1664.
Geog 4676.93 Correa, F. Relaçam do successo que teve o patacho chamado N. Sra. da Candelaria da ilha da Madeira. Lisboa, 1734.
Htn Geog 4677.35* Gomes de Brito, Bernardo. História trágico-marítima. Lisboa, 1735. 3v.
Geog 4677.48 Marsden, Peter Richard V. The wreck of the Amsterdam. London, 1974.
Geog 4677.50 Bailey, J. God's wonders in the great deep. N.Y., 1750.
Geog 4677.66.3 Harrison, David. Een droevig verhaal van de ongelukkige reis en wonderdaadige verlossinge. Wesel, 176-?
Geog 4677.82 A true account of the loss of the Halsewell. v.1-2. London, 1786.
Geog 4677.96 Stout, Benjamin. The total loss of the American ship Hercules. London, 17- ?
Geog 4677.99 Molen, Sytze Jan van der. Goud in de golven. 's-Gravenhage, 1965.
Geog 4678.00 Perils of the ocean. N.Y., 1800?
Geog 4678.03 Fellows, William D. Narrative of the loss of the Lady Hobart Packet. London, 1803.
Geog 4678.06 Duncan, A. The mariner's chronicle. Philadelphia, 1806. 4v.
Geog 4678.09 Eustace, Thomas. The adventures. London, 1820.
Htn Geog 4678.10* Larcom, Henry. Distressing narrative of the loss of the ship Margaret of Salem. Beverly, 1810.
Geog 4678.10.5 Narrative of calamitous and interesting shipwrecks. Philadelphia, 1810.
Geog 4678.16.2 Allen, S. (Mrs.). A narrative of the shipwreck and unparalleled sufferings of Mrs. S. Allen. 2. ed. Boston, 1816.
Geog 4678.24 Stanford, John. Aetna; a discourse. N.Y., 1824.
Htn Geog 4678.25* Collins, Daniel. Narrative of the shipwreck of the brig Betsey of Wiscasset, Maine...Dec. 1824. Wiscasset, 1825.
Geog 4678.30 Burrows, Silas E. Russia and America. v.1-2. Hartford, 1865.
Htn Geog 4678.30.5* Smith, John. Narrative of the shipwreck and sufferings of the crew of the English brig Neptune. N.Y., 1830.
Geog 4678.32 Meredith, J.C. The tattooed man. 1. American ed. N.Y., 1959.
Geog 4678.33.5 Melancholy shipwreck and...divine providence. n.p., 1834.
Geog 4678.34 Duncan, A. The mariner's chronicle...shipwrecks, storms. New Haven, 1834.
Geog 4678.35 Mariners' chronicle. New Haven, 1835.
Geog 4678.36 Holden, H. Narrative of the shipwreck. Boston, 1836.
Geog 4678.36.5 Holden, H. Narrative of the shipwreck. Cooperstown, N.Y., 1841.
Geog 4678.36.10 Kendall, J. A sermon delivered February 14, 1836. Plymouth, 1836.
Htn Geog 4678.37* Palmer, John. Awful shipwreck...crew of the ship Francis Spaight. Boston, 1837.
Geog 4678.37.9 Smyth, Thomas. The voice of God in calamity, or Reflections on the loss of the steamboat "Home"...sermon. Charleston, 1837.
Geog 4678.37.15 Thomas, R. Interesting and authentic narratives of the most remarkable shipwrecks, fires. Hartford, 1837.
Geog 4678.40 A full and particular act of all the circumstances attending the loss of the steamboat Lexington. Providence, 1840.
Geog 4678.40.3 Cleveland, H.R. A letter to Daniel Webster. Boston, 1840.
Geog 4678.40.5 Stone, John S. A sermon occasioned by burning of steamer Lexington. Boston, 1840.
Htn Geog 4678.41.6* Slight, Julian. A narrative of the loss of the Royal George. 6. ed. Portsea, 1843.
Geog 4678.42 Guerrazzi, F.D. Replica...in causa di abbordaggio. Livorno, 1842.
Geog 4678.46 United States. Congress. House. Committee on Commerce. Owners and crew of ship Chandler Price. n.p., 1848.
Geog 4678.47 Greig, A.M. Fate of Blenden Hall, East Indiaman. N.Y., 1847.
Geog 4678.47.5 Lockhart, John Gilbert. Blenden Hall; the true story of a shipwreck. London, 1930.
Geog 4678.48.5 En route from Honolulu. Washington, 1931?
Geog 4678.51.2 Gilly, W.O.S. Narratives of shipwrecks. 2. ed. London, 1851.
Geog 4678.52 Addison, A.C. A deathless story. London, 1906.
Geog 4678.54 Smalley, Elam. A sermon. Troy, N.Y., 1854.
Geog 4678.54.5 Loss of the Arctic. n.p., 1854.
Geog 4678.54.10 Cola, S. de. Difesa pel cav F. Miceli...scontro dei...Ercolano e Sicilia. Messina, 1858.

Geog 4675 - 4680 Ocean life in general - Shipwrecks and other marine disasters - Individual wrecks, etc. (By date of event) - cont.

Geog 4678.54.15 Brown, A.C. Women and children last; the loss of the steamship. London, 1962.
Geog 4678.55 Hinckley, F. Wrecked on a reef in the China Sea. Boston, 1908.
Geog 4678.55.20 Schilling, Nikolai. Seeoffizier des Zaren. Köln, 1971.
Geog 4678.58 Corrao, Mario. Difesa del capitano ed armatore del Sicilia contro il cap. ed armatore dell'Ercolano. Palermo, 1858.
Geog 4678.58.5 Knowles, J.N. Crusoes of Pitcairn Island. Los Angeles, 1957.
Geog 4678.59 Mudie, Ian. Wreck of the Admella. London, 1967.
Geog 4678.62 United States. Navy Department. Wreck of steamer Governor and search for U.S. ship Vermont. Washington, 1862?
Geog 4678.65 Bribery and piracy...loss of S.S. Shooting Star. N.Y., 1870.
Geog 4678.65.10 Weiss, N. Naufrage de la Ville-du-Havre et du Losh-Earn. Paris, 1874.
Geog 4678.66 Comettant, O. Le naufrage de L'Evening Star. Paris, 1866.
Geog 4678.67 Fernández Duro, C. Naufragios de la armada española. Photoreproduction. Madrid, 1867.
Geog 4678.68 Cubbin, T. The wreck of the Serica. London, 1950.
Geog 4678.70 Read, George H. The last cruise of the Saginaw. Boston, 1912.
Geog 4678.72.15 Keating, Laurence J. The great Mary Celeste hoax. London, 1929.
Geog 4678.73 Weiss, Nathanael. Personal recollections of wreck of...Ville-du-Havre. N.Y., 1875.
Geog 4678.75F Campisi, Orazio. Petition and claim for damages to the ship Immacolata. n.p., 1879.
Geog 4678.87 Wreck of the Rainier; a sailors narrative. Portland, 1887.
Geog 4678.89 Forbes, Robert B. Notes on some few of the wrecks and rescues during the present century. Boston, 1889.
VGeog 4678.91 Henderson Brothers, Glasgow, defendants. Difesa dei signori Henderson, armatori contro i danneggiati del naufragio dell'Utopia. Napoli, 1895.
Geog 4679.12A Beesley, Lawrence. The loss of the Titanic. Boston, 1912.
Geog 4679.12B Beesley, Lawrence. The loss of the Titanic. Boston, 1912.
Geog 4679.12.2 Beesley, Lawrence. The loss of the S.S. Titanic. Boston, 1912.
Geog 4679.12.3 Gracie, A. The truth about the Titanic. N.Y., 1913.
Geog 4679.12.5 United States. Congress. Senate. Titanic disaster. Washington, 1912.
Geog 4679.12.7 Pelz von Felman, J. Titanic. Berlin, 1939.
Geog 4679.12.10 Thayer, J.B. The sinking of the S.S. Titanic, April 14-15, 1912. Philadelphia, 1940.
Geog 4679.12.13 Marshall, L. Sinking of the Titanic and great sea disasters. Philadelphia, 1912.
Geog 4679.12.17A Lord, Walter. A night to remember. N.Y., 1955.
Geog 4679.12.17B Lord, Walter. A night to remember. N.Y., 1955.
Geog 4679.12.18 Lord, Walter. A night to remember. London, 1956.
Geog 4679.12.20 O'Connor, Richard. Down to eternity. N.Y., 1956.
Geog 4679.12.25 Padfield, Peter. The Titanic and the Californian. London, 1965.
Geog 4679.12.30 Marcus, Geoffrey. The maiden voyage (Titanic, steamship). London, 1969.
Geog 4679.12.35 The Titanic commutator. 1,1912. Indian Orchard, Mass. Reprint ed. Indian Orchard, Mass. 1,1963+
Geog 4679.12.41 Bullock, Shan F. A Titanic hero, Thomas Andrews, shipbuilder. Riverside, 1973.
Geog 4679.14 Canada. Commission of Inquiry into the Loss of the British Steamship "Empress of Ireland". Report and evidence. Ottawa, 1914.
Geog 4679.17 Benson, N.P. The log of the El Dorado. San Francisco, 1917.
Geog 4679.23 Lockwood, C.A. Tragedy at Honda. Philadelphia, 1960.
Geog 4679.23.6 Foster, Cecil. 1700 miles in open boats; the story of the loss of the S.S. Trivessa in the Indian Ocean. London, 1952.
Geog 4679.25 Marine Research Society, Salem, Mass. The sea, the ship, and the sailor. Salem, Mass., 1925.
Htn Geog 4679.26* Chapin, H.M. Biblioteca Titanicana. Metuchen, 1926.
Geog 4679.26.2 Chapin, H.M. Biblioteca Titanicana. Metuchen, 1926.
Geog 4679.27 Lockhart, John Gilbert. A great sea mystery. London, 1927.
Geog 4679.30 Hadfield, R.L. Sea-toll of our time. London, 1930.
Geog 4679.30.5 Shaw, F.H. Full fathom five. N.Y., 1930.
Geog 4679.33 Tambs, E. The cruise of the Teddy. London, 1950.
Htn Geog 4679.38* Knowles, Josiah N. The crusoes of Pitcairn's Islands. n.p., 1938.
Geog 4679.38.5 Slocum, Victor. Castaway boats. N.Y., 1938.
Geog 4679.39 Swan, E.W. The first commission of H.M.S. Calliope. Newcastle-upon-Tyne, 1939.
Geog 4679.41 Hanson, Earl P. Highroad to adventure. N.Y., 1941.
Geog 4679.41.5 Swinson, A. Scotch on the rocks. London, 1963.
Geog 4679.42 Bryan, George S. Mystery ship. The Mary Celeste in fancy and fact. Philadelphia, 1942.
Geog 4679.42.5 Fay, Charles E. Mary Celeste; the odyssey of an abandoned ship. Salem, 1942.
Geog 4679.42.10 Trumbull, Robert. The raft. N.Y., 1942.
Geog 4679.44.5 Hardy, Alfred C. Wreck - S.O.S. London, 1944.
Geog 4679.47 Ainslie, Kenneth. Pacific ordeal. 1. American ed. N.Y., 1956.
Geog 4679.47.5 Meier, F. Hurricane warning. N.Y., 1947.
Geog 4679.49 Lederer, William J. The last cruise. N.Y., 1950.
Geog 4679.52 Bevan, David. Drums of the Birkenhead. Capetown, 1972.
Geog 4679.56 Armstrong, W. Last voyage. London, 1956.
Geog 4679.56.5 Villiers, A.J. Posted missing. N.Y., 1956.
Geog 4679.56.7 Villiers, A.J. Posted missing. London, 1975.
Geog 4679.59 Moscow, Alvin. Collision course: the Andrea Doria and the Stockholm. N.Y., 1959.
Geog 4679.59.5 Gallagher, Thomas M. Fire at sea. N.Y., 1959.
Geog 4679.60.5 Ziganskin, Askhat. 49 dnei v okeane. Kuibyshev, 1960.
Geog 4679.61 Mowat, Farley. The serpent's coil. 1. United States ed. Boston, 1961.
Geog 4679.62 Noyce, Wilfrid. They survived. London, 1962.
Geog 4679.63 Troebst, C.C. Auf Wunder ist kein Verlass. Düsseldorf, 1963.
Geog 4679.63.1 Troebst, C.C. The art of survival. Garden City, 1975.
Geog 4679.63.5 Marchbanks, David. The painted ship. London, 1964.
Geog 4679.64 Ruhen, Olaf. Minerva Reef. 1. American ed. Boston, 1964.
Geog 4679.67 Gill, Crispin. The wreck of the Torrey Canyon. Newton Abbot, 1967.
Geog 4679.67.10 Mabire, Jean. La marée noire du Torrey Canyon. Paris, 1967.
Geog 4679.67.20 Du Pontavice, Emmanuel. La pollution des mers..."Torrey Canyon". Paris, 1968.

Classified Listing

Geog 4675 - 4680 Ocean life in general - Shipwrecks and other marine disasters - Individual wrecks, etc. (By date of event) - cont.

Geog 4679.67.25 Petrow, Richard. The black tide: in the wake of the Torrey Canyon. London, 1968.
Geog 4679.67.30 Committee of Scientists on the Scientific and Technological Aspects of the Torrey Canyon Disaster. The Torrey Canyon; report. London, 1967.
Geog 4679.68 Lambert, Max. The Wahine disaster. Wellington, 1968.
Geog 4679.69.10 Doak, Wade. The Elingamite and its treasure. London, 1969.
Geog 4679.71 Fenn, Charles. Journal of a voyage to nowhere. London, 1971.
Geog 4679.72 Vignes, Jacques. La rage de survivre, quand il ne reste que la vie. Paris, 1973.
Geog 4679.72.1 Vignes, Jacques. The rage to survive. N.Y., 1976.
Geog 4679.73 Robertson, Dougal. Survive the savage sea. London, 1973.
Geog 4679.73.6 Bailey, Maurice. 117 days adrift. Lymington, 1974.

Geog 4706 - 4710 Ocean life in general - Piracy and treasure trove in general - General works (By date)

Geog 4707.23.2F Rhode Island (Colony). Court of Admiralty. Tryals of thirty-six persons for piracy. Boston, 1723.
Geog 4707.24A Seybold, R.F. Captured by pirates. Two diaries of 1724-1725. Boston, 1929.
Geog 4707.24B Seybold, R.F. Captured by pirates. Two diaries of 1724-1725. Boston, 1929.
Geog 4707.26 Jedre, J.B. The trials of five persons for piracy. Boston, 1726.
Geog 4707.26.5F Massachusetts (Colony). Court of Admiralty. The tryals of sixteen persons for piracy. London, 1726.
Geog 4708.06 Phillips, James D. Loss of the ship Essex in 1806. n.p., 1941.
Geog 4708.12.4 United States. Circuit Court (1st Circuit, Mass.). The trial of Samuel Tulley and John Dalton...piracy and murder committed Jan. 21, 1812. 4. ed. Boston, 1813.
Htn Geog 4708.16.15* Onion, S.B. Narrative of the mutiny on board the schooner Plattsburg. Boston, 1819.
Htn Geog 4708.16.20* The pirates. Boston, 181-
Htn Geog 4708.19* Lives and confessions of John Williams, Francis Frederick, John P. Roy and P. Peterson. 2. ed. Boston, 1819. 2 pam.
Htn Geog 4708.19.2* Lives and confessions of John Williams, Francis Frederick, John P. Roy and P. Peterson. Boston, 1819.
Geog 4708.21 Lincoln, B. Narrative of the capture...of Captain B. Lincoln...by piratical schooner. Boston, 1822.
Htn Geog 4708.31* Mutiny and murder. Confession of Charles Gibbs. Providence, 1831.
Geog 4708.34 Pamphlet vol. Piracy on brig Mexican of Salem, report of trial. 4 pam.
Geog 4708.34.2 A report of the trial of Pedro Gilbert...piracy. Boston, 1834.
Geog 4708.36 Lives and bloody exploits of most noted pirates. Hartford, Conn., 1836.
Htn Geog 4708.37* The pirates own book. Portland, 1837.
Geog 4708.61 Trial of officers and crew of the privateer Savannah on charge of piracy. N.Y., 1862.
Geog 4708.68 Bollo, G.A. Petizione...relativa alla catastrofe toccata alla nave Teresa. n.p., n.d. 2 pam.
Geog 4708.74 Galleon Treasure Company, N.Y. Narrative of the circumstances of and attending the sinking of the Spanish galleons. N.Y., 1874.
Geog 4709.7 Knight, E.F. The cruise of the "Alerte". London, 1907.
Geog 4709.11 Paine, R.D. The book of buried treasure. London, 1911.
Geog 4709.22 French, J.L. Great pirate stories. N.Y., 1922.
Geog 4709.22.2 French, J.L. Great pirate stories. 2. ser. N.Y., 1925.
Geog 4709.23 Hill, Samuel C. Notes on piracy in eastern waters. London, 1923-28.
Geog 4709.24A The pirates own book. Salem, 1924.
Geog 4709.24B The pirates own book. Salem, 1924.
Geog 4709.25 Seitz, Don Carlos. Under the black flag. N.Y., 1925.
Geog 4709.26 Gosse, Philip. My pirate library. London, 1926.
Geog 4709.27 Partridge, Eric. Pirates, highwaymen, and adventurers. London, 1927.
Geog 4709.27.5 Seitz, Don Carlos. Under the black flag. London, 1927.
Geog 4709.28A Wycherley, George. Buccaneers of the Pacific. Indianapolis, 1928.
Geog 4709.28B Wycherley, George. Buccaneers of the Pacific. Indianapolis, 1928.
Geog 4709.28.5 Besson, Maurice. Les frères de la coste, flibustiers et corsaires. Paris, 1928.
Geog 4709.28.10 Gallomb, J. Pirates, old and new. N.Y., 1928.
Geog 4709.30A Dobie, J. Frank. Coronado's children. Dallas, 1930.
Geog 4709.30B Dobie, J. Frank. Coronado's children. Dallas, 1930.
Geog 4709.32 Gosse, Philip. The history of piracy. N.Y., 1946.
Geog 4709.33 Grey, Charles. Pirates of the Eastern seas, 1618-1723. London, 1933.
Geog 4709.34 Magre, M. Pirates, flibustiers, negriers. 9. éd. Paris, 1934.
Geog 4709.39 Dobie, J.F. Apache gold and yaqui silver. Boston, 1939.
Geog 4709.40 Hawes, Hildreth Gilman. The Bellamy treasure. Augusta, Maine, 1940.
Geog 4709.40.5 Wilkins, H.T. Panorama of treasure hunting. N.Y., 1940.
Geog 4709.42 Santschi, R.J. Treasure trails. Glen Ellyn, Ill., 1942.
Geog 4709.43 Taylor, James. Gold from the sea; the epic story of the "Niagara's" bullion. London, 1943.
Geog 4709.48 Wilkins, H.T. A modern treasure hunter. London, 1948.
Geog 4709.52 Franchi, A. Storia della pirateria nel mondo. Milano, 1952. 2v.
Geog 4709.53 Pringle, P. Jolly Roger. N.Y., 1953.
Geog 4709.53.2 Pringle, P. Jolly Roger. London, 1953.
Geog 4709.53.5 Cabal, Juan. Historias de pirates. 1. ed. Barcelona, 1953.
Geog 4709.55 Quarrell, Charles. Buried treasure. London, 1955.
Geog 4709.57 Freminville, René Marie de La Poix de. Tels etaient corsaires et flibustiers. Paris, 1957.
Geog 4709.57.5 Monti, Mario. I pirati. Milano, 1957.
Geog 4709.57.10 Carse, Robert. The age of piracy. N.Y., 1957.
Geog 4709.57.12 Carse, Robert. The age of piracy. N.Y., 1965.
Geog 4709.59 Maséa de Ros, Angeles. Historia general de la piratería. Barcelona, 1959.
Geog 4709.59.5 Leip, Hans. Bordbuch des Satans. München, 1959.
Geog 4709.65 Cochran, Hamilton. Freebooters of the Red Sea: pirates, politicians and pieces of eight. Indianapolis, 1965.
Geog 4709.65.5 Clausen, Carl J. A 1715 Spanish treasure ship. Gainesville, 1965.
Geog 4709.66 Wagner, Kip. Pieces of eight; recovering the riches of a lost Spanish treasure fleet. 1. ed. N.Y., 1966.

Geog 4706 - 4710 Ocean life in general - Piracy and treasure trove in general - General works (By date) - cont.

Geog 4709.66.5 Course, Alfred George. Pirates of the Eastern seas. London, 1966.
Geog 4709.68 Skriagin, Lev N. Sokrovishcha pogibshikh korablei. Moskva, 1968.
Geog 4709.69 Morris, Roland. Island treasure: the search for Sir Cloudesley Shovell's flagship "Association". London, 1969.
Geog 4709.69.5 Blond, Georges. Histoire de la flibuste. Paris, 1969.
Geog 4709.70 Earle, Peter. Corsairs of Malta and Barbary. London, 1970.
Geog 4709.71 Cousteau, Jacques Yves. Diving for sunken treasure. London, 1971.
Geog 4709.71.5 Sokol, Hans Hugo. Unter der Flagge mit dem Totenkopf. Herford, 1971.

Geog 4750 Tropics in general - Bibliographies

Geog 4750.2 International Geographical Union. Special Commission on the Humid Tropics. A select annotated bibliography of the humid tropics. Montreal, 1960.
Geog 4750.4 Texas Instruments, Inc. An inventory of geographic research of the humid tropic environment. Dallas, Texas, 1967? 2v.

Geog 4756 - 4760 Tropics in general - General works (By date)

Geog 4758.24 Denis, F. Scènes de la nature sous les tropiques. Paris, 1824.
Htn Geog 4758.24.2* Denis, F. Scènes de la nature sous les tropiques. Paris, 1824.
Htn Geog 4758.24.3* Denis, F. Scènes de la nature sous les tropiques. Paris, 1824.
Geog 4759.30 Waugh, Alec. The coloured countries. London, 1930.
Geog 4759.30.5A Waugh, Alec. Hot countries. N.Y., 1930.
Geog 4759.30.5B Waugh, Alec. Hot countries. N.Y., 1930.
Geog 4759.30.7 Waugh, Alec. Hot countries. N.Y., 1930.
Geog 4759.39 Price, A.G. White settlers in the tropics. N.Y., 1939.
Geog 4759.47 Gourou, P. Les pays tropicaux. Paris, 1947.
Geog 4759.47.3 Gourou, Pierre. Les pays tropicaux, principes d'une géographie humaine et économique. 4. éd. Paris, 1966.
Geog 4759.47.4 Gourou, Pierre. The tropical world. N.Y., 1966.
Geog 4759.52 Bates, M. Where winter never comes. N.Y., 1952.
Geog 4759.53 Gourou, Pierre. The tropical world. 2. ed. London, 1958.
Geog 4759.59 Troll, Carl. Die tropischen Gebirge. Bonn, 1959.
Geog 4759.64 Steel, R.W. Geographers and the tropics. London, 1964.
Geog 4759.71.5 Gourou, Pierre. Leçons de géographie tropicale. Paris, 1971.
Geog 4759.71.10 Benchétrit, Maurice. Géographie zonale des régions chaudes. Paris, 1971.
Geog 4759.72 Études de géographie tropicale offertes à Pierre Gourou. Paris, 1972.
Geog 4759.73 Daveau, Suzanne. La zone intertropicale humid. Paris, 1973.
Geog 4759.73.5 Birot, Pierre. Géographie physique générale de la zone intertropicale. Paris, 1973.

Geog 5050 Arctic and Antarctic Regions in general - Bibliographies

Geog 5050.2 Boston Public Library. Arctic regions and Antarctic regions. Boston, 1894.
Geog 5050.3 Chavanne, J. Die Literatur über die Polar-Regionen. Wien, 1878.
Geog 5050.5 United States. Works Progress Administration, N.Y.C. Annotated bibliography of the polar regions. Series B. pt. 1. N.Y., 1938.
Geog 5050.7 Zavatti, S. Saggio di bibliografia polare. 2. ed. Roma, 1952.
Geog 5050.8 Scotland. National Library, Edinburgh. Shelf-catalogue of the Wordie collection of polar exploration. Boston, 1964.
Geog 5050.9 Expéditions Polaires Francaises, 1948. Terre Adélie, Greenland, 1947-55; bibliographie. Grenoble, 1956.
Geog 5050.10 David Davis Polar Library. Catalog of the David Davis Polar Library. Menlo Park, 1969.
Geog 5050.12 Recent polar literature. Cambridge, Eng. 1973+

Geog 5055 Arctic and Antarctic Regions in general - Pamphlet volumes

Geog 5055.1 Pamphlet vol. Geography. Polar regions. Arctic and Antarctic in general.

Geog 5057 Arctic and Antarctic Regions in general - Periodicals and Societies

Geog 5057.5 The polar times. N.Y. 1,1935+ 4v.

Geog 5060 Arctic and Antarctic Regions in general - History of polar exploration

Geog 5060.5 Richardson, J. The polar regions. Edinburgh, 1861.
Geog 5060.7 Victor, Paul Émile. Man and the conquest of the poles. N.Y., 1963.
Geog 5060.7.1 Victor, Paul Émile. L'homme à la conquête des pôles. Evreux, 1971.
Geog 5060.7.6 Victor, Paul Émile. Terres polaires, terres tragiques. Paris, 1973.
Geog 5060.8 Marshall, Logan. The story of polar conquest. n.p., 1913.
Geog 5060.10A Joerg, W.L.G. Brief history of polar exploration since the introduction of flying. N.Y., 1930.
Geog 5060.10B Joerg, W.L.G. Brief history of polar exploration since the introduction of flying. N.Y., 1930.
Geog 5060.10C Joerg, W.L.G. Brief history of polar exploration since the introduction of flying. N.Y., 1930.
Geog 5060.11 Toreoja y Miret, José M. Las republicas hispano-americanos y la exploracion de las regiones polares. Madrid, 193-
Geog 5060.15 Rasmussen, K. Heldenbuch der Arktis. Leipzig, 1933.
Geog 5060.16 Zeidler, Paul G. Helden im ewigen Eis; im Kampf um den Nord- und Südpol. Leipzig, 1936.
Geog 5060.17 Kirwan, Laurence P. A history of polar exploration. N.Y., 1960.
Geog 5060.18 Mountfield, David. A history of polar exploration. London, 1974.

Geog 5070 Arctic and Antarctic Regions in general - General treatises

Geog 5070.5 M'Cormick, R. Voyages of discovery in Arctic and Antarctic. London, 1884. 2v.
Geog 5070.9 Markham, C.R. The lands of silence. Cambridge, Eng., 1921.
Geog 5070.11 Hassert, Kurt. Die Polarforschung. 3. Aufl. Leipzig, 1914.
Geog 5070.11.5 Hassert, Kurt. Die Polarforschung. München, 1956.
Geog 5070.15 Löwenberg, J. Die Entdeckungs- und Forschungsreisen im dem Leiden Polarzonen. Leipzig, 1886.
Geog 5070.20 Brown, R.N.R. A naturalist at the poles. London, 1923.
Geog 5070.20.5 Brown, R.N.R. The polar regions. London, 1927.

Classified Listing

Geog 5070   Arctic and Antarctic Regions in general - General treatises - cont.
- Geog 5070.25   Congrès International pour l'Étude des Régions Polaires, 1st, Brussels, 1906. Rapport d'ensemble. Bruxelles, 1906.
- Geog 5070.30   Greely, A.W. The polar regions in the twentieth century. Boston, 1928.
- Geog 5070.35A   American Geographic Society, N.Y. Problems of polar research. N.Y., 1928.
- Geog 5070.35B   American Geographic Society, N.Y. Problems of polar research. N.Y., 1928.
- Geog 5070.36A   Nordenskjöld, Otto. The geography of the polar regions. N.Y., 1928.
- Geog 5070.36B   Nordenskjöld, Otto. The geography of the polar regions. N.Y., 1928.
- Geog 5070.37   Moreno Fuentes, José. Las regiones heladas de los polos norte y sur. Madrid, 1884.
- Geog 5070.38   Debenham, F. The polar regions. London, 1930.
- Geog 5070.44   Zeidler, P.G. Polarfahrten. Berlin, 1927.
- Geog 5070.46   Brown, R.N.R. Some problems of polar geography. Washington, 1929.
- Geog 5070.55   Ellsworth, L. Beyond horizons. N.Y., 1938.
- Geog 5070.57   Courtauld, A. From the ends of the earth; an anthology of polar writings. London, 1958.
- Geog 5070.62   Croft, Andrew. Polar exploration. London, 1939.
- Geog 5070.62.5   Croft, Andrew. Polar exploration. 2. ed. London, 1947.
- Geog 5070.65   Riabinin, A.K. Iz Arktiki v Antarktiku. Murmansk, 1959.
- Geog 5070.66   Poliarnyi krug. Moskva, 1974.
- Geog 5070.68   Dieck, Herman. The marvelous wonders of the polar world. Philadelphia, 1965?
- Geog 5070.70   Tilman, Harold William. Mostly mischief: voyages to the Arctic and to the Antarctic. London, 1966.
- Geog 5070.70.5   Tilman, Harold William. In mischief's wake. London, 1971.
- Geog 5070.72   Kosack, Hans P. Die Poterforschung; ein Datenbuch über die Natur-, Kultur-, Wirtschaftsverhältnisse und die Erforschungsgeschichte der Polarregionen. Braunschweig, 1967.

Geog 5080   Arctic and Antarctic Regions in general - Special topics - Sovereignty
- Geog 5080.5   Smedal, Gustav. Aquisition of sovereignty over polar areas. Oslo, 1931.
- Geog 5080.5.5   Smedal, Gustav. Souveränitätsfragen der Polargebiete. Oslo? 1943.

Geog 5100   Arctic and Antarctic Regions in general - Special topics - Essays, addresses
- Geog 5100.5   Prentiss, H.M. The great polar current. Cambridge, 1897.
- Geog 5100.10   Gould, Laurence McKinley. The polar regions in their relation to human affairs. N.Y., 1958.

Geog 5110   Arctic and Antarctic Regions in general - Special topics - Miscellaneous speculations
- Geog 5110.3   Rud, William. The phantom of the poles. N.Y., 1906.
- Geog 5110.5   Bronner, Finn E. The polar regions. Santa Barbara, 1958.

Geog 5120   Arctic and Antarctic Regions in general - Special topics - Instructions for explorers
- Geog 5120.1   Bertram, Colin. Arctic and Antartic. Cambridge, 1939.

Geog 5130   Arctic and Antarctic Regions in general - Special topics - Aerial exploration
- Geog 5130.2   Grierson, J. Challenge to the poles. Hamden, Conn., 1964.
- Geog 5130.5   Domaas, K. Polarskipet frani. Oslo, 1954.

Geog 5150   Arctic Regions - Bibliographies
- Geog 5150.2   Toronto. Public Library. The Northwest passage, 1534-1859. Toronto, 1963.
- Geog 5150.5   Dutilly, A. Bibliographies on the Arctic. Washington, 1945.

Geog 5153   Arctic Regions - Periodicals and Societies
- Geog 5153.5   Danish Arctic research. Charlottenlund. 1,1955+
- Geog 5153.10   Polar notes. Hanover, N.H. 1-14,1959-1975// 2v.
- Geog 5153.15   Inter-nord. Paris. 4+ 6v.
- Geog 5153.20   The musk-ox; a journal on the north. Saskatoon. 1,1967+ 2v.

Geog 5155   Arctic Regions - Pamphlet volumes
- Geog 5155.1   Pamphlet box. Geography. Arctic regions. North Polar regions.
- Geog 5155.1.2   Pamphlet vol. Geography. Arctic regions. North Polar regions.
- Geog 5155.1.5   Pamphlet vol. Geography. Arctic regions. North Polar regions. German dissertations.
- Geog 5155.3   Pamphlet vol. Geography. Arctic exploration. 22 pam.
- Geog 5155.5   Gerritsz, H. Detectio freti Hudsoni. Amsterdam, 1878.
- Geog 5155.8   Pamphlet box. Stefansson, V. Minor works on Arctic exploration.

Geog 5160   Arctic Regions - General works, geography and description
- Geog 5160.3   Scoresby, William. An account of the Arctic regions. Edinburgh, 1820. 2v.
- Geog 5160.5F   The Arctic world; its plants, animals and natural phenomena. London, 1876?
- Geog 5160.7A   Stefansson, V. The northward course of empire. N.Y., 1922.
- Geog 5160.7B   Stefansson, V. The northward course of empire. N.Y., 1922.
- Geog 5160.8   Kersting, Rudolf. The white world. N.Y., 1902.
- Geog 5160.10F   Hartwig, G. The Arctic regions. London, n.d.
- Geog 5160.11   Pamphlet box. Stefansson, V. Minor publications.
- Geog 5160.11.5   Stefansson, V. Unsolved mysteries of the Arctic. N.Y., 1939.
- Geog 5160.11.10   Stefansson, V. Ultima Thule. N.Y., 1940.
- Geog 5160.11.12   Stefansson, V. Ultima Thule. Reykjavik, 1942.
- Geog 5160.11.15   Stefansson, V. Arctic manual. N.Y., 1944.
- Geog 5160.12   United States. Office of Chief of Air Corps (War Department). Arctic manual. v.1-2. Washington, 1940.
- Geog 5160.13.5   Brower, Charles D. Fifty years below zero. N.Y., 1942.
- Geog 5160.14   Ross, Colin. Mit Kind und Kegel in die Arktis. Leipzig, 1934.
- Geog 5160.15   Stefansson, Evelyn Schwartz Baird. Within the circle. N.Y., 1945.
- Geog 5160.16   Rodahl, Kåre. North. 1. ed. N.Y., 1953.
- Geog 5160.20   Kimble, G.H.T. Geography of the northlands. N.Y., 1955.
- Geog 5160.25   Grigor'ev, A.A. Subarktika. 2. Izd. Moskva, 1956.
- Geog 5160.30   Collins, H.B. Arctic area. México, 1954.
- Geog 5160.35   Calder, Ritchie. Men against the frozen north. N.Y., 1957.

Geog 5160   Arctic Regions - General works, geography and description - cont.
- Geog 5160.40   Stanwell-Fletcher, T.M. Clear lands and icy seas. N.Y., 1958.
- Geog 5160.47.5   Giaever, John. Hardbalne polarkarer. 5. Oppl. Oslo, 1973.
- Geog 5160.55   Liakh, N.N. Zapiski poliarnika. Novosibirsk, 1961.
- Geog 5160.56   Arctic Institute of North America. The Arctic basin. Washington, 1963.
- Geog 5160.57   Santillan de Andres, Selva E. La vida humana en el subecumene ártico. Tucuman, 1962.
- Geog 5160.58   Hantschel, A. Weltgeschehen am Rande des Polarmeeres. Würzburg, 1964.
- Geog 5160.59   Macdonald, Ronald. The Arctic frontier. Toronto, 1966.
- Geog 5160.60   Stefansson, Evelyn Schwartz Baird. Here is the far north. N.Y., 1957.
- Geog 5160.61F   Bruemmer, Fred. The Arctic. Scarborough, Ont., 1974.

Geog 5170   Arctic Regions - Special topics - Theory of North Polar exploration
- Geog 5170.5   Howgate, H.W. Polar colonization. Washington, 1878.
- Htn   Geog 5170.7*   Barrington, D. Probability of reaching the North Pole. London, 1775.
- Htn   Geog 5170.9*   Barrington, D. Possibility of approaching the North Pole. London, 1818.
- Geog 5170.11   A letter to John Barrow...polar expedition. London, 1819.
- Geog 5170.13   Remarks on the voyage of the ships "Resolution". London, 1780.
- Geog 5170.15   Projet pour tenter la decouverte. La Haye, 1772.
- Geog 5170.17   Seidenfaden, G. Modern Arctic exploration. Boston, 1939.

Geog 5180   Arctic Regions - Special topics - Miscellaneous essays, etc.
- Geog 5180.5   Heilprin, A. The Arctic problem. Philadelphia, 1893.
- Geog 5180.6   Symposium on Circumpolar Problems, Luleå, Sweden and Tromsø, Norway, 1969. Circumpolar problems. 1. ed. Oxford, 1973.
- Geog 5180.7   De Croy, D. Memoire sur le passage par le nord. Paris, 1782.
- Geog 5180.8   Balch, E.S. The North Pole and Bradley land. Philadelphia, 1913.
- Geog 5180.9   Weyer, E.M. The Eskimos. New Haven, 1932.
- Geog 5180.10   Byhan, A. Die Polarvölker. Leipzig, 1909.
- Geog 5180.11   Marquis, R. La conquête du pôle nord en ballon. Paris, 19- .
- Geog 5180.12   Hatt, G. Arktiske skinddragter i Eurasian og Amerika. Kjøbenhavn, 1914.
- Geog 5180.12.5   Hatt, G. Moccasins and their relation to Arctic footwear. Lancaster, 1915?
- Geog 5180.16   Stefansson, V. The Arctic in fact and fable. N.Y., 1945.
- Geog 5180.18   Smith, I. Norman. The unbelievable land; 29 experts bring us closer to the Arctic. Ottawa, 1964.
- Geog 5180.35   Hopper, Bruce C. Sovereignty in the Arctic. N.Y., 1937.
- Geog 5180.37   Nas razdeliaet okean. Magadan, 1969.

Geog 5182   Arctic Regions - Special topics - Eskimos - General works
- Geog 5182.4   Le peuple esquimau aujourd'hui et demain. Paris, 1973.
- Geog 5182.5   Fainberg, Lev A. Ocherki etnicheskoi istorii zarubezhnogo Severa. Moskva, 1971.
- Geog 5182.6   Gilbertson, Albert N. Some ethical phases of Eskimo culture. Worcester, 1914.
- Geog 5182.7   Steensby, Hans P. An anthropological study of the origin of the Eskimo culture. Kjøbenhavn, 1916.
- Geog 5182.8.3   Birket-Smith, Kaj. Eskimoerne. 3. Udg. Kjøbenhavn, 1971.
- Geog 5182.8.5   Birket-Smith, Kaj. The Eskimoes. London, 1936.

Geog 5208 - 5210   Arctic Regions - History of exploration and discoveries - General works
- Geog 5208.18   Barrow, John. Chronological history of voyages into Arctic regions. London, 1818.
- Geog 5208.32A   Tytler, P.F. Historical view of progress of discovery. Edinburgh, 1832.
- Geog 5208.32B   Tytler, P.F. Historical view of progress of discovery. Edinburgh, 1832.
- Geog 5208.33   Tytler, P.F. Historical view of progress of discovery. Edinburgh, 1833.
- Geog 5208.46   Barrow, John. Voyages of discovery and research within the Arctic regions, from the year 1818 to present. London, 1846.
- Geog 5208.46.3   Barrow, John. Voyages of discovery and research within the Arctic regions, from the year 1818 to present. London, 1846.
- Geog 5208.51   Shillinglaw, J.J. A narrative of Arctic discovery. 2. ed. London, 1851.
- Geog 5208.53   Northern coasts of America. London, 1853.
- Geog 5208.57   Sargent, E. Arctic adventure by sea and land. Boston, 1857.
- Geog 5208.68   Hayes, I.I. Progress of Arctic discovery. N.Y., 1868.
- Geog 5208.68.2   Hayes, I.I. Progress of Arctic discovery. N.Y., 1868.
- Geog 5208.69.15   Hartwig, G. The polar world. London, 1886.
- Geog 5208.73.2   Markham, C.R. The threshold of the unknown region. London, 1873.
- Geog 5208.73.4   Markham, C.R. The threshold of the unknown region. 4. ed. London, 1876.
- Geog 5208.79   Markham, A.H. Northward ho! London, 1879.
- Geog 5208.80   Rezzadore, P. I viaggi polari. Roma, 1880.
- Geog 5208.81   Hellwald, F. von. Im ewigen Eis. Stuttgart, 1881.
- Geog 5208.81.5   Hellwald, F. von. Voblasti vechnago l'da. Sankt Peterburg, 1881.
- Geog 5208.82.2   Perry, R. Jeannette, and a complete and authentic narrative encyclopedia of all voyages. San Francisco, 1883.
- Geog 5208.89   Smith, C.C. Arctic explorations in 18th and 19th centuries. Boston, 1889.
- Geog 5208.95   Greely, A.W. Handbook of arctic discoveries. Boston, 1896.
- Geog 5208.95.3   Greely, A.W. Handbook of polar discoveries. 3. ed. Boston, 1906.
- Geog 5208.95.7   Greely, A.W. Handbook of polar discoveries. 5th ed. Boston, 1910.
- Geog 5209.03   Thompson, G.M. Rannsóknarferđ i kafli. Gimli, 1903.
- Geog 5209.06   Hoare, J.D. Arctic exploration. N.Y., 1906.
- Geog 5209.06.2   Hoare, J.D. Arctic exploration. London, 1906.
- Geog 5209.07   Gordon, W.J. Round about the North Pole. N.Y., 1907.
- Geog 5209.10   Wright, H.S.S. (Mrs.). The great white North. N.Y., 1910.
- Geog 5209.10.3   Edwards, D.M. The toll of the arctic seas. N.Y., 1910.
- Geog 5209.23.3   Rouch, J. Le pôle nord. Paris, 1923.
- Geog 5209.29   Einarsson, S. Norđr um höf. Reykjavik, 1924.
- Geog 5209.31   Samoilovich, R. Der Weg nach dem Pol. Biehefeld, 1931.
- Geog 5209.32   Houhen, H.H. The call of the North. London, 1932.
- Geog 5209.34   Mirsky, J. To the North: The story of arctic exploration from earliest times to the present. N.Y., 1934.

Classified Listing

Geog 5208 - 5210 Arctic Regions - History of exploration and discoveries - General works - cont.

Geog 5209.34.2 Mirsky, J. To the Arctic! N.Y., 1948.
Geog 5209.39 Segal, Louis. The conquest of the Arctic. London, 1939.
Geog 5209.43 Einarsson, S. Sudr um höf. Reykjavik, 1943.
Geog 5209.47 Crouse, Nellis M. The search for the North Pole. N.Y., 1947.
Geog 5209.53 Mountevans, C.R.G.R.E. Arctic solitudes. N.Y., 1953.
Geog 5209.57 Förster, Hans A. Der weisse Weg; Forscher erobern die Arktis. Leipzig, 1957.
Geog 5209.60 Mendeleev, D.I. Nauchnyi arkhiv: osvoenie krainego severa. Leningrad, 1960.
Geog 5209.60.5 Dainelli, G. La gara verso il Polo Nord. Torino, 1960.
Geog 5209.67 Mowat, Farley. The polar passion: the quest for the North Pole. Toronto, 1967.

Geog 5220 Arctic Regions - History of exploration and discoveries - Special periods - Before 1800

Geog 5220.2 Dresden, Germany. Armen-Realschule. Programm...Offentlichen Prüfung. Dresden, 1873.
Geog 5220.5 Forster, J.R. History of voyages...in the North. London, 1786.
Geog 5220.7 Forster, J.R. Histoire des decouvertes...dans le Nord. Paris, 1788. 2v.
Geog 5220.9 Leslie, J. Narrative of discovery...polar seas. N.Y., 1833.
Geog 5220.9.3 Leslie, J. Narrative of discovery...polar seas. N.Y., 1836.
Geog 5220.9.6 Leslie, J. Narrative of discovery...polar seas. Edinburgh, 1845.
Geog 5220.11 Capel, R. Norden oder zu Wasser und Lande. Hamburg, 1678.
Htn Geog 5220.13* L'Isle, J.N. de. Explication de la carte des...decouvertes. Paris, 1752. 3 pam.
Geog 5220.15 L'Isle, J.N. de. Erklarung der Charte...Entdeckungen. Berlin, 1753.
Htn Geog 5220.17* Letter from Russian sea-officer. London, 1754.
Htn Geog 5220.19* Müller, G.F. Voyages from Asia to America. London, 1761.
Htn Geog 5220.21* Buache, P. Considerations geographiques. Paris, 1753.
Geog 5220.22 Nansen, F. Nord i tåkeheimen. Kristiania, 1911.
Geog 5220.22.5 Nansen, F. In northern mists, Arctic exploration in early times. London, 1911. 2v.
Geog 5220.22.10 Nansen, F. In northern mists. N.Y., 1911. 2v.
Geog 5220.23 Ahlenius, K. Fornskandinaviska upptäcktsfärder i Nordatlantiska hafvet. Stockholm, 1837.

Geog 5225 Arctic Regions - History of exploration and discoveries - Special periods - 19th century

Geog 5225.5 Force, P. Grinnell land. Washington, 1852.
Geog 5225.7 Smucker, Samuel. Arctic explorations and discoveries. N.Y., 1857.
Geog 5225.8 Mangles, James. Papers and dispatches...Arctic searching expeditions. 2. ed. London, 1852.
Geog 5225.10 Bruun, Daniel. Kampen om nordpolen. Kjøbenhavn, 1902.

Geog 5235 Arctic Regions - History of exploration and discoveries - Special periods - 20th century

Geog 5235.5 American Geographical Society, N.Y. Physical map of the Arctic. N.Y., 1930.
Geog 5235.10 Hayes, J.G. The conquest of the North Pole. N.Y., 1934.
Geog 5235.15 Ushakov, G.A. Po nekhoshenoi zemle. Moskva, 1953.
Geog 5235.20 Euller, John. Arctic world. London, 1958.
Geog 5235.22 Irvine, Tom A. The ice was all between. Toronto, 1959.
Geog 5235.24 Weems, John E. Race for the pole. N.Y., 1960.
Geog 5235.25 Wright, Theon. The big nail; the story of the Cook-Peary feud. N.Y., 1970.

Geog 5300 Arctic Regions - History of exploration and discoveries - By special countries

Geog 5300.5 Campen, S.R. The Dutch in the arctic seas. London, 1877.
Geog 5300.6 Cronheim, P. Fahrten und Forschunger der Holländer in die Polargebieten. Leipzig, 1913.
Geog 5300.7 DePeyster, J.W. The Dutch at the North Pole. N.Y., 1857.
Geog 5300.9 Nourse, J.E. American explorations in the ice zones. Boston, 1884.
Geog 5300.11 Brontman, L.K. On top of the world. N.Y., 1938.
Geog 5300.25 Muratov. M.V. Pervye razvedchiki velikogo puti. Moskva, 1943.
Geog 5300.30 Burkhanov, V.F. Novye sovetskie issledovannia v Arktike. Moskva, 1955.
Geog 5300.35 Caswell, J.E. Arctic frontiers. 1. ed. Norman, 1956.
Geog 5300.40 Zavalti, Silvio. Pionieri italiani nelle regioni polari. Brescia, 1952.
Geog 5300.45 Armstrong, Terence. The Russians in the Arctic. London, 1958.
Geog 5300.46 Akkuratov, Valentin I. Pravo na risk. Moskva, 1974.
Geog 5300.47 Gordienko, Pavel A. Die Polarforschung der Sowjetunion. Düsseldorf, 1967.
Geog 5300.48 Pastskii, Vasilii M. Arkticheskie puteshevstviia rossiian. Moskva, 1974.
Geog 5300.50 Vodop'ianov, M.V. Puti otvazhnykh. Moskva, 1958.
Geog 5300.52 National Research Council. Committee on Polar Research. Science in the Arctic Ocean basin. Washington, 1963.
Geog 5300.56 Pasetskii, Vasilii M. Eestist pärit Arktika-uurijad. Tallinn, 1970.

Geog 5311 - 5336 Arctic Regions - Biographies of explorers - Individual (A-Z by person)

Geog 5311.1.3 Amundsen, Roald. My life as an explorer. Garden City, N.Y., 1928.
Geog 5311.3.5 Arnesen, Odd. Roald Amundsen som hon var. 4. Oppl. Oslo, 1946.
Geog 5311.3.10 Partridge, B. Amundsen. London, 1953.
Geog 5311.3.15 Amundsen, Roald. Apdagelsesreiser. Oslo, 1928-30. 4v.
Geog 5311.3.20 Hovdenak, Gunnar. Roald Amundsen siste fere. Oslo, 1934.
Geog 5311.3.25 Malfatti, Alberto. Roald Amundsen. Roma, 1959.
Geog 5311.3.30 Centkiewicz, Alina. Człowiek o Którego upomniało się morze. Warszawa, 1966.
Geog 5312.2A Bartlett, Robert A. The log of Bob Bartlett. N.Y., 1928.
Geog 5312.2B Bartlett, Robert A. The log of Bob Bartlett. N.Y., 1928.
Geog 5312.2.5 Putnam, George. Mariner of the North. N.Y., 1947.
Geog 5312.3 Pamphlet vol. Geography. Arctic regions. Biography. Richard E. Byrd.
Geog 5312.3.5 Foster, C. Rear Admiral Byrd and the polar expeditions. N.Y., 1930.
Geog 5312.3.10 Hoyt, Edwin Palmer. The last explorer; the adventures of Admiral Byrd. N.Y., 1968.
Geog 5312.3.15A Murphy, Charles J.V. Struggle; the life and exploits of Commander Richard E. Byrd. N.Y., 1928.

Geog 5311 - 5336 Arctic Regions - Biographies of explorers - Individual (A-Z by person) - cont.

Geog 5312.3.15B Murphy, Charles J.V. Struggle; the life and exploits of Commander Richard E. Byrd. N.Y., 1928.
Geog 5312.4 Bolotnekov, N. Nikifor Belichev. 2. izd. Moskva, 1954.
Geog 5312.6 Blosseville, B. Ernest Poret. Jules de Blosseville. Évreux, 1954.
Geog 5312.7 Lauridsen, Peter. Vitus J. Bering og de russiske opdagelsesrejser fra 1725-43. Kjøbenhavn, 1885.
Geog 5312.7.5 Lauridsen, Peter. Vitus Bering; the discoverer of Bering Strait. Chicago, 1889.
Geog 5312.7.10A Golder, Frank A. Bering's voyages. N.Y., 1922-25. 2v.
Geog 5312.7.10B Golder, Frank A. Bering's voyages. N.Y., 1922-25. 2v.
Geog 5312.7.10C Golder, Frank A. Bering's voyages. v.2. N.Y., 1922-25.
Geog 5312.7.15 Goodhue, Cornelia. Journey into the fog; the story of Vitus Bering and the Bering Sea. Garden City, N.Y., 1944.
Geog 5312.7.20 Pokrovskii, Aleksei A. Ekspeditsiia Beringa. Moskva, 1941.
Geog 5312.7.25 Murphy, Robert William. The haunted journey. 1. ed. Garden City, 1961.
Geog 5313.2 Eames, Hugh. Winner lose all; Dr. Cook and the theft of the North Pole. Boston, 1973.
Geog 5316.1A Freuchen, P. Arctic adventure; my life in the frozen North. N.Y., 1935.
Geog 5316.1B Freuchen, P. Arctic adventure; my life in the frozen North. N.Y., 1935.
Geog 5316.1.5A Freuchen, P. Vagrant Viking. N.Y., 1953.
Geog 5316.1.5B Freuchen, P. Vagrant Viking. N.Y., 1953.
Geog 5316.1.6 Freuchen, P. Vagrant Viking. London, 1954.
Geog 5316.1.8 Freuchen, P. Min gronlandske ungdom. Kjøbenhavn, 1959.
Geog 5316.1.10 Freuchen, P. It's all adventure. London, 1938.
Geog 5316.2 Fraser, Robert J. Arctic adventurer. Lincoln, Ont., 1972.
Geog 5316.3 Baetzkes, Ottile G. Sir Martin Frobisher's search for the Northwest Passage. N.Y., 1964.
Geog 5317.5 Johann Georg, 1709-1755. München, 1911.
Geog 5317.10 Greely, A.W. Reminiscences of adventure and service. N.Y., 1927.
Geog 5317.10.5 Mitchell, William. General Greely; the story of a great American. N.Y., 1936.
Geog 5317.15 Gal'perin, I.M. Poliarnye zori; zapiski zhurnaliste. Moskva, 1956.
Geog 5318.1 Hendrik, Hans. Memoirs of Hans Hendrik, the Arctic traveller. London, 1878.
Geog 5318.2 Miller, F. Ahdoolo: the biography of M.A. Henson. 1. ed. N.Y., 1963.
Geog 5318.2.5 Robinson, Bradley. Dark companion. N.Y., 1947.
Geog 5318.2.10 Henson, Matthew Alexander. A Negro explorer at the North Pole. N.Y., 1969.
Geog 5321.1 United States. Naval Observatory. Reports on medals to Arctic explorers, Kane, Hayes. Washington, 1876.
Geog 5321.1.3 Elder, William. Biography of Elisha Kent Kane. Philadelphia, 1858.
Geog 5321.1.4 Elder, William. Biography of Elisha Kent Kane. Philadelphia, 1858.
Geog 5321.1.5 Shields, C.W. Funeral eulogy at obsequies of Dr. Kane. Philadelphia, 1857.
Geog 5321.1.6 Corner, George Washington. Doctor Kane of the Arctic seas. Philadelphia, 1972.
Geog 5321.1.7 Allen, J.H. Elisha Kent Kane. Bangor, 1857.
Geog 5321.1.8 The love life of Dr. Kane. N.Y., 1866.
Geog 5321.1.9 Kane Monument. Act of incorporation, preamble. N.Y., 1859.
Geog 5321.1.11 Kane, Elisha Kent. Access to an open polar sea. N.Y., 1853.
Geog 5321.1.15 Andrews, E.W. Memoir and eulogy of Dr. Elisha K. Kane. N.Y., 1857.
Geog 5321.1.17 Mirsky, J. Elisha Kent Kane and the seafaring frontier. 1. ed. Boston, 1954.
Geog 5321.2 Klengenberg, C. Klengenberg of the Arctic. London, 1932.
Geog 5322.1 Perevalov, V. Lomonosov i Arktika. Moskva, 1949.
Geog 5323.1.2 Markham, C. Life of admiral Sir Leopold McClintock. London, 1909.
Geog 5323.2 Mikkelsen, Ejnar. Mirage in the Arctic. London, 1955.
Geog 5323.2.5 Mikkelsen, Ejnar. Ukendt mand til ukendt land. Kjøbenhavn, 1954.
Geog 5323.3 Woel, Cai Magens. Hilsen til Ejnar Mikkelsen. Kjøbenhavn, 1930.
Geog 5323.5 Hansen, T. Jens Munk. 4. Opl. Kjøbenhavn, 1966.
Geog 5324.1.2 Leslie, A. Arctic voyages of Adolf Erik Nordenskiöld. London, 1879.
Geog 5324.1.5 Anderson, G. A.E. Nordenskiöld. In memoriam. Stockholm, 1901.
Geog 5324.1.10 Ramsey, H. Nordenskiöld. Stockholm, 1950.
Geog 5324.1.15 Kish, George. North-east passage: Adolf Erik Nordenskiöld, his life and times. Amsterdam, 1973.
Geog 5324.2 Enzberg, E. von. Fridtjof Nansen. Dresden, 1898.
Geog 5324.2.5 Bull, Jacob B. Fridtjof Nansen, a book for the young. Boston, 1903.
Geog 5324.2.10 Sörensen, Jon. The saga of Fridtjof Nansen. N.Y., 1932.
Geog 5324.2.15 Reynolds, E.E. Nansen. London, 1932.
Geog 5324.2.20 Turley, C. Nansen of Norway. London, 1933.
Geog 5324.2.25 Risteheuber, René. La double aventure de Fridtjof Nansen. Montréal, 1945.
Geog 5324.2.30 Byström, D.G.V. Fridtjof Nansen. Stockholm, 1940.
Geog 5324.2.37 Wetterfors, Paul. Fridtjof Nansen. Uppsala, 1932.
Geog 5324.2.40 Nansen, Fridtjof. Nansens røst. Oslo, 1945. 3v.
Geog 5324.2.45 Dolman, Frederick. Dr. Nansen. London, 1897.
Geog 5324.2.50 Øverås, Asbjørn. Fridtjof Nansen. Stavanger, 1946.
Geog 5324.2.55 Sponsel, H. Fridtjof Nansen. Nürnberg, 1952.
Geog 5324.2.60 Whitehouse, J.H. Nansen. London, 1930.
Geog 5324.2.65 Nockher, L. Fridtjof Nansen. Stuttgart, 1955.
Geog 5324.2.69 Hoeyer, Liv Nansen. Eva og Fridtjof Nansen. Oslo, 1954.
Geog 5324.2.70 Hoeyer, Liv Nansen. Eva og Fridtjof Nansen. 5. Oppl. Oslo, 1955.
Geog 5324.2.75 Hoeyer, Liv Nansen. Nansen og verden. 2. Oppl. Oslo, 1955.
Geog 5324.2.80 Gal'perin, I.M. On byl pervym. Moskva, 1958.
Geog 5324.2.85 Knuth, Eigil. Fridtjof Nansen og Knud Rasmussen. Kjøbenhavn, 1948.
Geog 5324.2.90 Nansen, Fridtjof. Brev utgitt for Nansenfondet av Steinar Kjaerheim. Oslo, 1961- 4v.
Geog 5324.2.100 Fridtjof Nansen minneforelesninger. Oslo.
Geog 5324.2.105 Greve, Tim. Fridtjof Nansen. Oslo, 1973-74. 2v.
Geog 5326.15.5 Parry A. Parry of the Arctic. London, 1963.
Geog 5326.21A Green, Fitzhugh. Peary; the man who refused to fail. N.Y., 1926.

Classified Listing

Geog 5311 - 5336 Arctic Regions - Biographies of explorers - Individual (A-Z by person) - cont.

Geog 5326.21B Green, Fitzhugh. Peary; the man who refused to fail. N.Y., 1926.
Geog 5326.21.5 Hobbs, William H. Peary. N.Y., 1936.
Geog 5326.21.10 Buck, M. Memorial book of Robert Edwin Peary. n.p., 1937.
Geog 5326.21.15 Hayes, J.G. Robert Edwin Peary, a record of his explorations, 1886-1909. London, 1929.
Geog 5326.21.20 Weems, John Edward. Peary, the explorer and the man, based on his personal papers. Boston, 1967.
Geog 5328.3 Rasmussen, Knud. Mindeudgave. Kjøbenhavn, 1934-35. 3v.
Geog 5328.5 Birket-Smith, Kaj. Knud Rasmussens saga. Kjøbenhavn, 1941.
Geog 5328.5.5 Rasmussen, Knud. In der Heimat des Polarmenschen. Leipzig, 1922.
Geog 5328.10 Freuchen, Peter. I sailed with Rasmussen. N.Y., 1958.
Geog 5328.15 Bogen om Knud. Skrevet af hans venner. Kjøbenhavn, 1943.
Geog 5328.17 Vejlager, Johannes. Knud Rasmussen. Kjøbenhavn, 1934.
Geog 5328.20 Arnesen, Odd. Eskimoenes venn Knud Rasmussen. Oslo, 1944.
Geog 5328.27 Hare, Kirsten. Kender du Knud? 2. Udg. Kjøbenhavn, 1971.
Geog 5328.50 Dodge, Ernest S. The polar Rosses: John and James Clark Ross and their explorations. London, 1973.
Geog 5329.1 Scoresby, R.E. Life of William Scoresby. London, 1861.
Geog 5329.3A Vilhjalmur Stefansson. N.Y., 1925.
Geog 5329.3B Vilhjalmur Stefansson. N.Y., 1925.
Geog 5329.3.2 Vilhjalmur Stefansson. N.Y., 1929.
Geog 5329.3.5 Finnbogason, G. Vilhjalmur Stefansson. Akureyri, 1927.
Geog 5329.3.7 Ol'khina, Evgeniia A. Vil'iamur Stefanson. Moskva, 1970.
Geog 5329.3.10 Stefansson, Vilhjalmur. Discovery. 1. ed. N.Y., 1964.
Geog 5330.1 Vittenburg, P.V. Zhizn' i nauchnaia deiatel'nost' E.V. Tollia. Leningrad, 1960.
Geog 5332.1 Davydov, Iurii V. Ferdinand Vrangel'. Moskva, 1959.
Geog 5333.5 Scott, J.M. Gino Watkins. London, 1935.
Geog 5333.15 Grierson, John. Sir Hubert Wilkins. London, 1960.
Geog 5333.20 Thomas, Lowell. Sir Hubert Wilkins. N.Y., 1961.

Geog 5340 Arctic Regions - Collected Arctic voyages

Geog 5340.5F In mezzo ai ghiacci. Milano, 1880.
Htn Geog 5340.7* Drie voyagien gedaen na Groenlandt. Amsterdam, n.d.
Htn Geog 5340.8F* Scoresby, William. Seven log-books concerning the Arctic voyages. N.Y., 1917. 8v.

Geog 5350 Arctic Regions - Northeast and Northwest Passages together

Geog 5350.5 Alexander, P.F. The north-west and north-east passages. Cambridge, 1915.

Geog 5365 - 5370 Arctic Regions - Eastern hemisphere - Northeast Passage - General works (By date)

Htn Geog 5367.61.3* Müller, G.F. Voyages from Asia to America. 2. ed. London, 1764.
Htn Geog 5367.65* Engel, S. Memoires...geographiques...pays septentrionaux. Lausanne, 1765.
Geog 5367.65.3 Engel, S. Geographische...Nachrichten. Mietau, 1772.
Geog 5367.65.5 Engel, S. Neuer Versuch über...Schriften. Basel, 1777.
Geog 5367.65.7 Engel, S. Extraits...des voyages...de l'Asie. Lausanne, 1779.
Geog 5368.80 Stuxberg, A. Nordostpassagens historia. Stockholm, 1880.
Geog 5369.52 Armstrong, T. The northern sea route; Soviet exploitation of the North East Passage. Cambridge, Eng., 1952.
Geog 5369.56 Istoriia otkrytiia i osvoeniia...morskogo poeti. v.1-3. Moskva, 1956- 4v.
Geog 5369.70 Sovetskaia Arktika; moria i ostrava Severnogo Ledovitogo okeana. Moskva, 1970.

Geog 5375 - 5380 Arctic Regions - Eastern hemisphere - Northeast Passage - Special expeditions (By date of voyage)

Htn Geog 5375.72* Röslin, H. Mitternächtige Schiffarth von den Herrn Staden. Oppenheim, 1611.
Htn Geog 5375.94* Veer, G. de. Oost-indische ende west-indische voyagien. Amsterdam, 1619.
Htn Geog 5375.94.3F* Veer, G. de. Vraye description de trois voyages de mer tres admirables. Amsterdam, 1600.
Htn Geog 5375.94.4* Veer, G. de. Les trois navigations admirables. Paris, 1610.
Geog 5375.94.5 Veer, G. de. Reizen van Willem Barents. 's-Gravenhage, 1917. 2v.
Geog 5375.94.11 Veer, G. de. The true and perfect description of three voyages by the ships of Holland and Zeland. Facsimile. Amsterdam, 1970.
Geog 5377.74 Journal of a voyage...by the Honorable Commodore Phipps, and Captain Lutwidge. London, 1774. 3 pam.
Htn Geog 5377.95* Broughton, William. A voyage of discovery...North Pacific Ocean. London, 1804.
Htn Geog 5377.95.2* Broughton, William. A voyage of discovery...North Pacific Ocean. London, 1804.
Geog 5378.18 Beechey, F.W. A voyage of discovery towards the North Pole in H.M. ships Dorothea and Trent...1818. London, 1843.
Geog 5378.80 Nordenskiöld, A.E. Voyage of the Vega round Asia and Europe. London, 1881. 2v.
Geog 5378.80.2 Nordenskiöld, A.E. The voyage of the Vega round Asia and Europe. N.Y., 1882.
Geog 5378.80.3 Hovgaard, A.P. Nordenskiöld's Voyage round Asia. London, 1882.
Geog 5378.80.4 Hovgaard, A.P. Nordenskiöld's Rejse omkring Asien og Europa. Kjøbenhavn, 1881.
Geog 5378.80.5 Nordenskiöld, A.E. Vegas färd rking Asien och Europa. Stockholm, 1880-81. 2v.
Geog 5378.80.7 Nordenskiöld, A.E. Studien und Forschungen...Reisen. Leipzig, 1885.
Geog 5378.80.9 Nordenskiöld, A.E. Lettres...du passage du pôle nord. Paris, 188-.
Geog 5378.80.11 Nordenskiöld, A.E. Nordenskiölds Vegafahrt um Asien und Europa. Leipzig, 1886.
Geog 5378.80.13 Nordenskiöld, A.E. Shvedskaia poliarnaia ekspeditsiia, 1878-79 g. Sankt Peterburg, 1880.
Geog 5378.80.20 Skoog, Gösta. Vega. En aktualisering av händelserna kring Vega-expeditionen 1878-1880. Göteborg, 1965.
Geog 5379.18 Amundsen, Roald. Nordostpassagen. Kristiania, 1921.
Geog 5379.35 Cheliuskin Expedition, 1933-34. The voyage of the Chelyuskin. N.Y., 1935.
Geog 5379.35.3 Kak my spasali chelnoskiktsev. Moskva, 1934.
Geog 5379.35.5 Khmyznikov, P. Na Chelnoskike. Leningrad, 1936.
Geog 5379.67 Petrow, Richard. Across the top of Russia. N.Y., 1967.
Geog 5379.67.5 Karalt, Charles. To the top of the world. 1. ed. N.Y., 1968.

Geog 5395 - 5400 Arctic Regions - Eastern hemisphere - Polar expeditions (By date of expedition)

Htn Geog 5397.73* Phipps, J. Journal of voyage...North Pole. London, 1774.
Geog 5397.73.3 Phipps, J. Voyage...North Pole...1773. London, 1774.
Geog 5397.73.5 Phipps, J. Voyages au pôle boréal. Paris, 1775.
Geog 5397.73.7 Phipps, J. Reise nach dem Nordpol. Berlin, 1777.
Geog 5398.20.3 Wrangell, F. von. Reise des...Flotten-Lieutenants. Berlin, 1839. 2v.
Geog 5398.20.5 Wrangell, F. von. Narrative of an expedition to the Polar Sea. London, 1840.
Geog 5398.20.7 Wrangell, F. von. Narrative of an expedition to the Polar Sea. London, 1844.
Geog 5398.20.10 Wrangell, F. von. Statistische...Nachrichten über...Besetzungen. St. Petersburg, 1839.
Geog 5398.20.12 Wrangell, F. von. Ferdinand von Wrangell und seine Reiselängs der Nordküste von Sibirien. Leipzig, 1885.
Geog 5398.20.20 Wrangell, F. von. Le nord de la Sibérie. Paris, 1843. 2v.
Geog 5398.20.25 Wrangell, F. von. Le nord de la Sibérie. Limoges, 188-?
Geog 5398.25 Beechey, F.W. Narrative of voyage to the Pacific. London, 1831. 2v.
Htn Geog 5398.50F* Browne, W.H. Ten coloured views during the Arctic expedition. London, 1850.
Geog 5398.56 Dufferin and Ava, F.T. Blackwood. Un voyage en yacht, lettres de hautes latitudes. Montreal, 1876.
Geog 5398.56.5 Dufferin and Ava, F.T. Blackwood. A yacht voyage; letters from high latitudes. N.Y., 1878.
Geog 5398.67 Wheildon, W.W. The new Arctic continent. Cambridge, 1869.
Geog 5398.70 Heuglin, M.T. Reisen nach dem Nordpolarmeer. Braunschweig, 1872-74. 3v.
Geog 5398.71 Brynjulfson, A. Have de gamle Nordboer havt Kjendskab. Kjøbenhavn, 1871.
Geog 5398.72 Payer, J. New Lands within the Arctic Circle. London, 1876. 2v.
Geog 5398.72.3 Payer, J. Die österreichisch-ungarische Nordpol-Expedition. Wien, 1876.
Geog 5398.72.5 Payer, J. New lands within the Arctic Circle. N.Y., 1877.
Geog 5398.72.7 Payer, J. L'expédition du Tegetthoff. Paris, 1878.
Geog 5398.72.11 Krisch, Otto. Das Tagebuch des Maschinisten Otto Krisch. Graz, 1973.
Geog 5398.72.12 Hundert Jahre Franz Josefs-Land. Zur Erinnerung an die Entdeckungsreise die österreichisch-ungarische Nordpol-Expedition, 1872-1874. Wien, 1973.
Geog 5398.75 Stuxberg, A. Einringar fran Svenska Expeditionerna. Stockholm, 1877.
Geog 5398.75.3 Young, Allen. The two voyages of the Pandora in 1875-76. London, 1879.
Geog 5398.79 Markham, A.H. Polar reconnaissance. London, 1881.
Geog 5398.79.5 Danenhower. Narrative of the "Jeannette". Boston, 1882.
Geog 5398.79.7 De Long, G.W. Voyage of the "Jeannette". Boston, 1883. 2v.
Geog 5398.79.9F N.Y. Herald. The Jeannette in the Arctic regions. N.Y., 1882.
Geog 5398.79.11 Soley, J.R. Address...at unveiling of Jeannette monument. Baltimore, 1891.
Geog 5398.79.15 United States. Navy Department. Letter from the Secretary of the Navy. Washington, 1884.
Geog 5398.79.20 History of the adventurous voyage and terrible shipwreck of the United States steamer "Jeannette" in polar seas. N.Y., 1882.
Geog 5398.79.25 Bliss, R.W. Our lost explorers; a narrative of the Jeannette Arctic expedition. Hartford, 1882.
Geog 5398.79.30A Ellsberg, Edward. Hell on ice; the saga of the "Jeannette". N.Y., 1938.
Geog 5398.79.30B Ellsberg, Edward. Hell on ice; the saga of the "Jeannette". N.Y., 1938.
Geog 5398.79.35 Hoehling, Adolph A. The Jeannette expedition, an ill fated journey to the Arctic. London, 1969.
Geog 5398.79.40 Geslin, Jules. L'expédition de la Jeannette au pôle nord, racontée par tous les membres de l'expédition. Paris, 1883? 2v.
Geog 5398.80 Markham, C.R. Voyage of the 'Eira'. London, 1881.
Geog 5398.81 Gilder, W.H. Ice-pack and tundra. London, 1883.
Geog 5398.81.5 Melville, G.W. In the Lena Delta. Boston, 1892.
Geog 5398.82 Hovgaard, A.P. Dijmphna's expeditionen, 1882-83. Kjøbenhavn, 1884.
Geog 5398.90 Clutterbuck, W.J. The skipper in Arctic seas. London, 1890.
Geog 5398.91 Keely, Robert. In Arctic seas...voyage of the "Kite". Philadelphia, 1892.
Geog 5398.93 Nansen, Fridtjof. Farthest north. Westminster, 1897. 2v.
Geog 5398.93.1 Nansen, Fridtjof. Farthest north. N.Y., 1897. 2v.
Geog 5398.93.3 Johansen, H. With Nansen in the north. London, 1899.
Geog 5398.93.5 Nansen, Fridtjof. In Nacht undEis. Leipzig, 1897. 2v.
Geog 5398.93.15 Plass, F. Reise-Erinnerungen...Aug. 1893..."Admiral". Hamburg, 1894.
Geog 5398.97 Lachambre, Henri. Andrée; au pôle nord en ballon. Paris, 1897.
Geog 5398.97.3 Fonvielle, W. La vérité sur l'expédition Andrée. Strasbourg, 1897.
Geog 5398.97.4 Svenska Sällskapet för Antropologi och Geografi. Med örnen mot polen av S.A. Andrée. Stockholm, 1930.
Geog 5398.97.5 Andrée, S.A. Andrée's story; the complete record of his polar flight, 1897. N.Y., 1930.
Geog 5398.97.6 Sundman, Per Olaf. Ingen fruktan, intet hopp. Stockholm, 1968.
Geog 5398.97.10 Putnam, George Palmer. Andrée. N.Y., 1930.
Geog 5398.98 Sverdrup, O. Neues Land, vier Jahre in arktischen Regionen. Leipzig, 1903. 2v.
Geog 5398.98.2 Sverdrup, O. New land, four years in Arctic regions. London, 1904. 2v.
Geog 5398.98.5 Nathorst, A.G. Tva somrar, norra ishafvet. Stockholm, 1900. 2v.
Geog 5398.99 Amadeus, Louis. Farther north than Nansen. London, 1901.
Geog 5398.99.3 Amadeus, Louis. On the "Polar Star". N.Y., 1903. 2v.
Geog 5398.99.5 Amadeus, Louis. Die Stella Polare im Eismeer. Leipzig, 1903.
Geog 5399.01 Makarov, S.O. S.O. Makarov v zavosvanie Ariuku. Moskva, 1943.
Geog 5399.03 Fiala, A. Fighting the polar ice. N.Y., 1906.
Geog 5399.03.2 Fiala, A. Fighting the polar ice. 2. ed. N.Y., 1907.
Geog 5399.06 Holmes, B.F. The log of the "Laura" in polar seas. Cambridge, 1907.
Geog 5399.06.9A Stefansson, V. Hunters of the great North. N.Y., 1922.
Geog 5399.06.9B Stefansson, V. Hunters of the great North. N.Y., 1922.

Classified Listing 65

Geog 5395 - 5400 Arctic Regions - Eastern hemisphere - Polar expeditions (By date of expedition) - cont.

Geog 5399.08F The Norwegian Aurora polaris expedition. Christiania, 1908.
Geog 5399.12 Al'banov, V.A. Au pays de la mort blanche. Paris, 1928.
Geog 5399.12.5 Al'banov, V.A. Mezhdu zhuznbiu v Smertbiu. Berlin, 1925.
Geog 5399.12.10 Al'banov, V.A. Podvig shturmaka V.I. Al'banova. Moskva, 1953.
Geog 5399.14 Stefansson, V. The friendly Arctic. N.Y. 1922.
Geog 5399.21 Stefansson, V. The friendly Arctic. N.Y., 1921.
Geog 5399.21.2 Stefansson, V. The friendly Arctic. N.Y., 1943.
Geog 5399.22 Ashton, James M. Ice-bound, a trader's adventure in the Siberian Arctic. N.Y., 1928.
Geog 5399.25 Stefansson, V. The adventure of Wrangel Island. N.Y., 1925.
Geog 5399.25.5 Ellsworth, L. The hop-off. N.Y., 1925.
Geog 5399.25.9 Amundsen, R. Amundsen-Ellsworths polflyvning, 1925. Oslo, 1925.
Geog 5399.25.10 Amundsen, R. Our polar flight. N.Y., 1925.
Geog 5399.25.19 Amundsen, R. Den første flukt over Polhavet. Oslo, 1926.
Geog 5399.25.20A Amundsen, R. First crossing of the Polar Sea. N.Y., 1927.
Geog 5399.25.20B Amundsen, R. First crossing of the Polar Sea. N.Y., 1927.
Geog 5399.25.21 Amundsen, R. First crossing of the Polar Sea. Garden City, 1928.
Geog 5399.25.35 Rasmussen, K. Fra Grønland til Stillehavet. København, 1925-26. 2v.
Geog 5399.27.3 Worsley, F.A. Under sail in the frozen north. London, 1927.
Geog 5399.28 Lee, Herbert P. Policing the top of the world. London, 1928.
Geog 5399.28.5 Giudici, Davide. The tragedy of the Italian with the rescuers to the Red tent. N.Y., 1929.
Geog 5399.28.10 Nobile, Umberto. L'"Italia" al Polo Nord. Milano, 1930.
Geog 5399.28.15 Nobile, Umberto. With the "Italia" to the North Pole. London, 1930.
Geog 5399.28.20 Nobile, Umberto. Met de "Italia" naar de Noordpool. Baarn, 1931.
Geog 5399.28.25 Běhounek, Franz. Sieben Wochen auf der Eisscholle. Leipzig, 1929.
Geog 5399.28.30 Viglieri, Alfredo. Quarantotto giorni sul pack. Milano, 1929.
Geog 5399.28.35 Hogg, Garry. Airship over the Pole; the story of the Italia. London, 1969.
Geog 5399.28.40 Katz, Otto. Neun Männer im Eis. Berlin, 1929.
Geog 5399.28.45 Cross, Wilbur. Ghost ship of the Pole. N.Y., 1960.
Geog 5399.28.52 Tomaselli, Francesco. L'inferno bianco. 4. ed. Milano, 1929.
Geog 5399.31 Meyer, W. Der Kampf um Nobile. Berlin, 1931.
Geog 5399.31.2 Itin, Vivian A. Vykhod k moriu. 2. izd. Novosibirsk, 1935.
Geog 5399.31.5 Sieburg, F. Die rote Arktis. Frankfurt, 1932.
Geog 5399.32 Kalmár, G. Kuz delmek a fehér halál országában. Budapest, 1932-33. 2v.
Geog 5399.32.5 Vize, Vladimir I. Na "Sibiriakove" v "Litke"...1932 i 1934. Moskva, 1946.
Geog 5399.32.6 Vize, Vladimir I. Ne "Sibirakove" v Tikhii okean. Leningrad, 1934.
Geog 5399.32.8 Grigor'ev, Origonii K. Dorogi vedut v Arktiku. Moskva, 1969.
Geog 5399.32.10 Shneiderov, V.A. Velikim Severnym. 2. izd. Moskva, 1963.
Geog 5399.34.5 Morozov, S. Put' skvoz l'dy. Moskva, 1934.
Geog 5399.35 Glen, A.R. Under the Pole star. London, 1937.
Geog 5399.36 Vize, Vladimir I. Moria Sovetskoi Arktiki. Leningrad, 1936.
Geog 5399.37 Laktionov, A.F. Severnyi polius. Arkhangelsk, 1939.
Geog 5399.37.5 Ekspeditsiia SSSR na Severnyi Polius, 1937. Trudy. Leningrad, 1940.
Geog 5399.37.7 Laktionov, A.F. Severnyi polius. Moskva, 1955.
Geog 5399.37.10 Kreukel', E.T. Chetyre tovarishcha. Moskva, 1940.
Geog 5399.46 Mineev, A.I. Ostrov Vrangelia. Moskva, 1946.
Geog 5399.48 Vibe, Christian. Laugthen og nordpaa. København, 1948.
Geog 5399.53 Hansen, Leo. I Knuds sloedespor. København, 1953.
Geog 5399.56 Centkiewicz, Alina. Zanaevanie Arktiki. Moskva, 1956.
Geog 5399.57.5 Iatsun, E.P. Na l'dine cherez polius. Moskva, 1957.
Geog 5399.57.10 Ruzov, L.V. Na sushe i na more v Arktike. Moskva, 1957.
Geog 5399.57.15 Sushkina, N.N. Dva leta v Arktike. Moskva, 1957.
Geog 5399.57.20 Freuchen, Peter. Fra Thule til Rio. København, 1957.
Geog 5399.57.25 Belov, M.I. Severnyi morskoi put. Leningrad, 1957.
Geog 5399.58 Rasmussen, Knud. Den store slaederejse. København, 1958.
Geog 5399.58.10 Problemy Severa. Moskva. 1-18 5v.
Geog 5399.60.5 Kudenko, O.I. Teplaia Arktika. Moskva, 1960.
Geog 5399.61 Gromov, L.V. Ostrov Vrangelia. Magadan, 1961.
Geog 5399.61.5 Reut, V.F. Moskva - Severnyi Polius. Moskva, 1961.
Geog 5399.64 Report from the Arctic; foreign and Soviet correspondents on their trip aboard the Soviet atomic icebreaker. Moscow, 1964.
Geog 5399.64.1 Report from the Arctic; foreign and Soviet correspondents on their trip aboard the Soviet atomic icebreaker. Moscow, 1968.
Geog 5399.72 Chilingarov, A.N. Pod nogami ostrov ledianoi. Moskva, 1972.
Geog 5399.73 Krenkel', Ernst T. RAEM - moi pozyvnye. Moskva, 1973.

Geog 5505 - 5510 Arctic Regions - Western hemisphere - Northwest passage - General works (By date)

Geog 5506.35.1 Fox, Luke. North-west Fox. n.p., 1965.
Htn Geog 5507.45* Description of coast...in Button's Bay. London, 1745?
Geog 5507.45.6 North West Committee. Articles of agreement. London, 1745.
Htn Geog 5507.49* Reasons...navigable passage to west American oceans. London, 1749.
Htn Geog 5507.82* Pickersgill, R. Concise account of voyages...discovery. London, 1782.
Geog 5507.87 LaRoche, F.C. de. Sendschreiben...des alten Grönlands. Kopenhagen, 1787.
Htn Geog 5507.93* Goldson, W. Observations on passage...Atlantic and Pacific. Portsmouth, 1793.
Geog 5507.98F Disertaciones sobre la navegación. Isla de León, 1798.
Htn Geog 5508.29* Northern regions. Boston, 1829.
Geog 5508.31 Snelling, W.J. Polar regions of the western continent. Boston, 1831.
Geog 5508.36 Tytler, P.F. Historical view of discovery...of America. N.Y., 1836.
Geog 5508.38 Hülstett, G.K.A. Über die nordwestliche Durchfahrt. Düsseldorf, 1838.
Geog 5508.51 Simmonds, P.L. Sir John Franklin and Arctic regions. London, 1851.

Geog 5505 - 5510 Arctic Regions - Western hemisphere - Northwest passage - General works (By date) - cont.

Geog 5508.51.5 Simmonds, P.L. Sir John Franklin and Arctic regions. 3. ed. London, 1853.
Geog 5509.07 Randall, Harry. The conquest of the Northwest Passage. Minneapolis, 1907.
Geog 5509.15 Beyl, Hermann. Der lange Zeit angenomme Zusammenhang der Hudsonbai mit dem Stille Ozean. Würzburg, 1915.
Geog 5509.34 Crouse, N.M. The search for the Northwest Passage. N.Y., 1934.
Geog 5509.58 Neatby, Leslie Hamilton. In quest of the North West Passage. London, 1958.
Geog 5509.58.5 Stefansson, V. Northwest to fortune. N.Y., 1958.
Geog 5509.61 Dodge, Ernest S. Northwest by sea. N.Y., 1961.
Geog 5509.61.5 Dainelli, Giotto. Il passaggio di nord-ouest. Torino, 1961.
Geog 5509.62 Williams, G. The British search for the Northwest Passage in the 18th century. London, 1962.
Geog 5509.66 Neatby, Leslie Hamilton. Conquest of the last frontier. Toronto, 1966.
Geog 5509.70 Keating, Bern. The Northwest Passage: frm the Mathew to Manhattan, 1497 to 1969. Chicago, 1970.
Geog 5509.71 Wilkinson, Doug. Arctic fever. Toronto, 1971.
Geog 5509.75 Thomson, George Malcolm. The North-west passage. London, 1975.

Geog 5515 - 5520 Arctic Regions - Western hemisphere - Northwest passage - Special expeditions (By date of voyage)

Htn Geog 5515.77* Settle, D. True report of Martin Frobisher's voyage, 1577. London, 1577.
Htn Geog 5515.77.5* Best, George. De Martini Forbisseri Angli. n.p., 1580.
Geog 5515.77.15 Best, George. The three voyages of Martin Frobisher in search of a passage to Cathay...1576-78. London, 1938. 2v.
Geog 5515.78 Parks, G.B. Frobisher's third voyage, 1578. n.p., 1935.
Geog 5515.88 Maldonado, L.F. Voyage de la mer atlantique...pacifique. Plaisance, 1812.
Geog 5516.19 Munk, J. Navigatio septentrionalis. Kjøbenhavn, 1723.
Geog 5516.19.3 Munk, J. Navigatio septentrionalis. Kjøbenhavn, 1883.
Geog 5516.19.5 Voyage of Captain John Monk. Glasgow, 1792.
Geog 5516.19.15 Hofrenning, Bernt Monson. Navigatio septentrionalis. Fergus Falls, Minn., 1942.
Htn Geog 5516.31* James, T. Strange and dangerous voyage...N.W. Passage. London, 1633.
Htn Geog 5516.31.2* James, T. Strange and dangerous voyage...N.W. Passage. London, 1633.
Geog 5516.31.3 James, T. The strange and dangerous voyage. Toronto, 1975.
Geog 5516.31.5A Bodilly, R.B. The voyage of Captain Thomas James...1631. London, 1928.
Geog 5516.31.5B Bodilly, R.B. The voyage of Captain Thomas James...1631. London, 1928.
Htn Geog 5516.31.9* Fox, Luke. North-west Fox...voyages of Cabot, Frobisher. London, 1635.
Geog 5516.88 Maldonado, Lorenzo F. Viaggio dal mare atlantico al pacifico. n.p., 1810.
Geog 5517.46.3 Ellis, H. Reise nach Hudsons Meerbusen. Göttingen, 1750.
Geog 5517.46.4 Ellis, H. Voyage à la baye de Hudson. Leide, 1750.
Geog 5517.46.5 Ellis, H. Voyage de la baye de Hudson. Paris, 1749.
Htn Geog 5517.46.10* Drage, T.S. Account of voyage...Northwest Passage. London, 1748-49. 2v.
Geog 5517.46.15 Eavenson, Howard N. Two early works on Arctic exploration. Pittsburgh, 1946.
Geog 5518.16 Thirty years in the Arctic regions. N.Y., 1859.
Geog 5518.18 Ross, J. Voyage of discovery...Baffin's Bay. London, 1819.
Geog 5518.18.3 Ross, J. Voyage vers le Pole Arctique. Paris, 1819.
Geog 5518.18.4 Sabine, Edward. Remarks on the account of the late voyage of discovery to Baffin's Bay. 2. ed. London, 1819.
Geog 5518.18.5 Fisher, Alexander. Journal of a voyage of discovery to the Arctic regions. London, 1819.
Geog 5518.18.7 Fisher, Alexander. Journal of a voyage of discovery to the Arctic regions. London, 1968.
Htn Geog 5518.19* Franklin, John. Narrative of journey to...Polar Sea. London, 1823.
Htn Geog 5518.19.2* Franklin, John. Narrative of journey to...Polar Sea. London, 1823. 2v.
Geog 5518.19.3 Franklin, John. Narrative of journey to...Polar Sea. 2. ed. London, 1824. 2v.
Geog 5518.19.4 Franklin, John. Narrative of journey to...Polar Sea. 3. ed. London, 1824. 2v.
Geog 5518.19.5 Franklin, John. Narrative of journey to...Polar Sea. Philadelphia, 1824.
Geog 5518.19.6 Parry, W.E. The North Georgia Gazette and Winter Chronicle. No.1-20. London, 1821.
Geog 5518.19.6.2 Parry, W.E. Journal of voyage...Northwest Passage and North Georgia Gazette. Philadelphia, 1821.
Geog 5518.19.7 Parry, W.E. Journal of voyage...Northwest Passage and North Georgia Gazette. London, 1821.
Geog 5518.19.8 Parry, W.E. Supplement to the appendix of Capt. Parry's voyage. London, 1824.
Geog 5518.19.10 Fisher, Alexander. Journal of a voyage of discovery. London, 1821.
Geog 5518.19.12 Fisher, Alexander. Journal of a voyage of discovery. London, 1821.
Geog 5518.19.14 Franklin, John. Narrative of a journey to the shores of the Polar Sea in the years 1819-1922. Rutland, 1970.
Geog 5518.19.20 Hood, Robert. To the Arctic by canoe, 1819-1821. Montreal, 1974.
Geog 5518.21 Parry, W.E. Journal of second voyage for discovery. London, 1824.
Geog 5518.21.2 Parry, W.E. Appendix to...Parry's Journal of second voyage. London, 1825.
Geog 5518.21.5 Lyon, G.F. Private journal of Captain G.F. Lyon. Boston, 1824.
Geog 5518.24 Parry, W.E. Journal of third voyage...Northwest Passage. Philadelphia, 1826.
Geog 5518.24.2 Parry, W.E. Journal of third voyage...Northwest Passage. London, 1826.
Geog 5518.24.7 Lyon, G.F. Brief narrative of attempt to reach Repulse Bay. London, 1825.
Geog 5518.25 Franklin, John. Narrative of second expedition to Polar Sea. Philadelphia, 1828.
Htn Geog 5518.25.2F* Franklin, John. Narrative of a second expedition to the shores of the Polar Sea, in the year 1825-1827. London, 1828.

Classified Listing

Geog 5515 - 5520 Arctic Regions - Western hemisphere - Northwest passage - Special expeditions (By date of voyage) - cont.

	Geog 5518.25.5	Franklin, John. Narrative of a second expedition to the shores of the Polar Sea in the years 1825-1827. Rutland, 1971.
	Geog 5518.27	Parry, W.E. Narrative of attempt to reach the North Pole. London, 1828.
	Geog 5518.29	Ross, J. Narrative of 2nd voyage...Northwest Passage. London, 1835.
	Geog 5518.29.2	Ross, J. Appendix to Narrative of second voyage. London, 1835.
	Geog 5518.29.3	Ross, J. Narrative of a second voyage in search of a northwest passage. Philadelphia, 1835.
	Geog 5518.29.5	Huish, Robert. The last voyage of Captain Sir John Ross, R.N., to the Arctic regions. London, 1835.
	Geog 5518.33	Back, George. Narrative of Arctic land expedition. London, 1836.
	Geog 5518.33.3	Back, George. Narrative of Arctic land expedition. Philadelphia, 1836.
	Geog 5518.36	Simpson, Thomas. Narrative of discoveries on north coast. London, 1843.
	Geog 5518.36.2	Simpson, Thomas. Narrative of the discoveries on the north coast of America. 2. ed. Toronto, 1970. 2v.
	Geog 5518.36.3	Simpson, A. Life and travels of Thomas Simpson. London, 1845.
	Geog 5518.36.4	Simpson, A. The life and travels of Thomas Simpson. Toronto, 1963.
	Geog 5518.36.7	Back, George. Narrative of an expedition in H.M.S. Terror. London, 1838.
	Geog 5518.36.9	Back, George. Narrative of the Arctic land expedition to the mouth of the Great Fish River. 2. ed. Philadelphia, 1837.
Htn	Geog 5518.46*	Rae, J. Narrative of expedition to Arctic Sea. London, 1850.
	Geog 5518.50.4	Osborn, S. Discovery of the Northwest Passage. Edinburgh, 1865.
	Geog 5518.50.10	M'Clure, R. Captain M'Clure's despatches from her majesty's discovery ship "Investigator". London, 1852.
	Geog 5518.50.15	Miertsching, Johann August. Frozen ships; the Arctic diary of Johann Miertsching, 1850-1854. N.Y., 1967.
	Geog 5518.92	Kallstenius, Alfhild. Tragedin i Smiths sund; Björling-Kallstenius expeditionen, 1892. Kalmar, 1966.
	Geog 5519.03	Amundsen, R. The North West Passage. London, 1908. 2v.
	Geog 5519.03.1	Amundsen, R. Roald Amundsen's "The North West Passage". N.Y., 1908. 2v.
	Geog 5519.39	Poncins, G. The ghost voyage. 1. ed. Garden City, 1954.
	Geog 5519.40.5	Tranter, G.J. Plowing the Arctic. London, 1944.
	Geog 5519.69	Smith, William D. Northwest Passage. N.Y., 1970.

Geog 5530 Arctic Regions - Western hemisphere - Northwest passage - Sir John Franklin's voyage of 1845-1846 - General works; Biography of Franklin

	Geog 5530.5	Markham, A.H. Life of Sir John Franklin. London, 1891.
	Geog 5530.5.2	Markham, A.H. Life of Sir John Franklin. N.Y., 1891.
	Geog 5530.7	Beesly, A.H. Sir John Frankin. N.Y., 1881.
	Geog 5530.9	Wright, A. Versus Tennysonianos Franklini. Cantabrigiae, 1882.
	Geog 5530.13	Bell, Benjamin. Lieut. John Irving, of HMS "Terror". Edinburgh, 1881.
	Geog 5530.14	Osborn, Sherard. The career, last voyage. London, 1860.
	Geog 5530.15	Franklin, Jane G. The life, diaries and correspondence...1792-1875. London, 1923.
	Geog 5530.15.5	Woodward, F.J. Portrait of Jane. London, 1951.
	Geog 5530.18	Traill, Henry Duff. The life of Sir John Franklin, R.N. London, 1896.
	Geog 5530.75	Gell, E.M.B. (Mrs.). John Franklin's, Eleanor Anne Porden. London, 1930.
	Geog 5530.80	Lamb, G.F. Franklin. London, 1956.
	Geog 5530.82	Nanton, Paul. Arctic breakthrough: Franklin's expeditions, 1819-1847. London, 1971.

Geog 5535 Arctic Regions - Western hemisphere - Northwest passage - Sir John Franklin's voyage of 1845-1846 - Search for Franklin - History

	Geog 5535.10.2	Brown, John. The Northwest Passage, and the plans for the search for Sir John Franklin. 2. ed. London, 1860.
	Geog 5535.15	Brandes, Karl. Sir John Franklin; die Unternehmungen für seine Rettung und die Nordwestliche Durchfahrt. Berlin, 1854.
	Geog 5535.20	Neatby, Leslie Hamilton. The search for Franklin. London, 1970.

Geog 5538 Arctic Regions - Western hemisphere - Northwest passage - Sir John Franklin's voyage of 1845-1846 - Search for Franklin - Special voyages (By date of voyage)

	Geog 5538.48	Richardson, J. Arctic searching expedition. N.Y., 1852.
	Geog 5538.48.5	Richardson, J. Arctic searching expedition. N.Y., 1854.
	Geog 5538.50.2	Osborn, S. Stray leaves from an Arctic journal. N.Y., 1852.
	Geog 5538.50.4	Osborn, S. Stray leaves from an Arctic journal. Edinburgh, 1865.
	Geog 5538.50.10	Osborn, S. The Polar regions...Franklin's expedition. N.Y., 1854.
	Geog 5538.50.15	Collinson, R. Journal of H.M.S. Enterprise...1850-55. London, 1889.
	Geog 5538.50.17	Snow, W.P. Voyage of Prince Albert...Sir John Franklin. London, 1851.
	Geog 5538.50.19	Arctic miscellanies. London, 1852.
	Geog 5538.50.25	Armstrong, Alexander. A personal narrative of the discovery of the Northwest Passage...while in search of the expedition under Sir John Franklin. London, 1857.
	Geog 5538.51	Kane, E.K. United States Grinnell expedition. N.Y., 1854.
	Geog 5538.51.3	Kane, E.K. United States Grinnell expedition. N.Y., 1857.
	Geog 5538.51.5	Kane, E.K. Access to an open Polar Sea. N.Y., 1853.
	Geog 5538.51.7	Bellot, J.R. Journal d'un voyage aux mers polaires...en 1851 et 1852. Paris, 1854.
	Geog 5538.51.9	Bellot, J.R. Memoirs of and journal of voyage...search of Sir John Franklin. London, 1855. 2v.
	Geog 5538.51.15	Kennedy, William. A short narrative of the second voyage of the Prince Albert in search of Sir John Franklin. London, 1853.
	Geog 5538.52	Belcher, E. The last of the Arctic voyages. London, 1855. 2v.
	Geog 5538.52.5	M'Dougall, G.F. Eventful voyage of H.M. discovery ship "Resolute". London, 1857.
	Geog 5538.52.10	Inglefield, Edward A. A summer search for Sir John Franklin. London, 1853.
	Geog 5538.53A	Kane, E.K. Arctic explorations. Philadelphia, 1856. 2v.

Geog 5538 Arctic Regions - Western hemisphere - Northwest passage - Sir John Franklin's voyage of 1845-1846 - Search for Franklin - Special voyages (By date of voyage) - cont.

	Geog 5538.53B	Kane, E.K. Arctic explorations. Philadelphia, 1856. 2v.
	Geog 5538.53.1	Kane, E.K. Arctic explorations. Philadelphia, 1856. 2v.
	Geog 5538.53.2	Kane, E.K. Arctic explorations. Philadelphia, 1856. 2v.
	Geog 5538.53.3	Kane, E.K. Arctic explorations. Hartford, 1881.
	Geog 5538.53.4	Kane, E.K. Arctic voyage. N.Y., 1857.
	Geog 5538.53.5	Sonntag, A. Professor Sonntag's thrilling narrative. Philadelphia, 1857.
	Geog 5538.53.6	Sonntag, A. Professor Sonntag's thrilling narrative. Philadelphia, 1857.
	Geog 5538.53.7	Strangers in Greenland. N.Y., 18- ?
	Geog 5538.54.2	Hayes, I.I. An Arctic boat journey. Boston, 1860.
	Geog 5538.54.5	Hayes, I.I. An Arctic boat journey, in the autumn of 1854. Boston, 1867.
	Geog 5538.54.8	Hayes, I.I. Arctic boat journey, in the autumn of 1854. Boston, 1883.
	Geog 5538.57	M'Clintock, F.L. Voyage of "Fox"...narrative...of fate of Sir John Franklin. Boston, 1860.
	Geog 5538.57.2	M'Clintock, F.L. The voyage of the "Fox" in the Arctic seas. London, 1859.
	Geog 5538.57.3	M'Clintock, F.L. The voyage of the "Fox" in the Arctic seas. Rutland, 1972.
	Geog 5538.57.5	Petersen, Carl. Den sidste Franklin expedition med "Fox". Kjøbenhavn, 1860.
	Geog 5538.78	Klutschak, H.W. Als Eskimo unter den Eskimos; eine Schilderung der Erlebnisse der Schwatka'schen Franklin-Aufsuchungs-Expedition in den Jahren 1878-80. Wien, 1881.
	Geog 5538.78.5	Gilder, W.H. Schwatka's search...Franklin records. N.Y., 1881.

Geog 5555 - 5560 Arctic Regions - Western hemisphere - Polar expeditions (By date of expedition)

	Geog 5557.25	Dall, W.H. Critical review of Bering's first expedition, 1725-30. Washington, 1890.
	Geog 5557.25.3	Dall, W.H. Early expeditions to the Bering Sea and Strait. Washington, 1891.
Htn	Geog 5557.25.5*	Berkh, V.N. Pervoe morskoe puteshestvie. Sankt Peterburg, 1823.
	Geog 5557.33	Ostrovskii, B.G. Veliki severnaia ekspeditsiia, 1733-43. Arkhangel'sk, 1935.
	Geog 5557.33.5	Davidson, George. The tracks and landfalls of Bering and Chirikof on the northwest coast of America. San Francisco, 1901.
	Geog 5558.17	O'Reilly, B. Greenland...Northwest Passage. N.Y., 1818.
	Geog 5558.17.3	O'Reilly, B. Greenland...Northwest Passage. London, 1818.
	Geog 5558.18	Parry, W.E. Journal of a voyage of discovery. London, 1820. 2 pam.
Htn	Geog 5558.21*	Manby, G.W. Journal of voyage to Greenland...1821. London, 1822.
	Geog 5558.22	Scoresby, William. Journal of voyage to the northern whale-fishery. Edinburgh, 1823.
	Geog 5558.26	Duncan, D. Arctic regions. Voyage to Davis Strait. London, 1827.
	Geog 5558.50	Miertsching, J.A. Reise-Tagebuch des Missionars. Gnadan, 1855.
	Geog 5558.53	Hayes, I.I. An Arctic boat journey...1854. Boston, 1868.
	Geog 5558.60	Hayes, I.I. The open Polar Sea. N.Y., 1867.
	Geog 5558.60.3	American Geographical and Statistical Society. The polar exploring expedition. N.Y., 1860.
	Geog 5558.64	Nourse, J.E. Narrative of 2nd Arctic expedition. Washington, 1879.
	Geog 5558.69	Verein für die Deutsche Nordpolarfahrt. Leipzig, 1873-74. 4v.
	Geog 5558.69.2	Koldewey, K. German Arctic expedition of 1869-70. London, 1874.
	Geog 5558.69.5	Verein in die Deutsche Nordpolarfahrt. Die zweite deutsche Nordpolarfahrt. Leipzig, 1883.
	Geog 5558.71	Davis, C.H. Narrative of north polar expedition. Washington, 1876.
	Geog 5558.71.5	Blake, E.V. Arctic experiences. N.Y., 1874.
	Geog 5558.75F	Nares, G.S. Arctic expedition 1875-76, journal. London, 1877.
	Geog 5558.75.3	Nares, G.S. Narrative of voyage to Polar Sea...1875-76. London, 1878. 2v.
	Geog 5558.75.7	Markham, A.H. The great frozen sea...1875-76. London, 1878.
	Geog 5558.75.9	MacGahan, J.A. Under the Northern Lights. London, 1876.
	Geog 5558.75.11	Johnston, R. The Arctic expedition of 1875-76. London, 1877.
	Geog 5558.75.13	Kan, C.M. De jongste engelsche pooltocht ende expedities der toekomst. Utrecht, 1877.
	Geog 5558.75.15PF	Moss, Edward L. Shores of the Polar Sea. London, 1878.
	Geog 5558.81F	Unites States. Expedition to Lady Franklin Bay. Report on proceedings. Washington, 1888. 2v.
	Geog 5558.81.3	Greely, A.W. Three years of Arctic service...1881-84. N.Y., 1886. 2v.
	Geog 5558.81.5	United States. Board of Officers. Expedition...Relief of Lieutenant Greely. Report. Washington, 1884.
	Geog 5558.81.7A	Schley, W.S. The rescue of Greely. N.Y., 1885.
	Geog 5558.81.7B	Schley, W.S. The rescue of Greely. N.Y., 1885.
	Geog 5558.81.8F	Schley, W.S. Report...Greely relief expedition of 1884. Washington, 1887.
	Geog 5558.81.9F	Rosse, I.C. Cruise of revenue-steamer Corwin. Washington, 1883.
	Geog 5558.81.10	Greely, A.W. The Greely Arctic expedition. Philadelphia, 1884.
	Geog 5558.81.15	McGinley, W.A. Greely relief expedition. Washington, 1884.
	Geog 5558.81.20	Brainard, David L. The outpost of the lost. Indianapolis, 1929.
	Geog 5558.81.25	Brainard, David L. Six came back; the Arctic adventure. Indianapolis, 1940.
	Geog 5558.81.55	Muir, John. The cruise of the Corwin. Boston, 1917.
	Geog 5558.81.60	Powell, T. The long rescue. 1. ed. Garden City, 1960.
	Geog 5558.81.65	Gauroy, Pierre. Les affamés de la banquise. Paris, 1964.
	Geog 5558.82	Nansen, F. Hunting and adventure in the Arctic. N.Y., 1925.
	Geog 5558.84	Lindsay, D.M. A voyage to the Arctic in the whaler Aurora. Boston, 1911.
	Geog 5558.84.5	Todd, Alden. Abandoned. 1. ed. N.Y., 1961.
	Geog 5558.86	Peary, R.E. Northward over the "Great Ice"...1886. N.Y., 1898. 2v.

Classified Listing

Geog 5555 - 5560 Arctic Regions - Western hemisphere - Polar expeditions (By date of expedition) - cont.

Geog 5558.91A Peary, Josephine D. My Arctic journal. N.Y., 1893.
Geog 5558.91B Peary, Josephine D. My Arctic journal. N.Y., 1893.
Geog 5558.91.3 Astrup, E. With Peary near the Pole. London, 1894.
Geog 5558.91.4 Astrup, E. Blandt Nordpolens naboer. Kristiania, 1895.
Geog 5558.91.5 Keely, R.N. In Arctic seas...with Peary expedition. Philadelphia, 1892.
Geog 5558.94 Walsh, H.C. Last cruise of the Miranda. N.Y., 1896.
Geog 5558.96 Hoppin, Benjamin. A diary kept while with the Peary Arctic expedition of 1896. New Haven? 1897?
Geog 5558.97 Berens, S.L. The "Fram" expedition. Nansen in the frozen world. Philadelphia, 1897.
Geog 5559.03 Low, A.P. Report on dominion government expedition. Ottawa, 1906.
Geog 5559.05 Peary, R.E. Nearest the Pole...1905-06. N.Y., 1907.
Geog 5559.06 Mikkelsen, E. Conquering the Arctic ice. Philadelphia, 1909.
Geog 5559.06.5 Canada. Department of Marine and Fisheries. Report on the dominion government expedition to Arctic islands and the Hudson Strait on board the C.G.S. "Arctic", 1906-07. Ottawa, 1909.
Geog 5559.08 Cook, F.A. Meine Eroberung des Nordpols. Hamburg, 1912.
Geog 5559.08.10 Cook, F.A. Return from the Pole. N.Y., 1951.
Geog 5559.09A Peary, R.E. North Pole; its discovery in 1909. N.Y., 1910.
Geog 5559.09B Peary, R.E. North Pole; its discovery in 1909. N.Y., 1910.
Geog 5559.09.3 Peary, R.E. The North Pole. 2. ed. N.Y., 1910.
Geog 5559.09.8 Moore, J.H. Peary's discovery of the North Pole. Washington, 1910.
Geog 5559.09.15 MacMillan, D.B. How Peary reached the pole. Boston, 1934.
Geog 5559.09.20 Ingersoll, Ernest. The conquest of the North. N.Y., 1909.
Geog 5559.09.25 Rawlins, Dennis. Peary at the North Pole; fact or fiction? Washington, 1973.
Geog 5559.09.100 Mikkelsen, Ejnar. Lost in the Arctic; being the story of the "Alabama" expedition, 1909-1912. London, 1913.
Geog 5559.09.101 Mikkelsen, Ejnar. Lost in the Arctic; being the story of the "Alabama" expedition, 1909-1912. N.Y., 1913.
Geog 5559.09.106 Mikkelsen, Ejnar. Tre aar paa Grønlands østkyst. 2. udg. Kjøbenhavn, 1914.
Geog 5559.09.110 Mikkelsen, Ejnar. Two against the ice. London, 1957.
Geog 5559.10 Canada. Department of Marine and Fisheries. Report...dominion government expedition...1910. Ottawa? 1911.
Geog 5559.10.5 Tremblay, Alfred. Cruise of the Minnie Maud; Arctic seas and Hudson Bay, 1910-1913. Quebec, 1921.
Geog 5559.11 Bruce, William S. Polar exploration. London, 1911.
Geog 5559.11.5 Borup, George. A tenderfoot with Peary. London, 1911.
Geog 5559.12 Cook, F.A. Im Kampfe mit Bär und Walross. Braunschweig, 192-.
Geog 5559.13A Bartlett, Robert A. The last voyage of the Karluk. Boston, 1916.
Geog 5559.13B Bartlett, Robert A. The last voyage of the Karluk. Boston, 1916.
Geog 5559.13.5A MacMillan, D.B. Four years in the white north. N.Y., 1918.
Geog 5559.13.5B MacMillan, D.B. Four years in the white north. N.Y., 1918.
Geog 5559.14 Nagornyi, S. Sedov. Moskva, 1939.
Geog 5559.14.5 Seleznev, Stepan A. Pervaia russkaia ekspeditsiia k severnomy poliucu. Arkhangel'sk, 1964.
Geog 5559.17A Peary, R.E. Secrets of polar travel. N.Y., 1917.
Geog 5559.17B Peary, R.E. Secrets of polar travel. N.Y., 1917.
Geog 5559.17.5 Hall, Thomas F. Has the North Pole been discovered? Boston, 1917.
Geog 5559.21 Rasmussen, Knud. Across Arctic America. N.Y., 1927.
Geog 5559.23 Mittelholzer, W. Im Flugzeug dem Nordpol entgegen. Zürich, 1924.
Geog 5559.27 Putnam, D.B. Davial goes to Baffinland. N.Y., 1927.
Geog 5559.28 Wilkins, George H. Flying the Arctic. N.Y., 1928.
Geog 5559.28.5 Passarello, Gaetano. Un giornalista al Polo Nord. Roma, 1965.
Geog 5559.30 Chapman, F.S. Northern lights. London, 1932.
Geog 5559.31 Wilkins, H. Under the North Pole. N.Y., 1931.
Geog 5559.31.5 Wilkins, H. Under the North Pole. N.Y., 1931.
Geog 5559.32 Polunin, N. The isle of auks. London, 1932.
Geog 5559.34 Rich, E.G. Hans the Eskimo. Boston, 1934.
Geog 5559.34.5 Shackleton, Edward. Arctic journeys; the story of the Oxford University Ellesmire land expedition, 1934-35. London, 1937.
Geog 5559.37 Papanin, Ivan D. Zhizn' na l'dine. Moskva, 1940.
Geog 5559.37.10 Severn Iulius Zavoevan Bolbshevik. Moskva, 1937.
Geog 5559.37.20 Papanin, Ivan D. Life on an icefloe. London, 1947.
Geog 5559.37.25 Deviat' mesiatsev na Dreifuiushchei stantsii severnyi polius. Moskva, 1938.
Geog 5559.38 Manning, E.W. Igloo for the night. London, 1943.
Geog 5559.42 Herrmann, Ernst. Mit dem Fieseler-Storch ins Nordpolarmeer. Berlin, 1942.
Geog 5559.47 Robinson, Bradley. Dark companion. 1. ed. N.Y., 1947.
Geog 5559.48 MacMillan, M. Green seas and white ice. N.Y., 1948.
Geog 5559.56 Morozov, S.T. U poslednikh parallelei. Moskva, 1956.
Geog 5559.56.5 Vodopianov, Mikhail V. Wings over the Arctic. Moscow, 1956?
Geog 5559.56.6 Vodopianov, Mikhail V. Na kryliakh v Arktiku. Moskva, 1955.
Geog 5559.57.5 Volovich, V.G. God na poliuse. Moskva, 1957.
Geog 5559.57.10 Agranat, G.A. Zarubezhmyi Sever. Moskva, 1957.
Geog 5559.58 Anderson, William R. Nautilus 90 north. Cleveland, 1959.
Geog 5559.68 Herbert, Wally. Across the top of the world. N.Y., 1971.
Geog 5559.69 Simpson, Myrtle. Due north. London, 1970.

Geog 5600 Arctic Regions - Greenland - Bibliographies

Geog 5600.1 Hagerups, H. Boghandel. Litteratur om Grønland. København, 1932.
Geog 5600.2 Vartdal, Hroar. Bibliographie des ouvrages norvégiens relatifs au Groenland. Oslo, 1935.
Geog 5600.7 Groenlandsk avis og tidsskrift index. Godthåb. 1962//

Geog 5603 Arctic Regions - Greenland - Pamphlet volumes

Geog 5603.1 Pamphlet box. Geography. Greenland.

Geog 5606 Arctic Regions - Greenland - Periodicals and Societies

Geog 5606.5 Grønland; tidsskrift for dunsk-grønlandsk samvirhe. Hellerup. 1953+ 16v.
Geog 5606.5.5 Grønland. Inholdsfortegnelse, 1953-62. n.p., n.d.
Geog 5606.6 Grønland; turist foreningen for Danmark årbog. Ringkjøbing. 1952-1953

Geog 5606 Arctic Regions - Greenland - Periodicals and Societies - cont.

Geog 5606.8 Greenland. Landsraadet. Kalatalet-nunãne landsrådip okalokatigîssutai. 1956-1958 2v.
Geog 5606.10 Denmark. Udvalget for samfundsforskning i Grønland. Skrifter. 1-9,1961-1963// 2v.
Geog 5606.12 Det Groenlandske samfund. København. 1,1967//

Geog 5610 Arctic Regions - Greenland - Collected source materials

Geog 5610.5 Bobé, Louis. Diplomatarium Groenlandicum, 1492-1814. København, 1936.
Geog 5610.5.2 Bobé, Louis. Opdagelsesrejser til Grønland, 1473-1806. København, 1936.

Geog 5636 - 5640 Arctic Regions - Greenland - History - General works (By date)

Geog 5638.99 Jonsson, Finnur. Um Graenland ad fornu og nýju. Kaupmannahöfn, 1899.
Geog 5639.05 Först, J. Geschichte der Entdeckung Grönlands von den ältesten Zeit bis zum Anfang des 19. Jahrhunderts. Worms, 1906.
Geog 5639.28 Berlin, Knud. Grønlands betydning. København, 1928.
Geog 5639.36 Oldendow, Knud. Traek af Grønlands politiske historie. København, 1936.
Geog 5639.36.4 Bobé, Louis. Den grønlanske handels og kolonisations historie indtil 1870. København, 1936.
Geog 5639.46 Rosing, Christian. Ostgronlaenderne tunuamuit. København, 1946.
Geog 5639.46.5 Gad, Finn. Grønlands historie. København, 1946.
Geog 5639.50 Birket-Smith, K. Grønlands bogen. København, 1950. 2v.
Geog 5639.50.5 Birket-Smith, K. Grenlandiia: sbornik statei. Moskva, 1953.
Geog 5639.57 Mikkelsen, Ejnar. Fra fribytler til embedsmand. København, 1957.
Geog 5639.60 Mikkelsen, Ejnar. Svundne tider i Østgrønland. København, 1960.
Geog 5639.67 Gad, Finn. Grønlands historie. København, 1967- 2v.
Geog 5639.67.2 Gad, Finn. The history of Greenland. London, 1970- 2v.
Geog 5639.70 Greenland past and present. Copenhagen, 1970.
Geog 5639.71 Wedin, Bertil. Tusen år på Grönland. Stockholm, 1971.

Geog 5645 Arctic Regions - Greenland - History - General special

Geog 5645.5 Berlin, Knud. Denmark's to Greenland. London, 1932.
Geog 5645.5.5 Skeie, Jon. Greenland; the dispute between Norway and Denmark. London, 1932.
Geog 5645.5.10 Dúason, Jón. Die koloniale Stellung Grönlands. Göttingen, 1955.
Geog 5645.5.15 Lidegaard, Mads. Ligestilling uden lighed (Grønland og Danmark). København, 1973.
Geog 5645.5.20 Norway and Greenland; a short survey. Oslo, 1931.

Geog 5650 Arctic Regions - Greenland - History - Special periods - Before 1600

Geog 5650.5 Jonsson, Finnur. Graenlendinga saga. Kaupmannahöfn, 1899.
Geog 5650.7 Grönlands historiske mindesmaerker. Kjobenhavn, 1838-45. 3v.
Geog 5650.9 Giesecke, C.L. Norwegian settlements on...east coast of Greenland. Dublin, 1824.
Geog 5650.10 Bruun, Daniel. Erik den Rode og nordhokolonierne il Gronland. Kristiania, 1915.
Geog 5650.10.3 Bruun, Daniel. De gamle Nordbokdonieri Grønland. København, 1905.
Geog 5650.10.5 Bruun, Daniel. The Icelandic colonization of Greenland. København, 1918.
Geog 5650.10.9 Bruun, Daniel. Oversigt over Nordboruiner i Godhaat. København, 1917.
Geog 5650.10.11 Bruun, Daniel. Godthaabsegnen og den gamle Vesterbygd. Odense, 1908.
Geog 5650.11 Hovgaard, William. The Norsemen in Greenland; recent discoveries at Herjolfones. N.Y., 1925.
Geog 5650.12 Dúason, Jón. Grønlands statsretslige stilling i Midelalderen. Oslo, 1928.
Geog 5650.12.5 Dúason, Jón. Nybygd i Grønland. n.p., 192-.
Geog 5650.13 Larusson, O. Rettarstad Graenlands adornu. Kaupmannahöfn, 1924.
Geog 5650.15 Bárdrson, I. Det gamle Grønlands beskrivelse. København, 1930.
Geog 5650.16 Hermansson, H. Landafundir og sjóferd i nordrhofum. Winnipeg, 19- .
Geog 5650.17 Nørlund, P. Viking settlers in Greenland. London, 1936.
Geog 5650.18 Holtved, Erik. Archeological investigation in the Thule district. n.p., n.d.
Geog 5650.19 Noerlund, P. De gambe nordbobygder ved verdens ende. 3. opl. København, 1942.
Geog 5650.20 Mathiassen, T. Skraelingerne i Grønland. København, 1935.
Geog 5650.25 Krogh, Knud J. Erik den Rødes Grønland. København, 1967.
Geog 5650.25.2 Krogh, Knud J. Viking Greenland with a supplement of saga. København, 1967.
Geog 5650.30F Aron, of Kangeg. K'avdlunãtsianik. Godthåt, 1969.
Geog 5650.35 Meldgaard, Jørgen. Nordboerne i Grønland. Købwnhavn, 1965.

Geog 5660 Arctic Regions - Greenland - History - Special periods - 1600-1900

Geog 5660.15 Dahl, F. Andrae og Østgrønland. København, 1932.
Geog 5660.25 Castberg, Frede. Østgrönlandsavtalen. Kristiania, 1924.
Geog 5660.26A Stefansson, V. Greenland. 1. ed. Garden City, 1942.
Geog 5660.26B Stefansson, V. Greenland. 1. ed. Garden City, 1942.
Geog 5660.30 Ostermann, H. Traekaf kolonien Jakobshavns. Kjøbenhavn, 1941.
Geog 5660.37 Smedal, G. Grönland und der Norden. Oslo? 1942?
Geog 5660.42 Gulloev, H.C. Igdlucrúnerit. København, 1971.

Geog 5670 Arctic Regions - Greenland - History - Special periods - 1900-

Geog 5670.2 Betaenkning afgivet af dit i december maaned 1920 nedsatte udvalgtel drøftelse af de Grønlandski anliggender. København, 1921.
Geog 5670.3 Bendixen, Ole. Grønlandstraktaten, belyst gennem de i dagspressen fremkomme artikler. København, 1924.
Geog 5670.4 Jensen, Bent. En livsform ved horsvejen. København, 1971.
Geog 5670.5 Blom, Ida. Kamper om Eirik Raudes land. Oslo, 1973.

Geog 5700 Arctic Regions - Greenland - Religion, Missions

Geog 5700.2 Egede, H. Omstaendelig...grønlandste missions. Kjøbenhavn, 1738.
Geog 5700.4 Egede, H. Ausführliche...Nachricht...Grönland. Hamburg, 1740.

Classified Listing

Geog 5700   Arctic Regions - Greenland - Religion, Missions - cont.
Htn   Geog 5700.4.5*   Egede, H. Det gamle Grønlands...naturel-historie. Kjøbenhavn, 1741.
Geog 5700.4.8   Bobé, Louis. Hans Egede. Copenhagen, 1952.
Geog 5700.4.9   Bobé, Louis. Hans Egede og Grønland. København, 1941.
Geog 5700.4.10   Dalager, Lars. Grønlandske relationer. København, 1715.
VGeog 5700.4.15   Petersen, Niels Matthias. Hans Egedes levnet. København, 1839.
Geog 5700.4.20   Fenger, Niels. Palase, Hans Egede i Grønland. København, 1971.
Geog 5700.5   Crantz, D. History of Greenland. London, 1767. 2v.
Geog 5700.5.3A   Crantz, D. Historie von Grönland...mission. Barby, 1770. 3v.
Geog 5700.5.3B   Crantz, D. Historie von Grönland...mission. Barby, 1770. 2v.
Geog 5700.5.10   Crantz, D. History of Greenland...account of mission. London, 1820. 2v.
Geog 5700.6   Religious Tract Society. Missionary records; northern countries. London, 1839.
Geog 5700.7   Fenger, H.M. Bidrag til Hans Egede...missions. Kjøbenhavn, 1879.
Geog 5700.8   Vormbaum, R. Hans Egede, der Prediger des Evangeliums. Elberfeld, 1861.
Geog 5700.8.5   Strangers in Greenland. n.p., n.d.
Geog 5700.9   Beauvois, E. Origines et fondation du plus ancien évêché du nouveau monde. Paris, 1878.
Geog 5700.10   Monnier, Edouard. Histoire de la mission chrétienne au Groënland. Thèse. Strasbourg, 1853.
Geog 5700.11   Thorhallason, Egiel. Beskrivelse over missionerne i Grønlands Søndre Distrikt. København, 1914.
Geog 5700.12   Glahn, Henrik C. Missionaer i Grønland, Henric Christopher Glahns. København, 1921.
Geog 5700.13   Lund, Jacob Johan. Første missionair paa Grønland. København, 1778.

Geog 5835 - 5840   Arctic Regions - Greenland - Geography, description and travel (By date)
Geog 5835.85   Davys, John. Tre rejser til Grønland i aarene 1585-87. København, 1930.
Geog 5837.03   Vidalin, Arngrimur Thorkilsson. Den tredie part af det saa kaldede gamle og mye Grønlands beskrifvelse. København, 1971.
Geog 5837.06   Torfaeo, T. Gronlandia antiqua. Havniae, 1706.
Htn   Geog 5837.06.2*   Torfaeo, T. Gronlandia antiqua. Havniae, 1706.
Geog 5837.20   Zorgdrager, C.G. Groenlandsche visschery. Amsterdam, 1720.
Geog 5837.20.3   Zorgdrager, C.G. Bloeijende opkomst...visschery. 's-Gravenhage, 1727.
Geog 5837.20.5   Zorgdrager, C.G. Beschreibung des...Wallfischfangs. Nürnberg, 1750.
Htn   Geog 5837.20.10*   Mesange, P. de. La vie...et le voyage de Groenland. v.1-2. Amsterdam, 1720.
Geog 5837.39   Walløe, Peder Olsen. Peder Olsen Walløes dagbøger. Kjøbenhavn, 1927.
Geog 5837.40F   Mauricius, Jan J. Naleesing over Groenland voor de historie van den noorweeschen Erik. pt.1-5. Hamburg? 1740.
Geog 5837.63   Egede, H. Beschreibung und Natur-Geschichte. Berlin, 1763.
Geog 5837.63.5   Egede, H. Description of Greenland. London, 1818.
Geog 5837.75   Stauning, J. Kort beskrivelse over Grønland. v.1-3. Viborg, 1775.
Geog 5837.76   Thorhallesen, E. Esterretning om rudera...Grønlands. Kiøbenhavn, 1776.
Geog 5837.77   Groot, J.J. Beknopt en getrouw verhaal. Amsterdam, 1779.
Geog 5837.78   Saabye, Hans E. Greenland...extracts...journal. London, 1818.
Geog 5837.78.5   Saabye, Hans E. Fragmentes af en dagbok hållen i Grönland. Stockholm, 1811.
Geog 5837.92   Eggers, H.P. Om Grönlands osterbygds. Kjøbenhavn, 1794.
Geog 5838.11   Zimmermann, E.A.H. Die Erde und ihre Bewohner. Leipzig, 1811.
Geog 5838.25   Graah, W.A. Beskrivelse...Grønland. Kjøbenhavn, 1825.
Geog 5838.28   Graah, W.A. Undersogelses-reise...Grønland. Kjøbenhavn, 1832.
Geog 5838.28.3   Graah, W.A. Narrative of expedition to...Greenland. London, 1837.
Geog 5838.36   Breidfjörd, S. Fra Graenlandi. Kaupmannahöfn, 1836.
Geog 5838.45   Rafn, C.C. Amerikas...gamle geographie. Kjøbenhavn, 1845.
Geog 5838.49   Janssen, Carl Emil. En grønlandspraests optegnelser, 1844-49. København, 1913.
Geog 5838.57   Rink, Hinrich. Grønland geographisk. Kjøbenhavn, 1857. 2v.
Geog 5838.57.3   Rink, Hinrich. Grønland geographisch...beschreiben. Stuttgart, 1860.
Geog 5838.68   Peirce, B.M. Report on the resources of Iceland and Greenland. Washington, 1868.
Geog 5838.70F   Nordenskjölds Reise nach Westgrönland im Jahre 1870. n.p., 187-?
Geog 5838.71   Hayes, I.I. The land of desolation...Greenland. London, 1871.
Geog 5838.72   Fries, T.M. Grönland dess natur och innevanare. Upsala, 1872.
Geog 5838.72.5   Oldendow, K. Grønlaendervennen Hinrich Rink. København, 1955.
Geog 5838.75   Jones, T.R. Manual of the natural history, geology and physics of Greenland. London, 1875.
Geog 5838.77.1   Rink, Henrik. Danish Greenland, its people and products. Montreal, 1974.
Geog 5838.82   Rink, Hinrich. Om Grønlaenderne, deres fremtid. Kjøbenhavn, 1882.
Geog 5838.83   Nordenskiold, A.E. Den andra Dicksonska Expedition. Stockholm, 1885.
Geog 5838.83.3   Nordenskiold, A.E. Grönland; seine Eiswüsten im Innern. Leipzig, 1886.
Geog 5838.83.5   Maps. Greenland, 1380-1482. Trois cartes. Copenhagen, 1883.
Geog 5838.83.10   Holm, G. Den Danske Konebaads-Expedition. Kjøbenhavn, 1887.
Geog 5838.84   Hansen, Johannes. Den grønlandske Kateket Hansêraks Dagbog. København, 1933.
Geog 5838.88   Nansen, F. Paa ski over Grønland...1888-90. Kristiania, 1890.
Geog 5838.88.3A   Nansen, F. The first crossing of Greenland. London, 1890. 2v.
Geog 5838.88.3B   Nansen, F. The first crossing of Greenland. London, 1890. 2v.
Geog 5838.88.15   Fridiksson, O. Fra vestfjördm til vestribydr för F. Nansen. Reykjavik, 1927.

Geog 5835 - 5840   Arctic Regions - Greenland - Geography, description and travel (By date) - cont.
Geog 5838.88.16   Fridiksson, O. Fra vestfjördm til vestribydr för F. Nansen. Reykjavik, 1927.
Geog 5838.90   Carstensen, A.R. Two summers in Greenland. London, 1890.
Geog 5838.94   Bruun, Daniel. Mellem fangere og jaegere. Kjøbenhavn, 1897.
Geog 5838.96   Wright, G.F. Greenland icefields. N.Y., 1896.
Geog 5838.98   Jonsson, F. Grønlands gamle topografi. Kjøbenhavn, 1899.
Geog 5839.03   Mylius, L. Grønland...1903-04. Kjøbenhavn, 1906.
Geog 5839.09   Friis, A. Danmark expeditionen. Kjøbenhavn, 1909.
Geog 5839.09.5   Quervain, Alfred. Durch Grönlands Eiswüste. 2. Aufl. Strassburg, 1911.
Geog 5839.09.10   Steenstrup, K. Geologiske og antivariske Iagttageker. København, 1909.
Geog 5839.10   Whitney, H. Hunting with the Eskimos. London, 1910.
Geog 5839.12.2   Rasmussen, Knud. Min rejsedagbog. 2. ugd. Kjøbenhavn, 1921.
Geog 5839.16   Rasmussen, Knud. Grønland langs Polhavet. Kristiania, 1919.
Geog 5839.16.5   Olrik, Harald. Forslag om at bebygge Scoresby Sund-Egnen i Østgrønland ved vestgrønlandske saelforgere. København, 1916.
Geog 5839.21   Rasmussen, Knud. Greenland by the Polar Sea. London, 1921.
Geog 5839.23   Dahl, Kai R. The "Teddy" expedition. N.Y., 1925.
Geog 5839.24   Nissen, L.W. Die sudwestgrönländische Landschaft und das Siedlungsgebiet der Normannen. Hamburg, 1924.
Geog 5839.24.5   Koht, Halvdan. Del Grønland vi miste og de vi ikkje miste. Kristiania, 1924.
Geog 5839.26   Rawson, K.L. A boy's-eye view of the Arctic. N.Y., 1926.
Geog 5839.27   MacMillan, D.B. Etah and beyond. Boston, 1927.
Geog 5839.27.5   Bendixen, Ole. Grønland som Nybyggerland. København, 1927.
Geog 5839.28   Denmark. Kommittee for Ledelsen af Degeoliske og Geografiske Undersøgelser i Grønland. Greenland...invest in Greenland. Copenhagen, 1928. 3v.
Geog 5839.29   Arnason, A. Graenlandsför 1929. Reykjavik, 1929.
Geog 5839.29.5   Charcot, J.B. La mer du Groenland, croisières du "Pourquoi pas?". Paris, 1929.
Geog 5839.30A   Kent, Rockwell. N. by E. N.Y., 1931.
Geog 5839.30B   Kent, Rockwell. N. by E. N.Y., 1931.
Geog 5839.30.5   Wegener, E. Greenland journey; the story of Wegener's German expedition...1930-31. London, 1939.
Geog 5839.30.6   Wegener, A. Tagebücher, Briefe. Wiesbaden, 1960.
Geog 5839.30.10F   Krabbe, Thomas N. Greenland; its natures, inhabitants and history. Copenhagen, 1930.
Geog 5839.30.15   Soctt, J.M. Portrait of an ice cap. London, 1953.
Geog 5839.30.20   Foreniger Det ry Grønland. Det ry Grønland. pt.3-5. København, 1930.
Geog 5839.35   Georgi, J. Mid-ice. N.Y., 1935.
Geog 5839.35.7   Kent, Rockwell. Greeland journal. N.Y., 1962.
Geog 5839.35.15   Ingstad, H.M. East of the great glacier. N.Y., 1937.
Geog 5839.35.20A   Boyd, S.A. The fiord region of east Greenland. N.Y., 1935. 2v.
Geog 5839.35.20B   Boyd, S.A. The fiord region of east Greenland. N.Y., 1935. 2v.
Geog 5839.35.25   Oldendow, Knud. Naturfredning i Grønland. København, 1935.
Geog 5839.37   Boyd, Louise A. The coast of northeast Greenland. N.Y., 1948.
Geog 5839.38   Knuth, Eigil. Under det nordligste danne broa; beretning om dansk Nordøstgrønlands Ekspedition, 1938-39. København, 1940.
Geog 5839.38.5   Gitz-Johansen, Aage. Skitzebogsblade fra Angmagssalik. København, 1938.
Geog 5839.39   Haig-Thomas, D. Tracks in the snow. N.Y., 1939.
Geog 5839.39.5   Koch, Lange. Fra Lissabon til Peary Land. København, 1939.
Geog 5839.44   Draatrup, E. Grønlandsfaerd. København, 1944.
Geog 5839.46   Rodahl, Karl. The ice capped island: Greenland. London, 1946.
Geog 5839.46.10   Lindsay, M. Three got through. London, 1946.
Geog 5839.48   Boegvad, R. Orientering og faerdsee i terraenet. København, 1948.
Geog 5839.48.5   Tschaen, Louis. Groenland 1948-1949-1950: astronomie, nivellement géodésique sur l'Inlandsis. Paris, 1959.
Geog 5839.50   Bomholt, Julius. Grønland foran en ny epoke. København, 1950.
Geog 5839.52   Denmark. Udenrigsministeriet. Greenland. Ringkjøbing, 1952?
Geog 5839.52.5   Banks, Mike. High Arctic; the story of the British North Greenland Expedition. London, 1957.
Geog 5839.52.10   Hamilton, Richard A. Venture to the Arctic. Harmondsworth, 1958.
Geog 5839.53   Williamson, G. Changing Greenland. London, 1953.
Geog 5839.54   Freuchen, P. Ice floes and flaming water. N.Y., 1954.
Geog 5839.59   Ingstad, Helge Marcus. Landet under leidarstjernen. Oslo, 1959.
Geog 5839.59.2   Ingstad, Helge Marcus. Land under the pole star: a voyage to the Norse settlements of Greenland and the saga of the people that vanished. London, 1966.
Geog 5839.59.5   Munch, Ebbe. Strejftog i Nord. København, 1959.
Geog 5839.68   Boucht, Christer. Grönland tvärs. Helsingfors, 1968.
Geog 5839.71   Sandstroem, Lennart. Leva på Grønland. Stockholm, 1971.
Geog 5839.74   Bechmann, Elke. Mennesker og miljøer i Grønland. København, 1974.
Geog 5839.75   Banks, Mike. Greenland. Newton Abbot, 1975.

Geog 5845   Arctic Regions - Greenland - General economic conditions
Geog 5845.2   Rasmussen, Holger. Grønland og dets problemer. København, 1947.
Geog 5845.3   Boserup, Mogens. Økonomisk politik i Grønland. København, 1963.
Geog 5845.4   Denmark. Ministeriet for Grønland. Perspektivplan for Grønland 1971-85. København, 1971.
Geog 5845.5   Hoejlund, Niels. Krise uden alternativ. København, 1972.
Geog 5845.6   Oldendow, Knud. Grønlands fremtid. København, 1945.

Geog 5850   Arctic Regions - Greenland - Civilization, Social life and conditions
Geog 5850.5   Trebitsch, R. Bei den Eskimos in Westgrønland. Berlin, 1910.
Geog 5850.7   Wallem, F.B. Universitetets Eskimoiske samlinger. Christiania, 1911.
Geog 5850.9   Nansen, F. Eskimoliv. Kristiania, 1891.
Geog 5850.12   Wedin, Bertil. Aktion och reaktion på Grönland. Stockholm, 1971.

Classified Listing

Geog 5855 Arctic Regions - Greenland - Races

Geog 5855.3 Rasmussen, Knud. Neue Menschen...des Nordpols. Bern, 1907.

Geog 5855.5 Rasmussen, Knud. The people of the Polar North. London, 1908.

Geog 5855.5.5 Rasmussen, Knud. The people of the Polar North. Philadelphia, 1908.

Geog 5855.7 Nansen, F. Eskimo life. London, 1893.

Geog 5855.9 Solberg, O. Beiträge zur Vorgeschichte. Christiania, 1907.

Geog 5855.10 Elgström, Ossian. Moderna eskimåer. Stockholm, 1916.

Geog 5855.11 Ekblaw, W.E. The material response of the polar eskimo to their far Arctic environment. Diss. N.Y., 1928.

Geog 5855.12 Porsild, N.P. Studies on the material culture of the Eskimo in West Greenland. København, 1915.

Geog 5855.13 Lorm, A.J. Kunstzin der Eskimos. 's-Gravenhage, 1945.

Geog 5855.15 Ostermann, Hother. Danske i Grønland i det 18. aarhundrede. København, 1945.

Geog 5855.16 Freuchen, Peter. Fangstmaend i Melvillebugten. København, 1956.

Geog 5855.17 Malaurie, Jean. The last kings of Thule. London, 1956.

Geog 5861 - 5886 Arctic Regions - Greenland - Local history, etc. (A-Z by place)

Geog 5882.5 Bjoergmose, Rasmus. Stensnaes - den glemte kirke. Odense, 1967.

Geog 5896 Arctic Regions - Greenland - Biographies - Collected

Geog 5896.2 Ostermann, Hother. Nordmaend paa Grønland, 1721-1814. Oslo, 1940. 2v.

Geog 5898 Arctic Regions - Greenland - Biographies - Individual (299 scheme, A-Z by person)

Geog 5898.241 Rosing, Jens. Isimardik. København, 1960.

Geog 5900 Antarctic Regions - Bibliographies

Geog 5900.5 International Antarctic bibliography. Braunschweig. 1958

Geog 5900.10 Hayton, Robert D. National interests in Antarctica; an annotated bibliography. Washington, 1959-60.

Geog 5900.15 Denucé, J. Bibliographie antarctique. Bruxelles, 1913.

Geog 5900.16 Barkov, N.I. Desiat' let sovetskikh issledovanii v Antarktike. Leningrad, 1968.

Geog 5900.18 United States. Naval Photographic Interpretation Center. Antarctic bibliography. Washington, 1951.

Geog 5905 Antarctic Regions - Pamphlet volumes

Geog 5905.1 Pamphlet box. Antarctic regions.

Geog 5907 Antarctic Regions - Periodicals and Societies

Geog 5907.5 Instituto Antartico Argentino. Boletin. Buenos Aires.

Geog 5907.8 British Antarctic Survey. Bulletin. London. 1-27,1963-1972 4v.

Geog 5907.15 United States. Naval Support Force, Antarctica. History and Research Division. Monograph. Washington. 1,1971+

Geog 5918 - 5920 Antarctic Regions - General works, geography and description (By date)

Geog 5918.98 Fricker, Karl. Antarktis. Berlin, 1898.

Geog 5918.98.3A Fricker, Karl. The Antarctic regions. London, 1900.

Geog 5918.98.3B Fricker, Karl. The Antarctic regions. London, 1900.

Geog 5919.01A Neumayer, G. Auf zum Südpol! Berlin, 1901.

Geog 5919.01B Neumayer, G. Auf zum Südpol! Berlin, 1901.

Geog 5919.01.5 Murray, G. The Antarctic manual. London, 1901.

Geog 5919.08 Gourdon, Ernest. Géographie physique, glaciologie, pétrographie des régions visitées par l'Expédition antarctique française...1903-05. Paris, 1908.

Geog 5919.18 Wright, H.S. (Mrs.). The seventh continent. Boston, 1918.

Geog 5919.28 Hayes, James G. Antarctica; a treatise on the southern continent. London, 1928.

Geog 5919.30 Taylor, Griffith. Antarctic adventure and research. N.Y., 1930.

Geog 5919.38 Ommaney, Francis D. Below the roaring forties. London, 1938.

Geog 5919.38.2 Ommaney, Francis D. South latitude. London, 1947.

Geog 5919.45 Cordovez Madariga, Enrique. La Antártida sudamericana. Santiago, 1945.

Geog 5919.48 Weetman, Charles. All about Antarctica. Melbourne, 1948.

Geog 5919.50 Henry, T. The white continent. N.Y., 1950.

Geog 5919.51 Christie, E.W.H. The Antarctic problem. London, 1951.

Geog 5919.52 Simpson, F.A. The Antarctic today. Wellington, 1952.

Geog 5919.55 Kosack, Hans P. Die Antarktis; eine Länderkunde. Heidelberg, 1955.

Geog 5919.57 Sullivan, Walter. Quest for a continent. N.Y., 1957.

Geog 5919.57.5 Machowski, J. Antarktyka. Wyd. 2. Warszawa, 1957.

Geog 5919.57.10 Lebedev, Vladimir L. Antarktika. Moskva, 1957.

Geog 5919.58 Liverridge, Douglas. The last continent. London, 1958.

Geog 5919.58.5 Bertram, George Colin L. Antarctica today and tomorrow. Cambridge, 1958.

Geog 5919.58.10 Knapp, Willem H.C. Antarctica; de geschiedenis van het geheimzinnige Zuidland. Haarlem, 1958.

Geog 5919.59 Steinitz, Hans. Der 7. Kontinent. Bern, 1959.

Geog 5919.62 Caras, Roger A. Antarctica. 1st ed. Philadelphia, 1962.

Geog 5919.65 Lewis, Richard S. A continent for science; the Antarctic adventure. N.Y., 1965.

Geog 5919.65.5 Hatherton, Trevor. Antarctica. Wellington, 1965.

Geog 5919.69 King, H.G.R. The Antarctic. London, 1969.

Geog 5925 Antarctic Regions - Special topics

Geog 5925.5 London, England. Royal Geographical Society. Antarctic exploration. London, 1898.

Geog 5925.10 Weddell, James. Observations on South Pole. London, 1826.

Geog 5925.15 Rouch, J. Le pôle sud. Paris, 1921.

Geog 5925.20 Kollback, K. Der Südpol. Bielefeld, 1911.

Geog 5925.25 Hayward, Walter B. The last continent of adventure. N.Y., 1930.

Geog 5925.30 Shackleton, E. South. N.Y., 1920.

Geog 5925.45 Gould, Lawrence M. Antarctica in world affairs. N.Y., 1958.

Geog 5925.50 Pinochet de la Barra, Oscar. Chilean sovereignty in Antarctica. Santiago, 1955.

Geog 5925.55 Rodriguez, Juan C. La Republica Argentina y las adquisiciones territoriales en el continente antartico. Buenos Aires, 1941.

Geog 5925.55.6 Argentina. Comisión Nacional del Antártico. Soberanía argentina en la Antártida. 2. ed. Buenos Aires, 1948.

Geog 5925.55.10 Sampay, Arturo E. La soberania argentina sobre la Antártida. La Plata, 1950.

Geog 5925.60 Diaz, Emilio. Relatos antarticos. Buenos Aires, 1958.

Geog 5925 Antarctic Regions - Special topics - cont.

Geog 5925.65 Slevich, S.B. Antarktika dolzhna stat' zonoi mira. Leningrad, 1960.

Geog 5925.75 Dralkin, A.G. V mire kholoda. Moskva, 1961.

Geog 5925.89 Conference on Antarctica, Washington, D.C., 1959. Conference documents. Washington, 1960.

Geog 5925.91 Vsesoiuznoe Soveshchenie po Izucheniiu Antarktiki, Moscow, 1966. Osnovnye itogi izucheniia Antarktiki za 10 let. Moskva, 1967.

Geog 5925.92 Markov, Konstantin K. Geografiia Antarktidy. Moskva, 1968.

Geog 5925.93 Potter, Neal. Natural resource potentials of the Antarctic. N.Y., 1969.

Geog 5925.94 Treshnika, Aleksei F. Vokrug Antarktidy. Leningrad, 1970.

Geog 5925.95 Vedenskii, Anatolii A. Vsnegakh Krainego Iuga. Leningrad, 1972.

Geog 5938 - 5940 Antarctic Regions - History of exploration and discoveries - General works (By date)

Geog 5938.80 Cardon, F. Le regioni polari antartiche. Roma, 1880.

Geog 5938.89 Fonville, W. de. Le pôle sud. Paris, 1889.

Geog 5939.02 Balch, E.S. Antarctica. Philadelphia, 1902.

Geog 5939.05 Mill, H.R. Siege of the South Pole. London, 1905.

Geog 5939.07 Murray, J. Antarctic days. London, 1913.

Geog 5939.07.5 Chun, Karl. Die Erforschung der Antarktis. Leipzig, 1907.

Geog 5939.25 Hurley, Frank. Argonauts of the South. N.Y., 1925.

Geog 5939.29 Stefansson, V. The theoretical continent. N.Y., 1929.

Geog 5939.32 Hayes, J.G. The conquest of the South Pole. London, 1932.

Geog 5939.38.3A Byrd, Richard E. Alone. N.Y., 1938.

Geog 5939.38.3B Byrd, Richard E. Alone. N.Y., 1938.

Geog 5939.38.3C Byrd, Richard E. Alone. N.Y., 1938.

Geog 5939.41 Owen, Russell. The Antarctic ocean. N.Y., 1941.

Geog 5939.46 Bezemer, K.W.L. Der Kampf um den Südpol. 2. Aufl. Zürich, 1952.

Geog 5939.50 Mountevans, E. The desolate Antarctic. London, 1950.

Geog 5939.51 Arsen'ev, V.A. V straie kitov i pingvinov. Moskva, 1951.

Geog 5939.55 Mountevans, E. The Antarctic challenged. London, 1955.

Geog 5939.55.5 Kearns, W.H. The silent continent. 1st ed. N.Y., 1955.

Geog 5939.56.5 Burkhanov, V.F. K beregam Antarktidy. Moskva, 1956.

Geog 5939.57 Kemp, Norman. The conquest of the Antarctic. N.Y., 1957.

Geog 5939.57.5 Markov, K.K. Puteshestvie v Antarktidu. Moskva, 1957.

Geog 5939.58.5 Pavlovskii, E.N. Antarktika. Moskva, 1958.

Geog 5939.58.15 Zavatti, Silvio. L'esplorazione dell'Antartide. Torino, 1958.

Geog 5939.59 Debenham, Frank. Antarctica. London, 1959.

Geog 5939.59.10 Pasetskii, V.M. Na samoi iuzhnoi zemle. Moskva, 1959.

Geog 5939.59.15 Reboux, Michael. Demain l'Antarchtique. Paris, 1959.

Geog 5939.59.20 Smuul, Juhan. Ledovaia kniga. Moskva, 1959.

Geog 5939.59.22 Smuul, Juhan. Antarctica ahoy! The icebook. Moscow, 196-?

Geog 5939.59.24 Smuul, Juhan. Jaine raamat. Tallinn, 1962.

Geog 5939.62 Ignatov, V.S. God na poliuse kholoda. Moskva, 1962.

Geog 5939.62.5 Fitte, Ernesto. El descubrimiento de la Antártida. Buenos Aires, 1962.

Geog 5939.63 Treshnikov, Aleksei G. Istoriia otkrytiia i issledovaniia Antarktidy. Moskva, 1963.

Geog 5939.64.5 Astapenko, Pavel D. Puteshestvie k ostrovu chetyrekh vulkanov. Moskva, 1964.

Geog 5939.73 Kutuzov, Pavel S. Antarkticheskii dnevnik. Moskva, 1973.

Geog 5939.73.5 Slevick, Solomon B. Osnovnye problemy osvoeniia Antarktiki. Leningrad, 1973.

Geog 5970 Antarctic Regions - History of exploration and discoveries - By special countries

Geog 5970.10 Buenos Aires. Universidad. Cronologia de los viajes a las regiones australes. Buenos Aires, 1950.

Geog 5970.15 Mitterling, Philip. America in the Antarctic to 1840. Urbana, 1959.

Geog 5970.15.20 Bertrand, Kenneth John. Americans in Antarctica, 1775-1948. N.Y., 1971.

Geog 5970.16 Siple, Paul. 90 degrees south; the story of the American South Pole conquest. N.Y., 1959.

Geog 5970.30 Swan, Robert A. Australia in the Antarctic. Victoria, 1961.

Geog 5970.36 Law, Phillip G. Australia and the Antarctic. St. Lucia, 1962.

Geog 5970.37 Billing, Graham. South: man and nature in Antarctica. Wellington, 1964.

Geog 5970.38 United States. Antarctic Projects Office. The United States in the Antarctic, 1820-1962. Washington, 1962?

Geog 5970.38.5 National Science Foundation. Office of Antarctic Programs. USARP; United States Antarctic Research Program, National Science Foundation. Washington, 1963.

Geog 5970.40 Takhariev, Vasil I. Bulgarin na Antarktida. Sofiia, 1971.

Geog 5970.42 Quartermain, Leslie B. New Zealand and the Antarctic. Wellington, 1971.

Geog 5970.44 Gasev, Aleksandr M. Ot El'brusa do Antarktidy. Moskva, 1972.

Geog 5970.44.5 Semontovskii, Vladimir N. Russkie otkrytiia v Antarktike w 1819-1820-1821 godakh. Moskva, 1951.

Geog 5970.46 Priestley, Raymond E. Antarctic research; a review of British scientific achievement in Antarctica. London, 1964.

Geog 5985 Antarctic Regions - Biographies of explorers - Individual (363 scheme, A-Z by person)

Geog 5985.34 Seaver, George. "Birdie" Bowers of the Antarctic. London, 1938.

Geog 5985.55 Davis, John K. High latitude [autobiographical]. Parkville, Victoria, 1962.

Geog 5985.122 Herbert, Wally. A world of men: exploration in Antarctica. London, 1968.

Geog 5985.187 Mawson, Paquita. Mawson of the Antarctic; the life of Sir Douglas Mawson. London, 1964.

Geog 5985.187.5 Saziumov, Evgenii M. A life given to the Antarctic; Douglas Mawson - Antarctic explorer. Adelaide, 1968.

Geog 5985.253 Gwynn, Stephen. Captain Scott. N.Y., 1930.

Geog 5985.253.5 Seaver, George. Scott of the Antarctic. London, 1940.

Geog 5985.253.10 Pound, Reginald. Scott of the Antarctic. London, 1966.

Geog 5985.253.15 Brent, Peter L. Captain Scott and the Antarctic tragedy. London, 1974.

Geog 5985.258A Mill, Hugh R. The life of Sir Ernest Shackleton. Boston, 1923.

Geog 5985.258B Mill, Hugh R. The life of Sir Ernest Shackleton. Boston, 1923.

Geog 5985.258.5 Worsley, Frank A. Endurance; an epic of polar adventure. London, 1931.

Geog 5985.258.10 Fisher, Margery. Shackleton. London, 1957.

70

Classified Listing

Geog 5985 Antarctic Regions - Biographies of explorers - Individual (363 scheme, A-Z by person) - cont.

Geog 5985.258.15 Lansing, Alfred. Endurance; Shackleton's incredible voyage. N.Y., 1959.
Geog 5985.325 Seaver, George. Edward Wilson of the Antarctic, naturalist and friend. N.Y., 1937.
Geog 5985.325.5 Seaver, George. The faith of Edward Wilson. London, 1948.

Geog 6005 Antarctic Regions - Collected expeditions

Geog 6005.5 Neider, Charles. Antarctica; authentic accounts of life and exploration. N.Y., 1972.

Geog 6008 - 6010 Antarctic Regions - Special expeditions (By date of voyage)

Geog 6008.19 Leningrad. Arkticheskii i Antarkticheskii Nauchno-Issledovatel'skii Institut. Pervaia russkaia antarkticheskaia ekspeditsiia, 1819-1821 gg. i ee otchetnaia navigatsionnaia karta. Leningrad, 1963.
Geog 6008.38.5 Ross, Frank E. The Antarctic explorations of Charles Wilkes, 1838-42. n.p., 1935.
Geog 6008.39 Ross, J.C. Voyage...southern and Antarctic regions. London, 1847. 2v.
Geog 6008.41 Dumont, J.S.C. Voyage au Pole Sud. Paris, 1841. 10v.
Geog 6008.92 Burn Murdoch, W.G. From Edinburgh to the Antarctic. London, 1894.
Geog 6008.97 Gerlache de Gomery, Adrien de. Voyage de la Belgica, quinze mois dans l'Antarctique. Paris, 1902.
Geog 6008.97.2 Gerlache de Gomery, Adrien de. Quinze mois dans l'Antarctique. Bruxelles, 1902.
Geog 6008.97.3 Lecointe, G. Au pays des manchots. Bruxelles, 1910.
Geog 6008.97.5 Gerlache de Gomery, Adrien de. Victoire sur la nuit antarctique. Tournai, 1960.
Geog 6008.97.10 Dobrowolski, A.B. Dziennik wyprawy na Antarktydę. Wrocław, 1962.
Geog 6008.98 Borchgrevink, C.E. First on the Antarctic continent. London, 1901.
Geog 6008.98.3 Cook, F.A. Through the first Antarctic night. N.Y., 1900.
Geog 6009.01 Nordenskjöld, Otto. Antarctic...zwei Jahre. Berlin, 1904. 2v.
Geog 6009.01.3 Nordenskjöld, Otto. Antarctica, or Two years...South Pole. London, 1905.
Geog 6009.01.7A Scott, R.F. Voyage of the 'Discovery'. N.Y., 1905. 2v.
Geog 6009.01.7B Scott, R.F. Voyage of the 'Discovery'. N.Y., 1905. 2v.
Htn Geog 6009.01.8* Scott, R.F. Voyage of the 'Discovery'. London, 1905. 2v.
Geog 6009.01.9 Armitage, A.B. Two years in the Antarctic. London, 1905.
Geog 6009.01.15 Scott, R.F. The voyages of Captain Scott retold from The voyage of the 'Discovery'. N.Y., 1915.
Geog 6009.01.20 Wilson, Edward Adrian. Diary of the 'Discovery' expedition to the Antarctic regions, 1901-1904. London, 1966.
Geog 6009.01.25 Hardy, Alister C. Great waters; a voyage of natural history to study whales, plankton and the waters of the southern ocean. N.Y., 1967.
Geog 6009.01.30 Lashly, William. Under Scott's command: Lashly's Antarctic diaries. London, 1969.
Geog 6009.02 Brown, R.N.R. Voyage of the "Scotia". Edinburgh, 1906.
Geog 6009.02.9 Doorly, G.S. The voyages of the "Morning". London, 1916.
Geog 6009.02.12 Doorly, G.S. The songs of the "Morning". Melbourne, 1943.
Geog 6009.03 Charcot, J.B. Le "Francais" au Pole Sud. Paris, 1906.
Geog 6009.03.5 Argentine Republic. La Argentina en los mares antarticos. Buenos Aires, 1903.
Geog 6009.04 Maveroff, J.O. Por los mares antarticos. Buenos Aires, 1954.
Geog 6009.07A Shackleton, Ernest H. Heart of the Antarctic. Philadelphia, 1909. 2v.
Geog 6009.07B Shackleton, Ernest H. Heart of the Antarctic. Philadelphia, 1909. 2v.
Geog 6009.07.5 Shackleton, Ernest H. The heart of the Antarctic. London, 1911.
Geog 6009.08 Charcot, Jean B. The voyage of the 'Why not?' in the Atlantic. London, n.d.
Geog 6009.08.5 Charcot, Jean B. Autour du pôle sud; expédition du "Pourquoi pas?" 1908-1910. London, 1937.
Geog 6009.10A Amundsen, R. The South Pole...Norwegian Antarctic expedition, 1910-1912. London, 1912. 2v.
Geog 6009.10B Amundsen, R. The South Pole...Norwegian Antarctic expedition, 1910-1912. London, 1912. 2v.
Geog 6009.10.3 Amundsen, R. Au Pôle Sud, éxpedition du "Fram", 1910-1912. Paris, 1913.
Geog 6009.10.5 Amundsen, R. Sydpolen...med Fram, 1910-1912. Kristiania, 1912. 2v.
Geog 6009.10.14 Scott, R.F. Scott's last expedition. 2d ed. London, 1964.
Geog 6009.10.15A Scott, R.F. Scott's last expedition. N.Y., 1913. 2v.
Geog 6009.10.15B Scott, R.F. Scott's last expedition. N.Y., 1913. 2v.
Geog 6009.10.15.2 Scott, R.F. Scott's last expedition. London, 1913. 2v.
Geog 6009.10.16 Like English gentlemen. London, 19- .
Geog 6009.10.19 Taylor, G. With Scott; the silver lining. London, 1916.
Geog 6009.10.20 Simpson, G.C. Soctt's polar journey and the weather. Oxford, 1926.
Geog 6009.10.21 Ponting, H.G. The great white South. London, 1922.
Geog 6009.10.23 Evans, E.R.G.R. South with Scott. London, 1925.
Geog 6009.10.30 Cherry-Garrard, Apsley G.B. The worst journey in the world; Antarctic 1910-13. N.Y., 1930.
Geog 6009.10.32 Cherry-Garrard, Apsley G.B. The worst journey in the world; Antarctic 1910-13. v.1-2. N.Y., 1937.
Geog 6009.10.40 Lindsay, M. The epic of Captain Scott. N.Y., 1934.
Geog 6009.10.50 Wilson, Edward Adrian. Diary of the "Terra Nova" expedition to the Antarctic, 1910-12. London, 1972.
Geog 6009.10.60 Ponting, Herbert George. Scott's last voyage. London, 1974.
Geog 6009.11 Priestly, R.E. Antarctic adventure; Scott's northern party. London, 1914.
Geog 6009.11.5 Davis, John King. With the "Aurora" in the Antarctic, 1911-14. London, 1919.
Geog 6009.12.5 Murphy, R.C. Logbook for Grace. N.Y., 1947.
Geog 6009.14 Shackleton, Ernest H. South; the story of Shackleton's last expedition. London, 1919.
Geog 6009.14.15 Joyce, Ernest. The South Pole trail. London, 1929.
Geog 6009.14.20 Mawson, Douglas. The home of the blizzard. Philadelphia, 1914. 2v.
Geog 6009.14.25 Laseron, Charles F. South with Mawson. 2d ed. Sydney, 1957.

Geog 6008 - 6010 Antarctic Regions - Special expeditions (By date of voyage) - cont.

Geog 6009.14.35 Richards, R.W. The Ross Sea Shore Party, 1914-17. Cambridge, 1962.
Geog 6009.16F Davis, John King. 1918 "Aurora" relief expedition. Melbourne, 1918.
Geog 6009.20 Bagshawe, T.W. Two men in the Antarctic; an expedition to Graham Land, 1920-22. N.Y., 1939.
Geog 6009.21 Wild, Frank. Shackleton's last voyage; the story of the Quest. London, 1923.
Geog 6009.28 Byrd, Richard E. Into the home of the blizzard. N.Y., 1928.
Geog 6009.28.5 Joerg, W.L.G. The work of the Byrd Antarctic expedition. N.Y., 1930.
Geog 6009.28.8A Byrd, Richard E. Little America. 1st ed. N.Y., 1930.
Geog 6009.28.8B Byrd, Richard E. Little America. 1st ed. N.Y., 1930.
Geog 6009.28.10 Byrd, Richard E. Little America. N.Y., 1930.
Geog 6009.28.25 Gould, L.M. Cold, the record of an Antarctic sledge journey. N.Y., 1931.
Geog 6009.28.30 Owen, Russell. South of the sun. N.Y., 1934.
Geog 6009.28.31 Owen, Russell. South of the sun. N.Y., 1934.
Geog 6009.29 Mawson, Douglas. The B.A.N.Z. Antarctic research expedition, 1929-31. London, 1932.
Geog 6009.29.5 British, Australian and New Zealand Research Expedition, 1929-1931. The winning of Australian Antarctica. Sydney, 1962.
Geog 6009.33A Byrd, Richard E. Discovery; the story of the second Byrd Antarctic expedition. N.Y., 1935.
Geog 6009.33B Byrd, Richard E. Discovery; the story of the second Byrd Antarctic expedition. N.Y., 1935.
Geog 6009.33.5 Walden, Jane Brevoort. The long whip; the story of a great husky. N.Y., 1936.
Geog 6009.33.10 Olsen, Magnus L. Saga of the white horizon. Lymington, 1972.
Geog 6009.34 Rymill, John. Southern lights; the official account of the British Graham Land Expedition, 1934-37. London, 1938.
Geog 6009.35 Joerg, W.L.G. The topographical results of Ellsworth's trans-Antarctic flight of 1935. N.Y., 1936.
Geog 6009.36 Christensen, Lars. My last expedition to the Antarctic, 1936-37. Oslo, 1938.
Geog 6009.46 Menster, William J. Strong men south. Milwaukee, 1949.
Geog 6009.47 Orrego Vicuña, Eugenio. Terra australis. Santiago, 1948.
Geog 6009.49 Giaener, John. The white desert. London, 1954.
Geog 6009.52 Marret, M. Seven men among the penguins. 1st American ed. N.Y., 1955.
Geog 6009.53 Solianik, A.N. Slava v Antarktike. Moskva, 1954.
Geog 6009.53.5 Solianik, A.N. Cruising in the Antarctic. Moscow, 1956.
Geog 6009.54 Dufek, George John. Operation deepfreeze. N.Y., 1957.
Geog 6009.55.5F Royal Society of London. The Royal Society International Geophysical Year. London, 1960. 4v.
Geog 6009.55.10 Fuchs, Vivian. The crossing of Antarctica; the Commonwealth Antarctic Expedition, 1955-58. London, 1958.
Geog 6009.55.15 Hillary, Edmund. No latitude for error. London, 1961.
Geog 6009.55.20 Barber, Noël. The white desert. N.Y., 1958.
Geog 6009.55.25 Trans-Antarctic Expedition, 1955-1958. Antarctica. Wellington, 1964.
Geog 6009.55.50 Sovetskaia Antarkticheskaia Ekspeditsiia, 1955-1958. Soviet Antarctic expedition; information bulletin. Amsterdam, N.Y., 1964-65. 3v.
Geog 6009.57 Gerlache de Gomery, Gaston de. Retour dans l'Antarctique; récit de l'Expédition antarctique belge, 1957-58. Tournai, 1960.
Geog 6009.62 Thomson, Robert Baden. The coldest place on earth. Wellington, 1965.
Geog 6009.63 Peskov, Vasilii M. Belye suy. Moskva, 1965.
Geog 6009.63.5 Nudel'man, Aizik V. Sovetskie ekspeditsii v Antarktiku, 1961-63 gg. Moskva, 1965.
Geog 6009.64 Vojtěch, Vaclav. Námořníkem, topičem a psovodem za jižním polárním kruhem. Vyd. 1. Praha, 1968.
Geog 6009.64.5 Hayter, Adrian. The year of the quiet sun: one year at Scott Base, Antarctica. London, 1968.

Geog 6100 Antarctic Regions - Local history, description, etc. (99 scheme, A-Z by place)

Geog 6100.2 Pamphlet vol. Terre Adélie, 1958-60. 2 pam.
Geog 6100.2.10 Expéditions Polaires Françaises. Terre Adélie, 1959-61. Paris, 1963.
Geog 6100.2.12 Expéditions Polaires Françaises. Terre Adélie, 1960-62; raport d'activités. Paris, 1965.
Geog 6100.5 Law, Phillip G. Anare: Australia's Antarctic outposts. Melbourne, 1957.
Geog 6100.37 Scholes, A. Fourteen men. London, 1951.
Geog 6100.37.10 Temple, Philip. The sea and the snow; the South Indian Ocean Expedition to Heard Island. Melbourne, 1966.
Geog 6100.44 Bossière, R.E. Nouvelle notice sur les îles Kerquelen. Paris, 1907.
Geog 6100.77 New Zealand. Geographic Board. Provisional gazetteer of the Ross Dependency. Wellington, 1958.
Geog 6100.77.5 Quartermain, Leslie B. South to the Pole: the early history of the Ross Sea sector, Antarctica. London, 1967.

An 5 Anthropology and ethnology in general - Bibliographies

An 5.1 Ripley, W.Z. Selected bibliography of the anthropology and ethnology of Europe. Boston, 1899.
An 5.2 Riccardi, P. Saggio di un catalogo bibliografico antropologico italiano. Modena, 1883.
An 5.3 Société d'Anthropologie, Paris. Catalogue de la bibliothèque...1890. Paris, 1891.
An 5.4 Steinmetz, S.R. Essai d'une bibliographie ssytématique de l'ethnologie...1911. Bruxelles, 1911.
An 5.7 Bibliography of anthropology and folk-lore. London.
An 5.8 Gercke, Achim. Die Rasse im Schrifttum. Berlin, 1933.
An 5.9 Royal Empire Society, London. Library. Select bibliography of recent publications in the library. London, 1926.
An 5.10 Conklin, Harold C. Folk classification: a topically arranged bibliography of contemporary and background references through 1971. New Haven, 1972.
An 5.15 Keesing, Felix. Culture change; an analysis and bibliography of anthropological sources to 1952. Stanford, Calif., 1953.
An 5.18 Thomas, William L. International directory of anthropological institutions. N.Y., 1953.
An 5.20 Biennial review of anthropology. Stanford. 1,1959+ 6v.
An 5.21F Tozzer Library. Catalogue. Authors. Boston, 1963. 26v.
An 5.21.1F Tozzer Library. Catalogue. Authors. 1st supplement. Boston, 1970. 6v.
An 5.21.2F Tozzer Library. Catalogue. Authors. 2d supplement. Boston, 1971. 2v.

Classified Listing

An 5   Anthropology and ethnology in general - Bibliographies - cont.

An 5.21.3F	Tozzer Library. Catalogue. Authors. 3d supplement. Boston, 1975. 3v.
An 5.22F	Tozzer Library. Catalogue. Subjects. Boston, 1963. 27v.
An 5.22.01F	Tozzer Library. Catalogue. Index to subject headings. Boston, 1963.
An 5.22.1F	Tozzer Library. Catalogue. Subjects. 1st supplement. Boston, 1970. 6v.
An 5.22.2F	Tozzer Library. Catalogue. Subjects. 2d supplement. Boston, 1971. 3v.
An 5.22.3F	Tozzer Library. Catalogue. Subjects. 3d supplement. Boston, 1975. 4v.
An 5.25	Vasconcellos-Abreu, Guilherme de. O critério nomolójico. Photoreproduction. Lisboa, 1887.
An 5.26	Library-Anthropology Resource Group. Serial publications in anthropology. Chicago, 1973.
An 5.30	Sándor, István. A Magyar néprajztudomány bibliográfiája 1945-1954. Budapest, 1965.
An 5.31	Antropoľsko Društvo Jugoslavije. Antropološka bibliografija o Jugoslaviji. Beograd, 1963.
An 5.32	Česnys, Gintautas. Lietuvos antropologijos bibliografija, 1470-1970. Vilnius, 1974.
An 5.33	Murdock, George Peter. Ethnographic atlas. Pittsburg, 1967.
An 5.34	Dokládal, Milan. Československá antropologická bibliografie, 1955-1964. Brno, 1966-67.
An 5.35	Pas, H.T. Economic anthropology 1946-1972. Oosterhout, 1973.

An 36 - 40   Anthropology and ethnology in general - Dictionaries (By date of issue)

An 39.08	Matsumura, A. A gazetteer of ethnology. Tokyo, 1908.
An 39.58	Vuorela, Taivo. Kansatieteen sanasto. Helsinki, 1958.
An 39.60	International dictionary of regional Europen ethnology and folklore. Copenhagen, 1960- 2v.
An 39.72	Davies, David Michael. A dictionary of anthropology. N.Y., 1973.

An 96 - 100   Anthropology and ethnology in general - History (By date of issue)

An 98.47	Bartlett, J.R. The progress of ethnology. 2. ed. N.Y., 1847.
An 98.67	Quatrefages de Bréaux, A. de. Rapport sur la progrès de l'anthropologie. Paris, 1867.
An 98.81	Bastian, A. Vorgeschichte der Ethnologie. Berlin, 1881.
An 98.89	Achelis, Thomas. Die Entwickelung der modernen Ethnologie. Berlin, 1889.
An 98.93	Topinard, P. L'anthropologie aux États-Unis. Paris, 1893.
An 98.96	Achelis, Thomas. Moderne Völkerlunde. Stuttgart, 1896.
An 99.33	Mitra Panchanan. A history of American anthropology. Calcutta, 1933.
An 99.35	Penniman, T.K. A hundred years of anthropology. London, 1935.
An 99.37.1	Lowie, Robert Harry. The history of ethnological theory. N.Y., 1937.
An 99.57	Estena-Fabregat, Claudio. La antropología contemporánea. Madrid, 1957. 2 pam.
An 99.58	Hays, H.R. From ape to angel. 1. ed. N.Y., 1958.
An 99.58.5	Rudolph, Wolfgang. Das Problem der kulturellen Werte in den Arbeiten der neueren amerikanischen Ethnologie. Berlin, 1958.
An 99.60	Levin, M.G. Ocherki po istorii antropologiia v Rossii. Moskva, 1960.
An 99.61A	Kluckhohn, Clyde. Anthropology and the classics. Providence, 1961.
An 99.61B	Kluckhohn, Clyde. Anthropology and the classics. Providence, 1961.
An 99.64A	Hodgen, Margaret. Early anthropology in the sixteenth and seventeenth centuries. Philadelphia, 1964.
An 99.64B	Hodgen, Margaret. Early anthropology in the sixteenth and seventeenth centuries. Philadelphia, 1964.
An 99.66	Tokarev, Sergei A. Istoriia russkoi etnografii, dook tiabr'skii period. Moskva, 1966.
An 99.66.5	Mercier, Paul. Histoire de l'anthropologie. Paris, 1966.
VAn 99.66.10	Moliński, Bogdan. Historia, osobowość, sztuka. Wyd. 1. Warszawa, 1966.
An 99.68	Harris, Marvin. The rise of anthropological theory. N.Y., 1968.
An 99.68.5	One hundred years of anthropology. Cambridge, Mass., 1968.
An 99.70	Moravia, Sergio. La scienza dell'uomo nel settecento. Bari, 1970.
An 99.71	Bausinger, Hermann. Volkskunde: von der Altertumsforschung zur Kulturanalyse. Berlin, 1971.
An 99.72.5	Leclerc, Gérard. Anthropologie et colonialisme. Paris, 1972.
An 99.73	Kuper, Adam. Anthropologists and anthropology. N.Y., 1973.
An 99.73.5	Duvignaud, Jean. Le langage perdu; essai sur la différence anthropologique. 1. éd. Paris, 1973.
An 99.74	De Waal Malefijt, Annemarie. Images of man; a history of anthropological thought. 1. ed. N.Y., 1974.
An 99.75	Voget, Fred W. A history of ethnology. N.Y., 1975.

An 124   Anthropology and ethnology in general - Biographies of anthropologists - Collected

An 124.1	National Research Council. International directory of anthropologists. Washington, 1938.
An 124.5	National Research Council. International directory of anthropologists. 3. ed. Washington, 1950.
An 124.11	Kardiner, Abram. They studied man. N.Y., 1963.
An 124.15	Golde, Peggy. Women in the field; anthropological experience. Chicago, 1970.

An 125 - 150   Anthropology and ethnology in general - Biographies of anthropologists - Individual (A-Z by person)

An 126.1	American Philosophical Society. Brinton memorial meeting. Philadelphia, 1900.
An 126.2	Boas anniversary volume. N.Y., 1906.
An 126.2.5	Herskovits, Melville Jean. Franz Boas. N.Y., 1953.
An 126.2.6	Herskovits, Melville Jean. Frank Boas: the science of man in the making. Clifton, 1973.
An 126.4	Bertillon, Suzanne. Vie d'Alphonse Bertillon, inventeur de l'anthropométrie. Paris? 1941.
An 126.5	Bandelier, Adolph Francis. Pioneers in American anthropology: the Bandelier-Morgan letters, 1873-1883. Albuquerque, 1940. 2v.

An 125 - 150   Anthropology and ethnology in general - Biographies of anthropologists - Individual (A-Z by person) - cont.

An 126.6	Viking Fund. Ruth Fulton Benedict; a memorial. N.Y., 1949.
An 126.6.5A	Benedict, Ruth (Fulton). An anthropologist at work. Boston, 1959.
An 126.6.5B	Benedict, Ruth (Fulton). An anthropologist at work. Boston, 1959.
An 126.8	Whitehill, Walter Muir. A memoir of John Otis Brew. Boston, 1968.
An 126.10	Salter, Elizabeth. Daisy Bates; "the great white queen of the never never". Sydney, 1971.
An 127.5	Currelly, Charles T. I brought the ages home. 1. ed. Toronto, 1967.
An 127.10	Boissel, Jean. Victor Courtet, 1813-1867. Paris, 1972.
An 130.1	Hough, Walter. Biographical memoir of J.W. Fawkes, 1850-1930. Washington, 1932.
An 130.5	Field, Henry. The track of man. 1. ed. Garden City, 1953.
An 130.10	Downie, Robert Angus. James George Frazer. London, 1940.
An 130.10.5	Downie, Robert Angus. Frazer and the golden bough. London, 1970.
An 130.15	Marett, Robert R. James George Frazer 1854-1941. London, 1941?
An 130.20	Hanke, Lewis. Gilberto Freyne. N.Y., 1939.
An 132.1	Reinach, Theodore. Notice sur la vie et les travaux de M. Ernest Hamy. Paris, 1910.
An 132.2	Rubín de la Borbolla, D. Bibliografía del Dr. Ales Hrdlicka. México, 193-.
An 132.3A	Quiggin, A. Hingston (Mrs.). Hadden, the head hunter. Cambridge, 1942.
An 132.3B	Quiggin, A. Hingston (Mrs.). Hadden, the head hunter. Cambridge, 1942.
An 132.4	Simpson, George Eaton. Melville J. Herskovits. N.Y., 1973.
An 132.4.5	Northwestern University, Evanston, Ill. Program of African Studies. Bibliography of Melville J. Herskovits, 1920-1962. Evanston, Ill., 1962?
An 132.5	LeRoux, Robert. L'anthropologie comparée de Guillaume de Humboldt. Paris, 1958.
An 132.6	Hanson, F. Allan. Meaning in culture. London, 1975.
An 135.1	Essays in anthropology. Kroeber. Berkeley, 1936.
An 135.1.5	Kroeber, Theodora. Alfred Kroeber: a personal configuration. Berkeley, 1970.
An 135.3	Keith, Arthur. An autobiography. London, 1950.
An 135.5	Feher, Geza. Felikcě F. Kanitsě. Sofiia, 1936.
An 136.5	Paz, Octavio. Claude Lévi-Strauss; an introduction. Ithaca, 1970.
An 136.5.5	Leach, Edmund Ronald. Claude Lévi-Strauss. N.Y., 1970.
An 136.5.10	Simonis, Yvan. Claude Lévi-Strauss ou la Passion de l'inceste. Paris, 1968.
An 136.5.17	Paz, Octavio. Claude Lévi-Strauss o el nuevo festin de Esopo. 2. ed. Mexico, 1969.
An 136.5.20	Moravio, Sergio. La ragione nascosta. Firenze, 1969.
An 136.5.25	Lévi-Strauss, Claude. Leçon inaugurale faite le mardi. Paris, 1966.
An 136.5.30	Dumasy, Annegret. Restloses. Erkenhen; die Diskussione über den Strukturalismus de Claude Lévi-Strauss in Frankreich. Berlin, 1972.
An 136.5.35	Lepenies, Wolf. Orte des wilden Denkens. 1. Aufl. Frankfurt am Main, 1970.
An 136.5.40	Remotti, Francesco. Lévi-Strauss. Torino, 1971.
An 136.5.45	Scardnelli, Pietro. L'analisi strutturali dei miti. Milano, 1971.
An 136.5.50	Lima, Luiz Costa. O estruturalismo de Lévi-Strauss. Petrópolis, 1968.
An 136.5.60	Boon, James A. From symbolism to structuralism. N.Y., 1972.
An 136.5.61	Boon, James A. From symbolism to structuralism: Lévi-Strauss in a literary tradition. Oxford, 1972.
An 136.5.65	Ipola, Emilio Rafael de. Le structuralisme, ou L'histoire en exil. Thèse. Nantorre? 1969.
An 136.5.70	Montes, Santiago. Claude Lévi-Strauss. 1. ed. San Salvador, 1971.
An 136.5.75	Makarius, Raoul. Structuralisme ou ethnologie. Paris, 1973.
An 136.5.80	Marc-Lipiansky, Mireille. Le structuralisme de Lévi-Strauss. Paris, 1973.
An 136.5.87	Leach, Edmund Ronald. Claude Lévi-Strauss. N.Y., 1974.
An 136.20	California. University. Robert H. Lowie Museum of Anthropology. The complete bibliography of Robert H. Lowie. Berkeley, 1966.
An 136.25	Murphy, Robert Francis. Robert H. Lowie. N.Y., 1972.
An 137.1	Putnam, F.W. Sketch of...Lewis H. Morgan. Boston, 1882.
An 137.1.5	Stern, B.J. Lewis Henry Morgan, social evolutionist. Chicago, 1931.
An 137.2	Meigs, C.D. Memoir of Samuel George Morton, M.D. Philadephia, 1851. 2 pam.
An 137.3	Marett, Robert R. A Jerseyman at Oxford. N.Y., 1941.
An 137.3.5	Ruse, H.J. Robert Ranulph Marett. London, 1944?
An 137.4	Association of Polish University Professors and Lecturers in Great Britain. Professor Bronislaw Malinowski. London, 1943.
An 137.4.5.4	Firth, Raymond W. Man and culture. 4. ed. London, 1963.
An 137.6	Wotte, Herbert. Kaaram Tamo Mann vom Mond; Leben undd Reisen Miklucho-Makleis. Leipzig, 1973.
An 143.2	Gagen-Torn, Nina I. Lev Iakovlevich Shternberg. Moskva, 1975.
An 147.5	In memoriam Karl Weule. Leipzig, 1929.

An 170   Anthropology and ethnology in general - Collected essays, etc. - Several authors (299 scheme, A-Z)

An 170.111	Anthropological Society of Washington. The Saturday lectures. Washington, D.C., 1882.
An 170.117	Athayde, Alfredo. Introdução a antropología tropical. Lisboa, 1962.
An 170.129	Beiträge zur Gesellungs- und Völkerwissenschaft. Berlin, 1950.
An 170.135.10	Biosocial anthropology. N.Y., 1975.
An 170.142	Braidwood, Robert John. Courses toward urban life. Chicago, 1962.
An 170.144	Antropologiia i genogerafiia. Moskva, 1974.
An 170.157	Chicago. University. Seminar of Racial and Cultural Contacts. Proceedings, 1935-36. n.p., n.d.
An 170.161	Cohen, Yehudi A. Man in adaptation. Chicago, 1968. 2v.
An 170.163	Conference on New Approaches in Social Anthropology, Jesus College, Cambridge, Eng., 1963. The social anthropology of complex societies. London, 1966.

Classified Listing

An 170 Anthropology and ethnology in general - Collected essays, etc. - Several authors (299 scheme, A-Z) - cont.

An 170.163.5	Congresso de Etnografia e Folelore, 1st, Braga, 1956. Actas...promovido pela Câmarã municipal de Braga. Lisboa, 1963. 3v.
An 170.163.15	Conference on New Approaches in Social Anthropology, Jesus College, Cambridge, Eng., 1963. The social anthropology of complex societies. London, 1969.
An 170.165A	Count, Carl W. This is race. N.Y., 1950.
An 170.165B	Count, Carl W. This is race. N.Y., 1950.
An 170.173.5	Diamond, Stanley. Primitive views of the world. N.Y., 1964.
An 170.174	Dole, Gertrude Evelyn. Essays in the science of culture. N.Y., 1960.
An 170.174.5	Douglas, Mary T. Implicit meanings. London, 1975.
An 170.189	Essays in anthropology. Lucknow, 194-.
An 170.190	Etnoniny. Moskva, 1970.
An 170.191	The translation of culture; essays to E.E. Evans-Pritchard. London, 1973.
An 170.200	Social organization; essays presented to Raymond Firth. London, 1967.
An 170.202	Ford, Clellan Stearns. Cross-cultural approaches. New Haven, 1967.
An 170.203	Fortes, Meyer. Social structure. Oxford, 1949.
An 170.204	Forschungen zur Völkerpsychologie. Leipzig. 1-14,1925-1935 14v.
An 170.206	Leo Frobenius. Leipzig, 1933.
An 170.210	Mexico (City). Universidad Nacional. Estudios antropologicos publicados en homenaje al Dr. Manuel Gamio. México, 1956.
An 170.211	Garn, Stanley Marion. Readings on race. Springfield, Ill., 1960.
An 170.212	Antropologicheskaia rekonstruktsiia i problemy paleoetnografii. Moskva, 1973.
An 170.221	Göttingen. Universität. Institut für Völkerkunde. Göttinger völkerkundliche Studien. Leipzig, 1939.
An 170.227	Hammond, Peter B. Cultural and social anthropology. N.Y., 1964.
An 170.227.5	Festschrift Eduard Hahn zum LX. Geburtstag. Stuttgart, 1917.
An 170.228	Kontakte und Grenzen. Göttingen, 1969.
An 170.230	Homenaje a Don Luis de Hoyos Sainz. Madrid, 1949-50. 2v.
An 170.230.5	Kongres Československých anthropologů, 10th, Prague and Humpolie, 1969. Anthropological congress dedicated to Aleš Hrdlička. Prague, 1971.
An 170.230.10	L'homme, hier et aujourd'hui. Paris, 1973.
An 170.238	International Congress of Anthropological and Ethnological Sciences, 5th, Philadelphia, 1960. Selected papers. Philadelphia, 1960.
An 170.238.10	International Congress of Anthropological and Ethnological Sciences, 7th, Moscow, 1964. VII Mezhdunarodnyi Kongress antropologicheskikh i etnograficheskikh nauk. Trudy. v.1-3,5-10; photoreproduction. Moskva, 1967- 13v.
An 170.246	Haberland, Eike. Festschrift für A.E. Jensen. München, 1964. 2v.
An 170.251.100	Koppers, W. Kultur und Sprache. Wien, 1952.
An 170.252	Kraeber, A.L. Source book in anthropology. N.Y., 1931.
An 170.257	Échanges et communications; mélanges offerts à Claude Lévi-Strauss. The Hague, 1970. 2v.
An 170.264	Madan, T.N. Indian anthropology. Bombay, 1962.
An 170.266	Manners, Robert A. Process and pattern in culture; essays in honor of Julian H. Steward. Chicago, 1964.
An 170.268	Marxist analyses and social anthropology. N.Y., 1975.
An 170.270	Mead, M. Primitive heritage. N.Y., 1953.
An 170.270.5	Mead, M. The golden age of American anthropology. N.Y., 1960.
An 170.274	Moore, Frank. Readings in cross-cultural methodology. New Haven, 1966.
An 170.275.5	Montagu, Ashley. The concept of race. N.Y., 1964.
An 170.275.10	Montagu, Ashley. Culture: man's adaptive dimension. N.Y., 1968.
An 170.278	Mühlemann, Wilhelm Emil. Kulturanthropologie. Köln, 1966.
An 170.279	Goodenough, Ward Hunt. Explorations in cultural anthropology; essays in honor of George Peter Murdock. N.Y., 1964.
An 170.283	Neue Anthropologie. Stuttgart, 1972-75. 7v.
An 170.296	Miscelanea de estudios dedicados a Fernando Ortiz por sus discipulos. La Habana, 1955-57. 3v.
An 170.312	Mélanges Pittard. Brive, 1957.
An 170.321	Diamond, Stanley. Culture in history; essays in honor of Paul Radin. N.Y., 1960.
An 170.323	Rethinking modernization; anthropological perspectives. Westport, 1974.
An 170.324	The interpretation of ritual; essays in honour of A.I. Richards. London, 1972.
An 170.328.8	Royal Anthropological Institute of Great Britain and Ireland. Notes and queries on anthropology. 6. ed. London, 1971.
An 170.335	Schmitz, Carl August. Historische Völkerkunde. Frankfurt am Main, 1967.
An 170.340	Simpozium Antropologiia 70-kh Godov, Moscow, 1972. Simpozium "Antropologiia 70-kh godov". Moskva, 1972.
An 170.341	Slotkin, James Sydney. Reading in early anthropology. N.Y., 1965.
An 170.341.5A	Grafton Elliot Smith: the man and his work. Portland, Ore., 1974.
An 170.341.5B	Grafton Elliot Smith: the man and his work. Portland, Ore., 1974.
An 170.345	Spiro, Melford E. Context and meaning in cultural anthropology. N.Y., 1965.
An 170.347	Festschrift Alfred Steinmann. Bern, 1972.
An 170.349	Symposium on Community Studies in Anthropology. Symposium on community studies in anthropology. Seattle, 1964.
An 170.351	Tax, Sol. Horizons of anthropology. Chicago, 1964.
An 170.360	Tyler, Stephen A. Cognitive anthropology; readings. N.Y., 1969.
An 170.361	The interpretation of symbolism. N.Y., 1975.
An 170.367	UNESCO. Le racisme devant la science. Paris, 1960.
An 170.386	Festschrift für Robert Wildhaber zum 70. Geburtstag am 3. August 1972. Basel, 1973.
An 170.396	Festschrift...Otto Zerries. Bern, 1974.

An 175 - 200 Anthropology and ethnology in general - Collected essays, etc. - Individual authors (A-Z)

An 175.1	Andree, R. Ethnographische Parallelen und Vergleiche. v.1-2. Leipzig, 1878-89.
An 175.2	Sbornik" v chest' semidesiatiletiia...D.N. Anuchina. Moskva, 1913.
An 176.1	Blumenbach, J.F. The anthropological treatises. London, 1865.
An 176.6A	Boas, Franz. Race, language and culture. N.Y., 1940.
An 176.6B	Boas, Franz. Race, language and culture. N.Y., 1940.
An 176.7	Bandelier, A.F. The scientist on the trail. Berkeley, 1949.
An 177.1	Capitan, J.L. Notice sur les travaux scientifiques. n.p., 1911.
An 177.2	Carias, Rafael. La indecision. Caracas, 1970.
An 179.1A	Eiseley, L.C. The immense journey. N.Y., 1957.
An 179.1B	Eiseley, L.C. The immense journey. N.Y., 1957.
An 180.1	Force, M.F. Pre-historic man. Cincinnati, 1873.
An 180.2	Festschrift der deutscher anthropologischer Gesellschaft zur XXVI allgemeinen Versammlung. Cassel, 1895.
An 180.3	Frobenius, Leo. Erlebte Erdteile; Ergebnisse eines deutschen Forscherlebens. Frankfurt am Main, 1925-29. 7v.
An 180.4	Evans-Pritchard, Edward Evans. The position of women in primitive societies. N.Y., 1965.
An 180.5	Fortes, Meyer. Time and social structure and other essays. London, 1970.
An 180.6	Fox, Robin. Encounter with anthropology. Harmondsworth, 1975.
An 181.1.2	Goldenweiser, Alexander A. History, psychology, and culture. Gloucester, 1968.
An 182.2	Humboldt, Wilhelm von. Philosophische Anthropologie und Theorie der Menschenkenntnis. Halle, 1929.
An 182.3	Harris, Marvin. Cows, pigs, wars and witches. 1. ed. N.Y., 1974.
An 182.4	Hodgen, Margaret. Anthropology, history and cultural change. Tucson, 1974.
An 182.8	Herskovits, Melville Jean. Cultural relativism. 1. ed. N.Y., 1972.
An 184.1F	Jagor, Fedor. Aus Fedor Jogor's Nachlass. Berlin, 1914.
An 185.1.2	Kluckhohn, Clyde. Mirror for man. N.Y., 1949.
An 185.2	Kluckhohn, Clyde. Culture and behaviour. N.Y., 1962.
An 186.2.5	Charbonnier, Georges. Conservations with Claude Lévi-Strauss. London, 1969.
An 187.1.2	Pamphlet box. Macdonald, A. Anthropology.
An 187.5.2	Pamphlet vol. MacCurdy. Anthropology and archaeology. Papers 1907-1913. 18 pam.
An 187.5.3	Pamphlet vol. MacCurdy. Anthropology and archaeology. Papers, 1914-1920. 17 pam.
An 187.5.25	Malinowski, B. A scientific theory of culture and other essays. Chapel Hill, 1944.
An 187.6	Mendes Corrèa, A.A. Da biologia à história. Porto, 1934.
An 187.8	Mauss, Marcel. Oeuvres. Paris, 1961- 3v.
An 187.9	Montagu, Ashley. Anthropology and human nature. Boston, 1957.
An 190.1	Pratt, Orson. Wonders of the universe. Salt Lake City, 1937.
An 190.2.2	Peacock, James Lowe. The human direction. 2. ed. N.Y., 1973.
An 192.1	Redfield, Robert. Human nature and the study of society. Chicago, 1962.
An 193.1	Sergi, Giuseppe. Problemi di scienza contemporanea. Palermo, 1904.
An 193.1.5	Sergi, Giuseppe. Problemi di scienza contemporanea. Torino, 1916.
An 197.1	Wilson, D. The lost Atlantis. N.Y., 1892.
An 197.5	Ward, D.J.H. Letters to future ages. n.p., 1955.
An 197.6	White, Leslie A. The concept of cultural systems. N.Y., 1975.

An 205 Anthropology and ethnology in general - General pamphlet volumes

An 205.1	Jomard, E.F. Lettre à M.P.F. de Siebold. Paris, 1845-69. 3 pam.
An 205.2	Pamphlet vol. Anthropology. Miscellaneous tracts. 19 pam.
An 205.4	Dewey, Chester. Examination of some reasonings against unity of mankind. n.p., 1862. 2 pam.

An 346 - 350 Anthropology and ethnology in general - General treatises - Folios [Discontinued]

An 348.43F	Prichard, J.C. Ethnographical maps to the natural history of man. London, 1843.
An 348.45	Slack, David B. An essay on the human color. Providence, 1845.
An 348.61	Prichard, J.C. Explanatory notice of the ethnographical map. London, 1861.
An 348.87	Hamy, E.T. Etudes ethnographiques et archéologiques sur l'exposition coloniale. Paris, 1887.
An 349.24A	Thomson, John Arthur. What is man? N.Y., 1924.
An 349.24B	Thomson, John Arthur. What is man? N.Y., 1924.
An 349.36	Numelin, R. The wandering spirit. Copenhagen, 1936.
An 349.48	Gillin, John. The ways of men. N.Y., 1948.

An 355 - 360 Anthropology and ethnology in general - General treatises - Monographs (By date of issue)

	An 355.85	Lascovius, P.M. De homine magno illo in rerum natura. Libro II. Witebergae, 1585.
Htn	An 356.77*	Pechlin, J.N. De habitu et colore Aethiopum. Kiloni, 1677.
	An 357.76	Demeunier, J.N. L'esprit des usages et des coutumes des différens peuples. Londres, 1776. 3v.
	An 357.95	Blumenbach, J.F. De generis humani varietate nativa. Gottingae, 1795.
	An 358.00.5	Girando, J.M. Considérations sur les diverses méthodes à suivre dans l'observation des peuples sauvages. Paris, 1800.
	An 358.01	Richard, Jerôme. Voyages chez les peuples sauvages. Paris, 1801. 3v.
	An 358.10.3	Smith, Samuel Stanhope. An essay on the causes of the variety of complexion and figure in the human species. 2. ed. N.Y., 1810.
	An 358.10.5	Smith, Samuel Stanhope. An essay on the causes of the variety of complexion and figure in the human species. Cambridge, 1965.
	An 358.11	Meinere, C.M. Untersuchungen über die Verschiedenheiten der Menschennaturen. Tübingen, 1811-15. 3v.
	An 358.24	Virey, Julien J. Histoire naturale du genre humain. Paris, 1824. 3v.

Classified Listing

An 355 - 360 Anthropology and ethnology in general - General treatises - Monographs (By date of issue) - cont.

	An 358.26	Prichard, J.C. Researches into the physical history of mankind. London, 1826. 2v.
	An 358.26.5	Prichard, J.C. Eastern origin of the Celtic nations...forming a supplement to Researches into the physical history of mankind. Oxford, 1831.
	An 358.28	Lawrence, William. Lectures on physiology, zoology and natural history of man. Salem, 1828.
	An 358.28.3	Leupoldt, J.M. Die Diätetik des physischen und psychischen Menschenlebens. Berlin, 1828.
	An 358.36	Prichard, J.C. Researches into the physical history of mankind. 3. ed. London, 1836-47. 6v.
	An 358.37	Brotonne, F. de. Histoire de la filiation et migrations des peuples. v.1-2. Paris, 1837.
	An 358.39	Kinmont, A. Twelve lectures on the natural history of man. Cincinnati, 1839.
	An 358.41	Prichard, J.C. Researches into the physical history of mankind. v.1-2, 4. ed.; v.3, 3. ed. London, 1837-41. 3v.
	An 358.41.2	Prichard. J.C. Researches into the physical history of mankind. v.1,5, 4. ed; v.2-4, 3. ed. London, 1837-47. 5v.
	An 358.41.5	Azaïs, H. Question philosophique de première importance: Quelle est dans l'univers, la destinée du genre humain? Paris, 1841.
	An 358.43	Prichard, J.C. The natural history of man. London, 1843.
	An 358.43.3	Bond, Thomas E. A dissertation on the varieties of mankind. Middletown, 1843.
	An 358.43.5	Virey, Julien J. Historia natural del jénero humano. 3. ed. Barcelona, 1842-46. 2v.
	An 358.44	Nott, Josiah C. Two lectures on the natural history of the Caucasian and Negro races. Mobile, 1844.
	An 358.45	Prichard, J.C. The natural history of man. 2. ed. London, 1845.
	An 358.45.8F	Schinz, Heinrich Rudolf. Naturgeschichte und Abbildungen des Menschen der verschiedenen Rassen. 3. Aufl. Zürich, 1848.
	An 358.48	Van Amringe, W.F. An investigation of the theories of the natural history of man. N.Y., 1848.
	An 358.49.3	Smith, Ashbel. An oration pronounced before the Connecticut Alpha of the Phi Beta Kappa at Yale College. New Haven, 1849.
	An 358.50	Knox, R. The races of men. Philadelphia, 1850.
	An 358.50.4	Pickering, Charles. The races of men. London, 1850.
	An 358.50.6	Bachman, J. Doctrine of unity of human race examined on principles of science. Charleston, S.C., 1850.
	An 358.51	Prichard, J.C. Researches into the physical history of mankind. v.1-2,5, 4. ed.; v.3-4, 3. ed. London, 1841-51. 5v.
	An 358.51.3	Smith, C.H. The natural history of the human species. Boston, 1851.
	An 358.52	Smith, C.H. The natural history of the human species. London, 1852.
	An 358.52.2	Latham, R.G. Man and his migrations. N.Y., 1852.
	An 358.52.4	Frankenheim, M.L. Völkerkunde. Breslau, 1852.
	An 358.54	Nott, Josiah C. Types of mankind. 2. ed. Philadelphia, 1854.
	An 358.54.2	Nott, Josiah C. Types of mankind. 2. ed. Philadelphia, 1854.
	An 358.54.5	Pickering, Charles. The races of man; and their geographical distribution. London, 1854.
	An 358.55	Prichard, J.C. The natural history of man. 4. ed. London, 1855. 2v.
	An 358.57	Nott, Josiah C. Indigenous races of the earth. Philadelphia, 1857.
	An 358.57.3	Baldwin, Samuel D. Dominion; or Unity and trinity of the human race. Nashville, 1857.
	An 358.59	Latham, R.G. Descriptive ethnology. London, 1859. 2v.
	An 358.59.3	Perty, M. Grundzüge der Ethnographie. Leipzig, 1859.
	An 358.59.5	Waitz, T. Anthropologie der Naturvölker. Leipzig, 1859. 6v.
	An 358.60	Clavel, A. Races humaines. Paris, 1860.
Htn	An 358.60.3*	Reid, Mayne. Odd people. N.Y., 1860.
	An 358.61	Reid, Mayne. Odd people. Boston, 1861.
	An 358.61.5	Quatrefages de Bréau, A. de. Unité de l'espèce humaine. Paris, 1861.
	An 358.63	Brace, C.L. The races of the old world. N.Y., 1863.
	An 358.63.2	Waitz, T. Introduction to anthropology. London, 1863.
	An 358.63.4	Pickering, Charles. Races of man; and their geographical distrbution. London, 1863.
	An 358.64	Vogt, C. Lecture on man: his place in creation and in the history of the earth. London, 1864.
	An 358.64.3	Diefenbach, L. Vorschule der Völkerkunde. Frankfurt am Main, 1864.
	An 358.64.5	Bonstetten, K.V. von. The man of the north and man of the south. N.Y., 1864.
	An 358.64.7	Reid, Mayne. Odd people. Boston, 1864.
	An 358.64.9	Pouchet, Georges. The plurality of the human race. 2. ed. London, 1864.
	An 358.65	Debay, A. Histoire naturelle de l'homme et de la femme. 12. éd. Paris, 1865.
	An 358.66	Matériaux pour l'histoire...de l'homme. 2. sér. Toulouse. 1872-1877 2v.
	An 358.66.5	Tuttle, Hudson. The origin and antiquity of physical man scientifically considered. 2. ed. Boston, 1866?
	An 358.68	Lesley, J.P. Man's origin and destiny. Philadelphia, 1868.
	An 358.68.5	Bastian, A. Der Beständige in den Menschenrassen. Berlin, 1868.
Htn	An 358.69*	Stratton, C.S. Sketch of life, personal appearance, character and manners. N.Y., 1869.
	An 358.69.3	Lesley, J.P. Man's origin and destiny sketched from the platform of the sciences. Philadelphia, 1869.
	An 358.70	Wood, J.G. The uncivilized races of men in all countries of the world. Hartford, 1870. 2v.
	An 358.70.5	Wood, J.G. The uncivilized races of men in all countries of the world. Hartford, 1876. 2v.
	An 358.71	Nott, Josiah C. Types of mankind. 10. ed. Philadelphia, 1871.
	An 358.71.3	Broca, Paul. Mémoires d'anthropologie. v.5. Paris, 1888.
	An 358.71.7	Bastian, A. Ethnologische Forschungen und Sammlung von Material für die selben. Jena, 1871-73. 2v.
	An 358.72	Figuier, Louis. Les races humaines. Paris, 1872.
	An 358.75	Quatrefages de Bréau, A. de. Natural history of man. N.Y., 1875.
	An 358.75.3	Gerland, Georg. Anthropologische Beiträge. Halle, 1875.
	An 358.76	Topinard, P. L'anthropologie. Paris, 1876.
	An 358.76.2	Topinard, P. L'anthropologie. Paris, 1876.

An 355 - 360 Anthropology and ethnology in general - General treatises - Monographs (By date of issue) - cont.

	An 358.76.3	Peschel, Oscar F. The races of man. N.Y., 1876.
	An 358.76.5	Peschel, Oscar F. Völkerkunde. 3. Aufl. Leipzig, 1876.
	An 358.77	Waitz, T. Anthropologie der Naturvölker. 2. Aufl. Leipzig, 1860.
	An 358.78	Oberländer, R. Der Mensch vormals und heute. Leipzig, 1878.
	An 358.78.3A	Morgan, L.H. Ancient society. N.Y., 1878.
	An 358.78.3B	Morgan, L.H. Ancient society. N.Y., 1878.
	An 358.78.5	Topinard, P. Anthropology. London, 1878.
	An 358.78.7	Congrès International des Sciences Anthropologiques, Paris, 1878. Comptes rendus sténographiques. Paris, 1880.
	An 358.79	Quatrefages de Bréau, A. de. The human species. N.Y., 1879.
	An 358.79.2	Farrer, J.A. Primitive manners and customs. London, 1879.
	An 358.79.5	Müller, F. Allgemeine Ethnographie. Wien, 1879.
	An 358.80	Antropologia. Lisboa, 1880.
	An 358.81	Girard de Rialle, J. Les peuples de l'Asie et de l'Europe. Paris, 1881.
	An 358.81.4	Tylor, Edward B. Anthropology, introduction to the study of man and civilization. N.Y., 1881.
	An 358.81.5	Lesley, J.P. Man's origin and destiny. Boston, 1881.
	An 358.81.7	Bastian, W.A. Der Völkergedanke. Berlin, 1881.
	An 358.81.9	Hovelacque, Abel. Les débuts de l'humanité. Paris, 1881.
	An 358.82	Hovelacque, Abel. Les races humaines. Paris, 1882.
	An 358.82.3	Bertillon, A. Les races sauvages. Paris, 1882.
	An 358.82.5	Hellwald, Friedrich. Naturgeschichte des Menschen. Stuttgart, 1882-85. 2v.
	An 358.83F	Oberländer, R. Fremde Völker ethnographische Schilderungen aus der alten und neuen Welt. Leipzig, 1883.
	An 358.84	Quatrefages de Bréau, A. de. Hommes fossiles et hommes sauvages. Paris, 1884.
	An 358.84.3	Lind, G.D. Man; embracing his origin. Chicago, 1884.
	An 358.84.10	Fontaine, Edward. How the world was peopled. N.Y., 1884.
	An 358.85	Schneider, W. Die Naturwölker. v.1-2. Paderborn, 1885-86.
	An 358.85.3	Peschel, O.F. Völkerkunde. Leipzig, 1885.
	An 358.85.7	Spencer, Herbert. La especie humana. Madrid, 1885.
	An 358.87	Quatrefages de Bréau, A. de. Histoire générale des races humaines. Paris, 1887-89. 2v.
	An 358.87.3	Ratzel, F. Völkerkunde. Leipzig, 1887-88. 3v.
	An 358.87.5	Featherman, A. Social history of the races of mankind. v.1-5. Boston, 1881-91. 7v.
	An 358.87.6	Featherman, A. Social history of the races of mankind. v.2,5. London, 1881-91. 2v.
	An 358.87.7	Lorenzini, C. Il regale istruttivo. Roma, 1887.
	An 358.87.9	Hovelacque, Abel. Précis d'anthropologie. Paris, 1887.
	An 358.88	Tylor, Edward B. Anthropology. N.Y., 1888.
NEDL	An 358.88.3	Quatrefages de Bréau, A. de. The human species. N.Y., 1888.
	An 358.88.5	Morris, Charles. The Aryan race: its origin and its achievements. Chicago, 1888.
	An 358.89	Tylor, Edward B. Anthropology. N.Y., 1889.
X Cg	An 358.89.3	Buel, J.W. The story of man. Philadelphia, 1889.
	An 358.90	Brinton, D.G. Races and peoples. N.Y., 1890.
	An 358.90.3	Peschel, O.F. The races of man and their geographical distribution. N.Y., 1890.
	An 358.91	Reclus, E. Primitive folk. London, 1891.
	An 358.91.3	Tylor, Edward B. Anthropology. N.Y., 1891.
	An 358.91.15	Kinmont, A. The natural history of man. 2. ed. Philadelphia, 1891.
	An 358.92	Lombroso, C. L'uomo bianco e l'uomo di colore. Firenze, 1892.
	An 358.92.3	Ranke, Johannes. Beiträge zur physischen Anthropologie der Bayern. München, 1892.
	An 358.92.5	Carmichael, Charles H.E. L'anthropologie et les origines de la Société...l'Orient et l'Occident. Lisbonne, 1892.
	An 358.92.7	Haliburton, Robert G. Survivals of prehistoric races in Mt. Atlas and Pyrenees. Lisbon, 1892.
	An 358.93	Schurtz, H. Katechismus der Völkerkunde. Leipzig, 1893.
	An 358.93.1	Beddoe, J. Anthropological history of Europe. London, 1893.
	An 358.94	Ranke, Johannes. Der Mensch. Leipzig, 1894. 2v.
	An 358.94.3	Tylor, Edward B. Anthropology. N.Y., 1894.
	An 358.94.5	Lavrov, P.L. Antropologicheskaia zhizne. Zheneva, 1894. 2v.
	An 358.95	Babington, W.D. Fallacies of race theories. London, 1895.
	An 358.96A	Keane, A.H. Ethnology. Cambridge, Eng., 1896.
	An 358.96B	Keane, A.H. Ethnology. Cambridge, Eng., 1896.
	An 358.96.2	Keane, A.H. Ethnology. 2. ed. Cambridge, Eng., 1896.
	An 358.96.3A	Ratzel, F. The history of mankind. London, 1896-98. 3v.
	An 358.96.3B	Ratzel, F. The history of mankind. London, 1896-98. 3v.
	An 358.96.5	Piette, Édouard. Étude d'éthnographie préhistorique. no.2-3,6-8. Paris, 1896-1925
	An 358.97	Munro, R. Prehistoric problems. Edinburgh, 1897.
	An 358.98A	Haddon, A.C. The study of man. N.Y., 1898.
	An 358.98B	Haddon, A.C. The study of man. N.Y., 1898.
	An 358.98.3	Tenishef, V. L'activité de l'homme. Paris, 1898.
	An 358.99A	Keane, A.H. Man - past and present. Cambridge, Eng., 1899.
	An 358.99B	Keane, A.H. Man - past and present. Cambridge, Eng., 1899.
	An 358.99.3	Beck, G. Der Urmensch. Basel, 1899.
	An 358.99.5	Herbertson, A.J. Man and his work. London, 1899.
	An 358.99.6	Aranzadi, G. de. Etnologia. Madrid, 1899.
	An 358.99.7A	Ripley, W.Z. The races of Europe. N.Y., 1899.
	An 358.99.7B	Ripley, W.Z. The races of Europe. N.Y., 1899.
	An 358.99.8A	Ripley, W.Z. The races of Europe. Supplement. N.Y., 1899.
	An 358.99.8B	Ripley, W.Z. The races of Europe. Supplement. N.Y., 1899.
	An 358.99.8C	Ripley, W.Z. The races of Europe. Supplement. N.Y., 1899.
	An 359.00.3	Deniker, J. Les races et les peuples de la terre. Paris, 1900.
	An 359.00.4	Deniker, J. The races of man. London, 1900.
	An 359.01	Demolins, E. Les grandes routes des peuples. Paris, 1901. 2v.
	An 359.01.3	Sergi, G. The Mediterranean race. London, 1901.
	An 359.01.5	Setourneau, C.J.E. La psychologie ethnique. Paris, 1901.
	An 359.01.7	Kirchoff, A. Mensch und Erde. Leipzig, 1901.
	An 359.02A	Brinton, D.G. Basis of social relations. N.Y., 1902.
	An 359.02B	Brinton, D.G. Basis of social relations. N.Y., 1902.
	An 359.02.3	Bryce, J. The relationship of the advanced and backward races of mankind. Oxford, 1902.
	An 359.02.5	Steenstrup, J. Ethnografien. Kjøbenhavn, 1902.
	An 359.02.6	Pogodin, A.L. Sbornik" statei po arkheologii i etnografii. Sankt Peterburg, 1902.

74

Classified Listing

An 355 - 360 Anthropology and ethnology in general - General treatises - Monographs (By date of issue) - cont.

An 359.03 Clevenger, S.V. The evolution of man and his mind. Chicago, 1903.
An 359.03.3 Ward, D.J.H. The human races. n.p., 1903.
An 359.03.5 Reches, Elie. Les primitifs, étude d'ethnologie. Paris, 1903.
An 359.03.7 Schurtz, H. Völkerkunde. Leipzig, 1903.
An 359.03.9 Bryce, J. Relationship of the advanced and backward races of mankind. 2. ed. Oxford, 1903.
An 359.04 Yertz, Friedrich O. Moderne Rassentheorien. Wien, 1904.
An 359.04.3 Tylor, Edward B. Anthropology. N.Y., 1904.
An 359.05 Koropchevskago, A.A. Zhachenie "geograficheskikh provintsie". Sankt Peterburg, 1905.
An 359.05.3 Rotzell, W.E. Man: an introduction to anthropology. 2. ed. Philadelphia, 1905.
An 359.06 Balch, E.S. Comparative art. Philadelphia, 1906.
An 359.06.3 Finot, Jean. Le préjugé des races. Paris, 1906.
An 359.06.4 Finot, Jean. Race prejudice. London, 1906.
An 359.06.4.1 Finot, Jean. Race prejudice. N.Y., 1907.
An 359.06.4.2 Finot, Jean. Race prejudice. Miami, 1969.
An 359.06.5 Chalikiopoulos, L. Landschafte...Kulturtypen. Leipzig, 1906.
An 359.06.7 Toro, E. Antropologia general. Caracas, 1906.
An 359.06.12 Maldonado, S.D. Defensa de la Antropologia general. Caracas, 1906.
An 359.07 Lankester, E.R. Kingdom of man. N.Y., 1907.
An 359.08 Dieserud, J. The scope and content of the science of anthropology. Chicago, 1908.
An 359.08.3 Coffin, E.W. On the education of backward races. Worcester, 1908.
An 359.08.5 British Museum (Natural History). Department of Zoology. Guide...specimens illustrating the races of mankind. London, 1908.
An 359.08.7A Gehring, A. Racial contrasts. N.Y., 1908.
An 359.08.7B Gehring, A. Racial contrasts. N.Y., 1908.
An 359.08.9 Keane, A.H. The world's peoples. London, 1908.
An 359.08.10 Keane, A.H. The world's peoples. N.Y., 1908.
An 359.09 Frobenius, Leo. The childhood of man. London, 1909.
An 359.09.3 Haddon, A.C. Races of man and their distribution. London, 1909.
An 359.09.5 Buschan, Georg H.T. Menschenkunde. Stuttgart, 1909.
An 359.09.8 Herbertson, A.J. Man and his work. London, 1909.
An 359.09.11 Tylor, Edward B. Anthropology. N.Y., 1909.
An 359.09.12 Tylor, Edward B. Anthropology. N.Y., 1909.
An 359.09.15 Morgan, L.H. Ancient society. Chicago, 1909.
An 359.10 Schmidt, W. Die Stellung der Pygmäenvölker in der Entwicklungsgeschichte des Menschen. Stuttgart, 1910.
An 359.10.3 Ripley, W.Z. Races of Europe. N.Y., 1910.
An 359.10.5 Simpson, B.P. The conflict of colour. N.Y., 1910.
An 359.10.7 Buschan, Georg H.T. Illustrierte Völkerkunde. Stuttgart, 1910.
An 359.11 Boas, Franz. Curso de antropologia general. Mexico, 1911.
An 359.11.3 Savorgnan, Franco. Gli indici di omogamia delle razze e delle nazionalità. Cagliari, 1911.
An 359.11.5 Marett, Robert R. Anthropology. N.Y., 1911.
An 359.12.2 Deniker, J. The races of man. London, 1912.
An 359.12.3 Marett, Robert R. Anthropology. London, 1912.
An 359.12.5 Munro, Robert. Palaeolithic man. Edinburgh, 1912.
An 359.12.7 Grow, O. The antagonism of races. Waterloo, 1912.
An 359.12.9 Bradley, R.N. Malta and the Mediterranean race. London, 1912.
An 359.12.11 British Museum (Natural History). Department of Zoology. Guide to the specimens...races of mankind. 2. ed. London, 1912.
An 359.12.13A Thomas, W.I. Source book for social origins. Chicago, 1912.
An 359.12.13B Thomas, W.I. Source book for social origins. Chicago, 1912.
An 359.12.15 Montalto de Jesus, C.A. Oriente modernisado. Lisboa, 1912.
An 359.12.17 Brunhes, Jean. La géographie humaine. 2. éd. Paris, 1912.
An 359.12.21 Giuffrida-Ruggeri, V. L'uomo come specie collettiva. Napoli, 1912.
An 359.12.23 Marett, Robert R. Anthropology. N.Y., 1912.
An 359.13 Rivers, W.H.R. Reports upon the present condition and future needs of the science of anthropology. Washington, 1913.
An 359.14 Wirth, Albrecht. Rasse und Volk. Halle, 1914.
An 359.14.3 Van Waters, Miriam. The adolescent girl among primitive peoples. Thesis. Worcester, 1914.
An 359.15 Smith, George E. The migrations of early culture. Manchester, 1915.
An 359.15.3 Latcham, R.E. Conferencias sobre antropología, etnología. Santiago, 1915-
An 359.15.5 Arldt, Theodor. Die Stammesgeschichte der Primaten und die Entwicklung der Menschrassen. Berlin, 1915.
An 359.15.7 Hertz, Friedrich. Rasse und Kultur. 2. Aufl. Leipzig, 1915.
An 359.15.10 Gobineau, A. The inequality of human races. N.Y., 1915.
An 359.16 Blackford, K.M.H. Blondes and brunets. Photoreproduction. N.Y., 1916.
An 359.16.5 Myres, John L. The influence of anthropology on the course of political science. Berkeley, 1916.
An 359.17 Restrepo-Hernandez, J. Lecciones de antropologia. Bogotá, 1917.
An 359.17.3 Humphrey, Seth King. Mankind; racial values and racial prospect. N.Y., 1917.
An 359.17.5 Nordenstreng, Rolf. Europas människoraser. 2. uppl. Stockholm, 1917-20.
An 359.17.7 Haberlandt, M. Völkerkunde. 3. Aufl. Berlin, 1917. 2v.
An 359.18 Numelin, R. Orsakerna till fokevandringerna. Helsingfors, 1918.
An 359.19 Marett, Robert R. Anthropology. London, 1919.
An 359.19.3 Gattefossé, R.M. Adam, l'homme tertiaire. Lyon, 1919.
An 359.20 Humphrey, Seth King. The racial prospect. N.Y., 1920.
An 359.20.3 Lowie, Robert H. Primitive society. N.Y., 1920.
An 359.20.5A Johnston, Harry. The backward peoples and our relations with them. London, 1920.
An 359.20.5B Johnston, Harry. The backward peoples and our relations with them. London, 1920.
An 359.20.7 Roy, Sarat Chandra. Principles and methods of physical anthropology. Patna, 1920.
An 359.20.9 Klaatsch, Hermann. Der Werdegang der Menschheit und die Entstehung der Kultur. Berlin, 1920.
An 359.20.11 Keane, W.H. Man past and present. Cambridge, 1920.
An 359.21A Korzybski, Alfred. Manhood of humanity. N.Y., 1921.

An 355 - 360 Anthropology and ethnology in general - General treatises - Monographs (By date of issue) - cont.

An 359.21B Korzybski, Alfred. Manhood of humanity. N.Y., 1921.
An 359.21C Korzybski, Alfred. Manhood of humanity. N.Y., 1923.
An 359.22 Armitage, F.P. Diet and race. London, 1922.
An 359.22.3A Carr-Saunders, A.M. The population problem. Oxford, 1922.
An 359.22.3B Carr-Saunders, A.M. The population problem. Oxford, 1922.
An 359.23A Dixon, Roland B. The racial history of man. N.Y., 1923.
An 359.23B Dixon, Roland B. The racial history of man. N.Y., 1923.
An 359.23C Dixon, Roland B. The racial history of man. N.Y., 1923.
An 359.23D Dixon, Roland B. The racial history of man. N.Y., 1923.
An 359.23.3A Wissler, Clark. Man and culture. N.Y., 1923.
An 359.23.3B Wissler, Clark. Man and culture. N.Y., 1923.
An 359.23.5 Bradley, R.N. Duality; a study in the psycho-analysis of race. N.Y., 1923.
An 359.23.7 Pape, A.G. Is there a new race type? Edinburgh, 1923.
An 359.23.9 Kroeber, A.L. Anthropology. N.Y., 1923.
An 359.23.11 Shirokogorov, S.M. Etnos". Shankhai, 1923.
An 359.24A Shirokogorov, S.M. Ethnical unit and milieu. Shanghai, 1924.
An 359.24B Shirokogorov, S.M. Ethnical unit and milieu. Shanghai, 1924.
An 359.24.3 Paudler, F. Die hellfarbigen Rassen und ihre Sprachstamme, Kulturen und Urkeimaten. Heidelberg, 1924.
An 359.24.5A Smith, G.E. Evolution of man. London, 1924.
An 359.24.5B Smith, G.E. Evolution of man. London, 1924.
An 359.24.7 Smith, G.E. The evolution of man. 2. ed. London, 1927.
An 359.24.10 Mathews, Basil. The clash of colour. 7. ed. London, 1924.
An 359.24.15 Mathews, Basil. The clash of color. 2. ed. N.Y., 1924.
An 359.24.18 Mathews, Basil. The clash of colour. Port Washington, 1973.
An 359.24.20 Schmidt, Max. Völkerkunde. Berlin, 1924.
An 359.24.25 Marett, Robert R. Mannfraedi. Reykjavik, 1924.
An 359.24.27 Schmidt, W. Völker und Kulturen. Regensburg, 1924.
An 359.25.1 Schiller, F.C.S. Tantalus; or The future of man. N.Y., 1925.
An 359.25.1.3 Schiller, F.C.S. Tantalus; or The future of man. London, 1924.
An 359.25.2 Schiller, F.C.S. Tantalus; or The future of man. N.Y., 1925.
An 359.25.4 Gregory, John W. The menace of colour. 2. ed. London, 1925.
An 359.25.5 Scheidt, W. Allgemeine Rassenkunde. München, 1925.
An 359.25.7 Hertz, F.O. Rasse und Kultur. 3. Aufl. Leipzig, 1925.
An 359.25.9 Anantha, Krishna Iyer L. Lectures on ethnogrpahy. Calcutta, 1925.
An 359.25.11 Haddon, A.C. Races of man and their distribution. N.Y., 1925.
An 359.26 Schütte, G. Vor folkegruppe Gottjod. Kjøbenhavn, 1926.
An 359.26.4A Hawkins, Frank H. The racial basis of civilization. N.Y., 1926.
An 359.26.4B Hawkins, Frank H. The racial basis of civilization. N.Y., 1926.
An 359.26.4C Hawkins, Frank H. The racial basis of civilization. N.Y., 1926.
An 359.26.10 Schmidt, Max. The primitive races of mankind. London, 1926.
An 359.26.15 Mathews, Basil. The clash of colour. 16. ed. London, 1926.
An 359.27 Wolff, K.F. Rassenlehre. Leipzig, 1927.
An 359.27.3 Marett, Robert R. The diffusion of culture. Cambridge, Eng., 1927.
An 359.27.4 Marett, Robert R. Man in the making. London, 1927.
An 359.27.4.3 Marett, Robert R. Man in the making. Garden City, 1928.
An 359.27.5 Gohier, W.D. Cassandre, ou La folie des blancs. Paris, 1927.
An 359.27.7A Worrell, W.H. A study of races in the ancient Near East. N.Y., 1927.
An 359.27.7B Worrell, W.H. A study of races in the ancient Near East. N.Y., 1927.
An 359.27.9A Pitt-Rivers, G.H.L.F. The clash of culture and the contact of races. London, 1927.
An 359.27.9B Pitt-Rivers, G.H.L.F. The clash of culture and the contact of races. London, 1927.
An 359.27.11A Muntz, Earl E. Race contact. N.Y., 1927.
An 359.27.11B Muntz, Earl E. Race contact. N.Y., 1927.
An 359.27.13 Penniston, J.B. Racial old age, being further adventures in philosophy. Seattle, 1927.
An 359.28A Dixon, Roland B. The building of cultures. N.Y., 1928.
An 359.28B Dixon, Roland B. The building of cultures. N.Y., 1928.
An 359.28.3 Hertz, F.O. Race and civilization. N.Y., 1928.
An 359.28.5 Hertz, F.O. Race and civilization. N.Y., 1928.
An 359.28.7 Scheler, Max. Die Stellung des Menschen im Kosmos. Darmstadt, 1928.
An 359.28.9 Hildebrandt, K. Staat und Rasse. Breslau, 1928.
An 359.28.11 Schemann, L. Die Rasse in den Geisteswissenschaften. München, 1928-31. 3v.
An 359.29 Morand, Paul. Black magic. N.Y., 1929.
An 359.29.3 Schütte, Gudmund. Our forefathers, the Gothonic nations. Cambridge, 1929-33. 2v.
An 359.29.5 Smith, G.E. Migrations of early culture. Manchester, 1929.
An 359.29.7 Günther, H.F.K. Rassenkunde Europas. 3. Aufl. München, 1929.
An 359.29.9 Duncan, H. Race and population problems. 1. ed. N.Y., 1929.
An 359.29.12 Foster, T.S. Travels and settlements of early man. N.Y., 1929.
An 359.30 Tylor, Edward B. Anthropology; an introduction to the study of man and civilization. London, 1930. 2v.
An 359.31.3 Gregory, John W. Race as a political factor. London, 1931.
An 359.31.5 Spiller, Gustav. The origin and nature of man. London, 1931.
An 359.31.7 Keith, Arthur. The place of prejudice in modern civilization. N.Y., 1931.
An 359.32 Muckermann, H. Rassenforschung und Volk der Zukunft. Berlin, 1932.
An 359.32.5 Bean, R.B. The races of man. N.Y., 1932.
An 359.33 Radin, Paul. The method and theory of ethnology. N.Y., 1933.
An 359.33.3 Ritter, J. Über den Sinn und die Grenze der Lehre vom Menschen. Potsdam, 1933.
An 359.33.5 Cornelius, William J.J. L'homme d'après la religion et la science. Thèse. Londres, 1933.
An 359.33.7 Voegelin, Erich. Die Rassenidee in der Geistesgeschichte Ray bis Carus. Berlin, 1933.

75

Classified Listing

An 355 - 360 Anthropology and ethnology in general - General treatises - Monographs (By date of issue) - cont.

An 359.33.9 Mendes Correia, A.A. Introdução i antropobiologia. Coimbra, 1933.
An 359.33.11 Montandon, Georges. La race, les races. Paris, 1933.
An 359.34A Radin, Paul. The racial myth. N.Y., 1934.
An 359.34B Radin, Paul. The racial myth. N.Y., 1934.
An 359.34.3 Poisson, Georges. Les aryens. Paris, 1934.
An 359.34.5 Reuter, E.B. Race and culture contacts. N.Y., 1934.
An 359.34.7 Kryzwicki, L. Primitive society and its vital statistics. London, 1934.
An 359.34.9 Clauss, Ludwig F. Rasse und Seele; eine Einführung in den Sinn der leiblichen Gestalt. München, 1934.
An 359.34.10 Menghin, O. Geist und Blut. Wien, 1934.
An 359.35 Orton, E.F. Links with past ages. Cambridge, Eng., 1935.
An 359.35.3 Spiller, Gustav. The origin and nature of man. 2. ed. London, 1935.
An 359.35.4A Carrell, Alexis. Man, the unknown. N.Y., 1935.
An 359.35.4B Carrell, Alexis. Man, the unknown. N.Y., 1935.
An 359.35.5A Carrel, Alexis. Man, the unknown. 2. ed. N.Y., 1935.
An 359.35.5B Carrel, Alexis. Man, the unknown. 19. ed. N.Y., 1935.
An 359.35.5.8A Carrel, Alexis. Man, the unknown. 11. ed. N.Y., 1935.
An 359.35.5.8B Carrel, Alexis. Man, the unknown. 3. ed. N.Y., 1935.
An 359.35.5.10 Carrel, Alexis. Man, the unknown. 25. ed. N.Y., 1935.
An 359.35.6 Carrel, Alexis. Man, the unknown. 50. ed. N.Y., 1935.
An 359.35.7 Huxley, Julian S. We Europeans. London, 1935.
An 359.35.7.10 Huxley, Julian S. We Europeans. 1. ed. N.Y., 1936.
An 359.35.9 Weigner, K. Die Gleichwertigkeit der europäischen Rassen und die Wege zu ihrer Vervollkommnung. Prag, 1935.
An 359.36 White, C.L. Geography. N.Y., 1936.
An 359.36.3 Linton, Ralph. The study of man. N.Y., 1936.
An 359.36.5 Die farbige Front; hinter den Kulissen der Weltpolitik. Leipzig, 1936.
An 359.36.10 Schmelzle, K. Rassengeschichte und Vorgeschichte. 2. Aufl. Bamberg, 1936.
An 359.37 Dover, C. Half-caste. London, 1937.
An 359.37.3 Preuss, K.T. Lehrbuch der Völkerkunde. Stuttgart, 1937.
An 359.37.3.5 Preuss, K.T. Lehrbuch der Völkerkunde. 2. Aufl. Stuttgart, 1939.
An 359.37.5A Hooton, E.A. Apes, man, and morons. N.Y., 1937.
An 359.37.5B Hooton, E.A. Apes, man, and morons. N.Y., 1937.
X Cg An 359.37.7 Eickstedt, E. Rassenkunde und Rassengeschichte der Menschheit. 2. Aufl. v.1, pt.1-15. Stuttgart, 1937-43.
An 359.37.9 Baschmakoff, A.A. Cinquante siècles d'évolution ethnique autour de la Mer noire. Paris, 1937.
An 359.37.11 Clauss, Ludwig F. Rasse ist Gestalt. München, 1937.
An 359.37.13 Krzywicki, L. Społeczeństwo pierwotne. Warszawa, 1937.
An 359.38 Barzun, Jacques. Race; a study in modern superstition. London, 1938.
An 359.38.3A Demiashkevich, Michael John. The national mind. N.Y., 1938.
An 359.38.3B Demiashkevich, Michael John. The national mind. N.Y., 1938.
An 359.38.3C Demiashkevich, Michael John. The national mind. N.Y., 1938.
An 359.38.5 Schmidt, Wilhelm. Razza e nazione. n.p., 1938.
An 359.38.7 Sombart, W. Von Menschen; Versuch einer geistwissenschaftlichen Anthropologie. Berlin, 1938.
An 359.38.11 Mühlmann, W.E. Methodik der Völkerkunde. Stuttgart, 1938.
An 359.38.13 Bang, Paul. Die farbige Gefahr. 2. Aufl. Göttingen, 1938.
An 359.38.15 Nauka o rasakh i rasizm. Moskva, 1938.
An 359.38.20 Boas, Franz. General anthropology. Boston, 1938.
An 359.38.26 Herskovits, Melville Jean. Acculturation. Gloucester, 1958.
An 359.39A Coon, Carleton Stevens. The races of Europe. N.Y., 1939.
An 359.39B Coon, Carleton Stevens. The races of Europe. N.Y., 1939.
An 359.39.3 Frazer, J.G. The native races of Asia and Europe. London, 1939.
An 359.39.5A Hooton, E.A. Twilight of man. N.Y., 1939.
An 359.39.5B Hooton, E.A. Twilight of man. N.Y., 1939.
An 359.39.7 Seligmann, H.J. Race against man. N.Y., 1939.
An 359.39.15 Schmidt, Wilhelm. The culture historical method of ethnology. N.Y., 1939.
An 359.40 Benedict, Ruth (Fulton). Race: science and politics. N.Y., 1940.
An 359.40.3A Hooton, E.A. Why men behave like apes and vice versa, or Body behavior. Princeton, 1940.
An 359.40.3B Hooton, E.A. Why men behave like apes and vice versa, or Body behavior. Princeton, 1940.
An 359.40.3C Hooton, E.A. Why men behave like apes and vice versa, or Body behavior. Princeton, 1940.
An 359.40.10 Fyfe, Henry H. The illusion of national character. London, 1940.
An 359.40.25 Wiese und Kaiserwaldau, Leopold M. von. Homo sum, Gedanken zu einer zusammenfassenden Anthropologie. Jena, 1940.
An 359.41 Estabrooks, G.H. Man, the mechanical misfit. N.Y., 1941.
An 359.41.10 Scientific aspects of the race problem. Washington, 1941.
An 359.42A Chapple, Eliot D. Principles of anthropology. N.Y., 1942.
An 359.42B Chapple, Eliot D. Principles of anthropology. N.Y., 1942.
An 359.42.5 Webster, Hutton. Taboo, a sociological study. Stanford, 1942.
An 359.42.10A Montagu, Ashley. Man's most dangerous myth: the fallacy of race. N.Y., 1942.
An 359.42.10B Montagu, Ashley. Man's most dangerous myth: the fallacy of race. N.Y., 1942.
An 359.42.12A Montagu, Ashley. Man's most dangerous myth. 2. ed. N.Y., 1945.
An 359.42.12B Montagu, Ashley. Man's most dangerous myth. 2. ed. N.Y., 1945.
An 359.42.15 Laidler, H.W. The role of the races in our future civilization. N.Y., 1942.
An 359.43 Dover, C. Hell in the sunshine. London, 1943.
An 359.43.5 Posnansky, A. Qué es raza. La Paz, 1943.
An 359.43.10 Claessens, A. Race prejudice. N.Y., 1943.
An 359.43.20 Murphy, J.D. Lamps of anthropology. Manchester, Eng., 1943.
An 359.43.25 Benedict, Ruth (Fulton). Race; science and politics. N.Y., 1943.
An 359.43.30 Nicholson, James E. Anthropos; or, The problem of man. London, 1943.
An 359.44 Harvard University. Department of Psychology. ABC's of scapegoating. Chicago, 1943.
An 359.44.5A Roback, A.A. A dictionary of the international slurs. Cambridge, Mass., 1944.
An 359.44.5B Roback, A.A. A dictionary of the international slurs. Cambridge, Mass., 1944.
An 359.44.10 Leiper, H.S. Blind spots. N.Y., 1944.
An 359.44.15 Mendes Correia, A.A. Gérmen e cultura. Pôrto, 1944.

An 355 - 360 Anthropology and ethnology in general - General treatises - Monographs (By date of issue) - cont.

An 359.44.25A Kluckhohn, Clyde. The concept of culture. n.p., 194-.
An 359.44.25B Kluckhohn, Clyde. The concept of culture. n.p., 194-.
An 359.45 Linton, R. The science of man in the world crisis. N.Y., 1945.
An 359.45.10A Hooton, E.A. "Young man, you are normal". N.Y., 1945.
An 359.45.10B Hooton, E.A. "Young man, you are normal". N.Y., 1945.
An 359.45.10C Hooton, E.A. "Young man, you are normal". N.Y., 1945.
An 359.45.15 Andrews, R.C. Meet your ancestors, a biography of primitive man. N.Y., 1945.
An 359.45.20 Boas, Franz. Race and democratic society. N.Y., 1945.
An 359.45.25 Montagu, Ashley. An introduction to physical anthropology. Springfield, 1945.
An 359.46 Clark, Grahame. From savagery to civilization. London, 1946.
An 359.46.5 Baruch, Dorothy W. Glasshouse of prejudice. N.Y., 1946.
An 359.46.10A Stewart, George R. Man: an autobiography. N.Y., 1946.
An 359.46.10B Stewart, George R. Man: an autobiography. N.Y., 1946.
An 359.46.17A Schmidt, Wilhelm. Rassen und Völker. 3. Aufl. Luzern, 1946-49. 3v.
An 359.46.17B Schmidt, Wilhelm. Rassen und Völker. v.3. 3. Aufl. Luzern, 1946-49.
An 359.48A Cox, Oliver Cromwell. Caste, class and race. 1. ed. Garden City, N.Y., 1948.
An 359.48B Cox, Oliver Cromwell. Caste, class and race. 1. ed. Garden City, N.Y., 1948.
An 359.48.5 Embree, Edwin R. Peoples of the earth. N.Y., 1948.
An 359.48.10 Bendict, Ruth (Fulton). In Henry's backyard. N.Y., 1947.
An 359.48.15 Coon, Carleton Stevens. A reader in general anthropology. N.Y., 1948.
An 359.49.1 Murdock, George Peter. Social structure. N.Y., 1965.
An 359.49.5 Price, A.G. White settlers and native peoples. Melbourne, 1949.
An 359.49.10 Herskovits, Melville Jean. Man and his works. N.Y., 1949.
An 359.49.15 Caro Baroja, Julio. Análisis de la cultura; etnología, historia, folklore. Barcelona, 1949.
An 359.49.20 Scheler, Max. Die Stellung des Menschen in Kosmos. München, 1949.
An 359.50 Park, R.E. Race and culture. Glencoe, Ill., 1950.
An 359.50.5 Schwidetzky, Ilse. Grundzäge der Völkerbiologie. Stuttgart, 1950.
An 359.50.10 Dempf, Alois. Theoretische Anthropologie. Bonn, 1950.
An 359.50.15 Korzybski, Alfred. Manhood of humanity. 2. ed. Lakeville, Conn., 1950.
An 359.51.5 Evans-Pritchard, Edward Evan. Social anthropology, and other essays. N.Y., 1964.
An 359.52 UNESCO. The race concept. Paris, 1952.
An 359.52.5 Morant, G.M. The significance of racial differences. Paris, 1952.
An 359.52.10A Kroeber, Alfred L. The nature of culture. Chicago, 1952.
An 359.52.10B Kroeber, Alfred L. The nature of culture. Chicago, 1952.
An 359.52.15 Campbell, Byram. American race theorists, a critique of their thoughts and methods. Boston, 1952.
An 359.52.20 Herskovits, Melville Jean. Man and his works: the science of cultural anthropology. N.Y., 1952.
An 359.53 Kroeber, Alfred L. Anthropology today. Chicago, 1953.
An 359.53.5 L'évolution, l'homme, la race. Paris, 1953?
An 359.54 LaBarre, W. The human animal. Chicago, 1954.
An 359.54.5 Mason, Philip. An essay on racial tension. London, 1954.
An 359.54.10 Missenard, André. A la recherche de l'homme. Paris, 1954.
An 359.54.15 Schwidetzky, Ilse. Das Problem des Völkertodes. Stuttgart, 1954.
An 359.54.20 Flatz, Josef. Die Kultur. Linz, 1954.
An 359.55 Hawkes, J. Man on earth. 1. American ed. N.Y., 1955.
An 359.55.5 Troise, Emilio. Racismo. Buenos Aires, 1955.
An 359.56 Bavinck, J.H. Het rassenvraagstuk. Kampen, 1956.
An 359.56.5 LaFarge, John. The Catholic viewpoint on race relations. 1. ed. Garden City, N.Y., 1956.
An 359.56.10 Chicago Urban League. Staff report on a scientist's report on race differences. Chicago, 1956.
An 359.56.15 UNESCO. The race question in modern science. Paris, 1957.
An 359.56.20 Gjessing, Gutorm. Socio-culture. v.1,4. Oslo, 1956- 2v.
An 359.56.25 White, Lynn T. Frontiers of knowledge in the study of man. 1. ed. N.Y., 1956.
An 359.57 Frazier, E.F. Race and culture contacts in the modern world. 1. ed. N.Y., 1957.
An 359.57.5 Lebeuf, Jean Paul. Application de l'ethnologie à l'assistance sanitaire. Bruxelles, 1957.
An 359.57.10 Drexler, John Paul. Die Front der Farbigen. München, 1957.
An 359.57.15 Griaule, Marcel. Méthode de l'ethnographie. Paris, 1957.
An 359.57.20 Akademiia Nauk SSSR. Institut Etnografii. Ocherki obshchei etnografii. v.1-4. Moskva, 1957- 5v.
An 359.57.30 Missenard, André. In search of man. N.Y., 1957.
An 359.58 Miroglio, Abel. La psychologie des peuples. Paris, 1958.
An 359.58.5 Silva, Mello. Estudos sôbre o negro. Rio de Janeiro, 1958.
An 359.58.10 Campbell, Byram. Race and social revolution. N.Y., 1958.
An 359.58.15 Boer, Wolfgang de. Das Problem des Menschen und die Kultur. Bonn, 1958.
An 359.58.20 Lévi-Strauss, Claude. Anthropologie structurale. Paris, 1958-1973. 2v.
An 359.59 Haselden, Kyle. The racial problem in Christian perspective. 1. ed. N.Y., 1959.
An 359.59.5 Goldschmidt, W.R. Man's way. Cleveland, 1959.
An 359.59.10 Bibby, Cyril. Race, prejudice and education. London, 1959.
An 359.59.15 Bertram, Colin. Adam's brood. London, 1959.
An 359.59.20 Cox, Oliver Cromwell. Caste, class, and race. N.Y., 1959.
An 359.59.25 Bunker, Robert. The first look at strangers. New Brunswick, N.J., 1959.
An 359.59.35 Škerlj, Božo. Antropologija i etnologija. Beograd, 1959.
An 359.59.40 Finlay, William G. Races in chaos. Johannesburg, 1959.
An 359.59.45 Benedict, Ruth (Fulton). Race: science and politics. N.Y., 1959.
An 359.59.50 Benedict, Ruth (Fulton). Race and racism. London, 1959.
An 359.60 Medawar, Peter Brian. The future of man. London, 1960.
An 359.60.5 Royal Anthropological Institute of Great Britain and Ireland. Man, race and Darwin. London, 1960.
An 359.60.15 Ghent. Université. Het lever; eer serie lezingen gehouden in de aula van de Rijksuniversität te Gent. Gent, 1960.
An 359.61 Hsu, Francis L.K. Psychological anthropology. Homewood, Ill., 1961.
An 359.61.5 Thompson, L. Toward a science of mankind. N.Y., 1961.
An 359.61.10 Garn, S.M. Human races. Springfield, Ill., 1961.

## Classified Listing

An 355 - 360 Anthropology and ethnology in general - General treatises - Monographs (By date of issue) - cont.

An 359.61.15	Gehlen, Arnold. Anthropologische Forschung. Reinbek, 1961.
An 359.61.20	Grieger, Paul. La caracterologie ethnique. Paris, 1961.
An 359.61.25	Contemporary raciology and racism. Bloomington, 1961.
An 359.61.30	Dias, Jorge. Ensaios etnológicas. Lisboa, 1961.
An 359.61.35	Fried, Morton Herbert. Readings in anthropology. N.Y., 1961-64. 2v.
An 359.61.40	Montagu, Ashley. Man in process. 1. ed. Cleveland, 1961.
An 359.61.45	Evans-Pritchard, Edward Evan. Anthropology and history. Manchester, Eng., 1961.
An 359.61.50	Scheler, Max. Man's place in nature. Boston, 1961.
An 359.61.55	UNESCO. The race question in modern science; race and science. N.Y., 1961.
An 359.62	Coon, Carleton Stevens. The origin of races. N.Y., 1962.
An 359.62.5	Haste, Hans. Rasisner. Stockholm, 1962.
An 359.62.10	Mason, Philip. Prospero's magic. London, 1962.
An 359.62.15	Evans-Pritchard, Edward Evan. Essays in social anthropology. London, 1962.
An 359.62.20	Fiezefontaine, J. Du racisme à l'universalisme. Paris, 1962.
An 359.62.25	Snyder, L.L. The idea of racialism. Princeton, 1962.
An 359.62.30	George, W.C. The biology of the race problem. N.Y., 1962.
An 359.63	Jennings, J.D. Anthropology and the world of science. Salt Lake City, 1963.
An 359.63.5	Weyl, N. The geography of intellect. Chicago, 1963.
An 359.63.10	Hsu, Francis L.K. Clan, caste, club. Princeton, 1963.
An 359.63.15	Ciba Foundation. Man and his future. London, 1963.
An 359.63.20	Akademiia Nauk SSSR. Sovremennaia amerikanskaia etnografiia. Moskva, 1963.
An 359.63.25	Tazerout, M. Manifeste contre le racisme. Rodez, 1963.
An 359.63.30	Spelling, K. Miljøets indflydelse på intelligensvaviklingen. København, 1963.
An 359.63.35	Barabas, Jenő. Kartografiai modszer a neprajzban. Budapest, 1963.
An 359.63.40A	Lévi-Strauss, Claude. Structural anthropology. N.Y., 1963-76. 2v.
An 359.63.40B	Lévi-Strauss, Claude. Structural anthropology. N.Y., 1963-76. 2v.
An 359.64	Beattie, John. Other cultures. N.Y., 1964.
An 359.64.5	Lienhardt, G. Social anthropology. London, 1964.
An 359.64.10	Moskovskoe Obshchestvo Ispytatelei Prirody. Sovremennaia antropologiia. Moskva, 1964.
An 359.64.15	Köbben, A.J.F. Van primitieven tot medeburgers. Assen, 1964.
An 359.64.20	Isä, Ali Ahmad. Social anthropology in theory and practice. Cairo, 1964.
An 359.64.25	Wolf, Eric Robert. Anthropology. Englewood Cliffs, N.J., 1964.
An 359.64.30	Harris, Marvin. The nature of cultural things. N.Y., 1964.
An 359.64.35	Grottanelli, Virigi L. L'etnologia e le leggi della condotta umana. Roma, 1964.
An 359.64.40	Morgan, Lewis H. Ancient society. Cambridge, 1964.
An 359.64.45	Montagu, Ashley. Man's most dangerous myth. 4. ed. Cleveland, 1964.
An 359.65	Montagu, Ashley. The idea of race. Lincoln, 1965.
An 359.65.5	Montagu, Ashley. The human revolution. Cleveland, 1965.
An 359.65.10	Jürgens, Hans W. Beiträge zur menschlichen Typenkunde. Stuttgart, 1965.
An 359.65.15	Coon, Carleton Stevens. The living races of man. 1. ed. N.Y., 1965.
An 359.65.20	Murdock, George Peter. Culture and society. Pittsburgh, 1965.
An 359.65.25	Barzun, Jacques. Race; a study in superstition. N.Y., 1965.
An 359.66	Melady, Thomas Patrick. The revolution of color. N.Y., 1966.
An 359.66.5	Podol'nyi Roman G. Predki i my. Moskva, 1966.
An 359.67	Lévi-Strauss, Claude. The scope of anthropology. London, 1967.
An 359.67.5	Race and modern science. N.Y., 1967.
VAn 359.68	Tokarev, Sergei A. Oskovy etnografii. Moskva, 1968.
An 359.68.5A	American Association for the Advancement of Science. Science and the concept of race. N.Y., 1968.
An 359.68.5B	American Association for the Advancement of Science. Science and the concept of race. N.Y., 1968.
An 359.68.10	Poirier, Jean. Ethnologie générale. Paris, 1968.
An 359.68.15	Stocking, George W. Race, culture, and evolution. N.Y., 1968.
An 359.68.16	Stocking, George W. Race, culture, and evolution. N.Y., 1971.
An 359.68.25	Bates, Marston. Gluttons and libertines; human problems of being natural. N.Y., 1968.
An 359.68.30	Rassengeschichte der Menschheit. München, 1968- 3v.
An 359.69	Leiris, Michel. Cinq études d'ethnologie. Paris, 1969.
An 359.69.10	Blacking, John. Process and product in human society. Johannesburg, 1969.
An 359.69.15	Hiernaux, Jean. Égalité ou inégalité des races? Paris, 1969.
An 359.69.20	Etnologia. Wyd. 1. Warszawa, 1969.
An 359.69.25	Vuorela, Toivo. Kansatieteen periaateoppia. Helsinki, 1969.
An 359.69.30	Montagu, Ashley. Man, his first two million years. N.Y., 1969.
An 359.70	Korolev, Stanislav I. Voprosy etnopsikhologii v vabotakh zanebezhnykh avtarov. Moskva, 1970.
An 359.70.5	Lévi-Strauss, Claude. Claude Lévi-Strauss; ou, La structure et le malheur. Paris, 1970.
An 359.70.10	Kovalev, Sergei M. O chloveke, ego poraboshchenii i osvdsozhdenii. Moskva, 1970.
An 359.70.15	Stiglmayr, Engelbert. Ganzheitliche Ethnologie. Wien, 1970.
An 359.70.20	Costanzo, Giorgio. La costruzione dell'uomo. Roma, 1970.
An 359.70.25	Rimet, Michel. Contracts, interférences ethniques et culturelles. Montpellier, 1970.
An 359.70.30	Schweppe, John S. Man: a remarkable animal. Chicago, 1970.
An 359.71	Harris, Marvin. Culture, man, and nature. N.Y., 1971.
An 359.71.5	Goldsby, Richard A. Race and races. N.Y., 1971.
An 359.71.10	Wolf, Josef. Integral anthropology. Praha, 1971.
An 359.71.15	Cheboksarov, Nikolai N. Narody, rasy, kul'tury. Moskva, 1971.
An 359.71.20	Alland, Alexander. Human diversity. N.Y., 1971.
An 359.71.25	Murphy, Robert Francis. The dialectics of social life: alarms and excursions in anthropological theory. N.Y., 1971.

An 355 - 360 Anthropology and ethnology in general - General treatises - Monographs (By date of issue) - cont.

An 359.71.30	Dzhandil'din, Nurymbet. Priroda natsional'noi psikholgii. Alma-Ata, 1971.
An 359.71.40	Poliahov, Leon. La mythe aryen. Paris, 1971.
An 359.72	Bicchieri, M.G. Hunters and gatherers today. N.Y., 1972.
An 359.72.10	Alland, Alexander. The human imperative. N.Y., 1972.
An 359.72.15	Mendel, Gérard. Vers une anthropologie sociopsychanalytique. 1. éd. Paris, 1972.
An 359.72.20	Ansari, G. Recent trends in cultural anthropology. Wien, 1972.
An 359.72.25	Bezerra, Felte. Antropologia sociocultural. Brasilia, 1972.
An 359.73	Morin, Edgar. La paradigme perdu: la nature humaine. Paris, 1973.
An 359.73.5	Goertz, Clifford. The interpretation of cultures. N.Y., 1973.
An 359.73.10	Bromlei, Iulian V. Etnos i etnografiia. Moskva, 1973.
An 359.73.15	Etnologicheskie issledovaniia za rubezhom. Moskva, 1973.
An 359.73.20	Sunderland, Eric. Elements of human and social geography. 1. ed. Oxford, 1973.
An 359.73.25	Waligórski, Andrzej. Antropologiczna koncepcja człowieka. Wyd. 1. Warszawa, 1973.
An 359.73.30	Jaulin, Robert. Gens du soi, gens de l'autre. Paris, 1973.
An 359.73.35	Bradfield, Maitland. A natural history of associations. London, 1973. 2v.
An 359.74	Baker, John Randal. Race. N.Y., 1974.
An 359.74.5	Diamond, Stanley. In search of the primitive. N.Y., 1974.
An 359.74.10	Edgerton, Robert B. Methods and styles in the study of culture. San Francisco, 1974.
An 359.74.15	Montagu, Ashley. Man's most dangerous myth. 5. ed. N.Y., 1974.
An 359.74.20	Szyfelbejn-Sokolewicz, Zofia. Wprowadzenie do etnologii. Wyd. 1. Warszawa, 1974.
An 359.74.25	Langness, Lewis L. The study of culture. San Francisco, 1974.
An 359.74.30	Rasogeneticheskie protesy v etnicheskoi istorii. Moskva, 1974.
An 359.75	Wagner, Roy. The invention of culture. Englewood Cliffs, 1975.
An 359.75.5	Banton, Michael P. The race concept. Newton Abbot, 1975.
An 359.75.10	Goodall, Vanne Morris. The quest for man. N.Y., 1975.
An 359.75.15	Racial variation in man: proceedings of a symposium held at the Royal Geographical Society, London, on 19 and 20 September 1974. London, 1975.

An 375 Anthropology and ethnology in general - Anthropogeography - Pamphlet volumes

An 375.5	Urabayen, Leoncio. Geografia umana. Madrid, 1934. 2 pam.

An 376 - 380 Anthropology and ethnology in general - Anthropogeography - General works (By date of issue)

An 378.24	Bonstetten, K.V. von. L'homme du midi et l'homme du nord. Genève, 1924.
An 378.24.5	Gioja, M. Riflessioni...su l'opera "L'homme du midi". Milano, 1830.
An 378.58	Rathlef, C. Die welthistorische Bedeuting dei Meere insbesondere des Mittelmeers. Dorpat, 1858.
An 378.82	Ratzel, F. Anthropo-Geographie. Stuttgart, 1882. 2v.
An 378.84	Bordier, A. La géographie médicale. Paris, 1884.
An 378.99	Ratzel, Friedrich. Anthropogeographie. v.2. Stuttgart, 1899.
An 378.99.1	Ratzel, Friedrich. Anthropogeographie. Darmstadt, 1975.
An 379.00	Matteweuzzi, A. Les facteurs de l'évolution des peuples. Bruxelles, 1900.
An 379.08	Richthofen, Ferdinand von. Vorlesungen. Berlin, 1908.
An 379.11	Semple, E.M. Influences of geographic environment on the basis of Ratzel's system. N.Y., 1911.
An 379.12	Vallaux, C. Géographie sociale. La mer. Paris, 1908.
An 379.12.3	Vallaux, C. Géographie sociale. Le sol et l'état. Paris, 1911.
An 379.12.5	Maranelli, C. La geografia umana di Jean Brunhes. Firenze, 1912.
An 379.14	Kirchhoff, Alfred. Man and earth. N.Y., 1914.
An 379.18A	Koller, A.H. The theory of environment. Menasha, Wis., 1918.
An 379.18B	Koller, A.H. The theory of environment. Menasha, Wis., 1918.
An 379.19	Schulz, George J. Geography. College Park, Md., 1919.
An 379.20.3	Brunhes, Jean. Human geography. Chicago, 1920.
An 379.21A	Huntington, Ellsworth. Principles of human geography. N.Y., 1921.
An 379.21B	Huntington, Ellsworth. Principles of human geography. N.Y., 1921.
An 379.21.3	McDougall, William. National welfare and national decay. London, 1921.
An 379.21.5	Krebs, Norbert. Die Verbreitung des Menschen auf der Erdoberfläche (Anthropogeographie). Leipzig, 1921.
An 379.22	Vidal de la Blache, P. Principes de géographie humaine. Paris, 1922.
An 379.22.3A	The evolution of man. New Haven, 1922.
An 379.22.3B	The evolution of man. New Haven, 1922.
An 379.22.4	Baitsell, George A. The evolution of man. New Haven, 1923.
An 379.22.5	Von Engeln, Oscar D. Inheriting the earth. N.Y., 1922.
An 379.22.7	Huntington, Ellsworth. Principles of human geography. 2. ed. N.Y., 1922.
An 379.22.9	Kaltenbach, Ernst. Beiträge zur Anthropographie des Bodenseegebiets. Inaug. Diss. Basel, 1922.
An 379.24	Huntington, Ellsworth. Character of races as influenced by...environment. N.Y., 1924.
An 379.24.3	Jefferson, M.S.W. Man in Europe, here and there. Ypsilanti, Mich., 1924.
An 379.24.5	Huntington, Ellsworth. Principles of human geography. 3. ed. N.Y., 1924.
An 379.25	Brunhes, Jean. La géographie humaine. 3. éd. Paris, 1925. 3v.
An 379.26A	Vidal de la Blache, P. Principles of human nature. N.Y., 1926.
An 379.26B	Vidal de la Blache, P. Principles of human nature. N.Y., 1926.
An 379.26C	Vidal de la Blache, P. Principles of human nature. N.Y., 1926.
An 379.26D	Vidal de la Blache, P. Principles of human nature. N.Y., 1926.
An 379.26.5	Günther, Hans F.K. Rasse und Stil. München, 1926.

Classified Listing

An 376 - 380 Anthropology and ethnology in general - Anthropogeography - General works (By date of issue) - cont.

An 379.26.7	Günther, Hans F.K. Rasse und Stil. 2. Aufl. München, 1927.
An 379.27A	Taylor, Griffith. Environment and race. London, 1927.
An 379.27B	Taylor, Griffith. Environment and race. London, 1927.
An 379.27.3	Huntington, Ellsworth. The human habitat. N.Y., 1927.
An 379.28	Fleure, H.J. The races of mankind. London, 1928.
An 379.28.3	Huntington, Ellsworth. The human habitat. London, 1928.
An 379.28.5	Günther, Hans F.K. The racial elements of European history. N.Y., 1928.
An 379.30	Hauemeyer, Loomis. Ethnography. Boston, 1929.
An 379.30.3	Huntington, C.C. Environmental basis of social geography. N.Y., 1930.
An 379.30.5	Huntington, Ellsworth. The human habitat. London, 1930.
An 379.30.7	Peate, I.C. Studies in regional consciousness and environment. London, 1930.
An 379.30.9	Conference on Regional Phenomena, Washington, D.C., 1930. Conference...Apr. 11-12, 1930. Washington, 1930.
An 379.31	Penrose, R.A.F. Geology as an agent in human welfare. N.Y., 1931.
An 379.32	Hubbard, George D. Geographic conditions. n.p., 1932.
An 379.33A	McIverny, A.J. The rôle of the deserts. London, 1933.
An 379.33B	McIverny, A.J. The rôle of the deserts. London, 1933.
An 379.33.5	Bryan, P.W. Man's adaptation of nature. London, 1933.
An 379.34	Forde, C.D. Habitat, economy and society. London, 1934.
An 379.34.2	Forde, C.D. Habitat, economy and society. N.Y., 1937.
An 379.34.3	Brunhes, Jean. La géographie humaine. 4. éd. Paris, 1934. 3v.
An 379.35.3	Bews, J.W. Human ecology. London, 1935.
An 379.36	Marett, J.R. de la H. Race, sex, and environment. London, 1936.
An 379.36.3	Lester, P. Les races humaines. Paris, 1936.
An 379.37	Machin, A. Darwin's theory applied to mankind. London, 1937.
An 379.37.3A	Taylor, G. Environment, race and migration. Chicago, 1937.
An 379.37.3B	Taylor, G. Environment, race and migration. Chicago, 1937.
An 379.38	McIverny, A.J. Through the great arid filter (man's drift to Europe). London? 1938.
An 379.38.5	Sanchez, Pedro C. Enseñanzas fundamentales de la geografia humana. Tacubaya, 1938.
An 379.39.5	Deffontaines, P. Problèmes de géographie humaine. Paris, 1939.
An 379.39.10	Hardy, Georges. La géographie psychologique. Paris, 1939.
An 379.41	Pasada, J. de la C. Geografia humana. Medellín, 1941.
An 379.41.5	Vermooten, Willem H. De mens in de geografie. Assen, 1941.
An 379.42	Cools, R.H.A. De geographische gedechte bij Jean Brunhes. Utrecht, 194-.
An 379.42.5	Brunhes, Jean. La géographie humaine. Paris, 1942.
An 379.43	Sorre, Maximilian. Les fondements biologiques de la géographie. v.1-3. Paris, 1943-52. 4v.
An 379.46	Taylor, G. Our evolving civilization. Toronto, 1946.
An 379.46.5	Taylor, G. Environment, race and migration. 2. ed. Chicago, 1946.
An 379.46.10	Amorim Girão, Aristides de. Geografia humana. Porto, 1946.
An 379.47	Hettner, Alfred. Allgemeine Geographie des Menschen. Stuttgart, 1947.
An 379.47.5	Huntington, Ellsworth. Principles of human geography. 5. ed. N.Y., 1947.
An 379.47.10	Sorre, Maximilien. Les fondements de la géographie humaine. 2. éd. Paris, 1947-
An 379.49	Le Lannou, M. La géographie humaine. Paris, 1949.
An 379.50F	Childe, Vere Gordon. Prehistoric migrations in Europe. Cambridge, 1950.
An 379.51	Huntington, Ellsworth. Principles of human geography. 6. ed. N.Y., 1951.
An 379.52	Brunhes, Jean. Human geography. London, 1952.
An 379.52.5F	Gutkind, E.A. Our world from the air. London, 1952.
An 379.52.10	Dickinson, Robert Eric. City, region and regionalism. London, 1952.
An 379.53	Baxter, William J. Today's revolution in weather. N.Y., 1953.
An 379.53.5	Kimble, G.H.T. The way of the world. N.Y., 1953.
An 379.54	Brown, H.S. The challenge of man's future. N.Y., 1954.
An 379.56.1	International Symposium on Man's Role in Changing the Face of the Earth, Princeton, N.J., 1955. Man's role in changing the face of the earth. Chicago, 1971.
An 379.57	Sears, Paul. The ecology of man. Eugene, 1957.
An 379.60	Dickinson, Robert Eric. Some problems of human geography. Leeds, 1960.
An 379.61	Maas, Walther. Probleme der Sozialgeographie. Berlin, 1961.
An 379.61.5	Sorre, Maximilien. L'homme sur la terre. Paris, 1961.
An 379.61.10	Otto, Klaus. Das Aufkommen sozialgeographischer Betrachtungsweisen in der deutschen länderkundlichen Literatur seit Beginn des 20. Jahrhunderts. Inaug. Diss. Münster, 1961.
An 379.63	Mills, C.A. World power and shifting climates. Boston, 1963.
An 379.64	Troëng, Ivan. Kulturer före istiden. Uppsala, 1964.
An 379.67	Glacken, Clarence J. Traces on the Rhodian shore. Berkeley, 1967.
An 379.67.5	Russell, William Moy Stratton. Man, nature and history. London, 1967.
An 379.67.10	Lundman, Bertil. Geographische Anthropologie. Stuttgart, 1967.
An 379.68	Priroda i obshchestvo. Moskva, 1968.
An 379.68.5	Zum Standort der Sozialgeographie. Kallmünz, 1968.
An 379.68.10	Gerling, Walter. Die Problematik der Sozialgeographie. Würzburg, 1968.
An 379.68.15	Heinemeyer, W.F. De sociale geografie in de rij van de sociale wetenschappen. Meppel, 1968.
An 379.68.20	Vermooten, Willem Hendrik. Sociografie en sociale geografie in Nederland. Assen, 1968.
An 379.69	Storkebaum, Werner. Sozialgeographie. Darmstadt, 1969.
An 379.70	Maršík, Miroslav. Natural environment and society in the theory of geographical determinism. Praha, 1970.
An 379.71	Chisholm, Michael. Research in human geography. London, 1971.
An 379.72	Cox, Kevin R. Man, location, and behavior: an introduction to human geography. N.Y., 1972.
An 379.72.5	Lopatina, Elena B. Otsenka prirodnykh uslovii zhizni naseleniia. Moskva, 1972.

An 376 - 380 Anthropology and ethnology in general - Anthropogeography - General works (By date of issue) - cont.

An 379.72.10	Niemeier, Georg. Siedlungsgeographie. 3. Aufl. Braunschweig, 1972.
An 379.73	Glacken, Clarence J. Traces on the Rhodian shore. Berkeley, 1973.
An 379.73.5	Johnston, Ronald John. Spatial structures. London, 1973.
An 379.73.10	Milkov, Fedor N. Chelovek in landshafty. Moskva, 1973.
An 379.73.15	Gourou, Jierre. Pour une géographie humaine. Paris, 1973.
An 379.73.20	Studies in human geography. London, 1973.
An 379.74	Alekseev, Valerii P. Geografiia chelovecheskikh ras. Moskva, 1974.

An 606 - 610 Anthropology and ethnology in general - Anthropological museums (By date of issue)

An 608.62	Jomard, E.F. Classification méthodique des produits de l'industrie. Paris, 1862.
An 608.70	Steinhauer, C.L. Kort vriledning i det kgl. ethnographiske museum. København, 1870.
An 608.78	Victoria, Australia. Public Library, Museum and National Galery. Catalogue of the objects of ethnotypical art. Melbourne, 1878.
An 608.81	Hamburg, Germany. Museum. Die ethnographisch-anthropologische AbtheilungMZ Godeffroy. Hamburg, 1881.
An 608.87	Moscow. Dashkovskii Etnograficheskii Muzei. Sistematicheskoe opisanie kollektsii Dashkovskago etnograficheskago muzeia. v.4. Moskva, 1895.
An 608.609.07	Nielsen, Y. Universitets ethnografiske samlinger, 1857-1907. Christiania, 1907.
An 609.07.2	Neilsen, Y. Universitets lappiske samlinger, 1857-1911. Christiania, 1911.
An 609.10	British Museum. Handbook to ethnographical collections. Oxford, 1910.
An 609.21	British Museum (Natural History). Department of Zoology. Guide to the specimens illustrating the races of mankind. 4. ed. London, 1921.
An 609.22	British Museum (Natural History). Department of Geology. Guide to the fossil remains of man. 3. ed. London, 1922.
An 609.25	British Museum. Department of Ancient and Mediaeval Antiquities and Ethnography. Handbook of the ethnographical collections. 2. ed. London, 1925.
An 609.28	Kölner Anthropologische Gesellschaft. 25 Jahre Kölner anthropologische Gesellschaft und städtisches Museum für Vor- und Frühgeschichte 1903-28. Festschrift. Köln, 1928.
An 609.67	Zagreb. Etnografski Muzej. Daleki sujtovi naših putnika i pomoraea. Zagreb, 1967.
An 609.69	Leipzig. Staedtisches Museum für Völkerkunde. Zum hundertjährigen Bestehen, 1869-1969. Berlin, 1969.
An 609.70A	Abel, Herbert. Vom Raritätenkabinett zum Bremener Überseemuseum. Bremen, 1970.
An 609.70B	Abel, Herbert. Vom Raritätenkabinett zum Bremener Überseemuseum. Bremen, 1970.
An 609.72	Materialy po rabote i istorii etnograficheskikh muzeev i vystavok. Moskva, 1972.

An 626 - 630 Anthropology and ethnology in general - Research methods and apparatus (By date of issue)

An 628.68	Gibbs, G. Instructions for research...ethnology and philology of America. Washington, 1863-68. 3 pam.
An 628.92	Brinton, D.G. Anthropology. Philadelphia, 1892.
An 629.03	Keller, A.G. Queries in ethnography. N.Y., 1903.
An 629.07	Frazer, J.G. Questions on customs, beliefs and languages of savages. Cambridge, 1907.
An 629.11	Graebner, Fritz. Methode der Ethnologie. Heidelberg, 1911.
An 629.36	Lévi-Strauss, D. Instruções praticas para pesquisas de antropologia fisica e cultural. pt.1. São Paulo, 1936-
An 629.37	Schultz, Bruno K. Taschenbuch der rassenkundlichen Messtechnik. München, 1937.
An 629.38	Balfour, Henry. Spinners and weavers in anthropological research. Oxford, 1938.
An 629.50	Yale University. Institute of Human Relations. Outline of cultural material. 3. ed. New Haven, 1950.
An 629.53	Mead, M. The study of culture at a distance. Chicago, 1953.
An 629.60	Inverarity, Robert Bruce. Visual files coding index. Bloomington, 1960.
An 629.61	Lindzey, G. Projective techniques and cross-cultural research. N.Y., 1961.
An 629.61.5	Jones, W.T. The romantic syndrome. The Hague, 1961.
An 629.63	Evans-Pritchard, E.E. The comparative method in social anthropology. London, 1963.
An 629.64	Jarvie, Ian C. The revolution in anthropology. London, 1964.
An 629.65	Conference on New Approaches in Social Anthropology. The relevance of models for social anthropology. London, 1965.
An 629.67	Jongnians, Douve Geert. Anthropologists in the field. Assen, 1967.
VAn 629.67.5	Pešić-Golutovic, Zagorka. Antropologia kao drustveno nauka. Beograd, 1967.
An 629.68	Research in social anthropology. London, 1968.
An 629.68.5	Pouwer, Jan. Translation at sight; the job of a social anthropologist. Wellington, 1968.
An 629.70	Akademiia Nauk SSSR. Institut Etnografii. Voprosy metodiki etnograficheskikh i etno-sotsiologicheskih issledovanii. Moskva, 1970.
An 629.70.5	Sarmela, Matti. Perinnerineiston kvantitatiivisesta tutkimuksesta. Helsinki, 1970.
An 629.71	Tennekes, J. Anthropology, relativism and method. Assen, 1971.
An 629.74	Vogt, Evon Zartman. Aerial photography in anthropological field research. Cambridge, 1974.

An 2052 Prehistoric and primitive man - Bibliographies

An 2052.5	Divale, William T. Warfare in primitive societies; a selected bibliography. Los Angeles, 1971.

An 2056 - 2060 Prehistoric and primitive man - Antiquity and origin of man (By date of issue)

Htn	An 2056.55*	La Peyrère, I. de. Systema theologicum...Praeadamitae. n.p., 1655.
Htn	An 2056.56*	Peyrère, I. de. Men before Adam: discourse on Romans 5-12,13,14. London, 1656.
	An 2056.56.5	Romano, E. Animad versiones in librum prae-adamitarum. Paris, 1656.
	An 2056.61	Peyrère, I. de. Praeadamiten. Jaer, 1661.

Classified Listing

An 2056 - 2060 Prehistoric and primitive man - Antiquity and origin of man (By date of issue) - cont.

An 2058.63 Lyell, C. The geological evidence of the antiquity of man. Philadelphia, 1863.
An 2058.63.2 Lyell, C. The geological evidence of the antiquity of man. 2d American ed. Philadelphia, 1863.
An 2058.65 Thioly, F. Débris de l'industrie humaine...caverne de Bossey. Genève, 1865.
An 2058.65.3 Gastaldi, B. Lake habitations and pre-historic remains...north, centre Italy. London, 1865.
An 2058.65.5 Pereira da Costa, F.A. Da existencia do homem em epochas remotas no valle do tejo. Lisboa, 1865.
An 2058.68 Saporta, G. da. La paléontologie appliquée a l'étude des races humaines. n.p., 1868.
An 2058.69 Bromby, J.E. Pre-historic man. Melbourne, 1869.
An 2058.73 Lyell, C. The geological evidence of the antiquity of man. 4. ed. London, 1873.
An 2058.74 Ratzel, F. Vorgeschichte des europäischen Menschen. München, 1874.
An 2058.74.5 Whitmore, J.H. Evidences of the antiquity of man. Rochester, N.Y., 1874.
An 2058.75 Southall, J.C. The recent origin of man. Philadelphia, 1875.
An 2058.75.5 Lester, A. Hoyle. The pre-adamite. Philadelphia, 1875.
An 2058.76 Rau, C. Early man in Europe. N.Y., 1876.
An 2058.76.3 Meunier, Victor. Los antepasados de Adan. Madrid, 1876.
An 2058.77 Dawkins, W.B. Address on the antiquity of man. Salford, 1877.
An 2058.77.3 Jones, T.R. Lecture on the antiquity of man. London, 1877.
An 2058.77.5 Day, J.V. The prehistoric use of iron and steel. London, 1877.
An 2058.78 Winchell, A. Adamites and preadamites. Syracuse, 1878.
An 2058.78.3 Southall, J.C. Address on man's age in the world. Richmond, 1878.
An 2058.78.5 Southall, J.C. The epoch of the Mammoth. Philadelphia, 1878.
An 2058.79 Piette, Edouard. Nomenclature des temps anthropiques primitifs. Laon, 1879.
An 2058.80 Winchell, A. Preadamites. Chicago, 1880.
An 2058.81 Evans, J. A few words on tertiary man. n.p., 1881.
An 2058.81.5 Sunderland, J.T. Dr. Winchell's "preadamites". n.p., 1881.
An 2058.83 Rawlinson, G. The antiquity of man historically considered. N.Y., 1883.
An 2058.83.5 Girard de Rialle, J. Nos ancêtres. Paris, 1883.
An 2058.84 Morse, E.S. Man in the tertiaries. Salem, 1884.
An 2058.85.3 Topinard, Paul. Eléments d'anthropologie générale. Paris, 1885.
An 2058.87 Burge, Lorenzo. Pre-glacial man and the Aryan race. 2. ed. Boston, 1887.
An 2058.92 Doughty, F.W. Evidences of man in the drift. N.Y., 1892.
An 2058.92.3 Wright, G.F. The antiquity and origin of the human race. Boston, 1892.
An 2059.01 Chantre, E. L'homme quaternaire. Paris, 1901.
An 2059.03 Hoernes, Moriz. Der diluviale Mensch in Europa. Braunschweig, 1903.
An 2059.03.5 Joby, N. Man before metals. N.Y., 1903.
An 2059.03.7 Mortillet, G. de. Musée prehistorique. 2. éd. Paris, 1903.
An 2059.05 Maccurdy, G.G. The eolithic problem. Lancaster, 1905.
An 2059.07 Lapparent, A. Les silex taillés et l'ancienneté de l'homme. Paris, 1907.
An 2059.11 Sollas, William J. Ancient hunters. London, 1911.
An 2059.12A Duckworth, W.L.H. Prehistoric man. Cambridge, 1912.
An 2059.12B Duckworth, W.L.H. Prehistoric man. Cambridge, 1912.
An 2059.12.3 Jacob, K.H. Der diluviale Mensch und seine Zeitgenossen aus dem Tierreiche. Leipzig, 1912.
An 2059.12.5A Keith, Arthur. Ancient types of man. 2. ed. London, 1912.
An 2059.12.5B Keith, Arthur. Ancient types of man. 2. ed. London, 1912.
An 2059.14 Geike, James. The antiquity of man in Europe. N.Y., 1914.
An 2059.14.5 Lyell, C. The geological evidence of the antiquity of man. London, 1914.
An 2059.15 British Museum (Natural History). Department of Geology. A guide to the fossil remains of man in the department. London, 1915.
An 2059.22 Soergel, W. Die Jagd der Vorzeit. Jena, 1922.
An 2059.23 Boule, Marcellin. Les hommes fossiles. 2. éd. Paris, 1923.
An 2059.24 Sollas, William J. Ancient hunters and their modern representatives. 3. ed. N.Y., 1924.
An 2059.24.1 Sollas, William J. Ancient hunters and their modern representatives. 3. ed. London, 1924.
An 2059.25 Black, Davidson. Asia and the dispersal of primates. Peking, 1925.
An 2059.25.5 Vulliamy, C.E. Our prehistoric forerunners. N.Y., 1925.
An 2059.27 Lyell, C. The antiquity of man. London, 1927.
An 2059.29 Hrdlička, A. The neanderthal phase of man. Washington, 1929.
An 2059.29.5 Wissler, Clark. An introduction to social anthropology. N.Y., 1929.
An 2059.30 Keith, Arthur. New discoveries relating to the antiquity of man. N.Y., 1930.
An 2059.31 Duncan, G.S. Prehistoric man. Boston, 1931.
An 2059.33 Black, Davidson. Fossil man in China. Peiping, 1933.
An 2059.37 International Symposium on Early Man. Early man. Philadelphia, 1937.
An 2059.41 Mattos, Anibal. A raça de Lagôa Santa. São Paulo, 1941.
An 2059.51 Coates, Adrian. Prelude to history. London, 1951.
An 2059.53 Leakey, Louis S. Adam's ancestors. 4. ed. London, 1953.
An 2059.54 Masiker, Reuben. The Australopithecinae. Cape Town, 1954.
An 2059.55 Vere, Francis. The Piltdown fantasy. London, 1955.
An 2059.55.5 Weiner, Joseph S. The Piltdown forgery. London, 1955.
An 2059.59 Lange, Hurt. Fremdling zwischen Tier und Gott. Gütersloh, 1959.
An 2059.59.5 Dart, Raymond. Adventures with the missing link. N.Y., 1959.
An 2059.59.10 Howells, William. Mankind in the making. Garden City, 1959.
An 2059.59.15 Anthropological Society of Washington. Evolution and anthropology. Washington, 1959.
An 2059.59.20 Nougier, L.R. Geographie humaine prehistorique. Paris, 1959.
An 2059.59.25 Overhage, Paul. Um des Erscheinungsbild der ersten Menschen. Basel, 1959.
An 2059.61 Ardrey, Robert. African genesis. N.Y., 1961.
An 2059.61.5 Overhage, Paul. Das Problem der Hominisation. Freiburg, 1961.

An 2056 - 2060 Prehistoric and primitive man - Antiquity and origin of man (By date of issue) - cont.

An 2059.61.10A Leakey, Louis S. The progress and evolution of man in Africa. London, 1961.
An 2059.61.10B Leakey, Louis S. The progress and evolution of man in Africa. London, 1961.
An 2059.63 Muschalek, H. Urmensch-Adam. Berlin, 1963.
An 2059.65 Atlas de prehistoire. Paris, 1965. 3v.
An 2059.66.5 Semenov, Iurii I. Kak vozniklo chelovechestvo. Moskva, 1966.
An 2059.67 Zinevich, Galina P. Ocherki paleoantropologii Ukrainy. Kiev, 1967.
An 2059.68.5 Nantevil, Hugues de. Sur les traces d'Adam; nouvel aperçu sur les origines de l'homme. Paris, 1968.
An 2059.69 Pfeiffer, John. E. The emergence of man. N.Y., 1969.
An 2059.69.5 Overhage, Paul. Menschenformen im Eiszeitalter. 1. Aufl. Frankfurt, 1969.
An 2059.69.10 Altner, Günter. Kreatur Mensch. München, 1969.
An 2059.72 Millar, Ronald William. The Piltdown men. London, 1972.
An 2059.72.5 Matiushin, Geral'd N. U kolybeli istorii. Moskva, 1972.

An 2106 - 2110 Prehistoric and primitive man - Primitive customs and institutions - General works (By date of issue)

An 2108.45 Brotonne, F. de. Civilisation primitive. Paris, 1845.
An 2108.62 Wilson, D. Prehistoric man. Cambridge, 1862. 2v.
An 2108.68 Bastian, A. Beiträge zur vergleichenden Psychologie. Berlin, 1868.
An 2108.69 Lubbock, J. Pre-historic times. 2. ed. London, 1869.
An 2108.69.3 Campbell, G.D. Primeval man. N.Y., 1869.
An 2108.69.4 Campbell, G.D. Primeval man. N.Y., 1869.
An 2108.69.5 Campbell, G.D. Primeval man. London, 1869.
An 2108.69.7 Bourlot, J. Histoire de l'homme préhistorique anté et postdiluvien. Paris, 1869.
An 2108.70 Lubbock, J. The origin of civilisation and the primitive condition of man. 2. ed. London, 1870.
An 2108.70.2 Lubbock, J. The origin of civilisation and the primitive condition of man. London, 1870.
An 2108.70.5 Figuier, L.G. L'homme primitif. 2. ed. Paris, 1870.
An 2108.70.6 Figuier, L.G. Primitive man. N.Y., 1870.
An 2108.70.7 Figuier, L.G. Primitive man. N.Y., 1870.
An 2108.70.8 Tylor, Edward Burnett. Researches into the early history of mankind. London, 1870.
An 2108.71.1 Tylor, Edward Burnett. Primitive culture: researches into the development of mythology. N.Y., 1974. 2v.
An 2108.71.3 Avebury, J.L. The origin of civilization and the primitive condition of man. N.Y., 1871.
An 2108.73 Tylor, Edward Burnett. Die Anfänge der Cultur. v.1-2. Leipzig, 1873.
An 2108.73.3 Clodd, Edward. The childhood of the world. 2. ed. London, 1873.
An 2108.73.10 Pireteau, Jean. Origine et destinée de l'homme. Paris, 1973.
An 2108.73.15 Cahen, Daniel. Un site tshitolien sur le plateau des Bateke. Tervuren, 1973.
An 2108.74 Tylor, Edward Burnett. Primitive culture. Boston, 1874. 2v.
An 2108.74.3 Dawkins, W.B. Cave hinting. London, 1874.
An 2108.74.5 Clodd, Edward. The childhood of the world. Boston, 1874.
An 2108.74.7 Zaborowski-Moindron, S. De l'ancienneté de l'homme. v.1-2. Paris, 1874.
An 2108.75 Clodd, Edward. The childhood of the world. 3. ed. London, 1875.
An 2108.75.3 Avebury, J.L. Die Entstehung der Civilisation. 3e Aufl. Jena, 1875.
An 2108.75.5 Avebury, J.L. The origin of civilisation. 3. ed. London, 1875.
An 2108.76 Wilson, D. Prehistoric man. 3. ed. London, 1876. 2v.
An 2108.77 Caspari, O. Die Urgeschichte der Menschheit. Leipzig, 1877. 2v.
An 2108.77.3 Tylor, Edward Burnett. Primitive culture. N.Y., 1877. 2v.
An 2108.78 Zaborowski-Moindron, S. L'homme préhistorique. Paris, 1878.
An 2108.78.3 Argyll, G.D.C. Primeval man. N.Y., 1878.
An 2108.79.5 Greenwood, James. The wild man at home, or Pictures of life in savage lands. London, 1879.
An 2108.80 Geiger, E.L. Contribution to the history of the development of the human race. London, 1880.
An 2108.80.3 Clodd, Edward. Childhood of the world. London, 1880.
An 2108.82 Avebury, J.L. The origin of civilisation and the primitive condition of man. London, 1882.
An 2108.82.2 Avebury, J.L. The origin of civilisation and the primitive condition of man. 4. ed. London, 1882.
An 2108.82.5 Avebury, J.L. The origin of civilisation and the primitive condion of man. 4. ed. N.Y., 1882.
An 2108.83 Dawson, J.W. Fossil men and their modern representatives. 2. ed. London, 1883.
An 2108.84 Van Overloop, E. Sur une méthode a suivre dans les études dites préhistoriques. Bruxelles, 1884.
An 2108.84.5F Römer, F. The bone caves of Ojcow in Poland. London, 1884.
An 2108.88 Debierre, C. L'homme avant l'histoire. Paris, 1888.
An 2108.89 Keary, C.F. The dawn of history. N.Y., 1889.
An 2108.91A Tylor, Edward Burnett. Primitive culture. 3. ed. London, 1891. 2v.
An 2108.91B Tylor, Edward Burnett. Primitive culture. 3. ed. London, 1891. 2v.
An 2108.91.3 Taylor, I. The prehistoric races of Italy. Washington, 1891.
An 2108.92 Hoernes, Moriz. Die Urgeschichte des Menschen. Wien, 1892.
An 2108.94 Hewitt, J.F. The ruling races of prehistoric times. Westminster, 1894-95. 2v.
An 2108.95 Starr, F. Some first steps in human progress. Meadville, Pa., 1895.
An 2108.95.3 Clodd, Edward. The story of "primitive" man. N.Y., 1895.
An 2108.95.5 Mason, O.T. Origins of invention. London, 1895.
An 2108.97 Hoernes, Moriz. Urgeschichte der Menschheit. 2e Aufl. Leipzig, 1897.
An 2109.00 Avebury, J.L. Prehistoric times. London, 1900.
An 2109.00.2 Avebury, J.L. Prehistoric times. 6. ed. N.Y., 1900.
An 2109.00.3 Schurtz, H. Urgeschichte der Kultur. Leipzig, 1900.
An 2109.01.3 Frobenius, L. Aus dem Flegel Jahren der Menschheit. Hannover, 1901.
An 2109.01.5 Bastian, A. Der Menschheitsgedanke durch Raum und Zeit. Berlin, 1901.
An 2109.01.7 Finot, J. Race prejudice. N.Y., 1907.

Classified Listing

An 2106 - 2110 Prehistoric and primitive man - Primitive customs and institutions - General works (By date of issue) - cont.

	An 2109.01.9	Starr, F. Some first steps in human progress. Cleveland, 1901.
	An 2109.01.15	Radliński, Ignacy. Przeszłość w teraźniejszości. Warszawa, 1901.
	An 2109.02	Schurtz, H. Altersklassen und Männerbünde. Berlin, 1902.
	An 2109.03A	Tylor, Edward Burnett. Primitive culture, researches. London, 1903. 2v.
	An 2109.03B	Tylor, Edward Burnett. Primitive culture, researches. London, 1903. 2v.
	An 2109.03.5	Hellwig, A. Das Asylrecht der Naturvölker. Berlin, 1903.
	An 2109.06	Pitt-Rivers, A.L.F. The evolution of culture and other essays. Oxford, 1906.
	An 2109.06.3	Zuccarelli, A. Gli uomini primitivi delle selci e delle caverne. Napoli, 1906.
	An 2109.08	Webster, H. Primitive secret societies. N.Y., 1908.
	An 2109.09A	Thomas, W.I. Source book for social origins. Chicago, 1909.
	An 2109.09B	Thomas, W.I. Source book for social origins. Chicago, 1909.
	An 2109.09.5	Thomas, W.I. Source book for social origins. 6th ed. Boston, 1909.
	An 2109.09.7A	Crawley, A.E. The idea of the soul. London, 1909.
	An 2109.09.7B	Crawley, A.E. The idea of the soul. London, 1909.
	An 2109.09.9	Somló, Felix. Der Güterverkehr in der Urgesellschaft. Bruxelles, 1909.
	An 2109.09.11	Bölsche, W. Der Mensch der Vorzeit. v.1-2. Stuttgart, 1909-11.
	An 2109.10A	Lévy-Bruhl, L. Les fonctions mentales dans les sociétées inférieures. Paris, 1910.
	An 2109.10B	Lévy-Bruhl, L. Les fonctions mentales dans les sociétées inférieures. Paris, 1910.
	An 2109.11	Avebury, J.L. Marriage, totenism, and relggion. London, 1911.
	An 2109.11.3A	Boas, Franz. The mind of primitive man. N.Y., 1911.
	An 2109.11.3B	Boas, Franz. The mind of primitive man. N.Y., 1911.
	An 2109.11.7	Visscher, H. Religion und soziales Leben bei Naturvölkern. Bonn, 1911. 2v.
	An 2109.12	Classen, K. Die Völker Europas zur jüngeren Steinzeit. Stuttgart, 1912.
	An 2109.13	Avebury, J.L. Prehistoric times. 7. ed. London, 1913.
	An 2109.13.10	Buttel-Reepen, H. Man and his forerunners. London, 1913.
	An 2109.14	Boas, Franz. Kultur und Rasse. Leipzig, 1914.
	An 2109.15	Hobhouse, L.T. The material culture of the simpler peoples. London, 1915.
	An 2109.15.3	Elliot, G.F.S. Prehistoric man and his story. London, 1915.
	An 2109.15.4	Elliot, G.F.S. Prehistoric man and his story. Philadelphia, 1915.
	An 2109.15.5A	Osborn, H.F. Men of the old stone age. N.Y., 1915.
	An 2109.15.5B	Osborn, H.F. Men of the old stone age. N.Y., 1915.
	An 2109.16	Smith, G.E. Primitive man. London, 1916.
	An 2109.17	Koppers, W. Die ethnologische Wirtschaftsforschung. Wien, 1917.
	An 2109.18.3	Aberg, Nils. Das nordische Kultur-Gebiet in Mitteleuropa. v.1-2. Uppsala, 1918.
	An 2109.20	Tylor, Edward Burnett. Primitive culture. 6. ed. London, 1920. 2v.
	An 2109.20.2	Tylor, Edward Burnett. Primitive culture. 1st American ed. Boston, 1874. 2v.
	An 2109.21.3	Fehlinger, H. Sexual life of primitive people. London, 1921.
	An 2109.21.5	Boelsche, Wilhelm. Der Mensch der Vorzeit. v.1-2. 28e Aufl. Stuttgart, 1921.
	An 2109.21.10	Osborn, Henry F. Men of the old stone age. 3. ed. N.Y., 1921.
	An 2109.22	Lévy-Bruhl, L. La mentalité primitive. Paris, 1922.
	An 2109.22.3A	Goldenweiser, A.A. Early civilization. N.Y., 1922.
	An 2109.22.3B	Goldenweiser, A.A. Early civilization. N.Y., 1922.
	An 2109.23.3	Osborn, H.F. Men of the old stone age. 3. ed. N.Y., 1923.
	An 2109.23.5	Perry, W.J. The children of the sun. N.Y., 1923.
	An 2109.23.6	Bartlett, F.C. Psychology and primitive culture. N.Y., 1923.
	An 2109.23.7A	Bartlett, F.C. Psychology and primitive culture. Cambridge, 1923.
	An 2109.23.7B	Bartlett, F.C. Psychology and primitive culture. Cambridge, 1923.
	An 2109.23.9	Goldenweiser, A.A. Early civilization. N.Y., 1923.
	An 2109.23.11	Quellen zur ethnologischen Rechtsforschung von Nordafrika, Asien und Australien. Stuttgart, 1923.
	An 2109.23.15	Lévy-Bruhl, L. Primitive mentality. N.Y., 1923.
	An 2109.23.20	Preuss, Konrad Theodor. Die geistige Kultur der Naturvölker. 2e Aufl. Leipzig, 1923.
	An 2109.24A	MacCurdy, G.G. Human origins, a manual of prehistory. N.Y., 1924. 2v.
	An 2109.24B	MacCurdy, G.G. Human origins, a manual of prehistory. N.Y., 1924. 2v.
	An 2109.24.3	Rivers, W.H.R. Social organization. London, 1924.
	An 2109.24.5	Danzel, T.W. Kultur undReligion des primitiven Menschen. Stuttgart, 1924.
	An 2109.24.7	Capitan, Louis. L'humanité primitive dans la région des Eyzies. Paris, 1924.
	An 2109.24.9	Graebner, F. Das Weltbild der Primitiven. München, 1924.
	An 2109.24.10	Tylor, Edward Burnett. Primitive culture. 7. ed. v.1-2. N.Y., 1924.
	An 2109.24.15	Whitnall, H.O. The dawn of mankind. Boston, 1924.
	An 2109.25A	Tozzer, A.M. Social origins and social continuities. N.Y., 1925.
	An 2109.25B	Tozzer, A.M. Social origins and social continuities. N.Y., 1925.
	An 2109.25.3A	Childe, Vere Gordon. The dawn of European civilization. London, 1925.
	An 2109.25.3B	Childe, Vere Gordon. The dawn of European civilization. London, 1925.
	An 2109.25.4	Childe, Vere Gordon. The dawn of European civilization. N.Y., 1925.
	An 2109.25.5F	Frobenius, Leo. Hádschra Máktuba. München, 1925.
X Cg	An 2109.25.7	Lévy-Bruhl, L. How natives think. N.Y., 1927.
	An 2109.25.9	Leroy, Olivier. Essai d'introduction critique à l'étude de l'economie primitive. Paris, 1925.
	An 2109.25.19	Bumüller, J. Die Urzeit des Menschen. 4e Aufl. Augsburg, 1925.
	An 2109.26.3	Hambly, W.D. Tribal dancing and social development. London, 1926.
	An 2109.26.5	Lévy-Bruhl, L. How natives think. London, 1926.
	An 2109.26.7	Hambly, W.D. Origins of education among primitive peoples. London, 1926.

An 2106 - 2110 Prehistoric and primitive man - Primitive customs and institutions - General works (By date of issue) - cont.

	An 2109.27.3A	Malinowski, B. Sex and repression in savage society. London, 1927.
	An 2109.27.3B	Malinowski, B. Sex and repression in savage society. London, 1927.
	An 2109.27.4	Malinowski, B. Sex and repression in savage society. N.Y., 1927.
	An 2109.27.7	Murphy, John. Primitive man, his essential quest. London, 1927.
	An 2109.27.9	Osborn, H.F. Man rises to Parnassus. Princeton, 1927.
	An 2109.27.11	Boyle, Mary E. In search of our ancestors. London, 1927.
	An 2109.27.13	Lévy-Bruhl, L. L'ame primitive. 3e éd. Paris, 1927.
	An 2109.27.15	Ellwood, Charles A. Cultural evolution; a study of social origins and development. N.Y., 1927.
	An 2109.27.17	Esanielsertier, Daniel. Les formes inférieures de l'explication. Paris, 1927.
	An 2109.27.18	Essertier, Daniel. Les formes inférieures de l'explication. Thèse. Paris, 1927.
	An 2109.27.19	Fahrenfort, J.J. Het hoogste wezen der primitieven. Groningen, 1927.
	An 2109.27.21	Fahl, Anton. Ein Beitrag zur Wirtschaftsgeschichte der Vorzeit. Inaug. Diss. Halle, 1927.
	An 2109.27.23	Allier, Raoul. La non-civilisé et nous. Paris, 1927.
	An 2109.27.25A	Leroy, Olivier. La raison primitive. Paris, 1927.
	An 2109.27.25B	Leroy, Olivier. La raison primitive. Paris, 1927.
	An 2109.27.27	Marazzi, A. Fra i selvaggi e fra i civilizzate. Milano, 1927.
	An 2109.27.30	Cunow, Heinrich. Technik und Wirtschaft des europäischen Urmenschen. Berlin, 1927.
	An 2109.27.35	Henderson, Keith. Prehistoric man. N.Y., 1927.
	An 2109.28.3	Osborn, H.F. Man rises to Parnassus. Princeton, 1928.
	An 2109.28.5A	Miller, Nathan. The child in primitive society. N.Y., 1928.
	An 2109.28.5B	Miller, Nathan. The child in primitive society. N.Y., 1928.
	An 2109.28.7.1	Lévy-Bruhl, L. The "soul" of the primitive. N.Y., 1966.
	An 2109.28.9	Hankin, E.H. The cave man's legacy. London, 1928.
	An 2109.28.11	Leeuw, Gerardus. La structure de la mentalité primitive. Strasbourg, 1928.
	An 2109.28.13	Smith, Grafton Eliot. In the beginning; the origin of civilization. London, 1928.
	An 2109.28.20	Tozzer, Alfred M. Social origins and social communities. N.Y., 1928.
Htn	An 2109.29*	Crawley, A.E. Studies of savages and sex. London, 1929.
X Cg	An 2109.29.5	Malinowski, B. The sexual life of savages in north west Melanesia. London, 1929.
	An 2109.29.7	Allier, Raoul. Mind of the savage. London, 1929.
	An 2109.29.9	Butt-Thompson, F.W. West African secret societies. London, 1929.
	An 2109.29.11	Rivers, W.H.R. Social organization. N.Y., 1929.
	An 2109.29.15	Boas, Franz. The mind of primitive man. N.Y., 1929.
	An 2109.30	Danzel, Theodor. Gefüge und Fundamente der Kultur vom Standpunkte der Ethnologie. Hamburg, 1930.
	An 2109.30.5	Gandert, O.F. Forschungen zur Geschichte des Hauskundes. Leipzig, 1930.
	An 2109.31	Aldrich, C.R. The primitive mind and modern civilization. London, 1931.
X Cg	An 2109.31.3	Crawley, A.E. Dress, drinks and drums. London, 1931.
	An 2109.31.5	Lévy-Bruhl, L. La mentalité primitive. Oxford, 1931.
	An 2109.31.5.5	Lévy-Bruhl, L. La mentalité primitive. 15. ed. Paris, 1960.
	An 2109.31.7	Georg, Eugen. The adventure of mankind. N.Y., 1931.
	An 2109.31.9	Early man, his origin, development and culture. London, 1931.
	An 2109.31.11	Lévy-Bruhl, L. Le surnaturel et la nature. Paris, 1931.
	An 2109.31.13	Winthius, Josef. Einführung in die Vorstellungswelt primitiver Völker. Leipzig, 1931.
	An 2109.32	Thurnwald, R. Economics in primitive communities. London, 1932.
	An 2109.32.3	Westermarck, E. Early beliefs and their social influence. London, 1932.
	An 2109.32.9A	Radin, Paul. Social anthropology. 1st ed. N.Y., 1932.
	An 2109.32.9B	Radin, Paul. Social anthropology. 1st ed. N.Y., 1932.
	An 2109.33	Wisse, Jakob. Selbstmord und Todesfurcht bei den Naturvölkern. Proefschrift. Zutpher, 1933.
	An 2109.33.3	Hofstra, Sjoerd. Differenzierungserscheinungen in einigen afrikanischen Gruppen. Proefschrift. Amsterdam, 1933.
	An 2109.34.3A	Róheim, Géza. The riddle of the sphinx, or Human origins. London, 1934.
	An 2109.34.3B	Róheim, Géza. The riddle of the sphinx, or Human origins. London, 1934.
	An 2109.34.5A	Benedict, Ruth (Fulton). Patterns of culture. Boston, 1934.
	An 2109.34.5B	Benedict, Ruth (Fulton). Patterns of culture. Boston, 1934.
	An 2109.34.10	Osborn, H.F. Men of the old stone age. 3. ed. N.Y., 1934.
	An 2109.35	Brelsford, Vernon. Primitive philosophy. London, 1935.
	An 2109.35.5	Marett, Robert R. Head, heart and hands in human evolution. London, 1935.
	An 2109.36	Viljoen, S. The economics of primitive peoples. London, 1936.
	An 2109.36.3	Schmidt, R.R. The dawn of the human mind. London, 1936.
	An 2109.36.5	Bateson, G. Naven; a survey of the problems suggested by a composite picture of the culture of a New Guinea tribe. Cambridge, Eng., 1936.
	An 2109.36.7	Lévy-Bruhl, L. Morceaux choisis. 2e éd. Paris, 1936.
	An 2109.36.9	Olbrechts, F.M. Ethnologie...primitiere beschaving. Zutphen, 1936.
	An 2109.37	Portens, S.D. Primitive intelligence and endowment. N.Y., 1937.
	An 2109.37.3	Montagu, Ashley. Coming into being among the Australian aborigines. London, 1937.
	An 2109.37.5	Mead, M. Cooperation and competition among primitive peoples. N.Y., 1937.
	An 2109.37.7	Laubscher, B.J.F. Sex, custom and psychopathology. London, 1937.
	An 2109.37.9	Hildebrand, E. Die Geheimbünde Westafrikas als Problem der Religionwissenschrift. Inaug. Diss. Leipzig, 1937.
	An 2109.37.11	Thomas, W.I. Primitive behavior. 1. ed. N.Y., 1937.
	An 2109.38	Landtman, G. The origin of the inequality of the social classes. London, 1938.
	An 2109.38.3	Page, J.W. Primitive races of to-day. London, 1938.
	An 2109.38.5	Boas, Franz. The mind of primitive man. N.Y., 1938.
	An 2109.38.7	Lévy-Bruhl, L. L'expérience mystique et les symboles chez les primitifs. Paris, 1938.
	An 2109.38.9	Hutton, J.H. A primitive philosophy of life. Oxford, 1938.

Classified Listing

An 2106 - 2110 Prehistoric and primitive man - Primitive customs and institutions - General works (By date of issue) - cont.

An 2109.38.11 Dunbar, George. Other men's lives...primitive peoples. London, 1938.
An 2109.39 Wright, W.B. Tools and the man. London, 1939.
An 2109.39.3 Childe, Vere Gordon. Dawn of European civilization. N.Y., 1939.
An 2109.39.3.6 Childe, Vere Gordon. The dawn of European civilization. 6. ed. London, 1957.
An 2109.39.5 Goodfellow, D.M. Principles of economic sociology. Philadelphia, 1939.
An 2109.39.6 Goodfellow, D.M. Principles of economic sociology. London, 1939.
An 2109.39.7 Kardiner, A. The individual and his society. N.Y., 1939.
An 2109.39.15 Mead, Margaret. From the South Seas. N.Y., 1939.
An 2109.40 Herskovits, M.J. The economic life of primitive peoples. N.Y., 1940.
An 2109.40.3 Lowie, R.H. An introduction to cultural anthropology. N.Y., 1940.
An 2109.40.5 Cooper, G. I searched the world for death. London, 1940.
An 2109.40.10 Cole, M.C. The story of primitive man. Chicago, 1940.
An 2109.40.15 Gattefossé, Jean. L'Hyperborée et les migrations néolithiques. Droguignan, 1940.
An 2109.40.20 Bawmann, Evert Dirk. Historische Betrachtungen über das Koitus-Konzeption Problem. Arnhem, 1940.
An 2109.41.5 Sieber, Sylvester A.M. The social life of primitive man. St. Louis, 1941.
An 2109.43 Murray, R.W. Man's unknown ancestors; the story of prehistoric man. Milwaukee, 1943.
An 2109.43.5 Kelsen, Hans. Society and nature; a sociological inquiry. Chicago, 1943.
An 2109.44 Sommerfelt, A. Is there a fundamental mental difference between primitive man and the civilized European. n.p., 1944.
An 2109.45 Simmons, L.W. The role of the aged in primitive society. New Haven, 1945.
An 2109.45.1 Simmons, L.W. The role of the aged in primitive society. Hamden, Ct., 1970.
An 2109.46 Goldenweiser, Alexander A. Anthropology; au introduction to primitive culture. N.Y., 1946.
An 2109.46.5 Boas, Franz. The mind of primitive man. N.Y., 1946.
An 2109.46.10 Kelsen, Hans. Society and nature; a sociological inquiry. London, 1946.
An 2109.49 Breuil, H. Beyond the bonds of history. London, 1949.
An 2109.49.5 Koppers, W. Der Urmensch und sein Weltbild. Wien, 1949.
An 2109.49.10 Hoernes, Moriz. Vorgeschichte Europas. 7. Aufl. Berlin, 1949.
An 2109.49.15 Murdock, George P. Our primitive contemporaries. N.Y., 1949.
An 2109.50 Numelin, R. The beginnings of diplomacy. London, 1950.
An 2109.50.5 Eildermann, H. Die Urgesellschaft. Berlin, 1950.
An 2109.51 Childe, V.G. Social evolution. London, 1951.
An 2109.51.2 Childe, V.G. Social evolution. N.Y., 1951.
An 2109.51.10 Davison, D. The story of prehistoric civilizations. London, 1951.
An 2109.51.15 Weinert, H. Der geistige Aufstieg der Menschheit. 2. Aufl. Stuttgart, 1951.
An 2109.52 Radcliffe-Brown, A.R. Structure and function in primitive society. London, 1952.
An 2109.52.2 Radcliffe-Brown, A.R. Structure and function in primitive society. London, 1956.
An 2109.52.5 Sanders, I.T. Societies around the world. v.4. Lexington, 1952.
An 2109.52.10 Herskovits, M.J. Economic anthropology; a study in comparative economics. 2d ed. N.Y., 1952.
An 2109.53 Sanders, I.T. Societies around the world. N.Y., 1953. 2v.
An 2109.53.5 Radin, Paul. The world of primitive man. N.Y., 1953.
An 2109.53.10 Kern, Fritz. Der Beginn der Weltgeschichte. Bern, 1953.
An 2109.53.15 Heuse, Georges A. La psychologie ethnique. Paris, 1953.
An 2109.53.20 Wormington, Hannah Marie. Origins, indigenous period. Mexico, 1953.
An 2109.53.27 Kuehn, Herbert. Auf den Spuren des Eiszeitmenschen. 2. Aufl. Wiesbaden, 1953.
An 2109.54 Howells, William. Back of history. 1. ed. Garden City, 1954.
An 2109.54.6 Values of Primitive Society (Radio Program). The institutions of primitive society. Oxford, 1967.
An 2109.54.10 Hocart, A.M. Social origins. London, 1954.
An 2109.56 Barrière, Claude. Les civilisations tardenoisiennes en Europe occidentale. Bordeaux, 1956.
An 2109.56.5 Gegeshidze, M.K. Gruzinskii narodnyi transport. Tbilisi, 1956.
An 2109.56.10 Kuehn, Herbert. Das Erwachen der Menschheit. Frankfurt, 1956.
An 2109.57 Kosven, M.O. Abriss der Geschichte und Kultur der Urgesellschaft. Berlin, 1957.
An 2109.57.5 Radin, Paul. Primitive man as philosopher. 2. ed. N.Y., 1957.
An 2109.58 Hibben, F.C. Prehistoric man in Europe. 1. ed. Norman, Okla., 1958.
An 2109.58.5 Gómez-Moreno, Manuel. Adam y la prehistoria. Madrid, 1958.
An 2109.58.10 Bergoonlour, F.M. La préhistoire et ses problèmes. Paris, 1958.
An 2109.58.15 Maluquer de Motes, Juan. La humanidad prehistórica. Barcelona, 1958.
An 2109.59 Symposium on the Evolution, Chicago, 1957. The evolution of man's capacity for culture. Detroit, 1959.
An 2109.59.5 Cantoni, Remo. Il pensiero dei primitivi. Milano, 1959.
An 2109.59.10 Eggers, H.J. Einführung in die Vorgeschichte. München, 1959.
An 2109.59.15 Varagnac, André. L'homme avant l'écriture. Paris, 1959.
An 2109.59.20 Udy, Stanley H. Organization of work. New Haven, 1959.
An 2109.60 Cornelius, Friedrich. Geistesgeschichte der Frühzeit. v.1-2. Leiden, 1960.
An 2109.60.5 Birket-Smith, Kaj. Primitive man and his ways. London, 1960.
An 2109.60.10 Smolla, G. Neolithische Kulturerscheinungen. Bonn, 1960.
An 2109.60.15 Grand, P.M. Découverte de la préhistoire. Paris, 1960.
An 2109.60.20 Kern, Fritz. The wildbooters. Edinburgh, 1960.
An 2109.60.25 Mauss, Marcel. Sociologie et anthropologie. 2. ed. Paris, 1960.
An 2109.61 Lissher, Ivar. Man, God, and magic. N.Y., 1961.
An 2109.61.5 Clark, J.G.D. World prehistory. Cambridge, Eng., 1961.
An 2109.61.10 Real y Ramos, C.A. del. Sociología pre y protohistórica. Madrid, 1961.

An 2106 - 2110 Prehistoric and primitive man - Primitive customs and institutions - General works (By date of issue) - cont.

An 2109.61.15F Spivack, M.R. La danse cosmique de Lascaux. Montignac, 1961.
An 2109.61.20 Mead, Margaret. Cooperation and competition among primitive peoples. Boston, 1961.
An 2109.62 Bowra, Cecil M. Primitive song. London, 1962.
An 2109.62.5 Lévi-Strauss, Claude. La pensée sauvage. Paris, 1962.
An 2109.62.6 Lévi-Strauss, Claude. The savage mind. Chicago, 1966.
An 2109.62.7 Lévi-Strauss, Claude. The savage mind. Chicago, 1968.
An 2109.62.10 Gluckman, Max. Essays on the ritual of social relations. Manchester, 1962.
An 2109.62.15 Domanskii, Ia.V. Po besovym sledam. Leningrad, 1962.
An 2109.62.20 Malinowski, B. Geschlechtstrieb und Verdrängung bei den Primitiven. Reinbek bei Hamburg, 1962.
An 2109.63 Barnouw, V. Culture and personality. Homewood, Ill., 1963.
An 2109.63.5 Scotti, Pietro. La vita sociale dei popoli primitivi. Brescia, 1963.
An 2109.63.10 Howells, William. Back of history, the story of our origins. Garden City, N.Y., 1963.
An 2109.63.15 Durkheim, Emile. Primitive classification. Chicago, 1963.
An 2109.64.5 Cohen, Y.A. The transition from childhood to adolescence. Chicago, 1964.
An 2109.64.15 Zolotarev, Aleksandr M. Rodovoi stroi i pervobytnaia mifologiia. Moskva, 1964.
An 2109.64.20 Forcieri, Luigi. Le role des croyances magiques et religieuses dans les économies primitives. Bordeaux, 1964.
An 2109.65 Gluckman, Max. Politics, law and ritual in tribal society. Oxford, 1965.
An 2109.65.5 De Beer, Gavin Rylands. Genetics and prehistory. Cambridge, Eng., 1965.
An 2109.65.10 Clark, John Grahame Douglas. Prehistoric soctieties. 1st American ed. N.Y., 1965.
An 2109.65.15 Herskovits, M.J. Economic anthropology; the economic life of primitive peoples. N.Y., 1965.
An 2109.66 Anisimov, Arkadii F. Dukhovnaia zhizn' pervobytnogo obshchestva. Leningrad, 1966.
An 2109.66.5 Cotlow, Lewis N. In search of the primitive. 1st ed. Boston, 1966.
An 2109.66.10 Nash, Manning. Primitive and peasant economic systems. San Francisco, 1966.
An 2109.66.15 Narr, Karl J. Handbuch der Urgeschichte. Bern, 1966. 2v.
An 2109.66.20 Kennedy, Kenneth A.R. Human skeletal remains from Chalcolithic and Indo Roman levels from Nevasa: an anthropometric and comparative analysis. Poona, 1966.
An 2109.66.25 Kuehn, Herbert. Erwachen und Aufstieg der Menschheit. Frankfurt, 1966.
An 2109.66.30 Bessaignet, Pierre. Principes de l'ethnologie économique. Paris, 1966.
An 2109.66.35 Symposium on Man the Hunter, Chicago, 1966. Man the hunter. Chicago, 1969.
An 2109.66.40 Lévy-Bruhl, Lucien. How natives think. N.Y., 1966.
An 2109.67 Clark, John Grahame Douglas. The Stone Age hunters. London, 1967.
An 2109.67.5 Goremykina, Vera I. Istoriia pervobytnogo obshchestva. Minsk, 1967.
An 2109.67.15 Firth, Raymond William. Themes in economic anthropology. London, 1967.
An 2109.67.20 Sigrist, Christian. Regulierte Anarchie. Olten, 1967.
An 2109.67.35F Laet, Sigfried Jan de. De voorgeschiedenis van Europa. Hasselt, 1967.
An 2109.67.40 Fried, Morton Herbert. The evolution of political society: an essay in political anthropology. N.Y., 1967.
An 2109.68 Montagu, Ashley. The concept of the primitive. N.Y., 1968.
An 2109.68.5 Birket Smith, Kaj. Strejftog. Arktiske, tropiske og midt im ellem. København, 1968.
An 2109.68.10 Semenov, Sergei A. Razvitie tekhniki v kameunom veke. Leningrad, 1968.
An 2109.68.15 Pershits, Abram I. Istoriia pervobytnogo obshchestva. Moskva, 1968.
An 2109.68.17 Pershits, Abram I. Istoriia pervobytnogo obshchestva. Izd. 2. Moskva, 1974.
An 2109.68.20 Razlozhenie rodovogo stroia i formirovanie klassovogo obshchestva. Moskva, 1968.
An 2109.68.25 Association of Social Anthropologists of the Commonwealth. History and social anthropology. London, 1968.
An 2109.68.30.2 Varagnac, André. L'homme avant l'écriture. 2e éd. Paris, 1968.
An 2109.68.35 Forno, Mario. L'acculturazione dei popoli primitivi. Roma, 1968.
An 2109.70 Clark, John Grahame Douglas. Aspects of prehistory. Berkeley, 1970.
An 2109.70.5 Nougier, Louis René. L'économie préhistorique. Paris, 1970.
An 2109.70.10 Ramseyer, Urs. Soziale Bezüge des Musizierens in Naturvolkkulturen. Bern, 1970.
An 2109.71.1 Marshack, Alexander. The roots of civilization. 1st ed. N.Y., 1971.
An 2109.72 Sahlins, Marshall David. Stone age economics. Chicago, 1972.
An 2109.72.5 Okhotniki, sobirateli, rybolovy. Leningrad, 1972.
An 2109.72.10 Cocchiara, Giuseppe. L'eterno selvaggio. Palermo, 1972.
An 2109.72.15 Rovse, Irving. Introduction to prehistory; a systematic approach. N.Y., 1972.
An 2109.72.20 Quilici, Folco. Primitive societies. 1st ed. London, 1972.
An 2109.72.25 Brain, Robert. Into the primitive environment: survival on the edge of our civilization. London, 1972.
An 2109.72.32 Terray, Emmanuel. Le Marxisme devant les sociétés primitives. 2. éd. Paris, 1972.
An 2109.72.35 Terrey, Emmanuel. Marxism and "primitive" societies. N.Y., 1972.
An 2109.72.40 Jelinek, Jan. Das grosse Bilderlexikon des Menschen in der Vorzeit. Gütersloh, 1972.
An 2109.72.45 Horken, K. Ex nocte lux. Tübingen, 1972.
An 2109.73 Knyshenko, Iurii V. Istoriia pervobytnogo obshchestvo. Rostov-na-Donu, 1973.
An 2109.73.5 Perry, William James. The primordial ocean; an introductory contribution to social psychology. London, 1973.
An 2109.73.10 Research Seminar in Archaeology and Related Subjects, University of Sheffield, 1971. The explanation of culture change; models in prehistory. Pittsburgh, 1973.
An 2109.73.15 American Ethnological Society. Learning and culture. Seattle, 1973.

Classified Listing

An 2106 - 2110		Prehistoric and primitive man - Primitive customs and institutions - General works (By date of issue) - cont.
	An 2109.73.20	Felgenhauer, Fritz. Einführung in die Urgeschichtsforschung. 1. Aufl. Breisgau, 1973.
	An 2109.73.25	Social and cultural identity; problems of persistence and change. Athens, 1974.
	An 2109.73.30	Enzyklopädie der Technikgeschichte: über 7000 Jahre frühe technische Kultur. Stuttgart, 1973.
	An 2109.74	Schneider, Harold K. Economic man; the anthropology of economics. N.Y., 1974.
	An 2109.74.5	Balandier, Georges. Anthropo-logiques. 1e éd. Paris, 1974.
	An 2109.74.10	Hawkes, Jacquetta Hopkins. Atlas of ancient archaeology. N.Y., 1974.
	An 2109.74.15	Godelier, Maurice. Un domaine contesté: l'anthropologie économique. Paris, 1974.
	An 2109.75	Bodley, John H. Victims of progress. Menlo Park, 1975.
	An 2109.75.5	Clarke, Robin. The challenge of the primitives. London, 1975.
	An 2109.75.10	Pervobytnoe obshchestvo. Moskva, 1975.
	An 2109.75.15	Borer, Mary Irene Cathcart. Background to archaeology. London, 1975.

An 2156 - 2160		Prehistoric and primitive man - Primitive customs and institutions - Science and technology (By date of issue)
	An 2159.64	Semenov, Sergei A. Prehistoric technology. London, 1964.
	An 2159.73	Hawkins, Gerald Stanley. Beyond Stonehenge. 1st ed. N.Y., 1973.
	An 2159.73.15	Gur'ev, Dmitrii V. Stanovlenie obshchestvennogo proizvodstva. Moskva, 1973.
	An 2159.73.20	Moreau, Marcel. Les civilisations des étoiles. Paris, 1973.
	An 2159.73.25F	Feustel, Rudolf. Technik der Steinzeit. Weimar, 1973.
	An 2159.74	Frolov, Boris A. Chisla v grafike paleolita. Novosibirsk, 1974.

An 2206 - 2210		Prehistoric and primitive man - Primitive customs and institutions - Medicine (By date of issue)
	An 2208.83	Maturi, R. La medicina preistorica. Napoli, 1883.
	An 2208.93	Bartels, Max. Die Medicin der Naturvölker. Leipzig, 1893.
	An 2209.23	Maddox, J.L. The medicine man. N.Y., 1923.
	An 2209.24	Rivers, W.H.R. Medicine, magic and religion. London, 1924.
	An 2209.30	Laufer, B. Geophagy. Chicago, 1930.
	An 2209.32	Paudler, F. Scheitelnarbensitte, Anschwellungsglaube und Kulturkreislehre. Brunn, 1932.
	An 2209.46	Elkin, A.P. Aboriginal men of high degree. Sydney, 1946.
	An 2209.71	Ackevknecht, Erwin Heinz. Medicine and ethnology. Baltimore, 1971.

An 2306 - 2310		Prehistoric and primitive man - Primitive customs and institutions - Burial (By date of issue)
	An 2308.92	Olshausen, O. Leichenverbrennung. Berlin, 1892.

An 2336 - 2340		Prehistoric and primitive man - Primitive customs and institutions - Arms and armor, tools (By date of issue)
	An 2338.85	Morse, E.S. Ancient and modern methods of arrow-release. Salem, 1885.
	An 2338.85.2	Morse, E.S. Additional notes on arrow-release. Salem, 1922.
	An 2338.95	Hough, Walter. Primitive American armor. Washington, 1895.
	An 2338.95.5	Meyer, Hermann. Bogen und Pfeil in Central Brasilien. Leipzig, 1895.
	An 2339.01	Frobenius, L. Die Bogen der Oceanier. Berlin, 1901.
	An 2339.02	Frobenius, L. Menschenjagden und Zweikämpfe. Jena, 1902.
	An 2339.11	Balfour, Henry. The origin of West African crossbows. Washington, 1911.
	An 2339.12	Dieck, Alfred. Die Waffen der Naturvölker Süd-Amerikas. Ställuponen, 1912.
	An 2339.17	Churchill, William. Club-types of nuclear Polynesia. Washington, 1917.
	An 2339.49	Turney-High, H. Primitive war. Columbia, 1949.
	An 2339.65	Rust, Alfred. Uber Waffen- und Werkzeugtechnik des Altmenschen. Neumünster, 1965.

An 2356 - 2360		Prehistoric and primitive man - Primitive customs and institutions - Cannibalism (By date of issue)
	An 2358.86	Darling, C.W. Anthropophagy. Utica, 1886.
	An 2359.75	Tannahill, Reay. Flesh and blood. London, 1975.

An 2406 - 2410		Prehistoric and primitive man - Primitive customs and institutions - Law (By date of issue)
	An 2408.72	Bastian, P.W.A. Die Rechtsverhältnisse. Berlin, 1872.

An 2600		Prehistoric and primitive man - Wild men, Wolf children
	An 2600.1	Koenig, H.C. De hominum inter feras educatorum. Hanover, 1730.
	An 2600.2	Itard, E.M. An historical account of...education of a savage man. London, 1802.
	An 2600.2.8	Itard, Jean Marc. The wild boy of Aveyron. N.Y., 1932.
	An 2600.3	Ausführliches Leben und besondere Schiksale eines wilden Knaben. Frankfurt, 1759.
	An 2600.4	Gesell, Arnold. Wolf child and human child. N.Y., 1941.
	An 2600.5	Singh, Joseph. Wolf children and feral man. N.Y., 1942.
	An 2600.5.2	Singh, Joseph. Wolf-children and feral man. Hamden, 1966.
	An 2600.6.1	Hecquet, (Mme.). Histoire d'une jeune fille sauvage trouvée dans les bois à l'âge de dix ans. Bordeaux, 1970.
	An 2600.7.1	Malson, Lucien. Wolf children. London, 1972.

An 3206 - 3210		Somatology, Physical anthropology - General works (By date of issue)
	An 3206.47	Otto, Andreas. Anthroposcopia. Regiomonti, 1647.
	An 3207.85	Sommering, S.T. Ueber...Verschiedenheit des Negers vom Europäer. Frankfurt, 1785.
Htn	An 3207.91*	Camper, P. Dissertation physique...traits du visage. Utrecht, 1791. 2 pam.
	An 3208.22	Lawrence, W. Lectures on physiology, zoology and the natural history of man. London, 1822.
	An 3208.22.2	Lawrence, W. Lectures on physiology, zoology and the natural history of man. London, 1822.
	An 3208.38	Mudie, D. Man, in his physical structure and adaptations. London, 1838.
	An 3208.50	Browne, P.A. The classification of mankind by the hair. Philadelphia, 1850.
	An 3208.80	Otis, George A. List of the specimens in the anatomical section of the United States Army Medical Museum. Washington, 1880.

An 3206 - 3210 of issue) - cont.		Somatology, Physical anthropology - General works (By date
	An 3208.93	Ammon, Alto. Die natürliche Auslese bein Menschen. Jena, 1893.
	An 3208.97	Ehrenreich, P. Anthropologische Studien über die Urbewohner Brasiliens. Braunschweig, 1897.
	An 3208.99	Karlsruhe, Germany. Altertumsverein. Zur Anthropologie der Badener. Jena, 1899.
	An 3209.01	Russell, Frank. Laboratory outlines in somatology. N.Y., 1901.
	An 3209.12	Loth, E. Beiträge zur...der Negerweichteile. Stuttgart, 1912.
	An 3209.33	Hildén, Kaarlo. Maapallon esihistorialliset ja. Helsinki, 1933.
	An 3209.60	Montagu, Ashley. An introduction to physical anthropology. 3. ed. Springfield, 1960.
	An 3209.61	Notschaele, Lucien Aimé. Een somatometrisch onderzoek in de Noord-oost Polder. Amsterdam, 1961.
VAn	3209.67	Czekanowski, Jan. Człowiek w czasie i przestrzeni. 3. wyd. Warszawa, 1967.
	An 3209.68	Zubov, Aleksandr A. Problemy evoliutsii cheloveka i ego ras. Moskva, 1968.
	An 3209.68.5	Contributions to the physical anthropology of central Asia and the Caucasus. Cambridge, 1968.
	An 3209.71	Young, John Z. An introduction to the study of man. Oxford, 1971.
	An 3209.72	Alekseev, Valerii P. V poishakh predkov. Moskva, 1972.

An 3305 - 3310 of issue)		Somatology, Physical anthropology - Anthropometry (By date
Htn	An 3305.63*	Elsholt, J.S. Anthropometria. Francofurti, 1563.
	An 3308.62F	Liharzih, F. Gesetz des Wachstumes...Proportionslehre. Wien, 1862.
	An 3308.76	Pagliani, L. Sopra alcuni fattori dello suiluppo umano. Torino, 1876.
	An 3308.78.2	Roberts, C. A manual of anthropometry. London, 1878. 2 pam.
	An 3308.79	Bowditch, J.P. The growth of children. Boston, 1879.
	An 3308.84	Galton, Francis. Life history album. London, 1884.
	An 3308.85	Bertillon, A. Identification anthropometrique. Melun, 1885.
	An 3308.87	Amherst College. The anthrometric manual 1887. Amherst, 1887.
	An 3308.87.5	Hitchcock, E. Need of anthropometry. Brooklyn, 1887.
	An 3308.89	Amherst College. Anthropometric manual 1889. Amherst, 1889.
	An 3308.89.3	Anuchin, D.N. O geograficheskom' raspred'nenii rosta muzhskago naseleniia Rossii. Sankt Peterburg, 1889.
	An 3308.90	Seaver, J.W. Anthropometry and physical examination. New Haven, 1890.
	An 3308.93	Livi, Ridolfo. Antropometria militare. Roma, 1893-1903. 2v.
	An 3308.95	Farmer, J.B. Extract from Science progress. Boston, 1895.
	An 3308.96	Seaver, J.W. Anthropometry and physical examination. 2. ed. New Haven, 1896.
	An 3309.00	Barros Ovalle, P.N. Manual de antropometria. Santiago de Chile, 1900.
	An 3309.02	Hastings, William W. Manual for physical measurements. Springfield, 1902.
	An 3309.05	Montessori, Maria. Caratteri fisici delle Giovani donne del Lazio. Roma, 1905.
	An 3309.08	New South Wales. Department of Public Instruction. Report - physical condition of children. Sydney, 1908.
	An 3309.08.5	Tocher, James F. Pigmentation survey of school children in Scotland. Aberdeen, 1908.
	An 3309.09	Bertillon, A. Anthropologie métrique. Paris, 1909.
	An 3309.11	Weissenberg, S. Das Wachstum des Menschen. Stuttgart, 1911.
	An 3309.12	Bresciani, C. La correlazione fra la statura e l'indice cefalico. Palermo, 1912.
	An 3309.13	Rocca, Pierre. Les corses devant l'anthropologie. Paris, 1913.
	An 3309.14	Friedenthal, Hans. Allgemeine und spezielle Physiologie des Menschenwachstums. Berlin, 1914.
	An 3309.14.3F	Stevenson, B.L. Constancy or variability in Scandinavian type. Leiden, 1914.
	An 3309.16	Stevenson, B.L. Socio-anthropometry. Boston, 1916.
	An 3309.20	Wilder, Harris H. A laboratory manual of anthropometry. Philadelphia, 1920.
	An 3309.21	Baldwin, Bird T. The physical growth of children from birth to maturity. Iowa City, 1921.
	An 3309.21.3	Dreyer, George. The assessment of physical fitness. N.Y., 1921.
	An 3309.24	Tocher, James F. Anthropometric observations on samples of the civil populations of Aberdeenshire, Banffshire, and Kincardineshire. Edinburgh, 1924.
	An 3309.24.3	Martin, Rudolf. Richtlineen für Körpermessungen. München, 1924.
	An 3309.25	Hrdlička, A. The old Americans. Baltimore, 1925.
	An 3309.27	Davenport, C.B. Guide to physical anthropometry and anthroposcopy. Baltimore, 1927.
	An 3309.28	Sullivan, Louis R. Essentials of anthropometry. N.Y., 1928.
	An 3309.29	Davenport, C.B. Race crossing in Jamaica. Washington, 1929.
	An 3309.29.3	Puccioni, Nello. Affrica nord-orientale e Arabia. Pavia, 1929.
	An 3309.30	Hoaton, E.A. The Indians of Pecos Pueblo. New Haven, 1930.
	An 3309.30.5	Measurement of man. Minneapolis, 1930.
	An 3309.30.10	Herskovits, M.J. The anthropometry of the American Negro. N.Y., 1930.
	An 3309.32	Bowles, G.T. New types of old Americans at Harvard and at eastern women's colleges. Cambridge, 1932.
	An 3309.32.3F	Frassetto, F. Note antropologiche sulla popolagione del Bolognese. Bologna, 1932.
	An 3309.32.5	Biedermann, Ernst. Körperform und Leistung sechzehnjähriger Lehslinge. Zürich, 1932.
	An 3309.35	Boyd, Edith. The growth of the surface area of the human body. Minneapolis, 1935.
	An 3309.36	Simon, Carleton. The retinal method of identification. n.p., 1936.
	An 3309.37	Chattapadkyay, Bojra K. Les affinités somatiques des Brahmines Moithils et Kanarijias de Bihar. Thèse. Paris, 1937?
	An 3309.38	Simon, Carleton. La identificación personal por la retina. La Habana, 1938.

Classified Listing

An 3305 - 3310 Somatology, Physical anthropology - Anthropometry (By date of issue) - cont.

An 3309.40	Sheldon, W.H. The varieties of human physique. 1st ed. N.Y., 1940.
An 3309.43	Goldstein, M.S. Demographic and bodily changes in descendants of Mexican immigrants. Austin, 1943.
An 3309.43.5	Riggs, F.B. Tall men have their problems too. Cambridge, 1943.
An 3309.45	Hoaton, E.A. A survey in seating. Cambridge, 1945.
An 3309.54F	Sheldon, W.H. Atlas of men. 1st ed. N.Y., 1954.
An 3309.63.5	Henzel, Tadeusz. Badania struktury rasowej ludności Afryki Srotkowej. Łódz, 1963.
An 3309.64	Sarkar, Sasanka Sekhar. Ancient races of Baluchistan. 1st ed. Calcutta, 1964.
An 3309.65.5	Łódzskic Towarzystwo Naukowe. Wydział III, Nauk Matematyczno-Przyrodniczych. Studia afrykanistyczne. Łódź, 1965.
An 3309.65.10	Joint Arabic-Polish Anthropological Expedition, 1958. Publications. Warsaw, 1965.
An 3309.67	Chai, Chen Kang. Taiwan aborigines. Cambridge, 1967.
An 3309.69	Nurse, G.T. Height and history in Malawi. Malawi, 1969.
An 3309.70.1	Debets, Georgii Frantsevich. Physical anthropology of Afghanistan. v.1-2. Cambridge, 1970.
An 3309.73	Zinevich, Galina P. Antropologicheskie materialy srednevekovykh mogil'nikov dugo-zapadnogo kryma. Kiev, 1973.
An 3309.73.5	Konduktorova, Tamara S. Antropologiia naseleniia Ukrainy mezolita, neolita, i epokki bronzy. Moskva, 1973.

An 3404 - 3410 Somatology, Physical anthropology - Fingerprints (By date of issue)

	An 3404.01	Pamphlet box. Finger prints.
Htn	An 3408.92*	Galton, F. Finger prints. London, 1892.
Htn	An 3408.93*	Galton, F. Decipherment of blurred fingerprints. London, 1893.
	An 3408.95	Galton, F. Fingerprint directories. London, 1895.
	An 3409.09	Reyna Almandos, Luis. Dactiloscopia argentina. La Plata, 1909.
	An 3409.10	Brayley, F.A. Arrangement of finger prints identification and their uses. Boston, 1910.
	An 3409.12	Reyna Almandos, Luis. Origen e influencia jurídico-social del sistema dactiloscopico argentino. La Plata, 1912.
	An 3409.13	Ortiz, F. La indentificación dactiloscópica. Habana, 1913.
	An 3409.13.5	Henry, Edward R. Classification and uses of fingerprints. 4. ed. London, 1913.
	An 3409.16	Unites States. Bureau of Navigation. Navy Department. How to obtain good fingerprints. 2. ed. Washington, 1916.
	An 3409.16.3	Herschel, W.J. The origin of finger-printing. London, 1916.
	An 3409.16.5	Kuhne, F. The finger print instructor. N.Y., 1916.
	An 3409.16.10	Kuhne, F. The finger print instructor. 3. ed. N.Y., 1942.
	An 3409.23	Crosskey, Walter C.S. The single finger print identification system. San Francisco, 1923.
	An 3409.30	Miranda Pinto, O. Contribution à la morphologie comparée des crêtes papillaires. Thèse. Lyon, 1930.
	An 3409.31	Unites States. Federal Bureau of Investigation. How to take fingerprints. Washington, 1931.
	An 3409.31.3	Battley, Harry. Single fingerprints. New Haven, 1931.
	An 3409.31.5	Forest, H.P. de. The evolution of dactyloscopy in the United States. Youngstown, 1931.
	An 3409.32	United States. Federal Bureau of Investigation. How to take fingerprints. Washington, 1932.
	An 3409.34	United States. Federal Bureau of Investigation. Fingerprints. Washington, 1934.
	An 3409.34.3	Pacheco, F. Vucetich e Reyna Abmandos. Rio de Janeiro, 1934.
	An 3409.35	Unites States. Federal Bureau of Investigation. Fingerprints. Washington, 1935.
	An 3409.35.5	Castellanos, I. Dactiloscopia clinica. Habana, 1935.
	An 3409.36	United States. Federal Bureau of Investigation. Classification of fingerprints. J.E. Hoover, director. Washington, 1936.
	An 3409.37	United States. Federal Bureau of Investigation. Fingerprints. Washington, 1937.
	An 3409.37.3	Borges Badell, A. Clasificación anatómica-matemática del dactilograma. La Habana, 1937?
	An 3409.41	Chapel, Charles E. Fingerprinting; a manual of identification. N.Y., 1941.
	An 3409.42	Bridges, B.C. Practical fingerprinting. N.Y., 1942.
	An 3409.43	Cummins, H. Fingerprints, palms and soles. Philadelphia, 1943.

An 3506 - 3510 Somatology, Physical anthropology - Osteology, Skeleton (By date of issue)

An 3508.82	Harkness, H.W. Footprints found at the Carson State Prison. n.p., n.d.
An 3508.84	Studley, C.A. Notes upon human remains from the caves of Coahuila, Mexico. Salem, 1884.
An 3508.93	Matthews, W. The human bones...Hemingway collection...United States Medical Museum at Washington. n.p., 1893.
An 3509.07	Hrdlička, A. Skeletal remains...early man...North America. Washington, 1907.
An 3509.12	Hrdlička, A. Early man in South America. Washington, 1912.
An 3509.31	Reynolds, E. The evolution of the human pelvis in relation to the mechanics of the erect posture. Cambridge, 1931.
An 3509.31.10	Schreiner, Kristian Emil. Zur Osteologie der Lappen. Oslo, 1931-35. 2v.
An 3509.34	Cameron, John. The skeleton of British neolithic man. London, 1934.
An 3509.35	Martin, Cecil P. Prehistoric man in Ireland. London, 1935.
An 3509.37	Buyssens, Paul. Le pithécanthrope était-il un Pygmée? Bruxelles, 1937.
An 3509.38	Bryan, Kirk. Discovery of Sauk Valley man of Minnesota. Abilene, Texas, 1938.
An 3509.66	Alekseev, Valerii P. Osteometriia. Moskva, 1966.
An 3509.70F	Nielsen, Ole Vagn. The Nubian skeleton through 4000 years. Thesis. København, 1970.
An 3509.72	Aron, Jean Paul. Anthropologie du conscrit français d'après les comptes numeriques et sommaires du recrutement de l'année, 1819-1826. Paris, 1972.

An 3538 - 3539 Somatology, Physical anthropology - Craniometry, Skull - Folios [Discontinued]

An 3538.39	Morton, S.G. Crania Americana. Philadelphia, 1839.
An 3538.44	Morton, S.G. Crania Aegyptiaca. Philadelphia, 1844.
An 3538.65	Davis, J.B. Crania Britannica. London, 1865. 2v.
An 3538.80	Carr, L. Notes on the Crania of New England Indians. Boston, 1880.
An 3538.80.2	Carr, L. Notes on the Crania of New England Indians. Boston, 1880.
An 3538.82	Quatrefages de Bréau, A. de. Crania ethnica. Les crânes des races humaines. Paris, 1873-81. 2v.
An 3538.92	Virchow, R. Crania ethnica americana. Berlin, 1892.
An 3538.96	Allen, Harrison. Crania from the mounds of the St. John's. Philadelphia, 1896.
An 3539.01	Koeze, G.A. Crania ethnica Philippinica. Haarlem, 1901-04.
An 3539.01.3	Randall-Maciver, D. The earliest inhabitants of Abydos. Oxford, 1901.
An 3539.07	Oetteking, B. Kraniologische Studien...Altägypten. Braunschweig, 1907.
An 3539.09F	Pittard, E. Anthropologie de la Suisse. Genève, 1909-10.
An 3539.10PF	Düben, G. Crania Lapponica. Holmiae, 1910. 2v.

An 3556 - 3560 Somatology, Physical anthropology - Craniometry, Skull - Monographs (By date of issue)

	An 3558.46	Zeune, August. Über Schädelbildung zur festern Begründung der Menschenrassen. Berlin, 1846.
	An 3558.67	Davis, J.B. Thesaurus Craniorum. London, 1867.
	An 3558.67.2	Davis, J.B. Thesaurus Craniorum. Supplement. London, 1875.
	An 3558.68	Wyman, J. Observations on Crania. Boston, 1868.
	An 3558.77	Virchow, R. Beitraege zur physischen Anthropologie der Deutschen. Berlin, 1877.
	An 3558.81	Bessel Hagen, F.K. Zur Kritik und Verbesserung der Winkelmessungen am Kopfe. Königsberg, 1881.
	An 3558.90	Török, A. Grundz. einer systematischen Kraniometrie. Stuttgart, 1890.
	An 3558.94	Olóriz y Aguilera, Frederico. Distribución geográfica del'índice cefálico en España. Madrid, 1894.
	An 3558.96	Barth, J. Norrønaskaller. Crania antiqua. Christiania, 1896.
	An 3558.98	Silva Boasto, A.J. de. Indices cephálicos dos Portuguêses. Coimbra, 1898.
	An 3559.06	Hauser, K. Das kraniologische der New Guinea Expedition. Berlin, 1906.
	An 3559.08	Channing, Walter. The hard palate in normal and feeble-minded individuals. N.Y., 1908.
	An 3559.09	Frizzi, E. Ein Beitrag zur Anthropologie des "Homo Alpinus Tirolensis". Wien, 1909.
	An 3559.12	Paul-Boncour, G. Anthropologie anatomique, crane-face. Paris, 1912.
	An 3559.12.3	Sergi, Sergio. Crania Habessinica contributo all'antropologia dell'Africa Orientale. Roma, 1912.
	An 3559.13	Dillenius, J.A. Craneometría comparativa de los antiguos habitantes de la isla y del pukara de Tilcara. Buenos Aires, 1913.
	An 3559.15	Mendes Correa, A.A. Sobre três crânios de negros Mossumles. Porto, 1915.
	An 3559.17	Mendes Correa, A.A. Sôbre algunos crânios da India Portuguêsa. Porto, 1917.
	An 3559.27	Nyèssen, D.J.H. The passing of the Frisians. 's-Gravenhage, 1927.
X Cg	An 3559.29	Nyèssen, D.J.H. Somatical investigation of the Javanese, 1929. Bandoeng, 1929.
	An 3559.30F	Black, Davidson. On an adolescent skull of "Sinanthropus pekinensis". Peiping, 1930.
	An 3559.33	Aguilar y R.F. Origen castellano del prognatismo. Madrid, 1933.
	An 3559.34	Mascarenhas, Constância. Os povos de Angola. Bastorá, 1934.
	An 3559.38	Froe, A. de. Meethare variabelen van den menschlijken schedel en hun an derlinge. Amsterdam, 1938.
	An 3559.64.5	Alekseev, Valerii P. Kraniometriia. Moskva, 1964.
	An 3559.66	Abdushelishvili, Malkhaz G. K kraniologii drevnego i sovremennogo naseleniia Kavkaza. Tbilisi, 1966.
	An 3559.68	Zubov, Aleksander A. Odontologiia. Moskva, 1968.
	An 3559.68.5	Zinevich, Galina P. Antropologichna kharakteristika davn'oho naselennia teritorii Ukrainy. Kyiv, 1968.
	An 3559.69	Alekseev, Valerii P. Proiskhozhdenie narodov Vostochnoi Evropy. Moskva, 1969.
	An 3559.70	Ismagulov, Orazak. Naselenie Kazakhstana ot epokhi bronzy do sovremennosti. Alma-Ata, 1970.
	An 3559.72	Konduktorova, Tamara S. Antropologiia drevnego nasteniia Ukrainy. Moskva, 1972.
	An 3559.72.10	Kruts, Svitlana O. Naselenie territorii Ukrainy epokhi medi-bronzy. Kyiv, 1972.
	An 3559.73	Zubov, Aleksandr A. Etnicheskaia odontologiia. Moskva, 1973.

WIDENER LIBRARY SHELFLIST, 60

GEOGRAPHY
AND
ANTHROPOLOGY

CHRONOLOGICAL LISTING

Chronological Listing

No date

	Geog 3.35.2	Acta geographica. Tables générales, anciennes séries, 1947-1969. Paris, n.d.
	Geog 13.4.4	American Geographical Society of New York. Index to the Bulletin of the American Geographical Society. 1852-1910. N.Y., n.d.
	Geog 3240.20	Andriani, G. Giacomo Bracelli nella storia della geografia. Pontremoli, n.d.
	Geog 4208.70.5	Around the world. n.p., n.d.
	Geog 13.9.5	Association of American Geographers. Annals. Index for volumes 1-25. Lancaster, Pa.? n.d.
	Geog 4206.54	Bäckhoff. Anhang zwer Reisen. Berlin, n.d.
	Geog 3055.27	Baour-Lormian, Pierre Marie F.L. L'Atlantide. Paris, n.d.
Htn	Geog 816.40*	Bertius, Petrus. Livre premier [-septiesme] des tables geographiques auquel est traité du monde en general. n.p., n.d.
	Geog 3055.9	Block, R. de. Quelques mots sur l'Atlantide. n.p., n.d.
	Geog 3240.9	Boek, C.P. Lettres...intitulé: liber guidonis. n.p., n.d.
	Geog 4708.68	Bollo, G.A. Petizione...relativa alla catastrofe toccata alla nave Teresa. n.p., n.d. 2 pam.
	Geog 4672.7	Book of shipwrecks and adventures on the ocean. N.Y., n.d.
	Geog 577.60F	Brice, A. Grand gazetteer, or topographical dictionary. n.p., n.d.
	Geog 4307.03	Broekhuizen, G. Die nieuwe Béreisde wereld. Amsterdam, n.d. 2 pam.
	Geog 3235.29.3	Buchon, J.A.C. Notices et extraits des manuscrits. Paris, n.d.
	Geog 6009.08	Charcot, Jean B. The voyage of the 'Why not?' in the Atlantic. London, n.d.
	An 170.157	Chicago. University. Seminar of Racial and Cultural Contacts. Proceedings, 1935-36. n.p., n.d.
	Geog 4208.40	Cleveland, R.J. Voyages and commercial enterprises. N.Y., n.d.
	Geog 3018.56	Cortambert, E. Coup d'oeil...sur...les progrès de géographie. Langny, n.d.
	Geog 4209.20	Curle, Richard. Wanderings. n.p., n.d.
	Geog 4181.91	Debenham, Frank. Voyage of Captain Bellingshausen to Antarctic seas, 1819-21. London, n.d. 2v.
	Geog 3070.24	Desimoni, C. Elenco di carte ed Atlanti nautici. Genova, n.d.
Htn	Geog 5340.7*	Drie voyagien gedaen na Groenland. Amsterdam, n.d.
	Geog 665.26	Famous islands. Boston, n.d.
	Geog 85.379.2	Geographia Polonica. Index, 1-32,1964-1975. Warszawa, n.d.
	Geog 85.201.10	Geographical review. Index, 1-45, 1926-1957. N.Y., n.d. 4v.
	Geog 5606.5.5	Grønland. Inholdsfortegnelse, 1953-62. n.p., n.d.
	An 3508.82	Harkness, H.W. Footprints found at the Carson State Prison. n.p., n.d.
	Geog 5160.10F	Hartwig, G. The Arctic regions. London, n.d.
	Geog 4268.31.7	Hofland, B. (Mrs.). Panorama of Europe. 7th ed. London, n.d.
	Geog 5650.18	Holtved, Erik. Archeological investigation in the Thule district. n.p., n.d.
	Geog 4168.50	Hombron, Bernard. Aventures...des voyageurs. Paris, n.d.
	Geog 608.2	Hubbard, Gardiner Greene. Contents of box in corner stone of Hubbard Memorial. n.p., n.d.
	Geog 3025.7	International Geographical Congress. Uber...Herstellung...Erdkarte im Mafestabe. Wien, n.d.
Htn	Geog 755.19*	Introductio in Ptolomei. n.p., n.d.
	Geog 112.5	Istanbul. Universite. Cografya Ensitüsü. Coğrafi araştirmalar. n.p., n.d.
	28.4	Jonard, E.F. Notice sur l'etablissement géographique de Bruxelles. v.1-4. Paris, n.d.
	Geog 4206.65.9	Journal des voyages de Monsieur de Monconys. Lyon, n.d.
	Geog 4348.18	Journal of Jonathan Russell, 1818-1819. n.p., n.d.
	Geog 142.3.2	The journal of tropical geography. Index. 1-39,1953-1974. Singapore, n.d. 2v.
	Geog 4219.20F	Kincaid, Earle H. History and cruises of the United States ship Whipple. Constantinople, n.d.
Htn	Geog 4308.78.5*	King, Edward. A collection of newspaper articles, describing his experiences in Europe, 1878-1880. n.p., n.d.
	Geog 139.2	London. Royal Geographical Society. Journal. Index. v.1-20,21-40,41-50. London, n.d. 3v.
NEDL	Geog 4158.33	Montemont, A. Histoire...des voyages. Paris, n.d. 46v.
	Geog 151.6.5	Norois. Table décennale, 1954-73. Poitiers, n.d. 2v.
	Geog 4417.64	Orleans, G. de. Travels of Father William Orleans a Jesuit. London, n.d.
	Geog 4219.04	Pearson, D.G. Trip to the Phillipines...in 1904. n.p., n.d.
	Geog 959.00	Photograph album containing views of Alaska about 1900, and mountaineering in the Swiss Alps. n.p., n.d.
Htn	Geog 4237.52.5*	Pierce, Nathaniel. Account of great dangers...and remarkable deliverance of N. Pierce. n.p., n.d.
	Geog 4308.72.9F	Pressed flowers and leaves; souvenirs of travel. n.p., n.d.
	Geog 4110.9.2	Rand, McNally and Co. Pocket atlas of the world. Chicago, n.d.
	Geog 190.20.2	Regio Basiliensis. Hefte für jurassische und oberheinische Landeskunde. Register. 1-10,1959-1969. Basel, n.d.
	Geog 4306.87	Regnard, J.F. Voyages. n.p., n.d.
	Geog 192.5.5	Revue géographique des pyrénées et du sud-ouest. Table décennale, 1950-1959. Toulouse, n.d.
	Geog 4308.91.3F	Richardson, W.L. Cornwall-Devon, 1891; [Collection of photographic views...of Germany, Holland, Switzerland, Paris and England]. n.p., n.d.
	Geog 4308.67.11F	Richardson, W.L. Europe, 1867-1869. n.p., n.d.
	Geog 3235.47.9	Santarem. Examen des assertions contenues...des monuments de la géographie. n.p., n.d.
	Geog 200.1.3	Scottish geographical magazine. Index. v.1-50, 1885-1934. Edinburgh, n.d.
NEDL	Geog 212.100.2	Societa Geografica Italiana. Bollettino. Indice generale della serie II-III. Roma, n.d.
	Geog 212.203.5	Société de Géographie. Bulletin. Table, series V-VII. Paris, n.d.
NEDL	Geog 212.200.5	Société Royale Belge de Géographie. Compte rendu. Tables des matières des v.1-25, 1876-1901. Bruxelles, n.d.
	Geog 5700.8.5	Strangers in Greenland. n.p., n.d.
	Geog 226.2.5	Tour du monde. Table alphabetique, 1860-1910. Paris, n.d.
	Geog 4300.31.8	Touring Club Suisse. Europa touring; guide automobile d'Europe. Berne, n.d.
	Geog 3055.21	Unger, F.X. The sunken island of Atlantis. n.p., n.d.
	Geog 243.15.5	L'universo. Index, 1920-40. Firenze, n.d.
	Geog 817.96	Walker, J. Elements of geography. London, n.d.
	Geog 3251.19	Wieser, F. Der Portulan des Infanten. n.p., n.d.

No date - cont.

	Geog 275.2	Ymer; tidskrift utgifven af svenska sällskapet för antropologi och geografi. Person-och ämnesregister; 1-70, 1881-1950. n.p., n.d. 2v.

1510-1519

Htn	Geog 4415.11F*	Varthema, L. de. Itinerarium aethio piae. Milan, 1511.
Htn	Geog 815.19*	Denciso, M.F. Suma de geographia. Seville, 1519.

1520-1529

Htn	Geog 755.27*	Glareanus, H. De geographia liber unus. Basileae, 1527.

1530-1539

Htn	Geog 755.27.2*	Glareanus, H. De geographia liber unus. Friburgum, 1530.
Htn	Geog 3205.34F*	Franck, S. Weltbuch: Spiegel...in Asiam, Aphrica, Europam und America. n.p., 1534.
Htn	Geog 815.34*	Watt, J. von. Epitome trium terrae. Tiguri, 1534.
Htn	Geog 815.38*	Rithaymer, Georg. Georgii Rithaymeri De orbis terrarvm sitr compendium. Norimbergae, 1538.

1540-1549

Htn	Geog 3205.40.5*	Boehme, J. Omnium gentium, mores, leges. Friburgi, 1540.
Htn	Geog 3205.40.7*	Boehme, J. Omnium gentium, mores, leges. Antverpiae, 1542.

1550-1559

Htn	Geog 3205.40.25*	Boehme, J. Fardle of facious...ancient manners. London, 1555. 3v.

1560-1569

Htn	Geog 3205.40.21*	Boehme, J. Gli costumi, le leggi et lusanze. Venetia, 1560.
Htn	Geog 4265.61*	Barreiros, G. Chorographia. Coimbra, 1561.
Htn	Geog 3205.40.11*	Boehme, J. Mores, leges et ritus omnium gentium. Lugduni, 1561.
Htn	Geog 4105.12*	Gratarolo, G. De regimine iter agentium. Basileae, 1561.
	An 3305.63*	Elsholt, J.S. Anthropometria. Francofurti, 1563.
Htn	Geog 755.63.2*	Postel, Guillaume. De universitate liber. 2a ed. pt.1-2. Parisiis, 1563.
Htn	Geog 805.67F*	Franck, S. Erst Theil dieses Weltbuchs von neuen Erfundnen Landtschafften Warhafftige Beschreibunge aller Theil der Welt. Franckfurt, 1567.

1570-1579

Htn	Geog 755.71*	Nores, Jason de. Breve trattato del mondo e delle sue parti. Venetia, 1571.
Htn	Geog 4341.73*	Montano, B.A. Itinerarium beniamini tudelemsis. Antwerpiae, 1575.
Htn	Geog 805.75F*	Thevet, A. La cosmographie universelle. Paris, 1575. 2v.
Htn	Geog 3205.40.15*	Boehme, J. Mores, leges et ritus omnium gentium. Lugduni, 1576.
Htn	Geog 4265.76*	Verstegen, R. The post of the world. London, 1576.
Htn	Geog 5515.77*	Settle, D. True report of Martin Frobisher's voyage, 1577. London, 1577.

1580-1589

Htn	Geog 5515.77.5*	Best, George. De Martini Forbisseri Angli. n.p., 1580.
Htn	Geog 3205.40.17*	Boehme, J. Mores, leges et ritus omnium gentium. Lugduni, 1582.
	An 355.85	Lascovius, P.M. De homine magno illo in rerum natura. Libro II. Witebergae, 1585.
Htn	Geog 4215.86*	Neaudro, M. Orbis terrae partium. Lipsiae, 1586.
Htn	Geog 3520.10F*	Hakluyt, Richard. Principall navigations, voiages. London, 1589.

1590-1599

Htn	Geog 4125.91*	De arte peregrinandi. Libri II. Noribergae, 1591.
Htn	Geog 3205.95*	Romanus, A. Parvum theatrum urbium. Frankoforti, 1595.
Htn	Geog 815.59.3*	Botero, G. Le relationi universali...divise in quattro parti. Venetia, 1597.
Htn	Geog 4205.98.2F*	Linschoten, Jan H. van. Discours of voyages...East and West Indies. London, 1598.
Htn	Geog 4205.98F*	Linschoten, Jan H. van. Discours of voyages...East and West Indies. London, 1598.
Htn	Geog 3520.10.5F*	Hakluyt, Richard. Principal navigations, voyages. v.1-3. London, 1599. 2v.
Htn	Geog 4205.99F*	Linschoten, Jan H. van. Navigatio ac itinerarium. Hagae-Comitis, 1599.

1600-1609

Htn	Geog 816.00*	Abbott, George. A briefe description of the whole world. London, 1600.
Htn	Geog 4324.59*	Ehingen, Georg von. Itinerarium. Augsburg, 1600.
Htn	Geog 5375.94.3F*	Veer, G. de. Vraye description de trois voyages de mer tres admirables. Amsterdam, 1600.
Htn	Geog 815.59.5*	Botero, G. The travellers breviat, or An historical description of the most famous kingdoms in the world. London, 1601.
Htn	Geog 4676.02*	Godinho Cardozo, M. Relaçam do naufragio da nao Santiago e itinerario da gente que delle se salvou. Lisboa, 1602. 3 pam.
Htn	Geog 815.50*	Hondius, Jodocus. Thresor de chartes contenant les tableaux de tous les pays du monde. Franckfort? 1602.
Htn	Geog 4105.19*	Palmer, Thomas. An essay of the meanes how to make...travailes. London, 1606.
	Geog 4345.99.2	Nixon, Anthony. The three English brothers. London, 1607.
Htn	Geog 816.00.3*	Abbott, George. A briefe description of the whole world. 3d ed. London, 1608. 2 pam.
	Geog 4266.09	Ens, Gaspar. Deliciarum germaniae...viatorius...itinera ad omnes ciirtates. Coloniae, 1608.

Chronological Listing

1610-1619

Htn	Geog 4326.10	Ens, K. Deliciae transmari. Coloniae Agrippinae, 1610.
Htn	Geog 5375.94.4*	Veer, G. de. Les trois navigations admirables. Paris, 1610.
Htn	Geog 5375.72*	Röslin, H. Mitternächtige Schiffarth von den Herrn Staden. Oppenheim, 1611.
Htn	Geog 4305.96*	Hentznero, P. Itinerarium Germaniae, Galliae. Norinbergae, 1612.
Htn	Geog 4206.13F*	Purchas, Samuel. Purchas, his pilgrimage. London, 1613.
Htn	Geog 4126.14*	Lithgow, W. Peregrination from Scotland. Lond, 1614.
	Geog 4306.17.7	Küsel Saxon, S. Itinerarium Germaniae, Siciliae. Erphordiae, 1617.
Htn	Geog 4306.17F*	Moryson, Fynes. An itinerary. London, 1617.
Htn	Geog 4206.13.5F*	Purchas, Samuel. His pilgrimage, or Relations of the world. 3rd ed. London, 1617.
Htn	Geog 4206.10.5*	Linschoten, Jan H. van. Histoire de la navigation. 2e éd. v.1-3. Amsterdam, 1619.
Htn	Geog 5375.94*	Veer, G. de. Oost-indische ende west-indische voyagien. Amsterdam, 1619.

1620-1629

Htn	Geog 816.00.5*	Abbott, George. A briefe description of the whole world. 5th ed. London, 1620.
Htn	Geog 815.59.7*	Botero, G. Le relationi universali...divise in quattro parti. Venetia, 1622.
Htn	Geog 4206.13.9F*	Purchas, Samuel. His pilgrimes. London, 1625- 4v.
Htn	Geog 816.61.16*	Cluverius, P. Introductionis in universam geographiam. Lugdunum Batavorum, 1627.
Htn	Geog 756.28*	Velazquez Minaya, F. Esfera forma del mundo. Madrid, 1628.

1630-1639

	Geog 4206.13.25	Amman, H.J. Reiss ins gelobte Land...Servian...Aegypten. Zürich, 1630.
	Geog 816.61.19	Cluverius, P. Introduction à la geographie universelle. Paris, 1631.
Htn	Geog 4266.31*	Gölnitz, A. Ulysses belgico-gallicus...per Belgium hispan. Lugdunum Batavorum, 1631.
Htn	Geog 4326.32*	Lithgow, William. The totall discourse of...adventures. Lyon, 1632.
	Geog 4341.73.3*	Benjamin ben Jonah, of Tudela. Itinerarium Benjaminis. Lugdunum Batavorum, 1633.
Htn	Geog 4126.33*	Essex, Robert. Profitable instrucions. London, 1633.
Htn	Geog 816.33.5*	Heylyn, Peter. Mikrokosmos; little description. Oxford, 1633.
Htn	Geog 5516.31*	James, T. Strange and dangerous voyage...N.W. Passage. London, 1633.
Htn	Geog 5516.31.2*	James, T. Strange and dangerous voyage...N.W. Passage. London, 1633.
Htn	Geog 816.35*	Carpenter, N. Geographie delineated forth in two books. Oxford, 1635.
Htn	Geog 5516.31.9*	Fox, Luke. North-west Fox...voyages of Cabot, Frobisher. London, 1635.
	Geog 816.36	Carvalho Da Costa, Antonio. Compendio geographico. Lisboa, 1636.
Htn	Geog 816.21.7*	Heylyn, Peter. Mikrokosmos: a little description of the great world. 7th ed. Oxford, 1636.
	Geog 4266.36	Mervla, P. Cosmographiae generalis libri tres. Amsterdam, 1636.
Htn	Geog 756.36*	Postelli, E. Cosmographica disciplina. Lugduni, 1636.
Htn	Geog 4416.26*	Herbert, Thomas. Some years travels into...Asia and Afrique. London, 1638.
Htn	Geog 4206.10.6F*	Linschoten, Jan H. van. Histoire de la navigation. Amsterdam, 1638.
Htn	Geog 816.33.8*	Heylyn, Peter. Mikrokosmos; a little description of the great world. 8th ed. Oxford, 1639.
Htn	Geog 816.33.7*	Heylyn, Peter. Mikrokosmos; little description. Oxford, 1639.

1640-1649

Htn	Geog 4326.32.3*	Lithgow, William. The totall discourse of...adventures. London, 1640.
Htn	Geog 4126.43*	Neale, Thomas. A treatise of direction, how to travell safely, and profitably into forraigne countries. London, 1643.
	Geog 756.44	Herigone, P. Cursus mathematicus. v.4. n.p., 1644.
	Geog 4205.96.9F	Linschoten, Jan H. van. Itinerarium oste schip-vaert naer. Amsterdam, 1644.
Htn	Geog 4206.45*	Mocquet, J. Voyages en Afrique, Asie. Rouen, 1645.
	VGeog 4306.00	Rohan de. Voyage...en l'an 1600. Amsterdam, 1646.
	An 3206.47	Otto, Andreas. Anthroposcopia. Regiomonti, 1647.
Htn	Geog 3055.37*	Tomasi, T. La spinalba antica historia. Venetia, 1647.
Htn	Geog 3106.48*	Brietio, O. Parallela geographiae. Paris, 1648-49. 3v.

1650-1659

Htn	Geog 807.51F*	Sanson, N. La France,...les Isles Britanniques. Paris, 1651.
Htn	Geog 816.52*	Francois, Jean. La science de la geographie divisée en trois parties. Rennes, 1652.
	Geog 816.46.13	Labbé, Philippe. La geographie royale. 2e éd. Paris, 1653.
Htn	Geog 4206.43.5*	La Boullaye, C. Goux. Les voyages et observations. Paris, 1653.
	Geog 4416.54	LeBlanc, V. De vermaarde reizen. Amsterdam, 1654.
Htn	An 2056.55*	La Peyrère, I. de. Systema theologicum...Praeadamitae. n.p., 1655.
Htn	Geog 816.55*	Linda, Lucas de. Descriptio orbis et omniumejus rerumpublicarum. Lugdunum Batavorum, 1655.
	Geog 4206.13.12	Purchas, Samuel. Samuel Purchas pelgrimagir. Amsterdam, 1655.
	Geog 4306.56*	Ogier, Charles. Ephemerides sive iter Danicum, Suecicum, Polonicum. Lutetiae Parisiorum, 1656.
Htn	An 2056.56*	Peyrère, I. de. Men before Adam: discourse on Romans 5-12,13,14. London, 1656.
	An 2056.56.5	Romano, E. Animad versiones in librum prae-adamitarum. Paris, 1656.
NEDL	Geog 816.61.3	Cluverius, P. Introduction into geography. Oxford, 1657.
	Geog 4416.54.5	LeBlanc, V. Les voyages fameux. Paris, 1658.
Htn	Geog 816.61.4*	Cluverius, P. Introductionis in universam geographiam. Amsterdam, 1659.

1660-1669

Htn	Geog 4206.60*	LeBlanc, V. The world surveyed. London, 1660.
Htn	Geog 3057.3*	Lipenius, M. Navigatio Salomonis Ophirifica. n.p., 1660.
	Geog 816.61	Cluverius, P. Introductionis in...geographiam. Amsterdam, 1661.
	An 2056.61	Peyrère, I. de. Praeadamiten. Jaer, 1661.
Htn	Geog 4306.60.2*	Loménie, Louis Henri. Itinerarium. Paris, 1662.
	Geog 4206.62	Valle, Pietro. Les fameux voyages de Pietro della Valle. Paris, 1663-70. 4v.
	Geog 4676.61	Kerckhoven, J. van. Wijtloopig...beschrijvinge van d. onzel voyage van't schip Arnhem, 1661. Amsterdam, 1664.
	Geog 4166.64F	Thevenot, M. de. Relations de divers voyages. Paris, 1664-66. 2v.
	Geog 4206.58	Welsch, Hier. Warhafftige Reiss-Beschreibung. v.1-2. Stuttgart, 1664.
Htn	Geog 4206.65*	Desboys du Chastelet, R. L'odyssee ou diversite d'avantures...en Europe, Asie et Afrique. Fleche, 1665.
Htn	Geog 4266.65*	Gerbier, B. Subsidium peregrinatibus. Oxford, 1665.
	Geog 816.65	Linda, Lucas de. Descriptio orbis. Amsterdam, 1665.
	Geog 4306.63.3	Payen. Les voyages de Monsieur Payen. 2e éd. Paris, 1667.
	Geog 576.67	Poyares, P. de. Diccionario lusitanico-latino de nomes proprios de regioens, reinos, provincias, cidades. Lisboa, 1667.
Htn	Geog 4306.68*	Birken, S. von. Hoch fürstlicher Brandenburgischer Ulysses. Bayreuth, 1668.
Htn	Geog 3060.9*	Das verdächtiger Pineser-Eyland. Hamburg, 1668.

1670-1679

	Geog 576.70F	Ferrari, Filippo. Lexicon geographicum. Parisiis, 1670.
	Geog 816.70	Joosten, J. De kleyne wonderlijcke werelt. Amsterdam, 1670.
Htn	Geog 816.57.5*	Clarke, Samuel. A geographical description of all the countries in the known world. London, 1671.
Htn	Geog 4476.71*	Schooten, Henry (pseud.). The hairy-giants. London, 1671.
	Geog 816.61.5	Cluverius, P. Introductionis in...geographiam. Brunsvigae, 1672.
Htn	Geog 4306.73*	Brown, E. Brief account of some travels. London, 1673.
	Geog 4326.74	Smith, T. Epistulae...moribus...turcarum. Oxonii, 1674.
Htn	Geog 4476.75*	Hamilton, William (pseud.). O-Brazile. London, 1675.
Htn	Geog 4136.76*	Bosch, L. Leeven en daden...zee-helden. Amsterdam, 1676.
	Geog 4306.71.5	Patin, Charles. Relations historiques...de voyages en Allemagne. 2e éd. Lyon, 1676.
	Geog 4416.26.5F	Herbert, Thomas. Some years travels into...Africa and Asia. London, 1677.
Htn	An 356.77*	Pechlin, J.N. De habitu et colore Aethiopum. Kiloni, 1677.
Htn	Geog 4266.41.7F*	Zeiller, Martin. Sämtliche...Topographias. v.1-31. Frankfurt, 1677-1736. 10v.
	Geog 4206.13.27	Amman, H.J. Reiss in das gelobte Land. Berlegung, 1678.
	Geog 5220.11	Capel, R. Norden oder zu Wasser und Lande. Hamburg, 1678.
	Geog 4166.78.5	Dyck, J. Seer gedenckwaerdige voyagien. Amsterdam, 1678.
Htn	Geog 956.78*	Meissner, D. Sciagraphia cosmica. v.1-8. Nürnberg, 1678. 2v.

1680-1689

	Geog 576.80	Fondeur, F. Urbium insularum regionum. Londuni, 1680.
	Geog 576.82F	Baudrand, M.A. Parisini geographia. Parisiis, 1681-82. 2v.
	Geog 4136.76.5	Bosch, L. Leben...der...See-Helden. Nürnberg, 1681.
	Geog 4206.81	Melton, E. (pseud.). Eduward Meltons, engelsch edelmans, zeldzaame en gedenkwaardige zee- en land reizen. Amsterdam, 1681.
Htn	Geog 4346.76.5*	Struys, Jan J. Les voyages de Jean Struys. Amsterdam, 1681.
	Geog 816.82	Duval, Pierre. La geographie du temps. pt.1-2. Paris, 1682. 2v.
Htn	Geog 4206.82*	Glanius. A new voyage to the East Indies. 2d ed. London, 1682.
Htn	Geog 4306.73.5F*	Brown, E. Brief account of some travels. London, 1685.
	Geog 4306.85	Pacichelli, G.B. Memorie di viaggi per l'Europa. Napoli, 1685. 3v.
Htn	Geog 816.85*	Suval, P. Geographia universalis. London, 1685.
Htn	Geog 4306.86*A	Burnet, Gilbert. Some letters...Switzerland, Italy. Rotterdam, 1686.
	Geog 816.61.6	Cluverius, P. Introductionis in...geographiam. Amsterdam, 1686.
	Geog 4326.86	Neitzschitz, G.C. Sieben-Jährige...Welt-Beschauung. Nurnberg, 1686.
Htn	Geog 4306.86.3*	Burnet, Gilbert. Dr. Burnet's travels or letters. Amsterdam, 1687.
	Geog 806.87	Happelius, E.G. Mundus mirabilis tripartitus. Ulm, 1687. 3v.
Htn	Geog 4260.5F*	Zeiller, M. Topographia Italiae. Franckfurt, 1688.

1690-1699

Htn	Geog 756.91*	Abraham ben Mordecai Farissol. Itinera mundi. v.1-2. Oxonii, 1691.
Htn	Geog 4476.92*	Les avantures de Jaques Sadeur. Paris, 1692.
Htn	Geog 4306.92*	Voyages historiques de l'Europe. v.1-8. Paris, 1692. 2v.
	Geog 4346.85	Avril, P. Voyage en divers etats d'Europe et d'Asie. Paris, 1693.
Htn	Geog 576.93F	Bohun, E. Geographical dictionary. London, 1693.
Htn	Geog 4476.92.2*	Foigny, Gabriel de. A new discovery of terra Incognita Australis. London, 1693.
	Geog 4346.93	Ray, John. Collection of curious travels and voyages...eastern countries. London, 1693.
	Geog 816.93	Sanson, N. Introduction à la geographie. Paris, 1693.
Htn	Geog 4206.93*	Account of several late voyages and discoveries to south and north. London, 1694.
Htn	Geog 816.94*	Seller, John. A new system of geography. n.p., 1694?
	Geog 4306.95	Patin, Charles. Relations historiques...de voyages. Amsterdam, 1695.
	Geog 816.96	El atlas abreviado ô compendiosa geografia. Amberes, 1696.
	Geog 4206.45.5	Mocquet, J. Travels and voyages into Africa. London, 1696.
	Geog 4166.64.5F	Thevenot, M. de. Relation de divers voyages. Paris, 1696. 2v.
	Geog 816.61.7	Cluverius, P. Introductionis in universam geographiam. Amsterdam, 1697.
	Geog 576.97F	Ferrari, P. Novum lexicon geographicum. Patavii, 1697.
	Geog 4216.99	Careri, G. Giro del mondo. Napoli, 1699-1700. 6v.
Htn	Geog 4206.99*	Hacke, William. A collection of original voyages. London, 1699.

Chronological Listing

1690-1699 - cont.

	Geog 3057.7	Der wohleingerichtete Staat des Bishero von vielen gesuchten aber nicht gefundenen Königreichs Ophir. Photoreproduction. Leipzig, 1699. 2v.

17-

	Geog 3520.7F	Aa, P. van der. De wijd beroemde voyagien...der Engelsen. Leyden, 17- . 2v.
	Geog 757.64F	Fenning, Daniel. A new system of geography: or A general description of the world. n.p., 17- . 2v.
	Geog 807.00F	Middleton, C.T. System of geography. v.1-2. London, 17- 4v.
	Geog 3060.15	Relação que trata de como em cincoenta e oito gráos do sul fay descuberta huma ilha. pt.2. Lisboa, 17- ?
	Geog 4677.96	Stout, Benjamin. The total loss of the American ship Hercules. London, 17- ?

1700-1709

Htn	Geog 817.00.5*	Bion. L'usages des globes celestes et terrestes. Adam, 1700.
Htn	Geog 816.80.4*	Morden, Robert. Geography rectified. 4th ed. London, 1700.
Htn	Geog 817.00*	Sanson, N. Description de tout universe. Amsterdam, 1700.
	Geog 577.01	Baudrand, M.A. Dictionaire geographique universal. Amsterdam, 1701.
Htn	Geog 817.01F*	Moll, H. System of geography. London, 1701.
	Geog 4157.04F	Churchill, J. Collection of voyages and travels. London, 1704-07. 8v.
	Geog 817.04	Gordon, P. Geography anatomized. 4th ed. London, 1704.
	Geog 4306.92.6	Bromley, William. Remarks in the grand tour of France and Italy. 2d ed. London, 1705.
Htn	Geog 4157.05.2F*	Harris, J. Navigantium...compleat collection. London, 1705.
	Geog 4127.05	Schroeter, J.C. Diatriba...peregrinationum eruditarum. Jenae, 1705.
	Geog 5837.06	Torfaeo, T. Gronlandia antiqua. Havniae, 1706.
Htn	Geog 5837.06.2*	Torfaeo, T. Gronlandia antiqua. Havniae, 1706.
NEDL	Geog 4157.07	Aa, P.V. Naaukeurige...der zee en land-reysen. v.1-28. Leyden, 1707. 29v.
Htn	Geog 3570.11*	Barrós, J. de. De alder erste scheepo-tog ten der Portugesen. Leyden, 1707?
	Geog 4217.03	Funnell, William. Voyage round the world. London, 1707.
Htn	Geog 577.04*	The gazetteer's or newsman's interpreter. pt.2. London, 1707.
	Geog 4167.07.2	Bellegarde, J.B.M. General history of all voyages...old and new world. London, 1708.
	Geog 4306.86.5	Burnet, Gilbert. Some letters...Switzerland, Italy. n.p., 1708.
Htn	Geog 4477.08*	Posos, Juan de (pseud.). Beschryoinge van het magtig. Amsterdam, 1708.
Htn	Geog 577.09*	Eachard, Laurence. Gazeteer's geographical index of Europe. London, 1709.
	Geog 4477.09	Relation...du voyage du Prince de Montberaud. Merinde, 1709.

1710-1719

	Geog 807.03	Scherer, R.P.H. Atlas novus exhibens orbem. pt.1-7. Augustae, 1710. 4v.
Htn	Geog 4477.10*	Voyages et avantures de Jaques Masse. Bourdeaux, 1710.
	Geog 4206.59.5	Account of several late voyages and discoveries to north and south. London, 1711.
Htn	Geog 4347.01.5F*	Bruyn, C. de. Reizen over Muskovie door Persie. Amsterdam, 1711.
Htn	Geog 816.61.9*	Cluverius, P. Introductionis in...geographiam. Londini, 1711.
Htn	Geog 817.04.6*	Gordon, P. Geography anatomized. 6th ed. London, 1711.
Htn	Geog 4167.11*	Stevens, J. New collection of voyages. London, 1711. 2v.
	Geog 577.13F	Savonarola, R. Universus terrarum orbis. Patavii, 1713. 2v.
	Geog 5700.4.10	Dalager, Lars. Grønlandske relationer. København, 1715.
	Geog 807.16F	Notoras, C. Introductio ad geographia et spheram. Paris, 1716.
Htn	Geog 4347.01.8F*	Bruyn, C. de. Voyages...en Perse et aux Indes Orient. Amsterdam, 1718. 2v.
	Geog 4267.18	Pockh, J.J. Der politische catholische Passagier. Augspurg, 1718.
Htn	Geog 4347.18*	Struys, Jan J. Les voyages...en Moscovie. v.1-3. Amsterdam, 1718.
	Geog 4300.48	Vidari, Giovanni Maria. Il viaggio in pratica. Venezia, 1718.
	Geog 817.04.8	Gordon, P. Geography anatomized. 8th ed. London, 1719.

1720-1729

Htn	Geog 5837.20.10*	Mesange, P. de. La vie...et le voyage de Groenland. v.1-2. Amsterdam, 1720.
	Geog 5837.20	Zorgdrager, C.G. Groenlandsche visschery. Amsterdam, 1720.
	Geog 4477.21	Relation d'un voyage du Pole Arctique au Pole Antarctique. Amsterdam, 1721.
	Geog 4307.22	Careri, G. Viaggi per Europa. Napoli, 1722. 2v.
	Geog 4306.91.1	Retcher, Wilhelm. Der sächsische Robinson. Leipzig, 1722.
	Geog 817.01.5F	Moll, H. The compleat geographer. London, 1723.
	Geog 5516.19	Munk, J. Navigatio septentrionalis. Kjøbenhavn, 1723.
	Geog 4707.23.2F	Rhode Island (Colony). Court of Admiralty. Tryals of thirty-six persons for piracy. Boston, 1723.
Htn	Geog 4347.01.6F*	Bruyn, C. de. Voyage au Levant. Paris, 1725. 5v.
Htn	Geog 4477.26*	Boyle, R. (pseud.). Voyages and adventures of Captain Boyle. London, 1726.
	Geog 4267.26	Breval, John D. Remarks on several parts of Europe. v.1-2. London, 1726.
	Geog 577.26F	Bruzen de la Martiniere, A.A. Le grand dictionnaire géographique. Haye, 1726-39. 10v.
	Geog 4707.26	Jedre, J.B. The trials of five persons for piracy. Boston, 1726.
	Geog 817.26	Kolb, P.G. Compendium totius orbis. Rottwike, 1726.
	Geog 4707.26.5F	Massachusetts (Colony). Court of Admiralty. The tryals of sixteen persons for piracy. London, 1726.
	Geog 3107.26	Wells, Edward. Treatise of antient and present geography. London, 1726.
Htn	Geog 4266.41.8F*	Zeiller, Martin. Sämtliche...Topographias. Haupt-Register. Frankfurt, 1726.

1720-1729 - cont.

	Geog 4347.27	La Montraye, A. de. Voyage du...en Europe, Asie et Afrique. Haye, 1727. 2v.
	Geog 5837.20.3	Zorgdrager, C.G. Bloeijende opkomst...visschery. 's-Gravenhage, 1727.
	Geog 4217.28	La Barbinais le Gentil. Nouveau voyage autour du monde. Amsterdam, 1728. 2v.
	Geog 4307.28	Remarques d'un voyageur. Haye, 1728.
	Geog 807.29	Berckenmeier, P. Le curieux antiquaire. Leide, 1729. 3v.
Htn	Geog 816.61.13*	Cluverius, P. Introductionis in universam geographiam. Amsterdam, 1729.
	Geog 4157.29	Dampier, W. Collection of voyages. London, 1729. 4v.

1730-1739

	Geog 817.04.12	Gordon, P. Geography anatomized. 12th ed. London, 1730.
	An 2600.1	Koenig, H.C. De hominum inter ferar educatorum. Hanover, 1730.
	Geog 4157.31	Bernard, I.F. Recueil de voyages au nord. Amsterdam, 1731-37. 9v.
	Geog 3107.31	Cellarius, C. Notitia orbis antiqui. Lipsiae, 1731-32. 2v.
	Geog 577.31	Hederich, M.B. Reales Schul-Lexicon. Leipzig, 1731.
	Geog 4347.01.7F	Bruyn, C. de. Voyages...au Levant. Haye, 1732. 5v.
	Geog 817.82	Guthrie, William. A new geographical, historical, and commercial grammar and present state of several kingdoms of the world. 7th ed. London, 1732.
	Geog 4347.27.3F	La Montraye, A. de. Voyage en Anglois et en François. Haye, 1732.
Htn	Geog 4347.27.5*	La Montraye, A. de. Voyages and travels, Prussia, Russia and Poland. London, 1732. 3v.
	Geog 4676.93	Correa, F. Relaçam do successo que teve o patacho chamado N. Sra. da Candelaria da ilha da Madeira. Lisboa, 1734.
	Geog 4477.21.3	Relation un voyage du Pole Arctique au Pole Antarctique. 2. ed. La Haye, 1734.
Htn	Geog 4677.35*	Gomes de Brito, Bernardo. História trágico-marítima. Lisboa, 1735. 3v.
	Geog 4307.35.3	Pöllnitz, K.L. von. Memoires de Charles-Louis, Baron de Pöllnitz. Amsterdam, 1735. 2v.
	Geog 4217.37	Behrens, E.F. Sud-Lander und um die Balt. Leipzig, 1737.
	Geog 4347.01.10FA	Bruyn, C. de. Travels into Muscovy, Persia, E. Indies. London, 1737. 2v.
	Geog 4307.37	Pöllnitz, K.L. von. Memoirs of...travels. London, 1737. 4v.
	Geog 4307.35.5	Pöllnitz, K.L. von. Nouveaux memoires du Baron de Pöllnitz. Amsterdam, 1737. 2v.
	Geog 4267.38	Breval, John D. Remarks on several parts of Europe. v.1-2. London, 1738.
	Geog 5700.2	Egede, H. Omstaendelig...grønlandste missions. Kjøbenhavn, 1738.
	Geog 4346.73.2	Ray, John. Travels thro' Low Countries, Germany. 2nd ed. London, 1738.
NEDL	Geog 807.40	Salmon, Thomas. Lo stato presente di tutti i paesi e popoli del mondo naturale, politico e morale. 2a ed. v.1-26. Venezia, 1738-66. 27v.
	Geog 4417.38F	Shaw, Thomas. Travels or observations. Oxford, 1738.
Htn	Geog 4327.39*	Campbell, J. Travels and adventures of Eduard Brown. London, 1739.
	Geog 4307.37.2	Pöllnitz, K.L. von. Memoirs of Charles-Lewis, Baron de Pöllnitz. 2nd ed. London, 1739. 4v.
	Geog 4477.29	Les voyages de Glantzby. Paris, 1739.

1740-1749

	Geog 5700.4	Egede, H. Ausführliche...Nachricht...Grönland. Hamburg, 1740.
	Geog 5837.40F	Mauricius, Jan J. Naleesing over Groenland voor de historie van den noorweeschen Erik. pt.1-5. Hamburg? 1740.
	Geog 4477.26.5	Boyle, R. (pseud.). The voyages and adventures of Captain Boyle. Dublin, 1741.
Htn	Geog 5700.4.5*	Egede, H. Det gamle Grønlands...naturel-historie. Kjøbenhavn, 1741.
	Geog 4207.41	Kuhns, J.M. Lebens und Reise Beschreibung. Gotha, 1741.
	Geog 807.42	Lenglet, N. Methode pour etudier la geographie. Paris, 1742. 7v.
Htn	Geog 757.42*	Maupertuis, Pierre L. Elements de geographie. Paris, 1742.
	Geog 4307.43	Blainville, J. de. Travels through Holland, Germany, Switzerland, and other parts of Europe. London, 1743-1745. 3v.
	Geog 4327.22	Shaw, Thomas. Voyages...dans plusieurs provinces de la Barbarie et du Levant. La Haye, 1743. 2v.
	Geog 4157.04.3F	Churchill, J. Collection of voyages and travels. 3. ed. London, 1744. 6v.
	Geog 4157.05.10F	Harris, J. Navigantium...compleat collection. London, 1744. 2v.
	Geog 4417.44	Lade, R. Voyages du capitaine...Lade. Paris, 1744. 2v.
	Geog 807.39.5F	Salmon, Thomas. Modern history. 3rd ed. London, 1744-1746. 3v.
	Geog 4327.44	Thompson, C. Travels. Reading, 1744. 2v.
Htn	Geog 5507.45*	Description of coast...in Button's Bay. London, 1745?
	Geog 4157.45	Green, J. New collection of voyages and travels. London, 1745-47. 4v.
	Geog 5507.45.6	North West Committee. Articles of agreement. London, 1745.
	Geog 4157.46	Prevost, A.T. Histoire generale des voyages. Paris, 1746-89. 20v.
	Geog 4157.47	Allgemeine Historie der Reisen. v.1-19,21. Leipzig, 1747-74. 20v.
	Geog 815.59.10	Botero, G. Descripcion de todas las provincias. Gerona, 1748.
Htn	Geog 5517.46.10*	Drage, T.S. Account of voyage...Northwest Passage. London, 1748-49. 2v.
	Geog 5517.46.5	Ellis, H. Voyage de la baye de Hudson. Paris, 1749.
	Geog 817.04.19	Gordon, P. Geography anatomized. 19th ed. London, 1749.
Htn	Geog 5507.49*	Reasons...navigable passage to west American oceans. London, 1749.
	Geog 817.51.2	Salmon, T. A new geographical...grammar. London, 1749.

1750-1759

	Geog 4677.50	Bailey, J. God's wonders in the great deep. N.Y., 1750.
	Geog 4306.86.10	Burnet, Gilbert. Travels through France, Italy. London, 1750.
	Geog 4307.50	Clancy, Michael. The memoirs. v.1-2. Dublin, 1750.

Chronological Listing

1750-1759 - cont.

	Geog 5517.46.3	Ellis, H. Reise nach Hudsons Meerbusen. Göttingen, 1750.
	Geog 5517.46.4	Ellis, H. Voyage à la baye de Hudson. Leide, 1750.
	Geog 5837.20.5	Zorgdrager, C.G. Beschreibung des...Wallfischfangs. Nürnberg, 1750.
	Geog 577.09.15	Eachard, Laurence. The gazetteer's or news-man's interpreter. London, 1751.
	Geog 4477.51	Geschichte des Gaudenti di Lucca. Frankfurt, 1751.
	Geog 4167.51	Reisen nach Peru, Acadien und Egypten. Göttingen, 1751.
	Geog 817.51.3	Salmon, T. New geographical...grammar. London, 1751.
Htn	Geog 5220.13*	L'Isle, J.N. de. Explication de la carte des...decouvertes. Paris, 1752. 3 pam.
Htn	Geog 5220.21*	Buache, P. Considerations geographiques. Paris, 1753.
Htn	Geog 5220.15	L'Isle, J.N. de. Erklarung der Charte...Entdeckungen. Berlin, 1753.
Htn	Geog 4307.53*	Uffenbach, Z.C. Merkwürdige Reisen. Ulm, 1753-1754. 3v.
	Geog 4327.44.3F	Drummond, A. Travels thro' different cities of Germany, Italy, Greece and parts of Asia. London, 1754.
	Geog 817.04.20	Gordon, P. Geography anatomized. 20th ed. London, 1754.
Htn	Geog 5220.17*	Letter from Russian sea-officer. London, 1754.
	Geog 3017.55	Robert de Vaugondy, D. Essai sur l'histoire de géographie. Paris, 1755.
	Geog 757.55	Vaissete, Joseph. Géographie historique, ecclesiastique et civile. Paris, 1755. 4v.
	Geog 4307.56	Nugent, T. Grand tour...through Netherlands. London, 1756. 4v.
	Geog 4237.56	Pierce, Nathaniel. Account of the great dangers...of Captain N. Pierce. Boston, 1756.
	Geog 4477.57	Histoire d'un peuple nouveau. Londres, 1757.
Htn	Geog 3060.16*	Noticia certa do descobrimento de huma nova terra. Lisboa, 1757.
	Geog 4207.53.3	Hanway, Jonas. Reize van London, door Rusland, nae en in Persie. Amsterdam, 1758. 2v.
	Geog 577.58	Salmon, T. Modern gazeteer. London, 1758.
	An 2600.3	Ausführliches Leben und merkwürdige Schiksale eines wilden Knaben. Frankfurt, 1759.
	Geog 4477.59	Brachfelds, J.M. Begebenheisen...Sudlandern. Eisenach, 1759.
	Geog 577.59F	New geographical dictionary of the known world. London, 1759.
	Geog 577.58.2	Salmon, T. Modern gazeteer. London, 1759.
	Geog 4157.59	World displayed. London, 1759-61. 20v.

1760-1769

	Geog 4677.66.3	Harrison, David. Een droevig verhaal van de ongelukkige reis en wonderdaadige verlossinge. Wesel, 176-?
	Geog 577.58.4	Salmon, T. Modern gazeteer. London, 176-.
	Geog 577.60.3	The universal gazetteer. 2. ed. London, 1760.
	Geog 4157.59.3	World displayed. v.3-4,5-6,8,15-16,17-18. 3.-4. ed. London, 1760-88. 5v.
	Geog 4237.61	An account of the voyages and cruizes of Captain Walker. Boston, 1761.
	Geog 4477.61	Listonai. Le voyageur philosophe. Amsterdam, 1761. 2v.
Htn	Geog 5220.19*	Müller, G.F. Voyages from Asia to America. London, 1761.
Htn	Geog 3017.62*	Bollan, William. Colonae anglicanae illustratae. Londini, 1762.
	Geog 577.62	Brookes, R. The general gazeteer. London, 1762.
	Geog 807.62	Büsching, Anton Friedrich. New system of geography. London, 1762. 6v.
	Geog 4207.53.5F	Hanway, Jonas. Historical account of British trade. 3. ed. London, 1762. 2v.
	Geog 4477.51.15	Berington, S. The adventures of Gaudentio di Lucca. London, 1763.
	Geog 5837.63	Egede, H. Beschreibung und Natur-Geschichte. Berlin, 1763.
	Geog 4267.63	Expilly, J.J. d'. Le geographe manuel. 6e éd. Paris, 1763.
	Geog 557.62	Succinta descrizione. Venezia, 1763.
	Geog 4157.05.15F	Harris, J. Navigantium...voyages and travels. London, 1764. 2v.
Htn	Geog 5367.61.3*	Müller, G.F. Voyages from Asia to America. 2. ed. London, 1764.
	Geog 817.51.5	Salmon, T. New geographical...grammar. London, 1764.
	Geog 4167.65	Barrow, J. A collection of authentic...voyages. v.2. London, 1765.
Htn	Geog 5367.65*	Engel, S. Memoires...geographiques...pays septentrionaux. Lausanne, 1765.
	Geog 4417.38.5	Shaw, Thomas. Reisen der Barbery und der Levante. Leipzig, 1765.
	Geog 4157.66	Barrow, J. Abrégé...histoire des decouvertes. Paris, 1766. 3v.
	Geog 4157.66.5	Callander, J. Terra Australis cognita, or Voyages. Edinburgh, 1766. 3v.
	Geog 500.13	Hagers, J.G. Geographischer Buchersaal. Chemnitz, 1766-78. 3v.
	Geog 817.51.7	Salmon, T. New geographical...grammar. London, 1766.
	Geog 5700.5	Crantz, D. History of Greenland. London, 1767. 2v.
	Geog 4157.67	Knox, J. New collection of voyages. London, 1767. 7v.
	Geog 577.68F	Bruzen de la Martiniere, A.A. Le grand dictionnaire géographique. Paris, 1768. 6v.
	Geog 4217.64.10	Byron, John V. Viaggio intorno al mondo. Firenze, 1768.
Htn	Geog 4477.35.3*	Godwin, Francis. The strange voyage and adventures of Domingo Gonsales. 2. ed. London, 1768.
	Geog 4417.68	Poivre, Pierre. Voyages d'un philosophe. Yverdon, 1768.
	Geog 4207.68	Viaggi da Parma in varie partie del mondo. Parma, 1768.
	Geog 807.68	The wonders of nature and art. 2nd ed. London, 1768. 6v.
	Geog 4217.64.7	Ortega, D.C. de. Viage del Comandante Byron al rededor del mundo. Madrid, 1769.
	Geog 817.51.8	Salmon, T. New geographical...grammar. London, 1769.

1770-1779

	Geog 757.70	Buy de Mornas, C. Cosmographie methodique. Paris, 1770.
	Geog 5700.5.3A	Crantz, D. Historie von Grønland...mission. Barby, 1770. 3v.
	Geog 577.09.18	Ladvocat, J.B. Dictionnaire géographique portatif. Paris, 1770.
	Geog 4217.66	Bougainville, L. de. Voyage autour du monde. Paris, 1771.
	Geog 4217.66.11	Bougainville, L. de. Reis rondom de weereldt. Dordrecht, 1772.
	Geog 4217.66.5	Bougainville, L. de. Voyage autour du monde. Paris, 1772. 3v.
	Geog 4217.66.3A	Bougainville, L. de. Voyage round the world...1766. London, 1772.
	Geog 5367.65.3	Engel, S. Geographische...Nachrichten. Mietau, 1772.

1770-1779 - cont.

	Geog 4307.72	Marshall, J. Travels through Holland. London, 1772. 3v.
	Geog 5170.15	Projet pour tenter la decouverte. La Haye, 1772.
	Geog 4157.74	Henry, David. Historical account of voyages. London, 1773-74. 4v.
	Geog 817.73	Jones, E. The young geographer and astronomer's best companion. London, 1773.
Htn	Geog 577.73*	Macbean, A. Dictionary of ancient geography. London, 1773.
	Geog 4306.97F	Sheremt'ev, B.P. Zapiska puteshestviia. Moskva, 1773.
	Geog 4477.51.7	The adventures of Signore Gaudentio di Lucca. London, 1774.
Htn	Geog 817.74*	André, Noël. Description et usages de la Mappemonde. Paris, 1774.
	Geog 5377.74	Journal of a voyage...by the Honorable Commodore Phipps, and Captain Lutwidge. London, 1774. 3 pam.
Htn	Geog 5397.73*	Phipps, J. Journal of voyage...North Pole. London, 1774.
	Geog 5397.73.3	Phipps, J. Voyage...North Pole...1773. London, 1774.
Htn	Geog 5170.7*	Barrington, D. Probability of reaching the North Pole. London, 1775.
	Geog 4167.75	Dalrymple, A. A collection of voyages, chiefly in the S. Atlantic. London, 1775.
	Geog 4207.55	Morris, Drake. Travels of...voyage at sea. London, 1775.
	Geog 5397.73.5	Phipps, J. Voyages au pôle boréal. Paris, 1775.
	Geog 5837.75	Staunning, J. Kort beskrivelse over Grønland. v.1-3. Viborg, 1775.
	An 357.76	Demeunier, J.N. L'esprit des usages et des coutumes des différens peuples. Londres, 1776. 3v.
	Geog 577.76	Johnson, R. New gazeteer. London, 1776.
	Geog 5837.76	Thorhallesen, E. Esterretning om rudera...Grønlands. Kiøbenhavn, 1776.
	Geog 5367.65.5	Engel, S. Neuer Versuch über...Schriften. Basel, 1777.
	Geog 5397.73.7	Phipps, J. Reise nach dem Nordpol. Bern, 1777.
	Geog 4237.78	Alexandre e Silva, Elias. Relação, ou Noticia particular da infeliz viajem da nas de Nassa Senhora da Ajuda. Lisboa, 1778.
	Geog 5700.13	Lund, Jacob Johan. Første missionair paa Grønland. København, 1778.
Htn	Geog 3055.33*	Bailly, J.S. Lettres sur l'Atlantide de Platon. Paris, 1779.
	Geog 817.79F	Carver, J. New universal traveller. London, 1779.
	Geog 5367.65.7	Engel, S. Extraits...des voyages...de l'Asie. Lausanne, 1779.
	Geog 5837.77	Groot, J.J. Beknopt en getrouw verhaal. Amsterdam, 1779.
	Geog 4267.79	Raff, G.C. Geographie für Kinder. Tübingen, 1779.

1780-1789

	Geog 4307.80	Björnståhls, J.J. Briefe auf seinem Auslandischen Reisen. Leipzig, 1780-1783. 6v.
	Geog 4417.80	Irwin, E. Series of adventures...Red Sea, Arabia, Egypt. Dublin, 1780.
	Geog 4157.80	LaHarpe, J.F. Abrégé de l'histoire generale des voyages. v.1-32, atlas. Paris, 1780-1801. 33v.
	Geog 4427.80	Le Gentil de la Galaisière, G.J.H.J.B. Voyage dans les mers de l'Inde. v.1-2. Suisse, 1780.
	Geog 5170.13	Remarks on the voyage of the ships "Resolution". London, 1780.
Htn	Geog 149.1*	Pallas, Peter. Neue nordische Beyträge. St. Petersburg, 1781. 4v.
	Geog 4477.64.5	Seriman, Zaccaria. Viages de Enrique Wanton. Madrid, 1781-85. 4v.
	Geog 4307.82	Bruce, Peter Henry. Memoirs of Peter Henry Bruce, Esq. London, 1782.
	Geog 5180.7	De Croy, D. Memoire sur le passage par le nord. Paris, 1782.
	Geog 4267.71.3	Gutierrez de la Hacera, P.R. Descripción general de la Europa. Madrid, 1782. 2v.
	Geog 665.20	Historisch-statistisch-geographische Belustigungen. Leipzig, 1782.
	Geog 4427.82	Kerguélen. Relation de deux voyages. Paris, 1782.
	Geog 4217.67	Pagès, P.M.F. de. Voyage autour du monde. Paris, 1782. 2v.
Htn	Geog 5507.82*	Pickersgill, R. Concise account of voyages...discovery. London, 1782.
	Geog 577.58.6	Salmon, T. Modern gazeteer. London, 1782.
	Geog 4307.83.2	The American wanderer...Europe. Dublin, 1783.
	Geog 4307.83	The American wanderer...Europe. London, 1783.
	Geog 817.83	Guthrie, William. A new geographical, historical, and commercial grammar and present state of the several kingdoms of the world. 8th ed. London, 1783.
	Geog 4357.81	Rooke, Henry. Travels to the coast of Arabia Felix, and from thence by the Red Sea and Egypt to Europe. London, 1783.
	Geog 4307.76	Sander, Heinrich. Beschreibung seiner Reisen durch Frankreich. Pt.1. Leipzig, 1783.
	Geog 807.87.2	Büsching, Anton Friedrich. Grosse Erdbeschreibung. Troppau, 1784-1787. 24v.
	Geog 4207.79	Evers, Journal kept on a journey from Bassora to Bagdad...1779. Horsham, 1784.
	Geog 4427.35	Schwartz, G.L. Utdrag af en Ost-Indisk rese-bescrifning. Westerås, 1784.
	Geog 4217.72.3	Sparrman, Anders. Reise nach dem Vorgebirge der guten Hoffnung. Berlin, 1784.
	Geog 500.29	Stuck, G.H. Verzeichnis von...Land und Reisebeschreibungen. Halle, 1784.
	Geog 807.85	Büsching, Anton Friedrich. Auszug aus seiner Erdbeschreibung. 6. Aufl. Hamburg, 1785.
	Geog 807.85.5	Handbuch der alten Erdbeschreibung zum Gebrauch der Eilf. v.1-2, pt.1-2. Nürnberg, 1785-1793. 5v.
	Geog 4324.89	Illustrazioni in un anonimo Viaggiatore (1489). Livorno, 1785.
	Geog 3017.85	Pluche, N.A. Concorde de la géographie des différents ages. Paris, 1785.
	Geog 4557.85	Pontoppidam, C. Hval og robbesangsten. Risbenhavn, 1785.
	Geog 4307.85.7	Ponz, A. Viage fuera de Espana. Madrid, 1785. 2v.
	Geog 817.51.9	Salmon, T. Geographical and astronomical grammar. London, 1785.
	An 3207.85	Sommering, S.T. Ueber...Verschiedenheit des Negers vom Europäer. Frankfurt, 1785.
Htn	Geog 4347.85F*	Witsen, N. Noord en Oost Tartaryen. Amsterdam, 1785. 2v.
	Geog 5220.5	Forster, J.R. History of voyages...in the North. London, 1786.
	Geog 4307.86.8	Moore, J. View of society and manners. 6th ed. London, 1786. 2v.

Chronological Listing

1780-1789 - cont.

	Geog 4677.82	A true account of the loss of the Halsewell. v.1-2. London, 1786.
	Geog 807.87.5	Bankes, T. A new, royal, authentic and complete system of universal geography. London, 1787-1810? 2v.
	Geog 4207.80.2	Baranshchikov, V. Neshchastiia prikliucheniia...Amerike, Azii i Evrope s" 1780 ro 1787 god". 2. izd. Sankt Peterburg, 1787.
	Geog 807.87	Büsching, Anton Friedrich. Erdbeschreibung. Hamburg, 1787-1792. 10v.
	Geog 4417.80.3	Irwin, E. Series of adventures...Red Sea, Arabia, Egypt. 3. ed. London, 1787. 2v.
	Geog 5507.87	LaRoche, F.C. de. Sendschreiben...des alten Grönlands. Kopenhagen, 1787.
	Geog 4217.72	Sparrman, Anders. Voyage au Cap de Bonne-Esperance. Paris, 1787.
	Geog 4477.87.7	Voyages imaginaires, songes. v.1-31,36. Amsterdam, 1787-89. 32v.
	Geog 4477.87	Wollap, G. Mémoires. v.1-4. Londres, 1787. 2v.
	Geog 4157.88	Berenger, J.P. Collection de tous les voyages. Lausanne, 1788-91. 9v.
	Geog 4268.83.6	Dutens, Louis. Itinéraire des routes les plus fréquentées. 6. éd. Paris, 1788.
	Geog 5220.7	Forster, J.R. Histoire des decouvertes...dans le Nord. Paris, 1788. 3v.
	Geog 817.88	Guthrie, William. A new geographical, historical and commercial grammar. 11th ed. London, 1788.
	Geog 4327.88	Jardine, A. Letters from Barbary, France. London, 1788. 2v.
	Geog 4307.86.13	Ansbach, Elizabeth Craven. Journey thro' Crimea to Constantinople. London, 1789.
Htn	Geog 4307.86.12*	Ansbach, Elizabeth Craven. Journey thro' Crimea to Constantinople. London, 1789. 2 pam.
	Geog 817.89	Gordon, William. New geographical grammar. Edinburgh, 1789.
	Geog 4227.89	Langstedt, F.L. Reisen nach Sudamerika, Asien. Hildesheim, 1789.
Htn	Geog 4217.85.3F*	Portlock, N. Voyage round the world. London, 1789.
	Geog 4217.72.5	Sparrman, Anders. Voyage to Cape of Good Hope. Perth, 1789.

1790

	Geog 4672.21.2	Deperthes, Jean L.H.S. Histoire des naufrages, ou Recueil des relations les plus intéressantes des naufrages. Paris, 1790. 3v.
	Geog 500.31	Ersch, J.S. Repertorium über...allgemeinern...Journale. Lemgo, 1790. 3v.

1791

	Geog 4327.88.10	Bisani, Alessandro. Lettres sur divers endroits de l'Europe. Londres, 1791.
	Geog 577.91	Brookes, R. General gazeteer. London, 1791.
Htn	An 3207.91*	Camper, P. Dissertation physique...traits du visage. Utrecht, 1791. 2 pam.
	Geog 4217.67.5	Pagès, P.M.F. de. Travels round the world. London, 1791. 3v.
	Geog 4307.87.5	Walker, Adam. Bemerkungen auf einer Reise durch Flandern. Berlin, 1791.

1792

	Geog 4167.92	Adams, J. Flowers of modern travels. London, 1792. 2v.
	Geog 817.92	Guthrie, William. A new geographical, historical and commercial grammar. 13th ed. London, 1792.
	Geog 4267.92	Meusel, J.G. Lehrbuch der Statistik. Leipzig, 1792.
	Geog 4307.86.6	Moore, J. View of society and manners. Boston, 1792.
	Geog 5516.19.5	Voyage of Captain John Monk. Glasgow, 1792.

1793

	Geog 4327.93	Bisani, Alessandro. Picturesque tour thro'...Europe, Asia. London, 1793.
Htn	Geog 5507.93*	Goldson, W. Observations on passage...Atlantic and Pacific. Portsmouth, 1793.
	Geog 4307.86.7	Moore, J. View of society and manners. 5th ed. Dublin, 1793. 2v.
	Geog 4307.86	Smith, J.E. Sketch of a tour on the continent. London, 1793. 3v.
	Geog 4417.70	Thunberg, Karl P. Travels in Europe, Africa and Asia. London, 1793? 4v.

1794

	Geog 3107.94	Adam, A. Summary of geography and history. Edinburgh, 1794.
	Geog 5837.92	Eggers, H.P. Om Grönlands osterbygds. Kjøbenhavn, 1794.
	Geog 817.94	Guthrie, William. New system of modern geography. Philadelphia, 1794-1795. 2v.
	Geog 4307.91	Stolberg, F.L.G. Reise in Deutschland, der Schweiz. Königsberg, 1794.
	Geog 4417.70.2	Thunberg, Karl P. Voyage en Afrique et en Asie. 1770-1779. Paris, 1794.
	Geog 4307.87.2	Watkins, T. Travels through Switzerland, Italy. London, 1794. 2v.

1795

	Geog 3107.94.5	Adam, A. Geographical index. Edinburgh, 1795.
	An 357.95	Blumenbach, J.F. De generis humani varietate nativa. Gottingae, 1795.
	Geog 757.93	Brookes, R. The general gazetteer. London, 1795.
	Geog 817.94.5F	Guthrie, William. New system of modern geography. London, 1795.
	Geog 4307.94.2	Radcliffe, A. (Mrs.). Journey...through Holland. 2. ed. London, 1795. 2v.
Htn	Geog 4307.94*	Radcliffe, A. (Mrs.). A journey made in summer of 1794. London, 1795.

1796

	Geog 4477.26.3	Boyle, R. (pseud.). The voyages and adventures of Captain Boyle. 1st American ed. Cooperstown, N.Y., 1796.
	Geog 577.91.3	Brookes, R. General gazeteer. London, 1796.

1796 - cont.

	Geog 817.94.10	Guthrie, William. New geographical, historical...grammar. London, 1796.
	Geog 4157.96	Mavor, William. Historical account of...voyages. v.1-2,4-22. London, 1796. 19v.
	Geog 4307.91.2	Stolberg, F.L.G. Travels through Germany. London, 1796. 2v.

1797

Htn	Geog 4167.92.3*	Adams, J. Flowers of modern travels. Boston, 1797. 2v.
	Geog 4167.97	Heron, Robert. A collection of late voyages and travels. Edinburgh, 1797.
Htn	Geog 577.97*	Malham, J. Naval gazetteer. Boston, 1797. 2v.
Htn	Geog 4217.97*	Milet-Mureau, M.L.A. Voyage de La Pérouse. Paris, 1797. 4v.
Htn	Geog 4217.97PF*	Milet-Mureau, M.L.A. Voyage de La Pérouse. Atlas. Paris, 1797.
	Geog 4217.88	Pagès, P.M.F. de. Nouveau voyage...en 1788. Paris, 1797. 3v.
	Geog 4307.97.2	Pratt, S.J. Gleanings through Wales. Dublin, 1797. 2v.
	Geog 4307.92	Sketches and observations...tour. London, 1797.
	Geog 4307.91.4	Stolberg, F.L.G. Travels through Germany, Switzerland, Italy. 2nd ed. London, 1797. 4v.

1798

Htn	Geog 4557.98*	Colnett, J. Voyages to the South Atlantic. London, 1798.
	Geog 577.98	Cruttwell, C. New universal gazetteer. London, 1798. 3v.
	Geog 5507.98F	Disertaciones sobre la navegación. Isla de León, 1798.
	Geog 4217.90.7	Fleurieu, Charles P. Claret de. Voyage autour du monde...1790-1792. Paris, 1798-1800. 4v.
	Geog 4347.98	Forster, G. Journey from Bengal to England. London, 1798.
	Geog 3127.98	Gosselin, P.F.J. Recherches sur la géographie...des anciens. Paris, 1798. 2v.
	Geog 4427.99	Henry, Pierre. Route de l'Inde. Paris, 1798-99.
	Geog 4417.98F	Pennant, T. The view of Hindoostan. London, 1798-1800. 4v.
Htn	Geog 4217.90*	Vancouver, G. Voyage of discovery...in 1790-1792. London, 1798. 3v.

1799

	Geog 3055.33.10	L'antiquité dévoilée par les principes de la magie naturelle. Paris? 1799-1800.
	Geog 4347.99.5	Jackson, J. Journey from India. London, 1799.
	Geog 4307.85.5	Matthisson, F. Letters...from various parts of the continent. London, 1799.
	Geog 4327.38	Montague, J. Voyage...round the Mediterranean. London, 1799.
	Geog 4347.99.7	Taylor, J. Travels from England to India. London, 1799. 2v.

18-

	Geog 4168.75	Adventures of famous travellers in many lands. N.Y., 18-
Htn	Geog 958.10*	Künstliche Erdkugel zur Uerbreitung gemeinnütziger Kentnisse über die Eintheilung. n.p., 18- ?
	Geog 5538.53.7	Strangers in Greenland. N.Y., 18- ?

1800

	Geog 4268.00F	Boetticher, J.G. Geographical, historical...description. London, 1800.
	An 358.00.5	Girando, J.M. Considérations sur les diverses méthodes à suivre dans l'observation des peuples sauvages. Paris, 1800. 2v.
	Geog 818.00.5	LaCroix, L.A.N. Geographie moderne et universelle. Paris, 1800. 2v.
	Geog 4678.00	Perils of the ocean. N.Y., 1800?

1801

	Geog 4217.90.18	Fleurieu, Charles P. Claret de. Die neuste Reise um die Welt in den Jahren 1790, 1791 und 1792 von Etienne Marchand. Leipzig, 1801? 2v.
	Geog 4217.90.15	Fleurieu, Charles P. Claret de. A voyage round the world. London, 1801. 3v.
	An 358.01	Richard, Jerôme. Voyages chez les peuples sauvages. Paris, 1801. 3v.
	Geog 4217.90.3	Vancouver, G. A voyage of discovery of the North Pacific Ocean. London, 1801. 6v.
	Geog 4308.01.5	Wakefield, Priscilla (Bell). The juvenile travellers. London, 1801.

1802

	Geog 818.02	Adam, A. A summary of geography and history. 3rd ed. London, 1802.
	Geog 807.87.3	Bohn, C.E. Ankündigung einer neuen Ausgabe von Büschings Erdbeschreibung welche auf Pränumeration gedruckt wird. Hamburg, 1802.
	An 2600.2	Itard, E.M. An historical account of...education of a savage man. London, 1802.
	Geog 4157.96.3	Mavor, William. Historical account of...voyages...Columbus to present. Philadelphia, 1802.
	Geog 4157.96.5	Mavor, William. Historical account of most celebrated voyages, travels, and discoveries from time of Columbus. New Haven, 1802-03. 24v.
Htn	Geog 578.02*	Morse, Judith. A new gazetteer of the Eastern continent. Charlestown, 1802.
	Geog 818.02.90	Pinkerton, J. Modern geography. London, 1802. 2v.
	Geog 4327.98F	Willyams, Cooper. A voyage up the Mediterranean. London, 1802.

1803

	Geog 4158.03F	Burney, J. Chronological history...discoveries..South Sea. London, 1803-1817. 5v.
	Geog 4218.03.11	Camus, A.G. Rapport à l'Institut national. Paris, 1803.
	Geog 3018.03	Clarke, J.S. Progress of maritime discovery. London, 1803.
	Geog 4678.03	Fellows, William D. Narrative of the loss of the Lady Hobart Packet. London, 1803.

Chronological Listing

1803 - cont.

| | Geog 4307.86.10 | Moore, J. View of society and manners. Paris, 1803. 2v. |
| Htn | Geog 3550.10F* | Morelli, Iacopo. Dissertazione intorno ad alcuni viaggiatori eruditi veneziani. Venezia, 1803. |

1804

	Geog 3055.33.5	Bailly, J.S. Lettres sur l'Atlantide de Platon. Paris, 1804.
Htn	Geog 5377.95*	Broughton, William. A voyage of discovery...North Pacific Ocean. London, 1804.
Htn	Geog 5377.95.2*	Broughton, William. A voyage of discovery...North Pacific Ocean. London, 1804.
	Geog 4308.04	Kotzebue, A. Travels from Berlin thro' Switzerland. London, 1804. 3v.
	Geog 4267.92.5	Meusel, J.G. Lehrbuch der Statistik. Leipzig, 1804.
	Geog 818.04	Pinkerton, J. Modern geography. Philadelphia, 1804. 2v.

1805

Htn	Geog 4558.05*	Burton, C. Journal of voyage from London. London, 1805.
Htn	Geog 818.05*	Davies, Benjamin. A new system of modern geography. Philadelphia, 1805.
	Geog 4328.05	Griffiths, J. Travels in Europe, Asia Minor, and Arabia. London, 1805.
	Geog 4158.05	Phillips, R. Collection of...voyages and travels. London, 1805-08. 7v.
	Geog 4158.05.5	Pouqueville, F.C.H. Voyage en Morée. v.1,2-3. Paris, 1805. 2v.

1806

	Geog 4308.06	Cruz y Bahamonde, N. Viage de Espana, Francia. v.1-10,12,14. Madrid, 1806-1813. 8v.
	Geog 4678.06	Duncan, A. The mariner's chronicle. Philadelphia, 1806. 4v.
	Geog 4208.06	General history of voyages and travels. Glasgow, 1806.
	Geog 4308.06.5	Wakefield, Priscilla (Bell). Juvenile travellers. London, 1806.

1807

	Geog 818.07	Aikin, J. Geographical delineations. Philadelphia, 1807.
	Geog 808.04.3	Blomfield, E. A general view of the world. Bungay, 1807. 2v.
	Geog 4348.07	Campbell, D. Journey over land to India. Philadelphia, 1807.
	Geog 4168.07	Campe, Joachim Heinrich. Voyages anecdotiques. Paris, 1807.
	Geog 758.07.2	Dicuil. Liber de mensura orbis terrae. Paris, 1807.
	Geog 758.07	Dicuil. Liber de mensura orbis terrae. Paris, 1807.
	Geog 3128.07	Dureau de la Malle, Adolphe. Géographie physique de la Mer Noire, de l'intérieur de l'Afrique et de la Méditerranée. Paris, 1807.
	Geog 4307.86.2	Smith, J.E. Sketch of a tour on the continent. 2nd ed. London, 1807.
	Geog 808.07	System of geography. Glasgow, 1807. 4v.

1808

	Geog 500.11	Beckmann, J. Literatur der...Reisebeschreibungen. Göttingen, 1808-09. 2v.
	Geog 500.7	Boucher, G. Bibliothèque universelle des voyages. Paris, 1808. 6v.
	Geog 4417.80.5	Parsons, Abraham. Travels in Asia and Africa. London, 1808.
	Geog 3235.17	Pezzana, A. De l'ancienneté de la mappemonde. Genes, 1808.
	Geog 4158.08	Pinkerton, John. General collection of voyages. London, 1808-14. 17v.
	Geog 808.08	Playfair, J. System of geography. Edinburgh, 1808. 6v.
Htn	Geog 4308.08.2*	Salvo, C. Travels in...1806 from Italy to England. Troy, N.Y., 1808.
	Geog 4308.08	Salvo, C. Travels in the year 1806, from Italy to England. Troy, N.Y., 1808.
	Geog 4417.38.3	Shaw, Thomas. Travels or observations...Barbary...Levant. Edinburgh, 1808. 2v.

1809

	Geog 4418.02F	Annesley, G. Voyages and travels to India. London, 1809. 3v.
	Geog 818.09	Guthrie, William. A new geographical, historical and commercial grammar. 1st American ed. Philadelphia, 1809.
	Geog 3520.10.10F	Hakluyt, Richard. Collection of early voyages, travels. London, 1809. 5v.

181-

| Htn | Geog 4708.16.20* | The pirates. Boston, 181- |

1810

	Geog 4347.99	Abu Taleb Khan. Travels...in Asia,...Europe. London, 1810. 2v.
	Geog 818.10.5	Evans, J. New system of geography. London, 1810. 2v.
	Geog 817.94.13	Guthrie, William. New geographical, historical...grammar. Montreal, 1810.
	Geog 4417.44.5	Lade, R. Voyages du capitaine...Lade. Paris, 1810.
Htn	Geog 4678.10*	Larcom, Henry. Distressing narrative of the loss of the ship Margaret of Salem. Beverly, 1810.
	Geog 5516.88	Maldonado, Lorenzo F. Viaggio dal mare atlantico al pacifico. n.p., 1810.
	Geog 4678.10.5	Narrative of calamitous and interesting shipwrecks. Philadelphia, 1810.
	Geog 4158.05.2	Phillips, R. Collection of...voyages and travels. London, 1810. 3v.
	Geog 818.10	Phillips, R. General view of manners...of nations. Philadelphia, 1810. 2v.
	An 358.10.3	Smith, Samuel Stanhope. An essay on the causes of the variety of complexion and figure in the human species. 2. ed. N.Y., 1810.

1811

	Geog 808.11	Bigland, J. Geographical...view of the world. Boston, 1811. 5v.
	Geog 4348.10.5	Clarke, E.D. Travels in various countries of Europe, Asia and Africa. Edinburgh, 1811. 2v.
	Geog 4158.11	Kerr, R. General history...of voyages. Edinburgh, 1811-24. 18v.
Htn	Geog 4218.03.3*	Krusenstern, A.J. von. Reise um die Welt in den Jahren 1803-1806 auf Befehl seiner Kaisere. v.1-2. Berlin, 1811-12. 3v.
	An 358.11	Meinere, C.M. Untersuchungen über die Verschiedenheiten der Menschennaturen. Tübingen, 1811-15. 3v.
	Geog 818.04.5	Pinkerton, J. Modern geography. London, 1811. 2v.
NEDL	Geog 4158.10F	Pinkerton, John. General collection of voyages. Philadelphia, 1811-14. 17v.
	Geog 5838.11	Zimmermann, E.A.H. Die Erde und ihre Bewohner. Leipzig, 1811.

1812

	Geog 4313.5	Catteau-Calleville, J.P. Tableau de la Mer Baltique. Paris, 1812. 2v.
	Geog 665.15	Geographical, commercial...essays. London, 1812.
	Geog 578.10.2	Ladvocat, J.B. Dictionnaire géographique. 2. ed. Paris, 1812.
Htn	Geog 4218.12*	Langsdorff, G.H. von. Bemerkungen auf einer Reise um die Welt in den Jahren 1803-1807. Frankfurt, 1812. 3v.
	Geog 5515.88	Maldonado, L.F. Voyage de la mer atlantique...pacifique. Plaisance, 1812.
	Geog 808.12.5	Malte-Brun, Conrad. Précis de la geographie universelle. Paris, 1812-1829. 8v.
	Geog 808.12.5F	Malte-Brun, Conrad. Précis de la geographie universelle. Atlas. Paris, 1812.
	Geog 4672.22	Shipwrecks and disasters at sea, or Historical narratives of the most noted calamities. Edinburgh, 1812. 3v.
	Geog 4308.05.2	Silliman, B. Journal of travels. Boston, 1812. 2v.

1813

	Geog 4417.99	Abu Taleb Khan. Reise...durch Asien, Africa. Wien, 1813.
	Geog 4348.13.5	Clarke, E.D. Travels in various countries of Europe. N.Y., 1813. 4v.
	Geog 818.13	Dickinson, R. Elements of geography. Boston, 1813.
	Geog 3140.15	Gosselin, P.Z.J. De l'évaluation et de l'emploi des mesures itinéraires grecques et romaines. Paris, 1813.
Htn	Geog 4218.03.5*	Krusenstern, A.J. Voyage around the world...1803-1806. London, 1813. 2v.
Htn	Geog 4218.03*	Langsdorff, G.H. von. Voyages and travels...in 1803-1805. London, 1813. 2v.
	Geog 4672.4	Remarkable shipwrecks. Hartford, 1813.
	Geog 4247.96	Smith, William. Journal of a voyage in the Duff to Pacific. N.Y., 1813.
	Geog 4708.12.4	United States. Circuit Court (1st Circuit, Mass.). The trial of Samuel Tulley and John Dalton...piracy and murder committed Jan. 21, 1812. 4. ed. Boston, 1813.

1814

	Geog 4418.03	Badia y Leblich. Voyages d'Ali Bey el Abbassi en Afrique et en Asie...1803...1807. v.1-3, atlas. Paris, 1814. 4v.
Htn	Geog 4218.03.9*	Lisianskii, I. Voyage round the world...1803-06. London, 1814.
	Geog 4326.32.5	Lithgow, William. Travels and voyages thro' Europe, Asia and Africa. 12th ed. Leith, 1814.

1815

	Geog 3060.14	Carta em resposta a hum amigo. Lisboa, 1815.
Htn	Geog 4348.10.10*	Clarke, E.D. Travels in various countries of Europe. 1st American ed. v.3-4. N.Y., 1815. 2v.
	Geog 4328.10	Cockburn, G. A voyage to Cadiz and Gibraltar, up the Mediterranean, to Sicily and Malta, in 1810 and 1811. London, 1815. 2v.

1816

NEDL	Geog 818.16	Adam, A. Summary of geography and history. London, 1816.
	Geog 4678.16.2	Allen, S. (Mrs.). A narrative of the shipwreck and unparalleled sufferings of Mrs. S. Allen. 2. ed. Boston, 1816.
	Geog 577.91.5	Brookes, R. General gazetteer. Boston, 1816.
Htn	Geog 4218.06*	Campbell, A. Voyage round the world from 1806-12. Edinburgh, 1816.
	Geog 4348.13.10	Clarke, E.D. Travels in...Europe, Asia and Africa. London, 1816-1824. 11v.
	Geog 4308.16	Raumer, F.L.G. Die Herbstreise nach Benedig. Berlin, 1816.

1817

	Geog 808.12.7	Malte-Brun, Conrad. Précis de la geographie universelle. v.5. Paris, 1817.
Htn	Geog 4248.08*	Patterson, S. Narrative of adventures...of S. Patterson, experienced in Pacific Ocean. Palmer, 1817.
	Geog 808.17	Ritter, Carl. Die Erkunde. Berlin, 1817-1818. 2v.
	Geog 5837.78.5	Saabye, Hans E. Fragmentes af en dagbok hållen i Grönland. Stockholm, 1817.
	Geog 578.17	Worcester, J.E. Geographical dictionary or universal gazetteer. Andover, 1817. 2v.

1818

Htn	Geog 5170.9*	Barrington, D. Possibility of approaching the North Pole. London, 1818.
	Geog 5208.18	Barrow, John. Chronological history of voyages into Arctic regions. London, 1818.
	Geog 5837.63.5	Egede, H. Description of Greenland. London, 1818.
	Geog 4348.17	Johnson, John. A journey from India to England thro' Persia, Georgia, Russia, Poland and Prussia, in 1817. London, 1818.
	Geog 4308.18	Neale, A. Travels thro'...Germany, Poland. London, 1818.
	Geog 5558.17.3	O'Reilly, B. Greenland...Northwest Passage. London, 1818.
	Geog 5558.17	O'Reilly, B. Greenland...Northwest Passage. N.Y., 1818.
	Geog 4308.17	Raffles, T. Letters during tour thro'...France. Liverpool, 1818.
	Geog 5837.78	Saabye, Hans E. Greenland...extracts...journal. London, 1818.

Chronological Listing

1819

	Geog 4308.18.5	Baillie, M. First impressions on tour upon continent. London, 1819.
	Geog 4328.01	Collins, F. Voyages...from 1796-1801. London, 1819.
	Geog 5518.18.5	Fisher, Alexander. Journal of a voyage of discovery to the Arctic regions. London, 1819.
	Geog 4208.17F	Fitzclarence. Journal of route across India...to England. London, 1819.
	Geog 4328.19.10	James, John Thomas. Journal of a tour in Germany, Russia, Poland in 1813-1814. 3rd ed. London, 1819.
	Geog 5170.11	A letter to John Barrow...polar expedition. London, 1819.
Htn	Geog 4708.19*	Lives and confessions of John Williams, Francis Frederick, John P. Roy and P. Peterson. 2. ed. Boston, 1819. 2 pam.
Htn	Geog 4708.19.2*	Lives and confessions of John Williams, Francis Frederick, John P. Roy and P. Peterson. Boston, 1819.
	Geog 4328.19	Noah, M.M. Travels in England, France and Spain. N.Y., 1819.
Htn	Geog 4708.16.15*	Onion, S.B. Narrative of the mutiny on board the schooner Plattsburg. Boston, 1819.
	Geog 4158.19	Phillips, R. New voyages and travels. London, 1819-20. 9v.
	Geog 818.19	Riise, J. Haanbog i geographien. Kjobenhavn, 1819-1820. 2v.
	Geog 5518.18	Ross, J. Voyage of discovery...Baffin's Bay. London, 1819.
	Geog 5518.18.3	Ross, J. Voyage vers le Pole Arctique. Paris, 1819.
	Geog 5518.18.4	Sabine, Edward. Remarks on the account of the late voyage of discovery to Baffin's Bay. 2. ed. London, 1819.

182-

	Geog 818.20	The traveller; or, An entertaining journey round the habitable globe. 3rd ed. London, 182-.

1820

	Geog 4328.13.2	Bramsen, John. Travels in Egypt, Syria, Cyprus. 2nd ed. London, 1820. 2v.
	Geog 4217.85.5	Burney, J. Memoir on the voyage...in search of La Pérouse. London, 1820.
	Geog 5700.5.10	Crantz, D. History of Greenland...account of mission. London, 1820. 2v.
	Geog 4307.65	Defeller. Itinéraire ou voyages. Liege, 1820. 2v.
	Geog 4558.20.3	Durand, J.R. The life and adventures. Rochester, 1820.
	Geog 4678.09	Eustace, Thomas. The adventures. London, 1820.
	Geog 4308.17.3	Matthews, H. The diary of an invalid. 2nd ed. London, 1820.
	Geog 4328.20	Monson, W.J.M. Extracts from a journal. London, 1820.
	Geog 5558.18	Parry, W.E. Journal of a voyage of discovery. London, 1820. 2 pam.
	Geog 5160.3	Scoresby, William. An account of the Arctic regions. Edinburgh, 1820. 2v.
	Geog 4311.7	Senkel, K.F. De istri ostiis dissertatio historico-geographica. Wratislaviae, 1820.
	Geog 4308.05.4	Silliman, B. Journal of travels. 3rd ed. New Haven, 1820. 3v.

1821

	Geog 4348.14	Almerté, T. Voyages de sa majesté la reine d'Angleterre. Paris, 1821.
	Geog 4307.99	A concise narrative of a tour through some parts of England, France, Holland, Switzerland, and Italy in the years 1799, 1800, 1801, and 1802. Philadelphia, 1821.
	Geog 5518.19.12	Fisher, Alexander. Journal of a voyage of discovery. London, 1821.
	Geog 5518.19.10	Fisher, Alexander. Journal of a voyage of discovery. London, 1821.
Htn	Geog 4218.03.7*	Krusenstern, A.J. Voyage autour du monde...1803-06. Paris, 1821. 2v.
Htn	Geog 4218.03.8F*	Krusenstern, A.J. Voyage autour du monde. Atlas. Paris, 1821.
	Geog 4228.16	Montulé, E. de. Voyage en Amerique. Paris, 1821. 2v.
	Geog 4228.17	Montulé, E. de. A voyage to North America and West Indies in 1817. London, 1821.
	Geog 578.21	Morse, Jedidiah. New universal gazetteer of the known world. 3. ed. New Haven, Maine.
	Geog 818.21	Oddsson, G. Almenn landaskipunarfraede. Kaupmannahøfn, 1821-1827. 2v.
	Geog 5518.19.6.2	Parry, W.E. Journal of voyage...Northwest Passage and North Georgia Gazette. Philadelphia, 1821.
	Geog 5518.19.7	Parry, W.E. Journal of voyage...Northwest Passage and North Georgia Gazette. London, 1821.
	Geog 5518.19.6	Parry, G.F. The North Georgia Gazette and Winter Chronicle. No.1-20. London, 1821.
	Geog 665.22	Phillips, R. The hundred wonders of the world. 1st American ed. New Haven, 1821.
	Geog 4308.21.5	Scott, J. Sketches of manners...in French provinces. London, 1821.

1822

	Geog 578.22	Edinburgh gazetteer. v.2-6. Edinburgh, 1822. 4v.
	Geog 4158.22	Eyries, J.B.B. Abrégé des voyages modernes. Paris, 1822-24. 14v.
	An 3208.22.2	Lawrence, W. Lectures on physiology, zoology and the natural history of man. London, 1822.
	An 3208.22	Lawrence, W. Lectures on physiology, zoology and the natural history of man. London, 1822.
	Geog 4708.21	Lincoln, B. Narrative of the capture...of Captain B. Lincoln...by piratical schooner. Boston, 1822.
	Geog 4208.19	Lumsden, T. A journey from Merut in India to London. London, 1822.
Htn	Geog 5558.21*	Manby, G.W. Journal of voyage to Greenland...1821. London, 1822.
	Geog 4308.17.3.5	Matthews, H. The diary of an invalid. 3rd ed. London, 1822. 2v.
	Geog 818.22	Morse, S.E. New system of modern geography. Boston, 1822.
	Geog 4308.20	Niemeyer, A.H. Beobachtungen auf Reisen in und ausser Deutschland. v.2-4. Halle, 1822-1826. 4v.
	Geog 818.21.3	Oddsson, G. Almenn jardarfraedi og landskipun edur geographia. Kaupmannahøfn, 1822. 4v.
	Geog 4208.22	Rezo, José Luiz do. Viagens à China. Porto, 1822.
	Geog 808.17.5	Ritter, Carl. Die Erkunde. v.1-19. Berlin, 1822-1859. 21v.

1823

	Geog 4218.17	Arago, J. Narrative of a voyage round the world. 1823.
Htn	Geog 5557.25.5*	Berkh, V.N. Pervoe morskoe puteshestvie. Sankt Peterburg, 1823.
	Geog 577.91.4	Brookes, R. General gazetteer. London, 1823.
	Geog 4181.54	Cieza de Leon, P. de. The war of Las Salinas. London, 1823.
	Geog 4168.23	Collection de relations de voyages. Paris, 1823.
	Geog 578.23.5	Dictionaire géographique. Paris, 1823. 10v.
Htn	Geog 5518.19*	Franklin, John. Narrative of journey to...Polar Sea. London, 1823.
Htn	Geog 5518.19.2*	Franklin, John. Narrative of journey to...Polar Sea. London, 1823. 2v.
	Geog 4308.18.3	Griscom, J. A year in Europe. N.Y., 1823.
	Geog 4228.23	Landolphe, Jean François. Mémoires du capitaine Landolphe. Paris, 1823. 2v.
	Geog 612.1	Materialien zu einer Biographie des Fr. J.M. Liechtenstern. Schneeberg, 1823.
	Geog 4248.12	Porter, David. Voyage in the South Seas in...1812-1814. London, 1823.
	Geog 4218.16	Roquefeuil, C. de. Voyage autour du monde. Paris, 1823. 2v.
	Geog 5558.22	Scoresby, William. Journal of voyage to the northern whale-fishery. Edinburgh, 1823.
	Geog 578.17.2	Worcester, J.E. Geographical dictionary or universal gazetteer. 2. ed. Boston, 1823. 2v.
	Geog 818.23	Worcester, J.E. Sketches of the earth. Boston, 1823. 2v.

1824

	Geog 4758.24	Denis, F. Scènes de la nature sous les tropiques. Paris, 1824.
Htn	Geog 4758.24.2*	Denis, F. Scènes de la nature sous les tropiques. Paris, 1824.
Htn	Geog 4758.24.3*	Denis, F. Scènes de la nature sous les tropiques. Paris, 1824.
Htn	Geog 818.24*	Engelmann, G. Porte-feuille géographique et ethnographique. 2e éd. v.1-2, Atlas. Mulhouse, 1824. 2v.
	Geog 5518.19.5	Franklin, John. Narrative of journey to...Polar Sea. Philadelphia, 1824.
	Geog 5518.19.3	Franklin, John. Narrative of journey to...Polar Sea. 2. ed. London, 1824. 2v.
	Geog 5518.19.4	Franklin, John. Narrative of journey to...Polar Sea. 3. ed. London, 1824. 2v.
	Geog 5650.9	Giesecke, C.L. Norwegian settlements on...east coast of Greenland. Dublin, 1824.
	Geog 4308.18.4	Griscom, J. A year in Europe. 2nd ed. N.Y., 1824. 2v.
	Geog 5518.21.5	Lyon, G.F. Private journal of Captain G.F. Lyon. Boston, 1824.
	Geog 808.12.9	Malte-Brun, Conrad. Universal geography. Boston, 1824. 8v.
	Geog 808.12.10	Malte-Brun, Conrad. Universal geography. v.1-14. Boston, 1824-1829. 7v.
	Geog 4218.24	Memoires du Capitaine Péron. Paris, 1824. 2v.
	Geog 4158.24	Paris. Société de Geographie. Recueil de voyages et de memoires. Paris, 1824-1864. 7v.
	Geog 5518.21	Parry, W.E. Journal of second voyage for discovery. London, 1824.
	Geog 5518.19.8	Parry, W.E. Supplement to the appendix of Capt. Parry's voyage. London, 1824.
	Geog 4678.24	Stanford, John. Aetna; a discourse. N.Y., 1824.
	Geog 4308.21	Tennant, C. Tour thro'...Netherlands, Holland. London, 1824. 2v.
	An 358.24	Virey, Julien J. Histoire naturale du genre humain. Paris, 1824. 3v.
	Geog 4308.23.2	Wilson, D. Letters from an absent brother. London, 1824.

1825

	Geog 3108.25	Butler, S. Sketch of modern and ancient geography. London, 1825.
	Geog 808.25	Casado Giraldes, J.P.C. Tratado completo de cosmographia. Paris, 1825-1828. 4v.
Htn	Geog 4678.25*	Collins, Daniel. Narrative of the shipwreck of the brig Betsey of Wiscasset, Maine...Dec. 1824. Wiscasset, 1825.
	Geog 4308.22.5	Duppa, Richard. Miscellaneous observations and opinions on the continent. London, 1825.
	Geog 585.5	A geographical instructor. London, 1825.
	Geog 5838.25	Graah, W.A. Beskrivelse...Grønland. Kjøbenhavn, 1825.
	Geog 4208.25	Howison, W. Foreign scenes and travel recreations. 2. ed. Edinburgh, 1825. 2v.
	Geog 5518.24.7	Lyon, G.F. Brief narrative of attempt to reach Repulse Bay. London, 1825.
	Geog 4308.17.4	Matthews, H. The diary of an invalid. London, 1825.
	Geog 5518.21.2	Parry, W.E. Appendix to...Parry's Journal of second voyage. London, 1825.
NEDL	Geog 4308.25.10	Pennington, Thomas. A journey into various parts of Europe. London, 1825. 2v.
	Geog 4345.99	Sherley, A. The three brothers or travels. London, 1825.
	Geog 4326.75	Teonge, Henry. The diary of H. Teonge...1675-1679. London, 1825.

1826

	Geog 818.26	Blake, J.L. Geographical, chronological...atlas. N.Y., 1826.
	Geog 4029.2	Mämpel, J.C. The young rifleman's comrade. London, 1826.
	Geog 5518.24.2	Parry, W.E. Journal of third voyage...Northwest Passage. London, 1826.
	Geog 5518.24	Parry, W.E. Journal of third voyage...Northwest Passage. Philadelphia, 1826.
	Geog 818.10.3	Phillips, R. Geographical view of the world. N.Y., 1826.
	An 358.26	Prichard, J.C. Researches into the physical history of mankind. London, 1826. 2v.
	Geog 4300.18.5	Starke, M. Information and directions for travellers. Paris, 1826.
	Geog 5925.10	Weddell, James. Observations on South Pole. London, 1826.

1827

	Geog 4308.25	Carter, N.H. Letters from Europe. v.2. N.Y., 1827.
	Geog 578.27	Darby, William. Darby's universal gazetteer. 2. ed. Philadelphia, 1827.

Chronological Listing

1827 - cont.

	Geog 4308.27	Delestre, P. De Paris a Varsovie. Paris, 1827.
	Geog 5558.26	Duncan, D. Arctic regions. Voyage to Davis Strait. London, 1827.
	Geog 4300.25.3	Engelmann, J.B. Manuel pour voyageurs en Allemagne et dans les Payslimitrophes. 3.ed. Francfort, 1827.
	Geog 578.27.3	Hawkes, P. The American companion. Philadelphia, 1827.
	Geog 4105.14	Kitchiner, William. The travellers' oracle. 2. ed. London, 1827. 2v.
	Geog 808.12.11	Malte-Brun, Conrad. Universal geography. Philadelphia, 1827-1832. 6v.
	Geog 4304.90	Martyr. Relation d'un voyage fait en Europe. Paris, 1827.
	Geog 4208.10	Swaving, J.G. Swavings reizen en logevallen, doorhem. v.1-2. Dordrecht, 1827.

1828

	Geog 4300.22	Audin, J.M.V. Guide classique du voyageur. Paris, 1828-1829. 2v.
	Geog 4168.29	Bennet, Roelof Gabriel. Nederlandsch zeeseizen in het laatst der z estiende. v.1-4. Wijk, 1828-29. 3v.
	Geog 585.7	Coulier, P.J. Tables des principales positions geonomiques. Paris, 1828.
	Geog 4672.21.4	Deperthes, Jean L.H.S. Histoire des naufrages, ou Recueil des relations les plus intéressantes des naufrages. Paris, 1828. 3v.
Htn	Geog 5518.25.2F*	Franklin, John. Narrative of a second expedition to the shores of the Polar Sea, in the year 1825-1827. London, 1828.
	Geog 5518.25	Franklin, John. Narrative of second expedition to Polar Sea. Philadelphia, 1828.
	Geog 3018.28	Larenaudière. Histoire abrégée de l'origine...de géographie. Paris, 1828.
	Geog 4218.24.5PF	La Touanne, E. de. Album pittoresque de la frégate la Thétis et de la corvette l'Esperance. Paris, 1828.
	An 358.28	Lawrence, William. Lectures on physiology, zoology and natural history of man. Salem, 1828.
	An 358.28.3	Leupoldt, J.M. Die Diätetik des psychischen und physischen Menschenlebens. Berlin, 1828.
	Geog 5518.27	Parry, W.E. Narrative of attempt to reach the North Pole. London, 1828.
	Geog 4207.76A	Sparks, Jared. The life of John Ledyard. Cambridge, 1828.
	Geog 4308.28.5	Walter, Weever. Letters from the continent. Edinburgh, 1828.

1829

	Geog 578.29	Bischoff, F.H.T. Vergleichendes Wörterbuch. Gotha, 1829.
	Geog 4308.25.2	Carter, N.H. Letters from Europe. N.Y., 1829. 2v.
	Geog 4413.25.5A	Ibn Batoutah. Travels of Ibn Batuta. London, 1829.
	Geog 4328.29	Jones, George. Sketches of naval life...on Mediterranean. New Haven, 1829. 2v.
	Geog 4418.29	Lushington, C. (Mrs.). Narrative of journey from Calcutta. London, 1829.
Htn	Geog 5508.29*	Northern regions. Boston, 1829.
	Geog 4207.76.12	Sparks, Jared. Leben des...Americanischen Reisenden J. Ledyard. Leipzig, 1829.
	Geog 4207.76.5	Sparks, Jared. The life of John Ledyard. 2. ed. Cambridge, 1829.
	Geog 818.28	Venning, I.A. (Mrs.). A geographical present. 1st American ed. N.Y., 1829.

1830

	Geog 4208.30.5	Ames, N. A mariner's sketches. Providence, 1830.
NEDL	Geog 520.25	Conder, Josiah. The modern traveller. London, 1830. 30v.
	Geog 3018.30	Cooley, W.D. The history of maritime and inland discovery. London, 1830-31. 3v.
	An 378.24.5	Gioja, M. Riflessioni...su l'opera "L'homme du midi". Milano, 1830.
	Geog 818.30	Hale, N. Epitome of universal geography. Boston, 1830.
	Geog 4208.30	Holbrook, S.P. Sketches by a traveller. Boston, 1830.
Htn	Geog 4678.30.5*	Smith, John. Narrative of the shipwreck and sufferings of the crew of the English brig Neptune. N.Y., 1830.

1831

	Geog 4168.31	Adams, W. The modern voyager and traveller. London, 1831. 4v.
	Geog 5398.25	Beechey, F.W. Narrative of voyage to the Pacific. London, 1831. 2v.
Htn	Geog 4208.14*	Bunnell, D.C. Travels and adventures. Palmyra, N.Y., 1831.
	Geog 4208.28	Butterworth, William. Three years adventures of a minor in England Africa, South Carolina and Georgia. Leeds, 1831.
	Geog 4228.31	Compendious view of...travels. London, 1831.
	Geog 665.38	A geographical present. N.Y., 1831.
	Geog 4208.31	Hall, B. Fragments of voyages and travels. Philadelphia, 1831. 2v.
	Geog 4308.30	Inglis, H.D. Switzerland...France and Pyrenees. Edinburgh, 1831. 2v.
	Geog 4308.31.2A	Johnson, J. Change of air, or The diary. London, 1831.
	Geog 4138.31	Lives and voyages of Drake. Edinburgh, 1831.
	Geog 578.31	Müller, Johan Wilhelm. Lexicon manuale. Lipsiae, 1831.
Htn	Geog 4708.31*	Mutiny and murder. Confession of Charles Gibbs. Providence, 1831.
	An 358.26.5	Prichard, J.C. Eastern origin of the Celtic nations...forming a supplement to Researches into the physical history of mankind. Oxford, 1831.
	Geog 5508.31	Snelling, W.J. Polar regions of the western continent. Boston, 1831.

1832

	Geog 4208.21.5	Ames, Nathaniel. Nautical reminiscences. Providence, 1832.
	Geog 577.91.6	Brookes, R. A new universal gazetteer containing a description of the principal nations. N.Y., 1832.
	Geog 3530.11	Estancelin, L. Recherches sur les voyages et découvertes des...Normands. Paris, 1832.
	Geog 818.32	Goodrich, S.G. A system of universal geography. Boston, 1832.
	Geog 5838.28	Graah, W.A. Undersogelses-reise...Grønland. Kjøbenhavn, 1832.
	Geog 4208.31.2	Hall, B. Fragments of voyages and travels. v.1-2. 2d ed. Edinburgh, 1832.

1832 - cont.

	Geog 808.12.17	Malte-Brun, Conrad. Traité elementaire de géographie. Bruxelles, 1832.
	Geog 4208.31.5	Morrell, B. Narrative of four voyages. N.Y., 1832.
	Geog 4308.32.10	Ritchie, Leitch. Travelling sketches in the north of Italy, the Tyrol, and on the Rhine. London, 1832.
	Geog 4348.32	Stocqueler, J.H. Fifteen months pilgrimage. London, 1832. 2v.
	Geog 4418.32	Tyerman, David. Journal of voyages and travels. Boston, 1832. 3v.
	Geog 5208.32A	Tytler, P.F. Historical view of progress of discovery. Edinburgh, 1832.
Htn	Geog 4328.32*	Wines, E.C. Two years and a half...in Mediterranean and Levant. Philadelphia, 1832.

1833

	Geog 4308.19	Beskow, B. Wandrings minnen. Stockholm, 1833-1834. 2v.
	Geog 808.33.2	Blanc, L.G. Handbuch...der Natur und Geschichte der Erde. Halle, 1833-1834. 3v.
	Geog 4217.64	Byron, John V. Viaje al alrededor del mundo hecho en 1764, 65 y 66. Madrid, 1833.
	Geog 3018.30.5	Cooley, W.D. The history of maritime and inland discovery. London, 1833. 2v.
	Geog 818.33	Depping, G.B. Evening entertainments...manners and customs of nations. Philadelphia, 1833.
	Geog 818.33.5	Goodrich, S.G. A system of universal geography. 2d ed. Boston, 1833.
	Geog 4218.33.25	Laplace, Cyril P.T. Voyage autour du monde. Paris, 1833-39. 4v.
	Geog 5220.9	Leslie, J. Narrative of discovery...polar seas. N.Y., 1833.
Htn	Geog 4558.33*	Mariner's library. Boston, 1833.
	Geog 958.33F	Meyer, J. Meyer's Universum. v.1-6,9. Hildburghausen, 1833-. 3v.
	Geog 808.33	Montenegro Colon, Feliciano. Geografia general para el uso de la juventud de Venezuela. Caracás, 1833-1837. 4v.
	Geog 4308.24	Moore, John. A journey from London to Odessa. Paris, 1833.
	Geog 4208.31.7	Morrell, Abby J. Narrative of voyage to Ethiopia. N.Y., 1833.
NEDL	Geog 4418.22	Owen, William Fitz William. Narrative of voyages to explore the shores of Africa, Arabia, and Madagascar. London, 1833. 2v.
	Geog 4208.29.3	Prinsep, A. (Mrs.). The journal of a voyage from Calcutta. London, 1833.
	Geog 4658.33	Redding, Cyrus. A history of shipwrecks, and disasters at sea. London, 1833. 2v.
	Geog 4308.33	Ritchie, Leitch. Travelling sketches. London, 1833.
	Geog 4308.33.10	Strombeck, Friedrich Karl von. Darstellungen aus meinem Leben und aus meiner Zeit. Braunschweig, 1833-1840. 4v.
	Geog 5208.33	Tytler, P.F. Historical view of progress of discovery. Edinburgh, 1833.

1834

	Geog 4167.92.10	Adams, J. Flowers of celebrated travellers. Baltimore, 1834.
	Geog 4307.80.3	Beckford, William. Italy, with sketches of Spain and Portugal. Philadelphia, 1834.
	Geog 4308.34.5	Boddington, Mary. Slight reminiscences of the Rhine. London, 1834. 2v.
Htn	Geog 4218.26*	Duhaut-Cilly, A. Voyage autour du monde. Paris, 1834.
	Geog 4168.34	Dumont d'Urville, Jules. Voyage pittoresque autour du monde. Paris, 1834-35. 2v.
	Geog 4678.34	Duncan, A. The mariner's chronicle...shipwrecks, storms. New Haven, 1834.
	Geog 4308.31.5	McLellan, H.B. Journal of residence in Scotland and tour. Boston, 1834.
	Geog 808.12.12	Malte-Brun, Conrad. A system of universal geography. Boston, 1834. 3v.
	Geog 4678.33.5	Melancholy shipwreck and...divine providence. n.p., 1834.
	Geog 4218.30.5	Meyen, F.J.F. Reise um die Erde. Berlin, 1834-35. 2v.
	Geog 818.34	Murray, H. An encyclopaedia of geography. London, 1834.
	Geog 4208.34	Rapelje, George. A narrative of excursions, voyages and travels. N.Y., 1834.
	Geog 4300.20.8	Reichard, H.A.O. Reichard's Passagier. Berlin, 1834.
	Geog 4708.34.2	A report of the trial of Pedro Gilbert...piracy. Boston, 1834.
	Geog 4208.29.5	Roberts, Jane. Two years at sea. London, 1834.
	Geog 585.13	Tavole sinottiche di geografia. Livorno, 1834.
	Geog 578.34	Wright, G.N. New and comprehensive gazetteer. London, 1834-37. 4v.

1835

	Geog 3055.13.2	Baer, F.C. Essai sur l'Atlantique des anciens. Avignon, 1835.
	Geog 4228.26	Boelen, Jacobus. Reize naar de oost- en westkust van Zuid-Amerika. Amsterdam, 1835-36. 3v.
	Geog 4218.30	Holman, James. A voyage round the world. v.4. London, 1835.
	Geog 5518.29.5	Huish, Robert. The last voyage of Captain Sir John Ross, R.N., to the Arctic regions. London, 1835.
	Geog 4218.33.27PF	Laplace, Cyril P.T. Voyage autour du monde. Atlas. Paris, 1835.
	Geog 4678.35	Mariners' chronicle. New Haven, 1835.
	Geog 4308.35.10	Pückler-Muskau, H. Vorletzter Weltgang von Semilasso. Stuttgart, 1835. 3v.
	Geog 4218.31.2	Reynolds, J.N. Voyage of the United States frigate Potomac. N.Y., 1835.
	Geog 4218.31	Reynolds, J.N. Voyage of the United States frigate Potomac. N.Y., 1835.
	Geog 5518.29.2	Ross, J. Appendix to Narrative of second voyage. London, 1835.
	Geog 5518.29.3	Ross, J. Narrative of a second voyage in search of a northwest passage. Philadelphia, 1835.
	Geog 5518.29	Ross, J. Narrative of 2nd voyage...Northwest Passage. London, 1835.
	Geog 4328.35	Ship and shore...cruise to the Levant. N.Y., 1835.
	Geog 4207.69.5	Wallenberg, Jacob. Min son på galejan. Stockholm, 1835.
	Geog 4218.31.3	Warriner, F. Cruise of the United States frigate Potomac round the world. N.Y., 1835.

Chronological Listing

1836

	Geog 5518.33	Back, George. Narrative of Arctic land expedition. London, 1836.
	Geog 5518.33.3	Back, George. Narrative of Arctic land expedition. Philadelphia, 1836.
	Geog 4418.03.5	Badia y Leblich. Viajes de Ali Bay el Abbassi por Africa y Asia. Valencia, 1836. 3v.
	Geog 3018.36.2	Bajot, L.M. Abrégé historique...des...voyages. Paris, 1836.
	Geog 808.36	Bell, J. System of geography. Glasgow, 1836. 6v.
	Geog 3140.9	Blau, J. Memoirs sur deux monuments geographiques. Nancy, 1836.
	Geog 5838.36	Breidfjörd, S. Fra Graenlandi. Kaupmannahöfn, 1836.
	Geog 818.36	Cannabich, J.G.F. Lehrbuch der Geographie. 14e Aufl. Weimar, 1836.
	Geog 4308.32	Cooper, J.F. Residence in France with excursion up the Rhine to Switzerland. Paris, 1836.
	Geog 4308.33.2	Dewey, O. Old world and the new. N.Y., 1836. 2v.
	Geog 4110.21	Goodrich, Charles A. The universal traveller. 2. ed. Hartford, 1836.
	Geog 4308.36.10	Gross-Hoffinger, A.J. Empfindsame Reise eines expatriirten Schwärmers. Leipzig, 1836.
	Geog 4678.36	Holden, H. Narrative of the shipwreck. Boston, 1836.
	Geog 4678.36.10	Kendall, J. A sermon delivered February 14, 1836. Plymouth, 1836.
	Geog 5220.9.3	Leslie, J. Narrative of discovery...polar seas. N.Y., 1836.
	Geog 28.4.3	Lettre sur l'etablissement géographique de Bruxelles. Bruxelles, 1836.
	Geog 4708.36	Lives and bloody exploits of most noted pirates. Hartford, Conn., 1836.
	Geog 4138.36	Lives and voyages of Drake. N.Y., 1836.
	Geog 4308.17.5	Matthews, H. Diary of an invalid...journey...1817. Paris, 1836.
	An 358.36	Prichard, J.C. Researches into the physical history of mankind. 3. ed. London, 1836-47. 6v.
	Geog 5508.36	Tytler, P.F. Historical view of discovery...of America. N.Y., 1836.

1837

	Geog 5220.23	Ahlenius, K. Fornskandinaviska upptäcktsfärder i Nordatlantiska hafvet. Stockholm, 1837.
	Geog 5518.36.9	Back, George. Narrative of the Arctic land expedition to the mouth of the Great Fish River. 2. ed. Philadelphia, 1837.
	An 358.37	Brotonne, F. de. Histoire de la filiation et migrations des peuples. v.1-2. Paris, 1837.
	Geog 4218.37.5	Dumont d'Urville, J. Malerische Reise um die Welt. Leipzig, 1837.
	Geog 5838.28.3	Graah, W.A. Narrative of expedition to...Greenland. London, 1837.
	Geog 3018.37.2	Historical account of circumnavigation. N.Y., 1837.
	Geog 3018.37	Historical account of circumnavigation. N.Y., 1837.
	Geog 4308.35.15	Hoppus, John. The continent in 1835. N.Y., 1837.
	Geog 4138.37.5	Johnstone, C. (Mrs.). Lives and voyages of Drake. Edinburgh, 1837.
	Geog 4348.34	Marmont, Auguste Frédéric L. Voyage du...en Hongrie, en Transylvanie. Paris, 1837. 4v.
	Geog 808.37	Murray, H. Encyclopedia of geography. Philadelphia, 1837. 3v.
	Geog 1724.2	Murray, John, publisher, London. Handbook for...southern Germany. London, 1837.
Htn	Geog 4678.37*	Palmer, John. Awful shipwreck...crew of the ship Francis Spaight. Boston, 1837.
Htn	Geog 4708.37*	The pirates own book. Portland, 1837.
	An 358.41	Prichard, J.C. Researches into the physical history of mankind. v.1-2, 4. ed.; v.3, 3. ed. London, 1837-41. 3v.
	An 358.41.2	Prichard, J.C. Researches into the physical history of mankind. v.1,5, 4. ed; v.2-4, 3. ed. London, 1837-47. 5v.
	Geog 4208.29.7	Roberts, Jane. Two years at sea. 2. ed. London, 1837.
	Geog 4138.37	St. John, J.A. Lives of celebrated travellers. N.Y., 1837. 3v.
	Geog 3108.37.2	Schirlitz, S.C. Handbuch der alten Geographie. Halle, 1837.
	Geog 4678.37.9	Smyth, Thomas. The voice of God in calamity, or Reflections on the loss of the steamboat "Home"...sermon. Charleston, 1837.
	Geog 4678.37.15	Thomas, R. Interesting and authentic narratives of the most remarkable shipwrecks, fires. Hartford, 1837.

1838

	Geog 5518.36.7	Back, George. Narrative of an expedition in H.M.S. Terror. London, 1838.
	Geog 3235.29	Buchon, J.A.C. Notice sur un atlas en langue catalane. Paris, 1838.
	Geog 4308.37.3	Cavalier auf Reisen im...1837. Leipzig, 1838.
	Geog 4228.35	Ferrer, Antonio C. Pases por Europa y América, en 1835 y 1836. Madrid, 1838.
	Geog 4308.35.5	Fisk, Wilbur. Travels in Europe. 4th ed. N.Y., 1838.
	Geog 5650.7	Grönlands historiske mindesmaerker. Kjobenhavn, 1838-45. 3v.
	Geog 5508.38	Hülstett, G.K.A. Über die nordwestliche Durchfahrt. Düsseldorf, 1838.
	Geog 4308.37.5	Jewett, I.A. Paggages in foreign travel. Boston, 1838. 2v.
	Geog 4328.36.5	Maritime scraps, or Scenes in the frigate United States during a cruise in the Mediterranean. Boston, 1838.
Htn	Geog 4311.15F*	Meyer's Donau-Ansichten. pt.1-2,5,7-11. Hildburghausen, 1838.
	An 3208.38	Mudie, R. Man, in his physical structure and adaptations. London, 1838.
	Geog 4308.37	Mundt, T. Spaziergänge und Weltfahrten. Altona, 1838-1839. 3v.
	Geog 1735.1	Murray, John, publisher, London. Handbook for...Switzerland. London, 1838.
	Geog 1707.3	Murray, John, publisher, London. Handbook for travellers on the continent. 2. ed. London, 1838.
	Geog 4328.38	Records of travel. Boston, 1838.
	Geog 4308.35	Roby, John. Seven wekks in Belgium, Switzerland, Lombardy. London, 1838. 2v.
	Geog 4208.38.3	Rose, W.G. Three months' leave, or Military reminiscences. London, 1838.
	Geog 4218.35.3	Ruschenberger, W.S.W. A voyage round the world. Philadelphia, 1838.
	Geog 4418.38	Sherer, M. The imagery of foreign travel. London, 1838.

1838 - cont.

Htn	Geog 4308.38.8*A	Stephens, J.L. Incidents of travel in Greece, Turkey and Poland.7th ed. N.Y., 1838. 2v.
	Geog 4308.36	The tourist in Europe. N.Y.. 1838.
	Geog 4308.38	Wright, H.H. Desultory reminiscences of a tour. Boston, 1838.

1839

	Geog 4218.17.3	Arago, J. Souvenirs...voyage...du monde. Paris, 1839.
	Geog 3108.39.3	Arrowsmith, A. Compendium of ancient and modern geography. London, 1839.
	Geog 3108.39	Arrowsmith, A. Grammar of ancient geography. London, 1839.
	Geog 4348.13.20	Clarke, E.D. Travels in Russia, Tartary and Turkey. Edinburgh, 1839.
	An 358.39	Kinmont, A. Twelve lectures on the natural history of man. Cincinnati, 1839.
	Geog 4208.37	Lofé, G. Cartas á mis hijos. N.Y., 1839.
	Geog 4328.39	Maximilian, H. Wanderung nach dem Orient. München, 1839.
	Geog 818.39	Mitchell, S.A. Accompaniment to map of world. Philadelphia, 1839.
	An 3538.39	Morton, S.G. Crania Americana. Philadelphia, 1839.
	Geog 1735.3	Murray, John, publisher, London. Handbook for...Switzerland. London, 1839.
	Geog 818.36.10	Perkins, Samuel. The world as it is. 5th ed. n.p., 1839.
	VGeog 5700.4.15	Petersen, Niels Matthias. Hans Egedes levnet. København, 1839.
	Geog 5700.6	Religious Tract Society. Missionary records; northern countries. London, 1839.
	Geog 4308.39.7	Tuttolasso. Wanderungen durch Deutschland, Polen. Stuttgart, 1839.
	Geog 5398.20.3	Wrangell, F. von. Reise des...Flotten-Lieutenants. Berlin, 1839. 2v.
	Geog 5398.20.10	Wrangell, F. von. Statistische...Nachrichten über...Besetzungen. St. Petersburg, 1839.

184-

	Geog 4308.50.25	Crockett, Henry C. The American in Europe. London, 184-.
	Geog 958.40	Our Globe. Philadelphia, 184-.
	Geog 4314.6	Symonds, W. Extract from journal in the Black Sea. London, 184-?

1840

	Geog 4307.80.5	Beckford, William. Italy, Spain and Portugal. London, 1840.
	Geog 4341.73.5	Benjamin ben Jonah, of Tudela. The itinerary of Rabbi Benjamin of Tudela. pt.1-2. London, 1840.
	Geog 4678.40.3	Cleveland, H.R. A letter to Daniel Webster. Boston, 1840.
	Geog 4328.36.2	Cumming, W.F. Notes of a wanderer. London, 1840. 2v.
Htn	Geog 4558.40*	Dana, Richard Henry. Two years before the mast. N.Y., 1840.
	Geog 4558.40.5	Dana, Richard Henry. Two years before the mast. N.Y., 1840.
	Geog 4678.40	A full and particular act of all the circumstances attending the loss of the steamboat Lexington. Providence, 1840.
	Geog 818.40.4	Goodrich, S.G. A pictorial geography of the world. Boston, 1840.
	Geog 818.40.5	Goodrich, S.G. Pictorial geography of the world. 2nd ed. Boston, 1840. 2v.
	Geog 818.40.6	Goodrich, S.G. Pictorial geography of the world. 3rd ed. Boston, 1840.
	Geog 4218.40	Lafond, G. Quinze ans de voyages. Paris, 1840. 2v.
	Geog 578.40	Landmann, G. Universal gazeteer. London, 1840.
	Geog 4303.99	Lannoy, Guillebert de. Voyages et ambassades...1399-1450. Mons, 1840.
	Geog 818.40	Murray, H. Encyclopedia of geography. London, 1840.
	Geog 1782.1	Murray, John, publisher, London. Handbook for...Ionian islands. London, 1840.
	Geog 1707.5	Murray, John, publisher, London. Handbook the travellers on the continent. London, 1840.
	Geog 4218.38.4	Murrell, William M. Cruise frigate Columbia around the world...1838-1840. Boston, 1840.
	Geog 818.36.11	Perkins, Samuel. The world as it is. 5th ed. n.p., 1840.
	Geog 4678.40.5	Stone, John S. A sermon occasioned by burning of steamer Lexington. Boston, 1840.
	Geog 4168.40	Ternaux-Compans, H. Archives des voyages. pt.1-4. Paris, 1840. 4v.
	Geog 5398.20.5	Wrangell, F. von. Narrative of an expedition to the Polar Sea. London, 1840.
	Geog 4328.40.10	Wright, G.N. The shores and islands of the Mediterranean. London, 1840.

1841

	An 358.41.5	Azaïs, H. Question philosophique de première importance: Quelle est dans l'univers, la destinée du genre humain? Paris, 1841.
	Geog 585.11	Crump, William H. The world in a pocket-book. Philadelphia, 1841.
Htn	Geog 4558.40.7*	Dana, Richard Henry. Two years before the mast. London, 1841.
	Geog 6008.41	Dumont, J.S.C. Voyage au Pole Sud. Paris, 1841. 10v.
	Geog 4672.5	Ellms, Charles. The tragedy of the seas. Philadelphia, 1841.
	Geog 4308.39	Gibson, William. Rambles in Europe in 1839. Philadelphia, 1841.
	Geog 4308.40.3	Hall, Basil. Patchwork. London, 1841. 3v.
	Geog 4308.40.5	Hall, Basil. Patchwork. 2nd ed. London, 1841. 3v.
	Geog 4678.36.5	Holden, H. Narrative of the shipwreck. Cooperstown, N.Y., 1841.
	Geog 4308.38.5	Holmes, D. A ride to Florence thro' France. London, 1841. 2v.
	Geog 4105.2	Jackson, J.R. What to observe. London, 1841.
	Geog 4308.41	Lindeberg, A. Betraktelser under en resa. Stockholm, 1841.
	Geog 578.41	McCulloch, J.R. Dictionary geographical, statistical. London, 1841-42. 2v.
	Geog 808.12.15F	Malte-Brun, Conrad. Geographie universelle. v.1-6, Atlas. Paris, 1841-1847. 7v.
	Geog 4308.33.5	Marmier, X. Souvenirs de voyages. Paris, 1841.
	Geog 4248.39	Mercier, H.J. Life in a man-of-war...during cruise in Pacific. Philadelphia, 1841.

Chronological Listing

1841 - cont.

An 358.51		Prichard, J.C. Researches into the physical history of mankind. v.1-2,5, 4. ed.; v.3-4, 3. ed. London, 1841-51. 5v.
	Geog 808.17.6	Ritter, Carl. Namen- und Sach-Verzeichniss. Berlin, 1841-1849. 2v.
Htn	Geog 4308.39.5*	Sedgwick, C.M. Letters from abroad. N.Y., 1841. 2v.
	Geog 500.17	Ternaux-Compaus, H. Bibliothèque asiatique et africaine. Paris, 1841.
	Geog 500.17.2	Ternaux-Compaus, H. Bibliothèque asiatique et africaine. Notes. Paris, 1841.
Htn	Geog 4328.39.3*	Torrey, F.P. Journal of a cruise of the U.S.S. Ohio in the Mediterranean. Boston, 1841.

1842

	Geog 4308.42.5	Baruffi, G.F. Pellegrinazioni autunnali ed opuscoli. Torino, 1842.
Htn	Geog 4208.40.3*	Cleveland, R.J. Narrative of voyages and commercial enterprises. Cambridge, 1842. 2v.
	Geog 585.11.2	Crump, William H. The world in a pocket-book. Philadelphia, 1842.
	Geog 4328.42.5	Damer, G.L.D. Diary of a tour in Greece, Turkey. London, 1842.
Htn	Geog 4558.40.6*	Dana, Richard Henry. Two years before the mast. Glasgow, 1842.
	Geog 4105.15	Darde, J.B. The travellers' handbook. London, 1842.
	Geog 4218.26.5	Duhaut-Cilly, A. Viaggio intorno al globo. Napoli, 1842.
	Geog 4308.42	Faber, F.W. Sights and thoughts...foreign. Photoreproduction. London, 1842.
	Geog 3108.42	Forbiger, A. Handbuch der alten Geographie. Leipzig, 1842-48. 3v.
	Geog 4678.42	Guerrazzi, F.D. Replica...in causa di abbordaggio. Livorno, 1842.
	Geog 818.42	Laurie, J. System of universal geography. Edinburgh, 1842.
	Geog 818.42.5	Mitchell, S.A. An accompaniment to Mitchell's map. Philadelphia, 1842.
	Geog 4328.34	Mott, V. Travels in Europe and the East. N.Y., 1842.
	Geog 818.36.14	Perkins, Samuel. The world as it is. 6th ed. n.p., 1842.
	Geog 4324.59.5	Pfeiffer, F. Georg von Ehingen Reisen. Stuttgart, 1842.
	Geog 4308.42.10	Rives, J.P.W. Tales and souvenirs of a residence in Europe. Philadelphia, 1842.
	Geog 4328.42	Rockwell, C. Sketches of foreign travel. Boston, 1842. 2v.
	Geog 4158.42	Smith, William. Nouvelle bibliotheque des voyages. Paris, 1842. 12v.
	An 358.43.5	Virey, Julien J. Historia natural del jénero humano. 3. ed. Barcelona, 1842-46. 2v.
	Geog 4308.32.2	Willis, N.P. Pencillings by the way. London, 1842.

1843

	Geog 5378.18	Beechey, F.W. A voyage of discovery towards the North Pole in H.M. ships Dorothea and Trent...1818. London, 1843.
	Geog 4218.35	Belcher, Edward. Narrative of a voyage round the world...1836-42. London, 1843. 2v.
	An 358.43.3	Bond, Thomas E. A dissertation on the varieties of mankind. Middletown, 1843.
	Geog 578.41.2	McCulloch, J.R. Dictionary geographical, statistical. N.Y., 1843-44. 2v.
	Geog 818.43.5	Murray, Hugh. The encyclopaedia of geography. Philadelphia, 1843. 3v.
	Geog 1740.1	Murray, John, publisher, London. Handbook for...central Italy. London, 1843.
	Geog 1715.2	Murray, John, publisher, London. Handbook for travellers in France. London, 1843.
	An 348.43F	Prichard, J.C. Ethnographical maps to the natural history of man. London, 1843.
	An 358.43	Prichard, J.C. The natural history of man. London, 1843.
	Geog 4303.99.3	Sachet, E. Examen critique des voyages et ambassades de G. de Lannoy, 1399-1450. Bruxelles, 1843.
	Geog 818.43	St. Platou, L. Stutt landaskipunarfraedi. n.p., 1843.
	Geog 5518.36	Simpson, Thomas. Narrative of discoveries on north coast. London, 1843.
Htn	Geog 4678.41.6*	Slight, Julian. A narrative of the loss of the Royal George. 6. ed. Portsea, 1843.
	Geog 5398.20.20	Wrangell, F. von. Le nord de la Sibérie. Paris, 1843. 2v.

1844

	Geog 3235.52	Avezac, M.A.P. Deux notes sur l'anciennes cartes historiées. Paris, 1844.
	Geog 4311.17	Beattie, William. The Danube, its history, scenery. London, 1844.
Htn	Geog 4308.32.5*	Briggs, Charles F. Working a passage: or Life in a liner. N.Y., 1844.
	Geog 577.91.6.5	Brookes, R. A new universal gazetteer. Philadelphia, 1844.
	Geog 4308.44	Durbin, J.P. Observations in Europe. N.Y., 1844. 2v.
	Geog 4672.5.5	Ellms, Charles. Shipwrecks and disasters at sea. N.Y., 1844.
	Geog 4303.99.5	Lelewel, J. Guilbert de Lannoy et ses voyages en 1413, 1414 et 1421. Bruxelles, 1844.
	Geog 4558.44.2	Little, G. Life on the ocean. Boston, 1844.
	Geog 578.41.3	McCulloch, J.R. Dictionary geographical, statistical. N.Y., 1844-45. 2v.
	Geog 578.41.4	McCulloch, J.R. Dictionary geographical, statistical. N.Y., 1844-47. 2v.
	Geog 808.12.13	Malte-Brun, Conrad. System of universal geography. v.1-2. Boston, 1844.
	Geog 818.44	Mitchell, S.A. An accompaniment to Mitchell's map. Philadelphia, 1844.
	An 3538.44	Morton, S.G. Crania Aegyptiaca. Philadelphia, 1844.
	Geog 1715.2.5	Murray, John, publisher, London. Handbook for travellers in France. London, 1844.
	An 358.44	Nott, Josiah C. Two lectures on the natural history of the Caucasian and Negro races. Mobile, 1844.
	Geog 4418.39.4	Olin, S. Travels in Egypt, Arabia, Petraea and Holy Land. 4th ed. N.Y., 1844. 2v.
	Geog 4304.65	Schaschek. Des böhmischen Herrn Leo's von Rozmital...Reise...Abendlande 1465-1467. Stuttgart, 1844.
	Geog 4208.44.3	Smith, T.W. Narrative of life...of T.W. Smith. Boston, 1844.

1844 - cont.

	Geog 4328.40.3	Wilde, William R. Narrative of a voyage to Madeira, Teneriffe. 2nd ed. Dublin, 1844.
Htn	Geog 4308.32.3*	Willis, N.P. Pencillings by the way. 1st ed. N.Y., 1844.
	Geog 5398.20.7	Wrangell, F. von. Narrative of an expedition to the Polar Sea. London, 1844.

1845

	Geog 578.45	Baldwin, Thomas. Universal pronouncing gazetteer. Philadelphia, 1845.
	Geog 4348.45	Barber, J. Overland guide-book. London, 1845.
	Geog 4307.80.7	Beckford, William. Italy, Spain and Portugal. Pt.1-2. N.Y., 1845.
	Geog 4328.45	Borrer, D. Journey from Naples to Jerusalem. London, 1845.
	An 2108.45	Brotonne, F. de. Civilisation primitive. Paris, 1845.
	Geog 4208.45	Bustamante, J. Viaje. Lima, 1845.
	Geog 585.11.5A	Crump, William H. The world in a pocket-book. Philadelphia, 1845.
	Geog 4308.43	Dana, William C. A transatlantic tour. Philadelphia, 1845.
	Geog 4238.45	D'Avezac. Notice des découvertes...dans l'Ocean. Paris, 1845.
	Geog 4418.45	Frege, C.G. Erinnerungen aus dem Osten. Leipzig, 1845.
	Geog 4308.23.5	Glagolev, A.G. Zapiski russkago puteshestve norika [s 1823 po 1826 g.]. Sankt Peterburg, 1845. 4v.
	An 205.1	Jomard, E.F. Lettre à M.P.F. de Siebold. Paris, 1845-69. 3 pam.
	Geog 5220.9.6	Leslie, J. Narrative of discovery...polar seas. Edinburgh, 1845.
	Geog 818.45	Mitchell, S.A. An accompaniment to Mitchell's map. Philadelphia, 1845.
	Geog 808.34.10	Murray, Hugh. The encyclopaedia of geography. Philadelphia, 1845. 3v.
	Geog 1782.2	Murray, John, publisher, London. Handbook for...Ionian islands. London, 1845.
	Geog 1748.1	Murray, John, publisher, London. A handbook for travellers in Spain. London, 1845. 2v.
	Geog 1707.6	Murray, John, publisher, London. Handbook for travellers on the continent. 5. ed. London, 1845.
	An 358.45	Prichard, J.C. The natural history of man. 2. ed. London, 1845.
	Geog 3057.4	Quatremère. Memoire sur le pays d'Ophir. Paris, 1845.
	Geog 5838.45	Rafn, C.C. Amerikas...gamle geographie. Kjøbenhavn, 1845.
	Geog 5518.36.3	Simpson, A. Life and travels of Thomas Simpson. London, 1845.
	An 348.45	Slack, David B. An essay on the human color. Providence, 1845.
	Geog 4308.38.14	Stephens, J.L. Incidents of travel in Greece, Turkey and Poland. 17th ed. N.Y., 1845. 2v.
	Geog 4308.45.15	Talfourd, T.N. Vacation rambles and thoughts. London, 1845. 2v.
	Geog 4308.45.30	White, T.H. A pilgrim's reliquary. London, 1845.

1846

	Geog 1522.15	Baedeker, publishers. Handbuch für Reisende in Deutschland. 2. Aufl. Coblenz, 1846.
	Geog 5208.46	Barrow, John. Voyages of discovery and research within the Arctic regions, from the year 1818 to present. London, 1846.
	Geog 5208.46.3	Barrow, John. Voyages of discovery and research within the Arctic regions, from the year 1818 to present. London, 1846.
	Geog 585.12F	Borbstaedt, A. Allgemeine geographische und statistische Verhältnisse. Berlin, 1846.
	Geog 4208.44	Capobianco, R. Breve racconto delle cose. Napoli, 1846.
	Geog 4418.44	Davidson, G.F. Trade and travel in the Far East. London, 1846.
	Geog 3055.25	Jolibois. Dissertation sur l'Atlantide. Lyon, 1846.
	Geog 4308.46.2	Laing, S. Notes of a traveller. Philadelphia, 1846.
	Geog 4308.39.3	Locmaria. Souvenirs des voyages de...duc de Bordeaux. Paris, 1846. 2v.
	Geog 818.46	Mitchell, S.A. An accompaniment to Mitchell's map. Philadelphia, 1846.
	Geog 1735.7	Murray, John, publisher, London. Handbook for...Switzerland. 3. ed. London, 1846.
	Geog 4207.99	Nevens, William. Forty years at sea. Portland, 1846.
	Geog 4308.45	Pedestrian and other reminiscences. London, 1846.
NEDL	Geog 4218.46.3	Pfeiffer, I. A woman's journey round the world. London, 1846.
	Geog 3506.5	Saint-Genois, Jules de. Les voyageurs belges. v.1-2. Bruxelles, 1846-47?
	Geog 4328.46	Schroeder, F. Shores of the Mediterranean. N.Y., 1846. 2v.
	Geog 4308.46.5	Smith, J.J. Summer's jaunt across the water. v.1-2. Philadelphia, 1846.
	An 3558.46	Zeune, August. Über Schädelbildung zur festern Begründung der Menschenrassen. Berlin, 1846.

1847

	An 98.47	Bartlett, J.R. The progress of ethnology. 2. ed. N.Y., 1847.
	Geog 577.91.7	Brookes, R. A new universal gazetteer. Philadelphia, 1847.
	Geog 4208.47.5	Clark, J.G. Lights and shadows of sailor life. Boston, 1847.
	Geog 4180.2	Columbus, C. Letters...four voyages to new world. London, 1847.
	Geog 4558.40.9	Dana, Richard Henry. Two years before the mast. N.Y., 1847.
	Geog 4308.47.5	Eames, J.A. A budget of letters. Boston, 1847.
	Geog 4678.47	Greig, A.M. Fate of Blenden Hall, East Indiaman. N.Y., 1847.
	Geog 4180.1	Hawkins, R. Observations...in voyage...South Sea. London, 1847.
	Geog 4558.44.3	Little, G. Life on the ocean, or Twenty years at sea. Aberdeen, 1847.
	Geog 4348.47	Marmier, X. Du Rhin au Nil. Paris, 1847. 2v.
	Geog 1816.5	Murray, John, publisher, London. Handbook for...Egypt. London, 1847.
	Geog 1748.2	Murray, John, publisher, London. Handbook for travellers in Spain. 2. ed. London, 1847.
	Geog 578.47	Ritter, Karl. Geographisches-statistisches Lexikon. 3. Aufl. Leipzig, 1847.

Chronological Listing

1847 - cont.

	Geog 6008.39	Ross, J.C. Voyage...southern and Antarctic regions. London, 1847. 2v.
	Geog 4218.41.3	Simpson, George. Narrative of a journey round the world. London, 1847. 2v.
Htn	Geog 4218.41*	Simpson, George. Overland journey round the world. Philadelphia, 1847.
	Geog 4218.41.2	Simpson, George. Overland journey round the world. Philadelphia, 1847.
	Geog 4207.76.9	Sparks, Jared. Life of John Ledyard...American traveller. Boston, 1847.
	Geog 4218.38.3	Taylor, F.W. A voyage around the world. v.1-2. 9th ed. New Haven, 1847.

1848

	Geog 4208.48.5	Barbeida, Claudio Lagrange. Huma viagem de duas mil legoas. Nova-Goa, 1848.
	Geog 4308.48	Buckingham, J.S. Belgium, the Rhine, Switzerland. London, 1848. 2v.
	Geog 4208.47	Clark, J.G. Lights and shadows of sailor life. Boston, 1848.
	Geog 4268.48.5	Description of Bayne's gigantic panorama of a voyage to Europe. Boston, 1848.
	Geog 4558.48	Hart, J.C. Romance of yachting. N.Y., 1848.
	Geog 4308.48.5	Heinzelmann, F. Reisen durch Belgien, Holland und Grossbritannien. Leipzig, 1848.
	Geog 500.15	Jomard, E.F. De la collection geographique. Paris, 1848.
	Geog 4268.48	Maps. Europe. Description...of countries of Europe. Hartford, 1848.
	Geog 1715.3	Murray, John, publisher, London. Handbook for travellers in France. 3. ed. London, 1848.
	Geog 4180.3	Raleigh, W. Discovery of...empire of Guiana. London, 1848.
	Geog 758.48	Raumer, K.G. Lehrbuch der allgemeinen Geographie. Leipzig, 1848.
	An 358.45.8F	Schinz, Heinrich Rudolf. Naturgeschichte und Abbildungen des Menschen der verschiedenen Rassen. 3. Aufl. Zürich, 1848.
	Geog 4348.48	Terry, C. Scenes and thoughts in foreign lands. London, 1848.
	Geog 4678.46	United States. Congress. House. Committee on Commerce. Owners and crew of ship Chandler Price. n.p., 1848.
	An 358.48	Van Amringe, W.F. An investigation of the theories of the natural history of man. N.Y., 1848.

1849

	Geog 4328.43	Chenavard, Antoine N. Voyage en Grèce et dans le Levant fait en 1843-1844. Lyon, 1849.
	Geog 4308.49	Colman, H. European life and manners. Boston, 1849. 2v.
	Geog 1710.5	Cunningham, Peter. Handbook for London, past and present. London, 1849. 2v.
	Geog 4308.48.2	Description of Bayne's gigantic panorama of a voyage to Europe. Philadelphia, 1849.
	Geog 4208.49	Donald, C.S. Adventures of a Greek lady. London, 1849. 2v.
	Geog 4308.49.5	Kirkland, C.M. Holidays abroad. N.Y., 1849. 2v.
	Geog 578.41.5	McCulloch, J.R. Dictionary geographical, statistical. N.Y., 1849. 2v.
	Geog 4180.4	Maynarde, T. Sir Francis Drake, his voyage 1595. London, 1849.
	Geog 1706.5	Murray, John, publisher, London. Handbook for northern Europe. London, 1849. 2v.
	Geog 4332.8	Paton, A.A. Highlands and islands of the Adriatic. London, 1849. 2v.
	Geog 4348.49.2	Richardson, D.L. Anglo-Indian passage. London, 1849.
	Geog 4180.5	Rundall, T. Narratives of voyages...north-west. London, 1849.
	An 358.49.3	Smith, Ashbel. An oration pronounced before the Connecticut Alpha of the Phi Beta Kappa at Yale College. New Haven, 1849.
	Geog 4180.6	Strachey, W. Historie of travaile. London, 1849.
	Geog 4217.89	Viana, F.J. de. Diario del viaje explorador...en 1789-94. Cerrito de la Victoria, 1849.

185-

	Geog 1710.24	Cunningham, Peter. London, as it is. London, 185-.

1850

	Geog 3108.50	Anthon, C. System of ancient and mediaeval geography. N.Y., 1850.
	Geog 3235.49	Avezac, M.A.P. Note sur un atlas hydrographique. v.1-2. Paris, 1850
	An 358.50.6	Bachman, J. Doctrine of unity of human race examined on principles of science. Charleston, S.C., 1850.
	Geog 4308.49.50	Baxter, William E. Impressions of central and southern Europe. London, 1850.
	Geog 808.36.5	Bell, J. System of geography. London, 1850. 6v.
	Geog 4477.51.25	Berington, S. The adventures of Gaudentio di Lucca. London, 1850.
	Geog 577.91.10	Brookes, R. New universal gazetteer. Boston, 1850.
	Geog 4558.50	Browne, J.R. Etchings of a whaling cruise. N.Y., 1850.
	An 3208.50	Browne, P.A. The classification of mankind by the hair. Philadelphia, 1850.
Htn	Geog 5398.50F*	Browne, W.H. Ten coloured views during the Arctic expedition. London, 1850.
Htn	Geog 4228.50*	Bryant, W.C. Letters of a traveller. N.Y., 1850.
	Geog 4558.50.5	Cheever, H.T. The whale and its captors. N.Y., 1850.
	Geog 4208.40.5	Cleveland, R.J. Narrative of voyages and commercial enterprises. Boston, 1850.
	Geog 4308.49.2	Colman, H. European life and manners. Boston, 1850. 2v.
	Geog 1710.7	Cunningham, Peter. Handbook for London, past and present. London, 1850.
	Geog 4347.9	Dodge, Robert. Diary, sketches and reviews. N.Y., 1850.
	Geog 4208.13	Dunham, Jacob. Journal of voyages. N.Y., 1850.
	Geog 4180.7A	Hackluyt, R. Divers voyages...discovery of America. London, 1850.
	An 358.50	Knox, R. The races of men. Philadelphia, 1850.
	Geog 4308.46.3	Laing, S. Observations on social...state of European people. London, 1850.
	Geog 4268.50	Lavallée, T. Military topography of...Europe. London, 1850.

1850 - cont.

	Geog 4308.50	Lundeberg, A. La perle trouvée. Lausanne, 1850.
	Geog 4328.47	Margoliouth, Moses. A pilgrimage to the land of my fathers. London, 1850. 2v.
	Geog 958.33.3	Meyer, J. Meyer's Universum. v.1,4. N.Y., 1850. 2v.
	Geog 758.50	Milner, Thomas. A universal geography in four parts. London, 1850.
Htn	Geog 1724.5*	Murray, John, publisher, London. Handbook for...southern Germany. 5. ed. London, 1850.
	Geog 1707.7	Murray, John, publisher, London. Handbook for travellers on the continent. 7. ed. London, 1850.
	An 358.50.4	Pickering, Charles. The races of men. London, 1850.
Htn	Geog 5518.46*	Rae, J. Narrative of expedition to Arctic Sea. London, 1850.
	Geog 4658.50	Rastoul de Mongeot, Alphonse. Histoire des naufrages depuis les temps ancien jusqu'au 1850. v.1-2. Bruxelles, 1850.
	Geog 4218.38.2	Taylor, F.W. A voyage around the world. v.1-2. 9th ed. New Haven, 1850.
	Geog 4308.50.15	Taylor, J.B. Views a-foot or Europe. N.Y., 1850.
	Geog 4168.43.4	Voyages round the world from the death of Captain Cook to the present time. 4. ed. London, 1850.

1851

	Geog 578.49.9	Baldwin, Thomas. A pronouncing gazetteer. 9. ed. Philadelphia, 1851.
	Geog 4228.51	Bryant, W.C. Letters of a traveller. 3. ed. N.Y., 1851.
	Geog 4228.50.2	Bryant, W.C. The picturesque souvenir. N.Y., 1851.
	Geog 4208.51	Coggeshall, G. Voyages to various parts of the world...1799-1844. N.Y., 1851.
	Geog 4328.51	Colton, W. Ship and shore in Madeira. N.Y., 1851.
	Geog 4558.51	Congar, Obadiah. The autobiography and memorials of Captain Obadiah Congar. N.Y., 1851.
	Geog 4308.50.9	Conway, George. Running sketches of men and places. N.Y., 1851.
	Geog 1710.9A	Cunningham, Peter. Handbook for modern London. London, 1851.
	Geog 818.51	Geelmuyden, J. Loerebogi geografien. Christiania, 1851.
	Geog 4678.51.2	Gilly, W.O.S. Narratives of shipwrecks. 2. ed. London, 1851.
	Geog 4308.51.20	Greely, H. Glances at Europe. N.Y., 1851.
X Cg	Geog 4180.1	Herberstein, Sigmund. Notes upon Russia. London, 1851-52. 2v.
	Geog 578.51	Johnstone, A.K. Dictionary of geography. London, 1851.
	Geog 578.41.6	McCulloch, J.R. Dictionary geographical, statistical. London, 1851.
	Geog 4308.51.15	McFarland, A. The escape or loiterings amid the scenes. Boston, 1851.
X Cg	Geog 4308.50.5	McFarland, A. Five months abroad...1850. Concord, 1851.
	An 137.2	Meigs, C.D. Memoir of Samuel George Morton, M.D. Philadelphia, 1851. 2 pam.
	Geog 1724.5.5	Murray, John, publisher, London. Handbook for...southern Germany. 5. ed. London, 1851.
	Geog 1735.9	Murray, John, publisher, London. Handbook for...Switzerland. 4. ed. London, 1851.
	Geog 4328.51.5	Noel-Fearn, H. The shores and islands of the Mediterranean. London, 1851. 3v.
	Geog 4180.9	Relacam Verdadeira dos Trabalhos. Discovery...of Terra Florida by De Soto. London, 1851.
	Geog 4248.37	Rovings in the Pacific from 1837 to 1849. London, 1851.
	Geog 4208.44.5	Santos, Eusebio. Diario del viaje desde Madrid a Manila. Madrid, 1851.
	Geog 5208.51	Shillinglaw, J.J. A narrative of Arctic discovery. 2. ed. London, 1851.
	Geog 5508.51	Simmonds, P.L. Sir John Franklin and Arctic regions. London, 1851.
	An 358.51.3	Smith, C.H. The natural history of the human species. Boston, 1851.
	Geog 5538.50.17	Snow, W.P. Voyage of Prince Albert...Sir John Franklin. London, 1851.
	Geog 4308.45.19	Talfourd, T.N. Vacation rambles...1841-1843. 3rd ed. London, 1851.

1852

	Geog 4328.52.2	Aiton, John. The lands of the Messiah, Mahomet, and the pope. 2nd ed. London, 1852.
	Geog 5538.50.19	Arctic miscellanies. London, 1852.
	Geog 4308.52.3	Baxter, W.E. The Tagus and the Tiber. London, 1852. 2v.
	Geog 4308.50.7	Bullard, A.T.J. (Mrs.). Sights and scenes in Europe. St. Louis, 1852.
Htn	Geog 4308.52.7*	Calvert, G.H. Scenes and thoughts in Europe. 2d series. N.Y., 1852.
	Geog 4341.73.7	Carmoly, E. Notice historique sur Benjamin de Tudèle. Bruxelles, 1852.
	Geog 4308.52.10	Clarke, J.F. Eleven weeks in Europe. Boston, 1852.
	Geog 4180.11	Coats, W. Geography of Hudson's Bay. London, 1852.
	Geog 4208.48.2	Cöllen, Franz A. Reise Album vom 15ten bis zum 22ten Lebensjahre. Hamburg, 1852.
	Geog 4208.52.5	Colvocoresses, G.M. Four years in a government exploring expedition. N.Y., 1852.
	Geog 4308.51.7	Cox, Samuel S. A Buckeye abroad. N.Y., 1852.
	Geog 5225.5	Force, P. Grinnell land. Washington, 1852.
	An 358.52.4	Frankenheim, M.L. Völkerkunde. Breslau, 1852.
	Geog 4308.50.12	George, W.C. A year abroad. Boston, 1852.
	Geog 4268.52	Hughes, W. Manual of geography. pt.1. London, 1852.
	An 358.52.2	Latham, R.G. Man and his migrations. N.Y., 1852.
	Geog 4558.44.4	Little, G. Life on the ocean, or Twenty years at sea. 14. ed. N.Y., 1852.
	Geog 500.21	London. Royal Geographical Society. Catalogue of the library. London, 1852.
	Geog 5518.50.10	M'Clure, R. Captain M'Clure's despatches from her majesty's discovery ship "Investigator". London, 1852.
	Geog 578.41.7	McCulloch, J.R. Dictionary geographical, statistical. N.Y., 1852. 2v.
	Geog 5225.8	Mangles, James. Papers and dispatches...Arctic searching expeditions. 2. ed. London, 1852.
	Geog 958.33.5	Meyer, J. Meyer's Universum, or Views of...all contries. N.Y., 1852.
	Geog 1739.4	Murray, John, publisher, London. Handbook for...northern Italy. 4. ed. London, 1852.
	Geog 1735.11	Murray, John, publisher, London. Handbook for...Switzerland. 5. ed. London, 1852.
	Geog 5538.50.2	Osborn, S. Stray leaves from an Arctic journal. N.Y., 1852.

Chronological Listing

1852 - cont.

Geog 4218.46.5	Pfeiffer, I. A lady's voyage round the world. N.Y., 1852.
Geog 5538.48	Richardson, J. Arctic searching expedition. N.Y., 1852.
Geog 4308.52	Sewell, E.M. Diary...summer tour. N.Y., 1852.
An 358.52	Smith, C.H. The natural history of the human species. London, 1852.
Geog 4308.51.25	A summer's tour in Europe in 1851. Charleston, 1852.
Geog 665.13	Tagsberichte über die Forstschritte. Weimar, 1852.
Geog 4308.52.5	Tappan, H.P. A step from the new world to the old. N.Y., 1852. 2v.
Geog 3240.7	Thomassy, M.J.R. Les papes géographes. Paris, 1852.
Geog 4308.52.9	Trafton, M. Rambles in Europe. Boston, 1852.
Geog 4180.13	Veer, G. de. True description...voyages by the North-East. London, 1852-53.
Geog 4328.37.2	Wilde, William R. Narrative voyage to Madeira...Cyprus and Greece. Dublin, 1852.
Geog 4208.52	The world here and there. N.Y., 1852.

1853

Geog 1535.10	Baedeker, publishers. Die Schweiz. 5. Aufl. Coblenz, 1853.
Geog 4218.45	Bille, C.S.A. Beretning om corvetten Galathea's reise omkring jorden i 1845. 2. udg. Kjøbenhavn, 1853.
Geog 4208.53.2	Coggeshall, G. Voyages to various parts...1800-1831. N.Y., 1853.
Geog 578.53	Gazetteer of the world. London, 1853. 14v.
Geog 4218.50	Gerstaecker, F. Narrative of a journey round the world. N.Y., 1853.
Geog 4180.14	Gonzalez de Mendoza, J. History of...kingdom of China. London, 1853-54. 2v.
Geog 4238.53	Hauch, J.C. von. En Rejse til Amerika. Kjøbenhavn, 1853.
Geog 4558.53	Holcomb, E. A wonderful providence in many incidents at sea. 8. ed. Boston, 1853.
Geog 5538.52.10	Inglefield, Edward A. A summer search for Sir John Franklin. London, 1853.
Geog 4248.53	Jenkins, J.S. Recent exploring expeditions to the Pacific. London, 1853.
Geog 5538.51.5	Kane, E.K. Access to an open Polar Sea. N.Y., 1853.
Geog 5321.1.11	Kane, Elisha Kent. Access to an open polar sea. N.Y., 1853.
Geog 5538.51.15	Kennedy, William. A short narrative of the second voyage of the Prince Albert in search of Sir John Franklin. London, 1853.
Geog 4218.50.3	Marmier, Xavier. Voyage d'une femme autour du monde. v.1-2. Bruxelles, 1853.
Geog 5700.10	Monnier, Edouard. Histoire de la mission chrétienne au Groënland. Thèse. Strasbourg, 1853.
Geog 1740.3	Murray, John, publisher, London. Handbook for...central Italy. 3. ed. London, 1853. 2v.
Geog 1724.6	Murray, John, publisher, London. Handbook for...southern Germany. 6. ed. London, 1853.
Geog 1742.1	Murray, John, publisher, London. Handbook for...southern Italy. London, 1853.
Geog 1715.4	Murray, John, publisher, London. Handbook for travellers in France. 4. ed. London, 1853.
Geog 1707.9	Murray, John, publisher, London. Handbook for travellers on the continent. 9. ed. London, 1853.
Geog 4308.53	Murray, N. Men and things...in Europe. N.Y., 1853.
Geog 4308.45.9	Nicholson, Asenath. Loose papers...residence in Ireland. N.Y., 1853.
Geog 5208.53	Northern coasts of America. London, 1853.
Geog 4328.53.5	Paris. Services Maritimes. Les paquebots du Levant. Paris, 1853.
Geog 4208.50.3	Prince, Nancy. Narrative of life and travels. 2. ed. Boston, 1853.
Geog 818.53	Savage, C.C. The world; geographical, historical, statistical. N.Y., 1853.
Geog 665.21	Scenes in foreign lands. N.Y., 1853.
Geog 4218.45.5	Seemann, Berthold. Narrative of the voyage of H.M.S. Herald during the years 1845-51...being a circumnavigation of the globe. London, 1853.
Geog 4308.51	Silliman, B. Visit to Europe in 1851. N.Y., 1853. 2v.
Geog 5508.51.5	Simmonds, P.L. Sir John Franklin and Arctic regions. 3. ed. London, 1853.
Geog 4308.53.4	Tripp, A. Crests from the ocean-world. Boston, 1853.
Geog 4308.53.3	Tripp, A. Crests from the ocean-world. Boston, 1853.
Geog 4208.53.3	Ungewitter, D.H. Portfolio für Iländer und Völkerkunde. Pest, 1853.
Geog 4308.53.8	Wild oats sown abroad. Philadelphia, 1853.
Geog 4328.53	Willis, N.P. Summer cruise in the Mediterranean. Auburn, 1853.
Geog 3235.9	Wuttke, J.K.H. Über Erdkunde und Karten. Leipzig, 1853.
Geog 4208.43	Yvan, M. Voyages et recits. v.1-2. Bruxelles, 1853.

1854

Geog 1522.22	Baedeker, publishers. Handbuch für Reisende in Deutschland. 5. Aufl. Coblenz, 1854.
Geog 1535.13	Baedeker, publishers. Die Schweiz. 5. Aufl. Coblenz, 1854.
Geog 5538.51.7	Bellot, J.R. Journal d'un voyage aux mers polaires...en 1851 et 1852. Paris, 1854.
Geog 4218:40.10	Blok, G.K. Dva goda iz zhizni russkago moriaka. Sanktpeterburg, 1854. 2v.
Geog 5535.15	Brandes, Karl. Sir John Franklin; die Unternehmungen für seine Rettung und die Nordwestliche Durchfahrt. Berlin, 1854.
Geog 4138.54	Charton, E.T. Voyageurs anciens et modernes. Paris, 1854-57. 4v.
Geog 4168.54	Charton, Edouard T. Voyageurs anciens et modernes. v.1-4. Paris, 1854-57. 2v.
Geog 4308.54.25	Choules, J.O. Cruise of the steam yacht, North Star. Boston, 1854.
Geog 4308.54.20	Choules, J.O. Cruise of the steam yacht, North Star. Boston, 1854.
Geog 818.54.10	Condaminas, S. Nociones generales de geografía astronómica. Matanzas, 1854.
Geog 4558.40.8	Dana, Richard Henry. Two years before the mast. N.Y., 1854.
Geog 4348.54	Hoppin, J.M. Notes of a theological student. N.Y., 1854.
Geog 818.54	Ingerslev, C.F. Stutt kennslubok i landafroedinni. Reykjavik, 1854.
Geog 4208.38	Jenkins, J.S. United States exploring expeditions...voyage...1838-1842. New Orleans, 1854.
Geog 5538.51	Kane, E.K. United States Grinnell expedition. N.Y., 1854.

1854 - cont.

	Geog 578.54A	Knight, Charles. The English cyclopaedia. London, 1854-55. 4v.
	Geog 4308.54.4	Lippincott, S.J. Haps and mishaps...in Europe. Boston, 1854.
	Geog 4308.54.3	Lippincott, S.J. Haps and mishaps...in Europe. Boston, 1854.
	Geog 4308.54.2	Lippincott, S.J. Haps and mishaps...in Europe. Boston, 1854.
	Geog 4308.54	Lippincott, S.J. Haps and mishaps...in Europe. Boston, 1854.
	Geog 4678.54.5	Loss of the Arctic. n.p., 1854.
	Geog 4308.54.15	Maney, H. Memories over the water. Nashville, 1854.
	Geog 4418.54.3	Mortimer, F.L.B. (Mrs.). Far off, or Asia and Australia described with anecdotes and illustrations. N.Y., 1854.
	Geog 4308.54.10	Murray, E.C.G. The roving Englishman. London, 1854.
	Geog 1785.2	Murray, John, publisher, London. Handbook for...Greece. London, 1854.
	Geog 1735.13	Murray, John, publisher, London. Handbook for...Switzerland. 6. ed. London, 1854.
NEDL	Geog 1803.3	Murray, John, publisher, London. Handbook for...Turkey. London, 1854.
	Geog 1707.10	Murray, John, publisher, London. Handbook for travellers on the continent. 10. ed. London, 1854.
	Geog 4248.54	Niboyet, P. Les mondes nouveaux. Paris, 1854.
	Geog 3055.23	Norof, A.S. Die Atlantis. St. Petersburg, 1854.
	An 358.54	Nott, Josiah C. Types of mankind. 2. ed. Philadelphia, 1854.
	An 358.54.2	Nott, Josiah C. Types of mankind. 2. ed. Philadelphia, 1854.
	Geog 5538.50.10	Osborn, S. The Polar regions...Franklin's expedition. N.Y., 1854.
	Geog 4418.54	Osgood, J.B.F. Notes of travel or recollections. Salem, 1854.
	An 358.54.5	Pickering, Charles. The races of man; and their geographical distribution. London, 1854.
	Geog 5538.48.5	Richardson, J. Arctic searching expedition. N.Y., 1854.
	Geog 4208.54.15	Romance of travel...by an old traveler. Philadelphia, 1854.
	Geog 4314.5	Das Schwarze Meer. Leipzig, 1854.
	Geog 4308.51.4	Silliman, B. Visit to Europe in 1851. N.Y., 1854. 2v.
	Geog 4308.51.2	Silliman, B. Visit to Europe in 1851. N.Y., 1854. 2v.
	Geog 4678.54	Smalley, Elam. A sermon. Troy, N.Y., 1854.
	Geog 4328.54	Smyth, W.H. The Mediterranean. London, 1854.
Htn	Geog 4308.54.5*	Stowe, H.B. Sunny memories of foreign lands. Boston, 1854. 2v.
	Geog 4308.45.25	Talfourd, T.N. Supplement to "Vacation rambles". London, 1854.
	Geog 4308.46.35	Taylor, Bayard. Views a-foot. 17th ed. N.Y., 1854.

1855

	Geog 1558.10	Baedeker, publishers. Belgien. 5. Aufl. Coblenz, 1855.
	Geog 1522.23	Baedeker, publishers. Handbuch für Reisende in Deutschland. pt.1-2. 6. Aufl. Coblenz, 1855.
	Geog 4308.55.5	Bartol, C.A. Pictures of Europe...ideas. Boston, 1855.
	Geog 5538.52	Belcher, E. The last of the Arctic voyages. London, 1855. 2v.
	Geog 5538.51.9	Bellot, J.R. Memoirs of and journal of voyage...search of Sir John Franklin. London, 1855. 2v.
	Geog 4308.53.14	Boucher de Crèrecoeur de Perthes. Voyage à Constantinople par l'Italie. v.2. Paris, 1855.
	Geog 4228.50.5	Bryant, W.C. Letters of a traveller. 4. ed. N.Y., 1855.
NEDL	Geog 818.16.20	Cannabish, T.G.F. Lehrbuch der Geographie. 17e Aufl. Weimar, 1855.
	Geog 4208.52.6	Colvocoresses, G.M. Four years in the government exploring expedition. 5. ed. N.Y., 1855.
	Geog 4418.55	Eames, J.A. Another budget...in the East. Boston, 1855.
	Geog 4308.52.11	Horwitz, O. Brushwood, picked up on the continent. Philadelphia, 1855.
	Geog 4208.38.1	Jenkins, J.S. United States exploring expeditions. N.Y., 1855.
	Geog 578.55	Lippincott, J.B. and Co. Complete pronouncing gazetteer. Philadelphia, 1855.
	Geog 4180.18	Martens, F. Collection of documents on Spitzbergen and Greenland. London, 1855.
	Geog 4180.19	Middleton, H. Voyage...to Bantum and Maluco Islands. London, 1855.
	Geog 5558.50	Miertsching, J.A. Reise-Tagebuch des Missionars. Gnadan, 1855.
	Geog 1750.1	Murray, John, publisher, London. Handbook for...Portugal. London, 1855.
	Geog 1724.7	Murray, John, publisher, London. Handbook for...southern Germany. 7. ed. London, 1855.
	Geog 1724.7.2	Murray, John, publisher, London. Handbook for...southern Germany. 7. ed. London, 1855.
	Geog 1742.2	Murray, John, publisher, London. Handbook for...southern Italy. 2. ed. London, 1855.
	Geog 1748.3A	Murray, John, publisher, London. Handbook for...Spain. 3. ed. London, 1855. 2v.
	Geog 4558.55	Ocean scenes. N.Y., 1855.
	An 358.55	Prichard, J.C. The natural history of man. 4. ed. London, 1855. 2v.
	Geog 4308.53.12	Prime, Samuel I. Travels in Europe and the East. N.Y., 1855. 2v.
	Geog 578.55.15	Smith, John Calvin. Harper's statistical gazetteer of the world. N.Y., 1855.
	Geog 4218.52	Spalding, J.W. The Japan expedition. N.Y., 1855.
	Geog 4218.51	Stogman, C.D. Fregatten Eugenies resa omkring jorden, aren 1851-1853. v.1-2. Stockholm, 1855?
	Geog 4328.55	Taylor, B. Lands of the Saracew. N.Y., 1855.
	Geog 4308.55.3A	Wallace, H.B. Art, scenery and philosophy. Philadelphia, 1855.
	Geog 4308.53.9	Young, Samuel. A Wall-street bear in Europe. N.Y., 1855.

1856

	Geog 1525.15	Baedeker, publishers. Die Rheinlande. 9. Aufl. Coblenz, 1856.
	Geog 1535.16	Baedeker, publishers. Die Schweiz, die italienischen Seen. 6. Aufl. Coblenz, 1856.
	Geog 4308.55.7	Bartol, C.A. Pictures of Europe...ideas. Boston, 1856.
	Geog 4308.56	Channing, W. A physician's vacation. Boston, 1856.
	Geog 1710.15	Cunningham, Peter. London. London, 1856.

Chronological Listing

1856 - cont.

Geog 4308.56.3	Cuvillier-Heury, A.A. Voyages et voyageurs, 1837-1854. Paris, 1856.
Geog 4308.50.17	Dickinson, A. My first visit to Europe. 5th ed. N.Y., 1856.
Geog 4308.56.9	Eddy, D.C. Europa, or Scenes and society. Boston, 1856.
Geog 4128.56	Galton, Francis. The art of travel. London, 1856.
Geog 818.40.7	Goodrich, S.G. Pictorial geography of the world. Boston, 1856. 2v.
Geog 4308.56.15	Haskins, G.F. Travels in England, France. Boston, 1856.
Geog 5538.53.1	Kane, E.K. Arctic explorations. Philadelphia, 1856. 2v.
Geog 5538.53A	Kane, E.K. Arctic explorations. Philadelphia, 1856. 2v.
Geog 5538.53.2	Kane, E.K. Arctic explorations. Philadelphia, 1856. 2v.
Geog 1739.6	Murray, John, publisher, London. Handbook for...northern Italy. 6. ed. London, 1856. 2v.
Geog 1750.2	Murray, John, publisher, London. Handbook for...Portugal. 2. ed. London, 1856.
Geog 1735.15	Murray, John, publisher, London. Handbook for...Switzerland. 7. ed. London, 1856.
Geog 1709.37	Murray, John, publisher, London. Handbook for...Wiltshire, Dorsetshire and Somersetshire. London, 1856.
Geog 1707.11	Murray, John, publisher, London. Handbook for travellers on the continent. 11. ed. London, 1856.
Geog 4558.56	Nordhoff, C. Whaling and fishing. Cincinnati, 1856.
Geog 4218.55.5	Nordhoff, Charles. Man-of-war life. Cincinnati, 1856.
Geog 4228.56A	Ossoli, M.F. At home and abroad. 3. ed. Boston, 1856.
Geog 4218.46.7	Pfeiffer, I. A lady's second journey...world. N.Y., 1856.
Geog 4308.53.13	Prime, Samuel I. Travels in Europe and the East. N.Y., 1856. 2v.
Geog 4208.50.5	Prince, Nancy. Narrative of life and travels. 3. ed. Boston, 1856.
Geog 4308.56.13	Sigourney, L.H. Pleasant memories. Boston, 1856.
Geog 3018.56.5	Taylor, B. Cyclopedia of modern travel. Cincinnati, 1856.
Geog 4308.56.7	Vicuña Mackenna, Benjamin. Pajinas de mi diario. Santiago, 1856.

1857

Geog 5321.1.7	Allen, J.H. Elisha Kent Kane. Bangor, 1857.
Geog 5321.1.15	Andrews, E.W. Memoir and eulogy of Dr. Elisha K. Kane. N.Y., 1857.
Geog 5538.50.25	Armstrong, Alexander. A personal narrative of the discovery of the Northwest Passage...while in search of the expedition under Sir John Franklin. London, 1857.
Geog 1522.25	Baedeker, publishers. Deutschland. 7. Aufl. Coblenz, 1857.
An 358.57.3	Baldwin, Samuel D. Dominion; or Unity and trinity of the human race. Nashville, 1857.
Geog 5300.7	DePeyster, J.W. The Dutch at the North Pole. N.Y., 1857.
Geog 4308.57	Doré. A stroller in Europe. N.Y., 1857.
Geog 4308.56.17	Edwards, J.E. Random sketches...travel in 1856. N.Y., 1857.
Geog 4308.57.8	Fisk, Samuel. Mr. Dunn Browne's experiences in foreign parts. Boston, 1857.
Geog 4308.57.7	Fisk, Samuel. Mr. Dunn Browne's experiences in foreign parts. Boston, 1857.
Geog 5538.53.4	Kane, E.K. Arctic voyage. N.Y., 1857.
Geog 5538.51.3	Kane, E.K. United States Grinnell expedition. N.Y., 1857.
Geog 4308.57.3	LeVert, O.W. Souvenirs of travel. Mobile, 1857. 2v.
Geog 578.55.3	Lippincott, J.B. and Co. Lippincott's pronouncing gazetteer. Philadelphia, 1857.
Geog 5538.52.5	M'Dougall, G.F. Eventful voyage of H.M. discovery ship "Resolute". London, 1857.
Geog 4180.22	Major, R.H. India in the fifteenth century. London, 1857.
Geog 1740.4	Murray, John, publisher, London. Handbook for...central Italy. 4. ed. London, 1857.
Geog 4208.57	Nordhoff, Charles. Nine years a sailor. Cincinnati, 1857.
An 358.57	Nott, Josiah C. Indigenous races of the earth. Philadelphia, 1857.
Geog 5838.57	Rink, Hinrich. Grønland geographisk. Kjøbenhavn, 1857. 2v.
Geog 5208.57	Sargent, E. Arctic adventure by sea and land. Boston, 1857.
Geog 5321.1.5	Shields, C.W. Funeral eulogy at obsequies of Dr. Kane. Philadelphia, 1857.
Geog 5225.7	Smucker, Samuel. Arctic explorations and discoveries. N.Y., 1857.
Geog 5538.53.6	Sonntag, A. Professor Sonntag's thrilling narrative. Philadelphia, 1857.
Geog 5538.53.5	Sonntag, A. Professor Sonntag's thrilling narrative. Philadelphia, 1857.
Geog 4418.57	Train, G.F. An American merchant. N.Y., 1857.
Geog 4418.57.5	Train, G.F. Young American abroad. London, 1857.
Geog 4308.44.5	Volkov, M. Otryuki iz zagranichnykh pisem. Sankt Peterburg, 1857.
Geog 4328.57	Wise, H.A. Scampavias from Gibel Tarek to Stamboul. N.Y., 1857.

1858

Geog 1522.30	Baedeker, publishers. Deutschland. 8. Aufl. Coblenz, 1858.
Geog 4208.58.5	Coggeshall, G. Thirty-six voyages to various parts of the world, 1799-1841. 3. ed. N.Y., 1858.
Geog 4678.54.10	Cola, S. de. Difesa pel cav F. Miceli...scontro dei...Ercolano e Sicilia. Messina, 1858.
Geog 4678.58	Corrao, Mario. Difesa del capitano ed armatore del Sicilia contro il cap. ed armatore dell'Ercolano. Palermo, 1858.
Geog 4208.58	Davenport, H.E. Rovings on land and sea. Boston, 1858.
Geog 4218.58	Dorr, D.F. A colored man round the world. Cleveland? 1858.
Geog 5321.1.3	Elder, William. Biography of Elisha Kent Kane. Philadelphia, 1858.
Geog 5321.1.4	Elder, William. Biography of Elisha Kent Kane. Philadelphia, 1858.
Geog 500.19	Engelmann, W. Bibliotheca geographica. Leipzig, 1858.
Geog 3018.58	Goodrich, F.B. Man upon the sea. Philadelphia, 1858.
Geog 4308.47	Herzen, A.I. Tsis'ma iz Frantsii i Italii. London, 1858.
Geog 1763.3	Murray, John, publisher, London. Handbook for...Denmark, Norway, Sweden. 3. ed. London, 1858.
Geog 1816.10	Murray, John, publisher, London. Handbook for...Egypt. London, 1858.
Geog 1709.17	Murray, John, publisher, London. Handbook for...Kent and Sussex. London, 1858.

1858 - cont.

Geog 1724.8	Murray, John, publisher, London. Handbook for...southern Germany. 8. ed. London, 1858.
Geog 1742.3	Murray, John, publisher, London. Handbook for...southern Italy. 3. ed. London, 1858.
Geog 1709.33	Murray, John, publisher, London. Handbook for...Surrey, Hampshire. London, 1858.
Geog 1735.17	Murray, John, publisher, London. Handbook for...Switzerland. 8. ed. London, 1858.
Geog 1805.1	Murray, John, publisher, London. Handbook for...Syria and Palestine. London, 1858. 2v.
Geog 1707.12	Murray, John, publisher, London. Handbook for travellers on the continent. 12. ed. London, 1858.
Geog 1741.5	Murray, John, publisher, London. Handbook of Rome and its environs. 5. ed. London, 1858.
Geog 3248.58	Peschel, O.F. Geschichte des Zeitalters. Stuttgart, 1858.
Geog 4208.58.9	Petano y Mazariegos, Gorgonio. Viajes por Europa y América de Don Gorgonio Petano y Mazariegos. Paris, 1858.
An 378.58	Rathlef, C. Die welthistorische Bedeuting dei Meere insbesondere des Mittelmeers. Dorpat, 1858.
VGeog 4218.51.5	Svenska Vetenskaps-Akademien, Stockholm. Voyage autour du monde sur la frégate suédoise l'Eugénie executé pendant des années 1851-1853. Stockholm, 1858-74.

1859

Geog 4412.91.5	Avezac-Macaya. Expédition génoise des frères Vivaldi...route maritime des Indes orientales XIII. siècle. Paris, 1859.
Geog 4328.59.5	Böttger, C. Das Mittelmeer. Leipzig, 1859.
Geog 4105.10	Buffum, E.G. Pocket guide for Americans going to Europe. N.Y., 1859.
Geog 4180.23	Champlain, S. Narrative of voyages to West Indies and Mexico. London, 1859.
Geog 4180.26	Clavijo, R.G. de. Narrative of embassy...to court of Timar. London, 1859.
Geog 4308.59.3	Field, H.M. Summer pictures. N.Y., 1859.
Geog 4308.58.3	Glimpses of Europe. Cincinnati, 1859.
Geog 4208.59	Johnes, M. Boys' book of modern travel. N.Y., 1859.
Geog 5321.1.9	Kane Monument. Act of incorporation, preamble. N.Y., 1859.
Geog 808.59	Klöden, G.A. Handbuch der Erdkunde. Berlin, 1859-1862. 3v.
An 358.59	Latham, R.G. Descriptive ethnology. London, 1859. 2v.
Geog 4308.58	Letters from Europe...summer of 1858. Buffalo, 1859.
Geog 5538.57.2	M'Clintock, F.L. The voyage of the "Fox" in the Arctic seas. London, 1859.
Geog 4180.25	Major, R.H. Early voyages to Terra Australis. London, 1859.
Geog 808.12.16F	Malte-Brun, Conrad. A description of all parts of the world. Boston, 1859. 3v.
Geog 1807.5	Murray, John, publisher, London. Handbook for...India. London, 1859. 2v.
Geog 1709.37.2	Murray, John, publisher, London. Handbook for...Wiltshire, Dorsetshire and Somersetshire. London, 1859.
Geog 1715.7	Murray, John, publisher, London. Handbook for travellers in France. 7. ed. London, 1859.
Geog 818.52.5	Nicolay, Charles G. A manuel of geographical science. v.11. London, 1859.
An 358.59.3	Perty, M. Grundzüge der Ethnographie. Leipzig, 1859.
Geog 4412.91	Pertz, G.H. Der älteste Versuch zur Entdeckung des Seeweges nach Ostindien im Jahre 1291. Berlin, 1859.
Geog 818.59	Pütz, Wilhelm. Charakteristiken zur Erd und Völkerkunde. Köln, 1859-1860. 2v.
Geog 4308.59	Sweat, M.J.M. Highways of travel. Boston, 1859.
Geog 4308.59.5	Tait, J.R. European life, legend. Philadelphia, 1859.
Geog 4328.59	Taylor, Bayard. Travels in Greece and Russia. N.Y., 1859.
Geog 5518.16	Thirty years in the Arctic regions. N.Y., 1859.
An 358.59.5	Waitz, T. Anthropologie der Naturvölker. Leipzig, 1859. 6v.
Geog 4228.59	Warren, T.R. Dust and foam, or Three oceans and two continents. N.Y., 1859.
Geog 818.59.9	Young, Francis. Elementary geography. London, 1859.

1860

Geog 5558.60.3	American Geographical and Statistical Society. The polar exploring expedition. N.Y., 1860.
Geog 4180.27A	Asher, G.M. Henry Hudson the navigator. London, 1860.
Geog 500.33	Asher, George Michael. Viro venerabili Friderico Laurentio Hoffmann. Berolinenses, 1860.
Htn Geog 4308.60*	Benedict, E.C. A run through Europe. N.Y., 1860.
Geog 5535.10.2	Brown, John. The Northwest Passage, and the plans for the search for Sir John Franklin. 2. ed. London, 1860.
An 358.60	Clavel, A. Races humaines. Paris, 1860.
Geog 4208.60.5	Coggeshall, G. An historical sketch of commerce and navigation. N.Y., 1860.
Geog 585.11.10	Crump, William H. The world in a pocket-book, or Universal popular statistics. 12th ed. Phialdelphia, 1860.
Geog 4308.47.7	Eames, J.A. The budget closed. Boston, 1860.
Geog 4308.52.20	Eddy, Daniel C. Europa, or Scenes and society in England, France. 50th ed. Boston, 1860.
Geog 618.1.7	Guyot, A. Carl Ritter. Princeton, 1860.
Geog 5538.54.2	Hayes, I.I. An Arctic boat journey. Boston, 1860.
Geog 5538.57	M'Clintock, F.L. Voyage of "Fox"...narrative...of fate of Sir John Franklin. Boston, 1860.
Geog 3235.35	Matkovic, P.P. Alte handschriftliche Schifferkarten. Agram, 1860.
Geog 1709.1	Murray, John, publisher, London. Handbook for...Berks, Bucks and Oxfordshire. London, 1860.
Geog 1744.1	Murray, John, publisher, London. Handbook for...Corsica and Sardinia. London, 1860.
Geog 1739.8	Murray, John, publisher, London. Handbook for...northern Italy. 8. ed. London, 1860.
Geog 1707.13	Murray, John, publisher, London. Handbook for travellers on the continent. 13. ed. London, 1860.
Geog 5530.14	Osborn, Sherard. The career, last voyage. London, 1860.
Geog 4228.56.5	Ossoli, M.F. At home and abroad. Boston, 1860.
Geog 5538.57.5	Petersen, Carl. Den sidste Franklin expedition med "Fox". Kjøbenhavn, 1860.
Htn An 358.60.3*	Reid, Mayne. Odd people. N.Y., 1860.
Geog 5838.57.3	Rink, Hinrich. Grönland geographisch...beschreiben. Stuttgart, 1860.
Geog 3018.56.10	Taylor, B. Cyclopaedia of modern travel. N.Y., 1860. 2v.
Geog 3055.19	Unger, F.X. Die versunkene Insel Atlantis. Wien, 1860.

Chronological Listing

1860 - cont.

	An 358.77	Waitz, T. Anthropologie der Naturvölker. 2. Aufl. Leipzig, 1860.

1861

	Geog 4218.61.3	Ainsworth, W.F. All round the world. London, 1861. 2v.
	Geog 4308.61	Bremer, F. Two years in Switzerland and Italy. London, 1861. 2v.
	Geog 3550.9	Canale, M.G. Indicazioni di opere...sopra i viaggi, le navigazioni, le scoperte...degl'Italiani nel medio evo. Lucca, 1861.
Htn	Geog 4218.61*	D'Wolf, J. Voyage to the North Pacific. Cambridge, 1861.
	Geog 4158.60	Galton, F. Vacation tourists...in 1860, 1861, 1862, 1863. Cambridge, 1861-64. 3v.
Htn	Geog 578.61.5*	Graesse, J.G.T. Orbis latinus. Dresden, 1861.
	Geog 578.61	Graesse, J.G.T. Orbis latinus. Dresden, 1861.
	Geog 4308.61.5	Hale, E.E. Ninety days worth of Europe. Boston, 1861.
	Geog 4248.61	Jones. Life and adventures in the south Pacific. N.Y., 1861.
	Geog 4218.40.5F	Lafond, G. Fragments de voyages autour du monde. Paris, 1861.
	Geog 818.61	Mackay, A. Manual of modern geography. Edinburgh, 1861.
	Geog 1740.5	Murray, John, publisher, London. Handbook for...central Italy. 5. ed. London, 1861.
	Geog 4308.55	Ochoa, Eugenio de. Paris, Londres y Madrid. Paris, 1861.
	Geog 4332.8.5	Paton, A.A. Researches on the Danube and the Adriatic. Leipzig, 1861. 2v.
	An 348.61	Prichard, J.C. Explanatory notice of the ethnographical map. London, 1861.
	An 358.61.5	Quatrefages de Bréau, A. de. Unité de l'espèce humaine. Paris, 1861.
	Geog 4418.61.3	Rand, E.A. All aboard for sunrise lands. 2. ed. Boston, 1861.
	An 358.61	Reid, Mayne. Odd people. Boston, 1861.
	Geog 5060.5	Richardson, J. The polar regions. Edinburgh, 1861.
	Geog 3018.61	Ritter, C. Geschichte der Erdkunde. Berlin, 1861.
	Geog 665.9.3	Ritter, Karl. Geographical studies. Cincinnati, 1861.
	Geog 4218.57	Scherzer, Karl. Narrative of circumnavigation of globe...1857-58. London, 1861-63. 3v.
	Geog 5329.1	Scoresby, R.E. Life of William Scoresby. London, 1861.
	Geog 4418.61	Tilley, H.A. Japan, the Amoor and the Pacific. London, 1861.
	Geog 4308.53.6	Tripp, A. Crests from the ocean-world. Boston, 1861.
	Geog 5700.8	Vormbaum, R. Hans Egede, der Prediger des Evangeliums. Elberfeld, 1861.

1862

	Geog 3235.51	Avezac, M.A.P. Note sur la mappemonde historiée de la cathedral de Hereford. Paris, 1862.
	Geog 1522.33	Baedeker, publishers. Deutschland. v.2. 10. Aufl. Coblenz, 1862.
	Geog 4418.62	Brumund, J.F.G. Schetsen eener mail-reize van Batavia naar Maastricht op reis en thuis. Amsterdam, 1862.
	Geog 4308.62	Bucher, L. Unterwegs. v.1-2. Berlin, 1862.
	Geog 818.62.5	Cortambert, P.F.E. Cours de geographie. Paris, 1862.
	An 205.4	Dewey, Chester. Examination of some reasonings against unity of mankind. n.p., 1862. 2 pam.
	Geog 4180.30A	Galvano, A. Discoveries of the world. London, 1862.
	Geog 818.62	Harris, A. Geographical hand-book. Lancaster, Pa., 1862.
	An 608.62	Jomard, E.F. Classification méthodique des produits de l'industrie. Paris, 1862.
	Geog 4345.73	Krafft, Hans U. Ein deutscher Kaufmann. Göttingen, 1862.
	An 3308.62F	Liharzih, F. Gesetz des Wachstumes...Proportionslehre. Wien, 1862.
	Geog 1742.4	Murray, John, publisher, London. Handbook for...southern Italy. 4. ed. London, 1862.
	Geog 4308.58.5	Samper, J.M. Viajes de un Colombiano en Europa. Paris, 1862. 2v.
	Geog 4708.61	Trial of officers and crew of the privateer Savannah on charge of piracy. N.Y., 1862.
	Geog 4308.53.7	Tripp, A. Crests from the ocean-world. Boston, 1862.
	Geog 4678.62	United States. Navy Department. Wreck of steamer Governor and search for U.S. ship Vermont. Washington, 1862?
	An 2108.62	Wilson, D. Prehistoric man. Cambridge, 1862. 2v.

1863

	An 358.63	Brace, C.L. The races of the old world. N.Y., 1863.
	Geog 4308.52.8	Calvert, G.H. Scenes and thoughts in Europe. 1st-2d series. Boston, 1863. 2v.
	Geog 610.1	Cortambert, R. Notice sur la vie et les oeuvres de M. Jomard. Paris, 1863.
	Geog 3070.8	D'Avezac, M.A.P. Coup d'oeil historique sur projection. Paris, 1863.
	An 628.68	Gibbs, G. Instructions for research...ethnology and philology of America. Washington, 1863-68. 3 pam.
	Geog 4332.17	Goracuchi, J.A. Die Adria und ihre Küsten. Triest, 1863.
	Geog 4218.59	Hoffman, William. The Monitor; or, jottings of a N.Y. merchant during a trip around the globe. N.Y., 1863.
	An 2058.63	Lyell, C. The geological evidence of the antiquity of man. Philadelphia, 1863.
	An 2058.63.2	Lyell, C. The geological evidence of the antiquity of man. 2d American ed. Philadelphia, 1863.
	Geog 4208.63	Massett, S.C. Drifting about. N.Y., 1863.
	Geog 3070.22	Matkovic, P. Alte handschriftliche Schiffer-Karten. Wien, 1863.
	Geog 1739.9	Murray, John, publisher, London. Handbook for...northern Italy. 9. ed. London, 1863.
	Geog 1724.9	Murray, John, publisher, London. Handbook for...southern Germany. 9. ed. London, 1863.
	Geog 1709.8.5	Murray, John, publisher, London. Handbook for Devon and Cornwall. 5. ed. London, 1863.
	An 358.63.4	Pickering, Charles. Races of man; and their geographical distrbution. London, 1863.
	Geog 4268.63	Ritter, K. Europa. Berlin, 1863.
	Geog 665.9A	Ritter, Karl. Geographical studies. Boston, 1863.
	Geog 4208.63.7F	Tillotson, John. The overland route to India. London, 1863?
	Geog 4308.63	Travelling notes in France, Italy...of an invalid in search of health. Glasgow, 1863.
	Geog 4105.6	Verax, V. (pseud.). Cautions for the first tour. London, 1863.
	An 358.63.2	Waitz, T. Introduction to anthropology. London, 1863.

1864

	Geog 1525.20	Baedeker, publishers. A handbook for travellers on the Rhine. 2. ed. Coblenz, 1864.
	Geog 1519.10	Baedeker, publishers. Paris, Rouen, Havre, Dieppe. 5. Aufl. Coblenz, 1864.
	Geog 1535.24	Baedeker, publishers. Switzerland and the adjacent portions of Italy. 2. ed. Coblenz, 1864.
	An 358.64.5	Bonstetten, K.V. von. The man of the north and man of the south. N.Y., 1864.
	Geog 4180.33	Cieza de Leon, P. de. Travels of Piedro Cieza de Leon. London, 1864.
	Geog 4208.48.3	Cöllen, Franz A. Reisen und Dichtungen. Berlin, 1864.
	Geog 4228.28	Combier, C. Voyage au Golfe de Californie. Paris, 1864.
	An 358.64.3	Diefenbach, L. Vorschule der Völkerkunde. Frankfurt am Main, 1864.
	Geog 4228.57	Ennemoser, F.J. Reise vom Mittelrhein...nach dem Nordamerikanischen Freistaaten. Kaiserlautern, 1864.
	Geog 818.64	Milner, Thomas. The gallery of geography, a pictorial and descriptive tour of the world. London, 1864. 2v.
	Geog 1740.6	Murray, John, publisher, London. Handbook for...central Italy. 6. ed. London, 1864.
	Geog 1714.1.2	Murray, John, publisher, London. Handbook for...North Wales. 2. ed. London, 1864.
	Geog 1743.1	Murray, John, publisher, London. Handbook for...Sicily. London, 1864.
	Geog 1715.9	Murray, John, publisher, London. Handbook for travellers in France. 9. ed. London, 1864.
	Geog 1719.1	Murray, John, publisher, London. Handbook for visitors to Paris. London, 1864.
	Geog 1741.7	Murray, John, publisher, London. Handbook of Rome and its environs. 7. ed. London, 1864.
	Geog 1735.22	Murray, John, publisher, London. Knapsack guide for...Switzerland. London, 1864.
	Geog 4328.62	Phelps, S.D. Holy Land with glimpses of Europe and Egypt. N.Y., 1864.
	An 358.64.9	Pouchet, Georges. The plurality of the human race. 2. ed. London, 1864.
	An 358.64.7	Reid, Mayne. Odd people. Boaston, 1864.
	Geog 4418.64	Russell-Killough, H. Seize mille lieues a travers l'Asie et l'Oceanie. Paris, 1864.
	Geog 4207.76.10	Sparks, Jared. Life of John Ledyard, the American traveller. Boston, 1864.
	Geog 3070.26	Thomas, G.M. Der Periplus des Pontus Euxinus nach Münchener Handschriften. München, 1864.
	An 358.64	Vogt, C. Lecture on man: his place in creation and in the history of the earth. London, 1864.

1865

	Geog 4180.34	Andagoya, P. de. Narrative of proceedings of Pedrarias Davila. London, 1865.
	Geog 1522.35	Baedeker, publishers. Deutschland. 12. Aufl. Coblenz, 1865.
	Geog 1539.5	Baedeker, publishers. L'Italie...septentrionale. 3. éd. Coblenz, 1865.
	Geog 1519.13	Baedeker, publishers. Paris, Rouen, Havre, Dieppe. Coblenz, 1865.
	An 176.1	Blumenbach, J.F. The anthropological treatises. London, 1865.
	Geog 818.61.10	Bohn, Henry G. A pictorial hand-book of modern geography. 1st ed. London, 1865.
	Geog 4678.30	Burrows, Silas E. Russia and America. v.1-2. Hartford, 1865.
	Geog 600.7	Ciampi, I. Oltre l'Alpe e il mar. Roma, 1865.
	An 3538.65	Davis, J.B. Crania Britannica. London, 1865. 2v.
	An 358.65	Debay, A. Histoire naturelle de l'homme et de la femme. 12. éd. Paris, 1865.
	Geog 4308.65	Felton, C.C. Familiar letters from Europe. Boston, 1865.
	An 2058.65.3	Gastaldi, B. Lake habitations and pre-historic remains...north, centre Italy. London, 1865.
	Geog 3060.7	Hofmann, C. Setzungsberichte. n.p., 1865.
	Geog 500.21.3	London. Royal Geographical Society. Catalogue of the library. London, 1865.
	Geog 4208.64	Mompou, José. De la Habana a Madrid. Habana, 1865.
	Geog 1735.23A	Murray, John, publisher, London. Handbook for...Switzerland. 11. ed. London, 1865.
	Geog 1707.15	Murray, John, publisher, London. Handbook for travellers on the continent. 15. ed. London, 1865.
	Geog 5518.50.4	Osborn, S. Discovery of the Northwest Passage. Edinburgh, 1865.
	Geog 5538.50.4	Osborn, S. Stray leaves from an Arctic journal. Edinburgh, 1865.
	An 2058.65.5	Pereira da Costa, F.A. Da existencia do homem em epochas remotas no vale do tejo. Lisboa, 1865.
	Geog 818.59.6	Pütz, Wilhelm. Grundriss der Geographie und Geschichte der...Zeit. v.1-3. 10. Aufl, 12. Aufl. Koblenz, 1865-1867.
	Geog 4208.64.5	Root, Sidney. Exotic leaves, gathered by a wanderer. London, 1865.
	Geog 3140.5	Ruge, S. Der Chaldäer Seleukos. Dresden, 1865.
	An 2058.65	Thioly, F. Débris de l'industrie humaine...caverne de Bossey. Genève, 1865.
	Geog 4348.65	Ussher, John. Journey from London to Persepolis. London, 1865.
	Geog 4558.65	Waites, Alfred. My diary from England to India around Cape of Good Hope. Calcutta, 1865.

1866

	Geog 1555.10	Baedeker, publishers. Belgique et Hollande. 4. éd. Coblenz, 1866.
	Geog 1538.20	Baedeker, publishers. Italien. Coblenz, 1866-68. 2v.
	Geog 1525.22	Baedeker, publishers. Die Rheinlande von der Schweize zu holländisch Grenze. 14. Aufl. Coblenz, 1866.
	Geog 4180.35	Barbosa, D. Description of...East Africa and Malabar. London, 1866.
	Geog 3235.39	Berchet, G. Portolani, esistenti nelle...bibliotheque. Venezia, 1866.
	Geog 32.5.5	Bombay Geographical Society. Transactions. Index, v.1-17. Edinburgh, 1866.
	Geog 4308.66	Calvert, G.H. First years in Europe. Boston, 1866.
	Geog 4678.66	Comettant, O. Le naufrage de L'Evening Star. Paris, 1866.
	Geog 4208.66.3	Cruise of "The Wave". N.Y., 1866.
	Geog 808.66	Daniel, H.A. Handbuch der Geographie. Leipzig, 1866-1868. 4v.
	Geog 3235.7F	De Luca, G. Carte nautiche del medio evo. Napoli, 1866.
	Geog 4308.65.3	Felton, C.C. Familiar letters from Europe. Boston, 1866.
	Geog 4308.65.5	Haven, Gilbert. The pilgrim's wallet. N.Y., 1866.

Chronological Listing

1866 - cont.

Geog 5321.1.8	The love life of Dr. Kane. N.Y., 1866.
Geog 4308.66.5	Macgregor, J. Thousand miles in the Rob Roy canoe. London, 1866.
Geog 1739.10	Murray, John, publisher, London. Handbook for...northern Italy. 10. ed. London, 1866.
Geog 1713.2.5	Murray, John, publisher, London. Handbook for travellers in Ireland. 2. ed. London, 1866.
Geog 818.59.7	Pütz, Wilhelm. Grundriss der Geographie und Geschichte. Koblenz, 1866.
Geog 4328.66.5	Sleeper, M.G.Z. (Mrs.). The Mediterranean Islands. Boston, 1866.
An 358.66.5	Tuttle, Hudson. The origin and antiquity of physical man scientifically considered. 2. ed. Boston, 1866?
Geog 4308.66.12	Watt, Robert. Tgjennen Europa. København, 1866.
Geog 4180.36	Yule, H. Cathay and the way thether. London, 1866. 2v.

1867

Geog 1540.5	Baedeker, publishers. Italy. Pt. 2: Central Italy and Rome. Coblenz, 1867.
Geog 1542.5	Baedeker, publishers. Italy. Pt. 3: Southern Italy, Sicily. Coblenz, 1867.
Geog 1519.15	Baedeker, publishers. Paris and northern France. 2. ed. Coblenz, 1867.
Geog 1535.20	Baedeker, publishers. La Suisse. 7. éd. Coblenz, 1867.
Geog 1535.24.25	Baedeker, publishers. Switzerland and the adjacent portions of Italy. 3. ed. Coblenz, 1867.
Geog 4180.38	Best, G. Three voyages of Martin Frobisher. London, 1867.
An 3558.67	Davis, J.B. Thesaurus Craniorum. London, 1867.
Geog 3070.2	Desimoni, C. Atlante idrografico. Genova, 1867.
Geog 4678.67	Fernández Duro, C. Naufragios de la armada española. Photoreproduction. Madrid, 1867.
Geog 4308.67	Forney, J.W. Letters from Europe. Philadelphia, 1867.
Geog 618.1.5	Gage, W.L. Life of Carl Ritter. N.Y., 1867.
Geog 3550.4	Gubernatis, A. de. Memoria ai viaggiatori italiani. Firenze, 1867.
Geog 5538.54.5	Hayes, I.I. An Arctic boat journey, in the autumn of 1854. Boston, 1867.
Geog 5558.60	Hayes, I.I. The open Polar Sea. N.Y., 1867.
Geog 3570.10	Maix de Sori, A.F. Descobrimentos dos Portuguezes nos seculos XV e XVI. Lisboa, 1867.
Geog 3530.7	Margery, P. Les navigations françaises. Paris, 1867.
Geog 3530.7.2	Margery, P. Les navigations françaises. Paris, 1867.
Geog 4308.67.3	Meissner, A. Unterwegs. Leipzig, 1867.
Geog 4308.65.9	Morford, Henry. Over-sea. N.Y., 1867.
Geog 1740.7.5	Murray, John, publisher, London. Handbook for...central Italy. 7. ed. London, 1867.
Geog 1709.12	Murray, John, publisher, London. Handbook for...Gloucestershire, Worcestershire. London, 1867.
Geog 1739.11	Murray, John, publisher, London. Handbook for...northern Italy. 11. ed. London, 1867.
Geog 1712.1	Murray, John, publisher, London. Handbook for...Scotland. London, 1867.
Geog 1724.10	Murray, John, publisher, London. Handbook for...southern Germany. 10. ed. London, 1867.
Geog 1735.25	Murray, John, publisher, London. Handbook for...Switzerland. 12. ed. London, 1867.
Geog 1715.10	Murray, John, publisher, London. Handbook for travellers in France. 10. ed. London, 1867.
Geog 1719.3	Murray, John, publisher, London. Handbook for visitors to Paris. 3. ed. London, 1867.
Geog 1709.39	Murray, John, publisher, London. Handbook for Yorkshire. London, 1867.
Geog 1741.8	Murray, John, publisher, London. Handbook of Rome and its environs. 8. ed. London, 1867.
Geog 1733.5	Murray, John, publisher, London. The knapsack guide...Tyrol and the eastern Alps. London, 1867.
Geog 1735.26	Murray, John, publisher, London. Knapsack guide for...Switzerland. London, 1867.
Geog 4208.67.5	Pacific Mail Steamship Co. A sketch of the new route to China and Japan. San Francisco, 1867.
Geog 4110.22	Pacific Mail Steamship Co. A sketch of the route to California, China. San Francisco, 1867.
An 98.67	Quatrefages de Bréaux, A. de. Rapport sur la progrès de l'anthropologie. Paris, 1867.
Htn Geog 4308.67.5*	Sala, George A. From Waterloo to the Peninsula. London, 1867. 2v.
Htn Geog 4308.67.9*	Stone, H.S. From Cleveland to Russia. n.p., 1867.
Geog 4208.67	Watt, Robert. Kjøbenhavn. Kjøbenhavn, 1867.

1868

Geog 1539.8	Baedeker, publishers. Northern Italy, as far as Leghorn. Coblentz, 1868.
Geog 1523.15	Baedeker, publishers. Rhine and northern Germany. 3. ed. Coblenz, 1868.
An 2108.68	Bastian, A. Beiträge zur vergleichenden Psychologie. Berlin, 1868.
An 358.68.5	Bastian, A. Der Beständige in den Menschenrassen. Berlin, 1868.
Geog 578.68	Beeton, S.O. Dictionary of geography. London, 1868.
Geog 4308.68.7	Bellows, H.W. The old world in its new face. N.Y., 1868. 2v.
Geog 4428.68	Codine, J. Mémoire géographique sur mer des Indes. Paris, 1868.
Geog 4308.68.20	Coghill, James Henry. Abroad. N.Y., 1868.
Geog 4180.40	Cortes, H. Fifth letter of...Cortes to Emperor Charles V. London, 1868.
Htn Geog 4308.68.15*	Darley, F.O.C. Sketches abroad with pen and pencil. N.Y., 1868.
Geog 808.50.9	Grube, A.W. Geographische Charakterbilder in algerundeten Gemälden aus der Länder- und Völkerkunde. Leipzig, 1868. 3v.
Geog 4308.67.7	Haeseler, Charles H. Across the Atlantic. Philadelphia, 1868.
Geog 5558.53	Hayes, I.I. An Arctic boat journey...1854. Boston, 1868.
Geog 5208.68	Hayes, I.I. Progress of Arctic discovery. N.Y., 1868.
Geog 5208.68.2	Hayes, I.I. Progress of Arctic discovery. N.Y., 1868.
Htn Geog 4308.68.3*	Howe, Julia W. From the oak to the olive. Boston, 1868.
Geog 665.24	Ingwood of Westchester. Transatlantic souvenirs. N.Y., 1868.
Geog 818.68	Lavallée, T.S. Physical, historical and military geography. London, 1868.
An 358.68	Lesley, J.P. Man's origin and destiny. Philadelphia, 1868.

1868 - cont.

Geog 4308.66.7	Macgregor, J. Voyage alone in the yawl "Rob Roy". London, 1868.
Geog 4180.39	Morga, A. de. Phillipine Islands, Moluccas. London, 1868.
NEDL Geog 1709.17.3	Murray, John, publisher, London. Handbook for...Kent and Sussex. 3. ed. London, 1868.
Geog 1775.2	Murray, John, publisher, London. Handbook for...Russia, Poland and Finland. 2. ed. London, 1868.
Geog 1712.2	Murray, John, publisher, London. Handbook for...Scotland. 2. ed. London, 1868.
Geog 1742.6	Murray, John, publisher, London. Handbook for...southern Italy. 6. ed. London, 1868.
Geog 1707.16	Murray, John, publisher, London. Handbook for travellers on the continent. 16. ed. London, 1868.
Geog 4308.68	Peabody, A.P. Reminiscences of European travel. N.Y., 1868.
Geog 5838.68	Peirce, B.M. Report on the resources of Iceland and Greenland. Washington, 1868.
Geog 4417.45	Poivre, Pierre. Un manuscrit inédit de Pierre Poivre: Les mémoires d'un voyageur. Paris, 1868.
Geog 4300.17	Practical general continental guide. London, 1868.
An 2058.68	Saporta, G. da. La paléontologie appliquée à l'étude des races humaines. n.p., 1868.
Geog 4325.90.12	Webbe, Edward. Edward Webbe, chief master gunner, his travailes, 1590. London, 1868.
Geog 4325.90.10	Webbe, Edward. Edward Webbe, his travailes, 1590. Birmingham, 1868.
An 3558.68	Wyman, J. Observations on Crania. Boston, 1868.

1869

Geog 4205.03	Avezac, A. de. Campagne du navire l'Espoir de Honfleur, 1503-1505. Paris, 1869.
Geog 1555.13	Baedeker, publishers. Belgium and Holland. Coblenz, 1869.
Geog 1522.36	Baedeker, publishers. Deutschland und Österreich. 14. Aufl. Coblenz, 1869.
Geog 1540.10	Baedeker, publishers. Italy. Pt. 2: Central Italy and Rome. 2. ed. Coblenz, 1869.
Geog 1542.9	Baedeker, publishers. Italy. Pt. 3: Southern Italy, Sicily. 2. ed. Coblenz, 1869.
Geog 1542.10	Baedeker, publishers. Italy. Pt. 3: Southern Italy, Sicily. 2. ed. Coblenz, 1869.
Geog 1535.22	Baedeker, publishers. Die Schweiz. 12. Aufl. Coblenz, 1869.
Geog 1535.23	Baedeker, publishers. La Suisse. 8. éd. Coblenz, 1869.
Geog 1535.25	Baedeker, publishers. Switzerland. 3. ed. Coblenz, 1869.
Geog 1535.26	Baedeker, publishers. Switzerland. 4. ed. Coblenz, 1869.
An 2108.69.7	Bourlot, J. Histoire de l'homme préhistorique anté et postdiluvien. Paris, 1869.
Geog 613.1.11	Breusing, A. Gerhard Kremer...Mercator. Duisburg, 1869.
An 2058.69	Bromby, J.E. Pre-historic man. Melbourne, 1869.
Geog 4308.69.5	Buffum, E.G. Sights and sensations in France. N.Y., 1869.
An 2108.69.5	Campbell, G.D. Primeval man. London, 1869.
An 2108.69.4	Campbell, G.D. Primeval man. N.Y., 1869.
An 2108.69.3	Campbell, G.D. Primeval man. N.Y., 1869.
Geog 4218.69	Coffin, C.C. Our new way round the world. Boston, 1869.
Geog 4180.42	Correa, G. Three voyages of Vasco da Gama. Photoreproduction. London, 1869.
Geog 4328.69	Cox, S.S. Search for winter sunbeams. London, 1869.
Geog 4558.40.11.5	Dana, Richard Henry. Two years before the mast. Boston, 1869.
Htn Geog 4558.40.10*	Dana, Richard Henry. Two years before the mast. Boston, 1869.
Geog 4558.40.11	Dana, Richard Henry. Two years before the mast. Boston, 1869.
Geog 4308.69.3	Darley, F.O.C. Sketches abroad with pen and pencil. N.Y., 1869.
Geog 4218.49.5	Davis, R.C. Reminiscences of a voyage around the world. Ann Arbor, 1869.
Geog 3235.55	Desimoni, C. Nuovi studi sull'atlante luxoro. Geneva, 1869.
An 358.69.3	Lesley, J.P. Man's origin and destiny sketched from the platform of the sciences. Philadelphia, 1869.
An 2108.69	Lubbock, J. Pre-historic times. 2. ed. London, 1869.
Geog 4308.69.8	Macmillan, Hugh. Holidays on high lands. London, 1869.
Geog 4308.69.9	Montgomery, J.E. Our admiral's flag abroad. N.Y., 1869.
Geog 1715.11	Murray, John, publisher, London. Handbook for travellers in France. 11. ed. London, 1869.
Geog 1709.36.2	Murray, John, publisher, London. Handbook for Westmorland...and the Lakes. 2. ed. London, 1869.
Geog 1741.9	Murray, John, publisher, London. Handbook of Rome and its environs. 9. ed. London, 1869.
Geog 1710.20	Murray, John, publisher, London. Handbook to London. London, 1869?
Geog 613.1.3	Raemdonck, J. van. Gérard Mercator. St. Nicolas, 1869.
Htn An 358.69*	Stratton, C.S. Sketch of life, personal appearance, character and manners. N.Y., 1869.
Geog 4308.69	Taylor, B. By-ways of Europe. N.Y., 1869.
Htn Geog 4308.69.2*	Taylor, B. Byeways of Europe. London, 1869. 2v.
Geog 4308.69.7	Tousey, S. Papers from over the water. N.Y., 1869.
Geog 5398.67	Wheildon, W.W. The new Arctic continent. Cambridge, 1869.

187-

Geog 5838.70F	Nordenskjölds Reise nach Westgrönland im Jahre 1870. n.p., 187-?

1870

Geog 3235.45	Avezac, M.A.P. Un digression géographique...la mappemonde. Paris, 1870.
Geog 1539.10	Baedeker, publishers. L'Italie...septentrionale. 5. éd. Coblenz, 1870.
Geog 1539.12	Baedeker, publishers. Italy...northern Italy. 2. ed. Coblenz, 1870.
Geog 4418.70	Beauvior, L. Voyage round the world. London, 1870. 2v.
Geog 4328.61	Bennet, J.H. Winter and spring on the shores of the Mediterranean. 4th ed. London, 1870.
Geog 4308.70.10	Beste, John R.D. Nowadays. London, 1870. 2v.
Geog 4678.65	Bribery and piracy...loss of S.S. Shooting Star. N.Y., 1870.
Geog 4180.43	Columbus, C. Letters...four voyages to the new world. London, 1870.
Geog 578.70	Deschamps, P. Dictionnaire de géographie, ancienne et moderne. Paris, 1870.

Chronological Listing

1870 - cont.

	Geog 758.70	Dicuil. Liber de mensura orbis terra a Gustavo Parthey recognitus. Bercolini, 1870.
	An 2108.70.5	Figuier, L.G. L'homme primitif. 2. ed. Paris, 1870.
	An 2108.70.7	Figuier, L.G. Primitive man. N.Y., 1870.
	An 2108.70.6	Figuier, L.G. Primitive man. N.Y., 1870.
	Geog 4328.70	Fouché, William Grey (Mrs.). Dagbok...resatill Egypten, Konstantinopel. Stockholm, 1870.
	Geog 4308.70	Harding, W.M. Trans-Atlantic sketches. N.Y., 1870.
	An 2108.70.2	Lubbock, J. The origin of civilisation and the primitive condition of man. London, 1870.
	An 2108.70	Lubbock, J. The origin of civilisation and the primitive condition of man. 2. ed. London, 1870.
	Geog 1709.29	Murray, John, publisher, London. Handbook for Shropshire. London, 1870.
	Geog 1707.17	Murray, John, publisher, London. Handbook for travellers on the continent. 17. ed. London, 1870.
	Geog 1710.22	Murray, John, publisher, London. Handbook to London. London, 1870.
	Geog 4228.70	Pumpelly, Raphael. Across America and Asia. N.Y., 1870.
	Geog 613.1.5	Raemdonck, J. van. Gérard de Cremer ou Mercator. St. Nicolas, 1870.
	Geog 4300.12.2	Sargent, H.W. Skeleton tours. N.Y., 1870.
	Geog 4300.12	Sargent, H.W. Skeleton tours. N.Y., 1870.
	Geog 4308.70.5	Spender, E. Fjord, Isle and Tor. London, 1870.
	An 608.70	Steinhauer, C.L. Kort vriledning i det kgl. ethnographiske museum. København, 1870.
	An 2108.70.8	Tylor, Edward Burnett. Researches into the early history of mankind. London, 1870.
	An 358.70	Wood, J.G. The uncivilized races of men in all countries of the world. Hartford, 1870. 2v.

1871

	Geog 4218.71.15	Adams, N. A voyage around the world. Boston, 1871.
	Geog 3108.71	Anthon, C. System of ancient and mediaeval geography. N.Y., 1871.
	An 2108.71.3	Avebury, J.L. The origin of civilization and the primitive condition of man. N.Y., 1871.
	Geog 1555.15	Baedeker, publishers. Belgique et Hollande. 6. éd. Coblenz, 1871.
	Geog 1555.16	Baedeker, publishers. Belgium and Holland. 2. ed. Coblenz, 1871.
	Geog 1524.5A	Baedeker, publishers. Southern Germany and Austria. 2. ed. Coblenz, 1871.
	An 358.71.7	Bastian, A. Ethnologische Forschungen und Sammlung von Material für die selben. Jena, 1871-73. 2v.
	Geog 602.1	Baudet, P.J.H. Leven en werken van Williem Janszoon Blaeu. Utrecht, 1871.
	Geog 5398.71	Brynjulfson, A. Have de gamle Nordboer havt Kjendskab. Kjøbenhavn, 1871.
Htn	Geog 4558.40.13*	Dana, Richard Henry. Two years before the mast. Boston, 1871.
	Geog 4110.16	Dempsey, J.M. Our ocean highways. London, 1871.
	Geog 4308.71A	Guild, C. Over the ocean. Boston, 1871.
	Geog 5838.71	Hayes, I.I. The land of desolation...Greenland. London, 1871.
	Geog 4308.47.3	Herzen, A.I. Lettres de France et d'Italie (1847-1852). Genève, 1871.
	Geog 4308.71.9	McCray, R.H. Narrative of travels, scenes...in the Old World. 3rd ed. Philadelphia, 1871.
	Geog 1763.6	Murray, John, publisher, London. Handbook for...Denmark, Norway and Sweden. 3. ed. London, 1871.
	Geog 1724.11	Murray, John, publisher, London. Handbook for...southern Germany. 11. ed. London, 1871.
	Geog 1707.17.5	Murray, John, publisher, London. Handbook for travellers on the continent. 17. ed. London, 1871.
	Geog 1741.10	Murray, John, publisher, London. Handbook of Rome and its environs. 10. ed. London, 1871.
	Geog 1710.26	Murray, John, publisher, London. Handbook to London. London, 1871.
	An 358.71	Nott, Josiah C. Types of mankind. 10. ed. Philadelphia, 1871.
	Geog 4228.71.5	Pumpelly, Raphael. Across America and Asia. N.Y., 1871.
Htn	Geog 578.71*A	Rosser, William H. The Bijou gazetteer of the world. London, 1871.
	Geog 4300.12.3	Sargent, H.W. Skeleton tours. N.Y., 1871.
	Geog 4218.68	Smiles, S. Round the world. N.Y., 1871.

1872

	Geog 1540.12	Baedeker, publishers. Italy. Pt. 2: Central Italy and Rome. 3. ed. Coblenz, 1872.
	Geog 1519.21	Baedeker, publishers. Paris and northern France. 3. ed. Coblenz, 1872.
	Geog 1519.20	Baedeker, publishers. Paris and northern France. 3. ed. Coblenz, 1872.
	Geog 1519.21.3	Baedeker, publishers. Paris and northern France. 3. ed. Leipsic, 1872.
	Geog 1535.30	Baedeker, publishers. Switzerland...Italy, Savoy...Tyrol. 5. ed. Coblenz, 1872.
	Geog 4208.72.7	Barker, Mary Ann B. Travelling about over new and old ground. London, 1872.
	An 2408.72	Bastian, P.W.A. Die Rechtsverhältnisse. Berlin, 1872.
	Geog 4218.68.5	Bell, William M. Other countries. London, 1872. 2v.
	Geog 4218.72.15	Brooks, James. A seven month's run up and down and around the world. N.Y., 1872.
	Geog 4145.11.10	Fernandez de Navarrete, Eustaquio. Historia de Juan Sebastian del Cano. Vitoria, 1872.
	An 358.72	Figuier, Louis. Les races humaines. Paris, 1872.
	Geog 4418.72	Fogg, W.P. "Round the world", letters from Japan, China, India, and Egypt. Cleveland, 1872.
	Geog 5838.72	Fries, T.M. Grönland dess natur och innevanare. Upsala, 1872.
	Geog 4308.71.5	Haskins, George F. Six weeks abroad in Ireland, England and Belgium. Boston, 1872.
	Geog 5398.70	Heuglin, M.T. Reisen nach dem Nordpolarmeer. Braunschweig, 1872-74. 3v.
	Geog 4218.72.5	Hildebrandt, E.W. Reise um die Erde. 3. Aufl. Boston, 1872.
	Geog 4308.71.3	Hoyt, James M. Glances on the wing at foreign lands. Cleveland, 1872.
	Geog 110.1	International Geographical Congress. v.1-8. Anvers, 1872. 23v.
	Geog 4308.72	Jackson, H.H. Bits of travel. Boston, 1872.
	Geog 4328.66	Lycklama, A. Nijeholt. Voyage en Russie, au Caucase. Paris, 1872-1875. 4v.

1872 - cont.

	Geog 4300.27	Morford, Henry. Morford's short-trip guide to Europe. N.Y., 1872.
	Geog 1785.4	Murray, John, publisher, London. Handbook for...Greece. 4. ed. Lonon, 1872.
	Geog 1709.8.8	Murray, John, publisher, London. Handbook for Devon and Cornwall. 8. ed. London, 1872.
	Geog 1719.5	Murray, John, publisher, London. Handbook for visitors to Paris. 5. ed. London, 1872.
	Geog 4208.72	Revere, J.W. Keel and saddle, a retrospect. Boston, 1872.
	Geog 4308.72.3	Trafton, A. An American girl abroad. Boston, 1872.
	Geog 4308.72.6	Warner, Charles Dudley. Saunterings. Boston, 1872.
	Geog 4308.72.5	Warner, Charles Dudley. Saunterings. Boston, 1872.

1873

	Geog 1542.13	Baedeker, publishers. Italy. Pt. 3: Southern Italy, Sicily. 4. ed. Leipsic, 1873.
	Geog 1523.20	Baedeker, publishers. Northern Germany. 5. ed. Coblenz, 1873.
	Geog 1525.25.3	Baedeker, publishers. The Rhine from Rotterdam to Constance. 5. ed. Leipsic, 1873.
	Geog 1525.25	Baedeker, publishers. The Rhine from Rotterdam to Constance. 5. ed. Leipsic, 1873.
	Geog 1524.10	Baedeker, publishers. Southern Germany and Austria. 3. ed. Coblenz, 1873.
	Geog 1535.31	Baedeker, publishers. Switzerland...Italy, Savoy...Tyrol. 6. ed. Coblenz, 1873.
	Geog 4218.73.20	Bastian, Adolf. Geographische und ethnologische Bilder. Jena, 1873.
	Geog 3235.5	Bevan, W.L. Mediaeval geography. London, 1873.
	Geog 4418.73	Blyden, E.W. From West Africa to Palestine. Freetown, Sierra Leone, 1873.
	Geog 3550.7	Branca, G. Storia dei viaggiatori italiani. Roma, 1873.
	An 2108.73.3	Clodd, Edward. The childhood of the world. 2. ed. London, 1873.
	Geog 4218.73.15	Cook, Thomas. Letters from the sea and from foreign lands. London, 1873.
	Geog 4348.67	Creagh, J. A scamper to Sebastopol and Jerusalem 1867. London, 1873.
	Geog 5220.2	Dresden, Germany. Armen-Realschule. Programm...Offentlichen Prüfung. Dresden, 1873.
	Geog 4248.73	Eardley-Wilmot, S. Our journal in the Pacific. London, 1873.
	An 180.1	Force, M.F. Pre-historic man. Cincinnati, 1873.
	Geog 665.36	Freeman, Henry. Wonders of the world. Boston, 1873.
	Geog 4218.71.2	Hübner, J.A. Promenade autour du monde, 1871. 2e éd. Paris, 1873. 2v.
	Geog 4308.72.2	Jackson, H.H. Bits of travel. Boston, 1873.
	Geog 808.59.3	Klöden, G.A. Handbuch der Erdkunde. Berlin, 1873. 5v.
	Geog 4268.73.2	Levasseur, E. L'Europe...geographie. Paris, 1873.
	Geog 578.55.7	Lippincott, J.B. and Co. Lippincott's pronouncing gazetteer. Philadelphia, 1873.
	An 2058.73	Lyell, C. The geological evidence of the antiquity of man. 4. ed. London, 1873.
	Geog 5208.73.2	Markham, C.R. The threshold of the unknown region. London, 1873.
	Geog 3048.73	Moreau de Jonnes, A.C. L'ocean des anciens. Paris, 1873.
	Geog 1709.10.5	Murray, John, publisher, London. Handbook for...Durham and Northumberland. London, 1873.
	Geog 1816.15	Murray, John, publisher, London. Handbook for...Egypt. 4. ed. London, 1873.
	Geog 1724.12	Murray, John, publisher, London. Handbook for...southern Germany. 12. ed. London, 1873.
	Geog 1715.12	Murray, John, publisher, London. Handbook for travellers in France. 12. ed. London, 1873. 2v.
	Geog 1707.18	Murray, John, publisher, London. Handbook for travellers on the continent. 18. ed. London, 1873-74. 2v.
	Geog 1710.28	Murray, John, publisher, London. Handbook to London. London, 1873.
	Geog 4324.84	Neeffs, E. Un voyage au XVe siècle. Louvain, 1873.
	Geog 4418.73.5	Obligade, P.S. Viaje a Oriente. Paris, 1873.
	Geog 4308.73.5	Prime, Samuel I. The Alhambra and the Kremlin. N.Y., 1873.
	An 3538.82	Quatrefages de Bréau, A. de. Crania ethnica. Les crânes des races humaines. Paris, 1873-81. 2v.
	Geog 4208.72.2	Revere, J.W. Keel and saddle, a retrospect. Boston, 1873.
	Geog 4218.70.5	Seward, William H. William H. Seward's travels around the world. N.Y., 1873.
	Geog 4218.70.4	Seward, William H. William H. Seward's travels around the world. N.Y., 1873.
	An 2108.73	Tylor, Edward Burnett. Die Anfänge der Cultur. v.1-2. Leipzig, 1873.
	Geog 5558.69	Verein für die Deutsche Nordpolarfahrt. Leipzig, 1873-74. 4v.
	Geog 4208.70.2	Willis, G.R. The cruise of the Colorado. N.Y., 1873.
X Cg	Geog 4208.70	Willis, G.R. The cruise of the Colorado. N.Y., 1873.
	Geog 4558.73	Wogan, E. de. Du Far-West à Bornéo. Paris, 1873.
	Geog 4180.50	Zeno, N. Voyages of...Nicolo and Antonio Zeno. London, 1873.

1874

	Geog 500.22	Amat di San Filippo, Pietro. Bibliografia di viaggiatori italiani ordinata cronologicamenti. Roma, 1874.
	Geog 818.51.14	Arendts, Carl. Leitfaden für den ersten Wissenschaftlichen Unterricht in der Geographie. 14. Aufl. Regensburg, 1874.
	Geog 1555.17	Baedeker, publishers. Belgium and Holland. 3. ed. Leipsic, 1874.
	Geog 1539.17	Baedeker, publishers. Italy...northern Italy. 3. ed. Leipsic, 1874.
	Geog 1539.17.5	Baedeker, publishers. Italy...northern Italy. 3. ed. Leipsic, 1874.
	Geog 1519.23	Baedeker, publishers. Paris and its environs. 4. ed. Leipsic, 1874.
	Geog 5558.71.5	Blake, E.V. Arctic experiences. N.Y., 1874.
	An 2108.74.5	Clodd, Edward. The childhood of the world. Boston, 1874.
	Geog 4328.69.5	Cox, S.S. Search for winter sunbeams. N.Y., 1874.
	Geog 808.66.5	Daniel, H.A. Handbuch der Geographie. Leipzig, 1874. 4v.
	An 2108.74.3	Dawkins, W.B. Cave hinting. London, 1874.
	Geog 4708.74	Galleon Treasure Company, N.Y. Narrative of the circumstances of and attending the sinking of the Spanish galleons. N.Y., 1874.
	Geog 4218.71.3	Hübner, J.A. Ramble round the world. London, 1874. 2v.

Chronological Listing

1874 - cont.

Geog 4218.71	Hübner, J.A. Ramble round the world. N.Y., 1874.
Geog 4413.25.9	Ibn Batoutah. Voyages. Paris, 1874. 5v.
Geog 4558.74.3	Jewell, J.G. Among our sailors. N.Y., 1874.
Geog 5558.69.2	Koldewey, K. German Arctic expedition of 1869-70. London, 1874.
Geog 4218.70	Longworth, M.T. Teresina peregrina. London, 1874. 2v.
Geog 3108.74	Moreau de Jonnés, A. Estudios prehistóricos. Sevilla, 1874.
Geog 1709.7.2	Murray, John, publisher, London. Handbook for...Derbyshire, Nottinghamshire. 2. ed. London, 1874.
Geog 1735.30	Murray, John, publisher, London. Handbook for...Switzerland. 15. ed. London, 1874.
Geog 1770.5	Murray, John, publisher, London. Handbook for Norway. 5. ed. London, 1874.
Geog 1710.29	Murray, John, publisher. London. Handbook to London. London, 1874.
Geog 4558.74	Nordhoff, C. Life on the ocean. Cincinnati, 1874.
Geog 4105.4	Noyes, E.H. Steamship notes...a handbook. N.Y., 1874.
Geog 4218.74	Prime, E.D.C. Around the world, sketches of travel. N.Y., 1874.
An 2058.74	Ratzel, F. Vorgeschichte des europäischen Menschen. München, 1874.
Geog 4306.81.3	Regnard, J.F. Voyage de...en Flandre, Hollanke, Danemark. Paris, 1874.
Geog 578.47.9	Ritter, Karl. Geographisches-statistisches Lexikon. Leipzig, 1874. 2v.
Geog 3055.31	Roisel, G. de. Les Atlantes. Paris, 1874.
Geog 4218.70.6	Seward, William H. William H. Seward's travels around the world. N.Y., 1874.
Geog 4218.72	Simpson, William. Meeting the sun; journey all round the world. London, 1874.
Geog 4180.51	Stade, H. Captivity of Hans Stade of Hesse. London, 1874.
Geog 4180.52A	Stanley, E.J. First voyage round the world by Magellan. London, 1874.
Geog 3018.74	Tiele, Pieter Anton. De ontdekkingsreizen sedert de vijftiende eeuw. Leiden, 1874.
An 2108.74	Tylor, Edward Burnett. Primitive culture. Boston, 1874. 2v.
An 2109.20.2	Tylor, Edward Burnett. Primitive culture. 1st American ed. Boston, 1874. 2v.
Geog 4678.65.10	Weiss, N. Naufrage de la Ville-du-Havre et du Losh-Earn. Paris, 1874.
Geog 4238.72	Wells, Theodore. Narrative of life and adventures of Captain Wells...voyages. Biddeford, 1874.
An 2058.74.5	Whitmore, J.H. Evidences of the antiquity of man. Rochester, N.Y., 1874.
An 2108.74.7	Zaborowski-Moindron, S. De l'ancienneté de l'homme. v.1-2. Paris, 1874.

1875

Geog 4328.74.5	Aguilar, Federico C. Recuerdos de un viaje a Oriente en el año de 1874. Bogota, 1875.
An 2108.75.3	Avebury, J.L. Die Entstehung der Civilisation. 3e Aufl. Jena, 1875.
An 2108.75.5	Avebury, J.L. The origin of civilisation. 3. ed. London, 1875.
Geog 1555.18	Baedeker, publishers. Belgique et Hollande. 8. ed. Leipzig, 1875.
Geog 1525.27	Baedeker, publishers. Les bords du Rhin de la frontière suisse à la frontière de Hollande. 9. éd. Leipzig, 1875.
Geog 1540.17	Baedeker, publishers. Italie. Pt. 2: Italie centrale et Rome. 4. éd. Leipzig, 1875.
Geog 1542.14	Baedeker, publishers. Italie. Pt. 3: Italie du sud et la Sicile. 4. éd. Leipzig, 1875.
Geog 1540.15	Baedeker, publishers. Italy. Pt. 2: Central Italy and Rome. 4. ed. Leipsic, 1875.
Geog 1510.3	Baedeker, publishers. Londres, ses environs: le Sud de l'Angleterre. 3. éd. Leipzig, 1875.
Geog 4208.73.5	Chapman, Charles. The ocean waves. London, 1875.
An 2108.75	Clodd, Edward. The childhood of the world. 3. ed. London, 1875.
Geog 808.75F	Colange, Leo de. The picturesque world, or Scenes in many lands. pt.1,3-6,8-13,15-16,18,20,26. Boston, 1875.
Geog 3570.15	Colleccão de opusculos...relativos a historia das navegacões. pt.1-4. Lisboa, 1875.
An 3558.67.2	Davis, J.B. Thesaurus Craniorum. Supplement. London, 1875.
Geog 4418.75	Fogg, W.P. Arabistan, or Land of "Arabian Nights." London, 1875.
An 358.75.3	Gerland, Georg. Anthropologische Beiträge. Halle, 1875.
Geog 3550.5	Gubernatis, A. de. Storia dei viaggiatori italiani. Livorno, 1875.
Geog 4308.71.2	Guild, C. Over the ocean, or Sights and scenes in foreign lands. Boston, 1875.
Geog 4218.73.9	Guillemard, A.C. Over land and sea. London, 1875.
Htn Geog 4308.75*	James, H. Transatlantic sketches. Boston, 1875.
Geog 5838.75	Jones, T.R. Manual of the natural history, geology and physics of Greenland. London, 1875.
An 2058.75.5	Lester, A. Hoyle. The pre-adamite. Philadelphia, 1875.
Geog 1740.9	Murray, John, publisher, London. Handbook for...central Italy. 9. ed. London, 1875.
Geog 1765.4	Murray, John, publisher, London. Handbook for...Denmark. 4. ed. London, 1875.
Geog 1816.17	Murray, John, publisher, London. Handbook for...Egypt. 5. ed. London, 1875.
Geog 1750.3	Murray, John, publisher, London. Handbook for...Portugal. 3. ed. London, 1875.
Geog 1775.3	Murray, John, publisher, London. Handbook for...Russia, Poland and Finland. 3. ed. London, 1875.
Geog 1712.4	Murray, John, publisher, London. Handbook for...Scotland. 4. ed. London, 1875.
Geog 1773.4	Murray, John, publisher, London. Handbook for...Sweden. 4. ed. London, 1875.
Geog 1805.1.10	Murray, John, publisher, London. Handbook for...Syria and Palestine. London, 1875.
Geog 1709.11.2	Murray, John, publisher, London. Handbook for Essex, Suffolk and Norfolk. 2. ed. London, 1875.
Geog 1715.13	Murray, John, publisher, London. Handbook for travellers in France. 13. ed. London, 1875.
Geog 1707.19	Murray, John, publisher, London. Handbook for travellers on the continent. 19. ed. London, 1875. 2v.
Geog 1741.12	Murray, John, publisher, London. Handbook of Rome and its environs. 12. ed. London, 1875.
Geog 4268.75F	Picturesque Europe. N.Y., 1875-1879. 3v.

1875 - cont.

Geog 578.75	Post-Lexicon; ein Verzeichniss der wichtigeren Verkehrs Orte. Berlin, 1875.
An 358.75	Quatrefages de Bréau, A. de. Natural history of man. N.Y., 1875.
Geog 4208.72.9	Smith, G.A. Correspondence of Palestine tourists...while traveling in Europe, Asia and Africa. Salt Lake City, 1875.
An 2058.75	Southall, J.C. The recent origin of man. Philadelphia, 1875.
Geog 3550.3	Studi bibliografici e biografici sulla storia della geografia in Italia. Roma, 1875.
Geog 4678.73	Weiss, Nathanael. Personal recollections of wreck of...Ville-du-Havre. N.Y., 1875.
Geog 4208.73	Wood, C.F. Yachting cruise in the South Seas. London, 1875.

1876

Geog 5160.5F	The Arctic world; its plants, animals and natural phenomena. London, 1876?
Geog 1539.18	Baedeker, publishers. L'Italie...septentrionale. 7. éd. Leipzig, 1876.
Geog 1542.15	Baedeker, publishers. Italy. Pt. 3: Southern Italy, Sicily. 6. ed. Leipsic, 1876.
Geog 1605.5	Baedeker, publishers. Palestine and Syria. Leipzig, 1876.
Geog 1519.24	Baedeker, publishers. Paris, ses environs et les principaux itinéraires. 4. éd. Leipzig, 1876.
Geog 1519.26	Baedeker, publishers. Paris and its environs. 5. ed. Leipsic, 1876.
Geog 1519.25	Baedeker, publishers. Paris and its environs. 5. ed. Leipsic, 1876.
Geog 4218.76.9	Bedinello, Ugo. Diario del viaggio intorno al globo. Trieste, 1876.
Geog 3228.76	Beltrán y Rózpide, Ricardo. Viajes y descubrimientos, efectuados en la edad media. Madrid, 1876.
Geog 578.76	Blackie, W.G. The imperial gazetteer. London, 1876. 2v.
Geog 4218.76.5	Campbell, J.F. My circular notes. v.1-2. London, 1876.
Geog 4218.76	Curtis, B.R. Dottings round the circle. Boston, 1876.
Geog 5558.71	Davis, C.H. Narrative of north polar expedition. Washington, 1876.
Geog 5398.56	Dufferin and Ava, F.T. Blackwood. Un voyage en yacht, lettres de hautes latitudes. Montreal, 1876.
Geog 4308.74	Forney, John W. A centennial commissioner in Europe 1874-1876. Philadelphia, 1876.
Geog 3235.53F	Gaffarel, Paul. Étude sur un portolan inédit. Dijon, 1876.
Geog 4308.72.7	Kennedy, David. Colonial travel. London, 1876.
Geog 4558.76	Lindsay, W.S. Recollections of a sailor. London, 1876.
Geog 5558.75.9	MacGahan, J.A. Under the Northern Lights. London, 1876.
Geog 5208.73.4	Markham, C.R. The threshold of the unknown region. 4. ed. London, 1876.
An 2058.76.3	Meunier, Victor. Los antepasados de Adan. Madrid, 1876.
Geog 1719.8	Murray, John, publisher, London. Handbook for visitors to Paris. 8. ed. London, 1876.
Geog 1710.30	Murray, John, publisher, London. Handbook to London. London, 1876.
An 3308.76	Pagliani, L. Sopra alcuni fattori dello suiluppo umano. Torino, 1876.
Geog 5398.72	Payer, J. New Lands within the Arctic Circle. London, 1876. 2v.
Geog 5398.72.3	Payer, J. Die österreichisch-ungarische Nordpol-Expedition. Wien, 1876.
Geog 4218.70.3	Peebles, James. Around the world. 2. ed. Boston, 1876.
An 358.76.3	Peschel, Oscar F. The races of man. N.Y., 1876.
An 358.76.5	Peschel, Oscar F. Völkerkunde. 3. Aufl. Leipzig, 1876.
An 2058.76	Rau, C. Early man in Europe. N.Y., 1876.
NEDL Geog 808.76	Reclus, J.J.E. Nouvelle geographie universelle. Paris, 1876-1894. 19v.
Geog 500.9	Rome, Italy. Biblioteca Collegio Romano. Catalogo ragionato. Roma, 1876.
Geog 1711.1	Thorne, James. Handbook to environs of London. London, 1876. 2v.
An 358.76.2	Topinard, P. L'anthropologie. Paris, 1876.
An 358.76	Topinard, P. L'anthropologie. Paris, 1876.
Geog 4558.76.3	The tourist, a magazine of information for ocean travellers, 1876. Boston, 1876.
Geog 5321.1	United States. Naval Observatory. Reports on medals to Arctic explorers, Kane, Hayes. Washington, 1876.
Geog 4180.54	Veer, G. de. Three voyages of...William Barents. London, 1876.
Geog 4308.75.6	Waring, G.E. A farmer's vacation. Boston, 1876.
Geog 4218.66.5	Weppner, M. The North Star and the Southern Cross. Albany, 1876.
An 2108.76	Wilson, D. Prehistoric man. 3. ed. London, 1876. 2v.
An 358.70.5	Wood, J.G. The uncivilized races of men in all countries of the world. Hartford, 1876. 2v.

1877

Geog 1540.17.25	Baedeker, publishers. Italie. Pt. 2: Italie centrale et Rome. 5. éd. Leipzig, 1877.
Geog 1542.16	Baedeker, publishers. Italie. Pt. 3: Italie méridionale et la Sicile. 5. éd. Leipzig, 1877.
Geog 1523.25	Baedeker, publishers. Northern Germany. 6. ed. Leipsic, 1877.
Geog 1523.26	Baedeker, publishers. Northern Germany. 6. ed. Leipsic, 1877.
Geog 1535.33	Baedeker, publishers. Switzerland...Italy, Savoy...Tyrol. 7. ed. Leipsic, 1877.
Geog 808.77	Brown, R. The countries of the world. London, 1877-1880? 4v.
Geog 3018.76	Bruniatti, A. I progressi della generale e della geografia esploratrice in Europa. Vicenza, 1877.
Geog 5300.5	Campen, S.R. The Dutch in the arctic seas. London, 1877.
An 2108.77	Caspari, O. Die Urgeschichte der Menschheit. Leipzig, 1877. 2v.
Geog 3235.11	Cortambert, P.F.E. Tros...monuments geographiques. Paris, 1877.
Geog 4218.76.2	Curtis, B.R. Dottings round the circle. 3. ed. Boston, 1877.
Geog 4110.11F	Cuward Steamship co. Official guide and album. London, 1877.
An 2058.77	Dawkins, W.B. Address on the antiquity of man. Salford, 1877.

Chronological Listing

1877 - cont.

An 2058.77.5	Day, J.V. The prehistoric use of iron and steel. London, 1877.
Geog 4218.73.10	Dupuy de Lôme, E. De Madrid a Madrid. Madrid, 1877.
Geog 4268.77.5	Faucher, J. Vergleichende Kultur Gilder. Hannover, 1877.
Geog 4218.76.15	Fenzi, S. Gita intorno alla terra dal gennaio al settembre dell'anno 1876. Firenze, 1877.
Geog 4308.77.8	Field, Henry M. From Egypt to Japan. 13th ed. N.Y., 1877.
Geog 3108.42.5	Forbiger, A. Handbuch der alten Geographie. Hamburg, 1877. 3v.
Geog 500.25	Georg, Carl. Die Reiseliteratur Deutschlands. Leipzig, 1877.
Geog 4308.77A	Guild, C. Abroad again. Boston, 1877.
Geog 4110.19	Hall, E.H. The picturesque tourist. N.Y., 1877.
Geog 4218.71.7	Hübner, J.A. Promenade autour du monde, 1871. Paris, 1877. 2v.
Geog 578.77	Johnston, A.K. A general dictionary of geography. London, 1877.
Geog 5558.75.11	Johnston, R. The Arctic expedition of 1875-76. London, 1877.
An 2058.77.3	Jones, T.R. Lecture on the antiquity of man. London, 1877.
Geog 5558.75.13	Kan, C.M. De jongste engelsche pooltocht ende expedities der toekomst. Utrecht, 1877.
Geog 4208.30.7	Lefavor, William. Captain Lefavor's forty years' travels. Philadelphia, 1877.
Geog 3070.14	Mayer, E. Die Entwicklung der Seekarten. Wien, 1877.
Geog 1739.14	Murray, John, publisher, London. Handbook for...northern Italy. 14. ed. London, 1877.
Geog 1723.19	Murray, John, publisher, London. Handbook for north Germany. 19. ed. London, 1877.
Geog 1709.34.4	Murray, John, publisher, London. Handbook for Sussex. 4. ed. London, 1877.
Geog 1709.17.4	Murray, John, publisher, London. Handbook for travellers in Kent. 4. ed. London, 1877.
Geog 5558.75F	Nares, G.S. Arctic expedition 1875-76, journal. London, 1877.
Geog 3235.41	Odorici, F. Carte geografiche. Milano, 1877.
Geog 5398.72.5	Payer, J. New lands within the Arctic Circle. N.Y., 1877.
Geog 3018.77.2	Peschel, O. Geschichte der Erdkunde. München, 1877.
Geog 665.19	Peschel, O.F. Abhandlungen zur Erd- und Völkerkunde. v.1,2-3. Leipzig, 1877. 2v.
Geog 3248.58.5	Peschel, O.F. Geschichte des Zeitalters. Stuttgart, 1877.
Geog 818.73.3	Reclus, O. Geographie. La terre a vol d'oiseau. 3e ed. Paris, 1877. 2v.
Geog 4307.39F	Sagramoso, M.E. Lettera...al E. Ignazio Zanardi di Mantova. Verona, 1877.
Geog 4268.77F	Sherer, J. Europe...picturesque scenes. London, 1877. 2v.
Geog 4218.72.2	Simpson, William. Meeting the sun; journey all round the world. London, 1877.
Geog 5398.75	Stuxberg, A. Einringar fran Svenska Expeditionerna. Stockholm, 1877.
Geog 4238.77	Thomson, C.W. Voyage of the Challenger. The Atlantic. v.1-2, atlas. London, 1877. 3v.
An 2108.77.3	Tylor, Edward Burnett. Primitive culture. N.Y., 1877. 2v.
An 3558.77	Virchow, R. Beitraege zur physischen Anthropologie der Deutschen. Berlin, 1877.
Geog 4208.68	Vogel, H.W. Vom indischen Ocean bis zum Goldlande. Berlin, 1877.
Geog 4218.73.5	Walworth, E.H. An old world...travels around the world. N.Y., 1877.

1878

Geog 3235.31	Amat, P. Del planisferio. Roma, 1878.
An 175.1	Andree, R. Ethnographische Parallelen und Vergleiche. v.1-2. Leipzig, 1878-89.
An 2108.78.3	Argyll, G.D.C. Primeval man. N.Y., 1878.
Geog 1524.14	Baedeker, publishers. L'Allemagne, l'Autriche. 6. éd. Leipzig, 1878.
Geog 1555.21	Baedeker, publishers. Belgium and Holland. 5. ed. Leipsic, 1878.
Geog 1555.27	Baedeker, publishers. Belgium and Holland. 5. ed. Leipsic, 1878.
Geog 1555.20	Baedeker, publishers. Belgium and Holland. 5. ed. Leipsic, 1878.
Geog 1618.5	Baedeker, publishers. Egypt...lower Egypt. Leipsic, 1878.
Geog 1539.19	Baedeker, publishers. L'Italie...septentrionale. 8. éd. Leipzig, 1878.
Geog 1510.5	Baedeker, publishers. London and its environs. Leipsic, 1878. 2v.
Geog 1519.30	Baedeker, publishers. Paris and its environs. 6. ed. Leipsic, 1878.
Geog 1519.27	Baedeker, publishers. Paris et ses environs. 5. éd. Leipzig, 1878.
Geog 1525.29	Baedeker, publishers. The Rhine from Rotterdam to Constance. 6. ed. Leipsic, 1878.
Geog 1524.75	Baedeker, publishers. Südbaiern, Tirol und Salzburg. 18. Aufl. Leipzig, 1878.
Geog 4228.73	Barra de Cobo, Maipina de la. Mis impresiones y mis vicisitudes en mi viaje a Europa. Buenos Aires, 1878.
Geog 5700.9	Beauvois, E. Origines et fondation du plus ancien évêché du nouveau monde. Paris, 1878.
Geog 4238.78	Benjamin, S.G.W. Atlantic Islands. N.Y., 1878.
Geog 4218.78	Brassey, A. (Mrs.). Around the world in the "Sunbeam". N.Y., 1878.
Geog 5050.3	Chavanne, J. Die Literatur über die Polar-Regionen. Wien, 1878.
Geog 4308.69.4	Darley, F.O.C. Sketches abroad with pen and pencil. Boston, 1878.
Geog 5398.56.5	Dufferin and Ava, F.T. Blackwood. A yacht voyage; letters from high latitudes. N.Y., 1878.
Geog 818.78	Erslev, Ed. Agrip af landafraedi. Reykjavik, 1878.
Geog 4208.78	Forbes, R.L. Personal reminiscences. Boston, 1878.
Geog 5155.5	Gerritsz, H. Detectio freti Hudsoni. Amsterdam, 1878.
Geog 4308.77.5	Halsey, F.W. Two months abroad. Binghampton, 1878.
Geog 5318.1	Hendrik, Hans. Memoirs of Hans Hendrik, the Arctic traveller. London, 1878.
Geog 5170.5	Howgate, H.W. Polar colonization. Washington, 1878.
Geog 4308.74.5	Hutton, William. Twelve thousand miles over land and sea. Philadelphia, 1878.
Geog 3108.78	Kiepert, J.S.H. Lehrbuch der alten Geographie. Berlin, 1878.
Geog 4308.78	King, H. Sketches of travel...Europe. Washington, 1878.
Geog 4208.52.7	Lamson, Joseph. Round Cape Horn...in 1852. Bangor, 1878.

1878 - cont.

Geog 4303.99.2	Lannoy, Guillebert de. Oeuvres. Louvain, 1878.
Geog 5558.75.7	Markham, A.H. The great frozen sea...1875-76. London, 1878.
Geog 4180.57	Markham, C.R. The Hawkins voyages. London, 1878.
An 358.78.3A	Morgan, L.H. Ancient society. N.Y., 1878.
Geog 5558.75.15PF	Moss, Edward L. Shores of the Polar Sea. London, 1878.
Geog 1820.2	Murray, John, publisher, London. Handbook for...Algeria and Tunis. 2. ed. London, 1878.
Geog 1709.24	Murray, John, publisher, London. Handbook for...Northamptonshire. London, 1878.
Geog 1742.8	Murray, John, publisher, London. Handbook for...southern Italy. 8. ed. London, 1878.
Geog 1748.5	Murray, John, publisher, London. Handbook for...Spain. 5. ed. London, 1878.
Geog 1708.1	Murray, John, publisher, London. Handbook for England and Wales. London, 1878.
Geog 1713.4	Murray, John, publisher, London. Handbook for travellers in Ireland. 4. ed. London, 1878.
Geog 1803.4	Murray, John, publisher, London. Handbook for Turkey. 4. ed. London, 1878.
Geog 5558.75.3	Nares, G.S. Narrative of voyage to Polar Sea...1875-76. London, 1878. 2v.
An 358.78	Oberländer, R. Der Mensch vormals und heute. Leipzig, 1878.
Geog 5398.72.7	Payer, J. L'expédition du Tegetthoff. Paris, 1878.
Geog 4308.78.10	Prentis, N.L. A Kansan abroad. Topeka, Kansas, 1878.
An 3308.78.2	Roberts, C. A manual of anthropometry. London, 1878. 2 pam.
An 2058.78.3	Southall, J.C. Address on man's age in the world. Richmond, 1878.
An 2058.78.5	Southall, J.C. The epoch of the Mammoth. Philadelphia, 1878.
Geog 4238.73	Thomson, C.W. The Atlantic. N.Y., 1878. 2v.
An 358.78.5	Topinard, P. Anthropology. London, 1878.
An 608.78	Victoria, Australia. Public Library, Museum and National Galery. Catalogue of the objects of ethnotypical art. Melbourne, 1878.
Geog 4478.78	Voyage to the North Pole. Cambridge, 1878.
Geog 4208.78.7	Wakeman, Edgar. The log of an ancient marine. San Francisco, 1878.
Geog 4218.78.15	Wernich, Agathon. Geographisch-medicinische Studien. Berlin, 1878.
An 2058.78	Winchell, A. Adamites and preadamites. Syracuse, 1878.
Geog 4332.7F	Yriarte, C. Les bords de l'Adriatique. Paris, 1878.
An 2108.78	Zaborowski-Moindron, S. L'homme préhistorique. Paris, 1878.

1879

Geog 4558.79	Adams, Robert C. On board the "Rocket". Boston, 1879.
Geog 4218.27	Allyn, Gurdon L. The old sailor's story. Norwich, Conn., 1879.
Geog 1533.4	Baedeker, publishers. Eastern Alps. 4. ed. Leipsic, 1879.
Geog 1538.30	Baedeker, publishers. Italien. 9. Aufl. Leipzig, 1879-89. 2v.
Geog 1539.21	Baedeker, publishers. Italy...northern Italy. 5. ed. Leipsic, 1879.
Geog 1540.18	Baedeker, publishers. Italy. Pt. 2: Central Italy and Rome. 6. ed. Leipsic, 1879.
Geog 1510.8	Baedeker, publishers. London and its environs. 2. ed. Leipsic, 1879.
Geog 1568.5	Baedeker, publishers. Norway and Sweden. Leipzig, 1879.
Geog 1524.15	Baedeker, publishers. Süd-Deutschland und Oesterrreich. 8. Aufl. Leipzig, 1879.
Geog 1535.35	Baedeker, publishers. Switzerland...Italy, Savoy...Tyrol. 8. ed. Leipsic, 1879.
Geog 758.79	Bevan, W.L. Students' manual of modern geography mathematical, physical and descriptive. London, 1879.
An 3308.79	Bowditch, J.P. The growth of children. Boston, 1879.
Geog 4218.78.5	Brassey, A. (Mrs.). Around the world in the "Sunbeam". N.Y., 1879.
Geog 4218.78.3	Brassey, A. (Mrs.). A voyage in the "Sunbeam". London, 1879.
Geog 212.110.10	Brunialti, A. Relazioni sulla fondazione e sull'ordinamento della sezione de geografia commerciale della Societa italiana. Roma, 1879.
Geog 4678.75F	Campisi, Orazio. Petition and claim for damages to the ship Immacolata. n.p., 1879.
Geog 4218.78.11	Carnegie, Andrew. Notes of a trip round the world. N.Y., 1879.
Geog 4418.79	Farnham, J.M.W. Homeward or travels. Schenectady, 1879.
An 358.79.2	Farrer, J.A. Primitive manners and customs. London, 1879.
Geog 5700.7	Fenger, H.M. Bidrag til Hans Egede...missions. Kjobenhavn, 1879.
An 2108.79.5	Greenwood, James. The wild man at home, or Pictures of life in savage lands. London, 1879.
Geog 4218.71.9F	Hübner, J.A. Passeggiata intorno al mondo. Milano, 1879.
Geog 4328.79	Jeréz Perchét, Augusto. El mediterráneo. 3a ed. Madrid, 1879.
Geog 3235.47	Jomard, E.F. Introduction à l'atlas des monuments de la géographie. Paris, 1879.
Geog 3018.79.5	Jurien de la Gravière, J.B.E. Les marins du XVe et du XVIe siècle. Paris, 1879. 2v.
Geog 5324.1.2	Leslie, A. Arctic voyages of Adolf Erik Nordenskiöld. London, 1879.
Geog 4418.79.5	Ludwig, Salvator. Die Karawanen-Strasse von Aegypten nach Syrien. Prag, 1879.
Geog 5208.79	Markham, A.H. Northward ho! London, 1879.
Geog 4238.79	Moulton, Francis D. At sea in the Celtic from January 23rd to February 11th, 1879. n.p., 1879.
An 358.79.5	Müller, F. Allgemeine Ethnographie. Wien, 1879.
Geog 1709.29.3	Murray, John, publisher, London. Handbook for Shropshire and Cheshire. London, 1879.
Geog 1719.10	Murray, John, publisher, London. Handbook for visitors to Paris. 10. ed. London, 1879.
Geog 1809.55	Murray, John, publisher, London. Handbook of the Madras presidency. London, 1879.
Geog 1710.35	Murray, John, publishers, London. Handbook to London. London, 1879.
Geog 5558.64	Nourse, J.E. Narrative of 2nd Arctic expedition. Washington, 1879.
An 2058.79	Piette, Edouard. Nomenclature des temps anthropiques primitifs. Laon, 1879.
An 358.79	Quatrefages de Bréau, A. de. The human species. N.Y., 1879.

Chronological Listing

1879 - cont.

	Geog 4180.58A	Schiltberger, J. Bondage and travels of J. Schiltberger. London, 1879.
	Geog 3018.79	Verne, Jules. Exploration of the world. N.Y., 1879-81. 3v.
	Geog 3018.79.3	Verne, Jules. Jardens op da gelseshistorie. Kristiania, 1879-83.
	Geog 578.79F	Vivien de St. Martin, L. Nouveau dictionnaire de géographie universelle contenant...la géographie physique. Paris, 1879-95. 7v.
	Geog 578.79.3F	Vivien de St. Martin, L. Nouveau dictionnaire de géographie universelle contenant...la géographie physique. Paris, 1879-95. 7v.
NEDL	Geog 4218.79	Wylie, A.H. Chatty letters from the East and West. London, 1879.
	Geog 5398.75.3	Young, Allen. The two voyages of the Pandora in 1875-76. London, 1879.

188-

	Geog 4228.79	MacMichael, M. A landlubber's log of a voyage round the "Horn". Philadelphia, 188-?
	Geog 5378.80.9	Nordenskiöld, A.E. Lettres...du passage du pôle nord. Paris, 188-.
	Geog 4300.33.3	Palmer, J.E. Palmer's European pocket guide. N.Y., 188-.
	Geog 5398.20.25	Wrangell, F. von. Le nord de la Sibérie. Limoges, 188-?

1880

	Geog 4180.60	Acosta, J. de. National and moral history of the Indies. v.2. London, 1880.
	An 358.80	Antropologia. Lisboa, 1880.
	Geog 1539.22	Baedeker, publishers. L'Italie...septentrionale. 9. éd. Leipzig, 1880.
	Geog 1525.30	Baedeker, publishers. The Rhine from Rotterdam to Constance. 7. ed. Leipsic, 1880.
	Geog 1524.30A	Baedeker, publishers. Southern Germany and Austria. 4. ed. Leipsic, 1880.
	Geog 3235.27F	Berchet, G. Il planisfero di Giovanni Leardo. Venezia, 1880.
	Geog 4208.80.5	Boyle, Frederick. Chronioles of no-man's land. London, 1880.
	Geog 4208.78.9	Browne, Ernest A. Round the north hemisphere. London, 1880.
	Geog 3228.80.5	Bullo, C. La vera patria di Nicolò de Conti e di Giovanni Caboto. Chioggia, 1880.
	Geog 4308.80	Bush, R. Personal impressions...of European travel. Photoreproduction. N.Y., 1880?
	Geog 818.80	Cañas Pinochet, A. El estudio de la jeografía por el dibujo de las cartas jeográficas. Santiago de Chile, 1880.
	Geog 5938.80	Cardon, F. Le regioni polari antartiche. Roma, 1880.
	An 3538.80	Carr, L. Notes on the Crania of New England Indians. Boston, 1880.
	An 3538.80.2	Carr, L. Notes on the Crania of New England Indians. Boston, 1880.
	Geog 622.1	Ciampi, Ignatius. Della vita e...opere di Pietro della Valle il Pellegrino. Roma, 1880.
	An 2108.80.3	Clodd, Edward. Childhood of the world. London, 1880.
	An 358.78.7	Congrès International des Sciences Anthropologiques, Paris, 1878. Comptes rendus sténographiques. Paris, 1880.
	Geog 4180.59	Davis, J. Voyages and works of J. Davis. London, 1880.
	An 2108.80	Geiger, E.L. Contribution to the history of the development of the human race. London, 1880.
	Geog 3228.80	Gravier, S. La cosmographie avant la decouverte de l'Amerique. Paris, 1880.
Htn	Geog 4558.26F*	Hunnewell, J. Journal of voyage of "Missionary Packet". Charlestown, 1880.
	Geog 5340.5F	In mezzo ai ghiacci. Milano, 1880.
	Geog 818.80.5	Johnston, Alexander K. A physical, historical, political, and descriptive geography. London, 1880.
	Geog 578.55.9	Lippincott, J.B. and Co. Lippincott's gazetteer of the world. Philadelphia, 1880.
	Geog 4308.66.10	Macgregor, J. Voyage alone in the yawl "Rob Roy". London, 1880.
	Geog 4180.59.3	Maps (1600). Map of the world A.D. 1600. London, 1880.
	Geog 1709.18.3	Murray, John, publisher, London. Handbook for Lancashire. London, 1880.
	Geog 1770.7	Murray, John, publisher, London. Handbook for Norway. 7. ed. London, 1880.
	Geog 5378.80.13	Nordenskiöld, A.E. Shvedskaia poliarnaia ekspeditsiia, 1878-79 g. Sankt Peterburg, 1880.
	Geog 5378.80.5	Nordenskiöld, A.E. Vegas färd rking Asien och Europa. Stockholm, 1880-81. 2v.
	An 3208.80	Otis, George A. List of the specimens in the anatomical section of the United States Army Medical Museum. Washington, 1880.
	Geog 4218.80.7	Packard, J.F. Grant's tour around the world. Cincinnati, 1880.
	Geog 613.1.13	Raemdonck, J. van. Relations commerciales entre G. Mercator et C. Plantin. Anvers, 1880.
	Geog 5208.80	Rezzadore, P. I viaggi polari. Roma, 1880.
	Geog 4308.78.15	Stokes, F.A. College tramps. N.Y., 1880.
	Geog 5368.80	Stuxberg, A. Nordostpassagens historia. Stockholm, 1880.
	Geog 4218.80.5	Vetromile, E. A tour in both hemispheres. N.Y., 1880.
	Geog 4218.66	Weppner, M. The North Star and the Southern Cross. 3. American ed. Albany, 1880.
	An 2058.80	Winchell, A. Preadamites. Chicago, 1880.

1881

	Geog 4180.64	Alvares, F. Narrative of Portuguese Embassy. London, 1881.
	Geog 1555.30	Baedeker, publishers. Belgium and Holland. 6. ed. Leipsic, 1881.
	Geog 1540.18.25	Baedeker, publishers. Italy. Pt. 2: Central Italy and Rome. 7. ed. Leipsic, 1881.
	Geog 1510.10	Baedeker, publishers. London and its environs. 3. ed. Leipsic, 1881.
	Geog 1523.30	Baedeker, publishers. Northern Germany. 7. ed. Leipsic, 1881.
	Geog 1519.33.2	Baedeker, publishers. Paris and its invirons. 7. ed. Leipsic, 1881.
	Geog 1519.33	Baedeker, publishers. Paris et ses environs. 7. éd. Leipzig, 1881.
	Geog 1535.40	Baedeker, publishers. Switzerland...Italy, Savoy...Tyrol. 9. ed. Leipsic, 1881.
	An 98.81	Bastian, A. Vorgeschichte der Ethnologie. Berlin, 1881.

1881 - cont.

	An 358.81.7	Bastian, W.A. Der Völkergedanke. Berlin, 1881.
	Geog 5530.7	Beesly, A.H. Sir John Frankin. N.Y., 1881.
	Geog 5530.13	Bell, Benjamin. Lieut. John Irving, of HMS "Terror". Edinburgh, 1881.
	An 3558.81	Bessel Hagen, F.K. Zur Kritik und Verbesserung der Winkelmessungen am Kopfe. Königsberg, 1881.
	Geog 4218.78.8	Brassey, A. (Mrs.). A voyage in the "Sunbeam". Chicago, 1881.
	Geog 4308.81.5	Butterworth, Hezekiah. Zigzag journeys in classic lands. Boston, 1881.
	Geog 4308.81	Colton, G. A Maryland editor abroad. Annapolis, 1881.
	Geog 658.81	Cora, G. Cenni intorno all'attuale. Torino, 1881.
	Geog 4348.81	Cuyler, T.L. From the Nile to Norway and homeward. N.Y., 1881.
	An 2058.81	Evans, J. A few words on tertiary man. n.p., 1881.
	An 358.87.5	Featherman, A. Social history of the races of mankind. v.1-5. Boston, 1881-91. 7v.
	An 358.87.6	Featherman, A. Social history of the races of mankind. v.2,5. London, 1881-91. 2v.
	Geog 623.1	Gebhard, J.F. Het leven van Mr. Nicolaas C. Witsen. Utrecht, 1881. 3v.
	Geog 5538.78.5	Gilder, W.H. Schwatka's search...Franklin records. N.Y., 1881.
	An 358.81	Girard de Rialle, J. Les peuples de l'Asie et de l'Europe. Paris, 1881.
	Geog 4218.81.5	Glass, C. The world, round it and over it. Toronto, 1881.
	An 608.81	Hamburg, Germany. Museum. Die ethnographisch-anthropologische AbtheilungMZ Godeffroy. Hamburg, 1881.
	Geog 5208.81	Hellwald, F. von. Im ewigen Eis. Stuttgart, 1881.
	Geog 5208.81.5	Hellwald, F. von. Voblasti vechnago l'da. Sankt Peterburg, 1881.
	Geog 4218.76.3	Hendrix, E.R. Around the world. St. Louis, 1881.
	An 358.81.9	Hovelacque, Abel. Les débuts de l'humanité. Paris, 1881.
	Geog 5378.80.4	Hovgaard, A.P. Nordenskiöld's Rejse omkring Asien og Europa. Kjøbenhavn, 1881.
	Geog 500.27A	Jackson, J. Liste provisoire de bibliographies géographiques. Paris, 1881.
	Geog 500.27.2A	Jackson, J. Liste provisoire de bibliographies géographiques. Paris, 1881.
	Geog 818.80.6	Johnston, Alexander K. A physical, historical, political, and descriptive geography. 2nd ed. London, 1881.
	Geog 5538.53.3	Kane, E.K. Arctic explorations. Hartford, 1881.
	Geog 5538.78	Klutschak, H.W. Als Eskimo unter den Eskimos; eine Schilderung der Erlebnisse der Schwatka'schen Franklin-Aufsuchungs-Expedition in den Jahren 1878-80. Wien, 1881.
	Geog 4128.80.3	Knox, Thomas W. How to travel. N.Y., 1881.
	An 358.81.5	Lesley, J.P. Man's origin and destiny. Boston, 1881.
	Geog 3018.81	Low, C.R. Maritime discovery. London, 1881. 2v.
	Geog 4218.81	Ludwig, S. Um die Welt ohne zu wollen. Prag, 1881.
	Geog 5398.79	Markham, A.H. Polar reconnaissance. London, 1881.
	Geog 5398.80	Markham, C.R. Voyage of the 'Eira'. London, 1881.
	Geog 4308.76	Moulton, Louise. Random rambles. Boston, 1881.
	Geog 1724.14	Murray, John, publisher, London. Handbook for...south Germany and Austria. 14. ed. London, 1881.
	Geog 1809.30	Murray, John, publisher, London. Handbook of the Bombay presidency. London, 1881.
	Geog 5378.80	Nordenskiöld, A.E. Voyage of the Vega round Asia and Europe. London, 1881. 2v.
	Geog 4218.78.9	Parry, S.H. Jones. My journey round the world. London, 1881. 2v.
	Geog 3235.37	Pezzana, A. Estratto di una nota pesta a.f. 365-66...la carta nautica. Berlin, 1881.
	Geog 4508.81	Reclus, Elisee. The history of a mountain. N.Y., 1881.
	Geog 3248.81	Ruge, Sophus. Geschichte des Zeitalters der Entdeckungen. Berlin, 1881.
	An 2058.81.5	Sunderland, J.T. Dr. Winchell's "preadamites". n.p., 1881.
	An 358.81.4	Tylor, Edward B. Anthropology, introduction to the study of man and civilization. N.Y., 1881.

1882

	An 170.111	Anthropological Society of Washington. The Saturday lectures. Washington, D.C., 1882.
	An 2108.82.5	Avebury, J.L. The origin of civilisation and the primitive condion of man. 4. ed. London, 1882.
	An 2108.82	Avebury, J.L. The origin of civilisation and the primitive condition of man. London, 1882.
	An 2108.82.2	Avebury, J.L. The origin of civilisation and the primitive condition of man. 4. ed. London, 1882.
	Geog 1525.37	Baedeker, publishers. Les bords du Rhin. 12. éd. Leipzig, 1882.
	Geog 1539.23	Baedeker, publishers. Italy...northern Italy. 6. ed. Leipsic, 1882.
	Geog 1525.35	Baedeker, publishers. The Rhine from Rotterdam to Constance. 8. ed. Leipsic, 1882.
	Geog 1525.35.3	Baedeker, publishers. The Rhine from Rotterdam to Constance. 8. ed. Leipsic, 1882.
	Geog 1524.32	Baedeker, publishers. Süd-Deutschland und Oesterreich. 19. Aufl. Leipzig, 1882.
	Geog 4218.81.9	Bennett, D.M. A truth seeker around the world. N.Y., 1882. 4v.
	An 358.82.3	Bertillon, A. Les races sauvages. Paris, 1882.
	Geog 5398.79.25	Bliss, R.W. Our lost explorers; a narrative of the Jeannette Arctic expedition. Hartford, 1882.
	Geog 4180.65	Butler, N. Historye of the Bermudaes. London, 1882.
	Geog 4308.81.6	Butterworth, Hezekiah. Zigzag journeys in classic lands. Boston, 1882.
	Geog 3248.82	Cat, E. Les grands decouvertes maritimes. Paris, 1882.
	Geog 4218.80	Coop, Timothy. A trip around the world. Cincinnati, 1882.
	Geog 4208.82.6	Coote, W. Wanderings, south and east. London, 1882.
	Geog 5398.79.5	Danenhower. Narrative of the "Jeannette". Boston, 1882.
	Geog 3240.17	Devic, L. Marcel. Coup d'oeil sur la litterature géographique arabe au moyen âge. Paris, 1882.
	Geog 3055.39.10	Donnelly, Ignatius. Atlantis, the antediluvian world. 7th ed. N.Y., 1882.
	Geog 3018.82.5	Dussieux, L.E. Les grands faits de l'histoire de géographie. Paris, 1882-83. 5v.
	Geog 4138.82.7	Embacher, F. Lexikon der Reisen und Entdeckungen. Leipzig, 1882.
	Geog 3550.6	Fischer, Theobald. Über italienischen Seekarten und Kartographen des Mittelalters. Berlin, 1882.
Htn	Geog 4208.78.3*	Forbes, R.B. Personal reminiscences. 2. ed. Boston, 1882.
X Cg	Geog 4208.78.2	Forbes, R.L. Personal reminiscences. 2. ed. Boston, 1882.

Chronological Listing

1882 - cont.

Geog 4311.5	Götz, W. Das Donaugebeit mit Rücksicht. Stuttgart, 1882.
Geog 818.82.5	Guthe, H. Lehrbuch der Geographie. Hannover, 1882-1883. 2v.
Geog 3108.82	Hahn, H. Leitfaden der alten Geographie. Leipzig, 1882.
Geog 4308.82.5	Hale, E.E. A family flight thro' France, Germany, Norway and Switzerland. Boston, 1882.
Geog 520.1	Harrisse, H. Jean et Sébastien Cabot. Paris, 1882.
An 358.82.5	Hellwald, Friedrich. Naturgeschichte des Menschen. Stuttgart, 1882-85. 2v.
Geog 4208.82	Helms, L.V. Pioneering in the Far East. London, 1882.
Geog 5398.79.20	History of the adventurous voyage and terrible shipwreck of the United States steamer "Jeannette" in polar seas. N.Y., 1882.
An 358.82	Hovelacque, Abel. Les races humaines. Paris, 1882.
Geog 5378.80.3	Hovgaard, A.P. Nordenskiöld's Voyage round Asia. London, 1882.
Geog 4308.51.10	Huidskoper, A. Glimpses of Europe in 1851 and 1867-1868. Meadville, Pa., 1882.
Geog 3018.82.2	Kingsley, H. Tales of old travel. London, 1882.
Geog 3228.82	Marinelli, G. La geografia. Roma, 1882.
Htn Geog 4208.82.3*	Meyer, Hans. Blätter aus mienem Reisetagbuch 1881-1883. Leipzig, 1882.
Geog 1709.1.3	Murray, John, publisher, London. Handbook for...Berks, Bucks and Oxfordshire. 3. ed. London, 1882.
Geog 1709.5.10	Murray, John, publisher, London. Handbook for...Cornwall. 10. ed. London, 1882.
Geog 1748.6	Murray, John, publisher, London. Handbook for...Spain. 6. ed. London, 1882. 2v.
Geog 1709.37.10	Murray, John, publisher, London. Handbook for...Wiltshire, Dorsetshire and Somersetshire. 4. ed. London, 1882.
Geog 1709.39.3	Murray, John, publisher, London. Handbook for...Yorkshire. 3. ed. London, 1882.
Geog 1715.16	Murray, John, publisher, London. Handbook for travellers in France. 16. ed. London, 1882-84. 2v.
Geog 1809.5	Murray, John, publisher, London. Handbook of the Bengal presidency. London, 1882.
Geog 5398.79.9F	N.Y. Herald. The Jeannette in the Arctic regions. N.Y., 1882.
Geog 5378.80.2	Nordenskiöld, A.E. The voyage of the Vega round Asia and Europe. N.Y., 1882.
Geog 4300.8	Osgood, J.R. Osgood's pocket guide to Europe. Boston, 1882.
Geog 4308.82	Pitman, M.J. European breezes. Boston, 1882.
An 137.1	Putnam, F.W. Sketch of...Lewis H. Morgan. Boston, 1882.
An 378.82	Ratzel, F. Anthropo-Geographie. Stuttgart, 1882. 2v.
Geog 5838.82	Rink, Hinrich. Om Grønlaenderne, deres fremtid. Kjøbenhavn, 1882.
Geog 520.2	Le voyage de la Sainte Cyté de Hierusalem. Paris, 1882.
Geog 3140.7	Warren, W.F. True key to ancient cosmology. Boston, 1882.
Geog 4138.82.5	Werner, R. Berühmte Seeleute. Berlin, 1882.
Geog 5530.9	Wright, A. Versus Tennysonianos Franklini. Cantabrigiae, 1882.

1883

X Cg Geog 4308.83.3	Aldrich, T.B. From Ponkapog to Pesth. Boston, 1883.
Geog 1585.5	Baedeker, publishers. Griechenland. Leipzig, 1883.
Geog 1540.19	Baedeker, publishers. Italy. Pt. 2: Central Italy and Rome. 8. ed. Leipsic, 1883.
Geog 1540.19.5	Baedeker, publishers. Italy. Pt. 2: Central Italy and Rome. 8. ed. Leipsic, 1883.
Geog 1542.20	Baedeker, publishers. Italy. Pt. 3: Southern Italy, Sicily. 8. ed. Leipsic, 1883.
Geog 1542.21	Baedeker, publishers. Italy. Pt. 3: Southern Italy, Sicily. 8. ed. Leipsic, 1883.
Geog 1510.15	Baedeker, publishers. London and its invirons. 4. ed. Leipsic, 1883.
Geog 1524.33	Baedeker, publishers. Southern Germany and Austria. 5. ed. Leipsic, 1883.
Geog 1535.42	Baedeker, publishers. Switzerland...Italy, Savoy...Tyrol. 10. ed. Leipsic, 1883.
Geog 4348.60.10	Beluze, X. Les pérégrinations en Orient et en Occident. 21e éd. Citeaux, 1883. 2v.
Geog 3070.4	Breusing, A. Leitfaden durch das Wiegenalter. Frankfurt, 1883.
Geog 4218.83.5	Bridges, F.D. Journal of a lady's travels. London, 1883.
Geog 3018.83.10	Crotambert, R. Nouvelle histoire des voyages. Paris, 1883-84.
Geog 4558.40.12	Dana, Richard Henry. Two years before the mast. Boston, 1883.
An 2108.83	Dawson, J.W. Fossil men and their modern representatives. 2. ed. London, 1883.
Geog 5398.79.7	De Long, G.W. Voyage of the "Jeannette". Boston, 1883. 2v.
Geog 4228.40	Fitch, F.Y. Life, travels...of an American wanderer...Alonzo P. De Milt. N.Y., 1883.
Geog 3060.5	Gaffarel, P. Les isles fantastiques de l'Atlantique. n.p., 1883.
Geog 5398.79.40	Geslin, Jules. L'expédition de la Jeannette au pôle nord, racontée par tous les membres de l'expédition. Paris, 1883? 2v.
Geog 5398.81	Gilder, W.H. Ice-pack and tundra. London, 1883.
An 2058.83.5	Girard de Rialle, J. Nos ancêtres. Paris, 1883.
Geog 520.3	Harrisse, H. Les Corte-Real et leurs voyages. Paris, 1883.
Geog 520.3.5	Harrisse, H. Gaspar Corte-Real. Paris, 1883.
Geog 5538.54.8	Hayes, I.I. Arctic boat journey, in the autumn of 1854. Boston, 1883.
Geog 4218.83.10	Lambert, C. Voyage of the Wanderer. London, 1883.
Geog 578.55.11	Lippincott, J.B. and Co. Supplementary tables of population. Philadelphia, 1883.
Geog 3070.29	Mager, Henri. De la lecture des cartes étrangères. Paris, 1883.
Geog 5838.83.5	Maps. Greenland, 1380-1482. Trois cartes. Copenhagen, 1883.
Geog 808.83	Marinelli, G. La terra. Milano, 1883-1885. 7v.
An 2208.83	Maturi, R. La medicina preistorica. Napoli, 1883.
Geog 5516.19.3	Munk, J. Navigatio septentrionalis. Kjøbenhavn, 1883.
Geog 1773.6	Murray, John, publisher, London. Handbook for...Sweden. 6. ed. London, 1883.
Geog 1809.80	Murray, John, publisher, London. Handbook of the Punjab, Kashmir. London, 1883.
Geog 4218.55.9	Nordhoff, Charles. Man-of-war life. N.Y., 1883.
An 358.83F	Oberländer, R. Fremde Völker ethnographische Schilderungen aus der alten und neuen Welt. Leipzig, 1883.

1883 - cont.

Geog 4300.8.5A	Osgood, J.R. Osgood's pocket guide to Europe. Boston, 1883.
Geog 520.4	Parmentier, J. Le discours de la navigation. Paris, 1883.
Geog 5208.82.2	Perry, R. Jeannette, and a complete and authentic narrative encyclopedia of all voyages. San Francisco, 1883.
Geog 4218.83.2	Pidgeon, D. An engineer's holiday...trip from 0 degrees to 0 degrees. 2. ed. London, 1883.
An 2058.83	Rawlinson, G. The antiquity of man historically considered. N.Y., 1883.
An 5.2	Riccardi, P. Saggio di un catalogo bibliografico antropologico italiano. Modena, 1883.
Geog 5558.81.9F	Rosse, I.C. Cruise of revenue-steamer Corwin. Washington, 1883.
Geog 3018.83	Rubiner, W. Die Entdeckungsreisen. Glogau, 1883.
Geog 4332.5	Schweiger-Lerchenfeld, A.F. Die Adria. Wien, 1883.
Geog 4308.83.5	Stokes, F.A. A jolly summer. N.Y., 1883.
Geog 4308.83	Thacher, S.O. What I saw in Europe. Topeka, Kansas, 1883.
Geog 5558.69.5	Verein in die Deutsche Nordpolarfahrt. Die zweite deutsche Nordpolarfahrt. Leipzig, 1883.
Geog 4208.83.10	Wilkinson, H. Sunny lands and seas. London, 1883.

1884

Geog 1555.36	Baedeker, publishers. Belgium and Holland. 7. ed. Leipsic, 1884.
Geog 1555.35	Baedeker, publishers. Belgium and Holland. 7. ed. Leipsic, 1884.
Geog 1516.3	Baedeker, publishers. Le Nord de la France. Leipzig, 1884.
Geog 1523.36	Baedeker, publishers. Northern Germany. 8. ed. Leipsic, 1884.
Geog 1523.35	Baedeker, publishers. Northern Germany. 8. ed. Leipsic, 1884.
Geog 1530.18	Baedeker, publishers. Oesterreich-Ungarn. 20. Aufl. Leipzig, 1884.
Geog 1519.36	Baedeker, publishers. Paris and its environs. 8. ed. Leipsic, 1884.
Geog 1519.36.2	Baedeker, publishers. Paris et ses environs. 7. éd. Leipzig, 1884.
Geog 4218.84	Ballou, M.M. Due west, or Round the world. Boston, 1884.
Geog 4218.84.2	Ballou, M.M. Due west, or Round the world in ten months. 2. ed. Boston, 1884.
Geog 3025.5	Barbier, J.V. Rapport sur les travaux cartographiques. Nancy, 1884.
An 378.84	Bordier, A. La géographie médicale. Paris, 1884.
Geog 4238.83.5	Brassey, Thomas. 11,506 knots in the "Sunbeam". London, 1884.
Geog 4168.84.5	Buel, James W. The worlds wonders. St. Louis, 1884.
An 358.84.10	Fontaine, Edward. How the world was peopled. N.Y., 1884.
An 3308.84	Galton, Francis. Life history album. London, 1884.
Geog 5558.81.10	Greely, A.W. The Greely Arctic expedition. Philadelphia, 1884.
Geog 520.6	Harrisse, H. Christophe Colomb. Paris, 1884. 2v.
Geog 3055.11	Hoernes, M. Atlantis. Wien, 1884.
Geog 5398.82	Hovgaard, A.P. Dijmphna's expeditionen, 1882-83. Kjøbenhavn, 1884.
Htn Geog 4308.84*	James, H. Portraits of places. Boston, 1884.
Geog 4308.84	James, H. Portraits of places. Boston, 1884.
An 358.84.3	Lind, G.D. Man; embracing his origin. Chicago, 1884.
Geog 5070.5	M'Cormick, R. Voyages of discovery in Arctic and Antarctic. London, 1884. 2v.
Geog 5558.81.15	McGinley, W.A. Greely relief expedition. Washington, 1884.
Geog 3199.5	Manitius, M. Anonymi de situ orbis. Stuttgardiae, 1884.
Geog 3228.82.3	Marinelli, G. Die Erdkunde bei den Kirchenvätern. Leipzig, 1884.
Geog 5070.37	Moreno Fuentes, José. Las regiones heladas de los polos norte y sur. Madrid, 1884.
An 2058.84	Morse, E.S. Man in the tertiaries. Salem, 1884.
Geog 1709.12.3	Murray, John, publisher, London. Handbook for...Gloucestershire, Worcestershire. 3. ed. London, 1884.
Geog 1785.5	Murray, John, publisher, London. Handbook for...Greece. 5. ed. London, 1884. 2v.
Geog 3251.7	Nordenskiold, A.E. Om en märklig globkarta. Stockholm, 1884.
Geog 5300.9	Nourse, J.E. American explorations in the ice zones. Boston, 1884.
An 358.84	Quatrefages de Bréau, A. de. Hommes fossiles et hommes sauvages. Paris, 1884.
Geog 4558.84	Rideing, William H. Boys coastwise, or All along the shore. N.Y., 1884.
An 2108.84.5F	Römer, F. The bone caves of Ojcow in Poland. London, 1884.
Geog 3570.5.3F	Sonsa Viterbo, F.M. Trabalhos nauticos dos Portuguezes. Lisboa, 1884-1900. 2v.
Geog 4235.70.2	Sousa, F. de. Tratado das ilhas novas. Ponta Delgada, 1884.
An 3508.84	Studley, C.A. Notes upon human remains from the caves of Coahuila, Mexico. Salem, 1884.
Geog 4208.83.6	Tangye, Richard. Reminiscences of travel in Australia, America and Egypt. 2. ed. London, 1884.
Geog 520.5	Thenault, J. Le voyage d'Outremer. Paris, 1884.
Geog 5558.81.5	United States. Board of Officers. Expedition...Relief of Lieutenant Greely. Report. Washington, 1884.
Geog 5398.79.15	United States. Navy Department. Letter from the Secretary of the Navy. Washington, 1884.
An 2108.84	Van Overloop, E. Sur une méthode a suivre dans les études dites préhistoriques. Bruxelles, 1884.
Geog 4308.84.7	Warner, C.D. A roundabout journey. Boston, 1884.

1885

Geog 3580.10	Anghiera, P.M. d'. Lettres relatives aux découvertes maritimes des Espagnols. Paris, 1885.
Geog 4218.85.10	Arme-Webb, A. (Mrs.). A glimpse of two hemispheres. Hertford, 1885.
Geog 1555.38	Baedeker, pbulishers. Belgium and Holland. 8. ed. Leipsic, 1885.
Geog 1618.10A	Baedeker, publishers. Egypt...lower Egypt. 2. ed. Leipsic, 1885.
Geog 1510.20	Baedeker, publishers. London and its invirons. 5. ed. Leipsic, 1885.
Geog 1568.15	Baedeker, publishers. Norway and Sweden. 3. ed. Leipsic, 1885.

Chronological Listing

1885 - cont.

	Geog 1568.18A	Baedeker, publishers. Schweden und Norwegen. 3. Aufl. Leipzig, 1885.
	Geog 1535.44	Baedeker, publishers. La Suisse. 14. éd. Leipzig, 1885.
	Geog 1535.43	Baedeker, publishers. Switzerland...Italy, Savoy...Tyrol. 11. ed. Leipsic, 1885.
	Geog 1568.16	Baedekr, publishers. Norway and Sweden. 3. ed. Leipsic, 1885.
	Geog 4218.84.5	Ballou, M.M. Due west, or Round the world. 3. ed. Boston, 1885.
	Geog 4228.85	Beehler, W.H. The cruise of the Brooklyn. Philadelphia, 1885.
	An 3308.85	Bertillon, A. Identification anthropometrique. Melun, 1885.
NEDL	Geog 4238.83.7	Brassey, Anna. 14,000 miles in the Sunbeam in 1883. London, 1885.
NEDL	Geog 39.2.5	Club Alpino Italiano. Indice generale dei cinquanta primi numeri (dal 1865 al 1884). Torino, 1885.
	Geog 4308.85.20	Evans, M.L. Glimpses by sea and land. Philadelphia, 1885.
	Geog 4208.85.3	Forbes, Archibald. Souvenirs of some continents. London, 1885.
	Geog 4558.11	Forbes, R.B. Voyage of the Niedas, 1811. Boston, 1885.
	Geog 4208.85	Fracheboud, J.J. Monséjour en France...mes pelerinages a Rome. Fribourg, 1885.
	Geog 4308.85.9	Gjellerup, Karl. Vandreaaret. København, 1885.
	Geog 3520.10.20	Hakluyt, Richard. Principal navigations, voyages. v.1-12,14-16. Edinburgh, 1885-90. 15v.
NEDL	Geog 3520.10.20	Hakluyt, Richard. Principal navigations, voyages. v.13. Edinburgh, 1885-90.
	Geog 4308.85	Johnson, E.C. On the track of the crescent. London, 1885.
	Geog 4308.85.15	Johnston, Richard M. Two gray tourists. Baltimore, 1885.
	Geog 5312.7	Lauridsen, Peter. Vitus J. Bering og de russiske opdagelsesrejser fra 1725-43. København, 1885.
X Cg	Geog 4180.70	Linschoten, J.H. Voyage of Linschoten to East Indies. London, 1885. 2v.
	Geog 4208.83	Lucy, H.W. East by West. London, 1885. 2v.
	Geog 3140.13	Mer, A. Memoire sur le Periple d'Hannon. Paris, 1885.
	An 2338.85	Morse, E.S. Ancient and modern methods of arrow-release. Salem, 1885.
	Geog 1714.1.5	Murray, John. publisher, London. Handbook for...North Wales. 5. ed. London, 1885.
	Geog 3055.29	Nicaise, A. Les terres disparues. Chalons sur Marne, 1885.
	Geog 5378.80.7	Nordenskiöld, A.E. Studien und Forschungen...Reisen. Leipzig, 1885.
	Geog 5838.83	Nordenskiold, A.E. Den andra Dicksonska Expedition. Stockholm, 1885.
	An 358.85.3	Peschel, O.F. Völkerkunde. Leipzig, 1885.
	Geog 4206.65.5	Piolin, Paul. René Desboys du Chastelet. n.p., 1885.
	Geog 808.76.5	Reclus, J.J.E. Earth and its inhabitants - Europe. N.Y., 1885. 5v.
	Geog 4268.85	Rudler, F.W. Europe. London, 1885.
Htn	Geog 4308.85.7*	Sala, George A. A journey due south, travels. London, 1885.
	Geog 5558.81.7A	Schley, W.S. The rescue of Greely. N.Y., 1885.
	An 358.85	Schneider, W. Die Naturwölker. v.1-2. Paderborn, 1885-86.
	An 358.85.7	Spencer, Herbert. La especie humana. Madrid, 1885.
	An 2058.85.3	Topinard, Paul. Eléments d'anthropologie générale. Paris, 1885.
	Geog 4308.85.3	Tyler, I. Waymarks, or Sola in Europe. Chicago, 1885.
	Geog 110.2.3	United States. War Department. Corps of Engineers. Report of 3rd International Geographical Congress and Exhibition at Venice, Italy, 1881. Washington, 1885.
	Geog 4208.54.5	Warren, E. A doctor's experiences in three continents. Baltimore, 1885.
	Geog 4325.90.6	Webbe, Edward. Edward Webbe, chief master gunner, his travailes. Edinburgh, 1885.
Htn	Geog 4325.90.5*	Webbe, Edward. Edward Webbe, his travailes. Edinburgh, 1885.
	Geog 5398.20.12	Wrangell, F. von. Ferdinand von Wrangell und seine Reiselängs der Nordküste von Sibirien. Leipzig, 1885.

1886

	Geog 1539.25	Baedeker, publishers. Italy...northern Italy. 7. ed. Leipsic, 1886.
	Geog 1540.20	Baedeker, publishers. Italy. Pt. 2: Central Italy and Rome. 9. ed. Leipsic, 1886.
	Geog 1540.21	Baedeker, publishers. Italy. Pt. 2: Central Italy and Rome. 9. ed. Leipsic, 1886.
	Geog 1523.41	Baedeker, publishers. Northern Germany. 9. ed. Leipsic, 1886.
	Geog 1523.40	Baedeker, publishers. Northern Germany. 9. ed. Leipsic, 1886.
	Geog 1525.40A	Baedeker, publishers. The Rhine from Rotterdam to Constance. 10. ed. Leipsic, 1886.
	Geog 1524.77	Baedeker, publishers. Südbaiern, Tirol und Salzburg. 22. Aufl. Leipzig, 1886.
	Geog 4308.86.20	Bartlett, David L. Letters from Europe. Baltimore, 1886.
	Geog 4208.86	Böckmann, W. Reise nach Japan aus Briefen und Tagebüchern zusammengestellt. Berlin, 1886.
	Geog 4238.83.8F	Brassey, Anna. In the trades, the tropics, and the 'roaring forties. London, 1886.
	Geog 3251.9	Brenner, O.K. Die ächte Karte des Olaus Magnus vom Jahre 1539. Christiana, 1886.
	Geog 4308.86.10	Buckley, James M. The midnight sun; the tsar and the nihilist. Boston, 1886.
	Geog 4208.40.10	Cleveland, R.J. Voyages of a merchant navigator. N.Y., 1886.
	An 2358.86	Darling, C.W. Anthropophagy. Utica, 1886.
	Geog 4168.86	Documentos para la historia de la nautica en Chile. Santiago de Chile, 1886.
	Geog 4308.77.7	Field, Henry M. From Egypt to Japan. 13th ed. N.Y., 1886.
	Geog 3070.6	Fischer, T. Sammlung mittelalterlicher Welt. Venedig, 1886.
	Geog 5558.81.3	Greely, A.W. Three years of Arctic service...1881-84. N.Y., 1886. 2v.
	Geog 4248.86	Guillemard, F.H.H. Cruise of the Marchesa. London, 1886.
	Geog 5208.69.15	Hartwig, G. The polar world. London, 1886.
	Geog 611.2	Heinrich Kiepert bei seiner Rückkehr November 1886 von Freunden gewidmet. Berlin, 1886.
	Geog 958.98F	Hirt, F. Geographische Bildertafeln. v.1-3. Breslau, 1886.
	Geog 665.34	Hoel. Geographische Charakter-Bilder für Schule und Haus. pt.1-10. Supplement. Wien, 1886.

1886 - cont.

	Geog 4208.78.5	Holden, J.W. A wizard's wanderings from China to Peru. London, 1886.
	Geog 5070.15	Löwenberg, J. Die Entdeckungs- und Forschungsreisen im dem Leiden Polarzonen. Leipzig, 1886.
	Geog 4228.86	Molinari, M.G. Au Canada et aux montagnes. Paris, 1886.
	Geog 1735.35	Murray, John, publisher, London. Handbook for...Switzerland. 17. ed. London, 1886.
	Geog 5378.80.11	Nordenskiölds, A.E. Nordenskiölds Vegafahrt um Asien und Europa. Leipzig, 1886.
	Geog 5838.83.3	Nordenskiold, A.E. Grönland; seine Eiswüsten im Innern. Leipzig, 1886.
	Geog 818.86	Pequeña geografia. Asunción, 1886.
	Geog 500.35	Petherick, E.A. Catalogue of the York Gate Library. London, 1886.
	Geog 4308.86.5	Preston, M.J. (Mrs.) A handfull of monographs, continental and English. N.Y., 1886.
	Geog 4218.86	Raum, G.E. A tour around the world. N.Y., 1886.
	Geog 808.76.7	Reclus, J.J.E. Earth and its inhabitants - Africa. N.Y., 1886-1890. 4v.
	Geog 3235.43	Schweder, E. Über die Weltkarte des Kosmographen. Kiel, 1886.

1887

	An 3308.87	Amherst College. The anthropometric manual 1887. Amherst, 1887.
	Geog 1508.5	Baedeker, publishers. Great Britain. Leipsic, 1887.
	Geog 1542.25	Baedeker, publishers. Italy. Pt. 3: Southern Italy, Sicily. 9. ed. Leipsic, 1887.
	Geog 1510.25	Baedeker, publishers. London and its environs. 6. ed. Leipsic, 1887.
	Geog 1516.4	Baedeker, publishers. Le Nord de la France. 2. éd. Leipzig, 1887.
	Geog 1519.37	Baedeker, publishers. Paris et ses environs. 8. éd. Leipzig, 1887.
	Geog 1524.35	Baedeker, publishers. Southern Germany and Austria. 6. ed. Leipsic, 1887.
	Geog 1524.36	Baedeker, publishers. Southern Germany and Austria. 6. ed. Leipsic, 1887.
	Geog 1535.45	Baedeker, publishers. Switzerland...Italy, Savoy...Tyrol. 12. ed. Leipsic, 1887.
	An 2058.87	Burge, Lorenzo. Pre-glacial man and the Aryan race. 2. ed. Boston, 1887.
	Geog 520.8	Chesneau, J. Le voyage de M. d'Aramon. Paris, 1887.
	Geog 4168.87.2F	Colange, Leo de. Voyages and travels. Boston, 1887.
	Geog 4168.87F	Colange, Leo de. Voyages and travels. Boston, 1887. 2v.
	Geog 4308.87.7	Cust, R.H.H. How I spent my summer holidays. Eton, 1887.
	Geog 4558.31	Davis, George. Recollections of a sea-wanderer's life.
	Geog 4218.87.20	Floyd Jones. Letters from the Far East. N.Y., 1887.
	Geog 4228.87	Francis, Harriet E. Across the meridians. N.Y., 1887.
	An 348.87	Hamy, E.T. Etudes ethnographiques et archéologiques sur l'exposition coloniale. Paris, 1887.
	Geog 3235.33	Hany, E.T. La mappemonde. Paris, 1887.
	Geog 4180.74	Hedges, W. Diary of William Hedges...during agency in Bengal. London, 1887-89. 3v.
	An 3308.87.5	Hitchcock, E. Need of anthropometry. Brooklyn, 1887.
	Geog 5838.83.10	Holm, G. Den Danske Konebaads-Expedition. København, 1887.
	An 358.87.9	Hovelacque, Abel. Précis d'anthropologie. Paris, 1887.
	Geog 808.87	Kirchoff, A. Landerkunde des Erdteils Europa. v.1-3. Wien, 1887-1907. 5v.
	An 358.87.7	Lorenzini, C. Il regale istruttivo. Roma, 1887.
	Geog 4208.84.5	McCarthy, W.J. Cruise of the U.S. flagship Lancaster. n.p., 1887.
	Geog 4348.87	McKenzie, A. Some things abroad. Boston, 1887.
Htn	Geog 3235.57*	Marcel, G. Note sur une carte catalane de Dulceri. Paris, 1887.
	Geog 4308.87	Meriwether, Lee. A tramp trip. N.Y., 1887.
	Geog 4308.87.1	Meriwether, Lee. A tramp trip. 5th ed. N.Y., 1887.
	Geog 4308.87.3	Morrison, Leonard A. Rambles in Europe. Boston, 1887.
	Geog 4308.87.2	Morrison, Leonard A. Rambles in Europe. Boston, 1887.
	Geog 1709.8.10A	Murray, John, publisher, London. Handbook for...Devonshire. 10. ed. London, 1887.
	Geog 4208.41.2	Oliphant, L. Episodes in a life of adventure. Edinburgh, 1887.
	Geog 4208.41	Oliphant, L. Episodes in a life of adventure. N.Y., 1887.
	Geog 4208.41.4	Oliphant, L. Episodes in a life of adventure. 4. ed. Edinburgh, 1887.
	Geog 4208.87.10	Palgrove, William G. Ulysses; or Scenes and studies in many lands. London, 1887.
	Geog 4218.82.3	Pitt, George. The collected remarkable travels of G. Pitt. 3 ed. Glasgow, 1887.
NEDL	Geog 4180.76	Pyrard, F. Voyage of F. Pyrard...to East Indies. v.1-2. London, 1887. 3v.
	An 358.87	Quatrefages de Bréau, A. de. Histoire générale des races humaines. Paris, 1887-89. 2v.
	Geog 4110.9	Rand, McNally and Co. Pocket atlas of the world. N.Y., 1887.
	An 358.87.3	Ratzel, F. Völkerkunde. Leipzig, 1887-88. 3v.
	Geog 4558.87A	Samuels, J. From the forecastle to the cabin. N.Y., 1887.
	Geog 5558.81.8F	Schley, W.S. Report...Greely relief expedition of 1884. Washington, 1887.
	Geog 4314.8	Schweiger-Lerchenfeld, Amand von. Zwischen Donau und Kaukasus. Wien, 1887.
	Geog 4110.23	Sherriff's illustrated route charts and travellers' hand book. v.1-4. London, 1887.
	Geog 4218.87	Stevens, Thomas. Around the world on a bicycle. N.Y., 1887-88. 2v.
	An 5.25	Vasconcellos-Abreu, Guilherme de. O critério nomolójico. Photoreproduction. Lisboa, 1887.
	Geog 4206.65.11	Les voyages de Balthasar de Monconys. Paris, 1887.
	Geog 4678.87	Wreck of the Rainier; a sailors narrative. Portland, 1887.

1888

	Geog 3140.16	Antichan, P.N. Grands voyages de découvertes des anciens. Paris, 1888.
	Geog 4418.03.9	Badia y Leblich. Viatjes de Ali Bey el Abbassi per Africa y Asia. v.1-3. Barcelona, 1888.
	Geog 1524.38	Baedeker, publishers. Allemagne du sud et Autriche. 9. éd. Leipzig, 1888.
	Geog 1555.43	Baedeker, publishers. Belgien und Holland. 18. Aufl. Leipzig, 1888.

Chronological Listing

1888 - cont.

	Geog 1555.40	Baedeker, publishers. Belgium and Holland. 9. ed. Leipsic, 1888.
	Geog 1533.6	Baedeker, publishers. Eastern Alps. 6. ed. Leipsic, 1888.
	Geog 1519.39	Baedeker, publishers. Paris and its environs. 9. ed. Leipsic, 1888.
	Geog 1524.19	Baedeker, publishers. Süd-Deutschland. 22. Aufl. Leipzig, 1888.
	An 358.71.3	Broca, Paul. Mémoires d'anthropologie. v.5. Paris, 1888.
	An 2108.88	Debierre, C. L'homme avant l'histoire. Paris, 1888.
	Geog 4208.88	Howells, W.D. Library of universal adventure by sea and land. N.Y., 1888.
	Geog 4208.88.5	Keane, J.F. Three years of a wanderer's life. London, 1888.
	Geog 4308.88	Kingston, W.B. A wanderer's notes. London, 1888. 2v
	Geog 4128.80.5	Knox, Thomas W. How to travel. N.Y., 1888.
	Geog 4268.86.4	Lanier, L. L'Europe...choix de lectures de geographie. 4. ed. Paris, 1888.
	Geog 4218.88	Lewis, I.N. Pleasant hours in sunny lands. Boston, 1888.
	Geog 4218.85.5	Mitford, R.C.W.M. Orient and Occident. London, 1888.
	An 358.88.5	Morris, Charles. The Aryan race: its origin and its achievements. Chicago, 1888.
	Geog 1816.19	Murray, John, publisher, London. Handbook for...Egypt. 7. ed. London, 1888.
	Geog 1709.33.4	Murray, John, publisher, London. Handbook for...Surrey, Hampshire. 4. ed. London, 1888.
NEDL	An 358.88.3	Quatrefages de Bréau, A. de. The human species. N.Y., 1888.
	Geog 4218.85	Richardson, D.N. A girdle round the earth. Chicago, 1888.
	Geog 3018.88	Ruge, S. Abhandlungen...zur Geschichte der Erdkunde. Dresden, 1888.
	Geog 4558.88	Spear, P.S. The old sailor's story of his life. Portland, 1888.
	Geog 665.7	Strachey, R. Lectures on geography. London, 1888.
	An 358.88	Tylor, Edward B. Anthropology. N.Y., 1888.
	Geog 5558.81F	Unites States. Expedition to Lady Franklin Bay. Report on proceedings. Washington, 1888. 2v.
	Geog 520.9	Varthema, L. di. Les voyages de L. di Varthema. Paris, 1888.

1889

	An 98.89	Achelis, Thomas. Die Entwickelung der modernen Ethnologie. Berlin, 1889.
	Geog 4558.89	Aldrich, H.L. Arctic Alaska and Siberia. Chicago, 1889.
	An 3308.89	Amherst College. Anthropometric manual 1889. Amherst, 1889.
	An 3308.89.3	Anuchin, D.N. O geograficheskom' raspred'nenii rosta muzhskago naseleniia Rossii. Sankt Peturburg, 1889.
	Geog 1585.10	Baedeker, publishers. Greece. Leipsic, 1889.
	Geog 1539.11	Baedeker, publishers. L'Italie...septentrionale. 12. éd. Leipzig, 1889.
	Geog 1539.30	Baedeker, publishers. Italy...northern Italy. 8. ed. Leipsic, 1889.
	Geog 1510.30	Baedeker, publishers. London and its environs. 7. ed. Leipsic, 1889.
	Geog 1516.5	Baedeker, publishers. Northern France. Leipsic, 1889.
	Geog 1519.42	Baedeker, publishers. Paris et ses environs. Leipzig, 1889.
	Geog 1525.41	Baedeker, publishers. Die Rheinlande von der Schweize zu holländisch Grenze. 28. Aufl. Leipzig, 1889.
	Geog 1525.41.3	Baedeker, publishers. The Rhine from Rotterdam to Constance. 11. ed. Leipsic, 1889.
	Geog 1535.46	Baedeker, publishers. Switzerland...Italy, Savoy...Tyrol. 13. ed. Leipsic, 1889.
	Geog 1535.47	Baedeker, publishers. Switzerland...Italy, Savoy...Tyrol. 13. ed. Leipsic, 1889.
	Geog 4218.84.14	Ballou, M.M. Foot-prints of travel. Boston, 1889.
	Geog 4218.84.13	Ballou, M.M. Foot-prints of travel. Boston, 1889.
	Geog 4228.89	Barneby, W.H. New Far West and old Far East. London, 1889.
	Geog 4418.89A	Brassey, A. The last voyage. London, 1889.
X Cg	An 358.89.3	Buel, J.W. The story of man. Philadelphia, 1889.
	Geog 4218.87.12	Cecil, Evelyn. Notes of my journey round the world. London, 1889.
	Geog 4208.87.15	Chapin, J.H. From Japan to Granada. N.Y., 1889.
	Geog 4308.89	Child, T. Summer holidays...Europe. N.Y., 1889.
	Geog 5538.50.15	Collinson, R. Journal of H.M.S. Enterprise...1850-55. London, 1889.
	Geog 4228.89.5	Faucher de Sant-Maurice, N.H.E. Loin du Pays. Souvenirs d'Europe, d'Afrique, et d'Amérique. Quebec, 1889. 2v.
	Geog 4508.89	Fiorio, I pericoli dell'Alpinismo e noime per vitarli. Torino, 1889.
	Geog 5938.89	Fonville, W. de. Le pôle sud. Paris, 1889.
	Geog 4678.89	Forbes, Robert B. Notes on some few of the wrecks and rescues during the present century. Boston, 1889.
	Geog 3560.5	Great Britain. India Office. Map of the world, commonly known as the second Borgian map. n.p., 1889.
	Geog 4248.86.5	Guillemard, F.H.H. Cruise of the Marchesa to Kamschatka and New Guinea. 2nd ed. London, 1889.
	Geog 4180.79	Hues, R. Tractatus de globis et eorum usu. London, 1889.
	Geog 3590.72	Kaulbars, N.A. Aperçu des travaux géographiques en Russie. St. Pétersbourg, 1889.
	An 2108.89	Keary, C.F. The dawn of history. N.Y., 1889.
	Geog 5312.7.5	Lauridsen, Peter. Vitus Bering; the discoverer of Bering Strait. Chicago, 1889.
	Geog 1740.11	Murray, John, publisher, London. Handbook for...central Italy. 11. ed. pt.2. London, 1889.
	Geog 1709.6.25	Murray, John, publisher, London. Handbook to the English lakes...Cumberland, Westmorland, and Lancashire. London, 1889.
	Geog 4308.85.5	Ninde, Mary L. We two-alone in Europe. Chicago, 1889.
	Geog 199.1.18	Schweizer Alpenclub. Die ersten 25 Jahre des Schweizer Alpenclub. Glarus, 1889.
	Geog 5208.89	Smith, C.C. Arctic explorations in 18th and 19th centuries. Boston, 1889.
Htn	Geog 4308.89.10*	Stockton, F.R. Personally conducted. N.Y., 1889.
	Geog 4168.89	Travel, adventure and sport. v.2-4. N.Y., 1889.
	An 358.89	Tylor, Edward B. Anthropology. N.Y., 1889.
	Geog 4308.89.3	Walker, B. Aboard and abroad. Lowell, 1889.
	Geog 4218.75	Wieting, M.E. Prominent incidents in life of John M. Wieting...around the world. N.Y., 1889.

189-

	Geog 4268.73.4	Levasseur, E. Precis de la geographie...de l'Europe. v.1, atlas. Paris, 189-. 2v.
	Geog 958.90F	La panorama, merveilles de France, Belgique. Paris, 189-?
	Geog 958.92.5F	Stoddard, John L. Portfolio of photographs of famous scenes, cities and paintings. Chicago, 189-?
	Geog 4268.89.5	Vidal de la Blache, Paul. Etats et nations de l'Europe, autour de la France. 4. éd. Paris, 189-?

1890

	Geog 4558.90.5	Amicis, E. de. Sull'oceano. Milano, 1890.
	Geog 1508.10	Baedeker, publishers. Great Britain. 2. ed. Leipsic, 1890.
	Geog 1508.11	Baedeker, publishers. Great Britain. 2. ed. Leipsic, 1890.
	Geog 1540.25	Baedeker, publishers. Italy. Pt. 2: Central Italy and Rome. 10. ed. Leipsic, 1890.
	Geog 1540.26	Baedeker, publishers. Italy. Pt. 2: Central Italy and Rome. 10. ed. Leipsic, 1890.
	Geog 1542.27	Baedeker, publishers. Italy. Pt. 3: Southern Italy, Sicily. 10. ed. Leipsic, 1890.
	Geog 1523.45	Baedeker, publishers. Northern Germany. 10. ed. Leipsic, 1890.
	Geog 1523.46	Baedeker, publishers. Northern Germany. 10. ed. Leipsic, 1890.
	Geog 1524.19.2	Baedeker, publishers. Süd-Deutschland. 23. Aufl. Leipzig, 1890.
	An 358.90	Brinton, D.G. Races and peoples. N.Y., 1890.
	Geog 5838.90	Carstensen, A.R. Two summers in Greenland. London, 1890.
	Geog 5398.90	Clutterbuck, W.J. The skipper in Arctic seas. London, 1890.
	Geog 4218.90.6	Cotes, Sara J.D. A social departure. London, 1890.
	Geog 4218.90.5	Cotes, Sara J.D. A social departure. N.Y., 1890.
	Geog 5557.25	Dall, W.H. Critical review of Bering's first expedition, 1725-30. Washington, 1890.
	Geog 808.90F	DePuy, W.H. The universal guide and gazetteer. N.Y., 1890.
	Geog 4218.38	Erskine, C. Twenty years before the mast. Boston, 1890.
	Geog 755.27.10	Fritzsche, O.F. Glarean sein Leben und seine Schriften. Frauenfeld, 1890.
	Geog 606.1	Gallois, L. De Crontio finaeo gallico geographo. Paris, 1890.
	Geog 3540.7	Gallois, Lucien. Les géographes allemands de la renaissance. Paris, 1890.
	Geog 4168.90	Griswold, W.M. Travel, series of narratives of...visits. v.1-106. Cambridge, 1890. 2v.
	Geog 758.90	Günther, S. Handbuch der Mathematischen Geographie. Stuttgart, 1890.
	Geog 3520.5	Jurien de la Gravière, J.B.E. Les Anglais et les Hollandais. Paris, 1890. 2v.
	Geog 4218.90	Leland, Lillian. Travelling alone, a woman's journey. N.Y., 1890.
	Geog 4110.13	Loftie, W.J. Orient line guide. London, 1890.
	Geog 4218.87.7	M'Collester, S.H. Round the globe. 3. ed. Boston, 1890.
	Geog 4358.90	Maupassant, G. de. La vie errante. 8. éd. Paris, 1890.
	Geog 1714.2.4	Murray, John, publisher, London. Handbook for...South Wales. 4. ed. London, 1890.
	Geog 1748.7	Murray, John, publisher, London. Handbook for...Spain. 7. ed. London, 1890. 2v.
	Geog 1708.15	Murray, John, publisher, London. Handbook for England and Wales. 2. ed. London, 1890.
	Geog 1709.20	Murray, John, publisher, London. Handbook for Lincolnshire. London, 1890.
	Geog 5838.88.3A	Nansen, F. The first crossing of Greenland. London, 1890. 2v.
	Geog 5838.88	Nansen, F. Paa ski over Grønland...1888-90. Kristiania, 1890.
	Geog 4110.20	Peninsular and Oriental Steam Navigation Company. Pocket book, 1890. London, 1890.
	An 358.90.3	Peschel, O.F. The races of man and their geographical distribution. N.Y., 1890.
	Geog 4208.90	Platen, Carl von. Impressions of travels. Stockholm, 1890.
	Geog 520.11	Possot, D. Le voyage de la Terre Sainte. Paris, 1890.
	Geog 4308.90.5	Potter, W.W. To Europe on a stretcher. N.Y., 1890.
	Geog 808.76.9	Reclus, J.J.E. Earth and its inhabitants - North America. N.Y., 1890-1893. 3v.
	Geog 808.76.8	Reclus, J.J.E. Earth and its inhabitants - Oceanica. N.Y., 1890.
	Geog 4308.90	Robinson, L.B. A bundle of letters from over the sea. Boston, 1890.
	An 3308.90	Seaver, J.W. Anthropometry and physical examination. New Haven, 1890.
	Geog 4304.65.10	Slavík, František A. Cesta pana Lva z Rožmitála po západní Evropĕ roku 1465-1467. Telči, 1890.
Htn	Geog 3570.5*	Sonsa Viterbo, F.M. Trabalhos nauticos dos Portuguezes nos seculos XVI e XVII. Lisboa, 1890.
	Geog 4418.90	Tchihatchef, P. de. Etudes de géographie et d'histoire naturelle. Florence, 1890.
	An 3558.90	Török, A. Grundz. einer systematischen Kraniometrie. Stuttgart, 1890.
	Geog 4658.90	Vidal Gormaz, Francisco. Algunos naufragios occuridos en las costas chilenes. Valparaiso, 1890.

1891

	Geog 1555.45	Baedeker, publishers. Belgium and Holland. 10. ed. Leipsic, 1891.
	Geog 1526.10	Baedeker, publishers. Berlin und Umgebungen. 7. Aufl. Leipzig, 1891.
	Geog 1533.7	Baedeker, publishers. Eastern Alps. 7. ed. Leipsic, 1891.
	Geog 1519.46.2	Baedeker, publishers. Paris and its environs. 10. ed. Leipsic, 1891.
	Geog 1519.46	Baedeker, publishers. Paris and its environs. 10. ed. Leipsic, 1891.
	Geog 1519.45	Baedeker, publishers. Paris et ses environs. 10. éd. Leipzig, 1891.
	Geog 1535.23.9	Baedeker, publishers. Die Schweiz. 24. Aufl. Leipzig, 1891.
	Geog 1524.41	Baedeker, publishers. Southern Germany and Austria. 7. ed. Leipsic, 1891.
	Geog 1524.40	Baedeker, publishers. Southern Germany and Austria. 7. ed. Leipsic, 1891.
	Geog 1535.48	Baedeker, publishers. Switzerland...Italy, Savoy...Tyrol. 14. ed. Leipsic, 1891.

Chronological Listing

1891 - cont.

Geog 4218.84.9	Ballou, M.M. Due west, or Round the world. 9. ed. Boston, 1891.
Geog 4308.91	Bates, J.H. Notes of foreign travel. N.Y., 1891.
Geog 4218.87.13	Caine, William S. A trip round the world in 1887-88. London, 1891.
Geog 4208.91.5	Carrasco, Gabriel. Del Atlántico al Pacífico y un Argentino en Europa. 2a ed. Buenos Aires, 1891.
Geog 1811.2	Chamberlain, B.H. Handbook for...Japan. 3. ed. London, 1891.
Geog 4218.91	Cotes, Sara J.D. A social departure. How Orthodocia and I went round the world by ourselves. N.Y., 1891.
Geog 5557.25.3	Dall, W.H. Early expeditions to the Bering Sea and Strait. Washington, 1891.
Geog 758.45.25	Daniel, H.A. Lehrbuch der Geographie für höhere Unterrichtsanstalten. Halle, 1891.
Geog 4180.81	Dominguez, L.L. Conquest of the river plate. London, 1891.
Geog 135.2	Gesellschaft für Erdkunde zu Leipzig. Beiträge zur Geographie des festen wassers. v.1. Leipzig, 1891.
Geog 4218.89.5	Gillis, C.J. Around the world in seven months. N.Y., 1891.
Geog 4328.91	Groome, P.L. Rambles...in three continents. Greensboro, 1891.
Geog 3506.7	Hennequin, Emile. Etude historique sur l'exécution de la carte de Ferraris et l'évolution de la cartographie topographique en Belgique. Bruxelles, 1891.
Geog 4128.91	Hunt, R. Steamship lines of the world. N.Y., 1891.
Geog 4228.86.10	Jackson, Helen F.H. Glimpses of three coasts. Boston, 1891.
Geog 3240.26	Jacob, Georg. Studien in arabischen Geographen. Berlin, 1891-92.
An 358.91.15	Kinmont, A. The natural history of man. 2. ed. Philadelphia, 1891.
Geog 4180.82	Leguat, F. Voyage of F. Leguat...to Rodriguez. London, 1891. 2v.
Geog 5530.5	Markham, A.H. Life of Sir John Franklin. London, 1891.
Geog 5530.5.2	Markham, A.H. Life of Sir John Franklin. N.Y., 1891.
Geog 4306.83.2	Melle, Jakob von. Beschreibung einer Reise durch das Nordwestliche Deutschland nach den Niederländen. Lübeck, 1891.
Geog 4308.87.5	Morrison, Leonard A. Among the Scotch-Irish. Boston, 1891.
Geog 1820.4	Murray, John, publisher, London. Handbook for...Algeria and Tunis. 4. ed. London, 1891.
Geog 1816.20A	Murray, John, publisher, London. Handbook for...Egypt. 8. ed. London, 1891.
Geog 5850.9	Nansen, F. Eskimoliv. Kristiania, 1891.
Geog 520.10	Odoric de Pardenone. Les voyages en Asie. Paris, 1891.
Geog 603.2	Partsch, J. Philipp Clüver der Begründer der historischen Landerkunde. Wien, 1891.
Geog 4308.89.5	Un paseo por Europa. Habana, 1891.
Geog 4131.5	Rae, W.F. The business of travel. London, 1891.
An 358.91	Reclus, E. Primitive folk. London, 1891.
Geog 808.76.6	Reclus, J.J.E. Earth and its inhabitants - Asia. N.Y., 1891. 4v.
Geog 958.91	Shepp, J.W. Shepp's photographs of the world. Philadelphia, 1891.
An 5.3	Société d'Anthropologie, Paris. Catalogue de la bibliothèque...1890. Paris, 1891.
Geog 5398.79.11	Soley, J.R. Address...at unveiling of Jeannette monument. Baltimore, 1891.
An 2108.91.3	Taylor, I. The prehistoric races of Italy. Washington, 1891.
An 358.91.3	Tylor, Edward B. Anthropology. N.Y., 1891.
An 2108.91A	Tylor, Edward Burnett. Primitive culture. 3. ed. London, 1891. 2v.
Geog 4218.91.4	Vassar, J.G. Twenty years around the world. N.Y., 1891.
Geog 4218.89.7	Wetmore, E.B. A flying trip around the world. N.Y, 1891.

1892

Geog 4218.92.15	Aubertin, J.J. Wanderings and wonderings. London, 1892.
Geog 4418.03.10	Badia y Leblich. Atlas dels viatjes d'Ali Bey el Abbassi. v.1-3. Barcelona, 1892.
Geog 1617.5	Baedeker, publishers. Egypt. Leipsic, 1892.
Geog 1618.15	Baedeker, publishers. Egypt. Leipsic, 1892-95. 2v.
Geog 1539.35	Baedeker, publishers. Italy...northern Italy. 9. ed. Leipsic, 1892.
Geog 1539.36	Baedeker, publishers. Italy...northern Italy. 9. ed. Leipsic, 1892.
Geog 1510.36	Baedeker, publishers. London and its environs. 8. ed. Leipsic, 1892.
Geog 1510.35	Baedeker, publishers. London and its environs. 8. ed. Leipsic, 1892.
Geog 1568.24	Baedeker, publishers. Norway, Sweden and Denmark. 5. ed. Leipsic, 1892.
Geog 1525.45.5	Baedeker, publishers. Die Rheinlande von der Schweize. 26. Aufl. Leipzig, 1892.
Geog 1525.45.2	Baedeker, publishers. The Rhine from Rotterdam to Constance. 12. ed. Leipsic, 1892.
Geog 1525.45	Baedeker, publishers. The Rhine from Rotterdam to Constance. 12. ed. Leipsic, 1892.
Geog 4218.84.15	Ballou, M.M. Foot-prints of travel. Boston, 1892.
Geog 4311.9	Bigelow, P. Paddles and politics down the Danube. London, 1892.
Geog 4358.92	Blackburn, Henry. Artistic travel in Normandy, Brittany, the Pyrenees, Spain and Algeria. London, 1892.
An 628.92	Brinton, D.G. Anthropology. Philadelphia, 1892.
Geog 4218.87.14	Caine, W.S. A trip round the world in 1887-88. London, 1892.
Geog 500.23	Cardon, F. Publicazioni geografiche. Roma, 1892.
An 358.92.5	Carmichael, Charles H.E. L'anthropologie et les origines de la Société...l'Orient et l'Occident. Lisbonne, 1892.
Htn Geog 4558.40.14*	Dana, Richard Henry. Two years before the mast. N.Y., 1892.
An 2058.92	Doughty, F.W. Evidences of man in the drift. N.Y., 1892.
Geog 4218.92	Esmonde, T.H.G. Round the world with the Irish delegates. Dublin, 1892.
Htn An 3408.92*	Galton, F. Finger prints. London, 1892.
Geog 4208.92	Great streets of the world. London, 1892.
An 358.92.7	Haliburton, Robert G. Survivals of prehistoric races in Mt. Atlas and Pyrenees. Lisbon, 1892.
Geog 4308.92	Hayward, H.C. From Finland to Greece, or Three seasons in Eastern Europe. N.Y., 1892.
An 2108.92	Hoernes, Moriz. Die Urgeschichte des Menschen. Wien, 1892.

1892 - cont.

Geog 5558.91.5	Keely, R.N. In Arctic seas...with Peary expedition. Philadelphia, 1892.
Geog 5398.91	Keely, Robert. In Arctic seas...voyage of the "Kite". Philadelphia, 1892.
Geog 520.12	La Broquière, Bertrandon de. Le voyage d'Outremer. Photoreproduction. Paris, 1892.
Geog 3590.10	Lialina, M.A. Russkie moreplavateli, arkticheckie krulosvetnye. Sankt Peterburg, 1892.
An 358.92	Lombroso, C. L'uomo bianco e l'uomo di colore. Firenze, 1892.
Geog 4218.92.20	MacGregor, John. Toil and travel. London, 1892.
Geog 5398.81.5	Melville, G.W. In the Lena Delta. Boston, 1892.
Geog 4328.92	Meriwether, Lee. Afloat and ashore on the Mediterranean. N.Y., 1892.
Geog 1807.10	Murray, John, publisher, London. Handbook for...India and Ceylon. London, 1892.
Geog 1770.8	Murray, John, publisher, London. Handbook for...Norway. 8. ed. London, 1892.
Geog 1742.9	Murray, John, publisher, London. Handbook for...southern Italy and Sicily. 9. ed. London, 1892.
Geog 1805.2	Murray, John, publisher, London. Handbook for...Syria and Palestine. London, 1892.
Geog 1715.18	Murray, John, publisher, London. Handbook for travellers in France. 18. ed. London, 1892. 2v.
Geog 1795.3	Murray, John, publisher, London. Handbook to the Mediterranean. 3. ed. London, 1892. 2v.
An 2308.92	Olshausen, O. Leichenverbrennung. Berlin, 1892.
Geog 3205.50F	Pacheco Pereira. Esmeraldo de situ orbis. Lisboa, 1892.
Geog 4110.18	Phillips, Morris. Abroad and at home. N.Y., 1892.
Geog 4218.92.5F	Radde, G.F.R. 23.000 mil'na iakhte "Tamara". Sanktpeterburg, 1892-93. 2v.
An 358.92.3	Ranke, Johannes. Beiträge zur physischen Anthropologie der Bayern. München, 1892.
Geog 818.92	Reclus, O. A bird's-eye view of the world. Boston, 1892.
Geog 607.20	Schweizinscher Zofingverein. Souvenir de l'inauguration du monument élevé à Arnold Guyot par la Société de Zofingue à l'Académie de Neuchâtel le 6 mai 1892. Neuchâtel, 1892.
Geog 4308.86.17	Smith, F.H. Well-worn roads of Spain, Holland and Italy. Boston, 1892.
Geog 4558.61.5	Stevens, Charles. A sailor boy's experience. Napanee, Ont., 1892.
Geog 958.92F	Stoddard, John L. Glimpses of the world. Chicago, 1892.
Geog 4300.26	Union Steamship Company, Ltd. Homeward bound. 3rd ed. London, 1892.
Geog 4180.84A	Valle, P. della. Travels of...Valle in India. London, 1892. 2v.
An 3538.92	Virchow, R. Crania ethnica americana. Berlin, 1892.
An 197.1	Wilson, D. The lost Atlantis. N.Y., 1892.
An 2058.92.3	Wright, G.F. The antiquity and origin of the human race. Boston, 1892.

1893

An 3208.93	Ammon, Alto. Die natürliche Auslese bein Menschen. Jena, 1893.
Geog 1540.31A	Baedeker, publishers. Italy. Pt. 2: Central Italy and Rome. 11. ed. Leipzig, 1893.
Geog 1540.30	Baedeker, publishers. Italy. Pt. 2: Central Italy and Rome. 11. ed. Leipzig, 1893.
Geog 1542.30	Baedeker, publishers. Italy. Pt. 3: Southern Italy, Sicily. 11. ed. Leipsic, 1893.
Geog 1523.50	Baedeker, publishers. Northern Germany. 11. ed. Leipsic, 1893.
Geog 1535.50	Baedeker, publishers. Switzerland...Italy, Savoy...Tryol. 15. ed. Leipsic, 1893.
Geog 1645.5.5	Baedeker, publishers. United States...Mexico. Leipsic, 1893.
Htn Geog 1645.5*	Baedeker, publishers. United States...Mexico. Leipsic, 1893.
Geog 4508.93	Balch, E.S. Mountain exploration. Philadelphia, 1893.
Geog 4208.49.5	Barra, E.I. A tale of two oceans. San Francisco, 1893.
An 2208.93	Bartels, Max. Die Medicin der Naturvölker. Leipzig, 1893.
An 358.93.1	Beddoe, J. Anthropological history of Europe. London, 1893.
Geog 4180.87A	Bent, J.T. Early voyages and travels in the Levant. London, 1893.
Geog 4308.93.3	Bishop, W.H. A house-hunter in Europe. N.Y., 1893.
Htn Geog 1811.3*	Chamberlain, B.H. Handbook for...Japan. N.Y., 1893.
Geog 4308.93.5	D'Apery, Tello J. Europe seen through a boy's eyes. N.Y., 1893.
Geog 4278.93	Fernandez Duro, Cesario. Viajes regios por mar en el trancurso de quinientos años. Madrid, 1893.
Htn An 3408.93*	Galton, F. Decipherment of blurred fingerprints. London, 1893.
Geog 818.93	Gilbert, Frank. The world, historical and actual. Chicago, 1893.
Geog 4138.93	Greely, A.W. Explorers and travellers. N.Y., 1893.
Geog 5180.5	Heilprin, A. The Arctic problem. Philadelphia, 1893.
Geog 4308.84.5	James, H. Portraits of places. 3rd ed. Boston, 1893.
Geog 4110.5	King, M. Where to stop. Boston, 1893.
Geog 3055.15	Knötel, A.F.R. Atlantis und das Volk der Atlanten. Leipzig, 1893.
Geog 578.55.13A	Lippincott, J.B. and Co. Gazetteer of the world. Philadelphia, 1893.
An 3308.93	Livi, Ridolfo. Antropometria militare. Roma, 1893-1903. 2v.
Geog 4658.93.2	Lussich, Antonio D. Naufragios célebres en el cabo Polonio, el banco Inglés, y el océano Atlántico. 2. ed. Montevideo, 1893.
Geog 4180.86A	Markham, C.R. Journal of Christopher Columbus. London, 1893.
An 3508.93	Matthews, W. The human bones...Hemingway collection...United States Medical Museum at Washington. n.p., 1893.
Geog 4508.93.10	Meurer, Julius. Der Bergsteiger in Hochgebirge. Wien, 1893.
Geog 4308.34	Michelet, J. Sur les chemins de l'Europe. Paris, 1893.
Geog 1828.1	Murray, John, publisher, London. Handbook for...New Zealand. London, 1893.
Geog 1775.5	Murray, John, publisher, London. Handbook for...Russia, Poland and Finland. 5. ed. London, 1893.
Geog 5855.7	Nansen, F. Eskimo life. London, 1893.
Geog 3570.7	Oliveira Marlins, J.P. de. Les explorations des Portugais. Paris, 1893.
Geog 4308.93.10	Papa, Dario. Viaggi. v.2. Milano, 1893.
Geog 5558.91A	Peary, Josephine D. My Arctic journal. N.Y., 1893.

Chronological Listing

1893 - cont.

Geog 4308.93	Pennell, J. Our sentimental journey thro' France. London, 1893.
Geog 3018.93	Rainaud, A. Le continent austral. Paris, 1893.
Geog 3140.11F	Rylands, T.G. Geography of Ptolemy elucidated. Dublin, 1893.
An 358.93	Schurtz, H. Katechismus der Völkerkunde. Leipzig, 1893.
Geog 4348.93	Scollard, Clinton. On sunny shores. N.Y., 1893.
Geog 4218.93	Thompson, F.D. In the track of the sun. N.Y., 1893.
An 98.93	Topinard, P. L'anthropologie aux États-Unis. Paris, 1893.
Geog 4238.92	Ward, Artemus. Columbus outdone, an exact narrative of the voyage of the Yankee skipper Captain N.A. Andrews. N.Y., 1893.
Geog 4508.93.5	Wilson, Claude. Mountaineering. London, 1893.

1894

Geog 585.15.3	Albrecht, Theodor. Formeln und Hülfstafeln für geographische Ortsbestemmungen. 3. Aufl. Leipzig, 1894.
Geog 4208.94	Arnold, E. Wandering words. London, 1894.
Geog 5558.91.3	Astrup, E. With Peary near the Pole. London, 1894.
Geog 1555.50	Baedeker, publishers. Belgium and Holland. 11. ed. Leipsic, 1894.
Geog 1555.51	Baedeker, publishers. Belgium and Holland. 11. ed. Leipsic, 1894.
Geog 1526.12	Baedeker, publishers. Berlin und Umgebungen. 8. Aufl. Leipzig, 1894.
Geog 1640.5	Baedeker, publishers. Dominion of Canada. Leipzig, 1894.
Geog 1508.13	Baedeker, publishers. Great Britain. 3. ed. Leipsic, 1894.
NEDL Geog 1508.12	Baedeker, publishers. Great Britain. 3. ed. Leipsic, 1894.
Geog 1585.15	Baedeker, publishers. Greece. 2. ed. Leipsic, 1894.
Geog 1538.35	Baedeker, publishers. Italien. 14. Aufl. Leipzig, 1894.
Geog 1510.40	Baedeker, publishers. London and its environs. 9. ed. Leipsic, 1894.
Geog 1510.41	Baedeker, publishers. London and its environs. 9. ed. Leipsic, 1894.
Geog 1510.45	Baedeker, publishers. London and its environs. 10. ed. Leipsic, 1894.
Geog 1510.42	Baedeker, publishers. London und Umgebungen. 11. Aufl. Leipzig, 1894.
Geog 1516.10.2	Baedeker, publishers. Northern France. 2. ed. Leipsic, 1894.
Geog 1516.10	Baedeker, publishers. Northern France. 2. ed. Leipsic, 1894.
Geog 1605.10	Baedeker, publishers. Palestine and Syria. 2. ed. Leipsic, 1894.
Geog 1519.48	Baedeker, publishers. Paris and its environs. 11. ed. Leipsic, 1894.
Geog 1519.49	Baedeker, publishers. Paris et ses environs. 11. éd. Leipzig, 1894.
Geog 5050.2	Boston Public Library. Arctic regions and Antarctic regions. Boston, 1894.
Geog 6008.92	Burn Murdoch, W.G. From Edinburgh to the Antarctic. London, 1894.
Geog 3251.13	Ceradini, G. A proposito dei due globi mercatoriani. Milano, 1894.
Geog 578.94	Chamber's consise gazetteer of the world, topographical, statistical, historical. London, 1894.
Geog 4180.88	Christy, M. Voyages of Captain Luke Foxe and Captain T. James. London, 1894. 2v.
Geog 613.1.7	Cologne, Germany. Stadtbibliothek. Katalog einer Mercator-Ausstellung. Koeln, 1894.
Geog 4328.94A	Davis, R.H. Rulers of the Mediterranean. N.Y., 1894.
Geog 613.1.15	Dinse, Paul. Zum Gedächtnis Gerhard Mercator's. Berlin, 1894.
Geog 4218.94	Dunn, S.H. The world's highway. London, 1894.
Geog 4218.93.10	Egerton, Alice A. Glimpses of four continents. London, 1894.
An 2108.94	Hewitt, J.F. The ruling races of prehistoric times. Westminster, 1894-95. 2v.
Geog 4328.94.5	Hoyt, W.M. A cruise on the Mediterranean. Chicago, 1894.
Geog 3530.13	Jolly, Raoul. Les missions françaises; causeries géographiques. Paris, 1894-96. 2v.
Geog 4110.5.2	King, M. Where to stop. Boston, 1894.
An 358.94.5	Lavrov, P.L. Antropologicheskaia zhizne. Zheneva, 1894. 2v.
Geog 4180.90	Markham, C.R. Letters of Amerigo Vespucci. London, 1894.
Geog 4218.57.3	Miller, Thomas. Over five seas and oceans. N.Y., 1894.
Geog 1709.27	Murray, John, publisher, London. Handbook for...Oxfordshire. London, 1894.
Geog 1709.38.4	Murray, John, publisher, London. Handbook for...Worcestershire. 4. ed. London, 1894.
An 3558.94	Olóriz y Aguilera, Frederico. Distribución geográfica del índice cefálico en España. Madrid, 1894.
Geog 4268.94	Philippson, A. Europa...eine...Landeskunde. Leipzig, 1894.
Geog 958.94	Photographic views of the world. Boston, 1894.
Geog 5398.93.15	Plass, F. Reise-Erinnerungen...Aug. 1893..."Admiral". Hamburg, 1894.
An 358.94	Ranke, Johannes. Der Mensch. Leipzig, 1894. 2v.
Geog 808.76.10	Reclus, J.J.E. Earth and its inhabitants - South America. N.Y., 1894-1895. 2v.
Geog 4328.94.7PF	Rothschild, N. Skizzer aus dem Süden. Wien, 1894.
Geog 4218.92.7	Scott, Clement. Pictures of the world. London, 1894.
Geog 4218.93.5	Stephens, A.A. A Queenslander's travel notes. Sydney, 1894.
An 358.94.3	Tylor, Edward B. Anthropology. N.Y., 1894.

1895

Geog 5558.91.4	Astrup, E. Blandt Nordpolens naboer. Kristiania, 1895.
An 358.95	Babington, W.D. Fallacies of race theories. London, 1895.
Geog 1533.8	Baedeker, publishers. Eastern Alps. 8. ed. Leipsic, 1895.
Geog 1538.37	Baedeker, publishers. Italien. v.3. 11. Aufl. Leipzig, 1895.
Geog 1538.45	Baedeker, publishers. Italien von den Alpen bis Neapel. 3. Aufl. Leipzig, 1895.
Geog 1539.38	Baedeker, publishers. Italy...northern Italy. 10. ed. Leipsic, 1895.
Geog 1518.55	Baedeker, publishers. Le Nord-est de la France. 5. éd. Leipsic, 1895.
Geog 1518.80	Baedeker, publishers. Le Nord-ouest de la France. 5. éd Leipsic, 1895.
Geog 1568.25A	Baedeker, publishers. Norway, Sweden and Denmark. 6. ed. Leipsic, 1895.

1895 - cont.

Geog 1518.26	Baedeker, publishers. South-western France from the Loire and the Rhone to the Spanish frontier. 2. ed. Leipsic, 1895.
Geog 1524.46	Baedeker, publishers. Southern Germany. 8. ed. Leipsic, 1895.
Geog 1524.45	Baedeker, publishers. Southern Germany. 8. ed. Leipsic, 1895.
Geog 1535.55	Baedeker, publishers. Switzerland...Italy, Savoy...Tyrol. 16. ed. Leipsic, 1895.
Geog 4208.95	Ballou, M.M. Foot-prints of travel. Boston, 1895.
Geog 4328.93	Barber, J.L. Mediterranean mosaics. N.Y., 1895.
Geog 3240.5	Bernard, A. De Adamo Bremensi geographo. Parisiis, 1895.
Geog 4168.95	Bonnaffé, E. Voyages et voyageurs de la renaissance. Paris, 1895.
Geog 4218.95.5	Brewster, F. Carroll. From Independence Hall around the world. Philadelphia, 1895.
Geog 4328.95	Brooks, N. The Mediterranean trip. N.Y., 1895.
Geog 4328.88	Buckley, J.M. Travels in three continents. N.Y., 1895.
Geog 3055.40	Buelua, E. La Atlantida y la ultima tule. Mexico, 1895.
Geog 4208.91	Cimon, Henri. Impressions de voyage. Québec, 1895-1902. 2v.
An 2108.95.3	Clodd, Edward. The story of "primitive" man. N.Y., 1895.
Geog 4300.32	Cook, T., firm. Tourist handbook for Holland, Belgium, the Rhine and Black Forest. London, 1895.
Geog 4558.40.15	Dana, Richard Henry. Two years before the mast. Cambridge, 1895.
Geog 808.66.7	Daniel, H.A. Handbuch der Geographie. Leipzig, 1895. 4v.
An 3308.95	Farmer, J.B. Extract from Science progress. Boston, 1895.
An 180.2	Festschrift der deutscher anthropologischer Gesellschaft zur XXVI allgemeinen Versammlung. Cassel, 1895.
An 3408.95	Galton, F. Fingerprint directories. London, 1895.
VGeog 4678.91	Henderson Brothers, Glasgow, defendants. Difesa dei signori Henderson, armatori contro i danneggiati del naufragio dell'Utopia. Napoli, 1895.
An 2338.95	Hough, Walter. Primitive American armor. Washington, 1895.
Geog 500.21.5	London. Royal Geographical Society. Catalogue of the library. London, 1895.
Geog 4308.95	Magness, E. Tramp tales of Europe. Buffalo, 1895.
Geog 618.2	Markham, C.R. Major Jame Rennell. London, 1895.
Geog 618.2.2	Markham, C.R. Major James Rennell. N.Y., 1895.
An 2108.95.5	Mason, O.T. Origins of invention. London, 1895.
An 2338.95.5	Meyer, Hermann. Bogen und Pfeil in Central Brasilien. Leipzig, 1895.
An 608.87	Moscow. Dashkovskii Etnograficheskii Muzei. Sistematicheskoe opisanie kollektsii Dashkovskago etnograficheskago muzeia. v.4. Moskva, 1895.
Geog 1820.5	Murray, John, publisher, London. Handbook for...Algeria and Tunis. 5. ed. London, 1895.
Geog 1709.12.4	Murray, John, publisher, London. Handbook for...Gloucestershire. 4. ed. London, 1895.
Geog 1801.1	Murray, John, publisher, London. Handbook for Asia Minor. London, 1895.
Geog 1709.15	Murray, John, publisher, London. Handbook for Hertfordshire, Bedfordshire. London, 1895.
Geog 4208.95.5.5	Nordhoff, C. The merchant vessel. N.Y., 1895.
Geog 4208.95.5	Nordhoff, C. The merchant vessel. N.Y., 1895.
Geog 135.2.2	Ratzel, Friedrich. Anthropogeographische Beiträge. v.2. Leipzig, 1895.
Geog 4218.95	Raum, G.E. A tour around the world. N.Y., 1895.
Geog 3618.3	Saellskapit för Finlands Geografi, Helsingfors. Exposé des travaux géographiques executés en Finlande jusqu'en 1895. Helsingfors, 1895.
Geog 4180.91	Sarmiento de Gamboa, Pedro. Narrative of voyages of...Sarmiento. London, 1895.
Geog 4308.39.9	Sreznevskii, I.I. Putevyia pis'ma...iz slavianskikh zemel, 1839-1842. Sankt Peterburg, 1895.
An 2108.95	Starr, F. Some first steps in human progress. Meadville, Pa., 1895.
Geog 4558.94	Stewart, J.A.E. En mer. n.p., 1895?
Geog 4418.95	Tiffany, F. This goodly frame, the earth. Boston, 1895.
Geog 3070.12	Wauwermans, H. Histoire de l'ecole cartographique. Bruxelles, 1895. 2v.

1896

An 98.96	Achelis, Thomas. Moderne Völkerlunde. Stuttgart, 1896.
An 3538.96	Allen, Harrison. Crania from the mounds of the St. John's. Philadelphia, 1896.
Geog 4208.96	Arnold, E. East and West. London, 1896.
Geog 4180.95	Azurara, G.E. de. Chronicle of discovery...of Guinea. London, 1896-99. 2v.
Geog 1530.8	Baedeker, publishers. Austria. 8. ed. Leipsic, 1896.
Geog 1542.33	Baedeker, publishers. Italy. Pt. 3: Southern Italy, Sicily. 12. ed. Leipsic, 1896.
Geog 1510.46	Baedeker, publishers. London and its environs. 10. ed. Leipsic, 1896.
Geog 1523.75	Baedeker, publishers. Nordost-Deutschland. 25. Aufl. Leipzig, 1896.
Geog 1519.50	Baedeker, publishers. Paris and its environs. 12. ed. Leipsic, 1896.
Geog 1519.50.2	Baedeker, publishers. Paris and its environs. 12. ed. Leipsic, 1896.
Geog 1519.51	Baedeker, publishers. Paris et ses environs. 12. éd. Leipsic, 1896.
Geog 1525.47	Baedeker, publishers. The Rhine from Rotterdam to Constance. 13. ed. Leipsic, 1896.
An 3558.96	Barth, J. Norrønaskaller. Crania antiqua. Christiania, 1896.
Geog 135.2.3	Baumann, Oskar. Der Sansibar-Archipel. v.3. Leipzig, 1896.
Geog 4248.13	Corney, Peter. Voyages in the northern Pacific, 1813-1818. Honolulu, 1896.
Geog 958.96	Dubois, M. Album géographique. Paris, 1896-1906. 5v.
Geog 4308.96	Dudley, L.B. Letters to Ruth. N.Y., 1896.
Geog 5208.95	Greely, A.W. Handbook of arctic discoveries. Boston, 1896.
Geog 3018.96	Hamy, J.T.E. Études historiques et géographiques. Paris, 1896.
An 358.96A	Keane, A.H. Ethnology. Cambridge, Eng., 1896.
An 358.96.2	Keane, A.H. Ethnology. 2. ed. Cambridge, Eng., 1896.
Geog 808.96	Kerp, H. Methodisches Lehrbuch...Erdkund. Bonn, 1896-1904. 3v.

Chronological Listing

1896 - cont.

Geog 4180.92	Leo Africanus, J. History and description of Africa. London, 1896. 3v.
Geog 520.13	Leone, G. Description de l'Afrique. v.2-3. Paris, 1896-98. 2v.
Geog 139.3.5	London. Royal Geographical Society. Proceedings. Index. 1879-1892. London, 1896.
Geog 4179.5	Markham, C.R. Richard Hakluyt: his life and work. London, 1896.
Geog 4308.71.12	Morris, Caspar. Letters of travel. Philadelphia, 1896. 2v.
Geog 500.32	Muller, F. Topographie ancienne - catalogue a prix marqués de cartes anciennes. Amsterdam, 1896.
Geog 1785.6	Murray, John, publisher, London. Handbook for...Greece. 6. ed. London, 1896. 2v.
Geog 1811.4	Murray, John, publisher, London. Handbook for...Japan. 4. ed. London, 1896.
Geog 1713.5	Murray, John, publisher, London. Handbook for travellers in Ireland. 5. ed. London, 1896.
An 358.96.5	Piette, Edouard. Etude d'éthnographie préhistorique. no.2-3,6-8. Paris, 1896-1925.
An 358.96.3A	Ratzel, F. The history of mankind. London, 1896-98. 3v.
Geog 4311.11	Schweiger-Lerchenfeld, Amand von. Die Donau als Völkeweg. Wien, 1896.
An 3308.96	Seaver, J.W. Anthropometry and physical examination. 2. ed. New Haven, 1896.
Geog 4418.95.3	Tiffany, F. This goodly frame, the earth. Boston, 1896.
Geog 4418.95.2	Tiffany, F. This goodly frame, the earth. Boston, 1896.
Geog 5530.18	Traill, Henry Duff. The life of Sir John Franklin, R.N. London, 1896.
Geog 5558.94	Walsh, H.C. Last cruise of the Miranda. N.Y., 1896.
Geog 4328.96	Woolson, C.F. Mentone, Cairo and Corfu. N.Y., 1896.
Geog 5838.96	Wright, G.F. Greenland icefields. N.Y., 1896.
Geog 4308.96.3	Zorrilla de San Martin, J. Resonancias del camino. Paris, 1896.

1897

Geog 4558.90.7	Amicis, E. de. On blue water. N.Y., 1897.
Geog 1555.55	Baedeker, publishers. Belgium and Holland. 12. ed. Leipsic, 1897.
Geog 1508.16	Baedeker, publishers. Great Britain. 4. ed. Leipsic, 1897.
Geog 1508.15	Baedeker, publishers. Great Britain. 4. ed. Leipsic, 1897.
Geog 1540.37	Baedeker, publishers. Italy. Pt. 2: Central Italy and Rome. 12. ed. Leipsic, 1897.
Geog 1523.55	Baedeker, publishers. Northern Germany. 12. ed. Leipsic, 1897.
Geog 1575.5	Baedeker, publishers. La Russie. 2. éd. Leipzig, 1897.
Geog 1517.5	Baedeker, publishers. Southern France...including Corsica. Leipsic, 1897.
Geog 1535.57	Baedeker, publishers. Switzerland...Italy, Savoy...Tyrol. 17. ed. Leipsic, 1897.
Geog 4218.97	Barrows, John D. A world-pilgrimage. Chicago, 1897.
Geog 3228.97A	Beazley, C.R. Dawn of modern geography. London, 1897. 3v.
Geog 3228.97.2	Beazley, C.R. Dawn of modern geography. London, 1897-1906. 3v.
Geog 5558.97	Berens, S.L. The "Fram" expedition. Nansen in the frozen world. Philadelphia, 1897.
Geog 5838.94	Bruun, Daniel. Mellem fangere og jaegere. Kjøbenhavn, 1897.
Geog 4308.97.3	Cassell and Company, Inc. Complete pocket guide to Europe. London, 1897.
Geog 4180.98A	Cosmas. Christian topography of Cosmas. London, 1897.
Geog 4208.97	Davis, R.H. A year from a reporter's note-book. N.Y., 1897.
Geog 5324.2.45	Dolman, Frederick. Dr. Nansen. London, 1897.
Geog 520.16	Du Fresne-Canayl, P. Le voyage du Levant. Paris, 1897.
An 3208.97	Ehrenreich, P. Anthropologische Studien über die Urbewohner Brasiliens. Braunschweig, 1897.
Geog 5398.97.3	Fonvielle, W. La vérité sur l'expédition Andrée. Strasbourg, 1897.
Geog 4180.96A	Gosch, C.C.A. Danish Arctic expeditions, 1605-20. London, 1897. 2v.
Geog 818.97.5	Hellwald, F. von. Die Erde und ihre Völker. Stuttgart, 1897.
An 2108.97	Hoernes, Moriz. Urgeschichte der Menschheit. 2e Aufl. Leipzig, 1897.
Geog 5558.96	Hoppin, Benjamin. A diary kept while with the Peary Arctic expedition of 1896. New Haven? 1897?
Geog 121.1	The journal of school geography. Lancaster, 1897. 5v.
Geog 5398.97	Lachambre, Henri. Andrée; au pôle nord en ballon. Paris, 1897.
Geog 3240.6	Lönborg, S. Adam af Bremen, skildring af Nordeuropas länder. Uppsala, 1897.
Geog 4128.97	Ludwig, Friedrich. Untersuchungen...Reise...Itineraire der deutsch Königer. Inaug. Diss. Berlin, 1897.
Geog 4128.97.1	Ludwig, Friedrich. Untersuchungen über die Reise und Marschgesehwindigkeit in XII. und XIII. Jahrhundert. Berlin, 1897.
Geog 614.5F	Marinelli, G. Cristoforo Negri. Torino, 1897.
An 358.97	Munro, R. Prehistoric problems. Edinburgh, 1897.
Geog 1709.29.5	Murray, John, publisher, London. Handbook for Shropshire and Cheshire. 3. ed. London, 1897.
Geog 5398.93.1	Nansen, Fridtjof. Farthest north. N.Y., 1897. 2v.
Geog 5398.93	Nansen, Fridtjof. Farthest north. Westminster, 1897. 2v.
Geog 5398.93.5	Nansen, Fridtjof. In Nacht undEis. Leipzig, 1897. 2v.
Geog 4218.92.3	Peters, G.H. Impressions of a journey round the world. London, 1897.
Geog 5100.5	Prentiss, H.M. The great polar current. Cambridge, 1897.
Geog 4208.97.7	Schanz, M. Ein Zug nach Osten. Hamburg, 1897. 2v.
Geog 4216.19	Schouten, W.C. Relación diaria del viaje de J. Le Maire y G.C. Shouten. Santiago de Chile, 1897.
Geog 578.79.4F	Vivien de St. Martin, L. Nouveau dictionnaire de géographie universelle contenant...la géographie physique. Supplement. Paris, 1897-1900. 2v.
Geog 4308.97	Widmann, J.V. Sommerwanderungen und Winterfahrten. Frauenfeld, 1897.
Geog 4508.97	Wilson, E.L. Mountain climbing. N.Y., 1897.
Geog 3018.97	Wisotzki, E. Zeitströmungen in der Geographie. Leipsic, 1897.

1898

Geog 4218.98	Albrey, J. d'. Du Tonkin au Havre. Paris, 1898.
Geog 1616.15	Baedeker, publishers. Egypt. 4. ed. Leipsic, 1898.
Geog 1616.18	Baedeker, publishers. Egypt. 4. ed. Leipsic, 1898.
Geog 1510.47	Baedeker, publishers. London and its environs. 11. ed. Leipsic, 1898.
Geog 1518.3	Baedeker, publishers. South eastern France. 3. ed. Leipsic, 1898.
Geog 1545.5	Baedeker, publishers. Spain and Portugal. Leipsic, 1898.
Geog 4208.96.10	Bond, C. Goldfields and chrysantemums. London, 1898.
Geog 4558.98.10	Bullen, F.T. Cruise of the Cachalot. London, 1898.
Geog 4208.85.5	Burton, R.G. Tropics and snows. London, 1898.
Geog 114.7.85	Coën, A. Venticinque anni di Lavoro dell'Istituto Geographico Militare. Firenze, 1898.
Geog 3530.14	Deschamps, Léon. De Rasiliiis Gabriel, Isaac et Claudio proenominatis Richelii adjutoribus. Paris, 1898.
Geog 4328.98	Domsthorpe, Wordsworth. Down the stream of civilization. London, 1898.
Geog 5324.2	Enzberg, E. von. Fridtjof Nansen. Dresden, 1898.
Geog 4308.98	Ferro, Adersón. Minhas viagens. Ceara, 1898.
Geog 5918.98	Fricker, Karl. Antarktis. Berlin, 1898.
Geog 4180.99A	Gama, V. da. Journal of first voyage of V. da Gama. London, 1898.
Geog 578.98	Garollo, G. Dizionario geografico universale. 4. ed. Milano, 1898.
An 358.98A	Haddon, A.C. The study of man. N.Y., 1898.
Geog 5925.5	London, England. Royal Geographical Society. Antarctic exploration. London, 1898.
Geog 4208.98.18	Mackin, S.M.A. (Mrs.). A society woman on two continents. N.Y., 1898.
Geog 1709.13.5	Murray, John, publisher, London. Handbook for...Hampshire. 5. ed. London, 1898.
Geog 1807.13	Murray, John, publisher, London. Handbook for...India, Burma and Ceylon. 3. ed. London, 1898.
Geog 1709.33.5	Murray, John, publisher, London. Handbook for travellers in Surrey. 5. ed. London, 1898.
Geog 5558.86	Peary, R.E. Northward over the "Great Ice"...1886. N.Y., 1898. 2v.
Geog 139.5.5	Playfair, R.L. Supplement to the bibliography of Algeria, 1895. London, 1898.
Geog 665.23	Sherwood, M.E. Here and there and everywhere. Chicago, 1898.
An 3558.98	Silva Boasto, A.J. de. Indices cephálicos dos Portuguêses. Coimbra, 1898.
Geog 4558.98	Stevenson, P.E. A deep-water voyage. Philadelphia, 1898.
An 358.98.3	Tenishef, V. L'activité de l'homme. Paris, 1898.
Geog 665.4	Umlauft, F. Die Pflege der Erdkunde in Oesterreich...Festschrift...Franz Josef I. Wien, 1898.

1899

Geog 4208.99.2	Antin, M. From Plotzk to Boston. Boston, 1899.
An 358.99.6	Aranzadi, G. de. Etnologia. Madrid, 1899.
Geog 1539.40	Baedeker, publishers. Italy...northern Italy. 11. ed. Leipsic, 1899.
Geog 1523.80	Baedeker, publishers. Nordwest-Deutschland. 26. Aufl. Leipzig, 1899.
Geog 1516.15.3	Baedeker, publishers. Northern France. 3. ed. Leipsic, 1899.
Geog 1516.15	Baedeker, publishers. Northern France. 3. ed. Leipsic, 1899.
Geog 1568.30	Baedeker, publishers. Norway, Sweden and Denmark. 7. ed. Leipzig, 1899.
Geog 1535.58	Baedeker, publishers. Switzerland and the adjacent portions of Italy, Savoy, and Tyrol. 18. ed. Leipsic, 1899.
Geog 1645.7.5	Baedeker, publishers. United States...Mexico. 2. ed. Leipsic, 1899.
Geog 1645.7	Baedeker, publishers. United States...Mexico. 2. ed. Leipsic, 1899.
An 358.99.3	Beck, G. Der Urmensch. Basel, 1899.
Geog 819.00	Beltrán y Rózpide, R. La geografía en 1898. Madrid, 1899.
Geog 4558.98.12	Bullen, F.T. Cruise of the Cachalot. N.Y., 1899.
Geog 4558.98.11	Bullen, F.T. Cruise of the Cachalot. N.Y., 1899.
Geog 4558.98.13	Bullen, F.T. Cruise of the Cachalot. N.Y., 1899.
Geog 4558.98.14	Bullen, F.T. Cruise of the Cachalot. N.Y., 1899.
Geog 4558.99A	Bullen, F.T. The log of a sea-waif. N.Y., 1899.
Geog 4268.85.5A	Chisholm, G.G. Europe. London, 1899-1902. 2v.
Geog 4181.3	Dudley, R. Voyage of R. Dudley...to West Indies. London, 1899.
Geog 4218.96	Fraser, J.F. Round the world on a wheel. N.Y., 1899.
Geog 4508.99A	Gribble, F. The early mountaineers. London, 1899.
An 358.99.5	Herbertson, A.J. Man and his work. London, 1899.
Geog 4328.99	Honeyman, Abraham Van Doren. From America to the Orient. Plainfield, 1899.
Geog 110.2.17	International Geographical Congress, 7th, Berlin. Miscellaneous papers. v.1-6. Berlin, 1899.
Geog 3018.99	Jacobs, J. Story of geographical discovery. London, 1899.
Geog 5398.93.3	Johansen, H. With Nansen in the north. London, 1899.
Geog 5838.98	Jonsson, F. Grønlands gamle topografi. Kjøbenhavn, 1899.
Geog 5650.5	Jonsson, Finnur. Graenlendinga saga. Kaupmannahöfn, 1899.
Geog 5638.99	Jonsson, Finnur. Um Graenland ad fornu og nýju. Kaupmannahöfn, 1899.
An 3208.99	Karlsruhe, Germany. Altertumsverein. Zur Anthropologie der Badener. Jena, 1899.
An 358.99A	Keane, A.H. Man - past and present. Cambridge, Eng., 1899.
Geog 3018.99.15	Keane, John. Evolution of geography. London, 1899.
Geog 4208.87A	Kipling, Rudyard. From sea to sea. N.Y., 1899. 2v.
Geog 1709.37.12	Murray, John, publisher, London. Handbook for...Wilts and Dorset. 5. ed. London, 1899.
Geog 1811.5	Murray, John, publisher, London. Handbook for Japan. 5. ed. London, 1899.
Geog 1709.30.5	Murray, John, publisher, London. Handbook for Somerset. 5. ed. London, 1899.
Geog 1709.35	Murray, John, publisher, London. Handbook of Warwickshire. London, 1899.
Geog 3018.99.5	Nystrom, J.F. Geografiens och de...historia. Stockholm, 1899.
Geog 3018.99.10	Partsch, J. Die geographische Arbeit des 19. Jahrhunderts. Breslau, 1899.
An 378.99	Ratzel, Friedrich. Anthropogeographie. v.2. Stuttgart, 1899.
Geog 135.2.4	Ratzel, Friedrich. Beiträge zur Geographie des mittleren Deutschland. v.4. Leipzig, 1899.
An 358.99.7A	Ripley, W.Z. The races of Europe. N.Y., 1899.
An 358.99.8A	Ripley, W.Z. The races of Europe. Supplement. N.Y., 1899.

Chronological Listing

1899 - cont.

An 5.1	Ripley, W.Z. Selected bibliography of the anthropology and ethnology of Europe. Boston, 1899.
Geog 4228.99	Riseis, G. de. Dagli stati uniti alle Indie. Roma, 1899.
Geog 4181.1	Roe, T. Embassy of Sir T. Roe to court of great mogul. London, 1899. 2v.
Geog 3570.33	Sa, Ayres de. Frei Goncalo Velho. Lisboa, 1899-1900. 2v.
Geog 4208.98.5	Schweitzer, Georg. Eine Reise um die Welt. 2e Aufl. Berlin, 1899.
Geog 4558.98.5	Stevenson, P.E. By way of Cape Horn. Philadelphia, 1899.
Geog 4208.97.3	Stoddard, J.L. Lectures. Boston, 1899-1900. 10v.
Geog 4208.98.11	Temple, E.L. Old world memories. Boston, 1899. 2v.
Geog 4328.70.5	Woerl, Leo. Erzherzog Ludwig Salvator aus dem Osterreichischen Kaiserhause als Forscher des Mittelmeeres. Leipzig, 1899.
Geog 4508.99.5	Zurbriggen, M. From the Alps to the Andes. London, 1899.

19-

Geog 4219.08.10	Bertrand, A. Quelques notes sur les conférences de Tokyo et de Shanghai. Genève, 19- .
Geog 613.5	Borodajkez, Taras. Konrad Millers Lebenswerk. Salzburg, 19- .
Geog 4237.83	Burall, Paul. Cornwall to America in 1783 from the journal of Paul Burall (1755-1826). n.p., 19- .
Geog 4502.5.16	Hager, F. Das Chiemgaubuch. München, 19- .
Geog 5650.16	Hermansson, H. Landafundir og sjóferdir i nordrhofum. Winnipeg, 19- .
Geog 6009.10.16	Like English gentlemen. London, 19- .
Geog 5180.11	Marquis, R. La conquête du pôle nord en ballon. Paris, 19- .
Geog 4502.5.21	Montis, R. Kampf um den Berg. München, 19- .
Geog 4218.87.15	Paterson, Florence E. Reminiscences of a skipper's wife. London, 19- .
Geog 4502.5.22	Rohrer, M. Der Feuerberg. München, 19- .
Geog 4502.5.23	Schmitt, F. Mensch, Berg und Tod. München, 19- .
Geog 4509.34.5	Spencer, S. Mountaineering. London, 19- .
Geog 4502.5.20	Stephen, L. Der Tummelplatz Europas. München, 19-

190-

Geog 4308.96.5	Zorrilla de San Martin, J. Resonancias del camino. Barcelona, 190-?

1900

Geog 4329.00	Acker, F. Pen sketches. Philadelphia, 1900.
An 126.1	American Philosophical Society. Brinton memorial meeting. Philadelphia, 1900.
An 2109.00	Avebury, J.L. Prehistoric times. London, 1900.
An 2109.00.2	Avebury, J.L. Prehistoric times. 6. ed. N.Y., 1900.
Geog 1530.9	Baedeker, publishers. Austria. 9. ed. Leipsic, 1900.
Geog 1530.9.5	Baedeker, publishers. Austria. 9. ed. Leipsic, 1900.
Geog 1526.17	Baedeker, publishers. Berlin und Umgebungen. 11. Aufl. Leipzig, 1900.
Geog 1640.11	Baedeker, publishers. Dominion of Canada. 2. ed. Leipsic, 1900.
Geog 1540.40	Baedeker, publishers. Italy. Pt. 2: Central Italy and Rome. 13. ed. Leipsic, 1900.
Geog 1542.35	Baedeker, publishers. Italy. Pt. 3: Southern Italy, Sicily. 13. ed. Leipsic, 1900.
Geog 1510.47.12	Baedeker, publishers. London and its environs. 12. ed. Leipsic, 1900.
Geog 1523.60	Baedeker, publishers. Northern Germany. 13. ed. Leipsic, 1900.
Geog 1519.54	Baedeker, publishers. Paris and its environs. 14. ed. Leipsic, 1900.
Geog 1519.55	Baedeker, publishers. Paris and its environs. 14. ed. Leipsic, 1900.
Geog 1525.50.2	Baedeker, publishers. The Rhine from Rotterdam to Constance. 14. ed. Leizsic, 1900.
Geog 1525.50	Baedeker, publishers. The Rhine from Rotterdam to Constance. 14. ed. Leizsic, 1900.
An 3309.00	Barros Ovalle, P.N. Manual de antropometria. Santiago de Chile, 1900.
Geog 4209.00	Boyer, Bessie B. Chief Officer Brown. Chicago? 1900.
Geog 4308.69.15	Byers, S.H.M. Twenty years in Europe. Chicago, 1900.
Geog 6008.98.3	Cook, F.A. Through the first Antarctic night. N.Y., 1900.
Geog 4300.24	Cyclist's touring club handbook and guide. London, 1900.
Geog 3530.5	Dahlgren, E.W. De franska sjöfärderna. Stockholm, 1900.
An 359.00.3	Deniker, J. Les races et les peuples de la terre. Paris, 1900.
An 359.00.4	Deniker, J. The races of man. London, 1900.
Geog 5918.98.3A	Fricker, Karl. The Antarctic regions. London, 1900.
Geog 3019.00	Johnson, W.H. The world's discoveries. Boston, 1900.
Geog 4218.98.5	McMahon, William. Journey with the sun around the world. Cleveland, 1900.
An 379.00	Matteweuzzi, A. Les facteurs de l'évolution des peuples. Bruxelles, 1900.
Geog 819.01.10	Mill, H.R. International geography. 2nd ed. N.Y., 1900.
Geog 3235.19	Miller, K. Die Ebstorfkarte. Stuttgart, 1900.
Geog 1804.1A	Murray, John, publisher, London. Handbook for...Constantinople. London, 1900.
Geog 1785.7	Murray, John, publisher, London. Handbook for...Greece. 7. ed. London, 1900.
Geog 5398.98.5	Nathorst, A.G. Tva somrar, norra ishafvet. Stockholm, 1900. 2v.
Geog 4181.4	Rubruquis, G. de. Journey of William of Rubruck...and John of Pian de Carpine. London, 1900.
Geog 4181.5	Saris, J. Voyage of Captain John Saris to Japan, 1613. London, 1900.
Geog 4309.00.10	Schück, Henrik. Ur en resandes antecknigar. v.1-3. Stockholm, 1900-1909.
An 2109.00.3	Schurtz, H. Urgeschichte der Kultur. Leipzig, 1900.
Geog 665.37	Singleton, E. Wonders of nature. N.Y., 1900.
Geog 4219.00.5	Slocum, Joshua. Sailing alone around the world. N.Y., 1900.
Geog 4219.00	Slocum, Joshua. Sailing alone around the world. N.Y., 1900.
Geog 4309.00	Steevens, G.W. Glimpses of three nations. N.Y., 1900.
Geog 4309.00.15	Taylor, Charles M. Odd bits of travel with brush and camera. Philadelphia, 1900.
Geog 4308.72.5.2	Warner, Charles Dudley. Saunterings. Boston, 1900.
Geog 4521.23.5	Whymper, Edward. A letter addressed to the...Alpine Club. London, 1900.

1901

Geog 5398.99	Amadeus, Louis. Farther north than Nansen. London, 1901.
Geog 4181.7	Amherst of Hackney. Discovery of Solomon Islands by...Mendaña. London, 1901. 2v.
Geog 5324.1.5	Anderson, G. A.E. Nordenskiöld. In memoriam. Stockholm, 1901.
Geog 1555.60	Baedeker, publishers. Belgium and Holland. 13. ed. Leipsic, 1901.
Geog 1508.20	Baedeker, publishers. Great Britain. 5. ed. Leipsic, 1901.
Geog 1575.10	Baedeker, publishers. Russland. Handbuch für Reisende. 5. Aufl. Leipzig, 1901.
Geog 1545.10	Baedeker, publishers. Spain and Portugal. 2. ed. Leipsic, 1901.
Geog 1545.11	Baedeker, publishers. Spain and Protugal. 2. ed. Leipsic, 1901.
Geog 1535.59	Baedeker, publishers. Switzerland and the adjacent portions of Italy, Savoy, and Tyrol. 19. ed. Leipsic, 1901.
An 2109.01.5	Bastian, A. Der Menschheitsgedanke durch Raum und Zeit. Berlin, 1901.
Geog 4181.6	Battell, A. Strange adventures of A. Battell. London, 1901.
Geog 6008.98	Borchgrevink, C.E. First on the Antarctic continent. London, 1901.
Geog 4209.01.5A	Burton, Richard F. Wanderings in three continents. N.Y., 1901.
An 2059.01	Chantre, E. L'homme quaternaire. Paris, 1901.
Geog 4209.01	Creelman, J. On the great highway. Boston, 1901.
Geog 5557.33.5	Davidson, George. The tracks and landfalls of Bering and Chirikof on the northwest coast of America. San Francisco, 1901.
An 359.01	Demolins, E. Les grandes routes des peuples. Paris, 1901. 2v.
Geog 4309.01.5	Dudley, L.B. (Mrs.). A royal journey. N.Y., 1901.
An 2109.01.3	Frobenius, L. Aus dem Flegel Jahren der Menschheit. Hannover, 1901.
An 2339.01	Frobenius, L. Die Bogen der Oceanier. Berlin, 1901.
Geog 3057.6	Keane, Augustus Henry. The gold of Ophir. London, 1901.
An 359.01.7	Kirchoff, A. Mensch und Erde. Leipzig, 1901.
An 3539.01	Koeze, G.A. Crania ethnica Philippinica. Haarlem, 1901-04.
Geog 4239.01	Lindsey, N. Allen. Cruising in the Madiana. Boston, 1901.
Geog 520.17	Maurand, J. Itineraire de Jerome Maurand. Paris, 1901.
Geog 819.01	Mill, H.R. The international geography. N.Y., 1901.
Geog 5919.01.5	Murray, G. The Antarctic manual. London, 1901.
Geog 1811.6	Murray, John, publisher, London. Handbook for...Japan. 6. ed. London, 1901.
Geog 1709.24.2	Murray, John, publisher, London. Handbook for...Northamptonshire. 2. ed. London, 1901.
Geog 5919.01A	Neumayer, G. Auf zum Südpol! Berlin, 1901.
Geog 4216.99.10	Nunnari, F.A. Un viaggiatore calabrese. Mesina, 1901.
Geog 4309.01.9	Proctor, Rachel. To the land of the midnight sun. Utica, N.Y., 1901.
An 2109.01.15	Radliński, Ignacy. Przeszłość w teraźniejszości. Warszawa, 1901.
An 3539.01.3	Randall-Maciver, D. The earliest inhabitants of Abydos. Oxford, 1901.
An 3209.01	Russell, Frank. Laboratory outlines in somatology. N.Y., 1901.
An 359.01.3	Sergi, G. The Mediterranean race. London, 1901.
An 359.01.5	Setourneau, C.J.E. La psychologie ethnique. Paris, 1901.
Geog 4249.01	Silva, Abeillard. A través da Malasia. Coimbra, 1901.
Geog 4169.01	Singleton, Lesther. Romantic castles and palaces, as seen and described by famous writers. N.Y., 1901.
An 2109.01.9	Starr, F. Some first steps in human progress. Cleveland, 1901.
Geog 4208.97.4	Stoddard, J.L. Lectures. Supplement. n.p., 1901-05. 4v.
Geog 135.2.5	Ule, Willi. Der Würmsee (Starnbergersee) in Oberbayern. v.5. Leipzig, 1901.
Geog 4658.90.2	Vidal Gormaz, Francisco. Algunas naufragios occuridos en las costas chilenes. Santiago de Chile, 1901.
Geog 520.18	Vignaud, Henry. La lettre et la carte. Paris, 1901.
Geog 4309.01	Vivian, O.W.H. (Mrs.). The romance of religion. N.Y., 1901.

1902

Geog 14.200.15	Annales de géographie. Table décennale, 1891-1901. Paris, 1902- 3v.
Geog 1616.20	Baedeker, publishers. Egypt. 5. ed. Leipsic, 1902.
Geog 1510.49	Baedeker, publishers. London and its environs. 13. ed. Leipsic, 1902.
Geog 1510.48	Baedeker, publishers. London and its environs. 13. ed. Leipsic, 1902.
Geog 1510.50	Baedeker, publishers. London and its environs. 13. ed. Leipsic, 1902. 2v.
Geog 1523.81	Baedeker, publishers. Nordwestdeutschland. 27. Aufl. Leipzig, 1902.
Geog 1517.17	Baedeker, publishers. Southern France. 4. ed. Leipsic, 1902.
Geog 1524.51	Baedeker, publishers. Southern Germany. 9. ed. Leipsic, 1902.
Geog 5939.02	Balch, E.S. Antarctica. Philadelphia, 1902.
Geog 4309.02.20	Bell, L.L. Abroad with the Jimmies. Boston, 1902.
Geog 4309.02.4	Belloc, H. The path to Rome. N.Y., 1902.
An 359.02A	Brinton, D.G. Basis of social relations. N.Y., 1902.
Geog 5225.10	Bruun, Daniel. Kampen om nordpolen. Kjøbenhavn, 1902.
An 359.02.3	Bryce, J. The relationship of the advanced and backward races of mankind. Oxford, 1902.
Geog 4269.02	Carpenter, F.G. Europe. N.Y., 1902.
Geog 4181.10	Castanhoso. Portugese expedition to Abyssinia. London, 1902.
Geog 4209.02.3	Colquhoun, E.C. (Mrs.). Two on their travels. London, 1902.
Geog 4209.02.5	Colquhoun, E.C. (Mrs.). Two on their travels. N.Y., 1902.
An 2339.02	Frobenius, L. Menschenjagden und Zweikämpfe. Jena, 1902.
Geog 6008.97.2	Gerlache de Gomery, Adrien de. Quinze mois dans l'Antarctique. Bruxelles, 1902.
Geog 6008.97	Gerlache de Gomery, Adrien de. Voyage de la Belgica, quinze mois dans l'Antarctique. Paris, 1902.
Geog 616.1.3	Gravier, G. Notice sur Jean Parmentier. Rouen, 1902.
Geog 3019.02	Günther, S. Entdeckungsgeschichte. Berlin, 1902.
An 3309.02	Hastings, William W. Manual for physical measurements. Springfield, 1902.
Geog 5160.8	Kersting, Rudolf. The white world. N.Y., 1902.

Chronological Listing

1902 - cont.

Geog 4419.02	Krausz, S. Zu Land und See im Orient. Chicago, 1902.
Geog 4308.99	Letchworth, J. Home letters from the continent. N.Y., 1902.
Geog 4559.02	Lubbock, A.B. Round the Horn before the mast. N.Y., 1902.
Geog 4329.02	McCready, R.H. Cruise of the Celtic...Mediterranean. N.Y., 1902.
Geog 1709.1.10	Murray, John, publisher, London. Handbook for Berkshire. London, 1902.
Geog 1713.6	Murray, John, publisher, London. Handbook for travellers in Ireland. 6. ed. London, 1902.
An 359.02.6	Pogodin, A.L. Sbornik" statei po arkheologii i etnografii. Sankt Peterburg, 1902.
Geog 4209.02	Prado, Eduardo. Viagens. v.1, 2a ed; v.2, 1a ed. São Paulo, 1902. 2v.
Geog 959.12F	Raymond, E.L. Marvelous scenes of the world. Chicago, 1902.
Geog 3055.49	Rosmy, Léon de. L'Atlantide historique. Paris, 1902.
An 2109.02	Schurtz, H. Altersklassen und Männerbünde. Berlin, 1902.
Geog 819.02	Seydlitz, Ernst von. Grosses Lehrbuch der Geographie. Breslau, 1902.
An 359.02.5	Steenstrup, J. Ethnografien. Kjøbenhavn, 1902.
Geog 4181.9	Teixeira, P. Travels of Teixeira with his "kings of Harmuz". London, 1902.
Geog 4169.02A	Voyages and travels, 16th and 17th centuries. N.Y., 1902. 2v.

1903

Geog 5398.99.3	Amadeus, Louis. On the "Polar Star". N.Y., 1903. 2v.
Geog 5398.99.5	Amadeus, Louis. Die Stella Polare im Eismeer. Leipzig, 1903.
Geog 6009.03.5	Argentine Republic. La Argentina en los mares antarticos. Buenos Aires, 1903.
Geog 1526.20	Baedeker, publishers. Berlin and its environs. Leipsic, 1903.
Geog 1533.10	Baedeker, publishers. Eastern Alps. 10. ed. Leipsic, 1903.
Geog 1539.45	Baedeker, publishers. Italy...northern Italy. 12. ed. Leipsic, 1903.
Geog 1539.46	Baedeker, publishers. Italy...northern Italy. 12. ed. Leipsic, 1903.
Geog 1568.35	Baedeker, publishers. Norway, Sweden and Denmark. 8. ed. Leipsic, 1903.
Geog 1525.52.5	Baedeker, publishers. The Rhine from Rotterdam to Constance. 15. ed. Leipzig, 1903.
Geog 1535.59.50	Baedeker, publishers. Switzerland and the adjacent portions of Italy, Savoy, and Tyrol. 20. ed. Leipsic, 1903.
Geog 4341.73.6	Benjamin ben Jonah, of Tudela. Die Reisebeschreibungen. Jerusalem, 1903.
Geog 4309.03	Bolton, C.E. Travels in Europe. N.Y., 1903.
An 359.03.9	Bryce, J. Relationship of the advanced and backward races of mankind. 2. ed. Oxford, 1903.
Geog 5324.2.5	Bull, Jacob B. Fridtjof Nansen, a book for the young. Boston, 1903.
An 359.03	Clevenger, S.V. The evolution of man and his mind. Chicago, 1903.
Geog 520.19	Codex Ramirez. Histoire de l'origine des Indiens. Paris, 1903.
Geog 3235.21	Crivellari, G. Alcuni cimeli della cartografia. Firenze, 1903.
VGeog 4309.03.15	Deák, Imre. Zarandoklat Rómába. Fajsz? 1903.
Geog 3070.16	Dröber, W. Kartographie bei den Naturvölkern. Erlangen, 1903.
Geog 500.40	Gesellschaft für Erdkunde zu Berlin. Bibliothek. Katalog der Bibliothek der Gesellschaft für Erdkunde zu Berlin. Berlin, 1903.
Geog 3030.3F	Guénin, E. La route de l'Inde. Paris, 1903.
Geog 4184.1A	Hakluyt, R. Principal navigations...voyages traffiques. Glasgow, 1903-05. 12v.
Geog 4184.13	Hakluyt, R. Texts and versions of...Carpini and...Rubruquis. London, 1903.
An 2109.03.5	Hellwig, A. Das Asylrecht der Naturvölker. Berlin, 1903.
An 2059.03	Hoernes, Moriz. Der diluviale Mensch in Europa. Braunschweig, 1903.
Geog 3249.03	Hugues, Luigi. Cronologia delle scoperte e delle esplorazioni geografiche dall'anno 1492 a tutto il secolo XIX. Milano, 1903.
An 2059.03.5	Joby, N. Man before metals. N.Y., 1903.
An 629.03	Keller, A.G. Queries in ethnography. N.Y., 1903.
Geog 4219.01	Loewenbach, Lottraire. Promenade autour du globe. Paris, 1903.
An 2059.03.7	Mortillet, G. de. Musée prehistorique. 2. éd. Paris, 1903.
Geog 4306.17.3	Moryson, Fynes. Shakespeare's Europe. London, 1903.
Geog 1712.7.25	Murray, John, publisher, London. Handbook for...Scotland. 8. ed. London, 1903.
Geog 1709.2	Murray, John, publisher, London. Handbook for Buckinghamshire. London, 1903.
Geog 1709.20.2	Murray, John, publisher, London. Handbook for Lincolnshire. 2. ed. London, 1903.
Geog 4269.03	Partsch, J. Central Europe. London, 1903.
Geog 4308.19.5	Plessis, Joseph O. Journal d'un voyage en Europe. Québec, 1903.
An 359.03.5	Reches, Elie. Les primitifs, étude d'ethnologie. Paris, 1903.
Geog 3550.14	Revelli, P. La casa di Savoia e gli studii geografici. Milano, 1903.
An 359.03.7	Schurtz, H. Völkerkunde. Leipzig, 1903.
Geog 819.03	Seydlitz, Ernst von. Kleines Lehrbuch der Geographie. Breslau, 1903.
Geog 4129.03	Shand, A.I. Old-time travel. London, 1903.
Geog 4209.03	Simpson, William. Autobiography. London, 1903.
Geog 212.109	Societa Geografica Italiana. Catalogo della biblioteca sociale. Roma, 1903.
Geog 4309.02.5	Some account of...travels of myself. N.Y., 1903.
Geog 4307.77	Staszic, S. Dziennik podrózy, 1777-1791. v.1-2. Warszawa, 1903.
Geog 5398.98	Sverdrup, O. Neues Land, vier Jahre in arktischen Regionen. Leipzig, 1903. 2v.
Geog 4269.03.5	Symons, A. Cities. London, 1903.
Geog 5209.03	Thompson, G.M. Rannsóknarferð i kafli. Gimli, 1903.
An 2109.03A	Tylor, Edward Burnett. Primitive culture, researches. London, 1903. 2v.

1903 - cont.

Geog 818.82.7	Wagner, H. Lehrbuch der Geographie. Hannover, 1903. 2v.
An 359.03.3	Ward, D.J.H. The human races. n.p., 1903.
Geog 4309.02	Woodward, M. His last log. Chicago, 1903.

1904

Geog 4229.04	Armstrong, W.N. Around the world with a king. London, 1904.
Geog 4169.04	Autour du monde. Paris, 1904.
Geog 1555.29	Baedeker, publishers. Belgien und Holland. 23. Aufl. Leipzig, 1904.
Geog 1540.45	Baedeker, publishers. Italy. Pt. 2: Central Italy and Rome. 14. ed. Leipzig, 1904.
Geog 1538.50	Baedeker, publishers. Italy from the Alps to Naples. Leipzig, 1904.
Geog 1645.11	Baedeker, publishers. Nordamerika. Die Vereinigten Staaten...Mexiko. 2. Aufl. Leipzig, 1904.
Geog 1523.66	Baedeker, publishers. Northern Germany. 14. ed. Leipzig, 1904.
Geog 1523.65	Baedeker, publishers. Northern Germany. 14. ed. Leipzig, 1904.
Geog 1605.18	Baedeker, publishers. Palästina und Syrien. 6. Aufl. Leipzig, 1904.
Geog 1519.58	Baedeker, publishers. Paris and its environs. 15. ed. Leipzig, 1904.
Geog 1645.10A	Baedeker, publishers. United States...Mexico. 3. ed. Leipzig, 1904.
Geog 135.2.6	Beiträge zur Biogeographie und Morphologie der Alpen. v.6. Leipzig, 1904.
Geog 4181.14	Belmonte Bermudez, L. de. Voyages of Pedro Fernandez de Quiros. London, 1904. 2v.
Geog 3019.04.3	Böhme, Max. Die grossen Reisesammlungen des 16. Jahrhunderts und ihre Bedeutung. Strassburg, 1904.
Geog 4181.11	Conway, W.M. Early Dutch and English voyages. London, 1904.
Geog 1616.20.2	Egyptian Museum, Cairo. Gratis supplement to the fifth edition of Baedeker's Egypt. Leipzig, 1904.
Geog 520.20	Fonteneau, J. La cosmographie. Paris, 1904.
Geog 618.2.5	Frenzel, C.A. Major James Rennell. Leipzig, 1904.
Geog 3019.04	Günther, S. Geschichte der Erdkunde. Leipzig, 1904.
Geog 4329.04	Loring, J.M. The old world thro' new world eyes. St. Louis, 1904.
Geog 4208.53.5	Mather, Frank J. A clipper ship and her commander. n.p., 1904.
Geog 1709.39.4	Murray, John, publisher, London. Handbook for...Yorkshire. 4. ed. London, 1904.
Geog 1709.7.3	Murray, John, publisher, London. Handbook for Derbyshire, Nottinghamshire. 3. ed. London, 1904.
Geog 1807.30	Murray, John, publisher, London. The Imperial guide to India. London, 1904.
Geog 6009.01	Nordenskjöld, Otto. Antarctic...zwei Jahre. Berlin, 1904. 2v.
Geog 665.5	Ratzel, F. Zu Friedrich Ratzels Gedächtnis. Leipzig, 1904.
An 193.1	Sergi, Giuseppe. Problemi di scienza contemporanea. Palermo, 1904.
Geog 5398.98.2	Sverdrup, O. New land, four years in Arctic regions. London, 1904. 2v.
An 359.04.3	Tylor, Edward B. Anthropology. N.Y., 1904.
An 359.04	Yertz, Friedrich O. Moderne Rassentheorien. Wien, 1904.

1905

Geog 6009.01.9	Armitage, A.B. Two years in the Antarctic. London, 1905.
Geog 3019.05	Athlenius, K. Landkonturer och hafsvidder. Stockholm, 1905.
Geog 1530.10A	Baedeker, publishers. Austria-Hungary. 10. ed. Leipzig, 1905.
Geog 1555.65	Baedeker, publishers. Belgium and Holland. 14. ed. Leipzig, 1905.
Geog 1526.25	Baedeker, publishers. Berlin and its environs. 2. ed. Leipsic, 1905.
Geog 1604.5	Baedeker, publishers. Konstaninopel und das westliche Kleinasien. Leipzig, 1905.
Geog 1510.55	Baedeker, publishers. London and its environs. 14. ed. Leipzig, 1905.
Geog 1510.2	Baedeker, publishers. London und Umgebung. 15. Aufl. Leipzig, 1905.
Geog 1516.20.4	Baedeker, publishers. Northern France. 4. ed. Leipzig, 1905.
Geog 1519.19	Baedeker, publishers. Paris, nebst einigen Routen. 16. Aufl. Leipzig, 1905.
Geog 1535.60	Baedeker, publishers. Switzerland and the adjacent portions of Italy, Savoy, and Tyrol. 21. ed. Leipsic, 1905.
Geog 1535.61	Baedeker, publishers. Switzerland and the adjacent portions of Italy, Savoy, and Tyrol. 21. ed. Leipsic, 1905.
Geog 3530.9	Barré, H. Voyageurs et explorateurs provençaux. Marseille, 1905.
Geog 4181.12	Bowrey, T. Geographical account...countries...Bengal. Cambridge, 1905.
Geog 5650.10.3	Bruun, Daniel. De gamle Nordbokdonieri Grønland. København, 1905.
Geog 4209.05	Herboso, F.J. Reminiscencias de viajes. Caracas, 1905- 3v.
Geog 4309.03.5	Higinbotham, John V. Three weeks in Europe. Chicago, 1905.
Geog 4181.16	Jourdain, J. Journal of John Jourdain, 1608-17. Cambridge, 1905.
An 359.05	Koropchevskago, A.A. Zhachenie "geograficheskikh provintsie". Sankt Peterburg, 1905.
Geog 585.9	Levasseur, E. Extrait de l'Annuaire in Bureau de Longitudes. n.p., 1905.
An 2059.05	Maccurdy, G.G. The eolithic problem. Lancaster, 1905.
Geog 4229.05	Macdonald, A. In search of El Dorado. London, 1905.
Geog 5939.05	Mill, H.R. Siege of the South Pole. London, 1905.
An 3309.05	Montessori, Maria. Caratteri fisici delle Giovani donne del Lazio. Roma, 1905.
Geog 1807.17	Murray, John, publisher, London. Handbook for...India, Burma and Ceylon. 5. ed. London, 1905.
Geog 1709.34.5	Murray, John, publisher, London. Handbook for Sussex. 5. ed. London, 1905.
Geog 6009.01.3	Nordenskjöld, Otto. Antarctica, or Two years...South Pole. London, 1905.

Chronological Listing

1905 - cont.

	Geog 3205.50.5	Pacheco Pereira. Esmeraldo de situ orbis. Lisboa, 1905.
	Geog 4184.14A	Purchas, Samuel. Hakluytus posthumus, or Purchas, his pilgrimes. Glasgow, 1905-07. 20v.
	Geog 618.1.9	Richter, O. Der teleologische Zug im Denken C. Ritters. Borna, 1905.
	An 359.05.3	Rotzell, W.E. Man: an introduction to anthropology. 2. ed. Philadelphia, 1905.
Htn	Geog 6009.01.8*	Scott, R.F. Voyage of the 'Discovery'. London, 1905. 2v.
	Geog 6009.01.7A	Scott, R.F. Voyage of the 'Discovery'. N.Y., 1905. 2v.
	Geog 4208.97.5	Stoddard, J.L. Lectures. v.8. Boston, 1905.
	Geog 4219.05.4	Treves, F. The other side of the lantern. London, 1905.
	Geog 4219.05.3	Treves, F. The other side of the lantern. London, 1905.
	Geog 500.65	Verein für Erdkunde, Dresden. Bibliothek. Bücherei-Verzeichnis des Vereins für Erdkunde zu Dresden. Dresden, 1905.
	Geog 4209.05.5	Williams, Archibald. The romance of modern exploration. Philadelphia, 1905.

1906

Geog 4678.52 — Addison, A.C. A deathless story. London, 1906.
Geog 16.1.2 — Appalachia. Index, v.1-10. Boston, 1906.
Geog 4209.04 — Arthur, Richard. Ten thousand miles in a yacht. N.Y., 1906.
Geog 1508.25 — Baedeker, publishers. Great Britain. 6. ed. Leipsic, 1906.
Geog 1539.50 — Baedeker, publishers. Italy...northern Italy. 13. ed. Leipsic, 1906-
Geog 1525.53 — Baedeker, publishers. The Rhine from Rotterdam to Constance. 16. ed. Leipzig, 1906.
An 359.06 — Balch, E.S. Comparative art. Philadelphia, 1906.
Geog 3251.21 — Behrmann, W. Uber die niederdeutschen Seebücher. Hamburg, 1906.
Geog 4309.02.7 — Bierbaum, Otto Julius. Mit der Kraft; Automobilia. Berlin, 1906.
An 126.2 — Boas anniversary volume. N.Y., 1906.
Geog 6009.02 — Brown, R.N.R. Voyage of the "Scotia". Edinburgh, 1906.
Geog 4309.05 — Cabrera, Raimundo. Cartas a Estevez. Habana, 1906.
An 359.06.5 — Chalikiopoulos, L. Landschafte...Kulturtypen. Leipzig, 1906.
Geog 6009.03 — Charcot, J.B. Le "Francais" au Pole Sud. Paris, 1906.
Geog 5070.25 — Congrès International pour l'Étude des Régions Polaires, 1st, Brussels, 1906. Rapport d'ensemble. Bruxelles, 1906.
Geog 5399.03 — Fiala, A. Fighting the polar ice. N.Y., 1906.
An 359.06.3 — Finot, Jean. Le préjugé des races. Paris, 1906.
An 359.06.4 — Finot, Jean. Race prejudice. London, 1906.
Geog 4329.06 — Fischer, T. Mittelmeerbilder. Leipzig, 1906.
Geog 5639.05 — Först, J. Geschichte der Entdeckung Grönlands von den ältesten Zeit bis zum Anfang des 19. Jahrhunderts. Worms, 1906.
Geog 4209.06 — Gambier, J.W. Links on my life on land and sea. N.Y., 1906.
Geog 618.3.5 — Greef, Guillaume J. de. Discours prononcé par Monsieur le recteur Guillaume de Greef. Gand, 1906.
Geog 5208.95.3 — Greely, A.W. Handbook of polar discoveries. 3. ed. Boston, 1906.
Geog 4300.16 — Guerber, H.A. How to prepare for Europe. N.Y., 1906.
An 3559.06 — Hauser, K. Das kraniologische der New Guinea Expedition. Berlin, 1906.
Geog 5209.06.2 — Hoare, J.D. Arctic exploration. London, 1906.
Geog 5209.06 — Hoare, J.D. Arctic exploration. N.Y., 1906.
Geog 4219.06 — Huber, Max. Tagebuchblätter aus Siberien, Japan. Zürich, 1906.
Geog 4321.83 — Ibn Gubáyer. Viaggio in Espana, Sicilia, Siria e Palestina. Roma, 1906.
Geog 4326.32.10A — Lithgow, William. Travels and voyages thro' Eruope, Asia and Africa. Glasgow, 1906.
Geog 139.4.5 — London. Royal Geographical Society. Geographical journal. Index. 1893-1902. London, 1906. 3v.
Geog 5559.03 — Low, A.P. Report on dominion government expedition. Ottawa, 1906.
Geog 4105.18.4 — Luce, Robert. Going abroad? 4. ed. Boston, 1906.
An 359.06.12 — Maldonado, S.D. Defensa de la Antropologia general. Caracas, 1906.
Geog 1807.17.5 — Murray, John, publisher, London. Handbook for...India, Burma and Ceylon. 5. ed. London, 1906.
Geog 1713.7 — Murray, John, publisher, London. Handbook for travellers in Ireland. 7. ed. London, 1906.
Geog 1713.6.75 — Murray, John, publisher, London. Handbook for travellers in Ireland. 7. ed. London, 1906.
Geog 1713.8 — Murray, John, publisher, London. Handbook for travellers in Ireland. 7. ed. London, 1906.
Geog 5839.03 — Mylius, L. Grønland...1903-04. København, 1906.
Geog 4268.94.5 — Philippson, A. Europa. Leipzig, 1906.
An 2109.06 — Pitt-Rivers, A.L.F. The evolution of culture and other essays. Oxford, 1906.
Geog 4129.06 — Raleigh, Walter. The English voyages of the sixteenth century. Glasgow, 1906.
Geog 3019.06 — Roberts, C.G.D. Discoveries and explorations in the century. Toronto, 1906.
Geog 5110.3 — Rud, William. The phantom of the poles. N.Y., 1906.
Geog 3229.06 — Sensburg, W. Poggio Bracciolini und Nicolò de Conti. Wien, 1906.
Geog 665.37.10 — Singleton, E. Greatest wonders of the world. N.Y., 1906.
Geog 4181.18 — Spillbergen, J. van. East and West Indian mirror. London, 1906.
Geog 4308.28A — Topliff, S. Topliff's travels, letters from abroad. Boston, 1906.
An 359.06.7 — Toro, E. Antropologia general. Caracas, 1906.
An 2109.06.3 — Zuccarelli, A. Gli uomini primitivi delle selci e delle caverne. Napoli, 1906.

1907

Geog 4219.07 — Aldao, C.A. A través del mundo. Buenos Aires, 1907.
Geog 520.21 — Anghiera, P.M. De orbe novo. Paris, 1907.
Geog 1640.15A — Baedeker, publishers. Dominion of Canada with Newfoundland and an excursion to Alaska. 3. ed. Leipzig, 1907.
Geog 1533.12 — Baedeker, publishers. Eastern Alps. 11. ed. Leipzig, 1907.
Geog 1519.60 — Baedeker, publishers. Paris and its environs. 16. ed. Leipzig, 1907.
Geog 1517.19.5 — Baedeker, publishers. Southern France. 5. ed. Leipzig, 1907.

1907 - cont.

Geog 1517.19 — Baedeker, publishers. Southern France. 5. ed. Leipzig, 1907.
Geog 1524.55 — Baedeker, publishers. Southern Germany. 10. ed. Leipzig, 1907.
Geog 1535.65 — Baedeker, publishers. Switzerland and the adjacent portions of Italy, Savoy and Tyrol. 22. ed. Leipzig, 1907.
Geog 4218.48 — Beck, Christian. Reise um die Welt. 11e Aufl. Dresden, 1907.
Geog 4341.73.10A — Benjamin ben Jonah, of Tudela. The itinerary of Benjamin of Tudela. London, 1907.
Geog 6100.44 — Bossière, R.E. Nouvelle notice sur les îles Kerquelen. Paris, 1907.
Geog 4309.07.5 — Bull, Edvard. Over lave og høie fjelde; breve fra en aeldre berre. Kristiania, 1907.
Geog 5939.07.5 — Chun, Karl. Die Erforschung der Antarktis. Leipzig, 1907.
Geog 4558.40.20A — Dana, Richard Henry. Two years before the mast. N.Y., 1907.
Geog 579.07 — Demangeon, A. Dictionnaire, manuel, illustré. Paris, 1907.
Geog 4181.21 — Espinosa, A. de. The guanches of Tenerife. London, 1907.
Geog 5399.03.2 — Fiala, A. Fighting the polar ice. 2. ed. N.Y., 1907.
An 2109.01.7 — Finot, Jean. Race prejudice. N.Y., 1907.
An 359.06.4.1 — Finot, Jean. Race prejudice. N.Y., 1907.
Geog 4300.10 — Frager, M.D. Practical European guide. Boston, 1907.
An 629.07 — Frazer, J.G. Questions on customs, beliefs and languages of savages. Cambridge, 1907.
Geog 5209.07 — Gordon, W.J. Round about the North Pole. N.Y., 1907.
Geog 959.07 — Grosvenor, G.H. Scenes from every land. Washington, D.C., 1907.
Geog 665.17 — Günther, S. Geographische Studien. Stuttgart, 1907.
Geog 3520.10.35 — Hakluyt, Richard. Principal navigations, voyages. London, 1907-13. 8v.
Geog 809.07 — Hettner, A. Grundzüge der Landerkunde. Leipzig, 1907-1925. 2v.
Geog 5399.06 — Holmes, B.F. The log of the "Laura" in polar seas. Cambridge, 1907.
An 3509.07 — Hrdlička, A. Skeletal remains...early man...North America. Washington, 1907.
Geog 4709.7 — Knight, E.F. The cruise of the "Alerte". London, 1907.
An 359.07 — Lankester, E.R. Kingdom of man. N.Y., 1907.
An 2059.07 — Lapparent, A. Les silex taillés et l'ancienneté de l'homme. Paris, 1907.
Geog 4329.07 — Letters from a Tarheel traveler. Oxford, 1907.
Geog 3235.13 — Longhena, M. Atlanti e carte nautiche. Parma, 1907.
Geog 4209.07 — Mallik, M.C. Impressions of a wanderer. London, 1907.
Geog 4110.27 — Meyer, H.J. Weltreise. Leipzig, 1907.
Geog 819.01.7 — Mill, H.R. The international geography. N.Y., 1907.
Geog 4305.91A — Moryson, Fynes. An itinerary containing his 10 yeeres travell. Glasgow, 1907. 4v.
Geog 4181.17 — Mundy, P. Travels of Peter Mundy in Europe and Asia, 1608-67. v.1-5. Cambridge, 1907-36. 6v.
Geog 1816.25 — Murray, John, publisher, London. Handbook for...Egypt and the Sudan. 11. ed. London, 1907.
Geog 1807.18 — Murray, John, publisher, London. Handbook for...India, Burma and Ceylon. 6. ed. London, 1907.
Geog 1811.8A — Murray, John, publisher, London. Handbook for...Japan. 8. ed. London, 1907.
Geog 1712.8 — Murray, John, publisher, London. Handbook for...Scotland. 8. ed. London, 1907.
An 608.609.07 — Nielsen, Y. Universitets ethnografiske samlinger, 1857-1907. Christiania, 1907.
An 3539.07 — Oetteking, B. Kraniologische Studien...Altägypten. Braunschweig, 1907.
Geog 5559.05 — Peary, R.E. Nearest the Pole...1905-06. N.Y., 1907.
Geog 5509.07 — Randall, Harry. The conquest of the Northwest Passage. Minneapolis, 1907.
Geog 5855.3 — Rasmussen, Knud. Neue Menschen...des Nordpols. Bern, 1907.
Geog 4209.07.5 — Russell, J.W. The romance of an old time shipmaster. N.Y., 1907.
Geog 4181.22A — Sarmiento de Gamboa, Pedro. History of the Incas and execution of the Inca Tupac Amaru. Cambridge, 1907.
Geog 4309.07 — Seitz, Don C. Discoveries in every-day Europe. N.Y., 1907.
Geog 212.9 — Sociedad Geográfica de la Paz. Estatutos de la "Sociedad geográfica de la Paz." 2. ed. La Paz, 1907.
Geog 4309.07.10 — Society recollections in Paris and Vienna, 1789-1904, by an English officer. London, 1907.
Geog 5855.9 — Solberg, O. Beiträge zur Vorgeschichte. Christiania, 1907.
Geog 3251.23 — Stevenson, E.L. Map of the world. N.Y., 1907.
Geog 4110.7 — Thorpe, D. Universal guide of standard routes. Boston, 1907.

1908

Geog 4509.07.1 — Abraham, G.D. The complete mountaineer. N.Y., 1908.
Geog 4509.07.2 — Abraham, G.D. The complete mountaineer. 2. ed. London, 1908.
Geog 5519.03 — Amundsen, R. The North West Passage. London, 1908. 2v.
Geog 5519.03.1 — Amundsen, R. Roald Amundsen's "The North West Passage". N.Y., 1908. 2v.
Geog 1526.30 — Baedeker, publishers. Berlin and its environs. 3. ed. Leipsic, 1908.
Geog 1616.21 — Baedeker, publishers. Egypt. 6. ed. Leipzig, 1908.
Geog 1539.51 — Baedeker, publishers. Italie septentrionale. 17. éd. Leipzig, 1908.
Geog 1510.56 — Baedeker, publishers. London and its environs. 15. ed. Leipzig, 1908.
Geog 1510.57 — Baedeker, publishers. London and its environs. 15. ed. Leipzig, 1908.
Geog 1542.37 — Baedeker, publishers. Southern Italy and Sicily. 15. ed. Leipzig, 1908.
Geog 1545.15 — Baedeker, publishers. Spain and Portugal. 3. ed. Leipzig, 1908.
An 359.08.5 — British Museum (Natural History). Department of Zoology. Guide...specimens illustrating the races of mankind. London, 1908.
Geog 5650.10.11 — Bruun, Daniel. Godthaabsegnen og den gamle Vesterbygd. Odense, 1908.
Geog 4219.08.5 — Carter, James. In the wake of the setting sun. London, 1908.
An 3559.08 — Channing, Walter. The hard palate in normal and feeble-minded individuals. N.Y., 1908.

Chronological Listing

1908 - cont.

	An 359.08.3	Coffin, E.W. On the education of backward races. Worcester, 1908.
	Geog 4218.49	Coffin, George. Pioneer voyage to California and round the world, 1849-1852. Chicago? 1908.
	Geog 4208.60.3	Colquohoun, A.R. Dan to Beersheba. London, 1908.
	Geog 4181.13	Corney, B.G. Voyage of Captain Don Felippe Gonzalez. Cambridge, 1908.
	Geog 4181.23A	Diaz del Castillo, B. True history of conquest of New Spain. London, 1908-10. 5v.
	An 359.08	Dieserud, J. The scope and content of the science of anthropology. Chicago, 1908.
	Geog 4329.06.2	Fischer, T. Mittelmeerbilder. Leipzig, 1908.
	Geog 613.2.5	Frenzel, R. Malthe Conrad Bruun. Crimmitschau, 1908.
	An 359.08.7A	Gehring, A. Racial contrasts. N.Y., 1908.
	Geog 5919.08	Gourdon, Ernest. Géographie physique, glaciologie, pétrographie des régions visitées par l'Expédition antarctique française...1903-05. Paris, 1908.
	Geog 4678.55	Hinckley, F. Wrecked on a reef in the China Sea. Boston, 1908.
X Cg	Geog 4269.07.7PF	Hunnewell, H.H. Nineteen thousand miles in 1907. Boston, 1908. 8v.
	An 359.08.9	Keane, A.H. The world's peoples. London, 1908.
	An 359.08.10	Keane, A.H. The world's peoples. N.Y., 1908.
	Geog 4215.19.5	Kölliker, O. Die erste Umseglung der Erde...1519-1522. München, 1908.
	Geog 520.22	Le Bouvier, Gilles. Le livre de la description des pays. Paris, 1908.
	An 39.08	Matsumura, A. A gazetteer of ethnology. Tokyo, 1908.
NEDL	Geog 144.1.2F	Mittheilungen. Berlin, 1908. 6v.
	Geog 4309.07.13	More society recollections, by an English officer. London, 1908.
	An 3309.08	New South Wales. Department of Public Instruction. Report - physical condition of children. Sydney, 1908.
	Geog 5399.08F	The Norwegian Aurora polaris expedition. Christiania, 1908.
	Geog 4309.08	Osborne, T.M. Adventures of a green dragon. Auburn, 1908.
	Geog 4110.20.3	Peninsular and Oriental Steam Navigation Company. The P. and O. pocket book. London, 1908.
	Geog 3057.5	Peters, Karl. Ophir nach dem neuesten Forschungen. Berlin, 1908.
	Geog 5855.5	Rasmussen, Knud. The people of the Polar North. London, 1908.
	Geog 5855.5.5	Rasmussen, Knud. The people of the Polar North. Philadelphia, 1908.
	An 379.08	Richthofen, Ferdinand von. Vorlesungen. Berlin, 1908.
	Geog 665.47	Scritti di geografia e di storia della geografia conserventi l'Italia pubblicati in onore di Giuseppe della Vedova. Firenze, 1908.
	Geog 3251.23.5	Stevenson, E.L. Marine world chart, 1502. N.Y., 1908.
	An 3309.08.5	Tocher, James F. Pigmentation survey of school children in Scotland. Aberdeen, 1908.
	Geog 4219.08.2	Treves, F. Other side of the lantern. London, 1908.
	An 379.12	Vallaux, C. Géographie sociale. La mer. Paris, 1908.
	An 2109.08	Webster, H. Primitive secret societies. N.Y., 1908.

1909

	Geog 4308.99.10	Allen, Grant. The European tour. N.Y., 1909.
	Geog 1540.48	Baedeker, publishers. Central Italy and Rome. 15. ed. Leipzig, 1909.
	Geog 1540.47	Baedeker, publishers. Central Italy and Rome. 15. ed. Leipzig, 1909.
	Geog 1585.26	Baedeker, publishers. Greece. 4. ed. Leipzig, 1909.
	Geog 1585.25	Baedeker, publishers. Greece. 4. ed. Leipzig, 1909.
	Geog 1538.56A	Baedeker, publishers. Italy from the Alps to Naples. 2. ed. Leipzig, 1909.
	Geog 1516.25	Baedeker, publishers. Northern France. 5. ed. Leipzig, 1909.
	Geog 1568.40	Baedeker, publishers. Norway, Sweden and Denmark. 9. ed. Leipzig, 1909.
	Geog 1535.68	Baedeker, publishers. Switzerland and the adjacent portions of Italy, Savoy, and Tyrol. 23. ed. Leipzig, 1909.
	Geog 1645.15A	Baedeker, publishers. United States...Mexico. 4. ed. Leipzig, 1909.
	Geog 4329.09	Bayne, S.G. A fantasy of Mediterranean travel. N.Y., 1909.
	An 3309.09	Bertillon, A. Anthropologie métrique. Paris, 1909.
	An 2109.09.11	Bölsche, W. Der Mensch der Vorzeit. v.1-2. Stuttgart, 1909-11.
	Geog 4209.09	Burge, C.O. Adventures of a civil engineer. London, 1909.
	An 359.09.5	Buschan, Georg H.T. Menschenkunde. Stuttgart, 1909.
	Geog 5180.10	Byhan, A. Die Polarvölker. Leipzig, 1909.
	Geog 5559.06.5	Canada. Department of Marine and Fisheries. Report on the dominion government expedition to Arctic islands and the Hudson Strait on board the C.G.S. "Arctic", 1906-07. Ottawa, 1909.
	An 2109.09.7A	Crawley, A.E. The idea of the soul. London, 1909.
	Geog 4304.65.5	Cust, Nina W.G. Gentlemen errant; being the journeys and adventures of four noblemen in Europe during the fifteenth and sixteenth centuries. London, 1909.
	Geog 5839.09	Friis, A. Danmark expeditionen. Kjøbenhavn, 1909.
	An 3559.09	Frizzi, E. Ein Beitrag zur Anthropologie des "Homo Alpinus Tirolensis". Wien, 1909.
	An 359.09	Frobenius, Leo. The childhood of man. London, 1909.
	Geog 4181.19	Fryer, J. New account of East India and Persia. London, 1909. 3v.
	Geog 959.07.5	Grosvenor, G.H. Scenes from every land. 2. ser. Washington, D.C., 1909.
	An 359.09.3	Haddon, A.C. Races of man and their distribution. London, 1909.
	Geog 3520.10.25A	Hakluyt, Richard. Voyages of Drake and Gilbert. Oxford, 1909.
	An 359.09.8	Herbertson, A.J. Man and his work. London, 1909.
	Geog 5559.09.20	Ingersoll, Ernest. The conquest of the North. N.Y., 1909.
	Geog 4208.87.3	Kipling, Rudyard. From sea to sea. N.Y., 1909.
	Geog 3235.50	Kretschmer, K. Die italienischen Portolane des Mittelalters. Berlin, 1909.
	Geog 5323.1.2	Markham, C. Life of admiral Sir Leopold McClintock. London, 1909.
	Geog 5559.06	Mikkelsen, E. Conquering the Arctic ice. Philadelphia, 1909.
	An 359.09.15	Morgan, L.H. Ancient society. Chicago, 1909.
	Geog 4182.1	Muller, S. De reis van Jan C. May. 's-Gravenhage, 1909.
	Geog 4329.09.10	Naumann, Friedrich. Sonnenfahrten. Berlin, 1909.
	Geog 4219.09.5	Pageot, G. A travers les pays jaunes. Paris, 1909.

1909 - cont.

	An 3539.09F	Pittard, E. Anthropologie de la Suisse. Genève, 1909-10.
	Geog 4209.08	Podmore, P.S.M. Rambles and adventures in Australasia. London, 1909.
	An 3409.09	Reyna Almandos, Luis. Dactiloscopia argentina. La Plata, 1909.
	Geog 4329.09.5	Roberts, William C. The boys' account of it...foreign travel. N.Y., 1909.
	Geog 819.09	Scobel, A. Geographisches Handbuch. Bielefeld, 1909-1910. 2v.
	Geog 6009.07A	Shackleton, Ernest H. Heart of the Antarctic. Philadelphia, 1909. 2v.
	An 2109.09.9	Somló, Felix. Der Güterverkehr in der Urgesellschaft. Bruxelles, 1909.
	Geog 5839.09.10	Steensrup, K. Geologiske og antivariske Iagttageker. København, 1909.
	Geog 4219.07.5	Stephens, Edwin William. Around the world; a narrative in letter form of a trip around the world, from October 1907 to July 1908. Columbia, 1909.
	An 2109.09A	Thomas, W.I. Source book for social origins. Chicago, 1909.
	An 2109.09.5	Thomas, W.I. Source book for social origins. 6th ed. Boston, 1909.
	An 359.09.12	Tylor, Edward B. Anthropology. N.Y., 1909.
	An 359.09.11	Tylor, Edward B. Anthropology. N.Y., 1909.

191-

	Geog 500.62	Stevenson, Edward L. Publications of Edward L. Stevenson. n.p., 191-.

1910

	Geog 4509.10	Abraham, G.D. Mountain adventures at home and abroad. London, 1910.
	Geog 4349.10	The all-rail route between the Far East and Europe. London, 1910.
	Geog 1555.70	Baedeker, publishers. Belgium and Holland. 15. ed. Leipzig, 1910.
	Geog 1508.30	Baedeker, publishers. Great Britain. 7. ed. Leipzig, 1910.
	Geog 1523.67A	Baedeker, publishers. Northern Germany. 15. ed. Leipzig, 1910.
	Geog 1519.63	Baedeker, publishers. Paris and its environs. 17. ed. Leipzig, 1910.
	Geog 1524.56	Baedeker, publishers. Southern Germany. 11. ed. Leipzig, 1910.
	Geog 4269.10.5	Bartholomew, J.G. Literary and historical atlas of Europe. London, 1910.
	Geog 4219.09	Bidwell, Daniel D. As far as the East is from the West. Hartford, 1910.
	An 3409.10	Brayley, F.A. Arrangement of finger prints identification and their uses. Boston, 1910.
	An 609.10	British Museum. Handbook to ethnographical collections. Oxford, 1910.
	An 359.10.7	Buschan, Georg H.T. Illustrierte Völkerkunde. Stuttgart, 1910.
	An 3539.10PF	Düben, G. Crania Lapponica. Holmiae, 1910. 2v.
	Geog 5209.10.3	Edwards, D.M. The toll of the arctic seas. N.Y., 1910.
	Geog 3229.10	Errera, C. L'epoca delle grandi scoperte geografiche. Milano, 1910.
	Geog 4219.10	Franck, H.A. Vagabond journey around the world. N.Y., 1910.
	Geog 85.100.5F	Geografisk tidskrift. Index, 1-20. Kjøbenhavn, 1910.
	Geog 4209.09.10	Gordon, F. Listy z podrozy. Chicago, 1910.
	Geog 5208.95.7	Greely, A.W. Handbook of polar discoveries. 5th ed. Boston, 1910.
	Geog 4419.07	Hochberg, Fritz. An eastern voyage; travels thro' the British Empire...in East and Japan. London, 1910. 2v.
	Geog 4129.10	Hopkins, A.A. Scientific American handbook of travel. N.Y., 1910.
	Geog 4139.10	Johnstone, C. (Mrs.). Buccaneers of America. Akron, 1910.
	Geog 809.10	Land og folk. Geografi i skildringer og livsbilleder. Kjøbenhavn, 1910. 2v.
	Geog 6008.97.3	Lecointe, G. Au pays des manchots. Bruxelles, 1910.
	An 2109.10A	Lévy-Bruhl, L. Les fonctions mentales dans les sociétés inférieures. Paris, 1910.
	Geog 4182.2	Linschoten, Jan H. van. Itinerario voyage. 's-Gravenhage, 1910. 5v.
	Geog 4309.10.10	Matto de Turner, C. Viaje de recrea. Valencia, 1910?
	Geog 5559.09.8	Moore, J.H. Peary's discovery of the North Pole. Washington, 1910.
	Geog 4159.10.10	Northcliffe, A.C.W.H. The world's greatest books. v.19. n.p., 1910.
	Geog 4309.09	O'Laughlin, John. From the jungle through Europe with Roosevelt. Boston, 1910.
	Geog 5559.09A	Peary, R.E. North Pole; its discovery in 1909. N.Y., 1910.
	Geog 5559.09.3	Peary, R.E. The North Pole. 2. ed. N.Y., 1910.
	Geog 4419.10	Pick, Emil G. Reisebriefe eines österreichischen Industriellen aus Abessinien, Indien und Ostasien. Prag, 1910?
	Geog 819.10F	Rand, McNally and Company. The world and its peoples photographed and described. Chicago, 1910.
	An 132.1	Reinach, Theodore. Notice sur la vie et les travaux de M. Ernest Hamy. Paris, 1910.
	An 359.10.3	Ripley, W.Z. Races of Europe. N.Y., 1910.
	Geog 665.30F	Santarem, M.F. de B. Opusculos e esparsos. Lisbon, 1910. 2v.
	An 359.10	Schmidt, W. Die Stellung der Pygmaenvölker in der Entwicklungsgeschichte des Menschen. Stuttgart, 1910.
	Geog 4559.10	Shaw, Frank H. The sea and its story from viking ship to submarine. London, 1910.
	An 359.10.5	Simpson, B.P. The conflict of colour. N.Y., 1910.
	Geog 4269.10	Sothriados, G. Chôrai kai ladi tēs Evrōpēs. Athēnai, 1910.
Htn	Geog 4238.03*	Sutherland, D. A diary kept by Reverend David Sutherland on a voyage from Greenock Scotland to New York. Woodsville, 1910.
	Geog 5850.5	Trebitsch, R. Bei den Eskimos in Westgrönland. Berlin, 1910.
	Geog 5839.10	Whitney, H. Hunting with the Eskimos. London, 1910.
	Geog 5209.10	Wright, H.S.S. (Mrs.). The great white North. N.Y., 1910.
	Geog 4558.81	Wyman, Walter. A cruise on the United States practice ship, S.P. Chase. N.Y., 1910.

Chronological Listing

1911

An 2109.11	Avebury, J.L. Marriage, totenism, and relggion. London, 1911.
Geog 1530.13	Baedeker, publishers. Austria-Hungary. 11. ed. Leipzig, 1911.
Geog 1533.13	Baedeker, publishers. Eastern Alps. 12. ed. Leipzig, 1911.
Geog 1510.59A	Baedeker, publishers. London and its invirons. 16. ed. Leipzig, 1911.
Geog 1595.5A	Baedeker, publishers. Mediterranean...Madeira, Canary Islands. Leipzig, 1911.
Geog 1525.55A	Baedeker, publishers. The Rhine including the Black Forest and Vosges. 17. ed. Leipzig, 1911.
Geog 1535.70	Baedeker, publishers. Switzerland and the adjacent portions of Italy, Savoy, and Tyrol. 24. ed. Leipzig, 1911.
An 2339.11	Balfour, Henry. The origin of West African crossbows. Washington, 1911.
Geog 4279.11A	Bates, Ernest S. Touring in 1600. Boston, 1911.
Geog 4209.10	Beltramelli, Antonio. Il diario di un viandante, dal deserto al mar glaciale. Milano, 1911.
Geog 4219.11.5	Bertram, C. A magician in many lands. N.Y., 1911.
An 359.11	Boas, Franz. Curso de antropologia general. Mexico, 1911.
An 2109.11.3A	Boas, Franz. The mind of primitive man. N.Y., 1911.
Geog 5559.11.5	Borup, George. A tenderfoot with Peary. London, 1911.
Geog 5559.11	Bruce, William S. Polar exploration. London, 1911.
Geog 819.11.3	Busson, Henri. Les principales puissances du monde. Paris, 1911.
Geog 5559.10	Canada. Department of Marine and Fisheries. Report...dominion government expedition...1910. Ottawa? 1911.
An 177.1	Capitan, J.L. Notice sur les travaux scientifiques. n.p., 1911.
Geog 4182.3	Colenbrander, H.T. Korte historiael...voyagiens...David P. de Vries. 's-Gravenhage, 1911.
Geog 4558.40.27	Dana, Richard Henry. Two years before the mast. Boston, 1911.
Geog 4558.40.28	Dana, Richard Henry. Two years before the mast. Boston, 1911.
Geog 4558.40.24A	Dana, Richard Henry. Two years before the mast. Boston, 1911.
Geog 4558.40.25	Dana, Richard Henry. Two years before the mast. Boston, 1911.
Geog 4558.40.22	Dana, Richard Henry. Two years before the mast. N.Y., 1911.
Geog 4219.11.7	Dittmar, J. Eine Fahrt um die Welt. Berlin, 1911.
Geog 3235.50.9	Errera, Carlo. I portolani italiani del medioevo secondo l'opera di K. Kretschmer. Firenze, 1911.
Geog 4219.10.2	Franck, H.A. A vagabond journey around the world. N.Y., 1911.
Geog 4209.10.6	Fraser, Mary Crawford. A diplomatist's wife in many lands. London, 1911. 2v.
Geog 4209.10.5	Fraser, Mary Crawford. A diplomatist's wife in many lands. N.Y., 1911. 2v.
Geog 759.11	Giamnitrapani, D. Geografia mathematica, geografia generale. Firenze, 1911.
An 629.11	Graebner, Fritz. Methode der Ethnologie. Heidelberg, 1911.
Geog 135.2.7	Hauthal, Rudolf. Reisen in Bolivien und Peru. v.7. Leipzig, 1911.
Geog 4311.12	Jerrold, Walter. The Danube. N.Y., 1911.
Geog 5317.5	Johann Georg, 1709-1755. München, 1911.
Geog 5925.20	Kollback, K. Der Südpol. Bielefeld, 1911.
Geog 5558.84	Lindsay, D.M. A voyage to the Arctic in the whaler Aurora. Boston, 1911.
Geog 4209.11.5	Lowther, H.C. From pillar to post. London, 1911.
An 359.11.5	Marett, Robert R. Anthropology. N.Y., 1911.
Geog 4179.5.5	Markham, C.R. Address...on fiftieth anniversary. London, 1911.
Geog 4181.28	Markham, C.R. Early Spanish voyages to the Strait of Magellan. v.28. London, 1911.
Geog 4309.11	Meriwether, Lee. Seeing Europe by automobile. N.Y., 1911.
Geog 4182.4	Mulert, F.E. De reis van Mr. Jacob Roggeveen. 's-Gravenhage, 1911.
Geog 1807.19	Murray, John, publisher, London. Handbook for...India, Burma and Ceylon. 8. ed. London, 1911.
Geog 5220.22.5	Nansen, F. In northern mists, Arctic exploration in early times. London, 1911. 2v.
Geog 5220.22.10	Nansen, F. In northern mists. N.Y., 1911. 2v.
Geog 5220.22	Nansen, F. Nord i tåkeheimen. Kristiania, 1911.
An 609.07.2	Neilsen, Y. Universitets lappiske samlinger, 1857-1911. Christiania, 1911.
Geog 819.11	Newbigin, Marion I. Modern geography. London, 1911.
Geog 819.11.1	Newbigin, Marion I. Modern geography. N.Y., 1911.
Geog 4709.11	Paine, R.D. The book of buried treasure. London, 1911.
Geog 3540.10	Pannwitz, Max. Deutsche Pfadfinder des 16. Jahrhunderts in Afrika, Asien und Südamerika. Stuttgart, 1911-12.
Geog 4300.28.7	Presbrey, F. Presbrey's information guide to transatlantic travelers. 7th ed. N.Y., 1911.
Geog 5839.09.5	Quervain, Alfred. Durch Grönlands Eiswüste. 2. Aufl. Strassburg, 1911.
Geog 618.3	Reclus, Élisée. Correspondance. v.1-2,3. Paris, 1911-25. 2v.
An 359.11.3	Savorgnan, Franco. Gli indici di omogamia delle razze e delle nazionalità. Cagliari, 1911.
An 379.11	Semple, E.M. Influences of geographic environment on the basis of Ratzel's system. N.Y., 1911.
Geog 6009.07.5	Shackleton, Ernest H. The heart of the Antarctic. London, 1911.
Geog 665.37.5	Singleton, E. Wonders of nature. N.Y., 1911.
Geog 142.2.2	Sociedad Geografico de Madrid. Repertorio, 1901-1910. Madrid, 1911.
An 2059.11	Sollas, William J. Ancient hunters. London, 1911.
An 5.4	Steinmetz, S.R. Essai d'une bibliographie ssytématique de l'ethnologie...1911. Bruxelles, 1911.
Geog 3235.56A	Stevenson, E.L. Portolan charts. N.Y., 1911.
Geog 4181.26	Storm van 's Gravesande, L. Rise of British Guiana. London, 1911. 2v.
An 379.12.3	Vallaux, C. Géographie sociale. Le sol et l'état. Paris, 1911.
An 2109.11.7	Visscher, H. Religion und soziales Leben bei Naturvölkern. Bonn, 1911.
Geog 5850.7	Wallem, F.B. Universitetets Eskimoiske samlinger. Christiania, 1911.
An 3309.11	Weissenberg, S. Das Wachstum des Menschen. Stuttgart, 1911.
Geog 616.25	Weller, E. August Petermann. Leipzig, 1911.

1912

Geog 6009.10A	Amundsen, R. The South Pole...Norwegian Antarctic expedition, 1910-1912. London, 1912. 2v.
Geog 6009.10.5	Amundsen, R. Sydpolen...med Fram, 1910-1912. Kristiania, 1912. 2v.
Geog 1526.35	Baedeker, publishers. Berlin and its environs. 5. ed. Leipzig, 1912.
Geog 1568.45A	Baedeker, publishers. Norway, Sweden and Denmark. 10. ed. Leipzig, 1912.
Geog 1605.25A	Baedeker, publishers. Palestine and Syria. 5. ed. Leipzig, 1912.
Geog 1525.56	Baedeker, publishers. Die Rheinland, Schwarzwald, Vogesen. 32. Aufl. Leipzig, 1912.
Geog 4679.12.2	Beesley, Lawrence. The loss of the S.S. Titanic. Boston, 1912.
Geog 4679.12A	Beesley, Lawrence. The loss of the Titanic. Boston, 1912.
Geog 4169.12	Boer, M.G. de. Van oude voyagien. Amsterdam, 1912-13. 3v.
An 359.12.9	Bradley, R.N. Malta and the Mediterranean race. London, 1912.
An 3309.12	Bresciani, C. La correlazione fra la statura e l'indice cefalico. Palermo, 1912.
An 359.12.11	British Museum (Natural History). Department of Zoology. Guide to the specimens...races of mankind. 2. ed. London, 1912.
An 359.12.17	Brunhes, Jean. La géographie humaine. 2. éd. Paris, 1912.
An 2109.12	Classen, K. Die Völker Europas zur jüngeren Steinzeit. Stuttgart, 1912.
Geog 3570.2	Consiglieri-Pedroso, L. Catalogo bibliographico das publicacões. Lisboa, 1912.
Geog 5559.08	Cook, F.A. Meine Eroberung des Nordpols. Hamburg, 1912.
Geog 4559.12	Crutchly, William C. My life at sea. London, 1912.
An 359.12.2	Deniker, J. The races of man. London, 1912.
Geog 3070.27	Denucé, J. Oud-Nederlandsches kaartmakers in betrekking met Plantijn. Antwerpe, 1912. 2v.
An 2339.12	Dieck, Alfred. Die Waffen der Naturvölker Süd-Amerikas. Ställuponen, 1912.
An 2059.12A	Duckworth, W.L.H. Prehistoric man. Cambridge, 1912.
Geog 4209.11.2	Fraser, Mary Crawford. Further reminiscences of a diplomat's wife. London, 1912.
Geog 4419.12A	Gardner, G.P. Chiefly the Orient. Boston, 1912.
Geog 4307.91.10	Gaultier, Pierre R.A. Séjour de mon grand-oncle, P. Gaultier, en Espagne...1791-1802. v.1-2. Angers, 1912.
An 359.12.21	Giuffrida-Ruggeri, V. L'uomo come specie collettiva. Napoli, 1912.
An 359.12.7	Grow, O. The antagonism of races. Waterloo, 1912.
Geog 4309.10A	Hale, Louise (Closser). Motor journeys. Chicago, 1912.
Geog 4309.12.10	Howell, Charles F. Around the clock in Europe. Boston, 1912.
An 3509.12	Hrdlička, A. Early man in South America. Washington, 1912.
An 2059.12.3	Jacob, K.H. Der diluviale Mensch und seine Zeitgenossen aus dem Tierreiche. Leipzig, 1912.
An 2059.12.5A	Keith, Arthur. Ancient types of man. 2. ed. London, 1912.
An 3209.12	Loth, E. Beiträge zur...der Negerweichteile. Stuttgart, 1912.
Geog 4209.11.6	Lowther, H.C. From pillar to post. London, 1912.
Geog 4325.73	Lubenau, R. Beschreibung der Reisen des Reinhold Lubenau. Königsberg, 1912-1920. 3v.
Geog 4309.12	Maraini, Y. A year of strangers. N.Y., 1912.
An 379.12.5	Maranelli, C. La geografia umana di Jean Brunhes. Firenze, 1912.
Geog 4182.5	Marees, P. de. Beschrijving...van het Gout Koninckrijek van Gunea. 's-Gravenhage, 1912.
An 359.12.3	Marett, Robert R. Anthropology. London, 1912.
An 359.12.23	Marett, Robert R. Anthropology. N.Y., 1912.
Geog 4181.29	Markham, C.R. Book of knowledge of all kingdoms, lands. London, 1912.
Geog 4679.12.13	Marshall, L. Sinking of the Titanic and great sea disasters. Philadelphia, 1912.
An 359.12.15	Montalto de Jesus, C.A. Oriente modernisado. Lisboa, 1912.
An 359.12.5	Munro, Robert. Palaeolithic man. Edinburgh, 1912.
An 3559.12	Paul-Boncour, G. Anthropologie anatomique, crane-face. Paris, 1912.
Geog 4309.12.5	Peçanha, Nilo. Impressões da Europa. Nice, 1912.
Geog 3025.8.2F	Portolan charts of XVth, XVIth, and XVIIth centuries. N.Y., 1912.
Geog 3025.8	Portolan charts of XVth, XVIth, XVIIth centuries collected by Dr. Theodore J.E. Hamy. N.Y., 1912.
Geog 4678.70	Read, George H. The last cruise of the Saginaw. Boston, 1912.
An 3409.12	Reyna Almandos, Luis. Origen e influencia jurídico-social del sistema dactiloscopico argentino. La Plata, 1912.
Geog 3550.13	Ricchieri, G. Il contributo degli Italiani alla conoscenza della terra ed agli studi cinquantennio. Roma, 1912.
Geog 612.3.10	Scott, Ernest. Laperouse. Sydney, 1912.
An 3559.12.3	Sergi, Sergio. Crania Habessinica contributo all'antropologia dell'Africa Orientale. Roma, 1912.
Geog 4309.10.5	Smith, Bertha W. Traveller's tales told in letters. N.Y., 1912.
Geog 3251.23.3	Stevenson, E.L. Genovese world map, 1457. Facsimile. N.Y., 1912.
An 359.12.13A	Thomas, W.I. Source book for social origins. Chicago, 1912.
Geog 4679.12.5	United States. Congress. Senate. Titanic disaster. Washington, 1912.
Geog 4309.12.15	Van Allen, William H. Travel-pictures. 2nd series. Milwaukee, 1912.
Geog 4219.12	Zobeltitz, Fedor von. Ein Bummel um die Welt. Berlin, 1912.

1913

Geog 4558.90.6	Amicis, E. de. Sull'oceano. Milano, 1913.
Geog 6009.10.3	Amundsen, R. Au Pôle Sud, éxpedition du "Fram", 1910-1912. Paris, 1913.
Geog 819.13	Andrews, A.W. A text-book of geography. London, 1913.
Geog 4305.17	Antonio de Beatis, Don. Voyage du Cardinal d'Aragon en Allemagne. Paris, 1913.
An 2109.13	Avebury, J.L. Prehistoric times. 7. ed. London, 1913.
Geog 4208.12	Bacardí Moreau, E. Hacia tierras viejas. Valencia, 1913.
Geog 1518.89	Baedeker, publishers. Le Nord-ouest de la France. 9. éd. Leipzig, 1913.
Geog 1523.70	Baedeker, publishers. Northern Germany. 16. ed. Leipzig, 1913.

Chronological Listing

1913 - cont.

Geog 1539.55	Baedeker, publishers. Northern Italy. 14. ed. Leipzig, 1913.
Geog 1530.15	Baedeker, publishers. Österreich-Ungarn. 29. Aufl. Leipzig, 1913.
Geog 1519.68	Baedeker, publishers. Paris and its environs. 18. ed. Leipzig, 1913.
Geog 1545.18A	Baedeker, publishers. Spain and Portugal. 4. ed. Leipzig, 1913.
Geog 1535.72	Baedeker, publishers. Switzerland and the adjacent portions of Italy, Savoy, and Tyrol. 25. ed. Leipzig, 1913.
Geog 5180.8	Balch, E.S. The North Pole and Bradley land. Philadelphia, 1913.
Geog 4208.51.5	Blaney, Henry. Journal of voyages to China and return. Boston, 1913.
Geog 4559.13	Burns, Walter. A year with a whaler. N.Y., 1913.
An 2109.13.10	Buttel-Reepen, H. Man and his forerunners. London, 1913.
Geog 4509.13	Casella, G. L'alpinisme. Paris, 1913.
Geog 4181.31	Cieza del Leon, P. de. The war of Quito. London, 1913.
Geog 4181.32	Corney, B.G. The quest and occupation of Tahiti by emissaries of Spain 1772-76. London, 1913-19. 3v.
Geog 5300.6	Cronheim, P. Fahrten und Forschunger der Holländer in die Polargebieten. Leipzig, 1913.
Geog 5900.15	Denucé, J. Bibliographie antarctique. Bruxelles, 1913.
An 3559.13	Dillenius, J.A. Craneometría comparativa de los antiguos habitantes de la isla y del pukara de Tilcara. Buenos Aires, 1913.
Geog 4309.13.5	Dreiser, Theodore. A traveler at forty. N.Y., 1913.
Geog 199.1.20	Dübi, Heinrich. Die ersten fünfzig Jahre des Schweizer Alpenclub. Bern, 1913.
Geog 4269.13.10	Fitch, G.H. Critic in the occident. San Francisco, 1913.
Geog 4209.10.9	Fraser, Mary Crawford. Reminiscences of a diplomat's wife. N.Y., 1913.
Geog 4679.12.3	Gracie, A. The truth about the Titanic. N.Y., 1913.
Geog 603.1	Graubner, Paul. Fr. Cannabich (1777-1859) sein Leben und sein Werke. Koniejeberg, 1913.
Geog 3235.65	Gross, Hans. Zur Entstehungs-Geschichte der Tabula Purtingeriana. Diss. Bonn, 1913.
An 3409.13.5	Henry, Edward R. Classification and uses of fingerprints. 4. ed. London, 1913.
Geog 3209.13	En Islandsk Vejviser for pilgrimme fra 12 årh. København, 1913.
Geog 4209.11.7	Jacobs, A.H. Reisbrieven uit Afrika en Azie. Almelo, 1913. 2v.
Geog 5838.49	Janssen, Carl Emil. En grønlandspraests optegnelser, 1844-49. København, 1913.
Geog 3019.13	Keltie, J.S. History of geography. N.Y., 1913.
Geog 4269.13	Lyde, L.W. The continent of Europe. London, 1913.
Geog 5060.8	Marshall, Logan. The story of polar conquest. n.p., 1913.
Geog 5559.09.100	Mikkelsen, Ejnar. Lost in the Arctic; being the story of the "Alabama" expedition, 1909-1912. London, 1913.
Geog 5559.09.101	Mikkelsen, Ejnar. Lost in the Arctic; being the story of the "Alabama" expedition, 1909-1912. N.Y., 1913.
Geog 5939.07	Murray, J. Antarctic days. London, 1913.
Geog 1811.9A	Murray, John, publisher, London. Handbook for...Japan. 9. ed. London, 1913.
Geog 1712.9A	Murray, John, publisher, London. Handbook for...Scotland. 9. ed. London, 1913.
Geog 4182.6	Naber, S.P. Toortse der zee-vaert...D. Ruiters 1623 en Samuel Brun's schiffarten (1624). 's-Gravenhage, 1913.
Geog 4558.77	Nelson, A.W. Yankee Swanson. N.Y., 1913.
An 3409.13	Ortiz, F. La indentificación dactiloscópica. Habana, 1913.
Geog 4307.37.9	Pöllnitz, K.L.. von. A vagabond courtier. London, 1913. 2v.
Geog 189.2.4	Quebec Geographical Society. Tables des matieres contenus dans le Bulletin. Quebec, 1913.
An 359.13	Rivers, W.H.R. Reports upon the present condition and future needs of the science of anthropology. Washington, 1913.
An 3309.13	Rocca, Pierre. Les corses devant l'anthropologie. Paris, 1913.
Gcog 819.13.5	Salisbury, Rollin D. Modern geography for high schools. N.Y., 1913.
An 175.2	Sbornik" v chest' semidesiatiletiia...D.N. Anuchina. Moskva, 1913.
Geog 6009.10.15.2	Scott, R.F. Scott's last expedition. London, 1913. 2v.
Geog 6009.10.15A	Scott, R.F. Scott's last expedition. N.Y., 1913. 2v.
Geog 3070.25	Stevenson, E.L. Maps reproduced as glass transparencies. N.Y., 1913.
Geog 4509.11	Turner, S.F.R. My climbing adventures in 4 continents. London, 1913.
Geog 4228.15A	Van Denburgh, L.D. My voyage in the U.S. filgate "Congress". N.Y., 1913.
Gcog 4309.13	Vasconcelos, C. de. Notas da Europa. Rio de Janeiro, 1913.
Geog 4208.98	Voss, J.C. Venturesome voyages of Captain Voss. Tokyo (Kanda), Japan. Yokohama, 1913.
Geog 4207.69.9	Wallenberg, Jacob. Min son på galejan. 2. uppl. Stockholm, 1913.
Geog 4181.33	Yule, Henry. Cathay and the way thither. London, 1913-14. 4v.

1914

Geog 4332.9	L'Adriatico. Milano, 1914.
Geog 4209.14	Aldao, Carlos A. A través del mundo. 5. ed. Buenos Aires, 1914.
Geog 1616.22	Baedeker, publishers. Egypt and the Sudan. 7. ed. Leipzig, 1914.
Geog 1604.15	Baedeker, publishers. Konstantinopel und Kleinasien, Archipel, Cypern. 2. Aufl. Leipzig, 1914.
Geog 1575.15A	Baedeker, publishers. Russia. N.Y., 1914.
Geog 1517.21A	Baedeker, publishers. Southern France. 6. ed. Leipzig, 1914.
An 2109.14	Boas, Franz. Kultur und Rasse. Leipzig, 1914.
Geog 4110.17	Burns, Philp and Co. Picturesque travel. Sydney, 1914.
Geog 809.19.4	Camena d'Almeida, P. Asia, India insular, Africa. Barcelona, 1914.
Geog 809.19.2	Camena d'Almeida, P. Europa. Barcelona, 1914.
Geog 4679.14	Canada. Commission of Inquiry into the Loss of the British Steamship "Empress of Ireland". Report and evidence. Ottawa, 1914.

1914 - cont.

Geog 4209.13	Casey, R.H. Notes made during a cruise around the world in 1913. N.Y., 1914.
Geog 4309.14.11A	Cobb, I.S. Europe revised. N.Y., 1914.
Geog 4308.91.14	Elson, L.C. European reminiscences. Philadelphia, 1914.
Geog 4209.09.6	Ewers, Hanns H. "Mit meinen Augen"; Fahrten durch die lateinische Welt. München, 1914.
An 3309.14	Friedenthal, Hans. Allgemeine und spezielle Physiologie des Menschenwachstums. Berlin, 1914.
An 2059.14	Geike, James. The antiquity of man in Europe. N.Y., 1914.
Geog 5182.6	Gilbertson, Albert N. Some ethical phases of Eskimo culture. Worcester, 1914.
Geog 5070.11	Hassert, Kurt. Die Polarforschung. 3. Aufl. Leipzig, 1914.
Geog 5180.12	Hatt, G. Arktiske skinddragter i Eurasian og Amerika. København, 1914.
Geog 4129.14	Howard, Clare. English travellers of the Renaissance. London, 1914.
An 184.1F	Jagor, Fedor. Aus Fedor Jogor's Nachlass. Berlin, 1914.
Geog 4329.14F	Januszewski, W. Dokoła morza śródziemnego. Wilnie, 1914.
An 379.14	Kirchhoff, Alfred. Man and earth. N.Y., 1914.
Geog 4182.8	Linschoten, Jan H. van. Reizen van...naar het noorden 1594-95. 's-Gravenhage, 1914.
An 2059.14.5	Lyell, C. The geological evidence of the antiquity of man. London, 1914.
Geog 3251.31	Magnagni, A. D'Anamia e Botero. Cirié, 1914.
Geog 6009.14.20	Mawson, Douglas. The home of the blizzard. Philadelphia, 1914. 2v.
Geog 4129.14.5	Mead, William E. The grand tour in the 18th century. Boston, 1914.
Geog 4309.14.15	Mencken, Henry L. Europe after 8:15. N.Y., 1914.
Geog 5559.09.106	Mikkelsen, Ejnar. Tre aar paa Grønlands østkyst. 2. udg. København, 1914.
Geog 4181.34	Mittall, Zelia. New light on Drake. London, 1914.
Geog 4327.94	Morritt, John B.S. The letters of...journeys in Europe and Asia Minor. London, 1914.
Geog 1590.25	Orlowicz, M. Illustrierter Führer durch Galizien. Wien, 1914.
Geog 3240.18	Ortroy, F.G. van. L'oeuvre cartographique de Gérard et de Corneille de Jode. Gand, 1914.
Geog 4219.11	Pearse, Albert W. Recent travel. Sydney, 1914.
Geog 6009.11	Priestly, R.E. Antarctic adventure; Scott's northern party. London, 1914.
Geog 4509.14	Rey, G. Alpinismo acrobatico. Torino, 1914.
Geog 665.39	Ricchieri, G. Dopo il viaggio d'istruzione. Firenze, 1914.
Geog 665.29F	Santarem, M.F. de B. Inéditos (miscellanea). Lisboa, 1914.
Geog 819.14	Seydlitz, Ernst von. Handbuch der Geographie. Breslau, 1914.
An 3309.14.3F	Stevenson, B.L. Constancy or variability in Scandinavian type. Leiden, 1914.
Geog 602.1.5	Stevenson, E.L. Willem Janszoon Blaeu. N.Y., 1914.
Geog 5700.11	Thorhallason, Egiel. Beskrivelse over missionerne i Grønlands Søndre Distrikt. København, 1914.
An 359.14.3	Van Waters, Miriam. The adolescent girl among primitive peoples. Thesis. Worcester, 1914.
Geog 4309.14	Vrooman, Carl S. The lure and the lore of travel. Boston, 1914.
Geog 4521.23	Whymper, Edward. Edward Whymper, alpinist of the heroic age. Nashville, Tenn., 1914.
An 359.14	Wirth, Albrecht. Rasse und Volk. Halle, 1914.
Geog 4349.14	Young, Ernest. From Russia to Siam, with a voyage down the Danube. London, 1914.

1915

Geog 4332.10	L'Adriatico; studio geografico. Milano, 1915.
Geog 5350.5	Alexander, P.F. The north-west and north-east passages. Cambridge, 1915.
An 359.15.5	Arldt, Theodor. Die Stammesgeschichte der Primaten und die Entwicklung der Menschrassen. Berlin, 1915.
Geog 5509.15	Beyl, Hermann. Der lange Zeit angenomme Zusammenhang der Hudsonbai mit dem Stille Ozean. Würzburg, 1915.
An 2059.15	British Museum (Natural History). Department of Geology. A guide to the fossil remains of man in the department. London, 1915.
Geog 5650.10	Bruun, Daniel. Erik den Rode og nordhokolonierne il Gronland. Kristiania, 1915.
Geog 4332.12	Cassi, Gellio. Il Mare Adriatico. Milano, 1915.
Geog 4309.06	Cestero, T.M. Hombres y piedras. Madrid, 1915.
Geog 4269.15	Couperus, Louis. Van en over alles en iedereen. pt.1-5. Amsterdam, 1915. 5v.
Geog 4558.40.30	Dana, Richard Henry. Two years before the mast. N.Y., 1915.
An 2109.15.3	Elliot, G.F.S. Prehistoric man and his story. London, 1915.
An 2109.15.4	Elliot, G.F.S. Prehistoric man and his story. Philadelphia, 1915.
Geog 4309.15	Gardiner, Sarah D. Pages in azure and gold. New Haven, 1915.
An 359.15.10	Gobineau, A. The inequality of human races. N.Y., 1915.
Geog 5180.12.5	Hatt, G. Moccasins and their relation to Arctic footwear. Lancaster, 1915?
An 359.15.7	Hertz, Friedrich. Rasse und Kultur. 2. Aufl. Leipzig, 1915.
An 2109.15	Hobhouse, L.T. The material culture of the simpler peoples. London, 1915.
Geog 4182.9	Ijzermann, J.W. Dirck Gerritsz Pomp. 's-Gravenhage, 1915.
Geog 4209.13.7	Kawabata, A. A hermit turned loose. 2. ed. Tokio, 1915.
An 359.15.3	Latcham, R.E. Conferencias sobre antropología, etnología. Santiago, 1915-
Geog 4208.80	Lopez, Lucio V. Recuerdos de viaje. Buenos Aires, 1915.
Geog 4209.15.10	Lubbock, A.B. Round the Horn before the most. London, 1915.
Geog 3530.16	Martonne, E. de. La science géographique. Paris, 1915.
An 3559.15	Mendes Correa, A.A. Sobre três crânios de negros Mossumles. Porto, 1915.
NEDL Geog 144.1.3	Mittheilungen. v.2. Berlin, 1915.
Geog 607.1	North, S.N.D. Henry Gannett, president of the National Geographic Society, 1910-1914. n.p., 1915.
An 2109.15.5A	Osborn, H.F. Men of the old stone age. N.Y., 1915.
Geog 5855.12	Porsild, N.P. Studies on the material culture of the Eskimo in West Greenland. København, 1915.
Geog 4182.10	Rogerius, Abraham. De open-deure tot het verborgen heydendom. 's-Gravenhagen, 1915.

Chronological Listing

1915 - cont.

Geog 4182.7	Rouffaer, G.P. De eerste schipvaart der nederlanders naar Ost-Indie. 's-Gravenhage, 1915-29. 3v.
Geog 6009.01.15	Scott, R.F. The voyages of Captain Scott retold from The voyage of the 'Discovery'. N.Y., 1915.
An 359.15	Smith, George E. The migrations of early culture. Manchester, 1915.
Geog 953.10	Stockholm. Biblioteket. Magnus Gabriel de la Gordie's samling af öldre stadsvger. Stockholm, 1915.

1916

Geog 4215.19	Alexander, Philip. Earliest voyages round the world, 1519-1617. Cambridge, 1916.
Geog 819.16	Andrews, A.W. A text book of geography. London, 1916.
Geog 4559.16.5	Andrews, R.C. Whale hunting with gun and camera. N.Y., 1916.
Geog 4303.08	Anonymi descriptio Europae orientalis. Cracoviae, 1916.
Geog 3070.30	Anthiaume, A. Cartes marines, constructions navales, 1500-1650. Paris, 1916. 2v.
Geog 5559.13A	Bartlett, Robert A. The last voyage of the Karluk. Boston, 1916.
Geog 4309.02.3A	Belloc, H. The path to Rome. 4th ed. London, 1916.
An 359.16	Blackford, K.M.H. Blondes and brunets. Photoreproduction. N.Y., 1916.
Geog 809.19.6	Blásquez y Delgado Aguilera, Antonio. América Meridional, Oceania. Barcelona, 1916.
Geog 809.19.5	Camena d'Almeida, P. América Septentrional, América Central, Las Antillas, Alaska, Canada, Estados Unidos. Barcelona, 1916.
Geog 6009.02.9	Doorly, G.S. The voyages of the "Morning". London, 1916.
Geog 5855.10	Elgström, Ossian. Moderna eskimåer. Stockholm, 1916.
Geog 4182.11	Godeé Molsbergen, E.C. Reizen in Zuid-Afrika. 's-Gravenhage, 1916-32. 4v.
An 3409.16.3	Herschel, W.J. The origin of finger-printing. London, 1916.
Geog 4500.5	Jeffers, LeRoy. Selected list of books on mountaineering. N.Y., 1916.
An 3409.16.5	Kuhne, F. The finger print instructor. N.Y., 1916.
An 359.16.5	Myres, John L. The influence of anthropology on the course of political science. Berkeley, 1916.
Geog 5839.16.5	Olrik, Harald. Forslag om at bebygge Scoresby Sund-Egnen i Østgrønland ved vestgrønlandske saelforgere. København, 1916.
Geog 4209.15.5	Rohrbach, P. Weltpolitisches Wanderbuch 1897-1915. Königstein, 1916.
Geog 4209.15	Rohrbach, P. Weltpolitisches Wanderbuch 1897-1915. Königstein, 1916.
An 193.1.5	Sergi, Giuseppe. Problemi di scienza contemporanea. Torino, 1916.
Geog 4269.16	Sievers, Wilhelm. Die geographischen Grenzen Mitteleuropas. Giessen, 1916.
An 2109.16	Smith, G.E. Primitive man. London, 1916.
Geog 5182.7	Steensby, Hans P. An anthropological study of the origin of the Eskimo culture. Kjøbenhavn, 1916.
An 3309.16	Stevenson, B.L. Socio-anthropometry. Boston, 1916.
Geog 4509.16	Taüber, C. Auf fremden Bergfpfaden. Zürich, 1916.
Geog 6009.10.19	Taylor, G. With Scott; the silver lining. London, 1916.
Geog 4559.16.3	The thrilling adventures of the whaler Alcyone. Peabody, 1916.
An 3409.16	Unites States. Bureau of Navigation. Navy Department. How to obtain good fingerprints. 2. ed. Washington, 1916.
Geog 3510.7	Van Loon, H.W. The golden book of the Dutch navigators. N.Y., 1916.
Geog 4559.16	Verrill, A.H. The real story of the whaler. N.Y., 1916.
Geog 4309.16	Villaverde, J.R. Desde lajos. Habana, 1916.

1917

Geog 4209.17	Anderson, I. (Mrs.). Odd corners. N.Y., 1917.
Geog 4679.17	Benson, N.P. The log of the El Dorado. San Francisco, 1917.
Geog 3570.9	Bersaude, J. Les légendes allemandes. Genève, 1917-20. 2v.
Geog 5650.10.9	Bruun, Daniel. Oversigt over Nordboruiner i Godhaat. København, 1917.
Geog 4208.81	Cané, Miguel. En viaje, 1881-82. Buenos Aires, 1917.
An 2339.17	Churchill, William. Club-types of nuclear Polynesia. Washington, 1917.
Geog 4329.17.5	Cimon, Henri H.M. Aux vieux pays. 3. ed. Montréal, 1917.
An 170.227.5	Festschrift Eduard Hahn zum LX. Geburtstag. Stuttgart, 1917.
Geog 4559.17	Gelett, Charles N. A life on the ocean; autobiography. Honolulu, 1917.
An 359.17.7	Haberlandt, M. Völkerkunde. 3. Aufl. Berlin, 1917. 2v.
Geog 5559.17.5	Hall, Thomas F. Has the North Pole been discovered? Boston, 1917.
Geog 4105.13A	Harvard Travel Club. Handbook of travel. Cambridge, 1917.
An 359.17.3	Humphrey, Seth King. Mankind; racial values and racial prospect. N.Y., 1917.
An 2109.17	Koppers, W. Die ethnologische Wirtschaftsforschung. Wien, 1917.
Geog 613.4.5	Markham, A.H. The life of Sir Clements R. Markham. London, 1917.
An 3559.17	Mendes Correa, A.A. Sôbre algunos crânios da India Portuguêsa. Porto, 1917.
Geog 5558.81.55	Muir, John. The cruise of the Corwin. Boston, 1917.
An 359.17.5	Nordenstreng, Rolf. Europas människoraser. 2. uppl. Stockholm, 1917-20.
Geog 5559.17A	Peary, R.E. Secrets of polar travel. N.Y., 1917.
An 359.17	Restrepo-Hernandez, J. Lecciones de antropologia. Bogotá, 1917.
Htn Geog 5340.8F*	Scoresby, William. Seven log-books concerning the Arctic voyages. N.Y., 1917. 8v.
Geog 3019.17	Teleki, Pal. A foldrajzi gondolat története. Budapest, 1917.
Geog 5375.94.5	Veer, G. de. Reizen van Willem Barents. 's-Gravenhage, 1917. 2v.
Geog 520.23	Vignaud, Henry. Americe Vespuce, 1451-1512. Paris, 1917.

1918

An 2109.18.3	Aberg, Nils. Das nordische Kultur-Gebiet in Mitteleuropa. v.1-2. Uppsala, 1918.
Geog 4181.36	Barbosa, Duarte. The book. London, 1918-21. 2v.

1918 - cont.

Geog 4341.73.15	Benjamin ben Jonah, of Tudela. Viajes de Benjamin de Tudela, 1160-1173. Madrid, 1918.
Geog 4218.78.7A	Brassey, Thomas. The "Sunbeam". London, 1918.
Geog 5650.10.5	Bruun, Daniel. The Icelandic colonization of Greenland. København, 1918.
Geog 4181.35	Cieza de Leon, P. de. The war of Chupas. London, 1918.
Geog 6009.16F	Davis, John King. 1918 "Aurora" relief expedition. Melbourne, 1918.
Geog 959.07.15	Grosvenor, G.H. Scenes from every land. Washington, D.C., 1918.
An 379.18A	Koller, A.H. The theory of environment. Menasha, Wis., 1918.
Geog 5559.13.5A	MacMillan, D.B. Four years in the white north. N.Y., 1918.
Geog 4269.18A	National Research Council. Division of Geology and Geography. The geography of Europe. New Haven, 1918.
Geog 500.55	Newark, N.J. Free Public Library. Foreign countries. Washington, 1918.
An 359.18	Numelin, R. Orsakerna till fokevandringerna. Helsingfors, 1918.
Geog 4182.12	Ottsen, Hendrick. Journael van de reis naar Zuid-Amerika. 's-Gravenhage, 1918.
Geog 4208.54A	Pumpelly, R. My reminiscences. N.Y., 1918. 2v.
Geog 4228.97	Riesenberg, Felix. Under sail. N.Y., 1918.
Geog 4332.14	Roncagli, Giovanni. Il problema militare dell'Adriatico spiegatto a tutti. Roma, 1918.
Geog 4308.18.6	Russell, Jonathan. Journal, 1818-1819. Boston, 1918.
Geog 819.18	Santa Cruz, Alonso de. Islario general de todas las isles del mundo. Atlas. Madrid, 1918. 2v.
Geog 4509.18	Schweizer Alpen Club. Le conseiller de l'ascensionniste. v.2. Genève, 1918.
Geog 520.30	Turistresor och forskningsfärder. v.3-12. Helsingfors, 1918. 5v.
Geog 5919.18	Wright, H.S. (Mrs.). The seventh continent. Boston, 1918.

1919

Geog 600.5	Abulhasan Mansur. Arabische Schriftsteller über die Geographie Indiens. Inaug. Diss. Berlin, 1919.
Geog 500.57	Almagia, Roberto. La geografia. Roma, 1919.
Geog 4206.13.30F	Amman, H.J. Hans Jakob Ammann genannt der Thalwyler Schärer und seineReise ins gelobte Land. Zürich, 1919.
Geog 665.27	Aufsätze Prof. Dr. Eugen Oberhummer gewidmet. Brünn, 1919.
Geog 819.19	Banse, E. Illustrierte Länderkunde. Berlin, 1919.
Geog 759.19	Beltran, Juan G. Lo inerte y lo vital. Buenos Aires, 1919.
Geog 4269.19	Beltrán y Rozpide, Ricardo. Nuevas nacionalidades en Europa. 2a ed. Madrid, 1919.
Geog 809.19	Camena d'Almeida, P. La tierra; geografia general. 2a ed. Barcelona, 1919.
Geog 6009.11.5	Davis, John King. With the "Aurora" in the Antarctic, 1911-14. London, 1919.
An 359.19.3	Gattefossé, R.M. Adam, l'homme tertiaire. Lyon, 1919.
Geog 4269.03.10	Herbertson, (Mrs.). Descriptive geography of Europe from original sources. London, 1919.
Geog 4209.19	Mackay, John. The ten islands and Ireland. Dublin, 1919.
An 359.19	Marett, Robert R. Anthropology. London, 1919.
Geog 613.3.5	Murray, John. John Murray III, 1808-1892; memoir. London, 1919.
Geog 4169.19A	Newbolt, Henry. The book of the long trail. N.Y., 1919.
Geog 5839.16	Rasmussen, Knud. Grønland langs Polhavet. Kristiania, 1919.
Htn Geog 4209.19.5*	Romero de Terreros, Juan. Apuntaciones de viaje en 1849. México, 1919.
Geog 659.19	Schrader, Franz. The foundations of geography in the 20th century. Oxford, 1919.
An 379.19	Schulz, George J. Geography. College Park, Md., 1919.
Geog 6009.14	Shackleton, Ernest H. South; the story of Shackleton's last expedition. London, 1919.

192-

Geog 4329.20	Bunge de Galvez, D. Tierras del mar Azul. Buenos Aires, 192-.
Geog 5559.12	Cook, F.A. Im Kampfe mit Bär und Walross. Braunschweig, 192-.
Geog 5650.12.5	Dúason, Jón. Nybygd i Grønland. n.p., 192-.
Geog 665.40F	Hachette, firm, publishers, France. Les merveilles du monde. Paris, 192-?
Geog 4515.201	Prusik, Karl. Gymnastik für Bergsteiger. München, 192-.

1920

Geog 1545.25	Baedeker, publishers. Espagne et Portugal. 3. éd. Leipzig, 1920.
Geog 4309.02.13	Bierbaum, Otto Julius. Die Yankeedoodle-Fahrt und andere Reisegeschichten. München, 1920.
Geog 4208.60.7	Bray, M.M. A sea trip in clipper ship days. Boston, 1920.
An 379.20.3	Brunhes, Jean. Human geography. Chicago, 1920.
Geog 4558.98.15	Bullen, F.T. Cruise of the Cachalot. N.Y., 1920.
Geog 4509.20.10	Conway, W.M.C. Mountain memories. N.Y., 1920.
Geog 3229.20	Dark, Richard. The quest of the Indies. N.Y., 1920.
Htn Geog 3060.8*	Ford, Worthington Chauncey. The isle of pines, 1668. Boston, 1920.
Geog 3060.8	Ford, Worthington Chauncey. The isle of pines, 1668. Boston, 1920.
Geog 4182.14	Hamel, Hendrik. Verhaal van het vergaan van het jacht De Sperwer. 's-Gravenhage, 1920.
An 359.20	Humphrey, Seth King. The racial prospect. N.Y., 1920.
An 359.20.5A	Johnston, Harry. The backward peoples and our relations with them. London, 1920.
An 359.20.11	Keane, W.H. Man past and present. Cambridge, 1920.
An 359.20.9	Klaatsch, Hermann. Der Werdegang der Menschheit und die Entstehung der Kultur. Berlin, 1920.
An 359.20.3	Lowie, Robert H. Primitive society. N.Y., 1920.
Geog 4269.13.2	Lyde, L.W. The international of Europe. London, 1920.
Geog 819.01.6	Mill, H.R. The international geography. N.Y., 1920.
Geog 4181.48	Montesinos, Fernando. Memorias antiguas historiales del Peru. London, 1920.
Geog 1807.21	Murray, John, publisher, London. Handbook for...India, Burma and Ceylon. 10. ed. London, 1920.
Geog 4509.20.5	Raeburn, H. Mountaineering art. N.Y., 1920.
An 359.20.7	Roy, Sarat Chandra. Principles and methods of physical anthropology. Patna, 1920.
Geog 5925.30	Shackleton, E. South. N.Y., 1920.
Geog 3019.20	Synge, M.B. A book of discovery. N.Y., 1920.

Chronological Listing

1920 - cont.

An 2109.20	Tylor, Edward Burnett. Primitive culture. 6. ed. London, 1920. 2v.
An 3309.20	Wilder, Harris H. A laboratory manual of anthropometry. Philadelphia, 1920.
Geog 4308.01	Wilmot, Catherine. An Irish peer on the continent. London, 1920.
Geog 4509.20	Young, G.W. Mountain craft. London, 1920.

1921

Geog 5379.18	Amundsen, Roald. Nordostpassagen. Kristiania, 1921.
Geog 809.21	Bader, G. Erläuterungen zu 938 ausgewählten Lichtbilder zur Länderkunde. Stuttgart, 1921. 3v.
Geog 1527.15	Baedeker, publishers. München. Leipzig, 1921.
An 3309.21	Baldwin, Bird T. The physical growth of children from birth to maturity. Iowa City, 1921.
Geog 5670.2	Betaenkning afgivet af dit i december maaned 1920 nedsatte udvalgtel drøftelse af de Grønlandski anliggender. København, 1921.
Geog 809.19.3	Blásquez y Delgado Aguilera, Antonio. Peninsula Ibérica. 2a ed. Barcelona, 1921.
An 2109.21.5	Boelsche, Wilhelm. Der Mensch der Vorzeit. v.1-2. 28e Aufl. Stuttgart, 1921.
Geog 819.21	Bowman, I. The new world. Yonkers-on-Hudson, 1921.
An 609.21	British Museum (Natural History). Department of Zoology. Guide to the specimens illustrating the races of mankind. 4. ed. London, 1921.
Geog 4219.21	Büchler, E. Rund um die Erde. Bern, 1921.
Geog 4558.40.35A	Dana, Richard Henry. Two years before the mast. N.Y., 1921.
An 3309.21.3	Dreyer, George. The assessment of physical fitness. N.Y., 1921.
An 2109.21.3	Fehlinger, H. Sexual life of primitive people. London, 1921.
Geog 3070.31	Fordham, Herbert G. Maps. Cambridge, Eng., 1921.
Geog 212.216.95	Foucart, G. La société Sultanich de géographie du Caire. Le Caire, 1921.
Geog 5700.12	Glahn, Henrik C. Missionaer i Grønland, Henric Christopher Glahns. København, 1921.
Geog 4209.21.3	Hamilton, Frederick. Here, there, and everywhere. London, 1921?
Geog 4209.21.4	Hamilton, Frederick. Here, there, and everywhere. N.Y., 1921.
Geog 759.21.5	Herbertson, Andrew J. The senior geography. 5th ed. Oxford, 1921.
An 379.21A	Huntington, Ellsworth. Principles of human geography. N.Y., 1921.
Geog 4182.19	Juet, Robert. Henry Hudson's reize onder Nederlander vlag. 's-Gravenhage, 1921.
An 359.21A	Korzybski, Alfred. Manhood of humanity. N.Y., 1921.
An 379.21.5	Krebs, Norbert. Die Verbreitung des Menschen auf der Erdoberfläche (Anthropogeographie). Leipzig, 1921.
Geog 4209.21	Lucas, E.V. Roving east and roving west. London, 1921.
Geog 4209.21.2	Lucas, E.V. Roving east and roving west. N.Y., 1921.
An 379.21.3	McDougall, William. National welfare and national decay. London, 1921.
Geog 5070.9	Markham, C.R. The lands of silence. Cambridge, Eng., 1921.
Geog 665.31	Miller, Émile. Pour qu'on aime la géographie. Montréal, 1921.
An 2109.21.10	Osborn, Henry F. Men of the old stone age. 3. ed. N.Y., 1921.
Geog 5839.21	Rasmussen, Knud. Greenland by the Polar Sea. London, 1921.
Geog 5839.12.2	Rasmussen, Knud. Min rejsedagbog. 2. udg. Kjøbenhavn, 1921.
Geog 4129.21	Roget, S.R. Travel in the two last centuries of three generations. N.Y., 1921.
Geog 5925.15	Rouch, J. Le pôle sud. Paris, 1921.
Geog 5399.21	Stefansson, V. The friendly Arctic. N.Y., 1921.
Geog 5559.10.5	Tremblay, Alfred. Cruise of the Minnie Maud; Arctic seas and Hudson Bay, 1910-1913. Quebec, 1921.
Geog 4309.21	Unstead, J.V. Europe of to-day. London, 1921.
Geog 819.21.3	Wilmore, Albert. The groundwork of modern geography. London, 1921.

1922

An 359.22	Armitage, F.P. Diet and race. London, 1922.
Geog 3060.6A	Babcock, William H. Legendary islands of the Atlantic. N.Y., 1922.
Geog 1640.15.2	Baedeker, publishers. The dominion of Canada. 4. ed. Leipzig, 1922.
Geog 1535.74	Baedeker, publishers. Switzerland together with Chamonix and the Italian lakes. 26. ed. Leipzig, 1922.
Geog 819.22	Bowman, I. The new world; problems in political geography. Yonkers-on-Hudson, 1922.
An 609.22	British Museum (Natural History). Department of Geology. Guide to the fossil remains of man. 3. ed. London, 1922.
Geog 4105.16	Brouwer, H.A. Practical hints to scientific travellers. Leyden, 1922-29. 6v.
An 359.22.3A	Carr-Saunders, A.M. The population problem. Oxford, 1922.
Geog 4558.40.26	Dana, Richard Henry. Two years before the mast. N.Y., 1922.
Geog 4228.28.5	Davis, S.H. Journal of Captain S.H. Davis, a Gloucester sea-captain, 1828-1846. Norwood, 1922.
Geog 4181.51	Edmundson, George. Journal of...Father Samuel Fritz. London, 1922.
An 379.22.3A	The evolution of man. New Haven, 1922.
Geog 4709.22	French, J.L. Great pirate stories. N.Y., 1922.
Geog 4209.22	Glass, James. Chats over a pipe; a tale of two brothers. London, 1922.
An 2109.22.3A	Goldenweiser, A.A. Early civilization. N.Y., 1922.
Geog 5312.7.10A	Golder, Frank A. Bering's voyages. N.Y., 1922-25. 2v.
Geog 4309.21.15A	Graham, S. Europe - whither bound? N.Y., 1922.
Geog 809.22F	Grauper, Ernest. Nouvelle géographie universelle. pt.1-10. Paris, 1922.
Geog 4509.22	Hamilton, H. Mountain madness. London, 1922.
An 379.22.7	Huntington, Ellsworth. Principles of human geography. 2. ed. N.Y., 1922.
Geog 3019.22	Ispizúa, Segundo de. Historia de la geografía y de la cosmografía. Madrid, 1922-26. 2v.
Geog 121.1.6	The journal of geography. Index. 1897-1956. N.Y., 1922-58. 2v.

1922 - cont.

An 379.22.9	Kaltenbach, Ernst. Beiträge zur Anthropogeographie des Bodenseegebiets. Inaug. Diss. Basel, 1922.
An 2109.22	Lévy-Bruhl, L. La mentalité primitive. Paris, 1922.
Geog 4329.22	Maurras, Charles. Anthénea. Paris, 1922.
An 2338.85.2	Morse, E.S. Additional notes on arrow-release. Salem, 1922.
Geog 4309.22	Morse, William I. Seeing Europe backwards. Boston, 1922.
Geog 4672.10	Paine, Ralph D. Lost ships and lonely seas. N.Y., 1922.
Geog 6009.10.21	Ponting, H.G. The great white South. London, 1922.
Geog 5328.5.5	Rasmussen, Knud. In der Heimat des Polarmenschen. Leipzig, 1922.
Geog 4558.66	Smith, C.E. From the deep of the sea. London, 1922.
An 2059.22	Soergel, W. Die Jagd der Vorzeit. Jena, 1922.
Geog 5399.14	Stefansson, V. The friendly Arctic. N.Y. 1922.
Geog 5399.06.9A	Stefansson, V. Hunters of the great North. N.Y., 1922.
Geog 5160.7A	Stefansson, V. The northward course of empire. N.Y., 1922.
Geog 809.22.15	Vahl, Martin. Jorden og menneskelivet. København, 1922-1927. 4v.
An 379.22	Vidal de la Blache, P. Principes de géographie humaine. Paris, 1922.
An 379.22.5	Von Engeln, Oscar D. Inheriting the earth. N.Y., 1922.

1923

Geog 4509.07.3	Abraham, G.D. The complete mountaineer. 3. ed. London, 1923.
Geog 500.56	Amsterdam. Universiteit. Bibliotheek. Catalogus geographie en reizen. Amsterdam, 1923.
Geog 4559.23	Angel, W.H. The clipper ship "Sheila". Boston, 1923.
Geog 1526.36	Baedeker, publishers. Berlin and its environs. 6. ed. Leipzig, 1923.
Geog 1510.62	Baedeker, publishers. London and its environs. 18. ed. Leipzig, 1923.
Geog 1510.61	Baedeker, publishers. London and its environs. 18. ed. Leipzig, 1923.
Geog 1531.3	Baedeker, publishers. Tirol: Vorarlberg und Teile von Salzburg und Kärnten. 37. Aufl. Leipzig, 1923.
An 379.22.4	Baitsell, George A. The evolution of man. New Haven, 1923.
Geog 4209.23.15	Balmont, K.D. Visions solaires Mexique, Égypte, Inde, Japon, Océanie. Paris, 1923.
Geog 4269.23	Bartholomew, J.G. A literary and historical atlas of Europe. London, 1923.
An 2109.23.7A	Bartlett, F.C. Psychology and primitive culture. Cambridge, 1923.
An 2109.23.6	Bartlett, F.C. Psychology and primitive culture. N.Y., 1923.
An 2059.23	Boule, Marcellin. Les hommes fossiles. 2. éd. Paris, 1923.
Geog 819.22.2	Bowman, I. Supplement to The new world. Yonkers-on-Hudson, 1923-1924. 2v.
An 359.23.5	Bradley, R.N. Duality; a study in the psycho-analysis of race. N.Y., 1923.
Geog 5070.20	Brown, R.N.R. A naturalist at the poles. London, 1923.
Geog 4209.23.20	Bryce, J.B. Memories of travel. N.Y., 1923.
Geog 4209.23	Buchan, John. The last secrets. London, 1923.
An 3409.23	Crosskey, Walter C.S. The single finger print identification system. San Francisco, 1923.
Geog 520.24	Denucé, J. Pigafetta; relation du premier voyage...par Magellan. Paris, 1923.
Geog 3055.45	Dévigné, R. Un continent disparu l'Atlantide. Paris, 1923.
An 359.23A	Dixon, Roland B. The racial history of man. N.Y., 1923.
Geog 4509.23	Dupin. Zum Wortschatze des Bergsteigers. Wien, 1923-
Geog 5530.15	Franklin, Jane G. The life, diaries and correspondence...1792-1875. London, 1923.
Geog 4509.23.10	Freshfield, D.M. Below the snow line. London, 1923.
Geog 3580.30	Garcia de Herreros, E. Quatre voyageurs espagnols à Alexandria d'Egypt. Alexandria, 1923.
Geog 3055.47	Gattefossé, R.M. La vérité sur l'Atlantide. Lyon, 1923.
An 2109.23.9	Goldenweiser, A.A. Early civilization. N.Y., 1923.
Geog 759.23	Gribandi, P. Il mondo e l'Italia. 3. éd. v.1-2. Torino, 1923.
Geog 4329.23	Hildebrand, A.S. Blue water. N.Y., 1923.
Geog 4709.23	Hill, Samuel C. Notes on piracy in eastern waters. London, 1923-28.
Geog 4129.23	Hungerford, E. Planning a trip abroad. N.Y., 1923.
Geog 4129.23.5	Hungerford, E. Planning a trip abroad. N.Y., 1923.
An 359.23.9	Kroeber, A.L. Anthropology. N.Y., 1923.
An 2109.23.10	Lévy-Bruhl, L. Primitive mentality. N.Y., 1923.
Geog 4181.53	Life of the Icelander Jón Ólafsson. London, 1923-31. 2v.
An 2209.23	Maddox, J.L. The medicine man. N.Y., 1923.
Geog 3055.41	Manzi, M. Le livre de l'Atlantide. Paris, 1923.
Geog 142.6	Marseilles. Exposition Coloniale Nationale, 1922. Semaine internationale des géographes, des explorateur et des ethnologues, 22-28 Sept. 1922. Marseilles, 1923.
Geog 5985.258A	Mill, Hugh R. The life of Sir Ernest Shackleton. Boston, 1923.
Geog 4219.23	Northcliffe, A.C.W.H. My journey round the world. Philadelphia, 1923.
An 2109.23.3	Osborn, H.F. Men of the old stone age. 3. ed. N.Y., 1923.
Geog 4309.23.5	Osborne, A.B. Finding the worth while in Europe. N.Y., 1923.
An 359.23.7	Pape, A.G. Is there a new race type? Edinburgh, 1923.
An 2109.23.5	Perry, W.J. The children of the sun. N.Y., 1923.
Geog 4209.23.25	Phelan, James D. Travel and comment. San Francisco, 1923.
An 2109.23.20	Preuss, Konrad Theodor. Die geistige Kultur der Naturvölker. 2e Aufl. Leipzig, 1923.
Geog 4329.21	Prioleau, John. The adventures of Imshi; a two-seater in search of the sun. Boston, 1923.
An 2109.23.11	Quellen zur ethnologischen Rechtsforschung von Nordafrika, Asien und Australien. Stuttgart, 1923.
Geog 5929.23.3	Rouch, J. Le pôle nord. Paris, 1923.
An 359.23.11	Shirokogorov, S.M. Etnos". Shankhai, 1923.
Geog 212.202.5	Société de Géographie de Lyon. Bulletin du cinquantaire, 1922-23. Lyon, 1923.
Geog 4309.23	Traz, Robert de. Dépaysements. Paris, 1923.
Geog 4182.21	Wieder, F.C. De reis van Mahu en de Cordes. 's-Gravenhage, 1923-25. 3v.
Geog 6009.21	Wild, Frank. Shackleton's last voyage; the story of the Quest. London, 1923.
An 359.23.3A	Wissler, Clark. Man and culture. N.Y., 1923.
Geog 500.175	Wright, John K. Aids to geographical research. N.Y., 1923.

Chronological Listing

1924

Geog 1519.70	Baedeker, publishers. Paris and its environs. 19. ed. Leipzig, 1924.
Geog 5670.3	Bendixen, Ole. Grønlandstraktaten, belyst gennem de i dagspressen fremkomme artikler. København, 1924.
Geog 4219.24.5	Blasco Ibáñez, Vicente. La vuelta al mundo de un novelista. Valencia, 1924-25. 3v.
An 378.24	Bonstetten, K.V. von. L'homme du midi et l'homme du nord. Genève, 1924.
Geog 4209.24.5A	Burton, Richard F. Selected papers on anthropology, travel and exploration. London, 1924.
Geog 819.11.5	Busson, Henri. Les principales puissances d'aujourd'hui. 5. éd. Paris, 1924.
Geog 3019.24	Capasso, C. Le scoperte geografiche e i viaggi d'esplorazione. Messina, 1924.
An 2109.24.7	Capitan, Louis. L'humanité primitive dans la région des Eyzies. Paris, 1924.
Geog 5660.25	Castberg, Frede. Østgrønlandsavtalen. Kristiania, 1924.
Geog 4559.24	Chatterton, E.K. Seamen all. Boston, 1924.
Geog 4219.24	Clements, Rex. A gipsy of the Horn. London, 1924.
Geog 4509.23.15	Collins, F.A. Mountain climbing. London, 1924.
Geog 4209.25A	Coolidge, J.G. Random letters from many countries. Boston, 1924.
An 2109.24.5	Danzel, T.W. Kultur und Religion des primitiven Menschen. Stuttgart, 1924.
Geog 3055.45.5	Dévigné, R. Un continent disparu l'Atlantide. Paris, 1924.
Geog 5209.29	Einarsson, S. Nordr um höf. Reykjavik, 1924.
Geog 4218.33.10A	Fanning, E. Voyages and discoveres in the South Seas, 1792-1832. Salem, 1924.
Geog 4509.24.6	Finch, G.L. The making of a mountaineer. London, 1924.
Geog 3060.10	Firestone, C.B. The coasts of illusion. N.Y., 1924.
Geog 4509.24	Flaig, W. Felsklettern in Bildern und Merkworten. Stuttgart, 1924.
Geog 619.1	Gedenkboek ter herinnering aan den 70sten verjaardag van R. Schuiling, 27 mei, 1924. Groningen, 1924.
Geog 4239.23	Gerbault, Alain. Seul a travers l'Atlantique. Paris, 1924.
An 2109.24.9	Graebner, F. Das Weltbild der Primitiven. München, 1924.
An 379.24	Huntington, Ellsworth. Character of races as influenced by...environment. N.Y., 1924.
An 379.24.5	Huntington, Ellsworth. Principles of human geography. 3. ed. N.Y., 1924.
An 379.24.3	Jefferson, M.S.W. Man in Europe, here and there. Ypsilanti, Mich., 1924.
Geog 5839.24.5	Koht, Halvdan. Del Grønland vi miste og de vi ikkje miste. Kristiania, 1924.
Geog 4209.24A	Landor, A.H.S. Everywhere; the memoirs of an explorer. N.Y., 1924. 2v.
Geog 5650.13	Larusson, O. Rettarstad Graenlands aðornu. Kaupmannahöfn, 1924.
Geog 4672.16.10	Lockhart, John Gilbert. Mysteries of the sea. London, 1924.
Geog 4269.13.4	Lyde, L.W. The continent of Europe. 2nd ed. London, 1924.
An 2109.24A	MacCurdy, G.G. Human origins, a manual of prehistory. N.Y., 1924. 2v.
An 359.24.25	Marett, Robert R. Mannfraedi. Reykjavik, 1924.
An 3309.24.3	Martin, Rudolf. Richtlineen für Körpermessungen. München, 1924.
An 359.24.5	Mathews, Basil. The clash of colour. N.Y., 1924.
An 359.24.10	Mathews, Basil. The clash of colour. 7. ed. London, 1924.
Geog 4512.561	Mayer, T.H. Der Führer; Novelle. München, 1924.
Geog 5559.23	Mittelholzer, W. Im Flugzeug dem Nordpol entgegen. Zürich, 1924.
Geog 3055.46	Moreux, T. L'Atlantide a-t-elle existé? Paris, 1924.
Geog 4309.24	Morse, William I. Twisting trails in the Auvergnes, Cevennes, Alps of Provence. Boston, 1924.
Geog 4182.23	Naber, S.P. Hessel Gerritsz, samoyeden land ten Spitsberghe. 's-Gravenhage, 1924.
Geog 5839.24	Nissen, N.W. Die sudwestgrönlandische Landschaft und das Siedlungsgebiet der Normannen. Hamburg, 1924.
Geog 613.4.20	Olivas, A. Contribución a la bibliotheca de C.R. Markham. Lima, 1924.
An 359.24.3	Paudler, F. Die hellfarbigen Rassen und ihre Sprachstamme, Kulturen und Urkeimaten. Heidelberg, 1924.
Geog 4709.24A	The pirates own book. Salem, 1924.
An 2209.24	Rivers, W.H.R. Medicine, magic and religion. London, 1924.
An 2109.24.3	Rivers, W.H.R. Social organization. London, 1924.
An 359.25.1.3	Schiller, F.C.S. Tantalus; or The future of man. London, 1924.
An 359.24.20	Schmidt, Max. Völkerkunde. Berlin, 1924.
An 359.24.27	Schmidt, W. Völker und Kulturen. Regensburg, 1924.
Geog 4509.24.10	Schwartz, M. Vers l'idéal par la montagne. Paris, 1924.
An 359.24A	Shirokogorov, S.M. Ethnical unit and milieu. Shanghai, 1924.
An 359.24.5A	Smith, G.E. Evolution of man. London, 1924.
An 2059.24.1	Sollas, William J. Ancient hunters and their modern representatives. 3. ed. London, 1924.
An 2059.24	Sollas, William J. Ancient hunters and their modern representatives. 3. ed. N.Y., 1924.
An 349.24A	Thomson, John Arthur. What is man? N.Y., 1924.
An 3309.24	Tocher, James F. Anthropometric observations on samples of the civil populations of Aberdeenshire, Banffshire, and Kincardineshire. Edinburgh, 1924.
An 2109.24.10	Tylor, Edward Burnett. Primitive culture. 7. ed. v.1-2. N.Y., 1924.
Geog 4209.23.10	Wehde, Albert. Seit ich die heimat Verliess. Berlin, 1924.
An 2109.24.15	Whitnall, H.O. The dawn of mankind. Boston, 1924.

1925

Geog 5399.12.5	Al'banov, V.A. Mezhdu zhuznbiu v Smertbiu. Berlin, 1925.
Geog 5399.25.9	Amundsen, R. Amundsen-Ellsworths polflyvning, 1925. Oslo, 1925.
Geog 5399.25.10	Amundsen, R. Our polar flight. N.Y., 1925.
An 359.25.9	Anantha, Krishna Iyer L. Lectures on ethnogrpahy. Calcutta, 1925.
Geog 1522.38	Baedeker, publishers. Deutschland in einem Bande. 4. Aufl. Leipzig, 1925.
Geog 1523.73	Baedeker, publishers. Northern Germany. 17. ed. Leipzig, 1925.
Geog 4139.25A	Beston, Henry B. The book of gallant vagabonds. N.Y., 1925.
An 2059.25	Black, Davidson. Asia and the dispersal of primates. Peking, 1925.

1925 - cont.

Geog 4209.25.5	Bridgman, H.B. Conquering the world. N.Y., 1925.
An 609.25	British Museum. Department of Ancient and Mediaeval Antiquities and Ethnography. Handbook of the ethnographical collections. 2. ed. London, 1925.
Geog 4559.25.15	Brown, Charles W. My ditty bag. Boston, 1925.
An 379.25	Brunhes, Jean. La géographie humaine. 3. éd. Paris, 1925. 3v.
An 2109.25.19	Bumüller, J. Die Urzeit des Menschen. 4e Aufl. Augsburg, 1925.
An 2109.25.3A	Childe, Vere Gordon. The dawn of European civilization. London, 1925.
An 2109.25.4	Childe, Vere Gordon. The dawn of European civilization. N.Y., 1925.
Geog 4219.24.2	Clements, Rex. A gipsy of the Horn. 2. ed. Boston, 1925.
Geog 5839.23	Dahl, Kai R. The "Teddy" expedition. N.Y., 1925.
Geog 4219.25	Dennison, C.S. Around the world with Texaco. Houston, 1925.
Geog 5399.25.5	Ellsworth, L. The hop-off. N.Y., 1925.
Geog 6009.10.23	Evans, E.R.G.R. South with Scott. London, 1925.
Geog 4515.105.2	Faes, Henry. Le manuel du skieur. 2e éd. Lausanne, 1925.
Geog 4509.25.6	Flaig, W. Eistechnik des Bergsteigers in Bildern und Merkworten. pt.1-4. 2. Aufl. Stuttgart, 1925.
Geog 4359.25	Flambeau, V. Red letter days in Europe. With a glimpse of Northern Africa. N.Y., 1925.
Geog 4709.22.2	French, J.L. Great pirate stories. 2. ser. N.Y., 1925.
An 180.3	Frobenius, Leo. Erlebte Erdteile; Ergebnisse eines deutschen Forscherlebens. Frankfurt am Main, 1925-29. 7v.
An 2109.25.5F	Frobenius, Leo. Hádschra Máktuba. München, 1925.
Geog 4129.25.2	Gorce, Denys. Les voyages l'hospitalité...dans le monde chrétien des IVe et Ve siècles. Thèse. Wépion-sur-Meuse, 1925.
Geog 4129.25	Gorce, Denys. Les voyages l'hospitalité...dans le monde chrétien des IVe et Ve siècles. Wépion-sur-Meuse, 1925.
An 359.25.4	Gregory, John W. The menace of colour. 2. ed. London, 1925.
An 359.25.11	Haddon, A.C. Races of man and their distribution. N.Y., 1925.
Geog 4219.25.15	Halliburton, Richard. The royal road to romance. Garden City, 1925.
Geog 4219.25.10A	Halliburton, Richard. The royal road to romance. Indianapolis, 1925.
Geog 4309.25.10	Hamilton, C.M. Wanderings. Garden City, 1925.
Geog 4181.56	Harlow, V.T. Colonizing expeditions to the West Indies and Guiana, 1623-67. London, 1925.
Geog 4209.25.10	Harper, H.H. Highlights of foreign travel. N.Y., 1925.
Geog 4559.25	Hayes, Bertram. Hull down. N.Y., 1925.
An 359.25.7	Hertz, F.O. Rasse und Kultur. 3. Aufl. Leipzig, 1925.
Geog 4515.102	Hoferer, Erwin. Winterliches Bergsteigen alpine Schilaustechnik. München, 1925.
Geog 5650.11	Hovgaard, William. The Norsemen in Greenland; recent discoveries at Herjolfones. N.Y., 1925.
An 3309.25	Hrdlička, A. The old Americans. Baltimore, 1925.
Geog 4419.25	Huntington, E. West of the Pacific. N.Y., 1925.
Geog 5939.25	Hurley, Frank. Argonauts of the South. N.Y., 1925.
Geog 4209.25.15	Jacques, N. Im Kaleidoskop der Weltteile. Berlin, 1925.
Geog 4557.95	Kelly, Samuel. Samuel Kelly, an 18th century seaman. N.Y., 1925.
Geog 4219.25.3	Knight, C. Trucking the sunset. Atlanta, 1925.
An 2109.25.9	Leroy, Olivier. Essai d'introduction critique à l'étude de l'economie primitive. Paris, 1925.
Geog 4181.52	Lockberry, William. The journal of William Lockerby. London, 1925.
Geog 4559.25.10F	Lubbock, A.B. Adventures by sea from art of old time. London, 1925.
Geog 4509.25	Lunn, A.H.M. The mountains of youth. London, 1925.
Geog 4512.537.5	Luther, C.J. Skunterhaltungen. München, 1925.
Geog 4512.537	Luther, C.J. Wil der Schneelauf nach Deutschland kam. München, 1925.
Geog 4679.25	Marine Research Society, Salem, Mass. The sea, the ship, and the sailor. Salem, Mass., 1925.
Geog 4181.57	Mortoft, F. Francis Mortoft; his book...1658-59. London, 1925.
Geog 5558.82	Nansen, F. Hunting and adventure in the Arctic. N.Y., 1925.
Geog 4300.36	Nederlandsch-Amerikaansche Stoomvaart Maatschappij, N.V. Guide for Europe. The Hague, 1925.
Geog 4309.23.10	Osborne, A.B. Finding the worth while in Europe. N.Y., 1925.
Geog 3249.25	Pereyra, Carlos. La conquête des routes océaniques d'Henri le navigateur à Magellan. Paris, 1925.
Geog 4559.25.3	Perry, F. Fair winds and foul. Boston, 1925.
Geog 5399.25.35	Rasmussen, K. Fra Grønland til Stillehavet. København, 1925-26. 2v.
Geog 4219.25.5	Reinhardt, Walther. Querweltein. Berlin, 1925.
An 359.25.5	Scheidt, W. Allgemeine Rassenkunde. München, 1925.
An 359.25.1	Schiller, F.C.S. Tantalus; or The future of man. N.Y., 1925.
An 359.25.2	Schiller, F.C.S. Tantalus; or The future of man. N.Y., 1925.
Geog 4515.301	Schmidkunz, W. Kochbuch für Bergsteiger. München, 1925.
Geog 4709.25	Seitz, Don Carlos. Under the black flag. N.Y., 1925.
Geog 3055.42	Spence, Lewis. Atlantis in America. London, 1925.
Geog 3055.64	Spence, Lewis. The problem of Atlantis. 2d ed. N.Y., 1925?
Geog 5399.25	Stefansson, V. The adventure of Wrangel Island. N.Y., 1925.
An 2109.25A	Tozzer, A.M. Social origins and social continuities. N.Y., 1925.
Geog 665.32	Vallaux, Camille. Les sciences géographiques. Paris, 1925.
Geog 5329.3A	Vilhjalmur Stefansson. N.Y., 1925.
An 2059.25.5	Vulliamy, C.E. Our prehistoric forerunners. N.Y., 1925.
Geog 4239.24	Wells, F. de W. The last cruise of the Shanghai. N.Y., 1925.
Geog 4182.26	Wieder, F.C. Die stichting van New York in Juli 1625. 's-Gravenhage, 1925.
Geog 3229.25A	Wright, John K. The geographical lore of the time of the Crusades. N.Y., 1925.

1926

Geog 4249.26	Akademiia Nauk, Petrograd. The Pacific Russian scientific investigation. Leningrad, 1926.
Geog 5399.25.19	Amundsen, R. Den første flukt over Polhavet. Oslo, 1926.
Geog 604.2	Anthiaume. Pierre Desceliers. Rouen, 1926.

Chronological Listing

1926 - cont.

	Geog 1525.66	Baedeker, publishers. The Rhine from the Dutch to the Alsatian frontier. 18. ed. Leipzig, 1926.
	Geog 3070.36	Barker, W.H. The history of cartography. Manchester, 1926.
	Geog 4559.26.20	Beaumont, J.C.H. Ships and people. N.Y., 1926.
	Geog 612.3.8	Bellessort, André. La Pérouse. Paris, 1926.
	Geog 4145.7	Bertrand, Alfred. Alfred Bertrand. London, 1926.
	Geog 4309.26.10	Blanco-Fombona, R. Por los caminos del mundo. Madrid, 1926.
	Geog 4559.26.15	Boughton, George P. Seafaring. London, 1926.
	Geog 4208.80.3	Briggs, L.V. Around Cape Horn to Honolulu on the bark "Amy Turner" 1880. Boston, 1926.
	Geog 4308.26	Carus, Karl Gustav. Reisen und Briefe. Leipzig, 1926. 2v.
Htn	Geog 4679.26*	Chapin, H.M. Biblioteca Titanicana. Metuchen, 1926.
	Geog 4679.26.2	Chapin, H.M. Biblioteca Titanicana. Metuchen, 1926.
	Geog 4559.26.10	Crane, Mannin. Yarns from a windjammer. Boston, 1926.
	Geog 4559.26.3	Farmer, H.F. The log of a shellback. N.Y., 1926.
	Geog 4309.21.10A	Futera, Count Y. The crown prince's European tour (March 3-September 3, 1921). Osaka, 1926.
	Geog 3055.48	Gattefossé, Jean. Bibliographie de l'Atlantide et des questions connexes. Lyon, 1926.
	Geog 4709.26	Gosse, Philip. My pirate library. London, 1926.
	Geog 5326.21A	Green, Fitzhugh. Peary; the man who refused to fail. N.Y., 1926.
	An 379.26.5	Günther, Hans F.K. Rasse und Stil. München, 1926.
	Geog 3520.10.15	Hakluyt, Richard. A selection of the principal voyages. N.Y., 1926.
	Geog 4209.26	Hall, James Norman. On the stream of travel. Boston, 1926.
	An 2109.26.7	Hambly, W.D. Origins of education among primitive peoples. London, 1926.
	An 2109.26.3	Hambly, W.D. Tribal dancing and social development. London, 1926.
	An 359.26.4A	Hawkins, Frank H. The racial basis of civilization. N.Y., 1926.
	Geog 4559.26	Hemy, Thomas M. Deep sea days. London, 1926.
	Geog 3070.33.2A	Holman, Louis A. Old maps and their makers. 2d ed. Boston, 1926.
	Geog 4229.26	Huxley, Aldous. Jesting pilate; an intellectual holiday. N.Y., 1926.
	Geog 4182.27	Ijzermann, J.W. De reis om de wereld door Olivier van Noort. 's-Gravehage, 1926. 2v.
	Geog 4672.12	Kitchin, Frederick H. Dead men's tales. Edinburgh, 1926.
	Geog 3055.43	Le Cour, Paul. A la recherche d'un monde perdu. Paris, 1926.
	An 2109.26.5	Lévy-Bruhl, L. How natives think. London, 1926.
	Geog 4672.16.1	Lockhart, John Gilbert. Strange adventures of the sea. N.Y., 1926.
	Geog 4515.103.2	Lunn, A.H.M. Alpine ski-ing at all heights and seasons. 2d ed. London, 1926.
	Geog 4329.25	McAllister, J.G. Borderlands of the Mediterranean. Richmond, 1926.
	An 359.26.15	Mathews, Basil. The clash of colour. 16. ed. London, 1926.
	Geog 4309.26A	Morse, William I. The diary of a musketeer. Boston, 1926.
	Geog 4205.98.10	Nijhoff, Wouter. Bibliographie...die voyagie om den geheelen werelt...door Olivier van Noort. 's-Gravenhage, 1926.
	Geog 4672.14	O'Donnell, Elliott. Strange sea mysteries. London, 1926.
	Geog 3019.26	Parlss, George B. The forerunners of Hakluyt. n.p., 1926?
	Geog 665.92	Plattie, Roderick. College geography. Boston, 1926.
	Geog 5839.26	Rawson, K.L. A boy's-eye view of the Arctic. N.Y., 1926.
	Geog 4309.26.25	Reynolds, Bruce. A cocktail continentale. N.Y., 1926.
	An 5.9	Royal Empire Society, London. Library. Select bibliography of recent publications in the library. London, 1926.
	An 359.26.10	Schmidt, Max. The primitive races of mankind. London, 1926.
	An 359.26	Schütte, G. Vor folkegruppe Gottjod. Kjøbenhavn, 1926.
	Geog 6009.10.20	Simpson, G.C. Soctt's polar journey and the weather. Oxford, 1926.
	Geog 3055.42.5	Spence, Lewis. The history of Atlantis. London, 1926.
	Geog 4344.35.12	Tafur, Pero. Travels and adventures, 1435-1439. London, 1926.
	Geog 4344.35.10	Tafur, Pero. Travels and adventures, 1435-1439. N.Y., 1926.
	Geog 579.26	Vergara y Martin, Gabriel Maria. Diccionario de voces y términos geográficos. Madrid, 1926.
	An 379.26A	Vidal de la Blache, P. Principles of human nature. N.Y., 1926.
	Geog 4219.26	Wells, Linton. Around the world in twenty-eight days. Boston, 1926.
	Geog 4329.26	Wilstach, Paul. Islands of the Mediterranean. London, 1926.

1927

	An 2109.27.23	Allier, Raoul. La non-civilisé et nous. Paris, 1927.
	Geog 5399.25.20A	Amundsen, R. First crossing of the Polar Sea. N.Y., 1927.
	Geog 604.1	Anthiaume. L'abbé Guillaume Denys de Dieppe, 1624-1689. Paris, 1927.
Htn	Geog 500.59F*	Atkinson, G. La litterature géographique française de la renaissance. Paris, 1927.
	Geog 1508.32	Baedeker, publishers. Great Britain. 8. ed. Leipzig, 1927.
	Geog 1560.67	Baedeker, publishers. Holland; Hanbuch für Reisende. 26. Aufl. Leipzig, 1927.
	Geog 1533.14A	Baedeker, publishers. Tyrol and the Dolomites including the Bavarian Alps. 13. ed. Leipzig, 1927.
	Geog 3070.34	Baulig, Henri. Exercices cartographiques. Paris, 1927.
	Geog 5839.27.5	Bendixen, Ole. Grønland som Nybyggerland. København, 1927.
	Geog 4268.30	Bertaut, Jules. Villégiatures romantiques. Paris, 1927.
	Geog 3055.44	Bjorkman, E. The search for Atlantis. N.Y., 1927.
	Geog 4307.65.10	Bousquet, (Mrs.). Mrs. Bousquet's diary, 1765. Norwich, 1927.
	Geog 4181.58	Bowry, T. The papers. London, 1927.
	An 2109.27.11	Boyle, Mary E. In search of our ancestors. London, 1927.
	Geog 5070.20.5	Brown, R.N.R. The polar regions. London, 1927.
	Geog 4307.70.10A	Burney, Charles. Continental travels. London, 1927.
	Geog 4559.27.15	Chatterton, E.K. The brotherhood of the sea. London, 1927.
	Geog 3580.40	Colbrecht, Jozsf. De vleminfen en de Spansche. Antwerp, 1927.
	Geog 4559.27.10	Cooper, F.G. Yarns of the seven seas. London, 1927.

1927 - cont.

	An 2109.27.30	Cunow, Heinrich. Technik und Wirtschaft des europäischen Urmenschen. Berlin, 1927.
	Geog 4348.08	Darwin, F.S. Travels in Spain and the East, 1808-1810. Cambridge, 1927.
	An 3309.27	Davenport, C.B. Guide to physical anthropometry and anthroposcopy. Baltimore, 1927.
	Geog 4500.7	Dreyer, Aloys. Bücherverzeichnis der Alpenvereinsbucherei, mit Verfasser- und Bergnamen-Verzeichnis. München, 1927.
	Geog 4309.27.15	Edschmid, Kasimir. Das grosse Reisebuch. Berlin, 1927.
	An 2109.27.15	Ellwood, Charles A. Cultural evolution; a study of social origins and development. N.Y., 1927.
	An 2109.27.17	Esanielsertier, Daniel. Les formes inférieures de l'explication. Paris, 1927.
	An 2109.27.18	Essertier, Daniel. Les formes inférieures de l'explication. Thèse. Paris, 1927.
	An 2109.27.21	Fahl, Anton. Ein Beitrag zur Wirtschaftsgeschichte der Vorzeit. Inaug. Diss. Halle, 1927.
	An 2109.27.19	Fahrenfort, J.J. Het hoogste wezen der primitieven. Groningen, 1927.
	Geog 5329.3.5	Finnbogason, G. Vilhjalmur Stefansson. Akureyri, 1927.
	Geog 5838.88.15	Friðiksson, O. Fra vestfjörðm til vestribyðr för F. Nansen. Reykjavik, 1927.
	Geog 5838.88.16	Friðiksson, O. Fra vestfjörðm til vestribyðr för F. Nansen. Reykjavik, 1927.
	Geog 809.27A	Geographie universelle. v.1-15. Paris, 1927-1946. 22v.
	An 359.27.5	Gohier, W.D. Cassandre, ou La folie des blancs. Paris, 1927.
	Geog 5317.10	Greely, A.W. Reminiscences of adventure and service. N.Y., 1927.
	An 379.26.7	Günther, Hans F.K. Rasse und Stil. 2. Aufl. München, 1927.
	Geog 3520.10.75A	Hakluyt, Richard. Fighting merchant men. Boston, 1927.
	Geog 3520.10.45	Hakluyt, Richard. Principal navigations, voyages...of the English nation. London, 1927. 8v.
	Geog 4329.27	Haliburton, R. The glorious adventure. Indianapolis, 1927.
	Geog 4208.27	Haus, Rosalie. Voyage of the Caroline, 1827-28. London, 1927.
	An 2109.27.35	Henderson, Keith. Prehistoric man. N.Y., 1927.
	Geog 819.27.5	Hettner, A. Die Geographie; ihre Geschichte. Breslau, 1927.
	Geog 4209.27	Hoare, Samuel. India by air. London, 1927.
	Geog 4300.30.4	Hungerford, E. Planning a trip abroad. N.Y., 1927.
	An 379.27.3	Huntington, Ellsworth. The human habitat. N.Y., 1927.
	Geog 4239.27	Ihering, H. von. Die Geschichte des Atlantischen Ozeans. Jena, 1927.
	Geog 3019.27.5	Iorga, N. Les voyageurs orientaux en France. Paris, 1927.
	Geog 618.3.3	Ishill, Joseph. Élisée and Elie Reclus, in memoriam. Berkeley Heights, N.J., 1927.
	Geog 4672.12.6	Kitchen, Frederick H. Tales of S.O.S. and T.T.T. Edinburgh, 1927.
	An 2109.27.25A	Leroy, Olivier. La raison primitive. Paris, 1927.
	An 2109.27.13	Lévy-Bruhl, L. L'ame primitive. 3e éd. Paris, 1927.
X Cg	An 2109.25.7	Lévy-Bruhl, L. How natives think. N.Y., 1927.
	Geog 4679.27	Lockhart, John Gilbert. A great sea mystery. London, 1927.
	Geog 4329.05.20	Lorenz, I.E. The new Mediterranean traveller. 13th ed. N.Y., 1927.
	Geog 4181.59A	Luard, C.E. Travels of Fray Sebanstian Manrique. London, 1927. 2v.
	Geog 4329.27.15	Ludwig, Emil. Am Mittelmeer. Berlin, 1927.
	Geog 4219.27	Lunn, Henry. Round the world with a dictaphone. London, 1927.
	An 2059.27	Lyell, C. The antiquity of man. London, 1927.
	Geog 5839.27	MacMillan, D.B. Etah and beyond. Boston, 1927.
	An 2109.27.3A	Malinowski, B. Sex and repression in savage society. London, 1927.
	An 2109.27.4	Malinowski, B. Sex and repression in savage society. N.Y., 1927.
	An 2109.27.27	Marazzi, A. Fra i selvaggi e fra i civilizzate. Milano, 1927.
	An 359.27.3	Marett, Robert R. The diffusion of culture. Cambridge, Eng., 1927.
	An 359.27.4	Marett, Robert R. Man in the making. London, 1927.
	Geog 4219.27.5	Martin, Lillien J. Round the world with a psychologist. San Francisco, 1927.
	Geog 4329.27.20	Meier-Graefe, J. Pyramide und Tempel. Berlin, 1927.
	Geog 500.58	Milan. Bibliotheca. Catalogo ragionato della geografica. Milano, 1927.
Htn	Geog 4329.27.10*	Morse, William I. Sicilian days and other journeys round the Mediterranean and Adriatic. Boston, 1927.
	An 359.27.11A	Muntz, Earl E. Race contact. N.Y., 1927.
	An 2109.27.7	Murphy, John. Primitive man, his essential quest. London, 1927.
	Geog 4515.101	Niederl, Franz. Das gehen auf Eis und Schnee. 2. Aufl. München, 1927.
	An 3559.27	Nyèssen, D.J.H. The passing of the Frisians. 's-Gravenhage, 1927.
	Geog 4329.27.5	Oppenheim, E.P. The quest for winter sunshine. Boston, 1927.
	An 2109.27.9	Osborn, H.F. Man rises to Parnassus. Princeton, 1927.
	Geog 819.27.10	Paquet, A. Städte, Landschaften und ewige Bewegung. Hamburg, 1927.
	Geog 4709.27	Partridge, Eric. Pirates, highwaymen, and adventurers. London, 1927.
	An 359.27.13	Penniston, J.B. Racial old age, being further adventures in philosophy. Seattle, 1927.
	An 359.27.9A	Pitt-Rivers, G.H.L.F. The clash of culture and the contact of races. London, 1927.
	Geog 5559.27	Putnam, D.B. Davial goes to Baffinland. N.Y., 1927.
	Geog 3620.2	Radoščić, Nikola. Geografsko znanje o Srbiji početkom 19 vcka. Beograd, 1927.
	Geog 5559.21	Rasmussen, Knud. Across Arctic America. N.Y., 1927.
	Geog 4349.23	Ross, Halford. By devious ways. London, 1927.
	Geog 4300.29	Schoonmaker, F. Through Europe on two dollars a day. N.Y., 1927.
	Geog 4709.27.5	Seitz, Don Carlos. Under the black flag. London, 1927.
	Geog 4559.27.5A	Senior, William. Naval history in the law courts. London, 1927.
	Geog 4559.27	Smith, Cicely Fox. A sea chest; an anthology of ships and sailormen. London, 1927.
	An 359.24.7	Smith, G.E. The evolution of man. 2. ed. London, 1927.
	Geog 3055.42.7	Spence, Lewis. The history of Atlantis. 4th ed. London, 1927.
	Geog 4105.17.5	Tatchell, Frank. The happy traveller. 5. ed. London, 1927.

Chronological Listing

1927 - cont.

An 379.27A	Taylor, Griffith. Environment and race. London, 1927.
Geog 4326.75.5A	Teonge, Henry. The diary of H. Teonge, chaplain on board H.M.'s ships Assistance, Bristol and Royal Oak, 1675-1679. London, 1927.
Geog 4309.27	Thomas, Lowell. European skyways. Boston, 1927.
Geog 4182.29	Van Nouhuys, J.W. De eerste nederlandsche. 's-Gravenhage, 1927-51. 2v.
Geog 5837.39	Wallóe, Peder Olsen. Peder Olsen Wallóes dagbóger. Kjøbenhavn, 1927.
Geog 4145.25.5	Wilson, P.W. An explorer of changing horizons. N.Y., 1927.
An 359.27	Wolff, K.F. Rassenlehre. Leipzig, 1927.
An 359.27.7A	Worrell, W.H. A study of races in the ancient Near East. N.Y., 1927.
Geog 5399.27.3	Worsley, F.A. Under sail in the frozen north. London, 1927.
Geog 5070.44	Zeidler, P.G. Polarfahrten. Berlin, 1927.

1928

Geog 5399.12	Al'banov, V.A. Au pays de la mort blanche. Paris, 1928.
Geog 809.28	Allgemeine Länderkunde der Erdteile. v.3-4,6. Hannover, 1928-1935.
Geog 5070.35A	American Geographic Society, N.Y. Problems of polar research. N.Y., 1928.
Geog 4559.28.15	Amicis, E. de. Sull'oceano. Milano, 1928.
Geog 5399.25.21	Amundsen, R. First crossing of the Polar Sea. Garden City, 1928.
Geog 5311.3.15	Amundsen, Roald. Apdagelsesreiser. Oslo, 1928-30. 4v.
Geog 5311.1.3	Amundsen, Roald. My life as an explorer. Garden City, N.Y., 1928.
Geog 5399.22	Ashton, James M. Ice-bound, a trader's adventure in the Siberian Arctic. N.Y., 1928.
Geog 1558.20	Baedeker, publishers. Belgique et Luxembourg. 20. éd. Leipzig, 1928.
Geog 1538.57A	Baedeker, publishers. Italy from the Alps to Naples. 3. ed. Leipzig, 1928.
Geog 1535.76	Baedeker, publishers. Switzerland together with Chamonix. 27. ed. Leipzig, 1928.
Geog 3070.40	Bagrow, Leo. A. Ortelii Catalogus cartographorum bearbeitet. v.1-2. Gotha, 1928-30.
Geog 5312.2A	Bartlett, Robert A. The log of Bob Bartlett. N.Y., 1928.
Geog 4307.80.15	Beckford, William. The travel-diares of William Beckford of Fonthill. Cambridge, 1928. 2v.
Geog 4359.28	Belloc, Hilaire. Many cities. London, 1928.
Geog 4209.28.10	Benson, Stella. Worlds within worlds. London, 1928.
Geog 5639.28	Berlin, Knud. Grønlands betydning. København, 1928.
Geog 4709.28.5	Besson, Maurice. Les frères de la coste, flibustiers et corsaires. Paris, 1928.
Geog 4209.28.25	Bloem, Walter. Weldgesicht; ein Buch von heutiger und kommender Menscheit. Leipzig, 1928.
Geog 5516.31.5A	Bodilly, R.B. The voyage of Captain Thomas James...1631. London, 1928.
Geog 4419.28	Bosch, Carl. Karawanen-Reisen...Ägypten, Mesopotamien, Persien und Abessinien. Berlin, 1928.
Geog 819.28.5A	Bowman, I. The new world. 4th ed. Yonkers-on-Hudson, 1928.
Geog 4309.28.5	Brandt, R. Das Gesicht Europas. Hamburg, 1928.
Geog 4349.28	Brennan, John W. The student abroad. Boston, 1928.
Geog 603.4	Briceno, Alfonso. Augustin Codazzi. Tesis. Caracas, 1928.
Geog 4028.4.2	Bridges, Thomas C. Heroes of modern adventure. Boston, 1928.
Geog 6009.28	Byrd, Richard E. Into the home of the blizzard. N.Y., 1928.
Geog 4209.28.20A	Cameron, J. John Cameron's odyssey. N.Y., 1928.
Geog 4169.28	Chatterton, E.K. Ventures and voyages. London, 1928.
Geog 4219.28	Cook, Thomas, firm, publishers, London. The supreme travel-adventure: around the world in the "Franconia" 1929. N.Y., 1928.
Geog 4219.28.10	Costes Dieudonné. La grande croisière de Costes et Le Brie. Paris, 1928.
Geog 604.3	Couto, Gustavo. O cosmografo Fernam Vaz Dourado, fronteiro da India e a sua obra. Lisboa, 1928.
Geog 4209.28.5	Cust, Nina. Wanderers. London, 1928.
Geog 4558.40.40	Dana, Richard Henry. Two years before the mast. N.Y., 1928.
Geog 5839.28	Denmark. Kommittee for Ledelsen af Degeoliske og Geografiske Undersøgelser i Grønland. Greenland...invest in Greenland. Copenhagen, 1928. 3v.
An 359.28A	Dixon, Roland B. The building of cultures. N.Y., 1928.
Geog 5650.12	Dúason, Jón. Grønlands statsretslige stilling i Midelalderen. Oslo, 1928.
Geog 611.1	Dupouy, A. Le Briton Yves de Kerguelen. Paris, 1928.
Geog 5855.11	Ekblaw, W.E. The material response of the polar eskimo to their far Arctic environment. Diss. N.Y., 1928.
Geog 4181.60	Farcourt, R. Relation of a voyage to Guiana. London, 1928.
Geog 4309.26.20	Farson, N. Sailing across Europe. London, 1928.
Geog 4216.21	Fernberger von Egenberg, C.M. Unfreiwillige Reise um die Welt, 1621-1628. Leipzig, 1928.
An 379.28	Fleure, H.J. The races of mankind. London, 1928.
Geog 4209.28	Forbes, Rosita. Adventure. Boston, 1928.
Geog 4709.28.10	Gallomb, J. Pirates, old and new. N.Y., 1928.
Geog 5070.30	Greely, A.W. The polar regions in the twentieth century. Boston, 1928.
An 379.28.5	Günther, Hans F.K. The racial elements of European history. N.Y., 1928.
An 2109.28.9	Hankin, E.H. The cave man's legacy. London, 1928.
Htn Geog 4559.28.6*	Harlow, Frederick Pease. The making of a sailor. Salem, Mass., 1928.
Geog 4559.28.5	Harlow, Frederick Pease. The making of a sailor. Salem, Mass., 1928.
Geog 5919.28	Hayes, James G. Antarctica; a treatise on the southern continent. London, 1928.
An 359.28.5	Hertz, F.O. Race and civilization. N.Y., 1928.
An 359.28.3	Hertz, F.O. Race and civilization. N.Y., 1928.
An 359.28.9	Hildebrandt, K. Staat und Rasse. Breslau, 1928.
An 379.28.3	Huntington, Ellsworth. The human habitat. London, 1928.
Geog 3019.27.10	Iorga, N. Une vingtaine de voyageurs dans l'Orient européen. Paris, 1928.
Geog 3019.27.2	Iorga, N. Les voyageurs français dans l'Orient européen. Paris, 1928.
Geog 4219.28.5	Kircheiss, C. Meine Weltumsegelung mit dem Fischkutter Hamburg. 2. Aufl. Berlin, 1928.
Geog 4229.25.5	Kleinschmitt, Edmund. Durch Werkstätten und Gassen dreier Erdteile. 2. Aufl. Hamburg, 1928.

1928 - cont.

An 609.28	Kölner Anthropologische Gesellschaft. 25 Jahre Kölner anthropologische Gesellschaft und städtisches Museum für Vor- und Frühgeschichte 1903-28. Festschrift. Köln, 1928.
Geog 4515.106.2	Kurz, Marcel. Alpinisme hivernal. Paris, 1928.
Geog 5399.28	Lee, Herbert P. Policing the top of the world. London, 1928.
An 2109.28.11	Leeuw, Gerardus. La structure de la mentalité primitive. Strasbourg, 1928.
Geog 4326.32.20	Lithgow, William. Rare adventures and painefull peregrinations. London, 1928.
Geog 4229.28	Log of the ketch Seven Bells. n.p., 1928?
Geog 4219.27.2	Lunn, Henry. Round the world with a dictaphone. London, 1928.
Geog 4309.25.11	MacDonald, J.R. Wanderings and excursions. Indianapolis, 1928.
An 359.27.4.3	Marett, Robert R. Man in the making. Garden City, 1928.
Geog 4309.28	Marín Vicuna, Santiago. Viajando. Santiago, 1928.
An 2109.28.5A	Miller, Nathan. The child in primitive society. N.Y., 1928.
Geog 819.28	Mitchell, J. Leslie. Hauno, or Future of exploration. London, 1928.
Geog 4309.29	Muller, Julius W. "Ever thine". N.Y., 1928.
Geog 5312.3.15A	Murphy, Charles J.V. Struggle; the life and exploits of Commander Richard E. Byrd. N.Y., 1928.
Geog 618.3.7	Nettlau, Max. Elisée Reclus. Berlin, 1928.
Geog 5070.36A	Nordenskjöld, Otto. The geography of the polar regions. N.Y., 1928.
Geog 3251.27F	Nunn, George E. World map of Francesco Roselli. Philadelphia, 1928.
An 2109.28.3	Osborn, H.F. Man rises to Parnassus. Princeton, 1928.
Geog 3520.11A	Parks, George Bruner. Richard Hakluyt and the English voyages. N.Y., 1928.
Geog 4502.5.2	Rohrer, M. Berglieder der Völker. München, 1928.
Geog 4329.28	Rydh, Hanna. Kring medelhavets stränder. Stockholm, 1928.
Geog 4502.5	Saussure, H.B. de. Relation abrégés d'un voyage a la cime du Mont-Blanc en aout 1787. Genève, 1928? 2 pam.
An 359.28.7	Scheler, Max. Die Stellung des Menschen im Kosmos. Darmstadt, 1928.
An 359.28.11	Schemann, L. Die Rasse in den Geisteswissenschaften. München, 1928-31. 3v.
Geog 4559.28.10	Smith, Cicely Fox. Ancient mariners. London, 1928.
An 2109.28.13	Smith, Grafton Eliot. In the beginning; the origin of civilization. London, 1928.
Geog 4559.28	Sorrell, George. The man before the mast. London, 1928.
Geog 4227.85F	Strange, James. Journal and narrative of the...expedition from Bombay to the N.W. coast of America. Madras, 1928.
An 3309.28	Sullivan, Louis R. Essentials of anthropometry. N.Y., 1928.
An 2109.28.20	Tozzer, Alfred M. Social origins and social communities. N.Y., 1928.
Geog 4219.05.10	Treves, F. The other side of the lantern. London, 1928.
Geog 5559.28	Wilkins, George H. Flying the Arctic. N.Y., 1928.
Geog 4182.30	Woard, C. de. Zeeuwsche expedite...Cornelis Evertsen. 's-Gravenhage, 1928.
Geog 4709.28A	Wycherley, George. Buccaneers of the Pacific. Indianapolis, 1928.
Geog 4309.28.10	Zozulia, Efim D. Iz Moskvy na Korsiku i Obratno. Leningrad, 1928.

1929

An 2109.29.7	Allier, Raoul. Mind of the savage. London, 1929.
Geog 500.70	Anderson, Ernst. Bok-katalog omfattande geografi och resor. v.1-2. Stockholm, 1929-31.
Geog 5839.29	Arnason, A. Graenlandsför 1929. Reykjavik, 1929.
Geog 1530.69	Baedeker, publishers. Austria, together with Budapest. 12. ed. Leipzig, 1929.
Geog 1616.24A	Baedeker, publishers. Egypt and the Sudan. 8. ed. Leipzig, 1929.
Geog 1524.69A	Baedeker, publishers. Southern Germany. 13. ed. Leipzig, 1929.
Geog 1531.4	Baedeker, publishers. Tirol, Vorarlberg, Etschland. 39. Aufl. Leipzig, 1929.
Geog 4219.29.5	Baus, Thomas J. 30,000 miles around the world. Philadelphia, 1929.
Geog 5399.28.25	Běhounek, Franz. Sieben Wochen auf der Eisscholle. Leipzig, 1929.
Geog 3550.11	Bertacchi, C. Geografi ed exploratori italiani contemporanei. Milano, 1929.
Geog 4515.104.5	Bilgeri, George. Colonel Bilgeri's handbook on mountain ski-ing. London, 1929.
An 2109.29.15	Boas, Franz. The mind of primitive man. N.Y., 1929.
Geog 5558.81.20	Brainard, David L. The outpost of the lost. Indianapolis, 1929.
Geog 5070.46	Brown, R.N.R. Some problems of polar geography. Washington, 1929.
Geog 4209.29.10	Bruce, Michael. Peaks of hazard. Indianapolis, 1929.
An 2109.29.9	Butt-Thompson, F.W. West African secret societies. London, 1929.
Geog 4209.29.5	Buxton, Noel. Travels and reflections. London, 1929.
Geog 4182.31	Caland, W. Die remonstrantie von W. Geleynssen de Jongh. 's-Gravenhage, 1929.
Geog 4181.63	Carruthers, D. The desert route to India. London, 1929.
Geog 5839.29.5	Charcot, J.B. La mer du Groenland, croisières du "Pourquoi pas?". Paris, 1929.
Geog 4559.29.15	Chatterton, E.K. On the high seas. Philadelphia, 1929.
Geog 110.3.10	Congrès des Géographes et Ethnographes Slaves, 2nd, Krakow, 1927. Pamiętnik II zjardu słowiańskich geografów i etnografów odbytego w Polsce w roku, 1927. Kraków, 1929-30. 2v.
Htn An 2109.29*	Crawley, A.E. Studies of savages and sex. London, 1929.
An 3309.29	Davenport, C.B. Race crossing in Jamaica. Washington, 1929.
Geog 4029.8	Dean, Harry. The Pedro Gorino. Boston, 1929.
An 359.29.9	Duncan, H. Race and population problems. 1. ed. N.Y., 1929.
Geog 4324.59.10	Ehingen, Georg von. Diary. London, 1929.
Geog 4349.29	Ellis, M.H. Express to Hindustan. London, 1929.
Geog 4239.29.5	England, George Allan. Isles of romance. N.Y., 1929.
An 359.29.12	Foster, T.S. Travels and settlements of early man. N.Y., 1929.
Geog 4029.4	Franco, Ramon. Aguilas y garras. Madrid, 1929.
Geog 5399.28.5	Giudici, Davide. The tragedy of the Italian with the rescuers to the Red tent. N.Y., 1929.
Geog 4209.29	Goldring, Douglas. People and places. Boston, 1929.

Chronological Listing

1929 - cont.

An 359.29.7 Günther, H.F.K. Rassenkunde Europas. 3. Aufl. München, 1929.
An 379.30 Hauemeyer, Loomis. Ethnography. Boston, 1929.
Geog 3070.35 Hayes, Gerald R. The production of an admiralty chart. London, 1929.
Geog 5326.21.15 Hayes, J.G. Robert Edwin Peary, a record of his explorations, 1886-1909. London, 1929.
Geog 3030.4 Honigmann, Ernst. Die sieben Klimata. Heidelberg, 1929.
An 2059.29 Hrdlička, A. The neanderthal phase of man. Washington, 1929.
An 182.2 Humboldt, Wilhelm von. Philosophische Anthropologie und Theorie der Menschenkenntnis. Halle, 1929.
An 147.5 In memoriam Karl Weule. Leipzig, 1929.
Geog 6009.14.15 Joyce, Ernest. The South Pole trail. London, 1929.
Geog 5399.28.40 Katz, Otto. Neun Männer im Eis. Berlin, 1929.
Geog 4678.72.15 Keating, Laurence J. The great Mary Celeste hoax. London, 1929.
Geog 4219.29.15 Lewis, Aswold. Because I've not been there before. London, 1929.
Geog 4672.16.5 Lockhart, John Gilbert. Strange tales of the seven seas. London, 1929.
Geog 4329.29.10 Ludwig, Emil. On Mediterranean shores. Boston, 1929.
Geog 4309.25.12 MacDonald, J.R. Wanderings and excursions. London, 1929.
Geog 4309.29.10 MacOrlan, Pierre. Villes: Rouen-Montmartre-Brest-Londres-Villes rhénanes-Rome. 4. éd. Paris, 1929.
Geog 500.60 Maggs Bros., London. Bibliotheca asiatica et africana. pt.4-5. London, 1929.
X Cg An 2109.29.5 Malinowski, B. The sexual life of savages in north west Melanesia. London, 1929.
Geog 4219.29.2 Mann, Erik. Rundherm. 3.-4. Aufl. Berlin, 1929.
An 359.29 Morand, Paul. Black magic. N.Y., 1929.
Geog 4559.29 Munro, D.J. The roaring forties and after. London, 1929.
Geog 1807.25 Murray, John, publisher, London. Handbook for...India, Burma and Ceylon. 13. ed. London, 1929.
Geog 1807.25.2 Murray, John, publisher, London. Handbook for...India, Burma and Ceylon. 13. ed. N.Y., 1929.
Geog 3229.29 Mžik, Hans von. Beiträge zur historischen Geographie, Kulturgeographie, Ethnographie und Kartographie. Leipzig, 1929.
Geog 618.3.8 Nettlau, Max. Eliseo Reclus, la vida de un sabio justo. Barcelona, 1929. 2v.
Geog 3060.11 Nunn, G.E. Origin of Strait of Asian concept. Philadelphia, 1929.
X Cg An 3559.29 Nyèssen, D.J.H. Somatical investigation of the Javanese, 1929. Bandoeng, 1929.
Geog 158.1.25 Österreichische Touristen-Zeitung. Sondernummer anläslich des 60 jähren Bestande. Innsbruck, 1929.
An 3309.29.3 Puccioni, Nello. Affrica nord-orientale e Arabia. Pavia, 1929.
An 2109.29.11 Rivers, W.H.R. Social organization. N.Y., 1929.
Geog 4502.5.4 Rombert, E. Das Murmeltier mit dem Halsband. München, 1929.
An 359.29.3 Schütte, Gudmund. Our forefathers, the Gothonic nations. Cambridge, 1929-33. 2v.
Geog 4707.24A Seybold, R.F. Captured by pirates. Two diaries of 1724-1725. Boston, 1929.
An 359.29.5 Smith, G.E. Migrations of early culture. Manchester, 1929.
Geog 4329.29 Spaini, Alberto. Viaggi di Bertoldo. Aquila, 1929.
Geog 4559.29.3 Stanton, William H. The journey of...pilot, of Deal. London, 1929.
Geog 5939.29 Stefansson, V. The theoretical continent. N.Y., 1929.
Geog 4502.5.3 Steinberger, S. Leben und Schriften. München, 1929.
Geog 5399.28.52 Tomaselli, Francesco. L'inferno bianco. 4. ed. Milano, 1929.
Htn Geog 4219.29.10F* Vanderbilt, William K. Taking one's own ship around the world. N.Y., 1929.
Geog 5399.28.30 Viglieri, Alfredo. Quarantotto giorni sul pack. Milano, 1929.
Geog 5329.3.2 Vilhjalmur Stefansson. N.Y., 1929.
Geog 4559.29.30 Villiers, A.J. Falmouth for orders...Cape Horn. Garden City, N.Y., 1929.
Geog 3055.50 Whishaw, M. Atlantis in Andalucia; a study of folk memory. London, 1929.
An 2059.29.5 Wissler, Clark. An introduction to social anthropology. N.Y., 1929.
Geog 4181.62 Wright, L.A. Spanish documents concerning English voyages to the Caribbean. n.p., 1929.

193-

Geog 819.30 Newbigin, Marion I. A new regional geography of the world. N.Y., 193-?
An 132.2 Rubín de la Borbolla, D. Bibliografía del Dr. Ales Hrdlicka. México, 193-.
Geog 5060.11 Toreoja y Miret, José M. Las republicas hispano-americanos y la exploracion de las regiones polares. Madrid, 193-

1930

Geog 4169.30 Adler, Elkan N. Jewish travellers. London, 1930.
Geog 5235.5 American Geographical Society, N.Y. Physical map of the Arctic. N.Y., 1930.
Geog 4329.30.7 Anderson, I.P. (Mrs.). A yacht in Mediterranean seas. Boston, 1930.
Geog 5398.97.5 Andrée, S.A. Andrée's story; the complete record of his polar flight, 1897. N.Y., 1930.
Geog 1539.60 Baedeker, publishers. Northern Italy. 15. ed. Leipzig, 1930.
Geog 1540.50 Baedeker, publishers. Rome and central Italy. 16. ed. N.Y., 1930.
Geog 1542.39 Baedeker, publishers. Southern Italy and Sicily. 17. ed. Leipzig, 1930.
Geog 5650.15 Bárdrson, I. Det gamle Grønlands beskrivelse. København, 1930.
Geog 4559.30.5 Barnes, William M. When ships were ships and not tin pots. N.Y., 1930.
Geog 4309.30.15 Benedict, C.W. (Mrs.). "The Benedicts abroad". London, 1930.
An 3559.30F Black, Davidson. On an adolescent skull of "Sinanthropus pekinensis". Peiping, 1930.
Geog 4209.30.10 Bremer, M. Memoirs of a Ceylon planter's travels, 1851 to 1921. London, 1930.
Geog 4138.30 Brendon, J.A. Great navigators and discoverers. N.Y., 1930.

1930 - cont.

Geog 6009.28.10 Byrd, Richard E. Little America. N.Y., 1930.
Geog 6009.28.8A Byrd, Richard E. Little America. 1st ed. N.Y., 1930.
Geog 4209.30.20 Chable, Jacques Edouard. Jazz, boomerang et kimonos. Paris, 1930.
Geog 6009.10.30 Cherry-Garrard, Apsley G.B. The worst journey in the world; Antarctic 1910-13. N.Y., 1930.
An 379.30.9 Conference on Regional Phenomena, Washington, D.C., 1930. Conference...Apr. 11-12, 1920. Washington, 1930.
Geog 4515.107 Couttet, Alfred. L'enchantement du ski. Paris, 1930.
An 2109.30 Danzel, Theodor. Gefüge und Fundamente der Kultur vom Standpunkte der Ethnologie. Hamburg, 1930.
Geog 5835.85 Davys, John. Tre rejser til Grønland i aarene 1585-87. København, 1930.
Geog 5070.38 Debenham, F. The polar regions. London, 1930.
Geog 4709.30A Dobie, J. Frank. Coronado's children. Dallas, 1930.
Geog 4309.13.6 Dreiser, Theodore. A traveler at forty. N.Y., 1930.
Geog 4209.30.5 Fairchild, David G. Exploring for plants. N.Y., 1930.
Geog 4309.30 Ferreira de Mira, M. Em viagem (na velha Europa). Lisboa, 1930.
Geog 4145.28.5 Filchner, Wilhelm. In China, auf Asiens Hochsteppen. Freiburg, 1930.
Geog 5839.30.20 Foreniger Det ry Grønland. Det ry Grønland. pt.3-5. København, 1930.
Geog 5312.3.5 Foster, C. Rear Admiral Byrd and the polar expeditions. N.Y., 1930.
Geog 4239.29 Franceschi, F. Odisea del yate "Mary" de Puerto Rico a España. Madrid, 1930.
Htn Geog 3550.12* Franciulli, G. I grandi navigatori italiani. Roma, 1930.
An 2109.30.5 Gandert, O.F. Forschungen zur Geschichte des Hauskundes. Leipzig, 1930.
Geog 5530.75 Gell, E.M.B. (Mrs.). John Franklin's, Eleanor Anne Porden. London, 1930.
Geog 5985.253 Gwynn, Stephen. Captain Scott. N.Y., 1930.
Geog 4679.30 Hadfield, R.L. Sea-toll of our time. London, 1930.
Geog 5925.25 Hayward, Walter B. The last continent of adventure. N.Y., 1930.
An 3309.30.10 Herskovits, M.J. The anthropometry of the American Negro. N.Y., 1930.
An 3309.30 Hoaton, E.A. The Indians of Pecos Pueblo. New Haven, 1930.
An 379.30.3 Huntington, C.C. Environmental basis of social geography. N.Y., 1930.
An 379.30.5 Huntington, Ellsworth. The human habitat. London, 1930.
Geog 3249.30 Jacome Correa, Ayres. Discussão historica das medidas geographicas no seculo XVI. Lisboa, 1930.
Geog 4181.65 Jane, Cecil. Select documents illustrating the 4 voyages of Columbus. London, 1930-33. 2v.
Geog 5060.10A Joerg, W.L.G. Brief history of polar exploration since the introduction of flying. N.Y., 1930.
Geog 6009.28.5 Joerg, W.L.G. The work of the Byrd Antarctic expedition. N.Y., 1930.
An 2059.30 Keith, Arthur. New discoveries relating to the antiquity of man. N.Y., 1930.
Geog 5839.30.10F Krabbe, Thomas N. Greenland; its natures, inhabitants and history. Copenhagen, 1930.
Geog 4309.30.30 Kushner, Boris A. 103 dnia na Zapade, 1924-1926 gg. 2. izd. Moskva, 1930.
An 2209.30 Laufer, B. Geophagy. Chicago, 1930.
Geog 4678.47.5 Lockhart, John Gilbert. Blenden Hall; the true story of a shipwreck. London, 1930.
Geog 139.10 London. Royal Geographical Society. Its foundation and history. London, 1930.
Geog 3240.19A Malone, Kemp. King Alfred's North. Cambridge, 1930.
Geog 4209.30.15 Mather, Norman C. Travel notes, 1929-30. Chicago, 1930.
An 3309.30.5 Measurement of man. Minneapolis, 1930.
Geog 139.25 Mill, Hugh R. The record of the Royal Geographical Society, 1830-1930. London, 1930.
An 3409.30 Miranda Pinto, O. Contribution à la morphologie comparée des crêtes papillaires. Thèse. Lyon, 1930.
Geog 4419.30 Monfreid, Henri de. Pearls, arms and hashish; pages from the life of a Red Sea navigator. N.Y., 1930.
Geog 4309.30.6 Morse, W.I. Nordic trails. Boston, 1930.
Htn Geog 4309.30.5* Morse, W.I. Nordic trails. Boston, 1930.
Geog 4219.30 Nelson, Robert. World beaters. Boston, 1930.
Geog 4209.30.2 Newton, A.E. A tourist in spite of himself. Boston, 1930.
Geog 5399.28.10 Nobile, Umberto. L'"Italia" al Polo Nord. Milano, 1930.
Geog 5399.28.15 Nobile, Umberto. With the "Italia" to the North Pole. London, 1930.
An 379.30.7 Peate, I.C. Studies in regional consciousness and environment. London, 1930.
Geog 4559.30.20 Perry, Joseph Malcolm. Cruising in many seas. Springfield, 1930.
Geog 500.71 Perthes, Justus, publishers. Wandkarten, Cottanten, Bücher, Zeitschriften für den geographischen Unterricht. Gotha, 1930.
Geog 5398.97.10 Putnam, George Palmer. Andrée. N.Y., 1930.
Geog 4559.30.10 Randell, Jack. I'm alone, by Captain Jack Randell, as told to Meigs O. Frost. 1. ed. Indianapolis, 1930.
Geog 4502.5.5 Rickmers, W.R. Querschnitt durch mich. München, 1930.
Geog 4502.5.6 Schmidkunz, W. Der Berg des Herzens. München, 1930.
Geog 4679.30.5 Shaw, F.H. Full fathom five. N.Y., 1930.
Geog 4181.64 Stevens, H.N. New light on the discovery of Australia. London, 1930.
Geog 5398.97.4 Svenska Sällskapet för Antropologi och Geografi. Med örnen mot polen av S.A. Andrée. Stockholm, 1930.
Geog 3520.17 Taylor, Eva G. Tudor geography, 1485-1583. London, 1930.
Geog 5919.30 Taylor, Griffith. Antarctic adventure and research. N.Y., 1930.
Geog 4309.30.20 Terán, Juan B. Lo gótico, signo de Europa. Buenos Aires, 1930?
Geog 4229.30 Toller, Ernest. Querdurch; Reisebilder und Reden. Berlin, 1930.
Geog 4309.30.25 Tully, J. Beggaro abroad. Garden City, 1930.
An 359.30 Tylor, Edward B. Anthropology; an introduction to the study of man and civilization. London, 1930. 2v.
Geog 4559.30.15 Valéry, Paul. Mer, marines, marins. Paris, 1930.
Geog 4559.30.3A Villiers, A.J. By way of Cape Horn. N.Y., 1930.
Geog 4182.33 Warnsinck, J.E.M. Reisen van Nicolaus de Graaff. 's-Gravenhage, 1930.
Geog 4759.30 Waugh, Alec. The coloured countries. London, 1930.
Geog 4759.30.7 Waugh, Alec. Hot countries. N.Y., 1930.
Geog 4759.30.5A Waugh, Alec. Hot countries. N.Y., 1930.
Geog 4329.30.3 Waugh, E. Labels; a Mediterranean journal. London, 1930.
Geog 5324.2.60 Whitehouse, J.H. Nansen. London, 1930.
Geog 5323.3 Woel, Cai Magens. Hilsen til Ejnar Mikkelsen. Kjøbenhavn, 1930.

Chronological Listing

1931

Geog 4509.33	Abraham, G.D. Modern mountaineering. London, 1931.
An 2109.31	Aldrich, C.R. The primitive mind and modern civilization. London, 1931.
Geog 1558.30	Baedeker, publishers. Belgium and Luxemburg. 16. ed. Leipzig, 1931.
Geog 1519.75	Baedeker, publishers. Paris und Umgebung und Supplement. 20. Aufl. Leipzig, 1931-37. 2v.
Geog 3019.31	Baker, John Norman Leonard. A history of geographical discovery and exploration. London, 1931.
Geog 3055.51	Barroso, Gustavo. Aquem da Atlantida. São Paulo, 1931.
An 3409.31.3	Battley, Harry. Single fingerprints. New Haven, 1931.
Geog 4359.27	Belloc, Hilaire. Towns of destiny. N.Y., 1931.
Geog 4169.31.15	Blossom, F.A. Told at the Explorers Club. N.Y., 1931.
Geog 4329.29.5	Chalmers, T.M. Under the olive trees. N.Y., 1931.
Geog 4559.31	Chatterton, E.K. Sailing the seas. London, 1931.
Geog 4209.31.20	Cornish, Naughan. The poetic impression of natural scenery. London, 1931.
X Cg An 2109.31.3	Crawley, A.E. Dress, drinks and drums. London, 1931.
Geog 4309.29.20	Davis, Robert H. With Bob Davis hither and yon. N.Y., 1931.
Geog 4209.31.5	Desmond, Alice C. Far horizons. N.Y., 1931.
Geog 4309.31.10	Duhamel, Georges. Mon Europe. Paris, 1931.
Geog 4169.31	Dulles, F.R. Eastward ho! London, 1931.
Geog 4169.31.2	Dulles, F.R. Eastward ho! The 1st English adventures to the Orient. Boston, 1931.
An 2059.31	Duncan, G.S. Prehistoric man. Boston, 1931.
An 2109.31.9	Early man, his origin, development and culture. London, 1931.
Geog 4309.31.15	Eddelbüttel, H.F. Schön ist die Welt. Berlin, 1931.
Geog 4678.48.5	En route from Honolulu. Washington, 1931?
An 3409.31.5	Forest, H.P. de. The evolution of dactyloscopy in the United States. Youngstown, 1931.
Geog 4181.67	Foster, William. Travels of John Sanderson in the Levant. London, 1931.
Geog 3580.35	Gavira, José. La ciencia geografica española del siglo XVI. Madrid, 1931.
An 2109.31.7	Georg, Eugen. The adventure of mankind. N.Y., 1931.
Geog 4509.31.10	Giussani, C. Chiacchiere di un alpinista. Milano, 1931.
Geog 6009.28.25	Gould, L.M. Cold, the record of an Antarctic sledge journey. N.Y., 1931.
An 359.31.3	Gregory, John W. Race as a political factor. London, 1931.
Geog 4500.16	Harvard University. Library. Mountaineering literature in the Harvard College Library. n.p., 1931.
Geog 4300.30.8	Hungerford, E. Planning a trip abroad. N.Y., 1931.
Geog 4209.31	Hyne, C.J.C. People and places. London, 1931.
Geog 110.1.13	International Geographical Congress, 13th, 1931. Livret-guide du congressiste. Paris, 1931.
Geog 110.1.13.5	International Geographical Congress. pt. A-B. Paris, 1931. 2v.
Geog 4309.31.25	Iraizoz y de Villar, Antonio. Apuntes de un turista tropical. Habana, 1931.
Geog 3055.53	Karst, Josef. Atlantis und der Liby-athiopische Kulturkreis. Heidelberg, 1931.
Geog 4209.31.25	Keilpflug, Erich R. An den Rändern dreier Erdteile. Berlin, 1931.
An 359.31.7	Keith, Arthur. The place of prejudice in modern civilization. N.Y., 1931.
Geog 5839.30A	Kent, Rockwell. N. by E. N.Y., 1931.
An 170.252	Kraeber, A.L. Source book in anthropology. N.Y., 1931.
Geog 4209.31.10	Leslie, L.A.D. Wilderness trails in three continents. London, 1931.
An 2109.31.5	Lévy-Bruhl, L. La mentalité primitive. Oxford, 1931.
An 2109.31.11	Lévy-Bruhl, L. Le surnaturel et la nature. Paris, 1931.
Geog 600.6	Literary record of Cleveland Abbe. Ithaca, 1931.
Geog 4672.16	Lockhart, John Gilbert. Strange adventures of the sea. N.Y., 1931.
Geog 4328.83	Lyautey, L.H.G. Lettres de jeunesse. Paris, 1931.
Geog 4269.31	Lyde, L.W. Peninsular Europe. London, 1931.
Geog 4309.31	McGeehan, William O. Trouble in the Balkans. N.Y., 1931.
Geog 5399.31	Meyer, W. Der Kampf um Nobile. Berlin, 1931.
Geog 4181.66	Moreland, W.H. Relations of Golconda in the early 17th century. London, 1931.
Geog 4182.34	Naber, S.P. Johannes de Laet. Iaerlyck Verhael van der verichtinghen der...Compagnie. 's-Gravenhage, 1931-37. 4v.
Geog 5399.28.20	Nobile, Umberto. Met de "Italia" naar de Noordpool. Baarn, 1931.
Geog 5645.5.20	Norway and Greenland; a short survey. Oslo, 1931.
Geog 4309.31.20	Osborne, A.B. Finding the worth while in Europe. N.Y., 1931.
An 379.31	Penrose, R.A.F. Geology as an agent in human welfare. N.Y., 1931.
Geog 4309.31.5	Ponten, J. Zwischen Rhone und Wolga. Leipzig, 1931.
Geog 4206.13.35	Purchas, Samuel. Narratives from Purchas, his pilgrimes. Cambridge, 1931.
Geog 4209.31.15	Rastron, A.H. Home from the sea. N.Y., 1931.
An 3509.31	Reynolds, E. The evolution of the human pelvis in relation to the mechanics of the erect posture. Cambridge, 1931.
Geog 4509.31	Robbins, L.H. Mountains and men. N.Y., 1931.
Geog 4238.32.10	Roebling, John A. Diary of my journey from Muehlhausen in Thurangia via Bremen to the United States...1831. Trenton, 1931.
Geog 4349.30	St.-Félix, Max de. A travers l'Orient (1930). Paris, 1931.
Geog 5209.31	Samoilovich, R. Der Weg nach dem Pol. Biehefeld, 1931.
An 3509.31.10	Schreiner, Kristian Emil. Zur Osteologie der Lappen. Oslo, 1931-35. 2v.
Geog 4313.6	Schulz, Bruno. Die deutsche Ostsee. Bielefeld, 1931.
Geog 4509.31.5	Schwartz, M. Et la montagne conquit l'homme. Paris, 1931.
Geog 4502.5.7	Simmler, J. De alpibus Commentarius; die Alpen. München, 1931.
Geog 5080.5	Smedal, Gustav. Aquisition of sovereignty over polar areas. Oslo, 1931.
Geog 4509.31.15	Smythe, F.S. Climbs and ski runs. Edinburgh, 1931.
An 359.31.5	Spiller, Gustav. The origin and nature of man. London, 1931.
Geog 4307.78	Staszic, S. Dziennik podrózy, 1789-1805. Krakow, 1931.
An 137.1.5	Stern, B.J. Lewis Henry Morgan, social evolutionist. Chicago, 1931.
Geog 4269.31.10	Suedost- und Südeuropa in Natur. v.1-7,8-14,18. Potsdam, 1931-1936. 3v.
Geog 4129.31	Titayna (pseud.). Mademoiselle against the world. N.Y., 1931.
Geog 4349.31	Toynbee, Arnold J. A journey to China; or, Things which are seen. London, 1931.

1931 - cont.

Geog 4515.25	Underhill, R.L.K. On the use and management of the rope in rock work. San Francisco, 1931.
An 3409.31	Unites States. Federal Bureau of Investigation. How to take fingerprints. Washington, 1931.
Geog 5559.31	Wilkins, H. Under the North Pole. N.Y., 1931.
Geog 5559.31.5	Wilkins, H. Under the North Pole. N.Y., 1931.
Geog 4329.31	Wilstach, Paul. Islands of the Mediterranean. N.Y., 1931.
An 2109.31.13	Winthius, Josef. Einführung in die Vorstellungswelt primitiver Völker. Leipzig, 1931.
Geog 5985.258.5	Worsley, Frank A. Endurance; an epic of polar adventure. London, 1931.

1932

Geog 4181.69	Barlow, Roger. A brief summe of geographie. London, 1932.
An 359.32.5	Bean, R.B. The races of man. N.Y., 1932.
Geog 5645.5	Berlin, Knud. Denmark's to Greenland. London, 1932.
Geog 3055.52	Bessmertny, A. Das Atlantisrätsel. Leipzig, 1932.
An 3309.32.5	Biedermann, Ernst. Körperform und Leistung sechzehnjähriger Lehslinge. Zürich, 1932.
An 3309.32	Bowles, G.T. New types of old Americans at Harvard and at eastern women's colleges. Cambridge, 1932.
Geog 91.1.10	Braun, Gustav. Geographische Gesellschaft zu Greifswald, 1882-1927. Greifswald, 1932.
Geog 3129.32A	Burton, H.E. The discovery of the ancient world. Cambridge, 1932.
Geog 759.32.5	Case, E.C. College geography. N.Y., 1932.
Geog 5559.30	Chapman, F.S. Northern lights. London, 1932.
Geog 5660.15	Dahl, F. Andrae og Østgrønland. København, 1932.
Geog 4219.32.15	Dinglreiter, Senta. Deutsches Mädel auf Fahrt um die Welt. Leipzig, 1932.
Geog 3530.15	Dupic, Jeanne. La Normandie exploratrice et colonisatrice du XVe au XVIIIe siècle. Rouen, 1932.
Geog 4329.32	Edschmid, K. Zauber und Grösse des Mittelmeers. Frankfurt, 1932.
Geog 4219.32	Fletcher, A.C.B. Keep moving. Chiago, 1932.
An 3309.32.3F	Frassetto, F. Note antropologiche sulla popolagione del Bolognese. Bologna, 1932.
Geog 3570.16	Frazão de Vasconcellos, J. Os pilotos dos seculos XV e XVI e a nobreza do reino. Lisboa, 1932.
Geog 4219.32.5	Gilmore, A.F. Yes, 'tis round. Boston, 1932.
Geog 192.4.5	Grenoble. Université. Institut de Géographie Alpine. Table décennale. v.2 (1923-32), v.5 (1953-62). Grenoble, 1932-63. 3v.
Geog 4309.32.5	Grisar, Erich. Mit Kamera und Schreibonaschine durch Europa. Berlin, 1932.
Geog 5600.1	Hagerups, H. Boghandel. Litteratur om Grønland. København, 1932.
Geog 4219.32.10A	Halliburton, Richard. The flying carpet. Indianapolis, 1932.
Geog 4269.32	Hausenstein, W. Europäische Hauptstädte. Erlenbach, 1932.
Geog 4209.32.20	Hauser, H. Fair winds and foul. N.Y., 1932.
Geog 5939.32	Hayes, J.G. The conquest of the South Pole. London, 1932.
Geog 3570.58	Heleno, Manuel Domingues. Colaboração portuguesa nos descobrimentos nauticos das outras nações. Lisboa, 1932.
Geog 4209.32.5	Hering, O.C. Down the world. N.Y., 1932.
Geog 4502.5.8	Hoek, H. Der denkende Wanderer. München, 1932.
An 130.1	Hough, Walter. Biographical memoir of J.W. Fawkes, 1850-1930. Washington, 1932.
Geog 5209.32	Houhen, H.H. The call of the North. London, 1932.
An 379.32	Hubbard, George D. Geographic conditions. n.p., 1932.
An 2600.2.8	Itard, Jean Marc. The wild boy of Aveyron. N.Y., 1932.
Geog 4209.32.25	Johnson, Irving. Round the Horn in a square rigger. Springfield, 1932.
Geog 5399.32	Kalmár, G. Kuz delmek a fehér halál országában. Budapest, 1932-33. 2v.
Geog 5321.2	Klengenberg, C. Klengenberg of the Arctic. London, 1932.
Geog 4502.5.9	Knorr, O. Der Grossvenediger in der Alpinismus. München, 1932.
Geog 4309.32	Landau, M.A. Eine unsentimentale Reise. München, 1932.
Geog 6009.29	Mawson, Douglas. The B.A.N.Z. Antarctic research expedition, 1929-31. London, 1932.
Geog 665.49	Mélanges géographiques offerts par ses élèves à Raoul Blanchard. Grenoble, 1932.
Geog 4209.32.15	Mordaunt, E. (pseud.). Rich tapestry. N.Y., 1932.
Geog 3019.32F	Morrison, E.R. Explorographs. Cleveland Heights, 1932.
An 359.32	Muckermann, H. Rassenforschung und Volk der Zukunft. Berlin, 1932.
Geog 4209.32	Murchie, Guy. Men on the horizon. Boston, 1932.
Geog 4332.20	Novak, Grga. Naše more. 2. izd. Zagreb, 1932.
An 2209.32	Paudler, F. Scheitelnarbensitte, Anschwellungsglaube und Kulturkreislehre. Brunn, 1932.
Geog 4219.32.35.5	Pidgeon, H. Around the world single-handed. N.Y., 1932.
Geog 5559.32	Polunin, N. The isle of auks. London, 1932.
Geog 4209.32.10	Powell, Edward A. Yonder lies adventure! N.Y., 1932.
An 2109.32.9A	Radin, Paul. Social anthropology. 1st ed. N.Y., 1932.
Geog 3055.57	Requena, Rafael. Vestigios de la Atlántida. Caracas, 1932.
Geog 5324.2.15	Reynolds, E.E. Nansen. London, 1932.
Geog 4502.5.10	Schuster, K. Weise Berge - schwarze Zelte. München, 1932.
Geog 815.35.4	Servetus, Michael. Descripciones geograficas del estado moderno de las regiones. Madrid, 1932.
Geog 815.35.5	Servetus, Michael. Descripciones geograficas del estado moderno de las regiones. Madrid, 1932.
Geog 5399.31.5	Sieburg, F. Die rote Arktis. Frankfurt, 1932.
Geog 5645.5.5	Skeie, Jon. Greenland; the dispute between Norway and Denmark. London, 1932.
Geog 5324.2.10	Sörensen, Jon. The saga of Fridtjof Nansen. N.Y., 1932.
Geog 4145.9	Thiery, Maurice. Bougainville. London, 1932.
An 2109.32	Thurnwald, R. Economics in primitive communities. London, 1932.
An 3409.32	United States. Federal Bureau of Investigation. How to take fingerprints. Washington, 1932.
Geog 665.50	Van Loon, H.W. Van Loon's geography, the story of the world we live in. N.Y., 1932.
Geog 604.4	Wagner, H.R. George Davidson, geographer of the northwest coast of America. n.p., 1932.
An 2109.32.3	Westermarck, E. Early beliefs and their social influence. London, 1932.
Geog 5324.2.37	Wetterfors, Paul. Fridtjof Nansen. Uppsala, 1932.
Geog 5180.9	Weyer, E.M. The Eskimos. New Haven, 1932.

Chronological Listing

1933

	An 3559.33	Aguilar y R.F. Origen castellano del prognatismo. Madrid, 1933.
	Geog 759.33	Allgemeine Geographie. Potsdam, 1933. 2v.
	Geog 4309.33	American Automobile Association. Foreign. Motering abroad. N.Y., 1933.
	Geog 3019.33.10	Banse, Ewald. Grosse Forschungsreisende. München, 1933.
	An 2059.33	Black, Davidson. Fossil man in China. Peiping, 1933.
	An 379.33.5	Bryan, P.W. Man's adaptation of nature. London, 1933.
	Geog 4181.72	Burnell, John. Bombay in the days of Queen Anne. London, 1933.
	Geog 4218.79.10A	Carnegie, Andrew. Round the world. Garden City, 1933.
	Geog 110.3	Congrès des Géographes et Ethnographes Slaves, 3rd, 1930. Zbornik radova III Kongresa slovenskih geografa i ethnografa u Kraljevini Jugoslaviji, 1930. Beograd, 1933.
	An 359.33.5	Cornelius, William J.J. L'homme d'après la religion et la science. Thèse. Londres, 1933.
	Geog 3019.33	Dickinson, R.E. The making of geography. Oxford, 1933.
	Geog 4502.5.14	Gallhuber, J. Das Gesäuse und seine Berge. 2. Aufl. München, 1933.
	An 5.8	Gercke, Achim. Die Rasse im Schrifttum. Berlin, 1933.
	Geog 4709.33	Grey, Charles. Pirates of the Eastern seas, 1618-1723. London, 1933.
	Geog 4309.33.3	Halévy, Daniel. Courrier d'Europe. Paris, 1933.
	Geog 4219.32.11	Halliburton, Richard. The flying carpet. London, 1933.
	Geog 5838.84	Hansen, Johannes. Den grønlandske Kateket Hansêraks Dagbog. København, 1933.
	An 3209.33	Hildén, Kaarlo. Maapallon esihistorialliset ja. Helsinki, 1933.
	An 2109.33.3	Hofstra, Sjoerd. Differenzierungserscheinungen in einigen afrikanischen Gruppen. Proefschrift. Amsterdam, 1933.
	Geog 4502.5.13	Lebwald, A. Damographia oder Gemsen-Beschreibung in zewy Theil Abgetheilet. Saltzburg, 1933.
	An 170.206	Leo Frobenius. Leipzig, 1933.
	Geog 4559.33	McCulloch, J.H. A million miles in sail. N.Y., 1933.
	An 379.33A	McIverny, A.J. The rôle of the deserts. London, 1933.
	Geog 4419.33	Makin, William J. Red Sea nights. N.Y., 1933.
	An 359.33.9	Mendes Correia, A.A. Introdução i antropobiologia. Coimbra, 1933.
	An 99.33	Mitra Panchanan. A history of American anthropology. Calcutta, 1933.
	An 359.33.11	Montandon, Georges. La race, les races. Paris, 1933.
	Geog 1807.27	Murray, John, publisher, London. Handbook for...India, Burma and Ceylon. 14. ed. London, 1933.
	Geog 4209.30.3	Newton, A.E. A tourist in spite of himself. Boston, 1933.
	Geog 4131.10	Ogilvie, F.W. The tourist movement. London, 1933.
	Geog 3019.29.5	Olsen, Ørjan. La conquête de la terre. Paris, 1933-36. 5v.
	Geog 4309.27.10	Petre, E.R. When you go to Europe. N.Y., 1933.
	Geog 4509.33.5A	Pickman, D.L. Some mountain views. Boston, 1933.
	Geog 3019.33.5	Plischke, Hans. Entdeckungsgeschichte von Altertum bis zur Neuzeit. Leipzig, 1933.
	Geog 3570.17	Prestage, Edgar. The Portuguese pioneers. London, 1933.
	An 359.33	Radin, Paul. The method and theory of ethnology. N.Y., 1933.
	Geog 5060.15	Rasmussen, K. Heldenbuch der Arktis. Leipzig, 1933.
	Geog 4182.38	De reis van Voris van Spilbergen naar Ceylon, Atjeh en Bantam, 1601-1604. 's-Gravenhage, 1933.
	Geog 4329.33	Rèpaci, L. Con la ciurma dell'Alessandro. Milano, 1933.
	An 359.33.3	Ritter, J. Über den Sinn und die Grenze der Lehre vom Menschen. Potsdam, 1933.
	Geog 4300.44	Ross, Janet. Budgetouring Europe. N.Y., 1933.
	Geog 3060.13.10	Spence, Lewis. The problem of Lemuria. Philadelphia, 1933.
	Geog 4502.5.12	Tscharner, J.B. von. Die Bernina. 1786. München, 1933.
	Geog 5324.2.20	Turley, C. Nansen of Norway. London, 1933.
Htn	Geog 4219.33F*	Vanderbilt, William K. West made East with the loss of a day. N.Y., 1933.
	Geog 4559.32	Villiers, A.J. Grain race. N.Y., 1933.
	An 359.33.7	Voegelin, Erich. Die Rassenidee in der Geistesgeschichte Ray bis Carus. Berlin, 1933.
	Geog 4169.33	Ward, Edward. Five travel scripts commonly attributed to Edward Ward. N.Y., 1933.
	An 2109.33	Wisse, Jakob. Selbstmord und Todesfurcht bei den Naturvölkern. Proefschrift. Zutpher, 1933.
	Geog 4181.71	Wright, I.A. Documents concerning English voyages to the Spanish Main, 1569-1580. London, 1933.

1934

	Geog 4219.34.5	Atkinson, B. The Cingalese prince. Garden City, 1934.
	Geog 1595.7	Baedeker, publishers. Mittelmeer. 2. Aufl. Leipzig, 1934.
	Geog 4556.59	Barlow, E. Barlow's journal of his life at sea in king's ships, East and West Indiamen. London, 1934. 2v.
	An 2109.34.5A	Benedict, Ruth (Fulton). Patterns of culture. Boston, 1934.
	Geog 4269.34	Bogardus, J.F. Europe; a geographical survey. N.Y., 1934.
	An 379.34.3	Brunhes, Jean. La géographie humaine. 4. éd. Paris, 1934. 3v.
	An 3509.34	Cameron, John. The skeleton of British neolithic man. London, 1934.
	Geog 3060.12.25	Churchward, J. The lost continent of Mu. N.Y., 1934.
	Geog 3060.12.45A	Churchward, J. The sacred symbols of Mu. N.Y., 1934.
	An 359.34.9	Clauss, Ludwig F. Rasse und Seele; eine Einführung in den Sinn der leiblichen Gestalt. München, 1934.
	Geog 4509.34	Club Alpin Francais. Manuel d'alpinisme. Chambéry, 1934. 2v.
	Geog 4209.34	Coyne, William M. A sailor's log of facts: not fables. Boston, 1934.
	Geog 5509.34	Crouse, N.M. The search for the Northwest Passage. N.Y., 1934.
	An 379.34	Forde, C.D. Habitat, economy and society. London, 1934.
	Geog 4181.75	Foster, William. The voyage of Thomas Best to the East Indies. London, 1934.
	Geog 4209.34.7	Fuchs, Hans. Heimkehr ins Dritte Reich. Dresden, 1934.
	Geog 4559.34	Grant, George H. Consigned to Davy Jones. Boston, 1934.
	Geog 4131.34	Grünthal, Adolf. Probleme der Fremdenverkehrsgeographie. Diss. Berlin, 1934.
	Geog 4209.34.10	Gutierrez Zamora, I. Contando un viaje. n.p., 1934.
	Geog 3520.18	Haender, Wilhelmina. De Engelsche geographie in de 20ste eeuw. Utrecht, 1934.
	Geog 5235.10	Hayes, J.G. The conquest of the North Pole. N.Y., 1934.
	Geog 5311.3.20	Hovdenak, Gunnar. Roald Amundsen siste fere. Oslo, 1934.
	Geog 110.1.14	International Geographical Congress, 14th, Warsaw, 1934. Comptes rendus. v.1-7. Varsovie, 1934.
	Geog 4181.73	Joyce, L.E. Elliott. A new voyage and description of the Isthmus of America. Oxford, 1934.

1934 - cont.

	Geog 3129.34	Kahlo, G. Die Keuntnis der Erde im Altertum. München, 1934.
	Geog 5379.35.3	Kak my spasali chelnoskiktsev. Moskva, 1934.
	Geog 819.34	Kalmár, Gusztáv. Négy Világrésg Földje és Népei. Budapest, 1934-1935.
	Geog 4309.34.10	Kisch, Egon E. Eintritt verboten. Paris, 1934.
	An 359.34.7	Kryzwicki, L. Primitive society and its vital statistics. London, 1934.
	Geog 3055.76	Lehmann, Einar. Atlantis. København, 1934.
	Geog 6009.10.40	Lindsay, M. The epic of Captain Scott. N.Y., 1934.
	Geog 5559.09.15	MacMillan, D.B. How Peary reached the pole. Boston, 1934.
	Geog 4709.34	Magre, M. Pirates, flibustiers, negriers. 9. éd. Paris, 1934.
	An 3559.34	Mascarenhas, Constância. Os povos de Angola. Bastorá, 1934.
	An 187.6	Mendes Corrèa, A.A. Da biologia à história. Porto, 1934.
	An 359.34.10	Menghin, O. Geist und Blut. Wien, 1934.
	Geog 5209.34	Mirsky, J. To the North: The story of arctic exploration from earliest times to the present. N.Y., 1934.
	Geog 3019.34.10	Mitchell, J.L. Earth conquerors. N.Y., 1934.
	Geog 4419.34.4	Monfreid, Henri de. La poursuite du "Kaipan". Paris, 1934.
	Geog 4419.34	Monfreid, Henri de. Secrets of the Red Sea. London, 1934.
	Geog 4181.74	Moreland, W.H. Peter Floris. London, 1934.
	Geog 5399.34.5	Morozov, S. Put' skvoz l'dy. Moskva, 1934.
	An 2109.34.10	Osborn, H.F. Men of the old stone age. 3. ed. N.Y., 1934.
	Geog 6009.28.30	Owen, Russell. South of the sun. N.Y., 1934.
	Geog 6009.28.31	Owen, Russell. South of the sun. N.Y., 1934.
	An 3409.34.3	Pacheco, F. Vucetich e Reyna Abmandos. Rio de Janeiro, 1934.
	Geog 4309.34	Patterson, Sara K. Out of the fog. Caldwell, 1934.
	An 359.34.3	Poisson, Georges. Les aryens. Paris, 1934.
	Geog 3570.18	Prestage, Edgar. Descobridores portugueses. Porto, 1934.
	An 359.34A	Radin, Paul. The racial myth. N.Y., 1934.
	Geog 5328.3	Rasmussen, Knud. Mindeudgave. Kjøbenhavn, 1934-35. 3v.
	An 359.34.5	Reuter, E.B. Race and culture contacts. N.Y., 1934.
	Geog 5559.34	Rich, E.G. Hans the Eskimo. Boston, 1934.
	Geog 4419.34.5	Rogers, C.K. (Mrs.). Journal letters from the Orient. Boston, 1934.
	Geog 4559.29.22	Rogers, S.R.H. Sea lore. London, 1934.
	An 2109.34.3A	Róheim, Géza. The riddle of the sphinx, or Human origins. London, 1934.
	Geog 5160.14	Ross, Colin. Mit Kind und Kegel in die Arktis. Leipzig, 1934.
	Geog 4313.7	Schlump, E. Die politisch-geographische Bedeutung der Ostsee. Inaug.-Diss. Königsberg, 1934.
X Cg	Geog 4502.5.15	Schmidkunz, W. Auf der Alm. Erfurt, 1934.
	Geog 4219.06.7	Sebok, I. Öt világrészun heresztül. Budapest, 1934.
	Geog 4309.34.5	Stackpole, Edward J. Land of the midnight sun, Scandinavian region, Russia, and Germany. Harrisburg, 1934.
	Geog 3019.34	Sykes, P.M. A history of exploration from the earliest times to the present day. N.Y., 1934.
	An 3409.34	United States. Federal Bureau of Investigation. Fingerprints. Washington, 1934.
	Geog 4219.34.10	Unwin, S. Two young men see the world. London, 1934.
	An 375.5	Urabayen, Leoncio. Geografia umana. Madrid, 1934. 2 pam.
	Geog 5328.17	Vejlager, Johannes. Knud Rasmussen. Kjøbenhavn, 1934.
	Geog 5399.32.6	Vize, Vladimir I. Ne "Sibirakove" v Tikhii okean. Leningrad, 1934.

1935

	Geog 4209.35.33	Aubert de la Rue, Edgar. L'homme et les iles. 2. ed. Paris, 1935.
	Geog 4329.35.3	Audisio, G. Jeunesse de la Méditerranée. Paris, 1935-1936.
	Geog 1527.2	Baedeker, publishers. München und Südbayern. 39. Aufl. Leipzig, 1935.
	Geog 1527.5	Baedeker, publishers. München und Umgebung. Augsburg. Leipzig, 1935.
	Geog 4219.35	Benn, Wedgwood. Beckoning horizon. London, 1935.
	An 379.35.3	Bews, J.W. Human ecology. London, 1935.
	Geog 4269.35	Blanchard, Raoul. A geography of Europe. N.Y., 1935.
	Geog 4329.35.10	Bloomfield, Paul. The Mediterranean; an anthology. London, 1935.
	An 3309.35	Boyd, Edith. The growth of the surface area of the human body. Minneapolis, 1935.
	Geog 5839.35.20A	Boyd, S.A. The fiord region of east Greenland. N.Y., 1935. 2v.
	An 2109.35	Brelsford, Vernon. Primitive philosophy. London, 1935.
	Geog 4559.35.10	Brown, Charles W. Journal and letters, 1876-84. n.p., 1935.
	Geog 148.1.8	Buxbaum, Edwin Clarence. Collecting national geographic magazines. Milwaukee, 1935.
	Geog 6009.33A	Byrd, Richard E. Discovery; the story of the second Byrd Antarctic expedition. N.Y., 1935.
	Geog 4229.35	Carré, J.M. Promenades dans trois continents. Paris, 1935.
	An 359.35.5A	Carrel, Alexis. Man, the unknown. 2. ed. N.Y., 1935.
	An 359.35.5.8A	Carrel, Alexis. Man, the unknown. 11. ed. N.Y., 1935.
	An 359.35.5.10	Carrel, Alexis. Man, the unknown. 25. ed. N.Y., 1935.
	An 359.35.6	Carrel, Alexis. Man, the unknown. 50. ed. N.Y., 1935.
	An 359.35.4A	Carrell, Alexis. Man, the unknown. N.Y., 1935.
	An 3409.35.5	Castellanos, I. Dactiloscopia clinica. Habana, 1935.
	Geog 5379.35	Cheliuskin Expedition, 1933-34. The voyage of the Chelyuskin. N.Y., 1935.
	Geog 4559.35.20	Devine, Eric. Midget Magellans; great cruises in small ships. N.Y., 1935.
	Geog 953.12	Ellis, Jessie (Croft). Travel through pictures; references to pictures, in books and periodicals. Boston, 1935.
	Geog 3019.35.20	Ellsworth, L. Exploring today. N.Y., 1935.
	Geog 4209.35.5	Foley, Arthur. Breezy adventure. Boston, 1935.
	Geog 5316.1A	Freuchen, P. Arctic adventure; my life in the frozen North. N.Y., 1935.
	Geog 5839.35	Georgi, J. Mid-ice. N.Y., 1935.
	Geog 4209.35.20A	Halliburton, Richard. Seven league boots. Indianapolis, 1935.
	Geog 4209.35.25	Harris, P.P. Peregrinations. v.2-3. n.p., 1935-37. 2v.
	Geog 4105.13.3	Harvard Travel Club. Handbook of travel. Cambridge, 1935.
	Geog 4309.35	Hull, M.L. The gangway to Europe. Boston, 1935.
	An 359.35.7	Huxley, Julian S. We Europeans. London, 1935.

Chronological Listing

1935 - cont.

	Geog 5399.31.2	Itin, Vivian A. Vykhod k moriu. 2. izd. Novosibirsk, 1935.
	Geog 819.35	James, Preston E. An outline of geography. Boston, 1935.
	Geog 4521.11	Kain, Conrad. Where the clouds can go. N.Y., 1935.
	Geog 4219.32.20	Katz, Richard. Ernte. Zürich, 1935.
	Geog 614.6	Kupferschmidt, F. Karl Neumann. Inaug. Diss. Leipzig, 1935.
	Geog 4329.35.15	Laughlin, C.E. So you're going to the Mediterranean. Boston, 1935.
	Geog 3251.32	Lehmann, Edgar. Alte deutsche Landkarten. Leipzig, 1935.
	An 2109.35.5	Marett, Robert R. Head, heart and hands in human evolution. London, 1935.
	An 3509.35	Martin, Cecil P. Prehistoric man in Ireland. London, 1935.
	Geog 5650.20	Mathiassen, T. Skraelingerne i Grønland. København, 1935.
	Geog 759.35	Mitteleuropa. Potsdam, 1935.
	Geog 4419.34.3	Monfreid, Henri de. Les secrets de la Mer Rouge. Paris, 1935.
	Geog 4309.35.15	Ofaire, Cilette. The San Luca. N.Y., 1935.
	Geog 5839.35.25	Oldendow, Knud. Naturfredning i Grønland. København, 1935.
Htn	Geog 3240.18.5*	Orion, booksellers, ltd., London. Description of a rare and precious atlas. London, 1935?
	An 359.35	Orton, E.F. Links with past ages. Cambridge, Eng., 1935.
	Geog 5557.33	Ostrovskii, B.G. Veliki severnaia ekspeditsiia, 1733-43. Arkhangel'sk, 1935.
	Geog 3019.35	Outhwaite, L. Unrolling the map. N.Y., 1935.
	Geog 500.63	Pan American Union. Columbus Memorial Library. Books and magazine articles on geography in the Columbus Memorial Library. Washington, 1935.
	Geog 4309.35.10	Paquet, Alfons. Fluggast über Europa. München, 1935.
	Geog 5515.78	Parks, G.B. Frobisher's third voyage, 1578. n.p., 1935.
	An 99.35	Penniman, T.K. A hundred years of anthropology. London, 1935.
	Geog 4509.35.5	Richards, D.P. (Mrs.). Climbing days. London, 1935.
	Geog 3019.35.10	Rosh, J.H. Man and the sea. Cambridge, Eng., 1935.
	Geog 6008.38.5	Ross, Frank E. The Antarctic explorations of Charles Wilkes, 1838-42. n.p., 1935.
	Geog 3019.35.25	Sanchez, Pedro C. Evolución de la geografía. México, 1935.
	Geog 5333.5	Scott, J.M. Gino Watkins. London, 1935.
	Geog 4559.35.15	Sheridan, Richard B. Heavenly hell. London, 1935.
	Geog 3019.35.5	Spilhaus, M.N. (Mrs.). The background of geography. Philadelphia, 1935.
	An 359.35.3	Spiller, Gustav. The origin and nature of man. 2. ed. London, 1935.
	Geog 4559.35.5	Stevers, M.D. Sea lanes; man's conquest of the ocean. N.Y., 1935.
	Geog 4346.76.25	Struys, Jan J. Tri puteshestviia. Moskva, 1935.
	Geog 4181.76	Taylor, E.G.R. The original writings and correspondance of the two Richard Hakluyts. London, 1935. 2v.
	An 3409.35	Unites States. Federal Bureau of Investigation. Fingerprints. Washington, 1935.
	Geog 4559.35	Van Loon, H.W. Ships and how they sailed the seven seas. (5000 B.C.-A.D. 1935). N.Y., 1935.
	Geog 4269.35.10A	Van Valkenburg, S. Europe. N.Y., 1935.
	Geog 5600.2	Vartdal, Hroar. Bibliographie des ouvrages norvégiens relatifs au Groenland. Oslo, 1935.
	An 359.35.9	Weigner, K. Die Gleichwertigkeit der europäischen Rassen und die Wege zu ihrer Vervollkommnung. Prag, 1935.

1936

	Geog 4309.36.5	Abella Caprile, Margarita. Geografias (notas de viaje). . Buenos Aires, 1936.
	Geog 4249.36	Allen, Edward Weber. North Pacific: Japan, Siberia, Alaska, Canada. N.Y., 1936.
Htn	Geog 500.59.2F*	Atkinson, G. La litterature géographique française de la renaissance. Supplement. Paris, 1936.
	Geog 1522.42	Baedeker, publishers. Das Deutsche Reich. 6. Aufl. Leipzig, 1936.
	Geog 1525.55.15	Baedeker, publishers. Schwarzwald. 3. Aufl. Leipzig, 1936.
	An 2109.36.5	Bateson, G. Naven; a survey of the problems suggested by a composite picture of the culture of a New Guinea tribe. Cambridge, Eng., 1936.
	Geog 3540.11	Beck, Carl. Deutsches Reisen im Wandel. Berlin, 1936.
	Geog 5182.8.5	Birket-Smith, Kaj. The Eskimoes. London, 1936.
	Geog 4269.35.4	Blanchard, Raoul. Géographie de l'Europe. Paris, 1936.
	Geog 5610.5	Bobé, Louis. Diplomatarium Groenlandicum, 1492-1814. København, 1936.
	Geog 5639.36.4	Bobé, Louis. Den grønlanske handels og kolonisations historie indtil 1870. København, 1936.
	Geog 5610.5.2	Bobé, Louis. Opdagelsesrejser til Grønland, 1473-1806. København, 1936.
	Geog 500.67	Curtiss, Frederic H. A little book on travel books. Boston, 1936.
	Geog 4209.36.5	Davis, R.H. People, people, everywhere. N.Y., 1936.
	Geog 4209.36.30	Durtain, L. Le globe sous le bras. Paris, 1936.
	Geog 4329.36.2	Ervine, St. J.G. A journey to Jerusalem. London, 1936.
	Geog 613.6	Espinosa Cordero, N. Pedro Vicente Maldonado, y la Misión geodésica del siglo XVIII. Cuenca, 1936.
	An 135.1	Essays in anthropology. Kroeber. Berkeley, 1936.
	An 359.36.5	Die farbige Front; hinter den Kulissen der Weltpolitik. Leipzig, 1936.
	An 135.5	Feher, Geza. Felikcě F. Kanitsě. Sofiia, 1936.
	Geog 4672.20.4	Gomes de Brito, Bernardo. História trágico-marítima. v.1-5. Pôrto, 1936-37.
	Geog 4309.36	Harris, J.P. Traveling light (Europe between boats). Hutchinson, Kansas, 1936.
	Geog 4169.36	Hennig, R. Terrae incognitae. Leiden, 1936-39. 4v.
	Geog 5326.21.5	Hobbs, William H. Peary. N.Y., 1936.
	Geog 4209.36.10	Howland, Charles. Travel essays. n.p., 1936?
	Geog 4219.36.10	Hurja, E.E. Westward ho - fare paid. Juneau, Alaska, 1936.
	An 359.35.7.10	Huxley, Julian S. We Europeans. 1. ed. N.Y., 1936.
	Geog 4169.36.5A	Ingram, B.S. Three sea journals of Stuart times. London, 1936.
	Geog 4659.36	Ingram, Charles W.N. Shipwrecks; New Zealand disasters, 1795-1936. Dunedin, 1936.
	Geog 500.68	International Committee of Historical Sciences. Committee on the History of Great Voyages and Great Discoveries. Travaux de la Commission pour l'histoire des grands voyages et des grandes decouvertes. Paris, 1936.
	Geog 603.5	Jean-Baptiste Charcot, 1867-1936. Paris, 1936.

1936 - cont.

	Geog 6009.35	Joerg, W.L.G. The topographical results of Ellsworth's trans-Antarctic flight of 1935. N.Y., 1936.
	Geog 4219.36.5	Johnson, Irving. Westward bound in the schooner Yankee. N.Y., 1936.
	Geog 3019.35.15	Kábmár, G. Régi népak, ujvilagok. Budapest, 1936.
	Geog 4219.32.25	Katz, Richard. Ein Bummel um die Welt. Zürich, 1936.
	Geog 5379.35.5	Khmyznikov, P. Na Chelnoskike. Leningrad, 1936.
	Geog 4208.55	King, Paul. Voyaging to China in 1855 and 1904. London, 1936.
	Geog 4502.5.19	Kobell, F. von. Wildonger. München, 1936.
	An 379.36.3	Lester, P. Les races humaines. Paris, 1936.
	An 629.36	Lévi-Strauss, D. Instruções praticas para pesquisas de antropologia fisica e cultural. pt.1. São Paulo, 1936-
	An 2109.36.7	Lévy-Bruhl, L. Morceaux choisis. 2e éd. Paris, 1936.
	An 359.36.3	Linton, Ralph. The study of man. N.Y., 1936.
	Geog 4209.36.35	Lissner, Ivar. Völker und Kontinente. Hamburg, 1936.
	Geog 139.4.13	London. Royal Geographical Society. Recent geographical literature, maps. Index. 1-4, 1918-1932. London, 1936.
	Geog 4209.36	Macdonald, J.R. At home and abroad; essays. London, 1936.
	Geog 4502.5.17	Maduschka, L. Junger Mensch im Gebirg. München, 1936.
	An 379.36	Marett, J.R. de la H. Race, sex, and environment. London, 1936.
	Geog 4659.36.10	Masters, David. S.O.S., a book of sea adventure. N.Y., 1936.
	Geog 4209.35.15	Matisse, Maarten. Wanderer from sea to sea. N.Y., 1936.
	Geog 665.35	Mélanges de géographie offerts par ses collègues et amis de l'étranger à M. Václav Svambera. Praha, 1936.
	Geog 4219.36	Merrick, Lesley. A good time. N.Y., 1936.
	Geog 5317.10.5	Mitchell, William. General Greely; the story of a great American. N.Y., 1936.
	Geog 4349.36.3	Morand, Paul. La route des Indes. Paris, 1936.
	Geog 4329.36.6	Morton, H.V. In the steps of St. Paul. London, 1936.
	Geog 4329.36.7	Morton, H.V. In the steps of St. Paul. N.Y., 1936.
	Geog 4349.36	Nichols, Beverley. No place like home. London, 1936.
	Geog 4349.36.2	Nichols, Beverley. No place like home. 1st ed. Garden City, 1936.
	Geog 5650.17	Nørlund, P. Viking settlers in Greenland. London, 1936.
	An 349.36	Numelin, R. The wandering spirit. Copenhagen, 1936.
	An 2109.36.9	Olbrechts, F.M. Ethnologie...primitiere beschaving. Zutphen, 1936.
	Geog 5639.36	Oldendow, Knud. Traek af Grønlands politiske historie. København, 1936.
	Geog 809.36	Ozonf, R. (Mme.). Lectures géographiques. v.1-2. Paris, 1936-1938. 4v.
	Geog 4329.36.10	Parain, Charles. La Méditerranée; les hommes et leurs travaux. 4e éd. Paris, 1936.
	Geog 4502.10.2	Petrarca, F. Des F. Petrarca Sendschreiben die Besteigung des Mont Ventoux betreffend. München, 1936.
	Geog 4269.36	Price, Lucien. We Northmen. Boston, 1936.
	Geog 4502.5.18	Rudatis, D. Das Letzte im Fels. München, 1936.
	Geog 3129.36	Saa, Mario. Evudania. Lisboa, 1936.
	An 359.36.10	Schmelzle, K. Rassengeschichte und Vorgeschichte. 2. Aufl. Bamberg, 1936.
	An 2109.36.3	Schmidt, R.R. The dawn of the human mind. London, 1936.
	An 3309.36	Simon, Carleton. The retinal method of identification. n.p., 1936.
	Geog 4237.00	Stearns, R.P. The course of Captain Edmond Halley in the year 1700. n.p., 1936.
	An 3409.36	United States. Federal Bureau of Investigation. Classification of fingerprints. J.E. Hoover, director. Washington, 1936.
	Geog 4308.56.7.10	Vicuña Mackenna, Benjamin. Páginas de mi diario durante tres años de viaje. Santiago de Chile, 1936. 2v.
	An 2109.36	Viljoen, S. The economics of primitive peoples. London, 1936.
	Geog 5399.36	Vize, Vladimir I. Moria Sovetskoi Arktiki. Leningrad, 1936.
	Geog 6009.33.5	Walden, Jane Brevoort. The long whip; the story of a great husky. N.Y., 1936.
	An 359.36	White, C.L. Geography. N.Y., 1936.
	Geog 4209.36.15	Wøller, Johan. Zest for life; recallections of a philosophic traveller. London, 1936.
	Geog 5060.16	Zeidler, Paul G. Helden im ewigen Eis; im Kampf um den Nord- und Südpol. Leipzig, 1936.

1937

	Geog 4558.56.5	Abbey, Charles A. Before the mast in clippers...1856 to 1860. N.Y., 1937.
	Geog 1508.34	Baedeker, publishers. Great Britain. N.Y., 1937.
	Geog 4129.37	Barraud, G. Touristes de jadis. Paris, 1937.
	An 359.37.9	Baschmakoff, A.A. Cinquante siècles d'évolution ethnique autour de la Mer noire. Paris, 1937.
	Geog 4219.37	Belfrage, C. Away from it all. N.Y., 1937.
	Geog 613.4.15	Bernstein, H. Sir Clements R. Markham as a translator. n.p., 1937?
	An 3409.37.3	Borges Badell, A. Clasificación anatómica-matematica del dactilograma. La Habana, 1937?
	Geog 5326.21.10	Buck, M. Memorial book of Robert Edwin Peary. n.p., 1937.
	An 3509.37	Buyssens, Paul. Le pithécanthrope était-il un Pygmée? Bruxelles, 1937.
	Geog 6009.08.5	Charcot, Jean B. Autour du pôle sud; expédition du "Pourquoi pas?" 1908-1910. London, 1937.
	An 3309.37	Chattapadkyay, Bojra K. Les affinités somatiques des Brahmines Moithils et Kanarijias de Bihar. Thèse. Paris, 1937?
	Geog 6009.10.32	Cherry-Garrard, Apsley G.B. The worst journey in the world; Antarctic 1910-13. v.1-2. N.Y., 1937.
	Geog 665.53	Chicago. University. Norman Wait Harris Foundation. Reports of round tables, 1937. Geographic aspects of international relations. n.p., 1937.
	Geog 4502.10.3	Christ, Fritz. Die erste Ersteigung des Totenkirches durch den Christ Fick-Kamin. München, 1937.
	Geog 3060.12.50	Churchward, J. The children of Mu. N.Y., 1937.
	An 359.37.11	Clauss, Ludwig F. Rasse ist Gestalt. München, 1937.
	Geog 4181.80	Crone, G.R. The voyages of Cadamosto. London, 1937.
Htn	Geog 4559.37*	Dana, Richard Henry. Cruelty to seamen; being the case of Nichols and Couch. Berkeley, 1937.
	Geog 4309.37.5A	Dorgelès, R. Vive la liberté. Paris, 1937.
	An 359.37	Dover, C. Half-caste. London, 1937.
X Cg	An 359.37.7	Eickstedt, E. Rassenkunde und Rassengeschichte der Menschheit. 2. Aufl. v.1, pt.1-15. Stuttgart, 1937-43.
	Geog 665.43	Errera, Carlo. Scritti geografici scelti e ordinati a cura del Comitato nazionale. Bologna, 1937.
	An 379.34.2	Forde, C.D. Habitat, economy and society. N.Y., 1937.
	Geog 5399.35	Glen, A.R. Under the Pole star. London, 1937.

Chronological Listing

1937 - cont.

Geog 3135.5A Heidel, William A. The frame of the ancient Greek maps. N.Y., 1937.
An 2109.37.9 Hildebrand, E. Die Geheimbünde Westafrikas als Problem der Religionwissenschrift. Inaug. Diss. Leipzig, 1937.
Geog 4306.56.5 Hiltebrandt, C.J. Dreifache schwedische Gesandtschaftsreise nach Siebenbürgen. Leiden, 1937.
An 359.37.5A Hooton, E.A. Apes, man, and morons. N.Y., 1937.
Geog 5180.35 Hopper, Bruce C. Sovereignty in the Arctic. N.Y., 1937.
Geog 4269.37 Hubbard, George D. The geography of Europe. N.Y., 1937.
Geog 4329.37F Hürlimann, M. Das Mittelmeer. Zürich, 1937.
Geog 4309.37A Ichikawa, H. Japanese lady in Europe. N.Y., 1937.
Geog 3199.10 al-Idrisi, Muhammad ibn Muhammad. Deutschland und seine Naehbarländer nach der grossen Geographie des Idrisi. Stuttgart, 1937.
Geog 5839.35.15 Ingstad, H.M. East of the great glacier. N.Y., 1937.
An 2059.37 International Symposium on Early Man. Early man. Philadelphia, 1937.
Geog 3070.39 Jervis, W.W. The world in maps. N.Y., 1937.
An 359.37.13 Krzywicki, L. Społeczeństwo pierwotne. Warszawa, 1937.
Htn Geog 4209.37F* LaCombe, Jean de. A compendium of the East; being an account of voyages to the Grand Indies. London, 1937.
An 2109.37.7 Laubscher, B.J.F. Sex, custom and psychopathology. London, 1937.
Geog 4249.37 Lissner, Ivar. Menschen und Mächte am Pazifik. Hamburg, 1937.
An 99.37.1 Lowie, Robert Harry. The history of ethnological theory. N.Y., 1937.
An 379.37 Machin, A. Darwin's theory applied to mankind. London, 1937.
Geog 4559.37.5 Making, V.L. In sail and steam...merchant service, 1902-1927. London, 1937.
Geog 612.2 Mazuel, J. L'oeuvre géographique de Linant de Bellefonds. Thèse. Le Caire, 1937.
An 2109.37.5 Mead, M. Cooperation and competition among primitive peoples. N.Y., 1937.
Geog 611.3 Meier-Lemgo, K. Engelbert Kampfer, der erste deutsche Forschungsreisende, 1651-1716. Stuttgart, 1937.
Geog 665.60 Mélanges de géographie et d'orientalisme offerts à E.-F. Gautier. Tours, 1937.
An 2109.37.3 Montagu, Ashley. Coming into being among the Australian aborigines. London, 1937.
Geog 148.1.6 National geographic magazine. Cumulative index, 1899-1936. Washington, 1937.
Geog 4208.22.6 Nicol, John. The life and adventures of John Nicol, mariner. London, 1937.
Geog 3019.37 Olschki, L. Storia letteraria delle scoperte geografiche. Firenze, 1937.
Geog 603.5.5 Oulié, M. Jean Charcot. 14. éd. Paris, 1937.
Geog 4181.79 Pacheco Pereira, Duarte. Esmeraldo de situ orbis. London, 1937.
An 2109.37 Portens, S.D. Primitive intelligence and endowment. N.Y., 1937.
An 190.1 Pratt, Orson. Wonders of the universe. Salt Lake City, 1937.
An 359.37.3 Preuss, K.T. Lehrbuch der Völkerkunde. Stuttgart, 1937.
An 629.37 Schultz, Bruno K. Taschenbuch der rassenkundlichen Messtechnik. München, 1937.
Geog 5985.325 Seaver, George. Edward Wilson of the Antarctic, naturalist and friend. N.Y., 1937.
Geog 5559.37.10 Severn Iulius Zavoevan Bolbshevik. Moskva, 1937.
Geog 5559.34.5 Shackleton, Edward. Arctic journeys; the story of the Oxford University Ellesmire land expedition, 1934-35. London, 1937.
An 379.37.3A Taylor, G. Environment, race and migration. Chicago, 1937.
An 2109.37.11 Thomas, W.I. Primitive behavior. 1. ed. N.Y., 1937.
An 3409.37 United States. Federal Bureau of Investigation. Fingerprints. Washington, 1937.
Geog 4219.37.10 Villiers, A.J. Cruise of the Conrad...1934-1936. London, 1937.
Geog 4182.41 Vogel, J.P. Journal van J.J. Ketelaar's hofreis...1711-1713. 's-Gravenhage, 1937.
Geog 4209.37.5 Vrázová, Vlasta. Život a cesty E. St. Vráze. Praha, 1937.
Geog 4559.37.3 Wead, Frank W. Gales, ice and men. N.Y., 1937.
Geog 4209.37.10 Wechsberg, J. Die grosse Mauer; das Buch einer Weltreise. Leipzig, 1937.

1938

Geog 1531.5 Baedeker, publishers. Tirol, Vorarlberg, westliche Salzburg, Hochkärnten. 40. Aufl. Leipzig, 1938.
An 629.38 Balfour, Henry. Spinners and weavers in anthropological research. Oxford, 1938.
An 359.38.13 Bang, Paul. Die farbige Gefahr. 2. Aufl. Göttingen, 1938.
An 359.38 Barzun, Jacques. Race, a study in modern superstition. London, 1938.
Geog 4209.38.21 Beddington, C. We sailed from Brixham. London, 1938.
Geog 5515.77.15 Best, George. The three voyages of Martin Frobisher in search of a passage to Cathay...1576-78. London, 1938. 2v.
An 359.38.20 Boas, Franz. General anthropology. Boston, 1938.
An 2109.38.5 Boas, Franz. The mind of primitive man. N.Y., 1938.
Geog 4209.38.35 Bonnard, A. Le bouquet du monde. Paris, 1938.
Geog 4228.61 Bradford, Ruth. Maskee! The journal and letters of Ruth Bradford, 1861-1872. Hartford, 1938.
Geog 3055.54A Bramwell, James. Lost Atlantis. N.Y., 1938.
Geog 5300.11 Brontman, L.K. On top of the world. N.Y., 1938.
An 3509.38 Bryan, Kirk. Discovery of Sauk Valley man of Minnesota. Abilene, Texas, 1938.
Geog 5939.38.3A Byrd, Richard E. Alone. N.Y., 1938.
Geog 6009.36 Christensen, Lars. My last expedition to the Antarctic, 1936-37. Oslo, 1938.
Geog 4169.38 Compton, Ray. The open road. N.Y., 1938.
Geog 110.3.25 Congrès des Géographes et Ethnographes Slaves, 4th, Sofia, 1936. Sbornik na IV Kongress na slavianskite geografi i etnografi. Sofiia, 1938.
Geog 4209.38.30 Craig, John D. Danger is my business. N.Y., 1938.
An 359.38.3A Demiashkevich, Michael John. The national mind. N.Y., 1938.
Geog 5559.37.25 Deviat' mesiatsev na Dreifuiushchei stantsii severnyi polius. Moskva, 1938.
Geog 4209.38.10 Dighy, G. Goose feathers. N.Y., 1938.
Geog 4209.38.16 Dos Passos, John. Journeys between wars. London, 1938.
Geog 4209.38.15A Dos Passos, John. Journeys between wars. N.Y., 1938.
Geog 4502.5.25 Dreyer, A. Geschichte der Alpinen Literatur. München, 1938.

1938 - cont.

An 2109.38.11 Dunbar, George. Other men's lives...primitive peoples. London, 1938.
Geog 4329.38.5 Duran, Faik Sabri. Akdenirde leir yor gerintisi. Istanbul, 1938.
Geog 5398.79.30A Ellsberg, Edward. Hell on ice; the saga of the "Jeannette". N.Y., 1938.
Geog 5070.55 Ellsworth, L. Beyond horizons. N.Y., 1938.
Geog 4249.38 Fahnestvak, B. Stars to windward. N.Y., 1938.
Geog 4309.38.5 Fellman, A. Voyage en Orient du roi Erik Ejegod. Helsinki, 1938.
Geog 4309.38.15 Fish, Helen D. Invitation to travel. N.Y., 1938.
Geog 3570.31.5 Fontoura da Costa, Abel. Descobrimentos maritimos africanos dos Portugueses com D. Henrique. Lisboa, 1938.
Geog 5316.1.10 Freuchen, P. It's all adventure. London, 1938.
An 3559.38 Froe, A. de. Meethare variabelen van den menschlijken schedel en hun an derlinge. Amsterdam, 1938.
Geog 4219.33.15 Gezork, H. So sah ich die Welt. 12e Aufl. Kassel, 1938?
Geog 5839.38.5 Gitz-Johansen, Aage. Skitzebogsblade fra Angmagssalik. København, 1938.
Geog 819.38F Globus geograficheskii eksegodnik dlia detei. Moskva, 1938.
Geog 4181.81 Greenlee, William B. The voyage of Pedro Alvares Cabral to Brazil and India. London, 1938.
Geog 4209.38.5A Hallet, R.M. The rolling world. Boston, 1938.
Geog 4309.38 Hamburg-American Line. Your trip to Europe. N.Y., 1938.
Geog 4209.38 Hart, M.R. Who called that lady a skipper? N.Y., 1938.
An 2109.38.9 Hutton, J.H. A primitive philosophy of life. Oxford, 1938.
Geog 4219.38 Hypes, J.L. Knights of the road. Washington, 1938.
Geog 110.1.15 International Geographical Congress, 15th, Amsterdam, 1938. Comptes rendus du Congrès international de géographie, Amsterdam, 1938. v.1-5,21. Leiden, 1938. 6v.
Geog 4512.1 Irving, R.L.G. The mountain way. London, 1938.
Geog 4502.5.26 Javelle, E. Erinnerungen eines Bergsteigers. München, 1938.
Geog 3070.39.5 Jervis, W.W. The world in maps. 2d ed. N.Y., 1938.
Geog 4182.42 Keuning, J. De tweede schipvaart der Nederlanders. v.1-5. 's-Gravenhage, 1938-51. 8v.
Geog 3019.38 Key, Charles E. The story of twentieth-century exploration. N.Y., 1938.
Geog 3229.38 Kimble, George H.T. Geography in the Middle Ages. London, 1938.
Htn Geog 4679.38* Knowles, Josiah N. The crusoes of Pitcairn's Islands. n.p., 1938.
An 2109.38 Landtman, G. The origin of the inequality of the social classes. London, 1938.
Geog 3019.38.5F La Roncière, Charles. Histoire de la découverte de la terre. Paris, 1938.
Geog 4131.38.5 Leveillé-Nizerolle, Claude. Le tourisme dans l'economie contemporaine. Thèse. Paris, 1938.
An 2109.38.7 Lévy-Bruhl, L. L'expérience mystique et les symboles chez les primitifs. Paris, 1938.
Geog 4658.93.4 Lussich, Antonio D. Naufragios célebres en el cabo Polonio, banco Inglés, y océano Atlántico. Montevideo, 1938.
An 379.38 McIverny, A.J. Through the great arid filter (man's drift to Europe). London? 1938.
Geog 4329.38 Miller, M.S. (Mrs.). Cruising the Mediterranean. N.Y., 1938.
Geog 4419.38.15 Monfreid, Henri de. Le trésor du pélerin; roman. 12. ed. Paris, 1938.
Geog 4419.38 Morton, H.V. Through the lands of the Bible. London, 1938.
Geog 4239.38 Moyne, W.E.G. Atlantic circle. London, 1938.
An 359.38.11 Mühlmann, W.E. Methodik der Völkerkunde. Stuttgart, 1938.
An 124.1 National Research Council. International directory of anthropologists. Washington, 1938.
An 359.38.15 Nauka o rasakh i rasizm. Moskva, 1938.
Geog 5919.38 Ommaney, Francis D. Below the roaring forties. London, 1938.
Geog 603.5.10 Oulié, M. Charcot of the Antartic. London, 1938.
An 2109.38.3 Page, J.W. Primitive races of to-day. London, 1938.
Geog 4345.99.4 Penrose, B. The Sherleian odyssey, being a record of the travels of three famous brothers. Taunton, 1938.
Geog 3410.5 Pfeifer, Gottfried. Regional geography in the United States since the war. N.Y., 1938.
Geog 4309.38.10 Pinochet, T. Viaje, plebeys por Europa. 2. ed. Santiago, 1938.
Geog 4228.10 Reynolds, Stephen. The voyage of the New Hazard...1810-1813. Salem, 1938.
Geog 6009.34 Rymill, John. Southern lights; the official account of the British Graham Land Expedition, 1934-37. London, 1938.
Geog 3019.38.3 St. Croix de la Roncière, G. À la conquête des mers. Paris, 1938.
An 379.38.5 Sanchez, Pedro C. Enseñanzas fundamentales de la geografia humana. Tacubaya, 1938.
Geog 579.36A Schmidt, A.J. Kleines deutsch-portugiesisches...Verzeichnis geographischer Eigennamen. Rio de Janeiro, 1938.
An 359.38.5 Schmidt, Wilhelm. Razza e nazione. n.p., 1938.
Geog 5985.34 Seaver, George. "Birdie" Bowers of the Antarctic. London, 1938.
Geog 4209.38.40 Sigvaldson, J. Ferdasaga Fritz Liebig. Reykjavik, 1938.
An 3309.38 Simon, Carleton. La identificación personal por la retina. La Habana, 1938.
Geog 4679.38.5 Slocum, Victor. Castaway boats. N.Y., 1938.
An 359.38.7 Sombart, W. Von Menschen; Versuch einer geistwissenschaftlichen Anthropologie. Berlin, 1938.
Geog 4559.35.7 Stevers, M.D. Sea lanes; man's conquest of the ocean. Garden City, N.Y., 1938.
Geog 4559.38 Tompkins, W.M. Fifty south to fifty south...around Cape Horn...Wander Bird. N.Y., 1938.
Geog 4131.38 Trimbach, André. Le tourisme international. Thèse. Paris, 1938.
Geog 5050.5 United States. Works Progress Administration, N.Y.C. Annotated bibliography of the polar regions. Series B. pt. 1. N.Y., 1938.
Geog 4502.10.4 Vierthaler, Franz Michael. Die Reise auf den Grossglockner, 1800. München, 1938.
Geog 4559.38.5 Wead, F.W. Gales, ice and men; a biography of the Steam Barkentine Bear. London, 1938.
Geog 759.38 West- und Nordeuropa in Natur. Potsdam, 1938.

Chronological Listing

1939

Geog 4219.39	Alford, B. Around the world on $100.00 a month. Emmaus, 1939.
Geog 953.5FA	Bachmann, Friedrich. Die alten Städtebilder. Leipzig, 1939.
Geog 1522.39.5	Baedeker, publishers. Autoführer, Deutsches Reich (Grossdeutschland). 2. Aufl. Leipzig, 1939.
Geog 6009.20	Bagshawe, T.W. Two men in the Antarctic; an expedition to Graham Land, 1920-22. N.Y., 1939.
Geog 5120.1	Bertram, Colin. Arctic and Antarctic. Cambridge, 1939.
Geog 4169.12.5	Boer, M.G. de. Van oude voyagien. 3. druk. Amsterdam, 1939.
Geog 4219.38.6	Bradshaw, M.J. Third class world. Alliance, 1939.
An 2109.39.3	Childe, Vere Gordon. Dawn of European civilization. N.Y., 1939.
An 359.39A	Coon, Carleton Stevens. The races of Europe. N.Y., 1939.
Geog 4110.25	Coon, Horace. 100 vacations costing from $50.00 to $500.00. N.Y., 1939.
Geog 5070.62	Croft, Andrew. Polar exploration. London, 1939.
An 379.39.5	Deffontaines, P. Problèmes de géographie humaine. Paris, 1939.
Geog 4709.39	Dobie, J.F. Apache gold and yaqui silver. Boston, 1939.
Geog 4209.39	Forbes, Rosita T. These are real people. N.Y., 1939.
Geog 4181.82	Foster, William. The voyage of Nicholas Downton to the East Indies, 1614-15. London, 1939.
An 359.39.3	Frazer, J.G. The native races of Asia and Europe. London, 1939.
Geog 4269.39	Gandy, Roxana Smith. Are people more a like than different. Cape May Court House, 1939.
An 170.221	Göttingen. Universität. Institut für Völkerkunde. Göttinger völkerkundliche Studien. Leipzig, 1939.
An 2109.39.6	Goodfellow, D.M. Principles of economic sociology. London, 1939.
An 2109.39.5	Goodfellow, D.M. Principles of economic sociology. Philadelphia, 1939.
Geog 4502.5.27	Graber, A. Der Weg zum Berg. München, 1939?
Geog 5839.39	Haig-Thomas, D. Tracks in the snow. N.Y., 1939.
Geog 4308.36.5	Hall, Fanny W. Rambles in Europe. N.Y., 1939. 2v.
An 130.20	Hanke, Lewis. Gilberto Freyne. N.Y., 1939.
An 379.39.10	Hardy, Georges. La géographie psychologique. Paris, 1939.
Geog 4209.39.20A	Harris, N.D. Moving on; the romance of travel. Chicago, 1939.
An 359.39.5A	Hooton, E.A. Twilight of man. N.Y., 1939.
An 2109.39.7	Kardiner, A. The individual and his society. N.Y., 1939.
Geog 5839.39.5	Koch, Lange. Fra Lissabon til Peary Land. København, 1939.
Geog 665.45	Kühn, Arthur. Die Neugestaltung der deutschen Geographie im 18. Jahrhundert. Leipzig, 1939.
Geog 5399.37	Laktionov, A.F. Severnyi polius. Arkhangelsk, 1939.
Geog 4182.05	Linschoten. Vereeniging. Werken. Register. v.1-25, 26-50. 's-Gravenhage, 1939-57. 2v.
Geog 4209.38.1	Long, Dwight. Seven seas on a shoestring. N.Y., 1939.
Geog 4209.39.10	Maury, Richard. The saga of "Cimba". N.Y., 1939.
An 2109.39.15	Mead, Margaret. From the South Seas. N.Y., 1939.
Geog 4207.76.15	Munford, K. John Ledyard: an American Marco Polo. Portland, 1939.
Geog 5559.14	Nagornyi, S. Sedov. Moskva, 1939.
Geog 4559.38.12	Nixon, Laurence A. Vagabond voyaging; the story of freighter travel. Boston, 1939.
Geog 4209.39.5	O'Malley, C.J. It was news to me. Boston, 1939.
Geog 4679.12.7	Pelz von Felman, J. Titanic. Berlin, 1939.
Geog 4559.40.10	Pierrefeu, François de. Les confessions de Tatibouet. Paris, 1939.
An 359.37.3.5	Preuss, K.T. Lehrbuch der Völkerkunde. 2. Aufl. Stuttgart, 1939.
Geog 4759.39	Price, A.G. White settlers in the tropics. N.Y., 1939.
Geog 4209.39.15	Reinius, I. Journal hållen på resan till Canton i China. Helsingfors, 1939.
Geog 3229.39	Rohr, Heinz. Die Entwicklung des Kartenbildes. Inaug. Diss. Borna, 1939.
Geog 4238.47	Rosenbaum, S.E. A voyage to America 90 years ago...Hamburg to New York in 1847. N.Y., 1939.
An 359.39.15	Schmidt, Wilhelm. The culture historical method of ethnology. N.Y., 1939.
Geog 5209.39	Segal, Louis. The conquest of the Arctic. London, 1939.
Geog 5170.17	Seidenfaden, G. Modern Arctic exploration. Boston, 1939.
Geog 4219.39.5	Seligman, A. The voyage of the Cap Pilar. London, 1939.
An 359.39.7	Seligmann, H.J. Race against man. N.Y., 1939.
Geog 5160.11.5	Stefansson, V. Unsolved mysteries of the Arctic. N.Y., 1939.
Geog 4502.5.28	Stöger-Ostin, Georg. Georg Jennerwein, der Wildschütz. München, 1939?
Geog 4679.39	Swan, E.W. The first commission of H.M.S. Calliope. Newcastle-upon-Tyne, 1939.
Geog 4559.39.5	Tompkins, W.M. Two sailors and their voyage around Cape Horn. N.Y., 1939.
Geog 5839.30.5	Wegener, E. Greenland journey; the story of Wegener's German expedition...1930-31. London, 1939.
An 2109.39	Wright, W.B. Tools and the man. London, 1939.

194-

Geog 4209.47	Colam, Lance. Death over my shoulder. London, 194-.
An 379.42	Cools, R.H.A. De geographische gedechte bij Jean Brunhes. Utrecht, 194-.
Geog 759.40	Costa Pereira, J.V. da. Geographia humana. Rio de Janeiro, 194-.
An 170.189	Essays in anthropology. Lucknow, 194-.
Geog 4129.11	Hedin, Sven. Van pool tot pool. Amsterdam, 194-?
An 359.44.25A	Kluckhohn, Clyde. The concept of culture. n.p., 194-.

1940

Geog 4309.40	Abbe, P. No place like home. N.Y., 1940.
Geog 4209.40.15	Baarslag, K. Islands of adventure. N.Y., 1940.
Geog 3019.40	Balen, Willem Julius van. Pioniers (De ontdekking van de wereld). Amsterdam, 1940. 2v.
An 126.5	Bandelier, Adolph Francis. Pioneers in American anthropology: the Bandelier-Morgan letters, 1873-1883. Albuquerque, 1940. 2v.
An 2109.40.20	Bawmann, Evert Dirk. Historische Betrachtungen über das Koitus-Konzeption Problem. Arnhem, 1940.
An 359.40	Benedict, Ruth (Fulton). Race: science and politics. N.Y., 1940.
Geog 4559.40.9	Bisschop, E. de. The voyage of the Kaimiloa. London, 1940.
Geog 4559.40.7	Blain, W. Home is the sailor. N.Y., 1940.

1940 - cont.

An 176.6A	Boas, Franz. Race, language and culture. N.Y., 1940.
Geog 3055.56	Bragbine, A. The shadow of Atlantis. N.Y., 1940.
Geog 5558.81.25	Brainard, David L. Six came back; the Arctic adventure. Indianapolis, 1940.
Geog 5324.2.30	Byström, D.G.V. Fridtjof Nansen. Stockholm, 1940.
An 2109.40.10	Cole, M.C. The story of primitive man. Chicago, 1940.
Geog 4247.89	Colnett, J. The journal of Captain James Colnett aboard the Argonaut from Apr. 26, 1789 to Nov. 3, 1791. Toronto, 1940.
An 2109.40.5	Cooper, G. I searched the world for death. London, 1940.
Geog 4419.40	Cutting, S. The fire ox and other years. N.Y., 1940.
Geog 3019.40.5	Dainville, François de. La géographie des humanistes. Paris, 1940.
Geog 4209.40.10	Davis, R.H. Let's go with Bob Davis to India, Ceylon, Venezuela. N.Y., 1940.
Geog 4209.40.25	Dennis, W.H. Glittering horizons. N.Y., 1940.
An 130.10	Downie, Robert Angus. James George Frazer. London, 1940.
Geog 5399.37.5	Ekspeditsiia SSSR na Severnyi Polius, 1937. Trudy. Leningrad, 1940.
Geog 4502.5.29	Flaig, Walther. Das Silvretta-Buch. München, 1940.
Geog 3570.31	Fontoura da Costa, Abel. Roteiros portugueses ineditos. Lisboa, 1940.
Geog 4209.40.3	Forsyth, W.J. Journal. Boston, 1940.
Geog 4181.85	Foster, William. The voyages of Sir James Lancaster to Brazil and the East Indies, 1591-1603. London, 1940.
An 359.40.10	Fyfe, Henry H. The illusion of national character. London, 1940.
Geog 4559.40.3	Gainard, J.A. Yankee skipper; the life story of Joseph A. Gainard. N.Y., 1940.
An 2109.40.15	Gattefossé, Jean. L'Hyperborée et les migrations néolithiques. Droguignan, 1940.
Geog 4502.5.31	Geissler, Paul. Um den Montblanc. München, 1940?
Geog 4209.40.20	Halliburton, Richard. Richard Halliburton, his story of his life's adventures as told in letters to his mother and father. Indianapolis, 1940.
Geog 4209.40.45	Haven, V.S. Many ports of call. N.Y., 1940.
Geog 4709.40	Hawes, Hildreth Gilman. The Bellamy treasure. Augusta, Maine, 1940.
An 2109.40	Herskovits, M.J. The economic life of primitive peoples. N.Y., 1940.
Geog 4209.40.40	Hogben, L. Author in transit. N.Y., 1940.
An 359.40.3A	Hooton, E.A. Why men behave like apes and vice versa, or Body behavior. Princeton, 1940.
Geog 4209.40.30	Hurst, Ida. Dare to live. London, 1940.
Geog 4509.40.5	Irving, R.L.G. Ten great mountains. N.Y., 1940.
Geog 4509.40	Jahn, A. Exrusionismo y alpinismo; historia de su evolución. Caracas, 1940.
Geog 4559.40.12	Jónsson, Gisli. Frekjan; aefintýralegt...frá Danmoëku. Reykjavik, 1940.
Geog 5839.38	Knuth, Eigil. Under det nordligste danne broa; beretning om dansk Nordøstgrønlands Ekspedition, 1938-39. København, 1940.
Geog 5399.37.10	Kreukel', E.T. Chetyre tovarishcha. Moskva, 1940.
An 2109.40.3	Lowie, R.H. An introduction to cultural anthropology. N.Y., 1940.
Geog 4209.40.60	Martyr, W. The wandering years. N.Y., 1940.
Geog 4502.5.30	Mason, Alfred E.W. Das Gesetz der Berge. München, 1940?
Geog 4219.40.10	Michael, Rudolf. Roman einer Weltreise. Hamburg, 1940.
Geog 4359.40	Newman, Bernard. Savoy! Corsica! Tunis! Mussolini's dream lands. London, 1940.
Geog 4559.40	Nordhoff, C. In Yankee windjammers. N.Y., 1940.
Geog 5896.2	Ostermann, Hother. Nordmaend paa Grønland, 1721-1814. Oslo, 1940. 2v.
Geog 5559.37	Papanin, Ivan D. Zhizn' na l'dine. Moskva, 1940.
Geog 3560.10	Pereyra, C. La conquista de las rutas oceánicas. Madrid, 1940.
Geog 4329.40	Pfuhl, E. Ostgriechische Reisen; Kleinasien, Kypros und Syrien. Basel, 1940.
Geog 4181.83	Quinn, David B. The voyages...of Sir Humphrey Gilbert. London, 1940. 2v.
Geog 5985.253.5	Seaver, George. Scott of the Antarctic. London, 1940.
An 3309.40	Sheldon, W.H. The varieties of human physique. 1st ed. N.Y., 1940.
Geog 4209.40.50	Simon, C.R. Gran'ma goes by freight. Placerville, Calif., 1940.
Geog 4559.40.5	Smale, R. There go the ships. Caldwell, 1940.
Geog 5160.11.10	Stefansson, V. Ultima Thule. N.Y., 1940.
Geog 4209.40.55	Tangye, Derek. The time was mine. London, 1940.
Geog 4679.12.10	Thayer, J.B. The sinking of the S.S. Titanic, April 14-15, 1912. Philadelphia, 1940.
Geog 3570.45	Tracey, Hugh. Antonio Fernandes. Lourenco Marques, 1940.
Geog 5160.12	United States. Office of Chief of Air Corps (War Department). Arctic manual. v.1-2. Washington, 1940.
Geog 665.50.5	Van Loon, H.W. Van Loon's geography. Garden City, N.Y., 1940.
Geog 4429.40	Villiers, Alan J. Sons of Sinbad. N.Y., 1940.
Geog 4209.40.35	Weidman, J. Letter of credit. N.Y., 1940.
An 359.40.25	Wiese und Kaiserwaldau, Leopold M. von. Homo sum, Gedanken zu einer zusammenfassenden Anthropologie. Jena, 1940.
Geog 4709.40.5	Wilkins, H.T. Panorama of treasure hunting. N.Y., 1940.
Geog 4209.40	Woodward, C. Lanterns alight; journeys to far places. Chicago, 1940.
Geog 4559.40.11	Wossidlo, R. Reise, Quartier in Gottesnaam. Seestadt Rostock, 1940.

1941

Geog 4182.45	Adrichem, D. van. Journaal van Dircq Van Adrichem hofreis. 's-Gravenhage, 1941.
Geog 612.3	Allen, Edward W. Jean François Galaup de Lapérouse. San Francisco, 1941.
Geog 4209.41.5	Baerlein, Henry. Travels without a passport. London, 1941.
Geog 4309.41.10	Belloc, Hilaire. Places. N.Y., 1941.
An 126.4	Bertillon, Suzanne. Vie d'Alphonse Bertillon, inventeur de l'anthropométrie. Paris? 1941.
Geog 5328.5	Birket-Smith, Kaj. Knud Rasmussens saga. Kjøbenhavn, 1941.
Geog 5700.4.9	Bobé, Louis. Hans Egede og Grønland. København, 1941.
Geog 579.41	Bonacker, Wilhelm. Karten-Wörterbuch eine Verdeutschung fremdsprachiger Kartensignatur-Bezeichnungen. Berlin, 1941.
Geog 603.9	Brown, Lloyd A. Jean Domenique Cassini and his world map of 1696. Ann Arbor, 1941.
Geog 4209.41.3	Buck, F. All in a lifetime. N.Y., 1941.

Chronological Listing

1941 - cont.

An 3409.41	Chapel, Charles E. Fingerprinting; a manual of identification. N.Y., 1941.
Geog 4208.54.10	Coffin, Robert. The last of the Logan. Ithaca, 1941.
Geog 4209.41.10	DeValda, F.W. Adventure is my business. London, 1941.
An 359.41	Estabrooks, G.H. Man, the mechanical misfit. N.Y., 1941.
Geog 4169.41	Explorers Club, N.Y. Thru hell and high water. N.Y., 1941.
Geog 4219.41.10	Faber, kurt. Tausend und ein Abenteuer. Berlin, 1941.
An 2600.4	Gesell, Arnold. Wolf child and human child. N.Y., 1941.
Geog 607.16	Gunchev, Guncho S. Guncho Gunchev. Sofiia, 1941.
Geog 4209.35.23	Halliburton, Richard. Seven league boots. 3. ed. London, 1941.
Geog 4679.41	Hanson, Earl P. Highroad to adventure. N.Y., 1941.
Geog 4209.41.15	Low, G.W. Gold rush by sea. Philadelphia, 1941.
Geog 4209.41	Lynch, Kathleen M. Travellers must be content. N.Y., 1941.
Geog 4309.41.5	McGee, A.E. Black America abraod. Boston, 1941.
An 130.15	Marett, Robert R. James George Frazer 1854-1941. London, 1941?
An 137.3	Marett, Robert R. A Jerseyman at Oxford. N.Y., 1941.
An 2059.41	Mattos, Anibal. A raça de Lagôa Santa. São Paulo, 1941.
Geog 4209.41.20	Maury, A.F. Intimate Virginiana. Richmond, 1941.
Geog 4309.41	Moody, M.N. Trails of two travelers. N.Y., 1941.
Geog 4218.78.20	Nichols, E.P. The Ocean Chronicle, 1878-1891. N.Y., 1941.
Geog 5660.30	Ostermann, H. Traekaf kolonien Jakobshavns. Kjøbenhavn, 1941.
Geog 5939.41	Owen, Russell. The Antarctic ocean. N.Y., 1941.
Geog 4219.41	Parsons, C. Vagabondage. London, 1941.
An 379.41	Pasada, J. de la C. Geografia humana. Medellín, 1941.
Geog 4509.41.5	Peacocke, T.A.H. Mountaineering. London, 1941.
Geog 616.3.5	Peattie, R. The incurable romantic. N.Y., 1941.
Geog 4708.06	Phillips, James D. Loss of the ship Essex in 1806. n.p., 1941.
Geog 5312.7.20	Pokrovskii, Aleksei A. Ekspeditsiia Beringa. Moskva, 1941.
Geog 5925.55	Rodriguez, Juan C. La Republica Argentina y las adquisiciones territoriales en el continente antartico. Buenos Aires, 1941.
Geog 4509.41.10	Rondet, C. Montagne: des terrasses aux arêtes. Paris, 1941.
An 359.41.10	Scientific aspects of the race problem. Washington, 1941.
An 2109.41.5	Sieber, Sylvester A.M. The social life of primitive man. St. Louis, 1941.
Geog 4132.6	Trantina, V. Staří Čechové na cestách. Praha, 1941.
Geog 4509.41	Ullman, J.R. High conquest, the story of mountaineering. Philadelphia, 1941.
An 379.41.5	Vermooten, Willem H. De mens in de geografie. Assen, 1941.
Geog 4219.41.5	Wiener, P. Last man around the world. N.Y., 1941.

1942

Geog 608.5	Almagià, Roberto. L'opera geografica di Luca Holstenio. Citta del Vaticano, 1942.
Geog 4209.42.10	Barrett, C.L. On the Wallaby. Melbourne, 1942.
Geog 4209.42.15	Birnbaum, M. Vanishing Eden. N.Y., 1942.
Geog 4181.86A	Blake, John W. Europeans in West Africa, 1450-1560. London, 1942. 2v.
Geog 4309.41.15	Blanco-Fombona, R. Dos años y medio de Inquietad. Caracas, 1942.
Geog 4217.21	Bree, Levinus W. de. Jacob Roggeveen en zijn reis naar het zuidland, 1721-1722. Amsterdam, 1942.
An 3409.42	Bridges, B.C. Practical fingerprinting. N.Y., 1942.
Geog 5160.13.5	Brower, Charles D. Fifty years below zero. N.Y., 1942.
An 379.42.5	Brunhes, Jean. La géographie humaine. Paris, 1942.
Geog 4679.42	Bryan, George S. Mystery ship. The Mary Celeste in fancy and fact. Philadelphia, 1942.
An 359.42A	Chapple, Eliot D. Principles of anthropology. N.Y., 1942.
Geog 659.42	Cholley, André. Guide de l'étudiant en géographie. 1. éd. Paris, 1942.
Geog 4208.78.4A	Connolly, J.B. Canton captain. Garden City, N.Y., 1942.
Geog 4209.42.5	Considine, J.J. Across a world. Toronto, 1942.
Geog 4309.08.5	Darling, Kenneth G. My early life in California and my first American and European tours (1890-1908). Cambridge, Mass.? 1942?
Geog 819.42.2	Davis, D.H. The earth and man. N.Y., 1942.
Geog 4209.42	Domville-Fife, C.W. I tell of the seven seas. London, 1942.
Geog 4229.42	Faber, Kurt. Der göttliche Vagabund. Stuttgart, 1942.
Geog 3560.15	Farinelli, Arturo. Viajes por España y Portugal. Roma, 1942-44. 3v.
Geog 4679.42.5	Fay, Charles E. Mary Celeste; the odyssey of an abandoned ship. Salem, 1942.
Geog 3570.16.5	Frazão de Vasconcellos, J. Pilobas das navegacões portuguesas das seculos XVI e XVII. Lisboa, 1942.
Geog 4129.11.5	Hedin, Sven. Von Pol zu Pol. 81. Aufl. Leipzig, 1942
Geog 4509.42	Henderson, K.A. The American Alpine Club's handbook of American mountaineering. Boston, 1942.
Geog 5559.42	Herrmann, Ernst. Mit dem Fieseler-Storch ins Nordpolarmeer. Berlin, 1942.
Geog 5516.19.15	Hofrenning, Bernt Monson. Navigatio septentrionalis. Fergus Falls, Minn., 1942.
An 3409.16.10	Kuhne, F. The finger print instructor. 3. ed. N.Y., 1942.
An 359.42.15	Laidler, H.W. The role of the races in our future civilization. N.Y., 1942.
Geog 759.42F	Lawrence, C.H. New world horizions; geography for the air age. N.Y., 1942.
Geog 4209.42.20	Maillart, E.K. Gypsy afloat. London, 1942.
An 359.42.10A	Montagu, Ashley. Man's most dangerous myth: the fallacy of race. N.Y., 1942.
Geog 4304.95	Muenzer, Hieronymus. Voyage aux Pays-Bas, 1495. Bruxelles, 1942.
Geog 5630.19	Noerlund, P. De gambe nordbobygder ved verdens ende. 3. opl. København, 1942.
Geog 4139.42	Penrose, Boies. Urbane travelers, 1591-1635. Philadelphia, 1942.
An 132.3A	Quiggin, A. Hingston (Mrs.). Hadden, the head hunter. Cambridge, 1942.
Geog 4709.42	Santschi, R.J. Treasure trails. Glen Ellyn, Ill., 1942.
An 2600.5	Singh, Joseph. Wolf children and feral man. N.Y., 1942.
Geog 5660.37	Smedal, G. Grönland und der Norden. Oslo? 1942?
Geog 4509.41.17	Smythe, F.S. The mountain vision. London, 1942.
Geog 5660.26A	Stefansson, V. Greenland. 1. ed. Garden City, 1942.
Geog 5160.11.12	Stefansson, V. Ultima Thule. Reykjavik, 1942.
Geog 4679.42.10	Trumbull, Robert. The raft. N.Y., 1942.

1942 - cont.

Geog 212.30	Urteaga, H.H. Memoria del presidente de la Sociedad Geográfica de Lima. Lima, 1942.
Geog 4209.42.25	Vejarano, J.R. Rutas del mundo. Bogota, 1942.
An 359.42.5	Webster, Hutton. Taboo, a sociological study. Stanford, 1942.

1943

An 137.4	Association of Polish University Professors and Lecturers in Great Britain. Professor Bronislaw Malinowski. London, 1943.
Geog 1580.5	Baedeker, publishers. Das General gouvernement. Leipzig, 1943.
Geog 1531.6	Baedeker, publishers. Tirol, Vorarlberg, westliche Salzburg, Hochkärnten. 41. Aufl. Leipzig, 1943.
Geog 1531.15	Baedeker, publishers. Wien und Niederdonau. Leipzig, 1943.
Geog 4209.41.6	Baerlein, Henry. Travels without a passport. London, 1943.
Geog 3580.20	Ballesteros Gaibrois, Manuel. España en los mares. 2. ed. Madrid, 1943.
Geog 3235.58	Beans, G.H. A collection of maps. Jenkintown, Pa., 1943.
An 359.43.25	Benedict, Ruth (Fulton). Race; science and politics. N.Y., 1943.
Geog 5328.15	Bogen om Knud. Skrevet af hans venner. Kjøbenhavn, 1943.
Geog 500.66	Bolles, E.C. The literature of sea travel since the introduction of steam, 1830-1930. Philadelphia, 1943.
An 359.43.10	Claessens, A. Race prejudice. N.Y., 1943.
Geog 4209.43.5	Collins, Grace. Chin takes and talks, an absurd travel diary. Chicago, 1943.
Geog 4145.17	Connolly, J.B. Master mariner...Amasa Delano. Garden City, 1943.
An 3409.43	Cummins, H. Fingerprints, palms and soles. Philadelphia, 1943.
Geog 6009.02.12	Doorly, G.S. The songs of the "Morning". Melbourne, 1943.
An 359.43	Dover, C. Hell in the sunshine. London, 1943.
Geog 5209.43	Einarsson, S. Suðr um höf. Reykjavik, 1943.
Geog 819.43	Engelhardt, N.L. Toward new frontiers of our global world. N.Y., 1943.
Geog 4181.88	Foster, William. The voyage of Sir Henry Middleton to the Moluccas. London, 1943.
An 3309.43	Goldstein, M.S. Demographic and bodily changes in descendants of Mexican immigrants. Austin, 1943.
An 359.44	Harvard University. Department of Psychology. ABC's of scapegoating. Chicago, 1943.
Geog 759.43	Howe, E.L. Air world. Denver, 1943.
Geog 4308.07	Irving, Peter. Peter Irving's journals. N.Y., 1943.
An 2109.43.5	Kelsen, Hans. Society and nature; a sociological inquiry. Chicago, 1943.
Geog 579.43	Kosack, H.P. Wörterverzeichnis für russische Karten. Berlin, 1943.
Geog 4512.535	Lunn, A. Mountain jubilee. London, 1943.
Geog 5399.01	Makarov, S.O. S.O. Makarov v zavosvanie Ariuku. Moskva, 1943.
Geog 5559.38	Manning, E.W. Igloo for the night. London, 1943.
Geog 4209.43.10	Medern, W.E.A.A. Blick in die weite Welt. Berlin, 1943.
Geog 4219.43	Metzelaar, A.C. Europa ahoy! De geschiedenis van een zeilschip. Amsterdam, 1943.
Geog 3019.43	Mexico (City). Biblioteca Benjamin Franklin. La era de las exploraciones. México, 1943.
Geog 5300.25	Muratov. M.V. Pervye razvedchiki velikogo puti. Moskva, 1943.
An 359.43.20	Murphy, J.D. Lamps of anthropology. Manchester, Eng., 1943.
An 2109.43	Murray, R.W. Man's unknown ancestors; the story of prehistoric man. Milwaukee, 1943.
An 359.43.30	Nicholson, James E. Anthropos; or, The problem of man. London, 1943.
Geog 3570.20	Peres, Damião. Historia dos descobrimentos portugueses. pt.1-15. Lisboa, 1943-45.
Geog 3570.27	Pina Manique, Luiz da. Subsidios para a história de cartografia portuguesa. Lisboa, 1943.
An 359.43.5	Posnansky, A. Qué es raza. La Paz, 1943.
Geog 3570.18.5	Prestage, Edgar. Descobridores portugueses. 2. ed. Lisboa, 1943.
An 3309.43.5	Riggs, F.B. Tall men have their problems too. Cambridge, 1943.
Geog 3070.45	Salishahev, K.A. Osnovy kartovesheniia. Izd. 2. Moskva, 1943.
Geog 4509.43.5	Shipton, Eric E. Upon that mountain. London, 1943.
Geog 5080.5.5	Smedal, Gustav. Souveränitätsfragen der Polargebiete. Oslo? 1943.
An 379.43	Sorre, Maximilian. Les fondements biologiques de la géographie. v.1 3. Paris, 1943-52. 4v.
Geog 4182.47	Spilbergen, J. van. De reis om de wereld. 's-Gravenhage, 1943. 2v.
Geog 5399.21.2	Stefansson, V. The friendly Arctic. N.Y., 1943.
Geog 4709.43	Taylor, James. Gold from the sea; the epic story of the "Niagara's" bullion. London, 1943.
Geog 4209.43	Thatcher, T.C. Travel letters. Yarmouth Port, 1943.
Geog 4229.00.5	Winter, James M. New York to Alaska...May to July, 1900. Middletown, N.Y, 1943.

1944

Geog 4209.44	Allan, D. Lightning strikes once. N.Y., 1944.
Geog 4559.45.10	Anson, P.F. Harbour Head. London, 1944.
Geog 603.6	Armao, Ermanno. Vincenzo Coronelli. Firenze, 1944.
Geog 4209.44.5	Armstrong, W. Saltwater tramp. London, 1944.
Geog 5328.20	Arnesen, Odd. Eskimoenes venn Knud Rasmussen. Oslo, 1944.
Geog 4209.44.10	Baerlein, Henry. The caravan rolls on. London, 1944.
Geog 4309.44.5	Baretti, Giuseppe. Oga Magoga. Milano, 1944.
Geog 579.44	Bargilliot, A. Vocabulaire pratique anglais-français. Paris, 1944.
Geog 4309.44	Cornette, Arthur H. Van Toledo tot Budapest. Antwerpen, 1944.
Geog 4181.89	Cortesão, Armando. Suma Oriental of Tomé Pires...book of Francisco Rodrigues. London, 1944. 2v.
Geog 3570.24	Costa Brochado. Historia de uma polemica. Lisboa, 1944.
Geog 5839.44	Draatrup, E. Grønlandsfaerd. København, 1944.
Geog 4329.44	Fateili, A. Dea senza volto. Milano, 1944.
Geog 4209.44.15	Gagnon, H.J. Blanc et noir. Montréal, 1944.
Geog 4672.20.10	Gomes de Brito, Bernardo. Quadros da História trágico-marítima. Lisboa, 1944.
Geog 5312.7.15	Goodhue, Cornelia. Journey into the fog; the story of Vitus Bering and the Bering Sea. Garden City, N.Y., 1944.

Chronological Listing

1944 - cont.

Geog 4559.44.5 — Green, Alfred. Jottings from a cruise. Seattle, 1944.
Geog 819.44.5F — Hankins, G.C. Our global world. N.Y., 1944.
Geog 4679.44.5 — Hardy, Alfred C. Wreck - S.O.S. London, 1944.
Geog 4169.36.3 — Hennig, R. Terrae incognitae. Leiden, 1944- 4v.
Geog 4309.44.10 — Hiett, Helen. No matter where. N.Y., 1944.
An 359.44.10 — Leiper, H.S. Blind spots. N.Y., 1944.
Geog 3520.15 — Lynam, Edward. British maps and map-makers. London, 1944.
An 187.5.25 — Malinowski, B. A scientific theory of culture and other essays. Chapel Hill, 1944.
Geog 4249.44A — Medrs, Eliot G. Pacific Ocean handbook. Stanford, Calif., 1944.
An 359.44.15 — Mendes Correia, A.A. Gérmen e cultura. Pôrto, 1944.
Geog 819.44 — Renner, G.T. Global geography. N.Y., 1944.
An 359.44.5A — Roback, A.A. A dictionary of the international slurs. Cambridge, Mass., 1944.
An 137.3.5 — Rose, H.J. Robert Ranulph Marett. London, 1944?
Geog 4559.44 — Snow, A.R. Log of a sea captain's daughter. Boston, 1944.
Geog 212.20 — Sociedad Geográfica de Lima. La reorganización de la Sociedad Geográfica de Lima. Lima, 1944.
An 2109.44 — Sommerfelt, A. Is there a fundamental mental difference between primitive man and the civilized European. n.p., 1944.
Geog 4217.72.10F — Sparrman, Anders. A voyage round the world with Captain James Cook. London, 1944.
Geog 5160.11.15 — Stefansson, V. Arctic manual. N.Y., 1944.
Geog 5519.40.5 — Tranter, G.J. Plowing the Arctic. London, 1944.
Geog 3070.42 — Wroth, Lawrence C. The early cartography of the Pacific. N.Y., 1944.

1945

Geog 4209.45.10 — Allan, Doug. Gamblers with fate. 1. ed. N.Y., 1945.
An 359.45.15 — Andrews, R.C. Meet your ancestors, a biography of primitive man. N.Y., 1945.
Geog 4145.89 — Arteche, J. de. Urdaneta. Madrid, 1945.
Geog 4169.45 — Bailey, Leslie. Travellers' tales, a series of BBC programmes broadcast throughout the world. London, 1945.
An 359.45.20 — Boas, Franz. Race and democratic society. N.Y., 1945.
Geog 3249.28.3 — Bullon y Fernández, Eloy. Miguel Servet y la geographia del renacimiento. 3. ed. Madrid, 1945.
Geog 4209.45 — Caldwell, E.N. Satin skirts of commerce. N.Y., 1945.
Geog 4559.45.5 — Carter, G.G. Looming lights. London, 1945.
Geog 5919.45 — Cordovez Madariga, Enrique. La Antártida sudamericana. Santiago, 1945.
Geog 4556.94 — Coxere, Edward. Adventures by sea of Edward Coxere. Oxford, 1945.
Geog 5150.5 — Dutilly, A. Bibliographies on the Arctic. Washington, 1945.
Geog 603.5.8 — Emmanuel, Marthe. J.B. Charcot. Paris, 1945.
Geog 3580.5.5 — Fernandes de Navaretti, Martin. Coleccion de los viajes y descubrimientos. Buenos Aires, 1945-46. 5v.
Geog 4209.45.15 — Galimir, Mosco. Half a century of world travel; impressions and reflections. N.Y., 1945.
Geog 3570.22 — Goncalões Niana, M. As viagens terrestres dos Portugueses. Porto, 1945.
An 3309.45 — Hoaton, E.A. A survey in seating. Cambridge, 1945.
An 359.45.10A — Hooton, E.A. "Young man, you are normal". N.Y., 1945.
Geog 4303.99.20A — Klimas, Petras. Ghillebert de Lannoy in medieval Lithuania. N.Y., 1945.
Geog 4559.45 — Klitgaard, Kaj. Oil and deep water. Chapel Hill, 1945.
Geog 4182.49 — LeMaire, Jacob. De ontdekkingsreis van Jacob le Maire. 's-Gravenhage, 1945. 2v.
An 359.45 — Linton, R. The science of man in the world crisis. N.Y., 1945.
Geog 5855.13 — Lorm, A.J. Kunstzin der Eskimos. 's-Gravenhage, 1945.
Geog 500.80 — Migliorini, Elio. Giuda bibliografica allo studio della geografia. Napoli, 1945.
An 359.45.25 — Montagu, Ashley. An introduction to physical anthropology. Springfield, 1945.
An 359.42.12A — Montagu, Ashley. Man's most dangerous myth. 2. ed. N.Y., 1945.
Geog 5324.2.40 — Nansen, Fridtjof. Nansens røst. Oslo, 1945. 3v.
Geog 3140.17 — Ninck, M. Die Entdeckung von Europa durch die Griechen. Basel, 1945.
Geog 5845.6 — Oldendow, Knud. Grønlands fremtid. København, 1945.
Geog 4209.45.5 — Orcutt, R. Merchant of alphabets. Garden City, N.Y., 1945.
Geog 5855.15 — Ostermann, Hother. Danske i Grønland i det 18. aarhundrede. København, 1945.
Geog 819.45 — Pickles, Thomas. The work of men. 1st ed. London, 1945.
Geog 4419.45 — Ponder, S.E.G. A wanderer in khaki. London, 1945.
Geog 4209.45.20 — Retuerto, Marcial. Sur les traces de Magellan. Paris, 1945.
Geog 5324.2.25 — Risteheuber, René. La double aventure de Fridtjof Nansen. Montréal, 1945.
An 2109.45 — Simmons, L.W. The role of the aged in primitive society. New Haven, 1945.
Geog 5160.15 — Stefansson, Evelyn Schwartz Baird. Within the circle. N.Y., 1945.
Geog 5180.16 — Stefansson, V. The Arctic in fact and fable. N.Y., 1945.
Geog 665.42 — Ward, F.K. Modern exploration. London, 1945.

1946

Geog 759.45 — Almagia, R. Fondamenti di geografia generale. v.1-2. Roma, 1946-48.
An 379.46.10 — Amorim Girão, Aristides de. Geografia humana. Porto, 1946.
Geog 4269.46 — Antonovych, Roman. Burlats'kym shliakom. Vynnypeg, 1946.
Geog 5311.3.5 — Arnesen, Odd. Roald Amundsen som hon var. 4. Oppl. Oslo, 1946.
Geog 4207.76.20 — Augur, Helen. Passage to glory; John Ledyard's America. 1. ed. Garden City, 1946.
Geog 4209.46 — Baerlein, Henry. Leaves in the wind. London, 1946.
Geog 4209.46.2 — Baerlein, Henry. So many roads. London, 1946?
Geog 3570.26 — Bandeira Ferreira, F. As viagens de descobrimento de iniciativa particular no tempo. Lisboa, 1946.
An 359.46.5 — Baruch, Dorothy W. Glasshouse of prejudice. N.Y., 1946.
Geog 3510.11 — Belen, Willem Julius. Nederlands voorhoede. 2. druk. Amsterdam, 1946.
Geog 3590.15 — Berg, L.S. Ocherki po istorii russk. geogr. otkrytii. Moskva, 1946.
An 2109.46.5 — Boas, Franz. The mind of primitive man. N.Y., 1946.
Geog 3570.25 — Castro Soromenho. A maravilhosa viagem dos exploradores portugueses. v.1-12. Lisboa, 1946.
Geog 4559.46 — Chaveriat, R. Ceux de la voile. Paris, 1946.

1946 - cont.

An 359.46 — Clark, Grahame. From savagery to civilization. London, 1946.
Htn Geog 4558.40.50* — Dana, Richard Henry. Two years before the mast. London, 1946?
Htn Geog 4558.40.52* — Dana, Richard Henry. Two years before the mast. Sydney, 1946.
Geog 3019.46 — Darby, Henry C. The theory and practice of geography. Liverpool, 1946.
Geog 5517.46.15 — Eavenson, Howard N. Two early works on Arctic exploration. Pittsburgh, 1946.
An 2209.46 — Elkin, A.P. Aboriginal men of high degree. Sydney, 1946.
Geog 4209.46.15 — Etherton, P.T. All over the world. London, 1946.
Geog 607.5 — Fraerman, R.I. Zhizn' i prikl. K.L. Golovnina. Moskva, 1946.
Geog 5639.46.5 — Gad, Finn. Grønlands historie. København, 1946.
Geog 4209.46.10 — George, Albert J. The cap'n's wife. Syracuse, 1946.
An 2109.46 — Goldenweiser, Alexander A. Anthropology; au introduction to primitive culture. N.Y., 1946.
Geog 4709.32 — Gosse, Philip. The history of piracy. N.Y., 1946.
Geog 4179.10 — Hakluyt Society, London. A list of the publications of the Hakluyt Society. London, 1946.
Geog 4181.94 — Harff, Arnold. The pilgrimage of Arnold von Harff. London, 1946.
Geog 607.5.10 — Ivashchenko, M.M. Admiral Golovnin. Moskva, 1946.
Geog 4309.46 — Jaunes, Elly. Mánnishor därute. Stockholm, 1946.
An 2109.46.10 — Kelsen, Hans. Society and nature; a sociological inquiry. London, 1946.
Geog 3570.28 — Lima, Manuel C. Deux voyages portuguèses de découverte dans l'Atlantique occidental. Lisbonne, 1946.
Geog 5839.46.10 — Lindsay, M. Three got through. London, 1946.
Geog 4181.93 — Lynam, E. Richard Hakluyt and his successors. London, 1946.
Geog 612.3.5 — Maine, René. Lapérouse. Paris, 1946.
Geog 4139.46 — Majó Framis, R. Vida de los navegantes y conquistadores españoles del siglo XVI. 1. ed. Madrid, 1946.
Geog 5399.46 — Mineev, A.I. Ostrov Vrangelia. Moskva, 1946.
Geog 5324.2.50 — Øverås, Asbjdrn. Fridtjof Nansen. Stavanger, 1946.
Geog 4249.46 — Robson, R.W. Where the trade-winds blow. Sydney, 1946.
Geog 5839.46 — Rodahl, Karl. The ice capped island: Greenland. London, 1946.
Geog 5639.46 — Rosing, Christian. Ostgronlaenderne tunuamuit. København, 1946.
An 359.46.17A — Schmidt, Wilhelm. Rassen und Völker. 3. Aufl. Luzern, 1946-49. 3v.
Geog 4209.46.5 — Shaw, Frank H. White sails and spendthrift. London, 1946.
An 359.46.10A — Stewart, George R. Man: an autobiography. N.Y., 1946.
Geog 4169.45.5 — Stood, Frederick T. Modern travel. London, 1946.
An 379.46.5 — Taylor, G. Environment, race and migration. 2. ed. Chicago, 1946.
An 379.46 — Taylor, G. Our evolving civilization. Toronto, 1946.
Geog 4509.46 — Tilman, H.W. When man and mountains meet. Cambridge, Eng., 1946.
Geog 3580.25A — Vicens Vives, J. Rumbos oceanicos. Barcelona, 1946.
Geog 5399.32.5 — Vize, Vladimir I. Na "Sibiriakove" v "Litke"...1932 i 1934. Moskva, 1946.
Geog 759.46 — Ward, Francis K. About this earth. London, 1946.
Geog 4209.46.28 — Waugh, Evelyn. When the going was good. London, 1946.
Geog 4209.46.20 — Wever, Jan. Journal der merkwaardige reizen van Jan Wever. 's-Gravenhage, 1946.

1947

Geog 3240.22 — Ahmad, Nafia. Muslim contribution to geography. Lahore, 1947.
Geog 665.46 — Arden-Close, Charles. Geographical by-ways and some other geographical essays. London, 1947.
Geog 3019.47 — Beckman, Leif. Vår väg genom världen. Stockholm, 1947-51. 3v.
An 359.48.10 — Bendict, Ruth (Fulton). In Henry's backyard. N.Y., 1947.
Geog 3590.25 — Bodnarskii, M.S. Ocherkii po istorii russk. zemleved. Moskva, 1947.
Geog 4181.95 — Carre. The travels of the Abbé Carre in India and the Near East. London, 1947-48. 3v.
Geog 4129.47 — Carrington, D. The traveller's eye. N.Y., 1947.
Geog 4300.35 — Continental touring, 1947-1948 for the British motorist. Leeds, 1947.
Geog 4129.47.5 — Cook, H.K. Over the hills and far away. London, 1947.
Geog 5070.62.5 — Croft, Andrew. Polar exploration. 2. ed. London, 1947.
Geog 5209.47 — Crouse, Nellis M. The search for the North Pole. N.Y., 1947.
Geog 819.47 — Davis, D.H. The earth and man. N.Y., 1947.
Geog 4145.20 — Day, George. Dumont d'Urville. Paris, 1947.
Geog 4239.47 — Durand-Couppel de St. Trent, M.N.P. Wind aloft, wind alone. N.Y., 1947.
Geog 4419.47 — Goolid-Adams, R. Middle East journey. 1. ed. London, 1947.
Geog 4759.47 — Gourou, P. Les pays tropicaux. Paris, 1947.
Geog 4145.34.2 — Grum-Grzhimailo, A.G. Dela i dni G.E. Grum-Grzhimailo. Moskva, 1947.
An 379.47 — Hettner, Alfred. Allgemeine Geographie des Menschen. Stuttgart, 1947.
An 379.47.5 — Huntington, Ellsworth. Principles of human geography. 5. ed. N.Y., 1947.
Geog 819.47.5 — Kinkead, Eugene. Our own Baedeker, from the New Yorker. N.Y., 1947.
Geog 4169.47.5 — Ley, Charles D. Portuguese voyages, 1498-1663. London, 1947.
Geog 4218.03.10 — Lisianskii, I. Putesh. vokrug sveta na kor. ""Neva", 1803-1806. Moskva, 1947.
Geog 4679.47.5 — Meier, F. Hurricane warning. N.Y., 1947.
Geog 6009.12.5 — Murphy, R.C. Logbook for Grace. N.Y., 1947.
Geog 4329.47 — Newman, B. Middle eastern journey. London, 1947.
Geog 3590.5 — Nozikov, N. Russian voyages round the world. London, 1947.
Geog 3590.5.5 — Nozikov, N. Russkie krugosvetiye moreplavateli. Izd. 2. Moskva, 1947.
Geog 4145.71 — Obruchev, V.A. Grigorii Nikolaenin Potanin; zhizn' i deiatel'nost'. Moskva, 1947.
Geog 5919.38.2 — Ommaney, Francis D. South latitude. London, 1947.
Geog 5559.37.20 — Papanin, Ivan D. Life on an icefloe. London, 1947.
Geog 5312.2.5 — Putnam, George. Mariner of the North. N.Y., 1947.
Geog 4269.47A — Rahv, Philip. Discovery of Europe. Boston, 1947.
Geog 5845.2 — Rasmussen, Holger. Grønland og dets problemer. København, 1947.
Geog 5318.2.5 — Robinson, Bradley. Dark companion. N.Y., 1947.
Geog 5559.47 — Robinson, Bradley. Dark companion. 1. ed. N.Y., 1947.

Chronological Listing

1947 - cont.

Geog 3055.58	Rodriguez Prampolini, Ida. La Atlántida de Platón en los cronistas del siglo XVI. Mexico, 1947.
Geog 4509.47	Smythe, F.S. The mountain top. London, 1947.
An 379.47.10	Sorre, Maximilien. Les fondements de la géographie humaine. 2. éd. Paris, 1947-
Geog 4169.47	Stefansson, V. Great adventures and explorations. N.Y., 1947.
Geog 4309.47.5	Thompson, R.W. Devil at my heels. London, 1947.
Geog 4229.47	Vargas Ugarte, Ruben. Relaciones de viajes. Lima, 1947.
Geog 809.27.2	Vidal de La Blache, P. Géographie universelle. 2. éd. v.6, pt.1. Paris, 1947.
Geog 4209.47.5	Wethered, Herbert Newton. The four paths of pilgrimage. London, 1947.
Geog 4309.47A	Wilson, Edmund. Europe without Baedeker. 1st ed. Garden City, 1947.

1948

Geog 5925.55.6	Argentina. Comissión Nacional del Antártico. Soberanía argentina en la Antártida. 2. ed. Buenos Aires, 1948.
Geog 1527.50	Baedeker, publishers. Leipzig. Leipzig, 1948.
Geog 4309.48.5F	Bemelmans, Ludwig. The best of times, an account of Europe revisited. N.Y., 1948.
Geog 4311.20	Blairy, Jean. Le beau Danube gris. Paris, 1948.
Geog 5839.48	Boegvad, R. Orientering og faerdsee i terraenet. København, 1948.
Geog 5839.37	Boyd, Louise A. The coast of northeast Greenland. N.Y., 1948.
An 359.48.15	Coon, Carleton Stevens. A reader in general anthropology. N.Y., 1948.
An 359.48A	Cox, Oliver Cromwell. Caste, class and race. 1. ed. Garden City, N.Y., 1948.
Geog 3590.35.5	Efimov, Aleksei V. Iz istorii russkikh aksped. na Tikhom okeane. Moskva, 1948.
An 359.48.5	Embree, Edwin R. Peoples of the earth. N.Y., 1948.
Geog 612.4.4	Fradkin, Naum G. Puteshestviia I.I. Lepekhina, N. Ia. Ozeretskovskogo, V.F. Zueva. Moskva, 1948.
An 349.48	Gillin, John. The ways of men. N.Y., 1948.
Geog 3530.17	Julien, C.A. Les voyages de découverte et les premiers établissements. 1. éd. Paris, 1948.
Geog 5324.2.85	Knuth, Eigil. Fridtjof Nansen og Knud Rasmussen. Kjøbenhavn, 1948.
Geog 4509.48.5	Kurz, Marcel. Fremde Berge-Ferne Ziele. Bern, 1948.
Geog 4209.39.25	Larigaudie, Guy de. La route aux aventures. Paris, 1948.
Geog 4218.26.17	Luetke, Fedor P. Puteshestvne vokrug sveda na volunom shliupe "Seniavin" 1826-1829. 2. izd. Moskva, 1948.
Geog 4209.48	Mackenzie, C. All over the place. London, 1948.
Geog 5559.48	MacMillan, M. Green seas and white ice. N.Y., 1948.
Geog 5209.34.2	Mirsky, J. To the Arctic! N.Y., 1948.
Geog 4559.48	Öngör, Sami. Deniz cağrafyasj. Ankara, 1948.
Geog 6009.47	Orrego Vicuña, Eugenio. Terra australis. Santiago, 1948.
Geog 3590.20	Pallas, P.S. Bering's successors, 1745-1780. Seattle, 1948.
Geog 4307.65.15	Pennant, T. Tour on the continent. London, 1948.
Geog 3560.20	Perez Emlied, F. Los descubrimientos en el Atlántico y la rivalidad castellano-portuguesa hosta el tratado de Tordesillas. 1. ed. Sevilla, 1948.
Geog 4181.98A	Robertson, George. The discovery of Tahiti. London, 1948.
Geog 3019.48	Schwarz, G. Die Entwicklung der geographischen Wissenschaft seit dem 18. Jahrhunderts. Berlin, 1948.
Geog 5985.325.5	Seaver, George. The faith of Edward Wilson. London, 1948.
Geog 4219.00.10	Slocum, Joshua. Sailing alone around the world. London, 1948.
VGeog 4308.65.15	Smith, Amy G. Letters from Europe, 1865-1866. Washington, 1948.
Geog 4145.83	Stark, Freya. Perseus in the wind. London, 1948.
Geog 3129.48A	Thomson, J.O. History of ancient geography. Cambridge, 1948.
Geog 4182.51	Unger, W. De oudste reizen van de Zieuwen naar Oost-Indie. 's-Gravenhage, 1948.
Geog 5399.48	Vibe, Christian. Laugthen og nordpaa. København, 1948.
Geog 5919.48	Weetman, Charles. All about Antarctica. Melbourne, 1948.
Geog 4709.48	Wilkins, H.T. A modern treasure hunter. London, 1948.
Geog 4131.48	Winble, Ernest W. European recovery, 1948-1951, and the tourist industry. London, 1948.

1949

Geog 4509.49	Allain, Pierre. Alpinisme et compétition. Grenoble, 1949.
Geog 3129.49	Almagià, Roberto. Storia dell'esplorazione e della scienza geografica: l'eta greca. Roma, 1949.
Geog 1525.70	Baedeker, publishers. Schleswig-Holstein und Hamburg. Hamburg, 1949.
An 176.7	Bandelier, A.F. The scientist on the trail Berkeley, 1949.
Geog 500.105	Barras de Aragón, F. Los ultimos escritores de Indias. Madrid, 1949.
Geog 3060.18	Barreto, Costa. A lenda das Sete Cidades. Porto, 1949.
Geog 3590.15.2	Berg, L.S. Ocherki po istorii russk. geogr. otkrytii. 2. izd. Moskva, 1949. 2v.
An 2109.49	Breuil, H. Beyond the bonds of history. London, 1949.
Geog 3070.48	Brown, Lloyd A. The story of maps. 1st ed. Boston, 1949.
Geog 4502.5.11.5	Bühler, H. Alpine Bibliographie. Gesamtregister, 1931-1938. München, 1949.
Geog 4208.81.5	Cané, Miguel. En viaje. Buenos Aires, 1949.
An 359.49.15	Caro Baroja, Julio. Análisis de la cultura; etnología, historia, folklore. Barcelona, 1949.
Geog 4145.70	Collis, Maurice. The grand peregrination; being the life of...Fernão Mendes Pinto. London, 1949.
Geog 3055.39.20	Donnelly, Ignatius. Atlantis, the antediluvian world. 1st ed. N.Y., 1949.
Geog 616.5	Eavenson, H.N. Map maker and Indian traders...John Patten. Pittsburgh, 1949.
Geog 3590.35.3	Efimov, Aleksei V. Iz istorii velikikh russkikh geograf. otkrytii. Moskva, 1949.
Geog 613.5.5	Festschrift zum 70...Konrad Miller. Bremen, 1949.
An 170.203	Fortes, Meyer. Social structure. Oxford, 1949.
Geog 4419.49	Foster, William. The Red Sea and adjacent countries...seventeenth century. London, 1949.
Geog 4181.100	Foster, William. The Red Sea and adjacent countries of the close of the 17th century as described by Joseph Pitts. London, 1949.
Geog 607.10	Galorm'a, R.M. Sochineniia. Moskva, 1949.
Geog 3019.49A	Hanson, Earl P. New worlds emerging. 1. ed. N.Y., 1949.
Geog 3542.5	Hassinger, H. Österreichs Anteil an der Erforschung der Erde. Wien, 1949?

1949 - cont.

An 359.49.10	Herskovits, Melville Jean. Man and his works. N.Y., 1949.
An 2109.49.10	Hoernes, Moriz. Vorgeschichte Europas. 7. Aufl. Berlin, 1949.
An 170.230	Homenaje a Don Luis de Hoyos Sainz. Madrid, 1949-50. 2v.
Geog 759.49	James, P.E. A geography of man. Boston, 1949.
An 185.1.2	Kluckhohn, Clyde. Mirror for man. N.Y., 1949.
An 2109.49.5	Koppers, W. Der Urmensch und sein Weltbild. Wien, 1949.
Geog 3590.30	Lebedev, Dimitrii M. Geogr. v Rossii XVII v. ocherkii. Moskva, 1949.
An 379.49	Le Lannou, M. La géographie humaine. Paris, 1949.
Geog 4228.49	Lewis, Oscar. Sea routes to the gold fields. 1st ed. N.Y., 1949.
Geog 665.54	Livre jubilaire offert a Maurice Zimmermann. Lyon, 1949.
Geog 6009.46	Menster, William J. Strong men south. Milwaukee, 1949.
An 2109.49.15	Murdock, George P. Our primitive contemporaries. N.Y., 1949.
Geog 1807.29	Murray, John, publisher, London. Handbook for...India, Pakistan, Burma and Ceylon. London, 1949.
Geog 4329.49	Newman, B. Mediterranean background. London, 1949.
Geog 4219.23.5	O'Brien, Conor. Across three oceans. London, 1949.
Geog 5322.1	Perevalov, V. Lomonosov i Arktika. Moskva, 1949.
An 359.49.5	Price, A.G. White settlers and native peoples. Melbourne, 1949.
An 359.49.20	Scheler, Max. Die Stellung des Menschen in Kosmos. München, 1949.
Geog 4309.25.20	Shotwell, J.T. A Balkan mission. N.Y., 1949.
An 2339.49	Turney-High, H. Primitive war. Columbia, 1949.
Geog 4415.02	Varthema, L. de. Itinerário. Lisboa, 1949.
An 126.6	Viking Fund. Ruth Fulton Benedict; a memorial. N.Y., 1949.
Geog 4559.49	Villiers, A.J. The set of the sails. N.Y., 1949.
Geog 4208.98.2	Voss, J.C. The venturesome voyages of Captain Voss. 2. ed. London, 1949.
Geog 4239.49	Wightman, Frank Armstrong. The wind is free. N.Y., 1949.
Geog 4145.53	Willers, U. Xavier Marmier och Sverige. Stockholm, 1949.

195-

Geog 13.2.10	American Geographical Society of New York. The role of geography in the modern world. N.Y., 195-.
Geog 3570.23F	Cortesão, Jaime. Os descrobrimentos portugueses. Lisboa, 195-? 2v.
Geog 4219.37.12	Marine Historical Association, Mystic, Conn. The Joseph Conrad. Mystic, 195-.
Geog 190.5.5	Ricossa, J.A. Pueblos. Buenos Aires, 195-? 6v.

1950

Geog 3590.40	Adamov, Arkadii. Pervye russkie issledovateli Aliaski. Moskva, 1950.
Geog 1527.10	Baedeker, publishers. Munich and its environs. Hamburg, 1950.
Geog 4249.50.5	Barrett, C.L. The Pacific. Melbourne, 1950?
An 170.129	Beiträge zur Gesellungs- und Völkerwissenschaft. Berlin, 1950.
Geog 3590.15.10	Berg, L.S. Belikie russkie iute estvenniki. Moskva, 1950.
Geog 3570.9.5	Bersaude, J. The attacks against Portuguese history. Lisbon, 1950.
Geog 5639.50	Birket-Smith, K. Grønlands bogen. København, 1950. 2v.
Geog 5839.50	Bomholt, Julius. Grønland foran en ny epoke. København, 1950.
Geog 4182.52	Broecke, Pieter van den. Reizen naar West-Afrika van Pieter van den Broecke. 's-Gravenhage, 1950.
Geog 3070.48.5	Brown, Lloyd A. The story of maps. Boston, 1950.
Geog 5970.10	Buenos Aires. Universidad. Cronologia de los viajes a las regiones australes. Buenos Aires, 1950.
An 379.50F	Childe, Vere Gordon. Prehistoric migrations in Europe. Cambridge, 1950.
An 170.165A	Count, Carl W. This is race. N.Y., 1950.
Geog 4678.68	Cubbin, T. The wreck of the Serica. London, 1950.
Geog 3019.50	Dainelli, G. La conquista della terra. Torino, 1950.
An 359.50.10	Dempf, Alois. Theoretische Anthropologie. Bonn, 1950.
Geog 3590.35	Efimov, Aleksei V. Iz istorii russkii geogr. otkrytii v sever ledov i tiloke okeanov. Moskva, 1950.
An 2109.50.5	Eildermann, H. Die Urgesellschaft. Berlin, 1950.
Geog 612.4.1	Fradkin, Naum G. Akademik I.I. Lepekhin i ego puteshestviia po Rossii v 1768-1773 gg. Moskva, 1950.
Geog 4269.50	Gottmann, J. A geography of Europe. N.Y., 1950.
Geog 500.90	Harris, Chauncy Donnison. A union list of geographical serials. 2. ed. Chicago, 1950.
Geog 3570.30A	Hart, Henry H. Sea road to the Indies. N.Y., 1950.
Geog 5919.50	Henry, T. The white continent. N.Y., 1950.
Geog 4249.50.10A	Hesselberg, E. Kon-Tiki and I. N.Y., 1950.
Geog 4249.50	Heyerdahl, T. Kon-Tiki. Chicago, 1950.
Geog 110.1.16	International Geographical Congress. Comptes rendus du Congrès international de géographie. Lisbonne, 1950. 2v.
Geog 4323.35	Jacopo da Verona. Liber peregrinationis. Roma, 1950.
An 135.3	Keith, Arthur. An autobiography. London, 1950.
An 359.50.15	Korzybski, Alfred. Manhood of humanity. 2. ed. Lakeville, Conn., 1950.
Geog 579.50F	Lagoa, J.A.M.J. Glossário taponimico da antiga historiografia. Lisboa, 1950- 4v.
Geog 3590.30.5	Lebedev, Dimitri M. Geog. v Rossii petrovsk. vremeni. Moskva, 1950.
Geog 4679.49	Lederer, William J. The last cruise. N.Y., 1950.
Geog 4129.50	Michael, M. Traveller's quest. London, 1950.
Geog 4209.38.45	Mielche, H. Let's see if the world is round. London, 1950.
Geog 603.32	Mogey, J.M. The study of geography. London, 1950.
Geog 5939.50	Mountevans, E. The desolate Antarctic. London, 1950.
An 124.5	National Research Council. International directory of anthropologists. 3. ed. Washington, 1950.
An 2109.50	Numelin, R. The beginnings of diplomacy. London, 1950.
Geog 4306.56.10	Ogier, Charles. Ozjennik podrózy do Polski, 1635-1636. Gdańsk, 1950-1953. 2v.
An 359.50	Park, R.E. Race and culture. Glencoe, Ill., 1950.
Geog 5324.1.10	Ramsey, H. Nordenskiöld. Stockholm, 1950.
Geog 4209.50	Riddell, J. Flight of fancy. London, 1950.
Geog 5925.55.10	Sampay, Arturo E. La soberania argentina sobre la Antártida. La Plata, 1950.
An 359.50.5	Schwidetzky, Ilse. Grundzäge der Völkerbiologie. Stuttgart, 1950.
Geog 4145.81	Slocum, Victor. Captain Joshua Slocum. N.Y., 1950.

Chronological Listing

1950 - cont.

Geog 4145.84	Stark, Freya. Traveller's prelude. London, 1950.
Geog 4679.33	Tambs, E. The cruise of the Teddy. London, 1950.
Geog 4145.87	Tschiffely, A. Bohemia junction. London, 1950.
An 629.50	Yale University. Institute of Human Relations. Outline of cultural material. 3. ed. New Haven, 1950.
Geog 4145.71.500	Zariu, V.M. Puteshchestviia A.V. Potaninoi. Moskva, 1950.

1951

Geog 3570.21	Amzalah, Moses. La Méditerranée et les découvertes maritimes des Portugais. Lisbonne, 1951.
Geog 5939.51	Arsen'ev, V.A. V straie kitov i pingvinov. Moskva, 1951.
Geog 1510.62.2	Baedeker, publishers. London and its environs. 20. ed. Hamburg, 1951.
Geog 1524.72	Baedeker, publishers. Northern Bavaria. Hamburg, 1951.
Geog 3070.40.5	Bagrow, Leo. Die Geschichte der Kartographie. Berlin, 1951.
Geog 4209.51.10	Barreto, Paulo. A alma encantadora das ruas. Rio de Janeiro, 1951.
Geog 4239.50	Barton, H.D.E. Westward crossing. 1st American ed. N.Y., 1951.
Geog 4239.51.5	Buhler, Jean. Sur les routes de l'Atlantique. Lausanne, 1951.
Geog 4419.51	Calder, Ritchie. Med against the desert. London, 1951.
Geog 4521.26	Chapman, F.S. Memories of a mountaineer. London, 1951.
Geog 4308.17.6	Charleville, H.C.B. A journey to Florence in 1817. London, 1951.
An 2109.51	Childe, V.G. Social evolution. London, 1951.
An 2109.51.2	Childe, V.G. Social evolution. N.Y., 1951.
Geog 659.42.5	Cholley, André. La géographie. 2. éd. Paris, 1951.
Geog 5919.51	Christie, E.W.H. The Antarctic problem. London, 1951.
Geog 4249.51	Clune, Frank. Hands across the Pacific. Sydney, 1951.
An 2059.51	Coates, Adrian. Prelude to history. London, 1951.
Geog 5559.08.10	Cook, F.A. Return from the Pole. N.Y., 1951.
Geog 4110.37	Croft-Cooke, Rupert. Cities. London, 1951.
An 2109.51.10	Davison, D. The story of prehistoric civilizations. London, 1951.
Geog 3560.25	Dória, A.A. Los descubrimientos en el Atlántico y la rivalidad castellano-portuguesa. Braga, 1951.
Geog 3590.35.10	Efimov, Aleksei V. Otkrytiia russk. zemleprov. na ser. Vost. Asii. Moskva, 1951.
Geog 665.68	Geographische Studien. Wien, 1951.
Geog 4418.03.15	Gonzalez y Rodriguez de la Peña, Hipolito. Viajes Ali Bey el Abbassi. Madrid, 1951.
Geog 809.51	Gutersohn, H. Die Erde. v.1-13. Bern, 1951. 2v.
Geog 4209.51	Harris, F.R. Itchin' feet. Greenfield, Ohio, 1951.
Geog 4309.51.5	Hausenstein, Wilhelm. Abendländische Wanderungen. München, 1951.
Geog 4209.33	Howard, S. Thames to Tahiti. London, 1951.
Geog 4269.52.5F	Hürlimann, Martin. Europe in photographs. London, 1951.
An 379.51	Huntington, Ellsworth. Principles of human geography. 6. ed. N.Y., 1951.
Geog 4500.10	Istituto di Bibliografia Alpina, Biella. Opere antiche e rare sulla montagna. Biella, 1951.
Geog 665.62	Karper, Kurt. Landschaft und Lund. Remagen, 1951.
Geog 819.51	Krebs, N. Vergleichende Landerkunde. Stuttgart, 1951.
Geog 3590.30.10	Lebedev, Dimitrii M. Plavanie A.I. Chirukova. Moskva, 1951.
Geog 4218.26.10	Le Netrel, Edmond. Voyage of the Lerós. Los Angeles, 1951.
Geog 613.7	Mill, H.R. An autobiography. London, 1951.
Geog 4228.17.5	Montulé, E. de. Travels in America. 1st ed. Bloomington, 1951.
Geog 4219.51.10	Mulhauser, G.H.P. The cruise of the Amaryllis. London, 1951.
Geog 1707.8	Murray, John, publisher, London. Handbook for travellers on the continent. 8. ed. London, 1951.
Geog 4218.03.12	Nevskii, Vladimir V. Perv. putesh. rossii. vokrug sveta, 1803-1806. Moskva, 1951.
Geog 500.73	Ossa Varela, P. Catálogo alfabético de algunos géografos y exploradores. Bogotá, 1951.
Geog 4509.48.13	Pause, Walter. Mit gleichlichen Augen. 3. Aufl. München, 1951.
Geog 4145.72	Price, W. I cannot rest from travel. N.Y., 1951.
Geog 4209.51.5	Rosenberg, Halger. Jorden rundt med Halger Rosenberg. København, 1951.
Geog 6100.37	Scholes, A. Fourteen men. London, 1951.
Geog 5970.44.5	Sementovskii, Vladimir N. Russkie otkrytiia v Antarktike w 1819-1820-1821 godakh. Moskva, 1951.
Geog 4219.51	Skolov, A.V. Tri prugosvetiykh plavaniia M.P. Lazareva. Moskva, 1951.
Geog 4181.99	Spain. Archivo General de Indias, Seville. Further English voyages to Spanish America. London, 1951.
Geog 665.56	Stamp, L. London essays in geography. Cambridge, 1951.
Geog 4145.84.5	Stark, Freya. Beyond Euphrates. 1. ed. London, 1951.
Geog 665.58A	Taylor, Griffith. Geography in the twentieth century; a story of growth, fields, techniques, aims and trends. N.Y., 1951.
Geog 4329.51	Thompson, D. The phoenix in the desert. London, 1951.
Geog 4219.51.5	Tregaskis, R.W. Seven leagues to paradise. 1. ed. Garden City, 1951.
Geog 4125.75.1	Turler, H. The traveller, 1575. Gainesville, Fla., 1951.
Geog 5900.18	United States. Naval Photographic Interpretation Center. Antarctic bibliography. Washington, 1951.
Geog 4239.51	Uriburu, E.C. Seagoing Gaucho. N.Y., 1951.
Geog 4169.51	Van Thal, H. Victoria's subjects travelled. London, 1951.
An 2109.51.15	Weinert, H. Der geistige Aufstieg der Menschheit. 2. Aufl. Stuttgart, 1951.
Geog 5530.15.5	Woodward, F.J. Portrait of Jane. London, 1951.
Geog 3019.51	Wooldridge, S. William. The spirit and purpose of geography. London, 1951.
Geog 616.30	Wyder, Samuel. Die Schaffhauser Karten von Hauptmann Heinrich Peyer (1621-1690). Zürich, 1951.

1952

Geog 500.72	American School of Classic Studies at Athens. Voyages and travels in the Near East made during the nineteenth century. Princeton, N.J., 1952.
Geog 602.1.10	Amsterdam. Nederlansh Historisch Schlepvaart Museum. De Blaeu's beschrijvers van land. Amsterdam, 1952.
Geog 5369.52	Armstrong, T. The northern sea route; Soviet exploitation of the North East Passage. Cambridge, Eng., 1952.
Geog 4759.52	Bates, M. Where winter never comes. N.Y., 1952.
Geog 4209.52.5	Beonio-Brocchieri, V. Il Marcopolo. Milano, 1952.

1952 - cont.

Geog 5939.46	Bezemer, K.W.L. Der Kampf um den Südpol. 2. Aufl. Zürich, 1952.
Geog 5700.4.8	Bobé, Louis. Hans Egede. Copenhagen, 1952.
Geog 665.70	Bologna. Università. Istituto di Geografia. Studi geografici in onore di Antonio Renato Toniolo. Milano, 1952.
Geog 4182.54	Bontekae, W.Y. Journalen van de gedenckenaerdige reijsen. 's-Gravenhage, 1952.
An 379.52	Brunhes, Jean. Human geography. London, 1952.
An 359.52.15	Campbell, Byram. American race theorists, a critique of their thoughts and methods. Boston, 1952.
Geog 3070.56	Codazzi, Angela. Storia delle carte geografiche. Milano, 1952-
Geog 579.52.5F	The Columbia Lippincott gazetteer of the world. N.Y., 1952.
Geog 606.2	Coma Soley, V. Jaime Ferrer...y el descubimiento de America. 1. ed. Barcelona, 1952.
Geog 665.64	Dardel, E. L'homme et la terre. Paris, 1952.
Geog 3049.52	DeCamp, L.S. Lands beyond. N.Y., 1952.
Geog 665.88	Demangeon, Albert. Problèmes de géographie humaine. 4. éd. Paris, 1952.
Geog 5839.52	Denmark. Udenrigsministeriet. Greenland. Ringkjøbing, 1952?
An 379.52.10	Dickinson, Robert Eric. City, region and regionalism. London, 1952.
Geog 3070.47.5	Durand, D.B. The Vienna-Klosterneuberg map corpus of the 15th century. Leiden, 1952.
Geog 4679.23.6	Foster, Cecil. 1700 miles in open boats; the story of the loss of the S.S. Trivessa in the Indian Ocean. London, 1952.
Geog 4709.52	Franchi, A. Storia della pirateria nel mondo. Milano, 1952. 2v.
An 379.52.5F	Gutkind, E.A. Our world from the air. London, 1952.
Geog 3019.52	Herrmann, Paul. Sieben vorbei und acht verweht. 2. Aufl. Hamburg, 1952.
An 2109.52.10	Herskovits, M.J. Economic anthropology; a study in comparative economics. 2d ed. N.Y., 1952.
An 359.52.20	Herskovits, Melville Jean. Man and his works: the science of cultural anthropology. N.Y., 1952.
Geog 4269.52F	Hürlimann, Martin. Europe in photographs. London, 1952.
An 170.251.100	Koppers, W. Kultur und Sprache. Wien, 1952.
An 359.52.10A	Kroeber, Alfred L. The nature of culture. Chicago, 1952.
Geog 4019.52	Lamb, Geoffrey F. Modern action and adventure. London, 1952.
Geog 3530.18	Lauga, Henri. De la banquise à la jungle. Paris, 1952.
Geog 4309.52	Lins do Rêgo, José. Bota de sete léguas. Rio de Janeiro, 1952.
Geog 4672.16.15	Lockhart, John Gilbert. The "Mary Celeste" and other strange tales of the sea. London, 1952.
Geog 4218.98.10	Mikhailovskii, N.G. Iz dnevnikov krugosvetnogo puteshestviia. Moskva, 1952.
An 359.52.5	Morant, G.M. The significance of racial differences. Paris, 1952.
Geog 4509.52	Negri, Carlo. Alpinismo. 3. ed. Milano, 1952.
Geog 4672.18	Neider, Charles. Great shipwrecks and castaways; authentic accounts of adventures at sea. N.Y., 1952.
Geog 4429.52.5	Ommanney, F.D. The shoals of Capricorn. London, 1952.
Geog 3249.52A	Penrose, B. Travel and discovery in the Renaissance. Cambridge, 1952.
An 2109.52	Radcliffe-Brown, A.R. Structure and function in primitive society. London, 1952.
Geog 4559.52	Rasmussen, A.H. Sea fever. London, 1952.
An 2109.52.5	Sanders, I.T. Societies around the world. v.4. Lexington, 1952.
Geog 4145.98	Seaver, G. Francis Younghusband. London, 1952.
Geog 4145.81.50	Siegfried, André. Geographie poétique des cinq continents. Paris, 1952.
Geog 5919.52	Simpson, F.A. The Antarctic today. Wellington, 1952.
Geog 4659.52	Snow, Edgar R. Great gales and dire disasters. N.Y., 1952.
Geog 5324.2.55	Sponsel, H. Fridtjof Nansen. Nürnberg, 1952.
Geog 4279.52	Stoye, J.W. English travellers abroad. London, 1952.
Geog 4209.52.10	Szos Kies, Henryk J. Your world and mine. N.Y., 1952.
Geog 809.52	Teran, M. de. Imago mundi. Madrid, 1952. 2v.
An 359.52	UNESCO. The race concept. Paris, 1952.
Geog 559.52	United States. National Archives. Geographical exploration and topographic mapping by the United States government. Washington, 1952.
Geog 4239.52	Veedam, V. Sailing to freedom. N.Y., 1952.
Geog 4429.52A	Villiers, Alan J. Monsoon seas. N.Y., 1952.
Geog 819.52F	Visintin, L. Continenti e poesi. Novara, 1952.
Geog 4209.52	Warden, W.R. Vale enchanting. London, 1952.
Geog 13.4.20	Wright, J.K. Geography in the making; the American Geographical Society, 1851-1951. N.Y., 1952.
Geog 5300.40	Zavalti, Silvio. Pionieri italiani nelle regioni polari. Brescia, 1952.
Geog 5050.7	Zavatti, S. Saggio di bibliografia polare. 2. ed. Roma, 1952.
Geog 579.52	Zavatti, Silvio. Dizionario geografico. Catania, 1952.

1953

Geog 5399.12.10	Al'banov, V.A. Podvig shturmaka V.I. Al'banova. Moskva, 1953.
Geog 500.72.5	American School of Classical Studies at Athens. Voyages and travels in Greece. Princeton, N.J., 1953.
Geog 4239.29.13	Andersen, Lis. Lis sails the Atlantic. London, 1953.
Geog 4029.10	Anderton, Russ. Tic-polonga. 1. ed. Garden City, 1953.
Geog 1527.70	Baedeker, publishers. Hamburg und die Niederelbe. Hamburg, 1953.
Geog 1525.72	Baedeker, publishers. Köln und der Rheinland zwischen Köln und Mainz. Hamburg, 1953.
Geog 1524.73	Baedeker, publishers. Südbayern: Alpenvorland, Alpen, österreichische Gunzgebiete. 41. Aufl. Hamburg, 1953.
An 379.53	Baxter, William J. Today's revolution in weather. N.Y., 1953.
Geog 4219.36.15	Bernicot, Louis. The voyage of Anahita; single-handed round the world. London, 1953.
Geog 5639.50.5	Birket-Smith, K. Grenlandiia: sbornik statei. Moskva, 1953.
Geog 4329.53.10	Birot, Pierre. La Méditerranée et le Moyen-Orient. Paris, 1953. 2v.
Geog 622.1.5	Blunt, Wilfred. Pietro's pilgrimage. London, 1953.
Geog 3019.53	Bonse, Ewald. Entwicklung und Aufgabe der Geographie. Stuttgart, 1953.

Chronological Listing

1953 - cont.

Geog 4181.106	Boxer, C.R. South China in the sixteenth century. London, 1953.
Geog 4709.53.5	Cabal, Juan. Historias de pirates. 1. ed. Barcelona, 1953.
Geog 4145.15.2	Chapman, F.S. Living dangerously. London, 1953.
Geog 4145.15	Chapman, F.S. Living dangerously. N.Y., 1953.
Geog 4509.53.5	Chatellus, A. de. Alpiniste. Grenoble, 1953.
Geog 4509.53.15	Club Alpino Italiano. Alpinismo italiano nel mondo. Milano, 1953.
Geog 4329.53.15	Cordan, W. Das Mittelmeer. Düsseldorf, 1953.
Geog 4209.53.5	Davenport, P. The voyage of Waltzing Matilda. London, 1953.
Geog 3590.30.15	Divin, Vasilii A. Belikii russkii moreplavateli A.I. Chirukov. Moskva, 1953.
Geog 4429.53A	Doyle, A.C. Heaven has claws. N.Y., 1953.
Geog 4329.53.5F	Dunand, M. De l'Amanus au Sinai. Beyrouth, 1953.
An 359.53.5	L'évolution, l'homme, la race. Paris, 1953?
An 130.5	Field, Henry. The track of man. 1. ed. Garden City, 1953.
Geog 4145.28	Filchner, Wilhelm. Ein Forscherleben. 3. Aufl. Wiesbaden, 1953.
Geog 3019.53.10	Fochler-Hanke, Gustav. Introducción a la historia de la geografía. Tucumán, 1953.
Geog 4145.29	Fournier-Aubry, Fernand. Mon metier l'aventure. Paris, 1953.
Geog 612.4.2	Fradkin, Naum G. Akademik I.I. Lepekhin i ego puteshestviia po Rossii v 1768-1773 gg. 2. izd. Moskva, 1953.
Geog 5316.1.5A	Freuchen, P. Vagrant Viking. N.Y., 1953.
Geog 4559.53.10	Freuchen, Peter. Vagrant viking. N.Y., 1953.
Geog 3590.67	Gesellschaft für Deutsch-Sowjetische Freundschaft. Beiträge aus der sowjetische Kartographie. Berlin, 1953.
Geog 5399.53	Hansen, Leo. I Knuds sloedespor. København, 1953.
Geog 4559.53.5	Hayet, Armand. Us et coutumes à bord des long-courriers. Paris, 1953.
An 126.2.5	Herskovits, Melville Jean. Franz Boas. N.Y., 1953.
An 2109.53.15	Heuse, Georges A. La psychologie ethnique. Paris, 1953.
Geog 604.5	Hoff, Bert van. Jacob van Deventer. 's-Gravenhage, 1953.
Geog 4509.53.20	Jerschak, Federico. L'alpinismo a Cortina dai suoi primordi ai giorni nostri. Roma, 1953.
Geog 819.50	Jones, S.B. Geography and world affairs. Chicago, 1953.
An 5.15	Keesing, Felix. Culture change; an analysis and bibliography of anthropological sources to 1952. Stanford, Calif., 1953.
An 2109.53.10	Kern, Fritz. Der Beginn der Weltgeschichte. Bern, 1953.
An 379.53.5	Kimble, G.H.T. The way of the world. N.Y., 1953.
An 359.53	Kroeber, Alfred L. Anthropology today. Chicago, 1953.
An 2109.53.27	Kuehn, Herbert. Auf den Spuren des Eiszeitmenschen. 2. Aufl. Wiesbaden, 1953.
Geog 4329.53	Landery, C.F. Whistling for a wind. 1st American ed. N.Y., 1953.
An 2059.53	Leakey, Louis S. Adam's ancestors. 4. ed. London, 1953.
Geog 4309.53	Lockley, R.M. Travels with a tent in western Europe. London, 1953.
Geog 4218.03.13	Lupach, V.S. I.F. Kruzenshtern. Moskva, 1953.
Geog 3590.45	Lupach, V.S. Russkie moraplavateli. Moskva, 1953.
Geog 3070.50A	Lynam, E. The mapmaker's art. London, 1953.
Geog 4181.101	Mandeville, J. Mandeville's travels. London, 1953. 2v.
An 170.270	Mead, M. Primitive heritage. N.Y., 1953.
An 629.53	Mead, M. The study of culture at a distance. Chicago, 1953.
Geog 4329.53.20	Méditerranée orientale. Paris, 1953.
Geog 4209.53	Meiss-Teuffen, Hans. Wanderlust. N.Y., 1953.
Geog 4509.53.10	Michel, Aime. Montagnes héroïques. Tours, 1953.
Geog 4309.53.5	Middleton, E.E. The cruise of the Kate. London, 1953.
Geog 618.7	Mil'kov, Fedor N. P.I. Rychkov. Moskva, 1953.
Geog 4559.53	Mitchell, C. Beyond horizons. 1. ed. London, 1953.
Geog 5209.53	Mountevans, C.R.G.R.E. Arctic solitudes. N.Y., 1953.
Geog 614.7	Nałkowska, Z. Moj ojciec. Warszawa, 1953.
Geog 4218.03.14	Nevskii, Vladimir V. Vokrug sveta pod russkim flagom. Moskva, 1953.
Geog 4145.71.5	Obruchev, V.A. Puteshchestviia Potanina. Moskva, 1953.
Geog 5311.3.10	Partridge, B. Amundsen. London, 1953.
Geog 4509.41.7	Peacocke, T.A.H. Mountaineering. 3. ed. London, 1953.
Geog 4219.47	Petterson, H. Westward ho with the Albatross. 1. ed. N.Y., 1953.
Geog 4509.53	Petzoldt, P. On top of the world. N.Y., 1953.
Geog 4709.53.2	Pringle, P. Jolly Roger. London, 1953.
Geog 4709.53	Pringle, P. Jolly Roger. N.Y., 1953.
Geog 4131.53	Pudney, J. The Thomas Cook story. London, 1953.
An 2109.53.5	Radin, Paul. The world of primitive man. N.Y., 1953.
Geog 4182.55	Ratelband, K. Vijf dagregisters van het Kasteel São Jorge da Maria. 's-Gravenhage, 1953.
Geog 5160.16	Rodahl, Kåre. North. 1. ed. N.Y., 1953.
Geog 3055.65	Saint-Michel, Léonard. Aux sources de l'Atlantide. Bourges, 1953.
An 2109.53	Sanders, I.T. Societies around the world. N.Y., 1953. 2v.
Geog 815.35.10	Servetus, Michael. Michael Servetus, a translation of his geographical, medical, and astrological writings. Philadelphia, 1953.
Geog 659.53	Sestini, Aldo. Avviamento allo studio della geografia. 1. ed. Firenze, 1953.
Geog 5839.30.15	Soctt, J.M. Portrait of an ice cap. London, 1953.
Geog 659.53.5	Spain. Consejo Superior de Investigaciones Cientificas. Iniciación a la geografia local. Zaragoza, 1953.
Geog 3055.60	Spanuth, Jürgen. Das enträtselte Atlantis. Stuttgart, 1953.
Geog 4145.84.10	Stark, Freya. The coast of incense. 1. ed. London, 1953.
Geog 4181.103	Strachey, William. The historie of travel into Virginia Britania. London, 1953.
Geog 665.58.2	Taylor, Griffith. Geography in the twentieth century. 2nd ed. N.Y., 1953.
An 5.18	Thomas, William L. International directory of anthropological institutions. N.Y., 1953.
Geog 3019.38.10	Toschi, Umberto. Schemi e notizie di storia delle esplorazione geografiche. 5. ed. Firenze, 1953?
Geog 5235.15	Ushakov, G.A. Po nekhoshenoi zemle. Moskva, 1953.
Geog 579.53	Villalba y Rubio, F. Diccionario geografico universal. Madrid, 1953.
Geog 3055.62	Weyl, R. Atlantis enträtselt? Kiel, 1953.
Geog 5839.53	Williamson, G. Changing Greenland. London, 1953.
An 2109.53.20	Wormington, Hannah Marie. Origins, indigenous period. Mexico, 1953.

1954

Geog 12.2.3	Alpine journal. Index. v.39-58, 1927-52. London, 1954.
Geog 500.75A	Baranskii, N.N. Istoricheskii obzor uchebniko geografii, 1876-1934. Moskva, 1954.
Geog 4181.107	Beckingham, C.F. Some records of Ethiopia. London, 1954.
Geog 3590.15.15	Berg, L.S. Geschichte der russischer geographischer Entdeckungen. Leipzig, 1954.
Geog 4559.54.5	Berge, Victor. Danger is my life. London, 1954.
Geog 5312.6	Blosseville, B. Ernest Poret. Jules de Blosseville. Évreux, 1954.
Geog 579.54A	Bodnarskii, M.S. Slovar' geograficheskikh nazvanii. Moskva, 1954.
Geog 5312.4	Bolotnekov, N. Nikifor Belichev. 2. izd. Moskva, 1954.
An 379.54	Brown, H.S. The challenge of man's future. N.Y., 1954.
Geog 4509.54.10	Buhl, H. Achttausend, drüber und drunter. München, 1954.
Geog 3060.13.5	Cervé, W.S. Lemuria. 6th ed. San Jose, Calif., 1954.
Geog 5160.30	Collins, H.B. Arctic area. México, 1954.
Geog 3070.80	Conseil Scientifique pour l'Afrique au Sud du Sahara. Mapping and surveying of Africa south of the Sahara. London, 1954.
Geog 665.156	Davis, William Morris. Geographical essays. N.Y., 1954.
Geog 3055.72	DeCamp, L.S. Lost continents; the Atlantis theme in history, science and literature. 1st ed. N.Y., 1954.
Geog 5130.5	Domaas, K. Polarskipet frani. Oslo, 1954.
Geog 4559.54.20	Eaddy, P.A. Sails beneath the Southern Cross. Wellington, 1954.
Geog 4309.53.11	Edschmid, Kasimir. Europäisches Reisebuch. Hamburg, 1954.
Geog 4559.54.10	Edwards, H.W. Under four flags. London, 1954.
Geog 4145.23	Ely, Edward. The wanderings of Edward Ely. N.Y., 1954.
An 359.54.20	Flatz, Josef. Die Kultur. Linz, 1954.
Geog 4269.54.10	Fremantle, Anne (Jackson). Europe: a journey with pictures. N.Y., 1954.
Geog 5839.54	Freuchen, P. Ice floes and flaming water. N.Y., 1954.
Geog 5316.1.6	Freuchen, P. Vagrant Viking. London, 1954.
Geog 4269.54.5	George, Pierre. L'Europe centrale. 1. éd. v.1-2. Paris, 1954.
Geog 6009.49	Giaener, John. The white desert. London, 1954.
Geog 3019.52.2	Herrmann, Paul. Conquest by man. N.Y., 1954.
An 2109.54.10	Hocart, A.M. Social origins. London, 1954.
Geog 5324.2.69	Hoeyer, Liv Nansen. Eva og Fridtjof Nansen. Oslo, 1954.
An 2109.54	Howells, William. Back of history. 1. ed. Garden City, 1954.
Geog 500.100	Istituto Veneto di Scienza, Lettre ed Arte. Carte geografiche cinquecentesche a stampa della Biblioteca Marciana e della Biblioteca del museo correr di Venezia. Venezia, 1954.
Geog 3400.5A	James, Preston E. American geography; inventory and prospect. Syracuse, 1954.
Geog 4509.54	Jolis Felisart, A. La conquista de la montaña. Barcelona, 1954.
Geog 4209.54.5	Kivikoski, Olavi. Yksin yli Atlantin. Helsinki, 1954.
An 359.54	LaBarre, W. The human animal. Chicago, 1954.
Geog 4559.54.15	Learmont, J.S. Master in sail. 2d ed. London, 1954.
Geog 3019.54.5	Le Gentil, Georges. Découverte du monde. Paris, 1954.
Geog 604.4.5	Lewis, Oscar. George Davidson. California, 1954.
Geog 4659.54	Mackenzie, Margaret E. Shipwrecks; being the historical account of shipwrecks along the Victorian Coast, 1813-1914. Melbourne, 1954.
An 2059.54	Masiker, Reuben. The Australopithecinae. Cape Town, 1954.
An 359.54.5	Mason, Philip. An essay on racial tension. London, 1954.
Geog 6009.04	Maveroff, J.O. Por los mares antarticos. Buenos Aires, 1954.
Geog 4329.54	Méditerranée occidentale. Paris, 1954.
Geog 665.72	Mélanges géographiques offerts au Doyen ernest Benevent. Gap, 1954.
Geog 5323.2.5	Mikkelsen, Ejnar. Ukendt mand til ukendt land. Kjøbenhavn, 1954.
Geog 5321.1.17	Mirsky, J. Elisha Kent Kane and the seafaring frontier. 1. ed. Boston, 1954.
An 359.54.10	Missenard, André. A la recherche de l'homme. Paris, 1954.
Geog 4239.54	Mitchell, Carleton. Passage east. London, 1954.
Geog 3055.67	Muck, O.H. Atlantis-gefunden. Stuttgart, 1954.
Geog 4559.54	Neider, C. Man against nature. 1. ed. N.Y., 1954.
Geog 3249.54	Nowell, C.E. The great discoveries and the first colonial empires. Ithaca, 1954.
Geog 4209.54	Osborn, C. He lived for adventure. Washington, 1954.
Geog 4300.37	Pastene, J. Auto guide to Europe. N.Y., 1954.
Geog 5519.39	Poncins, G. The ghost voyage. 1. ed. Garden City, 1954.
Geog 4182.56	Quast, Mathijs H. De reis van Mathijs Hendriksz. 's-Gravenhage, 1954.
Geog 3055.66	Saurat, Denis. L'Atlantide et la règne des géants. Paris, 1954.
An 359.54.15	Schwidetzky, Ilse. Das Problem des Völkertodes. Stuttgart, 1954.
An 3309.54F	Sheldon, W.H. Atlas of men. 1st ed. N.Y., 1954.
Geog 6009.53	Solianik, A.N. Slava v Antarktike. Moskva, 1954.
Geog 4509.54.5	Ullman, J.R. The age of mountaineering. Philadelphia, 1954.
Geog 4300.47	Vanderbilt, C. European travel directory. N.Y., 1954.
Geog 3590.47	Zubov, N.N. Otechestvennye moraplavateli-issledovanii morel i okeanov. Moskva, 1954.

1955

Geog 4429.55	Auber, Jacques. Histoire de l'Ocean Indien. Tananarive, 1955.
Geog 4672.24	Baldwin, Hanson W. Sea fights and shipwrecks. Garden City, 1955.
Geog 4521.2	Banks, Mike. Commando climber. London, 1955.
Geog 4239.55	Barton, H.D.E. Atlantic adventurers. N.Y., 1955?
Geog 4145.39	Baum, Jiří. Holub a Mašukulumbové. Praha, 1955.
Geog 3590.15.40	Berg, R.L. Po ozeram Sibiri i Srednei Azii. Moskva, 1955.
Geog 5300.30	Burkhanov, V.F. Novye sovetskie issledovaniia v Arktike. Moskva, 1955.
Geog 3129.55	Codazzi, Angela. La geografia dei Greci e dei Romani. 2. ed. Milano, 1955.
Geog 4169.55	Cooper, G. Forbidden lands. N.Y., 1955.
Geog 4309.55.5	Cox, Alfred B. The other half. Adelaide, 1955?
Geog 4559.55.5	Cross, Beverley. Mars in Capricorn. 1. ed. Boston, 1955.
Geog 5645.5.10	Dúason, Jón. Die koloniale Stellung Grönlands. Göttingen, 1955.
Geog 4659.55A	Duffy, James E. Shipwrecks and empire; being an account of Portuguese maritime disasters in a century of decline. Cambridge, Mass., 1955.
Geog 4672.25	Dyall, Valentine. Famous sea tragedies. London, 1955.
Geog 4309.55	Gibbings, R. Trumpets from Montparnasse. London, 1955.
Geog 4329.55	Golding, Louis. Good-bye to Ithaca. London, 1955.

Chronological Listing

1955 - cont.

Geog 500.78	Hanover, S. Geographie. Hannover, 1955.
An 359.55	Hawkes, J. Man on earth. 1. American ed. N.Y., 1955.
Geog 5324.2.70	Hoeyer, Liv Nansen. Eva og Fridtjof Nansen. 5. Oppl. Oslo, 1955.
Geog 5324.2.75	Hoeyer, Liv Nansen. Nansen og verden. 2. Oppl. Oslo, 1955.
Geog 4131.55	Ignacio de Arrillaga, José. Sistema de política turistíca. Madrid, 1955.
Geog 4521.10	Jackson, John. More than mountains. London, 1955.
Geog 659.55	Jong, Guben. Het karakter van de geografische totaliteit. Groningen, 1955.
Geog 5939.55.5	Kearns, W.H. The silent continent. 1st ed. N.Y., 1955.
Geog 5160.20	Kimble, G.H.T. Geography of the northlands. N.Y., 1955.
Geog 5919.55	Kosack, Hans P. Die Antarktis; eine Länderkunde. Heidelberg, 1955.
Geog 500.98F	Kosack, Hans P. Die Kartographie, 1943-1954. Lahr, 1955.
Geog 5399.37.7	Laktionov, A.F. Severnyi polius. Moskva, 1955.
Geog 3070.90	Lauf, G.B. The origin and development of cartography. Johannesburg, 1955.
Geog 4269.55	Lehmann, H. Europa. 16. Aufl. Frankfurt, 1955.
Geog 3019.53.5	Leithaeuser, J.G. Worlds beyond the horizon. 1. American ed. N.Y., 1955.
Geog 4219.53	Le Toumelin, J.I. Kurun around the world. N.Y., 1955.
Geog 4182.57	Linschoten, Jan H. van. Itinerario. 's-Gravenhage, 1955-57. 3v.
Geog 4679.12.17A	Lord, Walter. A night to remember. N.Y., 1955.
Geog 6009.52	Marret, M. Seven men among the penguins. 1st American ed. N.Y., 1955.
Geog 4239.55.5	Mastrangelo, F. Atlantica dall'intuizione alla scoperta del nuovo mondo. Napoli, 1955.
Geog 5323.2	Mikkelsen, Ejnar. Mirage in the Arctic. London, 1955.
An 170.296	Miscelanea de estudios dedicados a Fernando Ortiz por sus discipulos. La Habana, 1955-57. 3v.
Geog 3590.51	Moscow. Gosudarstvennyi Biblioteka SSSR Imeni V.I. Lenin. Russkie geografii i puteshestvenniki. Moskva, 1955.
Geog 5939.55	Mountevans, E. The Antarctic challenged. London, 1955.
Geog 614.8	Nałkowska, Z. Moja ojciec. Wyd. 2. Warszawa, 1955.
Geog 5324.2.65	Nockher, L. Fridtjof Nansen. Stuttgart, 1955.
Geog 5838.72.5	Oldendow, K. Grønlaendervennen Hinrich Rink. København, 1955.
Geog 4659.55.5	O'May, Harry. Wrecks in Tasmanian waters, 1797-1950. Tasmania, 1955?
Geog 3019.55	Parias, L.H. Histoire universelle des explorations. Paris, 1955- 4v.
Geog 4515.108	Patani, Osvaldo. Soli con le montagne. Milano, 1955.
Geog 4209.55.15	Pineda Yañez, R. La isla y Colón. Buenos Aires, 1955.
Geog 5925.50	Pinochet de la Barra, Oscar. Chilean sovereignty in Antarctica. Santiago, 1955.
Geog 4209.55.10	Praz, Mario. Viaggi in Occidente. Firenze, 1955.
Geog 4709.55	Quarrell, Charles. Buried treasure. London, 1955.
Geog 4181.104	Quinn, D.B. The Roanoke voyages, 1584-1590. London, 1955. 2v.
Geog 3602.5	Richter, Søren. Great Norwegian expeditions. Oslo, 1955.
Geog 3055.70	Rousseau-Liessens, A. Les colonnes d'Hercule et l'Atlantide. Bruxelles, 1955.
Geog 3019.55.5	Samhaber, Ernst. Knaurs Geschichte der Entdeckungsreisen. München, 1955.
Geog 759.55	Schmieder, Oscar. Geografía del viejo mundo. México, 1955.
Geog 4500.15	Smith, J.A. Mountaineering. Cambridge, 1955.
Geog 4209.55.5	Starobin, J.R. Paris to Peking. 1. ed. N.Y., 1955.
Geog 4559.55.10	Stenhouse, J.R. Cracker hash. London, 1955.
Geog 4559.55	Thomas, L.J. Great true adventures. N.Y., 1955.
An 359.55.5	Troise, Emilio. Racismo. Buenos Aires, 1955.
An 2059.55	Vere, Francis. The Piltdown fantasy. London, 1955.
Geog 4219.55	Vocino, Michele. Marinai italiani e iberici sulle vie delle Indie. Roma, 1955.
Geog 5559.56.6	Vodopianov, Mikhail V. Na kryliakh v Arktiku. Moskva, 1955.
Geog 579.55A	Volostnova, M.B. Slovar' russkoi...geograficheskikh nazvanii. Moskva, 1955. 2v.
An 197.5	Ward, D.J.H. Letters to future ages. n.p., 1955.
Geog 579.55.5	Webster's geographical dictionary. Springfield, 1955.
An 2059.55.5	Weiner, Joseph S. The Piltdown forgery. London, 1955.
Geog 4249.54	Willis, William. The gods were kind. 1st ed. N.Y., 1955.
Geog 4209.55	Woodcock, A.W.W. To Hong Kong and return. Boston, 1955.
Geog 3590.49	Zabrodskaia, M.R. Russkie puteshestvenniki po Afrike. Moskva, 1955.

1956

Geog 4679.47	Ainslie, Kenneth. Pacific ordeal. 1. American ed. N.Y., 1956.
Geog 4679.56	Armstrong, W. Last voyage. London, 1956.
Geog 1527.12	Baedeker, publishers. Munich and its environs. 2. ed. Hamburg, 1956.
Geog 1525.55.16	Baedeker, publishers. Schwarzwald. 4. Aufl. Malente, 1956.
Geog 1525.75	Baedeker, publishers. Wiesbaden, Mainz, Rheingau, Rheinhessen. Malente, 1956.
An 2109.56	Barrière, Claude. Les civilisations tardenoisiennes en Europe occidentale. Bordeaux, 1956.
An 359.56	Bavinck, J.H. Het rassenvraagstuk. Kampen, 1956.
Geog 3590.15.20	Berg, L.S. Izbrannye trudy. Moskva, 1956. 5v.
Geog 5939.56.5	Burkhanov, V.F. K beregam Antarktidy. Moskva, 1956.
Geog 5300.35	Caswell, J.E. Arctic frontiers. 1. ed. Norman, 1956.
Geog 5399.56	Centkiewicz, Alina. Zanaevanie Arktiki. Moskva, 1956.
An 359.56.10	Chicago Urban League. Staff report on a scientist's report on race differences. Chicago, 1956.
Geog 3019.56.5	Colamonico, Carmelo. Compendio di storia della geografia e delle esplorazioni geografiche. Napoli, 1956.
Geog 4145.53.5	Davydov, I.V. V moriakh i stranstviakh. Moskva, 1956.
Geog 4415.05	Den Öchten Weg Auss zu Faren von Lissbona gen Kallakuth. From Lisbon to Calicut. Minneapolis, 1956.
Geog 4209.56.5	Elmberg, J.E.O. Islands of to-morrow. London, 1956.
Geog 4509.56.15	Evans, Charles. On climbing. London, 1956.
Geog 5050.9	Expéditions Polaires Francaises, 1948. Terre Adélie, Greenland, 1947-55; bibliographie. Grenoble, 1956.
Geog 179.2.5	France. Centre de Documentation Cartographique et Géographique. Centre de documentation cartographique et géographique. Paris, 1956?
Geog 5855.16	Freuchen, Peter. Fangstmaend i Melvillebugten. København, 1956.
Geog 5317.15	Gal'perin, I.M. Poliarnye zori; zapiski zhurnaliste. Moskva, 1956.
Geog 4509.56	Garobbio, A. Uomini del sesto grado. Milano, 1956.

1956 - cont.

Geog 4209.56.30	Gaudio, Attilio. A la recherche des îles ignorees. Paris, 1956.
An 2109.56.5	Gegeshidze, M.K. Gruzinskii narodnyi transport. Tbilisi, 1956.
Geog 665.73	Geograficheskoe Obshchestvo SSSR. Essais de géegraphie. Moscou, 1956.
Geog 4248.82	Gilbay, Bernard. A voyage of pleasure. Cambridge, 1956.
An 359.56.20	Gjessing, Gutorm. Socio-culture. v.1,4. Oslo, 1956- 2v.
Geog 4182.59	Goens, Rijklof van. De vijf gezantschapsreizen van Rijklof van Goens naar Hethof van Mataram. 's-Gravenhage, 1956.
Geog 4209.56.10	Gouy, Robert. Terres de l'amitié. Neuchâtel, 1956.
Geog 5160.25	Grigor'ev, A.A. Subarktika. 2. Izd. Moskva, 1956.
Geog 579.56	Grigson, G. Places. N.Y., 1956?
Geog 3070.95	Guarnieri, Gino. Geografia e cartografia nautica nella loro evoluzione storica e scientifica. Genova, 1956.
Geog 5070.11.5	Hassert, Kurt. Die Polarforschung. München, 1956.
Geog 4309.56	Haya de la Torre. Mensaje de la Europa nordica. Buenos Aires, 1956.
Geog 3019.56	Herrmann, Paul. Zeigt mir Adams Testament. Hamburg, 1956.
Geog 4509.56.10	Herzog, Maurice. La montagne. Paris, 1956.
Geog 4332.19	Hodgkinson, H. The Adriatic Sea. N.Y., 1956.
Geog 110.1.18	International Geographical Congress, 18th, Rio de Janeiro. Abstracts of papers. Rio de Janeiro, 1956.
Geog 5369.56	Istoriia otkrytiia i osvoeniia...morskogo poeti. v.1-3. Moskva, 1956- 4v.
Geog 4559.56.10	Jones, William Herbert Sidney. The Cape Horn breed. N.Y., 1956.
Geog 4300.40	Joseph, Richard. Guide to Europe and the Mediterranean. 1st ed. Garden City, N.Y., 1956.
An 2109.56.10	Kuehn, Herbert. Das Erwachen der Menschheit. Frankfurt, 1956.
An 359.56.5	LaFarge, John. The Catholic viewpoint on race relations. 1. ed. Garden City, N.Y., 1956.
Geog 5530.80	Lamb, G.F. Franklin. London, 1956.
Geog 3590.55	Lebedev, Dmitrii M. Ocherki po istorii geografii v Rossii XV i XVI vekov. Moskva, 1956.
Geog 4679.12.18	Lord, Walter. A night to remember. London, 1956.
Geog 5855.17	Malaurie, Jean. The last kings of Thule. London, 1956.
An 170.210	Mexico (City). Universidad Nacional. Estudios antropologicos publicados en homenaje al Dr. Manuel Gamio. México, 1956.
Geog 5559.56	Morozov, S.T. U poslednikh parallelei. Moskva, 1956.
Geog 500.85	Moscow. Gosudarstvennyi Biblioteka SSSR imeni V.I. Lenin. Glazami sovetskikh liudel. Moskva, 1956.
Geog 3590.60	Murator, M.V. Navatrechu apasnostiam. Moskva, 1956.
Geog 4300.42	Nagel, publishers. Europe. Geneva, 1956.
Geog 4559.56	Newby, Eric. The last grain race. Boston, 1956.
Geog 4559.56.15	Noble, Arthur H. Three ships came sailing. London, 1956.
Geog 4679.12.20	O'Connor, Richard. Down to eternity. N.Y., 1956.
Geog 4300.46	Olson, H.S. Aboard and abroad. 3d ed. Philadelphia, 1956.
An 2109.52.2	Radcliffe-Brown, A.R. Structure and function in primitive society. London, 1956.
Geog 3070.85	Raisz, E.J. Mapping the world. N.Y., 1956.
Geog 4249.56	Raitt, Helen. Exploring the deep Pacific. 1st ed. N.Y., 1956.
Geog 4509.56.20	Rébuffat, Gaston. Mont Blanc to Everest. London, 1956.
Geog 4559.56.5	Rutzebeck, Hjalmar. Mad sea. N.Y., 1956.
Geog 4512.767	Salacrow, Armand. A pied, au-dessus des nuages. Paris, 1956.
Geog 4520.5	Ségogne, Henry de. Les alpinistes célébres. Paris, 1956.
Geog 622.2	Shostiu, N.A. M.P. Vronchenko. Moskva, 1956.
Geog 4672.26	Snow, Edward Rowe. The vengeful sea. N.Y., 1956.
Geog 6009.53.5	Solianik, A.N. Cruising in the Antarctic. Moscow, 1956.
Geog 4209.56.15	Svenson, Sven. Varlden ar så stor, så stor. Stockholm, 1956.
Geog 579.56.5	Swayne, James Colin. A concise glossary of geographical terms. London, 1956.
Geog 4145.81.5	Teller, Walter Magnes. The search for Captain Slocum. N.Y., 1956.
Geog 4559.56.20	Tunstall-Behrens, Hilary. Pamir. London, 1956.
Geog 4209.56.25	Uberti, Roberto degli. Soli attraverso gli oceani. 2. ed. Brescia, 1956.
Geog 4509.56.25	Underhill, Miriam. Give me the hills. London, 1956.
Geog 606.3	Uzbekistan. Tsentral'nyi Gosudarstvennyi Arkhiv. Otdel Dorevoliutsionnykh Fondov. A.P. Fedchenko. Tashkent, 1956.
Geog 4679.56.5	Villiers, A.J. Posted missing. N.Y., 1956.
Geog 5559.56.5	Vodopianov, Mikhail V. Wings over the Arctic. Moscow, 1956?
An 359.56.25	White, Lynn T. Frontiers of knowledge in the study of man. 1. ed. N.Y., 1956.
Geog 4209.56A	Wilson, Edmund. Red, black, blond, and olive. N.Y., 1956.
Geog 4219.56	Wollschläger, A. Grosse Weltreise mit A.E Johann (pseud.). 1. Aufl. Gutersloh, 1956.
Geog 3590.42	Zogosken, L.A. Puteshestviia i issledovaniu...v Russkoi Amerike v 1802-04 gg. Moskva, 1956.

1957

Geog 5559.57.10	Agranat, G.A. Zarubezhmyi Sever. Moskva, 1957.
An 359.57.20	Akademiia Nauk SSSR. Institut Etnografii. Ocherki obshchei etnografii. v.1-4. Moskva, 1957- 5v.
Geog 5839.52.5	Banks, Mike. High Arctic; the story of the British North Greenland Expedition. London, 1957.
Geog 600.10	Baranskii, N.N. Otechestvennye ekonomiko-geografy XVIII-XX vv. Moskva, 1957.
Geog 5399.57.25	Belov, M.I. Severnyi morskoi put. Leningrad, 1957.
Geog 5160.35	Calder, Ritchie. Men against the frozen north. N.Y., 1957.
Geog 4709.57.10	Carse, Robert. The age of piracy. N.Y., 1957.
Geog 4426.08	Castro Daire, A. de Ataide. Viagens de Reino para a India e da India para o Reino. Lisboa, 1957-58. 3v.
Geog 603.6.5	Catalogo dei globi antici conservati in Italia. pt.1-2. Firenze, 1957.
An 2109.39.3.6	Childe, Vere Gordon. The dawn of European civilization. 6. ed. London, 1957.
Geog 3055.75	Cordeau, Catherine L. Poséidones. Paris, 1957.
Geog 500.97	Coronelli, Marco V. Il catalogo degli autori; una biobibliografia geografica del '600. Firenze, 1957.
Geog 4219.55.5	Crove, Bill. Heaven, hell and salt water. London, 1957.
An 359.57.10	Drexler, John Paul. Die Front der Farbigen. München, 1957.
Geog 6009.54	Dufek, George John. Operation deepfreeze. N.Y., 1957.
An 179.1A	Eiseley, L.C. The immense journey. N.Y., 1957.
Geog 4305.93.5	Esprinchard, Jacques. Vie. Paris, 1957.

Chronological Listing

1957 - cont.

An 99.57	Estena-Fabregat, Claudio. La antropología contemporánea. Madrid, 1957. 2 pam.
Geog 659.57	Filchner, W. Route-mapping and position locating in unexplored regions. Basel, 1957.
Geog 759.36.4	Finch, V.C. Elements of geography. 4th ed. N.Y., 1957.
Geog 5985.258.10	Fisher, Margery. Shackleton. London, 1957.
Geog 759.57	Fiziko-geograficheskie raionirovanie Kitaia; sbornik statei. Moskva, 1957.
Geog 5209.57	Förster, Hans A. Der weisse Weg; Forscher erobern die Arktis. Leipzig, 1957.
Geog 4110.30	Ford, Norman D. Bargain paradises of the world. 4. ed. Greenlawn, N.Y., 1957.
Geog 4110.30.5	Ford, Norman D. How to travel without being rich. Greenlawn, N.Y., 1957.
Geog 500.88	Fossati Bellani, Luigi Vittorio. I libri di viaggio e le guide della raccolta Luigi Vittorio. Roma, 1957. 3v.
Geog 607.5.5	Fraerman, R.I. Zhizn' i neobyknovennye prikl. K.L. Golovnina. Moskva, 1957.
An 359.57	Frazier, E.F. Race and culture contacts in the modern world. 1. ed. N.Y., 1957.
Geog 4709.57	Freminville, René Marie de La Poix de. Tels etaient corsaires et flibustiers. Paris, 1957.
Geog 4559.57	Freuchen, Peter. Book of the seven seas. N.Y., 1957.
Geog 5399.57.20	Freuchen, Peter. Fra Thule til Rio. København, 1957.
Geog 3240.25	Garcia Franco. Le legua nautica en la edad media. Madrid, 1957.
Geog 819.57	Geografia y Atlas Universal. Geografia y atlas universal. Barcelona, 1957.
Geog 665.69	Geographische Gesellschaft in Wien. Festschrift zur Hundertjahrfeier der geographischen Gesellschaft in Wien, 1856-1956. Wien, 1957.
Geog 4659.57	Gibbs, James A. Shipwrecks of the Pacific Coast. Portland, Ore., 1957.
Geog 4209.57	Gosset, R.P. Retour du bout du monde. Paris, 1957.
An 359.57.15	Griaule, Marcel. Méthode de l'ethnographie. Paris, 1957.
Geog 3590.65	Gvozdetskii, Nikolai Andreevich. Sorok let issledovanii i otkrytii. Moskva, 1957.
Geog 665.40.5F	Hachette, firm, publishers, France. Les merveilles du monde. Paris, 1957.
Geog 4559.57.5	Houter, F. den. Neptuania, maritieme curiosa. Amsterdam, 1957.
Geog 4129.57	Hughes, Spike. The art of coarse travel. London, 1957.
Geog 5399.57.5	Iatsun, E.P. Na l'dine cherez polius. Moskva, 1957.
Geog 4209.57.10	Jara Peralta, José. La ciudad de Toledo. Madrid, 1957.
Geog 5939.57	Kemp, Norman. The conquest of the Antarctic. N.Y., 1957.
Geog 4678.58.5	Knowles, J.N. Crusoes of Pitcairn Island. Los Angeles, 1957.
Geog 665.80	Koegel, Ludwig. Geographische Plaudereien. Bonn, 1957.
An 2109.57	Kosven, M.O. Abriss der Geschichte und Kultur der Urgesellschaft. Berlin, 1957.
Geog 6009.14.25	Laseron, Charles F. South with Mawson. 2d ed. Sydney, 1957.
Geog 6100.5	Law, Phillip G. Anare: Australia's Antarctic outposts. Melbourne, 1957.
Geog 3590.55.5	Lebedev, Dmitrii M. Ocherki po istorii geografii v Rossii XVIII veka. Moskva, 1957.
Geog 5919.57.10	Lebedev, Vladimir L. Antarktika. Moskva, 1957.
An 359.57.5	Lebeuf, Jean Paul. Application de l'ethnologie à l'assistance sanitaire. Bruxelles, 1957.
Geog 4181.108	Letts, M.H.I. The travels of Leo of Rozmital through Germany. Cambridge, 1957.
Geog 500.83	L'Information Geographique. La géographie française au milieu du XX. siècle. Paris, 1957.
Geog 4145.14	Lopes, Francisco Fernandes. The brothers Corte Real. Lisboa, 1957.
Geog 5919.57.5	Machowski, J. Antarktyka. Wyd. 2. Warszawa, 1957.
Geog 3019.57.5	Magidovich, I.P. Ocherki po istorii geograficheskii otkrytii. Moskva, 1957.
Geog 5939.57.5	Markov, K.K. Puteshestvie v Antarktidu. Moskva, 1957.
An 170.312	Mélanges Pittard. Brive, 1957.
Geog 5639.57	Mikkelsen, Ejnar. Fra fribytler til embedsmand. København, 1957.
Geog 5559.09.110	Mikkelsen, Ejnar. Two against the ice. London, 1957.
An 359.57.30	Missenard, André. In search of man. N.Y., 1957.
An 187.9	Montagu, Ashley. Anthropology and human nature. Boston, 1957.
Geog 4709.57.5	Monti, Mario. I pirati. Milano, 1957.
Geog 4309.57	Nothomb, Pierre. Pélerinages européens. Paris, 1957.
Geog 4269.57	Ogilvie, A.G. Europe and its borderlands. Edinburgh, 1957.
Geog 4239.57.5	Outhwaite, L. The Atlantic. N.Y., 1957.
Geog 3129.58	Paassen, Christian van. The classical tradition of geography. Groningen, 1957.
Geog 3129.57	Paassen, Christian van. The classical tradition of geography. Groningen, 1957.
Geog 4145.11	Peres, Damião. Diogo Cão. Lisboa, 1957.
Geog 3594.5	Pertek, Jerzy. Polacy ner szlakach morskich świata. Gdańsk, 1957.
An 2109.57.5	Radin, Paul. Primitive man as philosopher. 2. ed. N.Y., 1957.
Geog 4209.57.5	Raitchewitch, M. Biographien und Autogramme beruhmter Staatsmanner und anderer Personlichkeiten des 20. Jahrhundert. Bielefeld, 1957.
Geog 3570.34	Rogers, F.M. Valentim Fernandes. Lisboa, 1957?
Geog 5399.57.10	Ruzov, L.V. Na sushe i na more v Arktike. Moskva, 1957.
An 379.57	Sears, Paul. The ecology of man. Eugene, 1957.
Geog 4209.56.20	Slessor, Tim. First overland. London, 1957.
Geog 665.82	Sorre, Maximilien. Rencontres de la géographie et de la sociologie. Paris, 1957.
Geog 5160.60	Stefansson, Evelyn Schwartz Baird. Here is the far north. N.Y., 1957.
Geog 5919.57	Sullivan, Walter. Quest for a continent. N.Y., 1957.
Geog 5399.57.15	Sushkina, N.N. Dva leta v Arktike. Moskva, 1957.
Geog 665.58.3	Taylor, Griffith. Geography in the twentieth century. 3rd ed. N.Y., 1957.
Geog 4329.57	Todd, Roberto. Viajando por Europa. Madrid, 1957.
An 359.56.15	UNESCO. The race question in modern science. Paris, 1957.
Geog 3550.2	Venice. Biblioteca Nazionale Marciana. Mostra dei navigatori veneti del quattrocento e del cinquecento. Venezia, 1957.
Geog 4239.57	Villiers, Alan John. Wild ocean. N.Y., 1957.
Geog 5559.57.5	Volovich, V.G. God na poliuse. Moskva, 1957.
Geog 4019.57	Wie sie entkamen; mit einer Einleitung von Kasimir Edschmid. Düsseldorf, 1957.
Geog 4509.57.5	Young, G.W. The influence of mountains upon the development of human intelligence. Glasgow, 1957.

1958

Geog 659.58	Ackerman, Ed. Geography as a fundamental research discipline. Chicago, 1958.
Geog 3019.58	Albertini, Renzo. Storia delle esplorazioni geografiche. Venezia, 1958.
Geog 5300.45	Armstrong, Terence. The Russians in the Arctic. London, 1958.
Geog 6009.55.20	Barber, Noël. The white desert. N.Y., 1958.
Geog 4309.58	Bell, Robert. By road to Turkey. London, 1958.
Geog 4329.58.10F	Beny, Roloff. Merveilles de la Méditerranée. Paris, 1958.
Geog 4329.58F	Beny, Roloff. The thrones of earth and heaven. London, 1958.
An 2109.58.10	Bergoonlour, F.M. La préhistoire et ses problèmes. Paris, 1958.
Geog 5919.58.5	Bertram, George Colin L. Antarctica today and tomorrow. Cambridge, 1958.
Geog 579.54.5	Bodnarskii, M.S. Slovar' geograficheskikh nazvanii. Moskva, 1958.
An 359.58.15	Boer, Wolfgang de. Das Problem des Menschen und die Kultur. Bonn, 1958.
Geog 622.5	Bonapace, Umberto. Luigi Visintin. Novara, 1958.
Geog 5110.5	Bronner, Finn E. The polar regions. Santa Barbara, 1958.
An 359.58.10	Campbell, Byram. Race and social revolution. N.Y., 1958.
Geog 4269.58	Chabot, Georges. L'Europe du Nord et du Nord-Ouest. Paris, 1958. 3v.
Geog 3070.56.5	Codazzi, Angela. Storia delle carte geografiche. Milano, 1958.
Geog 4515.14	Colò, Carlo. Attrezzature per soccorso alpino. Trento, 1958.
Geog 4239.58	Costa Brochado, José de. Descobrimento do Atlântico. Lisboa, 1958.
Geog 5070.57	Courtauld, A. From the ends of the earth; an anthology of polar writings. London, 1958.
Geog 4181.109	Crawford, O.G.S. Ethiopian itineraries circa 1400-1524. Cambridge, Eng., 1958.
Geog 5925.60	Diaz, Emilio. Relatos antarticos. Buenos Aires, 1958.
Geog 4269.58.15	Egli, Emil. Flugbild Europas. Zürich, 1958.
Geog 5235.20	Euller, John. Arctic world. London, 1958.
Geog 3580.5	Fernandes de Navaretti, Martin. Coleccion de los viajes y descubrimientos. 2. ed. v.1-3. Madrid, 1958. 5v.
Geog 5328.10	Freuchen, Peter. I sailed with Rasmussen. N.Y., 1958.
Geog 6009.55.10	Fuchs, Vivian. The crossing of Antarctica; the Commonwealth Antarctic Expedition, 1955-58. London, 1958.
Geog 5324.2.80	Gal'perin, I.M. On byl pervym. Moskva, 1958.
Geog 4145.68.5	Gnevusheva, E.I. Zaky tyi pute Shestvennik. Moskva, 1958.
An 2109.58.5	Gómez-Moreno, Manuel. Adam y la prehistoria. Madrid, 1958.
Geog 5100.10	Gould, Laurence McKinley. The polar regions in their relation to human affairs. N.Y., 1958.
Geog 5925.45	Gould, Lawrence M. Antarctica in world affairs. N.Y., 1958.
Geog 4759.53	Gourou, Pierre. The tropical world. 2. ed. London, 1958.
Geog 4129.58	Greenen, E. Reisen seit Anno dazumal. Hamburg, 1958.
Geog 4209.58	Hadley, Leila. Give me the world. N.Y., 1958.
Geog 5839.52.10	Hamilton, Richard A. Venture to the Arctic. Harmondsworth, 1958.
An 99.58	Hays, H.R. From ape to angel. 1. ed. N.Y., 1958.
Geog 3019.56.2	Herrmann, Paul. The great age of discovery. N.Y., 1958.
An 359.38.26	Herskovits, Melville Jean. Acculturation. Gloucester, 1958.
An 2109.58	Hibben, F.C. Prehistoric man in Europe. 1. ed. Norman, Okla., 1958.
Geog 4181.110A	Ibn Batuta. The travels of Ibn Battuta. Cambridge, 1958. 2v.
Geog 3070.115	Internationaler Kurz für Kartendruck. Fortschrittsberichte auf dem Gebiet des Kartendrucks. Hamburg, 1958.
Geog 5919.58.10	Knapp, Willem H.C. Antarctica; de geschiedenis van het geheimzinnige Zuidland. Haarlem, 1958.
Geog 4209.58.25	Knies, Donald. Walk the wide world. N.Y., 1958.
Geog 4269.58.5	Koeppen, Wolfgang. Nach Russland und anderswohin. Stuttgart, 1958.
Geog 4209.58.5	Korshunov, V. Khodili my pokhodami. Moskva, 1958.
Geog 4209.58.20	La Croix, Robert de. Mysteres des îles. Paris, 1958.
Geog 819.58.5	Larousse, firm, publishers. Geographie universelle Larousse. Paris, 1958. 3v.
Geog 3070.105	Leithäuser, J.G. Mappae Mundi. Berlin, 1958.
An 132.5	LeRoux, Robert. L'anthropologie comparée de Guillaume de Humboldt. Paris, 1958.
An 359.58.20	Lévi-Strauss, Claude. Anthropologie structurale. Paris, 1958-1973. 2v.
Geog 5919.58	Liverridge, Douglas. The last continent. London, 1958.
An 2109.58.15	Maluquer de Motes, Juan. La humanidad prehistórica. Barcelona, 1958.
Geog 4159.58	Les marins à la découverte. Paris, 1958. 12v.
An 359.58	Miroglio, Abel. La psychologie des peuples. Paris, 1958.
Geog 4215.19.10	Mitchell, Mairin. Elcano. London, 1958.
Geog 4509.58	Moravec, Fritz. Weisse Berge. Wien, 1958.
Geog 4145.68	Mueller, Martin. Julius von Payer. Stuttgart, 1958.
Geog 5509.58	Neatby, Leslie Hamilton. In quest of the North West Passage. London, 1958.
Geog 6100.77	New Zealand. Geographic Board. Provisional gazetteer of the Ross Dependency. Wellington, 1958.
Geog 4209.07.10	Nicholson, T.R. Adventurer's road. N.Y., 1958.
Geog 4145.61	Noailles, Loise de. Souvenirs de quatre horizons. Fribourg, 1958.
Geog 611.5	Novlianskaia, Mariia G. I.K. Kirilov i ego. Atlas Vserossisskoi imperii. Moskva, 1958.
Geog 4559.58.5	Noyce, Wilfrid. The springs of adventure. Cleveland, 1958.
Geog 4209.58.15	Olszewicz, B. Ze wspomnień podróżników. Warszawa, 1958.
Geog 5939.58.5	Pavlovskii, E.N. Antarktika. Moskva, 1958.
Geog 4129.58.5	Randall, C.B. International travel. Washington, 1958.
Geog 5399.58	Rasmussen, Knud. Den store slaederejse. København, 1958.
Geog 759.58	Rebagliato, F. Geografia universal. Barcelona, 1958.
Geog 510.20	Reden over geografie. v.1-2. Groningen, 1958.
Geog 4559.58	Rehm, Arnold. Das fröhliche Logbuch. Bremerhaven, 1958.
An 99.58.5	Rudolph, Wolfgang. Das Problem der kulturellen Werte in den Arbeiten der neueren amerikanischen Ethnologie. Berlin, 1958.
Geog 4145.80	Schindler, Fritz. Meine schönste Autoreise. Wien, 1958.
An 359.58.5	Silva, Mello. Estudos sôbre o negro. Rio de Janeiro, 1958.
Geog 4145.81.10	Slocum, Joshua. The voyages of Joshua Slocum. New Brunswick, 1958.
Geog 616.2	Spano, Benito. Gli atlanti corografici del cavaliere C.G. Pocelli. Bari, 1958.

134

Chronological Listing

1958 - cont.

Geog 5160.40 Stanwell-Fletcher, T.M. Clear lands and icy seas. N.Y., 1958.
Geog 5509.58.5 Stefansson, V. Northwest to fortune. N.Y., 1958.
Geog 4209.58.10 Toynbee, A.J. East to west. N.Y., 1958.
Geog 4131.58 Tusci, Leonida. Elementi di tecnica professionale turistica. Roma, 1958.
Geog 4219.60.5 Uittenbogaard, Leo. De wereld. Den Haag, 1958-60. 3v.
Geog 819.58F Unsere Erde. Heidelberg, 1958.
Geog 4329.58.5 Van Sinderen, Adrian. A voyage through the azure seas. N.Y., 1958.
Geog 5300.50 Vodop'ianov, M.V. Puti otvazhnykh. Moskva, 1958.
An 39.58 Vuorela, Taivo. Kansatieteen sanasto. Helsinki, 1958.
Geog 3520.8 Warmer, Oliver. English maritime writing. London, 1958.
Geog 5939.58.15 Zavatti, Silvio. L'esplorazione dell'Antartide. Torino, 1958.
Geog 4314.7 Zenkovich, V.P. Berega Chennogo i Azovskego morei. Moskva, 1958.

1959

Geog 612.3.2 Allen, Edward W. The vanishing Frenchman. Rutland, 1959.
Geog 5559.58 Anderson, William R. Nautilus 90 north. Cleveland, 1959.
Geog 4181.111 Andrews, Kenneth. English privateering voyages to the West Indies. Cambridge, 1959.
An 2059.59.15 Anthropological Society of Washington. Evolution and anthropology. Washington, 1959.
Geog 4249.58 Baker, DeVere. The raft Lehi IV. Long Beach, 1959.
Geog 600.11 Baranskii, N.N. Otechestvennye fiziko-geografy i puteshestvenniki. Moskva, 1959.
An 126.6.5A Benedict, Ruth (Fulton). An anthropologist at work. Boston, 1959.
An 359.59.45 Benedict, Ruth (Fulton). Race: science and politics. N.Y., 1959.
An 359.59.50 Benedict, Ruth (Fulton). Race and racism. London, 1959.
An 359.59.15 Bertram, Colin. Adam's brood. London, 1959.
An 359.59.10 Bibby, Cyril. Race, prejudice and education. London, 1959.
Geog 659.59.5 Birot, Pierre. Précis de géographie physique générale. Paris, 1959.
Geog 4249.58.5 Bisschop, Eric de. Tahiti-Nui. London, 1959.
Geog 607.15 Bonasera, F. Un Gobo Terrestre di Matteo Greuter conservato nella Biblioteca. Camarino? 1959.
Geog 4559.59.5 Bradford, G. Yonder is the sea. Barre, Mass., 1959.
Geog 665.84 Bulgarska Akademiia na Naukite, Sofia. Geografski Institut. Sbornik v chest na akademik Anastas Stoianov Beshkov. Sofiia, 1959.
An 359.59.25 Bunker, Robert. The first look at strangers. New Brunswick, N.J., 1959.
Geog 4208.45.2 Bustamante, J. Viaje al antiguo mundo. 2. ed. Lima, 1959.
NEDL Geog 3019.59.15 Cabal, Juan. Grandes exploradores, en la mar. Barcelona, 1959.
An 2109.59.5 Cantoni, Remo. Il pensiero dei primitivi. Milano, 1959.
Geog 600.15 Carcie, Giuseppe. Tre Fiorentini del Rinascimento. v.7. Roma, 1959?
Geog 4559.59 Carfi, Francesco. Geografia economica e sociale del mare. Livorno, 1959.
Geog 4521.3 Clark, Ronald. An eccentric in the Alps. London, 1959.
Geog 3019.59.10 Codazzi, Angelo. Storia della geografia. Milano, 1959.
Geog 3049.59 Correa-Calderon, E. Floria de la Atlantida y otras historias fabulosas. Madrid, 1959.
An 359.59.20 Cox, Oliver Cromwell. Caste, class, and race. N.Y., 1959.
An 2059.59.5 Dart, Raymond. Adventures with the missing link. N.Y., 1959.
Geog 5332.1 Davydov, Iurii V. Ferdinand Vrangel'. Moskva, 1959.
Geog 5939.59 Debenham, Frank. Antarctica. London, 1959.
Geog 4169.59 Dillon, Richard. Embarcadero. N.Y., 1959.
Geog 759.59 Doerr, Arthur H. Principles of geography. N.Y., 1959.
Geog 612.3.15 Dondo, M.M. La Perouse in Maui. Wailuku, 1959.
An 2109.59.10 Eggers, H.J. Einführung in die Vorgeschichte. München, 1959.
Geog 4269.59.5 Egli, Emil. Europe from the air. London, 1959.
An 359.59.40 Finlay, William G. Races in chaos. Johannesburg, 1959.
Geog 4309.59 Fortin Magaña, Romeo. Ajo otros cielos. San Salvador, 1959.
Geog 540.15 Frabetti, Pietro. La collezione della antiche carte geografiche, a cura di Pietro Frabetti, Il museo delle navi, a cura di Amedio Rizzi. Bologna, 1959.
Geog 3019.59 Fradkin, N.G. Roshidenie karty. Moskva, 1959.
Geog 5316.1.8 Freuchen, P. Min gronlandske ungdom. Kjøbenhavn, 1959.
Geog 4679.59.5 Gallagher, Thomas M. Fire at sea. N.Y., 1959.
An 359.59.5 Goldschmidt, W.R. Man's way. Cleveland, 1959.
Geog 4181.112A Gomes de Brito, Bernardo. The tragic history of the sea. Cambridge, Eng., 1959.
Geog 659.39.5A Hartshorne, Richard. Perspective on the nature of geography. Chicago, 1959.
An 359.59 Haselden, Kyle. The racial problem in Christian perspective. 1. ed. N.Y., 1959.
Geog 5900.10 Hayton, Robert D. National interests in Antarctica; an annotated bibliography. Washington, 1959-60.
Geog 4182.61 Hein, Pieter P. De Westafrikaanse reis van Piet Heyn. 's-Gravenhage, 1959.
Geog 3019.59.5 Herrmann, Paul. Traumen, Wagen und Vollbringen. Hamburg, 1959.
An 2059.59.10 Howells, William. Mankind in the making. Garden City, 1959.
Geog 5839.59 Ingstad, Helge Marcus. Landet under leidarstjernen. Oslo, 1959.
Geog 110.1.18.5 International Geographical Congress, 18th, Rio de Janeiro, 1956. Comptes rendus du XVIIIe congrès international de géographie. Rio de Janeiro, 1959. 4v.
Geog 665.81 International Geographical Union. Regional Conference in Japan, Tokyo and Nara, 1957. Proceedings of the IGU Regional Conference in Japan, August 28-September 3, 1957. Tokyo, 1959.
Geog 5235.22 Irvine, Tom A. The ice was all between. Toronto, 1959.
Geog 809.59.10F Istituto Geografico de Agostini. Il milione; enciclopedia di geografia. Novara, 1959-1965. 15v.
Geog 759.49.2A James, P.E. A geography of man. 2nd ed. Boston, 1959.
Geog 3590.84 Krempol'skii, Viktor F. Istoriia razvitiia kartoizdaniia v Rossii i v SSSR. Moskva, 1959.
An 2059.59 Lange, Hurt. Fremdling zwischen Tier und Gott. Gütersloh, 1959.
Geog 5985.258.15 Lansing, Alfred. Endurance; Shackleton's incredible voyage. N.Y., 1959.
Geog 4709.59.5 Leip, Hans. Bordbuch des Satans. München, 1959.

1959 - cont.

Geog 4521.25 Lowe, George. Because it is there. London, 1959.
Geog 5311.3.25 Malfatti, Alberto. Roald Amundsen. Roma, 1959.
Geog 4709.59 Maséa de Ros, Angeles. Historia general de la piratería. Barcelona, 1959.
Geog 613.1.18 Mercator, Gerardus. Correspondance mercatorienne. Anvers, 1959.
Geog 4678.32 Meredith, J.C. The tattooed man. 1. American ed. N.Y., 1959.
Geog 665.86 Miller, Ronald. Geographical essays in memory of Alan G. Ogilivie. London, 1959.
Geog 5970.15 Mitterling, Philip. America in the Antarctic to 1840. Urbana, 1959.
Geog 4269.59.10 Monkhouse, Francis. A regional geography of western Europe. London, 1959.
Geog 4679.59 Moscow, Alvin. Collision course: the Andrea Doria and the Stockholm. N.Y., 1959.
Geog 5839.59.5 Munch, Ebbe. Strejftog i Nord. København, 1959.
An 2059.59.20 Nougier, L.R. Geographie humaine prehistorique. Paris, 1959.
Geog 809.59 Obst, Erich. Lehrbuch der allgemeinen Geographie. v.1-4,6-8,10-11. Berlin, 1959. 9v.
Geog 579.59 Öngör, Sami. Coğrafya sözlüğü. Fasc.1-5. İstanbul, 1959.
An 2059.59.25 Overhage, Paul. Um des Erscheinungsbild der ersten Menschen. Basel, 1959.
Geog 5939.59.10 Pasetskii, V.M. Na samoi iuzhnoi zemle. Moskva, 1959.
Geog 3570.20.2 Peres, Damião. Historia dos descobrimentos portugueses. Lisboa, 1959.
Geog 4209.59 Pizzinelli, Corrado. Viaggio nel mondo. Bologna, 1959.
Geog 4309.31.30 Powell, Gertrude C. The quiet side of Europe. Los Angeles, 1959.
Geog 5939.59.15 Reboux, Michael. Demain l'Antarchtique. Paris, 1959.
Geog 3570.36 Renault-Roulier, Gilbert. The caravels of Christ. N.Y., 1959.
Geog 5070.65 Riabinin, A.K. Iz Arktiki v Antarktiku. Murmansk, 1959.
Geog 4145.79 Sack, John. Report from practically nowhere. N.Y., 1959.
Geog 3070.155 Salinari Emiliani, M. Nozioni di cartografia. Roma, 1959.
Geog 4515.401 Saysse-Tobiczyk, K. W skałach i lodach swiata. Warszawa, 1959- 2v.
Geog 4129.59 Schadendorf, Wulf. Zu Pferde, im Wagen, zu Fuss. München, 1959.
Geog 4131.59 Schweizerischer Fremdenverkehrsverband. Festschrift für Walter Hienzeker zum 60. Geburtstag. Bern, 1959.
Geog 659.59 Scotti, Pietro. Elementi di geografia. Genova, 1959.
Geog 4419.59 Serstevens, Albert. Les précurseurs de Marco Polo. Paris, 1959.
Geog 5970.16 Siple, Paul. 90 degrees south; the story of the American South Pole conquest. N.Y., 1959.
An 359.59.35 Škerlj, Božo. Antropologija i etnologija. Beograd, 1959.
Geog 5939.59.20 Smuul, Juhan. Ledovaia kniga. Moskva, 1959.
Geog 4169.59.5 Spagnol, Mario. Avventure e viaggi di mare. Milano, 1959.
Geog 3055.60.5 Spanuth, Jürgen. Und doch. Stuttgart, 1959.
Geog 5919.59 Steinitz, Hans. Der 7. Kontinent. Bern, 1959.
An 2109.59 Symposium on the Evolution, Chicago, 1957. The evolution of man's capacity for culture. Detroit, 1959.
Geog 4181.113 Taylor, Eva. The troublesome voyage of Captain Edward Fenton. Cambridge, Eng., 1959.
Geog 4209.59.10 Tey, José M. Hong Kong-Barcelona en el junco Rubia. Barcelona, 1959.
Geog 4145.78 Touring Club de France. De l'Himalaya aux Pyrénées. Paris? 1959.
Geog 4759.59 Troll, Carl. Die tropischen Gebirge. Bonn, 1959.
Geog 4209.59.5 Trolliet, Héli. Sur les routes du monde. Lausanne, 1959.
Geog 5839.48.5 Tschaen, Louis. Groenland 1948-1949-1950: astronomie, nivellement géodésique sur l'Inlandsis. Paris, 1959.
An 2109.59.20 Udy, Stanley H. Organization of work. New Haven, 1959.
Geog 4249.59 Van Sinderen, Adrian. The other half of the earth. N.Y., 1959. 2v.
An 2109.59.15 Varagnac, André. L'homme avant l'écriture. Paris, 1959.
Geog 4239.57.10 Villiers, Alan John. The new Mayflower. Leicester, 1959.
Geog 809.59.5 Vooys, Adriaan. Panorama der Wireld. Roermond, 1959. 3v.

196-

Geog 5939.59.22 Smuul, Juhan. Antarctica ahoy! The icebook. Moscow, 196-?

1960

Geog 3590.81 Akademiia Nauk SSSR. Sovetskaia geografiia. Moskva, 1960.
Geog 608.15 Alfred Hettner. 6.8.1859 Gedenkschrift zum 100. Geburtstag. Heidelberg, 1960.
Geog 1527.60 Baedeker, publishers. Köln und Umgebung. 2. Aufl. Freiburg, 1960.
Geog 1527.18 Baedeker, publishers. München und Umgebung. 4. Aufl. Freiburg, 1960.
Geog 4209.60.15 Bajsen-Moeller, Axel. Globetrotter og hovedjaeger. Vordingborg, 1960.
Geog 4209.60.50 Basso, Hamilton. A quota of seaweed. Garden City, 1960.
Geog 4219.60 Baty, Eben Neal. Citizen abroad. N.Y., 1960.
Geog 3019.60.5F Bettex, A.W. Welten der Entdecker. München, 1960.
An 2109.60.5 Birket-Smith, Kaj. Primitive man and his ways. London, 1960.
Geog 3070.135 Brown, Lloyd. Map making; the art that became a science. 1st ed. Boston, 1960.
Geog 4209.60.20 Burden, W.D. Look to the wilderness. Boston, 1960.
Geog 36.3.5 Canadian geographical journal. Regional index of articles, 1930-59. Ottawa, 1960.
Geog 3060.12.55 Churchward, J. The sacred symbols of Mu. London, 1960.
Geog 5925.89 Conference on Antarctica, Washington, D.C., 1959. Conference documents. Washington, 1960.
Geog 3570.34.5 Congresso Internacional de Historia dos Descobrimentos. Actas. Lisboa, 1960- 6v.
An 2109.60 Cornelius, Friedrich. Geistesgeschichte der Frühzeit. v.1-2. Leiden, 1960.
Geog 4239.58.5 Costa Brochado, José de. The discovery of the Atlantic. Lisboa, 1960.
Geog 5399.28.45 Cross, Wilbur. Ghost ship of the Pole. N.Y., 1960.
Geog 5209.60.5 Dainelli, G. La gara verso il Polo Nord. Torino, 1960.
An 170.321 Diamond, Stanley. Culture in history; essays in honor of Paul Radin. N.Y., 1960.
An 379.60 Dickinson, Robert Eric. Some problems of human geography. Leeds, 1960.
An 170.174 Dole, Gertrude Evelyn. Essays in the science of culture. N.Y., 1960.
Geog 4145.14.5 Fernandes, José dos Santos. Miguel Corte Real. Coimbra, 1960.

Chronological Listing

1960 - cont.

Geog 4131.60.5	Fuss, K. Geschichte der Reisebüros. Darmstadt, 1960.
An 170.211	Garn, Stanley Marion. Readings on race. Springfield, Ill., 1960.
Geog 6008.97.5	Gerlache de Gomery, Adrien de. Victoire sur la nuit antarctique. Tournai, 1960.
Geog 6009.57	Gerlache de Gomery, Gaston de. Retour dans l'Antarctique; récit de l'Expédition antarctique belge, 1957-58. Tournai, 1960.
An 359.60.15	Ghent. Université. Het lever; eer serie lezingen gehouden in de aula van de Rijksuniversität te Gent. Gent, 1960.
An 2109.60.15	Grand, P.M. Découverte de la préhistoire. Paris, 1960.
Geog 5333.15	Grierson, John. Sir Hubert Wilkins. London, 1960.
Geog 4521.8	Hackel, Heinrich. Meine Berge, nuen Leben. Salzburg, 1960.
Geog 4209.60	Hammond-Innes, Ralph. Harvest of journeys. N.Y., 1960.
Geog 4328.38.5	Happoldt, Christopher. The Christopher Happoldt journal. Charleston, 1960.
Geog 500.91	Harris, Chauncy Donnison. International list of geographical serials. Chicago, 1960.
An 170.238	International Congress of Anthropological and Ethnological Sciences, 5th, Philadelphia, 1960. Selected papers. Philadelphia, 1960.
An 39.60	International dictionary of regional Europen ethnology and folklore. Copenhagen, 1960- 2v.
Geog 4750.2	International Geographical Union. Special Commission on the Humid Tropics. A select annotated bibliography of the humid tropics. Montreal, 1960.
An 629.60	Inverarity, Robert Bruce. Visual files coding index. Bloomington, 1960.
Geog 4209.60.40	Jelling Kristoffersen, Frode. Jorden rundt på tommelfinger. Sirius, 1960.
Geog 4311.19	Jonák, Ján. 2200 kilo-metrov po Dunaiu. Moskva, 1960.
An 2109.60.20	Kern, Fritz. The wildbooters. Edinburgh, 1960.
Geog 5060.17	Kirwan, Laurence P. A history of polar exploration. N.Y., 1960.
Geog 4131.60	Knebel, Hans J. Soziologische Strukturwandlungen im Mardernen. Stuttgart, 1960.
Geog 579.60	Kratkaia geograficheskaia entsiklopecha. Moskva, 1960. 5v.
Geog 5399.60.5	Kudenko, O.I. Teplaia Arktika. Moskva, 1960.
Geog 4269.60	Kuleshov, A.P. 500,000 [i.e. Piat'sot tysiach] kolometrov v puti. Moskva, 1960.
An 99.60	Levin, M.G. Ocherki po istorii antropologiia v Rossii. Moskva, 1960.
An 2109.31.5.5	Lévy-Bruhl, L. La mentalité primitive. 15. ed. Paris, 1960.
Geog 4679.23	Lockwood, C.A. Tragedy at Honda. Philadelphia, 1960.
An 2109.60.25	Mauss, Marcel. Sociologie et anthropologie. 2. ed. Paris, 1960.
An 170.270.5	Mead, M. The golden age of American anthropology. N.Y., 1960.
An 359.60	Medawar, Peter Brian. The future of man. London, 1960.
Geog 5209.60	Mendeleev, D.I. Nauchnyi arkhiv: osvoenie krainego severa. Leningrad, 1960.
Geog 4019.60	Mier, Waldo de. Escogieron la inquietud. Madrid, 1960.
Geog 5639.60	Mikkelsen, Ejnar. Svundne tider i Østgrønland. København, 1960.
Geog 4209.60.25	Moncharmont, Simone. La croisière du Ly-Kau. Paris, 1960. 2v.
An 3209.60	Montagu, Ashley. An introduction to physical anthropology. 3. ed. Springfield, 1960.
Geog 665.95	Mori, Assunto. Scritti geografici, scelti e ordinati. Pisa, 1960.
Geog 4269.60.5	Nejedlý, O. Malířovy toulky po Evropě, Eejlonu a Indii. Praha, 1960.
Geog 4209.60.5	Nicholson, Timothy. Five roads to danger. London, 1960.
Geog 4209.60.10	Nikhailov, Nikolai I. From pole to pole. Moscow, 1960.
Geog 4559.60.5	O'Daniel, John. The nation that refused to starve. N.Y., 1960.
Geog 258.1.5	Österreichische Geographische Gesellschaft. Registerband, 1908-59. Wien, 1960.
Geog 3570.20.4	Peres, Damião. A history of the Portuguese. Lisbon, 1960.
Geog 601.5	Pertish, E.N. K.I. Arsen'ev i ego raboty po Raiomirovamiiu Rossii. Moskva, 1960.
Geog 659.60	Phlipponneau, Michel. Geographie et action. Paris, 1960.
Geog 4209.60.35	Portes Gil, Emilio. El mundo a través de sus grandes estadistas. México, 1960.
Geog 500.110	Powell, Lawrence. Around the world in sixty books. Los Angeles, 1960.
Geog 5558.81.60	Powell, T. The long rescue. 1. ed. Garden City, 1960.
Geog 4209.37.7	Prague. Městská Lidová Knihovna. Život a délo E. St. Vráze. Praha, 1960.
Geog 4659.60	Ratigan, William. Great Lakes shipwrecks and survivals. Grand Rapids, 1960.
Geog 618.4	Romer, E. Wybór prac. Warszawa, 1960. 3v.
Geog 5898.241	Rosing, Jens. Isimardik. København, 1960.
Geog 4209.60.30	Roy, Claude. Le journal des voyages. Paris, 1960.
An 359.60.5	Royal Anthropological Institute of Great Britain and Ireland. Man, race and Darwin. London, 1960.
Geog 6009.55.5F	Royal Society of London. The Royal Society International Geophysical Year. London, 1960. 4v.
Geog 4266.41.80	Schuchhard, C. Die Zeiller-Merianschen Topographien bibliographisch Beschrieben. Hamburg, 1960.
Geog 3604.5	Selander, Sten. Linnélärjungar i främmande länder. Stockholm, 1960.
Geog 4249.60	Sharp, C.A. The discovery of the Pacific Islands. Oxford, 1960.
Geog 619.5	Shokal'skaia, Z. Iu. Zhiznennyi put' Iu. M. Shokal'skogo. Moskva, 1960.
Geog 5925.65	Slevich, S.B. Antarktika dolzhna stat' zonoi mira. Leningrad, 1960.
An 2109.60.10	Smolla, G. Neolithische Kulturerscheinungen. Bonn, 1960.
Geog 3590.86	Tel', Sergei E. Kartografiia Rossii XVIII veka. Moskva, 1960.
An 170.367	UNESCO. Le racisme devant la science. Paris, 1960.
Geog 5330.1	Vittenburg, P.V. Zhizn' i nauchnaia deiatel'nost' E.V. Tollia. Leningrad, 1960.
Geog 579.49	Webster's geographical dictionary. Springfield, Mass., 1960.
Geog 5235.24	Weems, John E. Race for the pole. N.Y., 1960.
Geog 5839.30.6	Wegener, A. Tagebücher, Briefe. Wiesbaden, 1960.
Geog 4559.60	Whipple, Addison. Tall ships and great captains. N.Y., 1960.
Geog 579.60.5	The worldmark encyclopedia of the nations. N.Y., 1960.
Geog 4266.41.10F	Zeiller, Martin. Topographia Sueviae. Kassel, 1960.
Geog 4679.60.5	Ziganskin, Askhat. 49 dnei v okeane. Kuibyshev, 1960.

1961

Geog 3019.61.5	Almagia, Roberto. Scritti geografici. Roma, 1961.
Geog 4181.114	Alvares, Francisco. The Prester John of the Indies. Cambridge, 1961. 2v.
An 2059.61	Ardrey, Robert. African genesis. N.Y., 1961.
Geog 4428.01	Avine, Grégoire. Les voyages du chirurgien Avine à l'Ile de France et dans la mer de Indes au début du XIXe siècle. Paris, 1961.
Geog 1538.61	Baedeker, publishers. Italy, including Sicily and Sardinia. Freiburg, 1961.
Geog 1531.10	Baedeker, publishers. Tyrol and Salzburg. 14. ed. Freiburg, 1961.
Geog 4309.61.15	Beadle, Muriel. These ruins are inhabited. N.Y., 1961.
Geog 602.3	Blanchard, Raoul. Ma jeunese sous l'aile de Péguy. Paris, 1961.
Geog 4145.9.5	Bodrick, Alan Houghton. Casual change. London, 1961.
Geog 579.61	British Association for the Advancement of Science. A glossary of geographical terms. London, 1961.
Geog 4131.61	Cheechi and Company, Washington, D.C. The future of tourism in the Pacific and Far East. Washington, 1961.
An 2109.61.5	Clark, J.G.D. World prehistory. Cambridge, Eng., 1961.
An 359.61.25	Contemporary raciology and racism. Bloomington, 1961.
Geog 5509.61.5	Dainelli, Giotto. Il passaggio di nord-ouest. Torino, 1961.
Geog 4181.116	Davies, J. The history of the Tahitian Mission, 1799-1830. Cambridge, Eng., 1961.
An 359.61.30	Dias, Jorge. Ensaios etnológicas. Lisboa, 1961.
Geog 5509.61	Dodge, Ernest S. Northwest by sea. N.Y., 1961.
Geog 4209.61.10	Drachev, B.S. K beregam Vostoka. Moskva, 1961.
Geog 5925.75	Dralkin, A.G. V mire kholoda. Moskva, 1961.
Geog 4313.8	Egor'eva, A.V. Baltiiskoe more. Moskva, 1961.
An 359.61.45	Evans-Pritchard, Edward Evan. Anthropology and history. Manchester, Eng., 1961.
Geog 4219.58.5	Fenyvessy, J. Mein Traum wurde Wirklichkeit. 10. Aufl. Köln, 1961.
An 359.61.35	Fried, Morton Herbert. Readings in anthropology. N.Y., 1961-64. 2v.
An 359.61.10	Garn, S.M. Human races. Springfield, Ill., 1961.
An 359.61.15	Gehlen, Arnold. Anthropologische Forschung. Reinbek, 1961.
Geog 613.8	Gilbert, Edmund W. Sir Halford MacKinder, 1861-1947, an appreciation of his life and work. London, 1961.
Geog 4309.61.10	Giloteaux, Paulin. Voyage à Moscou et au-dela. Paris, 1961.
Geog 4217.85.15	Gordon, Mona. The mystery of La Pérouse. Christchurch, N.Z., 1961.
Geog 4269.50.5	Gottmann, J. A geography of Europe. N.Y., 1961.
An 359.61.20	Grieger, Paul. La caracterologie ethnique. Paris, 1961.
Geog 5399.61	Gromov, L.V. Ostrov Vrangelia. Magadan, 1961.
Geog 665.96	Hafemann, D. Mainzer geographische Studien. Braunschweig, 1961.
Geog 500.116	Harris, Chauncy Donnison. Geographic bibliography. Chicago, 1961.
Geog 4209.61.15	Hentzel, R. Aventyr jorden runt. Stockholm, 1961.
Geog 6009.55.15	Hillary, Edmund. No latitude for error. London, 1961.
Geog 4249.61	Hoever, Otto. Alt-Asiaten unter Segel in Indischen und Pazifischen. Braunschweig, 1961.
Geog 4269.61.10	Hoffman, George W. A geography of Europe including Asiatic USSR. N.Y., 1961.
An 359.61	Hsu, Francis L.K. Psychological anthropology. Homewood, Ill., 1961.
Geog 4659.36.3	Ingram, Charles W.N. New Zealand shipwrecks, 1795-1960. 3. ed. Wellington, 1961.
An 629.61.5	Jones, W.T. The romantic syndrome. The Hague, 1961.
Geog 4209.61	Kahn, E.J. A reporter here and there. N.Y., 1961.
Geog 4209.61.5	Kazakova, M.S. Goroda i vstrechi. Riga, 1961.
An 99.61A	Kluckhohn, Clyde. Anthropology and the classics. Providence, 1961.
Geog 4139.61	Kunský, J. Čeští cestovatelé. Praha, 1961. 2v.
Geog 4145.48	Lantzsch, W. Die Welt in allen Zonen. München, 1961.
Geog 819.58.10	Larousse encyclopedia of geography. N.Y., 1961.
An 2059.61.10A	Leakey, Louis S. The progress and evolution of man in Africa. London, 1961.
Geog 5160.55	Liakh, N.N. Zapiski poliarnika. Novosibirsk, 1961.
An 629.61	Lindzey, G. Projective techniques and cross-cultural research. N.Y., 1961.
An 2109.61	Lissher, Ivar. Man, God, and magic. N.Y., 1961.
An 379.61	Maas, Walther. Probleme der Sozialgeographie. Berlin, 1961.
Geog 4309.61.5	Malinowksi, P. Der synthetische Bazar. Wien, 1961.
Geog 819.61	Manley, Gordon. Geography; our planet, its peoples and resources. London, 1961.
An 187.8	Mauss, Marcel. Oeuvres. Paris, 1961- 3v.
An 2109.61.20	Mead, Margaret. Cooperation and competition among primitive peoples. Boston, 1961.
An 359.61.40	Montagu, Ashley. Man in process. 1. ed. Cleveland, 1961.
Geog 4679.61	Mowat, Farley. The serpent's coil. 1. United States ed. Boston, 1961.
Geog 5312.7.25	Murphy, Robert William. The haunted journey. 1. ed. Garden City, 1961.
Geog 4269.61.5	Mutton, Alice. Central Europe; a regional and human geography. London, 1961.
Geog 5324.2.90	Nansen, Fridtjof. Brev utgitt for Nansenfondet av Steinar Kjaerheim. Oslo, 1961- 4v.
An 3209.61	Notschaele, Lucien Aimé. Een somatometrisch onderzoek in de Noord-oost Polder. Amsterdam, 1961.
An 379.61.10	Otto, Klaus. Das Aufkommen sozialgeographischer Betrachtungsweisen in der deutschen länderkundlichen Literatur seit Beginn des 20. Jahrhunderts. Inaug. Diss. Münster, 1961.
An 2059.61.5	Overhage, Paul. Das Problem der Hominisation. Freiburg, 1961.
Geog 3520.11.2	Parks, George Bruner. Richard Hakluyt and the English voyages. 2. ed. N.Y., 1961.
An 2109.61.10	Real y Ramos, C.A. del. Sociología pre y protohistórica. Madrid, 1961.
Geog 5399.61.5	Reut, V.F. Moskva - Severnyi Polius. Moskva, 1961.
Geog 3229.61	Roux, Jean Paul. Les explorateurs au Moyen Âge. Paris, 1961.
Geog 4182.62	Ruyter, M.A. De reis van Michiel Adriaanszoom de Ruyter in 1664-1665. 's-Gravenhage, 1961.
An 359.61.50	Scheler, Max. Man's place in nature. Boston, 1961.
Geog 3570.51	O seculo dos descobrimentos. São Paulo, 1961.
An 379.61.5	Sorre, Maximilien. L'homme sur la terre. Paris, 1961.
An 2109.61.15F	Spivack, M.R. La danse cosmique de Lascaux. Montignac, 1961.

Chronological Listing

1961 - cont.

Geog 85.385.5	Spomenica o pedecetogodišnjici Srpskog geógrafskog društva, 1910-1960. Beograd, 1961.
Geog 4145.84.15	Stark, Freya. Dust in the lion's paw. London, 1961.
Geog 5970.30	Swan, Robert A. Australia in the Antarctic. Victoria, 1961.
Geog 5333.20	Thomas, Lowell. Sir Hubert Wilkins. N.Y., 1961.
An 359.61.5	Thompson, L. Toward a science of mankind. N.Y., 1961.
Geog 3540.13	Timpte, Helmut. Typologische Studien zur historischen Kartographie in Westfalen. Düsseldorf, 1961.
Geog 5558.84.5	Todd, Alden. Abandoned. 1. ed. N.Y., 1961.
Geog 4239.47.5	Turen, T. The Tuntsa. Chicago, 1961.
An 359.61.55	UNESCO. The race question in modern science; race and science. N.Y., 1961.
Geog 4249.61.5	United States. National Archives. United States scientific geographical exploration of the Pacific Basin, 1783-1899. Washington, 1961.
Geog 618.5	Wanklyn, H.G. Friedrich Ratzel. Cambridge, Eng., 1961.
Geog 4477.64	White, D.M. Zaccaria Seriman, 1709-1784, and the Viaggi di Enrico Wanton. Manchester, 1961.

1962

Geog 3060.19	Adams, Percy G. Travellers and travel liars, 1660-1800. Berkeley, 1962.
Geog 3590.71.3	Akademiia Nauk SSSR. Soviet geography, accomplishments and tasks. N.Y., 1962.
Geog 3570.50	Albuquerque, Louis. Introducão a historia dos descobrimentos. Coimbra, 1962.
Geog 4309.62.5	Alimzhanov, An. Piat'desiat tysiach nil' po vode i sushe. Alma-Ata, 1962.
Geog 500.115.2F	American Geographical Society of New York. Research catalogue. Map supplement. Boston, 1962.
Geog 500.115F	American Geographical Society of New York. Research catalogue of the American Geographical Society. Boston, 1962. 15v.
An 170.117	Athayde, Alfredo. Introdução a antropología tropical. Lisboa, 1962.
Geog 618.5.5F	Babicz, J. Nauka o ludakh Fryderyka Ratzla. Wrocław, 1962.
Geog 4218.17.10	Bassett, M.M. Realms and islands. London, 1962.
An 2109.62	Bowra, Cecil M. Primitive song. London, 1962.
Geog 4559.62	Bradford, G. In with the sea wind. Barre, 1962.
An 170.142	Braidwood, Robert John. Courses toward urban life. Chicago, 1962.
Geog 6009.29.5	British, Australian and New Zealand Research Expedition, 1929-1931. The winning of Australian Antarctica. Sydney, 1962.
Geog 4182.63	Broecke, Pieter van den. Pieter van den Broecke in Azie. 's-Gravenhage, 1962. 2v.
Geog 4678.54.15	Brown, A.C. Women and children last; the loss of the steamship. London, 1962.
Geog 148.1.9	Buxbaum, Edwin Clarence. Collectors guide to the national geographic magazine. Wilmington, 1962.
Geog 5919.62	Caras, Roger A. Antarctica. 1st ed. Philadelphia, 1962.
An 359.62	Coon, Carleton Stevens. The origin of races. N.Y., 1962.
Geog 3049.62	Cozzi, Piero. Los paises legendarios de la mitologia. Buenos Aires, 1962.
Geog 4559.62.10	Dane, Peter. The seamen are down below. Ilfracombe, 1962.
Geog 5985.55	Davis, John K. High latitude [autobiographical]. Parkville, Victoria, 1962.
Geog 6008.97.10	Dobrowolski, A.B. Dziennik wyprawy na Antarktydę. Wrocław, 1962.
An 2109.62.15	Domanskii, Ia.V. Po besovym sledam. Leningrad, 1962.
An 359.62.15	Evans-Pritchard, Edward Evan. Essays in social anthropology. London, 1962.
An 359.62.20	Fiezefontaine, J. Du racisme à l'universalisme. Paris, 1962.
Geog 5939.62.5	Fitte, Ernesto. El descubrimiento de la Antártida. Buenos Aires, 1962.
Geog 665.93	France. Centre National de la Recherche Scientifique. Colloque national de géographie appliquée, Strasbourg, 20-22 avril 1961. Paris, 1962.
Geog 3019.62.5	Freeman, Thomas. A hundred years of geograpy. Chicago, 1962.
Geog 4219.62	Gandon, Yves. A la recherche de l'Eden. Paris, 1962.
An 359.62.30	George, W.C. The biology of the race problem. N.Y., 1962.
An 2109.62.10	Gluckman, Max. Essays on the ritual of social relations. Manchester, 1962.
Geog 3019.62	Grenville, J.A.S. The coming of the Europeans. London, 1962.
Geog 600.20	Harms, Hans. Künstler des Kartenbildes. Oldenburg, 1962.
An 359.62.5	Haste, Hans. Rasisner. Stockholm, 1962.
Geog 3049.62.5	Hutin, Serge. Les civilisations inconnues; mythes ou réalités. Paris, 1962.
Geog 5939.62	Ignatov, V.S. God na poliuse kholoda. Moskva, 1962.
Geog 4309.62	Johnson, Irving. Yankee sails across Europe. N.Y., 1962.
Geog 659.62	Jong, Guben. Chonological differentiation as the fundamental principle of geography. Groningen, 1962.
Geog 5839.35.7	Kent, Rockwell. Greeland journal. N.Y., 1962.
An 185.2	Kluckhohn, Clyde. Culture and behaviour. N.Y., 1962.
Geog 4269.62.10	Koen, Albert. Putishta i spirki. Sofiia, 1962.
Geog 4209.62.5	Koroteev, V.I. Ia eto videl; ocherki raznykh let. Moskva, 1962.
Geog 4269.62	Kukanov, A.A. Doroga k serdtsu; reportazh. Tallin, 1962.
Geog 5970.36	Law, Phillip G. Australia and the Antarctic. St. Lucia, 1962.
Geog 665.97	Leidlmar, A. Herman von Wissmann-Festschrift. Tübingen, 1962.
An 2109.62.5	Lévi-Strauss, Claude. La pensée sauvage. Paris, 1962.
Geog 4145.4	Lewis, Warren. Levantine adventurer...Chevalier d'Arvieux. London, 1962.
Geog 665.94	McCashill, Murray. Land and livelihood. Christchurch, 1962.
Geog 4217.08.10	MacLiesh, Archibald F. The privateers. N.Y., 1962.
An 170.264	Madan, T.N. Indian anthropology. Bombay, 1962.
Geog 500.118	Maggs Bros., London. Voyages and travels. v.1,4-5. London, 1962- 3v.
An 2109.62.20	Malinowski, B. Geschlechtstrieb und Verdrängung bei den Primitiven. Reinbek bei Hamburg, 1962.
An 359.62.10	Mason, Philip. Prospero's magic. London, 1962.
Geog 4181.118	Navarrete, Domingo Fernández de. The travels and controversies of Friar Domingo Navarrete. Cambridge, 1962. 2v.
Geog 4239.62	Nikiforovskii, V.A. Ekspeditsiia na "Sedove" v Atlanticheskii okean. Moskva, 1962.

1962 - cont.

An 132.4.5	Northwestern University, Evanston, Ill. Program of African Studies. Bibliography of Melville J. Herskovits, 1920-1962. Evanston, Ill., 1962?
Geog 4679.62	Noyce, Wilfrid. They survived. London, 1962.
Geog 614.9	Olszewicz, B. Wacław Nałkowski. Warszawa, 1962.
Geog 3520.12	Penrose, Boies. Tudor and early Stuart voyaging. Washington, 1962.
Geog 4131.62.5	Poeschl, A.E. Fremdenverkehr und Fremdenverkehrspolitik. Berlin, 1962.
An 192.1	Redfield, Robert. Human nature and the study of society. Chicago, 1962.
Geog 6009.14.35	Richards, R.W. The Ross Sea Shore Party, 1914-17. Cambridge, 1962.
Geog 3520.16	Robinson, Adrian. Marine cartography in Britain. Leicester, 1962.
Geog 4269.62.5	Salvadori, M. Western ports in Europe. N.Y., 1962.
Geog 5160.57	Santillan de Andres, Selva E. La vida humana en el subecumene ártico. Tucuman, 1962.
Geog 4127.77.2	Schloezer, August Ludwig von. Vorlesungen über Land- und Seereisen. Göttingen, 1962.
Geog 4209.62	Shakirianov, N.A. Atlanticheskii dnevnik. Riga, 1962.
Geog 665.37.6	Singleton, E. The wonders of nature as seen and described by Alexandre Dumas. Washington, 1962.
Geog 5939.59.24	Smuul, Juhan. Jaine raamat. Tallinn, 1962.
An 359.62.25	Snyder, L.L. The idea of racialism. Princeton, 1962.
Geog 3590.70	Sovetskie ekspeditisii god. 1959. Moskva, 1962.
Geog 579.56.7	Swayne, James Colin. A concise glossary of geographical terms. 2. ed. London, 1962.
Geog 4209.62.15	Thorlacius, Sigriðm. Feroabók. Akureyri, 1962.
Geog 5970.38	United States. Antarctic Projects Office. The United States in the Antarctic, 1820-1962. Washington, 1962?
Geog 4559.62.5	Villiers, A.J. Men, ships, and the sea. Washington, 1962.
Geog 4169.62	Vincenti, Leonello. Viaggiatori del Settecento. Torino, 1962.
Geog 579.49.5	Webster's geographical dictionary. Springfield, Mass., 1962.
Geog 5509.62	Williams, G. The British search for the Northwest Passage in the 18th century. London, 1962.
Geog 4181.120	Williamson, J.A. The Cabot voyages and Bristol discovery under Henry VII. Cambridge, 1962.

1963

Geog 4209.63.15	Ajala, O. An African abroad. London, 1963.
An 359.63.20	Akademiia Nauk SSSR. Sovremennaia amerikanskaia etnografiia. Moskva, 1963.
An 5.31	Antropološko Društvo Jugoslavije. Antropološka bibliografija o Jugoslaviji. Beograd, 1963.
Geog 5160.56	Arctic Institute of North America. The Arctic basin. Washington, 1963.
Geog 4559.63	Asgeirsson, Rikki. Alltaf má fá annad skip. Reykjavik, 1963.
Geog 3019.63.5	Baker, J.N.L. The history of geography. Oxford, 1963.
An 359.63.35	Barabas, Jenó. Kartografiai modszer a neprajzban. Budapest, 1963.
An 2109.63	Barnouw, V. Culture and personality. Homewood, Ill., 1963.
Geog 602.3.5	Blanchard, Raoul. Je découvre l'université. Paris, 1963.
Geog 5845.3	Boserup, Mogens. Økonomisk politik i Grønland. København, 1963.
Geog 4181.121	Bourne, William. A regiment for the sea. Cambridge, Eng., 1963.
Geog 4209.63.30	Bouvier, Nicolas. L'usage du monde. Genève, 1963.
Geog 4415.02.5	Bracciolini, Poggio. A traveler in disguise. Cambridge, 1963.
An 359.63.15	Ciba Foundation. Man and his future. London, 1963.
Geog 4309.63	Cluzel, M.E. Des brumes de la Mer du Nord. Paris, 1963.
An 170.163.5	Congresso de Etnografia e Folelore, 1st, Braga, 1956. Actas...promovido pela Câmarã municipal de Braga. Lisboa, 1963. 3v.
Geog 4209.63.20	Diez del Corval, L. Del nuevo al viejo mundo. Madrid, 1963.
An 2109.63.15	Durkheim, Emile. Primitive classification. Chicago, 1963.
An 629.63	Evans-Pritchard, E.E. The comparative method in social anthropology. London, 1963.
Geog 6100.2.10	Expéditions Polaires Françaises. Terre Adélie, 1959-61. Paris, 1963.
Geog 4145.27	Fink, O. Auf dem Kirs der Raben. Hamburg, 1963.
An 137.4.5.4	Firth, Raymond W. Man and culture. 4. ed. London, 1963.
Geog 4571.2	Garland, Joseph E. Lone voyager. 1. ed. Boston, 1963.
Geog 4269.63	Glushchenk, I. Strany, strechi, uchenye. Moskva, 1963.
Geog 4209.63.35	Green, Lawrence George. A decent fellow doesn't work. Cape Town, 1963.
Geog 4509.63	Gudhjohnsen, P. Endurminningar fjallgöngumanns. Reykjavik, 1963.
Geog 3520.10.50.2	Hakluyt, Richard. Voyages and documents. London, 1963.
An 3309.63.5	Henzel, Tadeusz. Badania struktury rasowej ludności Afryki Srotkowej. Łódz, 1963.
Geog 4180.10A	Herberstein, Sigmund. Notes upon Russia. N.Y., 1963? 2v.
Geog 4219.59	Hiscock, Eric C. Beyond the west horizon. London, 1963.
An 2109.63.10	Howells, William. Back of history, the story of our origins. Garden City, N.Y., 1963.
An 359.63.10	Hsu, Francis L.K. Clan, caste, club. Princeton, 1963.
Geog 579.63	Hustich, I. Tämän päivän maailmaa. 4. ed. Helsinki, 1963.
An 359.63	Jennings, J.D. Anthropology and the world of science. Salt Lake City, 1963.
An 124.11	Kardiner, Abram. They studied man. N.Y., 1963.
Geog 6008.19	Leningrad. Arkticheskii i Antarkticheskii Nauchno-Issledovatel'skii Institut. Pervaia russkaia antarkticheskaia ekspeditsiia, 1819-1821 gg. i ee otchetnaia navigatsionnaia karta. Leningrad, 1963.
Geog 4309.63.5	Levi, P. La tregua. Torino, 1963.
An 359.63.40A	Lévi-Strauss, Claude. Structural anthropology. N.Y., 1963-76. 2v.
Geog 3570.8	Marcondes de Souza, T.O. Novas achegos à historia dos descobrimentos maritimos. São Paulo, 1963.
Geog 4239.63	Marx, R.F. The voyage of the Niña II. 1st ed. Cleveland, 1963.
Geog 3618.10	Mead, William R. The geographical tradition in Finland. London, 1963.
Geog 5318.2	Miller, F. Ahdoolo: the biography of M.A. Henson. 1. ed. N.Y., 1963.
An 379.63	Mills, C.A. World power and shifting climates. Boston, 1963.
Geog 4329.63	Morpurgo, J.E. The road to Athens. London, 1963.

Chronological Listing

1963 - cont.

Geog 4209.63.10	Morris, J. Cities. London, 1963.
Geog 3590.82	Moscow. Universitet. Sovetskaia geografiia v period stroitel'stva kommunizma. Moskva, 1963.
An 2059.63	Muschalek, H. Urmensch-Adam. Berlin, 1963.
Geog 4209.63	Nasedkin, V.N. Piatnadtsat' let skitanii po zemnomu sharu. Moskva, 1963.
Geog 4019.63	National Geographic Society, Washington. Great adventures with NationalGeographic. Washington, 1963.
Geog 5300.52	National Research Council. Committee on Polar Research. Science in the Arctic Ocean basin. Washington, 1963.
Geog 5970.38.5	National Science Foundation. Office of Antarctic Programs. USARP; United States Antarctic Research Program, National Science Foundation. Washington, 1963.
Geog 3019.63	Parry, John Horace. The age of reconnaissance. London, 1963.
Geog 3019.63.1	Parry, John Horace. The age of reconnaissance. 1. ed. Cleveland, 1963.
Geog 5326.15.5	Parry A. Parry of the Arctic. London, 1963.
An 2109.63.5	Scotti, Pietro. La vita sociale dei popoli primitivi. Brescia, 1963.
Geog 3019.63.10	Sharaf, A. Torayah. A short history of geographical discovery. Alexandria, 1963.
Geog 5399.32.10	Shneiderov, V.A. Velikim Severnym. 2. izd. Moskva, 1963.
Geog 5518.36.4	Simpson, A. The life and travels of Thomas Simpson. Toronto, 1963.
An 359.63.30	Spelling, K. Miljøets indflydelse på intelligensvaviklingen. København, 1963.
Geog 4679.41.5	Swinson, A. Scotch on the rocks. London, 1963.
An 359.63.25	Tazerout, M. Manifeste contre le racisme. Rodez, 1963.
Geog 5150.2	Toronto. Public Library. The Northwest passage, 1534-1859. Toronto, 1963.
An 5.21F	Tozzer Library. Catalogue. Authors. Boston, 1963. 26v.
An 5.22.01F	Tozzer Library. Catalogue. Index to subject headings. Boston, 1963.
An 5.22F	Tozzer Library. Catalogue. Subjects. Boston, 1963. 27v.
Geog 5939.63	Treshnikov, Aleksei G. Istoriia otkrytiia i issledovaniia Antarktidy. Moskva, 1963.
Geog 4679.63	Troebst, C.C. Auf Wunder ist kein Verlass. Düsseldorf, 1963.
Geog 4209.63.25	Turri, Eugenio. Viaggio a Samacanda. Novara, 1963.
Geog 5060.7	Victor, Paul Emile. Man and the conquest of the poles. N.Y., 1963.
An 359.63.5	Weyl, N. The geography of intellect. Chicago, 1963.
Geog 4309.63.10	Wićaz, Jurij. Z Kamjenskim nosom; što dóžiwi serbski nowinar we swěće. 2. wyd. Budyšin, 1963.
Geog 4209.63.5	Zhukov, Iu.A. Eti semnadtsat' let. Moskva, 1963.

1964

Geog 4180.60.2	Acosta, J. de. National and moral history of the Indies. N.Y., 1964? 2v.
Geog 665.99	Akademia Nauk SSSR. Institut Geografii. Razvitie i preobrazo vanie geografii predy. Moskva, 1964.
Geog 4180.53	Alboquerque, A. d'. Commentaries of Dalboquerque. N.Y., 1964?
An 3559.64.5	Alekseev, Valerii P. Kraniometriia. Moskva, 1964.
Geog 5939.64.5	Astapenko, Pavel D. Puteshestvie k ostrovu chetyrekh vulkanov. Moskva, 1964.
Geog 1526.40	Baedeker, publishers. Berlin. 23. Aufl. Freiburg, 1964.
Geog 5316.3	Baetzkes, Ottile G. Sir Martin Frobisher's search for the Northwest Passage. N.Y., 1964.
Geog 4180.63	Baffin, W. Voyages of William Baffin 1612-22. N.Y., 1964?
Geog 3070.157	Bagrow, Leo. History of cartography. Cambridge, 1964.
Geog 4180.49	Barbaro, J. Travels to Tana and Persia. N.Y., 1964?
An 359.64	Beattie, John. Other cultures. N.Y., 1964.
Geog 4209.64.10	Benchley, P. Time and a ticket. Boston, 1964.
Geog 4180.21	Benzoni, G. History of the new world. N.Y., 1964?
Geog 5970.37	Billing, Graham. South: man and nature in Antarctica. Wellington, 1964.
Geog 4329.53.12	Birot, Pierre. La Méditerranée et le Moyen-Orient. 2. éd. Paris, 1964.
Geog 4269.64.15	Bonasera, Francesco. Gli stati d'Europa. Palermo, 1964.
Geog 4180.46	Bontier, P. The Canarien...conquest and conversion. N.Y., 1964?
Geog 4181.123	Bovill, E.W. Missions to the Niger. Cambridge, Eng., 1964. 4v.
Geog 4209.64.15	Brooks, P. Roadles area. 1. ed. N.Y., 1964.
Geog 4311.21	Buettner, R. Burgen und Schlösser and der Donau. Wein, 1964.
Geog 665.98	Bulgarska Akademiia na Naukite, Sofia. Geografski Institut. Sbornik v chestna chlen-korespondent Iordan Zakhariev. Sofiia, 1964.
Geog 4181.122	Byron, John. Byron's journal of his circum navigation, 1764-1766. Cambridge, 1964.
Geog 4180.68	Cieza de León, P. de. Second part of chronicle of Peru. N.Y., 1964?
Geog 39.2.11	Club Alpino Italiano. I cento anni del club alpino italiano. 2. ed. Milano, 1964.
Geog 4309.64.5	Cluzel, M.E. Impressions sur l'Europe. Paris, 1964.
Geog 4180.66	Cocks, R. Diary of Richard Cocks...1615-22. N.Y., 1964. 2v.
An 2109.64.5	Cohen, Y.A. The transition from childhood to adolescence. Chicago, 1964.
Geog 4558.40.60	Dana, Richard Henry. Two years before the mast. Los Angeles, 1964. 2v.
An 170.173.5	Diamond, Stanley. Primitive views of the world. N.Y., 1964.
Geog 4279.64	Dulles, F.R. Americans abroad. Ann Arbor, 1964.
Geog 4209.64.35	Eckert, Gerhard. Richtig Reisen. Bergisch Gladbach, 1964.
Geog 3590.73	Esakov, V.A. Russkie geograficheskie issledovaniia Evropeiskoi Rossii i Vrala v XIX-nachale XX v. Moskva, 1964.
An 359.51.5	Evans-Pritchard, Edward Evan. Social anthropology, and other essays. N.Y., 1964.
Geog 4180.16	Feltcher, F. The world encompassed by Francis Drake. N.Y., 1964?
Geog 4209.64.40A	Fleming, Ian. Thrilling cities. N.Y., 1964.
Geog 4180.20	Fletcher, G. Russia at close of sixteenth century. N.Y., 1964?
An 2109.64.20	Forcieri, Luigi. Le role des croyances magiques et religieuses dans les économies primitives. Bordeaux, 1964.
Geog 5558.81.65	Gauroy, Pierre. Les affamés de la banquise. Paris, 1964.
Geog 4269.64.10	George, Pierre. Géographie de l'Europe centrale slave et danubienne. Paris, 1964.

1964 - cont.

Geog 819.45.10	George, Pierre. Géographie sociale du monde. 6. éd. Paris, 1964.
Geog 3249.64	Goldstein, Thomas. Fifteenth century geography against the background of medieval science. Salem, Mass., 1964.
An 170.279	Goodenough, Ward Hunt. Explorations in cultural anthropology; essays in honor of George Peter Murdock. N.Y., 1964.
Geog 5130.2	Grierson, J. Challenge to the poles. Hamden, Conn., 1964.
An 359.64.35	Grottanelli, Virigi L. L'etnologia e le leggi della condotta umana. Roma, 1964.
An 170.246	Haberland, Eike. Festschrift für A.E. Jensen. München, 1964. 2v.
Geog 3520.10.55	Hakluyt, Richard. They told Mr. Hakluyt. London, 1964.
An 170.227	Hammond, Peter B. Cultural and social anthropology. N.Y., 1964.
Geog 5160.58	Hantschel, A. Weltgeschehen am Rande des Polarmeeres. Würzburg, 1964.
Geog 500.116.5	Harris, Chauncy Donnison. Annotated world list of selected current geographical serials in English. 2. ed. Chicago, 1964.
An 359.64.30	Harris, Marvin. The nature of cultural things. N.Y., 1964.
Geog 4249.50.2	Heyerdahl, T. Kon-Tiki; across the Pacific by raft. Chicago, 1964.
An 99.64A	Hodgen, Margaret. Early anthropology in the sixteenth and seventeenth centuries. Philadelphia, 1964.
Geog 110.1.20	International Geographical Congress, 20th, London, 1964. Abstracts of papers. London, 1964.
Geog 110.1.20.2	International Geographical Congress, 20th, London, 1964. Abstracts of papers. Supplement. London, 1964.
Geog 110.1.20.5	International Geographical Congress, 20th, London, 1964. Congress programme. London, 1964.
Geog 110.1.25	International Geographical Congress, 20th, London, 1964. Sovremennye problemy geografii. London, 1964.
An 359.64.20	Isā, Alī Ahmad. Social anthropology in theory and practice. Cairo, 1964.
Geog 4169.64	Jahn, Janheinz. Wir nannten sie Wilde. München, 1964.
An 629.64	Jarvie, Ian C. The revolution in anthropology. London, 1964.
Geog 4180.72	Jenkinson, A. Early voyages...to Russia and Persia. N.Y., 1964? 2v.
Geog 4180.31	Jordanus de Saxonia. Mirabilia descripta...wonders of the East. N.Y., 1964?
Geog 500.117	Kaufman, Isaak M. Geograficheskie slovari; bibliografiia. Moskva, 1964.
Geog 4209.64	Knowles, J. Double vision. N.Y., 1964.
An 359.64.15	Köbben, A.J.F. Van primitieven tot medeburgers. Assen, 1964.
Geog 3235.61	Koeman, Cornelis. The history of Lucas Janszoon Wazehnaer and his Spieghel der Zeevaerdt. Lausanne, 1964.
Geog 4209.64.30	Kozlovs'kyi, Mykola F. Cherez 15 morei i 2 okeana; puteshestviia. Kiev, 1964.
Geog 3019.64	Landström, B. The quest for India. London, 1964.
Geog 4209.62.10	Lattmann, D. Mit einem deutschem Pass. München, 1964.
Geog 4427.80.5	Le Gentil de la Galaisière, G.J.H.J.B. A voyage to the Indian Seas. Manila, 1964.
Geog 4329.64	Le Lannou, Maurice. Problèmes géographiques de la Méditerranée européenne. Paris, 1964.
Geog 4180.92.2	Leo Africanus, J. History and description of Africa. N.Y., 1964? 3v.
An 359.64.5	Lienhardt, G. Social anthropology. London, 1964.
Geog 4309.64.25	Linhares, Temístocles. Jornal da Europe. Brasil, 1964.
Geog 4180.70	Linschoten, J.H. Voyage of Linschoten to East Indies. N.Y., 1964? 2v.
An 170.266	Manners, Robert A. Process and pattern in culture; essays in honor of Julian H. Steward. Chicago, 1964.
Geog 4679.63.5	Marchbanks, David. The painted ship. London, 1964.
Geog 4180.24	Markham, C.R. Expeditions into valley of the Amazons. N.Y., 1964.
Geog 4180.48	Markham, C.R. Narrative of the rites and laws of the Yncas. N.Y., 1964?
Geog 4180.47	Markham, C.R. Reports on the discovery of Peru. N.Y., 1964.
Geog 4180.56	Markham, C.R. Voyages of Sir James Lancaster. N.Y., 1964?
Geog 5985.187	Mawson, Paquita. Mawson of the Antarctic; the life of Sir Douglas Mawson. London, 1964.
Geog 4139.64	Mirsky, J. The great Chinese travelers, an anthology. N.Y., 1964.
Geog 4145.89.5	Mitchell, Mairin. Friar Andrés de Urdaneta, O.S.A. London, 1964.
An 170.275.5	Montagu, Ashley. The concept of race. N.Y., 1964.
An 359.64.45	Montagu, Ashley. Man's most dangerous myth. 4. ed. Cleveland, 1964.
An 359.64.40	Morgan, Lewis H. Ancient society. Cambridge, 1964.
An 359.64.10	Moskovskoe Obshchestvo Ispytatelei Prirody. Sovremennaia antropologiia. Moskva, 1964.
Geog 4209.64.20	Muhlethader, J. Toutes voiles dehors. Genève, 1964.
Geog 665.104	Neue Fragen der allgemeinen Geographie. Wurzburg, 1964.
Geog 4659.64	Noall, Cyril. Wreck and rescue round the Cornish coast. Truro, 1964-65. 3v.
Geog 611.5.5	Novlianskaia, Mariia G. Ivan Kirilovich Kirilov, geograf XVIII veka. London, 1964.
Geog 4180.17	Orleans, P.J. d'. History of...tartar conquerors of China. N.Y., 1964?
Geog 4145.50	Parr, Charles. Jan van Linschaten. N.Y., 1964.
Geog 4269.64.20	Phloros, Paulos. Staurodromia tēs Eurōpēs. Athēnai, 1964.
Geog 5970.46	Priestley, Raymond E. Antarctic research; a review of British scientific achievement in Antarctica. London, 1964.
Geog 4309.64.10	Pritchett, V.S. Foreign faces. London, 1964.
Geog 4309.64	Pritchett, V.S. The offensive traveller. 1. ed. N.Y., 1964.
Geog 4180.76	Pyrard, F. Voyage of F. Pyrard...to East Indies. v.1-2. N.Y., 1964? 3v.
Geog 4182.65	De reis om de wereld van de nassausche vloot, 1623-1926. 's-Gravenhage, 1964.
Geog 5399.64	Report from the Arctic; foreign and Soviet correspondents on their trip aboard the Soviet atomic icebreaker. Moscow, 1964.
Geog 619.6	Rowley, V.M. J. Russell Smith, geographer, educator and conservationist. Philadelphia, 1964.
Geog 4679.64	Ruhen, Olaf. Minerva Reef. 1. American ed. Boston, 1964.
Geog 4180.8	Rundall, T. Memorials of the empire of Japan. N.Y., 1964?
Geog 4209.64.25	Sakhnin, Arkadii Ia. Vot liudi. Moskva, 1964.
Geog 4180.44	Salîl-Ibn-Razîk. History of the Inâms...of Omân. N.Y., 1964?

Chronological Listing

1964 - cont.

An 3309.64	Sarkar, Sasanka Sekhar. Ancient races of Baluchistan. 1st ed. Calcutta, 1964.
Geog 500.135	Schmidt, Rolf Dietrich. Verzeichnis der geographischen Zeitschriften, periodischen Veröffentlichungen und Schriftreihen Deutschlands. Bad Godesberg, 1964.
Geog 619 7	Schultén, N.G. Levnadsteckning. Helsingfors, 1964.
Geog 4500.17	Scotland. National Library, Edinburgh. Shelf-catalogue of the Lloyd collection of Alpine books. Boston, 1964.
Geog 5050.8	Scotland. National Library, Edinburgh. Shelf-catalogue of the Wordie collection of polar exploration. Boston, 1964.
Geog 6009.10.14	Scott, R.F. Soctt's last expedition. 2d ed. London, 1964.
Geog 5559.14.5	Seleznev, Stepan A. Pervaia russkaia ekspeditsiia k severnomy poliucu. Arkhangel'sk, 1964.
Geog 4309.64.15	Selucký, Radoslav. Západ je Západ. Praha, 1964.
An 2159.64	Semenov, Sergei A. Prehistoric technology. London, 1964.
Geog 4110.36	Sheraton, Mimi. City portraits; a guide to 60 of the world's great cities. 1. ed. N.Y., 1964.
Geog 4180.28	Simon, P. Expedition of...Ursua and...Aguirre. N.Y., 1964?
Geog 5180.18	Smith, I. Norman. The unbelievable land; 29 experts bring us closer to the Arctic. Ottawa, 1964.
Geog 6009.55.50	Sovetskaia Antarkticheskaia Ekspeditsiia, 1955-1958. Soviet Antarctic expedition; information bulletin. Amsterdam, N.Y., 1964-65. 3v.
Geog 4759.64	Steel, R.W. Geographers and the tropics. London, 1964.
Geog 5329.3.10	Stefansson, Vilhjalmur. Discovery. 1. ed. N.Y., 1964.
Geog 4309.64.20	Streeter, Edward. Along the ridge; from northwestern Spain to southern Yugoslavia. N.Y., 1964.
An 170.349	Symposium on Community Studies in Anthropology. Symposium on community studies in anthropology. Seattle, 1964.
An 170.351	Tax, Sol. Horizons of anthropology. Chicago, 1964.
Geog 6009.55.25	Trans-Antarctic Expedition, 1955-1958. Antarctica. Wellington, 1964.
An 379.64	Troëng, Ivan. Kulturer före istiden. Uppsala, 1964.
Geog 4180.32	Varthema, S. di. Travels of Varthema. N.Y., 1964?
Geog 4180.41	Vega, G. de la. First part of royal commentaries of the Yncas. N.Y., 1964. 2v.
Geog 4208.39	Were, J.B. Voyage from Plymonth to Melbourne in 1839. Melbourne, 1964.
An 359.64.25	Wolf, Eric Robert. Anthropology. Englewood Cliffs, N.J., 1964.
An 2109.64.15	Zolotarev, Aleksandr M. Rodovoi stroi i pervobytnaia mifologiia. Moskva, 1964.

1965

An 2059.65	Atlas de prehistoire. Paris, 1965. 3v.
Geog 953.5.2F	Bachmann, Friedrich. Die alten Städtebilder. 2. Aufl. Stuttgart, 1965.
Geog 614.15	Barcinski, Florian. Stanislaw Nowakowski. Warszawa, 1965.
An 359.65.25	Barzun, Jacques. Race; a study in superstition. N.Y., 1965.
Geog 3019.65	Belov, Mikhail I. Puteshestviia i geograficheskia otkrytiia v XV-XIX vv. Leningrad, 1965.
Geog 4300.50	Bertelsmann, C. Ferien einmal anders. Gütersloh, 1965.
Geog 4509.65	Boutron, Michel. La montagne et ses hommes. Paris, 1965.
Geog 4145.14.10	Brazäo, Eduardo. Os Corte Reais e o novo mundo. Lisboa, 1965.
Geog 4709.57.12	Carse, Robert. The age of piracy. N.Y., 1965.
Geog 4181.124	Carteret, Philip. Carteret's voyage round the world, 1766-69. Cambridge, 1965. 2v.
Geog 4309.65.5	Cione, Edmondo. Questa Europa. Milano, 1965.
An 2109.65.10	Clark, John Grahame Douglas. Prehistoric soctieties. 1st American ed. N.Y., 1965.
Geog 4709.65.5	Clausen, Carl J. A 1715 Spanish treasure ship. Gainesville, 1965.
Geog 4709.65	Cochran, Hamilton. Freebooters of the Red Sea: pirates, politicians and pieces of eight. Indianapolis, 1965.
An 629.65	Conference on New Approaches in Social Anthropology. The relevance of models for social anthropology. London, 1965.
An 359.65.15	Coon, Carleton Stevens. The living races of man. 1. ed. N.Y., 1965.
Geog 4249.65	Dainelli, Giotto. L'esplorazione del Grande Oceano. Torino, 1965.
An 2109.65.5	De Beer, Gavin Rylands. Genetics and prehistory. Cambridge, Eng., 1965.
Geog 3580.45	Diaz-Trechuelo Spinola, Maria Lourdes. Navegantes y conquistadores vascos. Madrid, 1965.
Geog 5070.68	Dieck, Herman. The marvelous wonders of the polar world. Philadelphia, 1965?
Geog 4209.65.10	Efimov, Gerontii V. Strany i liudi. Leningrad, 1965.
Geog 600.12	Ekonomicheskaia geografiia v SSSR, istoriia i sovremennoe razvitie. Moskva, 1965.
Geog 60.3.5	Erdkunde; Archiv für wissenschaftliche Geographie. Gesamtregister, v.1-17, 1947-63. Bonn, 1965.
An 180.4	Evans-Pritchard, Edward Evans. The position of women in primitive societies. N.Y., 1965.
Geog 6100.2.12	Expéditions Polaires Françaises. Terre Adélie, 1960-62; raport d'activités. Paris, 1965.
Geog 665.138	Festkolloquim: 100 Jahre Geographie in Giessen. Giessen, 1965.
Geog 4269.65	Filipenko, Aloiz A. Na chuzhikh ulitsakh. Moskva, 1965.
Geog 5506.35.1	Fox, Luke. North-west Fox. n.p., 1965.
An 2109.65	Gluckman, Max. Politics, law and ritual in tribal society. Oxford, 1965.
Geog 618.6	Gol'denberg, Leonid A. Semen Ul'ianovich Remezov. Moskva, 1965.
Geog 4228.17.10	Golovnin, Vasilii M. Puteshestvie vokrug sveta. Moskva, 1965.
Geog 759.65	Grigor'ev, Andrei A. Razvitie teoretichekikh problem sovetskoi fizicheskoi geografii, 1917-1934 gg. Moskva, 1965.
Geog 3520.10.45.5	Hakluyt, Richard. The principal navigations. Cambridge, 1965. 2v.
Geog 5919.65.5	Hatherton, Trevor. Antarctica. Wellington, 1965.
Geog 3019.12.2	Heawood, Edward A. History of geographical discovery in the seventeenth and eighteenth centuries. N.Y., 1965.
An 2109.65.15	Herskovits, M.J. Economic anthropology; the economic life of primitive peoples. N.Y., 1965.
Geog 665.103	House, John. The frontiers of geography. New Castle upon Tyne, 1965.
Geog 500.180	International Committee of Historical Sciences. Commission Internationale d'Histoire Maritime. Compte rendu des travaux de la commission internationale d'histoire maritime, 1965. Paris? 1965?

1965 - cont.

An 3309.65.10	Joint Arabic-Polish Anthropological Expedition, 1958. Publications. Warsaw, 1965.
An 359.65.10	Jürgens, Hans W. Beiträge zur menschlichen Typenkunde. Stuttgart, 1965.
Geog 4311.23	Kazaka, Tenin. Plavashchi domovc. Plovdiv, 1965.
Geog 4145.87.5	Kharitanovskii, Aleksandr A. Chelovek s zheleznym olenem. Moskva, 1965.
Geog 624.1	Konnyüi, Leslie. John Xantus. Köln, 1965.
Geog 4217.85.13	Lapérouse, Jean François de Galaup. Voyage de Lapérouse autour du monde. Paris, 1965.
Geog 665.126	Leipziger geographische Beiträge. Text and Atlas. Leipzig, 1965. 2v.
Geog 5919.65	Lewis, Richard S. A continent for science; the Antarctic adventure. N.Y., 1965.
An 3309.65.5	Łódzskie Towarzystwo Naukowe. Wydział III, Nauk Matematyczno-Przyrodniczych. Studia afrykanistyczne. Łódź, 1965.
Geog 612.4.5	Lukina, Tat'iana A. Ivan Ivanovich Lepekhin. Leningrad, 1965.
Geog 4309.65	MacShane, Frank. The American in Europe. 1st ed. N.Y., 1965.
Geog 5650.35	Meldgaard, Jørgen. Nordboerne i Grønland. Kǿbnhavn, 1965.
Geog 4677.99	Molen, Sytze Jan van der. Goud in de golven. 's-Gravenhage, 1965.
An 359.65.5	Montagu, Ashley. The human revolution. Cleveland, 1965.
An 359.65	Montagu, Ashley. The idea of race. Lincoln, 1965.
An 359.65.20	Murdock, George Peter. Culture and society. Pittsburgh, 1965.
An 359.49.1	Murdock, George Peter. Social structure. N.Y., 1965.
Geog 665.100	National Research Council. Ad Hoc Committee on Geography. The science of geography. Washington, 1965.
Geog 4309.47.12	Noguera Mora, Neftali. Alegría y llanto de Europe; verano de 1964. 2. ed. Mérida, 1965.
Geog 6009.63.5	Nudel'man, Aizik V. Sovetskie ekspeditsyi v Antarktiku, 1961-63 gg. Moskva, 1965.
Geog 659.65	Overbeck, Hermann. Kulturlandschaftsforschung und Landeskunde. Heidelberg, 1965.
Geog 4679.12.25	Padfield, Peter. The Titanic and the Californian. London, 1965.
Geog 5559.28.5	Passarello, Gaetano. Un giornalista al Polo Nord. Roma, 1965.
Geog 6009.63	Peskov, Vasilii M. Belye suy. Moskva, 1965.
Geog 4209.65.15	Pleshakov, Leonid P. Vokrug sveta s Zarei. Moskva, 1965.
Geog 500.120	Polska Akademia Nauk. Institut Geografii. Katalog rękopisów geograficznych w zbiorach polskich. Warszawa, 1965- 2v.
Geog 4209.65	Rand, Christopher. Mountains and water. N.Y., 1965.
Geog 4029.16	Root, Jonathan. Halliburton, the magnificient myth; a biography. N.Y., 1965.
An 2339.65	Rust, Alfred. Über Waffen- und Werkzeugtechnik des Altmenschen. Neumünster, 1965.
An 5.30	Sándor, István. A Magyar néprajztudomány bibliográfiája 1945-1954. Budapest, 1965.
Geog 665.115	Sauer, Carl Ortwin. Land and life. Berkeley, 1965.
Geog 665.102	Schickel, Joachim. Terra incognita. Bergisch Gladbach, 1965.
Geog 4309.64.16	Selucký, Radoslav. Západ je Západ. 2. vyd. Praha, 1965.
Geog 4269.34.7	Shackleton, Margaret Reid. Europe: a regional geography. 7th ed. N.Y., 1965.
Geog 4209.65.5	Shelekhov, Grigorii I. Puteshestvie s 1783 po 1790 god iz okhotska po vostoch. okeanu. Ann Arbor, 1965. 2v.
Geog 5378.80.20	Skoog, Gösta. Vega. En aktualisering av händelserna kring Vega-expeditionen 1878-1880. Göteborg, 1965.
An 170.341	Slotkin, James Sydney. Reading in early anthropology. N.Y., 1965.
An 358.10.5	Smith, Samuel Stanhope. An essay on the causes of the variety of complexion and figure in the human species. Cambridge, 1965.
Geog 4332.21	Soljan, Antun. The thousand islands of the Adriatic. Beograd, 1965.
Geog 3055.60.10	Spanuth, Jürgen. Atlantis; Heimat. Tübingen, 1965.
An 170.345	Spiro, Melford E. Context and meaning in cultural anthropology. N.Y., 1965.
Geog 665.130	Steering Committee for Celebration of the Sixtieth Year of Prof. S.P. Chatterjee. Essays in geography. Calcutta, 1965.
Geog 579.65.5	Stolitsy stran mira. Moskva, 1965.
Geog 6009.62	Thomson, Robert Baden. The coldest place on earth. Wellington, 1965.
Geog 4559.65	West, Ellsworth Luce. Captain's papers. Barre, Mass., 1965.
Geog 665.101	Whittow, John Byron. Essays in geography for Austin Miller. Reading, Eng., 1965.
Geog 500.140	Wystawa pt. Rozwój Historyczny Geografii Polskiej i Pismicnictwo Polskie o Zakresu Historii Geografii, Warsaw, 1965. Catalogue of literature on the history of geography at the exposition. Warsaw, 1965.

1966

An 3559.66	Abdushelishvili, Malkhaz G. K kraniologii drevnego i sovremennogo naseleniia Kavkaza. Tbilisi, 1966.
Geog 3229.66	Alavi, S.M. Ziauddin. Geography in the Middle Ages. 1st ed. Delhi, 1966.
Geog 3590.74	Alekseev, Aleksandr. Kolumby rosskie. Magadan, 1966.
An 3509.66	Alekseev, Valerii P. Osteometriia. Moskva, 1966.
Geog 4309.66.5	Andersch, Alfred. Aus einem römischen Winter. Olten, 1966.
Geog 665.107	Andrews, John. Frontiers and men; a volume in memory of Griffith Taylor (1880-1963). Melbourne, 1966.
Geog 4238.80	Andrews, William Albert. Dangerous voyages of Captain William Andrews. N.Y., 1966.
An 2109.66	Anisimov, Arkadii F. Dukhovnaia zhizn' pervobytnogo obshchestva. Leningrad, 1966.
Geog 602.3.10	Association des Amis de l'Université de Grenoble. In memoriam Raoul Blanchard, 1877-1965. Grenoble, 1966.
Geog 623.3	Babicz, Józef. Teoria Moritza Wagnera o powstawaniu gatunków. Wrocław, 1966.
Geog 4309.66	Basnett, Fred. Travels of a capitalist lackey. South Brunswick, 1966.
An 2109.66.30	Bessaignet, Pierre. Principes de l'ethnologie économique. Paris, 1966.
Geog 4029.6	Borden, Norman E. Dear Sarah. Freeport, Me., 1966.
Geog 4324.80	Brasca, Santo. Viaggio in Terrasanta di Santo Brasca, 1480. Milano, 1966.

Chronological Listing

1966 - cont.

An 136.20	California. University. Robert H. Lowie Museum of Anthropology. The complete bibliography of Robert H. Lowie. Berkeley, 1966.
Geog 3570.55	Campos, Viriato. Viagens de Diogo Cão e de Bartolomeu Dias. Lisboa, 1966.
Geog 4308.30.5	Candolle, Augustin Pyramus de. L'Europe de 1830 vue à travers la correspondence de Augustin Pyremus de Candolle et Madame de Cércourt. Genève, 1966.
Geog 5311.3.30	Centkicwicz, Alina. Człowick o Którego upomniało się morze. Warszawa, 1966.
Geog 665.110	Chile. Universidad, Santiago. Facultad de Filosofia y Educacion. Estudios geográficos. Homenaje de la Facultad de Filosofia y Educacion a Don Huberto Fuenzalida Villegas. Santiago, 1966.
Geog 500.119	Church, Martha. A basic geographical library. Washington, 1966.
An 170.163	Conference on New Approaches in Social Anthropology, Jesus College, Cambridge, Eng., 1963. The social anthropology of complex societies. London, 1966.
Geog 3570.52	Cortesão, Jaime. Os descobrimentos pre-colombinos dos Portugueses. Lisboa, 1966.
An 2109.66.5	Cotlow, Lewis N. In search of the primitive. 1st ed. Boston, 1966.
Geog 4709.66.5	Course, Alfred George. Pirates of the Eastern seas. London, 1966.
An 5.34	Dokládal, Milan. Československá anthropologická bibliografie, 1955-1964. Brno, 1966-67.
Geog 4145.24	Eschels, Jeus Jacob. Das obenteuerliche Leben des Jeus Jacob Eschels. Hamburg, 1966.
Geog 4659.66	Farr, Grahame E. Wreck and rescue in the Bristol Channel. Truro, 1966-67.
Geog 4311.18	Feuerstein, Ernst. Deudonex. Dichtung. n.p., 1966.
Geog 619.9	Gol'denberg, Leonid A. Fedor Ivanovich Soimonov, 1692-1780. Moskva, 1966.
Geog 4759.47.3	Gourou, Pierre. Les pays tropicaux, principes d'une géographie humaine et économique. 4. éd. Paris, 1966.
Geog 4759.47.4	Gourou, Pierre. The tropical world. London, 1966.
Geog 4559.64.3	Hanke, Helmut. Männer, Planken, Ozeane. 3. Aufl. Leipzig, 1966.
Geog 5323.5	Hansen, T. Jens Munk. 4. Opl. Kjøbenhavn, 1966.
Geog 4652.5	Heden, Karl E. Direction of shipwrecks of the Great Lakes. Boston, 1966.
Geog 4209.02.10	Hellen, Gustav. Aus meiner Wanderzeit, 1902-1906 und 1908-1909. Hamburg, 1966.
Geog 4145.39.5	Hoyt, Jo Wasson. For the love of Mike. N.Y., 1966.
Geog 5839.59.2	Ingstad, Helge Marcus. Land under the pole star: a voyage to the Norse settlements of Greenland and the saga of the people that vanished. London, 1966.
Geog 759.66.5	Journaux, André. Géographie générale. Paris, 1966.
Geog 5518.92	Kallstenius, Alfhild. Tragedin i Smiths sund; Björling-Kallstenius expeditionen, 1892. Kalmar, 1966.
Geog 4269.65.12	Kares, Olavi. Kallaveden rannalta, päiväkirja valpurista 1964 valpuriin 1965. 3. painos. Porvoo, 1966.
Geog 4559.66	Karlsson, Elis. Pully-Haul: the story of a voyage. London, 1966.
Geog 4181.126	Kelly, Celsus. La Austrialia del Espiritu Santo. Cambridge, 1966. 2v.
An 2109.66.20	Kennedy, Kenneth A.R. Human skeletal remains from Chalcolithic and Indo Roman levels from Nevasa: an anthropometric and comparative analysis. Poona, 1966.
Geog 953.14	Kinauer, Rudolf. Lexikon geographischer Bilbände. Wien, 1966.
Geog 612.5	Kondracki, Jerzy. Stanisław Lencewicz. Warszawa, 1966.
Geog 4209.66.5	Kovanov, Vladimir V. Meridiany, sobytiia, vstrechi. Moskva, 1966.
An 2109.66.25	Kuehn, Herbert. Erwachen und Aufstieg der Menschheit. Frankfurt, 1966.
An 136.5.25	Lévi-Strauss, Claude. Leçon inaugurale faite le mardi. Paris, 1966.
An 2109.62.6	Lévi-Strauss, Claude. The savage mind. Chicago, 1966.
An 2109.28.7.1	Lévy-Bruhl, L. The "soul" of the primitive. N.Y., 1966.
An 2109.66.40	Lévy-Bruhl, Lucien. How natives think. N.Y., 1966.
Geog 579.66	Longman's dictionary of geography. London, 1966.
Geog 5160.59	Macdonald, Ronald. The Arctic frontier. Toronto, 1966.
Geog 4309.29.11	MacOrlan, Pierre. Villes, mémoires: Montmartre, Rouen, souvenirs de Picardie et d'Artois, Brest, Londres, Villes rhénanes. Paris, 1966.
Geog 4239.65	Manry, Robert. Tinkerbelle. N.Y., 1966.
Geog 4219.66	Marti-Ibáñez, Félix. Journey around myself. N.Y., 1966.
Geog 618.4.5	Mazurkiewicz-Herzowa, Kucja. Eugeniusz Romer. Warszawa, 1966.
An 359.66	Melady, Thomas Patrick. The revolution of color. N.Y., 1966.
Geog 4219.64	Menon, Kumara P.S. Journey round the world. Bombay, 1966.
An 99.66.5	Mercier, Paul. Histoire de l'anthropologie. Paris, 1966.
Geog 4209.66	Mikhailov, Nikolai N. Š planetoi vmestve. Moskva, 1966.
VAn 99.66.10	Moliński, Bogdan. Historia, osobowość, sztuka. Wyd. 1. Warszawa, 1966.
Geog 4209.66.15	Monteil, Vincent. Soldat de fortune. Paris, 1966.
An 170.274	Moore, Frank. Readings in cross-cultural methodology. New Haven, 1966.
Geog 3520.13	Moorehead, Alan. The fatal impact. London, 1966.
An 170.278	Mühlemann, Wilhelm Emil. Kulturanthropologie. Köln, 1966.
An 2109.66.15	Narr, Karl J. Handbuch der Urgeschichte. Bern, 1966. 2v.
An 2109.66.10	Nash, Manning. Primitive and peasant economic systems. San Francisco, 1966.
Geog 110.1.20.15	Natsional'nyi Komitet Sovetskikh Geografov. XX mezhdunarodnyi geograficheskii kongress, London, iiul 1964 gg. Moskva, 1966.
Geog 5509.66	Neatby, Leslie Hamilton. Conquest of the last frontier. Toronto, 1966.
Geog 619.8	Novlianskaia, Mariia G. Filipp Iogann Stralenberg. Leningrad, 1966.
Geog 759.59.7	Obst, Erich. Lehrbuch der allegemeinen Geographie. 3. Aufl. v.1,6,7. Berlin, 1966- 3v.
Geog 3520.14	Pennington, Loren E. Hakluytus posthumus; Samuel Purches and the promotion of English overseas expansion. Emporia, 1966.
An 359.66.5	Podol'nyi Roman G. Predki i my. Moskva, 1966.
Geog 4209.66.10	Pogačnik, Bogdan. Povsod so ljudje. Maribor, 1966.
Geog 5985.253.10	Pound, Reginald. Scott of the Antarctic. London, 1966.
Geog 4209.66.20	Quilici, Folco. Giramare. Roma, 1966.
Geog 4559.66.5	Randier, Jean. Hommes et navires au Cap Horn, 1616-1939. Paris, 1966.
Geog 618.5.10	Ratzel, Fridrich. Jugenderinnerungen. München, 1966.
Geog 4131.66	Ritter, Wigand. Fremdenverkehr in Europa. Leiden, 1966.
Geog 4309.64.31	Sansom, William. Away to it all. N.Y., 1966.
Geog 4209.66.25	Scott, Jack Denton. Passport to adventure. N.Y., 1966.
An 2059.66.5	Semenov, Iurii I. Kak vozniklo chelovechestvo. Moskva, 1966.
Geog 4311.22	Sergeeva, Nelli A. Dunai-reka druzhby. Odessa, 1966.
An 2600.5.2	Singh, Joseph. Wolf-children and feral man. Hamden, 1966.
Geog 4239.66	Snaith, William. Across the western ocean. 1st ed. N.Y., 1966.
Geog 579.66.5	Soto Mora, Consuelo. Glosario de términos geográficos. 1. ed. México, 1966.
Geog 6100.37.10	Temple, Philip. The sea and the snow; the South Indian Ocean Expedition to Heard Island. Melbourne, 1966.
Geog 619.12	Termer, Franz. Karl Theodor Sapper, 1866-1945. Leipzig, 1966.
Geog 5070.70	Tilman, Harold William. Mostly mischief: voyages to the Arctic and to the Antarctic. London, 1966.
An 99.66	Tokarev, Sergei A. Istoriia russkoi etnografii, dook tiabr'skii period. Moskva, 1966.
Geog 4429.61.5	Toussaint, Auguste. History of the Indian Ocean. Chicago, 1966.
Geog 4429.61	Toussaint, Auguste. History of the Indian Ocean. London, 1966.
Geog 4309.66.10	Utchenko, Sergei L. Glazami istorika. Moskva, 1966.
Geog 4709.66	Wagner, Kip. Pieces of eight; recovering the riches of a lost Spanish treasure fleet. 1. ed. N.Y., 1966.
Geog 4218.57.10	Wallisch, Friedrich. Sein Schiff hiess Novara. Wien, 1966.
Geog 665.105	Weigt, Ernst. Angewandte Geographie; Festschrift für Professor Dr. Erwin Scheu. Nürnberg, 1966.
Geog 4309.47.2	Wilson, Edmund. Europe without Baedeker. 2nd ed. N.Y., 1966.
Geog 6009.01.20	Wilson, Edward Adrian. Diary of the 'Discovery' expedition to the Antarctic regions, 1901-1904. London, 1966.
Geog 4182.66	Witsen, Nicolaas. Moscovische reyse 1664-1665. 's-Gravenhage, 1966-67. 3v.
Geog 623.2	Wright, John K. Human nature in geography. Cambridge, 1966.
Geog 4181.33.1	Yule, Henry. Cathay and the way thither. v.1-4. Taipei, 1966. 2v.

1967

Geog 4145.3	Allen, Katharine M. Foreign service diary. Washington, 1967.
Geog 4309.60	Anjaneyulu, D. Window to the West. 1st ed. Madras, 1967.
Geog 3019.67	Baker, John Norman Leonard. A history of geographical discovery and exploration. N.Y., 1967.
Geog 3019.31.2	Baker, John Norman Leonard. A history of geographical discovery and exploration. N.Y., 1967.
Geog 4219.58.10	Bedford, Jimmy. Around the world on a nickel. New Delhi, 1967.
Geog 4341.73.16	Benjamin ben Jonah, of Tudela. Beniamini Tudelensis Itinerarium; ex versione Benedicti Ariae Montani. Bononiae, 1967.
Geog 4181.131	Bishop, Charles. The journal and letters of Captain Charles Bishop on the north-west of America. Cambridge, 1967.
Geog 5882.5	Bjoergmose, Rasmus. Stensnaes - den glemte kirke. Odense, 1967.
Geog 4145.14.10.5	Brazão, Eduardo. Les Corte-Real et le nouveau monde. Lisbonne, 1967.
Geog 4659.67	Burman, Jose. Great shipwrecks off the coast of southern Africa. Cape Town, 1967.
An 3309.67	Chai, Chen Kang. Taiwan aborigines. Cambridge, 1967.
Geog 4219.67	Chichester, Francis Charles. Gipsy Moth circles the world. N.Y., 1967.
An 2109.67	Clark, John Grahame Douglas. The Stone Age hunters. London, 1967.
Geog 3019.42.4	Clozier, René. Histoire de la géographie. 4. éd. Paris, 1967.
Geog 4679.67.30	Committee of Scientists on the Scientific and Technological Aspects of the Torrey Canyon Disaster. The Torrey Canyon; report. London, 1967.
Geog 4145.56	Cordier, Stéphane. Balthazar de Monconys. Bruxelles, 1967.
An 127.5	Currelly, Charles T. I brought the ages home. 1. ed. Toronto, 1967.
VAn 3209.67	Czekanowski, Jan. Człowiek w czasie i przestrzeni. 3. wyd. Warszawa, 1967.
Geog 4145.52	Dainelli, Giotto. Il duca degli Abruzzi. Torino, 1967.
Geog 4345.99.10	Davies, David William. Elizabethans errant; the strange fortunes of Sir Thomas Sherley and his three sons. Ithaca, N.Y., 1967.
Geog 4332.22	Denham, Henry Mangles. The Adriatic; a sea guide to its coasts and islands. London, 1967.
Geog 4209.67.10	Díaz Plaja, Guillermo. Con variado rumbo. 1. ed. Barcelona, 1967.
Geog 758.07.5	Dicuil. Liber de mensura orbis terrae. Dublin, 1967.
Geog 614.10	Dontsova, Zoia N. Sergei Semenovich Neustruev, 1874-1928. Moskva, 1967.
Geog 4249.67.5	Downs, Hugh. A shoal of stars. Garden City, 1967.
Geog 3650.5	Dube, Beehan. Geographical concepts in ancient India. Varanasi, 1967.
Geog 603.5.8.5	Emmanuel, Marthe. Tel fut Charcot, 1867-1936. Paris, 1967.
Geog 4209.67.15	Fiore, Ilario. L'italiano di Ponte Cayumba. Firenze, 1967.
Geog 4209.60.45	Firsé, Adolf. Reise-Journal. Gütersloh, 1967.
An 2109.67.15	Firth, Raymond William. Themes in economic anthropology. London, 1967.
An 170.202	Ford, Clellan Stearns. Cross-cultural approaches. New Haven, 1967.
An 2109.67.40	Fried, Morton Herbert. The evolution of political society: an essay in political anthropology. N.Y., 1967.
Geog 4249.67	Friis, Herman Ralph. The Pacific Basin; a history of its geographical exploration. N.Y., 1967.
Geog 5639.67	Gad, Finn. Grønlands historie. København, 1967- 2v.
Geog 4679.67	Gill, Crispin. The wreck of the Torrey Canyon. Newton Abbot, 1967.
An 379.67	Glacken, Clarence J. Traces on the Rhodian shore. Berkeley, 1967.
Geog 5300.47	Gordienko, Pavel A. Die Polarforschung der Sowjetunion. Düsseldorf, 1967.

Chronological Listing

1967 - cont.

An 2109.67.5 Goremykina, Vera I. Istoriia pervobytnogo obshchestva. Minsk, 1967.

Geog 3590.75 Gvozdetskii, Nikolai Andreevich. Sovetskie geograficheskie issledovaniia i otkrytiia. Moskva, 1967.

Geog 6009.01.25 Hardy, Alister C. Great waters; a voyage of natural history to study whales, plankton and the waters of the southern ocean. N.Y., 1967.

Geog 4559.67.10 Heikkinen, Helge. Runt Kap Horn med Herzogin Cecilie. Ekenäs, 1967.

VGeog 3019.56.4 Herrmann, Paul. Historia de los descubrimientos geográficos. 2. ed. Barcelona, 1967.

Geog 4329.67 Houston, James M. The western Mediterranean world. N.Y., 1967.

Geog 4269.67.10 Houter, F. den. Vluchtige verlenningen. Laven, 1967.

Geog 4559.67.15A Hugill, Stan. Sailortown. London, 1967.

An 170.238.10 International Congress of Anthropological and Ethnological Sciences, 7th, Moscow, 1964. VII Mezhdunarodnyi Kongress antropologicheskikh i ethnograficheskikh nauk. Trudy. v.1-3,5-10; photoreproduction. Moskva, 1967- 13v.

Geog 110.1.20.10 International Geographical Congress, 20th, London, 1964. Congress proceedings. London, 1967.

Geog 4145.37 Jacoby, Arnold. Señor Kon-Tiki; the biography of Thor Heyerdahl. Chicago, 1967.

Geog 665.118 Johnston, William. Dynamic relationships in physical geography. Christchurch, 1967.

An 629.67 Jongnians, Douve Geert. Anthropologists in the field. Assen, 1967.

Geog 4559.67.5 Jupp, Ursula. Home port Victoria. Victoria, 1967.

Geog 5070.72 Kosack, Hans P. Die Poterforschung; ein Datenbuch über die Natur-, Kultur-, Wirtschaftsverhältnisse und die Erforschungsgeschichte der Polarregionen. Braunschweig, 1967.

Geog 5650.25 Krogh, Knud J. Erik den Rødes Grønland. København, 1967.

Geog 5650.25.2 Krogh, Knud J. Viking Greenland with a supplement of saga. København, 1967.

Geog 3594.10 Kuźmiński, Bolesław. Polskie nazwy na mapie śceiata. Wyd. 1. Warszawa, 1967.

Geog 819.59.4 Länder der Erde. 4. Aufl. Berlin, 1967.

An 2109.67.35F Laet, Sigfried Jan de. De voorgeschiedenis van Europa. Hasselt, 1967.

Geog 4269.67.5 Le Lannou, Maurice. Le déménagement du territoire. Paris, 1967.

An 359.67 Lévi-Strauss, Claude. The scope of anthropology. London, 1967.

Geog 4309.67 Lorck, Carl von. Europa Privat. Frankfurt, 1967.

An 379.67.10 Lundman, Bertil. Geographische Anthropologie. Stuttgart, 1967.

Geog 4679.67.10 Mabire, Jean. La marée noire du Torrey Canyon. Paris, 1967.

Geog 4269.67 Maevskii, Viktor. V. Evropa bez dzhentl'menov. Moskva, 1967.

Geog 4217.91.2 Marshall, James S. Vancouver's voyage. 2. ed. Vancouver, 1967.

Geog 665.142 Mass, Walther Gerhard Eduard. Menschen und Landschaften. Hildesheim, 1967.

Geog 665.152 Mélanges de géographie physique. Gembloux, 1967. 2v.

Geog 5518.50.15 Miertsching, Johann August. Frozen ships; the Arctic diary of Johann Miertsching, 1850-1854. N.Y., 1967.

Geog 4209.67.5 Mikosha, Vladislav V. Gody i strany. Moskva, 1967.

Geog 3240.26.5 Miquel, André. La géographie humaine du monde musalman jusqu'au milieu du 11e siècle. Paris, 1967.

Geog 4306.17.4 Moryson, Fynes. Shakespeare's Europe: a survey of the condition of Europe at the end of the 16th century. 2nd ed. N.Y., 1967.

Geog 5209.67 Mowat, Farley. The polar passion: the quest for the North Pole. Toronto, 1967.

Geog 4678.59 Mudie, Ian. Wreck of the Admella. London, 1967.

An 5.33 Murdock, George Peter. Ethnographic atlas. Pittsburg, 1967.

Geog 3590.83 Novokshanova, Zinaida K. Kartograficheskie i geodezicheskie raboty v Rossii v XIX-nachale XX v. Moskva, 1967.

Geog 600.21 Olszewicz, Bolesław. Dziewięć wieków geografii polskiej. Warszawa, 1967.

Geog 4308.48.11 Ow, Max von. Mit dem jüngsten Sohn König Ludwigs I. als Reisebegleiter nach Spanien 1848-1849. München, 1967.

VAn 629.67.5 Pešič-Golutovic, Zagorka. Antropologia kao drustveno nauka. Beograd, 1967.

Geog 5379.67 Petrow, Richard. Across the top of Russia. N.Y., 1967.

Geog 4307.84 Piozzi, Hester Lynch Thrale. Observations and reflections made in the course of a journey through France, Italy and Germany. Ann Arbor, 1967.

Geog 4209.67 Pochivalov, Leonid V. My za granitsei. Moskva, 1967.

Geog 3410.7 Problems and trends in American geography. N.Y., 1967.

VGeog 4559.67 Proface, Bruno. Nebo, more i ponekad kopno. Zagreb, 1967.

Geog 6100.77.5 Quartermain, Leslie B. South to the Pole: the early history of the Ross Sea sector, Antarctica. London, 1967.

An 359.67.5 Race and modern science. N.Y., 1967.

Geog 4239.66.5 Ridgway, John M. A fighting chance. London, 1967.

An 379.67.5 Russell, William Moy Stratton. Man, nature and history. London, 1967.

Geog 4249.67.10 Sandvad, Jørgen. Rejsernes rytme. Kronikker. København, 1967.

An 170.335 Schmitz, Carl August. Historische Völkerkunde. Frankfurt am Main, 1967.

VGeog 4131.67.5 Seminàr o Ekonomickej Efektívnosti Iuvesticii Cestovného Ruchu, Pieštány, 1966. Efektivnost' investicí cestovného ruchu. Bratislava, 1967.

An 2109.67.20 Sigrist, Christian. Regulierte Anarchie. Olten, 1967.

Geog 4219.66.5 Simpson, Colin. Sir Francis Chichester: voyage of the century. London, 1967.

An 170.200 Social organization; essays presented to Raymond Firth. London, 1967.

Geog 4131.67 Spain. Comisaría del Plan de Desarrollo Economico y Social. Comisión de Turismo. Turismo. Madrid, 1967?

Geog 659.67 Storkebaum, Werner. Zum Gegenstand und zer Methode der Geographie. Darmstadt, 1967.

Geog 665.124 Symposium über Fragen der Naturräumlichen Gliederung. Probleme der landschaftsökölogischen Erkundung und naturräumlichen Gliederung. Leipzig, 1967.

Geog 4145.83.5 Szumańska-Grossowa, Hanna. Podróze Stefana Srolca Rogozinskiego. Warszawa, 1967.

Geog 4750.4 Texas Instruments, Inc. An inventory of geographic research of the humid tropic environment. Dallas, Texas, 1967? 2v.

Geog 4279.67 Trease, Geoffrey. The grand tour. London, 1967.

1967 - cont.

An 2109.54.6 Values of Primitive Society (Radio Program). The institutions of primitive society. Oxford, 1967.

Geog 500.182 Vinge, Clarence L. United States government publications for research and teaching in geography. Totowa, N.J., 1967.

Geog 4239.67 Volovich, Vitalii G. Tridtsatyi meridian. Moskva, 1967.

Geog 5925.91 Vsesoiuznoe Soveshchenie po Izucheniiu Antarktiki, Moscow, 1966. Osnovnye itogi izucheniia Antarktiki za 10 let. Moskva, 1967.

Geog 5326.21.20 Weems, John Edward. Peary, the explorer and the man, based on his personal papers. Boston, 1967.

Geog 4659.63.4 Williams, Peter J. Shipwrecks at Port Phillips Heads since 1840. Melbourne, 1967.

Geog 4145.94 Willis, William. The hundred lives of an ancient mariner. London, 1967.

Geog 3590.76 Wotte, Herbert. Kurs auf Unerforscht. Leipzig, 1967.

An 609.67 Zagreb. Etnografski Muzej. Daleki sujtovi naših putnika i pomoraea. Zagreb, 1967.

An 2059.67 Zinevich, Galina P. Ocherki paleoantropologii Ukrainy. Kiev, 1967.

1968

Geog 4131.68.10 Aeschlimann, Jean Louis. Structure et tâches d'un organisme national de tourisme. Inaug. Diss. Bern, 1968.

An 359.68.5A American Association for the Advancement of Science. Science and the concept of race. N.Y., 1968.

VGeog 4131.68 Anan'ev, Mikhail A. Mezhdunarodruji turizm. Moskva, 1968.

Geog 4559.68.5 Armstrong, Richard. A history of seafaring. N.Y., 1968-69. 3v.

An 2109.68.25 Association of Social Anthropologists of the Commonwealth. History and social anthropology. London, 1968.

Geog 5900.16 Barkov, N.I. Desiat' let sovetskikh issledovanii v Antarktike. Leningrad, 1968.

An 359.68.25 Bates, Marston. Gluttons and libertines; human problems of being natural. N.Y., 1968.

An 2109.68.5 Birket Smith, Kaj. Strejftog. Arktiske, tropiske og midt im ellem. København, 1968.

Geog 4182.70 Booy, A. de. De derde reis van de V.O.C. naar Oost-Indië onder het beleid van Admiraal van Caerden. 's-Gravenhage, 1968.

Geog 5839.68 Boucht, Christer. Grönland tvärs. Helsingfors, 1968.

Geog 665.120 Bowen, Emzys George. Geography at Aberystwyth: essays written on the occasion of the departmental jubilee 1917-1918-1967-1968. Cardiff, 1968.

VGeog 4349.68 Bregman, Aleksander. Rubiezè woluorci. Londyn, 1968.

Geog 4239.68 Cassidy, Vincent H. The sea around them; the Atlantic Ocean, A.D. 1250. Baton Rouge, 1968.

Geog 3055.78 Cayce, Edgar Evans. Edgar Cayce in Atlantis. N.Y., 1968.

An 170.161 Cohen, Yehudi A. Man in adaptation. Chicago, 1968. 2v.

Geog 4209.68.5 Cole, Jean. Trimaran against the trades. Wellington, 1968.

Geog 759.68 Cole, John Peter. Quantitative geography. London, 1968.

Geog 665.112 Colloque International de Géographie Appliquée, Liège, 1967. Comptes rendus. Liège, 1968.

An 3209.68.5 Contributions to the physical anthropology of central Asia and the Caucasus. Cambridge, 1968.

Geog 4131.68.20 Corna Pellegrini, Giacomo. Studi e ricerche sulla regione turistica. Milano, 1968.

Geog 607.5.15 Davydov, Iurii Vl. Golóvnin. Moskva, 1968.

Geog 619.10 Dobrowolska, Maria. Ludomir Sawicki. Warszawa, 1968.

Geog 4679.67.20 Du Pontavice, Emmanuel. La pollution des mers..."Torrey Canyon". Paris, 1968.

Geog 4659.68.5 Farr, Grahame E. Wreck and rescue on the coast of Devon. Truro, 1968.

Geog 665.122 Festschrift für Hans Kinzl zum siebzigsten Geburtstag. Innsbruck, 1968.

Geog 5518.18.7 Fisher, Alexander. Journal of a voyage of discovery to the Arctic regions. London, 1968.

An 2109.68.35 Forno, Mario. L'acculturazione dei popoli primitivi. Roma, 1968.

Geog 665.134 Geographers in government: a series of papers given at meetings of the Geography Section of the AAAS in New York N.Y., 1968.

An 379.68.10 Gerling, Walter. Die Problematik der Sozialgeographie. Würzburg, 1968.

An 181.1.2 Goldenweiser, Alexander A. History, psychology, and culture. Gloucester, 1968.

Geog 4181.132 Gomes de Brito, Bernardo. Further selections from the tragic history of the sea, 1559-1565. Cambridge, 1968.

Geog 4419.68 Gotev, Goran. Probuzhdane to na Shekherazada. Sofiia, 1968.

Geog 4209.68.10 Gritskevich, Valentin P. Puteshestviia nashikh zemliakov. Minsk, 1968.

Geog 3129.68 Guarnieri, Giuseppe Gino. Le correnti del pensiero geografico nell'antichità classica e il lorocontributo alla cartografia nautica medioevale. Pisa, 1968-69. 2v.

Geog 3249.68 Hale, John Higby. Renaissance exploration. London, 1968.

An 99.68 Harris, Marvin. The rise of anthropological theory. N.Y., 1968.

Geog 4239.68.15 Haward, Peter. High latitude crossing. London, 1968.

Geog 6009.64.5 Hayter, Adrian. The year of the quiet sun: one year at Scott Base, Antarctica. London, 1968.

An 379.68.15 Heinemeyer, W.F. De sociale geografie in de rij van de sociale wetenschappen. Meppel, 1968.

Geog 5985.122 Herbert, Wally. A world of men: exploration in Antarctica. London, 1968.

Geog 4239.65.5 Hiscock, Eric C. Atlantic cruise in Wanderer III. London, 1968.

Geog 5312.3.10 Hoyt, Edwin Palmer. The last explorer; the adventures of Admiral Byrd. N.Y., 1968.

Geog 110.1.21 International Geographical Congress, 21st, Delhi, 1968. Abstracts of papers. Supplement. Calcutta, 1968.

Geog 4131.68.15 Internationale Informationstagung zur Geographie des Fremdenverkehrs. Dresden, 1965. Probleme der Geographie des Fremdenverkehrs der Deutschen Demokratischen Republik und anderer Staaten. Leipzig, 1968.

Geog 4218.68.10 Jentsch, August. Ein Mann und ein Traktor auf Weltreise. v.2. Wien, 1968-69.

Geog 4346.83 Kaempfer, Engelbert. Die Reisetagebücher. Wiesbaden, 1968.

Geog 579.68 Kalesnik, Stanislaw W. Entsiklopedicheskii slovar' geograficheskikh terminov. Moskva, 1968.

Geog 5379.67.5 Karalt, Charles. To the top of the world. 1. ed. N.Y., 1968.

Chronological Listing

1968 - cont.

VGeog 4269.68	Lah, Avguštin. Naše sosedne države. Ljubljana, 1968.
Geog 4679.68	Lambert, Max. The Wahine disaster. Wellington, 1968.
Geog 4181.133	Leichhardt, Ludwig. The letters of F.W. Ludwig Leichhardt. Cambridge, Eng., 1968. 3v.
An 2109.62.7	Lévi-Strauss, Claude. The savage mind. Chicago, 1968.
An 136.5.50	Lima, Luiz Costa. O estruturalismo de Lévi-Strauss. Petrópolis, 1968.
Geog 4309.68	Lisakovskii, Igor' N. Dalnie sosedi. Odessa, 1968.
Geog 759.56	Lobeck, Armin Kohl. Things maps don't tell us; an adventure into map interpretation. N.Y., 1968.
Geog 4659.68.10	Loney, Jack Kenneth. Wrecks on the Gippsland coast. Geelong, 1968.
Geog 665.116	Lulovac, Milisav. Cvijićev zbornik u spomen 100. godišnjice njegovog rodjeuja. Beograd, 1968.
Geog 4209.68	Maevskii, Viktor V. Polmilliona kilometrov pozadi. Moskva, 1968.
Geog 4659.68	Malster, Robert. Wreck and rescue on the Essex coast. Truro, 1968.
Geog 5925.92	Markov, Konstantin K. Geografiia Antarktidy. Moskva, 1968.
Geog 610.5	Martin, Geoffrey J. Mark Jefferson, geographer. Ypsilanti, Mich., 1968.
Geog 4131.68.5	Meinke, Hans. Tourismus und wirtschaftliche Entwicklung. Göttingen, 1968.
An 2109.68	Montagu, Ashley. The concept of the primitive. N.Y., 1968.
An 170.275.10	Montagu, Ashley. Culture: man's adaptive dimension. N.Y., 1968.
An 2059.68.5	Nantevil, Hugues de. Sur les traces d'Adam; nouvel aperçu sur les origines de l'homme. Paris, 1968.
Geog 4239.68.10	Naydler, Merton. The penance way; the mystery of Puffin's Atlantic voyage. London, 1968.
Geog 4209.68.15	Ogni dalekikh gorodov. Moskva, 1968.
Geog 616.31	Olszewicz, Bolesław. Stanisław Pawłowski. Warszawa, 1968.
An 99.68.5	One hundred years of anthropology. Cambridge, Mass., 1968.
Geog 4300.51	Owen, Charles. Britons abroad; a report on the package tour. London, 1968.
Geog 4418.22	Owen, William Fitz William. Narrative of voyages to explore the shores of Africa, Arabia, and Madagascar. Farnborough, 1968. 2v.
An 2109.68.15	Pershits, Abram I. Istoriia pervobytnogo obshchestva. Moskva, 1968.
Geog 4019.68	Petro, W. Triple commission. London, 1968.
Geog 4679.67.25	Petrow, Richard. The black tide: in the wake of the Torrey Canyon. London, 1968.
Geog 4559.68	Phillips-Burt, Douglas H.C. Reflections in the sea. Lymington, 1968.
Geog 4269.68.5	Phloros, Paulos. Eurōpaikē suphmōnia taxidia. Athēnai, 1968.
An 359.68.10	Poirier, Jean. Ethnologie générale. Paris, 1968.
Geog 185.8	Polskie Towarzystwo Geograficzne. Polskie towarzystwo geograficzne w 50 rocznicę dzialalności. Warszawa, 1968.
Geog 185.15	Polsko-Czeskie Seminarium Geograficzne, 3rd, Warsaw, 1967. Polsko-czeskie seminarium geograficzne. Wyd. 1. Warszawa, 1968.
Geog 4509.68	Potočnik, Miha. Srečanja z gorami. Ljubljana, 1968.
An 629.68.5	Pouwer, Jan. Translation at sight; the job of a social anthropologist. Wellington, 1968.
An 379.68	Priroda i obshchestvo. Moskva, 1968.
Geog 3520.11.10	Quinn, David Beers. A study of the facsimile edition of Richard Hakluyt's Divers voyages. Amsterdam, 1968. 2v.
An 359.68.30	Rassengeschichte der Menschheit. München, 1968- 3v.
Geog 4312.5	Raynes, Rozelle. North in a nutshell. Lymington, 1968.
An 2109.68.20	Razlozhenie rodovogo stroia i formirovanie klassovogo obshchestva. Moskva, 1968.
Geog 5399.64.1	Report from the Arctic: foreign and Soviet correspondents on their trip aboard the Soviet atomic icebreaker. Moscow, 1968.
An 629.68	Research in social anthropology. London, 1968.
Geog 4209.68.25	Rodin, Leonid Efimovich. Po iuzhnym stranam. Moskva, 1968.
Geog 4219.66.10	Rowland, John. Lone adventurer: the story of Sir Francis Chichester. London, 1968.
Geog 4239.68.5	Sauer, Carl Ortwin. Northern mists. Berkeley, 1968.
Geog 5985.187.5	Saziumov, Evgenii M. A life given to the Antarctic; Douglas Mawson - Antarctic explorer. Adelaide, 1968.
Geog 3550.16	Scarin, Maria Luisa. Viaggi ed esplorazioni di capitani marittimi della Riviera di Levante nella prima metà del secolo XIV. Genova, 1968.
Geog 4182.69	Schagen, Adriaen. Reijse gadaen bij Adriaen Schagen. 's-Gravenhage, 1968.
An 2109.68.10	Semenov, Sergei A. Razvitie tekhniki v kameunom veke. Leningrad, 1968.
An 136.5.10	Simonis, Yvan. Claude Lévi-Strauss ou la Passion de l'inceste. Paris, 1968.
Geog 4709.68	Skriagin, Lev N. Sokrovishcha pogibshikh korablei. Moskva, 1968.
An 359.68.15	Stocking, George W. Race, culture, and evolution. N.Y., 1968.
Geog 5398.97.6	Sundman, Per Olaf. Ingen fruktan, intet hopp. Stockholm, 1968.
Geog 3520.17.6	Taylor, Eva G. Late Tudor and early Stuart geography, 1583-1650. N.Y., 1968.
Geog 3590.85	Tikhomirov, Georgii S. Bibliograficheskii ocherk istorii geografii v Rossii XVIII veka. Moskva, 1968.
VAn 359.68	Tokarev, Sergei A. Oskovy etnografii. Moskva, 1968.
Geog 3594.15	Turley, Tomasz J. Polacy badacze Ameryki. Chicago, 1968.
Geog 4521.27	Ullman, James Ramsey. Straight up; the life and death of John Harlin. Garden City, N.Y., 1968.
An 2109.68.30.2	Varagnac, André. L'homme avant l'écriture. 2e éd. Paris, 1968.
An 379.68.20	Vermooten, Willem Hendrik. Sociografie en sociale geografie in Nederland. Assen, 1968.
Geog 6009.64	Vojtěch, Vaclav. Námořníkem, topičem a psovodem za jižním polárním kruhem. Vyd. 1. Praha, 1968.
Geog 4139.68	Wertheim, Willem Frederik. Ketters en kwezels, regenten en rebellen. Drachten, 1968.
Geog 579.68.5	Westermann Lexikon der Geographie. Braunschweig, 1968-72. 5v.
An 126.8	Whitehill, Walter Muir. A memoir of John Otis Brew. Boston, 1968.
An 3559.68.5	Zinevich, Galina P. Antropologichna kharakteristika davn'oho naselennia teritorii Ukrainy. Kyiv, 1968.
An 3559.68	Zubov, Aleksander A. Odontologiia. Moskva, 1968.
An 3209.68	Zubov, Aleksandr A. Problemy evoliutsii cheloveka i ego ras. Moskva, 1968.

1968 - cont.

An 379.68.5	Zum Standort der Sozialgeographie. Kallmünz, 1968.

1969

An 3559.69	Alekseev, Valerii P. Proiskhozhdenie narodov Vostochnoi Evropy. Moskva, 1969.
An 2059.69.10	Altner, Günter. Kreatur Mensch. München, 1969.
Geog 4279.69	Anderson, Patrick. Over the Alps. London, 1969.
Geog 5650.30F	Aron, of Kangeg. K'avdlunätsianik. Godthåt, 1969.
Geog 4181.136	Barbour, Philip L. The Jamestown voyages under the first charter, 1606-1609. Cambridge, Eng., 1969. 2v.
Geog 3055.82	Berlitz, Charles Frambach. The mystery of Atlantis. N.Y., 1969.
An 359.69.10	Blacking, John. Process and product in human society. Johannesburg, 1969.
Geog 4709.69.5	Blond, Georges. Histoire de la flibuste. Paris, 1969.
Geog 4209.69.20	Boldyrev, Ivan I. Na raznykh shirotakh. Riga, 1969.
Geog 4652.8	Boswell, David B. Loss list of Grimsby vessels, 1800-1960. Grimsby, 1969.
Geog 4559.69	Bradley, Wendell P. They live by the wind. 1. ed. N.Y., 1969.
Geog 4239.69.5	Castelbajac, Bertrand de. De Plymouth à Newport par le sud de Nantucket. Paris, 1969.
An 186.2.5	Charbonnier, Georges. Conservations with Claude Lévi-Strauss. London, 1969.
Geog 3019.69	Chaunu, Pierre. L'expansion européene du XIIIe au XVe siècle. 1. éd. Paris, 1969.
An 170.163.15	Conference on New Approaches in Social Anthropology, Jesus College, Cambridge, Eng., 1963. The social anthropology of complex societies. London, 1969.
Geog 659.69.5	Conference on Quantitative Methods in Geography, New York, 1969. Quantitative methods in geography; a symposium. N.Y., 1969.
Geog 603.7	Constantini, Otto. Leben und Wirken eines österreichischen Geographieprofessors und Erwachsenenbildners. Linz, 1969.
Geog 665.116.5	Cvijić, Jovan. Opšta geografija; antropogeogratija. Beograd, 1969.
Geog 4479.69	Darvère, Pierre. Le Baitoto. Paris, 1969.
Geog 5050.10	David Davis Polar Library. Catalog of the David Davis Polar Library. Menlo Park, 1969.
Geog 4679.69.10	Doak, Wade. The Elingamite and its treasure. London, 1969.
Geog 4309.69	Duffus, Robert Luther. The polar route to time gone by. 1st ed. N.Y., 1969.
Geog 4247.69	Dunmore, John. The fateful voyage of the St. Jean Baptiste. Christchurch, 1969.
An 359.69.20	Etnologia. Wyd. 1. Warszawa, 1969.
An 359.06.4.2	Finot, Jean. Race prejudice. Miami, 1969.
Geog 4559.69.5	Fréminville, René Marie de La Poix de. La vie quotidienne des marins au Moyen Age. Paris, 1969.
Geog 4131.69	Frentrup, Klaus. Die ökonomische Bedeutung des internationalen Tourismus fur die Ehtwicklungsländer. Hamburg, 1969.
Geog 3019.69.5	Fuson, Robert Henderson. A geography of geography; origins and development of the discipline. Dubuque, Iowa, 1969.
Geog 3055.80	Galanopoulos, Angelos Georgiou. Atlantis, the truth behind the legend. Indianapolis, 1969.
Geog 3055.87	Gleich, Sigismund von. Der Mensch der Eiszeit und Atlantis. Stuttgart, 1969.
Geog 665.146	Gottmann, Jean. The renewal of the geographic environment: an inaugural lecture delivered before the University of Oxford on 11 February 1969. Oxford, 1969.
Geog 5399.32.8	Grigor'ev, Origonii K. Dorogi vedut v Arktiku. Moskva, 1969.
Geog 4110.38	Gunther, John. Twelve cities. 1. ed. N.Y., 1969.
Geog 4145.34.25	Guzanov, Vitalii G. Odissei s Beloi Rusi. Minsk, 1969.
Geog 4208.81.10	Haeckel, Ernst. Tropenfahrten Reiseschilderungen aus Ceylon, Java und den Mittelmeergebieten. Leipzig, 1969.
Geog 5318.2.10	Henson, Matthew Alexander. A Negro explorer at the North Pole. N.Y., 1969.
Geog 4279.69.5	Hibbert, Christopher. The grand tour. London, 1969.
An 359.69.15	Hiernaux, Jean. Egalité ou inégalité des races? Paris, 1969.
Geog 4139.69	Histoire générale des grands aventuriers de la mer. v.1,4-15,17-18. Paris, 1969. 16v.
Geog 4652.7	Hocking, Charles. Dictionary of disasters at sea during the age of steam, 1824-1962. London, 1969. 2v.
Geog 5398.79.35	Hoehling, Adolph A. The Jeannette expedition, an ill fated journey to the Arctic. London, 1969.
Geog 5399.28.35	Hogg, Garry. Airship over the Pole; the story of the Italia. London, 1969.
An 136.5.65	Ipola, Emilio Rafael de. Le structuralisme, ou L'histoire en exil. Thèse. Nantorre? 1969.
VGeog 4209.69.5	Karlin, Alma M. Samotno potovanje. Ljubljana, 1969.
Geog 5919.69	King, H.G.R. The Antarctic. London, 1969.
Geog 659.69	King, Leslie J. Statistical analysis in geography. Englewood Cliffs, 1969.
Geog 4219.69.5	Knox-Johnston, Robin. A world of my own: the single-handed, non-stop circumnavigation of the world in Suahili. London, 1969.
An 170.228	Kontakte und Grenzen. Göttingen, 1969.
Geog 4659.69.5	Larn, Richard. Cornish shipwrecks. Newton Abbot, 1969- 2v.
Geog 6009.01.30	Lashly, William. Under Scott's command: Lashly's Antarctic diaries. London, 1969.
An 609.69	Leipzig. Staedtisches Museum für Völkerkunde. Zum hundertjährigen Bestehen, 1869-1969. Berlin, 1969.
An 359.69	Leiris, Michel. Cinq études d'ethnologie. Paris, 1969.
Geog 4219.64.5	Lewis, David Henry. Children of three oceans. London, 1969.
Geog 4309.29.12	MacOrlan, Pierre. Villes, mémoires: Montmartre, Rouen, souvenirs de Picardie et d'Artois. 1. éd. Evreux, 1969.
Geog 4679.12.30	Marcus, Geoffrey. The maiden voyage (Titanic, steamship). London, 1969.
Geog 3530.20	Meynier, André. Histoire de la pensée géographique en France, 1872-1969. Paris, 1969.
An 359.69.30	Montagu, Ashley. Man, his first two million years. N.Y., 1969.
An 136.5.20	Moravio, Sergio. La ragione nascosta. Firenze, 1969.
Geog 4709.69	Morris, Roland. Island treasure: the search for Sir Cloudesley Shovell's flagship "Association". London, 1969.
Geog 5180.37	Nas razdeliaet okean. Magadan, 1969.
Geog 4328.73	Nāsir, al-Dīn Shāh. Ein Harem in Bismarcks Reich. Tübingen, 1969.
Geog 4419.69	Nolan, Cynthia. A sight of China. London, 1969.
An 3309.69	Nurse, G.T. Height and history in Malawi. Malawi, 1969.

Chronological Listing

1969 - cont.

VGeog 4219.69.10	Ogrin, Miran. Na juga sveta. Ljubljana, 1969.
VGeog 4209.69	Ogrin, Miran. Širine sveta. Ljubljana, 1969.
An 2059.69.5	Overhage, Paul. Menschenformen im Eiszeitalter. 1. Aufl. Frankfurt, 1969.
Geog 4209.69.15	Paleckis, Justas. Kelioniu, Knyga. Vilnius, 1969.
Geog 4145.91	Parr, Charles. The voyages of David de Vries. N.Y., 1969.
Geog 4659.69	Parry, Henry. Wreck and rescue on the coast of Wales. Truro, 1969-73. 2v.
An 136.5.17	Paz, Octavio. Claude Lévi-Strauss o el nuevo festin de Esopo. 2. ed. Mexico, 1969.
Geog 3057.5.5	Peters, Karl. King Solomon's golden Ophir. N.Y., 1969.
An 2059.69	Pfeiffer, John. E. The emergence of man. N.Y., 1969.
Geog 4312.10	Pilkington, Roger. Small boat to Elsinore. N.Y., 1969.
VGeog 4509.69	Planinsko berilo. Ljubljana, 1969.
Geog 5925.93	Potter, Neal. Natural resource potentials of the Antarctic. N.Y., 1969.
Geog 4219.69	Price, Willard. Odd way around the world. N.Y., 1969.
Geog 4659.60.2	Ratigan, William. Great Lakes shipwrecks and survivals. 2. ed. Grand Rapids, 1969.
Geog 500.170	Schwickerath, Hildegard. Inhaltsverzeichnis der Festschriften zur Ehrung und Würdigung deutscher. Bad Godesberg, 1969.
Geog 665.132	Settlement and encounter; geographical studies presented to Sir Grenfell Price. Melbourne, 1969.
Geog 500.155	Smith, Harold F. American travellers abroad. Carbondale, 1969.
Geog 189.2.6	Société de Géographie de Québec. Bulletin. Index. 1880-1934. Québec, 1969.
Geog 4559.69.10	Spiers, George. The Wavertree; being an account of an ocean wanderer. N.Y., 1969.
Geog 4347.91.1	Steuke, Johann Kaspar. Von Amsterdam nach Temiswar. Berlin, 1969.
Geog 4209.69.25	Stoianovich, Ivan. Bregovete na khorata. Sofiia, 1969.
An 379.69	Storkebaum, Werner. Sozialgeographie. Darmstadt, 1969.
An 2109.66.35	Symposium on Man the Hunter, Chicago, 1966. Man the hunter. Chicago, 1969.
Geog 4209.69.10	Sytin, Viktor A. Puteshestviia. Moskva, 1969.
Geog 500.165	Trecento tesi di laurea in geografia. Padova, 1969.
Geog 665.144	Trends in geography. 1st ed. Oxford, 1969.
Geog 4228.98	Trevelyan, Charles Philips. Letters from North America and the Pacific, 1898. London, 1969.
An 170.360	Tyler, Stephen A. Cognitive anthropology; readings. N.Y., 1969.
Geog 3060.12.70	Vincent, Louis Claude. Le paradis perdu de Mu. Paris? 1969. 2v.
An 359.69.25	Vuorela, Toivo. Kansatieteen periaateoppia. Helsinki, 1969.
Geog 4239.69	Wharran, James. Two girls, two catamarans. London, 1969.

1970

An 609.70A	Abel, Herbert. Vom Raritätenkabinett zum Bremener Überseemuseum. Bremen, 1970.
An 629.70	Akademiia Nauk SSSR. Institut Etnografii. Voprosy metodiki etnograficheskikh i etno-sotsiologicheskikh issledovanii. Moskva, 1970.
Geog 612.6	Alekseev, Aleksandr G. R. Fedor Petrovich Litke. Moskva, 1970.
Geog 3590.78	Alekseev, Aleksandr Ivanovich. Syny otvazhnye Rossii. Magadan, 1970.
Geog 4181.138	Allen, William Edward David. Russian embassies to the Georgian kings, 1589-1605. Cambridge, 1970. 2v.
Geog 4129.70	Anderson, John Richard Lane. The Ulysses factor: the exploring instinct in man. London, 1970.
Geog 665.150	Argumenta geographica; Festschrift Carl Troll zum 70. Geburtstag. Bonn, 1970.
Geog 500.160	Arnim, Hlmuth. Bibliographie der geographischen Literatur in deutscher Sprache. 1. Aufl. Baden Baden, 1970.
Geog 4227.35.1	Atkins, John. A voyage to Guinea, Brazil, and the West Indies. London, 1970.
Geog 1575.16	Baedeker, publishers. Handbook for travellers: Russia. N.Y., 1970.
Geog 1523.17	Baedeker, publishers. Rhine and northern Germany. 4. ed. Coblenz, 1970.
Geog 4349.07	Barzini, Luigi. Da Pechino a Parigi in sessanta giorni. Milano, 1970.
Geog 665.148	Beiträge zur Geographie der Tropen und Subtropen; Festschrift zum 60. Geburtstag von Herbert Wilhelmy. Tübingen, 1970.
Geog 4559.70	Bertino, Serge. Guide de la mer mystérieuse. Paris, 1970.
Geog 759.70	Birot, Pierre. Les régions naturelles du globe. Paris, 1970.
Geog 4359.70	Brandi, Cesare. A passo d'uomo. Milano, 1970.
Geog 611.1.5	Brossard, Maurice Raymond de. Kerguelen, le découvreur et ses îles. Paris, 1970-1971. 2v.
Geog 500.150	Browning, Clyde Eugene. A bibliography of dissertations in geography, 1901 to 1969; American and Canadian universities. Chapel Hill, 1970.
Geog 4307.82.3	Bruce, Peter Henry. Memoirs of Peter Henry Bruce, Esq. Reprint. London, 1970.
An 177.2	Carias, Rafael. La indecision. Caracas, 1970.
Geog 4219.20.5	Chevalier, Haakon Maurice. The last voyage of the schooner Rosamond. London, 1970.
Geog 659.65.7	Chorley, Richard J. Frontiers in geographical teaching. 2nd ed. London, 1970.
An 2109.70	Clark, John Grahame Douglas. Aspects of prehistory. Berkeley, 1970.
Geog 4269.70	Coghill, Ian G. Western Europe: geographical studies. London, 1970.
An 359.70.20	Costanzo, Giorgio. La costruzione dell'uomo. Roma, 1970.
An 3309.70.1	Debets, Georgii Frantsevich. Physical anthropology of Afghanistan. v.1-2. Cambridge, 1970.
Geog 4218.17.8	Delano, Amasa. Narrative of voyages and travels. N.Y., 1970.
Geog 535.15	Denis, Jacques. Guide de la recherche géographique en Belgique. Namur, 1970.
Geog 579.70	Dictionnaire de la géographie. Paris, 1970.
Geog 3055.39.25	Donnelly, Ignatius. Atlantis, the antediluvian world. London, 1970.
Geog 3048.83.1	Donnelly, Ignatius. Ragnarok: the age of fire and gravel. N.Y., 1970.
An 130.10.5	Downie, Robert Angus. Frazer and the golden bough. London, 1970.
Geog 4709.70	Earle, Peter. Corsairs of Malta and Barbary. London, 1970.

1970 - cont.

An 170.257	Échanges et communications; mélanges offerts à Claude Lévi-Strauss. The Hague, 1970. 2v.
Geog 665.154	Eesti Geograafia Selts. Orazvitii geografii v Estonskoi SSR 1960-1968. Tallin, 1970.
An 170.190	Etnoniny. Moskva, 1970.
Geog 4209.68.20	Expédition France-Inde. Expédition France-Inde; mission culturelle 1968-1969 sous le patronage de la Fédération régionale spéléologique, Nancy. Saint Max, 1970.
An 180.5	Fortes, Meyer. Time and social structure and other essays. London, 1970.
Geog 5518.19.14	Franklin, John. Narrative of a journey to the shores of the Polar Sea in the years 1819-1922. Rutland, 1970.
Geog 5639.67.2	Gad, Finn. The history of Greenland. London, 1970- 2v.
Geog 3590.77	Geograficheskoe Obshchestvo SSSR. Geograficheskoe obshchestvo za 125 let. Leningrad, 1970.
Geog 665.136	Geographical essays in honour of K.C. Edwards. Nottingham, 1970.
Geog 85.303.2	Geographische Zeitschrift. Register. v.1-50, 1895-1944. Wiesbaden, 1970. 2v.
Geog 4227.46	Goelet, Francis. The voyages and travels of Francis Goelet, 1746-1758. N.Y., 1970.
An 124.15	Golde, Peggy. Women in the field; anthropological experience. Chicago, 1970.
Geog 4028.2	Green, Timothy. The adventures; four profiles of contemporary travellers. London, 1970.
Geog 5639.70	Greenland past and present. Copenhagen, 1970.
Geog 4180.29	Guzman, A.E. de. Life and acts of...Guzman. N.Y., 1970.
Geog 659.70	Harvey, David. Explanation in geography. N.Y., 1970.
An 2600.6.1	Hecquet, (Mme.). Histoire d'une jeune fille sauvage trouvée dans les bois à l'âge de dix ans. Bordeaux, 1970.
Geog 4239.70	Hills, Lawrence Donegan. Lands of the morning. London, 1970.
An 3559.70	Ismagulov, Orazak. Naselenie Kazakhstana ot epokhi bronzy do sovremennosti. Alma-Ata, 1970.
Geog 5509.70	Keating, Bern. The Northwest Passage: frm the Mathew to Manhattan, 1497 to 1969. Chicago, 1970.
Geog 4209.70.5	Kiknadze, Aleksandr Vasil'evich. Ot Madrida do Tokio. Baku, 1970.
Geog 3049.70	Kohlenberg, Karl Friedrich. Enträtselte Vorzeit. München, 1970.
Geog 4332.26	Kolar, Vladimir. Jadran - more mira. Ljubljana, 1970.
An 359.70	Korolev, Stanislav I. Voprosy etnopsikhologii v vabotakh zanebezhnykh avtarov. Moskva, 1970.
An 359.70.10	Kovalev, Sergei M. O chloveke, ego poraboshchenii i osvdsozhdenii. Moskva, 1970.
Geog 4209.70.10	Kozima, Marjan. Veseli potopisi. Maribor, 1970.
Geog 3070.165	Kremling, Helmut. Die Beziehungsgrundlage in thematischen Karten in ihrem Verhältnis zum Kartengegenstand. München, 1970.
An 135.1.5	Kroeber, Theodora. Alfred Kroeber: a personal configuration. Berkeley, 1970.
An 136.5.5	Leach, Edmund Ronald. Claude Lévi-Strauss. N.Y., 1970.
An 136.5.35	Lepenies, Wolf. Orte des wilden Denkens. 1. Aufl. Frankfurt am Main, 1970.
An 359.70.5	Lévi-Strauss, Claude. Claude Lévi-Strauss; ou, La structure et le malheur. Paris, 1970.
Geog 4309.70	Lottman, Herbert R. Detours from the grand tour. Englewood Cliffs, N.J., 1970.
Geog 3030.7	Magidovich, I.P. Istoriia otkrytiia i issledovaniia Evropy. Moskva, 1970.
Geog 3570.60	Marjay, Frederico Pedro. Navegadores portugueses, herois do mar. Lisboa, 1970.
An 379.70	Maršík, Miroslav. Natural environment and society in the theory of geographical determinism. Praha, 1970.
An 99.70	Moravia, Sergio. La scienza dell'uomo nel settecento. Bari, 1970.
Geog 1775.10	Murray, John, publisher, London. Hand-book for northern Europe: Finland and Russia. N.Y., 1970.
Geog 5535.20	Neatby, Leslie Hamilton. The search for Franklin. London, 1970.
An 3509.70F	Nielsen, Ole Vagn. The Nubian skeleton through 4000 years. Thesis. København, 1970.
An 2109.70.5	Nougier, Louis René. L'économie préhistorique. Paris, 1970.
Geog 613.9	Novlianskaia, Mariia G. Daniil Gotlib Messerschmidt i ego raboty po issledovaniiu Sibiri. Leningrad, 1970.
Geog 5329.3.7	Ol'khina, Evgeniia A. Vil'iamur Stefanson. Moskva, 1970.
Geog 5300.56	Pasetskii, Vasilii M. Eestist pärit Arktika-uurijad. Tallinn, 1970.
An 136.5	Paz, Octavio. Claude Lévi-Strauss; an introduction. Ithaca, 1970.
Geog 665.167	Prague. Universita Karlova. Sborník prací Geografických kateder UK k 75. narozeninám prof. dr. Jaromíra Korčáka, DrSc. Praha, 1970.
An 2109.70.10	Ramseyer, Urs. Soziale Bezüge des Musizierens in Naturvolkkulturen. Bern, 1970.
An 359.70.25	Rimet, Michel. Contracts, interférences ethniques et culturelles. Montpellier, 1970.
Geog 665.158	Riva, Ambrogio. Una piccola biblioteca. Milano, 1970.
Geog 194.1.6	Rivista di geografia italiana. Firenze. Reprint ed. Amsterdam, 1970. 8v.
Geog 4209.70.15	Rubissow, Helen. Ozni na dorogakh chetyrekh chastei sveta. Parizh, 1970.
Geog 4131.70	Ruppert, Karl. Zur Geographie des Freizeitverhaltens. Kallmünz, 1970.
An 629.70.5	Sarmela, Matti. Perinnerineiston kvantitatiivisesta tutkimuksesta. Helsinki, 1970.
Geog 3019.70	Schmithuesen, Josef. Geschichte der geographischen Wissenschaft von den ersten Anfägen bis zur Ende des 18. Jahrhunderts. Mannheim, 1970.
An 359.70.30	Schweppe, John S. Man: a remarkable animal. Chicago, 1970.
An 2109.45.1	Simmons, L.W. The role of the aged in primitive society. Hamden, Ct., 1970.
Geog 5559.69	Simpson, Myrtle. Due north. London, 1970.
Geog 5518.36.2	Simpson, Thomas. Narrative of the discoveries on the north coast of America. 2. ed. Toronto, 1970. 2v.
Geog 5519.69	Smith, William D. Northwest Passage. N.Y., 1970.
Geog 5369.70	Sovetskaia Arktika; moria i ostrava Severnogo Ledovitogo okeana. Moskva, 1970.
An 359.70.15	Stiglmayr, Engelbert. Ganzheitliche Ethnologie. Wien, 1970.
Geog 3240.28	Strzelczyk, Jerzy. Gerwazy z Tilbury. Wrocław, 1970.
Geog 4239.68.25	Svirin, Vladimir P. Vzemliakh blizkikh i dal'nikh. Stavropol', 1970.

Chronological Listing

1970 - cont.

Geog 4239.68.20	Tomalin, Nichols. The strange voyage of Donald Crowhurst. London, 1970.
An 5.21.1F	Tozzer Library. Catalogue. Authors. 1st supplement. Boston, 1970. 6v.
An 5.22.1F	Tozzer Library. Catalogue. Subjects. 1st supplement. Boston, 1970. 6v.
Geog 5925.94	Treshnika, Aleksei F. Vokrug Antarktidy. Leningrad, 1970.
Geog 5375.94.11	Veer, G. de. The true and perfect description of three voyages by the ships of Holland and Zeland. Facsimile. Amsterdam, 1970.
Geog 759.70.5	Voprovy geografii. Kaliningrad, 1970.
Geog 4209.70	Wollschaeger, Alfred. Weltreise auf den Spuren der Unruhe. Gütersloh, 1970.
Geog 3249.70	Wright, Louis Booker. Gold, glory, and the gospel. 1. ed. N.Y., 1970.
Geog 5235.25	Wright, Theon. The big nail; the story of the Cook-Peary feud. N.Y., 1970.
Geog 4419.70	Zemlia dobrykh liudei. Tashkent, 1970.

1971

An 2209.71	Ackevknecht, Erwin Heinz. Medicine and ethnology. Baltimore, 1971.
An 359.71.20	Alland, Alexander. Human diversity. N.Y., 1971.
Geog 4309.71	Babiński, Gan. Noc w Soho. Wyd. 1. Łódź, 1971.
Geog 4129.71	Bauer, Hans. Wenn einer eine Reise tat. Leipzig, 1971.
An 99.71	Bausinger, Hermann. Volkskunde: von der Altertumsforschung zur Kulturanalyse. Berlin, 1971.
Geog 659.71.15	Beaujeu-Garnier, Jacqueline. La géographie: méthods et perspectives. Paris, 1971.
Geog 4218.70.8	Beaumont, Thomas E. Pencilling by the way: a constitutional voyage round the world: 1870 and 1871. London, 1971.
Geog 4759.71.10	Benchétrit, Maurice. Géographie zonale des régions chaudes. Paris, 1971.
Geog 3055.84	Bergquist, Nils Olof. Ymdogat-Atlantis. Solna, 1971.
Geog 5970.15.20	Bertrand, Kenneth John. Americans in Antarctica, 1775-1948. N.Y., 1971.
Geog 5182.8.3	Birket-Smith, Kaj. Eskimoerne. 3. Udg. Kjøbenhavn, 1971.
Geog 4219.70	Blyth, Chay. The impossible voyage. London, 1971.
Geog 4279.71	Bocca, Geoffrey. The great resorts. N.Y., 1971.
Geog 4559.53.16	Bombard, Alain. The voyage of the Hérétique. Ann Arbor, Mich., 1971.
Geog 4182.72	Briel, Hans Jurgen. De expeditie van Anthonio Hurdt. 's-Gravenhage, 1971.
Geog 4239.71.5	Brinnin, John M. The sway of the grand saloon: a social history of the North Atlantic. N.Y., 1971.
Geog 142.12.5	Brown, Theodore Nigel Leslie. The history of the Manchester Geographical Society, 1884-1950. Manchester, 1971.
An 359.71.15	Cheboksarov, Nikolai N. Narody, rasy, kul'tury. Moskva, 1971.
Geog 4329.71	Chernishev, Slaveho. Po more. Sofiia, 1971.
Geog 4239.70.10	Chichester, Francis Charles. The romantic challenge. London, 1971.
An 379.71	Chisholm, Michael. Research in human geography. London, 1971.
Geog 659.71.5	Chorley, Richard J. Models in geography. London, 1971.
Geog 4709.71	Cousteau, Jacques Yves. Diving for sunken treasure. London, 1971.
Geog 4239.21	Craig, Gavin. Boy aloft. Lymington, 1971.
Geog 5845.4	Denmark. Ministeriet for Grønland. Perspektivplan for Grønland 1971-85. København, 1971.
Geog 4309.71.5	Derruau, Max. L'Europe. Paris, 1971.
An 2052.5	Divale, William T. Warfare in primitive societies; a selected bibliography. Los Angeles, 1971.
Geog 3590.79	Divin, Vasilii A. Russkie moreplavaniia na Tikhom okeane r XVIII neke. Moskva, 1971.
Geog 3055.39.1	Donnelly, Ignatius. Atlantis: the antediluvian world. Blauvelt, N.Y., 1971.
An 359.71.30	Dzhandil'din, Nurymbet. Priroda natsional'noi psikholgii. Alma-Ata, 1971.
Geog 3590.35.2	Efimov, Aleksei V. Iz istorii velikikh russkikh geograf. otkrytii. Moskva, 1971.
Geog 5182.5	Fainberg, Lev A. Ocherki etnicheskoi istorii zarubezhnogo Severa. Moskva, 1971.
Geog 5700.4.20	Fenger, Niels. Palase, Hans Egede i Grønland. København, 1971.
Geog 4679.71	Fenn, Charles. Journal of a voyage to nowhere. London, 1971.
Geog 579.71	Fochler-Hauze, Gustav. Allgemeine Geographie. Frankfurt am Main, 1971.
Geog 665.165	Forschungen zur allgemeinen und regionalen Geographie; Festschrift für Kurt Kayser. Wiesbaden, 1971.
Geog 5518.25.5	Franklin, John. Narrative of a second expedition to the shores of the Polar Sea in the years 1825-1827. Rutland, 1971.
Geog 4476.38.1	Godwin, Francis. The man in the moon, 1638. London, 1971.
An 359.71.5	Goldsby, Richard A. Race and races. N.Y., 1971.
Geog 4209.71.5	Górnicki, Wiesław. Opowieści zdyszane. 1. wyd. Warszawa, 1971.
Geog 4759.71.5	Gourou, Pierre. Leçons de géographie tropicale. Paris, 1971.
Geog 659.63.2	Gregory, Stanley. Statistical methods and the geographer. 2nd ed. London, 1971.
Geog 5660.42	Gulloev, H.C. Igdlucrúnerit. København, 1971.
Geog 659.71	Hampl, Martin. Teorie komplexity a diferenciace světa se zolášťnim zřetelem na diferenciaci geografickṡu. 1. vyd. Praha, 1971.
Geog 5328.27	Hare, Kirsten. Kender du Knud? 2. Udg. Kjøbenhavn, 1971.
An 359.71	Harris, Marvin. Culture, man, and nature. N.Y., 1971.
Geog 5559.68	Herbert, Wally. Across the top of the world. N.Y., 1971.
Geog 4239.70.6	Heyerdahl, Thor. The Ra expeditions. London, 1971.
Geog 4309.71.15	Holmqvist, Lasse. Vitt och brett. Stockholm, 1971.
Geog 4227.90	Ingraham, Joseph. Journal of the brigantine Hope on the voyage to the northwest coast of North America, 1790-92. Barre, 1971.
An 379.56.1	International Symposium on Man's Role in Changing the Face of the Earth, Princeton, N.J., 1955. Man's role in changing the face of the earth. Chicago, 1971.
Geog 3019.71	Isachenko, Anatolii G. Razvitie geograficheskikh idei. Moskva, 1971.
Geog 665.164	James, Preston Everett. On geography: Selected writings of Preston E. James. 1st ed. Syracuse, 1971.
Geog 5670.4	Jensen, Bent. En livsform ved horsvejen. København, 1971.

1971 - cont.

Geog 4209.71.35	Kearns, Des. World wanderer; 100,000 miles under sail. Sydney, 1971.
Geog 4209.71.40	Knudsen, Kim. Grenseløs. Oslo, 1971.
An 170.230.5	Kongres Československých anthropologů, 10th, Prague and Humpolie, 1969. Anthropological congress dedicated to Aleš Hrdlička. Prague, 1971.
Geog 4332.24	Košutić, Ivo. Gusari Jadranskoga mora. Zagreb, 1971.
Geog 4309.71.10	Krueger, Horst. Fremde Vaterländer. München, 1971.
Geog 4209.71.25	Kuleshov, Aleksandr P. Shest' gorodov piati kontinentov. Moskva, 1971.
Geog 3590.80	Lebedev, Dmitrii M. Russkie geograficheskie otkrytiia i issledovaniia s drevnikh vremen do 1917 goda. Moskva, 1971.
Geog 3055.86	Le Cour, Paul. L'Atlantide atlantique. Bordeaux, 1971.
Geog 4182.15	Ledyard, Gari. The Dutch come to Korea. Seoul, 1971.
Geog 4326.32.30	Lithgow, William. Discourse of a peregrination in Europe, Asia and Affricke. Facsimile. N.Y., 1971.
Geog 4416.21	Lobo, Jeronymo. Itinefario e outros escritos inéditos. Porto, 1971.
Geog 4029.12	Loeffler, Johann Friedrich. Abenteur in drei Erdteilen. Heidenheim, 1971.
Geog 4239.69.10	McClean, Tom. I had to dare; rouring the Atlantic in seventy days. London, 1971.
Geog 4269.66.5	Markovic, Slobodan. Putovanja grešnog Elefterija. Beograd, 1971.
An 2109.71.1	Marshack, Alexander. The roots of civilization. 1st ed. N.Y., 1971.
Geog 600.22	Matveeva, T.P. Russkie geografy i puteshestvenniki. Leningrad, 1971.
An 136.5.70	Montes, Santiago. Claude Lévi-Strauss. 1. ed. San Salvador, 1971.
Geog 4181.140	Morga, Antonio de. Sucesos de las Islas Filipinas. Cambridge, Eng., 1971.
Geog 189.2.8	Morissonneau, Christian. La société de géographie de Québec, 1877-1970. Québec, 1971.
Geog 4306.17.1F	Moryson, Fynes. An itinerary. Amsterdam, 1971.
An 359.71.25	Murphy, Robert Francis. The dialectics of social life: alarms and excursions in anthropological theory. N.Y., 1971.
Geog 1709.8.4	Murray, John, publisher, London. Handbook for Devon and Cornwall. Newton Abbot, Eng., 1971.
Geog 5530.82	Nanton, Paul. Arctic breakthrough: Franklin's expeditions, 1819-1847. London, 1971.
Geog 150.5.2	New Zealand geographer. Cumulative index, v.1-25, 1945-1969. Wellington, 1971.
Geog 4209.71.30	Nikolaev, Vladimir D. Maiak Zemli. Moskva, 1971.
Geog 4308.59.11	Pas Soldan y Unánue, Pedro. Memories de un viajero peruano; apuntes y recuerdos de Europa y Oriente (1859-1863). Lima, 1971.
Geog 4131.71	Patev, Iliia. Statistika na turizma. Varne, 1971.
An 359.71.40	Poliahov, Leon. La mythe aryen. Paris, 1971.
Geog 759.71.5	Prirodnye resursy i kulturnye landshafty materikov. Moskva, 1971.
Geog 5970.42	Quartermain, Leslie B. New Zealand and the Antarctic. Wellington, 1971.
An 136.5.40	Remotti, Francesco. Lévi-Strauss. Torino, 1971.
Geog 4029.14	Ridgway, John M. Journey to Ardmore. London, 1971.
Geog 4559.71.10	Robinson, Cyril. Men against the sea. Windsor, N.S., 1971.
An 170.328.8	Royal Anthropological Institute of Great Britain and Ireland. Notes and queries on anthropology. 6. ed. London, 1971.
An 126.10	Salter, Elizabeth. Daisy Bates; "the great white queen of the never never". Sydney, 1971.
Geog 5839.71	Sandstroem, Lennart. Leva på Grönland. Stockholm, 1971.
Geog 659.71.10	Santos, Milton. Le métier de géographe en pays sous-développé. Paris, 1971.
Geog 3560.26	Sanz, Carlos. La huella de España en el mundo. Madrid, 1971-73. 3v.
Geog 4265.95.1	Saur, Abraham. Theatrum urbium. Unterschneidheim, 1971.
An 136.5.45	Scardnelli, Pietro. L'analisi strutturali dei miti. Milano, 1971.
Geog 4678.55.20	Schilling, Nikolai. Seeoffizier des Zaren. Köln, 1971.
Geog 3540.15	Schulte-Althoff, Franz Josef. Studien zur politischen Wissenschaftsgeschichte der deutschen Geographie im Zeitalter des Imperialismus. Paderborn, 1971.
Geog 4209.71.20	Shelekhov, Grigorii I. Rossiiskogo kuptsa Grigoriia Shelekhova stranstvoraniia iz Okholska po Vostochnomu okeanu k amerikanskiam beregam. Khabarovsk, 1971.
Geog 4229.71	Shur, Leonid A. K beregam Novogo Sveta. Moskva, 1971.
Geog 3510.13	Smet, Antoine de. La cartographie hollandaise. Bruxelles, 1971.
Geog 4709.71.5	Sokol, Hans Hugo. Unter der Flagge mit dem Totenkopf. Herford, 1971.
Geog 4209.71.10	Šteinbergs, Valentins Filosofiia i dzhontol'mony. Riga, 1971.
An 359.68.16	Stocking, George W. Race, culture, and evolution. N.Y., 1971.
Geog 4209.71	Šulentić, Zeatko. Ljudi krajevi Beskraj. Karlovac, 1971.
Geog 5970.40	Takhariev, Vasil I. Bulgarin na Antartida. Sofiia, 1971.
Geog 4145.81.6	Teller, Walter Magnes. Joshua Slocum. New Brunswick, 1971.
An 629.71	Tennekes, J. Anthropology, relativism and method. Assen, 1971.
VGeog 759.71	Thuchkevich, Vadim A. Geografiia v tsifrakh i sravneniiakh. Minsk, 1971.
Geog 5070.70.5	Tilman, Harold William. In mischief's wake. London, 1971.
Geog 500.171	Tolchinskaia, L.I. Geograficheskaia literatura. Moskva, 1971.
An 5.21.2F	Tozzer Library. Catalogue. Authors. 2d supplement. Boston, 1971. 2v.
An 5.22.2F	Tozzer Library. Catalogue. Subjects. 2d supplement. Boston, 1971. 3v.
Geog 4205.96.1	Ultzheimer, Andreas J. Warhaffte Beschreibung ettlicher Reisen in Europa, Africa, Asien und America 1596-1610. Tübingen, 1971.
Geog 5060.7.1	Victor, Paul Émile. L'homme à la conquête des pôles. Evreux, 1971.
Geog 5837.03	Vidalin, Arngrimur Thorkilsson. Den tredie part af det saa kaldede gamle og mye Grønlands beskrifvelse. København, 1971.
Geog 4239.71	Vihlen, Hugo. April Fool or, How I sailed from Casablanca to Florida in a six-foot boat. Chicago, 1971.
Geog 4559.71.5	Villiers, A.J. The war with Cape Horn. N.Y., 1971.
Geog 3235.66	Vinland Map Conference, Smithsonian Institution. Proceedings. Chicago, 1971.

Chronological Listing

1971 - cont.

Geog 607.11	Vlora, Gribaudi. L'uomo e lo studioso. Bari, 1971.
Geog 5850.12	Wedin, Bertil. Aktion och reaktion på Grönland. Stockholm, 1971.
Geog 5639.71	Wedin, Bertil. Tusen år på Grönland. Stockholm, 1971.
Geog 5509.71	Wilkinson, Doug. Arctic fever. Toronto, 1971.
An 359.71.10	Wolf, Josef. Integral anthropology. Praha, 1971.
Geog 4239.69.15	Woolass, Peter. Stelda, George and I. London, 1971.
An 3209.71	Young, John Z. An introduction to the study of man. Oxford, 1971.

1972

Geog 3019.72.25	Aleksandrovskaia, Ol'ga A. Frantsvzskaia geograficheskaia shkola kontsa deviatnadtsatogo nachala dvadtsatogo veka. Moskva, 1972.
An 3209.72	Alekseev, Valerii P. V poishakh predkov. Moskva, 1972.
An 359.72.10	Alland, Alexander. The human imperative. N.Y., 1972.
Geog 4181.142A	Andrews, Kenneth Raymond. The last voyage of Drake and Hawkins. Cambridge, 1972.
An 359.72.20	Ansari, G. Recent trends in cultural anthropology. Wien, 1972.
Geog 3019.72.15	Anuchin, V.A. Teoreticheskie osnovy geografii. Moskva, 1972.
Geog 4029.9	Aresty, Miguel de. Los papeles de Juan de Aresty. Barcelona, 1972.
An 3509.72	Aron, Jean Paul. Anthropologie du conscrit français d'après les comptes numeriques et sommaires du recrutement de l'année, 1819-1826. Paris, 1972.
Geog 3030.8	Baker, Alan R.H. Progress in historical geography. N.Y., 1972.
Geog 4209.72.15	Bamans, Godfried. Op reis rond de wereld en op Rottumerplaat. Amsterdam, 1972.
Geog 4679.52	Bevan, David. Drums of the Birkenhead. Capetown, 1972.
An 359.72.25	Bezerra, Felte. Antropologia sociocultural. Brasilia, 1972.
An 359.72	Bicchieri, M.G. Hunters and gatherers today. N.Y., 1972.
An 127.10	Boissel, Jean. Victor Courtet, 1813-1867. Paris, 1972.
An 136.5.60	Boon, James A. From symbolism to structuralism. N.Y., 1972.
An 136.5.61	Boon, James A. From symbolism to structuralism: Lévi-Strauss in a literary tradition. Oxford, 1972.
An 2109.72.25	Brain, Robert. Into the primitive environment: survival on the edge of our civilization. London, 1972.
Geog 4306.86.6	Burnet, Gilbert. Some letters containing an account of what seemed most remarkable in Switzerland. Menston, Eng., 1972.
Geog 4559.72.10	Chapman, Gifford Desmond. Kangaroo Island shipwrecks. Canberra, 1972.
Geog 5399.72	Chilingarov, A.N. Pod nogami ostrov ledianoi. Moskva, 1972.
Geog 3019.72.35	Claral, Paul. La pensée géographique. Paris, 1972.
An 2109.72.10	Cocchiara, Giuseppe. L'eterno selvaggio. Palermo, 1972.
An 5.10	Conklin, Harold C. Folk classification: a topically arranged bibliography of contemporary and background references through 1971. New Haven, 1972.
Geog 5321.1.6	Corner, George Washington. Doctor Kane of the Arctic seas. Philadelphia, 1972.
An 379.72	Cox, Kevin R. Man, location, and behavior: an introduction to human geography. N.Y., 1972.
Geog 3570.53	Crone, Gerald. The discovery of the East. London, 1972.
Geog 3019.72.30	Deschamps, Hubert J. Les Européens hors d'Europe de 1434 à 1815. 1. éd. Paris, 1972.
Geog 45.12	Deutscher Geographentag, 38th, Erlangen and Nuremberg, 1971. Tagungsbericht und wissenschaftliche Abhandlungen. Wiesbaden, 1972.
An 136.5.30	Dumasy, Annegret. Restloses. Erkenhen; die Diskussione über den Strukturalismus de Claude Lévi-Strauss in Frankreich. Berlin, 1972.
Geog 4759.72	Études de géographie tropicale offertes à Pierre Gourou. Paris, 1972.
Geog 3570.61	Ferro, Gaetano. Le conoscenze geografiche del Medioevo. Genova, 1972.
An 170.347	Festschrift Alfred Steinmann. Bern, 1972.
Geog 4145.29.5	Fournier-Aubry, Fernand. Don Fernando. Paris, 1972.
Geog 3019.72.5	Fradkin, Naum G. Geografickeskie otkrytiia i nauchnoe poznanie Zemli. Moskva, 1972.
Geog 5316.2	Fraser, Robert J. Arctic adventurer. Lincoln, Ont., 1972.
Geog 5970.44	Gasev, Aleksandr M. Ot El'brusa do Antarktidy. Moskva, 1972.
Geog 4239.70.15	Genovés Tarazaga, Santiago. Ra, una balsa de papyrus a través del Atlantico. 1. ed. Mexico, 1972.
Geog 85.195.2F	The geographical magazine. Index, 1950-1972. London, 1972.
Geog 600.24	Gilbert, Edmund William. British pioneers in geography. Newton Abbot, 1972.
Geog 4219.65	Graham, Robin Lee. Dove. 1. ed. N.Y., 1972.
Geog 759.72	Haggeh, Peter. Geography: a modern synthesis. N.Y., 1972.
Geog 4209.72	Hammond Innes, Dorothy. Occasions. London, 1972.
Geog 4559.72.1	Hassloef, Olof. Ships and shipyards, sailors and fishermen. Copenhagen, 1972.
An 182.8	Herskovits, Melville Jean. Cultural relativism. 1. ed. N.Y., 1972.
Geog 4309.72	Hillaby, John D. Journey through Europe. London, 1972.
Geog 5845.5	Hoejlund, Niels. Krise uden alternativ. København, 1972.
An 2109.72.45	Horken, K. Ex nocte lux. Tübingen, 1972.
Geog 110.1.22	International Geographical Congress, 22nd, Montreal, 1972. Doklady k XXII mezhdunarodnomu geograficheskomu kongress (Kanada, avgust, 1972). Leningrad, 1972.
An 170.324	The interpretation of ritual; essays in honour of A.I. Richards. London, 1972.
Geog 3019.72	James, Preston Everett. All possible worlds; a history of geographical ideas. Indianapolis, 1972.
An 2109.72.40	Jelinek, Jan. Das grosse Bilderlexikon des Menschen in der Vorzeit. Gütersloh, 1972.
Geog 4313.9	Kiecksee, Heinz. Die Ostsee-Sturmflut 1872. Heide, 1972.
Geog 603.8	Kleopov, Igor' L'. Aleksandr Lavrent'evich Chekanovskii, 1833-1876. Leningrad, 1972.
An 3559.72	Konduktorova, Tamara S. Antropologiia drevnego nasteniia Ukrainy. Moskva, 1972.
Geog 3249.72	Krämer, Walter. Neue Horizonte. 1. Aufl. Leipzig, 1972.
An 3559.72.10	Kruts, Svitlana O. Naselenie territorii Ukrainy epokhi medi-bronzy. Kyiv, 1972.
Geog 3019.72.20	Langley, Michael. When the pole star shone. London, 1972.
An 99.72.5	Leclerc, Gérard. Anthropologie et colonialisme. Paris, 1972.
Geog 606.3.5	Leonov, Nikolai I. Aleksei Pavlovich Fedchenko, 1844-1873. Moskva, 1972.

1972 - cont.

Geog 4329.69	Lewis, Cecil. Turn right for Corfu. London, 1972.
Geog 4269.72	Lisakovskii, Igor' N. Gamlet ne prishel na svidanie. Odessa, 1972.
An 379.72.5	Lopatina, Elena B. Otsenka prirodnykh uslovii zhizni naseleniia. Moskva, 1972.
Geog 5538.57.3	M'Clintock, F.L. The voyage of the "Fox" in the Arctic seas. Rutland, 1972.
An 2600.7.1	Malson, Lucien. Wolf children. London, 1972.
An 609.72	Materialy po rabote i istorii etnograficheskikh muzeev i vystavok. Moskva, 1972.
An 2059.72.5	Matiushin, Geral'd N. U kolybeli istorii. Moskva, 1972.
An 359.72.15	Mendel, Gérard. Vers une anthropologie sociopsychanalytique. 1. éd. Paris, 1972.
An 2059.72	Millar, Ronald William. The Piltdown men. London, 1972.
Geog 4209.72.5	Mon'ko, Aleksei M. Potoratiny. Volgograd, 1972.
Geog 4209.72.10	Morris, James. Places. London, 1972.
An 136.25	Murphy, Robert Francis. Robert H. Lowie. N.Y., 1972.
Geog 6005.5	Neider, Charles. Antarctica; authentic accounts of life and exploration. N.Y., 1972.
An 170.283	Neue Anthropologie. Stuttgart, 1972-75. 7v.
An 379.72.10	Niemeier, Georg. Siedlungsgeographie. 3. Aufl. Braunschweig, 1972.
An 2109.72.5	Okhotniki, sobirateli, rybolovy. Leningrad, 1972.
Geog 6009.33.10	Olsen, Magnus L. Saga of the white horizon. Lymington, 1972.
Geog 4349.72	Ovcharenko, Aleksandr I. V etom bushuiushchem mire. Moskva, 1972.
Geog 665.166	La pensée géographique française contemporaine; Mélanges offerts à André Meynier. Saint-Brieuc, 1972.
Geog 759.72.5	Predsrazhenskii, Vladimir S. Besedy v sovremennoi fisicheskoi geografii. Moskva, 1972.
Geog 3580.46	Prieto, Carlos. El Oceano pacifico. Navegantes española del siglo XVI. Madrid, 1972.
An 2109.72.20	Quilici, Folco. Primitive societies. 1st ed. London, 1972.
Geog 665.162	Raeumliche und zeitliche und Bewegungen. Würzburg, 1972.
An 2109.72.15	Rovse, Irving. Introduction to prehistory; a systematic approach. N.Y., 1972.
An 2109.72	Sahlins, Marshall David. Stone age economics. Chicago, 1972.
Geog 619.11	Schmieder, Oskar. Lebenserinnerungen und Tagebuchblätter eines Geographen. Kiel, 1972.
Geog 4132.5	Silverberg, Robert. The longest voyage; circumnavigators in the age of discovery. Indianapolis, 1972.
An 170.340	Simpozium Antropologiia 70-kh Godov, Moscow, 1972. Simpozium "Antropologiia 70-kh godov". Moskva, 1972.
Geog 209.1	Slutskaia, Raisa D. Geograficheskoe obshchestvo globus. Moskva, 1972.
Geog 659.28.1	Spethman, Hans. Dynamische Länderkunde. Kiel, 1972.
Geog 4229.72	Struve, Wolfgang. Unglaubliche Wirklichkeit. Salzburg, 1972.
Geog 4429.72	Tendiuk, Leonid M. Al'batros-clukach moriv. Kyïv, 1972.
An 2109.72.32	Terray, Emmanuel. Le Marxisme devant les sociétés primitives. 2. éd. Paris, 1972.
An 2109.72.35	Terrey, Emmanuel. Marxism and "primitive" societies. N.Y., 1972.
Geog 4559.72.5	Throner, William R. Life at sea in the age of sail. London, 1972.
Geog 5925.95	Vedenskii, Anatolii A. Vsnegakh Krainego Iuga. Leningrad, 1972.
Geog 4238.48	Whaley, Thomas. Consignments to El Dorado. 1st ed. N.Y., 1972.
Geog 4229.11	Wilmore, C. Ray. Square rigger round the Horn. Camden, Me., 1972.
Geog 6009.10.50	Wilson, Edward Adrian. Diary of the "Terra Nova" expedition to the Antarctic, 1910-12. London, 1972.

1973

An 2109.73.15	American Ethnological Society. Learning and culture. Seattle, 1973.
An 170.212	Antropologicheskaia rekonstruktsiia i problemy paleoetnografii. Moskva, 1973.
Geog 3019.73.15	Beck, Hanno. Geographie; europäische Entwicklung in Texten und Erläuterungen. Freiburg, 1973.
Geog 4759.73.5	Birot, Pierre. Géographie physique générale de la zone intertropicale. Paris, 1973.
Geog 5670.5	Blom, Ida. Kamper om Eirik Raudes land. Oslo, 1973.
An 359.73.35	Bradfield, Maitland. A natural history of associations. London, 1973. 2v.
Geog 4207.55.5	Brelin, Johan. En äfventyrlig resa til och ifrån Ost-Indien, Södra America och en del af Europa åren 1755, 56 och 57. Stockholm, 1973.
An 359.73.10	Bromlei, Iulian V. Etnos i etnografiia. Moskva, 1973.
Geog 4269.73	Budrewicz, Olgierd. Scieżkami Starego Świata. Wyd. 1. Warszawa, 1973.
Geog 3019.73.10	Büttner, Manfred. Die Geographie generalis vor Varenius. Wiesbaden, 1973.
Geog 4679.12.41	Bullock, Shan F. A Titanic hero, Thomas Andrews, shipbuilder. Riverside, 1973.
Geog 4145.10	Burkov, Boris S. Ustrechi na piati kontinentakh. Moskva, 1973.
An 2108.73.15	Cahen, Daniel. Un site tshitolien sur le plateau des Bateke. Tervuren, 1973.
Geog 4219.72.5	Cluzel, Magdeleine. Au fil de l'eau, autour de la terre. Paris, 1973.
Geog 4229.73	Cousteau, Jacques Yves. Trois aventures de la Calypso: Galapago, Titicaca, Trous Bleus. Paris, 1973.
Geog 4759.73	Daveau, Suzanne. La zone intertropicale humid. Paris, 1973.
An 39.72	Davies, David Michael. A dictionary of anthropology. N.Y., 1973.
Geog 3070.166	Ditmar, Andrei B. Rubezh oikumeny. Moskva, 1973.
Geog 5328.50	Dodge, Ernest S. The polar Rosses: John and James Clark Ross and their explorations. London, 1973.
An 99.73.5	Duvignaud, Jean. Le langage perdu; essai sur la différence anthropologique. 1. éd. Paris, 1973.
Geog 5313.2	Eames, Hugh. Winner lose all; Dr. Cook and the theft of the North Pole. Boston, 1973.
Geog 579.73	Entsiklopedicheskii slovar' geograficheskikh nazvanii. Moskva, 1973.
Geog 659.73.10	Entwicklungstendenzen der Geographie. Berlin, 1973.
An 2109.73.30	Enzyklopädie der Technikgeschichte: über 7000 Jahre frühe technische Kultur. Stuttgart, 1973.
Geog 759.73	Eramov, R.A. Fizicheskaia geografiia zarubezhnoi Evropy. Moskva, 1973.

Chronological Listing

1973 - cont.

An 359.73.15	Etnologicheskie issledovaniia za rubezhom. Moskva, 1973.
An 2109.73.20	Felgenhauer, Fritz. Einführung in die Urgeschichtsforschung. 1. Aufl. Breisgau, 1973.
An 170.386	Festschrift für Robert Wildhaber zum 70. Geburtstag am 3. August 1972. Basel, 1973.
An 2159.73.25F	Feustel, Rudolf. Technik der Steinzeit. Weimar, 1973.
Geog 3055.68	Gadow, Gerhard. Der Atlantis-Streit. Frankfurt am Main, 1973.
Geog 4309.68.5	Garrido, Felipe. Viejo continente. 1. ed. México, 1973.
Geog 85.225.2	Geographische Gesellschaft in Hamburg. Mittheilungen. Register. 1-60, 1873-1972. Hamburg, 1973.
Geog 5160.47.5	Giaever, John. Hardbalne polarkarer. 5. Oppl. Oslo, 1973.
An 379.73	Glacken, Clarence J. Traces on the Rhodian shore. Berkeley, 1973.
An 359.73.5	Goertz, Clifford. The interpretation of cultures. N.Y., 1973.
An 379.73.15	Gourou, Jierre. Pour une géographie humaine. Paris, 1973.
Geog 5324.2.105	Greve, Tim. Fridtjof Nansen. Oslo, 1973-74. 2v.
An 2159.73.15	Gur'ev, Dmitrii V. Stanovlenie obshchestvennego proizvodstva. Moskva, 1973.
Geog 659.73.15	Hard, Gerhard. Die Geographie. Berlin, 1973.
An 2159.73	Hawkins, Gerald Stanley. Beyond Stonehenge. 1st ed. N.Y., 1973.
Geog 4217.40.5	Heaps, Leo. Log of the centurion. London, 1973.
An 126.2.6	Herskovits, Melville Jean. Frank Boas: the science of man in the making. Clifton, 1973.
An 170.230.10	L'homme, hier et aujourd'hui. Paris, 1973.
Geog 4209.65.20	Horvat, Joža. Besa-brodski dnevnik. Zagreb, 1973.
Geog 5398.72.12	Hundert Jahre Franz Josefs-Land. Zur Erinnerung an die Entdeckungsreise die österreichisch-ungarische Nordpol-Expedition, 1872-1874. Wien, 1973.
Geog 4329.73	Isnard, Hildebert. Pays et paysages mediterranéens. 1. éd. Paris, 1973.
An 359.73.30	Jaulin, Robert. Gens du soi, gens de l'autre. Paris, 1973.
An 379.73.5	Johnston, Ronald John. Spatial structures. London, 1973.
Geog 500.121	Josuweit, Werner. Studienbibliographie Geographie. Wiesbaden, 1973.
Geog 4131.73	Keller, Peter. Soziologische Probleme in modernen Tourismus. Bern, 1973.
Geog 5324.1.15	Kish, George. North-east passage: Adolf Erik Nordenskiöld, his life and times. Amsterdam, 1973.
An 2109.73	Knyshenko, Iurii V. Istoriia pervobytnogo obschchestvo. Rostov-na-Donu, 1973.
An 3309.73.5	Konduktorova, Tamara S. Antropologiia naseleniiu Ukrainy mezolita, neolita, i epokki bronzy. Moskva, 1973.
Geog 5399.73	Krenkel', Ernst T. RAEM - moi pozyvnye. Moskva, 1973.
Geog 5398.72.11	Krisch, Otto. Das Tagebuch des Maschinisten Otto Krisch. Graz, 1973.
An 99.73	Kuper, Adam. Anthropologists and anthropology. N.Y., 1973.
Geog 5939.73	Kutuzov, Pavel S. Antarkticheskii dnevnik. Moskva, 1973.
Geog 4139.73.5	Kuźmiński, Bolesław. Przygody polskich obieżyświatów na morzach i lądach. 1. wyd. Gdańsk, 1973.
An 5.26	Library-Anthropology Resource Group. Serial publications in anthropology. Chicago, 1973.
Geog 5645.5.15	Lidegaard, Mads. Ligestilling uden lighed (Grønland og Danmark). København, 1973.
Geog 819.73	McCormick, Donald. How to buy an island. N.Y., 1973.
An 136.5.75	Makarius, Raoul. Structuralisme ou ethnologie. Paris, 1973.
An 136.5.80	Marc-Lipiansky, Mireille. Le structuralisme de Lévi-Strauss. Paris, 1973.
Geog 613.10	Markov, Konstantin K. Vospominaniia i razmyshleniia geografa. Moskva, 1973.
Geog 608.10	Martin, Geoffrey J. Ellsworth Huntington; his life and thought. Hamden, Conn., 1973.
An 359.24.18	Mathews, Basil. The clash of colour. Port Washington, 1973.
Geog 4209.73	Matveev, Vikentii A. Sud'ia-vremia. Moskva, 1973.
An 379.73.10	Milkov, Fedor N. Chelovek in landshafty. Moskva, 1973.
An 2159.73.20	Moreau, Marcel. Les civilisations des étoiles. Paris, 1973.
An 359.73	Morin, Edgar. La paradigme perdu: la nature humaine. Paris, 1973.
An 5.35	Pas, H.T. Economic anthropology 1946-1972. Oosterhout, 1973.
An 190.2.2	Peacock, James Lowe. The human direction. 2. ed. N.Y., 1973.
Geog 4181.143	Peard, George. To the Pacific and Arctic with Beechey. Cambridge, 1973.
An 2109.73.5	Perry, William James. The primordial ocean; an introductory contribution to social psychology. London, 1973.
Geog 5182.4	Le peuple esquimau aujourd'hui et demain. Paris, 1973.
Geog 579.73.5	Pietkiewicz, Stanisław. Słownik pojęć geograficznych. Wyd. 1. Warszawa, 1973.
Geog 4309.73	Piontek, Heinz. Helle Tage anderswo. München, 1973.
An 2108.73.10	Pireteau, Jean. Origine et destinée de l'homme. Paris, 1973.
Geog 659.73	Racine, Jean Bernard. L'analyse quantitative en géographie. 1. éd. Paris, 1973.
Geog 5559.09.25	Rawlins, Dennis. Peary at the North Pole; fact or fiction? Washington, 1973.
An 2109.73.10	Research Seminar in Archaeology and Related Subjects, University of Sheffield, 1971. The explanation of culture change; models in prehistory. Pittsburgh, 1973.
Geog 4679.73	Robertson, Dougal. Survive the savage sea. London, 1973.
Geog 4029.17	Seering, Ruth. Mein tödliches Risiko. Bergisch Gladbach, 1973.
Geog 4239.73	Senkevich, Iurii A. Na "RA" cherez Atlantiku. Leningrad, 1973.
Geog 4515.80	Serebrianye lyzhi. Uzhgorod, 1973.
Geog 659.73.5	Simpozium po teoreticheskim problemam geografii, Riga, 1973. Teoreticheskaia geografiia. Riga, 1973.
An 132.4	Simpson, George Eaton. Melville J. Herskovits. N.Y., 1973.
Geog 4139.73	Słabezyński, Wacław. Polscy podróżnicy i odkrywcy. 1. wyd. Warszawa, 1973.
Geog 5939.73.5	Slevick, Solomon B. Osnovnye problemy osvoeniia Antarktiki. Leningrad, 1973.
Geog 3019.73	Studia z dziejów geografii i kartografii. Wrocław, 1973.
An 379.73.20	Studies in human geography. London, 1973.
An 359.73.20	Sunderland, Eric. Elements of human and social geography. 1. ed. Oxford, 1973.

1973 - cont.

Geog 5180.6	Symposium on Circumpolar Problems, Luleå, Sweden and Tromsø, Norway, 1969. Circumpolar problems. 1. ed. Oxford, 1973.
An 170.191	The translation of culture; essays to E.E. Evans-Pritchard. London, 1973.
Geog 5060.7.6	Victor, Paul Émile. Terres polaires, terres tragiques. Paris, 1973.
Geog 4679.72	Vignes, Jacques. La rage de survivre, quand il ne reste que la vie. Paris, 1973.
Geog 3596.2	Vitásek, František. Výroj moravské geografie. 1. vyd. Praha, 1973.
Geog 3019.73.5	Voprosy istoricheskoi geografii i istorii geografii. Moskva, 1973.
An 359.73.25	Waligórski, Andrzej. Antropologiczna koncepcja człowieka. Wyd. 1. Warszawa, 1973.
Geog 4145.95	Wollschläger, Alfred. Menschen an meinen Wegen. München, 1973.
An 137.6	Wotte, Herbert. Kaaram Tamo Mann vom Mond; Leben undd Reisen Miklucho-Makleis. Leipzig, 1973.
An 3309.73	Zinevich, Galina P. Antropologicheskie materialy srednevekovykh mogil'nikov dugo-zapadnogo kryma. Kiev, 1973.
An 3559.73	Zubov, Aleksandr A. Etnicheskaia odontologiia. Moskva, 1973.

1974

Geog 4269.74	Agapov, Boris N. Shest' zagranits. Moskva, 1974.
Geog 5300.46	Akkuratov, Valentin I. Pravo na risk. Moskva, 1974.
An 379.74	Alekseev, Valerii P. Geografiia chelovecheskikh ras. Moskva, 1974.
Geog 500.115.1F	American Geographical Society of New York. Research catalogue. First supplement. Boston, 1974. 2v.
An 170.144	Antropologiia i genogeografiia. Moskva, 1974.
Geog 4679.73.6	Bailey, Maurice. 117 days adrift. Lymington, 1974.
An 359.74	Baker, John Randal. Race. N.Y., 1974.
Geog 659.74.10	Baker, Laurie. A selection of geographical computer programs. London, 1974.
An 2109.74.5	Balandier, Georges. Anthropo-logiques. 1e éd. Paris, 1974.
Geog 5839.74	Bechmann, Elke. Mennesker og miljøer i Grønland. København, 1974.
Geog 4659.74	Berlitz, Charles F. The Bermuda triangle. 1st ed. Garden City, 1974.
Geog 5985.253.15	Brent, Peter L. Captain Scott and the Antarctic tragedy. London, 1974.
Geog 5160.61F	Bruemmer, Fred. The Arctic. Scarborough, Ont., 1974.
Geog 4028.3	Cartier, Jean Pierre. Explorateurs et explorations. Paris, 1974.
An 5.32	Česnys, Gintautas. Lietuvos antropologijos bibliografija, 1470-1970. Vilnius, 1974.
An 99.74	De Waal Malefijt, Annemarie. Images of man; a history of anthropological thought. 1. ed. N.Y., 1974.
An 359.74.5	Diamond, Stanley. In search of the primitive. N.Y., 1974.
An 359.74.10	Edgerton, Robert B. Methods and styles in the study of culture. San Francisco, 1974.
Geog 4309.74.5F	Europas Hovedstaeder. Billeder og stemninger oplevet af Eva Bendix. København, 1974.
Geog 3055.85	Falk, Bertil. Atlantis och svenskarna. Stockholm, 1974.
An 170.396	Festschrift...Otto Zerries. Bern, 1974.
Geog 4145.29.6	Fournier-Aubry, Fernand. Don Fernando. 1st American ed. N.Y., 1974.
Geog 3019.74	Fradkin, Naum G. Obraz Zemli. Moskva, 1974.
Geog 4329.74	Fries, Carl. Europas morgen. Stockholm, 1974.
An 2159.74	Frolov, Boris A. Chisla v grafike paleolita. Novosibirsk, 1974.
An 2109.74.15	Godelier, Maurice. Un domaine contesté: l'anthropologie économique. Paris, 1974.
Geog 665.168	Gould, Peter R. Mental maps. Harmondsworth, 1974.
An 170.341.5A	Grafton Elliot Smith: the man and his work. Portland, Ore., 1974.
Geog 3590.75.1	Gvozdetskii, Nikolai Andreevich. Soviet geographical explorations and discoveries. Moscow, 1974.
Geog 659.74	Hammond, Robert. Quantitative techniques in geography: an introduction. Oxford, 1974.
Geog 665.169	Hans Graul-Festschrift. Heidelberg, 1974.
An 182.3	Harris, Marvin. Cows, pigs, wars and witches. 1. ed. N.Y., 1974.
Geog 4131.74.5	Haulot, Arthur. Tourisme et environnement. Verviers, 1974.
An 2109.74.10	Hawkes, Jacquetta Hopkins. Atlas of ancient archaeology. N.Y., 1974.
An 182.4	Hodgen, Margaret. Anthropology, history and cultural change. Tucson, 1974.
Geog 5518.19.20	Hood, Robert. To the Arctic by canoe, 1819-1821. Montreal, 1974.
Geog 4672.23	Huntress, Keith G. Narrative of shipwrecks and disasters, 1586-1860. Ames, 1974.
Geog 4131.74	International Geographical Union. Working Group, Geography of Tourism and Recreation. Studies in the geography of tourism. Frankfurt, 1974.
Geog 759.74	Juillard, Etienne. La région, contributions à une géographie générale des espaces régionaux. Paris, 1974.
Geog 3049.68	Kolosimo, Peter. Timeless earth. Secaucus, N.J., 1974.
Geog 3590.88	Krupenikov, Igor' A. Istoriia geograficheskoi mysli v Moldavii. Kishinev, 1974.
Geog 4309.74	Kujundžić, Miodrag. Tudje avlije. Novi Sad, 1974.
An 359.74.25	Langness, Lewis L. The study of culture. San Francisco, 1974.
An 136.5.87	Leach, Edmund Ronald. Claude Lévi-Strauss. N.Y., 1974.
Geog 3030.7.1	Magidovich, I.P. Historiia poznania Europy. Wyd 1. Warszawa, 1974.
Geog 4677.48	Marsden, Peter Richard V. The wreck of the Amsterdam. London, 1974.
Geog 4029.15	Monfreid, Henri de. Le feu de Saint-Elme. Paris, 1974.
An 359.74.15	Montagu, Ashley. Man's most dangerous myth. 5. ed. N.Y., 1974.
Geog 5060.18	Mountfield, David. A history of polar exploration. London, 1974.
Geog 659.74.5	Mukitanov, Naurzbai K. Problema tselostnosti v fizicheskoi geografii. Alma-Ata, 1974.
Geog 4129.74	Parry, John Horace. The discovery of the sea. N.Y., 1974.
Geog 4145.46	Pasetskii, Vasilii M. Ivan Fedorovich Kruzenshtern. Moskva, 1974.
Geog 5300.48	Pastskii, Vasilii M. Arkticheskie putushestviia rossiian. Moskva, 1974.

Chronological Listing

1974 - cont.

An 2109.68.17	Pershits, Abram I. Istoriia pervobytnogo obshchestva. Izd. 2. Moskva, 1974.
Geog 5070.66	Poliarnyi krug. Moskva, 1974.
Geog 6009.10.60	Ponting, Herbert George. Scott's last voyage. London, 1974.
An 359.74.30	Rasogeneticheskie protsessy v etnicheskoi istorii. Moskva, 1974.
An 170.323	Rethinking modernization; anthropological perspectives. Westport, 1974.
Geog 5838.77.1	Rink, Henrik. Danish Greenland, its people and products. Montreal, 1974.
An 2109.74	Schneider, Harold K. Economic man; the anthropology of economics. N.Y., 1974.
Geog 3506.8	Smet, Antoine de. Album Antoine de Smet. Bruxelles, 1974.
An 2109.73.25	Social and cultural identity; problems of persistence and change. Athens, 1974.
Geog 4145.84.20	Stark, Freya. Letters. Compton Chamber, 1974-
Geog 4209.71.45	Swale, Rosie. Children of Cape Horn. London, 1974.
Geog 4145.14.100	Swinglehurst, Edmund. The romantic journey. London, 1974.
An 359.74.20	Szyfelbejn-Sokolewicz, Zofia. Wprowadzenie do etnologii. Wyd. 1. Warszawa, 1974.
Geog 616.32	Talyzin, Fedor F. Puteshestviia za nevidimym vragom. Moskva, 1974.
An 2108.71.1	Tylor, Edward Burnett. Primitive culture: researches into the development of mythology. N.Y., 1974. 2v.
Geog 4209.65.25	Vinogradov, Aleksandr A. Gde shumiat chuzhie goroda. Moskva, 1974.
An 629.74	Vogt, Evon Zartman. Aerial photography in anthropological field research. Cambridge, 1974.
Geog 3540.12	Wotte, Herbert. In blaver Ferne lag Amerika. 3. Aufl. Leipzig, 1974.
Geog 4019.74	Zweig, Paul. The adventure. London, 1974.

1975

Geog 659.75	Amedeo, Douglas. An introduction to scientific reasoning in geography. N.Y., 1975.
Geog 5839.75	Banks, Mike. Greenland. Newton Abbot, 1975.
An 359.75.5	Banton, Michael P. The race concept. Newton Abbot, 1975.
An 170.135.10	Biosocial anthropology. N.Y., 1975.
An 2109.75	Bodley, John H. Victims of progress. Menlo Park, 1975.
An 2109.75.15	Borer, Mary Irene Cathcart. Background to archaeology. London, 1975.
An 2109.75.5	Clarke, Robin. The challenge of the primitives. London, 1975.
An 170.174.5	Douglas, Mary T. Implicit meanings. London, 1975.
Geog 608.11	Drake, Fred W. China charts the world. Hsu, Chi-Yü and his geography of 1848. Cambridge, 1975.
An 180.6	Fox, Robin. Encounter with anthropology. Harmondsworth, 1975.
An 143.2	Gagen-Torn, Nina I. Lev Iakovlevich Shternberg. Moskva, 1975.
Geog 759.75	Géographie régionale. Paris, 1975-
An 359.75.10	Goodall, Vanne Morris. The quest for man. N.Y., 1975.
An 132.6	Hanson, F. Allan. Meaning in culture. London, 1975.
Geog 759.75.5	Hart, John Fraser. The look of the land. Englewood Cliffs, 1975.
An 170.361	The interpretation of symbolism. N.Y., 1975.
Geog 5516.31.3	James, T. The strange and dangerous voyage. Toronto, 1975.
Geog 4238.49F	Jenkins, F.H. Journal of a voyage to San Francisco, 1849. Northridge? 1975.
Geog 4182.76	Kreekel, Willem. De reis van Z.M. De Vlieg. 's-Gravenhage, 1975.
Geog 4659.75	Kusche, Lawrence D. The Bermuda triangle mystery - solved. N.Y., 1975.
Geog 4652.9	Marx, Robert F. Shipwrecks of the Western Hemisphere, 1492-1825. N.Y., 1975.
An 170.268	Marxist analyses and social anthropology. N.Y., 1975.
Geog 3550.17	Miscellanea di storia delle esplorazioni. Genova, 1975.
Geog 3019.75F	Newby, Eric. The Mitchell Beazley world atlas of exploration. London, 1975.
Geog 579.59.5	Öngör, Sami. Coğrafya terimleri sözlüğü. İstanbul, 1975.
An 2109.75.10	Pervobytnoe obshchestvo. Moskva, 1975.
Geog 4182.75	Pijnacker, Cornelis. Historysch verhael van den steden Thunes. 's-Gravenhage, 1975.
Geog 185.15.10	Polsko-Czeskie Seminarium Geograficzne, 5th, Warsaw, 1972. V. Czesko-Polskie seminarium geograficzne. Wyd. 1. Warszawa, 1975.
An 359.75.15	Racial variation in man: proceedings of a symposium held at the Royal Geographical Society, London, on 19 and 20 September 1974. London, 1975.
An 378.99.1	Ratzel, Friedrich. Anthropogeographie. Darmstadt, 1975.
Geog 3019.34.2	Sykes, P.M. A history of exploration. Westport, Conn., 1975.
An 2359.75	Tannahill, Reay. Flesh and blood. London, 1975.
Geog 5509.75	Thomson, George Malcolm. The North-west passage. London, 1975.
An 5.21.3F	Tozzer Library. Catalogue. Authors. 3d supplement. Boston, 1975. 3v.
An 5.22.3F	Tozzer Library. Catalogue. Subjects. 3d supplement. Boston, 1975. 4v.
Geog 4679.63.1	Troebst, C.C. The art of survival. Garden City, 1975.
Geog 4679.56.7	Villiers, A.J. Posted missing. London, 1975.
An 99.75	Voget, Fred W. A history of ethnology. N.Y., 1975.
An 359.75	Wagner, Roy. The invention of culture. Englewood Cliffs, 1975.
An 197.6	White, Leslie A. The concept of cultural systems. N.Y., 1975.

1976

Geog 4679.72.1	Vignes, Jacques. The rage to survive. N.Y., 1976.

WIDENER LIBRARY SHELFLIST, 60

GEOGRAPHY
AND
ANTHROPOLOGY

AUTHOR AND TITLE LISTING

Author and Title Listing

Geog 3070.40 A. Ortelii Catalogus cartographorum bearbeitet. v.1-2. (Bagrow, Leo.) Gotha, 1928-30.
Geog 5324.1.5 A.E. Nordenskiöld. In memoriam. (Anderson, G.) Stockholm, 1901.
Geog 606.3 A.P. Fedchenko. (Uzbekistan. Tsentral'nyi Gosudarstvennyi Arkhiv. Otdel Dorevoliutsionnykh Fondov.) Tashkent, 1956.
Geog 3019.38.3 À la conquête des mers. (St. Croix de la Roncière, G.) Paris, 1938.
Geog 4219.62 A la recherche de l'Eden. (Gandon, Yves.) Paris, 1962.
An 359.54.10 A la recherche de l'homme. (Missenard, André.) Paris, 1954.
Geog 4209.56.30 A la recherche des îles ignorees. (Gaudio,,Attilio.) Paris, 1956.
Geog 3055.43 A la recherche d'un monde perdu. (Le Cour, Paul.) Paris, 1926.
Geog 4359.70 A passo d'uomo. (Brandi, Cesare.) Milano, 1970.
Geog 4512.767 A pied, au-dessus des nuages. (Salacrow, Armand.) Paris, 1956.
Geog 3251.13 A proposito dei due globi mercatoriani. (Ceradini, G.) Milano, 1894.
Geog 4219.09.5 À travers les pays jaunes. (Pageot, G.) Paris, 1909.
Geog 4349.30 A travers l'Orient (1930). (St.-Félix, Max de.) Paris, 1931.
Geog 4219.07 A través del mundo. (Aldao, C.A.) Buenos Aires, 1907.
Geog 4209.14 A través del mundo. 5. ed. (Aldao, Carlos A.) Buenos Aires, 1914.
Geog 3520.7F Aa, P. van der. De wijd beroemde voyagien...der Engelsen. Leyden, 17- . 2v.
NEDL Geog 4157.07 Aa, P.V. Naaukeurige...der zee en land-reysen. v.1-28. Leyden, 1707. 29v.
Geog 91.4 Aarskrift. (Grønlandske Selskab.) Kjøbenhavn. 1910-1952 10v.
Geog 5558.84.5 Abandoned. 1. ed. (Todd, Alden.) N.Y., 1961.
Geog 4309.40 Abbe, P. No place like home. N.Y., 1940.
Geog 604.1 L'abbé Guillaume Denys de Dieppe, 1624-1689. (Anthiaume.) Paris, 1927.
Geog 4558.56.5 Abbey, Charles A. Before the mast in clippers...1856 to 1860. N.Y., 1937.
Htn Geog 816.00* Abbott, George. A briefe description of the whole world. London, 1600.
Htn Geog 816.00.3* Abbott, George. A briefe description of the whole world. 3d ed. London, 1608. 2 pam.
Htn Geog 816.00.5* Abbott, George. A briefe description of the whole world. 5th ed. London, 1620.
An 359.44 ABC's of scapegoating. (Harvard University. Department of Psychology.) Chicago, 1943.
An 3559.66 Abdushelishvili, Malkhaz G. K kraniologii drevnego i sovremennogo naseleniia Kavkaza. Tbilisi, 1966.
An 609.70A Abel, Herbert. Vom Raritätenkabinett zum Bremener Überseemuseum. Bremen, 1970.
Geog 4309.36.5 Abella Caprile, Margarita. Geografías (notas de viaje). Buenos Aires, 1936.
Geog 4309.51.5 Abendländische Wanderungen. (Hausenstein, Wilhelm.) München, 1951.
Geog 4029.12 Abenteur in drei Erdteilen. (Loeffler, Johann Friedrich.) Heidenheim, 1971.
An 2109.18.3 Aberg, Nils. Das nordische Kultur-Gebiet in Mitteleuropa. v.1-2. Uppsala, 1918.
Geog 3018.88 Abhandlungen...zur Geschichte der Erdkunde. (Ruge, S.) Dresden, 1888.
Geog 28.3.40 Abhandlungen. (Berlin. Freie Universität. Geographisches Institut.) 1-16 8v.
Geog 258.2 Abhandlungen. (Vienna. K.K. Geographische Gesellschaft.) Wien. 1-18,1899-1959 11v.
Geog 665.19 Abhandlungen zur Erd- und Völkerunde. v.1,2-3. (Peschel, O.F.) Leipzig, 1877. 2v.
Geog 4308.89.3 Aboard and abroad. (Walker, B.) Lowell, 1889.
Geog 4300.46 Aboard and abroad. 3d ed. (Olson, H.S.) Philadelphia, 1956.
An 2209.46 Aboriginal men of high degree. (Elkin, A.P.) Sydney, 1946.
Geog 759.46 About this earth. (Ward, Francis K.) London, 1946.
Geog 4509.07.1 Abraham, G.D. The complete mountaineer. N.Y., 1908.
Geog 4509.07.2 Abraham, G.D. The complete mountaineer. 2. ed. London, 1908.
Geog 4509.07.3 Abraham, G.D. The complete mountaineer. 3. ed. London, 1923.
Geog 4509.33 Abraham, G.D. Modern mountaineering. London, 1931.
Geog 4509.10 Abraham, G.D. Mountain adventures at home and abroad. London, 1910.
Htn Geog 756.91* Abraham ben Mordecai Farissol. Itinera mundi. v.1-2. Oxonii, 1691.
Geog 4157.66 Abrégé...histoire des decouvertes. (Barrow, J.) Paris, 1766. 12v.
Geog 4157.80 Abrégé de l'histoire generale des voyages. v.1-32, atlas. (LaHarpe, J.F.) Paris, 1780-1801. 33v.
Geog 4158.22 Abrégé des voyages modernes. (Eyries, J.B.B.) Paris, 1822-24. 14v.
Geog 3018.36.2 Abrégé historique...des...voyages. (Bajot, L.M.) Paris, 1836.
An 2109.57 Abriss der Geschichte und Kultur der Urgesellschaft. (Kosven, M.O.) Berlin, 1957.
Geog 4308.68.20 Abroad. (Coghill, James Henry.) N.Y., 1868.
Geog 4308.77A Abroad again. (Guild, C.) Boston, 1877.
Geog 4110.18 Abroad and at home. (Phillips, Morris.) N.Y., 1892.
Geog 4309.02.20 Abroad with the Jimmies. (Bell, L.L.) Boston, 1902.
Geog 110.1.18 Abstracts of papers. (International Geographical Congress, 18th, Rio de Janeiro.) Rio de Janeiro, 1956.
Geog 110.1.20 Abstracts of papers. (International Geographical Congress, 20th, London, 1964.) London, 1964.
Geog 110.1.20.2 Abstracts of papers. Supplement. (International Geographical Congress, 20th, London, 1964.) London, 1964.
Geog 110.1.21 Abstracts of papers. Supplement. (International Geographical Congress, 21st, Delhi, 1968.) Calcutta, 1968.
Geog 4417.99 Abu Taleb Khan. Reise...durch Asien, Africa. Wien, 1813.
Geog 4347.99 Abu Taleb Khan. Travels...in Asia,...Europe. London, 1810. 2v.
Geog 600.5 Abulhasan Mansur. Arabische Schriftsteller über die Geographie Indiens. Inaug. Diss. Berlin, 1919.
Geog 18.10 Academia...Buenos Aires. Academia Argentina de Geografia, Buenos Aires. Anales. 1-6,1957-1962// 3v.
Geog 5538.51.5 Access to an open Polar Sea. (Kane, E.K.) N.Y., 1853.
Geog 5321.1.11 Access to an open polar sea. (Kane, Elisha Kent.) N.Y., 1853.
Geog 818.39 Accompaniment to map of world. (Mitchell, S.A.) Philadelphia, 1839.
Geog 818.42.5 An accompaniment to Mitchell's map. (Mitchell, S.A.) Philadelphia, 1842.
Geog 818.44 An accompaniment to Mitchell's map. (Mitchell, S.A.) Philadelphia, 1844.
Geog 818.45 An accompaniment to Mitchell's map. (Mitchell, S.A.) Philadelphia, 1845.
Geog 818.46 An accompaniment to Mitchell's map. (Mitchell, S.A.) Philadelphia, 1846.
Htn Geog 4237.52.5* Account of great dangers...and remarkable deliverance of N. Pierce. (Pierce, Nathaniel.) n.p., n.d.
Geog 4206.59.5 Account of several late voyages and discoveries to north and south. London, 1711.
Htn Geog 4206.59* Account of several late voyages and discoveries to south and north. London, 1694.
Geog 5160.3 An account of the Arctic regions. (Scoresby, William.) Edinburgh, 1820. 2v.
Geog 4237.56 Account of the great dangers...of Captain N. Pierce. (Pierce, Nathaniel.) Boston, 1756.
Geog 4237.61 An account of the voyages and cruizes of Captain Walker. Boston, 1761.
Htn Geog 5517.46.10* Account of voyage...Northwest Passage. (Drage, T.S.) London, 1748-49. 2v.
An 359.38.26 Acculturation. (Herskovits, Melville Jean.) Gloucester, 1958.
An 2109.68.35 L'acculturazione dei popoli primitivi. (Forno, Mario.) Roma, 1968.
An 98.89 Achelis, Thomas. Die Entwickelung der modernen Ethnologie. Berlin, 1889.
An 98.96 Achelis, Thomas. Moderne Völkerlunde. Stuttgart, 1896.
Geog 4509.54.10 Achttausend, drüber und drunter. (Buhl, H.) München, 1954.
Geog 4329.00 Acker, F. Pen sketches. Philadelphia, 1900.
Geog 659.58 Ackerman, Ed. Geography as a fundamental research discipline. Chicago, 1958.
An 2209.71 Ackevknecht, Erwin Heinz. Medicine and ethnology. Baltimore, 1971.
Geog 4180.60.2 Acosta, J. de. National and moral history of the Indies. N.Y., 1964? 2v.
Geog 4180.60 Acosta, J. de. National and moral history of the Indies. v.2. London, 1880.
Geog 4209.42.5 Across a world. (Considine, J.J.) Toronto, 1942.
Geog 4228.70 Across America and Asia. (Pumpelly, Raphael.) N.Y., 1870.
Geog 4228.71.5 Across America and Asia. (Pumpelly, Raphael.) N.Y., 1871.
Geog 5559.21 Across Arctic America. (Rasmussen, Knud.) N.Y., 1927.
Geog 4308.67.7 Across the Atlantic. (Haeseler, Charles H.) Philadelphia, 1868.
Geog 4228.87 Across the meridians. (Francis, Harriet E.) N.Y., 1887.
Geog 5379.67 Across the top of Russia. (Petrow, Richard.) N.Y., 1967.
Geog 5559.68 Across the top of the world. (Herbert, Wally.) N.Y., 1971.
Geog 4239.66 Across the western ocean. 1st ed. (Snaith, William.) N.Y., 1966.
Geog 4219.23.5 Across three oceans. (O'Brien, Conor.) London, 1949.
Geog 5321.1.9 Act of incorporation, preamble. (Kane Monument.) N.Y., 1859.
Geog 3.1 Acta arctica. København. 1-10,1943-1958 3v.
Geog 85.135 Acta geographica. (Geografiska Sällskspet i Finland, Helsingfors.) Helsinki. 1-27,1927-1972// 11v.
Geog 3.25 Acta geographica. Kaapstad. 1,1967+
Geog 3.35 Acta geographica. Paris. 1970+ 2v.
Geog 3.35.2 Acta geographica. Tables générales, anciennes séries, 1947-1969. Paris, n.d.
Geog 3.30 Acta geographica Lovaniensia. Louvain. 1,1961+ 9v.
An 170.163.5 Actas...promovido pela Câmarã municipal de Braga. (Congresso de Etnografia e Folelore, 1st, Braga, 1956.) Lisboa, 1963. 3v.
Geog 3570.34.5 Actas. (Congresso Internacional de Historia dos Descobrimentos.) Lisboa, 1960- 6v.
Geog 81.1.3 Actes du congrès national des sociétés savantes. (France. Comité des Travaux Historiques et Scientifiques. Section de Géographie.) Paris. 84,1959+ 11v.
An 358.98.3 L'activité de l'homme. (Tenishef, V.) Paris, 1898.
Geog 3107.94.5 Adam, A. Geographical index. Edinburgh, 1795.
Geog 3107.94 Adam, A. Summary of geography and history. Edinburgh, 1794.
NEDL Geog 818.16 Adam, A. Summary of geography and history. London, 1816.
Geog 818.02 Adam, A. A summary of geography and history. 3rd ed. London, 1802.
An 359.19.3 Adam, l'homme tertiaire. (Gattefossé, R.M.) Lyon, 1919.
Geog 3240.6 Adam af Bremen...skildring af Nordeuropas länder. (Lönborg, S.) Uppsala, 1897.
An 2109.58.5 Adam y la prehistoria. (Gómez-Moreno, Manuel.) Madrid, 1958.
An 2058.78 Adamites and preadamites. (Winchell, A.) Syracuse, 1878.
Geog 3590.40 Adamov, Arkadii. Pervye russkie issledovateli Aliaski. Moskva, 1950.
Geog 4167.92.10 Adams, J. Flowers of celebrated travellers. Baltimore, 1834.
Htn Geog 4167.92.3* Adams, J. Flowers of modern travels. Boston, 1797. 2v.
Geog 4167.92 Adams, J. Flowers of modern travels. London, 1792. 2v.
Geog 4218.71.15 Adams, N. A voyage around the world. Boston, 1871.
Geog 3060.19 Adams, Percy G. Travellers and travel liars, 1660-1800. Berkeley, 1962.
Geog 4558.79 Adams, Robert C. On board the "Rocket". Boston, 1879.
Geog 4168.31 Adams, W. The modern voyager and traveller. London, 1831. 4v.
An 2059.53 Adam's ancestors. 4. ed. (Leakey, Louis S.) London, 1953.
An 359.59.15 Adam's brood. (Bertram, Colin.) London, 1959.
Geog 4678.52 Addison, A.C. A deathless story. London, 1906.
An 2338.85.2 Additional notes on arrow-release. (Morse, E.S.) Salem, 1922.
Geog 5398.79.11 Address...at unveiling of Jeannette monument. (Soley, J.R.) Baltimore, 1891.
Geog 4179.5.5 Address...on fiftieth anniversary. (Markham, C.R.) London, 1911.
Geog 139.7 Address at anniversary meeting of the Royal Geographical Society. (Smyth, W.H.) London. 1851-1868
An 2058.78.3 Address on man's age in the world. (Southall, J.C.) Richmond, 1878.
An 2058.77 Address on the antiquity of man. (Dawkins, W.B.) Salford, 1877.
Geog 4169.30 Adler, Elkan N. Jewish travellers. London, 1930.
Geog 607.5.10 Admiral Golovnin. (Ivashchenko, M.M.) Moskva, 1946.
An 359.14.3 The adolescent girl among primitive peoples. Thesis. (Van Waters, Miriam.) Worcester, 1914.
Geog 4332.5 Die Adria. (Schweiger-Lerchenfeld, A.F.) Wien, 1883.
Geog 4332.17 Die Adria und ihre Küsten. (Goracuchi, J.A.) Triest, 1863.

Author and Title Listing

Geog 4332.22 The Adriatic; a sea guide to its coasts and islands. (Denham, Henry Mangles.) London, 1967.
Geog 4332.19 The Adriatic Sea. (Hodgkinson, H.) N.Y., 1956.
Geog 4332.10 L'Adriatico; studio geografico. Milano, 1915.
Geog 4332.9 L'Adriatico. Milano, 1914.
Geog 4182.45 Adrichem, D. van. Journaal van Dircq Van Adrichem hofreis. 's-Gravenhage, 1941.
Geog 4209.28 Adventure. (Forbes, Rosita.) Boston, 1928.
Geog 4019.74 The adventure. (Zweig, Paul.) London, 1974.
Geog 4209.41.10 Adventure is my business. (DeValda, F.W.) London, 1941.
An 2109.31.7 The adventure of mankind. (Georg, Eugen.) N.Y., 1931.
Geog 5399.25 The adventure of Wrangel Island. (Stefansson, V.) N.Y., 1925.
Geog 4209.07.10 Adventurer's road. (Nicholson, T.R.) N.Y., 1958.
Geog 4028.2 The adventures; four profiles of contemporary travellers. (Green, Timothy.) London, 1970.
Geog 4678.09 The adventures. (Eustace, Thomas.) London, 1820.
Geog 4559.25.10F Adventures by sea from art of old time. (Lubbock, A.B.) London, 1925.
Geog 4556.94 Adventures by sea of Edward Coxere. (Coxere, Edward.) Oxford, 1945.
Geog 4209.09 Adventures of a civil engineer. (Burge, C.O.) London, 1909.
Geog 4208.49 Adventures of a Greek lady. (Donald, C.S.) London, 1849. 2v.
Geog 4309.08 Adventures of a green dragon. (Osborne, T.M.) Auburn, 1908.
Geog 4168.75 Adventures of famous travellers in many lands. N.Y., 18-
Geog 4477.51.15 The adventures of Gaudentio di Lucca. (Berington, S.) London, 1763.
Geog 4477.51.25 The adventures of Gaudentio di Lucca. (Berington, S.) London, 1850.
Geog 4329.21 The adventures of Imshi; a two-seater in search of the sun. (Prioleau, John.) Boston, 1923.
Geog 4477.51.7 The adventures of Signore Gaudentio di Lucca. London, 1774.
An 2059.59.5 Adventures with the missing link. (Dart, Raymond.) N.Y., 1959.
Geog 3251.9 Die ächte Karte des Olaus Magnus vom Jahre 1539. (Brenner, O.K.) Christiania, 1886.
Geog 4207.55.5 En äfventyrlig resa til och ifrån Ost-Indien, Södra America och en del af Europa åren 1755, 56 och 57. (Brelin, Johan.) Stockholm, 1973.
Geog 4412.91 Der älteste Versuch zur Entdeckung des Seeweges nach Ostindien im Jahre 1291. (Pertz, G.H.) Berlin, 1859.
An 629.74 Aerial photography in anthropological field research. (Vogt, Evon Zartman.) Cambridge, 1974.
Geog 4131.68.10 Aeschlimann, Jean Louis. Structure et tâches d'un organisme national de tourisme. Inaug. Diss. Bern, 1968.
Geog 4678.24 Aetna; a discourse. (Stanford, John.) N.Y., 1824.
Geog 5558.81.65 Les affamés de la banquise. (Gauroy, Pierre.) Paris, 1964.
An 3309.37 Les affinités somatiques des Brahmines Moithils et Kanarijias de Bihar. Thèse. (Chattapadkyay, Bojra K.) Paris, 1937?
An 3309.29.3 Affrica nord-orientale e Arabia. (Puccioni, Nello.) Pavia, 1929.
Geog 4328.92 Afloat and ashore on the Mediterranean. (Meriwether, Lee.) N.Y., 1892.
Geog 4209.63.15 An African abroad. (Ajala, O.) London, 1963.
An 2059.61 African genesis. (Ardrey, Robert.) N.Y., 1961.
Geog 4269.74 Agapov, Boris N. Shest' zagranits. Moskva, 1974.
Geog 4509.54.5 The age of mountaineering. (Ullman, J.R.) Philadelphia, 1954.
Geog 4709.57.10 The age of piracy. (Carse, Robert.) N.Y., 1957.
Geog 4709.57.12 The age of piracy. (Carse, Robert.) N.Y., 1965.
Geog 3019.63 The age of reconnaissance. (Parry, John Horace.) London, 1963.
Geog 3019.63.1 The age of reconnaissance. 1. ed. (Parry, John Horace.) Cleveland, 1963.
Geog 5559.57.10 Agranat, G.A. Zarubezhmyi Sever. Moskva, 1957.
Geog 818.78 Agrip af landafraedi. (Erslev, Ed.) Reykjavik, 1878.
Geog 4328.74.5 Aguilar, Federico C. Recuerdos de un viaje a Oriente en el año de 1874. Bogota, 1875.
An 3559.33 Aguilar y R.F. Origen castellano del prognatismo. Madrid, 1933.
Geog 4029.4 Aguilas y garras. (Franco, Ramon.) Madrid, 1929.
Geog 5318.2 Ahdoolo: the biography of M.A. Henson. 1. ed. (Miller, F.) N.Y., 1963.
Geog 5220.23 Ahlenius, K. Fornskandinaviska upptäcktsfärder i Nordatlantiska hafvet. Stockholm, 1837.
Geog 3240.22 Ahmad, Nafia. Muslim contribution to geography. Lahore, 1947.
Geog 500.175 Aids to geographical research. (Wright, John K.) N.Y., 1923.
Geog 818.07 Aikin, J. Geographical delineations. Philadelphia, 1807.
Geog 4679.47 Ainslie, Kenneth. Pacific ordeal. 1. American ed. N.Y., 1956.
Geog 4218.61.3 Ainsworth, W.F. All round the world. London, 1861. 2v.
Geog 759.43 Air world. (Howe, E.L.) Denver, 1943.
Geog 5399.28.35 Airship over the Pole; the story of the Italia. (Hogg, Garry.) London, 1969.
Geog 4328.52.2 Aiton, John. The lands of the Messiah, Mahomet, and the pope. 2nd ed. London, 1852.
Geog 4209.63.15 Ajala, O. An African abroad. London, 1963.
Geog 4309.59 Ajo otros cielos. (Fortin Magaña, Romeo.) San Salvador, 1959.
Geog 665.99 Akademia Nauk SSSR. Institut Geografii. Razvitie i preobrazo vanie geografii predy. Moskva, 1964.
Geog 4249.26 Akademiia Nauk, Petrograd. The Pacific Russian scientific investigation. Leningrad, 1926.
Geog 3590.81 Akademiia Nauk SSSR. Sovetskaia geografiia. Moskva, 1960.
Geog 3590.71.3 Akademiia Nauk SSSR. Soviet geography, accomplishments and tasks. N.Y., 1962.
An 359.63.20 Akademiia Nauk SSSR. Sovremennaia amerikanskaia etnografiia. Moskva, 1963.
An 359.57.20 Akademiia Nauk SSSR. Institut Etnografii. Ocherki obshchei etnografii. v.1-4. Moskva, 1957- 5v.
An 629.70 Akademiia Nauk SSSR. Institut Etnografii. Voprosy metodiki etnograficheskikh i etno-sotsiologicheskikh issledovanii. Moskva, 1970.
Geog 10.6 Akademiia Nauk SSSR. Institut Geografii. Geograficheskie soobshcheniia. Moskva. 2,1961+
Geog 10.2 Akademiia Nauk SSSR. Institut Nauchnoi Informatsii. Gidrologiia rushi. Moskva. 1963+
Geog 10.4 Akademiia Nauk SSSR. Institut Nauchnoi Informatsii. Itogi nauki: geografiia SSSR. Moskva. 1,1965+ 4v.
Geog 612.4.1 Akademik I.I. Lepekhin i ego puteshestviia po Rossii v 1768-1773 gg. (Fradkin, Naum G.) Moskva, 1950.
Geog 612.4.2 Akademik I.I. Lepekhin i ego puteshestviia po Rossii v 1768-1773 gg. 2. izd. (Fradkin, Naum G.) Moskva, 1953.
Geog 4329.38.5 Akdenirde leir yor gerintisi. (Duran, Faik Sabri.) Istanbul, 1938.
Geog 5300.46 Akkuratov, Valentin I. Pravo na risk. Moskva, 1974.
Geog 5850.12 Aktion och reaktion på Grönland. (Wedin, Bertil.) Stockholm, 1971.
Geog 3229.66 Alavi, S.M. Ziauddin. Geography in the Middle Ages. 1st ed. Delhi, 1966.
Geog 5399.12 Al'banov, V.A. Au pays de la mort blanche. Paris, 1928.
Geog 5399.12.5 Al'banov, V.A. Mezhdu zhuznbiu v Smertbiu. Berlin, 1925.
Geog 5399.12.10 Al'banov, V.A. Podvig shturmaka V.I. Al'banova. Moskva, 1953.
Geog 4429.72 Al'batros-clukach moriv. (Tendiuk, Leonid M.) Kyïv, 1972.
Geog 3019.58 Albertini, Renzo. Storia delle esplorazioni geografiche. Venezia, 1958.
Geog 4180.53 Alboquerque, A. d'. Commentaries of Dalboquerque. N.Y., 1964?
Geog 585.15.3 Albrecht, Theodor. Formeln und Hülfstafeln für geographische Ortsbestemmungen. 3. Aufl. Leipzig, 1894.
Geog 4218.98 Albrey, J. d'. Du Tonkin au Havre. Paris, 1898.
Geog 3506.8 Album Antoine de Smet. (Smet, Antoine de.) Bruxelles, 1974.
Geog 958.96 Album géographique. (Dubois, M.) Paris, 1896-1906. 5v.
Geog 4218.24.5PF Album pittoresque de la frégate la Thétis et de la corvette l'Esperance. (La Touanne, E. de.) Paris, 1828.
Geog 3570.50 Albuquerque, Louis. Introducão a historia dos descobrimentos. Coimbra, 1962.
Geog 3235.21 Alcuni cimeli della cartografia. (Crivellari, G.) Firenze, 1903.
Geog 4219.07 Aldao, C.A. A través del mundo. Buenos Aires, 1907.
Geog 4209.14 Aldao, Carlos A. A través del mundo. 5. ed. Buenos Aires, 1914.
Htn Geog 3570.11* De alder erste scheepo-tog ten der Portugesen. (Barrós, J. de.) Leyden, 1707?
An 2109.31 Aldrich, C.R. The primitive mind and modern civilization. London, 1931.
Geog 4558.89 Aldrich, H.L. Arctic Alaska and Siberia. Chicago, 1889.
Geog 4308.83.3 Aldrich, T.B. From Ponkapog to Pesth. Boston, 1883.
Geog 4309.47.12 Alegría y llanto de Europe; verano de 1946. 2. ed. (Noguera Mora, Neftali.) Mérida, 1965.
Geog 603.8 Aleksandr Lavrent'evich Chekanovskii, 1833-1876. (Kleopov, Igor' L'.) Leningrad, 1972.
Geog 3019.72.25 Aleksandrovskaia, Ol'ga A. Frantsvzskaia geograficheskaia shkola kontsa deviatnadtsatogo nachala dvadtsatogo veka. Moskva, 1972.
Geog 3590.74 Alekseev, Aleksandr. Kolumby rosskie. Magadan, 1966.
Geog 612.6 Alekseev, Aleksandr G. R. Fedor Petrovich Litke. Moskva, 1970.
Geog 3590.78 Alekseev, Aleksandr Ivanovich. Syny otvazhnye Rossii. Magadan, 1970.
An 379.74 Alekseev, Valerii P. Geografiia chelovecheskikh ras. Moskva, 1974.
An 3559.64.5 Alekseev, Valerii P. Kraniometriia. Moskva, 1964.
An 3509.66 Alekseev, Valerii P. Osteometriia. Moskva, 1966.
An 3559.69 Alekseev, Valerii P. Proiskhozhdenie narodov Vostochnoi Evropy. Moskva, 1969.
An 3209.72 Alekseev, Valerii P. V poishakh predkov. Moskva, 1972.
Geog 606.3.5 Aleksei Pavlovich Fedchenko, 1844-1873. (Leonov, Nikolai I.) Moskva, 1972.
Geog 5350.5 Alexander, P.F. The north-west and north-east passages. Cambridge, 1915.
Geog 4215.19 Alexander, Philip. Earliest voyages round the world, 1519-1617. Cambridge, 1916.
Geog 4237.78 Alexandre e Silva, Elias. Relação, ou Noticia particular da infeliz viajem da nas de Nassa Senhora da Ajuda. Lisboa, 1778.
Geog 4219.39 Alford, B. Around the world on $100.00 a month. Emmaus, 1939.
Geog 4145.7 Alfred Bertrand. (Bertrand, Alfred.) London, 1926.
Geog 608.15 Alfred Hettner. 6.8.1859 Gedenkschrift zum 100. Geburtstag. Heidelberg, 1960.
An 135.1.5 Alfred Kroeber: a personal configuration. (Kroeber, Theodora.) Berkeley, 1970.
Geog 4658.90.2 Algunas naufragios occuridos en las costas chilenes. (Vidal Gormaz, Francisco.) Santiago de Chile, 1901.
Geog 4658.90 Algunos naufragios occuridos en las costas chilenes. (Vidal Gormaz, Francisco.) Valparaiso, 1890.
Geog 4308.73.5 The Alhambra and the Kremlin. (Prime, Samuel I.) N.Y., 1873.
Geog 4309.62.5 Alimzhanov, An. Piat'desiat tysiach nil' po vode i sushe. Alma-Ata, 1962.
Geog 4418.61.3 All aboard for sunrise lands. 2. ed. (Rand, E.A.) Boston, 1861.
Geog 5919.48 All about Antarctica. (Weetman, Charles.) Melbourne, 1948.
Geog 4209.41.3 All in a lifetime. (Buck, F.) N.Y., 1941.
Geog 4209.48 All over the place. (Mackenzie, C.) London, 1948.
Geog 4209.46.15 All over the world. (Etherton, P.T.) London, 1946.
Geog 3019.72 All possible worlds; a history of geographical ideas. (James, Preston Everett.) Indianapolis, 1972.
Geog 4349.10 The all-rail route between the Far East and Europe. London, 1910.
Geog 4218.61.3 All round the world. (Ainsworth, W.F.) London, 1861. 2v.
Geog 4509.49 Allain, Pierre. Alpinisme et compétition. Grenoble, 1949.
Geog 4209.44 Allan, D. Lightning strikes once. N.Y., 1944.
Geog 4209.45.10 Allan, Doug. Gamblers with fate. 1. ed. N.Y., 1945.
An 359.71.20 Alland, Alexander. Human diversity. N.Y., 1971.
An 359.72.10 Alland, Alexander. The human imperative. N.Y., 1972.
Geog 1524.14 L'Allemagne, l'Autriche. 6. éd. (Baedeker, publishers.) Leipzig, 1878.
Geog 1524.38 Allemagne du sud et Autriche. 9. éd. (Baedeker, publishers.) Leipzig, 1888.
Geog 612.3 Allen, Edward W. Jean François Galaup de Lapérouse. San Francisco, 1941.
Geog 612.3.2 Allen, Edward W. The vanishing Frenchman. Rutland, 1959.
Geog 4249.36 Allen, Edward Weber. North Pacific: Japan, Siberia, Alaska, Canada. N.Y., 1936.
Geog 4308.99.10 Allen, Grant. The European tour. N.Y., 1909.
An 3538.96 Allen, Harrison. Crania from the mounds of the St. John's. Philadelphia, 1896.
Geog 5321.1.7 Allen, J.H. Elisha Kent Kane. Bangor, 1857.
Geog 4145.3 Allen, Katharine M. Foreign service diary. Washington, 1967.

Author and Title Listing

Call Number	Entry
Geog 4678.16.2	Allen, S. (Mrs.). A narrative of the shipwreck and unparalleled sufferings of Mrs. S. Allen. 2. ed. Boston, 1816.
Geog 4181.138	Allen, William Edward David. Russian embassies to the Georgian kings, 1589-1605. Cambridge, 1970. 2v.
An 358.79.5	Allgemeine Ethnographie. (Müller, F.) Wien, 1879.
Geog 579.71	Allgemeine Geographie. (Fochler-Hauze, Gustav.) Frankfurt am Main, 1971.
Geog 759.33	Allgemeine Geographie. Potsdam, 1933. 2v.
An 379.47	Allgemeine Geographie des Menschen. (Hettner, Alfred.) Stuttgart, 1947.
NEDL Geog 12.1	Allgemeine geographische Ephemeriden. Weimar. 1-51 51v.
Geog 585.12F	Allgemeine geographische und statistische Verhältnisse. (Borbstaedt, A.) Berlin, 1846.
Geog 4157.47	Allgemeine Historie der Reisen. v.1-19,21. Leipzig, 1747-74. 20v.
Geog 809.28	Allgemeine Länderkunde der Erdteile. v.3-4,6. Hannover, 1928-1935.
An 359.25.5	Allgemeine Rassenkunde. (Scheidt, W.) München, 1925.
An 3309.14	Allgemeine und spezielle Physiologie des Menschenwachstums. (Friedenthal, Hans.) Berlin, 1914.
An 2109.29.7	Allier, Raoul. Mind of the savage. London, 1929.
An 2109.27.23	Allier, Raoul. La non-civilisé et nous. Paris, 1927.
Geog 4559.63	Alltaf má fá annad skip. (Asgeirsson, Rikki.) Reykjavik, 1963.
Geog 4218.27	Allyn, Gurdon L. The old sailor's story. Norwich, Conn., 1879.
Geog 4209.51.10	A alma encantadora das ruas. (Barreto, Paulo.) Rio de Janeiro, 1951.
Geog 759.45	Almagia, R. Fondamenti di geografia generale. v.1-2. Roma, 1946-48.
Geog 500.57	Almagia, Roberto. La geografia. Roma, 1919.
Geog 608.5	Almagià, Roberto. L'opera geografica di Luca Holstenio. Citta del Vaticano, 1942.
Geog 3019.61.5	Almagia, Roberto. Scritti geografici. Roma, 1961.
Geog 3129.49	Almagià, Roberto. Storia dell'esplorazione e della scienza geografica: l'eta greca. Roma, 1949.
Geog 12.6	Almanacco del turista. Roma. 1951-1960 4v.
X Cg Geog 12.5	Almanach géographique, ou Petit atlas élémentaire. Paris. 1770
Geog 818.21.3	Almenn jardarfrädi og landskipun edur geographia. (Oddsson, G.) Kaupmannahøfn, 1822. 4v.
Geog 818.21	Almenn landaskipunarfraede. (Oddsson, G.) Kaupmannahøfn, 1821-1827. 2v.
Geog 4348.14	Almerté, T. Voyages de sa majesté la reine d'Angleterre. Paris, 1821.
Geog 5939.38.3A	Alone. (Byrd, Richard E.) N.Y., 1938.
Geog 4309.64.20	Along the ridge; from northwestern Spain to southern Yugoslavia. (Streeter, Edward.) N.Y., 1964.
Geog 12.4.5	Die Alpen; Zeitschrift des SAC. Bern. 33,1957+ 14v.
Geog 12.4	Die Alpen. Monatsschrift des Schweitzer Alpenclub. Bern. 1,1915+ 50v.
Geog 199.3	Alpina; Mitteilungen des Schweizer Alpen-Club. Zürich. 1-32 18v.
Geog 4502.5.11	Alpine Bibliographie. (Bühler, H.) München. 1932
Geog 4502.5.11.5	Alpine Bibliographie. Gesamtregister, 1931-1938. (Bühler, H.) München, 1949.
Geog 12.3	Alpine Club of Canada. The gazette. Banff. 1,1921
Geog 12.2	Alpine journal. London. 1,1863+ 46v.
Geog 12.2.3	Alpine journal. Index. v.39-58, 1927-52. London, 1954.
Geog 4515.103.2	Alpine ski-ing at all heights and seasons. 2d ed. (Lunn, A.H.M.) London, 1926.
Geog 4509.13	L'alpinisme. (Casella, G.) Paris, 1913.
Geog 4509.49	Alpinisme et compétition. (Allain, Pierre.) Grenoble, 1949.
Geog 4515.106.2	Alpinisme hivernal. (Kurz, Marcel.) Paris, 1928.
Geog 4509.52	Alpinismo. 3. ed. (Negri, Carlo.) Milano, 1952.
Geog 4509.53.20	L'alpinismo a Cortina dai suoi primordi ai giorni nostri. (Jerschak, Federico.) Roma, 1953.
Geog 4509.14	Alpinismo acrobatico. (Rey, G.) Torino, 1914.
Geog 4509.53.15	Alpinismo italiano nel mondo. (Club Alpino Italiano.) Milano, 1953.
Geog 4509.53.5	Alpiniste. (Chatellus, A. de.) Grenoble, 1953.
Geog 4520.5	Les alpinistes célèbres. (Ségogne, Henry de.) Paris, 1956.
Geog 5538.78	Als Eskimo unter den Eskimos; eine Schilderung der Erlebnisse der Schwatka'schen Franklin-Aufsuchungs-Expedition in den Jahren 1878-80. (Klutschak, H.W.) Wien, 1881.
Geog 4249.61	Alt-Asiaten unter Segel in Indischen und Pazifischen. (Hoever, Otto.) Braunschweig, 1961.
Geog 3251.32	Alte deutsche Landkarten. (Lehmann, Edgar.) Leipzig, 1935.
Geog 3070.22	Alte handschriftliche Schiffer-Karten (Matkovic, P.) Wien, 1863.
Geog 3235.35	Alte handschriftliche Schifferkarten. (Matkovic, P.P.) Agram, 1860.
Geog 953.5FA	Die alten Städtebilder. (Bachmann, Friedrich.) Leipzig, 1939.
Geog 953.5.2F	Die alten Städtebilder. 2. Aufl. (Bachmann, Friedrich.) Stuttgart, 1965.
An 2109.02	Altersklassen und Männerbünde. (Schurtz, H.) Berlin, 1902.
An 2059.69.10	Altner, Günter. Kreatur Mensch. München, 1969.
Geog 4180.64	Alvares, F. Narrative of Portuguese Embassy. London, 1881.
Geog 4181.114	Alvares, Francisco. The Prester John of the Indies. Cambridge, 1961. 2v.
Geog 4329.27.15	Am Mittelmeer. (Ludwig, Emil.) Berlin, 1927.
Geog 5398.99	Amadeus, Louis. Farther north than Nansen. London, 1901.
Geog 5398.99.3	Amadeus, Louis. On the "Polar Star". N.Y., 1903. 2v.
Geog 5398.99.5	Amadeus, Louis. Die Stella Polare im Eismeer. Leipzig, 1903.
Geog 3235.31	Amat, P. Del planisferio. Roma, 1878.
Geog 500.22	Amat di San Filippo, Pietro. Bibliografia di viaggiatori italiani ordinata cronologicamenti. Roma, 1874.
An 2109.27.13	L'ame primitive. 3e éd. (Lévy-Bruhl, L.) Paris, 1927.
Geog 659.75	Amedeo, Douglas. An introduction to scientific reasoning in geography. N.Y., 1975.
Geog 5970.15	America in the Antarctic to 1840. (Mitterling, Philip.) Urbana, 1959.
Geog 809.19.6	América Meridional, Oceania. (Blásquez y Delgado Aguilera, Antonio.) Barcelona, 1916.
Geog 809.19.5	América Septentrional, América Central, Las Antillas, Alaska, Canada, Estados Unidos. (Camena d'Almeida, P.) Barcelona, 1916.
Geog 4509.42	The American Alpine Club's handbook of American mountaineering. (Henderson, K.A.) Boston, 1942.
Geog 4502.12	The American alpine journal. N.Y. 7,1948+ 13v.
An 359.68.5A	American Association for the Advancement of Science. Science and the concept of race. N.Y., 1968.
Geog 4309.33	American Automobile Association. Foreign. Motering abroad. N.Y., 1933.
Geog 13.12	American Bureau of Geography. Bulletin. Winona. 1900-1901
Geog 578.27.3	The American companion. (Hawkes, P.) Philadelphia, 1827.
An 2109.73.15	American Ethnological Society. Learning and culture. Seattle, 1973.
Geog 5300.9	American explorations in the ice zones. (Nourse, J.E.) Boston, 1884.
Geog 5070.35A	American Geographic Society, N.Y. Problems of polar research. N.Y., 1928.
Geog 5558.60.3	American Geographical and Statistical Society. The polar exploring expedition. N.Y., 1860.
Geog 5235.5	American Geographical Society, N.Y. Physical map of the Arctic. N.Y., 1930.
Geog 13.1	American Geographical Society of New York. Bulletin. 1-2,1852-1856
Geog 13.2.5	American Geographical Society of New York. Charter, by-laws and list of members. N.Y. 1870
Geog 13.2.3	American Geographical Society of New York. Charter and by-laws. 1857
Geog 13.4.4	American Geographical Society of New York. Index to the Bulletin of the American Geographical Society. 1852-1910. N.Y., n.d.
Geog 13.4	American Geographical Society of New York. Journal. 1-47,1895-1915 45v.
Geog 13.3	American Geographical Society of New York. Proceedings. 1-2
Geog 500.115.1F	American Geographical Society of New York. Research catalogue. First supplement. Boston, 1974. 2v.
Geog 500.115.2F	American Geographical Society of New York. Research catalogue. Map supplement. Boston, 1962.
Geog 500.115F	American Geographical Society of New York. Research catalogue of the American Geographical Society. Boston, 1962. 15v.
Geog 13.2.10	American Geographical Society of New York. The role of geography in the modern world. N.Y., 195-.
Geog 13.2	Pamphlet vol. American Geographical Society of New York. Charter and by-laws. 15 pam.
Geog 3400.5A	American geography; inventory and prospect. (James, Preston E.) Syracuse, 1954.
Geog 4308.72.3	An American girl abroad. (Trafton, A.) Boston, 1872.
Geog 4308.50.25	The American in Europe. (Crockett, Henry C.) London, 184-.
Geog 4309.65	The American in Europe. 1st ed. (MacShane, Frank.) N.Y., 1965.
Geog 4418.57	An American merchant. (Train, G.F.) N.Y., 1857.
An 126.1	American Philosophical Society. Brinton memorial meeting. Philadelphia, 1900.
An 359.52.15	American race theorists, a critique of their thoughts and methods. (Campbell, Byram.) Boston, 1952.
Geog 500.72	American School of Classic Studies at Athens. Voyages and travels in the Near East made during the nineteenth century. Princeton, N.Y., 1954.
Geog 500.72.5	American School of Classical Studies at Athens. Voyages and travels in Greece. Princeton, N.Y., 1953.
Geog 500.155	American travellers abroad. (Smith, Harold F.) Carbondale, 1969.
Geog 4307.83.2	The American wanderer...Europe. Dublin, 1783.
Geog 4307.83	The American wanderer...Europe. London, 1783.
Geog 4279.64	Americans in Egypt. (Dulles, F.R.) Ann Arbor, 1964.
Geog 5970.15.20	Americans in Antarctica, 1775-1948. (Bertrand, Kenneth John.) N.Y., 1971.
Geog 520.23	Americe Vespuce, 1451-1512. (Vignaud, Henry.) Paris, 1917.
Geog 5838.45	Amerikas...gamle geographie. (Rafn, C.C.) Kjøbenhavn, 1845.
Geog 4208.30.5	Ames, N. A mariner's sketches. Providence, 1830.
Geog 4208.21.5	Ames, Nathaniel. Nautical reminiscences. Providence, 1832.
An 3308.89	Amherst College. Anthropometric manual 1889. Amherst, 1889.
An 3308.87	Amherst College. The anthropometric manual 1887. Amherst, 1887.
Geog 4181.7	Amherst of Hackney. Discovery of Solomon Islands by...Mendaña. London, 1901. 2v.
Geog 4558.90.7	Amicis, E. de. On blue water. N.Y., 1897.
Geog 4558.90.5	Amicis, E. de. Sull'oceano. Milano, 1890.
Geog 4558.90.6	Amicis, E. de. Sull'oceano. Milano, 1913.
Geog 4559.28.15	Amicis, E. de. Sull'oceano. Milano, 1928.
Geog 4206.13.30F	Amman, H.J. Hans Jakob Ammann genannt der Thalwyler Schärer und seineReise ins gelobte Land. Zürich, 1919.
Geog 4206.13.27	Amman, H.J. Reiss in das gelobte Land. Berlegung, 1678.
Geog 4206.13.25	Amman, H.J. Reiss ins gelobte Land...Servian...Aegypten. Zürich, 1630.
An 3208.93	Ammon, Alto. Die natürliche Auslese bein Menschen. Jena, 1893.
Geog 4558.74.3	Among our sailors. (Jewell, J.G.) N.Y., 1874.
Geog 4308.87.5	Among the Scotch-Irish. (Morrison, Leonard A.) Boston, 1891.
An 379.46.10	Amorim Girão, Aristides de. Geografia humana. Porto, 1946.
Geog 602.1.10	Amsterdam. Nederlansh Historisch Schlepvaart Museum. De Blaeu's beschrijvers van land. Amsterdam, 1952.
Geog 500.56	Amsterdam. Universiteit. Bibliotheek. Catalogus geographie en reizen. Amsterdam, 1923.
Geog 5399.25.9	Amundsen, R. Amundsen-Ellsworths polflyvning, 1925. Oslo, 1925.
Geog 6009.10.3	Amundsen, R. Au Pôle Sud, éxpedition du "Fram", 1910-1912. Paris, 1913.
Geog 5399.25.21	Amundsen, R. First crossing of the Polar Sea. Garden City, 1928.
Geog 5399.25.20A	Amundsen, R. First crossing of the Polar Sea. N.Y., 1927.
Geog 5399.25.19	Amundsen, R. Den første flukt over Polhavet. Oslo, 1926.
Geog 5519.03	Amundsen, R. The North West Passage. London, 1908. 2v.
Geog 5399.25.10	Amundsen, R. Our polar flight. N.Y., 1925.
Geog 5519.03.1	Amundsen, R. Roald Amundsen's "The North West Passage". N.Y., 1908. 2v.

Author and Title Listing

Geog 6009.10A Amundsen, R. The South Pole...Norwegian Antarctic expedition, 1910-1912. London, 1912. 2v.

Geog 6009.10.5 Amundsen, R. Sydpolen...med Fram, 1910-1912. Kristiania, 1912. 2v.

Geog 5311.3.15 Amundsen, Roald. Apdagelsesreiser. Oslo, 1928-30. 4v.

Geog 5311.1.3 Amundsen, Roald. My life as an explorer. Garden City, N.Y., 1928.

Geog 5379.18 Amundsen, Roald. Nordostpassagen. Kristiania, 1921.

Geog 5311.3.10 Amundsen. (Partridge, B.) London, 1953.

Geog 5399.25.9 Amundsen-Ellsworths polflyvning, 1925. (Amundsen, R.) Oslo, 1925.

Geog 3570.21 Amzalah, Moses. La Méditerranée et les découvertes maritimes des Portugais. Lisbonne, 1951.

Geog 4209.31.25 An den Rändern dreier Erdteile. (Keilpflug, Erich R.) Berlin, 1931.

Geog 19.3 Anais. (Associação dos Geografos Brasileiros.) 1,1945+ 8v.

Geog 18.10 Anales. (Academia...Buenos Aires. Academia Argentina de Geografía, Buenos Aires.) 1-6,1957-1962// 3v.

An 136.5.45 L'analisi strutturali dei miti. (Scardnelli, Pietro.) Milano, 1971.

An 359.49.15 Análisis de la cultura; etnología, historia, folklore. (Caro Baroja, Julio.) Barcelona, 1949.

Geog 659.73 L'analyse quantitative en géographie. 1. éd. (Racine, Jean Bernard.) Paris, 1973.

VGeog 4131.68 Anan'ev, Mikhail A. Mezhdunarodruji turizm. Moskva, 1968.

An 359.25.9 Anantha, Krishna Iyer L. Lectures on ethnogrpahy. Calcutta, 1925.

Geog 6100.5 Anare: Australia's Antarctic outposts. (Law, Phillip G.) Melbourne, 1957.

An 2338.85 Ancient and modern methods of arrow-release. (Morse, E.S.) Salem, 1885.

An 2059.11 Ancient hunters. (Sollas, William J.) London, 1911.

An 2059.24.1 Ancient hunters and their modern representatives. 3. ed. (Sollas, William J.) London, 1924.

An 2059.24 Ancient hunters and their modern representatives. 3. ed. (Sollas, William J.) N.Y., 1924.

Geog 4559.28.10 Ancient mariners. (Smith, Cicely Fox.) London, 1928.

An 3309.64 Ancient races of Baluchistan. 1st ed. (Sarkar, Sasanka Sekhar.) Calcutta, 1964.

An 359.09.15 Ancient society. (Morgan, L.H.) Chicago, 1909.

An 358.78.3A Ancient society. (Morgan, L.H.) N.Y., 1878.

An 359.64.40 Ancient society. (Morgan, Lewis H.) Cambridge, 1964.

An 2059.12.5A Ancient types of man. 2. ed. (Keith, Arthur.) London, 1912.

Geog 4180.34 Andagoya, P. de. Narrative of proceedings of Pedrarias Davila. London, 1865.

Geog 4309.66.5 Andersch, Alfred. Aus einem römischen Winter. Olten, 1966.

Geog 4239.29.13 Andersen, Lis. Lis sails the Atlantic. London, 1953.

Geog 500.70 Anderson, Ernst. Bok-katalog omfattande geografi och resor. v.1-2. Stockholm, 1929-31.

Geog 5324.1.5 Anderson, G. A.E. Nordenskiöld. In memoriam. Stockholm, 1901.

Geog 4209.17 Anderson, I. (Mrs.). Odd corners. N.Y., 1917.

Geog 4329.30.7 Anderson, I.P. (Mrs.). A yacht in Mediterranean seas. Boston, 1930.

Geog 4129.70 Anderson, John Richard Lane. The Ulysses factor: the exploring instinct in man. London, 1970.

Geog 4279.69 Anderson, Patrick. Over the Alps. London, 1969.

Geog 5559.58 Anderson, William R. Nautilus 90 north. Cleveland, 1959.

Geog 4029.10 Anderton, Russ. Tic-polonga. 1. ed. Garden City, 1953.

Geog 5838.83 Den andra Dicksonska Expedition. (Nordenskiold, A.E.) Stockholm, 1885.

Geog 5660.15 Andrae og Østgrønland. (Dahl, F.) København, 1932.

Htn Geog 817.74* André, Noël. Description et usages de la Mappemonde. Paris, 1774.

Geog 5398.97 Andrée; au pôle nord en ballon. (Lachambre, Henri.) Paris, 1897.

An 175.1 Andree, R. Ethnographische Parallelen und Vergleiche. v.1-2. Leipzig, 1878-89.

Geog 5398.97.5 Andrée, S.A. Andrée's story; the complete record of his polar flight, 1897. N.Y., 1930.

Geog 5398.97.10 Andrée. (Putnam, George Palmer.) N.Y., 1930.

Geog 5398.97.5 Andrée's story; the complete record of his polar flight, 1897. (Andrée, S.A.) N.Y., 1930.

Geog 819.13 Andrews, A.W. A text-book of geography. London, 1913.

Geog 819.16 Andrews, A.W. A text book of geography. London, 1916.

Geog 5321.1.15 Andrews, E.W. Memoir and eulogy of Dr. Elisha K. Kane. N.Y., 1857.

Geog 665.107 Andrews, John. Frontiers and men; a volume in memory of Griffith Taylor (1880-1963). Melbourne, 1966.

Geog 4181.111 Andrews, Kenneth. English privateering voyages to the West Indies. Cambridge, 1959.

Geog 4181.142A Andrews, Kenneth Raymond. The last voyage of Drake and Hawkins. Cambridge, 1972.

An 359.45.15 Andrews, R.C. Meet your ancestors, a biography of primitive man. N.Y., 1945.

Geog 4559.16.5 Andrews, R.C. Whale hunting with gun and camera. N.Y., 1916.

Geog 4238.80 Andrews, William Albert. Dangerous voyages of Captain William Andrews. N.Y., 1966.

Geog 3240.20 Andriani, G. Giacomo Bracelli nella storia della geografia. Pontremoli, n.d.

An 2108.73 Die Anfänge der Cultur. v.1-2. (Tylor, Edward Burnett.) Leipzig, 1873.

Geog 4559.23 Angel, W.H. The clipper ship "Sheila". Boston, 1923.

Geog 665.105 Angewandte Geographie; Festschrift für Professor Dr. Erwin Scheu. (Weigt, Ernst.) Nürnberg, 1966.

Geog 520.21 Anghiera, P.M. De orbe novo. Paris, 1907.

Geog 3580.10 Anghiera, P.M. d'. Lettres relatives aux découvertes maritimes des Espagnols. Paris, 1885.

Geog 3520.5 Les Anglais et les Hollandais. (Jurien de la Gravière, J.B.E.) Paris, 1890. 2v.

Geog 4348.49.2 Anglo-Indian passage. (Richardson, D.L.) London, 1849.

Geog 4206.54 Anhang zwer Reisen. (Bäckhoff.) Berlin, n.d.

An 2056.56.5 Animad versiones in librum prae-adamitarum. (Romano, E.) Paris, 1656.

An 2109.66 Anisimov, Arkadii F. Dukhovnaia zhizn' pervobytnogo obshchestva. Leningrad, 1966.

Geog 4309.60 Anjaneyulu, D. Window to the West. 1st ed. Madras, 1967.

Geog 807.87.3 Ankündigung einer neuen Ausgabe von Büschings Erdbeschreibung welche auf Pränumeration gedruckt wird. (Bohn, C.E.) Hamburg, 1802.

Geog 185.1 Annaes da commissão central permanente de geographia. (Portugal. Ministerio dos Negocios da Marinha e Ultramar.) Lisboa. 1,1876

Geog 32.90 Annales. Sectio geographica. (Budapest. Tudomány-Egyetem.) Budapest. 1,1965+ 2v.

Geog 14.200 Annales de géographie. Paris. 1,1892+ 76v.

Geog 14.200.15 Annales de géographie. Table décennale, 1891-1901. Paris, 1902- 3v.

NEDL Geog 14.207 Annales des voyages. Paris. 1808-1814 25v.

NEDL Geog 14.208 Annales des voyages. Table. Paris. 1813

NEDL Geog 14.201 Annales maritimes. 99v.

Geog 14.10 Annali di richerche e studi di geografia. Genova. 5,1949+ 11v.

Geog 13.9 Annals. (Association of American Geographers.) Albany, N.Y. 1+ 47v.

Geog 13.9.5 Annals. Index for volumes 1-25. (Association of American Geographers.) Lancaster, Pa.? n.d.

Geog 14.500 L'année cartographique. 1-23

Geog 14.501 L'année geographique. 1862-1878 16v.

Geog 4418.02F Annesley, G. Voyages and travels to India. London, 1809. 3v.

Geog 5050.5 Annotated bibliography of the polar regions. Series B. pt. 1. (United States. Works Progress Administration, N.Y.C.) N.Y., 1938.

Geog 500.116.5 Annotated world list of selected current geographical serials in English. 2. ed. (Harris, Chauncy Donnison.) Chicago, 1964.

NEDL Geog 39.1 Annuaire. (Club Alpin Française.) Paris. 1-30,1874-1903 30v.

NEDL Geog 39.1.3 Annuaire. Table générale. (Club Alpin Française.) Paris. 1874-1888

Geog 14.550 Annuaire des voyages et de la géographie. Paris. 1844-1845 2v.

Geog 16.1.10 Annual report. (Appalachian Mountain Club.)

Geog 39.2.18 Annuario. (Centro Alpinistico Italiano. Sezione di Biella.) Biella.

Geog 39.2.20 Annuario. (Club Alpino Italiano. Sezione Antonio Locatelli, Bergamo.) 1951+ 4v.

Geog 39.2.16 Annuario. (Club Alpino Italiano. Sezione di Roma.) Roma. 1-3,1886-1889

Geog 114.5 Annuario dell'Istituto. (Istituto Cartografico Italiano, Rome.) Roma. 1-4,1884-1889

Geog 3199.5 Anonymi de situ orbis. (Manitius, M.) Stuttgardiae, 1884.

Geog 4303.08 Anonymi descriptio Europae orientalis. Cracoviae, 1916.

Geog 4418.55 Another budget...in the East. (Eames, J.A.) Boston, 1855.

An 359.72.20 Ansari, G. Recent trends in cultural anthropology. Wien, 1972.

Geog 4307.86.13 Ansbach, Elizabeth Craven. Journey thro' Crimea to Constantinople. London, 1789.

Htn Geog 4307.86.12* Ansbach, Elizabeth Craven. Journey thro' Crimea to Constantinople. London, 1789. 2 pam.

Geog 4559.45.10 Anson, P.F. Harbour Head. London, 1944.

An 359.12.7 The antagonism of races. (Grow, O.) Waterloo, 1912.

Geog 15.2 Antarctic; a news bulletin published quarterly by the New Zealand Antarctic Society. Wellington. 1,1956+ 5v.

Geog 6009.01 Antarctic...zwei Jahre. (Nordenskjöld, Otto.) Berlin, 1904. 2v.

Geog 5919.69 The Antarctic. (King, H.G.R.) London, 1969.

Geog 6009.11 Antarctic adventure; Scott's northern party. (Priestly, R.E.) London, 1914.

Geog 5900.18 Antarctic bibliography. (United States. Naval Photographic Interpretation Center.) Washington, 1951.

Geog 5939.55 The Antarctic challenged. (Mountevans, E.) London, 1955.

Geog 5939.07 Antarctic days. (Murray, J.) London, 1913.

Geog 5925.5 Antarctic exploration. (London, England. Royal Geographical Society.) London, 1898.

Geog 6008.38.5 The Antarctic explorations of Charles Wilkes, 1838-42. (Ross, Frank E.) n.p., 1935.

Geog 5919.01.5 The Antarctic manual. (Murray, G.) London, 1901.

Geog 5939.41 The Antarctic ocean. (Owen, Russell.) N.Y., 1941.

Geog 5919.51 The Antarctic problem. (Christie, E.W.H.) London, 1951.

Geog 5905.1 Pamphlet box. Antarctic regions.

Geog 5918.98.3A The Antarctic regions. (Fricker, Karl.) London, 1900.

Geog 5970.46 Antarctic research; a review of British scientific achievement in Antarctica. (Priestley, Raymond E.) London, 1964.

Geog 5919.52 The Antarctic today. (Simpson, F.A.) Wellington, 1952.

Geog 5919.28 Antarctica; a treatise on the southern continent. (Hayes, James G.) London, 1928.

Geog 6005.5 Antarctica; authentic accounts of life and exploration. (Neider, Charles.) N.Y., 1972.

Geog 5919.58.10 Antarctica; de geschiedenis van het geheimzinnige Zuidland. (Knapp, Willem H.C.) Haarlem, 1958.

Geog 6009.01.3 Antarctica, or Two years...South Pole. (Nordenskjöld, Otto.) London, 1905.

Geog 15.5 Antarctica, the last frontier; the annual report of the officer in charge, United States Antarctic Programs. Washington.

Geog 5939.02 Antarctica. (Balch, E.S.) Philadelphia, 1902.

Geog 5939.59 Antarctica. (Debenham, Frank.) London, 1959.

Geog 5919.65.5 Antarctica. (Hatherton, Trevor.) Wellington, 1965.

Geog 6009.55.25 Antarctica. (Trans-Antarctic Expedition, 1955-1958.) Wellington, 1964.

Geog 5919.62 Antarctica. 1st ed. (Caras, Roger A.) Philadelphia, 1962.

Geog 5939.59.22 Antarctica ahoy! The icebook. (Smuul, Juhan.) Moscow, 196-?

Geog 5925.45 Antarctica in world affairs. (Gould, Lawrence M.) N.Y., 1958.

Geog 5919.58.5 Antarctica today and tomorrow. (Bertram, George Colin L.) Cambridge, 1958.

Geog 5939.73 Antarkticheskii dnevnik. (Kutuzov, Pavel S.) Moskva, 1973.

Geog 5919.57.10 Antarktika. (Lebedev, Vladimir L.) Moskva, 1957.

Geog 5939.58.5 Antarktika. (Pavlovskii, E.N.) Moskva, 1958.

Geog 5925.65 Antarktika dolzhna stat' zonoi mira. (Slevich, S.B.) Leningrad, 1960.

Geog 5919.55 Die Antarktis; eine Länderkunde. (Kosack, Hans P.) Heidelberg, 1955.

Geog 5918.98 Antarktis. (Fricker, Karl.) Berlin, 1898.

Geog 5919.57.5 Antarktyka. Wyd. 2. (Machowski, J.) Warszawa, 1957.

Geog 5919.45 La Antártida sudamericana. (Cordovez Madariga, Enrique.) Santiago, 1945.

An 2058.76.3 Los antepasados de Adan. (Meunier, Victor.) Madrid, 1876.

Geog 4329.22 Anthénea. (Maurras, Charles.) Paris, 1922.

Geog 3070.30 Anthiaume, A. Cartes marines, constructions navales, 1500-1650. Paris, 1916. 2v.

Geog 604.1 Anthiaume. L'abbé Guillaume Denys de Dieppe, 1624-1689. Paris, 1927.

Geog 604.2 Anthiaume. Pierre Desceliers. Rouen, 1926.

Geog 3108.50 Anthon, C. System of ancient and mediaeval geography. N.Y., 1850.

Author and Title Listing

Call Number	Entry
Geog 3108.71	Anthon, C. System of ancient and mediaeval geography. N.Y., 1871.
An 378.82	Anthropo-Geographie. (Ratzel, F.) Stuttgart, 1882. 2v.
An 2109.74.5	Anthropo-logiques. 1e éd. (Balandier, Georges.) Paris, 1974.
An 378.99.1	Anthropogeographie. (Ratzel, Friedrich.) Darmstadt, 1975.
An 378.99	Anthropogeographie. v.2. (Ratzel, Friedrich.) Stuttgart, 1899.
Geog 135.2.2	Anthropogeographische Beiträge. v.2. (Ratzel, Friedrich.) Leipzig, 1895.
An 170.230.5	Anthropological congress dedicated to Aleš Hrdlička. (Kongres Ceskolovenských anthropologû, 10th, Prague and Humpolie, 1969.) Prague, 1971.
An 358.93.1	Anthropological history of Europe. (Beddoe, J.) London, 1893.
An 2059.59.15	Anthropological Society of Washington. Evolution and anthropology. Washington, 1959.
An 170.111	Anthropological Society of Washington. The Saturday lectures. Washington, D.C., 1882.
Geog 5182.7	An anthropological study of the origin of the Eskimo culture. (Steensby, Hans P.) Kjøbenhavn, 1916.
An 176.1	The anthropological treatises. (Blumenbach, J.F.) London, 1865.
An 358.76	L'anthropologie. (Topinard, P.) Paris, 1876.
An 358.76.2	L'anthropologie. (Topinard, P.) Paris, 1876.
An 3559.12	Anthropologie anatomique, crane-face. (Paul-Boncour, G.) Paris, 1912.
An 98.93	L'anthropologie aux États-Unis. (Topinard, P.) Paris, 1893.
An 132.5	L'anthropologie comparée de Guillaume de Humboldt. (LeRoux, Robert.) Paris, 1958.
An 3539.09F	Anthropologie de la Suisse. (Pittard, E.) Genève, 1909-10.
An 358.59.5	Anthropologie der Naturvölker. (Waitz, T.) Leipzig, 1859. 6v.
An 358.77	Anthropologie der Naturvölker. 2. Aufl. (Waitz, T.) Leipzig, 1860.
An 3509.72	Anthropologie du conscrit français d'après les comptes numeriques et sommaires du recrutement de l'année, 1819-1826. (Aron, Jean Paul.) Paris, 1972.
An 99.72.5	Anthropologie et colonialisme. (Leclerc, Gérard.) Paris, 1972.
An 358.92.5	L'anthropologie et les origines de la Société...l'Orient et l'Occident. (Carmichael, Charles H.E.) Lisbonne, 1892.
An 3309.09	Anthropologie métrique. (Bertillon, A.) Paris, 1909.
An 359.58.20	Anthropologie structurale. (Lévi-Strauss, Claude.) Paris, 1958-1973. 2v.
An 358.75.3	Anthropologische Beiträge. (Gerland, Georg.) Halle, 1875.
An 359.61.15	Anthropologische Forschung. (Gehlen, Arnold.) Reinbek, 1961.
An 3208.97	Anthropologische Studien über die Urbewohner Brasiliens. (Ehrenreich, P.) Braunschweig, 1897.
An 126.6.5A	An anthropologist at work. (Benedict, Ruth (Fulton).) Boston, 1959.
An 99.73	Anthropologists and anthropology. (Kuper, Adam.) N.Y., 1973.
An 629.67	Anthropologists in the field. (Jongnians, Douve Geert.) Assen, 1967.
An 359.30	Anthropology; an introduction to the study of man and civilization. (Tylor, Edward B.) London, 1930. 2v.
An 2109.46	Anthropology; au introduction to primitive culture. (Goldenweiser, Alexander A.) N.Y., 1946.
An 182.4	Anthropology, history and cultural change. (Hodgen, Margaret.) Tucson, 1974.
An 358.81.4	Anthropology, introduction to the study of man and civilization. (Tylor, Edward B.) N.Y., 1881.
An 629.71	Anthropology, relativism and method. (Tennekes, J.) Assen, 1971.
An 628.92	Anthropology. (Brinton, D.G.) Philadelphia, 1892.
An 359.23.9	Anthropology. (Kroeber, A.L.) N.Y., 1923.
An 359.12.3	Anthropology. (Marett, Robert R.) London, 1912.
An 359.19	Anthropology. (Marett, Robert R.) London, 1919.
An 359.11.5	Anthropology. (Marett, Robert R.) N.Y., 1911.
An 359.12.23	Anthropology. (Marett, Robert R.) N.Y., 1912.
An 358.78.5	Anthropology. (Topinard, P.) London, 1878.
An 358.88	Anthropology. (Tylor, Edward B.) N.Y., 1888.
An 358.89	Anthropology. (Tylor, Edward B.) N.Y., 1889.
An 358.91.3	Anthropology. (Tylor, Edward B.) N.Y., 1891.
An 358.94.3	Anthropology. (Tylor, Edward B.) N.Y., 1894.
An 359.04.3	Anthropology. (Tylor, Edward B.) N.Y., 1904.
An 359.09.12	Anthropology. (Tylor, Edward B.) Chicago, 1909.
An 359.09.11	Anthropology. (Tylor, Edward B.) N.Y., 1909.
An 359.64.25	Anthropology. (Wolf, Eric Robert.) Englewood Cliffs, N.J., 1964.
An 205.2	Pamphlet vol. Anthropology. Miscellaneous tracts. 19 pam.
An 359.61.45	Anthropology and history. (Evans-Pritchard, Edward Evan.) Manchester, Eng., 1961.
An 187.9	Anthropology and human nature. (Montagu, Ashley.) Boston, 1957.
An 99.61A	Anthropology and the classics. (Kluckhohn, Clyde.) Providence, 1961.
An 359.63	Anthropology and the world of science. (Jennings, J.D.) Salt Lake City, 1963.
An 359.53	Anthropology today. (Kroeber, Alfred L.) Chicago, 1953.
Htn An 3305.63*	Anthropometria. (Elsholt, J.S.) Francofurti, 1563.
An 3308.89	Anthropometric manual 1889. (Amherst College.) Amherst, 1889.
An 3309.24	Anthropometric observations on samples of the civil populations of Aberdeenshire, Banffshire, and Kincardineshire. (Tocher, James F.) Edinburgh, 1924.
An 3308.90	Anthropometry and physical examination. (Seaver, J.W.) New Haven, 1890.
An 3308.96	Anthropometry and physical examination. 2. ed. (Seaver, J.W.) New Haven, 1896.
An 3309.30.10	The anthropometry of the American Negro. (Herskovits, M.J.) N.Y., 1930.
An 2358.86	Anthropophagy. (Darling, C.W.) Utica, 1886.
An 359.43.30	Anthropos; or, The problem of man. (Nicholson, James E.) London, 1943.
An 3206.47	Anthroposcopia. (Otto, Andreas.) Regiomonti, 1647.
An 3308.87	The anthroprometric manual 1887. (Amherst College.) Amherst, 1887.
Geog 3140.16	Antichan, P.N. Grands voyages de découvertes des anciens. Paris, 1888.
Geog 4208.99.2	Antin, M. From Plotzk to Boston. Boston, 1899.
Geog 3055.33.10	L'antiquité dévoilée par les principles de la magie naturelle. Paris? 1799-1800.
An 2058.92.3	The antiquity and origin of the human race. (Wright, G.F.) Boston, 1892.
An 2059.27	The antiquity of man. (Lyell, C.) London, 1927.
An 2058.83	The antiquity of man historically considered. (Rawlinson, G.) N.Y., 1883.
An 2059.14	The antiquity of man in Europe. (Geike, James.) N.Y., 1914.
Geog 4305.17	Antonio de Beatis, Don. Voyage du Cardinal d'Aragon en Allemagne. Paris, 1913.
Geog 3570.45	Antonio Fernandes. (Tracey, Hugh.) Lourenco Marques, 1940.
Geog 4269.46	Antonovych, Roman. Burlats'kym shliakom. Vynnypeg, 1946.
An 358.80	Antropologia. Lisboa, 1880.
An 99.57	La antropología contemporánea. (Estena-Fabregat, Claudio.) Madrid, 1957. 2 pam.
An 359.06.7	Antropologia general. (Toro, E.) Caracas, 1906.
VAn 629.67.5	Antropologia kao drustveno nauka. (Pešić-Golutovic, Zagorka.) Beograd, 1967.
An 359.72.25	Antropologia sociocultural. (Bezerra, Felte.) Brasilia, 1972.
An 170.212	Antropologicheskaia rekonstruktsiia i problemy paleoetnografii. Moskva, 1973.
An 358.94.5	Antropologicheskaia zhizne. (Lavrov, P.L.) Zhenëva, 1894. 2v.
An 3309.73	Antropologicheskie materialy srednevekovykh mogil'nikov dugo-zapadnogo kryma. (Zinevich, Galina P.) Kiev, 1973.
An 3559.68.5	Antropologichna kharakteristika davn'oho naselennia teritorii Ukrainy. (Zinevich, Galina P.) Kyiv, 1968.
An 359.73.25	Antropologiczna koncepcja człowieka. Wyd. 1. (Waligórski, Andrzej.) Warszawa, 1973.
An 3559.72	Antropologiia drevnego nasteniia Ukrainy. (Konduktorova, Tamara S.) Moskva, 1972.
An 170.144	Antropologiia i genogeografiia. Moskva, 1974.
An 3309.73.5	Antropologiia naseleniia Ukrainy mezolita, neolita, i epokki bronzy. (Konduktorova, Tamara S.) Moskva, 1973.
An 359.59.35	Antropologija i etnologija. (Škerlj, Božo.) Beograd, 1959.
An 5.31	Antropološka bibliografija o Jugoslaviji. (Antropolško Društvo Jugoslavije.) Beograd, 1963.
An 5.31	Antropolško Društvo Jugoslavije. Antropološka bibliografija o Jugoslaviji. Beograd, 1963.
An 3308.93	Antropometria militare. (Livi, Ridolfo.) Roma, 1893-1903. 2v.
Geog 32.50	Anuário. (Brazil. Diretoria do Serviço Geográfico do Exército.) 1962+
Geog 143.5	Anuario de geografía. (México (City). Universidad Nacional. Facultad de Filosofiá y Letras.) 1,1961 4v.
An 3308.89.3	Anuchin, D.N. O geograficheskom' raspred'nenii rosta muzhskago naseleniia Rossii. Sankt Peterburg, 1889.
Geog 3019.72.15	Anuchin, V.A. Teoreticheskie osnovy geografii. Moskva, 1972.
Geog 4709.39	Apache gold and yaqui silver. (Dobie, J.F.) Boston, 1939.
Geog 5311.3.15	Apdagelsesreiser. (Amundsen, Roald.) Oslo, 1928-30. 4v.
Geog 3590.72	Aperçu des travaux géographiques en Russie. (Kaulbars, N.A.) St. Pétersbourg, 1889.
An 359.37.5A	Apes, man, and morons. (Hooton, E.A.) N.Y., 1937.
Geog 16.1	Appalachia. Boston. 1,1879+ 39v.
Geog 16.1.2	Appalachia. Index, v.1-10. Boston, 1906.
Geog 16.1.10	Appalachian Mountain Club. Annual report.
Geog 16.1.7	Appalachian Mountain Club. Bulletin. 1-27,1907-1934 27v.
Geog 16.1.8	Appalachian Mountain Club. Bulletin. 1935+ 26v.
Geog 16.1.5	Appalachian Mountain Club. Register. 1888-1955 13v.
Geog 16.1.9.5F	Pamphlet vol. Appalachian Mountain Club.
Geog 16.1.9	Pamphlet vol. Appalachian Mountain Club.
Geog 16.2	Appalachian Trail Conference. Publication. Washington. 4,1942 3v.
Geog 16.2.9	Pamphlet vol. Appalachian Trail Conference. Washington. 2 pam.
Geog 16.2.5	Appalachian trailway news. Washington. 25,1964+ 4v.
Geog 5518.21.2	Appendix to...Parry's Journal of second voyage. (Parry, W.E.) London, 1825.
Geog 5518.29.2	Appendix to Narrative of second voyage. (Ross, J.) London, 1835.
An 359.57.5	Application de l'ethnologie à l'assistance sanitaire. (Lebeuf, Jean Paul.) Bruxelles, 1957.
Geog 4239.71	April Fool or, How I sailed from Casablanca to Florida in a six-foot boat. (Vihlen, Hugo.) Chicago, 1971.
Htn Geog 4209.19.5*	Apuntaciones de viaje en 1849. (Romero de Terreros, Juan.) México, 1919.
Geog 4309.31.25	Apuntes de un turista tropical. (Iraizoz y de Villar, Antonio.) Habana, 1931.
Geog 3055.51	Aquem da Atlantida. (Barroso, Gustavo.) São Paulo, 1931.
Geog 5080.5	Aquisition of sovereignty over polar areas. (Smedal, Gustav.) Oslo, 1931.
Geog 600.5	Arabische Schriftsteller über die Geographie Indiens. Inaug. Diss. (Abulhasan Mansur.) Berlin, 1919.
Geog 4418.75	Arabistan, or Land of "Arabian Nights." (Fogg, W.P.) London, 1875.
Geog 4218.17	Arago, J. Narrative of a voyage round the world. London, 1823.
Geog 4218.17.3	Arago, J. Souvenirs...voyage...du monde. Paris, 1839.
An 358.99.6	Aranzadi, G. de. Etnologia. Madrid, 1899.
Geog 197.10	Arbeiten. (Saarbruecken. Universität des Saarlandes. Geographisches Institut.) 1,1956+ 4v.
Geog 5650.18	Archeological investigation in the Thule district. (Holtved, Erik.) n.p., n.d.
NEDL Geog 18.1	Archiv für Geographie, Historie. Wien. 1-19,1810-1828 19v.
Geog 4168.40	Archives des voyages. pt.1-4. (Ternaux-Compans, H.) Paris, 1840. 4v.
Geog 18.155	Arctic; journal of the Arctic Institute of North America. Montreal. 1,1948+ 21v.
Geog 5160.61F	The Arctic. (Bruemmer, Fred.) Scarborough, Ont., 1974.
Geog 5316.1A	Arctic adventure; my life in the frozen North. (Freuchen, P.) N.Y., 1935.
Geog 5208.57	Arctic adventure by sea and land. (Sargent, E.) Boston, 1857.
Geog 5316.2	Arctic adventurer. (Fraser, Robert J.) Lincoln, Ont., 1972.
Geog 4558.89	Arctic Alaska and Siberia. (Aldrich, H.L.) Chicago, 1889.
Geog 5120.1	Arctic and Antarctic. (Bertram, Colin.) Cambridge, 1939.
Geog 5160.30	Arctic area. (Collins, H.B.) México, 1954.

Author and Title Listing

Geog 5160.56 The Arctic basin. (Arctic Institute of North America.) Washington, 1963.
Geog 5538.54.5 An Arctic boat journey, in the autumn of 1854. (Hayes, I.I.) Boston, 1867.
Geog 5538.54.8 Arctic boat journey, in the autumn of 1854. (Hayes, I.I.) Boston, 1883.
Geog 5558.53 An Arctic boat journey...1854. (Hayes, I.I.) Boston, 1868.
Geog 5538.54.2 An Arctic boat journey. (Hayes, I.I.) Boston, 1860.
Geog 5530.82 Arctic breakthrough: Franklin's expeditions, 1819-1847. (Nanton, Paul.) London, 1971.
Geog 5558.75.11 The Arctic expedition of 1875-76. (Johnston, R.) London, 1877.
Geog 5558.75F Arctic expedition 1875-76, journal. (Nares, G.S.) London, 1877.
Geog 5558.71.5 Arctic experiences. (Blake, E.V.) N.Y., 1874.
Geog 5209.06.2 Arctic exploration. (Hoare, J.D.) London, 1906.
Geog 5209.06 Arctic exploration. (Hoare, J.D.) N.Y., 1906.
Geog 5538.53.3 Arctic explorations. (Kane, E.K.) Hartford, 1881.
Geog 5538.53.2 Arctic explorations. (Kane, E.K.) Philadelphia, 1856. 2v.
Geog 5538.53.1 Arctic explorations. (Kane, E.K.) Philadelphia, 1856. 2v.
Geog 5538.53A Arctic explorations. (Kane, E.K.) Philadelphia, 1856. 2v.
Geog 5225.7 Arctic explorations and discoveries. (Smucker, Samuel.) N.Y., 1857.
Geog 5208.89 Arctic explorations in 18th and 19th centuries. (Smith, C.C.) Boston, 1889.
Geog 5509.71 Arctic fever. (Wilkinson, Doug.) Toronto, 1971.
Geog 5160.59 The Arctic frontier. (Macdonald, Ronald.) Toronto, 1966.
Geog 5300.35 Arctic frontiers. 1. ed. (Caswell, J.E.) Norman, 1956.
Geog 5180.16 The Arctic in fact and fable. (Stefansson, V.) N.Y., 1945.
Geog 5160.56 Arctic Institute of North America. The Arctic basin. Washington, 1963.
Geog 18.150 Arctic Institute of North America. Bulletin. Montreal.
Geog 18.160 Arctic Institute of North America. Special publication. Washington. 1-4,1952-1962//? 3v.
Geog 5559.34.5 Arctic journeys; the story of the Oxford University Ellesmire land expedition, 1934-35. (Shackleton, Edward.) London, 1937.
Geog 5160.11.15 Arctic manual. (Stefansson, V.) N.Y., 1944.
Geog 5160.12 Arctic manual. v.1-2. (United States. Office of Chief of Air Corps (War Department).) Washington, 1940.
Geog 5538.50.19 Arctic miscellanies. London, 1852.
Geog 5180.5 The Arctic problem. (Heilprin, A.) Philadelphia, 1893.
Geog 5160.10F The Arctic regions. (Hartwig, G.) London, n.d.
Geog 5558.26 Arctic regions. Voyage to Davis Strait. (Duncan, D.) London, 1827.
Geog 5050.2 Arctic regions and Antarctic regions. (Boston Public Library.) Boston, 1894.
Geog 5538.48 Arctic searching expedition. (Richardson, J.) N.Y., 1852.
Geog 5538.48.5 Arctic searching expedition. (Richardson, J.) N.Y., 1854.
Geog 5209.53 Arctic solitudes. (Mountevans, C.R.G.R.E.) N.Y., 1953.
Geog 5538.53.4 Arctic voyage. (Kane, E.K.) N.Y., 1857.
Geog 5324.1.2 Arctic voyages of Adolf Erik Nordenskiöld. (Leslie, A.) London, 1879.
Geog 5160.5F The Arctic world; its plants, animals and natural phenomena. London, 1876?
Geog 5235.20 Arctic world. (Euller, John.) London, 1958.
Geog 665.46 Arden-Close, Charles. Geographical by-ways and some other geographical essays. London, 1947.
An 2059.61 Ardrey, Robert. African genesis. N.Y., 1961.
Geog 4269.39 Are people more a like than different. (Gandy, Roxana Smith.) Cape May Court House, 1939.
Geog 818.51.14 Arendts, Carl. Leitfaden für den ersten Wissenschaftlichen Unterricht in der Geographie. 14. Aufl. Regensburg, 1874.
Geog 4029.9 Aresty, Miguel de. Los papeles de Juan de Aresty. Barcelona, 1972.
Geog 5925.55.6 Argentina. Comissión Nacional del Antártico. Soberanía argentina en la Antártida. 2. ed. Buenos Aires, 1948.
Geog 6009.03.5 La Argentina en los mares antarticos. (Argentine Republic.) Buenos Aires, 1903.
Geog 6009.03.5 Argentine Republic. La Argentina en los mares antarticos. Buenos Aires, 1903.
Geog 5939.25 Argonauts of the South. (Hurley, Frank.) N.Y., 1925.
Geog 665.150 Argumenta geographica; Festschrift Carl Troll zum 70. Geburtstag. Bonn, 1970.
An 2108.78.3 Argyll, G.D.C. Primeval man. N.Y., 1878.
Geog 5300.48 Arkticheskie puteshevstviia rossiian. (Pastskii, Vasilii M.) Moskva, 1974.
Geog 18.5 Arktis; Vierteljahrsschrift der Internationalen Studiengesellschaft zur Erforschung der Arktie. Gotha. 1-4,1928-1931 2v.
Geog 5180.12 Arktiske skinddragter i Eurasian og Amerika. (Hatt, G.) Kjøbenhavn, 1914.
An 359.15.5 Arldt, Theodor. Die Stammesgeschichte der Primaten und die Entwicklung der Menschrassen. Berlin, 1915.
Geog 603.6 Armao, Ermanno. Vincenzo Coronelli. Firenze, 1944.
Geog 4218.85.10 Arme-Webb, A. (Mrs.). A glimpse of two hemispheres. Hertford, 1885.
Geog 6009.01.9 Armitage, A.B. Two years in the Antarctic. London, 1905.
An 359.22 Armitage, F.P. Diet and race. London, 1922.
Geog 5538.50.25 Armstrong, Alexander. A personal narrative of the discovery of the Northwest Passage...while in search of the expedition under Sir John Franklin. London, 1857.
Geog 4559.68.5 Armstrong, Richard. A history of seafaring. N.Y., 1968-69. 3v.
Geog 5369.52 Armstrong, T. The northern sea route; Soviet exploitation of the North East Passage. Cambridge, Eng., 1952.
Geog 5300.45 Armstrong, Terence. The Russians in the Arctic. London, 1958.
Geog 4679.56 Armstrong, W. Last voyage. London, 1956.
Geog 4209.44.5 Armstrong, W. Saltwater tramp. London, 1944.
Geog 4229.04 Armstrong, W.N. Around the world with a king. London, 1904.
Geog 5839.29 Arnason, A. Graenlandsför 1929. Reykjavik, 1929.
Geog 5328.20 Arnesen, Odd. Eskimoenes venn Knud Rasmussen. Oslo, 1944.
Geog 5311.3.5 Arnesen, Odd. Roald Amundsen som hon var. 4. Oppl. Oslo, 1946.
Geog 500.160 Arnim, Hlmuth. Bibliographie der geographischen Literatur in deutscher Sprache. 1. Aufl. Baden Baden, 1970.
Geog 4208.96 Arnold, E. East and West. London, 1896.
Geog 4208.94 Arnold, E. Wandering words. London, 1894.
An 3509.72 Aron, Jean Paul. Anthropologie du conscrit français d'après les comptes numeriques et sommaires du recrutement de l'année, 1819-1826. Paris, 1972.
Geog 5650.30F Aron, of Kangeg. K'avdlunätsianik. Godthåt, 1969.
Geog 4208.80.3 Around Cape Horn to Honolulu on the bark "Amy Turner" 1880. (Briggs, L.V.) Boston, 1926.
Geog 4309.12.10 Around the clock in Europe. (Howell, Charles F.) Boston, 1912.
Geog 4219.07.5 Around the world; a narrative in letter form of a trip around the world, from October 1907 to July 1908. (Stephens, Edwin William.) Columbia, 1909.
Geog 4218.74 Around the world, sketches of travel. (Prime, E.D.C.) N.Y., 1874.
Geog 4218.76.3 Around the world. (Hendrix, E.R.) St. Louis, 1881.
Geog 4208.70.5 Around the world. n.p., n.d.
Geog 4218.70.3 Around the world. 2. ed. (Peebles, James.) Boston, 1876.
Geog 4218.89.5 Around the world in seven months. (Gillis, C.J.) N.Y., 1891.
Geog 500.110 Around the world in sixty books. (Powell, Lawrence.) Los Angeles, 1960.
Geog 4218.78 Around the world in the "Sunbeam". (Brassey, A. (Mrs.).) N.Y., 1878.
Geog 4218.78.5 Around the world in the "Sunbeam". (Brassey, A. (Mrs.).) N.Y., 1879.
Geog 4219.26 Around the world in twenty-eight days. (Wells, Linton.) Boston, 1926.
Geog 4219.39 Around the world on $100.00 a month. (Alford, B.) Emmaus, 1939.
Geog 4218.87 Around the world on a bicycle. (Stevens, Thomas.) N.Y., 1887-88. 2v.
Geog 4219.58.10 Around the world on a nickel. (Bedford, Jimmy.) New Delhi, 1967.
Geog 4219.32.35.5 Around the world single-handed. (Pidgeon, H.) N.Y., 1932.
Geog 4229.04 Around the world with a king. (Armstrong, W.N.) London, 1904.
Geog 4219.25 Around the world with Texaco. (Dennison, C.S.) Houston, 1925.
An 3409.10 Arrangement of finger prints identification and their uses. (Brayley, F.A.) Boston, 1910.
Geog 3108.39.3 Arrowsmith, A. Compendium of ancient and modern geography. London, 1839.
Geog 3108.39 Arrowsmith, A. Grammar of ancient geography. London, 1839.
Geog 5939.51 Arsen'ev, V.A. V straie kitov i pingvinov. Moskva, 1951.
Geog 4308.55.3A Art, scenery and philosophy. (Wallace, H.B.) Philadelphia, 1855.
Geog 4129.57 The art of coarse travel. (Hughes, Spike.) London, 1957.
Geog 4679.63.1 The art of survival. (Troebst, C.C.) Garden City, 1975.
Geog 4128.56 The art of travel. (Galton, Francis.) London, 1856.
Geog 4145.89 Arteche, J. de. Urdaneta. Madrid, 1945.
Geog 4209.04 Arthur, Richard. Ten thousand miles in a yacht. N.Y., 1906.
Geog 5507.45.6 Articles of agreement. (North West Committee.) London, 1745.
Geog 4358.92 Artistic travel in Normandy, Brittany, the Pyrenees, Spain and Algeria. (Blackburn, Henry.) London, 1892.
An 358.88.5 The Aryan race: its origin and its achievements. (Morris, Charles.) Chicago, 1888.
An 359.34.3 Les aryens. (Poisson, Georges.) Paris, 1934.
Geog 4219.09 As far as the East is from the West. (Bidwell, Daniel D.) Hartford, 1910.
Geog 4559.63 Asgeirsson, Rikki. Alltaf má fá annad skip. Reykjavik, 1963.
Geog 4180.27A Asher, G.M. Henry Hudson the navigator. London, 1860.
Geog 500.33 Asher, George Michael. Viro venerabili Friderico Laurentio Hoffmann. Berolinenses, 1860.
Geog 5399.22 Ashton, James M. Ice-bound, a trader's adventure in the Siberian Arctic. N.Y., 1928.
Geog 809.19.4 Asia, India insular, Africa. (Camena d'Almeida, P.) Barcelona, 1914.
An 2059.25 Asia and the dispersal of primates. (Black, Davidson.) Peking, 1925.
An 2109.70 Aspects of prehistory. (Clark, John Grahame Douglas.) Berkeley, 1970.
An 3309.21.3 The assessment of physical fitness. (Dreyer, George.) N.Y., 1921.
Geog 19.3 Associação dos Geografos Brasileiros. Anais. 1,1945+ 8v.
Geog 19.2 Associated Mountaineering Clubs of North America. Bulletin. N.Y. 1918-1921
Geog 602.3.10 Association des Amis de l'Université de Grenoble. In memoriam Raoul Blanchard, 1877-1965. Grenoble, 1966.
Geog 13.9 Association of American Geographers. Annals. Albany, N.Y. 1+ 47v.
Geog 13.9.5 Association of American Geographers. Annals. Index for volumes 1-25. Lancaster, Pa.? n.d.
Geog 13.10 Association of American Geographers. Handbook-directory. Washington. 1956
Geog 19.5 Association of Indian Geographers. Bulletin. New Delhi. 1-2,1956-1957
Geog 19.5.10 Association of Indian Geographers. The Indian geographer. New Delhi. 1957
An 137.4 Association of Polish University Professors and Lecturers in Great Britain. Professor Bronislaw Malinowski. London, 1943.
An 2109.68.25 Association of Social Anthropologists of the Commonwealth. History and social anthropology. London, 1968.
Geog 5939.64.5 Astapenko, Pavel D. Puteshestvie k ostrovu chetyrekh vulkanov. Moskva, 1964.
Geog 5558.91.4 Astrup, E. Blandt Nordpolens naboer. Kristiania, 1895.
Geog 5558.91.3 Astrup, E. With Peary near the Pole. London, 1894.
An 2109.03.5 Das Asylrecht der Naturvölker. (Hellwig, A.) Berlin, 1903.
Geog 4209.36 At home and abroad; essays. (Macdonald, J.R.) London, 1936.
Geog 4228.56.5 At home and abroad. (Ossoli, M.F.) Boston, 1860.
Geog 4228.56A At home and abroad. 3. ed. (Ossoli, M.F.) Boston, 1856.
Geog 4238.79 At sea in the Celtic from January 23rd to February 11th, 1879. (Moulton, Francis D.) n.p., 1879.
An 170.117 Athayde, Alfredo. Introdução a antropología tropical. Lisboa, 1962.
Geog 3019.05 Athlenius, K. Landkonturer och hafsvidder. Stockholm, 1905.
Geog 4227.35.1 Atkins, John. A voyage to Guinea, Brazil, and the West Indies. London, 1970.
Geog 4219.34.5 Atkinson, B. The Cingalese prince. Garden City, 1934.
Htn Geog 500.59F* Atkinson, G. La litterature géographique française de la renaissance. Paris, 1927.
Htn Geog 500.59.2F* Atkinson, G. La litterature géographique française de la renaissance. Supplement. Paris, 1936.
Geog 3070.2 Atlante idrografico. (Desimoni, C.) Genova, 1867.

Author and Title Listing

	Geog 3055.31	Les Atlantes. (Roisel, G. de.) Paris, 1874.
	Geog 616.2	Gli atlanti corografici del cavaliere C.G. Pocelli. (Spano, Benito.) Bari, 1958.
	Geog 3235.13	Atlanti e carte nautiche. (Longhena, M.) Parma, 1907.
	Geog 4239.57.5	The Atlantic. (Outhwaite, L.) N.Y., 1957.
	Geog 4238.73	The Atlantic. (Thomson, C.W.) N.Y., 1878. 2v.
	Geog 4239.55	Atlantic adventurers. (Barton, H.D.E.) N.Y., 1955?
	Geog 4239.38	Atlantic circle. (Moyne, W.E.G.) London, 1938.
	Geog 4239.65.5	Atlantic cruise in Wanderer III. (Hiscock, Eric C.) London, 1968.
	Geog 4238.78	Atlantic Islands. (Benjamin, S.G.W.) N.Y., 1878.
	Geog 4239.55.5	Atlantica dall'intuizione alla scoperta del nuovo mondo. (Mastrangelo, F.) Napoli, 1955.
	Geog 4209.62	Atlanticheskii dnevnik. (Shakirianov, N.A.) Riga, 1962.
	Geog 3055.58	La Atlántida de Platón en los cronistas del siglo XVI. (Rodriguez Prampolini, Ida.) Mexico, 1947.
	Geog 3055.40	La Atlantida y la ultima tule. (Buelua, E.) Mexico, 1895.
	Geog 3055.27	L'Atlantide. (Baour-Lormian, Pierre Marie F.L.) Paris, n.d.
	Geog 3055.46	L'Atlantide a-t-elle existé? (Moreux, T.) Paris, 1924.
	Geog 3055.86	L'Atlantide atlantique. (Le Cour, Paul.) Bordeaux, 1971.
	Geog 3055.66	L'Atlantide et la règne des géants. (Saurat, Denis.) Paris, 1954.
	Geog 3055.49	L'Atlantide historique. (Rosmy, Léon de.) Paris, 1902.
	Geog 3055.60.10	Atlantis; Heimat. (Spanuth, Jürgen.) Tübingen, 1965.
	Geog 3055.39.25	Atlantis, the antediluvian world. (Donnelly, Ignatius.) London, 1970.
	Geog 3055.39.20	Atlantis, the antediluvian world. 1st ed. (Donnelly, Ignatius.) N.Y., 1949.
	Geog 3055.39.10	Atlantis, the antediluvian world. 7th ed. (Donnelly, Ignatius.) N.Y., 1882.
	Geog 3055.80	Atlantis, the truth behind the legend. (Galanopoulos, Angelos Georgiou.) Indianapolis, 1969.
	Geog 3055.11	Atlantis. (Hoernes, M.) Wien, 1884.
	Geog 3055.76	Atlantis. (Lehmann, Einar.) København, 1934.
	Geog 3055.23	Die Atlantis. (Norof, A.S.) St. Petersburg, 1854.
	Geog 3055.39.1	Atlantis: the antediluvian world. (Donnelly, Ignatius.) Blauvelt, N.Y., 1971.
	Geog 3055.62	Atlantis enträtselt? (Weyl, R.) Kiel, 1953.
	Geog 3055.67	Atlantis-gefunden. (Muck, O.H.) Stuttgart, 1954.
	Geog 3055.42	Atlantis in America. (Spence, Lewis.) London, 1925.
	Geog 3055.50	Atlantis in Andalucia; a study of folk memory. (Whishaw, M.) London, 1929.
	Geog 3055.85	Atlantis och svenskarna. (Falk, Bertil.) Stockholm, 1974.
	Geog 3055.68	Der Atlantis-Streit. (Gadow, Gerhard.) Frankfurt am Main, 1973.
	Geog 3055.15	Atlantis und das Volk der Atlanten. (Knötel, A.F.R.) Leipzig, 1893.
	Geog 3055.53	Atlantis und der Liby-athiopische Kulturkreis. (Karst, Josef.) Heidelberg, 1931.
	Geog 3055.52	Das Atlantisrätsel. (Bessmertny, A.) Leipzig, 1932.
	An 2059.65	Atlas de prehistoire. Paris, 1965. 3v.
	Geog 4418.03.10	Atlas dels viatjes d'Ali Bey el Abbassi. v.1-3. (Badia y Leblich.) Barcelona, 1892.
	Geog 807.03	Atlas novus exhibens orbem. pt.1-7. (Scherer, R.P.H.) Augustae, 1710. 4v.
	An 2109.74.10	Atlas of ancient archaeology. (Hawkes, Jacquetta Hopkins.) N.Y., 1974.
	An 3309.54F	Atlas of men. 1st ed. (Sheldon, W.H.) N.Y., 1954.
	Geog 4168.36	Atlas zur Kunde fremder Weltheile. Leipzig. 1-2 2v.
	Geog 3570.9.5	The attacks against Portuguese history. (Bersaude, J.) Lisbon, 1950.
	Geog 40.2	Atti. (Congresso Geografico Italiano.) Genova. 1892-1907 39v.
	Geog 4515.14	Attrezzature per soccorso alpino. (Colò, Carlo.) Trento, 1958.
	Geog 4228.86	Au Canada et aux montagnes. (Molinari, M.G.) Paris, 1886.
	Geog 4219.72.5	Au fil de l'eau, autour de la terre. (Cluzel, Magdeleine.) Paris, 1973.
	Geog 5399.12	Au pays de la mort blanche. (Al'banov, V.A.) Paris, 1928.
	Geog 6008.97.3	Au pays des manchots. (Lecointe, G.) Bruxelles, 1910.
	Geog 6009.10.3	Au Pôle Sud, éxpedition du "Fram", 1910-1912. (Amundsen, R.) Paris, 1913.
	Geog 4429.55	Auber, Jacques. Histoire de l'Ocean Indien. Tananarive, 1955.
	Geog 4209.35.33	Aubert de la Rue, Edgar. L'homme et les iles. 2. ed. Paris, 1935.
	Geog 4218.92.15	Aubertin, J.J. Wanderings and wonderings. London, 1892.
	Geog 4300.22	Audin, J.M.V. Guide classique du voyageur. Paris, 1828-1829. 2v.
	Geog 4329.35.3	Audisio, G. Jeunesse de la Méditerranée. Paris, 1935-1936.
	Geog 4145.27	Auf dem Kirs der Raben. (Fink, O.) Hamburg, 1963.
	An 2109.53.27	Auf den Spuren des Eiszeitmenschen. 2. Aufl. (Kuehn, Herbert.) Wiesbaden, 1953.
X Cg	Geog 4502.5.15	Auf der Alm. (Schmidkunz, W.) Erfurt, 1934.
	Geog 4509.16	Auf fremden Bergpfaden. (Taüber, C.) Zürich, 1916.
	Geog 4679.63	Auf Wunder ist kein Verlass. (Troebst, C.C.) Düsseldorf, 1963.
	Geog 5919.01A	Auf zum Südpol! (Neumayer, G.) Berlin, 1901.
	An 379.61.10	Das Aufkommen sozialgeographischer Betrachtungsweisen in der deutschen länderkundlichen Literatur seit Beginn des 20. Jahrhunderts. Inaug. Diss. (Otto, Klaus.) Münster, 1961.
	Geog 665.27	Aufsätze Prof. Dr. Eugen Oberhummer gewidmet. Brünn, 1919.
	Geog 4207.76.20	Augur, Helen. Passage to glory; John Ledyard's America. 1. ed. Garden City, 1946.
	Geog 616.25	August Petermann. (Weller, E.) Leipzig, 1911.
	Geog 603.4	Augustin Codazzi. Tesis. (Briceno, Alfonso.) Caracas, 1928.
	An 2109.01.3	Aus dem Flegel Jahren der Menschheit. (Frobenius, L.) Hannover, 1901.
	Geog 4309.66.5	Aus einem römischen Winter. (Andersch, Alfred.) Olten, 1966.
	An 184.1F	Aus Fedor Jagor's Nachlass. (Jagor, Fedor.) Berlin, 1914.
	Geog 4209.02.10	Aus meiner Wanderzeit, 1902-1906 und 1908-1909. (Hellen, Gustav.) Hamburg, 1966.
	Geog 5700.4	Ausführliche...Nachricht...Grönland. (Egede, H.) Hamburg, 1740.
	An 2600.3	Ausführliches Leben und besondere Schiksale eines wilden Knaben. Frankfurt, 1759.
	Geog 5970.36	Australia and the Antarctic. (Law, Phillip G.) St. Lucia, 1962.
	Geog 5970.30	Australia in the Antarctic. (Swan, Robert A.) Victoria, 1961.
	Geog 21.15	The Australian geographer. Sydney. 8,1960+
	Geog 21.25	The Australian geographical record. Armidale. 1-7
	Geog 21.26	Australian geographical studies. Melbourne. 1,1963+ 5v.
	An 2059.54	The Australopithecinae. (Masiker, Reuben.) Cape Town, 1954.
	Geog 1530.69	Austria, together with Budapest. 12. ed. (Baedeker, publishers.) Leipzig, 1929.
	Geog 1530.8	Austria. 8. ed. (Baedeker, publishers.) Leipsic, 1896.
	Geog 1530.9	Austria. 9. ed. (Baedeker, publishers.) Leipsic, 1900.
	Geog 1530.9.5	Austria. 9. ed. (Baedeker, publishers.) Leipsic, 1900.
	Geog 1530.10A	Austria-Hungary. 10. ed. (Baedeker, publishers.) Leipzig, 1905.
	Geog 1530.13	Austria-Hungary. 11. ed. (Baedeker, publishers.) Leipzig, 1911.
	Geog 4181.126	La Austrialia del Espiritu Santo. (Kelly, Celsus.) Cambridge, 1966. 2v.
	Geog 807.85	Auszug aus seiner Erdbeschreibung. 6. Aufl. (Büsching, Anton Friedrich.) Hamburg, 1785.
	Geog 4209.40.40	Author in transit. (Hogben, L.) N.Y., 1940.
	Geog 4300.37	Auto guide to Europe. (Pastene, J.) N.Y., 1954.
	An 135.3	An autobiography. (Keith, Arthur.) London, 1950.
	Geog 613.7	An autobiography. (Mill, H.R.) London, 1951.
	Geog 4209.03	Autobiography. (Simpson, William.) London, 1903.
	Geog 4558.51	The autobiography and memorials of Captain Obadiah Congar. (Congar, Obadiah.) N.Y., 1851.
	Geog 1522.39.5	Autoführer, Deutsches Reich (Grossdeutschland). 2. Aufl. (Baedeker, publishers.) Leipzig, 1939.
	Geog 4169.04	Autour du monde. Paris, 1904.
	Geog 6009.08.5	Autour du pôle sud; expédition du "Pourquoi pas?" 1908-1910. (Charcot, Jean B.) London, 1937.
	Geog 3055.65	Aux sources de l'Atlantide. (Saint-Michel, Léonard.) Bourges, 1953.
	Geog 4329.17.5	Aux vieux pays. 3. ed. (Cimon, Henri H.M.) Montréal, 1917.
Htn	Geog 4476.92*	Les avantures de Jaques Sadeur. Paris, 1692.
	An 2108.75.3	Avebury, J.L. Die Entstehung der Civilisation. 3e Aufl. Jena, 1875.
	An 2109.11	Avebury, J.L. Marriage, totenism, and relggion. London, 1911.
	An 2108.75.5	Avebury, J.L. The origin of civilisation. 3. ed. London, 1875.
	An 2108.82.5	Avebury, J.L. The origin of civilisation and the primitive condion of man. 4. ed. N.Y., 1882.
	An 2108.82	Avebury, J.L. The origin of civilisation and the primitive condition of man. London, 1882.
	An 2108.82.2	Avebury, J.L. The origin of civilisation and the primitive condition of man. 4. ed. London, 1882.
	An 2108.71.3	Avebury, J.L. The origin of civilization and the primitive condition of man. N.Y., 1871.
	An 2109.00	Avebury, J.L. Prehistoric times. London, 1900.
	An 2109.00.2	Avebury, J.L. Prehistoric times. 6. ed. N.Y., 1900.
	An 2109.13	Avebury, J.L. Prehistoric times. 7. ed. London, 1913.
	Geog 4168.50	Aventures...des voyageurs. (Hombron, Bernard.) Paris, n.d.
	Geog 4209.61.15	Aventyr jorden runt. (Hentzel, R.) Stockholm, 1961.
	Geog 4205.03	Avezac, A. de. Campagne du navire l'Espoir de Honfleur, 1503-1505. Paris, 1869.
	Geog 3235.52	Avezac, M.A.P. Deux notes sur l'anciennes cartes historiées. Paris, 1844.
	Geog 3235.45	Avezac, M.A.P. Un digression géographique...la mappemonde. Paris, 1870.
	Geog 3235.51	Avezac, M.A.P. Note sur la mappemonde historiée de la cathedral de Hereford. Paris, 1862.
	Geog 3235.49	Avezac, M.A.P. Note sur un atlas hydrographique. v.1-2. Paris, 1850.
	Geog 4412.91.5	Avezac-Macaya. Expédition génoise des frères Vivaldi...route maritime des Indes orientales XIII. siècle. Paris, 1859.
	Geog 4428.01	Avine, Grégoire. Les voyages du chirurgien Avine à l'Ile de France et dans la mer de Indes au début du XIXe siècle. Paris, 1961.
	Geog 4346.85	Avril, P. Voyage en divers etats d'Europe et d'Asie. Paris, 1693.
	Geog 4169.59.5	Avventure e viaggi di mare. (Spagnol, Mario.) Milano, 1959.
	Geog 659.53	Avviamento allo studio della geografia. 1. ed. (Sestini, Aldo.) Firenze, 1953.
	Geog 4219.37	Away from it all. (Belfrage, C.) N.Y., 1937.
	Geog 4309.64.31	Away to it all. (Sansom, William.) N.Y., 1966.
Htn	Geog 4678.37*	Awful shipwreck...crew of the ship Francis Spaight. (Palmer, John.) Boston, 1837.
	An 358.41.5	Azaïs, H. Question philosophique de première importance: Quelle est dans l'univers, la destinée du genre humain? Paris, 1841.
	Geog 4180.95	Azurara, G.E. de. Chronicle of discovery...of Guinea. London, 1896-99. 2v.
	Geog 6009.29	The B.A.N.Z. Antarctic research expedition, 1929-31. (Mawson, Douglas.) London, 1932.
	Geog 4209.40.15	Baarslag, K. Islands of adventure. N.Y., 1940.
	Geog 3060.6A	Babcock, William H. Legendary islands of the Atlantic. N.Y., 1922.
	Geog 618.5.5F	Babicz, J. Nauka o ludakh Fryderyka Ratzla. Wrocław, 1962.
	Geog 623.3	Babicz, Józef. Teoria Moritza Wagnera o powstawaniu gatunków. Wrocław, 1966.
	An 358.95	Babington, W.D. Fallacies of race theories. London, 1895.
	Geog 4309.71	Bąbiński, Gan. Noc w Soho. Wyd. 1. Łódź, 1971.
	Geog 4208.12	Bacardí Moreau, E. Hacia tierras viejas. Valencia, 1913.
	An 358.50.6	Bachman, J. Doctrine of unity of human race examined on principles of science. Charleston, S.C., 1850.
	Geog 953.5FA	Bachmann, Friedrich. Die alten Städtebilder. Leipzig, 1939.
	Geog 953.5.2F	Bachmann, Friedrich. Die alten Städtebilder. 2. Aufl. Stuttgart, 1965.
	Geog 5518.36.7	Back, George. Narrative of an expedition in H.M.S. Terror. London, 1838.
	Geog 5518.33	Back, George. Narrative of Arctic land expedition. London, 1836.
	Geog 5518.33.3	Back, George. Narrative of Arctic land expedition. Philadelphia, 1836.
	Geog 5518.36.9	Back, George. Narrative of the Arctic land expedition to the mouth of the Great Fish River. 2. ed. Philadelphia, 1837.
	An 2109.63.10	Back of history, the story of our origins. (Howells, William.) Garden City, N.Y., 1963.
	An 2109.54	Back of history. 1. ed. (Howells, William.) Garden City, 1954.
	Geog 3019.35.5	The background of geography. (Spilhaus, M.N. (Mrs.).) Philadelphia, 1935.
	An 2109.75.15	Background to archaeology. (Borer, Mary Irene Cathcart.) London, 1975.

Author and Title Listing

Call number	Entry
An 359.20.5A	The backward peoples and our relations with them. (Johnston, Harry.) London, 1920.
An 3309.63.5	Badania struktury rasowej ludności Afryki Środkowej. (Henzel, Tadeusz.) Łódz, 1963.
Geog 809.21	Bader, G. Erläuterungen zu 938 ausgewählten Lichtbilder zur Länderkunde. Stuttgart, 1921. 3v.
Geog 4418.03.10	Badia y Leblich. Atlas dels viatjes d'Ali Bey el Abbassi. v.1-3. Barcelona, 1892.
Geog 4418.03.5	Badia y Leblich. Viajes de Ali Bay el Abbassi por Africa y Asia. Valencia, 1836. 3v.
Geog 4418.03.9	Badia y Leblich. Viatjes de Ali Bey el Abbassi per Africa y Asia. v.1-3. Barcelona, 1888.
Geog 4418.03	Badia y Leblich. Voyages d'Ali Bey el Abbassi en Afrique et en Asie...1803...1807. v.1-3, atlas. Paris, 1814. 4v.
Geog 4206.54	Bäckhoff. Anhang zwer Reisen. Berlin, n.d.
Geog 1555.38	Baedeker, pbulishers. Belgium and Holland. 8. ed. Leipsic, 1885.
Geog 1524.14	Baedeker, publishers. L'Allemagne, l'Autriche. 6. éd. Leipzig, 1878.
Geog 1524.38	Baedeker, publishers. Allemagne du sud et Autriche. 9. éd. Leipzig, 1888.
Geog 1530.69	Baedeker, publishers. Austria, together with Budapest. 12. ed. Leipzig, 1929.
Geog 1530.8	Baedeker, publishers. Austria. 8. ed. Leipsic, 1896.
Geog 1530.9	Baedeker, publishers. Austria. 9. ed. Leipsic, 1900.
Geog 1530.9.5	Baedeker, publishers. Austria. 9. ed. Leipsic, 1900.
Geog 1530.10A	Baedeker, publishers. Austria-Hungary. 10. ed. Leipzig, 1905.
Geog 1530.13	Baedeker, publishers. Austria-Hungary. 11. ed. Leipzig, 1911.
Geog 1522.39.5	Baedeker, publishers. Autoführer, Deutsches Reich (Grossdeutschland). 2. Aufl. Leipzig, 1939.
Geog 1558.10	Baedeker, publishers. Belgien. 5. Aufl. Coblenz, 1855.
Geog 1555.43	Baedeker, publishers. Belgien und Holland. 18. Aufl. Leipzig, 1888.
Geog 1555.29	Baedeker, publishers. Belgien und Holland. 23. Aufl. Leipzig, 1904.
Geog 1555.10	Baedeker, publishers. Belgique et Hollande. 4. éd. Coblenz, 1866.
Geog 1555.15	Baedeker, publishers. Belgique et Hollande. 6. éd. Coblenz, 1871.
Geog 1555.18	Baedeker, publishers. Belgique et Hollande. 8. ed. Leipzig, 1875.
Geog 1558.20	Baedeker, publishers. Belgique et Luxembourg. 20. éd. Leipzig, 1928.
Geog 1555.13	Baedeker, publishers. Belgium and Holland. Coblenz, 1869.
Geog 1555.16	Baedeker, publishers. Belgium and Holland. 2. ed. Coblenz, 1871.
Geog 1555.17	Baedeker, publishers. Belgium and Holland. 3. ed. Leipsic, 1874.
Geog 1555.21	Baedeker, publishers. Belgium and Holland. 5. ed. Leipsic, 1878.
Geog 1555.27	Baedeker, publishers. Belgium and Holland. 5. ed. Leipsic, 1878.
Geog 1555.20	Baedeker, publishers. Belgium and Holland. 5. ed. Leipsic, 1878.
Geog 1555.30	Baedeker, publishers. Belgium and Holland. 6. ed. Leipsic, 1881.
Geog 1555.36	Baedeker, publishers. Belgium and Holland. 7. ed. Leipsic, 1884.
Geog 1555.35	Baedeker, publishers. Belgium and Holland. 7. ed. Leipsic, 1884.
Geog 1555.40	Baedeker, publishers. Belgium and Holland. 9. ed. Leipsic, 1888.
Geog 1555.45	Baedeker, publishers. Belgium and Holland. 10. ed. Leipsic, 1891.
Geog 1555.50	Baedeker, publishers. Belgium and Holland. 11. ed. Leipsic, 1894.
Geog 1555.51	Baedeker, publishers. Belgium and Holland. 11. ed. Leipsic, 1894.
Geog 1555.55	Baedeker, publishers. Belgium and Holland. 12. ed. Leipsic, 1897.
Geog 1555.60	Baedeker, publishers. Belgium and Holland. 13. ed. Leipsic, 1901.
Geog 1555.65	Baedeker, publishers. Belgium and Holland. 14. ed. Leipzig, 1905.
Geog 1555.70	Baedeker, publishers. Belgium and Holland. 15. ed. Leipzig, 1910.
Geog 1558.30	Baedeker, publishers. Belgium and Luxembourg. 16. ed. Leipzig, 1931.
Geog 1526.40	Baedeker, publishers. Berlin. 23. Aufl. Freiburg, 1964.
Geog 1526.20	Baedeker, publishers. Berlin and its environs. Leipsic, 1903.
Geog 1526.25	Baedeker, publishers. Berlin and its environs. 2. ed. Leipsic, 1905.
Geog 1526.30	Baedeker, publishers. Berlin and its environs. 3. ed. Leipsic, 1908.
Geog 1526.35	Baedeker, publishers. Berlin and its environs. 5. ed. Leipsic, 1912.
Geog 1526.36	Baedeker, publishers. Berlin and its environs. 6. ed. Leipzig, 1923.
Geog 1526.10	Baedeker, publishers. Berlin und Umgebungen. 7. Aufl. Leipzig, 1891.
Geog 1526.12	Baedeker, publishers. Berlin und Umgebungen. 8. Aufl. Leipzig, 1894.
Geog 1526.17	Baedeker, publishers. Berlin und Umgebungen. 11. Aufl. Leipzig, 1900.
Geog 1525.37	Baedeker, publishers. Les bords du Rhin. 12. éd. Leipzig, 1882.
Geog 1525.27	Baedeker, publishers. Les bords du Rhin de la frontière suisse à la frontière de Hollande. 9. éd. Leipzig, 1875.
Geog 1540.47	Baedeker, publishers. Central Italy and Rome. 15. ed. Leipzig, 1909.
Geog 1540.48	Baedeker, publishers. Central Italy and Rome. 15. ed. Leipzig, 1909.
Geog 1522.42	Baedeker, publishers. Das Deutsche Reich. 6. Aufl. Leipzig, 1936.
Geog 1522.33	Baedeker, publishers. Deutschland. v.2. 10. Aufl. Coblenz, 1862.
Geog 1522.25	Baedeker, publishers. Deutschland. 7. Aufl. Coblenz, 1857.
Geog 1522.30	Baedeker, publishers. Deutschland. 8. Aufl. Coblenz, 1858.
Geog 1522.35	Baedeker, publishers. Deutschland. 12. Aufl. Coblenz, 1865.
Geog 1522.38	Baedeker, publishers. Deutschland in einem Bande. 4. Aufl. Leipzig, 1925.
Geog 1522.36	Baedeker, publishers. Deutschland und Österreich. 14. Aufl. Coblenz, 1869.
Geog 1640.5	Baedeker, publishers. Dominion of Canada. Leipzig, 1894.
Geog 1640.11	Baedeker, publishers. Dominion of Canada. 2. ed. Leipsic, 1900.
Geog 1640.15.2	Baedeker, publishers. The dominion of Canada. 4. ed. Leipzig, 1922.
Geog 1640.15A	Baedeker, publishers. Dominion of Canada with Newfoundland and an excursion to Alaska. 3. ed. Leipzig, 1907.
Geog 1533.4	Baedeker, publishers. Eastern Alps. 4. ed. Leipsic, 1879.
Geog 1533.6	Baedeker, publishers. Eastern Alps. 6. ed. Leipsic, 1888.
Geog 1533.7	Baedeker, publishers. Eastern Alps. 7. ed. Leipsic, 1891.
Geog 1533.8	Baedeker, publishers. Eastern Alps. 8. ed. Leipsic, 1895.
Geog 1533.10	Baedeker, publishers. Eastern Alps. 10. ed. Leipsic, 1903.
Geog 1533.12	Baedeker, publishers. Eastern Alps. 11. ed. Leipzig, 1907.
Geog 1533.13	Baedeker, publishers. Eastern Alps. 12. ed. Leipzig, 1911.
Geog 1618.5	Baedeker, publishers. Egypt...lower Egypt. Leipsic, 1878.
Geog 1618.10A	Baedeker, publishers. Egypt...lower Egypt. 2. ed. Leipsic, 1885.
Geog 1617.5	Baedeker, publishers. Egypt. Leipsic, 1892.
Geog 1618.15	Baedeker, publishers. Egypt. Leipsic, 1892-95. 2v.
Geog 1616.15	Baedeker, publishers. Egypt. 4. ed. Leipsic, 1898.
Geog 1616.18	Baedeker, publishers. Egypt. 4. ed. Leipsic, 1898.
Geog 1616.20	Baedeker, publishers. Egypt. 5. ed. Leipsic, 1902.
Geog 1616.21	Baedeker, publishers. Egypt. 6. ed. Leipsic, 1908.
Geog 1616.22	Baedeker, publishers. Egypt and the Sudan. 7. ed. Leipzig, 1914.
Geog 1616.24A	Baedeker, publishers. Egypt and the Sudan. 8. ed. Leipzig, 1929.
Geog 1545.25	Baedeker, publishers. Espagne et Portugal. 3. éd. Leipzig, 1920.
Geog 1580.5	Baedeker, publishers. Das General gouvernement. Leipzig, 1943.
Geog 1508.5	Baedeker, publishers. Great Britain. Leipsic, 1887.
Geog 1508.34	Baedeker, publishers. Great Britain. N.Y., 1937.
Geog 1508.10	Baedeker, publishers. Great Britain. 2. ed. Leipsic, 1890.
Geog 1508.11	Baedeker, publishers. Great Britain. 2. ed. Leipsic, 1890.
NEDL Geog 1508.12	Baedeker, publishers. Great Britain. 3. ed. Leipsic, 1894.
Geog 1508.13	Baedeker, publishers. Great Britain. 3. ed. Leipsic, 1894.
Geog 1508.15	Baedeker, publishers. Great Britain. 4. ed. Leipsic, 1897.
Geog 1508.16	Baedeker, publishers. Great Britain. 4. ed. Leipsic, 1897.
Geog 1508.20	Baedeker, publishers. Great Britain. 5. ed. Leipsic, 1901.
Geog 1508.25	Baedeker, publishers. Great Britain. 6. ed. Leipsic, 1906.
Geog 1508.30	Baedeker, publishers. Great Britain. 7. ed. Leipzig, 1910.
Geog 1508.32	Baedeker, publishers. Great Britain. 8. ed. Leipzig, 1927.
Geog 1585.10	Baedeker, publishers. Greece. Leipsic, 1889.
Geog 1585.15	Baedeker, publishers. Greece. 2. ed. Leipsic, 1894.
Geog 1585.25	Baedeker, publishers. Greece. 4. ed. Leipzig, 1909.
Geog 1585.26	Baedeker, publishers. Greece. 4. ed. Leipzig, 1909.
X Cg Geog 1585.5	Baedeker, publishers. Griechenland. Leipzig, 1883.
Geog 1527.70	Baedeker, publishers. Hamburg und die Niederelbe. Hamburg, 1953.
Geog 1575.16	Baedeker, publishers. Handbook for travellers: Russia. N.Y., 1970.
Geog 1525.20	Baedeker, publishers. A handbook for travellers on the Rhine. 2. ed. Coblenz, 1864.
Geog 1522.23	Baedeker, publishers. Handbuch für Reisende in Deutschland. pt.1-2. 6. Aufl. Coblenz, 1855.
Geog 1522.15	Baedeker, publishers. Handbuch für Reisende in Deutschland. 2. Aufl. Coblenz, 1846.
Geog 1522.22	Baedeker, publishers. Handbuch für Reisende in Deutschland. 5. Aufl. Coblenz, 1854.
Geog 1560.67	Baedeker, publishers. Holland; Hanbuch für Reisende. 26. Aufl. Leipzig, 1927.
Geog 1539.5	Baedeker, publishers. L'Italie...septentrionale. 3. éd. Coblenz, 1865.
Geog 1539.10	Baedeker, publishers. L'Italie...septentrionale. 5. éd. Coblenz, 1870.
Geog 1539.18	Baedeker, publishers. L'Italie...septentrionale. 7. éd. Leipsic, 1876.
Geog 1539.19	Baedeker, publishers. L'Italie...septentrionale. 8. éd. Leipsic, 1878.
Geog 1539.22	Baedeker, publishers. L'Italie...septentrionale. 9. éd. Leipsic, 1880.
Geog 1539.11	Baedeker, publishers. L'Italie...septentrionale. 12. éd. Leipsic, 1889.
Geog 1540.17	Baedeker, publishers. Italie. Pt. 2: Italie centrale et Rome. 4. éd. Leipzig, 1875.
Geog 1540.17.25	Baedeker, publishers. Italie. Pt. 2: Italie centrale et Rome. 5. éd. Leipzig, 1877.
Geog 1542.14	Baedeker, publishers. Italie. Pt. 3: Italie du sud et la Sicile. 4. éd. Leipzig, 1875.
Geog 1542.16	Baedeker, publishers. Italie. Pt. 3: Italie méridionale et la Sicile. 5. éd. Leipzig, 1877.
Geog 1539.51	Baedeker, publishers. Italie septentrionale. 17. éd. Leipzig, 1908.
Geog 1538.20	Baedeker, publishers. Italien. Coblenz, 1866-68. 2v.
Geog 1538.37	Baedeker, publishers. Italien. v.3. 11. Aufl. Leipzig, 1895.
Geog 1538.30	Baedeker, publishers. Italien. 9. Aufl. Leipzig, 1879-89. 2v.
Geog 1538.35	Baedeker, publishers. Italien. 14. Aufl. Leipzig, 1894.
Geog 1538.45	Baedeker, publishers. Italien von den Alpen bis Neapel. 3. Aufl. Leipzig, 1895.
Geog 1538.61	Baedeker, publishers. Italy, including Sicily and Sardinia. Freiburg, 1961.
Geog 1539.12	Baedeker, publishers. Italy...northern Italy. 2. ed. Coblenz, 1870.
Geog 1539.17.5	Baedeker, publishers. Italy...northern Italy. 3. ed. Leipsic, 1874.
Geog 1539.17	Baedeker, publishers. Italy...northern Italy. 3. ed. Leipzig, 1874.
Geog 1539.21	Baedeker, publishers. Italy...northern Italy. 5. ed. Leipsic, 1879.

Author and Title Listing

Call Number	Entry
Geog 1539.23	Baedeker, publishers. Italy...northern Italy. 6. ed. Leipsic, 1882.
Geog 1539.25	Baedeker, publishers. Italy...northern Italy. 7. ed. Leipsic, 1886.
Geog 1539.30	Baedeker, publishers. Italy...northern Italy. 8. ed. Leipsic, 1889.
Geog 1539.35	Baedeker, publishers. Italy...northern Italy. 9. ed. Leipsic, 1892.
Geog 1539.36	Baedeker, publishers. Italy...northern Italy. 9. ed. Leipsic, 1892.
Geog 1539.38	Baedeker, publishers. Italy...northern Italy. 10. ed. Leipsic, 1895.
Geog 1539.40	Baedeker, publishers. Italy...northern Italy. 11. ed. Leipsic, 1899.
Geog 1539.46	Baedeker, publishers. Italy...northern Italy. 12. ed. Leipsic, 1903.
Geog 1539.45	Baedeker, publishers. Italy...northern Italy. 12. ed. Leipsic, 1903.
Geog 1539.50	Baedeker, publishers. Italy...northern Italy. 13. ed. Leipsic, 1906-
Geog 1540.5	Baedeker, publishers. Italy. Pt. 2: Central Italy and Rome. Coblenz, 1867.
Geog 1540.10	Baedeker, publishers. Italy. Pt. 2: Central Italy and Rome. 2. ed. Coblenz, 1869.
Geog 1540.12	Baedeker, publishers. Italy. Pt. 2: Central Italy and Rome. 3. ed. Coblenz, 1872.
Geog 1540.15	Baedeker, publishers. Italy. Pt. 2: Central Italy and Rome. 4. ed. Leipsic, 1875.
Geog 1540.18	Baedeker, publishers. Italy. Pt. 2: Central Italy and Rome. 6. ed. Leipsic, 1879.
Geog 1540.18.25	Baedeker, publishers. Italy. Pt. 2: Central Italy and Rome. 7. ed. Leipsic, 1881.
Geog 1540.19	Baedeker, publishers. Italy. Pt. 2: Central Italy and Rome. 8. ed. Leipsic, 1883.
Geog 1540.19.5	Baedeker, publishers. Italy. Pt. 2: Central Italy and Rome. 8. ed. Leipsic, 1883.
Geog 1540.20	Baedeker, publishers. Italy. Pt. 2: Central Italy and Rome. 9. ed. Leipsic, 1886.
Geog 1540.21	Baedeker, publishers. Italy. Pt. 2: Central Italy and Rome. 9. ed. Leipsic, 1886.
Geog 1540.25	Baedeker, publishers. Italy. Pt. 2: Central Italy and Rome. 10. ed. Leipsic, 1890.
Geog 1540.26	Baedeker, publishers. Italy. Pt. 2: Central Italy and Rome. 10. ed. Leipsic, 1890.
Geog 1540.31A	Baedeker, publishers. Italy. Pt. 2: Central Italy and Rome. 11. ed. Leipzig, 1893.
Geog 1540.30	Baedeker, publishers. Italy. Pt. 2: Central Italy and Rome. 11. ed. Leipzig, 1893.
Geog 1540.37	Baedeker, publishers. Italy. Pt. 2: Central Italy and Rome. 12. ed. Leipsic, 1897.
Geog 1540.40	Baedeker, publishers. Italy. Pt. 2: Central Italy and Rome. 13. ed. Leipsic, 1900.
Geog 1540.45	Baedeker, publishers. Italy. Pt. 2: Central Italy and Rome. 14. ed. Leipzig, 1904.
Geog 1542.5	Baedeker, publishers. Italy. Pt. 3: Southern Italy, Sicily. Coblenz, 1867.
Geog 1542.9	Baedeker, publishers. Italy. Pt. 3: Southern Italy, Sicily. 2. ed. Coblenz, 1869.
Geog 1542.10	Baedeker, publishers. Italy. Pt. 3: Southern Italy, Sicily. 2. ed. Coblenz, 1869.
Geog 1542.13	Baedeker, publishers. Italy. Pt. 3: Southern Italy, Sicily. 4. ed. Leipsic, 1873.
Geog 1542.15	Baedeker, publishers. Italy. Pt. 3: Southern Italy, Sicily. 6. ed. Leipsic, 1876.
Geog 1542.21	Baedeker, publishers. Italy. Pt. 3: Southern Italy, Sicily. 8. ed. Leipsic, 1883.
Geog 1542.20	Baedeker, publishers. Italy. Pt. 3: Southern Italy, Sicily. 8. ed. Leipsic, 1883.
Geog 1542.25	Baedeker, publishers. Italy. Pt. 3: Southern Italy, Sicily. 9. ed. Leipsic, 1887.
Geog 1542.27	Baedeker, publishers. Italy. Pt. 3: Southern Italy, Sicily. 10. ed. Leipsic, 1890.
Geog 1542.30	Baedeker, publishers. Italy. Pt. 3: Southern Italy, Sicily. 11. ed. Leipsic, 1893.
Geog 1542.33	Baedeker, publishers. Italy. Pt. 3: Southern Italy, Sicily. 12. ed. Leipsic, 1896.
Geog 1542.35	Baedeker, publishers. Italy. Pt. 3: Southern Italy, Sicily. 13. ed. Leipsic, 1900.
Geog 1538.50	Baedeker, publishers. Italy from the Alps to Naples. Leipzig, 1904.
Geog 1538.56A	Baedeker, publishers. Italy from the Alps to Naples. 2. ed. Leipzig, 1909.
Geog 1538.57A	Baedeker, publishers. Italy from the Alps to Naples. 3. ed. Leipzig, 1928.
Geog 1525.72	Baedeker, publishers. Köln und der Rheinland zwischen Köln und Mainz. Hamburg, 1953.
Geog 1527.60	Baedeker, publishers. Köln und Umgebung. 2. Aufl. Freiburg, 1960.
Geog 1604.5	Baedeker, publishers. Konstantinopel und das westliche Kleinasien. Leipzig, 1905.
Geog 1604.15	Baedeker, publishers. Konstantinopel und Kleinasien, Archipel, Cypern. 2. Aufl. Leipzig, 1914.
Geog 1527.50	Baedeker, publishers. Leipzig. Leipzig, 1948.
Geog 1510.5	Baedeker, publishers. London and its environs. Leipsic, 1878. 2v.
Geog 1510.8	Baedeker, publishers. London and its environs. 2. ed. Leipsic, 1879.
Geog 1510.10	Baedeker, publishers. London and its environs. 3. ed. Leipsic, 1881
Geog 1510.25	Baedeker, publishers. London and its environs. 6. ed. Leipsic, 1887.
Geog 1510.30	Baedeker, publishers. London and its environs. 7. ed. Leipsic, 1889.
Geog 1510.35	Baedeker, publishers. London and its environs. 8. ed. Leipsic, 1892.
Geog 1510.36	Baedeker, publishers. London and its environs. 8. ed. Leipsic, 1892.
Geog 1510.41	Baedeker, publishers. London and its environs. 9. ed. Leipsic, 1894.
Geog 1510.40	Baedeker, publishers. London and its environs. 9. ed. Leipsic, 1894.
Geog 1510.45	Baedeker, publishers. London and its environs. 10. ed. Leipsic, 1894.
Geog 1510.46	Baedeker, publishers. London and its environs. 10. ed. Leipsic, 1896.
Geog 1510.47	Baedeker, publishers. London and its environs. 11. ed. Leipsic, 1898.
Geog 1510.47.12	Baedeker, publishers. London and its environs. 12. ed. Leipsic, 1900.
Geog 1510.50	Baedeker, publishers. London and its environs. 13. ed. Leipsic, 1902.
Geog 1510.49	Baedeker, publishers. London and its environs. 13. ed. Leipsic, 1902.
Geog 1510.48	Baedeker, publishers. London and its environs. 13. ed. Leipsic, 1902. 2v.
Geog 1510.55	Baedeker, publishers. London and its environs. 14. ed. Leipsic, 1905.
Geog 1510.56	Baedeker, publishers. London and its environs. 15. ed. Leipsic, 1908.
Geog 1510.57	Baedeker, publishers. London and its environs. 15. ed. Leipzig, 1908.
Geog 1510.61	Baedeker, publishers. London and its environs. 18. ed. Leipzig, 1923.
Geog 1510.62	Baedeker, publishers. London and its environs. 18. ed. Leipzig, 1923.
Geog 1510.62.2	Baedeker, publishers. London and its environs. 20. ed. Hamburg, 1951.
Geog 1510.15	Baedeker, publishers. London and its invirons. 4. ed. Leipsic, 1883.
Geog 1510.20	Baedeker, publishers. London and its invirons. 5. ed. Leipsic, 1885.
Geog 1510.59A	Baedeker, publishers. London and its invirons. 16. ed. Leipzig, 1911.
Geog 1510.2	Baedeker, publishers. London und Umgebung. 15. Aufl. Leipzig, 1905.
Geog 1510.42	Baedeker, publishers. London und Umgebungen. 11. Aufl. Leipsic, 1894.
Geog 1510.3	Baedeker, publishers. Londres, ses environs: le Sud de l'Angleterre. 3. éd. Leipzig, 1875.
Geog 1595.5A	Baedeker, publishers. Mediterranean...Madeira, Canary Islands. Leipzig, 1911.
Geog 1595.7	Baedeker, publishers. Mittelmeer. 2. Aufl. Leipzig, 1934.
Geog 1527.15	Baedeker, publishers. München. Leipzig, 1921.
Geog 1527.2	Baedeker, publishers. München und Südbayern. 39. Aufl. Leipzig, 1935.
Geog 1527.5	Baedeker, publishers. München und Umgebung. Augsburg. Leipzig, 1935.
Geog 1527.18	Baedeker, publishers. München und Umgebung. 4. Aufl. Freiburg, 1960.
Geog 1527.10	Baedeker, publishers. Munich and its environs. Hamburg, 1950.
Geog 1527.12	Baedeker, publishers. Munich and its environs. 2. ed. Hamburg, 1956.
Geog 1516.3	Baedeker, publishers. Le Nord de la France. Leipsic, 1884.
Geog 1516.4	Baedeker, publishers. Le Nord de la France. 2. éd. Leipsic, 1887.
Geog 1518.55	Baedeker, publishers. Le Nord-est de la France. 5. éd. Leipsic, 1895.
Geog 1518.80	Baedeker, publishers. Le Nord-ouest de la France. 5. éd. Leipsic, 1895.
Geog 1518.89	Baedeker, publishers. Le Nord-ouest de la France. 9. éd. Leipsic, 1913.
Geog 1645.11	Baedeker, publishers. Nordamerika. Die Vereinigten Staaten...Mexiko. 2. Aufl. Leipzig, 1904.
Geog 1523.75	Baedeker, publishers. Nordost-Deutschland. 25. Aufl. Leipsic, 1896.
Geog 1523.80	Baedeker, publishers. Nordwest-Deutschland. 26. Aufl. Leipsic, 1899.
Geog 1523.81	Baedeker, publishers. Nordwestdeutschland. 27. Aufl. Leipsic, 1902.
Geog 1524.72	Baedeker, publishers. Northern Bavaria. Hamburg, 1951.
Geog 1516.5	Baedeker, publishers. Northern France. Leipsic, 1889.
Geog 1516.10.2	Baedeker, publishers. Northern France. 2. ed. Leipsic, 1894.
Geog 1516.10	Baedeker, publishers. Northern France. 2. ed. Leipsic, 1894.
Geog 1516.15	Baedeker, publishers. Northern France. 3. ed. Leipsic, 1899.
Geog 1516.15.3	Baedeker, publishers. Northern France. 3. ed. Leipsic, 1899.
Geog 1516.20.4	Baedeker, publishers. Northern France. 4. ed. Leipzig, 1905.
Geog 1516.25	Baedeker, publishers. Northern France. 5. ed. Leipzig, 1909.
Geog 1523.20	Baedeker, publishers. Northern Germany. 5. ed. Coblenz, 1873.
Geog 1523.25	Baedeker, publishers. Northern Germany. 6. ed. Leipsic, 1877.
Geog 1523.26	Baedeker, publishers. Northern Germany. 6. ed. Leipsic, 1877.
Geog 1523.30	Baedeker, publishers. Northern Germany. 7. ed. Leipsic, 1881.
Geog 1523.36	Baedeker, publishers. Northern Germany. 8. ed. Leipsic, 1884.
Geog 1523.35	Baedeker, publishers. Northern Germany. 8. ed. Leipsic, 1884.
Geog 1523.41	Baedeker, publishers. Northern Germany. 9. ed. Leipsic, 1886.
Geog 1523.40	Baedeker, publishers. Northern Germany. 9. ed. Leipsic, 1886.
Geog 1523.46	Baedeker, publishers. Northern Germany. 10. ed. Leipsic, 1890.
Geog 1523.45	Baedeker, publishers. Northern Germany. 10. ed. Leipsic, 1890.
Geog 1523.50	Baedeker, publishers. Northern Germany. 11. ed. Leipsic, 1893.
Geog 1523.55	Baedeker, publishers. Northern Germany. 12. ed. Leipsic, 1897.
Geog 1523.60	Baedeker, publishers. Northern Germany. 13. ed. Leipsic, 1900.
Geog 1523.65	Baedeker, publishers. Northern Germany. 14. ed. Leipzig, 1904.
Geog 1523.66	Baedeker, publishers. Northern Germany. 14. ed. Leipzig, 1904.
Geog 1523.67A	Baedeker, publishers. Northern Germany. 15. ed. Leipzig, 1910.
Geog 1523.70	Baedeker, publishers. Northern Germany. 16. ed. Leipzig, 1913.
Geog 1523.73	Baedeker, publishers. Northern Germany. 17. ed. Leipzig, 1925.
Geog 1539.8	Baedeker, publishers. Northern Italy, as far as Leghorn. Coblentz, 1868.
Geog 1539.55	Baedeker, publishers. Northern Italy. 14. ed. Leipzig, 1913.

Author and Title Listing

Call Number	Entry
Geog 1539.60	Baedeker, publishers. Northern Italy. 15. ed. Leipzig, 1930.
Geog 1568.24	Baedeker, publishers. Norway, Sweden and Denmark. 5. ed. Leipsic, 1892.
Geog 1568.25A	Baedeker, publishers. Norway, Sweden and Denmark. 6. ed. Leipsic, 1895.
Geog 1568.30	Baedeker, publishers. Norway, Sweden and Denmark. 7. ed. Leipsic, 1899.
Geog 1568.35	Baedeker, publishers. Norway, Sweden and Denmark. 8. ed. Leipsic, 1903.
Geog 1568.40	Baedeker, publishers. Norway, Sweden and Denmark. 9. ed. Leipzig, 1909.
Geog 1568.45A	Baedeker, publishers. Norway, Sweden and Denmark. 10. ed. Leipzig, 1912.
Geog 1568.5	Baedeker, publishers. Norway and Sweden. Leipzig, 1879.
Geog 1568.15	Baedeker, publishers. Norway and Sweden. 3. ed. Leipsic, 1885.
Geog 1530.18	Baedeker, publishers. Oesterreich-Ungarn. 20. Aufl. Leipzig, 1884.
Geog 1530.15	Baedeker, publishers. Österreich-Ungarn. 29. Aufl. Leipzig, 1913.
Geog 1605.18	Baedeker, publishers. Palästina und Syrien. 6. Aufl. Leipzig, 1904.
Geog 1605.5	Baedeker, publishers. Palestine and Syria. Leipzig, 1876.
Geog 1605.10	Baedeker, publishers. Palestine and Syria. 2. ed. Leipzig, 1894.
Geog 1605.25A	Baedeker, publishers. Palestine and Syria. 5. ed. Leipzig, 1912.
Geog 1519.19	Baedeker, publishers. Paris, nebst einigen Routen. 16. Aufl. Leipzig, 1905.
Geog 1519.13	Baedeker, publishers. Paris, Rouen, Havre, Dieppe. Coblenz, 1865.
Geog 1519.10	Baedeker, publishers. Paris, Rouen, Havre, Dieppe. 5. Aufl. Coblenz, 1864.
Geog 1519.24	Baedeker, publishers. Paris, ses environs et les principaux itinéraires. 4. éd. Leipzig, 1876.
Geog 1519.23	Baedeker, publishers. Paris and its environs. 4. ed. Leipsic, 1874.
Geog 1519.26	Baedeker, publishers. Paris and its environs. 5. ed. Leipsic, 1876.
Geog 1519.25	Baedeker, publishers. Paris and its environs. 5. ed. Leipsic, 1876.
Geog 1519.30	Baedeker, publishers. Paris and its environs. 6. ed. Leipsic, 1878.
Geog 1519.36	Baedeker, publishers. Paris and its environs. 8. ed. Leipsic, 1884.
Geog 1519.39	Baedeker, publishers. Paris and its environs. 9. ed. Leipsic, 1888.
Geog 1519.46.2	Baedeker, publishers. Paris and its environs. 10. ed. Leipsic, 1891.
Geog 1519.46	Baedeker, publishers. Paris and its environs. 10. ed. Leipsic, 1891.
Geog 1519.48	Baedeker, publishers. Paris and its environs. 11. ed. Leipsic, 1894.
Geog 1519.50	Baedeker, publishers. Paris and its environs. 12. ed. Leipsic, 1896.
Geog 1519.50.2	Baedeker, publishers. Paris and its environs. 12. ed. Leipsic, 1896.
Geog 1519.54	Baedeker, publishers. Paris and its environs. 14. ed. Leipsic, 1900.
Geog 1519.55	Baedeker, publishers. Paris and its environs. 14. ed. Leipsic, 1900.
Geog 1519.58	Baedeker, publishers. Paris and its environs. 15. ed. Leipsic, 1904.
Geog 1519.60	Baedeker, publishers. Paris and its environs. 16. ed. Leipsic, 1907.
Geog 1519.63	Baedeker, publishers. Paris and its environs. 17. ed. Leipzig, 1910.
Geog 1519.68	Baedeker, publishers. Paris and its environs. 18. ed. Leipzig, 1913.
Geog 1519.70	Baedeker, publishers. Paris and its environs. 19. ed. Leipzig, 1924.
Geog 1519.33.2	Baedeker, publishers. Paris and its invirons. 7. ed. Leipsic, 1881.
Geog 1519.15	Baedeker, publishers. Paris and northern France. 2. ed. Coblenz, 1867.
Geog 1519.20	Baedeker, publishers. Paris and northern France. 3. ed. Coblenz, 1872.
Geog 1519.21	Baedeker, publishers. Paris and northern France. 3. ed. Coblenz, 1872.
Geog 1519.21.3	Baedeker, publishers. Paris and northern France. 3. ed. Leipsic, 1872.
Geog 1519.27	Baedeker, publishers. Paris et ses environs. 5. éd. Leipzig, 1878.
Geog 1519.33	Baedeker, publishers. Paris et ses environs. 7. éd. Leipzig, 1881.
Geog 1519.36.2	Baedeker, publishers. Paris et ses environs. 7. éd. Leipzig, 1884.
Geog 1519.37	Baedeker, publishers. Paris et ses environs. 8. éd. Leipzig, 1887.
Geog 1519.42	Baedeker, publishers. Paris et ses environs. 9. éd. Leipzig, 1889.
Geog 1519.45	Baedeker, publishers. Paris et ses environs. 10. éd. Leipzig, 1891.
Geog 1519.49	Baedeker, publishers. Paris et ses environs. 11. éd. Leipzig, 1894.
Geog 1519.51	Baedeker, publishers. Paris et ses environs. 12. éd. Leipzig, 1896.
Geog 1519.75	Baedeker, publishers. Paris und Umgebung und Supplement. 20. Aufl. Leipzig, 1931-37. 2v.
Geog 1525.56	Baedeker, publishers. Die Rheinland, Schwarzwald, Vogesen. 32. Aufl. Leipzig, 1912.
Geog 1525.15	Baedeker, publishers. Die Rheinlande. 9. Aufl. Coblenz, 1856.
Geog 1525.45.5	Baedeker, publishers. Die Rheinlande von der Schweize. 26. Aufl. Leipzig, 1892.
Geog 1525.22	Baedeker, publishers. Die Rheinlande von der Schweize zu holländisch Grenze. 14. Aufl. Leipzig, 1866.
Geog 1525.41	Baedeker, publishers. Die Rheinlande von der Schweize zu holländisch Grenze. 28. Aufl. Leipzig, 1889.
Geog 1523.15	Baedeker, publishers. Rhine and northern Germany. 3. ed. Coblenz, 1868.
Geog 1523.17	Baedeker, publishers. Rhine and northern Germany. 4. ed. Coblenz, 1970.
Geog 1525.25.3	Baedeker, publishers. The Rhine from Rotterdam to Constance. 5. ed. Leipsic, 1873.
Geog 1525.25	Baedeker, publishers. The Rhine from Rotterdam to Constance. 5. ed. Leipsic, 1873.
Geog 1525.29	Baedeker, publishers. The Rhine from Rotterdam to Constance. 6. ed. Leipsic, 1878.
Geog 1525.30	Baedeker, publishers. The Rhine from Rotterdam to Constance. 7. ed. Leipsic, 1880.
Geog 1525.35	Baedeker, publishers. The Rhine from Rotterdam to Constance. 8. ed. Leipsic, 1882.
Geog 1525.35.3	Baedeker, publishers. The Rhine from Rotterdam to Constance. 8. ed. Leipsic, 1882.
Geog 1525.40A	Baedeker, publishers. The Rhine from Rotterdam to Constance. 10. ed. Leipsic, 1886.
Geog 1525.41.3	Baedeker, publishers. The Rhine from Rotterdam to Constance. 11. ed. Leipsic, 1889.
Geog 1525.45	Baedeker, publishers. The Rhine from Rotterdam to Constance. 12. ed. Leipsic, 1892.
Geog 1525.45.2	Baedeker, publishers. The Rhine from Rotterdam to Constance. 12. ed. Leipsic, 1892.
Geog 1525.47	Baedeker, publishers. The Rhine from Rotterdam to Constance. 13. ed. Leipsic, 1896.
Geog 1525.50	Baedeker, publishers. The Rhine from Rotterdam to Constance. 14. ed. Leizsic, 1900.
Geog 1525.50.2	Baedeker, publishers. The Rhine from Rotterdam to Constance. 14. ed. Leizsic, 1900.
Geog 1525.52.5	Baedeker, publishers. The Rhine from Rotterdam to Constance. 15. ed. Leipzig, 1903.
Geog 1525.53	Baedeker, publishers. The Rhine from Rotterdam to Constance. 16. ed. Leipzig, 1906.
Geog 1525.66	Baedeker, publishers. The Rhine from the Dutch to the Alsatian frontier. 18. ed. Leipzig, 1926.
Geog 1525.55A	Baedeker, publishers. The Rhine including the Black Forest and Vosges. 17. ed. Leipzig, 1911.
Geog 1540.50	Baedeker, publishers. Rome and central Italy. 16. ed. N.Y., 1930.
Geog 1575.15A	Baedeker, publishers. Russia. N.Y., 1914.
Geog 1575.5	Baedeker, publishers. La Russie. 2. éd. Leipzig, 1897.
Geog 1575.10	Baedeker, publishers. Russland. Handbuch für Reisende. 5. Aufl. Leipzig, 1901.
Geog 1525.70	Baedeker, publishers. Schleswig-Holstein und Hamburg. Hamburg, 1949.
Geog 1525.55.15	Baedeker, publishers. Schwarzwald. 3. Aufl. Leipzig, 1936.
Geog 1525.55.16	Baedeker, publishers. Schwarzwald. 4. Aufl. Malente, 1956.
Geog 1568.18A	Baedeker, publishers. Schweden und Norwegen. 3. Aufl. Leipzig, 1885.
Geog 1535.16	Baedeker, publishers. Die Schweiz, die italienischen Seen. 6. Aufl. Coblenz, 1856.
Geog 1535.10	Baedeker, publishers. Die Schweiz. 5. Aufl. Coblenz, 1853.
Geog 1535.13	Baedeker, publishers. Die Schweiz. 5. Aufl. Coblenz, 1854.
Geog 1535.22	Baedeker, publishers. Die Schweiz. 12. Aufl. Coblenz, 1869.
Geog 1535.23.9	Baedeker, publishers. Die Schweiz. 24. Aufl. Leipzig, 1891.
Geog 1518.3	Baedeker, publishers. South eastern France. 3. ed. Leipsic, 1898.
Geog 1518.26	Baedeker, publishers. South-western France from the Loire and the Rhone to the Spanish frontier. 2. ed. Leipsic, 1895.
Geog 1517.5	Baedeker, publishers. Southern France...including Corsica. Leipsic, 1897.
Geog 1517.17	Baedeker, publishers. Southern France. 4. ed. Leipsic, 1902.
Geog 1517.19	Baedeker, publishers. Southern France. 5. ed. Leipzig, 1907.
Geog 1517.19.5	Baedeker, publishers. Southern France. 5. ed. Leipzig, 1907.
Geog 1517.21A	Baedeker, publishers. Southern France. 6. ed. Leipzig, 1914.
Geog 1524.46	Baedeker, publishers. Southern Germany. 8. ed. Leipsic, 1895.
Geog 1524.45	Baedeker, publishers. Southern Germany. 8. ed. Leipsic, 1895.
Geog 1524.51	Baedeker, publishers. Southern Germany. 9. ed. Leipzig, 1902.
Geog 1524.55	Baedeker, publishers. Southern Germany. 10. ed. Leipzig, 1907.
Geog 1524.56	Baedeker, publishers. Southern Germany. 11. ed. Leipzig, 1910.
Geog 1524.69A	Baedeker, publishers. Southern Germany. 13. ed. Leipzig, 1929.
Geog 1524.5A	Baedeker, publishers. Southern Germany and Austria. 2. ed. Coblenz, 1871.
Geog 1524.10	Baedeker, publishers. Southern Germany and Austria. 3. ed. Coblenz, 1873.
Geog 1524.30A	Baedeker, publishers. Southern Germany and Austria. 4. ed. Leipsic, 1880.
Geog 1524.33	Baedeker, publishers. Southern Germany and Austria. 5. ed. Leipsic, 1883.
Geog 1524.35	Baedeker, publishers. Southern Germany and Austria. 6. ed. Leipsic, 1887.
Geog 1524.36	Baedeker, publishers. Southern Germany and Austria. 6. ed. Leipsic, 1887.
Geog 1524.40	Baedeker, publishers. Southern Germany and Austria. 7. ed. Leipsic, 1891.
Geog 1524.41	Baedeker, publishers. Southern Germany and Austria. 7. ed. Leipsic, 1891.
Geog 1542.37	Baedeker, publishers. Southern Italy and Sicily. 15. ed. Leipzig, 1908.
Geog 1542.39	Baedeker, publishers. Southern Italy and Sicily. 17. ed. Leipzig, 1930.
Geog 1545.5	Baedeker, publishers. Spain and Portugal. Leipsic, 1898.
Geog 1545.10	Baedeker, publishers. Spain and Portugal. 2. ed. Leipsic, 1901.
Geog 1545.15	Baedeker, publishers. Spain and Portugal. 3. ed. Leipsic, 1908.
Geog 1545.18A	Baedeker, publishers. Spain and Portugal. 4. ed. Leipzig, 1913.
Geog 1545.11	Baedeker, publishers. Spain and Prutugal. 2. ed. Leipsic, 1901.
Geog 1524.19	Baedeker, publishers. Süd-Deutschland. 22. Aufl. Leipzig, 1888.
Geog 1524.19.2	Baedeker, publishers. Süd-Deutschland. 23. Aufl. Leipzig, 1890.
Geog 1524.32	Baedeker, publishers. Süd-Deutschland und Oesterreich. 19. Aufl. Leipzig, 1882.
Geog 1524.15	Baedeker, publishers. Süd-Deutschland und Oesterrreich. 8. Aufl. Leipzig, 1879.

Author and Title Listing

Geog 1524.75 Baedeker, publishers. Südbaiern, Tirol und Salzburg. 18. Aufl. Leipzig, 1878.

Geog 1524.77 Baedeker, publishers. Südbaiern, Tirol und Salzburg. 22. Aufl. Leipzig, 1886.

Geog 1524.73 Baedeker, publishers. Südbayern: Alpenvorland, Alpen, österreichische Gunzgebiete. 41. Aufl. Hamburg, 1953.

Geog 1535.20 Baedeker, publishers. La Suisse. 7. éd. Coblenz, 1867.

Geog 1535.23 Baedeker, publishers. La Suisse. 8. éd. Coblenz, 1869.

Geog 1535.44 Baedeker, publishers. La Suisse. 14. éd. Leipzig, 1885.

Geog 1535.50 Baedeker, publishers. Switzerland...Italy, Savoy...Tryol. 15. ed. Leipsic, 1893.

Geog 1535.30 Baedeker, publishers. Switzerland...Italy, Savoy...Tyrol. 5. ed. Coblenz, 1872.

Geog 1535.31 Baedeker, publishers. Switzerland...Italy, Savoy...Tyrol. 6. ed. Coblenz, 1873.

Geog 1535.33 Baedeker, publishers. Switzerland...Italy, Savoy...Tyrol. 7. ed. Leipsic, 1877.

Geog 1535.35 Baedeker, publishers. Switzerland...Italy, Savoy...Tyrol. 8. ed. Leipsic, 1879.

Geog 1535.40 Baedeker, publishers. Switzerland...Italy, Savoy...Tyrol. 9. ed. Leipsic, 1881.

Geog 1535.42 Baedeker, publishers. Switzerland...Italy, Savoy...Tyrol. 10. ed. Leipsic, 1883.

Geog 1535.43 Baedeker, publishers. Switzerland...Italy, Savoy...Tyrol. 11. ed. Leipsic, 1885.

Geog 1535.45 Baedeker, publishers. Switzerland...Italy, Savoy...Tyrol. 12. ed. Leipsic, 1887.

Geog 1535.46 Baedeker, publishers. Switzerland...Italy, Savoy...Tyrol. 13. ed. Leipsic, 1889.

Geog 1535.47 Baedeker, publishers. Switzerland...Italy, Savoy...Tyrol. 13. ed. Leipsic, 1889.

Geog 1535.48 Baedeker, publishers. Switzerland...Italy, Savoy...Tyrol. 14. ed. Leipsic, 1891.

Geog 1535.55 Baedeker, publishers. Switzerland...Italy, Savoy...Tyrol. 16. ed. Leipsic, 1895.

Geog 1535.57 Baedeker, publishers. Switzerland...Italy, Savoy...Tyrol. 17. ed. Leipsic, 1897.

Geog 1535.25 Baedeker, publishers. Switzerland. 3. ed. Coblenz, 1869.

Geog 1535.26 Baedeker, publishers. Switzerland. 4. ed. Coblenz, 1869.

Geog 1535.58 Baedeker, publishers. Switzerland and the adjacent portions of Italy, Savoy, and Tyrol. 18. ed. Leipsic, 1899.

Geog 1535.59 Baedeker, publishers. Switzerland and the adjacent portions of Italy, Savoy, and Tyrol. 19. ed. Leipsic, 1901.

Geog 1535.59.50 Baedeker, publishers. Switzerland and the adjacent portions of Italy, Savoy, and Tyrol. 20. ed. Leipsic, 1903.

Geog 1535.60 Baedeker, publishers. Switzerland and the adjacent portions of Italy, Savoy, and Tyrol. 21. ed. Leipsic, 1905.

Geog 1535.61 Baedeker, publishers. Switzerland and the adjacent portions of Italy, Savoy, and Tyrol. 21. ed. Leipsic, 1905.

Geog 1535.68 Baedeker, publishers. Switzerland and the adjacent portions of Italy, Savoy, and Tyrol. 23. ed. Leipzig, 1909.

Geog 1535.70 Baedeker, publishers. Switzerland and the adjacent portions of Italy, Savoy, and Tyrol. 24. ed. Leipzig, 1911.

Geog 1535.72 Baedeker, publishers. Switzerland and the adjacent portions of Italy, Savoy, and Tyrol. 25. ed. Leipzig, 1913.

Geog 1535.65 Baedeker, publishers. Switzerland and the adjacent portions of Italy, Savoy and Tyrol. 22. ed. Leipzig, 1907.

Geog 1535.24 Baedeker, publishers. Switzerland and the adjacent portions of Italy. 2. ed. Coblenz, 1864.

Geog 1535.24.25 Baedeker, publishers. Switzerland and the adjacent portions of Italy. 3. ed. Coblenz, 1867.

Geog 1535.76 Baedeker, publishers. Switzerland together with Chamonix. 27. ed. Leipzig, 1928.

Geog 1535.74 Baedeker, publishers. Switzerland together with Chamonix and the Italian lakes. 26. ed. Leipzig, 1922.

Geog 1531.4 Baedeker, publishers. Tirol, Vorarlberg, Etschland. 39. Aufl. Leipzig, 1929.

Geog 1531.5 Baedeker, publishers. Tirol, Vorarlberg, westliche Salzburg, Hochkärnten. 40. Aufl. Leipzig, 1938.

Geog 1531.6 Baedeker, publishers. Tirol, Vorarlberg, westliche Salzburg, Hochkärnten. 41. Aufl. Leipzig, 1943.

Geog 1531.3 Baedeker, publishers. Tirol: Vorarlberg und Teile von Salzburg und Kärnten. 37. Aufl. Leipzig, 1923.

Geog 1531.10 Baedeker, publishers. Tyrol and Salzburg. 14. ed. Freiburg, 1961.

Geog 1533.14A Baedeker, publishers. Tyrol and the Dolomites including the Bavarian Alps. 13. ed. Leipzig, 1927.

Htn Geog 1645.5* Baedeker, publishers. United States...Mexico Leipsic, 1893.

Geog 1645.5.5 Baedeker, publishers. United States...Mexico. Leipsic, 1893.

Geog 1645.7.5 Baedeker, publishers. United States...Mexico. 2. ed. Leipsic, 1899.

Geog 1645.7 Baedeker, publishers. United States...Mexico. 2. ed. Leipsic, 1899.

Geog 1645.10A Baedeker, publishers. United States...Mexico. 3. ed. Leipzig, 1904.

Geog 1645.15A Baedeker, publishers. United States...Mexico. 4. ed. Leipzig, 1909.

Geog 1531.15 Baedeker, publishers. Wien und Niederdonau. Leipzig, 1943.

Geog 1525.75 Baedeker, publishers. Wiesbaden, Mainz, Rheingau, Rheinhessen. Malente, 1956.

Geog 1568.16 Baedekr, publishers. Norway and Sweden. 3. ed. Leipsic, 1885.

Geog 3055.13.2 Baer, F.C. Essai sur l'Atlantique des anciens. Avignon, 1835.

Geog 4209.44.10 Baerlein, Henry. The caravan rolls on. London, 1944.

Geog 4209.46 Baerlein, Henry. Leaves in the wind. London, 1946.

Geog 4209.46.2 Baerlein, Henry. So many roads. London, 1946?

Geog 4209.41.5 Baerlein, Henry. Travels without a passport. London, 1941.

Geog 4209.41.6 Baerlein, Henry. Travels without a passport. London, 1943.

Geog 5316.3 Baetzkes, Ottile G. Sir Martin Frobisher's search for the Northwest Passage. N.Y., 1964.

Geog 4180.63 Baffin, W. Voyages of William Baffin 1612-22. N.Y., 1964?

Geog 3070.40 Bagrow, Leo. A. Ortelii Catalogus cartographorum bearbeitet. v.1-2. Gotha, 1928-30.

Geog 3070.40.5 Bagrow, Leo. Die Geschichte der Kartographie. Berlin, 1951.

Geog 3070.157 Bagrow, Leo. History of cartography. Cambridge, 1964.

Geog 6009.20 Bagshawe, T.W. Two men in the Antarctic; an expedition to Graham Land, 1920-22. N.Y., 1939.

Geog 4677.50 Bailey, J. God's wonders in the great deep. N.Y., 1750.

Geog 4169.45 Bailey, Leslie. Travellers' tales, a series of BBC programmes broadcast throughout the world. London, 1945.

Geog 4679.73.6 Bailey, Maurice. 117 days adrift. Lymington, 1974.

Geog 4308.18.5 Baillie, M. First impressions on tour upon continent. London, 1819.

Htn Geog 3055.33* Bailly, J.S. Lettres sur l'Atlantide de Platon. Paris, 1779.

Geog 3055.33.5 Bailly, J.S. Lettres sur l'Atlantide de Platon. Paris, 1804.

Geog 4479.69 Le Baitoto. (Darvère, Pierre.) Paris, 1969.

An 379.22.4 Baitsell, George A. The evolution of man. New Haven, 1923.

Geog 3018.36.2 Bajot, L.M. Abrégé historique...des...voyages. Paris, 1836.

Geog 4209.60.15 Bajsen-Moeller, Axel. Globetrotter og hovedjaeger. Vordingborg, 1960.

Geog 3030.8 Baker, Alan R.H. Progress in historical geography. N.Y., 1972.

Geog 4249.58 Baker, DeVere. The raft Lehi IV. Long Beach, 1959.

Geog 3019.63.5 Baker, J.N.L. The history of geography. Oxford, 1963.

Geog 3019.31 Baker, John Norman Leonard. A history of geographical discovery and exploration. London, 1931.

Geog 3019.67 Baker, John Norman Leonard. A history of geographical discovery and exploration. N.Y., 1967.

Geog 3019.31.2 Baker, John Norman Leonard. A history of geographical discovery and exploration. N.Y., 1967.

An 359.74 Baker, John Randal. Race. N.Y., 1974.

Geog 659.74.10 Baker, Laurie. A selection of geographical computer programs. London, 1974.

An 2109.74.5 Balandier, Georges. Anthropo-logiques. 1e éd. Paris, 1974.

Geog 5939.02 Balch, E.S. Antarctica. Philadelphia, 1902.

An 359.06 Balch, E.S. Comparative art. Philadelphia, 1906.

Geog 4508.93 Balch, E.S. Mountain exploration. Philadelphia, 1893.

Geog 5180.8 Balch, E.S. The North Pole and Bradley land. Philadelphia, 1913.

An 3309.21 Baldwin, Bird T. The physical growth of children from birth to maturity. Iowa City, 1921.

Geog 4672.24 Baldwin, Hanson W. Sea fights and shipwrecks. Garden City, 1955.

An 358.57.3 Baldwin, Samuel D. Dominion; or Unity and trinity of the human race. Nashville, 1857.

Geog 578.49.9 Baldwin, Thomas. A pronouncing gazetteer. 9. ed. Philadelphia, 1851.

Geog 578.45 Baldwin, Thomas. Universal pronouncing gazetteer. Philadelphia, 1845.

Geog 3019.40 Balen, Willem Julius van. Pioniers (De ontdekking van de wereld). Amsterdam, 1940. 2v.

An 2339.11 Balfour, Henry. The origin of West African crossbows. Washington, 1911.

An 629.38 Balfour, Henry. Spinners and weavers in anthropological research. Oxford, 1938.

Geog 4309.25.20 A Balkan mission. (Shotwell, J.T.) N.Y., 1949.

Geog 3580.20 Ballesteros Gaibrois, Manuel. España en los mares. 2. ed. Madrid, 1943.

Geog 4218.84 Ballou, M.M. Due west, or Round the world. Boston, 1884.

Geog 4218.84.5 Ballou, M.M. Due west, or Round the world. 3. ed. Boston, 1885.

Geog 4218.84.9 Ballou, M.M. Due west, or Round the world. 9. ed. Boston, 1891.

Geog 4218.84.2 Ballou, M.M. Due west, or Round the world in ten months. 2. ed. Boston, 1884.

Geog 4218.84.13 Ballou, M.M. Foot-prints of travel. Boston, 1889.

Geog 4218.84.14 Ballou, M.M. Foot-prints of travel. Boston, 1889.

Geog 4218.84.15 Ballou, M.M. Foot-prints of travel. Boston, 1892.

Geog 4208.95 Ballou, M.M. Foot-prints of travel. Boston, 1895.

Geog 4209.23.15 Balmont, K.D. Visions solaires Mexique, Égypte, Inde, Japon, Océanie. Paris, 1923.

Geog 4145.56 Balthazar de Monconys. (Cordier, Stéphane.) Bruxelles, 1967.

Geog 4313.8 Baltiiskoe more. (Egor'eva, A.V.) Moskva, 1961.

Geog 4209.72.15 Bamans, Godfried. Op reis rond de wereld en op Rottumerplaat. Amsterdam, 1972.

Geog 3570.26 Bandeira Ferreira, F. As viagens de descobrimento de iniciativa particular no tempo. Lisboa, 1946.

An 176.7 Bandelier, A.F. The scientist on the trail. Berkeley, 1949.

An 126.5 Bandelier, Adolph Francis. Pioneers in American anthropology: the Bandelier-Morgan letters, 1873-1883. Albuquerque, 1940. 2v.

An 359.38.13 Bang, Paul. Die farbige Gefahr. 2. Aufl. Göttingen, 1938

Geog 807.87.5 Bankes, T. A new, royal, authentic and complete system of universal geography. London, 1787-1810? 2v.

Geog 4521.2 Banks, Mike. Commando climber. London, 1955.

Geog 5839.75 Banks, Mike. Greenland. Newton Abbot, 1975.

Geog 5839.52.5 Banks, Mike. High Arctic; the story of the British North Greenland Expedition. London, 1957.

Geog 819.19 Banse, E. Illustrierte Länderkunde. Berlin, 1919.

Geog 3019.33.10 Banse, Ewald. Grosse Forschungsreisende. München, 1933.

An 359.75.5 Banton, Michael P. The race concept. Newton Abbot, 1975.

Geog 3055.27 Baour-Lormian, Pierre Marie F.L. L'Atlantide. Paris, n.d.

An 359.63.35 Barabas, Jenó. Kartografiai modszei a neprajzban. Budapest, 1963.

Geog 4207.80.2 Baranshchikov, V. Neshchastiia prikliucheniia...Amerike, Azii i Evrope s" 1780 ro 1787 god". 2. izd. Sankt Peterburg, 1787.

Geog 500.75A Baranskii, N.N. Istoricheskii obzor uchebniko geografii, 1876-1934. Moskva, 1954.

Geog 600.10 Baranskii, N.N. Otechestvennye ekonomiko-geografy XVIII-XX vv. Moskva, 1957.

Geog 600.11 Baranskii, N.N. Otechestvennye fiziko-geografy i puteshestvenniki. Moskva, 1959.

Geog 4180.49 Barbaro, J. Travels to Tana and Persia. N.Y., 1964?

Geog 4208.48.5 Barbeida, Claudio Lagrange. Huma viagem de duas mil legoas. Nova-Goa, 1848.

Geog 4348.45 Barber, J. Overland guide-book. London, 1845.

Geog 4328.93 Barber, J.L. Mediterranean mosaics. N.Y., 1895.

Geog 6009.55.20 Barber, Noël. The white desert. N.Y., 1958.

Geog 3025.5 Barbier, J.V. Rapport sur les travaux cartographiques. Nancy, 1884.

Geog 4180.35 Barbosa, D. Description of...East Africa and Malabar. London, 1866.

Author and Title Listing

Geog 4181.36 Barbosa, Duarte. The book. London, 1918-21. 2v.
Geog 4181.136 Barbour, Philip L. The Jamestown voyages under the first charter, 1606-1609. Cambridge, Eng., 1969. 2v.
Geog 614.15 Barcinski, Florian. Stanislaw Nowakowski. Warszawa, 1965.
Geog 5650.15 Bárdrson, I. Det gamle Grønlands beskrivelse. København, 1930.
Geog 4309.44.5 Baretti, Giuseppe. Oga Magoga. Milano, 1944.
Geog 4110.30 Bargain paradises of the world. 4. ed. (Ford, Norman D.) Greenlawn, N.Y., 1957.
Geog 579.44 Bargilliot, A. Vocabulaire pratique anglais-français. Paris, 1944.
Geog 4208.72.7 Barker, Mary Ann B. Travelling about over new and old ground. London, 1872.
Geog 3070.36 Barker, W.H. The history of cartography. Manchester, 1926.
Geog 5900.16 Barkov, N.I. Desiat' let sovetskikh issledovanii v Antarktike. Leningrad, 1968.
Geog 4556.59 Barlow, E. Barlow's journal of his life at sea in king's ships, East and West Indiamen. London, 1934. 2v.
Geog 4181.69 Barlow, Roger. A brief summe of geographie. London, 1932.
Geog 4556.59 Barlow's journal of his life at sea in king's ships, East and West Indiamen. (Barlow, E.) London, 1934. 2v.
Geog 4228.89 Barneby, W.H. New Far West and old Far East. London, 1889.
Geog 4559.30.5 Barnes, William M. When ships were ships and not tin pots. N.Y., 1930.
An 2109.63 Barnouw, V. Culture and personality. Homewood, Ill., 1963.
Geog 4208.49.5 Barra, E.I. A tale of two oceans. San Francisco, 1893.
Geog 4228.73 Barra de Cobo, Maipina de la. Mis impresiones y mis vicisitudes en mi viaje a Europa. Buenos Aires, 1878.
Geog 500.105 Barras de Aragón, F. Los ultimos escritores de Indias. Madrid, 1949.
Geog 4129.37 Barraud, G. Touristes de jadis. Paris, 1937.
Geog 3530.9 Barré, H. Voyageurs et explorateurs provençaux. Marseille, 1905.
Htn Geog 4265.61* Barreiros, G. Chorographia. Coimbra, 1561.
Geog 3060.18 Barreto, Costa. A lenda das Sete Cidades. Porto, 1949.
Geog 4209.51.10 Barreto, Paulo. A alma encantadora das ruas. Rio de Janeiro, 1951.
Geog 4209.42.10 Barrett, C.L. On the Wallaby. Melbourne, 1942.
Geog 4249.50.5 Barrett, C.L. The Pacific. Melbourne, 1950?
An 2109.56 Barrière, Claude. Les civilisations tardenoisiennes en Europe occidentale. Bordeaux, 1956.
Htn Geog 5170.9* Barrington, D. Possibility of approaching the North Pole. London, 1818.
Htn Geog 5170.7* Barrington, D. Probability of reaching the North Pole. London, 1775.
Htn Geog 3570.11* Barrós, J. de. De alder erste scheepo-tog ten der Portugesen. Leyden, 1707?
An 3309.00 Barros Ovalle, P.N. Manual de antropometria. Santiago de Chile, 1900.
Geog 3055.51 Barroso, Gustavo. Aquem da Atlantida. São Paulo, 1931.
Geog 4157.66 Barrow, J. Abrégé...histoire des decouvertes. Paris, 1766. 12v.
Geog 4167.65 Barrow, J. A collection of authentic...voyages. v.2. London, 1765.
Geog 5208.18 Barrow, John. Chronological history of voyages into Arctic regions. London, 1818.
Geog 5208.46.3 Barrow, John. Voyages of discovery and research within the Arctic regions, from the year 1818 to present. London, 1846.
Geog 5208.46 Barrow, John. Voyages of discovery and research within the Arctic regions, from the year 1818 to present. London, 1846.
Geog 4218.97 Barrows, John D. A world-pilgrimage. Chicago, 1897.
An 2208.93 Bartels, Max. Die Medicin der Naturvölker. Leipzig, 1893.
An 3558.96 Barth, J. Norrønaskaller. Crania antiqua. Christiania, 1896.
Geog 4269.10.5 Bartholomew, J.G. Literary and historical atlas of Europe. London, 1910.
Geog 4269.23 Bartholomew, J.G. A literary and historical atlas of Europe. London, 1923.
Geog 4308.86.20 Bartlett, David L. Letters from Europe. Baltimore, 1886.
An 2109.23.7A Bartlett, F.C. Psychology and primitive culture. Cambridge, 1923.
An 2109.23.6 Bartlett, F.C. Psychology and primitive culture. N.Y., 1923.
An 98.47 Bartlett, J.R. The progress of ethnology. 2. ed. N.Y., 1847.
Geog 5559.13A Bartlett, Robert A. The last voyage of the Karluk. Boston, 1916.
Geog 5312.2A Bartlett, Robert A. The log of Bob Bartlett. N.Y., 1928.
Geog 4308.55.5 Bartol, C.A. Pictures of Europe...ideas. Boston, 1855.
Geog 4308.55.7 Bartol, C.A. Pictures of Europe...ideas. Boston, 1856.
Geog 4239.55 Barton, H.D.E. Atlantic adventurers. N.Y., 1955?
Geog 4239.50 Barton, H.D.E. Westward crossing. 1st American ed. N.Y., 1951.
An 359.46.5 Baruch, Dorothy W. Glasshouse of prejudice. N.Y., 1946.
Geog 4308.42.5 Baruffi, G.F. Pellegrinazioni autunnali ed opuscoli. Torino, 1842.
Geog 4349.07 Barzini, Luigi. Da Pechino a Parigi in sessanta giorni. Milano, 1970.
An 359.38 Barzun, Jacques. Race; a study in modern superstition. London, 1938.
An 359.65.25 Barzun, Jacques. Race; a study in superstition. N.Y., 1965.
An 359.37.9 Baschmakoff, A.A. Cinquante siècles d'évolution ethnique autour de la Mer noire. Paris, 1937.
Geog 500.119 A basic geographical library. (Church, Martha.) Washington, 1966.
An 359.02A Basis of social relations. (Brinton, D.G.) N.Y., 1902.
Geog 4309.66 Basnett, Fred. Travels of a capitalist lackey. South Brunswick, 1966.
Geog 4218.17.10 Bassett, M.M. Realms and islands. London, 1962.
Geog 4209.60.50 Basso, Hamilton. A quota of seaweed. Garden City, 1960.
An 2108.68 Bastian, A. Beiträge zur vergleichenden Psychologie. Berlin, 1868.
An 358.68.5 Bastian, A. Der Beständige in den Menschenrassen. Berlin, 1868.
An 358.71.7 Bastian, A. Ethnologische Forschungen und Sammlung von Material für die selben. Jena, 1871-73. 2v.
An 2109.01.5 Bastian, A. Der Menschheitsgedanke durch Raum und Zeit. Berlin, 1901.
An 98.81 Bastian, A. Vorgeschichte der Ethnologie. Berlin, 1881.
Geog 4218.73.20 Bastian, Adolf. Geographische und ethnologische Bilder. Jena, 1873.
An 2408.72 Bastian, P.W.A. Die Rechtsverhältnisse. Berlin, 1872.
An 358.81.7 Bastian, W.A. Der Völkergedanke. Berlin, 1881.
Geog 4279.11A Bates, Ernest S. Touring in 1600. Boston, 1911.
Geog 4308.91 Bates, J.H. Notes of foreign travel. N.Y., 1891.
Geog 4759.52 Bates, M. Where winter never comes. N.Y., 1952.
An 359.68.25 Bates, Marston. Gluttons and libertines; human problems of being natural. N.Y., 1968.
An 2109.36.5 Bateson, G. Naven; a survey of the problems suggested by a composite picture of the culture of a New Guinea tribe. Cambridge, Eng., 1936.
Geog 4181.6 Battell, A. Strange adventures of A. Battell. London, 1901.
An 3409.31.3 Battley, Harry. Single fingerprints. New Haven, 1931.
Geog 4219.60 Baty, Eben Neal. Citizen abroad. N.Y., 1960.
Geog 602.1 Baudet, P.J.H. Leven en werken van Williem Janszoon Blaeu. Utrecht, 1871.
Geog 577.01 Baudrand, M.A. Dictionaire geographique universal. Amsterdam, 1701.
Geog 576.82F Baudrand, M.A. Parisini geographia. Parisiis, 1681-82. 2v.
Geog 4129.71 Bauer, Hans. Wenn einer eine Reise tat. Leipzig, 1971.
Geog 3070.34 Baulig, Henri. Exercices cartographiques. Paris, 1927.
Geog 4145.39 Baum, Jiří. Holub a Mašukulumbové. Praha, 1955.
Geog 135.2.3 Baumann, Oskar. Der Sansibar-Archipel. v.3. Leipzig, 1896.
Geog 4219.29.5 Baus, Thomas J. 30,000 miles around the world. Philadelphia, 1929.
An 99.71 Bausinger, Hermann. Volkskunde: von der Altertumsforschung zur Kulturanalyse. Berlin, 1971.
An 359.56 Bavinck, J.H. Het rassenvraagstuk. Kampen, 1956.
An 2109.40.20 Bawmann, Evert Dirk. Historische Betrachtungen über das Koitus-Konzeption Problem. Arnhem, 1940.
Geog 4308.52.3 Baxter, W.E. The Tagus and the Tiber. London, 1852. 2v.
Geog 4308.49.10 Baxter, William E. Impressions of central and southern Europe. London, 1850.
An 379.53 Baxter, William J. Today's revolution in weather. N.Y., 1953.
Geog 4329.09 Bayne, S.G. A fantasy of Mediterranean travel. N.Y., 1909.
Geog 4309.61.15 Beadle, Muriel. These ruins are inhabited. N.Y., 1961.
An 359.32.5 Bean, R.B. The races of man. N.Y., 1932.
Geog 3235.58 Beans, G.H. A collection of maps. Jenkintown, Pa., 1943.
An 359.64 Beattie, John. Other cultures. N.Y., 1964.
Geog 4311.17 Beattie, William. The Danube, its history, scenery. London, 1844?
Geog 4311.20 Le beau Danube gris. (Blairy, Jean.) Paris, 1948.
Geog 659.71.15 Beaujeu-Garnier, Jacqueline. La géographie: méthods et perspectives. Paris, 1971.
Geog 4559.26.20 Beaumont, J.C.H. Ships and people. N.Y., 1926.
Geog 4218.70.8 Beaumont, Thomas E. Pencilling by the way: a constitutional voyage round the world: 1870 and 1871. London, 1971.
Geog 4418.70 Beauvior, L. Voyage round the world. London, 1870. 2v.
Geog 5700.9 Beauvois, E. Origines et fondation du plus ancien évêché du nouveau monde. Paris, 1878.
Geog 3228.97A Beazley, C.R. Dawn of modern geography. London, 1897. 3v.
Geog 3228.97.2 Beazley, C.R. Dawn of modern geography. London, 1897-1906. 3v.
Geog 4521.25 Because it is there. (Lowe, George.) London, 1959.
Geog 4219.29.15 Because I've not been there before. (Lewis, Aswold.) London, 1929.
Geog 5839.74 Bechmann, Elke. Mennesker og miljøer i Grønland. København, 1974.
Geog 3540.11 Beck, Carl. Deutsches Reisen im Wandel. Berlin, 1936.
Geog 4218.48 Beck, Christian. Reise um die Welt. 11e Aufl. Dresden, 1907.
An 358.99.3 Beck, G. Der Urmensch. Basel, 1899.
Geog 3019.73.15 Beck, Hanno. Geographie; europäische Entwicklung in Texten und Erläuterungen. Freiburg, 1973.
Geog 4307.80.5 Beckford, William. Italy, Spain and Portugal. London, 1840.
Geog 4307.80.7 Beckford, William. Italy, Spain and Portugal. Pt.1-2. N.Y., 1845.
Geog 4307.80.3 Beckford, William. Italy, with sketches of Spain and Portugal. Philadelphia, 1834.
Geog 4307.80.15 Beckford, William. The travel-diares of William Beckford of Fonthill. Cambridge, 1928. 2v.
Geog 4181.107 Beckingham, C.F. Some records of Ethiopia. London, 1954.
Geog 3019.47 Beckman, Leif. Vår väg genom världen. Stockholm, 1947-51. 3v.
Geog 500.11 Beckmann, J. Literatur der...Reisebeschreibungen. Göttingen, 1808-09. 2v.
Geog 4219.35 Beckoning horizon. (Benn, Wedgwood.) London, 1935.
Geog 4209.38.21 Beddington, C. We sailed from Brixham. London, 1938.
An 358.93.1 Beddoe, J. Anthropological history of Europe. London, 1893.
Geog 4219.58.10 Bedford, Jimmy. Around the world on a nickel. New Delhi, 1967.
Geog 4218.76.9 Bedinello, Ugo. Diario del viaggio intorno al globo. Trieste, 1876.
Geog 5398.25 Beechey, F.W. Narrative of voyage to the Pacific. London, 1831. 2v.
Geog 5378.18 Beechey, F.W. A voyage of discovery towards the North Pole in H.M. ships Dorothea and Trent...1818. London, 1843.
Geog 4228.85 Beehler, W.H. The cruise of the Brooklyn. Philadelphia, 1885.
Geog 4679.12.2 Beesley, Lawrence. The loss of the S.S. Titanic. Boston, 1912.
Geog 4679.12A Beesley, Lawrence. The loss of the Titanic. Boston, 1912.
Geog 5530.7 Beesly, A.H. Sir John Frankin. N.Y., 1881.
Geog 578.68 Beeton, S.O. Dictionary of geography. London, 1868.
Geog 4558.56.5 Before the mast in clippers...1856 to 1860. (Abbey, Charles A.) N.Y., 1937.
Geog 4477.59 Begebenheisen...Sudlandern. (Brachfelds, J.M.) Eisenach, 1759.
Geog 4309.30.25 Beggaro abroad. (Tully, J.) Garden City, 1930.
An 2109.53.10 Der Beginn der Weltgeschichte. (Kern, Fritz.) Bern, 1953.
An 2109.50 The beginnings of diplomacy. (Numelin, R.) London, 1950.
Geog 5399.28.25 Běhounek, Franz. Sieben Wochen auf der Eisscholle. Leipzig, 1929.
Geog 4217.37 Behrens, E.F. Sud-Lander und um die Balt. Leipzig, 1737.
Geog 3251.21 Behrmann, W. Über die niederdeutschen Seebücher. Hamburg, 1906.
Geog 5850.5 Bei den Eskimos in Westgrönland. (Trebitsch, R.) Berlin, 1910.

Author and Title Listing

Call Number	Entry
Geog 199.1.3	Beilagen zum Jahrbuch. (Schweizer Alpenclub.) Bern. 1-46 46v.
Geog 3590.67	Beiträge aus der sowjetische Kartographie. (Gesellschaft für Deutsch-Sowjetische Freundschaft.) Berlin, 1953.
An 3209.12	Beiträge zur...der Negerweichteile. (Loth, E.) Stuttgart, 1912.
An 379.22.9	Beiträge zur Anthropogeographie des Bodenseegebiets. Inaug. Diss. (Kaltenbach, Ernst.) Basel, 1922.
Geog 135.2.6	Beiträge zur Biogeographie und Morphologie der Alpen. v.6. Leipzig, 1904.
Geog 665.148	Beiträge zur Geographie der Tropen und Subtropen; Festschrift zum 60. Geburtstag von Herbert Wilhelmy. Tübingen, 1970.
Geog 135.2	Beiträge zur Geographie des festen wassers. v.1. (Gesellschaft für Erdkunde zu Leipzig.) Leipzig, 1891.
Geog 135.2.4	Beiträge zur Geographie des mittleren Deutschland. v.4. (Ratzel, Friedrich.) Leipzig, 1899.
An 170.129	Beiträge zur Gesellungs- und Völkerwissenschaft. Berlin, 1950.
Geog 3229.29	Beiträge zur historischen Geographie, Kulturgeographie, Ethnographie und Kartographie. (Mžik, Hans von.) Leipzig, 1929.
An 359.65.10	Beiträge zur menschlichen Typenkunde. (Jürgens, Hans W.) Stuttgart, 1965.
An 358.92.3	Beiträge zur physischen Anthropologie der Bayern. (Ranke, Johannes.) München, 1892.
An 3558.77	Beitraege zur physischen Anthropologie der Deutschen. (Virchow, R.) Berlin, 1877.
An 2108.68	Beiträge zur vergleichenden Psychologie. (Bastian, A.) Berlin, 1868.
Geog 5855.9	Beiträge zur Vorgeschichte. (Solberg, O.) Christiania, 1907.
An 3559.09	Ein Beitrag zur Anthropologie des "Homo Alpinus Tirolensis". (Frizzi, E.) Wien, 1909.
An 2109.27.21	Ein Beitrag zur Wirtschaftsgeschichte der Vorzeit. Inaug. Diss. (Fahl, Anton.) Halle, 1927.
Geog 5837.77	Beknopt en getrouw verhaal. (Groot, J.J.) Amsterdam, 1779.
Geog 5538.52	Belcher, E. The last of the Arctic voyages. London, 1855. 2v.
Geog 4218.35	Belcher, Edward. Narrative of a voyage round the world...1836-42. London, 1843. 2v.
Geog 3510.11	Belen, Willem Julius. Nederlands voorhoede. 2. druk. Amsterdam, 1946.
Geog 4219.37	Belfrage, C. Away from it all. N.Y., 1937.
Geog 1558.10	Belgien. 5. Aufl. (Baedeker, publishers.) Coblenz, 1855.
Geog 1555.43	Belgien und Holland. 18. Aufl. (Baedeker, publishers.) Leipzig, 1888.
Geog 1555.29	Belgien und Holland. 23. Aufl. (Baedeker, publishers.) Leipzig, 1904.
Geog 1555.10	Belgique et Hollande. 4. éd. (Baedeker, publishers.) Coblenz, 1866.
Geog 1555.15	Belgique et Hollande. 6. éd. (Baedeker, publishers.) Coblenz, 1871.
Geog 1555.18	Belgique et Hollande. 8. ed. (Baedeker, publishers.) Leipzig, 1875.
Geog 1558.20	Belgique et Luxembourg. 20. éd. (Baedeker, publishers.) Leipzig, 1928.
Geog 4308.48	Belgium, the Rhine, Switzerland. (Buckingham, J.S.) London, 1848. 2v.
Geog 1555.13	Belgium and Holland. (Baedeker, publishers.) Coblenz, 1869.
Geog 1555.16	Belgium and Holland. 2. ed. (Baedeker, publishers.) Coblenz, 1871.
Geog 1555.17	Belgium and Holland. 3. ed. (Baedeker, publishers.) Leipsic, 1874.
Geog 1555.27	Belgium and Holland. 5. ed. (Baedeker, publishers.) Leipsic, 1878.
Geog 1555.21	Belgium and Holland. 5. ed. (Baedeker, publishers.) Leipsic, 1878.
Geog 1555.20	Belgium and Holland. 5. ed. (Baedeker, publishers.) Leipsic, 1878.
Geog 1555.30	Belgium and Holland. 6. ed. (Baedeker, publishers.) Leipsic, 1881.
Geog 1555.36	Belgium and Holland. 7. ed. (Baedeker, publishers.) Leipsic, 1884.
Geog 1555.35	Belgium and Holland. 7. ed. (Baedeker, publishers.) Leipsic, 1884.
Geog 1555.38	Belgium and Holland. 8. ed. (Baedeker, pbulishers.) Leipsic, 1885.
Geog 1555.40	Belgium and Holland. 9. ed. (Baedeker, publishers.) Leipsic, 1888.
Geog 1555.45	Belgium and Holland. 10. ed. (Baedeker, publishers.) Leipsic, 1891.
Geog 1555.50	Belgium and Holland. 11. ed. (Baedeker, publishers.) Leipsic, 1894.
Geog 1555.51	Belgium and Holland. 11. ed. (Baedeker, publishers.) Leipsic, 1894.
Geog 1555.55	Belgium and Holland. 12. ed. (Baedeker, publishers.) Leipsic, 1897.
Geog 1555.60	Belgium and Holland. 13. ed. (Baedeker, publishers.) Leipsic, 1901.
Geog 1555.65	Belgium and Holland. 14. ed. (Baedeker, publishers.) Leipzig, 1905.
Geog 1555.70	Belgium and Holland. 15. ed. (Baedeker, publishers.) Leipzig, 1910.
Geog 1558.30	Belgium and Luxemburg. 16. ed. (Baedeker, publishers.) Leipzig, 1931.
Geog 85.377	Belgrade. Geografski Institut. Zbornik radova. 18+ 5v.
Geog 85.384	Belgrade. Univerzitet. Prirodno-Matematički Fakultet. Geografski Zavod. Zbornik radova. Beograd. 13,1966+ 2v.
Geog 3590.15.10	Belikie russkie iute estvenniki. (Berg, L.S.) Moskva, 1950.
Geog 3590.30.15	Belikii russkii moreplavateli A.I. Chirukov. (Divin, Vasilii A.) Moskva, 1953.
Geog 5530.13	Bell, Benjamin. Lieut. John Irving, of HMS "Terror". Edinburgh, 1881.
Geog 808.36	Bell, J. System of geography. Glasgow, 1836. 6v.
Geog 808.36.5	Bell, J. System of geography. London, 1850. 6v.
Geog 4309.02.20	Bell, L.L. Abroad with the Jimmies. Boston, 1902.
Geog 4309.58	Bell, Robert. By road to Turkey. London, 1958.
Geog 4218.68.5	Bell, William M. Other countries. London, 1872. 2v.
Geog 4709.40	The Bellamy treasure. (Hawes, Hildreth Gilman.) Augusta, Maine, 1940.
Geog 4167.07.2	Bellegarde, J.B.M. General history of all voyages...old and new world. London, 1708.
Geog 612.3.8	Bellessort, André. La Pérouse. Paris, 1926.
Geog 4309.02.4	Belloc, H. The path to Rome. N.Y., 1902.
Geog 4309.02.3A	Belloc, H. The path to Rome. 4th ed. London, 1916.
Geog 4359.28	Belloc, Hilaire. Many cities. London, 1928.
Geog 4309.41.10	Belloc, Hilaire. Places. N.Y., 1941.
Geog 4359.27	Belloc, Hilaire. Towns of destiny. N.Y., 1931.
Geog 5538.51.7	Bellot, J.R. Journal d'un voyage aux mers polaires...en 1851 et 1852. Paris, 1854.
Geog 5538.51.9	Bellot, J.R. Memoirs of and journal of voyage...search of Sir John Franklin. London, 1855. 2v.
Geog 4308.68.7	Bellows, H.W. The old world in its new face. N.Y., 1868. 2v.
Geog 4181.14	Belmonte Bermudez, L. de. Voyages of Pedro Fernandez de Quiros. London, 1904. 2v.
Geog 5399.57.25	Belov, M.I. Severnyi morskoi put. Leningrad, 1957.
Geog 3019.65	Belov, Mikhail I. Puteshestviia i geograficheskia otkrytiia v XV-XIX vv. Leningrad, 1965.
Geog 5919.38	Below the roaring forties. (Ommaney, Francis D.) London, 1938.
Geog 4509.23.10	Below the snow line. (Freshfield, D.M.) London, 1923.
Geog 4209.10	Beltramelli, Antonio. Il diario di un viandante, dal deserto al mar glaciale. Milano, 1911.
Geog 759.19	Beltran, Juan G. Lo inerte y lo vital. Buenos Aires, 1919.
Geog 819.00	Beltrán y Rózpide, R. La geografía en 1898. Madrid, 1899.
Geog 4269.19	Beltrán y Rózpide, Ricardo. Nuevas nacionalidades en Europa. 2a ed. Madrid, 1919.
Geog 3228.76	Beltrán y Rózpide, Ricardo. Viajes y descubrimientos, efectuados en la edad media. Madrid, 1918.
Geog 4348.60.10	Beluze, X. Les pérégrinations en Orient et en Occident. 21e éd. Citeaux, 1883. 2v.
Geog 6009.63	Belye suy. (Peskov, Vasilii M.) Moskva, 1965.
Geog 4309.48.5F	Bemelmans, Ludwig. The best of times, an account of Europe revisited. N.Y., 1948.
Geog 4307.87.5	Bemerkungen auf einer Reise durch Flandern. (Walker, Adam.) Berlin, 1791.
Htn Geog 4218.12*	Bemerkungen auf einer Reise um die Welt in den Jahren 1803-1807. (Langsdorff, G.H. von.) Frankfurt, 1812. 3v.
Geog 4759.71.10	Benchétrit, Maurice. Géographie zonale des régions chaudes. Paris, 1971.
Geog 4209.64.10	Benchley, P. Time and a ticket. Boston, 1964.
An 359.48.10	Bendict, Ruth (Fulton). In Henry's backyard. N.Y., 1947.
Geog 5839.27.5	Bendixen, Ole. Grønland som Nybyggerland. København, 1927.
Geog 5670.3	Bendixen, Ole. Grønlandstraktaten, belyst gennem de i dagspressen fremkomke artikler. København, 1924.
Geog 4309.30.15	Benedict, C.W. (Mrs.). "The Benedicts abroad". London, 1930.
Htn Geog 4308.60*	Benedict, E.C. A run through Europe. N.Y., 1860.
An 126.6.5A	Benedict, Ruth (Fulton). An anthropologist at work. Boston, 1959.
An 2109.34.5A	Benedict, Ruth (Fulton). Patterns of culture. Boston, 1934.
An 359.43.25	Benedict, Ruth (Fulton). Race; science and politics. N.Y., 1943.
An 359.40	Benedict, Ruth (Fulton). Race: science and politics. N.Y., 1940.
An 359.59.45	Benedict, Ruth (Fulton). Race: science and politics. N.Y., 1959.
An 359.59.50	Benedict, Ruth (Fulton). Race and racism. London, 1959.
Geog 4309.30.15	"The Benedicts abroad". (Benedict, C.W. (Mrs.).) London, 1930.
Geog 4341.73.16	Beniamini Tudelensis Itinerarium; ex versione Benedicti Ariae Montani. (Benjamin ben Jonah, of Tudela.) Bononiae, 1967.
Geog 4238.78	Benjamin, S.G.W. Atlantic Islands. N.Y., 1878.
Geog 4341.73.16	Benjamin ben Jonah, of Tudela. Beniamini Tudelensis Itinerarium; ex versione Benedicti Ariae Montani. Bononiae, 1967.
Htn Geog 4341.73.3*	Benjamin ben Jonah, of Tudela. Itinerarium Benjaminis. Lugdunum Batavorum, 1633.
Geog 4341.73.10A	Benjamin ben Jonah, of Tudela. The itinerary of Benjamin of Tudela. London, 1907.
Geog 4341.73.5	Benjamin ben Jonah, of Tudela. The itinerary of Rabbi Benjamin of Tudela. pt.1-2. London, 1840.
Geog 4341.73.6	Benjamin ben Jonah, of Tudela. Die Reisebeschreibungen. Jerusalem, 1903.
Geog 4341.73.15	Benjamin ben Jonah, of Tudela. Viajes de Benjamin de Tudela, 1160-1173. Madrid, 1918.
Geog 4219.35	Benn, Wedgwood. Beckoning horizon. London, 1935.
Geog 4328.61	Bennet, J.H. Winter and spring on the shores of the Mediterranean. 4th ed. London, 1870.
Geog 4168.29	Bennet, Roelof Gabriel. Nederlandsch zeeseizen in het laatst der z estiende. v.1-4. Wijk, 1828-29. 3v.
Geog 4218.81.9	Bennett, D.M. A truth seeker around the world. N.Y., 1882. 4v.
Geog 4679.17	Benson, N.P. The log of the El Dorado. San Francisco, 1917.
Geog 4209.28.10	Benson, Stella. Worlds within worlds. London, 1928.
Geog 4180.87A	Bent, J.T. Early voyages and travels in the Levant. London, 1893.
Geog 4329.58.10F	Beny, Roloff. Merveilles de la Méditerranée. Paris, 1958.
Geog 4329.58F	Beny, Roloff. The thrones of earth and heaven. London, 1958.
Geog 4180.21	Benzoni, G. History of the new world. N.Y., 1964?
Geog 4308.20	Beobachtungen auf Reisen in und ausser Deutschland. v.2-4. (Niemeyer, A.H.) Halle, 1822-1826 4v
Geog 4209.52.5	Beonio-Brocchieri, V. Il Marcopolo. Milano, 1952.
Geog 3235.27F	Berchet, G. Il planisfero di Giovanni Leardo. Venezia, 1880.
Geog 3235.39	Berchet, G. Portolani, esistenti nelle...bibliotheque. Venezia, 1866.
Geog 807.29	Berckenmeier, P. Le curieux antiquaire. Leide, 1729. 3v.
Geog 4314.7	Berega Chennogo i Azovskogo morei. (Zenkovich, V.P.) Moskva, 1958.
Geog 4157.88	Berenger, J.P. Collection de tous les voyages. Lausanne, 1788-91. 9v.
Geog 5558.97	Berens, S.L. The "Fram" expedition. Nansen in the frozen world. Philadelphia, 1897.
Geog 4218.45	Beretning om corvetten Galathea's reise omkring jorden i 1845. 2. udg. (Bille, C.S.A.) Kjøbenhavn, 1853.
Geog 3590.15.10	Berg, L.S. Belikie russkie iute estvenniki. Moskva, 1950.
Geog 3590.15.15	Berg, L.S. Geschichte der russischer geographischer Entdeckungen. Leipzig, 1954.
Geog 3590.15.20	Berg, L.S. Izbrannye trudy. Moskva, 1956. 5v.

Author and Title Listing

	Geog 3590.15	Berg, L.S. Ocherki po istorii russk. geogr. otkrytii. Moskva, 1946.	Htn	Geog 4477.08*	Beschryoinge van het magtig. (Posos, Juan de (pseud.).) Amsterdam, 1708.

Geog 3590.15 — Berg, L.S. Ocherki po istorii russk. geogr. otkrytii. Moskva, 1946.
Geog 3590.15.2 — Berg, L.S. Ocherki po istorii russk. geogr. otkrytii. 2. izd. Moskva, 1949. 2v.
Geog 3590.15.40 — Berg, R.L. Po ozeram Sibiri i Srednei Azii. Moskva, 1955.
Geog 4502.5.6 — Der Berg des Herzens. (Schmidkunz, W.) München, 1930.
Geog 28.6 — Berg und Buch. München. 1,1928
Geog 4559.54.5 — Berge, Victor. Danger is my life. London, 1954.
Geog 4502.10.15 — Berge der Welt. Zürich. 1946-1957 10v.
Geog 4502.5.2 — Berglieder der Völker. (Rohrer, M.) München, 1928.
An 2109.58.10 — Bergoonlour, F.M. La préhistoire et ses problèmes. Paris, 1958.
Geog 3055.84 — Bergquist, Nils Olof. Ymdogat-Atlantis. Solna, 1971.
Geog 4508.93.10 — Der Bergsteiger in Hochgebirge. (Meurer, Julius.) Wien, 1893.
Geog 3590.20 — Bering's successors, 1745-1780. (Pallas, P.S.) Seattle, 1948.
Geog 5312.7.10A — Bering's voyages. (Golder, Frank A.) N.Y., 1922-25. 2v.
Geog 4477.51.15 — Berington, S. The adventures of Gaudentio di Lucca. London, 1763.
Geog 4477.51.25 — Berington, S. The adventures of Gaudentio di Lucca. London, 1850.
Htn Geog 5557.25.5* — Berkh, V.N. Pervoe morskoe puteshestvie. Sankt Peterburg, 1823.
Geog 5645.5 — Berlin, Knud. Denmark's to Greenland. London, 1932.
Geog 5639.28 — Berlin, Knud. Grønlands betydning. København, 1928.
Geog 28.3.40 — Berlin. Freie Universität. Geographisches Institut. Abhandlungen. 1-16 8v.
NEDL Geog 28.2.5 — Berlin. Gesellschaft für Erdkunde. Übersicht der Aufsatze...in den Monatsberichten über die Verhandlungen. Berlin. 1863
NEDL Geog 28.3 — Berlin. Gesellschaft für Erdkunde. Übersicht der Aufsatze. Zeitschrift. Berlin. 1884-1921 15v.
NEDL Geog 28.2 — Berlin. Gesellschaft für Erdkunde. Verhandlungen. Berlin. 1-27,1873-1896 25v.
Geog 28.3.50 — Berlin. Universität. Verein der Studierenden der Geographie. Mitteilungen. Berlin. 1-2,1915-1918
Geog 1526.40 — Berlin. 23. Aufl. (Baedeker, publishers.) Freiburg, 1964.
Geog 1526.20 — Berlin and its environs. (Baedeker, publishers.) Leipsic, 1903.
Geog 1526.25 — Berlin and its environs. 2. ed. (Baedeker, publishers.) Leipsic, 1905.
Geog 1526.30 — Berlin and its environs. 3. ed. (Baedeker, publishers.) Leipsic, 1908.
Geog 1526.35 — Berlin and its environs. 5. ed. (Baedeker, publishers.) Leipzig, 1912.
Geog 1526.36 — Berlin and its environs. 6. ed. (Baedeker, publishers.) Leipzig, 1923.
Geog 1526.10 — Berlin und Umgebungen. 7. Aufl. (Baedeker, publishers.) Leipzig, 1891.
Geog 1526.12 — Berlin und Umgebungen. 8. Aufl. (Baedeker, publishers.) Leipzig, 1894.
Geog 1526.17 — Berlin und Umgebungen. 11. Aufl. (Baedeker, publishers.) Leipzig, 1900.
Geog 28.3.60 — Berliner geographische Arbeiter. Berlin.
Geog 4659.74 — Berlitz, Charles F. The Bermuda triangle. 1st ed. Garden City, 1974.
Geog 3055.82 — Berlitz, Charles Frambach. The mystery of Atlantis. N.Y., 1969.
Geog 4659.74 — The Bermuda triangle. 1st ed. (Berlitz, Charles F.) Garden City, 1974.
Geog 4659.75 — The Bermuda triangle mystery - solved. (Kusche, Lawrence D.) N.Y., 1975.
Geog 3240.5 — Bernard, A. De Adamo Bremensi geographo. Parisiis, 1895.
Geog 4157.31 — Bernard, I.F. Recueil de voyages au nord. Amsterdam, 1731-37. 9v.
Geog 28.1 — Berne. Geographische Gesellschaft. Jahresbericht. Bern. 1,1879+ 14v.
Geog 4219.36.15 — Bernicot, Louis. The voyage of Anahita; single-handed round the world. London, 1953.
Geog 4502.5.12 — Die Bernina. 1786. (Tscharner, J.B. von.) München, 1933.
Geog 613.4.15 — Bernstein, H. Sir Clements R. Markham as a translator. n.p., 1937?
Geog 3570.9.5 — Bersaude, J. The attacks against Portuguese history. Lisbon, 1950.
Geog 3570.9 — Bersaude, J. Les légendes allemandes. Genève, 1917-20. 2v.
Geog 28.7 — Bersteiger Almanach. München.
Geog 3550.11 — Bertacchi, C. Geografi ed exploratori italiani contemporanei. Milano, 1929.
Geog 4268.30 — Bertaut, Jules. Villégiatures romantiques. Paris, 1927.
Geog 4300.50 — Bertelsmann, C. Ferien einmal anders. Gütersloh, 1965.
An 3309.09 — Bertillon, A. Anthropologie métrique. Paris, 1909.
An 3308.85 — Bertillon, A. Identification anthropometrique. Melun, 1885.
An 358.82.3 — Bertillon, A. Les races sauvages. Paris, 1882.
An 126.4 — Bertillon, Suzanne. Vie d'Alphonse Bertillon, inventeur de l'anthropométrie. Paris? 1941.
Geog 4559.70 — Bertino, Serge. Guide de la mer mystérieuse. Paris, 1970.
Htn Geog 816.40* — Bertius, Petrus. Livre premier [-septiesme] des tables geographiques auquel est traité du monde en general. n.p., n.d.
Geog 4219.11.5 — Bertram, C. A magician in many lands. N.Y., 1911.
An 359.59.15 — Bertram, Colin. Adam's brood. London, 1959.
Geog 5120.1 — Bertram, Colin. Arctic and Antarctic. Cambridge, 1939.
Geog 5919.58.5 — Bertram, George Colin L. Antarctica today and tomorrow. Cambridge, 1958.
Geog 4219.08.10 — Bertrand, A. Quelques notes sur les conférences de Tokyo et de Shanghai. Genève, 19- .
Geog 4145.7 — Bertrand, Alfred. Alfred Bertrand. London, 1926.
Geog 5970.15.20 — Bertrand, Kenneth John. Americans in Antarctica, 1775-1948. N.Y., 1971.
Geog 4138.82.5 — Berühmte Seeleute. (Werner, R.) Berlin, 1882.
Geog 4209.65.20 — Besa-brodski dnevnik. (Horvat, Joža.) Zagreb, 1973.
Geog 4325.73 — Beschreiben der Reisen des Reinhold Lubenau. (Lubenau, R.) Königsberg, 1912-1920. 3v.
Geog 5837.20.5 — Beschreibung des...Wallfischfangs. (Zorgdrager, C.G.) Nürnberg, 1750.
Geog 4306.83.2 — Beschreibung einer neue Reise durch das Nordwestliche Deutschland nach den Niederländen. (Melle, Jakob von.) Lübeck, 1891.
Geog 4307.76 — Beschreibung seiner Reisen durch Frankreich. Pt.1. (Sander, Heinrich.) Leipzig, 1783.
Geog 5837.63 — Beschreibung und Natur-Geschichte. (Egede, H.) Berlin, 1763.
Geog 4182.5 — Beschrijving...van het Gout Koninckrijk van Gunea. (Marees, P. de.) 's-Gravenhage, 1912.

Htn Geog 4477.08* — Beschryoinge van het magtig. (Posos, Juan de (pseud.).) Amsterdam, 1708.
Geog 759.72.5 — Besedy v sovremennoi fisicheskoi geografii. (Predsrazhenskii, Vladimir S.) Moskva, 1972.
Geog 4308.19 — Beskow, B. Wandrings minnen. Stockholm, 1833-1834. 2v.
Geog 5838.25 — Beskrivelse...Grønland. (Graah, W.A.) Kjøbenhavn, 1825.
Geog 5700.11 — Beskrivelse over missionerne i Grønlands Søndre Distrikt. (Thorhallason, Egiel.) København, 1914.
An 2109.66.30 — Bessaignet, Pierre. Principes de l'ethnologie économique. Paris, 1966.
An 3558.81 — Bessel Hagen, F.K. Zur Kritik und Verbesserung der Winkelmessungen am Kopfe. Königsberg, 1881.
Geog 3055.52 — Bessmertny, A. Das Atlantisrätsel. Leipzig, 1932.
Geog 4709.28.5 — Besson, Maurice. Les frères de la coste, flibustiers et corsaires. Paris, 1928.
Geog 4180.38 — Best, G. Three voyages of Martin Frobisher. London, 1867.
Htn Geog 5515.77.5* — Best, George. De Martini Forbisseri Angli. n.p., 1580.
Geog 5515.77.15 — Best, George. The three voyages of Martin Frobisher in search of a passage to Cathay...1576-78. London, 1938. 2v.
Geog 4309.48.5F — The best of times, an account of Europe revisited. (Bemelmans, Ludwig.) N.Y., 1948.
An 358.68.5 — Der Beständige in den Menschenrassen. (Bastian, A.) Berlin, 1868.
Geog 4308.70.10 — Beste, John R.D. Nowadays. London, 1870. 2v.
Geog 4139.25A — Beston, Henry B. The book of gallant vagabonds. N.Y., 1925.
Geog 5670.2 — Betaenkning afgivet af dit i december maaned 1920 nedsatte udvalgtel drøftelse af de Grønlandski anliggender. København, 1921.
Geog 4308.41 — Betraktelser under en resa. (Lindeberg, A.) Stockholm, 1841.
Geog 3019.60.5F — Bettex, A.W. Welten der Entdecker. München, 1960.
Geog 4679.52 — Bevan, David. Drums of the Birkenhead. Capetown, 1972.
Geog 3235.5 — Bevan, W.L. Mediaeval geography. London, 1873.
Geog 758.79 — Bevan, W.L. Students' manual of modern geography mathematical, physical and descriptive. London, 1879.
An 379.35.3 — Bews, J.W. Human ecology. London, 1935.
Geog 5509.15 — Beyl, Hermann. Der lange Zeit angenomme Zusammenhang der Hudsonbai mit dem Stille Ozean. Würzburg, 1915.
Geog 4145.84.5 — Beyond Euphrates. 1. ed. (Stark, Freya.) London, 1951.
Geog 5070.55 — Beyond horizons. (Ellsworth, L.) N.Y., 1938.
Geog 4559.53 — Beyond horizons. 1. ed. (Mitchell, C.) London, 1953.
An 2159.73 — Beyond Stonehenge. 1st ed. (Hawkins, Gerald Stanley.) N.Y., 1973.
An 2109.49 — Beyond the bonds of history. (Breuil, H.) London, 1949.
Geog 4219.59 — Beyond the west horizon. (Hiscock, Eric C.) London, 1963.
Geog 5939.46 — Bezemer, K.W.L. Der Kampf um den Südpol. 2. Aufl. Zürich, 1952.
An 359.72.25 — Bezerra, Felte. Antropologia sociocultural. Brasilia, 1972.
Geog 3070.165 — Die Beziehungsgrundlage in thematischen Karten in ihrem Verhältnis zum Kartengegenstand. (Kremling, Helmut.) München, 1970.
An 359.59.10 — Bibby, Cyril. Race, prejudice and education. London, 1959.
An 132.2 — Bibliografía del Dr. Ales Hrdlicka. (Rubín de la Borbolla, D.) México, 193-.
Geog 500.22 — Bibliografia di viaggiatori italiani ordinata cronologicamenti. (Amat di San Filippo, Pietro.) Roma, 1874.
Geog 500.95 — Bibliografia geografii polskiej. Warszawa. 1936+ 8v.
Geog 3590.85 — Bibliograficheskii ocherk istorii geografii v Rossii XVIII veka. (Tikhomirov, Georgii S.) Moskva, 1968.
Geog 4205.98.10 — Bibliographie...die voyagie om den geheelen werelt...door Olivier van Noort. (Nijhoff, Wouter.) 's-Gravenhage, 1926.
Geog 5900.15 — Bibliographie antarctique. (Denucé, J.) Bruxelles, 1913.
Geog 3055.48 — Bibliographie de l'Atlantide et des questions connexes. (Gattefossé, Jean.) Lyon, 1926.
Geog 500.160 — Bibliographie der geographischen Literatur in deutscher Sprache. 1. Aufl. (Arnim, Hlmuth.) Baden Baden, 1970.
Geog 5600.2 — Bibliographie des ouvrages norvégiens relatifs au Groenland. (Vartdal, Hroar.) Oslo, 1935.
Geog 5150.5 — Bibliographies on the Arctic. (Dutilly, A.) Washington, 1945.
An 5.7 — Bibliography of anthropology and folk-lore. London.
Geog 500.150 — A bibliography of dissertations in geography, 1901 to 1969; American and Canadian universities. (Browning, Clyde Eugene.) Chapel Hill, 1970.
An 132.4.5 — Bibliography of Melville J. Herskovits, 1920-1962. (Northwestern University, Evanston, Ill. Program of African Studies.) Evanston, Ill., 1962?
Htn Geog 4679.26* — Biblioteca Titanicana. (Chapin, H.M.) Metuchen, 1926.
Geog 4679.26.2 — Biblioteca Titanicana. (Chapin, H.M.) Metuchen, 1926.
Geog 500.60 — Bibliotheca asiatica et africana. pt.4-5. (Maggs Bros., London.) London, 1929.
Geog 500.19 — Bibliotheca geographica. (Engelmann, W.) Leipzig, 1858.
Geog 30.2 — Bibliotheca geographica. (Paschin, O.) Berlin. 1-19,1891-1912 19v.
Geog 500.17 — Bibliothèque asiatique et africaine. (Ternaux-Compaus, H.) Paris, 1841.
Geog 500.17.2 — Bibliothèque asiatique et africaine. Notes. (Ternaux-Compaus, H.) Paris, 1841.
Geog 500.7 — Bibliothèque universelle des voyages. (Boucher, G.) Paris, 1808. 6v.
An 359.72 — Bicchieri, M.G. Hunters and gatherers today. N.Y., 1972.
Geog 5700.7 — Bidrag til Hans Egede...missions. (Fenger, H.M.) Kjøbenhavn, 1879.
Geog 4219.09 — Bidwell, Daniel D. As far as the East is from the West. Hartford, 1910.
An 3309.32.5 — Biedermann, Ernst. Körperform und Leistung sechzehnjähriger Lehslinge. Zürich, 1932.
An 5.20 — Biennial review of anthropology. Stanford. 1,1959+ 6v.
Geog 4309.02.7 — Bierbaum, Otto Julius. Mit der Kraft; Automobilia. Berlin, 1906.
Geog 4309.02.13 — Bierbaum, Otto Julius. Die Yankeedoodle-Fahrt und andere Reisegeschichten. München, 1920.
Geog 5235.25 — The big nail; the story of the Cook-Peary feud. (Wright, Theon.) N.Y., 1970.
Geog 4311.9 — Bigelow, P. Paddles and politics down the Danube. London, 1892.
Geog 808.11 — Bigland, J. Geographical...view of the world. Boston, 1811. 5v.
Geog 13.201 — Bijbladen. (Nederlandsch Aardrijkskundig Genootschap, Amsterdam.) Amsterdam. 1879-1883 3v.

Author and Title Listing

Htn	Geog 578.71*A	The Bijou gazetteer of the world. (Rosser, William H.) London, 1871.
	Geog 4515.104.5	Bilgeri, George. Colonel Bilgeri's handbook on mountain ski-ing. London, 1929.
	Geog 4218.45	Bille, C.S.A. Beretning om corvetten Galathea's reise omkring jorden i 1845. 2. udg. Kjøbenhavn, 1853.
	Geog 4309.74.5F	Billeder og stemninger oplevet af Eva Bendix. (Europas Hovedstaeder.) København, 1974.
	Geog 5970.37	Billing, Graham. South: man and nature in Antarctica. Wellington, 1964.
	An 130.1	Biographical memoir of J.W. Fawkes, 1850-1930. (Hough, Walter.) Washington, 1932.
	Geog 4209.57.5	Biographien und Autogramme beruhmter Staatsmanner und anderer Personlichkeiten des 20. Jahrhundert. (Raitchewitch, M.) Bielefeld, 1957.
	Geog 5321.1.3	Biography of Elisha Kent Kane. (Elder, William.) Philadelphia, 1858.
	Geog 5321.1.4	Biography of Elisha Kent Kane. (Elder, William.) Philadelphia, 1858.
	An 359.62.30	The biology of the race problem. (George, W.C.) N.Y., 1962.
Htn	Geog 817.00.5*	Bion. L'usages des globes celestes et terrestes. Adam, 1700.
	An 170.135.10	Biosocial anthropology. N.Y., 1975.
	Geog 5985.34	"Birdie" Bowers of the Antarctic. (Seaver, George.) London, 1938.
	Geog 818.92	A bird's-eye view of the world. (Reclus, O.) Boston, 1892.
Htn	Geog 4306.68*	Birken, S. von. Hoch fürstlicher Brandenburgischer Ulysses. Bayreuth, 1668.
	Geog 5639.50.5	Birket-Smith, K. Grenlandiia: sbornik statei. Moskva, 1953.
	Geog 5639.50	Birket-Smith, K. Grønlands bogen. København, 1950. 2v.
	Geog 5182.8.3	Birket-Smith, Kaj. Eskimoerne. 3. Udg. Kjøbenhavn, 1971.
	Geog 5182.8.5	Birket-Smith, Kaj. The Eskimoes. London, 1936.
	Geog 5328.5	Birket-Smith, Kaj. Knud Rasmussens saga. Kjøbenhavn, 1941.
	An 2109.60.5	Birket-Smith, Kaj. Primitive man and his ways. London, 1960.
	An 2109.68.5	Birket Smith, Kaj. Strejftog. Arktiske, tropiske og midt im ellem. København, 1968.
	Geog 4209.42.15	Birnbaum, M. Vanishing Eden. N.Y., 1942.
	Geog 4759.73.5	Birot, Pierre. Géographie physique générale de la zone intertropicale. Paris, 1973.
	Geog 4329.53.10	Birot, Pierre. La Méditerranée et le Moyen-Orient. Paris, 1953. 2v.
	Geog 4329.53.12	Birot, Pierre. La Méditerranée et le Moyen-Orient. 2. éd. Paris, 1964.
	Geog 659.59.5	Birot, Pierre. Précis de géographie physique générale. Paris, 1959.
	Geog 759.70	Birot, Pierre. Les régions naturelles du globe. Paris, 1970.
	Geog 4327.88.10	Bisani, Alessandro. Lettres sur divers endroits de l'Europe. Londres, 1791.
	Geog 4327.93	Bisani, Alessandro. Picturesque tour thro'...Europe, Asia. London, 1793.
	Geog 578.29	Bischoff, F.H.T. Vergleichendes Wörterbuch. Gotha, 1829.
	Geog 4181.131	Bishop, Charles. The journal and letters of Captain Charles Bishop on the north-west of America. Cambridge, 1967.
	Geog 4308.93.3	Bishop, W.H. A house-hunter in Europe. N.Y., 1893.
	Geog 4559.40.9	Bisschop, E. de. The voyage of the Kaimiloa. London, 1940.
	Geog 4249.58.5	Bisschop, Eric de. Tahiti-Nui. London, 1959.
	Geog 4308.72	Bits of travel. (Jackson, H.H.) Boston, 1872.
	Geog 4308.72.2	Bits of travel. (Jackson, H.H.) Boston, 1873.
	Geog 5882.5	Bjoergmose, Rasmus. Stensnaes - den glemte kirke. Odense, 1967.
	Geog 4307.80	Björnstähls, J.J. Briefe auf seinem Auslandischen Reisen. Leipzig, 1780-1783. 6v.
	Geog 3055.44	Bjorkman, E. The search for Atlantis. N.Y., 1927.
	An 2059.25	Black, Davidson. Asia and the dispersal of primates. Peking, 1925.
	An 2059.33	Black, Davidson. Fossil man in China. Peiping, 1933.
	An 3559.30F	Black, Davidson. On an adolescent skull of "Sinanthropus pekinensis". Peiping, 1930.
	Geog 4309.41.5	Black America abraod. (McGee, A.E.) Boston, 1941.
	An 359.29	Black magic. (Morand, Paul.) N.Y., 1929.
	Geog 4679.67.25	The black tide: in the wake of the Torrey Canyon. (Petrow, Richard.) London, 1968.
	Geog 4358.92	Blackburn, Henry. Artistic travel in Normandy, Brittany, the Pyrenees, Spain and Algeria. London, 1892.
	An 359.16	Blackford, K.M.H. Blondes and brunets. Photoreproduction. N.Y., 1916.
	Geog 578.76	Blackie, W.G. The imperial gazetteer. London, 1876 2v.
	An 359.69.10	Blacking, John. Process and product in human society. Johannesburg, 1969.
Htn	Geog 4208.82.3*	Blätter aus mienem Reisetagbuch 1881-1883. (Meyer, Hans.) Leipzig, 1882.
	Geog 602.1.10	De Blaeu's beschrijvers van land. (Amsterdam. Nederlansh Historisch Schlepvaart Museum.) Amsterdam, 1952.
	Geog 4559.40.7	Blain, W. Home is the sailor. N.Y., 1940.
	Geog 4307.43	Blainville, J. de. Travels through Holland, Germany, Switzerland, and other parts of Europe. London, 1743-1745. 3v.
	Geog 4311.20	Blairy, Jean. Le beau Danube gris. Paris, 1948.
	Geog 5558.71.5	Blake, E.V. Arctic experiences. N.Y., 1874.
	Geog 818.26	Blake, J.L. Geographical, chronological...atlas. N.Y., 1826.
	Geog 4181.86A	Blake, John W. Europeans in West Africa, 1450-1560. London, 1942. 2v.
	Geog 808.33.2	Blanc, L.G. Handbuch der Natur und Geschichte der Erde. Halle, 1833-1834. 3v.
	Geog 4209.44.15	Blanc et noir. (Gagnon, H.J.) Montréal, 1944.
	Geog 4269.35.4	Blanchard, Raoul. Géographie de l'Europe. Paris, 1936.
	Geog 4269.35	Blanchard, Raoul. A geography of Europe. N.Y., 1935.
	Geog 602.3.5	Blanchard, Raoul. Je découvre l'université. Paris, 1963.
	Geog 602.3	Blanchard, Raoul. Ma jeunese sous l'aile de Péguy. Paris, 1961.
	Geog 4309.41.15	Blanco-Fombona, R. Dos años y medio de Inquietad. Caracas, 1942.
	Geog 4309.26.10	Blanco-Fombona, R. Por los caminos del mundo. Madrid, 1926.
	Geog 5558.91.4	Blandt Nordpolens naboer. (Astrup, E.) Kristiania, 1895.
	Geog 4208.51.5	Blaney, Henry. Journal of voyages to China and return. Boston, 1913.
	Geog 4219.24.5	Blasco Ibáñez, Vicente. La vuelta al mundo de un novelista. Valencia, 1924-25. 3v.
	Geog 809.19.6	Blásquez y Delgado Aguilera, Antonio. América Meridional, Oceania. Barcelona, 1916.
	Geog 809.19.3	Blásquez y Delgado Aguilera, Antonio. Peninsula Ibérica. 2a ed. Barcelona, 1921.
	Geog 3140.9	Blau, J. Memoirs sur deux monuments geographiques. Nancy, 1836.
	Geog 4678.47.5	Blenden Hall; the true story of a shipwreck. (Lockhart, John Gilbert.) London, 1930.
	Geog 4209.43.10	Blick in die weite Welt. (Medern, W.E.A.A.) Berlin, 1943.
	An 359.44.10	Blind spots. (Leiper, H.S.) N.Y., 1944.
	Geog 5398.79.25	Bliss, R.W. Our lost explorers; a narrative of the Jeannette Arctic expedition. Hartford, 1882.
	Geog 3055.9	Block, R. de. Quelques mots sur l'Atlantide. n.p., n.d.
	Geog 5837.20.3	Bloeijende opkomst...visschery. (Zorgdrager, C.G.) 's-Gravenhage, 1727.
	Geog 4209.28.25	Bloem, Walter. Weldgesicht; ein Buch von heutiger und kommender Menscheit. Leipzig, 1928.
	Geog 4218.40.10	Blok, G.K. Dva goda iz zhizni russkago moriaka. Sanktpeterburg, 1854. 2v.
	Geog 5670.5	Blom, Ida. Kamper om Eirik Raudes land. Oslo, 1973.
	Geog 808.04.3	Blomfield, E. A general view of the world. Bungay, 1807. 2v.
	Geog 4709.69.5	Blond, Georges. Histoire de la flibuste. Paris, 1969.
	An 359.16	Blondes and brunets. Photoreproduction. (Blackford, K.M.H.) N.Y., 1916.
	Geog 4329.35.10	Bloomfield, Paul. The Mediterranean; an anthology. London, 1935.
	Geog 5312.6	Blosseville, B. Ernest Poret. Jules de Blosseville. Evreux, 1954.
	Geog 4169.31.15	Blossom, F.A. Told at the Explorers Club. N.Y., 1931.
	Geog 4329.23	Blue water. (Hildebrand, A.S.) N.Y., 1923.
	An 176.1	Blumenbach, J.F. The anthropological treatises. London, 1865.
	An 357.95	Blumenbach, J.F. De generis humani varietate nativa. Gottingae, 1795.
	Geog 622.1.5	Blunt, Wilfred. Pietro's pilgrimage. London, 1953.
	Geog 4418.73	Blyden, E.W. From West Africa to Palestine. Freetown, Sierra Leone, 1873.
	Geog 4219.70	Blyth, Chay. The impossible voyage. London, 1971.
	An 359.11	Boas, Franz. Curso de antropologia general. Mexico, 1911.
	An 359.38.20	Boas, Franz. General anthropology. Boston, 1938.
	An 2109.14	Boas, Franz. Kultur und Rasse. Leipzig, 1914.
	An 2109.11.3A	Boas, Franz. The mind of primitive man. N.Y., 1911.
	An 2109.29.15	Boas, Franz. The mind of primitive man. N.Y., 1929.
	An 2109.38.5	Boas, Franz. The mind of primitive man. N.Y., 1938.
	An 2109.46.5	Boas, Franz. The mind of primitive man. N.Y., 1946.
	An 176.6A	Boas, Franz. Race, language and culture. N.Y., 1940.
	An 359.45.20	Boas, Franz. Race and democratic society. N.Y., 1945.
	An 126.2	Boas anniversary volume. N.Y., 1906.
	Geog 5610.5	Bobé, Louis. Diplomatarium Groenlandicum, 1492-1814. København, 1936.
	Geog 5639.36.4	Bobé, Louis. Den grønlanske handels og kolonisations historie indtil 1870. København, 1936.
	Geog 5700.4.8	Bobé, Louis. Hans Egede. Copenhagen, 1952.
	Geog 5700.4.9	Bobé, Louis. Hans Egede og Grønland. København, 1941.
	Geog 5610.5.2	Bobé, Louis. Opdagelsesrejser til Grønland, 1473-1806. København, 1936.
	Geog 4279.71	Bocca, Geoffrey. The great resorts. N.Y., 1971.
	Geog 4308.34.5	Boddington, Mary. Slight reminiscences of the Rhine. London, 1834. 2v.
	Geog 5516.31.5A	Bodilly, R.B. The voyage of Captain Thomas James...1631. London, 1928.
	An 2109.75	Bodley, John H. Victims of progress. Menlo Park, 1975.
	Geog 3590.25	Bodnarskii, M.S. Ocherkii po istorii russk. zemleved. Moskva, 1947.
	Geog 579.54A	Bodnarskii, M.S. Slovar' geograficheskikh nazvanii. Moskva, 1954.
	Geog 579.54.5	Bodnarskii, M.S. Slovar' geograficheskikh nazvanii. Moskva, 1958.
	Geog 4145.9.5	Bodrick, Alan Houghton. Casual change. London, 1961.
	Geog 4208.86	Böckmann, W. Reise nach Japan aus Briefen und Tagebüchern zusammengestellt. Berlin, 1886.
	Geog 5839.48	Boegvad, R. Orientering og faerdsee i terraenet. København, 1948.
Htn	Geog 3205.40.21*	Boehme, J. Gli costumi, le leggi et lusanze. Venetia, 1560.
Htn	Geog 3205.40.25*	Boehme, J. Fardle of facious...ancient manners. London, 1555. 3v.
Htn	Geog 3205.40.11*	Boehme, J. Mores, leges et ritus omnium gentium. Lugduni, 1561.
Htn	Geog 3205.40.15*	Boehme, J. Mores, leges et ritus omnium gentium. Lugduni, 1576.
Htn	Geog 3205.40.17*	Boehme, J. Mores, leges et ritus omnium gentium. Lugduni, 1582.
Htn	Geog 3205.40.7*	Boehme, J. Omnium gentium, mores, leges. Antverpiae, 1542.
Htn	Geog 3205.40.5*	Boehme, J. Omnium gentium, mores, leges. Friburgi, 1540.
	Geog 3019.04.3	Böhme, Max. Die grossen Reisesammlungen des 16. Jahrhunderts und ihre Bedeutung. Strassburg, 1904.
	Geog 3240.9	Boek, C.P. Lettres...intitulé: liber guidonis. n.p., n.d.
	Geog 4228.26	Boelen, Jacobus. Reize naar de oost- en westkust van Zuid-Amerika. Amsterdam, 1835-36. 3v.
	An 2109.09.11	Bölsche, W. Der Mensch der Vorzeit. v.1-2. Stuttgart, 1909-11.
	An 2109.21.5	Boelsche, Wilhelm. Der Mensch der Vorzeit. v.1-2. 28e Aufl. Stuttgart, 1921.
	Geog 4169.12	Boer, M.G. de. Van oude voyagien. Amsterdam, 1912-13. 3v.
	Geog 4169.12.5	Boer, M.G. de. Van oude voyagien. 3. druk. Amsterdam, 1939.
	An 359.58.15	Boer, Wolfgang de. Das Problem des Menschen und die Kultur. Bonn, 1958.
	Geog 4328.59.5	Böttger, C. Das Mittelmeer. Leipzig, 1859.
	Geog 4268.00F	Boetticher, J.G. Geographical, historical...description. London, 1800.
	Geog 4269.34	Bogardus, J.F. Europe; a geographical survey. N.Y., 1934.
	An 2339.01	Die Bogen der Oceanier. (Frobenius, L.) Berlin, 1901.
	Geog 5328.15	Bogen om Knud. Skrevet af hans venner. Kjøbenhavn, 1943.
	An 2338.95.5	Bogen und Pfeil in Central Brasilien. (Meyer, Hermann.) Leipzig, 1895.
	Geog 4145.87	Bohemia junction. (Tschiffely, A.) London, 1950.
	Geog 807.87.3	Bohn, C.E. Ankündigung einer neuen Ausgabe von Büschings Erdbeschreibung welche auf Pränumeration gedruckt wird. Hamburg, 1802.
	Geog 818.61.10	Bohn, Henry G. A pictorial hand-book of modern geography. 1st ed. London, 1865.

164
Author and Title Listing

Htn Geog 576.93F* Bohun, E. Geographical dictionary. London, 1693.
An 127.10 Boissel, Jean. Victor Courtet, 1813-1867. Paris, 1972.
Geog 500.70 Bok-katalog omfattande geografi och resor. v.1-2. (Anderson, Ernst.) Stockholm, 1929-31.
Geog 4209.69.20 Boldyrev, Ivan I. Na raznykh shirotakh. Riga, 1969.
Geog 186.5 Boletim do Departamento de Geografia. (Presidente Prudente, Brazil. Faculdade de Filosofia, Ciências e Letras. Departamento de Geografia.) Presidente Prudente. 4,1972+
Geog 5907.5 Boletin. (Instituto Antartico Argentino.) Buenos Aires.
NEDL Geog 137.2 Boletin. (Lisbon. Sociedade de Geographia.) 1-17,1877-1899
Geog 137.2 Boletin. (Lisbon. Sociedade de Geographia.) 1900+ 48v.
Geog 143.10.5 Boletín. (México (City). Universidad Nacional. Instututo de Geografía.) México. 1,1969+ 2v.
Geog 212.40 Boletin. (Sociedad Geográfica de Colombia.) Bogotá. 14,1956+ 5v.
Geog 212.1 Boletin. (Sociedad Geográfica de la Paz.) La Paz. 27-66,1909-1943 3v.
NEDL Geog 142.2 Boletin. (Sociedad Geografico de Madrid.) Madrid. 1-44,1876-1900
Geog 142.2 Boletin. (Sociedad Geografico de Madrid.) Madrid. 1876-1901 32v.
Geog 142.2.3 Boletin. Revista geografica colonial y mercantil. Actas. (Sociedad Geografico de Madrid.) Madrid. 1-21,1899-1924 19v.
Geog 30.5 Boletin de estudios geograficos. Mendoza. 2,1950+ 6v.
Geog 30.7 Boletin paulista de geografia. Sao Paulo, Brazil. 1,1949+ 8v.
Htn Geog 3017.62* Bollan, William. Colonae anglicanae illustratae. Londini, 1762.
Geog 500.66 Bolles, E.C. The literature of sea travel since the introduction of steam, 1830-1930. Philadelphia, 1943.
NEDL Geog 39.2 Bollettino. (Club Alpino Italiano.) Torino. 1-41 13v.
Geog 85.309F Bollettino. (Genoa. Civico Istituto Colombiano.) 1-4,1953-1956
NEDL Geog 212.100 Bollettino. (Societa Geografica Italiana.) Firenze. 1868-1899 12v.
Geog 212.100 Bollettino. (Societa Geografica Italiana.) Firenze. 1900+ 82v.
NEDL Geog 212.100.2 Bollettino. Indice generale della serie II-III. (Societa Geografica Italiana.) Roma, n.d.
Geog 4708.68 Bollo, G.A. Petizione...relativa alla catastrofe toccata alla nave Teresa. n.p., n.d. 2 pam.
Geog 665.70 Bologna. Università. Istituto di Geografia. Studi geografici in onore di Antonio Renato Toniolo. Milano, 1952.
Geog 5312.4 Bolotnekov, N. Nikifor Belichev. 2. izd. Moskva, 1954.
Geog 4309.03 Bolton, C.E. Travels in Europe. N.Y., 1903.
Geog 4559.53.16 Bombard, Alain. The voyage of the Hérétique. Ann Arbor, Mich., 1971.
Geog 32.10 Bombay geographical magazine. Bombay.
Geog 32.5 Bombay Geographical Society. Transactions. Bombay. 1-19,1836-1873 19v.
Geog 32.5.5 Bombay Geographical Society. Transactions. Index, v.1-17. Edinburgh, 1866.
Geog 4181.72 Bombay in the days of Queen Anne. (Burnell, John.) London, 1933.
Geog 5839.50 Bomholt, Julius. Grønland foran en ny epoke. København, 1950.
Geog 579.41 Bonacker, Wilhelm. Karten-Wörterbuch eine Verdeutschung fremdsprachiger Kartensignatur-Bezeichnungen. Berlin, 1941.
Geog 622.5 Bonapace, Umberto. Luigi Visintin. Novara, 1958.
Geog 607.15 Bonasera, F. Un Gobo Terrestre di Matteo Greuter conservato nella Biblioteca. Camarino? 1959.
Geog 4269.64.15 Bonasera, Francesco. Gli stati d'Europa. Palermo, 1964.
Geog 4208.96.10 Bond, C. Goldfields and chrysantemums. London, 1898.
An 358.43.3 Bond, Thomas E. A dissertation on the varieties of mankind. Middletown, 1843.
Geog 4180.58A Bondage and travels of J. Schiltberger. (Schiltberger, J.) London, 1879.
An 2108.84.5F The bone caves of Ojcow in Poland. (Römer, F.) London, 1884.
Geog 4168.95 Bonnaffé, E. Voyages et voyageurs de la renaissance. Paris, 1895.
Geog 4209.38.35 Bonnard, A. Le bouquet du monde. Paris, 1938.
Geog 3019.53 Bonse, Ewald. Entwicklung und Aufgabe der Geographie. Stuttgart, 1953.
An 378.24 Bonstetten, K.V. von. L'homme du midi et l'homme du nord. Genève, 1924.
An 358.64.5 Bonstetten, K.V. von. The man of the north and man of the south. N.Y., 1864.
Geog 4182.54 Bontekae, W.Y. Journalen van de gedenckenaerdige reijsen. 's-Gravenhage, 1952.
Geog 4180.46 Bontier, P. The Canarien...conquest and conversion. N.Y., 1964?
Geog 4181.36 The book. (Barbosa, Duarte.) London, 1918-21. 2v.
Geog 4709.11 The book of buried treasure. (Paine, R.D.) London, 1911.
Geog 3019.20 A book of discovery. (Synge, M.B.) N.Y., 1920.
Geog 4139.25A The book of gallant vagabonds. (Beston, Henry B.) N.Y., 1925.
Geog 4181.29 Book of knowledge of all kingdoms, lands. (Markham, C.R.) London, 1912.
Geog 4672.7 Book of shipwrecks and adventures on the ocean. N.Y., n.d.
Geog 4169.19A The book of the long trail. (Newbolt, Henry.) N.Y., 1919.
Geog 4559.57 Book of the seven seas. (Freuchen, Peter.) N.Y., 1957.
Geog 500.63 Books and magazine articles on geography in the Columbus Memorial Library. (Pan American Union. Columbus Memorial Library.) Washington, 1935.
An 136.5.60 Boon, James A. From symbolism to structuralism. N.Y., 1972.
An 136.5.61 Boon, James A. From symbolism to structuralism: Lévi-Strauss in a literary tradition. Oxford, 1972.
Geog 4182.70 Booy, A. de. De derde reis van de V.O.C. naar Oost-Indië onder het beleid van Admiraal van Caerden. 's-Gravenhage, 1968.
Geog 585.12F Borbstaedt, A. Allgemeine geographische und statistische Verhältnisse. Berlin, 1846.
Geog 6008.98 Borchgrevink, C.E. First on the Antarctic continent. London, 1901.
Geog 4709.59.5 Bordbuch des Satans. (Leip, Hans.) München, 1959.
Geog 4029.6 Borden, Norman E. Dear Sarah. Freeport, Me., 1966.
Geog 4329.25 Borderlands of the Mediterranean. (McAllister, J.G.) Richmond, 1926.
An 378.84 Bordier, A. La géographie médicale. Paris, 1884.
Geog 4332.7F Les bords de l'Adriatique. (Yriarte, C.) Paris, 1878.
Geog 1525.37 Les bords du Rhin. 12. éd. (Baedeker, publishers.) Leipzig, 1882.
Geog 1525.27 Les bords du Rhin de la frontière suisse à la frontière de Hollande. 9. éd. (Baedeker, publishers.) Leipzig, 1875.
An 2109.75.15 Borer, Mary Irene Cathcart. Background to archaeology. London, 1975.
An 3409.37.3 Borges Badell, A. Clasificación anatómica-matematica del dactilograma. La Habana, 1937?
Geog 613.5 Borodajkez, Taras. Konrad Millers Lebenswerk. Salzburg, 19- .
Geog 4328.45 Borrer, D. Journey from Naples to Jerusalem. London, 1845.
Geog 5559.11.5 Borup, George. A tenderfoot with Peary. London, 1911.
Geog 4419.28 Bosch, Carl. Karawanen-Reisen...Ägypten, Mesopotamien, Persien und Abessinien. Berlin, 1928.
Geog 4136.76.5 Bosch, L. Leben...der...See-Helden. Nürnberg, 1681.
Htn Geog 4136.76* Bosch, L. Leeven en daden...zee-helden. Amsterdam, 1676.
Geog 5845.3 Boserup, Mogens. Økonomisk politik i Grønland. København, 1963.
Geog 6100.44 Bossière, R.E. Nouvelle notice sur les îles Kerquelen. Paris, 1907.
Geog 5050.2 Boston Public Library. Arctic regions and Antarctic regions. Boston, 1894.
Geog 4652.8 Boswell, David B. Loss list of Grimsby vessels, 1800-1960. Grimsby, 1969.
Geog 4309.52 Bota de sete léguas. (Lins do Rêgo, José.) Rio de Janeiro, 1952.
Geog 815.59.10 Botero, G. Descripcion de todas las provincias. Gerona, 1748.
Htn Geog 815.59.3* Botero, G. Le relationi universali...divise in quattro parti. Venetia, 1597.
Htn Geog 815.59.7* Botero, G. Le relationi universali...divise in quattro parti. Venetia, 1622.
Htn Geog 815.59.5* Botero, G. The travellers breviat, or An historical description of the most famous kingdoms in the world. London, 1601.
Geog 500.7 Boucher, G. Bibliothèque universelle des voyages. Paris, 1808. 6v.
Geog 4308.53.14 Boucher de Crèrecoeur de Perthes. Voyage à Constantinople par l'Italie. v.2. Paris, 1855.
Geog 5839.68 Boucht, Christer. Grönland tvärs. Helsingfors, 1968.
Geog 4217.66.11 Bougainville, L. de. Reis rondom de weereldt. Dordrecht, 1772.
Geog 4217.66 Bougainville, L. de. Voyage autour du monde. Paris, 1771.
Geog 4217.66.5 Bougainville, L. de. Voyage autour du monde. Paris, 1772. 3v.
Geog 4217.66.3A Bougainville, L. de. Voyage round the world...1766. London, 1772.
Geog 4145.9 Bougainville. (Thiery, Maurice.) London, 1932.
Geog 4559.26.15 Boughton, George P. Seafaring. London, 1926.
An 2059.23 Boule, Marcellin. Les hommes fossiles. 2. éd. Paris, 1923.
Geog 4209.38.35 Le bouquet du monde. (Bonnard, A.) Paris, 1938.
An 2108.69.7 Bourlot, J. Histoire de l'homme préhistorique anté et postdiluvien. Paris, 1869.
Geog 4181.121 Bourne, William. A regiment for the sea. Cambridge, Eng., 1963.
Geog 4307.65.10 Bousquet, (Mrs.). Mrs. Bousquet's diary, 1765. Norwich, 1927.
Geog 4509.65 Boutron, Michel. La montagne et ses hommes. Paris, 1965.
Geog 4209.63.30 Bouvier, Nicolas. L'usage du monde. Genève, 1963.
Geog 4181.123 Bovill, E.W. Missions to the Niger. Cambridge, Eng., 1964. 4v.
An 3308.79 Bowditch, J.P. The growth of children. Boston, 1879.
Geog 665.120 Bowen, Emzys George. Geography at Aberystwyth: essays written on the occasion of the departmental jubilee 1917-1918-1967-1968. Cardiff, 1968.
An 3309.32 Bowles, G.T. New types of old Americans at Harvard and at eastern women's colleges. Cambridge, 1932.
Geog 819.22 Bowman, I. The new world; problems in political geography. Yonkers-on-Hudson, 1922.
Geog 819.21 Bowman, I. The new world. Yonkers-on-Hudson, 1921.
Geog 819.28.5A Bowman, I. The new world. 4th ed. Yonkers-on-Hudson, 1928.
Geog 819.22.2 Bowman, I. Supplement to The new world. Yonkers-on-Hudson, 1923-1924. 2v.
An 2109.62 Bowra, Cecil M. Primitive song. London, 1962.
Geog 4181.12 Bowrey, T. Geographical account...countries...Bengal. Cambridge, 1905.
Geog 4181.58 Bowry, T. The papers. London, 1927.
Geog 4181.106 Boxer, C.R. South China in the sixteenth century. London, 1953.
Geog 4239.21 Boy aloft. (Craig, Gavin.) Lymington, 1971.
An 3309.35 Boyd, Edith. The growth of the surface area of the human body. Minneapolis, 1935.
Geog 5839.37 Boyd, Louise A. The coast of northeast Greenland. N.Y., 1948.
Geog 5839.35.20A Boyd, S.A. The fiord region of east Greenland. N.Y., 1935. 2v.
Geog 4209.00 Boyer, Bessie B. Chief Officer Brown. Chicago? 1900.
Geog 4208.80.5 Boyle, Frederick. Chronioles of no-man's land. London, 1880.
An 2109.27.11 Boyle, Mary E. In search of our ancestors. London, 1927.
Geog 4477.26.5 Boyle, R. (pseud.). The voyages and adventures of Captain Boyle. Dublin, 1741.
Htn Geog 4477.26* Boyle, R. (pseud.). Voyages and adventures of Captain Boyle. London, 1726.
Geog 4477.26.3 Boyle, R. (pseud.). The voyages and adventures of Captain Boyle. 1st American ed. Cooperstown, N.Y., 1796.
Geog 4329.09.5 The boys' account of it...foreign travel. (Roberts, William C.) N.Y., 1909.
Geog 4208.59 Boys' book of modern travel. (Johnes, M.) N.Y., 1859.
Geog 4558.84 Boys coastwise, or All along the shore. (Rideing, William H.) N.Y., 1884.
Geog 5839.26 A boy's-eye view of the Arctic. (Rawson, K.L.) N.Y., 1926.
Geog 4415.02.5 Bracciolini, Poggio. A traveler in disguise. Cambridge, 1963.
An 358.63 Brace, C.L. The races of the old world. N.Y., 1863.
Geog 4477.59 Brachfelds, J.M. Begebenheisen...Sudlandern. Eisenach, 1759.
An 359.73.35 Bradfield, Maitland. A natural history of associations. London, 1973. 2v.
Geog 4559.62 Bradford, G. In with the sea wind. Barre, 1962.
Geog 4559.59.5 Bradford, G. Yonder is the sea. Barre, Mass., 1959.
Geog 4228.61 Bradford, Ruth. Maskee! The journal and letters of Ruth Bradford, 1861-1872. Hartford, 1938.

Author and Title Listing

	An 359.23.5	Bradley, R.N. Duality; a study in the psycho-analysis of race. N.Y., 1923.
	An 359.12.9	Bradley, R.N. Malta and the Mediterranean race. London, 1912.
	Geog 4559.69	Bradley, Wendell P. They live by the wind. 1. ed. N.Y., 1969.
	Geog 4219.38.6	Bradshaw, M.J. Third class world. Alliance, 1939.
	Geog 3055.56	Bragbine, A. The shadow of Atlantis. N.Y., 1940.
	An 170.142	Braidwood, Robert John. Courses toward urban life. Chicago, 1962.
	An 2109.72.25	Brain, Robert. Into the primitive environment: survival on the edge of our civilization. London, 1972.
	Geog 5558.81.20	Brainard, David L. The outpost of the lost. Indianapolis, 1929.
	Geog 5558.81.25	Brainard, David L. Six came back; the Arctic adventure. Indianapolis, 1940.
	Geog 4328.13.2	Bramsen, John. Travels in Egypt, Syria, Cyprus. 2nd ed. London, 1820. 2v.
	Geog 3055.54A	Bramwell, James. Lost Atlantis. N.Y., 1938.
	Geog 3550.7	Branca, G. Storia dei viaggiatori italiani. Roma, 1873.
	Geog 5535.15	Brandes, Karl. Sir John Franklin; die Unternehmungen für seine Rettung und die Nordwestliche Durchfahrt. Berlin, 1854.
	Geog 4359.70	Brandi, Cesare. A passo d'uomo. Milano, 1970.
	Geog 4309.28.5	Brandt, R. Das Gesicht Europas. Hamburg, 1928.
	Geog 4324.80	Brasca, Santo. Viaggio in Terrasanta di Santo Brasca, 1480. Milano, 1966.
	Geog 4418.89A	Brassey, A. The last voyage. London, 1889.
	Geog 4218.78	Brassey, A. (Mrs.). Around the world in the "Sunbeam". N.Y., 1878.
	Geog 4218.78.5	Brassey, A. (Mrs.). Around the world in the "Sunbeam". N.Y., 1879.
	Geog 4218.78.8	Brassey, A. (Mrs.). A voyage in the "Sunbeam". Chicago, 1881.
	Geog 4218.78.3	Brassey, A. (Mrs.). A voyage in the "Sunbeam". London, 1879.
	Geog 4238.83.8F	Brassey, Anna. In the trades, the tropics, and the 'roaring forties. London, 1886.
NEDL	Geog 4238.83.7	Brassey, Anna. 14,000 miles in the Sunbeam in 1883. London, 1885.
	Geog 4218.78.7A	Brassey, Thomas. The "Sunbeam". London, 1918.
	Geog 4238.83.5	Brassey, Thomas. 11,506 knots in the "Sunbeam". London, 1884.
	Geog 91.1.10	Braun, Gustav. Geographische Gesellschaft zu Greifswald, 1882-1927. Greifswald, 1932.
	Geog 4208.60.7	Bray, M.M. A sea trip in clipper ship days. Boston, 1920.
	An 3409.10	Brayley, F.A. Arrangement of finger prints identification and their uses. Boston, 1910.
	Geog 4145.14.10	Brazão, Eduardo. Os Corte Reais e o novo mundo. Lisboa, 1965.
	Geog 4145.14.10.5	Brazão, Eduardo. Les Corte-Real et le nouveau monde. Lisbonne, 1967.
	Geog 32.50	Brazil. Diretoria do Serviço Geográfico do Exército. Anuário. 1962+
	Geog 4217.21	Bree, Levinus W. de. Jacob Roggeveen en zijn reis naar het zuidland, 1721-1722. Amsterdam, 1942.
	Geog 4209.35.5	Breezy adventure. (Foley, Arthur.) Boston, 1935.
	VGeog 4349.68	Bregman, Aleksander. Rubiezè woluorci. Londyn, 1968.
	Geog 4209.69.25	Bregovete na khorata. (Stoianovich, Ivan.) Sofiia, 1969.
	Geog 5838.36	Breidfjörd, S. Fra Graenlandi. Kaupmannahöfn, 1836.
	Geog 4207.55.5	Brelin, Johan. En äfventyrlig resa til och ifrån Ost-Indien, Södra America och en del af Europa åren 1755, 56 och 57. Stockholm, 1973.
	An 2109.35	Brelsford, Vernon. Primitive philosophy. London, 1935.
	Geog 4308.61	Bremer, F. Two years in Switzerland and Italy. London, 1861. 2v.
	Geog 4209.30.10	Bremer, M. Memoirs of a Ceylon planter's travels, 1851 to 1921. London, 1930.
	Geog 4138.30	Brendon, J.A. Great navigators and discoverers. N.Y., 1930.
	Geog 4349.28	Brennan, John W. The student abroad. Boston, 1928.
	Geog 3251.9	Brenner, O.K. Die ächte Karte des Olaus Magnus vom Jahre 1539. Christiana, 1886.
	Geog 5985.253.15	Brent, Peter L. Captain Scott and the Antarctic tragedy. London, 1974.
	An 3309.12	Bresciani, C. La correlazione fra la statura e l'indice cefalico. Palermo, 1912.
	An 2109.49	Breuil, H. Beyond the bonds of history. London, 1949.
	Geog 613.1.11	Breusing, A. Gerhard Kremer...Mercator. Duisburg, 1869.
	Geog 3070.4	Breusing, A. Leitfaden durch das Wiegenalter. Frankfurt, 1883.
	Geog 5324.2.90	Brev utgitt for Nansenfondet av Steinar Kjaerheim. (Nansen, Fridtjof.) Oslo, 1961- 4v.
	Geog 4267.26	Breval, John D. Remarks on several parts of Europe. v.1-2. London, 1726.
	Geog 4267.38	Breval, John D. Remarks on several parts of Europe. v.1-2. London, 1738.
	Geog 4208.44	Breve racconto delle cose. (Capobianco, R.) Napoli, 1846.
Htn	Geog 755.71*	Breve trattato del mondo e delle sue parti. (Nores, Jason de.) Venetia, 1571.
	Geog 4218.95.5	Brewster, F. Carroll. From Independence Hall around the world. Philadelphia, 1895.
	Geog 4678.65	Bribery and piracy...loss of S.S. Shooting Star. N.Y., 1870.
	Geog 577.60F	Brice, A. Grand gazetteer, or topographical dictionary. n.p., n.d.
	Geog 603.4	Briceno, Alfonso. Augustin Codazzi. Tesis. Caracas, 1928.
	An 3409.42	Bridges, B.C. Practical fingerprinting. N.Y., 1942.
	Geog 4218.83.5	Bridges, F.D. Journal of a lady's travels. London, 1883.
	Geog 4028.4.2	Bridges, Thomas C. Heroes of modern adventure. Boston, 1928.
	Geog 4209.25.5	Bridgman, H.B. Conquering the world. N.Y., 1925.
Htn	Geog 4306.73*	Brief account of some travels. (Brown, E.) London, 1673.
Htn	Geog 4306.73.5F*	Brief account of some travels. (Brown, E.) London, 1685.
	Geog 5060.10A	Brief history of polar exploration since the introduction of flying. (Joerg, W.L.G.) N.Y., 1930.
	Geog 5518.24.7	Brief narrative of attempt to reach Repulse Bay. (Lyon, G.F.) London, 1825.
	Geog 4181.69	A brief summe of geographie. (Barlow, Roger.) London, 1932.
	Geog 4307.80	Briefe auf seinem Auslandischen Reisen. (Björnstähls, J.J.) Leipzig, 1780-1783. 6v.
Htn	Geog 816.00*	A briefe description of the whole world. (Abbott, George.) London, 1600.
Htn	Geog 816.00.3*	A briefe description of the whole world. 3d ed. (Abbott, George.) London, 1608. 2 pam.
Htn	Geog 816.00.5*	A briefe description of the whole world. 5th ed. (Abbott, George.) London, 1620.
	Geog 4182.72	Briel, Johan Jurgen. De expeditie van Anthonio Hurdt. 's-Gravenhage, 1971.
Htn	Geog 3106.48*	Brietio, O. Parallela geographiae. Paris, 1648-49. 3v.
Htn	Geog 4308.32.5*	Briggs, Charles F. Working a passage: or Life in a liner. N.Y., 1844.
	Geog 4208.80.3	Briggs, L.V. Around Cape Horn to Honolulu on the bark "Amy Turner" 1880. Boston, 1926.
	Geog 4239.71.5	Brinnin, John M. The sway of the grand saloon: a social history of the North Atlantic. N.Y., 1971.
	An 628.92	Brinton, D.G. Anthropology. Philadelphia, 1892.
	An 359.02A	Brinton, D.G. Basis of social relations. N.Y., 1902.
	An 358.90	Brinton, D.G. Races and peoples. N.Y., 1890.
	An 126.1	Brinton memorial meeting. (American Philosophical Society.) Philadelphia, 1900.
	Geog 6009.29.5	British, Australian and New Zealand Research Expedition, 1929-1931. The winning of Australian Antarctica. Sydney, 1962.
	Geog 5907.8	British Antarctic Survey. Bulletin. London. 1-27,1963-1972 4v.
	Geog 579.61	British Association for the Advancement of Science. A glossary of geographical terms. London, 1961.
	Geog 32.60	British Columbia geographical series. Vancouver. 3,1965+ 2v.
	Geog 3520.15	British maps and map-makers. (Lynam, Edward.) London, 1944.
	An 609.10	British Museum. Handbook to ethnographical collections. Oxford, 1910.
	An 609.25	British Museum. Department of Ancient and Mediaeval Antiquities and Ethnography. Handbook of the ethnographical collections. 2. ed. London, 1925.
	An 2059.15	British Museum (Natural History). Department of Geology. A guide to the fossil remains of man in the department. London, 1915.
	An 609.22	British Museum (Natural History). Department of Geology. Guide to the fossil remains of man. 3. ed. London, 1922.
	An 359.08.5	British Museum (Natural History). Department of Zoology. Guide...specimens illustrating the races of mankind. London, 1908.
	An 359.12.11	British Museum (Natural History). Department of Zoology. Guide to the specimens...races of mankind. 2. ed. London, 1912.
	An 609.21	British Museum (Natural History). Department of Zoology. Guide to the specimens illustrating the races of mankind. 4. ed. London, 1921.
	Geog 600.24	British pioneers in geography. (Gilbert, Edmund William.) Newton Abbot, 1972.
	Geog 5509.62	The British search for the Northwest Passage in the 18th century. (Williams, G.) London, 1962.
	Geog 611.1	Le Briton Yves de Kerguelen. (Dupouy, A.) Paris, 1928.
	Geog 4300.51	Britons abroad; a report on the package tour. (Owen, Charles.) London, 1968.
	An 358.71.3	Broca, Paul. Mémoires d'anthropologie. v.5. Paris, 1888.
	Geog 4182.63	Broecke, Pieter van den. Pieter van den Broecke in Azie. 's-Gravenhage, 1962. 2v.
	Geog 4182.52	Broecke, Pieter van den. Reizen naar West-Afrika van Pieter van den Broecke. 's-Gravenhage, 1950.
	Geog 4307.03	Broekhuizen, G. Die nieuwe Béreisde wereld. Amsterdam, n.d. 2 pam.
	An 2058.69	Bromby, J.E. Pre-historic man. Melbourne, 1869.
	An 359.73.10	Bromlei, Iulian V. Etnos i etnografiia. Moskva, 1973.
	Geog 4306.92.6	Bromley, William. Remarks in the grand tour of France and Italy. 2d ed. London, 1705.
	Geog 5110.5	Bronner, Finn E. The polar regions. Santa Barbara, 1958.
	Geog 5300.11	Brontman, L.K. On top of the world. N.Y., 1938.
	Geog 577.91.5	Brookes, R. General gazeteer. Boston, 1816.
	Geog 577.62	Brookes, R. The general gazeteer. London, 1762.
	Geog 577.91	Brookes, R. General gazeteer. London, 1791.
	Geog 577.91.3	Brookes, R. General gazeteer. London, 1796.
	Geog 577.91.4	Brookes, R. General gazeteer. London, 1823.
	Geog 757.93	Brookes, R. The general gazetteer. London, 1795.
	Geog 577.91.10	Brookes, R. New universal gazeteer. Boston, 1850.
	Geog 577.91.6.5	Brookes, R. A new universal gazetteer. Philadelphia, 1844.
	Geog 577.91.7	Brookes, R. A new universal gazetteer. Philadelphia, 1847.
	Geog 577.91.6	Brookes, R. A new universal gazetteer containing a description of the principal nations. N.Y., 1832.
	Geog 4218.72.15	Brooks, James. A seven month's run up and down and around the world. N.Y., 1872.
	Geog 4328.95	Brooks, N. The Mediterranean trip. N.Y., 1895.
	Geog 4209.64.15	Brooks, P. Roadles area. 1. ed. N.Y., 1964.
	Geog 611.1.5	Brossard, Maurice Raymond de. Kerguelen, le découvreur et ses îles. Paris, 1970-1971. 2v.
	Geog 4559.27.15	The brotherhood of the sea. (Chatterton, E.K.) London, 1927.
	Geog 4145.14	The brothers Corte Real. (Lopes, Francisco Fernandes.) Lisboa, 1957.
	An 2108.45	Brotonne, F. de. Civilisation primitive. Paris, 1845.
	An 358.37	Brotonne, F. de. Histoire de la filiation et migrations des peuples. v.1-2. Paris, 1837.
Htn	Geog 5377.95*	Broughton, William. A voyage of discovery...North Pacific Ocean. London, 1804.
Htn	Geog 5377.95.2*	Broughton, William. A voyage of discovery...North Pacific Ocean. London, 1804.
	Geog 4105.16	Brouwer, H.A. Practical hints to scientific travellers. Leyden, 1922-29. 6v.
	Geog 5160.13.5	Brower, Charles D. Fifty years below zero. N.Y., 1942.
	Geog 4678.54.15	Brown, A.C. Women and children last; the loss of the steamship. London, 1962.
	Geog 4559.35.10	Brown, Charles W. Journal and letters, 1876-84. n.p., 1935.
	Geog 4559.25.15	Brown, Charles W. My ditty bag. Boston, 1925.
Htn	Geog 4306.73*	Brown, E. Brief account of some travels. London, 1673.
Htn	Geog 4306.73.5F*	Brown, E. Brief account of some travels. London, 1685.
	An 379.54	Brown, H.S. The challenge of man's future. N.Y., 1954.
	Geog 5535.10.2	Brown, John. The Northwest Passage, and the plans for the search for Sir John Franklin. 2. ed. London, 1860.
	Geog 3070.135	Brown, Lloyd. Map making; the art that became a science. 1st ed. Boston, 1960.
	Geog 603.9	Brown, Lloyd A. Jean Domenique Cassini and his world map of 1696. Ann Arbor, 1941.
	Geog 3070.48.5	Brown, Lloyd A. The story of maps. Boston, 1950.
	Geog 3070.48	Brown, Lloyd A. The story of maps. 1st ed. Boston, 1949.
	Geog 808.77	Brown, R. The countries of the world. London, 1877-1880? 4v.
	Geog 5070.20	Brown, R.N.R. A naturalist at the poles. London, 1923.
	Geog 5070.20.5	Brown, R.N.R. The polar regions. London, 1927.

Author and Title Listing

Geog 5070.46 Brown, R.N.R. Some problems of polar geography. Washington, 1929.
Geog 6009.02 Brown, R.N.R. Voyage of the "Scotia". Edinburgh, 1906.
Geog 142.12.5 Brown, Theodore Nigel Leslie. The history of the Manchester Geographical Society, 1884-1950. Manchester, 1971.
Geog 4208.78.9 Browne, Ernest A. Round the north hemisphere. London, 1880.
Geog 4558.50 Browne, J.R. Etchings of a whaling cruise. N.Y., 1850.
An 3208.50 Browne, P.A. The classification of mankind by the hair. Philadelphia, 1850.
Htn Geog 5398.50F* Browne, W.H. Ten coloured views during the Arctic expedition. London, 1850.
Geog 500.150 Browning, Clyde Eugene. A bibliography of dissertations in geography, 1901 to 1969; American and Canadian universities. Chapel Hill, 1970.
Geog 4209.29.10 Bruce, Michael. Peaks of hazard. Indianapolis, 1929.
Geog 4307.82 Bruce, Peter Henry. Memoirs of Peter Henry Bruce, Esq. London, 1782.
Geog 4307.82.3 Bruce, Peter Henry. Memoirs of Peter Henry Bruce, Esq. Reprint. London, 1970.
Geog 5559.11 Bruce, William S. Polar exploration. London, 1911.
Geog 5160.61F Bruemmer, Fred. The Arctic. Scarborough, Ont., 1974.
Geog 34.15 Bruenn. Universita. Přérodovědecká Fakulta. Geographia. Praha. 5,1971+
Geog 4418.62 Brumund, J.F.G. Schetsen eener mail-reize van Batavia naar Maastricht op reis en thuis. Amsterdam, 1862.
An 379.42.5 Brunhes, Jean. La géographie humaine. Paris, 1942.
An 359.12.17 Brunhes, Jean. La géographie humaine. 2. éd. Paris, 1912.
An 379.25 Brunhes, Jean. La géographie humaine. 3. éd. Paris, 1925. 3v.
An 379.34.3 Brunhes, Jean. La géographie humaine. 4. éd. Paris, 1934. 3v.
An 379.20.3 Brunhes, Jean. Human geography. Chicago, 1920.
An 379.52 Brunhes, Jean. Human geography. London, 1952.
Geog 212.110.10 Brunialti, A. Relazioni sulla fondazione e sull'ordinamento della sezione de geografia commerciale della Societa italiana. Roma, 1879.
Geog 3018.76 Bruniatti, A. I progressi della generale e della geografia esploratrice in Europa. Vicenza, 1877.
Geog 4308.52.11 Brushwood, picked up on the continent. (Horwitz, O.) Philadelphia, 1855.
Geog 28.4.5 Brussels. Université Nouvelle. Institut Géographique. Publication. Bruxelles.
Geog 5650.10 Bruun, Daniel. Erik den Rode og nordhokolonierne il Gronland. Kristiania, 1915.
Geog 5650.10.3 Bruun, Daniel. De gamle Nordbokdonieri Grønland. København, 1905.
Geog 5650.10.11 Bruun, Daniel. Godthaabsegnen og den gamle Vesterbygd. Odense, 1908.
Geog 5650.10.5 Bruun, Daniel. The Icelandic colonization of Greenland. København, 1918.
Geog 5225.10 Bruun, Daniel. Kampen om nordpolen. Kjøbenhavn, 1902.
Geog 5838.94 Bruun, Daniel. Mellem fangere og jaegere. Kjøbenhavn, 1897.
Geog 5650.10.9 Bruun, Daniel. Oversigt over Nordboruiner i Godhaat. København, 1917.
Htn Geog 4347.01.5F* Bruyn, C. de. Reizen over Muskovie door Persie. Amsterdam, 1711.
Geog 4347.01.10FA Bruyn, C. de. Travels into Muscovy, Persia, E. Indies. London, 1737. 2v.
Htn Geog 4347.01.6F* Bruyn, C. de. Voyage au Levant. Paris, 1725. 5v.
Geog 4347.01.7F Bruyn, C. de. Voyages...au Levant. Haye, 1732. 5v.
Htn Geog 4347.01.8F* Bruyn, C. de. Voyages...en Perse et aux Indes Orient. Amsterdam, 1718. 2v.
Geog 577.26F Bruzen de la Martiniere, A.A. Le grand dictionnaire géographique. Haye, 1726-39. 10v.
Geog 577.68F Bruzen de la Martiniere, A.A. Le grand dictionnaire géographique. Paris, 1768. 6v.
Geog 4679.42 Bryan, George S. Mystery ship. The Mary Celeste in fancy and fact. Philadelphia, 1942.
An 3509.38 Bryan, Kirk. Discovery of Sauk Valley man of Minnesota. Abilene, Texas, 1938.
An 379.33.5 Bryan, P.W. Man's adaptation of nature. London, 1933.
Htn Geog 4228.50* Bryant, W.C. Letters of a traveller. N.Y., 1850.
Geog 4228.51 Bryant, W.C. Letters of a traveller. 3. ed. N.Y., 1851.
Geog 4228.50.5 Bryant, W.C. Letters of a traveller. 4. ed. N.Y., 1855.
Geog 4228.50.2 Bryant, W.C. The picturesque souvenir. N.Y., 1851.
An 359.02.3 Bryce, J. The relationship of the advanced and backward races of mankind. Oxford, 1902.
An 359.03.9 Bryce, J. Relationship of the advanced and backward races of mankind. 2. ed. Oxford, 1903.
Geog 4209.23.20 Bryce, J.B. Memories of travel. N.Y., 1923.
Geog 5398.71 Brynjulfson, A. Have de gamle Nordboer havt Kjendskab. Kjøbenhavn, 1871.
Htn Geog 5220.21* Buache, P. Considerations geographiques. Paris, 1753.
Geog 4139.10 Buccaneers of America. (Johnstone, C. (Mrs.).) Akron, 1910.
Geog 4709.28A Buccaneers of the Pacific. (Wycherley, George.) Indianapolis, 1928.
Geog 4209.23 Buchan, John. The last secrets. London, 1923.
Geog 4308.62 Bucher, L. Unterwegs. v.1-2. Berlin, 1862.
Geog 3235.29 Buchon, J.A.C. Notice sur un atlas en langue catalane. Paris, 1838.
Geog 3235.29.3 Buchon, J.A.C. Notices et extraits des manuscrits. Paris, n.d.
Geog 4209.41.3 Buck, F. All in a lifetime. N.Y., 1941.
Geog 5326.21.10 Buck, M. Memorial book of Robert Edwin Peary. n.p., 1937.
Geog 4308.51.7 A Buckeye abroad. (Cox, Samuel S.) N.Y., 1852.
Geog 4308.48 Buckingham, J.S. Belgium, the Rhine, Switzerland. London, 1848. 2v.
Geog 4328.88 Buckley, J.M. Travels in three continents. N.Y., 1895.
Geog 4308.86.10 Buckley, James M. The midnight sun; the tsar and the nihilist. Boston, 1886.
Geog 32.90 Budapest. Tudomány-Egyetem. Annales. Sectio geographica. Budapest. 1,1965+ 2v.
Geog 4308.47.7 The budget closed. (Eames, J.A.) Boston, 1860.
Geog 4308.47.5 A budget of letters. (Eames, J.A.) Boston, 1847.
Geog 4300.44 Budgetouring Europe. (Ross, Janet.) N.Y., 1933.
Geog 4269.73 Budrewicz, Olgierd. Ścieżkami Starego Świáta. Wyd. 1. Warszawa, 1973.
Geog 500.65 Bücherei-Verzeichnis des Vereins für Erdkunde zu Dresden. (Verein für Erdkunde, Dresden. Bibliothek.) Dresden, 1905.
Geog 4500.7 Bücherverzeichnis der Alpenvereinsbucherei, mit Verfasser- und Bergnamen-Verzeichnis. (Dreyer, Aloys.) München, 1927.
Geog 4219.21 Büchler, E. Rund um die Erde. Bern, 1921.
Geog 4502.5.11 Bühler, H. Alpine Bibliographie. München. 1932
Geog 4502.5.11.5 Bühler, H. Alpine Bibliographie. Gesamtregister, 1931-1938. München, 1949.
X Cg An 358.89.3 Buel, J.W. The story of man. Philadelphia, 1889.
Geog 4168.84.5 Buel, James W. The worlds wonders. St. Louis, 1884.
Geog 3055.40 Buelua, E. La Atlantida y la ultima tule. Mexico, 1895.
Geog 5970.10 Buenos Aires. Universidad. Cronologia de los viajes a las regiones australes. Buenos Aires, 1950.
Geog 807.85 Büsching, Anton Friedrich. Auszug aus seiner Erdbeschreibung. 6. Aufl. Hamburg, 1785.
Geog 807.87 Büsching, Anton Friedrich. Erdbeschreibung. Hamburg, 1787-1792. 10v.
Geog 807.87.2 Büsching, Anton Friedrich. Grosse Erdbeschreibung. Troppau, 1784-1787. 24v.
Geog 807.62 Büsching, Anton Friedrich. New system of geography. London, 1762. 6v.
Geog 3019.73.10 Büttner, Manfred. Die Geographie generalis vor Varenius. Wiesbaden, 1973.
Geog 4311.21 Buettner, R. Burgen und Schlösser and der Donau. Wein, 1964.
Geog 4105.10 Buffum, E.G. Pocket guide for Americans going to Europe. N.Y., 1859.
Geog 4308.69.5 Buffum, E.G. Sights and sensations in France. N.Y., 1869.
Geog 4509.54.10 Buhl, H. Achttausend, drüber und drunter. München, 1954.
Geog 4239.51.5 Buhler, Jean. Sur les routes de l'Atlantique. Lausanne, 1951.
An 359.28A The building of cultures. (Dixon, Roland B.) N.Y., 1928.
Geog 5970.40 Bulgarin na Antarktida. (Takhariev, Vasil I.) Sofiia, 1971.
Geog 33.10 Bulgarska Akademiia na Naukite, Sofia. Geografska Institut. Izvestiia. 1,1951+ 8v.
Geog 665.84 Bulgarska Akademiia na Naukite, Sofia. Geografski Institut. Sbornik v chest na akademik Anastas Stoianov Beshkov. Sofiia, 1959.
Geog 665.98 Bulgarska Akademiia na Naukite, Sofia. Geografski Institut. Sbornik v chestna chlen-korespondent Iordan Zakhariev. Sofiia, 1964.
Geog 34.5 Bulgarsko Geografsko Druzhestvo, Sofia. Izvestiia. 1,1933+ 11v.
Geog 4309.07.5 Bull, Edvard. Over lave og høie fjelde; breve fra en aeldre berre. Kristiania, 1907.
Geog 5324.2.5 Bull, Jacob B. Fridtjof Nansen, a book for the young. Boston, 1903.
Geog 4308.50.7 Bullard, A.T.J. (Mrs.). Sights and scenes in Europe. St. Louis, 1852.
Geog 4558.98.10 Bullen, F.T. Cruise of the Cachalot. London, 1898.
Geog 4558.98.11 Bullen, F.T. Cruise of the Cachalot. N.Y., 1899.
Geog 4558.98.12 Bullen, F.T. Cruise of the Cachalot. N.Y., 1899.
Geog 4558.98.13 Bullen, F.T. Cruise of the Cachalot. N.Y., 1899.
Geog 4558.98.14 Bullen, F.T. Cruise of the Cachalot. N.Y., 1899.
Geog 4558.98.15 Bullen, F.T. Cruise of the Cachalot. N.Y., 1920.
Geog 4558.99A Bullen, F.T. The log of a sea-waif. N.Y., 1899.
Geog 13.12 Bulletin. (American Bureau of Geography.) Winona. 1900-1901
Geog 13.1 Bulletin. (American Geographical Society of New York.) 1-2,1852-1856
Geog 16.1.7 Bulletin. (Appalachian Mountain Club.) 1-27,1907-1934 27v.
Geog 16.1.8 Bulletin. (Appalachian Mountain Club.) 1935+ 26v.
Geog 18.150 Bulletin. (Arctic Institute of North America.) Montreal.
Geog 19.2 Bulletin. (Associated Mountaineering Clubs of North America.) N.Y. 1918-1921
Geog 19.5 Bulletin. (Association of Indian Geographers.) New Delhi. 1-2,1956-1957
Geog 5907.8 Bulletin. (British Antarctic Survey.) London. 1-27,1963-1972 4v.
Geog 35.3 Bulletin. (California. Geographical Society.) 1-2
Geog 85.315 Bulletin. (Geographical Society of Ireland.) Dublin. 1,1944+ 5v.
Geog 85.206 Bulletin. (Geographical Society of the Pacific.) San Francisco. 1905
Geog 148.5 Bulletin. (National Geographical Society of India.) Benares.
Geog 182.1 Bulletin. (Philadelphia Geographical Club.) Philadelphia. 1-36 16v.
Geog 189.2.3 Bulletin. (Quebec Geographical Society.) Quebec. 3-23,1908-1929 10v.
NEDL Geog 199.10 Bulletin. (Schweizer Alpenclub. Section Genevoise.) Genève.
Geog 212.200.20 Bulletin. (Société Belge d'Études Coloniales.) Bruxelles. 1894-1925 22v.
NEDL Geog 212.203 Bulletin. (Société de Géographie.) Paris. 1-20 73v.
NEDL Geog 212.215 Bulletin. (Société de Géographie de l'Est.) Nancy. 1879-1912 34v.
NEDL Geog 212.235 Bulletin. (Société de Géographie de Lille.) Lille. 1-32,1882-1899 17v.
Geog 212.235 Bulletin. (Société de Géographie de Lille.) Lille. 1900-1962 20v.
NEDL Geog 212.202 Bulletin. (Société de Géographie de Lyon.) Lyon. 1-16,1875-1900 11v.
Geog 212.202 Bulletin. (Société de Géographie de Lyon.) Lyon. 1901-1929 6v.
Geog 189.2.5 Bulletin. (Société de Géographie de Québec.) 2v.
NEDL Geog 212.225 Bulletin. (Société de Géographie de Toulouse.) Toulouse. 1-18,1882-1899 18v.
Geog 212.225 Bulletin. (Société de Géographie de Toulouse.) Toulouse. 19-50 18v.
Geog 212.218 Bulletin. (Société de Géographie du Maroc.) Casablanca. 1916-1933 6v.
NEDL Geog 212.245 Bulletin. (Société Languedocieme de Géographie, Montpellier.) Montpellier. 1-34,1878-1911 33v.
Geog 152.1 Bulletin. (Société neuchateloise de géographie.) Neuchatel. 1,1885+ 16v.
NEDL Geog 212.201 Bulletin. (Société Normande de Géographie.) Rouen. 1-21,1879-1899 21v.
Geog 212.201 Bulletin. (Société Normande de Géographie.) Rouen. 22-43,1900-1928 14v.
NEDL Geog 212.200 Bulletin. (Société Royale Belge de Géographie.) Bruxelles. 1-23,1877-1899 23v.
Geog 212.200 Bulletin. (Société Royale Belge de Géographie.) Bruxelles. 24,1900+ 46v.
NEDL Geog 14.600 Bulletin. (Société Royale de Géographie d'Anvers.) Anvers. 1-23,1877-1899 23v.
Geog 14.600 Bulletin. (Société Royale de Géographie d'Anvers.) Anvers. 1-75,1877-1964 30v.

Author and Title Listing

Geog 212.216.10 Bulletin. (Société Royale de Géographie d'Egypte.) Le Caire. 2+ 27v.

Geog 243.2 Bulletin. (Union Geographique du Nord de la France.) Lille. 1-34,1880-1913 17v.

Geog 189.2.6 Bulletin. Index. 1880-1934. (Société de Géographie de Québec.) Québec, 1969.

Geog 212.203.5 Bulletin. Table, series V-VII. (Société de Géographie.) Paris, n.d.

Geog 212.216.11 Bulletin. Tables. (Société Royale de Géographie d'Egypte.) 16-30,1928-1957

Geog 212.240 Bulletin année. (Société Geographique de Liège.) Liège. 1,1965+ 2v.

Geog 81.1 Bulletin de géographie historique et descriptive. (France. Ministère de l'Instruction Publique.) Paris. 1886+ 58v.

Geog 33.1 Bulletin des sciences géographiques. Paris. 1-28 28v.

Geog 212.202.5 Bulletin du cinquantenaire, 1922-23. (Société de Géographie de Lyon.) Lyon, 1923.

Geog 3228.80.5 Bullo, C. La vera patria di Nicolò de Conti e di Giovanni Caboto. Chioggia, 1880.

Geog 4679.12.41 Bullock, Shan F. A Titanic hero, Thomas Andrews, shipbuilder. Riverside, 1973.

Geog 3249.28.3 Bullon y Fernández, Eloy. Miguel Servet y la geographia del renacimiento. 3. ed. Madrid, 1945.

Geog 4219.32.25 Ein Bummel um die Welt. (Katz, Richard.) Zürich, 1936.

Geog 4219.12 Ein Bummel um die Welt. (Zobeltitz, Fedor von.) Berlin, 1912.

An 2109.25.19 Bumüller, J. Die Urzeit des Menschen. 4e Aufl. Augsburg, 1925.

Geog 4308.90 A bundle of letters from over the sea. (Robinson, L.B.) Boston, 1890.

Geog 4329.20 Bunge de Galvez, D. Tierras del mar Azul. Buenos Aires, 192-.

An 359.59.25 Bunker, Robert. The first look at strangers. New Brunswick, N.J., 1959.

Htn Geog 4208.14* Bunnell, D.C. Travels and adventures. Palmyra, N.Y., 1831.

Geog 4237.83 Burall, Paul. Cornwall to America in 1783 from the journal of Paul Burall (1755-1826). n.p., 19- .

Geog 4209.60.20 Burden, W.D. Look to the wilderness. Boston, 1960.

Geog 4209.09 Burge, C.O. Adventures of a civil engineer. London, 1909.

An 2058.87 Burge, Lorenzo. Pre-glacial man and the Aryan race. 2. ed. Boston, 1887.

Geog 4311.21 Burgen und Schlösser and der Donau. (Buettner, R.) Wein, 1964.

Geog 4709.55 Buried treasure. (Quarrell, Charles.) London, 1955.

Geog 5939.56.5 Burkhanov, V.F. K beregam Antarktidy. Moskva, 1956.

Geog 5300.30 Burkhanov, V.F. Novye sovetskie issledovannia v Arktike. Moskva, 1955.

Geog 4145.10 Burkov, Boris S. Ustrechi na piati kontinentakh. Moskva, 1973.

Geog 4269.46 Burlats'kym shliakom. (Antonovych, Roman.) Vynnypeg, 1946.

Geog 4659.67 Burman, Jose. Great shipwrecks off the coast of southern Africa. Cape Town, 1967.

Geog 6008.92 Burn Murdoch, W.G. From Edinburgh to the Antarctic. London, 1894.

Geog 4181.72 Burnell, John. Bombay in the days of Queen Anne. London, 1933.

Htn Geog 4306.86.3* Burnet, Gilbert. Dr. Burnet's travels or letters. Amsterdam, 1687.

Geog 4306.86.5 Burnet, Gilbert. Some letters...Switzerland, Italy. n.p., 1708.

Htn Geog 4306.86*A Burnet, Gilbert. Some letters...Switzerland, Italy. Rotterdam, 1686.

Geog 4306.86.6 Burnet, Gilbert. Some letters containing an account of what seemed most remarkable in Switzerland. Menston, Eng., 1972.

Geog 4306.86.10 Burnet, Gilbert. Travels through France, Italy. London, 1750.

Geog 4307.70.10A Burney, Charles. Continental travels. London, 1927.

Geog 4158.03F Burney, J. Chronological history...discoveries..South Sea. London, 1803-1817. 5v.

Geog 4217.85.5 Burney, J. Memoir on the voyage...in search of La Pérouse. London, 1820.

Geog 4110.17 Burns, Philp and Co. Picturesque travel. Sydney, 1914.

Geog 4559.13 Burns, Walter. A year with a whaler. N.Y., 1913.

Geog 4678.30 Burrows, Silas E. Russia and America. v.1-2. Hartford, 1865.

Htn Geog 4558.05* Burton, C. Journal of voyage from London. London, 1805.

Geog 3129.32A Burton, H.E. The discovery of the ancient world. Cambridge, 1932.

Geog 4208.85.5 Burton, R.G. Tropics and snows. London, 1898.

Geog 4209.24.5A Burton, Richard F. Selected papers on anthropology, travel and exploration. London, 1924.

Geog 4209.01.5A Burton, Richard F. Wanderings in three continents. N.Y., 1901.

An 359.10.7 Buschan, Georg H.T. Illustrierte Völkerkunde. Stuttgart, 1910.

An 359.09.5 Buschan, Georg H.T. Menschenkunde. Stuttgart, 1909.

Geog 4308.80 Bush, R. Personal impressions...of European travel. Photoreproduction. N.Y., 1880?

Geog 4131.5 The business of travel. (Rae, W.F.) London, 1891.

Geog 819.11.5 Busson, Henri. Les principales puissances d'aujourd'hui. 5. éd. Paris, 1924.

Geog 819.11.3 Busson, Henri. Les principales puissances du monde. Paris, 1911.

Geog 4208.45 Bustamante, J. Viaje. Lima, 1845.

Geog 4208.45.2 Bustamante, J. Viaje al antiguo mundo. 2. ed. Lima, 1959.

Geog 4180.65 Butler, N. Historye of the Bermudaes. London, 1882.

Geog 3108.25 Butler, S. Sketch of modern and ancient geography. London, 1825.

An 2109.29.9 Butt-Thompson, F.W. West African secret societies. London, 1929.

An 2109.13.10 Buttel-Reepen, H. Man and his forerunners. London, 1913.

Geog 4308.81.5 Butterworth, Hezekiah. Zigzag journeys in classic lands. Boston, 1881.

Geog 4308.81.6 Butterworth, Hezekiah. Zigzag journeys in classic lands. Boston, 1882.

Geog 4208.28 Butterworth, William. Three years adventures of a minor in England Africa, South Carolina and Georgia. Leeds, 1831.

Geog 148.1.8 Buxbaum, Edwin Clarence. Collecting national geographic magazines. Milwaukee, 1935.

Geog 148.1.9 Buxbaum, Edwin Clarence. Collectors guide to the national geographic magazine. Wilmington, 1962.

Geog 4209.29.5 Buxton, Noel. Travels and reflections. London, 1929.

Geog 757.70 Buy de Mornas, C. Cosmographie methodique. Paris, 1770.

An 3509.37 Buyssens, Paul. Le pithécanthrope était-il un Pygmée? Bruxelles, 1937.

Geog 4349.23 By devious ways. (Ross, Halford.) London, 1927.

Geog 4309.58 By road to Turkey. (Bell, Robert.) London, 1958.

Geog 4558.98.5 By way of Cape Horn. (Stevenson, P.E.) Philadelphia, 1899.

Geog 4559.30.3A By way of Cape Horn. (Villiers, A.J.) N.Y., 1930.

Geog 4308.69 By-ways of Europe. (Taylor, B.) N.Y., 1869.

Geog 4308.69.15 Byers, S.H.M. Twenty years in Europe. Chicago, 1900.

Htn Geog 4308.69.2* Byeways of Europe. (Taylor, B.) London, 1869. 2v.

Geog 5180.10 Byhan, A. Die Polarvölker. Leipzig, 1909.

Geog 5939.38.3A Byrd, Richard E. Alone. N.Y., 1938.

Geog 6009.33A Byrd, Richard E. Discovery; the story of the second Byrd Antarctic expedition. N.Y., 1935.

Geog 6009.28 Byrd, Richard E. Into the home of the blizzard. N.Y., 1928.

Geog 6009.28.10 Byrd, Richard E. Little America. N.Y., 1930.

Geog 6009.28.8A Byrd, Richard E. Little America. 1st ed. N.Y., 1930.

Geog 4181.122 Byron, John. Byron's journal of his circum navigation, 1764-1766. Cambridge, 1964.

Geog 4217.64.10 Byron, John V. Viaggio intorno al mondo. Firenze, 1768.

Geog 4217.64 Byron, John V. Viaje al rededor del mundo hecho en 1764, 65 y 66. Madrid, 1833.

Geog 4181.122 Byron's journal of his circum navigation, 1764-1766. (Byron, John.) Cambridge, 1964.

Geog 5324.2.30 Byström, D.G.V. Fridtjof Nansen. Stockholm, 1940.

NEDL Geog 3019.59.15 Cabal, Juan. Grandes exploradores, en la mar. Barcelona, 1959.

Geog 4709.53.5 Cabal, Juan. Historias de pirates. 1. ed. Barcelona, 1953.

Geog 4181.120 The Cabot voyages and Bristol discovery under Henry VII. (Williamson, J.A.) Cambridge, 1962.

Geog 4309.05 Cabrera, Raimundo. Cartas a Estevez. Habana, 1906.

An 2108.73.15 Cahen, Daniel. Un site tshitolien sur le plateau des Bateke. Tervuren, 1973.

Geog 35.8 Cahiers de géographie; publication de l'institut d'histoire et de géographie, Université Laval. Québec.

Geog 35.9 Cahiers de géographie de Québec. Québec. 1,1955+ 13v.

Geog 35.6 Les cahiers d'Outre-mer. Bordeaux. 1,1948+ 21v.

Geog 4218.87.14 Caine, W.S. A trip round the world in 1887-88. London, 1892.

Geog 4218.87.13 Caine, William S. A trip round the world in 1887-88. London, 1891.

Geog 4182.31 Caland, W. Die remonstrantie von W. Geleynssen de Jongh. 's-Gravenhage, 1929.

Geog 4419.51 Calder, Ritchie. Med against the desert. London, 1951.

Geog 5160.35 Calder, Ritchie. Men against the frozen north. N.Y., 1957.

Geog 4209.45 Caldwell, E.N. Satin skirts of commerce. N.Y., 1945.

Geog 35.3 California. Geographical Society. Bulletin. 1-2

Geog 35.4 California. University. Publications in geography. Berkeley, Calif. 1,1913+ 19v.

An 136.20 California. University. Robert H. Lowie Museum of Anthropology. The complete bibliography of Robert H. Lowie. Berkeley, 1966.

Geog 5209.32 The call of the North. (Houhen, H.H.) London, 1932.

Geog 4157.66.5 Callander, J. Terra Australis cognita, or Voyages. Edinburgh, 1766. 3v.

Geog 4308.66 Calvert, G.H. First years in Europe. Boston, 1866.

Geog 4308.52.8 Calvert, G.H. Scenes and thoughts in Europe. 1st-2d series. Boston, 1863. 2v.

Htn Geog 4308.52.7* Calvert, G.H. Scenes and thoughts in Europe. 2d series. N.Y., 1852.

Geog 809.19.5 Camena d'Almeida, P. América Septentrional, América Central, Las Antillas, Alaska, Canada, Estados Unidos. Barcelona, 1916.

Geog 809.19.4 Camena d'Almeida, P. Asia, India insular, Africa. Barcelona, 1914.

Geog 809.19.2 Camena d'Almeida, P. Europa. Barcelona, 1914.

Geog 809.19 Camena d'Almeida, P. La tierra; geografia general. 2a ed. Barcelona, 1919.

Geog 4209.28.20A Cameron, J. John Cameron's odyssey. N.Y., 1928.

An 3509.34 Cameron, John. The skeleton of British neolithic man. London, 1934.

Geog 4205.03 Campagne du navire l'Espoir de Honfleur, 1503-1505. (Avezac, A. de.) Paris, 1869.

Htn Geog 4218.06* Campbell, A. Voyage round the world from 1806-12. Edinburgh, 1816.

An 359.52.15 Campbell, Byram. American race theorists, a critique of their thoughts and methods. Boston, 1952.

An 359.58.10 Campbell, Byram. Race and social revolution. N.Y., 1958.

Geog 4348.07 Campbell, D. Journey over land to India. Philadelphia, 1807.

An 2108.69.5 Campbell, G.D. Primeval man. London, 1869.

An 2108.69.3 Campbell, G.D. Primeval man. N.Y., 1869.

An 2108.69.4 Campbell, G.D. Primeval man. N.Y., 1869.

Htn Geog 4327.39* Campbell, J. Travels and adventures of Eduard Brown. London, 1739.

Geog 4218.76.5 Campbell, J.F. My circular notes. v.1-2. London, 1876.

Geog 4168.07 Campe, Joachim Heinrich. Voyages anecdotiques. Paris, 1807.

Geog 5300.5 Campen, S.R. The Dutch in the arctic seas. London, 1877.

Htn An 3207.91* Camper, P. Dissertation physique...traits du visage. Utrecht, 1791. 2 pam.

Geog 4678.75F Campisi, Orazio. Petition and claim for damages to the ship Immacolata. n.p., 1879.

Geog 3570.55 Campos, Viriato. Viagens de Diogo Cão e de Bartolomeu Dias. Lisboa, 1966.

Geog 4218.03.11 Camus, A.G. Rapport à l'Institut national. Paris, 1803.

Geog 4679.14 Canada. Commission of Inquiry into the Loss of the British Steamship "Empress of Ireland". Report and evidence. Ottawa, 1914.

Geog 5559.10 Canada. Department of Marine and Fisheries. Report...dominion government expedition...1910. Ottawa? 1911.

Geog 5559.06.5 Canada. Department of Marine and Fisheries. Report on the dominion government expedition to Arctic islands and the Hudson Strait on board the C.G.S. "Arctic", 1906-07. Ottawa, 1909.

Geog 36.2 The Canadian Alpine journal. Banff. 1,1907+ 23v.

Geog 36.25.2 Canadian Association of Geographers. Canadien geographer. Author and subject index. Montreal. 1951+

Geog 36.26 Canadian Association of Geographers. Occasional papers in geography. Vancouver. 1-7,1960-1965//

Geog 36.25 Canadian geographer. Ottawa? 1,1951+ 13v.

Geog 36.3 Canadian geographical journal. Ottawa. 1,1930+ 46v.

Author and Title Listing

Geog 36.3.5 Canadian geographical journal. Regional index of articles, 1930-59. Ottawa, 1960.

Geog 36.15 Pamphlet box. Canadian Geographical Society.

Geog 36.25.2 Canadien geographer. Author and subject index. (Canadian Association of Geographers.) Montreal. 1951+

Geog 3550.9 Canale, M.G. Indicazioni di opere...sopra i viaggi, le navigazioni, le scoperte...degl'Italiani nel medio evo. Lucca, 1861.

Geog 4180.46 The Canarien...conquest and conversion. (Bontier, P.) N.Y., 1964?

Geog 818.80 Cañas Pinochet, A. El estudio de la jeografía por el dibujo de las cartas jeográficas. Santiago de Chile, 1880.

Geog 4308.30.5 Candolle, Augustin Pyramus de. L'Europe de 1830 vue à travers la correspondence de Augustin Pyremus de Candolle et Madame de Cércourt. Genève, 1966.

Geog 4208.81 Cané, Miguel. En viaje, 1881-82. Buenos Aires, 1917.

Geog 4208.81.5 Cané, Miguel. En viaje. Buenos Aires, 1949.

Geog 818.36 Cannabich, J.G.F. Lehrbuch der Geographie. 14e Aufl. Weimar, 1836.

NEDL Geog 818.16.20 Cannabich, T.G.F. Lehrbuch der Geographie. 17e Aufl. Weimar, 1855.

Geog 4208.78.4A Canton captain. (Connolly, J.B.) Garden City, N.Y., 1942.

An 2109.59.5 Cantoni, Remo. Il pensiero dei primitivi. Milano, 1959.

Geog 3019.24 Capasso, C. Le scoperte geografiche e i viaggi d'esplorazione. Messina, 1924.

Geog 4559.56.10 The Cape Horn breed. (Jones, William Herbert Sidney.) N.Y., 1956.

Geog 5220.11 Capel, R. Norden oder zu Wasser und Lande. Hamburg, 1678.

An 177.1 Capitan, J.L. Notice sur les travaux scientifiques. n.p., 1911.

An 2109.24.7 Capitan, Louis. L'humanité primitive dans la région des Eyzies. Paris, 1924.

Geog 4209.46.10 The cap'n's wife. (George, Albert J.) Syracuse, 1946.

Geog 4208.44 Capobianco, R. Breve racconto delle cose. Napoli, 1846.

Geog 4145.81 Captain Joshua Slocum. (Slocum, Victor.) N.Y., 1950.

Geog 4208.30.7 Captain Lefavor's forty years' travels. (Lefavor, William.) Philadelphia, 1877.

Geog 5518.50.10 Captain M'Clure's despatches from her majesty's discovery ship "Investigator". (M'Clure, R.) London, 1852.

Geog 5985.253 Captain Scott. (Gwynn, Stephen.) N.Y., 1930.

Geog 5985.253.15 Captain Scott and the Antarctic tragedy. (Brent, Peter L.) London, 1974.

Geog 4559.65 Captain's papers. (West, Ellsworth Luce.) Barre, Mass., 1965.

Geog 4180.51 Captivity of Hans Stade of Hesse. (Stade, H.) London, 1874.

Geog 4707.24A Captured by pirates. Two diaries of 1724-1725. (Seybold, R.F.) Boston, 1929.

An 359.61.20 La caracterologie ethnique. (Grieger, Paul.) Paris, 1961.

Geog 5919.62 Caras, Roger A. Antarctica. 1st ed. Philadelphia, 1962.

An 3309.05 Caratteri fisici delle Giovani donne del Lazio. (Montessori, Maria.) Roma, 1905.

Geog 4209.44.10 The caravan rolls on. (Baerlein, Henry.) London, 1944.

Geog 3570.36 The caravels of Christ. (Renault-Roulier, Gilbert.) N.Y., 1959.

Geog 600.15 Carcie, Giuseppe. Tre Fiorentini del Rinascimento. v.7. Roma, 1959?

Geog 500.23 Cardon, F. Publicazioni geografiche. Roma, 1892.

Geog 5938.80 Cardon, F. Le regioni polari antartiche. Roma, 1880.

Geog 5530.14 The career, last voyage. (Osborn, Sherard.) London, 1860.

Geog 4216.99 Careri, G. Giro del mondo. Napoli, 1699-1700. 6v.

Geog 4307.22 Careri, G. Viaggi per Europa. Napoli, 1722. 2v.

Geog 4559.59 Carfì, Francesco. Geografia economica e sociale del mare. Livorno, 1959.

An 177.2 Carias, Rafael. La indecision. Caracas, 1970.

Geog 618.1.7 Carl Ritter. (Guyot, A.) Princeton, 1860.

An 358.92.5 Carmichael, Charles H.E. L'anthropologie et les origines de la Société...l'Orient et l'Occident. Lisbonne, 1892.

Geog 4341.73.7 Carmoly, E. Notice historique sur Benjamin de Tudèle. Bruxelles, 1852.

Geog 4218.78.11 Carnegie, Andrew. Notes of a trip round the world. N.Y., 1879.

Geog 4218.79.10A Carnegie, Andrew. Round the world. Garden City, 1933.

An 359.49.15 Caro Baroja, Julio. Análisis de la cultura; etnología, historia, folklore. Barcelona, 1949.

Geog 4269.02 Carpenter, F.G. Europe. N.Y., 1902.

Htn Geog 816.35* Carpenter, N. Geographie delineated forth in two books. Oxford, 1635.

An 3538.80.2 Carr, L. Notes on the Crania of New England Indians. Boston, 1880.

An 3538.80 Carr, L. Notes on the Crania of New England Indians. Boston, 1880.

An 359.22.3A Carr-Saunders, A.M. The population problem. Oxford, 1922.

Geog 4208.91.5 Carrasco, Gabriel. Del Atlántico al Pacífico y un Argentino en Europa. 2a ed. Buenos Aires, 1891.

Geog 4229.35 Carré, J.M. Promenades dans trois continents. Paris, 1935.

Geog 4181.95 Carre. The travels of the Abbé Carre in India and the Near East. London, 1947-48. 3v.

An 359.35.5A Carrel, Alexis. Man, the unknown. 2. ed. N.Y., 1935.

An 359.35.5.8A Carrel, Alexis. Man, the unknown. 11. ed. N.Y., 1935.

An 359.35.5.10 Carrel, Alexis. Man, the unknown. 25. ed. N.Y., 1935.

An 359.35.6 Carrel, Alexis. Man, the unknown. 50. ed. N.Y., 1935.

An 359.35.4A Carrell, Alexis. Man, the unknown. N.Y., 1935.

Geog 4129.47 Carrington, D. The traveller's eye. N.Y., 1947.

Geog 4181.63 Carruthers, D. The desert route to India. London, 1929.

Geog 4709.57.10 Carse, Robert. The age of piracy. N.Y., 1957.

Geog 4709.57.12 Carse, Robert. The age of piracy. N.Y., 1965.

Geog 5838.90 Carstensen, A.R. Two summers in Greenland. London, 1890.

Geog 3060.14 Carta em resposta a hum amigo. Lisboa, 1815.

Geog 4309.05 Cartas a Estevez. (Cabrera, Raimundo.) Habana, 1906.

Geog 4208.37 Cartas á mis hijos. (Lofé, G.) N.Y., 1839.

Geog 3235.41 Carte geografiche. (Odorici, F.) Milano, 1877.

Geog 500.100 Carte geografiche cinquecentesche a stampa della Biblioteca Marciana e della Biblioteca del museo correr di Venezia. (Istituto Veneto di Scienza, Lettre ed Arte.) Venezia, 1954.

Geog 3235.7F Carte nautiche del medio evo. (De Luca, G.) Napoli, 1866.

Geog 4559.45.5 Carter, G.G. Looming lights. London, 1945.

Geog 4219.08.5 Carter, James. In the wake of the setting sun. London, 1908.

Geog 4308.25.2 Carter, N.H. Letters from Europe. N.Y., 1829. 2v.

Geog 4308.25 Carter, N.H. Letters from Europe. v.2. N.Y., 1827.

Geog 4181.124 Carteret, Philip. Carteret's voyage round the world, 1766-69. Cambridge, 1965. 2v.

Geog 4181.124 Carteret's voyage round the world, 1766-69. (Carteret, Philip.) Cambridge, 1965. 2v.

Geog 3070.30 Cartes marines, constructions navales, 1500-1650. (Anthiaume, A.) Paris, 1916. 2v.

Geog 4028.3 Cartier, Jean Pierre. Explorateurs et explorations. Paris, 1974.

Geog 3510.13 La cartographie hollandaise. (Smet, Antoine de.) Bruxelles, 1971.

Geog 3070.01 Pamphlet vol. Cartography.

Geog 4308.26 Carus, Karl Gustav. Reisen und Briefe. Leipzig, 1926. 2v.

Geog 816.36 Carvalho Da Costa, Antonio. Compendio geographico. Lisboa, 1636.

Geog 817.79F Carver, J. New universal traveller. London, 1779.

Geog 3550.14 La casa di Savoia e gli studii geografici. (Revelli, P.) Milano, 1903.

Geog 808.25 Casado Giraldes, J.P.C. Tratado completo de cosmographia. Paris, 1825-1828. 4v.

Geog 759.32.5 Case, E.C. College geography. N.Y., 1932.

Geog 4509.13 Casella, G. L'alpinisme. Paris, 1913.

Geog 4209.13 Casey, R.H. Notes made during a cruise around the world in 1913. N.Y., 1914.

An 2108.77 Caspari, O. Die Urgeschichte der Menschheit. Leipzig, 1877. 2v.

An 359.27.5 Cassandre, ou La folie des blancs. (Gohier, W.D.) Paris, 1927.

Geog 4308.97.3 Cassell and Company, Inc. Complete pocket guide to Europe. London, 1897.

Geog 4332.12 Cassi, Gellio. Il Mare Adriatico. Milano, 1915.

Geog 4239.68 Cassidy, Vincent H. The sea around them; the Atlantic Ocean, A.D. 1250. Baton Rouge, 1968.

Geog 4181.10 Castanhoso. Portugese expedition to Abyssinia. London, 1902.

Geog 4679.38.5 Castaway boats. (Slocum, Victor.) N.Y., 1938.

Geog 5660.25 Castberg, Frede. Østgrönlandsavtalen. Kristiania, 1924.

An 359.59.20 Caste, class, and race. (Cox, Oliver Cromwell.) N.Y., 1959.

An 359.48A Caste, class and race. 1. ed. (Cox, Oliver Cromwell.) Garden City, N.Y., 1948.

Geog 4239.69.5 Castelbajac, Bertrand de. De Plymouth à Newport par le sud de Nantucket. Paris, 1969.

An 3409.35.5 Castellanos, I. Dactiloscopia clinica. Habana, 1935.

Geog 4426.08 Castro Daire, A. de Ataide. Viagens de Reino para a India e da India para o Reino. Lisboa, 1957-58. 3v.

Geog 3570.25 Castro Soromenho. A maravilhosa viagem dos exploradores portugueses. v.1-12. Lisboa, 1946.

Geog 4145.9.5 Casual change. (Bodrick, Alan Houghton.) London, 1961.

Geog 5300.35 Caswell, J.E. Arctic frontiers. 1. ed. Norman, 1956.

Geog 3248.82 Cat, E. Les grands decouvertes maritimes. Paris, 1882.

Geog 5050.10 Catalog of the David Davis Polar Library. (David Davis Polar Library.) Menlo Park, 1969.

Geog 500.73 Catálogo alfabético de algunos géografos y exploradores. (Ossa Varela, P.) Bogotá, 1951.

Geog 3570.2 Catalogo bibliographico das publicacões. (Consiglieri-Pedroso, L.) Lisboa, 1912.

Geog 137.2.12 Catalogo de Vendas. (Lisbon. Sociedade de Geographia.) 1900-1916

Geog 500.97 Il catalogo degli autori; una biobibliografia geografica del '600. (Coronelli, Marco V.) Firenze, 1957.

Geog 603.6.5 Catalogo dei globi antici conservati in Italia. pt.1-2. Firenze, 1957.

Geog 212.109 Catalogo della biblioteca sociale. (Societa Geografica Italiana.) Roma, 1903.

Geog 500.9 Catalogo ragionato. (Rome, Italy. Bibliotheca Collegio Romano.) Roma, 1876.

Geog 500.58 Catalogo ragionato della geografica. (Milan. Bibliotheca.) Milano, 1927.

An 5.21F Catalogue. Authors. (Tozzer Library.) Boston, 1963. 26v.

An 5.21.1F Catalogue. Authors. 1st supplement. (Tozzer Library.) Boston, 1970. 6v.

An 5.21.2F Catalogue. Authors. 2d supplement. (Tozzer Library.) Boston, 1971. 2v.

An 5.21.3F Catalogue. Authors. 3d supplement. (Tozzer Library.) Boston, 1975. 3v.

An 5.22.01F Catalogue. Index to subject headings. (Tozzer Library.) Boston, 1963.

An 5.22F Catalogue. Subjects. (Tozzer Library.) Boston, 1963. 27v.

An 5.22.1F Catalogue. Subjects. 1st supplement. (Tozzer Library.) Boston, 1970. 6v.

An 5.22.2F Catalogue. Subjects. 2d supplement. (Tozzer Library.) Boston, 1971. 3v.

An 5.22.3F Catalogue. Subjects. 3d supplement. (Tozzer Library.) Boston, 1975. 4v.

An 5.3 Catalogue de la bibliothèque...1890. (Société d'Anthropologie, Paris.) Paris, 1891.

Geog 500.140 Catalogue of literature on the history of geography at the exposition. (Wystawa pt. Rozwój Historyczny Geografii Polskiej i Pismicnnictwo Polskie o Zakresu Historii Geografii, Warsaw, 1965.) Warsaw, 1965.

Geog 500.21 Catalogue of the library. (London. Royal Geographical Society.) London, 1852.

Geog 500.21.3 Catalogue of the library. (London. Royal Geographical Society.) London, 1865.

Geog 500.21.5 Catalogue of the library. (London. Royal Geographical Society.) London, 1895.

An 608.78 Catalogue of the objects of ethnotypical art. (Victoria, Australia. Public Library, Museum and National Galery.) Melbourne, 1878.

Geog 500.35 Catalogue of the York Gate Library. (Petherick, E.A.) London, 1886.

Geog 500.56 Catalogus geographie en reizen. (Amsterdam. Universiteit. Bibliotheek.) Amsterdam, 1923.

Geog 4180.36 Cathay and the way thether. (Yule, H.) London, 1866. 2v.

Geog 4181.33 Cathay and the way thither. (Yule, Henry.) London, 1913-14. 4v.

Geog 4181.33.1 Cathay and the way thither. v.1-4. (Yule, Henry.) Taipei, 1966. 2v.

An 359.56.5 The Catholic viewpoint on race relations. 1. ed. (LaFarge, John.) Garden City, N.Y., 1956.

Geog 4313.5 Catteau-Calleville, J.P. Tableau de la Mer Baltique. Paris, 1812. 2v.

Geog 4105.6 Cautions for the first tour. (Verax, V. (pseud.).) London, 1863.

Geog 4308.37.3 Cavalier auf Reisen im...1837. Leipzig, 1838.

An 2108.74.3 Cave hinting. (Dawkins, W.B.) London, 1874.

An 2109.28.9 The cave man's legacy. (Hankin, E.H.) London, 1928.

Geog 3055.78 Cayce, Edgar Evans. Edgar Cayce in Atlantis. N.Y., 1968.

Author and Title Listing

Geog 4218.87.12 Cecil, Evelyn. Notes of my journey round the world. London, 1889.
Geog 3107.31 Cellarius, C. Notitia orbis antiqui. Lipsiae, 1731-32. 2v.
Geog 658.81 Cenni intorno all'attuale. (Cora, G.) Torino, 1881.
Geog 4308.74 A centennial commissioner in Europe 1874-1876. (Forney, John W.) Philadelphia, 1876.
Geog 5311.3.30 Centkiewicz, Alina. Człowick o Którego upomniało się morze. Warszawa, 1966.
Geog 5399.56 Centkiewicz, Alina. Zanaevanie Arktiki. Moskva, 1956.
Geog 39.2.11 I cento anni del club alpino italiano. 2. ed. (Club Alpino Italiano.) Milano, 1964.
Geog 4269.61.5 Central Europe; a regional and human geography. (Mutton, Alice.) London, 1961.
Geog 4269.03 Central Europe. (Partsch, J.) London, 1903.
Geog 1540.47 Central Italy and Rome. 15. ed. (Baedeker, publishers.) Leipzig, 1909.
Geog 1540.48 Central Italy and Rome. 15. ed. (Baedeker, publishers.) Leipzig, 1909.
Geog 179.2.5 Centre de documentation cartographique et géographique. (France. Centre de Documentation Cartographique et Géographique.) Paris, 1956?
Geog 39.2.18 Centro Alpinistico Italiano. Sezione di Biella. Annuario. Biella.
Geog 3251.13 Ceradini, G. A proposito dei due globi mercatoriani. Milano, 1894.
Geog 3060.13.5 Çervé, W.S. Lemuria. 6th ed. San Jose, Calif., 1954.
An 5.34 Ceskoslovenská anthropologická bibliografie, 1955-1964. (Dokládal, Milan.) Brno, 1966-67.
An 5.32 Česnys, Gintautas. Lietuvos antropologijos bibliografija, 1470-1970. Vilnius, 1974.
Geog 4304.65.10 Cesta pana Lva z Rožmitála po západní Europě roku 1465-1467. (Slavík, František A.) Telči, 1890.
Geog 4309.06 Çestero, T.M. Hombres y piedras. Madrid, 1915.
Geog 4139.61 Čeští cestovatelé. (Kunský, J.) Praha, 1961. 2v.
Geog 4559.46 Ceux de la voile. (Chaveriat, R.) Paris, 1946.
Geog 4209.30.20 Chable, Jacques Edouard. Jazz, boomerang et kimonos. Paris, 1930.
Geog 4269.58 Chabot, Georges. L'Europe du Nord et du Nord-Ouest. Paris, 1958. 3v.
An 3309.67 Chai, Chen Kang. Taiwan aborigines. Cambridge, 1967.
Geog 3140.5 Der Chaldäer Seleukos. (Ruge, S.) Dresden, 1865.
An 359.06.5 Chalikiopoulos, L. Landschafte...Kulturtypen. Leipzig, 1906.
An 379.54 The challenge of man's future. (Brown, H.S.) N.Y., 1954.
An 2109.75.5 The challenge of the primitives. (Clarke, Robin.) London, 1975.
Geog 5130.2 Challenge to the poles. (Grierson, J.) Hamden, Conn., 1964.
Geog 4329.29.5 Chalmers, T.M. Under the olive trees. N.Y., 1931.
Htn Geog 1811.3* Chamberlain, B.H. Handbook for...Japan. N.Y., 1893.
Geog 1811.2 Chamberlain, B.H. Handbook for...Japan. 3. ed. London, 1891.
Geog 578.94 Chamber's consise gazetteer of the world, topographical, statistical, historical. London, 1894.
Geog 4180.23 Champlain, S. Narrative of voyages to West Indies and Mexico. London, 1859.
Geog 4308.31.2A Change of air, or The diary. (Johnson, J.) London, 1831.
Geog 5839.53 Changing Greenland. (Williamson, G.) London, 1953.
Geog 4308.56 Channing, W. A physician's vacation. Boston, 1856.
An 3559.08 Channing, Walter. The hard palate in normal and feeble-minded individuals. N.Y., 1908.
An 2059.01 Chantre, E. L'homme quaternaire. Paris, 1901.
An 3409.41 Chapel, Charles E. Fingerprinting; a manual of identification. N.Y., 1941.
Htn Geog 4679.26* Chapin, H.M. Biblioteca Titanicana. Metuchen, 1926.
Geog 4679.26.2 Chapin, H.M. Biblioteca Titanicana. Metuchen, 1926.
Geog 4208.87.15 Chapin, J.H. From Japan to Granada. N.Y., 1889.
Geog 4208.73.5 Chapman, Charles. The ocean waves. London, 1875.
Geog 4145.15.2 Chapman, F.S. Living dangerously. London, 1953.
Geog 4145.15 Chapman, F.S. Living dangerously. N.Y., 1953.
Geog 4521.26 Chapman, F.S. Memories of a mountaineer. London, 1951.
Geog 5559.30 Chapman, F.S. Northern lights. London, 1932.
Geog 4559.72.10 Chapman, Gifford Desmond. Kangaroo Island shipwrecks. Canberra, 1972.
An 359.42A Chapple, Eliot D. Principles of anthropology. N.Y., 1942.
An 379.24 Character of races as influenced by...environment. (Huntington, Ellsworth.) N.Y., 1924.
Geog 818.59 Charakteristiken zur Erd und Völkerkunde. (Pütz, Wilhelm.) Köln, 1859-1860. 2v.
An 186.2.5 Charbonnier, Georges. Conservations with Claude Lévi-Strauss. London, 1969.
Geog 6009.03 Charcot, J.B. Le "Francais" au Pole Sud. Paris, 1906.
Geog 5839.29.5 Charcot, J.B. La mer du Groenland, croisières du "Pourquoi pas?". Paris, 1929.
Geog 6009.08.5 Charcot, Jean B. Autour du pole sud; expédition du "Pourquoi pas?" 1908-1910. London, 1937.
Geog 6009.08 Charcot, Jean B. The voyage of the 'Why not?' in the Atlantic. London, n.d.
Geog 603.5.10 Charcot of the Antartic. (Oulié, M.) London, 1938.
Geog 4308.17.6 Charleville, H.C.B. A journey to Florence in 1817. London, 1951.
Geog 13.2.5 Charter, by-laws and list of members. (American Geographical Society of New York.) N.Y. 1870
Geog 182.2 Charter. (Philadelphia Geographical Society.) Philadelphia.
Geog 13.2.3 Charter and by-laws. (American Geographical Society of New York.) 1857
Geog 4138.54 Charton, E.T. Voyageurs anciens et modernes. Paris, 1854-57. 4v.
Geog 4168.54 Charton, Edouard T. Voyageurs anciens et modernes. v.1-4. Paris, 1854-57. 2v.
Geog 4509.53.5 Chatellus, A. de. Alpiniste. Grenoble, 1953.
Geog 4209.22 Chats over a pipe; a tale of two brothers. (Glass, James.) London, 1922.
An 3309.37 Chattapadkyay, Bojra K. Les affinités somatiques des Brahmines Moithils et Kanarijias de Bihar. Thèse. Paris, 1937?
Geog 4559.27.15 Chatterton, E.K. The brotherhood of the sea. London, 1927.
Geog 4559.29.15 Chatterton, E.K. On the high seas. Philadelphia, 1929.
Geog 4559.31 Chatterton, E.K. Sailing the seas. London, 1931.
Geog 4559.24 Chatterton, E.K. Seamen all. Boston, 1924.
Geog 4169.28 Chatterton, E.K. Ventures and voyages. London, 1928.
NEDL Geog 4218.79 Chatty letters from the East and West. (Wylie, A.H.) London, 1879.
Geog 3019.69 Chaunu, Pierre. L'expansion européene du XIIIe au XVe siècle. 1. éd. Paris, 1969.
Geog 5050.3 Chavanne, J. Die Literatur über die Polar-Regionen. Wien, 1878.
Geog 4559.46 Chaveriat, R. Ceux de la voile. Paris, 1946.
An 359.71.15 Cheboksarov, Nikolai N. Narody, rasy, kul'tury. Moskva, 1971.
Geog 4131.61 Cheechi and Company, Washington, D.C. The future of tourism in the Pacific and Far East. Washington, 1961.
Geog 4558.50.5 Cheever, H.T. The whale and its captors. N.Y., 1850.
Geog 5379.35 Cheliuskin Expedition, 1933-34. The voyage of the Chelyuskin. N.Y., 1935.
An 379.73.10 Chelovek in landshafty. (Milkov, Fedor N.) Moskva, 1973.
Geog 4145.87.5 Chelovek s zheleznym olenem. (Kharitanovskii, Aleksandr A.) Moskva, 1965.
Geog 4328.43 Chenavard, Antoine N. Voyage en Grèce et dans le Levant fait en 1843-1844. Lyon, 1849.
Geog 4209.64.30 Cherez 15 morei i 2 okeana; puteshestviia. (Kozlovs'kyi, Mykola F.) Kiev, 1964.
Geog 4329.71 Chernishev, Slaveho. Po more. Sofiia, 1971.
Geog 6009.10.30 Cherry-Garrard, Apsley G.B. The worst journey in the world; Antarctic 1910-13. N.Y., 1930.
Geog 6009.10.32 Cherry-Garrard, Apsley G.B. The worst journey in the world; Antarctic 1910-13. v.1-2. N.Y., 1937.
Geog 520.8 Chesneau, J. Le voyage de M. d'Aramon. Paris, 1887.
Geog 5399.37.10 Chetyre tovarishcha. (Kreukel', E.T.) Moskva, 1940.
Geog 4219.20.5 Chevalier, Haakon Maurice. The last voyage of the schooner Rosamond. London, 1970.
Geog 4509.31.10 Chiacchiere di un alpinista. (Giussani, C.) Milano, 1931.
Geog 665.53 Chicago. University. Norman Wait Harris Foundation. Reports of round tables, 1937. Geographic aspects of international relations. n.p., 1937.
An 170.157 Chicago. University. Seminar of Racial and Cultural Contacts. Proceedings, 1935-36. n.p., n.d.
An 359.56.10 Chicago Urban League. Staff report on a scientist's report on race differences. Chicago, 1956.
Geog 4219.67 Chichester, Francis Charles. Gipsy Moth circles the world. N.Y., 1967.
Geog 4239.70.10 Chichester, Francis Charles. The romantic challenge. London, 1971.
Geog 4209.00 Chief Officer Brown. (Boyer, Bessie B.) Chicago? 1900.
Geog 4419.12A Chiefly the Orient. (Gardner, G.P.) Boston, 1912.
Geog 4502.5.16 Das Chiemgaubuch. (Hager, F.) München, 19- .
Geog 4308.89 Child, T. Summer holidays...Europe. N.Y., 1889.
An 2109.28.5A The child in primitive society. (Miller, Nathan.) N.Y., 1928.
An 2109.51 Childe, V.G. Social evolution. London, 1951.
An 2109.51.2 Childe, V.G. Social evolution. N.Y., 1951.
An 2109.25.3A Childe, Vere Gordon. The dawn of European civilization. London, 1925.
An 2109.25.4 Childe, Vere Gordon. The dawn of European civilization. N.Y., 1925.
An 2109.39.3 Childe, Vere Gordon. Dawn of European civilization. N.Y., 1939.
An 2109.39.3.6 Childe, Vere Gordon. The dawn of European civilization. 6. ed. London, 1957.
An 379.50F Childe, Vere Gordon. Prehistoric migrations in Europe. Cambridge, 1950.
An 359.09 The childhood of man. (Frobenius, Leo.) London, 1909.
An 2108.74.5 The childhood of the world. (Clodd, Edward.) Boston, 1874.
An 2108.80.3 Childhood of the world. (Clodd, Edward.) London, 1880.
An 2108.73.3 The childhood of the world. 2. ed. (Clodd, Edward.) London, 1873.
An 2108.75 The childhood of the world. 3. ed. (Clodd, Edward.) London, 1875.
Geog 4209.71.45 Children of Cape Horn. (Swale, Rosie.) London, 1974.
Geog 3060.12.50 The children of Mu. (Churchward, J.) N.Y., 1937.
An 2109.23.5 The children of the sun. (Perry, W.J.) N.Y., 1923.
Geog 4219.64.5 Children of three oceans. (Lewis, David Henry.) London, 1969.
Geog 665.110 Chile. Universidad, Santiago. Facultad de Filosofia y Educacion. Estudios geográficos. Homenaje de la Facultad de Filosofia y Educacion a Don Huberto Fuenzalida Villegas. Santiago, 1966.
Geog 5925.50 Chilean sovereignty in Antarctica. (Pinochet de la Barra, Oscar.) Santiago, 1955.
Geog 5399.72 Chilingarov, A.N. Pod nogami ostrov ledianoi. Moskva, 1972.
Geog 4209.43.5 Chin takes and talks, an absurd travel diary. (Collins, Grace.) Chicago, 1943.
Geog 608.11 China charts the world. Hsu, Chi-Yü and his geography of 1848. (Drake, Fred W.) Cambridge, 1975.
Geog 4268.85.5A Chisholm, G.G. Europe. London, 1899-1902. 2v.
An 379.71 Chisholm, Michael. Research in human geography. London, 1971.
An 2159.74 Chisla v grafike paleolita. (Frolov, Boris A.) Novosibirsk, 1974.
Geog 659.42.5 Cholley, André. La géographie. 2. éd. Paris, 1951.
Geog 659.42 Cholley, André. Guide de l'étudiant en géographie. 1. éd. Paris, 1942.
Geog 659.62 Chonological differentiation as the fundamental principle of geography. (Jong, Guben.) Groningen, 1962.
Geog 4269.10 Chōrai kai ladi tēs Evrōpēs. (Sothriados, G.) Athēnai, 1910.
Geog 659.65.7 Chorley, Richard J. Frontiers in geographical teaching. 2nd ed. London, 1970.
Geog 659.71.5 Chorley, Richard J. Models in geography. London, 1971.
Htn Geog 4265.61* Chorographia. (Barreiros, G.) Coimbra, 1561.
Geog 4308.54.25 Choules, J.O. Cruise of the steam yacht, North Star. Boston, 1854.
Geog 4308.54.20 Choules, J.O. Cruise of the steam yacht, North Star. Boston, 1854.
Geog 4502.10.3 Christ, Fritz. Die erste Ersteigung des Totenkirches durch den Christ Fick-Kamin. München, 1937.
Geog 6009.36 Christensen, Lars. My last expedition to the Antarctic, 1936-37. Oslo, 1938.
Geog 4180.98A Christian topography of Cosmas. (Cosmas.) London, 1897.
Geog 5919.51 Christie, E.W.H. The Antarctic problem. London, 1951.
Geog 520.6 Christophe Colomb. (Harrisse, H.) Paris, 1884. 2v.
Geog 4328.38.5 The Christopher Happoldt journal. (Happoldt, Christopher.) Charleston, 1960.
Geog 4180.88 Christy, M. Voyages of Captain Luke Foxe and Captain T. James. London, 1894. 2v.
Geog 4180.95 Chronicle of discovery...of Guinea. (Azurara, G.E. de.) London, 1896-99. 2v.
Geog 4208.80.5 Chronioles of no-man's land. (Boyle, Frederick.) London, 1880.

170

Author and Title Listing

	Geog 4158.03F	Chronological history...discoveries..South Sea. (Burney, J.) London, 1803-1817. 5v.
	Geog 5208.18	Chronological history of voyages into Arctic regions. (Barrow, John.) London, 1818.
	Geog 5939.07.5	Chun, Karl. Die Erforschung der Antarktis. Leipzig, 1907.
	Geog 500.119	Church, Martha. A basic geographical library. Washington, 1966.
	Geog 4157.04F	Churchill, J. Collection of voyages and travels. London, 1704-07. 8v.
	Geog 4157.04.3F	Churchill, J. Collection of voyages and travels. 3. ed. London, 1744. 6v.
	An 2339.17	Churchill, William. Club-types of nuclear Polynesia. Washington, 1917.
	Geog 3060.12.50	Churchward, J. The children of Mu. N.Y., 1937.
	Geog 3060.12.25	Churchward, J. The lost continent of Mu. N.Y., 1934.
	Geog 3060.12.55	Churchward, J. The sacred symbols of Mu. London, 1960.
	Geog 3060.12.45A	Churchward, J. The sacred symbols of Mu. N.Y., 1934.
	Geog 600.7	Ciampi, I. Oltre l'Alpe e il mare. Roma, 1865.
	Geog 622.1	Ciampi, Ignatius. Della vita e...opere di Pietro della Valle il Pellegrino. Roma, 1880.
	An 359.63.15	Ciba Foundation. Man and his future. London, 1963.
	Geog 3580.35	La ciencia geografica española del siglo XVI. (Gavira, José.) Madrid, 1931.
	Geog 4180.68	Cieza de Leon, P. de. Second part of chronicle of Peru. N.Y., 1964?
	Geog 4180.33	Cieza de Leon, P. de. Travels of Piedro Cieza de Leon. London, 1864.
	Geog 4181.35	Cieza de Leon, P. de. The war of Chupas. London, 1918.
	Geog 4181.54	Cieza de Leon, P. de. The war of Las Salinas. London, 1823.
	Geog 4181.31	Cieza del Leon, P. de. The war of Quito. London, 1913.
	Geog 4208.91	Cimon, Henri. Impressions de voyage. Québec, 1895-1902. 2v.
	Geog 4329.17.5	Cimon, Henri H.M. Aux vieux pays. 3. ed. Montréal, 1917.
	Geog 4219.34.5	The Cingalese prince. (Atkinson, B.) Garden City, 1934.
	An 359.69	Cinq études d'ethnologie. (Leiris, Michel.) Paris, 1969.
	An 359.37.9	Cinquante siècles d'évolution ethnique autour de la Mer noire. (Baschmakoff, A.A.) Paris, 1937.
	Geog 4309.65.5	Cione, Edmondo. Questa Europa. Milano, 1965.
	Geog 5180.6	Circumpolar problems. 1. ed. (Symposium on Circumpolar Problems, Luleå, Sweden and Tromsø, Norway, 1969.) Oxford, 1973.
	Geog 4110.37	Cities. (Croft-Cooke, Rupert.) London, 1951.
	Geog 4209.63.10	Cities. (Morris, J.) London, 1963.
	Geog 4269.03.5	Cities. (Symons, A.) London, 1903.
	Geog 4219.60	Citizen abroad. (Baty, Eben Neal.) N.Y., 1960.
	An 379.52.10	City, region and regionalism. (Dickinson, Robert Eric.) London, 1952.
	Geog 4110.36	City portraits; a guide to 60 of the world's great cities. 1. ed. (Sheraton, Mimi.) N.Y., 1964.
	Geog 4209.57.10	La ciudad de Toledo. (Jara Peralta, José.) Madrid, 1957.
	An 2108.45	Civilisation primitive. (Brotonne, F. de.) Paris, 1845.
	An 2159.73.20	Les civilisations des étoiles. (Moreau, Marcel.) Paris, 1973.
	Geog 3049.62.5	Les civilisations inconnues; mythes ou réalités. (Hutin, Serge.) Paris, 1962.
	An 2109.56	Les civilisations tardenoisiennes en Europe occidentale. (Barrière, Claude.) Bordeaux, 1956.
	An 359.43.10	Claessens, A. Race prejudice. N.Y., 1943.
	An 359.63.10	Clan, caste, club. (Hsu, Francis L.K.) Princeton, 1963.
	Geog 4307.50	Clancy, Michael. The memoirs. v.1-2. Dublin, 1750.
	Geog 3019.72.35	Claral, Paul. La pensée géographique. Paris, 1972.
	An 359.46	Clark, Grahame. From savagery to civilization. London, 1946.
	Geog 4208.47.5	Clark, J.G. Lights and shadows of sailor life. Boston, 1847.
	Geog 4208.47	Clark, J.G. Lights and shadows of sailor life. Boston, 1848.
	An 2109.61.5	Clark, J.G.D. World prehistory. Cambridge, Eng., 1961.
	An 2109.70	Clark, John Grahame Douglas. Aspects of prehistory. Berkeley, 1970.
	An 2109.65.10	Clark, John Grahame Douglas. Prehistoric soctieties. 1st American ed. N.Y., 1965.
	An 2109.67	Clark, John Grahame Douglas. The Stone Age hunters. London, 1967.
	Geog 4521.3	Clark, Ronald. An eccentric in the Alps. London, 1959.
	Geog 4348.13.10	Clarke, E.D. Travels in...Europe, Asia and Africa. London, 1816-1824. 11v.
	Geog 4348.13.20	Clarke, E.D. Travels in Russia, Tartary and Turkey. Edinburgh, 1839.
	Geog 4348.10.5	Clarke, E.D. Travels in various countries of Europe, Asia and Africa. Edinburgh, 1811. 2v.
	Geog 4348.13.5	Clarke, E.D. Travels in various countries of Europe. N.Y., 1813. 4v.
Htn	Geog 4348.10.10*	Clarke, E.D. Travels in various countries of Europe. 1st American ed. v.3-4. N.Y., 1815. 2v.
	Geog 4308.52.10	Clarke, J.F. Eleven weeks in Europe. Boston, 1852.
	Geog 3018.03	Clarke, J.S. Progress of maritime discovery. London, 1803.
	An 2109.75.5	Clarke, Robin. The challenge of the primitives. London, 1975.
Htn	Geog 816.57.5*	Clarke, Samuel. A geographical description of all the countries in the known world. London, 1671.
	An 359.24.15	The clash of color. 2. ed. (Mathews, Basil.) N.Y., 1924.
	An 359.24.18	The clash of colour. (Mathews, Basil.) Port Washington, 1973.
	An 359.24.10	The clash of colour. 7. ed. (Mathews, Basil.) London, 1924.
	An 359.26.15	The clash of colour. 16. ed. (Mathews, Basil.) London, 1926.
	An 359.27.9A	The clash of culture and the contact of races. (Pitt-Rivers, G.H.L.F.) London, 1927.
	An 3409.37.3	Clasificación anatómica-matematica del dactilograma. (Borges Badell, A.) La Habana, 1937?
	An 2109.12	Classen, K. Die Völker Europas zur jüngeren Steinzeit. Stuttgart, 1912.
	Geog 3129.58	The classical tradition of geography. (Paassen, Christian van.) Groningen, 1957.
	Geog 3129.57	The classical tradition of geography. (Paassen, Christian van.) Groningen, 1957.
	An 3409.13.5	Classification and uses of fingerprints. 4. ed. (Henry, Edward R.) London, 1913.
	An 608.62	Classification méthodique des produits de l'industrie. (Jomard, E.F.) Paris, 1862.
	An 3409.36	Classification of fingerprints. J.E. Hoover, director. (United States. Federal Bureau of Investigation.) Washington, 1936.
	An 3208.50.	The classification of mankind by the hair. (Browne, P.A.) Philadelphia, 1850.
	An 136.5	Claude Lévi-Strauss; an introduction. (Paz, Octavio.) Ithaca, 1970.
	An 359.70.5	Claude Lévi-Strauss; ou, La structure et le malheur. (Lévi-Strauss, Claude.) Paris, 1970.
	An 136.5.5	Claude Lévi-Strauss. (Leach, Edmund Ronald.) N.Y., 1970.
	An 136.5.87	Claude Lévi-Strauss. (Leach, Edmund Ronald.) N.Y., 1974.
	An 136.5.70	Claude Lévi-Strauss. 1. ed. (Montes, Santiago.) San Salvador, 1971.
	An 136.5.17	Claude Lévi-Strauss o el nuevo festin de Esopo. 2. ed. (Paz, Octavio.) Mexico, 1969.
	An 136.5.10	Claude Lévi-Strauss ou la Passion de l'inceste. (Simonis, Yvan.) Paris, 1968.
	Geog 4709.65.5	Clausen, Carl J. A 1715 Spanish treasure ship. Gainesville, 1965.
	An 359.37.11	Clauss, Ludwig F. Rasse ist Gestalt. München, 1937.
	An 359.34.9	Clauss, Ludwig F. Rasse und Seele; eine Einführung in den Sinn der leiblichen Gestalt. München, 1934.
	An 358.60	Clavel, A. Races humaines. Paris, 1860.
	Geog 4180.26	Clavijo, R.G. de. Narrative of embassy...to court of Timar. London, 1859.
	Geog 5160.40	Clear lands and icy seas. (Stanwell-Fletcher, T.M.) N.Y., 1958.
	Geog 4219.24	Clements, Rex. A gipsy of the Horn. London, 1924.
	Geog 4219.24.2	Clements, Rex. A gipsy of the Horn. 2. ed. Boston, 1925.
	Geog 4678.40.3	Cleveland, H.R. A letter to Daniel Webster. Boston, 1840.
	Geog 4208.40.5	Cleveland, R.J. Narrative of voyages and commercial enterprises. Boston, 1850.
Htn	Geog 4208.40.3*	Cleveland, R.J. Narrative of voyages and commercial enterprises. Cambridge, 1842. 2v.
	Geog 4208.40	Cleveland, R.J. Voyages and commercial enterprises. N.Y., n.d.
	Geog 4208.40.10	Cleveland, R.J. Voyages of a merchant navigator. N.Y., 1886.
	An 359.03	Clevenger, S.V. The evolution of man and his mind. Chicago, 1903.
	Geog 4509.35.5	Climbing days. (Richards, D.P. (Mrs.).) London, 1935.
	Geog 4509.31.15	Climbs and ski runs. (Smythe, F.S.) Edinburgh, 1931.
	Geog 4208.53.5	A clipper ship and her commander. (Mather, Frank J.) n.p., 1904.
	Geog 4559.23	The clipper ship "Sheila". (Angel, W.H.) Boston, 1923.
	An 2108.74.5	Clodd, Edward. The childhood of the world. Boston, 1874.
	An 2108.80.3	Clodd, Edward. Childhood of the world. London, 1880.
	An 2108.73.3	Clodd, Edward. The childhood of the world. 2. ed. London, 1873.
	An 2108.75	Clodd, Edward. The childhood of the world. 3. ed. London, 1875.
	An 2108.95.3	Clodd, Edward. The story of "primitive" man. N.Y., 1895.
	Geog 3019.42.4	Clozier, René. Histoire de la géographie. 4. éd. Paris, 1967.
	Geog 4509.34	Club Alpin Francais. Manuel d'alpinisme. Chambéry, 1934. 2v.
NEDL	Geog 39.1	Club Alpin Française. Annuaire. Paris. 1-30,1874-1903 30v.
NEDL	Geog 39.1.3	Club Alpin Française. Annuaire. Table générale. Paris. 1874-1888
	Geog 39.1.5	Pamphlet vol. Club Alpin Française. 2 pam.
	Geog 4509.53.15	Club Alpino Italiano. Alpinismo italiano nel mondo. Milano, 1953.
NEDL	Geog 39.2	Club Alpino Italiano. Bollettino. Torino. 1-41 13v.
	Geog 39.2.11	Club Alpino Italiano. I cento anni del club alpino italiano. 2. ed. Milano, 1964.
NEDL	Geog 39.2.5	Club Alpino Italiano. Indice generale dei cinquanta primi numeri (dal 1865 al 1884). Torino, 1885.
	Geog 39.2.10	Club Alpino Italiano. Revista mensile publicata per cura de consiglio direttivo della sede centrale. Torino. 1-60,1882-1941 37v.
NEDL	Geog 39.2.10	Club Alpino Italiano. Revista mensile publicata per cura de consiglio direttivo della sede centrale. Torino. 1-18,1882-1899 18v.
	Geog 39.2.20	Club Alpino Italiano. Sezione Antonio Locatelli, Bergamo. Annuario. 1951+ 4v.
	Geog 39.2.16	Club Alpino Italiano. Sezione di Roma. Annuario. Roma. 1-3,1886-1889
	An 2339.17	Club-types of nuclear Polynesia. (Churchill, William.) Washington, 1917.
	Geog 4249.51	Clune, Frank. Hands across the Pacific. Sydney, 1951.
	Geog 5398.90	Clutterbuck, W.J. The skipper in Arctic seas. London, 1890.
	Geog 816.61.19	Cluverius, P. Introduction à la geographie universelle. Paris, 1631.
NEDL	Geog 816.61.3	Cluverius, P. Introduction into geography. Oxford, 1657.
	Geog 816.61	Cluverius, P. Introductionis in...geographiam. Amsterdam, 1661.
	Geog 816.61.6	Cluverius, P. Introductionis in...geographiam. Amsterdam, 1686.
	Geog 816.61.5	Cluverius, P. Introductionis in...geographiam. Brunsvigae, 1672.
Htn	Geog 816.61.9*	Cluverius, P. Introductionis in...geographiam. Londini, 1711.
Htn	Geog 816.61.4*	Cluverius, P. Introductionis in universam geographiam. Amsterdam, 1659.
	Geog 816.61.7	Cluverius, P. Introductionis in universam geographiam. Amsterdam, 1697.
Htn	Geog 816.61.13*	Cluverius, P. Introductionis in universam geographiam. Amsterdam, 1729.
Htn	Geog 816.61.16*	Cluverius, P. Introductionis in universam geographiam. Lugdunum Batavorum, 1627.
	Geog 4309.63	Cluzel, M.E. Des brumes de la Mer du Nord. Paris, 1963.
	Geog 4309.64.5	Cluzel, M.E. Impressions sur l'Europe. Paris, 1964.
	Geog 4219.72.5	Cluzel, Magdeleine. Au fil de l'eau, autour de la terre. Paris, 1973.
	Geog 4145.84.10	The coast of incense. 1. ed. (Stark, Freya.) London, 1953.
	Geog 5839.37	The coast of northeast Greenland. (Boyd, Louise A.) N.Y., 1948.
	Geog 3060.10	The coasts of illusion. (Firestone, C.B.) N.Y., 1924.
	An 2059.51	Coates, Adrian. Prelude to history. London, 1951.
	Geog 4180.11	Coats, W. Geography of Hudson's Bay. London, 1852.
	Geog 4309.14.11A	Cobb, I.S. Europe revised. N.Y., 1914.
	An 2109.72.10	Cocchiara, Giuseppe. L'eterno selvaggio. Palermo, 1972.
	Geog 4709.65	Cochran, Hamilton. Freebooters of the Red Sea: pirates, politicians and pieces of eight. Indianapolis, 1965.
	Geog 4328.10	Cockburn, G. A voyage to Cadiz and Gibraltar, up the Mediterranean, to Sicily and Malta, in 1810 and 1811. London, 1815. 2v.

Author and Title Listing

Geog 4180.66 Cocks, R. Diary of Richard Cocks...1615-22. N.Y., 1964. 2v.

Geog 4309.26.25 A cocktail continentale. (Reynolds, Bruce.) N.Y., 1926.

Geog 3129.55 Codazzi, Angela. La geografia dei Greci e dei Romani. 2. ed. Milano, 1955.

Geog 3070.56 Codazzi, Angela. Storia delle carte geografiche. Milano, 1952-

Geog 3070.56.5 Codazzi, Angela. Storia delle carte geografiche. Milano, 1958.

Geog 3019.59.10 Codazzi, Angelo. Storia della geografia. Milano, 1959.

Geog 520.19 Codex Ramirez. Histoire de l'origine des Indiens. Paris, 1903.

Geog 4428.68 Codine, J. Mémoire géographique sur mer des Indes. Paris, 1868.

Geog 4208.48.2 Cöllen, Franz A. Reise Album vom 15ten bis zum 22ten Lebensjahre. Hamburg, 1852.

Geog 4208.48.3 Cöllen, Franz A. Reisen und Dichtungen. Berlin, 1864.

Geog 114.7.85 Coën, A. Venticinque anni di Lavoro dell'Istituto Geographico Militare. Firenze, 1898.

Geog 4218.69 Coffin, C.C. Our new way round the world. Boston, 1869.

An 359.08.3 Coffin, E.W. On the education of backward races. Worcester, 1908.

Geog 4218.49 Coffin, George. Pioneer voyage to California and round the world, 1849-1852. Chicago? 1908.

Geog 4208.54.10 Coffin, Robert. The last of the Logan. Ithaca, 1941.

Geog 4208.60.5 Coggeshall, G. An historical sketch of commerce and navigation. N.Y., 1860.

Geog 4208.58.5 Coggeshall, G. Thirty-six voyages to various parts of the world, 1799-1841. 3. ed. N.Y., 1858.

Geog 4208.53.2 Coggeshall, G. Voyages to various parts...1800-1831. N.Y., 1853.

Geog 4208.51 Coggeshall, G. Voyages to various parts of the world...1799-1844. N.Y., 1851.

Geog 4269.70 Coghill, Ian G. Western Europe: geographical studies. London, 1970.

Geog 4308.68.20 Coghill, James Henry. Abroad. N.Y., 1868.

An 170.360 Cognitive anthropology; readings. (Tyler, Stephen A.) N.Y., 1969.

Geog 112.5 Coğrafi araştirmalar. (Istanbul. Üniversite. Cografya Ensitüsü.) n.p., n.d.

Geog 579.59 Coğrafya sözlüğü. Fasc.1-5. (Öngör, Sami.) Istanbul, 1959.

Geog 579.59.5 Coğrafya terimleri sözlüğü. (Öngör, Sami.) Istanbul, 1975.

An 2109.64.5 Cohen, Y.A. The transition from childhood to adolescence. Chicago, 1964.

An 170.161 Cohen, Yehudi A. Man in adaptation. Chicago, 1968. 2v.

Geog 4678.54.10 Cola, S. de. Difesa pel cav F. Miceli...scontro dei...Ercolano e Sicilia. Messina, 1858.

Geog 3570.58 Colaboração portuguesa nos descobrimentos nauticos das outras nações. (Heleno, Manuel Domingues.) Lisboa, 1932.

Geog 4209.47 Colam, Lance. Death over my shoulder. London, 194-.

Geog 3019.56.5 Colamonico, Carmelo. Compendio di storia della geografia e delle esplorazioni geografiche. Napoli, 1956.

Geog 808.75F Colange, Leo de. The picturesque world, or Scenes in many lands. pt.1,3-6,8-13,15-16,18,20,26. Boston, 1875.

Geog 4168.87F Colange, Leo de. Voyages and travels. Boston, 1887. 2v.

Geog 4168.87.2F Colange, Leo de. Voyages and travels. Boston, 1887.

Geog 3580.40 Colbrecht, Jozsf. De vleminfen en de Spansche. Antwerp, 1927.

Geog 6009.28.25 Cold, the record of an Antarctic sledge journey. (Gould, L.M.) N.Y., 1931.

Geog 6009.62 The coldest place on earth. (Thomson, Robert Baden.) Wellington, 1965.

Geog 4209.68.5 Cole, Jean. Trimaran against the trades. Wellington, 1968.

Geog 759.68 Cole, John Peter. Quantitative geography. London, 1968.

An 2109.40.10 Cole, M.C. The story of primitive man. Chicago, 1940.

Geog 3580.5.5 Coleccion de los viajes y descubrimientos. (Fernandes de Navaretti, Martin.) Buenos Aires, 1945-46. 5v.

Geog 3580.5 Coleccion de los viajes y descubrimientos. 2. ed. v.1-3. (Fernandes de Navaretti, Martin.) Madrid, 1958. 5v.

Geog 4182.3 Colenbrander, H.T. Korte historiael...voyagiens...David P. de Vries. 's-Gravenhage, 1911.

Geog 3570.15 Collecção de opuscolos...relativos a historia das navegacões. pt.1-4. Lisboa, 1875.

Geog 4218.82.3 The collected remarkable travels of G. Pitt. 3. ed. (Pitt, George.) Glasgow, 1887.

Geog 148.1.8 Collecting national geographic magazines. (Buxbaum, Edwin Clarence.) Milwaukee, 1935.

Geog 4168.23 Collection de relations de voyages. Paris, 1823.

Geog 4157.88 Collection de tous les voyages. (Berenger, J.P.) Lausanne, 1788-91. 9v.

Geog 4158.05 Collection of...voyages and travels. (Phillips, R.) London, 1805-08. 7v.

Geog 4158.05.2 Collection of...voyages and travels. (Phillips, R.) London, 1810. 3v.

Geog 4167.65 A collection of authentic...voyages. v.2. (Barrow, J.) London, 1765.

Htn Geog 4346.93* Collection of curious travels and voyages...eastern countries. (Ray, John.) London, 1693.

Geog 4180.18 Collection of documents on Spitzbergen and Greenland. (Martens, F.) London, 1855.

Geog 3520.10.10F Collection of early voyages, travels. (Hakluyt, Richard.) London, 1809. 5v.

Geog 4167.97 A collection of late voyages and travels. (Heron, Robert.) Edinburgh, 1797.

Geog 3235.58 A collection of maps. (Beans, G.H.) Jenkintown, Pa., 1943.

Htn Geog 4308.78.5* A collection of newspaper articles, describing his experiences in Europe, 1878-1880. (King, Edward.) n.p., n.d.

Htn Geog 4206.99* A collection of original voyages. (Hacke, William.) London, 1699.

Geog 4167.75 A collection of voyages, chiefly in the S. Atlantic. (Dalrymple, A.) London, 1775.

Geog 4157.29 Collection of voyages. (Dampier, W.) London, 1729. 4v.

Geog 4157.04F Collection of voyages and travels. (Churchill, J.) London, 1704-07. 8v.

Geog 4157.04.3F Collection of voyages and travels. 3. ed. (Churchill, J.) London, 1744. 6v.

Geog 148.1.9 Collectors guide to the national geographic magazine. (Buxbaum, Edwin Clarence.) Wilmington, 1962.

Geog 759.32.5 College geography. (Case, E.C.) N.Y., 1932.

Geog 665.92 College geography. (Plattie, Roderick.) Boston, 1926.

Geog 4308.78.15 College tramps. (Stokes, F.A.) N.Y., 1880.

Geog 540.15 La collezione della antiche carte geografiche, a cura di Pietro Frabetti, Il museo delle navi, a cura di Amedio Rizzi. (Frabetti, Pietro.) Bologna, 1959.

Htn Geog 4678.25* Collins, Daniel. Narrative of the shipwreck of the brig Betsey of Wiscasset, Maine...Dec. 1824. Wiscasset, 1825.

Geog 4328.01 Collins, F. Voyages...from 1796-1801. London, 1819.

Geog 4509.23.15 Collins, F.A. Mountain climbing. London, 1924.

Geog 4209.43.5 Collins, Grace. Chin takes and talks, an absurd travel diary. Chicago, 1943.

Geog 5160.30 Collins, H.B. Arctic area. México, 1954.

Geog 5538.50.15 Collinson, R. Journal of H.M.S. Enterprise...1850-55. London, 1889.

Geog 4145.70 Collis, Maurice. The grand peregrination; being the life of...Fernão Mendes Pinto. London, 1949.

Geog 4679.59 Collision course: the Andrea Doria and the Stockholm. (Moscow, Alvin.) N.Y., 1959.

Geog 665.112 Colloque International de Géographie Appliquée, Liège, 1967. Comptes rendus. Liège, 1968.

Geog 665.93 Colloque national de géographie appliquée, Strasbourg, 20-22 avril 1961. (France. Centre National de la Recherche Scientifique.) Paris, 1962.

Geog 4308.49 Colman, H. European life and manners. Boston, 1849. 2v.

Geog 4308.49.2 Colman, H. European life and manners. Boston, 1850. 2v.

Geog 4247.89 Colnett, J. The journal of Captain James Colnett aboard the Argonaut from Apr. 26, 1789 to Nov. 3, 1791. Toronto, 1940.

Htn Geog 4557.98* Colnett, J. Voyages to the South Atlantic. London, 1798.

Geog 4515.14 Colò, Carlo. Attrezzature per soccorso alpino. Trento, 1958.

Geog 613.1.7 Cologne, Germany. Stadtbibliothek. Katalog einer Mercator-Ausstellung. Koeln, 1894.

Htn Geog 3017.62* Colonae anglicanae illustratae. (Bollan, William.) Londini, 1762.

Geog 4515.104.5 Colonel Bilgeri's handbook on mountain ski-ing. (Bilgeri, George.) London, 1929.

Geog 4308.72.7 Colonial travel. (Kennedy, David.) London, 1876.

Geog 4181.56 Colonizing expeditions to the West Indies and Guiana, 1623-67. (Harlow, V.T.) London, 1925.

Geog 3055.70 Les colonnes d'Hercule et l'Atlantide. (Rousseau-Liessens, A.) Bruxelles, 1955.

Geog 4218.58 A colored man round the world. (Dorr, D.F.) Cleveland? 1858.

Geog 4759.30 The coloured countries. (Waugh, Alec.) London, 1930.

Geog 4209.02.3 Colquhoun, E.C. (Mrs.). Two on their travels. London, 1902.

Geog 4209.02.5 Colquhoun, E.C. (Mrs.). Two on their travels. N.Y., 1902.

Geog 4208.60.3 Colquohoun, A.R. Dan to Beersheba. London, 1908.

Geog 4308.81 Colton, G. A Maryland editor abroad. Annapolis, 1881.

Geog 4328.51 Colton, W. Ship and shore in Madeira. N.Y., 1851.

Geog 40.15 Colton's journal of geography and collateral sciences. N.Y.

Geog 579.52.5F The Columbia Lippincott gazetteer of the world. N.Y., 1952.

Geog 4180.2 Columbus, C. Letters...four voyages to new world. London, 1847.

Geog 4180.43 Columbus, C. Letters...four voyages to the new world. London, 1870.

Geog 4238.92 Columbus outdone, an exact narrative of the voyage of the Yankee skipper Captain N.A. Andrews. (Ward, Artemus.) N.Y., 1893.

Geog 4208.52.5 Colvocoresses, G.M. Four years in a government exploring expedition. N.Y., 1852.

Geog 4208.52.6 Colvocoresses, G.M. Four years in the government exploring expedition. 5. ed. N.Y., 1855.

Geog 606.2 Coma Soley, V. Jaime Ferrer...y el descubimiento de America. 1. ed. Barcelona, 1952.

Geog 4228.28 Combier, C. Voyage au Golfe de Californie. Paris, 1864.

Geog 4678.66 Comettant, O. Le naufrage de L'Evening Star. Paris, 1866.

An 2109.37.3 Coming into being among the Australian aborigines. (Montagu, Ashley.) London, 1937.

Geog 3019.62 The coming of the Europeans. (Grenville, J.A.S.) London, 1962.

Geog 4521.2 Commando climber. (Banks, Mike.) London, 1955.

Geog 4180.53 Commentaries of Dalboquerque. (Alboquerque, A. d'.) N.Y., 1964?

Geog 4679.67.30 Committee of Scientists on the Scientific and Technological Aspects of the Torrey Canyon Disaster. The Torrey Canyon; report. London, 1967.

Geog 85.350 Communicazione dell...la geografia. (Istituto Geografico de Agostini.) Novara. 1-18,1912-1930 9v.

An 359.06 Comparative art. (Balch, E.S.) Philadelphia, 1906.

An 629.63 The comparative method in social anthropology. (Evans-Pritchard, E.E.) London, 1963.

Geog 3019.56.5 Compendio di storia della geografia e delle esplorazioni geografiche. (Colamonico, Carmelo.) Napoli, 1956.

Geog 816.36 Compendio geographico. (Carvalho Da Costa, Antonio.) Lisboa, 1636.

Geog 4228.31 Compendious view of...travels. London, 1831.

Geog 3108.39.3 Compendium of ancient and modern geography. (Arrowsmith, A.) London, 1839.

Htn Geog 4209.37F* A compendium of the East; being an account of voyages to the Grand Indies. (LaCombe, Jean de.) London, 1937.

Geog 817.26 Compendium totius orbis. (Kolb, P.G.) Rottwike, 1726.

Geog 817.01.5F The compleat geographer. (Moll, H.) London, 1723.

An 136.20 The complete bibliography of Robert H. Lowie. (California. University. Robert H. Lowie Museum of Anthropology.) Berkeley, 1966.

Geog 4509.07.1 The complete mountaineer. (Abraham, G.D.) N.Y., 1908.

Geog 4509.07.2 The complete mountaineer. 2. ed. (Abraham, G.D.) London, 1908.

Geog 4509.07.3 The complete mountaineer. 3. ed. (Abraham, G.D.) London, 1923.

Geog 4308.97.3 Complete pocket guide to Europe. (Cassell and Company, Inc.) London, 1897.

Geog 578.55 Complete pronouncing gazetteer. (Lippincott, J.B. and Co.) Philadelphia, 1855.

NEDL Geog 212.200.5 Compte rendu. Tables des matières des v.1-25, 1876-1901. (Société Royale Belge de Géographie.) Bruxelles, n.d.

Geog 500.180 Compte rendu des travaux de la commission internationale d'histoire maritime, 1965. (International Committee of Historical Sciences. Commission Internationale d'Histoire Maritime.) Paris? 1965?

Geog 665.112 Comptes rendus. (Colloque International de Géographie Appliquée, Liège, 1967.) Liège, 1968.

Author and Title Listing

NEDL Geog 212.204 Comptes rendus. (Société de Géographie.) Paris. 1882-1899 18v.
Geog 110.1.14 Comptes rendus. v.1-7. (International Geographical Congress, 14th, Warsaw, 1934.) Varsovie, 1934.
Geog 110.1.15 Comptes rendus du Congrès international de géographie, Amsterdam, 1938. v.1-5,21. (International Geographical Congress, 15th, Amsterdam, 1938.) Leiden, 1938. 6v.
Geog 110.1.16 Comptes rendus du Congrès international de géographie. (International Geographical Congress.) Lisbonne, 1950. 2v.
Geog 110.1.18.5 Comptes rendus du XVIIIe congrès international de géographie. (International Geographical Congress, 18th, Rio de Janeiro, 1956.) Rio de Janeiro, 1959. 4v.
An 358.78.7 Comptes rendus sténographiques. (Congrès International des Sciences Anthropologiques, Paris, 1878.) Paris, 1880.
Geog 4169.38 Compton, Ray. The open road. N.Y., 1938.
Geog 4329.33 Con la ciurma dell'Alessandro. (Rèpaci, L.) Milano, 1933.
Geog 4209.67.10 Con variado rumbo. 1. ed. (Díaz Plaja, Guillermo.) Barcelona, 1967.
An 197.6 The concept of cultural systems. (White, Leslie A.) N.Y., 1975.
An 359.44.25A The concept of culture. (Kluckhohn, Clyde.) n.p., 194-.
An 170.275.5 The concept of race. (Montagu, Ashley.) N.Y., 1964.
An 2109.68 The concept of the primitive. (Montagu, Ashley.) N.Y., 1968.
Htn Geog 5507.82* Concise account of voyages...discovery. (Pickersgill, R.) London, 1782.
Geog 579.56.5 A concise glossary of geographical terms. (Swayne, James Colin.) London, 1956.
Geog 579.56.7 A concise glossary of geographical terms. 2. ed. (Swayne, James Colin.) London, 1962.
Geog 4307.99 A concise narrative of a tour through some parts of England, France, Holland, Switzerland, and Italy in the years 1799, 1800, 1801, and 1802. Philadelphia, 1821.
Geog 3017.85 Concorde de la géographie des différents ages. (Pluche, N.A.) Paris, 1785.
Geog 818.54.10 Condaminas, S. Nociones generales de geografía astronómica. Matanzas, 1854.
NEDL Geog 520.25 Conder, Josiah. The modern traveller. London, 1830. 30v.
An 379.30.9 Conference...Apr. 11-12, 1920. (Conference on Regional Phenomena, Washington, D.C., 1930.) Washington, 1930.
Geog 5925.89 Conference documents. (Conference on Antarctica, Washington, D.C., 1959.) Washington, 1960.
Geog 5925.89 Conference on Antarctica, Washington, D.C., 1959. Conference documents. Washington, 1960.
An 170.163 Conference on New Approaches in Social Anthropology, Jesus College, Cambridge, Eng., 1963. The social anthropology of complex societies. London, 1966.
An 170.163.15 Conference on New Approaches in Social Anthropology, Jesus College, Cambridge, Eng., 1963. The social anthropology of complex societies. London, 1969.
An 629.65 Conference on New Approaches in Social Anthropology. The relevance of models for social anthropology. London, 1965.
Geog 659.69.5 Conference on Quantitative Methods in Geography, New York, 1969. Quantitative methods in geography; a symposium. N.Y., 1969.
An 379.30.9 Conference on Regional Phenomena, Washington, D.C., 1930. Conference...Apr. 11-12, 1920. Washington, 1930.
An 359.15.3 Conferencias sobre antropología, etnología. (Latcham, R.E.) Santiago, 1915-
Geog 4559.40.10 Les confessions de Tatibouet. (Pierrefeu, François de.) Paris, 1939.
An 359.10.5 The conflict of colour. (Simpson, B.P.) N.Y., 1910.
Geog 4558.51 Congar, Obadiah. The autobiography and memorials of Captain Obadiah Congar. N.Y., 1851.
Geog 110.3.10 Congrès des Géographes et Ethnographes Slaves, 2nd, Krakow, 1927. Pamiętnik II zjardu słowiańskich geografów i etnografów odbytego w Polsce w roku, 1927. Kraków, 1929-30. 2v.
Geog 110.3 Congrès des Géographes et Ethnographes Slaves, 3rd, 1930. Zbornik radova III Kongresa slovenskih geografa i ethnografa u Kraljevini Jugoslaviji, 1930. Beograd, 1933.
Geog 110.3.25 Congrès des Géographes et Ethnographes Slaves, 4th, Sofia, 1936. Sbornik na IV Kongress na slavianskite geografi i etnografi. Sofiia, 1938.
An 358.78.7 Congrès International des Sciences Anthropologiques, Paris, 1878. Comptes rendus sténographiques. Paris, 1880.
Geog 5070.25 Congrès International pour l'Étude des Régions Polaires, 1st, Brussels, 1906. Rapport d'ensemble. Bruxelles, 1906.
Geog 81.5 Congrès national compte-rendu. (Union Geographique du France.) 1879-1904 22v.
Geog 110.1.20.10 Congress proceedings. (International Geographical Congress, 20th, London, 1964.) London, 1967.
Geog 110.1.20.5 Congress programme. (International Geographical Congress, 20th, London, 1964.) London, 1964.
Geog 40.10 Congresso Brasileiro de Geografia. 4 annaes. Pernambuco. 1-2 3v.
Geog 40.9 Congresso Brasiliero de Geografia. Miscellaneous publications.
An 170.163.5 Congresso de Etnografia e Folelore, 1st, Braga, 1956. Actas...promovido pela Câmarã municipal de Braga. Lisboa, 1963. 3v.
Geog 40.2 Congresso Geografico Italiano. Atti. Genova. 1892-1907 39v.
Geog 3570.34.5 Congresso Internacional de Historia dos Descobrimentos. Actas. Lisboa, 1960- 6v.
An 5.10 Conklin, Harold C. Folk classification: a topically arranged bibliography of contemporary and background references through 1971. New Haven, 1972.
Geog 4208.78.4A Connolly, J.B. Canton captain. Garden City, N.Y., 1942.
Geog 4145.17 Connolly, J.B. Master mariner...Amasa Delano. Garden City, 1943.
Geog 3570.61 Le conoscenze geografiche del Medioevo. (Ferro, Gaetano.) Genova, 1972.
Geog 5559.06 Conquering the Arctic ice. (Mikkelsen, E.) Philadelphia, 1909.
Geog 4209.25.5 Conquering the world. (Bridgman, H.B.) N.Y., 1925.
Geog 3019.52.2 Conquest by man. (Herrmann, Paul.) N.Y., 1954.
Geog 5939.57 The conquest of the Antarctic. (Kemp, Norman.) N.Y., 1957.
Geog 5209.39 The conquest of the Arctic. (Segal, Louis.) London, 1939.
Geog 5509.66 Conquest of the last frontier. (Neatby, Leslie Hamilton.) Toronto, 1966.
Geog 5559.09.20 The conquest of the North. (Ingersoll, Ernest.) N.Y., 1909.
Geog 5235.10 The conquest of the North Pole. (Hayes, J.G.) N.Y., 1934.
Geog 5509.07 The conquest of the Northwest Passage. (Randall, Harry.) Minneapolis, 1907.
Geog 4180.81 Conquest of the river plate. (Dominguez, L.L.) London, 1891.
Geog 5939.32 The conquest of the South Pole. (Hayes, J.G.) London, 1932.
Geog 3019.29.5 La conquête de la terre. (Olsen, Ørjan.) Paris, 1933-36. 5v.
Geog 3249.25 La conquête des routes océaniques d'Henri le navigateur à Magellan. (Pereyra, Carlos.) Paris, 1925.
Geog 5180.11 La conquête du pôle nord en ballon. (Marquis, R.) Paris, 19- .
Geog 4509.54 La conquista de la montaña. (Jolis Felisart, A.) Barcelona, 1954.
Geog 3560.10 La conquista de las rutas oceánicas. (Pereyra, C.) Madrid, 1940.
Geog 3019.50 La conquista della terra. (Dainelli, G.) Torino, 1950.
Geog 3070.80 Conseil Scientifique pour l'Afrique au Sud du Sahara. Mapping and surveying of Africa south of the Sahara. London, 1954.
Geog 4509.18 Le conseiller de l'ascensionniste. v.2. (Schweizer Alpen Club.) Genève, 1918.
An 186.2.5 Conservations with Claude Lévi-Strauss. (Charbonnier, Georges.) London, 1969.
Htn Geog 5220.21* Considerations geographiques. (Buache, P.) Paris, 1753.
An 358.00.5 Considérations sur les diverses méthodes à suivre dans l'observation des peuples sauvages. (Girando, J.M.) Paris, 1800.
Geog 4209.42.5 Considine, J.J. Across a world. Toronto, 1942.
Geog 3570.2 Consiglieri-Pedroso, L. Catalogo bibliographico das publicacões. Lisboa, 1912.
Geog 4559.34 Consigned to Davy Jones. (Grant, George H.) Boston, 1934.
Geog 4238.48 Consignments to El Dorado. 1st ed. (Whaley, Thomas.) N.Y., 1972.
An 3309.14.3F Constancy or variability in Scandinavian type. (Stevenson, B.L.) Leiden, 1914.
Geog 603.7 Constantini, Otto. Leben und Wirken eines österreichischen Geographieprofessors und Erwachsenenbildners. Linz, 1969
Geog 4209.34.10 Contando un viaje. (Gutierrez Zamora, I.) n.p., 1934.
An 359.61.25 Contemporary raciology and racism. Bloomington, 1961.
Geog 608.2 Contents of box in corner stone of Hubbard Memorial. (Hubbard, Gardiner Greene.) n.p., n.d.
An 170.345 Context and meaning in cultural anthropology. (Spiro, Melford E.) N.Y., 1965.
Geog 3018.93 Le continent austral. (Rainaud, A.) Paris, 1893.
Geog 3055.45 Un continent disparu l'Atlantide. (Dévigné, R.) Paris, 1923.
Geog 3055.45.5 Un continent disparu l'Atlantide. (Dévigné, R.) Paris, 1924.
Geog 5919.65 A continent for science; the Antarctic adventure. (Lewis, Richard S.) N.Y., 1965.
Geog 4308.35.15 The continent in 1835. (Hoppus, John.) N.Y., 1837.
Geog 4269.13 The continent of Europe. (Lyde, L.W.) London, 1913.
Geog 4269.13.2 The continent of Europe. (Lyde, L.W.) London, 1920.
Geog 4269.13.4 The continent of Europe. 2nd ed. (Lyde, L.W.) London, 1924.
Geog 4300.35 Continental touring, 1947-1948 for the British motorist. Leeds, 1947.
Geog 4307.70.10A Continental travels. (Burney, Charles.) London, 1927.
Geog 819.52F Continenti e poesi. (Visintin, L.) Novara, 1952.
An 359.70.25 Contracts, interférences ethniques et culturelles. (Rimet, Michel.) Montpellier, 1970.
Geog 613.4.20 Contribución a la bibliotheca de C.R. Markham. (Olivas, A.) Lima, 1924.
Geog 148.1.15 Contributed technical papers. Katmai series. (National Geographic Society.) Washington. 1,1923
Geog 148.1.18 Contributed technical papers. Stratosphere series. (National Geographic Society.) Washington.
An 3409.30 Contribution à la morphologie comparée des crêtes papillaires. Thèse. (Miranda Pinto, O.) Lyon, 1930.
An 2108.80 Contribution to the history of the development of the human race. (Geiger, E.L.) London, 1880.
Geog 161.1 Contributions in geographical exploration. (Ohio State University.) Columbus. 1,1920
An 3209.68.5 Contributions to the physical anthropology of central Asia and the Caucasus. Cambridge, 1968.
Geog 3550.13 Il contributo degli Italiani alla conoscenza della terra ed agli studi cinquatennio. (Ricchieri, G.) Roma, 1912.
Geog 4308.50.9 Conway, George. Running sketches of men and places. N.Y., 1851.
Geog 4181.11 Conway, W.M. Early Dutch and English voyages. London, 1904.
Geog 4509.20.10 Conway, W.M.C. Mountain memories. N.Y., 1920.
Geog 5559.12 Cook, F.A. Im Kampfe mit Bär und Walross. Braunschweig, 192-.
Geog 5559.08 Cook, F.A. Meine Eroberung des Nordpols. Hamburg, 1912.
Geog 5559.08.10 Cook, F.A. Return from the Pole. N.Y., 1951.
Geog 6008.98.3 Cook, F.A. Through the first Antarctic night. N.Y., 1900.
Geog 4129.47.5 Cook, H.K. Over the hills and far away. London, 1947.
Geog 4300.32 Cook, T., firm. Tourist handbook for Holland, Belgium, the Rhine and Black Forest. London, 1895.
Geog 4219.28 Cook, Thomas, firm, publishers, London. The supreme travel-adventure: around the world in the "Franconia" 1929. N.Y., 1928.
Geog 4218.73.15 Cook, Thomas. Letters from the sea and from foreign lands. London, 1873.
Geog 40.17.5F Cook's excursionist. American edition. N.Y.
Geog 40.17F Cook's excursionist. English edition. London.
Geog 3018.30 Cooley, W.D. The history of maritime and inland discovery. London, 1830-31. 3v.
Geog 3018.30.5 Cooley, W.D. The history of maritime and inland discovery. London, 1833. 2v.
Geog 4209.25A Coolidge, J.G. Random letters from many countries. Boston, 1924.
An 379.42 Cools, R.H.A. De geographische gedechte bij Jean Brunhes. Utrecht, 194-.
An 359.65.15 Coon, Carleton Stevens. The living races of man. 1. ed. N.Y., 1965.
An 359.62 Coon, Carleton Stevens. The origin of races. N.Y., 1962.
An 359.39A Coon, Carleton Stevens. The races of Europe. N.Y., 1939.
An 359.48.15 Coon, Carleton Stevens. A reader in general anthropology. N.Y., 1948.
Geog 4110.25 Coon, Horace. 100 vacations costing from $50.00 to $500.00. N.Y., 1939.
Geog 4218.80 Coop, Timothy. A trip around the world. Cincinnati, 1882.
Geog 4559.27.10 Cooper, F.G. Yarns of the seven seas. London, 1927.
Geog 4169.55 Cooper, G. Forbidden lands. N.Y., 1955.
An 2109.40.5 Cooper, G. I searched the world for death. London, 1940.

Author and Title Listing

Geog 4308.32 Cooper, J.F. Residence in France with excursion up the Rhine to Switzerland. Paris, 1836.

An 2109.37.5 Cooperation and competition among primitive peoples. (Mead, M.) N.Y., 1937.

An 2109.61.20 Cooperation and competition among primitive peoples. (Mead, Margaret.) Boston, 1961.

Geog 4208.82.6 Coote, W. Wanderings, south and east. London, 1882.

Geog 658.81 Cora, G. Cenni intorno all'attuale. Torino, 1881.

Geog 4329.53.15 Cordan, W. Das Mittelmeer. Düsseldorf, 1953.

Geog 3055.75 Cordeau, Catherine L. Poséidones. Paris, 1957.

Geog 4145.56 Cordier, Stéphane. Balthazar de Monconys. Bruxelles, 1967.

Geog 5919.45 Cordovez Madariga, Enrique. La Antártida sudamericana. Santiago, 1945.

Geog 4131.68.20 Corna Pellegrini, Giacomo. Studi e ricerche sulla regione turistica. Milano, 1968.

An 2109.60 Cornelius, Friedrich. Geistesgeschichte der Frühzeit. v.1-2. Leiden, 1960.

An 359.33.5 Cornelius, William J.J. L'homme d'après la religion et la science. Thèse. Londres, 1933.

Geog 5321.1.6 Corner, George Washington. Doctor Kane of the Arctic seas. Philadelphia, 1972.

Geog 4309.44 Cornette, Arthur H. Van Toledo tot Budapest. Antwerpen, 1944.

Geog 4181.32 Corney, B.G. The quest and occupation of Tahiti by emissaries of Spain 1772-76. London, 1913-19. 3v.

Geog 4181.13 Corney, B.G. Voyage of Captain Don Felippe Gonzalez. Cambridge, 1908.

Geog 4248.13 Corney, Peter. Voyages in the northern Pacific, 1813-1818. Honolulu, 1896.

Geog 4209.31.20 Cornish, Naughan. The poetic impression of natural scenery. London, 1931.

Geog 4659.69.5 Cornish shipwrecks. (Larn, Richard.) Newton Abbot, 1969- 2v.

Geog 4308.91.3F Cornwall-Devon, 1891; [Collection of photographic views...of Germany, Holland, Switzerland, Paris and England]. (Richardson, W.L.) n.p., n.d.

Geog 4237.83 Cornwall to America in 1783 from the journal of Paul Burall (1755-1826). (Burall, Paul.) n.p., 19- .

Geog 4709.30A Coronado's children. (Dobie, J. Frank.) Dallas, 1930.

Geog 500.97 Coronelli, Marco V. Il catalogo degli autori; una biobibliografia geografica del '600. Firenze, 1957.

Geog 4678.58 Corrao, Mario. Difesa del capitano ed armatore del Sicilia contro il cap. ed armatore dell'Ercolano. Palermo, 1858.

Geog 4676.93 Correa, F. Relaçam do successo que teve o patacho chamado N. Sra. da Candelaria da ilha da Madeira. Lisboa, 1734.

Geog 4180.42 Correa, G. Three voyages of Vasco da Gama. Photoreproduction. London, 1869.

Geog 3049.59 Correa-Calderon, E. Floria de la Atlantida y otras historias fabulosas. Madrid, 1959.

An 3309.12 La correlazione fra la statura e l'indice cefalico. (Bresciani, C.) Palermo, 1912.

Geog 3129.68 Le correnti del pensiero geografico nell'antichità classica e il lorocontributo alla cartografia nautica medioevale. (Guarnieri, Giuseppe Gino.) Pisa, 1968-69. 2v.

Geog 618.3 Correspondance. v.1-2,3. (Reclus, Élisée.) Paris, 1911-25. 2v.

Geog 613.1.18 Correspondance mercatorienne. (Mercator, Gerardus.) Anvers, 1959.

Geog 4208.72.9 Correspondence of Palestine tourists...while traveling in Europe, Asia and Africa. (Smith, G.A.) Salt Lake City, 1875.

Geog 4709.70 Corsairs of Malta and Barbary. (Earle, Peter.) London, 1970.

An 3309.13 Les corses devant l'anthropologie. (Rocca, Pierre.) Paris, 1913.

Geog 3018.56 Cortambert, E. Coup d'oeil...sur...les progrès de géographie. Langny, n.d.

Geog 818.62.5 Cortambert, P.F.E. Cours de geographie. Paris, 1862.

Geog 3235.11 Cortambert, P.F.E. Tros...monuments geographiques. Paris, 1877.

Geog 610.1 Cortambert, R. Notice sur la vie et les oeuvres de M. Jomard. Paris, 1863.

Geog 4145.14.10 Os Corte Reais e o novo mundo. (Brazão, Eduardo.) Lisboa, 1965.

Geog 4145.14.10.5 Les Corte-Real et le nouveau monde. (Brazão, Eduardo.) Lisbonne, 1967.

Geog 520.3 Les Corte-Real et leurs voyages. (Harrisse, H.) Paris, 1883.

Geog 4180.40 Cortes, H. Fifth letter of...Cortes to Emperor Charles V. London, 1868.

Geog 4181.89 Cortesão, Armando. Suma Oriental of Tomé Pires...book of Francisco Rodrigues. London, 1944. 2v.

Geog 3570.52 Cortesão, Jaime. Os descobrimentos pre-colombinos dos Portugueses. Lisboa, 1966.

Geog 3570.23F Cortesão, Jaime. Os descrobrimentos portugueses. Lisboa, 195-? 2v.

Geog 4180.98A Cosmas. Christian topography of Cosmas. London, 1897.

Geog 604.3 O cosmografo Fernam Vaz Dourado, fronteiro da India e a sua obra. (Couto, Gustavo.) Lisboa, 1928.

Geog 4266.36 Cosmographiae generalis libri tres. (Mervla, P.) Amsterdami, 1636.

Htn Geog 756.36* Cosmographica disciplina. (Postelli, G.) Lugduni, 1636.

Geog 520.20 La cosmographie. (Fonteneau, J.) Paris, 1904.

Geog 3228.80 La cosmographie avant la decouverte de l'Amerique. (Gravier, S.) Paris, 1880.

Geog 757.70 Cosmographie methodique. (Buy de Mornas, C.) Paris, 1770.

Htn Geog 805.75F* La cosmographie universelle. (Thevet, A.) Paris, 1575. 2v.

Geog 4239.58 Costa Brochado, José de. Descobrimento do Atlântico. Lisboa, 1958.

Geog 4239.58.5 Costa Brochado, José de. The discovery of the Atlantic. Lisboa, 1960.

Geog 3570.24 Costa Brochado. Historia de uma polemica. Lisboa, 1944.

Geog 759.40 Costa Pereira, J.V. da. Geographia humana. Rio de Janeiro, 194-.

An 359.70.20 Costanzo, Giorgio. La costruzione dell'uomo. Roma, 1970.

Geog 4219.28.10 Costes Dieudonné. La grande croisière de Costes et Le Brie. Paris, 1928.

An 359.70.20 La costruzione dell'uomo. (Costanzo, Giorgio.) Roma, 1970.

Htn Geog 3205.40.21* Gli costumi, le leggi et lusanze. (Boehme, J.) Venetia, 1560.

Geog 4218.90.6 Cotes, Sara J.D. A social departure. London, 1890.

Geog 4218.90.5 Cotes, Sara J.D. A social departure. N.Y., 1890.

Geog 4218.91 Cotes, Sara J.D. A social departure. How Orthodocia and I went round the world by ourselves. N.Y., 1891.

An 2109.66.5 Cotlow, Lewis N. In search of the primitive. 1st ed. Boston, 1966.

Geog 585.7 Coulier, P.J. Tables des principales positions geonomiques. Paris, 1828.

An 170.165A Count, Carl W. This is race. N.Y., 1950.

Geog 808.77 The countries of the world. (Brown, R.) London, 1877-1880? 4v.

Geog 3018.56 Coup d'oeil...sur...les progrès de géographie. (Cortambert, E.) Langny, n.d.

Geog 3070.8 Coup d'oeil historique sur projection. (D'Avezac, M.A.P.) Paris, 1863.

Geog 3240.17 Coup d'oeil sur la litterature géographique arabe au moyen âge. (Devic, L. Marcel.) Paris, 1882.

Geog 4269.14 Couperus, Louis. Van en over alles en iedereen. pt.1-5. Amsterdam, 1915. 5v.

Geog 4309.33.3 Courrier d'Europe. (Halévy, Daniel.) Paris, 1933.

Geog 818.62.5 Cours de geographie. (Cortambert, P.F.E.) Paris, 1862.

Geog 4709.66.5 Course, Alfred George. Pirates of the Eastern seas. London, 1966.

Geog 4237.00 The course of Captain Edmond Halley in the year 1700. (Stearns, R.P.) n.p., 1936.

An 170.142 Courses toward urban life. (Braidwood, Robert John.) Chicago, 1962.

Geog 5070.57 Courtauld, A. From the ends of the earth; an anthology of polar writings. London, 1958.

Geog 4709.71 Cousteau, Jacques Yves. Diving for sunken treasure. London, 1971.

Geog 4229.73 Cousteau, Jacques Yves. Trois aventures de la Calypso: Galapago, Titicaca, Trous Bleus. Paris, 1973.

Geog 604.3 Couto, Gustavo. O cosmografo Fernam Vaz Dourado, fronteiro da India e a sua obra. Lisboa, 1928.

Geog 4515.107 Couttet, Alfred. L'enchantement du ski. Paris, 1930.

An 182.3 Cows, pigs, wars and witches. 1. ed. (Harris, Marvin.) N.Y., 1974.

Geog 4309.55.5 Cox, Alfred B. The other half. Adelaide, 1955?

An 379.72 Cox, Kevin R. Man, location, and behavior: an introduction to human geography. N.Y., 1972.

An 359.59.20 Cox, Oliver Cromwell. Caste, class, and race. N.Y., 1959.

An 359.48A Cox, Oliver Cromwell. Caste, class and race. 1. ed. Garden City, N.Y., 1948.

Geog 4328.69 Cox, S.S. Search for winter sunbeams. London, 1869.

Geog 4328.69.5 Cox, S.S. Search for winter sunbeams. N.Y., 1874.

Geog 4308.51.7 Cox, Samuel S. A Buckeye abroad. N.Y., 1852.

Geog 4556.94 Coxere, Edward. Adventures by sea of Edward Coxere. Oxford, 1945.

Geog 4209.34 Coyne, William M. A sailor's log of facts: not fables. Boston, 1934.

Geog 3049.62 Cozzi, Piero. Los paises legendarios de la mitologia. Buenos Aires, 1962.

Geog 4559.55.10 Cracker hash. (Stenhouse, J.R.) London, 1955.

Geog 4239.21 Craig, Gavin. Boy aloft. Lymington, 1971.

Geog 4209.38.30 Craig, John D. Danger is my business. N.Y., 1938.

Geog 4559.26.10 Crane, Mannin. Yarns from a windjammer. Boston, 1926.

An 3559.13 Craneometría comparativa de los antiguos habitantes de la isla y del pukara de Tilcara. (Dillenius, J.A.) Buenos Aires, 1913.

An 3538.44 Crania Aegyptiaca. (Morton, S.G.) Philadelphia, 1844.

An 3538.39 Crania Americana. (Morton, S.G.) Philadelphia, 1839.

An 3538.65 Crania Britannica. (Davis, J.B.) London, 1865. 2v.

An 3538.82 Crania ethnica. Les crânes des races humaines. (Quatrefages de Bréau, A. de.) Paris, 1873-81. 2v.

An 3538.92 Crania ethnica americana. (Virchow, R.) Berlin, 1892.

An 3539.01 Crania ethnica Philippinica. (Koeze, G.A.) Haarlem, 1901-04.

An 3538.96 Crania from the mounds of the St. John's. (Allen, Harrison.) Philadelphia, 1896.

An 3559.12.3 Crania Habessinica contributo all'antropologia dell'Africa Orientale. (Sergi, Sergio.) Roma, 1912.

An 3539.10PF Crania Lapponica. (Düben, G.) Holmiae, 1910. 2v.

Geog 5700.5.3A Crantz, D. Historie von Grønland...mission. Barby, 1770. 3v.

Geog 5700.5.10 Crantz, D. History of Greenland...account of mission. London, 1820. 2v.

Geog 5700.5 Crantz, D. History of Greenland. London, 1767. 2v.

Geog 4181.109 Crawford, O.G.S. Ethiopian itineraries circa 1400-1524. Cambridge, Eng., 1958.

X Cg An 2109.31.3 Crawley, A.E. Dress, drinks and drums. London, 1931.

An 2109.09.7A Crawley, A.E. The idea of the soul. London, 1909.

Htn An 2109.29* Crawley, A.E. Studies of savages and sex. London, 1929.

Geog 4348.67 Creagh, J. A scamper to Sebastopol and Jerusalem 1867. London, 1873.

Geog 4209.01 Creelman, J. On the great highway. Boston, 1901.

Geog 4308.53.4 Crests from the ocean-world. (Tripp, A.) Boston, 1853.

Geog 4308.53.3 Crests from the ocean-world. (Tripp, A.) Boston, 1853.

Geog 4308.53.6 Crests from the ocean-world. (Tripp, A.) Boston, 1861.

Geog 4308.53.7 Crests from the ocean-world. (Tripp, A.) Boston, 1862.

Geog 614.5F Cristoforo Negri. (Marinelli, G.) Torino, 1897.

An 5.25 O critério nomolójico. Photoreproductión. (Vasconcellos-Abreu, Guilherme de.) Lisboa, 1887.

Geog 4269.13.10 Critic in the occident. (Fitch, G.H.) San Francisco, 1913.

Geog 5557.25 Critical review of Bering's first expedition, 1725-30. (Dall, W.H.) Washington, 1890.

Geog 3235.21 Crivellari, G. Alcuni cimeli della cartografia. Firenze, 1903.

Geog 4308.50.25 Crockett, Henry C. The American in Europe. London, 184-.

Geog 5070.62 Croft, Andrew. Polar exploration. London, 1939.

Geog 5070.62.5 Croft, Andrew. Polar exploration. 2. ed. London, 1947.

Geog 4110.37 Croft-Cooke, Rupert. Cities. London, 1951.

Geog 4209.60.25 La croisière du Ly-Kau. (Moncharmont, Simone.) Paris, 1960. 2v.

Geog 4181.80 Crone, G.R. The voyages of Cadamosto. London, 1937.

Geog 3570.53 Crone, Gerald. The discovery of the East. London, 1972.

Geog 5300.6 Cronheim, P. Fahrten und Forschunger der Holländer in die Polargebieten. Leipzig, 1913.

Geog 5970.10 Cronologia de los viajes a las regiones australes. (Buenos Aires. Universidad.) Buenos Aires, 1950.

Geog 3249.03 Cronologia delle scoperte e delle esplorazioni geografiche dall'anno 1492 a tutto il secolo XIX. (Hugues, Luigi.) Milano, 1903.

Geog 4559.55.5 Cross, Beverley. Mars in Capricorn. 1. ed. Boston, 1955.

Geog 5399.28.45 Cross, Wilbur. Ghost ship of the Pole. N.Y., 1960.

An 170.202 Cross-cultural approaches. (Ford, Clellan Stearns.) New Haven, 1967.

Geog 6009.55.10 The crossing of Antarctica; the Commonwealth Antarctic Expedition, 1955-58. (Fuchs, Vivian.) London, 1958.

An 3409.23 Crosskey, Walter C.S. The single finger print identification system. San Francisco, 1923.

Author and Title Listing

Geog 3018.83.10 — Crotambert, R. Nouvelle histoire des voyages. Paris, 1883-84.
Geog 5509.34 — Crouse, N.M. The search for the Northwest Passage. N.Y., 1934.
Geog 5209.47 — Crouse, Nellis M. The search for the North Pole. N.Y., 1947.
Geog 4219.55.5 — Crove, Bill. Heaven, hell and salt water. London, 1957.
Geog 4309.21.10A — The crown prince's European tour (March 3-September 3, 1921). (Futera, Count Y.) Osaka, 1926.
Htn Geog 4559.37* — Cruelty to seamen; being the case of Nichols and Couch. (Dana, Richard Henry.) Berkeley, 1937.
Geog 4218.38.4 — Cruise frigate Columbia around the world...1838-1840. (Murrell, William M.) Boston, 1840.
Geog 5558.81.9F — Cruise of revenue-steamer Corwin. (Rosse, I.C.) Washington, 1883.
Geog 4709.7 — The cruise of the "Alerte". (Knight, E.F.) London, 1907.
Geog 4219.51.10 — The cruise of the Amaryllis. (Mulhauser, G.H.P.) London, 1951.
Geog 4228.85 — The cruise of the Brooklyn. (Beehler, W.H.) Philadelphia, 1885.
Geog 4558.98.10 — Cruise of the Cachalot. (Bullen, F.T.) London, 1898.
Geog 4558.98.14 — Cruise of the Cachalot. (Bullen, F.T.) N.Y., 1899.
Geog 4558.98.12 — Cruise of the Cachalot. (Bullen, F.T.) N.Y., 1899.
Geog 4558.98.11 — Cruise of the Cachalot. (Bullen, F.T.) N.Y., 1899.
Geog 4558.98.13 — Cruise of the Cachalot. (Bullen, F.T.) N.Y., 1899.
Geog 4558.98.15 — Cruise of the Cachalot. (Bullen, F.T.) N.Y., 1920.
Geog 4329.02 — Cruise of the Celtic...Mediterranean. (McCready, R.H.) N.Y., 1902.
X Cg Geog 4208.70 — The cruise of the Colorado. (Willis, G.R.) N.Y., 1873.
Geog 4208.70.2 — The cruise of the Colorado. (Willis, G.R.) N.Y., 1873.
Geog 4219.37.10 — Cruise of the Conrad...1934-1936. (Villiers, A.J.) London, 1937.
Geog 5558.81.55 — The cruise of the Corwin. (Muir, John.) Boston, 1917.
Geog 4309.53.5 — The cruise of the Kate. (Middleton, E.E.) London, 1953.
Geog 4248.86 — Cruise of the Marchesa. (Guillemard, F.H.H.) London, 1886. 2v.
Geog 4248.86.5 — Cruise of the Marchesa to Kamschatka and New Guinea. 2nd ed. (Guillemard, F.H.H.) London, 1889.
Geog 5559.10.5 — Cruise of the Minnie Maud; Arctic seas and Hudson Bay, 1910-1913. (Tremblay, Alfred.) Quebec, 1921.
Geog 4308.54.20 — Cruise of the steam yacht, North Star. (Choules, J.O.) Boston, 1854.
Geog 4308.54.25 — Cruise of the steam yacht, North Star. (Choules, J.O.) Boston, 1854.
Geog 4679.33 — The cruise of the Teddy. (Tambs, E.) London, 1950.
Geog 4208.84.5 — Cruise of the U.S. flagship Lancaster. (McCarthy, W.J.) n.p., 1887.
Geog 4218.31.3 — Cruise of the United States frigate Potomac round the world. (Warriner, F.) N.Y., 1835.
Geog 4208.66.3 — Cruise of "The Wave". N.Y., 1866.
Geog 4328.94.5 — A cruise on the Mediterranean. (Hoyt, W.M.) Chicago, 1894.
Geog 4558.81 — A cruise on the United States practice ship, S.P. Chase. (Wyman, Walter.) N.Y., 1910.
Geog 4559.30.20 — Cruising in many seas. (Perry, Joseph Malcolm.) Springfield, 1930.
Geog 6009.53.5 — Cruising in the Antarctic. (Solianik, A.N.) Moscow, 1956.
Geog 4239.01 — Cruising in the Madiana. (Lindsey, N. Allen.) Boston, 1901.
Geog 4329.38 — Cruising the Mediterranean. (Miller, M.S. (Mrs.).) N.Y., 1938.
Geog 585.11.10 — Crump, William H. The world in a pocket-book, or Universal popular statistics. 12th ed. Phialdelphia, 1860.
Geog 585.11 — Crump, William H. The world in a pocket-book. Philadelphia, 1841.
Geog 585.11.2 — Crump, William H. The world in a pocket-book. Philadelphia, 1842.
Geog 585.11.5A — Crump, William H. The world in a pocket-book. Philadelphia, 1845.
Geog 4678.58.5 — Crusoes of Pitcairn Island. (Knowles, J.N.) Los Angeles, 1957.
Htn Geog 4679.38* — The crusoes of Pitcairn's Islands. (Knowles, Josiah N.) n.p., 1938.
Geog 4559.12 — Crutchley, William C. My life at sea. London, 1912.
Geog 577.98 — Cruttwell, C. New universal gazetteer. London, 1798. 3v.
Geog 4308.06 — Cruz y Bahamonde, N. Viage de Espana, Francia. v.1-10,12,14. Madrid, 1806-1813. 8v.
Geog 4678.68 — Cubbin, T. The wreck of the Serica. London, 1950.
An 170.227 — Cultural and social anthropology. (Hammond, Peter B.) N.Y., 1964.
An 2109.27.15 — Cultural evolution; a study of social origins and development. (Ellwood, Charles A.) N.Y., 1927.
An 182.8 — Cultural relativism. 1. ed. (Herskovits, Melville Jean.) N.Y., 1972.
An 359.71 — Culture, man, and nature. (Harris, Marvin.) N.Y., 1971.
An 170.275.10 — Culture: man's adaptive dimension. (Montagu, Ashley.) N.Y., 1968.
An 185.2 — Culture and behaviour. (Kluckhohn, Clyde.) N.Y., 1962.
An 2109.63 — Culture and personality. (Barnouw, V.) Homewood, Ill., 1963.
An 359.65.20 — Culture and society. (Murdock, George Peter.) Pittsburgh, 1965.
An 5.15 — Culture change; an analysis and bibliography of anthropological sources to 1952. (Keesing, Felix.) Stanford, Calif., 1953.
An 359.39.15 — The culture historical method of ethnology. (Schmidt, Wilhelm.) N.Y., 1939.
An 170.321 — Culture in history; essays in honor of Paul Radin. (Diamond, Stanley.) N.Y., 1960.
Geog 4328.36.2 — Cumming, W.F. Notes of a wanderer. London, 1840. 2v.
An 3409.43 — Cummins, H. Fingerprints, palms and soles. Philadelphia, 1943.
Geog 1710.5 — Cunningham, Peter. Handbook for London, past and present. London, 1849. 2v.
Geog 1710.7 — Cunningham, Peter. Handbook for London, past and present. London, 1850.
Geog 1710.9A — Cunningham, Peter. Handbook for modern London. London, 1851.
Geog 1710.24 — Cunningham, Peter. London, as it is. London, 185-.
Geog 1710.15 — Cunningham, Peter. London. London, 1856.
An 2109.27.30 — Cunow, Heinrich. Technik und Wirtschaft des europäischen Urmenschen. Berlin, 1927.
Geog 807.29 — Le curieux antiquaire. (Berckenmeier, P.) Leide, 1729. 3v.
Geog 4209.20 — Curle, Richard. Wanderings. n.p., n.d.
An 127.5 — Currelly, Charles T. I brought the ages home. 1. ed. Toronto, 1967.
Geog 500.64 — Current geographical publications. N.Y. 1,1938+ 35v.
An 359.11 — Curso de antropologia general. (Boas, Franz.) Mexico, 1911.
Geog 756.44 — Cursus mathematicus. v.4. (Herigone, P.) n.p., 1644.
Geog 4218.76 — Curtis, B.R. Dottings round the circle. Boston, 1876.
Geog 4218.76.2 — Curtis, B.R. Dottings round the circle. 3. ed. Boston, 1877.
Geog 500.67 — Curtiss, Frederic H. A little book on travel books. Boston, 1936.
Geog 4209.28.5 — Cust, Nina. Wanderers. London, 1928.
Geog 4304.65.5 — Cust, Nina W.G. Gentlemen errant; being the journeys and adventures of four noblemen in Europe during the fifteenth and sixteenth centuries. London, 1909.
Geog 4308.87.7 — Cust, R.H.H. How I spent my summer holidays. Eton, 1887.
Geog 4419.40 — Cutting, S. The fire ox and other years. N.Y., 1940.
Geog 4308.56.3 — Cuvillier-Heury, A.A. Voyages et voyageurs, 1837-1854. Paris, 1856.
Geog 4110.11F — Cuward Steamship co. Official guide and album. London, 1877.
Geog 4348.81 — Cuyler, T.L. From the Nile to Norway and homeward. N.Y., 1881.
Geog 665.116.5 — Cvijić, Jovan. Opšta geografija; antropogeogratija. Beograd, 1969.
Geog 665.116 — Cvijićev zbornik u spomen 100. godišnjice njegovog rodjeuja. (Lulovac, Milisav.) Beograd, 1968.
Geog 4300.24 — Cyclist's touring club handbook and guide. London, 1900.
Geog 3018.56.10 — Cyclopaedia of modern travel. (Taylor, B.) N.Y., 1860. 2v.
Geog 3018.56.5 — Cyclopedia of modern travel. (Taylor, B.) Cincinnati, 1856.
VAn 3209.67 — Czekanowski, Jan. Czlowiek w czasie i przestrzeni. 3. wyd. Warszawa, 1967.
Geog 5311.3.30 — Czlowick o Którego upomnialo się morze. (Centkicwicz, Alina.) Warszawa, 1966.
VAn 3209.67 — Czlowiek w czasie i przestrzeni. 3. wyd. (Czekanowski, Jan.) Warszawa, 1967.
An 187.6 — Da biologia à história. (Mendes Corrèa, A.A.) Porto, 1934.
An 2058.65.5 — Da existencia do homem em epochas remotas no valle do tejo. (Pereira da Costa, F.A.) Lisboa, 1865.
Geog 4349.07 — Da Pechino a Parigi in sessanta giorni. (Barzini, Luigi.) Milano, 1970.
An 3409.09 — Dactiloscopia argentina. (Reyna Almandos, Luis.) La Plata, 1909.
An 3409.35.5 — Dactiloscopia clinica. (Castellanos, I.) Habana, 1935.
Geog 4328.70 — Dagbok...resatill Egypten, Konstantinopel. (Fouché, William Grey (Mrs.).) Stockholm, 1870.
Geog 4228.99 — Dagli stati uniti alle Indie. (Riseis, G. de.) Roma, 1899.
Geog 5660.15 — Dahl, F. Andrae og Østgrønland. København, 1932.
Geog 5839.23 — Dahl, Kai R. The "Teddy" expedition. N.Y., 1925.
Geog 3530.5 — Dahlgren, E.W. De franska sjöfärderna. Stockholm, 1900.
Geog 3019.50 — Dainelli, G. La conquista della terra. Torino, 1950.
Geog 5209.60.5 — Dainelli, G. La gara verso il Polo Nord. Torino, 1960.
Geog 4145.52 — Dainelli, Giotto. Il duca degli Abruzzi. Torino, 1967.
Geog 4249.65 — Dainelli, Giotto. L'esplorazione del Grande Oceano. Torino, 1965.
Geog 5509.61.5 — Dainelli, Giotto. Il passaggio di nord-ouest. Torino, 1961.
Geog 3019.40.5 — Dainville, François de. La géographie des humanistes. Paris, 1940.
An 126.10 — Daisy Bates; "the great white queen of the never never". (Salter, Elizabeth.) Sydney, 1971.
Geog 5700.4.10 — Dalager, Lars. Grønlandske relationer. København, 1715.
An 609.67 — Daleki sujtovi naših putnika i pomoraea. (Zagreb. Etnografski Muzej.) Zagreb, 1967.
Geog 5557.25 — Dall, W.H. Critical review of Bering's first expedition, 1725-30. Washington, 1890.
Geog 5557.25.3 — Dall, W.H. Early expeditions to the Bering Sea and Strait. Washington, 1891.
Geog 4309.68 — Dalnie sosedi. (Lisakovskii, Igor' N.) Odessa, 1968.
Geog 4167.75 — Dalrymple, A. A collection of voyages, chiefly in the S. Atlantic. London, 1775.
Geog 4328.42.5 — Damer, G.L.D. Diary of a tour in Greece, Turkey. London, 1842.
Geog 4502.5.13 — Damographia oder Gemsen-Beschreibung in zewy Theil Abgetheilet. (Lebwald, A.) Saltzburg, 1933.
Geog 4157.29 — Dampier, W. Collection of voyages. London, 1729. 4v.
Geog 4208.60.3 — Dan to Beersheba. (Colquohoun, A.R.) London, 1908.
Htn Geog 4559.37* — Dana, Richard Henry. Cruelty to seamen; being the case of Nichols and Couch. Berkeley, 1937.
Geog 4558.40.11 — Dana, Richard Henry. Two years before the mast. Boston, 1869.
Htn Geog 4558.40.10* — Dana, Richard Henry. Two years before the mast. Boston, 1869.
Geog 4558.40.11.5 — Dana, Richard Henry. Two years before the mast. Boston, 1869.
Htn Geog 4558.40.13* — Dana, Richard Henry. Two years before the mast. Boston, 1871.
Geog 4558.40.12 — Dana, Richard Henry. Two years before the mast. Boston, 1883.
Geog 4558.40.24A — Dana, Richard Henry. Two years before the mast. Boston, 1911.
Geog 4558.40.28 — Dana, Richard Henry. Two years before the mast. Boston, 1911.
Geog 4558.40.25 — Dana, Richard Henry. Two years before the mast. Boston, 1911.
Geog 4558.40.27 — Dana, Richard Henry. Two years before the mast. Boston, 1911.
Geog 4558.40.15 — Dana, Richard Henry. Two years before the mast. Cambridge, 1895.
Htn Geog 4558.40.6* — Dana, Richard Henry. Two years before the mast. Glasgow, 1842.
Htn Geog 4558.40.7* — Dana, Richard Henry. Two years before the mast. London, 1841.
Htn Geog 4558.40.50* — Dana, Richard Henry. Two years before the mast. London, 1946?
Geog 4558.40.60 — Dana, Richard Henry. Two years before the mast. Los Angeles, 1964. 2v.
Geog 4558.40.5 — Dana, Richard Henry. Two years before the mast. N.Y., 1840.
Htn Geog 4558.40* — Dana, Richard Henry. Two years before the mast. N.Y., 1840.
Geog 4558.40.9 — Dana, Richard Henry. Two years before the mast. N.Y., 1847.
Geog 4558.40.8 — Dana, Richard Henry. Two years before the mast. N.Y., 1854.

Author and Title Listing

Htn Geog 4558.40.14* Dana, Richard Henry. Two years before the mast. N.Y., 1892.
Geog 4558.40.20A Dana, Richard Henry. Two years before the mast. N.Y., 1907.
Geog 4558.40.22 Dana, Richard Henry. Two years before the mast. N.Y., 1911.
Geog 4558.40.30 Dana, Richard Henry. Two years before the mast. N.Y., 1915.
Geog 4558.40.35A Dana, Richard Henry. Two years before the mast. N.Y., 1921.
Geog 4558.40.26 Dana, Richard Henry. Two years before the mast. N.Y., 1922.
Geog 4558.40.40 Dana, Richard Henry. Two years before the mast. N.Y., 1928.
Htn Geog 4558.40.52* Dana, Richard Henry. Two years before the mast. Sydney, 1946.
Geog 4308.43 Dana, William C. A transatlantic tour. Philadelphia, 1845.
Geog 3251.31 D'Anamia e Botero. (Magnagni, A.) Ciriè, 1914.
Geog 4559.62.10 Dane, Peter. The seamen are down below. Ilfracombe, 1962.
Geog 5398.79.5 Danenhower. Narrative of the "Jeannette". Boston, 1882.
Geog 4209.38.30 Danger is my business. (Craig, John D.) N.Y., 1938.
Geog 4559.54.5 Danger is my life. (Berge, Victor.) London, 1954.
Geog 4238.80 Dangerous voyages of Captain William Andrews. (Andrews, William Albert.) N.Y., 1966.
Geog 808.66 Daniel, H.A. Handbuch der Geographie. Leipzig, 1866-1868. 4v.
Geog 808.66.5 Daniel, H.A. Handbuch der Geographie. Leipzig, 1874. 4v.
Geog 808.66.7 Daniel, H.A. Handbuch der Geographie. Leipzig, 1895. 4v.
Geog 758.45.25 Daniel, H.A. Lehrbuch der Geographie für höhere Unterrichtsanstalten. Halle, 1891.
Geog 613.9 Daniil Gotlib Messershmidt i ego raboty po issledovaniiu Sibiri. (Novlianskaia, Mariia G.) Leningrad, 1970.
Geog 4180.96A Danish Arctic expeditions, 1605-20. (Gosch, C.C.A.) London, 1897. 2v.
Geog 5153.5 Danish Arctic research. Charlottenlund. 1,1955+
Geog 5838.77.1 Danish Greenland, its people and products. (Rink, Henrik.) Montreal, 1974.
Geog 5839.09 Danmark expeditionen. (Friis, A.) Kjøbenhavn, 1909.
An 2109.61.15F La danse cosmique de Lascaux. (Spivack, M.R.) Montignac, 1961.
Geog 5855.15 Danske i Grønland i det 18. aarhundrede. (Ostermann, Hother.) København, 1945.
Geog 5838.83.10 Den Danske Konebaads-Expedition. (Holm, G.) Kjøbenhavn, 1887.
Geog 4311.17 The Danube, its history, scenery. (Beattie, William.) London, 1844?
Geog 4311.12 The Danube. (Jerrold, Walter.) N.Y., 1911.
An 2109.24.5 Danzel, T.W. Kultur undReligion des primitiven Menschen. Stuttgart, 1924.
An 2109.30 Danzel, Theodor. Gefüge und Fundamente der Kultur vom Standpunkte der Ethnologie. Hamburg, 1930.
Geog 290.5 Danzig. Universytet. Wydział Biologii i Nauk o Ziemi. Zeszyty naukowe. Geografia. Gdańsk. 1,1970+
Geog 290.4 Danzig. Wyższa Szkoła Pedagogiczna. Wydział Geograficzny. Zeszyty geograficzne. 1-11,1959-1969// 3v.
Geog 4308.93.5 D'Apery, Tello J. Europe seen through a boy's eyes. N.Y., 1893.
Geog 3019.46 Darby, Henry C. The theory and practice of geography. Liverpool, 1946.
Geog 578.27 Darby, William. Darby's universal gazetteer. 2. ed. Philadelphia, 1827.
Geog 578.27 Darby's universal gazetteer. 2. ed. (Darby, William.) Philadelphia, 1827.
Geog 4105.15 Darde, J.B. The travellers' handbook. London, 1842.
Geog 665.64 Dardel, E. L'homme et la terre. Paris, 1952.
Geog 4209.40.30 Dare to live. (Hurst, Ida.) London, 1940.
Geog 3229.20 Dark, Richard. The quest of the Indies. N.Y., 1920.
Geog 5318.2.5 Dark companion. (Robinson, Bradley.) N.Y., 1947.
Geog 5559.47 Dark companion. 1. ed. (Robinson, Bradley.) N.Y., 1947.
Geog 4308.69.4 Darley, F.O.C. Sketches abroad with pen and pencil. Boston, 1878.
Htn Geog 4308.68.15* Darley, F.O.C. Sketches abroad with pen and pencil. N.Y., 1868.
Geog 4308.69.3 Darley, F.O.C. Sketches abroad with pen and pencil. N.Y., 1869.
An 2358.86 Darling, C.W. Anthropophagy. Utica, 1886.
Geog 4309.08.5 Darling, Kenneth G. My early life in California and my first American and European tours (1890-1908). Cambridge, Mass.? 1942?
Geog 4308.33.10 Darstellungen aus meinem Leben und aus meiner Zeit. (Strombeck, Friedrich Karl von.) Braunschweig, 1833-1840. 4v.
An 2059.59.5 Dart, Raymond. Adventures with the missing link. N.Y., 1959.
Geog 4479.69 Darvère, Pierre. Le Baitoto. Paris, 1969.
Geog 4348.08 Darwin, F.S. Travels in Spain and the East, 1808-1810. Cambridge, 1927.
An 379.37 Darwin's theory applied to mankind. (Machin, A.) London, 1937.
Geog 4759.73 Daveau, Suzanne. La zone intertropicale humid. Paris, 1973.
An 3309.27 Davenport, C.B. Guide to physical anthropometry and anthroposcopy. Baltimore, 1927.
An 3309.29 Davenport, C.B. Race crossing in Jamaica. Washington, 1929.
Geog 4208.58 Davenport, H.E. Rovings on land and sea. Boston, 1858.
Geog 4209.53.5 Davenport, P. The voyage of Waltzing Matilda. London, 1953.
Geog 3070.8 D'Avezac, M.A.P. Coup d'oeil historique sur projection. Paris, 1863.
Geog 4238.45 D'Avezac. Notice des découvertes...dans l'Ocean. Paris, 1845.
Geog 5559.27 Davial goes to Baffinland. (Putnam, D.B.) N.Y., 1927.
Geog 5050.10 David Davis Polar Library. Catalog of the David Davis Polar Library. Menlo Park, 1969.
Geog 4418.44 Davidson, G.F. Trade and travel in the Far East. London, 1846.
Geog 5557.33.5 Davidson, George. The tracks and landfalls of Bering and Chirikof on the northwest coast of America. San Francisco, 1901.
Htn Geog 818.05* Davies, Benjamin. A new system of modern geography. Philadelphia, 1805.
An 39.72 Davies, David Michael. A dictionary of anthropology. N.Y., 1973.
Geog 4345.99.10 Davies, David William. Elizabethans errant; the strange fortunes of Sir Thomas Sherley and his three sons. Ithaca, N.Y., 1967.
Geog 4181.116 Davies, J. The history of the Tahitian Mission, 1799-1830. Cambridge, Eng., 1961.
Geog 5558.71 Davis, C.H. Narrative of north polar expedition. Washington, 1876.
Geog 819.42.2 Davis, D.H. The earth and man. N.Y., 1942.
Geog 819.47 Davis, D.H. The earth and man. N.Y., 1947.
Geog 4558.31 Davis, George. Recollections of a sea-wanderer's life. N.Y., 1887.
Geog 4180.59 Davis, J. Voyages and works of J. Davis. London, 1880.
An 3538.65 Davis, J.B. Crania Britannica. London, 1865. 2v.
An 3558.67 Davis, J.B. Thesaurus Craniorum. London, 1867.
An 3558.67.2 Davis, J.B. Thesaurus Craniorum. Supplement. London, 1875.
Geog 5985.55 Davis, John K. High latitude [autobiographical]. Parkville, Victoria, 1962.
Geog 6009.11.5 Davis, John King. With the "Aurora" in the Antarctic, 1911-14. London, 1919.
Geog 6009.16F Davis, John King. 1918 "Aurora" relief expedition. Melbourne, 1918.
Geog 4218.49.5 Davis, R.C. Reminiscences of a voyage around the world. Ann Arbor, 1869.
Geog 4209.40.10 Davis, R.H. Let's go with Bob Davis to India, Ceylon, Venezuela. N.Y., 1940.
Geog 4209.36.5 Davis, R.H. People, people, everywhere. N.Y., 1936.
Geog 4328.94A Davis, R.H. Rulers of the Mediterranean. N.Y., 1894.
Geog 4208.97 Davis, R.H. A year from a reporter's note-book. N.Y., 1897.
Geog 4309.29.20 Davis, Robert H. With Bob Davis hither and yon. N.Y., 1931.
Geog 4228.28.5 Davis, S.H. Journal of Captain S.H. Davis, a Gloucester sea-captain, 1828-1846. Norwood, 1922.
Geog 665.156 Davis, William Morris. Geographical essays. N.Y., 1954.
An 2109.51.10 Davison, D. The story of prehistoric civilizations. London, 1951.
Geog 4145.53.5 Davydov, I.V. V moriakh i stranstviakh. Moskva, 1956.
Geog 5332.1 Davydov, Iurii V. Ferdinand Vrangel'. Moskva, 1959.
Geog 607.5.15 Davydov, Iurii Vl. Golóvnin. Moskva, 1968.
Geog 5835.85 Davys, John. Tre rejser til Grønland i aarene 1585-87. København, 1930.
An 2058.77 Dawkins, W.B. Address on the antiquity of man. Salford, 1877.
An 2108.74.3 Dawkins, W.B. Cave hinting. London, 1874.
An 2109.25.3A The dawn of European civilization. (Childe, Vere Gordon.) London, 1925.
An 2109.25.4 The dawn of European civilization. (Childe, Vere Gordon.) N.Y., 1925.
An 2109.39.3 Dawn of European civilization. (Childe, Vere Gordon.) N.Y., 1939.
An 2109.39.3.6 The dawn of European civilization. 6. ed. (Childe, Vere Gordon.) London, 1957.
An 2108.89 The dawn of history. (Keary, C.F.) N.Y., 1889.
An 2109.24.15 The dawn of mankind. (Whitnall, H.O.) Boston, 1924.
Geog 3228.97A Dawn of modern geography. (Beazley, C.R.) London, 1897. 3v.
Geog 3228.97.2 Dawn of modern geography. (Beazley, C.R.) London, 1897-1906. 3v.
An 2109.36.3 The dawn of the human mind. (Schmidt, R.R.) London, 1936.
An 2108.83 Dawson, J.W. Fossil men and their modern representatives. 2. ed. London, 1883.
Geog 4145.20 Day, George. Dumont d'Urville. Paris, 1947.
An 2058.77.5 Day, J.V. The prehistoric use of iron and steel. London, 1877.
Geog 3240.5 De Adamo Bremensi geographo. (Bernard, A.) Parisiis, 1895.
Geog 4502.5.7 De alpibus Commentarius; die Alpen. (Simmler, J.) München, 1931.
Htn Geog 4125.91* De arte peregrinandi. Libri II. Noribergae, 1591.
Geog 606.1 De Crontio finaeo gallico geographo. (Gallois, L.) Paris, 1890.
An 357.95 De generis humani varietate nativa. (Blumenbach, J.F.) Gottingae, 1795.
Htn Geog 755.27* De geographia liber unus. (Glareanus, H.) Basileae, 1527.
Htn Geog 755.27.2* De geographia liber unus. (Glareanus, H.) Friburgum, 1530.
Htn An 356.77* De habitu et colore Aethiopum. (Pechlin, J.N.) Kiloni, 1677.
An 355.85 De homine magno illo in rerum natura. Libro II. (Lascovius, P.M.) Witebergae, 1585.
An 2600.1 De hominum inter ferar educatorum. (Koenig, H.C.) Hanover, 1730.
Geog 4311.7 De istri ostiis dissertatio historico-geographica. (Senkel, K.F.) Wratislaviae, 1820.
Geog 3530.18 De la banquise à la jungle. (Lauga, Henri.) Paris, 1952.
Geog 500.15 De la collection geographique. (Jomard, E.F.) Paris, 1848.
Geog 4208.64 De la Habana a Madrid. (Mompou, José.) Habana, 1865.
Geog 3070.29 De la lecture des cartes étrangères. (Mager, Henri.) Paris, 1883.
Geog 4329.53.5F De l'Amanus au Sinai. (Dunand, M.) Beyrouth, 1953.
Geog 3235.17 De l'anciennete de la mappemonde. (Pezzana, A.) Genes, 1808.
An 2108.74.7 De l'ancienneté de l'homme. v.1-2. (Zaborowski-Moindron, S.) Paris, 1874.
Geog 3140.15 De l'évaluation et de l'emploi des mesures itinéraires grecques et romaines. (Gosselin, P.Z.J.) Paris, 1813.
Geog 4145.78 De l'Himalaya aux Pyrénées. (Touring Club de France.) Paris? 1959.
Geog 4218.73.10 De Madrid a Madrid. (Dupuy de Lôme, E.) Madrid, 1877.
Htn Geog 5515.77.5* De Martini Forbisseri Angli. (Best, George.) n.p., 1580.
Geog 520.21 De orbe novo. (Anghiera, P.M.) Paris, 1907.
Geog 4308.27 De Paris a Varsovie. (Delestre, P.) Paris, 1827.
Geog 4239.69.5 De Plymouth à Newport par le sud de Nantucket. (Castelbajac, Bertrand de.) Paris, 1969.
Geog 3530.14 De Rasilliis Gabriel, Isaac et Claudio proenominatis Richelii adjutoribus. (Deschamps, Léon.) Paris, 1898.
Htn Geog 4105.12* De regimine iter agentium. (Gratarolo, G.) Basileae, 1561.
Htn Geog 755.63.2* De universitate liber. 2a ed. pt.1-2. (Postel, Guillaume.) Parisiis, 1563.
Geog 4329.44 Dea senza volto. (Fateili, A.) Milano, 1944.
Geog 4672.12 Dead men's tales. (Kitchin, Frederick H.) Edinburgh, 1926.
VGeog 4309.03.15 Deák, Imre. Zarandoklat Rómába. Fajsz? 1903.
Geog 4029.8 Dean, Harry. The Pedro Gorino. Boston, 1929.
Geog 4029.6 Dear Sarah. (Borden, Norman E.) Freeport, Me., 1966.

176

Author and Title Listing

	Geog 4209.47	Death over my shoulder. (Colam, Lance.) London, 194-.
	Geog 4678.52	A deathless story. (Addison, A.C.) London, 1906.
	An 358.65	Debay, A. Histoire naturelle de l'homme et de la femme. 12. éd. Paris, 1865.
	An 2109.65.5	De Beer, Gavin Rylands. Genetics and prehistory. Cambridge, Eng., 1965.
	Geog 5070.38	Debenham, F. The polar regions. London, 1930.
	Geog 5939.59	Debenham, Frank. Antarctica. London, 1959.
	Geog 4181.91	Debenham, Frank. Voyage of Captain Bellingshausen to Antarctic seas, 1819-21. London, n.d. 2v.
	An 3309.70.1	Debets, Georgii Frantsevich. Physical anthropology of Afghanistan. v.1-2. Cambridge, 1970.
	An 2108.88	Debierre, C. L'homme avant l'histoire. Paris, 1888.
	An 2058.65	Débris de l'industrie humaine...caverne de Bossey. (Thioly, F.) Genève, 1865.
	An 358.81.9	Les débuts de l'humanité. (Hovelacque, Abel.) Paris, 1881.
	Geog 3049.52	DeCamp, L.S. Lands beyond. N.Y., 1952.
	Geog 3055.72	DeCamp, L.S. Lost continents; the Atlantis theme in history, science and literature. 1st ed. N.Y., 1954.
	Geog 4209.63.35	A decent fellow doesn't work. (Green, Lawrence George.) Cape Town, 1963.
Htn	An 3408.93*	Decipherment of blurred fingerprints. (Galton, F.) London, 1893.
	An 2109.60.15	Découverte de la préhistoire. (Grand, P.M.) Paris, 1960.
	Geog 3019.54.5	Découverte du monde. (Le Gentil, Georges.) Paris, 1954.
	Geog 5180.7	De Croy, D. Memoire sur le passage par le nord. Paris, 1782.
	Geog 4559.26	Deep sea days. (Hemy, Thomas M.) London, 1926.
	Geog 4558.98	A deep-water voyage. (Stevenson, P.E.) Philadelphia, 1898.
	Geog 4307.65	Defeller. Itinéraire ou voyages. Liege, 1820. 2v.
	An 359.06.12	Defensa de la Antropologia general. (Maldonado, S.D.) Caracas, 1906.
	An 379.39.5	Deffontaines, P. Problèmes de géographie humaine. Paris, 1939.
	Geog 4208.91.5	Del Atlántico al Pacífico y un Argentino en Europa. 2a ed. (Carrasco, Gabriel.) Buenos Aires, 1891.
	Geog 5839.24.5	Del Grønland vi miste og de vi ikkje miste. (Koht, Halvdan.) Kristiania, 1924.
	Geog 4209.63.20	Del nuevo al viejo mundo. (Diez del Corval, L.) Madrid, 1963.
	Geog 3235.31	Del planisferio. (Amat, P.) Roma, 1878.
	Geog 4145.34.2	Dela i dni G.E. Grum-Grzhimailo. (Grum-Grzhimailo, A.G.) Moskva, 1947.
	Geog 4218.17.8	Delano, Amasa. Narrative of voyages and travels. N.Y., 1970.
	Geog 4308.27	Delestre, P. De Paris a Varsovie. Paris, 1827.
	Geog 4326.10	Deliciae transmari. (Ens, K.) Coloniae Agrippinae, 1610.
	Geog 4266.09	Deliciarum germaniae...viatorius...itinera ad omnes ciirtates. (Ens, Gaspar.) Coloniae, 1608.
	Geog 622.1	Della vita e...opere di Pietro della Valle il Pellegrino. (Ciampi, Ignatius.) Roma, 1880.
	Geog 5398.79.7	De Long, G.W. Voyage of the "Jeannette". Boston, 1883. 2v.
	Geog 3235.7F	De Luca, G. Carte nautiche del medio evo. Napoli, 1866.
	Geog 5939.59.15	Demain l'Antarchtique. (Reboux, Michael.) Paris, 1959.
	Geog 579.07	Demangeon, A. Dictionnaire, manuel, illustré. Paris, 1907.
	Geog 665.88	Demangeon, Albert. Problèmes de géographie humaine. 4. éd. Paris, 1952.
	Geog 4269.67.5	Le déménagement du territoire. (Le Lannou, Maurice.) Paris, 1967.
	An 357.76	Demeunier, J.N. L'esprit des usages et des coutumes des différens peuples. Londres, 1776. 3v.
	An 359.38.3A	Demiashkevich, Michael John. The national mind. N.Y., 1938.
	An 3309.43	Demographic and bodily changes in descendants of Mexican immigrants. (Goldstein, M.S.) Austin, 1943.
	An 359.01	Demolins, E. Les grandes routes des peuples. Paris, 1901. 2v.
	An 359.50.10	Dempf, Alois. Theoretische Anthropologie. Bonn, 1950.
	Geog 4110.16	Dempsey, J.M. Our ocean highways. London, 1871.
Htn	Geog 815.19*	Denciso, M.F. Suma de geographia. Seville, 1519.
	Geog 4332.22	Denham, Henry Mangles. The Adriatic; a sea guide to its coasts and islands. London, 1967.
	An 359.00.3	Deniker, J. Les races et les peuples de la terre. Paris, 1900.
	An 359.00.4	Deniker, J. The races of man. London, 1900.
	An 359.12.2	Deniker, J. The races of man. London, 1912.
Htn	Geog 4758.24.3*	Denis, F. Scènes de la nature sous les tropiques. Paris, 1824.
Htn	Geog 4758.24.2*	Denis, F. Scènes de la nature sous les tropiques. Paris, 1824.
	Geog 4758.24	Denis, F. Scènes de la nature sous les tropiques. Paris, 1824.
	Geog 535.15	Denis, Jacques. Guide de la recherche géographique en Belgique. Namur, 1970.
	Geog 4559.48	Deniz cağrafyasi. (Öngör, Sami.) Ankara, 1948.
	Geog 4502.5.8	Der denkende Wanderer. (Hoek, H.) München, 1932.
	Geog 5839.28	Denmark. Kommittee for Ledelsen af Degeoliske og Geografiske Undersøgelser i Grønland. Greenland...invest in Greenland. Copenhagen, 1928. 3v.
	Geog 5845.4	Denmark. Ministeriet for Grønland. Perspektivplan for Grønland 1971-85. København, 1971.
	Geog 5839.52	Denmark. Udenrigsministeriet. Greenland. Ringkjøbing, 1952?
	Geog 5606.10	Denmark. Udvalget for samfundsforskning i Grønland. Skrifter. 1-9,1961-1963// 2v.
	Geog 5645.5	Denmark's to Greenland. (Berlin, Knud.) London, 1932.
	Geog 4209.40.25	Dennis, W.H. Glittering horizons. N.Y., 1940.
	Geog 4219.25	Dennison, C.S. Around the world with Texaco. Houston, 1925.
	Geog 4415.05	Den Öchten Weg Auss zu Faren von Lissbona gen Kallakuth. From Lisbon to Calicut. Minneapolis, 1956.
	Geog 5900.15	Denucé, J. Bibliographie antarctique. Bruxelles, 1913.
	Geog 3070.27	Denucé, J. Oud-Nederlandsches kaartmakers in betrekking met Plantijn. Antwerpe, 1912. 2v.
	Geog 520.24	Denucé, J. Pigafetta; relation du premier voyage...par Magellan. Paris, 1923.
	Geog 4309.23	Dépaysements. (Traz, Robert de.) Paris, 1923.
	Geog 4672.21.2	Deperthes, Jean L.H.S. Histoire des naufrages, ou Recueil des relations les plus intéressantes des naufrages. Paris, 1790. 3v.
	Geog 4672.21.4	Deperthes, Jean L.H.S. Histoire des naufrages, ou Recueil des relations les plus intéressantes des naufrages. Paris, 1828. 3v.
	Geog 5300.7	DePeyster, J.W. The Dutch at the North Pole. N.Y., 1857.
	Geog 818.33	Depping, G.B. Evening entertainments...manners and customs of nations. Philadelphia, 1833.
	Geog 808.90F	DePuy, W.H. The universal guide and gazetteer. N.Y., 1890.
	Geog 4182.70	De derde reis van de V.O.C. naar Oost-Indië onder het beleid van Admiraal van Caerden. (Booy, A. de.) 's-Gravenhage, 1968.
	Geog 4309.71.5	Derruau, Max. L'Europe. Paris, 1971.
	Geog 4304.65	Des böhmischen Herrn Leo's von Rozmital...Reise...Abendlande 1465-1467. (Schaschek.) Stuttgart, 1844.
	Geog 4309.63	Des brumes de la Mer du Nord. (Cluzel, M.E.) Paris, 1963.
	Geog 4502.10.2	Des F. Petrarca Sendschreiben die Besteigung des Mont Ventoux betreffend. (Petrarca, F.) München, 1936.
Htn	Geog 4206.65*	Desboys du Chastelet, R. L'odyssee ou diversite d'avantures...en Europe, Asie et Afrique. Fleche, 1665.
	Geog 3019.72.30	Deschamps, Hubert J. Les Européens hors d'Europe de 1434 à 1815. 1. éd. Paris, 1972.
	Geog 3530.14	Deschamps, Léon. De Rasilliis Gabriel, Isaac et Claudio proenominatis Richelii adjutoribus. Paris, 1898.
	Geog 578.70	Deschamps, P. Dictionnaire de géographie, ancienne et moderne. Paris, 1870.
	Geog 3570.18	Descobridores portugueses. (Prestage, Edgar.) Porto, 1934.
	Geog 3570.18.5	Descobridores portugueses. 2. ed. (Prestage, Edgar.) Lisboa, 1943.
	Geog 4239.58	Descobrimento do Atlântico. (Costa Brochado, José de.) Lisboa, 1958.
	Geog 3570.10	Descobrimentos dos Portuguezes nos seculos XV e XVI. (Maix de Sori, A.F.) Lisboa, 1867.
	Geog 3570.31.5	Descobrimentos maritimos africanos dos Portugueses com D. Henrique. (Fontoura da Costa, Abel.) Lisboa, 1938.
	Geog 3570.52	Os descobrimentos pre-colombinos dos Portugueses. (Cortesão, Jaime.) Lisboa, 1966.
	Geog 815.59.10	Descripcion de todas las provincias. (Botero, G.) Gerona, 1748.
	Geog 4267.71.3	Descripción general de la Europa. (Gutierrez de la Hacera, P.R.) Madrid, 1782. 2v.
	Geog 815.35.5	Descripciones geograficas del estado moderno de las regiones. (Servetus, Michael.) Madrid, 1932.
	Geog 815.35.4	Descripciones geograficas del estado moderno de las regiones. (Servetus, Michael.) Madrid, 1932.
	Geog 816.65	Descriptio orbis. (Linda, Lucas de.) Amsterdam, 1665.
Htn	Geog 816.55*	Descriptio orbis et omniumejus rerumpublicarum. (Linda, Lucas de.) Lugdunum Batavorum, 1655.
	Geog 4268.48	Description...of countries of Europe. (Maps. Europe.) Hartford, 1848.
	Geog 520.13	Description de l'Afrique. v.2-3. (Leone, G.) Paris, 1896-98. 2v.
Htn	Geog 817.00*	Description de tout universe. (Sanson, N.) Amsterdam, 1700.
Htn	Geog 817.74*	Description et usages de la Mappemonde. (André, Noël.) Paris, 1774.
	Geog 4180.35	Description of...East Africa and Malabar. (Barbosa, D.) London, 1866.
Htn	Geog 3240.18.5*	Description of a rare and precious atlas. (Orion, booksellers, ltd., London.) London, 1935?
	Geog 808.12.16F	A description of all parts of the world. (Malte-Brun, Conrad.) Boston, 1859. 3v.
	Geog 4268.48.5	Description of Bayne's gigantic panorama of a voyage to Europe. Boston, 1848.
	Geog 4308.48.2	Description of Bayne's gigantic panorama of a voyage to Europe. Philadelphia, 1849.
Htn	Geog 5507.45*	Description of coast...in Button's Bay. London, 1745?
	Geog 5837.63.5	Description of Greenland. (Egede, H.) London, 1818.
	An 358.59	Descriptive ethnology. (Latham, R.G.) London, 1859. 2v.
	Geog 953.1	Pamphlet box. Descriptive geographies. Views. Bibliographies.
	Geog 4269.03.10	Descriptive geography of Europe from original sources. (Herbertson, (Mrs.).) London, 1919.
	Geog 3570.23F	Os descrobrimentos portugueses. (Cortesão, Jaime.) Lisboa, 195-? 2v.
	Geog 5939.62.5	El descubrimiento de la Antártida. (Fitte, Ernesto.) Buenos Aires, 1962.
	Geog 3560.25	Los descubrimientos en el Atlántico y la rivalidad castellano-portuguesa. (Dória, A.A.) Braga, 1951.
	Geog 3560.20	Los descubrimientos en el Atlántico y la rivalidad castellano-portuguesa hosta el tratado de Tordesillas. 1. ed. (Perez Emlied, F.) Sevilla, 1948.
	Geog 4309.16	Desde lajos. (Villaverde, J.R.) Habana, 1916.
	Geog 4181.63	The desert route to India. (Carruthers, D.) London, 1929.
	Geog 5900.16	Desiat' let sovetskikh issledovanii v Antarktike. (Barkov, N.I.) Leningrad, 1968.
	Geog 3070.2	Desimoni, C. Atlante idrografico. Genova, 1867.
	Geog 3070.24	Desimoni, C. Elenco di carte ed Atlanti nautici. Genova, n.d.
	Geog 3235.55	Desimoni, C. Nuovi studi sull'atlante luxoro. Geneva, 1869.
	Geog 4209.31.5	Desmond, Alice C. Far horizons. N.Y., 1931.
	Geog 5939.50	The desolate Antarctic. (Mountevans, E.) London, 1950.
	Geog 4308.38	Desultory reminiscences of a tour. (Wright, H.H.) Boston, 1838.
	Geog 5155.5	Detectio freti Hudsoni. (Gerritsz, H.) Amsterdam, 1878.
	Geog 4309.70	Detours from the grand tour. (Lottman, Herbert R.) Englewood Cliffs, N.J., 1970.
	Geog 4311.18	Deudonex. Dichtung. (Feuerstein, Ernst.) n.p., 1966.
	Geog 45.1	Deutsche Geographentage. Verhandlungen. Berlin. 1-39 25v.
	Geog 45.2	Deutsche geographische Blätter. Bremen. 13
	Geog 4313.6	Die deutsche Ostsee. (Schulz, Bruno.) Bielefeld, 1931.
	Geog 3540.10	Deutsche Pfadfinder des 16. Jahrhunderts in Afrika, Asien und Südamerika. (Pannwitz, Max.) Stuttgart, 1911-12.
	Geog 1522.42	Das Deutsche Reich. 6. Aufl. (Baedeker, publishers.) Leipzig, 1936.
	Geog 45.12	Deutscher Geographentag, 38th, Erlangen and Nuremberg, 1971. Tagungsbericht und wissenschaftliche Abhandlungen. Wiesbaden, 1972.
	Geog 4345.73	Ein deutscher Kaufmann. (Krafft, Hans U.) Göttingen, 1862.
NEDL	Geog 45.5	Deutscher und Österreichischer Alpenvereins. Mittheilungen. Salzburg. 1-25,1875-1899 25v.
	Geog 45.5	Deutscher und Österreichischer Alpenvereins. Mittheilungen. Salzburg. 1875-1929 24v.
	Geog 45.10	Deutscher und Österreichischer Alpenvereins. Register zu den Vereinsschriften. Innsbruck. 1906
NEDL	Geog 45.6	Deutscher und Österreichischer Alpenvereins. Zeitschrift. Salzburg. 1-30,1869-1899 30v.

Author and Title Listing

	Geog 45.6	Deutscher und Österreichischer Alpenvereins. Zeitschrift. Salzburg. 1-74,1869-1949 40v.	Geog 3019.33	Dickinson, R.E. The making of geography. Oxford, 1933.
	Geog 45.6.7	Deutscher und Österreichischer Alpenvereins. Zeitschrift. Beilagen. 1869-1910	An 379.52.10	Dickinson, Robert Eric. City, region and regionalism. London, 1952.
	Geog 45.9	Pamphlet box. Deutscher und Oesterreichischer Alpenvereins.	An 379.60	Dickinson, Robert Eric. Some problems of human geography. Leeds, 1960.
	Geog 4219.32.15	Deutsches Mädel auf Fahrt um die Welt. (Dinglreiter, Senta.) Leipzig, 1932.	Geog 578.23.5	Dictionaire géographique. Paris, 1823. 10v.
	Geog 3540.11	Deutsches Reisen im Wandel. (Beck, Carl.) Berlin, 1936.	Geog 577.01	Dictionaire geographique universal. (Baudrand, M.A.) Amsterdam, 1701.
	Geog 1522.33	Deutschland. v.2. 10. Aufl. (Baedeker, publishers.) Coblenz, 1862.	Geog 578.41	Dictionary geographical, statistical. (McCulloch, J.R.) London, 1841-42. 2v.
	Geog 1522.25	Deutschland. 7. Aufl. (Baedeker, publishers.) Coblenz, 1857.	Geog 578.41.6	Dictionary geographical, statistical. (McCulloch, J.R.) London, 1851. 2v.
	Geog 1522.30	Deutschland. 8. Aufl. (Baedeker, publishers.) Coblenz, 1858.	Geog 578.41.2	Dictionary geographical, statistical. (McCulloch, J.R.) N.Y., 1843-44. 2v.
	Geog 1522.35	Deutschland. 12. Aufl. (Baedeker, publishers.) Coblenz, 1865.	Geog 578.41.3	Dictionary geographical, statistical. (McCulloch, J.R.) N.Y., 1844-45. 2v.
	Geog 1522.38	Deutschland in einem Bande. 4. Aufl. (Baedeker, publishers.) Leipzig, 1925.	Geog 578.41.4	Dictionary geographical, statistical. (McCulloch, J.R.) N.Y., 1844-47. 2v.
	Geog 1522.36	Deutschland und Österreich. 14. Aufl. (Baedeker, publishers.) Coblenz, 1869.	Geog 578.41.5	Dictionary geographical, statistical. (McCulloch, J.R.) N.Y., 1849. 2v.
	Geog 3199.10	Deutschland und seine Naehbarländer nach der grossen Geographie des Idrisi. (al-Idrisi, Muhammad ibn Muhammad.) Stuttgart, 1937.	Geog 578.41.7	Dictionary geographical, statistical. (McCulloch, J.R.) N.Y., 1852. 2v.
	Geog 3235.52	Deux notes sur l'anciennes cartes historiées. (Avezac, M.A.P.) Paris, 1844.	Htn Geog 577.73*	Dictionary of ancient geography. (Macbean, A.) London, 1773.
	Geog 3570.28	Deux voyages portuguèses de découverte dans l'Atlantique occidental. (Lima, Manuel C.) Lisbonne, 1946.	An 39.72	A dictionary of anthropology. (Davies, David Michael.) N.Y., 1973.
	Geog 4209.41.10	DeValda, F.W. Adventure is my business. London, 1941.	Geog 4652.7	Dictionary of disasters at sea during the age of steam, 1824-1962. (Hocking, Charles.) London, 1969. 2v.
	Geog 5559.37.25	Deviat' mesiatsev na Dreifuiushchei stantsii severnyi polius. Moskva, 1938.	Geog 578.68	Dictionary of geography. (Beeton, S.O.) London, 1868.
	Geog 3240.17	Devic, L. Marcel. Coup d'oeil sur la litterature géographique arabe au moyen âge. Paris, 1882.	Geog 578.51	Dictionary of geography. (Johnstone, A.K.) London, 1851.
	Geog 3055.45	Dévigné, R. Un continent disparu l'Atlantide. Paris, 1923.	An 359.44.5A	A dictionary of the international slurs. (Roback, A.A.) Cambridge, Mass., 1944.
	Geog 3055.45.5	Dévigné, R. Un continent disparu l'Atlantide. Paris, 1924.	Geog 579.07	Dictionnaire, manuel, illustré. (Demangeon, A.) Paris, 1907.
	Geog 4309.47.5	Devil at my heels. (Thompson, R.W.) London, 1947.	Geog 578.70	Dictionnaire de géographie, ancienne et moderne. (Deschamps, P.) Paris, 1870.
	Geog 4559.35.20	Devine, Eric. Midget Magellans; great cruises in small ships. N.Y., 1935.	Geog 579.70	Dictionnaire de la géographie. Paris, 1970.
	An 99.74	De Waal Malefijt, Annemarie. Images of man; a history of anthropological thought. 1. ed. N.Y., 1974.	Geog 578.10.2	Dictionnaire géographique. 2. ed. (Ladvocat, J.B.) Paris, 1812.
	An 205.4	Dewey, Chester. Examination of some reasonings against unity of mankind. n.p., 1862. 2 pam.	Geog 577.09.18	Dictionnaire géographique portatif. (Ladvocat, J.B.) Paris, 1770.
	Geog 4308.33.2	Dewey, O. Old world and the new. N.Y., 1836. 2v.	Geog 758.70	Dicuil. Liber de mensura orbis terra a Gustavo Parthey recognitus. Bercolini, 1870.
	An 358.28.3	Die Diätetik des physischen und psychischen Menschenlebens. (Leupoldt, J.M.) Berlin, 1828.	Geog 758.07.5	Dicuil. Liber de mensura orbis terrae. Dublin, 1967.
	An 359.71.25	The dialectics of social life: alarms and excursions in anthropological theory. (Murphy, Robert Francis.) N.Y., 1971.	Geog 758.07.2	Dicuil. Liber de mensura orbis terrae. Paris, 1807.
			Geog 758.07	Dicuil. Liber de mensura orbis terrae. Paris, 1807.
			An 2339.12	Dieck, Alfred. Die Waffen der Naturvölker Süd-Amerikas. Ställuponen, 1912.
	An 170.321	Diamond, Stanley. Culture in history; essays in honor of Paul Radin. N.Y., 1960.	Geog 5070.68	Dieck, Herman. The marvelous wonders of the polar world. Philadelphia, 1965?
	An 359.74.5	Diamond, Stanley. In search of the primitive. N.Y., 1974.	An 358.64.3	Diefenbach, L. Vorschule der Völkerkunde. Frankfurt am Main, 1864.
	An 170.173.5	Diamond, Stanley. Primitive views of the world. N.Y., 1964.	An 359.08	Dieserud, J. The scope and content of the science of anthropology. Chicago, 1908.
	Geog 4218.76.9	Diario del viaggio intorno al globo. (Bedinello, Ugo.) Trieste, 1876.	An 359.22	Diet and race. (Armitage, F.P.) London, 1922.
	Geog 4208.44.5	Diario del viaje desde Madrid a Manila. (Santos, Eusebio.) Madrid, 1851.	Geog 4209.63.20	Diez del Corval, L. Del nuevo al viejo mundo. Madrid, 1963.
	Geog 4217.89	Diario del viaje explorador...en 1789-94. (Viana, F.J. de.) Cerrito de la Victoria, 1849.	VGeog 4678.91	Difesa dei signori Henderson, armatori contro i danneggiati del naufragio dell'Utopia. (Henderson Brothers, Glasgow, defendants.) Napoli, 1895.
	Geog 4209.10	Il diario di un viandante, dal deserto al mar glaciale. (Beltramelli, Antonio.) Milano, 1911.	Geog 4678.58	Difesa del capitano ed armatore del Sicilia contro il cap. ed armatore dell'Ercolano. (Corrao, Mario.) Palermo, 1858.
	Geog 4308.47.9	Diary, sketches and reviews. (Dodge, Robert.) N.Y., 1850.		
	Geog 4308.52	Diary...summer tour. (Sewell, E.M.) N.Y., 1852.		
	Geog 4324.59.10	Diary. (Ehingen, Georg von.) London, 1929.	Geog 4678.54.10	Difesa pel cav F. Miceli...scontro dei...Ercolano e Sicilia. (Cola, S. de.) Messina, 1858.
Htn	Geog 4238.03*	A diary kept by Reverend David Sutherland on a voyage from Greenock Scotland to New York. (Sutherland, D.) Woodsville, 1910.	An 2109.33.3	Differenzierungserscheinungen in einigen afrikanischen Gruppen. Proefschrift. (Hofstra, Sjoerd.) Amsterdam, 1933.
	Geog 5558.96	A diary kept while with the Peary Arctic expedition of 1896. (Hoppin, Benjamin.) New Haven? 1897?	An 359.27.3	The diffusion of culture. (Marett, Robert R.) Cambridge, Eng., 1927.
	Geog 4309.26A	The diary of a musketeer. (Morse, William I.) Boston, 1926.	Geog 4209.38.10	Dighy, G. Goose feathers. N.Y., 1938.
	Geog 4328.42.5	Diary of a tour in Greece, Turkey. (Damer, G.L.D.) London, 1842.	Geog 3235.45	Un digression géographique...la mappemonde. (Avezac, M.A.P.) Paris, 1870.
	Geog 4308.17.5	Diary of an invalid...journey...1817. (Matthews, H.) Paris, 1836.	Geog 5398.82	Dijmphna's expeditionen, 1882-83. (Hovgaard, A.P.) Kjøbenhavn, 1884.
	Geog 4308.17.4	The diary of an invalid. (Matthews, H.) Paris, 1825.	An 3559.13	Dillenius, J.A. Craneometría comparativa de los antiguos habitantes de la isla y del pukara de Tilcara. Buenos Aires, 1913.
	Geog 4308.17.3	The diary of an invalid. 2nd ed. (Matthews, H.) London, 1820.		
	Geog 4308.17.3.5	The diary of an invalid. 3rd ed. (Matthews, H.) London, 1822. 2v.	Geog 4169.59	Dillon, Richard. Embarcadero. N.Y., 1959.
			An 2059.03	Der diluviale Mensch in Europa. (Hoernes, Moriz.) Braunschweig, 1903.
	Geog 4326.75.5A	The diary of H. Teonge, chaplain on board H.M.'s ships Assistance, Bristol and Royal Oak, 1675-1679. (Teonge, Henry.) London, 1927.	An 2059.12.3	Der diluviale Mensch und seine Zeitgenossen aus dem Tierreiche. (Jacob, K.H.) Leipzig, 1912.
	Geog 4326.75	The diary of H. Teonge...1675-1679. (Teonge, Henry.) London, 1825.	Geog 4219.32.15	Dinglreiter, Senta. Deutsches Mädel auf Fahrt um die Welt. Leipzig, 1932.
	Geog 4238.32.10	Diary of my journey from Muehlhausen in Thurangia via Bremen to the United States...1831. (Roebling, John A.) Trenton, 1931.	Geog 613.1.15	Dinse, Paul. Zum Gedächtnis Gerhard Mercator's. Berlin, 1894.
			Geog 4145.11	Diogo Cão. (Peres, Damião.) Lisboa, 1957.
	Geog 4180.66	Diary of Richard Cocks...1615-22. (Cocks, R.) N.Y., 1964. 2v.	Geog 5610.5	Diplomatarium Groenlandicum, 1492-1814. (Bobé, Louis.) København, 1936.
	Geog 6009.01.20	Diary of the 'Discovery' expedition to the Antarctic regions, 1901-1904. (Wilson, Edward Adrian.) London, 1966.	Geog 4209.10.6	A diplomatist's wife in many lands. (Fraser, Mary Crawford.) London, 1911. 2v.
	Geog 6009.10.50	Diary of the "Terra Nova" expedition to the Antarctic, 1910-12. (Wilson, Edward Adrian.) London, 1972.	Geog 4209.10.5	A diplomatist's wife in many lands. (Fraser, Mary Crawford.) N.Y., 1911. 2v.
	Geog 4180.74	Diary of William Hedges...during agency in Bengal. (Hedges, W.) London, 1887-89. 3v.	Geog 4182.9	Dirck Gerritsz Pomp. (Ijzermann, J.W.) 's-Gravenhage, 1915.
	An 359.61.30	Dias, Jorge. Ensaios etnológicas. Lisboa, 1961.	Geog 4652.5	Direction of shipwrecks of the Great Lakes. (Heden, Karl E.) Boston, 1966.
	Geog 4127.05	Diatriba...peregrinationum eruditarum. (Schroeter, J.C.) Jenae, 1705.	Geog 520.4	Le discours de la navigation. (Parmentier, J.) Paris, 1883.
	Geog 5925.60	Diaz, Emilio. Relatos antarticos. Buenos Aires, 1958.	Htn Geog 4205.98F*	Discours of voyages...East and West Indies. (Linschoten, Jan H. van.) London, 1598.
	Geog 4181.23A	Diaz del Castillo, B. True history of conquest of New Spain. London, 1908-10. 5v.	Htn Geog 4205.98.2F*	Discours of voyages...East and West Indies. (Linschoten, Jan H. van.) London, 1598.
	Geog 4209.67.10	Díaz Plaja, Guillermo. Con variado rumbo. 1. ed. Barcelona, 1967.	Geog 618.3.5	Discours prononcé par Monsieur le recteur Guillaume de Greef. (Greef, Guillaume J. de.) Gand, 1906.
	Geog 3580.45	Diaz-Trechuelo Spinola, Maria Lourdes. Navegantes y conquistadores vascos. Madrid, 1965.	Geog 4326.32.30	Discourse of a peregrination in Europe, Asia and Affricke. Facsimile. (Lithgow, William.) N.Y., 1971.
	Geog 579.26	Diccionario de voces y términos geográficos. (Vergara y Martin, Gabriel Maria.) Madrid, 1926.	Geog 3019.06	Discoveries and explorations in the century. (Roberts, C.G.D.) Toronto, 1906.
	Geog 579.53	Diccionario geografico universal. (Villalba y Rubio, F.) Madrid, 1953.	Geog 4309.07	Discoveries in every-day Europe. (Seitz, Don C.) N.Y., 1907.
	Geog 576.67	Diccionario lusitanico-latino de nomes proprios de regioens, reinos, provincias, cidades. (Poyares, P. de.) Lisboa, 1667.	Geog 4180.30A	Discoveries of the world. (Galvano, A.) London, 1862.
			Geog 6009.33A	Discovery; the story of the second Byrd Antarctic expedition. (Byrd, Richard E.) N.Y., 1935.
	Geog 4308.50.17	Dickinson, A. My first visit to Europe. 5th ed. N.Y., 1856.	Geog 4180.9	Discovery...of Terra Florida by De Soto. (Relacam Verdadeira dos Trabalhos.) London, 1851.
	Geog 818.13	Dickinson, R. Elements of geography. Boston, 1813.	Geog 5329.3.10	Discovery. 1. ed. (Stefansson, Vilhjalmur.) N.Y., 1964.

Author and Title Listing

Geog 4180.3 Discovery of...empire of Guiana. (Raleigh, W.) London, 1848.

Geog 4269.47A Discovery of Europe. (Rahv, Philip.) Boston, 1947.

An 3509.38 Discovery of Sauk Valley man of Minnesota. (Bryan, Kirk.) Abilene, Texas, 1938.

Geog 4181.7 Discovery of Solomon Islands by...Mendaña. (Amherst of Hackney.) London, 1901. 2v.

Geog 4181.98A The discovery of Tahiti. (Robertson, George.) London, 1948.

Geog 3129.32A The discovery of the ancient world. (Burton, H.E.) Cambridge, 1932.

Geog 4239.58.5 The discovery of the Atlantic. (Costa Brochado, José de.) Lisboa, 1960.

Geog 3570.53 The discovery of the East. (Crone, Gerald.) London, 1972.

Geog 5518.50.4 Discovery of the Northwest Passage. (Osborn, S.) Edinburgh, 1865.

Geog 4249.60 The discovery of the Pacific Islands. (Sharp, C.A.) Oxford, 1960.

Geog 4129.74 The discovery of the sea. (Parry, John Horace.) N.Y., 1974.

Geog 3249.30 Discussão historica das medidas geographicas no seculo XVI. (Jacome Correa, Ayres.) Lisboa, 1930.

Geog 5507.98F Disertaciones sobre la navegación. Isla de León, 1798.

An 358.43.3 A dissertation on the varieties of mankind. (Bond, Thomas E.) Middletown, 1843.

Htn An 3207.91* Dissertation physique...traits du visage. (Camper, P.) Utrecht, 1791. 2 pam.

Geog 3055.25 Dissertation sur l'Atlantide. (Jolibois.) Lyon, 1846.

Htn Geog 3550.10F* Dissertazione intorno ad alcuni viaggiatori eruditi veneziani. (Morelli, Iacopo.) Venezia, 1803.

Htn Geog 4678.10* Distressing narrative of the loss of the ship Margaret of Salem. (Larcom, Henry.) Beverly, 1810.

An 3558.94 Distribución geográfica del'índice cefálico en España. (Olóriz y Aguilera, Frederico.) Madrid, 1894.

Geog 3070.166 Ditmar, Andrei B. Rubezh oikumeny. Moskva, 1973.

Geog 4219.11.7 Dittmar, J. Eine Fahrt um die Welt. Berlin, 1911.

An 2052.5 Divale, William T. Warfare in primitive societies; a selected bibliography. Los Angeles, 1971.

Geog 4180.7A Divers voyages...discovery of America. (Hackluyt, R.) London, 1850.

Geog 3590.30.15 Divin, Vasilii A. Belikii russkii moreplavateli A.I. Chirukov. Moskva, 1953.

Geog 3590.79 Divin, Vasilii A. Russkie moreplavaniia na Tikhom okeane r XVIII neke. Moskva, 1971.

Geog 4709.71 Diving for sunken treasure. (Cousteau, Jacques Yves.) London, 1971.

An 359.28A Dixon, Roland B. The building of cultures. N.Y., 1928.

An 359.23A Dixon, Roland B. The racial history of man. N.Y., 1923.

Geog 579.52 Dizionario geografico. (Zavatti, Silvio.) Catania, 1952.

Geog 578.98 Dizionario geografico universale. 4. ed. (Garollo, G.) Milano, 1898.

Geog 4679.69.10 Doak, Wade. The Elingamite and its treasure. London, 1969.

Geog 4709.30A Dobie, J. Frank. Coronado's children. Dallas, 1930.

Geog 4709.39 Dobie, J.F. Apache gold and yaqui silver. Boston, 1939.

Geog 619.10 Dobrowolska, Maria. Ludomir Sawicki. Warszawa, 1968.

Geog 6008.97.10 Dobrowolski, A.B. Dziennik wyprawy na Antarktydę. Wrocław, 1962.

Geog 5321.1.6 Doctor Kane of the Arctic seas. (Corner, George Washington.) Philadelphia, 1972.

Geog 4208.54.5 A doctor's experiences in three continents. (Warren, E.) Baltimore, 1885.

An 358.50.6 Doctrine of unity of human race examined on principles of science. (Bachman, J.) Charleston, S.C., 1850.

Geog 500.145 Documentatio geographica; geographische Zeitschriften- und Serien-Literatur. Jahresband. Berlin. 1966+ 9v.

Geog 4168.86 Documentos para la historia de la nautica en Chile. Santiago de Chile, 1886.

Geog 4181.71 Documents concerning English voyages to the Spanish Main, 1569-1580. (Wright, I.A.) London, 1933.

Geog 5509.61 Dodge, Ernest S. Northwest by sea. N.Y., 1961.

Geog 5328.50 Dodge, Ernest S. The polar Rosses: John and James Clark Ross and their explorations. London, 1973.

Geog 4308.47.9 Dodge, Robert. Diary, sketches and reviews. N.Y., 1850.

Geog 759.59 Doerr, Arthur H. Principles of geography. N.Y., 1959.

An 5.34 Dokládal, Milan. Československá anthropologická bibliografie, 1955-1964. Brno, 1966-67.

Geog 110.1.22 Doklady k XXII mezhdunarodnomu geograficheskomu kongressu (Kanada, avgust, 1972). (International Geographical Congress, 22nd, Montreal, 1972.) Leningrad, 1972.

Geog 3590.15.5 Doklady na ezhegodnykh chteniiakh pamiati L.S. Berga. Moskva. 1-7

Geog 3590.17 Doklady na ezhegodnykh chteniiakh pamiati V.A. Obrucheva. Moskva. 1-5

Geog 4329.14F Dokoła morza śródziemnego. (Januszewski, W.) Wilnie, 1914.

An 170.174 Dole, Gertrude Evelyn. Essays in the science of culture. N.Y., 1960.

Geog 5324.2.45 Dolman, Frederick. Dr. Nansen. London, 1897.

Geog 5130.5 Domaas, K. Polarskipet frani. Oslo, 1954.

An 2109.74.15 Un domaine contesté: l'anthropologie économique. (Godelier, Maurice.) Paris, 1974.

An 2109.62.15 Domanskii, Ia.V. Po besovym sledam. Leningrad, 1962.

Geog 4180.81 Dominguez, L.L. Conquest of the river plate. London, 1891.

An 358.57.3 Dominion; or Unity and trinity of the human race. (Baldwin, Samuel D.) Nashville, 1857.

Geog 1640.5 Dominion of Canada. (Baedeker, publishers.) Leipzig, 1894.

Geog 1640.11 Dominion of Canada. 2. ed. (Baedeker, publishers.) Leipsic, 1900.

Geog 1640.15.2 The dominion of Canada. 4. ed. (Baedeker, publishers.) Leipzig, 1922.

Geog 1640.15A Dominion of Canada with Newfoundland and an excursion to Alaska. 3. ed. (Baedeker, publishers.) Leipzig, 1907.

Geog 4328.98 Domsthorpe, Wordsworth. Down the stream of civilization. London, 1898.

Geog 4209.42 Domville-Fife, C.W. I tell of the seven seas. London, 1942.

Geog 205.1 Don; the journal of Sheffield University Geographical Society. Sheffield, Eng. 1,1957+

Geog 4145.29.5 Don Fernando. (Fournier-Aubry, Fernand.) Paris, 1972.

Geog 4145.29.6 Don Fernando. 1st American ed. (Fournier-Aubry, Fernand.) N.Y., 1974.

Geog 4208.49 Donald, C.S. Adventures of a Greek lady. London, 1849. 2v.

Geog 4311.11 Die Donau als Völkeweg. (Schweiger-Lerchenfeld, Amand von.) Wien, 1896.

Geog 4311.5 Das Donaugebeit mit Rücksicht. (Götz, W.) Stuttgart, 1882.

Geog 612.3.15 Dondo, M.M. La Perouse in Maui. Wailuku, 1959.

Geog 3055.39.25 Donnelly, Ignatius. Atlantis, the antediluvian world. London, 1970.

Geog 3055.39.20 Donnelly, Ignatius. Atlantis, the antediluvian world. 1st ed. N.Y., 1949.

Geog 3055.39.10 Donnelly, Ignatius. Atlantis, the antediluvian world. 7th ed. N.Y., 1882.

Geog 3055.39.1 Donnelly, Ignatius. Atlantis: the antediluvian world. Blauvelt, N.Y., 1971.

Geog 3048.83.1 Donnelly, Ignatius. Ragnarok: the age of fire and gravel. N.Y., 1970.

Geog 614.10 Dontsova, Zoia N. Sergei Semenovich Neustruev, 1874-1928. Moskva, 1967.

Geog 6009.02.12 Doorly, G.S. The songs of the "Morning". Melbourne, 1943.

Geog 6009.02.9 Doorly, G.S. The voyages of the "Morning". London, 1916.

Geog 665.39 Dopo il viaggio d'istruzione. (Ricchieri, G.) Firenze, 1914.

Geog 4308.57 Doré. A stroller in Europe. N.Y., 1857.

Geog 4309.37.5A Dorgelès, R. Vive la liberté. Paris, 1937.

Geog 3560.25 Dória, A.A. Los descubrimientos en el Atlántico y la rivalidad castellano-portuguesa. Braga, 1951.

Geog 4269.62 Doroga k serdtsu; reportazh. (Kukanov, A.A.) Tallin, 1962.

Geog 5399.32.8 Dorogi vedut v Arktiku. (Grigor'ev, Origonii K.) Moskva, 1969.

Geog 4218.58 Dorr, D.F. A colored man round the world. Cleveland? 1858.

Geog 4309.41.15 Dos años y medio de Inquietad. (Blanco-Fombona, R.) Caracas, 1942.

Geog 4209.38.16 Dos Passos, John. Journeys between wars. London, 1938.

Geog 4209.38.15A Dos Passos, John. Journeys between wars. N.Y., 1938.

Geog 4218.76 Dottings round the circle. (Curtis, B.R.) Boston, 1876.

Geog 4218.76.2 Dottings round the circle. 3. ed. (Curtis, B.R.) Boston, 1877.

Geog 5324.2.25 La double aventure de Fridtjof Nansen. (Risteheuber, René.) Montréal, 1945.

Geog 4209.64 Double vision. (Knowles, J.) N.Y., 1964.

An 2058.92 Doughty, F.W. Evidences of man in the drift. N.Y., 1892.

An 170.174.5 Douglas, Mary T. Implicit meanings. London, 1975.

Geog 4219.65 Dove. 1. ed. (Graham, Robin Lee.) N.Y., 1972.

An 359.37 Dover, C. Half-caste. London, 1937.

An 359.43 Dover, C. Hell in the sunshine. London, 1943.

Geog 4328.98 Down the stream of civilization. (Domsthorpe, Wordsworth.) London, 1898.

Geog 4209.32.5 Down the world. (Hering, O.C.) N.Y., 1932.

Geog 4679.12.20 Down to eternity. (O'Connor, Richard.) N.Y., 1956.

An 130.10.5 Downie, Robert Angus. Frazer and the golden bough. London, 1970.

An 130.10 Downie, Robert Angus. James George Frazer. London, 1940.

Geog 4249.67.5 Downs, Hugh. A shoal of stars. Garden City, 1967.

Geog 4429.53A Doyle, A.C. Heaven has claws. N.Y., 1953.

Htn Geog 4306.86.3* Dr. Burnet's travels or letters. (Burnet, Gilbert.) Amsterdam, 1687.

Geog 5324.2.45 Dr. Nansen. (Dolman, Frederick.) London, 1897.

An 2058.81.5 Dr. Winchell's "preadamites". (Sunderland, J.T.) n.p., 1881.

Geog 5839.44 Draatrup, E. Grønlandsfaerd. København, 1944.

Geog 4209.61.10 Drachev, B.S. K beregam Vostoka. Moskva, 1961.

Htn Geog 5517.46.10* Drage, T.S. Account of voyage...Northwest Passage. London, 1748-49. 2v.

Geog 608.11 Drake, Fred W. China charts the world. Hsu, Chi-Yü and his geography of 1848. Cambridge, 1975.

Geog 5925.75 Dralkin, A.G. V mire kholoda. Moskva, 1961.

Geog 4306.56.5 Dreifache schwedische Gesandtschaftsreise nach Siebenbürgen. (Hiltebrandt, C.J.) Leiden, 1937.

Geog 4309.13.5 Dreiser, Theodore. A traveler at forty. N.Y., 1913.

Geog 4309.13.6 Dreiser, Theodore. A traveler at forty. N.Y., 1930.

Geog 5220.2 Dresden, Germany. Armen-Realschule. Programm...Offentlichen Prüfung. Dresden, 1873.

Geog 48.1.9 Dresden. Vereins für Erdkunde. Festschrift. Dresden. 1888

Geog 48.1 Dresden. Vereins für Erdkunde. Jahresbericht. Dresden. 1-27 12v.

Geog 48.1.7 Dresden. Vereins für Erdkunde. Mitgleider-Verzeichnis. Dresden.

Geog 48.1.3 Dresden. Vereins für Erdkunde. Mitteilungen. Dresden. 1905-1935 5v.

X Cg An 2109.31.3 Dress, drinks and drums. (Crawley, A.E.) London, 1931.

An 359.57.10 Drexler, John Paul. Die Front der Farbigen. München, 1957.

Geog 4502.5.25 Dreyer, A. Geschichte der Alpinen Literatur. München, 1938.

Geog 4500.7 Dreyer, Aloys. Bücherverzeichnis der Alpenvereinsbucherei, mit Verfasser- und Bergnamen-Verzeichnis. München, 1927.

An 3309.21.3 Dreyer, George. The assessment of physical fitness. N.Y., 1921.

Htn Geog 5340.7* Drie voyagien gedaen na Groenlandt. Amsterdam, n.d.

Geog 4208.63 Drifting about. (Massett, S.C.) N.Y., 1863.

Geog 3070.16 Dröber, W. Kartographie bei den Naturvölkern. Erlangen, 1903.

Geog 4677.66.3 Een droevig verhaal van de ongelukkige reis en wonderdaadige verlossinge. (Harrison, David.) Wesel, 176-?

Geog 4327.44.3F Drummond, A. Travels thro' different cities of Germany, Italy, Greece and parts of Asia. London, 1754.

Geog 4679.52 Drums of the Birkenhead. (Bevan, David.) Capetown, 1972.

Geog 4558.73 Du Far-West à Bornéo. (Wogan, E. de.) Paris, 1873.

An 359.62.20 Du racisme à l'universalisme. (Fiezefontaine, J.) Paris, 1962.

Geog 4348.47 Du Rhin au Nil. (Marmier, X.) Paris, 1847. 2v.

Geog 4218.98 Du Tonkin au Havre. (Albrey, J. d'.) Paris, 1898.

An 359.23.5 Duality; a study in the psycho-analysis of race. (Bradley, R.N.) N.Y., 1923.

Geog 5650.12 Dúason, Jón. Grønlands statsretslige stilling i Midelalderen. Oslo, 1928.

Geog 5645.5.10 Dúason, Jón. Die koloniale Stellung Grönlands. Göttingen, 1955.

Geog 5650.12.5 Dúason, Jón. Nybygd i Grønland. n.p., 192-.

Geog 3650.5 Dube, Beehan. Geographical concepts in ancient India. Varanasi, 1967.

Geog 958.96 Dubois, M. Album géographique. Paris, 1896-1906. 5v.

Geog 4145.52 Il duca degli Abruzzi. (Dainelli, Giotto.) Torino, 1967.

An 2059.12A Duckworth, W.L.H. Prehistoric man. Cambridge, 1912.

Geog 4308.96 Dudley, L.B. Letters to Ruth. N.Y., 1896.

Geog 4309.01.5 Dudley, L.B. (Mrs.). A royal journey. N.Y., 1901.

Author and Title Listing

	Geog 4181.3	Dudley, R. Voyage of R. Dudley...to West Indies. London, 1899.
	Geog 5559.69	Due north. (Simpson, Myrtle.) London, 1970.
	Geog 4218.84	Due west, or Round the world. (Ballou, M.M.) Boston, 1884.
	Geog 4218.84.5	Due west, or Round the world. 3. ed. (Ballou, M.M.) Boston, 1885.
	Geog 4218.84.9	Due west, or Round the world. 9. ed. (Ballou, M.M.) Boston, 1891.
	Geog 4218.84.2	Due west, or Round the world in ten months. 2. ed. (Ballou, M.M.) Boston, 1884.
	An 3539.10PF	Düben, G. Crania Lapponica. Holmiae, 1910. 2v.
	Geog 199.1.20	Dübi, Heinrich. Die ersten fünfzig Jahre des Schweizer Alpenclub. Bern, 1913.
	Geog 6009.54	Dufek, George John. Operation deepfreeze. N.Y., 1957.
	Geog 5398.56	Dufferin and Ava, F.T. Blackwood. Un voyage en yacht, lettres de hautes latitudes. Montreal, 1876.
	Geog 5398.56.5	Dufferin and Ava, F.T. Blackwood. A yacht voyage; letters from high latitudes. N.Y., 1878.
	Geog 4309.69	Duffus, Robert Luther. The polar route to time gone by. 1st ed. N.Y., 1969.
	Geog 4659.55A	Duffy, James E. Shipwrecks and empire; being an account of Portuguese maritime disasters in a century of decline. Cambridge, Mass., 1955.
	Geog 520.16	Du Fresne-Canayl, P. Le voyage du Levant. Paris, 1897.
	Geog 4309.31.10	Duhamel, Georges. Mon Europe. Paris, 1931.
	Geog 4218.26.5	Duhaut-Cilly, A. Viaggio intorno al globo. Napoli, 1842.
Htn	Geog 4218.26*	Duhaut-Cilly, A. Voyage autour du monde. Paris, 1834.
	An 2109.66	Dukhovnaia zhizn' pervobytnogo obshchestva. (Anisimov, Arkadii F.) Leningrad, 1966.
	Geog 4279.64	Dulles, F.R. Americans abroad. Ann Arbor, 1964.
	Geog 4169.31	Dulles, F.R. Eastward ho! London, 1931.
	Geog 4169.31.2	Dulles, F.R. Eastward ho! The 1st English adventures to the Orient. Boston, 1931.
	An 136.5.30	Dumasy, Annegret. Restloses. Erkenhen; die Diskussione über den Strukturalismus de Claude Lévi-Strauss in Frankreich. Berlin, 1972.
	Geog 6008.41	Dumont, J.S.C. Voyage au Pole Sud. Paris, 1841. 10v.
	Geog 4218.37.5	Dumont d'Urville, J. Malerische Reise um die Welt. Leipzig, 1837.
	Geog 4168.34	Dumont d'Urville, Jules. Voyage pittoresque autour du monde. Paris, 1834-35. 2v.
	Geog 4145.20	Dumont d'Urville. (Day, George.) Paris, 1947.
	Geog 4311.22	Dunai-reka druzhby. (Sergeeva, Nelli A.) Odessa, 1966.
	Geog 4329.53.5F	Dunand, M. De l'Amanus au Sinai. Beyrouth, 1953.
	An 2109.38.11	Dunbar, George. Other men's lives...primitive peoples. London, 1938.
	Geog 4678.34	Duncan, A. The mariner's chronicle...shipwrecks, storms. New Haven, 1834.
	Geog 4678.06	Duncan, A. The mariner's chronicle. Philadelphia, 1806. 4v.
	Geog 5558.26	Duncan, D. Arctic regions. Voyage to Davis Strait. London, 1827.
	An 2059.31	Duncan, G.S. Prehistoric man. Boston, 1931.
	An 359.29.9	Duncan, H. Race and population problems. 1. ed. N.Y., 1929.
	Geog 4208.13	Dunham, Jacob. Journal of voyages. N.Y., 1850.
	Geog 4247.69	Dunmore, John. The fateful voyage of the St. Jean Baptiste. Christchurch, 1969.
	Geog 4218.94	Dunn, S.H. The world's highway. London, 1894.
	Geog 3530.15	Dupic, Jeanne. La Normandie exploratrice et colonisatrice du XVe au XVIIIe siècle. Rouen, 1932.
	Geog 4509.23	Dupin. Zum Wortschatze des Bergsteigers. Wien, 1923-
	Geog 4679.67.20	Du Pontavice, Emmanuel. La pollution des mers..."Torrey Canyon". Paris, 1968.
	Geog 611.1	Dupouy, A. Le Briton Yves de Kerguelen. Paris, 1928.
	Geog 4308.22.5	Duppa, Richard. Miscellaneous observations and opinions on the continent. London, 1825.
	Geog 4218.73.10	Dupuy de Lôme, E. De Madrid a Madrid. Madrid, 1877.
	Geog 4329.38.5	Duran, Faik Sabri. Akdenirde leir yor gerintisi. Istanbul, 1938.
	Geog 3070.47.5	Durand, D.B. The Vienna-Klosterneuberg map corpus of the 15th century. Leiden, 1952.
	Geog 4558.20.3	Durand, J.R. The life and adventures. Rochester, 1820.
	Geog 4239.47	Durand-Couppel de St. Trent, M.N.P. Wind aloft, wind alone. N.Y., 1947.
	Geog 4308.44	Durbin, J.P. Observations in Europe. N.Y., 1844. 2v.
	Geog 5839.09.5	Durch Grönlands Eiswüste. 2. Aufl. (Quervain, Alfred.) Strassburg, 1911.
	Geog 4229.25.5	Durch Werkstätten und Gassen dreier Erdteile. 2. Aufl. (Kleinschmitt, Edmund.) Hamburg, 1928.
	Geog 3128.07	Dureau de la Malle, Adolphe. Géographie physique de la Mer Noire, de l'intérieur de l'Afrique et de la Méditerranée. Paris, 1807.
	Geog 51.5.10	Durham, England. University. Department of Geography. Occasional publications. Durham. 1,1973+
	Geog 51.5	Durham, England. University. Durham Colleges. Department of Geography. Occasional papers series.
	An 2109.63.15	Durkheim, Emile. Primitive classification. Chicago, 1963.
	Geog 4209.36.30	Durtain, L. Le globe sous le bras. Paris, 1936.
	Geog 3018.82.5	Dussieux, L.E. Les grands faits de l'histoire de géographie. Paris, 1882-83. 5v.
	Geog 4228.59	Dust and foam, or Three oceans and two continents. (Warren, T.R.) N.Y., 1859.
	Geog 4145.84.15	Dust in the lion's paw. (Stark, Freya.) London, 1961.
	Geog 5300.7	The Dutch at the North Pole. (DePeyster, J.W.) N.Y., 1857.
	Geog 4182.15	The Dutch come to Korea. (Ledyard, Gari.) Seoul, 1971.
	Geog 5300.5	The Dutch in the arctic seas. (Campen, S.R.) London, 1877.
	Geog 4268.83.6	Dutens, Louis. Itinéraire des routes les plus fréquentées. 6. éd. Paris, 1788.
	Geog 5150.5	Dutilly, A. Bibliographies on the Arctic. Washington, 1945.
	Geog 816.82	Duval, Pierre. La geographie du temps. pt.1-2. Paris, 1682. 2v.
	An 99.73.5	Duvignaud, Jean. Le langage perdu; essai sur la différence anthropologique. 1. éd. Paris, 1973.
	Geog 4218.40.10	Dva goda iz zhizni russkago moriaka. (Blok, G.K.) Sanktpeterburg, 1854. 2v.
	Geog 5399.57.15	Dva leta v Arktike. (Sushkina, N.N.) Moskva, 1957.
Htn	Geog 4218.61*	D'Wolf, J. Voyage to the North Pacific. Cambridge, 1861.
	Geog 4672.25	Dyall, Valentine. Famous sea tragedies. London, 1955.
	Geog 4166.78.5	Dyck, J. Seer gedenckwaerdige voyagien. Amsterdam, 1678.
	Geog 665.118	Dynamic relationships in physical geography. (Johnston, William.) Christchurch, 1967.
	Geog 659.28.1	Dynamische Länderkunde. (Spethman, Hans.) Kiel, 1972.
	An 359.71.30	Dzhandil'din, Nurymbet. Priroda natsional'noi psikholgii. Alma-Ata, 1971.
	Geog 4307.77	Dziennik podrózy, 1777-1791. v.1-2. (Staszic, S.) Warszawa, 1903.
	Geog 4307.78	Dziennik podrózy, 1789-1805. (Staszic, S.) Krakow, 1931.
	Geog 6008.97.10	Dziennik wyprawy na Antarktydę. (Dobrowolski, A.B.) Wrocław, 1962.
	Geog 600.21	Dziewięć wieków geografii polskiej. (Olszewicz, Bolesław.) Warszawa, 1967.
Htn	Geog 577.09*	Eachard, Laurence. Gazeteer's geographical index of Europe. London, 1709.
	Geog 577.09.15	Eachard, Laurence. The gazetteer's or news-man's interpreter. London, 1751.
	Geog 4559.54.20	Eaddy, P.A. Sails beneath the Southern Cross. Wellington, 1954.
	Geog 5313.2	Eames, Hugh. Winner lose all; Dr. Cook and the theft of the North Pole. Boston, 1973.
	Geog 4418.55	Eames, J.A. Another budget...in the East. Boston, 1855.
	Geog 4308.47.7	Eames, J.A. The budget closed. Boston, 1860.
	Geog 4308.47.5	Eames, J.A. A budget of letters. Boston, 1847.
	Geog 4248.73	Eardley-Wilmot, S. Our journal in the Pacific. London, 1873.
	Geog 4709.70	Earle, Peter. Corsairs of Malta and Barbary. London, 1970.
	An 3539.01.3	The earliest inhabitants of Abydos. (Randall-Maciver, D.) Oxford, 1901.
	Geog 4215.19	Earliest voyages round the world, 1519-1617. (Alexander, Philip.) Cambridge, 1916.
	An 99.64A	Early anthropology in the sixteenth and seventeenth centuries. (Hodgen, Margaret.) Philadelphia, 1964.
	An 2109.32.3	Early beliefs and their social influence. (Westermarck, E.) London, 1932.
	Geog 3070.42	The early cartography of the Pacific. (Wroth, Lawrence C.) N.Y., 1944.
	An 2109.22.3A	Early civilization. (Goldenweiser, A.A.) N.Y., 1922.
	An 2109.23.9	Early civilization. (Goldenweiser, A.A.) N.Y., 1923.
	Geog 4181.11	Early Dutch and English voyages. (Conway, W.M.) London, 1904.
	Geog 5557.25.3	Early expeditions to the Bering Sea and Strait. (Dall, W.H.) Washington, 1891.
	An 2109.31.9	Early man, his origin, development and culture. London, 1931.
	An 2059.37	Early man. (International Symposium on Early Man.) Philadelphia, 1937.
	An 2058.76	Early man in Europe. (Rau, C.) N.Y., 1876.
	An 3509.12	Early man in South America. (Hrdlička, A.) Washington, 1912.
	Geog 4508.99A	The early mountaineers. (Gribble, F.) London, 1899.
	Geog 4181.28	Early Spanish voyages to the Strait of Magellan. v.28. (Markham, C.R.) London, 1911.
	Geog 4180.72	Early voyages...to Russia and Persia. (Jenkinson, A.) N.Y., 1964? 2v.
	Geog 4180.87A	Early voyages and travels in the Levant. (Bent, J.T.) London, 1893.
	Geog 4180.25	Early voyages to Terra Australis. (Major, R.H.) London, 1859.
	Geog 808.76.7	Earth and its inhabitants - Africa. (Reclus, J.J.E.) N.Y., 1886-1890. 4v.
	Geog 808.76.6	Earth and its inhabitants - Asia. (Reclus, J.J.E.) N.Y., 1891. 4v.
	Geog 808.76.5	Earth and its inhabitants - Europe. (Reclus, J.J.E.) N.Y., 1885. 5v.
	Geog 808.76.9	Earth and its inhabitants - North America. (Reclus, J.J.E.) N.Y., 1890-1893. 3v.
	Geog 808.76.8	Earth and its inhabitants - Oceanica. (Reclus, J.J.E.) N.Y., 1890.
	Geog 808.76.10	Earth and its inhabitants - South America. (Reclus, J.J.E.) N.Y., 1894-1895. 2v.
	Geog 819.42.2	The earth and man. (Davis, D.H.) N.Y., 1942.
	Geog 819.47	The earth and man. (Davis, D.H.) N.Y., 1947.
	Geog 3019.34.10	Earth conquerors. (Mitchell, J.L.) N.Y., 1934.
	Geog 4208.96	East and West. (Arnold, E.) London, 1896.
	Geog 4181.18	East and West Indian mirror. (Spillbergen, J. van.) London, 1906.
	Geog 4208.83	East by West. (Lucy, H.W.) London, 1885. 2v.
	Geog 5839.35.15	East of the great glacier. (Ingstad, H.M.) N.Y., 1937.
	Geog 4209.58.10	East to west. (Toynbee, A.J.) N.Y., 1958.
	Geog 1533.4	Eastern Alps. 4. ed. (Baedeker, publishers.) Leipsic, 1879.
	Geog 1533.6	Eastern Alps. 6. ed. (Baedeker, publishers.) Leipsic, 1888.
	Geog 1533.7	Eastern Alps. 7. ed. (Baedeker, publishers.) Leipsic, 1891.
	Geog 1533.8	Eastern Alps. 8. ed. (Baedeker, publishers.) Leipsic, 1895.
	Geog 1533.10	Eastern Alps. 10. ed. (Baedeker, publishers.) Leipsic, 1903.
	Geog 1533.12	Eastern Alps. 11. ed. (Baedeker, publishers.) Leipzig, 1907.
	Geog 1533.13	Eastern Alps. 12. ed. (Baedeker, publishers.) Leipzig, 1911.
	An 358.26.5	Eastern origin of the Celtic nations...forming a supplement to Researches into the physical history of mankind. (Prichard, J.C.) Oxford, 1831.
	Geog 4419.07	An eastern voyage; travels thro' the British Empire...in East and Japan. (Hochberg, Fritz.) London, 1910. 2v.
	Geog 4169.31	Eastward ho! (Dulles, F.R.) London, 1931.
	Geog 4169.31.2	Eastward ho! The 1st English adventures to the Orient. (Dulles, F.R.) Boston, 1931.
	Geog 616.5	Eavenson, H.N. Map maker and Indian traders...John Patten. Pittsburgh, 1949.
	Geog 5517.46.15	Eavenson, Howard N. Two early works on Arctic exploration. Pittsburgh, 1946.
	Geog 3235.19	Die Ebstorfkarte. (Miller, K.) Stuttgart, 1900.
	Geog 4521.3	An eccentric in the Alps. (Clark, Ronald.) London, 1959.
	An 170.257	Échanges et communications; mélanges offerts à Claude Lévi-Strauss. The Hague, 1970. 2v.
	Geog 4209.64.35	Eckert, Gerhard. Richtig Reisen. Bergisch Gladbach, 1964.
	An 379.57	The ecology of man. (Sears, Paul.) Eugene, 1957.
	An 2109.52.10	Economic anthropology; a study in comparative economics. 2d ed. (Herskovits, M.J.) N.Y., 1952.
	An 2109.65.15	Economic anthropology; the economic life of primitive peoples. (Herskovits, M.J.) N.Y., 1965.
	An 5.35	Economic anthropology 1946-1972. (Pas, H.T.) Oosterhout, 1973.
	An 2109.40	The economic life of primitive peoples. (Herskovits, M.J.) N.Y., 1940.

180

Author and Title Listing

	An 2109.74	Economic man; the anthropology of economics. (Schneider, Harold K.) N.Y., 1974.
	An 2109.32	Economics in primitive communities. (Thurnwald, R.) London, 1932.
	An 2109.36	The economics of primitive peoples. (Viljoen, S.) London, 1936.
	An 2109.70.5	L'économie préhistorique. (Nougier, Louis René.) Paris, 1970.
	Geog 4309.31.15	Eddelbüttel, H.F. Schön ist die Welt. Berlin, 1931.
	Geog 4308.56.9	Eddy, D.C. Europa, or Scenes and society. Boston, 1856.
	Geog 4308.52.20	Eddy, Daniel C. Europa, or Scenes and society in England, France. 50th ed. Boston, 1860.
	Geog 3055.78	Edgar Cayce in Atlantis. (Cayce, Edgar Evans.) N.Y., 1968.
	An 359.74.10	Edgerton, Robert B. Methods and styles in the study of culture. San Francisco, 1974.
	Geog 55.1	Edinburgh. University. Department of Geography. Papers. 1960-1962
	Geog 578.22	Edinburgh gazetteer. v.2-6. Edinburgh, 1822. 4v.
	Geog 4181.51	Edmundson, George. Journal of...Father Samuel Fritz. London, 1922.
	Geog 4329.32	Edschmid, K. Zauber und Grösse des Mittelmeers. Frankfurt, 1932.
	Geog 4309.53.11	Edschmid, Kasimir. Europäisches Reisebuch. Hamburg, 1954.
	Geog 4309.27.15	Edschmid, Kasimir. Das grosse Reisebuch. Berlin, 1927.
	Geog 4206.81	Eduward Meltons, engelsch edelmans, zeldzaame en gedenkwaardige zee- en land reizen. (Melton, E. (pseud.).) Amsterdam, 1681.
	Geog 4325.90.12	Edward Webbe, chief master gunner, his travailes, 1590. (Webbe, Edward.) London, 1868.
	Geog 4325.90.6	Edward Webbe, chief master gunner, his travailes. (Webbe, Edward.) Edinburgh, 1885.
	Geog 4325.90.10	Edward Webbe, his travailes, 1590. (Webbe, Edward.) Birmingham, 1868.
Htn	Geog 4325.90.5*	Edward Webbe, his travailes. (Webbe, Edward.) Edinburgh, 1885.
	Geog 4521.23	Edward Whymper, alpinist of the heroic age. (Whymper, Edward.) Nashville, Tenn., 1914.
	Geog 5985.325	Edward Wilson of the Antarctic, naturalist and friend. (Seaver, George.) N.Y., 1937.
	Geog 5209.10.3	Edwards, D.M. The toll of the arctic seas. N.Y., 1910.
	Geog 4559.54.10	Edwards, H.W. Under four flags. London, 1954.
	Geog 4308.56.17	Edwards, J.E. Random sketches...travel in 1856. N.Y., 1857.
	Geog 4182.29	De eerste nederlandsche. (Van Nouhuys, J.W.) 's-Gravenhage, 1927-51. 2v.
	Geog 4182.7	De eerste schipvaart der nederlanders naar Ost-Indie. (Rouffaer, G.P.) 's-Gravenhage, 1915-29. 3v.
	Geog 665.154	Eesti Geograafia Selts. Orazvitii geografii v Estonskoi SSR 1960-1968. Tallin, 1970.
	Geog 5300.56	Eestist pärit Arktika-uurijad. (Pasetskii, Vasilii M.) Tallinn, 1970.
	VGeog 4131.67.5	Efektivnost' investicí cestovného ruchu. (Seminàr o Ekonomickej Efektívnosti Iuvesticii Cestovného Ruchu, Pieštány, 1966.) Bratislava, 1967.
	Geog 3590.35	Efimov, Aleksei V. Iz istorii russkii geogr. otkrytii v sever ledov i tiloke okeanov. Moskva, 1950.
	Geog 3590.35.5	Efimov, Aleksei V. Iz istorii russkikh aksped. na Tikhom okeane. Moskva, 1948.
	Geog 3590.35.3	Efimov, Aleksei V. Iz istorii velikikh russkikh geograf. otkrytii. Moskva, 1949.
	Geog 3590.35.2	Efimov, Aleksei V. Iz istorii velikikh russkikh geograf. otkrytii. Moskva, 1971.
	Geog 3590.35.10	Efimov, Aleksei V. Otkrytiia russk. zemleprov. na ser. Vost. Asii. Moskva, 1951.
	Geog 4209.65.10	Ęfimov, Gerontii V. Strany i liudi. Leningrad, 1965.
	An 359.69.15	Égalité ou inégalité des races? (Hiernaux, Jean.) Paris, 1969.
	Geog 5700.4	Egede, H. Ausführliche...Nachricht...Grönland. Hamburg, 1740.
	Geog 5837.63	Egede, H. Beschreibung und Natur-Geschichte. Berlin, 1763.
	Geog 5837.63.5	Egede, H. Description of Greenland. London, 1818.
Htn	Geog 5700.4.5*	Egede, H. Det gamle Grønlands...naturel-historie. Kjobenhavn, 1741.
	Geog 5700.2	Egede, H. Omstaendelig...grønlandste missions. Kjøbenhavn, 1738.
	Geog 4218.93.10	Egerton, Alice A. Glimpses of four continents. London, 1894.
	An 2109.59.10	Eggers, H.J. Einführung in die Vorgeschichte. München, 1959.
	Geog 5837.92	Eggers, H.P. Om Grönlands osterbygds. Kjøbenhavn, 1794.
	Geog 4269.59.5	Egli, Emil. Europe from the air. London, 1959.
	Geog 4269.58.15	Egli, Emil. Flugbild Europas. Zürich, 1958.
	Geog 4313.8	Egor'eva, A.V. Baltiiskoe more. Moskva, 1961.
	Geog 1618.5	Egypt...lower Egypt. (Baedeker, publishers.) Leipsic, 1878.
	Geog 1618.10A	Egypt...lower Egypt. 2. ed. (Baedeker, publishers.) Leipsic, 1885.
	Geog 1617.5	Egypt. (Baedeker, publishers.) Leipsic, 1892.
	Geog 1618.15	Egypt. (Baedeker, publishers.) Leipsic, 1892-95. 2v.
	Geog 1616.15	Egypt. 4. ed. (Baedeker, publishers.) Leipsic, 1898.
	Geog 1616.18	Egypt. 4. ed. (Baedeker, publishers.) Leipsic, 1898.
	Geog 1616.20	Egypt. 5. ed. (Baedeker, publishers.) Leipsic, 1902.
	Geog 1616.21	Egypt. 6. ed. (Baedeker, publishers.) Leipzig, 1908.
	Geog 1616.22	Egypt and the Sudan. 7. ed. (Baedeker, publishers.) Leipzig, 1914.
	Geog 1616.24A	Egypt and the Sudan. 8. ed. (Baedeker, publishers.) Leipzig, 1929.
	Geog 1616.20.2	Egyptian Museum, Cairo. Gratis supplement to the fifth edition of Baedeker's Egypt. Leipzig, 1904.
	Geog 4324.59.10	Ehingen, Georg von. Diary. London, 1929.
Htn	Geog 4324.59*	Ehingen, Georg von. Itinerarium. Augsburg, 1600.
	An 3208.97	Ehrenreich, P. Anthropologische Studien über die Urbewohner Brasiliens. Braunschweig, 1897.
X Cg	An 359.37.7	Eickstedt, E. Rassenkunde und Rassengeschichte der Menschheit. 2. Aufl. v.1, pt.1-15. Stuttgart, 1937-43.
	An 2109.50.5	Eildermann, H. Die Urgesellschaft. Berlin, 1950.
	Geog 5209.29	Einarsson, S. Nordr um höf. Reykjavik, 1924.
	Geog 5209.43	Einarsson, S. Sudr um höf. Reykjavik, 1943.
	An 2109.73.20	Einführung in die Urgeschichtsforschung. 1. Aufl. (Felgenhauer, Fritz.) Breisgau, 1973.
	An 2109.59.10	Einführung in die Vorgeschichte. (Eggers, H.J.) München, 1959.
	An 2109.31.13	Einführung in die Vorstellungswelt primitiver Völker. (Winthius, Josef.) Leipzig, 1931.
	Geog 5398.75	Einringar fran Svenska Expeditionerna. (Stuxberg, A.) Stockholm, 1877.
	Geog 4309.34.10	Eintritt verboten. (Kisch, Egon E.) Paris, 1934.
	An 179.1A	Eiseley, L.C. The immense journey. N.Y., 1957.
	Geog 4509.25.6	Eistechnik des Bergsteigers in Bildern und Merkworten. pt.1-4. 2. Aufl. (Flaig, W.) Stuttgart, 1925.
	Geog 5855.11	Ekblaw, W.E. The material response of the polar eskimo to their far Arctic environment. Diss. N.Y., 1928.
	Geog 600.12	Ekonomicheskaia geografiia v SSSR, istoriia i sovremennoe razvitie. Moskva, 1965.
	Geog 5312.7.20	Ekspeditsiia Beringa. (Pokrovskii, Aleksei A.) Moskva, 1941.
	Geog 4239.62	Ekspeditsiia na "Sedove" v Atlanticheskii okean. (Nikiforovskii, V.A.) Moskva, 1962.
	Geog 5399.37.5	Ekspeditsiia SSSR na Severnyi Polius, 1937. Trudy. Leningrad, 1940.
	Geog 816.96	El atlas abreviado ô compendiosa geografia. Amberes, 1696.
	Geog 4215.19.10	Elcano. (Mitchell, Mairin.) London, 1958.
	Geog 5321.1.3	Elder, William. Biography of Elisha Kent Kane. Philadelphia, 1858.
	Geog 5321.1.4	Elder, William. Biography of Elisha Kent Kane. Philadelphia, 1858.
	Geog 818.59.9	Elementary geography. (Young, Francis.) London, 1859.
	Geog 659.59	Elementi di geografia. (Scotti, Pietro.) Genova, 1959.
	Geog 4131.58	Elementi di tecnica professionale turistica. (Tusci, Leonida.) Roma, 1958.
	An 2058.85.3	Eléments d'anthropologie générale. (Topinard, Paul.) Paris, 1885.
Htn	Geog 757.42*	Elements de geographie. (Maupertius, Pierre L.) Paris, 1742.
	Geog 818.13	Elements of geography. (Dickinson, R.) Boston, 1813.
	Geog 817.96	Elements of geography. (Walker, J.) London, n.d.
	Geog 759.36.4	Elements of geography. 4th ed. (Finch, V.C.) N.Y., 1957.
	An 359.73.20	Elements of human and social geography. 1. ed. (Sunderland, Eric.) Oxford, 1973.
	Geog 3070.24	Elenco di carte ed Atlanti nautici. (Desimoni, C.) Genova, n.d.
	Geog 4308.52.10	Eleven weeks in Europe. (Clarke, J.F.) Boston, 1852.
	Geog 5855.10	Elgström, Ossian. Moderna eskimåer. Stockholm, 1916.
	Geog 4679.69.10	The Elingamite and its treasure. (Doak, Wade.) London, 1969.
	Geog 618.3.3	Élisée and Elie Reclus, in memoriam. (Ishill, Joseph.) Berkeley Heights, N.J., 1927.
	Geog 618.3.7	Elisée Reclus. (Nettlau, Max.) Berlin, 1928.
	Geog 618.3.8	Eliseo Reclus, la vida de un sabio justo. (Nettlau, Max.) Barcelona, 1929. 2v.
	Geog 5321.1.7	Elisha Kent Kane. (Allen, J.H.) Bangor, 1857.
	Geog 5321.1.17	Elisha Kent Kane and the seafaring frontier. 1. ed. (Mirsky, J.) Boston, 1954.
	Geog 4345.99.10	Elizabethans errant; the strange fortunes of Sir Thomas Sherley and his three sons. (Davies, David William.) Ithaca, N.Y., 1967.
	An 2209.46	Elkin, A.P. Aboriginal men of high degree. Sydney, 1946.
	An 2109.15.3	Elliot, G.F.S. Prehistoric man and his story. London, 1915.
	An 2109.15.4	Elliot, G.F.S. Prehistoric man and his story. Philadelphia, 1915.
	Geog 5517.46.3	Ellis, H. Reise nach Hudsons Meerbusen. Göttingen, 1750.
	Geog 5517.46.4	Ellis, H. Voyage à la baye de Hudson. Leide, 1750.
	Geog 5517.46.5	Ellis, H. Voyage de la baye de Hudson. Paris, 1749.
	Geog 953.12	Ellis, Jessie (Croft). Travel through pictures; references to pictures, in books and periodicals. Boston, 1935.
	Geog 4349.29	Ellis, M.H. Express to Hindustan. London, 1929.
	Geog 4672.5.5	Ellms, Charles. Shipwrecks and disasters at sea. N.Y., 1844.
	Geog 4672.5	Ellms, Charles. The tragedy of the seas. Philadelphia, 1841.
	Geog 5398.79.30A	Ellsberg, Edward. Hell on ice; the saga of the "Jeannette". N.Y., 1938.
	Geog 5070.55	Ellsworth, L. Beyond horizons. N.Y., 1938.
	Geog 3019.35.20	Ellsworth, L. Exploring today. N.Y., 1935.
	Geog 5399.25.5	Ellsworth, L. The hop-off. N.Y., 1925.
	Geog 608.10	Ellsworth Huntington; his life and thought. (Martin, Geoffrey J.) Hamden, Conn., 1973.
	An 2109.27.15	Ellwood, Charles A. Cultural evolution; a study of social origins and development. N.Y., 1927.
	Geog 4209.56.5	Elmberg, J.E.O. Islands of to-morrow. London, 1956.
Htn	An 3305.63*	Elsholt, J.S. Anthropometria. Francofurti, 1563.
	Geog 4308.91.14	Elson, L.C. European reminiscences. Philadelphia, 1914.
	Geog 4145.23	Ely, Edward. The wanderings of Edward Ely. N.Y., 1954.
	Geog 4309.30	Em viagem (na velha Europa). (Ferreira de Mira, M.) Lisboa, 1930.
	Geog 4138.82.7	Embacher, F. Lexikon der Reisen und Entdeckungen. Leipzig, 1882.
	Geog 4169.59	Embarcadero. (Dillon, Richard.) N.Y., 1959.
	Geog 4181.1	Embassy of Sir T. Roe to court of great mogul. (Roe, T.) London, 1899. 2v.
	An 359.48.5	Embree, Edwin R. Peoples of the earth. N.Y., 1948.
	An 2059.69	The emergence of man. (Pfeiffer, John. E.) N.Y., 1969.
	Geog 603.5.8	Emmanuel, Marthe. J.B. Charcot. Paris, 1945.
	Geog 603.5.8.5	Emmanuel, Marthe. Tel fut Charcot, 1867-1936. Paris, 1967.
	Geog 4308.36.10	Empfindsame Reise eines expatriirten Schwärmers. (Gross-Hoffinger, A.J.) Leipzig, 1836.
	Geog 4558.94	En mer. (Stewart, J.A.E.) n.p., 1895?
	Geog 4678.48.5	En route from Honolulu. Washington, 1931?
	Geog 4208.81	En viaje, 1881-82. (Cané, Miguel.) Buenos Aires, 1917.
	Geog 4208.81.5	En viaje. (Cané, Miguel.) Buenos Aires, 1949.
	Geog 4515.107	L'enchantement du ski. (Couttet, Alfred.) Paris, 1930.
	An 180.6	Encounter with anthropology. (Fox, Robin.) Harmondsworth, 1975.
	Geog 818.34	An encyclopaedia of geography. (Murray, H.) London, 1834.
	Geog 818.43.5	The encyclopaedia of geography. (Murray, Hugh.) Philadelphia, 1843. 3v.
	Geog 808.34.10	The encyclopaedia of geography. (Murray, Hugh.) Philadelphia, 1845. 3v.
	Geog 818.40	Encyclopedia of geography. (Murray, H.) London, 1840.
	Geog 808.37	Encyclopedia of geography. (Murray, H.) Philadelphia, 1837. 3v.
	Geog 5985.258.5	Endurance; an epic of polar adventure. (Worsley, Frank A.) London, 1931.
	Geog 5985.258.15	Endurance; Shackleton's incredible voyage. (Lansing, Alfred.) N.Y., 1959.
	Geog 4509.63	Endurminningar fjallgöngumanns. (Gudhjohnsen, P.) Reykjavik, 1963.
	Geog 5367.65.7	Engel, S. Extraits...des voyages...de l'Asie. Lausanne, 1779.
	Geog 5367.65.3	Engel, S. Geographische...Nachrichten. Mietau, 1772.
Htn	Geog 5367.65*	Engel, S. Memoires...geographiques...pays septentrionaux. Lausanne, 1765.

Author and Title Listing

	Geog 5367.65.5	Engel, S. Neuer Versuch über...Schriften. Basel, 1777.
	Geog 611.3	Engelbert Kampfer, der erste deutsche Forschungsreisende, 1651-1716. (Meier-Lemgo, K.) Stuttgart, 1937.
	Geog 819.43	Engelhardt, N.L. Toward new frontiers of our global world. N.Y., 1943.
Htn	Geog 818.24*	Engelmann, G. Porte-feuille géographique et ethonographique. 2e éd. v.1-2, Atlas. Mulhouse, 1824. 2v.
	Geog 4300.25.3	Engelmann, J.B. Manuel pour voyageurs en Allemagne et dans les Payslimitrophes. 3.ed. Francfort, 1827.
	Geog 500.19	Engelmann, W. Bibliotheca geographica. Leipzig, 1858.
	Geog 3520.18	De Engelsche geographie in de 20ste eeuw. (Haender, Wilhelmina.) Utrecht, 1934.
	Geog 4218.83.2	An engineer's holiday...trip from 0 degrees to 0 degrees. 2. ed. (Pidgeon, D.) London, 1883.
	Geog 4239.29.5	England, George Allan. Isles of romance. N.Y., 1929.
	Geog 578.54A	The English cyclopaedia. (Knight, Charles.) London, 1854-55. 4v.
	Geog 3520.8	English maritime writing. (Warmer, Oliver.) London, 1958.
	Geog 4181.111	English privateering voyages to the West Indies. (Andrews, Kenneth.) Cambridge, 1959.
	Geog 4279.52	English travellers abroad. (Stoye, J.W.) London, 1952.
	Geog 4129.14	English travellers of the Renaissance. (Howard, Clare.) London, 1914.
	Geog 4129.06	The English voyages of the sixteenth century. (Raleigh, Walter.) Glasgow, 1906.
	Geog 4228.57	Ennemoser, F.J. Reise vom Mittelrhein...nach dem Nordamerikanischen Freistaaten. Kaiserlautern, 1864.
	Geog 4266.09	Ens, Gaspar. Deliciarum germaniae...viatorius...itinera ad omnes ciirtates. Coloniae, 1608.
	Geog 4326.10	Ens, K. Deliciae transmari. Coloniae Agrippinae, 1610.
	An 359.61.30	Ensaios etnológicas. (Dias, Jorge.) Lisboa, 1961.
	An 379.38.5	Enseñanzas fundamentales de la geografia humana. (Sanchez, Pedro C.) Tacubaya, 1938.
	Geog 3140.17	Die Entdeckung von Europa durch die Griechen. (Ninck, M.) Basel, 1945.
	Geog 5070.15	Die Entdeckungs- und Forschungsreisen im dem Leiden Polarzonen. (Löwenberg, J.) Leipzig, 1886.
	Geog 3019.02	Entdeckungsgeschichte. (Günther, S.) Berlin, 1902.
	Geog 3019.33.5	Entdeckungsgeschichte von Altertum bis zur Neuzeit. (Plischke, Hans.) Leipzig, 1933.
	Geog 3018.83	Die Entdeckungsreisen. (Rubiner, W.) Glogau, 1883.
	Geog 3055.60	Das enträtselte Atlantis. (Spanuth, Jürgen.) Stuttgart, 1953.
	Geog 3049.70	Enträtselte Vorzeit. (Kohlenberg, Karl Friedrich.) München, 1970.
	Geog 579.73	Entsiklopedicheskii slovar' geograficheskikh nazvanii. Moskva, 1973.
	Geog 579.68	Entsiklopedicheskii slovar' geograficheskikh terminov. (Kalesnik, Stanislaw W.) Moskva, 1968.
	An 2108.75.3	Die Entstehung der Civilisation. 3e Aufl. (Avebury, J.L.) Jena, 1875.
	An 98.89	Die Entwickelung der modernen Ethnologie. (Achelis, Thomas.) Berlin, 1889.
	Geog 3019.48	Die Entwicklung der geographischen Wissenschaft seit dem 18. Jahrhunderts. (Schwarz, G.) Berlin, 1948.
	Geog 3070.14	Die Entwicklung der Seekarten. (Mayer, E.) Wien, 1877.
	Geog 3229.39	Die Entwicklung des Kartenbildes. Inaug. Diss. (Rohr, Heinz.) Borna, 1939.
	Geog 3019.53	Entwicklung und Aufgabe der Geographie. (Bonse, Ewald.) Stuttgart, 1953.
	Geog 659.73.10	Entwicklungstendenzen der Geographie. Berlin, 1973.
	An 379.37.3A	Environment, race and migration. (Taylor, G.) Chicago, 1937.
	An 379.46.5	Environment, race and migration. 2. ed. (Taylor, G.) Chicago, 1946.
	An 379.27A	Environment and race. (Taylor, Griffith.) London, 1927.
	An 379.30.3	Environmental basis of social geography. (Huntington, C.C.) N.Y., 1930.
	Geog 5324.2	Enzberg, E. von. Fridtjof Nansen. Dresden, 1898.
	An 2109.73.30	Enzyklopädie der Technikgeschichte: über 7000 Jahre frühe technische Kultur. Stuttgart, 1973.
	An 2059.05	The eolithic problem. (Maccurdy, G.G.) Lancaster, 1905.
Htn	Geog 4306.56*	Ephemerides sive iter Danicum, Suecicum, Polonicum. (Ogier, Charles.) Lutetiae Parisiorum, 1656.
	Geog 6009.10.40	The epic of Captain Scott. (Lindsay, M.) N.Y., 1934.
	Geog 4208.41.2	Episodes in a life of adventure. (Oliphant, L.) Edinburgh, 1887.
	Geog 4208.41	Episodes in a life of adventure. (Oliphant, L.) N.Y., 1887.
	Geog 4208.41.4	Episodes in a life of adventure. 4. ed. (Oliphant, L.) Edinburgh, 1887.
	Geog 4326.74	Epistulae...moribus...turcarum. (Smith, T.) Oxonii, 1674.
	Geog 818.30	Epitome of universal geography. (Hale, N.) Boston, 1830.
Htn	Geog 815.34*	Epitome trium terrae. (Watt, J. von.) Tiguri, 1534.
	Geog 3229.10	L'epoca delle grandi scoperte geografiche. (Errera, C.) Milano, 1910.
	An 2058.78.5	The epoch of the Mammoth. (Southall, J.C.) Philadelphia, 1878.
	Geog 3019.43	La era de las exploraciones. (Mexico (City). Biblioteca Benjamin Franklin.) México, 1943.
	Geog 759.73	Eramov, R.A. Fizicheskaia geografiia zarubezhnoi Evropy. Moskva, 1973.
	Geog 807.87	Erdbeschreibung. (Büsching, Anton Friedrich.) Hamburg, 1787-1792. 10v.
	Geog 42.1	Die Erde; Zeitschrift der Gesellschaft für Erdkunde zu Berlin. Berlin. 1,1949+ 20v.
	Geog 809.51	Die Erde. v.1-13. (Gutersohn, H.) Bern, 1951. 2v.
	Geog 5838.11	Die Erde und ihre Bewohner. (Zimmermann, E.A.H.) Leipzig, 1811.
	Geog 818.97.5	Die Erde und ihre Völker. (Hellwald, F. von.) Stuttgart, 1897.
	Geog 60.3	Erdkunde; Archiv für wissenschaftliche Geographie. Bonn. 5,1951+ 17v.
	Geog 60.3.5	Erdkunde; Archiv für wissenschaftliche Geographie. Gesamtregister, v.1-17, 1947-63. Bonn, 1965.
	Geog 3228.82.3	Die Erdkunde bei den Kirchenvätern. (Marinelli, G.) Leipzig, 1884.
	Geog 85.302.3	Erdkundliches Wissen. Wiesbaden. 1-19 9v.
	Geog 5939.07.5	Die Erforschung der Antarktis. (Chun, Karl.) Leipzig, 1907.
	Geog 5650.10	Erik den Rode og nordhokolonierne il Gronland. (Bruun, Daniel.) Kristiania, 1915.
	Geog 5650.25	Erik den Rødes Grønland. (Krogh, Knud J.) København, 1967.
	Geog 4418.45	Erinnerungen aus dem Osten. (Frege, C.G.) Leipzig, 1845.
	Geog 4502.5.26	Erinnerungen eines Bergsteigers. (Javelle, E.) München, 1938.
	Geog 5220.15	Erklarung der Charte...Entdeckungen. (L'Isle, J.N. de.) Berlin, 1753.
	Geog 808.17	Die Erkunde. (Ritter, Carl.) Berlin, 1817-1818. 2v.
	Geog 808.17.5	Die Erkunde. v.1-19. (Ritter, Carl.) Berlin, 1822-1859. 21v.
	Geog 809.21	Erläuterungen zu 938 ausgewählten Lichtbilder zur Länderkunde. (Bader, G.) Stuttgart, 1921. 3v.
	An 180.3	Erlebte Erdteile; Ergebnisse eines deutschen Forscherlebens. (Frobenius, Leo.) Frankfurt am Main, 1925-29. 7v.
	Geog 4219.32.20	Ernte. (Katz, Richard.) Zürich, 1935.
	Geog 3229.10	Errera, C. L'epoca delle grandi scoperte geografiche. Milano, 1910.
	Geog 3235.50.9	Errera, Carlo. I portolani italiani del medioevo secondo l'opera di K. Kretschmer. Firenze, 1911.
	Geog 665.43	Errera, Carlo. Scritti geografici scelti e ordinati a cura del Comitato nazionale. Bologna, 1937.
	Geog 500.31	Ersch, J.S. Repertorium über...allgemeinern...Journale. Lemgo, 1790. 3v.
	Geog 4218.38	Erskine, C. Twenty years before the mast. Boston, 1890.
	Geog 818.78	Erslev, Ed. Agrip af landafraedi. Reykjavik, 1878.
Htn	Geog 805.67F*	Erst Theil dieses Weltbuchs von neuen Erfundnen Landtschafften Warhafftige Beschreibunge aller Theil der Welt. (Franck, S.) Franckfurt, 1567.
	Geog 4502.10.3	Die erste Ersteigung des Totenkirches durch den Christ Fick-Kamin. (Christ, Fritz.) München, 1937.
	Geog 4215.19.5	Die erste Umseglung der Erde...1519-1522. (Kölliker, O.) München, 1908.
	Geog 199.1.20	Die ersten fünfzig Jahre des Schweizer Alpenclub. (Dübi, Heinrich.) Bern, 1913.
	Geog 199.1.18	Die ersten 25 Jahre des Schweizer Alpenclub. (Schweizer Alpenclub.) Glarus, 1889.
	Geog 4329.36.2	Ervine, St. J.G. A journey to Jerusalem. London, 1936.
	An 2109.56.10	Das Erwachen der Menschheit. (Kuehn, Herbert.) Frankfurt, 1956.
	An 2109.66.25	Erwachen und Aufstieg der Menschheit. (Kuehn, Herbert.) Frankfurt, 1966.
	Geog 3070.02	Pamphlet vol. Erwin Raisz.
	Geog 4328.70.5	Erzherzog Ludwig Salvator aus dem Osterreichischen Kaiserhause als Forscher des Mittelmeeres. (Woerl, Leo.) Leipzig, 1899.
	Geog 3590.73	Esakov, V.A. Russkie geograficheskie issledovaniia Evropeiskoi Rossii i Vrala v XIX-nachale XX v. Moskva, 1964.
	An 2109.27.17	Esanielsertier, Daniel. Les formes inférieures de l'explication. Paris, 1927.
	Geog 4308.51.15	The escape or loiterings amid the scenes. (McFarland, A.) Boston, 1851.
	Geog 4145.24	Eschels, Jeus Jacob. Das obenteuerliche Leben des Jeus Jacob Eschels. Hamburg, 1966.
	Geog 4019.60	Escogieron la inquietud. (Mier, Waldo de.) Madrid, 1960.
Htn	Geog 756.28*	Esfera forma del mundo. (Velazquez Minaya, F.) Madrid, 1628.
	Geog 5855.7	Eskimo life. (Nansen, F.) London, 1893.
	Geog 5328.20	Eskimoenes venn Knud Rasmussen. (Arnesen, Odd.) Oslo, 1944.
	Geog 5182.8.3	Eskimoerne. 3. Udg. (Birket-Smith, Kaj.) Kjøbenhavn, 1971.
	Geog 5182.8.5	The Eskimoes. (Birket-Smith, Kaj.) London, 1936.
	Geog 5850.9	Eskimoliv. (Nansen, F.) Kristiania, 1891.
	Geog 5180.9	The Eskimos. (Weyer, E.M.) New Haven, 1932.
	Geog 4181.79	Esmeraldo de situ orbis. (Pacheco Pereira, Duarte.) London, 1937.
	Geog 3205.50F	Esmeraldo de situ orbis. (Pacheco Pereira.) Lisboa, 1892.
	Geog 3205.50.5	Esmeraldo de situ orbis. (Pacheco Pereira.) Lisboa, 1905.
	Geog 4218.92	Esmonde, T.H.G. Round the world with the Irish delegates. Dublin, 1892.
	Geog 1545.25	Espagne et Portugal. 3. éd. (Baedeker, publishers.) Leipzig, 1920.
	Geog 3580.20	España en los mares. 2. ed. (Ballesteros Gaibrois, Manuel.) Madrid, 1943.
	An 358.85.7	La especie humana. (Spencer, Herbert.) Madrid, 1885.
	Geog 4181.21	Espinosa, A. de. The guanches of Tenerife. London, 1907.
	Geog 613.6	Espinosa Cordero, N. Pedro Vicente Maldonado, y la Misión geodésica del siglo XVIII. Cuenca, 1936.
	Geog 62.2	L'esploratore. Milano. 1-10,1876-1886 10v.
	Geog 62.3	Esplorazione commerciale. Milano. 2-32,1887-1917 31v.
	Geog 4249.65	L'esplorazione del Grande Oceano. (Dainelli, Giotto.) Torino, 1965.
	Geog 5939.58.15	L'esplorazione dell'Antartide. (Zavatti, Silvio.) Torino, 1958.
	Geog 4305.93.5	Esprinchard, Jacques. Vie. Paris, 1957.
	An 357.76	L'esprit des usages et des coutumes des différens peuples. (Demeunier, J.N.) Londres, 1776. 3v.
	An 2109.25.9	Essai d'introduction critique à l'étude de l'economie primitive. (Leroy, Olivier.) Paris, 1925.
	An 5.4	Essai d'une bibliographie ssytématique de l'ethnologie...1911. (Steinmetz, S.R.) Bruxelles, 1911.
	Geog 3055.13.2	Essai sur l'Atlantique des anciens. (Baer, F.C.) Avignon, 1835.
	Geog 3017.55	Essai sur l'histoire de géographie. (Robert de Vaugondy, D.) Paris, 1755.
	Geog 665.73	Essais de géegraphie. (Geograficheskoe Obshchestvo SSSR.) Moscou, 1956.
Htn	Geog 4105.19*	An essay of the meanes how to make...travailes. (Palmer, Thomas.) London, 1606.
	An 359.54.5	An essay on racial tension. (Mason, Philip.) London, 1954.
	An 358.10.5	An essay on the causes of the variety of complexion and figure in the human species. (Smith, Samuel Stanhope.) Cambridge, 1965.
	An 358.10.3	An essay on the causes of the variety of complexion and figure in the human species. 2. ed. (Smith, Samuel Stanhope.) N.Y., 1810.
	An 348.45	An essay on the human color. (Slack, David B.) Providence, 1845.
	An 170.189	Essays in anthropology. Lucknow, 194-.
	An 135.1	Essays in anthropology. Kroeber. Berkeley, 1936.
	Geog 665.130	Essays in geography. (Steering Committee for Celebration of the Sixtieth Year of Prof. S.P. Chatterjee.) Calcutta, 1965.
	Geog 665.101	Essays in geography for Austin Miller. (Whittow, John Byron.) Reading, Eng., 1965.
	An 359.62.15	Essays in social anthropology. (Evans-Pritchard, Edward Evan.) London, 1962.
	An 170.174	Essays in the science of culture. (Dole, Gertrude Evelyn.) N.Y., 1960.

182

Author and Title Listing

An 2109.62.10 — Essays on the ritual of social relations. (Gluckman, Max.) Manchester, 1962.
An 3309.28 — Essentials of anthropometry. (Sullivan, Louis R.) N.Y., 1928.
An 2109.27.18 — Essertier, Daniel. Les formes inférieures de l'explication. Thèse. Paris, 1927.
Htn Geog 4126.33* — Essex, Robert. Profitable instrucions. London, 1633.
An 359.41 — Estabrooks, G.H. Man, the mechanical misfit. N.Y., 1941.
Geog 3530.11 — Estancelin, L. Recherches sur les voyages et découvertes des...Normands. Paris, 1832.
Geog 212.9 — Estatutos de la "Sociedad geográfica de la Paz." 2. ed. (Sociedad Geográfica de la Paz.) La Paz, 1907.
An 99.57 — Estena-Fabregat, Claudio. La antropología contemporánea. Madrid, 1957. 2 pam.
Geog 5837.76 — Esterretning om rudera...Grønlands. (Thorhallesen, E.) Kiøbenhavn, 1776.
Geog 3235.37 — Estratto di una nota pesta a.f. 365-66...la carta nautica. (Pezzana, A.) Berlin, 1881.
An 136.5.50 — O estruturalismo de Lévi-Strauss. (Lima, Luiz Costa.) Petrópolis, 1968.
Geog 818.80 — El estudio de la jeografía por el dibujo de las cartas jeográficas. (Cañas Pinochet, A.) Santiago de Chile, 1880.
An 170.210 — Estudios antropologicos publicados en homenaje al Dr. Manuel Gamio. (Mexico (City). Universidad Nacional.) México, 1956.
Geog 63.5 — Estudios geograficos. Madrid. 1,1940+ 30v.
Geog 63.5.2 — Estudios geograficos. Madrid. 22-77,1950-1961 2v.
Geog 665.110 — Estudios geográficos. Homenaje de la Facultad de Filosofia y Educacion a Don Huberto Fuenzalida Villegas. (Chile. Universidad, Santiago. Facultad de Filosofia y Educacion.) Santiago, 1966.
Geog 3108.74 — Estudios prehistóricos. (Moreau de Jonnés, A.) Sevilla, 1874.
An 359.58.5 — Estudos sôbre o negro. (Silva, Mello.) Rio de Janeiro, 1958.
Geog 4509.31.5 — Et la montagne conquit l'homme. (Schwartz, M.) Paris, 1931.
Geog 5839.27 — Etah and beyond. (MacMillan, D.B.) Boston, 1927.
Geog 4268.89.5 — Etats et nations de l'Europe, autour de la France. 4. éd. (Vidal de la Blache, Paul.) Paris, 189-?
Geog 4558.50 — Etchings of a whaling cruise. (Browne, J.R.) N.Y., 1850.
An 2109.72.10 — L'eterno selvaggio. (Cocchiara, Giuseppe.) Palermo, 1972.
Geog 4209.46.15 — Etherton, P.T. All over the world. London, 1946.
Geog 4181.109 — Ethiopian itineraries circa 1400-1524. (Crawford, O.G.S.) Cambridge, Eng., 1958.
An 359.24A — Ethnical unit and milieu. (Shirokogorov, S.M.) Shanghai, 1924.
An 359.02.5 — Ethnografien. (Steenstrup, J.) Kjøbenhavn, 1902.
An 5.33 — Ethnographic atlas. (Murdock, George Peter.) Pittsburg, 1967.
An 348.43F — Ethnographical maps to the natural history of man. (Prichard, J.C.) London, 1843.
An 608.81 — Die ethnographisch-anthropologische AbtheilungMZ Godeffroy. (Hamburg, Germany. Museum.) Hamburg, 1881.
An 175.1 — Ethnographische Parallelen und Vergleiche. v.1-2. (Andree, R.) Leipzig, 1878-89.
An 379.30 — Ethnography. (Hauemeyer, Loomis.) Boston, 1929.
An 2109.36.9 — Ethnologie...primitiere beschaving. (Olbrechts, F.M.) Zutphen, 1936.
An 359.68.10 — Ethnologie générale. (Poirier, Jean.) Paris, 1968.
An 358.71.7 — Ethnologische Forschungen und Sammlung von Material für die selben. (Bastian, A.) Jena, 1871-73. 2v.
An 2109.17 — Die ethnologische Wirtschaftsforschung. (Koppers, W.) Wien, 1917.
An 358.96A — Ethnology. (Keane, A.H.) Cambridge, Eng., 1896.
An 358.96.2 — Ethnology. 2. ed. (Keane, A.H.) Cambridge, Eng., 1896.
Geog 4209.63.5 — Eti semnadtsat' let. (Zhukov, Iu.A.) Moskva, 1963.
An 3559.73 — Etnicheskaia odontologiia. (Zubov, Aleksandr A.) Moskva, 1973.
An 358.99.6 — Etnologia. (Aranzadi, G. de.) Madrid, 1899.
An 359.69.20 — Etnologia. Wyd. 1. Warszawa, 1969.
An 359.64.35 — L'etnologia e le leggi della condotta umana. (Grottanelli, Virigi L.) Roma, 1964.
An 359.73.15 — Etnologicheskie issledovaniia za rubezhom. Moskva, 1973.
An 170.190 — Etnoniny. Moskva, 1970.
An 359.23.11 — Etnos". (Shirokogorov, S.M.) Shankhai, 1923.
An 359.73.10 — Etnos i etnografiia. (Bromlei, Iulian V.) Moskva, 1973.
Htn Geog 535.7* — Étrennes intéressants des quatre parties du monde. Paris. 1788
An 358.96.5 — Étude d'éthnographie préhistorique. no.2-3,6-8. (Piette, Édouard.) Paris, 1896-1925.
Geog 3506.7 — Etude historique sur l'exécution de la carte de Ferraris et l'évolution de la cartographie topographique en Belgique. (Hennequin, Emile.) Bruxelles, 1891.
Geog 3235.53F — Étude sur un portolan inédit. (Gaffarel, Paul.) Dijon, 1876.
Geog 4418.90 — Etudes de géographie et d'histoire naturelle. (Tchihatchef, P. de.) Florence, 1890.
Geog 4759.72 — Études de géographie tropicale offertes à Pierre Gourou. Paris, 1972.
An 348.87 — Études ethnographiques et archéologiques sur l'exposition coloniale. (Hamy, E.T.) Paris, 1887.
Geog 3018.96 — Études historiques et géographiques. (Hamy, J.T.E.) Paris, 1896.
Geog 618.4.5 — Eugeniusz Romer. (Mazurkiewicz-Herzowa, Kucja.) Warszawa, 1966.
Geog 5235.20 — Euller, John. Arctic world. London, 1958.
Geog 4308.56.9 — Europa, or Scenes and society. (Eddy, D.C.) Boston, 1856.
Geog 4308.52.20 — Europa, or Scenes and society in England, France. 50th ed. (Eddy, Daniel C.) Boston, 1860.
Geog 4268.94 — Europa...eine...Landeskunde. (Philippson, A.) Leipzig, 1894.
Geog 809.19.2 — Europa. (Camena d'Almeida, P.) Barcelona, 1914.
Geog 4268.94.5 — Europa. (Philippson, A.) Leipzig, 1906.
Geog 4268.63 — Europa. (Ritter, K.) Berlin, 1863.
Geog 64.2 — Europa. Kjøbenhavn. 1-3,1895-1902 3v.
Geog 4269.55 — Europa. 16. Aufl. (Lehmann, H.) Frankfurt, 1955.
Geog 4219.43 — Europa ahoy! De geschiedenis van een zeilschip. (Metzelaar, A.C.) Amsterdam, 1943.
Geog 4309.67 — Europa Privat. (Lorck, Carl von.) Frankfurt, 1967.
Geog 4300.31.8 — Europa touring; guide automobile d'Europe. (Touring Club Suisse.) Berne, n.d.
Geog 4269.32 — Europäische Hauptstädte. (Hausenstein, W.) Erlenbach, 1932.
Geog 4309.53.11 — Europäisches Reisebuch. (Edschmid, Kasimir.) Hamburg, 1954.
Geog 4269.68.5 — Eurōpaikē suphmōnia taxidia. (Phloros, Paulos.) Athēnai, 1968.
Geog 4309.74.5F — Europas Hovedstaeder. Billeder og stemninger oplevet af Eva Bendix. København, 1974.
An 359.17.5 — Europas människoraser. 2. uppl. (Nordenstreng, Rolf.) Stockholm, 1917-20.
Geog 4329.74 — Europas morgen. (Fries, Carl.) Stockholm, 1974.
Geog 4269.34 — Europe; a geographical survey. (Bogardus, J.F.) N.Y., 1934.
Geog 4308.67.11F — Europe, 1867-1869. (Richardson, W.L.) n.p., n.d.
Geog 4268.86.4 — L'Europe...choix de lectures de geographie. 4. ed. (Lanier, L.) Paris, 1888.
Geog 4268.73.2 — L'Europe...geographie. (Levasseur, E.) Paris, 1873.
Geog 4268.77F — Europe...picturesque scenes. (Sherer, J.) London, 1877. 2v.
Geog 4260.3 — Pamphlet vol. Europe. 11 pam.
Geog 4269.02 — Europe. (Carpenter, F.G.) N.Y., 1902.
Geog 4268.85.5A — Europe. (Chisholm, G.G.) London, 1899-1902. 2v.
Geog 4309.71.5 — L'Europe. (Derruau, Max.) Paris, 1971.
Geog 4300.42 — Europe. (Nagel, publishers.) Geneva, 1956.
Geog 4268.85 — Europe. (Rudler, F.W.) London, 1885.
Geog 4269.35.10A — Europe. (Van Valkenburg, S.) N.Y., 1935.
Geog 4269.54.10 — Europe: a journey with pictures. (Fremantle, Anne (Jackson).) N.Y., 1954.
Geog 4269.34.7 — Europe: a regional geography. 7th ed. (Shackleton, Margaret Reid.) N.Y., 1965.
Geog 4309.21.15A — Europe - whither bound? (Graham, S.) N.Y., 1922.
Geog 4309.14.15 — Europe after 8:15. (Mencken, Henry L.) N.Y., 1914.
Geog 4269.57 — Europe and its borderlands. (Ogilvie, A.G.) Edinburgh, 1957.
Geog 4269.54.5 — L'Europe centrale. 1. éd. v.1-2. (George, Pierre.) Paris, 1954.
Geog 4308.30.5 — L'Europe de 1830 vue à travers la correspondence de Augustin Pyremus de Candolle et Madame de Cércourt. (Candolle, Augustin Pyramus de.) Genève, 1966.
Geog 4269.58 — L'Europe du Nord et du Nord-Ouest. (Chabot, Georges.) Paris, 1958. 3v.
Geog 4269.59.5 — Europe from the air. (Egli, Emil.) London, 1959.
Geog 4269.52.5F — Europe in photographs. (Hürlimann, Martin.) London, 1951.
Geog 4269.52F — Europe in photographs. (Hürlimann, Martin.) London, 1952.
Geog 4309.21 — Europe of to-day. (Unstead, J.V.) London, 1921.
Geog 4309.14.11A — Europe revised. (Cobb, I.S.) N.Y., 1914.
Geog 4308.93.5 — Europe seen through a boy's eyes. (D'Apery, Tello J.) N.Y., 1893.
Geog 4309.47A — Europe without Baedeker. 1st ed. (Wilson, Edmund.) Garden City, 1947.
Geog 4309.47.2 — Europe without Baedeker. 2nd ed. (Wilson, Edmund.) N.Y., 1966.
Geog 4308.82 — European breezes. (Pitman, M.J.) Boston, 1882.
Geog 4260.8 — Pamphlet vol. European geography.
Geog 4308.59.5 — European life, legend. (Tait, J.R.) Philadelphia, 1859.
Geog 4308.49 — European life and manners. (Colman, H.) Boston, 1849. 2v.
Geog 4308.49.2 — European life and manners. (Colman, H.) Boston, 1850. 2v.
Geog 4131.48 — European recovery, 1948-1951, and the tourist industry. (Winble, Ernest W.) London, 1948.
Geog 4308.91.14 — European reminiscences. (Elson, L.C.) Philadelphia, 1914.
Geog 4309.27 — European skyways. (Thomas, Lowell.) Boston, 1927.
Geog 4308.99.10 — The European tour. (Allen, Grant.) N.Y., 1909.
Geog 4300.47 — European travel directory. (Vanderbilt, C.) N.Y., 1954.
Geog 4181.86A — Europeans in West Africa, 1450-1560. (Blake, John W.) London, 1942. 2v.
Geog 3019.72.30 — Les Européens hors d'Europe de 1434 à 1815. 1. éd. (Deschamps, Hubert J.) Paris, 1972.
Geog 4678.09 — Eustace, Thomas. The adventures. London, 1820.
Geog 5324.2.69 — Eva og Fridtjof Nansen. (Hoeyer, Liv Nansen.) Oslo, 1954.
Geog 5324.2.70 — Eva og Fridtjof Nansen. 5. Oppl. (Hoeyer, Liv Nansen.) Oslo, 1955.
Geog 4509.56.15 — Evans, Charles. On climbing. London, 1956.
Geog 6009.10.23 — Evans, E.R.G.R. South with Scott. London, 1925.
An 2058.81 — Evans, J. A few words on tertiary man. n.p., 1881.
Geog 818.10.5 — Evans, J. New system of geography. London, 1810. 2v.
Geog 4308.85.20 — Evans, M.L. Glimpses by sea and land. Philadelphia, 1885.
An 629.63 — Evans-Pritchard, E.E. The comparative method in social anthropology. London, 1963.
An 359.61.45 — Evans-Pritchard, Edward Evan. Anthropology and history. Manchester, Eng., 1961.
An 359.62.15 — Evans-Pritchard, Edward Evan. Essays in social anthropology. London, 1962.
An 359.51.5 — Evans-Pritchard, Edward Evan. Social anthropology, and other essays. N.Y., 1964.
An 180.4 — Evans-Pritchard, Edward Evans. The position of women in primitive societies. N.Y., 1965.
Geog 818.33 — Evening entertainments...manners and customs of nations. (Depping, G.B.) Philadelphia, 1833.
Geog 5538.52.5 — Eventful voyage of H.M. discovery ship "Resolute". (M'Dougall, G.F.) London, 1857.
Geog 4309.29 — "Ever thine". (Muller, Julius W.) N.Y., 1928.
Geog 4207.79 — Evers. Journal kept on a journey from Bassora to Bagdad...1779. Horsham, 1784.
Geog 4209.24A — Everywhere; the memoirs of an explorer. (Landor, A.H.S.) N.Y., 1924. 2v.
An 2058.92 — Evidences of man in the drift. (Doughty, F.W.) N.Y., 1892.
An 2058.74.5 — Evidences of the antiquity of man. (Whitmore, J.H.) Rochester, N.Y., 1874.
Geog 3019.35.25 — Evolución de la geografía. (Sanchez, Pedro C.) México, 1935.
An 359.53.5 — L'évolution, l'homme, la race. Paris, 1953?
An 2059.59.15 — Evolution and anthropology. (Anthropological Society of Washington.) Washington, 1959.
An 2109.06 — The evolution of culture and other essays. (Pitt-Rivers, A.L.F.) Oxford, 1906.
An 3409.31.5 — The evolution of dactyloscopy in the United States. (Forest, H.P. de.) Youngstown, 1931.
Geog 3018.99.15 — Evolution of geography. (Keane, John.) London, 1899.
An 379.22.4 — The evolution of man. (Baitsell, George A.) New Haven, 1923.
An 359.24.5A — Evolution of man. (Smith, G.E.) London, 1924.
An 379.22.3A — The evolution of man. New Haven, 1922.
An 359.24.7 — The evolution of man. 2. ed. (Smith, G.E.) London, 1927.
An 359.03 — The evolution of man and his mind. (Clevenger, S.V.) Chicago, 1903.
An 2109.59 — The evolution of man's capacity for culture. (Symposium on the Evolution, Chicago, 1957.) Detroit, 1959.
An 2109.67.40 — The evolution of political society: an essay in political anthropology. (Fried, Morton Herbert.) N.Y., 1967.

Author and Title Listing

An 3509.31 The evolution of the human pelvis in relation to the mechanics of the erect posture. (Reynolds, E.) Cambridge, 1931.

Geog 4269.67 Evropa bez dzhentl'menov. (Maevskii, Viktor. V.) Moskva, 1967.

Geog 3129.36 Evudania. (Saa, Mario.) Lisboa, 1936.

Geog 4209.09.6 Ewers, Hanns H. "Mit meinen Augen"; Fahrten durch die lateinische Welt. München, 1914.

An 2109.72.45 Ex nocte lux. (Horken, K.) Tübingen, 1972.

Geog 4303.99.3 Examen critique des voyages et ambassades de G. de Lannoy, 1399-1450. (Sachet, E.) Bruxelles, 1843.

Geog 3235.47.9 Examen des assertions contenues...des monuments de la géographie. (Santarem.) n.p., n.d.

An 205.4 Examination of some reasonings against unity of mankind. (Dewey, Chester.) n.p., 1862. 2 pam.

Geog 3070.34 Exercices cartographiques. (Baulig, Henri.) Paris, 1927.

Geog 4208.64.5 Exotic leaves, gathered by a wanderer. (Root, Sidney.) London, 1865.

Geog 3019.69 L'expansion europèene du XIIIe au XVe siècle. 1. éd. (Chaunu, Pierre.) Paris, 1969.

Geog 4182.72 De expeditie van Anthonio Hurdt. (Briel, Johan Jurgen.) 's-Gravenhage, 1971.

Geog 5398.79.40 L'expédition de la Jeannette au pôle nord, racontée par tous les membres de l'expédition. (Geslin, Jules.) Paris, 1883? 2v.

Geog 5398.72.7 L'expédition du Tegetthoff. (Payer, J.) Paris, 1878.

Geog 4209.68.20 Expédition France-Inde; mission culturelle 1968-1969 sous le patronage de la Fédération régionale speleologique, Nancy. (Expédition France-Inde.) Saint Max, 1970.

Geog 4209.68.20 Expédition France-Inde. Expédition France-Inde; mission culturelle 1968-1969 sous le patronage de la Fédération régionale speleologique, Nancy. Saint Max, 1970.

Geog 4412.91.5 Expédition génoise des frères Vivaldi...route maritime des Indes orientales XIII. siècle. (Avezac-Macaya.) Paris, 1859.

Geog 4180.28 Expedition of...Ursua and...Aguirre. (Simon, P.) N.Y., 1964?

Geog 4180.24 Expeditions into valley of the Amazons. (Markham, C.R.) N.Y., 1964.

Geog 5050.9 Expéditions Polaires Francaises, 1948. Terre Adélie, Greenland, 1947-55; bibliographie. Grenoble, 1956.

Geog 6100.2.10 Expéditions Polaires Françaises. Terre Adélie, 1959-61. Paris, 1963.

Geog 6100.2.12 Expéditions Polaires Françaises. Terre Adélie, 1960-62; raport d'activités. Paris, 1965.

An 2109.38.7 L'expérience mystique et les symboles chez les primitifs. (Lévy-Bruhl, L.) Paris, 1938.

Geog 4267.63 Expilly, J.J. d'. Le geographe manuel. 6e éd. Paris, 1763.

Geog 659.70 Explanation in geography. (Harvey, David.) N.Y., 1970.

An 2109.73.10 The explanation of culture change; models in prehistory. (Research Seminar in Archaeology and Related Subjects, University of Sheffield, 1971.) Pittsburgh, 1973.

An 348.61 Explanatory notice of the ethnographical map. (Prichard, J.C.) London, 1861.

Htn Geog 5220.13* Explication de la carte des...decouvertes. (L'Isle, J.N. de.) Paris, 1752. 3 pam.

Geog 64.4F L'explorateur. Paris. 1-4,1875-1876 4v.

Geog 3229.61 Les explorateurs au Moyen Âge. (Roux, Jean Paul.) Paris, 1961.

Geog 4028.3 Explorateurs et explorations. (Cartier, Jean Pierre.) Paris, 1974.

Geog 64.3 L'exploration. Paris. 1-18,1876-1884 16v.

Geog 3018.79 Exploration of the world. (Verne, Jules.) N.Y., 1879-81. 3v.

Geog 3570.7 Les explorations des Portugais. (Oliveira Marlins, J.P. de.) Paris, 1893.

An 170.279 Explorations in cultural anthropology; essays in honor of George Peter Murdock. (Goodenough, Ward Hunt.) N.Y., 1964.

Geog 4145.25.5 An explorer of changing horizons. (Wilson, P.W.) N.Y., 1927.

Geog 4138.93 Explorers and travellers. (Greely, A.W.) N.Y., 1893.

Geog 4169.41 Explorers Club, N.Y. Thru hell and high water. N.Y., 1941.

Geog 64.5 The explorers' journal. N.Y. 36,1958+ 9v.

Geog 4209.30.5 Exploring for plants. (Fairchild, David G.) N.Y., 1930.

Geog 4249.56 Exploring the deep Pacific. 1st ed. (Raitt, Helen.) N.Y., 1956.

Geog 3019.35.20 Exploring today. (Ellsworth, L.) N.Y., 1935.

Geog 3019.32F Explorographs. (Morrison, E.R.) Cleveland Heights, 1932.

Geog 3618.3 Exposé des travaux géographiques executés en Finlande jusqu'en 1895. (Saellskapit för Finlands Geografi, Helsingfors.) Helsingfors, 1895.

Geog 4349.29 Express to Hindustan. (Ellis, M.H.) London, 1929.

Geog 4509.40 Exrusionismo y alpinismo; historia de su evolución. (Jahn, A.) Caracas, 1940.

Geog 4314.6 Extract from journal in the Black Sea. (Symonds, W.) London, 184-?

An 3308.95 Extract from Science progress. (Farmer, J.B.) Boston, 1895.

Geog 4328.20 Extracts from a journal. (Monson, W.J.M.) London, 1820.

Geog 585.9 Extrait de l'Annuaire in Bureau de Longitudes. (Levasseur, E.) n.p., 1905.

Geog 5367.65.7 Extraits...des voyages...de l'Asie. (Engel, S.) Lausanne, 1779.

Geog 4158.22 Eyries, J.B.B. Abrégé des voyages modernes. Paris, 1822-24. 14v.

Geog 4308.42 Faber, F.W. Sights and thoughts...foreign. Photoreproduction. London, 1842.

Geog 4229.42 Faber, Kurt. Der göttliche Vagabund. Stuttgart, 1942.

Geog 4219.41.10 Faber, kurt. Tausend und ein Abenteuer. Berlin, 1941.

An 379.00 Les facteurs de l'évolution des peuples. (Matteweuzzi, A.) Bruxelles, 1900.

Geog 4515.105.2 Faes, Henry. Le manuel du skieur. 2e éd. Lausanne, 1925.

An 2109.27.21 Fahl, Anton. Ein Beitrag zur Wirtschaftsgeschichte der Vorzeit. Inaug. Diss. Halle, 1927.

Geog 4249.38 Fahnestvak, B. Stars to windward. N.Y., 1938.

An 2109.27.19 Fahrenfort, J.J. Het hoogste wezen der primitieven. Groningen, 1927.

Geog 4219.11.7 Eine Fahrt um die Welt. (Dittmar, J.) Berlin, 1911.

Geog 5300.6 Fahrten und Forschunger der Holländer in die Polargebieten. (Cronheim, P.) Leipzig, 1913.

Geog 5182.5 Fainberg, Lev A. Ocherki etnicheskoi istorii zarubezhnogo Severa. Moskva, 1971.

Geog 4209.32.20 Fair winds and foul. (Hauser, H.) N.Y., 1932.

Geog 4559.25.3 Fair winds and foul. (Perry, F.) Boston, 1925.

Geog 4209.30.5 Fairchild, David G. Exploring for plants. N.Y., 1930.

Geog 5985.325.5 The faith of Edward Wilson. (Seaver, George.) London, 1948.

Geog 3055.85 Falk, Bertil. Atlantis och svenskarna. Stockholm, 1974.

An 358.95 Fallacies of race theories. (Babington, W.D.) London, 1895.

Geog 4559.29.30 Falmouth for orders...Cape Horn. (Villiers, A.J.) Garden City, N.Y., 1929.

Geog 4206.62 Les fameux voyages de Pietro della Valle. (Valle, Pietro.) Paris, 1663-70. 4v.

Geog 4308.65.3 Familiar letters from Europe. (Felton, C.C.) Boston, 1866.

Geog 4308.65 Familiar letters from Europe. (Felton,C.C.) Boston, 1865.

Geog 4308.82.5 A family flight thro' France, Germany, Norway and Switzerland. (Hale, E.E.) Boston, 1882.

Geog 665.26 Famous islands. Boston, n.d.

Geog 4672.25 Famous sea tragedies. (Dyall, Valentine.) London, 1955.

Geog 5855.16 Fangstmaend i Melvillebugten. (Freuchen, Peter.) København, 1956.

Geog 4218.33.10A Fanning, E. Voyages and discoveres in the South Seas, 1792-1832. Salem, 1924.

Geog 4329.09 A fantasy of Mediterranean travel. (Bayne, S.G.) N.Y., 1909.

Geog 4209.31.5 Far horizons. (Desmond, Alice C.) N.Y., 1931.

Geog 4418.54.3 Far off, or Asia and Australia described with anecdotes and illustrations. (Mortimer, F.L.B. (Mrs.).) N.Y., 1854.

An 359.36.5 Die farbige Front; hinter den Kulissen der Weltpolitik. Leipzig, 1936.

An 359.38.13 Die farbige Gefahr. 2. Aufl. (Bang, Paul.) Göttingen, 1938.

Geog 4181.60 Farcourt, R. Relation of a voyage to Guiana. London, 1928.

Htn Geog 3205.40.25* Fardle of facious...ancient manners. (Boehme, J.) London, 1555. 3v.

Geog 3560.15 Farinelli, Arturo. Viajes por España y Portugal. Roma, 1942-44. 3v.

Geog 4559.26.3 Farmer, H.F. The log of a shellback. N.Y., 1926.

An 3308.95 Farmer, J.B. Extract from Science progress. Boston, 1895.

Geog 4308.75.6 A farmer's vacation. (Waring, G.E.) Boston, 1876.

Geog 4418.79 Farnham, J.M.W. Homeward or travels. Schenectady, 1879.

Geog 4659.66 Farr, Grahame E. Wreck and rescue in the Bristol Channel. Truro, 1966-67. 2v.

Geog 4659.68.5 Farr, Grahame E. Wreck and rescue on the coast of Devon. Truro, 1968.

An 358.79.2 Farrer, J.A. Primitive manners and customs. London, 1879.

Geog 4309.26.20 Farson, N. Sailing across Europe. London, 1928.

Geog 5398.99 Farther north than Nansen. (Amadeus, Louis.) London, 1901.

Geog 5398.93.1 Farthest north. (Nansen, Fridtjof.) N.Y., 1897. 2v.

Geog 5398.93 Farthest north. (Nansen, Fridtjof.) Westminster, 1897. 2v.

Geog 3520.13 The fatal impact. (Moorehead, Alan.) London, 1966.

Geog 4678.47 Fate of Blenden Hall, East Indiaman. (Greig, A.M.) N.Y., 1847.

Geog 4247.69 The fateful voyage of the St. Jean Baptiste. (Dunmore, John.) Christchurch, 1969.

Geog 4329.44 Fateili, A. Dea senza volto. Milano, 1944.

Geog 4268.77.5 Faucher, J. Vergleichende Kultur Gilder. Hannover, 1877.

Geog 4228.89.5 Faucher de Sant-Maurice, N.H.E. Loin du Pays. Souvenirs d'Europe, d'Afrique, et d'Amérique. Quebec, 1889. 2v.

Geog 4679.42.5 Fay, Charles E. Mary Celeste; the odyssey of an abandoned ship. Salem, 1942.

An 358.87.5 Featherman, A. Social history of the races of mankind. v.1-5. Boston, 1881-91. 7v.

An 358.87.6 Featherman, A. Social history of the races of mankind. v.2,5. London, 1881-91. 2v.

Geog 619.9 Fedor Ivanovich Soimonov, 1692-1780. (Gol'denberg, Leonid A.) Moskva, 1966.

An 135.5 Feher, Geza. Felikcě F. Kanitsě. Sofiia, 1936.

An 2109.21.3 Fehlinger, H. Sexual life of primitive people. London, 1921.

An 2109.73.20 Felgenhauer, Fritz. Einführung in die Urgeschichtsforschung. 1. Aufl. Breisgau, 1973.

An 135.5 Felikcě F. Kanitsě. (Feher, Geza.) Sofiia, 1936.

Geog 4309.38.5 Fellman, A. Voyage en Orient du roi Erik Ejegod. Helsinki, 1938.

Geog 4678.03 Fellows, William D. Narrative of the loss of the Lady Hobart Packet. London, 1803.

Geog 68.2F Fels und Firn. München. 1925

Geog 4509.24 Felsklettern in Bildern und Merkworten. (Flaig, W.) Stuttgart, 1924.

Geog 4180.16 Feltcher, F. The world encompassed by Francis Drake. N.Y., 1964?

Geog 4308.65.3 Felton, C.C. Familiar letters from Europe. Boston, 1866.

Geog 4308.65 Felton,C.C. Familiar letters from Europe. Boston, 1865.

Geog 5700.7 Fenger, H.M. Bidrag til Hans Egede...missions. Kjobenhavn, 1879.

Geog 5700.4.20 Fenger, Niels. Palase, Hans Egede i Grønland. København, 1971.

Geog 4679.71 Fenn, Charles. Journal of a voyage to nowhere. London, 1971.

Geog 78.1 Fennia. 1+ 52v.

Geog 78.1F Fennia. 16-66 3v.

Geog 78.1PF Fennia. 48,1929

Geog 757.64F Fenning, Daniel. A new system of geography: or A general description of the world. n.p., 17- . 2v.

Geog 4219.58.5 Fenyvessy, J. Mein Traum wurde Wirklichkeit. 10. Aufl. Köln, 1961.

Geog 4218.76.15 Fenzi, S. Gita intorno alla terra dal gennaio al settembre dell'anno 1876. Firenze, 1877.

Geog 4209.38.40 Ferdasaga Fritz Liebig. (Sigvaldson, J.) Reykjavik, 1938.

Geog 5398.20.12 Ferdinand von Wrangell und seine Reiselängs der Nordküste von Sibirien. (Wrangell, F. von.) Leipzig, 1885.

Geog 5332.1 Ferdinand Vrangel'. (Davydov, Iurii V.) Moskva, 1959.

Geog 4300.50 Ferien einmal anders. (Bertelsmann, C.) Gütersloh, 1965.

Geog 4145.14.5 Fernandes, José dos Santos. Miguel Corte Real. Coimbra, 1960.

Geog 3580.5.5 Fernandes de Navaretti, Martin. Coleccion de los viajes y descubrimientos. Buenos Aires, 1945-46. 5v.

Geog 3580.5 Fernandes de Navaretti, Martin. Coleccion de los viajes y descubrimientos. 2. ed. v.1-3. Madrid, 1958. 5v.

Geog 4145.11.10 Fernandez de Navarrete, Eustaquio. Historia de Juan Sebastian del Cano. Vitoria, 1872.

Geog 4678.67 Fernández Duro, C. Naufragios de la armada española. Photoreproduction. Madrid, 1867.

Geog 4278.93 Fernandez Duro, Cesario. Viajes regios por mar en el trancurso de quimentos años. Madrid, 1893.

184

Author and Title Listing

	Geog 4216.21	Fernberger von Egenberg, C.M. Unfreiwillige Reise um die Welt, 1621-1628. Leipzig, 1928.
	Geog 4209.62.15	Feroabók. (Thorlacius, Sigriðm.) Akureyri, 1962.
	Geog 576.70F	Ferrari, Filippo. Lexicon geographicum. Parisiis, 1670.
	Geog 576.97F	Ferrari, P. Novum lexicon geographicum. Patavii, 1697.
	Geog 4309.30	Ferreira de Mira, M. Em viagem (na velha Europa). Lisboa, 1930.
	Geog 4228.35	Ferrer, Antonio C. Pases por Europa y América, en 1835 y 1836. Madrid, 1838.
	Geog 4308.98	Ferro, Adersón. Minhas viagens. Ceara, 1898.
	Geog 3570.61	Ferro, Gaetano. Le conoscenze geografiche del Medioevo. Genova, 1972.
	Geog 665.138	Festkolloquim: 100 Jahre Geographie in Giessen. Giessen, 1965.
	An 170.396	Festschrift...Otto Zerries. Bern, 1974.
	An 170.347	Festschrift Alfred Steinmann. Bern, 1972.
	An 180.2	Festschrift der deutscher anthropologischer Gesellschaft zur XXVI allgemeinen Versammlung. Cassel, 1895.
	An 170.227.5	Festschrift Eduard Hahn zum LX. Geburtstag. Stuttgart, 1917.
	An 170.246	Festschrift für A.E. Jensen. (Haberland, Eike.) München, 1964. 2v.
	Geog 665.122	Festschrift für Hans Kinzl zum siebzigsten Geburtstag. Innsbruck, 1968.
	An 170.386	Festschrift für Robert Wildhaber zum 70. Geburtstag am 3. August 1972. Basel, 1973.
	Geog 4131.59	Festschrift für Walter Hienzeker zum 60. Geburtstag. (Schweizerischer Fremdenverkehrsverband.) Bern, 1959.
	Geog 613.5.5	Festschrift zum 70...Konrad Miller. Bremen, 1949.
	Geog 665.69	Festschrift zur Hundertjahrfeier der geographischen Gesellschaft in Wien, 1856-1956. (Geographische Gesellschaft in Wien.) Wien, 1957.
	Geog 4300.5	Fetridge, W.P. Harper's handbook for...Europe. N.Y. 1-17,1862-1878 7v.
	Geog 4029.15	Le feu de Saint-Elme. (Monfreid, Henri de.) Paris, 1974.
	Geog 4502.5.22	Der Feuerberg. (Rohrer, M.) München, 19- .
	Geog 4311.18	Feuerstein, Ernst. Deudonex. Dichtung. n.p., 1966.
	An 2159.73.25F	Feustel, Rudolf. Technik der Steinzeit. Weimar, 1973.
	An 2058.81	A few words on tertiary man. (Evans, J.) n.p., 1881.
	Geog 5399.03	Fiala, A. Fighting the polar ice. N.Y., 1906.
	Geog 5399.03.2	Fiala, A. Fighting the polar ice. 2. ed. N.Y., 1907.
	Geog 4308.59.3	Field, H.M. Summer pictures. N.Y., 1859.
	An 130.5	Field, Henry. The track of man. 1. ed. Garden City, 1953.
	Geog 4308.77.8	Field, Henry M. From Egypt to Japan. 13th ed. N.Y., 1877.
	Geog 4308.77.7	Field, Henry M. From Egypt to Japan. 13th ed. N.Y., 1886.
	An 359.62.20	Fiezefontaine, J. Du racisme à l'universalisme. Paris, 1962.
	Geog 4348.32	Fifteen months pilgrimage. (Stocqueler, J.H.) London, 1832. 2v.
	Geog 3249.64	Fifteenth century geography against the background of medieval science. (Goldstein, Thomas.) Salem, Mass., 1964.
	Geog 4180.40	Fifth letter of...Cortes to Emperor Charles V. (Cortes, H.) London, 1868.
	Geog 4559.38	Fifty south to fifty south...around Cape Horn...Wander Bird. (Tompkins, W.M.) N.Y., 1938.
	Geog 5160.13.5	Fifty years below zero. (Brower, Charles D.) N.Y., 1942.
	Geog 4239.66.5	A fighting chance. (Ridgway, John M.) London, 1967.
	Geog 3520.10.75A	Fighting merchant men. (Hakluyt, Richard.) Boston, 1927.
	Geog 5399.03	Fighting the polar ice. (Fiala, A.) N.Y., 1906.
	Geog 5399.03.2	Fighting the polar ice. 2. ed. (Fiala, A.) N.Y., 1907.
	An 2108.70.5	Figuier, L.G. L'homme primitif. 2. ed. Paris, 1870.
	An 2108.70.7	Figuier, L.G. Primitive man. N.Y., 1870.
	An 2108.70.6	Figuier, L.G. Primitive man. N.Y., 1870.
	An 358.72	Figuier, Louis. Les races humaines. Paris, 1872.
	Geog 659.57	Filchner, W. Route-mapping and position locating in unexplored regions. Basel, 1957.
	Geog 4145.28	Filchner, Wilhelm. Ein Forscherleben. 3. Aufl. Wiesbaden, 1953.
	Geog 4145.28.5	Filchner, Wilhelm. In China, auf Asiens Hochsteppen. Freiburg, 1930.
	Geog 4269.65	Filipenko, Aloiz A. Na chuzhikh ulitsakh. Moskva, 1965.
	Geog 619.8	Filipp Iogann Stralenberg. (Novlianskaia, Mariia G.) Leningrad, 1966.
	Geog 4209.71.10	Filosofiia i dzhentel'meny. (Šteinbergs, Valentins.) Riga, 1971.
	Geog 4509.24.6	Finch, G.L. The making of a mountaineer. London, 1924.
	Geog 759.36.4	Finch, V.C. Elements of geography. 4th ed. N.Y., 1957.
	Geog 4309.23.5	Finding the worth while in Europe. (Osborne, A.B.) N.Y., 1923.
	Geog 4309.23.10	Finding the worth while in Europe. (Osborne, A.B.) N.Y., 1925.
	Geog 4309.31.20	Finding the worth while in Europe. (Osborne, A.B.) N.Y., 1931.
	An 3409.16.5	The finger print instructor. (Kuhne, F.) N.Y., 1916.
	An 3409.16.10	The finger print instructor. 3. ed. (Kuhne, F.) N.Y., 1942.
	An 3404.01	Pamphlet box. Finger prints.
Htn	An 3408.92*	Finger prints. (Galton, F.) London, 1892.
	An 3408.95	Fingerprint directories. (Galton, F.) London, 1895.
	An 3409.41	Fingerprinting; a manual of identification. (Chapel, Charles E.) N.Y., 1941.
	An 3409.43	Fingerprints, palms and soles. (Cummins, H.) Philadelphia, 1943.
	An 3409.34	Fingerprints. (United States. Federal Bureau of Investigation.) Washington, 1934.
	An 3409.37	Fingerprints. (United States. Federal Bureau of Investigation.) Washington, 1937.
	An 3409.35	Fingerprints. (Unites States. Federal Bureau of Investigation.) Washington, 1935.
	Geog 69.5	Finisterra. Lisboa. 1,1966+ 5v.
	Geog 4145.27	Fink, O. Auf dem Kirs der Raben. Hamburg, 1963.
	An 359.59.40	Finlay, William G. Races in chaos. Johannesburg, 1959.
	Geog 5329.3.5	Finnbogason, G. Vilhjalmur Stefansson. Akureyri, 1927.
	An 2109.01.7	Finot, J. Race prejudice. N.Y., 1907.
	An 359.06.3	Finot, Jean. Le préjugé des races. Paris, 1906.
	An 359.06.4	Finot, Jean. Race prejudice. London, 1906.
	An 359.06.4.2	Finot, Jean. Race prejudice. Miami, 1969.
	An 359.06.4.1	Finot, Jean. Race prejudice. N.Y., 1907.
	Geog 5839.35.20A	The fiord region of east Greenland. (Boyd, S.A.) N.Y., 1935. 2v.
	Geog 4209.67.15	Fiore, Ilario. L'italiano di Ponte Cayumba. Firenze, 1967.
	Geog 4508.89	Fiorio, C. I pericoli dell'Alpinismo e noime per vitarli. Torino, 1889.
	Geog 4679.59.5	Fire at sea. (Gallagher, Thomas M.) N.Y., 1959.
	Geog 4419.40	The fire ox and other years. (Cutting, S.) N.Y., 1940.
	Geog 3060.10	Firestone, C.B. The coasts of illusion. N.Y., 1924.
	Geog 4209.60.45	Firsé, Adolf. Reise-Journal. Gütersloh, 1967.
	Geog 4679.39	The first commission of H.M.S. Calliope. (Swan, E.W.) Newcastle-upon-Tyne, 1939.
	Geog 5838.88.3A	The first crossing of Greenland. (Nansen, F.) London, 1890. 2v.
	Geog 5399.25.21	First crossing of the Polar Sea. (Amundsen, R.) Garden City, 1928.
	Geog 5399.25.20A	First crossing of the Polar Sea. (Amundsen, R.) N.Y., 1927.
	Geog 4308.18.5	First impressions on tour upon continent. (Baillie, M.) London, 1819.
	An 359.59.25	The first look at strangers. (Bunker, Robert.) New Brunswick, N.J., 1959.
	Geog 6008.98	First on the Antarctic continent. (Borchgrevink, C.E.) London, 1901.
	Geog 4209.56.20	First overland. (Slessor, Tim.) London, 1957.
	Geog 4180.41	First part of royal commentaries of the Yncas. (Vega, G. de la.) N.Y., 1964. 2v.
	Geog 4180.52A	First voyage round the world by Magellan. (Stanley, E.J.) London, 1874.
	Geog 4308.66	First years in Europe. (Calvert, G.H.) Boston, 1866.
	An 137.4.5.4	Firth, Raymond W. Man and culture. 4. ed. London, 1963.
	An 2109.67.15	Firth, Raymond William. Themes in economic anthropology. London, 1967.
	Geog 4329.06	Fischer, T. Mittelmeerbilder. Leipzig, 1906.
	Geog 4329.06.2	Fischer, T. Mittelmeerbilder. Leipzig, 1908.
	Geog 3070.6	Fischer, T. Sammlung mittelalterlicher Welt. Venedig, 1886.
	Geog 3550.6	Fischer, Theobald. Über italienischen Seekarten und Kartographen des Mittelalters. Berlin, 1882.
	Geog 4309.38.15	Fish, Helen D. Invitation to travel. N.Y., 1938.
	Geog 5518.19.12	Fisher, Alexander. Journal of a voyage of discovery. London, 1821.
	Geog 5518.19.10	Fisher, Alexander. Journal of a voyage of discovery. London, 1821.
	Geog 5518.18.5	Fisher, Alexander. Journal of a voyage of discovery to the Arctic regions. London, 1819.
	Geog 5518.18.7	Fisher, Alexander. Journal of a voyage of discovery to the Arctic regions. London, 1968.
	Geog 5985.258.10	Fisher, Margery. Shackleton. London, 1957.
	Geog 4308.57.8	Fisk, Samuel. Mr. Dunn Browne's experiences in foreign parts. Boston, 1857.
	Geog 4308.57.7	Fisk, Samuel. Mr. Dunn Browne's experiences in foreign parts. Boston, 1857.
	Geog 4308.35.5	Fisk, Wilbur. Travels in Europe. 4th ed. N.Y., 1838.
	Geog 4228.40	Fitch, F.Y. Life, travels...of an American wanderer...Alonzo P. De Milt. N.Y., 1883.
	Geog 4269.13.10	Fitch, G.H. Critic in the occident. San Francisco, 1913.
	Geog 5939.62.5	Fitte, Ernesto. El descubrimiento de la Antártida. Buenos Aires, 1962.
	Geog 4208.17F	Fitzclarence. Journal of route across India...to England. London, 1819.
X Cg	Geog 4308.50.5	Five months abroad...1850. (McFarland, A.) Concord, 1851.
	Geog 4209.60.5	Five roads to danger. (Nicholson, Timothy.) London, 1960.
	Geog 4169.33	Five travel scripts commonly attributed to Edward Ward. (Ward, Edward.) N.Y., 1933.
	Geog 759.73	Fizicheskaia geografiia zarubezhnoi Evropy. (Eramov, R.A.) Moskva, 1973.
	Geog 759.57	Fiziko-geograficheskie raionirovanie Kitaia; sbornik statei. Moskva, 1957.
	Geog 4308.70.5	Fjord, Isle and Tor. (Spender, E.) London, 1870.
	Geog 4509.25.6	Flaig, W. Eistechnik des Bergsteigers in Bildern und Merkworten. pt.1-4. 2. Aufl. Stuttgart, 1925.
	Geog 4509.24	Flaig, W. Felsklettern in Bildern und Merkworten. Stuttgart, 1924.
	Geog 4502.5.29	Flaig, Walther. Das Silvretta-Buch. München, 1940.
	Geog 4359.25	Flambeau, V. Red letter days in Europe. With a glimpse of Northern Africa. N.Y., 1925.
	An 359.54.20	Flatz, Josef. Die Kultur. Linz, 1954.
	Geog 4209.64.40A	Fleming, Ian. Thrilling cities. N.Y., 1964.
	An 2359.75	Flesh and blood. (Tannahill, Reay.) London, 1975.
	Geog 4219.32	Fletcher, A.C.B. Keep moving. Chiago, 1932.
	Geog 4180.20	Fletcher, G. Russia at close of sixteenth century. N.Y., 1964?
	An 379.28	Fleure, H.J. The races of mankind. London, 1928.
	Geog 4217.90.18	Fleurieu, Charles P. Claret de. Die neuste Reise um die Welt in den Jahren 1790, 1791 und 1792 von Etienne Marchand. Leipzig, 1801? 2v.
	Geog 4217.90.7	Fleurieu, Charles P. Claret de. Voyage autour du monde...1790-1792. Paris, 1798-1800. 4v.
	Geog 4217.90.15	Fleurieu, Charles P. Claret de. A voyage round the world. London, 1801. 3v.
	Geog 4209.50	Flight of fancy. (Riddell, J.) London, 1950.
	Geog 3049.59	Floria de la Atlantida y otras historias fabulosas. (Correa-Calderon, E.) Madrid, 1959.
	Geog 70.2	Florida. University. Publication. Geography series. Gainesville.
	Geog 4167.92.10	Flowers of celebrated travellers. (Adams, J.) Baltimore, 1834.
Htn	Geog 4167.92.3*	Flowers of modern travels. (Adams, J.) Boston, 1797. 2v.
	Geog 4167.92	Flowers of modern travels. (Adams, J.) London, 1792. 2v.
	Geog 4218.87.20	Floyd Jones. Letters from the Far East. N.Y., 1887.
	Geog 4269.58.15	Flugbild Europas. (Egli, Emil.) Zürich, 1958.
	Geog 4309.35.10	Fluggast über Europa. (Paquet, Alfons.) München, 1935.
	Geog 4219.32.10A	The flying carpet. (Halliburton, Richard.) Indianapolis, 1932.
	Geog 4219.32.11	The flying carpet. (Halliburton, Richard.) London, 1933.
	Geog 5559.28	Flying the Arctic. (Wilkins, George H.) N.Y., 1928.
	Geog 4218.89.7	A flying trip around the world. (Wetmore, E.B.) N.Y, 1891.
	Geog 3019.53.10	Fochler-Hanke, Gustav. Introducción a la historia de la geografía. Tucumán, 1953.
	Geog 579.71	Fochler-Hauze, Gustav. Allgemeine Geographie. Frankfurt am Main, 1971.
	Geog 83.5	Focus, by the American Geographical Society. N.Y. 1,1950+ 3v.
	Geog 4300.4	Fodor's Europe. N.Y. 1967+ 3v.
	Geog 5639.05	Först, J. Geschichte der Entdeckung Grönlands von den ältesten Zeit bis zum Anfang des 19. Jahrhunderts. Worms, 1906.
	Geog 5399.25.19	Den første flukt over Polhavet. (Amundsen, R.) Oslo, 1926.
	Geog 5700.13	Første missionair paa Grønland. (Lund, Jacob Johan.) København, 1778.
	Geog 5209.57	Förster, Hans A. Der weisse Weg; Forscher erobern die Arktis. Leipzig, 1957.

Author and Title Listing

Geog 4418.75 Fogg, W.P. Arabistan, or Land of "Arabian Nights." London, 1875.

Geog 4418.72 Fogg, W.P. "Round the world", letters from Japan, China, India, and Egypt. Cleveland, 1872.

Htn Geog 4476.92.2* Foigny, Gabriel de. A new discovery of terra Incognita Australis. London, 1693.

Geog 3019.17 A foldrajzi gondolat története. (Teleki, Pal.) Budapest, 1917.

Geog 4209.35.5 Foley, Arthur. Breezy adventure. Boston, 1935.

Geog 79.2 Folia geographica. Series geographica-oeconomica. Kraków. 1,1968+ 2v.

An 5.10 Folk classification: a topically arranged bibliography of contemporary and background references through 1971. (Conklin, Harold C.) New Haven, 1972.

An 2109.10A Les fonctions mentales dans les sociétées inférieures. (Lévy-Bruhl, L.) Paris, 1910.

Geog 759.45 Fondamenti di geografia generale. v.1-2. (Almagia, R.) Roma, 1946-48.

An 379.43 Les fondements biologiques de la géographie. v.1-3. (Sorre, Maximilian.) Paris, 1943-52. 4v.

An 379.47.10 Les fondements de la géographie humaine. 2. éd. (Sorre, Maximilien.) Paris, 1947-

Geog 576.80 Fondeur, F. Urbium insularum regionum. Londuni, 1680.

An 358.84.10 Fontaine, Edward. How the world was peopled. N.Y., 1884.

Geog 520.20 Fonteneau, J. La cosmographie. Paris, 1904.

Geog 3570.31.5 Fontoura da Costa, Abel. Descobrimentos maritimos africanos dos Portugueses com D. Henrique. Lisboa, 1938.

Geog 3570.31 Fontoura da Costa, Abel. Roteiros portugueses ineditos. Lisboa, 1940.

Geog 5398.97.3 Fonvielle, W. La vérité sur l'expédition Andrée. Strasbourg, 1897.

Geog 5938.89 Fonville, W. de. Le pôle sud. Paris, 1889.

Geog 4218.84.13 Foot-prints of travel. (Ballou, M.M.) Boston, 1889.

Geog 4218.84.14 Foot-prints of travel. (Ballou, M.M.) Boston, 1889.

Geog 4218.84.15 Foot-prints of travel. (Ballou, M.M.) Boston, 1892.

Geog 4208.95 Foot-prints of travel. (Ballou, M.M.) Boston, 1895.

An 3508.82 Footprints found at the Carson State Prison. (Harkness, H.W.) n.p., n.d.

Geog 4145.39.5 For the love of Mike. (Hoyt, Jo Wasson.) N.Y., 1966.

Geog 4208.85.3 Forbes, Archibald. Souvenirs of some continents. London, 1885.

Htn Geog 4208.78.3* Forbes, R.B. Personal reminiscences. 2. ed. Boston, 1882.

Geog 4558.11 Forbes, R.B. Voyage of the Niedas, 1811. Boston, 1885.

Geog 4208.78 Forbes, R.L. Personal reminiscences. Boston, 1878.

X Cg Geog 4208.78.2 Forbes, R.L. Personal reminiscences. 2. ed. Boston, 1882.

Geog 4678.89 Forbes, Robert B. Notes on some few of the wrecks and rescues during the present century. Boston, 1889.

Geog 4209.28 Forbes, Rosita. Adventure. Boston, 1928.

Geog 4209.39 Forbes, Rosita T. These are real people. N.Y., 1939.

Geog 4169.55 Forbidden lands. (Cooper, G.) N.Y., 1955.

Geog 3108.42.5 Forbiger, A. Handbuch der alten Geographie. Hamburg, 1877. 3v.

Geog 3108.42 Forbiger, A. Handbuch der alten Geographie. Leipzig, 1842-48. 3v.

An 180.1 Force, M.F. Pre-historic man. Cincinnati, 1873.

Geog 5225.5 Force, P. Grinnell land. Washington, 1852.

An 2109.64.20 Forcieri, Luigi. Le role des croyances magiques et religieuses dans les économies primitives. Bordeaux, 1964.

An 170.202 Ford, Clellan Stearns. Cross-cultural approaches. New Haven, 1967.

Geog 4110.30 Ford, Norman D. Bargain paradises of the world. 4. ed. Greenlawn, N.Y., 1957.

Geog 4110.30.5 Ford, Norman D. How to travel without being rich. Greenlawn, N.Y., 1957.

Htn Geog 3060.8* Ford, Worthington Chauncey. The isle of pines, 1668. Boston, 1920.

Geog 3060.8 Ford, Worthington Chauncey. The isle of pines, 1668. Boston, 1920.

An 379.34 Forde, C.D. Habitat, economy and society. London, 1934.

An 379.34.2 Forde, C.D. Habitat, economy and society. N.Y., 1937.

Geog 3070.31 Fordham, Herbert G. Maps. Cambridge, Eng., 1921.

Geog 500.55 Foreign countries. (Newark, N.J. Free Public Library.) Washington, 1918.

Geog 4309.64.10 Foreign faces. (Pritchett, V.S.) London, 1964.

Geog 4208.25 Foreign scenes and travel recreations. 2. ed. (Howison, J.) Edinburgh, 1825. 2v.

Geog 4145.3 Foreign service diary. (Allen, Katharine M.) Washington, 1967.

Geog 5839.30.20 Foreniger Det ry Grønland. Det ry Grønland. pt.3-5. København, 1930.

Geog 3019.26 The forerunners of Hakluyt. (Parlss, George B.) n.p., 1926?

An 3409.31.5 Forest, H.P. de. The evolution of dactyloscopy in the United States. Youngstown, 1931.

Geog 585.15.3 Formeln und Hülfstafeln für geographische Ortsbestemmungen. 3. Aufl. (Albrecht, Theodor.) Leipzig, 1894.

An 2109.27.17 Les formes inférieures de l'explication. (Esanielsertier, Daniel.) Paris, 1927.

An 2109.27.18 Les formes inférieures de l'explication. Thèse. (Essertier, Daniel.) Paris, 1927.

Geog 4308.67 Forney, J.W. Letters from Europe. Philadelphia, 1867.

Geog 4308.74 Forney, John W. A centennial commissioner in Europe 1874-1876. Philadelphia, 1876.

An 2109.68.35 Forno, Mario. L'acculturazione dei popoli primitivi. Roma, 1968.

Geog 5220.23 Fornskandinaviska upptäcktsfärder i Nordatlantiska hafvet. (Ahlenius, K.) Stockholm, 1837.

Geog 4145.28 Ein Forscherleben. 3. Aufl. (Filchner, Wilhelm.) Wiesbaden, 1953.

Geog 665.165 Forschungen zur allgemeinen und regionalen Geographie; Festschrift für Kurt Kayser. Wiesbaden, 1971.

An 2109.30.5 Forschungen zur Geschichte des Hauskundes. (Gandert, O.F.) Leipzig, 1930.

Geog 72.5 Forschungen zur theoretischen Kartographie. Wien. 1,1971+ 2v.

An 170.204 Forschungen zur Völkerpsychologie. Leipzig. 1-14,1925-1935 14v.

Geog 5839.16.5 Forslag om at bebygge Scoresby Sund-Egnen i Østgrønland ved vestgrønlandske saelforgere. (Olrik, Harald.) København, 1916.

Geog 4347.98 Forster, G. Journey from Bengal to England. London, 1798.

Geog 5220.7 Forster, J.R. Histoire des decouvertes...dans le Nord. Paris, 1788. 2v.

Geog 5220.5 Forster, J.R. History of voyages...in the North. London, 1786.

Geog 4209.40.3 Forsyth, W.J. Journal. Boston, 1940.

An 170.203 Fortes, Meyer. Social structure. Oxford, 1949.

An 180.5 Fortes, Meyer. Time and social structure and other essays. London, 1970.

Geog 4309.59 Fortin Magaña, Romeo. Ajo otros cielos. San Salvador, 1959.

Geog 3070.115 Fortschrittsberichte auf dem Gebiet des Kartendrucks. (Internationaler Kurz für Kartendruck.) Hamburg, 1958.

Geog 4207.99 Forty years at sea. (Nevens, William.) Portland, 1846.

Geog 500.88 Fossati Bellani, Luigi Vittorio. I libri di viaggio e le guide della raccolta Luigi Vittorio. Roma, 1957. 3v.

An 2059.33 Fossil man in China. (Black, Davidson.) Peiping, 1933.

An 2108.83 Fossil men and their modern representatives. 2. ed. (Dawson, J.W.) London, 1883.

Geog 5312.3.5 Foster, C. Rear Admiral Byrd and the polar expeditions. N.Y., 1930.

Geog 4679.23.6 Foster, Cecil. 1700 miles in open boats; the story of the loss of the S.S. Trivessa in the Indian Ocean. London, 1952.

An 359.29.12 Foster, T.S. Travels and settlements of early man. N.Y., 1929.

Geog 4419.49 Foster, William. The Red Sea and adjacent countries...seventeenth century. London, 1949.

Geog 4181.100 Foster, William. The Red Sea and adjacent countries of the close of the 17th century as described by Joseph Pitts. London, 1949.

Geog 4181.67 Foster, William. Travels of John Sanderson in the Levant. London, 1931.

Geog 4181.82 Foster, William. The voyage of Nicholas Downton to the East Indies, 1614-15. London, 1939.

Geog 4181.88 Foster, William. The voyage of Sir Henry Middleton to the Moluccas. London, 1943.

Geog 4181.75 Foster, William. The voyage of Thomas Best to the East Indies. London, 1934.

Geog 4181.85 Foster, William. The voyages of Sir James Lancaster to Brazil and the East Indies, 1591-1603. London, 1940.

Geog 212.216.95 Foucart, G. La société Sultanich de géographie du Caire. Le Caire, 1921.

Geog 4328.70 Fouché, William Grey (Mrs.). Dagbok...resatill Egypten, Konstantinopel. Stockholm, 1870.

Geog 659.19 The foundations of geography in the 20th century. (Schrader, Franz.) Oxford, 1919.

Geog 4209.47.5 The four paths of pilgrimage. (Wethered, Herbert Newton.) London, 1947.

Geog 4208.52.5 Four years in a government exploring expedition. (Colvocoresses, G.M.) N.Y., 1852.

Geog 4208.52.6 Four years in the government exploring expedition. 5. ed. (Colvocoresses, G.M.) N.Y., 1855.

Geog 5559.13.5A Four years in the white north. (MacMillan, D.B.) N.Y., 1918.

Geog 4145.29.5 Fournier-Aubry, Fernand. Don Fernando. Paris, 1972.

Geog 4145.29.6 Fournier-Aubry, Fernand. Don Fernando. 1st American ed. N.Y., 1974.

Geog 4145.29 Fournier-Aubry, Fernand. Mon metier l'aventure. Paris, 1953.

Geog 6100.37 Fourteen men. (Scholes, A.) London, 1951.

Htn Geog 5516.31.9* Fox, Luke. North-west Fox...voyages of Cabot, Frobisher. London, 1635.

Geog 5506.35.1 Fox, Luke. North-west Fox. n.p., 1965.

An 180.6 Fox, Robin. Encounter with anthropology. Harmondsworth, 1975.

Geog 603.1 Fr. Cannabich (1777-1859) sein Leben und sein Werke. (Graubner, Paul.) Koniejeberg, 1913.

Geog 5639.57 Fra fribytler til embedsmand. (Mikkelsen, Ejnar.) København, 1957.

Geog 5838.36 Fra Graenlandi. (Breidfjórd, S.) Kaupmannahöfn, 1836.

Geog 5399.25.35 Fra Grønland til Stillehavet. (Rasmussen, K.) København, 1925-26. 2v.

An 2109.27.27 Fra i selvaggi e fra i civilizzate. (Marazzi, A.) Milano, 1927.

Geog 5839.39.5 Fra Lissabon til Peary Land. (Koch, Lange.) København, 1939.

Geog 5399.57.20 Fra Thule til Rio. (Freuchen, Peter.) København, 1957.

Geog 5838.88.16 Fra vestfjörðm til vestribyðr för F. Nansen. (Friðiksson, O.) Reykjavik, 1927.

Geog 5838.88.15 Fra vestfjörðm til vestribyðr för F. Nansen. (Friðiksson, O.) Reykjavik, 1927.

Geog 540.15 Frabetti, Pietro. La collezione della antiche carte geografiche, a cura di Pietro Frabetti, Il museo delle navi, a cura di Amedio Rizzi. Bologna, 1959.

Geog 4208.85 Frachebourd, J.J. Monséjour en France...mes pelerinages a Rome. Fribourg, 1885.

Geog 3019.59 Fradkin, N.G. Roshidenie karty. Moskva, 1959.

Geog 612.4.1 Fradkin, Naum G. Akademik I.I. Lepekhin i ego puteshestviia po Rossii v 1768-1773 gg. Moskva, 1950.

Geog 612.4.2 Fradkin, Naum G. Akademik I.I. Lepekhin i ego puteshestviia po Rossii v 1768-1773 gg. 2. izd. Moskva, 1953.

Geog 3019.72.5 Fradkin, Naum G. Geografickeskie otkrytiia i nauchnoe poznanie Zemli. Moskva, 1972.

Geog 3019.74 Fradkin, Naum G. Obraz Zemli. Moskva, 1974.

Geog 612.4.4 Fradkin, Naum G. Puteshestviia I.I. Lepekhina, N. Ia. Ozeretskovskogo, V.F. Zueva. Moskva, 1948.

Geog 607.5.5 Fraerman, R.I. Zhizn' i neobyknovennye prikl. K.L. Golovnina. Moskva, 1957.

Geog 607.5 Fraerman, R.I. Zhizn' i prikl. K.L. Golovnina. Moskva, 1946.

Geog 4300.10 Frager, M.D. Practical European guide. Boston, 1907.

Geog 5837.78.5 Fragmentes af en dagbok hållen i Grönland. (Saabye, Hans E.) Stockholm, 1817.

Geog 4218.40.5F Fragments de voyages autour du monde. (Lafond, G.) Paris, 1861.

Geog 4208.31 Fragments of voyages and travels. (Hall, B.) Philadelphia, 1831. 2v.

Geog 4208.31.2 Fragments of voyages and travels. v.1-2. 2d ed. (Hall, B.) Edinburgh, 1832.

Geog 5558.97 The "Fram" expedition. Nansen in the frozen world. (Berens, S.L.) Philadelphia, 1897.

Geog 3135.5A The frame of the ancient Greek maps. (Heidel, William A.) N.Y., 1937.

Geog 6009.03 Le "Francais" au Pole Sud. (Charcot, J.B.) Paris, 1906.

Htn Geog 807.51F* La France,...les Isles Britanniques. (Sanson, N.) Paris, 1651.

Geog 179.2.5 France. Centre de Documentation Cartographique et Géographique. Centre de documentation cartographique et géographique. Paris, 1956?

Geog 179.2 France. Centre de Documentation Cartographique et Géographique. Mémoires et documents. 1-10 22v.

186

Author and Title Listing

Geog 665.93 France. Centre National de la Recherche Scientifique. Colloque national de géographie appliquée, Strasbourg, 20-22 avril 1961. Paris, 1962.

Geog 81.1.3 France. Comité des Travaux Historiques et Scientifiques. Section de Géographie. Actes du congrès national des sociétés savantes. Paris. 84,1959+ 11v.

Geog 81.1 France. Ministère de l'Instruction Publique. Bulletin de géographie historique et descriptive. Paris. 1886+ 58v.

Geog 4239.29 Franceschi, F. Odisea del yate "Mary" de Puerto Rico a España. Madrid, 1930.

Geog 4709.52 Franchi, A. Storia della pirateria nel mondo. Milano, 1952. 2v.

Geog 4228.87 Francis, Harriet E. Across the meridians. N.Y., 1887.

Geog 4181.57 Francis Mortoft; his book...1658-59. (Mortoft, F.) London, 1925.

Geog 4145.98 Francis Younghusband. (Seaver, G.) London, 1952.

Htn Geog 3550.12* Franciulli, G. I grandi navigatori italiani. Roma, 1930.

Geog 4219.10 Franck, H.A. Vagabond journey around the world. N.Y., 1910.

Geog 4219.10.2 Franck, H.A. A vagabond journey around the world. N.Y., 1911.

Htn Geog 805.67F* Franck, S. Erst Theil dieses Weltbuchs von neuen Erfundnen Landtschafften Warhafftige Beschreibunge aller Theil der Welt. Franckfurt, 1567.

Htn Geog 3205.34F* Franck, S. Weltbuch: Spiegel...in Asiam, Aphrica, Europam und America. n.p., 1534.

Geog 4029.4 Franco, Ramon. Aguilas y garras. Madrid, 1929.

Htn Geog 816.52* Francois, Jean. La science de la geographie divisée en trois parties. Rennes, 1652.

An 126.2.6 Frank Boas: the science of man in the making. (Herskovits, Melville Jean.) Clifton, 1973.

An 358.52.4 Frankenheim, M.L. Völkerkunde. Breslau, 1852.

Geog 75.1.3 Frankfurter geographische Hefte. Frankfurt. 1,1927+ 14v.

Geog 75.1 Frankfurter Vereins. Jahresbericht - Geographie und Statistik. Frankfurt. 1905-1919

Geog 28.8 Frankische Geographische Gesellschaft. Mitteilungen. Erlangen. 1,1954+ 12v.

Geog 5530.15 Franklin, Jane G. The life, diaries and correspondence...1792-1875. London, 1923.

Geog 5518.19.14 Franklin, John. Narrative of a journey to the shores of the Polar Sea in the years 1819-1922. Rutland, 1970.

Htn Geog 5518.25.2F* Franklin, John. Narrative of a second expedition to the shores of the Polar Sea, in the year 1825-1827. London, 1828.

Geog 5518.25.5 Franklin, John. Narrative of a second expedition to the shores of the Polar Sea in the years 1825-1827. Rutland, 1971.

Htn Geog 5518.19.2* Franklin, John. Narrative of journey to...Polar Sea. London, 1823. 2v.

Htn Geog 5518.19* Franklin, John. Narrative of journey to...Polar Sea. London, 1823.

Geog 5518.19.5 Franklin, John. Narrative of journey to...Polar Sea. Philadelphia, 1824.

Geog 5518.19.3 Franklin, John. Narrative of journey to...Polar Sea. 2. ed. London, 1824. 2v.

Geog 5518.19.4 Franklin, John. Narrative of journey to...Polar Sea. 3. ed. London, 1824. 2v.

Geog 5518.25 Franklin, John. Narrative of second expedition to Polar Sea. Philadelphia, 1828.

Geog 5530.80 Franklin. (Lamb, G.F.) London, 1956.

Geog 3530.5 De franska sjöfärderna. (Dahlgren, E.W.) Stockholm, 1900.

Geog 3019.72.25 Frantsvzskaia geograficheskaia shkola kontsa deviatnadtsatogo nachala dvadtsatogo veka. (Aleksandrovskaia, Ol'ga A.) Moskva, 1972.

An 126.2.5 Franz Boas. (Herskovits, Melville Jean.) N.Y., 1953.

Geog 4218.96 Fraser, J.F. Round the world on a wheel. N.Y., 1899.

Geog 4209.10.6 Fraser, Mary Crawford. A diplomatist's wife in many lands. London, 1911. 2v.

Geog 4209.10.5 Fraser, Mary Crawford. A diplomatist's wife in many lands. N.Y., 1911. 2v.

Geog 4209.11.2 Fraser, Mary Crawford. Further reminiscences of a diplomat's wife. London, 1912.

Geog 4209.10.9 Fraser, Mary Crawford. Reminiscences of a diplomat's wife. N.Y., 1913.

Geog 5316.2 Fraser, Robert J. Arctic adventurer. Lincoln, Ont., 1972.

An 3309.32.3F Frassetto, F. Note antropologiche sulla popolagione del Bolognese. Bologna, 1932.

Geog 3570.16.5 Frazão de Vasconcellos, J. Pilobas das navegacões portuguesas das seculos XVI e XVII. Lisboa, 1942.

Geog 3570.16 Frazão de Vasconcellos, J. Os pilotos dos seculos XV e XVI e a nobreza do reino. Lisboa, 1932.

An 359.39.3 Frazer, J.G. The native races of Asia and Europe. London, 1939.

An 629.07 Frazer, J.G. Questions on customs, beliefs and languages of savages. Cambridge, 1907.

An 130.10.5 Frazer and the golden bough. (Downie, Robert Angus.) London, 1970.

An 359.57 Frazier, E.F. Race and culture contacts in the modern world. 1. ed. N.Y., 1957.

Geog 4709.65 Freebooters of the Red Sea: pirates, politicians and pieces of eight. (Cochran, Hamilton.) Indianapolis, 1965.

Geog 665.36 Freeman, Henry. Wonders of the world. Boston, 1873.

Geog 3019.62.5 Freeman, Thomas. A hundred years of geograpy. Chicago, 1962.

Geog 4218.51 Fregatten Eugenies resa omkring jorden, aren 1851-1853. v.1-2. (Stogman, C.D.) Stockholm, 1855?

Geog 4418.45 Frege, C.G. Erinnerungen aus dem Osten. Leipzig, 1845.

Geog 3570.33 Frei Goncalo Velho. (Sa, Ayres de.) Lisboa, 1899-1900. 2v.

Geog 4559.40.12 Frekjan; aefintýralegt...frá Danmoëku. (Jónsson, Gisli.) Reykjavik, 1940.

Geog 4269.54.10 Fremantle, Anne (Jackson). Europe: a journey with pictures. N.Y., 1954.

Geog 4509.48.5 Fremde Berge-Ferne Ziele. (Kurz, Marcel.) Bern, 1948.

Geog 4309.71.10 Fremde Vaterländer. (Krueger, Horst.) München, 1971.

An 358.83F Fremde Völker ethnographische Schilderungen aus der alten und neuen Welt. (Oberländer, R.) Leipzig, 1883.

Geog 4131.66 Fremdenverkehr in Europa. (Ritter, Wigand.) Leiden, 1966.

Geog 4131.62.5 Fremdenverkehr und Fremdenverkehrspolitik. (Poeschl, A.E.) Berlin, 1962.

An 2059.59 Fremdling zwischen Tier und Gott. (Lange, Hurt.) Gütersloh, 1959.

Geog 4709.57 Freminville, René Marie de La Poix de. Tels etaient corsaires et flibustiers. Paris, 1957.

Geog 4559.69.5 Fréminville, René Marie de La Poix de. La vie quotidienne des marins au Moyen Age. Paris, 1969.

Geog 4709.22 French, J.L. Great pirate stories. N.Y., 1922.

Geog 4709.22.2 French, J.L. Great pirate stories. 2. ser. N.Y., 1925.

Geog 4131.69 Frentrup, Klaus. Die ökonomische Bedeutung des internationalen Tourismus fur die Ehtwicklungsländer. Hamburg, 1969.

Geog 618.2.5 Frenzel, C.A. Major James Rennell. Leipzig, 1904.

Geog 613.2.5 Frenzel, R. Malthe Conrad Bruun. Crimmitschau, 1908.

Geog 4709.28.5 Les frères de la coste, flibustiers et corsaires. (Besson, Maurice.) Paris, 1928.

Geog 4509.23.10 Freshfield, D.M. Below the snow line. London, 1923.

Geog 5316.1A Freuchen, P. Arctic adventure; my life in the frozen North. N.Y., 1935.

Geog 5839.54 Freuchen, P. Ice floes and flaming water. N.Y., 1954.

Geog 5316.1.10 Freuchen, P. It's all adventure. London, 1938.

Geog 5316.1.8 Freuchen, P. Min gronlandske ungdom. Kjøbenhavn, 1959.

Geog 5316.1.6 Freuchen, P. Vagrant Viking. London, 1954.

Geog 5316.1.5A Freuchen, P. Vagrant Viking. N.Y., 1953.

Geog 4559.57 Freuchen, Peter. Book of the seven seas. N.Y., 1957.

Geog 5855.16 Freuchen, Peter. Fangstmaend i Melvillebugten. København, 1956.

Geog 5399.57.20 Freuchen, Peter. Fra Thule til Rio. København, 1957.

Geog 5328.10 Freuchen, Peter. I sailed with Rasmussen. N.Y., 1958.

Geog 4559.53.10 Freuchen, Peter. Vagrant viking. N.Y., 1953.

Geog 4145.89.5 Friar Andrés de Urdaneta, O.S.A. (Mitchell, Mairin.) London, 1964.

Geog 5918.98.3A Fricker, Karl. The Antarctic regions. London, 1900.

Geog 5918.98 Fricker, Karl. Antarktis. Berlin, 1898.

Geog 5838.88.15 Friðiksson, O. Fra vestfjörðm til vestribyðr för F. Nansen. Reykjavik, 1927.

Geog 5838.88.16 Friðiksson, O. Fra vestfjörðm til vestribyðr för F. Nansen. Reykjavik, 1927.

Geog 5324.2.5 Fridtjof Nansen, a book for the young. (Bull, Jacob B.) Boston, 1903.

Geog 5324.2.30 Fridtjof Nansen. (Byström, D.G.V.) Stockholm, 1940.

Geog 5324.2 Fridtjof Nansen. (Enzberg, E. von.) Dresden, 1898.

Geog 5324.2.105 Fridtjof Nansen. (Greve, Tim.) Oslo, 1973-74. 2v.

Geog 5324.2.65 Fridtjof Nansen. (Nockher, L.) Stuttgart, 1955.

Geog 5324.2.50 Fridtjof Nansen. (Øverås, Asbjdrn.) Stavanger, 1946.

Geog 5324.2.55 Fridtjof Nansen. (Sponsel, H.) Nürnberg, 1952.

Geog 5324.2.37 Fridtjof Nansen. (Wetterfors, Paul.) Uppsala, 1932.

Geog 5324.2.100 Fridtjof Nansen minneforelesninger. Oslo.

Geog 5324.2.85 Fridtjof Nansen og Knud Rasmussen. (Knuth, Eigil.) Kjøbenhavn, 1948.

An 2109.67.40 Fried, Morton Herbert. The evolution of political society: an essay in political anthropology. N.Y., 1967.

An 359.61.35 Fried, Morton Herbert. Readings in anthropology. N.Y., 1961-64. 2v.

An 3309.14 Friedenthal, Hans. Allgemeine und spezielle Physiologie des Menschenwachstums. Berlin, 1914.

Geog 618.5 Friedrich Ratzel. (Wanklyn, H.G.) Cambridge, Eng., 1961.

Geog 5399.21 The friendly Arctic. (Stefansson, V.) N.Y., 1921.

Geog 5399.21.2 The friendly Arctic. (Stefansson, V.) N.Y., 1943.

Geog 5399.14 The friendly Arctic. (Stefansson, V.) N.Y. 1922.

Geog 4329.74 Fries, Carl. Europas morgen. Stockholm, 1974.

Geog 5838.72 Fries, T.M. Grönland dess natur och innevanare. Upsala, 1872.

Geog 5839.09 Friis, A. Danmark expeditionen. Kjøbenhavn, 1909.

Geog 4249.67 Friis, Herman Ralph. The Pacific Basin; a history of its geographical exploration. N.Y., 1967.

Geog 755.27.10 Fritzsche, O.F. Glarean sein Leben und seine Schriften. Frauenfeld, 1890.

An 3559.09 Frizzi, E. Ein Beitrag zur Anthropologie des "Homo Alpinus Tirolensis". Wien, 1909.

An 2109.01.3 Frobenius, L. Aus dem Flegel Jahren der Menschheit. Hannover, 1901.

An 2339.01 Frobenius, L. Die Bogen der Oceanier. Berlin, 1901.

An 2339.02 Frobenius, L. Menschenjagden und Zweikämpfe. Jena, 1902.

An 359.09 Frobenius, Leo. The childhood of man. London, 1909.

An 180.3 Frobenius, Leo. Erlebte Erdteile; Ergebnisse eines deutschen Forscherlebens. Frankfurt am Main, 1925-29. 7v.

An 2109.25.5F Frobenius, Leo. Hádschra Máktuba. München, 1925.

Geog 5515.78 Frobisher's third voyage, 1578. (Parks, G.B.) n.p., 1935.

An 3559.38 Froe, A. de. Meethare variabelen van den menschlijken schedel en hun an derlinge. Amsterdam, 1938.

Geog 4559.58 Das fröhliche Logbuch. (Rehm, Arnold.) Bremerhaven, 1958.

An 2159.74 Frolov, Boris A. Chisla v grafike paleolita. Novosibirsk, 1974.

Geog 4328.99 From America to the Orient. (Honeyman, Abraham Van Doren.) Plainfield, 1899.

An 99.58 From ape to angel. 1. ed. (Hays, H.R.) N.Y., 1958.

Htn Geog 4308.67.9* From Cleveland to Russia. (Stone, H.S.) n.p., 1867.

Geog 6008.92 From Edinburgh to the Antarctic. (Burn Murdoch, W.G.) London, 1894.

Geog 4308.77.8 From Egypt to Japan. 13th ed. (Field, Henry M.) N.Y., 1877.

Geog 4308.77.7 From Egypt to Japan. 13th ed. (Field, Henry M.) N.Y., 1886.

Geog 4308.92 From Finland to Greece, or Three seasons in Eastern Europe. (Hayward, H.C.) N.Y., 1892.

Geog 4218.95.5 From Independence Hall around the world. (Brewster, F. Carroll.) Philadelphia, 1895.

Geog 4208.87.15 From Japan to Granada. (Chapin, J.H.) N.Y., 1889.

Geog 4415.05 From Lisbon to Calicut. (Den Öchten Weg Auss zu Faren von Lissbona gen Kallakuth.) Minneapolis, 1956.

Geog 4209.11.5 From pillar to post. (Lowther, H.C.) London, 1911.

Geog 4209.11.6 From pillar to post. (Lowther, H.C.) London, 1912.

Geog 4208.99.2 From Plotzk to Boston. (Antin, M.) Boston, 1899.

Geog 4209.60.10 From pole to pole. (Nikhailov, Nikolai I.) Moscow, 1960.

Geog 4308.83.3 From Ponkapog to Pesth. (Aldrich, T.B.) Boston, 1883.

Geog 4349.14 From Russia to Siam, with a voyage down the Danube. (Young, Ernest.) London, 1914.

An 359.46 From savagery to civilization. (Clark, Grahame.) London, 1946.

Geog 4208.87A From sea to sea. (Kipling, Rudyard.) N.Y., 1899. 2v.

Geog 4208.87.3 From sea to sea. (Kipling, Rudyard.) N.Y., 1909.

An 136.5.60 From symbolism to structuralism. (Boon, James A.) N.Y., 1972.

An 136.5.61 From symbolism to structuralism: Lévi-Strauss in a literary tradition. (Boon, James A.) Oxford, 1972.

Geog 4508.99.5 From the Alps to the Andes. (Zurbriggen, M.) London, 1899.

Geog 4558.66 From the deep of the sea. (Smith, C.E.) London, 1922.

Geog 5070.57 From the ends of the earth; an anthology of polar writings. (Courtauld, A.) London, 1958.

Geog 4558.87A From the forecastle to the cabin. (Samuels, S.) N.Y., 1887.

Author and Title Listing

Geog 4309.09 From the jungle through Europe with Roosevelt. (O'Laughlin, John.) Boston, 1910.
Geog 4348.81 From the Nile to Norway and homeward. (Cuyler, T.L.) N.Y., 1881.
Htn Geog 4308.68.3* From the oak to the olive. (Howe, Julia W.) Boston, 1868.
An 2109.39.15 From the South Seas. (Mead, Margaret.) N.Y., 1939.
Htn Geog 4308.67.5* From Waterloo to the Peninsula. (Sala, George A.) London, 1867. 2v.
Geog 4418.73 From West Africa to Palestine. (Blyden, E.W.) Freetown, Sierra Leone, 1873.
An 359.57.10 Die Front der Farbigen. (Drexler, John Paul.) München, 1957.
Geog 665.107 Frontiers and men; a volume in memory of Griffith Taylor (1880-1963). (Andrews, John.) Melbourne, 1966.
Geog 659.65.7 Frontiers in geographical teaching. 2nd ed. (Chorley, Richard J.) London, 1970.
Geog 665.103 The frontiers of geography. (House, John.) New Castle upon Tyne, 1965.
An 359.56.25 Frontiers of knowledge in the study of man. 1. ed. (White, Lynn T.) N.Y., 1956.
Geog 5518.50.15 Frozen ships; the Arctic diary of Johann Miertsching, 1850-1854. (Miertsching, Johann August.) N.Y., 1967.
Geog 4159.62 Fruehe Reisen und Seefahrten in Originalberichten. Graz. 1,1962+ 7v.
Geog 4181.19 Fryer, J. New account of East India and Persia. London, 1909. 3v.
Geog 4209.34.7 Fuchs, Hans. Heimkehr ins Dritte Reich. Dresden, 1934.
Geog 6009.55.10 Fuchs, Vivian. The crossing of Antarctica; the Commonwealth Antarctic Expedition, 1955-58. London, 1958.
Geog 4512.561 Der Führer; Novelle. (Mayer, T.H.) München, 1924.
Geog 4678.40 A full and particular act of all the circumstances attending the loss of the steamboat Lexington. Providence, 1840.
Geog 4679.30.5 Full fathom five. (Shaw, F.H.) N.Y., 1930.
Geog 5321.1.5 Funeral eulogy at obsequies of Dr. Kane. (Shields, C.W.) Philadelphia, 1857.
Geog 4217.03 Funnell, William. Voyage round the world. London, 1707.
Geog 4181.99 Further English voyages to Spanish America. (Spain. Archivo General de Indias, Seville.) London, 1951.
Geog 4209.11.2 Further reminiscences of a diplomat's wife. (Fraser, Mary Crawford.) London, 1912.
Geog 4181.132 Further selections from the tragic history of the sea, 1559-1565. (Gomes de Brito, Bernardo.) Cambridge, 1968.
Geog 3019.69.5 Fuson, Robert Henderson. A geography of geography; origins and development of the discipline. Dubuque, Iowa, 1969.
Geog 4131.60.5 Fuss, K. Geschichte der Reisebüros. Darmstadt, 1960.
Geog 4309.21.10A Futera, Count Y. The crown prince's European tour (March 3-September 3, 1921). Osaka, 1926.
An 359.60 The future of man. (Medawar, Peter Brian.) London, 1960.
Geog 4131.61 The future of tourism in the Pacific and Far East. (Cheechi and Company, Washington, D.C.) Washington, 1961.
An 359.40.10 Fyfe, Henry H. The illusion of national character. London, 1940.
Geog 5639.46.5 Gad, Finn. Grønlands historie. København, 1946.
Geog 5639.67 Gad, Finn. Grønlands historie. København, 1967- 2v.
Geog 5639.67.2 Gad, Finn. The history of Greenland. London, 1970- 2v.
Geog 3055.68 Gadow, Gerhard. Der Atlantis-Streit. Frankfurt am Main, 1973.
Geog 3060.5 Gaffarel, P. Les isles fantastiques de l'Atlantique. n.p., 1883.
Geog 3235.53F Gaffarel, Paul. Étude sur un portolan inédit. Dijon, 1876.
Geog 618.1.5 Gage, W.L. Life of Carl Ritter. N.Y., 1867.
An 143.2 Gagen-Torn, Nina I. Lev Iakovlevich Shternberg. Moskva, 1975.
Geog 4209.44.15 Gagnon, H.J. Blanc et noir. Montréal, 1944.
Geog 4559.40.3 Gainard, J.A. Yankee skipper; the life story of Joseph A. Gainard. N.Y., 1940.
Geog 3055.80 Galanopoulos, Angelos Georgiou. Atlantis, the truth behind the legend. Indianapolis, 1969.
Geog 4559.38.5 Gales, ice and men; a biography of the Steam Barkentine Bear. (Wead, F.W.) London, 1938.
Geog 4559.37.3 Gales, ice and men. (Wead, Frank W.) N.Y., 1937.
Geog 4209.45.15 Galimir, Mosco. Half a century of world travel; impressions and reflections. N.Y., 1945.
Geog 4679.59.5 Gallagher, Thomas M. Fire at sea. N.Y., 1959.
Geog 4708.74 Galleon Treasure Company, N.Y. Narrative of the circumstances of and attending the sinking of the Spanish galleons. N.Y., 1874.
Geog 818.64 The gallery of geography, a pictorial and descriptive tour of the world. (Milner, Thomas.) London, 1864. 2v.
Geog 4502.5.14 Gallhuber, J. Das Gesäuse und seine Berge. 2. Aufl. München, 1933.
Geog 606.1 Gallois, L. De Crontio finaeo gallico geographo. Paris, 1890.
Geog 3540.7 Gallois, Lucien. Les géographes allemands de la renaissance. Paris, 1890.
Geog 4709.28.10 Gallomb, J. Pirates, old and new. N.Y., 1928.
Geog 607.10 Galorm'a, R.M. Sochineniia. Moskva, 1949.
Geog 5324.2.80 Gal'perin, I.M. On byl pervym. Moskva, 1958.
Geog 5317.15 Gal'perin, I.M. Poliarnye zori; zapiski zhurnaliste. Moskva, 1956.
Htn An 3408.93* Galton, F. Decipherment of blurred fingerprints. London, 1893.
Htn An 3408.92* Galton, F. Finger prints. London, 1892.
An 3408.95 Galton, F. Fingerprint directories. London, 1895.
Geog 4158.60 Galton, F. Vacation tourists in 1860, 1861, 1862, 1863. Cambridge, 1861-64. 3v.
Geog 4128.56 Galton, Francis. The art of travel. London, 1856.
An 3308.84 Galton, Francis. Life history album. London, 1884.
Geog 4180.30A Galvano, A. Discoveries of the world. London, 1862.
Geog 4180.99A Gama, V. da. Journal of first voyage of V. da Gama. London, 1898.
Geog 5650.19 De gambe nordbobygder ved verdens ende. 3. opl. (Noerlund, P.) København, 1942.
Geog 4209.06 Gambier, J.W. Links on my life on land and sea. N.Y., 1906.
Geog 4209.45.10 Gamblers with fate. 1. ed. (Allan, Doug.) N.Y., 1945.
Htn Geog 5700.4.5* Det gamle Grønlands...naturel-historie. (Egede, H.) Kjobenhavn, 1741.
Geog 5650.15 Det gamle Grønlands beskrivelse. (Bárdrson, I.) København, 1930.
Geog 5650.10.3 De gamle Nordbokdonieri Grønland. (Bruun, Daniel.) København, 1905.
Geog 4269.72 Gamlet ne prishel na svidanie. (Lisakovskii, Igor' N.) Odessa, 1972.
An 2109.30.5 Gandert, O.F. Forschungen zur Geschichte des Hauskundes. Leipzig, 1930.
Geog 4219.62 Gandon, Yves. A la recherche de l'Eden. Paris, 1962.
Geog 4269.39 Gandy, Roxana Smith. Are people more a like than different. Cape May Court House, 1939.
Geog 4309.35 The gangway to Europe. (Hull, M.L.) Boston, 1935.
An 359.70.15 Ganzheitliche Ethnologie. (Stiglmayr, Engelbert.) Wien, 1970.
Geog 5209.60.5 La gara verso il Polo Nord. (Dainelli, G.) Torino, 1960.
Geog 3580.30 Garcia de Herreros, E. Quatre voyageurs espagnols à Alexandria d'Egypt. Alexandria, 1923.
Geog 3240.25 Garcia Franco. Le legua nautica en la edad media. Madrid, 1957.
Geog 4309.15 Gardiner, Sarah D. Pages in azure and gold. New Haven, 1915.
Geog 4419.12A Gardner, G.P. Chiefly the Orient. Boston, 1912.
Geog 4571.2 Garland, Joseph E. Lone voyager. 1. ed. Boston, 1963.
An 359.61.10 Garn, S.M. Human races. Springfield, Ill., 1961.
An 170.211 Garn, Stanley Marion. Readings on race. Springfield, Ill., 1960.
Geog 4509.56 Garobbio, A. Uomini del sesto grado. Milano, 1956.
Geog 578.98 Garollo, G. Dizionario geografico universale. 4. ed. Milano, 1898.
Geog 4309.68.5 Garrido, Felipe. Viejo continente. 1. ed. México, 1973.
Geog 5970.44 Gasev, Aleksandr M. Ot El'brusa do Antarktidy. Moskva, 1972.
Geog 520.3.5 Gaspar Corte-Real. (Harrisse, H.) Paris, 1883.
An 2058.65.3 Gastaldi, B. Lake habitations and pre-historic remains...north, centre Italy. London, 1865.
Geog 3055.48 Gattefossé, Jean. Bibliographie de l'Atlantide et des questions connexes. Lyon, 1926.
An 2109.40.15 Gattefossé, Jean. L'Hyperborée et les migrations néolithiques. Droguignan, 1940.
An 359.19.3 Gattefossé, R.M. Adam, l'homme tertiaire. Lyon, 1919.
Geog 3055.47 Gattefossé, R.M. La vérité sur l'Atlantide. Lyon, 1923.
Geog 4209.56.30 Gaudio, Attilio. A la recherche des îles ignorees. Paris, 1956.
Geog 4307.91.10 Gaultier, Pierre R.A. Séjour de mon grand-oncle, P. Gaultier, en Espagne...1791-1802. v.1-2. Angers, 1912.
Geog 5558.81.65 Gauroy, Pierre. Les affamés de la banquise. Paris, 1964.
Geog 3580.35 Gavira, José. La ciencia geografica española del siglo XVI. Madrid, 1931.
Htn Geog 577.09* Gazeteer's geographical index of Europe. (Eachard, Laurence.) London, 1709.
Geog 12.3 The gazette. (Alpine Club of Canada.) Banff. 1,1921
An 39.08 A gazetteer of ethnology. (Matsumura, A.) Tokyo, 1908.
Geog 578.55.13A Gazetteer of the world. (Lippincott, J.B. and Co.) Philadelphia, 1893.
Geog 578.53 Gazetteer of the world. London, 1853. 14v.
Geog 577.09.15 The gazetteer's or news-man's interpreter. (Eachard, Laurence.) London, 1751.
Htn Geog 577.04* The gazetteer's or newsman's interpreter. pt.2. London, 1707.
Geog 4209.65.25 Gde shumiat chuzhie goroda. (Vinogradov, Aleksandr A.) Moskva, 1974.
Geog 623.1 Gebhard, J.F. Het leven van Mr. Nicolaas C. Witsen. Utrecht, 1881. 3v.
Geog 619.1 Gedenkboek ter herinnering aan den 70sten verjaardag van R. Schuiling, 27 mei, 1924. Groningen, 1924.
Geog 818.51 Geelmuyden, J. Loerebogi geografien. Christiania, 1851.
An 2109.30 Gefüge und Fundamente der Kultur vom Standpunkte der Ethnologie. (Danzel, Theodor.) Hamburg, 1930.
An 2109.56.5 Gegeshidze, M.K. Gruzinskii narodnyi transport. Tbilisi, 1956.
An 2109.37.9 Die Geheimbünde Westafrikas als Problem der Religionwissenschrift. Inaug. Diss. (Hildebrand, E.) Leipzig, 1937.
Geog 4515.101 Das gehen auf Eis und Schnee. 2. Aufl. (Niederl, Franz.) München, 1927.
An 359.61.15 Gehlen, Arnold. Anthropologische Forschung. Reinbek, 1961.
An 359.08.7A Gehring, A. Racial contrasts. N.Y., 1908.
An 2108.80 Geiger, E.L. Contribution to the history of the development of the human race. London, 1880.
An 2059.14 Geike, James. The antiquity of man in Europe. N.Y., 1914.
Geog 4502.5.31 Geissler, Paul. Um den Montblanc. München, 1940?
An 359.34.10 Geist und Blut. (Menghin, O.) Wien, 1934.
An 2109.60 Geistesgeschichte der Frühzeit. v.1-2. (Cornelius, Friedrich.) Leiden, 1960.
An 2109.51.15 Der geistige Aufstieg der Menschheit. 2. Aufl. (Weinert, H.) Stuttgart, 1951.
An 2109.23.20 Die geistige Kultur der Naturvölker. 2e Aufl. (Preuss, Konrad Theodor.) Leipzig, 1923.
Geog 4559.17 Gelett, Charles N. A life on the ocean; autobiography. Honolulu, 1917.
Geog 5530.75 Gell, E.M.B. (Mrs.). John Franklin's, Eleanor Anne Porden. London, 1930.
An 359.38.20 General anthropology. (Boas, Franz.) Boston, 1938.
Geog 4158.08 General collection of voyages. (Pinkerton, John.) London, 1808-14. 17v.
NEDL Geog 4158.10F General collection of voyages. (Pinkerton, John.) Philadelphia, 1811-14. 17v.
Geog 578.77 A general dictionary of geography. (Johnston, A.K.) London, 1877.
Geog 577.91.5 General gazeteer. (Brookes, R.) Boston, 1816.
Geog 577.62 The general gazeteer. (Brookes, R.) London, 1762.
Geog 577.91 General gazeteer. (Brookes, R.) London, 1791.
Geog 577.91.3 General gazeteer. (Brookes, R.) London, 1796.
Geog 577.91.4 General gazeteer. (Brookes, R.) London, 1823.
Geog 757.93 The general gazetteer. (Brookes, R.) London, 1795.
Geog 1580.5 Das General gouvernement. (Baedeker, publishers.) Leipzig, 1943.
Geog 5317.10.5 General Greely; the story of a great American. (Mitchell, William.) N.Y., 1936.
Geog 4158.11 General history...of voyages. (Kerr, R.) Edinburgh, 1811-24. 18v.
Geog 4167.07.2 General history of all voyages...old and new world. (Bellegarde, J.B.M.) London, 1708.
Geog 4208.06 General history of voyages and travels. Glasgow, 1806.
Geog 818.10 General view of manners...of nations. (Phillips, R.) Philadelphia, 1810. 2v.
Geog 808.04.3 A general view of the world. (Blomfield, E.) Bungay, 1807. 2v.
An 2109.65.5 Genetics and prehistory. (De Beer, Gavin Rylands.) Cambridge, Eng., 1965.
Geog 85.1 Geneva. Société de Géographie. Mémoires. Paris. 1-104 43v.

Author and Title Listing

Geog 85.309F Genoa. Civico Istituto Colombiano. Bollettino. 1-4,1953-1956

Geog 84.5 Genoa. Università. Istituto di Scienze Geografiche. Pubblicazione. Genova. 7,1968+

Geog 4239.70.15 Genovés Tarazaga, Santiago. Ra, una balsa de papyrus a través del Atlantico. 1. ed. Mexico, 1972.

Geog 3251.23.3 Genovese world map, 1457. Facsimile. (Stevenson, E.L.) N.Y., 1912.

An 359.73.30 Gens du soi, gens de l'autre. (Jaulin, Robert.) Paris, 1973.

Geog 4304.65.5 Gentlemen errant; being the journeys and adventures of four noblemen in Europe during the fifteenth and sixteenth centuries. (Cust, Nina W.G.) London, 1909.

Geog 3590.30.5 Geog. v Rossii petrovsk. vremeni. (Lebedev, Dimitri M.) Moskva, 1950.

Geog 3590.30 Geogr. v Rossii XVII v. ocherkii. (Lebedev, Dimitrii M.) Moskva, 1949.

Geog 3550.11 Geografi ed exploratori italiani contemporanei. (Bertacchi, C.) Milano, 1929.

Geog 500.57 La geografia. (Almagia, Roberto.) Roma, 1919.

Geog 3228.82 La geografia. (Marinelli, G.) Roma, 1882.

Geog 197.5 Geografia. (Sao Paulo, Brazil. Universidade. Faculdade de Filosofia, Ciências e Letras.) 5-7

Geog 85.2 Geografia. Karachi. 3,1964+

Geog 3129.55 La geografia dei Greci e dei Romani. 2. ed. (Codazzi, Angela.) Milano, 1955.

Geog 759.55 Geografía del viejo mundo. (Schmieder, Oscar.) México, 1955.

Geog 3070.95 Geografia e cartografia nautica nella loro evoluzione storica e scientifica. (Guarnieri, Gino.) Genova, 1956.

Geog 4559.59 Geografia economica e sociale del mare. (Carfi, Francesco.) Livorno, 1959.

Geog 819.00 La geografía en 1898. (Beltrán y Rózpide, R.) Madrid, 1899.

Geog 808.33 Geografia general para el uso de la juventud de Venezuela. (Montenegro Colon, Feliciano.) Caracás, 1833-1837. 4v.

An 379.46.10 Geografia humana. (Amorim Girão, Aristides de.) Porto, 1946.

An 379.41 Geografia humana. (Pasada, J. de la C.) Medellín, 1941.

Geog 759.11 Geografia mathematica, geografia generale. (Giamnitrapani, D.) Firenze, 1911.

An 375.5 Geografia umana. (Urabayen, Leoncio.) Madrid, 1934. 2 pam.

An 379.12.5 La geografia umana di Jean Brunhes. (Maranelli, C.) Firenze, 1912.

Geog 759.58 Geografia universal. (Rebagliato, F.) Barcelona, 1958.

Geog 819.57 Geografia y Atlas Universal. Geografia y atlas universal. Barcelona, 1957.

Geog 819.57 Geografia y atlas universal. (Geografia y Atlas Universal.) Barcelona, 1957.

Geog 4309.36.5 Geografías (notas de viaje). (Abella Caprile, Margarita.) Buenos Aires, 1936.

Geog 500.171 Geograficheskaia literatura. (Tolchinskaia, L.I.) Moskva, 1971.

Geog 500.117 Geograficheskie slovari; bibliografiia. (Kaufman, Isaak M.) Moskva, 1964.

Geog 10.6 Geograficheskie soobshcheniia. (Akademiia Nauk SSSR. Institut Geografii.) Moskva. 2,1961+

Geog 85.386 Geograficheskii sbornik. Kazan. 2,1967+

Geog 85.388 Geograficheskii sbornik. L'vov. 7,1963+

Geog 85.387 Geograficheskii sbornik. Moskva. 1,1963+ 4v.

Geog 209.1 Geograficheskoe obshchestvo globus. (Slutskaia, Raisa D.) Moskva, 1972.

Geog 665.73 Geograficheskoe Obshchestvo SSSR. Essais de géegraphie. Moscou, 1956.

Geog 3590.77 Geograficheskoe Obshchestvo SSSR. Geograficheskoe obshchestvo za 125 let. Leningrad, 1970.

Geog 3590.77 Geograficheskoe obshchestvo za 125 let. (Geograficheskoe Obshchestvo SSSR.) Leningrad, 1970.

Geog 3019.72.5 Geografickeskie otkrytiia i nauchnoe poznanie Zemli. (Fradkin, Naum G.) Moskva, 1972.

Geog 510.25 Pamphlet vol. Geografie. 3 pam.

Geog 3018.99.5 Geografiens och de...historia. (Nystrom, J.F.) Stockholm, 1899.

Geog 5925.92 Geografiia Antarktidy. (Markov, Konstantin K.) Moskva, 1968.

An 379.74 Geografiia chelovecheskikh ras. (Alekseev, Valerii P.) Moskva, 1974.

VGeog 759.71 Geografiia v tsifrakh i sravneniiakh. (Thuchkevich, Vadim A.) Minsk, 1971.

Geog 85.100F Geografisk tidskrift. Kjøbenhavn. 1-28 28v.

Geog 85.100 Geografisk tidskrift. Kjøbenhavn. 29+ 20v.

Geog 85.100.5F Geografisk tidskrift. Index, 1-20. Kjøbenhavn, 1910.

Geog 85.105 Geografiska annaler. Stockholm. 1,1919+ 19v.

Geog 85.135 Geografiska Sällskspet i Finland, Helsingfors. Acta geographica. Helsinki. 1-27,1927-1972// 11v.

Geog 85.370 Geografski glasnik. Zagreb. 8-28 7v.

Geog 85.385 Geografski godišnjak. (Geografsko Društvo, Belgrade. Podnežnica, Kragujevac.) Kragujevac. 4,1968+

Geog 85.380 Geografski pregled. Sarajevo. 2,1958+ 3v.

Geog 85.383 Geografski razgledi. Skopje. 4,1966+ 5v.

Geog 85.382 Geografski zbornik. Ljubljana. 10,1967+ 6v.

Geog 85.385 Geografsko Društvo, Belgrade. Podnežnica, Kragujevac. Geografski godišnjak. Kragujevac. 4,1968+

Geog 3620.2 Geografsko znanje o Srbiji početkom 19 veka. (Radosčić, Nikola.) Beograd, 1927.

Geog 4267.63 Le geographe manuel. 6e éd. (Expilly, J.J. d'.) Paris, 1763.

Geog 535.5 Geographen Kalendar. Gotha. 1903-1914 12v.

Geog 105.15 The geographer. Aligarh, India.

Geog 4759.64 Geographers and the tropics. (Steel, R.W.) London, 1964.

Geog 665.134 Geographers in government: a series of papers given at meetings of the Geography Section of the AAAS in New York. N.Y., 1968.

Geog 3540.7 Les géographes allemands de la renaissance. (Gallois, Lucien.) Paris, 1890.

Geog 34.15 Geographia. (Bruenn. Universita. Přírodovědecká Fakulta.) Praha. 5,1971+

Geog 759.40 Geographia humana. (Costa Pereira, J.V. da.) Rio de Janeiro, 194-.

Geog 85.379 Geographia Polonica. Warszawa. 1,1964+ 13v.

Geog 85.379.2 Geographia Polonica. Index, 1-32,1964-1975. Warszawa, n.d.

Htn Geog 816.85* Geographia universalis. (Suval, P.) London, 1685.

Geog 500.116 Geographic bibliography. (Harris, Chauncy Donnison.) Chicago, 1961.

Geog 243.5 Geographic bulletin. (United States. Department of State. Office of the Geographer.)

An 379.32 Geographic conditions. (Hubbard, George D.) n.p., 1932.

Geog 85.78 Geographica; collana di sussidi didattici e bibliografici. Roma.

Geog 186.3 Geographica. (Prague. Universita Karlova. Acta Universitis Carolinae.) Praha. 1,1966+ 3v.

Geog 137.2.5 Geographica. Lisboa. 1,1965+ 8v.

Geog 85.80 Geographica. Uppsala. 1-38,1936-1968// 15v.

Geog 85.90 Geographica helvetica. Bern. 1,1946+ 17v.

Geog 85.95 Geographica slovaca. Bratislava. 1,1949

Geog 818.26 Geographical, chronological...atlas. (Blake, J.L.) N.Y., 1826.

Geog 665.15 Geographical, commercial...essays. London, 1812.

Geog 4268.00F Geographical, historical...description. (Boetticher, J.G.) London, 1800.

Geog 808.11 Geographical...view of the world. (Bigland, J.) Boston, 1811. 5v.

Geog 500.125 Geographical abstracts. A: Geomorphology. London. 1966+ 9v.

Geog 500.131 Geographical abstracts. Annual index. Norwich, Eng. 1972+ 4v.

Geog 500.126 Geographical abstracts. B: Biogeography, climatology and cartography. London. 1966-1974 10v.

Geog 500.127 Geographical abstracts. C: Economic geography. London. 1966+ 9v.

Geog 500.128 Geographical abstracts. D: Social geography. London. 1966+ 7v.

Geog 500.130 Geographical abstracts. F: Regional and community planning. Norwich, Eng. 1972+ 3v.

Geog 500.130.3 Geographical abstracts. G: Remote sensing and cartography. Norwich. 1974+

Geog 500.129 Geographical abstracts. Index: section A-D. London. 1966-1971// 6v.

Geog 4181.12 Geographical account...countries...Bengal. (Bowrey, T.) Cambridge, 1905.

Geog 817.51.9 Geographical and astronomical grammar. (Salmon, T.) London, 1785.

Geog 85.180.15 Pamphlet vol. Geographical Association, Great Britain.

Geog 665.46 Geographical by-ways and some other geographical essays. (Arden-Close, Charles.) London, 1947.

Geog 3650.5 Geographical concepts in ancient India. (Dube, Beehan.) Varanasi, 1967.

Geog 818.07 Geographical delineations. (Aikin, J.) Philadelphia, 1807.

Htn Geog 816.57.5* A geographical description of all the countries in the known world. (Clarke, Samuel.) London, 1671.

Htn Geog 576.93F* Geographical dictionary. (Bohun, E.) London, 1693.

Geog 578.17 Geographical dictionary or universal gazetteer. (Worcester, J.E.) Andover, 1817. 2v.

Geog 578.17.2 Geographical dictionary or universal gazetteer. 2. ed. (Worcester, J.E.) Boston, 1823. 2v.

Geog 85.375 The geographical digest. London. 1963+ 3v.

Geog 665.156 Geographical essays. (Davis, William Morris.) N.Y., 1954.

Geog 665.136 Geographical essays in honour of K.C. Edwards. Nottingham, 1970.

Geog 665.86 Geographical essays in memory of Alan G. Ogilivie. (Miller, Ronald.) London, 1959.

Geog 559.52 Geographical exploration and topographic mapping by the United States government. (United States. National Archives.) Washington, 1952.

Geog 818.62 Geographical hand-book. (Harris, A.) Lancaster, Pa., 1862.

Geog 3107.94.5 Geographical index. (Adam, A.) Edinburgh, 1795.

Geog 585.5 A geographical instructor. London, 1825.

Geog 139.4 Geographical journal. (London. Royal Geographical Society.) London. 1,1893+ 128v.

Geog 139.4.5 Geographical journal. Index. 1893-1902. (London. Royal Geographical Society.) London, 1906. 3v.

Geog 3229.25A The geographical lore of the time of the Crusades. (Wright, John K.) N.Y., 1925.

Geog 85.195 The geographical magazine. London. 1,1935+ 53v.

Geog 85.200 Geographical magazine. London. 1-5,1874-1878 5v.

Geog 85.195.2F The geographical magazine. Index, 1950-1972. London, 1972.

Geog 510.13 Pamphlet vol. Geographical papers. 20 pam.

Geog 85.389 Geographical papers. (Zagreb. Univerzitet. Geografski Institut.) Zagreb. 1,1970+

Geog 665.38 A geographical present. N.Y., 1831.

Geog 818.28 A geographical present. 1st American ed. (Venning, I.A. (Mrs.).) N.Y., 1829.

Geog 85.390 Geographical report. Umeå. 2,1971+

Geog 225.5 Geographical reports. (Tokyo Metropolitan University. Department of Geography.) Tokyo. 1,1966+ 2v.

Geog 85.201.10 Geographical review. Index, 1-45, 1926-1957. N.Y., n.d. 4v.

Geog 85.201 Geographical review published by the American Geographical Society. N.Y. 1,1916+ 55v.

Geog 85.315 Geographical Society of Ireland. Bulletin. Dublin. 1,1944+ 5v.

Geog 85.206 Geographical Society of the Pacific. Bulletin. San Francisco. 1905

Geog 85.205 Geographical Society of the Pacific. Transcriptions and proceedings. San Francisco. 1902

Geog 665.9A Geographical studies. (Ritter, Karl.) Boston, 1863.

Geog 665.9.3 Geographical studies. (Ritter, Karl.) Cincinnati, 1861.

Geog 85.365 Geographical studies. London. 1-5,1954-1958 3v.

Geog 85.202 Geographical Teacher. Geography; the magazine of the Geographical Association. London. 2,1902+ 41v.

Geog 85.203 Geographical teacher. Supplement. London. 1,1925

Geog 3618.10 The geographical tradition in Finland. (Mead, William R.) London, 1963.

Geog 818.10.3 Geographical view of the world. (Phillips, R.) N.Y., 1826.

Geog 212.205 La geographie; bulletin de la société de géographie. (Société de Géographie.) Paris. 1-72,1925-1939 60v.

Geog 3019.73.15 Geographie; europäische Entwicklung in Texten und Erläuterungen. (Beck, Hanno.) Freiburg, 1973.

Geog 819.27.5 Die Geographie; ihre Geschichte. (Hettner, A.) Breslau, 1927.

Geog 500.78 Geographie. (Hanover, S.) Hannover, 1955.

Geog 659.73.15 Die Geographie. (Hard, Gerhard.) Berlin, 1973.

Geog 818.73.3 Geographie. La terre a vol d'oiseau. 3e ed. (Reclus, O.) Paris, 1877. 2v.

Geog 659.71.15 La géographie: méthods et perspectives. (Beaujeu-Garnier, Jacqueline.) Paris, 1971.

Geog 659.42.5 La géographie. 2. éd. (Cholley, André.) Paris, 1951.

Geog 4269.35.4 Géographie de l'Europe. (Blanchard, Raoul.) Paris, 1936.

Geog 4269.64.10 Géographie de l'Europe centrale slave et danubienne. (George, Pierre.) Paris, 1964.

Author and Title Listing

Htn Geog 816.35* Geographie delineated forth in two books. (Carpenter, N.) Oxford, 1635.
Geog 3019.40.5 La géographie des humanistes. (Dainville, François de.) Paris, 1940.
Geog 816.82 La geographie du temps. pt.1-2. (Duval, Pierre.) Paris, 1682. 2v.
Geog 659.60 Geographie et action. (Phlipponneau, Michel.) Paris, 1960.
Geog 500.83 La géographie française au milieu du XX. siècle. (L'Information Géographique.) Paris, 1957.
Geog 4267.79 Geographie für Kinder. (Raff, G.C.) Tübingen, 1779.
Geog 759.66.5 Géographie générale. (Journaux, André.) Paris, 1966.
Geog 3019.73.10 Die Geographie generalis vor Varenius. (Büttner, Manfred.) Wiesbaden, 1973.
Geog 757.55 Géographie historique, ecclesiastique et civile. (Vaissete, Joseph.) Paris, 1755. 4v.
An 379.42.5 La géographie humaine. (Brunhes, Jean.) Paris, 1942.
An 379.49 La géographie humaine. (Le Lannou, M.) Paris, 1949.
An 359.12.17 La géographie humaine. 2. éd. (Brunhes, Jean.) Paris, 1912.
An 379.25 La géographie humaine. 3. éd. (Brunhes, Jean.) Paris, 1925. 3v.
An 379.34.3 La géographie humaine. 4. éd. (Brunhes, Jean.) Paris, 1934. 3v.
Geog 3240.26.5 La géographie humaine du monde musalman jusqu'au milieu du 11e siècle. (Miquel, André.) Paris, 1967.
An 2059.59.20 Geographie humaine prehistorique. (Nougier, L.R.) Paris, 1959.
An 378.84 La géographie médicale. (Bordier, A.) Paris, 1884.
Geog 818.00.5 Geographie moderne et universelle. (LaCroix, L.A.N.) Paris, 1800. 2v.
Geog 5919.08 Géographie physique, glaciologie, pétrographie des régions visitées par l'Expédition antarctique française...1903-05. (Gourdon, Ernest.) Paris, 1908.
Geog 3128.07 Géographie physique de la Mer Noire, de l'intérieur de l'Afrique et de la Méditerranée. (Dureau de la Malle, Adolphe.) Paris, 1807.
Geog 4759.73.5 Géographie physique générale de la zone intertropicale. (Birot, Pierre.) Paris, 1973.
Geog 4145.81.50 Geographie poétique des cinq continents. (Siegfried, André.) Paris, 1952.
An 379.39.10 La géographie psychologique. (Hardy, Georges.) Paris, 1939.
Geog 759.75 Géographie régionale. Paris, 1975-
Geog 816.46.13 La geographie royalle. 2e éd. (Labbé, Philippe.) Paris, 1653.
An 379.12 Géographie sociale. La mer. (Vallaux, C.) Paris, 1908.
An 379.12.3 Géographie sociale. Le sol et l'état. (Vallaux, C.) Paris, 1911.
Geog 819.45.10 Géographie sociale du monde. 6. éd. (George, Pierre.) Paris, 1964.
Geog 808.12.15F Geographie universelle. v.1-6, Atlas. (Malte-Brun, Conrad.) Paris, 1841-1847. 7v.
Geog 809.27A Geographie universelle. v.1-15. Paris, 1927-1946. 22v.
Geog 809.27.2 Géographie universelle. 2. éd. v.6, pt.1. (Vidal de La Blache, P.) Paris, 1947.
Geog 819.58.5 Geographie universelle Larousse. (Larousse, firm, publishers.) Paris, 1958. 3v.
Geog 4759.71.10 Géographie zonale des régions chaudes. (Benchétrit, Maurice.) Paris, 1971.
Geog 85.310 Geographisch-Ethnographische Gesellschaft in Basel. Mitteilungen der geographisch-Ethnographischen Gesellschaft in Basel. Basel. 1-8 4v.
Geog 4218.78.15 Geographisch-medicinische Studien. (Wernich, Agathon.) Berlin, 1878.
Geog 5367.65.3 Geographische...Nachrichten. (Engel, S.) Mietau, 1772.
Geog 85.301PF Geographische Abhandlungen. Wien. 6 2v.
Geog 85.301 Geographische Abhandlungen. Wien. 1886-1936 15v.
An 379.67.10 Geographische Anthropologie. (Lundman, Bertil.) Stuttgart, 1967.
Geog 3018.99.10 Die geographische Arbeit des 19. Jahrhunderts. (Partsch, J.) Breslau, 1899.
Geog 85.373 Geographische Berichte; Mitteilungen der geographischen Gesellschaft in der Deutschen Demokratischen Republik. Berlin. 1,1956+ 9v.
Geog 958.98F Geographische Bildertafeln. v.1-3. (Hirt, F.) Breslau, 1886-98. 5v.
Geog 665.34 Geographische Charakter-Bilder für Schule und Haus. pt.1-10. Supplement. (Hoel.) Wien, 1886.
Geog 808.50.9 Geographische Charakterbilder in algerundeten Gemälden aus der Länder- und Völkerkunde. (Grube, A.W.) Leipzig, 1868. 3v.
An 379.42 De geographische gedechte bij Jean Brunhes. (Cools, R.H.A.) Utrecht, 194-.
Geog 85.225 Geographische Gesellschaft in Hamburg. Mittheilungen. Hamburg. 1-61 47v.
Geog 85.225.2 Geographische Gesellschaft in Hamburg Mittheilungen. Register. 1-60, 1873-1972. Hamburg, 1973.
Geog 665.69 Geographische Gesellschaft in Wien. Festschrift zur Hundertjahrfeier der geographischen Gesellschaft in Wien, 1856-1956. Wien, 1957.
Geog 146.1 Geographische Gesellschaft Jahresbericht. (Munich.) 1869-1902 7v.
Geog 91.1.10 Geographische Gesellschaft zu Greifswald, 1882-1927. (Braun, Gustav.) Greifswald, 1932.
Geog 95 Geographische Gesellschaft zu Hannover. Jahrbuch. Hannover. 1926+ 8v.
Geog 95.5 Geographische Gesellschaft zu Hannover. Jahrbuch. Sonderheft. Hannover. 2,1968+ 4v.
Geog 180.1 Geographische Mittheilungen. 1+ 113v.
Geog 180.2 Geographische Mittheilungen. Ergänzungshefte. 2+ 73v.
Geog 180.3 Geographische Mittheilungen. Inhaltsverz. 1855-1934 6v.
Geog 665.80 Geographische Plaudereien. (Koegel, Ludwig.) Bonn, 1957.
Geog 665.17 Geographische Studien. (Günther, S.) Stuttgart, 1907.
Geog 665.68 Geographische Studien. Wien, 1951.
Geog 4218.73.20 Geographische und ethnologische Bilder. (Bastian, Adolf.) Jena, 1873.
Geog 85.302 Geographische Zeitschrift. Leipzig. 1,1895+ 55v.
Geog 85.303 Geographische Zeitschrift. Register. Leipzig. 1-20,1895-1914 2v.
Geog 85.303.2 Geographische Zeitschrift. Register. v.1-50, 1895-1944. Wiesbaden, 1970. 2v.
Geog 180.4 Geographischen Anzeiger. Gotha. 1-41 40v.
Geog 4269.16 Die geographischen Grenzen Mitteleuropas. (Sievers, Wilhelm.) Giessen, 1916.
Geog 500.13 Geographischer Buchersaal. (Hagers, J.G.) Chemnitz, 1766-78. 3v.
Geog 819.09 Geographisches Handbuch. (Scobel, A.) Bielefeld, 1909-1910. 2v.
Geog 85.228 Geographisches Jahrbuch. Gotha. 1-3,1850-1851
Geog 85.300 Geographisches Jahrbuch. Gotha. 1-61 60v.
Geog 578.47.9 Geographisches-statistisches Lexikon. (Ritter, Karl.) Leipzig, 1874. 2v.
Geog 578.47 Geographisches-statistisches Lexikon. 3. Aufl. (Ritter, Karl.) Leipzig, 1847.
Geog 819.61 Geography; our planet, its peoples and resources. (Manley, Gordon.) London, 1961.
Geog 85.202 Geography; the magazine of the Geographical Association. (Geographical Teacher.) London. 2,1902+ 41v.
Geog 510.5 Pamphlet vol. Geography. 3 pam.
Geog 510.1 Pamphlet vol. Geography. 10 pam.
Geog 510.3F Pamphlet vol. Geography.
Geog 510.6 Pamphlet vol. Geography. 4 pam.
Geog 510.7 Pamphlet vol. Geography. 7 pam.
An 379.19 Geography. (Schulz, George J.) College Park, Md., 1919.
An 359.36 Geography. (White, C.L.) N.Y., 1936.
Geog 759.72 Geography: a modern synthesis. (Haggeh, Peter.) N.Y., 1972.
Geog 3000.3 Pamphlet vol. Geography. Almagia. 10 pam.
Geog 5155.3 Pamphlet vol. Geography. Arctic exploration. 22 pam.
Geog 5312.3 Pamphlet vol. Geography. Arctic regions. Biography. Richard E. Byrd.
Geog 5155.1.2 Pamphlet vol. Geography. Arctic regions. North Polar regions.
Geog 5155.1 Pamphlet box. Geography. Arctic regions. North Polar regions.
Geog 5155.1.5 Pamphlet vol. Geography. Arctic regions. North Polar regions. German dissertations.
Geog 500.1 Pamphlet box. Geography. Bibliography.
Geog 500.5 Pamphlet vol. Geography. Bibliography. 15 pam.
Htn Geog 5510.50PF* Pamphlet box. Geography. Broadsides on geographical subjects.
Geog 5603.1 Pamphlet box. Geography. Greenland.
Geog 510.12 Pamphlet vol. Geography. Islands. 15 pam.
Geog 510.11 Pamphlet vol. Geography. Islands. 24 pam.
Geog 510.10 Pamphlet vol. Geography. Islands.
Geog 510.9 Pamphlet vol. Geography. Islands. 2 pam.
Geog 4672.6 Pamphlet vol. Geography. Ocean life. Shipwrecks.
Geog 4672.1 Pamphlet vol. Geography. Ocean life. Shipwrecks.
Geog 4672.3 Pamphlet vol. Geography. Ocean life. Shipwrecks. 6 pam.
Geog 5055.1 Pamphlet vol. Geography. Polar regions. Arctic and Antarctic in general.
Geog 817.04 Geography anatomized. 4th ed. (Gordon, P.) London, 1704.
Htn Geog 817.04.6* Geography anatomized. 6th ed. (Gordon, P.) London, 1711.
Geog 817.04.8 Geography anatomized. 8th ed. (Gordon, P.) London, 1719.
Geog 817.04.12 Geography anatomized. 12th ed. (Gordon, P.) London, 1730.
Geog 817.04.19 Geography anatomized. 19th ed. (Gordon, P.) London, 1749.
Geog 817.04.20 Geography anatomized. 20th ed. (Gordon, P.) London, 1754.
Geog 510.8 Pamphlet vol. Geography and maps. 5 pam.
Geog 819.50 Geography and world affairs. (Jones, S.B.) Chicago, 1953.
Geog 659.58 Geography as a fundamental research discipline. (Ackerman, Ed.) Chicago, 1958.
Geog 665.120 Geography at Aberystwyth: essays written on the occasion of the departmental jubilee 1917-1918-1967-1968. (Bowen, Emzys George.) Cardiff, 1968.
Geog 13.4.20 Geography in the making; the American Geographical Society, 1851-1951. (Wright, J.K.) N.Y., 1952.
Geog 3229.38 Geography in the Middle Ages. (Kimble, George H.T.) London, 1938.
Geog 3229.66 Geography in the Middle Ages. 1st ed. (Alavi, S.M. Ziauddin.) Delhi, 1966.
Geog 665.58A Geography in the twentieth century; a story of growth, fields, techniques, aims and trends. (Taylor, Griffith.) N.Y., 1951.
Geog 665.58.2 Geography in the twentieth century. 2nd ed. (Taylor, Griffith.) N.Y., 1953.
Geog 665.58.3 Geography in the twentieth century. 3rd ed. (Taylor, Griffith.) N.Y., 1957.
Geog 4269.35 A geography of Europe. (Blanchard, Raoul.) N.Y., 1935.
Geog 4269.50 A geography of Europe. (Gottmann, J.) N.Y., 1950.
Geog 4269.50.5 A geography of Europe. (Gottmann, J.) N.Y., 1961.
Geog 4269.37 The geography of Europe. (Hubbard, George D.) N.Y., 1937.
Geog 4269.18A The geography of Europe. (National Research Council. Division of Geology and Geography.) New Haven, 1918.
Geog 4269.61.10 A geography of Europe including Asiatic USSR. (Hoffman, George W.) N.Y., 1961.
Geog 3019.69.5 A geography of geography; origins and development of the discipline. (Fuson, Robert Henderson.) Dubuque, Iowa, 1969.
Geog 4180.11 Geography of Hudson's Bay. (Coats, W.) London, 1852.
An 359.63.5 The geography of intellect. (Weyl, N.) Chicago, 1963.
Geog 759.49 A geography of man. (James, P.E.) Boston, 1949.
Geog 759.49.2A A geography of man. 2nd ed. (James, P.E.) Boston, 1959.
Geog 3140.11F Geography of Ptolemy elucidated. (Rylands, T.G.) Dublin, 1893.
Geog 5160.20 Geography of the northlands. (Kimble, G.H.T.) N.Y., 1955.
Geog 5070.36A The geography of the polar regions. (Nordenskjöld, Otto.) N.Y., 1928.
Htn Geog 816.80.4* Geography rectified. 4th ed. (Morden, Robert.) London, 1700.
An 2059.14.5 The geological evidence of the antiquity of man. (Lyell, C.) London, 1914.
An 2058.63 The geological evidence of the antiquity of man. (Lyell, C.) Philadelphia, 1863.
An 2058.63.2 The geological evidence of the antiquity of man. 2d American ed. (Lyell, C.) Philadelphia, 1863.
An 2058.73 The geological evidence of the antiquity of man. 4. ed. (Lyell, C.) London, 1873.
Geog 5839.09.10 Geologiske og antivariske Iagttageker. (Steenstrup, K.) København, 1909.
An 379.31 Geology as an agent in human welfare. (Penrose, R.A.F.) N.Y., 1931.
An 2209.30 Geophagy. (Laufer, B.) Chicago, 1930.
Geog 500.25 Georg, Carl. Die Reiseliteratur Deutschlands. Leipzig, 1877.
An 2109.31.7 Georg, Eugen. The adventure of mankind. N.Y., 1931.
Geog 4502.5.28 Georg Jennerwein, der Wildschütz. (Stöger-Ostin, Georg.) München, 1939?
Geog 4324.59.5 Georg von Ehingen Reisen. (Pfeiffer, F.) Stuttgart, 1842.
Geog 4209.46.10 George, Albert J. The cap'n's wife. Syracuse, 1946.
Geog 4269.54.5 George, Pierre. L'Europe centrale. 1. éd. v.1-2. Paris, 1954.

Author and Title Listing

	Geog 4269.64.10	George, Pierre. Géographie de l'Europe centrale slave et danubienne. Paris, 1964.
	Geog 819.45.10	George, Pierre. Géographie sociale du monde. 6. éd. Paris, 1964.
	An 359.62.30	George, W.C. The biology of the race problem. N.Y., 1962.
	Geog 4308.50.12	George, W.C. A year abroad. Boston, 1852.
	Geog 604.4	George Davidson, geographer of the northwest coast of America. (Wagner, H.R.) n.p., 1932.
	Geog 604.4.5	George Davidson. (Lewis, Oscar.) California, 1954.
	Geog 5839.35	Georgi, J. Mid-ice. N.Y., 1935.
Htn	Geog 815.38*	Georgii Rithaymeri De orbis terrarvm sitr compendium. (Rithaymer, Georg.) Norimbergae, 1538.
	Geog 613.1.5	Gérard de Cremer ou Mercator. (Raemdonck, J. van.) St. Nicolas, 1870.
	Geog 613.1.3	Gérard Mercator. (Raemdonck, J. van.) St. Nicolas, 1869.
	Geog 613.1	Pamphlet vol. Gerardus Mercator.
	Geog 4239.23	Gerbault, Alain. Seul a travers l'Atlantique. Paris, 1924.
Htn	Geog 4266.65*	Gerbier, B. Subsidium peregrinatibus. Oxford, 1665.
	An 5.8	Gercke, Achim. Die Rasse im Schrifttum. Berlin, 1933.
	Geog 613.1.11	Gerhard Kremer...Mercator. (Breusing, A.) Duisburg, 1869.
	Geog 6008.97.2	Gerlache de Gomery, Adrien de. Quinze mois dans l'Antarctique. Bruxelles, 1902.
	Geog 6008.97.5	Gerlache de Gomery, Adrien de. Victoire sur la nuit antarctique. Tournai, 1960.
	Geog 6008.97	Gerlache de Gomery, Adrien de. Voyage de la Belgica, quinze mois dans l'Antarctique. Paris, 1902.
	Geog 6009.57	Gerlache de Gomery, Gaston de. Retour dans l'Antarctique; récit de l'Expédition antarctique belge, 1957-58. Tournai, 1960.
	An 358.75.3	Gerland, Georg. Anthropologische Beiträge. Halle, 1875.
	An 379.68.10	Gerling, Walter. Die Problematik der Sozialgeographie. Würzburg, 1968.
	Geog 5558.69.2	German Arctic expedition of 1869-70. (Koldewey, K.) London, 1874.
	An 359.44.15	Gérmen e cultura. (Mendes Correia, A.A.) Pôrto, 1944.
	Geog 5155.5	Gerritsz, H. Detectio freti Hudsoni. Amsterdam, 1878.
	Geog 4218.50	Gerstaecker, F. Narrative of a journey round the world. N.Y., 1853.
	Geog 3240.28	Gerwazy z Tilbury. (Strzelczyk, Jerzy.) Wrocław, 1970.
	Geog 4502.5.14	Das Gesäuse und seine Berge. 2. Aufl. (Gallhuber, J.) München, 1933.
	Geog 4502.5.25	Geschichte der Alpinen Literatur. (Dreyer, A.) München, 1938.
	Geog 5639.05	Geschichte der Entdeckung Grönlands von den ältesten Zeit bis zum Anfang des 19. Jahrhunderts. (Först, J.) Worms, 1906.
	Geog 3019.04	Geschichte der Erdkunde. (Günther, S.) Leipzig, 1904.
	Geog 3018.77.2	Geschichte der Erdkunde. (Peschel, O.) München, 1877.
	Geog 3018.61	Geschichte der Erdkunde. (Ritter, C.) Berlin, 1861.
	Geog 3019.70	Geschichte der geographischen Wissenschaft von den ersten Anfägen bis zur Ende des 18. Jahrhunderts. (Schmithuesen, Josef.) Mannheim, 1970.
	Geog 3070.40.5	Die Geschichte der Kartographie. (Bagrow, Leo.) Berlin, 1951.
	Geog 4131.60.5	Geschichte der Reisebüros. (Fuss, K.) Darmstadt, 1960.
	Geog 3590.15.15	Geschichte der russischer geographischer Entdeckungen. (Berg, L.S.) Leipzig, 1954.
	Geog 4239.27	Die Geschichte des Atlantischen Ozeans. (Ihering, H. von.) Jena, 1927.
	Geog 4477.51	Geschichte des Gaudenti di Lucca. Frankfurt, 1751.
	Geog 3248.58	Geschichte des Zeitalters. (Peschel, O.F.) Stuttgart, 1858.
	Geog 3248.58.5	Geschichte des Zeitalters. (Peschel, O.F.) Stuttgart, 1877.
	Geog 3248.81	Geschichte des Zeitalters der Entdeckungen. (Ruge, Sophus.) Berlin, 1881.
	An 2109.62.20	Geschlechtstrieb und Verdrängung bei den Primitiven. (Malinowski, B.) Reinbek bei Hamburg, 1962.
	An 2600.4	Gesell, Arnold. Wolf child and human child. N.Y., 1941.
	Geog 3590.67	Gesellschaft für Deutsch-Sowjetische Freundschaft. Beiträge aus der sowjetische Kartographie. Berlin, 1953.
	Geog 215.1	Gesellschaft für Erdkunde und Kolonialwesen zu Strassburg. Mitteilungen. Strassburg.
	Geog 500.40	Gesellschaft für Erdkunde zu Berlin. Bibliothek. Katalog der Bibliothek der Gesellschaft für Erdkunde zu Berlin. Berlin, 1903.
	Geog 135.2	Gesellschaft für Erdkunde zu Leipzig. Beiträge zur Geographie des festen wassers. v.1. Leipzig, 1891.
	Geog 135.1	Gesellschaft für Erdkunde zu Leipzig. Verein für Erdkunde. Jahresbericht. 1884-1941 24v.
	Geog 4502.5.30	Das Gesetz der Berge. (Mason, Alfred E.W.) München, 1940?
	An 3308.62F	Gesetz des Wachstumes...Proportionslehre. (Liharzih, F.) Wien, 1862.
	Geog 4309.28.5	Das Gesicht Europas. (Brandt, R.) Hamburg, 1928.
	Geog 5398.79.40	Geslin, Jules. L'expédition de la Jeannette au pôle nord, racontée par tous les membres de l'expédition. Paris, 1883? 2v.
	Geog 4219.33.15	Gezork, H. So sah ich die Welt. 12e Aufl. Kassel, 1938?
	An 359.60.15	Ghent. Université. Het lever; eer serie lezingen gehouden in de aula van de Rijksuniversität te Gent. Gent, 1960.
	Geog 4303.99.20A	Ghillebert de Lannoy in medieval Lithuania. (Klimas, Petras.) N.Y., 1945.
	Geog 5399.28.45	Ghost ship of the Pole. (Cross, Wilbur.) N.Y., 1960.
	Geog 5519.39	The ghost voyage. 1. ed. (Poncins, G.) Garden City, 1954.
	Geog 3240.20	Giacomo Bracelli nella storia della geografia. (Andriani, G.) Pontremoli, n.d.
	Geog 6009.49	Giaener, John. The white desert. London, 1954.
	Geog 5160.47.5	Giaever, John. Hardbalne polarkarer. 5. Oppl. Oslo, 1973.
	Geog 759.11	Giamnitrapani, D. Geografia mathematica, geografia generale. Firenze, 1911.
	Geog 4309.55	Gibbings, R. Trumpets from Montparnasse. London, 1955.
	An 628.68	Gibbs, G. Instructions for research...ethnology and philology of America. Washington, 1863-68. 3 pam.
	Geog 4659.57	Gibbs, James A. Shipwrecks of the Pacific Coast. Portland, Ore., 1957.
	Geog 4308.39	Gibson, William. Rambles in Europe in 1839. Philadelphia, 1841.
	Geog 10.2	Gidrologiia rushi. (Akademiia Nauk SSSR. Institut Nauchnoi Informatsii.) Moskva. 1963+
	Geog 5650.9	Giesecke, C.L. Norwegian settlements on...east coast of Greenland. Dublin, 1824.
	Geog 4248.82	Gilbay, Bernard. A voyage of pleasure. Cambridge, 1956.
	Geog 613.8	Gilbert, Edmund W. Sir Halford MacKinder, 1861-1947, an appreciation of his life and work. London, 1961.
	Geog 600.24	Gilbert, Edmund William. British pioneers in geography. Newton Abbot, 1972.
	Geog 818.93	Gilbert, Frank. The world, historical and actual. Chicago, 1893.
	An 130.20	Gilberto Freyne. (Hanke, Lewis.) N.Y., 1939.
	Geog 5182.6	Gilbertson, Albert N. Some ethical phases of Eskimo culture. Worcester, 1914.
	Geog 5398.81	Gilder, W.H. Ice-pack and tundra. London, 1883.
	Geog 5538.78.5	Gilder, W.H. Schwatka's search...Franklin records. N.Y., 1881.
	Geog 4679.67	Gill, Crispin. The wreck of the Torrey Canyon. Newton Abbot, 1967.
	An 349.48	Gillin, John. The ways of men. N.Y., 1948.
	Geog 4218.89.5	Gillis, C.J. Around the world in seven months. N.Y., 1891.
	Geog 4678.51.2	Gilly, W.O.S. Narratives of shipwrecks. 2. ed. London, 1851.
	Geog 4219.32.5	Gilmore, A.F. Yes, 'tis round. Boston, 1932.
	Geog 4309.61.10	Giloteaux, Paulin. Voyage à Moscou et au-dela. Paris, 1961.
	Geog 5333.5	Gino Watkins. (Scott, J.M.) London, 1935.
	An 378.24.5	Gioja, M. Riflessioni...su l'opera "L'homme du midi". Milano, 1830.
	Geog 88.25	Giornale popolare di viaggi. Weekly. Milano. 1-8,1871-1874 8v.
	Geog 5559.28.5	Un giornalista al Polo Nord. (Passarello, Gaetano.) Roma, 1965.
	Geog 4219.67	Gipsy Moth circles the world. (Chichester, Francis Charles.) N.Y., 1967.
	Geog 4219.24	A gipsy of the Horn. (Clements, Rex.) London, 1924.
	Geog 4219.24.2	A gipsy of the Horn. 2. ed. (Clements, Rex.) Boston, 1925.
	Geog 4209.66.20	Giramare. (Quilici, Folco.) Roma, 1966.
	An 358.00.5	Girando, J.M. Considérations sur les diverses méthodes à suivre dans l'observation des peuples sauvages. Paris, 1800.
	An 2058.83.5	Girard de Rialle, J. Nos ancêtres. Paris, 1883.
	An 358.81	Girard de Rialle, J. Les peuples de l'Asie et de l'Europe. Paris, 1881.
	Geog 4218.85	A girdle round the earth. (Richardson, D.N.) Chicago, 1888.
	Geog 4216.99	Giro del mondo. (Careri, G.) Napoli, 1699-1700. 6v.
	Geog 4218.76.15	Gita intorno alla terra dal gennaio al settembre dell'anno 1876. (Fenzi, S.) Firenze, 1877.
	Geog 5839.38.5	Gitz-Johansen, Aage. Skitzebogsblade fra Angmagssalik. København, 1938.
	Geog 500.80	Giuda bibliografica allo studio della geografia. (Migliorini, Elio.) Napoli, 1945.
	Geog 5399.28.5	Giudici, Davide. The tragedy of the Italian with the rescuers to the Red tent. N.Y., 1929.
	An 359.12.21	Giuffrida-Ruggeri, V. L'uomo come specie collettiva. Napoli, 1912.
	Geog 4509.31.10	Giussani, C. Chiacchiere di un alpinista. Milano, 1931.
	Geog 4509.56.25	Give me the hills. (Underhill, Miriam.) London, 1956.
	Geog 4209.58	Give me the world. (Hadley, Leila.) N.Y., 1958.
	Geog 4308.85.9	Gjellerup, Karl. Vandreaaret. Kjøbenhavn, 1885.
	An 359.56.20	Gjessing, Gutorm. Socio-culture. v.1,4. Oslo, 1956- 2v.
	An 379.67	Glacken, Clarence J. Traces on the Rhodian shore. Berkeley, 1967.
	An 379.73	Glacken, Clarence J. Traces on the Rhodian shore. Berkeley, 1973.
	Geog 4308.23.5	Glagolev, A.G. Zapiski russkago puteshestve norika [s 1823 po 1826 g.]. Sankt Peterburg, 1845. 4v.
	Geog 5700.12	Glahn, Henrik C. Missionaer i Grønland, Henric Christopher Glahns. København, 1921.
	Geog 4308.51.20	Glances at Europe. (Greely, H.) N.Y., 1851.
	Geog 4308.71.3	Glances on the wing at foreign lands. (Hoyt, James M.) Cleveland, 1872.
Htn	Geog 4206.82*	Glanius. A new voyage to the East Indies. 2d ed. London, 1682.
	Geog 755.27.10	Glarean sein Leben und seine Schriften. (Fritzsche, O.F.) Frauenfeld, 1890.
Htn	Geog 755.27*	Glareanus, H. De geographia liber unus. Basileae, 1527.
Htn	Geog 755.27.2*	Glareanus, H. De geographia liber unus. Friburgum, 1530.
	Geog 4218.81.5	Glass, C. The world, round it and over it. Toronto, 1881.
	Geog 4209.22	Glass, James. Chats over a pipe; a tale of two brothers. London, 1922.
	An 359.46.5	Glasshouse of prejudice. (Baruch, Dorothy W.) N.Y., 1946.
	Geog 4309.66.10	Glazami istorika. (Utchenko, Sergei L.) Moskva, 1966.
	Geog 500.85	Glazami sovetskikh liudel. (Moscow. Gosudarstvennyi Biblioteka SSSR imeni V.I. Lenin.) Moskva, 1956.
	Geog 4307.97.2	Gleanings through Wales. (Pratt, S.J.) Dublin, 1797. 2v.
	Geog 3055.87	Gleich, Sigismund von. Der Mensch der Eiszeit und Atlantis. Stuttgart, 1969.
	An 359.35.9	Die Gleichwertigkeit der europäischen Rassen und die Wege zu ihrer Vervollkommnung. (Weigner, K.) Prag, 1935.
	Geog 5399.35	Glen, A.R. Under the Pole star. London, 1937.
	Geog 4218.85.10	A glimpse of two hemispheres. (Arme-Webb, A. (Mrs.).) Hertford, 1885.
	Geog 4308.85.20	Glimpses by sea and land. (Evans, M.L.) Philadelphia, 1885.
	Geog 4308.58.3	Glimpses of Europe. Cincinnati, 1859.
	Geog 4308.51.10	Glimpses of Europe in 1851 and 1867-1868. (Huidskoper, A.) Meadville, Pa., 1882.
	Geog 4218.93.10	Glimpses of four continents. (Egerton, Alice A.) London, 1894.
	Geog 958.92F	Glimpses of the world. (Stoddard, John L.) Chicago, 1892.
	Geog 4228.86.10	Glimpses of three coasts. (Jackson, Helen F.H.) Boston, 1891.
	Geog 4309.00	Glimpses of three nations. (Steevens, G.W.) N.Y., 1900.
	Geog 4209.40.25	Glittering horizons. (Dennis, W.H.) N.Y., 1940.
	Geog 819.44	Global geography. (Renner, G.T.) N.Y., 1944.
	Geog 85.1.2	Le globe. Table des matières. Genève. 71-90,1932-1951
	Geog 4209.36.30	Le globe sous le bras. (Durtain, L.) Paris, 1936.
	Geog 4209.60.15	Globetrotter og hovedjaeger. (Bajsen-Moeller, Axel.) Vordingborg, 1960.
	Geog 89.1F	Globus; illustrirte Zeitschrift für Länder- und Völkerkunde. Braunschweig. 1-98,1862-1910 76v.
	Geog 819.38F	Globus geograficheskii eksegodnik dlia detei. Moskva, 1938.
	Geog 89.4	Der Globusfreund. Wien. 1-17 4v.
	Geog 4329.27	The glorious adventure. (Haliburton, R.) Indianapolis, 1927.
	Geog 579.66.5	Glosario de términos geográficos. 1. ed. (Soto Mora, Consuelo.) México, 1966.
	Geog 579.50F	Glossário taponimico da antiga historiografia. (Lagoa, J.A.M.J.) Lisboa, 1950- 4v.

Author and Title Listing

	Geog 579.61	A glossary of geographical terms. (British Association for the Advancement of Science.) London, 1961.
	An 2109.62.10	Gluckman, Max. Essays on the ritual of social relations. Manchester, 1962.
	An 2109.65	Gluckman, Max. Politics, law and ritual in tribal society. Oxford, 1965.
	Geog 4269.63	Glushchenk, I. Strany, strechi, uchenye. Moskva, 1963.
	An 359.68.25	Gluttons and libertines; human problems of being natural. (Bates, Marston.) N.Y., 1968.
	Geog 4145.68.5	Gnevusheva, E.I. Zaky tyi pute Shestvennik. Moskva, 1958.
	An 359.15.10	Gobineau, A. The inequality of human races. N.Y., 1915.
	Geog 607.15	Un Gobo Terrestre di Matteo Greuter conservato nella Biblioteca. (Bonasera, F.) Camarino? 1959.
	Geog 5559.57.5	God na poliuse. (Volovich, V.G.) Moskva, 1957.
	Geog 5939.62	God na poliuse kholoda. (Ignatov, V.S.) Moskva, 1962.
	Geog 4182.11	Godeé Molsbergen, E.C. Reizen in Zuid-Afrika. 's-Gravenhage, 1916-32. 4v.
	An 2109.74.15	Godelier, Maurice. Un domaine contesté: l'anthropologie économique. Paris, 1974.
Htn	Geog 4676.02*	Godinho Cardozo, M. Relaçam do naufragio da nao Santiago e itinerario da gente que delle se salvou. Lisboa, 1602. 3 pam.
	Geog 4249.54	The gods were kind. 1st ed. (Willis, William.) N.Y., 1955.
	Geog 4677.50	God's wonders in the great deep. (Bailey, J.) N.Y., 1750.
	Geog 5650.10.11	Godthaabsegnen og den gamle Vesterbygd. (Bruun, Daniel.) Odense, 1908.
	Geog 4476.38.1	Godwin, Francis. The man in the moon, 1638. London, 1971.
Htn	Geog 4477.35.3*	Godwin, Francis. The strange voyage and adventures of Domingo Gonsales. 2. ed. London, 1768.
	Geog 4209.67.5	Gody i strany. (Mikosha, Vladislav V.) Moskva, 1967.
	Geog 4227.46	Goelet, Francis. The voyages and travels of Francis Goelet, 1746-1758. N.Y., 1970.
Htn	Geog 4266.31*	Gölnitz, A. Ulysses belgico-gallicus...per Belgium hispan. Lugdunum Batavorum, 1631.
	Geog 4182.59	Goens, Rijklof van. De vijf gezantschapsreizen van Rijklof van Goens naar Hethof van Mataram. 's-Gravenhage, 1956.
	An 359.73.5	Goertz, Clifford. The interpretation of cultures. N.Y., 1973.
	An 170.221	Göttingen. Universität. Institut für Völkerkunde. Göttinger völkerkundliche Studien. Leipzig, 1939.
	Geog 90.15	Göttinger geographische Abhandlungen. Göttingen. 1,1948+ 19v.
	An 170.221	Göttinger völkerkundliche Studien. (Göttingen. Universität. Institut für Völkerkunde.) Leipzig, 1939.
	Geog 4229.42	Der göttliche Vagabund. (Faber, Kurt.) Stuttgart, 1942.
	Geog 4311.5	Götz, W. Das Donaugebeit mit Rücksicht. Stuttgart, 1882.
	An 359.27.5	Gohier, W.D. Cassandre, ou La folie des blancs. Paris, 1927.
	Geog 4105.18.4	Going abroad? 4. ed. (Luce, Robert.) Boston, 1906.
	Geog 3249.70	Gold, glory, and the gospel. 1. ed. (Wright, Louis Booker.) N.Y., 1970.
	Geog 4709.43	Gold from the sea; the epic story of the "Niagara's" bullion. (Taylor, James.) London, 1943.
	Geog 3057.6	The gold of Ophir. (Keane, Augustus Henry.) London, 1901.
	Geog 4209.41.15	Gold rush by sea. (Low, G.W.) Philadelphia, 1941.
	An 124.15	Golde, Peggy. Women in the field; anthropological experience. Chicago, 1970.
	An 170.270.5	The golden age of American anthropology. (Mead, M.) N.Y., 1960.
	Geog 3510.7	The golden book of the Dutch navigators. (Van Loon, H.W.) N.Y., 1916.
	Geog 619.9	Gol'denberg, Leonid A. Fedor Ivanovich Soimonov, 1692-1780. Moskva, 1966.
	Geog 618.6	Gol'denberg, Leonid A. Semen Ul'ianovich Remezov. Moskva, 1965.
	An 2109.22.3A	Goldenweiser, A.A. Early civilization. N.Y., 1922.
	An 2109.23.9	Goldenweiser, A.A. Early civilization. N.Y., 1923.
	An 2109.46	Goldenweiser, Alexander A. Anthropology; au introduction to primitive culture. N.Y., 1946.
	An 181.1.2	Goldenweiser, Alexander A. History, psychology, and culture. Gloucester, 1968.
	Geog 5312.7.10A	Golder, Frank A. Bering's voyages. N.Y., 1922-25. 2v.
	Geog 4208.96.10	Goldfields and chrysantemums. (Bond, C.) London, 1898.
	Geog 4329.55	Golding, Louis. Good-bye to Ithaca. London, 1955.
	Geog 4209.29	Goldring, Douglas. People and places. Boston, 1929.
	An 359.71.5	Goldsby, Richard A. Race and races. N.Y., 1971.
	An 359.59.5	Goldschmidt, W.R. Man's way. Cleveland, 1959.
Htn	Geog 5507.93*	Goldson, W. Observations on passage...Atlantic and Pacific. Portsmouth, 1793.
	An 3309.43	Goldstein, M.S. Demographic and bodily changes in descendants of Mexican immigrants. Austin, 1943.
	Geog 3249.64	Goldstein, Thomas. Fifteenth century geography against the background of medieval science. Salem, Mass., 1964.
	Geog 90.10	Goldthwaite's geographical magazine. N.Y. 2v.
	Geog 1228.17.10	Golovnin, Vasilii M. Puteshestvie vokrug sveta. Moskva, 1965.
	Geog 607.5.15	Golóvnin. (Davydov, Iurii Vl.) Moskva, 1968.
	Geog 4181.132	Gomes de Brito, Bernardo. Further selections from the tragic history of the sea, 1559-1565. Cambridge, 1968.
Htn	Geog 4677.35*	Gomes de Brito, Bernardo. História trágico-marítima. Lisboa, 1735. 3v.
	Geog 4672.20.4	Gomes de Brito, Bernardo. História trágico-marítima. v.1-5. Pôrto, 1936-37.
	Geog 4672.20.10	Gomes de Brito, Bernardo. Quadros da História trágico-marítima. Lisboa, 1944.
	Geog 4181.112A	Gomes de Brito, Bernardo. The tragic history of the sea. Cambridge, Eng., 1959.
	An 2109.58.5	Gómez-Moreno, Manuel. Adam y la prehistoria. Madrid, 1958.
	Geog 3570.22	Goncalões Niana, M. As viagens terrestres dos Portugueses. Porto, 1945.
	Geog 4180.14	Gonzalez de Mendoza, J. History of...kingdom of China. London, 1853-54. 2v.
	Geog 4418.03.15	Gonzalez y Rodriguez de la Peña, Hipolito. Viajes Ali Bey el Abbassi. Madrid, 1951.
	Geog 4329.55	Good-bye to Ithaca. (Golding, Louis.) London, 1955.
	Geog 4219.36	A good time. (Merrick, Lesley.) N.Y., 1936.
	An 359.75.10	Goodall, Vanne Morris. The quest for man. N.Y., 1975.
	An 170.279	Goodenough, Ward Hunt. Explorations in cultural anthropology; essays in honor of George Peter Murdock. N.Y., 1964.
	An 2109.39.6	Goodfellow, D.M. Principles of economic sociology. London, 1939.
	An 2109.39.5	Goodfellow, D.M. Principles of economic sociology. Philadelphia, 1939.
	Geog 5312.7.15	Goodhue, Cornelia. Journey into the fog; the story of Vitus Bering and the Bering Sea. Garden City, N.Y., 1944.
	Geog 4110.21	Goodrich, Charles A. The universal traveller. 2. ed. Hartford, 1836.
	Geog 3018.58	Goodrich, F.B. Man upon the sea. Philadelphia, 1858.
	Geog 818.40.4	Goodrich, S.G. A pictorial geography of the world. Boston, 1840.
	Geog 818.40.7	Goodrich, S.G. Pictorial geography of the world. Boston, 1856. 2v.
	Geog 818.40.5	Goodrich, S.G. Pictorial geography of the world. 2nd ed. Boston, 1840. 2v.
	Geog 818.40.6	Goodrich, S.G. Pictorial geography of the world. 3rd ed. Boston, 1840.
	Geog 818.32	Goodrich, S.G. A system of universal geography. Boston, 1832.
	Geog 818.33.5	Goodrich, S.G. A system of universal geography. 2d ed. Boston, 1833.
	Geog 4419.47	Goolid-Adams, R. Middle East journey. 1. ed. London, 1947.
	Geog 4209.38.10	Goose feathers. (Dighy, G.) N.Y., 1938.
	Geog 4332.17	Goracuchi, J.A. Die Adria und ihre Küsten. Triest, 1863.
	Geog 4129.25	Gorce, Denys. Les voyages l'hospitalité...dans le monde chrétien des IVe et Ve siècles. Wépion-sur-Meuse, 1925.
	Geog 4129.25.2	Gorce, Denys. Les voyages l'hospitalité...dans le monde chrétien des IVe et Ve siècles. Thèse. Wépion-sur-Meuse, 1925.
	Geog 5300.47	Gordienko, Pavel A. Die Polarforschung der Sowjetunion. Düsseldorf, 1967.
	Geog 4209.09.10	Gordon, F. Listy z podrozy. Chicago, 1910.
	Geog 4217.85.15	Gordon, Mona. The mystery of La Pérouse. Christchurch, N.Z., 1961.
	Geog 817.04	Gordon, P. Geography anatomized. 4th ed. London, 1704.
Htn	Geog 817.04.6*	Gordon, P. Geography anatomized. 6th ed. London, 1711.
	Geog 817.04.8	Gordon, P. Geography anatomized. 8th ed. London, 1719.
	Geog 817.04.12	Gordon, P. Geography anatomized. 12th ed. London, 1730.
	Geog 817.04.19	Gordon, P. Geography anatomized. 19th ed. London, 1749.
	Geog 817.04.20	Gordon, P. Geography anatomized. 20th ed. London, 1754.
	Geog 5209.07	Gordon, W.J. Round about the North Pole. N.Y., 1907.
	Geog 817.89	Gordon, William. New geographical grammar. Edinburgh, 1789.
	An 2109.67.5	Goremykina, Vera I. Istoriia pervobytnogo obshchestva. Minsk, 1967.
	Geog 4209.71.5	Górnicki, Wiesław. Opowieści zdyszane. 1. wyd. Warszawa, 1971.
	Geog 4209.61.5	Goroda i vstrechi. (Kazakova, M.S.) Riga, 1961.
	Geog 4180.96A	Gosch, C.C.A. Danish Arctic expeditions, 1605-20. London, 1897. 2v.
	Geog 4709.32	Gosse, Philip. The history of piracy. N.Y., 1946.
	Geog 4709.26	Gosse, Philip. My pirate library. London, 1926.
	Geog 3127.98	Gosselin, P.F.J. Recherches sur la géographie...des anciens. Paris, 1798. 2v.
	Geog 3140.15	Gosselin, P.Z.J. De l'évaluation et de l'emploi des mesures itinéraires grecques et romaines. Paris, 1813.
	Geog 4209.57	Gosset, R.P. Retour du bout du monde. Paris, 1957.
	Geog 4419.68	Gotev, Goran. Probuzhdane to na Shekherazada. Sofiia, 1968.
	Geog 4269.50	Gottmann, J. A geography of Europe. N.Y., 1950.
	Geog 4269.50.5	Gottmann, J. A geography of Europe. N.Y., 1961.
	Geog 665.146	Gottmann, Jean. The renewal of the geographic environment: an inaugural lecture delivered before the University of Oxford on 11 February 1969. Oxford, 1969.
	Geog 4677.99	Goud in de golven. (Molen, Sytze Jan van der.) 's-Gravenhage, 1965.
	Geog 6009.28.25	Gould, L.M. Cold, the record of an Antarctic sledge journey. N.Y., 1931.
	Geog 5100.10	Gould, Laurence McKinley. The polar regions in their relation to human affairs. N.Y., 1958.
	Geog 5925.45	Gould, Lawrence M. Antarctica in world affairs. N.Y., 1958.
	Geog 665.168	Gould, Peter R. Mental maps. Harmondsworth, 1974.
	Geog 5919.08	Gourdon, Ernest. Géographie physique, glaciologie, pétrographie des régions visitées par l'Expédition antarctique française...1903-05. Paris, 1908.
	An 379.73.15	Gourou, Jierre. Pour une géographie humaine. Paris, 1973.
	Geog 4759.47	Gourou, P. Les pays tropicaux. Paris, 1947.
	Geog 4759.71.5	Gourou, Pierre. Leçons de géographie tropicale. Paris, 1971.
	Geog 4759.47.3	Gourou, Pierre. Les pays tropicaux, principes d'une géographie humaine et économique. 4. éd. Paris, 1966.
	Geog 4759.47.4	Gourou, Pierre. The tropical world. N.Y., 1966.
	Geog 4759.53	Gourou, Pierre. The tropical world. 2. ed. London, 1958.
	Geog 4209.56.10	Gouy, Robert. Terres de l'amitié. Neuchâtel, 1956.
	Geog 5838.25	Graah, W.A. Beskrivelse...Grønland. Kjøbenhavn, 1825.
	Geog 5838.28.3	Graah, W.A. Narrative of expedition to...Greenland. London, 1837.
	Geog 5838.28	Graah, W.A. Undersogelses-reise...Grønland. Kjøbenhavn, 1832.
	Geog 4502.5.27	Graber, A. Der Weg zum Berg. München, 1939?
	Geog 4679.12.3	Gracie, A. The truth about the Titanic. N.Y., 1913.
	An 2109.24.9	Graebner, F. Das Weltbild der Primitiven. München, 1924.
	An 629.11	Graebner, Fritz. Methode der Ethnologie. Heidelberg, 1911.
	Geog 5839.29	Graenlandsför 1929. (Arnason, A.) Reykjavik, 1929.
	Geog 5650.5	Graenlendinga saga. (Jonsson, Finnur.) Kaupmannahöfn, 1899.
Htn	Geog 578.61.5*	Graesse, J.G.T. Orbis latinus. Dresden, 1861.
	Geog 578.61	Graesse, J.G.T. Orbis latinus. Dresden, 1861.
	An 170.341.5A	Grafton Elliot Smith: the man and his work. Portland, Ore., 1974.
	Geog 4219.65	Graham, Robin Lee. Dove. 1. ed. N.Y., 1972.
	Geog 4309.21.15A	Graham, S. Europe - whither bound? N.Y., 1922.
	Geog 4559.32	Grain race. (Villiers, A.J.) N.Y., 1933.
	Geog 3108.39	Grammar of ancient geography. (Arrowsmith, A.) London, 1839.
	An 2109.60.15	Grand, P.M. Découverte de la préhistoire. Paris, 1960.
	Geog 577.26F	Le grand dictionnaire géographique. (Bruzen de la Martiniere, A.A.) Haye, 1726-39. 10v.
	Geog 577.68F	Le grand dictionnaire géographique. (Bruzen de la Martiniere, A.A.) Paris, 1768. 6v.
	Geog 577.60F	Grand gazetteer, or topographical dictionary. (Brice, A.) n.p., n.d.
	Geog 4145.70	The grand peregrination; being the life of...Fernão Mendes Pinto. (Collis, Maurice.) London, 1949.
	Geog 4307.56	Grand tour...through Netherlands. (Nugent, T.) London, 1756. 4v.
	Geog 4279.69.5	The grand tour. (Hibbert, Christopher.) London, 1969.
	Geog 4279.67	The grand tour. (Trease, Geoffrey.) London, 1967.

Call Number	Entry
Geog 4129.14.5	The grand tour in the 18th century. (Mead, William E.) Boston, 1914.
Geog 4219.28.10	La grande croisière de Costes et Le Brie. (Costes Dieudonné.) Paris, 1928.
NEDL Geog 3019.59.15	Grandes exploradores, en la mar. (Cabal, Juan.) Barcelona, 1959.
An 359.01	Les grandes routes des peuples. (Demolins, E.) Paris, 1901. 2v.
Htn Geog 3550.12*	I grandi navigatori italiani. (Franciulli, G.) Roma, 1930.
Geog 3248.82	Les grands decouvertes maritimes. (Cat, E.) Paris, 1882.
Geog 3018.82.5	Les grands faits de l'histoire de géographie. (Dussieux, L.E.) Paris, 1882-83. 5v.
Geog 3140.16	Grands voyages de découvertes des anciens. (Antichan, P.N.) Paris, 1888.
Geog 4209.40.50	Gran'ma goes by freight. (Simon, C.R.) Placerville, Calif., 1940.
Geog 4559.34	Grant, George H. Consigned to Davy Jones. Boston, 1934.
Geog 4218.80.7	Grant's tour around the world. (Packard, J.F.) Cincinnati, 1880.
Htn Geog 4105.12*	Gratarolo, G. De regimine iter agientium. Basileae, 1561.
Geog 1616.20.2	Gratis supplement to the fifth edition of Baedeker's Egypt. (Egyptian Museum, Cairo.) Leipzig, 1904.
Geog 603.1	Graubner, Paul. Fr. Cannabich (1777-1859) sein Leben und sein Werke. Koniejeberg, 1913.
Geog 809.22F	Grauper, Ernest. Nouvelle géographie universelle. pt.1-10. Paris, 1922. 2v.
Geog 616.1.3	Gravier, G. Notice sur Jean Parmentier. Rouen, 1902.
Geog 3228.80	Gravier, S. La cosmographie avant la decouverte de l'Amerique. Paris, 1880.
Geog 4169.47	Great adventures and explorations. (Stefansson, V.) N.Y., 1947.
Geog 4019.63	Great adventures with NationalGeographic. (National Geographic Society, Washington.) Washington, 1963.
Geog 3019.56.2	The great age of discovery. (Herrmann, Paul.) N.Y., 1958.
Geog 1508.5	Great Britain. (Baedeker, publishers.) Leipsic, 1887.
Geog 1508.34	Great Britain. (Baedeker, publishers.) N.Y., 1937.
Geog 3560.5	Great Britain. India Office. Map of the world, commonly known as the second Borgian map. n.p., 1889.
Geog 1508.11	Great Britain. 2. ed. (Baedeker, publishers.) Leipsic, 1890.
Geog 1508.10	Great Britain. 2. ed. (Baedeker, publishers.) Leipsic, 1890.
NEDL Geog 1508.12	Great Britain. 3. ed. (Baedeker, publishers.) Leipsic, 1894.
Geog 1508.13	Great Britain. 3. ed. (Baedeker, publishers.) Leipsic, 1894.
Geog 1508.15	Great Britain. 4. ed. (Baedeker, publishers.) Leipsic, 1897.
Geog 1508.16	Great Britain. 4. ed. (Baedeker, publishers.) Leipsic, 1897.
Geog 1508.20	Great Britain. 5. ed. (Baedeker, publishers.) Leipsic, 1901.
Geog 1508.25	Great Britain. 6. ed. (Baedeker, publishers.) Leipsic, 1906.
Geog 1508.30	Great Britain. 7. ed. (Baedeker, publishers.) Leipzig, 1910.
Geog 1508.32	Great Britain. 8. ed. (Baedeker, publishers.) Leipzig, 1927.
Geog 4139.64	The great Chinese travelers, an anthology. (Mirsky, J.) N.Y., 1964.
Geog 3249.54	The great discoveries and the first colonial empires. (Nowell, C.E.) Ithaca, 1954.
Geog 5558.75.7	The great frozen sea...1875-76. (Markham, A.H.) London, 1878.
Geog 4659.52	Great gales and dire disasters. (Snow, Edgar R.) N.Y., 1952.
Geog 4659.60	Great Lakes shipwrecks and survivals. (Ratigan, William.) Grand Rapids, 1960.
Geog 4659.60.2	Great Lakes shipwrecks and survivals. 2. ed. (Ratigan, William.) Grand Rapids, 1969.
Geog 4678.72.15	The great Mary Celeste hoax. (Keating, Laurence J.) London, 1929.
Geog 4138.30	Great navigators and discoverers. (Brendon, J.A.) N.Y., 1930.
Geog 3602.5	Great Norwegian expeditions. (Richter, Sørea.) Oslo, 1955.
Geog 4709.22	Great pirate stories. (French, J.L.) N.Y., 1922.
Geog 4709.22.2	Great pirate stories. 2. ser. (French, J.L.) N.Y., 1925.
Geog 5100.5	The great polar current. (Prentiss, H.M.) Cambridge, 1897.
Geog 4279.71	The great resorts. (Bocca, Geoffrey.) N.Y., 1971.
Geog 4679.27	A great sea mystery. (Lockhart, John Gilbert.) London, 1927.
Geog 4672.18	Great shipwrecks and castaways; authentic accounts of adventures at sea. (Neider, Charles.) N.Y., 1952.
Geog 4659.67	Great shipwrecks off the coast of southern Africa. (Burman, Jose.) Cape Town, 1967.
Geog 4208.92	Great streets of the world. London, 1892.
Geog 4559.55	Great true adventures. (Thomas, L.J.) N.Y., 1955.
Geog 6009.01.25	Great waters; a voyage of natural history to study whales, plankton and the waters of the southern ocean. (Hardy, Alister C.) N.Y., 1967.
Geog 5209.10	The great white North. (Wright, H.S.S. (Mrs.).) N.Y., 1910.
Geog 6009.10.21	The great white South. (Ponting, H.G.) London, 1922.
Geog 665.37.10	Greatest wonders of the world. (Singleton, E.) N.Y., 1906.
Geog 1585.10	Greece. (Baedeker, publishers.) Leipsic, 1889.
Geog 1585.15	Greece. 2. ed. (Baedeker, publishers.) Leipzig, 1894.
Geog 1585.26	Greece. 4. ed. (Baedeker, publishers.) Leipzig, 1909.
Geog 1585.25	Greece. 4. ed. (Baedeker, publishers.) Leipzig, 1909.
Geog 618.3.5	Greef, Guillaume J. de. Discours prononcé par Monsieur le recteur Guillaume de Greef. Gand, 1906.
Geog 5839.35.7	Greeland journal. (Kent, Rockwell.) N.Y., 1962.
Htn Geog 665.28*	Pamphlet box. Greeley, A.W. Essays on geographical subjects.
Geog 4138.93	Greely, A.W. Explorers and travellers. N.Y., 1893.
Geog 5558.81.10	Greely, A.W. The Greely Arctic expedition. Philadelphia, 1884.
Geog 5208.95	Greely, A.W. Handbook of arctic discoveries. Boston, 1896.
Geog 5208.95.3	Greely, A.W. Handbook of polar discoveries. 3. ed. Boston, 1906.
Geog 5208.95.7	Greely, A.W. Handbook of polar discoveries. 5th ed. Boston, 1910.
Geog 5070.30	Greely, A.W. The polar regions in the twentieth century. Boston, 1928.
Geog 5317.10	Greely, A.W. Reminiscences of adventure and service. N.Y., 1927.
Geog 5558.81.3	Greely, A.W. Three years of Arctic service...1881-84. N.Y., 1886. 2v.
Geog 4308.51.20	Greely, H. Glances at Europe. N.Y., 1851.
Geog 5558.81.10	The Greely Arctic expedition. (Greely, A.W.) Philadelphia, 1884.
Geog 5558.81.15	Greely relief expedition. (McGinley, W.A.) Washington, 1884.
Geog 4559.44.5	Green, Alfred. Jottings from a cruise. Seattle, 1944.
Geog 5326.21A	Green, Fitzhugh. Peary; the man who refused to fail. N.Y., 1926.
Geog 4157.45	Green, J. New collection of voyages and travels. London, 1745-47. 4v.
Geog 4209.63.35	Green, Lawrence George. A decent fellow doesn't work. Cape Town, 1963.
Geog 4028.2	Green, Timothy. The adventurers; four profiles of contemporary travellers. London, 1970.
Geog 5559.48	Green seas and white ice. (MacMillan, M.) N.Y., 1948.
Geog 4129.58	Greenen, E. Reisen seit Anno dazumal. Hamburg, 1958.
Geog 5839.30.10F	Greenland; its natures, inhabitants and history. (Krabbe, Thomas N.) Copenhagen, 1930.
Geog 5645.5.5	Greenland; 1952?
Geog 5837.78	Greenland...extracts...journal. (Saabye, Hans E.) London, 1818.
Geog 5839.28	Greenland...invest in Greenland. (Denmark. Kommittee for Ledelsen af Degeoliske og Geografiske Undersøgelser i Grønland.) Copenhagen, 1928. 3v.
Geog 5558.17.3	Greenland...Northwest Passage. (O'Reilly, B.) London, 1818.
Geog 5558.17	Greenland...Northwest Passage. (O'Reilly, B.) N.Y., 1818.
Geog 5839.75	Greenland. (Banks, Mike.) Newton Abbot, 1975.
Geog 5839.52	Greenland. (Denmark. Udenrigsministeriet.) Ringkjøbing, 1952?
Geog 5606.8	Greenland. Landsraadet. Kalatalet-nunâne landsrådip okalokatigîssutai. 1956-1958 2v.
Geog 5660.26A	Greenland. 1. ed. (Stefansson, V.) Garden City, 1942.
Geog 5839.21	Greenland by the Polar Sea. (Rasmussen, Knud.) London, 1921.
Geog 5838.96	Greenland icefields. (Wright, G.F.) N.Y., 1896.
Geog 5839.30.5	Greenland journey; the story of Wegener's German expedition...1930-31. (Wegener, E.) London, 1939.
Geog 5639.70	Greenland past and present. Copenhagen, 1970.
Geog 4181.81	Greenlee, William B. The voyage of Pedro Alvares Cabral to Brazil and India. London, 1938.
An 2108.79.5	Greenwood, James. The wild man at home, or Pictures of life in savage lands. London, 1879.
An 359.25.4	Gregory, John W. The menace of colour. 2. ed. London, 1925.
An 359.31.3	Gregory, John W. Race as a political factor. London, 1931.
Geog 659.63.2	Gregory, Stanley. Statistical methods and the geographer. 2nd ed. London, 1971.
Geog 91.1	Greifswald. Geographische Gesellschaft. Jahresbericht. Greifswald. 1-60,1882-1942 16v.
Geog 4678.47	Greig, A.M. Fate of Blenden Hall, East Indiaman. N.Y., 1847.
Geog 5639.50.5	Grenlandiia: sbornik statei. (Birket-Smith, K.) Moskva, 1953.
Geog 192.4	Grenoble. Université. Institut de Géographie Alpine. Revue de géographie alpine. 1,1913+ 61v.
Geog 192.4.5	Grenoble. Université. Institut de Géographie Alpine. Table décennale. v.2 (1923-32), v.5 (1953-62) Grenoble, 1932-63. 3v.
Geog 4209.71.40	Grenseløs. (Knudsen, Kim.) Oslo, 1971.
Geog 3019.62	Grenville, J.A.S. The coming of the Europeans. London, 1962.
Geog 5324.2.105	Greve, Tim. Fridtjof Nansen. Oslo, 1973-74. 2v.
Geog 4709.33	Grey, Charles. Pirates of the Eastern seas, 1618-1723. London, 1933.
An 359.57.15	Griaule, Marcel. Méthode de l'ethnographie. Paris, 1957.
Geog 759.23	Gribandi, P. Il mondo e l'Italia. 3. éd. v.1-2. Torino, 1923.
Geog 4508.99A	Gribble, F. The early mountaineers. London, 1899.
X Cg Geog 1585.5	Griechenland. (Baedeker, publishers.) Leipzig, 1883.
An 359.61.20	Grieger, Paul. La caracterologie ethnique. Paris, 1961.
Geog 5130.2	Grierson, J. Challenge to the poles. Hamden, Conn., 1964.
Geog 5333.15	Grierson, John. Sir Hubert Wilkins. London, 1960.
Geog 4328.05	Griffiths, J. Travels in Europe, Asia Minor, and Arabia. London, 1805.
Geog 5160.25	Grigor'ev, A.A. Subarktika. 2. Izd. Moskva, 1956.
Geog 759.65	Grigor'ev, Andrei A. Razvitie teoreticheckikh problem sovetskoi fizicheskoi geografii, 1917-1934 gg. Moskva, 1965.
Geog 5399.32.8	Grigor'ev, Origonii K. Dorogi vedut v Arktiku. Moskva, 1969.
Geog 4145.71	Grigorii Nikolaenin Potanin; zhizn' i deiatel'nost'. (Obruchev, V.A.) Moskva, 1947.
Geog 579.56	Grigson, G. Places. N.Y., 1956?
Geog 5225.5	Grinnell land. (Fore, P.) Washington, 1852.
Geog 4309.32.5	Grisar, Erich. Mit Kamera und Schreibonaschine durch Europa. Berlin, 1932.
Geog 4308.18.3	Griscom, J. A year in Europe. N.Y., 1823.
Geog 4308.18.4	Griscom, J. A year in Europe. 2nd ed. N.Y., 1824. 2v.
Geog 4168.90	Griswold, W.M. Travel, series of narratives of...visits. v.1-106. Cambridge, 1890. 2v.
Geog 4209.68.10	Gritskevich, Valentin P. Puteshestviia nashikh zemliakov. Minsk, 1968.
Geog 5838.72.5	Grønlaendervennen Hinrich Rink. (Oldendow, K.) København, 1955.
Geog 5838.83.3	Grønland; seine Eiswüsten im Innern. (Nordenskiold, A.E.) Leipzig, 1886.
Geog 5606.5	Grønland; tidsskrift for dunsk-grønlandsk samvirhe. Hellerup. 1953+ 16v.
Geog 5606.6	Grønland; turist foreningen for Danmark årbog. Ringkjøbing. 1952-1953
Geog 5839.03	Grønland...1903-04. (Mylius, L.) Kjøbenhavn, 1906.
Geog 5606.5.5	Grønland. Inholdsfortegnelse, 1953-62. n.p., n.d.
Geog 5838.72	Grønland dess natur och innevanare. (Fries, T.M.) Upsala, 1870.
Geog 5839.50	Grønland foran en ny epoke. (Bomholt, Julius.) København, 1961.
Geog 5838.57.3	Grønland geographisch...beschreiben. (Rink, Hinrich.) Stuttgart, 1860.
Geog 5838.57	Grønland geographisk. (Rink, Hinrich.) Kjøbenhavn, 1857. 2v.

193

Author and Title Listing

Geog 5839.16 Grønland langs Polhavet. (Rasmussen, Knud.) Kristiania, 1919.
Geog 5845.2 Grønland og dets problemer. (Rasmussen, Holger.) København, 1947.
Geog 5839.27.5 Grønland som Nybyggerland. (Bendixen, Ole.) København, 1927.
Geog 5839.68 Grönland tvärs. (Boucht, Christer.) Helsingfors, 1968.
Geog 5660.37 Grönland und der Norden. (Smedal, G.) Oslo? 1942?
Geog 5839.48.5 Groenland 1948-1949-1950: astronomie, nivellement géodésique sur l'Inlandsis. (Tschaen, Louis.) Paris, 1959.
Geog 5639.28 Grønlands betydning. (Berlin, Knud.) København, 1928.
Geog 5639.50 Grønlands bogen. (Birket-Smith, K.) København, 1950. 2v.
Geog 5845.6 Grønlands fremtid. (Oldendow, Knud.) København, 1945.
Geog 5838.98 Grønlands gamle topografi. (Jonsson, F.) Kjøbenhavn, 1899.
Geog 5639.46.5 Grønlands historie. (Gad, Finn.) København, 1946.
Geog 5639.67 Grønlands historie. (Gad, Finn.) København, 1967- 2v.
Geog 5650.7 Grönlands historiske mindesmaerker. Kjobenhavn, 1838-45. 3v.
Geog 5650.12 Grønlands statsretslige stilling i Midelalderen. (Dúason, Jón.) Oslo, 1928.
Geog 5837.20 Groenlandsche visschery. (Zorgdrager, C.G.) Amsterdam, 1720.
Geog 5839.44 Grønlandsfaerd. (Draatrup, E.) København, 1944.
Geog 5600.7 Groenlandsk avis og tidsskrift index. Godthåb. 1962//
Geog 5838.84 Den grønlandske Kateket Hansêraks Dagbog. (Hansen, Johannes.) København, 1933.
Geog 5700.4.10 Grønlandske relationer. (Dalager, Lars.) København, 1715.
Geog 5606.12 Det Groenlandske samfund. København. 1,1967//
Geog 91.4 Grønlandske Selskab. Aarskrift. Kjøbenhavn. 1910-1952 10v.
Geog 5838.49 En grønlandspraests optegnelser, 1844-49. (Janssen, Carl Emil.) København, 1913.
Geog 5670.3 Grønlandstraktaten, belyst gennem de i dagspressen fremkomme artikler. (Bendixen, Ole.) København, 1924.
Geog 5639.36.4 Den grønlanske handels og kolonisations historie indtil 1870. (Bobé, Louis.) København, 1936.
Geog 5399.61 Gromov, L.V. Ostrov Vrangelia. Magadan, 1961.
Htn Geog 5837.06.2* Gronlandia antiqua. (Torfaeo, T.) Havniae, 1706.
Geog 5837.06 Gronlandia antiqua. (Torfaeo, T.) Havniae, 1706.
Geog 4328.91 Groome, P.L. Rambles...in three continents. Greensboro, 1891.
Geog 5837.77 Groot, J.J. Beknopt en getrouw verhaal. Amsterdam, 1779.
Geog 3235.65 Gross, Hans. Zur Entstehungs-Geschichte der Tabula Purtingeriana. Diss. Bonn, 1913.
Geog 4308.36.10 Gross-Hoffinger, A.J. Empfindsame Reise eines expatriirten Schwärmers. Leipzig, 1836.
An 2109.72.40 Das grosse Bilderlexikon des Menschen in der Vorzeit. (Jelinek, Jan.) Gütersloh, 1972.
Geog 807.87.2 Grosse Erdbeschreibung. (Büsching, Anton Friedrich.) Troppau, 1784-1787. 24v.
Geog 3019.33.10 Grosse Forschungsreisende. (Banse, Ewald.) München, 1933.
Geog 4209.37.10 Die grosse Mauer; das Buch einer Weltreise. (Wechsberg, J.) Leipzig, 1937.
Geog 4309.27.15 Das grosse Reisebuch. (Edschmid, Kasimir.) Berlin, 1927.
Geog 4219.56 Grosse Weltreise mit A.E Johann (pseud.). 1. Aufl. (Wollschläger, A.) Gutersloh, 1956.
Geog 3019.04.3 Die grossen Reisesammlungen des 16. Jahrhunderts und ihre Bedeutung. (Böhme, Max.) Strassburg, 1904.
Geog 819.02 Grosses Lehrbuch der Geographie. (Seydlitz, Ernst von.) Breslau, 1902.
Geog 4502.5.9 Der Grossvenediger in der Alpinismus. (Knorr, O.) München, 1932.
Geog 959.07 Grosvenor, G.H. Scenes from every land. Washington, D.C., 1907.
Geog 959.07.15 Grosvenor, G.H. Scenes from every land. Washington, D.C., 1918.
Geog 959.07.5 Grosvenor, G.H. Scenes from every land. 2. ser. Washington, D.C., 1909.
An 359.64.35 Grottanelli, Virigi L. L'etnologia e le leggi della condotta umana. Roma, 1964.
Geog 819.21.3 The groundwork of modern geography. (Wilmore, Albert.) London, 1921.
An 359.12.7 Grow, O. The antagonism of races. Waterloo, 1912.
An 3308.79 The growth of children. (Bowditch, J.P.) Boston, 1879.
An 3309.35 The growth of the surface area of the human body. (Boyd, Edith.) Minneapolis, 1935.
Geog 808.50.9 Grube, A.W. Geographische Charakterbilder in algerundeten Gemälden aus der Länder- und Völkerkunde. Leipzig, 1868. 3v.
Geog 4131.34 Grünthal, Adolf. Probleme der Fremdenverkehrsgeographie. Diss. Berlin, 1934.
Geog 4145.34.2 Grum-Grzhimailo, A.G. Dela i dni G.E. Grum-Grzhimailo. Moskva, 1947.
Geog 818.59.7 Grundriss der Geographie und Geschichte. (Pütz, Wilhelm.) Koblenz, 1866.
Geog 818.59.6 Grundriss der Geographie und Geschichte der...Zeit. v.1-3. 10. Aufl, 12. Aufl. (Pütz, Wilhelm.) Koblenz, 1865-1867.
An 3558.90 Grundz. einer systematischen Kraniometrie. (Török, A.) Stuttgart, 1890.
An 359.50.5 Grundzäge der Völkerbiologie. (Schwidetzky, Ilse.) Stuttgart, 1950.
An 358.59.3 Grundzüge der Ethnographie. (Perty, M.) Leipzig, 1859.
Geog 809.07 Grundzüge der Landerkunde. (Hettner, A.) Leipzig, 1907-1925. 2v.
An 2109.56.5 Gruzinskii narodnyi transport. (Gegeshidze, M.K.) Tbilisi, 1956.
Geog 4181.21 The guanches of Tenerife. (Espinosa, A. de.) London, 1907.
Geog 3070.95 Guarnieri, Gino. Geografia e cartografia nautica nella loro evoluzione storica e scientifica. Genova, 1956.
Geog 3129.68 Guarnieri, Giuseppe Gino. Le correnti del pensiero geografico nell'antichità classica e il lorocontributo alla cartografia nautica medioevale. Pisa, 1968-69. 2v.
Geog 3550.4 Gubernatis, A. de. Memoria ai viaggiatori italiani. Firenze, 1867.
Geog 3550.5 Gubernatis, A. de. Storia dei viaggiatori italiani. Livorno, 1875.
Geog 4509.63 Gudhjohnsen, P. Endurminningar fjallgöngumanns. Reykjavik, 1963.
Geog 3030.3F Guénin, E. La route de l'Inde. Paris, 1903.
An 359.29.7 Günther, H.F.K. Rassenkunde Europas. 3. Aufl. München, 1929.
An 379.28.5 Günther, Hans F.K. The racial elements of European history. N.Y., 1928.
An 379.26.5 Günther, Hans F.K. Rasse und Stil. München, 1926.
An 379.26.7 Günther, Hans F.K. Rasse und Stil. 2. Aufl. München, 1927.
Geog 3019.02 Günther, S. Entdeckungsgeschichte. Berlin, 1902.
Geog 665.17 Günther, S. Geographische Studien. Stuttgart, 1907.
Geog 3019.04 Günther, S. Geschichte der Erdkunde. Leipzig, 1904.
Geog 758.90 Günther, S. Handbuch der Mathematischen Geographie. Stuttgart, 1890.
Geog 4300.16 Guerber, H.A. How to prepare for Europe. N.Y., 1906.
Geog 4678.42 Guerrazzi, F.D. Replica...in causa di abbordaggio. Livorno, 1842.
An 2109.09.9 Der Güterverkehr in der Urgesellschaft. (Somló, Felix.) Bruxelles, 1909.
An 359.08.5 Guide...specimens illustrating the races of mankind. (British Museum (Natural History). Department of Zoology.) London, 1908.
Geog 4300.22 Guide classique du voyageur. (Audin, J.M.V.) Paris, 1828-1829. 2v.
Geog 4559.70 Guide de la mer mystérieuse. (Bertino, Serge.) Paris, 1970.
Geog 535.15 Guide de la recherche géographique en Belgique. (Denis, Jacques.) Namur, 1970.
Geog 659.42 Guide de l'étudiant en géographie. 1. éd. (Cholley, André.) Paris, 1942.
Geog 4300.36 Guide for Europe. (Nederlandsch-Amerikaansche Stoomvaart Maatschappij, N.V.) The Hague, 1925.
Geog 4300.34 Guide to Europe. (New York Herald Tribune. European Edition.) Paris?
Geog 4300.40 Guide to Europe and the Mediterranean. 1st ed. (Joseph, Richard.) Garden City, N.Y., 1956.
An 3309.27 Guide to physical anthropometry and anthroposcopy. (Davenport, C.B.) Baltimore, 1927.
An 609.22 Guide to the fossil remains of man. 3. ed. (British Museum (Natural History). Department of Geology.) London, 1922.
An 2059.15 A guide to the fossil remains of man in the department. (British Museum (Natural History). Department of Geology.) London, 1915.
An 359.12.11 Guide to the specimens...races of mankind. 2. ed. (British Museum (Natural History). Department of Zoology.) London, 1912.
An 609.21 Guide to the specimens illustrating the races of mankind. 4. ed. (British Museum (Natural History). Department of Zoology.) London, 1921.
Geog 4308.77A Guild, C. Abroad again. Boston, 1877.
Geog 4308.71.2 Guild, C. Over the ocean, or Sights and scenes in foreign lands. Boston, 1875.
Geog 4308.71A Guild, C. Over the ocean. Boston, 1871.
Geog 4303.99.5 Guillebert de Lannoy et ses voyages en 1413, 1414 et 1421. (Lelewel, J.) Bruxelles, 1844.
Geog 4218.73.9 Guillemard, A.C. Over land and sea. London, 1875.
Geog 4248.86 Guillemard, F.H.H. Cruise of the Marchesa. London, 1886. 2v.
Geog 4248.86.5 Guillemard, F.H.H. Cruise of the Marchesa to Kamschatka and New Guinea. 2nd ed. London, 1889.
Geog 5660.42 Gulloev, H.C. Igdlucrúnerit. København, 1971.
Geog 607.16 Gunchev, Guncho S. Guncho Gunchev. Sofiia, 1941.
Geog 607.16 Guncho Gunchev. (Gunchev, Guncho S.) Sofiia, 1941.
Geog 4110.38 Gunther, John. Twelve cities. 1. ed. N.Y., 1969.
An 2159.73.15 Gur'ev, Dmitrii V. Stanovlenie obshchestvennego proizvodstva. Moskva, 1973.
Geog 4332.24 Gusari Jadranskoga mora. (Košutić, Ivo.) Zagreb, 1971.
Geog 809.51 Gutersohn, H. Die Erde. v.1-13. Bern, 1951. 2v.
Geog 818.82.5 Guthe, H. Lehrbuch der Geographie. Hannover, 1882-1883. 2v.
Geog 817.82 Guthrie, William. A new geographical, historical, and commercial grammar and present state of several kingdoms of the world. 7th ed. London, 1732.
Geog 817.83 Guthrie, William. A new geographical, historical, and commercial grammar and present state of the several kingdoms of the world. 8th ed. London, 1783.
Geog 817.94.10 Guthrie, William. New geographical, historical...grammar. London, 1796.
Geog 817.94.13 Guthrie, William. New geographical, historical...grammar. Montreal, 1810.
Geog 818.09 Guthrie, William. A new geographical, historical and commercial grammar. 1st American ed. Philadelphia, 1809.
Geog 817.88 Guthrie, William. A new geographical, historical and commercial grammar. 11th ed. London, 1788.
Geog 817.92 Guthrie, William. A new geographical, historical and commercial grammar. 13th ed. London, 1792.
Geog 817.94.5F Guthrie, William. New system of modern geography. London, 1795.
Geog 817.94 Guthrie, William. New system of modern geography. Philadelphia, 1794-1795. 2v.
Geog 4267.71.3 Gutierrez de la Hacera, P.R. Descripción general de la Europa. Madrid, 1782. 2v.
Geog 4209.34.10 Gutierrez Zamora, I. Contando un viaje. n.p., 1934.
An 379.52.5F Gutkind, E.A. Our world from the air. London, 1952.
Geog 618.1.7 Guyot, A. Carl Ritter. Princeton, 1860.
Geog 4145.34.25 Guzanov, Vitalii G. Odissei s Beloi Rusi. Minsk, 1969.
Geog 4180.29 Guzman, A.E. de. Life and acts of...Guzman. N.Y., 1970.
Geog 3590.65 Gvozdetskii, Nikolai Andreevich. Sorok let issledovanii i otkrytii. Moskva, 1957.
Geog 3590.75 Gvozdetskii, Nikolai Andreevich. Sovetskie geograficheskie issledovaniia i otkrytiia. Moskva, 1967.
Geog 3590.75.1 Gvozdetskii, Nikolai Andreevich. Soviet geographical explorations and discoveries. Moscow, 1974.
Geog 5985.253 Gwynn, Stephen. Captain Scott. N.Y., 1930.
Geog 4515.201 Gymnastik fur Bergsteiger. (Prusik, Karl.) München, 192-.
Geog 4209.42.20 Gypsy afloat. (Maillart, E.K.) London, 1942.
Geog 818.19 Haanbog i geographien. (Riise, J.) Kjobenhavn, 1819-1820. 2v.
An 170.246 Haberland, Eike. Festschrift für A.E. Jensen. München, 1964. 2v.
An 359.17.7 Haberlandt, M. Völkerkunde. 3. Aufl. Berlin, 1917. 2v.
An 379.34 Habitat, economy and society. (Forde, C.D.) London, 1934.
An 379.34.2 Habitat, economy and society. (Forde, C.D.) N.Y., 1937.
Geog 665.40F Hachette, firm, publishers, France. Les merveilles du monde. Paris, 192-?
Geog 665.40.5F Hachette, firm, publishers, France. Les merveilles du monde. Paris, 1957.
Geog 4208.12 Hacia tierras viejas. (Bacardí Moreau, E.) Valencia, 1913.
Htn Geog 4206.99* Hacke, William. A collection of original voyages. London, 1699.
Geog 4521.8 Hackel, Heinrich. Meine Berge, nuen Leben. Salzburg, 1960.

194

Author and Title Listing

Geog 4180.7A Hackluyt, R. Divers voyages...discovery of America. London, 1850.
An 132.3A Hadden, the head hunter. (Quiggin, A. Hingston (Mrs.).) Cambridge, 1942.
An 359.09.3 Haddon, A.C. Races of man and their distribution. London, 1909.
An 359.25.11 Haddon, A.C. Races of man and their distribution. N.Y., 1925.
An 358.98A Haddon, A.C. The study of man. N.Y., 1898.
Geog 4679.30 Hadfield, R.L. Sea-toll of our time. London, 1930.
Geog 4209.58 Hadley, Leila. Give me the world. N.Y., 1958.
An 2109.25.5F Hádschra Máktuba. (Frobenius, Leo.) München, 1925.
Geog 4208.81.10 Haeckel, Ernst. Tropenfahrten Reiseschilderungen aus Ceylon, Java und den Mittelmeergebieten. Leipzig, 1969.
Geog 3520.18 Haender, Wilhelmina. De Engelsche geographie in de 20ste eeuw. Utrecht, 1934.
Geog 4308.67.7 Haeseler, Charles H. Across the Atlantic. Philadelphia, 1868.
Geog 665.96 Hafemann, D. Mainzer geographische Studien. Braunschweig, 1961.
Geog 4502.5.16 Hager, F. Das Chiemgaubuch. München, 19- .
Geog 500.13 Hagers, J.G. Geographischer Buchersaal. Chemnitz, 1766-78. 3v.
Geog 5600.1 Hagerups, H. Boghandel. Litteratur om Grønland. København, 1932.
Geog 759.72 Haggeh, Peter. Geography: a modern synthesis. N.Y., 1972.
Geog 3108.82 Hahn, H. Leitfaden der alten Geographie. Leipzig, 1882.
Geog 5839.39 Haig-Thomas, D. Tracks in the snow. N.Y., 1939.
Htn Geog 4476.71* The hairy-giants. (Schooten, Henry (pseud.).) London, 1671.
Geog 4184.1A Hakluyt, R. Principal navigations...voyages traffiques. Glasgow, 1903-05. 12v.
Geog 4184.13 Hakluyt, R. Texts and versions of...Carpini and...Rubruquis. London, 1903.
Geog 3520.10.10F Hakluyt, Richard. Collection of early voyages, travels. London, 1809. 5v.
Geog 3520.10.75A Hakluyt, Richard. Fighting merchant men. Boston, 1927.
Geog 3520.10.45 Hakluyt, Richard. Principal navigations, voyages...of the English nation. London, 1927. 8v.
Geog 3520.10.35 Hakluyt, Richard. Principal navigations, voyages. London, 1907-13. 8v.
Htn Geog 3520.10.5F* Hakluyt, Richard. Principal navigations, voyages. v.1-3. London, 1599. 2v.
Geog 3520.10.20 Hakluyt, Richard. Principal navigations, voyages. v.1-12,14-16. Edinburgh, 1885-90. 15v.
NEDL Geog 3520.10.20 Hakluyt, Richard. Principal navigations, voyages. v.13. Edinburgh, 1885-90.
Geog 3520.10.45.5 Hakluyt, Richard. The principal navigations. Cambridge, 1965. 2v.
Htn Geog 3520.10F* Hakluyt, Richard. Principall navigations, voiages. London, 1589.
Geog 3520.10.15 Hakluyt, Richard. A selection of the principal voyages. N.Y., 1926.
Geog 3520.10.55 Hakluyt, Richard. They told Mr. Hakluyt. London, 1964.
Geog 3520.10.50.2 Hakluyt, Richard. Voyages and documents. London, 1963.
Geog 3520.10.25A Hakluyt, Richard. Voyages of Drake and Gilbert. Oxford, 1909.
Geog 4179.10 Hakluyt Society, London. A list of the publications of the Hakluyt Society. London, 1946.
Geog 4179.7 Hakluyt Society, London. Prospectus and list of members. London. 1850-1918 2v.
Geog 4184.14A Hakluytus posthumus, or Purchas, his pilgrimes. (Purchas, Samuel.) Glasgow, 1905-07. 20v.
Geog 3520.14 Hakluytus posthumus; Samuel Purches and the promotion of English overseas expansion. (Pennington, Loren E.) Emporia, 1966.
Geog 4308.82.5 Hale, E.E. A family flight thro' France, Germany, Norway and Switzerland. Boston, 1882.
Geog 4308.61.5 Hale, E.E. Ninety days worth of Europe. Boston, 1861.
Geog 3249.68 Hale, John Higby. Renaissance exploration. London, 1968.
Geog 4309.10A Hale, Louise (Closser). Motor journeys. Chicago, 1912.
Geog 818.30 Hale, N. Epitome of universal geography. Boston, 1830.
Geog 4309.33.3 Halévy, Daniel. Courrier d'Europe. Paris, 1933.
Geog 4209.45.15 Half a century of world travel; impressions and reflections. (Galimir, Mosco.) N.Y., 1945.
An 359.37 Half-caste. (Dover, C.) London, 1937.
Geog 4329.27 Haliburton, R. The glorious adventure. Indianapolis, 1927.
An 358.92.7 Haliburton, Robert G. Survivals of prehistoric races in Mt. Atlas and Pyrenees. Lisbon, 1892.
Geog 4208.31 Hall, B. Fragments of voyages and travels. Philadelphia, 1831. 2v.
Geog 4208.31.2 Hall, B. Fragments of voyages and travels. v.1-2. 2d ed. Edinburgh, 1832.
Geog 4308.40.3 Hall, Basil. Patchwork. London, 1841. 3v.
Geog 4308.40.5 Hall, Basil. Patchwork. 2nd ed. London, 1841. 3v.
Geog 4110.19 Hall, E.H. The picturesque tourist. N.Y., 1877.
Geog 4308.36.5 Hall, Fanny W. Rambles in Europe. N.Y., 1939. 2v.
Geog 4209.26 Hall, James Norman. On the stream of travel. Boston, 1926.
Geog 5559.17.5 Hall, Thomas F. Has the North Pole been discovered? Boston, 1917.
Geog 4209.38.5A Hallet, R.M. The rolling world. Boston, 1938.
Geog 4219.32.10A Halliburton, Richard. The flying carpet. Indianapolis, 1932.
Geog 4219.32.11 Halliburton, Richard. The flying carpet. London, 1933.
Geog 4209.40.20 Halliburton, Richard. Richard Halliburton, his story of his life's adventures as told in letters to his mother and father. Indianapolis, 1940.
Geog 4219.25.15 Halliburton, Richard. The royal road to romance. Garden City, 1925.
Geog 4219.25.10A Halliburton, Richard. The royal road to romance. Indianapolis, 1925.
Geog 4209.35.20A Halliburton, Richard. Seven league boots. Indianapolis, 1935.
Geog 4209.35.23 Halliburton, Richard. Seven league boots. 3. ed. London, 1941.
Geog 4029.16 Halliburton, the magnificient myth; a biography. (Root, Jonathan.) N.Y., 1965.
Geog 4308.77.5 Halsey, F.W. Two months abroad. Binghampton, 1878.
An 2109.26.7 Hambly, W.D. Origins of education among primitive peoples. London, 1926.
An 2109.26.3 Hambly, W.D. Tribal dancing and social development. London, 1926.
An 608.81 Hamburg, Germany. Museum. Die ethnographisch-anthropologische AbtheilungMZ Godeffroy. Hamburg, 1881.
Geog 4309.38 Hamburg-American Line. Your trip to Europe. N.Y., 1938.
Geog 1527.70 Hamburg und die Niederelbe. (Baedeker, publishers.) Hamburg, 1953.
Geog 4182.14 Hamel, Hendrik. Verhaal van het vergaan van het jacht De Sperwer. 's-Gravenhage, 1920.
Geog 4309.25.10 Hamilton, C.M. Wanderings. Garden City, 1925.
Geog 4209.21.3 Hamilton, Frederick. Here, there, and everywhere. London, 1921?
Geog 4209.21.4 Hamilton, Frederick. Here, there, and everywhere. N.Y., 1921.
Geog 4509.22 Hamilton, H. Mountain madness. London, 1922.
Geog 5839.52.10 Hamilton, Richard A. Venture to the Arctic. Harmondsworth, 1958.
Htn Geog 4476.75* Hamilton, William (pseud.). O-Brazile. London, 1675.
An 170.227 Hammond, Peter B. Cultural and social anthropology. N.Y., 1964.
Geog 659.74 Hammond, Robert. Quantitative techniques in geography: an introduction. Oxford, 1974.
Geog 4209.72 Hammond Innes, Dorothy. Occasions. London, 1972.
Geog 4209.60 Hammond-Innes, Ralph. Harvest of journeys. N.Y., 1960.
Geog 659.71 Hampl, Martin. Teorie komplexity a diferenciace světa se zoláštnim zřetelem na diferenciaci géograficksu. 1. vyd. Praha, 1971.
An 348.87 Hamy, E.T. Études ethnographiques et archéologiques sur l'exposition çoloniale. Paris, 1887.
Geog 3018.96 Hamy, J.T.E. Études historiques et géographiques. Paris, 1896.
Geog 1775.10 Hand-book for northern Europe: Finland and Russia. (Murray, John, publisher, London.) N.Y., 1970.
Geog 13.10 Handbook-directory. (Association of American Geographers.) Washington. 1956
Geog 1820.2 Handbook for...Algeria and Tunis. 2. ed. (Murray, John, publisher, London.) London, 1878.
Geog 1820.4 Handbook for...Algeria and Tunis. 4. ed. (Murray, John, publisher, London.) London, 1891.
Geog 1820.5 Handbook for...Algeria and Tunis. 5. ed. (Murray, John, publisher, London.) London, 1895.
Geog 1709.1 Handbook for...Berks, Bucks and Oxfordshire. (Murray, John, publisher, London.) London, 1860.
Geog 1709.1.3 Handbook for...Berks, Bucks and Oxfordshire. 3. ed. (Murray, John, publisher, London.) London, 1882.
Geog 1740.1 Handbook for...central Italy. (Murray, John, publisher, London.) London, 1843.
Geog 1740.3 Handbook for...central Italy. 3. ed. (Murray, John, publisher, London.) London, 1853. 2v.
Geog 1740.4 Handbook for...central Italy. 4. ed. (Murray, John, publisher, London.) London, 1857.
Geog 1740.5 Handbook for...central Italy. 5. ed. (Murray, John, publisher, London.) London, 1861.
Geog 1740.6 Handbook for...central Italy. 6. ed. (Murray, John, publisher, London.) London, 1864.
Geog 1740.7.5 Handbook for...central Italy. 7. ed. (Murray, John, publisher, London.) London, 1867.
Geog 1740.9 Handbook for...central Italy. 9. ed. (Murray, John, publisher, London.) London, 1875.
Geog 1740.11 Handbook for...central Italy. 11. ed. pt.2. (Murray, John, publisher, London.) London, 1889.
Geog 1804.1A Handbook for...Constantinople. (Murray, John, publisher, London.) London, 1900.
Geog 1709.5.10 Handbook for...Cornwall. 10. ed. (Murray, John, publisher, London.) London, 1882.
Geog 1744.1 Handbook for...Corsica and Sardinia. (Murray, John, publisher, London.) London, 1860.
Geog 1763.3 Handbook for...Denmark, Norway, Sweden. 3. ed. (Murray, John, publisher, London.) London, 1858.
Geog 1763.6 Handbook for...Denmark, Norway and Sweden. 3. ed. (Murray, John, publisher, London.) London, 1871.
Geog 1765.4 Handbook for...Denmark. 4. ed. (Murray, John, publisher, London.) London, 1875.
Geog 1709.7.2 Handbook for...Derbyshire, Nottinghamshire. 2. ed. (Murray, John, publisher, London.) London, 1874.
Geog 1709.8.10A Handbook for...Devonshire. 10. ed. (Murray, John, publisher, London.) London, 1887.
Geog 1709.10.5 Handbook for...Durham and Northumberland. (Murray, John, publisher, London.) London, 1873.
Geog 1816.5 Handbook for...Egypt. (Murray, John, publisher, London.) London, 1847.
Geog 1816.10 Handbook for...Egypt. (Murray, John, publisher, London.) London, 1858.
Geog 1816.15 Handbook for...Egypt. 4. ed. (Murray, John, publisher, London.) London, 1873.
Geog 1816.17 Handbook for...Egypt. 5. ed. (Murray, John, publisher, London.) London, 1875.
Geog 1816.19 Handbook for...Egypt. 7. ed. (Murray, John, publisher, London.) London, 1888.
Geog 1816.20A Handbook for...Egypt. 8. ed. (Murray, John, publisher, London.) London, 1891.
Geog 1816.25 Handbook for...Egypt and the Sudan. 11. ed. (Murray, John, publisher, London.) London, 1907.
Geog 1709.12 Handbook for...Gloucestershire, Worcestershire. (Murray, John, publisher, London.) London, 1867.
Geog 1709.12.3 Handbook for...Gloucestershire, Worcestershire. 3. ed. (Murray, John, publisher, London.) London, 1884.
Geog 1709.12.4 Handbook for...Gloucestershire. 4. ed. (Murray, John, publisher, London.) London, 1895.
Geog 1785.2 Handbook for...Greece. (Murray, John, publisher, London.) London, 1854.
Geog 1785.4 Handbook for...Greece. 4. ed. (Murray, John, publisher, London.) Lonon, 1872.
Geog 1785.5 Handbook for...Greece. 5. ed. (Murray, John, publisher, London.) London, 1884. 2v.
Geog 1785.6 Handbook for...Greece. 6. ed. (Murray, John, publisher, London.) London, 1896. 2v.
Geog 1785.7 Handbook for...Greece. 7. ed. (Murray, John, publisher, London.) London, 1900.
Geog 1709.13.5 Handbook for...Hampshire. 5. ed. (Murray, John, publisher, London.) London, 1898.
Geog 1807.13 Handbook for...India, Burma and Ceylon. 3. ed. (Murray, John, publisher, London.) London, 1898.
Geog 1807.17 Handbook for...India, Burma and Ceylon. 5. ed. (Murray, John, publisher, London.) London, 1905.
Geog 1807.17.5 Handbook for...India, Burma and Ceylon. 5. ed. (Murray, John, publisher, London.) London, 1906.
Geog 1807.18 Handbook for...India, Burma and Ceylon. 6. ed. (Murray, John, publisher, London.) London, 1907.
Geog 1807.19 Handbook for...India, Burma and Ceylon. 8. ed. (Murray, John, publisher, London.) London, 1911.
Geog 1807.21 Handbook for...India, Burma and Ceylon. 10. ed. (Murray, John, publisher, London.) London, 1920.

195
Author and Title Listing

Geog 1807.25 Handbook for...India, Burma and Ceylon. 13. ed. (Murray, John, publisher, London.) London, 1929.
Geog 1807.25.2 Handbook for...India, Burma and Ceylon. 13. ed. (Murray, John, publisher, London.) N.Y., 1929.
Geog 1807.27 Handbook for...India, Burma and Ceylon. 14. ed. (Murray, John, publisher, London.) London, 1933.
Geog 1807.29 Handbook for...India, Pakistan, Burma and Ceylon. (Murray, John, publisher, London.) London, 1949.
Geog 1807.5 Handbook for...India. (Murray, John, publisher, London.) London, 1859. 2v.
Geog 1807.10 Handbook for...India and Ceylon. (Murray, John, publisher, London.) London, 1892.
Geog 1782.1 Handbook for...Ionian islands. (Murray, John, publisher, London.) London, 1840.
Geog 1782.2 Handbook for...Ionian islands. (Murray, John, publisher, London.) London, 1845.
Htn Geog 1811.3* Handbook for...Japan. (Chamberlain, B.H.) N.Y., 1893.
Geog 1811.2 Handbook for...Japan. 3. ed. (Chamberlain, B.H.) London, 1891.
Geog 1811.4 Handbook for...Japan. 4. ed. (Murray, John, publisher, London.) London, 1896.
Geog 1811.6 Handbook for...Japan. 6. ed. (Murray, John, publisher, London.) London, 1901.
Geog 1811.8A Handbook for...Japan. 8. ed. (Murray, John, publisher, London.) London, 1907.
Geog 1811.9A Handbook for...Japan. 9. ed. (Murray, John, publisher, London.) London, 1913.
Geog 1709.17 Handbook for...Kent and Sussex. (Murray, John, publisher, London.) London, 1858.
NEDL Geog 1709.17.3 Handbook for...Kent and Sussex. 3. ed. (Murray, John, publisher, London.) London, 1868.
Geog 1828.1 Handbook for...New Zealand. (Murray, John, publisher, London.) London, 1893.
Geog 1714.1.2 Handbook for...North Wales. 2. ed. (Murray, John, publisher, London.) London, 1864.
Geog 1714.1.5 Handbook for...North Wales. 5. ed. (Murray, John. publisher, London.) London, 1885.
Geog 1709.24 Handbook for...Northamptonshire. (Murray, John, publisher, London.) London, 1878.
Geog 1709.24.2 Handbook for...Northamptonshire. 2. ed. (Murray, John, publisher, London.) London, 1901.
Geog 1739.4 Handbook for...northern Italy. 4. ed. (Murray, John, publisher, London.) London, 1852.
Geog 1739.6 Handbook for...northern Italy. 6. ed. (Murray, John, publisher, London.) London, 1856. 2v.
Geog 1739.8 Handbook for...northern Italy. 8. ed. (Murray, John, publisher, London.) London, 1860.
Geog 1739.9 Handbook for...northern Italy. 9. ed. (Murray, John, publisher, London.) London, 1863.
Geog 1739.10 Handbook for...northern Italy. 10. ed. (Murray, John, publisher, London.) London, 1866.
Geog 1739.11 Handbook for...northern Italy. 11. ed. (Murray, John, publisher, London.) London, 1867.
Geog 1739.14 Handbook for...northern Italy. 14. ed. (Murray, John, publisher, London.) London, 1877.
Geog 1770.8 Handbook for...Norway. 8. ed. (Murray, John, publisher, London.) London, 1892.
Geog 1709.27 Handbook for...Oxfordshire. (Murray, John, publisher, London.) London, 1894.
Geog 1750.1 Handbook for...Portugal. (Murray, John, publisher, London.) London, 1855.
Geog 1750.2 Handbook for...Portugal. 2. ed. (Murray, John, publisher, London.) London, 1856.
Geog 1750.3 Handbook for...Portugal. 3. ed. (Murray, John, publisher, London.) London, 1875.
Geog 1775.2 Handbook for...Russia, Poland and Finland. 2. ed. (Murray, John, publisher, London.) London, 1868.
Geog 1775.3 Handbook for...Russia, Poland and Finland. 3. ed. (Murray, John, publisher, London.) London, 1875.
Geog 1775.5 Handbook for...Russia, Poland and Finland. 5. ed. (Murray, John, publisher, London.) London, 1893.
Geog 1712.1 Handbook for...Scotland. (Murray, John, publisher, London.) London, 1867.
Geog 1712.2 Handbook for...Scotland. 2. ed. (Murray, John, publisher, London.) London, 1868.
Geog 1712.4 Handbook for...Scotland. 4. ed. (Murray, John, publisher, London.) London, 1875.
Geog 1712.7.25 Handbook for...Scotland. 8. ed. (Murray, John, publisher, London.) London, 1903.
Geog 1712.8 Handbook for...Scotland. 8. ed. (Murray, John, publisher, London.) London, 1907.
Geog 1712.9A Handbook for...Scotland. 9. ed. (Murray, John, publisher, London.) London, 1913.
Geog 1743.1 Handbook for...Sicily. (Murray, John, publisher, London.) London, 1864.
Geog 1724.14 Handbook for...south Germany and Austria. 14. ed. (Murray, John, publisher, London.) London, 1881.
Geog 1714.2.4 Handbook for...South Wales. 4. ed. (Murray, John, publisher, London.) London, 1890.
Geog 1724.2 Handbook for...southern Germany. (Murray, John, publisher, London.) London, 1837.
Htn Geog 1724.5* Handbook for...southern Germany. 5. ed. (Murray, John, publisher, London.) London, 1850.
Geog 1724.5.5 Handbook for...southern Germany. 5. ed. (Murray, John, publisher, London.) London, 1851.
Geog 1724.6 Handbook for...southern Germany. 6. ed. (Murray, John, publisher, London.) London, 1853.
Geog 1724.7 Handbook for...southern Germany. 7. ed. (Murray, John, publisher, London.) London, 1855.
Geog 1724.7.2 Handbook for...southern Germany. 7. ed. (Murray, John, publisher, London.) London, 1855.
Geog 1724.8 Handbook for...southern Germany. 8. ed. (Murray, John, publisher, London.) London, 1858.
Geog 1724.9 Handbook for...southern Germany. 9. ed. (Murray, John, publisher, London.) London, 1863.
Geog 1724.10 Handbook for...southern Germany. 10. ed. (Murray, John, publisher, London.) London, 1867.
Geog 1724.11 Handbook for...southern Germany. 11. ed. (Murray, John, publisher, London.) London, 1871.
Geog 1724.12 Handbook for...southern Germany. 12. ed. (Murray, John, publisher, London.) London, 1873.
Geog 1742.1 Handbook for...southern Italy. (Murray, John, publisher, London.) London, 1853.
Geog 1742.2 Handbook for...southern Italy. 2. ed. (Murray, John, publisher, London.) London, 1855.
Geog 1742.3 Handbook for...southern Italy. 3. ed. (Murray, John, publisher, London.) London, 1858.
Geog 1742.4 Handbook for...southern Italy. 4. ed. (Murray, John, publisher, London.) London, 1862.
Geog 1742.6 Handbook for...southern Italy. 6. ed. (Murray, John, publisher, London.) London, 1868.
Geog 1742.8 Handbook for...southern Italy. 8. ed. (Murray, John, publisher, London.) London, 1878.
Geog 1742.9 Handbook for...southern Italy and Sicily. 9. ed. (Murray, John, publisher, London.) London, 1892.
Geog 1748.3A Handbook for...Spain. 3. ed. (Murray, John, publisher, London.) London, 1855. 2v.
Geog 1748.5 Handbook for...Spain. 5. ed. (Murray, John, publisher, London.) London, 1878.
Geog 1748.6 Handbook for...Spain. 6. ed. (Murray, John, publisher, London.) London, 1882. 2v.
Geog 1748.7 Handbook for...Spain. 7. ed. (Murray, John, publisher, London.) London, 1890. 2v.
Geog 1709.33 Handbook for...Surrey, Hampshire. (Murray, John, publisher, London.) London, 1858.
Geog 1709.33.4 Handbook for...Surrey, Hampshire. 4. ed. (Murray, John, publisher, London.) London, 1888.
Geog 1773.4 Handbook for...Sweden. 4. ed. (Murray, John, publisher, London.) London, 1875.
Geog 1773.6 Handbook for...Sweden. 6. ed. (Murray, John, publisher, London.) London, 1883.
Geog 1735.1 Handbook for...Switzerland. (Murray, John, publisher, London.) London, 1838.
Geog 1735.3 Handbook for...Switzerland. (Murray, John, publisher, London.) London, 1839.
Geog 1735.7 Handbook for...Switzerland. 3. ed. (Murray, John, publisher, London.) London, 1846.
Geog 1735.9 Handbook for...Switzerland. 4. ed. (Murray, John, publisher, London.) London, 1851.
Geog 1735.11 Handbook for...Switzerland. 5. ed. (Murray, John, publisher, London.) London, 1852.
Geog 1735.13 Handbook for...Switzerland. 6. ed. (Murray, John, publisher, London.) London, 1854.
Geog 1735.15 Handbook for...Switzerland. 7. ed. (Murray, John, publisher, London.) London, 1856.
Geog 1735.17 Handbook for...Switzerland. 8. ed. (Murray, John, publisher, London.) London, 1858.
Geog 1735.23A Handbook for...Switzerland. 11. ed. (Murray, John, publisher, London.) London, 1865.
Geog 1735.25 Handbook for...Switzerland. 12. ed. (Murray, John, publisher, London.) London, 1867.
Geog 1735.30 Handbook for...Switzerland. 15. ed. (Murray, John, publisher, London.) London, 1874.
Geog 1735.35 Handbook for...Switzerland. 17. ed. (Murray, John, publisher, London.) London, 1886.
Geog 1805.1 Handbook for...Syria and Palestine. (Murray, John, publisher, London.) London, 1858. 2v.
Geog 1805.1.10 Handbook for...Syria and Palestine. (Murray, John, publisher, London.) London, 1875.
Geog 1805.2 Handbook for...Syria and Palestine. (Murray, John, publisher, London.) London, 1892.
NEDL Geog 1803.3 Handbook for...Turkey. (Murray, John, publisher, London.) London, 1854.
Geog 1709.37.12 Handbook for...Wilts and Dorset. 5. ed. (Murray, John, publisher, London.) London, 1899.
Geog 1709.37 Handbook for...Wiltshire, Dorsetshire and Somersetshire. (Murray, John, publisher, London.) London, 1856.
Geog 1709.37.2 Handbook for...Wiltshire, Dorsetshire and Somersetshire. (Murray, John, publisher, London.) London, 1859.
Geog 1709.37.10 Handbook for...Wiltshire, Dorsetshire and Somersetshire. 4. ed. (Murray, John, publisher, London.) London, 1882.
Geog 1709.38.4 Handbook for...Worcestershire. 4. ed. (Murray, John, publisher, London.) London, 1894.
Geog 1709.39.3 Handbook for...Yorkshire. 3. ed. (Murray, John, publisher, London.) London, 1882.
Geog 1709.39.4 Handbook for...Yorkshire. 4. ed. (Murray, John, publisher, London.) London, 1904.
Geog 1801.1 Handbook for Asia Minor. (Murray, John, publisher, London.) London, 1895.
Geog 1709.1.10 Handbook for Berkshire. (Murray, John, publisher, London.) London, 1902.
Geog 1709.2 Handbook for Buckinghamshire. (Murray, John, publisher, London.) London, 1903.
Geog 1709.7.3 Handbook for Derbyshire, Nottinghamshire. 3. ed. (Murray, John, publisher, London.) London, 1904.
Geog 1709.8.4 Handbook for Devon and Cornwall. (Murray, John, publisher, London.) Newton Abbot, Eng., 1971.
Geog 1709.8.5 Handbook for Devon and Cornwall. 5. ed. (Murray, John, publisher, London.) London, 1863.
Geog 1709.8.8 Handbook for Devon and Cornwall. 8. ed. (Murray, John, publisher, London.) London, 1872.
Geog 1708.1 Handbook for England and Wales. (Murray, John, publisher, London.) London, 1878.
Geog 1708.15 Handbook for England and Wales. 2. ed. (Murray, John, publisher, London.) London, 1890.
Geog 1709.11.2 Handbook for Essex, Suffolk and Norfolk. 2. ed. (Murray, John, publisher, London.) London, 1875.
Geog 1709.15 Handbook for Hertfordshire, Bedfordshire. (Murray, John, publisher, London.) London, 1895.
Geog 1811.5 Handbook for Japan. 5. ed. (Murray, John, publisher, London.) London, 1899.
Geog 1709.18.3 Handbook for Lancashire. (Murray, John, publisher, London.) London, 1880.
Geog 1709.20 Handbook for Lincolnshire. (Murray, John, publisher, London.) London, 1890.
Geog 1709.20.2 Handbook for Lincolnshire. 2. ed. (Murray, John, publisher, London.) London, 1903.
Geog 1710.5 Handbook for London, past and present. (Cunningham, Peter.) London, 1849. 2v.
Geog 1710.7 Handbook for London, past and present. (Cunningham, Peter.) London, 1850.
Geog 1710.9A Handbook for modern London. (Cunningham, Peter.) London, 1851.
Geog 1723.19 Handbook for north Germany. 19. ed. (Murray, John, publisher, London.) London, 1877.
Geog 1706.5 Handbook for northern Europe. (Murray, John, publisher, London.) London, 1849. 2v.
Geog 1770.5 Handbook for Norway. 5. ed. (Murray, John, publisher, London.) London, 1874.
Geog 1770.7 Handbook for Norway. 7. ed. (Murray, John, publisher, London.) London, 1880.
Geog 1709.29 Handbook for Shropshire. (Murray, John, publisher, London.) London, 1870.
Geog 1709.29.3 Handbook for Shropshire and Cheshire. (Murray, John, publisher, London.) London, 1879.

Author and Title Listing

Geog 1709.29.5	Handbook for Shropshire and Cheshire. 3. ed. (Murray, John, publisher, London.) London, 1897.
Geog 1709.30.5	Handbook for Somerset. 5. ed. (Murray, John, publisher, London.) London, 1899.
Geog 1709.34.4	Handbook for Sussex. 4. ed. (Murray, John, publisher, London.) London, 1877.
Geog 1709.34.5	Handbook for Sussex. 5. ed. (Murray, John, publisher, London.) London, 1905.
Geog 1575.16	Handbook for travellers: Russia. (Baedeker, publishers.) N.Y., 1970.
Geog 1715.2	Handbook for travellers in France. (Murray, John, publisher, London.) London, 1843.
Geog 1715.2.5	Handbook for travellers in France. (Murray, John, publisher, London.) London, 1844.
Geog 1715.3	Handbook for travellers in France. 3. ed. (Murray, John, publisher, London.) London, 1848.
Geog 1715.4	Handbook for travellers in France. 4. ed. (Murray, John, publisher, London.) London, 1853.
Geog 1715.7	Handbook for travellers in France. 7. ed. (Murray, John, publisher, London.) London, 1859.
Geog 1715.9	Handbook for travellers in France. 9. ed. (Murray, John, publisher, London.) London, 1864.
Geog 1715.10	Handbook for travellers in France. 10. ed. (Murray, John, publisher, London.) London, 1867.
Geog 1715.11	Handbook for travellers in France. 11. ed. (Murray, John, publisher, London.) London, 1869.
Geog 1715.12	Handbook for travellers in France. 12. ed. (Murray, John, publisher, London.) London, 1873. 2v.
Geog 1715.13	Handbook for travellers in France. 13. ed. (Murray, John, publisher, London.) London, 1875.
Geog 1715.16	Handbook for travellers in France. 16. ed. (Murray, John, publisher, London.) London, 1882-84. 2v.
Geog 1715.18	Handbook for travellers in France. 18. ed. (Murray, John, publisher, London.) London, 1892. 2v.
Geog 1713.2.5	Handbook for travellers in Ireland. 2. ed. (Murray, John, publisher, London.) London, 1866.
Geog 1713.4	Handbook for travellers in Ireland. 4. ed. (Murray, John, publisher, London.) London, 1878.
Geog 1713.5	Handbook for travellers in Ireland. 5. ed. (Murray, John, publisher, London.) London, 1896.
Geog 1713.6	Handbook for travellers in Ireland. 6. ed. (Murray, John, publisher, London.) London, 1902.
Geog 1713.6.75	Handbook for travellers in Ireland. 7. ed. (Murray, John, publisher, London.) London, 1906.
Geog 1713.7	Handbook for travellers in Ireland. 7. ed. (Murray, John, publisher, London.) London, 1906.
Geog 1713.8	Handbook for travellers in Ireland. 7. ed. (Murray, John, publisher, London.) London, 1906.
Geog 1709.17.4	Handbook for travellers in Kent. 4. ed. (Murray, John, publisher, London.) London, 1877.
Geog 1748.1	A handbook for travellers in Spain. (Murray, John, publisher, London.) London, 1845. 2v.
Geog 1748.2	Handbook for travellers in Spain. 2. ed. (Murray, John, publisher, London.) London, 1847.
Geog 1709.33.5	Handbook for travellers in Surrey. 5. ed. (Murray, John, publisher, London.) London, 1898.
Geog 1707.3	Handbook for travellers on the continent. 2. ed. (Murray, John, publisher, London.) London, 1838.
Geog 1707.6	Handbook for travellers on the continent. 5. ed. (Murray, John, publisher, London.) London, 1845.
Geog 1707.7	Handbook for travellers on the continent. 7. ed. (Murray, John, publisher, London.) London, 1850.
Geog 1707.8	Handbook for travellers on the continent. 8. ed. (Murray, John, publisher, London.) London, 1951.
Geog 1707.9	Handbook for travellers on the continent. 9. ed. (Murray, John, publisher, London.) London, 1853.
Geog 1707.10	Handbook for travellers on the continent. 10. ed. (Murray, John, publisher, London.) London, 1854.
Geog 1707.11	Handbook for travellers on the continent. 11. ed. (Murray, John, publisher, London.) London, 1856.
Geog 1707.12	Handbook for travellers on the continent. 12. ed. (Murray, John, publisher, London.) London, 1858.
Geog 1707.13	Handbook for travellers on the continent. 13. ed. (Murray, John, publisher, London.) London, 1860.
Geog 1707.15	Handbook for travellers on the continent. 15. ed. (Murray, John, publisher, London.) London, 1865.
Geog 1707.16	Handbook for travellers on the continent. 16. ed. (Murray, John, publisher, London.) London, 1868.
Geog 1707.17	Handbook for travellers on the continent. 17. ed. (Murray, John, publisher, London.) London, 1870.
Geog 1707.17.5	Handbook for travellers on the continent. 17. ed. (Murray, John, publisher, London.) London, 1871.
Geog 1707.18	Handbook for travellers on the continent. 18. ed. (Murray, John, publisher, London.) London, 1873-74. 2v.
Geog 1707.19	Handbook for travellers on the continent. 19. ed. (Murray, John, publisher, London.) London, 1875. 2v.
Geog 1525.20	A handbook for travellers on the Rhine. 2. ed. (Baedeker, publishers.) Coblenz, 1864.
Geog 1803.4	Handbook for Turkey. 4. ed. (Murray, John, publisher, London.) London, 1878.
Geog 1719.1	Handbook for visitors to Paris. (Murray, John, publisher, London.) London, 1864.
Geog 1719.3	Handbook for visitors to Paris. 3. ed. (Murray, John, publisher, London.) London, 1867.
Geog 1719.5	Handbook for visitors to Paris. 5. ed. (Murray, John, publisher, London.) London, 1872.
Geog 1719.8	Handbook for visitors to Paris. 8. ed. (Murray, John, publisher, London.) London, 1876.
Geog 1719.10	Handbook for visitors to Paris. 10. ed. (Murray, John, publisher, London.) London, 1879.
Geog 1709.36.2	Handbook for Westmorland...and the Lakes. 2. ed. (Murray, John, publisher, London.) London, 1869.
Geog 1709.39	Handbook for Yorkshire. (Murray, John, publisher, London.) London, 1867.
Geog 5208.95	Handbook of arctic discoveries. (Greely, A.W.) Boston, 1896.
Geog 5208.95.3	Handbook of polar discoveries. 3. ed. (Greely, A.W.) Boston, 1906.
Geog 5208.95.7	Handbook of polar discoveries. 5th ed. (Greely, A.W.) Boston, 1910.
Geog 1741.5	Handbook of Rome and its environs. 5. ed. (Murray, John, publisher, London.) London, 1858.
Geog 1741.7	Handbook of Rome and its environs. 7. ed. (Murray, John, publisher, London.) London, 1864.
Geog 1741.8	Handbook of Rome and its environs. 8. ed. (Murray, John, publisher, London.) London, 1867.
Geog 1741.9	Handbook of Rome and its environs. 9. ed. (Murray, John, publisher, London.) London, 1869.
Geog 1741.10	Handbook of Rome and its environs. 10. ed. (Murray, John, publisher, London.) London, 1871.
Geog 1741.12	Handbook of Rome and its environs. 12. ed. (Murray, John, publisher, London.) London, 1875.
Geog 1809.5	Handbook of the Bengal presidency. (Murray, John, publisher, London.) London, 1882.
Geog 1809.30	Handbook of the Bombay presidency. (Murray, John, publisher, London.) London, 1881.
An 609.25	Handbook of the ethnographical collections. 2. ed. (British Museum. Department of Ancient and Mediaeval Antiquities and Ethnography.) London, 1925.
Geog 1809.55	Handbook of the Madras presidency. (Murray, John, publisher, London.) London, 1879.
Geog 1809.80	Handbook of the Punjab...Kashmir. (Murray, John, publisher, London.) London, 1883.
Geog 4105.13A	Handbook of travel. (Harvard Travel Club.) Cambridge, 1917.
Geog 4105.13.3	Handbook of travel. (Harvard Travel Club.) Cambridge, 1935.
Geog 1709.35	Handbook of Warwickshire. (Murray, John, publisher, London.) London, 1899.
Geog 1707.5	Handbook the travellers on the continent. (Murray, John, publisher, London.) London, 1840.
Geog 1711.1	Handbook to environs of London. (Thorne, James.) London, 1876. 2v.
An 609.10	Handbook to ethnographical collections. (British Museum.) Oxford, 1910.
Geog 1710.20	Handbook to London. (Murray, John, publisher, London.) London, 1869?
Geog 1710.22	Handbook to London. (Murray, John, publisher, London.) London, 1870.
Geog 1710.26	Handbook to London. (Murray, John, publisher, London.) London, 1871.
Geog 1710.28	Handbook to London. (Murray, John, publisher, London.) London, 1873.
Geog 1710.30	Handbook to London. (Murray, John, publisher, London.) London, 1876.
Geog 1710.29	Handbook to London. (Murray, John, publisher. London.) London, 1874.
Geog 1710.35	Handbook to London. (Murray, John, publishers, London.) London, 1879.
Geog 1709.6.25	Handbook to the English lakes...Cumberland, Westmorland, and Lancashire. (Murray, John, publisher, London.) London, 1889.
Geog 1795.3	Handbook to the Mediterranean. 3. ed. (Murray, John, publisher, London.) London, 1892. 2v.
Geog 808.33.2	Handbuch...der Natur und Geschichte der Erde. (Blanc, L.G.) Halle, 1833-1834. 3v.
Geog 807.85.5	Handbuch der alten Erdbeschreibung zum Gebrauch der Eilf. v.1-2, pt.1-2. Nürnberg, 1785-1793. 3v.
Geog 3108.42.5	Handbuch der alten Geographie. (Forbiger, A.) Hamburg, 1877. 3v.
Geog 3108.42	Handbuch der alten Geographie. (Forbiger, A.) Leipzig, 1842-48. 3v.
Geog 3108.37.2	Handbuch der alten Geographie. (Schirlitz, S.C.) Halle, 1837.
Geog 808.59	Handbuch der Erdkunde. (Klöden, G.A.) Berlin, 1859-1862. 3v.
Geog 808.59.3	Handbuch der Erdkunde. (Klöden, G.A.) Berlin, 1873. 5v.
Geog 808.66	Handbuch der Geographie. (Daniel, H.A.) Leipzig, 1866-1868. 4v.
Geog 808.66.5	Handbuch der Geographie. (Daniel, H.A.) Leipzig, 1874. 4v.
Geog 808.66.7	Handbuch der Geographie. (Daniel, H.A.) Leipzig, 1895. 4v.
Geog 819.14	Handbuch der Geographie. (Seydlitz, Ernst von.) Breslau, 1914.
Geog 758.90	Handbuch der Mathematischen Geographie. (Günther, S.) Stuttgart, 1890.
An 2109.66.15	Handbuch der Urgeschichte. (Narr, Karl J.) Bern, 1966. 2v.
Geog 1522.23	Handbuch für Reisende in Deutschland. pt.1-2. 6. Aufl. (Baedeker, publishers.) Coblenz, 1855.
Geog 1522.15	Handbuch für Reisende in Deutschland. 2. Aufl. (Baedeker, publishers.) Coblenz, 1846.
Geog 1522.22	Handbuch für Reisende in Deutschland. 5. Aufl. (Baedeker, publishers.) Coblenz, 1854.
Geog 4308.86.5	A handfull of monographs, continental and English. (Preston, M.J. (Mrs.).) N.Y., 1886.
Geog 4249.51	Hands across the Pacific. (Clune, Frank.) Sydney, 1951.
Geog 4559.64.3	Hanke, Helmut. Männer, Planken, Ozeane. 3. Aufl. Leipzig, 1966.
An 130.20	Hanke, Lewis. Gilberto Freyne. N.Y., 1939.
An 2109.28.9	Hankin, E.H. The cave man's legacy. London, 1928.
Geog 819.44.5F	Hankins, G.C. Our global world. N.Y., 1944.
Geog 500.78	Hanover, S. Geographie. Hannover, 1955.
Geog 5700.8	Hans Egede, der Prediger des Evangeliums. (Vormbaum, R.) Elberfeld, 1861.
Geog 5700.4.8	Hans Egede. (Bobé, Louis.) Copenhagen, 1952.
Geog 5700.4.9	Hans Egede og Grønland. (Bobé, Louis.) København, 1941.
VGeog 5700.4.15	Hans Egedes levnet. (Petersen, Niels Matthias.) København, 1839.
Geog 665.169	Hans Graul-Festschrift. Heidelberg, 1974.
Geog 4206.13.30F	Hans Jakob Ammann genannt der Thalwyler Schärer und seine Reise ins gelobte Land. (Amman, H.J.) Zürich, 1919.
Geog 5559.34	Hans the Eskimo. (Rich, E.G.) Boston, 1934.
Geog 5838.84	Hansen, Johannes. Den grønlandske Kateket Hansêraks Dagbog. København, 1933.
Geog 5399.53	Hansen, Leo. I Knuds sloedespor. København, 1953.
Geog 5323.5	Hansen, T. Jens Munk. 4. Opl. Kjøbenhavn, 1966.
Geog 4679.41	Hanson, Earl P. Highroad to adventure. N.Y., 1941.
Geog 3019.49A	Hanson, Earl P. New worlds emerging. 1. ed. N.Y., 1949.
An 132.6	Hanson, F. Allan. Meaning in culture. London, 1975.
Geog 5160.58	Hantschel, A. Weltgeschehen am Rande des Polarmeeres. Würzburg, 1964.
Geog 4207.53.5F	Hanway, Jonas. Historical account of British trade. 3. ed. London, 1762. 2v.
Geog 4207.53.3	Hanway, Jonas. Reize van London, door Rusland, nae en in Persie. Amsterdam, 1758. 2v.
Geog 3235.33	Hany, E.T. La mappemonde. Paris, 1887.
Geog 806.87	Happelius, E.G. Mundus mirabilis tripartitus. Ulm, 1687. 3v.
Geog 4328.38.5	Happoldt, Christopher. The Christopher Happoldt journal. Charleston, 1960.
Geog 4105.17.5	The happy traveller. 5. ed. (Tatchell, Frank.) London, 1927.

Author and Title Listing

Geog 4308.54.4 Haps and mishaps...in Europe. (Lippincott, S.J.) Boston, 1854.
Geog 4308.54.2 Haps and mishaps...in Europe. (Lippincott, S.J.) Boston, 1854.
Geog 4308.54 Haps and mishaps...in Europe. (Lippincott, S.J.) Boston, 1854.
Geog 4308.54.3 Haps and mishaps...in Europe. (Lippincott, S.J.) Boston, 1854.
Geog 4559.45.10 Harbour Head. (Anson, P.F.) London, 1944.
Geog 659.73.15 Hard, Gerhard. Die Geographie. Berlin, 1973.
An 3559.08 The hard palate in normal and feeble-minded individuals. (Channing, Walter.) N.Y., 1908.
Geog 5160.47.5 Hardbalne polarkarer. 5. Oppl. (Giaever, John.) Oslo, 1973.
Geog 4308.70 Harding, W.M. Trans-Atlantic sketches. N.Y., 1870.
Geog 4679.44.5 Hardy, Alfred C. Wreck - S.O.S. London, 1944.
Geog 6009.01.25 Hardy, Alister C. Great waters; a voyage of natural history to study whales, plankton and the waters of the southern ocean. N.Y., 1967.
An 379.39.10 Hardy, Georges. La géographie psychologique. Paris, 1939.
Geog 5328.27 Hare, Kirsten. Kender du Knud? 2. Udg. Kjøbenhavn, 1971.
Geog 4328.73 Ein Harem in Bismarcks Reich. (Nāsir, al-Dīn Shāh.) Tübingen, 1969.
Geog 4181.94 Harff, Arnold. The pilgrimage of Arnold von Harff. London, 1946.
An 3508.82 Harkness, H.W. Footprints found at the Carson State Prison. n.p., n.d.
Htn Geog 4559.28.6* Harlow, Frederick Pease. The making of a sailor. Salem, Mass., 1928.
Geog 4559.28.5 Harlow, Frederick Pease. The making of a sailor. Salem, Mass., 1928.
Geog 4181.56 Harlow, V.T. Colonizing expeditions to the West Indies and Guiana, 1623-67. London, 1925.
Geog 600.20 Harms, Hans. Künstler des Kartenbildes. Oldenburg, 1962.
Geog 4209.25.10 Harper, H.H. Highlights of foreign travel. N.Y., 1925.
Geog 4300.5 Harper's handbook for...Europe. (Fetridge, W.P.) N.Y. 1-17,1862-1878 7v.
Geog 578.55.15 Harper's statistical gazetteer of the world. (Smith, John Calvin.) N.Y., 1855.
Geog 818.62 Harris, A. Geographical hand-book. Lancaster, Pa., 1862.
Geog 500.116.5 Harris, Chauncy Donnison. Annotated world list of selected current geographical serials in English. 2. ed. Chicago, 1964.
Geog 500.116 Harris, Chauncy Donnison. Geographic bibliography. Chicago, 1961.
Geog 500.91 Harris, Chauncy Donnison. International list of geographical serials. Chicago, 1960.
Geog 500.90 Harris, Chauncy Donnison. A union list of geographical serials. 2. ed. Chicago, 1950.
Geog 4209.51 Harris, F.R. Itchin' feet. Greenfield, Ohio, 1951.
Htn Geog 4157.05.2F* Harris, J. Navigantium...compleat collection. London, 1705.
Geog 4157.05.10F Harris, J. Navigantium...compleat collection. London, 1744. 2v.
Geog 4157.05.15F Harris, J. Navigantium...voyages and travels. London, 1764. 2v.
Geog 4309.36 Harris, J.P. Traveling light (Europe between boats). Hutchinson, Kansas, 1936.
An 182.3 Harris, Marvin. Cows, pigs, wars and witches. 1. ed. N.Y., 1974.
An 359.71 Harris, Marvin. Culture, man, and nature. N.Y., 1971.
An 359.64.30 Harris, Marvin. The nature of cultural things. N.Y., 1964.
An 99.68 Harris, Marvin. The rise of anthropological theory. N.Y., 1968.
Geog 4209.39.20A Harris, N.D. Moving on; the romance of travel. Chicago, 1939.
Geog 4209.35.25 Harris, P.P. Peregrinations. v.2-3. n.p., 1935-37. 2v.
Geog 4677.66.3 Harrison, David. Een droevig verhaal van de ongelukkige reis en wonderdaadige verlossinge. Wesel, 176-?
Geog 520.6 Harrisse, H. Christophe Colomb. Paris, 1884. 2v.
Geog 520.3 Harrisse, H. Les Corte-Real et leurs voyages. Paris, 1883.
Geog 520.3.5 Harrisse, H. Gaspar Corte-Real. Paris, 1883.
Geog 520.1 Harrisse, H. Jean et Sébastien Cabot. Paris, 1882.
Geog 3570.30A Hart, Henry H. Sea road to the Indies. N.Y., 1950.
Geog 4558.48 Hart, J.C. Romance of yachting. N.Y., 1848.
Geog 759.75.5 Hart, John Fraser. The look of the land. Englewood Cliffs, 1975.
Geog 4209.38 Hart, M.R. Who called that lady a skipper? N.Y., 1938.
Geog 659.39.5A Hartshorne, Richard. Perspective on the nature of geography. Chicago, 1959.
Geog 5160.10F Hartwig, G. The Arctic regions. London, n.d.
Geog 5208.69.15 Hartwig, G. The polar world. London, 1886.
Geog 97.2 Harvard mountaineering. Cambridge, Mass. 10-14,1951-1959
Geog 4105.13A Harvard Travel Club. Handbook of travel. Cambridge, 1917.
Geog 4105.13.3 Harvard Travel Club. Handbook of travel. Cambridge, 1935.
Geog 97.10 Harvard Travellers' Club. Yearbook.
Geog 97.12 Pamphlet vol. Harvard Travellers' Club.
An 359.44 Harvard University. Department of Psychology. ABC's of scapegoating. Chicago, 1943.
Geog 4500.16 Harvard University. Library. Mountaineering literature in the Harvard College Library. n.p., 1931.
Geog 4209.60 Harvest of journeys. (Hammond-Innes, Ralph.) N.Y., 1960.
Geog 659.70 Harvey, David. Explanation in geography. N.Y., 1970.
Geog 5559.17.5 Has the North Pole been discovered? (Hall, Thomas F.) Boston, 1917.
An 359.59 Haselden, Kyle. The racial problem in Christian perspective. 1. ed. N.Y., 1959.
Geog 4308.56.15 Haskins, G.F. Travels in England, France. Boston, 1856.
Geog 4308.71.5 Haskins, George F. Six weeks abroad in Ireland, England and Belgium. Boston, 1872.
Geog 5070.11.5 Hassert, Kurt. Die Polarforschung. München, 1956.
Geog 5070.11 Hassert, Kurt. Die Polarforschung. 3. Aufl. Leipzig, 1914.
Geog 3542.5 Hassinger, H. Österreichs Anteil an der Erforschung der Erde. Wien, 1949?
Geog 4559.72.1 Hassloef, Olof. Ships and shipyards, sailors and fishermen. Copenhagen, 1972.
An 359.62.5 Haste, Hans. Rasisner. Stockholm, 1962.
An 3309.02 Hastings, William W. Manual for physical measurements. Springfield, 1902.
Geog 5919.65.5 Hatherton, Trevor. Antarctica. Wellington, 1965.
Geog 5180.12 Hatt, G. Arktiske skinddragter i Eurasian og Amerika. Kjøbenhavn, 1914.
Geog 5180.12.5 Hatt, G. Moccasins and their relation to Arctic footwear. Lancaster, 1915?
Geog 4238.53 Hauch, J.C. von. En Rejse til Amerika. Kjøbenhavn, 1853.
An 379.30 Hauemeyer, Loomis. Ethnography. Boston, 1929.
Geog 4131.74.5 Haulot, Arthur. Tourisme et environnement. Verviers, 1974.
Geog 819.28 Hauno, or Future of exploration. (Mitchell, J. Leslie.) London, 1928.
Geog 5312.7.25 The haunted journey. 1. ed. (Murphy, Robert William.) Garden City, 1961.
Geog 4208.27 Haus, Rosalie. Voyage of the Caroline, 1827-28. London, 1927.
Geog 4269.32 Hausenstein, W. Europäische Hauptstädte. Erlenbach, 1932.
Geog 4309.51.5 Hausenstein, Wilhelm. Abendländische Wanderungen. München, 1951.
Geog 4209.32.20 Hauser, H. Fair winds and foul. N.Y., 1932.
An 3559.06 Hauser, K. Das kraniologische der New Guinea Expedition. Berlin, 1906.
Geog 135.2.7 Hauthal, Rudolf. Reisen in Bolivien und Peru. v.7. Leipzig, 1911.
Geog 5398.71 Have de gamle Nordboer havt Kjendskab. (Brynjulfson, A.) Kjøbenhavn, 1871.
Geog 4308.65.5 Haven, Gilbert. The pilgrim's wallet. N.Y., 1866.
Geog 4209.40.45 Haven, V.S. Many ports of call. N.Y., 1940.
Geog 4239.68.15 Haward, Peter. High latitude crossing. London, 1968.
Geog 4709.40 Hawes, Hildreth Gilman. The Bellamy treasure. Augusta, Maine, 1940.
An 359.55 Hawkes, J. Man on earth. 1. American ed. N.Y., 1955.
An 2109.74.10 Hawkes, Jacquetta Hopkins. Atlas of ancient archaeology. N.Y., 1974.
Geog 578.27.3 Hawkes, P. The American companion. Philadelphia, 1827.
An 359.26.4A Hawkins, Frank H. The racial basis of civilization. N.Y., 1926.
An 2159.73 Hawkins, Gerald Stanley. Beyond Stonehenge. 1st ed. N.Y., 1973.
Geog 4180.1 Hawkins, R. Observations...in voyage...South Sea. London, 1847.
Geog 4180.57 The Hawkins voyages. (Markham, C.R.) London, 1878.
Geog 4309.56 Haya de la Torre. Mensaje de la Europa nordica. Buenos Aires, 1956.
Geog 4559.25 Hayes, Bertram. Hull down. N.Y., 1925.
Geog 3070.35 Hayes, Gerald R. The production of an admiralty chart. London, 1929.
Geog 5538.54.5 Hayes, I.I. An Arctic boat journey, in the autumn of 1854. Boston, 1867.
Geog 5538.54.8 Hayes, I.I. Arctic boat journey, in the autumn of 1854. Boston, 1883.
Geog 5558.53 Hayes, I.I. An Arctic boat journey...1854. Boston, 1868.
Geog 5538.54.2 Hayes, I.I. An Arctic boat journey. Boston, 1860.
Geog 5838.71 Hayes, I.I. The land of desolation...Greenland. London, 1871.
Geog 5558.60 Hayes, I.I. The open Polar Sea. N.Y., 1867.
Geog 5208.68.2 Hayes, I.I. Progress of Arctic discovery. N.Y., 1868.
Geog 5208.68 Hayes, I.I. Progress of Arctic discovery. N.Y., 1868.
Geog 5235.10 Hayes, J.G. The conquest of the North Pole. N.Y., 1934.
Geog 5939.32 Hayes, J.G. The conquest of the South Pole. London, 1932.
Geog 5326.21.15 Hayes, J.G. Robert Edwin Peary, a record of his explorations, 1886-1909. London, 1929.
Geog 5919.28 Hayes, James G. Antarctica; a treatise on the southern continent. London, 1928.
Geog 4559.53.5 Hayet, Armand. Us et coutumes à bord des long-courriers. Paris, 1953.
An 99.58 Hays, H.R. From ape to angel. 1. ed. N.Y., 1958.
Geog 6009.64.5 Hayter, Adrian. The year of the quiet sun: one year at Scott Base, Antarctica. London, 1968.
Geog 5900.10 Hayton, Robert D. National interests in Antarctica; an annotated bibliography. Washington, 1959-60.
Geog 4308.92 Hayward, H.C. From Finland to Greece, or Three seasons in Eastern Europe. N.Y., 1892.
Geog 5925.25 Hayward, Walter B. The last continent of adventure. N.Y., 1930.
Geog 4209.54 He lived for adventure. (Osborn, C.) Washington, 1954.
An 2109.35.5 Head, heart and hands in human evolution. (Marett, Robert R.) London, 1935.
Geog 4217.40.5 Heaps, Leo. Log of the centurion. London, 1973.
Geog 6009.07.5 The heart of the Antarctic. (Shackleton, Ernest H.) London, 1911.
Geog 6009.07A Heart of the Antarctic. (Shackleton, Ernest H.) Philadelphia, 1909. 2v.
Geog 4219.55.5 Heaven, hell and salt water. (Crove, Bill.) London, 1957.
Geog 4429.53A Heaven has claws. (Doyle, A.C.) N.Y., 1953.
Geog 4559.35.15 Heavenly hell. (Sheridan, Richard B.) London, 1935.
Geog 3019.12.2 Heawood, Edward A. History of geographical discovery in the seventeenth and eighteenth centuries. N.Y., 1965.
An 2600.6.1 Hecquet, (Mme.). Histoire d'une jeune fille sauvage trouvée dans les bois à l'âge de dix ans. Bordeaux, 1970.
Geog 4652.5 Heden, Karl E. Direction of shipwrecks of the Great Lakes. Boston, 1966.
Geog 577.31 Hederich, M.B. Reales Schul-Lexicon. Leipzig, 1731.
Geog 4180.74 Hedges, W. Diary of William Hedges...during agency in Bengal. London, 1887-89. 3v.
Geog 4129.11 Hedin, Sven. Van pool tot pool. Amsterdam, 194-?
Geog 4129.11.5 Hedin, Sven. Von Pol zu Pol. 81. Aufl. Leipzig, 1942.
Geog 3135.5A Heidel, William A. The frame of the ancient Greek maps. N.Y., 1937.
An 3309.69 Height and history in Malawi. (Nurse, G.T.) Malawi, 1969.
Geog 4559.67.10 Heikkinen, Helge. Runt Kap Horn med Herzogin Cecilie. Ekenäs, 1967.
Geog 5180.5 Heilprin, A. The Arctic problem. Philadelphia, 1893.
Geog 4209.34.7 Heimkehr ins Dritte Reich. (Fuchs, Hans.) Dresden, 1934.
Geog 4182.61 Hein, Pieter P. De Westafrikaanse reis van Piet Heyn. 's-Gravenhage, 1959.
An 379.68.15 Heinemeyer, W.F. De sociale geografie in de rij van de sociale wetenschappen. Meppel, 1968.
Geog 611.2 Heinrich Kiepert bei seiner Rückkehr November 1886 von Freunden gewidmet. Berlin, 1886.
Geog 4308.48.5 Heinzelmann, F. Reisen durch Belgien, Holland und Grossbritannien. Leipzig, 1848.
Geog 5060.16 Helden im ewigen Eis; im Kampf um den Nord- und Südpol. (Zeidler, Paul G.) Leipzig, 1936.
Geog 5060.15 Heldenbuch der Arktis. (Rasmussen, K.) Leipzig, 1933.
Geog 3570.58 Heleno, Manuel Domingues. Colaboração portuguesa nos descobrimentos nauticos das outras nações. Lisboa, 1932.
An 359.43 Hell in the sunshine. (Dover, C.) London, 1943.
Geog 5398.79.30A Hell on ice; the saga of the "Jeannette". (Ellsberg, Edward.) N.Y., 1938.
Geog 4309.73 Helle Tage anderswo. (Piontek, Heinz.) München, 1973.
Geog 4209.02.10 Hellen, Gustav. Aus meiner Wanderzeit, 1902-1906 und 1908-1909. Hamburg, 1966.

Author and Title Listing

	Call number	Entry
	An 359.24.3	Die hellfarbigen Rassen und ihre Sprachstamme, Kulturen und Urkeimaten. (Paudler, F.) Heidelberg, 1924.
	Geog 818.97.5	Hellwald, F. von. Die Erde und ihre Völker. Stuttgart, 1897.
	Geog 5208.81	Hellwald, F. von. Im ewigen Eis. Stuttgart, 1881.
	Geog 5208.81.5	Hellwald, F. von. Voblasti vechnago l'da. Sankt Peterburg, 1881.
	An 358.82.5	Hellwald, Friedrich. Naturgeschichte des Menschen. Stuttgart, 1882-85. 2v.
	An 2109.03.5	Hellwig, A. Das Asylrecht der Naturvölker. Berlin, 1903.
	Geog 4208.82	Helms, L.V. Pioneering in the Far East. London, 1882.
	Geog 4559.26	Hemy, Thomas M. Deep sea days. London, 1926.
	Geog 4509.42	Henderson, K.A. The American Alpine Club's handbook of American mountaineering. Boston, 1942.
	An 2109.27.35	Henderson, Keith. Prehistoric man. N.Y., 1927.
	VGeog 4678.91	Henderson Brothers, Glasgow, defendants. Difesa dei signori Henderson, armatori contro i danneggiati del naufragio dell'Utopia. Napoli, 1895.
	Geog 5318.1	Hendrik, Hans. Memoirs of Hans Hendrik, the Arctic traveller. London, 1878.
	Geog 4218.76.3	Hendrix, E.R. Around the world. St. Louis, 1881.
	Geog 3506.7	Hennequin, Emile. Etude historique sur l'exécution de la carte de Ferraris et l'évolution de la cartographie topographique en Belgique. Bruxelles, 1891.
	Geog 4169.36	Hennig, R. Terrae incognitae. Leiden, 1936-39. 4v.
	Geog 4169.36.3	Hennig, R. Terrae incognitae. Leiden, 1944- 4v.
	Geog 4157.74	Henry, David. Historical account of voyages. London, 1773-74. 4v.
	An 3409.13.5	Henry, Edward R. Classification and uses of fingerprints. 4. ed. London, 1913.
	Geog 4427.99	Henry, Pierre. Route de l'Inde. Paris, 1798-99.
	Geog 5919.50	Henry, T. The white continent. N.Y., 1950.
	Geog 607.1	Henry Gannett, president of the National Geographic Society, 1910-1914. (North, S.N.D.) n.p., 1915.
	Geog 4180.27A	Henry Hudson the navigator. (Asher, G.M.) London, 1860.
	Geog 4182.19	Henry Hudson's reize onder Nederlander vlag. (Juet, Robert.) 's-Gravenhage, 1921.
	Geog 5318.2.10	Henson, Matthew Alexander. A Negro explorer at the North Pole. N.Y., 1969.
	Geog 4209.61.15	Hentzel, R. Aventyr jorden runt. Stockholm, 1961.
Htn	Geog 4305.96*	Hentznero, P. Itinerarium Germaniae, Galliae. Norinbergae, 1612.
	An 3309.63.5	Henzel, Tadeusz. Badania struktury rasowej ludności Afryki Srotkowej. Łódz, 1963.
	Geog 240.5	Heohrafichnyi zbirnyk. (Ukrains'ke Heohrafichne Tovarystvo.) Kyïv. 1-5 5v.
X Cg	Geog 4180.10	Herberstein, Sigmund. Notes upon Russia. London, 1851-52. 2v.
	Geog 4180.10A	Herberstein, Sigmund. Notes upon Russia. N.Y., 1963? 2v.
	Geog 4416.26.5F	Herbert, Thomas. Some years travels into...Africa and Asia. London, 1677.
Htn	Geog 4416.26*	Herbert, Thomas. Some years travels into...Asia and Afrique. London, 1638.
	Geog 5559.68	Herbert, Wally. Across the top of the world. N.Y., 1971.
	Geog 5985.122	Herbert, Wally. A world of men: exploration in Antarctica. London, 1968.
	Geog 4269.03.10	Herbertson, (Mrs.). Descriptive geography of Europe from original sources. London, 1919.
	An 358.99.5	Herbertson, A.J. Man and his work. London, 1899.
	An 359.09.8	Herbertson, A.J. Man and his work. London, 1909.
	Geog 759.21.5	Herbertson, Andrew J. The senior geography. 5th ed. Oxford, 1921.
	Geog 4209.05	Herboso, F.J. Reminiscencias de viajes. Caracas, 1905- 3v.
	Geog 4308.16	Die Herbstreise nach Benedig. (Raumer, F.L.G.) Berlin, 1816. 2v.
	Geog 4209.21.3	Here, there, and everywhere. (Hamilton, Frederick.) London, 1921?
	Geog 4209.21.4	Here, there, and everywhere. (Hamilton, Frederick.) N.Y., 1921.
	Geog 665.23	Here and there and everywhere. (Sherwood, M.E.) Chicago, 1898.
	Geog 5160.60	Here is the far north. (Stefansson, Evelyn Schwartz Baird.) N.Y., 1957.
	Geog 756.44	Herigone, P. Cursus mathematicus. v.4. n.p., 1644.
	Geog 4209.32.5	Hering, O.C. Down the world. N.Y., 1932.
	Geog 665.97	Herman von Wissmann-Festschrift. (Leidlmar, A.) Tübingen, 1962.
	Geog 5650.16	Hermansson, H. Landafundir og sjóferð i norðrhofum. Winnipeg, 19- .
	Geog 4209.13.7	A hermit turned loose. 2. ed. (Kawabata, A.) Tokio, 1915.
	Geog 4028.4.2	Heroes of modern adventure. (Bridges, Thomas C.) Boston, 1928.
	Geog 4167.97	Heron, Robert. A collection of late voyages and travels. Edinburgh, 1797.
	Geog 5559.42	Herrmann, Ernst. Mit dem Fieseler-Storch ins Nordpolarmeer. Berlin, 1942.
	Geog 3019.52.2	Herrmann, Paul. Conquest by man. N.Y., 1954.
	Geog 3019.56.2	Herrmann, Paul. The great age of discovery. N.Y., 1958.
	VGeog 3019.56.4	Herrmann, Paul. Historia de los descubrimientos geográficos. 2. ed. Barcelona, 1967.
	Geog 3019.52	Herrmann, Paul. Sieben vorbei und acht verweht. 2. Aufl. Hamburg, 1952.
	Geog 3019.59.5	Herrmann, Paul. Traumen, Wagen und Vollbringen. Hamburg, 1959.
	Geog 3019.56	Herrmann, Paul. Zeigt mir Adams Testament. Hamburg, 1956.
	An 3409.16.3	Herschel, W.J. The origin of finger-printing. London, 1916.
	An 3309.30.10	Herskovits, M.J. The anthropometry of the American Negro. N.Y., 1930.
	An 2109.52.10	Herskovits, M.J. Economic anthropology; a study in comparative economics. 2d ed. N.Y., 1952.
	An 2109.65.15	Herskovits, M.J. Economic anthropology; the economic life of primitive peoples. N.Y., 1965.
	An 2109.40	Herskovits, M.J. The economic life of primitive peoples. N.Y., 1940.
	An 359.38.26	Herskovits, Melville Jean. Acculturation. Gloucester, 1958.
	An 182.8	Herskovits, Melville Jean. Cultural relativism. 1. ed. N.Y., 1972.
	An 126.2.6	Herskovits, Melville Jean. Frank Boas: the science of man in the making. Clifton, 1973.
	An 126.2.5	Herskovits, Melville Jean. Franz Boas. N.Y., 1953.
	An 359.49.10	Herskovits, Melville Jean. Man and his works. N.Y., 1949.
	An 359.52.20	Herskovits, Melville Jean. Man and his works: the science of cultural anthropology. N.Y., 1952.
	An 359.28.5	Hertz, F.O. Race and civilization. N.Y., 1928.
	An 359.28.3	Hertz, F.O. Race and civilization. N.Y., 1928.
	An 359.25.7	Hertz, F.O. Rasse und Kultur. 3. Aufl. Leipzig, 1925.
	An 359.15.7	Hertz, Friedrich. Rasse und Kultur. 2. Aufl. Leipzig, 1915.
	Geog 4308.47.3	Herzen, A.I. Lettres de France et d'Italie (1847-1852). Genève, 1871.
	Geog 4308.47	Herzen, A.I. Tsis'ma iz Frantsii i Italii. London, 1858.
	Geog 4509.56.10	Herzog, Maurice. La montagne. Paris, 1956.
	Geog 4182.23	Hessel Gerritsz, samoyeden land ten Spitsberghe. (Naber, S.P.) 's-Gravenhage, 1924.
	Geog 4249.50.10A	Hesselberg, E. Kon-Tiki and I. N.Y., 1950.
	Geog 819.27.5	Hettner, A. Die Geographie; ihre Geschichte. Breslau, 1927.
	Geog 809.07	Hettner, A. Grundzüge der Landerkunde. Leipzig, 1907-1925. 2v.
	An 379.47	Hettner, Alfred. Allgemeine Geographie des Menschen. Stuttgart, 1947.
	Geog 5398.70	Heuglin, M.T. Reisen nach dem Nordpolarmeer. Braunschweig, 1872-74. 3v.
	An 2109.53.15	Heuse, Georges A. La psychologie ethnique. Paris, 1953.
	An 2108.94	Hewitt, J.F. The ruling races of prehistoric times. Westminster, 1894-95. 2v.
	Geog 4249.50.2	Heyerdahl, T. Kon-Tiki; across the Pacific by raft. Chicago, 1964.
	Geog 4249.50	Heyerdahl, T. Kon-Tiki. Chicago, 1950.
	Geog 4239.70.6	Heyerdahl, Thor. The Ra expeditions. London, 1971.
Htn	Geog 816.33.8*	Heylyn, Peter. Mikrokosmos; a little description of the great world. 8th ed. Oxford, 1639.
Htn	Geog 816.33.5*	Heylyn, Peter. Mikrokosmos; little description. Oxford, 1633.
Htn	Geog 816.33.7*	Heylyn, Peter. Mikrokosmos; little description. Oxford, 1639.
Htn	Geog 816.21.7*	Heylyn, Peter. Mikrokosmos: a little description of the great world. 7th ed. Oxford, 1636.
	An 2109.58	Hibben, F.C. Prehistoric man in Europe. 1. ed. Norman, Okla., 1958.
	Geog 4279.69.5	Hibbert, Christopher. The grand tour. London, 1969.
	An 359.69.15	Hiernaux, Jean. Égalité ou inégalité des races? Paris, 1969.
	Geog 4309.44.10	Hiett, Helen. No matter where. N.Y., 1944.
	Geog 5839.52.5	High Arctic; the story of the British North Greenland Expedition. (Banks, Mike.) London, 1957.
	Geog 4509.41	High conquest, the story of mountaineering. (Ullman, J.R.) Philadelphia, 1941.
	Geog 5985.55	High latitude [autobiographical]. (Davis, John K.) Parkville, Victoria, 1962.
	Geog 4239.68.15	High latitude crossing. (Haward, Peter.) London, 1968.
	Geog 4332.8	Highlands and islands of the Adriatic. (Paton, A.A.) London, 1849. 2v.
	Geog 4209.25.10	Highlights of foreign travel. (Harper, H.H.) N.Y., 1925.
	Geog 4679.41	Highroad to adventure. (Hanson, Earl P.) N.Y., 1941.
	Geog 4308.59	Highways of travel. (Sweat, M.J.M.) Boston, 1859.
	Geog 4309.03.5	Higinbotham, John V. Three weeks in Europe. Chicago, 1905.
	Geog 4329.23	Hildebrand, A.S. Blue water. N.Y., 1923.
	An 2109.37.9	Hildebrand, E. Die Geheimbünde Westafrikas als Problem der Religionwissenschrift. Inaug. Diss. Leipzig, 1937.
	Geog 4218.72.5	Hildebrandt, E.W. Reise um die Erde. 3. Aufl. Boston, 1872.
	An 359.28.9	Hildebrandt, K. Staat und Rasse. Breslau, 1928.
	An 3209.33	Hildén, Kaarlo. Maapallon esihistorialliset ja. Helsinki, 1933.
	Geog 4709.23	Hill, Samuel C. Notes on piracy in eastern waters. London, 1923-28.
	Geog 4309.72	Hillaby, John D. Journey through Europe. London, 1972.
	Geog 6009.55.15	Hillary, Edmund. No latitude for error. London, 1961.
	Geog 4239.70	Hills, Lawrence Donegan. Lands of the morning. London, 1970.
	Geog 5323.3	Hilsen til Ejnar Mikkelsen. (Woel, Cai Magens.) Kjøbenhavn, 1930.
	Geog 4306.56.5	Hiltebrandt, C.J. Dreifache schwedische Gesandtschaftsreise nach Siebenbürgen. Leiden, 1937.
	Geog 4678.55	Hinckley, F. Wrecked on a reef in the China Sea. Boston, 1908.
	Geog 958.98F	Hirt, F. Geographische Bildertafeln. v.1-3. Breslau, 1886-98. 5v.
	Geog 4309.02	His last log. (Woodward, M.) Chicago, 1903.
Htn	Geog 4206.13.5F*	His pilgrimage, or Relations of the world. 3rd ed. (Purchas, Samuel.) London, 1617.
Htn	Geog 4206.13.9F*	His pilgrimes. (Purchas, Samuel.) London, 1625- 4v.
	Geog 4239.65.5	Hiscock, Eric C. Atlantic cruise in Wanderer III. London, 1968.
	Geog 4219.59	Hiscock, Eric C. Beyond the west horizon. London, 1963.
NEDL	Geog 4158.33	Histoire...des voyages. (Montemont, A.) Paris, n.d. 46v.
	Geog 3018.28	Histoire abrégée de l'origine...de géographie. (Larenaudière.) Paris, 1828.
	Geog 3019.38.5F	Histoire de la découverte de la terre. (La Roncière, Charles.) Paris, 1938.
	An 358.37	Histoire de la filiation et migrations des peuples. v.1-2. (Brotonne, F. de.) Paris, 1837.
	Geog 4709.69.5	Histoire de la flibuste. (Blond, Georges.) Paris, 1969.
	Geog 3019.42.4	Histoire de la géographie. 4. éd. (Clozier, René.) Paris, 1967.
	Geog 5700.10	Histoire de la mission chrétienne au Groënland. Thèse. (Monnier, Edouard.) Strasbourg, 1853.
Htn	Geog 4206.10.6F*	Histoire de la navigation. (Linschoten, Jan H. van.) Amsterdam, 1638.
Htn	Geog 4206.10.5*	Histoire de la navigation. 2e éd. v.1-3. (Linschoten, Jan H. van.) Amsterdam, 1619.
	Geog 3530.20	Histoire de la pensée géographique en France, 1872-1969. (Meynier, André.) Paris, 1969.
	An 99.66.5	Histoire de l'anthropologie. (Mercier, Paul.) Paris, 1966.
	Geog 3070.12	Histoire de l'ecole cartographique. (Wauwermans, H.) Bruxelles, 1895. 2v.
	An 2108.69.7	Histoire de l'homme préhistorique anté et postdiluvien. (Bourlot, J.) Paris, 1869.
	Geog 4429.55	Histoire de l'Ocean Indien. (Auber, Jacques.) Tananarive, 1955.
	Geog 520.19	Histoire de l'origine des Indiens. (Codex Ramirez.) Paris, 1903.
	Geog 5220.7	Histoire des decouvertes...dans le Nord. (Forster, J.R.) Paris, 1788. 2v.
	Geog 4672.21.2	Histoire des naufrages, ou Recueil des relations les plus intéressantes des naufrages. (Deperthes, Jean L.H.S.) Paris, 1790. 3v.

Author and Title Listing

Geog 4672.21.4 Histoire des naufrages, ou Recueil des relations les plus intéressantes des naufrages. (Deperthes, Jean L.H.S.) Paris, 1828. 3v.
Geog 4658.50 Histoire des naufrages depuis les temps ancien jusqu'au 1850. v.1-2. (Rastoul de Mongeot, Alphonse.) Bruxelles, 1850.
Geog 4477.57 Histoire d'un peuple nouveau. Londres, 1757.
An 2600.6.1 Histoire d'une jeune fille sauvage trouvée dans les bois à l'âge de dix ans. (Hecquet, (Mme.).) Bordeaux, 1970.
Geog 4139.69 Histoire générale des grands aventuriers de la mer. v.1,4-15,17-18. Paris, 1969. 16v.
An 358.87 Histoire générale des races humaines. (Quatrefages de Bréau, A. de.) Paris, 1887-89. 2v.
Geog 4157.46 Histoire generale des voyages. (Prevost, A.T.) Paris, 1746-89. 20v.
An 358.24 Histoire naturale du genre humain. (Virey, Julien J.) Paris, 1824. 3v.
An 358.65 Histoire naturelle de l'homme et de la femme. 12. éd. (Debay, A.) Paris, 1865.
Geog 3019.55 Histoire universelle des explorations. (Parias, L.H.) Paris, 1955- 4v.
VAn 99.66.10 Historia, osobowość, sztuka. Wyd. 1. (Moliński, Bogdan.) Warszawa, 1966.
Geog 4145.11.10 Historia de Juan Sebastian del Cano. (Fernandez de Navarrete, Eustaquio.) Vitoria, 1872.
Geog 3019.22 Historia de la geografía y de la cosmografía. (Ispizúa, Segundo de.) Madrid, 1922-26. 2v.
VGeog 3019.56.4 Historia de los descubrimientos geográficos. 2. ed. (Herrmann, Paul.) Barcelona, 1967.
Geog 3570.24 Historia de uma polemica. (Costa Brochado.) Lisboa, 1944.
Geog 3570.20.2 Historia dos descobrimentos portugueses. (Peres, Damião.) Lisboa, 1959.
Geog 3570.20 Historia dos descobrimentos portugueses. pt.1-15. (Peres, Damião.) Lisboa, 1943-45.
Geog 4709.59 Historia general de la piratería. (Maséa de Ros, Angeles.) Barcelona, 1959.
An 358.43.5 Historia natural del jénero humano. 3. ed. (Virey, Julien J.) Barcelona, 1842-46. 2v.
Htn Geog 4677.35* História trágico-marítima. (Gomes de Brito, Bernardo.) Lisboa, 1735. 3v.
Geog 4672.20.4 História trágico-marítima. v.1-5. (Gomes de Brito, Bernardo.) Pôrto, 1936-37.
Geog 4709.53.5 Historias de pirates. 1. ed. (Cabal, Juan.) Barcelona, 1953.
An 2600.2 An historical account of...education of a savage man. (Itard, E.M.) London, 1802.
Geog 4157.96.3 Historical account of...voyages...Columbus to present. (Mavor, William.) Philadelphia, 1802.
Geog 4157.96 Historical account of...voyages. v.1-2,4-22. (Mavor, William.) London, 1796. 19v.
Geog 4207.53.5F Historical account of British trade. 3. ed. (Hanway, Jonas.) London, 1762. 2v.
Geog 3018.37 Historical account of circumnavigation. N.Y., 1837.
Geog 3018.37.2 Historical account of circumnavigation. N.Y., 1837.
Geog 4157.96.5 Historical account of most celebrated voyages, travels, and discoveries from time of Columbus. (Mavor, William.) New Haven, 1802-03. 24v.
Geog 4157.74 Historical account of voyages. (Henry, David.) London, 1773-74. 4v.
Geog 4208.60.5 An historical sketch of commerce and navigation. (Coggeshall, G.) N.Y., 1860.
Geog 5508.36 Historical view of discovery...of America. (Tytler, P.F.) N.Y., 1836.
Geog 5208.32A Historical view of progress of discovery. (Tytler, P.F.) Edinburgh, 1832.
Geog 5208.33 Historical view of progress of discovery. (Tytler, P.F.) Edinburgh, 1833.
Geog 4180.6 Historie of travaile. (Strachey, W.) London, 1849.
Geog 4181.103 The historie of travel into Virginia Britania. (Strachey, William.) London, 1953.
Geog 5700.5.3A Historie von Grønland...mission. (Crantz, D.) Barby, 1770. 3v.
Geog 3030.7.1 Historiia poznania Europy. Wyd 1. (Magidovich, I.P.) Warszawa, 1974.
Geog 665.20 Historisch-statistisch-geographische Belustigungen. Leipzig, 1782.
An 2109.40.20 Historische Betrachtungen über das Koitus-Konzeption Problem. (Bawmann, Evert Dirk.) Arnhem, 1940.
An 170.335 Historische Völkerkunde. (Schmitz, Carl August.) Frankfurt am Main, 1967.
An 181.1.2 History, psychology, and culture. (Goldenweiser, Alexander A.) Gloucester, 1968.
Geog 4219.20F History and cruises of the United States ship Whipple. (Kincaid, Earle H.) Constantinople, n.d.
Geog 4180.92 History and description of Africa. (Leo Africanus, J.) London, 1896. 3v.
Geog 4180.92.2 History and description of Africa. (Leo Africanus, J.) N.Y., 1964? 3v.
An 2109.68.25 History and social anthropology. (Association of Social Anthropologists of the Commonwealth.) London, 1968.
Geog 4180.14 History of...kingdom of China. (Gonzalez de Mendoza, J.) London, 1853-54. 2v.
Geog 4180.17 History of...tartar conquerors of China. (Orleans, P.J. d'.) N.Y., 1964?
Geog 4508.81 The history of a mountain. (Reclus, Elisee.) N.Y., 1881.
An 99.33 A history of American anthropology. (Mitra Panchanan.) Calcutta, 1933.
Geog 3129.48A History of ancient geography. (Thomson, J.O.) Cambridge, 1948.
Geog 3055.42.5 The history of Atlantis. (Spence, Lewis.) London, 1926.
Geog 3055.42.7 The history of Atlantis. 4th ed. (Spence, Lewis.) London, 1927.
Geog 3070.157 History of cartography. (Bagrow, Leo.) Cambridge, 1964.
Geog 3070.36 The history of cartography. (Barker, W.H.) Manchester, 1926.
An 99.37.1 The history of ethnological theory. (Lowie, Robert Harry.) N.Y., 1937.
An 99.75 A history of ethnology. (Voget, Fred W.) N.Y., 1975.
Geog 3019.34.2 A history of exploration. (Sykes, P.M.) Westport, Conn., 1975.
Geog 3019.34 A history of exploration from the earliest times to the present day. (Sykes, P.M.) N.Y., 1934.
Geog 3019.31 A history of geographical discovery and exploration. (Baker, John Norman Leonard.) London, 1931.
Geog 3019.31.2 A history of geographical discovery and exploration. (Baker, John Norman Leonard.) N.Y., 1967.
Geog 3019.67 A history of geographical discovery and exploration. (Baker, John Norman Leonard.) N.Y., 1967.
Geog 3019.12.2 History of geographical discovery in the seventeenth and eighteenth centuries. (Heawood, Edward A.) N.Y., 1965.
Geog 3000.1 Pamphlet box. History of geography.
Geog 3000.1.2 Pamphlet box. History of geography.
Geog 3019.63.5 The history of geography. (Baker, J.N.L.) Oxford, 1963.
Geog 3019.13 History of geography. (Keltie, J.S.) N.Y., 1913.
Geog 5700.5.10 History of Greenland...account of mission. (Crantz, D.) London, 1820. 2v.
Geog 5700.5 History of Greenland. (Crantz, D.) London, 1767. 2v.
Geog 5639.67.2 The history of Greenland. (Gad, Finn.) London, 1970- 2v.
Geog 3235.61 The history of Lucas Janszoon Wazehnaer and his Spieghel der Zeevaerdt. (Koeman, Cornelis.) Lausanne, 1964.
An 358.96.3A The history of mankind. (Ratzel, F.) London, 1896-98. 3v.
Geog 3018.30 The history of maritime and inland discovery. (Cooley, W.D.) London, 1830-31. 3v.
Geog 3018.30.5 The history of maritime and inland discovery. (Cooley, W.D.) London, 1833. 2v.
Geog 4709.32 The history of piracy. (Gosse, Philip.) N.Y., 1946.
Geog 5060.17 A history of polar exploration. (Kirwan, Laurence P.) N.Y., 1960.
Geog 5060.18 A history of polar exploration. (Mountfield, David.) London, 1974.
Geog 4559.68.5 A history of seafaring. (Armstrong, Richard.) N.Y., 1968-69. 3v.
Geog 4658.33 A history of shipwrecks, and disasters at sea. (Redding, Cyrus.) London, 1833. 2v.
Geog 5398.79.20 History of the adventurous voyage and terrible shipwreck of the United States steamer "Jeannette" in polar seas. N.Y., 1882.
Geog 4180.44 History of the Inâms...of Omân. (Salîl-Ibn-Razîk.) N.Y., 1964?
Geog 4181.22A History of the Incas and execution of the Inca Tupac Amaru. (Sarmiento de Gamboa, Pedro.) Cambridge, 1907.
Geog 4429.61.5 History of the Indian Ocean. (Toussaint, Auguste.) Chicago, 1966.
Geog 4429.61 History of the Indian Ocean. (Toussaint, Auguste.) London, 1966.
Geog 142.12.5 The history of the Manchester Geographical Society, 1884-1950. (Brown, Theodore Nigel Leslie.) Manchester, 1971.
Geog 4180.21 History of the new world. (Benzoni, G.) N.Y., 1964?
Geog 3570.20.4 A history of the Portuguese. (Peres, Damião.) Lisbon, 1960.
Geog 4181.116 The history of the Tahitian Mission, 1799-1830. (Davies, J.) Cambridge, Eng., 1961.
Geog 5220.5 History of voyages...in the North. (Forster, J.R.) London, 1786.
Geog 4180.65 Historye of the Bermudaes. (Butler, N.) London, 1882.
Geog 4182.75 Historysch verhael van den steden Thunes. (Pijnacker, Cornelis.) 's-Gravenhage, 1975.
An 3308.87.5 Hitchcock, E. Need of anthropometry. Brooklyn, 1887.
Geog 5209.06.2 Hoare, J.D. Arctic exploration. London, 1906.
Geog 5209.06 Hoare, J.D. Arctic exploration. N.Y., 1906.
Geog 4209.27 Hoare, Samuel. India by air. London, 1927.
An 3309.30 Hoaton, E.A. The Indians of Pecos Pueblo. New Haven, 1930.
An 3309.45 Hoaton, E.A. A survey in seating. Cambridge, 1945.
Geog 5326.21.5 Hobbs, William H. Peary. N.Y., 1936.
An 2109.15 Hobhouse, L.T. The material culture of the simpler peoples. London, 1915.
An 2109.54.10 Hocart, A.M. Social origins. London, 1954.
Htn Geog 4306.68* Hoch fürstlicher Brandenburgischer Ulysses. (Birken, S. von.) Bayreuth, 1668.
Geog 4419.07 Hochberg, Fritz. An eastern voyage; travels thro' the British Empire...in East and Japan. London, 1910. 2v.
Geog 4652.7 Hocking, Charles. Dictionary of disasters at sea during the age of steam, 1824-1962. London, 1969. 2v.
An 182.4 Hodgen, Margaret. Anthropology, history and cultural change. Tucson, 1974.
An 99.64A Hodgen, Margaret. Early anthropology in the sixteenth and seventeenth centuries. Philadelphia, 1964.
Geog 4332.19 Hodgkinson, H. The Adriatic Sea. N.Y., 1956.
Geog 5398.79.35 Hoehling, Adolph A. The Jeannette expedition, an ill fated journey to the Arctic. London, 1969.
Geog 5845.5 Hoejlund, Niels. Krise uden alternativ. København, 1972.
Geog 4502.5.8 Hoek, H. Der denkende Wanderer. München, 1932.
Geog 665.34 Hoel. Geographische Charakter-Bilder für Schule und Haus. pt.1-10. Supplement. Wien, 1886.
Geog 3055.11 Hoernes, M. Atlantis. Wien, 1884.
An 2059.03 Hoernes, Moriz. Der diluviale Mensch in Europa. Braunschweig, 1903.
An 2108.97 Hoernes, Moriz. Urgeschichte der Menschheit. 2e Aufl. Leipzig, 1897.
An 2108.92 Hoernes, Moriz. Die Urgeschichte des Menschen. Wien, 1892.
An 2109.49.10 Hoernes, Moriz. Vorgeschichte Europas. 7. Aufl. Berlin, 1949.
Geog 4249.61 Hoever, Otto. Alt-Asiaten unter Segel in Indischen und Pazifischen. Braunschweig, 1961.
Geog 5324.2.69 Hoeyer, Liv Nansen. Eva og Fridtjof Nansen. Oslo, 1954.
Geog 5324.2.70 Hoeyer, Liv Nansen. Eva og Fridtjof Nansen. 5. Oppl. Oslo, 1955.
Geog 5324.2.75 Hoeyer, Liv Nansen. Nansen og verden. 2. Oppl. Oslo, 1955.
Geog 4515.102 Hoferer, Erwin. Winterliches Bergsteigen alpine Schilaustechnik. München, 1925.
Geog 604.5 Hoff, Bert van. Jacob van Deventer. 's-Gravenhage, 1953.
Geog 4269.61.10 Hoffman, George W. A geography of Europe including Asiatic USSR. N.Y., 1961.
Geog 4218.59 Hoffman, William. The Monitor; or, jottings of a N.Y. merchant during a trip around the globe. N.Y., 1863.
Geog 4268.31.7 Hofland, B. (Mrs.). Panorama of Europe. 7th ed. London, n.d.
Geog 3060.7 Hofmann, C. Setzungsberichte. n.p., 1865.
Geog 5516.19.15 Hofrenning, Bernt Monson. Navigatio septentrionalis. Fergus Falls, Minn., 1942.
An 2109.33.3 Hofstra, Sjoerd. Differenzierungserscheinungen in einigen afrikanischen Gruppen. Proefschrift. Amsterdam, 1933.
Geog 4209.40.40 Hogben, L. Author in transit. N.Y., 1940.
Geog 5399.28.35 Hogg, Garry. Airship over the Pole; the story of the Italia. London, 1969.
Geog 4208.30 Holbrook, S.P. Sketches by a traveller. Boston, 1830.
Geog 4558.53 Holcomb, E. A wonderful providence in many incidents at sea. 8. ed. Boston, 1853.
Geog 4678.36 Holden, H. Narrative of the shipwreck. Boston, 1836.

Author and Title Listing

Geog 4678.36.5 Holden, H. Narrative of the shipwreck. Cooperstown, N.Y., 1841.
Geog 4208.78.5 Holden, J.W. A wizard's wanderings from China to Peru. London, 1886.
Geog 4308.49.5 Holidays abroad. (Kirkland, C.M.) N.Y., 1849. 2v.
Geog 4308.69.8 Holidays on high lands. (Macmillan, Hugh.) London, 1869.
Geog 1560.67 Holland; Hanbuch für Reisende. 26. Aufl. (Baedeker, publishers.) Leipzig, 1927.
Geog 5838.83.10 Holm, G. Den Danske Konebaads-Expedition. Kjøbenhavn, 1887.
Geog 4218.30 Holman, James. A voyage round the world. v.4. London, 1835.
Geog 3070.33.2A Holman, Louis A. Old maps and their makers. 2d ed. Boston, 1926.
Geog 5399.06 Holmes, B.F. The log of the "Laura" in polar seas. Cambridge, 1907.
Geog 4308.38.5 Holmes, D. A ride to Florence thro' France. London, 1841. 2v.
Geog 4309.71.15 Holmqvist, Lasse. Vitt och brett. Stockholm, 1971.
Geog 5650.18 Holtved, Erik. Archeological investigation in the Thule district. n.p., n.d.
Geog 4145.39 Holub a Mašukulumbové. (Baum, Jiří.) Praha, 1955.
Geog 4328.62 Holy Land with glimpses of Europe and Egypt. (Phelps, S.D.) N.Y., 1864.
Geog 4309.06 Hombres y piedras. (Cestero, T.M.) Madrid, 1915.
Geog 4168.50 Hombron, Bernard. Aventures...des voyageurs. Paris, n.d.
Geog 4209.31.15 Home from the sea. (Rastron, A.H.) N.Y., 1931.
Geog 4559.40.7 Home is the sailor. (Blain, W.) N.Y., 1940.
Geog 4308.99 Home letters from the continent. (Letchworth, J.) N.Y., 1902.
Geog 6009.14.20 The home of the blizzard. (Mawson, Douglas.) Philadelphia, 1914. 2v.
Geog 4559.67.5 Home port Victoria. (Jupp, Ursula.) Victoria, 1967.
An 170.230 Homenaje a Don Luis de Hoyos Sainz. Madrid, 1949-50. 2v.
Geog 4300.26 Homeward bound. 3rd ed. (Union Steamship Company, Ltd.) London, 1892.
Geog 4418.79 Homeward or travels. (Farnham, J.M.W.) Schenectady, 1879.
An 170.230.10 L'homme, hier et aujourd'hui. Paris, 1973.
Geog 5060.7.1 L'homme à la conquête des pôles. (Victor, Paul Émile.) Evreux, 1971.
An 2109.59.15 L'homme avant l'écriture. (Varagnac, André.) Paris, 1959.
An 2109.68.30.2 L'homme avant l'écriture. 2e éd. (Varagnac, André.) Paris, 1968.
An 2108.88 L'homme avant l'histoire. (Debierre, C.) Paris, 1888.
An 359.33.5 L'homme d'après la religion et la science. Thèse. (Cornelius, William J.J.) Londres, 1933.
An 378.24 L'homme du midi et l'homme du nord. (Bonstetten, K.V. von.) Genève, 1924.
Geog 665.64 L'homme et la terre. (Dardel, E.) Paris, 1952.
Geog 4209.35.33 L'homme et les iles. 2. ed. (Aubert de la Rue, Edgar.) Paris, 1935.
An 2108.78 L'homme préhistorique. (Zaborowski-Moindron, S.) Paris, 1878.
An 2108.70.5 L'homme primitif. 2. ed. (Figuier, L.G.) Paris, 1870.
An 2059.01 L'homme quaternaire. (Chantre, E.) Paris, 1901.
An 379.61.5 L'homme sur la terre. (Sorre, Maximilien.) Paris, 1961.
Geog 4559.66.5 Hommes et navires au Cap Horn, 1616-1939. (Randier, Jean.) Paris, 1966.
An 2059.23 Les hommes fossiles. 2. éd. (Boule, Marcellin.) Paris, 1923.
An 358.84 Hommes fossiles et hommes sauvages. (Quatrefages de Bréau, A. de.) Paris, 1884.
An 359.40.25 Homo sum, Gedanken zu einer zusammenfassenden Anthropologie. (Wiese und Kaiserwaldau, Leopold M. von.) Jena, 1940.
Htn Geog 815.50* Hondius, Jodocus. Thresor de chartes contenant les tableaux de tous les pays du monde. Francksfort? 1602.
Geog 4328.99 Honeyman, Abraham Van Doren. From America to the Orient. Plainfield, 1899.
Geog 4209.59.10 Hong Kong-Barcelona en el junco Rubia. (Tey, José M.) Barcelona, 1959.
Geog 3030.4 Honigmann, Ernst. Die sieben Klimata. Heidelberg, 1929.
Geog 5518.19.20 Hood, Robert. To the Arctic by canoe, 1819-1821. Montreal, 1974.
An 2109.27.19 Het hoogste wezen der primitieven. (Fahrenfort, J.J.) Groningen, 1927.
An 359.37.5A Hooton, E.A. Apes, man, and morons. N.Y., 1937.
An 359.39.5A Hooton, E.A. Twilight of man. N.Y., 1939.
An 359.40.3A Hooton, E.A. Why men behave like apes and vice versa, or Body behavior. Princeton, 1940.
An 359.45.10A Hooton, E.A. "Young man, you are normal". N.Y., 1945.
Geog 5399.25.5 The hop-off. (Ellsworth, L.) N.Y., 1925.
Geog 4129.10 Hopkins, A.A. Scientific American handbook of travel. N.Y., 1910.
Geog 5180.35 Hopper, Bruce C. Sovereignty in the Arctic. N.Y., 1937.
Geog 5558.96 Hoppin, Benjamin. A diary kept while with the Peary Arctic expedition of 1896. New Haven? 1897?
Geog 4348.54 Hoppin, J.M. Notes of a theological student. N.Y., 1854.
Geog 4308.35.15 Hoppus, John. The continent in 1835. N.Y., 1837.
An 170.351 Horizons of anthropology. (Tax, Sol.) Chicago, 1964.
An 2109.72.45 Horken, K. Ex nocte lux. Tübingen, 1972.
Geog 4209.65.20 Horvat, Joža. Besa-brodski dnevnik. Zagreb, 1973.
Geog 4308.52.11 Horwitz, O. Brushwood, picked up on the continent. Philadelphia, 1855.
Geog 4759.30.5A Hot countries. (Waugh, Alec.) N.Y., 1930.
Geog 4759.30.7 Hot countries. (Waugh, Alec.) N.Y., 1930.
An 130.1 Hough, Walter. Biographical memoir of J.W. Fawkes, 1850-1930. Washington, 1932.
An 2338.95 Hough, Walter. Primitive American armor. Washington, 1895.
Geog 5209.32 Houhen, H.H. The call of the North. London, 1932.
Geog 665.103 House, John. The frontiers of geography. New Castle upon Tyne, 1965.
Geog 4308.93.3 A house-hunter in Europe. (Bishop, W.H.) N.Y., 1893.
Geog 4329.67 Houston, James M. The western Mediterranean world. N.Y., 1967.
Geog 4559.57.5 Houter, F. den. Neptuania, maritieme curiosa. Amsterdam, 1957.
Geog 4269.67.10 Houter, F. den. Vluchtige verlenningen. Laven, 1967.
Geog 5311.3.20 Hovdenak, Gunnar. Roald Amundsen siste fere. Oslo, 1934.
An 358.81.9 Hovelacque, Abel. Les débuts de l'humanité. Paris, 1881.
An 358.87.9 Hovelacque, Abel. Précis d'anthropologie. Paris, 1887.
An 358.82 Hovelacque, Abel. Les races humaines. Paris, 1882.
Geog 5398.82 Hovgaard, A.P. Dijmphna's expeditionen, 1882-83. Kjøbenhavn, 1884.
Geog 5378.80.4 Hovgaard, A.P. Nordenskiöld's Rejse omkring Asien og Europa. Kjøbenhavn, 1881.
Geog 5378.80.3 Hovgaard, A.P. Nordenskiöld's Voyage round Asia. London, 1882.
Geog 5650.11 Hovgaard, William. The Norsemen in Greenland; recent discoveries at Herjolfones. N.Y., 1925.
Geog 4308.87.7 How I spent my summer holidays. (Cust, R.H.H.) Eton, 1887.
An 2109.26.5 How natives think. (Lévy-Bruhl, L.) London, 1926.
X Cg An 2109.25.7 How natives think. (Lévy-Bruhl, L.) N.Y., 1927.
An 2109.66.40 How natives think. (Lévy-Bruhl, Lucien.) N.Y., 1966.
Geog 5559.09.15 How Peary reached the pole. (MacMillan, D.B.) Boston, 1934.
An 358.84.10 How the world was peopled. (Fontaine, Edward.) N.Y., 1884.
Geog 819.73 How to buy an island. (McCormick, Donald.) N.Y., 1973.
An 3409.16 How to obtain good fingerprints. 2. ed. (Unites States. Bureau of Navigation. Navy Department.) Washington, 1916.
Geog 4300.16 How to prepare for Europe. (Guerber, H.A.) N.Y., 1906.
An 3409.32 How to take fingerprints. (United States. Federal Bureau of Investigation.) Washington, 1932.
An 3409.31 How to take fingerprints. (Unites States. Federal Bureau of Investigation.) Washington, 1931.
Geog 4128.80.3 How to travel. (Knox, Thomas W.) N.Y., 1881.
Geog 4128.80.5 How to travel. (Knox, Thomas W.) N.Y., 1888.
Geog 4110.30.5 How to travel without being rich. (Ford, Norman D.) Greenlawn, N.Y., 1957.
Geog 4129.14 Howard, Clare. English travellers of the Renaissance. London, 1914.
Geog 4209.33 Howard, S. Thames to Tahiti. London, 1951.
Geog 759.43 Howe, E.L. Air world. Denver, 1943.
Htn Geog 4308.68.3* Howe, Julia W. From the oak to the olive. Boston, 1868.
Geog 4309.12.10 Howell, Charles F. Around the clock in Europe. Boston, 1912.
Geog 4208.88 Howells, W.D. Library of universal adventure by sea and land. N.Y., 1888.
An 2109.63.10 Howells, William. Back of history, the story of our origins. Garden City, N.Y., 1963.
An 2109.54 Howells, William. Back of history. 1. ed. Garden City, 1954.
An 2059.59.10 Howells, William. Mankind in the making. Garden City, 1959.
Geog 5170.5 Howgate, H.W. Polar colonization. Washington, 1878.
Geog 4208.25 Howison, J. Foreign scenes and travel recreations. 2. ed. Edinburgh, 1825. 2v.
Geog 4209.36.10 Howland, Charles. Travel essays. n.p., 1936?
Geog 5312.3.10 Hoyt, Edwin Palmer. The last explorer; the adventures of Admiral Byrd. N.Y., 1968.
Geog 4308.71.3 Hoyt, James M. Glances on the wing at foreign lands. Cleveland, 1872.
Geog 4145.39.5 Hoyt, Jo Wasson. For the love of Mike. N.Y., 1966.
Geog 4328.94.5 Hoyt, W.M. A cruise on the Mediterranean. Chicago, 1894.
An 3509.12 Hrdlička, A. Early man in South America. Washington, 1912.
An 2059.29 Hrdlička, A. The neanderthal phase of man. Washington, 1929.
An 3309.25 Hrdlička, A. The old Americans. Baltimore, 1925.
An 3509.07 Hrdlička, A. Skeletal remains...early man...North America. Washington, 1907.
An 359.63.10 Hsu, Francis L.K. Clan, caste, club. Princeton, 1963.
An 359.61 Hsu, Francis L.K. Psychological anthropology. Homewood, Ill., 1961.
Geog 608.2 Hubbard, Gardiner Greene. Contents of box in corner stone of Hubbard Memorial. n.p., n.d.
An 379.32 Hubbard, George D. Geographic conditions. n.p., 1932.
Geog 4269.37 Hubbard, George D. The geography of Europe. N.Y., 1937.
Geog 4219.06 Huber, Max. Tagebuchblätter aus Siberien, Japan. Zürich, 1906.
Geog 4218.71.9F Hübner, J.A. Passeggiata intorno al mondo. Milano, 1879.
Geog 4218.71.7 Hübner, J.A. Promenade autour du monde, 1871. Paris, 1877. 2v.
Geog 4218.71.2 Hübner, J.A. Promenade autour du monde, 1871. 2e éd. Paris, 1873. 2v.
Geog 4218.71.3 Hübner, J.A. Ramble round the world. London, 1874. 2v.
Geog 4218.71 Hübner, J.A. Ramble round the world. N.Y., 1874.
Geog 3560.26 La huella de España en el mundo. (Sanz, Carlos.) Madrid, 1971-73. 3v.
Geog 5508.38 Hülstett, G.K.A. Über die nordwestliche Durchfahrt. Düsseldorf, 1838.
Geog 4329.37F Hürlimann, M. Das Mittelmeer. Zürich, 1937.
Geog 4269.52.5F Hürlimann, Martin. Europe in photographs. London, 1951.
Geog 4269.52F Hürlimann, Martin. Europe in photographs. London, 1952.
Geog 4180.79 Hues, R. Tractatus de globis et eorum usu. London, 1889.
Geog 4129.57 Hughes, Spike. The art of coarse travel. London, 1957.
Geog 4268.52 Hughes, W. Manual of geography. pt.1. London, 1852.
Geog 4559.67.15A Hugill, Stan. Sailortown. London, 1967.
Geog 3249.03 Hugues, Luigi. Cronologia delle scoperte e delle esplorazioni geografiche dall'anno 1492 a tutto il secolo XIX. Milano, 1903.
Geog 4308.51.10 Huidskoper, A. Glimpses of Europe in 1851 and 1867-1868. Meadville, Pa., 1882.
Geog 5518.29.5 Huish, Robert. The last voyage of Captain Sir John Ross, R.N., to the Arctic regions. London, 1835.
Geog 4309.35 Hull, M.L. The gangway to Europe. Boston, 1935.
Geog 4559.25 Hull down. (Hayes, Bertram.) N.Y., 1925.
Geog 4208.48.5 Huma viagem de duas mil legoas. (Barbeida, Claudio Lagrange.) Nova-Goa, 1848.
An 359.54 The human animal. (LaBarre, W.) Chicago, 1954.
An 3508.93 The human bones...Hemingway collection...United States Medical Museum at Washington. (Matthews, W.) n.p., 1893.
An 190.2.2 The human direction. 2. ed. (Peacock, James Lowe.) N.Y., 1973.
An 359.71.20 Human diversity. (Alland, Alexander.) N.Y., 1971.
An 379.35.3 Human ecology. (Bews, J.W.) London, 1935.
An 379.20.3 Human geography. (Brunhes, Jean.) Chicago, 1920.
An 379.52 Human geography. (Brunhes, Jean.) London, 1952.
An 379.28.3 The human habitat. (Huntington, Ellsworth.) London, 1928.
An 379.30.5 The human habitat. (Huntington, Ellsworth.) London, 1930.
An 379.27.3 The human habitat. (Huntington, Ellsworth.) N.Y., 1927.
An 359.72.10 The human imperative. (Alland, Alexander.) N.Y., 1972.
An 192.1 Human nature and the study of society. (Redfield, Robert.) Chicago, 1962.
Geog 623.2 Human nature in geography. (Wright, John K.) Cambridge, 1966.
An 2109.24A Human origins, a manual of prehistory. (MacCurdy, G.G.) N.Y., 1924. 2v.
An 359.61.10 Human races. (Garn, S.M.) Springfield, Ill., 1961.
An 359.03.3 The human races. (Ward, D.J.H.) n.p., 1903.
An 359.65.5 The human revolution. (Montagu, Ashley.) Cleveland, 1965.

Author and Title Listing

	An 2109.66.20	Human skeletal remains from Chalcolithic and Indo Roman levels from Nevasa: an anthropometric and comparative analysis. (Kennedy, Kenneth A.R.) Poona, 1966.
	An 358.79	The human species. (Quatrefages de Bréau, A. de.) N.Y., 1879.
NEDL	An 358.88.3	The human species. (Quatrefages de Bréau, A. de.) N.Y., 1888.
	An 2109.58.15	La humanidad prehistórica. (Maluquer de Motes, Juan.) Barcelona, 1958.
	An 2109.24.7	L'humanité primitive dans la région des Eyzies. (Capitan, Louis.) Paris, 1924.
	An 182.2	Humboldt, Wilhelm von. Philosophische Anthropologie und Theorie der Menschenkenntnis. Halle, 1929.
	An 359.17.3	Humphrey, Seth King. Mankind; racial values and racial prospect. N.Y., 1917.
	An 359.20	Humphrey, Seth King. The racial prospect. N.Y., 1920.
	Geog 5398.72.12	Hundert Jahre Franz Josefs-Land. Zur Erinnerung an die Entdeckungsreise die österreichisch-ungarische Nordpol-Expedition, 1872-1874. Wien, 1973.
	Geog 4145.94	The hundred lives of an ancient mariner. (Willis, Wiliam.) London, 1967.
	Geog 665.22	The hundred wonders of the world. 1st American ed. (Phillips, R.) New Haven, 1821.
	An 99.35	A hundred years of anthropology. (Penniman, T.K.) London, 1935.
	Geog 3019.62.5	A hundred years of geograpy. (Freeman, Thomas.) Chicago, 1962.
	Geog 4129.23.5	Hungerford, E. Planning a trip abroad. N.Y., 1923.
	Geog 4129.23	Hungerford, E. Planning a trip abroad. N.Y., 1923.
	Geog 4300.30.4	Hungerford, E. Planning a trip abroad. N.Y., 1927.
	Geog 4300.30.8	Hungerford, E. Planning a trip abroad. N.Y., 1931.
X Cg	Geog 4269.07.7PF	Hunnewell, H.H. Nineteen thousand miles in 1907. Boston, 1908. 8v.
Htn	Geog 4558.26F*	Hunnewell, J. Journal of voyage of "Missionary Packet". Charlestown, 1880.
	Geog 4128.91	Hunt, R. Steamship lines of the world. N.Y., 1891.
	An 359.72	Hunters and gatherers today. (Bicchieri, M.G.) N.Y., 1972.
	Geog 5399.06.9A	Hunters of the great North. (Stefansson, V.) N.Y., 1922.
	Geog 5558.82	Hunting and adventure in the Arctic. (Nansen, F.) N.Y., 1925.
	Geog 5839.10	Hunting with the Eskimos. (Whitney, H.) London, 1910.
	An 379.30.3	Huntington, C.C. Environmental basis of social geography. N.Y., 1930.
	Geog 4419.25	Huntington, E. West of the Pacific. N.Y., 1925.
	An 379.24	Huntington, Ellsworth. Character of races as influenced by...environment. N.Y., 1924.
	An 379.28.3	Huntington, Ellsworth. The human habitat. London, 1928.
	An 379.30.5	Huntington, Ellsworth. The human habitat. London, 1930.
	An 379.27.3	Huntington, Ellsworth. The human habitat. N.Y., 1927.
	An 379.21A	Huntington, Ellsworth. Principles of human geography. N.Y., 1921.
	An 379.22.7	Huntington, Ellsworth. Principles of human geography. 2. ed. N.Y., 1922.
	An 379.24.5	Huntington, Ellsworth. Principles of human geography. 3. ed. N.Y., 1924.
	An 379.47.5	Huntington, Ellsworth. Principles of human geography. 5. ed. N.Y., 1947.
	An 379.51	Huntington, Ellsworth. Principles of human geography. 6. ed. N.Y., 1951.
	Geog 4672.23	Huntress, Keith G. Narrative of shipwrecks and disasters, 1586-1860. Ames, 1974.
	Geog 4219.36.10	Hurja, E.E. Westward ho - fare paid. Juneau, Alaska, 1936.
	Geog 5939.25	Hurley, Frank. Argonauts of the South. N.Y., 1925.
	Geog 4679.47.5	Hurricane warning. (Meier, F.) N.Y., 1947.
	Geog 4209.40.30	Hurst, Ida. Dare to live. London, 1940.
	Geog 579.63	Hustich, I. Tämän päivän maailmaa. 4. ed. Helsinki, 1963.
	Geog 3049.62.5	Hutin, Serge. Les civilisations inconnues; mythes ou réalités. Paris, 1962.
	An 2109.38.9	Hutton, J.H. A primitive philosophy of life. Oxford, 1938.
	Geog 4308.74.5	Hutton, William. Twelve thousand miles over land and sea. Philadelphia, 1878.
	Geog 4229.26	Huxley, Aldous. Jesting pilate; an intellectual holiday. N.Y., 1926.
	An 359.35.7	Huxley, Julian S. We Europeans. London, 1935.
	An 359.35.7.10	Huxley, Julian S. We Europeans. 1. ed. N.Y., 1936.
	Geog 4557.85	Hval og robbesangsten. (Pontoppidam, C.) Risbenhavn, 1785.
	Geog 4209.31	Hyne, C.J.C. People and places. London, 1931.
	An 2109.40.15	L'Hyperborée et les migrations néolithiques. (Gattefossé, Jean.) Droguignan, 1940.
	Geog 4219.38	Hypes, J.L. Knights of the road. Washington, 1938.
	Geog 4218.03.13	I.F. Kruzenshtern. (Lupach, V.S.) Moskva, 1953.
	Geog 611.5	I.K. Kirilov i ego. Atlas Vserossisskoi imperii. (Novlianskaia, Mariia G.) Moskva, 1958.
	An 127.5	I brought the ages home. 1. ed. (Currelly, Charles T.) Toronto, 1967.
	Geog 4145.72	I cannot rest from travel. (Price, W.) N.Y., 1951.
	Geog 4239.69.10	I had to dare; rouring the Atlantic in seventy days. (McClean, Tom.) London, 1971.
	Geog 5399.53	I Knuds sloedespor. (Hansen, Leo.) København, 1953.
	Geog 5328.10	I sailed with Rasmussen. (Freuchen, Peter.) N.Y., 1958.
	An 2109.40.5	I searched the world for death. (Cooper, G.) London, 1940.
	Geog 4209.42	I tell of the seven seas. (Domville-Fife, C.W.) London, 1942.
	Geog 4209.62.5	Ia eto videl; ocherki raznykh let. (Koroteev, V.I.) Moskva, 1962.
	Geog 5399.57.5	Iatsun, E.P. Na l'dine cherez polius. Moskva, 1957.
	Geog 4413.25.5A	Ibn Batoutah. Travels of Ibn Batuta. London, 1829.
	Geog 4413.25.9	Ibn Batoutah. Voyages. Paris, 1874. 5v.
	Geog 4181.110A	Ibn Batuta. The travels of Ibn Battuta. Cambridge, 1958. 2v.
	Geog 4321.83	Ibn Gubáyer. Viaggio in Espana, Sicilia, Siria e Palestina. Roma, 1906.
	Geog 5399.22	Ice-bound, a trader's adventure in the Siberian Arctic. (Ashton, James M.) N.Y., 1928.
	Geog 5839.46	The ice capped island: Greenland. (Rodahl, Karl.) London, 1946.
	Geog 5839.54	Ice floes and flaming water. (Freuchen, P.) N.Y., 1954.
	Geog 5398.81	Ice-pack and tundra. (Gilder, W.H.) London, 1883.
	Geog 5235.22	The ice was all between. (Irvine, Tom A.) Toronto, 1959.
	Geog 5650.10.5	The Icelandic colonization of Greenland. (Bruun, Daniel.) København, 1918.
	Geog 4309.37A	Ichikawa, H. Japanese lady in Europe. N.Y., 1937.
	An 359.65	The idea of race. (Montagu, Ashley.) Lincoln, 1965.
	An 359.62.25	The idea of racialism. (Snyder, L.L.) Princeton, 1962.
	An 2109.09.7A	The idea of the soul. (Crawley, A.E.) London, 1909.
	An 3309.38	La identificación personal por la retina. (Simon, Carleton.) La Habana, 1938.
	An 3308.85	Identification anthropometrique. (Bertillon, A.) Melun, 1885.
	Geog 3199.10	al-Idrisi, Muhammad ibn Muhammad. Deutschland und seine Naehbarländer nach der grossen Geographie des Idrisi. Stuttgart, 1937.
	Geog 5660.42	Igdlucrúnerit. (Gulloev, H.C.) København, 1971.
	Geog 5559.38	Igloo for the night. (Manning, E.W.) London, 1943.
	Geog 4131.55	Ignacio de Arrillaga, José. Sistema de política turística. Madrid, 1955.
	Geog 5939.62	Ignatov, V.S. God na poliuse kholoda. Moskva, 1962.
	Geog 113.5	The IGU newsletter. (International Geographical Union.) N.Y. 1-13,1950-1962
	Geog 4239.27	Ihering, H. von. Die Geschichte des Atlantischen Ozeans. Jena, 1927.
	Geog 4182.9	Ijzermann, J.W. Dirck Gerritsz Pomp. 's-Gravenhage, 1915.
	Geog 4182.27	Ijzermann, J.W. De reis om de wereld door Olivier van Noort. 's-Gravehage, 1926. 2v.
	Geog 809.59.10F	Il milione; enciclopedia di geografia. (Istituto Geografico de Agostini.) Novara, 1959-1965. 15v.
	An 359.40.10	The illusion of national character. (Fyfe, Henry H.) London, 1940.
	Geog 4300.23	Illustrated Europe. Zürich.
	Geog 4324.89	Illustrazioni in un anonimo Viaggiatore (1489). Livorno, 1785.
	Geog 819.19	Illustrierte Länderkunde. (Banse, E.) Berlin, 1919.
	An 359.10.7	Illustrierte Völkerkunde. (Buschan, Georg H.T.) Stuttgart, 1910.
	Geog 1590.25	Illustrierter Führer durch Galizien. (Orlowicz, M.) Wien, 1914.
	Geog 4559.30.10	I'm alone, by Captain Jack Randell, as told to Meigs O. Frost. 1. ed. (Randell, Jack.) Indianapolis, 1930.
	Geog 5208.81	Im ewigen Eis. (Hellwald, F. von.) Stuttgart, 1881.
	Geog 5559.23	Im Flugzeug dem Nordpol entgegen. (Mittelholzer, W.) Zürich, 1924.
	Geog 4209.25.15	Im Kaleidoskop der Weltteile. (Jacques, N.) Berlin, 1925.
	Geog 5559.12	Im Kampfe mit Bär und Walross. (Cook, F.A.) Braunschweig, 192-.
	Geog 4418.38	The imagery of foreign travel. (Sherer, M.) London, 1838.
	An 99.74	Images of man; a history of anthropological thought. 1. ed. (De Waal Malefijt, Annemarie.) N.Y., 1974.
	Geog 809.52	Imago mundi. (Teran, M. de.) Madrid, 1952. 2v.
	An 179.1A	The immense journey. (Eiseley, L.C.) N.Y., 1957.
	Geog 578.76	The imperial gazetteer. (Blackie, W.G.) London, 1876. 2v.
	Geog 1807.30	The Imperial guide to India. (Murray, John, publisher, London.) London, 1904.
	An 170.174.5	Implicit meanings. (Douglas, Mary T.) London, 1975.
	Geog 4219.70	The impossible voyage. (Blyth, Chay.) London, 1971.
	Geog 4208.91	Impressions de voyage. (Cimon, Henri.) Québec, 1895-1902. 2v.
	Geog 4218.92.3	Impressions of a journey round the world. (Peters, G.H.) London, 1897.
	Geog 4209.07	Impressions of a wanderer. (Mallik, M.C.) London, 1907.
	Geog 4308.49.10	Impressions of central and southern Europe. (Baxter, William E.) London, 1850.
	Geog 4208.90	Impressions of travels. (Platen, Carl von.) Stockholm, 1890.
	Geog 4309.64.5	Impressions sur l'Europe. (Cluzel, M.E.) Paris, 1964.
	Geog 4309.12.5	Impressões da Europa. (Peçanha, Nilo.) Nice, 1912.
	Geog 5398.91	In Arctic seas...voyage of the "Kite". (Keely, Robert.) Philadelphia, 1892.
	Geog 5558.91.5	In Arctic seas...with Peary expedition. (Keely, R.N.) Philadelphia, 1892.
	Geog 3540.12	In blaver Ferne lag Amerika. 3. Aufl. (Wotte, Herbert.) Leipzig, 1974.
	Geog 4145.28.5	In China, auf Asiens Hochsteppen. (Filchner, Wilhelm.) Freiburg, 1930.
	Geog 5328.5.5	In der Heimat des Polarmenschen. (Rasmussen, Knud.) Leipzig, 1922.
	An 359.48.10	In Henry's backyard. (Bendict, Ruth (Fulton).) N.Y., 1947.
	An 147.5	In memoriam Karl Weule. Leipzig, 1929.
	Geog 602.3.10	In memoriam Raoul Blanchard, 1877-1965. (Association des Amis de l'Université de Grenoble.) Grenoble, 1966.
	Geog 5340.5F	In mezzo ai ghiacci. Milano, 1880.
	Geog 5070.70.5	In mischief's wake. (Tilman, Harold William.) London, 1971.
	Geog 5398.93.5	In Nacht undEis. (Nansen, Fridtjof.) Leipzig, 1897. 2v.
	Geog 5220.22.5	In northern mists, Arctic exploration in early times. (Nansen, F.) London, 1911. 2v.
	Geog 5220.22.10	In northern mists. (Nansen, F.) N.Y., 1911. 2v.
	Geog 5509.58	In quest of the North West Passage. (Neatby, Leslie Hamilton.) London, 1958.
	Geog 4559.37.5	In sail and steam...merchant service, 1902-1927. (Making, V.L.) London, 1937.
	Geog 4229.05	In search of El Dorado. (Macdonald, A.) London, 1905.
	An 359.57.30	In search of man. (Missenard, André.) N.Y., 1957.
	An 2109.27.11	In search of our ancestors. (Boyle, Mary E.) London, 1927.
	An 359.74.5	In search of the primitive. (Diamond, Stanley.) N.Y., 1974.
	An 2109.66.5	In search of the primitive. 1st ed. (Cotlow, Lewis N.) Boston, 1966.
	An 2109.28.13	In the beginning; the origin of civilization. (Smith, Grafton Eliot.) London, 1928.
	Geog 5398.81.5	In the Lena Delta. (Melville, G.W.) Boston, 1892.
	Geog 4329.36.6	In the steps of St. Paul. (Morton, H.V.) London, 1936.
	Geog 4329.36.7	In the steps of St. Paul. (Morton, H.V.) N.Y., 1936.
	Geog 4218.93	In the track of the sun. (Thompson, F.D.) N.Y., 1893.
	Geog 4238.83.8F	In the trades, the tropics, and the 'roaring forties. (Brassey, Anna.) London, 1886.
	Geog 4219.08.5	In the wake of the setting sun. (Carter, James.) London, 1908.
	Geog 4559.62	In with the sea wind. (Bradford, G.) Barre, 1962.
	Geog 4559.40	In Yankee windjammers. (Nordhoff, C.) N.Y., 1940.
Htn	Geog 4308.38.8*A	Incidents of travel in Greece, Turkey and Poland.7th ed. (Stephens, J.L.) N.Y., 1838. 2v.
	Geog 4308.38.14	Incidents of travel in Greece, Turkey and Poland. 17th ed. (Stephens, J.L.) N.Y., 1845. 2v.
	Geog 616.3.5	The incurable romantic. (Peattie, R.) N.Y., 1941.
	An 177.2	La indecision. (Carias, Rafael.) Caracas, 1970.
	An 3409.13	La indentificación dactiloscópica. (Ortiz, F.) Habana, 1913.

202

Author and Title Listing

Geog 13.4.4 Index to the Bulletin of the American Geographical Society. 1852-1910. (American Geographical Society of New York.) N.Y., n.d.

Geog 4209.27 India by air. (Hoare, Samuel.) London, 1927.

Geog 4180.22 India in the fifteenth century. (Major, R.H.) London, 1857.

An 170.264 Indian anthropology. (Madan, T.N.) Bombay, 1962.

Geog 19.5.10 The Indian geographer. (Association of Indian Geographers.) New Delhi. 1957

An 3309.30 The Indians of Pecos Pueblo. (Hoaton, E.A.) New Haven, 1930.

Geog 3550.9 Indicazioni di opere...sopra i viaggi, le navigazioni, le scoperte...degl'Italiani nel medio evo. (Canale, M.G.) Lucca, 1861.

NEDL Geog 39.2.5 Indice generale dei cinquanta primi numeri (dal 1865 al 1884). (Club Alpino Italiano.) Torino, 1885.

An 3558.98 Indices cephálicos dos Portuguêses. (Silva Boasto, A.J. de.) Coimbra, 1898.

An 359.11.3 Gli indici di omogamia delle razze e delle nazionalità. (Savorgnan, Franco.) Cagliari, 1911.

An 358.57 Indigenous races of the earth. (Nott, Josiah C.) Philadelphia, 1857.

An 2109.39.7 The individual and his society. (Kardiner, A.) N.Y., 1939.

Geog 665.29F Inéditos (miscellanea). (Santarem, M.F. de B.) Lisboa, 1914.

An 359.15.10 The inequality of human races. (Gobineau, A.) N.Y., 1915.

Geog 5399.28.52 L'inferno bianco. 4. ed. (Tomaselli, Francesco.) Milano, 1929.

An 359.16.5 The influence of anthropology on the course of political science. (Myres, John L.) Berkeley, 1916.

Geog 4509.57.5 The influence of mountains upon the development of human intelligence. (Young, G.W.) Glasgow, 1957.

An 379.11 Influences of geographic environment on the basis of Ratzel's system. (Semple, E.M.) N.Y., 1911.

Geog 4300.18.5 Information and directions for travellers. (Starke, M.) Paris, 1826.

Geog 212.50 Informe sobre los trabajos cartográficos. (Sociedad Mexicana de Geográfica y Estadística.) México. 5-7,1938-1947 4v.

Geog 5398.97.6 Ingen fruktan, intet hopp. (Sundman, Per Olaf.) Stockholm, 1968.

Geog 818.54 Ingerslev, C.F. Stutt kennslubok i landafroedinni. Reykjavik, 1854.

Geog 5559.09.20 Ingersoll, Ernest. The conquest of the North. N.Y., 1909.

Geog 5538.52.10 Inglefield, Edward A. A summer search for Sir John Franklin. London, 1853.

Geog 4308.30 Inglis, H.D. Switzerland...France and Pyrenees. Edinburgh, 1831. 2v.

Geog 4227.90 Ingraham, Joseph. Journal of the brigantine Hope on the voyage to the northwest coast of North America, 1790-92. Barre, 1971.

Geog 4169.36.5A Ingram, B.S. Three sea journals of Stuart times. London, 1936.

Geog 4659.36.3 Ingram, Charles W.N. New Zealand shipwrecks, 1795-1960. 3. ed. Wellington, 1961.

Geog 4659.36 Ingram, Charles W.N. Shipwrecks; New Zealand disasters, 1795-1936. Dunedin, 1936.

Geog 5839.35.15 Ingstad, H.M. East of the great glacier. N.Y., 1937.

Geog 5839.59.2 Ingstad, Helge Marcus. Land under the pole star: a voyage to the Norse settlements of Greenland and the saga of the people that vanished. London, 1966.

Geog 5839.59 Ingstad, Helge Marcus. Landet under leidarstjernen. Oslo, 1959.

Geog 665.24 Ingwood of Westchester. Transatlantic souvenirs. N.Y., 1868.

Geog 500.170 Inhaltsverzeichnis der Festschriften zur Ehrung und Würdigung deutscher. (Schwickerath, Hildegard.) Bad Godesberg, 1969.

An 379.22.5 Inheriting the earth. (Von Engeln, Oscar D.) N.Y., 1922.

Geog 659.53.5 Iniciación a la geografía local. (Spain. Consejo Superior de Investigaciones Cientificas.) Zaragoza, 1953.

Geog 109.5 Institute of British Geographers. Publications. London. 1933-1975// 20v.

Geog 109.6 Institute of British Geographers. Transactions. New series. London. 1,1976+

An 2109.54.6 The institutions of primitive society. (Values of Primitive Society (Radio Program).) Oxford, 1967.

Geog 5907.5 Instituto Antartico Argentino. Boletin. Buenos Aires.

An 629.36 Instruções praticas para pesquisas de antropologia fisica e cultural. pt.1. (Lévi-Strauss, D.) São Paulo, 1936-

An 628.68 Instructions for research...ethnology and philology of America. (Gibbs, G.) Washington, 1863-68. 3 pam.

An 359.71.10 Integral anthropology. (Wolf, Josef.) Praha, 1971.

Geog 5153.15 Inter-nord. Paris. 4+ 6v.

Geog 4678.37.15 Interesting and authentic narratives of the most remarkable shipwrecks, fires. (Thomas, R.) Hartford, 1837.

Geog 5900.5 International Antarctic bibliography. Braunschweig. 1958

Geog 500.180 International Committee of Historical Sciences. Commission Internationale d'Histoire Maritime. Compte rendu des travaux de la commission internationale d'histoire maritime, 1965. Paris? 1965?

Geog 500.68 International Committee of Historical Sciences. Committee on the History of Great Voyages and Great Discoveries. Travaux de la Commission pour l'histoire des grands voyages et des grandes decouvertes. Paris, 1936.

An 170.238 International Congress of Anthropological and Ethnological Sciences, 5th, Philadelphia, 1960. Selected papers. Philadelphia, 1960.

An 170.238.10 International Congress of Anthropological and Ethnological Sciences, 7th, Moscow, 1964. VII Mezhdunarodnyi Kongress antropologicheskikh i ethnograficheskikh nauk. Trudy. v.1-3,5-10; photoreproduction. Moskva, 1967- 13v.

An 39.60 International dictionary of regional Europen ethnology and folklore. Copenhagen, 1960- 2v.

An 5.18 International directory of anthropological institutions. (Thomas, William L.) N.Y., 1953.

An 124.1 International directory of anthropologists. (National Research Council.) Washington, 1938.

An 124.5 International directory of anthropologists. 3. ed. (National Research Council.) Washington, 1950.

Geog 110.2.17 International Geographical Congress, 7th, Berlin. Miscellaneous papers. v.1-6. Berlin, 1899.

Geog 110.2.18 Pamphlet box. International Geographical Congress, 8th, Berlin.

Geog 110.01.8 Pamphlet vol. International Geographical Congress, 8th.

Geog 110.1.13 International Geographical Congress, 13th, 1931. Livret-guide du congressiste. Paris, 1931.

Geog 110.1.14 International Geographical Congress, 14th, Warsaw, 1934. Comptes rendus. v.1-7. Varsovie, 1934.

Geog 110.1.15 International Geographical Congress, 15th, Amsterdam, 1938. Comptes rendus du Congrès international de géographie, Amsterdam, 1938. v.1-5,21. Leiden, 1938. 6v.

Geog 110.1.18 International Geographical Congress, 18th, Rio de Janeiro. Abstracts of papers. Rio de Janeiro, 1956.

Geog 110.1.18.5 International Geographical Congress, 18th, Rio de Janeiro, 1956. Comptes rendus du XVIIIe congrès international de géographie. Rio de Janeiro, 1959. 4v.

Geog 110.1.20 International Geographical Congress, 20th, London, 1964. Abstracts of papers. London, 1964.

Geog 110.1.20.2 International Geographical Congress, 20th, London, 1964. Abstracts of papers. Supplement. London, 1964.

Geog 110.1.20.10 International Geographical Congress, 20th, London, 1964. Congress proceedings. London, 1967.

Geog 110.1.20.5 International Geographical Congress, 20th, London, 1964. Congress programme. London, 1964.

Geog 110.1.25 International Geographical Congress, 20th, London, 1964. Sovremennye problemy geografii. London, 1964.

Geog 110.1.21 International Geographical Congress, 21st, Delhi, 1968. Abstracts of papers. Supplement. Calcutta, 1968.

Geog 110.1.22 International Geographical Congress, 22nd, Montreal, 1972. Doklady k XXII mezhdunarodnomu geograficheskomu kongress (Kanada, avgust, 1972). Leningrad, 1972.

Geog 110.1.16 International Geographical Congress. Comptes rendus du Congrès international de géographie. Lisbonne, 1950. 2v.

Geog 3025.7 International Geographical Congress. Über...Herstellung...Erdkarte im Mafestabe. Wien, n.d.

Geog 110.1.500 Pamphlet vol. International Geographical Congress.

Geog 110.1.13.5 International Geographical Congress. pt. A-B. Paris, 1931. 2v.

Geog 110.1 International Geographical Congress. v.1-8. Anvers, 1872. 23v.

Geog 113.5 International Geographical Union. The IGU newsletter. N.Y. 1-13,1950-1962

Geog 665.81 International Geographical Union. Regional Conference in Japan, Tokyo and Nara, 1957. Proceedings of the IGU Regional Conference in Japan, August 28-September 3, 1957. Tokyo, 1959.

Geog 4750.2 International Geographical Union. Special Commission on the Humid Tropics. A select annotated bibliography of the humid tropics. Montreal, 1960.

Geog 4131.74 International Geographical Union. Working Group, Geography of Tourism and Recreation. Studies in the geography of tourism. Frankfurt, 1974.

Geog 819.01 The international geography. (Mill, H.R.) N.Y., 1901.

Geog 819.01.7 The international geography. (Mill, H.R.) N.Y., 1907.

Geog 819.01.6 The international geography. (Mill, H.R.) N.Y., 1920.

Geog 819.01.10 International geography. 2nd ed. (Mill, H.R.) N.Y., 1900.

Geog 500.91 International list of geographical serials. (Harris, Chauncy Donnison.) Chicago, 1960.

An 2059.37 International Symposium on Early Man. Early man. Philadelphia, 1937.

An 379.56.1 International Symposium on Man's Role in Changing the Face of the Earth, Princeton, N.J., 1955. Man's role in changing the face of the earth. Chicago, 1971.

Geog 4131.51 International Touring Association. Scientific Commission. Publication de la Commission scientifique de l'alliance internationale de tourisme. Berne. 1-6,1951-1956 3v.

Geog 4129.58.5 International travel. (Randall, C.B.) Washington, 1958.

Geog 4131.68.15 Internationale Informationstagung zur Geographie des Fremdenverkehrs. Dresden, 1965. Probleme der Geographie des Fremdenverkehrs der Deutschen Demokratischen Republik und anderer Staaten. Leipzig, 1968.

Geog 3070.115 Internationaler Kurz für Kartendruck. Fortschrittsberichte auf dem Gebiet des Kartendrucks. Hamburg, 1958.

An 359.73.5 The interpretation of cultures. (Goertz, Clifford.) N.Y., 1973.

An 170.324 The interpretation of ritual; essays in honour of A.I. Richards. London, 1972.

An 170.361 The interpretation of symbolism. N.Y., 1975.

Geog 4209.41.20 Intimate Virginiana. (Maury, A.F.) Richmond, 1941.

Geog 6009.28 Into the home of the blizzard. (Byrd, Richard E.) N.Y., 1928.

An 2109.72.25 Into the primitive environment: survival on the edge of our civilization. (Brain, Robert.) London, 1972.

An 170.117 Introdução a antropologia tropical. (Athayde, Alfredo.) Lisboa, 1962.

Geog 3570.50 Introducão a historia dos descobrimentos. (Albuquerque, Louis.) Coimbra, 1962.

An 359.33.9 Introdução i antropobiologia. (Mendes Correia, A.A.) Coimbra, 1933.

Geog 3019.53.10 Introducción a la historia de la geografía. (Fochler-Hanke, Gustav.) Tucumán, 1953.

Geog 807.16F Introductio ad geographia et spheram. (Notoras, C.) Paris, 1716.

Htn Geog 755.19* Introductio in Ptolomei. n.p., n.d.

Geog 816.93 Introduction à la geographie. (Sanson, N.) Paris, 1693.

Geog 816.61.19 Introduction à la geographie universelle. (Cluverius, P.) Paris, 1631.

Geog 3235.47 Introduction à l'atlas des monuments de la géographie. (Jomard, E.F.) Paris, 1879.

NEDL Geog 816.61.3 Introduction into geography. (Cluverius, P.) Oxford, 1657.

An 358.63.2 Introduction to anthropology. (Waitz, T.) London, 1863.

An 2109.40.3 An introduction to cultural anthropology. (Lowie, R.H.) N.Y., 1940.

An 359.45.25 An introduction to physical anthropology. (Montagu, Ashley.) Springfield, 1945.

An 3209.60 An introduction to physical anthropology. 3. ed. (Montagu, Ashley.) Springfield, 1960.

An 2109.72.15 Introduction to prehistory; a systematic approach. (Rovse, Irving.) N.Y., 1972.

Geog 659.75 An introduction to scientific reasoning in geography. (Amedeo, Douglas.) N.Y., 1975.

An 2059.29.5 An introduction to social anthropology. (Wissler, Clark.) N.Y., 1929.

An 3209.71 An introduction to the study of man. (Young, John Z.) Oxford, 1971.

Geog 816.61 Introductionis in...geographiam. (Cluverius, P.) Amsterdam, 1661.

Geog 816.61.6 Introductionis in...geographiam. (Cluverius, P.) Amsterdam, 1686.

Geog 816.61.5 Introductionis in...geographiam. (Cluverius, P.) Brunsvigae, 1672.

Author and Title Listing

Htn Geog 816.61.9* Introductionis in...geographiam. (Cluverius, P.) Londini, 1711.

Htn Geog 816.61.4* Introductionis in universam geographiam. (Cluverius, P.) Amsterdam, 1659.

Geog 816.61.7 Introductionis in universam geographiam. (Cluverius, P.) Amsterdam, 1697.

Htn Geog 816.61.13* Introductionis in universam geographiam. (Cluverius, P.) Amsterdam, 1729.

Htn Geog 816.61.16* Introductionis in universam geographiam. (Cluverius, P.) Lugdunum Batavorum, 1627.

An 359.75 The invention of culture. (Wagner, Roy.) Englewood Cliffs, 1975.

Geog 4750.4 An inventory of geographic research of the humid tropic environment. (Texas Instruments, Inc.) Dallas, Texas, 1967? 2v.

An 629.60 Inverarity, Robert Bruce. Visual files coding index. Bloomington, 1960.

An 358.48 An investigation of the theories of the natural history of man. (Van Amringe, W.F.) N.Y., 1848.

Geog 4309.38.15 Invitation to travel. (Fish, Helen D.) N.Y., 1938.

Geog 3019.27.10 Iorga, N. Une vingtaine de voyageurs dans l'Orient européen. Paris, 1928.

Geog 3019.27.2 Iorga, N. Les voyageurs français dans l'Orient européen. Paris, 1928.

Geog 3019.27.5 Iorga, N. Les voyageurs orientaux en France. Paris, 1927.

An 136.5.65 Ipola, Emilio Rafael de. Le structuralisme, ou L'histoire en exil. Thèse. Nantorre? 1969.

Geog 4309.31.25 Iraizoz y de Villar, Antonio. Apuntes de un turista tropical. Habana, 1931.

Geog 4308.01 An Irish peer on the continent. (Wilmot, Catherine.) London, 1920.

Geog 5235.22 Irvine, Tom A. The ice was all between. Toronto, 1959.

Geog 4308.07 Irving, Peter. Peter Irving's journals. N.Y., 1943.

Geog 4512.1 Irving, R.L.G. The mountain way. London, 1938.

Geog 4509.40.5 Irving, R.L.G. Ten great mountains. N.Y., 1940.

Geog 4417.80 Irwin, E. Series of adventures...Red Sea, Arabia, Egypt. Dublin, 1780.

Geog 4417.80.3 Irwin, E. Series of adventures...Red Sea, Arabia, Egypt. 3. ed. London, 1787. 2v.

An 2109.44 Is there a fundamental mental difference between primitive man and the civilized European. (Sommerfelt, A.) n.p., 1944.

An 359.23.7 Is there a new race type? (Pape, A.G.) Edinburgh, 1923.

An 359.64.20 Isā, Ali Ahmad. Social anthropology in theory and practice. Cairo, 1964.

Geog 3019.71 Isachenko, Anatolii G. Razvitie geograficheskikh idei. Moskva, 1971.

Geog 618.3.3 Ishill, Joseph. Élisée and Elie Reclus, in memoriam. Berkeley Heights, N.J., 1927.

Geog 5898.241 Isimardik. (Rosing, Jens.) København, 1960.

Geog 4709.69 Island treasure: the search for Sir Cloudesley Shovell's flagship "Association". (Morris, Roland.) London, 1969.

Geog 4209.40.15 Islands of adventure. (Baarslag, K.) N.Y., 1940.

Geog 4329.26 Islands of the Mediterranean. (Wilstach, Paul.) London, 1926.

Geog 4329.31 Islands of the Mediterranean. (Wilstach, Paul.) N.Y., 1931.

Geog 4209.56.5 Islands of to-morrow. (Elmberg, J.E.O.) London, 1956.

Geog 3209.13 En Islandsk Vejviser for pilgrimme fra 12 århundrede. København, 1913.

Geog 819.18 Islario general de todas las isles del mundo. Atlas. (Santa Cruz, Alonso de.) Madrid, 1918. 2v.

Geog 5559.32 The isle of auks. (Polunin, N.) London, 1932.

Htn Geog 3060.8* The isle of pines, 1668. (Ford, Worthington Chauncey.) Boston, 1920.

Geog 3060.8 The isle of pines, 1668. (Ford, Worthington Chauncey.) Boston, 1920.

Geog 3060.5 Les isles fantastiques de l'Atlantique. (Gaffarel, P.) n.p., 1883.

Geog 4239.29.5 Isles of romance. (England, George Allan.) N.Y., 1929.

An 3559.70 Ismagulov, Orazak. Naselenie Kazakhstana ot epokhi bronzy do sovremennosti. Alma-Ata, 1970.

Geog 4329.73 Isnard, Hildebert. Pays et paysages mediterranéens. 1. éd. Paris, 1973.

Geog 3019.22 Ispizúa, Segundo de. Historia de la geografía y de la cosmografía. Madrid, 1922-26. 2v.

Geog 112.5 Istanbul. Universite. Cografya Ensitüsü. Coğrafî araştirmalar. n.p., n.d.

Geog 114.5 Istituto Cartografico Italiano, Rome. Annuario dell'Istituto. Roma. 1-4,1884-1889

Geog 4500.10 Istituto di Bibliografia Alpina, Biella. Opere antiche e rare sulla montagna. Biella, 1951.

Geog 85.350 Istituto Geografico de Agostini. Communicazione dell...la geografia. Novara. 1-18,1912-1930 9v.

Geog 809.59.10F Istituto Geografico de Agostini. Il milione; enciclopedia di geografia. Novara, 1959-1965. 15v.

Geog 500.100 Istituto Veneto di Scienza, Lettre ed Arte. Carte geografiche cinquecentesche a stampa della Biblioteca Marciana e della Biblioteca del museo correr di Venezia. Venezia, 1954.

Geog 500.75A Istoricheskii obzor uchebniko geografii, 1876-1934. (Baranskii, N.N.) Moskva, 1954.

Geog 115.5 Istoriia geograficheskikh znanii i istoricheskaia geografiia. Etnografiia. Moskva. 2,1967+

Geog 3590.88 Istoriia geograficheskoi mysli v Moldavii. (Krupenikov, Igor' A.) Kishinev, 1974.

Geog 5939.63 Istoriia otkrytiia i issledovaniia Antarktidy. (Treshnikov, Aleksei G.) Moskva, 1963.

Geog 3030.7 Istoriia otkrytiia i issledovaniia Evropy. (Magidovich, I.P.) Moskva, 1970.

Geog 5369.56 Istoriia otkrytiia i osvoeniia...morskogo poeti. v.1-3. Moskva, 1956- 4v.

An 2109.73 Istoriia pervobytnogo obschchestvo. (Knyshenko, Iurii V.) Rostov-na-Donu, 1973.

An 2109.67.5 Istoriia pervobytnogo obshchestva. (Goremykina, Vera I.) Minsk, 1967.

An 2109.68.15 Istoriia pervobytnogo obshchestva. (Pershits, Abram I.) Moskva, 1968.

An 2109.68.17 Istoriia pervobytnogo obshchestva. Izd. 2. (Pershits, Abram I.) Moskva, 1974.

Geog 3590.84 Istoriia razvitiia kartoizdaniia v Rossii i v SSSR. (Krempol'skii, Viktor F.) Moskva, 1959.

An 99.66 Istoriia russkoi etnografii, dook tiabr'skii period. (Tokarev, Sergei A.) Moskva, 1966.

Geog 4209.39.5 It was news to me. (O'Malley, C.J.) Boston, 1939.

Geog 5399.28.10 L'"Italia" al Polo Nord. (Nobile, Umberto.) Milano, 1930.

Geog 4209.67.15 L'italiano di Ponte Cayumba. (Fiore, Ilario.) Firenze, 1967.

Geog 1539.5 L'Italie...septentrionale. 3. éd. (Baedeker, publishers.) Coblenz, 1865.

Geog 1539.10 L'Italie...septentrionale. 5. éd. (Baedeker, publishers.) Coblenz, 1870.

Geog 1539.18 L'Italie...septentrionale. 7. éd. (Baedeker, publishers.) Leipzig, 1876.

Geog 1539.19 L'Italie...septentrionale. 8. éd. (Baedeker, publishers.) Leipzig, 1878.

Geog 1539.22 L'Italie...septentrionale. 9. éd. (Baedeker, publishers.) Leipzig, 1880.

Geog 1539.11 L'Italie...septentrionale. 12. éd. (Baedeker, publishers.) Leipzig, 1889.

Geog 1540.17 Italie. Pt. 2: Italie centrale et Rome. 4. éd. (Baedeker, publishers.) Leipzig, 1875.

Geog 1540.17.25 Italie. Pt. 2: Italie centrale et Rome. 5. éd. (Baedeker, publishers.) Leipzig, 1877.

Geog 1542.14 Italie. Pt. 3: Italie du sud et la Sicile. 4. éd. (Baedeker, publishers.) Leipzig, 1875.

Geog 1542.16 Italie. Pt. 3: Italie méridionale et la Sicile. 5. éd. (Baedeker, publishers.) Leipzig, 1877.

Geog 1539.51 Italie septentrionale. 17. éd. (Baedeker, publishers.) Leipzig, 1908.

Geog 1538.20 Italien. (Baedeker, publishers.) Coblenz, 1866-68. 2v.

Geog 1538.37 Italien. v.3. 11. Aufl. (Baedeker, publishers.) Leipzig, 1895.

Geog 1538.30 Italien. 9. Aufl. (Baedeker, publishers.) Leipzig, 1879-89. 2v.

Geog 1538.35 Italien. 14. Aufl. (Baedeker, publishers.) Leipzig, 1894.

Geog 1538.45 Italien von den Alpen bis Neapel. 3. Aufl. (Baedeker, publishers.) Leipzig, 1895.

Geog 3235.50 Die italienischen Portolane des Mittelalters. (Kretschmer, K.) Berlin, 1909.

Geog 1538.61 Italy, including Sicily and Sardinia. (Baedeker, publishers.) Freiburg, 1961.

Geog 4307.80.5 Italy, Spain and Portugal. (Beckford, William.) London, 1840.

Geog 4307.80.7 Italy, Spain and Portugal. Pt.1-2. (Beckford, William.) N.Y., 1845.

Geog 4307.80.3 Italy, with sketches of Spain and Portugal. (Beckford, William.) Philadelphia, 1834.

Geog 1539.12 Italy...northern Italy. 2. ed. (Baedeker, publishers.) Coblenz, 1870.

Geog 1539.17 Italy...northern Italy. 3. ed. (Baedeker, publishers.) Leipsic, 1874.

Geog 1539.17.5 Italy...northern Italy. 3. ed. (Baedeker, publishers.) Leipsic, 1874.

Geog 1539.21 Italy...northern Italy. 5. ed. (Baedeker, publishers.) Leipsic, 1879.

Geog 1539.23 Italy...northern Italy. 6. ed. (Baedeker, publishers.) Leipsic, 1882.

Geog 1539.25 Italy...northern Italy. 7. ed. (Baedeker, publishers.) Leipsic, 1886.

Geog 1539.30 Italy...northern Italy. 8. ed. (Baedeker, publishers.) Leipsic, 1889.

Geog 1539.35 Italy...northern Italy. 9. ed. (Baedeker, publishers.) Leipsic, 1892.

Geog 1539.36 Italy...northern Italy. 9. ed. (Baedeker, publishers.) Leipsic, 1892.

Geog 1539.38 Italy...northern Italy. 10. ed. (Baedeker, publishers.) Leipsic, 1895.

Geog 1539.40 Italy...northern Italy. 11. ed. (Baedeker, publishers.) Leipsic, 1899.

Geog 1539.45 Italy...northern Italy. 12. ed. (Baedeker, publishers.) Leipsic, 1903.

Geog 1539.46 Italy...northern Italy. 12. ed. (Baedeker, publishers.) Leipsic, 1903.

Geog 1539.50 Italy...northern Italy. 13. ed. (Baedeker, publishers.) Leipsic, 1906-

Geog 1540.5 Italy. Pt. 2: Central Italy and Rome. (Baedeker, publishers.) Coblenz, 1867.

Geog 1540.10 Italy. Pt. 2: Central Italy and Rome. 2. ed. (Baedeker, publishers.) Coblenz, 1869.

Geog 1540.12 Italy. Pt. 2: Central Italy and Rome. 3. ed. (Baedeker, publishers.) Coblenz, 1872.

Geog 1540.15 Italy. Pt. 2: Central Italy and Rome. 4. ed. (Baedeker, publishers.) Leipsic, 1875.

Geog 1540.18 Italy. Pt. 2: Central Italy and Rome. 6. ed. (Baedeker, publishers.) Leipsic, 1879.

Geog 1540.18.25 Italy. Pt. 2: Central Italy and Rome. 7. ed. (Baedeker, publishers.) Leipsic, 1881.

Geog 1540.19.5 Italy. Pt. 2: Central Italy and Rome. 8. ed. (Baedeker, publishers.) Leipsic, 1883.

Geog 1540.19 Italy. Pt. 2: Central Italy and Rome. 8. ed. (Baedeker, publishers.) Leipsic, 1883.

Geog 1540.20 Italy. Pt. 2: Central Italy and Rome. 9. ed. (Baedeker, publishers.) Leipsic, 1886.

Geog 1540.21 Italy. Pt. 2: Central Italy and Rome. 9. ed. (Baedeker, publishers.) Leipsic, 1886.

Geog 1540.26 Italy. Pt. 2: Central Italy and Rome. 10. ed. (Baedeker, publishers.) Leipsic, 1890.

Geog 1540.25 Italy. Pt. 2: Central Italy and Rome. 10. ed. (Baedeker, publishers.) Leipsic, 1890.

Geog 1540.30 Italy. Pt. 2: Central Italy and Rome. 11. ed. (Baedeker, publishers.) Leipzig, 1893.

Geog 1540.31A Italy. Pt. 2: Central Italy and Rome. 11. ed. (Baedeker, publishers.) Leipzig, 1893.

Geog 1540.37 Italy. Pt. 2: Central Italy and Rome. 12. ed. (Baedeker, publishers.) Leipsic, 1897.

Geog 1540.40 Italy. Pt. 2: Central Italy and Rome. 13. ed. (Baedeker, publishers.) Leipsic, 1900.

Geog 1540.45 Italy. Pt. 2: Central Italy and Rome. 14. ed. (Baedeker, publishers.) Leipzig, 1904.

Geog 1542.5 Italy. Pt. 3: Southern Italy, Sicily. (Baedeker, publishers.) Coblenz, 1867.

Geog 1542.10 Italy. Pt. 3: Southern Italy, Sicily. 2. ed. (Baedeker, publishers.) Coblenz, 1869.

Geog 1542.9 Italy. Pt. 3: Southern Italy, Sicily. 2. ed. (Baedeker, publishers.) Coblenz, 1869.

Geog 1542.13 Italy. Pt. 3: Southern Italy, Sicily. 4. ed. (Baedeker, publishers.) Leipsic, 1873.

Geog 1542.15 Italy. Pt. 3: Southern Italy, Sicily. 6. ed. (Baedeker, publishers.) Leipsic, 1876.

Geog 1542.21 Italy. Pt. 3: Southern Italy, Sicily. 8. ed. (Baedeker, publishers.) Leipsic, 1883.

Geog 1542.20 Italy. Pt. 3: Southern Italy, Sicily. 8. ed. (Baedeker, publishers.) Leipsic, 1883.

Geog 1542.25 Italy. Pt. 3: Southern Italy, Sicily. 9. ed. (Baedeker, publishers.) Leipsic, 1887.

Author and Title Listing

Geog 1542.27 Italy. Pt. 3: Southern Italy, Sicily. 10. ed. (Baedeker, publishers.) Leipsic, 1890.
Geog 1542.30 Italy. Pt. 3: Southern Italy, Sicily. 11. ed. (Baedeker, publishers.) Leipsic, 1893.
Geog 1542.33 Italy. Pt. 3: Southern Italy, Sicily. 12. ed. (Baedeker, publishers.) Leipsic, 1896.
Geog 1542.35 Italy. Pt. 3: Southern Italy, Sicily. 13. ed. (Baedeker, publishers.) Leipsic, 1900.
Geog 1538.50 Italy from the Alps to Naples. (Baedeker, publishers.) Leipzig, 1904.
Geog 1538.56A Italy from the Alps to Naples. 2. ed. (Baedeker, publishers.) Leipzig, 1909.
Geog 1538.57A Italy from the Alps to Naples. 3. ed. (Baedeker, publishers.) Leipzig, 1928.
An 2600.2 Itard, E.M. An historical account of...education of a savage man. London, 1802.
An 2600.2.8 Itard, Jean Marc. The wild boy of Aveyron. N.Y., 1932.
Geog 4209.51 Itchin' feet. (Harris, F.R.) Greenfield, Ohio, 1951.
Geog 5399.31.2 Itin, Vivian A. Vykhod k moriu. 2. izd. Novosibirsk, 1935.
Htn Geog 756.91* Itinera mundi. v.1-2. (Abraham ben Mordecai Farissol.) Oxonii, 1691.
Geog 520.17 Itineraire de Jerome Maurand. (Maurand, J.) Paris, 1901.
Geog 4268.83.6 Itinéraire des routes les plus fréquentées. 6. éd. (Dutens, Louis.) Paris, 1788.
Geog 4307.65 Itinéraire ou voyages. (Defeller.) Liege, 1820. 2v.
Geog 4182.57 Itinerario. (Linschoten, Jan H. van.) 's-Gravenhage, 1955-57. 3v.
Geog 4415.02 Itinerário. (Varthema, L. de.) Lisboa, 1949.
Geog 4416.21 Itinerário e outros escritos inéditos. (Lobo, Jeronymo.) Porto, 1971.
Geog 4182.2 Itinerario voyage. (Linschoten, Jan H. van.) 's-Gravenhage, 1910. 5v.
Htn Geog 4324.59* Itinerarium. (Ehingen, Georg von.) Augsburg, 1600.
Htn Geog 4306.60.2* Itinerarium. (Loménie, Louis Henri.) Paris, 1662.
Htn Geog 4415.11F* Itinerarium aethio piae. (Varthema, L. de.) Milan, 1511.
Htn Geog 4341.73* Itinerarium beniamini tudelemsis. (Montano, B.A.) Antwerpiae, 1575.
Htn Geog 4341.73.3* Itinerarium Benjaminis. (Benjamin ben Jonah, of Tudela.) Lugdunum Batavorum, 1633.
Htn Geog 4305.96* Itinerarium Germaniae, Galliae. (Hentznero, P.) Norinbergae, 1612.
Geog 4306.17.7 Itinerarium Germaniae, Siciliae. (Küsel Saxon, S.) Erphordiae, 1617.
Geog 4205.96.9F Itinerarium oste schip-vaert naer. (Linschoten, Jan H. van.) Amsterdam, 1644.
Geog 4306.17.1F An itinerary. (Moryson, Fynes.) Amsterdam, 1971.
Htn Geog 4306.17F* An itinerary. (Moryson, Fynes.) London, 1617.
Geog 4305.91A An itinerary containing his 10 yeeres travell. (Moryson, Fynes.) Glasgow, 1907. 4v.
Geog 4341.73.10A The itinerary of Benjamin of Tudela. (Benjamin ben Jonah, of Tudela.) London, 1907.
Geog 4341.73.5 The itinerary of Rabbi Benjamin of Tudela. pt.1-2. (Benjamin ben Jonah, of Tudela.) London, 1840.
Geog 10.4 Itogi nauki: geografiia SSSR. (Akademiia Nauk SSSR. Institut Nauchnoi Informatsii.) Moskva. 1,1965+ 4v.
Geog 10.5 Itogi nauki: geomorfologiia. Moskva. 2,1971+ 3v.
Geog 220.10 Itogi nauki: Teoreticheskie i obshchie voprosy geografii. Moskva. 1,1974+
Geog 10.7 Itogi nauki i tekhniki. Geografiia zarubezhnykh stran. Moskva. 1,1972+
Geog 5316.1.10 It's all adventure. (Freuchen, P.) London, 1938.
Geog 139.10 Its foundation and history. (London. Royal Geographical Society.) London, 1930.
Geog 4145.46 Ivan Fedorovich Kruzenshtern. (Pasetskii, Vasilii M.) Moskva, 1974.
Geog 612.4.5 Ivan Ivanovich Lepekhin. (Lukina, Tat'iana A.) Leningrad, 1965.
Geog 611.5.5 Ivan Kirilovich Kirilov, geograf XVIII veka. (Novlianskaia, Mariia G.) London, 1964.
Geog 607.5.10 Ivashchenko, M.M. Admiral Golovnin. Moskva, 1946.
Geog 5070.65 Iz Arktiki v Antarktiku. (Riabinin, A.K.) Murmansk, 1959.
Geog 4218.98.10 Iz dnevnikov krugosvetnogo puteshestviia. (Mikhailovskii, N.G.) Moskva, 1952.
Geog 3590.35 Iz istorii russkii geogr. otkrytii v sever ledov i tiloke okeanov. (Efimov, Aleksei V.) Moskva, 1950.
Geog 3590.35.5 Iz istorii russkikh aksped. na Tikhom okeane. (Efimov, Aleksei V.) Moskva, 1948.
Geog 3590.35.3 Iz istorii velikikh russkikh geograf. otkrytii. (Efimov, Aleksei V.) Moskva, 1949.
Geog 3590.35.2 Iz istorii velikikh russkikh geograf. otkrytii. (Efimov, Aleksei V.) Moskva, 1971.
Geog 4309.28.10 Iz Moskvy na Korsiku i Obratno. (Zozulia, Efim D.) Leningrad, 1928.
Geog 3590.15.20 Izbrannye trudy. (Berg, L.S.) Moskva, 1956. 5v.
Geog 33.10 Izvestiia. (Bulgarska Akademiia na Naukite, Sofia. Geografska Institut.) 1,1951+ 8v.
Geog 34.5 Izvestiia. (Bulgarsko Geografsko Druzhestvo, Sofia.) 1,1933+ 11v.
Geog 619.6 J. Russell Smith, geographer, educator and conservationist. (Rowley, V.M.) Philadelphia, 1964.
Geog 603.5.8 J.B. Charcot. (Emmanuel, Marthe.) Paris, 1945.
Geog 137.5 Jaarveslag. (Linschoten. Vereinigung.) 1-31,1908-1938 2v.
Geog 4308.72 Jackson, H.H. Bits of travel. Boston, 1872.
Geog 4308.72.2 Jackson, H.H. Bits of travel. Boston, 1873.
Geog 4228.86.10 Jackson, Helen F.H. Glimpses of three coasts. Boston, 1891.
Geog 4347.99.5 Jackson, J. Journey from India. London, 1799.
Geog 500.27A Jackson, J. Liste provisoire de bibliographies geographiques. Paris, 1881.
Geog 500.27.2A Jackson, J. Liste provisoire de bibliographies geographiques. Paris, 1881.
Geog 4105.2 Jackson, J.R. What to observe. London, 1841.
Geog 4521.10 Jackson, John. More than mountains. London, 1955.
Geog 3240.26 Jacob, Georg. Studien in arabischen Geographen. Berlin, 1891-92.
An 2059.12.3 Jacob, K.H. Der diluviale Mensch und seine Zeitgenossen aus dem Tierreiche. Leipzig, 1912.
Geog 4217.21 Jacob Roggeveen en zijn reis naar het zuidland, 1721-1722. (Bree, Levinus W. de.) Amsterdam, 1942.
Geog 604.5 Jacob van Deventer. (Hoff, Bert van.) 's-Gravenhage, 1953.
Geog 4209.11.7 Jacobs, A.H. Reisbrieven uit Afrika en Azie. Almelo, 1913. 2v.
Geog 3018.99 Jacobs, J. Story of geographical discovery. London, 1899.
Geog 4145.37 Jacoby, Arnold. Señor Kon-Tiki; the biography of Thor Heyerdahl. Chicago, 1967.
Geog 3249.30 Jacome Correa, Ayres. Discussão historica das medidas geographicas no seculo XVI. Lisboa, 1930.
Geog 4323.35 Jacopo da Verona. Liber peregrinationis. Roma, 1950.
Geog 4209.25.15 Jacques, N. Im Kaleidoskop der Weltteile. Berlin, 1925.
Geog 4332.26 Jadran - more mira. (Kolar, Vladimir.) Ljubjana, 1970.
An 2059.22 Die Jagd der Vorzeit. (Soergel, W.) Jena, 1922.
An 184.1F Jagor, Fedor. Aus Fedor Jogor's Nachlass. Berlin, 1914.
Geog 4509.40 Jahn, A. Exrusionismo y alpinismo; historia de su evolución. Caracas, 1940.
Geog 4169.64 Jahn, Janheinz. Wir nannten sie Wilde. München, 1964.
Geog 95 Jahrbuch. (Geographische Gesellschaft zu Hannover.) Hannover. 1926+ 8v.
NEDL Geog 199.1 Jahrbuch. (Schweizer Alpenclub.) Bern. 1-35 23v.
Geog 199.1 Jahrbuch. (Schweizer Alpenclub.) Bern. 36-58,1900-1923 23v.
Geog 95.5 Jahrbuch. Sonderheft. (Geographische Gesellschaft zu Hannover.) Hannover. 2,1968+ 4v.
Geog 28.1 Jahresbericht. (Berne. Geographische Gesellschaft.) Bern. 1,1879+ 14v.
Geog 91.1 Jahresbericht. (Greifswald. Geographische Gesellschaft.) Greifswald. 1-60,1882-1942 16v.
Geog 75.1 Jahresbericht - Geographie und Statistik. (Frankfurter Vereins.) Frankfurt. 1905-1919
Geog 606.2 Jaime Ferrer...y el descubimiento de America. 1. ed. (Coma Soley, V.) Barcelona, 1952.
Geog 5939.59.24 Jaine raamat. (Smuul, Juhan.) Tallinn, 1962.
Geog 4308.84 James, H. Portraits of places. Boston, 1884.
Htn Geog 4308.84* James, H. Portraits of places. Boston, 1884.
Geog 4308.84.5 James, H. Portraits of places. 3rd ed. Boston, 1893.
Htn Geog 4308.75* James, H. Transatlantic sketches. Boston, 1875.
Geog 4328.19.10 James, John Thomas. Journal of a tour in Germany, Russia, Poland in 1813-1814. 3rd ed. London, 1819.
Geog 759.49 James, P.E. A geography of man. Boston, 1949.
Geog 759.49.2A James, P.E. A geography of man. 2nd ed. Boston, 1959.
Geog 3400.5A James, Preston E. American geography; inventory and prospect. Syracuse, 1954.
Geog 819.35 James, Preston E. An outline of geography. Boston, 1935.
Geog 3019.72 James, Preston Everett. All possible worlds; a history of geographical ideas. Indianapolis, 1972.
Geog 665.164 James, Preston Everett. On geography: Selected writings of Preston E. James. 1st ed. Syracuse, 1971.
Htn Geog 5516.31* James, T. Strange and dangerous voyage...N.W. Passage. London, 1633.
Htn Geog 5516.31.2* James, T. Strange and dangerous voyage...N.W. Passage. London, 1633.
Geog 5516.31.3 James, T. The strange and dangerous voyage. Toronto, 1975.
An 130.10 James George Frazer. (Downie, Robert Angus.) London, 1940.
An 130.15 James George Frazer 1854-1941. (Marett, Robert R.) London, 1941?
Geog 4181.136 The Jamestown voyages under the first charter, 1606-1609. (Barbour, Philip L.) Cambridge, Eng., 1969. 2v.
Geog 4145.50 Jan van Linschaten. (Parr, Charles.) N.Y., 1964.
Geog 4181.65 Jane, Cecil. Select documents illustrating the 4 voyages of Columbus. London, 1930-33. 2v.
Geog 5838.49 Janssen, Carl Emil. En grønlandspraests optegnelser, 1844-49. København, 1913.
Geog 4329.14F Januszewski, W. Dokoła morza śródziemnego. Wilnie, 1914.
Geog 4418.61 Japan, the Amoor and the Pacific. (Tilley, H.A.) London, 1861.
Geog 4218.52 The Japan expedition. (Spalding, J.W.) N.Y., 1855.
Geog 4309.37A Japanese lady in Europe. (Ichikawa, H.) N.Y., 1937.
Geog 4209.57.10 Jara Peralta, José. La ciudad de Toledo. Madrid, 1957.
Geog 3018.79.3 Jardens op da gelseshistorie. (Verne, Jules.) Kristiania, 1879-83.
Geog 4327.88 Jardine, A. Letters from Barbary, France. London, 1788. 2v.
An 629.64 Jarvie, Ian C. The revolution in anthropology. London, 1964.
An 359.73.30 Jaulin, Robert. Gens du soi, gens de l'autre. Paris, 1973.
Geog 4309.46 Jaunes, Elly. Mánnishor därute. Stockholm, 1946.
Geog 4502.5.26 Javelle, E. Erinnerungen eines Bergsteigers. München, 1938.
Geog 4209.30.20 Jazz, boomerang et kimonos. (Chable, Jacques Edouard.) Paris, 1930.
Geog 602.3.5 Je découvre l'université. (Blanchard, Raoul.) Paris, 1963.
Geog 603.5 Jean-Baptiste Charcot, 1867-1936. Paris, 1936.
Geog 603.5.5 Jean Charcot. 14. éd. (Oulié, M.) Paris, 1937.
Geog 603.9 Jean Domenique Cassini and his world map of 1696. (Brown, Lloyd A.) Ann Arbor, 1941.
Geog 520.1 Jean et Sébastien Cabot. (Harrisse, H.) Paris, 1882.
Geog 612.3 Jean François Galaup de Lapérouse. (Allen, Edward W.) San Francisco, 1941.
Geog 5208.82.2 Jeannette, and a complete and authentic narrative encyclopedia of all voyages. (Perry, R.) San Francisco, 1883.
Geog 5398.79.35 The Jeannette expedition, an ill fated journey to the Arctic. (Hoehling, Adolph A.) London, 1969.
Geog 5398.79.9F The Jeannette in the Arctic regions. (N.Y. Herald.) N.Y., 1882.
Geog 4707.26 Jedre, J.B. The trials of five persons for piracy. Boston, 1726.
Geog 4500.5 Jeffers, LeRoy. Selected list of books on mountaineering. N.Y., 1916.
An 379.24.3 Jefferson, M.S.W. Man in Europe, here and there. Ypsilanti, Mich., 1924.
An 2109.72.40 Jelinek, Jan. Das grosse Bilderlexikon des Menschen in der Vorzeit. Gütersloh, 1972.
Geog 4209.60.40 Jelling Kristoffersen, Frode. Jorden rundt på tommelfinger. Sirius, 1960.
Geog 4238.49F Jenkins, F.H. Journal of a voyage to San Francisco, 1849. Northridge? 1975.
Geog 4248.53 Jenkins, J.S. Recent exploring expeditions to the Pacific. London, 1853.
Geog 4208.38 Jenkins, J.S. United States exploring expeditions...voyage...1838-1842. New Orleans, 1854.
Geog 4208.38.1 Jenkins, J.S. United States exploring expeditions. N.Y., 1855.
Geog 4180.72 Jenkinson, A. Early voyages...to Russia and Persia. N.Y., 1964? 2v.
An 359.63 Jennings, J.D. Anthropology and the world of science. Salt Lake City, 1963.
Geog 5323.5 Jens Munk. 4. Opl. (Hansen, T.) Kjøbenhavn, 1966.
Geog 5670.4 Jensen, Bent. En livsform ved horsvejen. København, 1971.

Author and Title Listing

Geog 4218.68.10	Jentsch, August. Ein Mann und ein Traktor auf Weltreise. v.2. Wien, 1968-69.
Geog 4328.79	Jeréz Perchét, Augusto. El mediterráneo. 3a ed. Madrid, 1879.
Geog 4311.12	Jerrold, Walter. The Danube. N.Y., 1911.
Geog 4509.53.20	Jerschak, Federico. L'alpinismo a Cortina dai suoi primordi ai giorni nostri. Roma, 1953.
An 137.3	A Jerseyman at Oxford. (Marett, Robert R.) N.Y., 1941.
Geog 120.5	Jerusalem studies in geography. Jerusalem. 1,1970+
Geog 3070.39	Jervis, W.W. The world in maps. N.Y., 1937.
Geog 3070.39.5	Jervis, W.W. The world in maps. 2d ed. N.Y., 1938.
Geog 4229.26	Jesting pilate; an intellectual holiday. (Huxley, Aldous.) N.Y., 1926.
Geog 4329.35.3	Jeunesse de la Méditerranée. (Audisio, G.) Paris, 1935-1936.
Geog 4558.74.3	Jewell, J.G. Among our sailors. N.Y., 1874.
Geog 4308.37.5	Jewett, I.A. Paggages in foreign travel. Boston, 1838. 2v.
Geog 4110.35	Jewish travel guide. London, 1961+ 3v.
Geog 4169.30	Jewish travellers. (Adler, Elkan N.) London, 1930.
An 2059.03.5	Joby, N. Man before metals. N.Y., 1903.
Geog 5060.10A	Joerg, W.L.G. Brief history of polar exploration since the introduction of flying. N.Y., 1930.
Geog 6009.35	Joerg, W.L.G. The topographical results of Ellsworth's trans-Antarctic flight of 1935. N.Y., 1936.
Geog 6009.28.5	Joerg, W.L.G. The work of the Byrd Antarctic expedition. N.Y., 1930.
Geog 5317.5	Johann Georg, 1709-1755. München, 1911.
Geog 4182.34	Johannes de Laet. Iaerlyck Verhael van der verichtinghen der...Compagnie. (Naber, S.P.) 's-Gravenhage, 1931-37. 4v.
Geog 5398.93.3	Johansen, H. With Nansen in the north. London, 1899.
Geog 4209.28.20A	John Cameron's odyssey. (Cameron, J.) N.Y., 1928.
Geog 5530.75	John Franklin's, Eleanor Anne Porden. (Gell, E.M.B. (Mrs.).) London, 1930.
Geog 4207.76.15	John Ledyard: an American Marco Polo. (Munford, K.) Portland, 1939.
Geog 613.3.5	John Murray III, 1808-1892; memoir. (Murray, John.) London, 1919.
Geog 624.1	John Xantus. (Konnyü, Leslie.) Köln, 1965.
Geog 4208.59	Johnes, M. Boys' book of modern travel. N.Y., 1859.
Geog 4308.85	Johnson, E.C. On the track of the crescent. London, 1885.
Geog 4209.32.25	Johnson, Irving. Round the Horn in a square rigger. Springfield, 1932.
Geog 4219.36.5	Johnson, Irving. Westward bound in the schooner Yankee. N.Y., 1936.
Geog 4309.62	Johnson, Irving. Yankee sails across Europe. N.Y., 1962.
Geog 4308.31.2A	Johnson, J. Change of air, or The diary. London, 1831.
Geog 4348.17	Johnson, John. A journey from India to England thro' Persia, Georgia, Russia, Poland and Prussia, in 1817. London, 1818.
Geog 577.76	Johnson, R. New gazeteer. London, 1776.
Geog 3019.00	Johnson, W.H. The world's discoveries. Boston, 1900.
Geog 578.77	Johnston, A.K. A general dictionary of geography. London, 1877.
Geog 818.80.6	Johnston, Alexander K. A physical, historical, political, and descriptive geography. 2nd ed. London, 1881.
Geog 818.80.5	Johnston, Alexander K. A physical, historical, political, and descriptive geography. London, 1880.
An 359.20.5A	Johnston, Harry. The backward peoples and our relations with them. London, 1920.
Geog 5558.75.11	Johnston, R. The Arctic expedition of 1875-76. London, 1877.
Geog 4308.85.15	Johnston, Richard M. Two gray tourists. Baltimore, 1885.
An 379.73.5	Johnston, Ronald John. Spatial structures. London, 1973.
Geog 665.118	Johnston, William. Dynamic relationships in physical geography. Christchurch, 1967.
Geog 578.51	Johnstone, A.K. Dictionary of geography. London, 1851.
Geog 4139.10	Johnstone, C. (Mrs.). Buccaneers of America. Akron, 1910.
Geog 4138.37.5	Johnstone, C. (Mrs.). Lives and voyages of Drake. Edinburgh, 1837.
An 3309.65.10	Joint Arabic-Polish Anthropological Expedition, 1958. Publications. Warsaw, 1965.
Geog 3055.25	Jolibois. Dissertation sur l'Atlantide. Lyon, 1846.
Geog 4509.54	Jolis Felisart, A. La conquista de la montaña. Barcelona, 1954.
Geog 3530.13	Jolly, Raoul. Les missions françaises; causeries géographiques. Paris, 1894-96. 2v.
Geog 4709.53.2	Jolly Roger. (Pringle, P.) London, 1953.
Geog 4709.53	Jolly Roger. (Pringle, P.) N.Y., 1953.
Geog 4308.83.10	A jolly summer. (Stokes, F.A.) N.Y., 1883.
An 608.62	Jomard, E.F. Classification méthodique des produits de l'industrie. Paris, 1862.
Geog 500.15	Jomard, E.F. De la collection geographique. Paris, 1848.
Geog 3235.47	Jomard, E.F. Introduction à l'atlas des monuments de la géographie. Paris, 1879.
An 205.1	Jomard, E.F. Lettre à M.P.F. de Siebold. Paris, 1845-69. 3 pam.
Geog 4311.19	Jonák, Ján. 2200 kilo-metrov po Dunalu. Moskva, 1960.
Geog 28.4	Jonard, E.F. Notice sur l'etablissement géographique de Bruxelles. v.1-4. Paris, n.d.
Geog 817.73	Jones, E. The young geographer and astronomer's best companion. London, 1773.
Geog 4328.29	Jones, George. Sketches of naval life...on Mediterranean. New Haven, 1829. 2v.
Geog 819.50	Jones, S.B. Geography and world affairs. Chicago, 1953.
An 2058.77.3	Jones, T.R. Lecture on the antiquity of man. London, 1877.
Geog 5838.75	Jones, T.R. Manual of the natural history, geology and physics of Greenland. London, 1875.
An 629.61.5	Jones, W.T. The romantic syndrome. The Hague, 1961.
Geog 4559.56.10	Jones, William Herbert Sidney. The Cape Horn breed. N.Y., 1956.
Geog 4248.61	Jones. Life and adventures in the south Pacific. N.Y., 1861.
Geog 659.62	Jong, Guben. Chonological differentiation as the fundamental principle of geography. Groningen, 1962.
Geog 659.55	Jong, Guben. Het karakter van de geografische totaliteit. Groningen, 1955.
An 629.67	Jongnians, Douve Geert. Anthropologists in the field. Assen, 1967.
Geog 5558.75.13	De jongste engelsche pooltocht ende expedities der toekomst. (Kan, C.M.) Utrecht, 1877.
Geog 5838.98	Jonsson, F. Grønlands gamle topografi. Kjøbenhavn, 1899.
Geog 5650.5	Jonsson, Finnur. Graenlendinga saga. Kaupmannahöfn, 1899.
Geog 5638.99	Jonsson, Finnur. Um Graenland ad fornu og nýju. Kaupmannahöfn, 1899.
Geog 4559.40.12	Jónsson, Gisli. Frekjan; aefintýralegt...frá Danmoëku. Reykjavik, 1940.
Geog 816.70	Joosten, J. De kleyne wonderlijcke werelt. Amsterdam, 1670.
Geog 4180.31	Jordanus de Saxonia. Mirabilia descripta...wonders of the East. N.Y., 1964?
Geog 809.22.15	Jorden og menneskelivet. (Vahl, Martin.) København, 1922-1927. 4v.
Geog 4209.51.5	Jorden rundt med Halger Rosenberg. (Rosenberg, Halger.) København, 1951.
Geog 4209.60.40	Jorden rundt på tommelfinger. (Jelling Kristoffersen, Frode.) Sirius, 1960.
Geog 4309.64.25	Jornal da Europe. (Linhares, Temístocles.) Brasil, 1964.
Geog 4300.40	Joseph, Richard. Guide to Europe and the Mediterranean. 1st ed. Garden City, N.Y., 1956.
Geog 4219.37.12	The Joseph Conrad. (Marine Historical Association, Mystic, Conn.) Mystic, 195-.
Geog 4145.81.6	Joshua Slocum. (Teller, Walter Magnes.) New Brunswick, 1971.
Geog 500.121	Josuweit, Werner. Studienbibliographie Geographie. Wiesbaden, 1973.
Geog 4559.44.5	Jottings from a cruise. (Green, Alfred.) Seattle, 1944.
Geog 4181.16	Jourdain, J. Journal of John Jourdain, 1608-17. Cambridge, 1905.
Geog 4182.45	Journaal van Dircq Van Adrichem hofreis. (Adrichem, D. van.) 's-Gravenhage, 1941.
Geog 4182.12	Journael van de reis naar Zuid-Amerika. (Ottsen, Hendrick.) 's-Gravenhage, 1918.
Geog 4308.18.6	Journal, 1818-1819. (Russell, Jonathan.) Boston, 1918.
Geog 13.4	Journal. (American Geographical Society of New York.) 1-47,1895-1915 45v.
Geog 4209.40.3	Journal. (Forsyth, W.J.) Boston, 1940.
Geog 139.1	Journal. (London. Royal Geographical Society.) London. 1831-1880 50v.
Geog 142.12	Journal. (Manchester Geographical Society.) Manchester. 1-7 6v.
Geog 139.2	Journal. Index. v.1-20,21-40,41-50. (London. Royal Geographical Society.) London, n.d. 3v.
Geog 4559.35.10	Journal and letters, 1876-84. (Brown, Charles W.) n.p., 1935.
Geog 4181.131	The journal and letters of Captain Charles Bishop on the north-west of America. (Bishop, Charles.) Cambridge, 1967.
Geog 4227.85F	Journal and narrative of the...expedition from Bombay to the N.W. coast of America. (Strange, James.) Madras, 1928.
Geog 4209.46.20	Journal der merkwaardige reizen van Jan Wever. (Wever, Jan.) 's-Gravenhage, 1946.
Geog 4209.60.30	Le journal des voyages. (Roy, Claude.) Paris, 1960.
Geog 4206.65.9	Journal des voyages de Monsieur de Monconys. Lyon, n.d.
Geog 121.2F	Journal des voyages et des aventures de terre et de mer. Paris. 1877-1909 30v.
Geog 5538.51.7	Journal d'un voyage aux mers polaires...en 1851 et 1852. (Bellot, J.R.) Paris, 1854.
Geog 4308.19.5	Journal d'un voyage en Europe. (Plessis, Joseph O.) Québec, 1903.
Geog 4209.39.15	Journal hållen på resan till Canton i China. (Reinius, I.) Helsingfors, 1939.
Geog 4207.79	Journal kept on a journey from Bassora to Bagdad...1779. (Evers.) Horsham, 1784.
Geog 4419.34.5	Journal letters from the Orient. (Rogers, C.K. (Mrs.).) Boston, 1934.
Geog 4181.51	Journal of...Father Samuel Fritz. (Edmundson, George.) London, 1922.
Htn Geog 4328.39.3*	Journal of a cruise of the U.S.S. Ohio in the Mediterranean. (Torrey, F.P.) Boston, 1841.
Geog 4218.83.5	Journal of a lady's travels. (Bridges, F.D.) London, 1883.
Geog 4328.19.10	Journal of a tour in Germany, Russia, Poland in 1813-1814. 3rd ed. (James, John Thomas.) London, 1819.
Geog 5377.74	Journal of a voyage...by the Honorable Commodore Phipps, and Captain Lutwidge. London, 1774. 3 pam.
Geog 4208.29.3	The journal of a voyage from Calcutta. (Prinsep, A. (Mrs.).) London, 1833.
Geog 4247.96	Journal of a voyage in the Duff to Pacific. (Smith, William.) N.Y., 1813.
Geog 5518.19.12	Journal of a voyage of discovery. (Fisher, Alexander.) London, 1821.
Geog 5518.19.10	Journal of a voyage of discovery. (Fisher, Alexander.) London, 1821.
Geog 5558.18	Journal of a voyage of discovery. (Parry, W.E.) London, 1820. 2 pam.
Geog 5518.18.5	Journal of a voyage of discovery to the Arctic regions. (Fisher, Alexander.) London, 1819.
Geog 5518.18.7	Journal of a voyage of discovery to the Arctic regions. (Fisher, Alexander.) London, 1968.
Geog 4679.71	Journal of a voyage to nowhere. (Fenn, Charles.) London, 1971.
Geog 4238.49F	Journal of a voyage to San Francisco, 1849. (Jenkins, F.H.) Northridge? 1975.
Geog 4247.89	The journal of Captain James Colnett aboard the Argonaut from Apr. 26, 1789 to Nov. 3, 1791. (Colnett, J.) Toronto, 1940.
Geog 4228.28.5	Journal of Captain S.H. Davis, a Gloucester sea-captain, 1828-1846. (Davis, S.H.) Norwood, 1922.
Geog 4180.86A	Journal of Christopher Columbus. (Markham, C.R.) London, 1893.
Geog 4180.99A	Journal of first voyage of V. da Gama. (Gama, V. da.) London, 1898.
Geog 121.1.2	Journal of geography. Lancaster. 1,1902+ 63v.
Geog 121.1.6	The journal of geography. Index. 1897-1956. N.Y., 1922-58. 2v.
Geog 5538.50.15	Journal of H.M.S. Enterprise...1850-55. (Collinson, R.) London, 1889.
Geog 4181.16	Journal of John Jourdain, 1608-17. (Jourdain, J.) Cambridge, 1905.
Geog 4348.18	Journal of Jonathan Russell, 1818-1819. n.p., n.d.
Geog 4308.31.5	Journal of residence in Scotland and tour. (McLellan, H.B.) Boston, 1834.
Geog 4208.17F	Journal of route across India...to England. (Fitzclarence.) London, 1819.
Geog 121.1	The journal of school geography. Lancaster, 1897. 5v.
Geog 5518.21	Journal of second voyage for discovery. (Parry, W.E.) London, 1824.
Geog 4227.90	Journal of the brigantine Hope on the voyage to the northwest coast of North America, 1790-92. (Ingraham, Joseph.) Barre, 1971.

206

Author and Title Listing

Geog 5518.24.2 Journal of third voyage...Northwest Passage. (Parry, W.E.) London, 1826.
Geog 5518.24 Journal of third voyage...Northwest Passage. (Parry, W.E.) Philadelphia, 1826.
Geog 4308.05.2 Journal of travels. (Silliman, B.) Boston, 1812. 2v.
Geog 4308.05.4 Journal of travels. 3rd ed. (Silliman, B.) New Haven, 1820. 3v.
Geog 142.3.2 The journal of tropical geography. Index. 1-39,1953-1974. Singapore, n.d. 2v.
Htn Geog 5397.73* Journal of voyage...North Pole. (Phipps, J.) London, 1774.
Geog 5518.19.6.2 Journal of voyage...Northwest Passage and North Georgia Gazette. (Parry, W.E.) Philadelphia, 1821.
Geog 5518.19.7 Journal of voyage...Northwest Passage and North Georgia Gazette. (Parry, W.E.) London, 1821.
Htn Geog 4558.05* Journal of voyage from London. (Burton, C.) London, 1805.
Htn Geog 4558.26F* Journal of voyage of "Missionary Packet". (Hunnewell, J.) Charlestown, 1880.
Htn Geog 5558.21* Journal of voyage to Greenland...1821. (Manby, G.W.) London, 1822.
Geog 5558.22 Journal of voyage to the northern whale-fishery. (Scoresby, William.) Edinburgh, 1823.
Geog 4208.13 Journal of voyages. (Dunham, Jacob.) N.Y., 1850.
Geog 4418.32 Journal of voyages and travels. (Tyerman, David.) Boston, 1832. 3v.
Geog 4208.51.5 Journal of voyages to China and return. (Blaney, Henry.) Boston, 1913.
Geog 4181.52 The journal of William Lockerby. (Lockerby, William.) London, 1925.
Geog 4182.41 Journal van J.J. Ketelaar's hofreis...1711-1713. (Vogel, J.P.) 's-Gravenhage, 1937.
Geog 4182.54 Journalen van de gedenckenaerdige reijsen. (Bontekae, W.Y.) 's-Gravenhage, 1952.
Geog 759.66.5 Journaux, André. Géographie générale. Paris, 1966.
Geog 4307.94.2 Journey...through Holland. 2. ed. (Radcliffe, A. (Mrs.).) London, 1795. 2v.
Geog 4219.66 Journey around myself. (Marti-Ibáñez, Félix.) N.Y., 1966.
Htn Geog 4308.85.7* A journey due south, travels. (Sala, George A.) London, 1885.
Geog 4347.98 Journey from Bengal to England. (Forster, G.) London, 1798.
Geog 4347.99.5 Journey from India. (Jackson, J.) London, 1799.
Geog 4348.17 A journey from India to England thro' Persia, Georgia, Russia, Poland and Prussia, in 1817. (Johnson, John.) London, 1818.
Geog 4308.24 A journey from London to Odessa. (Moore, John.) Paris, 1833.
Geog 4348.65 Journey from London to Persepolis. (Ussher, John.) London, 1865.
Geog 4208.19 A journey from Merut in India to London. (Lumsden, T.) London, 1822.
Geog 4328.45 Journey from Naples to Jerusalem. (Borrer, D.) London, 1845.
Geog 5312.7.15 Journey into the fog; the story of Vitus Bering and the Bering Sea. (Goodhue, Cornelia.) Garden City, N.Y., 1944.
NEDL Geog 4308.25.10 A journey into various parts of Europe. (Pennington, Thomas.) London, 1825. 2v.
Htn Geog 4307.94* A journey made in summer of 1794. (Radcliffe, A. (Mrs.).) London, 1795.
Geog 4559.29.3 The journey of...pilot, of Deal. (Stanton, William H.) London, 1929.
Geog 4181.4 Journey of William of Rubruck...and John of Pian de Carpine. (Rubruquis, G. de.) London, 1900.
Geog 4348.07 Journey over land to India. (Campbell, D.) Philadelphia, 1807.
Geog 4219.64 Journey round the world. (Menon, Kumara P.S.) Bombay, 1966.
Geog 4307.86.13 Journey thro' Crimea to Constantinople. (Ansbach, Elizabeth Craven.) London, 1789.
Htn Geog 4307.86.12* Journey thro' Crimea to Constantinople. (Ansbach, Elizabeth Craven.) London, 1789. 2 pam.
Geog 4309.72 Journey through Europe. (Hillaby, John D.) London, 1972.
Geog 4029.14 Journey to Ardmore. (Ridgway, John M.) London, 1971.
Geog 4349.31 A journey to China; or, Things which are seen. (Toynbee, Arnold J.) London, 1931.
Geog 4308.17.6 A journey to Florence in 1817. (Charleville, H.C.B.) London, 1951.
Geog 4329.36.2 A journey to Jerusalem. (Ervine, St. J.G.) London, 1936.
Geog 4218.98.5 Journey with the sun around the world. (McMahon, William.) Cleveland, 1900.
Geog 4209.38.16 Journeys between wars. (Dos Passos, John.) London, 1938.
Geog 4209.38.15A Journeys between wars. (Dos Passos, John.) N.Y., 1938.
Geog 6009.14.15 Joyce, Ernest. The South Pole trail. London, 1929.
Geog 4181.73 Joyce, L.E. Elliott. A new voyage and description of the Isthmus of America. Oxford, 1934.
An 359.65.10 Jürgens, Hans W. Beiträge zur menschlichen Typenkunde. Stuttgart, 1965.
Geog 4182.19 Juet, Robert. Henry Hudson's reize onder Nederlander vlag. 's-Gravenhage, 1921.
Geog 618.5.10 Jugenderinnerungen. (Ratzel, Fridrich.) München, 1966.
Geog 759.74 Juillard, Étienne. La région, contributions à une géographie générale des espaces régionaux. Paris, 1974.
Geog 5312.6 Jules de Blosseville. (Blosseville, B. Ernest Poret.) Évreux, 1954.
Geog 3530.17 Julien, C.A. Les voyages de découverte et les premiers établissements. 1. éd. Paris, 1948.
Geog 4145.68 Julius von Payer. (Mueller, Martin.) Stuttgart, 1958.
Geog 4502.5.17 Junger Mensch im Gebirg. (Maduschka, L.) München, 1936.
Geog 4559.67.5 Jupp, Ursula. Home port Victoria. Victoria, 1967.
Geog 3520.5 Jurien de la Gravière, J.B.E. Les Anglais et les Hollandais. Paris, 1890. 2v.
Geog 3018.79.5 Jurien de la Gravière, J.B.E. Les marins du XVe et du XVIe siècle. Paris, 1879. 2v.
Geog 4308.01.5 The juvenile travellers. (Wakefield, Priscilla (Bell).) London, 1801.
Geog 4308.06.5 Juvenile travellers. (Wakefield, Priscilla (Bell).) London, 1806.
Geog 601.5 K.I. Arsen'ev i ego raboty po Raiomirovamiiu Rossii. (Pertish, E.N.) Moskva, 1960.
Geog 5939.56.5 K beregam Antarktidy. (Burkhanov, V.F.) Moskva, 1956.
Geog 4229.71 K beregam Novogo Sveta. (Shur, Leonid A.) Moskva, 1971.
Geog 4209.61.10 K beregam Vostoka. (Drachev, B.S.) Moskva, 1961.
An 3559.66 K kraniologii drevnego i sovremennogo naseleniia Kavkaza. (Abdushelishvili, Malkhaz G.) Tbilisi, 1966.
An 137.6 Kaaram Tamo Mann vom Mond; Leben undd Reisen Miklučho-Makleis. (Wotte, Herbert.) Leipzig, 1973.
Geog 3019.35.15 Kábmár, G. Régi népak, ujvilagok. Budapest, 1936.
Geog 4346.83 Kaempfer, Engelbert. Die Reisetagebücher. Wiesbaden, 1968.
Geog 3129.34 Kahlo, G. Die Keuntnis der Erde im Altertum. München, 1934.
Geog 4209.61 Kahn, E.J. A reporter here and there. N.Y., 1961.
Geog 4521.11 Kain, Conrad. Where the clouds can go. N.Y., 1935.
Geog 5379.35.3 Kak my spasali chelnoskiktsev. Moskva, 1934.
An 2059.66.5 Kak vozniklo chelovechestvo. (Semenov, Iurii I.) Moskva, 1966.
Geog 5606.8 Kalatalet-nunäne landsrådip okalokatigīssutai. (Greenland. Landsraadet.) 1956-1958 2v.
Geog 579.68 Kalesnik, Stanislaw W. Entsiklopedicheskii slovar' geograficheskikh terminov. Moskva, 1968.
Geog 4269.65.12 Kallaveden rannalta, päiväkirja valpurista 1964 valpuriin 1965. 3. painos. (Kares, Olavi.) Porvoo, 1966.
Geog 5518.92 Kallstenius, Alfhild. Tragedin i Smiths sund; Björling-Kallstenius expeditionen, 1892. Kalmar, 1966.
Geog 5399.32 Kalmár, G. Kuz delmek a fehér halál országában. Budapest, 1932-33. 2v.
Geog 819.34 Kalmár, Gusztáv. Négy Világrésg Földje és Népei. Budapest, 1934-1935.
An 379.22.9 Kaltenbach, Ernst. Beiträge zur Anthropogeographie des Bodenseegebiets. Inaug. Diss. Basel, 1922.
Geog 5225.10 Kampen om nordpolen. (Bruun, Daniel.) Kjøbenhavn, 1902.
Geog 5670.5 Kamper om Eirik Raudes land. (Blom, Ida.) Oslo, 1973.
Geog 4502.5.21 Kampf um den Berg. (Montis, R.) München, 19- .
Geog 5939.46 Der Kampf um den Südpol. 2. Aufl. (Bezemer, K.W.L.) Zürich, 1952.
Geog 5399.31 Der Kampf um Nobile. (Meyer, W.) Berlin, 1931.
Geog 5558.75.13 Kan, C.M. De jongste engelsche pooltocht ende expedities der toekomst. Utrecht, 1877.
Geog 5538.51.5 Kane, E.K. Access to an open Polar Sea. N.Y., 1853.
Geog 5538.53.3 Kane, E.K. Arctic explorations. Hartford, 1881.
Geog 5538.53A Kane, E.K. Arctic explorations. Philadelphia, 1856. 2v.
Geog 5538.53.1 Kane, E.K. Arctic explorations. Philadelphia, 1856. 2v.
Geog 5538.53.2 Kane, E.K. Arctic explorations. Philadelphia, 1856. 2v.
Geog 5538.53.4 Kane, E.K. Arctic voyage. N.Y., 1857.
Geog 5538.51 Kane, E.K. United States Grinnell expedition. N.Y., 1854.
Geog 5538.51.3 Kane, E.K. United States Grinnell expedition. N.Y., 1857.
Geog 5321.1.11 Kane, Elisha Kent. Access to an open polar sea. N.Y., 1853.
Geog 5321.1.9 Kane Monument. Act of incorporation, preamble. N.Y., 1859.
Geog 4559.72.10 Kangaroo Island shipwrecks. (Chapman, Gifford Desmond.) Canberra, 1972.
Geog 4308.78.10 A Kansan abroad. (Prentis, N.L.) Topeka, Kansas, 1878.
An 359.69.25 Kansatieteen periaateoppia. (Vuorela, Toivo.) Helsinki, 1969.
An 39.58 Kansatieteen sanasto. (Vuorela, Taivo.) Helsinki, 1958.
Geog 659.55 Het karakter van de geografische totaliteit. (Jong, Guben.) Groningen, 1955.
Geog 5379.67.5 Karalt, Charles. To the top of the world. 1. ed. N.Y., 1968.
Geog 4419.28 Karawanen-Reisen...Ägypten, Mesopotamien, Persien und Abessinien. (Bosch, Carl.) Berlin, 1928.
Geog 4418.79.5 Die Karawanen-Strasse von Aegypten nach Syrien. (Ludwig, Salvator.) Prag, 1879.
An 2109.39.7 Kardiner, A. The individual and his society. N.Y., 1939.
An 124.11 Kardiner, Abram. They studied man. N.Y., 1963.
Geog 4269.65.12 Kares, Olavi. Kallaveden rannalta, päiväkirja valpurista 1964 valpuriin 1965. 3. painos. Porvoo, 1966.
Geog 614.6 Karl Neumann. Inaug. Diss. (Kupferschmidt, F.) Leipzig, 1935.
Geog 619.12 Karl Theodor Sapper, 1866-1945. (Termer, Franz.) Leipzig, 1966.
VGeog 4209.69.5 Karlin, Alma M. Samotno potovanje. Ljubljana, 1969.
An 3208.99 Karlsruhe, Germany. Altertumsverein. Zur Anthropologie der Badener. Jena, 1899.
Geog 4559.66 Karlsson, Elis. Pully-Haul: the story of a voyage. London, 1966.
Geog 665.62 Karper, Kurt. Landschaft und Lund. Remagen, 1951.
Geog 3055.53 Karst, Josef. Atlantis und der Liby-athiopische Kulturkreis. Heidelberg, 1931.
Geog 579.41 Karten-Wörterbuch eine Verdeutschung fremdsprachiger Kartensignatur-Bezeichnungen. (Bonacker, Wilhelm.) Berlin, 1941.
An 359.63.35 Kartografiai modszer a neprajzban. (Barabas, Jenó.) Budapest, 1963.
Geog 3590.83 Kartograficheskie i geodezicheskie raboty v Rossii v XIX-nachale XX v. (Novokshanova, Zinaida K.) Moskva, 1967.
Geog 3590.86 Kartografiia Rossii XVIII veka. (Tel', Sergei E.) Moskva, 1960.
Geog 500.98F Die Kartographie, 1943-1954. (Kosack, Hans P.) Lahr, 1955.
Geog 3070.16 Kartographie bei den Naturvölkern. (Dröber, W.) Erlangen, 1903.
Geog 500.40 Katalog der Bibliothek der Gesellschaft für Erdkunde zu Berlin. (Gesellschaft für Erdkunde zu Berlin. Bibliothek.) Berlin, 1903.
Geog 613.1.7 Katalog einer Mercator-Ausstellung. (Cologne, Germany. Stadtbibliothek.) Koeln, 1894.
Geog 500.120 Katalog rękopisów geograficznych w zbiorach polskich. (Polska Akademia Nauk. Institut Geografii.) Warszawa, 1965- 2v.
An 358.93 Katechismus der Völkerkunde. (Schurtz, H.) Leipzig, 1893.
Geog 5399.28.40 Katz, Otto. Neun Männer im Eis. Berlin, 1929.
Geog 4219.32.25 Katz, Richard. Ein Bummel um die Welt. Zürich, 1936.
Geog 4219.32.20 Katz, Richard. Ernte. Zürich, 1935.
Geog 500.117 Kaufman, Isaak M. Geograficheskie slovari; bibliografiia. Moskva, 1964.
Geog 3590.72 Kaulbars, N.A. Aperçu des travaux géographiques en Russie. St. Pétersbourg, 1889.
Geog 5650.30F K'avdlunätsianik. (Aron, of Kangeg.) Godthåt, 1969.
Geog 4209.13.7 Kawabata, A. A hermit turned loose. 2. ed. Tokio, 1915.
Geog 4311.23 Kazaka, Tenin. Plavashchi domovc. Plovdiv, 1965.
Geog 4209.61.5 Kazakova, M.S. Goroda i vstrechi. Riga, 1961.
An 358.96A Keane, A.H. Ethnology. Cambridge, Eng., 1896.
An 358.96.2 Keane, A.H. Ethnology. 2. ed. Cambridge, Eng., 1896.
An 358.99A Keane, A.H. Man - past and present. Cambridge, Eng., 1899.
An 359.08.9 Keane, A.H. The world's peoples. London, 1908.
An 359.08.10 Keane, A.H. The world's peoples. N.Y., 1908.
Geog 3057.6 Keane, Augustus Henry. The gold of Ophir. London, 1901.

Author and Title Listing

Geog 4208.88.5   Keane, J.F. Three years of a wanderer's life. London, 1888.
Geog 3018.99.15   Keane, John. Evolution of geography. London, 1899.
An 359.20.11   Keane, W.H. Man past and present. Cambridge, 1920.
Geog 4209.71.35   Kearns, Des. World wanderer; 100,000 miles under sail. Sydney, 1971.
Geog 5939.55.5   Kearns, W.H. The silent continent. 1st ed. N.Y., 1955.
An 2108.89   Keary, C.F. The dawn of history. N.Y., 1889.
Geog 5509.70   Keating, Bern. The Northwest Passage: frm the Mathew to Manhattan, 1497 to 1969. Chicago, 1970.
Geog 4678.72.15   Keating, Laurence J. The great Mary Celeste hoax. London, 1929.
Geog 4208.72   Keel and saddle, a retrospect. (Revere, J.W.) Boston, 1872.
Geog 4208.72.2   Keel and saddle, a retrospect. (Revere, J.W.) Boston, 1873.
Geog 5558.91.5   Keely, R.N. In Arctic seas...with Peary expedition. Philadelphia, 1892.
Geog 5398.91   Keely, Robert. In Arctic seas...voyage of the "Kite". Philadelphia, 1892.
Geog 4219.32   Keep moving. (Fletcher, A.C.B.) Chiago, 1932.
An 5.15   Keesing, Felix. Culture change; an analysis and bibliography of anthropological sources to 1952. Stanford, Calif., 1953.
Geog 4209.31.25   Keilpflug, Erich R. An den Rändern dreier Erdteile. Berlin, 1931.
An 2059.12.5A   Keith, Arthur. Ancient types of man. 2. ed. London, 1912.
An 135.3   Keith, Arthur. An autobiography. London, 1950.
An 2059.30   Keith, Arthur. New discoveries relating to the antiquity of man. N.Y., 1930.
An 359.31.7   Keith, Arthur. The place of prejudice in modern civilization. N.Y., 1931.
Geog 4209.69.15   Kelioniu, Knyga. (Paleckis, Justas.) Vilnius, 1969.
An 629.03   Keller, A.G. Queries in ethnography. N.Y., 1903.
Geog 4131.73   Keller, Peter. Soziologische Probleme in modernen Tourismus. Bern, 1973.
Geog 4181.126   Kelly, Celsus. La Austrialia del Espiritu Santo. Cambridge, 1966. 2v.
Geog 4557.95   Kelly, Samuel. Samuel Kelly, an 18th century seaman. N.Y., 1925.
An 2109.43.5   Kelsen, Hans. Society and nature; a sociological inquiry. Chicago, 1943.
An 2109.46.10   Kelsen, Hans. Society and nature; a sociological inquiry. London, 1946.
Geog 3019.13   Keltie, J.S. History of geography. N.Y., 1913.
Geog 5939.57   Kemp, Norman. The conquest of the Antarctic. N.Y., 1957.
Geog 4678.36.10   Kendall, J. A sermon delivered February 14, 1836. Plymouth, 1836.
Geog 5328.27   Kender du Knud? 2. Udg. (Hare, Kirsten.) Kjøbenhavn, 1971.
Geog 4308.72.7   Kennedy, David. Colonial travel. London, 1876.
An 2109.66.20   Kennedy, Kenneth A.R. Human skeletal remains from Chalcolithic and Indo Roman levels from Nevasa: an anthropometric and comparative analysis. Poona, 1966.
Geog 5538.51.15   Kennedy, William. A short narrative of the second voyage of the Prince Albert in search of Sir John Franklin. London, 1853.
Geog 5839.35.7   Kent, Rockwell. Greeland journal. N.Y., 1962.
Geog 5839.30A   Kent, Rockwell. N. by E. N.Y., 1931.
Geog 4676.61   Kerckhoven, J. van. Wijtloopig...beschrijvinge van d. onzel voyage van't schip Arnhem, 1661. Amsterdam, 1664.
Geog 611.1.5   Kerguelen, le découvreur et ses îles. (Brossard, Maurice Raymond de.) Paris, 1970-1971. 2v.
Geog 4427.82   Kerguélen. Relation de deux voyages. Paris, 1782.
An 2109.53.10   Kern, Fritz. Der Beginn der Weltgeschichte. Bern, 1953.
An 2109.60.20   Kern, Fritz. The wildbooters. Edinburgh, 1960.
Geog 808.96   Kerp, H. Methodisches Lehrbuch...Erdkund. Bonn, 1896-1904. 3v.
Geog 4158.11   Kerr, R. General history...of voyages. Edinburgh, 1811-24. 18v.
Geog 5160.8   Kersting, Rudolf. The white world. N.Y., 1902.
Geog 4139.68   Ketters en kwezels, regenten en rebellen. (Wertheim, Willem Frederik.) Drachten, 1968.
Geog 4182.42   Keuning, J. De tweede schipvaart der Nederlanders. v.1-5. 's-Gravenhage, 1938-51. 8v.
Geog 3129.34   Die Keuntnis der Erde im Altertum. (Kahlo, G.) München, 1934.
Geog 3019.38   Key, Charles E. The story of twentieth-century exploration. N.Y., 1938.
Geog 4145.87.5   Kharitanovskii, Aleksandr A. Chelovek s zheleznym olenem. Moskva, 1965.
Geog 5379.35.5   Khmyznikov, P. Na Chelnoskike. Leningrad, 1936.
Geog 4209.58.5   Khodili my pokhodami. (Korshunov, V.) Moskva, 1958.
Geog 4313.9   Kiecksee, Heinz. Die Ostsee-Sturmflut 1872. Heide, 1972.
Geog 126.10   Kiel. Universität. Geographisches Institut. Schriften. Kiel. 1,1932+ 36v.
Geog 3108.78   Kiepert, J.S.H. Lehrbuch der alten Geographie. Berlin, 1878.
Geog 132.4   Kiev. Universitet. Visnyk. Ser. heohrafiï. Kiev. 9,1967+
Geog 4209.70.5   Kiknadze, Aleksandr Vasil'evich. Ot Madrida do Tokio. Baku, 1970.
Geog 5160.20   Kimble, G.H.T. Geography of the northlands. N.Y., 1955.
An 379.53.5   Kimble, G.H.T. The way of the world. N.Y., 1953.
Geog 3229.38   Kimble, George H.T. Geography in the Middle Ages. London, 1938.
Geog 953.14   Kinauer, Rudolf. Lexikon geographischer Bilbände. Wien, 1966.
Geog 4219.20F   Kincaid, Earle H. History and cruises of the United States ship Whipple. Constantinople, n.d.
Htn   Geog 4308.78.5*   King, Edward. A collection of newspaper articles, describing his experiences in Europe, 1878-1880. n.p., n.d.
Geog 4308.78   King, H. Sketches of travel...Europe. Washington, 1878.
Geog 5919.69   King, H.G.R. The Antarctic. London, 1969.
Geog 659.69   King, Leslie J. Statistical analysis in geography. Englewood Cliffs, 1969.
Geog 4110.5   King, M. Where to stop. Boston, 1893.
Geog 4110.5.2   King, M. Where to stop. Boston, 1894.
Geog 4208.55   King, Paul. Voyaging to China in 1855 and 1904. London, 1936.
Geog 3240.19A   King Alfred's North. (Malone, Kemp.) Cambridge, 1930.
Geog 3057.5.5   King Solomon's golden Ophir. (Peters, Karl.) N.Y., 1969.
An 359.07   Kingdom of man. (Lankester, E.R.) N.Y., 1907.
Geog 3018.82.2   Kingsley, H. Tales of old travel. London, 1882.
Geog 4308.88   Kingston, W.B. A wanderer's notes. London, 1888. 2v.
Geog 819.47.5   Kinkead, Eugene. Our own Baedeker, from the New Yorker. N.Y., 1947.

An 358.91.15   Kinmont, A. The natural history of man. 2. ed. Philadelphia, 1891.
An 358.39   Kinmont, A. Twelve lectures on the natural history of man. Cincinnati, 1839.
Geog 4208.87A   Kipling, Rudyard. From sea to sea. N.Y., 1899. 2v.
Geog 4208.87.3   Kipling, Rudyard. From sea to sea. N.Y., 1909.
Geog 4219.28.5   Kircheiss, C. Meine Weltumsegelung mit dem Fischkutter Hamburg. 2. Aufl. Berlin, 1928.
An 379.14   Kirchhoff, Alfred. Man and earth. N.Y., 1914.
Geog 808.87   Kirchoff, A. Landerkunde des Erdteils Europa. v.1-3. Wien, 1887-1907. 5v.
An 359.01.7   Kirchoff, A. Mensch und Erde. Leipzig, 1901.
Geog 4308.49.5   Kirkland, C.M. Holidays abroad. N.Y., 1849. 2v.
Geog 5060.17   Kirwan, Laurence P. A history of polar exploration. N.Y., 1960.
Geog 4309.34.10   Kisch, Egon E. Eintritt verboten. Paris, 1934.
Geog 5324.1.15   Kish, George. North-east passage: Adolf Erik Nordenskiöld, his life and times. Amsterdam, 1973.
Geog 4672.12.6   Kitchen, Frederick H. Tales of S.O.S. and T.T.T. Edinburgh, 1927.
Geog 4672.12   Kitchin, Frederick H. Dead men's tales. Edinburgh, 1926.
Geog 4105.14   Kitchiner, William. The travellers' oracle. 2. ed. London, 1827. 2v.
Geog 4209.54.5   Kivikoski, Olavi. Yksin yli Atlantin. Helsinki, 1954.
Geog 4208.67   Kjøbenhavn. (Watt, Robert.) Kjøbenhavn, 1867.
An 359.20.9   Klaatsch, Hermann. Der Werdegang der Menschheit und die Entstehung der Kultur. Berlin, 1920.
Geog 579.36A   Kleines deutsch-portugiesisches...Verzeichnis geographischer Eigennamen. (Schmidt, A.J.) Rio de Janeiro, 1938.
Geog 819.03   Kleines Lehrbuch der Geographie. (Seydlitz, Ernst von.) Breslau, 1900.
Geog 4229.25.5   Kleinschmitt, Edmund. Durch Werkstätten und Gassen dreier Erdteile. 2. Aufl. Hamburg, 1928.
Geog 5321.2   Klengenberg, C. Klengenberg of the Arctic. London, 1932.
Geog 5321.2   Klengenberg of the Arctic. (Klengenberg, C.) London, 1932.
Geog 603.8   Kleopov, Igor' L'. Aleksandr Lavrent'evich Chekanovskii, 1833-1876. Leningrad, 1972.
Geog 816.70   De kleyne wonderlijcke werelt. (Joosten, J.) Amsterdam, 1670.
Geog 4303.99.20A   Klimas, Petras. Ghillebert de Lannoy in medieval Lithuania. N.Y., 1945.
Geog 4559.45   Klitgaard, Kaj. Oil and deep water. Chapel Hill, 1945.
Geog 808.59   Klöden, G.A. Handbuch der Erdkunde. Berlin, 1859-1862. 3v.
Geog 808.59.3   Klöden, G.A. Handbuch der Erdkunde. Berlin, 1873. 5v.
An 99.61A   Kluckhohn, Clyde. Anthropology and the classics. Providence, 1961.
An 359.44.25A   Kluckhohn, Clyde. The concept of culture. n.p., 194-.
An 185.2   Kluckhohn, Clyde. Culture and behaviour. N.Y., 1962.
An 185.1.2   Kluckhohn, Clyde. Mirror for man. N.Y., 1949.
Geog 5538.78   Klutschak, H.W. Als Eskimo unter den Eskimos; eine Schilderung der Erlebnisse der Schwatka'schen Franklin-Aufsuchungs-Expedition in den Jahren 1878-80. Wien, 1881.
Geog 5919.58.10   Knapp, Willem H.C. Antarctica; de geschiedenis van het geheimzinnige Zuidland. Haarlem, 1958.
Geog 1733.5   The knapsack guide...Tyrol and the eastern Alps. (Murray, John, publisher, London.) London, 1867.
Geog 1735.22   Knapsack guide for...Switzerland. (Murray, John, publisher, London.) London, 1864.
Geog 1735.26   Knapsack guide for...Switzerland. (Murray, John, publisher, London.) London, 1867.
Geog 3019.55.5   Knaurs Geschichte der Entdeckungsreisen. (Samhaber, Ernst.) München, 1955.
Geog 4131.60   Knebel, Hans J. Soziologische Strukturwandlungen im Mardernen. Stuttgart, 1960.
Geog 4209.58.25   Knies, Donald. Walk the wide world. N.Y., 1958.
Geog 578.54A   Knight, Charles. The English cyclopaedia. London, 1854-55. 4v.
Geog 4709.7   Knight, E.F. The cruise of the "Alerte". London, 1907.
Geog 4219.25.3   Knight, L.L. Trucking the sunset. Atlanta, 1925.
Geog 4219.38   Knights of the road. (Hypes, J.L.) Washington, 1938.
Geog 3055.15   Knötel, A.F.R. Atlantis und das Volk der Atlanten. Leipzig, 1893.
Geog 4502.5.9   Knorr, O. Der Grossvenediger in der Alpinismus. München, 1932.
Geog 4209.64   Knowles, J. Double vision. N.Y., 1964.
Geog 4678.58.5   Knowles, J.N. Crusoes of Pitcairn Island. Los Angeles, 1957.
Htn   Geog 4679.38*   Knowles, Josiah N. The crusoes of Pitcairn's Islands. n.p., 1938.
Geog 4157.67   Knox, J. New collection of voyages. London, 1767. 7v.
An 358.50   Knox, R. The races of men. Philadelphia, 1850.
Geog 4128.80.3   Knox, Thomas W. How to travel. N.Y., 1881.
Geog 4128.80.5   Knox, Thomas W. How to travel. N.Y., 1888.
Geog 4219.69.5   Knox-Johnston, Robin. A world of my own: the single-handed, non-stop circumnavigation of the world in Suahili. London, 1969.
Geog 5328.17   Knud Rasmussen. (Vejlager, Johannes.) Kjøbenhavn, 1934.
Geog 5328.5   Knud Rasmussens saga. (Birket-Smith, Kaj.) Kjøbenhavn, 1941.
Geog 4209.71.40   Knudsen, Kim. Grenselös. Oslo, 1971.
Geog 5324.2.85   Knuth, Eigil. Fridtjof Nansen og Knud Rasmussen. Kjøbenhavn, 1948.
Geog 5839.38   Knuth, Eigil. Under det nordligste danne broa; beretning om dansk Nordøstgrønlands Ekspedition, 1938-39. København, 1940.
An 2109.73   Knyshenko, Iurii V. Istoriia pervobytnogo obschchestvo. Rostov-na-Donu, 1973.
Geog 4502.5.19   Kobell, F. von. Wildonger. München, 1936.
Geog 5839.39.5   Koch, Lange. Fra Lissabon til Peary Land. København, 1939.
Geog 4515.301   Kochbuch für Bergsteiger. (Schmidkunz, W.) München, 1925.
An 359.64.15   Köbben, A.J.F. Van primitieven tot medeburgers. Assen, 1964.
Geog 665.80   Koegel, Ludwig. Geographische Plaudereien. Bonn, 1957.
Geog 4215.19.5   Kölliker, O. Die erste Umsegelung der Erde...1519-1522. München, 1908.
Geog 1525.72   Köln und der Rheinland zwischen Köln und Mainz. (Baedeker, publishers.) Freiburg, 1953.
Geog 1527.60   Köln und Umgebung. 2. Aufl. (Baedeker, publishers.) Freiburg, 1960.
An 609.28   Kölner Anthropologische Gesellschaft. 25 Jahre Kölner anthropologische Gesellschaft und städtisches Museum für Vor- und Frühgeschichte 1903-28. Festschrift. Köln, 1928.

Author and Title Listing

	Geog 3235.61	Koeman, Cornelis. The history of Lucas Janszoon Wazehnaer and his Spieghel der Zeevaerdt. Lausanne, 1964.
	Geog 4269.62.10	Koen, Albert. Putishta i spirki. Sofiia, 1962.
	An 2600.1	Koenig, H.C. De hominum inter ferar educatorum. Hanover, 1730.
	Geog 4269.58.5	Koeppen, Wolfgang. Nach Russland und anderswohin. Stuttgart, 1958.
	An 3309.32.5	Körperform und Leistung sechzehnjähriger Lehslinge. (Biedermann, Ernst.) Zürich, 1932.
	An 3539.01	Koeze, G.A. Crania ethnica Philippinica. Haarlem, 1901-04.
	Geog 3049.70	Kohlenberg, Karl Friedrich. Enträtselte Vorzeit. München, 1970.
	Geog 5839.24.5	Koht, Halvdan. Del Grønland vi miste og de vi ikkje miste. Kristiania, 1924.
	Geog 4332.26	Kolar, Vladimir. Jadran - more mira. Ljubjana, 1970.
	Geog 817.26	Kolb, P.G. Compendium totius orbis. Rottwike, 1726.
	Geog 5558.69.2	Koldewey, K. German Arctic expedition of 1869-70. London, 1874.
	Geog 5925.20	Kollback, K. Der Südpol. Bielefeld, 1911.
	An 379.18A	Koller, A.H. The theory of environment. Menasha, Wis., 1918.
	Geog 5645.5.10	Die koloniale Stellung Grönlands. (Dúason, Jón.) Göttingen, 1955.
	Geog 3049.68	Kolosimo, Peter. Timeless earth. Secaucus, N.J., 1974.
	Geog 3590.74	Kolumby rosskie. (Alekseev, Aleksandr.) Magadan, 1966.
	Geog 4249.50.2	Kon-Tiki; across the Pacific by raft. (Heyerdahl, T.) Chicago, 1964.
	Geog 4249.50	Kon-Tiki. (Heyerdahl, T.) Chicago, 1950.
	Geog 4249.50.10A	Kon-Tiki and I. (Hesselberg, E.) N.Y., 1950.
	Geog 612.5	Kondracki, Jerzy. Stanisław Lencewicz. Warszawa, 1966.
	An 3559.72	Konduktorova, Tamara S. Antropologiia drevnego nasteniia Ukrainy. Moskva, 1972.
	An 3309.73.5	Konduktorova, Tamara S. Antropologiia naseleniia Ukrainy mezoliṭa, neolita, i epokki bronzy. Moskva, 1973.
	An 170.230.5	Kongres Československých anthropologă, 10th, Prague and Humpolie, 1969. Anthropological congress dedicated to Alěs Hrdličkă. Prague, 1971.
	Geog 128.2	Konigsberg. Universität. Geographisches Institut. Veröffentlichungen. Hamburg. 10v.
	Geog 128.2.5	Konigsberg. Universität. Geographisches Institut. Veröffentlichungen. N.F. Reihe Geographie. Konigsberg. 1-10,1931-1937 10v.
	Geog 128.2.8	Konigsberg. Universität. Geographisches Institut. Veröffentlichungen. N.F. Reihe Ethnographie. Neudamm. 1-2,1931-1932 2v.
	Geog 624.1	Konnyü, Leslie. John Xantus. Köln, 1965.
	Geog 613.5	Konrad Millers Lebenswerk. (Borodajkez, Taras.) Salzburg, 19- .
	Geog 1604.5	Konstaninopel und das westliche Kleinasien. (Baedeker, publishers.) Leipzig, 1905.
	Geog 1604.15	Konstantinopel und Kleinasien, Archipel, Cypern. 2. Aufl. (Baedeker, publishers.) Leipzig, 1914.
	An 170.228	Kontakte und Grenzen. Göttingen, 1969.
	An 2109.17	Koppers, W. Die ethnologische Wirtschaftsforschung. Wien, 1917.
	An 170.251.100	Koppers, W. Kultur und Sprache. Wien, 1952.
	An 2109.49.5	Koppers, W. Der Urmensch und sein Weltbild. Wien, 1949.
	An 359.70	Korolev, Stanislav I. Voprosy etnopsikhologii v vabotakh zanebezhnykh avtarov. Moskva, 1970.
	An 359.05	Koropchevskago, A.A. Zhachenie "geograficheskikh provintsie". Sankt Peterburg, 1905.
	Geog 4209.62.5	Koroteev, V.I. Ia eto videl; ocherki raznykh let. Moskva, 1962.
	Geog 4209.58.5	Korshunov, V. Khodili my pokhodami. Moskva, 1958.
	Geog 5837.75	Kort beskrivelse over Grønland. v.1-3. (Stauning, J.) Viborg, 1775.
	An 608.70	Kort vriledning i det kgl. ethnographiske museum. (Steinhauer, C.L.) København, 1870.
	Geog 4182.3	Korte historiael...voyagiens...David P. de Vries. (Colenbrander, H.T.) 's-Gravenhage, 1911.
	An 359.21A	Korzybski, Alfred. Manhood of humanity. N.Y., 1921.
	An 359.50.15	Korzybski, Alfred. Manhood of humanity. 2. ed. Lakeville, Conn., 1950.
	Geog 579.43	Kosack, H.P. Wörterverzeichnis für russische Karten. Berlin, 1943.
	Geog 5919.55	Kosack, Hans P. Die Antarktis; eine Länderkunde. Heidelberg, 1955.
	Geog 500.98F	Kosack, Hans P. Die Kartographie, 1943-1954. Lahr, 1955.
	Geog 5070.72	Kosack, Hans P. Die Poterforschung; ein Datenbuch über die Natur-, Kultur-, Wirtschaftsverhältnisse und die Erforschungsgeschichte der Polarregionen. Braunschweig, 1967.
	Geog 4332.24	Košutić, Ivo. Gusari Jadranskoga mora. Zagreb, 1971.
	An 2109.57	Kosven, M.O. Abriss der Geschichte und Kultur der Urgesellschaft. Berlin, 1957.
	Geog 4308.04	Kotzebue, A. Travels from Berlin thro' Switzerland. London, 1804. 3v.
	An 359.70.10	Kovalev, Sergei M. O chloveke, ego poraboshchenii i osvdsozhdenii. Moskva, 1970.
	Geog 4209.66.5	Kovanov, Vladimir V. Meridiany, sobytiia, vstrechi. Moskva, 1966.
	Geog 4209.70.10	Kozima, Marjan. Veseli potopisi. Maribor, 1970.
	Geog 4209.64.30	Kozlovs'kyi, Mykola F. Cherez 15 morei i 2 okeana; puteshestviia. Kiev, 1964.
	Geog 5839.30.10F	Krabbe, Thomas N. Greenland; its natures, inhabitants and history. Copenhagen, 1930.
	An 170.252	Kraeber, A.L. Source book in anthropology. N.Y., 1931.
	Geog 3249.72	Krämer, Walter. Neue Horizonte. 1. Aufl. Leipzig, 1972.
	Geog 4345.73	Krafft, Hans U. Ein deutscher Kaufmann. Göttingen, 1862.
	Geog 133.1	Krakow. Wyższa Szkola Pedagogiczna. Prace geograficzne. Kraków. 3,1964+ 2v.
	An 3559.06	Das kraniologische der New Guinea Expedition. (Hauser, K.) Berlin, 1906.
	An 3539.07	Kraniologische Studien...Altägypten. (Oetteking, B.) Braunschweig, 1907.
	An 3559.64.5	Kraniometriia. (Alekseev, Valerii P.) Moskva, 1964.
	Geog 579.60	Kratkaia geograficheskaia entsiklopecha. Moskva, 1960. 5v.
	Geog 4419.02	Krausz, S. Zu Land und See im Orient. Chicago, 1902.
	An 2059.69.10	Kreatur Mensch. (Altner, Günter.) München, 1969.
	Geog 819.51	Krebs, N. Vergleichende Landerkunde. Stuttgart, 1951.
	An 379.21.5	Krebs, Norbert. Die Verbreitung des Menschen auf der Erdoberfläche (Anthropogeographie). Leipzig, 1921.
	Geog 4182.76	Kreekel, Willem. De reis van Z.M. De Vlieg. 's-Gravenhage, 1975.
	Geog 3070.165	Kremling, Helmut. Die Beziehungsgrundlage in thematischen Karten in ihrem Verhältnis zum Kartengegenstand. München, 1970.
	Geog 3590.84	Krempol'skii, Viktor F. Istoriia razvitiia kartoizdaniia v Rossii i v SSSR. Moskva, 1959.
	Geog 5399.73	Krenkel', Ernst T. RAEM - moi pozyvnye. Moskva, 1973.
	Geog 3235.50	Kretschmer, K. Die italienischen Portolane des Mittelalters. Berlin, 1909.
	Geog 5399.37.10	Kreukel', E.T. Chetyre tovarishcha. Moskva, 1940.
	Geog 4329.28	Kring medelhavets stränder. (Rydh, Hanna.) Stockholm, 1928.
	Geog 5398.72.11	Krisch, Otto. Das Tagebuch des Maschinisten Otto Krisch. Graz, 1973.
	Geog 5845.5	Krise uden alternativ. (Hoejlund, Niels.) København, 1972.
	An 359.23.9	Kroeber, A.L. Anthropology. N.Y., 1923.
	An 359.53	Kroeber, Alfred L. Anthropology today. Chicago, 1953.
	An 359.52.10A	Kroeber, Alfred L. The nature of culture. Chicago, 1952.
	An 135.1.5	Kroeber, Theodora. Alfred Kroeber: a personal configuration. Berkeley, 1970.
	Geog 5650.25	Krogh, Knud J. Erik den Rødes Grønland. København, 1967.
	Geog 5650.25.2	Krogh, Knud J. Viking Greenland with a supplement of saga. København, 1967.
	Geog 4309.71.10	Krueger, Horst. Fremde Vaterländer. München, 1971.
	Geog 3590.88	Krupenikov, Igor' A. Istoriia geograficheskoi mysli v Moldavii. Kishinev, 1974.
Htn	Geog 4218.03.5*	Krusenstern, A.J. Voyage around the world...1803-1806. London, 1813. 2v.
Htn	Geog 4218.03.7*	Krusenstern, A.J. Voyage autour du monde...1803-06. Paris, 1821. 2v.
Htn	Geog 4218.03.8F*	Krusenstern, A.J. Voyage autour du monde. Atlas. Paris, 1821.
Htn	Geog 4218.03.3*	Krusenstern, A.J. von. Reise um die Welt in den Jahren 1803-1806 auf Befehl seiner Kaisere. v.1-2. Berlin, 1811-12. 3v.
	An 3559.72.10	Kruts, Svitlana O. Naselenie territorii Ukrainy epokhi medi-bronzy. Kyiv, 1972.
	Geog 129.5	Krymskii Garnyi Klub, Odessa. Zapiski. Odessa. 1895-1912 2v.
	An 359.34.7	Kryzwicki, L. Primitive society and its vital statistics. London, 1934.
	An 359.37.13	Krzywicki, L. Społeczeństwo pierwotne. Warszawa, 1937.
	Geog 5399.60.5	Kudenko, O.I. Teplaia Arktika. Moskva, 1960.
	Geog 665.45	Kühn, Arthur. Die Neugestaltung der deutschen Geographie im 18. Jahrhundert. Leipzig, 1939.
	An 2109.53.27	Kuehn, Herbert. Auf den Spuren des Eiszeitmenschen. 2. Aufl. Wiesbaden, 1953.
	An 2109.56.10	Kuehn, Herbert. Das Erwachen der Menschheit. Frankfurt, 1956.
	An 2109.66.25	Kuehn, Herbert. Erwachen und Aufstieg der Menschheit. Frankfurt, 1966.
	Geog 600.20	Künstler des Kartenbildes. (Harms, Hans.) Oldenburg, 1962.
Htn	Geog 958.10*	Künstliche Erdkugel zur Uerbreitung gemeinnütziger Kentnisse über die Eintheilung. n.p., 18- ?
	An 3409.16.5	Kuhne, F. The finger print instructor. N.Y., 1916.
	An 3409.16.10	Kuhne, F. The finger print instructor. 3. ed. N.Y., 1942.
	Geog 4207.41	Kuhns, J.M. Lebens und Reise Beschreibung. Gotha, 1741.
	Geog 4309.74	Kujundžić, Miodrag. Tudje avlije. Novi Sad, 1974.
	Geog 4269.62	Kukanov, A.A. Doroga k serdtsu; reportazh. Tallin, 1962.
	Geog 4269.60	Kuleshov, A.P. 500,000 [i.e. Piat'sot tysiach] kolometrov v puti. Moskva, 1960.
	Geog 4209.71.25	Kuleshov, Aleksandr P. Shest' gorodov piati kontinentov. Moskva, 1971.
	An 359.54.20	Die Kultur. (Flatz, Josef.) Linz, 1954.
	An 2109.14	Kultur und Rasse. (Boas, Franz.) Leipzig, 1914.
	An 2109.24.5	Kultur undReligion des primitiven Menschen. (Danzel, T.W.) Stuttgart, 1924.
	An 170.251.100	Kultur und Sprache. (Koppers, W.) Wien, 1952.
	An 170.278	Kulturanthropologie. (Mühlemann, Wilhelm Emil.) Köln, 1966.
	An 379.64	Kulturer före istiden. (Troëng, Ivan.) Uppsala, 1964.
	Geog 132.1	Kulturgeografi; tidsskrift. Kjøbenhavn. 5,1953+ 9v.
	Geog 659.65	Kulturlandschaftsforschung und Landeskunde. (Overbeck, Hermann.) Heidelberg, 1965.
	Geog 4139.61	Kunský, J. Čestí cestovatelé. Praha, 1961. 2v.
	Geog 5855.13	Kunstzin der Eskimos. (Lorm, A.J.) 's-Gravenhage, 1945.
	An 99.73	Kuper, Adam. Anthropologists and anthropology. N.Y., 1973.
	Geog 614.6	Kupferschmidt, F. Karl Neumann. Inaug. Diss. Leipzig, 1935.
	Geog 3590.76	Kurs auf Unerforscht. (Wotte, Herbert.) Leipzig, 1967.
	Geog 4219.53	Kurun around the world. (Le Toumelin, J.I.) N.Y., 1955.
	Geog 4515.106.2	Kurz, Marcel. Alpinisme hivernal. Paris, 1928.
	Geog 4509.48.5	Kurz, Marcel. Fremde Berge-Ferne Ziele. Bern, 1948.
	Geog 4659.75	Kusche, Lawrence D. The Bermuda triangle mystery - solved. N.Y., 1975.
	Geog 4306.17.7	Küsel Saxon, S. Itinerarium Germaniae, Siciliae. Erphordiae, 1617.
	Geog 4309.30.30	Kushner, Boris A. 103 dnia na Zapade, 1924-1926 gg. 2. izd. Moskva, 1930.
	Geog 5939.73	Kutuzov, Pavel S. Antarkticheskii dnevnik. Moskva, 1973.
	Geog 5399.32	Kuz delmek a fehér halál országában. (Kalmár, G.) Budapest, 1932-33. 2v.
	Geog 3594.10	Kuźmiński, Bolesław. Polskie nazwy na mapie śceiata. Wyd. 1. Warszawa, 1967.
	Geog 4139.73.5	Kuźmiński, Bolesław. Przygody polskich obieżyświatów na morzach i lądach. 1. wyd. Gdeńsk, 1973.
	Geog 4209.55.15	La isla y Colón. (Pineda Yañez, R.) Buenos Aires, 1955.
	Geog 612.3.15	La Perouse in Maui. (Dondo, M.M.) Wailuku, 1959.
	Geog 4217.28	La Barbinais le Gentil. Nouveau voyage autour du monde. Amsterdam, 1728. 2v.
	An 359.54	LaBarre, W. The human animal. Chicago, 1954.
	Geog 816.46.13	Labbé, Philippe. La geographie royalle. 2e éd. Paris, 1653.
	Geog 4329.30.3	Labels; a Mediterranean journal. (Waugh, E.) London, 1930.
	An 3309.20	A laboratory manual of anthropometry. (Wilder, Harris H.) Philadelphia, 1920.
	An 3209.01	Laboratory outlines in somatology. (Russell, Frank.) N.Y., 1901.
Htn	Geog 4206.43.5*	La Boullaye, C. Goux. Les voyages et observations. Paris, 1653.
	Geog 520.12	La Broquière, Bertrandon de. Le voyage d'Outremer. Photoreproduction. Paris, 1892.
	Geog 5398.97	Lachambre, Henri. Andrée; au pôle nord en ballon. Paris, 1897.

Author and Title Listing

Htn Geog 4209.37F* LaCombe, Jean de. A compendium of the East; being an account of voyages to the Grand Indies. London, 1937.

Geog 818.00.5 LaCroix, L.A.N. Geographie moderne et universelle. Paris, 1800. 2v.

Geog 4209.58.20 La Croix, Robert de. Mysteres des îles. Paris, 1958.

Geog 4417.44 Lade, R. Voyages du capitaine...Lade. Paris, 1744. 2v.

Geog 4417.44.5 Lade, R. Voyages du capitaine...Lade. Paris, 1810.

Geog 578.10.2 Ladvocat, J.B. Dictionnaire géographique. 2. ed. Paris, 1812.

Geog 577.09.18 Ladvocat, J.B. Dictionnaire géographique portatif. Paris, 1770.

Geog 4218.46.7 A lady's second journey...world. (Pfeiffer, I.) N.Y., 1856.

Geog 4218.46.5 A lady's voyage round the world. (Pfeiffer, I.) N.Y., 1852.

Geog 819.59.4 Länder der Erde. 4. Aufl. Berlin, 1967.

An 2109.67.35F Laet, Sigfried Jan de. De voorgeschiedenis van Europa. Hasselt, 1967.

An 359.56.5 LaFarge, John. The Catholic viewpoint on race relations. 1. ed. Garden City, N.Y., 1956.

Geog 4218.40.5F Lafond, G. Fragments de voyages autour du monde. Paris, 1861.

Geog 4218.40 Lafond, G. Quinze ans de voyages. Paris, 1840. 2v.

Geog 579.50F Lagoa, J.A.M.J. Glossário taponimico da antiga historiografia. Lisboa, 1950- 4v.

VGeog 4269.68 Lah, Avguštin. Naše sosedne države. Ljubljana, 1968.

Geog 4157.80 LaHarpe, J.F. Abrégé de l'histoire generale des voyages. v.1-32, atlas. Paris, 1780-1801. 33v.

An 359.42.15 Laidler, H.W. The role of the races in our future civilization. N.Y., 1942.

Geog 4308.46.2 Laing, S. Notes of a traveller. Philadelphia, 1846.

Geog 4308.46.3 Laing, S. Observations on social...state of European people. London, 1850.

An 2058.65.3 Lake habitations and pre-historic remains...north, centre Italy. (Gastaldi, B.) London, 1865.

Geog 5399.37 Laktionov, A.F. Severnyi polius. Arkhangelsk, 1939.

Geog 5399.37.7 Laktionov, A.F. Severnyi polius. Moskva, 1955.

Geog 5530.80 Lamb, G.F. Franklin. London, 1956.

Geog 4019.52 Lamb, Geoffrey F. Modern action and adventure. London, 1952.

Geog 4218.83.10 Lambert, C. Voyage of the Wanderer. London, 1883.

Geog 4679.68 Lambert, Max. The Wahine disaster. Wellington, 1968.

Geog 4347.27 La Montraye, A. de. Voyage du...en Europe, Asie et Afrique. Haye, 1727. 2v.

Geog 4347.27.3F La Montraye, A. de. Voyage en Anglois et en François. Haye, 1732.

Htn Geog 4347.27.5* La Montraye, A. de. Voyages and travels, Prussia, Russia and Poland. London, 1732. 3v.

An 359.43.20 Lamps of anthropology. (Murphy, J.D.) Manchester, Eng., 1943.

Geog 4208.52.7 Lamson, Joseph. Round Cape Horn...in 1852. Bangor, 1878.

Geog 665.115 Land and life. (Sauer, Carl Ortwin.) Berkeley, 1965.

Geog 665.94 Land and livelihood. (McCashill, Murray.) Christchurch, 1962.

Geog 5838.71 The land of desolation...Greenland. (Hayes, I.I.) London, 1871.

Geog 4309.34.5 Land of the midnight sun, Scandinavian region, Russia, and Germany. (Stackpole, Edward J.) Harrisburg, 1934.

Geog 809.10 Land og folk. Geografi i skildringer og livsbilleder. Kjøbenhavn, 1910. 2v.

Geog 5839.59.2 Land under the pole star: a voyage to the Norse settlements of Greenland and the saga of the people that vanished. (Ingstad, Helge Marcus.) London, 1966.

Geog 5650.16 Landafundir og sjóferðir i norðrhofum. (Hermansson, H.) Winnipeg, 19- .

Geog 4309.32 Landau, M.A. Eine unsentimentale Reise. München, 1932.

Geog 808.87 Landerkunde des Erdteils Europa. v.1-3. (Kirchoff, A.) Wien, 1887-1907. 5v.

Geog 4329.53 Landery, C.F. Whistling for a wind. 1st American ed. N.Y., 1953.

Geog 5839.59 Landet under leidarstjernen. (Ingstad, Helge Marcus.) Oslo, 1959.

Geog 3019.05 Landkonturer och hafsvidder. (Athlenius, K.) Stockholm, 1905.

Geog 4228.79 A landlubber's log of a voyage round the "Horn". (MacMichael, M.) Philadelphia, 188-?

Geog 578.40 Landmann, G. Universal gazeteer. London, 1840.

Geog 4228.23 Landolphe, Jean François. Mémoires du capitaine Landolphe. Paris, 1823. 2v.

Geog 4209.24A Landor, A.H.S. Everywhere; the memoirs of an explorer. N.Y., 1924. 2v.

Geog 3049.52 Lands beyond. (DeCamp, L.S.) N.Y., 1952.

Geog 5070.9 The lands of silence. (Markham, C.R.) Cambridge, Eng., 1921.

Geog 4328.52.2 The lands of the Messiah, Mahomet, and the pope. 2nd ed. (Aiton, John.) London, 1852.

Geog 4239.70 Lands of the morning. (Hills, Lawrence Donegan.) London, 1970.

Geog 4328.55 Lands of the Saracew. (Taylor, B.) N.Y., 1855.

Geog 665.62 Landschaft und Lund. (Karper, Kurt.) Remagen, 1951.

An 359.06.5 Landschafte...Kulturtypen. (Chalikiopoulos, L.) Leipzig, 1906.

Geog 3019.64 Landström, B. The quest for India. London, 1964.

An 2109.38 Landtman, G. The origin of the inequality of the social classes. London, 1938.

An 99.73.5 Le langage perdu; essai sur la différence anthropologique. 1. éd. (Duvignaud, Jean.) Paris, 1973.

An 2059.59 Lange, Hurt. Fremdling zwischen Tier und Gott. Gütersloh, 1959.

Geog 5509.15 Der lange Zeit angenomme Zusammenhang der Hudsonbai mit dem Stille Ozean. (Beyl, Hermann.) Würzburg, 1915.

Geog 3019.72.20 Langley, Michael. When the pole star shone. London, 1972.

An 359.74.25 Langness, Lewis L. The study of culture. San Francisco, 1974.

Htn Geog 4218.12* Langsdorff, G.H. von. Bemerkungen auf einer Reise um die Welt in den Jahren 1803-1807. Frankfurt, 1812. 3v.

Htn Geog 4218.03* Langsdorff, G.H. von. Voyages and travels...in 1803-1805. London, 1813. 2v.

Geog 4227.89 Langstedt, F.L. Reisen nach Sudamerika, Asien. Hildesheim, 1789.

Geog 4268.86.4 Lanier, L. L'Europe...choix de lectures de geographie. 4. ed. Paris, 1888.

An 359.07 Lankester, E.R. Kingdom of man. N.Y., 1907.

Geog 4303.99.2 Lannoy, Guillebert de. Oeuvres. Louvain, 1878.

Geog 4303.99 Lannoy, Guillebert de. Voyages et ambassades...1399-1450. Mons, 1840.

Geog 5985.258.15 Lansing, Alfred. Endurance; Shackleton's incredible voyage. N.Y., 1959.

Geog 4209.40 Lanterns alight; journeys to far places. (Woodward, C.) Chicago, 1940.

Geog 4145.48 Lantzsch, W. Die Welt in allen Zonen. München, 1961.

Geog 4217.85.13 Lapérouse, Jean François de Galaup. Voyage de Lapérouse autour du monde. Paris, 1965.

Geog 612.3.5 Lapérouse. (Maine, René.) Paris, 1946.

Geog 612.3.10 Laperouse. (Scott, Ernest.) Sydney, 1912.

Htn An 2056.55* La Peyrère, I. de. Systema theologicum...Praeadamitae. n.p., 1655.

Geog 4218.33.25 Laplace, Cyril P.T. Voyage autour du monde. Paris, 1833-39. 4v.

Geog 4218.33.27PF Laplace, Cyril P.T. Voyage autour du monde. Atlas. Paris, 1835.

An 2059.07 Lapparent, A. Les silex taillés et l'ancienneté de l'homme. Paris, 1907.

Htn Geog 4678.10* Larcom, Henry. Distressing narrative of the loss of the ship Margaret of Salem. Beverly, 1810.

Geog 3018.28 Larenaudière. Histoire abrégée de l'origine...de géographie. Paris, 1828.

Geog 4209.39.25 Larigaudie, Guy de. La route aux aventures. Paris, 1948.

Geog 4659.69.5 Larn, Richard. Cornish shipwrecks. Newton Abbot, 1969- 2v.

Geog 5507.87 LaRoche, F.C. de. Sendschreiben...des alten Grönlands. Kopenhagen, 1787.

Geog 3019.38.5F La Roncière, Charles. Histoire de la découverte de la terre. Paris, 1938.

Geog 819.58.5 Larousse, firm, publishers. Geographie universelle Larousse. Paris, 1958. 3v.

Geog 819.58.10 Larousse encyclopedia of geography. N.Y., 1961.

Geog 5650.13 Larusson, O. Rettarstad Graenlands aðornu. Kaupmannahöfn, 1924.

An 355.85 Lascovius, P.M. De homine magno illo in rerum natura. Libro II. Witebergae, 1585.

Geog 6009.14.25 Laseron, Charles F. South with Mawson. 2d ed. Sydney, 1957.

Geog 6009.01.30 Lashly, William. Under Scott's command: Lashly's Antarctic diaries. London, 1969.

Geog 5919.58 The last continent. (Liverridge, Douglas.) London, 1958.

Geog 5925.25 The last continent of adventure. (Hayward, Walter B.) N.Y., 1930.

Geog 4679.49 The last cruise. (Lederer, William J.) N.Y., 1950.

Geog 5558.94 Last cruise of the Miranda. (Walsh, H.C.) N.Y., 1896.

Geog 4678.70 The last cruise of the Saginaw. (Read, George H.) Boston, 1912.

Geog 4239.24 The last cruise of the Shanghai. (Wells, F. de W.) N.Y., 1925.

Geog 5312.3.10 The last explorer; the adventures of Admiral Byrd. (Hoyt, Edwin Palmer.) N.Y., 1968.

Geog 4559.56 The last grain race. (Newby, Eric.) Boston, 1956.

Geog 5855.17 The last kings of Thule. (Malaurie, Jean.) London, 1956.

Geog 4219.41.5 Last man around the world. (Wiener, P.) N.Y., 1941.

Geog 5538.52 The last of the Arctic voyages. (Belcher, E.) London, 1855. 2v.

Geog 4208.54.10 The last of the Logan. (Coffin, Robert.) Ithaca, 1941.

Geog 4209.23 The last secrets. (Buchan, John.) London, 1923.

Geog 4679.56 Last voyage. (Armstrong, W.) London, 1956.

Geog 4418.89A The last voyage. (Brassey, A.) London, 1889.

Geog 5518.29.5 The last voyage of Captain Sir John Ross, R.N., to the Arctic regions. (Huish, Robert.) London, 1835.

Geog 4181.142A The last voyage of Drake and Hawkins. (Andrews, Kenneth Raymond.) Cambridge, 1972.

Geog 5559.13A The last voyage of the Karluk. (Bartlett, Robert A.) Boston, 1916.

Geog 4219.20.5 The last voyage of the schooner Rosamond. (Chevalier, Haakon Maurice.) London, 1970.

An 359.15.3 Latcham, R.E. Conferencias sobre antropología, etnología. Santiago, 1915-

Geog 3520.17.6 Late Tudor and early Stuart geography, 1583-1650. (Taylor, Eva G.) N.Y., 1968.

An 358.59 Latham, R.G. Descriptive ethnology. London, 1859. 2v.

An 358.52.2 Latham, R.G. Man and his migrations. N.Y., 1852.

Geog 4218.24.5PF La Touanne, E. de. Album pittoresque de la frégate la Thétis et de la corvette l'Esperance. Paris, 1828.

Geog 4209.62.10 Lattmann, D. Mit einem deutschem Pass. München, 1964.

An 2109.37.7 Laubscher, B.J.F. Sex, custom and psychopathology. London, 1937.

Geog 3070.90 Lauf, G.B. The origin and development of cartography. Johannesburg, 1955.

An 2209.30 Laufer, B. Geophagy. Chicago, 1930.

Geog 3530.18 Lauga, Henri. De la banquise à la jungle. Paris, 1952.

Geog 4329.35.15 Laughlin, C.E. So you're going to the Mediterranean. Boston, 1935.

Geog 5399.48 Laugthen og nordpaa. (Vibe, Christian.) København, 1948.

Geog 5312.7.5 Lauridsen, Peter. Vitus Bering; the discoverer of Bering Strait. Chicago, 1889.

Geog 5312.7 Lauridsen, Peter. Vitus J. Bering og de russiske opdagelsesrejser fra 1725-43. Kjøbenhavn, 1885.

Geog 818.42 Laurie, J. System of universal geography. Edinburgh, 1842.

Geog 4268.50 Lavallée, T. Military topography of...Europe. London, 1850.

Geog 818.68 Lavallée, T.S. Physical, historical and military geography. London, 1868.

An 358.94.5 Lavrov, P.L. Antropologicheskaia zhizne. Zheneva, 1894. 2v.

Geog 6100.5 Law, Phillip G. Anare: Australia's Antarctic outposts. Melbourne, 1957.

Geog 5970.36 Law, Phillip G. Australia and the Antarctic. St. Lucia, 1962.

Geog 759.42F Lawrence, C.H. New world horizions; geography for the air age. N.Y., 1942.

An 3208.22.2 Lawrence, W. Lectures on physiology, zoology and the natural history of man. London, 1822.

An 3208.22 Lawrence, W. Lectures on physiology, zoology and the natural history of man. London, 1822.

An 358.28 Lawrence, William. Lectures on physiology, zoology and natural history of man. Salem, 1828.

Geog 3240.25 Le legua nautica en la edad media. (Garcia Franco.) Madrid, 1957.

An 136.5.5 Leach, Edmund Ronald. Claude Lévi-Strauss. N.Y., 1970.

An 136.5.87 Leach, Edmund Ronald. Claude Lévi-Strauss. N.Y., 1974.

An 2059.53 Leakey, Louis S. Adam's ancestors. 4. ed. London, 1953.

An 2059.61.10A Leakey, Louis S. The progress and evolution of man in Africa. London, 1961.

Geog 4559.54.15 Learmont, J.S. Master in sail. 2d ed. London, 1954.

Author and Title Listing

An 2109.73.15 Learning and culture. (American Ethnological Society.) Seattle, 1973.
Geog 4209.46 Leaves in the wind. (Baerlein, Henry.) London, 1946.
Geog 3590.30.5 Lebedev, Dimitri M. Geog. v Rossii petrovsk. vremeni. Moskva, 1950.
Geog 3590.30 Lebedev, Dimitrii M. Geogr. v Rossii XVII v. ocherkii. Moskva, 1949.
Geog 3590.30.10 Lebedev, Dimitrii M. Plavanie A.I. Chirukova. Moskva, 1951.
Geog 3590.55 Lebedev, Dmitrii M. Ocherki po istorii geografii v Rossii XV i XVI vekov. Moskva, 1956.
Geog 3590.55.5 Lebedev, Dmitrii M. Ocherki po istorii geografii v Rossii XVIII veka. Moskva, 1957.
Geog 3590.80 Lebedev, Dmitrii M. Russkie geograficheskie otkrytiia i issledovaniia s drevnikh vremen do 1917 goda. Moskva, 1971.
Geog 5919.57.10 Lebedev, Vladimir L. Antarktika. Moskva, 1957.
Geog 4136.76.5 Leben...der...See-Helden. (Bosch, L.) Nürnberg, 1681.
Geog 4207.76.12 Leben des...Americanischen Reisenden J. Ledyard. (Sparks, Jared.) Leipzig, 1829.
Geog 4502.5.3 Leben und Schriften. (Steinberger, S.) München, 1929.
Geog 603.7 Leben und Wirken eines österreichischen Geographieprofessors und Erwachsenenbildners. (Constantini, Otto.) Linz, 1969.
Geog 4207.41 Lebens und Reise Beschreibung. (Kuhns, J.M.) Gotha, 1741.
Geog 619.11 Lebenserinnerungen und Tagebuchblätter eines Geographen. (Schmieder, Oskar.) Kiel, 1972.
An 359.57.5 Lebeuf, Jean Paul. Application de l'ethnologie à l'assistance sanitaire. Bruxelles, 1957.
Geog 4416.54 LeBlanc, V. De vermaarde reizen. Amsterdam, 1654.
Geog 4416.54.5 LeBlanc, V. Les voyages fameux. Paris, 1658.
Htn Geog 4206.60* LeBlanc, V. The world surveyed. London, 1660.
Geog 520.22 Le Bouvier, Gilles. Le livre de la description des pays. Paris, 1908.
Geog 4502.5.13 Lebwald, A. Damographia oder Gemsen-Beschreibung in zewy Theil Abgetheilet. Saltzburg, 1933.
An 359.17 Lecciones de antropologia. (Restrepo-Hernandez, J.) Bogotá, 1917.
An 99.72.5 Leclerc, Gérard. Anthropologie et colonialisme. Paris, 1972.
Geog 6008.97.3 Lecointe, G. Au pays des manchots. Bruxelles, 1910.
An 136.5.25 Leçon inaugurale faite le mardi. (Lévi-Strauss, Claude.) Paris, 1966.
Geog 4759.71.5 Leçons de géographie tropicale. (Gourou, Pierre.) Paris, 1971.
Geog 3055.43 Le Cour, Paul. A la recherche d'un monde perdu. Paris, 1926.
Geog 3055.86 Le Cour, Paul. L'Atlantide atlantique. Bordeaux, 1971.
An 358.64 Lecture on man: his place in creation and in the history of the earth. (Vogt, C.) London, 1864.
An 2058.77.3 Lecture on the antiquity of man. (Jones, T.R.) London, 1877.
Geog 4208.97.3 Lectures. (Stoddard, J.L.) Boston, 1899-1900. 10v.
Geog 4208.97.4 Lectures. Supplement. (Stoddard, J.L.) n.p., 1901-05. 4v.
Geog 4208.97.5 Lectures. v.8. (Stoddard, J.L.) Boston, 1905.
Geog 809.36 Lectures géographiques. v.1-2. (Ozonf, R. (Mme.).) Paris, 1936-1938. 4v.
An 359.25.9 Lectures on ethnogrpahy. (Anantha, Krishna Iyer L.) Calcutta, 1925.
Geog 665.7 Lectures on geography. (Strachey, R.) London, 1888.
An 358.28 Lectures on physiology, zoology and natural history of man. (Lawrence, William.) Salem, 1828.
An 3208.22 Lectures on physiology, zoology and the natural history of man. (Lawrence, W.) London, 1822.
An 3208.22.2 Lectures on physiology, zoology and the natural history of man. (Lawrence, W.) London, 1822.
Geog 4679.49 Lederer, William J. The last cruise. N.Y., 1950.
Geog 5939.59.20 Ledovaia kniga. (Smuul, Juhan.) Moskva, 1959.
Geog 4182.15 Ledyard, Gari. The Dutch come to Korea. Seoul, 1971.
Geog 5399.28 Lee, Herbert P. Policing the top of the world. London, 1928.
An 2109.28.11 Leeuw, Gerardus. La structure de la mentalité primitive. Strasbourg, 1928.
Htn Geog 4136.76* Leeven en daden...zee-helden. (Bosch, L.) Amsterdam, 1676.
Geog 4208.30.7 Lefavor, William. Captain Lefavor's forty years' travels. Philadelphia, 1877.
Geog 3060.6A Legendary islands of the Atlantic. (Babcock, William H.) N.Y., 1922.
Geog 3570.9 Les légendes allemandes. (Bersaude, J.) Genève, 1917-20. 2v.
Geog 3019.54.5 Le Gentil, Georges. Découverte du monde. Paris, 1954.
Geog 4427.80 Le Gentil de la Galaisière, G.J.H.J.B. Voyage dans les mers de l'Inde. v.1-2. Suisse, 1780.
Geog 4427.80.5 Le Gentil de la Galaisière, G.J.H.J.B. A voyage to the Indian Seas. Manila, 1964.
Geog 4180.82 Leguat, F. Voyage of F. Leguat...to Rodriguez. London, 1891. 2v.
Geog 3251.32 Lehmann, Edgar. Alte deutsche Landkarten. Leipzig, 1935.
Geog 3055.76 Lehmann, Einar. Atlantis. København, 1934.
Geog 4269.55 Lehmann, H. Europa. 16. Aufl. Frankfurt, 1955.
Geog 759.59.7 Lehrbuch der allegemeinen Geographie. 3. Aufl. v.1,6,7. (Obst, Erich.) Berlin, 1966- 3v.
Geog 758.48 Lehrbuch der allgemeinen Geographie. (Raumer, K.G.) Leipzig, 1848.
Geog 809.59 Lehrbuch der allgemeinen Geographie. v.1-4,6-8,10-11. (Obst, Erich.) Berlin, 1959. 9v.
Geog 3108.78 Lehrbuch der alten Geographie. (Kiepert, J.S.H.) Berlin, 1878.
Geog 818.82.5 Lehrbuch der Geographie. (Guthe, H.) Hannover, 1882-1883. 2v.
Geog 818.82.7 Lehrbuch der Geographie. (Wagner, H.) Hannover, 1903. 2v.
Geog 818.36 Lehrbuch der Geographie. 14e Aufl. (Cannabich, J.G.F.) Weimar, 1836.
NEDL Geog 818.16.20 Lehrbuch der Geographie. 17e Aufl. (Cannabich, T.G.F.) Weimar, 1855.
Geog 758.45.25 Lehrbuch der Geographie für höhere Unterrichtsanstalten. (Daniel, H.A.) Halle, 1891.
Geog 4267.92 Lehrbuch der Statistik. (Meusel, J.G.) Leipzig, 1792.
Geog 4267.92.5 Lehrbuch der Statistik. (Meusel, J.G.) Leipzig, 1804.
An 359.37.3 Lehrbuch der Völkerkunde. (Preuss, K.T.) Stuttgart, 1937.
An 359.37.3.5 Lehrbuch der Völkerkunde. 2. Aufl. (Preuss, K.T.) Stuttgart, 1939.
An 2308.92 Leichenverbrennung. (Olshausen, O.) Berlin, 1892.
Geog 4181.133 Leichhardt, Ludwig. The letters of F.W. Ludwig Leichhardt. Cambridge, Eng., 1968. 3v.
Geog 665.97 Leidlmar, A. Herman von Wissmann-Festschrift. Tübingen, 1962.
Geog 4709.59.5 Leip, Hans. Bordbuch des Satans. München, 1959.
An 359.44.10 Leiper, H.S. Blind spots. N.Y., 1944.
Geog 1527.50 Leipzig. (Baedeker, publishers.) Leipzig, 1948.
An 609.69 Leipzig. Staedtisches Museum für Völkerkunde. Zum hundertjährigen Bestehen, 1869-1969. Berlin, 1969.
Geog 665.126 Leipziger geographische Beiträge. Text and Atlas. Leipzig, 1965. 2v.
An 359.69 Leiris, Michel. Cinq études d'ethnologie. Paris, 1969.
Geog 3108.82 Leitfaden der alten Geographie. (Hahn, H.) Leipzig, 1882.
Geog 3070.4 Leitfaden durch das Wiegenalter. (Breusing, A.) Frankfurt, 1883.
Geog 818.51.14 Leitfaden für den ersten Wissenschaftlichen Unterricht in der Geographie. 14. Aufl. (Arendts, Carl.) Regensburg, 1874.
Geog 3070.105 Leithäuser, J.G. Mappae Mundi. Berlin, 1958.
Geog 3019.53.5 Leithaeuser, J.G. Worlds beyond the horizon. 1. American ed. N.Y., 1955.
Geog 4218.90 Leland, Lillian. Travelling alone, a woman's journey. N.Y., 1890.
An 379.49 Le Lannou, M. La géographie humaine. Paris, 1949.
Geog 4269.67.5 Le Lannou, Maurice. Le déménagement du territoire. Paris, 1967.
Geog 4329.64 Le Lannou, Maurice. Problèmes géographiques de la Méditerranée européenne. Paris, 1964.
Geog 4303.99.5 Lelewel, J. Guillebert de Lannoy et ses voyages en 1413, 1414 et 1421. Bruxelles, 1844.
Geog 4182.49 LeMaire, Jacob. De ontdekkingsreis van Jacob le Maire. 's-Gravenhage, 1945. 2v.
Geog 3060.13.5 Lemuria. 6th ed. (Cervé, W.S.) San Jose, Calif., 1954.
Geog 3060.18 A lenda das Sete Cidades. (Barreto, Costa.) Porto, 1949.
Geog 4218.26.10 Le Netrel, Edmond. Voyage of the Lerós. Los Angeles, 1951.
Geog 807.42 Lenglet, N. Methode pour etudier la geographie. Paris, 1742. 7v.
Geog 6008.19 Leningrad. Arkticheskii i Antarkticheskii Nauchno-Issledovatel'skii Institut. Pervaia russkaia antarkticheskaia ekspeditsiia, 1819-1821 gg. i ee otchetnaia navigatsionnaia karta. Leningrad, 1963.
Geog 4180.92 Leo Africanus, J. History and description of Africa. London, 1896. 3v.
Geog 4180.92.2 Leo Africanus, J. History and description of Africa. N.Y., 1964? 3v.
An 170.206 Leo Frobenius. Leipzig, 1933.
Geog 520.13 Leone, G. Description de l'Afrique. v.2-3. Paris, 1896-98. 2v.
Geog 606.3.5 Leonov, Nikolai I. Aleksei Pavlovich Fedchenko, 1844-1873. Moskva, 1972.
An 136.5.35 Lepenies, Wolf. Orte des wilden Denkens. 1. Aufl. Frankfurt am Main, 1970.
An 132.5 LeRoux, Robert. L'anthropologie comparée de Guillaume de Humboldt. Paris, 1958.
An 2109.25.9 Leroy, Olivier. Essai d'introduction critique à l'étude de l'economie primitive. Paris, 1925.
An 2109.27.25A Leroy, Olivier. La raison primitive. Paris, 1927.
An 358.81.5 Lesley, J.P. Man's origin and destiny. Boston, 1881.
An 358.68 Lesley, J.P. Man's origin and destiny. Philadelphia, 1868.
An 358.69.3 Lesley, J.P. Man's origin and destiny sketched from the platform of the sciences. Philadelphia, 1869.
Geog 5324.1.2 Leslie, A. Arctic voyages of Adolf Erik Nordenskiöld. London, 1879.
Geog 5220.9.6 Leslie, J. Narrative of discovery...polar seas. Edinburgh, 1845.
Geog 5220.9 Leslie, J. Narrative of discovery...polar seas. N.Y., 1833.
Geog 5220.9.3 Leslie, J. Narrative of discovery...polar seas. N.Y., 1836.
Geog 4209.31.10 Leslie, L.A.D. Wilderness trails in three continents. London, 1931.
An 2058.75.5 Lester, A. Hoyle. The pre-adamite. Philadelphia, 1875.
An 379.36.3 Lester, P. Les races humaines. Paris, 1936.
Geog 4308.99 Letchworth, J. Home letters from the continent. N.Y., 1902.
Geog 4219.53 Le Toumelin, J.I. Kurun around the world. N.Y., 1955.
Geog 4110.42 Let's go II. Cambridge, Mass. 1968
Geog 4209.40.10 Let's go with Bob Davis to India, Ceylon, Venezuela. (Davis, R.H.) N.Y., 1940.
Geog 4209.38.45 Let's see if the world is round. (Mielche, H.) London, 1950.
Geog 4521.23.5 A letter addressed to the...Alpine Club. (Whymper, Edward.) London, 1900.
Htn Geog 5220.17* Letter from Russian sea-officer. London, 1754.
Geog 5398.79.15 Letter from the Secretary of the Navy. (United States. Navy Department.) Washington, 1884.
Geog 4209.40.35 Letter of credit. (Weidman, J.) N.Y., 1940.
Geog 4678.40.3 A letter to Daniel Webster. (Cleveland, H.R.) Boston, 1840.
Geog 5170.11 A letter to John Barrow...polar expedition. London, 1819.
Geog 4307.39F Lettera...al E. Ignazio Zanardi di Mantova. (Sagramoso, M.E.) Verona, 1877.
Geog 4180.2 Letters...four voyages to new world. (Columbus, C.) London, 1847.
Geog 4180.43 Letters...four voyages to the new world. (Columbus, C.) London, 1870.
Geog 4307.85.5 Letters...from various parts of the continent. (Matthisson, F.) London, 1799.
Geog 4145.84.20 Letters. (Stark, Freya.) Compton Chamber, 1974-
Geog 4308.17 Letters during tour thro'...France. (Raffles, T.) Liverpool, 1818.
Geog 4329.07 Letters from a Tarheel traveler. Oxford, 1907.
Htn Geog 4308.39.5* Letters from abroad. (Sedgwick, C.M.) N.Y., 1841. 2v.
Geog 4308.23.2 Letters from an absent brother. (Wilson, D.) London, 1824.
Geog 4327.88 Letters from Barbary, France. (Jardine, A.) London, 1788. 2v.
VGeog 4308.65.15 Letters from Europe, 1865-1866. (Smith, Amy G.) Washington, 1948.
Geog 4308.58 Letters from Europe...summer of 1858. Buffalo, 1859.
Geog 4308.86.20 Letters from Europe. (Bartlett, David L.) Baltimore, 1886.
Geog 4308.25.2 Letters from Europe. (Carter, N.H.) N.Y., 1829. 2v.
Geog 4308.67 Letters from Europe. (Forney, J.W.) Philadelphia, 1867.
Geog 4308.25 Letters from Europe. v.2. (Carter, N.H.) N.Y., 1827.

211

Author and Title Listing

	Geog 4228.98	Letters from North America and the Pacific, 1898. (Trevelyan, Charles Philips.) London, 1969.
	Geog 4308.28.5	Letters from the continent. (Walter, Weever.) Edinburgh, 1828.
	Geog 4218.87.20	Letters from the Far East. (Floyd Jones.) N.Y., 1887.
	Geog 4218.73.15	Letters from the sea and from foreign lands. (Cook, Thomas.) London, 1873.
	Geog 4327.94	The letters of...journeys in Europe and Asia Minor. (Morritt, John B.S.) London, 1914.
Htn	Geog 4228.50*	Letters of a traveller. (Bryant, W.C.) N.Y., 1850.
	Geog 4228.51	Letters of a traveller. 3. ed. (Bryant, W.C.) N.Y., 1851.
	Geog 4228.50.5	Letters of a traveller. 4. ed. (Bryant, W.C.) N.Y., 1855.
	Geog 4180.90	Letters of Amerigo Vespucci. (Markham, C.R.) London, 1894.
	Geog 4181.133	The letters of F.W. Ludwig Leichhardt. (Leichhardt, Ludwig.) Cambridge, Eng., 1968. 3v.
	Geog 4308.71.12	Letters of travel. (Morris, Caspar.) Philadelphia, 1896. 2v.
	An 197.5	Letters to future ages. (Ward, D.J.H.) n.p., 1955.
	Geog 4308.96	Letters to Ruth. (Dudley, L.B.) N.Y., 1896.
	An 205.1	Lettre à M.P.F. de Siebold. (Jomard, E.F.) Paris, 1845-69. 3 pam.
	Geog 520.18	La lettre et la carte. (Vignaud, Henry.) Paris, 1901.
	Geog 28.4.3	Lettre sur l'etablissement géographique de Bruxelles. Bruxelles, 1836.
	Geog 5378.80.9	Lettres...du passage du pôle nord. (Nordenskiöld, A.E.) Paris, 188-.
	Geog 3240.9	Lettres...intitulé: liber guidonis. (Boek, C.P.) n.p., n.d.
	Geog 4308.47.3	Lettres de France et d'Italie (1847-1852). (Herzen, A.I.) Genève, 1871.
	Geog 4328.83	Lettres de jeunesse. (Lyautey, L.H.G.) Paris, 1931.
	Geog 3580.10	Lettres relatives aux découvertes maritimes des Espagnols. (Anghiera, P.M. d'.) Paris, 1885.
	Geog 4327.88.10	Lettres sur divers endroits de l'Europe. (Bisani, Alessandro.) Londres, 1791.
Htn	Geog 3055.33*	Lettres sur l'Atlantide de Platon. (Bailly, J.S.) Paris, 1779.
	Geog 3055.33.5	Lettres sur l'Atlantide de Platon. (Bailly, J.S.) Paris, 1804.
	Geog 4181.108	Letts, M.H.I. The travels of Leo of Rozmital through Germany. Cambridge, 1957.
	Geog 4502.5.18	Das Letzte im Fels. (Rudatis, D.) München, 1936.
	An 358.28.3	Leupoldt, J.M. Die Diätetik des physischen und psychischen Menschenlebens. Berlin, 1828.
	An 143.2	Lev Iakovlevich Shternberg. (Gagen-Torn, Nina I.) Moskva, 1975.
	Geog 5839.71	Leva på Grönland. (Sandstroem, Lennart.) Stockholm, 1971.
	Geog 4145.4	Levantine adventurer...Chevalier d'Arvieux. (Lewis, Warren.) London, 1962.
	Geog 4268.73.2	Levasseur, E. L'Europe...geographie. Paris, 1873.
	Geog 585.9	Levasseur, E. Extrait de l'Annuaire in Bureau de Longitudes. n.p., 1905.
	Geog 4268.73.4	Levasseur, E. Precis de la geographie...de l'Europe. v.1, atlas. Paris, 189-. 2v.
	Geog 4131.38.5	Leveillé-Nizerolle, Claude. Le tourisme dans l'economie contemporaine. Thèse. Paris, 1938.
	Geog 602.1	Leven en werken van Williem Janszoon Blaeu. (Baudet, P.J.H.) Utrecht, 1871.
	Geog 623.1	Het leven van Mr. Nicolaas C. Witsen. (Gebhard, J.F.) Utrecht, 1881. 3v.
	An 359.60.15	Het lever; eer serie lezingen gehouden in de aula van de Rijksuniversität te Gent. (Ghent. Université.) Gent, 1960.
	Geog 4308.57.3	LeVert, O.W. Souvenirs of travel. Mobile, 1857. 2v.
	Geog 4309.63.5	Levi, P. La tregua. Torino, 1963.
	An 359.58.20	Lévi-Strauss, Claude. Anthropologie structurale. Paris, 1958-1973. 2v.
	An 359.70.5	Lévi-Strauss, Claude. Claude Lévi-Strauss; ou, La structure et le malheur. Paris, 1970.
	An 136.5.25	Lévi-Strauss, Claude. Leçon inaugurale faite le mardi. Paris, 1966.
	An 2109.62.5	Lévi-Strauss, Claude. La pensée sauvage. Paris, 1962.
	An 2109.62.6	Lévi-Strauss, Claude. The savage mind. Chicago, 1966.
	An 2109.62.7	Lévi-Strauss, Claude. The savage mind. Chicago, 1968.
	An 359.67	Lévi-Strauss, Claude. The scope of anthropology. London, 1967.
	An 359.63.40A	Lévi-Strauss, Claude. Structural anthropology. N.Y., 1963-76. 2v.
	An 629.36	Lévi-Strauss, D. Instruções praticas para pesquisas de antropologia fisica e cultural. pt.1. São Paulo, 1936-
	An 136.5.40	Lévi-Strauss. (Remotti, Francesco.) Torino, 1971.
	An 99.60	Levin, M.G. Ocherki po istorii antropologiia v Rossii. Moskva, 1960.
	Geog 619.7	Levnadsteckning. (Schultén, N.G.) Helsingfors, 1964.
	An 2109.27.13	Lévy-Bruhl, L. L'ame primitive. 3e éd. Paris, 1927.
	An 2109.38.7	Lévy-Bruhl, L. L'expérience mystique et les symboles chez les primitifs. Paris, 1938.
	An 2109.10A	Lévy-Bruhl, L. Les fonctions mentales dans les sociétées inférieures. Paris, 1910.
	An 2109.26.5	Lévy-Bruhl, L. How natives think. London, 1926.
X Cg	An 2109.25.7	Lévy-Bruhl, L. How natives think. N.Y., 1927.
	An 2109.31.5	Lévy-Bruhl, L. La mentalité primitive. Oxford, 1931.
	An 2109.22	Lévy-Bruhl, L. La mentalité primitive. Paris, 1922.
	An 2109.31.5.5	Lévy-Bruhl, L. La mentalité primitive. 15. ed. Paris, 1960.
	An 2109.36.7	Lévy-Bruhl, L. Morceaux choisis. 2e éd. Paris, 1936.
	An 2109.23.15	Lévy-Bruhl, L. Primitive mentality. N.Y., 1923.
	An 2109.28.7.1	Lévy-Bruhl, L. The "soul" of the primitive. N.Y., 1966.
	An 2109.31.11	Lévy-Bruhl, L. Le surnaturel et la nature. Paris, 1931.
	An 2109.66.40	Lévy-Bruhl, Lucien. How natives think. N.Y., 1966.
	Geog 4219.29.15	Lewis, Aswold. Because I've not been there before. London, 1929.
	Geog 4329.69	Lewis, Cecil. Turn right for Corfu. London, 1972.
	Geog 4219.64.5	Lewis, David Henry. Children of three oceans. London, 1969.
	Geog 4218.88	Lewis, I.N. Pleasant hours in sunny lands. Boston, 1888.
	Geog 604.4.5	Lewis, Oscar. George Davidson. California, 1954.
	Geog 4228.49	Lewis, Oscar. Sea routes to the gold fields. 1st ed. N.Y., 1949.
	Geog 5919.65	Lewis, Richard S. A continent for science; the Antarctic adventure. N.Y., 1965.
	Geog 4145.4	Lewis, Warren. Levantine adventurer...Chevalier d'Arvieux. London, 1962.
	An 137.1.5	Lewis Henry Morgan, social evolutionist. (Stern, B.J.) Chicago, 1931.
	Geog 576.70F	Lexicon geographicum. (Ferrari, Filippo.) Parisiis, 1670.
	Geog 578.31	Lexicon manuale. (Müller, Johan Wilhelm.) Lipsiae, 1831.
	Geog 4138.82.7	Lexikon der Reisen und Entdeckungen. (Embacher, F.) Leipzig, 1882.
	Geog 953.14	Lexikon geographischer Bilbände. (Kinauer, Rudolf.) Wien, 1966.
	Geog 4169.47.5	Ley, Charles D. Portuguese voyages, 1498-1663. London, 1947.
	Geog 5160.55	Liakh, N.N. Zapiski poliarnika. Novosibirsk, 1961.
	Geog 3590.10	Lialina, M.A. Russkie moreplavateli, arkticheckie krulosvetnye. Sankt Peterburg, 1892.
	Geog 758.70	Liber de mensura orbis terra a Gustavo Parthey recognitus. (Dicuil.) Bercolini, 1870.
	Geog 758.07.5	Liber de mensura orbis terrae. (Dicuil.) Dublin, 1967.
	Geog 758.07.2	Liber de mensura orbis terrae. (Dicuil.) Paris, 1807.
	Geog 758.07	Liber de mensura orbis terrae. (Dicuil.) Paris, 1807.
	Geog 4323.35	Liber peregrinationis. (Jacopo da Verona.) Roma, 1950.
	An 5.26	Library-Anthropology Resource Group. Serial publications in anthropology. Chicago, 1973.
	Geog 4208.88	Library of universal adventure by sea and land. (Howells, W.D.) N.Y., 1888.
	Geog 139.4.15F	Library series. (London. Royal Geographical Society. Library.)
	Geog 500.88	I libri di viaggio e le guide della raccolta Luigi Vittorio. (Fossati Bellani, Luigi Vittorio.) Roma, 1957. 3v.
	Geog 5645.5.15	Lidegaard, Mads. Ligestilling uden lighed (Grønland og Danmark). København, 1973.
	An 359.64.5	Lienhardt, G. Social anthropology. London, 1964.
	An 5.32	Lietuvos antropologijos bibliografija, 1470-1970. (Česnys, Gintautas.) Vilnius, 1974.
	Geog 5530.13	Lieut. John Irving, of HMS "Terror". (Bell, Benjamin.) Edinburgh, 1881.
	Geog 5530.15	The life, diaries and correspondence...1792-1875. (Franklin, Jane G.) London, 1923.
	Geog 4228.40	Life, travels...of an American wanderer...Alonzo P. De Milt. (Fitch, F.Y.) N.Y., 1883.
	Geog 4180.29	Life and acts of...Guzman. (Guzman, A.E. de.) N.Y., 1970.
	Geog 4558.20.3	The life and adventures. (Durand, J.R.) Rochester, 1820.
	Geog 4248.61	Life and adventures in the south Pacific. (Jones.) N.Y., 1861.
	Geog 4208.22.6	The life and adventures of John Nicol, mariner. (Nicol, John.) London, 1937.
	Geog 5518.36.3	Life and travels of Thomas Simpson. (Simpson, A.) London, 1845.
	Geog 5518.36.4	The life and travels of Thomas Simpson. (Simpson, A.) Toronto, 1963.
	Geog 4559.72.5	Life at sea in the age of sail. (Throner, William R.) London, 1972.
	Geog 5985.187.5	A life given to the Antarctic; Douglas Mawson - Antarctic explorer. (Saziumov, Evgenii M.) Adelaide, 1968.
	An 3308.84	Life history album. (Galton, Francis.) London, 1884.
	Geog 4248.39	Life in a man-of-war...during cruise in Pacific. (Mercier, H.J.) Philadelphia, 1841.
	Geog 5323.1.2	Life of admiral Sir Leopold McClintock. (Markham, C.) London, 1909.
	Geog 618.1.5	Life of Carl Ritter. (Gage, W.L.) N.Y., 1867.
	Geog 4207.76.10	Life of John Ledyard, the American traveller. (Sparks, Jared.) Boston, 1864.
	Geog 4207.76.9	Life of John Ledyard...American traveller. (Sparks, Jared.) Boston, 1847.
	Geog 4207.76A	The life of John Ledyard. (Sparks, Jared.) Cambridge, 1828.
	Geog 4207.76.5	The life of John Ledyard. 2. ed. (Sparks, Jared.) Cambridge, 1829.
	Geog 613.4.5	The life of Sir Clements R. Markham. (Markham, A.H.) London, 1917.
	Geog 5985.258A	The life of Sir Ernest Shackleton. (Mill, Hugh R.) Boston, 1923.
	Geog 5530.18	The life of Sir John Franklin, R.N. (Traill, Henry Duff.) London, 1896.
	Geog 5530.5	Life of Sir John Franklin. (Markham, A.H.) London, 1891.
	Geog 5530.5.2	Life of Sir John Franklin. (Markham, A.H.) N.Y., 1891.
	Geog 4181.53	Life of the Icelander Jón Ólafsson. London, 1923-31. 2v.
	Geog 5329.1	Life of William Scoresby. (Scoresby, R.E.) London, 1861.
	Geog 5559.37.20	Life on an icefloe. (Papanin, Ivan D.) London, 1947.
	Geog 4559.17	A life on the ocean; autobiography. (Gelett, Charles N.) Honolulu, 1917.
	Geog 4558.44.3	Life on the ocean, or Twenty years at sea. (Little, G.) Aberdeen, 1847.
	Geog 4558.44.4	Life on the ocean, or Twenty years at sea. 14. ed. (Little, G.) N.Y., 1852.
	Geog 4558.44.2	Life on the ocean. (Little, G.) Boston, 1844.
	Geog 4558.74	Life on the ocean. (Nordhoff, C.) Cincinnati, 1874.
	Geog 5645.5.15	Ligestilling uden lighed (Grønland og Danmark). (Lidegaard, Mads.) København, 1973.
	Geog 4209.44	Lightning strikes once. (Allan, D.) N.Y., 1944.
	Geog 4208.47.5	Lights and shadows of sailor life. (Clark, J.G.) Boston, 1847.
	Geog 4208.47	Lights and shadows of sailor life. (Clark, J.G.) Boston, 1848.
	An 3308.62F	Liharzih, F. Gesetz des Wachstumes...Proportionslehre. Wien, 1862.
	Geog 6009.10.16	Like English gentlemen. London, 19- .
	An 136.5.50	Lima, Luiz Costa. O estruturalismo de Lévi-Strauss. Petrópolis, 1968.
	Geog 3570.28	Lima, Manuel C. Deux voyages portuguèses de découverte dans l'Atlantique occidental. Lisbonne, 1946.
	Geog 4708.21	Lincoln, B. Narrative of the capture...of Captain B. Lincoln...by piratical schooner. Boston, 1822.
	An 358.84.3	Lind, G.D. Man; embracing his origin. Chicago, 1884.
	Geog 816.65	Linda, Lucas de. Descriptio orbis. Amsterdam, 1665.
Htn	Geog 816.55*	Linda, Lucas de. Descriptio orbis et omniumejus rerumpublicarum. Lugdunum Batavorum, 1655.
	Geog 4308.41	Lindeberg, A. Betraktelser under en resa. Stockholm, 1841.
	Geog 5558.84	Lindsay, D.M. A voyage to the Arctic in the whaler Aurora. Boston, 1911.
	Geog 6009.10.40	Lindsay, M. The epic of Captain Scott. N.Y., 1934.
	Geog 5839.46.10	Lindsay, M. Three got through. London, 1946.
	Geog 4558.76	Lindsay, W.S. Recollections of a sailor. London, 1876.
	Geog 4239.01	Lindsey, N. Allen. Cruising in the Madiana. Boston, 1901.
	An 629.61	Lindzey, G. Projective techniques and cross-cultural research. N.Y., 1961.
	Geog 500.83	L'Information Geographique. La géographie française au milleu du XX. siècle. Paris, 1957.
	Geog 4309.64.25	Linhares, Temístocles. Jornal da Europe. Brasil, 1964.
	Geog 4209.06	Links on my life on land and sea. (Gambier, J.W.) N.Y., 1906.

Author and Title Listing

	An 359.35	Links with past ages. (Orton, E.F.) Cambridge, Eng., 1935.
	Geog 3604.5	Linnélärjungar i främmande länder. (Selander, Sten.) Stockholm, 1960.
	Geog 4309.52	Lins do Rêgo, José. Bota de sete léguas. Rio de Janeiro, 1952.
X Cg	Geog 4180.70	Linschoten, J.H. Voyage of Linschoten to East Indies. London, 1885. 2v.
	Geog 4180.70	Linschoten, J.H. Voyage of Linschoten to East Indies. N.Y., 1964? 2v.
Htn	Geog 4205.98.2F*	Linschoten, Jan H. van. Discours of voyages...East and West Indies. London, 1598.
Htn	Geog 4205.98F*	Linschoten, Jan H. van. Discours of voyages...East and West Indies. London, 1598.
Htn	Geog 4206.10.6F*	Linschoten, Jan H. van. Histoire de la navigation. Amsterdam, 1638.
Htn	Geog 4206.10.5*	Linschoten, Jan H. van. Histoire de la navigation. 2e éd. v.1-3. Amsterdam, 1619.
	Geog 4182.57	Linschoten, Jan H. van. Itinerario. 's-Gravenhage, 1955-57. 3v.
	Geog 4205.96.9F	Linschoten, Jan H. van. Itinerarium oste schip-vaert naer. Amsterdam, 1644.
Htn	Geog 4205.99F*	Linschoten, Jan H. van. Navigatio ac itinerarium. Hagae-Comitis, 1599.
	Geog 4182.8	Linschoten, Jan H. van. Reizen van...naar het noorden 1594-95. 's-Gravenhage, 1914.
	Geog 4182.2	Linschoten, Jan H. van. Itinerario voyage. 's-Gravenhage, 1910. 5v.
	Geog 4182.05	Linschoten. Vereeniging. Werken. Register. v.1-25, 26-50. 's-Gravenhage, 1939-57. 2v.
	Geog 137.5	Linschoten. Vereinigung. Jaarveslag. 1-31,1908-1938 2v.
	Geog 137.10	Pamphlet vol. Linschoten. Vereinigung. Minor publications.
	An 359.45	Linton, R. The science of man in the world crisis. N.Y., 1945.
	An 359.36.3	Linton, Ralph. The study of man. N.Y., 1936.
Htn	Geog 3057.3*	Lipenius, M. Navigatio Salomonis Ophirifica. n.p., 1660.
	Geog 578.55	Lippincott, J.B. and Co. Complete pronouncing gazetteer. Philadelphia, 1855.
	Geog 578.55.13A	Lippincott, J.B. and Co. Gazetteer of the world. Philadelphia, 1893.
	Geog 578.55.9	Lippincott, J.B. and Co. Lippincott's gazetteer of the world. Philadelphia, 1880.
	Geog 578.55.3	Lippincott, J.B. and Co. Lippincott's pronouncing gazetteer. Philadelphia, 1857.
	Geog 578.55.7	Lippincott, J.B. and Co. Lippincott's pronouncing gazetteer. Philadelphia, 1873.
	Geog 578.55.11	Lippincott, J.B. and Co. Supplementary tables of population. Philadelphia, 1883.
	Geog 4308.54.4	Lippincott, S.J. Haps and mishaps...in Europe. Boston, 1854.
	Geog 4308.54.3	Lippincott, S.J. Haps and mishaps...in Europe. Boston, 1854.
	Geog 4308.54.2	Lippincott, S.J. Haps and mishaps...in Europe. Boston, 1854.
	Geog 4308.54	Lippincott, S.J. Haps and mishaps...in Europe. Boston, 1854.
	Geog 578.55.9	Lippincott's gazetteer of the world. (Lippincott, J.B. and Co.) Philadelphia, 1880.
	Geog 578.55.3	Lippincott's pronouncing gazetteer. (Lippincott, J.B. and Co.) Philadelphia, 1857.
	Geog 578.55.7	Lippincott's pronouncing gazetteer. (Lippincott, J.B. and Co.) Philadelphia, 1873.
	Geog 4239.29.13	Lis sails the Atlantic. (Andersen, Lis.) London, 1953.
	Geog 4309.68	Lisakovskii, Igor' N. Dalnie sosedi. Odessa, 1968.
	Geog 4269.72	Lisakovskii, Igor' N. Gamlet ne prishel na svidanie. Odessa, 1972.
NEDL	Geog 137.2	Lisbon. Sociedade de Geographia. Boletin. 1-17,1877-1899
	Geog 137.2	Lisbon. Sociedade de Geographia. Boletin. 1900+ 48v.
	Geog 137.2.12	Lisbon. Sociedade de Geographia. Catalogo de Vendas. 1900-1916
	Geog 4218.03.10	Lisianskii, I. Putesh. vokrug sveta na kor. ""Neva", 1803-1806. Moskva, 1947.
Htn	Geog 4218.03.9*	Lisianskii, I. Voyage round the world...1803-06. London, 1814.
	Geog 5220.15	L'Isle, J.N. de. Erklarung der Charte...Entdeckungen. Berlin, 1753.
Htn	Geog 5220.13*	L'Isle, J.N. de. Explication de la carte des...decouvertes. Paris, 1752. 3 pam.
	An 2109.61	Lissher, Ivar. Man, God, and magic. N.Y., 1961.
	Geog 4249.37	Lissner, Ivar. Menschen und Mächte am Pazifik. Hamburg, 1937.
	Geog 4209.36.35	Lissner, Ivar. Völker und Kontinente. Hamburg, 1936.
	Geog 4179.10	A list of the publications of the Hakluyt Society. (Hakluyt Society, London.) London, 1946.
	An 3208.80	List of the specimens in the anatomical section of the United States Army Medical Museum. (Otis, George A.) Washington, 1880.
	Geog 212.206	Liste des membres. (Société de Géographie.) Paris. 1868-1897 11v.
	Geog 500.27A	Liste provisoire de bibliographies geographiques. (Jackson, J.) Paris, 1881.
	Geog 500.27.2A	Liste provisoire de bibliographies geographiques. (Jackson, J.) Paris, 1881.
	Geog 4477.61	Listonai. Le voyageur philosophe. Amsterdam, 1761. 2v.
	Geog 4209.09.10	Listy z podrozy. (Gordon, F.) Chicago, 1910.
	Geog 4269.10.5	Literary and historical atlas of Europe. (Bartholomew, J.G.) London, 1910.
	Geog 4269.23	A literary and historical atlas of Europe. (Bartholomew, J.G.) London, 1923.
	Geog 600.6	Literary record of Cleveland Abbe. Ithaca, 1931.
	Geog 500.11	Literatur der...Reisebeschreibungen. (Beckmann, J.) Göttingen, 1808-09. 2v.
	Geog 5050.3	Die Literatur über die Polar-Regionen. (Chavanne, J.) Wien, 1878.
	Geog 500.66	The literature of sea travel since the introduction of steam, 1830-1930. (Bolles, E.C.) Philadelphia, 1943.
Htn	Geog 4126.14*	Lithgow, W. Peregrination from Scotland. Lond, 1614.
	Geog 4326.32.30	Lithgow, William. Discourse of a peregrination in Europe, Asia and Affricke. Facsimile. N.Y., 1971.
	Geog 4326.32.20	Lithgow, William. Rare adventures and painefull peregrinations. London, 1928.
Htn	Geog 4326.32.3*	Lithgow, William. The totall discourse of...adventures. London, 1640.
Htn	Geog 4326.32*	Lithgow, William. The totall discourse of...adventures. Lyon, 1632.
	Geog 4326.32.10A	Lithgow, William. Travels and voyages thro' Eruope, Asia and Africa. Glasgow, 1906.
	Geog 4326.32.5	Lithgow, William. Travels and voyages thro' Europe, Asia and Africa. 12th ed. Leith, 1814.
	Geog 5600.1	Litteratur om Grønland. (Hagerups, H. Boghandel.) København, 1932.
Htn	Geog 500.59F*	La litterature géographique française de la renaissance. (Atkinson, G.) Paris, 1927.
Htn	Geog 500.59.2F*	La litterature géographique française de la renaissance. Supplement. (Atkinson, G.) Paris, 1936.
	Geog 4558.44.3	Little, G. Life on the ocean, or Twenty years at sea. Aberdeen, 1847.
	Geog 4558.44.4	Little, G. Life on the ocean, or Twenty years at sea. 14. ed. N.Y., 1852.
	Geog 4558.44.2	Little, G. Life on the ocean. Boston, 1844.
	Geog 6009.28.10	Little America. (Byrd, Richard E.) N.Y., 1930.
	Geog 6009.28.8A	Little America. 1st ed. (Byrd, Richard E.) N.Y., 1930.
	Geog 500.67	A little book on travel books. (Curtiss, Frederic H.) Boston, 1936.
	Geog 5919.58	Liverridge, Douglas. The last continent. London, 1958.
	Geog 4708.36	Lives and bloody exploits of most noted pirates. Hartford, Conn., 1836.
Htn	Geog 4708.19.2*	Lives and confessions of John Williams, Francis Frederick, John P. Roy and P. Peterson. Boston, 1819.
Htn	Geog 4708.19*	Lives and confessions of John Williams, Francis Frederick, John P. Roy and P. Peterson. 2. ed. Boston, 1819. 2 pam.
	Geog 4138.37.5	Lives and voyages of Drake. (Johnstone, C. (Mrs.).) Edinburgh, 1837.
	Geog 4138.31	Lives and voyages of Drake. Edinburgh, 1831.
	Geog 4138.36	Lives and voyages of Drake. N.Y., 1836.
	Geog 4138.37	Lives of celebrated travellers. (St. John, J.A.) N.Y., 1837. 3v.
	An 3308.93	Livi, Ridolfo. Antropometria militare. Roma, 1893-1903. 2v.
	Geog 4145.15.2	Living dangerously. (Chapman, F.S.) London, 1953.
	Geog 4145.15	Living dangerously. (Chapman, F.S.) N.Y., 1953.
	An 359.65.15	The living races of man. 1. ed. (Coon, Carleton Stevens.) N.Y., 1965.
	Geog 520.22	Le livre de la description des pays. (Le Bouvier, Gilles.) Paris, 1908.
	Geog 3055.41	Le livre de l'Atlantide. (Manzi, M.) Paris, 1923.
	Geog 665.54	Livre jubilaire offert a Maurice Zimmermann. Lyon, 1949.
Htn	Geog 816.40*	Livre premier [-septiesme] des tables geographiques auquel est traité du monde en general. (Bertius, Petrus.) n.p., n.d.
	Geog 110.1.13	Livret-guide du congressiste. (International Geographical Congress, 13th, 1931.) Paris, 1931.
	Geog 5670.4	En livsform ved horsvejen. (Jensen, Bent.) København, 1971.
	Geog 4209.71	Ljudi krajevi Beskraj. (Šulentić, Zeatko.) Karlovac, 1971.
	Geog 4309.30.20	Lo gótico, signo de Europa. (Terán, Juan B.) Buenos Aires, 1930?
	Geog 759.19	Lo inerte y lo vital. (Beltran, Juan G.) Buenos Aires, 1919.
	Geog 759.56	Lobeck, Armin Kohl. Things maps don't tell us; an adventure into map interpretation. N.Y., 1968.
	Geog 4416.21	Lobo, Jeronymo. Itinefario e outros escritos inéditos. Porto, 1971.
	Geog 4181.52	Lockerby, William. The journal of William Lockerby. London, 1925.
	Geog 4678.47.5	Lockhart, John Gilbert. Blenden Hall; the true story of a shipwreck. London, 1930.
	Geog 4679.27	Lockhart, John Gilbert. A great sea mystery. London, 1927.
	Geog 4672.16.15	Lockhart, John Gilbert. The "Mary Celeste" and other strange tales of the sea. London, 1952.
	Geog 4672.16.10	Lockhart, John Gilbert. Mysteries of the sea. London, 1924.
	Geog 4672.16.1	Lockhart, John Gilbert. Strange adventures of the sea. N.Y., 1926.
	Geog 4672.16	Lockhart, John Gilbert. Strange adventures of the sea. N.Y., 1931.
	Geog 4672.16.5	Lockhart, John Gilbert. Strange tales of the seven seas. London, 1929.
	Geog 4309.53	Lockley, R.M. Travels with a tent in western Europe. London, 1953.
	Geog 4679.23	Lockwood, C.A. Tragedy at Honda. Philadelphia, 1960.
	Geog 4308.39.3	Locmaria. Souvenirs des voyages de...duc de Bordeaux. Paris, 1846. 2v.
	An 3309.65.5	Łódzskic Towarzystwo Naukowe. Wydział III, Nauk Matematyczno-Przyrodniczych. Studia afrykanistyczne. Łódź, 1965.
	Geog 4029.12	Loeffler, Johann Friedrich. Abenteur in drei Erdteilen. Heidenheim, 1971.
	Geog 3240.6	Lönborg, S. Adam af Bremen...skildring af Nordeuropas länder. Uppsala, 1897.
	Geog 818.51	Loerebogi geografien. (Geelmuyden, J.) Christiania, 1851.
	Geog 4219.01	Loewenbach, Lottraire. Promenade autour du globe. Paris, 1903.
	Geog 5070.15	Löwenberg, J. Die Entdeckungs- und Forschungsreisen im dem Leiden Polarzonen. Leipzig, 1886.
	Geog 4208.37	Lofé, G. Cartas á mis hijos. N.Y., 1839.
	Geog 4110.13	Loftie, W.J. Orient line guide. London, 1890.
	Geog 4559.44	Log of a sea captain's daughter. (Snow, A.R.) Boston, 1944.
	Geog 4558.99A	The log of a sea-waif. (Bullen, F.T.) N.Y., 1899.
	Geog 4559.26.3	The log of a shellback. (Farmer, H.F.) N.Y., 1926.
	Geog 4208.78.7	The log of an ancient marine. (Wakeman, Edgar.) San Francisco, 1878.
	Geog 5312.2A	The log of Bob Bartlett. (Bartlett, Robert A.) N.Y., 1928.
	Geog 4217.40.5	Log of the centurion. (Heaps, Leo.) London, 1973.
	Geog 4679.17	The log of the El Dorado. (Benson, N.P.) San Francisco, 1917.
	Geog 4229.28	Log of the ketch Seven Bells. n.p., 1928?
	Geog 5399.06	The log of the "Laura" in polar seas. (Holmes, B.F.) Cambridge, 1907.
	Geog 6009.12.5	Logbook for Grace. (Murphy, R.C.) N.Y., 1947.
	Geog 4228.89.5	Loin du Pays. Souvenirs d'Europe, d'Afrique, et d'Amérique. (Faucher de Sant-Maurice, N.H.E.) Quebec, 185-. 2v.
	An 358.92	Lombroso, C. L'uomo bianco e l'uomo di colore. Firenze, 1892.
Htn	Geog 4306.60.2*	Loménie, Louis Henri. Itinerarium. Paris, 1662.
	Geog 5322.1	Lomonosov i Arktika. (Perevalov, V.) Moskva, 1949.
	Geog 1710.24	London, as it is. (Cunningham, Peter.) London, 185-.
	Geog 5925.5	London, England. Royal Geographical Society. Antarctic exploration. London, 1898.
	Geog 1710.15	London. (Cunningham, Peter.) London, 1856.

Author and Title Listing

Geog 500.21	London. Royal Geographical Society. Catalogue of the library. London, 1852.
Geog 500.21.3	London. Royal Geographical Society. Catalogue of the library. London, 1865.
Geog 500.21.5	London. Royal Geographical Society. Catalogue of the library. London, 1895.
Geog 139.4	London. Royal Geographical Society. Geographical journal. London. 1,1893+ 128v.
Geog 139.4.5	London. Royal Geographical Society. Geographical journal. Index. 1893-1902. London, 1906. 3v.
Geog 139.10	London. Royal Geographical Society. Its foundation and history. London, 1930.
Geog 139.1	London. Royal Geographical Society. Journal. London. 1831-1880 50v.
Geog 139.2	London. Royal Geographical Society. Journal. Index. v.1-20,21-40,41-50. London, n.d. 3v.
Geog 139.3	London. Royal Geographical Society. Proceedings. London. 1-14,1879-1894 14v.
Geog 139.2.9	London. Royal Geographical Society. Proceedings. London. 2-22 16v.
Geog 139.3.5	London. Royal Geographical Society. Proceedings. Index. 1879-1892. London, 1896.
Geog 139.4.12	London. Royal Geographical Society. Recent geographical literature, maps. London. 19-41,1926-1932 11v.
Geog 139.4.13	London. Royal Geographical Society. Recent geographical literature, maps. Index. 1-4, 1918-1932. London, 1936.
Geog 139.5	London. Royal Geographical Society. Supplementary papers. v.1-3. Photoreproduction. 1951-1955 6v.
Geog 139.9	London. Royal Geographical Society. Technical series. London. 1-5,1920-1929
Geog 139.6	London. Royal Geographical Society. Yearbook record. London. 1898-1905 2v.
Geog 139.4.15F	London. Royal Geographical Society. Library. Library series.
Geog 139.30	London. University. Queen Mary College. Department of Geography. Occasional papers. London. 1,1974+
Geog 1510.5	London and its environs. (Baedeker, publishers.) Leipsic, 1878. 2v.
Geog 1510.8	London and its environs. 2. ed. (Baedeker, publishers.) Leipsic, 1879.
Geog 1510.10	London and its environs. 3. ed. (Baedeker, publishers.) Leipsic, 1881.
Geog 1510.25	London and its environs. 6. ed. (Baedeker, publishers.) Leipsic, 1887.
Geog 1510.30	London and its environs. 7. ed. (Baedeker, publishers.) Leipsic, 1889.
Geog 1510.35	London and its environs. 8. ed. (Baedeker, publishers.) Leipsic, 1892.
Geog 1510.36	London and its environs. 8. ed. (Baedeker, publishers.) Leipsic, 1892.
Geog 1510.40	London and its environs. 9. ed. (Baedeker, publishers.) Leipsic, 1894.
Geog 1510.41	London and its environs. 9. ed. (Baedeker, publishers.) Leipsic, 1894.
Geog 1510.45	London and its environs. 10. ed. (Baedeker, publishers.) Leipsic, 1894.
Geog 1510.46	London and its environs. 10. ed. (Baedeker, publishers.) Leipsic, 1896.
Geog 1510.47	London and its environs. 11. ed. (Baedeker, publishers.) Leipsic, 1898.
Geog 1510.47.12	London and its environs. 12. ed. (Baedeker, publishers.) Leipsic, 1900.
Geog 1510.48	London and its environs. 13. ed. (Baedeker, publishers.) Leipsic, 1902. 2v.
Geog 1510.49	London and its environs. 13. ed. (Baedeker, publishers.) Leipsic, 1902.
Geog 1510.50	London and its environs. 13. ed. (Baedeker, publishers.) Leipsic, 1902.
Geog 1510.55	London and its environs. 14. ed. (Baedeker, publishers.) Leipzig, 1905.
Geog 1510.57	London and its environs. 15. ed. (Baedeker, publishers.) Leipzig, 1908.
Geog 1510.56	London and its environs. 15. ed. (Baedeker, publishers.) Leipzig, 1908.
Geog 1510.61	London and its environs. 18. ed. (Baedeker, publishers.) Leipzig, 1923.
Geog 1510.62	London and its environs. 18. ed. (Baedeker, publishers.) Leipzig, 1923.
Geog 1510.62.2	London and its environs. 20. ed. (Baedeker, publishers.) Hamburg, 1951.
Geog 1510.15	London and its invirons. 4. ed. (Baedeker, publishers.) Leipsic, 1883.
Geog 1510.20	London and its invirons. 5. ed. (Baedeker, publishers.) Leipsic, 1885.
Geog 1510.59A	London and its invirons. 16. ed. (Baedeker, publishers.) Leipzig, 1911.
Geog 665.56	London essays in geography. (Stamp, L.) Cambridge, 1951.
Geog 1510.2	London und Umgebung. 15. Aufl. (Baedeker, publishers.) Leipzig, 1905.
Geog 1510.42	London und Umgebungen. 11. Aufl. (Baedeker, publishers.) Leipzig, 1894.
Geog 1510.3	Londres, ses environs: le Sud de l'Angleterre. 3. éd. (Baedeker, publishers.) Leipzig, 1875.
Geog 4219.66.10	Lone adventurer: the story of Sir Francis Chichester. (Rowland, John.) London, 1968.
Geog 4571.2	Lone voyager. 1. ed. (Garland, Joseph E.) Boston, 1963.
Geog 4659.68.10	Loney, Jack Kenneth. Wrecks on the Gippsland coast. Geelong, 1968.
Geog 4209.38.1	Long, Dwight. Seven seas on a shoestring. N.Y., 1939.
Geog 5558.81.60	The long rescue. 1. ed. (Powell, T.) Garden City, 1960.
Geog 91.5	The long trail news. Brandon, Vt. 2v.
Geog 6009.33.5	The long whip; the story of a great husky. (Walden, Jane Brevoort.) N.Y., 1936.
Geog 4132.5	The longest voyage; circumnavigators in the age of discovery. (Silverberg, Robert.) Indianapolis, 1972.
Geog 3235.13	Longhena, M. Atlanti e carte nautiche. Parma, 1907.
Geog 579.66	Longman's dictionary of geography. London, 1966.
Geog 4218.70	Longworth, M.T. Teresina peregrina. London, 1874. 2v.
Geog 759.75.5	The look of the land. (Hart, John Fraser.) Englewood Cliffs, 1975.
Geog 4209.60.20	Look to the wilderness. (Burden, W.D.) Boston, 1960.
Geog 4559.45.5	Looming lights. (Carter, G.G.) London, 1945.
Geog 4308.45.9	Loose papers...residence in Ireland. (Nicholson, Asenath.) N.Y., 1853.
An 379.72.5	Lopatina, Elena B. Otsenka prirodnykh uslovii zhizni naseleniia. Moskva, 1972.
Geog 4145.14	Lopes, Francisco Fernandes. The brothers Corte Real. Lisboa, 1957.
Geog 4208.80	Lopez, Lucio V. Recuerdos de viaje. Buenos Aires, 1915.
Geog 4309.67	Lorck, Carl von. Europa Privat. Frankfurt, 1967.
Geog 4679.12.18	Lord, Walter. A night to remember. London, 1956.
Geog 4679.12.17A	Lord, Walter. A night to remember. N.Y., 1955.
Geog 4329.05.20	Lorenz, I.E. The new Mediterranean traveller. 13th ed. N.Y., 1927.
An 358.87.7	Lorenzini, C. Il regale istruttivo. Roma, 1887.
Geog 4329.04	Loring, J.M. The old world thro' new world eyes. St. Louis, 1904.
Geog 5855.13	Lorm, A.J. Kunstzin der Eskimos. 's-Gravenhage, 1945.
Geog 4652.8	Loss list of Grimsby vessels, 1800-1960. (Boswell, David B.) Grimsby, 1969.
Geog 4678.54.5	Loss of the Arctic. n.p., 1854.
Geog 4679.12.2	The loss of the S.S. Titanic. (Beesley, Lawrence.) Boston, 1912.
Geog 4708.06	Loss of the ship Essex in 1806. (Phillips, James D.) n.p., 1941.
Geog 4679.12A	The loss of the Titanic. (Beesley, Lawrence.) Boston, 1912.
Geog 3055.54A	Lost Atlantis. (Bramwell, James.) N.Y., 1938.
An 197.1	The lost Atlantis. (Wilson, D.) N.Y., 1892.
Geog 3060.12.25	The lost continent of Mu. (Churchward, J.) N.Y., 1934.
Geog 3055.72	Lost continents; the Atlantis theme in history, science and literature. 1st ed. (DeCamp, L.S.) N.Y., 1954.
Geog 5559.09.100	Lost in the Arctic; being the story of the "Alabama" expedition, 1909-1912. (Mikkelsen, Ejnar.) London, 1913.
Geog 5559.09.101	Lost in the Arctic; being the story of the "Alabama" expedition, 1909-1912. (Mikkelsen, Ejnar.) N.Y., 1913.
Geog 4672.10	Lost ships and lonely seas. (Paine, Ralph D.) N.Y., 1922.
An 3209.12	Loth, E. Beiträge zur...der Negerweichteile. Stuttgart, 1912.
Geog 4309.70	Lottman, Herbert R. Detours from the grand tour. Englewood Cliffs, N.J., 1970.
Geog 5321.1.8	The love life of Dr. Kane. N.Y., 1866.
Geog 5559.03	Low, A.P. Report on dominion government expedition. Ottawa, 1906.
Geog 3018.81	Low, C.R. Maritime discovery. London, 1881. 2v.
Geog 4209.41.15	Low, G.W. Gold rush by sea. Philadelphia, 1941.
Geog 4521.25	Lowe, George. Because it is there. London, 1959.
An 2109.40.3	Lowie, R.H. An introduction to cultural anthropology. N.Y., 1940.
An 359.20.3	Lowie, Robert H. Primitive society. N.Y., 1920.
An 99.37.1	Lowie, Robert Harry. The history of ethnological theory. N.Y., 1937.
Geog 4209.11.5	Lowther, H.C. From pillar to post. London, 1911.
Geog 4209.11.6	Lowther, H.C. From pillar to post. London, 1912.
Geog 4181.59A	Luard, C.E. Travels of Fray Sebanstian Manrique. London, 1927. 2v.
Geog 4559.25.10F	Lubbock, A.B. Adventures by sea from art of old time. London, 1925.
Geog 4559.02	Lubbock, A.B. Round the Horn before the mast. N.Y., 1902.
Geog 4209.15.10	Lubbock, A.B. Round the Horn before the most. London, 1915.
An 2108.70.2	Lubbock, J. The origin of civilisation and the primitive condition of man. London, 1870.
An 2108.70	Lubbock, J. The origin of civilisation and the primitive condition of man. 2. ed. London, 1870.
An 2108.69	Lubbock, J. Pre-historic times. 2. ed. London, 1869.
Geog 4325.73	Lubenau, R. Beschreiben der Reisen des Reinhold Lubenau. Königsberg, 1912-1920. 3v.
Geog 4209.21	Lucas, E.V. Roving east and roving west. London, 1921.
Geog 4209.21.2	Lucas, E.V. Roving east and roving west. N.Y., 1921.
Geog 4105.18.4	Luce, Robert. Going abroad? 4. ed. Boston, 1906.
Geog 4208.83	Lucy, H.W. East by West. London, 1885. 2v.
Geog 619.10	Ludomir Sawicki. (Dobrowolska, Maria.) Warszawa, 1968.
Geog 4329.27.15	Ludwig, Emil. Am Mittelmeer. Berlin, 1927.
Geog 4329.29.10	Ludwig, Emil. On Mediterranean shores. Boston, 1929.
Geog 4128.97	Ludwig, Friedrich. Untersuchungen...Reise...Itineraire der deutsech Königer. Inaug. Diss. Berlin, 1897.
Geog 4128.97.1	Ludwig, Friedrich. Untersuchungen über die Reise und Marschegesehwindigkeit in XII. und XIII. Jahrhundert. Berlin, 1897.
Geog 4218.81	Ludwig, S. Um die Welt ohne zu wollen. Prag, 1881.
Geog 4418.79.5	Ludwig, Salvator. Die Karawanen-Strasse von Aegypten nach Syrien. Prag, 1879.
Geog 4218.26.17	Luetke, Fedor P. Puteshestvne vokrug sveda na volunom shliupe "Seniavin" 1826-1829. 2. izd. Moskva, 1948.
Geog 622.5	Luigi Visintin. (Bonapace, Umberto.) Novara, 1958.
Geog 612.4.5	Lukina, Tat'iana A. Ivan Ivanovich Lepekhin. Leningrad, 1965.
Geog 665.116	Lulovac, Milisav. Cvijićev zbornik u spomen 100. godišnjice njegovog rodjeuja. Beograd, 1968.
Geog 4208.19	Lumsden, T. A journey from Merut in India to London. London, 1822.
Geog 5700.13	Lund, Jacob Johan. Første missionair paa Grønland. København, 1778.
Geog 141.5F	Lund. Universitet Geografiska Institution. Meddelanden. Avhandlingar. Lund.
Geog 141.5	Lund. Universitet Geografiska Institution. Meddelanden. Avhandlingar. Lund. 1-60 32v.
Geog 141.10	Lund studies in geography. Series A: Physical geography. Lund. 1-37 4v.
Geog 141.10.2	Lund studies in geography. Series B: Human geography. Lund. 1-31 6v.
Geog 141.10.4	Lund studies in geography. Series C: General and mathematical geography. Lund. 1-4
Geog 4308.50	Lundeberg, A. La perle trouvée. Lausanne, 1850.
An 379.67.10	Lundman, Bertil. Geographische Anthropologie. Stuttgart, 1967.
Geog 4512.535	Lunn, A. Mountain jubilee. London, 1943.
Geog 4515.103.2	Lunn, A.H.M. Alpine ski-ing at all heights and seasons. 2d ed. London, 1926.
Geog 4509.25	Lunn, A.H.M. The mountains of youth. London, 1925.
Geog 4219.27	Lunn, Henry. Round the world with a dictaphone. London, 1927.
Geog 4219.27.2	Lunn, Henry. Round the world with a dictaphone. London, 1928.
Geog 4218.03.13	Lupach, V.S. I.F. Kruzenshtern. Moskva, 1953.
Geog 3590.45	Lupach, V.S. Russkie moraplavateli. Moskva, 1953.
Geog 4309.14	The lure and the lore of travel. (Vrooman, Carl S.) Boston, 1914.
Geog 4418.29	Lushington, C. (Mrs.). Narrative of journey from Calcutta. London, 1829.
Geog 4658.93.4	Lussich, Antonio D. Naufragios célebres en el cabo Polonio, banco Inglés, y océano Atlántico. Montevideo, 1938.

	Geog 4658.93.2	Lussich, Antonio D. Naufragios célebres en el cabo Polonio, el banco Inglés, y el océano Atlántico. 2. ed. Montevideo, 1893.
	Geog 4512.537.5	Luther, C.J. Skunterhaltungen. München, 1925.
	Geog 4512.537	Luther, C.J. Wil der Schneelauf nach Deutschland kam. München, 1925.
	Geog 4328.83	Lyautey, L.H.G. Lettres de jeunesse. Paris, 1931.
	Geog 4328.66	Lycklama, A. Nijeholt. Voyage en Russie, au Caucase. Paris, 1872-1875. 4v.
	Geog 4269.13	Lyde, L.W. The continent of Europe. London, 1913.
	Geog 4269.13.2	Lyde, L.W. The continent of Europe. London, 1920.
	Geog 4269.13.4	Lyde, L.W. The continent of Europe. 2nd ed. London, 1924.
	Geog 4269.31	Lyde, L.W. Peninsular Europe. London, 1931.
	An 2059.27	Lyell, C. The antiquity of man. London, 1927.
	An 2059.14.5	Lyell, C. The geological evidence of the antiquity of man. London, 1914.
	An 2058.63	Lyell, C. The geological evidence of the antiquity of man. Philadelphia, 1863.
	An 2058.63.2	Lyell, C. The geological evidence of the antiquity of man. 2d American ed. Philadelphia, 1863.
	An 2058.73	Lyell, C. The geological evidence of the antiquity of man. 4. ed. London, 1873.
	Geog 3070.50A	Lynam, E. The mapmaker's art. London, 1953.
	Geog 4181.93	Lynam, E. Richard Hakluyt and his successors. London, 1946.
	Geog 3520.15	Lynam, Edward. British maps and map-makers. London, 1944.
	Geog 4209.41	Lynch, Kathleen M. Travellers must be content. N.Y., 1941.
	Geog 5518.24.7	Lyon, G.F. Brief narrative of attempt to reach Repulse Bay. London, 1825.
	Geog 5518.21.5	Lyon, G.F. Private journal of Captain G.F. Lyon. Boston, 1824.
	Geog 622.2	M.P. Vronchenko. (Shostiu, N.A.) Moskva, 1956.
	Geog 602.3	Ma jeunese sous l'aile de Péguy. (Blanchard, Raoul.) Paris, 1961.
	An 3209.33	Maapallon esihistorialliset ja. (Hildén, Kaarlo.) Helsinki, 1933.
	An 379.61	Maas, Walther. Probleme der Sozialgeographie. Berlin, 1961.
	Geog 4679.67.10	Mabire, Jean. La marée noire du Torrey Canyon. Paris, 1967.
	Geog 4329.25	McAllister, J.G. Borderlands of the Mediterranean. Richmond, 1926.
Htn	Geog 577.73*	Macbean, A. Dictionary of ancient geography. London, 1773.
	Geog 4208.84.5	McCarthy, W.J. Cruise of the U.S. flagship Lancaster. n.p., 1887.
	Geog 665.94	McCashill, Murray. Land and livelihood. Christchurch, 1962.
	Geog 4239.69.10	McClean, Tom. I had to dare; rouring the Atlantic in seventy days. London, 1971.
	Geog 5538.57	M'Clintock, F.L. Voyage of "Fox"...narrative...of fate of Sir John Franklin. Boston, 1860.
	Geog 5538.57.2	M'Clintock, F.L. The voyage of the "Fox" in the Arctic seas. London, 1859.
	Geog 5538.57.3	M'Clintock, F.L. The voyage of the "Fox" in the Arctic seas. Rutland, 1972.
	Geog 5518.50.10	M'Clure, R. Captain M'Clure's despatches from her majesty's discovery ship "Investigator". London, 1852.
	Geog 4218.87.7	M'Collester, S.H. Round the globe. 3. ed. Boston, 1890.
	Geog 819.73	McCormick, Donald. How to buy an island. N.Y., 1973.
	Geog 5070.5	M'Cormick, R. Voyages of discovery in Arctic and Antarctic. London, 1884. 2v.
	Geog 4308.71.9	McCray, R.H. Narrative of travels, scenes...in the Old World. 3rd ed. Philadelphia, 1871.
	Geog 4329.02	McCready, R.H. Cruise of the Celtic...Mediterranean. N.Y., 1902.
	Geog 4559.33	McCulloch, J.H. A million miles in sail. N.Y., 1933.
	Geog 578.41	McCulloch, J.R. Dictionary geographical, statistical. London, 1841-42. 2v.
	Geog 578.41.6	McCulloch, J.R. Dictionary geographical, statistical. London, 1851. 2v.
	Geog 578.41.2	McCulloch, J.R. Dictionary geographical, statistical. N.Y., 1843-44. 2v.
	Geog 578.41.3	McCulloch, J.R. Dictionary geographical, statistical. N.Y., 1844-45. 2v.
	Geog 578.41.4	McCulloch, J.R. Dictionary geographical, statistical. N.Y., 1844-47. 2v.
	Geog 578.41.5	McCulloch, J.R. Dictionary geographical, statistical. N.Y., 1849. 2v.
	Geog 578.41.7	McCulloch, J.R. Dictionary geographical, statistical. N.Y., 1852. 2v.
	An 2059.05	Maccurdy, G.G. The eolithic problem. Lancaster, 1905.
	An 2109.24A	MacCurdy, G.G. Human origins, a manual of prehistory. N.Y., 1924. 2v.
	An 187.5.3	Pamphlet vol. MacCurdy. Anthropology and archaeology. Papers, 1914-1920. 17 pam.
	An 187.5.2	Pamphlet vol. MacCurdy. Anthropology and archaeology. Papers 1907-1913. 18 pam.
	Geog 4229.05	Macdonald, A. In search of El Dorado. London, 1905.
	An 187.1.2	Pamphlet box. Macdonald, A. Anthropology.
	Geog 4209.36	Macdonald, J.R. At home and abroad; essays. London, 1936.
	Geog 4309.25.11	MacDonald, J.R. Wanderings and excursions. Indianapolis, 1928.
	Geog 4309.25.12	MacDonald, J.R. Wanderings and excursions. London, 1929.
	Geog 5160.59	Macdonald, Ronald. The Arctic frontier. Toronto, 1966.
	Geog 5538.52.5	M'Dougall, G.F. Eventful voyage of H.M. discovery ship "Resolute". London, 1857.
	An 379.21.3	McDougall, William. National welfare and national decay. London, 1921.
	Geog 4308.51.15	McFarland, A. The escape or loiterings amid the scenes. Boston, 1851.
X Cg	Geog 4308.50.5	McFarland, A. Five months abroad...1850. Concord, 1851.
	Geog 5558.75.9	MacGahan, J.A. Under the Northern Lights. London, 1876.
	Geog 4309.41.5	McGee, A.E. Black America abraod. Boston, 1941.
	Geog 4309.31	McGeehan, William O. Trouble in the Balkans. N.Y., 1931.
	Geog 5558.81.15	McGinley, W.A. Greely relief expedition. Washington, 1884.
	Geog 4308.66.5	Macgregor, J. Thousand miles in the Rob Roy canoe. London, 1866.
	Geog 4308.66.7	Macgregor, J. Voyage alone in the yawl "Rob Roy". London, 1868.
	Geog 4308.66.10	Macgregor, J. Voyage alone in the yawl "Rob Roy". London, 1880.
	Geog 4218.92.20	MacGregor, John. Toil and travel. London, 1892.
	An 379.37	Machin, A. Darwin's theory applied to mankind. London, 1937.
	Geog 5919.57.5	Machowski, J. Antarktyka. Wyd. 2. Warszawa, 1957.
	An 379.33A	McIverny, A.J. The rôle of the deserts. London, 1933.
	An 379:38	McIverny, A.J. Through the great arid filter (man's drift to Europe). London? 1938.
	Geog 818.61	Mackay, A. Manual of modern geography. Edinburgh, 1861.
	Geog 4209.19	Mackay, John. The ten islands and Ireland. Dublin, 1919.
	Geog 4348.87	McKenzie, A. Some things abroad. Boston, 1887.
	Geog 4209.48	Mackenzie, C. All over the place. London, 1948.
	Geog 4659.54	Mackenzie, Margaret E. Shipwrecks; being the historical account of shipwrecks along the Victorian Coast, 1813-1914. Melbourne, 1954.
	Geog 4208.98.18	Mackin, S.M.A. (Mrs.). A society woman on two continents. N.Y., 1898.
	Geog 4308.31.5	McLellan, H.B. Journal of residence in Scotland and tour. Boston, 1834.
	Geog 4217.08.10	MacLiesh, Archibald F. The privateers. N.Y., 1962.
	Geog 4218.98.5	McMahon, William. Journey with the sun around the world. Cleveland, 1900.
	Geog 4228.79	MacMichael, M. A landlubber's log of a voyage round the "Horn". Philadelphia, 188-?
	Geog 5839.27	MacMillan, D.B. Etah and beyond. Boston, 1927.
	Geog 5559.13.5A	MacMillan, D.B. Four years in the white north. N.Y., 1918.
	Geog 5559.09.15	MacMillan, D.B. How Peary reached the pole. Boston, 1934.
	Geog 4308.69.8	Macmillan, Hugh. Holidays on high lands. London, 1869.
	Geog 5559.48	MacMillan, M. Green seas and white ice. N.Y., 1948.
	Geog 4309.29.11	MacOrlan, Pierre. Villes, mémoires: Montmartre, Rouen, souvenirs de Picardie et d'Artois, Brest, Londres, Villes rhénanes. Paris, 1966.
	Geog 4309.29.12	MacOrlan, Pierre. Villes, mémoires: Montmartre, Rouen, souvenirs de Picardie et d'Artois. 1. éd. Evreux, 1969.
	Geog 4309.29.10	MacOrlan, Pierre. Villes: Rouen-Montmartre-Brest-Londres-Villes rhénanes-Rome. 4. éd. Paris, 1929.
	Geog 4309.65	MacShane, Frank. The American in Europe. 1st ed. N.Y., 1965.
	Geog 4559.56.5	Mad sea. (Rutzebeck, Hjalmar.) N.Y., 1956.
	An 170.264	Madan, T.N. Indian anthropology. Bombay, 1962.
	An 2209.23	Maddox, J.L. The medicine man. N.Y., 1923.
	Geog 4129.31	Mademoiselle against the world. (Titayna (pseud.).) N.Y., 1931.
NEDL	Geog 142.1	Madrid. Instituto de Geográfico y Estadistico. Memorias. Madrid. 1-11,1875-1899 11v.
	Geog 142.1	Madrid. Instituto de Geográfico y Estadistico. Memorias. Madrid. 1925-1927 4v.
	Geog 142.2.10F	Madrid. Museo Naval. Publicaciones. Madrid. 1,1932
	Geog 4502.5.17	Maduschka, L. Junger Mensch im Gebirg. München, 1936.
	Geog 4029.2	Mämpel, J.C. The young rifleman's comrade. London, 1826.
	Geog 4559.64.3	Männer, Planken, Ozeane. 3. Aufl. (Hanke, Helmut.) Leipzig, 1966.
	Geog 4269.67	Maevskii, Viktor. V. Evropa bez dzhentl'menov. Moskva, 1967.
	Geog 4209.68	Maevskii, Viktor V. Polmilliona kilometrov pozadi. Moskva, 1968.
	Geog 152.5	Magazine. (Nottingham, Eng. University. Geographical Society.) 1-5,1963-1967//?
	Geog 3070.29	Mager, Henri. De la lecture des cartes étrangères. Paris, 1883.
	Geog 500.60	Maggs Bros., London. Bibliotheca asiatica et africana. pt.4-5. London, 1929.
	Geog 500.118	Maggs Bros., London. Voyages and travels. v.1,4-5. London, 1962- 3v.
	Geog 4219.11.5	A magician in many lands. (Bertram, C.) N.Y., 1911.
	Geog 3030.7.1	Magidovich, I.P. Historiia poznania Europy. Wyd 1. Warszawa, 1974.
	Geog 3030.7	Magidovich, I.P. Istoriia otkrytiia i issledovaniia Evropy. Moskva, 1970.
	Geog 3019.57.5	Magidovich, I.P. Ocherki po istorii geograficheskii otkrytii. Moskva, 1957.
	Geog 3251.31	Magnagni, A. D'Anamia e Botero. Ciriè, 1914.
	Geog 4308.95	Magness, E. Tramp tales of Europe. Buffalo, 1895.
	Geog 953.10	Magnus Gabriel de la Gordie's samling af öldre stadsvger. (Stockholm. Biblioteket.) Stockholm, 1915.
	Geog 4709.34	Magre, M. Pirates, flibustiers, negriers. 9. éd. Paris, 1934.
	An 5.30	A Magyar néprajztudomány bibliográfiája 1945-1954. (Sándor, István.) Budapest, 1965.
	Geog 4209.71.30	Maiak Zemli. (Nikolaev, Vladimir D.) Moskva, 1971.
	Geog 4679.12.30	The maiden voyage (Titanic, steamship). (Marcus, Geoffrey.) London, 1969.
	Geog 4209.42.20	Maillart, E.K. Gypsy afloat. London, 1942.
	Geog 612.3.5	Maine, René. Lapérouse. Paris, 1946.
	Geog 665.96	Mainzer geographische Studien. (Hafemann, D.) Braunschweig, 1961.
	Geog 3570.10	Maix de Sori, A.F. Descobrimentos dos Portuguezes nos seculos XV e XVI. Lisboa, 1867.
	Geog 4139.46	Majó Framis, R. Vida de los navegantes y conquistadores españoles del siglo XVI. 1. ed. Madrid, 1946.
	Geog 4180.25	Major, R.H. Early voyages to Terra Australis. London, 1859.
	Geog 4180.22	Major, R.H. India in the fifteenth century. London, 1857.
	Geog 618.2	Major Jame Rennell. (Markham, C.R.) London, 1895.
	Geog 618.2.5	Major James Rennell. (Frenzel, C.A.) Leipzig, 1904.
	Geog 618.2.2	Major James Rennell. (Markham, C.R.) N.Y., 1895.
	An 136.5.75	Makarius, Raoul. Structuralisme ou ethnologie. Paris, 1973.
	Geog 5399.01	Makarov, S.O. S.O. Makarov v zavosvanie Ariuku. Moskva, 1943.
	Geog 4419.33	Makin, William J. Red Sea nights. N.Y., 1933.
	Geog 4559.37.5	Making, V.L. In sail and steam...merchant service, 1902-1927. London, 1937.
	Geog 4509.24.6	The making of a mountaineer. (Finch, G.L.) London, 1924.
	Geog 4559.28.5	The making of a sailor. (Harlow, Frederick Pease.) Salem, Mass., 1928.
Htn	Geog 4559.28.6*	The making of a sailor. (Harlow, Frederick Pease.) Salem, Mass., 1928.
	Geog 3019.33	The making of geography. (Dickinson, R.E.) Oxford, 1933.
	Geog 5855.17	Malaurie, Jean. The last kings of Thule. London, 1956.
	Geog 142.3	Malayan journal of tropical geography. Singapore. 1,1953+ 10v.
	Geog 5515.88	Maldonado, L.F. Voyage de la mer atlantique...pacifique. Plaisance, 1812.
	Geog 5516.88	Maldonado, Lorenzo F. Viaggio dal mare atlantico al pacifico. n.p., 1810.
	An 359.06.12	Maldonado, S.D. Defensa de la Antropologia general. Caracas, 1906.
	Geog 4218.37.5	Malerische Reise um die Welt. (Dumont d'Urville, J.) Leipzig, 1837.
	Geog 5311.3.25	Malfatti, Alberto. Roald Amundsen. Roma, 1959.

Author and Title Listing

Htn	Geog 577.97*	Malham, J. Naval gazetteer. Boston, 1797. 2v.
	Geog 4309.61.5	Malinowksi, P. Der synthetische Bazar. Wien, 1961.
	An 2109.62.20	Malinowski, B. Geschlechtstrieb und Verdrängung bei den Primitiven. Reinbek bei Hamburg, 1962.
	An 187.5.25	Malinowski, B. A scientific theory of culture and other essays. Chapel Hill, 1944.
	An 2109.27.3A	Malinowski, B. Sex and repression in savage society. London, 1927.
	An 2109.27.4	Malinowski, B. Sex and repression in savage society. N.Y., 1927.
X Cg	An 2109.29.5	Malinowski, B. The sexual life of savages in north west Melanesia. London, 1929.
	Geog 4269.60.5	Malířovy toulky po Evropě, Eejlonu a Indii. (Nejedlý, O.) Praha, 1960.
	Geog 4209.07	Mallik, M.C. Impressions of a wanderer. London, 1907.
	Geog 3240.19A	Malone, Kemp. King Alfred's North. Cambridge, 1930.
	An 2600.7.1	Malson, Lucien. Wolf children. London, 1972.
	Geog 4659.68	Malster, Robert. Wreck and rescue on the Essex coast. Truro, 1968.
	An 359.12.9	Malta and the Mediterranean race. (Bradley, R.N.) London, 1912.
	Geog 808.12.16F	Malte-Brun, Conrad. A description of all parts of the world. Boston, 1859. 3v.
	Geog 808.12.15F	Malte-Brun, Conrad. Geographie universelle. v.1-6, Atlas. Paris, 1841-1847. 7v.
	Geog 808.12.5	Malte-Brun, Conrad. Précis de la geographie universelle. Paris, 1812-1829. 8v.
	Geog 808.12.5F	Malte-Brun, Conrad. Précis de la geographie universelle. Atlas. Paris, 1812.
	Geog 808.12.7	Malte-Brun, Conrad. Précis de la geographie universelle. v.5. Paris, 1817.
	Geog 808.12.12	Malte-Brun, Conrad. A system of universal geography. Boston, 1834. 3v.
	Geog 808.12.13	Malte-Brun, Conrad. System of universal geography. v.1-2. Boston, 1844.
	Geog 808.12.17	Malte-Brun, Conrad. Traité elementaire de géographie. Bruxelles, 1832.
	Geog 808.12.9	Malte-Brun, Conrad. Universal geography. Boston, 1824. 8v.
	Geog 808.12.11	Malte-Brun, Conrad. Universal geography. Philadelphia, 1827-1832. 6v.
	Geog 808.12.10	Malte-Brun, Conrad. Universal geography. v.1-14. Boston, 1824-1829. 7v.
	Geog 613.2.5	Malthe Conrad Bruun. (Frenzel, R.) Crimmitschau, 1908.
	An 2109.58.15	Maluquer de Motes, Juan. La humanidad prehistórica. Barcelona, 1958.
	An 358.84.3	Man; embracing his origin. (Lind, G.D.) Chicago, 1884.
	An 2109.61	Man, God, and magic. (Lissher, Ivar.) N.Y., 1961.
	An 359.69.30	Man, his first two million years. (Montagu, Ashley.) N.Y., 1969.
	An 3208.38	Man, in his physical structure and adaptations. (Mudie, R.) London, 1838.
	An 379.72	Man, location, and behavior: an introduction to human geography. (Cox, Kevin R.) N.Y., 1972.
	An 379.67.5	Man, nature and history. (Russell, William Moy Stratton.) London, 1967.
	An 359.60.5	Man, race and Darwin. (Royal Anthropological Institute of Great Britain and Ireland.) London, 1960.
	An 359.41	Man, the mechanical misfit. (Estabrooks, G.H.) N.Y., 1941.
	An 359.35.4A	Man, the unknown. (Carrell, Alexis.) N.Y., 1935.
	An 359.35.5A	Man, the unknown. 2. ed. (Carrel, Alexis.) N.Y., 1935.
	An 359.35.5.8A	Man, the unknown. 11. ed. (Carrel, Alexis.) N.Y., 1935.
	An 359.35.5.10	Man, the unknown. 25. ed. (Carrel, Alexis.) N.Y., 1935.
	An 359.35.6	Man, the unknown. 50. ed. (Carrel, Alexis.) N.Y., 1935.
	An 359.70.30	Man: a remarkable animal. (Schweppe, John S.) Chicago, 1970.
	An 359.46.10A	Man: an autobiography. (Stewart, George R.) N.Y., 1946.
	An 359.05.3	Man: an introduction to anthropology. 2. ed. (Rotzell, W.E.) Philadelphia, 1905.
	An 358.99A	Man - past and present. (Keane, A.H.) Cambridge, Eng., 1899.
	Geog 4559.54	Man against nature. 1. ed. (Neider, C.) N.Y., 1954.
	An 359.23.3A	Man and culture. (Wissler, Clark.) N.Y., 1923.
	An 137.4.5.4	Man and culture. 4. ed. (Firth, Raymond W.) London, 1963.
	An 379.14	Man and earth. (Kirchhoff, Alfred.) N.Y., 1914.
	An 2109.13.10	Man and his forerunners. (Buttel-Reepen, H.) London, 1913.
	An 359.63.15	Man and his future. (Ciba Foundation.) London, 1963.
	An 358.52.2	Man and his migrations. (Latham, R.G.) N.Y., 1852.
	An 358.99.5	Man and his work. (Herbertson, A.J.) London, 1899.
	An 359.09.8	Man and his work. (Herbertson, A.J.) London, 1909.
	An 359.49.10	Man and his works. (Herskovits, Melville Jean.) N.Y., 1949.
	An 359.52.20	Man and his works: the science of cultural anthropology. (Herskovits, Melville Jean.) N.Y., 1952.
	Geog 5060.7	Man and the conquest of the poles. (Victor, Paul Émile.) N.Y., 1963.
	Geog 3019.35.10	Man and the sea. (Rosh, J.H.) Cambridge, Eng., 1935.
	An 2059.03.5	Man before metals. (Joby, N.) N.Y., 1903.
	Geog 4559.28	The man before the mast. (Sorrell, George.) London, 1928.
	An 170.161	Man in adaptation. (Cohen, Yehudi A.) Chicago, 1968. 2v.
	An 379.24.3	Man in Europe, here and there. (Jefferson, M.S.W.) Ypsilanti, Mich., 1924.
	An 359.61.40	Man in process. 1. ed. (Montagu, Ashley.) Cleveland, 1961.
	An 359.27.4.3	Man in the making. (Marett, Robert R.) Garden City, 1928.
	An 359.27.4	Man in the making. (Marett, Robert R.) London, 1927.
	Geog 4476.38.1	The man in the moon, 1638. (Godwin, Francis.) London, 1971.
	An 2058.84	Man in the tertiaries. (Morse, E.S.) Salem, 1884.
	An 358.64.5	The man of the north and man of the south. (Bonstetten, K.V. von.) N.Y., 1864.
	Geog 4218.55.5	Man-of-war life. (Nordhoff, Charles.) Cincinnati, 1856.
	Geog 4218.55.9	Man-of-war life. (Nordhoff, Charles.) N.Y., 1883.
	An 359.55	Man on earth. 1. American ed. (Hawkes, J.) N.Y., 1955.
	An 359.20.11	Man past and present. (Keane, W.H.) Cambridge, 1920.
	An 2109.27.9	Man rises to Parnassus. (Osborn, H.F.) Princeton, 1927.
	An 2109.28.3	Man rises to Parnassus. (Osborn, H.F.) Princeton, 1928.
	An 2109.66.35	Man the hunter. (Symposium on Man the Hunter, Chicago, 1966.) Chicago, 1969.
	Geog 3018.58	Man upon the sea. (Goodrich, F.B.) Philadelphia, 1858.
Htn	Geog 5558.21*	Manby, G.W. Journal of voyage to Greenland...1821. London, 1822.
	Geog 142.12	Manchester Geographical Society. Journal. Manchester. 1-7 6v.
	Geog 4181.101	Mandeville, J. Mandeville's travels. London, 1953. 2v.
	Geog 4181.101	Mandeville's travels. (Mandeville, J.) London, 1953. 2v.
	Geog 4308.54.15	Maney, H. Memories over the water. Nashville, 1854.
	Geog 5225.8	Mangles, James. Papers and dispatches...Arctic searching expeditions. 2. ed. London, 1852.
	An 359.21A	Manhood of humanity. (Korzybski, Alfred.) N.Y., 1921.
	An 359.50.15	Manhood of humanity. 2. ed. (Korzybski, Alfred.) Lakeville, Conn., 1950.
	An 359.63.25	Manifeste contre le racisme. (Tazerout, M.) Rodez, 1963.
	Geog 3199.5	Manitius, M. Anonymi de situ orbis. Stuttgardiae, 1884.
	An 359.17.3	Mankind; racial values and racial prospect. (Humphrey, Seth King.) N.Y., 1917.
	An 2059.59.10	Mankind in the making. (Howells, William.) Garden City, 1959.
	Geog 819.61	Manley, Gordon. Geography; our planet, its peoples and resources. London, 1961.
	Geog 4219.29.2	Mann, Erik. Rundherm. 3.-4. Aufl. Berlin, 1929.
	Geog 4218.68.10	Ein Mann und ein Traktor auf Weltreise. v.2. (Jentsch, August.) Wien, 1968-69.
	An 170.266	Manners, Robert A. Process and pattern in culture; essays in honor of Julian H. Steward. Chicago, 1964.
	An 359.24.25	Mannfraedi. (Marett, Robert R.) Reykjavik, 1924.
	Geog 5559.38	Manning, E.W. Igloo for the night. London, 1943.
	Geog 4309.46	Mánnishor därute. (Jaunes, Elly.) Stockholm, 1946.
	Geog 4239.65	Manry, Robert. Tinkerbelle. N.Y., 1966.
	An 379.33.5	Man's adaptation of nature. (Bryan, P.W.) London, 1933.
	An 359.42.10A	Man's most dangerous myth: the fallacy of race. (Montagu, Ashley.) N.Y., 1942.
	An 359.42.12A	Man's most dangerous myth. 2. ed. (Montagu, Ashley.) N.Y., 1945.
	An 359.64.45	Man's most dangerous myth. 4. ed. (Montagu, Ashley.) Cleveland, 1964.
	An 359.74.15	Man's most dangerous myth. 5. ed. (Montagu, Ashley.) N.Y., 1974.
	An 358.81.5	Man's origin and destiny. (Lesley, J.P.) Boston, 1881.
	An 358.68	Man's origin and destiny. (Lesley, J.P.) Philadelphia, 1868.
	An 358.69.3	Man's origin and destiny sketched from the platform of the sciences. (Lesley, J.P.) Philadelphia, 1869.
	An 359.61.50	Man's place in nature. (Scheler, Max.) Boston, 1961.
	An 379.56.1	Man's role in changing the face of the earth. (International Symposium on Man's Role in Changing the Face of the Earth, Princeton, N.J., 1955.) Chicago, 1971.
	An 2109.43	Man's unknown ancestors; the story of prehistoric man. (Murray, R.W.) Milwaukee, 1943.
	An 359.59.5	Man's way. (Goldschmidt, W.R.) Cleveland, 1959.
	An 3309.00	Manual de antropometria. (Barros Ovalle, P.N.) Santiago de Chile, 1900.
	An 3309.02	Manual for physical measurements. (Hastings, William W.) Springfield, 1902.
	An 3308.78.2	A manual of anthropometry. (Roberts, C.) London, 1878. 2 pam.
	Geog 4268.52	Manual of geography. pt.1. (Hughes, W.) London, 1852.
	Geog 818.61	Manual of modern geography. (Mackay, A.) Edinburgh, 1861.
	Geog 5838.75	Manual of the natural history, geology and physics of Greenland. (Jones, T.R.) London, 1875.
	Geog 4509.34	Manuel d'alpinisme. (Club Alpin Francais.) Chambéry, 1934. 2v.
	Geog 4515.105.2	Le manuel du skieur. 2e éd. (Faes, Henry.) Lausanne, 1925.
	Geog 818.52.5	A manuel of geographical science. v.11. (Nicolay, Charles G.) London, 1859.
	Geog 4300.25.3	Manuel pour voyageurs en Allemagne et dans les Payslimitrophes. 3.ed. (Engelmann, J.B.) Francfort, 1827.
	Geog 4417.45	Un manuscrit inédit de Pierre Poivre: Les mémoires d'un voyageur. (Poivre, Pierre.) Paris, 1868.
	Geog 4359.28	Many cities. (Belloc, Hilaire.) London, 1928.
	Geog 4209.40.45	Many ports of call. (Haven, V.S.) N.Y., 1940.
	Geog 3055.41	Manzi, M. Le livre de l'Atlantide. Paris, 1923.
	Geog 616.5	Map maker and Indian traders...John Patten. (Eavenson, H.N.) Pittsburgh, 1949.
	Geog 3070.135	Map making; the art that became a science. 1st ed. (Brown, Lloyd.) Boston, 1960.
	Geog 3560.5	Map of the world, commonly known as the second Borgian map. (Great Britain. India Office.) n.p., 1889.
	Geog 3251.23	Map of the world. (Stevenson, E.L.) N.Y., 1907.
	Geog 4180.59.3	Map of the world A.D. 1600. (Maps (1600).) London, 1880.
	Geog 3070.50A	The mapmaker's art. (Lynam, E.) London, 1953.
	Geog 3070.105	Mappae Mundi. (Leithäuser, J.G.) Berlin, 1958.
	Geog 3235.33	La mappemonde. (Hany, E.T.) Paris, 1887.
	Geog 3070.80	Mapping and surveying of Africa south of the Sahara. (Conseil Scientifique pour l'Afrique au Sud du Sahara.) London, 1954.
	Geog 3070.05	Mapping the world. (Raisz, E.J.) N.Y., 1936.
	Geog 3070.31	Maps. (Fordham, Herbert G.) Cambridge, Eng., 1921.
	Geog 4268.48	Maps. Europe. Description...of countries of Europe. Hartford, 1848.
	Geog 5838.83.5	Maps. Greenland, 1380-1482. Trois cartes. Copenhagen, 1883.
	Geog 4180.59.3	Maps (1600). Map of the world A.D. 1600. London, 1880.
	Geog 3070.25	Maps reproduced as glass transparencies. (Stevenson, E.L.) N.Y., 1913.
	Geog 4309.12	Maraini, Y. A year of strangers. N.Y., 1912.
	An 379.12.5	Maranelli, C. La geografia umana di Jean Brunhes. Firenze, 1912.
	Geog 3570.25	A maravilhosa viagem dos exploradores portugueses. v.1-12. (Castro Soromenho.) Lisboa, 1946.
	An 2109.27.27	Marazzi, A. Fra i selvaggi e fra i civilizzate. Milano, 1927.
	Geog 142.4	Marburger geographischer Schrifter. Marburg. 1+ 43v.
	An 136.5.80	Marc-Lipiansky, Mireille. Le structuralisme de Lévi-Strauss. Paris, 1973.
Htn	Geog 3235.57*	Marcel, G. Note sur une carte catalane de Dulceri. Paris, 1887.
	Geog 4679.63.5	Marchbanks, David. The painted ship. London, 1964.
	Geog 142.7F	Marco Polo; turismo scolastico del touring club italiano. Milano. 5-10,1954-1959 7v.
	Geog 3570.8	Marcondes de Souza, T.O. Novas achegos à historia dos descobrimentos maritimos. São Paulo, 1963.
	Geog 4209.52.5	Il Marcopolo. (Beonio-Brocchieri, V.) Milano, 1952.
	Geog 4679.12.30	Marcus, Geoffrey. The maiden voyage (Titanic, steamship). London, 1969.
	Geog 4332.12	Il Mare Adriatico. (Cassi, Gellio.) Milano, 1915.
	Geog 4679.67.10	La marée noire du Torrey Canyon. (Mabire, Jean.) Paris, 1967.

Author and Title Listing

Geog 4182.5 Marees, P. de. Beschrijving...van het Gout Koninckrijek van Gunea. 's-Gravenhage, 1912.
An 379.36 Marett, J.R. de la H. Race, sex, and environment. London, 1936.
An 359.12.3 Marett, Robert R. Anthropology. London, 1912.
An 359.19 Marett, Robert R. Anthropology. London, 1919.
An 359.11.5 Marett, Robert R. Anthropology. N.Y., 1911.
An 359.12.23 Marett, Robert R. Anthropology. N.Y., 1912.
An 359.27.3 Marett, Robert R. The diffusion of culture. Cambridge, Eng., 1927.
An 2109.35.5 Marett, Robert R. Head, heart and hands in human evolution. London, 1935.
An 130.15 Marett, Robert R. James George Frazer 1854-1941. London, 1941?
An 137.3 Marett, Robert R. A Jerseyman at Oxford. N.Y., 1941.
An 359.27.4.3 Marett, Robert R. Man in the making. Garden City, 1928.
An 359.27.4 Marett, Robert R. Man in the making. London, 1927.
An 359.24.25 Marett, Robert R. Mannfraedi. Reykjavik, 1924.
Geog 3530.7 Margery, P. Les navigations françaises. Paris, 1867.
Geog 3530.7.2 Margery, P. Les navigations françaises. Paris, 1867.
Geog 4328.47 Margoliouth, Moses. A pilgrimage to the land of my fathers. London, 1850. 2v.
Geog 4309.28 Marín Vicuna, Santiago. Viajando. Santiago, 1928.
Geog 4219.55 Marinai italiani e iberici sulle vie delle Indie. (Vocino, Michele.) Roma, 1955.
Geog 3520.16 Marine cartography in Britain. (Robinson, Adrian.) Leicester, 1962.
Geog 4219.37.12 Marine Historical Association, Mystic, Conn. The Joseph Conrad. Mystic, 195-.
Geog 4679.25 Marine Research Society, Salem, Mass. The sea, the ship, and the sailor. Salem, Mass., 1925.
Geog 3251.23.5 Marine world chart, 1502. (Stevenson, E.L.) N.Y., 1908.
Geog 614.5F Marinelli, G. Cristoforo Negri. Torino, 1897.
Geog 3228.82.3 Marinelli, G. Die Erdkunde bei den Kirchenvätern. Leipzig, 1884.
Geog 3228.82 Marinelli, G. La geografia. Roma, 1882.
Geog 808.83 Marinelli, G. La terra. Milano, 1883-1885. 7v.
Geog 5312.2.5 Mariner of the North. (Putnam, George.) N.Y., 1947.
Geog 4678.34 The mariner's chronicle...shipwrecks, storms. (Duncan, A.) New Haven, 1834.
Geog 4678.06 The mariner's chronicle. (Duncan, A.) Philadelphia, 1806. 4v.
Geog 4678.35 Mariners' chronicle. New Haven, 1835.
Htn Geog 4558.33* Mariner's library. Boston, 1833.
Geog 4208.30.5 A mariner's sketches. (Ames, N.) Providence, 1830.
Geog 4159.58 Les marins à la découverte. Paris, 1958. 12v.
Geog 3018.79.5 Les marins du XVe et du XVIe siècle. (Jurien de la Gravière, J.B.E.) Paris, 1879. 2v.
Geog 3018.81 Maritime discovery. (Low, C.R.) London, 1881. 2v.
Geog 4328.36.5 Maritime scraps, or Scenes in the frigate United States during a cruise in the Mediterranean. Boston, 1838.
Geog 3570.60 Marjay, Frederico Pedro. Navegadores portugueses, herois do mar. Lisboa, 1970.
Geog 610.5 Mark Jefferson, geographer. (Martin, Geoffrey J.) Ypsilanti, Mich., 1968.
Geog 5558.75.7 Markham, A.H. The great frozen sea...1875-76. London, 1878.
Geog 613.4.5 Markham, A.H. The life of Sir Clements R. Markham. London, 1917.
Geog 5530.5 Markham, A.H. Life of Sir John Franklin. London, 1891.
Geog 5530.5.2 Markham, A.H. Life of Sir John Franklin. N.Y., 1891.
Geog 5208.79 Markham, A.H. Northward ho! London, 1879.
Geog 5398.79 Markham, A.H. Polar reconnaissance. London, 1881.
Geog 5323.1.2 Markham, C. Life of admiral Sir Leopold McClintock. London, 1909.
Geog 4179.5.5 Markham, C.R. Address...on fiftieth anniversary. London, 1911.
Geog 4181.29 Markham, C.R. Book of knowledge of all kingdoms, lands. London, 1912.
Geog 4181.28 Markham, C.R. Early Spanish voyages to the Strait of Magellan. v.28. London, 1911.
Geog 4180.24 Markham, C.R. Expeditions into valley of the Amazons. N.Y., 1964.
Geog 4180.57 Markham, C.R. The Hawkins voyages. London, 1878.
Geog 4180.86A Markham, C.R. Journal of Christopher Columbus. London, 1893.
Geog 5070.9 Markham, C.R. The lands of silence. Cambridge, Eng., 1921.
Geog 4180.90 Markham, C.R. Letters of Amerigo Vespucci. London, 1894.
Geog 618.2 Markham, C.R. Major Jame Rennell. London, 1895.
Geog 618.2.2 Markham, C.R. Major James Rennell. N.Y., 1895.
Geog 4180.48 Markham, C.R. Narrative of the rites and laws of the Yncas. N.Y., 1964?
Geog 4180.47 Markham, C.R. Reports on the discovery of Peru. N.Y., 1964.
Geog 4179.5 Markham, C.R. Richard Hakluyt: his life and work. London, 1896.
Geog 5208.73.2 Markham, C.R. The threshold of the unknown region. London, 1873.
Geog 5208.73.4 Markham, C.R. The threshold of the unknown region. 4. ed. London, 1876.
Geog 5398.80 Markham, C.R. Voyage of the 'Eira'. London, 1881.
Geog 4180.56 Markham, C.R. Voyages of Sir James Lancaster. N.Y., 1964?
Geog 5939.57.5 Markov, K.K. Puteshestvie v Antarktidu. Moskva, 1957.
Geog 5925.92 Markov, Konstantin K. Geografiia Antarktidy. Moskva, 1968.
Geog 613.10 Markov, Konstantin K. Vospominaniia i razmyshleniia geografa. Moskva, 1973.
Geog 4269.66.5 Markovic, Slobodan. Putovanja grešnog Elefterija. Beograd, 1971.
Geog 4348.47 Marmier, X. Du Rhin au Nil. Paris, 1847. 2v.
Geog 4308.33.5 Marmier, X. Souvenirs de voyages. Paris, 1841.
Geog 4218.50.3 Marmier, Xavier. Voyage d'une femme autour du monde. v.1-2. Bruxelles, 1853.
Geog 4348.34 Marmont, Auguste Frédéric L. Voyage du...en Hongrie, en Transylvanie. Paris, 1837. 5v.
Geog 5180.11 Marquis, R. La conquête du pôle nord en ballon. Paris, 19- .
Geog 6009.52 Marret, M. Seven men among the penguins. 1st American ed. N.Y., 1955.
An 2109.11 Marriage, totenism, and relgion. (Avebury, J.L.) London, 1911.
Geog 4559.55.5 Mars in Capricorn. 1. ed. (Cross, Beverley.) Boston, 1955.
Geog 4677.48 Marsden, Peter Richard V. The wreck of the Amsterdam. London, 1974.
NEDL Geog 142.5 Marseilles. Société de géographie. Marseilles. 1-23,1877-1899 23v.
Geog 142.5 Marseilles. Société de géographie. Marseilles. 1-65,1877-1954 18v.
Geog 142.6 Marseilles. Exposition Coloniale Nationale, 1922. Semaine internationale des géographes, des explorateur et des ethnologues, 22-28 Sept. 1922. Marseilles, 1923.
An 2109.71.1 Marshack, Alexander. The roots of civilization. 1st ed. N.Y., 1971.
Geog 4307.72 Marshall, J. Travels through Holland. London, 1772. 3v.
Geog 4217.91.2 Marshall, James S. Vancouver's voyage. 2. ed. Vancouver, 1967.
Geog 4679.12.13 Marshall, L. Sinking of the Titanic and great sea disasters. Philadelphia, 1912.
Geog 5060.8 Marshall, Logan. The story of polar conquest. n.p., 1913.
An 379.70 Maršík, Miroslav. Natural environment and society in the theory of geographical determinism. Praha, 1970.
Geog 4180.18 Martens, F. Collection of documents on Spitzbergen and Greenland. London, 1855.
Geog 4219.66 Marti-Ibáñez, Félix. Journey around myself. N.Y., 1966.
An 3509.35 Martin, Cecil P. Prehistoric man in Ireland. London, 1935.
Geog 608.10 Martin, Geoffrey J. Ellsworth Huntington; his life and thought. Hamden, Conn., 1973.
Geog 610.5 Martin, Geoffrey J. Mark Jefferson, geographer. Ypsilanti, Mich., 1968.
Geog 4219.27.5 Martin, Lillien J. Round the world with a psychologist. San Francisco, 1927.
An 3309.24.3 Martin, Rudolf. Richtlineen für Körpermessungen. München, 1924.
Geog 3530.16 Martonne, E. de. La science géographique. Paris, 1915.
Geog 4209.40.60 Martyr, W. The wandering years. N.Y., 1940.
Geog 4304.90 Martyr. Relation d'un voyage fait en Europe. Paris, 1827.
Geog 959.12F Marvelous scenes of the world. (Raymond, E.L.) Chicago, 1902.
Geog 5070.68 The marvelous wonders of the polar world. (Dieck, Herman.) Philadelphia, 1965?
Geog 4239.63 Marx, R.F. The voyage of the Niña II. 1st ed. Cleveland, 1963.
Geog 4652.9 Marx, Robert F. Shipwrecks of the Western Hemisphere, 1492-1825. N.Y., 1975.
An 2109.72.35 Marxism and "primitive" societies. (Terrey, Emmanuel.) N.Y., 1972.
An 2109.72.32 Le Marxisme devant les sociétés primitives. 2. éd. (Terray, Emmanuel.) Paris, 1972.
An 170.268 Marxist analyses and social anthropology. N.Y., 1975.
Geog 4679.42.5 Mary Celeste; the odyssey of an abandoned ship. (Fay, Charles E.) Salem, 1942.
Geog 4672.16.15 The "Mary Celeste" and other strange tales of the sea. (Lockhart, John Gilbert.) London, 1952.
Geog 4308.81 A Maryland editor abroad. (Colton, G.) Annapolis, 1881.
An 3559.34 Mascarenhas, Constância. Os povos de Angola. Bastorá, 1934.
Geog 4709.59 Maséa de Ros, Angeles. Historia general de la piratería. Barcelona, 1959.
An 2059.54 Masiker, Reuben. The Australopithecinae. Cape Town, 1954.
Geog 4228.61 Maskee! The journal and letters of Ruth Bradford, 1861-1872. (Bradford, Ruth.) Hartford, 1938.
Geog 4502.5.30 Mason, Alfred E.W. Das Gesetz der Berge. München, 1940?
An 2108.95.5 Mason, O.T. Origins of invention. London, 1895.
An 359.54.5 Mason, Philip. An essay on racial tension. London, 1954.
An 359.62.10 Mason, Philip. Prospero's magic. London, 1962.
Geog 665.142 Mass, Walther Gerhard Eduard. Menschen und Landschaften. Hildesheim, 1967.
Geog 4707.26.5F Massachusetts (Colony). Court of Admiralty. The tryals of sixteen persons for piracy. London, 1726.
Geog 4208.63 Massett, S.C. Drifting about. N.Y., 1863.
Geog 4559.54.15 Master in sail. 2d ed. (Learmont, J.S.) London, 1954.
Geog 4145.17 Master mariner...Amasa Delano. (Connolly, J.B.) Garden City, 1943.
Geog 4659.36.10 Masters, David. S.O.S., a book of sea adventure. N.Y., 1936.
Geog 4239.55.5 Mastrangelo, F. Atlantica dall'intuizione alla scoperta del nuovo mondo. Napoli, 1955.
An 2109.15 The material culture of the simpler peoples. (Hobhouse, L.T.) London, 1915.
Geog 5855.11 The material response of the polar eskimo to their far Arctic environment. Diss. (Ekblaw, W.E.) N.Y., 1928.
Geog 612.1 Materialien zu einer Biographie des Fr. J.M. Liechtenstern. Schneeberg, 1823.
An 609.72 Materialy po rabote i istorii etnograficheskikh muzeev i vystavok. Moskva, 1972.
An 358.66 Matériaux pour l'histoire...de l'homme. 2. sér. Toulouse. 1872-1877 2v.
Geog 4208.53.5 Mather, Frank J. A clipper ship and her commander. n.p., 1904.
Geog 4209.30.15 Mather, Norman C. Travel notes, 1929-30. Chicago, 1930.
An 359.24.15 Mathews, Basil. The clash of color. 2. ed. N.Y., 1924.
An 359.24.18 Mathews, Basil. The clash of colour. Port Washington, 1973.
An 359.24.10 Mathews, Basil. The clash of colour. 7. ed. London, 1924.
An 359.26.15 Mathews, Basil. The clash of colour. 16. ed. London, 1926.
Geog 5650.20 Mathiassen, T. Skraelingerne i Grønland. København, 1935.
Geog 4209.35.15 Matisse, Maarten. Wanderer from sea to sea. N.Y., 1936.
An 2059.72.5 Matiushin, Geral'd N. U kolybeli istorii. Moskva, 1972.
Geog 3070.22 Matkovic, P. Alte handschriftliche Schiffer-Karten. Wien, 1863.
Geog 3235.35 Matkovic, P.P. Alte handschriftliche Schifferkarten. Agram, 1860.
An 39.08 Matsumura, A. A gazetteer of ethnology. Tokyo, 1908.
An 379.00 Matteweuzzi, A. Les facteurs de l'évolution des peuples. Bruxelles, 1900.
Geog 4308.17.5 Matthews, H. Diary of an invalid...journey...1817. Paris, 1836.
Geog 4308.17.4 Matthews, H. The diary of an invalid. Paris, 1825.
Geog 4308.17.3 Matthews, H. The diary of an invalid. 2nd ed. London, 1820.
Geog 4308.17.3.5 Matthews, H. The diary of an invalid. 3rd ed. London, 1822. 2v.
An 3508.93 Matthews, W. The human bones...Hemingway collection...United States Medical Museum at Washington. n.p., 1893.
Geog 4307.85.5 Matthisson, F. Letters...from various parts of the continent. London, 1799.
Geog 4309.10.10 Matto de Turner, C. Viaje de recrea. Valencia, 1910?
An 2059.41 Mattos, Anibal. A raça de Lagôa Santa. São Paulo, 1941.
An 2208.83 Maturi, R. La medicina preistorica. Napoli, 1883.
Geog 4209.73 Matveev, Vikentii A. Sud'ia-vremia. Moskva, 1973.

Author and Title Listing

Geog 600.22 — Matveeva, T.P. Russkie geografy i puteshestvenniki. Leningrad, 1971.
Geog 4358.90 — Maupassant, G. de. La vie errante. 13. éd. Paris, 1890.
Htn Geog 757.42* — Maupertius, Pierre L. Elements de geographie. Paris, 1742.
Geog 520.17 — Maurand, J. Itineraire de Jerome Maurand. Paris, 1901.
Geog 5837.40F — Mauricius, Jan J. Naleesing over Groenland voor de historie van den noorweeschen Erik. pt.1-5. Hamburg? 1740.
Geog 4329.22 — Maurras, Charles. Anthénea. Paris, 1922.
Geog 4209.41.20 — Maury, A.F. Intimate Virginiana. Richmond, 1941.
Geog 4209.39.10 — Maury, Richard. The saga of "Cimba". N.Y., 1939.
An 187.8 — Mauss, Marcel. Oeuvres. Paris, 1961- 3v.
An 2109.60.25 — Mauss, Marcel. Sociologie et anthropologie. 2. ed. Paris, 1960.
Geog 6009.04 — Maveroff, J.O. Por los mares antarticos. Buenos Aires, 1954.
Geog 4157.96.3 — Mavor, William. Historical account of...voyages...Columbus to present. Philadelphia, 1802.
Geog 4157.96 — Mavor, William. Historical account of...voyages. v.1-2,4-22. London, 1796. 19v.
Geog 4157.96.5 — Mavor, William. Historical account of most celebrated voyages, travels, and discoveries from time of Columbus. New Haven, 1802-03. 24v.
Geog 6009.29 — Mawson, Douglas. The B.A.N.Z. Antarctic research expedition, 1929-31. London, 1932.
Geog 6009.14.20 — Mawson, Douglas. The home of the blizzard. Philadelphia, 1914. 2v.
Geog 5985.187 — Mawson, Paquita. Mawson of the Antarctic; the life of Sir Douglas Mawson. London, 1964.
Geog 5985.187 — Mawson of the Antarctic; the life of Sir Douglas Mawson. (Mawson, Paquita.) London, 1964.
Geog 4328.39 — Maximilian, H. Wanderung nach dem Orient. München, 1839.
Geog 3070.14 — Mayer, E. Die Entwicklung der Seekarten. Wien, 1877.
Geog 4512.561 — Mayer, T.H. Der Führer; Novelle. München, 1924.
Geog 4180.4 — Maynarde, T. Sir Francis Drake, his voyage 1595. London, 1849.
Geog 612.2 — Mazuel, J. L'oeuvre géographique de Linant de Bellefonds. Thèse. Le Caire, 1937.
Geog 618.4.5 — Mazurkiewicz-Herzowa, Kucja. Eugeniusz Romer. Warszawa, 1966.
An 2109.37.5 — Mead, M. Cooperation and competition among primitive peoples. N.Y., 1937.
An 170.270.5 — Mead, M. The golden age of American anthropology. N.Y., 1960.
An 170.270 — Mead, M. Primitive heritage. N.Y., 1953.
An 629.53 — Mead, M. The study of culture at a distance. Chicago, 1953.
An 2109.61.20 — Mead, Margaret. Cooperation and competition among primitive peoples. Boston, 1961.
An 2109.39.15 — Mead, Margaret. From the South Seas. N.Y., 1939.
Geog 4129.14.5 — Mead, William E. The grand tour in the 18th century. Boston, 1914.
Geog 3618.10 — Mead, William R. The geographical tradition in Finland. London, 1963.
An 132.6 — Meaning in culture. (Hanson, F. Allan.) London, 1975.
An 3309.30.5 — Measurement of man. Minneapolis, 1930.
Geog 4419.51 — Med against the desert. (Calder, Ritchie.) London, 1951.
Geog 5398.97.4 — Med örnen mot polen av S.A. Andrée. (Svenska Sällskapet för Antropologi och Geografi.) Stockholm, 1930.
An 359.60 — Medawar, Peter Brian. The future of man. London, 1960.
Geog 141.5F — Meddelanden. Avhandlingar. (Lund. Universitet Geografiska Institution.) Lund.
Geog 141.5 — Meddelanden. Avhandlingar. (Lund. Universitet Geografiska Institution.) Lund. 1-60 32v.
Geog 4209.43.10 — Medern, W.E.A.A. Blick in die weite Welt. Berlin, 1943.
Geog 3235.5 — Mediaeval geography. (Bevan, W.L.) London, 1873.
An 2208.93 — Die Medicin der Naturvölker. (Bartels, Max.) Leipzig, 1893.
An 2208.83 — La medicina preistorica. (Maturi, R.) Napoli, 1883.
An 2209.24 — Medicine, magic and religion. (Rivers, W.H.R.) London, 1924.
An 2209.71 — Medicine and ethnology. (Ackevknecht, Erwin Heinz.) Baltimore, 1971.
An 2209.23 — The medicine man. (Maddox, J.L.) N.Y., 1923.
Geog 4329.35.10 — The Mediterranean; an anthology. (Bloomfield, Paul.) London, 1935.
Geog 1595.5A — Mediterranean...Madeira, Canary Islands. (Baedeker, publishers.) Leipzig, 1911.
Geog 4328.54 — The Mediterranean. (Smyth, W.H.) London, 1854.
Geog 4329.49 — Mediterranean background. (Newman, B.) London, 1949.
Geog 4328.66.5 — The Mediterranean Islands. (Sleeper, M.G.Z. (Mrs.).) Boston, 1866.
Geog 4328.93 — Mediterranean mosaics. (Barber, J.L.) N.Y., 1895.
An 359.01.3 — The Mediterranean race. (Sergi, G.) London, 1901.
Geog 4328.95 — The Mediterranean trip. (Brooks, N.) N.Y., 1895.
Geog 4329.36.10 — La Méditerranée; les hommes et leurs travaux. 4e éd. (Parain, Charles.) Paris, 1936.
Geog 4329.53.10 — La Méditerranée et le Moyen-Orient. (Birot, Pierre.) Paris, 1953. 2v.
Geog 4329.53.12 — La Méditerranée et le Moyen-Orient. 2. éd. (Birot, Pierre.) Paris, 1964.
Geog 3570.21 — La Méditerranée et les découvertes maritimes des Portugais. (Amzalah, Moses.) Lisbonne, 1951.
Geog 4329.54 — Méditerranée occidentale. Paris, 1954.
Geog 4329.53.20 — Méditerranée orientale. Paris, 1953.
Geog 4328.79 — El mediterráneo. 3a ed. (Jeréz Perchét, Augusto.) Madrid, 1879.
Geog 4249.44A — Medrs, Eliot G. Pacific Ocean handbook. Stanford, Calif., 1944.
An 359.45.15 — Meet your ancestors, a biography of primitive man. (Andrews, R.C.) N.Y., 1945.
An 3559.38 — Meethare variabelen van den menschlijken schedel en hun an derlinge. (Froe, A. de.) Amsterdam, 1938.
Geog 4218.72 — Meeting the sun; journey all round the world. (Simpson, William.) London, 1874.
Geog 4218.72.2 — Meeting the sun; journey all round the world. (Simpson, William.) London, 1877.
Geog 4679.47.5 — Meier, F. Hurricane warning. N.Y., 1947.
Geog 4329.27.20 — Meier-Graefe, J. Pyramide und Tempel. Berlin, 1927.
Geog 611.3 — Meier-Lemgo, K. Engelbert Kampfer, der erste deutsche Forschungsreisende, 1651-1716. Stuttgart, 1937.
An 137.2 — Meigs, C.D. Memoir of Samuel George Morton, M.D. Philadelphia, 1851. 2 pam.
Geog 4029.17 — Mein tödliches Risiko. (Seering, Ruth.) Bergisch Gladbach, 1973.
Geog 4219.58.5 — Mein Traum wurde Wirklichkeit. 10. Aufl. (Fenyvessy, J.) Köln, 1961.
Geog 4521.8 — Meine Berge, nuen Leben. (Hackel, Heinrich.) Salzburg, 1960.
Geog 5559.08 — Meine Eroberung des Nordpols. (Cook, F.A.) Hamburg, 1912.
Geog 4145.80 — Meine schönste Autoreise. (Schindler, Fritz.) Wien, 1958.
Geog 4219.28.5 — Meine Weltumsegelung mit dem Fischkutter Hamburg. 2. Aufl. (Kircheiss, C.) Berlin, 1928.
An 358.11 — Meinere, C.M. Untersuchungen über die Verschiedenheiten der Menschennaturen. Tübingen, 1811-15. 3v.
Geog 4131.68.5 — Meinke, Hans. Tourismus und wirtschaftliche Entwicklung. Göttingen, 1968.
Geog 4209.53 — Meiss-Teuffen, Hans. Wanderlust. N.Y., 1953.
Geog 4308.67.3 — Meissner, A. Unterwegs. Leipzig, 1867.
Htn Geog 956.78* — Meissner, D. Sciagraphia cosmica. v.1-8. Nürnberg, 1678. 2v.
An 359.66 — Melady, Thomas Patrick. The revolution of color. N.Y., 1966.
Geog 4678.33.5 — Melancholy shipwreck and...divine providence. n.p., 1834.
Geog 665.60 — Mélanges de géographie et d'orientalisme offerts à E.-F. Gautier. Tours, 1937.
Geog 665.35 — Mélanges de géographie offerts par ses collègues et amis de l'étranger à M. Václav Svambera. Praha, 1936.
Geog 665.152 — Mélanges de géographie physique. Gembloux, 1967. 2v.
Geog 665.72 — Mélanges géographiques offerts au Doyen ernest Benevent. Gap, 1954.
Geog 665.49 — Mélanges géographiques offerts par ses élèves à Raoul Blanchard. Grenoble, 1932.
An 170.312 — Mélanges Pittard. Brive, 1957.
Geog 5650.35 — Meldgaard, Jørgen. Nordboerne i Grønland. Købwnhavn, 1965.
Geog 4306.83.2 — Melle, Jakob von. Beschreibung einer Reise durch das Nordwestliche Deutschland nach den Niederländen. Lübeck, 1891.
Geog 5838.94 — Mellem fangere og jaegere. (Bruun, Daniel.) Kjøbenhavn, 1897.
Geog 4206.81 — Melton, E. (pseud.). Eduward Meltons, engelsch edelmans, zeldzaame en gedenkwaardige zee- en land reizen. Amsterdam, 1681.
Geog 5398.81.5 — Melville, G.W. In the Lena Delta. Boston, 1892.
An 132.4 — Melville J. Herskovits. (Simpson, George Eaton.) N.Y., 1973.
Geog 5321.1.15 — Memoir and eulogy of Dr. Elisha K. Kane. (Andrews, E.W.) N.Y., 1857.
An 126.8 — A memoir of John Otis Brew. (Whitehill, Walter Muir.) Boston, 1968.
An 137.2 — Memoir of Samuel George Morton, M.D. (Meigs, C.D.) Philadephia, 1851. 2 pam.
Geog 4217.85.5 — Memoir on the voyage...in search of La Pérouse. (Burney, J.) London, 1820.
Geog 195.2 — Memoire geografiche. Roma. 1-9 7v.
Geog 4428.68 — Mémoire géographique sur mer des Indes. (Codine, J.) Paris, 1868.
Geog 5180.7 — Memoire sur le passage par le nord. (De Croy, D.) Paris, 1782.
Geog 3057.4 — Memoire sur le pays d'Ophir. (Quatremère.) Paris, 1845.
Geog 3140.13 — Memoire sur le Periple d'Hannon. (Mer, A.) Paris, 1885.
Htn Geog 5367.65* — Memoires...geographiques...pays septentrionaux. (Engel, S.) Lausanne, 1765.
Geog 85.1 — Mémoires. (Geneva. Société de Géographie.) Paris. 1-104 43v.
Geog 212.216F — Mémoires. (Société Royale de Géographie d'Egypte.) Le Caire. 1,1919+ 17v.
Geog 4477.87 — Mémoires. v.1-4. (Wollap, G.) Londres, 1787. 2v.
An 358.71.3 — Mémoires d'anthropologie. v.5. (Broca, Paul.) Paris, 1888.
Geog 4307.35.3 — Memoires de Charles-Louis, Baron de Pöllnitz. (Pöllnitz, K.L. von.) Amsterdam, 1735. 2v.
Geog 4228.23 — Mémoires du capitaine Landolphe. (Landolphe, Jean François.) Paris, 1823. 2v.
Geog 4218.24 — Memoires du Capitaine Péron. Paris, 1824. 2v.
Geog 179.2 — Mémoires et documents. (France. Centre de Documentation Cartographique et Géographique.) 1-10 22v.
Geog 4307.50 — The memoirs. v.1-2. (Clancy, Michael.) Dublin, 1750.
Geog 4307.37 — Memoirs of...travels. (Pöllnitz, K.L. von.) London, 1737. 4v.
Geog 4209.30.10 — Memoirs of a Ceylon planter's travels, 1851 to 1921. (Bremer, M.) London, 1930.
Geog 5538.51.9 — Memoirs of and journal of voyage...search of Sir John Franklin. (Bellot, J.R.) London, 1855. 2v.
Geog 4307.37.2 — Memoirs of Charles-Lewis, Baron de Pöllnitz. 2nd ed. (Pöllnitz, K.L. von.) London, 1739. 4v.
Geog 5318.1 — Memoirs of Hans Hendrik, the Arctic traveller. (Hendrik, Hans.) London, 1878.
Geog 4307.82 — Memoirs of Peter Henry Bruce, Esq. (Bruce, Peter Henry.) London, 1782.
Geog 4307.82.3 — Memoirs of Peter Henry Bruce, Esq. Reprint. (Bruce, Peter Henry.) London, 1970.
Geog 3140.9 — Memoirs sur deux monuments geographiques. (Blau, J.) Nancy, 1836.
Geog 3550.4 — Memoria ai viaggiatori italiani. (Gubernatis, A. de.) Firenze, 1867.
Geog 212.30 — Memoria del presidente de la Sociedad Geográfica de Lima. (Urteaga, H.H.) Lima, 1942.
Geog 5326.21.10 — Memorial book of Robert Edwin Peary. (Buck, M.) n.p., 1937.
Geog 4180.8 — Memorials of the empire of Japan. (Rundall, T.) N.Y., 1964?
NEDL Geog 142.1 — Memorias. (Madrid. Instituto de Geográfico y Estadistico.) Madrid. 1-11,1875-1899 11v.
Geog 142.1 — Memorias. (Madrid. Instituto de Geográfico y Estadistico.) Madrid. 1925-1927 4v.
Geog 4181.48 — Memorias antiguas historiales del Peru. (Montesinos, Fernando.) London, 1920.
Geog 212.105 — Memorie. (Societa Geografica Italiana.) Roma. 1-30 27v.
Geog 4306.85 — Memorie de viaggi per l'Europa. (Pacichelli, G.B.) Napoli, 1685. 3v.
Geog 194.1.5 — Memorie geografiche. Firenze. 1-13,1907-1919 11v.
Geog 4308.59.11 — Memories de un viajero peruano; apuntes y recuerdos de Europa y Oriente (1859-1863). (Pas Soldan y Unánue, Pedro.) Lima, 1971.
Geog 4521.26 — Memories of a mountaineer. (Chapman, F.S.) London, 1951.
Geog 4209.23.20 — Memories of travel. (Bryce, J.B.) N.Y., 1923.
Geog 4308.54.15 — Memories over the water. (Maney, H.) Nashville, 1854.
Geog 4559.62.5 — Men, ships, and the sea. (Villiers, A.J.) Washington, 1962.
Geog 5160.35 — Men against the frozen north. (Calder, Ritchie.) N.Y., 1957.

Author and Title Listing

Geog 4559.71.10	Men against the sea. (Robinson, Cyril.) Windsor, N.S., 1971.
Geog 4308.53	Men and things...in Europe. (Murray, N.) N.Y., 1853.
Htn An 2056.56*	Men before Adam: discourse on Romans 5-12,13,14. (Peyrère, I. de.) London, 1656.
An 2109.15.5A	Men of the old stone age. (Osborn, H.F.) N.Y., 1915.
An 2109.23.3	Men of the old stone age. 3. ed. (Osborn, H.F.) N.Y., 1923.
An 2109.34.10	Men of the old stone age. 3. ed. (Osborn, H.F.) N.Y., 1934.
An 2109.21.10	Men of the old stone age. 3. ed. (Osborn, Henry F.) N.Y., 1921.
Geog 4209.32	Men on the horizon. (Murchie, Guy.) Boston, 1932.
An 359.25.4	The menace of colour. 2. ed. (Gregory, John W.) London, 1925.
Geog 4309.14.15	Mencken, Henry L. Europe after 8:15. N.Y., 1914.
An 359.72.15	Mendel, Gérard. Vers une anthropologie sociopsychanalytique. 1. éd. Paris, 1972.
Geog 5209.60	Mendeleev, D.I. Nauchnyi arkhiv: osvoenie krainego severa. Leningrad, 1960.
An 187.6	Mendes Corrèa, A.A. Da biologia à história. Porto, 1934.
An 3559.17	Mendes Correa, A.A. Sôbre algunos crânios da India Portuguêsa. Porto, 1917.
An 3559.15	Mendes Correa, A.A. Sobre três crânios de negros Mossumles. Porto, 1915.
An 359.44.15	Mendes Correia, A.A. Gérmen e cultura. Pôrto, 1944.
An 359.33.9	Mendes Correia, A.A. Introdução i antropobiologia. Coimbra, 1933.
An 359.34.10	Menghin, O. Geist und Blut. Wien, 1934.
Geog 5839.74	Mennesker og miljøer i Grønland. (Bechmann, Elke.) København, 1974.
Geog 4219.64	Menon, Kumara P.S. Journey round the world. Bombay, 1966.
An 379.41.5	De mens in de geografie. (Vermooten, Willem H.) Assen, 1941.
Geog 4309.56	Mensaje de la Europa nordica. (Haya de la Torre.) Buenos Aires, 1956.
Geog 4502.5.23	Mensch, Berg und Tod. (Schmitt, F.) München, 19- .
An 358.94	Der Mensch. (Ranke, Johannes.) Leipzig, 1894. 2v.
Geog 3055.87	Der Mensch der Eiszeit und Atlantis. (Gleich, Sigismund von.) Stuttgart, 1969.
An 2109.09.11	Der Mensch der Vorzeit. v.1-2. (Bölsche, W.) Stuttgart, 1909-11.
An 2109.21.5	Der Mensch der Vorzeit. v.1-2. 28e Aufl. (Boelsche, Wilhelm.) Stuttgart, 1921.
An 359.01.7	Mensch und Erde. (Kirchoff, A.) Leipzig, 1901.
An 358.78	Der Mensch vormals und heute. (Oberländer, R.) Leipzig, 1878.
Geog 4145.95	Menschen an meinen Wegen. (Wollschläger, Alfred.) München, 1973.
Geog 665.142	Menschen und Landschaften. (Mass, Walther Gerhard Eduard.) Hildesheim, 1967.
Geog 4249.37	Menschen und Mächte am Pazifik. (Lissner, Ivar.) Hamburg, 1937.
An 2059.69.5	Menschenformen im Eiszeitalter. 1. Aufl. (Overhage, Paul.) Frankfurt, 1969.
An 2339.02	Menschenjagden und Zweikämpfe. (Frobenius, L.) Jena, 1902.
An 359.09.5	Menschenkunde. (Buschan, Georg H.T.) Stuttgart, 1909.
An 2109.01.5	Der Menschheitsgedanke durch Raum und Zeit. (Bastian, A.) Berlin, 1901.
Geog 6009.46	Menster, William J. Strong men south. Milwaukee, 1949.
Geog 665.168	Mental maps. (Gould, Peter R.) Harmondsworth, 1974.
An 2109.31.5	La mentalité primitive. (Lévy-Bruhl, L.) Oxford, 1931.
An 2109.22	La mentalité primitive. (Lévy-Bruhl, L.) Paris, 1922.
An 2109.31.5.5	La mentalité primitive. 15. ed. (Lévy-Bruhl, L.) Paris, 1960.
Geog 4328.96	Mentone, Cairo and Corfu. (Woolson, C.F.) N.Y., 1896.
Geog 3140.13	Mer, A. Memoire sur le Periple d'Hannon. Paris, 1885.
Geog 4559.30.15	Mer, marines, marins. (Valéry, Paul.) Paris, 1930.
Geog 5839.29.5	La mer du Groenland, croisières du "Pourquoi pas?". (Charcot, J.B.) Paris, 1929.
Geog 613.1.18	Mercator, Gerardus. Correspondance mercatorienne. Anvers, 1859.
Geog 4209.45.5	Merchant of alphabets. (Orcutt, R.) Garden City, N.Y., 1945.
Geog 4208.95.5.5	The merchant vessel. (Nordhoff, C.) N.Y., 1895.
Geog 4208.95.5	The merchant vessel. (Nordhoff, C.) N.Y., 1895.
Geog 4248.39	Mercier, H.J. Life in a man-of-war...during cruise in Pacific. Philadelphia, 1841.
An 99.66.5	Mercier, Paul. Histoire de l'anthropologie. Paris, 1966.
Geog 4678.32	Meredith, J.C. The tattooed man. 1. American ed. N.Y., 1938.
Geog 4209.66.5	Meridiany, sobytiia, vstrechi. (Kovanov, Vladimir V.) Moskva, 1966.
Geog 4328.92	Meriwether, Lee. Afloat and ashore on the Mediterranean. N.Y., 1892.
Geog 4309.11	Meriwether, Lee. Seeing Europe by automobile. N.Y., 1911.
Geog 4308.87	Meriwether, Lee. A tramp trip. N.Y., 1887.
Geog 4308.87.1	Meriwether, Lee. A tramp trip. 5th ed. N.Y., 1887.
Htn Geog 4307.53*	Merkwürdige Reisen. (Uffenbach, Z.C.) Ulm, 1753-1754. 3v.
Geog 4219.36	Merrick, Lesley. A good time. N.Y., 1936.
Geog 4329.58.10F	Merveilles de la Méditerranée. (Beny, Roloff.) Paris, 1958.
Geog 665.40F	Les merveilles du monde. (Hachette, firm, publishers, France.) Paris, 192-?
Geog 665.40.5F	Les merveilles du monde. (Hachette, firm, publishers, France.) Paris, 1957.
Geog 4266.36	Mervla, P. Cosmographiae generalis libri tres. Amsterdami, 1636.
Htn Geog 5837.20.10*	Mesange, P. de. La vie...et le voyage de Groenland. v.1-2. Amsterdam, 1720.
Geog 5399.28.20	Met de "Italia" naar de Noordpool. (Nobile, Umberto.) Baarn, 1931.
An 359.33	The method and theory of ethnology. (Radin, Paul.) N.Y., 1933.
An 359.57.15	Méthode de l'ethnographie. (Griaule, Marcel.) Paris, 1957.
An 629.11	Methode der Ethnologie. (Graebner, Fritz.) Heidelberg, 1911.
Geog 807.42	Methode pour etudier la geographie. (Lenglet, N.) Paris, 1742. 7v.
An 359.38.11	Methodik der Völkerkunde. (Mühlmann, W.E.) Stuttgart, 1938.
Geog 808.96	Methodisches Lehrbuch...Erdkund. (Kerp, H.) Bonn, 1896-1904. 3v.
An 359.74.10	Methods and styles in the study of culture. (Edgerton, Robert B.) San Francisco, 1974.
Geog 659.71.10	Le métier de géographe en pays sous-développé. (Santos, Milton.) Paris, 1971.
Geog 4219.43	Metzelaar, A.C. Europa ahoy! De geschiedenis van een zeilschip. Amsterdam, 1943.
Geog 146.3	Meunchener geographische Hefte. Kallmunz. 1,1953+ 17v.
An 2058.76.3	Meunier, Victor. Los antepasados de Adan. Madrid, 1876.
Geog 4508.93.10	Meurer, Julius. Der Bergsteiger in Hochgebirge. Wien, 1893.
Geog 4267.92	Meusel, J.G. Lehrbuch der Statistik. Leipzig, 1792.
Geog 4267.92.5	Meusel, J.G. Lehrbuch der Statistik. Leipzig, 1804.
Geog 3019.43	Mexico (City). Biblioteca Benjamin Franklin. La era de las exploraciones. México, 1943.
An 170.210	Mexico (City). Universidad Nacional. Estudios antropologicos publicados en homenaje al Dr. Manuel Gamio. México, 1956.
Geog 143.5	México (City). Universidad Nacional. Facultad de Filosofiá y Letras. Anuario de geografía. 1,1961 4v.
Geog 143.10	México (City). Universidad Nacional. Instituto de Geografia. Publicaciones. México. 1,1965+
Geog 143.10.5	México (City). Universidad Nacional. Instituto de Geografia. Boletín. México. 1,1969+ 2v.
Geog 4218.30.5	Meyen, F.J.F. Reise um die Erde. Berlin, 1834-35. 2v.
Geog 4110.27	Meyer, H.J. Weltreise. Leipzig, 1907.
Htn Geog 4208.82.3*	Meyer, Hans. Blätter aus mienem Reisetagbuch 1881-1883. Leipzig, 1882.
An 2338.95.5	Meyer, Hermann. Bogen und Pfeil in Central Brasilien. Leipzig, 1895.
Geog 958.33.5	Meyer, J. Meyer's Universum, or Views of...all contries. N.Y., 1852.
Geog 958.33.3	Meyer, J. Meyer's Universum. v.1,4. N.Y., 1850. 2v.
Geog 958.33F	Meyer, J. Meyer's Universum. v.1-6,9. Hildburghausen, 1833- 3v.
Geog 5399.31	Meyer, W. Der Kampf um Nobile. Berlin, 1931.
Htn Geog 4311.15F*	Meyer's Donau-Ansichten. pt.1-2,5,7-11. Hildburghausen, 1838.
Geog 958.33.5	Meyer's Universum, or Views of...all contries. (Meyer, J.) N.Y., 1852.
Geog 958.33.3	Meyer's Universum. v.1,4. (Meyer, J.) N.Y., 1850. 2v.
Geog 958.33F	Meyer's Universum. v.1-6,9. (Meyer, J.) Hildburghausen, 1833- 3v.
Geog 3530.20	Meynier, André. Histoire de la pensée géographique en France, 1872-1969. Paris, 1969.
Geog 5399.12.5	Mezhdu zhuznbiu v Smertbiu. (Al'banov, V.A.) Berlin, 1925.
VGeog 4131.68	Mezhdunarodruji turizm. (Anan'ev, Mikhail A.) Moskva, 1968.
Geog 4129.50	Michael, M. Traveller's quest. London, 1950.
Geog 4219.40.10	Michael, Rudolf. Roman einer Weltreise. Hamburg, 1940.
Geog 815.35.10	Michael Servetus, a translation of his geographical, medical, and astrological writings. (Servetus, Michael.) Philadelphia, 1953.
Geog 4509.53.10	Michel, Aime. Montagnes héroïques. Tours, 1953.
Geog 4308.34	Michelet, J. Sur les chemins de l'Europe. Paris, 1893.
Geog 5839.35	Mid-ice. (Georgi, J.) N.Y., 1935.
Geog 4419.47	Middle East journey. 1. ed. (Goolid-Adams, R.) London, 1947.
Geog 4329.47	Middle eastern journey. (Newman, B.) London, 1947.
Geog 807.00F	Middleton, C.T. System of geography. v.1-2. London, 17- . 4v.
Geog 4309.53.5	Middleton, E.E. The cruise of the Kate. London, 1953.
Geog 4180.19	Middleton, H. Voyage...to Bantum and Maluco Islands. London, 1855.
Geog 4559.35.20	Midget Magellans; great cruises in small ships. (Devine, Eric.) N.Y., 1935.
Geog 4308.86.10	The midnight sun; the tsar and the nihilist. (Buckley, James M.) Boston, 1886.
Geog 4209.38.45	Mielche, H. Let's see if the world is round. London, 1950.
Geog 4019.60	Mier, Waldo de. Escogieron la inquietud. Madrid, 1960.
Geog 5558.50	Miertsching, J.A. Reise-Tagebuch des Missionars. Gnadan, 1855.
Geog 5518.50.15	Miertsching, Johann August. Frozen ships; the Arctic diary of Johann Miertsching, 1850-1854. N.Y., 1967.
Geog 500.80	Migliorini, Elio. Giuda bibliografica allo studio della geografia. Napoli, 1945.
An 359.29.5	Migrations of early culture. (Smith, G.E.) Manchester, 1929.
An 359.15	The migrations of early culture. (Smith, George E.) Manchester, 1915.
Geog 4145.14.5	Miguel Corte Real. (Fernandes, José dos Santos.) Coimbra, 1960.
Geog 3249.28.3	Miguel Servet y la geografia del renacimiento. 3. ed. (Bullon y Fernández, Eloy.) Madrid, 1945.
Geog 4209.66	Mikhailov, Nikolai N. S planetoi vmestve. Moskva, 1966.
Geog 4218.98.10	Mikhailovskii, N.G. Iz dnevnikov krugosvetnogo puteshestviia. Moskva, 1952.
Geog 5559.06	Mikkelsen, E. Conquering the Arctic ice. Philadelphia, 1909.
Geog 5639.57	Mikkelsen, Ejnar. Fra fribytter til embedsmand. København, 1957.
Geog 5559.09.101	Mikkelsen, Ejnar. Lost in the Arctic; being the story of the "Alabama" expedition, 1909-1912. N.Y., 1913.
Geog 5559.09.100	Mikkelsen, Ejnar. Lost in the Arctic; being the story of the "Alabama" expedition, 1909-1912. London, 1913.
Geog 5323.2	Mikkelsen, Ejnar. Mirage in the Arctic. London, 1955.
Geog 5639.60	Mikkelsen, Ejnar. Svundne tider i Østgrønland. København, 1960.
Geog 5559.09.106	Mikkelsen, Ejnar. Tre aar paa Grønlands østkyst. 2. udg. Kjøbenhavn, 1914.
Geog 5559.09.110	Mikkelsen, Ejnar. Two against the ice. London, 1957.
Geog 5323.2.5	Mikkelsen, Ejnar. Ukendt mand til ukendt land. Kjøbenhavn, 1954.
Geog 4209.67.5	Mikosha, Vladislav V. Gody i strany. Moskva, 1967.
Htn Geog 816.33.8*	Mikrokosmos; a little description of the great world. 8th ed. (Heylyn, Peter.) Oxford, 1639.
Htn Geog 816.33.5*	Mikrokosmos; little description. (Heylyn, Peter.) Oxford, 1633.
Htn Geog 816.33.7*	Mikrokosmos; little description. (Heylyn, Peter.) Oxford, 1639.
Htn Geog 816.21.7*	Mikrokosmos: a little description of the great world. 7th ed. (Heylyn, Peter.) Oxford, 1636.
Geog 500.58	Milan. Bibliotheca. Catalogo ragionato della geografica. Milano, 1927.
Geog 143.70	Milan. Università Cattolica del Sacro Cuore. Pubblicazione. Serie 10. Scienze geografico. Milano.

Author and Title Listing

Htn	Geog 4217.97*	Milet-Mureau, M.L.A. Voyage de La Pérouse. Paris, 1797. 4v.
Htn	Geog 4217.97PF*	Milet-Mureau, M.L.A. Voyage de La Pérouse. Atlas. Paris, 1797.
	Geog 4268.50	Military topography of...Europe. (Lavallée, T.) London, 1850.
	An 359.63.30	Miljøets indflydelse på intelligensvaviklingen. (Spelling, K.) København, 1963.
	An 379.73.10	Milkov, Fedor N. Chelovek in landshafty. Moskva, 1973.
	Geog 618.7	Mil'kov, Fedor N. P.I. Rychkov. Moskva, 1953.
	Geog 613.7	Mill, H.R. An autobiography. London, 1951.
	Geog 819.01	Mill, H.R. The international geography. N.Y., 1901.
	Geog 819.01.7	Mill, H.R. The international geography. N.Y., 1907.
	Geog 819.01.6	Mill, H.R. The international geography. N.Y., 1920.
	Geog 819.01.10	Mill, H.R. International geography. 2nd ed. N.Y., 1900.
	Geog 5939.05	Mill, H.R. Siege of the South Pole. London, 1905.
	Geog 5985.258A	Mill, Hugh R. The life of Sir Ernest Shackleton. Boston, 1923.
	Geog 139.25	Mill, Hugh R. The record of the Royal Geographical Society, 1830-1930. London, 1930.
	An 2059.72	Millar, Ronald William. The Piltdown men. London, 1972.
	Geog 665.31	Miller, Émile. Pour qu'on aime la géographie. Montréal, 1921.
	Geog 5318.2	Miller, F. Ahdoolo: the biography of M.A. Henson. 1. ed. N.Y., 1963.
	Geog 3235.19	Miller, K. Die Ebstorfkarte. Stuttgart, 1900.
	Geog 4329.38	Miller, M.S. (Mrs.). Cruising the Mediterranean. N.Y., 1938.
	An 2109.28.5A	Miller, Nathan. The child in primitive society. N.Y., 1928.
	Geog 665.86	Miller, Ronald. Geographical essays in memory of Alan G. Ogilvie. London, 1959.
	Geog 4218.57.3	Miller, Thomas. Over five seas and oceans. N.Y., 1894.
	Geog 4559.33	A million miles in sail. (McCulloch, J.H.) N.Y., 1933.
	An 379.63	Mills, C.A. World power and shifting climates. Boston, 1963.
	Geog 818.64	Milner, Thomas. The gallery of geography, a pictorial and descriptive tour of the world. London, 1864. 2v.
	Geog 758.50	Milner, Thomas. A universal geography in four parts. London, 1850.
	Geog 5316.1.8	Min gronlandske ungdom. (Freuchen, P.) Kjøbenhavn, 1959.
	Geog 5839.12.2	Min rejsedagbog. 2. ugd. (Rasmussen, Knud.) Kjøbenhavn, 1921.
	Geog 4207.69.5	Min son på galejan. (Wallenberg, Jacob.) Stockholm, 1835.
	Geog 4207.69.9	Min son på galejan. 2. uppl. (Wallenberg, Jacob.) Stockholm, 1913.
	An 2109.11.3A	The mind of primitive man. (Boas, Franz.) N.Y., 1911.
	An 2109.29.15	The mind of primitive man. (Boas, Franz.) N.Y., 1929.
	An 2109.38.5	The mind of primitive man. (Boas, Franz.) N.Y., 1938.
	An 2109.46.5	The mind of primitive man. (Boas, Franz.) N.Y., 1946.
	An 2109.29.7	Mind of the savage. (Allier, Raoul.) London, 1929.
	Geog 5328.3	Mindeudgave. (Rasmussen, Knud.) Kjøbenhavn, 1934-35. 3v.
	Geog 5399.46	Mineev, A.I. Ostrov Vrangelia. Moskva, 1946.
	Geog 4679.64	Minerva Reef. 1. American ed. (Ruhen, Olaf.) Boston, 1964.
	Geog 4308.98	Minhas viagens. (Ferro, Adersón.) Ceara, 1898.
	Geog 3240.26.5	Miquel, André. La géographie humaine du monde musalman jusqu'au milieu du 11e siècle. Paris, 1967.
	Geog 4180.31	Mirabilia descripta...wonders of the East. (Jordanus de Saxonia.) N.Y., 1964?
	Geog 5323.2	Mirage in the Arctic. (Mikkelsen, Ejnar.) London, 1955.
	An 3409.30	Miranda Pinto, O. Contribution à la morphologie comparée des crêtes papillaires. Thèse. Lyon, 1930.
	An 359.58	Miroglio, Abel. La psychologie des peuples. Paris, 1958.
	An 185.1.2	Mirror for man. (Kluckhohn, Clyde.) N.Y., 1949.
	Geog 5321.1.17	Mirsky, J. Elisha Kent Kane and the seafaring frontier. 1. ed. Boston, 1954.
	Geog 4139.64	Mirsky, J. The great Chinese travelers, an anthology. N.Y., 1964.
	Geog 5209.34.2	Mirsky, J. To the Arctic! N.Y., 1948.
	Geog 5209.34	Mirsky, J. To the North: The story of arctic exploration from earliest times to the present. N.Y., 1934.
	Geog 4228.73	Mis impresiones y mis vicisitudes en mi viaje a Europa. (Barra de Cobo, Maipina de la.) Buenos Aires, 1878.
	An 170.296	Miscelanea de estudios dedicados a Fernando Ortiz por sus discipulos. La Habana, 1955-57. 3v.
	Geog 3550.17	Miscellanea di storia delle esplorazioni. Genova, 1975.
	Geog 4308.22.5	Miscellaneous observations and opinions on the continent. (Duppa, Richard.) London, 1825.
	Geog 110.2.17	Miscellaneous papers. v.1-6. (International Geographical Congress, 7th, Berlin.) Berlin, 1899.
	Geog 40.9	Miscellaneous publications. (Congresso Brasiliero de Geografia.)
	An 359.54.10	Missenard, André. A la recherche de l'homme. Paris, 1954.
	An 359.57.30	Missenard, André. In search of man. N.Y., 1957.
	Geog 5700.12	Missionaer i Grønland, Henric Christopher Glahns. (Glahn, Henrik C.) København, 1921.
	Geog 5700.6	Missionary records; northern countries. (Religious Tract Society.) London, 1839.
	Geog 3530.13	Les missions françaises; causeries géographiques. (Jolly, Raoul.) Paris, 1894-96. 2v.
	Geog 4181.123	Missions to the Niger. (Bovill, E.W.) Cambridge, Eng., 1964. 4v.
	Geog 5559.42	Mit dem Fieseler-Storch ins Nordpolarmeer. (Herrmann, Ernst.) Berlin, 1942.
	Geog 4308.48.11	Mit dem jüngsten Sohn König Ludwigs I. als Reisebegleiter nach Spanien 1848-1849. (Ow, Max von.) München, 1967.
	Geog 4309.02.7	Mit der Kraft; Automobilia. (Bierbaum, Otto Julius.) Berlin, 1906.
	Geog 4209.62.10	Mit einem deutschem Pass. (Lattmann, D.) München, 1964.
	Geog 4509.48.13	Mit gleichlichen Augen. 3. Aufl. (Pause, Walter.) München, 1951.
	Geog 4309.32.5	Mit Kamera und Schreibonaschine durch Europa. (Grisar, Erich.) Berlin, 1932.
	Geog 5160.14	Mit Kind und Kegel in die Arktis. (Ross, Colin.) Leipzig, 1934.
	Geog 4209.09.6	"Mit meinen Augen"; Fahrten durch die lateinische Welt. (Ewers, Hanns H.) München, 1914.
	Geog 4559.53	Mitchell, C. Beyond horizons. 1. ed. London, 1953.
	Geog 4239.54	Mitchell, Carleton. Passage east. London, 1954.
	Geog 819.28	Mitchell, J. Leslie. Hauno, or Future of exploration. London, 1928.
	Geog 3019.34.10	Mitchell, J.L. Earth conquerors. N.Y., 1934.
	Geog 4215.19.10	Mitchell, Mairin. Elcano. London, 1958.
	Geog 4145.89.5	Mitchell, Mairin. Friar Andrés de Urdaneta, O.S.A. London, 1964.
	Geog 818.39	Mitchell, S.A. Accompaniment to map of world. Philadelphia, 1839.
	Geog 818.42.5	Mitchell, S.A. An accompaniment to Mitchell's map. Philadelphia, 1842.
	Geog 818.44	Mitchell, S.A. An accompaniment to Mitchell's map. Philadelphia, 1844.
	Geog 818.45	Mitchell, S.A. An accompaniment to Mitchell's map. Philadelphia, 1845.
	Geog 818.46	Mitchell, S.A. An accompaniment to Mitchell's map. Philadelphia, 1846.
	Geog 5317.10.5	Mitchell, William. General Greely; the story of a great American. N.Y., 1936.
	Geog 3019.75F	The Mitchell Beazley world atlas of exploration. (Newby, Eric.) London, 1975.
	Geog 4218.85.5	Mitford, R.C.W.M. Orient and Occident. London, 1888.
	An 99.33	Mitra Panchanan. A history of American anthropology. Calcutta, 1933.
	Geog 4181.34	Mittall, Zelia. New light on Drake. London, 1914.
	Geog 28.3.50	Mitteilungen. (Berlin. Universität. Verein der Studierenden der Geographie.) Berlin. 1-2,1915-1918
	Geog 28.8	Mitteilungen. (Frankische Geographische Gesellschaft.) Erlangen. 1,1954+ 12v.
	Geog 215.1	Mitteilungen. (Gesellschaft für Erdkunde und Kolonialwesen zu Strassburg.) Strassburg.
	Geog 85.310	Mitteilungen der geographisch-Ethnographischen Gesellschaft in Basel. (Geographisch-Ethnographische Gesellschaft in Basel.) Basel. 1-8 4v.
	Geog 759.35	Mitteleuropa. Potsdam, 1935.
	Geog 5559.23	Mittelholzer, W. Im Flugzeug dem Nordpol entgegen. Zürich, 1924.
	Geog 4328.59.5	Das Mittelmeer. (Böttger, C.) Leipzig, 1859.
	Geog 4329.53.15	Das Mittelmeer. (Cordan, W.) Düsseldorf, 1953.
	Geog 4329.37F	Das Mittelmeer. (Hürlimann, M.) Zürich, 1937.
	Geog 1595.7	Mittelmeer. 2. Aufl. (Baedeker, publishers.) Leipzig, 1934.
	Geog 4329.06	Mittelmeerbilder. (Fischer, T.) Leipzig, 1906.
	Geog 4329.06.2	Mittelmeerbilder. (Fischer, T.) Leipzig, 1908.
	Geog 5970.15	Mitterling, Philip. America in the Antarctic to 1840. Urbana, 1959.
Htn	Geog 5375.72*	Mitternächtige Schiffarth von den Herrn Staden. (Röslin, H.) Oppenheim, 1611.
NEDL	Geog 45.5	Mittheilungen. (Deutscher und Österreichischer Alpenvereins.) Salzburg. 1-25,1875-1899 25v.
	Geog 45.5	Mittheilungen. (Deutscher und Österreichischer Alpenvereins.) Salzburg. 1875-1929 24v.
	Geog 85.225	Mittheilungen. (Geographische Gesellschaft in Hamburg.) Hamburg. 1-61 47v.
	Geog 157.50	Mittheilungen. (Österreichischer Alpen-Verein, Vienna.) Wien. 1-2,1863-1864 2v.
	Geog 258.1	Mittheilungen. (Vienna. K.K. Geographische Gesellschaft.) Wien. 1,1857+ 81v.
NEDL	Geog 144.1.2F	Mittheilungen. Berlin, 1908. 6v.
NEDL	Geog 144.1	Mittheilungen. Berlin. 1-36,1888-1929 36v.
	Geog 85.225.2	Mittheilungen. Register. 1-60, 1873-1972. (Geographische Gesellschaft in Hamburg.) Hamburg, 1973.
NEDL	Geog 144.1.3	Mittheilungen. v.2. Berlin, 1915.
	Geog 146.1.5	Mittheilungen der geographische Gesellschaft in München. (Munich.) München. 1,1906+ 33v.
	Geog 5180.12.5	Moccasins and their relation to Arctic footwear. (Hatt, G.) Lancaster, 1915?
	Geog 4206.45.5	Mocquet, J. Travels and voyages into Africa. London, 1696.
Htn	Geog 4206.45*	Mocquet, J. Voyages en Afrique, Asie. Rouen, 1645.
	Geog 659.71.5	Models in geography. (Chorley, Richard J.) London, 1971.
	Geog 4019.52	Modern action and adventure. (Lamb, Geoffrey F.) London, 1952.
	Geog 5170.17	Modern Arctic exploration. (Seidenfaden, G.) Boston, 1939.
	Geog 665.42	Modern exploration. (Ward, F.K.) London, 1945.
	Geog 577.58	Modern gazeteer. (Salmon, T.) London, 1758.
	Geog 577.58.2	Modern gazeteer. (Salmon, T.) London, 1759.
	Geog 577.58.4	Modern gazeteer. (Salmon, T.) London, 176-.
	Geog 577.58.6	Modern gazeteer. (Salmon, T.) London, 1782.
	Geog 819.11	Modern geography. (Newbigin, Marion I.) London, 1911.
	Geog 819.11.1	Modern geography. (Newbigin, Marion I.) N.Y., 1911.
	Geog 818.02.90	Modern geography. (Pinkerton, J.) London, 1802. 2v.
	Geog 818.04.5	Modern geography. (Pinkerton, J.) London, 1811. 2v.
	Geog 818.04	Modern geography. (Pinkerton, J.) Philadelphia, 1804. 2v.
	Geog 819.13.5	Modern geography for high schools. (Salisbury, Rollin D.) N.Y., 1913.
	Geog 807.39.5F	Modern history. 3rd ed. (Salmon, Thomas.) London, 1744-1746. 3v.
	Geog 4509.33	Modern mountaineering. (Abraham, G.D.) London, 1931.
	Geog 4169.45.5	Modern travel. (Stood, Frederick T.) London, 1946.
NEDL	Geog 520.25	The modern traveller. (Conder, Josiah.) London, 1830 30v.
	Geog 4709.48	A modern treasure hunter. (Wilkins, H.T.) London, 1948.
	Geog 4168.31	The modern voyager and traveller. (Adams, W.) London, 1831. 4v.
	Geog 5855.10	Moderna eskimåer. (Elgström, Ossian.) Stockholm, 1916.
	An 359.04	Moderne Rassentheorien. (Yertz, Friedrich O.) Wien, 1904.
	An 98.96	Moderne Völkerlunde. (Achelis, Thomas.) Stuttgart, 1896.
	Geog 665.52	Mogey, J.M. The study of geography. London, 1950.
	Geog 614.7	Moj ojciec. (Nałkowska, Z.) Warszawa, 1953.
	Geog 614.8	Moja ojciec. Wyd. 2. (Nałkowska, Z.) Warszawa, 1955.
	Geog 4677.99	Molen, Sytze Jan van der. Goud in de golven. 's-Gravenhage, 1965.
	Geog 4228.86	Molinari, M.G. Au Canada et aux montagnes. Paris, 1886.
	VAn 99.66.10	Moliński, Bogdan. Historia, osobowość, sztuka. Wyd. 1. Warszawa, 1966.
	Geog 817.01.5F	Moll, H. The compleat geographer. London, 1723.
Htn	Geog 817.01F*	Moll, H. System of geography. London, 1701.
	Geog 4208.64	Mompou, José. De la Habana a Madrid. Habana, 1865.
	Geog 4309.31.10	Mon Europe. (Duhamel, Georges.) Paris, 1931.
	Geog 4145.29	Mon metier l'aventure. (Fournier-Aubry, Fernand.) Paris, 1953.
	Geog 4209.60.25	Moncharmont, Simone. La croisière du Ly-Kau. Paris, 1960. 2v.
	Geog 4248.54	Les mondes nouveaux. (Niboyet, P.) Paris, 1854.
	Geog 759.23	Il mondo e l'Italia. 3. éd. v.1-2. (Gribandi, P.) Torino, 1923.
	Geog 4029.15	Monfreid, Henri de. Le feu de Saint-Elme. Paris, 1974.
	Geog 4419.30	Monfreid, Henri de. Pearls, arms and hashish; pages from the life of a Red Sea navigator. N.Y., 1930.
	Geog 4419.34.4	Monfreid, Henri de. La poursuite du "Kaipan". Paris, 1934.

220

Author and Title Listing

Geog 4419.34.3 Monfreid, Henri de. Les secrets de la Mer Rouge. Paris, 1935.
Geog 4419.34 Monfreid, Henri de. Secrets of the Red Sea. London, 1934.
Geog 4419.38.15 Monfreid, Henri de. Le trésor du pélerin; roman. 12. ed. Paris, 1938.
Geog 4218.59 The Monitor; or, jottings of a N.Y. merchant during a trip around the globe. (Hoffman, William.) N.Y., 1863.
Geog 4269.59.10 Monkhouse, Francis. A regional geography of western Europe. London, 1959.
Geog 4209.72.5 Mon'ko, Aleksei M. Potoratiny. Volgograd, 1972.
Geog 5700.10 Monnier, Edouard. Histoire de la mission chrétienne au Groënland. Thèse. Strasbourg, 1853.
Geog 5907.15 Monograph. (United States. Naval Support Force, Antarctica. History and Research Division.) Washington. 1,1971+
Geog 145.4 Monographs in geography. Enfield, Eng. 1,1973//
Geog 4208.85 Monséjour en France...mes pelerinages a Rome. (Frachebourd, J.J.) Fribourg, 1885.
Geog 4328.20 Monson, W.J.M. Extracts from a journal. London, 1820.
Geog 4429.52A Monsoon seas. (Villiers, Alan J.) N.Y., 1952.
Geog 4509.56.20 Mont Blanc to Everest. (Rébuffat, Gaston.) London, 1956.
Geog 4509.56.10 La montagne. (Herzog, Maurice.) Paris, 1956.
Geog 39.1.10 La montagne. Paris. 1904+ 38v.
Geog 4509.41.10 Montagne: des terrasses aux arêtes. (Rondet, C.) Paris, 1941.
Geog 4509.65 La montagne et ses hommes. (Boutron, Michel.) Paris, 1965.
Geog 144.10 Montagnes du monde. Genève. 1,1953+
Geog 4509.53.10 Montagnes héroïques. (Michel, Aime.) Tours, 1953.
An 187.9 Montagu, Ashley. Anthropology and human nature. Boston, 1957.
An 2109.37.3 Montagu, Ashley. Coming into being among the Australian aborigines. London, 1937.
An 170.275.5 Montagu, Ashley. The concept of race. N.Y., 1964.
An 2109.68 Montagu, Ashley. The concept of the primitive. N.Y., 1968.
An 170.275.10 Montagu, Ashley. Culture: man's adaptive dimension. N.Y., 1968.
An 359.65.5 Montagu, Ashley. The human revolution. Cleveland, 1965.
An 359.65 Montagu, Ashley. The idea of race. Lincoln, 1965.
An 359.45.25 Montagu, Ashley. An introduction to physical anthropology. Springfield, 1945.
An 3209.60 Montagu, Ashley. An introduction to physical anthropology. 3. ed. Springfield, 1960.
An 359.69.30 Montagu, Ashley. Man, his first two million years. N.Y., 1969.
An 359.61.40 Montagu, Ashley. Man in process. 1. ed. Cleveland, 1961.
An 359.42.10A Montagu, Ashley. Man's most dangerous myth: the fallacy of race. N.Y., 1942.
An 359.42.12A Montagu, Ashley. Man's most dangerous myth. 2. ed. N.Y., 1945.
An 359.64.45 Montagu, Ashley. Man's most dangerous myth. 4. ed. Cleveland, 1964.
An 359.74.15 Montagu, Ashley. Man's most dangerous myth. 5. ed. N.Y., 1974.
Geog 4327.38 Montague, J. Voyage...round the Mediterranean. London, 1799.
An 359.12.15 Montalto de Jesus, C.A. Oriente modernisado. Lisboa, 1912.
An 359.33.11 Montandon, Georges. La race, les races. Paris, 1933.
Htn Geog 4341.73* Montano, B.A. Itinerarium beniamini tudelemsis. Antwerpiae, 1575.
Geog 4209.66.15 Monteil, Vincent. Soldat de fortune. Paris, 1966.
NEDL Geog 4158.33 Montemont, A. Histoire...des voyages. Paris, n.d. 46v.
Geog 808.33 Montenegro Colon, Feliciano. Geografia general para el uso de la juventud de Venezuela. Caracás, 1833-1837. 4v.
An 136.5.70 Montes, Santiago. Claude Lévi-Strauss. 1. ed. San Salvador, 1971.
Geog 4181.48 Montesinos, Fernando. Memorias antiguas historiales del Peru. London, 1920.
An 3309.05 Montessori, Maria. Caratteri fisici delle Giovani donne del Lazio. Roma, 1905.
Geog 4308.69.9 Montgomery, J.E. Our admiral's flag abroad. N.Y., 1869.
Geog 4709.57.5 Monti, Mario. I pirati. Milano, 1957.
Geog 4502.5.21 Montis, R. Kampf um den Berg. München, 19- .
Geog 4228.17.5 Montulé, E. de. Travels in America. 1st ed. Bloomington, 1951.
Geog 4228.16 Montulé, E. de. Voyage en Amerique. Paris, 1821. 2v.
Geog 4228.17 Montulé, E. de. A voyage to North America and West Indies in 1817. London, 1821.
Geog 4309.41 Moody, M.N. Trails of two travelers. N.Y., 1941.
An 170.274 Moore, Frank. Readings in cross-cultural methodology. New Haven, 1966.
Geog 4307.86.6 Moore, J. View of society and manners. Boston, 1792.
Geog 4307.86.10 Moore, J. View of society and manners. Paris, 1803. 2v.
Geog 4307.86.7 Moore, J. View of society and manners. 5th ed. Dublin, 1793. 2v.
Geog 4307.86.8 Moore, J. View of society and manners. 6th ed. London, 1786. 2v.
Geog 5559.09.8 Moore, J.H. Peary's discovery of the North Pole. Washington, 1910.
Geog 4308.24 Moore, John. A journey from London to Odessa. Paris, 1833.
Geog 3520.13 Moorehead, Alan. The fatal impact. London, 1966.
An 359.29 Morand, Paul. Black magic. N.Y., 1929.
Geog 4349.36.3 Morand, Paul. La route des Indes. Paris, 1936.
An 359.52.5 Morant, G.M. The significance of racial differences. Paris, 1952.
Geog 4509.58 Moravec, Fritz. Weisse Berge. Wien, 1958.
An 99.70 Moravia, Sergio. La scienza dell'uomo nel settecento. Bari, 1970.
An 136.5.20 Moravio, Sergio. La ragione nascosta. Firenze, 1969.
An 2109.36.7 Morceaux choisis. 2e éd. (Lévy-Bruhl, L.) Paris, 1936.
Geog 4209.32.15 Mordaunt, E. (pseud.). Rich tapestry. N.Y., 1932.
Htn Geog 816.80.4* Morden, Robert. Geography rectified. 4th ed. London, 1700.
Geog 4309.07.13 More society recollections, by an English officer. London, 1908.
Geog 4521.10 More than mountains. (Jackson, John.) London, 1955.
An 2159.73.20 Moreau, Marcel. Les civilisations des étoiles. Paris, 1973.
Geog 3108.74 Moreau de Jonnés, A. Estudios prehistóricos. Sevilla, 1874.
Geog 3048.73 Moreau de Jonnes, A.C. L'ocean des anciens. Paris, 1873.
Geog 4181.74 Moreland, W.H. Peter Floris. London, 1934.
Geog 4181.66 Moreland, W.H. Relations of Golconda in the early 17th century. London, 1931.
Htn Geog 3550.10F* Morelli, Iacopo. Dissertazione intorno ad alcuni viaggiatori eruditi veneziani. Venezia, 1803.
Geog 5070.37 Moreno Fuentes, José. Las regiones heladas de los polos norte y sur. Madrid, 1884.
Htn Geog 3205.40.11* Mores, leges et ritus omnium gentium. (Boehme, J.) Lugduni, 1561.
Htn Geog 3205.40.15* Mores, leges et ritus omnium gentium. (Boehme, J.) Lugduni, 1576.
Htn Geog 3205.40.17* Mores, leges et ritus omnium gentium. (Boehme, J.) Lugduni, 1582.
Geog 3055.46 Moreux, T. L'Atlantide a-t-elle existé? Paris, 1924.
Geog 4300.27 Morford, Henry. Morford's short-trip guide to Europe. N.Y., 1872.
Geog 4308.65.9 Morford, Henry. Over-sea. N.Y., 1867.
Geog 4300.27 Morford's short-trip guide to Europe. (Morford, Henry.) N.Y., 1872.
Geog 4180.39 Morga, A. de. Phillipine Islands, Moluccas. London, 1868.
Geog 4181.140 Morga, Antonio de. Sucesos de las Islas Filipinas. Cambridge, Eng., 1971.
An 359.09.15 Morgan, L.H. Ancient society. Chicago, 1909.
An 358.78.3A Morgan, L.H. Ancient society. N.Y., 1878.
An 359.64.40 Morgan, Lewis H. Ancient society. Cambridge, 1964.
Geog 665.95 Mori, Assunto. Scritti geografici, scelti e ordinati. Pisa, 1960.
Geog 5399.36 Moria Sovetskoi Arktiki. (Vize, Vladimir I.) Leningrad, 1936.
An 359.73 Morin, Edgar. La paradigme perdu: la nature humaine. Paris, 1973.
Geog 189.2.8 Morissonneau, Christian. La société de géographie de Québec, 1877-1970. Québec, 1971.
Geog 5399.34.5 Morozov, S. Put' skvoz l'dy. Moskva, 1934.
Geog 5559.56 Morozov, S.T. U poslednikh parallelei. Moskva, 1956.
Geog 4329.63 Morpurgo, J.E. The road to Athens. London, 1963.
Geog 4208.31.7 Morrell, Abby J. Narrative of voyage to Ethiopia. N.Y., 1833.
Geog 4208.31.5 Morrell, B. Narrative of four voyages. N.Y., 1832.
Geog 4308.71.12 Morris, Caspar. Letters of travel. Philadelphia, 1896. 2v.
An 358.88.5 Morris, Charles. The Aryan race: its origin and its achievements. Chicago, 1888.
Geog 4207.55 Morris, Drake. Travels of...voyage at sea. London, 1775.
Geog 4209.63.10 Morris, J. Cities. London, 1963.
Geog 4209.72.10 Morris, James. Places. London, 1972.
Geog 4709.69 Morris, Roland. Island treasure: the search for Sir Cloudesley Shovell's flagship "Association". London, 1969.
Geog 3019.32F Morrison, E.R. Explorographs. Cleveland Heights, 1932.
Geog 4308.87.5 Morrison, Leonard A. Among the Scotch-Irish. Boston, 1891.
Geog 4308.87.2 Morrison, Leonard A. Rambles in Europe. Boston, 1887.
Geog 4308.87.3 Morrison, Leonard A. Rambles in Europe. Boston, 1887.
Geog 4327.94 Morritt, John B.S. The letters of...journeys in Europe and Asia Minor. London, 1914.
An 2338.85.2 Morse, E.S. Additional notes on arrow-release. Salem, 1922.
An 2338.85 Morse, E.S. Ancient and modern methods of arrow-release. Salem, 1885.
An 2058.84 Morse, E.S. Man in the tertiaries. Salem, 1884.
Geog 578.21 Morse, Jedidiah. New universal gazeteer of the known world. 3. ed. New Haven, 1821.
Htn Geog 578.02* Morse, Judith. A new gazeteer of the Eastern continent. Charlestown, 1802.
Geog 818.22 Morse, S.E. New system of modern geography. Boston, 1822.
Htn Geog 4309.30.5* Morse, W.I. Nordic trails. Boston, 1930.
Geog 4309.30.6 Morse, W.I. Nordic trails. Boston, 1930.
Geog 4309.26A Morse, William I. The diary of a musketeer. Boston, 1926.
Geog 4309.22 Morse, William I. Seeing Europe backwards. Boston, 1922.
Htn Geog 4329.27.10* Morse, William I. Sicilian days and other journeys round the Mediterranean and Adriatic. Boston, 1927.
Geog 4309.24 Morse, William I. Twisting trails in the Auvergnes, Cevennes, Alps of Provence. Boston, 1924.
An 2059.03.7 Mortillet, G. de. Musée prehistorique. 2. éd. Paris, 1903.
Geog 4418.54.3 Mortimer, F.L.B. (Mrs.). Far off, or Asia and Australia described with anecdotes and illustrations. N.Y., 1854.
Geog 4181.57 Mortoft, F. Francis Mortoft; his book...1658-59. London, 1925.
Geog 4329.36.6 Morton, H.V. In the steps of St. Paul. London, 1936.
Geog 4329.36.7 Morton, H.V. In the steps of St. Paul. N.Y., 1936.
Geog 4419.38 Morton, H.V. Through the lands of the Bible. London, 1938.
An 3538.44 Morton, S.G. Crania Aegyptiaca. Philadelphia, 1844.
An 3538.39 Morton, S.G. Crania Americana. Philadelphia, 1839.
Geog 4306.17.1F Moryson, Fynes. An itinerary. Amsterdam, 1971.
Htn Geog 4306.17F* Moryson, Fynes. An itinerary. London, 1617.
Geog 4305.91A Moryson, Fynes. An itinerary containing his 10 yeeres travell. Glasgow, 1907. 4v.
Geog 4306.17.3 Moryson, Fynes. Shakespeare's Europe. London, 1903.
Geog 4306.17.4 Moryson, Fynes. Shakespeare's Europe: a survey of the condition of Europe at the end of the 16th century. 2nd ed. N.Y., 1967.
Geog 4182.66 Moscovische reyse 1664-1665. (Witsen, Nicolaas.) 's-Gravenhage, 1966-67. 3v.
Geog 4679.59 Moscow, Alvin. Collision course: the Andrea Doria and the Stockholm. N.Y., 1959.
An 608.87 Moscow. Dashkovskii Etnograficheskii Muzei. Sistematicheskoe opisanie kollektsii Dashkovskago etnograficheskago muzeia. v.4. Moskva, 1895.
Geog 500.85 Moscow. Gosudarstvennyi Biblioteka SSSR imeni V.I. Lenin. Glazami sovetskikh liudel. Moskva, 1956.
Geog 3590.51 Moscow. Gosudarstvennyi Biblioteka SSSR Imeni V.I. Lenin. Russkie geografii i puteshestvenniki. Moskva, 1955.
Geog 3590.82 Moscow. Universitet. Sovetskaia geografiia v period stroitel'stva kommunizma. Moskva, 1963.
An 359.64.10 Moskovskoe Obshchestvo Ispytatelei Prirody. Sovremennaia antropologiia. Moskva, 1964.
Geog 5399.61.5 Moskva - Severnyi Polius. (Reut, V.F.) Moskva, 1961.
Geog 5558.75.15PF Moss, Edward L. Shores of the Polar Sea. London, 1878.
Geog 5070.70 Mostly mischief: voyages to the Arctic and to the Antarctic. (Tilman, Harold William.) London, 1966.
Geog 3550.2 Mostra dei navigatori veneti del quattrocento e del cinquecento. (Venice. Biblioteca Nazionale Marciana.) Venezia, 1957.
Geog 4309.33 Motering abroad. (American Automobile Association. Foreign.) N.Y., 1933.
Geog 4309.10A Motor journeys. (Hale, Louise (Closser).) Chicago, 1912.
Geog 4328.34 Mott, V. Travels in Europe and the East. N.Y., 1842.

Author and Title Listing

Geog 4238.79 Moulton, Francis D. At sea in the Celtic from January 23rd to February 11th, 1879. n.p., 1879.
Geog 4308.76 Moulton, Louise. Random rambles. Boston, 1881.
Geog 4509.10 Mountain adventures at home and abroad. (Abraham, G.D.) London, 1910.
Geog 4509.23.15 Mountain climbing. (Collins, F.A.) London, 1924.
Geog 4508.97 Mountain climbing. (Wilson, E.L.) N.Y., 1897.
Geog 4509.20 Mountain craft. (Young, G.W.) London, 1920.
Geog 4508.93 Mountain exploration. (Balch, E.S.) Philadelphia, 1893.
Geog 4512.535 Mountain jubilee. (Lunn, A.) London, 1943.
Geog 4509.22 Mountain madness. (Hamilton, H.) London, 1922.
Geog 4509.20.10 Mountain memories. (Conway, W.M.C.) N.Y., 1920.
Geog 4509.47 The mountain top. (Smythe, F.S.) London, 1947.
Geog 4509.41.17 The mountain vision. (Smythe, F.S.) London, 1942.
Geog 4512.1 The mountain way. (Irving, R.L.G.) London, 1938.
Geog 144.9 The mountain world. London. 1,1953+ 9v.
Geog 145.1.10 The mountaineer.
Geog 145.1 The mountaineer. Seattle. 1-9,1907-1916 2v.
Geog 4509.41.5 Mountaineering. (Peacocke, T.A.H.) London, 1941.
Geog 4500.15 Mountaineering. (Smith, J.A.) Cambridge, 1955.
Geog 4509.34.5 Mountaineering. (Spencer, S.) London, 19- .
Geog 4508.93.5 Mountaineering. (Wilson, Claude.) London, 1893.
Geog 4509.41.7 Mountaineering. 3. ed. (Peacocke, T.A.H.) London, 1953.
Geog 4509.20.5 Mountaineering art. (Raeburn, H.) N.Y., 1920.
Geog 4500.16 Mountaineering literature in the Harvard College Library. (Harvard University. Library.) n.p., 1931.
Geog 4509.31 Mountains and men. (Robbins, L.H.) N.Y., 1931.
Geog 4209.65 Mountains and water. (Rand, Christopher.) N.Y., 1965.
Geog 4509.25 The mountains of youth. (Lunn, A.H.M.) London, 1925.
Geog 5209.53 Mountevans, C.R.G.R.E. Arctic solitudes. N.Y., 1953.
Geog 5939.55 Mountevans, E. The Antarctic challenged. London, 1955.
Geog 5939.50 Mountevans, E. The desolate Antarctic. London, 1950.
Geog 5060.18 Mountfield, David. A history of polar exploration. London, 1974.
Geog 145.2F Le mouvement geographique. Bruxelles. 1884-1905 23v.
Geog 4209.39.20A Moving on; the romance of travel. (Harris, N.D.) Chicago, 1939.
Geog 5209.67 Mowat, Farley. The polar passion: the quest for the North Pole. Toronto, 1967.
Geog 4679.61 Mowat, Farley. The serpent's coil. 1. United States ed. Boston, 1961.
Geog 4239.38 Moyne, W.E.G. Atlantic circle. London, 1938.
Geog 4308.57.7 Mr. Dunn Browne's experiences in foreign parts. (Fisk, Samuel.) Boston, 1857.
Geog 4308.57.8 Mr. Dunn Browne's experiences in foreign parts. (Fisk, Samuel.) Boston, 1857.
Geog 4307.65.10 Mrs. Bousquet's diary, 1765. (Bousquet, (Mrs.).) Norwich, 1927.
Geog 3055.67 Muck, O.H. Atlantis-gefunden. Stuttgart, 1954.
An 359.32 Muckermann, H. Rassenforschung und Volk der Zukunft. Berlin, 1932.
Geog 4678.59 Mudie, Ian. Wreck of the Admella. London, 1967.
An 3208.38 Mudie, R. Man, in his physical structure and adaptations. London, 1838.
An 170.278 Mühlemann, Wilhelm Emil. Kulturanthropologie. Köln, 1966.
An 359.38.11 Mühlmann, W.E. Methodik der Völkerkunde. Stuttgart, 1938.
An 358.79.5 Müller, F. Allgemeine Ethnographie. Wien, 1879.
Htn Geog 5220.19* Müller, G.F. Voyages from Asia to America. London, 1761.
Htn Geog 5367.61.3* Müller, G.F. Voyages from Asia to America. 2. ed. London, 1764.
Geog 578.31 Müller, Johan Wilhelm. Lexicon manuale. Lipsiae, 1831.
Geog 4145.68 Mueller, Martin. Julius von Payer. Stuttgart, 1958.
Geog 1527.15 München. (Baedeker, publishers.) Leipzig, 1921.
Geog 1527.2 München und Südbayern. 39. Aufl. (Baedeker, publishers.) Leipzig, 1935.
Geog 1527.5 München und Umgebung. Augsburg. (Baedeker, publishers.) Leipzig, 1935.
Geog 1527.18 München und Umgebung. 4. Aufl. (Baedeker, publishers.) Freiburg, 1960.
Geog 4304.95 Muenzer, Hieronymus. Voyage aux Pays-Bas, 1495. Bruxelles, 1942.
Geog 4209.64.20 Muhlethader, J. Toutes voiles dehors. Genève, 1964.
Geog 5558.81.55 Muir, John. The cruise of the Corwin. Boston, 1917.
Geog 659.74.5 Mukitanov, Naurzbai K. Problema tselostnosti v fizicheskoi geografii. Alma-Ata, 1974.
Geog 4182.4 Mulert, F.E. De reis van Mr. Jacob Roggeveen. 's-Gravenhage, 1911.
Geog 4219.51.10 Mulhauser, G.H.P. The cruise of the Amaryllis. London, 1951.
Geog 500.32 Muller, F. Topographie ancienne - catalogue a prix marquès de cartes anciennes. Amsterdam, 1896.
Geog 4309.29 Muller, Julius W. "Ever thine". N.Y., 1928.
Geog 4182.1 Muller, S. De reis van Jan C. May. 's-Gravenhage, 1909.
Geog 5839.59.5 Munch, Ebbe. Strejftog i Nord. København, 1959.
Geog 4209.60.35 El mundo a través de sus grandes estadistas. (Portes Gil, Emilio.) México, 1960.
Geog 4308.37 Mundt, T. Spaziergänge und Weltfahrten. Altona, 1838-1839. 3v.
Geog 806.87 Mundus mirabilis tripartitus. (Happelius, E.G.) Ulm, 1687. 3v.
Geog 4181.17 Mundy, P. Travels of Peter Mundy in Europe and Asia, 1608-67. v.1-5. Cambridge, 1907-36. 6v.
Geog 4207.76.15 Munford, K. John Ledyard: an American Marco Polo. Portland, 1939.
Geog 146.1 Munich. Geographische Gesellschaft Jahresbericht. 1869-1902 7v.
Geog 146.1.5 Munich. Mittheilungen der geographische Gesellschaft in München. München. 1,1906+ 33v.
Geog 1527.10 Munich and its environs. (Baedeker, publishers.) Hamburg, 1950.
Geog 1527.12 Munich and its environs. 2. ed. (Baedeker, publishers.) Hamburg, 1956.
Geog 5516.19 Munk, J. Navigatio septentrionalis. Kjøbenhavn, 1723.
Geog 5516.19.3 Munk, J. Navigatio septentrionalis. Kjøbenhavn, 1883.
Geog 4559.29 Munro, D.J. The roaring forties and after. London, 1929.
An 358.97 Munro, R. Prehistoric problems. Edinburgh, 1897.
An 359.12.5 Munro, Robert. Palaeolithic man. Edinburgh, 1912.
An 359.27.11A Muntz, Earl E. Race contact. N.Y., 1927.
Geog 3590.60 Murator, M.V. Navatrechu apasnostiam. Moskva, 1956.
Geog 5300.25 Muratov. M.V. Pervye razvedchiki velikogo puti. Moskva, 1943.
Geog 4209.32 Murchie, Guy. Men on the horizon. Boston, 1932.
An 2109.49.15 Murdock, George P. Our primitive contemporaries. N.Y., 1949.
An 359.65.20 Murdock, George Peter. Culture and society. Pittsburgh, 1965.
An 5.33 Murdock, George Peter. Ethnographic atlas. Pittsburg, 1967.
An 359.49.1 Murdock, George Peter. Social structure. N.Y., 1965.
Geog 4502.5.4 Das Murmeltier mit dem Halsband. (Rombert, E.) München, 1929.
Geog 5312.3.15A Murphy, Charles J.V. Struggle; the life and exploits of Commander Richard E. Byrd. N.Y., 1928.
An 359.43.20 Murphy, J.D. Lamps of anthropology. Manchester, Eng., 1943.
An 2109.27.7 Murphy, John. Primitive man, his essential quest. London, 1927.
Geog 6009.12.5 Murphy, R.C. Logbook for Grace. N.Y., 1947.
An 359.71.25 Murphy, Robert Francis. The dialectics of social life: alarms and excursions in anthropological theory. N.Y., 1971.
An 136.25 Murphy, Robert Francis. Robert H. Lowie. N.Y., 1972.
Geog 5312.7.25 Murphy, Robert William. The haunted journey. 1. ed. Garden City, 1961.
Geog 4308.54.10 Murray, E.C.G. The roving Englishman. London, 1854.
Geog 5919.01.5 Murray, G. The Antarctic manual. London, 1901.
Geog 818.34 Murray, H. An encyclopaedia of geography. London, 1834.
Geog 818.40 Murray, H. Encyclopedia of geography. London, 1840.
Geog 808.37 Murray, H. Encyclopedia of geography. Philadelphia, 1837. 3v.
Geog 818.43.5 Murray, Hugh. The encyclopaedia of geography. Philadelphia, 1843. 3v.
Geog 808.34.10 Murray, Hugh. The encyclopaedia of geography. Philadelphia, 1845. 3v.
Geog 5939.07 Murray, J. Antarctic days. London, 1913.
Geog 1775.10 Murray, John, publisher, London. Hand-book for northern Europe: Finland and Russia. N.Y., 1970.
Geog 1820.2 Murray, John, publisher, London. Handbook for...Algeria and Tunis. 2. ed. London, 1878.
Geog 1820.4 Murray, John, publisher, London. Handbook for...Algeria and Tunis. 4. ed. London, 1891.
Geog 1820.5 Murray, John, publisher, London. Handbook for...Algeria and Tunis. 5. ed. London, 1895.
Geog 1709.1 Murray, John, publisher, London. Handbook for...Berks, Bucks and Oxfordshire. London, 1860.
Geog 1709.1.3 Murray, John, publisher, London. Handbook for...Berks, Bucks and Oxfordshire. 3. ed. London, 1882.
Geog 1740.1 Murray, John, publisher, London. Handbook for...central Italy. London, 1843.
Geog 1740.3 Murray, John, publisher, London. Handbook for...central Italy. 3. ed. London, 1853. 2v.
Geog 1740.4 Murray, John, publisher, London. Handbook for...central Italy. 4. ed. London, 1857.
Geog 1740.5 Murray, John, publisher, London. Handbook for...central Italy. 5. ed. London, 1861.
Geog 1740.6 Murray, John, publisher, London. Handbook for...central Italy. 6. ed. London, 1864.
Geog 1740.7.5 Murray, John, publisher, London. Handbook for...central Italy. 7. ed. London, 1867.
Geog 1740.9 Murray, John, publisher, London. Handbook for...central Italy. 9. ed. London, 1875.
Geog 1740.11 Murray, John, publisher, London. Handbook for...central Italy. 11. ed. pt.2. London, 1889.
Geog 1804.1A Murray, John, publisher, London. Handbook for...Constantinople. London, 1900.
Geog 1709.5.10 Murray, John, publisher, London. Handbook for...Cornwall. 10. ed. London, 1882.
Geog 1744.1 Murray, John, publisher, London. Handbook for...Corsica and Sardinia. London, 1860.
Geog 1763.3 Murray, John, publisher, London. Handbook for...Denmark, Norway, Sweden. 3. ed. London, 1858.
Geog 1763.6 Murray, John, publisher, London. Handbook for...Denmark, Norway and Sweden. 3. ed. London, 1871.
Geog 1765.4 Murray, John, publisher, London. Handbook for...Denmark. 4. ed. London, 1875.
Geog 1709.7.2 Murray, John, publisher, London. Handbook for...Derbyshire, Nottinghamshire. 2. ed. London, 1874.
Geog 1709.8.10A Murray, John, publisher, London. Handbook for...Devonshire. 10. ed. London, 1887.
Geog 1709.10.5 Murray, John, publisher, London. Handbook for...Durham and Northumberland. London, 1873.
Geog 1816.5 Murray, John, publisher, London. Handbook for...Egypt. London, 1847.
Geog 1816.10 Murray, John, publisher, London. Handbook for...Egypt. London, 1858.
Geog 1816.15 Murray, John, publisher, London. Handbook for...Egypt. 4. ed. London, 1873.
Geog 1816.17 Murray, John, publisher, London. Handbook for...Egypt. 5. ed. London, 1875.
Geog 1816.19 Murray, John, publisher, London. Handbook for...Egypt. 7. ed. London, 1888.
Geog 1816.20A Murray, John, publisher, London. Handbook for...Egypt. 8. ed. London, 1891.
Geog 1816.25 Murray, John, publisher, London. Handbook for...Egypt and the Sudan. 11. ed. London, 1907.
Geog 1709.12 Murray, John, publisher, London. Handbook for...Gloucestershire, Worcestershire. London, 1867.
Geog 1709.12.3 Murray, John, publisher, London. Handbook for...Gloucestershire, Worcestershire. 3. ed. London, 1884.
Geog 1709.12.4 Murray, John, publisher, London. Handbook for...Gloucestershire. 4. ed. London, 1895.
Geog 1785.2 Murray, John, publisher, London. Handbook for...Greece. London, 1854.
Geog 1785.4 Murray, John, publisher, London. Handbook for...Greece. 4. ed. Lonon, 1872.
Geog 1785.5 Murray, John, publisher, London. Handbook for...Greece. 5. ed. London, 1884. 2v.
Geog 1785.6 Murray, John, publisher, London. Handbook for...Greece. 6. ed. London, 1896. 2v.
Geog 1785.7 Murray, John, publisher, London. Handbook for...Greece. 7. ed. London, 1900.
Geog 1709.13.5 Murray, John, publisher, London. Handbook for...Hampshire. 5. ed. London, 1898.
Geog 1807.13 Murray, John, publisher, London. Handbook for...India, Burma and Ceylon. 3. ed. London, 1898.
Geog 1807.17 Murray, John, publisher, London. Handbook for...India, Burma and Ceylon. 5. ed. London, 1905.
Geog 1807.17.5 Murray, John, publisher, London. Handbook for...India, Burma and Ceylon. 5. ed. London, 1906.
Geog 1807.18 Murray, John, publisher, London. Handbook for...India, Burma and Ceylon. 6. ed. London, 1907.
Geog 1807.19 Murray, John, publisher, London. Handbook for...India, Burma and Ceylon. 8. ed. London, 1911.
Geog 1807.21 Murray, John, publisher, London. Handbook for...India, Burma and Ceylon. 10. ed. London, 1920.

Author and Title Listing

	Geog 1807.25	Murray, John, publisher, London. Handbook for...India, Burma and Ceylon. 13. ed. London, 1929.
	Geog 1807.25.2	Murray, John, publisher, London. Handbook for...India, Burma and Ceylon. 13. ed. N.Y., 1929.
	Geog 1807.27	Murray, John, publisher, London. Handbook for...India, Burma and Ceylon. 14. ed. London, 1933.
	Geog 1807.29	Murray, John, publisher, London. Handbook for...India, Pakistan, Burma and Ceylon. London, 1949.
	Geog 1807.5	Murray, John, publisher, London. Handbook for...India. London, 1859. 2v.
	Geog 1807.10	Murray, John, publisher, London. Handbook for...India and Ceylon. London, 1892.
	Geog 1782.1	Murray, John, publisher, London. Handbook for...Ionian islands. London, 1840.
	Geog 1782.2	Murray, John, publisher, London. Handbook for...Ionian islands. London, 1845.
	Geog 1811.4	Murray, John, publisher, London. Handbook for...Japan. 4. ed. London, 1896.
	Geog 1811.6	Murray, John, publisher, London. Handbook for...Japan. 6. ed. London, 1901.
	Geog 1811.8A	Murray, John, publisher, London. Handbook for...Japan. 8. ed. London, 1907.
	Geog 1811.9A	Murray, John, publisher, London. Handbook for...Japan. 9. ed. London, 1913.
	Geog 1709.17	Murray, John, publisher, London. Handbook for...Kent and Sussex. London, 1858.
NEDL	Geog 1709.17.3	Murray, John, publisher, London. Handbook for...Kent and Sussex. 3. ed. London, 1868.
	Geog 1828.1	Murray, John, publisher, London. Handbook for...New Zealand. London, 1893.
	Geog 1714.1.2	Murray, John, publisher, London. Handbook for...North Wales. 2. ed. London, 1864.
	Geog 1709.24	Murray, John, publisher, London. Handbook for...Northamptonshire. London, 1878.
	Geog 1709.24.2	Murray, John, publisher, London. Handbook for...Northamptonshire. 2. ed. London, 1901.
	Geog 1739.4	Murray, John, publisher, London. Handbook for...northern Italy. 4. ed. London, 1852.
	Geog 1739.6	Murray, John, publisher, London. Handbook for...northern Italy. 6. ed. London, 1856. 2v.
	Geog 1739.8	Murray, John, publisher, London. Handbook for...northern Italy. 8. ed. London, 1860.
	Geog 1739.9	Murray, John, publisher, London. Handbook for...northern Italy. 9. ed. London, 1863.
	Geog 1739.10	Murray, John, publisher, London. Handbook for...northern Italy. 10. ed. London, 1866.
	Geog 1739.11	Murray, John, publisher, London. Handbook for...northern Italy. 11. ed. London, 1867.
	Geog 1739.14	Murray, John, publisher, London. Handbook for...northern Italy. 14. ed. London, 1877.
	Geog 1770.8	Murray, John, publisher, London. Handbook for...Norway. 8. ed. London, 1892.
	Geog 1709.27	Murray, John, publisher, London. Handbook for...Oxfordshire. London, 1894.
	Geog 1750.1	Murray, John, publisher, London. Handbook for...Portugal. London, 1855.
	Geog 1750.2	Murray, John, publisher, London. Handbook for...Portugal. 2. ed. London, 1856.
	Geog 1750.3	Murray, John, publisher, London. Handbook for...Portugal. 3. ed. London, 1875.
	Geog 1775.2	Murray, John, publisher, London. Handbook for...Russia, Poland and Finland. 2. ed. London, 1868.
	Geog 1775.3	Murray, John, publisher, London. Handbook for...Russia, Poland and Finland. 3. ed. London, 1875.
	Geog 1775.5	Murray, John, publisher, London. Handbook for...Russia, Poland and Finland. 5. ed. London, 1893.
	Geog 1712.1	Murray, John, publisher, London. Handbook for...Scotland. London, 1867.
	Geog 1712.2	Murray, John, publisher, London. Handbook for...Scotland. 2. ed. London, 1868.
	Geog 1712.4	Murray, John, publisher, London. Handbook for...Scotland. 4. ed. London, 1875.
	Geog 1712.7.25	Murray, John, publisher, London. Handbook for...Scotland. 8. ed. London, 1903.
	Geog 1712.8	Murray, John, publisher, London. Handbook for...Scotland. 8. ed. London, 1907.
	Geog 1712.9A	Murray, John, publisher, London. Handbook for...Scotland. 9. ed. London, 1913.
	Geog 1743.1	Murray, John, publisher, London. Handbook for...Sicily. London, 1864.
	Geog 1724.14	Murray, John, publisher, London. Handbook for...south Germany and Austria. 14. ed. London, 1881.
	Geog 1714.2.4	Murray, John, publisher, London. Handbook for...South Wales. 4. ed. London, 1890.
	Geog 1724.2	Murray, John, publisher, London. Handbook for...southern Germany. London, 1837.
Htn	Geog 1724.5*	Murray, John, publisher, London. Handbook for...southern Germany. 5. ed. London, 1850.
	Geog 1724.5.5	Murray, John, publisher, London. Handbook for...southern Germany. 5. ed. London, 1851.
	Geog 1724.6	Murray, John, publisher, London. Handbook for...southern Germany. 6. ed. London, 1853.
	Geog 1724.7	Murray, John, publisher, London. Handbook for...southern Germany. 7. ed. London, 1855.
	Geog 1724.7.2	Murray, John, publisher, London. Handbook for...southern Germany. 7. ed. London, 1855.
	Geog 1724.8	Murray, John, publisher, London. Handbook for...southern Germany. 8. ed. London, 1858.
	Geog 1724.9	Murray, John, publisher, London. Handbook for...southern Germany. 9. ed. London, 1863.
	Geog 1724.10	Murray, John, publisher, London. Handbook for...southern Germany. 10. ed. London, 1867.
	Geog 1724.11	Murray, John, publisher, London. Handbook for...southern Germany. 11. ed. London, 1871.
	Geog 1724.12	Murray, John, publisher, London. Handbook for...southern Germany. 12. ed. London, 1873.
	Geog 1742.1	Murray, John, publisher, London. Handbook for...southern Italy. London, 1853.
	Geog 1742.2	Murray, John, publisher, London. Handbook for...southern Italy. 2. ed. London, 1855.
	Geog 1742.3	Murray, John, publisher, London. Handbook for...southern Italy. 3. ed. London, 1858.
	Geog 1742.4	Murray, John, publisher, London. Handbook for...southern Italy. 4. ed. London, 1862.
	Geog 1742.6	Murray, John, publisher, London. Handbook for...southern Italy. 6. ed. London, 1868.
	Geog 1742.8	Murray, John, publisher, London. Handbook for...southern Italy. 8. ed. London, 1878.
	Geog 1742.9	Murray, John, publisher, London. Handbook for...southern Italy and Sicily. 9. ed. London, 1892.
	Geog 1748.3A	Murray, John, publisher, London. Handbook for...Spain. 3. ed. London, 1855. 2v.
	Geog 1748.5	Murray, John, publisher, London. Handbook for...Spain. 5. ed. London, 1878.
	Geog 1748.6	Murray, John, publisher, London. Handbook for...Spain. 6. ed. London, 1882. 2v.
	Geog 1748.7	Murray, John, publisher, London. Handbook for...Spain. 7. ed. London, 1890. 2v.
	Geog 1709.33	Murray, John, publisher, London. Handbook for...Surrey, Hampshire. London, 1858.
	Geog 1709.33.4	Murray, John, publisher, London. Handbook for...Surrey, Hampshire. 4. ed. London, 1888.
	Geog 1773.4	Murray, John, publisher, London. Handbook for...Sweden. 4. ed. London, 1875.
	Geog 1773.6	Murray, John, publisher, London. Handbook for...Sweden. 6. ed. London, 1883.
	Geog 1735.1	Murray, John, publisher, London. Handbook for...Switzerland. London, 1838.
	Geog 1735.3	Murray, John, publisher, London. Handbook for...Switzerland. London, 1839.
	Geog 1735.7	Murray, John, publisher, London. Handbook for...Switzerland. 3. ed. London, 1846.
	Geog 1735.9	Murray, John, publisher, London. Handbook for...Switzerland. 4. ed. London, 1851.
	Geog 1735.11	Murray, John, publisher, London. Handbook for...Switzerland. 5. ed. London, 1852.
	Geog 1735.13	Murray, John, publisher, London. Handbook for...Switzerland. 6. ed. London, 1854.
	Geog 1735.15	Murray, John, publisher, London. Handbook for...Switzerland. 7. ed. London, 1856.
	Geog 1735.17	Murray, John, publisher, London. Handbook for...Switzerland. 8. ed. London, 1858.
	Geog 1735.23A	Murray, John, publisher, London. Handbook for...Switzerland. 11. ed. London, 1865.
	Geog 1735.25	Murray, John, publisher, London. Handbook for...Switzerland. 12. ed. London, 1867.
	Geog 1735.30	Murray, John, publisher, London. Handbook for...Switzerland. 15. ed. London, 1874.
	Geog 1735.35	Murray, John, publisher, London. Handbook for...Switzerland. 17. ed. London, 1886.
	Geog 1805.1	Murray, John, publisher, London. Handbook for...Syria and Palestine. London, 1858. 2v.
	Geog 1805.1.10	Murray, John, publisher, London. Handbook for...Syria and Palestine. London, 1875.
	Geog 1805.2	Murray, John, publisher, London. Handbook for...Syria and Palestine. London, 1892.
NEDL	Geog 1803.3	Murray, John, publisher, London. Handbook for...Turkey. London, 1854.
	Geog 1709.37.12	Murray, John, publisher, London. Handbook for...Wilts and Dorset. 5. ed. London, 1899.
	Geog 1709.37	Murray, John, publisher, London. Handbook for...Wiltshire, Dorsetshire and Somersetshire. London, 1856.
	Geog 1709.37.2	Murray, John, publisher, London. Handbook for...Wiltshire, Dorsetshire and Somersetshire. London, 1859.
	Geog 1709.37.10	Murray, John, publisher, London. Handbook for...Wiltshire, Dorsetshire and Somersetshire. 4. ed. London, 1882.
	Geog 1709.38.4	Murray, John, publisher, London. Handbook for...Worcestershire. 4. ed. London, 1894.
	Geog 1709.39.3	Murray, John, publisher, London. Handbook for...Yorkshire. 3. ed. London, 1882.
	Geog 1709.39.4	Murray, John, publisher, London. Handbook for...Yorkshire. 4. ed. London, 1904.
	Geog 1801.1	Murray, John, publisher, London. Handbook for Asia Minor. London, 1895.
	Geog 1709.1.10	Murray, John, publisher, London. Handbook for Berkshire. London, 1902.
	Geog 1709.2	Murray, John, publisher, London. Handbook for Buckinghamshire. London, 1903.
	Geog 1709.7.3	Murray, John, publisher, London. Handbook for Derbyshire, Nottinghamshire. 3. ed. London, 1904.
	Geog 1709.8.4	Murray, John, publisher, London. Handbook for Devon and Cornwall. Newton Abbot, Eng., 1971.
	Geog 1709.8.5	Murray, John, publisher, London. Handbook for Devon and Cornwall. 5. ed. London, 1863.
	Geog 1709.8.8	Murray, John, publisher, London. Handbook for Devon and Cornwall. 8. ed. London, 1872.
	Geog 1708.1	Murray, John, publisher, London. Handbook for England and Wales. London, 1878.
	Geog 1708.15	Murray, John, publisher, London. Handbook for England and Wales. 2. ed. London, 1890.
	Geog 1709.11.2	Murray, John, publisher, London. Handbook for Essex, Suffolk and Norfolk. 2. ed. London, 1875.
	Geog 1709.15	Murray, John, publisher, London. Handbook for Hertfordshire, Bedfordshire. London, 1895.
	Geog 1811.5	Murray, John, publisher, London. Handbook for Japan. 5. ed. London, 1899.
	Geog 1709.18.3	Murray, John, publisher, London. Handbook for Lancashire. London, 1880.
	Geog 1709.20	Murray, John, publisher, London. Handbook for Lincolnshire. London, 1890.
	Geog 1709.20.2	Murray, John, publisher, London. Handbook for Lincolnshire. 2. ed. London, 1903.
	Geog 1723.19	Murray, John, publisher, London. Handbook for north Germany. 19. ed. London, 1877.
	Geog 1706.5	Murray, John, publisher, London. Handbook for northern Europe. London, 1849. 2v.
	Geog 1770.5	Murray, John, publisher, London. Handbook for Norway. 5. ed. London, 1874.
	Geog 1770.7	Murray, John, publisher, London. Handbook for Norway. 7. ed. London, 1880.
	Geog 1709.29	Murray, John, publisher, London. Handbook for Shropshire. London, 1870.
	Geog 1709.29.3	Murray, John, publisher, London. Handbook for Shropshire and Cheshire. London, 1879.
	Geog 1709.29.5	Murray, John, publisher, London. Handbook for Shropshire and Cheshire. 3. ed. London, 1897.
	Geog 1709.30.5	Murray, John, publisher, London. Handbook for Somerset. 5. ed. London, 1899.
	Geog 1709.34.4	Murray, John, publisher, London. Handbook for Sussex. 4. ed. London, 1877.
	Geog 1709.34.5	Murray, John, publisher, London. Handbook for Sussex. 5. ed. London, 1905.
	Geog 1715.2	Murray, John, publisher, London. Handbook for travellers in France. London, 1843.
	Geog 1715.2.5	Murray, John, publisher, London. Handbook for travellers in France. London, 1844.

Author and Title Listing

Geog 1715.3 Murray, John, publisher, London. Handbook for travellers in France. 3. ed. London, 1848.

Geog 1715.4 Murray, John, publisher, London. Handbook for travellers in France. 4. ed. London, 1853.

Geog 1715.7 Murray, John, publisher, London. Handbook for travellers in France. 7. ed. London, 1859.

Geog 1715.9 Murray, John, publisher, London. Handbook for travellers in France. 9. ed. London, 1864.

Geog 1715.10 Murray, John, publisher, London. Handbook for travellers in France. 10. ed. London, 1867.

Geog 1715.11 Murray, John, publisher, London. Handbook for travellers in France. 11. ed. London, 1869.

Geog 1715.12 Murray, John, publisher, London. Handbook for travellers in France. 12. ed. London, 1873. 2v.

Geog 1715.13 Murray, John, publisher, London. Handbook for travellers in France. 13. ed. London, 1875.

Geog 1715.16 Murray, John, publisher, London. Handbook for travellers in France. 16. ed. London, 1882-84. 2v.

Geog 1715.18 Murray, John, publisher, London. Handbook for travellers in France. 18. ed. London, 1892. 2v.

Geog 1713.2.5 Murray, John, publisher, London. Handbook for travellers in Ireland. 2. ed. London, 1866.

Geog 1713.4 Murray, John, publisher, London. Handbook for travellers in Ireland. 4. ed. London, 1878.

Geog 1713.5 Murray, John, publisher, London. Handbook for travellers in Ireland. 5. ed. London, 1896.

Geog 1713.6 Murray, John, publisher, London. Handbook for travellers in Ireland. 6. ed. London, 1902.

Geog 1713.6.75 Murray, John, publisher, London. Handbook for travellers in Ireland. 7. ed. London, 1906.

Geog 1713.7 Murray, John, publisher, London. Handbook for travellers in Ireland. 7. ed. London, 1906.

Geog 1713.8 Murray, John, publisher, London. Handbook for travellers in Ireland. 7. ed. London, 1906.

Geog 1709.17.4 Murray, John, publisher, London. Handbook for travellers in Kent. 4. ed. London, 1877.

Geog 1748.1 Murray, John, publisher, London. A handbook for travellers in Spain. London, 1845. 2v.

Geog 1748.2 Murray, John, publisher, London. Handbook for travellers in Spain. 2. ed. London, 1847.

Geog 1709.33.5 Murray, John, publisher, London. Handbook for travellers in Surrey. 5. ed. London, 1898.

Geog 1707.3 Murray, John, publisher, London. Handbook for travellers on the continent. 2. ed. London, 1838.

Geog 1707.6 Murray, John, publisher, London. Handbook for travellers on the continent. 5. ed. London, 1845.

Geog 1707.7 Murray, John, publisher, London. Handbook for travellers on the continent. 7. ed. London, 1850.

Geog 1707.8 Murray, John, publisher, London. Handbook for travellers on the continent. 8. ed. London, 1951.

Geog 1707.9 Murray, John, publisher, London. Handbook for travellers on the continent. 9. ed. London, 1853.

Geog 1707.10 Murray, John, publisher, London. Handbook for travellers on the continent. 10. ed. London, 1854.

Geog 1707.11 Murray, John, publisher, London. Handbook for travellers on the continent. 11. ed. London, 1856.

Geog 1707.12 Murray, John, publisher, London. Handbook for travellers on the continent. 12. ed. London, 1858.

Geog 1707.13 Murray, John, publisher, London. Handbook for travellers on the continent. 13. ed. London, 1860.

Geog 1707.15 Murray, John, publisher, London. Handbook for travellers on the continent. 15. ed. London, 1865.

Geog 1707.16 Murray, John, publisher, London. Handbook for travellers on the continent. 16. ed. London, 1868.

Geog 1707.17 Murray, John, publisher, London. Handbook for travellers on the continent. 17. ed. London, 1870.

Geog 1707.17.5 Murray, John, publisher, London. Handbook for travellers on the continent. 17. ed. London, 1871.

Geog 1707.18 Murray, John, publisher, London. Handbook for travellers on the continent. 18. ed. London, 1873-74. 2v.

Geog 1707.19 Murray, John, publisher, London. Handbook for travellers on the continent. 19. ed. London, 1875. 2v.

Geog 1803.4 Murray, John, publisher, London. Handbook for Turkey. 4. ed. London, 1878.

Geog 1719.1 Murray, John, publisher, London. Handbook for visitors to Paris. London, 1864.

Geog 1719.3 Murray, John, publisher, London. Handbook for visitors to Paris. 3. ed. London, 1867.

Geog 1719.5 Murray, John, publisher, London. Handbook for visitors to Paris. 5. ed. London, 1872.

Geog 1719.8 Murray, John, publisher, London. Handbook for visitors to Paris. 8. ed. London, 1876.

Geog 1719.10 Murray, John, publisher, London. Handbook for visitors to Paris. 10. ed. London, 1879.

Geog 1709.36.2 Murray, John, publisher, London. Handbook for Westmorland...and the Lakes. 2. ed. London, 1869.

Geog 1709.39 Murray, John, publisher, London. Handbook for Yorkshire. London, 1867.

Geog 1741.5 Murray, John, publisher, London. Handbook of Rome and its environs. 5. ed. London, 1858.

Geog 1741.7 Murray, John, publisher, London. Handbook of Rome and its environs. 7. ed. London, 1864.

Geog 1741.8 Murray, John, publisher, London. Handbook of Rome and its environs. 8. ed. London, 1867.

Geog 1741.9 Murray, John, publisher, London. Handbook of Rome and its environs. 9. ed. London, 1869.

Geog 1741.10 Murray, John, publisher, London. Handbook of Rome and its environs. 10. ed. London, 1871.

Geog 1741.12 Murray, John, publisher, London. Handbook of Rome and its environs. 12. ed. London, 1875.

Geog 1809.5 Murray, John, publisher, London. Handbook of the Bengal presidency. London, 1882.

Geog 1809.30 Murray, John, publisher, London. Handbook of the Bombay presidency. London, 1881.

Geog 1809.55 Murray, John, publisher, London. Handbook of the Madras presidency. London, 1879.

Geog 1809.80 Murray, John, publisher, London. Handbook of the Punjab...Kashmir. London, 1883.

Geog 1709.35 Murray, John, publisher, London. Handbook of Warwickshire. London, 1899.

Geog 1707.5 Murray, John, publisher, London. Handbook the travellers on the continent. London, 1840.

Geog 1710.20 Murray, John, publisher, London. Handbook to London. London, 1869?

Geog 1710.22 Murray, John, publisher, London. Handbook to London. London, 1870.

Geog 1710.26 Murray, John, publisher, London. Handbook to London. London, 1871.

Geog 1710.28 Murray, John, publisher, London. Handbook to London. London, 1873.

Geog 1710.30 Murray, John, publisher, London. Handbook to London. London, 1876.

Geog 1709.6.25 Murray, John, publisher, London. Handbook to the English lakes...Cumberland, Westmorland, and Lancashire. London, 1889.

Geog 1795.3 Murray, John, publisher, London. Handbook to the Mediterranean. 3. ed. London, 1892. 2v.

Geog 1807.30 Murray, John, publisher, London. The Imperial guide to India. London, 1904.

Geog 1733.5 Murray, John, publisher, London. The knapsack guide...Tyrol and the eastern Alps. London, 1867.

Geog 1735.22 Murray, John, publisher, London. Knapsack guide for...Switzerland. London, 1864.

Geog 1735.26 Murray, John, publisher, London. Knapsack guide for...Switzerland. London, 1867.

Geog 1710.29 Murray, John, publisher. London. Handbook to London. London, 1874.

Geog 1710.35 Murray, John, publishers, London. Handbook to London. London, 1879.

Geog 613.3.5 Murray, John. John Murray III, 1808-1892; memoir. London, 1919.

Geog 1714.1.5 Murray, John. publisher, London. Handbook for...North Wales. 5. ed. London, 1885.

Geog 4308.53 Murray, N. Men and things...in Europe. N.Y., 1853.

An 2109.43 Murray, R.W. Man's unknown ancestors; the story of prehistoric man. Milwaukee, 1943.

Geog 146.5 Murray Park College of Advanced Education. Department of Geography. Occasional papers in geography series. Magill. 1,1974+

Geog 4218.38.4 Murrell, William M. Cruise frigate Columbia around the world...1838-1840. Boston, 1840.

An 2059.63 Muschalek, H. Urmensch-Adam. Berlin, 1963.

An 2059.03.7 Musée prehistorique. 2. éd. (Mortillet, G. de.) Paris, 1903.

Geog 5153.20 The musk-ox; a journal on the north. Saskatoon. 1,1967+ 2v.

Geog 3240.22 Muslim contribution to geography. (Ahmad, Nafia.) Lahore, 1947.

Htn Geog 4708.31* Mutiny and murder. Confession of Charles Gibbs. Providence, 1831.

Geog 4269.61.5 Mutton, Alice. Central Europe; a regional and human geography. London, 1961.

Geog 5558.91A My Arctic journal. (Peary, Josephine D.) N.Y., 1893.

Geog 4218.76.5 My circular notes. v.1-2. (Campbell, J.F.) London, 1876.

Geog 4509.11 My climbing adventures in 4 continents. (Turner, S.F.R.) London, 1913.

Geog 4558.65 My diary from England to India around Cape of Good Hope. (Waites, Alfred.) Calcutta, 1865.

Geog 4559.25.15 My ditty bag. (Brown, Charles W.) Boston, 1925.

Geog 4309.08.5 My early life in California and my first American and European tours (1890-1908). (Darling, Kenneth G.) Cambridge, Mass.? 1942?

Geog 4308.50.17 My first visit to Europe. 5th ed. (Dickinson, A.) N.Y., 1856.

Geog 4219.23 My journey round the world. (Northcliffe, A.C.W.H.) Philadelphia, 1923.

Geog 4218.78.9 My journey round the world. (Parry, S.H. Jones.) London, 1881. 2v.

Geog 6009.36 My last expedition to the Antarctic, 1936-37. (Christensen, Lars.) Oslo, 1938.

Geog 5311.1.3 My life as an explorer. (Amundsen, Roald.) Garden City, N.Y., 1928.

Geog 4559.12 My life at sea. (Crutchley, William C.) London, 1912.

Geog 4709.26 My pirate library. (Gosse, Philip.) London, 1926.

Geog 4208.54A My reminiscences. (Pumpelly, R.) N.Y., 1918. 2v.

Geog 4228.45A My voyage in the U.S. frigate "Congress". (Van Denburgh, E.D.) N.Y., 1913.

Geog 4209.67 My za granitsei. (Pochivalov, Leonid V.) Moskva, 1967.

Geog 5839.03 Mylius, L. Grønland...1903-04. Kjøbenhavn, 1906.

An 359.16.5 Myres, John L. The influence of anthropology on the course of political science. Berkeley, 1916.

Geog 4209.58.20 Mysteres des îles. (La Croix, Robert de.) Paris, 1958.

Geog 4672.16.10 Mysteries of the sea. (Lockhart, John Gilbert.) London, 1924.

Geog 3055.82 The mystery of Atlantis. (Berlitz, Charles Frambach.) N.Y., 1969.

Geog 4217.85.15 The mystery of La Pérouse. (Gordon, Mona.) Christchurch, N.Z., 1961.

Geog 4679.42 Mystery ship. The Mary Celeste in fancy and fact. (Bryan, George S.) Philadelphia, 1942.

An 359.71.40 La mythe aryen. (Poliahov, Leon.) Paris, 1971.

Geog 3229.29 Mžik, Hans von. Beiträge zur historischen Geographie, Kulturgeographie, Ethnographie und Kartographie. Leipzig, 1929.

Geog 5839.30A N. by E. (Kent, Rockwell.) N.Y., 1931.

Geog 5398.79.9F N.Y. Herald. The Jeannette in the Arctic regions. N.Y., 1882.

Geog 5379.35.5 Na Chelnoskike. (Khmyznikov, P.) Leningrad, 1936.

Geog 4269.65 Na chuzhikh ulitsakh. (Filipenko, Aloiz A.) Moskva, 1965.

VGeog 4219.69.10 Na juga sveta. (Ogrin, Miran.) Ljubljana, 1969.

Geog 5559.56.6 Na kryliakh v Arktiku. (Vodopianov, Mikhail V.) Moskva, 1955.

Geog 5399.57.5 Na l'dine cherez polius. (Iatsun, E.P.) Moskva, 1957.

Geog 4239.73 Na "RA" cherez Atlantiku. (Senkevich, Iurii A.) Leningrad, 1973.

Geog 4209.69.20 Na raznykh shirotakh. (Boldyrev, Ivan I.) Riga, 1969.

Geog 5939.59.10 Na samoi iuzhnoi zemle. (Pasetskii, V.M.) Moskva, 1959.

Geog 5399.32.5 Na "Sibiriakove" v "Litke"...1932 i 1934. (Vize, Vladimir I.) Moskva, 1946.

Geog 5399.57.10 Na sushe i na more v Arktike. (Ruzov, L.V.) Moskva, 1957.

NEDL Geog 4157.07 Naaukeurige...der zee en land-reysen. v.1-28. (Aa, P.V.) Leyden, 1707. 29v.

Geog 4182.23 Naber, S.P. Hessel Gerritsz, samoyeden land ten Spitsberghe. 's-Gravenhage, 1924.

Geog 4182.34 Naber, S.P. Johannes de Laet. Iaerlyck Verhael van der verichtinghen der...Compagnie. 's-Gravenhage, 1931-37. 4v.

Geog 4182.6 Naber, S.P. Toortse der zee-vaert...D. Ruiters 1623 en Samuel Brun's schiffarten (1624). 's-Gravenhage, 1913.

Geog 4269.58.5 Nach Russland und anderswohin. (Koeppen, Wolfgang.) Stuttgart, 1958.

Geog 4300.42 Nagel, publishers. Europe. Geneva, 1956.

Geog 5559.14 Nagornyi, S. Sedov. Moskva, 1939.

Geog 5837.40F Naleesing over Groenland voor de historie van den noorweeschen Erik. pt.1-5. (Mauricius, Jan J.) Hamburg? 1740.

Author and Title Listing

	Geog 614.7	Nałkowska, Z. Moj ojciec. Warszawa, 1953.
	Geog 614.8	Nałkowska, Z. Moja ojciec. Wyd. 2. Warszawa, 1955.
	Geog 808.17.6	Namen- und Sach-Verzeichniss. (Ritter, Carl.) Berlin, 1841-1849. 2v.
	Geog 6009.64	Námořníkem, topičem a psovodem za jižním polárním kruhem. Vyd. 1. (Vojtěch, Vaclav.) Praha, 1968.
	Geog 5855.7	Nansen, F. Eskimo life. London, 1893.
	Geog 5850.9	Nansen, F. Eskimoliv. Kristiania, 1891.
	Geog 5838.88.3A	Nansen, F. The first crossing of Greenland. London, 1890. 2v.
	Geog 5558.82	Nansen, F. Hunting and adventure in the Arctic. N.Y., 1925.
	Geog 5220.22.5	Nansen, F. In northern mists, Arctic exploration in early times. London, 1911. 2v.
	Geog 5220.22.10	Nansen, F. In northern mists. N.Y., 1911. 2v.
	Geog 5220.22	Nansen, F. Nord i tåkeheimen. Kristiania, 1911.
	Geog 5838.88	Nansen, F. Paa ski over Grønland...1888-90. Kristiania, 1890.
	Geog 5324.2.90	Nansen, Fridtjof. Brev utgitt for Nansenfondet av Steinar Kjaerheim. Oslo, 1961- 4v.
	Geog 5398.93.1	Nansen, Fridtjof. Farthest north. N.Y., 1897. 2v.
	Geog 5398.93	Nansen, Fridtjof. Farthest north. Westminster, 1897. 2v.
	Geog 5398.93.5	Nansen, Fridtjof. In Nacht undEis. Leipzig, 1897. 2v.
	Geog 5324.2.40	Nansen, Fridtjof. Nansens røst. Oslo, 1945. 3v.
	Geog 5324.2.15	Nansen. (Reynolds, E.E.) London, 1932.
	Geog 5324.2.60	Nansen. (Whitehouse, J.H.) London, 1930.
	Geog 5324.2.20	Nansen of Norway. (Turley, C.) London, 1933.
	Geog 5324.2.75	Nansen og verden. 2. Oppl. (Hoeyer, Liv Nansen.) Oslo, 1955.
	Geog 5324.2.40	Nansens røst. (Nansen, Fridtjof.) Oslo, 1945. 3v.
	An 2059.68.5	Nantevil, Hugues de. Sur les traces d'Adam; nouvel aperçu sur les origines de l'homme. Paris, 1968.
	Geog 5530.82	Nanton, Paul. Arctic breakthrough: Franklin's expeditions, 1819-1847. London, 1971.
	Geog 5558.75F	Nares, G.S. Arctic expedition 1875-76, journal. London, 1877.
	Geog 5558.75.3	Nares, G.S. Narrative of voyage to Polar Sea...1875-76. London, 1878. 2v.
	An 359.71.15	Narody, rasy, kul'tury. (Cheboksarov, Nikolai N.) Moskva, 1971.
	An 2109.66.15	Narr, Karl J. Handbuch der Urgeschichte. Bern, 1966. 2v.
	Geog 4218.50	Narrative of a journey round the world. (Gerstaecker, F.) N.Y., 1853.
	Geog 4218.41.3	Narrative of a journey round the world. (Simpson, George.) London, 1847. 2v.
	Geog 5518.19.14	Narrative of a journey to the shores of the Polar Sea in the years 1819-1922. (Franklin, John.) Rutland, 1970.
Htn	Geog 5518.25.2F*	Narrative of a second expedition to the shores of the Polar Sea, in the year 1825-1827. (Franklin, John.) London, 1828.
	Geog 5518.25.5	Narrative of a second expedition to the shores of the Polar Sea in the years 1825-1827. (Franklin, John.) Rutland, 1971.
	Geog 5518.29.3	Narrative of a second voyage in search of a northwest passage. (Ross, J.) Philadelphia, 1835.
	Geog 4218.35	Narrative of a voyage round the world...1836-42. (Belcher, Edward.) London, 1843. 2v.
	Geog 4218.17	Narrative of a voyage round the world. (Arago, J.) London, 1823.
	Geog 4328.40.3	Narrative of a voyage to Madeira, Teneriffe. 2nd ed. (Wilde, William R.) Dublin, 1844.
Htn	Geog 4248.08*	Narrative of adventures...of S. Patterson, experienced in Pacific Ocean. (Patterson, S.) Palmer, 1817.
	Geog 5518.36.7	Narrative of an expedition in H.M.S. Terror. (Back, George.) London, 1838.
	Geog 5398.20.5	Narrative of an expedition to the Polar Sea. (Wrangell, F. von.) London, 1840.
	Geog 5398.20.7	Narrative of an expedition to the Polar Sea. (Wrangell, F. von.) London, 1844.
	Geog 5208.51	A narrative of Arctic discovery. 2. ed. (Shillinglaw, J.J.) London, 1851.
	Geog 5518.33	Narrative of Arctic land expedition. (Back, George.) London, 1836.
	Geog 5518.33.3	Narrative of Arctic land expedition. (Back, George.) Philadelphia, 1836.
	Geog 5518.27	Narrative of attempt to reach the North Pole. (Parry, W.E.) London, 1828.
	Geog 4678.10.5	Narrative of calamitous and interesting shipwrecks. Philadelphia, 1810.
	Geog 4218.57	Narrative of circumnavigation of globe...1857-58. (Scherzer, Karl.) London, 1861-63. 3v.
	Geog 5518.36	Narrative of discoveries on north coast. (Simpson, Thomas.) London, 1843.
	Geog 5220.9.6	Narrative of discovery...polar seas. (Leslie, J.) Edinburgh, 1845.
	Geog 5220.9	Narrative of discovery...polar seas. (Leslie, J.) N.Y., 1833.
	Geog 5220.9.3	Narrative of discovery...polar seas. (Leslie, J.) N.Y., 1836.
	Geog 4180.26	Narrative of embassy...to court of Timar. (Clavijo, R.G. de.) London, 1859.
	Geog 4208.34	A narrative of excursions, voyages and travels. (Rapelje, George.) N.Y., 1834.
	Geog 5838.28.3	Narrative of expedition to...Greenland. (Graah, W.A.) London, 1837.
Htn	Geog 5518.46*	Narrative of expedition to Arctic Sea. (Rae, J.) London, 1850.
	Geog 4208.31.5	Narrative of four voyages. (Morrell, B.) N.Y., 1832.
	Geog 4418.29	Narrative of journey from Calcutta. (Lushington, C. (Mrs.).) London, 1829.
Htn	Geog 5518.19*	Narrative of journey to...Polar Sea. (Franklin, John.) London, 1823.
Htn	Geog 5518.19.2*	Narrative of journey to...Polar Sea. (Franklin, John.) London, 1823. 2v.
	Geog 5518.19.5	Narrative of journey to...Polar Sea. (Franklin, John.) Philadelphia, 1824.
	Geog 5518.19.3	Narrative of journey to...Polar Sea. 2. ed. (Franklin, John.) London, 1824. 2v.
	Geog 5518.19.4	Narrative of journey to...Polar Sea. 3. ed. (Franklin, John.) London, 1824. 2v.
	Geog 4208.44.3	Narrative of life...of T.W. Smith. (Smith, T.W.) Boston, 1844.
	Geog 4238.72	Narrative of life and adventures of Captain Wells...voyages. (Wells, Theodore.) Biddeford, 1874.
	Geog 4208.50.3	Narrative of life and travels. 2. ed. (Prince, Nancy.) Boston, 1853.
	Geog 4208.50.5	Narrative of life and travels. 3. ed. (Prince, Nancy.) Boston, 1856.
	Geog 5558.71	Narrative of north polar expedition. (Davis, C.H.) Washington, 1876.
	Geog 4180.64	Narrative of Portuguese Embassy. (Alvares, F.) London, 1881.
	Geog 4180.34	Narrative of proceedings of Pedrarias Davila. (Andagoya, P. de.) London, 1865.
	Geog 5518.25	Narrative of second expedition to Polar Sea. (Franklin, John.) Philadelphia, 1828.
	Geog 4672.23	Narrative of shipwrecks and disasters, 1586-1860. (Huntress, Keith G.) Ames, 1974.
	Geog 5518.36.9	Narrative of the Arctic land expedition to the mouth of the Great Fish River. 2. ed. (Back, George.) Philadelphia, 1837.
	Geog 4708.21	Narrative of the capture...of Captain B. Lincoln...by piratical schooner. (Lincoln, B.) Boston, 1822.
	Geog 4708.74	Narrative of the circumstances of and attending the sinking of the Spanish galleons. (Galleon Treasure Company, N.Y.) N.Y., 1874.
	Geog 5518.36.2	Narrative of the discoveries on the north coast of America. 2. ed. (Simpson, Thomas.) Toronto, 1970. 2v.
	Geog 5398.79.5	Narrative of the "Jeannette". (Danenhower.) Boston, 1882.
	Geog 4678.03	Narrative of the loss of the Lady Hobart Packet. (Fellows, William D.) London, 1803.
Htn	Geog 4678.41.6*	A narrative of the loss of the Royal George. 6. ed. (Slight, Julian.) Portsea, 1843.
Htn	Geog 4708.16.15*	Narrative of the mutiny on board the schooner Plattsburg. (Onion, S.B.) Boston, 1819.
	Geog 4180.48	Narrative of the rites and laws of the Yncas. (Markham, C.R.) N.Y., 1964?
	Geog 4678.36	Narrative of the shipwreck. (Holden, H.) Boston, 1836.
	Geog 4678.36.5	Narrative of the shipwreck. (Holden, H.) Cooperstown, N.Y., 1841.
Htn	Geog 4678.30.5*	Narrative of the shipwreck and sufferings of the crew of the English brig Neptune. (Smith, John.) N.Y., 1830.
	Geog 4678.16.2	A narrative of the shipwreck and unparalleled sufferings of Mrs. S. Allen. 2. ed. (Allen, S. (Mrs.).) Boston, 1816.
Htn	Geog 4678.25*	Narrative of the shipwreck of the brig Betsey of Wiscasset, Maine...Dec. 1824. (Collins, Daniel.) Wiscasset, 1825.
	Geog 4218.45.5	Narrative of the voyage of H.M.S. Herald during the years 1845-51...being a circumnavigation of the globe. (Seemann, Berthold.) London, 1853.
	Geog 4308.71.9	Narrative of travels, scenes...in the Old World. 3rd ed. (McCray, R.H.) Philadelphia, 1871.
	Geog 4208.31.7	Narrative of voyage to Ethiopia. (Morrell, Abby J.) N.Y., 1833.
	Geog 5558.75.3	Narrative of voyage to Polar Sea...1875-76. (Nares, G.S.) London, 1878. 2v.
	Geog 5398.25	Narrative of voyage to the Pacific. (Beechey, F.W.) London, 1831. 2v.
	Geog 4208.40.5	Narrative of voyages and commercial enterprises. (Cleveland, R.J.) Boston, 1850.
Htn	Geog 4208.40.3*	Narrative of voyages and commercial enterprises. (Cleveland, R.J.) Cambridge, 1842. 2v.
	Geog 4218.17.8	Narrative of voyages and travels. (Delano, Amasa.) N.Y., 1970.
	Geog 4180.91	Narrative of voyages of...Sarmiento. (Sarmiento de Gamboa, Pedro.) London, 1895.
NEDL	Geog 4418.22	Narrative of voyages to explore the shores of Africa, Arabia, and Madagascar. (Owen, William Fitz William.) London, 1833. 2v.
	Geog 4418.22	Narrative of voyages to explore the shores of Africa, Arabia, and Madagascar. (Owen, William Fitz William.) Farnborough, 1968. 2v.
	Geog 4180.23	Narrative of voyages to West Indies and Mexico. (Champlain, S.) London, 1859.
	Geog 5558.64	Narrative of 2nd Arctic expedition. (Nourse, J.E.) Washington, 1879.
	Geog 5518.29	Narrative of 2nd voyage...Northwest Passage. (Ross, J.) London, 1835.
	Geog 4328.37.2	Narrative voyage to Madeira...Cyprus and Greece. (Wilde, William R.) Dublin, 1852.
	Geog 4206.13.35	Narratives from Purchas, his pilgrimes. (Purchas, Samuel.) Cambridge, 1931.
	Geog 4678.51.2	Narratives of shipwrecks. 2. ed. (Gilly, W.O.S.) London, 1851.
	Geog 4180.5	Narratives of voyages...north-west. (Rundall, T.) London, 1849.
	Geog 5180.37	Nas razdeliaet okean. Magadan, 1969.
	Geog 4332.20	Naše more. 2. izd. (Novak, Grga.) Zagreb, 1932.
	VGeog 4269.68	Naše sosedne države. (Lah, Avguštin.) Ljubljana, 1968.
	Geog 4209.63	Nasedkin, V.N. Piatnadtsat' let skitanii po zemnomu sharu. Moskva, 1963.
	An 3559.70	Naselenie Kazakhstana ot epokhi bronzy do sovremennosti. (Ismagulov, Orazak.) Alma-Ata, 1970.
	An 3559.72.10	Naselenie territorii Ukrainy epokhi medi-bronzy. (Kruts, Svitlana O.) Kyiv, 1972.
	An 2109.66.10	Nash, Manning. Primitive and peasant economic systems. San Francisco, 1966.
	Geog 4328.73	Nāsir, al-Din Shāh. Ein Harem in Bismarcks Reich. Tübingen, 1969.
	Geog 5398.98.5	Nathorst, A.G. Tva somrar, norra ishafvet. Stockholm, 1900. 2v.
	Geog 4559.60.5	The nation that refused to starve. (O'Daniel, John.) N.Y., 1960.
	Geog 4180.60.2	National and moral history of the Indies. (Acosta, J. de.) N.Y., 1964? 2v.
	Geog 4180.60	National and moral history of the Indies. v.2. (Acosta, J. de.) London, 1880.
	Geog 147.5	National Council for Geographic Education. Yearbook. Palo Alto, Calif. 1,1970+ 2v.
Htn	Geog 148.1*	National geographic magazine. 1-6 6v.
	Geog 148.1	National geographic magazine. 1915+ 170v.
	Geog 148.1.6	National geographic magazine. Cumulative index, 1899-1936. Washington, 1937.
	Geog 4019.63	National Geographic Society, Washington. Great adventures with NationalGeographic. Washington, 1963.
	Geog 148.1.15	National Geographic Society. Contributed technical papers. Katmai series. Washington. 1,1923
	Geog 148.1.18	National Geographic Society. Contributed technical papers. Stratosphere series. Washington.
	Geog 148.1.30	Pamphlet vol. National Geographic Society.
	Geog 148.5.5	National geographical journal of India. Benares. 1,1955+ 7v.

Author and Title Listing

Geog 148.5	National Geographical Society of India. Bulletin. Benares.
Geog 5900.10	National interests in Antarctica; an annotated bibliography. (Hayton, Robert D.) Washington, 1959-60.
An 359.38.3A	The national mind. (Demiashkevich, Michael John.) N.Y., 1938.
An 124.1	National Research Council. International directory of anthropologists. Washington, 1938.
An 124.5	National Research Council. International directory of anthropologists. 3. ed. Washington, 1950.
Geog 665.100	National Research Council. Ad Hoc Committee on Geography. The science of geography. Washington, 1965.
Geog 5300.52	National Research Council. Committee on Polar Research. Science in the Arctic Ocean basin. Washington, 1963.
Geog 4269.18A	National Research Council. Division of Geology and Geography. The geography of Europe. New Haven, 1918.
Geog 5970.38.5	National Science Foundation. Office of Antarctic Programs. USARP; United States Antarctic Research Program, National Science Foundation. Washington, 1963.
An 379.21.3	National welfare and national decay. (McDougall, William.) London, 1921.
An 359.39.3	The native races of Asia and Europe. (Frazer, J.G.) London, 1939.
Geog 110.1.20.15	Natsional'nyi Komitet Sovetskikh Geografov. XX mezhdunarodnyi geograficheskii kongress, London, iiul 1964 gg. Moskva, 1966.
An 3208.93	Die natürliche Auslese bein Menschen. (Ammon, Alto.) Jena, 1893.
An 379.70	Natural environment and society in the theory of geographical determinism. (Maršík, Miroslav.) Praha, 1970.
An 359.73.35	A natural history of associations. (Bradfield, Maitland.) London, 1973. 2v.
An 358.43	The natural history of man. (Prichard, J.C.) London, 1843.
An 358.75	Natural history of man. (Quatrefages de Bréau, A. de.) N.Y., 1875.
An 358.91.15	The natural history of man. 2. ed. (Kinmont, A.) Philadelphia, 1891.
An 358.45	The natural history of man. 2. ed. (Prichard, J.C.) London, 1845.
An 358.55	The natural history of man. 4. ed. (Prichard, J.C.) London, 1855. 2v.
An 358.51.3	The natural history of the human species. (Smith, C.H.) Boston, 1851.
An 358.52	The natural history of the human species. (Smith, C.H.) London, 1852.
Geog 5925.93	Natural resource potentials of the Antarctic. (Potter, Neal.) N.Y., 1969.
Geog 5070.20	A naturalist at the poles. (Brown, R.N.R.) London, 1923.
An 359.64.30	The nature of cultural things. (Harris, Marvin.) N.Y., 1964.
An 359.52.10A	The nature of culture. (Kroeber, Alfred L.) Chicago, 1952.
Geog 5839.35.25	Naturfredning i Grønland. (Oldendow, Knud.) København, 1935.
An 358.82.5	Naturgeschichte des Menschen. (Hellwald, Friedrich.) Stuttgart, 1882-85. 2v.
An 358.45.8F	Naturgeschichte und Abbildungen des Menschen der verschiedenen Rassen. 3. Aufl. (Schinz, Heinrich Rudolf.) Zürich, 1848.
An 358.85	Die Naturwölker. v.1-2. (Schneider, W.) Paderborn, 1885-86.
Geog 5209.60	Nauchnyi arkhiv: osvoenie krainego severa. (Mendeleev, D.I.) Leningrad, 1960.
Geog 4678.65.10	Naufrage de la Ville-du-Havre et du Losh-Earn. (Weiss, N.) Paris, 1874.
Geog 4678.66	Le naufrage de L'Evening Star. (Comettant, O.) Paris, 1866.
Geog 4658.93.4	Naufragios célebres en el cabo Polonio, banco Inglés, y océano Atlántico. (Lussich, Antonio D.) Montevideo, 1938.
Geog 4658.93.2	Naufragios célebres en el cabo Polonio, el banco Inglés, y el océano Atlántico. 2. ed. (Lussich, Antonio D.) Montevideo, 1893.
Geog 4678.67	Naufragios de la armada española. Photoreproduction. (Fernández Duro, C.) Madrid, 1867.
Geog 618.5.5F	Nauka o ludakh Fryderyka Ratzla. (Babicz, J.) Wrocław, 1962.
An 359.38.15	Nauka o rasakh i rasizm. Moskva, 1938.
Geog 4329.09.10	Naumann, Friedrich. Sonnenfahrten. Berlin, 1909.
Geog 4208.21.5	Nautical reminiscences. (Ames, Nathaniel.) Providence, 1832.
Geog 5559.58	Nautilus 90 north. (Anderson, William R.) Cleveland, 1959.
Htn Geog 577.97*	Naval gazetteer. (Malham, J.) Boston, 1797. 2v.
Geog 4559.27.5A	Naval history in the law courts. (Senior, William.) London, 1927.
Geog 4181.118	Navarrete, Domingo Fernández de. The travels and controversies of Friar Domingo Navarrete. Cambridge, 1962. 2v.
Geog 3590.60	Navatrechu apasnostiam. (Murator, M.V.) Moskva, 1956.
Geog 3570.60	Navegadores portugueses, herois do mar. (Marjay, Frederico Pedro.) Lisboa, 1970.
Geog 3580.45	Navegantes y conquistadores vascos. (Diaz-Trechuelo Spinola, Maria Lourdes.) Madrid, 1965.
An 2109.36.5	Naven; a survey of the problems suggested by a composite picture of the culture of a New Guinea tribe. (Bateson, G.) Cambridge, Eng., 1936.
Htn Geog 4157.05.2F*	Navigantium...compleat collection. (Harris, J.) London, 1705.
Htn Geog 4157.05.10F	Navigantium...compleat collection. (Harris, J.) London, 1744. 2v.
Geog 4157.05.15F	Navigantium...voyages and travels. (Harris, J.) London, 1764. 2v.
Htn Geog 4205.99F*	Navigatio ac itinerarium. (Linschoten, Jan H. van.) Hagae-Comitis, 1599.
Htn Geog 3057.3*	Navigatio Salomonis Ophirifica. (Lipenius, M.) n.p., 1660.
Geog 5516.19.15	Navigatio septentrionalis. (Hofrenning, Bernt Monson.) Fergus Falls, Minn., 1942.
Geog 5516.19	Navigatio septentrionalis. (Munk, J.) Kjøbenhavn, 1723.
Geog 5516.19.3	Navigatio septentrionalis. (Munk, J.) Kjøbenhavn, 1883.
Geog 3530.7.2	Les navigations françaises. (Margery, P.) Paris, 1867.
Geog 3530.7	Les navigations françaises. (Margery, P.) Paris, 1867.
Geog 4239.68.10	Naydler, Merton. The penance way; the mystery of Puffin's Atlantic voyage. London, 1968.
Geog 5399.32.6	Ne "Sibirakove" v Tikhii okean. (Vize, Vladimir I.) Leningrad, 1934.
Geog 4308.18	Neale, A. Travels thro'...Germany, Poland. London, 1818.
Htn Geog 4126.43*	Neale, Thomas. A treatise of direction, how to travell safely, and profitably into forraigne countries. London, 1643.
An 2059.29	The neanderthal phase of man. (Hrdlička, A.) Washington, 1929.
Geog 5559.05	Nearest the Pole...1905-06. (Peary, R.E.) N.Y., 1907.
Geog 5509.66	Neatby, Leslie Hamilton. Conquest of the last frontier. Toronto, 1966.
Geog 5509.58	Neatby, Leslie Hamilton. In quest of the North West Passage. London, 1958.
Geog 5535.20	Neatby, Leslie Hamilton. The search for Franklin. London, 1970.
Htn Geog 4215.86*	Neaudro, M. Orbis terrae partium. Lipsiae, 1586.
VGeog 4559.67	Nebo, more i ponekad kopno. (Proface, Bruno.) Zagreb, 1967.
Geog 3510.11	Nederlands voorhoede. 2. druk. (Belen, Willem Julius.) Amsterdam, 1946.
Geog 13.201	Nederlandsch Aardrijkskundig Genootschap, Amsterdam. Bijbladen. Amsterdam. 1879-1883 3v.
Geog 13.200.5	Nederlandsch Aardrijkskundig Genootschap, Amsterdam. Systematisch register. Leiden. 1876-1960 3v.
Geog 13.200	Nederlandsch Aardrijkskundig Genootschap, Amsterdam. Tijdschrift. Amsterdam. 1,1900+ 72v.
NEDL Geog 13.200	Nederlandsch Aardrijkskundig Genootschap, Amsterdam. Tijdschrift. Amsterdam. 1-17,1876-1899
Geog 4300.36	Nederlandsch-Amerikaansche Stoomvaart Maatschappij, N.V. Guide for Europe. The Hague, 1925.
Geog 4168.29	Nederlandsch zeeseizen in het laatst der z estiende. v.1-4. (Bennet, Roelof Gabriel.) Wijk, 1828-29. 3v.
An 3308.87.5	Need of anthropometry. (Hitchcock, E.) Brooklyn, 1887.
Geog 4324.84	Neeffs, E. Un voyage au XVe siècle. Louvain, 1873.
Geog 4509.52	Negri, Carlo. Alpinismo. 3. ed. Milano, 1952.
Geog 5318.2.10	A Negro explorer at the North Pole. (Henson, Matthew Alexander.) N.Y., 1969.
Geog 819.34	Négy Világrésg Földje és Népei. (Kalmár, Gusztáv.) Budapest, 1934-1935.
Geog 4559.54	Neider, C. Man against nature. 1. ed. N.Y., 1954.
Geog 6005.5	Neider, Charles. Antarctica; authentic accounts of life and exploration. N.Y., 1972.
Geog 4672.18	Neider, Charles. Great shipwrecks and castaways; authentic accounts of adventures at sea. N.Y., 1952.
An 609.07.2	Neilsen, Y. Universitets lappiske samlinger, 1857-1911. Christiania, 1911.
Geog 4326.86	Neitzschitz, G.C. Sieben-Jährige...Welt-Beschauung. Nurnberg, 1686.
Geog 4269.60.5	Nejedlý, O. Malířovy toulky po Evropě, Eejlonu a Indii. Praha, 1960.
Geog 4558.77	Nelson, A.W. Yankee Swanson. N.Y., 1913.
Geog 4219.30	Nelson, Robert. World beaters. Boston, 1930.
An 2109.60.10	Neolithische Kulturerscheinungen. (Smolla, G.) Bonn, 1960.
Geog 4559.57.5	Neptuania, maritieme curiosa. (Houter, F. den.) Amsterdam, 1957.
Geog 4207.80.2	Neshchastiia prikliucheniia...Amerike, Azii i Evrope s" 1780 ro 1787 god". 2. izd. (Baranshchikov, V.) Sankt Peterburg, 1787.
Geog 618.3.7	Nettlau, Max. Elisée Reclus. Berlin, 1928.
Geog 618.3.8	Nettlau, Max. Eliseo Reclus, la vida de un sabio justo. Barcelona, 1929. 2v.
Geog 150.1	Neue allgemeine geographische Ephemeriden. Weimar.
An 170.283	Neue Anthropologie. Stuttgart, 1972-75. 7v.
Geog 665.104	Neue Fragen der allgemeinen Geographie. Wurzburg, 1964.
Geog 3249.72	Neue Horizonte. 1. Aufl. (Krämer, Walter.) Leipzig, 1972.
Geog 5855.3	Neue Menschen...des Nordpols. (Rasmussen, Knud.) Bern, 1907.
Htn Geog 149.1*	Neue nordische Beyträge. (Pallas, Peter.) St. Petersburg, 1781. 4v.
Geog 5367.65.5	Neuer Versuch über...Schriften. (Engel, S.) Basel, 1777.
NEDL Geog 18.2	Neues Archiv für Geschichte. Wien. 1-2,1829-1830 2v.
Geog 5398.98	Neues Land, vier Jahre in arktischen Regionen. (Sverdrup, O.) Leipzig, 1903. 2v.
Geog 665.45	Die Neugestaltung der deutschen Geographie im 18. Jahrhundert. (Kühn, Arthur.) Leipzig, 1939.
Geog 5919.01A	Neumayer, G. Auf zum Südpol! Berlin, 1901.
Geog 5399.28.40	Neun Männer im Eis. (Katz, Otto.) Berlin, 1929.
Geog 4217.90.18	Die neuste Reise um die Welt in den Jahren 1790, 1791 und 1792 von Etienne Marchand. (Fleurieu, Charles P. Claret de.) Leipzig, 1801? 2v.
Geog 4207.99	Nevens, William. Forty years at sea. Portland, 1846.
Geog 4218.03.12	Nevskii, Vladimir V. Perv. putesh. rossii. vokrug sveta, 1803-1806. Moskva, 1951.
Geog 4218.03.14	Nevskii, Vladimir V. Vokrug sveta pod russkim flagom. Moskva, 1953.
Geog 807.87.5	A new, royal, authentic and complete system of universal geography. (Bankes, T.) London, 1787-1810? 2v.
Geog 4181.19	New account of East India and Persia. (Fryer, J.) London, 1909. 3v.
Geog 578.34	New and comprehensive gazetteer. (Wright, G.N.) London, 1834-37. 4v.
Geog 5398.67	The new Arctic continent. (Wheildon, W.W.) Cambridge, 1869.
Geog 4157.67	New collection of voyages. (Knox, J.) London, 1767. 7v.
Htn Geog 4167.11*	New collection of voyages. (Stevens, J.) London, 1711. 2v.
Geog 4157.45	New collection of voyages and travels. (Green, J.) London, 1745 47. 4v.
An 2059.30	New discoveries relating to the antiquity of man. (Keith, Arthur.) N.Y., 1930.
Htn Geog 4476.92.2*	A new discovery of terra Incognita Australis. (Foigny, Gabriel de.) London, 1693.
Geog 4228.89	New Far West and old Far East. (Barneby, W.H.) London, 1889.
Geog 577.76	New gazeteer. (Johnson, R.) London, 1776.
Htn Geog 578.02*	A new gazeteer of the Eastern continent. (Morse, Judith.) Charlestown, 1802.
Geog 817.82	A new geographical, historical, and commercial grammar and present state of several kingdoms of the world. 7th ed. (Guthrie, William.) London, 1732.
Geog 817.83	A new geographical, historical, and commercial grammar and present state of the several kingdoms of the world. 8th ed. (Guthrie, William.) London, 1783.
Geog 817.94.10	New geographical, historical...grammar. (Guthrie, William.) London, 1796.
Geog 817.94.13	New geographical, historical...grammar. (Guthrie, William.) Montreal, 1810.

Author and Title Listing

	Geog 818.09	A new geographical, historical and commercial grammar. 1st American ed. (Guthrie, William.) Philadelphia, 1809.
	Geog 817.88	A new geographical, historical and commercial grammar. 11th ed. (Guthrie, William.) London, 1788.
	Geog 817.92	A new geographical, historical and commercial grammar. 13th ed. (Guthrie, William.) London, 1792.
	Geog 817.51.2	A new geographical...grammar. (Salmon, T.) London, 1749.
	Geog 817.51.3	New geographical...grammar. (Salmon, T.) London, 1751.
	Geog 817.51.5	New geographical...grammar. (Salmon, T.) London, 1764.
	Geog 817.51.7	New geographical...grammar. (Salmon, T.) London, 1766.
	Geog 817.51.8	New geographical...grammar. (Salmon, T.) London, 1769.
	Geog 577.59F	New geographical dictionary of the known world. London, 1759. 2v.
	Geog 817.89	New geographical grammar. (Gordon, William.) Edinburgh, 1789.
	Geog 5398.98.2	New land, four years in Arctic regions. (Sverdrup, O.) London, 1904. 2v.
	Geog 5398.72	New Lands within the Arctic Circle. (Payer, J.) London, 1876. 2v.
	Geog 5398.72.5	New lands within the Arctic Circle. (Payer, J.) N.Y., 1877.
	Geog 4181.34	New light on Drake. (Mittall, Zelia.) London, 1914.
	Geog 4181.64	New light on the discovery of Australia. (Stevens, H.N.) London, 1930.
	Geog 4239.57.10	The new Mayflower. (Villiers, Alan John.) Leicester, 1959.
	Geog 4329.05.20	The new Mediterranean traveller. 13th ed. (Lorenz, I.E.) N.Y., 1927.
	Geog 819.30	A new regional geography of the world. (Newbigin, Marion I.) N.Y., 193-?
	An 3309.08	New South Wales. Department of Public Instruction. Report - physical condition of children. Sydney, 1908.
	Geog 807.62	New system of geography. (Büsching, Anton Friedrich.) London, 1762. 6v.
	Geog 818.10.5	New system of geography. (Evans, J.) London, 1810. 2v.
Htn	Geog 816.94*	A new system of geography. (Seller, John.) n.p., 1694?
	Geog 757.64F	A new system of geography: or A general description of the world. (Fenning, Daniel.) n.p., 17- . 2v.
Htn	Geog 818.05*	A new system of modern geography. (Davies, Benjamin.) Philadelphia, 1805.
	Geog 817.94.5F	New system of modern geography. (Guthrie, William.) London, 1795.
	Geog 817.94	New system of modern geography. (Guthrie, William.) Philadelphia, 1794-1795. 2v.
	Geog 818.22	New system of modern geography. (Morse, S.E.) Boston, 1822.
	An 3309.32	New types of old Americans at Harvard and at eastern women's colleges. (Bowles, G.T.) Cambridge, 1932.
	Geog 577.91.10	New universal gazeteer. (Brookes, R.) Boston, 1850.
	Geog 578.21	New universal gazeteer of the known world. 3. ed. (Morse, Jedidiah.) New Haven, 1821.
	Geog 577.91.6.5	A new universal gazetteer. (Brookes, R.) Philadelphia, 1844.
	Geog 577.91.7	A new universal gazetteer. (Brookes, R.) Philadelphia, 1847.
	Geog 577.98	New universal gazetteer. (Cruttwell, C.) London, 1798. 3v.
	Geog 577.91.6	A new universal gazetteer containing a description of the principal nations. (Brookes, R.) N.Y., 1832.
	Geog 817.79F	New universal traveller. (Carver, J.) London, 1779.
	Geog 4181.73	A new voyage and description of the Isthmus of America. (Joyce, L.E. Elliott.) Oxford, 1934.
Htn	Geog 4206.82*	A new voyage to the East Indies. 2d ed. (Glanius.) London, 1682.
	Geog 4158.19	New voyages and travels. (Phillips, R.) London, 1819-20. 9v.
	Geog 819.22	The new world; problems in political geography. (Bowman, I.) Yonkers-on-Hudson, 1922.
	Geog 819.21	The new world. (Bowman, I.) Yonkers-on-Hudson, 1921.
	Geog 819.28.5A	The new world. 4th ed. (Bowman, I.) Yonkers-on-Hudson, 1928.
	Geog 759.42F	New world horizions; geography for the air age. (Lawrence, C.H.) N.Y., 1942.
	Geog 3019.49A	New worlds emerging. 1. ed. (Hanson, Earl P.) N.Y., 1949.
	Geog 4300.34	New York Herald Tribune. European Edition. Guide to Europe. Paris?
	Geog 4229.00.5	New York to Alaska...May to July, 1900. (Winter, James M.) Middletown, N.Y, 1943.
	Geog 6100.77	New Zealand. Geographic Board. Provisional gazetteer of the Ross Dependency. Wellington, 1958.
	Geog 5970.42	New Zealand and the Antarctic. (Quartermain, Leslie B.) Wellington, 1971.
	Geog 150.5	New Zealand geographer. 3,1947+ 12v.
	Geog 150.5.2	New Zealand geographer. Cumulative index, v.1-25, 1945-1969. Wellington, 1971.
	Geog 150.10	New Zealand Geographical Society. Record of proceedings of the society and its branches. Christchurch. 1,1946+ 3v.
	Geog 4659.36.3	New Zealand shipwrecks, 1795-1960. 3. ed. (Ingram, Charles W.N.) Wellington, 1961.
	Geog 500.55	Newark, N.J. Free Public Library. Foreign countries. Washington, 1918.
	Geog 819.11	Newbigin, Marion I. Modern geography. London, 1911.
	Geog 819.11.1	Newbigin, Marion I. Modern geography. N.Y., 1911.
	Geog 819.30	Newbigin, Marion I. A new regional geography of the world. N.Y., 193-?
	Geog 4169.19A	Newbolt, Henry. The book of the long trail. N.Y., 1919.
	Geog 4559.56	Newby, Eric. The last grain race. Boston, 1956.
	Geog 3019.75F	Newby, Eric. The Mitchell Beazley world atlas of exploration. London, 1975.
	Geog 4329.49	Newman, B. Mediterranean background. London, 1949.
	Geog 4329.47	Newman, B. Middle eastern journey. London, 1947.
	Geog 4359.40	Newman, Bernard. Savoy! Corsica! Tunis! Mussolini's dream lands. London, 1940.
	Geog 4209.30.2	Newton, A.E. A tourist in spite of himself. Boston, 1930.
	Geog 4209.30.3	Newton, A.E. A tourist in spite of himself. Boston, 1933.
	Geog 4248.54	Niboyet, P. Les mondes nouveaux. Paris, 1854.
	Geog 3055.29	Nicaise, A. Les terres disparues. Chalons sur Marne, 1885.
	Geog 4349.36	Nichols, Beverley. No place like home. London, 1936.
	Geog 4349.36.2	Nichols, Beverley. No place like home. 1st ed. Garden City, 1936.
	Geog 4218.78.20	Nichols, E.P. The Ocean Chronicle, 1878-1891. N.Y., 1941.
	Geog 4308.45.9	Nicholson, Asenath. Loose papers...residence in Ireland. N.Y., 1853.
	An 359.43.30	Nicholson, James E. Anthropos; or, The problem of man. London, 1943.
	Geog 4209.07.10	Nicholson, T.R. Adventurer's road. N.Y., 1958.
	Geog 4209.60.5	Nicholson, Timothy. Five roads to danger. London, 1960.
	Geog 4208.22.6	Nicol, John. The life and adventures of John Nicol, mariner. London, 1937.
	Geog 818.52.5	Nicolay, Charles G. A manuel of geographical science. v.11. London, 1859.
	Geog 4515.101	Niederl, Franz. Das gehen auf Eis und Schnee. 2. Aufl. München, 1927.
	An 3509.70F	Nielsen, Ole Vagn. The Nubian skeleton through 4000 years. Thesis. København, 1970.
	An 608.609.07	Nielsen, Y. Universitets ethnografiske samlinger, 1857-1907. Christiania, 1907.
	An 379.72.10	Niemeier, Georg. Siedlungsgeographie. 3. Aufl. Braunschweig, 1972.
	Geog 4308.20	Niemeyer, A.H. Beobachtungen auf Reisen in und ausser Deutschland. v.2-4. Halle, 1822-1826. 4v.
	Geog 4307.03	Die nieuwe Béreisde wereld. (Broekhuizen, G.) Amsterdam, n.d. 2 pam.
	Geog 4679.12.18	A night to remember. (Lord, Walter.) London, 1956.
	Geog 4679.12.17A	A night to remember. (Lord, Walter.) N.Y., 1955.
	Geog 4205.98.10	Nijhoff, Wouter. Bibliographie...die voyagie om den geheelen werelt...door Olivier van Noort. 's-Gravenhage, 1926.
	Geog 4209.60.10	Nikhailov, Nikolai I. From pole to pole. Moscow, 1960.
	Geog 5312.4	Nikifor Belichev. 2. izd. (Bolotnekov, N.) Moskva, 1954.
	Geog 4239.62	Nikiforovskii, V.A. Ekspeditsiia na "Sedove" v Atlanticheskii okean. Moskva, 1962.
	Geog 4209.71.30	Nikolaev, Vladimir D. Maiak Zemli. Moskva, 1971.
	Geog 3140.17	Ninck, M. Die Entdeckung von Europa durch die Griechen. Basel, 1945.
	Geog 4308.85.5	Ninde, Mary L. We two-alone in Europe. Chicago, 1889.
	Geog 4208.57	Nine years a sailor. (Nordhoff, Charles.) Cincinnati, 1857.
X Cg	Geog 4269.07.7PF	Nineteen thousand miles in 1907. (Hunnewell, H.H.) Boston, 1908. 8v.
	Geog 4308.61.5	Ninety days worth of Europe. (Hale, E.E.) Boston, 1861.
	Geog 5839.24	Nissen, N.W. Die sudwestgrönländische Landschaft und das Siedlungsgebiet der Normannen. Hamburg, 1924.
	Geog 4345.99.2	Nixon, Anthony. The three English brothers. London, 1607.
	Geog 4559.38.12	Nixon, Laurence A. Vagabond voyaging; the story of freighter travel. Boston, 1939.
	Geog 6009.55.15	No latitude for error. (Hillary, Edmund.) London, 1961.
	Geog 4309.44.10	No matter where. (Hiett, Helen.) N.Y., 1944.
	Geog 4309.40	No place like home. (Abbe, P.) N.Y., 1940.
	Geog 4349.36	No place like home. (Nichols, Beverley.) London, 1936.
	Geog 4349.36.2	No place like home. 1st ed. (Nichols, Beverley.) Garden City, 1936.
	Geog 4328.19	Noah, M.M. Travels in England, France and Spain. N.Y., 1819.
	Geog 4145.61	Noailles, Loise de. Souvenirs de quatre horizons. Fribourg, 1958.
	Geog 4659.64	Noall, Cyril. Wreck and rescue round the Cornish coast. Truro, 1964-65. 3v.
	Geog 5399.28.10	Nobile, Umberto. L'"Italia" al Polo Nord. Milano, 1930.
	Geog 5399.28.20	Nobile, Umberto. Met de "Italia" naar de Noordpool. Baarn, 1931.
	Geog 5399.28.15	Nobile, Umberto. With the "Italia" to the North Pole. London, 1930.
	Geog 4559.56.15	Noble, Arthur H. Three ships came sailing. London, 1956.
	Geog 4309.71	Noc w Soho. Wyd. 1. (Bąbiński, Gan.) Łódź, 1971.
	Geog 818.54.10	Nociones generales de geografía astronómica. (Condaminas, S.) Matanzas, 1854.
	Geog 5324.2.65	Nockher, L. Fridtjof Nansen. Stuttgart, 1955.
	Geog 4328.51.5	Noel-Fearn, H. The shores and islands of the Mediterranean. London, 1851. 3v.
	Geog 5650.19	Noerlund, P. De gambe nordbobygder ved verdens ende. 3. opl. København, 1942.
	Geog 5650.17	Nørlund, P. Viking settlers in Greenland. London, 1936.
	Geog 4309.47.12	Noguera Mora, Neftali. Alegría y llanto de Europe; verano de 1946. 2. ed. Mérida, 1965.
	Geog 4419.69	Nolan, Cynthia. A sight of China. London, 1969.
	An 2058.79	Nomenclature des temps anthropiques primitifs. (Piette, Edouard.) Laon, 1879.
	An 2109.27.23	La non-civilisé et nous. (Allier, Raoul.) Paris, 1927.
Htn	Geog 4347.85F*	Noord en Oost Tartaryen. (Witsen, N.) Amsterdam, 1785. 2v.
	Geog 1516.3	Le Nord de la France. (Baedeker, publishers.) Leipzig, 1884.
	Geog 1516.4	Le Nord de la France. 2. éd. (Baedeker, publishers.) Leipzig, 1887.
	Geog 5398.20.25	Le nord de la Sibérie. (Wrangell, F. von.) Limoges, 188-?
	Geog 5398.20.20	Le nord de la Sibérie. (Wrangell, F. von.) Paris, 1843. 2v.
	Geog 1518.55	Le Nord-est de la France. 5. éd. (Baedeker, publishers.) Leipzig, 1895.
	Geog 5220.22	Nord i tåkeheimen. (Nansen, F.) Kristiania, 1911.
	Geog 1518.80	Le Nord-ouest de la France. 5. éd. (Baedeker, publishers.) Leipzig, 1895.
	Geog 1518.89	Le Nord-ouest de la France. 9. éd. (Baedeker, publishers.) Leipzig, 1913.
	Geog 1645.11	Nordamerika. Die Vereinigten Staaten...Mexiko. 2. Aufl. (Baedeker, publishers.) Leipzig, 1904.
	Geog 5650.35	Nordboerne i Grønland. (Meldgaard, Jørgen.) Købwnhavn, 1965.
	Geog 5220.11	Norden oder zu Wasser und Lande. (Capel, R.) Hamburg, 1678.
	Geog 5378.80.9	Nordenskiöld, A.E. Lettres...du passage du pôle nord. Paris, 188-.
	Geog 5378.80.11	Nordenskiöld, A.E. Nordenskiölds Vegafahrt um Asien und Europa. Leipzig, 1886.
	Geog 5378.80.13	Nordenskiöld, A.E. Shvedskaia poliarnaia ekspeditsiia, 1878-79 g. Sankt Peterburg, 1880.
	Geog 5378.80.7	Nordenskiöld, A.E. Studien und Forschungen...Reisen. Leipzig, 1885.
	Geog 5378.80.5	Nordenskiöld, A.E. Vegas färd rking Asien och Europa. Stockholm, 1880-81. 2v.
	Geog 5378.80	Nordenskiöld, A.E. Voyage of the Vega round Asia and Europe. London, 1881. 2v.
	Geog 5378.80.2	Nordenskiöld, A.E. The voyage of the Vega round Asia and Europe. N.Y., 1882.
	Geog 5324.1.10	Nordenskiöld. (Ramsey, H.) Stockholm, 1950.
	Geog 5378.80.4	Nordenskiöld's Rejse omkring Asien og Europa. (Hovgaard, A.P.) Kjøbenhavn, 1881.
	Geog 5378.80.11	Nordenskiölds Vegafahrt um Asien und Europa. (Nordenskiöld, A.E.) Leipzig, 1886.
	Geog 5378.80.3	Nordenskiöld's Voyage round Asia. (Hovgaard, A.P.) London, 1882.
	Geog 5838.83	Nordenskiold, A.E. Den andra Dicksonska Expedition. Stockholm, 1885.

Author and Title Listing

	Geog 5838.83.3	Nordenskiold, A.E. Grönland; seine Eiswüsten im Innern. Leipzig, 1886.
	Geog 3251.7	Nordenskiold, A.E. Om en märklig globkarta. Stockholm, 1884.
	Geog 6009.01	Nordenskjöld, Otto. Antarctic...zwei Jahre. Berlin, 1904. 2v.
	Geog 6009.01.3	Nordenskjöld, Otto. Antarctica, or Two years...South Pole. London, 1905.
	Geog 5070.36A	Nordenskjöld, Otto. The geography of the polar regions. N.Y., 1928.
	Geog 5838.70F	Nordenskjölds Reise nach Westgrönland im Jahre 1870. n.p., 187-?
	An 359.17.5	Nordenstreng, Rolf. Europas människoraser. 2. uppl. Stockholm, 1917-20.
	Geog 4559.40	Nordhoff, C. In Yankee windjammers. N.Y., 1940.
	Geog 4558.74	Nordhoff, C. Life on the ocean. Cincinnati, 1874.
	Geog 4208.95.5	Nordhoff, C. The merchant vessel. N.Y., 1895.
	Geog 4208.95.5.5	Nordhoff, C. The merchant vessel. N.Y., 1895.
	Geog 4558.56	Nordhoff, C. Whaling and fishing. Cincinnati, 1856.
	Geog 4218.55.5	Nordhoff, Charles. Man-of-war life. Cincinnati, 1856.
	Geog 4218.55.9	Nordhoff, Charles. Man-of-war life. N.Y., 1883.
	Geog 4208.57	Nordhoff, Charles. Nine years a sailor. Cincinnati, 1857.
	Geog 4309.30.6	Nordic trails. (Morse, W.I.) Boston, 1930.
Htn	Geog 4309.30.5*	Nordic trails. (Morse, W.I.) Boston, 1930.
	An 2109.18.3	Das nordische Kultur-Gebiet in Mitteleuropa. v.1-2. (Aberg, Nils.) Uppsala, 1918.
	Geog 5896.2	Nordmaend paa Grønland, 1721-1814. (Ostermann, Hother.) Oslo, 1940. 2v.
	Geog 1523.75	Nordost-Deutschland. 25. Aufl. (Baedeker, publishers.) Leipzig, 1896.
	Geog 5379.18	Nordostpassagen. (Amundsen, Roald.) Kristiania, 1921.
	Geog 5368.80	Nordostpassagens historia. (Stuxberg, A.) Stockholm, 1880.
	Geog 5209.29	Norðr um höf. (Einarsson, S.) Reykjavik, 1924.
	Geog 1523.80	Nordwest-Deutschland. 26. Aufl. (Baedeker, publishers.) Leipzig, 1899.
	Geog 1523.81	Nordwestdeutschland. 27. Aufl. (Baedeker, publishers.) Leipzig, 1902.
Htn	Geog 755.71*	Nores, Jason de. Breve trattato del mondo e delle sue parti. Venetia, 1571.
	Geog 3530.15	La Normandie exploratrice et colonisatrice du XVe au XVIIIe siècle. (Dupic, Jeanne.) Rouen, 1932.
	Geog 3055.23	Norof, A.S. Die Atlantis. St. Petersburg, 1854.
	Geog 151.6	Norois; revue géographique de l'Ouest et des pays de l'Atlantique nord. Poitiers. 1,1954+ 20v.
	Geog 151.6.5	Norois. Table décennale, 1954-73. Poitiers, n.d. 2v.
	An 3558.96	Norrønaskaller. Crania antiqua. (Barth, J.) Christiania, 1896.
	Geog 5650.11	The Norsemen in Greenland; recent discoveries at Herjolfones. (Hovgaard, William.) N.Y., 1925.
	Geog 151.1.2	Norsk geografisk tidsskrift. Oslo. 18,1961+ 18v.
	Geog 151.1.3	Norsk geografisk tidsskrift. Register. Oslo.
	Geog 151.1	Norske geografisk selskabs aarbog. 1-32 20v.
	Geog 151.2	Den Norske turistforenings årbog. Kristiania. 1868-1913 23v.
	Geog 607.1	North, S.N.D. Henry Gannett, president of the National Geographic Society, 1910-1914. n.p., 1915.
	Geog 5160.16	North. 1. ed. (Rodahl, Kåre.) N.Y., 1953.
	Geog 5324.1.15	North-east passage: Adolf Erik Nordenskiöld, his life and times. (Kish, George.) Amsterdam, 1973.
	Geog 5518.19.6	The North Georgia Gazette and Winter Chronicle. No.1-20. (Parry, W.E.) London, 1821.
	Geog 151.4F	North German Lloyd bulletin. N.Y.
	Geog 4312.5	North in a nutshell. (Raynes, Rozelle.) Lymington, 1968.
	Geog 4249.36	North Pacific: Japan, Siberia, Alaska, Canada. (Allen, Edward Weber.) N.Y., 1936.
	Geog 5559.09A	North Pole; its discovery in 1909. (Peary, R.E.) N.Y., 1910.
	Geog 5559.09.3	The North Pole. 2. ed. (Peary, R.E.) N.Y., 1910.
	Geog 5180.8	The North Pole and Bradley land. (Balch, E.S.) Philadelphia, 1913.
	Geog 4218.66.5	The North Star and the Southern Cross. (Weppner, M.) Albany, 1876. 2v.
	Geog 4218.66	The North Star and the Southern Cross. 3. American ed. (Weppner, M.) Albany, 1880.
	Geog 4350.5	The north-west and north-east passages. (Alexander, P.F.) Cambridge, 1915.
	Geog 5507.45.6	North West Committee. Articles of agreement. London, 1745.
Htn	Geog 5516.31.9*	North-west Fox...voyages of Cabot, Frobisher. (Fox, Luke.) London, 1635.
	Geog 5506.35.1	North-west Fox. (Fox, Luke.) n.p., 1965.
	Geog 5519.03	The North West Passage. (Amundsen, R.) London, 1908. 2v.
	Geog 5509.75	The North-west passage. (Thomson, George Malcolm.) London, 1975.
	Geog 151.5	North Western University studies in geography. Evanston. 1+ 11v.
	Geog 4219.23	Northcliffe, A.C.W.H. My journey round the world. Philadelphia, 1923.
	Geog 4159.10.10	Northcliffe, A.C.W.H. The world's greatest books. v.19. n.p., 1910.
	Geog 1524.72	Northern Bavaria. (Baedeker, publishers.) Hamburg, 1951.
	Geog 5208.53	Northern coasts of America. London, 1853.
	Geog 1516.5	Northern France. (Baedeker, publishers.) Leipsic, 1889.
	Geog 1516.10	Northern France. 2. ed. (Baedeker, publishers.) Leipsic, 1894.
	Geog 1516.10.2	Northern France. 2. ed. (Baedeker, publishers.) Leipsic, 1894.
	Geog 1516.15	Northern France. 3. ed. (Baedeker, publishers.) Leipsic, 1899.
	Geog 1516.15.3	Northern France. 3. ed. (Baedeker, publishers.) Leipsic, 1899.
	Geog 1516.20.4	Northern France. 4. ed. (Baedeker, publishers.) Leipzig, 1905.
	Geog 1516.25	Northern France. 5. ed. (Baedeker, publishers.) Leipzig, 1909.
	Geog 1523.20	Northern Germany. 5. ed. (Baedeker, publishers.) Coblentz, 1873.
	Geog 1523.26	Northern Germany. 6. ed. (Baedeker, publishers.) Leipsic, 1877.
	Geog 1523.25	Northern Germany. 6. ed. (Baedeker, publishers.) Leipsic, 1877.
	Geog 1523.30	Northern Germany. 7. ed. (Baedeker, publishers.) Leipsic, 1881.
	Geog 1523.35	Northern Germany. 8. ed. (Baedeker, publishers.) Leipsic, 1884.
	Geog 1523.36	Northern Germany. 8. ed. (Baedeker, publishers.) Leipsic, 1884.
	Geog 1523.40	Northern Germany. 9. ed. (Baedeker, publishers.) Leipsic, 1886.
	Geog 1523.41	Northern Germany. 9. ed. (Baedeker, publishers.) Leipsic, 1886.
	Geog 1523.46	Northern Germany. 10. ed. (Baedeker, publishers.) Leipsic, 1890.
	Geog 1523.45	Northern Germany. 10. ed. (Baedeker, publishers.) Leipsic, 1890.
	Geog 1523.50	Northern Germany. 11. ed. (Baedeker, publishers.) Leipsic, 1893.
	Geog 1523.55	Northern Germany. 12. ed. (Baedeker, publishers.) Leipsic, 1897.
	Geog 1523.60	Northern Germany. 13. ed. (Baedeker, publishers.) Leipsic, 1900.
	Geog 1523.66	Northern Germany. 14. ed. (Baedeker, publishers.) Leipzig, 1904.
	Geog 1523.65	Northern Germany. 14. ed. (Baedeker, publishers.) Leipzig, 1904.
	Geog 1523.67A	Northern Germany. 15. ed. (Baedeker, publishers.) Leipzig, 1910.
	Geog 1523.70	Northern Germany. 16. ed. (Baedeker, publishers.) Leipzig, 1913.
	Geog 1523.73	Northern Germany. 17. ed. (Baedeker, publishers.) Leipzig, 1925.
	Geog 1539.8	Northern Italy, as far as Leghorn. (Baedeker, publishers.) Coblentz, 1868.
	Geog 1539.55	Northern Italy. 14. ed. (Baedeker, publishers.) Leipzig, 1913.
	Geog 1539.60	Northern Italy. 15. ed. (Baedeker, publishers.) Leipzig, 1930.
	Geog 5559.30	Northern lights. (Chapman, F.S.) London, 1932.
	Geog 4239.68.5	Northern mists. (Sauer, Carl Ortwin.) Berkeley, 1968.
Htn	Geog 5508.29*	Northern regions. Boston, 1829.
	Geog 5369.52	The northern sea route; Soviet exploitation of the North East Passage. (Armstrong, T.) Cambridge, Eng., 1952.
	Geog 152.10	The northern universities' geographical journal. Leicester, Eng. 1-9,1960-1968//?
	Geog 5160.7A	The northward course of empire. (Stefansson, V.) N.Y., 1922.
	Geog 5208.79	Northward ho! (Markham, A.H.) London, 1879.
	Geog 5558.86	Northward over the "Great Ice"...1886. (Peary, R.E.) N.Y., 1898. 2v.
	Geog 5509.61	Northwest by sea. (Dodge, Ernest S.) N.Y., 1961.
	Geog 5535.10.2	The Northwest Passage, and the plans for the search for Sir John Franklin. 2. ed. (Brown, John.) London, 1860.
	Geog 5150.2	The Northwest passage, 1534-1859. (Toronto. Public Library.) Toronto, 1963.
	Geog 5519.69	Northwest Passage. (Smith, William D.) N.Y., 1970.
	Geog 5509.70	The Northwest Passage: frm the Mathew to Manhattan, 1497 to 1969. (Keating, Bern.) Chicago, 1970.
	Geog 5509.58.5	Northwest to fortune. (Stefansson, V.) N.Y., 1958.
	An 132.4.5	Northwestern University, Evanston, Ill. Program of African Studies. Bibliography of Melville J. Herskovits, 1920-1962. Evanston, Ill., 1962?
	Geog 1568.24	Norway, Sweden and Denmark. 5. ed. (Baedeker, publishers.) Leipsic, 1892.
	Geog 1568.25A	Norway, Sweden and Denmark. 6. ed. (Baedeker, publishers.) Leipsic, 1895.
	Geog 1568.30	Norway, Sweden and Denmark. 7. ed. (Baedeker, publishers.) Leipzig, 1899.
	Geog 1568.35	Norway, Sweden and Denmark. 8. ed. (Baedeker, publishers.) Leipsic, 1903.
	Geog 1568.40	Norway, Sweden and Denmark. 9. ed. (Baedeker, publishers.) Leipzig, 1909.
	Geog 1568.45A	Norway, Sweden and Denmark. 10. ed. (Baedeker, publishers.) Leipzig, 1912.
	Geog 5645.5.20	Norway and Greenland; a short survey. Oslo, 1931.
	Geog 1568.5	Norway and Sweden. (Baedeker, publishers.) Leipzig, 1879.
	Geog 1568.15	Norway and Sweden. 3. ed. (Baedeker, publishers.) Leipsic, 1885.
	Geog 1568.16	Norway and Sweden. 3. ed. (Baedekr, publishers.) Leipsic, 1885.
	Geog 5399.08F	The Norwegian Aurora polaris expedition. Christiania, 1908.
	Geog 5650.9	Norwegian settlements on...east coast of Greenland. (Giesecke, C.L.) Dublin, 1824.
	An 2058.83.5	Nos ancêtres. (Girard de Rialle, J.) Paris, 1883.
	Geog 4309.13	Notas da Europa. (Vasconcelos, C. de.) Rio de Janeiro, 1913.
	An 3309.32.3F	Note antropologiche sulla popolagione del Bolognese. (Frassetto, F.) Bologna, 1932.
	Geog 3235.51	Note sur la mappemonde historiée de la cathedral de Hereford. (Avezac, M.A.P.) Paris, 1862.
	Geog 3235.49	Note sur un atlas hydrographique. v.1-2. (Avezac, M.A.P.) Paris, 1850.
Htn	Geog 3235.57*	Note sur une carte catalane de Dulceri. (Marcel, G.) Paris, 1887.
	An 170.328.8	Notes and queries on anthropology. 6. ed. (Royal Anthropological Institute of Great Britain and Ireland.) London, 1971.
	Geog 4209.13	Notes made during a cruise around the world in 1913. (Casey, R.H.) N.Y., 1914.
	Geog 4348.54	Notes of a theological student. (Hoppin, J.M.) N.Y., 1854.
	Geog 4308.46.2	Notes of a traveller. (Laing, S.) Philadelphia, 1846.
	Geog 4218.78.11	Notes of a trip round the world. (Carnegie, Andrew.) N.Y., 1879.
	Geog 4328.36.2	Notes of a wanderer. (Cumming, W.F.) London, 1840. 2v.
	Geog 4308.91	Notes of foreign travel. (Bates, J.H.) N.Y., 1891.
	Geog 4218.87.12	Notes of my journey round the world. (Cecil, Evelyn.) London, 1889.
	Geog 4418.54	Notes of travel or recollections. (Osgood, J.B.F.) Salem, 1854.
	Geog 4709.23	Notes on piracy in eastern waters. (Hill, Samuel C.) London, 1923-28.
	Geog 4678.89	Notes on some few of the wrecks and rescues during the present century. (Forbes, Robert B.) Boston, 1889.
	An 3538.80	Notes on the Crania of New England Indians. (Carr, L.) Boston, 1880.
	An 3538.80.2	Notes on the Crania of New England Indians. (Carr, L.) Boston, 1880.
	An 3508.84	Notes upon human remains from the caves of Coahuila, Mexico. (Studley, C.A.) Salem, 1884.
X Cg	Geog 4180.10	Notes upon Russia. (Herberstein, Sigmund.) London, 1851-52. 2v.

Author and Title Listing

	Geog 4180.10A	Notes upon Russia. (Herberstein, Sigmund.) N.Y., 1963? 2v.
	Geog 4309.57	Nothomb, Pierre. Pélerinages européens. Paris, 1957.
Htn	Geog 3060.16*	Notica certa do descobrimento de huma nova terra. Lisboa, 1757.
	Geog 4238.45	Notice des découvertes...dans l'Ocean. (D'Avezac.) Paris, 1845.
	Geog 4341.73.7	Notice historique sur Benjamin de Tudèle. (Carmoly, E.) Bruxelles, 1852.
	Geog 616.1.3	Notice sur Jean Parmentier. (Gravier, G.) Rouen, 1902.
	Geog 610.1	Notice sur la vie et les oeuvres de M. Jomard. (Cortambert, R.) Paris, 1863.
	An 132.1	Notice sur la vie et les travaux de M. Ernest Hamy. (Reinach, Theodore.) Paris, 1910.
	An 177.1	Notice sur les travaux scientifiques. (Capitan, J.L.) n.p., 1911.
	Geog 28.4	Notice sur l'etablissement géographique de Bruxelles. v.1-4. (Jonard, E.F.) Paris, n.d.
	Geog 3235.29	Notice sur un atlas en langue catalane. (Buchon, J.A.C.) Paris, 1838.
	Geog 3235.29.3	Notices et extraits des manuscrits. (Buchon, J.A.C.) Paris, n.d.
	Geog 3107.31	Notitia orbis antiqui. (Cellarius, C.) Lipsiae, 1731-32. 2v.
	Geog 228.25	Notiziario. (Trieste. Università. Istituto di Geografia.)
	Geog 807.16F	Notoras, C. Introductio ad geographia et spheram. Paris, 1716.
	An 3209.61	Notschaele, Lucien Aimé. Een somatometrisch onderzoek in de Noord-oost Polder. Amsterdam, 1961.
	An 358.57	Nott, Josiah C. Indigenous races of the earth. Philadelphia, 1857.
	An 358.44	Nott, Josiah C. Two lectures on the natural history of the Caucasian and Negro races. Mobile, 1844.
	An 358.54.2	Nott, Josiah C. Types of mankind. 2. ed. Philadelphia, 1854.
	An 358.54	Nott, Josiah C. Types of mankind. 2. ed. Philadelphia, 1854.
	An 358.71	Nott, Josiah C. Types of mankind. 10. ed. Philadelphia, 1871.
	Geog 152.5	Nottingham, Eng. University. Geographical Society. Magazine. 1-5,1963-1967//?
	An 2059.59.20	Nougier, L.R. Geographie humaine prehistorique. Paris, 1959.
	An 2109.70.5	Nougier, Louis René. L'économie préhistorique. Paris, 1970.
	Geog 5300.9	Nourse, J.E. American explorations in the ice zones. Boston, 1884.
	Geog 5558.64	Nourse, J.E. Narrative of 2nd Arctic expedition. Washington, 1879.
	Geog 578.79F	Nouveau dictionnaire de géographie universelle contenant...la géographie physique. (Vivien de St. Martin, L.) Paris, 1879-95. 7v.
	Geog 578.79.3F	Nouveau dictionnaire de géographie universelle contenant...la géographie physique. (Vivien de St. Martin, L.) Paris, 1879-95. 7v.
	Geog 578.79.4F	Nouveau dictionnaire de géographie universelle contenant...la géographie physique. Supplement. (Vivien de St. Martin, L.) Paris, 1897-1900. 2v.
	Geog 4217.88	Nouveau voyage...en 1788. (Pagès, P.M.F. de.) Paris, 1797. 3v.
	Geog 4217.28	Nouveau voyage autour du monde. (La Barbinais le Gentil.) Amsterdam, 1728. 2v.
	Geog 4307.35.5	Nouveaux memoires du Baron de Pöllnitz. (Pöllnitz, K.L. von.) Amsterdam, 1737. 2v.
	Geog 4158.42	Nouvelle bibliotheque des voyages. (Smith, William.) Paris, 1842. 12v.
NEDL	Geog 808.76	Nouvelle geographie universelle. (Reclus, J.J.E.) Paris, 1876-1894. 19v.
	Geog 809.22F	Nouvelle géographie universelle. pt.1-10. (Grauper, Ernest.) Paris, 1922. 2v.
	Geog 3018.83.10	Nouvelle histoire des voyages. (Crotambert, R.) Paris, 1883-84.
	Geog 6100.44	Nouvelle notice sur les îles Kerquelen. (Bossière, R.E.) Paris, 1907.
NEDL	Geog 14.204	Nouvelles annales de la marine. 1849-1864 15v.
NEDL	Geog 14.210	Nouvelles annales des voyages, géographiques, historiques, et archéologiques. Paris. 1855-1870 42v.
NEDL	Geog 14.209	Nouvelles annales des voyages, géographiques, historiques, et archéologiques. Paris. 1819-1854 124v.
	Geog 4332.20	Novak, Grga. Naše more. 2. izd. Zagreb, 1932.
	Geog 3570.8	Novas achegos à historia dos descobrimentos maritimos. (Marcondes de Souza, T.O.) São Paulo, 1963.
	Geog 613.9	Novlianskaia, Mariia G. Daniil Gotlib Messershmidt i ego raboty po issledovaniiu Sibiri. Leningrad, 1970.
	Geog 619.8	Novlianskaia, Mariia G. Filipp Iogann Stralenberg. Leningrad, 1966.
	Geog 611.5	Novlianskaia, Mariia G. I.K. Kirilov i ego. Atlas Vserossisskoi imperii. Moskva, 1958.
	Geog 611.5.5	Novlianskaia, Mariia G. Ivan Kirilovich Kirilov, geograf XVIII veka. London, 1964.
	Geog 3590.83	Novokshanova, Zinaida K. Kartograficheskie i geodezicheskie raboty v Rossii v XIX-nachale XX v. Moskva, 1967.
	Geog 576.97F	Novum lexicon geographicum. (Ferrari, P.) Patavii, 1697.
	Geog 5300.30	Novye sovetskie issledovannia v Arktike. (Burkhanov, V.F.) Moskva, 1955.
	Geog 4308.70.10	Nowadays. (Beste, John R.D.) London, 1870. 2v.
	Geog 3249.54	Nowell, C.E. The great discoveries and the first colonial empires. Ithaca, 1954.
	Geog 4559.58.5	Noyce, Wilfrid. The springs of adventure. Cleveland, 1958.
	Geog 4679.62	Noyce, Wilfrid. They survived. London, 1962.
	Geog 4105.4	Noyes, E.H. Steamship notes...a handbook. N.Y., 1874.
	Geog 3590.5	Nozikov, N. Russian voyages round the world. London, 1947.
	Geog 3590.5.5	Nozikov, N. Russkie krugosvetiye moreplavateli. Izd. 2. Moskva, 1947.
	Geog 3070.155	Nozioni di cartografia. (Salinari Emiliani, M.) Roma, 1959.
	An 3509.70F	The Nubian skeleton through 4000 years. Thesis. (Nielsen, Ole Vagn.) København, 1970.
	Geog 6009.63.5	Nudel'man, Aizik V. Sovetskie ekspeditsii v Antarktiku, 1961-63 gg. Moskva, 1965.
	Geog 4269.19	Nuevas nacionalidades en Europa. 2a ed. (Beltrán y Rozpide, Ricardo.) Madrid, 1919.
	Geog 4307.56	Nugent, T. Grand tour...through Netherlands. London, 1756. 4v.
	An 2109.50	Numelin, R. The beginnings of diplomacy. London, 1950.
	An 359.18	Numelin, R. Orsakerna till fokevandringerna. Helsingfors, 1918.
	An 349.36	Numelin, R. The wandering spirit. Copenhagen, 1936.
	Geog 3060.11	Nunn, G.E. Origin of Strait of Asian concept. Philadelphia, 1929.
	Geog 3251.27F	Nunn, George E. World map of Francesco Roselli. Philadelphia, 1928.
	Geog 4216.99.10	Nunnari, F.A. Un viaggiatore calabrese. Mesina, 1901.
	Geog 3235.55	Nuovi studi sull'atlante luxoro. (Desimoni, C.) Geneva, 1869.
	An 3309.69	Nurse, G.T. Height and history in Malawi. Malawi, 1969.
	Geog 5650.12.5	Nybygd i Grønland. (Dúason, Jón.) n.p., 192-.
	An 3559.27	Nyèssen, D.J.H. The passing of the Frisians. 's-Gravenhage, 1927.
X Cg	An 3559.29	Nyèssen, D.J.H. Somatical investigation of the Javanese, 1929. Bandoeng, 1929.
	Geog 3018.99.5	Nystrom, J.F. Geografiens och de...historia. Stockholm, 1899.
Htn	Geog 4476.75*	O-Brazile. (Hamilton, William (pseud.).) London, 1675.
	An 359.70.10	O chloveke, ego poraboshchenii i osvdsozhdenii. (Kovalev, Sergei M.) Moskva, 1970.
	An 3308.89.3	O geograficheskom' raspred'nenii rosta muzhskago naseleniia Rossii. (Anuchin, D.N.) Sankt Peterburg, 1889.
	Geog 4145.24	Das obenteuerliche Leben des Jeus Jacob Eschels. (Eschels, Jeus Jacob.) Hamburg, 1966.
	An 358.83F	Oberländer, R. Fremde Völker ethnographische Schilderungen aus der alten und neuen Welt. Leipzig, 1883.
	An 358.78	Oberländer, R. Der Mensch vormals und heute. Leipzig, 1878.
	Geog 4418.73.5	Obligade, P.S. Viaje a Oriente. Paris, 1873.
	Geog 3019.74	Obraz Zemli. (Fradkin, Naum G.) Moskva, 1974.
	Geog 4219.23.5	O'Brien, Conor. Across three oceans. London, 1949.
	Geog 4145.71	Obruchev, V.A. Grigorii Nikolaenin Potanin; zhizn' i deiatel'nost'. Moskva, 1947.
	Geog 4145.71.5	Obruchev, V.A. Puteshchestviia Potanina. Moskva, 1953.
	Geog 4180.1	Observations...in voyage...South Sea. (Hawkins, R.) London, 1847.
	Geog 4307.84	Observations and reflections made in the course of a journey through France, Italy and Germany. (Piozzi, Hester Lynch Thrale.) Ann Arbor, 1967.
	Geog 4308.44	Observations in Europe. (Durbin, J.P.) N.Y., 1844. 2v.
	An 3558.68	Observations on Crania. (Wyman, J.) Boston, 1868.
Htn	Geog 5507.93*	Observations on passage...Atlantic and Pacific. (Goldson, W.) Portsmouth, 1793.
	Geog 4308.46.3	Observations on social...state of European people. (Laing, S.) London, 1850.
	Geog 5925.10	Observations on South Pole. (Weddell, James.) London, 1826.
	Geog 759.59.7	Obst, Erich. Lehrbuch der allegemeinen Geographie. 3. Aufl. v.1,6,7. Berlin, 1966- 3v.
	Geog 809.59	Obst, Erich. Lehrbuch der allgemeinen Geographie. v.1-4,6-8,10-11. Berlin, 1959. 9v.
	Geog 139.30	Occasional papers. (London. University. Queen Mary College. Department of Geography.) London. 1,1974+
	Geog 36.26	Occasional papers in geography. (Canadian Association of Geographers.) Vancouver. 1-7,1960-1965//
	Geog 146.5	Occasional papers in geography series. (Murray Park College of Advanced Education. Department of Geography.) Magill. 1,1974+
	Geog 51.5	Occasional papers series. (Durham, England. University. Durham Colleges. Department of Geography.)
	Geog 51.5.10	Occasional publications. (Durham, England. University. Department of Geography.) Durham. 1,1973+
	Geog 4209.72	Occasions. (Hammond Innes, Dorothy.) London, 1972.
	Geog 4218.78.20	The Ocean Chronicle, 1878-1891. (Nichols, E.P.) N.Y., 1941.
	Geog 3048.73	L'ocean des anciens. (Moreau de Jonnes, A.C.) Paris, 1873.
	Geog 156.1	Ocean highways. 1-12
	Geog 4558.55	Ocean scenes. N.Y., 1855.
	Geog 4208.73.5	The ocean waves. (Chapman, Charles.) London, 1875.
	Geog 3580.46	El Oceano pacifico. Navegantes española del siglo XVI. (Prieto, Carlos.) Madrid, 1972.
	Geog 5182.5	Ocherki etnicheskoi istorii zarubezhnogo Severa. (Fainberg, Lev A.) Moskva, 1971.
	An 359.57.20	Ocherki obshchei etnografii. v.1-4. (Akademiia Nauk SSSR. Institut Etnografii.) Moskva, 1957- 5v.
	An 2059.67	Ocherki paleoantropologii Ukrainy. (Zinevich, Galina P.) Kiev, 1967.
	Geog 160.5	Ocherki po fizicheskoi geografii. Riga. 6,1966+
	An 99.60	Ocherki po istorii antropologiia v Rossii. (Levin, M.G.) Moskva, 1960.
	Geog 3019.57.5	Ocherki po istorii geograficheskii otkrytii. (Magidovich, I.P.) Moskva, 1957.
	Geog 3590.55	Ocherki po istorii geografii v Rossii XV i XVI vekov. (Lebedev, Dmitrii M.) Moskva, 1956.
	Geog 3590.55.5	Ocherki po istorii geografii v Rossii XVIII veka. (Lebedev, Dmitrii M.) Moskva, 1957.
	Geog 3590.15	Ocherki po istorii russk. geogr. otkrytii. (Berg, L.S.) Moskva, 1946.
	Geog 3590.15.2	Ocherki po istorii russk. geogr. otkrytii. 2. izd. (Berg, L.S.) Moskva, 1949. 2v.
	Geog 3590.25	Ocherkii po istorii russk. zemleved. (Bodnarskii, M.S.) Moskva, 1947.
	Geog 4308.55	Ochoa, Eugenio de. Paris, Londres y Madrid. Paris, 1861.
	Geog 4679.12.20	O'Connor, Richard. Down to eternity. N.Y., 1956.
	Geog 4559.60.5	O'Daniel, John. The nation that refused to starve. N.Y., 1960.
	Geog 4309.00.15	Odd bits of travel with brush and camera. (Taylor, Charles M.) Philadelphia, 1900.
	Geog 4209.17	Odd corners. (Anderson, I. (Mrs.).) N.Y., 1917.
	An 358.64.7	Odd people. (Reid, Mayne.) Boaston, 1864.
	An 358.61	Odd people. (Reid, Mayne.) Boston, 1861.
Htn	An 358.60.3*	Odd people. (Reid, Mayne.) N.Y., 1860.
	Geog 4219.69	Odd way around the world. (Price, Willard.) N.Y., 1969.
	Geog 818.21.3	Oddsson, G. Almenn jardarfrädi og landskipun edur geographia. Kaupmannahøfn, 1822. 4v.
	Geog 818.21	Oddsson, G. Almenn landaskipunarfraede. Kaupmannahøfn, 1821-1827. 2v.
	Geog 4239.29	Odisea del yate "Mary" de Puerto Rico a España. (Franceschi, F.) Madrid, 1930.
	Geog 4145.34.25	Odissei s Beloi Rusi. (Guzanov, Vitalii G.) Minsk, 1969.
	Geog 4672.14	O'Donnell, Elliott. Strange sea mysteries. London, 1926.
	An 3559.68	Odontologiia. (Zubov, Aleksander A.) Moskva, 1968.
	Geog 520.10	Odoric de Pardenone. Les voyages en Asie. Paris, 1891.
	Geog 3235.41	Odorici, F. Carte geografiche. Milano, 1877.

Author and Title Listing

Htn Geog 4206.65* L'odyssee ou diversite d'avantures...en Europe, Asie et Afrique. (Desboys du Chastelet, R.) Fleche, 1665.

Geog 4131.69 Die ökonomische Bedeutung des internationalen Tourismus fur die Ehtwicklungsländer. (Frentrup, Klaus.) Hamburg, 1969.

Geog 5845.3 Økonomisk politik i Grønland. (Boserup, Mogens.) København, 1963.

Geog 1530.18 Oesterreich-Ungarn. 20. Aufl. (Baedeker, publishers.) Leipzig, 1884.

Geog 1530.15 Österreich-Ungarn. 29. Aufl. (Baedeker, publishers.) Leipzig, 1913.

Geog 5398.72.3 Die österreichisch-ungarische Nordpol-Expedition. (Payer, J.) Wien, 1876.

Geog 258.1.5 Österreichische Geographische Gesellschaft. Registerband, 1908-59. Wien, 1960.

NEDL Geog 158.1 Österreichische Touristen-Zeitung. Wien. 1-56,1881-1936 28v.

Geog 158.1.25 Österreichische Touristen-Zeitung. Sondernummer anläslich des 60 jähren Bestande. Innsbruck, 1929.

Geog 157.50 Österreichischer Alpen-Verein, Vienna. Mittheilungen. Wien. 1-2,1863-1864 2v.

Geog 18.3 Österreichisches Archiv für Geschichte. Wien. 1-3,1831-1833 3v.

Geog 18.4 Österreichisches Zeitschrift für Geschichte. Wien. 1-3,1835-1837 3v.

Geog 3542.5 Österreichs Anteil an der Erforschung der Erde. (Hassinger, H.) Wien, 1949?

Geog 5660.25 Østgrönlandsavtalen. (Castberg, Frede.) Kristiania, 1924.

An 3539.07 Oetteking, B. Kraniologische Studien...Altägypten. Braunschweig, 1907.

Geog 3240.18 L'oeuvre cartographique de Gérard et de Corneille de Jode. (Ortroy, F.G. van.) Gand, 1914.

Geog 612.2 L'oeuvre géographique de Linant de Bellefonds. Thèse. (Mazuel, J.) Le Caire, 1937.

Geog 4303.99.2 Oeuvres. (Lannoy, Guillebert de.) Louvain, 1878.

An 187.8 Oeuvres. (Mauss, Marcel.) Paris, 1961- 3v.

Geog 5324.2.50 Øverås, Asbjdrn. Fridtjof Nansen. Stavanger, 1946.

Geog 4309.35.15 Ofaire, Cilette. The San Luca. N.Y., 1935.

Geog 4309.64 The offensive traveller. 1. ed. (Pritchett, V.S.) N.Y., 1964.

Geog 4110.11F Official guide and album. (Cuward Steamship co.) London, 1877.

Geog 4309.44.5 Oga Magoga. (Baretti, Giuseppe.) Milano, 1944.

Htn Geog 4306.56* Ogier, Charles. Ephemerides sive iter Danicum, Suecicum, Polonicum. Lutetiae Parisiorum, 1656.

Geog 4306.56.10 Ogier, Charles. Oziennik podrózy do Polski, 1635-1636. Gdańsk, 1950-1953. 2v.

Geog 4269.57 Ogilvie, A.G. Europe and its borderlands. Edinburgh, 1957.

Geog 4131.10 Ogilvie, F.W. The tourist movement. London, 1933.

Geog 4209.68.15 Ogni dalekikh gorodov. Moskva, 1968.

VGeog 4219.69.10 Ogrin, Miran. Na juga sveta. Ljubljana, 1969.

VGeog 4209.69 Ogrin, Miran. Širine sveta. Ljubljana, 1969.

Geog 161.1 Ohio State University. Contributions in geographical exploration. Columbus. 1,1920

Geog 4559.45 Oil and deep water. (Klitgaard, Kaj.) Chapel Hill, 1945.

An 2109.72.5 Okhotniki, sobirateli, rybolovy. Leningrad, 1972.

VGeog 162.5 Okhrana prirody i vosproizvodstvo prirodnykh resursov. Moskva. 1,1968+

Geog 4309.09 O'Laughlin, John. From the jungle through Europe with Roosevelt. Boston, 1910.

An 2109.36.9 Olbrechts, F.M. Ethnologie...primitiere beschaving. Zutphen, 1936.

An 3309.25 The old Americans. (Hrdlička, A.) Baltimore, 1925.

Geog 3070.33.2A Old maps and their makers. 2d ed. (Holman, Louis A.) Boston, 1926.

Geog 4218.27 The old sailor's story. (Allyn, Gurdon L.) Norwich, Conn., 1879.

Geog 4558.88 The old sailor's story of his life. (Spear, P.S.) Portland, 1888.

Geog 4129.03 Old-time travel. (Shand, A.I.) London, 1903.

Geog 4218.73.5 An old world...travels around the world. (Walworth, E.H.) N.Y., 1877.

Geog 4308.33.2 Old world and the new. (Dewey, O.) N.Y., 1836. 2v.

Geog 4308.68.7 The old world in its new face. (Bellows, H.W.) N.Y., 1868. 2v.

Geog 4208.98.11 Old world memories. (Temple, E.L.) Boston, 1899. 2v.

Geog 4329.04 The old world thro' new world eyes. (Loring, J.M.) St. Louis, 1904.

Geog 5838.72.5 Oldendow, K. Grønlaendervennen Hinrich Rink. København, 1955.

Geog 5845.6 Oldendow, Knud. Grønlands fremtid. København, 1945.

Geog 5839.35.25 Oldendow, Knud. Naturfredning i Grønland. København, 1935.

Geog 5639.36 Oldendow, Knud. Traek af Grønlands politiske historie. København, 1936.

Geog 4418.39.4 Olin, S. Travels in Egypt, Arabia, Petraea and Holy Land 4th ed. N.Y., 1844. 2v.

Geog 4208.41.2 Oliphant, L. Episodes in a life of adventure. Edinburgh, 1887.

Geog 4208.41 Oliphant, L. Episodes in a life of adventure. N.Y., 1887.

Geog 4208.41.4 Oliphant, L. Episodes in a life of adventure. 4. ed. Edinburgh, 1887.

Geog 613.4.20 Olivas, A. Contribución a la bibliotheca de C.R. Markham. Lima, 1924.

Geog 3570.7 Oliveira Marlins, J.P. de. Les explorations des Portugais. Paris, 1893.

Geog 5329.3.7 Ol'khina, Evgeniia A. Vil'iamur Stefanson. Moskva, 1970.

An 3558.94 Olóriz y Aguilera, Frederico. Distribución geográfica del'índice cefálico en España. Madrid, 1894.

Geog 5839.16.5 Olrik, Harald. Forslag om at bebygge Scoresby Sund-Egnen i Østgrønland ved vestgrønlandske saelforgere. København, 1916.

Geog 3019.37 Olschki, L. Storia letteraria delle scoperte geografiche. Firenze, 1937.

Geog 6009.33.10 Olsen, Magnus L. Saga of the white horizon. Lymington, 1972.

Geog 3019.29.5 Olsen, Ørjan. La conquête de la terre. Paris, 1933-36. 5v.

An 2308.92 Olshausen, O. Leichenverbrennung. Berlin, 1892.

Geog 4300.46 Olson, H.S. Aboard and abroad. 3d ed. Philadelphia, 1956.

Geog 614.9 Olszewicz, B. Wacław Nałkowski. Warszawa, 1962.

Geog 4209.58.15 Olszewicz, B. Ze wspomnień podrózników. Warszawa, 1958.

Geog 600.21 Olszewicz, Bolesław. Dziewięć wieków geografii polskiej. Warszawa, 1967.

Geog 616.31 Olszewicz, Bolesław. Stanisław Pawłowski. Warszawa, 1968.

Geog 600.7 Oltre l'Alpe e il mare. (Ciampi, I.) Roma, 1865.

Geog 3251.7 Om en märklig globkarta. (Nordenskiold, A.E.) Stockholm, 1884.

Geog 5838.82 Om Grønlaenderne, deres fremtid. (Rink, Hinrich.) Kjøbenhavn, 1882.

Geog 5837.92 Om Grönlands osterbygds. (Eggers, H.P.) Kjøbenhavn, 1794.

Geog 4209.39.5 O'Malley, C.J. It was news to me. Boston, 1939.

Geog 4659.55.5 O'May, Harry. Wrecks in Tasmanian waters, 1797-1950. Tasmania, 1955?

Geog 5919.38 Ommaney, Francis D. Below the roaring forties. London, 1938.

Geog 5919.38.2 Ommaney, Francis D. South latitude. London, 1947.

Geog 4429.52.5 Ommanney, F.D. The shoals of Capricorn. London, 1952.

Htn Geog 3205.40.7* Omnium gentium, mores, leges. (Boehme, J.) Antverpiae, 1542.

Htn Geog 3205.40.5* Omnium gentium, mores, leges. (Boehme, J.) Friburgi, 1540.

Geog 5700.2 Omstaendelig...grønlandste missions. (Egede, H.) Kjøbenhavn, 1738.

An 3559.30F On an adolescent skull of "Sinanthropus pekinensis". (Black, Davidson.) Peiping, 1930.

Geog 4558.90.7 On blue water. (Amicis, E. de.) N.Y., 1897.

Geog 4558.79 On board the "Rocket". (Adams, Robert C.) Boston, 1879.

Geog 5324.2.80 On byl pervym. (Gal'perin, I.M.) Moskva, 1958.

Geog 4509.56.15 On climbing. (Evans, Charles.) London, 1956.

Geog 665.164 On geography: Selected writings of Preston E. James. 1st ed. (James, Preston Everett.) Syracuse, 1971.

Geog 4329.29.10 On Mediterranean shores. (Ludwig, Emil.) Boston, 1929.

Geog 4348.93 On sunny shores. (Scollard, Clinton.) N.Y., 1893.

An 359.08.3 On the education of backward races. (Coffin, E.W.) Worcester, 1908.

Geog 4209.01 On the great highway. (Creelman, J.) Boston, 1901.

Geog 4559.29.15 On the high seas. (Chatterton, E.K.) Philadelphia, 1929.

Geog 5398.99.3 On the "Polar Star". (Amadeus, Louis.) N.Y., 1903. 2v.

Geog 4209.26 On the stream of travel. (Hall, James Norman.) Boston, 1926.

Geog 4308.85 On the track of the crescent. (Johnson, E.C.) London, 1885.

Geog 4515.25 On the use and management of the rope in rock work. (Underhill, R.L.K.) San Francisco, 1931.

Geog 4209.42.10 On the Wallaby. (Barrett, C.L.) Melbourne, 1942.

Geog 5300.11 On top of the world. (Brontman, L.K.) N.Y., 1938.

Geog 4509.53 On top of the world. (Petzoldt, P.) N.Y., 1953.

An 99.68.5 One hundred years of anthropology. Cambridge, Mass., 1968.

Geog 579.59 Öngör, Sami. Coğrafya sözlüğü. Fasc.1-5. İstanbul, 1959.

Geog 579.59.5 Öngör, Sami. Coğrafya terimleri sözlüğü. İstanbul, 1975.

Geog 4559.48 Öngör, Sami. Deniz cağrafyasi. Ankara, 1948.

Htn Geog 4708.16.15* Onion, S.B. Narrative of the mutiny on board the schooner Plattsburg. Boston, 1819.

Geog 4182.49 De ontdekkingsreis van Jacob le Maire. (LeMaire, Jacob.) 's-Gravenhage, 1945. 2v.

Geog 3018.74 De ontdekkingsreizen sedert de vijftiende eeuw. (Tiele, Pieter Anton.) Leiden, 1874.

Htn Geog 5375.94* Oost-indische ende west-indische voyagien. (Veer, G. de.) Amsterdam, 1619.

Geog 4209.72.15 Op reis rond de wereld en op Rottumerplaat. (Bamans, Godfried.) Amsterdam, 1972.

Geog 5610.5.2 Opdagelsesrejser til Grønland, 1473-1806. (Bobé, Louis.) København, 1936.

Geog 4182.10 De open-deure tot het verborgen heydendom. (Rogerius, Abraham.) 's-Gravenhagen, 1915.

Geog 5558.60 The open Polar Sea. (Hayes, I.I.) N.Y., 1867.

Geog 4169.38 The open road. (Compton, Ray.) N.Y., 1938.

Geog 608.5 L'opera geografica di Luca Holstenio. (Almagià, Roberto.) Citta del Vaticano, 1942.

Geog 6009.54 Operation deepfreeze. (Dufek, George John.) N.Y., 1957.

Geog 4500.10 Opere antiche e rare sulla montagna. (Istituto di Bibliografia Alpina, Biella.) Biella, 1951.

Geog 3057.5 Ophir nach dem neuesten Forschungen. (Peters, Karl.) Berlin, 1908.

Geog 4209.71.5 Opowieści zdyszane. 1. wyd. (Górnicki, Wiesław.) Warszawa, 1971.

Geog 4329.27.5 Oppenheim, E.P. The quest for winter sunshine. Boston, 1927.

Geog 665.116.5 Opšta geografija; antropogeografija. (Cvijić, Jovan.) Beograd, 1969.

Geog 665.30F Opusculos e esparsos. (Santarem, M.F. de B.) Lisboa, 1910. 2v.

An 358.49.3 An oration pronounced before the Connecticut Alpha of the Phi Beta Kappa at Yale College. (Smith, Ashbel.) New Haven, 1849.

Geog 665.154 Orazvitii geografii v Estonskoi SSR 1960-1968. (Eesti Geograafia Selts.) Tallin, 1970.

Geog 535.10 Orbis geographicus; world directory of geography. Wiesbaden. 1952 3v.

Htn Geog 578.61.5* Orbis latinus. (Graesse, J.G.T.) Dresden, 1861.

Geog 578.61 Orbis latinus. (Graesse, J.G.T.) Dresden, 1861.

Htn Geog 1215.86* Orbis terrae partium. (Neaudro, M.) Lipsiae, 1586.

Geog 4209.45.5 Orcutt, R. Merchant of alphabets. Garden City, N.Y., 1945.

Geog 5558.17.3 O'Reilly, B. Greenland...Northwest Passage. London, 1818.

Geog 5558.17 O'Reilly, B. Greenland...Northwest Passage. N.Y., 1818.

Geog 4131.2 Organization for European Economic Cooperation. Trends in economic sectors: tourism in Europe. Paris. 1953-1961

An 2109.59.20 Organization of work. (Udy, Stanley H.) New Haven, 1959.

Geog 4218.85.5 Orient and Occident. (Mitford, R.C.W.M.) London, 1888.

Geog 4110.13 Orient line guide. (Loftie, W.J.) London, 1890.

An 359.12.15 Oriente modernisado. (Montalto de Jesus, C.A.) Lisboa, 1912.

Geog 5839.48 Orientering og faerdsee i terraenet. (Boegvad, R.) København, 1948.

An 3559.33 Origen castellano del prognatismo. (Aguilar y R.F.) Madrid, 1933.

An 3409.12 Origen e influencia jurídico-social del sistema dactiloscopico argentino. (Reyna Almandos, Luis.) La Plata, 1912.

An 358.66.5 The origin and antiquity of physical man scientifically considered. 2. ed. (Tuttle, Hudson.) Boston, 1866?

Geog 3070.90 The origin and development of cartography. (Lauf, G.B.) Johannesburg, 1955.

An 359.31.5 The origin and nature of man. (Spiller, Gustav.) London, 1931.

An 359.35.3 The origin and nature of man. 2. ed. (Spiller, Gustav.) London, 1935.

An 2108.75.5 The origin of civilisation. 3. ed. (Avebury, J.L.) London, 1875.

An 2108.82.5 The origin of civilisation and the primitive condion of man. 4. ed. (Avebury, J.L.) N.Y., 1882.

An 2108.82 The origin of civilisation and the primitive condition of man. (Avebury, J.L.) London, 1882.

Author and Title Listing

	An 2108.70.2	The origin of civilisation and the primitive condition of man. (Lubbock, J.) London, 1870.
	An 2108.70	The origin of civilisation and the primitive condition of man. 2. ed. (Lubbock, J.) London, 1870.
	An 2108.82.2	The origin of civilisation and the primitive condition of man. 4. ed. (Avebury, J.L.) London, 1882.
	An 2108.71.3	The origin of civilization and the primitive condition of man. (Avebury, J.L.) N.Y., 1871.
	An 3409.16.3	The origin of finger-printing. (Herschel, W.J.) London, 1916.
	An 359.62	The origin of races. (Coon, Carleton Stevens.) N.Y., 1962.
	Geog 3060.11	Origin of Strait of Asian concept. (Nunn, G.E.) Philadelphia, 1929.
	An 2109.38	The origin of the inequality of the social classes. (Landtman, G.) London, 1938.
	An 2339.11	The origin of West African crossbows. (Balfour, Henry.) Washington, 1911.
	Geog 4181.76	The original writings and correspondance of the two Richard Hakluyts. (Taylor, E.G.R.) London, 1935. 2v.
	An 2108.73.10	Origine et destinée de l'homme. (Pireteau, Jean.) Paris, 1973.
	Geog 5700.9	Origines et fondation du plus ancien évêché du nouveau monde. (Beauvois, E.) Paris, 1878.
	An 2109.53.20	Origins, indigenous period. (Wormington, Hannah Marie.) Mexico, 1953.
	An 2109.26.7	Origins of education among primitive peoples. (Hambly, W.D.) London, 1926.
	An 2108.95.5	Origins of invention. (Mason, O.T.) London, 1895.
Htn	Geog 3240.18.5*	Orion, booksellers, ltd., London. Description of a rare and precious atlas. London, 1935?
	Geog 4417.64	Orleans, G. de. Travels of Father William Orleans a Jesuit. London, n.d.
	Geog 4180.17	Orleans, P.J. d'. History of...tartar conquerors of China. N.Y., 1964?
	Geog 1590.25	Orlowicz, M. Illustrierter Führer durch Galizien. Wien, 1914.
	Geog 6009.47	Orrego Vicuña, Eugenio. Terra australis. Santiago, 1948.
	An 359.18	Orsakerna till fokevandringerna. (Numelin, R.) Helsingfors, 1918.
	An 136.5.35	Orte des wilden Denkens. 1. Aufl. (Lepenies, Wolf.) Frankfurt am Main, 1970.
	Geog 4217.64.7	Ortega, D.C. de. Viage del Comandante Byron al rededor del mundo. Madrid, 1769.
	An 3409.13	Ortiz, F. La indentificación dactiloscópica. Habana, 1913.
	An 359.35	Orton, E.F. Links with past ages. Cambridge, Eng., 1935.
	Geog 3240.18	Ortroy, F.G. van. L'oeuvre cartographique de Gérard et de Corneille de Jode. Gand, 1914.
	Geog 4209.54	Osborn, C. He lived for adventure. Washington, 1954.
	An 2109.27.9	Osborn, H.F. Man rises to Parnassus. Princeton, 1927.
	An 2109.28.3	Osborn, H.F. Man rises to Parnassus. Princeton, 1928.
	An 2109.15.5A	Osborn, H.F. Men of the old stone age. N.Y., 1915.
	An 2109.23.3	Osborn, H.F. Men of the old stone age. 3. ed. N.Y., 1923.
	An 2109.34.10	Osborn, H.F. Men of the old stone age. 3. ed. N.Y., 1934.
	An 2109.21.10	Osborn, Henry F. Men of the old stone age. 3. ed. N.Y., 1921.
	Geog 5518.50.4	Osborn, S. Discovery of the Northwest Passage. Edinburgh, 1865.
	Geog 5538.50.10	Osborn, S. The Polar regions...Franklin's expedition. N.Y., 1854.
	Geog 5538.50.4	Osborn, S. Stray leaves from an Arctic journal. Edinburgh, 1865.
	Geog 5538.50.2	Osborn, S. Stray leaves from an Arctic journal. N.Y., 1852.
	Geog 5530.14	Osborn, Sherard. The career, last voyage. London, 1860.
	Geog 4309.23.5	Osborne, A.B. Finding the worth while in Europe. N.Y., 1923.
	Geog 4309.23.10	Osborne, A.B. Finding the worth while in Europe. N.Y., 1925.
	Geog 4309.31.20	Osborne, A.B. Finding the worth while in Europe. N.Y., 1931.
	Geog 4309.08	Osborne, T.M. Adventures of a green dragon. Auburn, 1908.
	Geog 4418.54	Osgood, J.B.F. Notes of travel or recollections. Salem, 1854.
	Geog 4300.8	Osgood, J.R. Osgood's pocket guide to Europe. Boston, 1882.
	Geog 4300.8.5A	Osgood, J.R. Osgood's pocket guide to Europe. Boston, 1883.
	Geog 4300.8	Osgood's pocket guide to Europe. (Osgood, J.R.) Boston, 1882.
	Geog 4300.8.5A	Osgood's pocket guide to Europe. (Osgood, J.R.) Boston, 1883.
	VAn 359.68	Oskovy etnografii. (Tokarev, Sergei A.) Moskva, 1968.
	Geog 5925.91	Osnovnye itogi izucheniia Antarktiki za 10 let. (Vsesoiuznoe Soveshchenie po Izucheniiu Antarktiki, Moscow, 1966.) Moskva, 1967.
	Geog 5939.73.5	Osnovnye problemy osvoeniia Antarktiki. (Slevick, Solomon B.) Leningrad, 1973.
	Geog 3070.45	Osnovy kartovesheniia. Izd. 2. (Salishahev, K.A.) Moskva, 1943.
	Geog 500.73	Ossa Varela, P. Catálogo alfabético de algunos géografos y exploradores. Bogotá, 1951.
	Geog 4228.56.5	Ossoli, M.F. At home and abroad. Boston, 1860.
	Geog 4228.56A	Ossoli, M.F. At home and abroad. 3. ed. Boston, 1856.
	An 3509.66	Osteometriia. (Alekseev, Valerii P.) Moskva, 1966.
	Geog 5660.30	Ostermann, H. Traekaf kolonien Jakobshavns. Kjøbenhavn, 1941.
	Geog 5855.15	Ostermann, Hother. Danske i Grønland i det 18. aarhundrede. København, 1945.
	Geog 5896.2	Ostermann, Hother. Nordmaend paa Grønland, 1721-1814. Oslo, 1940. 2v.
	Geog 4329.40	Ostgriechische Reisen; Kleinasien, Kypros und Syrien. (Pfuhl, E.) Basel, 1940.
	Geog 5639.46	Ostgronlaenderne tunuamuit. (Rosing, Christian.) København, 1946.
	Geog 5399.61	Ostrov Vrangelia. (Gromov, L.V.) Magadan, 1961.
	Geog 5399.46	Ostrov Vrangelia. (Mineev, A.I.) Moskva, 1946.
	Geog 5557.33	Ostrovskii, B.G. Veliki severnaia ekspeditsiia, 1733-43. Arkhangel'sk, 1935.
	Geog 4313.9	Die Ostsee-Sturmflut 1872. (Kiecksee, Heinz.) Heide, 1972.
	Geog 5970.44	Ot El'brusa do Antarktidy. (Gasev, Aleksandr M.) Moskva, 1972.
	Geog 4209.70.5	Ot Madrida do Tokio. (Kiknadze, Aleksandr Vasil'evich.) Baku, 1970.
	Geog 4219.06.7	Öt világrészun heresztül. (Sebok, I.) Budapest, 1934.
	Geog 600.10	Otechestvennye ekonomiko-geografy XVIII-XX vv. (Baranskii, N.N.) Moskva, 1957.
	Geog 600.11	Otechestvennye fiziko-geografy i puteshestvenniki. (Baranskii, N.N.) Moskva, 1959.
	Geog 3590.47	Otechestvennye moraplavateli-issledovanii morel i okeanov. (Zubov, N.N.) Moskva, 1954.
	Geog 4218.68.5	Other countries. (Bell, William M.) London, 1872. 2v.
	An 359.64	Other cultures. (Beattie, John.) N.Y., 1964.
	Geog 4309.55.5	The other half. (Cox, Alfred B.) Adelaide, 1955?
	Geog 4249.59	The other half of the earth. (Van Sinderen, Adrian.) N.Y., 1959. 2v.
	An 2109.38.11	Other men's lives...primitive peoples. (Dunbar, George.) London, 1938.
	Geog 4219.05.3	The other side of the lantern. (Treves, F.) London, 1905.
	Geog 4219.05.4	The other side of the lantern. (Treves, F.) London, 1905.
	Geog 4219.08.2	Other side of the lantern. (Treves, F.) London, 1908.
	Geog 4219.05.10	The other side of the lantern. (Treves, F.) London, 1928.
	An 3208.80	Otis, George A. List of the specimens in the anatomical section of the United States Army Medical Museum. Washington, 1880.
	Geog 3590.35.10	Otkrytiia russk. zemleprov. na ser. Vost. Asii. (Efimov, Aleksei V.) Moskva, 1951.
	Geog 4308.44.5	Otryuki iz zagranichnykh pisem. (Volkov, M.) Sankt Peterburg, 1857.
	An 379.72.5	Otsenka prirodnykh uslovii zhizni naseleniia. (Lopatina, Elena B.) Moskva, 1972.
	Geog 170.3	Ottawa, Ont. University. Department of Geography and Regional Planning. Travaux du Département de géographie et d'aménagement régional, Université d'Ottawa. Ottawa. 1,1971+
	An 3206.47	Otto, Andreas. Anthroposcopia. Regiomonti, 1647.
	An 379.61.10	Otto, Klaus. Das Aufkommen sozialgeographischer Betrachtungsweisen in der deutschen länderkundlichen Literatur seit Beginn des 20. Jahrhunderts. Inaug. Diss. Münster, 1961.
	Geog 4182.12	Ottsen, Hendrick. Journael van de reis naar Zuid-Amerika. 's-Gravenhage, 1918.
	Geog 3070.27	Oud-Nederlandsches kaartmakers in betrekking met Plantijn. (Denucé, J.) Antwerpe, 1912. 2v.
	Geog 4182.51	De oudste reizen van de Zieuwen naar Oost-Indie. (Unger, W.) 's-Gravenhage, 1948.
	Geog 603.5.10	Oulié, M. Charcot of the Antartic. London, 1938.
	Geog 603.5.5	Oulié, M. Jean Charcot. 14. éd. Paris, 1937.
	Geog 4308.69.9	Our admiral's flag abroad. (Montgomery, J.E.) N.Y., 1869.
	An 379.46	Our evolving civilization. (Taylor, G.) Toronto, 1946.
	An 359.29.3	Our forefathers, the Gothonic nations. (Schütte, Gudmund.) Cambridge, 1929-33. 2v.
	Geog 819.44.5F	Our global world. (Hankins, G.C.) N.Y., 1944.
	Geog 958.40	Our Globe. Philadelphia, 184-.
	Geog 4248.73	Our journal in the Pacific. (Eardley-Wilmot, S.) London, 1873.
	Geog 5398.79.25	Our lost explorers; a narrative of the Jeannette Arctic expedition. (Bliss, R.W.) Hartford, 1882.
	Geog 4218.69	Our new way round the world. (Coffin, C.C.) Boston, 1869.
	Geog 4110.16	Our ocean highways. (Dempsey, J.M.) London, 1871.
	Geog 819.47.5	Our own Baedeker, from the New Yorker. (Kinkead, Eugene.) N.Y., 1947.
	Geog 5399.25.10	Our polar flight. (Amundsen, R.) N.Y., 1925.
	An 2059.25.5	Our prehistoric forerunners. (Vulliamy, C.E.) N.Y., 1925.
	An 2109.49.15	Our primitive contemporaries. (Murdock, George P.) N.Y., 1949.
	Geog 4308.93	Our sentimental journey thro' France. (Pennell, J.) London, 1893.
	An 379.52.5F	Our world from the air. (Gutkind, E.A.) London, 1952.
	Geog 4309.34	Out of the fog. (Patterson, Sara K.) Caldwell, 1934.
	Geog 4239.57.5	Outhwaite, L. The Atlantic. N.Y., 1957.
	Geog 3019.35	Outhwaite, L. Unrolling the map. N.Y., 1935.
	An 629.50	Outline of cultural material. 3. ed. (Yale University. Institute of Human Relations.) New Haven, 1950.
	Geog 819.35	An outline of geography. (James, Preston E.) Boston, 1935.
	Geog 5558.81.20	The outpost of the lost. (Brainard, David L.) Indianapolis, 1929.
	Geog 4349.72	Ovcharenko, Aleksandr I. V etom bushuiushchem mire. Moskva, 1972.
	Geog 4218.57.3	Over five seas and oceans. (Miller, Thomas.) N.Y., 1894.
	Geog 4218.73.9	Over land and sea. (Guillemard, A.C.) London, 1875.
	Geog 4309.07.5	Over lave og høie fjelde; breve fra en aeldre berre. (Bull, Edvard.) Kristiania, 1907.
	Geog 4308.65.9	Over-sea. (Morford, Henry.) N.Y., 1867.
	Geog 4279.69	Over the Alps. (Anderson, Patrick.) London, 1969.
	Geog 4129.47.5	Over the hills and far away. (Cook, H.K.) London, 1947.
	Geog 4308.71.2	Over the ocean, or Sights and scenes in foreign lands. (Guild, C.) Boston, 1875.
	Geog 4308.71A	Over the ocean. (Guild, C.) Boston, 1871.
	Geog 659.65	Overbeck, Hermann. Kulturlandschaftsforschung und Landeskunde. Heidelberg, 1965.
	An 2059.69.5	Overhage, Paul. Menschenformen im Eiszeitalter. 1. Aufl. Frankfurt, 1969.
	An 2059.61.5	Overhage, Paul. Das Problem der Hominisation. Freiburg, 1961.
	An 2059.59.25	Overhage, Paul. Um des Erscheinungsbild der ersten Menschen. Basel, 1959.
	Geog 4348.45	Overland guide-book. (Barber, J.) London, 1845.
Htn	Geog 4218.41*	Overland journey round the world. (Simpson, George.) Philadelphia, 1847.
	Geog 4218.41.2	Overland journey round the world. (Simpson, George.) Philadelphia, 1847.
	Geog 4208.63.7F	The overland route to India. (Tillotson, John.) London, 1863?
	Geog 5650.10.9	Oversigt over Nordboruiner i Godhaat. (Bruun, Daniel.) København, 1917.
	Geog 4308.48.11	Ow, Max von. Mit dem jüngsten Sohn König Ludwigs I. als Reisebegleiter nach Spanien 1848-1849. München, 1967.
	Geog 4300.51	Owen, Charles. Britons abroad; a report on the package tour. London, 1968.
	Geog 5939.41	Owen, Russell. The Antarctic ocean. N.Y., 1941.
	Geog 6009.28.30	Owen, Russell. South of the sun. N.Y., 1934.
	Geog 6009.28.31	Owen, Russell. South of the sun. N.Y., 1934.
NEDL	Geog 4418.22	Owen, William Fitz William. Narrative of voyages to explore the shores of Africa, Arabia, and Madagascar. London, 1833. 2v.
	Geog 4418.22	Owen, William Fitz William. Narrative of voyages to explore the shores of Africa, Arabia, and Madagascar. Farnborough, 1968. 2v.
	Geog 4678.46	Owners and crew of ship Chandler Price. (United States. Congress. House. Committee on Commerce.) n.p., 1848.

Author and Title Listing

Call Number	Entry
Geog 174.5F	Oxford. University. School of Geography. Research papers. Oxford. 1,1972+
Geog 4306.56.10	Oziennik podróży do Polski, 1635-1636. (Ogier, Charles.) Gdańsk, 1950-1953. 2v.
Geog 4209.70.15	Ozni na dorogakh chetyrekh chastei sveta. (Rubissow, Helen.) Parizh, 1970.
Geog 809.36	Ozonf, R. (Mme.). Lectures géographiques. v.1-2. Paris, 1936-1938. 4v.
Geog 4110.20.3	The P. and O. pocket book. (Peninsular and Oriental Steam Navigation Company.) London, 1908.
Geog 618.7	P.I. Rychkov. (Mil'kov, Fedor N.) Moskva, 1953.
Geog 5838.88	Paa ski over Grønland...1888-90. (Nansen, F.) Kristiania, 1890.
Geog 3129.57	Paassen, Christian van. The classical tradition of geography. Groningen, 1957.
Geog 3129.58	Paassen, Christian van. The classical tradition of geography. Groningen, 1957.
An 3409.34.3	Pacheco, F. Vucetich e Reyna Abmandos. Rio de Janeiro, 1934.
Geog 4181.79	Pacheco Pereira, Duarte. Esmeraldo de situ orbis. London, 1937.
Geog 3205.50F	Pacheco Pereira. Esmeraldo de situ orbis. Lisboa, 1892.
Geog 3205.50.5	Pacheco Pereira. Esmeraldo de situ orbis. Lisboa, 1905.
Geog 4306.85	Pacichelli, G.B. Memorie de viaggi per l'Europa. Napoli, 1685. 3v.
Geog 4249.50.5	The Pacific. (Barrett, C.L.) Melbourne, 1950?
Geog 4249.67	The Pacific Basin; a history of its geographical exploration. (Friis, Herman Ralph.) N.Y., 1967.
Geog 178.7.5	Pamphlet vol. Pacific Crest Trail Conference, Pasadena, California.
Geog 178.7	Pamphlet vol. Pacific Crest Trail Conference, Pasadena, California.
Geog 178.5	Pacific geographic magazine. Beverly Hills, Calif.
Geog 178.1	Pamphlet vol. Pacific Geographic Society.
Geog 4208.67.5	Pacific Mail Steamship Co. A sketch of the new route to China and Japan. San Francisco, 1867.
Geog 4110.22	Pacific Mail Steamship Co. A sketch of the route to California, China. San Francisco, 1867.
Geog 4249.44A	Pacific Ocean handbook. (Medrs, Eliot G.) Stanford, Calif., 1944.
Geog 4679.47	Pacific ordeal. 1. American ed. (Ainslie, Kenneth.) N.Y., 1956.
Geog 4249.26	The Pacific Russian scientific investigation. (Akademiia Nauk, Petrograd.) Leningrad, 1926.
Geog 4218.80.7	Packard, J.F. Grant's tour around the world. Cincinnati, 1880.
Geog 4311.9	Paddles and politics down the Danube. (Bigelow, P.) London, 1892.
Geog 4679.12.25	Padfield, Peter. The Titanic and the Californian. London, 1965.
An 2109.38.3	Page, J.W. Primitive races of to-day. London, 1938.
Geog 4219.09.5	Pageot, G. À travers les pays jaunes. Paris, 1909.
Geog 4217.88	Pagès, P.M.F. de. Nouveau voyage...en 1788. Paris, 1797. 3v.
Geog 4217.67.5	Pagès, P.M.F. de. Travels round the world. London, 1791. 3v.
Geog 4217.67	Pagès, P.M.F. de. Voyage autour du monde. Paris, 1782. 2v.
Geog 4309.15	Pages in azure and gold. (Gardiner, Sarah D.) New Haven, 1915.
Geog 4308.37.5	Paggages in foreign travel. (Jewett, I.A.) Boston, 1838. 2v.
Geog 4308.56.7.10	Páginas de mi diario durante tres años de viaje. (Vicuña Mackenna, Benjamin.) Santiago de Chile, 1936. 2v.
An 3308.76	Pagliani, L. Sopra alcuni fattori dello suiluppo umano. Torino, 1876.
Geog 4709.11	Paine, R.D. The book of buried treasure. London, 1911.
Geog 4672.10	Paine, Ralph D. Lost ships and lonely seas. N.Y., 1922.
Geog 4679.63.5	The painted ship. (Marchbanks, David.) London, 1964.
Geog 3049.62	Los paises legendarios de la mitologia. (Cozzi, Piero.) Buenos Aires, 1962.
Geog 4308.56.7	Pajinas de mi diario. (Vicuña Mackenna, Benjamin.) Santiago, 1856.
An 359.12.5	Palaeolithic man. (Munro, Robert.) Edinburgh, 1912.
Geog 1605.18	Palästina und Syrien. 6. Aufl. (Baedeker, publishers.) Leipzig, 1904.
Geog 5700.4.20	Palase, Hans Egede i Grønland. (Fenger, Niels.) København, 1971.
Geog 4209.69.15	Paleckis, Justas. Kelioniu, Knyga. Vilnius, 1969.
An 2058.68	La paléontologie appliquée a l'étude des races humaines. (Saporta, G. da.) n.p., 1868.
Geog 1605.5	Palestine and Syria. (Baedeker, publishers.) Leipzig, 1876.
Geog 1605.10	Palestine and Syria. 2. ed. (Baedeker, publishers.) Leipsic, 1894.
Geog 1605.25A	Palestine and Syria. 5. ed. (Baedeker, publishers.) Leipzig, 1912.
Geog 4208.87.10	Palgrave, William G. Ulysses; or Scenes and studies in many lands. London, 1887.
Geog 3590.20	Pallas, P.S. Bering's successors, 1745-1780. Seattle, 1948.
Htn Geog 149.1*	Pallas, Peter. Neue nordische Beyträge. St. Petersburg, 1781. 4v.
Geog 4300.33.3	Palmer, J.E. Palmer's European pocket guide. N.Y., 188-.
Htn Geog 4678.37*	Palmer, John. Awful shipwreck...crew of the ship Francis Spaight. Boston, 1837.
Htn Geog 4105.19*	Palmer, Thomas. An essay of the meanes how to make...travailes. London, 1606.
Geog 4300.33.3	Palmer's European pocket guide. (Palmer, J.E.) N.Y., 188-.
Geog 190.10	Pam American Institute of Geography and History. Revista geográfica del instituto panamericano de geográfica e historia. México. 9,1949+ 11v.
Geog 110.3.10	Pamiętnik II zjardu słowiańskich geografów i etnografów odbytego w Polsce w roku, 1927. (Congrès des Géographes et Ethnographes Slaves, 2nd, Krakow, 1927.) Kraków, 1929-30. 2v.
Geog 4559.56.20	Pamir. (Tunstall-Behrens, Hilary.) London, 1956.
Geog 175.5	Pan American Institute of Geography and History. Commission on Cartography. Progress report on the cartographic activities of the United States. St. Louis. 1946-1952 3v.
Geog 500.63	Pan American Union. Columbus Memorial Library. Books and magazine articles on geography in the Columbus Memorial Library. Washington, 1935.
Geog 3540.10	Pannwitz, Max. Deutsche Pfadfinder des 16. Jahrhunderts in Afrika, Asien und Südamerika. Stuttgart, 1911-12.
Geog 958.90F	La panorama, merveilles de France, Belgique. Paris, 189-?
Geog 809.59.5	Panorama der Wireld. (Vooys, Adriaan.) Roermond, 1959. 3v.
Geog 4268.31.7	Panorama of Europe. 7th ed. (Hofland, B. (Mrs.).) London, n.d.
Geog 4709.40.5	Panorama of treasure hunting. (Wilkins, H.T.) N.Y., 1940.
Geog 4260.7	Pamphlet vol. Panoramas. 7 pam.
Geog 4308.93.10	Papa, Dario. Viaggi. v.2. Milano, 1893.
Geog 5559.37.20	Papanin, Ivan D. Life on an icefloe. London, 1947.
Geog 5559.37	Papanin, Ivan D. Zhizn' na l'dine. Moskva, 1940.
An 359.23.7	Pape, A.G. Is there a new race type? Edinburgh, 1923.
Geog 4029.9	Los papeles de Juan de Aresty. (Aresty, Miguel de.) Barcelona, 1972.
Geog 4181.58	The papers. (Bowry, T.) London, 1927.
Geog 55.1	Papers. (Edinburgh. University. Department of Geography.) 1960-1962
Geog 5225.8	Papers and dispatches...Arctic searching expeditions. 2. ed. (Mangles, James.) London, 1852.
Geog 4308.69.7	Papers from over the water. (Tousey, S.) N.Y., 1869.
Geog 3240.7	Les papes géographes. (Thomassy, M.J.R.) Paris, 1852.
Geog 4328.53.5	Les paquebots du Levant. (Paris. Services Maritimes.) Paris, 1853.
Geog 819.27.10	Paquet, A. Städte, Landschaften und ewige Bewegung. Hamburg, 1927.
Geog 4309.35.10	Paquet, Alfons. Fluggast über Europa. München, 1935.
An 359.73	La paradigme perdu: la nature humaine. (Morin, Edgar.) Paris, 1973.
Geog 3060.12.70	Le paradis perdu de Mu. (Vincent, Louis Claude.) Paris? 1969. 2v.
Geog 4329.36.10	Parain, Charles. La Méditerranée; les hommes et leurs travaux. 4e éd. Paris, 1936.
Htn Geog 3106.48*	Parallela geographiae. (Brietio, O.) Paris, 1648-49. 3v.
Geog 3019.55	Parias, L.H. Histoire universelle des explorations. Paris, 1955- 4v.
Geog 4308.55	Paris, Londres y Madrid. (Ochoa, Eugenio de.) Paris, 1861.
Geog 1519.19	Paris, nebst einigen Routen. 16. Aufl. (Baedeker, publishers.) Leipzig, 1905.
Geog 1519.13	Paris, Rouen, Havre, Dieppe. (Baedeker, publishers.) Coblenz, 1865.
Geog 1519.10	Paris, Rouen, Havre, Dieppe. 5. Aufl. (Baedeker, publishers.) Coblenz, 1864.
Geog 1519.24	Paris, ses environs et les principaux itinéraires. 4. éd. (Baedeker, publishers.) Leipzig, 1876.
Geog 4328.53.5	Paris. Services Maritimes. Les paquebots du Levant. Paris, 1853.
Geog 4158.24	Paris. Société de Geographie. Recueil de voyages et de memoires. Paris, 1824-1864. 7v.
Geog 1519.23	Paris and its environs. 4. ed. (Baedeker, publishers.) Leipsic, 1874.
Geog 1519.26	Paris and its environs. 5. ed. (Baedeker, publishers.) Leipsic, 1876.
Geog 1519.25	Paris and its environs. 5. ed. (Baedeker, publishers.) Leipsic, 1876.
Geog 1519.30	Paris and its environs. 6. ed. (Baedeker, publishers.) Leipsic, 1878.
Geog 1519.36	Paris and its environs. 8. ed. (Baedeker, publishers.) Leipsic, 1884.
Geog 1519.39	Paris and its environs. 9. ed. (Baedeker, publishers.) Leipsic, 1888.
Geog 1519.46	Paris and its environs. 10. ed. (Baedeker, publishers.) Leipsic, 1891.
Geog 1519.46.2	Paris and its environs. 10. ed. (Baedeker, publishers.) Leipsic, 1891.
Geog 1519.48	Paris and its environs. 11. ed. (Baedeker, publishers.) Leipsic, 1894.
Geog 1519.50.2	Paris and its environs. 12. ed. (Baedeker, publishers.) Leipsic, 1896.
Geog 1519.50	Paris and its environs. 12. ed. (Baedeker, publishers.) Leipsic, 1896.
Geog 1519.54	Paris and its environs. 14. ed. (Baedeker, publishers.) Leipsic, 1900.
Geog 1519.55	Paris and its environs. 14. ed. (Baedeker, publishers.) Leipsic, 1900.
Geog 1519.58	Paris and its environs. 15. ed. (Baedeker, publishers.) Leipzig, 1904.
Geog 1519.60	Paris and its environs. 16. ed. (Baedeker, publishers.) Leipzig, 1907.
Geog 1519.63	Paris and its environs. 17. ed. (Baedeker, publishers.) Leipzig, 1910.
Geog 1519.68	Paris and its environs. 18. ed. (Baedeker, publishers.) Leipzig, 1913.
Geog 1519.70	Paris and its environs. 19. ed. (Baedeker, publishers.) Leipzig, 1924.
Geog 1519.33.2	Paris and its invirons. 7. ed. (Baedeker, publishers.) Leipsic, 1884.
Geog 1519.15	Paris and northern France. 2. ed. (Baedeker, publishers.) Coblenz, 1867.
Geog 1519.21	Paris and northern France. 3. ed. (Baedeker, publishers.) Coblenz, 1872.
Geog 1519.20	Paris and northern France. 3. ed. (Baedeker, publishers.) Coblenz, 1872.
Geog 1519.21.3	Paris and northern France. 3. ed. (Baedeker, publishers.) Leipsic, 1872.
Geog 1519.27	Paris et ses environs. 5. éd. (Baedeker, publishers.) Leipzig, 1876.
Geog 1519.33	Paris et ses environs. 7. éd. (Baedeker, publishers.) Leipzig, 1881.
Geog 1519.36.2	Paris et ses environs. 7. éd. (Baedeker, publishers.) Leipzig, 1884.
Geog 1519.37	Paris et ses environs. 8. éd. (Baedeker, publishers.) Leipzig, 1887.
Geog 1519.42	Paris et ses environs. 9. éd. (Baedeker, publishers.) Leipzig, 1889.
Geog 1519.45	Paris et ses environs. 10. éd. (Baedeker, publishers.) Leipzig, 1891.
Geog 1519.49	Paris et ses environs. 11. éd. (Baedeker, publishers.) Leipzig, 1894.
Geog 1519.51	Paris et ses environs. 12. éd. (Baedeker, publishers.) Leipzig, 1896.
Geog 4209.55.5	Paris to Peking. 1. ed. (Starobin, J.R.) N.Y., 1955.
Geog 1519.75	Paris und Umgebung und Supplement. 20. Aufl. (Baedeker, publishers.) Leipzig, 1931-37. 2v.
Geog 576.82F	Parisini geographia. (Baudrand, M.A.) Parisiis, 1681-82. 2v.
An 359.50	Park, R.E. Race and culture. Glencoe, Ill., 1950.
Geog 5515.78	Parks, G.B. Frobisher's third voyage, 1578. n.p., 1935.

Author and Title Listing

	Geog 3520.11A	Parks, George Bruner. Richard Hakluyt and the English voyages. N.Y., 1928.
	Geog 3520.11.2	Parks, George Bruner. Richard Hakluyt and the English voyages. 2. ed. N.Y., 1961.
	Geog 3019.26	Parlss, George B. The forerunners of Hakluyt. n.p., 1926?
	Geog 520.4	Parmentier, J. Le discours de la navigation. Paris, 1883.
	Geog 4145.50	Parr, Charles. Jan van Linschaten. N.Y., 1964.
	Geog 4145.91	Parr, Charles. The voyages of David de Vries. N.Y., 1969.
	Geog 4659.69	Parry, Henry. Wreck and rescue on the coast of Wales. Truro, 1969-73. 2v.
	Geog 3019.63	Parry, John Horace. The age of reconnaissance. London, 1963.
	Geog 3019.63.1	Parry, John Horace. The age of reconnaissance. 1. ed. Cleveland, 1963.
	Geog 4129.74	Parry, John Horace. The discovery of the sea. N.Y., 1974.
	Geog 4218.78.9	Parry, S.H. Jones. My journey round the world. London, 1881. 2v.
	Geog 5518.21.2	Parry, W.E. Appendix to...Parry's Journal of second voyage. London, 1825.
	Geog 5558.18	Parry, W.E. Journal of a voyage of discovery. London, 1820. 2 pam.
	Geog 5518.21	Parry, W.E. Journal of second voyage for discovery. London, 1824.
	Geog 5518.24.2	Parry, W.E. Journal of third voyage...Northwest Passage. London, 1826.
	Geog 5518.24	Parry, W.E. Journal of third voyage...Northwest Passage. Philadelphia, 1826.
	Geog 5518.19.6.2	Parry, W.E. Journal of voyage...Northwest Passage and North Georgia Gazette. Philadelphia, 1821.
	Geog 5518.19.7	Parry, W.E. Journal of voyage...Northwest Passage and North Georgia Gazette. London, 1821.
	Geog 5518.27	Parry, W.E. Narrative of attempt to reach the North Pole. London, 1828.
	Geog 5518.19.6	Parry, W.E. The North Georgia Gazette and Winter Chronicle. No.1-20. London, 1821.
	Geog 5518.19.8	Parry, W.E. Supplement to the appendix of Capt. Parry's voyage. London, 1824.
	Geog 5326.15.5	Parry A. Parry of the Arctic. London, 1963.
	Geog 5326.15.5	Parry of the Arctic. (Parry A.) London, 1963.
	Geog 4417.80.5	Parsons, Abraham. Travels in Asia and Africa. London, 1808.
	Geog 4219.41	Parsons, C. Vagabondage. London, 1941.
	Geog 5311.3.10	Partridge, B. Amundsen. London, 1953.
	Geog 4709.27	Partridge, Eric. Pirates, highwaymen, and adventurers. London, 1927.
	Geog 4269.03	Partsch, J. Central Europe. London, 1903.
	Geog 3018.99.10	Partsch, J. Die geographische Arbeit des 19. Jahrhunderts. Breslau, 1899.
	Geog 603.2	Partsch, J. Philipp Clüver der Begründer der historischen Landerkunde. Wien, 1891.
Htn	Geog 3205.95*	Parvum theatrum urbium. (Romanus, A.) Frankoforti, 1595.
	An 5.35	Pas, H.T. Economic anthropology 1946-1972. Oosterhout, 1973.
	Geog 4308.59.11	Pas Soldan y Unánue, Pedro. Memories de un viajero peruano; apuntes y recuerdos de Europa y Oriente (1859-1863). Lima, 1971.
	An 379.41	Pasada, J. de la C. Geografía humana. Medellín, 1941.
	Geog 30.2	Paschin, O. Bibliotheca geographica. Berlin. 1-19,1891-1912 19v.
	Geog 4308.89.5	Un paseo por Europa. Habana, 1891.
	Geog 4228.35	Pases por Europa y América, en 1835 y 1836. (Ferrer, Antonio C.) Madrid, 1838.
	Geog 5939.59.10	Pasetskii, V.M. Na samoi iuzhnoi zemle. Moskva, 1959.
	Geog 5300.56	Pasetskii, Vasilii M. Eestist pärit Arktika-uurijad. Tallinn, 1970.
	Geog 4145.46	Pasetskii, Vasilii M. Ivan Fedorovich Kruzenshtern. Moskva, 1974.
	Geog 4239.54	Passage east. (Mitchell, Carleton.) London, 1954.
	Geog 4207.76.20	Passage to glory; John Ledyard's America. 1. ed. (Augur, Helen.) Garden City, 1946.
	Geog 5509.61.5	Il passaggio di nord-ouest. (Dainelli, Giotto.) Torino, 1961.
	Geog 5559.28.5	Passarello, Gaetano. Un giornalista al Polo Nord. Roma, 1965.
	Geog 4218.71.9F	Passeggiata intorno al mondo. (Hübner, J.A.) Milano, 1879.
	An 3559.27	The passing of the Frisians. (Nyèssen, D.J.H.) 's-Gravenhage, 1927.
	Geog 4209.66.25	Passport to adventure. (Scott, Jack Denton.) N.Y., 1966.
	Geog 4300.37	Pastene, J. Auto guide to Europe. N.Y., 1954.
	Geog 5300.48	Pastskii, Vasilii M. Arkticheskie puteshevstviia rossiian. Moskva, 1974.
	Geog 4515.108	Patani, Osvaldo. Soli con le montagne. Milano, 1955.
	Geog 4308.40.3	Patchwork. (Hall, Basil.) London, 1841. 3v.
	Geog 4308.40.5	Patchwork. 2nd ed. (Hall, Basil.) London, 1841. 3v.
	Geog 4218.87.15	Paterson, Florence E. Reminiscences of a skipper's wife. London, 19- .
	Geog 4131.71	Patev, Iliia. Statistika na turizma. Varne, 1971.
	Geog 4309.02.4	The path to Rome. (Belloc, H.) N.Y., 1902.
	Geog 4309.02.3A	The path to Rome. 4th ed. (Belloc, H.) London, 1916.
	Geog 4306.95	Patin, Charles. Relations historiques...de voyages. Amsterdam, 1695.
	Geog 4306.71.5	Patin, Charles. Relations historiques...de voyages en Allemagne. 2e éd. Lyon, 1676.
	Geog 4332.8	Paton, A.A. Highlands and islands of the Adriatic. London, 1849. 2v.
	Geog 4332.8.5	Paton, A.A. Researches on the Danube and the Adriatic. Leipzig, 1861. 2v.
	An 2109.34.5A	Patterns of culture. (Benedict, Ruth (Fulton).) Boston, 1934.
Htn	Geog 4248.08*	Patterson, S. Narrative of adventures...of S. Patterson, experienced in Pacific Ocean. Palmer, 1817.
	Geog 4309.34	Patterson, Sara K. Out of the fog. Caldwell, 1934.
	An 359.24.3	Paudler, F. Die hellfarbigen Rassen und ihre Sprachstamme, Kulturen und Urkeimaten. Heidelberg, 1924.
	An 2209.32	Paudler, F. Scheitelnarbensitte, Anschwellungsglaube und Kulturkreislehre. Brunn, 1932.
	An 3559.12	Paul-Boncour, G. Anthropologie anatomique, crane-face. Paris, 1912.
	Geog 4509.48.13	Pause, Walter. Mit gleichlichen Augen. 3. Aufl. München, 1951.
	Geog 5939.58.5	Pavlovskii, E.N. Antarktika. Moskva, 1958.
	Geog 4306.63.3	Payen. Les voyages de Monsieur Payen. 2e éd. Paris, 1667.
	Geog 5398.72.7	Payer, J. L'expédition du Tegetthoff. Paris, 1878.
	Geog 5398.72	Payer, J. New Lands within the Arctic Circle. London, 1876. 2v.
	Geog 5398.72.5	Payer, J. New lands within the Arctic Circle. N.Y., 1877.
	Geog 5398.72.3	Payer, J. Die österreichisch-ungarische Nordpol-Expedition. Wien, 1876.
	Geog 4329.73	Pays et paysages mediterranéens. 1. éd. (Isnard, Hildebert.) Paris, 1973.
	Geog 4759.47.3	Les pays tropicaux, principes d'une géographie humaine et économique. 4. éd. (Gourou, Pierre.) Paris, 1966.
	Geog 4759.47	Les pays tropicaux. (Gourou, P.) Paris, 1947.
	An 136.5	Paz, Octavio. Claude Lévi-Strauss; an introduction. Ithaca, 1970.
	An 136.5.17	Paz, Octavio. Claude Lévi-Strauss o el nuevo festin de Esopo. 2. ed. Mexico, 1969.
	Geog 4308.68	Peabody, A.P. Reminiscences of European travel. N.Y., 1868.
	An 190.2.2	Peacock, James Lowe. The human direction. 2. ed. N.Y., 1973.
	Geog 4509.41.5	Peacocke, T.A.H. Mountaineering. London, 1941.
	Geog 4509.41.7	Peacocke, T.A.H. Mountaineering. 3. ed. London, 1953.
	Geog 4209.29.10	Peaks of hazard. (Bruce, Michael.) Indianapolis, 1929.
	Geog 4181.143	Peard, George. To the Pacific and Arctic with Beechey. Cambridge, 1973.
	Geog 4419.30	Pearls, arms and hashish; pages from the life of a Red Sea navigator. (Monfreid, Henri de.) N.Y., 1930.
	Geog 4219.11	Pearse, Albert W. Recent travel. Sydney, 1914.
	Geog 4219.04	Pearson, D.G. Trip to the Phillipines...in 1904. n.p., n.d.
	Geog 5558.91A	Peary, Josephine D. My Arctic journal. N.Y., 1893.
	Geog 5559.05	Peary, R.E. Nearest the Pole...1905-06. N.Y., 1907.
	Geog 5559.09A	Peary, R.E. North Pole; its discovery in 1909. N.Y., 1910.
	Geog 5559.09.3	Peary, R.E. The North Pole. 2. ed. N.Y., 1910.
	Geog 5558.86	Peary, R.E. Northward over the "Great Ice"...1886. N.Y., 1898. 2v.
	Geog 5559.17A	Peary, R.E. Secrets of polar travel. N.Y., 1917.
	Geog 5326.21.20	Peary, the explorer and the man, based on his personal papers. (Weems, John Edward.) Boston, 1967.
	Geog 5326.21A	Peary; the man who refused to fail. (Green, Fitzhugh.) N.Y., 1926.
	Geog 5326.21.5	Peary. (Hobbs, William H.) N.Y., 1936.
	Geog 5559.09.25	Peary at the North Pole; fact or fiction? (Rawlins, Dennis.) Washington, 1973.
	Geog 5559.09.8	Peary's discovery of the North Pole. (Moore, J.H.) Washington, 1910.
	An 379.30.7	Peate, I.C. Studies in regional consciousness and environment. London, 1930.
	Geog 616.3.5	Peattie, R. The incurable romantic. N.Y., 1941.
	Geog 4309.12.5	Peçanha, Nilo. Impressões da Europa. Nice, 1912.
Htn	An 356.77*	Pechlin, J.N. De habitu et colore Aethiopum. Kiloni, 1677.
	Geog 5837.39	Peder Olsen Walløes dagbøger. (Walløe, Peder Olsen.) Kjøbenhavn, 1927.
	Geog 4308.45	Pedestrian and other reminiscences. London, 1846.
	Geog 4029.8	The Pedro Gorino. (Dean, Harry.) Boston, 1929.
	Geog 613.6	Pedro Vicente Maldonado, y la Misión geodésica del siglo XVIII. (Espinosa Cordero, N.) Cuenca, 1936.
	Geog 4218.70.3	Peebles, James. Around the world. 2. ed. Boston, 1876.
	Geog 5838.68	Peirce, B.M. Report on the resources of Iceland and Greenland. Washington, 1868.
	Geog 4309.57	Pélerinages européens. (Nothomb, Pierre.) Paris, 1957.
	Geog 4308.42.5	Pellegrinazioni autunnali ed opuscoli. (Baruffi, G.F.) Torino, 1842.
	Geog 4679.12.7	Pelz von Felman, J. Titanic. Berlin, 1939.
	Geog 4329.00	Pen sketches. (Acker, F.) Philadelphia, 1900.
	Geog 4239.68.10	The penance way; the mystery of Puffin's Atlantic voyage. (Naydler, Merton.) London, 1968.
	Geog 4218.70.8	Pencilling by the way: a constitutional voyage round the world: 1870 and 1871. (Beaumont, Thomas E.) London, 1971.
	Geog 4308.32.2	Pencillings by the way. (Willis, N.P.) London, 1842.
Htn	Geog 4308.32.3*	Pencillings by the way. 1st ed. (Willis, N.P.) N.Y., 1844.
	Geog 809.19.3	Peninsula Ibérica. 2a ed. (Blásquez y Delgado Aguilera, Antonio.) Barcelona, 1921.
	Geog 4110.20.3	Peninsular and Oriental Steam Navigation Company. The P. and O. pocket book. London, 1908.
	Geog 4110.20	Peninsular and Oriental Steam Navigation Company. Pocket book, 1890. London, 1890.
	Geog 4269.31	Peninsular Europe. (Lyde, L.W.) London, 1931.
	Geog 4307.65.15	Pennant, T. Tour on the continent. London, 1948.
	Geog 4417.98F	Pennant, T. The view of Hindoostan. London, 1798-1800. 4v.
	Geog 4308.93	Pennell, J. Our sentimental journey thro' France. London, 1893.
	An 99.35	Penniman, T.K. A hundred years of anthropology. London, 1935.
	Geog 3520.14	Pennington, Loren E. Hakluytus posthumus; Samuel Purches and the promotion of English overseas expansion. Emporia, 1966.
NEDL	Geog 4308.25.10	Pennington, Thomas. A journey into various parts of Europe. London, 1825. 2v.
	An 359.27.13	Penniston, J.B. Racial old age, being further adventures in philosophy. Seattle, 1927.
	Geog 4345.99.4	Penrose, B. The Sherleian odyssey, being a record of the travels of three famous brothers. Taunton, 1938.
	Geog 3249.52A	Penrose, B. Travel and discovery in the Renaissance. Cambridge, 1952.
	Geog 3520.12	Penrose, Boies. Tudor and early Stuart voyaging. Washington, 1962.
	Geog 4139.42	Penrose, Boies. Urbane travelers, 1591-1635. Philadelphia, 1942.
	An 379.31	Penrose, R.A.F. Geology as an agent in human welfare. N.Y., 1931.
	Geog 3019.72.35	La pensée géographique. (Claral, Paul.) Paris, 1972.
	Geog 665.166	La pensée géographique française contemporaine; Mélanges offerts à André Meynier. Saint-Brieuc, 1972.
	An 2109.62.5	La pensée sauvage. (Lévi-Strauss, Claude.) Paris, 1962.
	An 2109.59.5	Il pensiero dei primitivi. (Cantoni, Remo.) Milano, 1959.
	Geog 4209.36.5	People, people, everywhere. (Davis, R.H.) N.Y., 1936.
	Geog 4209.29	People and places. (Goldring, Douglas.) Boston, 1929.
	Geog 4209.31	People and places. (Hyne, C.J.C.) London, 1931.
	Geog 5855.5	The people of the Polar North. (Rasmussen, Knud.) London, 1908.
	Geog 5855.5.5	The people of the Polar North. (Rasmussen, Knud.) Philadelphia, 1908.
	An 359.48.5	Peoples of the earth. (Embree, Edwin R.) N.Y., 1948.
	Geog 818.86	Pequeña geografía. Asunción, 1886.
Htn	Geog 4126.14*	Peregrination from Scotland. (Lithgow, W.) Lond, 1614.
	Geog 4209.35.25	Peregrinations. v.2-3. (Harris, P.P.) n.p., 1935-37. 2v.

Author and Title Listing

	Geog 4348.60.10	Les pérégrinations en Orient et en Occident. 21e éd. (Beluze, X.) Citeaux, 1883. 2v.
	An 2058.65.5	Pereira da Costa, F.A. Da existencia do homem em epochas remotas no valle do tejo. Lisboa, 1865.
	Geog 4145.11	Peres, Damião. Diogo Cão. Lisboa, 1957.
	Geog 3570.20.2	Peres, Damião. Historia dos descobrimentos portugueses. Lisboa, 1959.
	Geog 3570.20	Peres, Damião. Historia dos descobrimentos portugueses. pt.1-15. Lisboa, 1943-45.
	Geog 3570.20.4	Peres, Damião. A history of the Portuguese. Lisbon, 1960.
	Geog 5322.1	Perevalov, V. Lomonosov i Arktika. Moskva, 1949.
	Geog 3560.10	Pereyra, C. La conquista de las rutas oceánicas. Madrid, 1940.
	Geog 3249.25	Pereyra, Carlos. La conquête des routes océaniques d'Henri le navigateur à Magellan. Paris, 1925.
	Geog 3560.20	Perez Emlied, F. Los descubrimientos en el Atlántico y la rivalidad castellano-portuguesa hosta el tratado de Tordesillas. 1. ed. Sevilla, 1948.
	Geog 4508.89	I pericoli dell'Alpinismo e noime per vitarli. (Fiorio, C.) Torino, 1889.
	Geog 4678.00	Perils of the ocean. N.Y., 1800?
	An 629.70.5	Perinnerineiston kvantitatiivisesta tutkimuksesta. (Sarmela, Matti.) Helsinki, 1970.
	Geog 3070.26	Der Periplus des Pontus Euxinus nach Münchener Handschriften. (Thomas, G.M.) München, 1864.
	Geog 818.36.10	Perkins, Samuel. The world as it is. 5th ed. n.p., 1839.
	Geog 818.36.11	Perkins, Samuel. The world as it is. 5th ed. n.p., 1840.
	Geog 818.36.14	Perkins, Samuel. The world as it is. 6th ed. n.p., 1842.
	Geog 4308.50	La perle trouvée. (Lundeberg, A.) Lausanne, 1850.
	Geog 612.3.8	La Pérouse. (Bellessort, André.) Paris, 1926.
	Geog 4559.25.3	Perry, F. Fair winds and foul. Boston, 1925.
	Geog 4559.30.20	Perry, Joseph Malcolm. Cruising in many seas. Springfield, 1930.
	Geog 5208.82.2	Perry, R. Jeannette, and a complete and authentic narrative encyclopedia of all voyages. San Francisco, 1883.
	An 2109.23.5	Perry, W.J. The children of the sun. N.Y., 1923.
	An 2109.73.5	Perry, William James. The primordial ocean; an introductory contribution to social psychology. London, 1973.
	Geog 4145.83	Perseus in the wind. (Stark, Freya.) London, 1948.
	An 2109.68.15	Pershits, Abram I. Istoriia pervobytnogo obshchestva. Moskva, 1968.
	An 2109.68.17	Pershits, Abram I. Istoriia pervobytnogo obshchestva. Izd. 2. Moskva, 1974.
	Geog 4308.80	Personal impressions...of European travel. Photoreproduction. (Bush, R.) N.Y., 1880?
	Geog 5538.50.25	A personal narrative of the discovery of the Northwest Passage...while in search of the expedition under Sir John Franklin. (Armstrong, Alexander.) London, 1857.
	Geog 4678.73	Personal recollections of wreck of...Ville-du-Havre. (Weiss, Nathanael.) N.Y., 1875.
	Geog 4208.78	Personal reminiscences. (Forbes, R.L.) Boston, 1878.
Htn	Geog 4208.78.3*	Personal reminiscences. 2. ed. (Forbes, R.B.) Boston, 1882.
X Cg	Geog 4208.78.2	Personal reminiscences. 2. ed. (Forbes, R.L.) Boston, 1882.
Htn	Geog 4308.89.10*	Personally conducted. (Stockton, F.R.) N.Y., 1889.
	Geog 659.39.5A	Perspective on the nature of geography. (Hartshorne, Richard.) Chicago, 1959.
	Geog 180.10	Perspectives in geography. Dekalb. 1,1971+
	Geog 5845.4	Perspektivplan for Grønland 1971-85. (Denmark. Ministeriet for Grønland.) København, 1971.
	Geog 3594.5	Pertek, Jerzy. Polacy ner szlakach morskich świata. Gdańsk, 1957.
	Geog 500.71	Perthes, Justus, publishers. Wandkarten, Cottanten, Bücher, Zeitschriften für den geographischen Unterricht. Gotha, 1930.
	Geog 601.5	Pertish, E.N. K.I. Arsen'ev i ego raboty po Raiomirovamiiu Rossii. Moskva, 1960.
	An 358.59.3	Perty, M. Grundzüge der Ethnographie. Leipzig, 1859.
	Geog 4412.91	Pertz, G.H. Der älteste Versuch zur Entdeckung des Seeweges nach Ostindien im Jahre 1291. Berlin, 1859.
	Geog 4218.03.12	Perv. putesh. rossii. vokrug sveta, 1803-1806. (Nevskii, Vladimir V.) Moskva, 1951.
	Geog 6008.19	Pervaia russkaia antarkticheskaia ekspeditsiia, 1819-1821 gg. i ee otchetnaia navigatsionnaia karta. (Leningrad. Arkticheskii i Antarkticheskii Nauchno-Issledovatel'skii Institut.) Leningrad, 1963.
	Geog 5559.14.5	Pervaia russkaia ekspeditsiia k severnomy poliucu. (Seleznev, Stepan A.) Arkhangel'sk, 1964.
	An 2109.75.10	Pervobytnoe obshchestvo. Moskva, 1975.
Htn	Geog 5557.25.5*	Pervoe morskoe puteshestvie. (Berkh, V.N.) Sankt Peterburg, 1823.
	Geog 5300.25	Pervye razvedchiki velikogo puti. (Muratov. M.V.) Moskva, 1943.
	Geog 3590.40	Pervye russkie issledovateli Aliaski. (Adamov, Arkadii.) Moskva, 1950.
	Geog 3018.77.2	Peschel, O. Geschichte der Erdkunde. München, 1877.
	Geog 665.19	Peschel, O.F. Abhandlungen zur Erd- und Völkerunde. v.1,2-3. Leipzig, 1877. 2v.
	Geog 3248.58	Peschel, O.F. Geschichte des Zeitalters. Stuttgart, 1858.
	Geog 3248.58.5	Peschel, O.F. Geschichte des Zeitalters. Stuttgart, 1877.
	An 358.90.3	Peschel, O.F. The races of man and their geographical distribution. N.Y., 1890.
	An 358.85.3	Peschel, O.F. Völkerkunde. Leipzig, 1885.
	An 358.76.3	Peschel, Oscar F. The races of man. N.Y., 1876.
	An 358.76.5	Peschel, Oscar F. Völkerkunde. 3. Aufl. Leipzig, 1876.
	VAn 629.67.5	Pešiě-Golutovic, Zagorka. Antropologia kao drustveno nauka. Beograd, 1967.
	Geog 6009.63	Peskov, Vasilii M. Belye suy. Moskva, 1965.
	Geog 4208.58.9	Petano y Mazariegos, Gorgonio. Viajes por Europa y América de Don Gorgonio Petano y Mazariegos. Paris, 1858.
	Geog 4181.74	Peter Floris. (Moreland, W.H.) London, 1934.
	Geog 4308.07	Peter Irving's journals. (Irving, Peter.) N.Y., 1943.
	Geog 4218.92.3	Peters, G.H. Impressions of a journey round the world. London, 1897.
	Geog 3057.5.5	Peters, Karl. King Solomon's golden Ophir. N.Y., 1969.
	Geog 3057.5	Peters, Karl. Ophir nach dem neuesten Forschungen. Berlin, 1908.
	Geog 5538.57.5	Petersen, Carl. Den sidste Franklin expedition med "Fox". Kjøbenhavn, 1860.
	VGeog 5700.4.15	Petersen, Niels Matthias. Hans Egedes levnet. København, 1839.
	Geog 500.35	Petherick, E.A. Catalogue of the York Gate Library. London, 1886.
	Geog 4678.75F	Petition and claim for damages to the ship Immacolata. (Campisi, Orazio.) n.p., 1879.
	Geog 4708.68	Petizione...relativa alla catastrofe toccata alla nave Teresa. (Bollo, G.A.) n.p., n.d. 2 pam.
	Geog 4502.10.2	Petrarca, F. Des F. Petrarca Sendschreiben die Besteigung des Mont Ventoux betreffend. München, 1936.
	Geog 4309.27.10	Petre, E.R. When you go to Europe. N.Y., 1933.
	Geog 4019.68	Petro, W. Triple commission. London, 1968.
	Geog 5379.67	Petrow, Richard. Across the top of Russia. N.Y., 1967.
	Geog 4679.67.25	Petrow, Richard. The black tide: in the wake of the Torrey Canyon. London, 1968.
	Geog 4219.47	Petterson, H. Westward ho with the Albatross. 1. ed. N.Y., 1953.
	Geog 4509.53	Petzoldt, P. On top of the world. N.Y., 1953.
	Geog 5182.4	Le peuple esquimau aujourd'hui et demain. Paris, 1973.
	An 358.81	Les peuples de l'Asie et de l'Europe. (Girard de Rialle, J.) Paris, 1881.
Htn	An 2056.56*	Peyrère, I. de. Men before Adam: discourse on Romans 5-12,13,14. London, 1656.
	An 2056.61	Peyrère, I. de. Praeadamiten. Jaer, 1661.
	Geog 3235.17	Pezzana, A. De l'anciennete de la mappemonde. Genes, 1808.
	Geog 3235.37	Pezzana, A. Estratto di una nota pesta a.f. 365-66...la carta nautica. Berlin, 1881.
	Geog 3410.5	Pfeifer, Gottfried. Regional geography in the United States since the war. N.Y., 1938.
	Geog 4324.59.5	Pfeiffer, F. Georg von Ehingen Reisen. Stuttgart, 1842.
	Geog 4218.46.7	Pfeiffer, I. A lady's second journey...world. N.Y., 1856.
	Geog 4218.46.5	Pfeiffer, I. A lady's voyage round the world. N.Y., 1852.
NEDL	Geog 4218.46.3	Pfeiffer, I. A woman's journey round the world. London, 1846.
	An 2059.69	Pfeiffer, John. E. The emergence of man. N.Y., 1969.
	Geog 665.4	Die Pflege der Erdkunde in Oesterreich...Festschrift...Franz Josef I. (Umlauft, F.) Wien, 1898.
	Geog 4329.40	Pfuhl, E. Ostgriechische Reisen; Kleinasien, Kypros und Syrien. Basel, 1940.
	Geog 5110.3	The phantom of the poles. (Rud, William.) N.Y., 1906.
	Geog 4209.23.25	Phelan, James D. Travel and comment. San Francisco, 1923.
	Geog 4328.62	Phelps, S.D. Holy Land with glimpses of Europe and Egypt. N.Y., 1864.
	Geog 182.1	Philadelphia Geographical Club. Bulletin. Philadelphia. 1-36 16v.
	Geog 182.2	Philadelphia Geographical Society. Charter. Philadelphia.
	Geog 603.2	Philipp Clüver der Begründer der historischen Landerkunde. (Partsch, J.) Wien, 1891.
	Geog 183.2	Philippine geographical journal. Manila. 1,1953+ 4v.
	Geog 4268.94	Philippson, A. Europa...eine...Landeskunde. Leipzig, 1894.
	Geog 4268.94.5	Philippson, A. Europa. Leipzig, 1906.
	Geog 4180.39	Phillipine Islands, Moluccas. (Morga, A. de.) London, 1868.
	Geog 4708.06	Phillips, James D. Loss of the ship Essex in 1806. n.p., 1941.
	Geog 4110.18	Phillips, Morris. Abroad and at home. N.Y., 1892.
	Geog 4158.05	Phillips, R. Collection of...voyages and travels. London, 1805-08. 7v.
	Geog 4158.05.2	Phillips, R. Collection of...voyages and travels. London, 1810. 3v.
	Geog 818.10	Phillips, R. General view of manners...of nations. Philadelphia, 1810. 2v.
	Geog 818.10.3	Phillips, R. Geographical view of the world. N.Y., 1826.
	Geog 665.22	Phillips, R. The hundred wonders of the world. 1st American ed. New Haven, 1821.
	Geog 4158.19	Phillips, R. New voyages and travels. London, 1819-20. 9v.
	Geog 4559.68	Phillips-Burt, Douglas H.C. Reflections in the sea. Lymington, 1968.
	An 182.2	Philosophische Anthropologie und Theorie der Menschenkenntnis. (Humboldt, Wilhelm von.) Halle, 1929.
Htn	Geog 5397.73*	Phipps, J. Journal of voyage...North Pole. London, 1774.
	Geog 5397.73.7	Phipps, J. Reise nach dem Nordpol. Berlin, 1777.
	Geog 5397.73.3	Phipps, J. Voyage...North Pole...1773. London, 1774.
	Geog 5397.73.5	Phipps, J. Voyages au pôle boréal. Paris, 1775.
	Geog 659.60	Phlipponneau, Michel. Geographie et action. Paris, 1960.
	Geog 4269.68.5	Phloros, Paulos. Eurōpaikē suphmōnia taxidia. Athēnai, 1968.
	Geog 4269.64.20	Phloros, Paulos. Staurodromia tēs Eurōpēs. Athēnai, 1964.
	Geog 4329.51	The phoenix in the desert. (Thompson, D.) London, 1951.
	Geog 959.00	Photograph album containing views of Alaska about 1900, and mountaineering in the Swiss Alps. n.p., n.d.
	Geog 958.94	Photographic views of the world. Boston, 1894.
	Geog 818.80.5	A physical, historical, political, and descriptive geography. (Johnston, Alexander K.) London, 1880.
	Geog 818.80.6	A physical, historical, political, and descriptive geography. 2nd ed. (Johnston, Alexander K.) London, 1881.
	Geog 818.68	Physical, historical and military geography. (Lavallée, T.S.) London, 1868.
	An 3309.70.1	Physical anthropology of Afghanistan. v.1-2. (Debets, Georgii Frantsevich.) Cambridge, 1970.
	An 3309.21	The physical growth of children from birth to maturity. (Baldwin, Bird T.) Iowa City, 1921.
	Geog 5235.5	Physical map of the Arctic. (American Geographical Society, N.Y.) N.Y., 1930.
	Geog 4308.56	A physician's vacation. (Channing, W.) Boston, 1856.
	Geog 4309.62.5	Piat'desiat tysiach nil' po vode i sushe. (Alimzhanov, An.) Alma-Ata, 1962.
	Geog 4209.63	Piatnadtsat' let skitanii po zemnomu sharu. (Nasedkin, V.N.) Moskva, 1963.
	Geog 665.158	Una piccola biblioteca. (Riva, Ambrogio.) Milano, 1970.
	Geog 4419.10	Pick, Emil G. Reisebriefe eines österreichischen Industriellen aus Abessinien, Indien und Ostasien. Prag, 1910?
	An 358.63.4	Pickering, Charles. Races of man; and their geographical distrbution. London, 1863.
	An 358.54.5	Pickering, Charles. The races of man; and their geographical distribution. London, 1854.
	An 358.50.4	Pickering, Charles. The races of men. London, 1850.
Htn	Geog 5507.82*	Pickersgill, R. Concise account of voyages...discovery. London, 1782.
	Geog 819.45	Pickles, Thomas. The work of men. 1st ed. London, 1945.
	Geog 4509.33.5A	Pickman, D.L. Some mountain views. Boston, 1933.
	Geog 818.40.4	A pictorial geography of the world. (Goodrich, S.G.) Boston, 1840.
	Geog 818.40.7	Pictorial geography of the world. (Goodrich, S.G.) Boston, 1856. 2v.
	Geog 818.40.5	Pictorial geography of the world. 2nd ed. (Goodrich, S.G.) Boston, 1840. 2v.

Geog 818.40.6 Pictorial geography of the world. 3rd ed. (Goodrich, S.G.) Boston, 1840.
Geog 818.61.10 A pictorial hand-book of modern geography. 1st ed. (Bohn, Henry G.) London, 1865.
Geog 4308.55.5 Pictures of Europe...ideas. (Bartol, C.A.) Boston, 1855.
Geog 4308.55.7 Pictures of Europe...ideas. (Bartol, C.A.) Boston, 1856.
Geog 4218.92.7 Pictures of the world. (Scott, Clement.) London, 1894.
Geog 4268.75F Picturesque Europe. N.Y., 1875-1879. 3v.
Geog 4228.50.2 The picturesque souvenir. (Bryant, W.C.) N.Y., 1851.
Geog 4327.93 Picturesque tour thro'...Europe, Asia. (Bisani, Alessandro.) London, 1793.
Geog 4110.19 The picturesque tourist. (Hall, E.H.) N.Y., 1877.
Geog 4110.17 Picturesque travel. (Burns, Philp and Co.) Sydney, 1914.
Geog 808.75F The picturesque world, or Scenes in many lands. pt.1,3-6,8-13,15-16,18,20,26. (Colange, Leo de.) Boston, 1875.
Geog 4218.83.2 Pidgeon, D. An engineer's holiday...trip from 0 degrees to 0 degrees. 2. ed. London, 1883.
Geog 4219.32.35.5 Pidgeon, H. Around the world single-handed. N.Y., 1932.
Geog 4709.66 Pieces of eight; recovering the riches of a lost Spanish treasure fleet. 1. ed. (Wagner, Kip.) N.Y., 1966.
Htn Geog 4237.52.5* Pierce, Nathaniel. Account of great dangers...and remarkable deliverance of N. Pierce. n.p., n.d.
Geog 4237.56 Pierce, Nathaniel. Account of the great dangers...of Captain N. Pierce. Boston, 1756.
Geog 604.2 Pierre Desceliers. (Anthiaume.) Rouen, 1926.
Geog 4559.40.10 Pierrefeu, François de. Les confessions de Tatibouet. Paris, 1939.
Geog 4182.63 Pieter van den Broecke in Azie. (Broecke, Pieter van den.) 's-Gravenhage, 1962. 2v.
Geog 579.73.5 Pietkiewicz, Stanisław. Słownik pojęć geograficznych. Wyd. 1. Warszawa, 1973.
Geog 622.1.5 Pietro's pilgrimage. (Blunt, Wilfred.) London, 1953.
An 358.96.5 Piette, Édouard. Étude d'éthnographie préhistorique. no.2-3,6-8. Paris, 1896-1925.
An 2058.79 Piette, Edouard. Nomenclature des temps anthropiques primitifs. Laon, 1879.
Geog 520.24 Pigafetta; relation du premier voyage...par Magellan. (Denucé, J.) Paris, 1923.
An 3309.08.5 Pigmentation survey of school children in Scotland. (Tocher, James F.) Aberdeen, 1908.
Geog 4182.75 Pijnacker, Cornelis. Historysch verhael van den steden Thunes. 's-Gravenhage, 1975.
Geog 4181.94 The pilgrimage of Arnold von Harff. (Harff, Arnold.) London, 1946.
Geog 4328.47 A pilgrimage to the land of my fathers. (Margoliouth, Moses.) London, 1850. 2v.
Geog 4308.45.30 A pilgrim's reliquary. (White, T.H.) London, 1845.
Geog 4308.65.5 The pilgrim's wallet. (Haven, Gilbert.) N.Y., 1866.
Geog 4312.10 Pilkington, Roger. Small boat to Elsinore. N.Y., 1969.
Geog 3570.16.5 Pilobas das navegacões portuguesas das seculos XVI e XVII. (Frazão de Vasconcellos, J.) Lisboa, 1942.
Geog 3570.16 Os pilotos dos seculos XV e XVI e a nobreza do reino. (Frazão de Vasconcellos, J.) Lisboa, 1932.
An 2059.55 The Piltdown fantasy. (Vere, Francis.) London, 1955.
An 2059.55.5 The Piltdown forgery. (Weiner, Joseph S.) London, 1955.
An 2059.72 The Piltdown men. (Millar, Ronald William.) London, 1972.
Geog 3570.27 Pina Manique, Luiz da. Subsidios para a história de cartografia portuguesa. Lisboa, 1943.
Geog 4209.55.15 Pineda Yañez, R. La isla y Colón. Buenos Aires, 1955.
Geog 818.02.90 Pinkerton, J. Modern geography. London, 1802. 2v.
Geog 818.04.5 Pinkerton, J. Modern geography. London, 1811. 2v.
Geog 818.04 Pinkerton, J. Modern geography. Philadelphia, 1804. 2v.
Geog 4158.08 Pinkerton, John. General collection of voyages. London, 1808-14. 17v.
NEDL Geog 4158.10F Pinkerton, John. General collection of voyages. Philadelphia, 1811-14. 17v.
Geog 4309.38.10 Pinochet, T. Viaje, plebeys por Europa. 2. ed. Santiago, 1938.
Geog 5925.50 Pinochet de la Barra, Oscar. Chilean sovereignty in Antarctica. Santiago, 1955.
Geog 4206.65.5 Piolin, Paul. René Desboys du Chastelet. n.p., 1885.
Geog 4218.49 Pioneer voyage to California and round the world, 1849-1852. (Coffin, George.) Chicago? 1908.
Geog 4208.82 Pioneering in the Far East. (Helms, L.V.) London, 1882.
An 126.5 Pioneers in American anthropology: the Bandelier-Morgan letters, 1873-1883. (Bandelier, Adolph Francis.) Albuquerque, 1940. 2v.
Geog 5300.40 Pionieri italiani nelle regioni polari. (Zavalti, Silvio.) Brescia, 1952.
Geog 3019.40 Pioniers (De ontdekking van de wereld). (Balen, Willem Julius van.) Amsterdam, 1940. 2v.
Geog 4309.73 Piontek, Heinz. Helle Tage anderswo. München, 1973.
Geog 4307.84 Piozzi, Hester Lynch Thrale. Observations and reflections made in the course of a journey through France, Italy and Germany. Ann Arbor, 1967.
Geog 4708.34 Pamphlet vol. Piracy on brig Mexican of Salem, report of trial. 4 pam.
Geog 4709.34 Pirates, flibustiers, negriers. 9. éd. (Magre, M.) Paris, 1934.
Geog 4709.27 Pirates, highwaymen, and adventurers. (Partridge, Eric.) London, 1927.
Geog 4709.28.10 Pirates, old and new. (Gallomb, J.) N.Y., 1928.
Htn Geog 4708.16.20* The pirates. Boston, 181-
Geog 4709.33 Pirates of the Eastern seas, 1618-1723. (Grey, Charles.) London, 1933.
Geog 4709.66.5 Pirates of the Eastern seas. (Course, Alfred George.) London, 1966.
Htn Geog 4708.37* The pirates own book. Portland, 1837.
Geog 4709.24A The pirates own book. Salem, 1924.
Geog 4709.57.5 I pirati. (Monti, Mario.) Milano, 1957.
An 2108.73.10 Pireteau, Jean. Origine et destinée de l'homme. Paris, 1973.
An 3509.37 Le pithécanthrope était-il un Pygmée? (Buyssens, Paul.) Bruxelles, 1937.
Geog 4308.82 Pitman, M.J. European breezes. Boston, 1882.
Geog 4218.82.3 Pitt, George. The collected remarkable travels of G. Pitt. 3. ed. Glasgow, 1887.
An 2109.06 Pitt-Rivers, A.L.F. The evolution of culture and other essays. Oxford, 1906.
An 359.27.9A Pitt-Rivers, G.H.L.F. The clash of culture and the contact of races. London, 1927.
An 3539.09F Pittard, E. Anthropologie de la Suisse. Genève, 1909-10.
Geog 4209.59 Pizzinelli, Corrado. Viaggio nel mondo. Bologna, 1959.
An 359.31.7 The place of prejudice in modern civilization. (Keith, Arthur.) N.Y., 1931.
Geog 4309.41.10 Places. (Belloc, Hilaire.) N.Y., 1941.
Geog 579.56 Places. (Grigson, G.) N.Y., 1956?
Geog 4209.72.10 Places. (Morris, James.) London, 1972.
Geog 183.5 Places. Indiana, Pa. 1,1974+
VGeog 4509.69 Planinsko berilo. Ljubljana, 1969.
Geog 3235.27F Il planisfero di Giovanni Leardo. (Berchet, G.) Venezia, 1880.
Geog 4129.23 Planning a trip abroad. (Hungerford, E.) N.Y., 1923.
Geog 4129.23.5 Planning a trip abroad. (Hungerford, E.) N.Y., 1923.
Geog 4300.30.4 Planning a trip abroad. (Hungerford, E.) N.Y., 1927.
Geog 4300.30.8 Planning a trip abroad. (Hungerford, E.) N.Y., 1931.
Geog 5398.93.15 Plass, F. Reise-Erinnerungen...Aug. 1893..."Admiral". Hamburg, 1894.
Geog 4208.90 Platen, Carl von. Impressions of travels. Stockholm, 1890.
Geog 665.92 Plattie, Roderick. College geography. Boston, 1926.
Geog 3590.30.10 Plavanie A.I. Chirukova. (Lebedev, Dimitrii M.) Moskva, 1951.
Geog 4311.23 Plavashchi domovc. (Kazaka, Tenin.) Plovdiv, 1965.
Geog 808.08 Playfair, J. System of geography. Edinburgh, 1808. 6v.
Geog 139.5.5 Playfair, R.L. Supplement to the bibliography of Algeria, 1895. London, 1898.
Geog 4218.88 Pleasant hours in sunny lands. (Lewis, I.N.) Boston, 1888.
Geog 4308.56.13 Pleasant memories. (Sigourney, L.H.) Boston, 1856.
Geog 4209.65.15 Pleshakov, Leonid P. Vokrug sveta s Zarei. Moskva, 1965.
Geog 4308.19.5 Plessis, Joseph O. Journal d'un voyage en Europe. Québec, 1903.
Geog 3019.33.5 Plischke, Hans. Entdeckungsgeschichte von Altertum bis zur Neuzeit. Leipzig, 1933.
Geog 5519.40.5 Plowing the Arctic. (Tranter, G.J.) London, 1944.
Geog 3017.85 Pluche, N.A. Concorde de la géographie des différents ages. Paris, 1785.
An 358.64.9 The plurality of the human race. 2. ed. (Pouchet, Georges.) London, 1864.
An 2109.62.15 Po besovym sledam. (Domanskii, Ia.V.) Leningrad, 1962.
Geog 4209.68.25 Po iuzhnym stranam. (Rodin, Leonid Efimovich.) Moskva, 1968.
Geog 4329.71 Po more. (Chernishev, Slaveho.) Sofiia, 1971.
Geog 5235.15 Po nekhoshenoi zemle. (Ushakov, G.A.) Moskva, 1953.
Geog 3590.15.40 Po ozeram Sibiri i Srednei Azii. (Berg, R.L.) Moskva, 1955.
Geog 4209.67 Pochivalov, Leonid V. My za granitsei. Moskva, 1967.
Geog 4110.9.2 Pocket atlas of the world. (Rand, McNally and Co.) Chicago, n.d.
Geog 4110.9 Pocket atlas of the world. (Rand, McNally and Co.) N.Y., 1887.
Geog 4110.20 Pocket book, 1890. (Peninsular and Oriental Steam Navigation Company.) London, 1890.
Geog 4105.10 Pocket guide for Americans going to Europe. (Buffum, E.G.) N.Y., 1859.
Geog 4267.18 Pockh, J.J. Der politische catholische Passagier. Augsburg, 1718.
Geog 5399.72 Pod nogami ostrov ledianoi. (Chilingarov, A.N.) Moskva, 1972.
Geog 4209.08 Podmore, P.S.M. Rambles and adventures in Australasia. London, 1909.
An 359.66.5 Podol'nyi Roman G. Predki i my. Moskva, 1966.
Geog 4145.83.5 Podróze Stefana Srolca Rogozinskiego. (Szumańska-Grossowa, Hanna.) Warszawa, 1967.
Geog 5399.12.10 Podvig shturmaka V.I. Al'banova. (Al'banov, V.A.) Moskva, 1953.
Geog 4307.37.9 Pöllnitz, K.L.. von. A vagabond courtier. London, 1913. 2v.
Geog 4307.35.3 Pöllnitz, K.L. von. Memoires de Charles-Louis, Baron de Pöllnitz. Amsterdam, 1735. 2v.
Geog 4307.37 Pöllnitz, K.L. von. Memoirs of...travels. London, 1737. 4v.
Geog 4307.37.2 Pöllnitz, K.L. von. Memoirs of Charles-Lewis, Baron de Pöllnitz. 2nd ed. London, 1739. 4v.
Geog 4307.35.5 Pöllnitz, K.L. von. Nouveaux memoires du Baron de Pöllnitz. Amsterdam, 1737. 2v.
Geog 4131.62.5 Poeschl, A.E. Fremdenverkehr und Fremdenverkehrspolitik. Berlin, 1962.
Geog 4209.31.20 The poetic impression of natural scenery. (Cornish, Naughan.) London, 1931.
Geog 4209.66.10 Pogačnik, Bogdan. Povsod so ljudje. Maribor, 1966.
Geog 3229.06 Poggio Bracciolini und Nicolò de Conti. (Sensburg, W.) Wien, 1906.
An 359.02.6 Pogodin, A.L. Sbornik" statei po arkheologii i etnografii. Sankt Peterburg, 1902.
An 359.68.10 Poirier, Jean. Ethnologie générale. Paris, 1968.
An 359.34.3 Poisson, Georges. Les aryens. Paris, 1934.
Geog 4417.45 Poivre, Pierre. Un manuscrit inédit de Pierre Poivre: Les mémoires d'un voyageur. Paris, 1868.
Geog 4417.68 Poivre, Pierre. Voyages d'un philosophe. Yverdon, 1768.
Geog 5312.7.20 Pokrovskii, Aleksei A. Ekspeditsiia Beringa. Moskva, 1941.
Geog 3594.15 Polacy badacze Ameryki. (Turley, Tomasz J.) Chicago, 1968.
Geog 3594.5 Polacy ner szlakach morskich świata. (Pertek, Jerzy.) Gdańsk, 1957.
Geog 5170.5 Polar colonization. (Howgate, H.W.) Washington, 1878.
Geog 5559.11 Polar exploration. (Bruce, William S.) London, 1911.
Geog 5070.62 Polar exploration. (Croft, Andrew.) London, 1939.
Geog 5070.62.5 Polar exploration. 2. ed. (Croft, Andrew.) London, 1947.
Geog 5558.60.3 The polar exploring expedition. (American Geographical and Statistical Society.) N.Y., 1860.
Geog 5153.10 Polar notes. Hanover, N.H. 1-14,1959-1975// 2v.
Geog 5209.67 The polar passion: the quest for the North Pole. (Mowat, Farley.) Toronto, 1967.
Geog 5398.79 Polar reconnaissance. (Markham, A.H.) London, 1881.
Geog 184.1 The polar record. Cambridge. 3,1941+ 14v.
Geog 5538.50.10 The Polar regions...Franklin's expedition. (Osborn, S.) N.Y., 1854.
Geog 5110.5 The polar regions. (Bronner, Finn E.) Santa Barbara, 1958.
Geog 5070.20.5 The polar regions. (Brown, R.N.R.) London, 1927.
Geog 5070.38 The polar regions. (Debenham, F.) London, 1930.
Geog 5060.5 The polar regions. (Richardson, J.) Edinburgh, 1861.
Geog 5070.30 The polar regions in the twentieth century. (Greely, A.W.) Boston, 1928.
Geog 5100.10 The polar regions in their relation to human affairs. (Gould, Laurence McKinley.) N.Y., 1958.
Geog 5508.31 Polar regions of the western continent. (Snelling, W.J.) Boston, 1831.
Geog 5328.50 The polar Rosses: John and James Clark Ross and their explorations. (Dodge, Ernest S.) London, 1973.

Author and Title Listing

Geog 4309.69 The polar route to time gone by. 1st ed. (Duffus, Robert Luther.) N.Y., 1969.
Geog 5057.5 The polar times. N.Y. 1,1935+ 4v.
Geog 5208.69.15 The polar world. (Hartwig, G.) London, 1886.
Geog 5070.44 Polarfahrten. (Zeidler, P.G.) Berlin, 1927.
Geog 5070.11.5 Die Polarforschung. (Hassert, Kurt.) München, 1956.
Geog 5070.11 Die Polarforschung. 3. Aufl. (Hassert, Kurt.) Leipzig, 1914.
Geog 5300.47 Die Polarforschung der Sowjetunion. (Gordienko, Pavel A.) Düsseldorf, 1967.
Geog 5130.5 Polarskipet frani. (Domaas, K.) Oslo, 1954.
Geog 5180.10 Die Polarvölker. (Byhan, A.) Leipzig, 1909.
Geog 5209.23.3 Le pôle nord. (Rouch, J.) Paris, 1923.
Geog 5938.89 Le pôle sud. (Fonville, W. de.) Paris, 1889.
Geog 5925.15 Le pôle sud. (Rouch, J.) Paris, 1921.
An 359.71.40 Poliahov, Leon. La mythe aryen. Paris, 1971.
Geog 5317.15 Poliarnye zori; zapiski zhurnaliste. (Gal'perin, I.M.) Moskva, 1956.
Geog 5070.66 Poliarnyi krug. Moskva, 1974.
Geog 5399.28 Policing the top of the world. (Lee, Herbert P.) London, 1928.
An 2109.65 Politics, law and ritual in tribal society. (Gluckman, Max.) Oxford, 1965.
Geog 4313.7 Die politisch-geographische Bedeutung der Ostsee. Inaug.-Diss. (Schlump, E.) Königsberg, 1934.
Geog 4267.18 Der politische catholische Passagier. (Pockh, J.J.) Augsburg, 1718.
Geog 4679.67.20 La pollution des mers..."Torrey Canyon". (Du Pontavice, Emmanuel.) Paris, 1968.
Geog 4209.68 Polmilliona kilometrov pozadi. (Maevskii, Viktor V.) Moskva, 1968.
Geog 4139.73 Polscy podróznicy i odkrywey. 1. wyd. (Słabezyński, Wacław.) Warszawa, 1973.
Geog 500.120 Polska Akademia Nauk. Institut Geografii. Katalog rękopisów geograficznych w zbiorach polskich. Warszawa, 1965- 2v.
Geog 185.16 Polska Akademia Nauk. Institut Geografii. Streszczenia prac habiltacyjnych i dobtorskich. Warszawa. 1967+
Geog 185.6 Polska Akademia Nauk. Instytut Geografii. Prace geograficzne. Warszawa. 1,1954+ 40v.
Geog 185.5 Polska Akademia Umiejetności. Komisja Geograficzna. Prace. Kraków. 1-2 2v.
Geog 185.5PF Polska Akademia Umiejetności. Komisja Geograficzna. Prace. Kraków. 2
Geog 185.7 Polski przegląd kartograficzny. Lwów.
Geog 3594.10 Polskie nazwy na mapie śceiata. Wyd. 1. (Kuźmiński, Bolesław.) Warszawa, 1967.
Geog 185.8 Polskie Towarzystwo Geograficzne. Polskie towarzystwo geograficzne w 50 rocznicę dzialalności. Warszawa, 1968.
Geog 185.8 Polskie towarzystwo geograficzne w 50 rocznicę dzialalności. (Polskie Towarzystwo Geograficzne.) Warszawa, 1968.
Geog 185.15 Polsko-Czeskie Seminarium Geograficzne, 3rd, Warsaw, 1967. Polsko-czeskie seminarium geograficzne. Wyd. 1. Warszawa, 1968.
Geog 185.15.10 Polsko-Czeskie Seminarium Geograficzne, 5th, Warsaw, 1972. V. Czesko-Polskie seminarium geograficzne. Wyd. 1. Warszawa, 1975.
Geog 185.15 Polsko-czeskie seminarium geograficzne. Wyd. 1. (Polsko-Czeskie Seminarium Geograficzne, 3rd, Warsaw, 1967.) Warszawa, 1968.
Geog 5559.32 Polunin, N. The isle of auks. London, 1932.
Geog 5519.39 Poncins, G. The ghost voyage. 1. ed. Garden City, 1954.
Geog 4419.45 Ponder, S.E.G. A wanderer in khaki. London, 1945.
Geog 4309.31.5 Ponten, J. Zwischen Rhone und Wolga. Leipzig, 1931.
Geog 6009.10.21 Ponting, H.G. The great white South. London, 1922.
Geog 6009.10.60 Ponting, Herbert George. Scott's last voyage. London, 1974.
Geog 4557.85 Pontoppidam, C. Hval og robbesangsten. Risbenhavn, 1785.
Geog 4307.85.7 Ponz, A. Viage fuera de Espana. Madrid, 1785. 2v.
An 359.22.3A The population problem. (Carr-Saunders, A.M.) Oxford, 1922.
Geog 4309.26.10 Por los caminos del mundo. (Blanco-Fombona, R.) Madrid, 1926.
Geog 6009.04 Por los mares antarticos. (Maveroff, J.O.) Buenos Aires, 1954.
Geog 5855.12 Porsild, N.P. Studies on the material culture of the Eskimo in West Greenland. København, 1915.
Htn Geog 818.24* Porte-feuille géographique et ethonographique. 2e éd. v.1-2, Atlas. (Engelmann, G.) Mulhouse, 1824. 2v.
An 2109.37 Portens, S.D. Primitive intelligence and endowment. N.Y., 1937.
Geog 4248.12 Porter, David. Voyage in the South Seas in...1812-1814. London, 1823.
Geog 4209.60.35 Portes Gil, Emilio. El mundo a través de sus grandes estadistas. México, 1960.
Geog 1208.53.3 Portfolio für Länder und Völkerkunde. (Ungewitter, D.H.) Pest, 1853.
Geog 958.92.5F Portfolio of photographs of famous scenes, cities and paintings. (Stoddard, John L.) Chicago, 189-?
Htn Geog 4217.85.3F* Portlock, N. Voyage round the world. London, 1789.
Geog 3235.56A Portolan charts. (Stevenson, E.L.) N.Y., 1911.
Geog 3025.8.2F Portolan charts of XVth, XVIth, and XVIIth centuries. N.Y., 1912.
Geog 3025.8 Portolan charts of XVth, XVIth, XVIIth centuries collected by Dr. Theodore J.E. Hamy. N.Y., 1912.
Geog 3235.39 Portolani, esistenti nelle...bibliotheque. (Berchet, G.) Venezia, 1866.
Geog 3235.50.9 I portolani italiani del medioevo secondo l'opera di K. Kretschmer. (Errera, Carlo.) Firenze, 1911.
Geog 5839.30.15 Portrait of an ice cap. (Soctt, J.M.) London, 1953.
Geog 5530.15.5 Portrait of Jane. (Woodward, F.J.) London, 1951.
Htn Geog 4308.84* Portraits of places. (James, H.) Boston, 1884.
Geog 4308.84 Portraits of places. (James, H.) Boston, 1884.
Geog 4308.84.5 Portraits of places. 3rd ed. (James, H.) Boston, 1893.
Geog 185.12 Portugal. Agrupamento de Estudos de Cartografia Antiga. Secção de Coimbra. Publicacion. Ser. monografias. Lisboa. 1,1963+ 12v.
Geog 185.10 Portugal. Agrupamento de Estudos de Cartografia Antiga. Secção de Lisboa. Publicacion. Ser. separatas. Lisboa. 1-21 5v.
Geog 185.1 Portugal. Ministerio dos Negocios da Marinha e Ultramar. Annaes da commissão central permanente de geographia. Lisboa. 1,1876
Geog 4181.10 Portugese expedition to Abyssinia. (Castanhoso.) London, 1902.
Geog 3570.17 The Portuguese pioneers. (Prestage, Edgar.) London, 1933.
Geog 4169.47.5 Portuguese voyages, 1498-1663. (Ley, Charles D.) London, 1947.
Geog 3251.19 Der Portulan des Infanten. (Wieser, F.) n.p., n.d.
Geog 3055.75 Poséidones. (Cordeau, Catherine L.) Paris, 1957.
An 180.4 The position of women in primitive societies. (Evans-Pritchard, Edward Evans.) N.Y., 1965.
An 359.43.5 Posnansky, A. Qué es raza. La Paz, 1943.
Htn Geog 4477.08* Posos, Juan de (pseud.). Beschryoinge van het magtig. Amsterdam, 1708.
Htn Geog 5170.9* Possibility of approaching the North Pole. (Barrington, D.) London, 1818.
Geog 520.11 Possot, D. Le voyage de la Terre Sainte. Paris, 1890.
Geog 578.75 Post-Lexicon; ein Verzeichniss der wichtigeren Verkehrs Orte. Berlin, 1875.
Htn Geog 4265.76* The post of the world. (Verstegen, R.) London, 1576.
Geog 4679.56.7 Posted missing. (Villiers, A.J.) London, 1975.
Geog 4679.56.5 Posted missing. (Villiers, A.J.) N.Y., 1956.
Htn Geog 755.63.2* Postel, Guillaume. De universitate liber. 2a ed. pt.1-2. Parisiis, 1563.
Htn Geog 756.36* Postelli, G. Cosmographica disciplina. Lugduni, 1636.
Geog 5070.72 Die Poterforschung; ein Datenbuch über die Natur-, Kultur-, Wirtschaftsverhältnisse und die Erforschungsgeschichte der Polarregionen. (Kosack, Hans P.) Braunschweig, 1967.
Geog 4509.68 Potočnik, Miha. Srečanja z gorami. Ljubljana, 1968.
Geog 4209.72.5 Potoratiny. (Mon'ko, Aleksei M.) Volgograd, 1972.
Geog 5925.93 Potter, Neal. Natural resource potentials of the Antarctic. N.Y., 1969.
Geog 4308.90.5 Potter, V.M. To Europe on a stretcher. N.Y., 1890.
An 358.64.9 Pouchet, Georges. The plurality of the human race. 2. ed. London, 1864.
Geog 5985.253.10 Pound, Reginald. Scott of the Antarctic. London, 1966.
Geog 4158.05.5 Pouqueville, F.C.H. Voyage en Morée. v.1,2-3. Paris, 1805. 2v.
Geog 665.31 Pour qu'on aime la géographie. (Miller, Émile.) Montréal, 1921.
An 379.73.15 Pour une géographie humaine. (Gourou, Jierre.) Paris, 1973.
Geog 4419.34.4 La poursuite du "Kaipan". (Monfreid, Henri de.) Paris, 1934.
An 629.68.5 Pouwer, Jan. Translation at sight; the job of a social anthropologist. Wellington, 1968.
An 3559.34 Os povos de Angola. (Mascarenhas, Constância.) Bastorá, 1934.
Geog 4209.66.10 Povsod so ljudje. (Pogačnik, Bogdan.) Maribor, 1966.
Geog 4209.32.10 Powell, Edward A. Yonder lies adventure! N.Y., 1932.
Geog 4309.31.30 Powell, Gertrude C. The quiet side of Europe. Los Angeles, 1959.
Geog 500.110 Powell, Lawrence. Around the world in sixty books. Los Angeles, 1960.
Geog 5558.81.60 Powell, T. The long rescue. 1. ed. Garden City, 1960.
Geog 576.67 Poyares, P. de. Diccionario lusitanico-latino de nomes proprios de regioens, reinos, provincias, cidades. Lisboa, 1667.
Geog 185.5 Prace. (Polska Akademia Umiejetności. Komisja Geograficzna.) Kraków. 1-2 2v.
Geog 185.5PF Prace. (Polska Akademia Umiejetności. Komisja Geograficzna.) Kraków. 2
Geog 133.1 Prace geograficzne. (Krakow. Wyższa Szkola Pedagogiczna.) Kraków. 3,1964+ 2v.
Geog 185.6 Prace geograficzne. (Polska Akademia Nauk. Instytut Geografii.) Warszawa. 1,1954+ 40v.
Geog 265.3 Prace i studia. (Warsaw. Universytet. Instytut Geograficzny. Katedra Geografii Fizycznej.) 1,1967+
Geog 265.2 Prace i studia. (Warsaw. Universytet. Instytut Geograficzny. Katedra Klimatologii.) 1,1964+
Geog 4300.10 Practical European guide. (Frager, M.D.) Boston, 1907.
An 3409.42 Practical fingerprinting. (Bridges, B.C.) N.Y., 1942.
Geog 4300.17 Practical general continental guide. London, 1868.
Geog 4105.16 Practical hints to scientific travellers. (Brouwer, H.A.) Leyden, 1922-29. 6v.
Geog 4209.02 Prado, Eduardo. Viagens. v.1, 2a ed; v.2, 1a ed. São Paulo, 1902. 2v.
An 2056.61 Praeadamiten. (Peyrère, I. de.) Jaer, 1661.
Geog 4209.37.7 Prague. Městská Lidová Knihovna. Život a délo E. St. Vráze. Praha, 1960.
Geog 665.167 Prague. Universita Karlova. Sborník prací Geografických kateder UK k 75. narozeninám prof. dr. Jaromíra Korčáka, DrSc. Praha, 1970.
Geog 186.3 Prague. Universita Karlova. Acta Universitis Carolinae. Geographica. Praha. 1,1966+ 3v.
An 190.1 Pratt, Orson. Wonders of the universe. Salt Lake City, 1937.
Geog 4307.97.2 Pratt, S.J. Gleanings through Wales. Dublin, 1797. 2v.
Geog 5300.46 Pravo na risk. (Akkuratov, Valentin I.) Moskva, 1974.
Geog 4209.55.10 Praz, Mario. Viaggi in Occidente. Firenze, 1955.
An 2058.75.5 The pre-adamite. (Lester, A. Hoyle.) Philadelphia, 1875.
An 2058.87 Pre-glacial man and the Aryan race. 2. ed. (Burge, Lorenzo.) Boston, 1887.
An 2058.69 Pre-historic man. (Bromby, J.E.) Melbourne, 1869.
An 180.1 Pre-historic man. (Force, M.F.) Cincinnati, 1873.
An 2108.69 Pre-historic times. 2. ed. (Lubbock, J.) London, 1869.
An 2058.80 Preadamites. (Winchell, A.) Chicago, 1880.
An 358.87.9 Précis d'anthropologie. (Hovelacque, Abel.) Paris, 1887.
Geog 659.59.5 Précis de géographie physique générale. (Birot, Pierre.) Paris, 1959.
Geog 4268.73.4 Precis de la geographie...de l'Europe. v.1, atlas. (Levasseur, E.) Paris, 189-. 2v.
Geog 808.12.5 Précis de la geographie universelle. (Malte-Brun, Conrad.) Paris, 1812-1829. 8v.
Geog 808.12.5F Précis de la geographie universelle. Atlas. (Malte-Brun, Conrad.) Paris, 1812.
Geog 808.12.7 Précis de la geographie universelle. v.5. (Malte-Brun, Conrad.) Paris, 1817.
Geog 4419.59 Les précurseurs de Marco Polo. (Serstevens, Albert.) Paris, 1959.
An 359.66.5 Predki i my. (Podol'nyi Roman G.) Moskva, 1966.
Geog 759.72.5 Predsrazhenskii, Vladimir S. Besedy v sovremennoi fisicheskoi geografii. Moskva, 1972.
An 2109.58.10 La préhistoire et ses problèmes. (Bergoonlour, F.M.) Paris, 1958.
An 2059.12A Prehistoric man. (Duckworth, W.L.H.) Cambridge, 1912.
An 2059.31 Prehistoric man. (Duncan, G.S.) Boston, 1931.
An 2109.27.35 Prehistoric man. (Henderson, Keith.) N.Y., 1927.
An 2108.62 Prehistoric man. (Wilson, D.) Cambridge, 1862. 2v.
An 2108.76 Prehistoric man. 3. ed. (Wilson, D.) London, 1876. 2v.

Author and Title Listing

An 2109.15.3	Prehistoric man and his story. (Elliot, G.F.S.) London, 1915.
An 2109.15.4	Prehistoric man and his story. (Elliot, G.F.S.) Philadelphia, 1915.
An 2109.58	Prehistoric man in Europe. 1. ed. (Hibben, F.C.) Norman, Okla., 1958.
An 3509.35	Prehistoric man in Ireland. (Martin, Cecil P.) London, 1935.
An 379.50F	Prehistoric migrations in Europe. (Childe, Vere Gordon.) Cambridge, 1950.
An 358.97	Prehistoric problems. (Munro, R.) Edinburgh, 1897.
An 2108.91.3	The prehistoric races of Italy. (Taylor, I.) Washington, 1891.
An 2109.65.10	Prehistoric socties. 1st American ed. (Clark, John Grahame Douglas.) N.Y., 1965.
An 2159.64	Prehistoric technology. (Semenov, Sergei A.) London, 1964.
An 2109.00	Prehistoric times. (Avebury, J.L.) London, 1900.
An 2109.00.2	Prehistoric times. 6. ed. (Avebury, J.L.) N.Y., 1900.
An 2109.13	Prehistoric times. 7. ed. (Avebury, J.L.) London, 1913.
An 2058.77.5	The prehistoric use of iron and steel. (Day, J.V.) London, 1877.
An 359.06.3	Le préjugé des races. (Finot, Jean.) Paris, 1906.
An 2059.51	Prelude to history. (Coates, Adrian.) London, 1951.
Geog 4308.78.10	Prentis, N.L. A Kansan abroad. Topeka, Kansas, 1878.
Geog 5100.5	Prentiss, H.M. The great polar current. Cambridge, 1897.
Geog 4300.28.7	Presbrey, F. Presbrey's information guide to transatlantic travelers. 7th ed. N.Y., 1911.
Geog 4300.28.7	Presbrey's information guide to transatlantic travelers. 7th ed. (Presbrey, F.) N.Y., 1911.
Geog 186.5	Presidente Prudente, Brazil. Faculdade de Filosofía, Ciências e Letras. Departamento de Geografia. Boletim do Departamento de Geografia. Presidente Prudente. 4,1972+
Geog 4308.72.9F	Pressed flowers and leaves; souvenirs of travel. n.p., n.d.
Geog 3570.18	Prestage, Edgar. Descobridores portugueses. Porto, 1934.
Geog 3570.18.5	Prestage, Edgar. Descobridores portugueses. 2. ed. Lisboa, 1943.
Geog 3570.17	Prestage, Edgar. The Portuguese pioneers. London, 1933.
Geog 4181.114	The Prester John of the Indies. (Alvares, Francisco.) Cambridge, 1961. 2v.
Geog 4308.86.5	Preston, M.J. (Mrs.). A handfull of monographs, continental and English. N.Y., 1886.
An 359.37.3	Preuss, K.T. Lehrbuch der Völkerkunde. Stuttgart, 1937.
An 359.37.3.5	Preuss, K.T. Lehrbuch der Völkerkunde. 2. Aufl. Stuttgart, 1939.
An 2109.23.20	Preuss, Konrad Theodor. Die geistige Kultur der Naturvölker. 2e Aufl. Leipzig, 1923.
Geog 4157.46	Prevost, A.T. Histoire generale des voyages. Paris, 1746-89. 20v.
An 359.49.5	Price, A.G. White settlers and native peoples. Melbourne, 1949.
Geog 4759.39	Price, A.G. White settlers in the tropics. N.Y., 1939.
Geog 4269.36	Price, Lucien. We Northmen. Boston, 1936.
Geog 4145.72	Price, W. I cannot rest from travel. N.Y., 1951.
Geog 4219.69	Price, Willard. Odd way around the world. N.Y., 1969.
An 358.26.5	Prichard, J.C. Eastern origin of the Celtic nations...forming a supplement to Researches into the physical history of mankind. Oxford, 1831.
An 348.43F	Prichard, J.C. Ethnographical maps to the natural history of man. London, 1843.
An 348.61	Prichard, J.C. Explanatory notice of the ethnographical map. London, 1861.
An 358.43	Prichard, J.C. The natural history of man. London, 1843.
An 358.45	Prichard, J.C. The natural history of man. 2. ed. London, 1845.
An 358.55	Prichard, J.C. The natural history of man. 4. ed. London, 1855. 2v.
An 358.26	Prichard, J.C. Researches into the physical history of mankind. London, 1826. 2v.
An 358.41	Prichard, J.C. Researches into the physical history of mankind. v.1-2, 4. ed.; v.3, 3. ed. London, 1837-41. 3v.
An 358.51	Prichard, J.C. Researches into the physical history of mankind. v.1-2,5, 4. ed.; v.3-4, 3. ed. London, 1841-51. 5v.
An 358.36	Prichard, J.C. Researches into the physical history of mankind. 3. ed. London, 1836-47. 6v.
An 358.41.2	Prichard, J.C. Researches into the physical history of mankind. v.1,5, 4. ed; v.2-4, 3. ed. London, 1837-47. 5v.
Geog 5970.46	Priestley, Raymond E. Antarctic research; a review of British scientific achievement in Antarctica. London, 1964.
Geog 6009.11	Priestly, R.E. Antarctic adventure; Scott's northern party. London, 1914.
Geog 3580.46	Prieto, Carlos. El Oceano pacifico. Navegantes española del siglo XVI. Madrid, 1972.
Geog 4218.74	Prime, E.D.C. Around the world, sketches of travel. N.Y., 1874.
Geog 4308.73.5	Prime, Samuel I. The Alhambra and the Kremlin. N.Y., 1873.
Geog 4308.53.12	Prime, Samuel I. Travels in Europe and the East. N.Y., 1855. 2v.
Geog 4308.53.13	Prime, Samuel I. Travels in Europe and the East. N.Y., 1856. 2v.
An 2108.78.3	Primeval man. (Argyll, G.D.C.) N.Y., 1878.
An 2108.69.5	Primeval man. (Campbell, G.D.) London, 1869.
An 2108.69.4	Primeval man. (Campbell, G.D.) N.Y., 1869.
An 2108.69.3	Primeval man. (Campbell, G.D.) N.Y., 1869.
An 359.03.5	Les primitifs, étude d'ethnologie. (Reches, Elie.) Paris, 1903.
An 2338.95	Primitive American armor. (Hough, Walter.) Washington, 1895.
An 2109.66.10	Primitive and peasant economic systems. (Nash, Manning.) San Francisco, 1966.
An 2109.37.11	Primitive behavior. 1. ed. (Thomas, W.I.) N.Y., 1937.
An 2109.63.15	Primitive classification. (Durkheim, Emile.) Chicago, 1963.
An 2109.03A	Primitive culture, researches. (Tylor, Edward Burnett.) London, 1903. 2v.
An 2108.74	Primitive culture. (Tylor, Edward Burnett.) Boston, 1874. 2v.
An 2108.77.3	Primitive culture. (Tylor, Edward Burnett.) N.Y., 1877. 2v.
An 2108.71.1	Primitive culture: researches into the development of mythology. (Tylor, Edward Burnett.) N.Y., 1974. 2v.
An 2109.20.2	Primitive culture. 1st American ed. (Tylor, Edward Burnett.) Boston, 1874. 2v.
An 2108.91A	Primitive culture. 3. ed. (Tylor, Edward Burnett.) London, 1891. 2v.
An 2109.20	Primitive culture. 6. ed. (Tylor, Edward Burnett.) London, 1920. 2v.
An 2109.24.10	Primitive culture. 7. ed. v.1-2. (Tylor, Edward Burnett.) N.Y., 1924.
An 358.91	Primitive folk. (Reclus, E.) London, 1891.
An 170.270	Primitive heritage. (Mead, M.) N.Y., 1953.
An 2109.37	Primitive intelligence and endowment. (Portens, S.D.) N.Y., 1937.
An 2109.27.7	Primitive man, his essential quest. (Murphy, John.) London, 1927.
An 2108.70.6	Primitive man. (Figuier, L.G.) N.Y., 1870.
An 2108.70.7	Primitive man. (Figuier, L.G.) N.Y., 1870.
An 2109.16	Primitive man. (Smith, G.E.) London, 1916.
An 2109.60.5	Primitive man and his ways. (Birket-Smith, Kaj.) London, 1960.
An 2109.57.5	Primitive man as philosopher. 2. ed. (Radin, Paul.) N.Y., 1957.
An 358.79.2	Primitive manners and customs. (Farrer, J.A.) London, 1879.
An 2109.23.15	Primitive mentality. (Lévy-Bruhl, L.) N.Y., 1923.
An 2109.31	The primitive mind and modern civilization. (Aldrich, C.R.) London, 1931.
An 2109.35	Primitive philosophy. (Brelsford, Vernon.) London, 1935.
An 2109.38.9	A primitive philosophy of life. (Hutton, J.H.) Oxford, 1938.
An 359.26.10	The primitive races of mankind. (Schmidt, Max.) London, 1926.
An 2109.38.3	Primitive races of to-day. (Page, J.W.) London, 1938.
An 2109.08	Primitive secret societies. (Webster, H.) N.Y., 1908.
An 2109.72.20	Primitive societies. 1st ed. (Quilici, Folco.) London, 1972.
An 359.20.3	Primitive society. (Lowie, Robert H.) N.Y., 1920.
An 359.34.7	Primitive society and its vital statistics. (Kryzwicki, L.) London, 1934.
An 2109.62	Primitive song. (Bowra, Cecil M.) London, 1962.
An 170.173.5	Primitive views of the world. (Diamond, Stanley.) N.Y., 1964.
An 2339.49	Primitive war. (Turney-High, H.) Columbia, 1949.
An 2109.73.5	The primordial ocean; an introductory contribution to social psychology. (Perry, William James.) London, 1973.
Geog 4208.50.3	Prince, Nancy. Narrative of life and travels. 2. ed. Boston, 1853.
Geog 4208.50.5	Prince, Nancy. Narrative of life and travels. 3. ed. Boston, 1856.
Geog 3520.10.45	Principal navigations, voyages...of the English nation. (Hakluyt, Richard.) London, 1927. 8v.
Geog 3520.10.35	Principal navigations, voyages. (Hakluyt, Richard.) London, 1907-13. 8v.
Htn Geog 3520.10.5F*	Principal navigations, voyages. v.1-3. (Hakluyt, Richard.) London, 1599. 2v.
Geog 3520.10.20	Principal navigations, voyages. v.1-12,14-16. (Hakluyt, Richard.) Edinburgh, 1885-90. 15v.
NEDL Geog 3520.10.20	Principal navigations, voyages. v.13. (Hakluyt, Richard.) Edinburgh, 1885-90.
Geog 4184.1A	Principal navigations...voyages traffiques. (Hakluyt, R.) Glasgow, 1903-05. 12v.
Geog 3520.10.45.5	The principal navigations. (Hakluyt, Richard.) Cambridge, 1965. 2v.
Geog 819.11.5	Les principales puissances d'aujourd'hui. 5. éd. (Busson, Henri.) Paris, 1924.
Geog 819.11.3	Les principales puissances du monde. (Busson, Henri.) Paris, 1911.
Htn Geog 3520.10F*	Principall navigations, voiages. (Hakluyt, Richard.) London, 1589.
An 379.22	Principes de géographie humaine. (Vidal de la Blache, P.) Paris, 1922.
An 2109.66.30	Principes de l'ethnologie économique. (Bessaignet, Pierre.) Paris, 1966.
An 359.20.7	Principles and methods of physical anthropology. (Roy, Sarat Chandra.) Patna, 1920.
An 359.42A	Principles of anthropology. (Chapple, Eliot D.) N.Y., 1942.
An 2109.39.6	Principles of economic sociology. (Goodfellow, D.M.) London, 1939.
An 2109.39.5	Principles of economic sociology. (Goodfellow, D.M.) Philadelphia, 1939.
Geog 759.59	Principles of geography. (Doerr, Arthur H.) N.Y., 1959.
An 379.21A	Principles of human geography. (Huntington, Ellsworth.) N.Y., 1921.
An 379.22.7	Principles of human geography. 2. ed. (Huntington, Ellsworth.) N.Y., 1922.
An 379.24.5	Principles of human geography. 3. ed. (Huntington, Ellsworth.) N.Y., 1924.
An 379.47.5	Principles of human geography. 5. ed. (Huntington, Ellsworth.) N.Y., 1947.
An 379.51	Principles of human geography. 6. ed. (Huntington, Ellsworth.) N.Y., 1951.
An 379.26A	Principles of human nature. (Vidal de la Blache, P.) N.Y., 1926.
Geog 4709.53.2	Pringle, P. Jolly Roger. London, 1953.
Geog 4709.53	Pringle, P. Jolly Roger. N.Y., 1953.
Geog 4208.29.3	Prinsep, A. (Mrs.). The journal of a voyage from Calcutta. London, 1833.
Geog 4329.21	Prioleau, John. The adventures of Imshi; a two-seater in search of the sun. Boston, 1923.
An 379.68	Priroda i obshchestvo. Moskva, 1968.
An 359.71.30	Priroda natsional'noi psikhlgii. (Dzhandil'din, Nurymbet.) Alma-Ata, 1971.
Geog 759.71.5	Prirodnye resursy i kulturnye landshafty materikov. Moskva, 1971.
Geog 4309.64.10	Pritchett, V.S. Foreign faces. London, 1964.
Geog 4309.64	Pritchett, V.S. The offensive traveller. 1. ed. N.Y., 1964.
Geog 5518.21.5	Private journal of Captain G.F. Lyon. (Lyon, G.F.) Boston, 1824.
Geog 4217.08.10	The privateers. (MacLiesh, Archibald F.) N.Y., 1962.
Htn Geog 5170.7*	Probability of reaching the North Pole. (Barrington, D.) London, 1775.
An 2059.61.5	Das Problem der Hominisation. (Overhage, Paul.) Freiburg, 1961.
An 99.58.5	Das Problem der kulturellen Werte in den Arbeiten der neueren amerikanischen Ethnologie. (Rudolph, Wolfgang.) Berlin, 1958.
An 359.58.15	Das Problem des Menschen und die Kultur. (Boer, Wolfgang de.) Bonn, 1958.

Author and Title Listing

An 359.54.15 Das Problem des Völkertodes. (Schwidetzky, Ilse.) Stuttgart, 1954.
Geog 3055.64 The problem of Atlantis. 2d ed. (Spence, Lewis.) N.Y., 1925?
Geog 3060.13.10 The problem of Lemuria. (Spence, Lewis.) Philadelphia, 1933.
Geog 4332.14 Il problema militaire dell'Adriatico spiegatto a tutti. (Roncagli, Giovanni.) Roma, 1918.
Geog 659.74.5 Problema tselostnosti v fizicheskoi geografii. (Mukitanov, Naurzbai K.) Alma-Ata, 1974.
An 379.68.10 Die Problematik der Sozialgeographie. (Gerling, Walter.) Würzburg, 1968.
Geog 4131.34 Probleme der Fremdenverkehrsgeographie. Diss. (Grünthal, Adolf.) Berlin, 1934.
Geog 4131.68.15 Probleme der Geographie des Fremdenverkehrs der Deutschen Demokratischen Republik und anderer Staaten. (Internationale Informationstagung zur Geographie des Fremdenverkehrs. Dresden, 1965.) Leipzig, 1968.
Geog 665.124 Probleme der landschaftsökölogischen Erkundung und naturräumlichen Gliederung. (Symposium über Fragen der Naturräumlichen Gliederung.) Leipzig, 1967.
An 379.61 Probleme der Sozialgeographie. (Maas, Walther.) Berlin, 1961.
An 379.39.5 Problèmes de géographie humaine. (Deffontaines, P.) Paris, 1939.
Geog 665.88 Problèmes de géographie humaine. 4. éd. (Demangeon, Albert.) Paris, 1952.
Geog 4329.64 Problèmes géographiques de la Méditerranée européenne. (Le Lannou, Maurice.) Paris, 1964.
An 193.1 Problemi di scienza contemporanea. (Sergi, Giuseppe.) Palermo, 1904.
An 193.1.5 Problemi di scienza contemporanea. (Sergi, Giuseppe.) Torino, 1916.
Geog 186.20 Problemi na geografiiata. Sofiia. 1,1975+
Geog 186.15 Problemi na paleogeomorfolozhko to razvitie na Bulgariia. Sofiia. 1,1971+
Geog 3410.7 Problems and trends in American geography. N.Y., 1967.
Geog 5070.35A Problems of polar research. (American Geographic Society, N.Y.) N.Y., 1928.
An 3209.68 Problemy evoliutsii cheloveka i ego ras. (Zubov, Aleksandr A.) Moskva, 1968.
Geog 3590.87 Problemy heohrafichnoï nauky v Ukrains'kii RSR. Kyïv. 1,1972+ 2v.
Geog 5399.58.10 Problemy Severa. Moskva. 1-18 5v.
Geog 4419.68 Probuzhdane to na Shekherazada. (Gotev, Goran.) Sofiia, 1968.
An 170.157 Proceedings, 1935-36. (Chicago. University. Seminar of Racial and Cultural Contacts.) n.p., n.d.
Geog 13.3 Proceedings. (American Geographical Society of New York.) 1-2
Geog 139.3 Proceedings. (London. Royal Geographical Society.) London. 1-14,1879-1894 14v.
Geog 139.2.9 Proceedings. (London. Royal Geographical Society.) London. 2-22 16v.
Geog 3235.66 Proceedings. (Vinland Map Conference, Smithsonian Institution.) Chicago, 1971.
Geog 139.3.5 Proceedings. Index. 1879-1892. (London. Royal Geographical Society.) London, 1896.
Geog 21.1 Proceedings and transactions. (Royal Geographical Society of Australasia.) Queensland. 1-62,1885-1964 9v.
Geog 665.81 Proceedings of the IGU Regional Conference in Japan, August 28-September 3, 1957. (International Geographical Union. Regional Conference in Japan, Tokyo and Nara, 1957.) Tokyo, 1959.
An 170.266 Process and pattern in culture; essays in honor of Julian H. Steward. (Manners, Robert A.) Chicago, 1964.
An 359.69.10 Process and product in human society. (Blacking, John.) Johannesburg, 1969.
Geog 4309.01.9 Proctor, Rachel. To the land of the midnight sun. Utica, N.Y., 1901.
Geog 3070.35 The production of an admiralty chart. (Hayes, Gerald R.) London, 1929.
VGeog 4559.67 Proface, Bruno. Nebo, more i ponekad kopno. Zagreb, 1967.
Geog 186.2 The professional geographer. Washington. 1,1949+ 14v.
An 137.4 Professor Bronislaw Malinowski. (Association of Polish University Professors and Lecturers in Great Britain.) London, 1943.
Geog 5538.53.5 Professor Sonntag's thrilling narrative. (Sonntag, A.) Philadelphia, 1857.
Geog 5538.53.6 Professor Sonntag's thrilling narrative. (Sonntag, A.) Philadelphia, 1857.
Htn Geog 4126.33* Profitable instrucions. (Essex, Robert.) London, 1633.
Geog 5220.2 Programm...Öffentlichen Prüfung. (Dresden, Germany. Armen-Realschule.) Dresden, 1873.
An 2059.61.10A The progress and evolution of man in Africa. (Leakey, Louis S.) London, 1961.
Geog 186.10 Progress in geography; international reviews of current research. London. 1,1969+ 8v.
Geog 3030.8 Progress in historical geography. (Baker, Alan R.H.) N.Y., 1972.
Geog 5208.68.2 Progress of Arctic discovery. (Hayes, I.I.) N.Y., 1868.
Geog 5208.68 Progress of Arctic discovery. (Hayes, I.I.) N.Y., 1868.
An 98.47 The progress of ethnology. 2. ed. (Bartlett, J.R.) N.Y., 1847.
Geog 3018.03 Progress of maritime discovery. (Clarke, J.S.) London, 1803.
Geog 175.5 Progress report on the cartographic activities of the United States. (Pan American Institute of Geography and History. Commission on Cartography.) St. Louis. 1946-1952 3v.
Geog 3018.76 I progressi della generale e della geografia esploratrice in Europa. (Bruniatti, A.) Vicenza, 1877.
An 3559.69 Proiskhozhdenie narodov Vostochnoi Evropy. (Alekseev, Valerii P.) Moskva, 1969.
An 629.61 Projective techniques and cross-cultural research. (Lindzey, G.) N.Y., 1961.
Geog 5170.15 Projet pour tenter la decouverte. La Haye, 1772.
Geog 4219.01 Promenade autour du globe. (Loewenbach, Lottraire.) Paris, 1903.
Geog 4218.71.7 Promenade autour du monde, 1871. (Hübner, J.A.) Paris, 1877. 2v.
Geog 4218.71.2 Promenade autour du monde, 1871. 2e éd. (Hübner, J.A.) Paris, 1873. 2v.
Geog 4229.35 Promenades dans trois continents. (Carré, J.M.) Paris, 1935.
Geog 4218.75 Prominent incidents in life of John M. Wieting...around the world. (Wieting, M.E.) N.Y., 1889.
Geog 578.49.9 A pronouncing gazetteer. 9. ed. (Baldwin, Thomas.) Philadelphia, 1851.
Geog 4179.7 Prospectus and list of members. (Hakluyt Society, London.) London. 1850-1918 2v.
An 359.62.10 Prospero's magic. (Mason, Philip.) London, 1962.
Geog 6100.77 Provisional gazetteer of the Ross Dependency. (New Zealand. Geographic Board.) Wellington, 1958.
Geog 4515.201 Prusik, Karl. Gymnastik für Bergsteiger. München, 192-.
An 2109.01.15 Przeszłość w teraźniejszości. (Radliński, Ignacy.) Warszawa, 1901.
Geog 4139.73.5 Przygody polskich obieżyświatów na morzach i lądach. 1. wyd. (Kuźmiński, Bolesław.) Gdeńsk, 1973.
An 359.61 Psychological anthropology. (Hsu, Francis L.K.) Homewood, Ill., 1961.
An 359.58 La psychologie des peuples. (Miroglio, Abel.) Paris, 1958.
An 2109.53.15 La psychologie ethnique. (Heuse, Georges A.) Paris, 1953.
An 359.01.5 La psychologie ethnique. (Setourneau, C.J.E.) Paris, 1901.
An 2109.23.7A Psychology and primitive culture. (Bartlett, F.C.) Cambridge, 1923.
An 2109.23.6 Psychology and primitive culture. (Bartlett, F.C.) N.Y., 1923.
Geog 84.5 Pubblicazione. (Genoa. Università. Istituto di Scienze Geografiche.) Genova. 7,1968+
Geog 143.70 Pubblicazione. Serie 10. Scienze geografico. (Milan. Università Cattolica del Sacro Cuore.) Milano.
Geog 195.3 Pubblicazioni. Serie B (Geostorica). (Rome, Italy. Università. Istituto di Geografia.) Roma. 1,1969+
Geog 195.4 Pubblicazioni. Serie C (Miscellanea). (Rome, Italy. Università. Istituto di Geografia.) Roma. 2,1969+
Geog 185.12 Publicacion. Ser. monografias. (Portugal. Agrupamento de Estudos de Cartografia Antiga. Secção de Coimbra.) Lisboa. 1,1963+ 12v.
Geog 185.10 Publicacion. Ser. separatas. (Portugal. Agrupamento de Estudos de Cartografia Antiga. Secção de Lisboa.) Lisboa. 1-21 5v.
Geog 142.2.10F Publicaciones. (Madrid. Museo Naval.) Madrid. 1,1932
Geog 143.10 Publicaciones. (México (City). Universidad Nacional. Instituto de Geografía.) México. 1,1965+
Geog 212.15 Publicaciones. Serie B. (Sociedad Geográfica, Madrid.) Madrid. 455,1966+
Geog 16.2 Publication. (Appalachian Trail Conference.) Washington. 4,1942 3v.
Geog 28.4.5 Publication. (Brussels. Université Nouvelle. Institut Géographique.) Bruxelles.
Geog 70.2 Publication. Geography series. (Florida. University.) Gainesville.
Geog 4131.51 Publication de la Commission scientifique de l'alliance internationale de tourisme. (International Touring Association. Scientific Commission.) Berne. 1-6,1951-1956 3v.
Geog 109.5 Publications. (Institute of British Geographers.) London. 1933-1975// 20v.
An 3309.65.10 Publications. (Joint Arabic-Polish Anthropological Expedition, 1958.) Warsaw, 1965.
Geog 212.235.15 Publications. (Société de Géographie de Lille.) 1943-1953
Geog 35.4 Publications in geography. (California. University.) Berkeley, Calif. 1,1913+ 19v.
Geog 500.62 Publications of Edward L. Stevenson. (Stevenson, Edward L.) n.p., 191-.
Geog 500.23 Publicazioni geografiche. (Cardon, F.) Roma, 1892.
An 3309.29.3 Puccioni, Nello. Affrica nord-orientale e Arabia. Pavia, 1929.
Geog 4131.53 Pudney, J. The Thomas Cook story. London, 1953.
Geog 190.5.5 Pueblos. (Ricossa, J.A.) Buenos Aires, 195-? 6v.
Geog 4308.35.10 Pückler-Muskau, H. Vorletzter Weltgang von Semilasso. Stuttgart, 1835. 3v.
Geog 818.59 Pütz, Wilhelm. Charakteristiken zur Erd und Völkerkunde. Köln, 1859-1860. 2v.
Geog 818.59.7 Pütz, Wilhelm. Grundriss der Geographie und Geschichte. Koblenz, 1866.
Geog 818.59.6 Pütz, Wilhelm. Grundriss der Geographie und Geschichte der...Zeit. v.1-3. 10. Aufl, 12. Aufl. Koblenz, 1865-1867.
Geog 4559.66 Pully-Haul: the story of a voyage. (Karlsson, Elis.) London, 1966.
Geog 4208.54A Pumpelly, R. My reminiscences. N.Y., 1918. 2v.
Geog 4228.70 Pumpelly, Raphael. Across America and Asia. N.Y., 1870.
Geog 4228.71.5 Pumpelly, Raphael. Across America and Asia. N.Y., 1871.
Htn Geog 4206.13F* Purchas, his pilgrimage. (Purchas, Samuel.) London, 1613.
Geog 4184.14A Purchas, Samuel. Hakluytus posthumus, or Purchas, his pilgrimes. Glasgow, 1905-07. 20v.
Htn Geog 4206.13.5F* Purchas, Samuel. His pilgrimage, or Relations of the world. 3rd ed. London, 1617.
Htn Geog 4206.13.9F* Purchas, Samuel. His pilgrimes. London, 1625- 4v.
Geog 4206.13.35 Purchas, Samuel. Narratives from Purchas, his pilgrimes. Cambridge, 1931.
Htn Geog 4206.13F* Purchas, Samuel. Purchas, his pilgrimage. London, 1613.
Geog 4206.13.12 Purchas, Samuel. Samuel Purchas pelgrimagir. Amsterdam, 1655.
Geog 5399.34.5 Put' skvoz l'dy. (Morozov, S.) Moskva, 1934.
Geog 4218.03.10 Putesh. vokrug sveta na kor. ""Neva", 1803-1806. (Lisianskii, I.) Moskva, 1947.
Geog 4145.71.500 Puteshchestviia A.V. Potaninoi. (Zariu, V.M.) Moskva, 1950.
Geog 4145.71.5 Puteshchestviia Potanina. (Obruchev, V.A.) Moskva, 1953.
Geog 5939.64.5 Puteshestvie k ostrovu chetyrekh vulkanov. (Astapenko, Pavel D.) Moskva, 1964.
Geog 4209.65.5 Puteshestvie s 1783 po 1790 god iz okhotska po vostoch. okeanu. (Shelekhov, Grigorii I.) Ann Arbor, 1965. 2v.
Geog 5939.57.5 Puteshestvie v Antarktidu. (Markov, K.K.) Moskva, 1957.
Geog 4228.17.10 Puteshestvie vokrug sveta. (Golovnin, Vasilii M.) Moskva, 1965.
Geog 4209.69.10 Puteshestviia. (Sytin, Viktor A.) Moskva, 1969.
Geog 612.4.4 Puteshestviia I.I. Lepekhina, N. Ia. Ozeretskovskogo, V.F. Zueva. (Fradkin, Naum G.) Moskva, 1948.
Geog 3019.65 Puteshestviia i geograficheskia otkrytiia v XV-XIX vv. (Belov, Mikhail I.) Leningrad, 1965.
Geog 3590.42 Puteshestviia i issledovaniu...v Russkoi Amerike v 1802-04 gg. (Zogosken, L.A.) Moskva, 1956.
Geog 4209.68.10 Puteshestviia nashikh zemliakov. (Gritskevich, Valentin P.) Minsk, 1968.
Geog 616.32 Puteshestviia za nevidimym vragom. (Talyzin, Fedor F.) Moskva, 1974.
Geog 4218.26.17 Puteshestvne vokrug sveda na volunom shliupe "Seniavin" 1826-1829. 2. izd. (Luetke, Fedor P.) Moskva, 1948.

Author and Title Listing

Geog 194.5 Putevoditel' po Krymu, Kavkazu, i Blizhnemu Vostoku. (Russkoe Obshchestvo Parokhodstva i Torgovli.) Odessa. 1913

Geog 4308.39.9 Putevyia pis'ma...iz slavianskikh zemel, 1839-1842. (Sreznevskii, I.I.) Sankt Peterburg, 1895.

Geog 5300.50 Puti otvazhnykh. (Vodop'ianov, M.V.) Moskva, 1958.

Geog 4269.62.10 Putishta i spirki. (Koen, Albert.) Sofiia, 1962.

Geog 5559.27 Putnam, D.B. Davial goes to Baffinland. N.Y., 1927.

An 137.1 Putnam, F.W. Sketch of...Lewis H. Morgan. Boston, 1882.

Geog 5312.2.5 Putnam, George. Mariner of the North. N.Y., 1947.

Geog 5398.97.10 Putnam, George Palmer. Andrée. N.Y., 1930.

Geog 4269.66.5 Putovanja grešnog Elefterija. (Markovic, Slobodan.) Beograd, 1971.

Geog 4329.27.20 Pyramide und Tempel. (Meier-Graefe, J.) Berlin, 1927.

NEDL Geog 4180.76 Pyrard, F. Voyage of F. Pyrard...to East Indies. v.1-2. London, 1887. 3v.

Geog 4180.76 Pyrard, F. Voyage of F. Pyrard...to East Indies. v.1-2. N.Y., 1964? 3v.

Geog 187.1 Quaderni geografica. Roma. 1-11

Geog 187.5 Quaderns de geografia. Barcelona. 1,1949 2v.

Geog 4672.20.10 Quadros da História trágico-marítima. (Gomes de Brito, Bernardo.) Lisboa, 1944.

Geog 187.10 Quaestiones geographicae. Poznań. 1,1974+

Geog 759.68 Quantitative geography. (Cole, John Peter.) London, 1968.

Geog 659.69.5 Quantitative methods in geography; a symposium. (Conference on Quantitative Methods in Geography, New York, 1969.) N.Y., 1969.

Geog 659.74 Quantitative techniques in geography: an introduction. (Hammond, Robert.) Oxford, 1974.

Geog 5399.28.30 Quarantotto giorni sul pack. (Viglieri, Alfredo.) Milano, 1929.

Geog 4709.55 Quarrell, Charles. Buried treasure. London, 1955.

Geog 5970.42 Quartermain, Leslie B. New Zealand and the Antarctic. Wellington, 1971.

Geog 6100.77.5 Quartermain, Leslie B. South to the Pole: the early history of the Ross Sea sector, Antarctica. London, 1967.

Geog 4182.56 Quast, Mathijs H. De reis van Mathijs Hendriksz. 's-Gravenhage, 1954.

Geog 3580.30 Quatre voyageurs espagnols à Alexandria d'Egypt. (Garcia de Herreros, E.) Alexandria, 1923.

An 3538.82 Quatrefages de Bréau, A. de. Crania ethnica. Les crânes des races humaines. Paris, 1873-81. 2v.

An 358.87 Quatrefages de Bréau, A. de. Histoire générale des races humaines. Paris, 1887-89. 2v.

An 358.84 Quatrefages de Bréau, A. de. Hommes fossiles et hommes sauvages. Paris, 1884.

An 358.79 Quatrefages de Bréau, A. de. The human species. N.Y., 1879.

NEDL An 358.88.3 Quatrefages de Bréau, A. de. The human species. N.Y., 1888.

An 358.75 Quatrefages de Bréau, A. de. Natural history of man. N.Y., 1875.

An 358.61.5 Quatrefages de Bréau, A. de. Unité de l'espèce humaine. Paris, 1861.

An 98.67 Quatrefages de Bréaux, A. de. Rapport sur la progrès de l'anthropologie. Paris, 1867.

Geog 3057.4 Quatremère. Memoire sur le pays d'Ophir. Paris, 1845.

An 359.43.5 Qué es raza. (Posnansky, A.) La Paz, 1943.

Geog 189.2.3 Quebec Geographical Society. Bulletin. Quebec. 3-23,1908-1929 10v.

Geog 189.2.4 Quebec Geographical Society. Tables des matieres contenus dans le Bulletin. Quebec, 1913.

Geog 189.2 Quebec Geographical Society. Transactions. Quebec. 1880-1897 3v.

Geog 4218.93.5 A Queenslander's travel notes. (Stephens, A.A.) Sydney, 1894.

An 2109.23.11 Quellen zur ethnologischen Rechtsforschung von Nordafrika, Asien und Australien. Stuttgart, 1923.

Geog 3055.9 Quelques mots sur l'Atlantide. (Block, R. de.) n.p., n.d.

Geog 4219.08.10 Quelques notes sur les conférences de Tokyo et de Shanghai. (Bertrand, A.) Genève, 19- .

Geog 4229.30 Querdurch; Reisebilder und Reden. (Toller, Ernest.) Berlin, 1930.

An 629.03 Queries in ethnography. (Keller, A.G.) N.Y., 1903.

Geog 4502.5.5 Querschnitt durch mich. (Rickmers, W.R.) München, 1930.

Geog 5839.09.5 Quervain, Alfred. Durch Grönlands Eiswüste. 2. Aufl. Strassburg, 1911.

Geog 4219.25.5 Querweltein. (Reinhardt, Walther.) Berlin, 1925.

Geog 4181.32 The quest and occupation of Tahiti by emissaries of Spain 1772-76. (Corney, B.G.) London, 1913-19. 3v.

Geog 5919.57 Quest for a continent. (Sullivan, Walter.) N.Y., 1957.

Geog 3019.64 The quest for India. (Landström, B.) London, 1964.

An 359.75.10 The quest for man. (Goodall, Vanne Morris.) N.Y., 1975.

Geog 4329.27.5 The quest for winter sunshine. (Oppenheim, E.P.) Boston, 1927.

Geog 3229.20 The quest of the Indies. (Dark, Richard.) N.Y., 1920.

Geog 4309.65.5 Questa Europa. (Cione, Edmondo.) Milano, 1965.

An 358.41.5 Question philosophique de première importance: Quelle est dans l'univers, la destinée du genre humain? (Azaïs, H.) Paris, 1841.

An 629.07 Questions on customs, beliefs and languages of savages. (Frazer, J.G.) Cambridge, 1907.

Geog 4309.31.30 The quiet side of Europe. (Powell, Gertrude C.) Los Angeles, 1959.

An 132.3A Quiggin, A. Hingston (Mrs.). Hadden, the head hunter. Cambridge, 1942.

Geog 4209.66.20 Quilici, Folco. Giramare. Roma, 1966.

An 2109.72.20 Quilici, Folco. Primitive societies. 1st ed. London, 1972.

Geog 4181.104 Quinn, D.B. The Roanoke voyages, 1584-1590. London, 1955. 2v.

Geog 4181.83 Quinn, David B. The voyages...of Sir Humphrey Gilbert. London, 1940. 2v.

Geog 3520.11.10 Quinn, David Beers. A study of the facsimile edition of Richard Hakluyt's Divers voyages. Amsterdam, 1968. 2v.

Geog 4218.40 Quinze ans de voyages. (Lafond, G.) Paris, 1840. 2v.

Geog 6008.97.2 Quinze mois dans l'Antarctique. (Gerlache de Gomery, Adrien de.) Bruxelles, 1902.

Geog 4209.60.50 A quota of seaweed. (Basso, Hamilton.) Garden City, 1960.

Geog 612.6 R. Fedor Petrovich Litke. (Alekseev, Aleksandr G.) Moskva, 1970.

Geog 4239.70.15 Ra, una balsa de papyrus a través del Atlantico. 1. ed. (Genovés Tarazaga, Santiago.) Mexico, 1972.

Geog 4239.70.6 The Ra expeditions. (Heyerdahl, Thor.) London, 1971.

An 2059.41 A raça de Lagôa Santa. (Mattos, Anibal.) São Paulo, 1941.

An 359.38 Race; a study in modern superstition. (Barzun, Jacques.) London, 1938.

An 359.65.25 Race; a study in superstition. (Barzun, Jacques.) N.Y., 1965.

An 359.68.15 Race, culture, and evolution. (Stocking, George W.) N.Y., 1968.

An 359.68.16 Race, culture, and evolution. (Stocking, George W.) N.Y., 1971.

An 176.6A Race, language and culture. (Boas, Franz.) N.Y., 1940.

An 359.33.11 La race, les races. (Montandon, Georges.) Paris, 1933.

An 359.59.10 Race, prejudice and education. (Bibby, Cyril.) London, 1959.

An 359.43.25 Race; science and politics. (Benedict, Ruth (Fulton).) N.Y., 1943.

An 379.36 Race, sex, and environment. (Marett, J.R. de la H.) London, 1936.

An 359.74 Race. (Baker, John Randal.) N.Y., 1974.

An 359.40 Race: science and politics. (Benedict, Ruth (Fulton).) N.Y., 1940.

An 359.59.45 Race: science and politics. (Benedict, Ruth (Fulton).) N.Y., 1959.

An 359.39.7 Race against man. (Seligmann, H.J.) N.Y., 1939.

An 359.28.5 Race and civilization. (Hertz, F.O.) N.Y., 1928.

An 359.28.3 Race and civilization. (Hertz, F.O.) N.Y., 1928.

An 359.50 Race and culture. (Park, R.E.) Glencoe, Ill., 1950.

An 359.34.5 Race and culture contacts. (Reuter, E.B.) N.Y., 1934.

An 359.57 Race and culture contacts in the modern world. 1. ed. (Frazier, E.F.) N.Y., 1957.

An 359.45.20 Race and democratic society. (Boas, Franz.) N.Y., 1945.

An 359.67.5 Race and modern science. N.Y., 1967.

An 359.29.9 Race and population problems. 1. ed. (Duncan, H.) N.Y., 1929.

An 359.71.5 Race and races. (Goldsby, Richard A.) N.Y., 1971.

An 359.59.50 Race and racism. (Benedict, Ruth (Fulton).) London, 1959.

An 359.58.10 Race and social revolution. (Campbell, Byram.) N.Y., 1958.

An 359.31.3 Race as a political factor. (Gregory, John W.) London, 1931.

An 359.75.5 The race concept. (Banton, Michael P.) Newton Abbot, 1975.

An 359.52 The race concept. (UNESCO.) Paris, 1952.

An 359.27.11A Race contact. (Muntz, Earl E.) N.Y., 1927.

An 3309.29 Race crossing in Jamaica. (Davenport, C.B.) Washington, 1929.

Geog 5235.24 Race for the pole. (Weems, John E.) N.Y., 1960.

An 359.43.10 Race prejudice. (Claessens, A.) N.Y., 1943.

An 2109.01.7 Race prejudice. (Finot, J.) N.Y., 1907.

An 359.06.4 Race prejudice. (Finot, Jean.) London, 1906.

An 359.06.4.2 Race prejudice. (Finot, Jean.) Miami, 1969.

An 359.06.4.1 Race prejudice. (Finot, Jean.) N.Y., 1907.

An 359.61.55 The race question in modern science; race and science. (UNESCO.) N.Y., 1961.

An 359.56.15 The race question in modern science. (UNESCO.) Paris, 1957.

An 358.90 Races and peoples. (Brinton, D.G.) N.Y., 1890.

An 359.00.3 Les races et les peuples de la terre. (Deniker, J.) Paris, 1900.

An 358.60 Races humaines. (Clavel, A.) Paris, 1860.

An 358.72 Les races humaines. (Figuier, Louis.) Paris, 1872.

An 358.82 Les races humaines. (Hovelacque, Abel.) Paris, 1882.

An 379.36.3 Les races humaines. (Lester, P.) Paris, 1936.

An 359.59.40 Races in chaos. (Finlay, William G.) Johannesburg, 1959.

An 359.39A The races of Europe. (Coon, Carleton Stevens.) N.Y., 1939.

An 358.99.7A The races of Europe. (Ripley, W.Z.) N.Y., 1899.

An 359.10.3 Races of Europe. (Ripley, W.Z.) N.Y., 1910.

An 358.99.8A The races of Europe. Supplement. (Ripley, W.Z.) N.Y., 1899.

An 358.63.4 Races of man; and their geographical distrbution. (Pickering, Charles.) London, 1863.

An 358.54.5 The races of man; and their geographical distribution. (Pickering, Charles.) London, 1854.

An 359.32.5 The races of man. (Bean, R.B.) N.Y., 1932.

An 359.00.4 The races of man. (Deniker, J.) London, 1900.

An 359.12.2 The races of man. (Deniker, J.) London, 1912.

An 358.76.3 The races of man. (Peschel, Oscar F.) N.Y., 1876.

An 359.09.3 Races of man and their distribution. (Haddon, A.C.) London, 1909.

An 359.25.11 Races of man and their distribution. (Haddon, A.C.) N.Y., 1925.

An 358.90.3 The races of man and their geographical distribution. (Peschel, O.F.) N.Y., 1890.

An 379.28 The races of mankind. (Fleure, H.J.) London, 1928.

An 358.50 The races of men. (Knox, R.) Philadelphia, 1850.

An 358.50.4 The races of men. (Pickering, Charles.) London, 1850.

An 358.63 The races of the old world. (Brace, C.L.) N.Y., 1863.

An 358.82.3 Les races sauvages. (Bertillon, A.) Paris, 1882.

An 359.26.4A The racial basis of civilization. (Hawkins, Frank H.) N.Y., 1926.

An 359.08.7A Racial contrasts. (Gehring, A.) N.Y., 1908.

An 379.28.5 The racial elements of European history. (Günther, Hans F.K.) N.Y., 1928.

An 359.23A The racial history of man. (Dixon, Roland B.) N.Y., 1923.

An 359.34A The racial myth. (Radin, Paul.) N.Y., 1934.

An 359.27.13 Racial old age, being further adventures in philosophy. (Penniston, J.B.) Seattle, 1927.

An 359.59 The racial problem in Christian perspective. 1. ed. (Haselden, Kyle.) N.Y., 1959.

An 359.20 The racial prospect. (Humphrey, Seth King.) N.Y., 1920.

An 359.75.15 Racial variation in man: proceedings of a symposium held at the Royal Geographical Society, London, on 19 and 20 September 1974. London, 1975.

Geog 659.73 Racine, Jean Bernard. L'analyse quantitative en géographie. 1. éd. Paris, 1973.

An 170.367 Le racisme devant la science. (UNESCO.) Paris, 1960.

An 359.55.5 Racismo. (Troise, Emilio.) Buenos Aires, 1955.

Geog 4307.94.2 Radcliffe, A. (Mrs.). Journey...through Holland. 2. ed. London, 1795. 2v.

Htn Geog 4307.94* Radcliffe, A. (Mrs.). A journey made in summer of 1794. London, 1795.

An 2109.52 Radcliffe-Brown, A.R. Structure and function in primitive society. London, 1952.

An 2109.52.2 Radcliffe-Brown, A.R. Structure and function in primitive society. London, 1956.

Geog 4218.92.5F Radde, G.F.R. 23.000 mil'na iakhte "Tamara". Sanktpeterburg, 1892-93. 2v.

An 359.33 Radin, Paul. The method and theory of ethnology. N.Y., 1933.

An 2109.57.5 Radin, Paul. Primitive man as philosopher. 2. ed. N.Y., 1957.

Author and Title Listing

An 359.34A	Radin, Paul. The racial myth. N.Y., 1934.	Geog 5328.3	Rasmussen, Knud. Mindeudgave. København, 1934-35. 3v.
An 2109.32.9A	Radin, Paul. Social anthropology. 1st ed. N.Y., 1932.	Geog 5855.3	Rasmussen, Knud. Neue Menschen...des Nordpols. Bern, 1907.
An 2109.53.5	Radin, Paul. The world of primitive man. N.Y., 1953.	Geog 5855.5	Rasmussen, Knud. The people of the Polar North. London, 1908.
An 2109.01.15	Radliński, Ignacy. Przeszłość w teraźniejszości. Warszawa, 1901.	Geog 5855.5.5	Rasmussen, Knud. The people of the Polar North. Philadelphia, 1908.
Geog 3620.2	Radošćić, Nikola. Geografsko znanje o Srbiji početkom 19 veka. Beograd, 1927.	Geog 5399.58	Rasmussen, Knud. Den store slaedereise. København, 1958.
Geog 85.372	Radovi. Travaux. (Zagreb. Univerzitet. Geografski Institut.) 7,1968+	An 359.74.30	Rasogeneticheskie protsesy v etnicheskoi istorii. Moskva, 1974.
Htn Geog 5518.46*	Rae, J. Narrative of expedition to Arctic Sea. London, 1850.	An 5.8	Die Rasse im Schrifttum. (Gercke, Achim.) Berlin, 1933.
Geog 4131.5	Rae, W.F. The business of travel. London, 1891.	An 359.28.11	Die Rasse in den Geisteswissenschaften. (Schemann, L.) München, 1928-31. 3v.
Geog 4509.20.5	Raeburn, H. Mountaineering art. N.Y., 1920.	An 359.37.11	Rasse ist Gestalt. (Clauss, Ludwig F.) München, 1937.
Geog 5399.73	RAEM - moi pozyvnye. (Krenkel', Ernst T.) Moskva, 1973.	An 359.15.7	Rasse und Kultur. 2. Aufl. (Hertz, Friedrich.) Leipzig, 1915.
Geog 613.1.5	Raemdonck, J. van. Gérard de Cremer ou Mercator. St. Nicolas, 1870.	An 359.25.7	Rasse und Kultur. 3. Aufl. (Hertz, F.O.) Leipzig, 1925.
Geog 613.1.3	Raemdonck, J. van. Gérard Mercator. St. Nicolas, 1869.	An 359.34.9	Rasse und Seele; eine Einführung in den Sinn der leiblichen Gestalt. (Clauss, Ludwig F.) München, 1934.
Geog 613.1.13	Raemdonck, J. van. Relations commerciales entre G. Mercator et C. Plantin. Anvers, 1880.	An 379.26.5	Rasse und Stil. (Günther, Hans F.K.) München, 1926.
Geog 665.162	Raeumliche und zeitliche und Bewegungen. Würzburg, 1972.	An 379.26.7	Rasse und Stil. 2. Aufl. (Günther, Hans F.K.) München, 1927.
Geog 4267.79	Raff, G.C. Geographie für Kinder. Tübingen, 1779.	An 359.14	Rasse und Volk. (Wirth, Albrecht.) Halle, 1914.
Geog 4308.17	Raffles, T. Letters during tour thro'...France. Liverpool, 1818.	An 359.46.17A	Rassen und Völker. 3. Aufl. (Schmidt, Wilhelm.) Luzern, 1946-49. 3v.
Geog 5838.45	Rafn, C.C. Amerikas...gamle geographie. Kjøbenhavn, 1845.	An 359.32	Rassenforschung und Volk der Zukunft. (Muckermann, H.) Berlin, 1930.
Geog 4679.42.10	The raft. (Trumbull, Robert.) N.Y., 1942.	An 359.68.30	Rassengeschichte der Menschheit. München, 1968- 3v.
Geog 4249.58	The raft Lehi IV. (Baker, DeVere.) Long Beach, 1959.	An 359.36.10	Rassengeschichte und Vorgeschichte. 2. Aufl. (Schmelzle, K.) Bamberg, 1936.
Geog 4679.72	La rage de survivre, quand il ne reste que la vie. (Vignes, Jacques.) Paris, 1973.	An 359.33.7	Die Rassenidee in der Geistesgeschichte Ray bis Carus. (Voegelin, Erich.) Berlin, 1933.
Geog 4679.72.1	The rage to survive. (Vignes, Jacques.) N.Y., 1976.	An 359.29.7	Rassenkunde Europas. 3. Aufl. (Günther, H.F.K.) München, 1929.
An 136.5.20	La ragione nascosta. (Moravio, Sergio.) Firenze, 1969.	X Cg An 359.37.7	Rassenkunde und Rassengeschichte der Menschheit. 2. Aufl. v.1, pt.1-15. (Eickstedt, E.) Stuttgart, 1937-43.
Geog 3048.83.1	Ragnarok: the age of fire and gravel. (Donnelly, Ignatius.) N.Y., 1970.	An 359.27	Rassenlehre. (Wolff, K.F.) Leipzig, 1927.
Geog 4269.47A	Rahv, Philip. Discovery of Europe. Boston, 1947.	An 359.56	Het rassenvraagstuk. (Bavinck, J.H.) Kampen, 1956.
Geog 3018.93	Rainaud, A. Le continent austral. Paris, 1893.	An 4658.50	Rastoul de Mongeot, Alphonse. Histoire des naufrages depuis les temps ancien jusqu'au 1850. v.1-2. Bruxelles, 1859.
An 2109.27.25A	La raison primitive. (Leroy, Olivier.) Paris, 1927.	Geog 4209.31.15	Rastron, A.H. Home from the sea. N.Y., 1931.
Geog 3070.85	Raisz, E.J. Mapping the world. N.Y., 1956.	Geog 4182.55	Ratelband, K. Vijf dagregisters van het Kasteel São Jorge da Maria. 's-Gravenhage, 1953.
Geog 4209.57.5	Raitchewitch, M. Biographien und Autogramme beruhmter Staatsmanner und anderer Personlichkeiten des 20. Jahrhundert. Bielefeld, 1957.	An 378.58	Rathlef, C. Die welthistorische Bedeutung dei Meere insbesondere des Mittelmeers. Dorpat, 1858.
Geog 4249.56	Raitt, Helen. Exploring the deep Pacific. 1st ed. N.Y., 1956.	Geog 4659.60	Ratigan, William. Great Lakes shipwrecks and survivals. Grand Rapids, 1960.
Geog 4180.3	Raleigh, W. Discovery of...empire of Guiana. London, 1848.	Geog 4659.60.2	Ratigan, William. Great Lakes shipwrecks and survivals. 2. ed. Grand Rapids, 1969.
Geog 4129.06	Raleigh, Walter. The English voyages of the sixteenth century. Glasgow, 1906.	An 378.82	Ratzel, F. Anthropo-Geographie. Stuttgart, 1882. 2v.
Geog 4218.71.3	Ramble round the world. (Hübner, J.A.) London, 1874. 2v.	An 358.96.3A	Ratzel, F. The history of mankind. London, 1896-98. 3v.
Geog 4218.71	Ramble round the world. (Hübner, J.A.) N.Y., 1874.	An 358.87.3	Ratzel, F. Völkerkunde. Leipzig, 1887-88. 3v.
Geog 4328.91	Rambles...in three continents. (Groome, P.L.) Greensboro, 1891.	An 2058.74	Ratzel, F. Vorgeschichte des europäischen Menschen. München, 1874.
Geog 4209.08	Rambles and adventures in Australasia. (Podmore, P.S.M.) London, 1909.	Geog 665.5	Ratzel, F. Zu Friedrich Ratzels Gedächtnis. Leipzig, 1904.
Geog 4308.36.5	Rambles in Europe. (Hall, Fanny W.) N.Y., 1939. 2v.	Geog 618.5.10	Ratzel, Fridrich. Jugenderinnerungen. München, 1966.
Geog 4308.87.3	Rambles in Europe. (Morrison, Leonard A.) Boston, 1887.	An 378.99.1	Ratzel, Friedrich. Anthropogeographie. Darmstadt, 1975.
Geog 4308.87.2	Rambles in Europe. (Morrison, Leonard A.) Boston, 1887.	An 378.99	Ratzel, Friedrich. Anthropogeographie. v.2. Stuttgart, 1899.
Geog 4308.52.9	Rambles in Europe. (Trafton, M.) Boston, 1852.	Geog 135.2.2	Ratzel, Friedrich. Anthropogeographische Beiträge. v.2. Leipzig, 1899.
Geog 4308.18	Rambles in Europe in 1839. (Gibson, William.) Philadelphia, 1841.	Geog 135.2.4	Ratzel, Friedrich. Beiträge zur Geographie des mittleren Deutschland. v.4. Leipzig, 1899.
Geog 5324.1.10	Ramsey, H. Nordenskiöld. Stockholm, 1950.	An 2058.76	Rau, C. Early man in Europe. N.Y., 1876.
An 2109.70.10	Ramseyer, Urs. Soziale Bezüge des Musizierens in Naturvolkkulturen. Bern, 1970.	Geog 4218.86	Raum, G.E. A tour around the world. N.Y., 1886.
Geog 4209.65	Rand, Christopher. Mountains and water. N.Y., 1965.	Geog 4218.95	Raum, G.E. A tour around the world. N.Y., 1895.
Geog 4418.61.3	Rand, E.A. All aboard for sunrise lands. 2. ed. Boston, 1861.	Geog 4308.16	Raumer, F.L.G. Die Herbstreise nach Benedig. Berlin, 1816. 2v.
Geog 4110.9.2	Rand, McNally and Co. Pocket atlas of the world. Chicago, n.d.	Geog 758.48	Raumer, K.G. Lehrbuch der allgemeinen Geographie. Leipzig, 1848.
Geog 4110.9	Rand, McNally and Co. Pocket atlas of the world. N.Y., 1887.	Geog 5559.09.25	Rawlins, Dennis. Peary at the North Pole; fact or fiction? Washington, 1973.
Geog 819.10F	Rand, McNally and Company. The world and its peoples photographed and described. Chicago, 1910.	An 2058.83	Rawlinson, G. The antiquity of man historically considered. N.Y., 1883.
Geog 4129.58.5	Randall, C.B. International travel. Washington, 1958.	Geog 5839.26	Rawson, K.L. A boy's-eye view of the Arctic. N.Y., 1926.
Geog 5509.07	Randall, Harry. The conquest of the Northwest Passage. Minneapolis, 1907.	Htn Geog 4346.93*	Ray, John. Collection of curious travels and voyages...eastern countries. London, 1693.
An 3539.01.3	Randall-MacIver, D. The earliest inhabitants of Abydos. Oxford, 1901.	Geog 4346.73.2	Ray, John. Travels thro' Low Countries, Germany. 2nd ed. London, 1738.
Geog 4559.30.10	Randell, Jack. I'm alone, by Captain Jack Randell, as told to Meigs O. Frost. 1. ed. Indianapolis, 1930.	Geog 959.12F	Raymond, E.L. Marvelous scenes of the world. Chicago, 1902.
Geog 4559.66.5	Randier, Jean. Hommes et navires au Cap Horn, 1616-1939. Paris, 1966.	Geog 4312.5	Raynes, Rozelle. North in a nutshell. Lymington, 1968.
Geog 4209.25A	Random letters from many countries. (Coolidge, J.G.) Boston, 1924.	An 2109.68.20	Razlozhenie rodovogo stroia i formirovanie klassovogo obshchestva. Moskva, 1968.
Geog 4308.76	Random rambles. (Moulton, Louise.) Boston, 1881.	Geog 3019.71	Razvitie geograficheskikh idei. (Isachenko, Anatolii G.) Moskva, 1971.
Geog 4308.56.17	Random sketches...travel in 1856. (Edwards, J.E.) N.Y., 1857.	Geog 665.99	Razvitie i preobrazo vanie geografii predy. (Akademia Nauk SSSR. Institut Geografii.) Moskva, 1964.
An 358.92.3	Ranke, Johannes. Beiträge zur physischen Anthropologie der Bayern. München, 1892.	An 2109.68.10	Razvitie tekhniki v kameunom veke. (Semenov, Sergei A.) Leningrad, 1968.
An 358.94	Ranke, Johannes. Der Mensch. Leipzig, 1894. 2v.	Geog 759.65	Razvitie teoretichekikh problem sovetskoi fizicheskoi geografii, 1917-1934 gg. (Grigor'ev, Andrei A.) Moskva, 1965.
Geog 5209.03	Rannsóknarferð i kafli. (Thompson, G.M.) Gimli, 1903.	An 359.38.5	Razza e nazione. (Schmidt, Wilhelm.) n.p., 1938.
Geog 4208.34	Rapelje, George. A narrative of excursions, voyages and travels. N.Y., 1834.	Geog 4678.70	Read, George H. The last cruise of the Saginaw. Boston, 1912.
Geog 4218.03.11	Rapport à l'Institut national. (Camus, A.G.) Paris, 1803.	An 359.48.15	A reader in general anthropology. (Coon, Carleton Stevens.) N.Y., 1948.
Geog 5070.25	Rapport d'ensemble. (Congrès International pour l'Étude des Régions Polaires, 1st, Brussels, 1906.) Bruxelles, 1906.	An 170.341	Reading in early anthropology. (Slotkin, James Sydney.) N.Y., 1965.
An 98.67	Rapport sur la progrès de l'anthropologie. (Quatrefages de Bréaux, A. de.) Paris, 1867.	An 359.61.35	Readings in anthropology. (Fried, Morton Herbert.) N.Y., 1961-64. 2v.
Geog 3025.5	Rapport sur les travaux cartographiques. (Barbier, J.V.) Nancy, 1884.	An 170.274	Readings in cross-cultural methodology. (Moore, Frank.) New Haven, 1966.
Geog 4326.32.20	Rare adventures and painefull peregrinations. (Lithgow, William.) London, 1928.	An 170.16	Readings on race. (Garn, Stanley Marion.) Springfield, Ill., 1960.
An 359.62.5	Rasisner. (Haste, Hans.) Stockholm, 1962.	An 4559.16	The real story of the whaler. (Verrill, A.H.) N.Y., 1916.
Geog 4559.52	Rasmussen, A.H. Sea fever. London, 1952.	An 2109.61.10	Real y Ramos, C.A. del. Sociología pre y protohistórica. Madrid, 1961.
Geog 5845.2	Rasmussen, Holger. Grønland og dets problemer. København, 1947.	Geog 577.31	Reales Schul-Lexicon. (Hederich, M.B.) Leipzig, 1731.
Geog 5399.25.35	Rasmussen, K. Fra Grønland til Stillehavet. København, 1925-26. 2v.	Geog 4218.17.10	Realms and islands. (Bassett, M.M.) London, 1962.
Geog 5060.15	Rasmussen, K. Heldenbuch der Arktis. Leipzig, 1933.	Geog 5312.3.5	Rear Admiral Byrd and the polar expeditions. (Foster, C.) N.Y., 1930.
Geog 5559.21	Rasmussen, Knud. Across Arctic America. N.Y., 1927.		
Geog 5839.21	Rasmussen, Knud. Greenland by the Polar Sea. London, 1921.		
Geog 5839.16	Rasmussen, Knud. Grønland langs Polhavet. Kristiania, 1919.		
Geog 5328.5.5	Rasmussen, Knud. In der Heimat des Polarmenschen. Leipzig, 1922.		
Geog 5839.12.2	Rasmussen, Knud. Min rejsedagbog. 2. udg. Kjøbenhavn, 1921.		

Author and Title Listing

Htn	Geog 5507.49*	Reasons...navigable passage to west American oceans. London, 1749.
	Geog 759.58	Rebagliato, F. Geografia universal. Barcelona, 1958.
	Geog 5939.59.15	Reboux, Michael. Demain l'Antarchtique. Paris, 1959.
	Geog 4509.56.20	Rébuffat, Gaston. Mont Blanc to Everest. London, 1956.
	Geog 4248.53	Recent exploring expeditions to the Pacific. (Jenkins, J.S.) London, 1853.
	Geog 139.4.12	Recent geographical literature, maps. (London. Royal Geographical Society.) London. 19-41,1926-1932 11v.
	Geog 139.4.13	Recent geographical literature, maps. Index. 1-4, 1918-1932. (London. Royal Geographical Society.) London, 1936.
	An 2058.75	The recent origin of man. (Southall, J.C.) Philadelphia, 1875.
	Geog 5050.12	Recent polar literature. Cambridge, Eng. 1973+
	Geog 4219.11	Recent travel. (Pearse, Albert W.) Sydney, 1914.
	An 359.72.20	Recent trends in cultural anthropology. (Ansari, G.) Wien, 1972.
	Geog 3127.98	Recherches sur la géographie...des anciens. (Gosselin, P.F.J.) Paris, 1798. 2v.
	Geog 3530.11	Recherches sur les voyages et découvertes des...Normands. (Estancelin, L.) Paris, 1832.
	An 359.03.5	Reches, Elie. Les primitifs, étude d'ethnologie. Paris, 1903.
	An 2408.72	Die Rechtsverhältnisse. (Bastian, P.W.A.) Berlin, 1872.
	An 358.91	Reclus, E. Primitive folk. London, 1891.
	Geog 618.3	Reclus, Élisée. Correspondance. v.1-2,3. Paris, 1911-25. 2v.
	Geog 4508.81	Reclus, Elisee. The history of a mountain. N.Y., 1881.
	Geog 808.76.7	Reclus, J.J.E. Earth and its inhabitants - Africa. N.Y., 1886-1890. 4v.
	Geog 808.76.6	Reclus, J.J.E. Earth and its inhabitants - Asia. N.Y., 1891. 4v.
	Geog 808.76.5	Reclus, J.J.E. Earth and its inhabitants - Europe. N.Y., 1885. 5v.
	Geog 808.76.9	Reclus, J.J.E. Earth and its inhabitants - North America. N.Y., 1890-1893. 3v.
	Geog 808.76.8	Reclus, J.J.E. Earth and its inhabitants - Oceanica. N.Y., 1890.
	Geog 808.76.10	Reclus, J.J.E. Earth and its inhabitants - South America. N.Y., 1894-1895. 2v.
NEDL	Geog 808.76	Reclus, J.J.E. Nouvelle geographie universelle. Paris, 1876-1894. 19v.
	Geog 818.92	Reclus, O. A bird's-eye view of the world. Boston, 1892.
	Geog 818.73.3	Reclus, O. Geographie. La terre a vol d'oiseau. 3e ed. Paris, 1877. 2v.
	Geog 4558.76	Recollections of a sailor. (Lindsay, W.S.) London, 1876.
	Geog 4558.31	Recollections of a sea-wanderer's life. (Davis, George.) N.Y., 1887.
	Geog 150.10	Record of proceedings of the society and its branches. (New Zealand Geographical Society.) Christchurch. 1,1946+ 3v.
	Geog 139.25	The record of the Royal Geographical Society, 1830-1930. (Mill, Hugh R.) London, 1930.
	Geog 4328.38	Records of travel. Boston, 1838.
	Geog 4157.31	Recueil de voyages au nord. (Bernard, I.F.) Amsterdam, 1731-37. 9v.
	Geog 4158.24	Recueil de voyages et de memoires. (Paris. Société de Geographie.) Paris, 1824-1864. 7v.
	Geog 4328.74.5	Recuerdos de un viaje a Oriente en el año de 1874. (Aguilar, Federico C.) Bogota, 1875.
	Geog 4208.80	Recuerdos de viaje. (Lopez, Lucio V.) Buenos Aires, 1915.
	Geog 4209.56A	Red, black, blond, and olive. (Wilson, Edmund.) N.Y., 1956.
	Geog 4359.25	Red letter days in Europe. With a glimpse of Northern Africa. (Flambeau, V.) N.Y., 1925.
	Geog 4419.49	The Red Sea and adjacent countries...seventeenth century. (Foster, William.) London, 1949.
	Geog 4181.100	The Red Sea and adjacent countries of the close of the 17th century as described by Joseph Pitts. (Foster, William.) London, 1949.
	Geog 4419.33	Red Sea nights. (Makin, William J.) N.Y., 1933.
	Geog 4658.33	Redding, Cyrus. A history of shipwrecks, and disasters at sea. London, 1833. 2v.
	Geog 510.20	Reden over geografie. v.1-2. Groningen, 1958.
	An 192.1	Redfield, Robert. Human nature and the study of society. Chicago, 1962.
	Geog 4559.68	Reflections in the sea. (Phillips-Burt, Douglas H.C.) Lymington, 1968.
	An 358.87.7	Il regale istruttivo. (Lorenzini, C.) Roma, 1887.
	Geog 3019.35.15	Régi népak, ujvilagok. (Kábmár, G.) Budapest, 1936.
	Geog 4181.121	A regiment for the sea. (Bourne, William.) Cambridge, Eng., 1963.
	Geog 190.20	Regio Basiliensis. Hefte für jurassische und oberheinische Landeskunde. Basel. 1,1959+ 8v.
	Geog 190.20.2	Regio Basiliensis. Hefte für jurassische und oberheinische Landeskunde. Register. 1-10,1959-1969. Basel, n.d.
	Geog 759.74	La région, contributions à une géographie générale des espaces régionaux. (Juillard, Étienne.) Paris, 1974.
	Geog 3410.5	Regional geography in the United States since the war. (Pfeifer, Gottfried.) N.Y., 1938.
	Geog 4269.59.10	A regional geography of western Europe. (Monkhouse, Francis.) London, 1959.
	Geog 5070.37	Las regiones heladas de los polos norte y sur. (Moreno Fuentes, José.) Madrid, 1884.
	Geog 5938.80	Le regioni polari antartiche. (Cardon, F.) Roma, 1880.
	Geog 759.70	Les régions naturelles du globe. (Birot, Pierre.) Paris, 1970.
	Geog 16.1.5	Register. (Appalachian Mountain Club.) 1888-1955 13v.
	Geog 45.10	Register zu den Vereinsschriften. (Deutscher und Österreichischer Alpenvereins.) Innsbruck. 1906
	Geog 258.1.5	Registerband, 1908-59. (Österreichische Geographische Gesellschaft.) Wien, 1960.
	Geog 4306.81.3	Regnard, J.F. Voyage de...en Flandre, Hollanke, Danemark. Paris, 1874.
	Geog 4306.87	Regnard, J.F. Voyages. n.p., n.d.
	An 2109.67.20	Regulierte Anarchie. (Sigrist, Christian.) Olten, 1967.
	Geog 4559.58	Rehm, Arnold. Das fröhliche Logbuch. Bremerhaven, 1958.
	Geog 4300.20.8	Reichard, H.A.O. Reichard's Passagier. Berlin, 1834.
	Geog 4300.20.8	Reichard's Passagier. (Reichard, H.A.O.) Berlin, 1834.
	An 358.64.7	Reid, Mayne. Odd people. Boaston, 1864.
	An 358.61	Reid, Mayne. Odd people. Boston, 1861.
Htn	An 358.60.3*	Reid, Mayne. Odd people. N.Y., 1860.
	Geog 4182.69	Reijse gadaen bij Adriaen Schagen. (Schagen, Adriaen.) 's-Gravenhage, 1968.
	Geog 193.10	Reims. Université. Institut de Géographie. Travaux. Reims. 3,1970+ 3v.
	An 132.1	Reinach, Theodore. Notice sur la vie et les travaux de M. Ernest Hamy. Paris, 1910.
	Geog 4219.25.5	Reinhardt, Walther. Querweltein. Berlin, 1925.
	Geog 4209.39.15	Reinius, I. Journal hållen på resan till Canton i China. Helsingfors, 1939.
	Geog 4182.47	De reis om de wereld. (Spilbergen, J. van.) 's-Gravenhage, 1943. 2v.
	Geog 4182.27	De reis om de wereld door Olivier van Noort. (Ijzermann, J.W.) 's-Gravehage, 1926. 2v.
	Geog 4182.65	De reis om de wereld van de nassausche vloot, 1623-1926. 's-Gravenhage, 1964.
	Geog 4217.66.11	Reis rondom de weereldt. (Bougainville, L. de.) Dordrecht, 1772.
	Geog 4182.1	De reis van Jan C. May. (Muller, S.) 's-Gravenhage, 1909.
	Geog 4182.21	De reis van Mahu en de Cordes. (Wieder, F.C.) 's-Gravenhage, 1923-25. 3v.
	Geog 4182.56	De reis van Mathijs Hendriksz. (Quast, Mathijs H.) 's-Gravenhage, 1954.
	Geog 4182.62	De reis van Michiel Adriaanszoom de Ruyter in 1664-1665. (Ruyter, M.A.) 's-Gravenhage, 1961.
	Geog 4182.4	De reis van Mr. Jacob Roggeveen. (Mulert, F.E.) 's-Gravenhage, 1911.
	Geog 4182.38	De reis van Voris van Spilbergen naar Ceylon, Atjeh en Bantam, 1601-1604. 's-Gravenhage, 1933.
	Geog 4182.76	De reis van Z.M. De Vlieg. (Kreekel, Willem.) 's-Gravenhage, 1975.
	Geog 4209.11.7	Reisbrieven uit Afrika en Azie. (Jacobs, A.H.) Almelo, 1913. 2v.
	Geog 4559.40.11	Reise, Quartier in Gottesnaam. (Wossidlo, R.) Seestadt Rostock, 1940.
	Geog 4417.99	Reise...durch Asien, Africa. (Abu Taleb Khan.) Wien, 1813.
	Geog 4208.48.2	Reise Album vom 15ten bis zum 22ten Lebensjahre. (Cöllen, Franz A.) Hamburg, 1852.
	Geog 4502.10.4	Die Reise auf den Grossglockner, 1800. (Vierthaler, Franz Michael.) München, 1938.
	Geog 5398.20.3	Reise des...Flotten-Lieutenants. (Wrangell, F. von.) Berlin, 1839. 2v.
	Geog 5398.93.15	Reise-Erinnerungen...Aug. 1893..."Admiral". (Plass, F.) Hamburg, 1894.
	Geog 4307.91	Reise in Deutschland, der Schweiz. (Stolberg, F.L.G.) Königsberg, 1794. 4v.
	Geog 4209.60.45	Reise-Journal. (Firsé, Adolf.) Gütersloh, 1967.
	Geog 5397.73.7	Reise nach dem Nordpol. (Phipps, J.) Berlin, 1777.
	Geog 4217.72.3	Reise nach dem Vorgebirge der guten Hoffnung. (Sparrman, Anders.) Berlin, 1784.
	Geog 5517.46.3	Reise nach Hudsons Meerbusen. (Ellis, H.) Göttingen, 1750.
	Geog 4208.86	Reise nach Japan aus Briefen und Tagebüchern zusammengestellt. (Böckmann, W.) Berlin, 1886.
	Geog 5558.50	Reise-Tagebuch des Missionars. (Miertsching, J.A.) Gnadan, 1855.
	Geog 4218.30.5	Reise um die Erde. (Meyen, F.J.F.) Berlin, 1834-35. 2v.
	Geog 4218.72.5	Reise um die Erde. 3. Aufl. (Hildebrandt, E.W.) Boston, 1872.
	Geog 4208.98.5	Eine Reise um die Welt. 2e Aufl. (Schweitzer, Georg.) Berlin, 1899.
	Geog 4218.48	Reise um die Welt. 11e Aufl. (Beck, Christian.) Dresden, 1907.
Htn	Geog 4218.03.3*	Reise um die Welt in den Jahren 1803-1806 auf Befehl seiner Kaisere. v.1-2. (Krusenstern, A.J. von.) Berlin, 1811-12. 3v.
	Geog 4228.57	Reise vom Mittelrhein...nach dem Nordamerikanischen Freistaaten. (Ennemoser, F.J.) Kaiserlautern, 1864.
	Geog 4341.73.6	Die Reisebeschreibungen. (Benjamin ben Jonah, of Tudela.) Jerusalem, 1903.
	Geog 4419.10	Reisebriefe eines österreichischen Industriellen aus Abessinien, Indien und Ostasien. (Pick, Emil G.) Prag, 1910?
	Geog 500.25	Die Reiseliteratur Deutschlands. (Georg, Carl.) Leipzig, 1877.
	Geog 4417.38.5	Reisen der Barbery und der Levante. (Shaw, Thomas.) Leipzig, 1765.
	Geog 4308.48.5	Reisen durch Belgien, Holland und Grossbritannien. (Heinzelmann, F.) Leipzig, 1848.
	Geog 135.2.7	Reisen in Bolivien und Peru. v.7. (Hauthal, Rudolf.) Leipzig, 1911.
	Geog 5398.70	Reisen nach dem Nordpolarmeer. (Heuglin, M.T.) Braunschweig, 1872-74. 3v.
	Geog 4167.51	Reisen nach Peru, Acadien und Egypten. Göttingen, 1751.
	Geog 4227.89	Reisen nach Sudamerika, Asien. (Langstedt, F.L.) Hildesheim, 1789.
	Geog 4129.58	Reisen seit Anno dazumal. (Greenen, E.) Hamburg, 1958.
	Geog 4308.26	Reisen und Briefe. (Carus, Karl Gustav.) Leipzig, 1926. 2v.
	Geog 4208.48.3	Reisen und Dichtungen. (Cöllen, Franz A.) Berlin, 1864.
	Geog 4182.33	Reisen van Nicolaus de Graaff. (Warnsinck, J.E.M.) 's-Gravenhage, 1930.
	Geog 4346.83	Die Reisetagebücher. (Kaempfer, Engelbert.) Wiesbaden, 1968.
	Geog 4206.13.27	Reiss in das gelobte Land. (Amman, H.J.) Berlegung, 1678.
	Geog 4206.13.25	Reiss ins gelobte Land...Servian...Aegypten. (Amman, H.J.) Zürich, 1630.
	Geog 4228.26	Reize naar de oost- en westkust van Zuid-Amerika. (Boelen, Jacobus.) Amsterdam, 1835-36. 3v.
	Geog 4207.53.3	Reize van London, door Rusland, nae en in Persie. (Hanway, Jonas.) Amsterdam, 1758. 2v.
	Geog 4182.11	Reizen in Zuid-Afrika. (Godeé Molsbergen, E.C.) 's-Gravenhage, 1916-32. 4v.
	Geog 4182.52	Reizen naar West-Afrika van Pieter van den Broecke. (Broecke, Pieter van den.) 's-Gravenhage, 1950.
Htn	Geog 4347.01.5F*	Reizen over Muskovie door Persie. (Bruyn, C. de.) Amsterdam, 1711.
	Geog 4182.8	Reizen van...naar het noorden 1594-95. (Linschoten, Jan H. van.) 's-Gravenhage, 1914.
	Geog 5375.94.5	Reizen van Willem Barents. (Veer, G. de.) 's-Gravenhage, 1917. 2v.
	Geog 4238.53	En Rejse til Amerika. (Hauch, J.C. von.) Kjøbenhavn, 1853.
	Geog 4249.67.10	Rejsernes rytme. Kronikker. (Sandvad, Jørgen.) København, 1967.
Htn	Geog 4676.02*	Relaçam do naufragio da nao Santiago e itinerario da gente que delle se salvou. (Godinho Cardozo, M.) Lisboa, 1602. 3 pam.
	Geog 4676.93	Relaçam do successo que teve o patacho chamado N. Sra. da Candelaria da ilha da Madeira. (Correa, F.) Lisboa, 1734.

Author and Title Listing

Call Number	Entry
Geog 4180.9	Relacam Verdadeira dos Trabalhos. Discovery...of Terra Florida by De Soto. London, 1851.
Geog 4237.78	Relação, ou Noticia particular da infeliz viajem da nas de Nassa Senhora da Ajuda. (Alexandre e Silva, Elias.) Lisboa, 1778.
Geog 3060.15	Relação que trata de como em cincoenta e oito gráos do sul fay descuberta huma ilha. pt.2. Lisboa, 17- ?
Geog 4216.19	Relación diaria del viaje de J. Le Maire y G.C. Shouten. (Schouten, W.C.) Santiago de Chile, 1897.
Geog 4229.47	Relaciones de viajes. (Vargas Ugarte, Ruben.) Lima, 1947.
Geog 4477.09	Relation...du voyage du Prince de Montberaud. Merinde, 1709.
Geog 4502.5	Relation abrégés d'un voyage a la cime du Mont-Blanc en aout 1787. (Saussure, H.B. de.) Genève, 1928? 2 pam.
Geog 4427.82	Relation de deux voyages. (Kerguélen.) Paris, 1782.
Geog 4166.64.5F	Relation de divers voyages. (Thevenot, M. de.) Paris, 1696. 2v.
Geog 4477.21	Relation d'un voyage du Pole Arctique au Pole Antarctique. Amsterdam, 1721.
Geog 4304.90	Relation d'un voyage fait en Europe. (Martyr.) Paris, 1827.
Geog 4181.60	Relation of a voyage to Guiana. (Farcourt, R.) London, 1928.
Geog 4477.21.3	Relation un voyage du Pole Arctique au Pole Antarctique. 2. ed. La Haye, 1734.
Htn Geog 815.59.3*	Le relationi universali...divise in quattro parti. (Botero, G.) Venetia, 1597.
Htn Geog 815.59.7*	Le relationi universali...divise in quattro parti. (Botero, G.) Venetia, 1622.
Geog 613.1.13	Relations commerciales entre G. Mercator et C. Plantin. (Raemdonck, J. van.) Anvers, 1880.
Geog 4166.64F	Relations de divers voyages. (Thevenot, M. de.) Paris, 1664-66. 2v.
Geog 4306.95	Relations historiques...de voyages. (Patin, Charles.) Amsterdam, 1695.
Geog 4306.71.5	Relations historiques...de voyages en Allemagne. 2e éd. (Patin, Charles.) Lyon, 1676.
Geog 4181.66	Relations of Golconda in the early 17th century. (Moreland, W.H.) London, 1931.
An 359.02.3	The relationship of the advanced and backward races of mankind. (Bryce, J.) Oxford, 1902.
An 359.03.9	Relationship of the advanced and backward races of mankind. 2. ed. (Bryce, J.) Oxford, 1903.
Geog 5925.60	Relatos antarticos. (Diaz, Emilio.) Buenos Aires, 1958.
Geog 212.110.10	Relazioni sulla fondazione e sull'ordinamento della sezione de geografia commerciale della Societa italiana. (Brunialti, A.) Roma, 1879.
An 629.65	The relevance of models for social anthropology. (Conference on New Approaches in Social Anthropology.) London, 1965.
An 2109.11.7	Religion und soziales Leben bei Naturvölkern. (Visscher, H.) Bonn, 1911. 2v.
Geog 5700.6	Religious Tract Society. Missionary records; northern countries. London, 1839.
Geog 4672.4	Remarkable shipwrecks. Hartford, 1813.
Geog 4306.92.6	Remarks in the grand tour of France and Italy. 2d ed. (Bromley, William.) London, 1705.
Geog 4267.26	Remarks on several parts of Europe. v.1-2. (Breval, John D.) London, 1726.
Geog 4267.38	Remarks on several parts of Europe. v.1-2. (Breval, John D.) London, 1738.
Geog 5518.18.4	Remarks on the account of the late voyage of discovery to Baffin's Bay. 2. ed. (Sabine, Edward.) London, 1819.
Geog 5170.13	Remarks on the voyage of the ships "Resolution". London, 1780.
Geog 4307.28	Remarques d'un voyageur. Haye, 1728.
Geog 4209.10.9	Reminiscences of a diplomat's wife. (Fraser, Mary Crawford.) N.Y., 1913.
Geog 4218.87.15	Reminiscences of a skipper's wife. (Paterson, Florence E.) London, 19- .
Geog 4218.49.5	Reminiscences of a voyage around the world. (Davis, R.C.) Ann Arbor, 1869.
Geog 5317.10	Reminiscences of adventure and service. (Greely, A.W.) N.Y., 1927.
Geog 4308.68	Reminiscences of European travel. (Peabody, A.P.) N.Y., 1868.
Geog 4208.83.6	Reminiscences of travel in Australia, America and Egypt. 2. ed. (Tangye, Richard.) London, 1884.
Geog 4209.05	Reminiscencias de viajes. (Herboso, F.J.) Caracas, 1905- 3v.
Geog 4182.31	Die remonstrantie von W. Geleynssen de Jongh. (Caland, W.) 's-Gravenhage, 1929.
An 136.5.40	Remotti, Francesco. Lévi-Strauss. Torino, 1971.
Geog 3249.68	Renaissance exploration. (Hale, John Higby.) London, 1968.
Geog 3570.36	Renault-Roulier, Gilbert. The caravels of Christ. N.Y., 1959.
Geog 665.82	Rencontres de la géographie et de la sociologie. (Sorre, Maximilien.) Paris, 1957.
Geog 4206.65.5	René Desboys du Chastelet. (Piolin, Paul.) n.p., 1885.
Geog 665.146	The renewal of the geographic environment: an inaugural lecture delivered before the University of Oxford on 11 February 1969. (Gottmann, Jean.) Oxford, 1969.
Geog 819.44	Renner, G.T. Global geography. N.Y., 1944.
Geog 193.1	Rennes. Université. Laboratoire de Géographie. Travaux. Rennes. 1-6,1903-1909 2v.
Geog 212.20	La reorganización de la Sociedad Geográfica de Lima. (Sociedad Geográfica de Lima.) Lima, 1944.
Geog 4329.33	Rèpaci, L. Con la ciurma dell'Alessandro. Milano, 1933.
Geog 142.2.2	Repertorio, 1901-1910. (Sociedad Geografico de Madrid.) Madrid, 1911.
Geog 500.31	Repertorium über...allgemeinern...Journale. (Ersch, J.S.) Lemgo, 1790. 3v.
Geog 199.1.5	Repertorium und Ortsregister. (Schweizer Alpenclub.) Bern. 1-44,1886-1910 2v.
Geog 4678.42	Replica...in causa di abbordaggio. (Guerrazzi, F.D.) Livorno, 1842.
Geog 5559.10	Report...dominion government expedition...1910. (Canada. Department of Marine and Fisheries.) Ottawa? 1911.
Geog 5558.81.8F	Report...Greely relief expedition of 1884. (Schley, W.S.) Washington, 1887.
Geog 5558.81.5	Report. (United States. Board of Officers. Expedition...Relief of Lieutenant Greely.) Washington, 1884.
An 3309.08	Report - physical condition of children. (New South Wales. Department of Public Instruction.) Sydney, 1908.
Geog 4679.14	Report and evidence. (Canada. Commission of Inquiry into the Loss of the British Steamship "Empress of Ireland".) Ottawa, 1914.
Geog 4145.79	Report from practically nowhere. (Sack, John.) N.Y., 1959.
Geog 5399.64	Report from the Arctic; foreign and Soviet correspondents on their trip aboard the Soviet atomic icebreaker. Moscow, 1964.
Geog 5399.64.1	Report from the Arctic: foreign and Soviet correspondents on their trip aboard the Soviet atomic icebreaker. Moscow, 1968.
Geog 4708.34.2	A report of the trial of Pedro Gilbert...piracy. Boston, 1834.
Geog 110.2.3	Report of 3rd International Geographical Congress and Exhibition at Venice, Italy, 1881. (United States. War Department. Corps of Engineers.) Washington, 1885.
Geog 5559.03	Report on dominion government expedition. (Low, A.P.) Ottawa, 1906.
Geog 5558.81F	Report on proceedings. (Unites States. Expedition to Lady Franklin Bay.) Washington, 1888. 2v.
Geog 5559.06.5	Report on the dominion government expedition to Arctic islands and the Hudson Strait on board the C.G.S. "Arctic", 1906-07. (Canada. Department of Marine and Fisheries.) Ottawa, 1909.
Geog 5838.68	Report on the resources of Iceland and Greenland. (Peirce, B.M.) Washington, 1868.
Geog 4209.61	A reporter here and there. (Kahn, E.J.) N.Y., 1961.
Geog 665.53	Reports of round tables, 1937. Geographic aspects of international relations. (Chicago. University. Norman Wait Harris Foundation.) n.p., 1937.
Geog 5321.1	Reports on medals to Arctic explorers, Kane, Hayes. (United States. Naval Observatory.) Washington, 1876.
Geog 4180.47	Reports on the discovery of Peru. (Markham, C.R.) N.Y., 1964-66. 2v.
An 359.13	Reports upon the present condition and future needs of the science of anthropology. (Rivers, W.H.R.) Washington, 1913.
Geog 5925.55	La Republica Argentina y las adquisiciones territoriales en el continente antartico. (Rodriguez, Juan C.) Buenos Aires, 1941.
Geog 5060.11	Las republicas hispano-americanos y la exploracion de las regiones polares. (Toreoja y Miret, José M.) Madrid, 193-
Geog 3055.57	Requena, Rafael. Vestigios de la Atlántida. Caracas, 1932.
Geog 5558.81.7A	The rescue of Greely. (Schley, W.S.) N.Y., 1885.
Geog 500.115.1F	Research catalogue. First supplement. (American Geographical Society of New York.) Boston, 1974. 2v.
Geog 500.115.2F	Research catalogue. Map supplement. (American Geographical Society of New York.) Boston, 1962.
Geog 500.115F	Research catalogue of the American Geographical Society. (American Geographical Society of New York.) Boston, 1962. 15v.
An 379.71	Research in human geography. (Chisholm, Michael.) London, 1971.
An 629.68	Research in social anthropology. London, 1968.
Geog 174.5F	Research papers. (Oxford. University. School of Geography.) Oxford. 1,1972+
An 2109.73.10	Research Seminar in Archaeology and Related Subjects, University of Sheffield, 1971. The explanation of culture change; models in prehistory. Pittsburgh, 1973.
An 2108.70.8	Researches into the early history of mankind. (Tylor, Edward Burnett.) London, 1870.
An 358.26	Researches into the physical history of mankind. (Prichard, J.C.) London, 1826. 2v.
An 358.41.2	Researches into the physical history of mankind. v.1,5, 4. ed; v.2-4, 3. ed. (Prichard. J.C.) London, 1837-47. 5v.
An 358.41	Researches into the physical history of mankind. v.1-2, 4. ed.; v.3, 3. ed. (Prichard, J.C.) London, 1837-41. 3v.
An 358.51	Researches into the physical history of mankind. v.1-2,5, 4. ed.; v.3-4, 3. ed. (Prichard, J.C.) London, 1841-51. 5v.
An 358.36	Researches into the physical history of mankind. 3. ed. (Prichard, J.C.) London, 1836-47. 6v.
Geog 4332.8.5	Researches on the Danube and the Adriatic. (Paton, A.A.) Leipzig, 1861. 2v.
Geog 4308.32	Residence in France with excursion up the Rhine to Switzerland. (Cooper, J.F.) Paris, 1836.
Geog 4308.96.5	Resonancias del camino. (Zorrilla de San Martin, J.) Barcelona, 190-?
Geog 4308.96.3	Resonancias del camino. (Zorrilla de San Martin, J.) Paris, 1896.
An 136.5.30	Restloses. Erkenhen; die Diskussione über den Strukturalismus de Claude Lévi-Strauss in Frankreich. (Dumasy, Annegret.) Berlin, 1972.
An 359.17	Restrepo-Hernandez, J. Lecciones de antropologia. Bogotá, 1917.
Geog 4306.91.1	Retcher, Wilhelm. Der sächsische Robinson. Leipzig, 1722.
An 170.323	Rethinking modernization; anthropological perspectives. Westport, 1974.
An 3309.36	The retinal method of identification. (Simon, Carleton.) n.p., 1936.
Geog 6009.57	Retour dans l'Antarctique; récit de l'Expédition antarctique belge, 1957-58. (Gerlache de Gomery, Gaston de.) Tournai, 1960.
Geog 4209.57	Retour du bout du monde. (Gosset, R.P.) Paris, 1957.
Geog 5650.13	Rettarstad Graenlands adornu. (Larusson, O.) Kaupmannahöfn, 1914.
Geog 4209.45.20	Retuerto, Marcial. Sur les traces de Magellan. Paris, 1945.
Geog 5559.08.10	Return from the Pole. (Cook, F.A.) N.Y., 1951.
Geog 5399.61.5	Reut, V.F. Moskva - Severnyi Polius. Moskva, 1961.
An 359.34.5	Reuter, E.B. Race and culture contacts. N.Y., 1934.
Geog 3550.14	Revelli, P. La casa di Savoia e gli studii geografici. Milano, 1903.
Geog 4208.72	Revere, J.W. Keel and saddle, a retrospect. Boston, 1872.
Geog 4208.72.2	Revere, J.W. Keel and saddle, a retrospect. Boston, 1873.
Geog 190.5	Revista geográfica americana. Buenos Aires. 1-43,1933-1959 41v.
Geog 190.10	Revista geográfica del instituto panamericano de geográfica e historia. (Pam American Institute of Geography and History.) México. 9,1949+ 11v.
NEDL Geog 39.2.10	Revista mensile publicata per cura de consiglio direttivo della sede centrale. (Club Alpino Italiano.) Torino. 1-18,1882-1899 18v.

242

Author and Title Listing

	Geog 39.2.10	Revista mensile publicata per cura de consiglio direttivo della sede centrale. (Club Alpino Italiano.) Torino. 1-60,1882-1941 37v.
	Geog 190.15	Revista uruguaya de geografia. Montevideo. 1-6,1950-1952
	An 629.64	The revolution in anthropology. (Jarvie, Ian C.) London, 1964.
	An 359.66	The revolution of color. (Melady, Thomas Patrick.) N.Y., 1966.
	Geog 36.20	Revue canadienne de géographie, organe de la Société de géographie de Montréal et de l'Institut de géographie de l'université de Montréal. Montréal. 1,1947+ 13v.
	Geog 192.1	Revue de geographie. Annuelle. Paris. 1-12,1906-1924 65v.
	Geog 192.4	Revue de géographie alpine. (Grenoble. Université. Institut de Géographie Alpine.) 1,1913+ 61v.
	Geog 192.3	Revue française de l'étranger et des colonies et exploration. Paris. 1-39,1886-1914 39v.
	Geog 192.5	Revue géographique des pyrénées et du sud-ouest. Toulouse. 1,1930+ 33v.
	Geog 192.5.5	Revue géographique des pyrénées et du sud-ouest. Table décennale, 1950-1959. Toulouse, n.d.
	Geog 192.2	Revue geographique internationale. 89-112,1883-1885
NEDL	Geog 14.205	Revue maritime. 1861-1899
	Geog 14.205	Revue maritime. 1900-1971//
NEDL	Geog 14.206	Revue maritime. Table. 1861-1888 2v.
	Geog 226.3	Revue mensuelle. (Touring Club de France.) Paris. 36-47,1926-1937 3v.
	Geog 4509.14	Rey, G. Alpinismo acrobatico. Torino, 1914.
	An 3409.09	Reyna Almandos, Luis. Dactiloscopia argentina. La Plata, 1909.
	An 3409.12	Reyna Almandos, Luis. Origen e influencia jurídico-social del sistema dactiloscopico argentino. La Plata, 1912.
	Geog 4309.26.25	Reynolds, Bruce. A cocktail continentale. N.Y., 1926.
	An 3509.31	Reynolds, E. The evolution of the human pelvis in relation to the mechanics of the erect posture. Cambridge, 1931.
	Geog 5324.2.15	Reynolds, E.E. Nansen. London, 1932.
	Geog 4218.31.2	Reynolds, J.N. Voyage of the United States frigate Potomac. N.Y., 1835.
	Geog 4218.31	Reynolds, J.N. Voyage of the United States frigate Potomac. N.Y., 1835.
	Geog 4228.10	Reynolds, Stephen. The voyage of the New Hazard...1810-1813. Salem, 1938.
	Geog 4208.22	Rezo, José Luiz do. Viagens à China. Porto, 1822.
	Geog 5208.80	Rezzadore, P. I viaggi polari. Roma, 1880.
	Geog 1525.56	Die Rheinland, Schwarzwald, Vogesen. 32. Aufl. (Baedeker, publishers.) Leipzig, 1912.
	Geog 1525.15	Die Rheinlande. 9. Aufl. (Baedeker, publishers.) Coblenz, 1856.
	Geog 1525.45.5	Die Rheinlande von der Schweize. 26. Aufl. (Baedeker, publishers.) Leipzig, 1892.
	Geog 1525.22	Die Rheinlande von der Schweize zu holländisch Grenze. 14. Aufl. (Baedeker, publishers.) Coblenz, 1866.
	Geog 1525.41	Die Rheinlande von der Schweize zu holländisch Grenze. 28. Aufl. (Baedeker, publishers.) Leipzig, 1889.
	Geog 1523.15	Rhine and northern Germany. 3. ed. (Baedeker, publishers.) Coblenz, 1868.
	Geog 1523.17	Rhine and northern Germany. 4. ed. (Baedeker, publishers.) Coblenz, 1970.
	Geog 1525.25	The Rhine from Rotterdam to Constance. 5. ed. (Baedeker, publishers.) Leipsic, 1873.
	Geog 1525.25.3	The Rhine from Rotterdam to Constance. 5. ed. (Baedeker, publishers.) Leipsic, 1873.
	Geog 1525.29	The Rhine from Rotterdam to Constance. 6. ed. (Baedeker, publishers.) Leipsic, 1878.
	Geog 1525.30	The Rhine from Rotterdam to Constance. 7. ed. (Baedeker, publishers.) Leipsic, 1880.
	Geog 1525.35.3	The Rhine from Rotterdam to Constance. 8. ed. (Baedeker, publishers.) Leipsic, 1882.
	Geog 1525.35	The Rhine from Rotterdam to Constance. 8. ed. (Baedeker, publishers.) Leipsic, 1882.
	Geog 1525.40A	The Rhine from Rotterdam to Constance. 10. ed. (Baedeker, publishers.) Leipsic, 1886.
	Geog 1525.41.3	The Rhine from Rotterdam to Constance. 11. ed. (Baedeker, publishers.) Leipsic, 1889.
	Geog 1525.45.2	The Rhine from Rotterdam to Constance. 12. ed. (Baedeker, publishers.) Leipsic, 1892.
	Geog 1525.45	The Rhine from Rotterdam to Constance. 12. ed. (Baedeker, publishers.) Leipsic, 1892.
	Geog 1525.47	The Rhine from Rotterdam to Constance. 13. ed. (Baedeker, publishers.) Leipsic, 1896.
	Geog 1525.50	The Rhine from Rotterdam to Constance. 14. ed. (Baedeker, publishers.) Leizsic, 1900.
	Geog 1525.50.2	The Rhine from Rotterdam to Constance. 14. ed. (Baedeker, publishers.) Leizsic, 1900.
	Geog 1525.52.5	The Rhine from Rotterdam to Constance. 15. ed. (Baedeker, publishers.) Leipzig, 1903.
	Geog 1525.53	The Rhine from Rotterdam to Constance. 16. ed. (Baedeker, publishers.) Leipzig, 1906.
	Geog 1525.66	The Rhine from the Dutch to the Alsatian frontier. 18. ed. (Baedeker, publishers.) Leipzig, 1926.
	Geog 1525.55A	The Rhine including the Black Forest and Vosges. 17. ed. (Baedeker, publishers.) Leipzig, 1911.
	Geog 4707.23.2F	Rhode Island (Colony). Court of Admiralty. Tryals of thirty-six persons for piracy. Boston, 1723.
	Geog 5070.65	Riabinin, A.K. Iz Arktiki v Antarktiku. Murmansk, 1959.
	An 5.2	Riccardi, P. Saggio di un catalogo bibliografico antropologico italiano. Modena, 1883.
	Geog 3550.13	Ricchieri, G. Il contributo degli Italiani alla conoscenza della terra ed agli studi cinquatennio. Roma, 1912.
	Geog 665.39	Ricchieri, G. Dopo il viaggio d'istruzione. Firenze, 1914.
	Geog 5559.34	Rich, E.G. Hans the Eskimo. Boston, 1934.
	Geog 4209.32.15	Rich tapestry. (Mordaunt, E. (pseud.).) N.Y., 1932.
	An 358.01	Richard, Jerôme. Voyages chez les peuples sauvages. Paris, 1801. 3v.
	Geog 4179.5	Richard Hakluyt: his life and work. (Markham, C.R.) London, 1896.
	Geog 4181.93	Richard Hakluyt and his successors. (Lynam, E.) London, 1946.
	Geog 3520.11A	Richard Hakluyt and the English voyages. (Parks, George Bruner.) N.Y., 1928.
	Geog 3520.11.2	Richard Hakluyt and the English voyages. 2. ed. (Parks, George Bruner.) N.Y., 1961.
	Geog 4209.40.20	Richard Halliburton, his story of his life's adventures as told in letters to his mother and father. (Halliburton, Richard.) Indianapolis, 1940.
	Geog 4509.35.5	Richards, D.P. (Mrs.). Climbing days. London, 1935.
	Geog 6009.14.35	Richards, R.W. The Ross Sea Shore Party, 1914-17. Cambridge, 1962.
	Geog 4348.49.2	Richardson, D.L. Anglo-Indian passage. London, 1849.
	Geog 4218.85	Richardson, D.N. A girdle round the earth. Chicago, 1888.
	Geog 5538.48	Richardson, J. Arctic searching expedition. N.Y., 1852.
	Geog 5538.48.5	Richardson, J. Arctic searching expedition. N.Y., 1854.
	Geog 5060.5	Richardson, J. The polar regions. Edinburgh, 1861.
	Geog 4308.91.3F	Richardson, W.L. Cornwall-Devon, 1891; [Collection of photographic views...of Germany, Holland, Switzerland, Paris and England]. n.p., n.d.
	Geog 4308.67.11F	Richardson, W.L. Europe, 1867-1869. n.p., n.d.
	Geog 618.1.9	Richter, O. Der teleologische Zug im Denken C. Ritters. Borna, 1905.
	Geog 3602.5	Richter, Sørea. Great Norwegian expeditions. Oslo, 1955.
	An 379.08	Richthofen, Ferdinand von. Vorlesungen. Berlin, 1908.
	Geog 4209.64.35	Richtig Reisen. (Eckert, Gerhard.) Bergisch Gladbach, 1964.
	An 3309.24.3	Richtlineen für Körpermessungen. (Martin, Rudolf.) München, 1924.
	Geog 4502.5.5	Rickmers, W.R. Querschnitt durch mich. München, 1930.
	Geog 190.5.5	Ricossa, J.A. Pueblos. Buenos Aires, 195-? 6v.
	Geog 4209.50	Riddell, J. Flight of fancy. London, 1950.
	An 2109.34.3A	The riddle of the sphinx, or Human origins. (Róheim, Géza.) London, 1934.
	Geog 4308.38.5	A ride to Florence thro' France. (Holmes, D.) London, 1841. 2v.
	Geog 4558.84	Rideing, William H. Boys coastwise, or All along the shore. N.Y., 1884.
	Geog 4239.66.5	Ridgway, John M. A fighting chance. London, 1967.
	Geog 4029.14	Ridgway, John M. Journey to Ardmore. London, 1971.
	Geog 4228.97	Riesenberg, Felix. Under sail. N.Y., 1918.
	An 378.24.5	Riflessioni...su l'opera "L'homme du midi". (Gioja, M.) Milano, 1830.
	An 3309.43.5	Riggs, F.B. Tall men have their problems too. Cambridge, 1943.
	Geog 818.19	Riise, J. Haanbog i geographien. Kjobenhavn, 1819-1820. 2v.
	An 359.70.25	Rimet, Michel. Contracts, interférences ethniques et culturelles. Montpellier, 1970.
	Geog 5838.77.1	Rink, Henrik. Danish Greenland, its people and products. Montreal, 1974.
	Geog 5838.57.3	Rink, Hinrich. Grönland geographisch...beschreiben. Stuttgart, 1860.
	Geog 5838.57	Rink, Hinrich. Grønland geographisk. Kjøbenhavn, 1857. 2v.
	Geog 5838.82	Rink, Hinrich. Om Grønlaenderne, deres fremtid. Kjøbenhavn, 1882.
	An 358.99.7A	Ripley, W.Z. The races of Europe. N.Y., 1899.
	An 359.10.3	Ripley, W.Z. Races of Europe. N.Y., 1910.
	An 358.99.8A	Ripley, W.Z. The races of Europe. Supplement. N.Y., 1899.
	An 5.1	Ripley, W.Z. Selected bibliography of the anthropology and ethnology of Europe. Boston, 1899.
	An 99.68	The rise of anthropological theory. (Harris, Marvin.) N.Y., 1968.
	Geog 4181.26	Rise of British Guiana. (Storm van 's Gravesande, L.) London, 1911. 2v.
	Geog 4228.99	Riseis, G. de. Dagli stati uniti alle Indie. Roma, 1899.
	Geog 5324.2.25	Risteheuber, René. La double aventure de Fridtjof Nansen. Montréal, 1945.
	Geog 4308.33	Ritchie, Leitch. Travelling sketches. London, 1833.
	Geog 4308.32.10	Ritchie, Leitch. Travelling sketches in the north of Italy, the Tyrol, and on the Rhine. London, 1832.
Htn	Geog 815.38*	Rithaymer, Georg. Georgii Rithaymeri De orbis terrarvm sitr compendium. Norimbergae, 1538.
	Geog 3018.61	Ritter, C. Geschichte der Erdkunde. Berlin, 1861.
	Geog 808.17	Ritter, Carl. Die Erkunde. Berlin, 1817-1818. 2v.
	Geog 808.17.5	Ritter, Carl. Die Erkunde. v.1-19. Berlin, 1822-1859. 21v.
	Geog 808.17.6	Ritter, Carl. Namen- und Sach-Verzeichniss. Berlin, 1841-1849. 2v.
	An 359.33.3	Ritter, J. Über den Sinn und die Grenze der Lehre vom Menschen. Potsdam, 1933.
	Geog 4268.63	Ritter, K. Europa. Berlin, 1863.
	Geog 665.9A	Ritter, Karl. Geographical studies. Boston, 1863.
	Geog 665.9.3	Ritter, Karl. Geographical studies. Cincinnati, 1861.
	Geog 578.47.9	Ritter, Karl. Geographisches-statistisches Lexikon. Leipzig, 1874. 2v.
	Geog 578.47	Ritter, Karl. Geographisches-statistisches Lexikon. 3. Aufl. Leipzig, 1847.
	Geog 4131.66	Ritter, Wigand. Fremdenverkehr in Europa. Leiden, 1966.
	Geog 665.158	Riva, Ambrogio. Una piccola biblioteca. Milano, 1970.
	An 2209.24	Rivers, W.H.R. Medicine, magic and religion. London, 1924.
	An 359.13	Rivers, W.H.R. Reports upon the present condition and future needs of the science of anthropology. Washington, 1913.
	An 2109.24.3	Rivers, W.H.R. Social organization. London, 1924.
	An 2109.29.11	Rivers, W.H.R. Social organization. N.Y., 1929.
	Geog 4308.42.10	Rives, J.P.W. Tales and souvenirs of a residence in Europe. Philadelphia, 1842.
	Geog 194.2	Rivista di geografia didattica. Firenze. 1-13,1917-1933 3v.
	Geog 194.1.6	Rivista geografia italiana. Firenze. Reprint ed. Amsterdam, 1970. 8v.
	Geog 194.1	Rivista geografica italiana. Roma. 1,1894+ 37v.
	Geog 4329.63	The road to Athens. (Morpurgo, J.E.) London, 1963.
	Geog 4209.64.15	Roadles area. 1. ed. (Brooks, P.) N.Y., 1964.
	Geog 5311.3.25	Roald Amundsen. (Malfatti, Alberto.) Roma, 1959.
	Geog 5311.3.20	Roald Amundsen siste fere. (Hovdenak, Gunnar.) Oslo, 1934.
	Geog 5311.3.5	Roald Amundsen som hon var. 4. Oppl. (Arnesen, Odd.) Oslo, 1946.
	Geog 5519.03.1	Roald Amundsen's "The North West Passage". (Amundsen, R.) N.Y., 1908. 2v.
	Geog 4181.104	The Roanoke voyages, 1584-1590. (Quinn, D.B.) London, 1955. 2v.
	Geog 4559.29	The roaring forties and after. (Munro, D.J.) London, 1929.
	An 359.44.5A	Roback, A.A. A dictionary of the international slurs. Cambridge, Mass., 1944.
	Geog 4509.31	Robbins, L.H. Mountains and men. N.Y., 1931.
	Geog 3017.55	Robert de Vaugondy, D. Essai sur l'histoire de géographie. Paris, 1755.
	Geog 5326.21.15	Robert Edwin Peary, a record of his explorations, 1886-1909. (Hayes, J.G.) London, 1929.
	An 136.25	Robert H. Lowie. (Murphy, Robert Francis.) N.Y., 1972.
	An 137.3.5	Robert Ranulph Marett. (Rose, H.J.) London, 1944?

Author and Title Listing

	An 3308.78.2	Roberts, C. A manual of anthropometry. London, 1878. 2 pam.
	Geog 3019.06	Roberts, C.G.D. Discoveries and explorations in the century. Toronto, 1906.
	Geog 4208.29.5	Roberts, Jane. Two years at sea. London, 1834.
	Geog 4208.29.7	Roberts, Jane. Two years at sea. 2. ed. London, 1837.
	Geog 4329.09.5	Roberts, William C. The boys' account of it...foreign travel. N.Y., 1909.
	Geog 4679.73	Robertson, Dougal. Survive the savage sea. London, 1973.
	Geog 4181.98A	Robertson, George. The discovery of Tahiti. London, 1948.
	Geog 3520.16	Robinson, Adrian. Marine cartography in Britain. Leicester, 1962.
	Geog 5318.2.5	Robinson, Bradley. Dark companion. N.Y., 1947.
	Geog 5559.47	Robinson, Bradley. Dark companion. 1. ed. N.Y., 1947.
	Geog 4559.71.10	Robinson, Cyril. Men against the sea. Windsor, N.S., 1971.
	Geog 4308.90	Robinson, L.B. A bundle of letters from over the sea. Boston, 1890.
	Geog 4249.46	Robson, R.W. Where the trade-winds blow. Sydney, 1946.
	Geog 4308.35	Roby, John. Seven wekks in Belgium, Switzerland, Lombardy. London, 1838. 2v.
	An 3309.13	Rocca, Pierre. Les corses devant l'anthropologie. Paris, 1913.
	Geog 4328.42	Rockwell, C. Sketches of foreign travel. Boston, 1842. 2v.
	Geog 5160.16	Rodahl, Kåre. North. 1. ed. N.Y., 1953.
	Geog 5839.46	Rodahl, Karl. The ice capped island: Greenland. London, 1946.
	Geog 4209.68.25	Rodin, Leonid Efimovich. Po iuzhnym stranam. Moskva, 1968.
	An 2109.64.15	Rodovoi stroi i pervobytnaia mifologiia. (Zolotarev, Aleksandr M.) Moskva, 1964.
	Geog 5925.55	Rodriguez, Juan C. La Republica Argentina y las adquisiciones territoriales en el continente antartico. Buenos Aires, 1941.
	Geog 3055.58	Rodriguez Prampolini, Ida. La Atlántida de Platón en los cronistas del siglo XVI. Mexico, 1947.
	Geog 4181.1	Roe, T. Embassy of Sir T. Roe to court of great mogul. London, 1899. 2v.
	Geog 4238.32.10	Roebling, John A. Diary of my journey from Muehlhausen in Thurangia via Bremen to the United States...1831. Trenton, 1931.
	An 2108.84.5F	Römer, F. The bone caves of Ojcow in Poland. London, 1884.
Htn	Geog 5375.72*	Röslin, H. Mitternächtige Schiffarth von den Herrn Staden. Oppenheim, 1611.
	Geog 4182.10	Rogerius, Abraham. De open-deure tot het verborgen heydendom. 's-Gravenhagen, 1915.
	Geog 4419.34.5	Rogers, C.K. (Mrs.). Journal letters from the Orient. Boston, 1934.
	Geog 3570.34	Rogers, F.M. Valentim Fernandes. Lisboa, 1957?
	Geog 4559.29.22	Rogers, S.R.H. Sea lore. London, 1934.
	Geog 4129.21	Roget, S.R. Travel in the two last centuries of three generations. N.Y., 1921.
	VGeog 4306.00	Rohan de. Voyage...en l'an 1600. Amsterdam, 1646.
	An 2109.34.3A	Róheim, Géza. The riddle of the sphinx, or Human origins. London, 1934.
	Geog 3229.39	Rohr, Heinz. Die Entwicklung des Kartenbildes. Inaug. Diss. Borna, 1939.
	Geog 4209.15.5	Rohrbach, P. Weltpolitisches Wanderbuch 1897-1915. Königstein, 1916.
	Geog 4209.15	Rohrbach, P. Weltpolitisches Wanderbuch 1897-1915. Königstein, 1916.
	Geog 4502.5.2	Rohrer, M. Berglieder der Völker. München, 1928.
	Geog 4502.5.22	Rohrer, M. Der Feuerberg. München, 19- .
	Geog 3055.31	Roisel, G. de. Les Atlantes. Paris, 1874.
	An 2109.64.20	Le role des croyances magiques et religieuses dans les économies primitives. (Forcieri, Luigi.) Bordeaux, 1964.
	Geog 13.2.10	The role of geography in the modern world. (American Geographical Society of New York.) N.Y., 195-.
	An 2109.45.1	The role of the aged in primitive society. (Simmons, L.W.) Hamden, Ct., 1970.
	An 2109.45	The role of the aged in primitive society. (Simmons, L.W.) New Haven, 1945.
	An 379.33A	The rôle of the deserts. (McIverny, A.J.) London, 1933.
	An 359.42.15	The role of the races in our future civilization. (Laidler, H.W.) N.Y., 1942.
	Geog 4209.38.5A	The rolling world. (Hallet, R.M.) Boston, 1938.
	Geog 4219.40.10	Roman einer Weltreise. (Michael, Rudolf.) Hamburg, 1940.
	Geog 4209.07.5	The romance of an old time shipmaster. (Russell, J.W.) N.Y., 1907.
	Geog 4209.05.5	The romance of modern exploration. (Williams, Archibald.) Philadelphia, 1905.
	Geog 4309.01	The romance of religion. (Vivian, O.W.H. (Mrs.).) N.Y., 1901.
	Geog 4208.54.15	Romance of travel...by an old traveler. Philadelphia, 1854.
	Geog 4558.48	Romance of yachting. (Hart, J.C.) N.Y., 1848.
	An 2056.56.5	Romano, E. Animad versiones in librum prae-adamitarum. Paris, 1656.
	Geog 4169.01	Romantic castles and palaces, as seen and described by famous writers. (Singleton, Lesther.) N.Y., 1901.
	Geog 4239.70.10	The romantic challenge. (Chichester, Francis Charles.) London, 1971.
	Geog 4145.14.100	The romantic journey. (Swinglehurst, Edmund.) London, 1974.
	An 629.61.5	The romantic syndrome. (Jones, W.T.) The Hague, 1961.
Htn	Geog 3205.95*	Romanus, A. Parvum theatrum urbium. Frankoforti, 1595.
	Geog 4502.5.4	Rombert, E. Das Murmeltier mit dem Halsband. München, 1929.
	Geog 500.9	Rome, Italy. Bibliotheca Collegio Romano. Catalogo ragionato. Roma, 1876.
	Geog 195.3	Rome, Italy. Università. Istituto di Geografia. Pubblicazioni. Serie B (Geostorica). Roma. 1,1969+
	Geog 195.4	Rome, Italy. Università. Istituto di Geografia. Pubblicazioni. Serie C (Miscellanea). Roma. 2,1969+
	Geog 1540.50	Rome and central Italy. 16. ed. (Baedeker, publishers.) N.Y., 1930.
	Geog 618.4	Romer, E. Wybór prac. Warszawa, 1960. 3v.
Htn	Geog 4209.19.5*	Romero de Terreros, Juan. Apuntaciones de viaje en 1849. México, 1919.
	Geog 4332.14	Roncagli, Giovanni. Il problema militaire dell'Adriatico spiegatto a tutti. Roma, 1918.
	Geog 4509.41.10	Rondet, C. Montagne: des terrasses aux arêtes. Paris, 1941.
	Geog 4357.81	Rooke, Henry. Travels to the coast of Arabia Felix, and from thence by the Red Sea and Egypt to Europe. London, 1783.
	Geog 4029.16	Root, Jonathan. Halliburton, the magnificient myth; a biography. N.Y., 1965.
	Geog 4208.64.5	Root, Sidney. Exotic leaves, gathered by a wanderer. London, 1865.
	An 2109.71.1	The roots of civilization. 1st ed. (Marshack, Alexander.) N.Y., 1971.
	Geog 4218.16	Roquefeuil, C. de. Voyage autour du monde. Paris, 1823. 2v.
	An 137.3.5	Rose, H.J. Robert Ranulph Marett. London, 1944?
	Geog 4208.38.3	Rose, W.G. Three months' leave, or Military reminiscences. London, 1838.
	Geog 4238.47	Rosenbaum, S.E. A voyage to America 90 years ago...Hamburg to New York in 1847. N.Y., 1939.
	Geog 4209.51.5	Rosenberg, Halger. Jorden rundt med Halger Rosenberg. København, 1951.
	Geog 3019.35.10	Rosh, J.H. Man and the sea. Cambridge, Eng., 1935.
	Geog 3019.59	Roshidenie karty. (Fradkin, N.G.) Moskva, 1959.
	Geog 5639.46	Rosing, Christian. Ostgronlaenderne tunuamuit. København, 1946.
	Geog 5898.241	Rosing, Jens. Isimardik. København, 1960.
	Geog 3055.49	Rosmy, Léon de. L'Atlantide historique. Paris, 1902.
	Geog 5160.14	Ross, Colin. Mit Kind und Kegel in die Arktis. Leipzig, 1934.
	Geog 6008.38.5	Ross, Frank E. The Antarctic explorations of Charles Wilkes, 1838-42. n.p., 1935.
	Geog 4349.23	Ross, Halford. By devious ways. London, 1927.
	Geog 5518.29.2	Ross, J. Appendix to Narrative of second voyage. London, 1835.
	Geog 5518.29.3	Ross, J. Narrative of a second voyage in search of a northwest passage. Philadelphia, 1835.
	Geog 5518.29	Ross, J. Narrative of 2nd voyage...Northwest Passage. London, 1835.
	Geog 5518.18	Ross, J. Voyage of discovery...Baffin's Bay. London, 1819.
	Geog 5518.18.3	Ross, J. Voyage vers le Pole Arctique. Paris, 1819.
	Geog 6008.39	Ross, J.C. Voyage...southern and Antarctic regions. London, 1847. 2v.
	Geog 4300.44	Ross, Janet. Budgetouring Europe. N.Y., 1933.
	Geog 6009.14.35	The Ross Sea Shore Party, 1914-17. (Richards, R.W.) Cambridge, 1962.
	Geog 5558.81.9F	Rosse, I.C. Cruise of revenue-steamer Corwin. Washington, 1883.
Htn	Geog 578.71*A	Rosser, William H. The Bijou gazetteer of the world. London, 1871.
	Geog 4209.71.20	Rossiiskogo kuptsa Grigoriia Shelekhova stranstvoraniia iz Okholska po Vostochnomu okeanu k amerikanskiam beregam. (Shelekhov, Grigorii I.) Khabarovsk, 1971.
	Geog 5399.31.5	Die rote Arktis. (Sieburg, F.) Frankfurt, 1932.
	Geog 3570.31	Roteiros portugueses ineditos. (Fontoura da Costa, Abel.) Lisboa, 1940.
	Geog 4328.94.7PF	Rothschild, N. Skizzer aus dem Süden. Wien, 1894.
	An 359.05.3	Rotzell, W.E. Man: an introduction to anthropology. 2. ed. Philadelphia, 1905.
	Geog 5209.23.3	Rouch, J. Le pôle nord. Paris, 1923.
	Geog 5925.15	Rouch, J. Le pôle sud. Paris, 1921.
	Geog 4182.7	Rouffaer, G.P. De eerste schipvaart der nederlanders naar Ost-Indie. 's-Gravenhage, 1915-29. 3v.
	Geog 5209.07	Round about the North Pole. (Gordon, W.J.) N.Y., 1907.
	Geog 4208.52.7	Round Cape Horn...in 1852. (Lamson, Joseph.) Bangor, 1878.
	Geog 4218.87.7	Round the globe. 3. ed. (M'Collester, S.H.) Boston, 1890.
	Geog 4559.02	Round the Horn before the mast. (Lubbock, A.B.) N.Y., 1902.
	Geog 4209.15.10	Round the Horn before the most. (Lubbock, A.B.) London, 1915.
	Geog 4209.32.25	Round the Horn in a square rigger. (Johnson, Irving.) Springfield, 1932.
	Geog 4208.78.9	Round the north hemisphere. (Browne, Ernest A.) London, 1880.
	Geog 4418.72	"Round the world", letters from Japan, China, India, and Egypt. (Fogg, W.P.) Cleveland, 1872.
	Geog 4218.79.10A	Round the world. (Carnegie, Andrew.) Garden City, 1933.
	Geog 4218.68	Round the world. (Smiles, S.) N.Y., 1871.
	Geog 4218.96	Round the world on a wheel. (Fraser, J.F.) N.Y., 1899.
	Geog 4219.27	Round the world with a dictaphone. (Lunn, Henry.) London, 1927.
	Geog 4219.27.2	Round the world with a dictaphone. (Lunn, Henry.) London, 1928.
	Geog 4219.27.5	Round the world with a psychologist. (Martin, Lillien J.) San Francisco, 1927.
	Geog 4218.92	Round the world with the Irish delegates. (Esmonde, T.H.G.) Dublin, 1892.
	Geog 4308.84.7	A roundabout journey. (Warner, C.D.) Boston, 1884.
	Geog 3055.70	Rousseau-Liessens, A. Les colonnes d'Hercule et l'Atlantide. Bruxelles, 1955.
	Geog 4209.39.25	La route aux aventures. (Larigaudie, Guy de.) Paris, 1948.
	Geog 3030.3F	La route de l'Inde. (Guénin, E.) Paris, 1903.
	Geog 4427.99	Route de l'Inde. (Henry, Pierre.) Paris, 1798-99.
	Geog 4349.36.3	La route des Indes. (Morand, Paul.) Paris, 1936.
	Geog 659.57	Route-mapping and position locating in unexplored regions. (Filchner, W.) Basel, 1957.
	Geog 3229.61	Roux, Jean Paul. Les explorateurs au Moyen Âge. Paris, 1961.
	Geog 4209.21	Roving east and roving west. (Lucas, E.V.) London, 1921.
	Geog 4209.21.2	Roving east and roving west. (Lucas, E.V.) N.Y., 1921.
	Geog 4308.54.10	The roving Englishman. (Murray, E.C.G.) London, 1854.
	Geog 4248.37	Rovings in the Pacific from 1837 to 1849. London, 1851.
	Geog 4208.58	Rovings on land and sea. (Davenport, H.E.) Boston, 1858.
	An 2109.72.15	Rovse, Irving. Introduction to prehistory; a systematic approach. N.Y., 1972.
	Geog 4219.66.10	Rowland, John. Lone adventurer: the story of Sir Francis Chichester. London, 1968.
	Geog 619.6	Rowley, V.M. J. Russell Smith, geographer, educator and conservationist. Philadelphia, 1964.
	Geog 4209.60.30	Roy, Claude. Le journal des voyages. Paris, 1960.
	An 359.20.7	Roy, Sarat Chandra. Principles and methods of physical anthropology. Patna, 1920.
	An 359.60.5	Royal Anthropological Institute of Great Britain and Ireland. Man, race and Darwin. London, 1960.
	An 170.328.8	Royal Anthropological Institute of Great Britain and Ireland. Notes and queries on anthropology. 6. ed. London, 1971.
	An 5.9	Royal Empire Society, London. Library. Select bibliography of recent publications in the library. London, 1926.
	Geog 139.8	Pamphlet vol. Royal Geographical Society.
	Geog 21.1	Royal Geographical Society of Australasia. Proceedings and transactions. Queensland. 1-62,1885-1964 9v.

Author and Title Listing

Geog 21.1.40 — Royal Geographical Society of Australasia. Victorian Branch. Transactions. Melbourne. 15-19,1898-1901

Geog 4309.01.5 — A royal journey. (Dudley, L.B. (Mrs.).) N.Y., 1901.

Geog 4219.25.15 — The royal road to romance. (Halliburton, Richard.) Garden City, 1925.

Geog 4219.25.10A — The royal road to romance. (Halliburton, Richard.) Indianapolis, 1925.

Geog 6009.55.5F — The Royal Society International Geophysical Year. (Royal Society of London.) London, 1960. 4v.

Geog 6009.55.5F — Royal Society of London. The Royal Society International Geophysical Year. London, 1960. 4v.

Geog 3070.166 — Rubezh oikumeny. (Ditmar, Andrei B.) Moskva, 1973.

VGeog 4349.68 — Rubiezè woluorci. (Bregman, Aleksander.) Londyn, 1968.

An 132.2 — Rubín de la Borbolla, D. Bibliografía del Dr. Ales Hrdlicka. México, 193-.

Geog 3018.83 — Rubiner, W. Die Entdeckungsreisen. Glogau, 1883.

Geog 4209.70.15 — Rubissow, Helen. Ozni na dorogakh chetyrekh chastei sveta. Parizh, 1970.

Geog 4181.4 — Rubruquis, G. de. Journey of William of Rubruck...and John of Pian de Carpine. London, 1900.

Geog 5110.3 — Rud, William. The phantom of the poles. N.Y., 1906.

Geog 4502.5.18 — Rudatis, D. Das Letzte im Fels. München, 1936.

Geog 4268.85 — Rudler, F.W. Europe. London, 1885.

An 99.58.5 — Rudolph, Wolfgang. Das Problem der kulturellen Werte in den Arbeiten der neueren amerikanischen Ethnologie. Berlin, 1958.

Geog 3018.88 — Ruge, S. Abhandlungen...zur Geschichte der Erdkunde. Dresden, 1888.

Geog 3140.5 — Ruge, S. Der Chaldäer Seleukos. Dresden, 1865.

Geog 3248.81 — Ruge, Sophus. Geschichte des Zeitalters der Entdeckungen. Berlin, 1881.

Geog 4679.64 — Ruhen, Olaf. Minerva Reef. 1. American ed. Boston, 1964.

Geog 4328.94A — Rulers of the Mediterranean. (Davis, R.H.) N.Y., 1894.

An 2108.94 — The ruling races of prehistoric times. (Hewitt, J.F.) Westminster, 1894-95. 2v.

Geog 3580.25A — Rumbos oceanicos. (Vicens Vives, J.) Barcelona, 1946.

Htn Geog 4308.60* — A run through Europe. (Benedict, E.C.) N.Y., 1860.

Geog 4219.21 — Rund um die Erde. (Büchler, E.) Bern, 1921.

Geog 4180.8 — Rundall, T. Memorials of the empire of Japan. N.Y., 1964?

Geog 4180.5 — Rundall, T. Narratives of voyages...north-west. London, 1849.

Geog 4219.29.2 — Rundherm. 3.-4. Aufl. (Mann, Erik.) Berlin, 1929.

Geog 4308.50.9 — Running sketches of men and places. (Conway, George.) N.Y., 1851.

Geog 4559.67.10 — Runt Kap Horn med Herzogin Cecilie. (Heikkinen, Helge.) Ekenäs, 1967.

Geog 4131.70 — Ruppert, Karl. Zur Geographie des Freizeitverhaltens. Kallmünz, 1970.

Geog 4218.35.3 — Ruschenberger, W.S.W. A voyage round the world. Philadelphia, 1838.

An 3209.01 — Russell, Frank. Laboratory outlines in somatology. N.Y., 1901.

Geog 4209.07.5 — Russell, J.W. The romance of an old time shipmaster. N.Y., 1907.

Geog 4308.18.6 — Russell, Jonathan. Journal, 1818-1819. Boston, 1918.

An 379.67.5 — Russell, William Moy Stratton. Man, nature and history. London, 1967.

Geog 4418.64 — Russell-Killough, H. Seize mille lieues a travers l'Asie et l'Oceanie. Paris, 1864.

Geog 1575.15A — Russia. (Baedeker, publishers.) N.Y., 1914.

Geog 4678.30 — Russia and America. v.1-2. (Burrows, Silas E.) Hartford, 1865.

Geog 4180.20 — Russia at close of sixteenth century. (Fletcher, G.) N.Y., 1964?

Geog 4181.138 — Russian embassies to the Georgian kings, 1589-1605. (Allen, William Edward David.) Cambridge, 1970. 2v.

Geog 3590.5 — Russian voyages round the world. (Nozikov, N.) London, 1947.

Geog 5300.45 — The Russians in the Arctic. (Armstrong, Terence.) London, 1958.

Geog 1575.5 — La Russie. 2. éd. (Baedeker, publishers.) Leipzig, 1897.

Geog 3590.73 — Russkie geograficheskie issledovaniia Evropeiskoi Rossii i Vrala v XIX-nachale XX v. (Esakov, V.A.) Moskva, 1964.

Geog 3590.80 — Russkie geograficheskie otkrytiia i issledovaniia s drevnikh vremen do 1917 goda. (Lebedev, Dmitrii M.) Moskva, 1971.

Geog 3590.51 — Russkie geografii i puteshestvenniki. (Moscow. Gosudarstvennyi Biblioteka SSSR Imeni V.I. Lenin.) Moskva, 1955.

Geog 600.22 — Russkie geografy i puteshestvenniki. (Matveeva, T.P.) Leningrad, 1971.

Geog 3590.5.5 — Russkie krugosvetiye moreplavateli. Izd. 2. (Nozikov, N.) Moskva, 1947.

Geog 3590.45 — Russkie moraplavateli. (Lupach, V.S.) Moskva, 1953.

Geog 3590.79 — Russkie moreplavaniia na Tikhom okeane r XVIII neke. (Divin, Vasilii A.) Moskva, 1971.

Geog 3590.10 — Russkie moreplavateli, arkticheckie krulosvetnye. (Lialina, M.A.) Sankt Peterburg, 1892.

Geog 5970.44.5 — Russkie otkrytiia v Antarktike w 1819-1820-1821 godakh. (Sementovskii, Vladimir N.) Moskva, 1951.

Geog 3590.49 — Russkie puteshestvenniki po Afrike. (Zabrodskaia, M.R.) Moskva, 1955.

Geog 194.5 — Russkoe Obshchestvo Parokhodstva i Torgovli. Putevoditel' po Krymu, Kavkazu, i Blizhnemu Vostoku. Odessa. 1913

Geog 1575.10 — Russland. Handbuch für Reisende. 5. Aufl. (Baedeker, publishers.) Leipzig, 1901.

An 2339.65 — Rust, Alfred. Über Waffen- und Werkzeugtechnik des Altmenschen. Neumünster, 1965.

Geog 4209.42.25 — Rutas del mundo. (Vejarano, J.R.) Bogota, 1942.

An 126.6 — Ruth Fulton Benedict; a memorial. (Viking Fund.) N.Y., 1949.

Geog 4559.56.5 — Rutzebeck, Hjalmar. Mad sea. N.Y., 1956.

Geog 4182.62 — Ruyter, M.A. De reis van Michiel Adriaanszoom de Ruyter in 1664-1665. 's-Gravenhage, 1961.

Geog 5399.57.10 — Ruzov, L.V. Na sushe i na more v Arktike. Moskva, 1957.

Geog 5839.30.20 — Det ry Grønland. pt.3-5. (Foreniger Det ry Grønland.) København, 1930.

Geog 4329.28 — Rydh, Hanna. Kring medelhavets stränder. Stockholm, 1928.

Geog 3140.11F — Rylands, T.G. Geography of Ptolemy elucidated. Dublin, 1893.

Geog 6009.34 — Rymill, John. Southern lights; the official account of the British Graham Land Expedition, 1934-37. London, 1938.

Geog 5399.01 — S.O. Makarov v zavosvanie Ariuku. (Makarov, S.O.) Moskva, 1943.

Geog 4659.36.10 — S.O.S., a book of sea adventure. (Masters, David.) N.Y., 1936.

Geog 4209.66 — S planetoi vmestve. (Mikhailov, Nikolai N.) Moskva, 1966.

Geog 3570.33 — Sa, Ayres de. Frei Goncalo Velho. Lisboa, 1899-1900. 2v.

Geog 3129.36 — Saa, Mario. Evudania. Lisboa, 1936.

Geog 5837.78.5 — Saabye, Hans E. Fragmentes af en dagbok hållen i Grönland. Stockholm, 1817.

Geog 5837.78 — Saabye, Hans E. Greenland...extracts...journal. London, 1818.

Geog 197.10 — Saarbruecken. Universität des Saarlandes. Geographisches Institut. Arbeiten. 1,1956+ 4v.

Geog 5518.18.4 — Sabine, Edward. Remarks on the account of the late voyage of discovery to Baffin's Bay. 2. ed. London, 1819.

Geog 4303.99.3 — Sachet, E. Examen critique des voyages et ambassades de G. de Lannoy, 1399-1450. Bruxelles, 1843.

Geog 4145.79 — Sack, John. Report from practically nowhere. N.Y., 1959.

Geog 3060.12.55 — The sacred symbols of Mu. (Churchward, J.) London, 1960.

Geog 3060.12.45A — The sacred symbols of Mu. (Churchward, J.) N.Y., 1934.

Geog 4306.91.1 — Der sächsische Robinson. (Retcher, Wilhelm.) Leipzig, 1722.

Geog 3618.3 — Saellskapit för Finlands Geografi, Helsingfors. Exposé des travaux géographiques executés en Finlande jusqu'en 1895. Helsingfors, 1895.

Htn Geog 4266.41.8F* — Sämtliche...Topographias. Haupt-Register. (Zeiller, Martin.) Frankfurt, 1726.

Htn Geog 4266.41.7F* — Sämtliche...Topographias. v.1-31. (Zeiller, Martin.) Frankfurt, 1677-1736. 10v.

Geog 4209.39.10 — The saga of "Cimba". (Maury, Richard.) N.Y., 1939.

Geog 5324.2.10 — The saga of Fridtjof Nansen. (Sörensen, Jon.) N.Y., 1932.

Geog 6009.33.10 — Saga of the white horizon. (Olsen, Magnus L.) Lymington, 1972.

Geog 5050.7 — Saggio di bibliografia polare. 2. ed. (Zavatti, S.) Roma, 1952.

An 5.2 — Saggio di un catalogo bibliografico antropologico italiano. (Riccardi, P.) Modena, 1883.

Geog 4307.39F — Sagramoso, M.E. Lettera...al E. Ignazio Zanardi di Mantova. Verona, 1877.

An 2109.72 — Sahlins, Marshall David. Stone age economics. Chicago, 1972.

Geog 4309.26.20 — Sailing across Europe. (Farson, N.) London, 1928.

Geog 4219.00.10 — Sailing alone around the world. (Slocum, Joshua.) London, 1948.

Geog 4219.00 — Sailing alone around the world. (Slocum, Joshua.) N.Y., 1900.

Geog 4219.00.5 — Sailing alone around the world. (Slocum, Joshua.) N.Y., 1900.

Geog 4559.31 — Sailing the seas. (Chatterton, E.K.) London, 1931.

Geog 4239.52 — Sailing to freedom. (Veedam, V.) N.Y., 1952.

Geog 4558.61.5 — A sailor boy's experience. (Stevens, Charles.) Napanee, Ont., 1892.

Geog 4209.34 — A sailor's log of facts: not fables. (Coyne, William M.) Boston, 1934.

Geog 4559.67.15A — Sailortown. (Hugill, Stan.) London, 1967.

Geog 4559.54.20 — Sails beneath the Southern Cross. (Eaddy, P.A.) Wellington, 1954.

Geog 3019.38.3 — St. Croix de la Roncière, G. À la conquête des mers. Paris, 1938.

Geog 4349.30 — St.-Félix, Max de. A travers l'Orient (1930). Paris, 1931.

Geog 3506.5 — Saint-Genois, Jules de. Les voyageurs belges. v.1-2. Bruxelles, 1846-47?

Geog 4138.37 — St. John, J.A. Lives of celebrated travellers. N.Y., 1837. 3v.

Geog 3055.65 — Saint-Michel, Léonard. Aux sources de l'Atlantide. Bourges, 1953.

Geog 818.43 — St. Platou, L. Stutt landaskipunarfraedi. n.p., 1843.

Geog 4209.64.25 — Sakhnin, Arkadii Ia. Vot liudi. Moskva, 1964.

Htn Geog 4308.67.5* — Sala, George A. From Waterloo to the Peninsula. London, 1867. 2v.

Htn Geog 4308.85.7* — Sala, George A. A journey due south, travels. London, 1885.

Geog 4512.767 — Salacrow, Armand. A pied, au-dessus des nuages. Paris, 1956.

Geog 4180.44 — Salîl-Ibn-Razîk. History of the Inâms...of Omân. N.Y., 1964?

Geog 3070.155 — Salinari Emiliani, M. Nozioni di cartografia. Roma, 1959.

Geog 819.13.5 — Salisbury, Rollin D. Modern geography for high schools. N.Y., 1913.

Geog 3070.45 — Salishahev, K.A. Osnovy kartovesheniia. Izd. 2. Moskva, 1943.

Geog 817.51.9 — Salmon, T. Geographical and astronomical grammar. London, 1785.

Geog 577.58 — Salmon, T. Modern gazeteer. London, 1758.

Geog 577.58.2 — Salmon, T. Modern gazeteer. London, 1759.

Geog 577.58.4 — Salmon, T. Modern gazeteer. London, 176-.

Geog 577.58.6 — Salmon, T. Modern gazeteer. London, 1782.

Geog 817.51.2 — Salmon, T. A new geographical...grammar. London, 1749.

Geog 817.51.3 — Salmon, T. New geographical...grammar. London, 1751.

Geog 817.51.5 — Salmon, T. New geographical...grammar. London, 1764.

Geog 817.51.7 — Salmon, T. New geographical...grammar. London, 1766.

Geog 817.51.8 — Salmon, T. New geographical...grammar. London, 1769.

Geog 807.39.5F — Salmon, Thomas. Modern history. 3rd ed. London, 1744-1746. 3v.

NEDL Geog 807.40 — Salmon, Thomas. Lo stato presente di tutti i paesi e popoli del mondo naturale, politico e morale. 2a ed. v.1-26. Venezia, 1738-66. 27v.

An 126.10 — Salter, Elizabeth. Daisy Bates; "the great white queen of the never never". Sydney, 1971.

Geog 4209.44.5 — Saltwater tramp. (Armstrong, W.) London, 1944.

Geog 4269.62.5 — Salvadori, M. Western ports in Europe. N.Y., 1962.

Htn Geog 4308.08.2* — Salvo, C. Travels in...1806 from Italy to England. Troy, N.Y., 1808.

Geog 4308.08 — Salvo, C. Travels in the year 1806, from Italy to England. Troy, N.Y., 1808.

Geog 3019.55.5 — Samhaber, Ernst. Knaurs Geschichte der Entdeckungsreisen. München, 1955.

Geog 3070.6 — Sammlung mittelalterlicher Welt. (Fischer, T.) Venedig, 1886.

Geog 5209.31 — Samoilovich, R. Der Weg nach dem Pol. Biehefeld, 1931.

VGeog 4209.69.5 — Samotno potovanje. (Karlin, Alma M.) Ljubljana, 1969.

Geog 5925.55.10 — Sampay, Arturo E. La soberania argentina sobre la Antártida. La Plata, 1950.

Geog 4308.58.5 — Samper, J.M. Viajes de un Colombiano en Europa. Paris, 1862. 2v.

Geog 4557.95 — Samuel Kelly, an 18th century seaman. (Kelly, Samuel.) N.Y., 1925.

Geog 4206.13.12 — Samuel Purchas pelgrimagir. (Purchas, Samuel.) Amsterdam, 1655.

Geog 4558.87A — Samuels, S. From the forecastle to the cabin. N.Y., 1887.

Geog 4309.35.15 — The San Luca. (Ofaire, Cilette.) N.Y., 1935.

Author and Title Listing

An 379.38.5 Sanchez, Pedro C. Enseñanzas fundamentales de la geografía humana. Tacubaya, 1938.

Geog 3019.35.25 Sanchez, Pedro C. Evolución de la geografía. México, 1935.

Geog 4307.76 Sander, Heinrich. Beschreibung seiner Reisen durch Frankreich. Pt.1. Leipzig, 1783.

An 2109.53 Sanders, I.T. Societies around the world. N.Y., 1953. 2v.

An 2109.52.5 Sanders, I.T. Societies around the world. v.4. Lexington, 1952.

An 5.30 Sándor, István. A Magyar néprajztudomány bibliográfiája 1945-1954. Budapest, 1965.

Geog 5839.71 Sandstroem, Lennart. Leva på Grönland. Stockholm, 1971.

Geog 4249.67.10 Sandvad, Jørgen. Rejsernes rytme. Kronikker. København, 1967.

Geog 135.2.3 Der Sansibar-Archipel. v.3. (Baumann, Oskar.) Leipzig, 1896.

Geog 4309.64.31 Sansom, William. Away to it all. N.Y., 1966.

Htn Geog 817.00* Sanson, N. Description de tout universe. Amsterdam, 1700.

Htn Geog 807.51F* Sanson, N. La France,...les Isles Britanniques. Paris, 1651.

Geog 816.93 Sanson, N. Introduction à la geographie. Paris, 1693.

Geog 819.18 Santa Cruz, Alonso de. Islario general de todas las isles del mundo. Atlas. Madrid, 1918. 2v.

Geog 665.29F Santarem, M.F. de B. Inéditos (miscellanea). Lisboa, 1914.

Geog 665.30F Santarem, M.F. de B. Opusculos e esparsos. Lisboa, 1910. 2v.

Geog 3235.47.9 Santarem. Examen des assertions contenues...des monuments de la géographie. n.p., n.d.

Geog 5160.57 Santillan de Andres, Selva E. La vida humana en el subecumene ártico. Tucuman, 1962.

Geog 4208.44.5 Santos, Eusebio. Diario del viaje desde Madrid a Manila. Madrid, 1851.

Geog 659.71.10 Santos, Milton. Le métier de géographe en pays sous-développé. Paris, 1971.

Geog 4709.42 Santschi, R.J. Treasure trails. Glen Ellyn, Ill., 1942.

Geog 3560.26 Sanz, Carlos. La huella de España en el mundo. Madrid, 1971-73. 3v.

Geog 197.5 Sao Paulo, Brazil. Universidade. Faculdade de Filosofia, Ciências e Letras. Geografia. 5-7

An 2058.68 Saporta, G. da. La paléontologie appliquée a l'étude des races humaines. n.p., 1868.

Geog 5208.57 Sargent, E. Arctic adventure by sea and land. Boston, 1857.

Geog 4300.12 Sargent, H.W. Skeleton tours. N.Y., 1870.

Geog 4300.12.2 Sargent, H.W. Skeleton tours. N.Y., 1870.

Geog 4300.12.3 Sargent, H.W. Skeleton tours. N.Y., 1871.

Geog 4181.5 Saris, J. Voyage of Captain John Saris to Japan, 1613. London, 1900.

An 3309.64 Sarkar, Sasanka Sekhar. Ancient races of Baluchistan. 1st ed. Calcutta, 1964.

An 629.70.5 Sarmela, Matti. Perinnerineiston kvantitatiivisesta tutkimuksesta. Helsinki, 1970.

Geog 4181.22A Sarmiento de Gamboa, Pedro. History of the Incas and execution of the Inca Tupac Amaru. Cambridge, 1907.

Geog 4180.91 Sarmiento de Gamboa, Pedro. Narrative of voyages of...Sarmiento. London, 1895.

Geog 4300.6 Satchel guide for the vacation tourist in Europe. N.Y. 1872-1929 27v.

Geog 4209.45 Satin skirts of commerce. (Caldwell, E.N.) N.Y., 1945.

An 170.111 The Saturday lectures. (Anthropological Society of Washington.) Washington, D.C., 1882.

Geog 665.115 Sauer, Carl Ortwin. Land and life. Berkeley, 1965.

Geog 4239.68.5 Sauer, Carl Ortwin. Northern mists. Berkeley, 1968.

Geog 4308.72.6 Saunterings. (Warner, Charles Dudley.) Boston, 1872.

Geog 4308.72.5 Saunterings. (Warner, Charles Dudley.) Boston, 1872.

Geog 4308.72.5.2 Saunterings. (Warner, Charles Dudley.) Boston, 1900.

Geog 4265.95.1 Saur, Abraham. Theatrum urbium. Unterschneidheim, 1971.

Geog 3055.66 Saurat, Denis. L'Atlantide et la règne des géants. Paris, 1954.

Geog 4502.5 Saussure, H.B. de. Relation abrégés d'un voyage a la cime du Mont-Blanc en aout 1787. Genève, 1928? 2 pam.

Geog 818.53 Savage, C.C. The world; geographical, historical, statistical. N.Y., 1853.

An 2109.62.6 The savage mind. (Lévi-Strauss, Claude.) Chicago, 1966.

An 2109.62.7 The savage mind. (Lévi-Strauss, Claude.) Chicago, 1968.

Geog 577.13F Savonarola, R. Universus terrarum orbis. Patavii, 1713. 2v.

An 359.11.3 Savorgnan, Franco. Gli indici di omogamia delle razze e delle nazionalità. Cagliari, 1911.

Geog 4359.40 Savoy! Corsica! Tunis! Mussolini's dream lands. (Newman, Bernard.) London, 1940.

Geog 4515.401 Saysse-Tobiczyk, K. W skałach i lodach swiata. Warszawa, 1959- 2v.

Geog 5985.187.5 Saziumov, Evgenii M. A life given to the Antarctic, Douglas Mawson - Antarctic explorer. Adelaide, 1968.

Geog 110.3.25 Sbornik na IV Kongress na slavianskite geografi i etnografi. (Congrès des Géographes et Ethnographes Slaves, 4th, Sofia, 1936.) Sofiia, 1938.

Geog 665.167 Sborník prací Geografických kateder UK k 75. narozeninám prof. dr. Jaromíra Korčáka, DrSc. (Prague. Universita Karlova.) Praha, 1970.

An 359.02.6 Sbornik" statei po arkheologii i etnografii. (Pogodin, A.L.) Sankt Peterburg, 1902.

Geog 665.84 Sbornik v chest na akademik Anastas Stoianov Beshkov. (Bulgarska Akademiia na Naukite, Sofia. Geografski Institut.) Sofiia, 1959.

An 175.2 Sbornik" v chest' semidesiatiletiia...D.N. Anuchina. Moskva, 1913.

Geog 665.98 Sbornik v chestna chlen-korespondent Iordan Zakhariev. (Bulgarska Akademiia na Naukite, Sofia. Geografski Institut.) Sofiia, 1964.

Geog 4328.57 Scampavias from Gibel Tarek to Stamboul. (Wise, H.A.) N.Y., 1857.

Geog 4348.67 A scamper to Sebastopol and Jerusalem 1867. (Creagh, J.) London, 1873.

An 136.5.45 Scardnelli, Pietro. L'analisi strutturali dei miti. Milano, 1971.

Geog 3550.16 Scarin, Maria Luisa. Viaggi ed esplorazioni di capitani marittimi della Riviera di Levante nella prima metà del secolo XIV. Genova, 1968.

Geog 4308.52.8 Scenes and thoughts in Europe. 1st-2d series. (Calvert, G.H.) Boston, 1863. 2v.

Htn Geog 4308.52.7* Scenes and thoughts in Europe. 2d series. (Calvert, G.H.) N.Y., 1852.

Geog 4348.48 Scenes and thoughts in foreign lands. (Terry, C.) London, 1848.

Geog 4758.24 Scènes de la nature sous les tropiques. (Denis, F.) Paris, 1824.

Htn Geog 4758.24.3* Scènes de la nature sous les tropiques. (Denis, F.) Paris, 1824.

Htn Geog 4758.24.2* Scènes de la nature sous les tropiques. (Denis, F.) Paris, 1824.

Geog 959.07 Scenes from every land. (Grosvenor, G.H.) Washington, D.C., 1907.

Geog 959.07.15 Scenes from every land. (Grosvenor, G.H.) Washington, D.C., 1918.

Geog 959.07.5 Scenes from every land. 2. ser. (Grosvenor, G.H.) Washington, D.C., 1909.

Geog 665.21 Scenes in foreign lands. N.Y., 1853.

Geog 4129.59 Schadendorf, Wulf. Zu Pferde, im Wagen, zu Fuss. München, 1959.

Geog 616.30 Die Schaffhauser Karten von Hauptmann Heinrich Peyer (1621-1690). (Wyder, Samuel.) Zürich, 1951.

Geog 4182.69 Schagen, Adriaen. Reijse gadaen bij Adriaen Schagen. 's-Gravenhage, 1968.

Geog 4208.97.7 Schanz, M. Ein Zug nach Osten. Hamburg, 1897. 2v.

Geog 4304.65 Schaschek. Des böhmischen Herrn Leo's von Rozmital...Reise...Abendlande 1465-1467. Stuttgart, 1844.

An 359.25.5 Scheidt, W. Allgemeine Rassenkunde. München, 1925.

An 2209.32 Scheitelnarbensitte, Anschwellungsglaube und Kulturkreislehre. (Paudler, F.) Brunn, 1932.

An 359.61.50 Scheler, Max. Man's place in nature. Boston, 1961.

An 359.28.7 Scheler, Max. Die Stellung des Menschen im Kosmos. Darmstadt, 1928.

An 359.49.20 Scheler, Max. Die Stellung des Menschen in Kosmos. München, 1949.

An 359.28.11 Schemann, L. Die Rasse in den Geisteswissenschaften. München, 1928-31. 3v.

Geog 3019.38.10 Schemi e notizie di storia delle esplorazione geografiche. 5. ed. (Toschi, Umberto.) Firenze, 1953?

Geog 807.03 Scherer, R.P.H. Atlas novus exhibens orbem. pt.1-7. Augustae, 1710. 4v.

Geog 4218.57 Scherzer, Karl. Narrative of circumnavigation of globe...1857-58. London, 1861-63. 3v.

Geog 4418.62 Schetsen eener mail-reize van Batavia naar Maastricht op reis en thuis. (Brumund, J.F.G.) Amsterdam, 1862.

Geog 665.102 Schickel, Joachim. Terra incognita. Bergisch Gladbach, 1965.

An 359.25.1.3 Schiller, F.C.S. Tantalus; or The future of man. London, 1924.

An 359.25.1 Schiller, F.C.S. Tantalus; or The future of man. N.Y., 1925.

An 359.25.2 Schiller, F.C.S. Tantalus; or The future of man. N.Y., 1925.

Geog 4678.55.20 Schilling, Nikolai. Seeoffizier des Zaren. Köln, 1971.

Geog 4180.58A Schiltberger, J. Bondage and travels of J. Schiltberger. London, 1879.

Geog 4145.80 Schindler, Fritz. Meine schönste Autoreise. Wien, 1958.

An 358.45.8F Schinz, Heinrich Rudolf. Naturgeschichte und Abbildungen des Menschen der verschiedenen Rassen. 3. Aufl. Zürich, 1848.

Geog 3108.37.2 Schirlitz, S.C. Handbuch der alten Geographie. Halle, 1837.

Geog 1525.70 Schleswig-Holstein und Hamburg. (Baedeker, publishers.) Hamburg, 1949.

Geog 5558.81.8F Schley, W.S. Report...Greely relief expedition of 1884. Washington, 1887.

Geog 5558.81.7A Schley, W.S. The rescue of Greely. N.Y., 1885.

Geog 4127.77.2 Schloezer, August Ludwig von. Vorlesungen über Land- und Seereisen. Göttingen, 1962.

Geog 4313.7 Schlump, E. Die politisch-geographische Bedeutung der Ostsee. Inaug.-Diss. Königsberg, 1934.

An 359.36.10 Schmelzle, K. Rassengeschichte und Vorgeschichte. 2. Aufl. Bamberg, 1936.

X Cg Geog 4502.5.15 Schmidkunz, W. Auf der Alm. Erfurt, 1934.

Geog 4502.5.6 Schmidkunz, W. Der Berg des Herzens. München, 1930.

Geog 4515.301 Schmidkunz, W. Kochbuch für Bergsteiger. München, 1925.

Geog 579.36A Schmidt, A.J. Kleines deutsch-portugiesisches...Verzeichnis geographischer Eigennamen. Rio de Janeiro, 1938.

An 359.26.10 Schmidt, Max. The primitive races of mankind. London, 1926.

An 359.24.20 Schmidt, Max. Völkerkunde. Berlin, 1924.

An 2109.36.3 Schmidt, R.R. The dawn of the human mind. London, 1936.

Geog 500.135 Schmidt, Rolf Dietrich. Verzeichnis der geographischen Zeitschriften, periodischen Veröffentlichungen und Schriftreihen Deutschlands. Bad Godesberg, 1964.

An 359.10 Schmidt, W. Die Stellung der Pygmäenvölker in der Entwicklungsgeschichte des Menschen. Stuttgart, 1910.

An 359.24.27 Schmidt, W. Völker und Kulturen. Regensburg, 1924.

An 359.39.15 Schmidt, Wilhelm. The culture historical method of ethnology. N.Y., 1939.

An 359.46.17A Schmidt, Wilhelm. Rassen und Völker. 3. Aufl. Luzern, 1946-49 3v

An 359.38.5 Schmidt, Wilhelm. Razza e nazione. n.p., 1938.

Geog 759.55 Schmieder, Oscar. Geografía del viejo mundo. México, 1955.

Geog 619.11 Schmieder, Oskar. Lebenserinnerungen und Tagebuchblätter eines Geographen. Kiel, 1972.

Geog 3019.70 Schmithuesen, Josef. Geschichte der geographischen Wissenschaft von den ersten Anfägen bis zur Ende des 18. Jahrhunderts. Mannheim, 1970.

Geog 4502.5.23 Schmitt, F. Mensch, Berg und Tod. München, 19- .

An 170.335 Schmitz, Carl August. Historische Völkerkunde. Frankfurt am Main, 1967.

An 2109.74 Schneider, Harold K. Economic man; the anthropology of economics. N.Y., 1974.

An 358.85 Schneider, W. Die Naturwölker. v.1-2. Paderborn, 1885-86.

Geog 4309.31.15 Schön ist die Welt. (Eddelbüttel, H.F.) Berlin, 1931.

Geog 6100.37 Scholes, A. Fourteen men. London, 1951.

Geog 4300.29 Schoonmaker, F. Through Europe on two dollars a day. N.Y., 1927.

Htn Geog 4476.71* Schooten, Henry (pseud.). The hairy-giants. London, 1671.

Geog 4216.19 Schouten, W.C. Relación diaria del viaje de J. Le Maire y G.C. Shouten. Santiago de Chile, 1897.

Geog 659.19 Schrader, Franz. The foundations of geography in the 20th century. Oxford, 1919.

An 3509.31.10 Schreiner, Kristian Emil. Zur Osteologie der Lappen. Oslo, 1931-35. 2v.

Geog 126.10 Schriften. (Kiel. Universität. Geographisches Institut.) Kiel. 1,1932+ 36v.

Geog 4328.46 Schroeder, F. Shores of the Mediterranean. N.Y., 1846. 2v.

Author and Title Listing

	Geog 4127.05	Schroeter, J.C. Diatriba...peregrinationum eruditarum. Jenae, 1705.
	Geog 4266.41.80	Schuchhard, C. Die Zeiller-Merianschen Topographien bibliographisch Beschrieben. Hamburg, 1960.
	Geog 4309.00.10	Schück, Henrik. Ur en resandes antecknigar. v.1-3. Stockholm, 1900-1909.
	An 359.26	Schütte, G. Vor folkegruppe Gottjod. Kjøbenhavn, 1926.
	An 359.29.3	Schütte, Gudmund. Our forefathers, the Gothonic nations. Cambridge, 1929-33. 2v.
	Geog 3540.15	Schulte-Althoff, Franz Josef. Studien zur politischen Wissenschaftsgeschichte der deutschen Geographie im Zeitalter des Imperialismus. Paderborn, 1971.
	Geog 619.7	Schultén, N.G. Levnadsteckning. Helsingfors, 1964.
	An 629.37	Schultz, Bruno K. Taschenbuch der rassenkundlichen Messtechnik. München, 1937.
	Geog 4313.6	Schulz, Bruno. Die deutsche Ostsee. Bielefeld, 1931.
	An 379.19	Schulz, George J. Geography. College Park, Md., 1919.
	An 2109.02	Schurtz, H. Altersklassen und Männerbünde. Berlin, 1902.
	An 358.93	Schurtz, H. Katechismus der Völkerkunde. Leipzig, 1893.
	An 2109.00.3	Schurtz, H. Urgeschichte der Kultur. Leipzig, 1900.
	An 359.03.7	Schurtz, H. Völkerkunde. Leipzig, 1903.
	Geog 4502.5.10	Schuster, K. Weise Berge - schwarze Zelte. München, 1932.
	Geog 4427.35	Schwartz, G.L. Utdrag af en Ost-Indisk rese-bescrifning. Westerås, 1784.
	Geog 4509.31.5	Schwartz, M. Et la montagne conquit l'homme. Paris, 1931.
	Geog 4509.24.10	Schwartz, M. Vers l'idéal par la montagne. Paris, 1924.
	Geog 3019.48	Schwarz, G. Die Entwicklung der geographischen Wissenschaft seit dem 18. Jahrhunderts. Berlin, 1948.
	Geog 4314.5	Das Schwarze Meer. Leipzig, 1854.
	Geog 1525.55.15	Schwarzwald. 3. Aufl. (Baedeker, publishers.) Leipzig, 1936.
	Geog 1525.55.16	Schwarzwald. 4. Aufl. (Baedeker, publishers.) Malente, 1956.
	Geog 5538.78.5	Schwatka's search...Franklin records. (Gilder, W.H.) N.Y., 1881.
	Geog 1568.18A	Schweden und Norwegen. 3. Aufl. (Baedeker, publishers.) Leipzig, 1885.
	Geog 3235.43	Schweder, E. Über die Weltkarte des Kosmographen. Kiel, 1886.
	Geog 4332.5	Schweiger-Lerchenfeld, A.F. Die Adria. Wien, 1883.
	Geog 4311.11	Schweiger-Lerchenfeld, Amand von. Die Donau als Völkeweg. Wien, 1896.
	Geog 4314.8	Schweiger-Lerchenfeld, Amand von. Zwischen Donau und Kaukasus. Wien, 1887.
	Geog 4208.98.5	Schweitzer, Georg. Eine Reise um die Welt. 2e Aufl. Berlin, 1899.
	Geog 1535.16	Die Schweiz, die italienischen Seen. 6. Aufl. (Baedeker, publishers.) Coblenz, 1856.
	Geog 1535.10	Die Schweiz. 5. Aufl. (Baedeker, publishers.) Coblenz, 1853.
	Geog 1535.13	Die Schweiz. 5. Aufl. (Baedeker, publishers.) Coblenz, 1854.
	Geog 1535.22	Die Schweiz. 12. Aufl. (Baedeker, publishers.) Coblenz, 1869.
	Geog 1535.23.9	Die Schweiz. 24. Aufl. (Baedeker, publishers.) Leipzig, 1891.
	Geog 4509.18	Schweizer Alpen Club. Le conseiller de l'ascensionniste. v.2. Genève, 1918.
	Geog 199.2	Schweizer Alpen-Zeitung. Zürich. 1-11 3v.
	Geog 199.1.3	Schweizer Alpenclub. Beilagen zum Jahrbuch. Bern. 1-46 46v.
	Geog 199.1.18	Schweizer Alpenclub. Die ersten 25 Jahre des Schweizer Alpenclub. Glarus, 1889.
NEDL	Geog 199.1	Schweizer Alpenclub. Jahrbuch. Bern. 1-35 23v.
	Geog 199.1	Schweizer Alpenclub. Jahrbuch. Bern. 36-58,1900-1923 23v.
	Geog 199.1.5	Schweizer Alpenclub. Repertorium und Ortsregister. Bern. 1-44,1886-1910 2v.
	Geog 199.1.4	Pamphlet vol. Schweizer Alpenclub. Clubhütten. Ergänzungsblätter.
NEDL	Geog 199.10	Schweizer Alpenclub. Section Genevoise. Bulletin. Genève.
	Geog 199.20	Der Schweizer Geograph; Zeitschrift des Vereins schweizerischer Geographialehrer sowie der geographischen Besellschaften von Basel, Bern, St. Gallen und Zürich. Bern. 1-22,1923-1945 4v.
	Geog 4131.59	Schweizerischer Fremdenverkehrsverband. Festschrift für Walter Hienzeker zum 60. Geburtstag. Bern, 1959.
	Geog 607.20	Schweizinscher Zofinguverein. Souvenir de l'inauguration du monument élevé à Arnold Guyot par la Société de Zofingue à l'Académie de Neuchâtel le 6 mai 1892. Neuchâtel, 1892.
	An 359.70.30	Schweppe, John S. Man: a remarkable animal. Chicago, 1970.
	Geog 500.170	Schwickerath, Hildegard. Inhaltsverzeichnis der Festschriften zur Ehrung und Würdigung deutscher. Bad Godesberg, 1969.
	An 359.50.5	Schwidetzky, Ilse. Grundzäge der Völkerbiologie. Stuttgart, 1950.
	An 359.54.15	Schwidetzky, Ilse. Das Problem des Völkertodes. Stuttgart, 1954.
Htn	Geog 956.78*	Sciagraphia cosmica. v.1-8. (Meissner, D.) Nürnberg, 1678. 2v.
	An 359.68.5A	Science and the concept of race. (American Association for the Advancement of Science.) N.Y., 1968.
Htn	Geog 816.52*	La science de la geographie divisée en trois parties. (Francois, Jean.) Rennes, 1652.
	Geog 3530.16	La science géographique. (Martonne, E. de.) Paris, 1915.
	Geog 5300.52	Science in the Arctic Ocean basin. (National Research Council. Committee on Polar Research.) Washington, 1963.
	Geog 665.100	The science of geography. (National Research Council. Ad Hoc Committee on Geography.) Washington, 1965.
	An 359.45	The science of man in the world crisis. (Linton, R.) N.Y., 1945.
	Geog 665.32	Les sciences géographiques. (Vallaux, Camille.) Paris, 1925.
	Geog 4129.10	Scientific American handbook of travel. (Hopkins, A.A.) N.Y., 1910.
	An 359.41.10	Scientific aspects of the race problem. Washington, 1941.
	An 187.5.25	A scientific theory of culture and other essays. (Malinowski, B.) Chapel Hill, 1944.
	An 176.7	The scientist on the trail. (Bandelier, A.F.) Berkeley, 1949.
	An 99.70	La scienza dell'uomo nel settecento. (Moravia, Sergio.) Bari, 1970.
	Geog 4269.73	Ścieżkami Starego Świáta. Wyd. 1. (Budrewicz, Olgierd.) Warszawa, 1973.
	Geog 819.09	Scobel, A. Geographisches Handbuch. Bielefeld, 1909-1910. 2v.
	Geog 4348.93	Scollard, Clinton. On sunny shores. N.Y., 1893.
	An 359.08	The scope and content of the science of anthropology. (Dieserud, J.) Chicago, 1908.
	An 359.67	The scope of anthropology. (Lévi-Strauss, Claude.) London, 1967.
	Geog 3019.24	Le scoperte geografiche e i viaggi d'esplorazione. (Capasso, C.) Messina, 1924.
	Geog 5329.1	Scoresby, R.E. Life of William Scoresby. London, 1861.
	Geog 5160.3	Scoresby, William. An account of the Arctic regions. Edinburgh, 1820. 2v.
	Geog 5558.22	Scoresby, William. Journal of voyage to the northern whale-fishery. Edinburgh, 1823.
Htn	Geog 5340.8F*	Scoresby, William. Seven log-books concerning the Arctic voyages. N.Y., 1917. 8v.
	Geog 4679.41.5	Scotch on the rocks. (Swinson, A.) London, 1963.
	Geog 4500.17	Scotland. National Library, Edinburgh. Shelf-catalogue of the Lloyd collection of Alpine books. Boston, 1964.
	Geog 5050.8	Scotland. National Library, Edinburgh. Shelf-catalogue of the Wordie collection of polar exploration. Boston, 1964.
	Geog 4218.92.7	Scott, Clement. Pictures of the world. London, 1894.
	Geog 612.3.10	Scott, Ernest. Laperouse. Sydney, 1912.
	Geog 4308.21.5	Scott, J. Sketches of manners...in French provinces. London, 1821.
	Geog 5333.5	Scott, J.M. Gino Watkins. London, 1935.
	Geog 4209.66.25	Scott, Jack Denton. Passport to adventure. N.Y., 1966.
	Geog 6009.10.15.2	Scott, R.F. Scott's last expedition. London, 1913. 2v.
	Geog 6009.10.15A	Scott, R.F. Scott's last expedition. N.Y., 1913. 2v.
	Geog 6009.10.14	Scott, R.F. Soctt's last expedition. 2d ed. London, 1964.
Htn	Geog 6009.01.8*	Scott, R.F. Voyage of the 'Discovery'. London, 1905. 2v.
	Geog 6009.01.7A	Scott, R.F. Voyage of the 'Discovery'. N.Y., 1905. 2v.
	Geog 6009.01.15	Scott, R.F. The voyages of Captain Scott retold from The voyage of the 'Discovery'. N.Y., 1915.
	Geog 5985.253.10	Scott of the Antarctic. (Pound, Reginald.) London, 1966.
	Geog 5985.253.5	Scott of the Antarctic. (Seaver, George.) London, 1940.
	Geog 659.59	Scotti, Pietro. Elementi di geografia. Genova, 1959.
	An 2109.63.5	Scotti, Pietro. La vita sociale dei popoli primitivi. Brescia, 1963.
	Geog 200.1	Scottish geographical magazine. Edinburgh. 1+ 63v.
	Geog 200.1.3	Scottish geographical magazine. Index. v.1-50, 1885-1934. Edinburgh, n.d.
	Geog 6009.10.15.2	Scott's last expedition. (Scott, R.F.) London, 1913. 2v.
	Geog 6009.10.15A	Scott's last expedition. (Scott, R.F.) N.Y., 1913. 2v.
	Geog 6009.10.60	Scott's last voyage. (Ponting, Herbert George.) London, 1974.
	Geog 665.47	Scritti di geografia e di storia della geografia consernenti l'Italia pubblicati in onore di Giuseppe della Vedova. Firenze, 1908.
	Geog 665.95	Scritti geografici, scelti e ordinati. (Mori, Assunto.) Pisa, 1960.
	Geog 3019.61.5	Scritti geografici. (Almagia, Roberto.) Roma, 1961.
	Geog 665.43	Scritti geografici scelti e ordinati a cura del Comitato nazionale. (Errera, Carlo.) Bologna, 1937.
	Geog 4679.25	The sea, the ship, and the sailor. (Marine Research Society, Salem, Mass.) Salem, Mass., 1925.
	Geog 4559.10	The sea and its story from viking ship to submarine. (Shaw, Frank H.) London, 1910.
	Geog 6100.37.10	The sea and the snow; the South Indian Ocean Expedition to Heard Island. (Temple, Philip.) Melbourne, 1966.
	Geog 4239.68	The sea around them; the Atlantic Ocean, A.D. 1250. (Cassidy, Vincent H.) Baton Rouge, 1968.
	Geog 4559.27	A sea chest; an anthology of ships and sailormen. (Smith, Cicely Fox.) London, 1927.
	Geog 4559.52	Sea fever. (Rasmussen, A.H.) London, 1952.
	Geog 4672.24	Sea fights and shipwrecks. (Baldwin, Hanson W.) Garden City, 1955.
	Geog 4559.35.7	Sea lanes; man's conquest of the ocean. (Stevens, M.D.) Garden City, N.Y., 1938.
	Geog 4559.35.5	Sea lanes; man's conquest of the ocean. (Stevens, M.D.) N.Y., 1935.
	Geog 4559.29.22	Sea lore. (Rogers, S.R.H.) London, 1934.
	Geog 3570.30A	Sea road to the Indies. (Hart, Henry H.) N.Y., 1950.
	Geog 4228.49	Sea routes to the gold fields. 1st ed. (Lewis, Oscar.) N.Y., 1949.
	Geog 4679.30	Sea-toll of our time. (Hadfield, R.L.) London, 1930.
	Geog 4208.60.7	A sea trip in clipper ship days. (Bray, M.M.) Boston, 1920.
	Geog 4559.26.15	Seafaring. (Boughton, George P.) London, 1926.
	Geog 4239.51	Seagoing Gaucho. (Uriburu, E.C.) N.Y., 1951.
	Geog 4559.24	Seamen all. (Chatterton, E.K.) Boston, 1924.
	Geog 4559.62.10	The seamen are down below. (Dane, Peter.) Ilfracombe, 1962.
	Geog 3055.44	The search for Atlantis. (Bjorkman, E.) N.Y., 1927.
	Geog 4145.81.5	The search for Captain Slocum. (Teller, Walter Magnes.) N.Y., 1956.
	Geog 5535.20	The search for Franklin. (Neatby, Leslie Hamilton.) London, 1970.
	Geog 5209.47	The search for the North Pole. (Crouse, Nellis M.) N.Y., 1947.
	Geog 5509.34	The search for the Northwest Passage. (Crouse, N.M.) N.Y., 1934.
	Geog 4328.69	Search for winter sunbeams. (Cox, S.S.) London, 1869.
	Geog 4328.69.5	Search for winter sunbeams. (Cox, S.S.) N.Y., 1874.
	An 379.57	Sears, Paul. The ecology of man. Eugene, 1957.
	Geog 4145.98	Seaver, G. Francis Younghusband. London, 1952.
	Geog 5985.34	Seaver, George. "Birdie" Bowers of the Antarctic. London, 1938.
	Geog 5985.325	Seaver, George. Edward Wilson of the Antarctic, naturalist and friend. N.Y., 1937.
	Geog 5985.325.5	Seaver, George. The faith of Edward Wilson. London, 1948.
	Geog 5985.253.5	Seaver, George. Scott of the Antarctic. London, 1940.
	An 3308.90	Seaver, J.W. Anthropometry and physical examination. New Haven, 1890.
	An 3308.96	Seaver, J.W. Anthropometry and physical examination. 2. ed. New Haven, 1896.
	Geog 4219.06.7	Sebok, I. Öt világrészun heresztül. Budapest, 1934.
	Geog 4180.68	Second part of chronicle of Peru. (Cieza de Leon, P. de.) N.Y., 1964?
	Geog 4419.34.3	Les secrets de la Mer Rouge. (Monfreid, Henri de.) Paris, 1935.
	Geog 5559.17A	Secrets of polar travel. (Peary, R.E.) N.Y., 1917.
	Geog 4419.34	Secrets of the Red Sea. (Monfreid, Henri de.) London, 1934.

Author and Title Listing

Geog 3570.51 O seculo dos descobrimentos. São Paulo, 1961.
Htn Geog 4308.39.5* Sedgwick, C.M. Letters from abroad. N.Y., 1841. 2v.
Geog 5559.14 Sedov. (Nagornyi, S.) Moskva, 1939.
Geog 4309.22 Seeing Europe backwards. (Morse, William I.) Boston, 1922.
Geog 4309.11 Seeing Europe by automobile. (Meriwether, Lee.) N.Y., 1911.
Geog 4218.45.5 Seemann, Berthold. Narrative of the voyage of H.M.S. Herald during the years 1845-51...being a circumnavigation of the globe. London, 1853.
Geog 4678.55.20 Seeoffizier des Zaren. (Schilling, Nikolai.) Köln, 1971.
Geog 4166.78.5 Seer gedenckwaerdige voyagien. (Dyck, J.) Amsterdam, 1678.
Geog 4029.17 Seering, Ruth. Mein tödliches Risiko. Bergisch Gladbach, 1973.
Geog 5209.39 Segal, Louis. The conquest of the Arctic. London, 1939.
Geog 4520.5 Ségogne, Henry de. Les alpinistes célébres. Paris, 1956.
Geog 5170.17 Seidenfaden, G. Modern Arctic exploration. Boston, 1939.
Geog 4218.57.10 Sein Schiff hiess Novara. (Wallisch, Friedrich.) Wien, 1966.
Geog 4209.23.10 Seit ich die heimat Verliess. (Wehde, Albert.) Berlin, 1924.
Geog 4309.07 Seitz, Don C. Discoveries in every-day Europe. N.Y., 1907.
Geog 4709.27.5 Seitz, Don Carlos. Under the black flag. London, 1927.
Geog 4709.25 Seitz, Don Carlos. Under the black flag. N.Y., 1925.
Geog 4418.64 Seize mille lieues a travers l'Asie et l'Oceanie. (Russell-Killough, H.) Paris, 1864.
Geog 4307.91.10 Séjour de mon grand-oncle, P. Gaultier, en Espagne...1791-1802. v.1-2. (Gaultier, Pierre R.A.) Angers, 1912.
Geog 3604.5 Selander, Sten. Linnélärjungar i främmande länder. Stockholm, 1960.
An 2109.33 Selbstmord und Todesfurcht bei den Naturvölkern. Proefschrift. (Wisse, Jakob.) Zutpher, 1933.
Geog 4750.2 A select annotated bibliography of the humid tropics. (International Geographical Union. Special Commission on the Humid Tropics.) Montreal, 1960.
An 5.9 Select bibliography of recent publications in the library. (Royal Empire Society, London. Library.) London, 1926.
Geog 4181.65 Select documents illustrating the 4 voyages of Columbus. (Jane, Cecil.) London, 1930-33. 2v.
An 5.1 Selected bibliography of the anthropology and ethnology of Europe. (Ripley, W.Z.) Boston, 1899.
Geog 4500.5 Selected list of books on mountaineering. (Jeffers, LeRoy.) N.Y., 1916.
An 170.238 Selected papers. (International Congress of Anthropological and Ethnological Sciences, 5th, Philadelphia, 1960.) Philadelphia, 1960.
Geog 4209.24.5A Selected papers on anthropology, travel and exploration. (Burton, Richard F.) London, 1924.
Geog 659.74.10 A selection of geographical computer programs. (Baker, Laurie.) London, 1974.
Geog 3520.10.15 A selection of the principal voyages. (Hakluyt, Richard.) N.Y., 1926.
Geog 5559.14.5 Seleznev, Stepan A. Pervaia russkaia ekspeditsiia k severnomy poliucu. Arkhangel'sk, 1964.
Geog 4219.39.5 Seligman, A. The voyage of the Cap Pilar. London, 1939.
An 359.39.7 Seligmann, H.J. Race against man. N.Y., 1939.
Htn Geog 816.94* Seller, John. A new system of geography. n.p., 1694?
Geog 4309.64.15 Selucký, Radoslav. Západ je Západ. Praha, 1964.
Geog 4309.64.16 Selucký, Radoslav. Západ je Západ. 2. vyd. Praha, 1965.
Geog 142.6 Semaine internationale des géographes, des explorateur et des ethnologues, 22-28 Sept. 1922. (Marseilles. Exposition Coloniale Nationale, 1922.) Marseilles, 1923.
Geog 618.6 Semen Ul'ianovich Remezov. (Gol'denberg, Leonid A.) Moskva, 1965.
An 2059.66.5 Semenov, Iurii I. Kak vozniklo chelovechestvo. Moskva, 1966.
An 2159.64 Semenov, Sergei A. Prehistoric technology. London, 1964.
An 2109.68.10 Semenov, Sergei A. Razvitie tekhniki v kameunom veke. Leningrad, 1968.
Geog 5970.44.5 Sementovskii, Vladimir N. Russkie otkrytiia v Antarktike w 1819-1820-1821 godakh. Moskva, 1951.
VGeog 4131.67.5 Seminàr o Ekonomickej Efektívnosti Iuvesticii Cestovného Ruchu, Piešťany, 1966. Efektivnost' investicí cestovného ruchu. Bratislava, 1967.
An 379.11 Semple, E.M. Influences of geographic environment on the basis of Ratzel's system. N.Y., 1911.
Geog 5507.87 Sendschreiben...des alten Grönlands. (LaRoche, F.C. de.) Kopenhagen, 1787.
Geog 4559.27.5A Senior, William. Naval history in the law courts. London, 1927.
Geog 759.21.5 The senior geography. 5th ed. (Herbertson, Andrew J.) Oxford, 1921.
Geog 4311.7 Senkel, K.F. De istri ostiis dissertatio historico-geographica. Wratislaviae, 1820.
Geog 4239.73 Senkevich, Iurii A. Na "RA" cherez Atlantiku. Leningrad, 1973.
Geog 4145.37 Señor Kon-Tiki; the biography of Thor Heyerdahl. (Jacoby, Arnold.) Chicago, 1967.
Geog 3229.06 Sensburg, W. Poggio Bracciolini und Nicolò de Conti. Wien, 1906.
Geog 4515.80 Serebrianye lyzhi. Uzhgorod, 1973.
Geog 4311.22 Sergeeva, Nelli A. Dunai-reka druzhby. Odessa, 1966.
Geog 614.10 Sergei Semenovich Neustruev, 1874-1928. (Dontsova, Zoia N.) Moskva, 1967.
An 359.01.3 Sergi, G. The Mediterranean race. London, 1901.
An 193.1 Sergi, Giuseppe. Problemi di scienza contemporanea. Palermo, 1904.
An 193.1.5 Sergi, Giuseppe. Problemi di scienza contemporanea. Torino, 1916.
An 3559.12.3 Sergi, Sergio. Crania Habessinica contributo all'antropologia dell'Africa Orientale. Roma, 1912.
An 5.26 Serial publications in anthropology. (Library-Anthropology Resource Group.) Chicago, 1973.
Geog 4417.80 Series of adventures...Red Sea, Arabia, Egypt. (Irwin, E.) Dublin, 1780.
Geog 4417.80.3 Series of adventures...Red Sea, Arabia, Egypt. 3. ed. (Irwin, E.) London, 1787. 2v.
Geog 4477.64.5 Seriman, Zaccaria. Viages de Enrique Wanton. Madrid, 1781-85. 4v.
Geog 4678.54 A sermon. (Smalley, Elam.) Troy, N.Y., 1854.
Geog 4678.36.10 A sermon delivered February 14, 1836. (Kendall, J.) Plymouth, 1836.
Geog 4678.40.5 A sermon occasioned by burning of steamer Lexington. (Stone, John S.) Boston, 1840.
Geog 4679.61 The serpent's coil. 1. United States ed. (Mowat, Farley.) Boston, 1961.
Geog 4419.59 Serstevens, Albert. Les précurseurs de Marco Polo. Paris, 1959.
Geog 815.35.5 Servetus, Michael. Descripciones geograficas del estado moderno de las regiones. Madrid, 1932.
Geog 815.35.4 Servetus, Michael. Descripciones geograficas del estado moderno de las regiones. Madrid, 1932.
Geog 815.35.10 Servetus, Michael. Michael Servetus, a translation of his geographical, medical, and astrological writings. Philadelphia, 1953.
Geog 659.53 Sestini, Aldo. Avviamento allo studio della geografia. 1. ed. Firenze, 1953.
Geog 4559.49 The set of the sails. (Villiers, A.J.) N.Y., 1949.
An 359.01.5 Setourneau, C.J.E. La psychologie ethnique. Paris, 1901.
Htn Geog 5515.77* Settle, D. True report of Martin Frobisher's voyage, 1577. London, 1577.
Geog 665.132 Settlement and encounter; geographical studies presented to Sir Grenfell Price. Melbourne, 1969.
Geog 3060.7 Setzungsberichte. (Hofmann, C.) n.p., 1865.
Geog 4239.23 Seul a travers l'Atlantique. (Gerbault, Alain.) Paris, 1924.
Geog 4209.35.20A Seven league boots. (Halliburton, Richard.) Indianapolis, 1935.
Geog 4209.35.23 Seven league boots. 3. ed. (Halliburton, Richard.) London, 1941.
Geog 4219.51.5 Seven leagues to paradise. 1. ed. (Tregaskis, R.W.) Garden City, 1951.
Htn Geog 5340.8F* Seven log-books concerning the Arctic voyages. (Scoresby, William.) N.Y., 1917. 8v.
Geog 6009.52 Seven men among the penguins. 1st American ed. (Marret, M.) N.Y., 1955.
Geog 4218.72.15 A seven month's run up and down and around the world. (Brooks, James.) N.Y., 1872.
Geog 4209.38.1 Seven seas on a shoestring. (Long, Dwight.) N.Y., 1939.
Geog 4308.35 Seven wekks in Belgium, Switzerland, Lombardy. (Roby, John.) London, 1838. 2v.
Geog 5919.18 The seventh continent. (Wright, H.S. (Mrs.).) Boston, 1918.
Geog 5559.37.10 Severn Iulius Zavoevan Bolbshevik. Moskva, 1937.
Geog 5399.57.25 Severnyi morskoi put. (Belov, M.I.) Leningrad, 1957.
Geog 5399.37 Severnyi polius. (Laktionov, A.F.) Arkhangelsk, 1939.
Geog 5399.37.7 Severnyi polius. (Laktionov, A.F.) Moskva, 1955.
Geog 4218.70.6 Seward, William H. William H. Seward's travels around the world. N.Y, 1874.
Geog 4218.70.5 Seward, William H. William H. Seward's travels around the world. N.Y., 1873.
Geog 4218.70.4 Seward, William H. William H. Seward's travels around the world. N.Y., 1873.
Geog 4308.52 Sewell, E.M. Diary summer tour. N.Y. 1852.
An 2109.37.7 Sex, custom and psychopathology. (Laubscher, B.J.F.) London, 1937.
An 2109.27.3A Sex and repression in savage society. (Malinowski, B.) London, 1927.
An 2109.27.4 Sex and repression in savage society. (Malinowski, B.) N.Y., 1927.
An 2109.21.3 Sexual life of primitive people. (Fehlinger, H.) London, 1921.
X Cg An 2109.29.5 The sexual life of savages in north west Melanesia. (Malinowski, B.) London, 1929.
Geog 4707.24A Seybold, R.F. Captured by pirates. Two diaries of 1724-1725. Boston, 1929.
Geog 819.02 Seydlitz, Ernst von. Grosses Lehrbuch der Geographie. Breslau, 1902.
Geog 819.14 Seydlitz, Ernst von. Handbuch der Geographie. Breslau, 1914.
Geog 819.03 Seydlitz, Ernst von. Kleines Lehrbuch der Geographie. Breslau, 1903.
Geog 5925.30 Shackleton, E. South. N.Y., 1920.
Geog 5559.34.5 Shackleton, Edward. Arctic journeys; the story of the Oxford University Ellesmire land expedition, 1934-35. London, 1937.
Geog 6009.07.5 Shackleton, Ernest H. The heart of the Antarctic. London, 1911.
Geog 6009.07A Shackleton, Ernest H. Heart of the Antarctic. Philadelphia, 1909. 2v.
Geog 6009.14 Shackleton, Ernest H. South; the story of Shackleton's last expedition. London, 1919.
Geog 4269.34.7 Shackleton, Margaret Reid. Europe: a regional geography. 7th ed. N.Y., 1965.
Geog 5985.258.10 Shackleton. (Fisher, Margery.) London, 1957.
Geog 6009.21 Shackleton's last voyage; the story of the Quest. (Wild, Frank.) London, 1923.
Geog 3055.56 The shadow of Atlantis. (Bragbine, A.) N.Y., 1940.
Geog 4306.17.3 Shakespeare's Europe. (Moryson, Fynes.) London, 1903.
Geog 4306.17.4 Shakespeare's Europe: a survey of the condition of Europe at the end of the 16th century. 2nd ed. (Moryson, Fynes.) N.Y., 1967.
Geog 4209.62 Shakirianov, N.A. Atlanticheskii dnevnik. Riga, 1962.
Geog 4129.03 Shand, A.I. Old-time travel. London, 1903.
Geog 3019.63.10 Sharaf, A. Torayah. A short history of geographical discovery. Alexandria, 1963.
Geog 4249.60 Sharp, C.A. The discovery of the Pacific Islands. Oxford, 1960.
Geog 4679.30.5 Shaw, F.H. Full fathom five. N.Y., 1930.
Geog 4559.10 Shaw, Frank H. The sea and its story from viking ship to submarine. London, 1910.
Geog 4209.46.5 Shaw, Frank H. White sails and spendthrift. London, 1946.
Geog 4417.38.5 Shaw, Thomas. Reisen der Barbery und der Levante. Leipzig, 1765.
Geog 4417.38.3 Shaw, Thomas. Travels or observations...Barbary...Levant. Edinburgh, 1808. 2v.
Geog 4417.38F Shaw, Thomas. Travels or observations. Oxford, 1738.
Geog 4327.22 Shaw, Thomas. Voyages...dans plusieurs provinces de la Barbarie et du Levant. La Haye, 1743. 2v.
An 3309.54F Sheldon, W.H. Atlas of men. 1st ed. N.Y., 1954.
An 3309.40 Sheldon, W.H. The varieties of human physique. 1st ed. N.Y., 1940.
Geog 4209.65.5 Shelekhov, Grigorii I. Puteshestvie s 1783 po 1790 god iz okhotska po vostoch. okeanu. Ann Arbor, 1965. 2v.
Geog 4209.71.20 Shelekhov, Grigorii I. Rossiiskogo kuptsa Grigoriia Shelekhova stranstvoraniia iz Okholska po Vostochnomu okeanu k amerikanskiam beregam. Khabarovsk, 1971.
Geog 4500.17 Shelf-catalogue of the Lloyd collection of Alpine books. (Scotland. National Library, Edinburgh.) Boston, 1964.
Geog 5050.8 Shelf-catalogue of the Wordie collection of polar exploration. (Scotland. National Library, Edinburgh.) Boston, 1964.

Author and Title Listing

Geog 958.91 Shepp, J.W. Shepp's photographs of the world. Philadelphia, 1891.
Geog 958.91 Shepp's photographs of the world. (Shepp, J.W.) Philadelphia, 1891.
Geog 4110.36 Sheraton, Mimi. City portraits; a guide to 60 of the world's great cities. 1. ed. N.Y., 1964.
Geog 4306.97F Sheremt'ev, B.P. Zapiska puteshestviia. Moskva, 1773.
Geog 4268.77F Sherer, J. Europe...picturesque scenes. London, 1877. 2v.
Geog 4418.38 Sherer, M. The imagery of foreign travel. London, 1838.
Geog 4559.35.15 Sheridan, Richard B. Heavenly hell. London, 1935.
Geog 4345.99.4 The Sherleian odyssey, being a record of the travels of three famous brothers. (Penrose, B.) Taunton, 1938.
Geog 4345.99 Sherley, A. The three brothers or travels. London, 1825.
Geog 4110.23 Sherriff's illustrated route charts and travellers' hand book. v.1-4. London, 1887.
Geog 665.23 Sherwood, M.E. Here and there and everywhere. Chicago, 1898.
Geog 4209.71.25 Shest' gorodov piati kontinentov. (Kuleshov, Aleksandr P.) Moskva, 1971.
Geog 4269.74 Shest' zagranits. (Agapov, Boris N.) Moskva, 1974.
Geog 5321.1.5 Shields, C.W. Funeral eulogy at obsequies of Dr. Kane. Philadelphia, 1857.
Geog 5208.51 Shillinglaw, J.J. A narrative of Arctic discovery. 2. ed. London, 1851.
Geog 4328.35 Ship and shore...cruise to the Levant. N.Y., 1835.
Geog 4328.51 Ship and shore in Madeira. (Colton, W.) N.Y., 1851.
Geog 4559.35 Ships and how they sailed the seven seas. (5000 B.C.-A.D. 1935). (Van Loon, H.W.) N.Y., 1935.
Geog 4559.26.20 Ships and people. (Beaumont, J.C.H.) N.Y., 1926.
Geog 4559.72.1 Ships and shipyards, sailors and fishermen. (Hassloef, Olof.) Copenhagen, 1972.
Geog 4509.43.5 Shipton, Eric E. Upon that mountain. London, 1943.
Geog 4659.54 Shipwrecks; being the historical account of shipwrecks along the Victorian Coast, 1813-1914. (Mackenzie, Margaret E.) Melbourne, 1954.
Geog 4659.36 Shipwrecks; New Zealand disasters, 1795-1936. (Ingram, Charles W.N.) Dunedin, 1936.
Geog 4672.22 Shipwrecks and disasters at sea, or Historical narratives of the most noted calamities. Edinburgh, 1812. 3v.
Geog 4672.5.5 Shipwrecks and disasters at sea. (Ellms, Charles.) N.Y., 1844.
Geog 4659.55A Shipwrecks and empire; being an account of Portuguese maritime disasters in a century of decline. (Duffy, James E.) Cambridge, Mass., 1955.
Geog 4659.63.4 Shipwrecks at Port Phillips Heads since 1840. (Williams, Peter J.) Melbourne, 1967.
Geog 4659.57 Shipwrecks of the Pacific Coast. (Gibbs, James A.) Portland, Ore., 1957.
Geog 4652.9 Shipwrecks of the Western Hemisphere, 1492-1825. (Marx, Robert F.) N.Y., 1975.
An 359.24A Shirokogorov, S.M. Ethnical unit and milieu. Shanghai, 1924.
An 359.23.11 Shirokogorov, S.M. Etnos". Shankhai, 1923.
Geog 5399.32.10 Shneiderov, V.A. Velikim Severnym. 2. izd. Moskva, 1963.
Geog 4249.67.5 A shoal of stars. (Downs, Hugh.) Garden City, 1967.
Geog 4429.52.5 The shoals of Capricorn. (Ommanney, F.D.) London, 1952.
Geog 619.5 Shokal'skaia, Z. Iu. Zhiznennyi put' Iu. M. Shokal'skogo. Moskva, 1960.
Geog 4328.51.5 The shores and islands of the Mediterranean. (Noel-Fearn, H.) London, 1851. 3v.
Geog 4328.40.10 The shores and islands of the Mediterranean. (Wright, G.N.) London, 1840.
Geog 4328.46 Shores of the Mediterranean. (Schroeder, F.) N.Y., 1846. 2v.
Geog 5558.75.15PF Shores of the Polar Sea. (Moss, Edward L.) London, 1878.
Geog 3019.63.10 A short history of geographical discovery. (Sharaf, A. Torayah.) Alexandria, 1963.
Geog 5538.51.15 A short narrative of the second voyage of the Prince Albert in search of Sir John Franklin. (Kennedy, William.) London, 1853.
Geog 622.2 Shostiu, N.A. M.P. Vronchenko. Moskva, 1956.
Geog 4309.25.20 Shotwell, J.T. A Balkan mission. N.Y., 1949.
Geog 4229.71 Shur, Leonid A. K beregam Novogo Sveta. Moskva, 1971.
Geog 5378.80.13 Shvedskaia poliarnaia ekspeditsiia, 1878-79 g. (Nordenskiöld, A.E.) Sankt Peterburg, 1880.
Htn Geog 4329.27.10* Sicilian days and other journeys round the Mediterranean and Adriatic. (Morse, William I.) Boston, 1927.
Geog 5538.57.5 Den sidste Franklin expedition med "Fox". (Petersen, Carl.) Kjøbenhavn, 1860.
Geog 4326.86 Sieben-Jährige...Welt-Beschauung. (Neitzschitz, G.C.) Nurnberg, 1686.
Geog 3030.4 Die sieben Klimata. (Honigmann, Ernst.) Heidelberg, 1929.
Geog 3019.52 Sieben vorbei und acht verweht. 2. Aufl. (Herrmann, Paul.) Hamburg, 1952.
Geog 5399.28.25 Sieben Wochen auf der Eisscholle. (Běhounek, Franz.) Leipzig, 1929.
An 2109.41.5 Sieber, Sylvester A.M. The social life of primitive man. St. Louis, 1941.
Geog 5399.31.5 Sieburg, F. Die rote Arktis. Frankfurt, 1932.
An 379.72.10 Siedlungsgeographie. 3. Aufl. (Niemeier, Georg.) Braunschweig, 1972.
Geog 5939.05 Siege of the South Pole. (Mill, H.R.) London, 1905.
Geog 4145.81.50 Siegfried, André. Geographie poétique des cinq continents. Paris, 1952.
Geog 207.1 Sierra club bulletin. San Francisco. 2,1897+ 30v.
Geog 207.5 Sierra club circular.
Geog 4269.16 Sievers, Wilhelm. Die geographischen Grenzen Mitteleuropas. Giessen, 1916.
Geog 4419.69 A sight of China. (Nolan, Cynthia.) London, 1969.
Geog 4308.50.7 Sights and scenes in Europe. (Bullard, A.T.J. (Mrs.).) St. Louis, 1852.
Geog 4308.69.5 Sights and sensations in France. (Buffum, E.G.) N.Y., 1869.
Geog 4308.42 Sights and thoughts...foreign. Photoreproduction. (Faber, F.W.) London, 1842.
An 359.52.5 The significance of racial differences. (Morant, G.M.) Paris, 1952.
Geog 4308.56.13 Sigourney, L.H. Pleasant memories. Boston, 1856.
An 2109.67.20 Sigrist, Christian. Regulierte Anarchie. Olten, 1967.
Geog 4209.38.40 Sigvaldson, J. Ferdasaga Fritz Liebig. Reykjavik, 1938.
Geog 5939.55.5 The silent continent. 1st ed. (Kearns, W.H.) N.Y., 1955.
An 2059.07 Les silex taillés et l'ancienneté de l'homme. (Lapparent, A.) Paris, 1907.
Geog 4308.05.2 Silliman, B. Journal of travels. Boston, 1812. 2v.
Geog 4308.05.4 Silliman, B. Journal of travels. 3rd ed. New Haven, 1820. 3v.
Geog 4308.51 Silliman, B. Visit to Europe in 1851. N.Y., 1853. 2v.
Geog 4308.51.4 Silliman, B. Visit to Europe in 1851. N.Y., 1854. 2v.
Geog 4308.51.2 Silliman, B. Visit to Europe in 1851. N.Y., 1854. 2v.
Geog 4249.01 Silva, Abeillard. A través da Malasia. Coimbra, 1901.
An 359.58.5 Silva, Mello. Estudos sôbre o negro. Rio de Janeiro, 1958.
An 3558.98 Silva Boasto, A.J. de. Indices cephálicos dos Portuguêses. Coimbra, 1898.
Geog 4132.5 Silverberg, Robert. The longest voyage; circumnavigators in the age of discovery. Indianapolis, 1972.
Geog 4502.5.29 Das Silvretta-Buch. (Flaig, Walther.) München, 1940.
Geog 4502.5.7 Simmler, J. De alpibus Commentarius; die Alpen. München, 1931.
Geog 5508.51 Simmonds, P.L. Sir John Franklin and Arctic regions. London, 1851.
Geog 5508.51.5 Simmonds, P.L. Sir John Franklin and Arctic regions. 3. ed. London, 1853.
An 2109.45.1 Simmons, L.W. The role of the aged in primitive society. Hamden, Ct., 1970.
An 2109.45 Simmons, L.W. The role of the aged in primitive society. New Haven, 1945.
Geog 4209.40.50 Simon, C.R. Gran'ma goes by freight. Placerville, Calif., 1940.
An 3309.38 Simon, Carleton. La identificación personal por la retina. La Habana, 1938.
An 3309.36 Simon, Carleton. The retinal method of identification. n.p., 1936.
Geog 4180.28 Simon, P. Expedition of...Ursua and...Aguirre. N.Y., 1964?
An 136.5.10 Simonis, Yvan. Claude Lévi-Strauss ou la Passion de l'inceste. Paris, 1968.
An 170.340 Simpozium Antropologiia 70-kh Godov, Moscow, 1972. Simpozium "Antropologiia 70-kh godov". Moskva, 1972.
An 170.340 Simpozium "Antropologiia 70-kh godov". (Simpozium Antropologiia 70-kh Godov, Moscow, 1972.) Moskva, 1972.
Geog 659.73.5 Simpozium po teoreticheskim problemam geografii, Riga, 1973. Teoreticheskaia geografiia. Riga, 1973.
Geog 5518.36.3 Simpson, A. Life and travels of Thomas Simpson. London, 1845.
Geog 5518.36.4 Simpson, A. The life and travels of Thomas Simpson. Toronto, 1963.
An 359.10.5 Simpson, B.P. The conflict of colour. N.Y., 1910.
Geog 4219.66.5 Simpson, Colin. Sir Francis Chichester: voyage of the century. London, 1967.
Geog 5919.52 Simpson, F.A. The Antarctic today. Wellington, 1952.
Geog 6009.10.20 Simpson, G.C. Soctt's polar journey and the weather. Oxford, 1926.
Geog 4218.41.3 Simpson, George. Narrative of a journey round the world. London, 1847. 2v.
Geog 4218.41.2 Simpson, George. Overland journey round the world. Philadelphia, 1847.
Htn Geog 4218.41* Simpson, George. Overland journey round the world. Philadelphia, 1847.
An 132.4 Simpson, George Eaton. Melville J. Herskovits. N.Y., 1973.
Geog 5559.69 Simpson, Myrtle. Due north. London, 1970.
Geog 5518.36 Simpson, Thomas. Narrative of discoveries on north coast. London, 1843.
Geog 5518.36.2 Simpson, Thomas. Narrative of the discoveries on the north coast of America. 2. ed. Toronto, 1970. 2v.
Geog 4209.03 Simpson, William. Autobiography. London, 1903.
Geog 4218.72 Simpson, William. Meeting the sun; journey all round the world. London, 1874.
Geog 4218.72.2 Simpson, William. Meeting the sun; journey all round the world. London, 1877.
An 2600.5.2 Singh, Joseph. Wolf-children and feral man. Hamden, 1966.
An 2600.5 Singh, Joseph. Wolf children and feral man. N.Y., 1942.
An 3409.23 The single finger print identification system. (Crosskey, Walter C.S.) San Francisco, 1923.
An 3409.31.3 Single fingerprints. (Battley, Harry.) New Haven, 1931.
Geog 665.37.10 Singleton, E. Greatest wonders of the world. N.Y., 1906.
Geog 665.37 Singleton, E. Wonders of nature. N.Y., 1900.
Geog 665.37.5 Singleton, E. Wonders of nature. N.Y., 1911.
Geog 665.37.6 Singleton, E. The wonders of nature as seen and described by Alexandre Dumas. Washington, 1962.
Geog 4169.01 Singleton, Lesther. Romantic castles and palaces, as seen and described by famous writers. N.Y., 1901.
Geog 4679.12.10 The sinking of the S.S. Titanic, April 14-15, 1912. (Thayer, J.B.) Philadelphia, 1940.
Geog 4679.12.13 Sinking of the Titanic and great sea disasters. (Marshall, L.) Philadelphia, 1912.
Geog 5970.16 Siple, Paul. 90 degrees south; the story of the American South Pole conquest. N.Y., 1959.
Geog 613.4.15 Sir Clements R. Markham as a translator. (Bernstein, H.) n.p., 1937?
Geog 4219.66.5 Sir Francis Chichester: voyage of the century. (Simpson, Colin.) London, 1967.
Geog 4180.4 Sir Francis Drake, his voyage 1595. (Maynarde, T.) London, 1849.
Geog 613.8 Sir Halford MacKinder, 1861-1947, an appreciation of his life and work. (Gilbert, Edmund W.) London, 1961.
Geog 5333.15 Sir Hubert Wilkins. (Grierson, John.) London, 1960.
Geog 5333.20 Sir Hubert Wilkins. (Thomas, Lowell.) N.Y., 1961.
Geog 5530.7 Sir John Frankin. (Beesly, A.H.) N.Y., 1881.
Geog 5535.15 Sir John Franklin; die Unternehmungen für seine Rettung und die Nordwestliche Durchfahrt. (Brandes, Karl.) Berlin, 1854.
Geog 5508.51 Sir John Franklin and Arctic regions. (Simmonds, P.L.) London, 1851.
Geog 5508.51.5 Sir John Franklin and Arctic regions. 3. ed. (Simmonds, P.L.) London, 1853.
Geog 5316.3 Sir Martin Frobisher's search for the Northwest Passage. (Baetzkes, Ottile G.) N.Y., 1964.
VGeog 4209.69 Širine sveta. (Ogrin, Miran.) Ljubljana, 1969.
Geog 4131.55 Sistema de política turistíca. (Ignacio de Arrillaga, José.) Madrid, 1955.
An 608.87 Sistematicheskoe opisanie kollektsii Dashkovskago etnograficheskago muzeia. v.4. (Moscow. Dashkovskii Etnograficheskii Muzei.) Moskva, 1895.
An 2108.73.15 Un site tshitolien sur le plateau des Bateke. (Cahen, Daniel.) Tervuren, 1973.
Geog 5558.81.25 Six came back; the Arctic adventure. (Brainard, David L.) Indianapolis, 1940.
Geog 4308.71.5 Six weeks abroad in Ireland, England and Belgium. (Haskins, George F.) Boston, 1872.

Author and Title Listing

	Geog 5645.5.5	Skeie, Jon. Greenland; the dispute between Norway and Denmark. London, 1932.
	An 3509.07	Skeletal remains...early man...North America. (Hrdlička, A.) Washington, 1907.
	An 3509.34	The skeleton of British neolithic man. (Cameron, John.) London, 1934.
	Geog 4300.12	Skeleton tours. (Sargent, H.W.) N.Y., 1870.
	Geog 4300.12.2	Skeleton tours. (Sargent, H.W.) N.Y., 1870.
	Geog 4300.12.3	Skeleton tours. (Sargent, H.W.) N.Y., 1871.
	An 359.59.35	Škerlj, Božo. Antropologija i etnologija. Beograd, 1959.
	An 137.1	Sketch of...Lewis H. Morgan. (Putnam, F.W.) Boston, 1882.
	Geog 4307.86	Sketch of a tour on the continent. (Smith, J.E.) London, 1793. 3v.
	Geog 4307.86.2	Sketch of a tour on the continent. 2nd ed. (Smith, J.E.) London, 1807. 3v.
Htn	An 358.69*	Sketch of life, personal appearance, character and manners. (Stratton, C.S.) N.Y., 1869.
	Geog 3108.25	Sketch of modern and ancient geography. (Butler, S.) London, 1825.
	Geog 4208.67.5	A sketch of the new route to China and Japan. (Pacific Mail Steamship Co.) San Francisco, 1867.
	Geog 4110.22	A sketch of the route to California, China. (Pacific Mail Steamship Co.) San Francisco, 1867.
	Geog 4308.69.4	Sketches abroad with pen and pencil. (Darley, F.O.C.) Boston, 1878.
Htn	Geog 4308.68.15*	Sketches abroad with pen and pencil. (Darley, F.O.C.) N.Y., 1868.
	Geog 4308.69.3	Sketches abroad with pen and pencil. (Darley, F.O.C.) N.Y., 1869.
	Geog 4307.92	Sketches and observations...tour. London, 1797.
	Geog 4208.30	Sketches by a traveller. (Holbrook, S.P.) Boston, 1830.
	Geog 4328.42	Sketches of foreign travel. (Rockwell, C.) Boston, 1842. 2v.
	Geog 4308.21.5	Sketches of manners...in French provinces. (Scott, J.) London, 1821.
	Geog 4328.29	Sketches of naval life...on Mediterranean. (Jones, George.) New Haven, 1829. 2v.
	Geog 818.23	Sketches of the earth. (Worcester, J.E.) Boston, 1823. 2v.
	Geog 4308.78	Sketches of travel...Europe. (King, H.) Washington, 1878.
	Geog 5398.90	The skipper in Arctic seas. (Clutterbuck, W.J.) London, 1890.
	Geog 5839.38.5	Skitzebogsblade fra Angmagssalik. (Gitz-Johansen, Aage.) København, 1938.
	Geog 4328.94.7PF	Skizzer aus dem Süden. (Rothschild, N.) Wien, 1894.
	Geog 4219.51	Skolov, A.V. Tri prugosvetiykh plavaniia M.P. Lazareva. Moskva, 1951.
	Geog 5378.80.20	Skoog, Gösta. Vega. En aktualisering av händelserna kring Vega-expeditionen 1878-1880. Göteborg, 1965.
	Geog 5650.20	Skraelingerne i Grønland. (Mathiassen, T.) København, 1935.
	Geog 4709.68	Skriagin, Lev N. Sokrovishcha pogibshikh korablei. Moskva, 1968.
	Geog 5606.10	Skrifter. (Denmark. Udvalget for samfundsforskning i Grønland.) 1-9,1961-1963// 2v.
	Geog 4512.537.5	Skunterhaltungen. (Luther, C.J.) München, 1925.
	Geog 4139.73	Słabezyński, Wacław. Polscy podróznicy i odkrywey. 1. wyd. Warszawa, 1973.
	An 348.45	Slack, David B. An essay on the human color. Providence, 1845.
	Geog 6009.53	Slava v Antarktike. (Solianik, A.N.) Moskva, 1954.
	Geog 4304.65.10	Slavík, František A. Cesta pana Lva z Rožmitála po západní Europě roku 1465-1467. Telči, 1890.
	Geog 4328.66.5	Sleeper, M.G.Z. (Mrs.). The Mediterranean Islands. Boston, 1866.
	Geog 4209.56.20	Slessor, Tim. First overland. London, 1957.
	Geog 5925.65	Slevich, S.B. Antarktika dolzhna stat' zonoi mira. Leningrad, 1960.
	Geog 5939.73.5	Slevick, Solomon B. Osnovnye problemy osvoeniia Antarktiki. Leningrad, 1973.
Htn	Geog 4678.41.6*	Slight, Julian. A narrative of the loss of the Royal George. 6. ed. Portsea, 1843.
	Geog 4308.34.5	Slight reminiscences of the Rhine. (Boddington, Mary.) London, 1834. 2v.
	Geog 4219.00.10	Slocum, Joshua. Sailing alone around the world. London, 1948.
	Geog 4219.00.5	Slocum, Joshua. Sailing alone around the world. N.Y., 1900.
	Geog 4219.00	Slocum, Joshua. Sailing alone around the world. N.Y., 1900.
	Geog 4145.81.10	Slocum, Joshua. The voyages of Joshua Slocum. New Brunswick, 1958.
	Geog 4145.81	Slocum, Victor. Captain Joshua Slocum. N.Y., 1950.
	Geog 4679.38.5	Slocum, Victor. Castaway boats. N.Y., 1938.
	An 170.341	Slotkin, James Sydney. Reading in early anthropology. N.Y., 1965.
	Geog 579.54A	Slovar' geograficheskikh nazvanii. (Bodnarskii, M.S.) Moskva, 1954.
	Geog 579.54.5	Slovar' geograficheskikh nazvanii. (Bodnarskii, M.S.) Moskva, 1958.
	Geog 579.55A	Slovar' russkoi...geograficheskikh nazvanii. (Volostnova, M.B.) Moskva, 1955. 2v.
	Geog 579.73.5	Słownik pojęć geograficznych. Wyd. 1. (Pietkiewicz, Stanisław.) Warszawa, 1973.
	Geog 209.1	Slutskaia, Raisa D. Geograficheskoe obshchestvo globus. Moskva, 1972.
	Geog 4559.40.5	Smale, R. There go the ships. Caldwell, 1940.
	Geog 4312.10	Small boat to Elsinore. (Pilkington, Roger.) N.Y., 1969.
	Geog 4678.54	Smalley, Elam. A sermon. Troy, N.Y., 1854.
	Geog 5660.37	Smedal, G. Grönland und der Norden. Oslo? 1942?
	Geog 5080.5	Smedal, Gustav. Aquisition of sovereignty over polar areas. Oslo, 1931.
	Geog 5080.5.5	Smedal, Gustav. Souveränitätsfragen der Polargebiete. Oslo? 1943.
	Geog 3506.8	Smet, Antoine de. Album Antoine de Smet. Bruxelles, 1974.
	Geog 3510.13	Smet, Antoine de. La cartographie hollandaise. Bruxelles, 1971.
	Geog 4218.68	Smiles, S. Round the world. N.Y., 1871.
	VGeog 4308.65.15	Smith, Amy G. Letters from Europe, 1865-1866. Washington, 1948.
	An 358.49.3	Smith, Ashbel. An oration pronounced before the Connecticut Alpha of the Phi Beta Kappa at Yale College. New Haven, 1849.
	Geog 4309.10.5	Smith, Bertha W. Traveller's tales told in letters. N.Y., 1912.
	Geog 5208.89	Smith, C.C. Arctic explorations in 18th and 19th centuries. Boston, 1889.
	Geog 4558.66	Smith, C.E. From the deep of the sea. London, 1922.

	An 358.51.3	Smith, C.H. The natural history of the human species. Boston, 1851.
	An 358.52	Smith, C.H. The natural history of the human species. London, 1852.
	Geog 4559.28.10	Smith, Cicely Fox. Ancient mariners. London, 1928.
	Geog 4559.27	Smith, Cicely Fox. A sea chest; an anthology of ships and sailormen. London, 1927.
	Geog 4308.86.17	Smith, F.H. Well-worn roads of Spain, Holland and Italy. Boston, 1892.
	Geog 4208.72.9	Smith, G.A. Correspondence of Palestine tourists...while traveling in Europe, Asia and Africa. Salt Lake City, 1875.
	An 359.24.5A	Smith, G.E. Evolution of man. London, 1924.
	An 359.24.7	Smith, G.E. The evolution of man. 2. ed. London, 1927.
	An 359.29.5	Smith, G.E. Migrations of early culture. Manchester, 1929.
	An 2109.16	Smith, G.E. Primitive man. London, 1916.
	An 359.15	Smith, George E. The migrations of early culture. Manchester, 1915.
	An 2109.28.13	Smith, Grafton Eliot. In the beginning; the origin of civilization. London, 1928.
	Geog 500.155	Smith, Harold F. American travellers abroad. Carbondale, 1969.
	Geog 5180.18	Smith, I. Norman. The unbelievable land; 29 experts bring us closer to the Arctic. Ottawa, 1964.
	Geog 4500.15	Smith, J.A. Mountaineering. Cambridge, 1955.
	Geog 4307.86	Smith, J.E. Sketch of a tour on the continent. London, 1793. 3v.
	Geog 4307.86.2	Smith, J.E. Sketch of a tour on the continent. 2nd ed. London, 1807. 3v.
	Geog 4308.46.5	Smith, J.J. Summer's jaunt across the water. v.1-2. Philadelphia, 1846.
Htn	Geog 4678.30.5*	Smith, John. Narrative of the shipwreck and sufferings of the crew of the English brig Neptune. N.Y., 1830.
	Geog 578.55.15	Smith, John Calvin. Harper's statistical gazetteer of the world. N.Y., 1855.
	An 358.10.3	Smith, Samuel Stanhope. An essay on the causes of the variety of complexion and figure in the human species. 2. ed. N.Y., 1810.
	An 358.10.5	Smith, Samuel Stanhope. An essay on the causes of the variety of complexion and figure in the human species. Cambridge, 1965.
	Geog 4326.74	Smith, T. Epistulae...moribus...turcarum. Oxonii, 1674.
	Geog 4208.44.3	Smith, T.W. Narrative of life...of T.W. Smith. Boston, 1844.
	Geog 4247.96	Smith, William. Journal of a voyage in the Duff to Pacific. N.Y., 1813.
	Geog 4158.42	Smith, William. Nouvelle bibliotheque des voyages. Paris, 1842. 12v.
	Geog 5519.69	Smith, William D. Northwest Passage. N.Y., 1970.
	An 2109.60.10	Smolla, G. Neolithische Kulturerscheinungen. Bonn, 1960.
	Geog 5225.7	Smucker, Samuel. Arctic explorations and discoveries. N.Y., 1857.
	Geog 5939.59.22	Smuul, Juhan. Antarctica ahoy! The icebook. Moscow, 196-?
	Geog 5939.59.24	Smuul, Juhan. Jaine raamat. Tallinn, 1962.
	Geog 5939.59.20	Smuul, Juhan. Ledovaia kniga. Moskva, 1959.
	Geog 4678.37.9	Smyth, Thomas. The voice of God in calamity, or Reflections on the loss of the steamboat "Home"...sermon. Charleston, 1837.
	Geog 139.7	Smyth, W.H. Address at anniversary meeting of the Royal Geographical Society. London. 1851-1868
	Geog 4328.54	Smyth, W.H. The Mediterranean. London, 1854.
	Geog 4509.31.15	Smythe, F.S. Climbs and ski runs. Edinburgh, 1931.
	Geog 4509.47	Smythe, F.S. The mountain top. London, 1947.
	Geog 4509.41.17	Smythe, F.S. The mountain vision. London, 1942.
	Geog 4239.66	Snaith, William. Across the western ocean. 1st ed. N.Y., 1966.
	Geog 5508.31	Snelling, W.J. Polar regions of the western continent. Boston, 1831.
	Geog 4559.44	Snow, A.R. Log of a sea captain's daughter. Boston, 1944.
	Geog 4659.52	Snow, Edgar R. Great gales and dire disasters. N.Y., 1952.
	Geog 4672.26	Snow, Edward Rowe. The vengeful sea. N.Y., 1956.
	Geog 5538.50.17	Snow, W.P. Voyage of Prince Albert...Sir John Franklin. London, 1851.
	An 359.62.25	Snyder, L.L. The idea of racialism. Princeton, 1962.
	Geog 4209.46.2	So many roads. (Baerlein, Henry.) London, 1946?
	Geog 4219.33.15	So sah ich die Welt. 12e Aufl. (Gezork, H.) Kassel, 1938?
	Geog 4329.35.15	So you're going to the Mediterranean. (Laughlin, C.E.) Boston, 1935.
	Geog 5925.55.6	Soberanía argentina en la Antártida. 2. ed. (Argentina. Comissión Nacional del Antártico.) Buenos Aires, 1948.
	Geog 5925.55.10	La soberania argentina sobre la Antártida. (Sampay, Arturo E.) La Plata, 1950.
	An 3559.17	Sôbre algunos crânios da India Portuguêsa. (Mendes Correa, A.A.) Porto, 1917.
	An 3559.15	Sobre três crânios de negros Mossumles. (Mendes Correa, A.A.) Porto, 1915.
	Geog 607.10	Sochineniia. (Galorm'a, R.M.) Moskva, 1949.
	An 2109.73.25	Social and cultural identity; problems of persistence and change. Athens, 1974.
	An 359.51.5	Social anthropology, and other essays. (Evans-Pritchard, Edward Evan.) N.Y., 1964.
	An 359.64.5	Social anthropology. (Lienhardt, G.) London, 1964.
	An 2109.32.9A	Social anthropology. 1st ed. (Radin, Paul.) N.Y., 1932.
	An 359.64.20	Social anthropology in theory and practice. (Isā, Alī Ahmad.) Cairo, 1964.
	An 170.163	The social anthropology of complex societies. (Conference on New Approaches in Social Anthropology, Jesus College, Cambridge, Eng., 1963.) London, 1966.
	An 170.163.15	The social anthropology of complex societies. (Conference on New Approaches in Social Anthropology, Jesus College, Cambridge, Eng., 1963.) London, 1969.
	Geog 4218.90.6	A social departure. (Cotes, Sara J.D.) London, 1890.
	Geog 4218.90.5	A social departure. (Cotes, Sara J.D.) N.Y., 1890.
	Geog 4218.91	A social departure. How Orthodocia and I went round the world by ourselves. (Cotes, Sara J.D.) N.Y., 1891.
	An 2109.51	Social evolution. (Childe, V.G.) London, 1951.
	An 2109.51.2	Social evolution. (Childe, V.G.) N.Y., 1951.
	An 358.87.5	Social history of the races of mankind. v.1-5. (Featherman, A.) Boston, 1881-91. 7v.
	An 358.87.6	Social history of the races of mankind. v.2,5. (Featherman, A.) London, 1881-91. 2v.
	An 2109.41.5	The social life of primitive man. (Sieber, Sylvester A.M.) St. Louis, 1941.
	An 170.200	Social organization; essays presented to Raymond Firth. London, 1967.
	An 2109.24.3	Social organization. (Rivers, W.H.R.) London, 1924.

Author and Title Listing

An 2109.29.11	Social organization. (Rivers, W.H.R.) N.Y., 1929.
An 2109.54.10	Social origins. (Hocart, A.M.) London, 1954.
An 2109.28.20	Social origins and social communities. (Tozzer, Alfred M.) N.Y., 1928.
An 2109.25A	Social origins and social continuities. (Tozzer, A.M.) N.Y., 1925.
An 170.203	Social structure. (Fortes, Meyer.) Oxford, 1949.
An 359.49.1	Social structure. (Murdock, George Peter.) N.Y., 1965.
An 379.68.15	De sociale geografie in de rij van de sociale wetenschappen. (Heinemeyer, W.F.) Meppel, 1968.
Geog 212.15	Sociedad Geográfica, Madrid. Publicaciones. Serie B. Madrid. 455,1966+
Geog 212.40	Sociedad Geográfica de Colombia. Boletin. Bogotá. 14,1956+ 5v.
Geog 212.1	Sociedad Geográfica de la Paz. Boletin. La Paz. 27-66,1909-1943 3v.
Geog 212.9	Sociedad Geográfica de la Paz. Estatutos de la "Sociedad geográfica de la Paz." 2. ed. La Paz, 1907.
Geog 212.20	Sociedad Geográfica de Lima. La reorganización de la Sociedad Geográfica de Lima. Lima, 1944.
NEDL Geog 142.2	Sociedad Geografica de Madrid. Boletin. Madrid. 1-44,1876-1900
Geog 142.2	Sociedad Geografica de Madrid. Boletin. Madrid. 1876-1901 32v.
Geog 142.2.3	Sociedad Geografica de Madrid. Boletin. Revista geografica colonial y mercantil. Actas. Madrid. 1-21,1899-1924 19v.
Geog 142.2.2	Sociedad Geografica de Madrid. Repertorio, 1901-1910. Madrid, 1911.
Geog 142.2.5	Pamphlet vol. Sociedad Geografico de Madrid.
Geog 212.50	Sociedad Mexicana de Geográfica y Estadística. Informe sobre los trabajos cartográficos. México. 5-7,1938-1947 4v.
NEDL Geog 212.100	Societa Geografica Italiana. Bollettino. Firenze. 1868-1899 12v.
Geog 212.100	Societa Geografica Italiana. Bollettino. Firenze. 1900+ 82v.
NEDL Geog 212.100.2	Societa Geografica Italiana. Bollettino. Indice generale della serie II-III. Roma, n.d.
Geog 212.109	Societa Geografica Italiana. Catalogo della biblioteca sociale. Roma, 1903.
Geog 212.105	Societa Geografica Italiana. Memorie. Roma. 1-30 27v.
Geog 212.110	La Societa Geografica Italiana e l'opera sua nel secolo XIX. (Vedova, G.) Roma. 5-6,1940-1941 3v.
Geog 212.200.20	Société Belge d'Études Coloniales. Bulletin. Bruxelles. 1894-1925 22v.
An 5.3	Société d'Anthropologie, Paris. Catalogue de la bibliothèque...1890. Paris, 1891.
NEDL Geog 212.203	Société de Géographie. Bulletin. Paris. 1-20 73v.
Geog 212.203.5	Société de Géographie. Bulletin. Table, series V-VII. Paris, n.d.
NEDL Geog 212.204	Société de Géographie. Comptes rendus. Paris. 1882-1899 18v.
Geog 212.205	Société de Géographie. La geographie; bulletin de la société de géographie. Paris. 1-72,1925-1939 60v.
Geog 212.206	Société de Géographie. Liste des membres. Paris. 1868-1897 11v.
NEDL Geog 142.5	Société de géographie. (Marseilles.) Marseilles. 1-23,1877-1899 23v.
Geog 142.5	Société de géographie. (Marseilles.) Marseilles. 1-65,1877-1954 18v.
NEDL Geog 212.215	Société de Géographie de l'Est. Bulletin. Nancy. 1879-1912 34v.
NEDL Geog 212.235	Société de Géographie de Lille. Bulletin. Lille. 1-32,1882-1899 17v.
Geog 212.235	Société de Géographie de Lille. Bulletin. Lille. 1900-1962 20v.
Geog 212.235.15	Société de Géographie de Lille. Publications. 1943-1953
Geog 500.61.5	Société de Géographie de Lille. Bibliothèque. Supplément au catalogue...paru en décembre, 1887. 1,1889
NEDL Geog 212.202	Société de Géographie de Lyon. Bulletin. Lyon. 1-16,1875-1900 11v.
Geog 212.202	Société de Géographie de Lyon. Bulletin. Lyon. 1901-1929 6v.
Geog 212.202.5	Société de Géographie de Lyon. Bulletin du cinquantenaire, 1922-23. Lyon, 1923.
Geog 189.2.8	La société de géographie de Québec, 1877-1970. (Morissonneau, Christian.) Québec, 1971.
Geog 189.2.5	Société de Géographie de Québec. Bulletin. 2v.
Geog 189.2.6	Société de Géographie de Québec. Bulletin. Index. 1880-1934. Québec, 1969.
NEDL Geog 212.225	Société de Géographie de Toulouse. Bulletin. Toulouse. 1-18,1882-1899 18v.
Geog 212.225	Société de Géographie de Toulouse. Bulletin. Toulouse. 19-50 18v.
Geog 212.218	Société de Géographie du Maroc. Bulletin. Casablanca. 1916-1933 6v.
Geog 212.240	Société Geographique de Liège. Bulletin année. Liège. 1,1965+ 2v.
Geog 35.2	Pamphlet vol. Société Khédivienne de Géographie.
NEDL Geog 212.245	Société Languedocieme de Géographie, Montpellier. Bulletin. Montpellier. 1-34,1878-1911 33v.
Geog 152.1	Société neuchateloise de géographie. Bulletin. Neuchatel. 1,1885+ 16v.
NEDL Geog 212.201	Société Normande de Géographie. Bulletin. Rouen. 1-21,1879-1899 21v.
Geog 212.201	Société Normande de Géographie. Bulletin. Rouen. 22-43,1900-1928 14v.
NEDL Geog 212.200	Société Royale Belge de Géographie. Bulletin. Bruxelles. 1-23,1877-1899 23v.
Geog 212.200	Société Royale Belge de Géographie. Bulletin. Bruxelles. 24,1900+ 46v.
NEDL Geog 212.200.5	Société Royale Belge de Géographie. Compte rendu. Tables des matières des v.1-25, 1876-1901. Bruxelles, n.d.
NEDL Geog 14.600	Société Royale de Géographie d'Anvers. Bulletin. Anvers. 1-23,1877-1899 23v.
Geog 14.600	Société Royale de Géographie d'Anvers. Bulletin. Anvers. 1-75,1876-1964 30v.
Geog 212.216.10	Société Royale de Géographie d'Egypte. Bulletin. Le Caire. 2+ 27v.
Geog 212.216.11	Société Royale de Géographie d'Egypte. Bulletin. Tables. 16-30,1928-1957
Geog 212.216F	Société Royale de Géographie d'Egypte. Mémoires. Le Caire. 1,1919+ 17v.
Geog 212.216.95	La société Sultanich de géographie du Caire. (Foucart, G.) Le Caire, 1921.
An 2109.53	Societies around the world. (Sanders, I.T.) N.Y., 1953. 2v.
An 2109.52.5	Societies around the world. v.4. (Sanders, I.T.) Lexington, 1952.
An 2109.43.5	Society and nature; a sociological inquiry. (Kelsen, Hans.) Chicago, 1943.
An 2109.46.10	Society and nature; a sociological inquiry. (Kelsen, Hans.) London, 1946.
Geog 4309.07.10	Society recollections in Paris and Vienna, 1789-1904, by an English officer. London, 1907.
Geog 4208.98.18	A society woman on two continents. (Mackin, S.M.A. (Mrs.).) N.Y., 1898.
An 3309.16	Socio-anthropometry. (Stevenson, B.L.) Boston, 1916.
An 359.56.20	Socio-culture. v.1,4. (Gjessing, Gutorm.) Oslo, 1956- 2v.
An 379.68.20	Sociografie en sociale geografie in Nederland. (Vermooten, Willem Hendrik.) Assen, 1968.
An 2109.61.10	Sociología pre y protohistórica. (Real y Ramos, C.A. del.) Madrid, 1961.
An 2109.60.25	Sociologie et anthropologie. 2. ed. (Mauss, Marcel.) Paris, 1960.
Geog 5839.30.15	Soctt, J.M. Portrait of an ice cap. London, 1953.
Geog 6009.10.14	Soctt's last expedition. 2d ed. (Scott, R.F.) London, 1964.
Geog 6009.10.20	Soctt's polar journey and the weather. (Simpson, G.C.) Oxford, 1926.
Geog 5324.2.10	Sörensen, Jon. The saga of Fridtjof Nansen. N.Y., 1932.
An 2059.22	Soergel, W. Die Jagd der Vorzeit. Jena, 1922.
Geog 4709.71.5	Sokol, Hans Hugo. Unter der Flagge mit dem Totenkopf. Herford, 1971.
Geog 4709.68	Sokrovishcha pogibshikh korablei. (Skriagin, Lev N.) Moskva, 1968.
Geog 5855.9	Solberg, O. Beiträge zur Vorgeschichte. Christiania, 1907.
Geog 4209.66.15	Soldat de fortune. (Monteil, Vincent.) Paris, 1966.
Geog 5398.79.11	Soley, J.R. Address...at unveiling of Jeannette monument. Baltimore, 1891.
Geog 4209.56.25	Soli attraverso gli oceani. 2. ed. (Uberti, Roberto degli.) Brescia, 1956.
Geog 4515.108	Soli con le montagne. (Patani, Osvaldo.) Milano, 1955.
Geog 6009.53.5	Solianik, A.N. Cruising in the Antarctic. Moscow, 1956.
Geog 6009.53	Solianik, A.N. Slava v Antarktike. Moskva, 1954.
Geog 4332.21	Soljan, Antun. The thousand islands of the Adriatic. Beograd, 1965.
An 2059.11	Sollas, William J. Ancient hunters. London, 1911.
An 2059.24.1	Sollas, William J. Ancient hunters and their modern representatives. 3. ed. London, 1924.
An 2059.24	Sollas, William J. Ancient hunters and their modern representatives. 3. ed. N.Y., 1924.
X Cg An 3559.29	Somatical investigation of the Javanese, 1929. (Nyèssen, D.J.H.) Bandoeng, 1929.
An 3209.61	Een somatometrisch onderzoek in de Noord-oost Polder. (Notschaele, Lucien Aimé.) Amsterdam, 1961.
An 359.38.7	Sombart, W. Von Menschen; Versuch einer geistwissenschaftlichen Anthropologie. Berlin, 1938.
Geog 4309.02.5	Some account of...travels of myself. N.Y., 1903.
Geog 5182.6	Some ethical phases of Eskimo culture. (Gilbertson, Albert N.) Worcester, 1914.
An 2109.01.9	Some first steps in human progress. (Starr, F.) Cleveland, 1901.
An 2108.95	Some first steps in human progress. (Starr, F.) Meadville, Pa., 1895.
Geog 4306.86.5	Some letters...Switzerland, Italy. (Burnet, Gilbert.) n.p., 1708.
Htn Geog 4306.86*A	Some letters...Switzerland, Italy. (Burnet, Gilbert.) Rotterdam, 1686.
Geog 4306.86.6	Some letters containing an account of what seemed most remarkable in Switzerland. (Burnet, Gilbert.) Menston, Eng., 1972.
Geog 4509.33.5A	Some mountain views. (Pickman, D.L.) Boston, 1933.
An 379.60	Some problems of human geography. (Dickinson, Robert Eric.) Leeds, 1960.
Geog 5070.46	Some problems of polar geography. (Brown, R.N.R.) Washington, 1929.
Geog 4181.107	Some records of Ethiopia. (Beckingham, C.F.) London, 1954.
Geog 4348.87	Some things abroad. (McKenzie, A.) Boston, 1887.
Geog 4416.26.5F	Some years travels into...Africa and Asia. (Herbert, Thomas.) London, 1677.
Htn Geog 4416.26*	Some years travels into...Asia and Afrique. (Herbert, Thomas.) London, 1638.
An 2109.09.9	Somló, Felix. Der Güterverkehr in der Urgesellschaft. Bruxelles, 1909.
An 2109.44	Sommerfelt, A. Is there a fundamental mental difference between primitive man and the civilized European. n.p., 1944.
An 3207.85	Sommering, S.T. Ueber...Verschiedenheit des Negers vom Europäer. Frankfurt, 1785.
Geog 4308.97	Sommerwanderungen und Winterfahrten. (Widmann, J.V.) Frauenfeld, 1897.
Geog 6009.02.12	The songs of the "Morning". (Doorly, G.S.) Melbourne, 1943.
Geog 4329.09.10	Sonnenfahrten. (Naumann, Friedrich.) Berlin, 1909.
Geog 5538.53.6	Sonntag, A. Professor Sonntag's thrilling narrative. Philadelphia, 1857.
Geog 5538.53.5	Sonntag, A. Professor Sonntag's thrilling narrative. Philadelphia, 1857.
Geog 4429.40	Sons of Sinbad. (Villiers, Alan J.) N.Y., 1940.
Geog 3570.5.3F	Sonsa Viterbo, F.M. Trabalhos nauticos dos Portuguezes. Lisboa, 1884-1900. 2v.
Htn Geog 3570.5*	Sonsa Viterbo, F.M. Trabalhos nauticos dos Portuguezes nos seculos XVI e XVII. Lisboa, 1890.
An 3308.76	Sopra alcuni fattori dello suiluppo umano. (Pagliani, L.) Torino, 1876.
Geog 3590.65	Sorok let issledovanii i otkrytii. (Gvozdetskii, Nikolai Andreevich.) Moskva, 1957.
An 379.43	Sorre, Maximilian. Les fondements biologiques de la géographie. v.1-3. Paris, 1943-52. 4v.
An 379.47.10	Sorre, Maximilien. Les fondements de la géographie humaine. 2. éd. Paris, 1947-
An 379.61.5	Sorre, Maximilien. L'homme sur la terre. Paris, 1961.
Geog 665.82	Sorre, Maximilien. Rencontres de la géographie et de la sociologie. Paris, 1957.
Geog 4559.28	Sorrell, George. The man before the mast. London, 1928.
Geog 4269.10	Sothriados, G. Chōrai kai ladi tēs Evrōpēs. Athēnai, 1910.

Author and Title Listing

Geog 579.66.5	Soto Mora, Consuelo. Glosario de términos geográficos. 1. ed. México, 1966.
An 2109.28.7.1	The "soul" of the primitive. (Lévy-Bruhl, L.) N.Y., 1966.
An 2109.09A	Source book for social origins. (Thomas, W.I.) Chicago, 1909.
An 359.12.13A	Source book for social origins. (Thomas, W.I.) Chicago, 1912.
An 2109.09.5	Source book for social origins. 6th ed. (Thomas, W.I.) Boston, 1909.
An 170.252	Source book in anthropology. (Kraeber, A.L.) N.Y., 1931.
Geog 4235.70.2	Sousa, F. de. Tratado das ilhas novas. Ponta Delgada, 1884.
Geog 6009.14	South; the story of Shackleton's last expedition. (Shackleton, Ernest H.) London, 1919.
Geog 5925.30	South. (Shackleton, E.) N.Y., 1920.
Geog 5970.37	South: man and nature in Antarctica. (Billing, Graham.) Wellington, 1964.
Geog 4181.106	South China in the sixteenth century. (Boxer, C.R.) London, 1953.
Geog 1518.3	South eastern France. 3. ed. (Baedeker, publishers.) Leipsic, 1898.
Geog 213.5	The South Hampshire geographer. Portsmouth, Eng. 3,1970+
Geog 5919.38.2	South latitude. (Ommaney, Francis D.) London, 1947.
Geog 6009.28.30	South of the sun. (Owen, Russell.) N.Y., 1934.
Geog 6009.28.31	South of the sun. (Owen, Russell.) N.Y., 1934.
Geog 6009.10A	The South Pole...Norwegian Antarctic expedition, 1910-1912. (Amundsen, R.) London, 1912. 2v.
Geog 6009.14.15	The South Pole trail. (Joyce, Ernest.) London, 1929.
Geog 6100.77.5	South to the Pole: the early history of the Ross Sea sector, Antarctica. (Quartermain, Leslie B.) London, 1967.
Geog 1518.26	South-western France from the Loire and the Rhone to the Spanish frontier. 2. ed. (Baedeker, publishers.) Leipsic, 1895.
Geog 6009.14.25	South with Mawson. 2d ed. (Laseron, Charles F.) Sydney, 1957.
Geog 6009.10.23	South with Scott. (Evans, E.R.G.R.) London, 1925.
An 2058.78.3	Southall, J.C. Address on man's age in the world. Richmond, 1878.
An 2058.78.5	Southall, J.C. The epoch of the Mammoth. Philadelphia, 1878.
An 2058.75	Southall, J.C. The recent origin of man. Philadelphia, 1875.
Geog 1517.5	Southern France...including Corsica. (Baedeker, publishers.) Leipsic, 1897.
Geog 1517.17	Southern France. 4. ed. (Baedeker, publishers.) Leipsic, 1902.
Geog 1517.19	Southern France. 5. ed. (Baedeker, publishers.) Leipzig, 1907.
Geog 1517.19.5	Southern France. 5. ed. (Baedeker, publishers.) Leipzig, 1907.
Geog 1517.21A	Southern France. 6. ed. (Baedeker, publishers.) Leipzig, 1914.
Geog 1524.45	Southern Germany. 8. ed. (Baedeker, publishers.) Leipsic, 1895.
Geog 1524.46	Southern Germany. 8. ed. (Baedeker, publishers.) Leipsic, 1895.
Geog 1524.51	Southern Germany. 9. ed. (Baedeker, publishers.) Leipsic, 1902.
Geog 1524.55	Southern Germany. 10. ed. (Baedeker, publishers.) Leipzig, 1907.
Geog 1524.56	Southern Germany. 11. ed. (Baedeker, publishers.) Leipzig, 1910.
Geog 1524.69A	Southern Germany. 13. ed. (Baedeker, publishers.) Leipzig, 1929.
Geog 1524.5A	Southern Germany and Austria. 2. ed. (Baedeker, publishers.) Coblenz, 1871.
Geog 1524.10	Southern Germany and Austria. 3. ed. (Baedeker, publishers.) Coblenz, 1873.
Geog 1524.30A	Southern Germany and Austria. 4. ed. (Baedeker, publishers.) Leipsic, 1880.
Geog 1524.33	Southern Germany and Austria. 5. ed. (Baedeker, publishers.) Leipsic, 1883.
Geog 1524.35	Southern Germany and Austria. 6. ed. (Baedeker, publishers.) Leipsic, 1887.
Geog 1524.36	Southern Germany and Austria. 6. ed. (Baedeker, publishers.) Leipsic, 1887.
Geog 1524.41	Southern Germany and Austria. 7. ed. (Baedeker, publishers.) Leipsic, 1891.
Geog 1524.40	Southern Germany and Austria. 7. ed. (Baedeker, publishers.) Leipsic, 1891.
Geog 1542.37	Southern Italy and Sicily. 15. ed. (Baedeker, publishers.) Leipzig, 1908.
Geog 1542.39	Southern Italy and Sicily. 17. ed. (Baedeker, publishers.) Leipzig, 1930.
Geog 6009.34	Southern lights; the official account of the British Graham Land Expedition, 1934-37. (Rymill, John.) London, 1938.
Geog 607.20	Souvenir de l'inauguration du monument élevé à Arnold Guyot par la Société de Zofingue à l'Académie de Neuchâtel le 6 mai 1892. (Schweizinscher Zofinguverein.) Neuchâtel, 1892.
Geog 4218.17.3	Souvenirs...voyage...du monde. (Arago, J.) Paris, 1839.
Geog 4145.61	Souvenirs de quatre horizons. (Noailles, Loise de.) Fribourg, 1958.
Geog 4308.33.5	Souvenirs de voyages. (Marmier, X.) Paris, 1841.
Geog 4308.39.3	Souvenirs des voyages de...duc de Bordeaux. (Locmaria.) Paris, 1846. 2v.
Geog 4208.85.3	Souvenirs of some continents. (Forbes, Archibald.) London, 1885.
Geog 4308.57.3	Souvenirs of travel. (LeVert, O.W.) Mobile, 1857. 2v.
Geog 5080.5.5	Souveränitätsfragen der Polargebiete. (Smedal, Gustav.) Oslo? 1943.
Geog 5180.35	Sovereignty in the Arctic. (Hopper, Bruce C.) N.Y., 1937.
Geog 6009.55.50	Sovetskaia Antarkticheskaia Ekspeditsiia, 1955-1958. Soviet Antarctic expedition; information bulletin. Amsterdam, N.Y., 1964-65. 3v.
Geog 5369.70	Sovetskaia Arktika; moria i ostrava Severnogo Ledovitogo okeana. Moskva, 1970.
Geog 3590.81	Sovetskaia geografiia. (Akademiia Nauk SSSR.) Moskva, 1960.
Geog 3590.82	Sovetskaia geografiia v period stroitel'stva kommunizma. (Moscow. Universitet.) Moskva, 1963.
Geog 3590.70	Sovetskie ekspeditsii god. 1959. Moskva, 1962.
Geog 6009.63.5	Sovetskie ekspeditsii v Antarktiku, 1961-63 gg. (Nudel'man, Aizik N.) Moskva, 1965.
Geog 3590.75	Sovetskie geograficheskie issledovaniia i otkrytiia. (Gvozdetskii, Nikolai Andreevich.) Moskva, 1967.
Geog 6009.55.50	Soviet Antarctic expedition; information bulletin. (Sovetskaia Antarkticheskaia Ekspeditsiia, 1955-1958.) Amsterdam, N.Y., 1964-65. 3v.
Geog 3590.75.1	Soviet geographical explorations and discoveries. (Gvozdetskii, Nikolai Andreevich.) Moscow, 1974.
Geog 3590.71.3	Soviet geography, accomplishments and tasks. (Akademiia Nauk SSSR.) N.Y., 1962.
Geog 214.5	Soviet geography: review and translation. N.Y. 1,1960+ 9v.
An 359.63.20	Sovremennaia amerikanskaia etnografiia. (Akademiia Nauk SSSR.) Moskva, 1963.
An 359.64.10	Sovremennaia antropologiia. (Moskovskoe Obshchestvo Ispytatelei Prirody.) Moskva, 1964.
Geog 110.1.25	Sovremennye problemy geografii. (International Geographical Congress, 20th, London, 1964.) London, 1964.
An 2109.70.10	Soziale Bezüge des Musizierens in Naturvolkkulturen. (Ramseyer, Urs.) Bern, 1970.
An 379.69	Sozialgeographie. (Storkebaum, Werner.) Darmstadt, 1969.
Geog 4131.73	Soziologische Probleme in modernen Tourismus. (Keller, Peter.) Bern, 1973.
Geog 4131.60	Soziologische Strukturwandlungen im Marderen. (Knebel, Hans J.) Stuttgart, 1960.
Geog 4169.59.5	Spagnol, Mario. Avventure e viaggi di mare. Milano, 1959.
Geog 4181.99	Spain. Archivo General de Indias, Seville. Further English voyages to Spanish America. London, 1951.
Geog 4131.67	Spain. Comisaría del Plan de Desarrollo Economico y Social. Comisión de Turismo. Turismo. Madrid, 1967?
Geog 659.53.5	Spain. Consejo Superior de Investigaciones Cientificas. Iniciación a la geografía local. Zaragoza, 1953.
Geog 3580.15	Pamphlet box. Spain. Consejo Superior de Investigaciones Cientificas. Instituto Historico de Marina.
Geog 1545.5	Spain and Portugal. (Baedeker, publishers.) Leipsic, 1898.
Geog 1545.10	Spain and Portugal. 2. ed. (Baedeker, publishers.) Leipsic, 1901.
Geog 1545.15	Spain and Portugal. 3. ed. (Baedeker, publishers.) Leipzig, 1908.
Geog 1545.18A	Spain and Portugal. 4. ed. (Baedeker, publishers.) Leipzig, 1913.
Geog 1545.11	Spain and Protugal. 2. ed. (Baedeker, publishers.) Leipsic, 1901.
Geog 4329.29	Spaini, Alberto. Viaggi di Bertoldo. Aquila, 1929.
Geog 4218.52	Spalding, J.W. The Japan expedition. N.Y., 1855.
Geog 4181.62	Spanish documents concerning English voyages to the Caribbean. (Wright, L.A.) n.p., 1929.
Geog 616.2	Spano, Benito. Gli atlanti corografici del cavaliere C.G. Pocelli. Bari, 1958.
Geog 3055.60.10	Spanuth, Jürgen. Atlantis; Heimat. Tübingen, 1965.
Geog 3055.60	Spanuth, Jürgen. Das enträtselte Atlantis. Stuttgart, 1953.
Geog 3055.60.5	Spanuth, Jürgen. Und doch. Stuttgart, 1959.
Geog 4207.76.12	Sparks, Jared. Leben des...Americanischen Reisenden J. Ledyard. Leipzig, 1829.
Geog 4207.76.10	Sparks, Jared. Life of John Ledyard, the American traveller. Boston, 1864.
Geog 4207.76.9	Sparks, Jared. Life of John Ledyard...American traveller. Boston, 1847.
Geog 4207.76A	Sparks, Jared. The life of John Ledyard. Cambridge, 1828.
Geog 4207.76.5	Sparks, Jared. The life of John Ledyard. 2. ed. Cambridge, 1829.
Geog 4217.72.3	Sparrman, Anders. Reise nach dem Vorgebirge der guten Hoffnung. Berlin, 1784.
Geog 4217.72	Sparrman, Anders. Voyage au Cap de Bonne-Esperance. Paris, 1787.
Geog 4217.72.10F	Sparrman, Anders. A voyage round the world with Captain James Cook. London, 1944.
Geog 4217.72.5	Sparrman, Anders. Voyage to Cape of Good Hope. Perth, 1789.
An 379.73.5	Spatial structures. (Johnston, Ronald John.) London, 1973.
Geog 4308.37	Spaziergänge und Weltfarhrten. (Mundt, T.) Altona, 1838-1839. 3v.
Geog 4558.88	Spear, P.S. The old sailor's story of his life. Portland, 1888.
Geog 18.160	Special publication. (Arctic Institute of North America.) Washington. 1-4,1952-1962//? 3v.
Geog 250.5	Special publication. (Vereineging vir Aardrykskunde-Onderwys.) Stellenbosch.
An 359.63.30	Spelling, K. Miljøets indflydelse på intelligensvaviklingen. København, 1963.
Geog 3055.42	Spence, Lewis. Atlantis in America. London, 1925.
Geog 3055.42.5	Spence, Lewis. The history of Atlantis. London, 1926.
Geog 3055.42.7	Spence, Lewis. The history of Atlantis. 4th ed. London, 1927.
Geog 3055.64	Spence, Lewis. The problem of Atlantis. 2d ed. N.Y., 1925?
Geog 3060.13.10	Spence, Lewis. The problem of Lemuria. Philadelphia, 1933.
An 358.85.7	Spencer, Herbert. La especie humana. Madrid, 1885.
Geog 4509.34.5	Spencer, S. Mountaineering. London, 19- .
Geog 4308.70.5	Spender, E. Fjord, Isle and Tor. London, 1870.
Geog 659.28.1	Spethman, Hans. Dynamische Länderkunde. Kiel, 1972.
Geog 4559.69.10	Spiers, George. The Wavertree; being an account of an ocean wanderer. N.Y., 1969.
Geog 4182.47	Spilbergen, J. van. De reis om de wereld. 's-Gravenhage, 1943. 2v.
Geog 3019.35.5	Spilhaus, M.N. (Mrs.). The background of geography. Philadelphia, 1935.
Geog 4181.18	Spillbergen, J. van. East and West Indian mirror. London, 1906.
An 359.31.5	Spiller, Gustav. The origin and nature of man. London, 1931.
An 359.35.3	Spiller, Gustav. The origin and nature of man. 2. ed. London, 1935.
Htn Geog 3055.37*	La spinalba antica historia. (Tomasi, T.) Venetia, 1647.
An 629.38	Spinners and weavers in anthropological research. (Balfour, Henry.) Oxford, 1938.
Geog 3019.51	The spirit and purpose of geography. (Wooldridge, S. William.) London, 1951.
An 170.345	Spiro, Melford E. Context and meaning in cultural anthropology. N.Y., 1965.
An 2109.61.15F	Spivack, M.R. La danse cosmique de Lascaux. Montignac, 1961.
An 359.37.13	Społeczeństwo pierwotne. (Krzywicki, L.) Warszawa, 1937.
Geog 85.385.5	Spomenica o pedecetogodišnjici Srpskog geografskog društva, 1910-1960. Beograd, 1961.
Geog 5324.2.55	Sponsel, H. Fridtjof Nansen. Nürnberg, 1952.

252
Author and Title Listing

Geog 4559.58.5 The springs of adventure. (Noyce, Wilfrid.) Cleveland, 1958.
Geog 4229.11 Square rigger round the Horn. (Wilmore, C. Ray.) Camden, Me., 1972.
Geog 4509.68 Srečanja z gorami. (Potočnik, Miha.) Ljubljana, 1968.
Geog 4308.39.9 Sreznevskii, I.I. Putevyia pis'ma...iz slavianskikh zemel, 1839-1842. Sankt Peterburg, 1895.
An 359.28.9 Staat und Rasse. (Hildebrandt, K.) Breslau, 1928.
Geog 4309.34.5 Stackpole, Edward J. Land of the midnight sun, Scandinavian region, Russia, and Germany. Harrisburg, 1934.
Geog 4180.51 Stade, H. Captivity of Hans Stade of Hesse. London, 1874.
Geog 819.27.10 Städte, Landschaften und ewige Bewegung. (Paquet, A.) Hamburg, 1927.
An 359.56.10 Staff report on a scientist's report on race differences. (Chicago Urban League.) Chicago, 1956.
An 359.15.5 Die Stammesgeschichte der Primaten und die Entwicklung der Menschrassen. (Arldt, Theodor.) Berlin, 1915.
Geog 665.56 Stamp, L. London essays in geography. Cambridge, 1951.
Geog 4678.24 Stanford, John. Aetna; a discourse. N.Y., 1824.
Geog 612.5 Stanisław Lencewicz. (Kondracki, Jerzy.) Warszawa, 1966.
Geog 614.15 Stanislaw Nowakowski. (Barcinski, Florian.) Warszawa, 1965.
Geog 616.31 Stanisław Pawłowski. (Olszewicz, Bolesław.) Warszawa, 1968.
Geog 4180.52A Stanley, E.J. First voyage round the world by Magellan. London, 1874.
An 2159.73.15 Stanovlenie obshchestvennego proizvodstva. (Gur'ev, Dmitrii V.) Moskva, 1973.
Geog 4559.29.3 Stanton, William H. The journey of...pilot, of Deal. London, 1929.
Geog 5160.40 Stanwell-Fletcher, T.M. Clear lands and icy seas. N.Y., 1958.
Geog 4132.6 Staří Čechové na cestách. (Trantina, V.) Praha, 1941.
Geog 4145.84.5 Stark, Freya. Beyond Euphrates. 1. ed. London, 1951.
Geog 4145.84.10 Stark, Freya. The coast of incense. 1. ed. London, 1953.
Geog 4145.84.15 Stark, Freya. Dust in the lion's paw. London, 1961.
Geog 4145.84.20 Stark, Freya. Letters. Compton Chamber, 1974-
Geog 4145.83 Stark, Freya. Perseus in the wind. London, 1948.
Geog 4145.84 Stark, Freya. Traveller's prelude. London, 1950.
Geog 4300.18.5 Starke, M. Information and directions for travellers. Paris, 1826.
Geog 4209.55.5 Starobin, J.R. Paris to Peking. 1. ed. N.Y., 1955.
An 2109.01.9 Starr, F. Some first steps in human progress. Cleveland, 1901.
An 2108.95 Starr, F. Some first steps in human progress. Meadville, Pa., 1895.
Geog 4249.38 Stars to windward. (Fahnestvak, B.) N.Y., 1938.
Geog 4307.77 Staszic, S. Dziennik podrózy, 1777-1791. v.1-2. Warszawa, 1903.
Geog 4307.78 Staszic, S. Dziennik podrózy, 1789-1805. Krakow, 1931.
Geog 4269.64.15 Gli stati d'Europa. (Bonasera, Francesco.) Palermo, 1964.
Geog 659.69 Statistical analysis in geography. (King, Leslie J.) Englewood Cliffs, 1969.
Geog 659.63.2 Statistical methods and the geographer. 2nd ed. (Gregory, Stanley.) London, 1971.
Geog 4131.71 Statistika na turizma. (Patev, Iliia.) Varne, 1971.
Geog 5398.20.10 Statistische...Nachrichten über...Besetzungen. (Wrangell, F. von.) St. Petersburg, 1839.
NEDL Geog 807.40 Lo stato presente di tutti i paesi e popoli del mondo naturale, politico e morale. 2a ed. v.1-26. (Salmon, Thomas.) Venezia, 1738-66. 27v.
Geog 5837.75 Stauning, J. Kort beskrivelse over Grønland. v.1-3. Viborg, 1775.
Geog 4269.64.20 Staurodromia tēs Eurōpēs. (Phloros, Paulos.) Athēnai, 1964.
Geog 4128.91 Steamship lines of the world. (Hunt, R.) N.Y., 1891.
Geog 4105.4 Steamship notes...a handbook. (Noyes, E.H.) N.Y., 1874.
Geog 4237.00 Stearns, R.P. The course of Captain Edmond Halley in the year 1700. n.p., 1936.
Geog 4759.64 Steel, R.W. Geographers and the tropics. London, 1964.
Geog 5182.7 Steensby, Hans P. An anthropological study of the origin of the Eskimo culture. Kjøbenhavn, 1916.
An 359.02.5 Steenstrup, J. Ethnografien. Kjøbenhavn, 1902.
Geog 5839.09.10 Steenstrup, K. Geologiske og antivariske Iagttageker. København, 1909.
Geog 665.130 Steering Committee for Celebration of the Sixtieth Year of Prof. S.P. Chatterjee. Essays in geography. Calcutta, 1965.
Geog 4309.00 Steevens, G.W. Glimpses of three nations. N.Y., 1900.
Geog 5160.60 Stefansson, Evelyn Schwartz Baird. Here is the far north. N.Y., 1957.
Geog 5160.15 Stefansson, Evelyn Schwartz Baird. Within the circle. N.Y., 1945.
Geog 5399.25 Stefansson, V. The adventure of Wrangel Island. N.Y., 1925.
Geog 5180.16 Stefansson, V. The Arctic in fact and fable. N.Y., 1945.
Geog 5160.11.15 Stefansson, V. Arctic manual. N.Y., 1944.
Geog 5399.21 Stefansson, V. The friendly Arctic. N.Y., 1921.
Geog 5399.21.2 Stefansson, V. The friendly Arctic. N.Y., 1943.
Geog 5399.14 Stefansson, V. The friendly Arctic. N.Y. 1922.
Geog 4169.47 Stefansson, V. Great adventures and explorations. N.Y., 1947.
Geog 5660.26A Stefansson, V. Greenland. 1. ed. Garden City, 1942.
Geog 5399.06.9A Stefansson, V. Hunters of the great North. N.Y., 1922.
Geog 5160.7A Stefansson, V. The northward course of empire. N.Y., 1922.
Geog 5509.58.5 Stefansson, V. Northwest to fortune. N.Y., 1958.
Geog 5939.29 Stefansson, V. The theoretical continent. N.Y., 1929.
Geog 5160.11.10 Stefansson, V. Ultima Thule. N.Y., 1940.
Geog 5160.11.12 Stefansson, V. Ultima Thule. Reykjavik, 1942.
Geog 5160.11.5 Stefansson, V. Unsolved mysteries of the Arctic. N.Y., 1939.
Geog 5160.11 Pamphlet box. Stefansson, V. Minor publications.
Geog 5155.8 Pamphlet box. Stefansson, V. Minor works on Arctic exploration.
Geog 5329.3.10 Stefansson, Vilhjalmur. Discovery. 1. ed. N.Y., 1964.
Geog 4502.5.3 Şteinberger, S. Leben und Schriften. München, 1929.
Geog 4209.71.10 Šteinbergs, Valentins. Filosofiia i dzhentel'meny. Riga, 1971.
An 608.70 Steinhauer, C.L. Kort vriledning i det kgl. ethnographiske museum. København, 1870.
Geog 5919.59 Steinitz, Hans. Der 7. Kontinent. Bern, 1959.
An 5.4 Steinmetz, S.R. Essai d'une bibliographie ssytématique de l'ethnologie...1911. Bruxelles, 1911.
Geog 4239.69.15 Stelda, George and I. (Woolass, Peter.) London, 1971.
Geog 5398.99.5 Die Stella Polare im Eismeer. (Amadeus, Louis.) Leipzig, 1903.
An 359.10 Die Stellung der Pygmäenvölker in der Entwicklungsgeschichte des Menschen. (Schmidt, W.) Stuttgart, 1910.
An 359.28.7 Die Stellung des Menschen im Kosmos. (Scheler, Max.) Darmstadt, 1928.
An 359.49.20 Die Stellung des Menschen in Kosmos. (Scheler, Max.) München, 1949.
Geog 4559.55.10 Stenhouse, J.R. Cracker hash. London, 1955.
Geog 5882.5 Stensnaes - den glemte kirke. (Bjoergmose, Rasmus.) Odense, 1967.
Geog 4308.52.5 A step from the new world to the old. (Tappan, H.P.) N.Y., 1852. 2v.
Geog 4502.5.20 Stephen, L. Der Tummelplatz Europas. München, 19- .
Geog 4218.93.5 Stephens, A.A. A Queenslander's travel notes. Sydney, 1894.
Geog 4219.07.5 Stephens, Edwin William. Around the world; a narrative in letter form of a trip around the world, from October 1907 to July 1908. Columbia, 1909.
Htn Geog 4308.38.8*A Stephens, J.L. Incidents of travel in Greece, Turkey and Poland.7th ed. N.Y., 1838. 2v.
Geog 4308.38.14 Stephens, J.L. Incidents of travel in Greece, Turkey and Poland. 17th ed. N.Y., 1845. 2v.
An 137.1.5 Stern, B.J. Lewis Henry Morgan, social evolutionist. Chicago, 1931.
Geog 4347.91.1 Steuke, Johann Kaspar. Von Amsterdam nach Temiswar. Berlin, 1969.
Geog 4558.61.5 Stevens, Charles. A sailor boy's experience. Napanee, Ont., 1892.
Geog 4181.64 Stevens, H.N. New light on the discovery of Australia. London, 1930.
Htn Geog 4167.11* Stevens, J. New collection of voyages. London, 1711. 2v.
Geog 4218.87 Stevens, Thomas. Around the world on a bicycle. N.Y., 1887-88. 2v.
An 3309.14.3F Stevenson, B.L. Constancy or variability in Scandinavian type. Leiden, 1914.
An 3309.16 Stevenson, B.L. Socio-anthropometry. Boston, 1916.
Geog 3251.23.3 Stevenson, E.L. Genovese world map, 1457. Facsimile. N.Y., 1912.
Geog 3251.23 Stevenson, E.L. Map of the world. N.Y., 1907.
Geog 3070.25 Stevenson, E.L. Maps reproduced as glass transparencies. N.Y., 1913.
Geog 3251.23.5 Stevenson, E.L. Marine world chart, 1502. N.Y., 1908.
Geog 3235.56A Stevenson, E.L. Portolan charts. N.Y., 1911.
Geog 602.1.5 Stevenson, E.L. Willem Janszoon Blaeu. N.Y., 1914.
Geog 500.62 Stevenson, Edward L. Publications of Edward L. Stevenson. n.p., 191-.
Geog 4558.98.5 Stevenson, P.E. By way of Cape Horn. Philadelphia, 1899.
Geog 4558.98 Stevenson, P.E. A deep-water voyage. Philadelphia, 1898.
Geog 4559.35.7 Stevers, M.D. Sea lanes; man's conquest of the ocean. Garden City, N.Y., 1938.
Geog 4559.35.5 Stevers, M.D. Sea lanes; man's conquest of the ocean. N.Y., 1935.
An 359.46.10A Stewart, George R. Man: an autobiography. N.Y., 1946.
Geog 4558.94 Stewart, J.A.E. En mer. n.p., 1895?
Geog 4182.26 Die stichting van New York in Juli 1625. (Wieder, F.C.) 's-Gravenhage, 1925.
An 359.70.15 Stiglmayr, Engelbert. Ganzheitliche Ethnologie. Wien, 1970.
Geog 953.10 Stockholm. Biblioteket. Magnus Gabriel de la Gordie's samling af öldre stadsvger. Stockholm, 1915.
An 359.68.15 Stocking, George W. Race, culture, and evolution. N.Y., 1968.
An 359.68.16 Stocking, George W. Race, culture, and evolution. N.Y., 1971.
Htn Geog 4308.89.10* Stockton, F.R. Personally conducted. N.Y., 1889.
Geog 4348.32 Stocqueler, J.H. Fifteen months pilgrimage. London, 1832. 2v.
Geog 4208.97.3 Stoddard, J.L. Lectures. Boston, 1899-1900. 10v.
Geog 4208.97.4 Stoddard, J.L. Lectures. Supplement. n.p., 1901-05. 4v.
Geog 4208.97.5 Stoddard, J.L. Lectures. v.8. Boston, 1905.
Geog 958.92F Stoddard, John L. Glimpses of the world. Chicago, 1892.
Geog 958.92.5F Stoddard, John L. Portfolio of photographs of famous scenes, cities and paintings. Chicago, 189-?
Geog 4502.5.28 Stöger-Ostin, Georg. Georg Jennerwein, der Wildschütz. München, 1939?
Geog 4218.51 Stogman, C.D. Fregatten Eugenies resa omkring jorden, aren 1851-1853. v.1-2. Stockholm, 1855?
Geog 4209.69.25 Stoianovich, Ivan. Bregovete na khorata. Sofiia, 1969.
Geog 4308.78.15 Stokes, F.A. College tramps. N.Y., 1880.
Geog 4308.83.10 Stokes, F.A. A jolly summer. N.Y., 1883.
Geog 4307.91 Stolberg, F.L.G. Reise in Deutschland, der Schweiz. Königsberg, 1794. 4v.
Geog 4307.91.4 Stolberg, F.L.G. Travels through Germany, Switzerland, Italy. 2nd ed. London, 1797. 4v.
Geog 4307.91.2 Stolberg, F.L.G. Travels through Germany. London, 1796. 2v.
Geog 579.65.5 Stolitsy stran mira. Moskva, 1965.
Htn Geog 4308.67.9* Stone, H.S. From Cleveland to Russia. n.p., 1867.
Geog 4678.40.5 Stone, John S. A sermon occasioned by burning of steamer Lexington. Boston, 1840.
An 2109.72 Stone age economics. (Sahlins, Marshall David.) Chicago, 1972.
An 2109.67 The Stone Age hunters. (Clark, John Grahame Douglas.) London, 1967.
Geog 4169.45.5 Stood, Frederick T. Modern travel. London, 1946.
Geog 5399.58 Den store slaederejse. (Rasmussen, Knud.) København, 1958.
Geog 3550.7 Storia dei viaggiatori italiani. (Branca, G.) Roma, 1873.
Geog 3550.5 Storia dei viaggiatori italiani. (Gubernatis, A. de.) Livorno, 1875.
Geog 3019.59.10 Storia della geografia. (Codazzi, Angelo.) Milano, 1959.
Geog 4709.52 Storia della pirateria nel mondo. (Franchi, A.) Milano, 1952. 2v.
Geog 3070.56 Storia delle carte geografiche. (Codazzi, Angela.) Milano, 1952-
Geog 3070.56.5 Storia delle carte geografiche. (Codazzi, Angela.) Milano, 1958.
Geog 3019.58 Storia delle esplorazioni geografiche. (Albertini, Renzo.) Venezia, 1958.
Geog 3129.49 Storia dell'esplorazione e della scienza geografica: l'eta greca. (Almagià, Roberto.) Roma, 1949.
Geog 3019.37 Storia letteraria delle scoperte geografiche. (Olschki, L.) Firenze, 1937.
An 379.69 Storkebaum, Werner. Sozialgeographie. Darmstadt, 1969.
Geog 659.67 Storkebaum, Werner. Zum Gegenstand und zer Methode der Geographie. Darmstadt, 1967.

Author and Title Listing

Geog 4181.26 Storm van 's Gravesande, L. Rise of British Guiana. London, 1911. 2v.

Geog 3018.99 Story of geographical discovery. (Jacobs, J.) London, 1899.

X Cg An 358.89.3 The story of man. (Buel, J.W.) Philadelphia, 1889.

Geog 3070.48.5 The story of maps. (Brown, Lloyd A.) Boston, 1950.

Geog 3070.48 The story of maps. 1st ed. (Brown, Lloyd A.) Boston, 1949.

Geog 5060.8 The story of polar conquest. (Marshall, Logan.) n.p., 1913.

An 2109.51.10 The story of prehistoric civilizations. (Davison, D.) London, 1951.

An 2108.95.3 The story of "primitive" man. (Clodd, Edward.) N.Y., 1895.

An 2109.40.10 The story of primitive man. (Cole, M.C.) Chicago, 1940.

Geog 3019.38 The story of twentieth-century exploration. (Key, Charles E.) N.Y., 1938.

Geog 4677.96 Stout, Benjamin. The total loss of the American ship Hercules. London, 17- ?

Htn Geog 4308.54.5* Stowe, H.B. Sunny memories of foreign lands. Boston, 1854. 2v.

Geog 4279.52 Stoye, J.W. English travellers abroad. London, 1952.

Geog 665.7 Strachey, R. Lectures on geography. London, 1888.

Geog 4180.6 Strachey, W. Historie of travaile. London, 1849.

Geog 4181.103 Strachey, William. The historie of travel into Virginia Britania. London, 1953.

Geog 4521.27 Straight up; the life and death of John Harlin. (Ullman, James Ramsey.) Garden City, N.Y., 1968.

Geog 4227.85F Strange, James. Journal and narrative of the...expedition from Bombay to the N.W. coast of America. Madras, 1928.

Geog 4181.6 Strange adventures of A. Battell. (Battell, A.) London, 1901.

Geog 4672.16.1 Strange adventures of the sea. (Lockhart, John Gilbert.) N.Y., 1926.

Geog 4672.16 Strange adventures of the sea. (Lockhart, John Gilbert.) N.Y., 1931.

Htn Geog 5516.31* Strange and dangerous voyage...N.W. Passage. (James, T.) London, 1633.

Htn Geog 5516.31.2* Strange and dangerous voyage...N.W. Passage. (James, T.) London, 1633.

Geog 5516.31.3 The strange and dangerous voyage. (James, T.) Toronto, 1975.

Geog 4672.14 Strange sea mysteries. (O'Donnell, Elliott.) London, 1926.

Geog 4672.16.5 Strange tales of the seven seas. (Lockhart, John Gilbert.) London, 1929.

Htn Geog 4477.35.3* The strange voyage and adventures of Domingo Gonsales. 2. ed. (Godwin, Francis.) London, 1768.

Geog 4239.68.20 The strange voyage of Donald Crowhurst. (Tomalin, Nichols.) London, 1970.

Geog 5700.8.5 Strangers in Greenland. n.p., n.d.

Geog 5538.53.7 Strangers in Greenland. N.Y., 18- ?

Geog 4269.63 Strany, strechi, uchenye. (Glushchenk, I.) Moskva, 1963.

Geog 4209.65.10 Strany i liudi. (Efimov, Gerontii V.) Leningrad, 1965.

Htn An 358.69* Stratton, C.S. Sketch of life, personal appearance, character and manners. N.Y., 1869.

Geog 5538.50.4 Stray leaves from an Arctic journal. (Osborn, S.) Edinburgh, 1865.

Geog 5538.50.2 Stray leaves from an Arctic journal. (Osborn, S.) N.Y., 1852.

Geog 4309.64.20 Streeter, Edward. Along the ridge; from northwestern Spain to southern Yugoslavia. N.Y., 1964.

An 2109.68.5 Strejftog. Arktiske, tropiske og midt im ellem. (Birket Smith, Kaj.) København, 1968.

Geog 5839.59.5 Strejftog i Nord. (Munch, Ebbe.) København, 1959.

Geog 185.16 Streszczenia prac habiltacyjnych i dobtorskich. (Polska Akademia Nauk. Institut Geografii.) Warszawa. 1967+

Geog 4308.57 A stroller in Europe. (Doré.) N.Y., 1857.

Geog 4308.33.10 Strombeck, Friedrich Karl von. Darstellungen aus meinem Leben und aus meiner Zeit. Braunschweig, 1833-1840. 4v.

Geog 6009.46 Strong men south. (Menster, William J.) Milwaukee, 1949.

An 359.63.40A Structural anthropology. (Lévi-Strauss, Claude.) N.Y., 1963-76. 2v.

An 136.5.65 Le structuralisme, ou L'histoire en exil. Thèse. (Ipola, Emilio Rafael de.) Nantorre? 1969.

An 136.5.80 Le structuralisme de Lévi-Strauss. (Marc-Lipiansky, Mireille.) Paris, 1973.

An 136.5.75 Structuralisme ou ethnologie. (Makarius, Raoul.) Paris, 1973.

An 2109.52 Structure and function in primitive society. (Radcliffe-Brown, A.R.) London, 1952.

An 2109.52.2 Structure and function in primitive society. (Radcliffe-Brown, A.R.) London, 1956.

An 2109.28.11 La structure de la mentalité primitive. (Leeuw, Gerardus.) Strasbourg, 1928.

Geog 4131.68.10 Structure et tâches d'un organisme national de tourisme. Inaug. Diss. (Aeschlimann, Jean Louis.) Bern, 1968.

Geog 5312.3.15A Struggle; the life and exploits of Commander Richard E. Byrd. (Murphy, Charles J.V.) N.Y., 1928.

Geog 4229.72 Struve, Wolfgang. Unglaubliche Wirklichkeit. Salzburg, 1972.

Geog 4346.76.25 Struys, Jan J. Tri puteshestviia. Moskva, 1935.

Htn Geog 4347.18* Struys, Jan J. Les voyages...en Moscovie. v.1-3. Amsterdam, 1718.

Htn Geog 4346.76.5* Struys, Jan J. Les voyages de Jean Struys. Amsterdam, 1681.

Geog 3240.28 Strzelczyk, Jerzy. Gerwazy z Tilbury. Wrocław, 1970.

Geog 500.29 Stuck, G.H. Verzeichnis von...Land und Reisebeschreibungen. Halle, 1784.

Geog 4349.28 The student abroad. (Brennan, John W.) Boston, 1928.

Geog 758.79 Students' manual of modern geography mathematical, physical and descriptive. (Bevan, W.L.) London, 1879.

Geog 3550.3 Studi bibliografici e biografici sulla storia della geografia in Italia. Roma, 1875.

Geog 4131.68.20 Studi e ricerche sulla regione turistica. (Corna Pellegrini, Giacomo.) Milano, 1968.

Geog 665.70 Studi geografici in onore di Antonio Renato Toniolo. (Bologna. Università. Istituto di Geografia.) Milano, 1952.

An 3309.65.5 Studia afrykanistyczne. (Łódzskic Towarzystwo Naukowe. Wydział III, Nauk Matematyczno-Przyrodniczych.) Łódź, 1965.

Geog 214.10 Studia geograficzno-fizyczne z obszasu opolszocyzny. Opole. 1,1968+

Geog 214.25 Studia geographica. Brno. 1,1969+ 4v.

Geog 3019.73 Studia z dziejów geografii i kartografii. Wrocław, 1973.

Geog 3240.26 Studien in arabischen Geographen. (Jacob, Georg.) Berlin, 1891-92.

Geog 5378.80.7 Studien und Forschungen...Reisen. (Nordenskiöld, A.E.) Leipzig, 1885.

Geog 214.20 Studien zur Kartographie. Berlin. 1-2

Geog 3540.15 Studien zur politischen Wissenschaftsgeschichte der deutschen Geographie im Zeitalter des Imperialismus. (Schulte-Althoff, Franz Josef.) Paderborn, 1971.

Geog 500.121 Studienbibliographie Geographie. (Josuweit, Werner.) Wiesbaden, 1973.

An 379.73.20 Studies in human geography. London, 1973.

An 379.30.7 Studies in regional consciousness and environment. (Peate, I.C.) London, 1930.

Geog 4131.74 Studies in the geography of tourism. (International Geographical Union. Working Group, Geography of Tourism an Recreation.) Frankfurt, 1974.

Htn An 2109.29* Studies of savages and sex. (Crawley, A.E.) London, 1929.

Geog 5855.12 Studies on the material culture of the Eskimo in West Greenland. (Porsild, N.P.) København, 1915.

An 3508.84 Studley, C.A. Notes upon human remains from the caves of Coahuila, Mexico. Salem, 1884.

An 359.74.25 The study of culture. (Langness, Lewis L.) San Francisco, 1974.

An 629.53 The study of culture at a distance. (Mead, M.) Chicago, 1953.

Geog 665.52 The study of geography. (Mogey, J.M.) London, 1950.

An 358.98A The study of man. (Haddon, A.C.) N.Y., 1898.

An 359.36.3 The study of man. (Linton, Ralph.) N.Y., 1936.

An 359.27.7A A study of races in the ancient Near East. (Worrell, W.H.) N.Y., 1927.

Geog 3520.11.10 A study of the facsimile edition of Richard Hakluyt's Divers voyages. (Quinn, David Beers.) Amsterdam, 1968. 2v.

Geog 818.54 Stutt kennslubok i landafroedinni. (Ingerslev, C.F.) Reykjavik, 1854.

Geog 818.43 Stutt landaskipunarfraedi. (St. Platou, L.) n.p., 1843.

Geog 5398.75 Stuxberg, A. Einringar fran Svenska Expeditionerna. Stockholm, 1877.

Geog 5368.80 Stuxberg, A. Nordostpassagens historia. Stockholm, 1880.

Geog 5160.25 Subarktika. 2. Izd. (Grigor'ev, A.A.) Moskva, 1956.

Geog 3570.27 Subsidios para a história de cartografia portuguesa. (Pina Manique, Luiz da.) Lisboa, 1943.

Htn Geog 4266.65* Subsidium peregrinatibus. (Gerbier, B.) Oxford, 1665.

Geog 557.62 Succinta descrizione. Venezia, 1763.

Geog 4181.140 Sucesos de las Islas Filipinas. (Morga, Antonio de.) Cambridge, Eng., 1971.

Geog 4217.37 Sud-Lander und um die Balt. (Behrens, E.F.) Leipzig, 1737.

Geog 4209.73 Sud'ia-vremia. (Matveev, Vikentii A.) Moskva, 1973.

Geog 5209.43 Suđr um höf. (Einarsson, S.) Reykjavik, 1943.

Geog 5839.24 Die sudwestgrönländische Landschaft und das Siedlungsgebiet der Normannen. (Nissen, N.W.) Hamburg, 1924.

Geog 1524.19 Süd-Deutschland. 22. Aufl. (Baedeker, publishers.) Leipzig, 1888.

Geog 1524.19.2 Süd-Deutschland. 23. Aufl. (Baedeker, publishers.) Leipzig, 1890.

Geog 1524.32 Süd-Deutschland und Oesterreich. 19. Aufl. (Baedeker, publishers.) Leipzig, 1882.

Geog 1524.15 Süd-Deutschland und Oesterrreich. 8. Aufl. (Baedeker, publishers.) Leipzig, 1879.

Geog 1524.75 Südbaiern, Tirol und Salzburg. 18. Aufl. (Baedeker, publishers.) Leipzig, 1878.

Geog 1524.77 Südbaiern, Tirol und Salzburg. 22. Aufl. (Baedeker, publishers.) Leipzig, 1886.

Geog 1524.73 Südbayern: Alpenvorland, Alpen, österreichische Gunzgebiete. 41. Aufl. (Baedeker, publishers.) Hamburg, 1953.

Geog 4269.31.10 Suedost- und Südeuropa in Natur. v.1-7,8-14,18. Potsdam, 1931-1936. 3v.

Geog 5925.20 Der Südpol. (Kollback, K.) Bielefeld, 1911.

Geog 1535.20 La Suisse. 7. éd. (Baedeker, publishers.) Coblenz, 1867.

Geog 1535.23 La Suisse. 8. éd. (Baedeker, publishers.) Coblenz, 1869.

Geog 1535.44 La Suisse. 14. éd. (Baedeker, publishers.) Leipzig, 1885.

Geog 4209.71 Šulentić, Zeatko. Ljudi krajevi Beskraj. Karlovac, 1971.

An 3309.28 Sullivan, Louis R. Essentials of anthropometry. N.Y., 1928.

Geog 5919.57 Sullivan, Walter. Quest for a continent. N.Y., 1957.

Geog 4558.90.5 Sull'oceano. (Amicis, E. de.) Milano, 1890.

Geog 4558.90.6 Sull'oceano. (Amicis, E. de.) Milano, 1913.

Geog 4559.28.15 Sull'oceano. (Amicis, E. de.) Milano, 1928.

Htn Geog 815.19* Suma de geographia. (Denciso, M.F.) Seville, 1519.

Geog 4181.89 Suma Oriental of Tomé Pires...book of Francisco Rodrigues. (Cortesão, Armando.) London, 1944. 2v.

Geog 3107.94 Summary of geography and history. (Adam, A.) Edinburgh, 1794.

NEDL Geog 818.16 Summary of geography and history. (Adam, A.) London, 1816.

Geog 818.02 A summary of geography and history. 3rd ed. (Adam, A.) London, 1802.

Geog 4328.53 Summer cruise in the Mediterranean. (Willis, N.P.) Auburn, 1853.

Geog 4308.89 Summer holidays...Europe. (Child, T.) N.Y., 1889.

Geog 4308.59.3 Summer pictures. (Field, H.M.) N.Y., 1859.

Geog 5538.52.10 A summer search for Sir John Franklin. (Inglefield, Edward A.) London, 1853.

Geog 4308.46.5 Summer's jaunt across the water. v.1-2. (Smith, J.J.) Philadelphia, 1846.

Geog 4308.51.25 A summer's tour in Europe in 1851. Charleston, 1852.

Geog 4218.78.7A The "Sunbeam". (Brassey, Thomas.) London, 1918.

An 359.73.20 Sunderland, Eric. Elements of human and social geography. 1. ed. Oxford, 1973.

An 2058.81.5 Sunderland, J.T. Dr. Winchell's "preadamites". n.p., 1881.

Geog 5398.97.6 Sundman, Per Olaf. Ingen fruktan, intet hopp. Stockholm, 1968.

Geog 3055.21 The sunken island of Atlantis. (Unger, F.X.) n.p., n.d.

Geog 4208.83.10 Sunny lands and seas. (Wilkinson, H.) London, 1883.

Htn Geog 4308.54.5* Sunny memories of foreign lands. (Stowe, H.B.) Boston, 1854. 2v.

Geog 500.61.5 Supplément au catalogue...paru en décembre, 1887. (Société de Géographie de Lille. Bibliothèque.) 1,1889

Geog 5518.19.8 Supplement to the appendix of Capt. Parry's voyage. (Parry, W.E.) London, 1824.

Geog 139.5.5 Supplement to the bibliography of Algeria, 1895. (Playfair, R.L.) London, 1898.

Geog 819.22.2 Supplement to The new world. (Bowman, I.) Yonkers-on-Hudson, 1923-1924. 2v.

Geog 4308.45.25 Supplement to "Vacation rambles". (Talfourd, T.N.) London, 1854.

Author and Title Listing

	Geog 139.5	Supplementary papers. v.1-3. Photoreproduction. (London. Royal Geographical Society.) 1951-1955 6v.
	Geog 578.55.11	Supplementary tables of population. (Lippincott, J.B. and Co.) Philadelphia, 1883.
	Geog 4219.28	The supreme travel-adventure: around the world in the "Franconia" 1929. (Cook, Thomas, firm, publishers, London.) N.Y., 1928.
	Geog 4308.34	Sur les chemins de l'Europe. (Michelet, J.) Paris, 1893.
	Geog 4239.51.5	Sur les routes de l'Atlantique. (Buhler, Jean.) Lausanne, 1951.
	Geog 4209.59.5	Sur les routes du monde. (Trolliet, Héli.) Lausanne, 1959.
	An 2059.68.5	Sur les traces d'Adam; nouvel aperçu sur les origines de l'homme. (Nantevil, Hugues de.) Paris, 1968.
	Geog 4209.45.20	Sur les traces de Magellan. (Retuerto, Marcial.) Paris, 1945.
	An 2108.84	Sur une méthode a suivre dans les études dites préhistoriques. (Van Overloop, E.) Bruxelles, 1884.
	An 2109.31.11	Le surnaturel et la nature. (Lévy-Bruhl, L.) Paris, 1931.
	An 3309.45	A survey in seating. (Hoaton, E.A.) Cambridge, 1945.
	An 358.92.7	Survivals of prehistoric races in Mt. Atlas and Pyrenees. (Haliburton, Robert G.) Lisbon, 1892.
	Geog 4679.73	Survive the savage sea. (Robertson, Dougal.) London, 1973.
	Geog 5399.57.15	Sushkina, N.N. Dva leta v Arktike. Moskva, 1957.
Htn	Geog 4238.03*	Sutherland, D. A diary kept by Reverend David Sutherland on a voyage from Greenock Scotland to New York. Woodsville, 1910.
Htn	Geog 816.85*	Suval, P. Geographia universalis. London, 1685.
	Geog 218.2	Svensk geografisk årsbok. Lund. 39,1963+ 23v.
	Geog 218.2.2	Svensk geografisk årsbok. Register. Lund.
	Geog 5398.97.4	Svenska Sällskapet för Antropologi och Geografi. Med örnen mot polen av S.A. Andrée. Stockholm, 1930.
	VGeog 4218.51.5	Svenska Vetenskaps-Akademien, Stockholm. Voyage autour du monde sur la frégate suédoise l'Eugénie executé pendant des années 1851-1853. Stockholm, 1858-74.
	Geog 4209.56.15	Svenson, Sven. Varlden ar så stor, så stor. Stockholm, 1956.
	Geog 5398.98	Sverdrup, O. Neues Land, vier Jahre in arktischen Regionen. Leipzig, 1903. 2v.
	Geog 5398.98.2	Sverdrup, O. New land, four years in Arctic regions. London, 1904. 2v.
	Geog 4239.68.25	Svirin, Vladimir P. Vzemliakh blizkikh i dal'nikh. Stavropol', 1970.
	Geog 5639.60	Svundne tider i Østgrønland. (Mikkelsen, Ejnar.) København, 1960.
	Geog 4209.71.45	Swale, Rosie. Children of Cape Horn. London, 1974.
	Geog 4679.39	Swan, E.W. The first commission of H.M.S. Calliope. Newcastle-upon-Tyne, 1939.
	Geog 5970.30	Swan, Robert A. Australia in the Antarctic. Victoria, 1961.
	Geog 4208.10	Swaving, J.G. Swavings reizen en logevallen, doorhem. v.1-2. Dordrecht, 1827.
	Geog 4208.10	Swavings reizen en logevallen, doorhem. v.1-2. (Swaving, J.G.) Dordrecht, 1827.
	Geog 4239.71.5	The sway of the grand saloon: a social history of the North Atlantic. (Brinnin, John M.) N.Y., 1971.
	Geog 579.56.5	Swayne, James Colin. A concise glossary of geographical terms. London, 1956.
	Geog 579.56.7	Swayne, James Colin. A concise glossary of geographical terms. 2. ed. London, 1962.
	Geog 4308.59	Sweat, M.J.M. Highways of travel. Boston, 1859.
	Geog 4145.14.100	Swinglehurst, Edmund. The romantic journey. London, 1974.
	Geog 4679.41.5	Swinson, A. Scotch on the rocks. London, 1963.
	Geog 4308.30	Switzerland...France and Pyrenees. (Inglis, H.D.) Edinburgh, 1831. 2v.
	Geog 1535.50	Switzerland...Italy, Savoy...Tryol. 15. ed. (Baedeker, publishers.) Leipsic, 1893.
	Geog 1535.30	Switzerland...Italy, Savoy...Tyrol. 5. ed. (Baedeker, publishers.) Coblenz, 1872.
	Geog 1535.31	Switzerland...Italy, Savoy...Tyrol. 6. ed. (Baedeker, publishers.) Coblenz, 1873.
	Geog 1535.33	Switzerland...Italy, Savoy...Tyrol. 7. ed. (Baedeker, publishers.) Leipsic, 1877.
	Geog 1535.35	Switzerland...Italy, Savoy...Tyrol. 8. ed. (Baedeker, publishers.) Leipsic, 1879.
	Geog 1535.40	Switzerland...Italy, Savoy...Tyrol. 9. ed. (Baedeker, publishers.) Leipsic, 1881.
	Geog 1535.42	Switzerland...Italy, Savoy...Tyrol. 10. ed. (Baedeker, publishers.) Leipsic, 1883.
	Geog 1535.43	Switzerland...Italy, Savoy...Tyrol. 11. ed. (Baedeker, publishers.) Leipsic, 1885.
	Geog 1535.45	Switzerland...Italy, Savoy...Tyrol. 12. ed. (Baedeker, publishers.) Leipsic, 1887.
	Geog 1535.47	Switzerland...Italy, Savoy...Tyrol. 13. ed. (Baedeker, publishers.) Leipsic, 1889.
	Geog 1535.46	Switzerland...Italy, Savoy...Tyrol. 13. ed. (Baedeker, publishers.) Leipsic, 1889.
	Geog 1535.48	Switzerland...Italy, Savoy...Tyrol. 14. ed. (Baedeker, publishers.) Leipsic, 1891.
	Geog 1535.55	Switzerland...Italy, Savoy...Tyrol. 16. ed. (Baedeker, publishers.) Leipsic, 1895.
	Geog 1535.57	Switzerland...Italy, Savoy...Tyrol. 17. ed. (Baedeker, publishers.) Leipsic, 1897.
	Geog 1535.25	Switzerland. 3. ed. (Baedeker, publishers.) Coblenz, 1869.
	Geog 1535.26	Switzerland. 4. ed. (Baedeker, publishers.) Coblenz, 1869.
	Geog 1535.58	Switzerland and the adjacent portions of Italy, Savoy, and Tyrol. 18. ed. (Baedeker, publishers.) Leipsic, 1899.
	Geog 1535.59	Switzerland and the adjacent portions of Italy, Savoy, and Tyrol. 19. ed. (Baedeker, publishers.) Leipsic, 1901.
	Geog 1535.59.50	Switzerland and the adjacent portions of Italy, Savoy, and Tyrol. 20. ed. (Baedeker, publishers.) Leipsic, 1903.
	Geog 1535.60	Switzerland and the adjacent portions of Italy, Savoy, and Tyrol. 21. ed. (Baedeker, publishers.) Leipsic, 1905.
	Geog 1535.61	Switzerland and the adjacent portions of Italy, Savoy, and Tyrol. 21. ed. (Baedeker, publishers.) Leipsic, 1905.
	Geog 1535.68	Switzerland and the adjacent portions of Italy, Savoy, and Tyrol. 23. ed. (Baedeker, publishers.) Leipzig, 1909.
	Geog 1535.70	Switzerland and the adjacent portions of Italy, Savoy, and Tyrol. 24. ed. (Baedeker, publishers.) Leipzig, 1911.
	Geog 1535.72	Switzerland and the adjacent portions of Italy, Savoy, and Tyrol. 25. ed. (Baedeker, publishers.) Leipzig, 1913.
	Geog 1535.65	Switzerland and the adjacent portions of Italy, Savoy and Tyrol. 22. ed. (Baedeker, publishers.) Leipzig, 1907.
	Geog 1535.24	Switzerland and the adjacent portions of Italy. 2. ed. (Baedeker, publishers.) Coblenz, 1864.
	Geog 1535.24.25	Switzerland and the adjacent portions of Italy. 3. ed. (Baedeker, publishers.) Coblenz, 1867.
	Geog 1535.76	Switzerland together with Chamonix. 27. ed. (Baedeker, publishers.) Leipzig, 1928.
	Geog 1535.74	Switzerland together with Chamonix and the Italian lakes. 26. ed. (Baedeker, publishers.) Leipzig, 1922.
	Geog 6009.10.5	Sydpolen...med Fram, 1910-1912. (Amundsen, R.) Kristiania, 1912. 2v.
	Geog 3019.34.2	Sykes, P.M. A history of exploration. Westport, Conn., 1975.
	Geog 3019.34	Sykes, P.M. A history of exploration from the earliest times to the present day. N.Y., 1934.
	Geog 4314.6	Symonds, W. Extract from journal in the Black Sea. London, 184-?
	Geog 4269.03.5	Symons, A. Cities. London, 1903.
	Geog 5180.6	Symposium on Circumpolar Problems, Luleå, Sweden and Tromsø, Norway, 1969. Circumpolar problems. 1. ed. Oxford, 1973.
	An 170.349	Symposium on Community Studies in Anthropology. Symposium on community studies in anthropology. Seattle, 1964.
	An 170.349	Symposium on community studies in anthropology. (Symposium on Community Studies in Anthropology.) Seattle, 1964.
	An 2109.66.35	Symposium on Man the Hunter, Chicago, 1966. Man the hunter. Chicago, 1969.
	An 2109.59	Symposium on the Evolution, Chicago, 1957. The evolution of man's capacity for culture. Detroit, 1959.
	Geog 665.124	Symposium über Fragen der Naturräumlichen Gliederung. Probleme der landschaftsökologischen Erkundung und naturräumlichen Gliederung. Leipzig, 1967.
	Geog 3019.20	Synge, M.B. A book of discovery. N.Y., 1920.
	Geog 4309.61.5	Der synthetische Bazar. (Malinowksi, P.) Wien, 1961.
	Geog 3590.78	Syny otvazhnye Rossii. (Alekseev, Aleksandr Ivanovich.) Magadan, 1970.
	Geog 3108.50	System of ancient and mediaeval geography. (Anthon, C.) N.Y., 1850.
	Geog 3108.71	System of ancient and mediaeval geography. (Anthon, C.) N.Y., 1871.
	Geog 808.36	System of geography. (Bell, J.) Glasgow, 1836. 6v.
	Geog 808.36.5	System of geography. (Bell, J.) London, 1850. 6v.
Htn	Geog 817.01F*	System of geography. (Moll, H.) London, 1701.
	Geog 808.08	System of geography. (Playfair, J.) Edinburgh, 1808. 6v.
	Geog 808.07	System of geography. Glasgow, 1807. 4v.
	Geog 807.00F	System of geography. v.1-2. (Middleton, C.T.) London, 17- 4v.
	Geog 818.32	A system of universal geography. (Goodrich, S.G.) Boston, 1832.
	Geog 818.42	System of universal geography. (Laurie, J.) Edinburgh, 1842.
	Geog 808.12.12	A system of universal geography. (Malte-Brun, Conrad.) Boston, 1834. 3v.
	Geog 808.12.13	System of universal geography. v.1-2. (Malte-Brun, Conrad.) Boston, 1844.
	Geog 818.33.5	A system of universal geography. 2d ed. (Goodrich, S.G.) Boston, 1833.
Htn	An 2056.55*	Systema theologicum...Praeadamitae. (La Peyrère, I. de.) n.p., 1655.
	Geog 13.200.5	Systematisch register. (Nederlandsch Aardrijkskundig Genootschap, Amsterdam.) Leiden. 1876-1960 3v.
	Geog 4209.69.10	Sytin, Viktor A. Puteshestviia. Moskva, 1969.
	Geog 4209.52.10	Szos Kies, Henryk J. Your world and mine. N.Y., 1952.
	Geog 4145.83.5	Szumańska-Grossowa, Hanna. Podróze Stefana Srolca Rogozinskiego. Warszawa, 1967.
	An 359.74.20	Szyfelbejn-Sokolewicz, Zofia. Wprowadzenie do etnologii. Wyd. 1. Warszawa, 1974.
	Geog 192.4.5	Table décennale. v.2 (1923-32), v.5 (1953-62). (Grenoble. Université. Institut de Géographie Alpine.) Grenoble, 1932-63. 3v.
	Geog 4313.5	Tableau de la Mer Baltique. (Catteau-Calleville, J.P.) Paris, 1812. 2v.
	Geog 189.2.4	Tables des matieres contenus dans le Bulletin. (Quebec Geographical Society.) Quebec, 1913.
	Geog 585.7	Tables des principales positions geonomiques. (Coulier, P.J.) Paris, 1828.
	An 359.42.5	Taboo, a sociological study. (Webster, Hutton.) Stanford, 1942.
	Geog 4344.35.12	Tafur, Pero. Travels and adventures, 1435-1439. London, 1926.
	Geog 4344.35.10	Tafur, Pero. Travels and adventures, 1435-1439. N.Y., 1926.
	Geog 5398.72.11	Das Tagebuch des Maschinisten Otto Krisch. (Krisch, Otto.) Graz, 1973.
	Geog 4219.06	Tagebuchblätter aus Siberien, Japan. (Huber, Max.) Zürich, 1906.
	Geog 5839.30.6	Tagebücher, Briefe. (Wegener, A.) Wiesbaden, 1960.
	Geog 665.13	Tagsberichte über die Forstschritte. Weimar, 1852.
	Geog 45.12	Tagungsbericht und wissenschaftliche Abhandlungen. (Deutscher Geographentag, 38th, Erlangen and Nuremberg, 1971.) Wiesbaden, 1972.
	Geog 4308.52.3	The Tagus and the Tiber. (Baxter, W.E.) London, 1852. 2v.
	Geog 4249.58.5	Tahiti-Nui. (Bisschop, Eric de.) London, 1959.
	Geog 4308.59.5	Tait, J.R. European life, legend. Philadelphia, 1859.
	An 3309.67	Taiwan aborigines. (Chai, Chen Kang.) Cambridge, 1967.
	Geog 5970.40	Takhariev, Vasil I. Bulgarin na Antarktida. Sofiia, 1971.
Htn	Geog 4219.29.10F*	Taking one's own ship around the world. (Vanderbilt, William K.) N.Y., 1929.
	Geog 4208.49.5	A tale of two oceans. (Barra, E.I.) San Francisco, 1893.
	Geog 4308.42.10	Tales and souvenirs of a residence in Europe. (Rives, J.P.W.) Philadelphia, 1842.
	Geog 3018.82.2	Tales of old travel. (Kingsley, H.) London, 1882.
	Geog 4672.12.6	Tales of S.O.S. and T.T.T. (Kitchen, Frederick H.) Edinburgh, 1927.
	Geog 4308.45.25	Talfourd, T.N. Supplement to "Vacation rambles". London, 1854.
	Geog 4308.45.19	Talfourd, T.N. Vacation rambles...1841-1843. 3rd ed. London, 1851.
	Geog 4308.45.15	Talfourd, T.N. Vacation rambles and thoughts. London, 1845. 2v.
	An 3309.43.5	Tall men have their problems too. (Riggs, F.B.) Cambridge, 1943.
	Geog 4559.60	Tall ships and great captains. (Whipple, Addison.) N.Y., 1960.
	Geog 616.32	Talyzin, Fedor F. Puteshestviia za nevidimym vragom. Moskva, 1974.
	Geog 579.63	Tämän päivän maailmaa. 4. ed. (Hustich, I.) Helsinki, 1963.
	Geog 4679.33	Tambs, E. The cruise of the Teddy. London, 1950.

Author and Title Listing

Geog 4209.40.55 Tangye, Derek. The time was mine. London, 1940.

Geog 4208.83.6 Tangye, Richard. Reminiscences of travel in Australia, America and Egypt. 2. ed. London, 1884.

An 2359.75 Tannahill, Reay. Flesh and blood. London, 1975.

An 359.25.1.3 Tantalus; or The future of man. (Schiller, F.C.S.) London, 1924.

An 359.25.1 Tantalus; or The future of man. (Schiller, F.C.S.) N.Y., 1925.

An 359.25.2 Tantalus; or The future of man. (Schiller, F.C.S.) N.Y., 1925.

Geog 4308.52.5 Tappan, H.P. A step from the new world to the old. N.Y., 1852. 2v.

An 629.37 Taschenbuch der rassenkundlichen Messtechnik. (Schultz, Bruno K.) München, 1937.

Geog 4105.17.5 Tatchell, Frank. The happy traveller. 5. ed. London, 1927.

Geog 4678.32 The tattooed man. 1. American ed. (Meredith, J.C.) N.Y., 1959.

Geog 4509.16 Taüber, C. Auf fremden Bergfpfaden. Zürich, 1916.

Geog 4219.41.10 Tausend und ein Abenteuer. (Faber, kurt.) Berlin, 1941.

Geog 585.13 Tavole sinottiche di geografia. Livorno, 1834.

An 170.351 Tax, Sol. Horizons of anthropology. Chicago, 1964.

Geog 4308.69 Taylor, B. By-ways of Europe. N.Y., 1869.

Htn Geog 4308.69.2* Taylor, B. Byeways of Europe. London, 1869. 2v.

Geog 3018.56.10 Taylor, B. Cyclopaedia of modern travel. N.Y., 1860. 2v.

Geog 3018.56.5 Taylor, B. Cyclopedia of modern travel. Cincinnati, 1856.

Geog 4328.55 Taylor, B. Lands of the Saracew. N.Y., 1855.

Geog 4328.59 Taylor, Bayard. Travels in Greece and Russia. N.Y., 1859.

Geog 4308.46.35 Taylor, Bayard. Views a-foot. 17th ed. N.Y., 1854.

Geog 4309.00.15 Taylor, Charles M. Odd bits of travel with brush and camera. Philadelphia, 1900.

Geog 4181.76 Taylor, E.G.R. The original writings and correspondance of the two Richard Hakluyts. London, 1935. 2v.

Geog 4181.113 Taylor, Eva. The troublesome voyage of Captain Edward Fenton. Cambridge, Eng., 1959.

Geog 3520.17.6 Taylor, Eva G. Late Tudor and early Stuart geography, 1583-1650. N.Y., 1968.

Geog 3520.17 Taylor, Eva G. Tudor geography, 1485-1583. London, 1930.

Geog 4218.38.3 Taylor, F.W. A voyage around the world. v.1-2. 9th ed. New Haven, 1847.

Geog 4218.38.2 Taylor, F.W. A voyage around the world. v.1-2. 9th ed. New Haven, 1850.

An 379.37.3A Taylor, G. Environment, race and migration. Chicago, 1937.

An 379.46.5 Taylor, G. Environment, race and migration. 2. ed. Chicago, 1946.

An 379.46 Taylor, G. Our evolving civilization. Toronto, 1946.

Geog 6009.10.19 Taylor, G. With Scott; the silver lining. London, 1916.

An 379.27A Taylor, Griffith. Environment and race. London, 1927.

Geog 665.58A Taylor, Griffith. Geography in the twentieth century; a story of growth, fields, techniques, aims and trends. N.Y., 1951.

Geog 665.58.2 Taylor, Griffith. Geography in the twentieth century. 2nd ed. N.Y., 1953.

Geog 665.58.3 Taylor, Griffith. Geography in the twentieth century. 3rd ed. N.Y., 1957.

Geog 5919.30 Taylor, Griffith. Antarctic adventure and research. N.Y., 1930.

An 2108.91.3 Taylor, I. The prehistoric races of Italy. Washington, 1891.

Geog 4347.99.7 Taylor, J. Travels from England to India. London, 1799. 2v.

Geog 4308.50.15 Taylor, J.B. Views a-foot or Europe. N.Y., 1850.

Geog 4709.43 Taylor, James. Gold from the sea; the epic story of the "Niagara's" bullion. London, 1943.

An 359.63.25 Tazerout, M. Manifeste contre le racisme. Rodez, 1963.

Geog 4418.90 Tchihatchef, P. de. Etudes de géographie et d'histoire naturelle. Florence, 1890.

Geog 139.9 Technical series. (London. Royal Geographical Society.) London. 1-5,1920-1929

An 2159.73.25F Technik der Steinzeit. (Feustel, Rudolf.) Weimar, 1973.

An 2109.27.30 Technik und Wirtschaft des europäischen Urmenschen. (Cunow, Heinrich.) Berlin, 1927.

Geog 5839.23 The "Teddy" expedition. (Dahl, Kai R.) N.Y., 1925.

Geog 4181.9 Teixeira, P. Travels of Teixeira with his "kings of Harmuz". London, 1902.

Geog 3590.86 Tel', Sergei E. Kartografiia Rossii XVIII veka. Moskva, 1960.

Geog 603.5.8.5 Tel fut Charcot, 1867-1936. (Emmanuel, Marthe.) Paris, 1967.

Geog 3019.17 Teleki, Pal. A foldrajzi gondolat torténete. Budapest, 1917.

Geog 618.1.9 Der teleologische Zug im Denken C. Ritters. (Richter, O.) Borna, 1905.

Geog 4145.81.6 Teller, Walter Magnes. Joshua Slocum. New Brunswick, 1971.

Geog 4145.81.5 Teller, Walter Magnes. The search for Captain Slocum. N.Y., 1956.

Geog 4709.57 Tels etaient corsaires et flibustiers. (Freminville, René Marie de La Poix de.) Paris, 1957.

Geog 4208.98.11 Temple, E.L. Old world memories. Boston, 1899. 2v.

Geog 6100.37.10 Temple, Philip. The sea and the snow; the South Indian Ocean Expedition to Heard Island. Melbourne, 1966.

Htn Geog 5398.50F* Ten coloured views during the Arctic expedition. (Browne, W.H.) London, 1850.

Geog 4509.40.5 Ten great mountains. (Irving, R.L.G.) N.Y., 1940.

Geog 4209.19 The ten islands and Ireland. (Mackay, John.) Dublin, 1919.

Geog 4209.04 Ten thousand miles in a yacht. (Arthur, Richard.) N.Y., 1906.

Geog 5559.11.5 A tenderfoot with Peary. (Borup, George.) London, 1911.

Geog 4429.72 Tendiuk, Leonid M. Al'batros-clukach moriv. Kyïv, 1972.

An 358.98.3 Tenishef, V. L'activité de l'homme. Paris, 1898.

Geog 4308.21 Tennant, C. Tour thro'...Netherlands, Holland. London, 1824. 2v.

An 629.71 Tennekes, J. Anthropology, relativism and method. Assen, 1971.

Geog 4326.75.5A Teonge, Henry. The diary of H. Teonge, chaplain on board H.M.'s ships Assistance, Bristol and Royal Oak, 1675-1679. London, 1927.

Geog 4326.75 Teonge, Henry. The diary of H. Teonge...1675-1679. London, 1825.

Geog 659.73.5 Teoreticheskaia geografiia. (Simpozium po teoreticheskim problemam geografii, Riga, 1973.) Riga, 1973.

Geog 3019.72.15 Teoreticheskie osnovy geografii. (Anuchin, V.A.) Moskva, 1972.

Geog 623.3 Teoria Moritza Wagnera o powstawaniu gatunków. (Babicz, Józef.) Wrocław, 1966.

Geog 659.71 Teorie komplexity a diferenciace světa se zoláštnim zřetelem na diferenciaci géograficksu. 1. vyd. (Hampl, Martin.) Praha, 1971.

Geog 5399.60.5 Teplaia Arktika. (Kudenko, O.I.) Moskva, 1960.

Geog 4309.30.20 Terán, Juan B. Lo gótico, signo de Europa. Buenos Aires, 1930?

Geog 809.52 Teran, M. de. Imago mundi. Madrid, 1952. 2v.

Geog 4218.70 Teresina peregrina. (Longworth, M.T.) London, 1874. 2v.

Geog 619.12 Termer, Franz. Karl Theodor Sapper, 1866-1945. Leipzig, 1966.

Geog 4168.40 Ternaux-Compans, H. Archives des voyages. pt.1-4. Paris, 1840. 4v.

Geog 500.17 Ternaux-Compaus, H. Bibliothèque asiatique et africaine. Paris, 1841.

Geog 500.17.2 Ternaux-Compaus, H. Bibliothèque asiatique et africaine. Notes. Paris, 1841.

Geog 85.135.5 Terra; geografiska sällskapets i Finland tidskrift. Helsingfors. 1,1888+ 32v.

Geog 808.83 La terra. (Marinelli, G.) Milano, 1883-1885. 7v.

Geog 6009.47 Terra australis. (Orrego Vicuña, Eugenio.) Santiago, 1948.

Geog 4157.66.5 Terra Australis cognita, or Voyages. (Callander, J.) Edinburgh, 1766. 3v.

Geog 212.112 La terra e la vita. Roma. 2v.

Geog 665.102 Terra incognita. (Schickel, Joachim.) Bergisch Gladbach, 1965.

Geog 4169.36 Terrae incognitae. (Hennig, R.) Leiden, 1936-39. 4v.

Geog 4169.36.3 Terrae incognitae. (Hennig, R.) Leiden, 1944- 4v.

Geog 220.5 Terrae incognitae. Amsterdam. 1,1969+

An 2109.72.32 Terray, Emmanuel. Le Marxisme devant les sociétés primitives. 2. éd. Paris, 1972.

Geog 5050.9 Terre Adélie, Greenland, 1947-55; bibliographie. (Expéditions Polaires Francaises, 1948.) Grenoble, 1956.

Geog 6100.2 Pamphlet vol. Terre Adélie, 1958-60. 2 pam.

Geog 6100.2.10 Terre Adélie, 1959-61. (Expéditions Polaires Françaises.) Paris, 1963.

Geog 6100.2.12 Terre Adélie, 1960-62; raport d'activités. (Expéditions Polaires Françaises.) Paris, 1965.

Geog 4209.56.10 Terres de l'amitié. (Gouy, Robert.) Neuchâtel, 1956.

Geog 3055.29 Les terres disparues. (Nicaise, A.) Chalons sur Marne, 1885.

Geog 5060.7.6 Terres polaires, terres tragiques. (Victor, Paul Émile.) Paris, 1973.

An 2109.72.35 Terrey, Emmanuel. Marxism and "primitive" societies. N.Y., 1972.

Geog 4348.48 Terry, C. Scenes and thoughts in foreign lands. London, 1848.

Geog 4750.4 Texas Instruments, Inc. An inventory of geographic research of the humid tropic environment. Dallas, Texas, 1967? 2v.

Geog 819.13 A text-book of geography. (Andrews, A.W.) London, 1913.

Geog 819.16 A text book of geography. (Andrews, A.W.) London, 1916.

Geog 4184.13 Texts and versions of...Carpini and...Rubruquis. (Hakluyt, R.) London, 1903.

Geog 4209.59.10 Tey, José M. Hong Kong-Barcelona en el junco Rubia. Barcelona, 1959.

Geog 4308.66.12 Tgjennen Europa. (Watt, Robert.) København, 1866.

Geog 4308.83 Thacher, S.O. What I saw in Europe. Topeka, Kansas, 1883.

Geog 4209.33 Thames to Tahiti. (Howard, S.) London, 1951.

Geog 4209.43 Thatcher, T.C. Travel letters. Yarmouth Port, 1943.

Geog 4679.12.10 Thayer, J.B. The sinking of the S.S. Titanic, April 14-15, 1912. Philadelphia, 1940.

Geog 4265.95.1 Theatrum urbium. (Saur, Abraham.) Unterschneidheim, 1971.

An 2109.67.15 Themes in economic anthropology. (Firth, Raymond William.) London, 1967.

Geog 520.5 Thenaud, J. Le voyage d'Outremer. Paris, 1884.

Geog 5939.29 The theoretical continent. (Stefansson, V.) N.Y., 1929.

An 359.50.10 Theoretische Anthropologie. (Dempf, Alois.) Bonn, 1950.

Geog 3019.46 The theory and practice of geography. (Darby, Henry C.) Liverpool, 1946.

An 379.18A The theory of environment. (Koller, A.H.) Menasha, Wis., 1918.

Geog 4559.40.5 There go the ships. (Smale, R.) Caldwell, 1940.

An 3558.67 Thesaurus Craniorum. (Davis, J.B.) London, 1867.

An 3558.67.2 Thesaurus Craniorum. Supplement. (Davis, J.B.) London, 1875.

Geog 4209.39 These are real people. (Forbes, Rosita T.) N.Y., 1939.

Geog 4309.61.15 These ruins are inhabited. (Beadle, Muriel.) N.Y., 1961.

Geog 4166.64.5F Thevenot, M. de. Relation de divers voyages. Paris, 1696. 2v.

Geog 4166.64F Thevenot, M. de. Relations de divers voyages. Paris, 1664-66. 2v.

Htn Geog 805.75F* Thevet, A. La cosmographie universelle. Paris, 1575. 2v.

Geog 4559.69 They live by the wind. 1. ed. (Bradley, Wendell P.) N.Y., 1969.

An 124.11 They studied man. (Kardiner, Abram.) N.Y., 1963.

Geog 4679.62 They survived. (Noyce, Wilfrid.) London, 1962.

Geog 3520.10.55 They told Mr. Hakluyt. (Hakluyt, Richard.) London, 1964.

Geog 4145.9 Thiery, Maurice. Bougainville. London, 1932.

Geog 759.56 Things maps don't tell us; an adventure into map interpretation. (Lobeck, Armin Kohl.) N.Y., 1968.

An 2058.65 Thioly, F. Débris de l'industrie humaine...caverne de Bossey. Genève, 1865.

Geog 4219.38.6 Third class world. (Bradshaw, M.J.) Alliance, 1939.

Geog 4208.58.5 Thirty-six voyages to various parts of the world, 1799-1841. 3. ed. (Coggeshall, G.) N.Y., 1858.

Geog 5518.16 Thirty years in the Arctic regions. N.Y., 1859.

Geog 4418.95 This goodly frame, the earth. (Tiffany, F.) Boston, 1895.

Geog 4418.95.2 This goodly frame, the earth. (Tiffany, F.) Boston, 1896.

Geog 4418.95.3 This goodly frame, the earth. (Tiffany, F.) Boston, 1896.

An 170.165A This is race. (Count, Carl W.) N.Y., 1950.

Geog 3070.26 Thomas, G.M. Der Periplus des Pontus Euxinus nach Münchener Handschriften. München, 1864.

Geog 4559.55 Thomas, L.J. Great true adventures. N.Y., 1955.

Geog 4309.27 Thomas, Lowell. European skyways. Boston, 1927.

Geog 5333.20 Thomas, Lowell. Sir Hubert Wilkins. N.Y., 1961.

Geog 4678.37.15 Thomas, R. Interesting and authentic narratives of the most remarkable shipwrecks, fires. Hartford, 1837.

An 2109.37.11 Thomas, W.I. Primitive behavior. 1. ed. N.Y., 1937.

An 2109.09A Thomas, W.I. Source book for social origins. Chicago, 1909.

An 359.12.13A Thomas, W.I. Source book for social origins. Chicago, 1912.

Author and Title Listing

An 2109.09.5 — Thomas, W.I. Source book for social origins. 6th ed. Boston, 1909.
An 5.18 — Thomas, William L. International directory of anthropological institutions. N.Y., 1953.
Geog 4131.53 — The Thomas Cook story. (Pudney, J.) London, 1953.
Geog 3240.7 — Thomassy, M.J.R. Les papes géographes. Paris, 1852.
Geog 4327.44 — Thompson, C. Travels. Reading, 1744. 2v.
Geog 4329.51 — Thompson, D. The phoenix in the desert. London, 1951.
Geog 4218.93 — Thompson, F.D. In the track of the sun. N.Y., 1893.
Geog 5209.03 — Thompson, G.M. Rannsóknarferð i kafli. Gimli, 1903.
An 359.61.5 — Thompson, L. Toward a science of mankind. N.Y., 1961.
Geog 4309.47.5 — Thompson, R.W. Devil at my heels. London, 1947.
Geog 4238.73 — Thomson, C.W. The Atlantic. N.Y., 1878. 2v.
Geog 4238.77 — Thomson, C.W. Voyage of the Challenger. The Atlantic. v.1-2, atlas. London, 1877. 3v.
Geog 5509.75 — Thomson, George Malcolm. The North-west passage. London, 1975.
Geog 3129.48A — Thomson, J.O. History of ancient geography. Cambridge, 1948.
An 349.24A — Thomson, John Arthur. What is man? N.Y., 1924.
Geog 6009.62 — Thomson, Robert Baden. The coldest place on earth. Wellington, 1965.
Geog 5700.11 — Thorhallason, Egiel. Beskrivelse over missionerne i Grønlands Søndre Distrikt. København, 1914.
Geog 5837.76 — Thorhallesen, E. Esterretning om rudera...Grønlands. Kiøbenhavn, 1776.
Geog 4209.62.15 — Thorlacius, Sigriðm. Feroabók. Akureyri, 1962.
Geog 1711.1 — Thorne, James. Handbook to environs of London. London, 1876. 2v.
Geog 4110.7 — Thorpe, D. Universal guide of standard routes. Boston, 1907.
Geog 4332.21 — The thousand islands of the Adriatic. (Soljan, Antun.) Beograd, 1965.
Geog 4308.66.5 — Thousand miles in the Rob Roy canoe. (Macgregor, J.) London, 1866.
Geog 4345.99 — The three brothers or travels. (Sherley, A.) London, 1825.
Geog 4345.99.2 — The three English brothers. (Nixon, Anthony.) London, 1607.
Geog 5839.46.10 — Three got through. (Lindsay, M.) London, 1946.
Geog 4208.38.3 — Three months' leave, or Military reminiscences. (Rose, W.G.) London, 1838.
Geog 4169.36.5A — Three sea journals of Stuart times. (Ingram, B.S.) London, 1936.
Geog 4559.56.15 — Three ships came sailing. (Noble, Arthur H.) London, 1956.
Geog 4180.54 — Three voyages of...William Barents. (Veer, G. de.) London, 1876.
Geog 4180.38 — Three voyages of Martin Frobisher. (Best, G.) London, 1867.
Geog 5515.77.15 — The three voyages of Martin Frobisher in search of a passage to Cathay...1576-78. (Best, George.) London, 1938. 2v.
Geog 4180.42 — Three voyages of Vasco da Gama. Photoreproduction. (Correa, G.) London, 1869.
Geog 4309.03.5 — Three weeks in Europe. (Higinbotham, John V.) Chicago, 1905.
Geog 4208.28 — Three years adventures of a minor in England Africa, South Carolina and Georgia. (Butterworth, William.) Leeds, 1831.
Geog 4208.88.5 — Three years of a wanderer's life. (Keane, J.F.) London, 1888.
Geog 5558.81.3 — Three years of Arctic service...1881-84. (Greely, A.W.) N.Y., 1886. 2v.
Geog 5208.73.2 — The threshold of the unknown region. (Markham, C.R.) London, 1873.
Geog 5208.73.4 — The threshold of the unknown region. 4. ed. (Markham, C.R.) London, 1876.
Htn Geog 815.50* — Thresor de chartes contenant les tableaux de tous les pays du monde. (Hondius, Jodocus.) Franckfort? 1602.
Geog 4559.16.3 — The thrilling adventures of the whaler Alcyone. Peabody, 1916.
Geog 4209.64.40A — Thrilling cities. (Fleming, Ian.) N.Y., 1964.
Geog 4559.72.5 — Throner, William R. Life at sea in the age of sail. London, 1972.
Geog 4329.58F — The thrones of earth and heaven. (Beny, Roloff.) London, 1958.
Geog 4300.29 — Through Europe on two dollars a day. (Schoonmaker, F.) N.Y., 1927.
Geog 6008.98.3 — Through the first Antarctic night. (Cook, F.A.) N.Y., 1900.
An 379.38 — Through the great arid filter (man's drift to Europe). (McIverny, A.J.) London? 1938.
Geog 4419.38 — Through the lands of the Bible. (Morton, H.V.) London, 1938.
Geog 4169.41 — Thru hell and high water. (Explorers Club, N.Y.) N.Y., 1941.
VGeog 759.71 — Thuchkevich, Vadim A. Geografiia v tsifrakh i sravneniiakh. Minsk, 1971.
Geog 4417.70 — Thunberg, Karl P. Travels in Europe, Africa and Asia. London, 1793? 4v.
Geog 4417.70.2 — Thunberg, Karl P. Voyage en Afrique et en Asie. 1770-1779. Paris, 1794.
An 2109.32 — Thurnwald, R. Economics in primitive communities. London, 1932.
Geog 4029.10 — Tic-polonga. 1. ed. (Anderton, Russ.) Garden City, 1953.
Geog 3018.74 — Tiele, Pieter Anton. De ontdekkingsreizen sedert de vijftiende eeuw. Leiden, 1874.
Geog 809.19 — La tierra; geografia general. 2a ed. (Camena d'Almeida, P.) Barcelona, 1919.
Geog 4329.20 — Tierras del mar Azul. (Bunge de Galvez, D.) Buenos Aires, 192-.
Geog 4418.95 — Tiffany, F. This goodly frame, the earth. Boston, 1895.
Geog 4418.95.2 — Tiffany, F. This goodly frame, the earth. Boston, 1896.
Geog 4418.95.3 — Tiffany, F. This goodly frame, the earth. Boston, 1896.
NEDL Geog 13.200 — Tijdschrift. (Nederlandsch Aardrijkskundig Genootschap, Amsterdam.) Amsterdam. 1-17,1876-1899
Geog 13.200 — Tijdschrift. (Nederlandsch Aardrijkskundig Genootschap, Amsterdam.) Amsterdam. 1,1900+ 72v.
Geog 224.3F — Tijdschrift voor economische en sociale geografie. Rotterdam. 40,1935+ 21v.
Geog 3590.85 — Tikhomirov, Georgii S. Bibliograficheskii ocherk istorii geografii v Rossii XVIII veka. Moskva, 1968.
Geog 4418.61 — Tilley, H.A. Japan, the Amoor and the Pacific. London, 1861.
Geog 4208.63.7F — Tillotson, John. The overland route to India. London, 1863?
Geog 4509.46 — Tilman, H.W. When man and mountains meet. Cambridge, Eng., 1946.
Geog 5070.70.5 — Tilman, Harold William. In mischief's wake. London, 1971.
Geog 5070.70 — Tilman, Harold William. Mostly mischief: voyages to the Arctic and to the Antarctic. London, 1966.
Geog 4209.64.10 — Time and a ticket. (Benchley, P.) Boston, 1964.
An 180.5 — Time and social structure and other essays. (Fortes, Meyer.) London, 1970.
Geog 4209.40.55 — The time was mine. (Tangye, Derek.) London, 1940.
Geog 3049.68 — Timeless earth. (Kolosimo, Peter.) Secaucus, N.J., 1974.
Geog 3540.13 — Timpte, Helmut. Typologische Studien zur historischen Kartographie in Westfalen. Düsseldorf, 1961.
Geog 4239.65 — Tinkerbelle. (Manry, Robert.) N.Y., 1966.
Geog 1531.4 — Tirol, Vorarlberg, Etschland. 39. Aufl. (Baedeker, publishers.) Leipzig, 1929.
Geog 1531.5 — Tirol, Vorarlberg, westliche Salzburg, Hochkärnten. 40. Aufl. (Baedeker, publishers.) Leipzig, 1938.
Geog 1531.6 — Tirol, Vorarlberg, westliche Salzburg, Hochkärnten. 41. Aufl. (Baedeker, publishers.) Leipzig, 1943.
Geog 1531.3 — Tirol: Vorarlberg und Teile von Salzburg und Kärnten. 37. Aufl. (Baedeker, publishers.) Leipzig, 1923.
Geog 4679.12.7 — Titanic. (Pelz von Felman, J.) Berlin, 1939.
Geog 4679.12.25 — The Titanic and the Californian. (Padfield, Peter.) London, 1965.
Geog 4679.12.35 — The Titanic commutator. 1,1912. Indian Orchard, Mass. Reprint ed. Indian Orchard, Mass. 1,1963+
Geog 4679.12.5 — Titanic disaster. (United States. Congress. Senate.) Washington, 1912.
Geog 4679.12.41 — A Titanic hero, Thomas Andrews, shipbuilder. (Bullock, Shan F.) Riverside, 1973.
Geog 4129.31 — Titayna (pseud.). Mademoiselle against the world. N.Y., 1931.
Geog 4308.90.5 — To Europe on a stretcher. (Potter, V.M.) N.Y., 1890.
Geog 4209.55 — To Hong Kong and return. (Woodcock, A.W.W.) Boston, 1955.
Geog 5209.34.2 — To the Arctic! (Mirsky, J.) N.Y., 1948.
Geog 5518.19.20 — To the Arctic by canoe, 1819-1821. (Hood, Robert.) Montreal, 1974.
Geog 4309.01.9 — To the land of the midnight sun. (Proctor, Rachel.) Utica, N.Y., 1901.
Geog 5209.34 — To the North: The story of arctic exploration from earliest times to the present. (Mirsky, J.) N.Y., 1934.
Geog 4181.143 — To the Pacific and Arctic with Beechey. (Peard, George.) Cambridge, 1973.
Geog 5379.67.5 — To the top of the world. 1. ed. (Karalt, Charles.) N.Y., 1968.
An 3309.24 — Tocher, James F. Anthropometric observations on samples of the civil populations of Aberdeenshire, Banffshire, and Kincardineshire. Edinburgh, 1924.
An 3309.08.5 — Tocher, James F. Pigmentation survey of school children in Scotland. Aberdeen, 1908.
An 379.53 — Today's revolution in weather. (Baxter, William J.) N.Y., 1953.
Geog 5558.84.5 — Todd, Alden. Abandoned. 1. ed. N.Y., 1961.
Geog 4329.57 — Todd, Roberto. Viajando por Europa. Madrid, 1957.
An 3558.90 — Török, A. Grundz. einer systematischen Kraniometrie. Stuttgart, 1890.
Geog 4218.92.20 — Toil and travel. (MacGregor, John.) London, 1892.
An 99.66 — Tokarev, Sergei A. Istoriia russkoi etnografii, dook tiabr'skii period. Moskva, 1966.
VAn 359.68 — Tokarev, Sergei A. Oskovy etnografii. Moskva, 1968.
Geog 225.5 — Tokyo Metropolitan University. Department of Geography. Geographical reports. Tokyo. 1,1966+ 2v.
Geog 500.171 — Tolchinskaia, L.I. Geograficheskaia literatura. Moskva, 1971.
Geog 4169.31.15 — Told at the Explorers Club. (Blossom, F.A.) N.Y., 1931.
Geog 5209.10.3 — The toll of the arctic seas. (Edwards, D.M.) N.Y., 1910.
Geog 4229.30 — Toller, Ernest. Querdurch; Reisebilder und Reden. Berlin, 1930.
Geog 4239.68.20 — Tomalin, Nichols. The strange voyage of Donald Crowhurst. London, 1970.
Geog 5399.28.52 — Tomaselli, Francesco. L'inferno bianco. 4. ed. Milano, 1929.
Htn Geog 3055.37* — Tomasi, T. La spinalba antica historia. Venetia, 1647.
Geog 4559.38 — Tompkins, W.M. Fifty south to fifty south...around Cape Horn...Wander Bird. N.Y., 1938.
Geog 4559.39.5 — Tompkins, W.M. Two sailors and their voyage around Cape Horn. N.Y., 1939.
An 2109.39 — Tools and the man. (Wright, W.B.) London, 1939.
Geog 4182.6 — Toortse der zee-vaert...D. Ruiters 1623 en Samuel Brun's schiffarten (1624). (Naber, S.P.) 's-Gravenhage, 1913.
An 358.76.2 — Topinard, P. L'anthropologie. Paris, 1876.
An 358.76 — Topinard, P. L'anthropologie. Paris, 1876.
An 98.93 — Topinard, P. L'anthropologie aux États-Unis. Paris, 1893.
An 358.78.5 — Topinard, P. Anthropology. London, 1878.
An 2058.85.3 — Topinard, Paul. Eléments d'anthropologie générale. Paris, 1885.
Geog 4308.28A — Topliff, S. Topliff's travels, letters from abroad. Boston, 1906.
Geog 4308.28A — Topliff's travels, letters from abroad. (Topliff, S.) Boston, 1906.
Htn Geog 4260.5F* — Topographia Italiae. (Zeiller, M.) Franckfurt, 1688.
Geog 4266.41.10F — Topographia Sueviae. (Zeiller, Martin.) Kassel, 1960.
Geog 6009.35 — The topographical results of Ellsworth's trans-Antarctic flight of 1935. (Joerg, W.L.G.) N.Y., 1936.
Geog 500.32 — Topographie ancienne - catalogue a prix marquès de cartes anciennes. (Muller, F.) Amsterdam, 1896.
Geog 5060.11 — Toreoja y Miret, José M. Las republicas hispano-americanos y la exploracion de las regiones polares. Madrid, 193-
Geog 5837.06 — Torfaeo, T. Gronlandia antiqua. Havniae, 1706.
Htn Geog 5837.06.2* — Torfaeo, T. Gronlandia antiqua. Havniae, 1706.
An 359.06.7 — Toro, E. Antropologia general. Caracas, 1906.
Geog 5150.2 — Toronto. Public Library. The Northwest passage, 1534-1859. Toronto, 1963.
Htn Geog 4328.39.3* — Torrey, F.P. Journal of a cruise of the U.S.S. Ohio in the Mediterranean. Boston, 1841.
Geog 4679.67.30 — The Torrey Canyon; report. (Committee of Scientists on the Scientific and Technological Aspects of the Torrey Canyon Disaster.) London, 1967.
Geog 226.1 — Toscanelli. 1,1893
Geog 3019.38.10 — Toschi, Umberto. Schemi e notizie di storia delle esplorazione geografiche. 5. ed. Firenze, 1953?
Geog 4677.96 — The total loss of the American ship Hercules. (Stout, Benjamin.) London, 17- ?
Htn Geog 4326.32.3* — The totall discourse of...adventures. (Lithgow, William.) London, 1640.
Htn Geog 4326.32* — The totall discourse of...adventures. (Lithgow, William.) Lyon, 1632.
Geog 4218.86 — A tour around the world. (Raum, G.E.) N.Y., 1886.

Author and Title Listing

Geog 4218.95 A tour around the world. (Raum, G.E.) N.Y., 1895.
NEDL Geog 226.2F Tour du monde. Paris. 1861-1899
Geog 226.2F Tour du monde. Paris. 1900-1914 15v.
Geog 226.2.5 Tour du monde. Table alphabetique, 1860-1910. Paris, n.d.
Geog 4218.80.5 A tour in both hemispheres. (Vetromile, E.) N.Y., 1880.
Geog 4307.65.15 Tour on the continent. (Pennant, T.) London, 1948.
Geog 4308.21 Tour thro'...Netherlands, Holland. (Tennant, C.) London, 1824. 2v.
Geog 4145.78 Touring Club de France. De l'Himalaya aux Pyrénées. Paris? 1959.
Geog 226.3 Touring Club de France. Revue mensuelle. Paris. 36-47,1926-1937 3v.
Geog 4300.31.8 Touring Club Suisse. Europa touring; guide automobile d'Europe. Berne, n.d.
Geog 4279.11A Touring in 1600. (Bates, Ernest S.) Boston, 1911.
Geog 4131.4 Tourism in O.E.C.D. member countries. Paris. 1961+ 6v.
Geog 4131.38.5 Le tourisme dans l'economie contemporaine. Thèse. (Leveillé-Nizerolle, Claude.) Paris, 1938.
Geog 4131.74.5 Tourisme et environnement. (Haulot, Arthur.) Verviers, 1974.
Geog 4131.38 Le tourisme international. Thèse. (Trimbach, André.) Paris, 1938.
Geog 4131.68.5 Tourismus und wirtschaftliche Entwicklung. (Meinke, Hans.) Göttingen, 1968.
Geog 4558.76.3 The tourist, a magazine of information for ocean travellers, 1876. Boston, 1876.
Geog 4300.32 Tourist handbook for Holland, Belgium, the Rhine and Black Forest. (Cook, T., firm.) London, 1895.
Geog 4308.36 The tourist in Europe. N.Y.. 1838.
Geog 4209.30.2 A tourist in spite of himself. (Newton, A.E.) Boston, 1930.
Geog 4209.30.3 A tourist in spite of himself. (Newton, A.E.) Boston, 1933.
Geog 4131.10 The tourist movement. (Ogilvie, F.W.) London, 1933.
Geog 4129.37 Touristes de jadis. (Barraud, G.) Paris, 1937.
Geog 4308.69.7 Tousey, S. Papers from over the water. N.Y., 1869.
Geog 4429.61.5 Toussaint, Auguste. History of the Indian Ocean. Chicago, 1966.
Geog 4429.61 Toussaint, Auguste. History of the Indian Ocean. London, 1966.
Geog 4209.64.20 Toutes voiles dehors. (Muhlethader, J.) Genève, 1964.
An 359.61.5 Toward a science of mankind. (Thompson, L.) N.Y., 1961.
Geog 819.43 Toward new frontiers of our global world. (Engelhardt, N.L.) N.Y., 1943.
Geog 4359.27 Towns of destiny. (Belloc, Hilaire.) N.Y., 1931.
Geog 4209.58.10 Toynbee, A.J. East to west. N.Y., 1958.
Geog 4349.31 Toynbee, Arnold J. A journey to China; or, Things which are seen. London, 1931.
An 2109.25A Tozzer, A.M. Social origins and social continuities. N.Y., 1925.
An 2109.28.20 Tozzer, Alfred M. Social origins and social communities. N.Y., 1928.
An 5.21F Tozzer Library. Catalogue. Authors. Boston, 1963. 26v.
An 5.21.1F Tozzer Library. Catalogue. Authors. 1st supplement. Boston, 1970. 6v.
An 5.21.2F Tozzer Library. Catalogue. Authors. 2d supplement. Boston, 1971. 2v.
An 5.21.3F Tozzer Library. Catalogue. Authors. 3d supplement. Boston, 1975. 3v.
An 5.22.01F Tozzer Library. Catalogue. Index to subject headings. Boston, 1963.
An 5.22F Tozzer Library. Catalogue. Subjects. Boston, 1963. 27v.
An 5.22.1F Tozzer Library. Catalogue. Subjects. 1st supplement. Boston, 1970. 6v.
An 5.22.2F Tozzer Library. Catalogue. Subjects. 2d supplement. Boston, 1971. 3v.
An 5.22.3F Tozzer Library. Catalogue. Subjects. 3d supplement. Boston, 1975. 4v.
Geog 3570.5.3F Trabalhos nauticos dos Portuguezes. (Sonsa Viterbo, F.M.) Lisboa, 1884-1900. 2v.
Htn Geog 3570.5* Trabalhos nauticos dos Portuguezes nos seculos XVI e XVII. (Sonsa Viterbo, F.M.) Lisboa, 1890.
An 379.67 Traces on the Rhodian shore. (Glacken, Clarence J.) Berkeley, 1967.
An 379.73 Traces on the Rhodian shore. (Glacken, Clarence J.) Berkeley, 1973.
Geog 3570.45 Tracey, Hugh. Antonio Fernandes. Lourenco Marques, 1940.
An 130.5 The track of man. 1. ed. (Field, Henry.) Garden City, 1953.
Geog 5557.33.5 The tracks and landfalls of Bering and Chirikof on the northwest coast of America. (Davidson, George.) San Francisco, 1901.
Geog 5839.39 Tracks in the snow. (Haig-Thomas, D.) N.Y., 1939.
Geog 4180.79 Tractatus de globis et eorum usu. (Hues, R.) London, 1889.
Geog 4418.44 Trade and travel in the Far East. (Davidson, G.F.) London, 1846.
Geog 5639.36 Traek af Grønlands politiske historie. (Oldendow, Knud.) København, 1936.
Geog 5660.30 Traekaf kolonien Jakobshavns. (Ostermann, H.) Kjøbenhavn, 1941.
Geog 4308.72.3 Trafton, A. An American girl abroad. Boston, 1872.
Geog 4308.52.9 Trafton, M. Rambles in Europe. Boston, 1852.
Geog 5518.92 Tragedin i Smiths sund; Björling-Kallstenius expeditionen, 1892. (Kallstenius, Alfhild.) Kalmar, 1966.
Geog 4679.23 Tragedy at Honda. (Lockwood, C.A.) Philadelphia, 1960.
Geog 5399.28.5 The tragedy of the Italian with the rescuers to the Red tent. (Giudici, Davide.) N.Y., 1929.
Geog 4672.5 The tragedy of the seas. (Ellms, Charles.) Philadelphia, 1841.
Geog 4181.112A The tragic history of the sea. (Gomes de Brito, Bernardo.) Cambridge, Eng., 1959.
Geog 5530.18 Traill, Henry Duff. The life of Sir John Franklin, R.N. London, 1896.
Geog 4309.41 Trails of two travelers. (Moody, M.N.) N.Y., 1941.
Geog 4418.57 Train, G.F. An American merchant. N.Y., 1857.
Geog 4418.57.5 Train, G.F. Young American abroad. London, 1857.
Geog 808.12.17 Traité elementaire de géographie. (Malte-Brun, Conrad.) Bruxelles, 1832.
Geog 4308.95 Tramp tales of Europe. (Magness, E.) Buffalo, 1895.
Geog 4308.87 A tramp trip. (Meriwether, Lee.) N.Y., 1887.
Geog 4308.87.1 A tramp trip. 5th ed. (Meriwether, Lee.) N.Y., 1887.
Geog 6009.55.25 Trans-Antarctic Expedition, 1955-1958. Antarctica. Wellington, 1964.
Geog 4308.70 Trans-Atlantic sketches. (Harding, W.M.) N.Y., 1870.
Geog 32.5 Transactions. (Bombay Geographical Society.) Bombay. 1-19,1836-1873 19v.
Geog 189.2 Transactions. (Quebec Geographical Society.) Quebec. 1880-1897 3v.
Geog 21.1.40 Transactions. (Royal Geographical Society of Australasia. Victorian Branch.) Melbourne. 15-19,1898-1901
Geog 32.5.5 Transactions. Index, v.1-17. (Bombay Geographical Society.) Edinburgh, 1866.
Geog 109.6 Transactions. New series. (Institute of British Geographers.) London. 1,1976+
Htn Geog 4308.75* Transatlantic sketches. (James, H.) Boston, 1875.
Geog 665.24 Transatlantic souvenirs. (Ingwood of Westchester.) N.Y., 1868.
Geog 4308.43 A transatlantic tour. (Dana, William C.) Philadelphia, 1845.
Geog 85.205 Transcriptions and proceedings. (Geographical Society of the Pacific.) San Francisco. 1902
An 2109.64.5 The transition from childhood to adolescence. (Cohen, Y.A.) Chicago, 1964.
An 629.68.5 Translation at sight; the job of a social anthropologist. (Pouwer, Jan.) Wellington, 1968.
An 170.191 The translation of culture; essays to E.E. Evans-Pritchard. London, 1973.
Geog 5519.40.5 Tranter, G.J. Plowing the Arctic. London, 1944.
Geog 4132.6 Trantina, V. Staří Čechové na cestách. Praha, 1941.
Geog 808.25 Tratado completo de cosmographia. (Casado Giraldes, J.P.C.) Paris, 1825-1828. 4v.
Geog 4235.70.2 Tratado das ilhas novas. (Sousa, F. de.) Ponta Delgada, 1884.
Geog 3019.59.5 Traumen, Wagen und Vollbringen. (Herrmann, Paul.) Hamburg, 1959.
Geog 193.10 Travaux. (Reims. Université. Institut de Géographie.) Reims. 3,1970+ 3v.
Geog 193.1 Travaux. (Rennes. Université. Laboratoire de Géographie.) Rennes. 1-6,1903-1909 2v.
Geog 500.68 Travaux de la Commission pour l'histoire des grands voyages et des grandes decouvertes. (International Committee of Historical Sciences. Committee on the History of Great Voyages and Great Discoveries.) Paris, 1936.
Geog 170.3 Travaux du Département de géographie et d'aménagement régional, Université d'Ottawa. (Ottawa, Ont. University. Department of Geography and Regional Planning.) Ottawa. 1,1971+
Geog 4168.89 Travel, adventure and sport. v.2-4. N.Y., 1889.
Geog 4168.90 Travel, series of narratives of...visits. v.1-106. (Griswold, W.M.) Cambridge, 1890. 2v.
Geog 4100.4 Pamphlet vol. Travel. 9 pam.
Geog 4100.3 Pamphlet vol. Travel.
Geog 4100.2 Pamphlet vol. Travel. 3 pam.
Geog 4100.5 Pamphlet vol. Travel.
Geog 4100.01F Pamphlet vol. Travel.
Geog 4100.1 Pamphlet box. Travel.
Geog 228.5F Travel and camera. N.Y.
Geog 4209.23.25 Travel and comment. (Phelan, James D.) San Francisco, 1923.
Geog 3249.52A Travel and discovery in the Renaissance. (Penrose, B.) Cambridge, 1952.
Geog 228.1 Travel and exploration; monthly illustrated magazine. London. 1-4,1909-1910 4v.
Geog 4307.80.15 The travel-diares of William Beckford of Fonthill. (Beckford, William.) Cambridge, 1928. 2v.
Geog 4209.36.10 Travel essays. (Howland, Charles.) n.p., 1936?
Geog 4129.21 Travel in the two last centuries of three generations. (Roget, S.R.) N.Y., 1921.
Geog 4209.43 Travel letters. (Thatcher, T.C.) Yarmouth Port, 1943.
Geog 4209.30.15 Travel notes, 1929-30. (Mather, Norman C.) Chicago, 1930.
Geog 4309.12.15 Travel-pictures. 2nd series. (Van Allen, William H.) Milwaukee, 1912.
Geog 4110.40 Travel routes around the world. N.Y. 23,1957
Geog 953.12 Travel through pictures; references to pictures, in books and periodicals. (Ellis, Jessie (Croft).) Boston, 1935.
Geog 4309.13.5 A traveler at forty. (Dreiser, Theodore.) N.Y., 1913.
Geog 4309.13.6 A traveler at forty. (Dreiser, Theodore.) N.Y., 1930.
Geog 4415.02.5 A traveler in disguise. (Bracciolini, Poggio.) Cambridge, 1963.
Geog 4309.36 Traveling light (Europe between boats). (Harris, J.P.) Hutchinson, Kansas, 1936.
Geog 818.20 The traveller; or, An entertaining journey round the habitable globe. 3rd ed. London, 182-.
Geog 4125.75.1 The traveller, 1575. (Turler, H.) Gainesville, Fla., 1951.
Geog 3060.19 Travellers and travel liars, 1660-1800. (Adams, Percy G.) Berkeley, 1962.
Htn Geog 815.59.5* The travellers breviat, or An historical description of the most famous kingdoms in the world. (Botero, G.) London, 1601.
Geog 4129.47 The traveller's eye. (Carrington, D.) N.Y., 1947.
Geog 4105.15 The travellers' handbook. (Darde, J.B.) London, 1842.
Geog 4209.41 Travellers must be content. (Lynch, Kathleen M.) N.Y., 1941.
Geog 4105.14 The travellers' oracle. 2. ed. (Kitchiner, William.) London, 1827. 2v.
Geog 4145.84 Traveller's prelude. (Stark, Freya.) London, 1950.
Geog 4129.50 Traveller's quest. (Michael, M.) London, 1950.
Geog 4169.45 Travellers' tales, a series of BBC programmes broadcast throughout the world. (Bailey, Leslie.) London, 1945.
Geog 4309.10.5 Traveller's tales told in letters. (Smith, Bertha W.) N.Y., 1912.
Geog 4208.72.7 Travelling about over new and old ground. (Barker, Mary Ann B.) London, 1872.
Geog 4218.90 Travelling alone, a woman's journey. (Leland, Lillian.) N.Y., 1890.
Geog 4308.63 Travelling notes in France, Italy...of an invalid in search of health. Glasgow, 1863.
Geog 4308.33 Travelling sketches. (Ritchie, Leitch.) London, 1833.
Geog 4308.32.10 Travelling sketches in the north of Italy, the Tyrol, and on the Rhine. (Ritchie, Leitch.) London, 1832.
Geog 4347.99 Travels...in Asia,...Europe. (Abu Taleb Khan.) London, 1810. 2v.
Geog 4327.44 Travels. (Thompson, C.) Reading, 1744. 2v.
Geog 4344.35.12 Travels and adventures, 1435-1439. (Tafur, Pero.) London, 1926.
Geog 4344.35.10 Travels and adventures, 1435-1439. (Tafur, Pero.) N.Y., 1926.
Htn Geog 4208.14* Travels and adventures. (Bunnell, D.C.) Palmyra, N.Y., 1831.
Htn Geog 4327.39* Travels and adventures of Eduard Brown. (Campbell, J.) London, 1739.

Author and Title Listing

Geog 4181.118 The travels and controversies of Friar Domingo Navarrete. (Navarrete, Domingo Fernández de.) Cambridge, 1962. 2v.

Geog 4209.29.5 Travels and reflections. (Buxton, Noel.) London, 1929.

An 359.29.12 Travels and settlements of early man. (Foster, T.S.) N.Y., 1929.

Geog 4206.45.5 Travels and voyages into Africa. (Mocquet, J.) London, 1696.

Geog 4326.32.10A Travels and voyages thro' Eruope, Asia and Africa. (Lithgow, William.) Glasgow, 1906.

Geog 4326.32.5 Travels and voyages thro' Europe, Asia and Africa. 12th ed. (Lithgow, William.) Leith, 1814.

Geog 4308.04 Travels from Berlin thro' Switzerland. (Kotzebue, A.) London, 1804. 3v.

Geog 4347.99.7 Travels from England to India. (Taylor, J.) London, 1799. 2v.

Geog 4348.13.10 Travels in...Europe, Asia and Africa. (Clarke, E.D.) London, 1816-1824. 11v.

Htn Geog 4308.08.2* Travels in...1806 from Italy to England. (Salvo, C.) Troy, N.Y., 1808.

Geog 4228.17.5 Travels in America. 1st ed. (Montulé, E. de.) Bloomington, 1951.

Geog 4417.80.5 Travels in Asia and Africa. (Parsons, Abraham.) London, 1808.

Geog 4418.39.4 Travels in Egypt, Arabia, Petraea and Holy Land. 4th ed. (Olin, S.) N.Y., 1844. 2v.

Geog 4328.13.2 Travels in Egypt, Syria, Cyprus. 2nd ed. (Bramsen, John.) London, 1820. 2v.

Geog 4308.56.15 Travels in England, France. (Haskins, G.F.) Boston, 1856.

Geog 4328.19 Travels in England, France and Spain. (Noah, M.M.) N.Y., 1819.

Geog 4417.70 Travels in Europe, Africa and Asia. (Thunberg, Karl P.) London, 1793? 4v.

Geog 4328.05 Travels in Europe, Asia Minor, and Arabia. (Griffiths, J.) London, 1805.

Geog 4309.03 Travels in Europe. (Bolton, C.E.) N.Y., 1903.

Geog 4308.35.5 Travels in Europe. 4th ed. (Fisk, Wilbur.) N.Y., 1838.

Geog 4328.34 Travels in Europe and the East. (Mott, V.) N.Y., 1842.

Geog 4308.53.12 Travels in Europe and the East. (Prime, Samuel I.) N.Y., 1855. 2v.

Geog 4308.53.13 Travels in Europe and the East. (Prime, Samuel I.) N.Y., 1856. 2v.

Geog 4328.59 Travels in Greece and Russia. (Taylor, Bayard.) N.Y., 1859.

Geog 4348.13.20 Travels in Russia, Tartary and Turkey. (Clarke, E.D.) Edinburgh, 1839.

Geog 4348.08 Travels in Spain and the East, 1808-1810. (Darwin, F.S.) Cambridge, 1927.

Geog 4308.08 Travels in the year 1806, from Italy to England. (Salvo, C.) Troy, N.Y., 1808.

Geog 4328.88 Travels in three continents. (Buckley, J.M.) N.Y., 1895.

Geog 4348.10.5 Travels in various countries of Europe, Asia and Africa. (Clarke, E.D.) Edinburgh, 1811. 2v.

Geog 4348.13.5 Travels in various countries of Europe. (Clarke, E.D.) N.Y., 1813. 4v.

Htn Geog 4348.10.10* Travels in various countries of Europe. 1st American ed. v.3-4. (Clarke, E.D.) N.Y., 1815. 2v.

Geog 4347.01.10FA Travels into Muscovy, Persia, E. Indies. (Bruyn, C. de.) London, 1737. 2v.

Geog 4180.84A Travels of...Valle in India. (Valle, P. della.) London, 1892. 2v.

Geog 4207.55 Travels of...voyage at sea. (Morris, Drake.) London, 1775.

Geog 4309.66 Travels of a capitalist lackey. (Basnett, Fred.) South Brunswick, 1966.

Geog 4417.64 Travels of Father William Orleans a Jesuit. (Orleans, G. de.) London, n.d.

Geog 4181.59A Travels of Fray Sebanstian Manrique. (Luard, C.E.) London, 1927. 2v.

Geog 4181.110A The travels of Ibn Battuta. (Ibn Batuta.) Cambridge, 1958. 2v.

Geog 4413.25.5A Travels of Ibn Batuta. (Ibn Batoutah.) London, 1829.

Geog 4181.67 Travels of John Sanderson in the Levant. (Foster, William.) London, 1931.

Geog 4181.108 The travels of Leo of Rozmital through Germany. (Letts, M.H.I.) Cambridge, 1957.

Geog 4181.17 Travels of Peter Mundy in Europe and Asia, 1608-67. v.1-5. (Mundy, P.) Cambridge, 1907-36. 6v.

Geog 4180.33 Travels of Piedro Cieza de Leon. (Cieza de Leon, P. de.) London, 1864.

Geog 4181.9 Travels of Teixeira with his "kings of Harmuz". (Teixeira, P.) London, 1902.

Geog 4181.95 The travels of the Abbé Carre in India and the Near East. (Carre.) London, 1947-48. 3v.

Geog 4180.32 Travels of Varthema. (Varthema, S. di.) N.Y., 1964?

Geog 4417.38.3 Travels or observations...Barbary...Levant. (Shaw, Thomas.) Edinburgh, 1808. 2v.

Geog 4417.38F Travels or observations. (Shaw, Thomas.) Oxford, 1738.

Geog 4217.67.5 Travels round the world. (Pagès, P.M.F. de.) London, 1791. 3v.

Geog 4308.18 Travels thro'...Germany, Poland. (Neale, A.) London, 1818.

Geog 4327.44.3F Travels thro' different cities of Germany, Italy, Greece and parts of Asia. (Drummond, A.) London, 1754.

Geog 4346.73.2 Travels thro' Low Countries, Germany. 2nd ed. (Ray, John.) London, 1738.

Geog 4306.86.10 Travels through France, Italy. (Burnet, Gilbert.) London, 1750.

Geog 4307.91.4 Travels through Germany, Switzerland, Italy. 2nd ed. (Stolberg, F.L.G.) London, 1797. 4v.

Geog 4307.91.2 Travels through Germany. (Stolberg, F.L.G.) London, 1796. 2v.

Geog 4307.43 Travels through Holland, Germany, Switzerland, and other parts of Europe. (Blainville, J. de.) London, 1743-1745. 3v.

Geog 4307.72 Travels through Holland. (Marshall, J.) London, 1772. 3v.

Geog 4307.87.2 Travels through Switzerland, Italy. (Watkins, T.) London, 1794. 2v.

Geog 4180.49 Travels to Tana and Persia. (Barbaro, J.) N.Y., 1964?

Geog 4357.81 Travels to the coast of Arabia Felix, and from thence by the Red Sea and Egypt to Europe. (Rooke, Henry.) London, 1783.

Geog 4309.53 Travels with a tent in western Europe. (Lockley, R.M.) London, 1953.

Geog 4209.41.5 Travels without a passport. (Baerlein, Henry.) London, 1941.

Geog 4209.41.6 Travels without a passport. (Baerlein, Henry.) London, 1943.

Geog 4249.01 A través da Malasia. (Silva, Abeillard.) Coimbra, 1901.

Geog 4309.23 Traz, Robert de. Dépaysements. Paris, 1923.

Geog 5559.09.106 Tre aar paa Grønlands østkyst. 2. udg. (Mikkelsen, Ejnar.) Kjøbenhavn, 1914.

Geog 600.15 Tre Fiorentini del Rinascimento. v.7. (Carcie, Giuseppe.) Roma, 1959?

Geog 5835.85 Tre rejser til Grønland i aarene 1585-87. (Davys, John.) København, 1930.

Geog 4279.67 Trease, Geoffrey. The grand tour. London, 1967.

Geog 4709.42 Treasure trails. (Santschi, R.J.) Glen Ellyn, Ill., 1942.

Geog 3107.26 Treatise of antient and present geography. (Wells, Edward.) London, 1726.

Htn Geog 4126.43* A treatise of direction, how to travell safely, and profitably into forraigne countries. (Neale, Thomas.) London, 1643.

Geog 5850.5 Trebitsch, R. Bei den Eskimos in Westgrönland. Berlin, 1910.

Geog 500.165 Trecento tesi di laurea in geografia. Padova, 1969.

Geog 5837.03 Den tredie part af det saa kaldede gamle og mye Grønlands beskrifvelse. (Vidalin, Arngrimur Thorkilsson.) København, 1971.

Geog 4219.51.5 Tregaskis, R.W. Seven leagues to paradise. 1. ed. Garden City, 1951.

Geog 4309.63.5 La tregua. (Levi, P.) Torino, 1963.

Geog 5559.10.5 Tremblay, Alfred. Cruise of the Minnie Maud; Arctic seas and Hudson Bay, 1910-1913. Quebec, 1921.

Geog 4131.2 Trends in economic sectors: tourism in Europe. (Organization for European Economic Cooperation.) Paris. 1953-1961

Geog 665.144 Trends in geography. 1st ed. Oxford, 1969.

Geog 5925.94 Treshnika, Aleksei F. Vokrug Antarktidy. Leningrad, 1970.

Geog 5939.63 Treshnikov, Aleksei G. Istoriia otkrytiia i issledovaniia Antarktidy. Moskva, 1963.

Geog 4419.38.15 Le trésor du pélerin; roman. 12. ed. (Monfreid, Henri de.) Paris, 1938.

Geog 4228.98 Trevelyan, Charles Philips. Letters from North America and the Pacific, 1898. London, 1969.

Geog 4219.05.4 Treves, F. The other side of the lantern. London, 1905.

Geog 4219.05.3 Treves, F. The other side of the lantern. London, 1905.

Geog 4219.08.2 Treves, F. Other side of the lantern. London, 1908.

Geog 4219.05.10 Treves, F. The other side of the lantern. London, 1928.

Geog 4219.51 Tri prugosvetiykh plavaniia M.P. Lazareva. (Skolov, A.V.) Moskva, 1951.

Geog 4346.76.25 Tri puteshestviia. (Struys, Jan J.) Moskva, 1935.

Geog 4708.61 Trial of officers and crew of the privateer Savannah on charge of piracy. N.Y., 1862.

Geog 4708.12.4 The trial of Samuel Tulley and John Dalton...piracy and murder committed Jan. 21, 1812. 4. ed. (United States. Circuit Court (1st Circuit, Mass.).) Boston, 1813.

Geog 4707.26 The trials of five persons for piracy. (Jedre, J.B.) Boston, 1726.

An 2109.26.3 Tribal dancing and social development. (Hambly, W.D.) London, 1926.

Geog 227.2 Trident; magazine of the sea. London. 17-18,1955-1956 2v.

Geog 4239.67 Tridtsatyi meridian. (Volovich, Vitalii G.) Moskva, 1967.

Geog 228.25 Trieste. Università. Istituto di Geografia. Notiziario.

Geog 4209.68.5 Trimaran against the trades. (Cole, Jean.) Wellington, 1968.

Geog 4131.38 Trimbach, André. Le tourisme international. Thèse. Paris, 1938.

Geog 4218.80 A trip around the world. (Coop, Timothy.) Cincinnati, 1882.

Geog 4218.87.14 A trip round the world in 1887-88. (Caine, W.S.) London, 1892.

Geog 4218.87.13 A trip round the world in 1887-88. (Caine, William S.) London, 1891.

Geog 4219.04 Trip to the Phillipines...in 1904. (Pearson, D.G.) n.p., n.d.

Geog 4019.68 Triple commission. (Petro, W.) London, 1968.

Geog 4308.53.4 Tripp, A. Crests from the ocean-world. Boston, 1853.

Geog 4308.53.3 Tripp, A. Crests from the ocean-world. Boston, 1853.

Geog 4308.53.6 Tripp, A. Crests from the ocean-world. Boston, 1861.

Geog 4308.53.7 Tripp, A. Crests from the ocean-world. Boston, 1862.

Geog 4679.63.1 Troebst, C.C. The art of survival. Garden City, 1975.

Geog 4679.63 Troebst, C.C. Auf Wunder ist kein Verlass. Düsseldorf, 1963.

An 379.64 Troëng, Ivan. Kulturer före istiden. Uppsala, 1964.

Geog 4229.73 Trois aventures de la Calypso: Galapago, Titicaca, Trous Bleus. (Cousteau, Jacques Yves.) Paris, 1973.

Geog 5838.83.5 Trois cartes. (Maps. Greenland, 1380-1482.) Copenhagen, 1883.

Htn Geog 5375.94.4* Les trois navigations admirables. (Veer, G. de.) Paris, 1610.

An 359.55.5 Troise, Emilio. Racismo. Buenos Aires, 1955.

Geog 4759.50 Troll, Carl. Die tropischen Gebirge. Bonn, 1959.

Geog 4209.59.5 Trolliet, Héli. Sur les routes du monde. Lausanne, 1959.

Geog 4208.81.10 Tropenfahrten Reiseschilderungen aus Ceylon, Java und den Mittelmeergebieten. (Haeckel, Ernst.) Leipzig, 1969.

Geog 4759.47.4 The tropical world. (Gourou, Pierre.) N.Y., 1966.

Geog 4759.53 The tropical world. 2. ed. (Gourou, Pierre.) London, 1958.

Geog 4208.85.5 Tropics and snows. (Burton, R.G.) London, 1898.

Geog 4759.59 Die tropischen Gebirge. (Troll, Carl.) Bonn, 1959.

Geog 3235.11 Tros...monuments geographiques. (Cortambert, P.F.E.) Paris, 1877.

Geog 4309.31 Trouble in the Balkans. (McGeehan, William O.) N.Y., 1931.

Geog 4181.113 The troublesome voyage of Captain Edward Fenton. (Taylor, Eva.) Cambridge, Eng., 1959.

Geog 4219.25.3 Trucking the sunset. (Knight, L.L.) Atlanta, 1925.

Geog 5399.37.5 Trudy. (Ekspeditsiia SSSR na Severnyi Polius, 1937.) Leningrad, 1940.

Geog 4677.82 A true account of the loss of the Halsewell. v.1-2. London, 1786.

Geog 5375.94.11 The true and perfect description of three voyages by the ships of Holland and Zeland. Facsimile. (Veer, G. de.) Amsterdam, 1970.

Geog 4180.13 True description...voyages by the North-East. (Veer, G. de.) London, 1852-53.

Geog 4181.23A True history of conquest of New Spain. (Diaz del Castillo, B.) London, 1908-10. 5v.

Geog 3140.7 True key to ancient cosmology. (Warren, W.F.) Boston, 1882.

Htn Geog 5515.77* True report of Martin Frobisher's voyage, 1577. (Settle, D.) London, 1577.

Author and Title Listing

Geog 4679.42.10	Trumbull, Robert. The raft. N.Y., 1942.
Geog 4309.55	Trumpets from Montparnasse. (Gibbings, R.) London, 1955.
Geog 4679.12.3	The truth about the Titanic. (Gracie, A.) N.Y., 1913.
Geog 4218.81.9	A truth seeker around the world. (Bennett, D.M.) N.Y., 1882. 4v.
Geog 4707.26.5F	The tryals of sixteen persons for piracy. (Massachusetts (Colony). Court of Admiralty.) London, 1726.
Geog 4707.23.2F	Tryals of thirty-six persons for piracy. (Rhode Island (Colony). Court of Admiralty.) Boston, 1723.
Geog 5839.48.5	Tschaen, Louis. Groenland 1948-1949-1950: astronomie, nivellement géodésique sur l'Inlandsis. Paris, 1959.
Geog 4502.5.12	Tscharner, J.B. von. Die Bernina. 1786. München, 1933.
Geog 4145.87	Tschiffely, A. Bohemia junction. London, 1950.
Geog 4308.47	Tsis'ma iz Frantsii i Italii. (Herzen, A.I.) London, 1858.
Geog 4309.74	Tudje avlije. (Kujundžić, Miodrag.) Novi Sad, 1974.
Geog 3520.12	Tudor and early Stuart voyaging. (Penrose, Boies.) Washington, 1962.
Geog 3520.17	Tudor geography, 1485-1583. (Taylor, Eva G.) London, 1930.
Geog 4309.30.25	Tully, J. Beggaro abroad. Garden City, 1930.
Geog 4502.5.20	Der Tummelplatz Europas. (Stephen, L.) München, 19-
Geog 4559.56.20	Tunstall-Behrens, Hilary. Pamir. London, 1956.
Geog 4239.47.5	The Tuntsa. (Turen, T.) Chicago, 1961.
Geog 4239.47.5	Turen, T. The Tuntsa. Chicago, 1961.
Geog 4131.67	Turismo. (Spain. Comisaría del Plan de Desarrollo Económico y Social. Comisión de Turismo.) Madrid, 1967?
Geog 520.30	Turistresor och forskningsfärder. v.3-12. Helsingfors, 1918. 5v.
Geog 4125.75.1	Turler, H. The traveller, 1575. Gainesville, Fla., 1951.
Geog 5324.2.20	Turley, C. Nansen of Norway. London, 1933.
Geog 3594.15	Turley, Tomasz J. Polacy badacze Ameryki. Chicago, 1968.
Geog 4329.69	Turn right for Corfu. (Lewis, Cecil.) London, 1972.
Geog 4509.11	Turner, S.F.R. My climbing adventures in 4 continents. London, 1913.
An 2339.49	Turney-High, H. Primitive war. Columbia, 1949.
Geog 4209.63.25	Turri, Eugenio. Viaggio a Samacanda. Novara, 1963.
Geog 4131.58	Tusci, Leonida. Elementi di tecnica professionale turistica. Roma, 1958.
Geog 5639.71	Tusen år på Grönland. (Wedin, Bertil.) Stockholm, 1971.
An 358.66.5	Tuttle, Hudson. The origin and antiquity of physical man scientifically considered. 2. ed. Boston, 1866?
Geog 4308.39.7	Tuttolasso. Wanderungen durch Deutschland, Polen. Stuttgart, 1839.
Geog 5398.98.5	Tva somrar, norra ishafvet. (Nathorst, A.G.) Stockholm, 1900. 2v.
Geog 4182.42	De tweede schipvaart der Nederlanders. v.1-5. (Keuning, J.) 's-Gravenhage, 1938-51. 8v.
Geog 4110.38	Twelve cities. 1. ed. (Gunther, John.) N.Y., 1969.
An 358.39	Twelve lectures on the natural history of man. (Kinmont, A.) Cincinnati, 1839.
Geog 4308.74.5	Twelve thousand miles over land and sea. (Hutton, William.) Philadelphia, 1878.
Geog 4218.91.4	Twenty years around the world. (Vassar, J.G.) N.Y., 1891.
Geog 4218.38	Twenty years before the mast. (Erskine, C.) Boston, 1890.
Geog 4308.69.15	Twenty years in Europe. (Byers, S.H.M.) Chicago, 1900.
An 359.39.5A	Twilight of man. (Hooton, E.A.) N.Y., 1939.
Geog 4309.24	Twisting trails in the Auvergnes, Cevennes, Alps of Provence. (Morse, William I.) Boston, 1924.
Geog 5559.09.110	Two against the ice. (Mikkelsen, Ejnar.) London, 1957.
Geog 5517.46.15	Two early works on Arctic exploration. (Eavenson, Howard N.) Pittsburgh, 1946.
Geog 4239.69	Two girls, two catamarans. (Wharran, James.) London, 1969.
Geog 4308.85.15	Two gray tourists. (Johnston, Richard M.) Baltimore, 1885.
An 358.44	Two lectures on the natural history of the Caucasian and Negro races. (Nott, Josiah C.) Mobile, 1844.
Geog 6009.20	Two men in the Antarctic; an expedition to Graham Land, 1920-22. (Bagshawe, T.W.) N.Y., 1939.
Geog 4308.77.5	Two months abroad. (Halsey, F.W.) Binghampton, 1878.
Geog 4209.02.3	Two on their travels. (Colquhoun, E.C. (Mrs.).) London, 1902.
Geog 4209.02.5	Two on their travels. (Colquhoun, E.C. (Mrs.).) N.Y., 1902.
Geog 4559.39.5	Two sailors and their voyage around Cape Horn. (Tompkins, W.M.) N.Y., 1939.
Geog 5838.90	Two summers in Greenland. (Carstensen, A.R.) London, 1890.
Geog 5398.75.3	The two voyages of the Pandora in 1875-76. (Young, Allen.) London, 1879.
Htn Geog 4328.32*	Two years and a half...in Mediterranean and Levant. (Wines, E.C.) Philadelphia, 1832.
Geog 4208.29.5	Two years at sea. (Roberts, Jane.) London, 1834.
Geog 4208.29.7	Two years at sea. 2. ed. (Roberts, Jane.) London, 1837.
Htn Geog 4558.40.10*	Two years before the mast. (Dana, Richard Henry.) Boston, 1869.
Geog 4558.40.11	Two years before the mast. (Dana, Richard Henry.) Boston, 1869.
Geog 4558.40.11.5	Two years before the mast. (Dana, Richard Henry.) Boston, 1869.
Htn Geog 4558.40.13*	Two years before the mast. (Dana, Richard Henry.) Boston, 1871.
Geog 4558.40.12	Two years before the mast. (Dana, Richard Henry.) Boston, 1883.
Geog 4558.40.28	Two years before the mast. (Dana, Richard Henry.) Boston, 1911.
Geog 4558.40.24A	Two years before the mast. (Dana, Richard Henry.) Boston, 1911.
Geog 4558.40.27	Two years before the mast. (Dana, Richard Henry.) Boston, 1911.
Geog 4558.40.25	Two years before the mast. (Dana, Richard Henry.) Boston, 1911.
Geog 4558.40.15	Two years before the mast. (Dana, Richard Henry.) Cambridge, 1895.
Htn Geog 4558.40.6*	Two years before the mast. (Dana, Richard Henry.) Glasgow, 1842.
Htn Geog 4558.40.7*	Two years before the mast. (Dana, Richard Henry.) London, 1841.
Htn Geog 4558.40.50*	Two years before the mast. (Dana, Richard Henry.) London, 1946?
Geog 4558.40.60	Two years before the mast. (Dana, Richard Henry.) Los Angeles, 1964. 2v.
Htn Geog 4558.40*	Two years before the mast. (Dana, Richard Henry.) N.Y., 1840.
Geog 4558.40.5	Two years before the mast. (Dana, Richard Henry.) N.Y., 1840.
Geog 4558.40.9	Two years before the mast. (Dana, Richard Henry.) N.Y., 1847.
Geog 4558.40.8	Two years before the mast. (Dana, Richard Henry.) N.Y., 1854.
Htn Geog 4558.40.14*	Two years before the mast. (Dana, Richard Henry.) N.Y., 1892.
Geog 4558.40.20A	Two years before the mast. (Dana, Richard Henry.) N.Y., 1907.
Geog 4558.40.22	Two years before the mast. (Dana, Richard Henry.) N.Y., 1911.
Geog 4558.40.30	Two years before the mast. (Dana, Richard Henry.) N.Y., 1915.
Geog 4558.40.35A	Two years before the mast. (Dana, Richard Henry.) N.Y., 1921.
Geog 4558.40.26	Two years before the mast. (Dana, Richard Henry.) N.Y., 1922.
Geog 4558.40.40	Two years before the mast. (Dana, Richard Henry.) N.Y., 1928.
Htn Geog 4558.40.52*	Two years before the mast. (Dana, Richard Henry.) Sydney, 1946.
Geog 4308.61	Two years in Switzerland and Italy. (Bremer, F.) London, 1861. 2v.
Geog 6009.01.9	Two years in the Antarctic. (Armitage, A.B.) London, 1905.
Geog 4219.34.10	Two young men see the world. (Unwin, S.) London, 1934.
Geog 224.10	Tydskrif vir aardrykskunde; tydskrif van die vereniging vir aardrykskunde-onderwys. Stellenbosch. 2,1962+ 2v.
Geog 4418.32	Tyerman, David. Journal of voyages and travels. Boston, 1832. 3v.
Geog 4308.85.3	Tyler, I. Waymarks, or Sola in Europe. Chicago, 1885.
An 170.360	Tyler, Stephen A. Cognitive anthropology; readings. N.Y., 1969.
An 359.30	Tylor, Edward B. Anthropology; an introduction to the study of man and civilization. London, 1930. 2v.
An 358.81.4	Tylor, Edward B. Anthropology, introduction to the study of man and civilization. N.Y., 1881.
An 358.88	Tylor, Edward B. Anthropology. N.Y., 1888.
An 358.89	Tylor, Edward B. Anthropology. N.Y., 1889.
An 358.91.3	Tylor, Edward B. Anthropology. N.Y., 1891.
An 358.94.3	Tylor, Edward B. Anthropology. N.Y., 1894.
An 359.04.3	Tylor, Edward B. Anthropology. N.Y., 1904.
An 359.09.12	Tylor, Edward B. Anthropology. N.Y., 1909.
An 359.09.11	Tylor, Edward B. Anthropology. N.Y., 1909.
An 2108.73	Tylor, Edward Burnett. Die Anfänge der Cultur. v.1-2. Leipzig, 1873.
An 2109.03A	Tylor, Edward Burnett. Primitive culture, researches. London, 1903. 2v.
An 2108.74	Tylor, Edward Burnett. Primitive culture. Boston, 1874. 2v.
An 2108.77.3	Tylor, Edward Burnett. Primitive culture. N.Y., 1877. 2v.
An 2108.71.1	Tylor, Edward Burnett. Primitive culture: researches into the development of mythology. N.Y., 1974. 2v.
An 2109.20.2	Tylor, Edward Burnett. Primitive culture. 1st American ed. Boston, 1874. 2v.
An 2108.91A	Tylor, Edward Burnett. Primitive culture. 3. ed. London, 1891. 2v.
An 2109.20	Tylor, Edward Burnett. Primitive culture. 6. ed. London, 1920. 2v.
An 2109.24.10	Tylor, Edward Burnett. Primitive culture. 7. ed. v.1-2. N.Y., 1924.
An 2108.70.8	Tylor, Edward Burnett. Researches into the early history of mankind. London, 1870.
An 358.54	Types of mankind. 2. ed. (Nott, Josiah C.) Philadelphia, 1854.
An 358.54.2	Types of mankind. 2. ed. (Nott, Josiah C.) Philadelphia, 1854.
An 358.71	Types of mankind. 10. ed. (Nott, Josiah C.) Philadelphia, 1871.
Geog 3540.13	Typologische Studien zur historischen Kartographie in Westfalen. (Timpte, Helmut.) Düsseldorf, 1961.
Geog 1531.10	Tyrol and Salzburg. 14. ed. (Baedeker, publishers.) Freiburg, 1961.
Geog 1533.14A	Tyrol and the Dolomites including the Bavarian Alps. 13. ed. (Baedeker, publishers.) Leipzig, 1927.
Geog 5508.36	Tytler, P.F. Historical view of discovery...of America. N.Y., 1836.
Geog 5208.32A	Tytler, P.F. Historical view of progress of discovery. Edinburgh, 1832.
Geog 5208.33	Tytler, P.F. Historical view of progress of discovery. Edinburgh, 1833.
An 2059.72.5	U kolybeli istorii. (Matiushin, Geral'd N.) Moskva, 1972.
Geog 5559.56	U poslednikh parallelei. (Morozov, S.T.) Moskva, 1956.
Geog 4209.56.25	Uberti, Roberto degli. Soli attraverso gli oceani. 2. ed. Brescia, 1956.
An 2109.59.20	Udy, Stanley H. Organization of work. New Haven, 1959.
Geog 3025.7	Ueber...Herstellung...Erdkarte im Mafestabe. (International Geographical Congress.) Wien, n.d.
An 3207.85	Ueber...Verschiedenheit des Negers vom Europäer. (Sommering, S.T.) Frankfurt, 1785.
An 359.33.3	Über den Sinn und die Grenze der Lehre vom Menschen. (Ritter, J.) Potsdam, 1933.
Geog 3251.21	Über die niederdeutschen Seebücher. (Behrmann, W.) Hamburg, 1906.
Geog 5508.38	Über die nordwestliche Durchfahrt. (Hülstett, G.K.A.) Düsseldorf, 1838.
Geog 3235.43	Über die Weltkarte des Kosmographen. (Schweder, E.) Kiel, 1886.
Geog 3235.9	Über Erdkunde und Karten. (Wuttke, J.K.II.) Leipzig, 1853.
Geog 3550.6	Über italienischen Seekarten und Kartographen des Mittelalters. (Fischer, Theobald.) Berlin, 1882.
An 3558.46	Über Schädelbildung zur festern Begründung der Menschenrassen. (Zeune, August.) Berlin, 1846.
An 2339.65	Über Waffen- und Werkzeugtechnik des Altmenschen. (Rust, Alfred.) Neumünster, 1965.
NEDL Geog 28.2.5	Übersicht der Aufsatze...in den Monatsberichten über die Verhandlungen. (Berlin. Gesellschaft für Erdkunde.) Berlin. 1863
NEDL Geog 28.3	Übersicht der Aufsatze. Zeitschrift. (Berlin. Gesellschaft für Erdkunde.) Berlin. 1884-1921 15v.
Htn Geog 4307.53*	Uffenbach, Z.C. Merkwürdige Reisen. Ulm, 1753-1754. 3v.
Geog 4219.60.5	Uittenbogaard, Leo. De wereld. Den Haag, 1958-60. 3v.
Geog 5323.2.5	Ukendt mand til ukendt land. (Mikkelsen, Ejnar.) Kjøbenhavn, 1954.
Geog 240.5	Ukrains'ke Heohrafichne Tovarystvo. Heohrafichnyi zbirnyk. Kyiv. 1-5 5v.

Author and Title Listing

Geog 135.2.5 Ule, Willi. Der Würmsee (Starnbergersee) in Oberbayern. v.5. Leipzig, 1901.
Geog 4509.54.5 Ullman, J.R. The age of mountaineering. Philadelphia, 1954.
Geog 4509.41 Ullman, J.R. High conquest, the story of mountaineering. Philadelphia, 1941.
Geog 4521.27 Ullman, James Ramsey. Straight up; the life and death of John Harlin. Garden City, N.Y., 1968.
Geog 5160.11.10 Ultima Thule. (Stefansson, V.) N.Y., 1940.
Geog 5160.11.12 Ultima Thule. (Stefansson, V.) Reykjavik, 1942.
Geog 500.105 Los ultimos escritores de Indias. (Barras de Aragón, F.) Madrid, 1949.
Geog 4205.96.1 Ultzheimer, Andreas J. Warhaffte Beschreibung ettlicher Reisen in Europa, Africa, Asien und America 1596-1610. Tübingen, 1971.
Geog 4208.87.10 Ulysses; or Scenes and studies in many lands. (Palgrove, William G.) London, 1887.
Htn Geog 4266.31* Ulysses belgico-gallicus...per Belgium hispan. (Gölnitz, A.) Lugdunum Batavorum, 1631.
Geog 4129.70 The Ulysses factor: the exploring instinct in man. (Anderson, John Richard Lane.) London, 1970.
Geog 4502.5.31 Um den Montblanc. (Geissler, Paul.) München, 1940?
An 2059.59.25 Um des Erscheinungsbild der ersten Menschen. (Overhage, Paul.) Basel, 1959.
Geog 4218.81 Um die Welt ohne zu wollen. (Ludwig, S.) Prag, 1881.
Geog 5638.99 Um Graenland ad fornu og nýju. (Jonsson, Finnur.) Kaupmannahöfn, 1899.
Geog 665.4 Umlauft, F. Die Pflege der Erdkunde in Oesterreich...Festschrift...Franz Josef I. Wien, 1898.
Geog 5180.18 The unbelievable land; 29 experts bring us closer to the Arctic. (Smith, I. Norman.) Ottawa, 1964.
An 358.70 The uncivilized races of men in all countries of the world. (Wood, J.G.) Hartford, 1870. 2v.
An 358.70.5 The uncivilized races of men in all countries of the world. (Wood, J.G.) Hartford, 1876. 2v.
Geog 3055.60.5 Und doch. (Spanuth, Jürgen.) Stuttgart, 1959.
Geog 5839.38 Under det nordligste danne broa; beretning om dansk Nordøstgrønlands Ekspedition, 1938-39. (Knuth, Eigil.) København, 1940.
Geog 4559.54.10 Under four flags. (Edwards, H.W.) London, 1954.
Geog 4228.97 Under sail. (Riesenberg, Felix.) N.Y., 1918.
Geog 5399.27.3 Under sail in the frozen north. (Worsley, F.A.) London, 1927.
Geog 6009.01.30 Under Scott's command: Lashly's Antarctic diaries. (Lashly, William.) London, 1969.
Geog 4709.27.5 Under the black flag. (Seitz, Don Carlos.) London, 1927.
Geog 4709.25 Under the black flag. (Seitz, Don Carlos.) N.Y., 1925.
Geog 5559.31 Under the North Pole. (Wilkins, H.) N.Y., 1931.
Geog 5559.31.5 Under the North Pole. (Wilkins, H.) N.Y., 1931.
Geog 5558.75.9 Under the Northern Lights. (MacGahan, J.A.) London, 1876.
Geog 4329.29.5 Under the olive trees. (Chalmers, T.M.) N.Y., 1931.
Geog 5399.35 Under the Pole star. (Glen, A.R.) London, 1937.
Geog 4509.56.25 Underhill, Miriam. Give me the hills. London, 1956.
Geog 4515.25 Underhill, R.L.K. On the use and management of the rope in rock work. San Francisco, 1931.
Geog 5838.28 Undersogelses-reise...Grønland. (Graah, W.A.) Kjøbenhavn, 1832.
An 359.52 UNESCO. The race concept. Paris, 1952.
An 359.61.55 UNESCO. The race question in modern science; race and science. N.Y., 1961.
An 359.56.15 UNESCO. The race question in modern science. Paris, 1957.
An 170.367 UNESCO. Le racisme devant la science. Paris, 1960.
Geog 4216.21 Unfreiwillige Reise um die Welt, 1621-1628. (Fernberger von Egenberg, C.M.) Leipzig, 1928.
Geog 3055.21 Unger, F.X. The sunken island of Atlantis. n.p., n.d.
Geog 3055.19 Unger, F.X. Die versunkene Insel Atlantis. Wien, 1860.
Geog 4182.51 Unger, W. De oudste reizen van de Zieuwen naar Oost-Indie. 's-Gravenhage, 1948.
Geog 4208.53.3 Ungewitter, D.H. Portfolio für Iländer und Völkerkunde. Pest, 1853.
Geog 4229.72 Unglaubliche Wirklichkeit. (Struve, Wolfgang.) Salzburg, 1972.
Geog 81.5 Union Geographique du France. Congrès national compte-rendu. 1879-1904 22v.
Geog 243.2 Union Geographique du Nord de la France. Bulletin. Lille. 1-34,1880-1913 17v.
Geog 500.90 A union list of geographical serials. 2. ed. (Harris, Chauncy Donnison.) Chicago, 1950.
Geog 4300.26 Union Steamship Company, Ltd. Homeward bound. 3rd ed. London, 1892.
An 358.61.5 Unité de l'espèce humaine. (Quatrefages de Bréau, A. de.) Paris, 1861.
Htn Geog 1645.5* United States...Mexico. (Baedeker, publishers.) Leipsic, 1893.
Geog 1645.5.5 United States...Mexico. (Baedeker, publishers.) Leipsic, 1893.
Geog 1645.7.5 United States...Mexico. 2. ed. (Baedeker, publishers.) Leipsic, 1899.
Geog 1645.7 United States...Mexico. 2. ed. (Baedeker, publishers.) Leipsic, 1899.
Geog 1645.10A United States...Mexico. 3. ed. (Baedeker, publishers.) Leipzig, 1904.
Geog 1645.15A United States...Mexico. 4. ed. (Baedeker, publishers.) Leipzig, 1909.
Geog 5970.38 United States. Antarctic Projects Office. The United States in the Antarctic, 1820-1962. Washington, 1962?
Geog 5558.81.5 United States. Board of Officers. Expedition...Relief of Lieutenant Greely. Report. Washington, 1884.
Geog 4708.12.4 United States. Circuit Court (1st Circuit, Mass.). The trial of Samuel Tulley and John Dalton...piracy and murder committed Jan. 21, 1812. 4. ed. Boston, 1813.
Geog 4678.46 United States. Congress. House. Committee on Commerce. Owners and crew of ship Chandler Price. n.p., 1848.
Geog 4679.12.5 United States. Congress. Senate. Titanic disaster. Washington, 1912.
Geog 243.5 United States. Department of State. Office of the Geographer. Geographic bulletin.
An 3409.36 United States. Federal Bureau of Investigation. Classification of fingerprints. J.E. Hoover, director. Washington, 1936.
An 3409.34 United States. Federal Bureau of Investigation. Fingerprints. Washington, 1934.
An 3409.37 United States. Federal Bureau of Investigation. Fingerprints. Washington, 1937.
An 3409.32 United States. Federal Bureau of Investigation. How to take fingerprints. Washington, 1932.
Geog 559.52 United States. National Archives. Geographical exploration and topographic mapping by the United States government. Washington, 1952.
Geog 4249.61.5 United States. National Archives. United States scientific geographical exploration of the Pacific Basin, 1783-1899. Washington, 1961.
Geog 5321.1 United States. Naval Observatory. Reports on medals to Arctic explorers, Kane, Hayes. Washington, 1876.
Geog 5900.18 United States. Naval Photographic Interpretation Center. Antarctic bibliography. Washington, 1951.
Geog 5907.15 United States. Naval Support Force, Antarctica. History and Research Division. Monograph. Washington. 1,1971+
Geog 5398.79.15 United States. Navy Department. Letter from the Secretary of the Navy. Washington, 1884.
Geog 4678.62 United States. Navy Department. Wreck of steamer Governor and search for U.S. ship Vermont. Washington, 1862?
Geog 5160.12 United States. Office of Chief of Air Corps (War Department). Arctic manual. v.1-2. Washington, 1940.
Geog 110.2.3 United States. War Department. Corps of Engineers. Report of 3rd International Geographical Congress and Exhibition at Venice, Italy, 1881. Washington, 1885.
Geog 5050.5 United States. Works Progress Administration, N.Y.C. Annotated bibliography of the polar regions. Series B. pt. 1. N.Y., 1938.
Geog 4208.38 United States exploring expeditions...voyage...1838-1842. (Jenkins, J.S.) New Orleans, 1854.
Geog 4208.38.1 United States exploring expeditions. (Jenkins, J.S.) N.Y., 1855.
Geog 500.182 United States government publications for research and teaching in geography. (Vinge, Clarence L.) Totowa, N.J., 1967.
Geog 5538.51 United States Grinnell expedition. (Kane, E.K.) N.Y., 1854.
Geog 5538.51.3 United States Grinnell expedition. (Kane, E.K.) N.Y., 1857.
Geog 5970.38 The United States in the Antarctic, 1820-1962. (United States. Antarctic Projects Office.) Washington, 1962?
Geog 4249.61.5 United States scientific geographical exploration of the Pacific Basin, 1783-1899. (United States. National Archives.) Washington, 1961.
An 3409.16 Unites States. Bureau of Navigation. Navy Department. How to obtain good fingerprints. 2. ed. Washington, 1916.
Geog 5558.81F Unites States. Expedition to Lady Franklin Bay. Report on proceedings. Washington, 1888. 2v.
An 3409.35 Unites States. Federal Bureau of Investigation. Fingerprints. Washington, 1935.
An 3409.31 Unites States. Federal Bureau of Investigation. How to take fingerprints. Washington, 1931.
Geog 578.40 Universal gazeteer. (Landmann, G.) London, 1840.
Geog 577.60.3 The universal gazetteer. 2. ed. London, 1760.
Geog 808.12.9 Universal geography. (Malte-Brun, Conrad.) Boston, 1824. 8v.
Geog 808.12.11 Universal geography. (Malte-Brun, Conrad.) Philadelphia, 1827-1832. 6v.
Geog 808.12.10 Universal geography. v.1-14. (Malte-Brun, Conrad.) Boston, 1824-1829. 7v.
Geog 758.50 A universal geography in four parts. (Milner, Thomas.) London, 1850.
Geog 808.90F The universal guide and gazetteer. (DePuy, W.H.) N.Y., 1890.
Geog 4110.7 Universal guide of standard routes. (Thorpe, D.) Boston, 1907.
Geog 578.45 Universal pronouncing gazetteer. (Baldwin, Thomas.) Philadelphia, 1845.
Geog 4110.21 The universal traveller. 2. ed. (Goodrich, Charles A.) Hartford, 1836.
Geog 5850.7 Universitetets Eskimoiske samlinger. (Wallem, F.B.) Christiania, 1911.
An 608.609.07 Universitets ethnografiske samlinger, 1857-1907. (Nielsen, Y.) Christiania, 1907.
An 609.07.2 Universitets lappiske samlinger, 1857-1911. (Neilsen, Y.) Christiania, 1911.
Geog 243.15 L'universo. Firenze. 1,1920+ 62v.
Geog 243.15.5 L'universo. Index, 1920-40. Firenze, n.d.
Geog 577.13F Universus terrarum orbis. (Savonarola, R.) Patavii, 1713. 2v.
Geog 3019.35 Unrolling the map. (Outhwaite, L.) N.Y., 1935.
Geog 4309.32 Eine unsentimentale Reise. (Landau, M.A.) München, 1932.
Geog 819.58F Unsere Erde. Heidelberg, 1958.
Geog 5160.11.5 Unsolved mysteries of the Arctic. (Stefansson, V.) N.Y., 1939.
Geog 4309.21 Unstead, J.V. Europe of to-day. London, 1921.
Geog 4709.71.5 Unter der Flagge mit dem Totenkopf. (Sokol, Hans Hugo.) Herford, 1971.
Geog 4128.97 Untersuchungen...Reise...Itineraire der deutsech Königer. Inaug. Diss. (Ludwig, Friedrich.) Berlin, 1897.
Geog 4128.97.1 Untersuchungen über die Reise und Marschegeshwindigkeit in XII. und XIII. Jahrhundert. (Ludwig, Friedrich.) Berlin, 1897.
An 358.11 Untersuchungen über die Verschiedenheiten der Menschennaturen. (Meinere, C.M.) Tübingen, 1811-15. 3v.
Geog 4308.67.3 Unterwegs. (Meissner, A.) Leipzig, 1867.
Geog 4308.62 Unterwegs. v.1-2. (Bucher, L.) Berlin, 1862.
Geog 4219.34.10 Unwin, S. Two young men see the world. London, 1934.
Geog 4509.56 Uomini del sesto grado. (Garobbio, A.) Milano, 1956.
An 2109.06.3 Gli uomini primitivi delle selci e delle caverne. (Zuccarelli, A.) Napoli, 1906.
An 358.92 L'uomo bianco e l'uomo di colore. (Lombroso, C.) Firenze, 1892.
An 359.12.21 L'uomo come specie collettiva. (Giuffrida-Ruggeri, V.) Napoli, 1912.
Geog 607.11 L'uomo e lo studioso. (Vlora, Gribaudi.) Bari, 1971.
Geog 4509.43.5 Upon that mountain. (Shipton, Eric E.) London, 1943.
Geog 4309.00.10 Ur en resandes antecknigar. v.1-3. (Schück, Henrik.) Stockholm, 1900-1909.
An 375.5 Urabayen, Leoncio. Geografia umana. Madrid, 1934. 2 pam.
Geog 4139.42 Urbane travelers, 1591-1635. (Penrose, Boies.) Philadelphia, 1942.
Geog 576.80 Urbium insularum regionum. (Fondeur, F.) Londuni, 1680.
Geog 4145.89 Urdaneta. (Arteche, J. de.) Madrid, 1945.
An 2109.00.3 Urgeschichte der Kultur. (Schurtz, H.) Leipzig, 1900.
An 2108.77 Die Urgeschichte der Menschheit. (Caspari, O.) Leipzig, 1877. 2v.
An 2108.97 Urgeschichte der Menschheit. 2e Aufl. (Hoernes, Moriz.) Leipzig, 1897.

Author and Title Listing

An 2108.92 Die Urgeschichte des Menschen. (Hoernes, Moriz.) Wien, 1892.
An 2109.50.5 Die Urgesellschaft. (Eildermann, H.) Berlin, 1950.
Geog 4239.51 Uriburu, E.C. Seagoing Gaucho. N.Y., 1951.
An 358.99.3 Der Urmensch. (Beck, G.) Basel, 1899.
An 2059.63 Urmensch-Adam. (Muschalek, H.) Berlin, 1963.
An 2109.49.5 Der Urmensch und sein Weltbild. (Koppers, W.) Wien, 1949.
Geog 212.30 Urteaga, H.H. Memoria del presidente de la Sociedad Geográfica de Lima. Lima, 1942.
An 2109.25.19 Die Urzeit des Menschen. 4e Aufl. (Bumüller, J.) Augsburg, 1925.
Geog 4559.53.5 Us et coutumes à bord des long-courriers. (Hayet, Armand.) Paris, 1953.
Geog 4209.63.30 L'usage du monde. (Bouvier, Nicolas.) Genève, 1963.
Htn Geog 817.00.5* L'usages des globes celestes et terrestes. (Bion.) Adam, 1700.
Geog 5970.38.5 USARP; United States Antarctic Research Program, National Science Foundation. (National Science Foundation. Office of Antarctic Programs.) Washington, 1963.
Geog 5235.15 Ushakov, G.A. Po nekhoshenoi zemle. Moskva, 1953.
Geog 4348.65 Ussher, John. Journey from London to Persepolis. London, 1865.
Geog 4145.10 Ustrechi na piati kontinentakh. (Burkov, Boris S.) Moskva, 1973.
Geog 4309.66.10 Utchenko, Sergei L. Glazami istorika. Moskva, 1966.
Geog 4427.35 Utdrag af en Ost-Indisk rese-bescrifning. (Schwartz, G.L.) Westerås, 1784.
Geog 606.3 Uzbekistan. Tsentral'nyi Gosudarstvennyi Arkhiv. Otdel Dorevoliutsionnykh Fondov. A.P. Fedchenko. Tashkent, 1956.
Geog 185.15.10 V. Czesko-Polskie seminarium geograficzne. Wyd. 1. (Polsko-Czeskie Seminarium Geograficzne, 5th, Warsaw, 1972.) Warszawa, 1975.
Geog 4349.72 V etom bushuiushchem mire. (Ovcharenko, Aleksandr I.) Moskva, 1972.
Geog 5925.75 V mire kholoda. (Dralkin, A.G.) Moskva, 1961.
Geog 4145.53.5 V moriakh i stranstviakh. (Davydov, I.V.) Moskva, 1956.
An 3209.72 V poishakh predkov. (Alekseev, Valerii P.) Moskva, 1972.
Geog 5939.51 V straie kitov i pingvinov. (Arsen'ev, V.A.) Moskva, 1951.
Geog 3019.47 Vår väg genom världen. (Beckman, Leif.) Stockholm, 1947-51. 3v.
Geog 4308.45.19 Vacation rambles...1841-1843. 3rd ed. (Talfourd, T.N.) London, 1851.
Geog 4308.45.15 Vacation rambles and thoughts. (Talfourd, T.N.) London, 1845. 2v.
Geog 4158.60 Vacation tourists...in 1860, 1861, 1862, 1863. (Galton, F.) Cambridge, 1861-64. 3v.
Geog 4307.37.9 A vagabond courtier. (Pöllnitz, K.L.. von.) London, 1913. 2v.
Geog 4219.10 Vagabond journey around the world. (Franck, H.A.) N.Y., 1910.
Geog 4219.10.2 A vagabond journey around the world. (Franck, H.A.) N.Y., 1911.
Geog 4559.38.12 Vagabond voyaging; the story of freighter travel. (Nixon, Laurence A.) Boston, 1939.
Geog 4219.41 Vagabondage. (Parsons, C.) London, 1941.
Geog 5316.1.6 Vagrant Viking. (Freuchen, P.) London, 1954.
Geog 5316.1.5A Vagrant Viking. (Freuchen, P.) N.Y., 1953.
Geog 4559.53.10 Vagrant viking. (Freuchen, Peter.) N.Y., 1953.
Geog 809.22.15 Vahl, Martin. Jorden og menneskelivet. København, 1922-1927. 4v.
Geog 757.55 Vaissete, Joseph. Géographie historique, ecclesiastique et civile. Paris, 1755. 4v.
Geog 4209.52 Vale enchanting. (Warden, W.R.) London, 1952.
Geog 3570.34 Valentim Fernandes. (Rogers, F.M.) Lisboa, 1957?
Geog 4559.30.15 Valéry, Paul. Mer, marines, marins. Paris, 1930.
An 379.12 Vallaux, C. Géographie sociale. La mer. Paris, 1908.
An 379.12.3 Vallaux, C. Géographie sociale. Le sol et l'état. Paris, 1911.
Geog 665.32 Vallaux, Camille. Les sciences géographiques. Paris, 1925.
Geog 4180.84A Valle, P. della. Travels of...Valle in India. London, 1892. 2v.
Geog 4206.62 Valle, Pietro. Les fameux voyages de Pietro della Valle. Paris, 1663-70. 4v.
An 2109.54.6 Values of Primitive Society (Radio Program). The institutions of primitive society. Oxford, 1967.
Geog 4269.14 Van en over alles en iedereen. pt.1-5. (Couperus, Louis.) Amsterdam, 1915. 5v.
Geog 665.50 Van Loon's geography, the story of the world we live in. (Van Loon, H.W.) N.Y., 1932.
Geog 665.50.5 Van Loon's geography. (Van Loon, H.W.) Garden City, N.Y., 1940.
Geog 4169.12 Van oude voyagien. (Boer, M.G. de.) Amsterdam, 1912-13. 3v.
Geog 4169.12.5 Van oude voyagien. 3. druk. (Boer, M.G. de.) Amsterdam, 1939.
Geog 4129.11 Van pool tot pool. (Hedin, Sven.) Amsterdam, 194-?
An 359.64.15 Van primitieven tot medeburgers. (Köbben, A.J.F.) Assen, 1964.
Geog 4309.44 Van Toledo tot Budapest. (Cornette, Arthur H.) Antwerpen, 1944.
Geog 4309.12.15 Van Allen, William H. Travel-pictures. 2nd series. Milwaukee, 1912.
An 358.48 Van Amringe, W.F. An investigation of the theories of the natural history of man. N.Y., 1848.
Htn Geog 4217.90* Vancouver, G. Voyage of discovery...in 1790-1792. London, 1798. 3v.
Geog 4217.90.3 Vancouver, G. A voyage of discovery of the North Pacific Ocean. London, 1801. 6v.
Geog 4217.91.2 Vancouver's voyage. 2. ed. (Marshall, James S.) Vancouver, 1967.
Geog 4228.45A Van Denburgh, E.D. My voyage in the U.S. frigate "Congress". N.Y., 1913.
Geog 4300.47 Vanderbilt, C. European travel directory. N.Y., 1954.
Htn Geog 4219.29.10F* Vanderbilt, William K. Taking one's own ship around the world. N.Y., 1929.
Htn Geog 4219.33F* Vanderbilt, William K. West made East with the loss of a day. N.Y., 1933.
Geog 4308.85.9 Vandreaaret. (Gjellerup, Karl.) Kjøbenhavn, 1885.
Geog 4209.42.15 Vanishing Eden. (Birnbaum, M.) N.Y., 1942.
Geog 612.3.2 The vanishing Frenchman. (Allen, Edward W.) Rutland, 1959.
Geog 3510.7 Van Loon, H.W. The golden book of the Dutch navigators. N.Y., 1916.
Geog 4559.35 Van Loon, H.W. Ships and how they sailed the seven seas. (5000 B.C.-A.D. 1935). N.Y., 1935.
Geog 665.50 Van Loon, H.W. Van Loon's geography, the story of the world we live in. N.Y., 1932.
Geog 665.50.5 Van Loon, H.W. Van Loon's geography. Garden City, N.Y., 1940.
Geog 4182.29 Van Nouhuys, J.W. De eerste nederlandsche. 's-Gravenhage, 1927-51. 2v.
An 2108.84 Van Overloop, E. Sur une méthode a suivre dans les études dites préhistoriques. Bruxelles, 1884.
Geog 4249.59 Van Sinderen, Adrian. The other half of the earth. N.Y., 1959. 2v.
Geog 4329.58.5 Van Sinderen, Adrian. A voyage through the azure seas. N.Y., 1958.
Geog 4169.51 Van Thal, H. Victoria's subjects travelled. London, 1951.
Geog 4269.35.10A Van Valkenburg, S. Europe. N.Y., 1935.
An 359.14.3 Van Waters, Miriam. The adolescent girl among primitive peoples. Thesis. Worcester, 1914.
An 2109.59.15 Varagnac, André. L'homme avant l'écriture. Paris, 1959.
An 2109.68.30.2 Varagnac, André. L'homme avant l'écriture. 2e éd. Paris, 1968.
Geog 4229.47 Vargas Ugarte, Ruben. Relaciones de viajes. Lima, 1947.
An 3309.40 The varieties of human physique. 1st ed. (Sheldon, W.H.) N.Y., 1940.
Geog 4209.56.15 Varlden ar så stor, så stor. (Svenson, Sven.) Stockholm, 1956.
Geog 5600.2 Vartdal, Hroar. Bibliographie des ouvrages norvégiens relatifs au Groenland. Oslo, 1935.
Geog 4415.02 Varthema, L. de. Itinerário. Lisboa, 1949.
Htn Geog 4415.11F* Varthema, L. de. Itinerarium aethio piae. Milan, 1511.
Geog 520.9 Varthema, L. di. Les voyages de L. di Varthema. Paris, 1888.
Geog 4180.32 Varthema, S. di. Travels of Varthema. N.Y., 1964?
An 5.25 Vasconcellos-Abreu, Guilherme de. O critério nomolójico. Photoreproduction. Lisboa, 1887.
Geog 4309.13 Vasconcelos, C. de. Notas da Europa. Rio de Janeiro, 1913.
Geog 4218.91.4 Vassar, J.G. Twenty years around the world. N.Y., 1891.
Geog 5925.95 Vedenskii, Anatolii A. Vsnegakh Krainego Iuga. Leningrad, 1972.
Geog 212.110 Vedova, G. La Societa Geografica Italiana e l'opera sua nel secolo XIX. Roma. 5-6,1940-1941 3v.
Geog 4239.52 Veedam, V. Sailing to freedom. N.Y., 1952.
Htn Geog 5375.94* Veer, G. de. Oost-indische ende west-indische voyagien. Amsterdam, 1619.
Geog 5375.94.5 Veer, G. de. Reizen van Willem Barents. 's-Gravenhage, 1917. 2v.
Geog 4180.54 Veer, G. de. Three voyages of...William Barents. London, 1876.
Htn Geog 5375.94.4* Veer, G. de. Les trois navigations admirables. Paris, 1610.
Geog 5375.94.11 Veer, G. de. The true and perfect description of three voyages by the ships of Holland and Zeland. Facsimile. Amsterdam, 1970.
Geog 4180.13 Veer, G. de. True description...voyages by the North-East. London, 1852-53.
Htn Geog 5375.94.3F* Veer, G. de. Vraye description de trois voyages de mer tres admirables. Amsterdam, 1600.
Geog 4180.41 Vega, G. de la. First part of royal commentaries of the Yncas. N.Y., 1964. 2v.
Geog 5378.80.20 Vega. En aktualisering av händelserna kring Vega-expeditionen 1878-1880. (Skoog, Gösta.) Göteborg, 1965.
Geog 5378.80.5 Vegas färd rking Asien och Europa. (Nordenskiöld, A.E.) Stockholm, 1880-81. 2v.
Geog 4209.42.25 Vejarano, J.R. Rutas del mundo. Bogota, 1942.
Geog 5328.17 Vejlager, Johannes. Knud Rasmussen. Kjøbenhavn, 1934.
Htn Geog 756.28* Velazquez Minaya, F. Esfera forma del mundo. Madrid, 1628.
Geog 5557.33 Veliki severnaia ekspeditsiia, 1733-43. (Ostrovskii, B.G.) Arkhangel'sk, 1935.
Geog 5399.32.10 Velikim Severnym. 2. izd. (Shneiderov, V.A.) Moskva, 1963.
Geog 4672.26 The vengeful sea. (Snow, Edward Rowe.) N.Y., 1956.
Geog 3550.2 Venice. Biblioteca Nazionale Marciana. Mostra dei navigatori veneti del quattrocento e del cinquecento. Venezia, 1957.
Geog 818.28 Venning, I.A. (Mrs.). A geographical present. 1st American ed. N.Y., 1829.
Geog 114.7.85 Venticinque anni di Lavoro dell'Istituto Geographico Militare. (Coën, A.) Firenze, 1898.
VGeog 249.5 Venture. Des Moines. 3,1967+ 10v.
Geog 5839.52.10 Venture to the Arctic. (Hamilton, Richard A.) Harmondsworth, 1958.
Geog 4169.28 Ventures and voyages. (Chatterton, E.K.) London, 1928.
Geog 4208.98 Venturesome voyages of Captain Voss. Tokyo (Kanda), Japan. (Voss, J.C.) Yokohama, 1913.
Geog 4208.98.2 The venturesome voyages of Captain Voss. 2. ed. (Voss, J.C.) London, 1949.
Geog 3228.80.5 La vera patria di Nicolò de Conti e di Giovanni Caboto. (Bullo, C.) Chioggia, 1880.
Geog 4105.6 Verax, V. (pseud.). Cautions for the first tour. London, 1863.
An 379.21.5 Die Verbreitung des Menschen auf der Erdoberfläche (Anthropogeographie). (Krebs, Norbert.) Leipzig, 1921.
Htn Geog 3060.9* Das verdächtiger Pineser-Eyland. Hamburg, 1668.
An 2059.55 Vere, Francis. The Piltdown fantasy. London, 1955.
Geog 5558.69 Verein für die Deutsche Nordpolarfahrt. Leipzig, 1873-74. 4v.
Geog 500.65 Verein für Erdkunde, Dresden. Bibliothek. Büchcrei Verzeichnis des Vereins für Erdkunde zu Dresden. Dresden, 1905.
Geog 135.1 Verein für Erdkunde. Jahresbericht. (Gesellschaft für Erdkunde zu Leipzig.) 1884-1941 24v.
Geog 5558.69.5 Verein in die Deutsche Nordpolarfahrt. Die zweite deutsche Nordpolarfahrt. Leipzig, 1883.
Geog 48.1.9 Vereins für Erdkunde. Festschrift. (Dresden.) Dresden. 1888
Geog 48.1 Vereins für Erdkunde. Jahresbericht. (Dresden.) Dresden. 1-27 12v.
Geog 48.1.7 Vereins für Erdkunde. Mitgleider-Verzeichnis. (Dresden.) Dresden.
Geog 48.1.3 Vereins für Erdkunde. Mitteilungen. (Dresden.) Dresden. 1905-1935 5v.
Geog 250.5 Vereniging vir Aardrykskunde-Onderwys. Special publication. Stellenbosch.
Geog 579.26 Vergara y Martin, Gabriel Maria. Diccionario de voces y términos geográficos. Madrid, 1926.
Geog 4268.77.5 Vergleichende Kultur Gilder. (Faucher, J.) Hannover, 1877.

262

Author and Title Listing

	Geog 819.51	Vergleichende Landerkunde. (Krebs, N.) Stuttgart, 1951.
	Geog 578.29	Vergleichendes Wörterbuch. (Bischoff, F.H.T.) Gotha, 1829.
	Geog 4182.14	Verhaal van het vergaan van het jacht De Sperwer. (Hamel, Hendrik.) 's-Gravenhage, 1920.
NEDL	Geog 28.2	Verhandlungen. (Berlin. Gesellschaft für Erdkunde.) Berlin. 1-27,1873-1896 25v.
	Geog 45.1	Verhandlungen. (Deutsche Geographentage.) Berlin. 1-39 25v.
	Geog 3055.47	La vérité sur l'Atlantide. (Gattefossé, R.M.) Lyon, 1923.
	Geog 5398.97.3	La vérité sur l'expédition Andrée. (Fonvielle, W.) Strasbourg, 1897.
	Geog 4416.54	De vermaarde reizen. (LeBlanc, V.) Amsterdam, 1654.
	An 379.41.5	Vermooten, Willem H. De mens in de geografie. Assen, 1941.
	An 379.68.20	Vermooten, Willem Hendrik. Sociografie en sociale geografie in Nederland. Assen, 1968.
	Geog 3018.79	Verne, Jules. Exploration of the world. N.Y., 1879-81. 3v.
	Geog 3018.79.3	Verne, Jules. Jardens op da gelseshistorie. Kristiania, 1879-83.
	Geog 128.2	Veröffentlichungen. (Konigsberg. Universität. Geographisches Institut.) Hamburg. 10v.
	Geog 128.2.8	Veröffentlichungen. N.F. Reihe Ethnographie. (Konigsberg. Universität. Geographisches Institut.) Neudamm. 1-2,1931-1932 2v.
	Geog 128.2.5	Veröffentlichungen. N.F. Reihe Geographie. (Konigsberg. Universität. Geographisches Institut.) Konigsberg. 1-10,1931-1937 10v.
	Geog 4559.16	Verrill, A.H. The real story of the whaler. N.Y., 1916.
	Geog 4509.24.10	Vers l'idéal par la montagne. (Schwartz, M.) Paris, 1924.
	An 359.72.15	Vers une anthropologie sociopsychanalytique. 1. éd. (Mendel, Gérard.) Paris, 1972.
Htn	Geog 4265.76*	Verstegen, R. The post of the world. London, 1576.
	Geog 3055.19	Die versunkene Insel Atlantis. (Unger, F.X.) Wien, 1860.
	Geog 5530.9	Versus Tennysonianos Franklini. (Wright, A.) Cantabrigiae, 1882.
	Geog 500.135	Verzeichnis der geographischen Zeitschriften, periodischen Veröffentlichungen und Schriftreihen Deutschlands. (Schmidt, Rolf Dietrich.) Bad Godesberg, 1964.
	Geog 500.29	Verzeichnis von...Land und Reisebeschreibungen. (Stuck, G.H.) Halle, 1784.
	Geog 4209.70.10	Veseli potopisi. (Kozima, Marjan.) Maribor, 1970.
	Geog 3055.57	Vestigios de la Atlántida. (Requena, Rafael.) Caracas, 1932.
	Geog 78.2	Vetenskapliga. 1-11,1892-1920 6v.
	Geog 4218.80.5	Vetromile, E. A tour in both hemispheres. N.Y., 1880.
	Geog 4308.06	Viage de Espana, Francia. v.1-10,12,14. (Cruz y Bahamonde, N.) Madrid, 1806-1813. 8v.
	Geog 4217.64.7	Viage del Comandante Byron al rededor del mundo. (Ortega, D.C. de.) Madrid, 1769.
	Geog 4307.85.7	Viage fuera de Espana. (Ponz, A.) Madrid, 1785. 2v.
	Geog 4209.02	Viagens. v.1, 2a ed; v.2, 1a ed. (Prado, Eduardo.) São Paulo, 1902. 2v.
	Geog 4208.22	Viagens à China. (Rezo, José Luiz do.) Porto, 1822.
	Geog 3570.26	As viagens de descobrimento de iniciativa particular no tempo. (Bandeira Ferreira, F.) Lisboa, 1946.
	Geog 3570.55	Viagens de Diogo Cão e de Bartolomeu Dias. (Campos, Viriato.) Lisboa, 1966.
	Geog 4426.08	Viagens de Reino para a India e da India para o Reino. (Castro Daire, A. de Ataide.) Lisboa, 1957-58. 3v.
	Geog 3570.22	As viagens terrestres dos Portugueses. (Goncalões Niana, M.) Porto, 1945.
	Geog 4477.64.5	Viages de Enrique Wanton. (Seriman, Zaccaria.) Madrid, 1781-85. 4v.
	Geog 4308.93.10	Viaggi. v.2. (Papa, Dario.) Milano, 1893.
	Geog 4207.68	Viaggi da Parma in varie partie del mondo. Parma, 1768.
	Geog 4329.29	Viaggi di Bertoldo. (Spaini, Alberto.) Aquila, 1929.
	Geog 252.2	Viaggi e scoperte di navigatori ed esploratori italiani. Milano. 1-18,1929-1932 18v.
	Geog 3550.16	Viaggi ed esplorazioni di capitani marittimi della Riviera di Levante nella prima metà del secolo XIV. (Scarin, Maria Luisa.) Genova, 1968.
	Geog 252.3	Viaggi esplorazioni e scoperte. Milano. 1-11 11v.
	Geog 4209.55.10	Viaggi in Occidente. (Praz, Mario.) Firenze, 1955.
	Geog 4307.22	Viaggi per Europa. (Careri, G.) Napoli, 1722. 2v.
	Geog 5208.80	I viaggi polari. (Rezzadore, P.) Roma, 1880.
	Geog 4216.99.10	Un viaggiatore calabrese. (Nunnari, F.A.) Mesina, 1901.
	Geog 4169.62	Viaggiatori del Settecento. (Vincenti, Leonello.) Torino, 1962.
	Geog 4209.63.25	Viaggio a Samacanda. (Turri, Eugenio.) Novara, 1963.
	Geog 5516.88	Viaggio dal mare atlantico al pacifico. (Maldonado, Lorenzo F.) n.p., 1810.
	Geog 4321.83	Viaggio in Espana, Sicilia, Siria e Palestina. (Ibn Gubáyer.) Roma, 1906.
	Geog 4300.48	Il viaggio in pratica. (Vidari, Giovanni Maria.) Venezia, 1718.
	Geog 4324.80	Viaggio in Terrasanta di Santo Brasca, 1480. (Brasca, Santo.) Milano, 1966.
	Geog 4218.26.5	Viaggio intorno al globo. (Duhaut-Cilly, A.) Napoli, 1842.
	Geog 4217.64.10	Viaggio intorno al mondo. (Byron, John V.) Firenze, 1768.
	Geog 4209.59	Viaggio nel mondo. (Pizzinelli, Corrado.) Bologna, 1959.
	Geog 4309.28	Viajando. (Marín Vicuna, Santiago.) Santiago, 1928.
	Geog 4329.57	Viajando por Europa. (Todd, Roberto.) Madrid, 1957.
	Geog 4309.38.10	Viaje, plebeys por Europa. 2. ed. (Pinochet, T.) Santiago, 1938.
	Geog 4208.45	Viaje. (Bustamante, J.) Lima, 1845.
	Geog 4418.73.5	Viaje a Oriente. (Obligade, P.S.) Paris, 1873.
	Geog 4208.45.2	Viaje al antiguo mundo. 2. ed. (Bustamante, J.) Lima, 1959.
	Geog 4217.64	Viaje al rededor del mundo hecho en 1764, 65 y 66. (Byron, John V.) Madrid, 1833.
	Geog 4309.10.10	Viaje de recrea. (Matto de Turner, C.) Valencia, 1910?
	Geog 4418.03.15	Viajes Ali Bey el Abbassi. (Gonzalez y Rodriguez de la Peña, Hipolito.) Madrid, 1951.
	Geog 4159.21	Viajes clasicos. Madrid. 1-34 23v.
	Geog 4418.03.5	Viajes de Ali Bay el Abbassi por Africa y Asia. (Badia y Leblich.) Valencia, 1836. 3v.
	Geog 4341.73.15	Viajes de Benjamin de Tudela, 1160-1173. (Benjamin ben Jonah, of Tudela.) Madrid, 1918.
	Geog 4308.58.5	Viajes de un Colombiano en Europa. (Samper, J.M.) Paris, 1862. 2v.
	Geog 3560.15	Viajes por España y Portugal. (Farinelli, Arturo.) Roma, 1942-44. 3v.
	Geog 4208.58.9	Viajes por Europa y América de Don Gorgonio Petano y Mazariegos. (Petano y Mazariegos, Gorgonio.) Paris, 1858.
	Geog 4278.93	Viajes regios por mar en el trancurso de quimentos años. (Fernandez Duro, Cesario.) Madrid, 1893.
	Geog 3228.76	Viajes y descubrimientos, efectuados en la edad media. (Beltrán y Rózpide, Ricardo.) Madrid, 1876.
	Geog 4217.89	Viana, F.J. de. Diario del viaje explorador...en 1789-94. Cerrito de la Victoria, 1849.
	Geog 4418.03.9	Viatjes de Ali Bey el Abbassi per Africa y Asia. v.1-3. (Badia y Leblich.) Barcelona, 1888.
	Geog 5399.48	Vibe, Christian. Laugthen og nordpaa. København, 1948.
	Geog 3580.25A	Vicens Vives, J. Rumbos oceanicos. Barcelona, 1946.
	An 2109.75	Victims of progress. (Bodley, John H.) Menlo Park, 1975.
	Geog 6008.97.5	Victoire sur la nuit antarctique. (Gerlache de Gomery, Adrien de.) Tournai, 1960.
	Geog 5060.7.1	Victor, Paul Émile. L'homme à la conquête des pôles. Evreux, 1971.
	Geog 5060.7	Victor, Paul Émile. Man and the conquest of the poles. N.Y., 1963.
	Geog 5060.7.6	Victor, Paul Émile. Terres polaires, terres tragiques. Paris, 1973.
	An 127.10	Victor Courtet, 1813-1867. (Boissel, Jean.) Paris, 1972.
	An 608.78	Victoria, Australia. Public Library, Museum and National Galery. Catalogue of the objects of ethnotypical art. Melbourne, 1878.
	Geog 4169.51	Victoria's subjects travelled. (Van Thal, H.) London, 1951.
	Geog 4308.56.7.10	Vicuña Mackenna, Benjamin. Páginas de mi diario durante tres años de viaje. Santiago de Chile, 1936. 2v.
	Geog 4308.56.7	Vicuña Mackenna, Benjamin. Pajinas de mi diario. Santiago, 1856.
	Geog 4139.46	Vida de los navegantes y conquistadores españoles del siglo XVI. 1. ed. (Majó Framis, R.) Madrid, 1946.
	Geog 5160.57	La vida humana en el subecumene ártico. (Santillan de Andres, Selva E.) Tucuman, 1962.
	Geog 809.27.2	Vidal de La Blache, P. Géographie universelle. 2. éd. v.6, pt.1. Paris, 1947.
	An 379.22	Vidal de la Blache, P. Principes de géographie humaine. Paris, 1922.
	An 379.26A	Vidal de la Blache, P. Principles of human nature. N.Y., 1926.
	Geog 4268.89.5	Vidal de la Blache, Paul. Etats et nations de l'Europe, autour de la France. 4. éd. Paris, 189-?
	Geog 4658.90.2	Vidal Gormaz, Francisco. Algunas naufragios occuridos en las costas chilenes. Santiago de Chile, 1901.
	Geog 4658.90	Vidal Gormaz, Francisco. Algunos naufragios occuridos en las costas chilenes. Valparaiso, 1890.
	Geog 5837.03	Vidalin, Arngrimur Thorkilsson. Den tredie part af det saa kaldede gamle og mye Grønlands beskrifvelse. København, 1971.
	Geog 4300.48	Vidari, Giovanni Maria. Il viaggio in pratica. Venezia, 1718.
Htn	Geog 5837.20.10*	La vie...et le voyage de Groenland. v.1-2. (Mesange, P. de.) Amsterdam, 1720.
	Geog 4305.93.5	Vie. (Esprinchard, Jacques.) Paris, 1957.
	An 126.4	Vie d'Alphonse Bertillon, inventeur de l'anthropométrie. (Bertillon, Suzanne.) Paris? 1941.
	Geog 4358.90	La vie errante. 13. éd. (Maupassant, G. de.) Paris, 1890.
	Geog 4559.69.5	La vie quotidienne des marins au Moyen Age. (Fréminville, René Marie de La Poix de.) Paris, 1969.
	Geog 4309.68.5	Viejo continente. 1. ed. (Garrido, Felipe.) México, 1973.
	Geog 258.2	Vienna. K.K. Geographische Gesellschaft. Abhandlungen. Wien. 1-18,1899-1959 11v.
	Geog 258.1	Vienna. K.K. Geographische Gesellschaft. Mittheilungen. Wien. 1,1857+ 81v.
	Geog 3070.47.5	The Vienna-Klosterneuberg map corpus of the 15th century. (Durand, D.B.) Leiden, 1952.
	Geog 4502.10.4	Vierthaler, Franz Michael. Die Reise auf den Grossglockner, 1800. München, 1938.
	Geog 4417.98F	The view of Hindoostan. (Pennant, T.) London, 1798-1800. 4v.
	Geog 4307.86.6	View of society and manners. (Moore, J.) Boston, 1792.
	Geog 4307.86.10	View of society and manners. (Moore, J.) Paris, 1803. 2v.
	Geog 4307.86.7	View of society and manners. 5th ed. (Moore, J.) Dublin, 1793. 2v.
	Geog 4307.86.8	View of society and manners. 6th ed. (Moore, J.) London, 1786. 2v.
	Geog 4308.46.35	Views a-foot. 17th ed. (Taylor, Bayard.) N.Y., 1854.
	Geog 4308.50.15	Views a-foot or Europe. (Taylor, J.B.) N.Y., 1850.
	Geog 5399.28.30	Viglieri, Alfredo. Quarantotto giorni sul pack. Milano, 1929.
	Geog 520.23	Vignaud, Henry. Americe Vespuce, 1451-1512. Paris, 1917.
	Geog 520.18	Vignaud, Henry. La lettre et la carte. Paris, 1901.
	Geog 4679.72	Vignes, Jacques. La rage de survivre, quand il ne reste que la vie. Paris, 1973.
	Geog 4679.72.1	Vignes, Jacques. The rage to survive. N.Y., 1976.
	Geog 4239.71	Vihlen, Hugo. April Fool or, How I sailed from Casablanca to Florida in a six-foot boat. Chicago, 1971.
	An 170.238.10	VII Mezhdunarodnyi Kongress antropologicheskikh i ethnograficheskikh nauk. Trudy. v.1-3,5-10; photoreproduction. (International Congress of Anthropological and Ethnological Sciences, 7th, Moscow, 1964.) Moskva, 1967- 13v.
	Geog 4182.55	Vijf dagregisters van het Kasteel São Jorge da Maria. (Ratelband, K.) 's-Gravenhage, 1953.
	Geog 4182.59	De vijf gezantschapsreizen van Rijklof van Goens naar Hethof van Mataram. (Goens, Rijklof van.) 's-Gravenhage, 1956.
	An 126.6	Viking Fund. Ruth Fulton Benedict; a memorial. N.Y., 1949.
	Geog 5650.25.2	Viking Greenland with a supplement of saga. (Krogh, Knud J.) København, 1967.
	Geog 5650.17	Viking settlers in Greenland. (Nørlund, P.) London, 1936.
	Geog 5329.3.5	Vilhjalmur Stefansson. (Finnbogason, G.) Akureyri, 1927.
	Geog 5329.3A	Vilhjalmur Stefansson. N.Y., 1925.
	Geog 5329.3.2	Vilhjalmur Stefansson. N.Y., 1929.
	Geog 5329.3.7	Vil'iamur Stefanson. (Ol'khina, Evgeniia A.) Moskva, 1970.
	An 2109.36	Viljoen, S. The economics of primitive peoples. London, 1936.
	Geog 579.53	Villalba y Rubio, F. Diccionario geografico universal. Madrid, 1953.
	Geog 4309.16	Villaverde, J.R. Desde lajos. Habana, 1916.
	Geog 4268.30	Villégiatures romantiques. (Bertaut, Jules.) Paris, 1927.
	Geog 4309.29.11	Villes, mémoires: Montmartre, Rouen, souvenirs de Picardie et d'Artois, Brest, Londres, Villes rhénanes. (MacOrlan, Pierre.) Paris, 1966.
	Geog 4309.29.12	Villes, mémoires: Montmartre, Rouen, souvenirs de Picardie et d'Artois. 1. éd. (MacOrlan, Pierre.) Evreux, 1969.

Author and Title Listing

Geog 4309.29.10	Villes: Rouen-Montmartre-Brest-Londres-Villes rhénanes-Rome. 4. éd. (MacOrlan, Pierre.) Paris, 1929.
Geog 4559.30.3A	Villiers, A.J. By way of Cape Horn. N.Y., 1930.
Geog 4219.37.10	Villiers, A.J. Cruise of the Conrad...1934-1936. London, 1937.
Geog 4559.29.30	Villiers, A.J. Falmouth for orders...Cape Horn. Garden City, N.Y., 1929.
Geog 4559.32	Villiers, A.J. Grain race. N.Y., 1933.
Geog 4559.62.5	Villiers, A.J. Men, ships, and the sea. Washington, 1962.
Geog 4679.56.7	Villiers, A.J. Posted missing. London, 1975.
Geog 4679.56.5	Villiers, A.J. Posted missing. N.Y., 1956.
Geog 4559.49	Villiers, A.J. The set of the sails. N.Y., 1949.
Geog 4559.71.5	Villiers, A.J. The war with Cape Horn. N.Y., 1971.
Geog 4429.52A	Villiers, Alan J. Monsoon seas. N.Y., 1952.
Geog 4429.40	Villiers, Alan J. Sons of Sinbad. N.Y., 1940.
Geog 4239.57.10	Villiers, Alan John. The new Mayflower. Leicester, 1959.
Geog 4239.57	Villiers, Alan John. Wild ocean. N.Y., 1957.
Geog 3060.12.70	Vincent, Louis Claude. Le paradis perdu de Mu. Paris? 1969. 2v.
Geog 4169.62	Vincenti, Leonello. Viaggiatori del Settecento. Torino, 1962.
Geog 603.6	Vincenzo Coronelli. (Armao, Ermanno.) Firenze, 1944.
Geog 500.182	Vinge, Clarence L. United States government publications for research and teaching in geography. Totowa, N.J., 1967.
Geog 3019.27.10	Une vingtaine de voyageurs dans l'Orient européen. (Iorga, N.) Paris, 1928.
Geog 3235.66	Vinland Map Conference, Smithsonian Institution. Proceedings. Chicago, 1971.
Geog 4209.65.25	Vinogradov, Aleksandr A. Gde shumiat chuzhie goroda. Moskva, 1974.
An 3558.77	Virchow, R. Beitraege zur physischen Anthropologie der Deutschen. Berlin, 1877.
An 3538.92	Virchow, R. Crania ethnica americana. Berlin, 1892.
An 358.24	Virey, Julien J. Histoire naturale du genre humain. Paris, 1824. 3v.
An 358.43.5	Virey, Julien J. Historia natural del jénero humano. 3. ed. Barcelona, 1842-46. 2v.
Geog 500.33	Viro venerabili Friderico Laurentio Hoffmann. (Asher, George Michael.) Berolinenses, 1860.
Geog 819.52F	Visintin, L. Continenti e poesi. Novara, 1952.
Geog 4209.23.15	Visions solaires Mexique, Égypte, Inde, Japon, Océanie. (Balmont, K.D.) Paris, 1923.
Geog 4308.51	Visit to Europe in 1851. (Silliman, B.) N.Y., 1853. 2v.
Geog 4308.51.2	Visit to Europe in 1851. (Silliman, B.) N.Y., 1854. 2v.
Geog 4308.51.4	Visit to Europe in 1851. (Silliman, B.) N.Y., 1854. 2v.
Geog 132.4	Visnyk. Ser. heohrafiï. (Kiev. Universitet.) Kiev. 9,1967+
An 2109.11.7	Visscher, H. Religion und soziales Leben bei Naturvölkern. Bonn, 1911. 2v.
An 629.60	Visual files coding index. (Inverarity, Robert Bruce.) Bloomington, 1960.
An 2109.63.5	La vita sociale dei popoli primitivi. (Scotti, Pietro.) Brescia, 1963.
Geog 3596.2	Vitásek, František. Výroj moravské geografie. 1. vyd. Praha, 1973.
Geog 4309.71.15	Vitt och brett. (Holmqvist, Lasse.) Stockholm, 1971.
Geog 5330.1	Vittenburg, P.V. Zhizn' i nauchnaia deiatel'nost' E.V. Tollia. Leningrad, 1960.
Geog 5312.7.5	Vitus Bering; the discoverer of Bering Strait. (Lauridsen, Peter.) Chicago, 1889.
Geog 5312.7	Vitus J. Bering og de russiske opdagelsesrejser fra 1725-43. (Lauridsen, Peter.) Kjøbenhavn, 1885.
Geog 4309.37.5A	Vive la liberté. (Dorgelès, R.) Paris, 1937.
Geog 4309.01	Vivian, O.W.H. (Mrs.). The romance of religion. N.Y., 1901.
Geog 578.79.3F	Vivien de St. Martin, L. Nouveau dictionnaire de géographie universelle contenant...la géographie physique. Paris, 1879-95. 7v.
Geog 578.79F	Vivien de St. Martin, L. Nouveau dictionnaire de géographie universelle contenant...la géographie physique. Paris, 1879-95. 7v.
Geog 578.79.4F	Vivien de St. Martin, L. Nouveau dictionnaire de géographie universelle contenant...la géographie physique. Supplement. Paris, 1897-1900. 2v.
Geog 5399.36	Vize, Vladimir I. Moria Sovetskoi Arktiki. Leningrad, 1936.
Geog 5399.32.5	Vize, Vladimir I. Na "Sibiriakove" v "Litke"...1932 i 1934. Moskva, 1946.
Geog 5399.32.6	Vize, Vladimir I. Ne "Sibirakove" v Tikhii okean. Leningrad, 1934.
Geog 3580.40	De vlaminfen en de Spansche. (Colbrecht, Jozef.) Antwerp, 1927.
Geog 607.11	Vlora, Gribaudi. L'uomo e lo studioso. Bari, 1971.
Geog 4269.67.10	Vluchtige verlenningen. (Houter, F. den.) Laven, 1967.
Geog 5208.81.5	Voblasti vechnago l'da. (Hellwald, F. von.) Sankt Peterburg, 1881.
Geog 579.44	Vocabulaire pratique anglais-français. (Bargilliot, A.) Paris, 1944.
Geog 4219.55	Vocino, Michele. Marinai italiani e iberici sulle vie delle Indie. Roma, 1955.
Geog 5300.50	Vodop'ianov, M.V. Puti otvazhnykh. Moskva, 1958.
Geog 5559.56.6	Vodopianov, Mikhail V. Na kryliakh v Arktiku. Moskva, 1955.
Geog 5559.56.5	Vodopianov, Mikhail V. Wings over the Arctic. Moscow, 1956?
An 359.33.7	Voegelin, Erich. Die Rassenidee in der Geistesgeschichte Ray bis Carus. Berlin, 1933.
An 2109.12	Die Völker Europas zur jüngeren Steinzeit. (Classen, K.) Stuttgart, 1912.
Geog 4209.36.35	Völker und Kontinente. (Lissner, Ivar.) Hamburg, 1936.
An 359.24.27	Völker und Kulturen. (Schmidt, W.) Regensburg, 1924.
An 358.81.7	Der Völkergedanke. (Bastian, W.A.) Berlin, 1881.
An 358.52.4	Völkerkunde. (Frankenheim, M.L.) Breslau, 1852.
An 358.85.3	Völkerkunde. (Peschel, O.F.) Leipzig, 1885.
An 358.87.3	Völkerkunde. (Ratzel, F.) Leipzig, 1887-88. 3v.
An 359.24.20	Völkerkunde. (Schmidt, Max.) Berlin, 1924.
An 359.03.7	Völkerkunde. (Schurtz, H.) Leipzig, 1903.
An 359.17.7	Völkerkunde. 3. Aufl. (Haberlandt, M.) Berlin, 1917. 2v.
An 358.76.5	Völkerkunde. 3. Aufl. (Peschel, Oscar F.) Leipzig, 1876.
Geog 4208.68	Vogel, H.W. Vom indischen Ocean bis zum Goldlande. Berlin, 1877.
Geog 4182.41	Vogel, J.P. Journal van J.J. Ketelaar's hofreis...1711-1713. 's-Gravenhage, 1937.
An 99.75	Voget, Fred W. A history of ethnology. N.Y., 1975.
An 358.64	Vogt, C. Lecture on man: his place in creation and in the history of the earth. London, 1864.
An 629.74	Vogt, Evon Zartman. Aerial photography in anthropological field research. Cambridge, 1974.
Geog 4678.37.9	The voice of God in calamity, or Reflections on the loss of the steamboat "Home"...sermon. (Smyth, Thomas.) Charleston, 1837.
Geog 6009.64	Vojtěch, Vaclav. Námořníkem, topičem a psovodem za jižním polárním kruhem. Vyd. 1. Praha, 1968.
Geog 5925.94	Vokrug Antarktidy. (Treshnika, Aleksei F.) Leningrad, 1970.
Geog 4218.03.14	Vokrug sveta pod russkim flagom. (Nevskii, Vladimir V.) Moskva, 1953.
Geog 4209.65.15	Vokrug sveta s Zarei. (Pleshakov, Leonid P.) Moskva, 1965.
Geog 4308.44.5	Volkov, M. Otryuki iz zagranichnykh pisem. Sankt Peterburg, 1857.
An 99.71	Volkskunde: von der Altertumsforschung zur Kulturanalyse. (Bausinger, Hermann.) Berlin, 1971.
Geog 579.55A	Volostnova, M.B. Slovar' russkoi...geograficheskikh nazvanii. Moskva, 1955. 2v.
Geog 5559.57.5	Volovich, V.G. God na poliuse. Moskva, 1957.
Geog 4239.67	Volovich, Vitalii G. Tridtsatyi meridian. Moskva, 1967.
Geog 4208.68	Vom indischen Ocean bis zum Goldlande. (Vogel, H.W.) Berlin, 1877.
An 609.70A	Vom Raritätenkabinett zum Bremener Überseemuseum. (Abel, Herbert.) Bremen, 1970.
Geog 4347.91.1	Von Amsterdam nach Temiswar. (Steuke, Johann Kaspar.) Berlin, 1969.
An 359.38.7	Von Menschen; Versuch einer geistwissenschaftlichen Anthropologie. (Sombart, W.) Berlin, 1938.
Geog 4129.11.5	Von Pol zu Pol. 81. Aufl. (Hedin, Sven.) Leipzig, 1942.
An 379.22.5	Von Engeln, Oscar D. Inheriting the earth. N.Y., 1922.
An 2109.67.35F	De voorgeschiedenis van Europa. (Laet, Sigfried Jan de.) Hasselt, 1967.
Geog 809.59.5	Vooys, Adriaan. Panorama der Wireld. Roermond, 1959. 3v.
An 359.70	Voprosy etnopsikhologii v vabotakh zanebezhnykh avtarov. (Korolev, Stanislav I.) Moskva, 1970.
Geog 260.5	Voprosy geografii. Moskva. 1-98,1946-1975 68v.
Geog 3019.73.5	Voprosy istoricheskoi geografii i istorii geografii. Moskva, 1973.
An 629.70	Voprosy metodiki etnograficheskikh i etno-sotsiologicheskikh issledovanii. (Akademiia Nauk SSSR. Institut Etnografii.) Moskva, 1970.
Geog 759.70.5	Voprovy geografii. Kaliningrad, 1970.
An 359.26	Vor folkegruppe Gottjod. (Schütte, G.) Kjøbenhavn, 1926.
An 98.81	Vorgeschichte der Ethnologie. (Bastian, A.) Berlin, 1881.
An 2058.74	Vorgeschichte des europäischen Menschen. (Ratzel, F.) München, 1874.
An 2109.49.10	Vorgeschichte Europas. 7. Aufl. (Hoernes, Moriz.) Berlin, 1949.
An 379.08	Vorlesungen. (Richthofen, Ferdinand von.) Berlin, 1908.
Geog 4127.77.2	Vorlesungen über Land- und Seereisen. (Schloezer, August Ludwig von.) Göttingen, 1962.
Geog 4308.35.10	Vorletzter Weltgang von Semilasso. (Pückler-Muskau, H.) Stuttgart, 1835. 3v.
Geog 5700.8	Vormbaum, R. Hans Egede, der Prediger des Evangeliums. Elberfeld, 1861.
An 358.64.3	Vorschule der Völkerkunde. (Diefenbach, L.) Frankfurt am Main, 1864.
Geog 613.10	Vospominaniia i razmyshleniia geografa. (Markov, Konstantin K.) Moskva, 1973.
Geog 4208.98	Voss, J.C. Venturesome voyages of Captain Voss. Tokyo (Kanda), Japan. Yokohama, 1913.
Geog 4208.98.2	Voss, J.C. The venturesome voyages of Captain Voss. 2. ed. London, 1949.
Geog 4209.64.25	Vot liudi. (Sakhnin, Arkadii Ia.) Moskva, 1964.
VGeog 4306.00	Voyage...en l'an 1600. (Rohan de.) Amsterdam, 1646.
Geog 5397.73.3	Voyage...North Pole...1773. (Phipps, J.) London, 1774.
Geog 4327.38	Voyage...round the Mediterranean. (Montague, J.) London, 1799.
Geog 6008.39	Voyage...southern and Antarctic regions. (Ross, J.C.) London, 1847. 2v.
Geog 4180.19	Voyage...to Bantum and Maluco Islands. (Middleton, H.) London, 1855.
Geog 4308.53.14	Voyage à Constantinople par l'Italie. v.2. (Boucher de Crèrecoeur de Perthes.) Paris, 1855.
Geog 5517.46.4	Voyage à la baye de Hudson. (Ellis, H.) Leide, 1750.
Geog 4309.61.10	Voyage à Moscou et au-dela. (Giloteaux, Paulin.) Paris, 1961.
Geog 4308.66.7	Voyage alone in the yawl "Rob Roy". (Macgregor, J.) London, 1868.
Geog 4308.66.10	Voyage alone in the yawl "Rob Roy". (Macgregor, J.) London, 1880.
Htn Geog 4218.03.5*	Voyage around the world...1803-1806. (Krusenstern, A.J.) London, 1813. 2v.
Geog 4218.71.15	A voyage around the world. (Adams, N.) Boston, 1871.
Geog 4218.38.3	A voyage around the world. v.1-2. 9th ed. (Taylor, F.W.) New Haven, 1847.
Geog 4218.38.2	A voyage around the world. v.1-2. 9th ed. (Taylor, F.W.) New Haven, 1850.
Geog 4217.72	Voyage au Cap de Bonne-Esperance. (Sparrman, Anders.) Paris, 1787.
Geog 4228.28	Voyage au Golfe de Californie. (Combier, C.) Paris, 1864.
Htn Geog 4347.01.6F*	Voyage au Levant. (Bruyn, C. de.) Paris, 1725. 5v.
Geog 6008.41	Voyage au Pole Sud. (Dumont, J.S.C.) Paris, 1841. 10v.
Geog 4324.84	Un voyage au XVe siècle. (Neeffs, E.) Louvain, 1873.
Geog 4217.90.7	Voyage autour du monde...1790-1792. (Fleurieu, Charles P. Claret de.) Paris, 1798-1800. 4v.
Htn Geog 4218.03.7*	Voyage autour du monde...1803-06. (Krusenstern, A.J.) Paris, 1821. 2v.
Geog 4217.66	Voyage autour du monde. (Bougainville, L. de.) Paris, 1771.
Geog 4217.66.5	Voyage autour du monde. (Bougainville, L. de.) Paris, 1772. 3v.
Htn Geog 4218.26*	Voyage autour du monde. (Duhaut-Cilly, A.) Paris, 1834.
Geog 4218.33.25	Voyage autour du monde. (Laplace, Cyril P.T.) Paris, 1833-39. 4v.
Geog 4217.67	Voyage autour du monde. (Pagès, P.M.F. de.) Paris, 1782. 2v.
Geog 4218.16	Voyage autour du monde. (Roquefeuil, C. de.) Paris, 1823. 2v.
Htn Geog 4218.03.8F*	Voyage autour du monde. Atlas. (Krusenstern, A.J.) Paris, 1821.

Author and Title Listing

Geog 4218.33.27PF Voyage autour du monde. Atlas. (Laplace, Cyril P.T.) Paris, 1835.
VGeog 4218.51.5 Voyage autour du monde sur la frégate suédoise l'Eugénie executé pendant des années 1851-1853. (Svenska Vetenskaps-Akademien, Stockholm.) Stockholm, 1858-74.
Geog 4304.95 Voyage aux Pays-Bas, 1495. (Muenzer, Hieronymus.) Bruxelles, 1942.
Geog 4427.80 Voyage dans les mers de l'Inde. v.1-2. (Le Gentil de la Galaisière, G.J.H.J.B.) Suisse, 1780.
Geog 4306.81.3 Voyage de...en Flandre, Hollanke, Danemark. (Regnard, J.F.) Paris, 1874.
Geog 5517.46.5 Voyage de la baye de Hudson. (Ellis, H.) Paris, 1749.
Geog 6008.97 Voyage de la Belgica, quinze mois dans l'Antarctique. (Gerlache de Gomery, Adrien de.) Paris, 1902.
Geog 5515.88 Voyage de la mer atlantique...pacifique. (Maldonado, L.F.) Plaisance, 1812.
Htn Geog 4217.97* Voyage de La Pérouse. (Milet-Mureau, M.L.A.) Paris, 1797. 4v.
Htn Geog 4217.97PF* Voyage de La Pérouse. Atlas. (Milet-Mureau, M.L.A.) Paris, 1797.
Geog 520.2 Le voyage de la Sainte Cyté de Hierusalem. Paris, 1882.
Geog 520.11 Le voyage de la Terre Sainte. (Possot, D.) Paris, 1890.
Geog 4217.85.13 Voyage de Lapérouse autour du monde. (Lapérouse, Jean François de Galaup.) Paris, 1965.
Geog 520.8 Le voyage de M. d'Aramon. (Chesneau, J.) Paris, 1887.
Geog 520.5 Le voyage d'Outremer. (Thenaud, J.) Paris, 1884.
Geog 520.12 Le voyage d'Outremer. Photoreproduction. (La Broquière, Bertrandon de.) Paris, 1892.
Geog 4347.27 Voyage du...en Europe, Asie et Afrique. (La Montraye, A. de.) Haye, 1727. 2v.
Geog 4348.34 Voyage du...en Hongrie, en Transylvanie. (Marmont, Auguste Frédéric L.) Paris, 1837. 5v.
Geog 4305.17 Voyage du Cardinal d'Aragon en Allemagne. (Antonio de Beatis, Don.) Paris, 1913.
Geog 520.16 Le voyage du Levant. (Du Fresne-Canayl, P.) Paris, 1897.
Geog 4218.50.3 Voyage d'une femme autour du monde. v.1-2. (Marmier, Xavier.) Bruxelles, 1853.
Geog 4417.70.2 Voyage en Afrique et en Asie. 1770-1779. (Thunberg, Karl P.) Paris, 1794.
Geog 4228.16 Voyage en Amerique. (Montulé, E. de.) Paris, 1821. 2v.
Geog 4347.27.3F Voyage en Anglois et en François. (La Montraye, A. de.) Haye, 1732.
Geog 4346.85 Voyage en divers etats d'Europe et d'Asie. (Avril, P.) Paris, 1693.
Geog 4328.43 Voyage en Grèce et dans le Levant fait en 1843-1844. (Chenavard, Antoine N.) Lyon, 1849.
Geog 4158.05.5 Voyage en Morée. v.1,2-3. (Pouqueville, F.C.H.) Paris, 1805. 2v.
Geog 4309.38.5 Voyage en Orient du roi Erik Ejegod. (Fellman, A.) Helsinki, 1938.
Geog 4328.66 Voyage en Russie, au Caucase. (Lycklama, A. Nijeholt.) Paris, 1872-1875. 4v.
Geog 5398.56 Un voyage en yacht, lettres de hautes latitudes. (Dufferin and Ava, F.T. Blackwood.) Montreal, 1876.
Geog 4208.39 Voyage from Plymonth to Melbourne in 1839. (Were, J.B.) Melbourne, 1964.
Geog 4248.12 Voyage in the South Seas in...1812-1814. (Porter, David.) London, 1823.
Geog 4218.78.8 A voyage in the "Sunbeam". (Brassey, A. (Mrs.).) Chicago, 1881.
Geog 4218.78.3 A voyage in the "Sunbeam". (Brassey, A. (Mrs.).) London, 1879.
Geog 4219.36.15 The voyage of Anahita; single-handed round the world. (Bernicot, Louis.) London, 1953.
Geog 4181.91 Voyage of Captain Bellingshausen to Antarctic seas, 1819-21. (Debenham, Frank.) London, n.d. 2v.
Geog 4181.13 Voyage of Captain Don Felippe Gonzalez. (Corney, B.G.) Cambridge, 1908.
Geog 5516.19.5 Voyage of Captain John Monk. Glasgow, 1792.
Geog 4181.5 Voyage of Captain John Saris to Japan, 1613. (Saris, J.) London, 1900.
Geog 5516.31.5A The voyage of Captain Thomas James...1631. (Bodilly, R.B.) London, 1928.
Geog 5518.18 Voyage of discovery...Baffin's Bay. (Ross, J.) London, 1819.
Htn Geog 4217.90* Voyage of discovery...in 1790-1792. (Vancouver, G.) London, 1798. 3v.
Htn Geog 5377.95* A voyage of discovery...North Pacific Ocean. (Broughton, William.) London, 1804.
Htn Geog 5377.95.2* A voyage of discovery...North Pacific Ocean. (Broughton, William.) London, 1804.
Geog 4217.90.3 A voyage of discovery of the North Pacific Ocean. (Vancouver, G.) London, 1801. 6v.
Geog 5378.18 A voyage of discovery towards the North Pole in H.M. ships Dorothea and Trent...1818. (Beechey, F.W.) London, 1843.
Geog 4180.82 Voyage of F. Leguat...to Rodriguez. (Leguat, F.) London, 1891. 2v.
NEDL Geog 4180.76 Voyage of F. Pyrard...to East Indies. v.1-2. (Pyrard, F.) London, 1887. 3v.
Geog 4180.76 Voyage of F. Pyrard...to East Indies. v.1-2. (Pyrard, F.) N.Y., 1964? 3v.
Geog 5538.57 Voyage of "Fox"...narrative...of fate of Sir John Franklin. (M'Clintock, F.L.) Boston, 1860.
X Cg Geog 4180.70 Voyage of Linschoten to East Indies. (Linschoten, J.H.) London, 1885. 2v.
Geog 4180.70 Voyage of Linschoten to East Indies. (Linschoten, J.H.) N.Y., 1964? 2v.
Geog 4181.82 The voyage of Nicholas Downton to the East Indies, 1614-15. (Foster, William.) London, 1939.
Geog 4181.81 The voyage of Pedro Alvares Cabral to Brazil and India. (Greenlee, William B.) London, 1938.
Geog 4248.82 A voyage of pleasure. (Gilbay, Bernard.) Cambridge, 1956.
Geog 5538.50.17 Voyage of Prince Albert...Sir John Franklin. (Snow, W.P.) London, 1851.
Geog 4181.3 Voyage of R. Dudley...to West Indies. (Dudley, R.) London, 1899.
Geog 4181.88 The voyage of Sir Henry Middleton to the Moluccas. (Foster, William.) London, 1943.
Geog 4219.39.5 The voyage of the Cap Pilar. (Seligman, A.) London, 1939.
Geog 4208.27 Voyage of the Caroline, 1827-28. (Haus, Rosalie.) London, 1927.
Geog 4238.77 Voyage of the Challenger. The Atlantic. v.1-2, atlas. (Thomson, C.W.) London, 1877. 3v.
Geog 5379.35 The voyage of the Chelyuskin. (Cheliuskin Expedition, 1933-34.) N.Y., 1935.
Htn Geog 6009.01.8* Voyage of the 'Discovery'. (Scott, R.F.) London, 1905. 2v.
Geog 6009.01.7A Voyage of the 'Discovery'. (Scott, R.F.) N.Y., 1905. 2v.
Geog 5398.80 Voyage of the 'Eira'. (Markham, C.R.) London, 1881.
Geog 5538.57.2 The voyage of the "Fox" in the Arctic seas. (M'Clintock, F.L.) London, 1859.
Geog 5538.57.3 The voyage of the "Fox" in the Arctic seas. (M'Clintock, F.L.) Rutland, 1972.
Geog 4559.53.16 The voyage of the Hérétique. (Bombard, Alain.) Ann Arbor, Mich., 1971.
Geog 5398.79.7 Voyage of the "Jeannette". (De Long, G.W.) Boston, 1883. 2v.
Geog 4559.40.9 The voyage of the Kaimiloa. (Bisschop, E. de.) London, 1940.
Geog 4218.26.10 Voyage of the Lerós. (Le Netrel, Edmond.) Los Angeles, 1951.
Geog 4228.10 The voyage of the New Hazard...1810-1813. (Reynolds, Stephen.) Salem, 1938.
Geog 4558.11 Voyage of the Niedas, 1811. (Forbes, R.B.) Boston, 1885.
Geog 4239.63 The voyage of the Niña II. 1st ed. (Marx, R.F.) Cleveland, 1963.
Geog 6009.02 Voyage of the "Scotia". (Brown, R.N.R.) Edinburgh, 1906.
Geog 4218.31 Voyage of the United States frigate Potomac. (Reynolds, J.N.) N.Y., 1835.
Geog 4218.31.2 Voyage of the United States frigate Potomac. (Reynolds, J.N.) N.Y., 1835.
Geog 5378.80 Voyage of the Vega round Asia and Europe. (Nordenskiöld, A.E.) London, 1881. 2v.
Geog 5378.80.2 The voyage of the Vega round Asia and Europe. (Nordenskiöld, A.E.) N.Y., 1882.
Geog 4218.83.10 Voyage of the Wanderer. (Lambert, C.) London, 1883.
Geog 6009.08 The voyage of the 'Why not?' in the Atlantic. (Charcot, Jean B.) London, n.d.
Geog 4181.75 The voyage of Thomas Best to the East Indies. (Foster, William.) London, 1934.
Geog 4209.53.5 The voyage of Waltzing Matilda. (Davenport, P.) London, 1953.
Geog 4168.34 Voyage pittoresque autour du monde. (Dumont d'Urville, Jules.) Paris, 1834-35. 2v.
Geog 4217.66.3A Voyage round the world...1766. (Bougainville, L. de.) London, 1772.
Htn Geog 4218.03.9* Voyage round the world...1803-06. (Lisianskii, I.) London, 1814.
Geog 4418.70 Voyage round the world. (Beauvior, L.) London, 1870. 2v.
Geog 4217.90.15 A voyage round the world. (Fleurieu, Charles P. Claret de.) London, 1801. 3v.
Geog 4217.03 Voyage round the world. (Funnell, William.) London, 1707.
Htn Geog 4217.85.3F* Voyage round the world. (Portlock, N.) London, 1789.
Geog 4218.35.3 A voyage round the world. (Ruschenberger, W.S.W.) Philadelphia, 1838.
Geog 4218.30 A voyage round the world. v.4. (Holman, James.) London, 1835.
Htn Geog 4218.06* Voyage round the world from 1806-12. (Campbell, A.) Edinburgh, 1816.
Geog 4217.72.10F A voyage round the world with Captain James Cook. (Sparrman, Anders.) London, 1944.
Geog 4329.58.5 A voyage through the azure seas. (Van Sinderen, Adrian.) N.Y., 1958.
Geog 4238.47 A voyage to America 90 years ago...Hamburg to New York in 1847. (Rosenbaum, S.E.) N.Y., 1939.
Geog 4328.10 A voyage to Cadiz and Gibraltar, up the Mediterranean, to Sicily and Malta, in 1810 and 1811. (Cockburn, G.) London, 1815. 2v.
Geog 4217.72.5 Voyage to Cape of Good Hope. (Sparrman, Anders.) Perth, 1789.
Geog 4227.35.1 A voyage to Guinea, Brazil, and the West Indies. (Atkins, John.) London, 1970.
Geog 4228.17 A voyage to North America and West Indies in 1817. (Montulé, E. de.) London, 1821.
Geog 5558.84 A voyage to the Arctic in the whaler Aurora. (Lindsay, D.M.) Boston, 1911.
Geog 4427.80.5 A voyage to the Indian Seas. (Le Gentil de la Galaisière, G.J.H.J.B.) Manila, 1964.
Htn Geog 4218.61* Voyage to the North Pacific. (D'Wolf, J.) Cambridge, 1861.
Geog 4478.78 Voyage to the North Pole. Cambridge, 1878.
Geog 4327.98F A voyage up the Mediterranean. (Willyams, Cooper.) London, 1802.
Geog 5518.18.3 Voyage vers le Pole Arctique. (Ross, J.) Paris, 1819.
Geog 4347.01.7F Voyages...au Levant. (Bruyn, C. de.) Haye, 1732. 5v.
Geog 4327.22 Voyages...dans plusieurs provinces de la Barbarie et du Levant. (Shaw, Thomas.) La Haye, 1743. 2v.
Htn Geog 4347.18* Les voyages...en Moscovie. v.1-3. (Struys, Jan J.) Amsterdam, 1718.
Htn Geog 4347.01.8F* Voyages...en Perse et aux Indes Orient. (Bruyn, C. de.) Amsterdam, 1718. 2v.
Geog 4328.01 Voyages...from 1796-1801. (Collins, F.) London, 1819.
Geog 4181.83 The voyages...of Sir Humphrey Gilbert. (Quinn, David B.) London, 1940. 2v.
Geog 4413.25.9 Voyages. (Ibn Batoutah.) Paris, 1874. 5v.
Geog 4306.87 Voyages. (Regnard, J.F.) n.p., n.d.
Geog 4477.26.5 The voyages and adventures of Captain Boyle. (Boyle, R. (pseud.).) Dublin, 1741.
Htn Geog 4477.26* Voyages and adventures of Captain Boyle. (Boyle, R. (pseud.).) London, 1726.
Geog 4477.26.3 The voyages and adventures of Captain Boyle. 1st American ed. (Boyle, R. (pseud.).) Cooperstown, N.Y., 1796.
Geog 4208.40 Voyages and commercial enterprises. (Cleveland, R.J.) N.Y., n.d.
Geog 4218.33.10A Voyages and discoveres in the South Seas, 1792-1832. (Fanning, E.) Salem, 1924.
Geog 3520.10.50.2 Voyages and documents. (Hakluyt, Richard.) London, 1963.
Geog 4209.64.5 Pamphlet vol. Voyages and travel. 2 pam.
Htn Geog 4347.27.5* Voyages and travels, Prussia, Russia and Poland. (La Montraye, A. de.) London, 1732. 3v.
Geog 4169.02A Voyages and travels, 16th and 17th centuries. N.Y., 1902. 2v.
Htn Geog 4218.03* Voyages and travels...in 1803-1805. (Langsdorff, G.H. von.) London, 1813. 2v.
Geog 4209.71.15 Pamphlet vol. "Voyages and travels". 6 pam.
Geog 4168.87F Voyages and travels. (Colange, Leo de.) Boston, 1887. 2v.
Geog 4168.87.2F Voyages and travels. (Colange, Leo de.) Boston, 1887.
Geog 500.118 Voyages and travels. v.1,4-5. (Maggs Bros., London.) London, 1962- 3v.
Geog 500.72.5 Voyages and travels in Greece. (American School of Classical Studies at Athens.) Princeton, N.J., 1953.

Author and Title Listing

Call Number	Entry
Geog 500.72	Voyages and travels in the Near East made during the nineteenth century. (American School of Classic Studies at Athens.) Princeton, N.J., 1952.
Geog 4227.46	The voyages and travels of Francis Goelet, 1746-1758. (Goelet, Francis.) N.Y., 1970.
Geog 4418.02F	Voyages and travels to India. (Annesley, G.) London, 1809. 3v.
Geog 4180.59	Voyages and works of J. Davis. (Davis, J.) London, 1880.
Geog 4168.07	Voyages anecdotiques. (Campe, Joachim Heinrich.) Paris, 1807.
Geog 5397.73.5	Voyages au pôle boréal. (Phipps, J.) Paris, 1775.
An 358.01	Voyages chez les peuples sauvages. (Richard, Jerôme.) Paris, 1801. 3v.
Geog 4418.03	Voyages d'Ali Bey el Abbassi en Afrique et en Asie...1803...1807. v.1-3, atlas. (Badia y Leblich.) Paris, 1814. 4v.
Geog 4206.65.11	Les voyages de Balthasar de Monconys. Paris, 1887.
Geog 3530.17	Les voyages de découverte et les premiers établissements. 1. éd. (Julien, C.A.) Paris, 1948.
Geog 4477.29	Les voyages de Glantzby. Paris, 1739.
Htn Geog 4346.76.5*	Les voyages de Jean Struys. (Struys, Jan J.) Amsterdam, 1681.
Geog 520.9	Les voyages de L. di Varthema. (Varthema, L. di.) Paris, 1888.
Geog 4306.63.3	Les voyages de Monsieur Payen. 2e éd. (Payen.) Paris, 1667.
Geog 4348.14	Voyages de sa majesté la reine d'Angleterre. (Almerté, T.) Paris, 1821.
Geog 4417.44	Voyages du capitaine...Lade. (Lade, R.) Paris, 1744. 2v.
Geog 4417.44.5	Voyages du capitaine...Lade. (Lade, R.) Paris, 1810.
Geog 4428.01	Les voyages du chirurgien Avine à l'Ile de France et dans la mer de Indes au début du XIXe siècle. (Avine, Grégoire.) Paris, 1961.
Geog 4417.68	Voyages d'un philosophe. (Poivre, Pierre.) Yverdon, 1768.
Htn Geog 4206.45*	Voyages en Afrique, Asie. (Mocquet, J.) Rouen, 1645.
Geog 520.10	Les voyages en Asie. (Odoric de Pardenone.) Paris, 1891.
Geog 4303.99	Voyages et ambassades...1399-1450. (Lannoy, Guillebert de.) Mons, 1840.
Htn Geog 4477.10*	Voyages et avantures de Jaques Masse. Bourdeaux, 1710.
Htn Geog 4206.43.5*	Les voyages et observations. (La Boullaye, C. Goux.) Paris, 1653.
Geog 4208.43	Voyages et recits. v.1-2. (Yvan, M.) Bruxelles, 1853.
Geog 4308.56.3	Voyages et voyageurs, 1837-1854. (Cuvillier-Heury, A.A.) Paris, 1856.
Geog 4168.95	Voyages et voyageurs de la renaissance. (Bonnaffé, E.) Paris, 1895.
Geog 4416.54.5	Les voyages fameux. (LeBlanc, V.) Paris, 1658.
Htn Geog 5220.19*	Voyages from Asia to America. (Müller, G.F.) London, 1761.
Htn Geog 5367.61.3*	Voyages from Asia to America. 2. ed. (Müller, G.F.) London, 1764.
Htn Geog 4306.92*	Voyages historiques de l'Europe. v.1-8. Paris, 1692. 2v.
Geog 4477.87.7	Voyages imaginaires, songes. v.1-31,36. Amsterdam, 1787-89. 32v.
Geog 4248.13	Voyages in the northern Pacific, 1813-1818. (Corney, Peter.) Honolulu, 1896.
Geog 4129.25	Les voyages l'hospitalité...dans le monde chrétien des IVe et Ve siècles. (Gorce, Denys.) Wépion-sur-Meuse, 1925.
Geog 4129.25.2	Les voyages l'hospitalité...dans le monde chrétien des IVe et Ve siècles. Thèse. (Gorce, Denys.) Wépion-sur-Meuse, 1925.
Geog 4180.50	Voyages of...Nicolo and Antonio Zeno. (Zeno, N.) London, 1873.
Geog 4208.40.10	Voyages of a merchant navigator. (Cleveland, R.J.) N.Y., 1886.
Geog 4181.80	The voyages of Cadamosto. (Crone, G.R.) London, 1937.
Geog 4180.88	Voyages of Captain Luke Foxe and Captain T. James. (Christy, M.) London, 1894. 2v.
Geog 6009.01.15	The voyages of Captain Scott retold from The voyage of the 'Discovery'. (Scott, R.F.) N.Y., 1915.
Geog 4145.91	The voyages of David de Vries. (Parr, Charles.) N.Y., 1969.
Geog 5208.46	Voyages of discovery and research within the Arctic regions, from the year 1818 to present. (Barrow, John.) London, 1846.
Geog 5208.46.3	Voyages of discovery and research within the Arctic regions, from the year 1818 to present. (Barrow, John.) London, 1846.
Geog 5070.5	Voyages of discovery in Arctic and Antarctic. (M'Cormick, R.) London, 1884. 2v.
Geog 3520.10.25A	Voyages of Drake and Gilbert. (Hakluyt, Richard.) Oxford, 1909.
Geog 4145.81.10	The voyages of Joshua Slocum. (Slocum, Joshua.) New Brunswick, 1958.
Geog 4181.14	Voyages of Pedro Fernandez de Quiros. (Belmonte Bermudez, L. de.) London, 1904. 2v.
Geog 4180.56	Voyages of Sir James Lancaster. (Markham, C.R.) N.Y., 1964?
Geog 4181.85	The voyages of Sir James Lancaster to Brazil and the East Indies, 1591-1603. (Foster, William.) London, 1940.
Geog 6009.02.9	The voyages of the "Morning". (Doorly, G.S.) London, 1916.
Geog 4180.63	Voyages of William Baffin 1612-22. (Baffin, W.) N.Y., 1964?
Geog 4168.43.4	Voyages round the world from the death of Captain Cook to the present time. 4. ed. London, 1850.
Htn Geog 4557.98*	Voyages to the South Atlantic. (Colnett, J.) London, 1798.
Geog 4208.53.2	Voyages to various parts...1800-1831. (Coggeshall, G.) N.Y., 1853.
Geog 4208.51	Voyages to various parts of the world...1799-1844. (Coggeshall, G.) N.Y., 1851.
Geog 4477.61	Le voyageur philosophe. (Listonai.) Amsterdam, 1761. 2v.
Geog 4138.54	Voyageurs anciens et modernes. (Charton, E.T.) Paris, 1854-57. 4v.
Geog 4168.54	Voyageurs anciens et modernes. v.1-4. (Charton, Edouard T.) Paris, 1854-57. 2v.
Geog 3506.5	Les voyageurs belges. v.1-2. (Saint-Genois, Jules de.) Bruxelles, 1846-47?
Geog 3530.9	Voyageurs et explorateurs provençaux. (Barré, H.) Marseille, 1905.
Geog 3019.27.2	Les voyageurs français dans l'Orient européen. (Iorga, N.) Paris, 1928.
Geog 3019.27.5	Les voyageurs orientaux en France. (Iorga, N.) Paris, 1927.
Geog 4208.55	Voyaging to China in 1855 and 1904. (King, Paul.) London, 1936.
Htn Geog 5375.94.3F*	Vraye description de trois voyages de mer tres admirables. (Veer, G. de.) Amsterdam, 1600.
Geog 4209.37.5	Vrázová, Vlasta. Život a cesty E. St. Vráze. Praha, 1937.
Geog 4309.14	Vrooman, Carl S. The lure and the lore of travel. Boston, 1914.
Geog 5925.91	Vsesoiuznoe Soveshchenie po Izucheniiu Antarktiki, Moscow, 1966. Osnovnye itogi izucheniia Antarktiki za 10 let. Moskva, 1967.
Geog 5925.95	Vsnegakh Krainego Iuga. (Vedenskii, Anatolii A.) Leningrad, 1972.
An 3409.34.3	Vucetich e Reyna Abmandos. (Pacheco, F.) Rio de Janeiro, 1934.
Geog 4219.24.5	La vuelta al mundo de un novelista. (Blasco Ibáñez, Vicente.) Valencia, 1924-25. 3v.
An 2059.25.5	Vulliamy, C.E. Our prehistoric forerunners. N.Y., 1925.
An 39.58	Vuorela, Taivo. Kansatieteen sanasto. Helsinki, 1958.
An 359.69.25	Vuorela, Toivo. Kansatieteen periaateoppia. Helsinki, 1969.
Geog 5399.31.2	Vykhod k moriu. 2. izd. (Itin, Vivian A.) Novosibirsk, 1935.
Geog 3596.2	Výroj moravské geografie. 1. vyd. (Vitásek, František.) Praha, 1973.
Geog 4239.68.25	Vzemliakh blizkikh i dal'nikh. (Svirin, Vladimir P.) Stavropol', 1970.
Geog 4515.401	W skałach i lodach swiata. (Saysse-Tobiczyk, K.) Warszawa, 1959- 2v.
An 3309.11	Das Wachstum des Menschen. (Weissenberg, S.) Stuttgart, 1911.
Geog 614.9	Wacław Nałkowski. (Olszewicz, B.) Warszawa, 1962.
An 2339.12	Die Waffen der Naturvölker Süd-Amerikas. (Dieck, Alfred.) Ställuponen, 1912.
Geog 818.82.7	Wagner, H. Lehrbuch der Geographie. Hannover, 1903. 2v.
Geog 604.4	Wagner, H.R. George Davidson, geographer of the northwest coast of America. n.p., 1932.
Geog 4709.66	Wagner, Kip. Pieces of eight; recovering the riches of a lost Spanish treasure fleet. 1. ed. N.Y., 1966.
An 359.75	Wagner, Roy. The invention of culture. Englewood Cliffs, 1975.
Geog 4679.68	The Wahine disaster. (Lambert, Max.) Wellington, 1968.
Geog 4558.65	Waites, Alfred. My diary from England to India around Cape of Good Hope. Calcutta, 1865.
An 358.59.5	Waitz, T. Anthropologie der Naturvölker. Leipzig, 1859. 6v.
An 358.77	Waitz, T. Anthropologie der Naturvölker. 2. Aufl. Leipzig, 1860.
An 358.63.2	Waitz, T. Introduction to anthropology. London, 1863.
Geog 4308.01.5	Wakefield, Priscilla (Bell). The juvenile travellers. London, 1801.
Geog 4308.06.5	Wakefield, Priscilla (Bell). Juvenile travellers. London, 1806.
Geog 4208.78.7	Wakeman, Edgar. The log of an ancient marine. San Francisco, 1878.
Geog 6009.33.5	Walden, Jane Brevoort. The long whip; the story of a great husky. N.Y., 1936.
An 359.73.25	Waligórski, Andrzej. Antropologiczna koncepcja człowieka. Wyd. 1. Warszawa, 1973.
Geog 4209.58.25	Walk the wide world. (Knies, Donald.) N.Y., 1958.
Geog 4307.87.5	Walker, Adam. Bemerkungen auf einer Reise durch Flandern. Berlin, 1791.
Geog 4308.89.3	Walker, B. Aboard and abroad. Lowell, 1889.
Geog 817.96	Walker, J. Elements of geography. London, n.d.
Geog 4308.53.9	A Wall-street bear in Europe. (Young, Samuel.) N.Y., 1855.
Geog 4308.55.3A	Wallace, H.B. Art, scenery and philosophy. Philadelphia, 1855.
Geog 5850.7	Wallem, F.B. Universitetets Eskimoiske samlinger. Christiania, 1911.
Geog 4207.69.5	Wallenberg, Jacob. Min son på galejan. Stockholm, 1835.
Geog 4207.69.9	Wallenberg, Jacob. Min son på galejan. 2. uppl. Stockholm, 1913.
Geog 4218.57.10	Wallisch, Friedrich. Sein Schiff hiess Novara. Wien, 1966.
Geog 5837.39	Walløe, Peder Olsen. Peder Olsen Walløes dagbøger. Kjøbenhavn, 1927.
Geog 5558.94	Walsh, H.C. Last cruise of the Miranda. N.Y., 1896.
Geog 4308.28.5	Walter, Weever. Letters from the continent. Edinburgh, 1828.
Geog 4218.73.5	Walworth, E.H. An old world...travels around the world. N.Y., 1877.
Geog 4209.35.15	Wanderer from sea to sea. (Matisse, Maarten.) N.Y., 1936.
Geog 4419.45	A wanderer in khaki. (Ponder, S.E.G.) London, 1945.
Geog 4209.28.5	Wanderers. (Cust, Nina.) London, 1928.
Geog 4308.88	A wanderer's notes. (Kingston, W.B.) London, 1888. 2v.
An 349.36	The wandering spirit. (Numelin, R.) Copenhagen, 1936.
Geog 4208.94	Wandering words. (Arnold, E.) London, 1894.
Geog 4209.40.60	The wandering years. (Martyr, W.) N.Y., 1940.
Geog 4208.82.6	Wanderings, south and east. (Coote, W.) London, 1882.
Geog 4209.20	Wanderings. (Curle, Richard.) n.p., n.d.
Geog 4309.25.10	Wanderings. (Hamilton, C.M.) Garden City, 1925.
Geog 4309.25.11	Wanderings and excursions. (MacDonald, J.R.) Indianapolis, 1928.
Geog 4309.25.12	Wanderings and excursions. (MacDonald, J.R.) London, 1929.
Geog 4218.92.15	Wanderings and wonderings. (Aubertin, J.J.) London, 1892.
Geog 4209.01.5A	Wanderings in three continents. (Burton, Richard F.) N.Y., 1901.
Geog 4145.23	The wanderings of Edward Ely. (Ely, Edward.) N.Y., 1954.
Geog 4209.53	Wanderlust. (Meiss-Teuffen, Hans.) N.Y., 1953.
Geog 4328.39	Wanderung nach dem Orient. (Maximilian, H.) München, 1839.
Geog 4308.39.7	Wanderungen durch Deutschland, Polen. (Tuttolasso.) Stuttgart, 1839.
Geog 500.71	Wandkarten, Cottanten, Bücher, Zeitschriften für den geographischen Unterricht. (Perthes, Justus, publishers.) Gotha, 1930.
Geog 4308.19	Wandrings minnen. (Beskow, B.) Stockholm, 1833-1834. 2v.
Geog 618.5	Wanklyn, H.G. Friedrich Ratzel. Cambridge, Eng., 1961.
Geog 4181.35	The war of Chupas. (Cieza de Leon, P. de.) London, 1918.
Geog 4181.54	The war of Las Salinas. (Cieza de Leon, P. de.) London, 1823.
Geog 4181.31	The war of Quito. (Cieza del Leon, P. de.) London, 1913.
Geog 4559.71.5	The war with Cape Horn. (Villiers, A.J.) N.Y., 1971.

Author and Title Listing

Geog 4238.92 — Ward, Artemus. Columbus outdone, an exact narrative of the voyage of the Yankee skipper Captain N.A. Andrews. N.Y., 1893.
An 359.03.3 — Ward, D.J.H. The human races. n.p., 1903.
An 197.5 — Ward, D.J.H. Letters to future ages. n.p., 1955.
Geog 4169.33 — Ward, Edward. Five travel scripts commonly attributed to Edward Ward. N.Y., 1933.
Geog 665.42 — Ward, F.K. Modern exploration. London, 1945.
Geog 759.46 — Ward, Francis K. About this earth. London, 1946.
Geog 4209.52 — Warden, W.R. Vale enchanting. London, 1952.
An 2052.5 — Warfare in primitive societies; a selected bibliography. (Divale, William T.) Los Angeles, 1971.
Geog 4205.96.1 — Warhaffte Beschreibung ettlicher Reisen in Europa, Africa, Asien und America 1596-1610. (Ultzheimer, Andreas J.) Tübingen, 1971.
Geog 4206.58 — Warhafftige Reiss-Beschreibung. v.1-2. (Welsch, Hier.) Stuttgart, 1664.
Geog 4308.75.6 — Waring, G.E. A farmer's vacation. Boston, 1876.
Geog 3520.8 — Warmer, Oliver. English maritime writing. London, 1958.
Geog 4308.84.7 — Warner, C.D. A roundabout journey. Boston, 1884.
Geog 4308.72.6 — Warner, Charles Dudley. Saunterings. Boston, 1872.
Geog 4308.72.5 — Warner, Charles Dudley. Saunterings. Boston, 1872.
Geog 4308.72.5.2 — Warner, Charles Dudley. Saunterings. Boston, 1900.
Geog 4182.33 — Warnsinck, J.E.M. Reisen van Nicolaus de Graaff. 's-Gravenhage, 1930.
Geog 4208.54.5 — Warren, E. A doctor's experiences in three continents. Baltimore, 1885.
Geog 4228.59 — Warren, T.R. Dust and foam, or Three oceans and two continents. N.Y., 1859.
Geog 3140.7 — Warren, W.F. True key to ancient cosmology. Boston, 1882.
Geog 4218.31.3 — Warriner, F. Cruise of the United States frigate Potomac round the world. N.Y., 1835.
Geog 265.3 — Warsaw. Universytet. Instytut Geograficzny. Katedra Geografii Fizycznej. Prace i studia. 1,1967+
Geog 265.2 — Warsaw. Universytet. Instytut Geograficzny. Katedra Klimatologii. Prace i studia. 1,1964+
Geog 4307.87.2 — Watkins, T. Travels through Switzerland, Italy. London, 1794. 2v.
Htn Geog 815.34* — Watt, J. von. Epitome trium terrae. Tiguri, 1534.
Geog 4208.67 — Watt, Robert. Kjøbenhavn. Kjøbenhavn, 1867.
Geog 4308.66.12 — Watt, Robert. Tgjennen Europa. København, 1866.
Geog 4759.30 — Waugh, Alec. The coloured countries. London, 1930.
Geog 4759.30.7 — Waugh, Alec. Hot countries. N.Y., 1930.
Geog 4759.30.5A — Waugh, Alec. Hot countries. N.Y., 1930.
Geog 4329.30.3 — Waugh, E. Labels; a Mediterranean journal. London, 1930.
Geog 4209.46.28 — Waugh, Evelyn. When the going was good. London, 1946.
Geog 3070.12 — Wauwermans, H. Histoire de l'ecole cartographique. Bruxelles, 1895. 2v.
Geog 4559.69.10 — The Wavertree; being an account of an ocean wanderer. (Spiers, George.) N.Y., 1969.
An 379.53.5 — The way of the world. (Kimble, G.H.T.) N.Y., 1953.
Geog 4308.85.3 — Waymarks, or Sola in Europe. (Tyler, I.) Chicago, 1885.
An 349.48 — The ways of men. (Gillin, John.) N.Y., 1948.
An 359.35.7 — We Europeans. (Huxley, Julian S.) London, 1935.
An 359.35.7.10 — We Europeans. 1. ed. (Huxley, Julian S.) N.Y., 1936.
Geog 4269.36 — We Northmen. (Price, Lucien.) Boston, 1936.
Geog 4209.38.21 — We sailed from Brixham. (Beddington, C.) London, 1938.
Geog 4308.85.5 — We two-alone in Europe. (Ninde, Mary L.) Chicago, 1889.
Geog 4559.38.5 — Wead, F.W. Gales, ice and men; a biography of the Steam Barkentine Bear. London, 1938.
Geog 4559.37.3 — Wead, Frank W. Gales, ice and men. N.Y., 1937.
Geog 4325.90.12 — Webbe, Edward. Edward Webbe, chief master gunner, his travailes, 1590. London, 1868.
Geog 4325.90.6 — Webbe, Edward. Edward Webbe, chief master gunner, his travailes. Edinburgh, 1885.
Geog 4325.90.10 — Webbe, Edward. Edward Webbe, his travailes, 1590. Birmingham, 1868.
Htn Geog 4325.90.5* — Webbe, Edward. Edward Webbe, his travailes. Edinburgh, 1885.
An 2109.08 — Webster, H. Primitive secret societies. N.Y., 1908.
An 359.42.5 — Webster, Hutton. Taboo, a sociological study. Stanford, 1942.
Geog 579.49 — Webster's geographical dictionary. Springfield, Mass., 1960.
Geog 579.49.5 — Webster's geographical dictionary. Springfield, Mass., 1962.
Geog 579.55.5 — Webster's geographical dictionary. Springfield, 1955.
Geog 4209.37.10 — Wechsberg, J. Die grosse Mauer; das Buch einer Weltreise. Leipzig, 1937.
Geog 5925.10 — Weddell, James. Observations on South Pole. London, 1826.
Geog 5850.12 — Wedin, Bertil. Aktion och reaktion på Grönland. Stockholm, 1971.
Geog 5639.71 — Wedin, Bertil. Tusen år på Grönland. Stockholm, 1971.
Geog 5235.24 — Weems, John E. Race for the pole. N.Y., 1960.
Geog 5326.21.20 — Weems, John Edward. Peary, the explorer and the man, based on his personal papers. Boston, 1967.
Geog 5919.48 — Weetman, Charles. All about Antarctica. Melbourne, 1948.
Geog 5209.31 — Der Weg nach dem Pol. (Samoilovich, R.) Biehefeld, 1931.
Geog 4502.5.27 — Der Weg zum Berg. (Graber, A.) München, 1939?
Geog 5839.30.6 — Wegener, A. Tagebücher, Briefe. Wiesbaden, 1960.
Geog 5839.30.5 — Wegener, E. Greenland journey; the story of Wegener's German expedition...1930-31. London, 1939.
Geog 4209.23.10 — Wehde, Albert. Seit ich die heimat Verliess. Berlin, 1924.
Geog 4209.40.35 — Weidman, J. Letter of credit. N.Y., 1940.
An 359.35.9 — Weigner, K. Die Gleichwertigkeit der europäischen Rassen und die Wege zu ihrer Vervollkommnung. Prag, 1935.
Geog 665.105 — Weigt, Ernst. Angewandte Geographie; Festschrift für Professor Dr. Erwin Scheu. Nürnberg, 1966.
An 2059.55.5 — Weiner, Joseph S. The Piltdown forgery. London, 1955.
An 2109.51.15 — Weinert, H. Der geistige Aufstieg der Menschheit. 2. Aufl. Stuttgart, 1951.
Geog 4502.5.10 — Weise Berge - schwarze Zelte. (Schuster, K.) München, 1932.
Geog 4678.65.10 — Weiss, N. Naufrage de la Ville-du-Havre et du Losh-Earn. Paris, 1874.
Geog 4678.73 — Weiss, Nathanael. Personal recollections of wreck of...Ville-du-Havre. N.Y., 1875.
Geog 4509.58 — Weisse Berge. (Moravec, Fritz.) Wien, 1958.
Geog 5209.57 — Der weisse Weg; Forscher erobern die Arktis. (Förster, Hans A.) Leipzig, 1957.
An 3309.11 — Weissenberg, S. Das Wachstum des Menschen. Stuttgart, 1911.
Geog 4209.28.25 — Weldgesicht; ein Buch von heutiger und kommender Menscheit. (Bloem, Walter.) Leipzig, 1928.
Geog 4308.86.17 — Well-worn roads of Spain, Holland and Italy. (Smith, F.H.) Boston, 1892.
Geog 616.25 — Weller, E. August Petermann. Leipzig, 1911.
Geog 3107.26 — Wells, Edward. Treatise of antient and present geography. London, 1726.
Geog 4239.24 — Wells, F. de W. The last cruise of the Shanghai. N.Y., 1925.
Geog 4219.26 — Wells, Linton. Around the world in twenty-eight days. Boston, 1926.
Geog 4238.72 — Wells, Theodore. Narrative of life and adventures of Captain Wells...voyages. Biddeford, 1874.
Geog 4206.58 — Welsch, Hier. Warhafftige Reiss-Beschreibung. v.1-2. Stuttgart, 1664.
Geog 4145.48 — Die Welt in allen Zonen. (Lantzsch, W.) München, 1961.
An 2109.24.9 — Das Weltbild der Primitiven. (Graebner, F.) München, 1924.
Htn Geog 3205.34F* — Weltbuch: Spiegel...in Asiam, Aphrica, Europam und America. (Franck, S.) n.p., 1534.
Geog 3019.60.5F — Welten der Entdecker. (Bettex, A.W.) München, 1960.
Geog 5160.58 — Weltgeschehen am Rande des Polarmeeres. (Hantschel, A.) Würzburg, 1964.
An 378.58 — Die welthistorische Bedeuting dei Meere insbesondere des Mittelmeers. (Rathlef, C.) Dorpat, 1858.
Geog 4209.15 — Weltpolitisches Wanderbuch 1897-1915. (Rohrbach, P.) Königstein, 1916.
Geog 4209.15.5 — Weltpolitisches Wanderbuch 1897-1915. (Rohrbach, P.) Königstein, 1916.
Geog 4110.27 — Weltreise. (Meyer, H.J.) Leipzig, 1907.
Geog 4209.70 — Weltreise auf den Spuren der Unruhe. (Wollschaeger, Alfred.) Gütersloh, 1970.
Geog 4129.71 — Wenn einer eine Reise tat. (Bauer, Hans.) Leipzig, 1971.
Geog 4218.66.5 — Weppner, M. The North Star and the Southern Cross. Albany, 1876. 2v.
Geog 4218.66 — Weppner, M. The North Star and the Southern Cross. 3. American ed. Albany, 1880.
An 359.20.9 — Der Werdegang der Menschheit und die Entstehung der Kultur. (Klaatsch, Hermann.) Berlin, 1920.
Geog 4208.39 — Were, J.B. Voyage from Plymonth to Melbourne in 1839. Melbourne, 1964.
Geog 4219.60.5 — De wereld. (Uittenbogaard, Leo.) Den Haag, 1958-60. 3v.
Geog 4182.05 — Werken. Register. v.1-25, 26-50. (Linschoten. Vereeniging.) 's-Gravenhage, 1939-57. 2v.
Geog 4138.82.5 — Werner, R. Berühmte Seeleute. Berlin, 1882.
Geog 4218.78.15 — Wernich, Agathon. Geographisch-medicinische Studien. Berlin, 1878.
Geog 4139.68 — Wertheim, Willem Frederik. Ketters en kwezels, regenten en rebellen. Drachten, 1968.
Geog 4559.65 — West, Ellsworth Luce. Captain's papers. Barre, Mass., 1965.
Geog 759.38 — West- und Nordeuropa in Natur. Potsdam, 1938.
An 2109.29.9 — West African secret societies. (Butt-Thompson, F.W.) London, 1929.
Htn Geog 4219.33F* — West made East with the loss of a day. (Vanderbilt, William K.) N.Y., 1933.
Geog 4419.25 — West of the Pacific. (Huntington, E.) N.Y., 1925.
Geog 4182.61 — De Westafrikaanse reis van Piet Heyn. (Hein, Pieter P.) 's-Gravenhage, 1959.
Geog 579.68.5 — Westermann Lexikon der Geographie. Braunschweig, 1968-72. 5v.
An 2109.32.3 — Westermarck, E. Early beliefs and their social influence. London, 1932.
Geog 4269.70 — Western Europe: geographical studies. (Coghill, Ian G.) London, 1970.
Geog 4329.67 — The western Mediterranean world. (Houston, James M.) N.Y., 1967.
Geog 4269.62.5 — Western ports in Europe. (Salvadori, M.) N.Y., 1962.
Geog 266.2 — Westfälische geographische Studien. Münster. 1,1949+ 8v.
Geog 4219.36.5 — Westward bound in the schooner Yankee. (Johnson, Irving.) N.Y., 1936.
Geog 4239.50 — Westward crossing. 1st American ed. (Barton, H.D.E.) N.Y., 1951.
Geog 4219.36.10 — Westward ho - fare paid. (Hurja, E.E.) Juneau, Alaska, 1936.
Geog 4219.47 — Westward ho with the Albatross. 1. ed. (Petterson, H.) N.Y., 1953.
Geog 4209.47.5 — Wethered, Herbert Newton. The four paths of pilgrimage. London, 1947.
Geog 4218.89.7 — Wetmore, E.B. A flying trip around the world. N.Y, 1891.
Geog 5324.2.37 — Wetterfors, Paul. Fridtjof Nansen. Uppsala, 1932.
Geog 4209.46.20 — Wever, Jan. Journal der merkwaardige reizen van Jan Wever. 's-Gravenhage, 1946.
Geog 5180.9 — Weyer, E.M. The Eskimos. New Haven, 1932.
An 359.63.5 — Weyl, N. The geography of intellect. Chicago, 1963.
Geog 3055.62 — Weyl, R. Atlantis enträtselt? Kiel, 1953.
Geog 4558.50.5 — The whale and its captors. (Cheever, H.T.) N.Y., 1850.
Geog 4559.16.5 — Whale hunting with gun and camera. (Andrews, R.C.) N.Y., 1916.
Geog 4238.48 — Whaley, Thomas. Consignments to El Dorado. 1st ed. N.Y., 1972.
Geog 4558.56 — Whaling and fishing. (Nordhoff, C.) Cincinnati, 1856.
Geog 4239.69 — Wharran, James. Two girls, two catamarans. London, 1969.
Geog 4308.83 — What I saw in Europe. (Thacher, S.O.) Topeka, Kansas, 1883.
An 349.24A — What is man? (Thomson, John Arthur.) N.Y., 1924.
Geog 4105.2 — What to observe. (Jackson, J.R.) London, 1841.
Geog 5398.67 — Wheildon, W.W. The new Arctic continent. Cambridge, 1869.
Geog 4509.46 — When man and mountains meet. (Tilman, H.W.) Cambridge, Eng., 1946.
Geog 4559.30.5 — When ships were ships and not tin pots. (Barnes, William M.) N.Y., 1930.
Geog 4209.46.28 — When the going was good. (Waugh, Evelyn.) London, 1946.
Geog 3019.72.20 — When the pole star shone. (Langley, Michael.) London, 1972.
Geog 4309.27.10 — When you go to Europe. (Petre, E.R.) N.Y., 1933.
Geog 4521.11 — Where the clouds can go. (Kain, Conrad.) N.Y., 1935.
Geog 4249.46 — Where the trade-winds blow. (Robson, R.W.) Sydney, 1946.
Geog 4110.5 — Where to stop. (King, M.) Boston, 1893.
Geog 4110.5.2 — Where to stop. (King, M.) Boston, 1894.
Geog 4759.52 — Where winter never comes. (Bates, M.) N.Y., 1952.
Geog 4559.60 — Whipple, Addison. Tall ships and great captains. N.Y., 1960.
Geog 3055.50 — Whishaw, M. Atlantis in Andalucia; a study of folk memory. London, 1929.
Geog 4329.53 — Whistling for a wind. 1st American ed. (Landery, C.F.) N.Y., 1953.
An 359.36 — White, C.L. Geography. N.Y., 1936.
Geog 4477.64 — White, D.M. Zaccaria Seriman, 1709-1784, and the Viaggi di Enrico Wanton. Manchester, 1961.

267

Author and Title Listing

	An 197.6	White, Leslie A. The concept of cultural systems. N.Y., 1975.
	An 359.56.25	White, Lynn T. Frontiers of knowledge in the study of man. 1. ed. N.Y., 1956.
	Geog 4308.45.30	White, T.H. A pilgrim's reliquary. London, 1845.
	Geog 5919.50	The white continent. (Henry, T.) N.Y., 1950.
	Geog 6009.55.20	The white desert. (Barber, Noël.) N.Y., 1958.
	Geog 6009.49	The white desert. (Giaener, John.) London, 1954.
	Geog 4209.46.5	White sails and spendthrift. (Shaw, Frank H.) London, 1946.
	An 359.49.5	White settlers and native peoples. (Price, A.G.) Melbourne, 1949.
	Geog 4759.39	White settlers in the tropics. (Price, A.G.) N.Y., 1939.
	Geog 5160.8	The white world. (Kersting, Rudolf.) N.Y., 1902.
	An 126.8	Whitehill, Walter Muir. A memoir of John Otis Brew. Boston, 1968.
	Geog 5324.2.60	Whitehouse, J.H. Nansen. London, 1930.
	An 2058.74.5	Whitmore, J.H. Evidences of the antiquity of man. Rochester, N.Y., 1874.
	An 2109.24.15	Whitnall, H.O. The dawn of mankind. Boston, 1924.
	Geog 5839.10	Whitney, H. Hunting with the Eskimos. London, 1910.
	Geog 665.101	Whittow, John Byron. Essays in geography for Austin Miller. Reading, Eng., 1965.
	Geog 4209.38	Who called that lady a skipper? (Hart, M.R.) N.Y., 1938.
	An 359.40.3A	Why men behave like apes and vice versa, or Body behavior. (Hooton, E.A.) Princeton, 1940.
	Geog 4521.23	Whymper, Edward. Edward Whymper, alpinist of the heroic age. Nashville, Tenn., 1914.
	Geog 4521.23.5	Whymper, Edward. A letter addressed to the...Alpine Club. London, 1900.
	Geog 4309.63.10	Wićaz, Jurij. Z Kamjenskim nosom; što dóžiwi serbski nowinar we swěče. 2. wyd. Budyšin, 1963.
	Geog 4308.97	Widmann, J.V. Sommerwanderungen und Winterfahrten. Frauenfeld, 1897.
	Geog 4019.57	Wie sie entkamen; mit einer Einleitung von Kasimir Edschmid. Düsseldorf, 1957.
	Geog 4182.21	Wieder, F.C. De reis van Mahu en de Cordes. 's-Gravenhage, 1923-25. 3v.
	Geog 4182.26	Wieder, F.C. Die stichting van New York in Juli 1625. 's-Gravenhage, 1925.
	Geog 1531.15	Wien und Niederdonau. (Baedeker, publishers.) Leipzig, 1943.
	Geog 4219.41.5	Wiener, P. Last man around the world. N.Y., 1941.
	Geog 1525.75	Wiesbaden, Mainz, Rheingau, Rheinhessen. (Baedeker, publishers.) Malente, 1956.
	An 359.40.25	Wiese und Kaiserwaldau, Leopold M. von. Homo sum, Gedanken zu einer zusammenfassenden Anthropologie. Jena, 1940.
	Geog 3251.19	Wieser, F. Der Portulan des Infanten. n.p., n.d.
	Geog 4218.75	Wieting, M.E. Prominent incidents in life of John M. Wieting...around the world. N.Y., 1889.
	Geog 4239.49	Wightman, Frank Armstrong. The wind is free. N.Y., 1949.
	Geog 3520.7F	De wijd beroemde voyagien...der Engelsen. (Aa, P. van der.) Leyden, 17- . 2v.
	Geog 4676.61	Wijtloopig...beschrijvinge van d. onzel voyage van't schip Arnhem, 1661. (Kerckhoven, J. van.) Amsterdam, 1664.
	Geog 4512.537	Wil der Schneelauf nach Deutschland kam. (Luther, C.J.) München, 1925.
	Geog 6009.21	Wild, Frank. Shackleton's last voyage; the story of the Quest. London, 1923.
	An 2600.2.8	The wild boy of Aveyron. (Itard, Jean Marc.) N.Y., 1932.
	An 2108.79.5	The wild man at home, or Pictures of life in savage lands. (Greenwood, James.) London, 1879.
	Geog 4308.53.8	Wild oats sown abroad. Philadelphia, 1853.
	Geog 4239.57	Wild ocean. (Villiers, Alan John.) N.Y., 1957.
	An 2109.60.20	The wildbooters. (Kern, Fritz.) Edinburgh, 1960.
	Geog 4328.40.3	Wilde, William R. Narrative of a voyage to Madeira, Teneriffe. 2nd ed. Dublin, 1844.
	Geog 4328.37.2	Wilde, William R. Narrative voyage to Madeira...Cyprus and Greece. Dublin, 1852.
	An 3309.20	Wilder, Harris H. A laboratory manual of anthropometry. Philadelphia, 1920.
	Geog 4209.31.10	Wilderness trails in three continents. (Leslie, L.A.D.) London, 1931.
	Geog 4502.5.19	Wildonger. (Kobell, F. von.) München, 1936.
	Geog 5559.28	Wilkins, George H. Flying the Arctic. N.Y., 1928.
	Geog 5559.31.5	Wilkins, H. Under the North Pole. N.Y., 1931.
	Geog 5559.31	Wilkins, H. Under the North Pole. N.Y., 1931.
	Geog 4709.48	Wilkins, H.T. A modern treasure hunter. London, 1948.
	Geog 4709.40.5	Wilkins, H.T. Panorama of treasure hunting. N.Y., 1940.
	Geog 5509.71	Wilkinson, Doug. Arctic fever. Toronto, 1971.
	Geog 4208.83.10	Wilkinson, H. Sunny lands and seas. London, 1883.
	Geog 602.1.5	Willem Janszoon Blaeu. (Stevenson, E.L.) N.Y., 1914.
	Geog 4145.53	Willers, U. Xavier Marmier och Sverige. Stockholm, 1949.
	Geog 4218.70.6	William H. Seward's travels around the world. (Seward, William H.) N.Y., 1874.
	Geog 4218.70.4	William H. Seward's travels around the world. (Seward, William H.) N.Y., 1873.
	Geog 4218.70.5	William H. Seward's travels around the world. (Seward, William H.) N.Y., 1873.
	Geog 4209.05.5	Williams, Archibald. The romance of modern exploration. Philadelphia, 1905.
	Geog 5509.62	Williams, G. The British search for the Northwest Passage in the 18th century. London, 1962.
	Geog 4659.63.4	Williams, Peter J. Shipwrecks at Port Phillips Heads since 1840. Melbourne, 1967.
	Geog 5839.53	Williamson, G. Changing Greenland. London, 1953.
	Geog 4181.120	Williamson, J.A. The Cabot voyages and Bristol discovery under Henry VII. Cambridge, 1962.
	Geog 4208.70.2	Willis, G.R. The cruise of the Colorado. N.Y., 1873.
X Cg	Geog 4208.70	Willis, G.R. The cruise of the Colorado. N.Y., 1873.
	Geog 4308.32.2	Willis, N.P. Pencillings by the way. London, 1842.
Htn	Geog 4308.32.3*	Willis, N.P. Pencillings by the way. 1st ed. N.Y., 1844.
	Geog 4328.53	Willis, N.P. Summer cruise in the Mediterranean. Auburn, 1853.
	Geog 4145.94	Willis, William. The hundred lives of an ancient mariner. London, 1967.
	Geog 4249.54	Willis, William. The gods were kind. 1st ed. N.Y., 1955.
	Geog 4327.98F	Willyams, Cooper. A voyage up the Mediterranean. London, 1802.
	Geog 819.21.3	Wilmore, Albert. The groundwork of modern geography. London, 1921.
	Geog 4229.11	Wilmore, C. Ray. Square rigger round the Horn. Camden, Me., 1972.
	Geog 4308.01	Wilmot, Catherine. An Irish peer on the continent. London, 1920.
	Geog 4508.93.5	Wilson, Claude. Mountaineering. London, 1893.
	Geog 4308.23.2	Wilson, D. Letters from an absent brother. London, 1824.
	An 197.1	Wilson, D. The lost Atlantis. N.Y., 1892.
	An 2108.62	Wilson, D. Prehistoric man. Cambridge, 1862. 2v.
	An 2108.76	Wilson, D. Prehistoric man. 3. ed. London, 1876. 2v.
	Geog 4508.97	Wilson, E.L. Mountain climbing. N.Y., 1897.
	Geog 4309.47A	Wilson, Edmund. Europe without Baedeker. 1st ed. Garden City, 1947.
	Geog 4309.47.2	Wilson, Edmund. Europe without Baedeker. 2nd ed. N.Y., 1966.
	Geog 4209.56A	Wilson, Edmund. Red, black, blond, and olive. N.Y., 1956.
	Geog 6009.01.20	Wilson, Edward Adrian. Diary of the 'Discovery' expedition to the Antarctic regions, 1901-1904. London, 1966.
	Geog 6009.10.50	Wilson, Edward Adrian. Diary of the "Terra Nova" expedition to the Antarctic, 1910-12. London, 1972.
	Geog 4145.25.5	Wilson, P.W. An explorer of changing horizons. N.Y., 1927.
	Geog 4329.26	Wilstach, Paul. Islands of the Mediterranean. London, 1926.
	Geog 4329.31	Wilstach, Paul. Islands of the Mediterranean. N.Y., 1931.
	Geog 4131.48	Winble, Ernest W. European recovery, 1948-1951, and the tourist industry. London, 1948.
	An 2058.78	Winchell, A. Adamites and preadamites. Syracuse, 1878.
	An 2058.80	Winchell, A. Preadamites. Chicago, 1880.
	Geog 4239.47	Wind aloft, wind alone. (Durand-Couppel de St. Trent, M.N.P.) N.Y., 1947.
	Geog 4239.49	The wind is free. (Wightman, Frank Armstrong.) N.Y., 1949.
	Geog 4309.60	Window to the West. 1st ed. (Anjaneyulu, D.) Madras, 1967.
Htn	Geog 4328.32*	Wines, E.C. Two years and a half...in Mediterranean and Levant. Philadelphia, 1832.
	Geog 5559.56.5	Wings over the Arctic. (Vodopianov, Mikhail V.) Moscow, 1956?
	Geog 5313.2	Winner lose all; Dr. Cook and the theft of the North Pole. (Eames, Hugh.) Boston, 1973.
	Geog 6009.29.5	The winning of Australian Antarctica. (British, Australian and New Zealand Research Expedition, 1929-1931.) Sydney, 1962.
	Geog 4229.00.5	Winter, James M. New York to Alaska...May to July, 1900. Middletown, N.Y, 1943.
	Geog 4328.61	Winter and spring on the shores of the Mediterranean. 4th ed. (Bennet, J.H.) London, 1870.
	Geog 4515.102	Winterliches Bergsteigen alpine Schilaustechnik. (Hoferer, Erwin.) München, 1925.
	An 2109.31.13	Winthius, Josef. Einführung in die Vorstellungswelt primitiver Völker. Leipzig, 1931.
	Geog 4169.64	Wir nannten sie Wilde. (Jahn, Janheinz.) München, 1964.
	An 359.14	Wirth, Albrecht. Rasse und Volk. Halle, 1914.
	Geog 4328.57	Wise, H.A. Scampavias from Gibel Tarek to Stamboul. N.Y., 1857.
	Geog 3018.97	Wisotzki, E. Zeitströmungen in der Geographie. Leipzig, 1897.
	An 2109.33	Wisse, Jakob. Selbstmord und Todesfurcht bei den Naturvölkern. Proefschrift. Zutpher, 1933.
	An 2059.29.5	Wissler, Clark. An introduction to social anthropology. N.Y., 1929.
	An 359.23.3A	Wissler, Clark. Man and culture. N.Y., 1923.
	Geog 4309.29.20	With Bob Davis hither and yon. (Davis, Robert H.) N.Y., 1931.
	Geog 5398.93.3	With Nansen in the north. (Johansen, H.) London, 1899.
	Geog 5558.91.3	With Peary near the Pole. (Astrup, E.) London, 1894.
	Geog 6009.10.19	With Scott; the silver lining. (Taylor, G.) London, 1916.
	Geog 6009.11.5	With the "Aurora" in the Antarctic, 1911-14. (Davis, John King.) London, 1919.
	Geog 5399.28.15	With the "Italia" to the North Pole. (Nobile, Umberto.) London, 1930.
	Geog 5160.15	Within the circle. (Stefansson, Evelyn Schwartz Baird.) N.Y., 1945.
Htn	Geog 4347.85F*	Witsen, N. Noord en Oost Tartaryen. Amsterdam, 1785. 2v.
	Geog 4182.66	Witsen, Nicolaas. Moscovische reyse 1664-1665. 's-Gravenhage, 1966-67. 3v.
	Geog 4208.78.5	A wizard's wanderings from China to Peru. (Holden, J.W.) London, 1886.
	Geog 4182.30	Woard, C. de. Zeeuwsche expedite...Cornelis Evertsen. 's-Gravenhage, 1928.
	Geog 5323.3	Woel, Cai Magens. Hilsen til Ejnar Mikkelsen. Kjøbenhavn, 1930.
	Geog 4209.36.15	Wøller, Johan. Zest for life; recallections of a philosophic traveller. London, 1936.
	Geog 4328.70.5	Woerl, Leo. Erzherzog Ludwig Salvator aus dem Osterreichischen Kaiserhause als Forscher des Mittelmeeres. Leipzig, 1899.
	Geog 579.43	Wörterverzeichnis für russische Karten. (Kosack, H.P.) Berlin, 1943.
	Geog 4558.73	Wogan, E. de. Du Far-West à Bornéo. Paris, 1873.
	Geog 3057.7	Der wohleingerichtete Staat des Bishero von vielen gesuchten aber nicht gefundenen Königreichs Ophir. Photoreproduction. Leipzig, 1699. 2v.
	An 359.64.25	Wolf, Eric Robert. Anthropology. Englewood Cliffs, N.J., 1964.
	An 359.71.10	Wolf, Josef. Integral anthropology. Praha, 1971.
	An 2600.4	Wolf child and human child. (Gesell, Arnold.) N.Y., 1941.
	An 2600.7.1	Wolf children. (Malson, Lucien.) London, 1972.
	An 2600.5.2	Wolf-children and feral man. (Singh, Joseph.) Hamden, 1966.
	An 2600.5	Wolf children and feral man. (Singh, Joseph.) N.Y., 1942.
	An 359.27	Wolff, K.F. Rassenlehre. Leipzig, 1927.
	Geog 4477.87	Wollap, G. Mémoires. v.1-4. Londres, 1787. 2v.
	Geog 4209.70	Wollschlaeger, Alfred. Weltreise auf den Spuren der Unruhe. Gütersloh, 1970.
	Geog 4219.56	Wollschläger, A. Grosse Weltreise mit A.E Johann (pseud.). 1. Aufl. Gutersloh, 1956.
	Geog 4145.95	Wollschläger, Alfred. Menschen an meinen Wegen. München, 1973.
NEDL	Geog 4218.46.3	A woman's journey round the world. (Pfeiffer, I.) London, 1846.
	Geog 4678.54.15	Women and children last; the loss of the steamship. (Brown, A.C.) London, 1962.
	An 124.15	Women in the field; anthropological experience. (Golde, Peggy.) Chicago, 1970.
	Geog 4558.53	A wonderful providence in many incidents at sea. 8. ed. (Holcomb, E.) Boston, 1853.
	Geog 665.37	Wonders of nature. (Singleton, E.) N.Y., 1900.
	Geog 665.37.5	Wonders of nature. (Singleton, E.) N.Y., 1911.
	Geog 807.68	The wonders of nature and art. 2nd ed. London, 1768. 6v.
	Geog 665.37.6	The wonders of nature as seen and described by Alexandre Dumas. (Singleton, E.) Washington, 1962.
	An 190.1	Wonders of the universe. (Pratt, Orson.) Salt Lake City, 1937.

Author and Title Listing

	Geog 665.36	Wonders of the world. (Freeman, Henry.) Boston, 1873.
	Geog 4208.73	Wood, C.F. Yachting cruise in the South Seas. London, 1875.
	An 358.70	Wood, J.G. The uncivilized races of men in all countries of the world. Hartford, 1870. 2v.
	An 358.70.5	Wood, J.G. The uncivilized races of men in all countries of the world. Hartford, 1876. 2v.
	Geog 4209.55	Woodcock, A.W.W. To Hong Kong and return. Boston, 1955.
	Geog 4209.40	Woodward, C. Lanterns alight; journeys to far places. Chicago, 1940.
	Geog 5530.15.5	Woodward, F.J. Portrait of Jane. London, 1951.
	Geog 4309.02	Woodward, M. His last log. Chicago, 1903.
	Geog 4239.69.15	Woolass, Peter. Stelda, George and I. London, 1971.
	Geog 3019.51	Wooldridge, S. William. The spirit and purpose of geography. London, 1951.
	Geog 4328.96	Woolson, C.F. Mentone, Cairo and Corfu. N.Y., 1896.
	Geog 578.17	Worcester, J.E. Geographical dictionary or universal gazetteer. Andover, 1817. 2v.
	Geog 578.17.2	Worcester, J.E. Geographical dictionary or universal gazetteer. 2. ed. Boston, 1823. 2v.
	Geog 818.23	Worcester, J.E. Sketches of the earth. Boston, 1823. 2v.
	Geog 819.45	The work of men. 1st ed. (Pickles, Thomas.) London, 1945.
	Geog 6009.28.5	The work of the Byrd Antarctic expedition. (Joerg, W.L.G.) N.Y., 1930.
Htn	Geog 4308.32.5*	Working a passage: or Life in a liner. (Briggs, Charles F.) N.Y., 1844.
	Geog 818.53	The world; geographical, historical, statistical. (Savage, C.C.) N.Y., 1853.
	Geog 818.93	The world, historical and actual. (Gilbert, Frank.) Chicago, 1893.
	Geog 4218.81.5	The world, round it and over it. (Glass, C.) Toronto, 1881.
	Geog 819.10F	The world and its peoples photographed and described. (Rand, McNally and Company.) Chicago, 1910.
	Geog 818.36.10	The world as it is. 5th ed. (Perkins, Samuel.) n.p., 1839.
	Geog 818.36.11	The world as it is. 5th ed. (Perkins, Samuel.) n.p., 1840.
	Geog 818.36.14	The world as it is. 6th ed. (Perkins, Samuel.) n.p., 1842.
	Geog 4219.30	World beaters. (Nelson, Robert.) Boston, 1930.
	Geog 4157.59	World displayed. London, 1759-61. 20v.
	Geog 4157.59.3	World displayed. v.3-4,5-6,8,15-16,17-18. 3.-4. ed. London, 1760-88. 5v.
	Geog 4180.16	The world encompassed by Francis Drake. (Feltcher, F.) N.Y., 1964?
	Geog 4208.52	The world here and there. N.Y., 1852.
	Geog 585.11.10	The world in a pocket-book, or Universal popular statistics. 12th ed. (Crump, William H.) Phialdelphia, 1860.
	Geog 585.11	The world in a pocket-book. (Crump, William H.) Philadelphia, 1841.
	Geog 585.11.2	The world in a pocket-book. (Crump, William H.) Philadelphia, 1842.
	Geog 585.11.5A	The world in a pocket-book. (Crump, William H.) Philadelphia, 1845.
	Geog 3070.39	The world in maps. (Jervis, W.W.) N.Y., 1937.
	Geog 3070.39.5	The world in maps. 2d ed. (Jervis, W.W.) N.Y., 1938.
	Geog 3251.27F	World map of Francesco Roselli. (Nunn, George E.) Philadelphia, 1928.
	Geog 5985.122	A world of men: exploration in Antarctica. (Herbert, Wally.) London, 1968.
	Geog 4219.69.5	A world of my own: the single-handed, non-stop circumnavigation of the world in Suahili. (Knox-Johnston, Robin.) London, 1969.
	An 2109.53.5	The world of primitive man. (Radin, Paul.) N.Y., 1953.
	Geog 4218.97	A world-pilgrimage. (Barrows, John D.) Chicago, 1897.
	An 379.63	World power and shifting climates. (Mills, C.A.) Boston, 1963.
	An 2109.61.5	World prehistory. (Clark, J.G.D.) Cambridge, Eng., 1961.
Htn	Geog 4206.60*	The world surveyed. (LeBlanc, V.) London, 1660.
	Geog 265.1F	World traveler. N.Y.
	Geog 4209.71.35	World wanderer; 100,000 miles under sail. (Kearns, Des.) Sydney, 1971.
	Geog 579.60.5	The worldmark encyclopedia of the nations. N.Y., 1960.
	Geog 3019.53.5	Worlds beyond the horizon. 1. American ed. (Leithaeuser, J.G.) N.Y., 1955.
	Geog 3019.00	The world's discoveries. (Johnson, W.H.) Boston, 1900.
	Geog 4159.10.10	The world's greatest books. v.19. (Northcliffe, A.C.W.H.) n.p., 1910.
	Geog 4218.94	The world's highway. (Dunn, S.H.) London, 1894.
	An 359.08.9	The world's peoples. (Keane, A.H.) London, 1908.
	An 359.08.10	The world's peoples. (Keane, A.H.) N.Y., 1908.
	Geog 4209.28.10	Worlds within worlds. (Benson, Stella.) London, 1928.
	Geog 4168.84.5	The worlds wonders. (Buel, James W.) St. Louis, 1884.
	An 2109.53.20	Wormington, Hannah Marie. Origins, indigenous period. Mexico, 1953.
	An 359.27.7A	Worrell, W.H. A study of races in the ancient Near East. N.Y., 1927.
	Geog 5399.27.3	Worsley, F.A. Under sail in the frozen north. London, 1927.
	Geog 5985.258.5	Worsley, Frank A. Endurance; an epic of polar adventure. London, 1931.
	Geog 6009.10.30	The worst journey in the world; Antarctic 1910-13. (Cherry-Garrard, Apsley G.B.) N.Y., 1930.
	Geog 6009.10.32	The worst journey in the world; Antarctic 1910-13. v.1-2. (Cherry-Garrard, Apsley G.B.) N.Y., 1937.
	Geog 4559.40.11	Wossidlo, R. Reise, Quartier in Gottesnaam. Seestadt Rostock, 1940.
	Geog 3540.12	Wotte, Herbert. In blaver Ferne lag Amerika. 3. Aufl. Leipzig, 1974.
	An 137.6	Wotte, Herbert. Kaaram Tamo Mann vom Mond; Leben undd Reisen Miklučho-Makleis. Leipzig, 1973.
	Geog 3590.76	Wotte, Herbert. Kurs auf Unerforscht. Leipzig, 1967.
	An 359.74.20	Wprowadzenie do etnologii. Wyd. 1. (Szyfelbejn-Sokolewicz, Zofia.) Warszawa, 1974.
	Geog 5398.20.12	Wrangell, F. von. Ferdinand von Wrangell und seine Reiselängs der Nordküste von Sibirien. Leipzig, 1885.
	Geog 5398.20.5	Wrangell, F. von. Narrative of an expedition to the Polar Sea. London, 1840.
	Geog 5398.20.7	Wrangell, F. von. Narrative of an expedition to the Polar Sea. London, 1844.
	Geog 5398.20.25	Wrangell, F. von. Le nord de la Sibérie. Limoges, 188-?
	Geog 5398.20.20	Wrangell, F. von. Le nord de la Sibérie. Paris, 1843. 2v.
	Geog 5398.20.3	Wrangell, F. von. Reise des...Flotten-Lieutenants. Berlin, 1839. 2v.
	Geog 5398.20.10	Wrangell, F. von. Statistische...Nachrichten über...Besetzungen. St. Petersburg, 1839.
	Geog 4679.44.5	Wreck - S.O.S. (Hardy, Alfred C.) London, 1944.
	Geog 4659.66	Wreck and rescue in the Bristol Channel. (Farr, Grahame E.) Truro, 1966-67. 2v.
	Geog 4659.68.5	Wreck and rescue on the coast of Devon. (Farr, Grahame E.) Truro, 1968.
	Geog 4659.69	Wreck and rescue on the coast of Wales. (Parry, Henry.) Truro, 1969-73. 2v.
	Geog 4659.68	Wreck and rescue on the Essex coast. (Malster, Robert.) Truro, 1968.
	Geog 4659.64	Wreck and rescue round the Cornish coast. (Noall, Cyril.) Truro, 1964-65. 3v.
	Geog 4678.62	Wreck of steamer Governor and search for U.S. ship Vermont. (United States. Navy Department.) Washington, 1862?
	Geog 4678.59	Wreck of the Admella. (Mudie, Ian.) London, 1967.
	Geog 4677.48	The wreck of the Amsterdam. (Marsden, Peter Richard V.) London, 1974.
	Geog 4678.87	Wreck of the Rainier; a sailors narrative. Portland, 1887.
	Geog 4678.68	The wreck of the Serica. (Cubbin, T.) London, 1950.
	Geog 4679.67	The wreck of the Torrey Canyon. (Gill, Crispin.) Newton Abbot, 1967.
	Geog 4678.55	Wrecked on a reef in the China Sea. (Hinckley, F.) Boston, 1908.
	Geog 4659.55.5	Wrecks in Tasmanian waters, 1797-1950. (O'May, Harry.) Tasmania, 1955?
	Geog 4659.68.10	Wrecks on the Gippsland coast. (Loney, Jack Kenneth.) Geelong, 1968.
	Geog 5530.9	Wright, A. Versus Tennysonianos Franklini. Cantabrigiae, 1882.
	An 2058.92.3	Wright, G.F. The antiquity and origin of the human race. Boston, 1892.
	Geog 5838.96	Wright, G.F. Greenland icefields. N.Y., 1896.
	Geog 578.34	Wright, G.N. New and comprehensive gazeteer. London, 1834-37. 4v.
	Geog 4328.40.10	Wright, G.N. The shores and islands of the Mediterranean. London, 1840.
	Geog 4308.38	Wright, H.H. Desultory reminiscences of a tour. Boston, 1838.
	Geog 5919.18	Wright, H.S. (Mrs.). The seventh continent. Boston, 1918.
	Geog 5209.10	Wright, H.S.S. (Mrs.). The great white North. N.Y., 1910.
	Geog 4181.71	Wright, I.A. Documents concerning English voyages to the Spanish Main, 1569-1580. London, 1933.
	Geog 13.4.20	Wright, J.K. Geography in the making; the American Geographical Society, 1851-1951. N.Y., 1952.
	Geog 500.175	Wright, John K. Aids to geographical research. N.Y., 1923.
	Geog 3229.25A	Wright, John K. The geographical lore of the time of the Crusades. N.Y., 1925.
	Geog 623.2	Wright, John K. Human nature in geography. Cambridge, 1966.
	Geog 4181.62	Wright, L.A. Spanish documents concerning English voyages to the Caribbean. n.p., 1929.
	Geog 3249.70	Wright, Louis Booker. Gold, glory, and the gospel. 1. ed. N.Y., 1970.
	Geog 5235.25	Wright, Theon. The big nail; the story of the Cook-Peary feud. N.Y., 1970.
	An 2109.39	Wright, W.B. Tools and the man. London, 1939.
	Geog 3070.42	Wroth, Lawrence C. The early cartography of the Pacific. N.Y., 1944.
	Geog 135.2.5	Der Würmsee (Starnbergersee) in Oberbayern. v.5. (Ule, Willi.) Leipzig, 1901.
	Geog 3235.9	Wuttke, J.K.H. Über Erdkunde und Karten. Leipzig, 1853.
	Geog 618.4	Wybór prac. (Romer, E.) Warszawa, 1960. 3v.
	Geog 4709.28A	Wycherley, George. Buccaneers of the Pacific. Indianapolis, 1928.
	Geog 616.30	Wyder, Samuel. Die Schaffhauser Karten von Hauptmann Heinrich Peyer (1621-1690). Zürich, 1951.
NEDL	Geog 4218.79	Wylie, A.H. Chatty letters from the East and West. London, 1879.
	An 3558.68	Wyman, J. Observations on Crania. Boston, 1868.
	Geog 4558.81	Wyman, Walter. A cruise on the United States practice ship, S.P. Chase. N.Y., 1910.
	Geog 500.140	Wystawa pt. Rozwój Historyczny Geografii Polskiej i Pismiennictwo Polskie o Zakresu Historii Geografii, Warsaw, 1965. Catalogue of literature on the history of geography at the exposition. Warsaw, 1965.
	Geog 4145.53	Xavier Marmier och Sverige. (Willers, U.) Stockholm, 1949.
	Geog 110.1.20.15	XX mezhdunarodnyi geograficheskii kongress, London, iiul 1964 gg. (Natsional'nyi Komitet Sovetskikh Geografov.) Moskva, 1966.
	Geog 4329.30.7	A yacht in Mediterranean seas. (Anderson, I.P. (Mrs.).) Boston, 1930.
	Geog 5398.56.5	A yacht voyage; letters from high latitudes. (Dufferin and Ava, F.T. Blackwood.) N.Y., 1878.
	Geog 4208.73	Yachting cruise in the South Seas. (Wood, C.F.) London, 1875.
	An 629.50	Yale University. Institute of Human Relations. Outline of cultural material. 3. ed. New Haven, 1950.
	Geog 4309.62	Yankee sails across Europe. (Johnson, Irving.) N.Y., 1962.
	Geog 4559.40.3	Yankee skipper; the life story of Joseph A. Gainard. (Gainard, J.A.) N.Y., 1940.
	Geog 4558.77	Yankee Swanson. (Nelson, A.W.) N.Y., 1913.
	Geog 4309.02.13	Die Yankeedoodle-Fahrt und andere Reisegeschichten. (Bierbaum, Otto Julius.) München, 1920.
	Geog 4559.26.10	Yarns from a windjammer. (Crane, Mannin.) Boston, 1926.
	Geog 4559.27.10	Yarns of the seven seas. (Cooper, F.G.) London, 1927.
	Geog 4308.50.12	A year abroad. (George, W.C.) Boston, 1852.
	Geog 4208.97	A year from a reporter's note-book. (Davis, R.H.) N.Y., 1897.
	Geog 4308.18.3	A year in Europe. (Griscom, J.) N.Y., 1823.
	Geog 4308.18.4	A year in Europe. 2nd ed. (Griscom, J.) N.Y., 1824. 2v.
	Geog 4309.12	A year of strangers. (Maraini, Y.) N.Y., 1912.
	Geog 6009.64.5	The year of the quiet sun: one year at Scott Base, Antarctica. (Hayter, Adrian.) London, 1968.
	Geog 4559.13	A year with a whaler. (Burns, Walter.) N.Y., 1913.
	Geog 97.10	Yearbook. (Harvard Travellers' Club.)
	Geog 147.5	Yearbook. (National Council for Geographic Education.) Palo Alto, Calif. 1,1970+ 2v.
	Geog 139.6	Yearbook record. (London. Royal Geographical Society.) London. 1898-1905 2v.
	An 359.04	Yertz, Friedrich O. Moderne Rassentheorien. Wien, 1904.
	Geog 4219.32.5	Yes, 'tis round. (Gilmore, A.F.) Boston, 1932.
	Geog 4209.54.5	Yksin yli Atlantin. (Kivikoski, Olavi.) Helsinki, 1954.
	Geog 3055.84	Ymdogat-Atlantis. (Bergquist, Nils Olof.) Solna, 1971.

Author and Title Listing

Geog 275.1 Ymer; tidskrift utgifven af svenska sällskapet för antropologi och geografi. Stockholm. 1+ 61v.
Geog 275.2 Ymer; tidskrift utgifven af svenska sällskapet för antropologi och geografi. Person-och ämnesregister; 1-70, 1881-1950. n.p., n.d. 2v.
Geog 4559.59.5 Yonder is the sea. (Bradford, G.) Barre, Mass., 1959.
Geog 4209.32.10 Yonder lies adventure! (Powell, Edward A.) N.Y., 1932.
Geog 5398.75.3 Young, Allen. The two voyages of the Pandora in 1875-76. London, 1879.
Geog 4349.14 Young, Ernest. From Russia to Siam, with a voyage down the Danube. London, 1914.
Geog 818.59.9 Young, Francis. Elementary geography. London, 1859.
Geog 4509.57.5 Young, G.W. The influence of mountains upon the development of human intelligence. Glasgow, 1957.
Geog 4509.20 Young, G.W. Mountain craft. London, 1920.
An 3209.71 Young, John Z. An introduction to the study of man. Oxford, 1971.
Geog 4308.53.9 Young, Samuel. A Wall-street bear in Europe. N.Y., 1855.
Geog 4418.57.5 Young American abroad. (Train, G.F.) London, 1857.
Geog 817.73 The young geographer and astronomer's best companion. (Jones, E.) London, 1773.
An 359.45.10A "Young man, you are normal". (Hooton, E.A.) N.Y., 1945.
Geog 4029.2 The young rifleman's comrade. (Mämpel, J.C.) London, 1826.
Geog 4309.38 Your trip to Europe. (Hamburg-American Line.) N.Y., 1938.
Geog 4209.52.10 Your world and mine. (Szos Kies, Henryk J.) N.Y., 1952.
Geog 4332.7F Yriarte, C. Les bords de l'Adriatique. Paris, 1878.
Geog 4180.36 Yule, H. Cathay and the way thether. London, 1866. 2v.
Geog 4181.33 Yule, Henry. Cathay and the way thither. London, 1913-14. 4v.
Geog 4181.33.1 Yule, Henry. Cathay and the way thither. v.1-4. Taipei, 1966. 2v.
Geog 4208.43 Yvan, M. Voyages et recits. v.1-2. Bruxelles, 1853.
Geog 4309.63.10 Z Kamjenskim nosom; što dóživi serbski nowinar we swěče. 2. wyd. (Wićaz, Jurij.) Budyšin, 1963.
An 2108.74.7 Zaborowski-Moindron, S. De l'ancienneté de l'homme. v.1-2. Paris, 1874.
An 2108.78 Zaborowski-Moindron, S. L'homme préhistorique. Paris, 1878.
Geog 3590.49 Zabrodskaia, M.R. Russkie puteshestvenniki po Afrike. Moskva, 1955.
Geog 4477.64 Zaccaria Seriman, 1709-1784, and the Viaggi di Enrico Wanton. (White, D.M.) Manchester, 1961.
An 609.67 Zagreb. Etnografski Muzej. Daleki sujtovi naših putnika i pomoraea. Zagreb, 1967.
Geog 85.389 Zagreb. Univerzitet. Geografski Institut. Geographical papers. Zagreb. 1,1970+
Geog 85.372 Zagreb. Univerzitet. Geografski Institut. Radovi. Travaux. 7,1968+
Geog 4145.68.5 Zaky tyi pute Shestvennik. (Gnevusheva, E.I.) Moskva, 1958.
Geog 5399.56 Zanaevanie Arktiki. (Centkiewicz, Alina.) Moskva, 1956.
Geog 4309.64.15 Západ je Západ. (Selucký, Radoslav.) Praha, 1964.
Geog 4309.64.16 Západ je Západ. 2. vyd. (Selucký, Radoslav.) Praha, 1965.
Geog 4306.97F Zapiska puteshestviia. (Sheremt'ev, B.P.) Moskva, 1773.
Geog 129.5 Zapiski. (Krymskii Garnyi Klub, Odessa.) Odessa. 1895-1912 2v.
Geog 5160.55 Zapiski poliarnika. (Liakh, N.N.) Novosibirsk, 1961.
Geog 4308.23.5 Zapiski russkago puteshestve norika [s 1823 po 1826 g.]. (Glagolev, A.G.) Sankt Peterburg, 1845. 4v.
VGeog 4309.03.15 Zarandoklat Rómába. (Deák, Imre.) Fajsz? 1903.
Geog 4145.71.500 Zariu, V.M. Puteshchestviia A.V. Potaninoi. Moskva, 1950.
Geog 5559.57.10 Zarubezhmyi Sever. (Agranat, G.A.) Moskva, 1957.
Geog 4329.32 Zauber und Grösse des Mittelmeers. (Edschmid, K.) Frankfurt, 1932.
Geog 5300.40 Zavalti, Silvio. Pionieri italiani nelle regioni polari. Brescia, 1952.
Geog 5050.7 Zavatti, S. Saggio di bibliografia polare. 2. ed. Roma, 1952.
Geog 579.52 Zavatti, Silvio. Dizionario geografico. Catania, 1952.
Geog 5939.58.15 Zavatti, Silvio. L'esplorazione dell'Antartide. Torino, 1958.
Geog 85.377 Zbornik radova. (Belgrade. Geografski Institut.) 18+ 5v.
Geog 85.384 Zbornik radova. (Belgrade. Univerzitet. Prirodno-Matematički Fakultet. Geografski Zavod.) Beograd. 13,1966+ 2v.
Geog 110.3 Zbornik radova III Kongresa slovenskih geografa i ethnografa u Kraljevini Jugoslaviji, 1930. (Congrès des Géographes et Ethnographes Slaves, 3rd, 1930.) Beograd, 1933.
Geog 4209.58.15 Ze wspomnień podróżników. (Olszewicz, B.) Warszawa, 1958.
Geog 4182.30 Zeeuwsche expedite...Cornelis Evertsen. (Woard, C. de.) 's-Gravenhage, 1928.
Geog 5070.44 Zeidler, P.G. Polarfahrten. Berlin, 1927.
Geog 5060.16 Zeidler, Paul G. Helden im ewigen Eis; im Kampf um den Nord- und Südpol. Leipzig, 1936.
Geog 3019.56 Zeigt mir Adams Testament. (Herrmann, Paul.) Hamburg, 1956.
Htn Geog 4260.5F* Zeiller, M. Topographia Italiae. Franckfurt, 1688.
Htn Geog 4266.41.8F* Zeiller, Martin. Sämtliche...Topographias. Haupt-Register. Frankfurt, 1726.
Htn Geog 4266.41.7F* Zeiller, Martin. Sämtliche...Topographias. v.1-31. Frankfurt, 1677-1736. 10v.
Geog 4266.41.10F Zeiller, Martin. Topographia Sueviae. Kassel, 1960.
Geog 4266.41.80 Die Zeiller-Merianschen Topographien bibliographisch Beschrieben. (Schuchhard, C.) Hamburg, 1960.
NEDL Geog 45.6 Zeitschrift. (Deutscher und Österreichischer Alpenvereins.) Salzburg. 1-30,1869-1899 30v.
Geog 45.6 Zeitschrift. (Deutscher und Österreichischer Alpenvereins.) Salzburg. 1-74,1869-1949 40v.
Geog 45.6.7 Zeitschrift. Beilagen. (Deutscher und Österreichischer Alpenvereins.) 1869-1910
Geog 287.4 Zeitschrift für Erdkunde. Frankfurt am Main. 4
Geog 287.2 Zeitschrift für wissenschaftliche Geographie. Lahr. 1-3,1880-1882 3v.
Geog 3018.97 Zeitströmungen in der Geographie. (Wisotzki, E.) Leipzig, 1897.
Geog 4419.70 Zemlia dobrykh liudei. Tashkent, 1970.
Geog 290.2 Zemlia i liudi. Moskva. 9,1966+ 9v.
Geog 4314.7 Zenkovich, V.P. Berega Chennogo i Azovskego morei. Moskva, 1958.
Geog 4180.50 Zeno, N. Voyages of...Nicolo and Antonio Zeno. London, 1873.
Geog 4209.36.15 Zest for life; recallections of a philosophic traveller. (Wøller, Johan.) London, 1936.
Geog 290.4 Zeszyty geograficzne. (Danzig. Wyższa Szkoła Pedagogiczna. Wydział Geograficzny.) 1-11,1959-1969// 3v.
Geog 290.5 Zeszyty naukowe. Geografia. (Danzig. Universytet. Wydział Biologii i Nauk o Ziemi.) Gdańsk. 1,1970+
An 3558.46 Zeune, August. Über Schädelbildung zur festern Begründung der Menschenrassen. Berlin, 1846.
An 359.05 Zhachenie "geograficheskikh provintsie". (Koropchevskago, A.A.) Sankt Peterburg, 1905.
Geog 5330.1 Zhizn' i nauchnaia deiatel'nost' E.V. Tollia. (Vittenburg, P.V.) Leningrad, 1960.
Geog 607.5.5 Zhizn' i neobyknovennye prikl. K.L. Golovnina. (Fraerman, R.I.) Moskva, 1957.
Geog 607.5 Zhizn' i prikl. K.L. Golovnina. (Fraerman, R.I.) Moskva, 1946.
Geog 5559.37 Zhizn' na l'dine. (Papanin, Ivan D.) Moskva, 1940.
Geog 619.5 Zhiznennyi put' Iu. M. Shokal'skogo. (Shokal'skaia, Z. Iu.) Moskva, 1960.
Geog 4209.63.5 Zhukov, Iu.A. Eti semnadtsat' let. Moskva, 1963.
Geog 4679.60.5 Ziganskin, Askhat. 49 dnei v okeane. Kuibyshev, 1960.
Geog 4308.81.5 Zigzag journeys in classic lands. (Butterworth, Hezekiah.) Boston, 1881.
Geog 4308.81.6 Zigzag journeys in classic lands. (Butterworth, Hezekiah.) Boston, 1882.
Geog 5838.11 Zimmermann, E.A.H. Die Erde und ihre Bewohner. Leipzig, 1811.
An 3309.73 Zinevich, Galina P. Antropologicheskie materialy srednevekovykh mogil'nikov dugo-zapadnogo kryma. Kiev, 1973.
An 3559.68.5 Zinevich, Galina P. Antropologichna kharakteristika davn'oho naselennia teritorii Ukrainy. Kyiv, 1968.
An 2059.67 Zinevich, Galina P. Ocherki paleoantropologii Ukrainy. Kiev, 1967.
Geog 4209.37.5 Život a cesty E. St. Vráze. (Vrázová, Vlasta.) Praha, 1937.
Geog 4209.37.7 Život a délo E. St. Vráze. (Prague. Městská Lidová Knihovna.) Praha, 1960.
Geog 4219.12 Zobeltitz, Fedor von. Ein Bummel um die Welt. Berlin, 1912.
Geog 3590.42 Zogosken, L.A. Puteshestviia i issledovaniu...v Russkoi Amerike v 1802-04 gg. Moskva, 1956.
An 2109.64.15 Zolotarev, Aleksandr M. Rodovoi stroi i pervobytnaia mifologiia. Moskva, 1964.
Geog 4759.73 La zone intertropicale humid. (Daveau, Suzanne.) Paris, 1973.
Geog 5837.20.5 Zorgdrager, C.G. Beschreibung des...Wallfischfangs. Nürnberg, 1750.
Geog 5837.20.3 Zorgdrager, C.G. Bloeijende opkomst...visschery. 's-Gravenhage, 1727.
Geog 5837.20 Zorgdrager, C.G. Groenlandsche visschery. Amsterdam, 1720.
Geog 4308.96.5 Zorrilla de San Martin, J. Resonancias del camino. Barcelona, 190-?
Geog 4308.96.3 Zorrilla de San Martin, J. Resonancias del camino. Paris, 1896.
Geog 4309.28.10 Zozulia, Efim D. Iz Moskvy na Korsiku i Obratno. Leningrad, 1928.
Geog 665.5 Zu Friedrich Ratzels Gedächtnis. (Ratzel, F.) Leipzig, 1904.
Geog 4419.02 Zu Land und See im Orient. (Krausz, S.) Chicago, 1902.
Geog 4129.59 Zu Pferde, im Wagen, zu Fuss. (Schadendorf, Wulf.) München, 1959.
An 3559.68 Zubov, Aleksander A. Odontologiia. Moskva, 1968.
An 3559.73 Zubov, Aleksandr A. Etnicheskaia odontologiia. Moskva, 1973.
An 3209.68 Zubov, Aleksandr A. Problemy evoliutsii cheloveka i ego ras. Moskva, 1968.
Geog 3590.47 Zubov, N.N. Otechestvennye moraplavateli-issledovanii morei i okeanov. Moskva, 1954.
An 2109.06.3 Zuccarelli, A. Gli uomini primitivi delle selci e delle caverne. Napoli, 1906.
Geog 4208.97.7 Ein Zug nach Osten. (Schanz, M.) Hamburg, 1897. 2v.
Geog 613.1.15 Zum Gedächtnis Gerhard Mercator's. (Dinse, Paul.) Berlin, 1894.
Geog 659.67 Zum Gegenstand und zer Methode der Geographie. (Storkebaum, Werner.) Darmstadt, 1967.
An 609.69 Zum hundertjährigen Bestehen, 1869-1969. (Leipzig. Staedtisches Museum für Völkerkunde.) Berlin, 1969.
An 379.68.5 Zum Standort der Sozialgeographie. Kallmünz, 1968.
Geog 4509.23 Zum Wortschatze des Bergsteigers. (Dupin.) Wien, 1923-
An 3208.99 Zur Anthropologie der Badener. (Karlsruhe, Germany. Altertumsverein.) Jena, 1899.
Geog 3235.65 Zur Entstehungs-Geschichte der Tabula Purtingeriana. Diss. (Gross, Hans.) Bonn, 1913.
Geog 4131.70 Zur Geographie des Freizeitverhaltens. (Ruppert, Karl.) Kallmünz, 1970.
An 3558.81 Zur Kritik und Verbesserung der Winkelmessungen am Kopfe. (Bessel Hagen, F.K.) Königsberg, 1881.
An 3509.31.10 Zur Osteologie der Lappen. (Schreiner, Kristian Emil.) Oslo, 1931-35. 2v.
Geog 4508.99.5 Zurbriggen, M. From the Alps to the Andes. London, 1899.
Geog 4019.74 Zweig, Paul. The adventure. London, 1974.
Geog 5558.69.5 Die zweite deutsche Nordpolarfahrt. (Verein in die Deutsche Nordpolarfahrt.) Leipzig, 1883.
Geog 4314.8 Zwischen Donau und Kaukasus. (Schweiger-Lerchenfeld, Amand von.) Wien, 1887.
Geog 4309.31.5 Zwischen Rhone und Wolga. (Ponten, J.) Leipzig, 1931.
Geog 40.10 4 annaes. (Congresso Brasileiro de Geografia.) Pernambuco. 1-2 3v.
Geog 5919.59 Der 7. Kontinent. (Steinitz, Hans.) Bern, 1959.
Geog 4238.83.5 11,506 knots in the "Sunbeam". (Brassey, Thomas.) London, 1884.
NEDL Geog 4238.83.7 14,000 miles in the Sunbeam in 1883. (Brassey, Anna.) London, 1885.
Geog 4218.92.5F 23.000 mil'na iakhte "Tamara". (Radde, G.F.R.) Sanktpeterburg, 1892-93. 2v.
An 609.28 25 Jahre Kölner anthropologische Gesellschaft und städtisches Museum fürVor- und Frühgeschichte 1903-28. Festschrift. (Kölner Anthropologische Gesellschaft.) Köln, 1928.
Geog 4219.29.5 30,000 miles around the world. (Baus, Thomas J.) Philadelphia, 1929.
Geog 4679.60.5 49 dnei v okeane. (Ziganskin, Askhat.) Kuibyshev, 1960.
Geog 5970.16 90 degrees south; the story of the American South Pole conquest. (Siple, Paul.) N.Y., 1959.
Geog 4110.25 100 vacations costing from $50.00 to $500.00. (Coon, Horace.) N.Y., 1939.
Geog 4309.30.30 103 dnia na Zapade, 1924-1926 gg. 2. izd. (Kushner, Boris A.) Moskva, 1930.

Geog 4679.73.6	117 days adrift. (Bailey, Maurice.) Lymington, 1974.	
Geog 4679.23.6	1700 miles in open boats; the story of the loss of the S.S. Trivessa in the Indian Ocean. (Foster, Cecil.) London, 1952.	
Geog 4709.65.5	A 1715 Spanish treasure ship. (Clausen, Carl J.) Gainesville, 1965.	
Geog 6009.16F	1918 "Aurora" relief expedition. (Davis, John King.) Melbourne, 1918.	
Geog 4311.19	2200 kilo-metrov po Dunaiu. (Jonák, Ján.) Moskva, 1960.	
Geog 4269.60	500,000 [i.e. Piat'sot tysiach] kolometrov v puti. (Kuleshov, A.P.) Moskva, 1960.	

Z
6009
H37
1979

JUL 9 1979